TEACHER'S EDITION

Miller & Levine

Biology

FOUNDATION EDITION

Kenneth R. Miller, Ph.D.
Professor of Biology, Brown University
Providence, Rhode Island

Joseph S. Levine, Ph.D.
Science Writer and Producer
Concord, Massachusetts

PEARSON

Boston, Massachusetts • Chandler, Arizona • Glenview, Illinois • Upper Saddle River, New Jersey

Overview of Teacher's Edition

Front matter acknowledgments: T6 Danny Kerr. **T7** Jen-Yi Wu.

Additional acknowledgments begin on page C–1, which constitutes an extension of this copyright page.

PEARSON

ISBN-13: 978-0-13-361470-1
ISBN-10: 0-13-361470-0
4 5 6 7 8 9 10 V063 13 12 11

Miller & Levine Biology: A Comprehensive Approach

The Miller & Levine Biology program offers a variety of unique product options to support a range of teaching styles. Communicate your love of science to your students in a way that will engage them and offer support for all levels and all types of learners. Choose the product that suits your teaching style and the needs of your classroom:

A Comprehensive Approach to Teaching Biology

Miller & Levine Biology

The most complete, accurate, comprehensible and student-friendly biology program available.

Foundation Edition: *Miller & Levine Biology*

Foundations offers accessible content with embedded learning strategies and instructional support to address the needs of struggling students.

Core Edition: *Miller & Levine Biology*

A concise textbook accompanied by a complete online program covering the core biology curriculum.

Biology.com and Biology.com Plus

Biology.com is the next generation of digital instruction offering complete online student and teacher editions, digital lessons, animations, simulations and assessments and much more!

What customers are saying about *Miller & Levine Biology*

"The Foundation text is easy to comprehend by all learners in the classroom, but mainly targets those learners that are below grade level reading. Each chapter provides the same basic information as the regular text, but adds the vocabulary word origins that are so important in the biology classroom. The extra guidance with the text activties helps students complete the larger task at hand. We will continue to use the Foundation text as a supplement for our struggling readers and those students having difficulty comprehending traditional science textbooks. This textbook's narrative form also provides all learners a better understanding of science."

Kim Lewis, Biology Teacher
Wellston High School, Wellston, Ohio

A program to support struggling learners

Introducing the *Foundation Edition* of *Miller & Levine Biology.*

Authors Ken Miller and Joe Levine deliver the same, trusted content but with a lower reading level and embedded reading support to aid your struggling learners. The *Foundation Edition* makes biology come alive for struggling students with comprehensive strategies to help increase reading comprehension, vocabulary retention, and conceptual understanding in a fresh new way with:

Engaging and accessible content

brings the "Big Ideas" of biology to life for all students with reading strategies built into every lesson.

Biology.com

the latest in instructional technology helps struggling students master concepts.

A robust ancillary array

provides point-of-use correlation with the *Foundation* Student Edition at a lower reading level.

Teaching support

includes a Teacher's Edition that provides a roadmap for supporting struggling learners for successful mastery of content.

Build a solid foundation for every student.

The *Foundation Edition* of *Miller & Levine Biology* will connect your students to current science content and to the stories and adventures of our world. *Miller & Levine Biology*, the *Foundation Edition* covers relevant content in more accessible ways with embedded reading support.

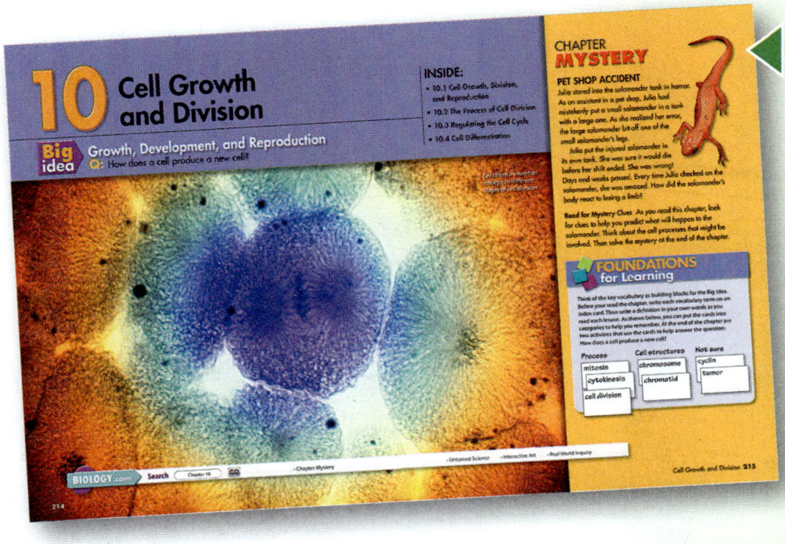

Big Ideas Introduce Chapters

Big Ideas and related questions set a framework for the chapter and pose direct questions that students will answer throughout the chapter in order to create a consistent learning-strategy strand.

Every chapter begins with an inquiry strand called a Chapter Mystery that aligns to Big Ideas and concepts introduced in the lessons. The mystery is solved at the end of the chapter.

Reading Strategies Embedded Throughout

Foundations for Learning is an added learning strategy to help students organize the key concepts of the chapter.

Build Understanding helps students build an outline of chapter concepts.

Build Connections helps students create a connection between the chapter concepts and their daily lives.

Build Vocabulary helps students master academic and scientific vocabulary from the chapter.

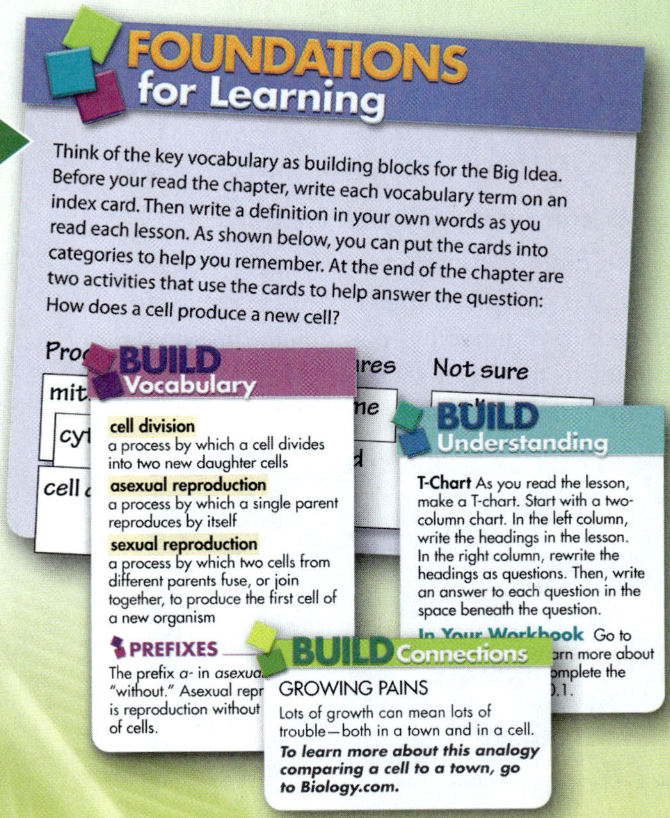

Student Support Ensures Success

Two levels of Study Workbooks offer leveled activities and strategies to enhance concepts learned in the textbook.

Study Workbook B

The adapted Study Workbook for the *Foundation Edition* of *Miller & Levine Biology* will help students:

- preview a chapter
- work with the key vocabulary
- practice strategies for learning the content
- review for chapter and unit tests
- learn strategies for taking standardized tests

Preview the Chapter

Before reading the chapter, students preview the main ideas using the concept map provided.

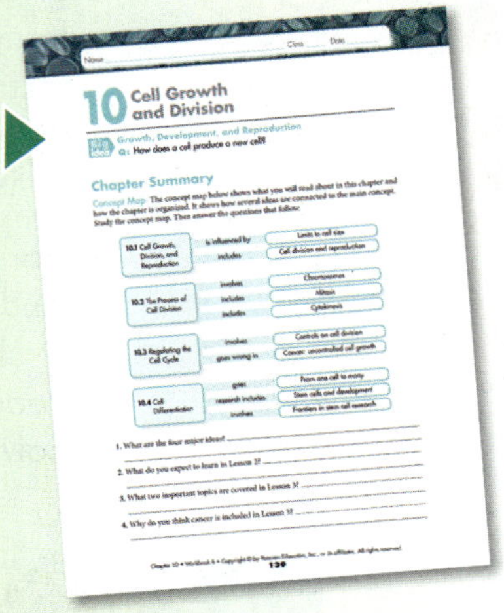

Build Vocabulary

Each lesson starts with a list of objectives and a chart that encourages students to learn key vocabulary. Have students highlight these key terms when they use them in the workbook.

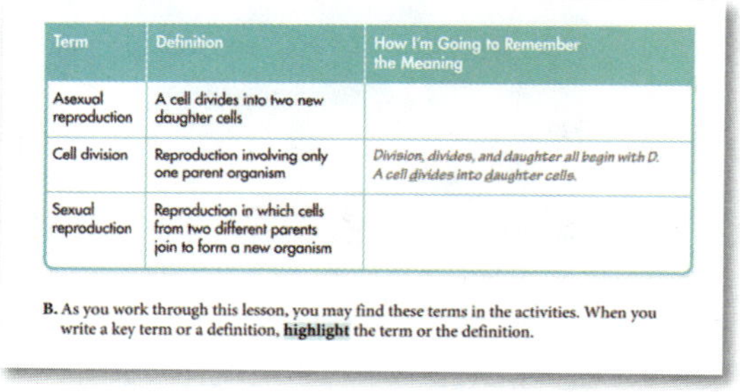

Term	Definition	How I'm Going to Remember the Meaning
Asexual reproduction	A cell divides into two new daughter cells	
Cell division	Reproduction involving only one parent organism	*Division, divides, and daughter all begin with D. A cell divides into daughter cells.*
Sexual reproduction	Reproduction in which cells from two different parents join to form a new organism	

B. As you work through this lesson, you may find these terms in the activities. When you write a key term or a definition, **highlight** the term or the definition.

Grant Wiggins and his "Big Idea"

Grant Wiggins, consultant and UbD creator

Miller & Levine Biology incorporates Grant Wiggins's *Understanding by Design* framework. It's a common sense approach for connecting curriculum, instruction, and assessment so that your students develop a deep understanding of the Big Ideas every day.

Understanding by Design

What's the Big Idea?

If deep understanding is the goal, how do you get your students there? *Understanding by Design* provides a number of ways to:

- Engage students in inquiry
- Promote a transfer of learning
- Provide a conceptual framework
- Uncover "big ideas"
- Develop appropriate assessments

Want to Learn More?

Check out Understanding by Design for yourself at Biology.com

Support for English Language Learners

English language learners are entering U.S. classrooms in steadily increasing numbers. The *Miller & Levine Biology* program is designed to help you address the needs of this growing and diverse population. Program materials have been developed according to the Pearson ELL Curriculum Framework, which incorporates five research-based essential principles within an Into/Through/Beyond structure. This framework establishes an effective instructional scaffold that will help you reach all your English language learners and prepare them for success.

In the Teacher's Edition, wrap-around notes include strategies for building vocabulary and helping students handle academic vocabulary.

Chapter planning guides provide built-in differentiated instruction to ensure you have the resources you need to reach all the learning levels in the classroom.

Wrap-around dialogue will help you emphasize the most important points of lessons and ask the questions that will reveal concepts that students struggle with. Suggestions for graphic organizers will help you guide students to organize information in meaningful ways.

Teacher's Edition

In addition, you will find a wealth of support for ELL instruction in this Teacher's Edition wrap-around. In particular, look for Focus on ELL strategies, which provide you with content-specific instructions for working with your ELL students.

Teacher's ELL Handbook

The ELL Handbook, *From Theory to Practice: Accelerating the Progress of English Language Learners,* is designed to help you scaffold and support instruction for your English language learners. With this support, English language learners can learn in ways that are comprehensible and meaningful. The handbook will enable you to promote the academic success and achievement of all your students.

Multilingual Glossary

The Multilingual Glossary contains all the highlighted terms from the *Foundation Edition* of *Miller & Levine Biology*. In addition to English-language definitions, the glossary provides translations of the terms in these languages: Arabic, Brazilian Portuguese, Chinese (both simplified and traditional), Haitian Creole, Hmong, Korean, Russian, Spanish, and Vietnamese.

"The knowledge base that research has generated about ELL students' academic trajectories shows clearly that ELL students must be understanding instruction and learning English across the curriculum if they are to catch up in time to meet graduation requirements. Teaching biology affords opportunities for extending ELL students' academic language proficiency. The Pearson ELL Curriculum Framework incorporates the essential elements that teachers need to implement."

—Jim Cummins, Ph.D.
University of Toronto

Deepen Understanding Through Inquiry

The *Foundation Edition* of *Miller & Levine Biology* encourages students to think like scientists, and enhances their understanding through participation in inquiry-based discovery. Multiple lab options are included for every chapter, providing you with the flexibility to address all types of learners and accommodate your class time and equipment requirements.

Inquiry into Scientific Thinking

Quick labs provide students the opportunity to discover science with a variety of investigations through guided inquiry. These labs deepen student understanding of Key Concepts and Big Ideas.

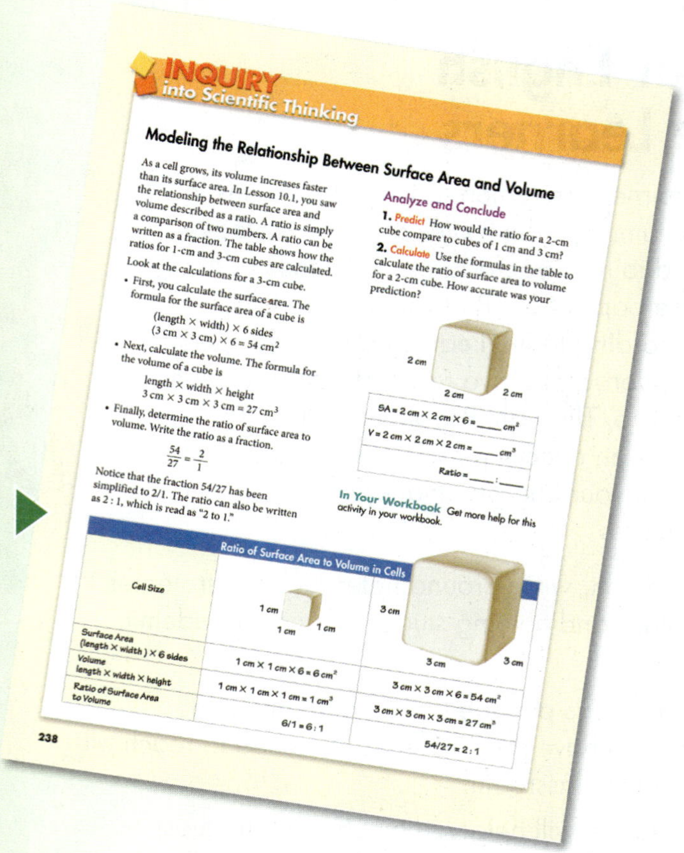

INQUIRY into Scientific Thinking

Modeling the Relationship Between Surface Area and Volume

As a cell grows, its volume increases faster than its surface area. In Lesson 10.1, you saw the relationship between surface area and volume described as a ratio. A ratio is simply a comparison of two numbers. A ratio can be written as a fraction. The table shows how the ratios for 1-cm and 3-cm cubes are calculated.

Look at the calculations for a 3-cm cube.

- First, you calculate the surface area. The formula for the surface area of a cube is
  (length × width) × 6 sides
  $(3\text{ cm} \times 3\text{ cm}) \times 6 = 54\text{ cm}^2$
- Next, calculate the volume. The formula for the volume of a cube is
  length × width × height
  $3\text{ cm} \times 3\text{ cm} \times 3\text{ cm} = 27\text{ cm}^3$
- Finally, determine the ratio of surface area to volume. Write the ratio as a fraction.
  $\frac{54}{27} = \frac{2}{1}$

Notice that the fraction 54/27 has been simplified to 2/1. The ratio can also be written as 2 : 1, which is read as "2 to 1."

Analyze and Conclude

1. Predict How would the ratio for a 2-cm cube compare to cubes of 1 cm and 3 cm?
2. Calculate Use the formulas in the table to calculate the ratio of surface area to volume for a 2-cm cube. How accurate was your prediction?

2 cm / 2 cm / 2 cm

$SA = 2\text{ cm} \times 2\text{ cm} \times 6 =$ ____ cm^2
$V = 2\text{ cm} \times 2\text{ cm} \times 2\text{ cm} =$ ____ cm^3
Ratio = ____ : ____

In Your Workbook Get more help for this activity in your workbook.

Ratio of Surface Area to Volume in Cells

Cell Size	1 cm × 1 cm × 1 cm	3 cm × 3 cm × 3 cm
Surface Area (length × width) × 6 sides	$1\text{ cm} \times 1\text{ cm} \times 6 = 6\text{ cm}^2$	$3\text{ cm} \times 3\text{ cm} \times 6 = 54\text{ cm}^2$
Volume length × width × height	$1\text{ cm} \times 1\text{ cm} \times 1\text{ cm} = 1\text{ cm}^3$	$3\text{ cm} \times 3\text{ cm} \times 3\text{ cm} = 27\text{ cm}^3$
Ratio of Surface Area to Volume	6/1 = 6 : 1	54/27 = 2 : 1

238

End-of-Chapter Labs

The pre-lab at the end of each chapter correlates to a lab in the Lab Manual, and connects the lab to the Big Idea of the chapter. Chapter Labs offer students a variety of inquiry investigations–including Skills Labs, Forensics Labs, Real-World Labs, and Design Your Own Labs.

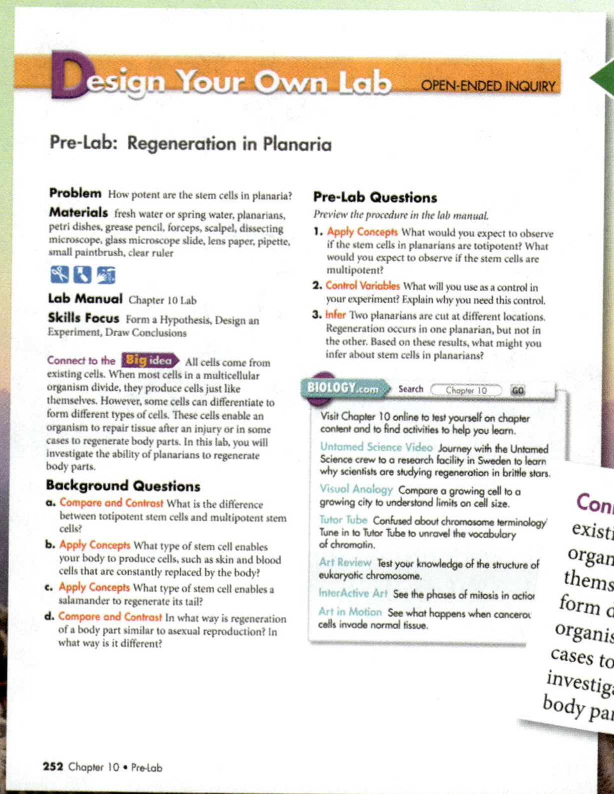

Design Your Own Lab OPEN-ENDED INQUIRY

Pre-Lab: Regeneration in Planaria

Problem How potent are the stem cells in planaria?

Materials fresh water or spring water, planarians, petri dishes, grease pencil, forceps, scalpel, dissecting microscope, glass microscope slide, lens paper, pipette, small paintbrush, clear ruler

Lab Manual Chapter 10 Lab

Skills Focus Form a Hypothesis, Design an Experiment, Draw Conclusions

Connect to the Big idea All cells come from existing cells. When most cells in a multicellular organism divide, they produce cells just like themselves. However, some cells can differentiate to form different types of cells. These cells enable an organism to repair tissue after an injury or in some cases to regenerate body parts. In this lab, you will investigate the ability of planarians to regenerate body parts.

Background Questions

a. Compare and Contrast What is the difference between totipotent stem cells and multipotent stem cells?
b. Apply Concepts What type of stem cell enables your body to produce cells, such as skin and blood cells that are constantly replaced by the body?
c. Apply Concepts What type of stem cell enables a salamander to regenerate its tail?
d. Compare and Contrast In what way is regeneration of a body part similar to asexual reproduction? In what way is it different?

Pre-Lab Questions

Preview the procedure in the lab manual.

1. Apply Concepts What would you expect to observe if the stem cells in planarians are totipotent? What would you expect to observe if the stem cells are multipotent?
2. Control Variables What will you use as a control in your experiment? Explain why you need this control.
3. Infer Two planarians are cut at different locations. Regeneration occurs in one planarian, but not in the other. Based on these results, what might you infer about stem cells in planarians?

BIOLOGY.com Search Chapter 10 GO

Visit Chapter 10 online to test yourself on chapter content and to find activities to help you learn.

Untamed Science Video Journey with the Untamed Science crew to a research facility in Sweden to learn why scientists are studying regeneration in brittle stars.

Visual Analogy Compare a growing cell to a growing city to understand limits on cell size.

Tutor Tube Confused about chromosome terminology? Tune in to Tutor Tube to unravel the vocabulary of chromatin.

Art Review Test your knowledge of the structure of eukaryotic chromosome.

InterActive Art See the phases of mitosis in actio

Art in Motion See what happens when cancerou cells invade normal tissue.

252 Chapter 10 • Pre-Lab

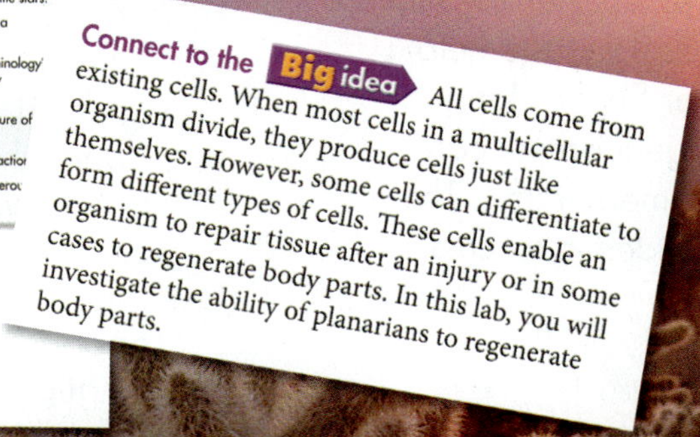
Connect to the Big idea All cells come from existing cells. When most cells in a multicellular organism divide, they produce cells just like themselves. However, some cells can differentiate to form different types of cells. These cells enable an organism to repair tissue after an injury or in some cases to regenerate body parts. In this lab, you will investigate the ability of planarians to regenerate body parts.

Inquiry Resources

Lab Manuals

The lab manuals offer leveled labs to accompany each chapter investigation. A variety of investigations are provided, including Design Your Own Lab, Real World, Skills, and Forensics Labs.

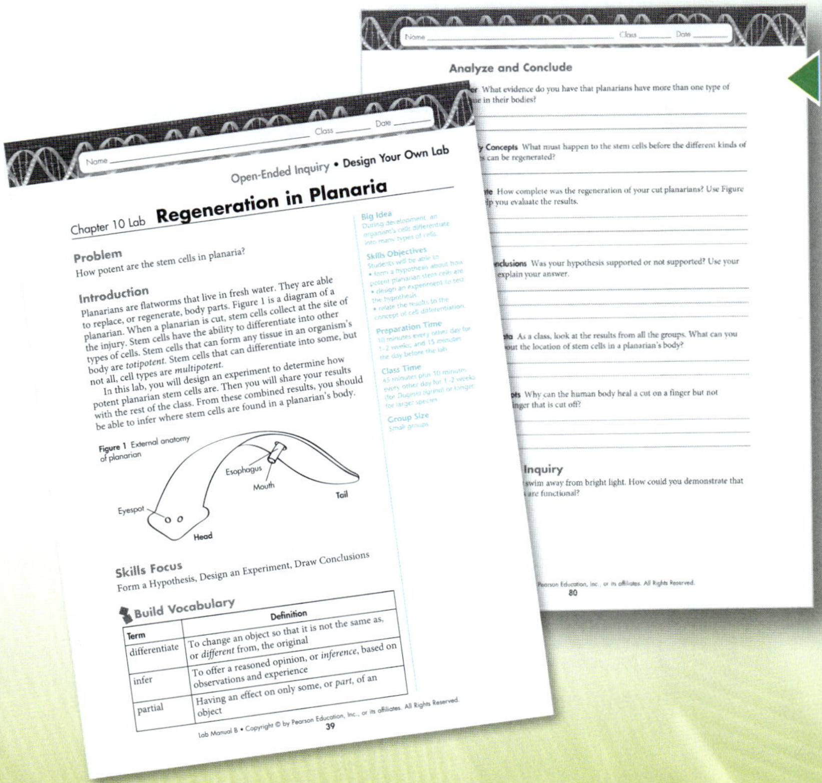

Name ______ Class ______ Date ______

Open-Ended Inquiry • Design Your Own Lab

Chapter 10 Lab **Regeneration in Planaria**

Problem
How potent are the stem cells in planaria?

Introduction
Planarians are flatworms that live in fresh water. They are able to replace, or regenerate, body parts. Figure 1 is a diagram of a planarian. When a planarian is cut, stem cells collect at the site of the injury. Stem cells have the ability to differentiate into other types of cells. Stem cells that can form any tissue in an organism's body are *totipotent*. Stem cells that can differentiate into some, but not all, cell types are *multipotent*.

In this lab, you will design an experiment to determine how potent planarian stem cells are. Then you will share your results with the rest of the class. From these combined results, you should be able to infer where stem cells are found in a planarian's body.

Figure 1 External anatomy of planarian

Skills Focus
Form a Hypothesis, Design an Experiment, Draw Conclusions

Build Vocabulary

Term	Definition
differentiate	To change an object so that it is not the same as, or *different* from, the original
infer	To offer a reasoned opinion, or *inference*, based on observations and experience
partial	Having an effect on only some, or *part*, of an object

Lab Manual B • Copyright © by Pearson Education, Inc., or its affiliates. All Rights Reserved.
39

Analyze and Conclude

Lab Manual B

This lab manual is aimed at students who require more support, including those who will be using the *Foundation Edition* of *Miller & Levine Biology* The 35 chapter labs are modified to reflect a lower reading level. Lab Manual B includes:

- 35 investigations introduced in the Student Edition Pre-Labs, made more accessible for students needing additional support
- a large selection of Data Analysis activities
- 24 additional Hands-On activities

Biology.com—the online habitat to engage struggling learners

Featuring the latest in instructional technology, Biology.com provides an engaging interface to bring biology to life for your struggling students. Visual, interactive, and differentiated components help reinforce content and provide a broad understanding of scientific concepts for students who may need a little extra support.

BIOLOGY.com offers:

For Students:

- Complete *Foundation* Student Edition with audio to help students read along online
- Interactive visuals, online tutorials, and simulations brings concepts come alive for students
- Scaffolded self-assessment prepares students for testing

For Teachers:

- Lesson planner with leveled activities crafts lessons to support struggling students
- Lesson overview presentations help present the Big Ideas of each lesson
- Comprehensive, auto-graded assessment and remediation for different levels of students
- Editable worksheets that allow you to cater your lessons to your class's needs

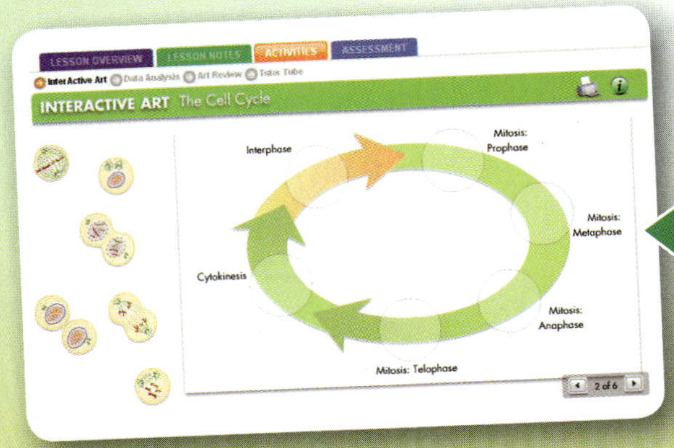

Online Activities

Vocabulary flash cards with audio, drag-and-drop activities, and engaging multimedia allow struggling students to interact with biology concepts and reinforce mastery of key ideas.

Biology.comPLUS offers you:

Virtual BioLab Online, an inquiry-based lab simulation environment.

BioAdventure Videos with Untamed Science

Join these backpacking "eco-geeks" as they take high school science students on real-life science adventures. Each video explores the concepts introduced in the Chapter Mystery out in the field.

Tutorials and Lesson Review

Video tutorials and Self-Assessments help students review key concepts in a safe and interactive way to help prepare for lesson and end-of-year testing.

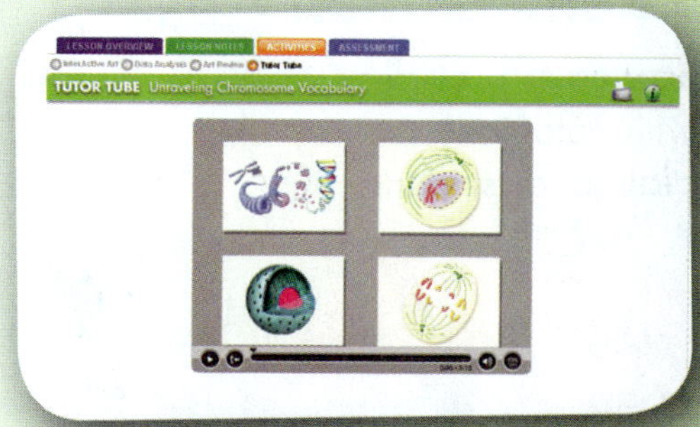

Note-taking and Graphic Organizers

Students are provided with their own online interface to take notes, organize their thoughts, and read lesson overviews to help ensure content mastery.

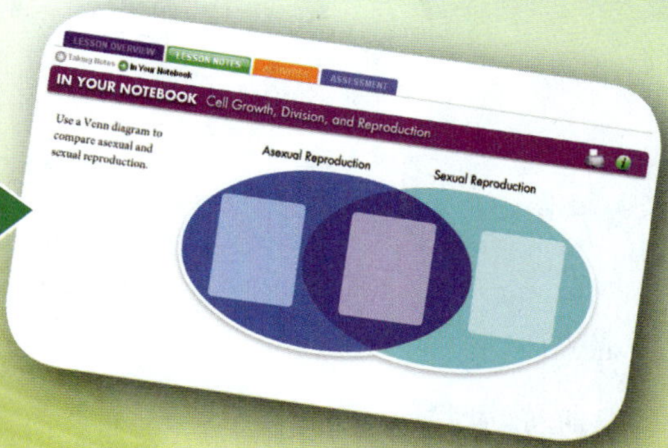

Assessment to Inform Instruction

A variety of assessment opportunities in the *Foundation Edition* of *Miller & Levine Biology* helps you monitor student progress, evaluate content mastery, and ensure student success on high-stakes tests.

Assessment Monitors Learning

Check Understanding at the end of every lesson calls on students to think critically, apply vocabulary, and use key concepts to solve the Chapter Mystery introduced on the Chapter Opener. Students are asked to summarize their understanding of topics covered in that chapter, and reach conclusions based on what they've learned.

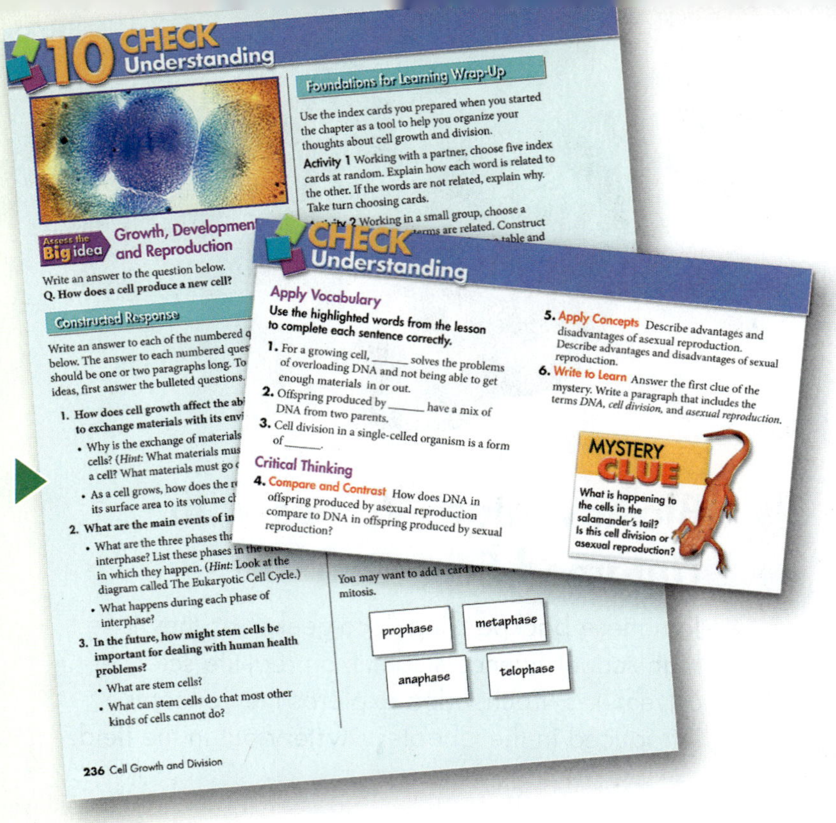

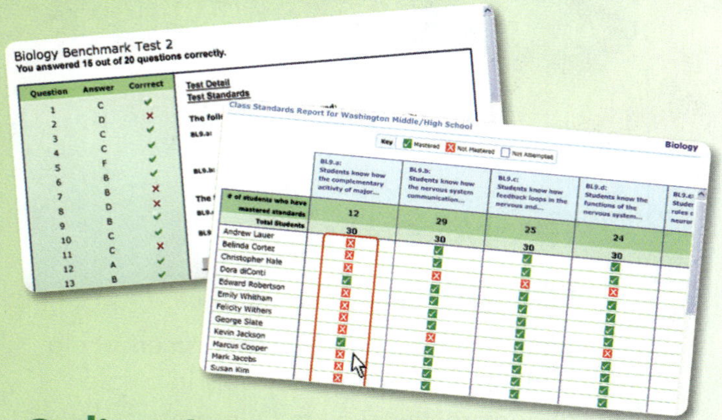

Online Assessment and Remediation

Biology.com provides Diagnostic and Benchmark tests with automatic scoring, class reports, and remediation for students.

Assessment Resource Book

Ensure your students are mastering content with this brand-new assessment program! Assessments include 80 leveled Unit Tests, 350 leveled Chapter Tests, and 130 Visual Quizzes to help you prepare your classroom for end-of-year testing. B-level tests provide extra support to aid struggling readers.

Exam*View*® *Assessment Suite* CD-ROM

Create and print custom tests in minutes that correlate to your state standards. This testbank includes over 4000 test questions and is compatible with QuickTake software.

National Science Education Standards

Standard	Chapter Lesson/Lab
Unifying Concepts and Processes	
I Systems, order, and organization	Ch. 1: 1.1, 1.3; Ch. 2: 2.1, 2.2, 2.3; Ch. 3: 3.1, 3.2, 3.3, 3.4, Ch. 3 Lab; Ch. 4: 4.1, 4.2, 4.3, 4.4; Ch. 5: 5.2; Ch. 6: 6.1, 6.2, 6.3, 6.4; Ch. 7: 7.1, 7.2, 7.4; Ch. 8: 8.1; Ch. 9: 9.1, 9.2, 9.3; Ch. 10: 10.1, 10.2, 10.3, 10.4; Ch. 11: 11.4; Ch. 12: 12.3, Ch. 12 Lab; Ch. 13: 13.4, Ch. 13 Lab; Ch. 14: 14.1, 14.3; Ch. 16: 16.2, 16.3, 16.4; Ch. 17: 17.1, 17.2; Ch. 18: 18.1, 18.2, 18.3, Ch. 18 Lab; Ch. 19: 19.1, 19.2, 19.3, Ch. 19 Lab; Ch. 20: 20.2; Ch. 21: 21.2, 21.3, 21.4; Ch. 22: 22.1, 22.2, 22.3, 22.4, Ch. 22 Lab; Ch. 23: 23.1, 23.2, 23.3, 23.4, 23.5; Ch. 24: 24.4; Ch. 25: 25.1, 25.2, Ch. 25 Lab; Ch. 26: 26.1, 26.2, 26.3; Ch. 27: 27.1, 27.2, 27.3, 27.4, Ch. 27 Lab; Ch. 28: 28.1, 28.2, 28.3, 28.4; Ch. 29: 29.1, 29.2; Ch. 30: 30.1, 30.3, 30.4; Ch. 31: 31.1, 31.2, 31.3, 31.4; Ch. 32: 32.1, 32.2, 32.3; Ch. 33: 33.1, 33.2, 33.3; Ch. 34: 34.1, 34.2, 34.3, 34.4, Ch. 34 Lab; Ch. 35: 35.1, 35.2, 35.4; Unit Projects: 2, 3, 6, 7, 8
II Evidence, models, and explanation	Ch. 1: 1.1, 1.2, 1.3; Ch. 2: 2.1, 2.4, Ch. 2 Lab; Ch. 3: 3.1, 3.3, 3.4; Ch. 4: 4.1, 4.2, 4.3; Ch. 5: 5.1, 5.3; Ch. 6: 6.2, 6.3, 6.4, Ch. 6 Lab; Ch. 7: Ch. 7 Lab; Ch. 8: 8.2; Ch. 10: 10.1, 10.2, 10.3, 10.4, Ch. 10 Lab; Ch. 11: 11.1, 11.2, 11.3, 11.4, Ch. 11 Lab; Ch. 12: 12.1, 12.2; Ch. 13: 13.1, 13.2, 13.3, 13.4, Ch. 13 Lab; Ch. 14: 14.1, 14.2, 14.3; Ch.15: 15.4; Ch. 16: 16.1, 16.2, 16.3, 16.4; Ch.17: 17.1, 17.2, 17.3, 17.4; Ch. 18: 18.2, 18.3; Ch. 19: 19.1, 19.2, 19.3, Ch. 19 Lab; Ch. 21: 21.1; Ch. 23: Ch. 23 Lab; Ch. 25: 25.1, 25.2; Ch. 26 26.1, 26.2, 26.3; Ch. 29: 29.1, 29.2, Ch. 29 Lab; Ch. 30: 30.1; Ch. 32: 32.1, 32.2, 32.3; Ch. 34; Ch. 35: 35.1, 35.3, Ch. 35 Lab; Unit Projects: 1, 2, 3, 4, 7, 8
III Change, constancy, and measurement	Ch. 1: 1.1, 1.3, Ch. 1 Lab; Ch. 2: 2.2, 2.4, Ch. 2 Lab; Ch. 3: 3.4; Ch. 4: 4.1, 4.2, 4.3, 4.4; Ch. 5: 5.1, 5.2, 5.3, Ch. 5 Lab; Ch. 6: 6.1, 6.2, 6.3, 6.4, Ch. 6 Lab; Ch. 7: 7.1, 7.3, 7.4; Ch. 8: 8.2, 8.3; Ch. 9: Ch. 9 Lab; Ch. 14: 14.3; Ch. 15: 15.1, 15.2, 15.3; Ch. 16: 16.4; Ch. 17: 17.1, 17.2, 17.3, 17.4; Ch. 19: 19.1, 19.2, 19.3; Ch. 23: Ch. 23 Lab; Ch. 24: Ch. 24 Lab; Ch. 26: 26.1, 26.2, 26.3, Ch. 26 Lab; Ch. 30: 30.4; Unit Projects: 1, 2
IV Evolution and equilibrium	Ch. 1: 1.3; Ch. 2: 2.4; Ch. 4: 4.3; Ch. 5: 5.1, 5.2, Ch. 5 Lab; Ch. 6: 6.3, 6.4; Ch. 7: 7.3, 7.4; Ch. 10: 10.1; Ch. 13: 13.3, 13.4; Ch. 14: 14.2; Ch. 15: 15.1; Ch. 16: 16.1, 16.2, 16.3, 16.4, Ch. 16 Lab; Ch. 17: 17.1, 17.2, 17.3, 17.4, 17.5, Ch. 17 Lab; Ch. 18: 18.1, 18.2, 18.3; Ch. 19: 19.1, 19.2, 19.3, Ch. 19 Lab; Ch. 21: 21.2, 21.3, 21.4; Ch. 22: 22.1, 22.2, 22.3, 22.4; Ch. 23: 23.4; Ch. 24: 24.3; Ch. 25: 25.1, 25.2; Ch. 26: 26.1, 26.2, 26.3; Ch. 27: 27.3; Ch. 28: 28.4; Ch. 29: 29.1, 29.2; Ch. 30: 30.1, 30.4; Ch. 34: 34.1, 34.2; Ch. 35: 35.3; Unit Projects: 5, 7
V Form and function	Ch. 1: 1.3; Ch. 2: 2.3, 2.4; Ch. 4: 4.4, 4.5; Ch. 7: 7.1, 7.2, 7.3, 7.4, Ch. 7 Lab; Ch. 8: 8.1, 8.3; Ch. 9: 9.1, 9.2; Ch. 12: 12.1, 12.2, 12.3; Ch. 13: 13.1, 13.2, 13.3, 13.4; Ch. 14: 14.2; Ch. 15: 15.1, 15.2, 15.3; Ch. 16: 16.2, 16.3, 16.4; Ch. 17: 17.3, 17.4; Ch. 19: 19.1, 19.2, 19.3; Ch. 20: 20.1, 20.2, 20.3; Ch. 21: 21.1, 21.2, 21.3, 21.4; Ch. 22: 22.2, 22.3, 22.4; Ch. 23: 23.1, 23.2, 23.3, 23.4, 23.5; Ch. 24: 24.1, 24.2; Ch. 25: 25.1, 25.2; Ch. 26: 26.2, 26.3; Ch. 27: 27.1, 27.2, 27.3, 27.4, Ch. 27 Lab; Ch. 28: 28.1, 28.2, 28.3, 28.4, Ch. 28 Lab; Ch. 30: 30.1, 30.2, 30.3, 30.4; Ch. 31: 31.1, 31.2, 31.3, 31.4, Ch. 31 Lab; Ch. 32: 32.1, 32.2, 32.3, Ch. 32 Lab; Ch. 33: 33.1, 33.2, 33.3, Ch. 33 Lab; Ch. 34: 34.1, 34.2, 34.3, 34.4; Ch. 35: 35.1, 35.2, 35.4; Unit Projects: 3, 7, 8

Standard	Chapter Lesson/Lab
CONTENT STANDARD A Science as Inquiry: A.1 Abilities Necessary to Do Scientific Inquiry	
A.1.a Identify questions and concepts that guide scientific investigations.	Ch. 6: Ch. 6 Lab; Ch. 8: 8.3; Ch. 10: 10.1; Ch. 11: 11.1; Ch. 13: 13.4; Ch. 16: 16.2; Ch. 21: 21.3, 21.4, Ch. 21 Lab; Ch. 22: 22.1; Ch. 23: 23.4, Ch. 23 Lab; Ch. 24; Ch. 25: 25.1; Ch. 26: Ch. 26 Lab; Ch. 28: 28.1, 28.4, Ch. 28 Lab; Ch. 29: 29.2, Ch. 29 Lab; Ch. 31: Ch. 31 Lab; Ch. 33: Ch. 33 Lab; Unit Project: 1
A.1.b Design and conduct scientific investigations.	Ch. 1: 1.2; Ch. 2: 2.1, 2.2, Ch. 2 Lab; Ch. 3: 3.1, 3.2; Ch. 4: 4.3; Ch. 5: 5.2; Ch. 6: 6.1, Ch. 6 Lab; Ch. 7: 7.1, 7.2; Ch. 8: 8.2; Ch. 9: 9.3; Ch. 10: 10.1, 10.2, Ch. 10 Lab; Ch. 11: 11.2; Ch. 12: 12.3, Ch. 12 Lab; Ch. 13: 13.2, 13.3; Ch. 14: 14.1, 14.3; Ch. 15: 15.2, 15.4; Ch. 16: 16.1, 16.2; Ch. 17: 17.2; Ch. 18: 18.2; Ch. 19: 19.1, Ch. 19 Lab; Ch. 20: 20.1, Ch. 20 Lab; Ch. 21: 21.1, 21.3, 21.4, Ch. 21 Lab; Ch. 22: 22.1, 22.4; Ch. 23: 23.1, 23.4, 23.5, Ch. 23 Lab; Ch. 24: 24.1, Ch. 24 Lab; Ch. 25: 25.1; Ch. 26: 26.3, Ch. 26 Lab; Ch. 27: 27.2, 27.4; Ch. 28: 28.1, 28.2; Ch. 29: 29.1, Ch. 29 Lab; Ch. 30: 30.1, 30.3; Ch. 31: 31.3; Ch. 32: 32.1, 32.2; Ch. 33: 33.1, 33.3, Ch. 33 Lab; Ch. 34: 34.3, 34.4; Ch. 35: 35.3; Unit Project: 1
A.1.c Use technology and mathematics to improve investigations and communications.	Ch. 1: 1.3, Ch. 1 Lab; Ch. 3: 3.3; Ch. 4: 4.1, 4.2, 4.3, 4.4; Ch. 5: 5.1, Ch. 5 Lab; Ch. 6: 6.2, 6.3; Ch. 7: 7.1, 7.2, 7.4, Ch. 7 Lab; Ch. 8: 8.3; Ch. 9: 9.1, 9.2, Ch. 9 Lab; Ch. 10: 10.1, 10.2, 10.3, 10.4; Ch. 11: 11.1, 11.2, 11.3, 11.4; Ch. 12: 12.2; Ch. 14: 14.1; Ch. 15: 15.3; Ch. 16: 16.2, 16.4; Ch. 17: 17.2; Ch. 19: 19.1, 19.3; Ch. 20: 20.1, 20.3; Ch. 21: 21.1, 21.3, 21.4; Ch. 22: 22.1, 22.2; Ch. 23: 23.3, 23.4; Ch. 24: 24.1, 24.2, 24.3, 24.4; Ch. 25: 25.2; Ch. 26: 26.3; Ch. 27: 27.1, Ch. 27 Lab; Ch. 28: 28.4; Ch. 29: 29.1, 29.2; Ch. 30: 30.1, 30.2, 30.4; Ch. 31: 31.4; Ch. 32: Ch 32.3; Ch. 33: 33.3, Ch. 33 Lab; Ch. 34: 34.4; Ch. 35: 35.2, 35.4, Ch. 35 Lab
A.1.d Formulate and revise scientific explanations and models using logic and evidence.	Ch. 2: 2.1; Ch. 3: 3.1, Ch. 3 Lab; Ch. 4: 4.3; Ch. 5: 5.1; Ch. 6: Ch. 6 Lab; Ch. 7: 7.2, 7.4; Ch. 10: 10.1; Ch. 11: Ch. 11 Lab; Ch. 12: 12.3; Ch. 13: 13.3, Ch. 13 Lab; Ch. 14: 14.1, 14.3; Ch. 15: 15.2; Ch. 18: Ch. 18 Lab; Ch. 20: 20.1; Ch. 21: 21.3; Ch. 22; Ch. 29: Ch. 29 Lab; Ch. 33: 33.1, Ch. 33 Lab
A.1.e Recognize and analyze alternative explanations and models.	Ch. 16: 16.2, Ch. 16 Lab; Ch. 17: 17.4; Ch. 19: 19.2; Ch. 34: Ch. 34 Lab
A.1.f Communicate and defend a scientific argument.	Ch. 4: 4.3; Ch. 23: Ch. 23 Lab; Unit Projects: 2, 5
A.2 Understandings About Scientific Inquiry	
A.2.a Scientists usually inquire about how physical, living, or designed systems function.	Ch. 1: 1.1, 1.3; Ch. 3: 3.1; Ch. 4: 4.2; Ch. 5: 5.1, 5.2, 5.3; Ch. 6: 6.1, 6.3, 6.4; Ch. 7: 7.1, 7.2, 7.3, 7.4; Ch. 8: 8.2, 8.3; Ch. 9: 9.1, 9.2, 9.3; Ch. 10: 10.1, 10.2, 10.3, 10.4; Ch. 11: 11.1, 11.2, 11.3, 11.4; Ch. 12: 12.1, 12.2; Ch. 13: 13.2, 13.4; Ch. 16: 16.1, 16.2, 16.3, 16.4; Ch. 17: 17.2, 17.3, Ch. 17 Lab; Ch. 19: 19.1, 19.2, 19.3; Ch. 20: 20.1, 20.2, 20.3, Ch. 20 Lab; Ch. 21: 21.1, 21.3, Ch. 21 Lab; Ch. 22: 22.1, 22.2, 22.3, 22.4; Ch. 23: 23.1, 23.2, 23.3, 23.4, 23.5; Ch. 24: 24.1, 24.2, 24.3; Ch. 25: 25.2; Ch. 27: 27.1, 27.2, 27.3, 27.4; Ch. 28: 28.1, 28.2, 28.3, 28.4; Ch. 29: 29.2, Ch. 29 Lab; Ch. 30: 30.1, 30.3, 30.4; Ch. 31: 31.1, 31.2, 31.3, 31.4; Ch. 32: 32.1, 32.2, 32.3; Ch. 33: 33.1, 33.2, 33.3; Ch. 34: 34.1, 34.2, 34.3, 34.4, Ch. 34 Lab; Ch. 35: 35.1, 35.2, 35.3, 35.4; Unit Projects: 1, 3, 7, 8

Standard	Chapter Lesson/Lab
A.2.b Scientists conduct investigations for a wide variety of reasons.	Ch. 1: 1.1, 1.2, 1.3; Ch. 3: 3.1; Ch. 8: Ch. 8 Lab; Ch. 11: 11.1, 11.2, 11.4; Ch. 12: 12.1, 12.2; Ch. 13: 13.4; Ch. 14: 14.3, Ch. 14 Lab; Ch. 15: 15.1; Ch. 16: 16.1, 16.4; Ch. 17: 17.3, Ch. 17 Lab; Ch. 18: 18.1; Ch. 19: 19.3; Ch. 21: 21.1, 21.4; Ch. 22: Ch. 22 Lab; Ch. 29: 29.1; Ch. 34: Ch. 34 Lab; Ch. 35: 35.3, Ch. 35 Lab
A.2.c Scientists rely on technology to enhance the gathering and manipulation of data.	Ch. 1: 1.1, 1.2, 1.3, Ch. 1 Lab; Ch. 7: 7.1; Ch. 8: Ch. 8 Lab; Ch. 9: Ch. 9 Lab; Ch. 12: 12.1, 12.2; Ch. 13: 13.4; Ch. 14: 14.3; Ch. 15: 15.2, 15.3; Ch. 16: 16.4; Ch. 20: 20.3; Ch. 35: Ch. 35 Lab
A.2.d Mathematics is essential in scientific inquiry.	Ch. 1: 1.1, 1.3, Ch. 1 Lab; Ch. 2: 2.4; Ch. 3: 3.1, 3.3; Ch. 4: 4.2, 4.4; Ch. 5: 5.1, 5.2, 5.3; Ch. 6: 6.2, 6.3; Ch. 9: 9.1; Ch. 11: 11.1, 11.2; Ch. 12: 12.2; Ch. 14: 14.1; Ch.16: 16.4; Ch. 17: 17.1, 17.2; Ch. 19: 19.1; Ch. 20: 20.3; Ch. 24: 24.4; Ch. 33: Ch. 33 Lab; Ch. 34: 34.4; Ch. 35: 35.4
A.2.e Scientific explanations must adhere to criteria such as: a proposed explanation must be logically consistent; it must abide by the rules of evidence; it must be open to questions and possible modification; and it must be based on historical and current scientific knowledge.	Ch. 1: 1.1, 1.2; Ch. 12: 12.1, 12.2; Ch. 14: 14.3; Ch. 15: 15.4; Ch. 16: 16.2, 16.3, 16.4; Ch. 17: 17.3, 17.4; Ch. 18: 18.3; Ch. 19: 19.3; Ch. 26: 26.3; Ch. 33: 33.2; Unit Projects: 2, 5
A.2.f Results of scientific inquiry—new knowledge and methods—emerge from different types of investigations and public communication among scientists. In communicating and defending the results of scientific inquiry, arguments must be logical and demonstrate connections between natural phenomena, investigations, and the historical body of scientific knowledge. In addition, the methods and procedures that scientists used to obtain evidence must be clearly reported to enhance opportunities for further investigation.	Ch. 1: 1.1, 1.2; Ch. 7: 7.1; Ch. 14: 14.3, Ch. 14 Lab; Ch. 15: 15.4; Ch. 16: 16.2, 16.3, 16.4; Ch. 18: 18.3; Ch. 17: 17.3, 17.4; Ch. 19: 19.3; Ch. 26: 26.3; Ch. 29 Lab; Ch. 33: 33.2; Ch. 35: 35.3
CONTENT STANDARD B Physical Science	
B.1 Structure of atoms	Ch. 2: 2.1, 2.2, 2.3; Ch. 19: 19.1
B.2 Structure and properties of matter	Ch. 2: 2.1, 2.2, 2.3; Ch. 3: 3.4; Ch. 13: 13.2; Ch. 14: 14.2, 14.3; Ch. 30: 30.2; Unit Project: 1
B.3 Chemical reactions	Ch. 2: 2.2, 2.4, Ch. 2 Lab; Ch. 8: 8.3; Ch. 9: 9.1, 9.2, 9.3, Ch. 9 Lab; Ch. 13: 13.1, 13.2, 13.4; Ch. 30: 30.2
B.4 Motions and forces	Ch. 2: 2.1, 2.2
B.5 Conservation of energy and the increase in disorder	Ch. 2: 2.4; Ch. 3: 3.3; Ch. 4: 4.1
B.6 Interactions of energy and matter	Ch. 4: 4.1; Ch. 8: 8.2, 8.3

Standard	Chapter Lesson/Lab
CONTENT STANDARD C Life Science: C.1 The Cell	
C.1.a Cells have particular structures that underlie their functions.	Ch. 1: 1.3; Ch. 2: 2.3, 2.4; Ch. 7: 7.1, 7.2, 7.3, 7.4, Ch. 7 Lab; Ch. 8: 8.3; Ch. 9: 9.1, 9.2, 9.3; Ch. 11: Ch. 11 Lab; Ch. 15: 15.2, 15.3; Ch. 19: 19.3; Ch. 21: 21.2; Ch. 23: 23.1; Ch. 27: 27.4; Ch. 30: 30.1; Ch. 31: 31.1; Ch. 32: 32.2; Ch. 33: 33.2; Ch. 34: 34.1; Ch. 35: 35.2, 35.3, 35.4; Unit Project: 3
C.1.b Most cell functions involve chemical reactions.	Ch. 2: 2.4; Ch. 7: 7.2, 7.3; Ch. 8: 8.1, 8.2, 8.3; Ch. 9: 9.1, 9.2, 9.3, Ch. 9 Lab; Ch. 20: 20.2; Ch. 27: 27.1; Ch. 30: 30.2, 30.3, Ch. 30 Lab; Ch. 32: 32.2; Unit Project: 3
C.1.c Cells store and use information to guide their functions.	Ch. 2: 2.3; Ch. 7: 7.2; Ch. 10: 10.1, 10.2; Ch. 12: 12.1, 12.2; Ch. 13: 13.2; Ch. 15: 15.2, 15.3, 15.4; Ch. 34: 34.1; Unit Project: 3
C.1.d Cell functions are regulated.	Ch. 2: 2.3, 2.4, Ch. 2 Lab; Ch. 7: 7.2, 7.4, Ch. 7 Lab; Ch. 10: 10.3, 10.4; Ch. 13: 13.4, Ch. 13 Lab; Ch. 14: 14.1, 14.2, 14.3; Ch. 15: 15.3; Ch. 24: 24.3, Ch. 24 Lab; Ch. 32: 32.2; Ch. 34: 34.1, 34.2; Ch. 35: 35.2; Unit Project: 3
C.1.e Plant cells contain chloroplasts, the site of photosynthesis.	Ch. 2: 2.4; Ch. 7: 7.2; Ch. 8: 8.2, 8.3, Ch. 8 Lab; Ch. 19: 19.3; Ch. 20: 20.2; Ch. 22: 22.1; Ch. 23: 23.4; Ch. 24: Ch. 24 Lab; Unit Project: 3
C.1.f Cells can differentiate, and complex multicellular organisms are formed as a highly organized arrangement of differentiated cells.	Ch. 7: 7.4; Ch. 10: 10.4; Ch. 13: 13.4; Ch. 15: 15.2; Ch. 16: 16.4; Ch. 23: 23.1, Ch. 23 Lab; Ch. 25: 25.1, 25.2, Ch. 25 Lab; Ch. 28: 28.1; Ch. 30: 30.1; Ch. 32: 32.1, 32.2, 32.3, Ch. 32 Lab; Ch. 34: 34.4; Unit Projects: 3, 8
C.2 The Molecular Basis of Heredity	
C.2.a In all organisms, the instructions for specifying the characteristics of the organism are carried in DNA, a large polymer formed from subunits of four kinds (A, G, C, and T).	Ch. 1: 1.3; Ch. 2: 2.3; Ch. 7: 7.2; Ch. 10: 10.1; Ch. 11: 11.1, 11.2, 11.3, 11.4; Ch. 12: 12.1, 12.2, 12.3, Ch. 12 Lab; Ch. 13: 13.1, 13.2; Ch. 14: 14.1, 14.2, 14.3; Ch. 15: 15.2, 15.3, 15.4, Ch. 15 Lab; Ch. 16: 16.4, Ch. 16 Lab; Ch. 17: 17.1, 17.4; Unit Project: 4
C.2.b Most of the cells in a human contain two copies of each of 22 different chromosomes.	Ch. 10: 10.2; Ch. 11: 11.1, 11.2, 11.3, 11.4; Ch. 12; Ch. 14: 14.1, 14.2, 14.3; Ch. 15: 15.1; Ch. 17: 17.1; Ch. 28: 28.3; Ch. 34: 34.3, 34.4
C.2.c Changes in DNA (mutations) occur spontaneously at low rates.	Ch. 12: 12.3; Ch. 13: 13.3; Ch. 14: 14.2, 14.3; Ch. 15: 15.1; Ch. 16: Ch. 16 Lab; Ch. 17: 17.1, 17.2, 17.3, 17.4
C.3 Biological Evolution	
C.3.a Species evolve over time.	Ch. 1: 1.3; Ch. 13: 13.3; Ch. 15: 15.1; Ch. 16: 16.1, 16.2, 16.3, 16.4; Ch. 17: 17.1, 17.2, 17.3, 17.4, Ch. 17 Lab; Ch. 18: 18.2, 18.3; Ch. 19: 19.1, 19.2, 19.3; Ch. 20: 20.2; Ch. 21: 21.1, 21.4; Ch. 22: 22.1; Ch. 25: 25.2; Ch. 26: 26.1, 26.2, 26.3; Ch. 27: 27.3; Ch. 28: 28.1, 28.3, 28.4; Ch. 29: 29.1, 29.2; Ch. 35: 35.3; Unit Project: 5
C.3.b The great diversity of organisms is the result of more than 3.5 billion years of evolution that has filled every available niche with life forms.	Ch. 16: 16.3; Ch. 17: 17.2, 17.3, 17.4; Ch. 18: 18.1, 18.2, 18.3; Ch. 19: 19.2, 19.3; Ch. 22: 22.1, 22.2, 22.3, 22.4; Ch. 24: 24.1; Ch. 25: 25.2; Ch. 26: 26.1, 26.2, 26.3; Ch. 29: 29.2; Unit Project: 5

Standard	Chapter Lesson/Lab
C.3.c Natural selection and its evolutionary consequences provide a scientific explanation for the fossil record of ancient life forms, as well as for the striking molecular similarities observed among the diverse species of living organisms.	Ch. 13: 13.4; Ch. 16: 16.3, 16.4; Ch. 17: 17.2, 17.4; Ch. 19: 19.1, 19.2, Ch. 19 Lab; Ch. 26: 26.1, 26.2, 26.3; Ch. 32: Ch. 32 Lab; Unit Project: 5
C.3.d The millions of different species of plants, animals, and microorganisms that live on Earth today are related by descent from common ancestors.	Ch. 16: 16.3, 16.4, Ch. 16 Lab; Ch. 17: 17.3, 17.4; Ch. 18: 18.2, 18.3; Ch. 19: 19.1, 19.2, 19.3; Ch. 21: 21.1; Ch. 22: 22.1, 22.2, 22.3, 22.4; Ch. 25: 25.1, 25.2; Ch. 26: 26.1, 26.2, 26.3; Unit Projects: 5, 6
C.3.e Biological classifications are based on how organisms are related.	Ch. 16: 16.3, Ch. 16 Lab; Ch. 17: 17.4; Ch. 18: 18.1, 18.2, 18.3, Ch. 18 Lab; Ch. 19: 19.2, 19.3; Ch. 20: 20.2; Ch. 21: 21.1; Ch. 22: 22.1, 22.2, 22.3, 22.4; Ch. 25: 25.1, 25.2; Ch. 26: 26.1, 26.2, 26.3, Ch. 26 Lab; Unit Project: 6
C.4 The Interdependence of Organisms	
C.4.a The atoms and molecules on the Earth cycle among the living and nonliving components of the biosphere.	Ch. 3: 3.3, 3.4, Ch. 3 Lab; Ch. 6: 6.4; Ch. 20: 20.2; Unit Project: 2
C.4.b Energy flows through ecosystems in one direction, from photosynthetic organisms to herbivores to carnivores and decomposers.	Ch. 3: 3.2, 3.3; Ch. 4: 4.2, 4.5; Ch. 6: 6.2; Ch. 8: 8.1; Ch. 20: 20.2; Ch. 21: 21.3, 21.4; Ch. 27: 27.1
C.4.c Organisms both cooperate and compete in ecosystems.	Ch. 4: 4.2, 4.3, 4.4, 4.5; Ch. 6: 6.3, 6.4; Ch. 17: Ch. 17 Lab; Ch. 21: 21.3, 21.4; Ch. 22: Ch. 22 Lab; Ch. 24: 24.2; Ch. 27: 27.1; Ch. 35: 35.1; Unit Project: 2
C.4.d Living organisms have the capacity to produce populations of infinite size, but environments and resources are finite.	Ch. 3: 3.4, Ch. 3 Lab; Ch. 4: 4.2; Ch. 5: 5.1, 5.2, 5.3, Ch. 5 Lab; Ch. 6: 6.1, 6.2, 6.3, 6.4; Ch. 16: 16.2, 16.3
C.4.e Human beings live within the world's ecosystems.	Ch. 3: 3.4, Ch. 3 Lab; Ch. 4: 4.3; Ch. 5: 5.1, 5.2, 5.3; Ch. 6: 6.1, 6.2, 6.3, 6.4; Ch. 22: Ch. 22 Lab; Ch. 24: 24.4; Ch. 35: 35.3; Unit Project: 2
C.5 Matter, Energy, and Organization in Living Systems	
C.5.a All matter tends toward more disorganized states.	Ch. 3: 3.2, 3.3; Ch. 8: 8.1, 8.2, 8.3; Ch. 9: 9.1; Ch. 30: 30.2
C.5.b The energy for life primarily derives from the sun.	Ch. 1: 1.3; Ch. 2: 2.3, 2.4; Ch. 3: 3.2, 3.3, Ch. 3 Lab; Ch. 4: 4.5; Ch. 8: 8.1, 8.2, 8.3, Ch. 8 Lab; Ch. 9: Ch. 9 Lab
C.5.c The chemical bonds of food molecules contain energy.	Ch. 2: 2.4; Ch. 8: 8.1; Ch. 9: 9.1, 9.2, 9.3; Ch. 30: 30.2
C.5.d The complexity and organization of organisms accommodate the need for obtaining, transforming, transporting, releasing, and eliminating the matter and energy used to sustain the organism.	Ch. 6: Ch. 6 Lab; Ch. 7: 7.4; Ch. 8: 8.1; Ch. 9: 9.1, 9.2, 9.3; Ch. 21; 21.3, 21.4, Ch. 21 Lab; Ch. 22: 22.1, 22.2, 22.3; Ch. 23: 23.1, 23.2, 23.3, 23.4, 23.5; Ch. 24: 24.3; Ch. 25: 25.1, 25.2, Ch. 25 Lab; Ch. 27: 27.1, 27.2, 27.3, 27.4, Ch. 27 Lab; Ch. 28: 28.1, 28.2, 28.3, 28.4, Ch. 28 Lab; Ch. 30: 30.1, 30.3, 30.4; Ch. 32: 32.1; Ch. 33: 33.1, 33.2, 33.3; Ch. 34: 34.2, 34.4; Unit Projects: 7, 8

Standard	Chapter Lesson/Lab
C.5.e The distribution and abundance of organisms and populations in ecosystems are limited by the availability of matter and energy and the ability of the ecosystem to recycle materials.	Ch. 3: 3.2, 3.3, 3.4, Ch. 3 Lab; Ch. 4: 4.2, 4.5; Ch. 6: 6.1, 6.2, 6.3, 6.4; Ch. 22: Ch. 22 Lab
C.5.f As matter and energy flow through different levels of organization of living systems—cells, organs, organisms, communities—and between living systems and the physical environment, chemical elements are recombined in different ways.	Ch. 3: 3.2, 3.3, 3.4
C.6 The Behavior of Organisms	
C.6.a Multicellular animals have nervous systems that generate behavior.	Ch. 25: 25.1; Ch. 26: 26.3; Ch. 28: 28.1; Ch. 29: 29.2; Ch. 31: 31.1, 31.2, 31.3, 31.4, Ch. 31 Lab; Unit Project: 7
C.6.b Organisms have behavioral responses to internal changes and to external stimuli.	Ch. 1: 1.3; Ch. 4: 4.4; Ch. 6: 6.4; Ch. 24: 24.3; Ch. 25: 25.1; Ch. 27: 27.1; Ch. 28: 28.1, 28.3; Ch. 29: 29.1, 29.2, Ch. 29 Lab; Ch. 31: 31.1, 31.2, 31.3, 31.4; Ch. 35: 35.1; Unit Project: 7
C.6.c Like other aspects of an organism's biology, behaviors have evolved through natural selection.	Ch. 16: 16.3; Ch. 17: 17.3, Ch. 17 Lab; Ch. 26: 26.3; Ch. 28: 28.3; Ch. 29: 29.1, 29.2
C.6.d Behavioral biology has implications for humans, as it provides links to psychology, sociology, and anthropology.	Ch. 26: 26.3; Ch. 29: 29.1
CONTENT STANDARD D Earth and Space Science	
D.1 Energy in the Earth system	Ch. 3: 3.2, 3.3; Ch. 4: 4.1, 4.4
D.2 Geochemical cycles	Ch. 3: 3.4
D.3 The origin and evolution of the Earth system	Ch. 16: 16.2, 16.4; Ch. 19: 19.1, 19.3, Ch. 19 Lab
D.4 The origin and evolution of the universe	Ch. 19: 19.3
CONTENT STANDARD E Science and Technology	
E.1 Abilities of technological design	Ch. 15: 15.1, 15.2, 15.3
E.2 Understandings about science and technology	Ch. 1: 1.2, 1.3; Ch. 2: 2.1; Ch. 13: 13.4; Ch. 15: 15.1, 15.2, 15.3, 15.4, Ch. 15 Lab; Ch. 16: 16.4

Standard	Chapter Lesson/Lab
CONTENT STANDARD F Science in Personal and Social Perspectives	
F.1 Personal and community health	Ch. 13: 13.3; Ch. 20: 20.1, 20.3, Ch. 20 Lab; Ch. 21: 21.3; Ch. 30: 30.2, 30.4, Ch. 30 Lab; Ch. 31: 31.2, Ch. 31 Lab; Ch. 32: 32.1, 32.2, 32.3; Ch. 33: 33.1, 33.2, 33.3, Ch. 33 Lab; Ch. 34: 34.2, 34.3, 34.4, Ch. 34 Lab; Ch. 35: 35.1, 35.2, 35.3, 35.4, Ch. 35 Lab; Unit Project: 1
F.2 Population growth	Ch. 5: 5.1, 5.2, 5.3, Ch. 5 Lab; Ch. 6: 6.3, 6.4; Ch. 16: 16.2
F.3 Natural resources	Ch. 3: 3.4; Ch. 4: Ch. 4 Lab; Ch. 5: 5.3; Ch. 6: 6.1, 6.2, 6.3, 6.4, Ch. 6 Lab; Ch. 16: 16.2; Ch. 24: 24.4; Unit Project: 2
F.4 Environmental quality	Ch. 3: 3.4, Ch. 3 Lab; Ch. 4: Ch. 4 Lab; Ch. 6: 6.1, 6.2, 6.3, 6.4, Ch. 6 Lab; Ch. 13: 13.3; Ch. 15: 15.4; Ch. 22: Ch. 22 Lab; Ch. 24: 24.4; Unit Project: 2
F.5 Natural and human-induced hazards	Ch. 3: Ch. 3 Lab; Ch. 4: 4.3; Ch. 6: 6.1, 6.2, 6.3, 6.4, Ch. 6 Lab; Ch. 24: 24.4; Ch. 35: 35.3, Ch. 35 Lab; Unit Project: 2
F.6 Science and technology in local, national, and global challenges	Ch. 1: 1.1, 1.2; Ch. 5: 5.3; Ch. 6: 6.1, 6.2, 6.3, 6.4; Ch. 14: 14.3; Ch. 15: 15.4; Ch. 24: 24.4; Ch. 35: 35.3; Unit Projects: 2, 4
CONTENT STANDARD G History and Nature of Science	
G.1 Science as a human endeavor	Ch. 1: 1.1, 1.2, 1.3; Ch. 3: 3.1; Ch. 6: 6.4; Ch. 11: 11.1; Ch. 12: 12.1, 12.2; Ch. 13: 13.4; Ch. 14: 14.3, Ch. 14 Lab; Ch. 15: 15.1, 15.2, 15.3; Ch. 16: 16.1, 16.2, 16.3, 16.4; Ch. 17: 17.3; Ch. 18: 18.1; Ch. 20: 20.3; Unit Project: 4
G.2 Nature of scientific knowledge	Ch. 1: 1.1, 1.2; Ch. 3: 3.1; Ch. 4: 4.3; Ch. 5: 5.3; Ch. 6: 6.4; Ch. 11: 11.2, 11.3; Ch. 12: 12.1, 12.2; Ch. 16: 16.1, 16.2, 16.3, 16.4; Ch. 18: 18.1; Ch. 26: 26.3
G.3 Historical perspectives	Ch. 1: 1.1, 1.2; Ch. 11: 11.1; Ch. 12: 12.1, 12.2; Ch. 13: 13.2; Ch. 15: 15.1, 15.2, 15.3; Ch. 16: 16.1, 16.2, 16.3, 16.4; Ch. 17: Ch. 17 Lab; Ch. 18: 18.1; Ch. 19: 19.3; Ch. 20: 20.1; Ch. 21: 21.1; Ch. 26: 26.3; Ch. 35: 35.3

Inquiry Into Scientific Thinking Materials List

Item	*Quantity	Lab
Apron, lab	1 per student	for all appropriate labs
Artichoke	1	23.1
Balance	1	27.4
Beans red white	 1 3	 14.1 14.1
Bell peppers, assorted colors	1	16.2
Blocks	10	1.2
Chicken wing, raw and bleached	1	32.2
Container	3	6.1
Coverslips	1 box per class	21.1, 24.1
Culture, live *Daphnia* Hydra Protist, mixed	 1 1 See investigation.	 25.1 25.1 21.1
Cups paper, large plastic, large and clear	 1 2 1 2	 19.1 5.2 8.2 14.1
Dropper pipette	1	21.1, 24.1, 25.1
Elodea plant, freshly cut	1	8.2
Flower	1	24.1
Forceps	1	24.1, 32.2
Fruits	See investigation.	22.4
Fruit juices, variety	See investigation.	2.2
Gloves, plastic	1 pair per student	for all appropriate labs
Goggles	1 pair per student	for all appropriate labs
Graduated cylinder 25-mL	1	27.4
Hand lens	1	22.4, 24.1
Index card	1	2.1
Knife	1	16.2, 23.1
Marker, black	1	5.2, 14.1
Methyl cellulose	1 drop	21.1

Item	*Quantity	Lab
Microscope compound	 1	 7.1, 21.1, 21.4, 23.4, 25.1, 33.3
Onion, raw	1	23.1
Paper, construction blue, gray, purple, green, red, yellow graph grid, 1-cm	 1 sheet of each 1 sheet 1 sheet	 12.3 33.3 19.1
Paper towels, white	1 roll per lab	32.2
Pencil, glass-marking	1	27.4
Petri dish	3–5	22.4
Petroleum jelly	1 jar per lab	33.3
Popcorn kernels	1 bag	2.1
Scalpel	1	22.4, 23.1, 24.1, 32.2
Scissors	1	12.3, 19.1, 32.2, 33.3
Seeds bean	 18	 5.2
Slides depression	1 5 1 box per class	21.1, 24.1, 33.3 25.1
Slides, prepared bacteria leaf, stem, or cross section nerve cells paramecia	 1 1 1 1	 7.1 7.1 7.1 7.1
Sodium bicarbonate solution (5g/L)	250 mL	8.2
Soil, potting	1 large bag per class	5.2
Sweet potato	1	23.1
Tape, transparent	1	12.3, 33.3
Test tube, large	1 2	8.2 27.4
Test-tube rack	1	27.4
Test-tube stopper, large	2	27.4
Trash, dry	See investigation.	6.1
Urea	2 grams	27.4
Uric acid	2 grams	27.4
Water pond	 1 drop	 21.1

***Quantities per group**

Live organisms and lab supplies are available from Science Kit. Order at www.sciencekit.com or call 800-828-7777.

Miller & Levine Biology

FOUNDATION EDITION

Kenneth R. Miller, Ph.D.

Professor of Biology, Brown University
Providence, Rhode Island

Joseph S. Levine, Ph.D.

Science Writer and Producer
Concord, Massachusetts

PEARSON

Boston, Massachusetts • Chandler, Arizona • Glenview, Illinois • Upper Saddle River, New Jersey

Print Components

Student Foundation Edition

Teacher's Foundation Edition

Study Workbook A

Study Workbook A, Teacher's Edition

Study Workbook B: Reading Foundations

Study Workbook B: Reading Foundations, Teacher's Edition

Laboratory Manual A

Laboratory Manual A, Teacher's Edition

Laboratory Manual B: Skill Foundations

Laboratory Manual B: Skill Foundations, Teacher's Edition

Probeware Lab Manual

Assessment Resources

Transparencies

Technology Components

Biology.com

Untamed Science™ Video Series: BioAdventures DVD

Classroom Resources DVD-ROM

ExamView CD-ROM

Virtual BioLab DVD-ROM with Lab Manual

English Language Learners

Teacher's ELL Handbook

Multilingual Glossary

Credits appear on pages C-1–C-3, which constitute an extension of this copyright page.

PEARSON

ISBN-13: 978-0-13-366961-9

ISBN-10: 0-13-366961-0

3 4 5 6 7 8 9 10 V057 14 13 12 11 10

About the Authors

Kenneth R. Miller grew up in Rahway, New Jersey, attended the local public schools, and graduated from Rahway High School in 1966. Miller attended Brown University on a scholarship and graduated with honors. He was awarded a National Defense Education Act fellowship for graduate study, and earned his Ph.D. in Biology at the University of Colorado. Miller is professor of Biology at Brown University in Providence, Rhode Island, where he teaches courses in general biology and cell biology.

Miller's research specialty is the structure of biological membranes. He has published more than 70 research papers in journals such as *CELL*, *Nature*, and *Scientific American*. He has also written the popular trade books *Finding Darwin's God* and *Only a Theory*. He is a fellow of the American Association for the Advancement of Science.

Miller lives with his wife, Jody, on a small farm in Rehoboth, Massachusetts. He is the father of two daughters, one a wildlife biologist and the other a high-school history teacher. He swims competitively in the masters' swimming program and umpires high school and collegiate softball.

Joseph S. Levine was born in Mount Vernon, New York, where he attended public schools. He earned a B.S. in Biology at Tufts University, a master's degree from the Boston University Marine Program, and a Ph.D. at Harvard University. His research has been published in scientific journals ranging from *Science* to *Scientific American*, and in several academic books. He has taught introductory biology, ecology, marine biology, neurobiology, and coral reef biology at Boston College and in the Boston University Marine Program. He has also co-taught a field biology course for high-school teachers entitled "Rainforests and Reefs" at the Organization for Tropical Studies in Costa Rica.

After receiving a Macy Fellowship in Science Broadcast Journalism at WGBH-TV, Levine dedicated himself to improving public understanding of science. His popular scientific writing has appeared in five trade books and in magazines such as *Smithsonian*, *GEO*, and *Natural History*. He has produced science features for National Public Radio and has designed exhibit programs for state aquarium projects in Texas, New Jersey, and Florida. Since 1987, Levine has served as scientific advisor at WGBH, where he worked on NOVA programs including *Judgment Day*, and on projects including the OMNI-MAX films *Cocos: Island of Sharks* and *Coral Reef Adventure*. He also served as science editor for the PBS series *The Secret of Life* and *The Evolution Project*.

Levine and his family live in Concord, Massachusetts, a short distance from Thoreau's Walden Pond.

Build a Solid Foundation

The Foundation Edition of *Miller & Levine Biology* is a new type of program. It offers the same content as *Miller & Levine Biology* made more accessible with:

- **Embedded reading support**
- **Inquiry** activities that promote scientific thinking
- **Engaging visuals** that enhance instruction
- **Adapted ancillary support**
- **Digital instruction for the next-generation,** built for 21st Century learners

Reading Foundations

The Foundation Edition provides enhanced reading support with embedded reading strategies to make content more accessible.

Build Understanding provides suggestions on how to use graphic organizers as a framework for learning.

Build Vocabulary boxes highlight academic and scientific vocabulary from the chapter.

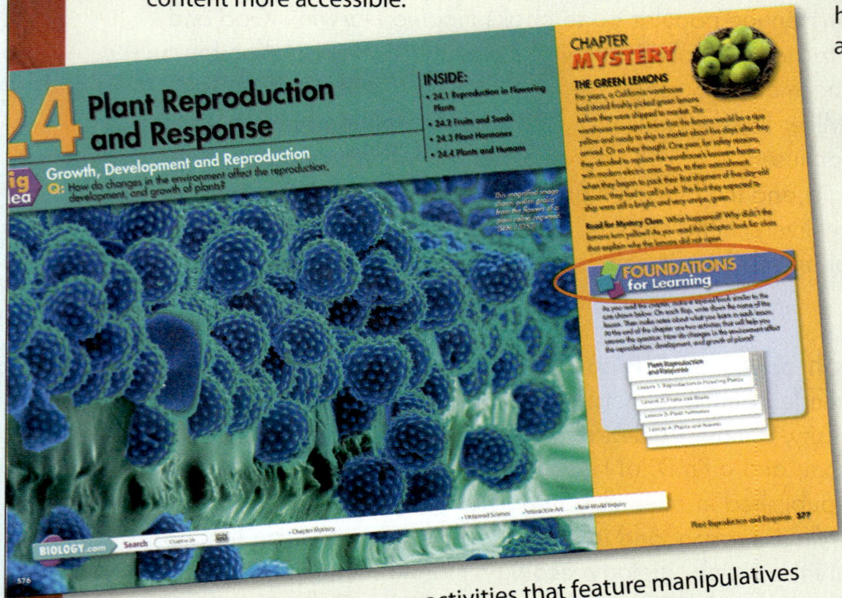

Foundations for Learning are activities that feature manipulatives you can use to organize key concepts as you read a chapter.

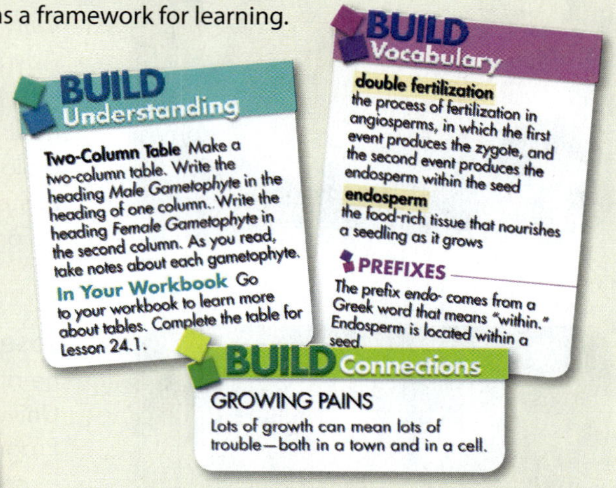

Build Connections are chapter features that help you create a connection between Big Ideas, putting science in a real-world context tied to a clear and easy-to-follow graphic.

Visuals Enhance Instruction

Stunning instructional visuals throughout each chapter offer a unique way to see what science looks like.

Visual analogies offer a way to relate key concepts to things you may see in your daily life. Visual summaries tie together the key concepts in a clear and easy-to-follow graphic. The magazine-style Visual Guide to the Diversity of Life provides a unique way to explore Earth's major organisms with key characteristics, habitats, behaviors, and other important facts.

Student Support Ensures Success

Biology is not a "one-size-fits-all" subject. *Miller & Levine Biology* provides content in ways that fit your unique learning style. The *Foundation* program offers workbooks that are written at an accessible reading level and work hand-in-hand with the *Foundation* Student Edition to help provide extra support to ensure content mastery. The accompanying ELL handbook also provides support for ELL classrooms.

The *Foundation Edition* supports all students with 2 levels of resources.

Study Workbook B

The adapted Study Workbook for the *Foundation Edition* of *Miller & Levine Biology* will help you:

- preview a chapter
- work with the key vocabulary
- practice strategies for learning the content
- review for chapter and unit tests
- learn strategies for taking standardized tests

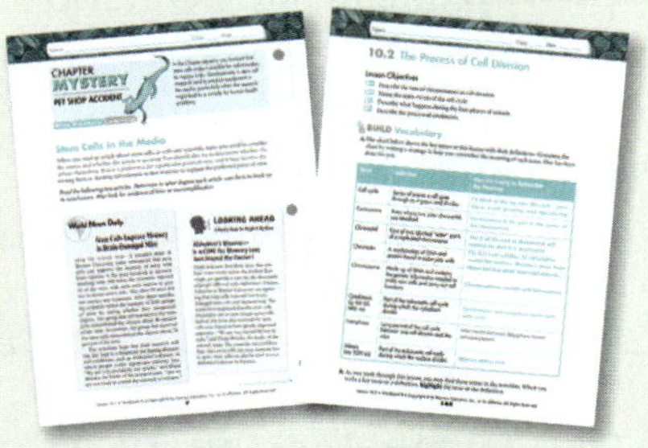

Lab Manual B

The adapted Lab Manual provides connections from the text to help you conduct lab experiments. You are encouraged to apply concepts, analyze data, and draw conclusions from lab activities. The *Extend Your Inquiry* feature encourages you to consider concepts outside of the step-by-step laboratory process.

Biology.com—Your Online Habitat for Learning

Featuring the latest in instructional technology, Biology.com is an engaging program that brings biology concepts to life. Visual, interactive, and differentiated components help reinforce content and provide a broad understanding of scientific concepts.

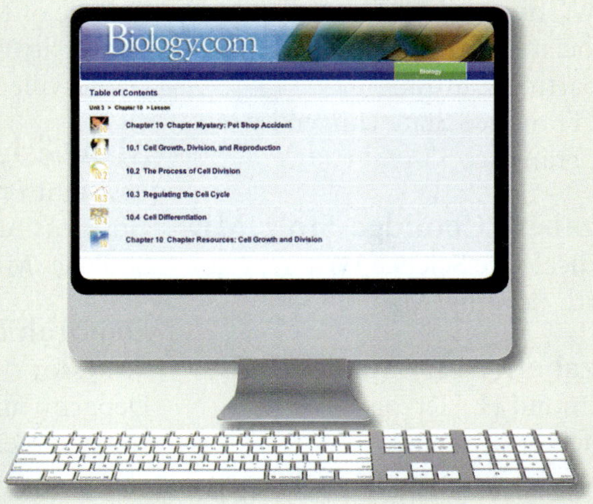

Biology.com offers:

For Students:

- Complete *Foundation* Student Edition with audio helps you read along online
- Interactive visuals, tutorials, and simulations help you build connections as concepts come alive online.
- Scaffolded self-assessment prepares you for content mastery and end-of-course testing.

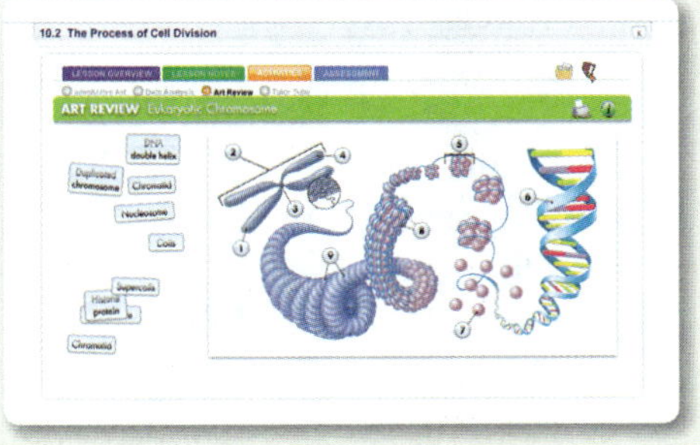

For Teachers:

- Lesson planner with leveled activities crafts lessons to support all levels of students.
- Lesson overview presentations help present the Big Ideas of each lesson
- Comprehensive, auto-graded assessment and remediation for different levels of students
- Editable worksheets that allow teachers to cater their lessons to your classroom's needs

BIOLOGY.com Search Lesson 22.4 GO • Art Review

Consultants/Reviewers

Grant Wiggins, E.d.D. is a co-author of Understanding by Design® (UbD), a philosophy of instructional design. UbD is a disciplined way of thinking about curriculum design, assessment, and instruction that moves teaching from covering the content to ensuring understanding.

Big idea Big Ideas are one of the core components of UbD in *Miller & Levine Biology.* These Big Ideas, such as Science as a Way of Knowing and the Cellular Basis of Life, establish a conceptual framework for the program. In the Student Edition, look for opportunities throughout each chapter to link back to the Big Ideas. And, since Understanding by Design is by nature a teaching tool, see the Teacher's Edition for additional applications of this philosophy.

Jim Cummins is Professor and Canada Research Chair in the Curriculum, Teaching and Learning department at the Ontario Institute for Studies in Education at the University of Toronto. His research focuses on literacy development in multilingual schools and the role of technology in promoting language and literacy development.

Program materials for *Miller & Levine Biology* incorporate research-based essential principles using Dr. Cummins's Into/Through/Beyond structure. You will find ample support for ELL instruction in the Teacher's Edition, Teacher's ELL Handbook, the Multilingual Glossary, as well as the Spanish components offered with this program.

Content Reviewers

Lily Chen
Associate Professor
Department of Biology
San Francisco State University
San Francisco, CA

Elizabeth Coolidge-Stolz, MD
Medical/Life Science Writer/Editor
North Reading, MA

Elizabeth A. De Stasio, Ph.D.
Raymond H. Herzog
Professor of Science
Associate Professor of Biology
Lawrence University
Appleton, WI

Jennifer C. Drew, Ph.D.
Lecturer/Scientist
University of Florida
Kennedy Space Center, FL

Donna H. Duckworth, Ph.D.
Professor Emeritus
College of Medicine
University of Florida
Gainesville, FL

Alan Gishlick, Ph.D.
Assistant Professor
Gustavus Adolphus College
St. Peter, MN

Deborah L. Gumucio, Ph.D.
Professor
Department of Cell and
Developmental Biology
University of Michigan
Ann Arbor, MI

Janet Lanza, Ph.D.
Professor of Biology
University of Arkansas
at Little Rock
Little Rock, AR

Charles F. Lytle, Ph.D.
Professor of Zoology
North Carolina State University
Raleigh, NC

Martha Newsome, DDS
Adjunct Instructor of Biology
Cy-Fair College, Fairbanks Center
Houston, TX

Jan A. Pechenik, Ph.D.
Professor of Biology
Tufts University
Medford, MA

Imara Y. Perera, Ph.D.
Research Assistant, Professor
Department of Plant Biology
North Carolina State University
Raleigh, NC

Daniel M. Raben, Ph.D.
Professor
Department of Biological
Chemistry
Johns Hopkins University
Baltimore, MD

Megan Rokop, Ph.D.
Educational Outreach Program Director
Broad Institute of MIT and Harvard
Cambridge, MA

Gerald P. Sanders
Former Biology Instructor
Grossmont College
Julian, CA

Ronald Sass, Ph.D.
Professor Emeritus
Rice University
Houston, TX

Linda Silveria, Ph.D.
Professor
University of Redlands
Redlands, CA

Richard K. Stucky, Ph.D.
Curator of Paleontology and Evolution
Denver Museum of Nature and Science
Denver, CO

Robert Thornton, Ph.D.
Senior Lecturer Emeritus
Department of Plant Biology
College of Biological Sciences
University of California at Davis
Davis, CA

Edward J. Zalisko, Ph.D.
Professor of Biology
Blackburn College
Carlinville, IL

ESL Lecturer

Nancy Vincent Montgomery, Ed.D.
Southern Methodist University
Dallas, TX

High-School Reviewers

Christine Bill
Sayreville War Memorial High School
Parlin, NJ

Jean T. (Caye) Boone
Central Gwinnett High School
Lawrenceville, GA

Samuel J. Clifford, Ph.D.
Biology Teacher
Round Rock High School
Round Rock, TX

Jennifer Collins, M.A.
South County Secondary School
Lorton, VA

Roy Connor, M.S.
Science Department Head
Muncie Central High School
Muncie, IN

Norm Dahm, Jr.
Belleville East High School
Belleville, IL

Cora Nadine Dickson
Science Department Chair
Jersey Village High School
Cypress Fairbanks ISD
Houston, TX

Dennis M. Dudley
Science Department Chair/Teacher
Shaler Area High School
Pittsburgh, PA

Mary K. Dulko
Sharon High School
Sharon, MA

Erica Everett, M.A.T., M.Ed.
Science Department Chair
Manchester-Essex Regional High School
Manchester, MA

Heather M. Gannon
Elisabeth Ann Johnson High School
Mt. Morris, MI

Virginia Glasscock
Science Teacher
California High School
Whittier, CA

Ruth Gleicher
Biology Teacher
Niles West High School
Skokie, IL

Lance Goodlock
Biology Teacher/Science Department Chairperson
Sturgis High School
Sturgis, MI

W. Tony Heiting, Ph.D.
State Science Supervisor (retired)
Iowa Department of Education
Panora, IA

Patricia Anne Johnson, M.S.
Biology Teacher
Ridgewood High School
Ridgewood, NJ

Judith Decherd Jones, M.A.T.
NBCT AYA Science
East Chapel Hill High School
Chapel Hill, NC

Shellie Jones
Science Teacher
California High School
Whittier, CA

Michelle Lauria, M.A.T.
Biology Teacher
Hopkinton High School
Hopkinton. MA

Kimberly Lewis
Science Department Chair
Wellston High School
Wellston, OH

Consultants *(continued)*

Lenora Lewis
Teacher
Creekview High School
Canton, GA

JoAnn Lindell-Overton, M.Ed.
Supervisor of Secondary Science
Chesapeake Public Schools
Chesapeake, VA

Lender Luse
H.W. Byers High School
Holly Springs, MS

Molly J. Markey, Ph.D.
Science Teacher
Newton Country Day School of the Sacred Heart
Newton, MA

Rebecca McLelland-Crawley
Biological Sciences Teacher
Piscataway, NJ

Mark L. Mettert, M.S. Ed.
Science Department Chair
New Haven High School
New Haven, IN

Jane Parker
Lewisville High School North
Lewisville, TX

Ian Pearce
Educator
Austin, TX

Jim Peters
Science Resource Teacher
Carroll County Public Schools
Westminster, MD

Michelle Phillips, M.A.T.
Secondary Science: Education
Science Teacher
Jordan High School
Durham, NC

Randy E. Phillips
Science Teacher/Department Chair
Green Bay East High School
Green Bay, WI

Nancy Richey
Educator
Longmont, CO

Linda Roberson
Department Chairman
Jenks Freshman Academy
Jenks, OK

Sharon D. Spencer
Assistant Principal
Bronx Center for Science and Math
Bronx, NY

Stephen David Wright, M.S.
Biology Teacher
Montgomery County Public Schools
Columbia, MD

Alan W. Zimroth, M.S.
Science Teacher/Department Chairperson
Hialeah-Miami Lakes High School
Hialeah, FL

Contents

BIOLOGY.com • Go Digital. See what awaits you at Biology.com.

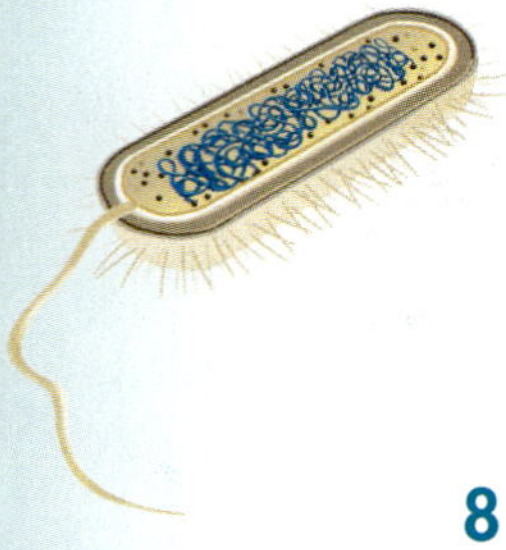

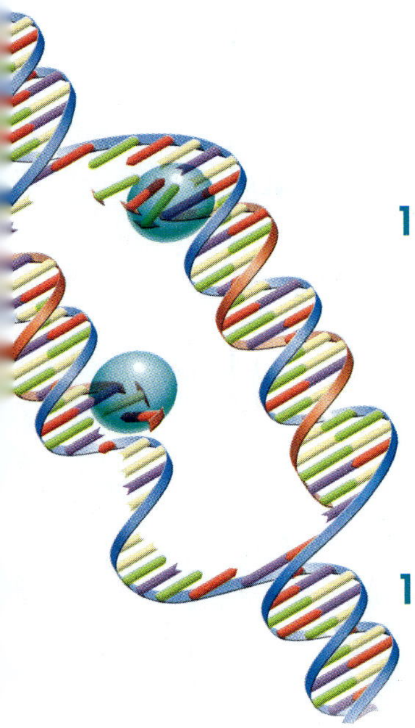

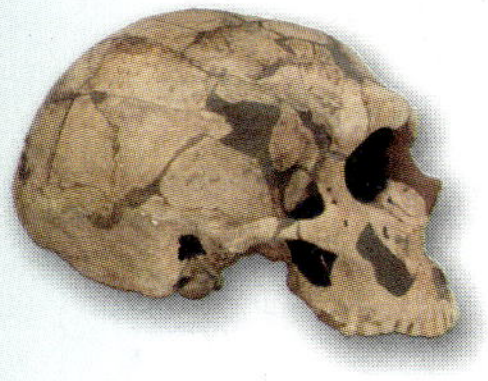

UNIT 6 From Microorganisms to Plants 477–602

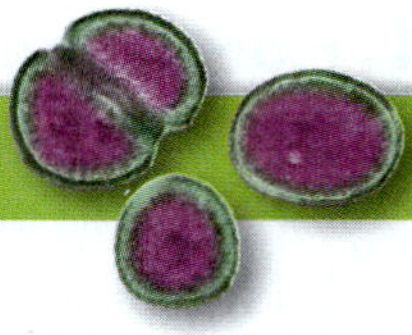

UNIT 7

Animals 603–710

BIOLOGY.com • Go Digital. See what awaits you at Biology.com.

BIOLOGY.com • Go Digital. See what awaits you at Biology.com.

Diversity of Life: A Visual Guide DOL • 1– DOL • 64

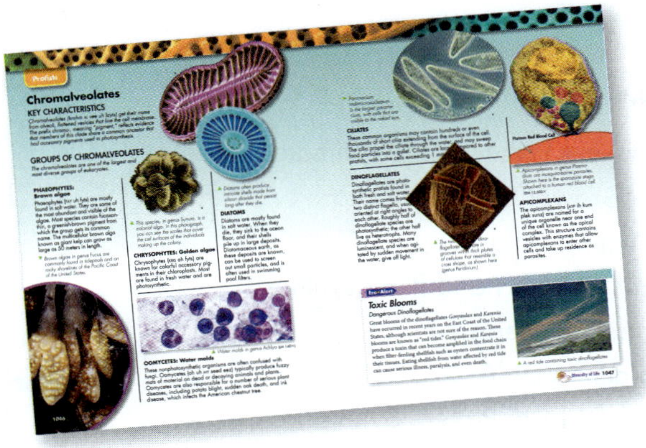

Labs and Activities

INQUIRY into Scientific Thinking

Labs and Activities *(continued)*

Design Your Own Lab

Forensics Lab

Real-World Lab

Skills Lab

BUILD Connections with Visual Analogies

Dear Student:

Welcome to our world—the endlessly fascinating world of biology.

I can guess what some of you are thinking right now. “Fascinating? Yeah, right. Totally.” Well, give us—and biology—a chance to show you that the study of the natural world really is more exciting, more fascinating, and more important to you personally than you’ve ever realized. In fact, biology is more important to our daily lives today than it has ever been.

Why? Three words: "We are one." This isn’t meant in a "touchy-feely" or "New Age" way. “We” includes all forms of life on Earth. And “are one” means that all of us are tied together more tightly, in more different ways, than anyone imagined until recently.

Both our “hardware” (body structures) and our “software” (genetic instructions and biochemical processes that program body functions) are incredibly similar to those of all other living things. Genetic instructions in our bodies are written in the same universal code as instructions in bacteria and palm trees. As biologists “read” and study that code, they find astonishingly similar processes in all of us. That’s why medical researchers can learn about human diseases that may strike you or your family by studying not only apes and pigs and mice, but even yeasts. We are one on the molecular level.

All organisms interact with one another and with the environment to weave our planet’s web of life. Organisms make rain forests and coral reefs, prairies and swamps—and farms and cities. We interact, too, with the winds and ocean currents that tie our planet together. Human activity is changing local and global environments in ways that we still don’t understand . . . and that affect our ability to produce food and protect ourselves from diseases. We are one ecologically with the rest of life on Earth.

All organisms evolve over time, adapting to their surroundings. If humans alter the environment, other organisms respond to that change. When we use antibiotics against bacteria, they develop resistance to our drugs. If we use pesticides against insects, they become immune to our poisons. We are one in our ability to evolve over time.

Those are the kinds of connections you will find in this book. Microscopic. Enormous. Amusing. Threatening. But always fascinating. That’s why—no matter where you start off in your attitude about biology—we think you are in for some surprises!

Sincerely,

Dear Student,

Biology is one of the subjects you're going to study this year, but I hope you'll realize from the very first pages of this book that biology is a lot more than just a "subject." Biology is what makes an eagle fly, a flower bloom, or a caterpillar turn into a butterfly. It's the study of ourselves—of how our bodies grow and change and respond to the outside world, and it's the study of our planet, a world transformed by the actions of living things. Of course, you might have known some of this already. But there's something more—you might call it a "secret" that makes biology unique.

That secret is that you've come along at just the right time. In all of human history, there has never been a moment like the present, a time when we stood so close to the threshold of answering the most fundamental questions about the nature of life. You belong to the first generation of students who can read the human genome almost as your parents might have read a book or a newspaper. You are the first students who will grow up in a world that has a chance to use that information for the benefit of humanity, and you are the very first to bear the burden of using that knowledge wisely.

If all of this seems like heavy stuff, it is. But there is another reason we wrote this book, and we hope that is not a secret at all. Science is fun! Biologists aren't a bunch of serious, grim-faced, middle-aged folks in lab coats who think of nothing but work. In fact, most of the people we know in science would tell you honestly, with broad grins on their faces, that they have the best jobs in the world. They would say there's nothing that compares to the excitement of doing scientific work, and that the beauty and variety of life make every day a new adventure.

We agree, and we hope that you'll keep something in mind as you begin the study of biology. You don't need a lab coat or a degree or a laboratory to be a scientist. What you do need is an inquiring mind, the patience to look at nature carefully, and the willingness to figure things out. We've filled this book with some of the latest and most important discoveries about living things, but we hope we've also filled it with something else: our wonder, our amazement, and our sheer delight in the variety of life itself. Come on in, and enjoy the journey!

Sincerely,

Ken Miller

The Nature of Life

UNIT 1

Chapters

INTRODUCE the Big ideas

- **Science as a Way of Knowing**
- **Matter and Energy**

"Science is 'a way of knowing'—a way of explaining the natural world through observations, questions, and experiments. But science isn't just dry old data, pressed between pages of this book like prom flowers in a school yearbook. Science is a living adventure story, aimed at understanding humans and the world around us. That story begins with the relationship between the matter that forms our bodies and the energy that powers life's processes."

Joe Levine

Dear Colleague,

Trying to cover all of biology in a single textbook is a terrifying, exhilarating, humbling, and sometimes overwhelming experience. But, if we had to pick a single word to describe our experience in writing this book, both of us would say "amazing."

Why? Because the more we learn about biology, the more we are astonished at how much science has progressed since our student days, and how much more we have yet to learn about the natural world. We wrote this book because we want to share with you and your students our amazement at scientific progress, our awe at the wonder of nature, and our delight with the way science works.

It's easy for students to get overwhelmed by the amount of information presented in an intro bio course. From molecules to the biosphere, there's an awful lot of stuff to learn. All this "stuff" can give students the impression that biology "is done"—that everything has been figured out. Scientists, a student once told us, are ashamed to admit that there's anything they don't know.

That student was dead wrong, and we wrote this book to help you correct that kind of misunderstanding. Scientists aren't interested in just memorizing what is known, but in using scientific methodology to explore what is not known. Beginning with Chapter 1, we've tried to present the "stuff" your students are required to know in a context that helps them share in the excitement that scientists feel as we investigate unsolved problems and unexplored territory in the living world.

If you find this book useful, we'll be happy. As teachers, you are the most important part of the scientific community. You are the nurturers of new talent, the caretakers of youthful curiosity. You can help invigorate the scientific enterprise with hope, energy, and vigor. We hope you'll share your thoughts, suggestions, and criticisms of this textbook with us, because we know we'll learn from them. And, most especially, we thank you for the honor of sharing your classroom with us.

Ken Miller

Joe Levine

The Science of Biology

Chapter Contents	NSES	Time	Core Resources
Chapter Preview			**Student Edition,** pp. 2–3 **Chapter Mystery,** p. 3 **Foundations for Learning,** p. 3
1.1 What Is Science? What Science Is and Is Not • Scientific Methodology • When Experiments Are Not Possible	I, II, III, A.2.a, A.2.b, A.2.c, A.2.d, A.2.e, F.6, G.1, G.2	1 period 1/2 block	**Student Edition,** pp. 4–8 **Study Workbook B** 1.1 Worksheets **Biology.com** *Art in Motion:* Experimental Design • *Art Review:* Revising Hypotheses • *InterActive Art:* Redi and Pasteur's Experiment • 1.1 Lesson Assessment
1.2 Science in Context Exploration and Discovery: Where Ideas Come From • Communicating Results: Reviewing and Sharing Ideas • Scientific Theories • Science and Society	II, A.1.b, A.2.b, A.2.c, A.2.e, A.2.f, E.2, F.6, G.1, G.2, G.3	1 period 1/2 block	**Student Edition,** pp. 9–12 **Inquiry Into Scientific Thinking,** p. 11 **Study Workbook B** 1.2 Worksheets **Biology.com** 1.2 Lesson Assessment
1.3 Studying Life Characteristics of Living Things • Big Ideas in Biology • Fields of Biology • Performing Biological Investigations	I, II, III, IV, V, A.2.a, A.2.b, A.2.c, A.2.d, C.1.a, C.2.a, C.3.a, C.5.b, C.6.b, E.2, G.1	1 period 1/2 block	**Student Edition,** pp. 13–19 **Build Connections,** pp. 14–15 **Study Workbook B** 1.3 Worksheets **Biology.com** 1.3 Lesson Assessment
Chapter Pre-Lab	III, A.1.c, A.2.c, A.2.d	1 period 1/2 block	**Student Edition,** p. 20 **Lab Manual B** *Using a Microscope to Estimate Size*

Tools for ELL Instruction

ELL Handbook describes teaching strategies to make *Miller & Levine Biology: Foundation Edition* more accessible to ELL students.

Spanish Study Workbook is a Spanish translation of Study Workbook A.

Multilingual Glossary is the *Miller & Levine Biology* glossary translated into ten languages.

Tools to Challenge Students

Study Workbook A includes more challenging worksheets and activities such as Chapter Mystery activities that reinforce 21st Century Skills.

Lab Manual A labs are geared toward on-level students. The Lab Manual includes fifteen labs in addition to the Chapter Labs.

Biology.com Data Analysis activities offer students a way to practice data gathering and analyzing skills.

Visual Quizzes in the Assessment Resources Book are geared toward on-level students and use visuals from *Miller & Levine Biology.*

Additional Resources

Biology.com Untamed Science Video • Flash Cards

Study Workbook A 1.1 Worksheets
Lab Manual B *Data Analysis:* Picturing Data • *Hands-On Activity:* Asking Questions Scientifically
Spanish Study Workbook 1.1 Worksheets
Assessment Resources Book Visual Quiz

Study Workbook A 1.2 Worksheets
Spanish Study Workbook 1.2 Worksheets

Study Workbook A 1.3 Worksheets
Lab Manual B *Data Analysis:* What's in a Diet?
Spanish Study Workbook 1.3 Worksheets
Biology.com *Data Analysis:* Adventures in Measurement
Assessment Resources Book Visual Quiz

Lab Manual A *Using a Microscope to Estimate Size*

Chapter Review

Student Edition Chapter Summary, p. 21
Study Workbook B Chapter 1 Review
Transparencies, pp. 1–12
Biology.com Untamed Science Video • Editable Worksheets of Study Workbooks A and B and Lab Manuals A and B • Chapter 1 Flash Cards and Crossword

Untamed Science DVD • Classroom Resources CD (includes lesson presentations and editable worksheets)

Chapter Assessment

Student Edition Check Understanding, pp. 22–25
Study Workbook B Chapter 1 Taking a Standardized Test
Assessment Resources Book Chapter 1 Test B
Biology.com Chapter 1 Assessment • Editable Worksheets of Chapter 1 Visual Quizzes* and Chapter 1 Tests A and B

Exam*View* *Assessment Suite* • Classroom Resources CD (includes lesson presentations and editable worksheets)

Time: 1 period, 1/2 block

* Visual Quizzes are based on visuals from *Miller & Levine Biology.* Review carefully to make sure they are appropriate for your class.

Pressed for Time?

Preview the Chapter Introduce the vocabulary for Lessons 1.1 and 1.3, and preview the Salt Marsh Experiment and Characteristics of Living Things visuals.

Cover the Chapter Quickly Have students read *Scientific Methodology* in Lesson 1.1, *Scientific Theories* in Lesson 1.2, and all of Lesson 1.3. Discuss the Characteristics of Living Things visual.

Assess Assign students Questions 1–3 and 5 in the 1.1 Check Understanding, Question 2 in the 1.2 Check Understanding, the 1.3 Check Understanding, and the Chapter 1 Standardized Test Prep, except Questions 1 and 3.

Get the Picture

Tell students that these paleontologists are using their senses to learn about a dinosaur that once lived on Earth. Explain that the word *observation* refers to the use of the senses to gather information.

Ask What senses are these paleontologists using in their observation of the dinosaur's skeleton? *(sight and touch)* What things do you notice about the dinosaur from looking at, or observing, this picture? *(Sample answers: It had many large, pointed teeth. It was bigger than a person.)*

Say Scientists gather facts about nature by observing. They also use prior knowledge and observation to make inferences. For example, paleontologists know from the study of zoology that animals with sharp pointed teeth eat meat. Therefore, we can infer that the dinosaur in this picture ate meat.

Ask What else, besides the dinosaur's diet, might be inferred from observations of its skull? *(Students should notice the large openings in the skull for eyes and nostrils, from which they might infer that the dinosaur probably had good eyesight and a good sense of smell.)* Help students realize that these senses would be advantages for finding animals to eat.

Activate Prior Knowledge

Ask What traits and skills do you think a good scientist should have? *(Sample answers: curiosity, asks questions, keen sense of observation, persistent, careful in doing work, can infer, good communication skills, adventurous)*

Discuss how these traits and skills are helpful to scientific work not only for scientists but for students studying science.

NATIONAL SCIENCE EDUCATION STANDARDS

UNIFYING CONCEPTS AND PROCESSES

I, II, III, IV, V

CONTENT

C.1.a, C.2.a, C.3.a, C.5.b, C.6.b, E.2, F.6, G.1, G.2, G.3

INQUIRY

A.1.b, A.1.c, A.2.a, A.2.b, A.2.c, A.2.d, A.2.e, A.2.f

1 The Science of Biology

Big idea Science as a Way of Knowing

Q: What role does science play in the study of life?

2

UbD Connect to the Big Idea

Invite a volunteer to read aloud the chapter number and the chapter title. Read the Big Idea and question. Emphasize the question: **What role does science play in the study of life?**

Set the Purpose

PREVIEW AND PREDICT Draw the graphic organizer on the board. Guide students to the INSIDE feature on the opposite page. Invite a volunteer to read each lesson number and title.

Ask How are the questions in the graphic organizer related to the Big Idea? *(Students' answers should reflect the idea that answering the questions helps you find specific information throughout the chapter about the study of life.)*

Encourage students to think about how biology relates to their lives as they read each lesson.

INSIDE:

- 1.1 What Is Science?
- 1.2 Science in Context
- 1.3 Studying Life

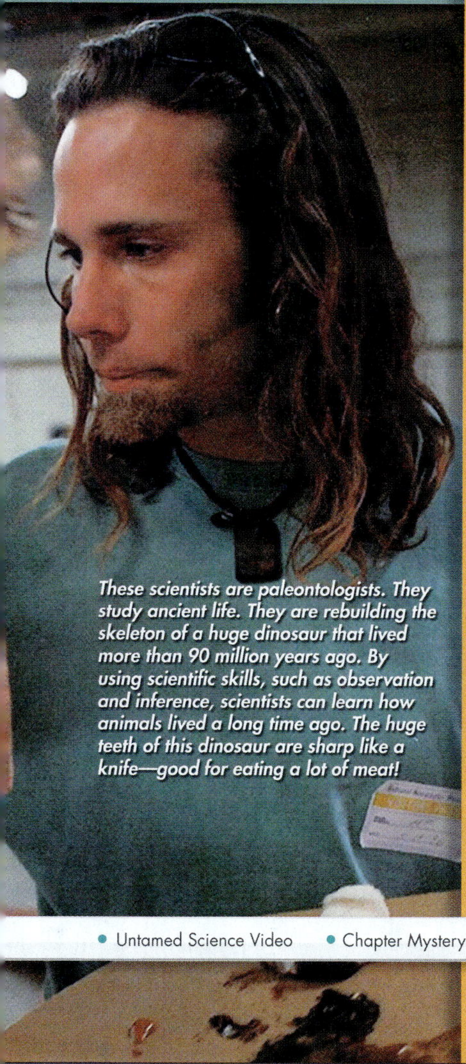

These scientists are paleontologists. They study ancient life. They are rebuilding the skeleton of a huge dinosaur that lived more than 90 million years ago. By using scientific skills, such as observation and inference, scientists can learn how animals lived a long time ago. The huge teeth of this dinosaur are sharp like a knife—good for eating a lot of meat!

- Untamed Science Video
- Chapter Mystery

CHAPTER MYSTERY

HEIGHT BY PRESCRIPTION

A doctor gives a shot of a powerful chemical to an eight-year-old boy named David. This boy is perfectly healthy. He shows no signs of being sick. The "problem" that he is being treated for is very common. He is short for his age.

The chemical David is being given is human growth hormone, or HGH. HGH, together with genes and diet, controls growth during childhood. Some people produce little or no HGH. These people are extremely short and may have other health problems. But David has a normal amount of HGH. He is short because his parents are both healthy, short people.

But if David isn't sick, why does his doctor prescribe HGH? Where does HGH come from? Is it safe? And what does its use in David's case say about science and society?

Read for Mystery Clues As you read this chapter, look for clues about the nature of science, the role of technology in our modern world, and the relationship between science and society. Then, solve the mystery.

FOUNDATIONS for Learning

All living things share certain characteristics. Biology is the study of these characteristics. Before you read the chapter, make an Inspired Shape Tree. Your tree should have a trunk and eight branches. Write "Characteristics of Living Things" on the trunk. Then, as you read Lesson 3, describe a different characteristic on each branch. At the end of the chapter are two activities that use your tree to help answer the question: What role does science play in the study of life?

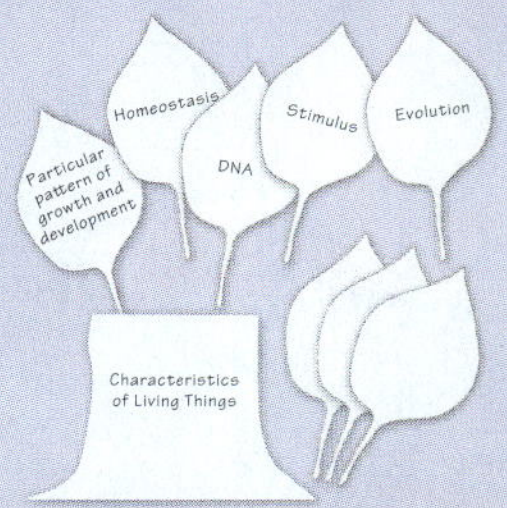

Foundations for Learning

Provide each student with a sheet of construction paper or card stock, scissors, and tape to use in making their Inspired Shape Trees. Explain that students will need a "tree trunk" and 8 thinner strips that will become the limbs or branches. Encourage students to be somewhat creative with the shapes of the trunk and branches, but be sure they understand that one or more sentences or phrases will need to be written on each branch.

Ask How do you think making an Inspired Shape Tree will help you learn the characteristics of living things? *(Sample answer: Thinking about something and writing it down will help me remember it. The tree also shows how things are related. The ideas on the branches are all characteristics of living things.)*

Encourage students to include as much detail as possible when they describe each characteristic. Tell students that their trees can be used as review and study aids.

Untamed Science Follow the Untamed Science crew in the video **What Do Biologists Look Like?** as they find the answers to the question "What is biology?" and dispel misconceptions about biologists.

CHAPTER MYSTERY Invite a volunteer to read aloud the Chapter Mystery.

Ask Do you think medicines such as HGH should be given to children who are not sick? Explain your answer. *(Accept all reasonable answers.)*

Have students predict how HGH affects human growth, and then have them write their prediction in their notebooks.

Tell students they will find clues in the chapter that will help them determine if their prediction is correct.

Digital Support

BIOLOGY.com Extend your reach by using the digital assets offered at **Biology.com.**

Chapter 1
Big Idea: Science as a Way of Knowing

↓

Chapter 1: What role does science play in the study of life?

- **1.1:** How do we find explanations for events in the natural world?
- **1.2:** How do the scientific community and society influence the process of science?
- **1.3:** What is biology?

LESSON 1.1

Pre-Reading

Objectives

1.1.1 State the goals of science.

1.1.2 Describe the steps used in scientific methodology.

Preview the Pages

1. Ask volunteers to locate and pronounce each highlighted vocabulary term in the lesson. Help students pronounce the words correctly.
2. Preview the series of marsh grass visuals. Invite volunteers to read the labels and captions.
3. Preview the steps in the flowchart Revising Hypotheses. Discuss the meanings of words in the steps, such as *analysis* and *interpretation,* that may be unfamiliar to students.

Activate Prior Knowledge

Ask How might a typical news report show that science is always changing? *(Sample answer: News reports often include stories about scientific discoveries.)*

Ask What patterns of whale movement might the scientists in the photograph observe? *(Sample answer: Whales swim in the same direction and about the same distance apart.)*

Build Understanding

FLOWCHART Refer students to the empty flowchart in Workbook B. Instruct students to complete the flowchart with information about performing a scientific investigation as they read the lesson.

NATIONAL SCIENCE EDUCATION STANDARDS

UNIFYING CONCEPTS AND PROCESSES

I, II, III

CONTENT

F.6, G.1, G.2

INQUIRY

A.2.a, A.2.b, A.2.c, A.2.d, A.2.e

For the corresponding lesson in *Miller & Levine Biology,* see pages 4–9.

1.1 What Is Science?

Key Questions

What are the goals of science?

What are the parts of scientific methodology?

BUILD Understanding

Flowchart As you read, create a flowchart showing the steps scientists use to answer questions about the natural world.

In Your Workbook Go to your workbook to learn more about making a flowchart.

What Science Is and Is Not

Where did plants and animals come from? How did I come to be? People have tried to answer those questions in different ways. Some ways of explaining the world have stayed the same over time. Science, however, is always changing.

This book has lots of facts and ideas about living things. But biology, or any science, is not just a collection of never-changing facts. Some of the facts in this book will change soon—if they have not changed already! What's more, science is not a lot of unchanging ideas about the world. Scientific ideas are open to testing, discussion, and change. If science is neither a list of facts nor a lot of unchanging ideas, what is it?

Science as a Way of Knowing **Science** is an organized way of gathering and analyzing evidence about the natural world. It is a way of looking, a way of thinking, and "a way of knowing" about the world. In other words, science is a *process*, not a "thing." The word *science* also refers to the body of knowledge that scientific studies have gathered over the years.

Science is different from other things people do in many ways. First, science deals only with the natural world. Second, scientists collect and organize information in an orderly way, looking for patterns and connections. Third, scientists suggest explanations that are based on evidence, not belief. Then they test those explanations with more evidence.

Science in Action This scientist is observing and recording information about the behavior of whales.

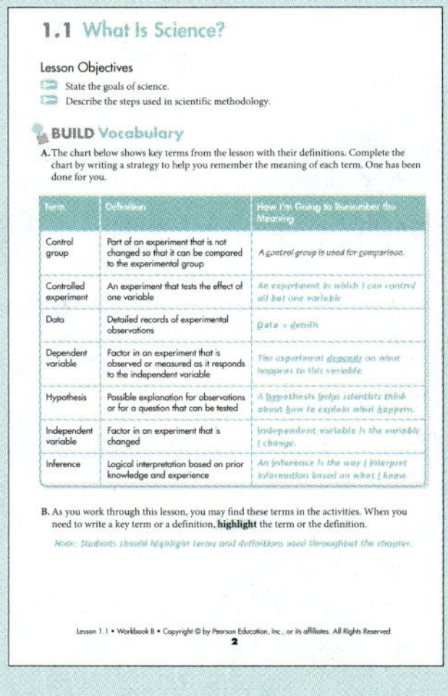

1.1 What Is Science?

Lesson Objectives

State the goals of science.

Describe the steps used in scientific methodology.

BUILD Vocabulary

A. The chart below shows key terms from the lesson with their definitions. Complete the chart by writing a strategy to help you remember the meaning of each term. One has been done for you.

Term	Definition	How I'm Going to Remember the Meaning
Control group	Part of an experiment that is not changed so that it can be compared to the experimental group	A control group is used for comparison.
Controlled experiment	An experiment that tests the effect of one variable	An experiment in which I can control all but one variable
Data	Detailed records of experimental observations	Data = details
Dependent variable	Factor in an experiment that is observed or measured as it responds to the independent variable	The experiment depends on what happens to this variable.
Hypothesis	Possible explanation for observations or for a question that can be tested	A hypothesis helps scientists think about how to explain what happens.
Independent variable	Factor in an experiment that is changed	Independent variable is the variable I change.
Inference	Logical interpretation based on prior knowledge and experience	An inference is the way I interpret information based on what I know.

B. As you work through this lesson, you may find these terms in the activities. When you need to write a key term or a definition, **highlight** the term or the definition.

Note: Students should highlight terms and definitions used throughout the chapter.

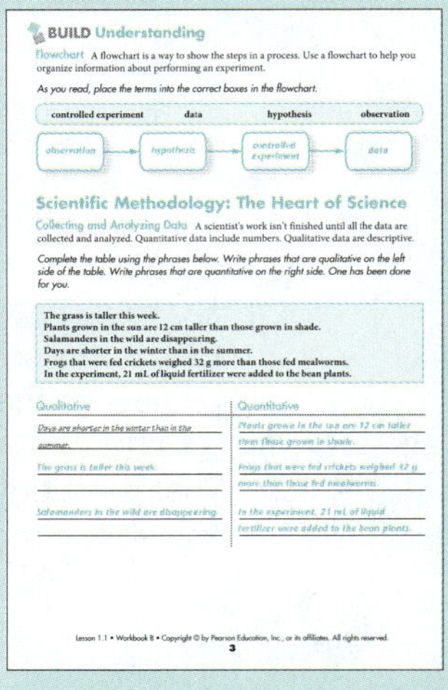

BUILD Understanding

Flowchart A flowchart is a way to show the steps in a process. Use a flowchart to help you organize information about performing an experiment.

As you read, place the terms into the correct boxes in the flowchart.

controlled experiment **data** **hypothesis** **observation**

observation → hypothesis → controlled experiment → data

Scientific Methodology: The Heart of Science

Collecting and Analyzing Data A scientist's work isn't finished until all the data are collected and analyzed. Quantitative data include numbers. Qualitative data are descriptive.

Complete the table using the phrases below. Write phrases that are qualitative on the left side of the table. Write phrases that are quantitative on the right side. One has been done for you.

The grass is taller this week.
Plants grown in the sun are 12 cm taller than those grown in shade.
Salamanders in the wild are disappearing.
Days are shorter in the winter than in the summer.
Frogs that were fed crickets weighed 32 g more than those fed mealworms.
In the experiment, 21 mL of liquid fertilizer were added to the bean plants.

Qualitative	Quantitative
Days are shorter in the winter than in the summer.	Plants grown in the sun are 12 cm taller than those grown in shade.
The grass is taller this week.	Frogs that were fed crickets weighed 32 g more than those fed mealworms.
Salamanders in the wild are disappearing.	In the experiment, 21 mL of liquid fertilizer were added to the bean plants.

The Goals of Science From a scientific view, objects in the universe, and all interactions among those objects, are ruled by natural laws. One goal of science is to use an understanding of those laws to give natural explanations for events in the natural world. Science also aims to use those explanations to understand patterns in nature and to make useful predictions about natural events.

Key Question What are the goals of science?
The goals of science are to give explanations for natural events, to understand patterns, and to make predictions.

Science, Change, and Uncertainty Scientists know a lot about nature. Yet, much of nature remains a mystery because science never stands still. This constant change does not mean science has failed. It shows that science is always improving.

Science rarely "proves" anything in absolute terms. Scientists aim for the best understanding that can be found. Uncertainty is a part of science and is part of what makes science exciting!

Scientific Methodology

There isn't any single "scientific method." But there is a general style of investigation that can be called scientific methodology. The parts of scientific methodology are described below.

Observing and Asking Questions Scientific studies begin with **observation,** the act of noticing and describing what is happening in an orderly way. Observation leads to new questions.

Inferring and Forming a Hypothesis After asking questions, scientists use further observations to make inferences. An inference is an idea based on what is already known. Inference, along with imagination, can lead to a hypothesis. A **hypothesis** is a scientific explanation for a set of observations that can be tested.

BUILD Vocabulary

science
an organized way of gathering and analyzing evidence about the natural world

observation
a process of noticing and describing events or processes in a careful, orderly way

hypothesis
a possible explanation for a set of observations or a possible answer to a scientific question

WORD ORIGINS
The word *science* comes from the Latin word *scientia,* which means "knowledge." Science represents knowledge that has been gathered over time.

Salt Marsh Experiment Scientists made observations about a salt marsh. Then they used scientific methodology to answer the questions that came from their observations.

Observing and Asking Questions

Scientists observed that marsh grass grows taller in some places than others. This observation led to a question: *Why do marsh grasses grow to different heights in different places?*

Inferring and Hypothesizing

The scientists inferred that something limits grass growth in some places. Temperature, sunlight, water, or nutrients could cause the difference. Based on their knowledge of salt marshes, they proposed a hypothesis: *Marsh grass growth is limited by available nitrogen.*

BIOLOGY.com Search Lesson 1.1 GO • Art in Motion • InterActive Art

ELL Focus on ELL: Access Content

ALL SPEAKERS Have students begin a T-chart in their notebooks with one column labeled Science and the other Not Science. Have volunteers read aloud the paragraphs under What Science Is and Is Not. After each paragraph is read, discuss examples of words and phrases to be recorded in each T-chart column.

For example, have a volunteer reread the last two sentences of the first paragraph.

Ask What are the two sentences about? *(ways of explaining the world)*

Ask How are the ways different? *(Some ways have stayed the same, and another way—science—is always changing.)*

Ask What is another way to say something has stayed the same? *(not changed, unchanged)*

Ask So, what should be the first two entries for the T-chart? *(The phrase* changing ways of explaining *could go under Science;* unchanging ways of explaining *could go under Not Science.)*

During Reading

KEY QUESTION Read aloud the Key Question and its answer.

Active Reading

SET PURPOSE FOR READING Have volunteers take turns reading aloud the four paragraphs on the previous page.

Say You have been introduced to the process of science. Now read The Goals of Science.

Ask What are two main goals of science? *(to explain events that happen in nature; to understand patterns in nature and use them to predict natural events)*

Say In this lesson, you will learn about specific kinds of actions scientists use to accomplish these goals.

DRAW CONCLUSIONS Help students see that the parts of scientific methodology are not formalized.

Ask Do the parts of scientific methodology always take place in the same order? *(no)* Explain. *(Parts may occur in different order, be repeated, or take place simultaneously.)*

USE VISUALS Have volunteers read the captions under the pictures of marsh grass. Make sure they understand that this is the first set of pictures in a series of pictures about a scientific investigation.

Ask What caused scientists to ask questions about the growth of marsh grass? *(seeing different heights of the grass in different places)*

Ask What did scientists hypothesize limited the growth of marsh grass? *(available nitrogen)*

Build Vocabulary

WORD ORIGINS Explain that many vocabulary words students will learn in biology come from either Latin or Greek.

Say *Hypothesis* comes from the Greek words *hupothe,* meaning "to suppose," and *hupothesis,* meaning "a proposal or supposition."

Explain that *hupothe* is a verb, shown by the word *to* in its definition, and *hupothesis* is a noun.

Ask How is the definition of *hypothesis* related to its Greek origin? *(Both involve possible explanations.)* Help students see the relationship by using the sentence frames "I propose that . . ." and "Suppose that . . ." and comparing them to "One hypothesis is that . . ."

Speed Bump

Ask Why does the text use *scientific methodology* instead of *scientific method*? *(because there is not just one method of conducting science but a number of methods that are used logically in investigations)*

LESSON 1.1

During Reading continued

Active Reading

SUMMARIZE Have students work in pairs to read the three paragraphs under Designing Controlled Experiments. Instruct each partner to write a sentence or two that summarize a paragraph after it is read. Then, have partners compare their summaries and discuss any differences. Tell partners that any disagreements that arise in the discussion of their summaries should be settled by rereading related text in a paragraph.

USE VISUALS Direct students' attention to the Designing Controlled Experiments visual. Make sure students understand that this is the second set of pictures in the series about a scientific investigation. Have a volunteer read the captions.

Ask What things are the same in both groups? *(numbers of plants, soil type, water supply, and height above the water level)* Are these things variables? *(yes)* Why do you say that? *(These things could change in some way.)*

Ask How do you know that nitrogen fertilizer is the independent variable in the scientists' experiment? *(Adding nitrogen fertilizer changes the makeup of the groups. The fertilizer is the only thing that is not in both groups.)*

Ask Why is the growth of marsh grass the dependent variable in the experiment? *(Based on their prior knowledge, scientists hypothesized that marsh grass growth changes in response to the independent variable, the nitrogen fertilizer.)*

MAKE INFERENCES Have students infer the uses of quantitative and qualitative data in the marsh grass experiment.

Ask What might quantitative data in the marsh grass experiment include? *(Sample answer: number of plants per plot, length and width of each blade of grass)*

Ask What might qualitative data include? *(Sample answer: how healthy the grass looks)*

Build Vocabulary

RELATED WORD FORMS Have students reread the last two sentences of the paragraph Collecting and Analyzing Data.

Ask How might you rewrite these sentences if they used the singular *datum* instead of the plural *data*? *(A quantitative datum is found by . . . A qualitative datum describes . . .)* Emphasize how the form of the subject changes the form of the verb.

BUILD Vocabulary

controlled experiment
an experiment in which only one variable is changed

control group
the group in an experiment that is exposed to the same conditions as the experimental group except for one independent variable

data
evidence; information gathered from observations

RELATED WORD FORMS

The word *data* is plural for *datum*. To help you remember that *data* is plural, you can think of data as being a lot of information and datum as just one piece of information.

Designing Controlled Experiments Testing a hypothesis often involves an experiment that keeps track of variables, or the things that can change. A few examples of variables include temperature, light, and time. Whenever possible, a hypothesis should be tested by an experiment in which only one variable is changed. All other variables should be kept unchanged, or controlled. An experiment in which only one variable is changed is called a **controlled experiment.**

▶ ***Controlling Variables*** Why are variables controlled? The reason is that if more than one variable is changed in an experiment, scientists cannot easily tell which variable caused the results. The variable that is changed is called the independent variable. The variable that is observed and that changes because of modifications to the independent variable is called the dependent variable.

▶ ***Control and Experimental Groups*** An experiment is usually divided into control and experimental groups. A **control group** is treated the same as the experimental group except for one independent variable. Scientists often repeat their experiments several times to see if they get the same results. This process is called replicating the experiment. So, scientists set up several sets of control and experimental groups, rather than just a single pair.

Collecting and Analyzing Data Scientists make detailed records of experimental observations by gathering information called **data.** There are two main kinds of data. Quantitative data are numbers found by counting or measuring. Qualitative data describe things that cannot be counted.

Designing Controlled Experiments

The scientists selected areas that had similar numbers of plants and similar soil type, water supply, and height above the water level. The areas were divided into control and experimental groups.

The scientists added nitrogen fertilizer (the independent variable) to the experimental groups. They then observed the growth of marsh grass (the dependent variable) in both experimental and control groups.

Focus on Academic Vocabulary

- variable
- independent
- dependent
- independent variable
- dependent variable

Write these words on the board. Tell students that all of these terms have particular meanings in scientific investigations.

Ask In the definition of a controlled experiment, what part of speech is the word *variable*? *(noun)* Which terms are used to modify variables? *(independent, dependent)* What part of speech are these terms? *(adjectives)*

Explain that a variable in an experiment is a factor that can change in some way.

Say Another name for an independent variable is *manipulated variable,* and another name for a dependent variable is *responding variable.* A dependent variable is one that responds to (depends on) changes in an independent variable.

▶ ***Research Tools*** Scientists pick the right tools for collecting and analyzing data. The tools may be simple, such as metersticks and calculators, or they may be complex, such as computers or robots. Charts and graphs are also tools that help scientists organize data.

▶ ***Sources of Error*** Scientists must avoid errors in data collection and analysis. Errors may happen when using tools. Tools have limited accuracy or can be read incorrectly. Data analysis and decisions about sample size must be carried out carefully to avoid errors. Sometimes experimental and control groups are very large. Why? The reason is because there are always differences among subjects in the groups. The larger the sample size, the more sure scientists are about their data analysis.

Drawing Conclusions Scientists use data to support or refute the hypothesis, to change the hypothesis, or to draw a valid conclusion. Hypotheses are often not shown to be completely right or wrong by an experiment. Rather, the data may show that the scientists have the right idea but are wrong about a few things. In that case, scientists change the first hypothesis, make new predictions, and design new experiments.

Key Question What are the seven parts of scientific methodology?
Scientific methodology involves observing and asking questions, making inferences and forming hypotheses, doing controlled experiments, collecting and analyzing data, and drawing conclusions.

Collecting and Analyzing Data

The scientists sampled all the areas during the growing season. They measured growth rates and plant sizes, and analyzed the chemicals in the leaves.

Drawing Conclusions

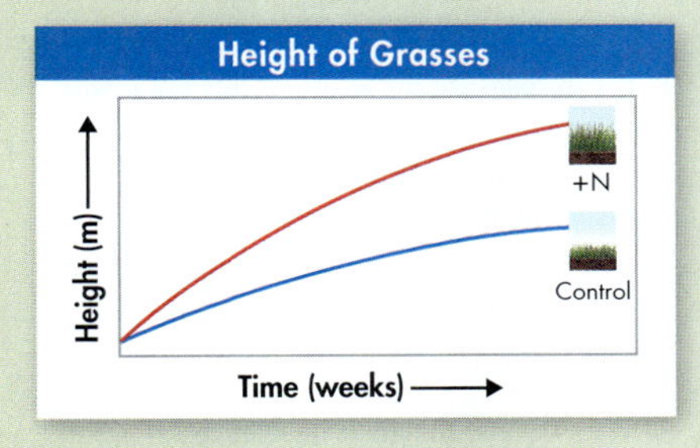

Data from all the areas were compared and evaluated. Data confirmed that marsh grasses with added nitrogen grow taller and larger than marsh grasses without added nitrogen. The hypothesis and its predictions were supported.

Active Reading

MAKE CHARTS Assign students to small groups. Have the groups draw a two-column table with rows on a sheet of lined paper, and title the columns Research Tools and Uses. Instruct the groups to list the tools mentioned in Research Tools in the left column and their uses in the right column. Encourage groups to brainstorm other tools that could be used in scientific research and to add the names and uses of these tools to their table.

USE VISUALS Direct students' attention to the Collecting and Analyzing Data and Drawing Conclusions visuals. Make sure students understand that this is the final set of images in the series about a scientific investigation. Have a volunteer read the captions.

Ask Compare the measurements of the dependent variable (growth of marsh grass) in the control group and experimental group. *(The grass in the experimental group is taller than the grass in the control group.)*

Say Go back to the second page of this lesson. Look at the first pictures in this series of pictures. Reread the captions.

Ask What hypothesis did scientists make about the growth of marsh grass? *(Marsh grass growth is limited by available nitrogen.)*

Ask Did the scientists hypothesize correctly? *(yes)* Explain. *(The independent variable, nitrogen fertilizer, added to the marsh plants in the experimental group, increased the height to which the plants would grow.)*

KEY QUESTION Invite a volunteer to read aloud the Key Question and answer.

Ask What is an inference? *(an idea based on what is already known)*

Speed Bump

Ask How many variables are changed in a controlled experiment? *(one)*

Ask Why should there be only one independent variable in an experiment? *(Scientists cannot easily tell which independent variable caused results if there is more than one in an experiment.)*

Ask Why do scientists set up several sets of control and experimental groups? *(They want to be sure that they get the same results when an experiment is done more than once.)*

Ask What can cause errors in data collection and analysis? *(Sample answer: the accuracy of scientific tools and the way the tools are used)*

Ask What happens if data does not support a scientist's hypothesis? *(The scientist must change the hypothesis, make new predictions, and conduct new experiments.)*

Hands-On Learning

TESTING A HYPOTHESIS Pair students, and provide each pair with 2 small paper cups and 2 grapes of equal size.

1. Place a labeled container of fresh water and one of salty water on a table in a central location.
2. Ask students if a grape soaked in salty water will be the same size after 10 minutes.
3. Have students answer by writing a hypothesis. *(Possible hypothesis: If a grape soaks in salty water, it will be smaller after 10 minutes.)*
4. Have students test their hypothesis by soaking one grape in fresh water and the other in salty water for 10 minutes.
5. Students should record observations of the grapes before *(about the same size)* and after soaking *(The grape in salty water got smaller.)*
6. **Ask** Was your hypothesis supported? *(Answers may vary.)* Is this a controlled experiment? *(yes)* What is the control? *(grape soaked in fresh water)* What is a possible source of error? *(measurement of the grapes)*

During Reading continued

Active Reading

INTERPRET DIAGRAMS Direct students' attention to the Revising Hypotheses flowchart. Have a volunteer read the caption. Then, have students trace the steps involved as you read aloud the labels.

Ask Which color arrows show the order of the steps in an experiment that supports the initial hypothesis? *(green)*

Ask Which color arrows show the order of the steps in an experiment when a new hypothesis is used? *(green and yellow)*

DRAW CONCLUSIONS Have small groups of students read and discuss When Experiments Are Not Possible.

Say Work with your group to make a list of examples of research in which hypotheses cannot be tested with experiments.

Say Develop a written answer to the following question. How is the ability to experiment an advantage in scientific research? *(Answers should reflect the idea that being able to experiment will result in more accurate results.)*

Speed Bump

Ask Can a scientific investigation be done without doing experiments? *(yes)* How? *(by making many observations of natural processes)*

Ask Why might a certain experiment not be done because it is not ethical? *(Sample answer: Some experiments might cause harm to people.)*

Post Reading

ASSESSMENT TIPS For Question 5, explain that the answer is not given directly in the text. Students need to evaluate the importance of hypotheses.

MYSTERY CLUE After students read the clue, have them suggest a possible control group *(children who have never taken HGH)* and an ethical experimental group *(children who must take HGH to be healthy)*.

Wrap-Up Activity

USE SCIENTIFIC METHODOLOGY Give small groups of students a "mystery box." Tell groups that they must develop a scientific method they can use to determine the contents of the box without opening it. Have groups list the steps of the method on paper. Then have groups share their methods.

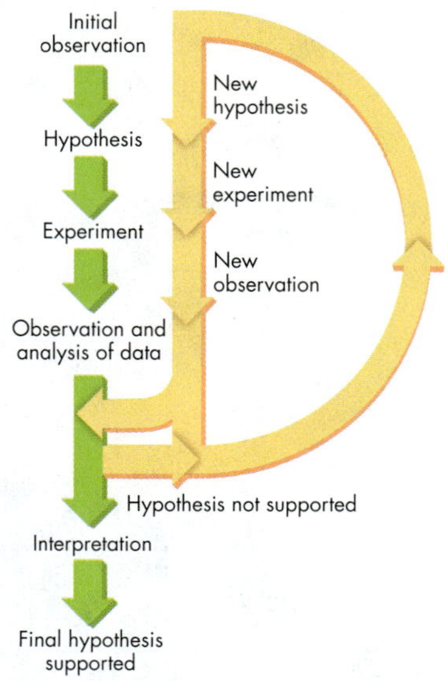

Revising Hypotheses During an investigation, scientists may have to change their hypotheses and redo experiments several times.

When Experiments Are Not Possible It is not always possible to test a hypothesis with an experiment. In some cases, scientists come up with hypotheses that can be tested by observations. For example, scientists who study how animals behave might want to learn how animal groups act in the wild. Studying this kind of natural behavior means that the scientists must go into the wild and watch the animals without bothering them. When scientists analyze data from these observations, they may come up with hypotheses that can be tested in different ways.

Sometimes, ethics keeps scientists from doing certain kinds of experiments. Ethics are beliefs about what is wrong or right. Some experiments on people are not ethical to do. For example, suppose that some scientists think that a chemical causes cancer in people who breathe it in. The scientists cannot make people breathe the chemical to see if they are correct! Instead, the scientists search for people who have already breathed in the chemical. Then, the scientists study people who have not breathed in the chemical.

When experiments are not possible, scientists still try to control as many variables as possible. They might not use people in their study who have serious health problems or known genetic conditions. Medical scientists often study large groups of subjects so that genetic differences between people do not give results that may be misleading.

CHECK Understanding

Apply Vocabulary

Use the highlighted words from the lesson to complete each sentence correctly.

1. If you measure the width of this book with a ruler, you are collecting ________.
2. A good experiment should have an experimental group and a(n) ________.
3. A(n) ________ is a possible explanation for a set of observations.

Critical Thinking

4. **Explain** List the goals of science.
5. **Explain** Why are hypotheses so important to controlled experiments?
6. **Write to Learn** Answer the first clue of the mystery. Be sure your answer identifies the independent and dependent variables.

MYSTERY CLUE

Describe a controlled experiment that would test the hypothesis that extra HGH helps children grow taller. What ethical issues can you imagine in actually carrying out such a study? (**Hint:** See p. 6.)

BIOLOGY.com Search Lesson 1.1 GO • Lesson Assessment • Art Review

Check Understanding

1. data
2. control group
3. hypothesis
4. Science gives explanations for events in the natural world, an understanding of patterns in nature, and predictions about natural events.
5. Hypotheses are important to controlled experiments because they are testable explanations for a set of observations.
6. Students should describe hypothetical experiments that include control groups and clearly identified independent and dependent variables. They should also mention the ethical problems involved in testing medical products in humans—for example, issues of safety, consent, and long-term health effects.

1.2 Science in Context

Exploration and Discovery: Where Ideas Come From

Scientific methodology is the heart of science. But that "heart" is only part of the full "body" of science. The full body of science is shown in the Process of Science at the bottom of this page. Part of science is exploration and discovery. How do exploration and discovery start? They often begin with scientific attitudes, practical problems, and new technology.

Scientific Attitudes Good scientists share scientific attitudes that lead them to exploration and discovery.

▶ ***Curiosity*** Scientists are curious and ask questions about what they observe. Results from previous studies also raise curiosity and lead to new questions.

▶ ***Skepticism*** Scientists are skeptics, which means that they question existing ideas and hypotheses. They also do not believe ideas without evidence.

▶ ***Open-Mindedness*** Scientists must be open-minded. They must be willing to accept new ideas that they may not agree with.

▶ ***Creativity*** Scientists need to think creatively to design experiments that provide good data.

Key Question What scientific attitudes help make new ideas? **Curiosity, skepticism, open-mindedness, and creativity help scientists come up with new ideas.**

Practical Problems Sometimes, ideas for scientific investigations come from practical problems. For example, people want cars that do not cause air pollution. Practical problems lead to questions, hypotheses, and experiments.

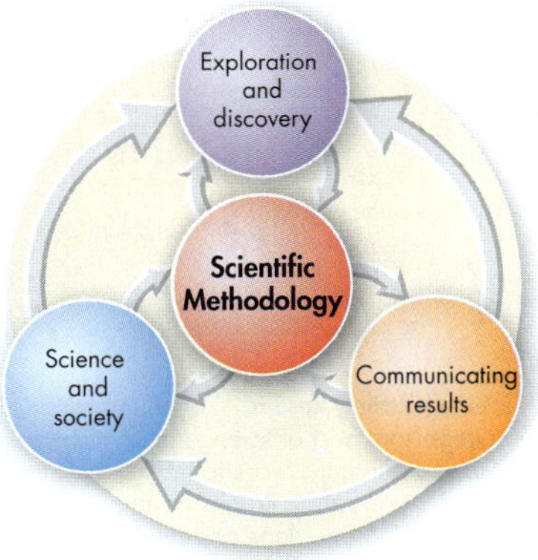

The Process of Science The arrows show that the different parts of science are connected in many ways. So, the process of science is flexible, unpredictable, and always changing.

Key Questions

- *What scientific attitudes help make new ideas?*
- *Why is peer review important?*
- *What is a scientific theory?*
- *How are science and society related?*

BUILD Understanding

Previewing Visuals Before you read, study the figure on the last page of this section. As you read, use the figure to describe the role that science plays in society.

In Your Workbook Go to your workbook to learn more about previewing visuals.

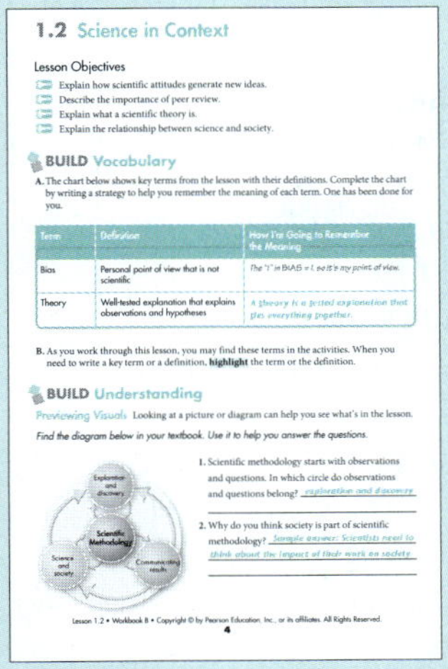

1.2 Science in Context

Lesson Objectives

- Explain how scientific attitudes generate new ideas.
- Describe the importance of peer review.
- Explain what a scientific theory is.
- Explain the relationship between science and society.

BUILD Vocabulary

A. The chart below shows key terms from the lesson with their definitions. Complete the chart by writing a strategy to help you remember the meaning of each term. One has been done for you.

Term	Definition	How I'm Going to Remember the Meaning
Bias	Personal point of view that is not scientific	The "I" in BIAS = I, so it's my point of view.
Theory	Well-tested explanation that explains observations and hypotheses	A theory is a tested explanation that ties everything together.

B. As you work through this lesson, you may find these terms in the activities. When you need to write a key term or a definition, **highlight** the term or the definition.

BUILD Understanding

Previewing Visuals Looking at a picture or diagram can help you see what's in the lesson.

Find the diagram below in your textbook. Use it to help you answer the questions.

1. Scientific methodology starts with observations and questions. In which circle do observations and questions belong? exploration and discovery
2. Why do you think society is part of scientific methodology? Sample answer: Scientists need to think about the impact of their work on society

Lesson 1.2 • Workbook B • Copyright © by Pearson Education, Inc., or its affiliates. All Rights Reserved.
4

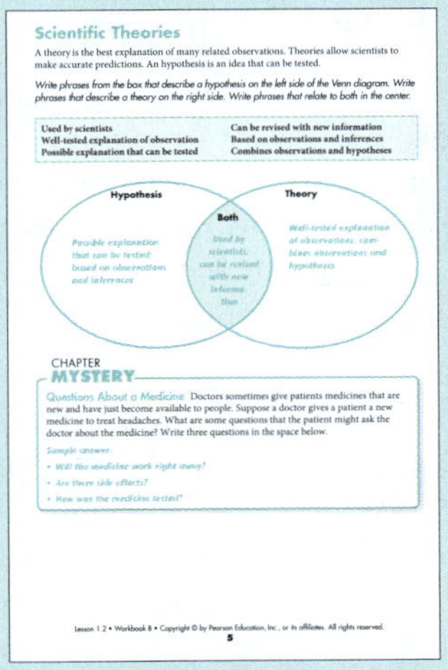

Scientific Theories

A theory is the best explanation of many related observations. Theories allow scientists to make accurate predictions. An hypothesis is an idea that can be tested.

Write phrases from the box that describe a hypothesis on the left side of the Venn diagram. Write phrases that describe a theory on the right side. Write phrases that relate to both in the center.

Used by scientists	Can be revised with new information
Well-tested explanation of observation	Based on observations and inferences
Possible explanation that can be tested	Combines observations and hypotheses

Hypothesis: Possible explanation that can be tested; based on observations and inferences
Both: Used by scientists; can be revised with new information
Theory: Well-tested explanation of observations; combines observations and hypotheses

CHAPTER MYSTERY

Questions About a Medicine Doctors sometimes give patients medicines that are new and have just become available to people. Suppose a doctor gives a patient a new medicine to treat headaches. What are some questions that the patient might ask the doctor about the medicine? Write three questions in the space below.

Sample answer:
- Will the medicine work right away?
- Are there side effects?
- How was the medicine tested?

Lesson 1.2 • Workbook B • Copyright © by Pearson Education, Inc., or its affiliates. All rights reserved.
5

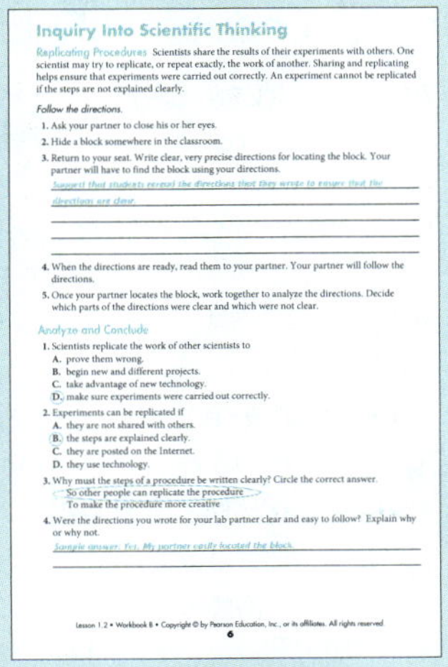

Inquiry Into Scientific Thinking

Replicating Procedures Scientists share the results of their experiments with others. One scientist may try to replicate, or repeat exactly, the work of another. Sharing and replicating helps ensure that experiments were carried out correctly. An experiment cannot be replicated if the steps are not explained clearly.

Follow the directions.

1. Ask your partner to close his or her eyes.
2. Hide a block somewhere in the classroom.
3. Return to your seat. Write clear, very precise directions for locating the block. Your partner will have to find the block using your directions.
Suggest that students reread the directions that they wrote to ensure that the directions are clear.
4. When the directions are ready, read them to your partner. Your partner will follow the directions.
5. Once your partner locates the block, work together to analyze the directions. Decide which parts of the directions were clear and which were not clear.

Analyze and Conclude

1. Scientists replicate the work of other scientists to
A. prove them wrong.
B. begin new and different projects.
C. take advantage of new technology.
D. make sure experiments were carried out correctly.
2. Experiments can be replicated if
A. they are not shared with others.
B. the steps are explained clearly.
C. they are posted on the Internet.
D. they use technology.
3. Why must the steps of a procedure be written clearly? Circle the correct answer.
So other people can replicate the procedure
To make the procedure more creative
4. Were the directions you wrote for your lab partner clear and easy to follow? Explain why or why not.
Sample answer: Yes. My partner easily located the block.

Lesson 1.2 • Workbook B • Copyright © by Pearson Education, Inc., or its affiliates. All rights reserved.
6

Pre-Reading

Objectives

1.2.1 Explain how scientific attitudes generate new ideas.

1.2.2 Describe the importance of peer review.

1.2.3 Explain what a scientific theory is.

1.2.4 Explain the relationship between science and society.

Preview the Pages

1. Preview the vocabulary words. Have students locate the highlighted words in the lesson and copy them into their notebooks. Help students with the pronunciation of the words.
2. Preview the visuals. Have volunteers read the captions and all the labels in each diagram. Point out that all of the diagrams in the lesson build on the diagram The Process of Science at the beginning of the lesson.

Build Understanding

PREVIEW VISUALS Direct students to Workbook B, where they will find a previewing visuals activity. Have students complete the activity as they read the first page of the lesson.

NATIONAL SCIENCE EDUCATION STANDARDS

UNIFYING CONCEPTS AND PROCESSES

II

CONTENT

E.2, F.6, G.1, G.2, G.3

INQUIRY

A.1.b, A.2.b, A.2.c, A.2.e, A.2.f

For the corresponding lesson in *Miller & Levine Biology,* see pages 10–15.

During Reading

Active Reading

KEY QUESTION Read aloud the Key Question and answer on the previous page. Discuss whether there is any contradiction between the needs for skepticism and open-mindedness. Explain that scientists are skeptical of new ideas without proper evidence, but they are open to new evidence that disputes their own hypotheses.

INTERPRET DIAGRAMS Direct students' attention to the diagram The Process of Science on the previous page.

Ask What is the center of science? *(scientific methodology)* What parts of science are linked to scientific methodology and to each other? *(exploration and discovery, communicating results, science and society)*

Point out the diagrams on this page.

Ask What do the diagrams show? *(details about how two major parts of science work: Exploration and Discovery and Communicating Results)*

Ask What actions lead to sharing data and ideas in the exploration of science? *(making observations, asking questions, exploring literature, finding inspiration)*

Ask What ways of communicating lead to new ideas in science? *(feedback and peer review, replication of results, discussion with colleagues, publication)*

USE SENTENCE FRAMES Write the following sentence frames on the board.

Sharing data and ideas ___(1)___ communication. ___(2)___ results leads to ___(3)___, which causes more ___(4)___ and discovery.

Say Read the words in the center oval of the diagrams, and think about what they mean. Then, complete the sentences by filling in the blanks. *(Sample answers: [1] requires, uses, involves; [2] Communicating; [3] new ideas; [4] exploration)*

KEY QUESTION Have a volunteer read aloud the Key Question and answer. Explain that a peer is a person of similar abilities or qualifications.

Ask Who are a scientist's peers? *(other scientists)*

Speed Bump

Ask How are new technologies helpful to scientists? *(New technologies provide ways for scientists to ask new questions and to gather data.)*

Ask Is an article about a scientific investigation in a newspaper or popular magazine likely to have been peer-reviewed? *(no)* Why? *(Peer-reviewed papers are published in scientific journals, and newspapers and popular magazines are not scientific journals.)*

Exploration and Discovery Ideas in science can turn up in many ways—from curiosity to the need to solve a certain problem. Scientists often begin investigations by making observations, asking questions, talking to other scientists, and reading about other experiments.

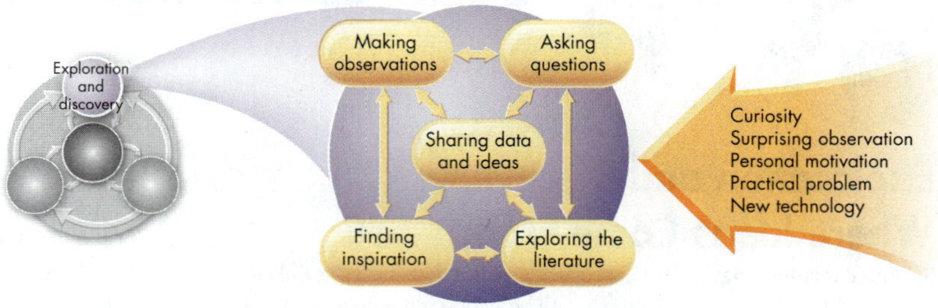

Adapted from *Understanding Science,* UC Berkeley, Museum of Paleontology

The Role of Technology Technology, science, and society are closely linked. Discoveries in science may lead to new technologies. Those technologies, in turn, enable scientists to ask new questions or to gather data in new ways. Technology can also have big impacts on daily life. For instance, it is now possible to make vitamins, antibiotics, and hormones that before were only available naturally.

Communicating Results: Reviewing and Sharing Ideas

Data collection and analysis can take a long time. Scientists may work on a single study for months or even years. Then, the scientists communicate results to other scientists.

Peer Review Scientists share their findings with other scientists by publishing papers that go through peer review. In peer review, papers are reviewed by other scientists. These reviewers read papers looking for mistakes and other problems. Peer review does not make sure that a piece of work is correct, but it does make sure that the work meets standards set by the scientific community. Publishing peer-reviewed papers in journals lets scientists share ideas and review each other's work.

Sharing Knowledge and New Ideas How do new findings fit into what is already known about science? Perhaps the findings lead to new questions. Each of those questions could lead to new hypotheses and new controlled experiments.

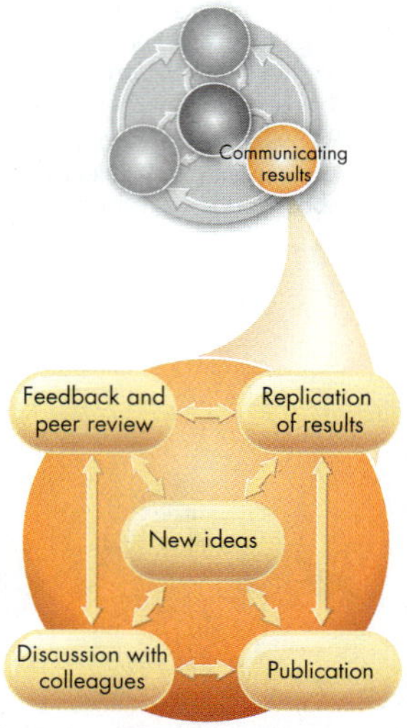

Adapted from *Understanding Science,* UC Berkeley, Museum of Paleontology

Communicating Results Communication is an important part of science. Scientists look over one another's work to make sure it meets scientific standards. Results from one study may lead to new ideas and new studies.

Key Question Why is peer review important? **Publishing peer-reviewed papers in journals lets scientists share ideas and review each other's work.**

10 Chapter 1 • Lesson 2

ELL Focus on ELL: Extend Language

ALL SPEAKERS Begin a word wall (see Workbook B, page A4 for a description of word walls) by posting the following terms from this lesson: *technology, technologies, explore, exploration, communicate, communication, publish, publication, replication, discover, discovery, discoveries, curious, curiosity, skepticism, skeptical.* Group words with the same root, such as *technology* and *technologies.* Invite students to add to the wall any other words from the lesson that they find difficult.

Have students identify the parts of speech for each word as it is used in the lesson, along with a definition for each word as it is used in the lesson. Encourage students to use a dictionary and context clues from the lesson text to develop their definitions. Invite volunteers to post their information about each word on the word wall.

Challenge students to creatively organize the information on the wall. For example, students may record the information on different colors or shapes of paper, indicating similarities such as part of speech, root word, or word origin.

INQUIRY into Scientific Thinking

Replicating Procedures

Scientists often repeat each other's experiments. They do this to make sure that the data and conclusions of the first scientist are correct. So, when scientists write papers to publish in journals, they must describe the procedures that they followed. The descriptions have to be good enough so that other scientists can repeat the first scientist's experiment.

1. Working with a partner behind a screen, arrange ten blocks into an unusual structure.
2. Write directions that others can use to build the same structure without seeing it.
3. Exchange directions with another team.
4. Build the other team's structure by following the directions.
5. Compare each new structure to the original. Identify which parts of the directions were clear and correct, and which parts were hard to understand or follow.

Analyze and Conclude

1. Evaluate How could you have written better directions?

2. Infer Why is it important that scientists write procedures that can be repeated?

In Your Workbook Get more help for this activity in your workbook.

Scientific Theories

In science, a **theory** is a well-tested explanation that accounts for a lot of observations and hypotheses and that lets scientists make good predictions. For example, Charles Darwin's early observations and hypotheses about change over time grew for years before he collected them into a theory of evolution by natural selection. Today, this theory is a central idea in biology.

A useful theory that has been tested and supported in many ways may become the most accepted view among scientists. But no theory is thought to be absolute truth. Science is always changing. As new evidence is found, a theory may be changed or replaced by a more useful explanation.

Key Question What is a scientific theory?
A scientific theory is a well-tested explanation that accounts for a lot of observations and hypotheses and that lets scientists make good predictions.

BUILD Vocabulary

theory
a well-tested explanation that unifies a broad range of observations and hypotheses, and enables scientists to make accurate predictions about new situations

bias
a particular preference or point of view that is personal, rather than scientific

MULTIPLE MEANINGS
The word *theory* is used both in science and in everyday life. In everyday life, when you say, "I have a theory," you may mean, "I have a hunch." But in science, a theory is much stronger than a hunch—it is a well-tested explanation.

Analyze and Conclude

1. Students should identify additional details or more precise language that would have improved the directions provided to their partners.
2. Scientists would not be able to verify findings if they were not able to repeat an experiment. Results of scientific studies must be replicable to be accepted. Replication would not be possible without carefully written directions.

Inquiry Into Scientific Thinking

MAKE CONNECTIONS Have students think of other situations in which a procedure must be clearly communicated. *(Sample answers: writing a recipe, giving directions to a lost motorist, teaching a play in sports)*

Ask In what ways might you "mess up" a recipe? *(Sample answers: using wrong ingredients, using wrong amounts of ingredients, leaving out a step or steps, not following the correct order of steps)*

Relate the importance of following a recipe or any set of instructions to the importance of repeating an experimental procedure to make it a fair test.

Lead a Discussion

Discuss experiences students may have had using directions to cook, find a location, build a model, or assemble something.

Ask Were you successful? Why or why not? *(Answers may vary. Reasons might include that directions were or were not clearly written and were or were not completely followed.)*

Active Reading

MAKE ANALOGIES Tell students that the word *theory* is often used in place of *hypothesis.* Explain that these words have different meanings in science. Write the following sentence frame on the board: A brick wall is to a brick as a _____ is to a _____.

Say This sentence represents an analogy, in which difficult concepts are likened to simpler concepts to make understanding easier. Put the words *theory* and *hypothesis* in the correct blanks to show how they are used in science. *(theory, hypothesis)*

KEY QUESTION Invite a volunteer to read aloud the Key Question and answer.

Ask What is a more powerful and complete explanation of something you observe—a hypothesis or a theory? *(theory)*

Build Vocabulary

MULTIPLE MEANINGS Tell students that the origin of *bias* is a French word that means "a slope or a slant." Have students look up *bias* in a dictionary.

Ask What are some meanings of *bias*? *(Sample answers: [1] a line that cuts diagonally across the weave of fabric [2] a mental leaning or inclination [3] an electronic signal that reduces distortion of recorded sound [4] an error that contributes to differences in samples taken for statistical analysis)*

Ask Which meaning is the same as the one in the text? *(#2)*

During Reading continued

Active Reading

INTERPRET DIAGRAMS Direct students' attention to the diagram.

Ask What does the diagram show? *(how science and society are related)*

Ask What can science and society working together do? *(develop technology, address societal issues, inform policy, solve everyday problems, satisfy curiosity, build knowledge)*

KEY QUESTION Read the Key Question and answer aloud. Then show students one or more recent newspaper or magazine articles about issues involving science. Read the articles aloud and discuss how the issues relate science and society. Point out any misinformation or bias as you discuss the articles.

Speed Bump

Ask What is an example of a question that science cannot address? *(Examples should involve ethical or moral issues.)*

Ask Why is it useful for citizens to know science, or be science literate? *(Sample answer: Science-literate citizens can tell if data are being used fairly when politicians debate a new law.)*

Post Reading

For Question 1, tell students that the word *personal* is a clue to the correct answer.

Tell students that Questions 3 and 4, which require them to explain, cannot be answered in only a few words. Encourage students to write answers that use at least two sentences.

MYSTERY CLUE After students read the clue, explain that technology enables people to produce artificial HGH. Before students answer the question, have the class discuss what an unlimited supply of a medicine would mean to the people who need it.

Wrap-Up Activity

BIAS REVIEW Have students look at newspapers, magazines, and online articles about life science, or give students copies of an article. Have small groups read the article and look for words that might suggest a bias or slanted point of view. Provide examples such as *should, better, worse.* Point out that phrases such as *my opinion* and *I think* also show bias, but the writer is letting the readers know that, so that they can form their own opinions.

Adapted from *Understanding Science,* UC Berkeley, Museum of Paleontology

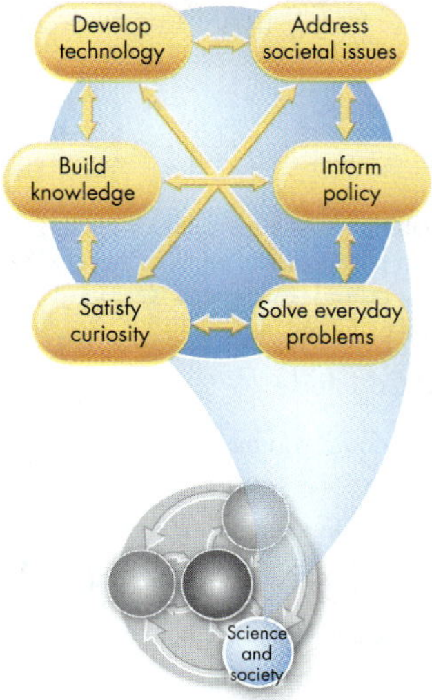

Science and Society Science can help society or change how people in society do things. But society also affects science. Problems and questions in society may lead scientists to do new experiments.

Science and Society

Many important questions can only be answered with the help of scientific information. But very few of these questions can be answered by science alone. These questions involve the society in which we live. Using science involves understanding its limitations and how it fits into society.

Science, Ethics, and Morality When scientists explain "why" something happens, their explanation is about only nature. Science does not include ethical or moral views. Scientists can try to explain what life is. But science cannot answer questions about why life exists or what the meaning of life is.

Avoiding Bias How science is used in society can be affected by bias. A **bias** is a certain preference or point of view that is personal, rather than scientific. Examples of biases include personal taste and liking one thing over another.

Scientists try to avoid bias. But scientific data can be used in the wrong way by people who want to prove a certain point. However, if enough of us understand science, we can help make sure that science is used in only helpful, correct ways.

Understanding and Using Science As you read this book, don't think of it as an encyclopedia. Don't memorize the scientific facts and ideas. Instead, try to understand how scientists developed those ideas. Understanding science will help you make decisions that also fit with society's values.

Key Question How are science and society related?
Using science involves understanding its limitations and how it fits into society.

CHECK Understanding

Apply Vocabulary

Use the highlighted words from the lesson to complete the sentence correctly.

1. A point of view that is personal rather than scientific is a ________.
2. Scientists can use a ________ to make good predictions.

Critical Thinking

3. **Explain** What does it mean to be skeptical? Why is skepticism a good quality in a scientist?
4. **Explain** Describe some of the limitations of science.
5. **Write to Learn** Write the answer to the second clue of the mystery.

MYSTERY CLUE

How does the ability to produce artificial HGH affect human life? (Hint: See p. 12.)

BIOLOGY.com Search Lesson 1.2 GO • Lesson Assessment

Check Understanding

1. bias
2. theory
3. A scientist is skeptical if he or she questions existing ideas and hypotheses. Skepticism is important in scientists because it means they refuse to accept explanations without evidence.
4. Sample answer: Science does not include ethical or moral viewpoints. It also may be influenced by bias.
5. Students should infer that the ability to produce artificial HGH increases its availability and the number of individuals who can receive treatment.

1.3 Studying Life

Characteristics of Living Things

Think about news stories that you have seen or heard. Bird flu spreads around the world killing thousand of birds and threatening an epidemic. People who use illegal drugs get permanent brain damage. These and many other stories are about biology—the study living things. (The Greek word *bios* means "life," and *-logy* means "study of.")

Biology is the study of life. But what is life? What is the difference between living things and nonliving matter? It is not as simple as you might think to describe what makes something alive. No single characteristic is enough to describe a living thing. Also, some nonliving things share one or more traits with living things. For example, automobiles and clouds (which are not alive) move around. However, mushrooms and trees (which are alive) stay in one spot.

Despite these difficulties, we can list characteristics that most living things have in common. Living things are made up of basic units called cells, are based on a universal genetic code, obtain and use materials and energy, grow and develop, reproduce, respond to their environment, maintain a stable internal environment, and change over time.

Key Question What characteristics do all living things share? **Living things are made up of basic units called cells, are based on a universal genetic code, obtain and use materials and energy, grow and develop, reproduce, respond to their environment, maintain a stable internal environment, and change over time.**

Key Questions

What characteristics do all living things share?

What are the big ideas of biology?

How do different fields of biology differ in the way they study life?

Why is the metric system important in science?

BUILD Understanding

Concept Map As you read, draw a concept map showing the big ideas in biology.

In Your Workbook Go to your workbook and finish the concept map for Lesson 1.3.

Is It Alive? The fish are clearly alive, but what about the colorful structure above them? Is it alive? Yes! It is an animal called elkhorn coral. Corals show all the characteristics of living things.

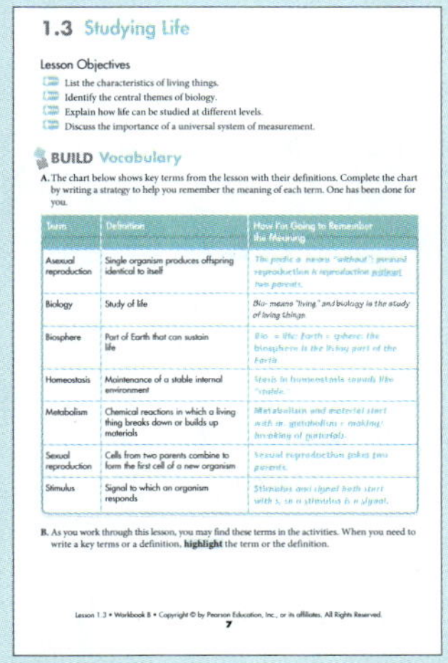

1.3 Studying Life

Lesson Objectives

- List the characteristics of living things.
- Identify the central themes of biology.
- Explain how life can be studied at different levels.
- Discuss the importance of a universal system of measurement.

BUILD Vocabulary

A. The chart below shows key terms from the lesson with their definitions. Complete the chart by writing a strategy to help you remember the meaning of each term. One has been done for you.

Term	Definition	How I'm Going to Remember the Meaning
Asexual reproduction	Single organism produces offspring identical to itself	
Biology	Study of life	*Bio- means "living," and biology is the study of living things.*
Biosphere	Part of Earth that can sustain life	
Homeostasis	Maintenance of a stable internal environment	
Metabolism	Chemical reactions in which a living thing breaks down or builds up materials	
Sexual reproduction	Cells from two parents combine to form the first cell of a new organism	
Stimulus	Signal to which an organism responds	

B. As you work through this lesson, you may find these terms in the activities. When you need to write a key terms or a definition, **highlight** the term or the definition.

Lesson 1.3 • Workbook B • Copyright © by Pearson Education, Inc., or its affiliates. All Rights Reserved.
7

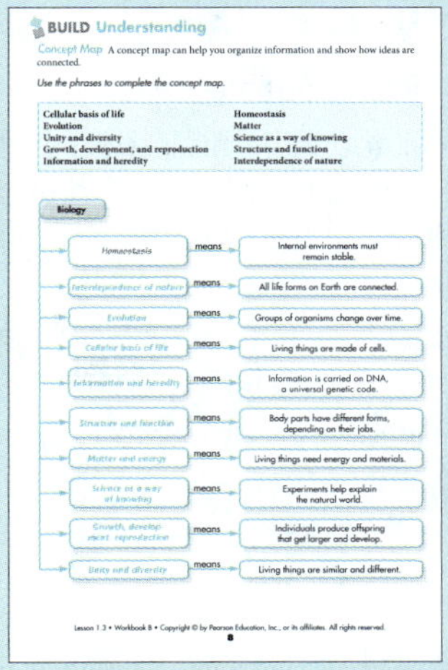

BUILD Understanding

Concept Map A concept map can help you organize information and show how ideas are connected.

Use the phrases to complete the concept map.

Cellular basis of life	**Homeostasis**
Evolution	**Matter**
Unity and diversity	**Science as a way of knowing**
Growth, development, and reproduction	**Structure and function**
Information and heredity	**Interdependence of nature**

Lesson 1.3 • Workbook B • Copyright © by Pearson Education, Inc., or its affiliates. All rights reserved.
8

Characteristics of Living Things

Scientists use certain characteristics to classify something as a living thing. Living things must be based on a universal genetic code and must have cells as the basic unit of life. They obtain and use energy. Living things also grow, develop, and reproduce. They maintain a stable internal environment. Over time, living things evolve. Look at the drawings. Label each thing as living or nonliving in the spaces provided.

In the chart, write yes or no next to the characteristics that apply to all living things.

Characteristics	Applies to living things?
Made of one or more cells	
Able to reproduce	
Stay warm	
Based on a genetic code	
Based on the metric system	
Grow and develop	
Obtain and use materials and energy	
Respond to the environment	
Maintain a stable environment	
Change over time	

Lesson 1.3 • Workbook B • Copyright © by Pearson Education, Inc., or its affiliates. All Rights Reserved.
9

LESSON 1.3

Pre-Reading

Objectives

1.3.1 List the characteristics of living things.

1.3.2 Identify the central themes of biology.

1.3.3 Explain how life can be studied at different levels.

1.3.4 Discuss the importance of a universal system of measurement.

Preview the Pages

1. Preview the vocabulary words by having students locate the highlighted terms in the text. Invite volunteers to pronounce the words. Help them with the words that are difficult to say.
2. Preview the Build Connections The Characteristics of Living Things. Have students identify the image in each text box and read aloud the boldfaced sentence.
3. Preview the Common Metric Units table. Have a volunteer model how each word in the table is pronounced.

Build Understanding

CONCEPT MAP Direct students to Workbook B, where they will find a partially filled-in concept map. Encourage students to complete the concept map as they read about the study of life.

NATIONAL SCIENCE EDUCATION STANDARDS

UNIFYING CONCEPTS AND PROCESSES

I, II, III, IV, V

CONTENT

C.1.a, C.2.a, C.3.a, C.5.b, C.6.b, E.2, G.1

INQUIRY

A.2.a, A.2.b, A.2.c A.2.d

For the corresponding lesson in *Miller & Levine Biology*, see pages 17–25.

During Reading

Active Reading

KEY QUESTION Have a volunteer read the Key Question and answer on the previous page aloud.

Ask What characteristic of most living things can you observe in some nonliving things? *(Sample answers: motion, increase in size)*

Build Connections

VISUAL SUMMARY Have a volunteer read the caption for the visual The Characteristics of Living Things. Point out that the visual extends across this page and the next. Be sure that students relate the pictures and text in the separate boxes to the picture of the apple tree, which is used as an example of a living thing.

Direct students to Workbook B, where they will find drawings and a table they can use to distinguish living and nonliving things. Encourage students to complete the labeling activity and the table as they read about the characteristics of life.

Ask What does the picture in the top box represent? *(DNA)* Where is the information carried by DNA found? *(in the genetic code)* How can the DNA molecule be described? *(Sample answer: two twisted strips held together by many crosspieces)* Tell students that the shape of a DNA molecule is often described as that of a twisted ladder.

Ask How is the young apple plant in the lower left box different from the apple tree? *(It is much smaller, has only 2 leaves, and lacks a woody trunk and flowers.)*

Ask What are some parts of the environment of the apple tree? *(air, light, space, soil, other plants, animals such as caterpillars)* Which stimulus from the environment is most important to the growth of plants such as the apple tree and grasses? *(light)*

BUILD Connections

THE CHARACTERISTICS OF LIVING THINGS

Apple trees share certain characteristics with other living things. How are the apple tree and the grass growing below similar? How are they different?

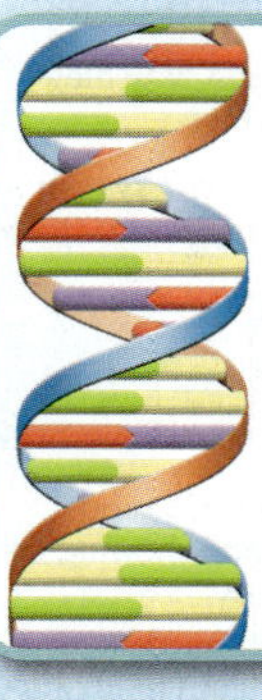

Living things are based on a universal genetic code. All organisms store the information they need to live, grow, and reproduce in a genetic code written in a molecule called **DNA.** That information is copied and passed from parent to offspring. With a few small differences, life's genetic code is almost the same in every living thing on Earth.

The information in an apple tree's DNA directs all of the tree's life processes.

Living things grow and develop. Every organism has a certain pattern of growth and development. During development, a single fertilized egg divides again and again. As these cells divide, they differentiate, which means they begin to look different from one another and to do different jobs.

An apple tree develops from a tiny seed.

Living things respond to their environment. Organisms notice and react to stimuli from their environment. A **stimulus** is a signal to which a living thing reacts.

Some plants can produce poisons to ward off caterpillars that feed on its leaves.

Focus on Academic Vocabulary

• universal • development • differentiate • environment • evolutionary

Write these words, which are all found in this lesson, on the board or post them on a word wall. Students may need help pronouncing and understanding words with multiple syllables. Have students look up the words in a dictionary and write a detailed definition of each one in their notebook.

Ask Which word has six syllables? *(evolutionary)* How many of the words have four syllables? *(three)* Which word has five syllables? *(differentiate)*

Encourage students to use a dictionary to break long words into syllables and to identify root words and word origins that will help them determine a word's meaning.

Living things reproduce. All living things make new similar living things. Most plants and animals engage in sexual reproduction. In sexual reproduction, cells from two parents come together to form the first cell of a new living thing. Other living things reproduce through asexual reproduction, in which a single living thing makes offspring exactly the same as itself.

Flowers form as part of the apple tree's reproductive cycle.

Living things maintain a stable internal environment. All living things need to keep conditions inside themselves as constant as possible, even when conditions outside of them change. When conditions inside organisms are kept within certain limits, this is called **homeostasis.**

These cells help leaves control gases that enter and leave the plant. SEM 1200×

Living things get and use material and energy. All living things must take in materials and energy to grow, develop, and reproduce. The chemical reactions through which a living thing builds up or breaks down materials are called **metabolism.**

Different metabolic reactions happen in leaves.

Living things are made up of cells. Living things are made up of one or more cells—the smallest units considered fully alive. Cells can grow, respond to their surroundings, and reproduce. Even though cells are small, they are complex and very organized.

One branch of a tree is made up of millions of cells. LM 800×

Taken as a group, living things evolve. Over generations, groups of living things evolve, or change over time. Evolutionary change links all forms of life to a common origin more than 3.5 billion years ago. Evidence of this shared history is found in all parts of living things and fossils, from body parts to proteins to information in DNA.

Signs of one of the first land plants are preserved in rock over 400 million years old.

The Science of Biology 15

Students Say

"I get more from the reading when the teacher hands us a web, and we fill it in as we read. Then I know that there are eight characteristics that all living things share."

Build Connections continued

VISUAL SUMMARY First, discuss with students the text in the box at the lower left of this page. Next, discuss the text in the boxes in the right column from top to bottom.

Ask What appears to be inside the plant cells shown in the picture? *(many smaller green structures or spheres)* What phrase in the text box at lower left does this photograph support? *(they are complex and very organized)*

Say Flowers help an apple tree reproduce by making seeds.

Ask Why do you think the apple tree has so many flowers? *(Sample answer: to make sure the tree produces some offspring)*

Say Look at the cells in the box that discusses homeostasis.

Ask Why do some of these cells have a dark slit? *(Sample answer: The slits are openings to let certain materials in and out.)* What scientific tool must have been used to take the photograph? *(a microscope)*

Ask What tiny structures make up the leaves of a plant? *(cells)* Is the leaf shown here what you expect a leaf to look like? *(yes)* Explain. *(Students should agree that most leaves are broad, flat, and thin.)*

Ask What feature of the tiny plant fossil in the picture is similar to a feature of the apple tree? *(The fossil plant has a main stem with branches, which is similar to the trunk and branches of the apple tree.)* How was this similarity transferred from the early plant to the apple tree? *(through evolution)*

Speed Bump

Ask What is the difference between sexual reproduction and asexual reproduction? *(Sexual reproduction occurs when cells from two parents come together. Asexual reproduction occurs when one organism makes offspring exactly like itself.)*

Ask What is homeostasis? *(When conditions inside living things are as constant as possible)*

Ask Which two characteristics of life are most obviously involved in making evolution possible? *(the genetic code in DNA and reproduction)*

During Reading continued

Active Reading

SET PURPOSE FOR READING Tell students that this page and the next introduce the Big Ideas, or main ideas, of biology.

Say Each chapter in this textbook will begin with one or more of these ideas, which will be followed by a question that ties the Big Idea to the content of the chapter.

Explain that students should not concentrate on memorizing the Big Ideas but, instead, should try to understand them and see how the ideas relate to the different topics they will study throughout the year.

VISUALIZE Ask students to describe images that come to mind as they read each paragraph. Direct students to one or two chapters that tie into each Big Idea so students can compare their images to the visuals in the chapters.

OUTLINE Encourage students to organize the Big Ideas of biology by making an outline for each one. Model an outline by writing the title of the first Big Idea on the board. Below that, list several important details from the paragraph that describes the Big Idea. Point out how you rephrased the information into your own words instead of copying it directly from the text.

Instruct students to copy this outline in their notebooks and to outline the remaining nine Big Ideas.

Alternatively, provide students with an outline shell on sheets of paper, with the Big Ideas written, each followed by several blank lines. Have students fill in the first outline as you model it on the board. Then they can outline the remaining Big Ideas.

Build Vocabulary

PREFIXES Explain that *meta-* is another prefix students will encounter in scientific vocabulary. Tell them that *meta-* means "a change."

Say Metabolism involves the changing of chemicals in living things to make other chemicals and to provide energy for processes that the organism needs to survive.

Ask What do you think *metabolic rate* means? *(how fast the changes of metabolism happen)*

BUILD Vocabulary

biology
the scientific study of life

DNA
the genetic material that organisms inherit from their parents

stimulus
a signal to which an organism responds

homeostasis
the relatively constant internal physical and chemical conditions that organisms maintain

metabolism
the combination of chemical reactions through which an organism builds up or breaks down materials

biosphere
the part of Earth in which life exists, including land, water, and air or atmosphere

PREFIXES
The prefix *homeo-* means "the same" or "similar." Homeostasis is how the body keeps conditions inside the same at all times.

Big Ideas in Biology

All parts of biology are tied together by big ideas. These big ideas overlap and interlock with one another. They will come up again and again throughout the book. Many of these big ideas overlap with the characteristics of life or the nature of science.

Big idea **Cellular Basis of Life** Living things are made of cells. Many living things are made up of only a single cell; they are called unicellular organisms. Plants and animals are multicellular. Cells in multicellular organisms display many different sizes, shapes, and functions.

Big idea **Information and Heredity** Living things carry information written in a universal genetic code in their DNA. The information coded in DNA forms an unbroken chain that stretches back roughly 3.5 billion years. Yet, the DNA inside your cells right now can affect your future, such as your risk of getting cancer and the color of your children's hair.

Big idea **Matter and Energy** Life needs matter that serves as nutrients to build body parts and energy that fuels the body. Some living things, such as plants, get energy from sunlight and take up nutrients from air, water, and soil. Other living things, including most animals, eat plants or other animals to get both nutrients and energy. The need for matter and energy ties all living things on Earth together into a living web of relationships.

Big idea **Growth, Development, and Reproduction** All living things reproduce, which means that they make more individuals. Offspring are almost always smaller than adults, so they grow and develop as they mature.

Big idea **Homeostasis** Living things keep a fairly stable internal environment. For most organisms, any problems with homeostasis may have serious or deadly consequences.

Big idea **Evolution** Evolutionary change ties all forms of life to a common beginning more than 3.5 billion years ago. Evolution is the central organizing idea of all biology.

Different But Similar This colorful bird is different from the plant it is sitting on. Yet, the two living things are similar on the molecular level. Unity and diversity of life is a big idea in biology.

Hands-On Learning

MAKE SPIDER MAPS Divide the class into 10 groups or pairs.

1. Have each group create a spider map of one of the Big Ideas. Randomly assign a Big Idea to each group so that all of the Big Ideas are covered. (See page A7 in Workbook B for a description of spider maps.)
2. Supply sheets of construction paper or card stock of different colors for students to use for developing their spider maps. Also, supply magazines as sources of images that they can use to illustrate concepts.
3. Post the complete maps around the classroom.

Big idea **Structure and Function** Each major group of living things has evolved its own set of body parts. These parts make certain functions possible. Living things have evolved into different forms as species have adapted to life in different places.

Big idea **Unity and Diversity of Life** Although life comes in a huge variety of forms, all living things are similar at the molecular level. All living things are made up of a common set of molecules, have DNA, and use proteins to build their body parts and carry out their functions. Evolutionary theory explains both this unity of life and its diversity.

Big idea **Interdependence in Nature** All forms of life on Earth are connected into a **biosphere,** which means "living planet." Within the biosphere, living things are linked to one another and to the land, water, and air around them. Relationships between living things and where they live depend on the cycling of matter and the flow of energy.

Big idea **Science as a Way of Knowing** Science is not a list of facts. The job of science is to use observations, questions, and experiments to explain the natural world. Good scientific research finds rules and patterns that can explain and predict at least some events in nature.

Key Question What are the big ideas of biology?
Biology's big ideas are the cellular basis of life; information and heredity; matter and energy; growth, development, and reproduction; homeostasis; evolution; structure and function; unity and diversity of life; interdependence in nature; and science as a way of knowing.

Fields of Biology

Biology is made up of many overlapping fields that use different tools to study life from the level of molecules to the whole planet. Here's a peek into a few of the many branches of biology.

Global Ecology The world is affected by everything that living things do. Global ecological studies using satellites and huge computers are letting us learn about people's global impact, which affects all life on Earth.

Biotechnology This field is based on the ability to read, write, and edit the genetic code. We may soon learn to correct or replace genes that cause diseases. But biotechnology also raises a lot of ethical, legal, and social questions.

Building the Tree of Life Biologists have found about 1.8 million different kinds of living things. Scientists want to use computers to gather what is known about all organisms to put all living things into a single "Tree of All Life."

Building the Tree of Life This paleontologist studies signs of ancient life—even fossilized dinosaur dung!

Focus on Academic Vocabulary

- tied together • overlap • interlock

Write these words on the board and point them out in the first paragraph on the previous page. Use your hands and fingers to demonstrate the differences in meaning among the words *interlocked, overlapped,* and *tied together*.

Ask If many objects are tied together, overlapped, or interlocked, what might happen if something happens to one of the objects? *(Sample answer: All of the objects might be affected in some way. For example, if one object moves, such as a train car connected to other cars, than the rest of the train cars will move.)*

Ask What do you think these words have to do with the Big Ideas of biology? *(Sample answer: The Big Ideas are all related to one another. For example, the Big Idea of information and heredity is related to cellular basis of life because DNA is in cells.)*

Discuss other examples of how the Big Ideas are related to one another.

Active Reading

SUMMARIZE Divide the class into small groups. Explain that each group member should read aloud at least one of the paragraphs about the Big Ideas while the other group members read along silently.

Say After you read a paragraph aloud, try to summarize it in your own words. Discuss the summary with others in your group. If something doesn't make sense, ask about it. If something important seems missing from the summary, mention it. When you all agree on a good summary, write it down.

Encourage students to reread a paragraph more than once to help clarify ideas. They can use their outlines, described on the previous page, if they have made them.

Have volunteers from each group read their summaries for each Big Idea paragraph. Compare summaries among groups and allow students to correct any misinterpretations and fill in important missing information.

KEY QUESTION Have a volunteer read aloud the Key Question and its answer.

Call on students to say one thing about each Big Idea, based on the details in each paragraph.

Speed Bump

Ask Which Big Ideas are about ways in which organisms change? *(growth, development, and reproduction; evolution)*

Ask Which Big Ideas are about how living things are put together? *(cellular basis of life, information and heredity, structure and function)*

Ask Which Big Ideas are about how living things are related to each other? *(unity and diversity of life, interdependence in nature; evolution)*

During Reading continued

Active Reading

USE GRAPHIC ORGANIZERS Have students make a Fishbone Map to organize the examples of the fields of biology described in the lesson. Give students the following as a pattern for their maps.

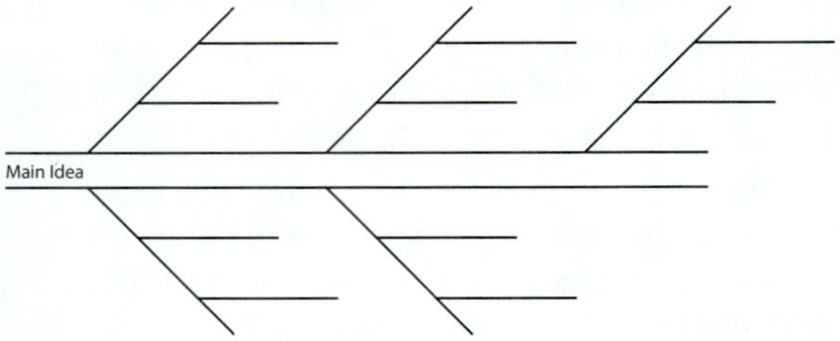

Explain that students should write the name of one field of biology in the lesson on each major branch and at least two details on the side branches.

KEY QUESTION Invite a volunteer to read aloud the Key Question and answer.

Ask Which field of biology is the most interesting to you? Why? *(Encourage students to verbalize why a certain field of study might interest them.)*

COMPARE AND CONTRAST Encourage students to look for the similarities and differences in metric units of measurement.

Ask What is a similarity? *(All use the same prefixes.)*

KEY QUESTION Read the question and answer.

Ask Why is the metric system easy to use? *(because its units are divided into multiples of 10, which makes conversion between different sizes of the units easy by moving the decimal point)*

Speed Bump

Ask What technologies have enabled global ecology to become a field of biology? *(satellites and computers)*

Ask Which other fields of biology rely on powerful computers? *(genomics and molecular biology)*

Ask Why is the study of the evolution of disease important? *(It is important because the germs that cause disease in animals can evolve and infect people. Knowing how this happens could help us fight these diseases.)*

Evolution of Diseases This wildlife biologist is studying baboons. Germs that cause diseases in wild animals can evolve and infect people.

Genomics and Molecular Biology A molecular biologist is analyzing a DNA sequence.

Ecology and Evolution of Diseases Over time, viruses, bacteria, and other organisms that cause disease evolve so that they still infect people, even as we try to fight them. Understanding how germs change and react to their surroundings is important for keeping people healthy.

Genomics and Molecular Biology Scientists are looking at the DNA of many different living things. Scientists analyze the data with powerful computers to learn about growth, development, and the history of life on Earth.

Key Question How do different fields of biology differ in the way they study life?
Biology is made up of many overlapping fields that use different tools to study life from the level of molecules to the whole planet.

Performing Biological Investigations

Scientists use a common system of measurement and practice safety when doing studies.

Scientific Measurement Most scientists use the metric system when gathering data and doing experiments. The metric system is a system of measurement in which units are based on multiples of 10. A revised version of the metric system is called the International System of Units, or SI. The metric system is easy to use because it's based on multiples of 10.

Safety Scientists are trained to be safe when doing investigations. Whenever you work in your biology lab, you must follow safety rules, too. Careful preparation is the key to staying safe while doing activities. Before doing any activity, be sure you understand the safety rules. Also, read all the steps in the activity and make sure that you understand them.

The most important safety rule is to always follow your teacher's instructions. You are responsible for your own safety and that of your teacher and classmates. If you are handling live animals, you are responsible for their safety, too.

Key Question Why is the metric system important in science?
Most scientists use the metric system when collecting data and performing experiments.

ELL Focus on ELL: Build Background

INTERMEDIATE AND ADVANCED SPEAKERS Have students make table tents that describe each field of biology mentioned in this lesson. Tell students that their table tents should be illustrated with images related to the field of biology featured. Provide magazines to use as sources of specific technologies. Encourage students to use their creativity to develop drawings and sketches.

Place the finished table tents around the classroom, and provide class time for a career fair in which students may circulate among the displays and consider possible careers in biology. Then, invite groups of students with similar interests to discuss fields of biology and careers of interest. Finally, have each student write a summary of the field of biology he or she is most interested in as a possible career choice.

Common Metric Units

Length	Mass
1 meter (m) = 100 centimeters (cm) 1 meter = 1000 millimeters (mm) 1000 meters = 1 kilometer (km)	1 kilogram (kg) = 1000 grams (g) 1 gram = 1000 milligrams (mg) 1000 kilograms = 1 metric ton (t)
Volume	**Temperature**
1 liter (L) = 1000 milliliters (mL) 1 liter = 1000 cubic centimeters (cm^3)	0°C = freezing point of water 100°C = boiling point of water

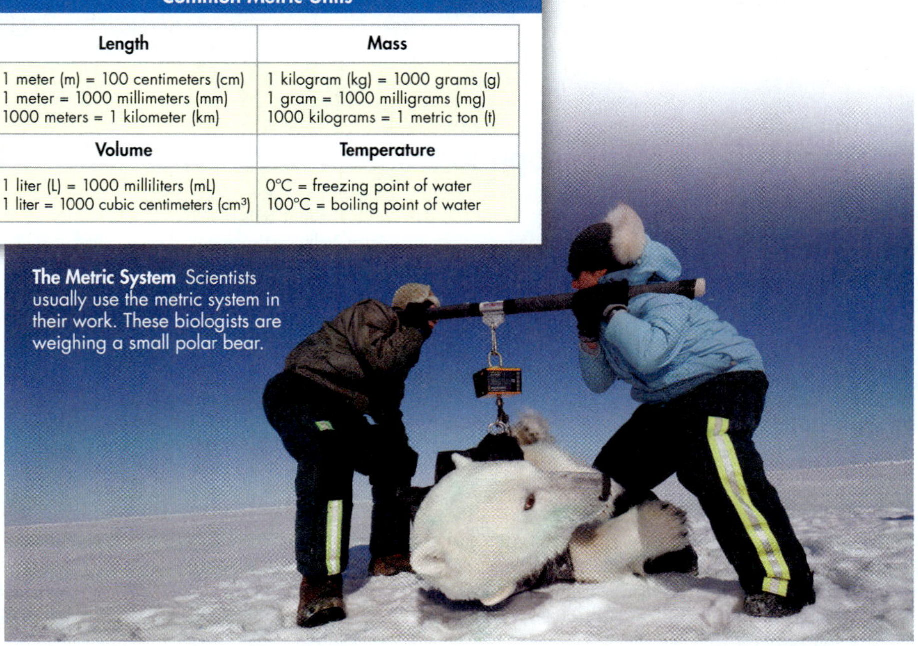

The Metric System Scientists usually use the metric system in their work. These biologists are weighing a small polar bear.

CHECK Understanding

Apply Vocabulary

Use the highlighted words from the lesson to complete the sentence correctly.

1. All the information that an animal needs to live is stored in its ________.
2. The part of Earth where living things are found is called the ________.
3. All the reactions by which a living thing builds up and breaks down matter is its ________.

Critical Thinking

4. **Applying Concepts** Suppose you feel hungry, so you reach for an apple in a fruit bowl. How do stimuli both outside and inside of your body lead you to take the apple?
5. **Relate Cause and Effect** Suppose that two scientists are doing an experiment using dangerous chemicals. How might their safety be affected by not using a common measurement?
6. **Write to Learn** Write the answer to the third clue of the mystery.

MYSTERY CLUE

What human values or biases are involved in the case of giving HGH to healthy children? What role does science play in this case?

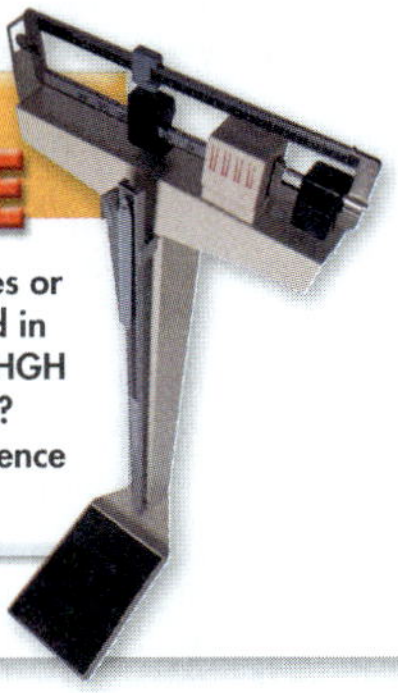

BIOLOGY.com Search Lesson 1.3 GO • Lesson Assessment

Check Understanding

1. DNA
2. biosphere
3. metabolism
4. Hunger is a stimulus inside your body and seeing the apple is a stimulus outside of your body. Feeling the hunger and seeing the apple leads you to take the apple.
5. Sample answer: If these scientists did not use a common system of measurement, they could mistakenly mix the wrong amount of chemicals, which could lead to dangerous consequences.
6. Students' response should reflect an understanding of the term *bias*. The responses should describe how human values and biases are part of the decision to give healthy children HGH. For example, parents may be concerned about a bias in society against short people. Students should understand that the role of pure science is to determine what can be done, not whether it should be done.

Active Reading

INTERPRET TABLES Review the information in the table. Demonstrate that the metric system is easy to use by having students calculate two things: the number of centimeters in 6 meters *(600 cm)* and the number of inches in 6 yards *(216 in.).*

Ask Which calculation was easier? Why? *(Sample answer: It was easier to convert from meters to centimeters because the conversion was based on multiples of 10.)*

Speed Bump

Ask What prefix indicates one thousand? *(kilo-)*

Ask What are the keys to staying safe in a biology lab? *(following safety rules and careful preparation)*

Post Reading

For Question 4, tell students that *apply* means "to use information previously learned to explain what happens in different but somewhat similar situations." Explain that the answer to this question is not found directly in the text.

For Question 5, explain that *relate* means "to make a connection" and that an effect is the outcome of a cause. Provide a simple safety example of a cause and effect, such as not wiping up a spill, which leads to a fall.

MYSTERY CLUE After students read the clue, remind them that values have to do with right and wrong or good and bad and that a bias is a personal opinion.

Wrap-Up Activity

SCIENTIFIC PURSUIT Have each student write one question about the information on each page of the lesson. Then, give each student seven index cards. Have students copy the question on one side of the card and the answer on the other side. Also, have them print the name of a category at the top of the question side of each card. Provide the following category names: Characteristics of Life, Big Ideas of Biology, Fields of Biology, and Performing Biological Investigations.

Sort the cards by category, and place them in separate envelopes. Then, have students play a game in which each competitor selects a category, and a game moderator draws a corresponding card and reads the question. If a competitor answers correctly, a point is awarded and the competitor plays again. If the answer is incorrect, the next competitor selects a category and answers a question. Students can compete individually or in teams of two or more.

CHAPTER LAB

Pre-Lab

Review the parts of a compound microscope and the instructions for using the microscope.

Ask Where are the lenses located in a compound microscope? *(in the ocular, or eyepiece, and in the objectives)* How many objective lenses does the compound microscope you will be using have? *(Answers may vary.)* How do you vary the total magnification of a compound microscope? *(by rotating the nosepiece to a higher-power or lower-power objective)* Which objective do you use when locating an object to be viewed and bringing it into focus? *(the low-power objective)*

Review the Background Questions and Pre-Lab Questions as a class.

Lab

Tell students they will perform the chapter lab *Using a Microscope to Estimate Size* described in **Lab Manual B.**

SAFETY Students should use caution when handling the microscope slides.

Post-Lab

LAYERED BOOK Have each lab group make a layered book that gives the instructions for estimating size with a microscope. Encourage students to use drawings along with clear and concise text to illustrate and explain the process.

Have lab groups exchange finished books and use them to estimate the size of another object.

Skills Lab

GUIDED INQUIRY

Pre-Lab: Using a Microscope to Estimate Size

Problem How can you use a microscope to estimate the size of an object?

Materials compound microscope, transparent 15-cm plastic ruler, prepared slide of plant root or stem, prepared slide of bacteria

Lab Manual Chapter 1 Lab

Skills Focus Observe, Measure, Calculate, Predict

Connect to the Big idea Science provides a way of knowing the world. The use of technology to gather data is a central part of modern science. In biology, the compound microscope is a vital tool. With a microscope, you can observe objects that are too tiny to see with the unaided eye. These objects include cells, which are the basis for all life.

In this lab, you will explore another important use of the microscope. You will use the microscope to estimate the size of cells.

Background Questions

a. **Explain** How did the invention of the microscope help scientists know the natural world?

b. **Explain** How can a microscope help a scientist use scientific methodology?

c. **Infer** List one important fact about life that scientists would not know without microscopes. *Hint:* Review the characteristics of living things.

Pre-Lab Questions

Preview the procedure in the lab manual.

1. **Review** Which lens provides more magnification—a low-power lens or a high-power lens?
2. **Infer** Which lens will provide the larger field of view—a low-power lens or a high-power lens?
3. **Calculate** Eight cells fit across a field of view of 160 μm. What is the width of each cell? MATH
4. **Predict** Which cell do you think will be larger, the plant cell or the bacterial cell? Give a reason for your answer.

BIOLOGY.com Search Chapter 1 GO

Visit Chapter 1 online to test yourself on chapter content and to find activities to help you learn.

Untamed Science Video Be prepared for some surprise answers as the Untamed Science crew hit the streets to ask people basic questions about science and biology.

Art in Motion Learn about the steps scientists use to solve problems. Change the variables, and watch what happens!

Art Review Review your understanding of the various steps of experimental processes.

InterActive Art Design your own experiment to test Redi's and Pasteur's spontaneous generation experiments.

Pre-Lab Answers

BACKGROUND QUESTIONS

a. Sample answer: Microscopes allowed scientists to gather information about parts of the natural world that they could not previously observe.

b. Most students will say that microscopes help scientists observe and collect data. Some may know that microscopes can be used to conduct experiments.

c. Sample answer: Living things are made up of units called cells.

PRE-LAB QUESTIONS

1. The high-power lens provides more magnification.
2. The low-power lens provides the larger field of view.
3. 20 μm
4. Accept any answer for which a student provides a reason. For example, some students may predict that plant cells are smaller because a plant has more than one cell.

NATIONAL SCIENCE EDUCATION STANDARDS

UCP III

INQUIRY A.1.c, A.2.c, A.2.d

1 CHAPTER Summary

1.1 What Is Science?

- The goals of science are to give natural explanations for natural events, to understand patterns, and to make predictions.
- Scientific methodology involves observing and asking questions, making inferences and forming hypotheses, doing controlled experiments, collecting and analyzing data, and drawing conclusions.

science (p. 4)
observation (p. 5)
hypothesis (p. 5)
controlled experiment (p. 6)
control group (p. 6)
data (p. 6)

1.2 Science in Context

- Curiosity, skepticism, open-mindedness, and creativity help scientists come up with new ideas.
- Publishing peer-reviewed papers in journals lets scientists share ideas and review each other's work.
- A scientific theory is a well-tested explanation that accounts for a lot of observations and hypotheses and that lets scientists make good predictions.
- Using science involves understanding its limitations and how it fits into society.

theory (p. 11)
bias (p. 12)

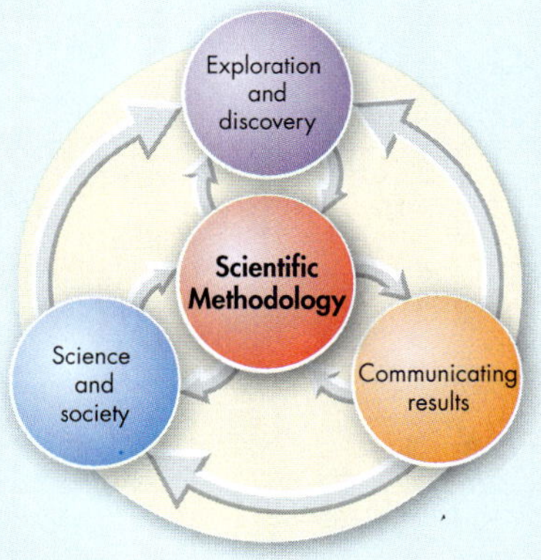

1.3 Studying Life

- Living things are made up of basic units called cells, are based on a universal genetic code, obtain and use materials and energy, grow and develop, reproduce, respond to their environment, maintain a stable internal environment, and change over time.
- Biology's big ideas are the cellular basis of life; information and heredity; matter and energy; growth, development, and reproduction; homeostasis; evolution; structure and function; unity and diversity of life; interdependence in nature; and science as a way of knowing.
- Biology is made up of many overlapping fields that use different tools to study life from the level of molecules to the whole planet.
- Most scientists use the metric system when collecting data and performing experiments.

biology (p. 13)
DNA (p. 14)
stimulus (p. 14)
homeostasis (p. 15)
metabolism (p. 15)
biosphere (p. 17)

UbD Transfer the Big Idea

TRANSFER Have students write an editorial for the school newspaper that explains the importance of studying science. In their editorials, students should explain how science knowledge helps citizens make better decisions about issues that involve science. They should also explain how science-related issues have become more and more important in recent years and will be important for the future. Encourage students to submit their editorials to the school paper for publication.

Chapter Summary

Think Visually

CONCEPT MAP Have students use information from the chapter to complete the following concept map:

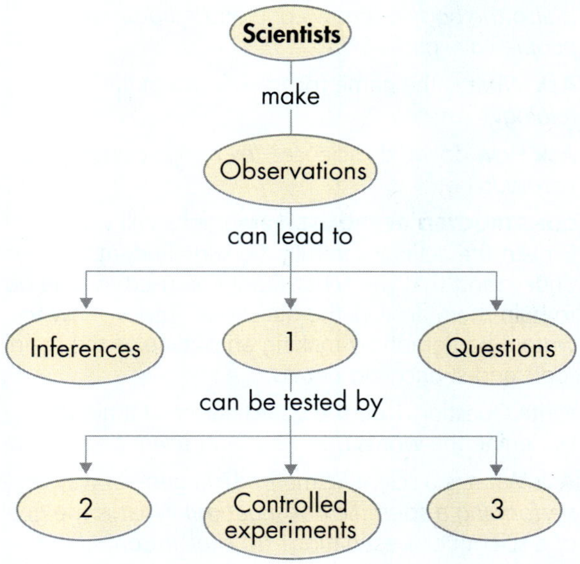

ANSWERS

1. Hypotheses **2.** Observations **3.** Field studies

BIOLOGY.com Study Online

REVIEW RESOURCES

Art in Motion Students can see the impact of changing variables.

InterActive Art Students explore historical experiments.

Data Analysis Students review strategies for obtaining difficult measurement.

Vocabulary Review The *Flash Cards* and *Crossword* are an interactive vocabulary review.

Editable Worksheets Editable pages of Study Workbook B, Lab Manual B, and the Assessment Resources Book are available online.

Lesson Overviews Lesson Overviews are based on Miller & Levine Biology. However, they are available on CD-ROM and can be edited for your class.

ASSESSMENT RESOURCES

Chapter Assessment Students can take an online version of the Chapter 1 Check Understanding.

Standardized Test Prep Students can take an online version of the Standardized Test Prep. You will receive their scores along with ideas for remediation.

Diagnostic and Benchmark Tests Use these tests to monitor your students' progress and supply remediation.

Assess

Big idea Invite a volunteer to read the Big Idea and the question. Write the words *science* and *biology* on the board.

Ask What are two definitions of science? *(Science is the way we go about learning about nature. Science is also the body of knowledge about nature that people have gathered.)*

Ask What is the name for the science of life? *(biology)*

Ask How do we do science? *(by using scientific methodology)*

CONSTRUCTED RESPONSE Paragraphs will vary. Explain the activity carefully. Be sure students understand that the *Hints* should be used as a guide to help them answer the questions. Review how to write a paragraph by making an outline of the main topic and supporting details.

Write Question 1 on the board and read it aloud. Underline the words *questions* and *scientific method.*

Ask What is a scientific method? *(a general way of performing a scientific investigation)* What is the goal of a scientific investigation? *(to explain something that happens in the natural world, such as a pattern in nature)*

Read the first hint.

Ask What do scientists do when they observe? *(They notice and describe what is happening in the natural world.)* What is an example of an observation? *(Sample answer: Some crabs in a population are brown, and others are black.)*

Write the sample observation on the board.

Say Observations lead directly to questions. What is a question that relates to the observation on the board? *(Sample answer: Why do crabs in a population have different colors?)*

Ask What do scientists do after asking a question? *(They make inferences and form a hypothesis, which is an explanation for the observations.)*

Model writing the first one or two sentences using the hints on the board and the text. *(Sample: The goal of a scientific investigation is to explain something that happens in the natural world. By asking questions, scientists identify things that they don't understand, things that need to be explained.)*

1 CHECK Understanding

Assess the Big idea: Science as a Way of Knowing

Write an answer to the question below.

Q: What role does science play in the study of life?

Constructed Response

Write an answer to each of the questions below. The answer to each question should be one or two paragraphs long. To help you begin, read the **Hints** below each of the questions.

1. **Why are questions so important in the scientific method?**

   Hint Scientific investigation begins with observation.

2. **Predict some problems that might happen if some scientists used SI units to measure objects and other scientists used inches.**

   Hint Scientists need to replicate each other's experiments.

   Hint The metric system is based on multiples of 10.

3. **Suppose that a classmate says that a wooden chair is alive. He says that the wood comes from a tree and that trees are alive. Explain why your classmate is right or wrong.**

   Hint If something does not have all the characteristics of living things, it is not alive.

Foundations for Learning Wrap-Up

Use the Inspired Shape Tree that you made when you started the chapter as a tool to help you learn the characteristics of living things.

Activity 1 Remove a branch from your Inspired Shape Tree. Write the characteristic on a sheet of paper. Then, write a brief description of what would happen to a tree if the tree did not have that characteristic. Continue until you have removed all the branches from your Inspired Shape Tree.

Activity 2 Remove the branches from your Inspired Shape Tree. Turn each branch over and write one or two words related to the characteristic on the front side of the branch. Mix up the branches. Have a partner read the words on one of the branches. Try to explain the characteristic of that branch. If your partner thinks that you explained the characteristic clearly, he or she will give you the branch to attach to your tree trunk. Continue until all the branches are picked. Take turns.

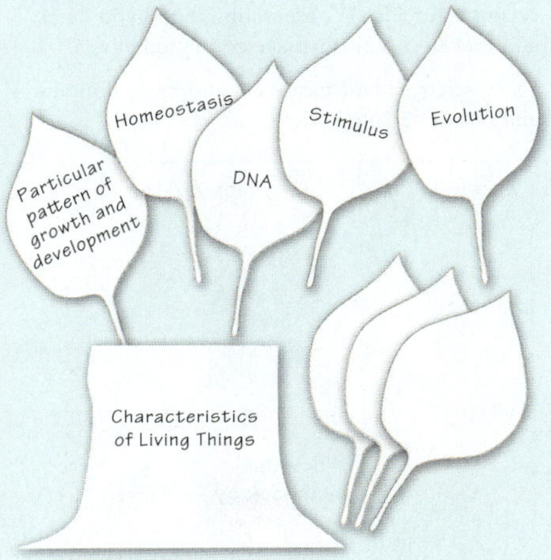

For Question 2, pair students and have them use the hints and your modeling to answer the question. Have students share their paragraphs with the class.

For Question 3, have students write the paragraphs independently for homework or as an anchor activity.

Foundations for Learning Wrap-Up

Have students take out their Inspired Shape Trees. Assign Activity 1 to be completed by individual students. Be sure that students have paper for the activity. Then, pair students and explain Activity 2. Have the pairs of students continue the activity until each member of the pairs has rebuilt his or her tree.

1.1 What Is Science?

Understand Key Concepts

1. Which of the following statements about a controlled experiment is true?
   a. All the variables must be kept the same.
   b. Only one variable is tested at a time.
   c. All hypotheses can be tested by setting up a controlled experiment.
   d. Controlled experiments cannot be performed on living things.

Test-Taking Tip

Read All the Answer Choices If you are not sure of the answer to a question, read all of the choices before picking one. Then begin to cross out incorrect answers. Read question 1 again. If all the variables are kept the same, the experiment would not show any changes in the dependent variable. So, **a** is not correct. Choices **c** and **d** are not correct. Some experiments cannot be tested with controlled experiments, and some controlled experiments can be done on living things. Therefore, **b** is the correct answer.

2. An inference is
   a. the same as an observation.
   b. a logical interpretation of an observation.
   c. a statement involving numbers.
   d. a way to avoid bias.
3. To be useful in science, a hypothesis must be
   a. measurable.
   b. observable.
   c. testable.
   d. correct.

Think Critically

4. **Apply Concepts** Suggest an experiment that would test the hypothesis that one food is better than another at speeding an animal's growth.
5. **Explain** Explain why you cannot draw a conclusion about the effect of one variable in an investigation when other key variables are not controlled.

1.2 Science in Context

Understand Key Concepts

6. A skeptical attitude in science
   a. prevents scientists from accepting new ideas.
   b. makes the acceptance of new ideas more likely.
   c. means a new idea will only be accepted if it is backed by evidence.
   d. is unimportant.
7. The purpose of peer review in science is to ensure that
   a. all scientific research is funded.
   b. the results of experiments are correct.
   c. all scientific results are published.
   d. published results meet standards set by the scientific community.
8. A scientific theory is
   a. the same as a hypothesis.
   b. a well-tested explanation that unifies a broad range of observations.
   c. the same as the conclusion of an experiment.
   d. the first step in a controlled experiment.

Think Critically

9. **Infer** How would having a scientific attitude help you in everyday activities—in trying to learn a new skill, for example?
10. **Apply Concepts** If you were one of the reviewers of a paper submitted for publication, what criteria would you use to determine whether or not the paper should be published?

1.3 Studying Life

Understand Key Concepts

11. The process in which two cells from different parents unite to produce the first cell of a new organism is called
   a. homeostasis.
   b. development.
   c. asexual reproduction.
   d. sexual reproduction.

Check Understanding **23**

Assessment Tips

Pair students and have them work together to answer the questions on this page. Questions 1–3, 6–8, and 11–12 require students to recall and comprehend what they learned about science and the study of life. Questions 4, 5, and 10 require students to apply what they have learned. Question 9 requires students to draw a conclusion based on knowledge and experience. Questions 13–15 require interpretation of a graph. Have students share their answers with the class and explain their reasoning for each answer.

Answers

Lesson 1.1

UNDERSTAND KEY CONCEPTS

1. b
2. b
3. c

THINK CRITICALLY

4. Sample answer: I'd find two groups of young animals of the same species whose weight is approximately the same. Feed each group a different food, and weigh the animals at intervals to find out which group of animals grows more quickly. For the most reliable results, this experiment would need to be done on many animals of the same size and species.

5. If variables are not controlled, there is no way of knowing which variable caused the observed results.

Lesson 1.2

UNDERSTAND KEY CONCEPTS

6. c
7. d
8. b

THINK CRITICALLY

9. Scientific attitudes include curiosity, skepticism, open-mindedness, and creativity. Each of these characteristics would be useful in learning a new skill. Curiosity—leads you to ask questions about the new skill; skepticism—keeps you from accepting explanations without evidence; open-mindedness—enables you to accept different ideas about how the skill could be learned; and creativity—helps you explore different ways the skill could be learned.

10. Sample answer: I would look to see whether the researchers who wrote it showed any bias or unfair influences in their conclusions. I would see if they made any mistakes in their techniques or in their reasoning. I also would make sure that their data supported their conclusions.

Lesson 1.3

UNDERSTAND KEY CONCEPTS

11. d

ASSESSMENT

THINK CRITICALLY

12. (1) This symbol warns that the lab includes glassware or other materials that could be hazardous if broken. (2) This symbol is used to indicate the possibility of electric shock. (3) This symbol means that the lab involves the use of sharp objects that could cause injury. (4) This symbol is a reminder to use heat-resistant gloves when handling hot materials.

Connecting Concepts

13. Graph 1 shows the number of organisms in the population increasing over time. Graph 2 shows an increase in the number of individuals in the population followed by a decrease, which results in no net change in the size of the population. Graph 3 shows several rapid increases in the number of individuals, each followed by a decrease in the number of individuals, resulting in no net change in the size of the population. Graph 4 shows a population that does not change in size over the time period represented in the graph.

14. Numerical values indicating time elapsed on the *x*-axis and number of individuals on the *y*-axis are needed before direct comparisons can be made.

15. Sample answer: The shape of Graph 1 could represent a chemical reaction in which the product accumulates, or it could represent the height of a plant over time.

CHAPTER MYSTERY After students have finished reading the chapter mystery, discuss the ethical and health issues presented.

Ask Why do some parents want their children to take HGH? *(because the children may grow taller and thus not experience the prejudice of society against short people)*

Ask Do you think that these reasons are moral and ethical? Why or why not? *(Answers may vary. Students should give reasons to support their opinions.)*

Ask Why did David's doctor agree to give him HGH? *(to avoid complaints from his parents)*

Ask Do you think the doctor's reason is moral and ethical? Why or why not? *(Answers may vary. Students should give reasons to support their opinions.)*

1 CHECK Understanding

Think Critically

12. Interpret Visuals Each of the following safety symbols might appear in a laboratory activity in this book. Describe what each symbol stands for. (**Hint:** Refer to Appendix B.)

Connecting Concepts

Use Science Graphs

The following graphs show the size of four different populations over a period of time. Use the graphs to answer questions 13–15.

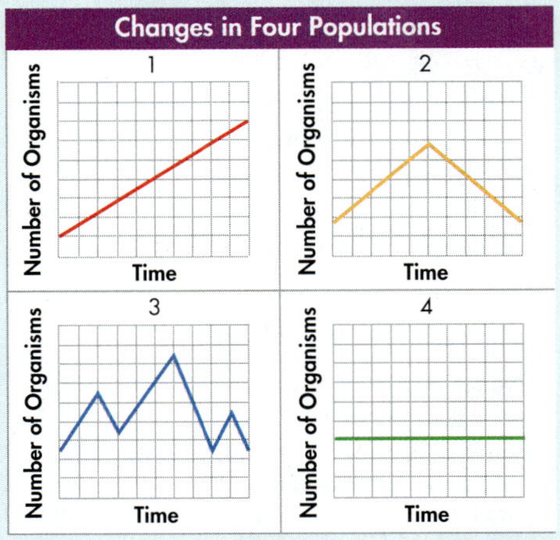

13. Analyze Data Write a sentence summarizing what each graph shows.

14. Interpret Graphs What information is missing from the graphs? (**Hint** Look at the *x* and *y* axes.)

15. Compare and Contrast Graphs of completely different events can look alike. Select one of the graphs and explain how the shape of the graph could apply to a different set of events.

solve the CHAPTER MYSTERY

HEIGHT BY PRESCRIPTION

Although scientific studies have not proved that HGH makes adults much taller, studies suggest that extra HGH may help some short kids grow taller sooner. David's doctor prescribed HGH so that David and his parents would not complain that HGH was not given as an option.

This situation is new. HGH used to be available only from cadavers, or dead bodies, and it was given only to people who had severe medical problems. Then, genetic engineering made it possible to make a lot of HGH—safe medicine for sick people.

Soon drug companies began trying to sell HGH to parents of healthy, short kids.

As David's case shows, science can change lives, but new scientific knowledge may raise more questions than they answer. Just because science makes something possible, does that mean it's right to do it? This question is difficult to answer. When thinking about how science should be applied, we must consider both its limitations and its context in society.

1. Relate Cause and Effect Search the Internet for the latest data on the use of HGH by healthy children.

2. Predict HGH was among the first products to come out of genetic engineering. Many more will follow. As products become available that could change other inherited traits, what challenges will society face?

Untamed Science **Never Stop Exploring Your World.** Finding the solution to the HGH mystery is only the beginning. Take a video field trip with the ecogeeks of Untamed Science to see where this mystery leads.

BIOLOGY.com Search Chapter 1 GO • Untamed Science Video • Chapter Mystery

CHAPTER MYSTERY ANSWERS

1. Responses may vary, depending on resources used. Students should try to find results from the latest studies possible. Remind them to consider who did the research—an HGH manufacturer or independent group?

2. Sample answer: As more genetically engineered products that can change inherited traits become available, society will need to figure out whether the new products are safe and whether they should be used when there is not a medical need.

Untamed Science Students can learn more about what it is like to be a scientist by watching the Untamed Science video **What Do Biologists Look Like?**

Standardized Test Prep

Multiple Choice

1. To ensure that a scientific work is free of bias and meets standards set by the scientific community, a research group's work is peer reviewed by
   A other scientists.
   B the general public.
   C the researchers' friends.
   D lawmakers.

2. The term for the chemical reactions in which organisms build up or break down materials is
   A biotechnology.
   B genetics.
   C metabolism.
   D genomics.

3. Which of the following is NOT an attitude that scientists need?
   A bias
   B skepticism
   C creativity
   D open-mindedness

4. A bird-watcher sees an unusual bird at a feeder. He takes careful notes on the bird's color, shape, and other physical features and then goes to a reference book to see if he can identify the species. What part of scientific thinking is most apparent in this situation?
   A observation
   B inference
   C hypothesis formation
   D controlled experimentation

5. Unlike sexual reproduction, asexual reproduction involves
   A two cells.
   B two parents.
   C one parent.
   D one nonliving thing.

6. One meter is equal to
   A 1000 millimeters.
   B 1 millimeter.
   C 10 kilometers.
   D 1 milliliter.

Questions 7–8

Once a month, a pet owner recorded the mass of her puppy in a table. When the puppy was 3 months old, she started to feed it a "special puppy food" she saw advertised on TV.

Change in a Puppy's Mass Over Time

Age (months)	Mass at Start of Month (kg)	Change in Mass per Month (kg)
2	5	–
3	8	+3
4	13	+5

7. According to the table, which statement is true?
   A The puppy's mass increased at the same rate for each month shown.
   B The puppy's mass was less than 5 kg at the start of the new diet.
   C The puppy gained 5 kg between age 3 and 4 months.
   D The puppy had gained 13 kg as a result of the new diet.

8. All of the following statements about the pet owner's study are true EXCEPT
   A The owner used the metric system.
   B The owner recorded data.
   C The owner could graph the data.
   D The owner conducted a controlled experiment.

Open-Ended Response

9. Explain how a controlled experiment works.

If You Have Trouble With . . .

Question	1	2	3	4	5	6	7	8	9
See Lesson	1.2	1.3	1.2	1.1	1.3	1.3	1.1	1.1	1.1

Test-Taking Tip

USE TIME WISELY

Advise students to skip to the next question if they get stuck on a difficult one. Suggest that they respond to all the questions they can answer easily first, and then go back to questions they have trouble answering. For these difficult questions, make sure students take the time to reread the entire question before attempting to answer it.

Before the Test

Review with students the Test-Taking Tip.

Point out the word *NOT* in Question 3 and the word *EXCEPT* in Question 8. Also point out that the answers to Questions 7 and 8 can be found by studying the data in the table.

Remind students that *open-ended response* means that they'll have to write a paragraph to answer the question in their notebooks. Have them paraphrase the question in their notebooks, and remind them about the steps in writing a paragraph.

Answers

1. A
2. C
3. A
4. A
5. C
6. A
7. C
8. D
9. Sample answer: In a controlled experiment, a scientific hypothesis is tested by keeping all variables unchanged except for one. The variable that is deliberately changed is called the independent variable, and it should be the only factor that is different between the control group and the experimental group.

CHAPTER PLANNER 2 The Chemistry of Life

Chapter Contents	NSES	Time	Core Resources
Chapter Preview			**Student Edition,** pp. 26–27 **Chapter Mystery,** p. 27 **Foundations for Learning,** p. 27
2.1 The Nature of Matter Atoms • Elements and Isotopes • Chemical Compounds • Chemical Bonds	I, II, A.1.b, A.1.d, B.1, B.2, B.4	1 period 1/2 block	**Student Edition,** pp. 28–32 **Inquiry Into Scientific Thinking,** p. 32 **Study Workbook B** 2.1 Worksheets **Biology.com** *Art Review:* Ionic and Covalent Bonding • 2.1 Lesson Assessment
2.2 Properties of Water The Water Molecule • Solutions and Suspensions • Acids, Bases, and pH	I, III, B.1, B.2, B.3, B.4	1 period 1/2 block	**Student Edition,** pp. 33–36 **Study Workbook B** 2.2 Worksheets **Biology.com** *Art in Motion:* A Salt Solution • 2.2 Lesson Assessment
2.3 Carbon Compounds The Chemistry of Carbon • Macromolecules	I, V, B.1, B.2, C.1.a, C.1.c, C.1.d, C.2.a	1 period 1/2 block	**Student Edition,** pp. 37–41 **Study Workbook B** 2.3 Worksheets **Biology.com** 2.3 Lesson Assessment
2.4 Chemical Reactions and Enzymes Chemical Reactions • Energy in Reactions • Enzymes	II, III, IV, V, B.3, B.5, C.1.a, C.1.b, C.1.d, C.1.e, C.5.b, C.5.c	1 period 1/2 block	**Student Edition,** pp. 42–45 **Build Connections,** p. 45 **Study Workbook B** 2.4 Worksheets **Biology.com** *Visual Analogy:* Lock and Key • 2.4 Lesson Assessment
Chapter Pre-Lab	I, II, III, A.1.b, A.2.a, B.3, C.1.d	1 period 1/2 block	**Student Edition,** p. 46 **Lab Manual B** *Temperature Affects Enzymes*

Tools for ELL Instruction

ELL Handbook describes teaching strategies to make *Miller & Levine Biology: Foundation Edition* more accessible to ELL students.

Spanish Study Workbook is a Spanish translation of Study Workbook A.

Multilingual Glossary is the *Miller & Levine Biology* glossary translated into ten languages.

Tools to Challenge Students

Study Workbook A includes more challenging worksheets and activities such as Chapter Mystery activities that reinforce 21st Century Skills.

Lab Manual A labs are geared toward on-level students. The Lab Manual includes fifteen labs in addition to the Chapter Labs.

Biology.com Data Analysis activities offer students a way to practice data gathering and analyzing skills.

Visual Quizzes in the Assessment Resources Book are geared toward on-level students and use visuals from *Miller & Levine Biology.*

Additional Resources

Biology.com Untamed Science Video • Flash Cards

Study Workbook A 2.1 Worksheets
Lab Manual B *Hands-On Activity:* Model an Ionic Compound
Spanish Study Workbook 2.1 Worksheets

Study Workbook A 2.2 Worksheets
Spanish Study Workbook 2.2 Worksheets
Biology.com *Data Analysis:* Acid Rain

Study Workbook A 2.3 Worksheets
Lab Manual B *Data Analysis:* Comparing Fatty Acids
Spanish Study Workbook 2.3 Worksheets
Assessment Resources Book Visual Quiz

Study Workbook A 2.4 Worksheets
Spanish Study Workbook 2.4 Worksheets
Assessment Resources Book Visual Quiz

Lab Manual A *Temperature Affects Enzymes* • *Making Models of Macromolecules*

Chapter Review

Student Edition Chapter Summary, p. 47 • Unit Project, p. 52
Study Workbook B Chapter 2 Review
Transparencies, pp. 13–25
Biology.com Untamed Science Video • You're the Director • Editable Worksheets of Study Workbooks A and B and Lab Manuals A and B • Chapter 2 Flash Cards and Match It

Untamed Science DVD • Classroom Resources CD (includes lesson presentations and editable worksheets)

Chapter Assessment

Student Edition Check Understanding, pp. 48–51
Study Workbook B Chapter 2 Taking a Standardized Test
Assessment Resources Book Chapter 2 Test B • Unit 1 Test B
Biology.com Chapter 2 Assessment • Editable Worksheets of Chapter 2 Visual Quizzes,* Chapter 2 Tests A and B, and Unit 1 Tests A and B

Exam*View* *Assessment Suite* • Classroom Resources CD (includes lesson presentations and editable worksheets)

Time: 1 period, 1/2 block

* Visual Quizzes are based on visuals from *Miller & Levine Biology*. Review carefully to make sure they are appropriate for your class.

Pressed for Time?

Preview the Chapter Introduce students to the Key Questions and vocabulary for Lessons 2.3 and 2.4.

Cover the Chapter Quickly Have students read *The Water Molecule* in Lesson 2.2 and all of Lesson 2.3, focusing on the Polymerization visual. Have them read all of Lesson 2.4 and go over the Effect of Enzymes graph.

Assess Assign Question 3 in the 2.2 Check Understanding, the 2.3 Check Understanding, the 2.4 Check Understanding, and Questions 7 and 9–19 in the Chapter 2 Check Understanding.

Get the Picture

Have students brainstorm some words that describe the environmental conditions in the picture. *(cold, icy, freezing)* Have students use their words in sentences that begin *This environment is . . .*

Ask What important chemical dominates this photograph? *(water)* What challenges do animals who live in this type of environment face? *(Sample answers: staying warm, having access to liquid water)*

Activate Prior Knowledge

Review with students the characteristics of living things described in Chapter 1. Then ask them to reread the caption on the next page.

Ask What characteristic of living things is being referenced in the caption? *(Living things obtain and use materials and energy.)*

Tell students that, over millions of years, populations have evolved adaptations to allow them to survive in their habitats. Polar bears have adaptations that allow them to survive on the Svalbard islands.

Explain that most life processes involve chemical reactions and that living things, such as the polar bear, use the energy from food to fuel these chemical reactions. Some of these reactions produce heat, which many living things use to warm their bodies.

Ask What are some chemical reactions or processes that produce heat? *(Sample answers: burning of wood, use of heat packs, and burning of fuel in car engines)*

NATIONAL SCIENCE EDUCATION STANDARDS

UNIFYING CONCEPTS AND PROCESSES

I, II, III, IV, V

CONTENT

B.1, B.2, B.3, B.4, B.5, C.1.a, C.1.b, C.1.c, C.1.d, C.1.e, C.2.a, C.5.b, C.5.c

INQUIRY

A.1.a, A.1.b, A.1.d, A.2.a

2 The Chemistry of Life

Big idea **Matter and Energy**

Q: What are the basic chemical principles that affect living things?

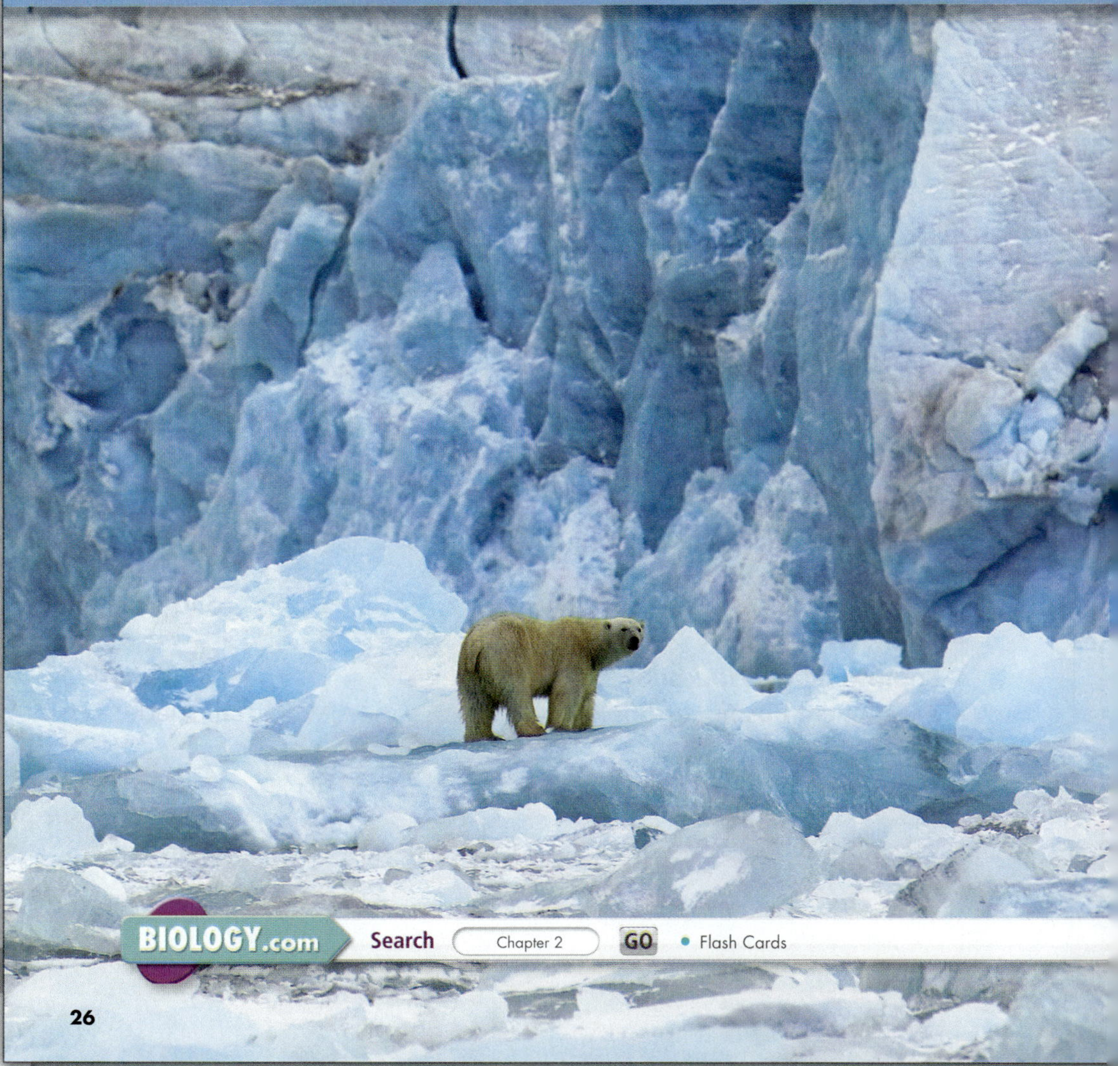

BIOLOGY.com Search Chapter 2 GO • Flash Cards

26

UbD Connect to the Big Idea

Invite a volunteer to read aloud the chapter number and the chapter title. Read the Big Idea and question. Emphasize the question, **What are the basic chemical principles that affect living things?**

Set the Purpose

PREVIEW AND PREDICT Draw the graphic organizer on the board. Invite a volunteer to read the questions.

Ask How does the chapter title relate to the lesson questions in the graphic organizer?

Explain that matter is anything that takes up space, and that chemistry is the study of matter. Tell students that living things are made of matter. Remind students to search for ways that chemical principles affect living things as they read the chapter.

INSIDE:

Polar bears live on the Svalbard islands of Norway. Living conditions on the islands are hard. The water is locked in ice. But some living things can still get the matter and energy that they need to survive.

- Untamed Science Video
- Chapter Mystery

CHAPTER MYSTERY

THE GHOSTLY FISH

Most fish have red blood, just like other vertebrates. Red blood cells carry oxygen, a gas needed for life. The cell's red color comes from an oxygen-binding protein called hemoglobin (HEE muh gloh bin).

But some fish don't have red blood cells. Their blood is clear. Because they live in cold antarctic waters and look ghostly, they are called "ice fish." How do ice fish live without red blood cells?

Read for Mystery Clues As you read this chapter, look for clues to help you explain the ice fish's clear blood. Think about the chemistry that might be involved. Then, solve the mystery.

FOUNDATIONS for Learning

Vocabulary words are important for learning the key ideas of this chapter. Before you read this chapter, make a vocabulary quiz sheet. Fold a sheet of paper into three equal parts lengthwise. As you read the chapter, write the vocabulary words in the first column. Write the definition for each word in the second column next to the word. Then, cut the third column to make tabs. The tabs will cover up the definitions. At the end of the chapter are two activities that use the vocabulary quiz sheet to answer the question: What are the basic chemical principles that affect living things?

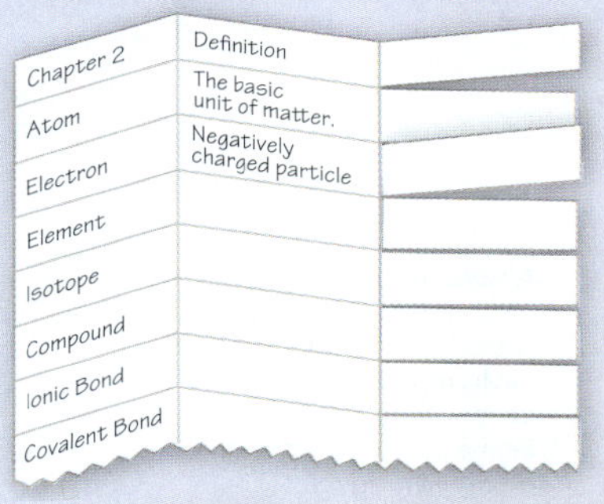

Chapter 2 Big Idea: Matter and Energy

Chapter 2: What are the basic chemical principles that affect living things?

- **2.1:** What is the matter in organisms made of?
- **2.2:** Why are the properties of water important to organisms?
- **2.3:** How do organisms use different types of carbon compounds?
- **2.4:** How do chemicals combine and break apart inside living things?

Foundations for Learning

Guide students as they make their Undercover Vocabulary Tool. Recommend that they use a sheet of lined paper. Remind students to space the vocabulary words in the first column so that they have room to write the definitions in the second column. Tell them to cut along corresponding lines to make tabs.

Ask Why is it important to know the meanings of vocabulary words? *(Knowing the meanings of vocabulary words helps me to understand the material in the chapter.)*

Ask How can you use the vocabulary sheet to study for a quiz? *(I can cover the definitions and ask myself what each word means. I can uncover the definitions one at a time to see if I was right.)*

Tell students that they will use the vocabulary quiz sheets as a review and study aid. Demonstrate how the sheets are used by covering the meaning of one word with a tab.

Untamed Science

Follow the ecogeeks of Untamed Science as they explore the unique chemistry of water in **Not a Drop to Drink.**

CHAPTER MYSTERY

Ask students if they know what a red blood cell is and what its function is. Invite a volunteer to read aloud the Chapter Mystery. Have students predict how the mystery is related to chemistry, red blood cells, and the chapter title.

Tell students that they will find clues in the chapter that answer the question, **How do ice fish live without red blood cells?** As they read the chapter, they should relate what they read to the function of red blood cells, oxygen, and the ice fish.

Digital Support

BIOLOGY.com Extend your reach by using the digital assets offered at **Biology.com.**

LESSON 2.1

Pre-Reading

Objectives

2.1.1 Identify the three subatomic particles found in atoms.

2.1.2 Explain how all of the isotopes of an element are similar and how they are different.

2.1.3 Explain how compounds are different from their component elements.

2.1.4 Describe the two main types of chemical bonds.

Preview the Pages

1. Preview the vocabulary words by asking students to find the words in the text. Ask if they have heard the words before. Have students define each word using context clues.
2. Point out the nucleus and three kinds of subatomic particles in the diagrams of atoms so students learn the location of each.

Activate Prior Knowledge

Ask In what other context have you heard the word *nucleus* while studying science? *(A cell has a nucleus.)*

Build Understanding

PREVIEWING VISUALS Direct students to Workbook B, where they will find a partially filled-in T-chart. As they preview the visuals in the lesson, they should write down questions they have. As they read, they should fill in answers.

NATIONAL SCIENCE EDUCATION STANDARDS

UNIFYING CONCEPTS AND PROCESSES

I, II

CONTENT

B.1, B.2, B.4

INQUIRY

A.1.d

For the corresponding lesson in *Miller & Levine Biology,* see pages 34–38.

2.1 The Nature of Matter

Key Questions

What three particles make up atoms?

How are all of the isotopes of an element similar?

How do compounds differ from the elements in them?

What are the main kinds of chemical bonds?

BUILD Understanding

Previewing Visuals Before you read the chapter, study the carbon atom figure below. Learn where in an atom you can find protons, neutrons, and electrons.

In Your Workbook Go to your workbook for a way to preview visuals.

Atoms

Nearly 2500 years ago, the Greek philosopher Democritus asked a simple question: If you break a stick of chalk in half, are both parts still chalk? The answer, of course, is yes. What happens if you break it in half again and again? Can you keep breaking it without changing it into something else? Democritus thought that there had to be a limit, and he was right. He called the smallest piece the atom. The **atom** is the basic unit of matter. The study of chemistry begins with atoms.

Atoms are very small. Placed side by side, 100 million atoms would make a row only about as wide as your little finger! Although they are small, atoms are made up of even smaller particles called protons, neutrons, and electrons.

Protons and Neutrons Protons and neutrons have about the same mass. However, protons are positively charged particles (+) and neutrons carry no charge at all. Strong forces hold protons and neutrons together to form the nucleus. The **nucleus** is the center of the atom.

Electrons An **electron** is a negatively charged particle (–). An electron has only 1/1840 the mass of a proton. Electrons are always moving around the nucleus. They are attracted to the positively charged nucleus. But electrons stay outside the nucleus because of the energy of their motion. Atoms have equal numbers of electrons and protons. So, the positive and negative charges balance and atoms have no charge.

Key Question What three particles make up atoms?
Protons, neutrons, and electrons are the particles that make up atoms.

A Carbon Atom A carbon atom has 6 protons and 6 electrons, in addition to the neutrons in its nucleus. Like all atoms, it has no charge, since the number of protons and electrons is equal.

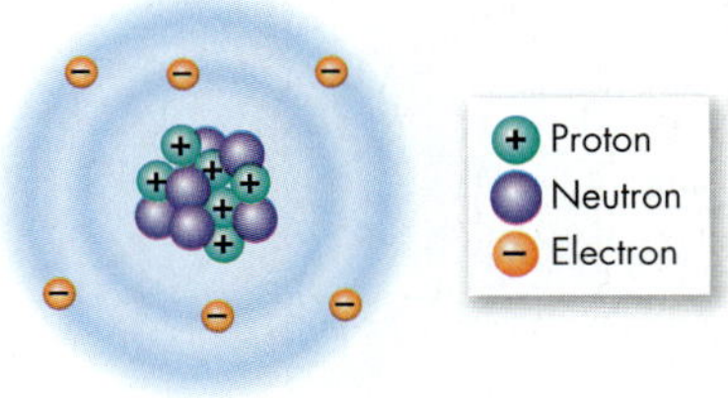

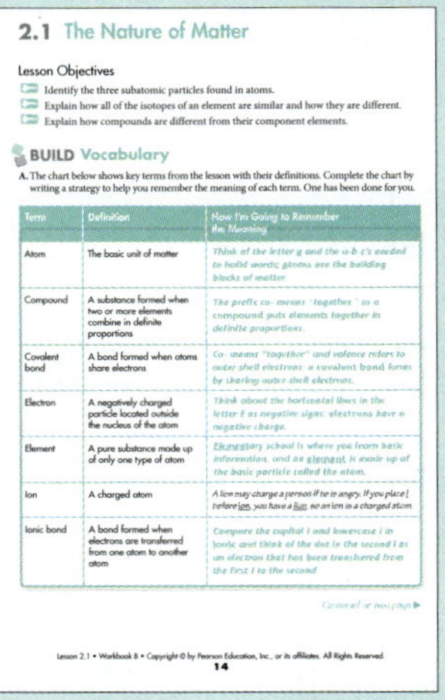

2.1 The Nature of Matter

Lesson Objectives

Identify the three subatomic particles found in atoms.
Explain how all of the isotopes of an element are similar and how they are different.
Explain how compounds are different from their component elements.

BUILD Vocabulary

A. The chart below shows key terms from the lesson with their definitions. Complete the chart by writing a strategy to help you remember the meaning of each term. One has been done for you.

Term	Definition	How I'm Going to Remember the Meaning
Atom	The basic unit of matter	Think of the letter a and the a-b-c's needed to build words; atoms are the building blocks of matter.
Compound	A substance formed when two or more elements combine in definite proportions	The prefix co- means "together," so a compound puts elements together in definite proportions.
Covalent bond	A bond formed when atoms share electrons	Co- means "together" and valence refers to outer shell electrons: a covalent bond forms by sharing outer shell electrons.
Electron	A negatively charged particle located outside the nucleus of the atom	Think about the horizontal lines in the letter E as negative signs: electrons have a negative charge.
Element	A pure substance made up of only one type of atom	Elementary school is where you learn basic information, and an element is made up of the basic particle called the atom.
Ion	A charged atom	A lion may charge a person if he is angry. If you place l before ion, you have a lion, so an ion is a charged atom.
Ionic bond	A bond formed when electrons are transferred from one atom to another atom	Compare the capital I and lowercase i in ionic and think of the dot in the second i as an electron that has been transferred from the first i to the second.

Continued on next page ▶

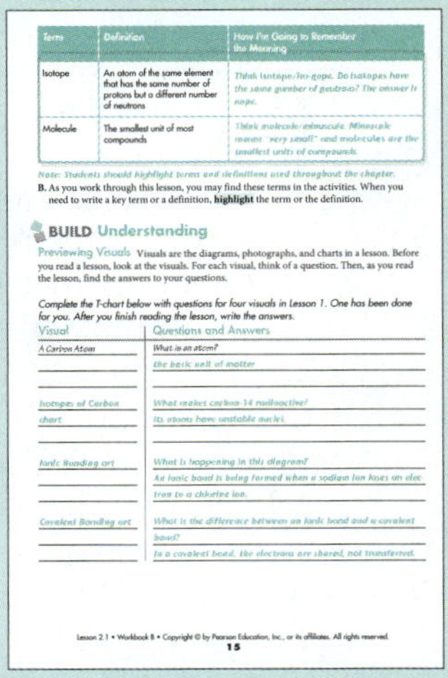

Term	Definition	How I'm Going to Remember the Meaning
Isotope	An atom of the same element that has the same number of protons but a different number of neutrons	Think isotope/iso-nope. Do isotopes have the same number of neutrons? The answer is nope.
Molecule	The smallest unit of most compounds	Think molecule/minuscule. Minuscule means "very small" and molecules are the smallest units of compounds.

Note: Students should highlight terms and definitions used throughout the chapter.

B. As you work through this lesson, you may find these terms in the activities. When you need to write a key term or a definition, **highlight** the term or the definition.

BUILD Understanding

Previewing Visuals Visuals are the diagrams, photographs, and charts in a lesson. Before you read a lesson, look at the visuals. For each visual, think of a question. Then, as you read the lesson, find the answers to your questions.

Complete the T-chart below with questions for four visuals in Lesson 1. One has been done for you. After you finish reading the lesson, write the answers.

Visual	Questions and Answers
A Carbon Atom	What is an atom? the basic unit of matter
Isotopes of Carbon chart	What makes carbon-14 radioactive? Its atoms have unstable nuclei.
Ionic Bonding art	What is happening in this diagram? An ionic bond is being formed when a sodium ion loses an electron to a chlorine ion.
Covalent Bonding art	What is the difference between an ionic bond and a covalent bond? In a covalent bond, the electrons are shared, not transferred.

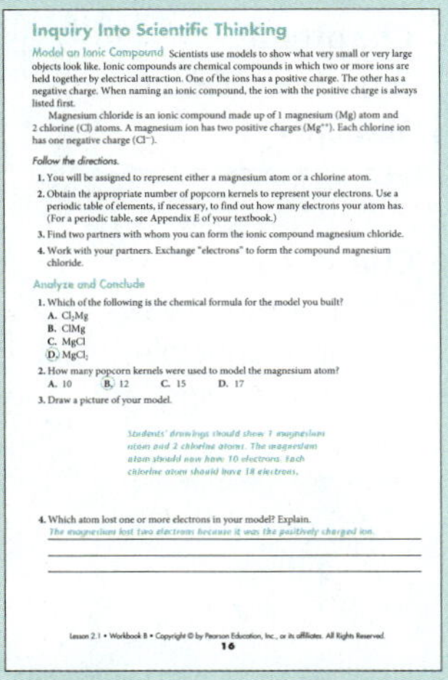

Inquiry Into Scientific Thinking

Model an Ionic Compound Scientists use models to show what very small or very large objects look like. Ionic compounds are chemical compounds in which two or more ions are held together by electrical attraction. One of the ions has a positive charge. The other has a negative charge. When naming an ionic compound, the ion with the positive charge is always listed first.

Magnesium chloride is an ionic compound made up of 1 magnesium (Mg) atom and 2 chlorine (Cl) atoms. A magnesium ion has two positive charges (Mg^{++}). Each chlorine ion has one negative charge (Cl^-).

Follow the directions.

1. You will be assigned to represent either a magnesium atom or a chlorine atom.
2. Obtain the appropriate number of popcorn kernels to represent your electrons. Use a periodic table of elements, if necessary, to find out how many electrons your atom has. (For a periodic table, see Appendix E of your textbook.)
3. Find two partners with whom you can form the ionic compound magnesium chloride.
4. Work with your partners. Exchange "electrons" to form the compound magnesium chloride.

Analyze and Conclude

1. Which of the following is the chemical formula for the model you built?
 A. Cl_2Mg
 B. ClMg
 C. MgCl
 D. $MgCl_2$
2. How many popcorn kernels were used to model the magnesium atom?
 A. 10 B. 12 C. 15 D. 17
3. Draw a picture of your model.
 Students' drawings should show 1 magnesium atom and 2 chlorine atoms. The magnesium atom should now have 10 electrons. Each chlorine atom should have 18 electrons.
4. Which atom lost one or more electrons in your model? Explain.
 The magnesium lost two electrons because it was the positively charged ion.

Elements and Isotopes

An **element** is a pure substance that is made up of only one kind of atom. Each element has a one- or two-letter symbol. For example, the symbol for carbon is C, and the symbol for sodium is Na. The number of protons in the atom of an element is the element's atomic number. Carbon's atomic number is 6 because carbon atoms have 6 protons. Carbon atoms also have 6 electrons.

Isotopes Atoms of an element may have different numbers of neutrons. For example, all atoms of carbon have 6 protons, but each atom may have 6, 7, or 8 neutrons. Atoms of the same element that have different numbers of neutrons are **isotopes** (EYE suh tohps). The total number of protons and neutrons in an atom is the atom's mass number. Isotopes are named using their mass numbers. A carbon atom with 6 neutrons has a mass number of 12 and is called carbon-12. The table below shows the isotopes of carbon.

Radioactive Isotopes Some isotopes are radioactive, which means that the nucleus of the atom breaks down at a constant rate over time. When these isotopes break down, they give off energy or particles called radiation. Radiation can be dangerous, but radioactive isotopes have important uses.

Radioactive isotopes can help determine the ages of rocks and fossils. Radiation can help find and treat cancer and kill bacteria. Radioactive isotopes can also serve as labels or "tracers" to follow how substances move in organisms.

Key Question How are all of the isotopes of an element similar?
Isotopes of an element all have the same number of protons.

BUILD Vocabulary

atom
the basic unit of matter

nucleus
the center of an atom, which contains the protons and neutrons

electron
a negatively charged particle; located in the space surrounding the nucleus

element
a pure substance that consists entirely of one type of atom

isotope
one of several forms of a single element, each of which contains the same number of protons but different numbers of neutrons

PREFIXES
The prefix *iso-* means "the same." Isotopes are atoms that have the same number of protons but different numbers of neutrons.

Isotopes of Carbon

Isotope	Number of Protons	Number of Electrons	Number of Neutrons
Carbon–12 (nonradioactive)	6	6	6
Carbon–13 (nonradioactive)	6	6	7
Carbon–14 (radioactive)	6	6	8

Carbon Isotopes Isotopes of carbon have the same number of protons but different numbers of neutrons. Isotopes are named by their mass number, which is the total number of protons and neutrons. So, an isotope that has 6 protons and 6 neutrons is called carbon-12.

During Reading

Active Reading

KEY QUESTION Read the Key Question on the previous page aloud. Have a volunteer read the answer.

Ask Atoms typically have equal numbers of electrons and protons. What does this tell you about their charge? *(They have no charge, because the + charges balance the – charges.)*

INTERPRET CHARTS AND TABLES Review the information in the table of carbon isotopes. Have students note the column heads. Point out that for each isotope, the numbers are identical in two columns but not in the third.

Say The table shows that carbon has three isotopes.

Ask Do the isotopes of carbon differ in their nucleus or in the space around the nucleus? *(In the nucleus. They have different numbers of neutrons, which are in the nucleus.)*

Ask If an isotope of carbon with 9 neutrons were discovered, what would its name be? *(carbon-15)*

KEY QUESTION Read aloud the Key Question and answer on this page.

Ask Nitrogen has an atomic number of 7. How many protons does a nitrogen atom have? *(7)* How many electrons does a nitrogen atom have? *(7)* How many neutrons does nitrogen-14 have? *(7)* How many neutrons does nitrogen-15 have? *(8)*

Build Vocabulary

PREFIXES Ask students if they have ever seen a weather map. Explain that the lines on the map are called isobars and isotherms. (*bar* = air pressure; *therm* = temperature)

Ask What do you know about places connected by the same isotherm on a weather map? *(They have the same temperature.)* What do you know about places that are not connected by the same isobar? *(They have different air pressures.)*

Speed Bump

Ask Where are electrons found in an atom? *(in the space around the nucleus)*

Ask What is an element? *(a pure substance that has only one kind of atom)*

Ask What happens when a radioactive isotope breaks down? *(It releases energy or particles called radiation.)*

Focus on Academic Language

- neutron
- nucleus

Point out to students that the first syllables of the words *neutron* and *nucleus* sound alike but contain different letter combinations. Explain that the spelling is different because the two words come from different root words. *Neutron* comes from the word *neutral.* The root for *nucleus* means "nut" or "center." Ask students if they can think of two other letter combinations that produce the same sound as *neu-* and *nu-*. (new- *and* noo-)

Make a four-column chart on the board, using these four letter combinations (*neu-, nu-, new-,* and *noo-*) as column heads. Have students work in pairs to think of other words that use one of the four letter combinations. Invite volunteers to write their words in the appropriate column.

Ask Do the different letter combinations have the same or different meanings? *(different)* Explain why. *(They come from root words that have different meanings.)*

During Reading continued

Active Reading

USE CONTEXT CLUES Have a volunteer read the second paragraph under Chemical Compounds. Help students use context clues to figure out the meaning of the phrase *physical and chemical properties.*

Ask What is a property? *(A property is what a substance looks like and what it does. A property is a characteristic.)* Does the paragraph give any examples of the physical properties of the elements that make up water? *(It says that hydrogen and oxygen are gases.)* Name a chemical property of these two elements. *(They combine to form water.)*

SUMMARIZE Have students read the paragraph headed Ionic Bonds. Have them write two or three sentences in their notebooks that explain in their own words what an ionic bond is.

Ask What holds the atoms together in an ionic bond? *(the attraction between the positive and negative ions)*

INTERPRET DIAGRAMS Have students examine the diagram of ionic bonding. Have volunteers identify the nucleus, protons, neutrons, and electrons in the sodium atom. Point out that one electron moves from a sodium atom to a chlorine atom.

Ask Does the nucleus take part in the formation of an ionic bond? *(No, only electrons take part.)*

KEY QUESTION Have a volunteer read the Key Question and answer.

Say Water is a compound.

Ask What elements make up water? *(hydrogen and oxygen)*

Ask Why isn't water an element? *(Water contains two different kinds of atoms. An element contains only one kind of atom.)*

Build Vocabulary

PREFIXES Have students brainstorm other words that have a *co-* prefix.

Ask What does it mean to cooperate? *(to work together)*

Speed Bump

Ask What is an ion? *(an atom that has gained or lost at least one electron, becoming electrically charged)*

Ask What kind of charge does an atom have when it gains electrons? *(a negative charge)* Why? *(The atom now has more negative electrons than positive protons.)*

BUILD Vocabulary

compound
a substance formed by the chemical combination of two or more elements in definite proportions

ionic bond
a chemical bond formed when one or more electrons are transferred from one atom to another

covalent bond
a type of bond between atoms in which the electrons are shared

molecule
the smallest unit of most compounds that displays all the properties of that compound

PREFIXES
The prefixes *com-* and *co-* mean "together." A compound is made up of elements joined together, and a covalent bond contains atoms that share electrons together.

Chemical Compounds

Most elements are not found as single atoms. Instead, they are joined with other elements in compounds. A **compound** is a substance formed by the chemical combination of two or more elements in definite amounts. The number of each element in a compound can be shown by chemical formulas. For example, water has 2 atoms of hydrogen and 1 atom of oxygen. So, the chemical formula for water is H_2O.

The physical and chemical properties of a compound are different from the properties of the elements that make up the compound. For example, hydrogen and oxygen are gases at room temperature. But they can combine explosively to form liquid water.

Key Question How do compounds differ from the elements in them?

Compounds have chemical and physical properties that are different from the properties of the elements in them.

Chemical Bonds

The atoms in compounds are held together by chemical bonds. The two main kinds of chemical bonds are ionic and covalent bonds.

Ionic Bonds An **ionic bond** is formed when one or more electrons are moved, or transferred, from one atom to another. An atom that loses electrons becomes positively charged. An atom that gains electrons has a negative charge. These positively and negatively charged atoms are known as ions. Oppositely charged ions are attracted to each other.

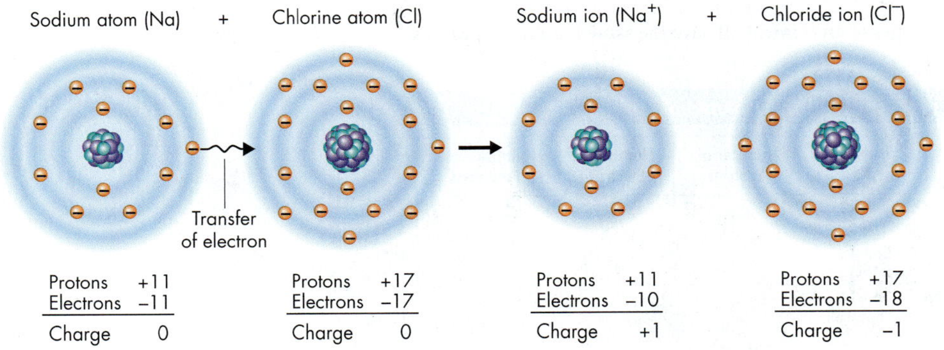

Ionic Bonding Table salt is made up of sodium ions and chloride ions. A sodium ion (Na^+) forms when a sodium atom loses 1 electron. A chloride ion (Cl^-) forms when a chlorine atom gains 1 electron. These oppositely charged ions attract each other and form an ionic bond.

ELL Focus on ELL: Access Content

ALL SPEAKERS Students are probably familiar with the word *attract,* but they may not understand that attraction in chemistry and physics involves forces. Point out that particles or objects that have unlike electrical charges are attracted to each other. Help students visualize this attraction by giving them small bar magnets. Let students manipulate them for a few minutes.

Ask Which poles of the magnets attract each other? *(unlike poles)*

Ask Which poles are not attracted to each other? *(like poles)*

Covalent Bonds Sometimes electrons are shared by atoms instead of being moved from one atom to another. What does it mean to share electrons? It means that the electrons travel around the nuclei of both atoms, forming a **covalent bond.** There are different types of covalent bonds. When atoms share 2 electrons, the bond is a single covalent bond. If atoms share 4 electrons, then the bond is a double bond. Atoms can also share 6 electrons and form a triple bond.

The structure that results when atoms are joined together by covalent bonds is called a molecule. The **molecule** is the smallest unit of most compounds. The bonds in a water molecule are covalent bonds. Each hydrogen atom in a water molecule is joined to the oxygen atom by a single covalent bond. Oxygen can also form double bonds. For example, the oxygen molecules that you breathe are 2 oxygen atoms joined by a double bond.

Key Question What are the main kinds of chemical bonds?
The main kinds of chemical bonds are ionic bonds and covalent bonds.

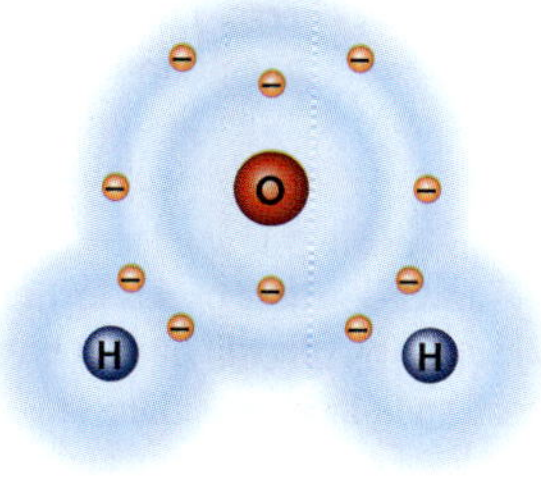

Water molecule (H_2O)

Covalent Bonding In a water molecule, each hydrogen atom shares 2 electrons with the oxygen atom. So, each hydrogen atom is joined to the oxygen atom by a covalent bond.

CHECK Understanding

Apply Vocabulary

Use the highlighted words from the lesson to complete each sentence correctly.

1. A(n) ______ is the basic unit of matter.
2. Sodium and chlorine form a(n) ______ when an electron moves from the sodium atom to the chlorine atom.
3. ______ are negatively charged and move around the nucleus of an atom.

Critical Thinking

4. **Infer** An atom of calcium has 20 protons. How many electrons does it have?
5. **Compare and Contrast** Compare carbon-12 and carbon-14 atoms. How are they the same, and how are they different?
6. **Apply Concepts** A potassium atom can easily lose 1 electron. What kind of bond will potassium form with chlorine?
7. **Write to Learn** Answer the first clue of the mystery. Write two or three sentences that include the words *atom* and *electron*.

MYSTERY CLUE

Fish do not break water molecules apart to get oxygen. Instead, they use oxygen gas that is dissolved in the water. How are the atoms in an oxygen molecule joined together? (Hint: See above.)

BIOLOGY.com Search Lesson 2.1 GO • Lesson Assessment

Check Understanding

1. atom
2. ionic bond
3. Electrons
4. A calcium atom has 20 electrons.
5. Carbon-12 and carbon-14 atoms each have 6 protons. But carbon-12 has 6 neutrons and carbon-14 has 8 neutrons.
6. Potassium will form an ionic bond with chlorine.
7. The atoms in an oxygen molecule share 4 electrons. These shared electrons form a double covalent bond.

Active Reading

USE VISUALIZATION Challenge students to think of an analogy for a covalent bond, in which two people or objects are held together by sharing something between them.

COMPARE AND CONTRAST Have students compare the two main kinds of chemical bonds.

Ask How are ionic bonds and covalent bonds alike and different? *(Both involve electrons of two atoms. In an ionic bond, the electron is actually transferred. In a covalent bond, the electron is shared.)*

USE GRAPHIC ORGANIZERS Draw a Venn diagram on the board. As students list ways that ionic and covalent bonds are alike and different, write them in the diagram.

KEY QUESTION Invite a volunteer to read the question and answer aloud.

Ask What is a chemical bond? *(the joining together of two atoms in a compound by sharing or transferring of electrons)*

Post Reading

ASSESSMENT TIPS Review the vocabulary words with students before they begin the assessment.

For Question 4, remind students that *infer* means that the answer will not be found directly in the text.

For Question 5, have student pairs make a graphic organizer to help them answer the question. Have them use complete sentences to write their answers.

MYSTERY CLUE Point out to students that various gases, including oxygen, are dissolved in water. Discuss whether these gases are bonded to water molecules and whether that affects how available the gases are to fish. *(The gas molecules are not bonded to water. They can easily be taken into the fishes' bodies.)*

Wrap-Up Activity

MODEL ISOTOPES Have students make models of the nuclei of carbon isotopes, using beads of two different colors to represent protons and neutrons. Have students make a pile of six "proton" beads. Then have them add six "neutron" beads. Explain that the pile represents the nucleus of a carbon-12 atom. Have students make models of nuclei of carbon-13 and carbon-14 isotopes in the same way.

Ask How many protons does each isotope have? *(six)* How many electrons? *(six)*

Have students add a proton bead to their model of carbon-14.

Ask Do the beads still model the nucleus of a carbon isotope? *(No, carbon atoms have six protons.)*

LESSON 2.1

Inquiry Into Scientific Thinking

DRAW A DIAGRAM Invite a volunteer to come to the board and draw a sodium atom. Invite another volunteer to draw a chlorine atom next to the sodium atom. Have a third volunteer draw an arrow to show the transfer of an electron.

Say An ionic bond forms when electrons are transferred from one atom to another.

Ask In which direction is an electron transferred? *(An electron moves from the sodium atom to the chlorine atom.)*

Ask What causes the sodium and chlorine atoms to stick together after the electron has been transferred? *(The two atoms are now electrically charged, and are called ions. The sodium and chloride ions that formed have opposite charges, and because of this they are attracted to each other.)*

MAKE CONNECTIONS Have students review the information in the chart.

Ask What is the atomic number of sodium? How do you know? *(The atomic number is the number of protons in an atom. For sodium, this number is 11.)*

Ask What is the mass number of chlorine-35? How do you know? *(The mass number is the number of protons plus neutrons. For chlorine, this number is 35.)*

Lead a Discussion

Discuss crystals with students. Explain that ions in table salt are arranged in an orderly, three-dimensional pattern called a crystal. The crystals formed by a substance are distinctive for that substance.

Say Sodium chloride crystals are white cubes.

Ask What holds the ions in place in a crystal of sodium chloride? *(ionic bonds)*

Say You have probably seen a diamond ring. A diamond is a crystal. The carbon bonds in a diamond are covalent bonds.

Ask What does a diamond look like? *(Answers may include descriptions of color, shape, size, and number of sides.)*

Say Sugar grains are crystals. These crystals are also formed by covalent bonds. You can see the characteristics of sugar crystals through a magnifying glass.

INQUIRY into Scientific Thinking GUIDED INQUIRY

Model an Ionic Compound

Ionic bonds are formed when electrons transfer from one atom to another. Sodium chloride, or table salt, is an ionic compound. To form table salt, sodium "gives" an electron to chlorine. The sodium atom then has more protons than electrons. It is an ion with a positive charge. The chlorine atom has more electrons than protons. It is an ion with a negative charge. Sodium and chloride ions form an orderly structure called a crystal.

In Lesson 2.1, you learned about ionic bonds and how these bonds form in table salt. In this activity, you will build a model of table salt.

1. Find a partner. One of you will make a model of a sodium atom, and the other will make a model of a chlorine atom.
2. Use the table below to find the correct number of electrons in your atom. Use popcorn kernels to represent the electrons in your atom.
3. Write the number of protons in your atom in the center of an index card and draw a circle around this number. This circle is the nucleus of your atom.
4. Arrange the correct number of electrons (popcorn kernels) on the index card around the nucleus.
5. Move your atom close to your partner's atom. Use the two atom models to form the ionic compound sodium chloride—table salt. (Form the ionic compound by moving an electron from one atom to the other.)
6. In table salt, the closely packed sodium and chloride ions form an orderly structure called a crystal. Using your compound models, work in a small group to model a sodium chloride crystal.

Analyze and Conclude

1. Relate Cause and Effect Describe how the popcorn kernels (electrons) moved as you formed the ionic bond. What electrical charges resulted from the movement?

2. Use Models How did you arrange the ions in the crystal model? Why did you choose this arrangement?

In Your Workbook Go to your workbook for more help with this activity.

Isotope	Number of Protons	Number of Electrons	Number of Neutrons
Sodium-23	11	11	12
Chrorline-35	17	17	18

Materials

bags of popcorn kernels, paper

Analyze and Conclude

1. One popcorn kernel was moved from the sodium atom to the chlorine atom. When this happened, the sodium became positively charged and the chlorine became negatively charged.
2. Sample answer: The crystal was arranged so that sodium ions alternated with chloride ions. This arrangement was chosen so that the positively charged sodium ions were always surrounded by negatively charged chloride ions.

2.2 Properties of Water

The Water Molecule

Water is a compound found over most of Earth's surface. In fact, just finding liquid water on a planet tells scientists that there could be life on that planet. What properties of water make it so important?

A Polar Molecule Like other molecules, water (H_2O) is neutral. The positive charges on its 10 protons balance out the negative charges on its 10 electrons. But an oxygen atom has 8 protons whereas a hydrogen atom has only 1. So, the electrons in a water molecule are pulled more toward the oxygen atom. They spend more time near the oxygen atom than near the hydrogen atoms. The oxygen atom is found on one end of the water molecule and the hydrogen atoms are on the other. As a result, the oxygen end of the molecule has a slight negative charge and the hydrogen end has a slight positive charge.

A molecule in which the charges are unevenly spread out is said to be "polar." The charges on a polar molecule are written in parentheses, (–) or (+), to show that they are weaker than the charges on ions such as Na^+ and Cl^-.

Hydrogen Bonds Because of their partial positive and negative charges, polar molecules such as water can attract each other. The pull between a partially positive hydrogen atom on one molecule and a partially negative oxygen atom on another is an example of a hydrogen bond. Hydrogen bonds form between the hydrogen atoms and oxygen atoms of different water molecules. Because water is a polar molecule, it can form many hydrogen bonds. Hydrogen bonds are not as strong as covalent or ionic bonds. But the hydrogen bonds do give water some special properties.

Key Questions

How does the structure of water lead to its special properties?

How does water's polarity affect its properties as a solvent?

Why is it important for cells to buffer solutions against rapid changes in pH?

BUILD Understanding

Venn Diagram As you read, draw a Venn diagram to compare solutions and suspensions. The diagram should show similarities and differences between the two.

In Your Workbook Go to your workbook to learn more about making a Venn diagram. Complete the Venn diagram for Lesson 2.2.

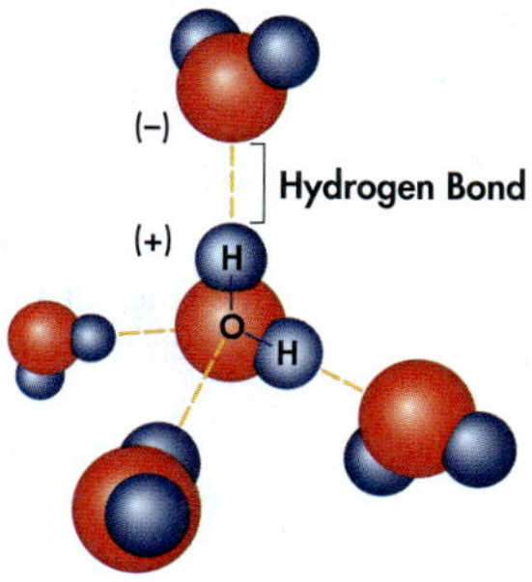

Hydrogen Bonding The molecules shown are water molecules. The red atoms are oxygen, and the blue atoms are hydrogen. The dashed lines show the hydrogen bonds that can form between water molecules.

Pre-Reading

Objectives

2.2.1 Discuss the unique properties of water.

2.2.2 Differentiate between solutions and suspensions.

2.2.3 Explain what acidic solutions and basic solutions are.

Preview the Pages

1. Preview the vocabulary words, pronouncing each word. Have students find each word in the text. Review the meaning of the prefix *co-* when you preview cohesion.
2. Polarity is often a difficult concept. Draw a diagram of a water molecule on the board. Show protons and electrons. Explain partial charges.
3. Preview the visuals and read the captions and labels. Be sure students understand what is happening in the diagram of buffer action.

Build Understanding

VENN DIAGRAM Remind students that details of the differences between solutions and suspensions go in the outer circles, and details of their similarities go where the circles overlap. Then, have students fill in the Venn diagram in Workbook B.

Speed Bump

Ask What is a hydrogen bond? *(the pull between a partially negative atom in one molecule and a partially positive atom in a different molecule)*

NATIONAL SCIENCE EDUCATION STANDARDS

UNIFYING CONCEPTS AND PROCESSES

I, III

CONTENT

B.1, B.2, B.3, B.4

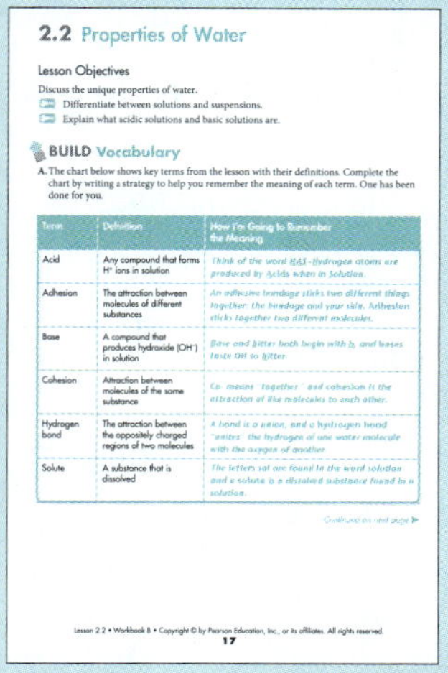

2.2 Properties of Water

Lesson Objectives

Discuss the unique properties of water.

Differentiate between solutions and suspensions.

Explain what acidic solutions and basic solutions are.

BUILD Vocabulary

A. The chart below shows key terms from the lesson with their definitions. Complete the chart by writing a strategy to help you remember the meaning of each term. One has been done for you.

Term	Definition	How I'm Going to Remember the Meaning
Acid	Any compound that forms H^+ ions in solution	*Think of the word HAS–Hydrogen atoms are produced by Acids when in Solution.*
Adhesion	The attraction between molecules of different substances	*An adhesive bandage sticks two different things together: the bandage and your skin. Adhesion sticks together two different molecules.*
Base	A compound that produces hydroxide (OH^-) in solution	*Base and bitter both begin with b, and bases taste OH so bitter.*
Cohesion	Attraction between molecules of the same substance	*Co- means "together," and cohesion is the attraction of like molecules to each other.*
Hydrogen bond	The attraction between the oppositely charged regions of two molecules	*A bond is a union, and a hydrogen bond "unites" the hydrogen of one water molecule with the oxygen of another.*
Solute	A substance that is dissolved	*The letters sol are found in the word solution and a solute is a dissolved substance found in a solution.*

Continued on next page ▶

17

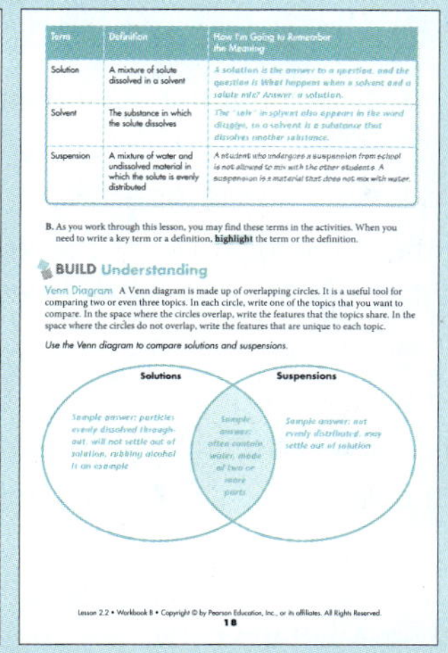

Term	Definition	How I'm Going to Remember the Meaning
Solution	A mixture of solute dissolved in a solvent	*A solution is the answer to a question, and the question is What happens when a solvent and a solute mix? Answer: a solution.*
Solvent	The substance in which the solute dissolves	*The "solv" in solvent also appears in the word dissolve, so a solvent is a substance that dissolves another substance.*
Suspension	A mixture of water and undissolved material in which the solute is evenly distributed	*A student who undergoes a suspension from school is not allowed to mix with the other students. A suspension is a material that does not mix with water.*

B. As you work through this lesson, you may find these terms in the activities. When you need to write a key term or a definition, **highlight** the term or the definition.

BUILD Understanding

Venn Diagram A Venn diagram is made up of overlapping circles. It is a useful tool for comparing two or even three topics. In each circle, write one of the topics that you want to compare. In the space where the circles overlap, write the features that the topics share. In the space where the circles do not overlap, write the features that are unique to each topic.

Use the Venn diagram to compare solutions and suspensions.

Solutions

Sample answer: particles evenly dissolved throughout, will not settle out of solution, rubbing alcohol is an example

Sample answer: often contain water, made of two or more parts

Suspensions

Sample answer: not evenly distributed, may settle out of solution

18

For the corresponding lesson in *Miller & Levine Biology,* see pages 40–44.

During Reading

Active Reading

MAKE CONNECTIONS Have students think about everyday examples that demonstrate cohesion and adhesion.

Ask What causes water to "climb" up a glass tube? *(adhesion—the attraction between water molecules and the molecules of the glass tube)*

Say Water beads up on a waxed car after it rains.

Ask Is cohesion or adhesion occurring when this happens? *(Both cohesion and adhesion are occurring. Cohesion causes the water to form a bead. Adhesion is the attraction of the bead to the wax. In this case, cohesion is stronger than adhesion.)*

KEY QUESTION Read aloud the Key Question and answer. Ask students to imagine what would happen if water had a heat capacity similar to that of aluminum or another metal.

Ask What would happen to ocean temperatures on a hot summer day? *(They would become very hot.)* On a cold night? *(They would become very cold.)* How would this affect organisms living in the ocean? *(Many would probably die. They might not be able to tolerate such extreme temperature changes.)*

Build Vocabulary

PREFIXES Explain that *ad-* means "to" or "toward."

Say I know that *adhesive tape* is used to stick a bandage onto skin.

Ask How does the prefix *ad-* relate to *adhesive tape*? *(Adhesive tape is sticky. It causes a bandage to stick to skin.)*

Ask How are *adhesive tape* and *adhesion* related? *(Adhesion causes water molecules to stick to other substances. Adhesive tape causes a bandage to stick to skin.)*

Speed Bump

Ask What is surface tension? *(a condition that results when cohesion pulls water molecules together, forming a "skin" on the surface of a body of water)*

Ask How does capillary action affect a plant? *(It causes water to move up the narrow tubes in a plant's roots to its stems and leaves.)*

Adhesion Adhesion between the molecules in the glass and the molecules in the water causes water to stick to the sides of this glass tube. The water climbs up the sides, and the water dips down in the middle. This curved water surface is called a meniscus.

BUILD Vocabulary

cohesion
the attraction between molecules of the same substance

adhesion
the force of attraction between different kinds of molecules

solution
a type of mixture in which all the components are evenly distributed

suspension
a mixture of water and nondissolved material

acid
a compound that releases hydrogen ions (H^+) in solution; a solution with a pH of less than 7

base
a compound that releases hydroxide ions (OH^-) in solution; a solution with a pH of more than 7

buffer
a compound that prevents sharp, sudden changes in pH

PREFIXES
The prefix *co-* means "together." Cohesion pulls molecules of the same substance together.

▶ ***Cohesion*** Because a water molecule can have as many as four hydrogen bonds at the same time, water is very cohesive. **Cohesion** (koh HEE zhun) is an attraction between molecules of the same substance. Cohesion pulls water molecules together. It causes surface tension, which enables some insects to walk on water's surface.

▶ ***Adhesion*** On the other hand, **adhesion** (ad HEE zhun) is an attraction between molecules of different substances. Adhesion between water and glass makes the surface of water in a tube curve upward along the sides of the tube. Adhesion also makes water rise in a narrow tube. This effect is called capillary action. Capillary action helps draw water out of the roots of a plant and up into its stems and leaves.

▶ ***Heat Capacity*** Another special property of water is its high heat capacity. Heat capacity is the amount of heat energy needed to increase a substance's temperature. In order for the temperature of water to rise, the molecules must move faster. The hydrogen bonds make it hard for water molecules to move faster. This property allows large bodies of water like oceans and lakes to absorb large amounts of energy without large changes in temperature.

Key Question How does the structure of water lead to its special properties?
Water is polar, so it can form hydrogen bonds. These bonds give water special properties, including cohesion, adhesion, and a high heat capacity.

Solutions and Suspensions

Water is often found as part of a mixture. A mixture is made up of elements or compounds that are combined but not bonded together. Living things are partly made up of water mixtures. Two kinds of mixtures that are made with water are solutions and suspensions.

Solutions If you put table salt in water, the ions on the salt's surface will be pulled toward the polar water molecules. Ions break away from the salt and are surrounded by water molecules. The ions spread out in the water to make a solution. A **solution** is a mixture in which the molecules of the mixed substances are evenly spread out. Solutions are made up of one or more solutes in a solvent. A solute is what is dissolved. A solvent is what does the dissolving. So, in a saltwater solution, salt is the solute and water is the solvent. Because water is polar, it can dissolve ionic compounds and other polar molecules.

ELL Focus on ELL: Build Background

BEGINNING SPEAKERS Remind students of the polar bear photograph in the chapter opener.

Ask Where do the polar bears live? *(near the North Pole; on the Svalbard islands)*

Display a world map and point out the locations of the North and South Poles. Explain that the regions around the North and South Poles are called polar regions.

Ask What do you think *polar* means in regards to regions? *(of or near the North or South Poles)*

Say *Pole* and *polar* have many meanings and some of them are related to opposites.

Ask What is opposite about a polar molecule? *(The charges on different parts of the molecule are opposite.)*

Have students draw a water molecule. Tell them to label the oxygen side of the molecule *Oxygen side (–)*. Have them label the hydrogen side *Hydrogen side (+)*. Add the caption: *Water is a polar molecule.*

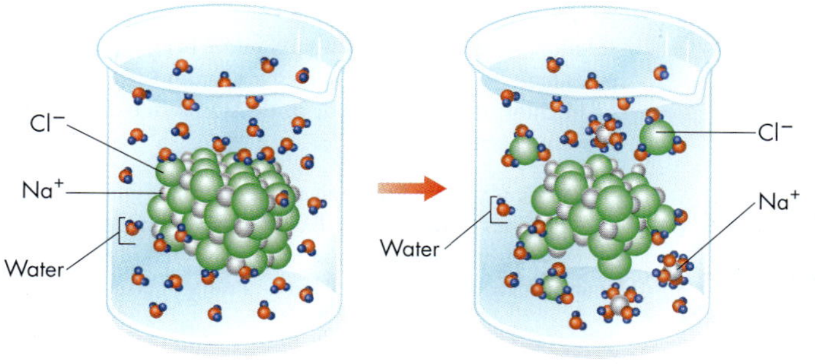

A Salt Solution An ionic compound, such as salt, dissolves in water. The water molecules surround the ions and pull them apart.

Suspensions Some things do not dissolve in water. But they may break into pieces so small that they do not settle out. Mixtures of water and nondissolved material that do not settle out are **suspensions.** Some important fluids in living things are both solutions and suspensions. Blood is mostly water. The water in the blood contains many dissolved compounds. But blood also has cells and other particles in suspension.

Key Question How does water's polarity affect its properties as a solvent?
Because water is polar, it can dissolve ionic compounds and other polar molecules.

Acids, Bases, and pH

Water molecules sometimes break apart to form ions. This reaction can be shown by a chemical equation. The double arrows show that the reaction can happen in either direction.

$$H_2O \rightleftharpoons H^+ + OH^-$$

water ⇌ hydrogen ion + hydroxide ion

The pH Scale Chemists use a measurement system called the pH scale to show the concentration of H^+ ions in solution. The pH scale ranges from 0 to 14. At a pH of 7, the concentration of H^+ ions and OH^- ions is equal. Pure water has a pH of 7. Solutions that have a pH below 7 are acidic. They have more H^+ ions than OH^- ions. The lower the pH is, the greater the acidity is. Solutions that have a pH above 7 are basic. They have more OH^- ions than H^+ ions. The higher the pH is, the more basic the solution is.

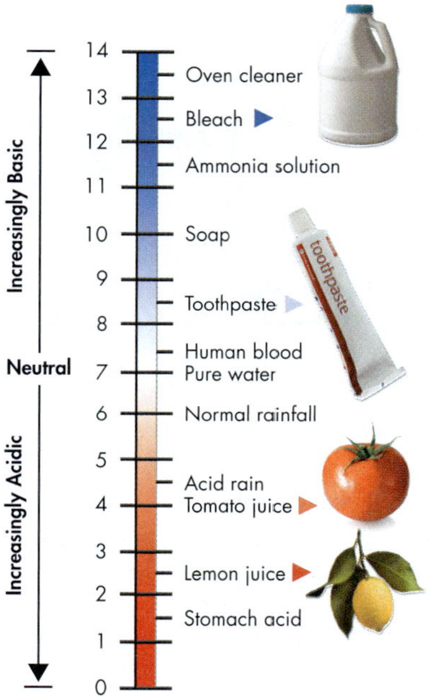

The pH Scale The concentration of H^+ ions determines whether a solution is acidic or basic.

Hands-On Learning

SURFACE TENSION Give each student group a paper towel, a dry penny, a dropper, and a cup with some water. Have them predict how many drops of water they can pile up on the penny before the water spills. Tell students to place the penny on the paper towel and count the number of drops they add. Allow students to practice using the dropper before they begin. The water pile is best viewed from the side.

Ask Were you able to add more water drops than you predicted? *(Students should be able to add about 40 drops of water to the penny before it spills.)*

Ask Why could you add so many drops to the penny? *(The water piles up because of surface tension. Cohesion holds the water drops together.)*

As an extension, have students add two drops of liquid dishwashing detergent to the water and repeat the activity. Explain that detergent decreases surface tension by interfering with hydrogen bonding between water molecules. *(Students should be able to pile up less than 15 drops of water before it spills.)*

Active Reading

DRAW A DIAGRAM Have students add to the salt solution diagram by drawing a third beaker in which all the salt crystals have dissolved.

Ask How are the dissolved ions distributed in your drawing? *(They are evenly distributed throughout the solution.)*

Ask How do the sodium and chloride ions appear in your drawing? *(Each Na+ and Cl– ion is surrounded by water molecules.)*

Ask Are there any salt crystals visible in your drawing? *(No, the crystals have all dissolved.)*

FIND THE MAIN IDEA Invite a volunteer to read aloud the paragraph headed Suspensions. Have students write a sentence or two in their notebooks identifying the main idea of the paragraph.

Ask How is blood both a solution and a suspension? *(it contains dissolved materials and materials that do not dissolve.)*

KEY QUESTION Read aloud the Key Question and answer.

Ask How does water dissolve polar substances? *(The partially charged atoms in water pull ions away from a polar substance. The positive hydrogen atoms in water are attracted to the negative ions in a polar substance, and the negative oxygen atoms are attracted to the positive ions.)*

Ask Why do you think water does not dissolve nonpolar substances? *(The partially charged atoms in water are not attracted to substances that have no charge.)*

Speed Bump

Ask What is formed when a water molecule breaks apart? *(a hydrogen ion and a hydroxide ion)*

Ask What are five types of molecules and ions found in a solution of salt and water? *(water molecules, sodium ions, and chloride ions, hydrogen ions, and hydroxide ions)*

Ask What is pH? *(a measure of the concentration of hydrogen ions in solution)*

During Reading continued

Active Reading

USE GRAPHIC ORGANIZERS Have students read the two paragraphs about acids and bases. Then have them make a T-chart to compare the characteristics of these kinds of compounds.

INTERPRET DIAGRAMS Have students write a short description explaining what is happening in each of the two diagrams.

Ask How are the two diagrams different? *(In the bottom diagram, a buffer has been added to the base. The buffer prevents a significant drop in pH when an acid is added.)*

KEY QUESTION Invite a volunteer to read the question and answer aloud.

Ask What is one way that buffers maintain homeostasis? *(They prevent blood pH from changing too much. Chemical reactions in cells can take place only in a narrow range of pH.)*

Speed Bump

Ask How would you describe a solution that has a pH of 13? *(It is a strong base.)*

Ask What do you know about the concentrations of hydrogen and hydroxide ions in a solution with a pH of 7? *(They are equal.)*

Post Reading

ASSESSMENT TIPS For Question 4, have students refer back to the Venn diagram they made at the beginning of the lesson.

For Question 5, remind students that *infer* means that the answer will not be found directly in the text. Tell students that they need to discuss how a serious drop in blood pH is prevented.

CHAPTER MYSTERY Discuss how oxygen dissolves in water. *(Water molecules are attracted to and surround the oxygen molecules.)* Have students infer why more oxygen dissolves at cold temperatures than at warm temperatures. *(At cold temperatures, oxygen molecules have less energy, move more slowly, and are less able to resist the attractions of water molecules.)*

Wrap-Up Activity

DETERMINE pH Provide students with some pH strips and small samples of common household liquids, such as soft drinks, orange juice, shampoo, and mouthwash. Have students determine the pH of each sample and classify it as acidic, basic, or neutral.

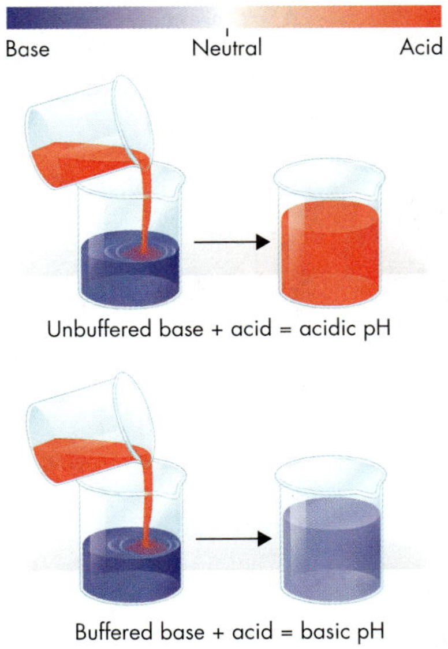

Buffers Adding an acid to a solution without a buffer causes the pH of the solution to drop. But if the solution has a buffer, adding an acid will cause only a small change in pH.

Acids An **acid** is any compound that releases H^+ ions in solution. Acidic solutions have pH values below 7. Strong acids tend to have pH values that range from 1 to 3. The hydrochloric acid (HCl) made by the stomach to digest food is a strong acid.

Bases A compound that releases hydroxide (OH^-) ions in solution is called a **base.** Basic solutions have pH values above 7. Strong bases, such as the lye (NaOH) used in soapmaking, tend to have pH values ranging from 11 to 14.

Buffers The pH in most cells in the human body must stay between 6.5 and 7.5. If the pH is lower or higher, it will affect the chemical reactions that take place in cells. So, keeping pH from changing is an important part of homeostasis (hoh mee oh STAY sis). Homeostasis is the process of keeping a stable internal environment inside living things. One way to control pH is with buffers. **Buffers** are weak acids or bases that can react with strong acids or bases to stop sharp, sudden changes in pH. Blood has a normal pH of 7.4. Sudden changes in blood pH are usually stopped by buffers. Buffers play an important role in keeping homeostasis in living things.

Key Question Why is it important for cells to buffer solutions against rapid changes in pH?
Buffers play an important role in keeping homeostasis in living things.

CHECK Understanding

Apply Vocabulary

Use the highlighted words from the lesson to complete each sentence correctly.

1. Cells do not dissolve in blood, but they form a _______ because they do not settle out of the blood either.
2. _______ stop sudden changes in pH.

Critical Thinking

3. **Explain** What property of water molecules allows them to form hydrogen bonds?
4. **Compare and Contrast** What is the difference between a solution and a suspension?
5. **Infer** During exercise many chemical changes happen in the body, including a drop in blood pH, which can be very serious. How is the body able to deal with such changes?
6. **Write to Learn** Answer the second clue of the mystery.

MYSTERY CLUE

The ability of a gas to dissolve increases as temperatures decrease. How might the temperature of antarctic water affect the amount of dissolved oxygen available for ice fish? (Hint: See p. 34.)

BIOLOGY.com Search Lesson 2.2 GO • Lesson Assessment

Check Understanding

1. suspension
2. Buffers
3. Water molecules can form hydrogen bonds becauses they are polar molecules.
4. A solution is a mixture of two or more substances in which the molecules of the substances are evenly spread out. A suspension is a mixture of water and nondissolved materials that do not settle out.
5. A drop in blood pH is minimized by buffers in the blood.
6. Antarctic water is very cold. The cold temperature increases the amount of dissolved oxygen in the water. There is more oxygen in the water for the ice fish to use.

2.3 Carbon Compounds

The Chemistry of Carbon

The chemistry of carbon is very important for living things. In fact, a whole branch of chemistry is set aside just to study carbon compounds. Why is carbon so interesting? There are two reasons. First, carbon atoms have four electrons available for bonding. These four electrons allow carbon atoms to form as many as four strong covalent bonds. Carbon can bond with many elements, including hydrogen, oxygen, phosphorus, sulfur, and nitrogen, to form the molecules of living things. Organisms are made up of molecules that consist of carbon and these other elements.

The second reason carbon compounds are so interesting is that carbon atoms can bond with each other. Bonding between carbon atoms gives carbon the ability to form chains that can be unlimited in length. These carbon-carbon bonds can be single, double, or triple covalent bonds. Chains of carbon atoms can even close up on themselves to form rings. Carbon can form millions of different large and complex structures. No other element comes close to matching carbon's ability to make so many different compounds.

Key Question What elements does carbon bond with to make up life's molecules? **Carbon bonds with hydrogen, oxygen, phosphorus, sulfur, and nitrogen to form life's molecules.**

Key Questions

What elements does carbon bond with to make up the molecules of living things?

What are the functions of each of the four groups of macromolecules?

BUILD Understanding

Compare/Contrast Table As you read, make a table that compares and contrasts the four groups of macromolecules.

In Your Workbook Go to your workbook to learn more about making a compare/contrast table. Fill out the compare/contrast table for Lesson 2.3.

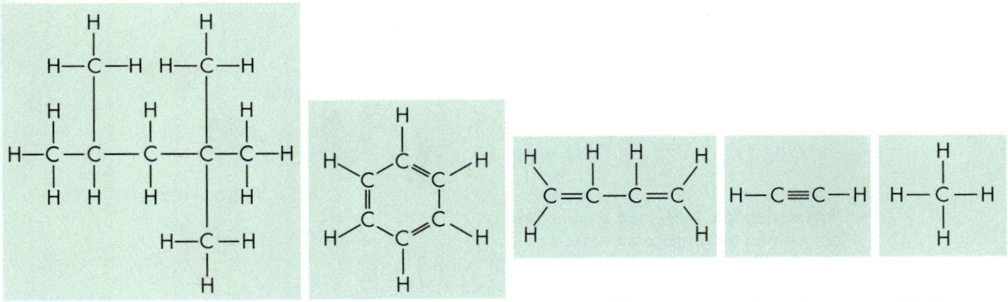

Carbon Structures Carbon can form single, double, or triple bonds with other carbon atoms. Each line between atoms in a molecular structure is one covalent bond. Carbon atoms can form straight chains, rings, or branched chains.

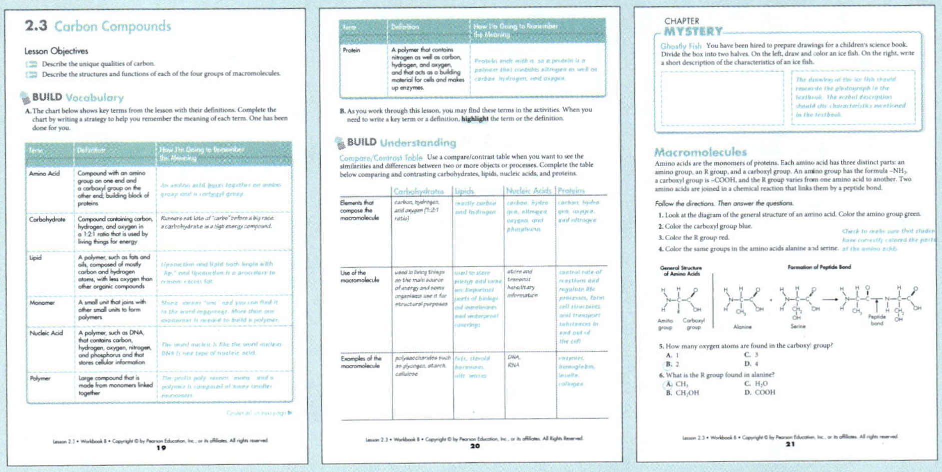
2.3 Carbon Compounds

Lesson Objectives

Describe the unique qualities of carbon.

Describe the structures and functions of each of the four groups of macromolecules.

BUILD Vocabulary

A. The chart below shows key terms from the lesson with their definitions. Complete the chart by writing a strategy to help you remember the meaning of each term. One has been done for you.

Term	Definition	How I'm Going to Remember the Meaning
Amino Acid	Compound with an amino group on one end and a carboxyl group on the other end; building block of proteins	An amino acid joins together an amino group and a carboxyl group.
Carbohydrate	Compound containing carbon, hydrogen, and oxygen in a 1:2:1 ratio that is used by living things for energy	Runners eat lots of "carbs" before a big race: a carbohydrate is a high energy compound.
Lipid	A polymer, such as fats and oils, composed of mostly carbon and hydrogen atoms, with less oxygen than other organic compounds	Liposuction and lipid both begin with "lip," and liposuction is a procedure to remove excess fat.
Monomer	A small unit that joins with other small units to form polymers	Mono means "one" and you can find it in the word [illegible]. More than one monomer is needed to build a polymer.
Nucleic Acid	A polymer, such as DNA, that contains carbon, hydrogen, oxygen, nitrogen, and phosphorus and that stores cellular information	The word nucleic is like the word nucleus. DNA is one type of nucleic acid.
Polymer	Large compound that is made from monomers linked together	The prefix poly means "many," and a polymer is composed of many smaller monomers.

Continued on next page ▶

Lesson 2.3 • Workbook B • Copyright © by Pearson Education, Inc., or its affiliates. All rights reserved.
19

Term	Definition	How I'm Going to Remember the Meaning
Protein	A polymer that contains nitrogen as well as carbon, hydrogen, and oxygen, and that acts as a building material for cells and makes up enzymes.	Protein ends with n, so a protein is a polymer that contains nitrogen as well as carbon, hydrogen, and oxygen.

B. As you work through this lesson, you may find these terms in the activities. When you need to write a key term or a definition, **highlight** the term or the definition.

BUILD Understanding

Compare/Contrast Table Use a compare/contrast table when you want to see the similarities and differences between two or more objects or processes. Complete the table below comparing and contrasting carbohydrates, lipids, nucleic acids, and proteins.

	Carbohydrates	Lipids	Nucleic Acids	Proteins
Elements that compose the macromolecule	carbon, hydrogen, and oxygen (1:2:1 ratio)	mostly carbon and hydrogen	carbon, hydrogen, nitrogen, oxygen, and phosphorus	carbon, hydrogen, oxygen, and nitrogen
Use of the macromolecule	used in living things as the main source of energy and some organisms use it for structural purposes	used to store energy and some are important parts of biological membranes and waterproof coverings	store and transmit hereditary information	control rate of reactions and regulate life processes, form cell structures, and transport substances in and out of the cell
Examples of the macromolecule	polysaccharides such as glycogen, starch, cellulose	fats, steroid hormones, oils, waxes	DNA, RNA	enzymes, hemoglobin, insulin, collagen

Lesson 2.3 • Workbook B • Copyright © by Pearson Education, Inc., or its affiliates. All Rights Reserved.
20

CHAPTER MYSTERY

Ghostly Fish You have been hired to prepare drawings for a children's science book. Divide the box into two halves. On the left, draw and color an ice fish. On the right, write a short description of the characteristics of an ice fish.

The drawing of the ice fish should resemble the photograph in the textbook. The verbal description should cite characteristics mentioned in the textbook.

Macromolecules

Amino acids are the monomers of proteins. Each amino acid has three distinct parts: an amino group, an R group, and a carboxyl group. An amino group has the formula $-NH_2$, a carboxyl group is $-COOH$, and the R group varies from one amino acid to another. Two amino acids are joined in a chemical reaction that links them by a peptide bond.

Follow the directions. Then answer the questions.

1. Look at the diagram of the general structure of an amino acid. Color the amino group green.
2. Color the carboxyl group blue.
3. Color the R group red.
4. Color the same groups in the amino acids alanine and serine.

Check to make sure that students have correctly colored the parts of the amino acids.

General Structure of Amino Acids — Amino group, Carboxyl group

Formation of Peptide Bond — Alanine, Serine, Peptide bond

5. How many oxygen atoms are found in the carboxyl group?
   A. 1 C. 3
   B. 2 D. 4
6. What is the R group found in alanine?
   A. CH_3 C. H_2O
   B. CH_2OH D. COOH

Lesson 2.3 • Workbook B • Copyright © by Pearson Education, Inc., or its affiliates. All rights reserved.
21

Pre-Reading

Objectives

2.3.1 Describe the unique qualities of carbon.

2.3.2 Describe the structures and functions of each of the four groups of macromolecules.

Preview the Pages

1. Preview the visuals, paying special attention to the bonding of carbon atoms in the molecular structure diagrams.
2. Point out the red arrow in the Polymerization diagram. Be sure students understand that the arrow signifies how polymer chains form.
3. Discuss the concept of levels of protein structure. Use analogies such as levels of organization in stores (by department and by type of products) or in a library (by category and by alphabetical order).
4. Preview the vocabulary words by having students find the words in the text. Ask students in what situations they have heard the words before. Point out that many of the words will be familiar to students who are interested in the components of the food they eat.

Build Understanding

COMPARE/CONTRAST TABLE Direct students to Workbook B where they will find a partially completed compare/contrast table. Tell students that they should fill out a column as they read about each kind of macromolecule. The completed table can be used later as a study guide.

NATIONAL SCIENCE EDUCATION STANDARDS

UNIFYING CONCEPTS AND PROCESSES

I, V

CONTENT

B.1, B.2, C.1.a, C.1.c, C.1.d, C.2.a

For the corresponding lesson in ***Miller & Levine Biology,*** see pages 45–49.

LESSON 2.3

During Reading

Active Reading

SCIENCE SUPPORT After students read the first paragraph of this lesson, make sure they understand the sentence "carbon atoms have four electrons available for bonding." Have students review the diagram of the carbon atom in Lesson 2.1. Point out that some of the electrons are farther from the nucleus than others. Explain that the outermost electrons are available for bonding with other atoms.

KEY QUESTION Have a student read the Key Question and answer on the previous page.

Ask What kind of chemical bond does carbon form with other elements? *(covalent bond)*

MAKE CONNECTIONS Have students read the paragraph about carbohydrates.

Say Often, the answer to a question is not given directly. You may need to infer from context.

Ask What does the sixth sentence say about glucose? *(Glucose is a sugar that releases energy when it is broken down.)*

Ask What does it mean to say that carbohydrates are a main energy source? *(Living things get much of the energy they need from breaking down sugars, such as glucose.)* If students have trouble making this inference, write the answer with blanks for students to fill in. Leave out key words such as *energy, sugars,* and *glucose*.

KEY QUESTION Have a student read the Key Question and answer on this page.

Ask What are some sources of carbohydrates? *(Sample answers: milk, fruits, bread)*

APPLY ACADEMIC VOCABULARY Have students read the caption on the visual and then close their book.

Ask What is polymerization? *(the process in which monomers join together to form polymers)*

Ask What do you think is formed when a polymer breaks down into its parts? *(Monomers are formed.)*

Build Vocabulary

WORD ORIGINS Explain that the general formula for a carbohydrate can be written as CH_2O.

Ask How does this formula reflect the prefix and word origin of *carbohydrate*? *(The formula shows that a carbohydrate is a combination of both carbon and water.)*

Speed Bump

Ask What are the four kinds of macromolecules? *(carbohydrates, lipids, nucleic acids, and proteins)*

BUILD Vocabulary

carbohydrate
a compound made up of carbon, hydrogen, and oxygen atoms; a type of nutrient that is the main source of energy for the body

lipid
a macromolecule made mostly from carbon and hydrogen atoms; includes fats, oils, and waxes

WORD ORIGINS
Carbohydrate is a combination of the prefix *carbo-*, which means "carbon," and the word *hydrate*, which means "water."

Macromolecules

Many of the carbon compounds in living cells are so large that they are known as macromolecules or "giant molecules." Macromolecules are made from thousands or even hundreds of thousands of smaller molecules. The smaller units, or monomers, join together to form polymers (PAHL uh murz). The monomers in a polymer may be the same, like links in a chain. Or, the monomers may be different, like different colored beads in a necklace.

Scientists sort the macromolecules found in living things into groups based on their chemical makeup. The four major groups of macromolecules found in living things are carbohydrates, lipids, nucleic acids, and proteins.

Carbohydrates Compounds made up of carbon, hydrogen, and oxygen atoms are called **carbohydrates.** There are usually twice as many hydrogen atoms as carbon or oxygen atoms in these molecules. Living things use carbohydrates as their main source of energy. Carbohydrates give plants, some animals, and other organisms structure. Sugars are carbohydrates. The breakdown of sugars, such as glucose, supplies energy for cell activities. Many living things store extra sugar as a carbohydrate called starch.

▶ ***Simple Sugars*** Single sugar molecules are called both monosaccharides (mahn oh SAK uh rydz) and simple sugars. Besides glucose, simple sugars include galactose, which is found in milk, and fructose, which is found in many fruits. Table sugar, sucrose, is made up of glucose and fructose. Since sucrose is made of two sugars it is called a disaccharide.

Key Question What are the functions of carbohydrates?
Carbohydrates provide energy and structural support for living things.

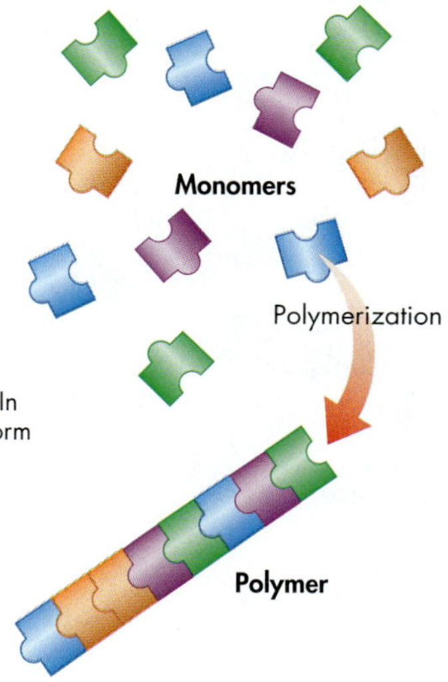

Polymerization Most macromolecules are formed by a process known as polymerization (pah lih mur ih ZAY shun). In this process, monomers join together to form polymers. Polymers can get very large.

ELL Focus on ELL: Extend Language

BEGINNING SPEAKERS Have students reread the paragraph about simple sugars.

Say *Saccharide* is another word for sugar. *Mono-* means "one, single, alone."

Ask Based on its word parts, what do you think *monosaccharide* means? *(one sugar)*

Show students pictures of a monorail and a monocle.

Ask What is a monorail? *(a train that runs on one track or rail)* What is a monocle? *(an eyeglass for one eye)* What is a monomer? *(a single unjoined molecule)*

Say *Di-* means "two, twice, double."

Ask Based on its word parts, what do you think *disaccharide* means? *(two sugars)*

Say Oxygen gas is a diatomic molecule.

Ask What do you think this statement means? *(A molecule of oxygen gas contains two atoms.)*

Carbohydrates Sugars and starches are carbohydrates. Starches form when sugars join together in long chains.

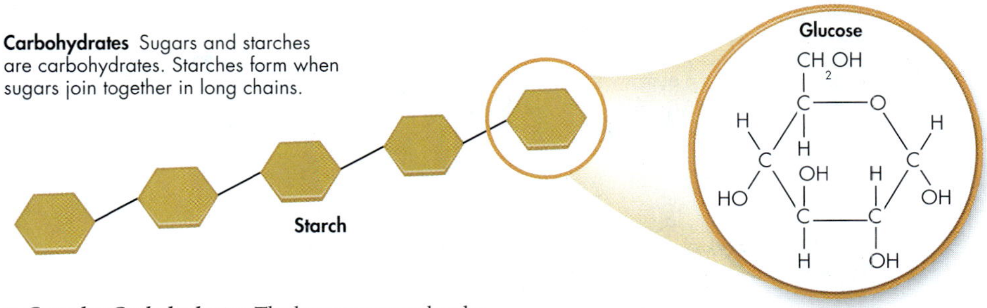

▶ ***Complex Carbohydrates*** The large macromolecules formed when simple sugars join together are called complex carbohydrates. Many animals store extra sugar in a complex carbohydrate called glycogen. When the amount of glucose in your blood runs low, glycogen is broken down into glucose, which then goes into the blood. The glycogen stored in your muscles gives energy to muscles so that you can move.

Plants use a different complex carbohydrate, called starch, to store extra sugar. Plants also make another complex carbohydrate called cellulose. Cellulose gives plants their strength and rigidity. Cellulose is the major part of wood and paper, so you are looking at cellulose as you read these words!

Lipids Fats, oils, and waxes are common lipids. **Lipids** are macromolecules that generally do not dissolve in water and are made mostly of carbon and hydrogen atoms. Some lipids store energy. Others form biological membranes, and some produce waterproof coverings on cells and tissues.

Many lipids are formed when a glycerol molecule joins with compounds called fatty acids. If each carbon atom in a lipid's fatty acids is joined to other carbon atoms by only single bonds, the lipid is called saturated. This means that the fatty acid has the most hydrogen atoms it can have. If there is at least one carbon-carbon double bond in a fatty acid, the fatty acid is called unsaturated. You have probably seen the terms *saturated* and *unsaturated* on food labels. Lipids that have unsaturated fatty acids are liquid at room temperature. Olive oil is one such lipid.

Key Question What are the functions of lipids? **Lipids are used to store energy and are parts of membranes and waterproof coverings.**

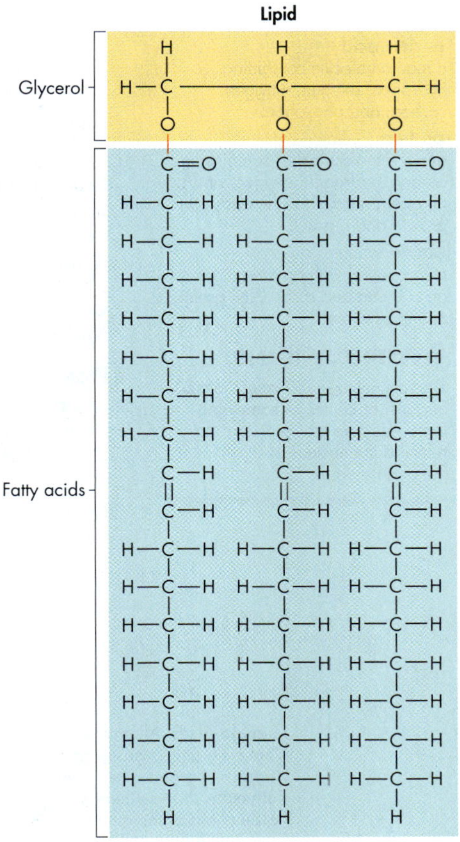

Lipids Lipids are made up of a glycerol molecule and fatty acids.

Active Reading

DRAW CONCLUSIONS Have volunteers read the two paragraphs about complex carbohydrates.

Ask How do plants and animals differ in the way they store extra sugar? *(Animals store sugar as glycogen. Plants store it as starch.)*

Ask What can happen when an animal needs extra energy? *(Some stored glycogen can be broken down into glucose, which is an energy source.)*

INTERPRET DIAGRAMS Guide students to the diagram of carbohydrates.

Ask What does each brown shape represent? *(a glucose molecule)*

Ask Is each brown shape a monomer or a polymer? Explain. *(It is a monomer, a single unit in a larger polymer.)*

Ask What atoms is carbon bonded to in a glucose molecule? *(other carbon atoms, oxygen, and hydrogen)*

SELF-QUESTION Have students read the section about lipids. Have them write a question they may have and place it in a box that you pass around. Read the questions aloud and call on volunteers to answer them.

COMPARE AND CONTRAST Have students reread the last paragraph on the page and examine the structure of a lipid.

Ask How are saturated and unsaturated lipids the same? *(Both have long chains of carbon and hydrogen atoms.)*

Ask How are they different? *(An unsaturated fatty acid contains at least one carbon-carbon double bond. A saturated fatty acid contains only single bonds.)*

Ask Is the lipid in the diagram saturated or unsaturated? *(unsaturated)*

KEY QUESTION Have a volunteer read the question and answer.

Ask What is a lipid? *(A lipid is a macromolecule made mostly of carbon and hydrogen atoms. Fats, oils, and waxes are lipids.)*

Focus on Academic Language

• saturated • unsaturated

Explain to students that the word *saturated* is used in several contexts. Ask if students have ever heard of a saturated solution. A saturated solution is one that holds as much solute as can be dissolved in the solvent. In a saturated lipid, each carbon atom is bonded to as many hydrogen atoms as it can be.

Say *Saturated* means "full."

Ask What does it mean to say that the ground is saturated after a heavy rainfall? *(The ground holds as much water as it can.)*

Explain that the prefix *un-* changes the meaning of a word. Ask students if they can figure out what the prefix *un-* means. *("not" or "opposite of")*

Ask What does it mean to say that the ground is unsaturated even though it has been raining? *(The ground can still hold more water.)*

Speed Bump

Ask What are two complex carbohydrates found in plants? *(starch and cellulose)*

Ask Which macromolecule often includes a glycerol molecule? *(lipid)*

Ask Glycogen is an example of what kind of macromolecule? *(carbohydrate)*

During Reading continued

Active Reading

USE DIAGRAMS Guide students to the diagram of a nucleotide.

Ask How is the sugar in nucleic acids different from glucose? *(The sugar in nucleic acids has five carbon atoms; glucose has six carbon atoms.)*

Ask What part of a nucleotide is a phosphate group bonded to? *(the 5-carbon sugar)*

CLARIFY WORD USAGE Have students read about nucleic acids.

Ask What do you think the word *nitrogenous* means? *(containing nitrogen)*

Say A nitrogenous base is part of a nucleotide.

Ask How would you describe a nitrogenous base? *(The base contains nitrogen and has the chemical properties of a base, meaning that it produces hydroxide ions in water.)*

KEY QUESTION Have a volunteer read the question aloud. Ask students to answer it in their own words.

Ask What is genetic information? *(the instructions for producing all the structures and chemicals in the body)*

Ask What does it mean to transmit information? *(to pass it on from one place to another)*

USE DIAGRAMS Guide students to the diagram of an amino acid.

Say An amino acid has a central carbon atom with four atoms or groups of atoms attached to it.

Ask What are the four atoms or groups? *(an amino group, a carboxyl group, a hydrogen atom, and a side chain)*

Build Vocabulary

WORD ORIGINS Explain to students that, like the root word *nucle-*, *nucleo-* refers to a nucleus.

Ask What are monomers that make up the nucleic acids in living cells known as? *(nucleotides)*

Speed Bump

Ask What kind of bonds hold the three parts of a nucleotide together? *(covalent bonds)*

Ask What is the monomer that makes up proteins? *(amino acids)*

Nucleotides Each nucleotide is made up of a 5-carbon sugar, a phosphate group, and a nitrogenous base. Nucleic acids are made up of many nucleotides joined together.

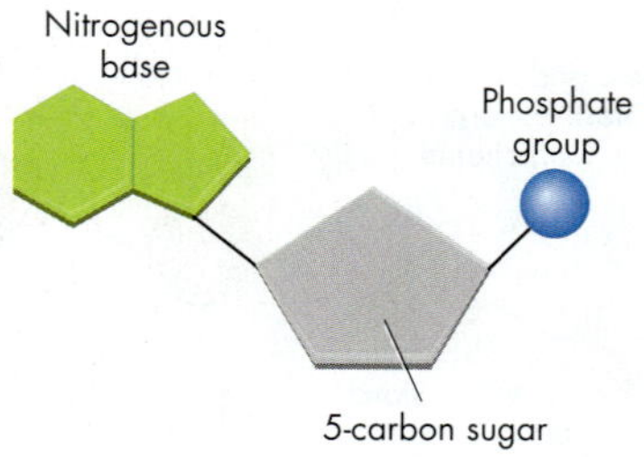

BUILD Vocabulary

nucleic acid
a macromolecule containing hydrogen, oxygen, nitrogen, carbon, and phosphorus

protein
a macromolecule that contains carbon, hydrogen, oxygen, and nitrogen; needed by the body for growth and repair

amino acid
a compound with an amino group on one end and a carboxyl group on the other end

WORD ORIGINS
The root word *nucle-* refers to a nucleus, or center, of something. In *nucleic acid* the root word refers to the nucleus of a cell, not of an atom.

Nucleic Acids **Nucleic** (noo KLEE ik) **acids** are macromolecules containing hydrogen, oxygen, nitrogen, carbon, and phosphorus. Nucleic acids are polymers. They are made up of monomers called nucleotides (NOO klee oh tydz). Nucleotides have three parts: a 5-carbon sugar, a phosphate group ($-PO_4$), and a nitrogenous base. Some nucleotides, including adenosine triphosphate (ATP), help in storing and transferring energy.

Single nucleotides can be linked by covalent bonds into long chains known as nucleic acids. Nucleic acids store and transmit genetic information. There are two kinds of nucleic acids: ribonucleic acid (RNA) and deoxyribonucleic acid (DNA). RNA contains the sugar ribose, and DNA contains the sugar deoxyribose.

Key Question What are the functions of nucleic acids?
Nucleic acids store and transmit genetic information.

Proteins Proteins are macromolecules that contain nitrogen as well as carbon, hydrogen, and oxygen. **Proteins** are polymers of molecules called amino acids. **Amino acids** are compounds with an amino group ($-NH_2$) on one end and a carboxyl group ($-COOH$) on the other end. Covalent bonds called peptide bonds link amino acids to form one or more polypeptides. A protein is a molecule made from one or more polypeptides. Some proteins control cell processes and the rate of reactions. Others form important parts in cells, while still others move substances into or out of cells or help to fight disease.

Amino Acids All amino acids have an amino group and a carboxyl group. The *R* in the structure shows where the side chain of the amino acid goes. Different side chains have different properties.

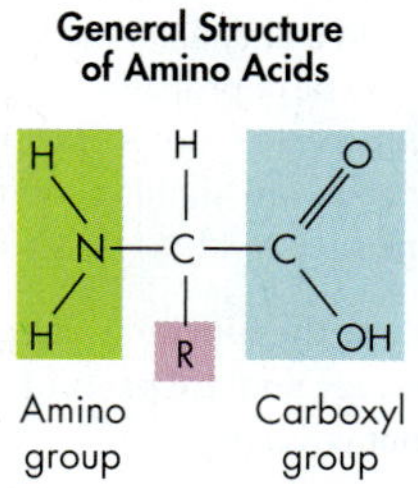

Focus on Academic Language

- nucleic

The letter combination *ei* is often confusing to readers. Usually, the *ei* letter combination is pronounced together as one sound. Ask students how the *ei* combination is pronounced in the word *height*. *(with a long* i *sound)* Then ask how the *ei* is pronounced in the word *seize*, meaning "to take." *(with a long* e *sound)* Explain that in the word *nucleic*, the *ei* is not used as a letter combination. Rather, the e and *i* are pronounced separately.

Ask How is the e pronounced in *nucleic*? *(with a long* a *sound)*

Ask How is the *i* pronounced? *(with a short* i *sound)*

Say The word *nucleic* has two parts—a root and a suffix.

Ask Do you think the *i* is part of the root or part of the suffix? *(part of the suffix)* Do you think the e is part of the root or part of the suffix? *(part of the root)*

▶ *Structure and Function* More than 20 different amino acids are found in nature. All amino acids are the same in the regions where they may be joined together by covalent bonds. This uniformity allows any amino acid to be joined to any other amino acid. Each amino acid has a different side chain called an R-group. The side chains have different properties. Some chains are acidic and some are basic. Some are polar, some are nonpolar, and some even contain large ring structures. Because the side chains are so different, proteins are diverse macromolecules.

▶ *Levels of Structure* Amino acids are joined in long chains according to instructions coded in DNA. Scientists describe proteins as having four levels of structure. A protein's primary structure is the order of its amino acids. Its secondary structure is the folding of the polypeptide chain. The third level of structure is the three-dimensional arrangement of a chain. Proteins that have more than one chain have a fourth level of structure. The fourth level describes how the different chains are placed next to each other. Hemoglobin, a protein in red blood cells that helps move oxygen in blood, has four levels of structure. The shape of a protein is kept by many forces, including ionic, covalent, and hydrogen bonds. The heme groups in the protein bind oxygen and also give hemoglobin its red color.

Key Question What are the functions of proteins?
Some proteins control cell processes and the rate of reactions. Others form important parts in cells, while still others move substances into or out of cells or help to fight disease.

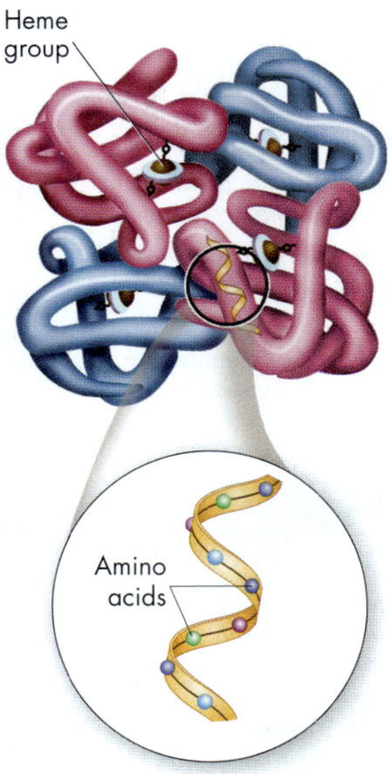

Protein Structure Hemoglobin is a protein made up of four polypeptide chains. Near the center of each polypeptide is an iron-containing heme group. An oxygen molecule is bound by each heme group.

CHECK Understanding

Apply Vocabulary

Use the highlighted words from the lesson to complete each sentence correctly.

1. Fats, oils, and waxes do not dissolve in water and are kinds of macromolecules called _______.
2. _______ store genetic information.
3. Compounds made up of carbon, hydrogen, and oxygen that are the main source of energy for the body are called _______.

Critical Thinking

4. **Relate Cause and Effect** Why can carbon form so many different structures?
5. **Explain** How do living things use lipids?
6. **Sequence** What are the four levels of organization in a protein?
7. **Write to Learn** Make a chart that shows the monomers that make up three of the four kinds of macromolecules.

BIOLOGY.com Search Lesson 2.3 GO • Lesson Assessment

Check Understanding

1. lipids
2. Nucleic acids
3. carbohydrates
4. Carbon atoms have four electrons available for bonding. These four electrons allow carbon atoms to form covalent bonds with other carbon atoms in long chains and rings.
5. Living things use lipids to store energy. Lipids are also parts of membranes and waterproof coverings.
6. the order of amino acids, the folding of the amino acid chain, the three-dimensional arrangement of the chain, and how the chains are placed next to each other
7. Students' charts should list the monomers that make up each of three macromolecules.

Active Reading

SELF-QUESTION Demonstrate how asking yourself questions helps to clarify information in the paragraph about levels of organization in proteins.

Say I know a protein has four levels of structure. I need to know what the levels are.

Say First, I will ask myself the following questions. What is the first level of structure in a protein? The second level? The third level? The fourth level?

Say Next, I will look back through the paragraph for information to answer my questions in order.

KEY QUESTION Have volunteers read the Key Question and answer.

Ask Suggest a reason why proteins have so many functions in cells. *(Because of the many possible side chains, huge numbers of different proteins with different functions can be made.)*

Speed Bump

Ask What are some of the properties of amino acid side chains? *(Side chains can be acidic, basic, polar, nonpolar, or made of ring structures.)*

Ask What kind of bonds hold amino acids together in a protein? *(covalent bonds)*

Post Reading

ASSESSMENT TIPS Review the highlighted vocabulary words in the lesson with students before they begin the assessment.

For Question 6, have student pairs make a graphic organizer that sequences the levels of organization in a protein. Then have them use the graphic organizer to answer the question.

Wrap-Up Activity

EXAMINE FOOD LABELS Have students bring in a food label from home and examine it for familiar words from this lesson. They should find the words *carbohydrates, sugar, fat, saturated,* and *protein.* Have students write a short paragraph explaining why macromolecules are an important part of their lives.

LESSON 2.4

Pre-Reading

Objectives

2.4.1 Explain how chemical reactions affect chemical bonds.

2.4.2 Describe how energy changes affect how easily a chemical reaction will occur.

2.4.3 Explain why enzymes are important to living things.

Preview the Pages

1. Preview the visuals, paying special attention to the graphs. Students may not be familiar with axes that are labeled with arrows rather than numbers.
2. Point out that the Key Questions that are listed on the first page of the lesson show main information students will learn in the lesson. Ask volunteers to read the three questions aloud. As a group, discuss what the questions are asking.
3. Be sure that students understand that energy comes in different forms and that light, heat, and sound all are forms of energy.
4. Preview the vocabulary words by having students find each word and its definition in the text. Be sure students are pronouncing each word correctly.

Build Understanding

CONCEPT MAP Direct students to Workbook B, where they will find a partially completed concept map. The concept map is quite complex. Suggest that students begin with the vertical path on the left side of the map.

NATIONAL SCIENCE EDUCATION STANDARDS

UNIFYING CONCEPTS AND PROCESSES

II, III, IV, V

CONTENT

B.3, B.5, C.1.a, C.1.b, C.1.d, C.1.e, C.5.b, C.5.c

For the corresponding lesson in ***Miller & Levine Biology,*** see pages 50–53.

2.4 Chemical Reactions and Enzymes

Key Questions

What happens to chemical bonds during chemical reactions?

How do energy changes affect whether a chemical reaction will happen?

What role do enzymes play in living things and what affects their function?

BUILD Understanding

Concept Map As you read, make a concept map that shows the relationship among the vocabulary words in this lesson.

In Your Workbook Go to your workbook for help in completing the concept map for Lesson 2.4.

Chemical Reactions

Living things are made up of chemical compounds. But chemistry is not just what life is made of—chemistry is also what life does. Everything that happens in living things is based on chemical reactions. A **chemical reaction** is a process that changes one set of chemicals into another. Chemical reactions also involve changes in energy. Some reactions release energy, and some use energy. This is also true for chemical reactions that happen in living things.

Some chemical reactions happen slowly, such as the reaction of iron and oxygen to form rust. Other reactions happen quickly. The elements or compounds that go into a chemical reaction are reactants. The elements or compounds that come out of a chemical reaction are products. During chemical reactions, some chemical bonds are broken and other bonds are formed. An important reaction in your bloodstream removes carbon dioxide from the body. That reaction is shown below.

Key Question What happens to chemical bonds during chemical reactions?

During chemical reactions, some bonds are broken and other bonds are formed.

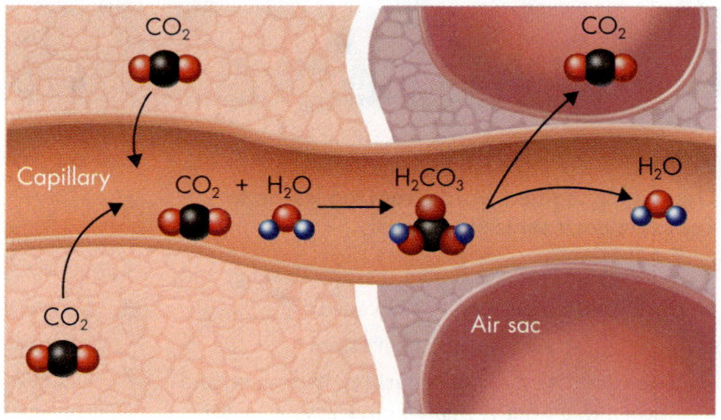

Reaction in the Bloodstream Carbon dioxide (CO_2) reacts with water (H_2O) to make carbonic acid (H_2CO_3). Your blood carries carbonic acid to your lungs. In your lungs, carbonic acid breaks apart to form water and carbon dioxide. You then breathe the carbon dioxide out.

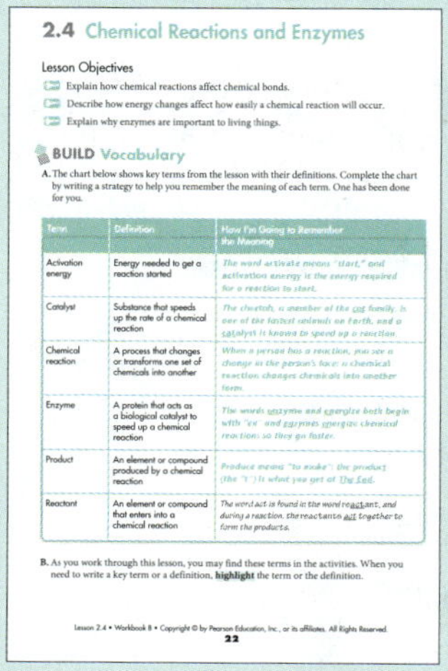

2.4 Chemical Reactions and Enzymes

Lesson Objectives

Explain how chemical reactions affect chemical bonds.

Describe how energy changes affect how easily a chemical reaction will occur.

Explain why enzymes are important to living things.

BUILD Vocabulary

A. The chart below shows key terms from the lesson with their definitions. Complete the chart by writing a strategy to help you remember the meaning of each term. One has been done for you.

Term	Definition	How I'm Going to Remember the Meaning
Activation energy	Energy needed to get a reaction started	The word activate means "start," and activation energy is the energy required for a reaction to start.
Catalyst	Substance that speeds up the rate of a chemical reaction	The cheetah, a member of the cat family, is one of the fastest animals on Earth, and a catalyst is known to speed up a reaction.
Chemical reaction	A process that changes or transforms one set of chemicals into another	When a person has a reaction, you see a change in the person's face; a chemical reaction changes chemicals into another form.
Enzyme	A protein that acts as a biological catalyst to speed up a chemical reaction	The words enzyme and energize both begin with "en" and enzymes energize chemical reactions so they go faster.
Product	An element or compound produced by a chemical reaction	Produce means "to make"; the product (the "t") is what you get at The End.
Reactant	An element or compound that enters into a chemical reaction	The word act is found in the word reactant, and during a reaction, the reactants act together to form the products.

B. As you work through this lesson, you may find these terms in the activities. When you need to write a key term or a definition, **highlight** the term or the definition.

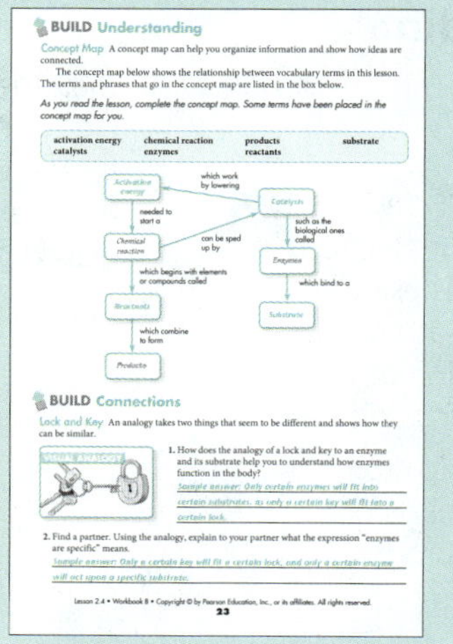

BUILD Understanding

Concept Map A concept map can help you organize information and show how ideas are connected.

The concept map below shows the relationship between vocabulary terms in this lesson. The terms and phrases that go in the concept map are listed in the box below.

As you read the lesson, complete the concept map. Some terms have been placed in the concept map for you.

activation energy **catalysts** **chemical reaction** **enzymes** **products** **reactants** **substrate**

BUILD Connections

Lock and Key An analogy takes two things that seem to be different and shows how they can be similar.

1. How does the analogy of a lock and key to an enzyme and its substrate help you to understand how enzymes function in the body?
Sample answer: Only certain enzymes will fit into certain substrates, as only a certain key will fit into a certain lock.

2. Find a partner. Using the analogy, explain to your partner what the expression "enzymes are specific" means.
Sample answer: Only a certain key will fit a certain lock, and only a certain enzyme will act upon a specific substrate.

Energy in Reactions

Energy is given off or taken in whenever chemical bonds are formed or broken. This means that chemical reactions involve changes in energy.

Energy Changes Energy changes are very important in determining whether a chemical reaction will happen. Chemical reactions that give off energy often happen on their own. Chemical reactions that take in energy will not happen without a source of energy. The burning of hydrogen gas gives off energy. Hydrogen reacts with oxygen to make water vapor.

$$2H_2 + O_2 \longrightarrow 2H_2O$$

If the hydrogen explodes, light and sound are given off in addition to the water vapor. The reverse reaction, in which water is changed into hydrogen and oxygen gas, requires so much energy that it doesn't happen by itself. In fact, the only way to reverse the reaction is to pass an electrical current through water to break the water down.

Energy Sources To stay alive, organisms need to carry out reactions that need added energy. Every living thing must have a source of energy to carry out these chemical reactions. Plants get that energy by trapping and storing the energy from sunlight in energy-rich compounds. Animals get their energy when they eat plants or other animals. Humans release energy needed to live when they break down the food they eat.

Activation Energy Chemical reactions that give off energy do not always happen on their own. If they did, the pages of this book might burst into flames! The cellulose in paper can burn and give off heat and light. However, paper burns only if you light it with a match, which gives enough energy to start the reaction. Chemists call the energy that is needed to get a reaction started **activation energy.** Energy is involved in all chemical reactions, whether the overall reaction gives off or takes in energy.

Key Question How do energy changes affect whether a chemical reaction will happen?
Chemical reactions that give off energy often happen on their own. Chemical reactions that take in energy will not happen without a source of energy.

BUILD Vocabulary

chemical reaction
the process that changes, or transforms, one set of chemicals into another set of chemicals

activation energy
the energy that is needed to get a reaction started

RELATED WORD FORMS
The verb *react* means "to act in response to something." The adjective *reactive* describes the tendency to react.

Energy-Absorbing Reaction

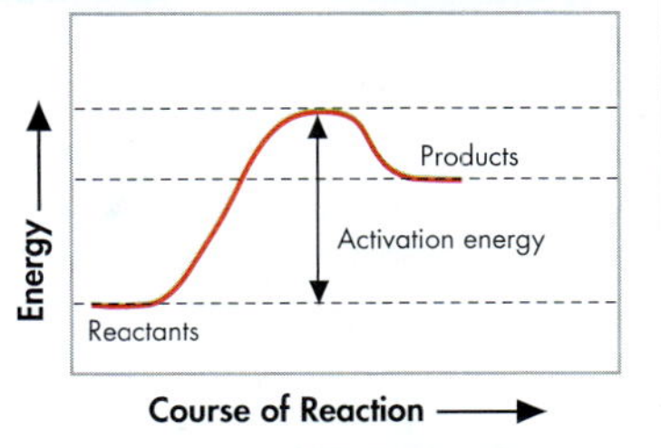

Energy-Releasing Reaction

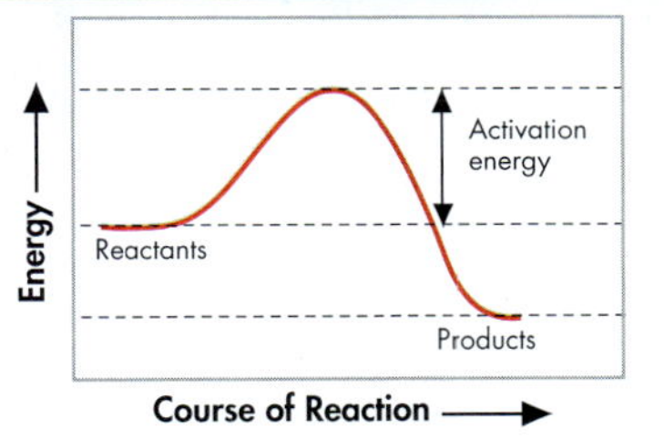

Activation Energy The high point of each graph shows the energy needed for the reaction to happen. The difference between this needed energy and the energy of the reactants is the activation energy.

Hands-On Learning

COLD PACKS AND HEAT PACKS Give student groups either a heat pack or a cold pack. Have students follow the directions on the product's label to activate the reaction. Then they should feel the outside of the pack. Have students draw a poster diagramming how energy moved when they activated the pack. Arrows and labels should indicate the flow of energy.

Ask Is the reaction that takes place in a heat pack energy-releasing or energy-absorbing? How do you know? *(Heat is produced. This means that the reaction in the pack is energy-releasing.)*

Ask Is the reaction that takes place in a cold pack energy-releasing or energy-absorbing? How do you know? *(The pack gets cold. This means that the reaction in the bag absorbed heat from the surroundings and the surroundings got colder.)*

During Reading

Active Reading

KEY QUESTION Have volunteers read aloud the Key Question and answer on the previous page.

Ask Are bonds of reactants broken or formed during a reaction? *(broken)* Do new bonds form in reactants or in products during a reaction? *(products)*

INTERPRET DIAGRAMS Guide students to the Reaction in the Bloodstream visual. Point out that carbon dioxide moves in opposite directions in body tissues and in lungs. Tell students carbon dioxide in body tissues comes from cellular respiration.

Ask What is one way that carbon dioxide is transported in the blood? *(as carbonic acid)*

Ask How does the body get rid of carbon dioxide? *(It is breathed out of the lungs.)*

INTERPRET GRAPHS Guide students to the two Activation Energy graphs. Point out the arrows that represent activation energy.

Ask In which type of reaction do the products have less energy than the reactants? *(energy-releasing reaction)*

Discuss the example from the text (paper being lit by a match and then giving off heat and light as it burns) as an example of an energy-releasing reaction. Have students trace the red line in the graph with their fingers as you describe how energy changes during the reaction.

MAKE CONNECTIONS Ask students which of the energy graphs could show the chemical reaction that occurs during photosynthesis. *(energy-absorbing graph)*

USE ANALOGIES Have students think of analogies that demonstrate activation energy. *(Sample answer: needing energy to get out of a chair before walking across a room)*

KEY QUESTION Invite volunteers to read the question and answer on this page.

Ask Does a reaction that gives off energy still need activation energy to start? Explain. *(Yes, the reaction still needs energy to get started before it can release energy.)*

Build Vocabulary

RELATED WORD FORMS Have students write a sentence in which *react* is used correctly. Have them do the same for *reactive.*

Ask What noun is related to the verb *react* and the adjective *reactive? (reaction)*

Have students write a sentence correctly using the noun *reaction.*

During Reading continued

Active Reading

INTERPRET GRAPHS Guide students to the graph showing the effect of enzymes. Point out the two curves, the labels, and the arrows. Emphasize the difference between the red and blue energy pathways.

Ask What does the shape of each curve tell you? *(changes in the amount of energy during the reaction)* What does the shorter height of the red curve tell you? *(Less activation energy is needed to start the reaction when an enzyme is added.)*

MAKE CONNECTIONS Have students read the paragraph Nature's Catalysts.

Ask What can you infer about the number of enzymes needed by the body? *(Because each enzyme works on only one reaction, there must be a large number of different enzymes in the body.)*

INTERPRET SYMBOLS Point out the chemical equation for the formation of carbonic acid.

Ask What does each chemical formula stand for? *(CO_2, carbon dioxide; H_2O, water; H_2CO_3, carbonic acid)*

Ask What does the plus sign mean? *(joins)*

Ask What does the arrow mean? *(reacts to form)*

PREDICT The pH of the stomach is acidic, but the pH of the small intestine is basic. Have students predict and explain whether a stomach enzyme will work in the small intestine. *(No, an enzyme will work only under specific conditions. A stomach enzyme will work only in acidic conditions.)*

Build Vocabulary

MULTIPLE MEANINGS Read the following sentence aloud to students: *Getting a bad grade on the test was a catalyst that made him study harder.* Have a student volunteer explain what the sentence means. Then have students write their own sentences using the word *catalyst* in a similar context.

Speed Bump

Ask What is a substrate? *(the reactant on which an enzyme works)*

Ask What does the word *site* mean? *(a place)*

Ask What is the active site of an enzyme? *(the place where the substrate binds to the enzyme)*

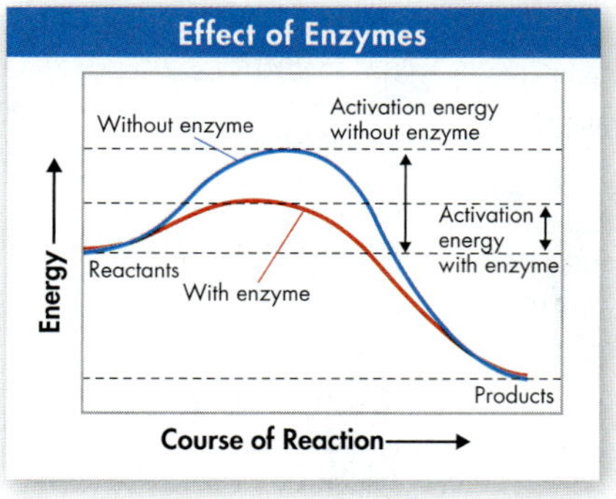

Effect of Enzymes Adding an enzyme lowers the activation energy of reactions. The enzyme speeds up reactions.

Enzymes

Some chemical reactions that make life possible are too slow to make them practical for living tissue. These chemical reactions need the help of catalysts. A **catalyst** is a substance that speeds up the rate of a chemical reaction. Catalysts work by lowering a reaction's activation energy.

Nature's Catalysts Many reactions in living things need catalysts. **Enzymes** are proteins that act as catalysts in living things. Enzymes speed up chemical reactions that take place in cells. Each enzyme generally works on only one chemical reaction in the body. Like other catalysts, enzymes lower activation energies. This has a big effect on how quickly a reaction happens. An enzyme affects the speed of a reaction in your blood that joins carbon dioxide and water to make carbonic acid, as shown below.

$$CO_2 + H_2O \longrightarrow H_2CO_3$$

Left to itself, this reaction is very slow. It is so slow, in fact, that dangerous levels of carbon dioxide could build up in your body. But the enzyme carbonic anhydrase makes the reaction happen immediately so that carbon dioxide is removed quickly.

Enzymes and Substrates How do enzymes do their jobs? For a chemical reaction to take place, the reactants must hit each other with enough energy for existing bonds to break and for new bonds to form. If the reactants do not have enough energy, they will not react.

An enzyme has a site where reactants can be brought together to react. This site lowers the energy needed for a reaction. The reactants of enzyme-catalyzed reactions are known as substrates. The substrates bind to a site on the enzyme called the active site. At the active site, the enzyme changes a substrate slightly so that a specific chemical bond is weakened. This weakened bond allows a chemical reaction to happen quickly. The active site and the substrates fit together like puzzle pieces. The fit is so precise that the active site and substrates are often compared to a lock and key.

Controlling How Enzymes Work Enzymes have important roles in controlling chemical reactions, making materials that cells need, giving off energy, and sending information. Enzymes are catalysts for reactions, so they can be affected by many things. Temperature, pH, and other molecules can affect how enzymes work.

BUILD Vocabulary

catalyst
a substance that speeds up the rate of a chemical reaction

enzyme
a protein catalyst that speeds up the rate of specific biological reactions

MULTIPLE MEANINGS
When the word *catalyst* is used in everyday language, it usually means "something that causes a change." This definition is related to the scientific meaning of the word. In science, a catalyst speeds up the rate of a chemical reaction. Change happens when the chemical reaction occurs.

Students Say

"I liked a game called Clue to learn vocabulary. The class was divided into four groups: A, B, C, D. Groups A and B went to the front of the room. Group A had to guess the vocabulary word while group B gave one-word clues. If group A couldn't guess the vocabulary word in one minute, then groups C and D would get a point. If group A got the word right, then A and B would get the point. After five words, we switched places, so that groups C and D went up to the front. If A or B broke the rules, such as giving more than one-word clues or guessing more than once for each clue, then C and D automatically got the point."

44 Chapter 2 • Lesson 4

Many enzymes are affected by changes in temperature. Enzymes made by human cells usually work best at around 37° Celsius, or normal human body temperature. Enzymes also work best at certain pH values. For example, enzymes in the stomach work best in acidic solutions. Also, most enzymes are controlled by molecules that switch the enzymes "on" or "off" as needed.

Key Question What role do enzymes play in living things and what affects their function? **Enzymes speed up chemical reactions that take place in cells. Temperature, pH, and other molecules can affect how enzymes work.**

BUILD Connections

UNLOCKING ENZYMES

The red molecule is a substrate and the blue molecule is the enzyme that it fits into. The gap in the enzyme is the active site. The fit between the enzyme and the substrate is like the fit between a lock and key.

In Your Workbook Go to your workbook to explain why a lock and key is a good analogy for a substrate and an enzyme.

CHECK Understanding

Apply Vocabulary

Use the highlighted words from the lesson to complete each sentence correctly.

1. During a(n) ________, some chemical bonds are broken and others are formed.
2. A(n) _______ is a protein that speeds up reactions in living things.

Critical Thinking

3. **Use Analogies** A change in pH can change the shape of a protein. How might a change in pH affect the function of an enzyme? (**Hint:** Think about the lock and key analogy.)
4. **Compare and Contrast** Describe the difference between a reaction that often happens on its own and one that does not.
5. **Explain** Explain how enzymes work, including the role of the active site.
6. **Review** What happens to chemical bonds during chemical reactions?
7. **Apply Concepts** Why is the melting of ice not a chemical reaction?
8. **Write to Learn** Answer the third clue of the mystery. Consider how temperature might affect chemical reactions.

MYSTERY CLUE

The chemical reactions of living things happen more slowly at lower temperatures. How would very cold antarctic waters affect the ice fish's need for oxygen? (Hint: See above.)

BIOLOGY.com Search Lesson 2.4 GO • Lesson Assessment • Visual Analogy

Check Understanding

1. chemical reaction
2. enzyme
3. If a change in pH changes the shape of an enzyme, it might result in the enzyme and substrates no longer fitting together properly. As a result, the enzyme would no longer be able to speed up the chemical reaction.
4. A reaction that can happen on its own gives off energy. A reaction that does not happen on its own requires energy to begin and continue.
5. An enzyme has an active site where reactants are brought together to react. This site lowers the activation energy needed for a reaction.
6. During a chemical reaction, some bonds are broken and other bonds are formed.
7. No bonds are broken when ice melts.
8. Very cold water would reduce the fish's need for oxygen because the fish's chemical reactions would happen more slowly at lower temperatures.

Build Connections

VISUAL ANALOGY Be sure students understand that the image is a visual analogy. It is comparing two different relationships: an enzyme and a substrate to a lock and key.

Ask Why is a lock and key a good analogy for an enzyme-substrate reaction? *(Both depend on a perfect fit between parts before they will work.)*

Ask Do you think the substrate will fit into one of the gaps on the outside of the enzyme? Explain. *(No, the shape of the substrate will not match the shape of any gap other than the active site.)*

Active Reading

KEY QUESTION Have a volunteer read the question and answer aloud.

Ask How do you think a change in pH or temperature might make an enzyme stop working? *(The shape of the active site may change.)*

Post Reading

ASSESSMENT TIPS Before students answer Question 4, have them make a graphic organizer to summarize the similarities and differences between reactions that happen on their own and those that do not. Then have them answer the question in complete sentences.

MYSTERY CLUE Remind students that temperature can affect the activity of enzymes and therefore the rate at which reactions take place. Guide them to conclude that the fish would need less oxygen if the reactions that require oxygen occur more slowly.

Wrap-Up Activity

COMPOSE A QUIZ Divide students into small groups. Have each group write a lesson quiz consisting of five multiple-choice questions or five fill-in-the-blank questions, as well as one constructed-response question. When the quizzes are finished, have groups switch quizzes and complete them. The group that wrote a test should correct it.

CHAPTER LAB

Pre-Lab

Review the meanings of *hypothesis, independent variable,* and *dependent variable.*

Ask students for definitions of *catalyst, enzyme,* and *substrate.* Have students describe how an enzyme works. Be sure they use the words *substrate* and *active site* in their descriptions.

Say Bonds are broken and new bonds are formed in a chemical reaction.

Ask Is a substrate a reactant or a product in a chemical reaction? *(a reactant)* Will bonds in a substrate break or will new bonds form? *(Bonds will break.)* What kind of macromolecule is an enzyme? *(a protein)*

Review the Background Questions and Pre-Lab Questions as a class.

Lab

Tell students they will perform the chapter lab *Temperature Affects Enzymes,* described in **Lab Manual B**. Explain that hydrogen peroxide, which is harmful to cells at high concentrations, is a product of many chemical reactions in cells. The liver produces an enzyme called catalase that breaks down hydrogen peroxide to form water and oxygen gas.

SAFETY Remind students to wear goggles, apron, and gloves during the lab. Hydrogen peroxide irritates skin and bleaches clothing. The liver is raw and may contain bacteria, so it is important that students wash their hands after handling it.

Post-Lab

EQUATION Have students write a chemical equation for the reaction they observe.

$2H_2O_2 \longrightarrow O_2 + 2H_2O$

NATIONAL SCIENCE EDUCATION STANDARDS

UCP I, II, III

CONTENT B.3, C.1.d

INQUIRY A.1.b, A.2.a

Design Your Own Lab

GUIDED INQUIRY

Pre-Lab: Temperature and Enzymes

Problem How does temperature affect the rate of an enzyme-catalyzed reaction?

Materials raw liver, forceps, petri dish, dropper pipette, 1% hydrogen peroxide solution, 25-mL graduated cylinder, 50-mL beakers, puréed liver, filter paper disks, paper towels, timer or clock with a second hand, water baths, thermometers, beaker tongs, graph paper

Lab Manual Chapter 2 Lab

Skills Focus Form a Hypothesis, Design an Experiment, Measure, Interpret Graphs

Connect to the Big idea Many chemical reactions in living organisms could not take place without enzymes. Enzymes catalyze the reactions that release energy from nutrients. They also catalyze the synthesis of the complex molecules that organisms need to grow and stay healthy. One factor that affects the action of enzymes is temperature. Think about why people store some foods in a refrigerator. The cold temperature limits the ability of enzymes to break down, or spoil, those foods.

Do high temperatures have the opposite effect on enzymes? Do they become more and more active as the temperature rises? In this lab, you will investigate the effect of temperature on an enzyme-catalyzed reaction.

Background Questions

a. **Review** Why do many reactions that occur in cells require enzymes? How do enzymes speed up chemical reactions?

b. **Review** Name three variables that can affect enzyme activity.

c. **Use Analogies** Use eggs and a frying pan on a stove as an analogy for reactants and an enzyme. Use the control knob on the stove burner as an analogy for how a variable can affect the action of an enzyme.

Pre-Lab Questions

Preview the procedure in the lab manual.

1. **Relate Cause and Effect** How will you know that a chemical reaction is taking place in Part A? How will you know in Part B?
2. **Control Variables** In Part B of the lab, which variable will you manipulate? Which variable is the dependent variable?
3. **Relate Cause and Effect** How is the time required for the filter-paper disk to float related to the activity of the enzyme?

BIOLOGY.com Search Chapter 2 GO

Visit Chapter 2 online to test yourself on chapter content and to find activities to help you learn.

Untamed Science Watch the Untamed Science crew find answers to the mystery of why water is such a special compound.

Art Review Learn about ionic and covalent bonding.

Art in Motion View an animation that shows the process of a salt crystal dissolving in water.

Visual Analogy Compare enzymes and substrates to a lock and key.

46 Chapter 2 • Pre-Lab

Pre-Lab Answers

BACKGROUND QUESTIONS

a. Reactions in cells are often too slow or require an activation energy that is not practical for living tissue. Enzymes lower the activation energy of reactions.

b. pH, temperature, and regulatory molecules

c. Sample answer: The frying pan is like an enzyme. It provides a location where the eggs can be cooked. The control knob allows the user to control the temperature of the frying pan, which affects how fast the eggs cook.

PRE-LAB QUESTIONS

1. I will observe bubbles of oxygen on the surface of the liver. The filter paper disk will rise to the top of the puree.
2. temperature; reaction time
3. Oxygen produced in the reaction causes the disk to float. The rate of the reaction that produces the oxygen depends on the activity of the enzyme. The more active the enzyme, the faster the oxygen is produced, and the quicker the disk will rise.

2 CHAPTER Summary

2.1 The Nature of Matter

- The particles that make up atoms are protons, neutrons, and electrons.
- All isotopes of an element have the same number of protons and the same chemical properties.
- The physical and chemical properties of a compound are usually very different from those of the elements that make it up.
- The main kinds of chemical bonds are ionic bonds and covalent bonds.

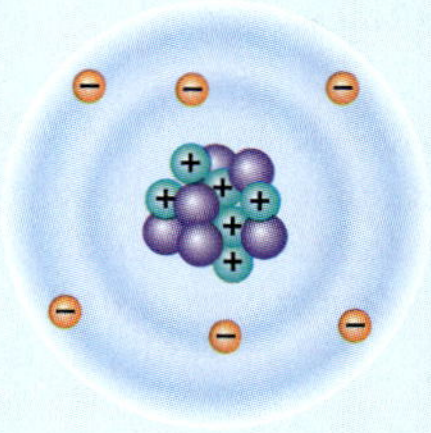

atom (p. 28)	**compound** (p. 30)
nucleus (p. 28)	**ionic bond** (p. 30)
electron (p. 28)	**covalent bond** (p. 31)
element (p. 29)	**molecule** (p. 31)
isotope (p. 29)	

2.2 Properties of Water

- Water is a polar molecule. It is able to form many hydrogen bonds, which makes water cohesive and adhesive.
- Water's polar nature gives it the ability to dissolve both ionic compounds and other polar molecules.
- Buffers play an important role in maintaining homeostasis in organisms.

cohesion (p. 34)	**acid** (p. 36)
adhesion (p. 34)	**base** (p. 36)
solution (p. 34)	**buffer** (p. 36)
suspension (p. 35)	

2.3 Carbon Compounds

- Carbon can bond with many elements, including hydrogen, oxygen, phosphorus, sulfur, and nitrogen, to form the molecules of life.
- Living things use carbohydrates as their main source of energy. Plants, some animals, and other living things also use carbohydrates for structural purposes.
- Lipids can be used to store energy. Some lipids are important parts of biological membranes and waterproof coverings.
- Nucleic acids store and transmit genetic information.
- Some proteins control cell processes and the rate of reactions. Others form important parts in cells, while still others move substances into or out of cells or help to fight disease.

carbohydrate (p. 38)	**protein** (p. 40)
lipid (p. 39)	**amino acid** (p. 40)
nucleic acid (p. 40)	

2.4 Chemical Reactions and Enzymes

- During chemical reactions, some bonds are broken and other bonds are formed.
- Chemical reactions that give off energy often happen on their own. Chemical reactions that take in energy will not happen without a source of energy.
- Enzymes speed up chemical reactions that take place in cells.
- Temperature, pH, and other molecules can affect how enzymes work.

chemical reaction (p. 42)	**catalyst** (p. 44)
activation energy (p. 43)	**enzyme** (p. 44)

BIOLOGY.com Search Chapter 2 GO • Match It • Chapter Assessment

UbD Transfer the Big Idea

TRANSFER Introduce the subject of genetic disorders by explaining that many genetic disorders are caused by a defective enzyme that is needed for a chemical reaction inside cells. Genes for these disorders are inherited from one or both parents. Explain that genetic disorders can cause serious health problems. Have students work in small groups to identify a particular genetic disorder—such as phenylketonuria (PKU) or Tay-Sachs—and produce a simple slide presentation, poster, or oral presentation about it. In their presentation, students should explain what causes the disorder, identify the body system that is affected, and tell how a defect in the enzyme affects health. Tell students that many other genetic disorders are caused by defective proteins that are not enzymes. Color blindness, sickle cell disease, and cystic fibrosis are examples.

Chapter Summary

Think Visually

USE GRAPHICS Have students create a table in which they compare the structures and functions of the following macromolecules: carbohydrates, lipids, proteins, and nucleic acids.

ANSWERS
Student tables should show that carbohydrates consist of simple sugars called monosaccharides or chains of sugars called complex carbohydrates; they provide energy or structure. Lipids consist of glycerol and fatty acids; they store energy or make membranes and waterproof coverings. Nucleic acids consist of chains of nucleotides; they store and transmit genetic information. Proteins consist of chains of amino acids; they control the rate of reactions, regulate cell processes, form cell structures, transport substances into or out of cells, or help fight disease.

BIOLOGY.com Study Online

REVIEW RESOURCES

Art Review This drag-and-drop activity lets students explore ionic and covalent bonds.

Art in Motion This animation shows the process of salt crystals dissolving in water.

Data Analysis Students can analyze data about the ecological impact of acid rain.

Visual Analogy Using this animation, students can further explore the lock-and-key analogy for an enzyme and its substrates.

Vocabulary Review The *Flash Cards* and *Match It* provide an interactive way to review chapter vocabulary.

Editable Worksheets Pages of Study Workbook B, Lab Manual B, and the Assessment Resources Book are available online. These documents can be easily edited using a word-processing program.

Lesson Overviews Lesson Overviews are based on Miller & Levine Biology. However, they are available on CD-ROM and can be edited for your class.

ASSESSMENT RESOURCES

Chapter Assessment Have students take an online version of the Chapter 2 Check Understanding.

Standardized Test Prep Students can take an online version of the Standardized Test Prep. You will receive their scores along with ideas for remediation.

Diagnostic and Benchmark Tests Use these tests to monitor your students' progress and supply remediation.

ASSESSMENT

Assess

Big idea Invite a volunteer to read the question. Write the terms *atoms, chemical bond,* and *chemical reaction* on the board. Ask students to write a single sentence that ties all three terms together. *(Sample sentence: A chemical bond forms between atoms and is broken during a chemical reaction.)*

Ask What is formed when a bond connects two different atoms? *(a compound)*

Ask What are the basic chemical principles that affect living things? *(Sample answer: Chemical bonds join together the molecules and compounds of life. Water and carbon compounds play important roles in living things, which carry out chemical reactions in their daily life processes.)*

Have students write their sentence and answers in their notebooks.

CONSTRUCTED RESPONSE Paragraphs may vary. Explain the activity carefully. Be sure students understand that the *Hints* should be used as a guide to help them answer the questions. Review how to use an outline as a prewriting tool.

Write Question 1 on the board. Underline the word *electrons*.

Say This question is about how the structure of atoms relates to chemical compounds. Remind students that electrons occupy space around an atom's nucleus.

Read the first hint.

Say When elements are *chemically joined,* they are forming a bond.

Ask How do elements chemically join? *(They may share electrons, or one element may take electrons away from another element.)*

Read the next hint. Emphasize the word *molecules*.

Ask What kind of bonds hold the atoms of a molecule together? *(covalent bonds)*

Model writing the answer using the hints on the board and in the text. *(Sample: Compounds are made when two or more elements join chemically. Atoms of elements join to make molecules by forming bonds. The electrons in atoms are involved in the formation of these bonds. In the covalent bonds of a molecule, electrons are shared between atoms. In the ionic bonds of an ionic compound, electrons move from one atom to another.)*

2 CHECK Understanding

Assess the Big idea Matter and Energy

Write an answer to the question below.

Q: What are the basic chemical principles that affect living things?

Constructed Response

Write an answer to each of the questions below. The answer to each question should be one or two paragraphs long. To help you begin, read the **Hints** below each of the questions.

1. **How are chemical compounds related to the electrons in atoms?**

   Hint Chemical compounds are made when two or more elements are chemically joined.

   Hint Molecules of compounds are formed when atoms of elements form bonds with each other.

2. **How does the structure of a water molecule help water dissolve ionic compounds?**

   Hint The atoms in water are joined by covalent bonds.

   Hint The electrons in water are not evenly spread out.

3. **Explain which of the four groups of macromolecules are polymers and which are not.**

   Hint A polymer is a large compound made by the joining of many monomers.

Foundations for Learning Wrap-Up

Use the vocabulary quiz sheet that you made when you started the chapter as a tool to help you learn the important words from this chapter. Do Activity 1 before you do Activity 2.

Activity 1 Compare your quiz sheet with a partner's quiz sheet. If your partner's sheet has words that you do not have, add them to your own sheet. Fold the tabs over the definitions. Then, swap sheets with your partner and take turns quizzing each other on the definitions of the words. Open each word's matching tab as you ask your partner what the word means.

Activity 2 Cut your vocabulary quiz sheet so that the words and definitions are on separate slips of paper. Cut off the tabs. Discard the tabs. Mix up the remaining slips of paper on your desk. Then, work to match up each vocabulary word with its definition.

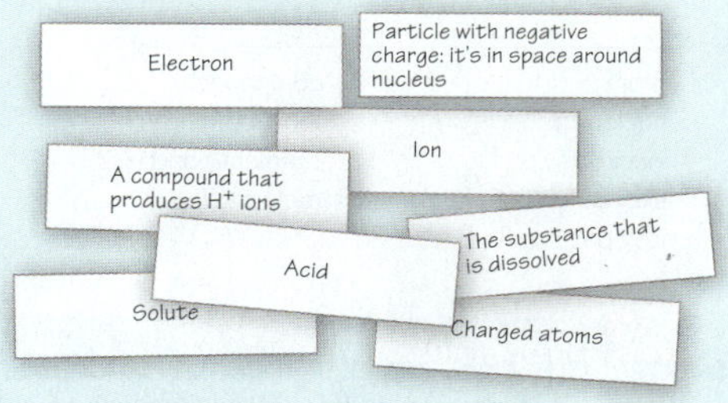

For Question 2, pair students and have them use the hints and your modeling to answer the question. Have students share their paragraphs with the class.

For Question 3, have students write the paragraphs independently for homework or as an anchor activity.

Foundations for Learning Wrap-Up

Assign students to pairs. Have students take out their vocabulary quiz sheets, and explain the activities. In Activity 1, tell students that if neither partner knows the definition, the pair should read the definition and then close the tab. Students can try the word again later.

2.1 The Nature of Matter

Understand Key Concepts

1. The positively charged particle in an atom is a(n)
   a. electron. c. neutron.
   b. ion. d. proton.
2. Two or more atoms are joined in definite amounts in any
   a. compound. c. isotope.
   b. element. d. symbol.
3. A covalent bond is formed by the
   a. gaining of electrons.
   b. losing of electrons.
   c. sharing of electrons.
   d. transfer of electrons.

Test-Taking Tip

Choose Among Similar Answers The answers for question 3 all end with "of electrons," so those words will not matter when picking an answer. You have to look at only the first word of each answer. When you do, you should realize that "gaining" and "losing" are part of a "transfer" of electrons. So, answers **a**, **b**, and **d** are all related and cannot all be correct. So, answer **c** is the right answer.

Think Critically

4. **Compare and Contrast** Describe how ionic bonds and covalent bonds are similar and how they are different.

2.2 Properties of Water

Understand Key Concepts

5. When you mix salt and water together in a beaker, you cause them to form a
   a. compound. c. mixture.
   b. solution. d. suspension.
6. A compound that releases hydrogen ions (H^+) in solution is a(n)
   a. acid. c. polymer.
   b. base. d. salt.

Think Critically

7. **Explain** Explain the properties of cohesion and adhesion. Then, give an example of each property.
8. **Apply Concepts** A student mixes a spoonful of sugar in a glass of water. Explain what the student has made using the terms *solution*, *solute*, and *solvent*.

2.3 Carbon Compounds

Understand Key Concepts

9. Proteins are polymers formed from
   a. amino acids.
   b. lipids.
   c. nucleic acids.
   d. simple sugars.
10. Carbohydrates are made up of all of the following elements EXCEPT
    a. carbon. c. oxygen.
    b. hydrogen. d. sodium.

Think Critically

11. **Apply Concepts** Explain the relationship between monomers and polymers using complex carbohydrates as an example.
12. **Explain** Describe the structure of a lipid and explain how to tell if the lipid is saturated or unsaturated.

2.4 Chemical Reactions and Enzymes

Understand Key Concepts

13. In a chemical reaction, a reactant binds to an enzyme at a place known as the
    a. active site. c. product.
    b. catalyst. d. substrate.

Assessment Tips

Read Question 1 and the options. Point out that the key words in the question are *positively charged.* Have students ask themselves which of the four answer choices is positively charged. Two of the options (*a* and *c*) can be eliminated. Ions can be either positive or negative. Ask students for the correct answer.

Read aloud the Test-Taking Tip. Make sure students understand it before they answer Question 3. Remind them that the tip is useful with any multiple-choice question. For Questions 5 and 8, have students review mixtures and solutions. Have students review monomers and polymers before answering Questions 9 and 11. Students must understand enzyme action to answer Questions 13–16 and basic concepts of graphing for Questions 17–19.

Answers

Lesson 2.1

UNDERSTAND KEY CONCEPTS

1. d

2. a

3. c

THINK CRITICALLY

4. Both ionic and covalent bonds form between atoms and both involve electrons. However, ionic bonds involve a transfer of electrons and covalent bonds involve sharing of electrons.

Lesson 2.2

UNDERSTAND KEY CONCEPTS

5. b

6. a

THINK CRITICALLY

7. Cohesion is an attraction between molecules of the same substance. Adhesion is an attraction between molecules of different substances. An example of cohesion is surface tension. An example of adhesion is the way water sticks to the side of a narrow tube and forms a meniscus.

8. The student has made a solution of sugar and water. In the solution, the sugar is the solute and water is the solvent.

Lesson 2.3

UNDERSTAND KEY CONCEPTS

9. a

10. d

THINK CRITICALLY

11. Polymers are made up of many monomers joined together. A complex carbohydrate is a polymer that is made up of many monomers called simple sugars.

12. A lipid is made up of a glycerol molecule that is joined to compounds called fatty acids. If each carbon atom in a lipid's fatty acids is joined to other carbon atoms only by single bonds, the lipid is said to be saturated. If there is at least one carbon-carbon double or triple bond in a fatty acid, the fatty acid is said to be unsaturated.

Lesson 2.4

UNDERSTAND KEY CONCEPTS

13. a

ASSESSMENT

14. c

THINK CRITICALLY

15. The temperature would lower the activity of the enzyme. Enzymes made by the human body work best when the temperature is around 37°C—the normal human body temperature.

16. The fit of an enzyme and its substrates at the enzyme's active site is so precise that the substrates are like a key and the enzyme is like a lock. Similar to a key in a lock, only substrates of a certain shape can fit into the active site of the enzyme.

Connecting Concepts

17. 35°C

18. The total product was doubled when the temperature of the reaction increased from 25°C to 35°C. It decreased to almost zero when the temperature was increased to 45°C. Enzymes work best at certain temperatures. The enzyme involved in this reaction works best at about 35°C. The enzyme's activity is lower at other temperatures.

19. about 1.5 mg

CHAPTER MYSTERY After students have read through the Chapter Mystery, discuss the concept of polarity. Review the structure of water as a polar compound. Review the topics of solutions and solubility.

Ask Why does more oxygen dissolve in cold water than in warm water? *(Oxygen molecules have less energy and move more slowly in cold water. Because they cannot resist the attraction of water molecules, more of them dissolve.)*

Ask How does the temperature of the water affect the fish's need for oxygen? *(The fish's life processes take place more slowly, and therefore require less oxygen, than the processes in organisms living in warmer temperatures.)*

2 CHECK Understanding

14. An enzyme speeds up a reaction by
a. giving off energy.
b. taking in energy.
c. lowering the activation energy.
d. raising the activation energy.

Think Critically

15. Infer An enzyme made by the human body is put in a solution in a lab. The temperature of the solution is 20°C. How would this temperature affect how the enzyme works? Explain your answer.

16. Use Analogies Explain why a lock and key are used to describe the way an enzyme works.

Connecting Concepts

Use Science Graphics

The following graph shows the total amount of product from a chemical reaction performed at three different temperatures. The same enzyme was involved in each case. Use the graph to answer questions 17–19.

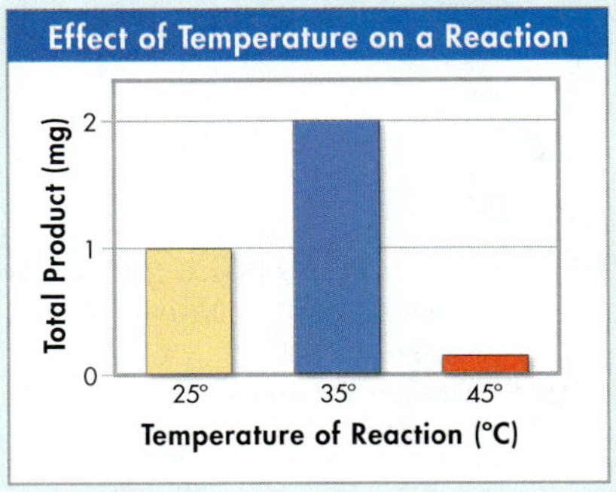

17. Interpret Graphs At which temperature was the greatest amount of product formed?

18. Draw Conclusions Describe the results of each reaction. How can you explain these results?

19. Predict A student performs the same chemical reaction at 30°C. About how much product can she expect to get?

solve the CHAPTER MYSTERY

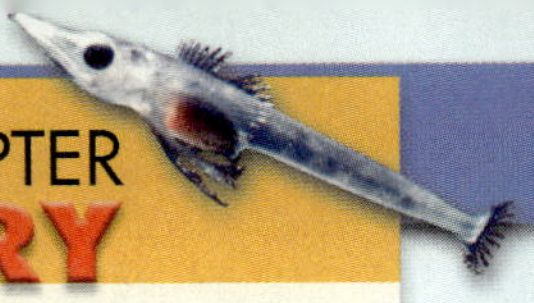

THE GHOSTLY FISH

The protein hemoglobin allows the blood of most fishes to carry nearly 50 times the oxygen it would without the protein. The ghostly white look of the antarctic ice fish comes from its clear blood—blood without hemoglobin. Ice fish can live without hemoglobin because of the properties of very cold water.

Oxygen dissolves in seawater, providing the oxygen that fishes need to live. Fishes take in dissolved oxygen directly through their gills, where it passes into their blood. More oxygen can dissolve in water when the water is cold. So, the icy cold antarctic waters have a lot of oxygen.

The large gills and scaleless skin of ice fish let them take in oxygen efficiently from the water. Compared to red-blooded fishes, ice fish have a higher blood volume, thinner blood, and larger hearts. Their blood can carry more dissolved oxygen. The large hearts can pump the thinner blood through the body faster. These and other physical features, along with the chemistry of oxygen in very cold water, enable ice fish to live where many other living things cannot.

1. Relate Cause and Effect Ice fish produce antifreeze proteins to keep their blood from freezing. Their body temperature stays below 0°C. How does low body temperature affect the blood's ability to carry dissolved oxygen?

2. Infer People living at high altitudes generally have more hemoglobin in their blood than people living at sea level. Why do you think this is so?

3. Predict If the antarctic oceans were to warm up, how might this affect ice fish?

Untamed Science™ Finding the solution to the fishy mystery is only the beginning. Take a video field trip with the ecogeeks of Untamed Science to see where this mystery leads.

CHAPTER MYSTERY ANSWERS

1. Low body temperature would increase the blood's ability to carry dissolved oxygen.

2. At high altitudes, less oxygen is available in the atmosphere than at low altitudes. Therefore, people (and other organisms) need more hemoglobin to ensure that enough oxygen is obtained to supply the tissues of the body.

3. Sample answer: There would be less dissolved oxygen available for the ice fish. At the same time, their need for oxygen would increase because warmer water temperatures would speed the fish's metabolism. The fish might migrate to colder waters farther south to avoid the temperature change.

Untamed Science™ Have students take a video field trip with the ecogeeks of Untamed Science to learn more about the unique properties of water in **Not a Drop to Drink.**

Standardized Test Prep

Multiple Choice

1. The elements or compounds that enter into a chemical reaction are called
   A products.
   B catalysts.
   C active sites.
   D reactants.

2. Chemical bonds that involve the total transfer of electrons from one atom or group of atoms to another are called
   A covalent bonds.
   B ionic bonds.
   C hydrogen bonds.
   D oxygen bonds.

3. Which of the following is NOT a macromolecule found in living organisms?
   A protein
   B nucleic acid
   C sodium chloride
   D lipid

4. Which combination of particle and charge is correct?
   A proton: positively charged
   B electron: positively charged
   C neutron: negatively charged
   D electron: no charge

5. In which of the following ways do isotopes of the same element differ?
   A in number of neutrons only
   B in number of protons only
   C in numbers of neutrons and protons
   D in number of neutrons and in mass

6. Which of the following molecules is made up of glycerol and fatty acids?
   A sugars
   B starches
   C lipids
   D nucleic acids

7. Nucleotides consist of a phosphate group, a nitrogenous base, and a
   A fatty acid.
   B lipid.
   C 5-carbon sugar.
   D 6-carbon sugar.

Questions 8–9

The enzyme catalase speeds up the chemical reaction that changes hydrogen peroxide into oxygen and water. The amount of oxygen given off is an indication of the rate of the reaction.

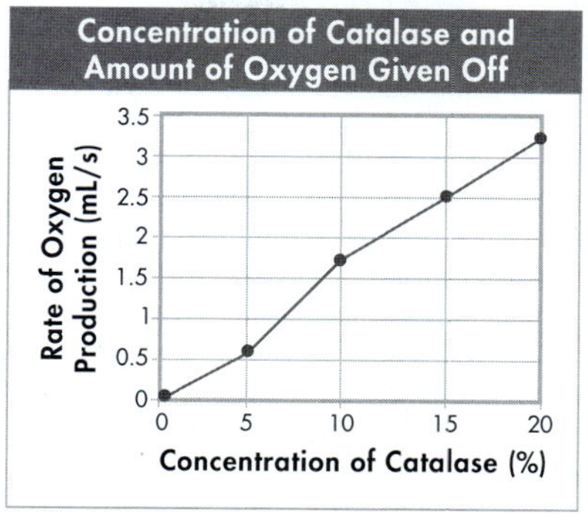

8. Based on the graph, what can you conclude about the relationship between enzyme concentration and reaction rate?
   A Reaction rate decreases with increasing enzyme concentration.
   B Reaction rate increases with decreasing enzyme concentration.
   C Reaction rate increases with increasing enzyme concentration.
   D The variables are indirectly proportional.

9. Which concentration of catalase will produce the fastest reaction rate?
   A 5%
   B 10%
   C 15%
   D 20%

Open-Ended Response

10. List some of the properties of water that make it such a unique substance.

If You Have Trouble With . . .

Question	1	2	3	4	5	6	7	8	9	10
See Lesson	2.4	2.1	2.3	2.1	2.1	2.3	2.3	2.4	2.4	2.2

ASSESSMENT

Before the Test

Review with students the Test-Taking Tip.

Point out the word *NOT* in Question 3.

Point out that each option in Question 4 is like a matching test. Students have to match each particle with the charge to determine the correct answer.

Point out the word *only* in options A and B of Question 5. This word affects which option is the correct answer.

Remind students that an open-ended response means that they'll have to write a paragraph to answer the question. Have them paraphrase the question in their notebooks, and remind them about modeling a paragraph.

Answers

1. D

2. B

3. C

4. A

5. D

6. C

7. C

8. C

9. D

10. Sample answer: Properties of water that make it such a unique substance include the polar nature of its molecules, which allows them to form hydrogen bonds with each other and with other molecules. Because it is a polar molecule, water exhibits cohesion, adhesion, and the ability to dissolve ionic compounds and other polar molecules.

Test-Taking Tip

READ ALL THE ANSWER CHOICES

Tell students to be sure to read all of the answer choices, even if the first choice seems to be correct. Remind them that a more complete or better answer choice may be present among the remaining choices. By reading all of the answer choices, they are more likely to choose the best one.

Plan Ahead

After students have read what their task will be in the Unit 1 Project, suggest they review how to design a controlled experiment and what independent and dependent variables are. Also, make sure students understand acids, bases, and pH. If students haven't already tested pH during a previous activity, consider demonstrating how pH paper is used. You may want to direct students to selected experiments in a lab manual to help familiarize them with experimental procedures.

Monitor the Project

Suggest students begin by writing a hypothesis for the first of their three experiments. Briefly check that each student's hypothesis can be tested with a controlled experiment. Then, as they design their experiments, ask individual students questions that will help them identify variables, write a procedure, and think of ways to collect and record data to test the hypothesis.

Ask What are the independent and dependent variables in your experiment?

Ask How will data be collected in this experiment to show whether the medication actually neutralizes stomach acid?

Project Assessment

Make sure students use the rubric and reflection questions to assess their work. Then, use the rubric to assign a final score. Note that it is important to value the creativity of students' work as well as the content when you score their projects. If desired, talk with students about any differences between their self-assessment scores and your assigned score.

UNIT 1 The Nature of Life

Unit Project

Design the Experiment

Did you ever wonder how a medication goes from the lab to your local drug store shelf? A lot of research and experimentation by scientists goes into testing a new medication to make sure it is safe and effective. Imagine you are a scientist working for a pharmaceutical company. Your current project is to test a new medication for heartburn. Heartburn is a painful condition in which acid inside the stomach backs up into the esophagus—the connection between your throat and stomach. This new medication helps neutralize stomach acid to prevent irritation.

Your Task Design *three* possible experiments to test the safety and effectiveness of the new heartburn medication. Before you begin, think about how you will know if the medication actually neutralizes stomach acid. Once you've written your procedures, you will propose the experiments to your company's Executive Board for Research and Development.

For each experiment,

- identify clear independent and dependent variables.
- identify a control.
- form a hypothesis—predict the results you'd expect to find if the medication worked.
- write a specific procedure that tests your hypothesis.

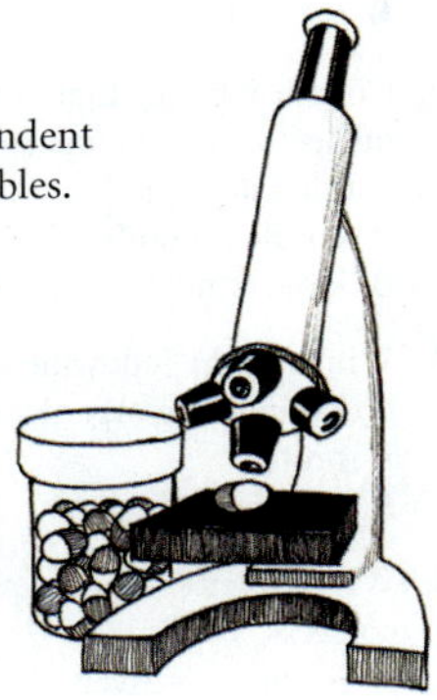

Reflection Questions

1. **Score your experimental designs using the rubric below. What score did you give yourself?**
2. **What did you do well in this project?**
3. **What about your designs needs improvement?**
4. **Are there any ethical dilemmas related to your experiments? Explain.**

Assessment Rubric

Score	Scientific Content	Quality of Experiments
4	Correctly and extensively applies knowledge and understanding of unit concepts (i.e., pH scale) to experimental designs and predictions.	Experimental designs are clever and effectively test the hypotheses. Experimental conditions are carefully controlled and variables are correctly identified.
3	Applies relevant knowledge and understanding of unit concepts (i.e., pH scale) to experimental designs and predictions.	Experimental designs are logical and test the hypotheses. Experimental conditions are controlled and variables are correctly identified.
2	Applies relevant knowledge and understanding of unit concepts (i.e., pH scale) incompletely to experimental designs and predictions.	Experimental designs need some revisions—some parts are unclear or do not fully test the hypotheses. Variables and controls need corrections.
1	Does not correctly apply knowledge and understanding of unit concepts (i.e., pH scale) to experimental designs and predictions.	Experimental designs are unclear and do not test the hypotheses. Variables and controls listed are incorrect or absent.

NATIONAL SCIENCE EDUCATION STANDARDS

UCP II, III

CONTENT B.2, F.1

INQUIRY A.1.a, A.1.b, A.2.a

21st Century Skills

To be successful in the 21st century, students need skills and learning experiences that extend beyond subject area mastery. The Unit 1 Project helps students build the following 21st Century Skills: *Critical Thinking and Systems Thinking; Problem Identification, Formulation, and Solution; Self-Direction;* and *Accountability and Adaptability.*

FOCUS ON COMMUNICATION Extend this Unit Project by having small groups of students design and create a magazine advertisement for a hypothetical heartburn medication. Suggest that the advertisement not only extol the virtues of the medication but also teach the consumer about heartburn medications by briefly describing an experiment that supports the medication's effectiveness. Have groups present their ads to the class.

For more practice building 21st Century Skills, see The Chapter Mystery pages in **Study Workbook A.**

Ecology

UNIT 2

Chapters

INTRODUCE the **Big ideas**

- **Matter and Energy**
- **Interdependence in Nature**

"Earth is a living planet on which all forms of life are linked to one another, and to land, water, and air. Through those links, energy flows and matter cycles in patterns that support life, including human society. We know enough about these patterns to realize that they are changing, due to human activity, in ways that we don't understand. Our challenge is to study our impact on the biosphere and plan for a healthy future."

Joe Levine

53

Dear Colleague,

Ever since I began helping my dad in the garden as a kid, I've loved plants. So, when my indoor jungle was first invaded by aphids, spider mites, and mealybugs a few years ago, I had a problem. It was too cold to take plants outside to hose them off. But, I wasn't comfortable using insecticides indoors, with my son and our dog running around. On a hunch, I searched the Web for "beneficial insects" and found several sites selling species that prey on plant-eating pests. The best sources supplied information on preferred foods and environmental requirements of each predatory species, so I could select the right ones for my conditions. Soon, my own little Integrated Pest Management system included ladybugs, lacewings, predatory mites, and a beetle named Cryptolaemus. *Those "good bugs" cost a little more than pesticides. But, I could rest easy, knowing that I could keep herbivore populations under control without exposing my son to potentially harmful chemicals.*

This wasn't rocket science. All forms of life on Earth are involved in interactions with one another and with their environment. Natural populations of insects and plants, predators and prey, parasites and hosts, grow and reproduce. They pass through cycles of increase and decrease, rarely going extinct, yet never taking over the planet, either.

Unfortunately, many Americans don't think about food chains or nutrient cycles except during food shortages, droughts, or floods. This level of ecological illiteracy is dangerous, because our species is Earth's most powerful force for change. Human activity now affects not only fish populations in the open ocean, but also vital life-support systems such as the ozone layer and the global greenhouse. I hope this unit helps inspire you to teach your students that human society is part of the biosphere; that biological diversity is a treasure; that clean air, water, and soil are invaluable resources; and that all life is connected. If your students learn these lessons, they and their descendants will enjoy a happier and healthier future.

Joe Levine

The Biosphere

Chapter Contents	NSES	Time	Core Resources
Chapter Preview			**Student Edition,** pp. 54–55 **Chapter Mystery,** p. 55 **Foundations for Learning,** p. 55
3.1 What Is Ecology? Studying Our Living Planet • Biotic and Abiotic Factors • Ecological Methods	I, II, A.2.a, A.2.b, A.2.d, G.1, G.2	1/2 period 1/4 block	**Student Edition,** pp. 56–59 **Study Workbook B** 3.1 Worksheets **Biology.com** *Art in Motion:* Levels of Organization • 3.1 Lesson Assessment
3.2 Energy, Producers, and Consumers Primary Producers • Consumers	I, C.4.b, C.5.a, C.5.b, C.5.e, C.5.f, D.1	1 period 1/2 block	**Student Edition,** pp. 60–62 **Study Workbook B** 3.2 Worksheets **Biology.com** *Art Review:* Producers and Consumers • *Tutor Tube:* Producers and Consumers • 3.2 Lesson Assessment
3.3 Energy Flow in Ecosystems Food Chains and Food Webs • Ecological Pyramids	I, II, A.1.c, A.2.d, B.5, C.4.a, C.4.b, C.5.a, C.5.b, C.5.e, C.5.f, D.1	1 period 1/2 block	**Student Edition,** pp. 63–67 **Build Connections,** p. 64 **Inquiry Into Scientific Thinking,** p. 66 **Study Workbook B** 3.3 Worksheets **Biology.com** *Visual Analogy:* Earth's Recycling Center • 3.3 Lesson Assessment
3.4 Cycles of Matter Recycling in the Biosphere • The Water Cycle • Nutrient Cycles • Nutrient Limitation	I, II, III, B.2, C.4.a, C.4.d, C.4.e, C.5.e, C.5.f, D.2, F.3, F.4	1 period 1/2 block	**Student Edition,** pp. 68–73 **Build Connections,** p. 68 **Study Workbook B** 3.4 Worksheets **Biology.com** *InterActive Art:* The Water Cycle • *Visual Analogy:* Interlocking Nutrient Cycles • 3.4 Lesson Assessment
Chapter Pre-Lab	II, A.1.b, A.2.a, C.4.a, C.4.d, C.4.e, C.5.b, C.5.e, F.4, F.5	1 period 1/2 block	**Student Edition,** p. 74 **Lab Manual B** *The Effect of Fertilizer on Algae*

Tools for ELL Instruction

ELL Handbook describes teaching strategies to make *Miller & Levine Biology: Foundation Edition* more accessible to ELL students.

Spanish Study Workbook is a Spanish translation of Study Workbook A.

Multilingual Glossary is the *Miller & Levine Biology* glossary translated into ten languages.

Tools to Challenge Students

Study Workbook A includes more challenging worksheets and activities such as Chapter Mystery activities that reinforce 21st Century Skills.

Lab Manual A labs are geared toward on-level students. The Lab Manual includes fifteen labs in addition to the Chapter Labs.

Biology.com Data Analysis activities offer students a way to practice data gathering and analyzing skills.

Visual Quizzes in the Assessment Resources Book are geared toward on-level students and use visuals from *Miller & Levine Biology.*

Additional Resources

Biology.com Untamed Science Video • Flash Cards

Study Workbook A 3.1 Worksheets
Lab Manual B *Hands-On Activity:* Abiotic Factors
Spanish Study Workbook 3.1 Worksheets
Biology.com *Data Analysis:* Counting on Nature

Study Workbook A 3.2 Worksheets
Spanish Study Workbook 3.2 Worksheets

Study Workbook A 3.3 Worksheets
Lab Manual B *Data Analysis:* The 10 Percent Rule
Spanish Study Workbook 3.3 Worksheets
Assessment Resources Book Visual Quiz

Study Workbook A 3.4 Worksheets
Spanish Study Workbook 3.4 Worksheets
Assessment Resources Book Visual Quiz

Lab Manual A *The Effect of Fertilizer on Algae* • *Enzymes in Detergents*

Chapter Review

Student Edition Chapter Summary, p. 75
Study Workbook B Chapter 3 Review
Transparencies, pp. 26–43
Biology.com Untamed Science Video • Editable Worksheets of Study Workbooks A and B and Lab Manuals A and B • Chapter 3 Flash Cards and Match It

Untamed Science DVD • Classroom Resources CD (includes lesson presentations and editable worksheets)

Chapter Assessment

Student Edition Check Understanding, pp. 76–79
Study Workbook B Chapter 3 Taking a Standardized Test
Assessment Resources Book Chapter 3 Test B
Biology.com Chapter 3 Assessment • Editable Worksheets of Chapter 3 Visual Quizzes* and Chapter 3 Tests A and B

Exam*View* *Assessment Suite* • Classroom Resources CD (includes lesson presentations and editable worksheets)

Time: 1 period, 1/2 block

* Visual Quizzes are based on visuals from *Miller & Levine Biology*. Review carefully to make sure they are appropriate for your class.

Pressed for Time?

Preview the Chapter Preview the Levels of Organization visual and the Consumers visual and introduce the vocabulary for Lesson 3.2.

Cover the Chapter Quickly Go over the Levels of Organization visual and have students read *Biotic and Abiotic Factors* in Lesson 3.1. Have students read all of Lesson 3.2, focusing on the Consumers visual. Briefly discuss the Food Web in the Everglades visual, assign students to read *Ecological Pyramids* in Lesson 3.3, and go over the Pyramid of Energy visual. Have students read *Recycling in the Biosphere* in Lesson 3.4.

Assess Assign Questions 3–5 in the 3.1 Check Understanding, the 3.2 Check Understanding, Questions 1 and 2 in the 3.3 Check Understanding, Question 3 in the 3.4 Check Understanding, and the Chapter 3 Standardized Test Prep except Question 5.

Get the Picture

Direct students' attention to the picture. Ask a volunteer to read aloud the caption.

Ask How would you describe the Everglades? *(Sample answers: has a lot of water, has many plants)*

Say The Everglades is sometimes called a "river of grass."

Ask What do you think that means? *(Sample answer: tall grass growing in shallow, flowing water)*

Say The Everglades is a marshland in southern Florida. Water moves slowly through the Everglades, like a slow-moving river.

Activate Prior Knowledge

Refer students to the chapter title. Use the picture as a springboard for discussion of what the biosphere is.

Ask What living things do you see in the picture? *(Sample answers: trees, ferns, bird)*

Ask What nonliving things do you see in the picture? *(Sample answers: water, rocks, light)*

Say The Everglades is part of the biosphere.

Ask What do you think the biosphere is made up of? *(living and nonliving things)*

3 The Biosphere

Big ideas Matter and Energy, Interdependence in Nature

Q: How do Earth's living and nonliving parts interact and affect the survival of organisms?

Great White Egret among some plants in the Florida Everglades

BIOLOGY.com Search Chapter 3 GO • Flash Cards

54

NATIONAL SCIENCE EDUCATION STANDARDS

UNIFYING CONCEPTS AND PROCESSES

I, II, IV

CONTENT

B.2, B.5, C.4.a, C.4.b, C.4.d, C.4.e, C.5.a, C.5.b, C.5.e, C.5.f, D.1, D.2, F.4, G.1, G.2

INQUIRY

A.1.a, A.1.b, A.1.c, A.2.a, A.2.d

UbD Connect to the Big Idea

Invite a volunteer to read aloud the chapter number and the chapter title. Read the Big Ideas and the question. Emphasize the question, **How do Earth's living and nonliving parts interact and affect the survival of organisms?**

Set the Purpose

PREVIEW AND PREDICT Draw the graphic organizer on the board. Guide students to the INSIDE feature on the next page. Invite a volunteer to read each lesson number and title.

Point out the root word *depend* in the phrase *Interdependence in Nature.*

Ask What do you think the phrase means? *(Sample answer: parts of nature depend on other parts)* What word in the question might be replaced with the phrase *depend on each other*? *(interact)*

INSIDE:
- 3.1 What Is Ecology?
- 3.2 Energy, Producers, and Consumers
- 3.3 Energy Flow in Ecosystems
- 3.4 Cycles of Matter

- Untamed Science Video
- Chapter Mystery

CHAPTER MYSTERY

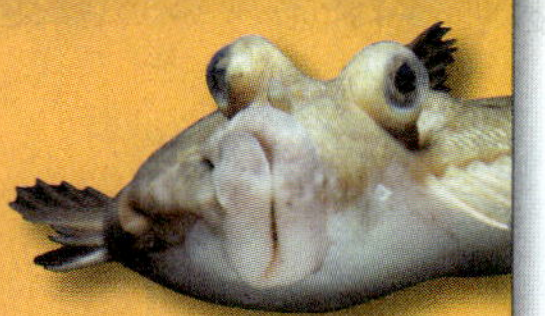

CHANGES IN THE BAY

Marine life in Rhode Island's Narragansett Bay is changing. One clue to those changes comes from fishing boat captains. These captains now boast about catching bluefish in November—a month after bluefish used to head south for winter. But catches of winter flounder are not as plentiful as they once were.

Other changes have also been observed in the bay. The annual spring increase in plant and animal growth is disappearing. Researchers working in the bay also report something puzzling. There are changes in the activities of bacteria living in mud on the bay floor.

What's going on? Farms, towns, and cities surround the bay. But direct human influence on the bay has not changed much lately. So why are there so many changes to the bay's plant and animal populations?

Read for Mystery Clues Could these changes be related to mud-dwelling bacteria? As you read the chapter, look for clues to help you understand the interactions of plants, animals, and bacteria in Narragansett Bay. Then, solve the mystery.

FOUNDATIONS for Learning

Make a layered book by stapling 4 sheets of paper together. Make sure that 2 to 3 cm at the bottom of each page is visible. Label the top page What Is Ecology? Label the remaining pages with the titles of Lessons 2, 3, and 4. Write the major headings from each lesson on the pages. When you finish reading a section, write a summary of the section in your layered book. Remember to use your vocabulary words in your summaries.

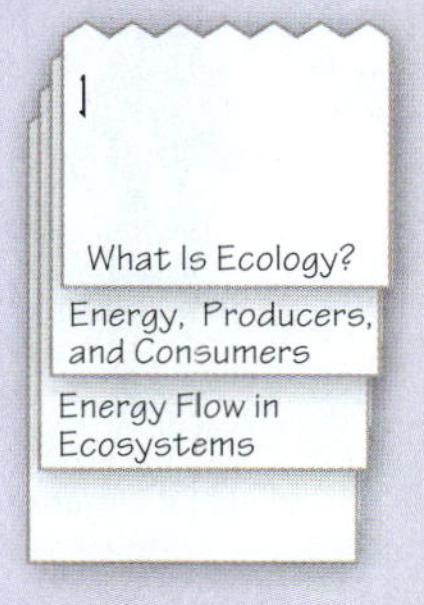

Chapter 3
Big Ideas: Matter and Energy, Interdependence in Nature

Chapter 3: How do Earth's living and nonliving parts interact and affect the survival of organisms?

- **3.1:** How do we study life?
- **3.2:** How do different organisms get the energy they need to survive?
- **3.3:** How does energy move through an ecosystem?
- **3.4:** Why is the cycling of matter important to life on Earth?

Foundations for Learning

Pass out four sheets of plain paper, and give students access to staplers for making their layered books. Help students understand the value of these summaries as a study tool.

Ask How do you write a summary of a section? *(Write in your own words the most important information in the section.)*

Ask Although the summary should be in your own words, why is it important to use the vocabulary words? *(to make sure you know the meanings of the words)*

Work through the first page together so that students have a model to follow. Remind students they will use their summaries as a study aid when they finish the chapter.

Untamed Science What a tangled web nature weaves! Follow the ecogeeks of Untamed Science as they unravel food-web relationships in **Nature's Tangled Web.**

CHAPTER MYSTERY Invite a volunteer to read aloud the Chapter Mystery. As a class, list the changes that hint at problems in Narragansett Bay. Write these changes on the board to revisit at the end of each lesson.

Have students predict how this mystery relates to the chapter title. Tell students they will find clues in the chapter to answer the question, **Could the changes in Narragansett Bay be related to mud-dwelling bacteria?**

Digital Support

BIOLOGY.com Extend your reach by using the digital assets offered at **Biology.com.**

LESSON 3.1

Pre-Reading

Objectives

3.1.1 Describe the study of ecology.

3.1.2 Explain how biotic and abiotic factors influence an ecosystem.

3.1.3 Describe the methods used to study ecology.

Preview the Pages

1. Preview the lesson vocabulary by having volunteers read each word and its definition aloud. Ask students to identify any words that are familiar and discuss their meanings.
2. Have students read the green headings in the lesson to preview the main ideas. Then, have them read the blue subheadings to find out the details they will study.
3. Preview the visuals. Have volunteers read aloud each caption. Emphasize the increasing numbers and types of organisms from left to right, as well as the increasing area, in the Levels of Organization visual.

Build Understanding

VENN DIAGRAM Direct students to Workbook B where they will find a partially filled-in Venn diagram. Remind students that the purpose of a Venn diagram is to identify characteristics that are different between two or more things and characteristics that are shared.

NATIONAL SCIENCE EDUCATION STANDARDS

UNIFYING CONCEPTS AND PROCESSES

I, II

CONTENT

G.1, G.2

INQUIRY

A.2.a, A.2.b, A.2.d

For the corresponding lesson in *Miller & Levine Biology,* see pages 64–68.

3.1 What Is Ecology?

Key Questions

What is ecology?

What are biotic and abiotic factors?

What methods are used in ecological studies?

BUILD Understanding

Venn Diagram Make a Venn diagram that shows how the environment consists of biotic factors and abiotic factors. The diagram should also show how some components are truly a mixture of both. Use examples from the lesson.

In Your Workbook Go to your workbook for help completing the Venn diagram for Lesson 3.1.

Studying Our Living Planet

Science writer Lewis Thomas once wrote: "Viewed from the distance of the moon, the astonishing thing about the earth . . . is that it is alive." That sounds good. But what does it mean? In a scientific sense, how is Earth a "living planet"? And how do we study it?

Biologists use the term *biosphere* when they talk about life on a global scale. The **biosphere** consists of all life on Earth and all parts of the Earth where life exists. This includes land, water, and the atmosphere. The biosphere contains every organism, from underground bacteria to giant rain forest trees, even humans. It extends from about 8 kilometers above Earth's surface to as far as 11 kilometers below the surface of the ocean.

The Science of Ecology Organisms in the biosphere interact with each other. But they also interact with their surroundings, or environment. The study of these interactions is called **ecology.** Ecology is the scientific study of the interactions between organisms and their physical environment.

Organisms respond to each other and to their environments. Organisms also affect each other and change their environments. These interactions produce an ever-changing, or dynamic, biosphere.

Individual Organism

A **population** is a group of individuals that belong to the same species and live in the same area.

An assemblage of different populations that live together in a defined area is called a **community.**

Levels of Organization The kinds of questions that ecologists ask about the living environment can vary. The questions depend on the level at which the ecologist works.

BIOLOGY.com Search Lesson 3.1 GO • Art in Motion

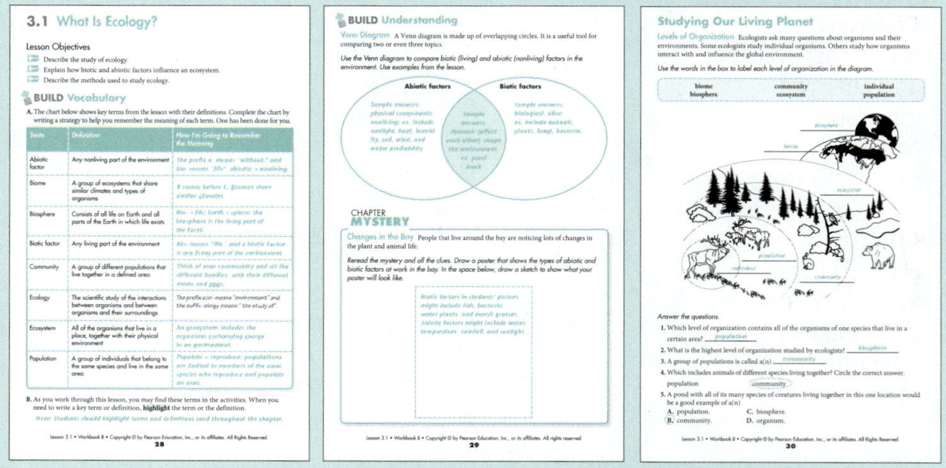
3.1 What Is Ecology?

BUILD Vocabulary

BUILD Understanding

CHAPTER MYSTERY

Studying Our Living Planet

Ecology and Economics Some people think that ecology doesn't really matter to people very much. But humans depend on ecological processes to produce many things we need, such as food and clean water. Food, water, and other things we need are often worth money. That's why ecology is linked to the field called economics. Economics is concerned with human interactions based on money or trade.

Levels of Organization Ecologists ask many questions about organisms and their environments. Some ecologists focus on the ecology of individual organisms. Others try to understand how interactions among organisms (including humans) influence our global environment. Ecological studies may focus on levels of organization that include those shown below.

Key Question What is ecology?
Ecology is the scientific study of interactions among organisms and between organisms and their physical environment.

Biotic and Abiotic Factors

Ecologists use the word *environment* to refer to all conditions, or factors, surrounding an organism. Environmental conditions include biotic factors and abiotic factors.

Biotic Factors A **biotic factor** is any living part of the environment. Biotic factors include animals, plants, mushrooms, and bacteria. For example, the insects that a bullfrog might eat are biotic factors. Another biotic factor relating to the bullfrog is the heron that might eat the bullfrog. Species that compete with the bullfrog for food or space are also biotic factors.

BUILD Vocabulary

biosphere
a part of Earth in which life exists, including land, water, and air, or atmosphere

ecology
the scientific study of interactions among organisms and between organisms and their environment

population
a group of individuals of the same species that live in the same area

community
a group of different populations that live together in a defined area

ecosystem
all the organisms that live in a place, together with their physical environment

biome
a group of ecosystems that share similar climates and typical organisms

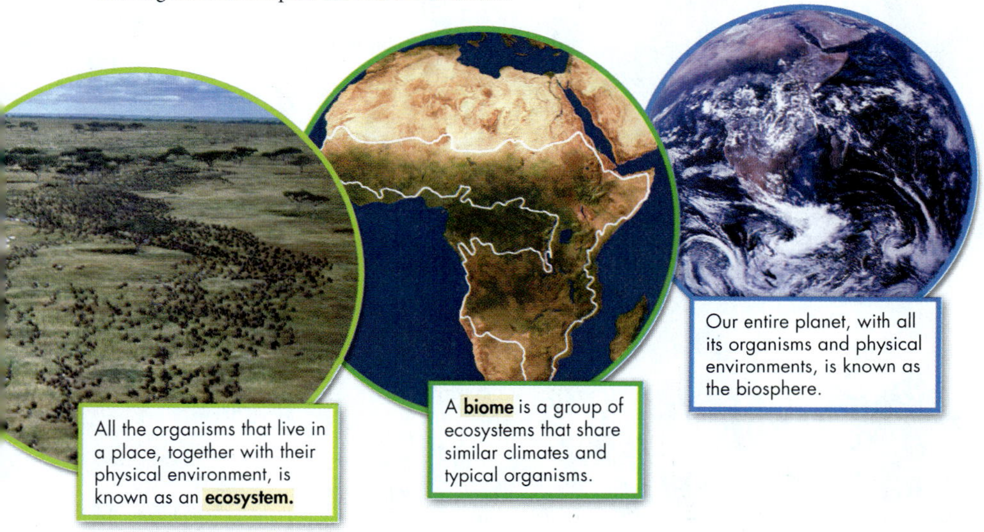

All the organisms that live in a place, together with their physical environment, is known as an **ecosystem.**

A **biome** is a group of ecosystems that share similar climates and typical organisms.

Our entire planet, with all its organisms and physical environments, is known as the biosphere.

ELL Focus on ELL: Extend Language

BEGINNING SPEAKERS It is important that students have a clear understanding of the words *interact* and *interdependence* for this chapter. Have students look up the meaning of the prefix *inter-* (between) and use it to phrase a definition of the words *interact* and *interdependence*. Then, have them look up other common words that use the prefix *inter-*, such as *interchange, intermingle, interstate,* and *interlibrary,* and phrase definitions for these words. Discuss students' definitions to clarify understanding.

During Reading

Active Reading

MAKE A GRAPHIC ORGANIZER Draw an inverted triangle on the board, and divide the triangle into six horizontal bands.

Ask How do the bands change in size? *(They become larger from bottom to top.)*

Ask How do the levels of organization in ecology change in size? *(They become larger from the individual to the planet Earth.)*

Have a volunteer label the bands with the levels of organization in order of size.

COMPARE AND CONTRAST Make sure students understand the meaning of *ecosystem,* as this word will be used often in the chapter.

Ask What is the difference between a community and an ecosystem? *(A community is all the populations that live in the same place. An ecosystem is all the populations that live in the same place together with all the nonliving things around them.)*

KEY QUESTION Invite a volunteer to read aloud the Key Question and answer.

Ask What are some ways organisms interact with each other? *(Sample answers: eating other organisms, competing for food, mating, taking care of offspring)*

Ask What are some ways organisms interact with nonliving things in their environment? *(Sample answers: breathing, drinking water, obtaining nutrients from the soil)*

Build Vocabulary

ROOT WORDS Write the word *biosphere* on the board and underline the root word *sphere.*

Say The root word *sphere* comes from the Greek word *sphaira,* meaning "ball."

Ask How is the biosphere similar to a ball? *(The biosphere consists of most of our planet, which is shaped like a ball.)*

Speed Bump

Say Imagine a grasshopper, all the grasshoppers in a vacant lot, and all the organisms in the vacant lot.

Ask Which is an individual, a population, and a community? *(grasshopper—individual, all grasshoppers in vacant lot—population, all organisms in vacant lot—community)*

LESSON 3.1

During Reading continued

Active Reading

INTERPRET VISUALS Direct students to the Biotic and Abiotic Factors visual. First, have them compare the two outer pictures.

Ask How do the two pictures differ? *(The picture on the left shows only living things. The picture on the right shows only nonliving things.)*

Ask What can you say about the picture in the middle? *(It is a combination of the two outer pictures. It shows both living and nonliving things.)*

Ask Which picture represents an ecosystem? *(the middle picture)*

KEY QUESTION Invite a volunteer to read aloud the question and answer. Point out a small environment near the school.

Ask What are the biotic factors? *(Students should list any type of living thing.)* What are the abiotic factors? *(Students should list nonliving things such as soil, air, and water.)*

Build Vocabulary

PREFIXES Have students compare the spellings and definitions of the terms *biotic factor* and *abiotic factor*.

Ask What do you think the prefix *a-* in the word *abiotic* means? *(not)*

Speed Bump

Ask When an organism interacts with a biotic factor, what is it interacting with? *(another organism)*

Ask What is an example of an abiotic factor? *(Sample answers: sunlight, water, air)*

BUILD Vocabulary

biotic factor
any living part of the environment with which an organism might interact

abiotic factor
a physical, or nonliving, factor that shapes an ecosystem

PREFIXES
The prefix *bio-* comes from the Greek word *bios*, which means "living tissue."

Abiotic Factors An **abiotic factor** is any nonliving part of the environment. Abiotic factors include things like sunlight, precipitation, and water currents. For example, a bullfrog might be affected by water availability, temperature, and humidity.

Biotic and Abiotic Factors Together The difference between biotic and abiotic factors is not always simple. Many physical factors can be strongly influenced by the activities of organisms. Remember the bullfrog? It lives in soft "muck" along the shores of ponds. This muck is a mix of biotic factors such as bacteria and abiotic factors such as sand. Trees and shrubs around the pond affect the amount of sunlight that reaches its shoreline. Plant roots determine how much soil is held in place and how much washes into the pond. This kind of dynamic mix of biotic and abiotic factors shapes every environment.

Key Question What are biotic and abiotic factors?
Biological influences on organisms are called biotic factors. Physical elements of an ecosystem are called abiotic factors.

Biotic and Abiotic Factors
This pond is affected by a combination of biotic and abiotic factors.

Focus on Academic Language

• influence • dynamic • approach • artificial

Write these words on the board and pronounce them. Ask several volunteers to read aloud the text on these pages. Note how these four terms are used. Have students look up the definitions of these words and then use them in a variety of sentences, to ensure they understand their meanings. Ask volunteers to share sentences and ask others to compare what the sentences say with the dictionary definitions of the words. Revise sentences as needed to clarify the correct usage of the words.

Ecological Methods

Ecologists study many different parts of the biosphere and often use different research tools. But all ecologists use three scientific approaches to their work: observation, experimentation, and modeling.

Observation Observation is often the first step in asking ecological questions. Some observations are simple: Which species live here? Other observations are more complex: How does an animal protect its young from predators? These types of questions may form the first step in designing experiments and models.

Experimentation Experiments can be used to test hypotheses. Suppose an ecologist wants to see how growing plants react to different temperatures. He or she may set up an artificial environment in a greenhouse to test the plants' responses. Other experiments carefully alter conditions in selected parts of natural ecosystems.

Modeling Many ecological events occur over long periods of time or over large distances. These events are difficult to study directly. Ecologists make models to help them understand these events. These models are usually based on data collected through observation and experimentation. Then more observations are made to test predictions based on those models.

Key Question What methods are used in ecological studies? **Modern ecologists use three methods in their work: observation, experimentation, and modeling.**

Ecology Field Work The three basic approaches to ecological research involve observing, experimenting, and modeling. This ecologist is measuring a giant *Rafflesia* flower in Borneo.

CHECK Understanding

Apply Vocabulary

Use the highlighted words from the lesson to complete each sentence correctly.

1. The _______ extends from about 8 kilometers above the Earth's surface to as far as 11 kilometers below ocean's surface.
2. The science that studies all the interactions between living and nonliving parts of an ecosystem is called _______.
3. A herd of wildebeests living on the Serengeti Plain would be classified as a _______ within an ecosystem.

Critical Thinking

4. **Explain** Is weather a biotic or abiotic factor? Explain your answer.
5. **Compare and Contrast** How are biotic and abiotic factors related? How are they different?
6. **Write to Learn** Answer the first clue of the mystery. Write a paragraph that explains why you think the three abiotic factors you chose are important.

MYSTERY CLUE

What are three examples of abiotic factors that might affect life in the Narragansett Bay? (**Hint:** See p. 58.)

BIOLOGY.com Search Lesson 3.1 GO • Lesson Assessment

Check Understanding

1. biosphere
2. ecology
3. population
4. Weather is an abiotic factor because it is a nonliving part of the environment.
5. Biotic and abiotic factors are both components of ecosystems. Biotic factors are the living components, and abiotic factors are the nonliving components.
6. Student answers may vary. Abiotic factors that could affect the bay include sunlight, heat, precipitation, wind currents, and water currents. Student explanations should take into account the ways his or her chosen factors would affect the bay.

Active Reading

COMPARE AND CONTRAST Make a three-column chart on the board with the headings Observation, Experimentation, and Modeling. Invite volunteers to fill in the chart by listing characteristics for each category. For example, under Observation, students might suggest *often first step, use senses, take notes, use tools, takes practice,* and *takes patience.*

Then, next to the chart, draw a Venn diagram showing three circles in a triangle formation with all three circles partly overlapping in the middle. Label the circles Observation, Experimentation, and Modeling, respectively.

As a class, determine whether each characteristic in the chart is unique to a particular scientific approach or applies to two or three of the approaches. Record the characteristic in the appropriate part of the Venn diagram.

KEY QUESTION Invite a volunteer to read aloud the question and answer.

Ask What does an ecological study do? *(It tries to answer questions about interactions among organisms and between organisms and the nonliving things in their environment.)*

Speed Bump

Ask How does an experiment differ from a model? *(Experiments test hypotheses in ways that enable scientists to see results directly. Models explain events that occur over long periods of time or great distances and are difficult to study directly.)*

Post Reading

ASSESSMENT TIPS Point out to students that Question 5 requires two answers.

For Question 6, remind students that a paragraph consists of a main idea sentence with supporting details.

MYSTERY CLUE Refer students to the section Abiotic Factors for general examples of abiotic factors. Then, have them think about abiotic factors particular to the bay environment.

Wrap-Up Activity

CLASSIFY Have students write the word *biotic* on one index card and *abiotic* on another index card. Show pictures from magazines, the Internet, or the textbook that depict a living thing or something nonliving. Have students hold up the appropriate index card to classify each object.

LESSON 3.2

Pre-Reading

Objectives

3.2.1 Define primary producers.

3.2.2 Describe how consumers obtain energy and nutrients.

Preview the Pages

1. Preview the vocabulary words, and help students pronounce them. Compare and contrast the words *autotroph* and *heterotroph*.
2. Preview the visuals. Have volunteers read aloud each caption. Point out and explain the chemical reaction that summarizes photosynthesis.
3. Draw students' attention to the Consumers image on the last page of the lesson. Connect the photographs to the information in the text boxes.

Activate Prior Knowledge

Tell students that all living things are either producers or consumers and that these words have specific meanings in biology.

Ask Are you a producer or a consumer? *(consumer)* Explain. *(Producers make their own food. Humans must consume food.)*

Build Understanding

CONCEPT MAP Remind students that a concept map shows the relationships among words and ideas. Have volunteers call out two vocabulary words and tell how they are related. Then, direct students to Workbook B to see how this relationship can be shown in a concept map.

NATIONAL SCIENCE EDUCATION STANDARDS

UNIFYING CONCEPTS AND PROCESSES

I

CONTENT

C.4.b, C.5.a, C.5.b, C.5.e, C.5.f, D.1

For the corresponding lesson in *Miller & Levine Biology,* see pages 69–72.

3.2 Energy, Producers, and Consumers

Key Questions

What are primary producers?

How do consumers obtain energy and nutrients?

BUILD Understanding

Concept Map As you read, use the highlighted vocabulary words to create a concept map that organizes the information in the lesson.

In Your Workbook Go to your workbook for help in completing your concept map.

Primary Producers

Organisms need energy for growth, reproduction, and their own metabolic processes. Without energy, there are no life functions!

No organism can create energy—organisms can only use energy from other sources. Where does that energy come from? Sunlight is the ultimate energy source for most life on Earth. But some organisms get their energy from chemical energy that is stored in inorganic chemical compounds.

Only algae, certain bacteria, and plants can capture energy from sunlight or chemicals. This energy is converted into forms that living cells can use. Such organisms are called **autotrophs.** They use solar or chemical energy to produce "food" by making organic molecules from inorganic compounds. Organisms that eat autotrophs, then use the energy stored in the bonds of these molecules. That's why autotrophs are also called **primary producers.** Primary producers are the first producers of energy-rich compounds that other organisms need to survive. Primary producers are essential to the flow of energy through the biosphere.

Energy From the Sun The most common primary producers capture solar energy through photosynthesis. Photosynthesis captures light energy and uses it to power chemical reactions. These reactions convert carbon dioxide and water into oxygen and energy-rich carbohydrates. Photosynthesis adds oxygen to the atmosphere and removes carbon dioxide. Without photosynthesis, there would not be enough oxygen for you to breathe! Plants are the main photosynthetic producers on land. Algae fill that role in freshwater ecosystems and in the sunlit upper layers of the ocean. Photosynthetic bacteria are the important primary producers in tidal flats and salt marshes.

Primary Producers Plants get energy from sunlight and turn it into nutrients. These nutrients can be eaten and used for energy by animals such as this caterpillar.

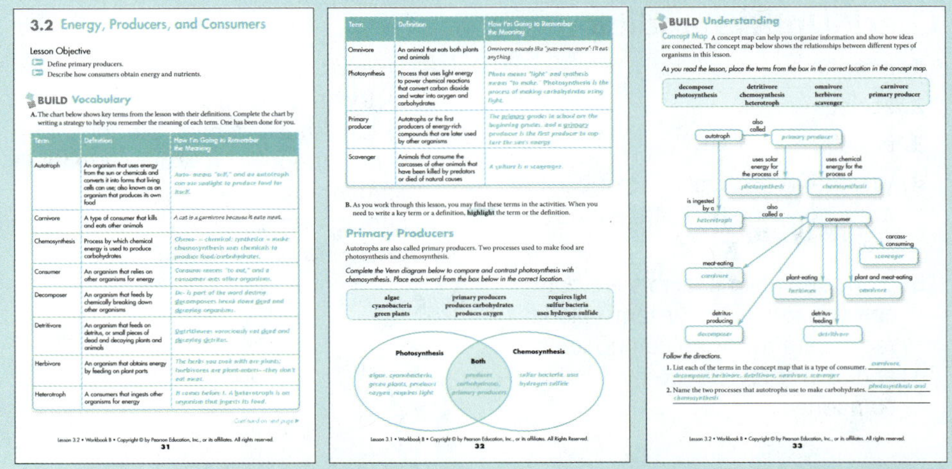

3.2 Energy, Producers, and Consumers

BUILD Vocabulary

Primary Producers

BUILD Understanding

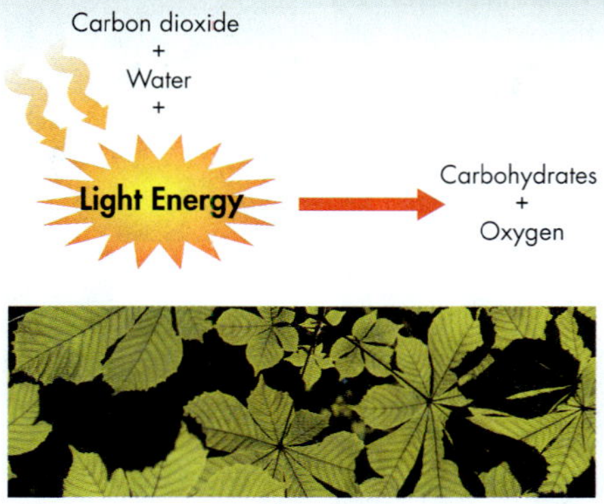

Photosynthesis
Plants use the energy from sunlight to carry out the process of photosynthesis. As a result, energy-rich carbohydrates are produced.

Photosynthesis

Life Without Light About 30 years ago, biologists found thriving ecosystems around deep-sea volcanic vents. Without light for photosynthesis, how did the primary producers capture energy? They used a process called chemosynthesis (kee moh SIN thuh sis) to harness energy from inorganic molecules. Chemosynthesis uses chemical energy to produce carbohydrates. Some chemosynthetic organisms live in hot springs while others live in tidal marshes.

Key Question What are primary producers?
Primary producers are the first producers of energy-rich compounds that are later used by other organisms.

Consumers

Animals, fungi, and many bacteria cannot capture energy directly from the environment. These organisms, called **heterotrophs** (HET uh roh trohfs), must get energy from other organisms by consuming them. So heterotrophs are also called **consumers.**

Types of Consumers Consumers are classified by the ways in which they acquire energy and nutrients. As you will see, the definition of *food* can vary quite a lot among different categories of consumers.

Beyond Consumer Categories The simple consumer categories often do not express how nature really works. For example, seeds and fruits are richer in nutrients and easier to digest than leaves. So herbivores that eat different plant parts need different types of digestive systems. Carnivores do not stay in their own categories, either. Hyenas are carnivores but will scavenge if they get a chance. This means it is important to understand how energy and nutrients move through ecosystems.

Key Question How do consumers obtain energy and nutrients?
Organisms that rely on other organisms for energy and nutrients are called consumers.

BUILD Vocabulary

autotroph
an organism that is able to capture energy from sunlight or chemicals and use it to produce its own food from inorganic compounds; also called a producer

primary producer
the first producer of energy-rich compounds that are later used by other organisms

heterotroph
an organism that obtains food by consuming other living things; also called a consumer

consumer
an organism that relies on other organisms for its energy and food supply; also called a heterotroph

PREFIXES
The prefix *auto-* means "by itself." The Greek word *trophikos* means "to feed." This means that an autotroph can be described as a "self feeder." It does not need to eat other organisms for food.

ELL Focus on ELL: Access Content

INTERMEDIATE AND ADVANCED SPEAKERS Have the class make a concept map comparing photosynthesis and chemosynthesis.

Ask What would be a good title for the concept map? *(Sample answers: Making Food, Primary Producers, Autotrophs)*

Ask What are two ways autotrophs make food? *(by photosynthesis and chemosynthesis)* Ask a volunteer to add these to the concept map.

Ask What is the energy source for each way of making food? *(light energy in photosynthesis, chemical energy in chemosynthesis)* Have a volunteer add these to the appropriate places on the concept map.

Work with the class to finish the comparison by adding the reactants and products of each process to the concept map.

During Reading

Active Reading

FIND THE MAIN IDEA Help students convert the sections Energy from the Sun and Life Without Light to an outline. Use the blue subheads for Roman numerals I and II. Then write the letter *A* under each subhead and fill it in with the main idea of each paragraph. Challenge students to help you list the supporting details under each main idea. Model the outlining process for the first subhead.

Ask What is the main idea? *(Most primary producers capture solar energy through photosynthesis.)*

Have students write the main idea in their notebooks.

Ask What is one supporting detail? *(Photosynthesis captures light energy.)* Have students add this detail as a numbered point under the main idea.

Continue asking students for supporting details and having them add bullet points under the main idea.

DETERMINE MEANING On the board, break the words *photosynthesis* and *chemosynthesis* into the root word *synthesis* and the prefixes *photo-* and *chemo-*. Have students suggest what the word parts mean. *(*Photo- *means "light."* Chemo- *means "chemical."* Synthesis *means "a combination or putting together.")*

KEY QUESTION Invite a volunteer to read aloud the first Key Question and answer.

Ask What are some examples of primary producers? *(Sample answers: algae, some bacteria, plants)*

KEY QUESTION Read the next question and answer.

Ask What are some examples of consumers? *(Sample answers: animals, fungi, many bacteria)*

Build Vocabulary

PREFIXES Have students review the prefix and root word of the term *autotroph*.

Ask Which other vocabulary word has the same root word? *(heterotroph)*

Say The prefix *hetero-* means "other."

Ask How does the prefix explain the meaning of the word *heterotroph*? *(A heterotroph is an "other feeder." It must get energy from other organisms.)*

Speed Bump

Ask What are two ways autotrophs make their own food? *(by photosynthesis and chemosynthesis)*

Ask What do all heterotrophs have in common? *(They consume other organisms.)*

LESSON 3.2

During Reading continued

Active Reading

MAKE A GRAPHIC ORGANIZER Have students work in pairs to make a table that summarizes the information about consumers provided in the visual.

Ask What would be a good title for your table? *(Sample answers: Consumers, Organisms that Feed On Other Organisms)*

Say Look at the text boxes. What two kinds of information are provided for each type of consumer? *(what the consumer eats and examples of the type of consumer)*

Ask How many columns should your table have? *(three)* What should its headings be? *(Type of Consumer, What It Eats, Examples)*

Have students complete their table.

Speed Bump

Ask What do herbivores eat? *(plant parts)*

Ask Which type of consumer feeds by chemically breaking down organic matter and absorbing nutrients? *(decomposer)*

Post Reading

ASSESSMENT TIPS For Question 4, remind students to give an example of each kind of consumer. For Question 5, point out that explaining a relationship means telling how the things are connected.

MYSTERY CLUE Have students review the descriptions of detritivores and decomposers on this page. They should then link this information to the bacterial communities in Narragansett Bay.

Ask Where do the bacteria that concern scientists live in the bay? *(in the mud at the bottom)* What else do you think is part of the mud? *(dead organisms and detritus)*

Wrap-Up Activity

SENTENCE GAME Divide the class into teams of four. Write on separate slips of paper the vocabulary words and other important words from this lesson, such as *photosynthesis, chemosynthesis, carnivore, herbivore,* and so on. Make one set of words for each team, and place the sets in separate containers. Have the teams compete with each other by drawing two slips of paper and using the two words to form a scientifically accurate sentence. Allow four turns per team. The team with the greatest number of accurate sentences wins.

CHECK Understanding

Apply Vocabulary

Use the highlighted words from the lesson to complete each sentence correctly.

1. Consumers are classified as ______ because they are unable to make their own food.

2. Autotrophs are also known as ______ because they are the first organisms to produce energy-rich compounds in an ecosystem.

Critical Thinking

3. Explain How do consumers obtain energy?

4. Compare and Contrast How are detritivores different from decomposers? Give an example of each.

5. Write to Learn Answer the second clue of the mystery. Write a paragraph explaining the relationship between detritivores and decomposers in an ecosystem.

MYSTERY CLUE

Bacteria are important members of the living community in Narragansett Bay. How do you think the bacterial communities on the floor of the bay might be linked to its producers and consumers? (Hint: See above.)

BIOLOGY.com Search Lesson 3.2 GO • Lesson Assessment • Art Review

62 Chapter 3 • Lesson 2

Check Understanding

1. heterotrophs

2. primary producers

3. Consumers obtain energy by eating other organisms.

4. Detritivores eat detritus particles, which often have decomposers living on or in them. Detritivores include worms, mites, snails, shrimp, and crabs. Decomposers are bacteria or fungi that feed by chemically breaking down organic matter. Their feeding activities break organic matter down into detritus.

5. Student answers may vary. Bacterial communities on the floor of the bay serve as the ecosystem's decomposers. Decomposers break down organic matter from dead plants (producers) and animals (consumers) in the ecosystem into detritus. The detritus, along with any attached bacteria, is consumed by the bay's detritivores.

3.3 Energy Flow in Ecosystems

Food Chains and Food Webs

What happens to energy stored in body tissues when one organism eats another? It moves from the "eaten" to the "eater." In every ecosystem, primary producers and consumers are linked through feeding relationships. There are many different feeding relationships in an ecosystem. But energy flows through the ecosystem in a one-way stream, from primary producers to various consumers.

Food Chains You can think of energy as passing through an ecosystem in a series of steps. This series of steps is called a **food chain.** In a food chain, organisms transfer energy by eating and being eaten. Food chains can vary in length. Look at the aquatic food chain below. The primary producers are a mixture of floating algae (called phytoplankton) and attached algae. These primary producers may be eaten by small fishes, such as flagfish. Larger fishes, like the largemouth bass, eat the small fishes. The bass are preyed upon by large wading birds, such as the anhinga. In the end, the anhinga may be eaten by an alligator. Therefore there are four steps in this food chain. So the top carnivore is four steps removed from the primary producer.

Food Webs In most ecosystems, feeding relationships are much more complicated than a simple chain can describe. One reason for this is that many animals eat more than one kind of food. Take Africa's Serengeti Plain, for example. Herbivores such as zebras, gazelles, and buffaloes often graze upon several different species of grasses. Several predators such as lions, hyenas, and leopards, in turn, often prey upon those herbivores! Ecologists call this network of feeding interactions a **food web.**

Key Questions

How does energy flow through ecosystems?

What do the three types of ecological pyramids illustrate?

BUILD Understanding

Preview Visuals Before you read, look at the diagrams called Food Chains and Food Web in the Everglades. Note how the diagrams are similar and how they are different.

In Your Workbook Complete the chart in Lesson 3.3 by noting the similarities and differences between the two diagrams.

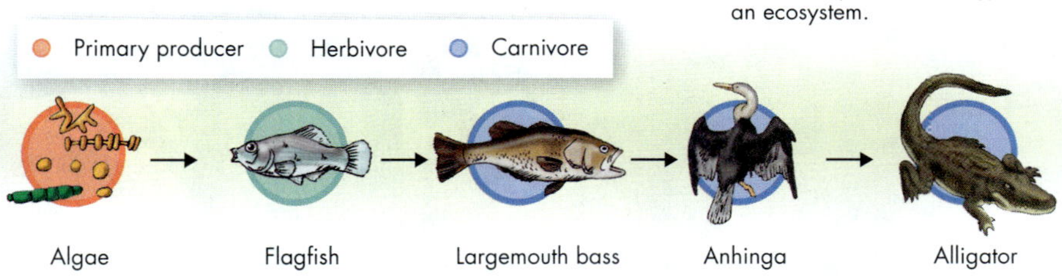

Food Chains Food chains show the one-way flow of energy in an ecosystem.

LESSON 3.3

Pre-Reading

Objectives

3.3.1 Trace the flow of energy through living systems.

3.3.2 Identify the three types of ecological pyramids.

Preview the Pages

1. Preview the vocabulary words by having students find the words in the text and read each aloud.
2. Preview the lesson visuals. Make sure students understand that the arrows in the food chain and food web diagrams show the direction in which energy flows. Point out the color-coding used to designate the primary producers and the different kinds of consumers.
3. Ask students to describe what they see in the Earth's Recycling Center visual. Tell them the picture is an analogy. It compares decomposers to a recycling center. Tell students the analogy will help them understand information in the text.

Build Understanding

PREVIEW VISUALS Have a volunteer read the names of the organisms in the food chain figure. Have another volunteer repeat the names with the phrase "gets eaten by" after each name. Then ask students to find this food chain in the visual Food Web in the Everglades later in the lesson. Finally, have them complete the chart in Workbook B.

NATIONAL SCIENCE EDUCATION STANDARDS

UNIFYING CONCEPTS AND PROCESSES

I, II

CONTENT

B.5, C.4.a, C.4.b, C.5.a, C.5.b, C.5.e, C.5.f, D.1

INQUIRY

A.1.c, A.2.d

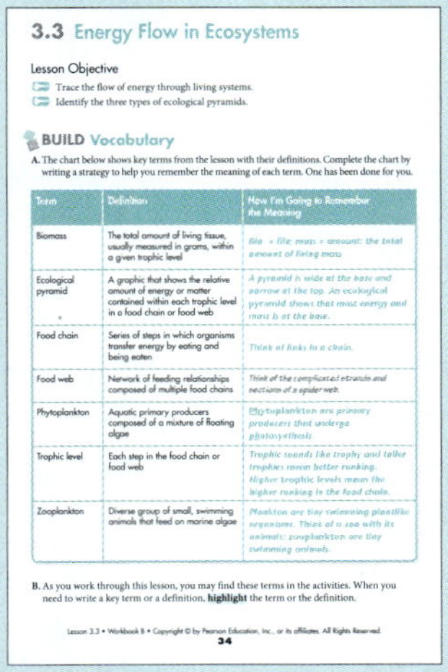

3.3 Energy Flow in Ecosystems

Lesson Objective

Trace the flow of energy through living systems.

Identify the three types of ecological pyramids.

BUILD Vocabulary

A. The chart below shows key terms from the lesson with their definitions. Complete the chart by writing a strategy to help you remember the meaning of each term. One has been done for you.

Term	Definition	How I'm Going to Remember the Meaning
Biomass	The total amount of living tissue, usually measured in grams, within a given trophic level	Bio = life; mass = amount: the total amount of living mass
Ecological pyramid	A graphic that shows the relative amount of energy or matter contained within each trophic level in a food chain or food web	A pyramid is wide at the base and narrow at the top. An ecological pyramid shows that most energy and mass is at the base.
Food chain	Series of steps in which organisms transfer energy by eating and being eaten	Think of links in a chain.
Food web	Network of feeding relationships composed of multiple food chains	Think of the complicated strands and sections of a spider web.
Phytoplankton	Aquatic primary producers composed of a mixture of floating algae	Phytoplankton are primary producers that undergo photosynthesis.
Trophic level	Each step in the food chain or food web	Trophic sounds like trophy and taller trophies mean better ranking. Higher trophic levels mean the higher ranking in the food chain.
Zooplankton	Diverse group of small, swimming animals that feed on marine algae	Plankton are tiny swimming plantlike organisms. Think of a zoo with its animals; zooplankton are tiny swimming animals.

B. As you work through this lesson, you may find these terms in the activities. When you need to write a key term or a definition, **highlight** the term or the definition.

Lesson 3.3 • Workbook B • Copyright © by Pearson Education, Inc., or its affiliates. All Rights Reserved.
34

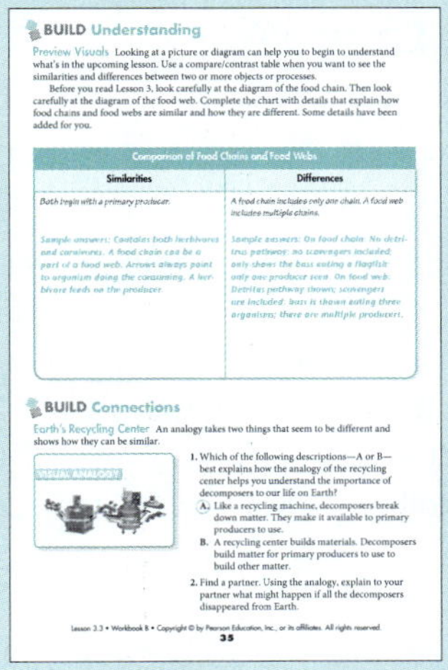

BUILD Understanding

Preview Visuals Looking at a picture or diagram can help you to begin to understand what's in the upcoming lesson. Use a compare/contrast table when you want to see the similarities and differences between two or more objects or processes.

Before you read Lesson 3, look carefully at the diagram of the food chain. Then look carefully at the diagram of the food web. Complete the chart with details that explain how food chains and food webs are similar and how they are different. Some details have been added for you.

Comparison of Food Chains and Food Webs	
Similarities	Differences
Both begin with a primary producer.	A food chain includes only one chain. A food web includes multiple chains.
Sample answers: Contains both herbivores and carnivores. A food chain can be a part of a food web. Arrows always point to organism doing the consuming. A herbivore feeds on the producer.	Sample answers: On food chain: No detritus pathway; no scavengers included; only shows the bass eating a flagfish; only one producer seen. On food web: Detritus pathway shown; scavengers are included; bass is shown eating three organisms; there are multiple producers.

BUILD Connections

Earth's Recycling Center An analogy takes two things that seem to be different and shows how they can be similar.

1. Which of the following descriptions—A or B—best explains how the analogy of the recycling center helps you understand the importance of decomposers to our life on Earth?
   A. Like a recycling machine, decomposers break down matter. They make it available to primary producers to use.
   B. A recycling center builds materials. Decomposers build matter for primary producers to use to build other matter.
2. Find a partner. Using the analogy, explain to your partner what might happen if all the decomposers disappeared from Earth.

Lesson 3.3 • Workbook B • Copyright © by Pearson Education, Inc., or its affiliates. All rights reserved.
35

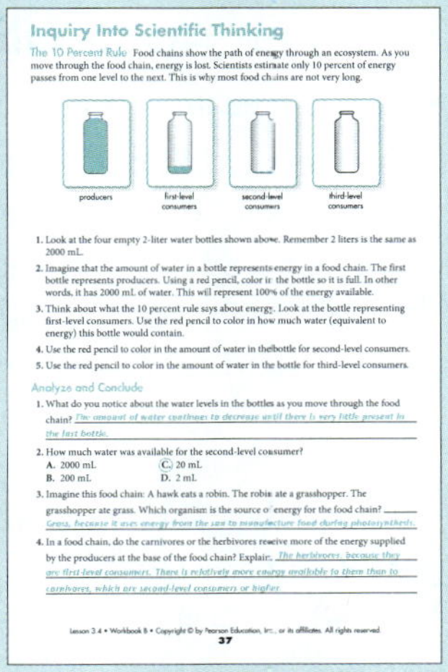

Inquiry Into Scientific Thinking

The 10 Percent Rule Food chains show the path of energy through an ecosystem. As you move through the food chain, energy is lost. Scientists estimate only 10 percent of energy passes from one level to the next. This is why most food chains are not very long.

producers | first-level consumers | second-level consumers | third-level consumers

1. Look at the four empty 2-liter water bottles shown above. Remember 2 liters is the same as 2000 mL.
2. Imagine that the amount of water in a bottle represents energy in a food chain. The first bottle represents producers. Using a red pencil, color in the bottle so it is full. In other words, it has 2000 mL of water. This will represent 100% of the energy available.
3. Think about what the 10 percent rule says about energy. Look at the bottle representing first-level consumers. Use the red pencil to color in how much water (equivalent to energy) this bottle would contain.
4. Use the red pencil to color in the amount of water in the bottle for second-level consumers.
5. Use the red pencil to color in the amount of water in the bottle for third-level consumers.

Analyze and Conclude

1. What do you notice about the water levels in the bottles as you move through the food chain? The amount of water continues to decrease until there is very little present in the last bottle.
2. How much water was available for the second-level consumer?
   A. 2000 mL
   B. 200 mL
   C. 20 mL
   D. 2 mL
3. Imagine this food chain: A hawk eats a robin. The robin ate a grasshopper. The grasshopper ate grass. Which organism is the source of energy for the food chain? Grass, because it uses energy from the sun to manufacture food during photosynthesis.
4. In a food chain, do the carnivores or the herbivores receive more of the energy supplied by the producers at the base of the food chain? Explain. The herbivores, because they are first-level consumers. There is relatively more energy available to them than to carnivores, which are second-level consumers or higher.

Lesson 3.4 • Workbook B • Copyright © by Pearson Education, Inc., or its affiliates. All rights reserved.
37

For the corresponding lesson in *Miller & Levine Biology,* see pages 73–78.

During Reading

Active Reading

INTERPRET SYMBOLS Have students study the food web on the next page. Be sure they connect the colored circles in the key to the colored circles behind each organism.

Ask What kind of organism is a bobcat? *(carnivore)* What kind of organism are algae? *(primary producer)* Which organisms can be classified more than one way? *(grass shrimp, worms, and Everglades crayfish)*

Point out the three kinds of arrows in the key.

Say The solid lines show the flow of energy from a producer to a consumer or from prey to a predator.

Ask What is the difference between the dashed arrows and the dotted arrows? *(The dashed arrows show the flow of energy from a dead organism to a scavenger. The dotted arrows show the flow of energy from dead material to detritus.)*

KEY QUESTION Invite a volunteer to read aloud the question and answer.

Ask How can you tell from the food chain shown on the previous page that energy flows in one direction only in an ecosystem? *(All the arrows point in one direction only.)*

Build Vocabulary

ACADEMIC WORDS Refer students to the Pyramid of Energy diagram later in the chapter to discuss the meaning of the term *ecological pyramid.* Then, have a volunteer read aloud the definition of *trophic level.*

Ask How many trophic levels make up the food chain on the previous page? *(five)*

Point out to students that each layer of an ecological pyramid represents a different trophic level.

Build Connections

VISUAL ANALOGY Help students understand the analogy portrayed in the visual.

Ask What is going into the machine labeled *decomposers*? *(an item made up of building blocks)* What does the item represent? *(a dead organism)* What is coming out of this machine? *(individual building blocks)* What do these represent? *(nutrients)*

Ask What is going into the machine labeled *primary producers*? *(the building blocks)* What is coming out of this machine? *(different items made from the building blocks)* What do these items represent? *(structures within the primary producer)*

Refer students to Workbook B to read more about this analogy.

BUILD Vocabulary

food chain
a series of steps in an ecosystem in which organisms transfer energy by eating and being eaten

food web
a network of complex interactions formed by the feeding relationships among the various organisms in an ecosystem

trophic level
each step in a food chain or food web

ecological pyramid
an illustration of the relative amounts of energy or matter contained within each trophic level in a given food chain or food web

ACADEMIC WORDS
An ecological pyramid looks more like a triangle than a pyramid. But ecologists call it a pyramid because it has layers that get smaller as you go toward the top—just like an Egyptian pyramid.

BUILD Connections

EARTH'S RECYCLING CENTER

Decomposers break down dead and decaying matter and release nutrients that can be reused by primary producers.

In Your Workbook How are decomposers like a recycling center? Go to your workbook to explain.

▶ ***Food Chains Within Food Webs*** The Everglades are a complex marshland ecosystem in southern Florida, with many overlapping feeding relationships. These relationships have been simplified —many species have been left out—and represented in the food web to the right. Many different paths can be taken from a primary producer to a predator. Each path is a food chain. So, a food web links together all of the food chains in an ecosystem. Now, you can begin to appreciate how complicated food webs are!

▶ ***Decomposers and Detritivores in Food Webs*** Decomposers and detritivores are as important in most food webs as other consumers are. Look again at the Everglades web. White-tailed deer, moorhens, raccoons, grass shrimp, crayfish, and flagfish feed at least partly on primary producers. But most producers die without being eaten, so decomposers convert that dead material to detritus. The detritus is eaten by detritivores, such as crayfish, grass shrimp, and worms. At the same time, the decomposition process releases nutrients that can be used by primary producers. Decomposers are nature's recyclers. Without decomposers, nutrients would remain locked in dead organisms.

Food Webs and Disturbance Relationships in food webs are not simple, so a disturbance can have dramatic effects. Look again at the food web. What might happen to the feeding relationships in it after a disturbance? Suppose an oil spill caused a major decrease in the number of decomposers. How might that affect the populations of detritivores? Would a change in those populations affect how pig frogs feed? How might a change in frog feeding behavior affect other species the frogs eat? As you can see, a decrease in decomposers would affect the whole web.

Key Question How does energy flow through ecosystems?
Energy flows through an ecosystem in a one-way stream, from primary producers to various consumers.

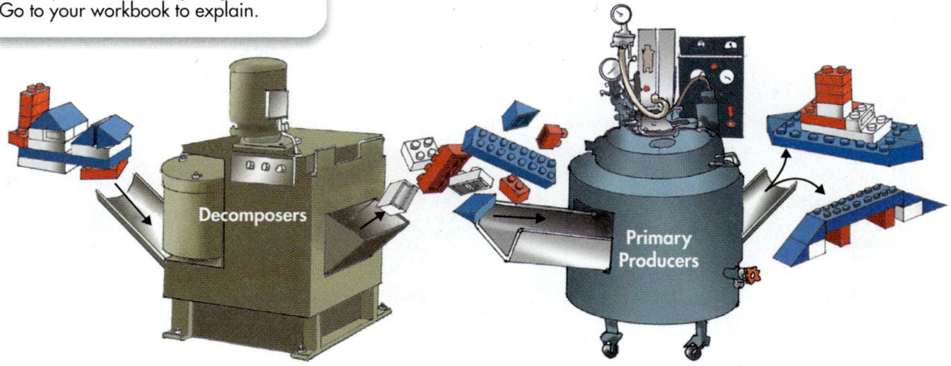

Focus on Academic Language

• relate • related • relation • relationship

Write these words on the board. Use the term *feeding relationships* from the text as a springboard to compare and contrast these words.

Ask What is a feeding relationship? *(a connection between two or more organisms that shows how one feeds on another)*

Say The word *relationship* is a noun. A relationship is a connection between things.

Ask What part of speech is the word *relate? (verb)* What does it mean? *(to make a connection between things)* Then ask a volunteer to use the word *relate* in a sentence. Elicit from the class whether the word was used correctly.

Repeat with *related (adjective; describes something as having a connection)* and *relation (noun; a relative, kin).*

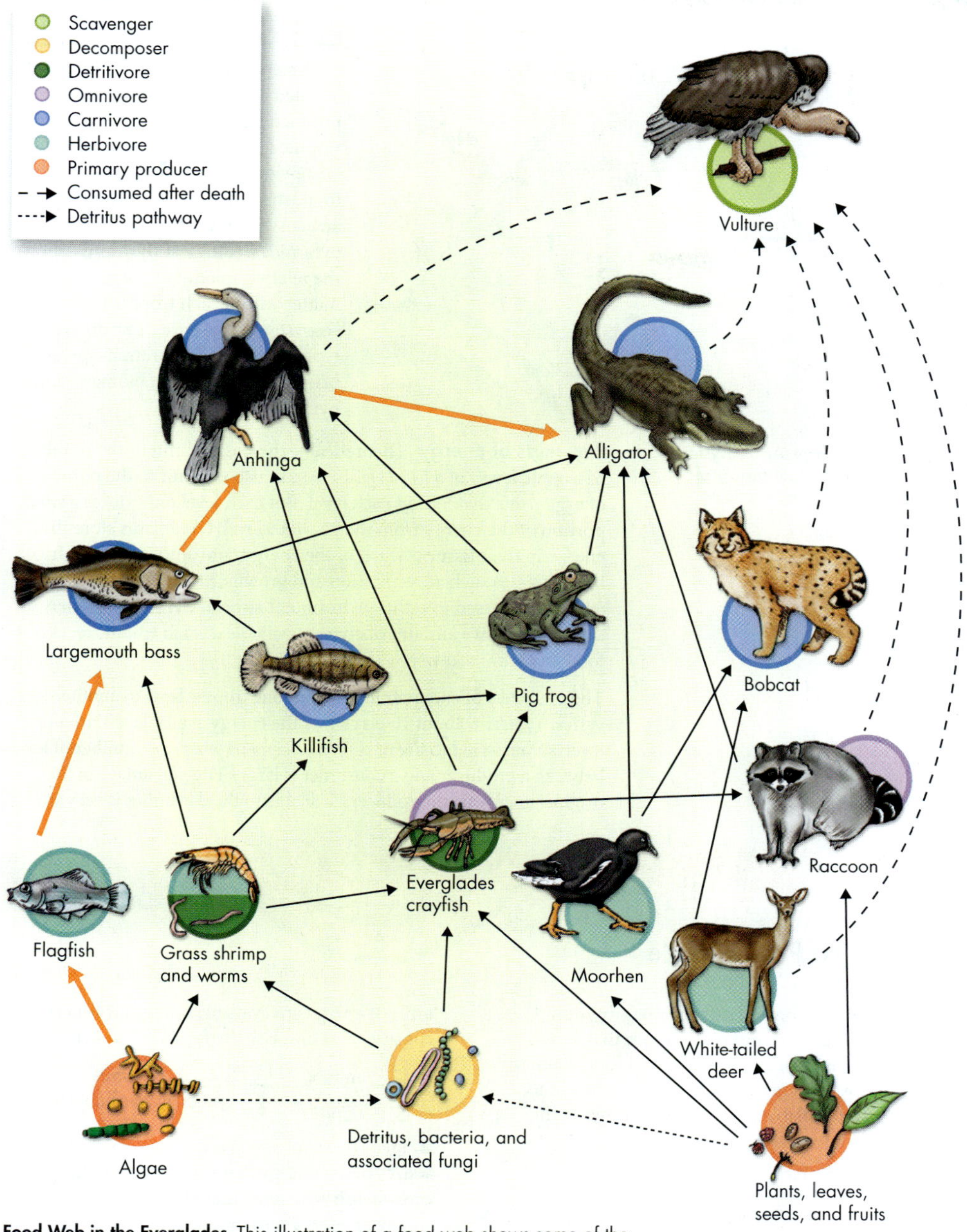

Food Web in the Everglades This illustration of a food web shows some of the feeding relationships within the Florida Everglades. The orange-highlighted aquatic food chain is one of many chains that make up this food web.

Active Reading

USE DIAGRAMS Point out the arrows surrounding the Everglades crayfish.

Ask From which organisms does energy flow to the crayfish? *(from detritus, bacteria, and associated fungi; grass shrimp and worms; and plants, leaves, seeds, and fruits)*

Ask To which organisms does the energy in the crayfish flow? *(Anhinga, pig frog, and raccoon)*

Say Look at the food chain we saw earlier in the lesson. Use your finger to trace that food chain in this food web.

Invite students to trace other food chains in this web. Ask volunteers to name the organisms, in sequence, as they trace a food chain.

Ask Do you think this food web is complete? *(no)* Be sure students understand that this food web shows only representative organisms in a Florida Everglades food web. An actual food web is much more complex and that, even in a small part of the Everglades, the web would include at least hundreds of different kinds of organisms.

DRAW CONCLUSIONS Focus students' attention on the vulture.

Ask What kind of consumer is a vulture? *(scavenger)* What happens to the parts of the dead organisms that the vulture doesn't eat? *(They will be broken down by detritivores and decomposers.)*

Ask What pathways can be added to the diagram? *(Pathways from any of the organisms that are not fully consumed to the decomposers.)*

Speed Bump

Ask What is the relationship between a food chain and a food web? *(A food chain shows one path of energy flow in an ecosystem. A food web links many food chains in an ecosystem.)*

Hands-On Learning

FOOD CHAIN LINE-UP Give each pair of students 16 index cards or sticky notes. Tell students to write the name of each organism shown in the Everglades food web on a separate card or sticky note. Instruct students to line up the names of organisms that make up one of the food chains in the food web. Then, have them choose one organism from that food chain and assemble another food chain to which that organism belongs. Students should repeat the process until all the organism's food chains have been assembled.

Have pairs share with the class the organism they chose and the number of food chains they assembled that included that organism.

During Reading continued

Active Reading

USE CONTEXT CLUES Point out context clues in the first and third sentences of the Pyramids of Energy paragraph.

Say The word *theoretically* at the beginning of the first sentence means the sentence is true based on an idea. The sentence may not be true in real life.

Have a volunteer read aloud the first three sentences.

Say Notice that the third sentence begins with *but.*

Ask What does the word *but* usually mean? *(except)* After reading the first three sentences, do you think there is no limit to the number of trophic levels or number of organisms on each level? *(No, there is a limit.)*

Speed Bump

Ask What percentage of energy available in the second trophic level of an ecosystem is transferred to the third trophic level? *(about 10 percent)*

Inquiry Into Scientific Thinking

MAKE CONNECTIONS Use the value of a dollar to review percentages with students.

Say Consider that one dollar is 100 percent. It is made up of 100 cents, or pennies.

Ask How much is 10 percent of a dollar? *(ten cents)* How much is 10 percent of ten cents? *(one cent)*

Lead a Discussion

Discuss how the body uses, loses, and stores energy.

Ask Where do you get energy? *(from food)*

Ask How does your body use energy every day? *(Sample answers: by moving, sleeping, thinking, undergoing cell processes)*

Ask What happens to the energy that your body doesn't use and does not release as heat? *(It is stored.)*

Tell students that the energy is stored in the form of chemical bonds in the body's molecules.

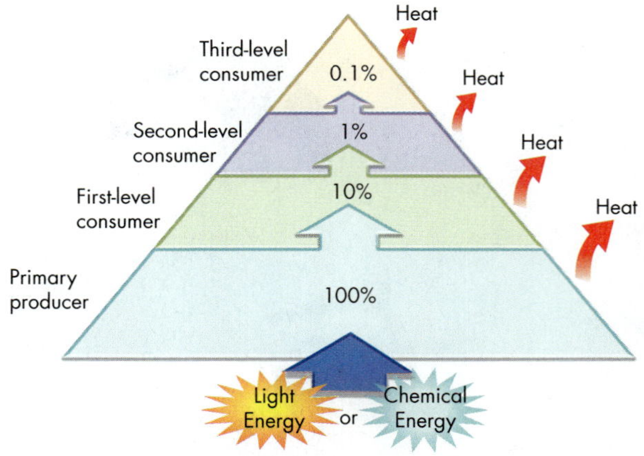

Pyramid of Energy An ecosystem must have a constant supply of energy from photosynthetic or chemosynthetic producers.

Ecological Pyramids

Each step in a food chain or food web is called a **trophic level.** Primary producers always make up the first trophic level. Various consumers occupy every other level. One way to illustrate the trophic levels in an ecosystem is with an ecological pyramid. **Ecological pyramids** show the relative amount of energy or matter within each trophic level of an ecosystem. There are three different types of ecological pyramids: pyramids of energy, pyramids of biomass, and pyramids of numbers.

Pyramids of Energy Theoretically, there is no limit to the number of trophic levels in a food web. There is also no limit to the number of organisms that live on each level. But each level only stores a small portion of the energy from the previous level. Organisms store this energy in their tissues. Much of the energy consumed is spent on life processes such as respiration and growth. Most of the remaining energy is released into the environment as heat. Pyramids of energy show the relative amount of energy available at each trophic level of a food chain or food web.

The efficiency of energy transfer from one trophic level to another varies. Typically, about 10 percent of the energy available within one level is transferred to the next. What happens when the number of levels between a producer and a consumer is large? The percentage of the original energy from producers available to the consumer is very small.

INQUIRY into Scientific Thinking — GUIDED INQUIRY

The 10 Percent Rule

The figure above shows an energy pyramid. An energy pyramid is a diagram that shows the transfer of energy through a food chain or food web. In general, only 10 percent of the energy available in one level is stored in the level above.

Analyze and Conclude

1. Calculate Suppose there are 1000 units of energy available at the producer level of the energy pyramid. Approximately how many units of energy are available to the third-level consumer? (Hint: Use the formula below.)

______ units × ______% = ______ units of energy available

2. Interpret Diagrams What is the original source of the energy that flows through the ecosystem? Why must there be a continual supply of energy into the ecosystem?

In Your Workbook Get more help for this activity in your workbook.

Analyze and Conclude

1. 1000 units × 0.1% = 1 unit of energy available
2. Light energy or chemical energy is the original source of energy in most ecosystems. There must be a continual flow of energy into the ecosystem because energy travels in only one direction through the ecosystem. Energy that is converted to heat is lost to the environment. More energy must enter the ecosystem to replace it.

Pyramids of Biomass and Numbers The total amount of living tissue within a trophic level is called its **biomass.** Biomass is usually measured in grams of organic matter per unit area. The amount of energy available to a trophic level helps determine the amount of biomass it can support. A pyramid of biomass illustrates the relative amount of living tissue available at each trophic level in an ecosystem.

Ecologists interested in the number of organisms at each trophic level use a pyramid of numbers. A pyramid of numbers shows the relative number of individual organisms at each trophic level in an ecosystem. Usually, the shapes of the pyramids of biomass and numbers are similar for an ecosystem. In this shape, the numbers of individuals on each level decrease from the level below it. The result might look something like the pyramid to the right.

In some cases, consumers are much smaller than the organisms they feed upon. For example, thousands of insects may graze on a single tree. Countless mosquitoes can feed off a few deer. Both the tree and deer have a lot of biomass, but they each represent only one organism. In such cases, the pyramid of numbers may be turned upside down, but the pyramid of biomass is right side up.

Pyramids of Biomass and Numbers In most cases, pyramids of biomass and numbers follow the same general pattern. In the field modeled here, there are more individual primary producers than first-level consumers. And the primary producers collectively have more mass. With each step to a higher trophic level, biomass and numbers decrease.

Key Question What do the three types of ecological pyramids illustrate? **A pyramid of energy shows the relative amounts of energy available at each trophic level of an ecosystem. A pyramid of biomass illustrates the relative amount of living organic matter available at each trophic level. A pyramid of numbers shows the relative number of organisms at each trophic level.**

BUILD Vocabulary

biomass the total amount of living tissue within a given trophic level

ACADEMIC WORDS
Mass is the amount of matter a substance or object has. The term *biomass* identifies a quantity of matter that is alive.

CHECK Understanding

Apply Vocabulary

Use the highlighted words from the lesson to complete each sentence correctly.

1. A(n) ________ is a way of representing the feeding relationships within an ecosystem.
2. A step in a food chain or food web is called a ________.

Critical Thinking

3. **Write to Learn** Answer the mystery clue.

MYSTERY CLUE

Zooplankton in Narragansett Bay graze on floating algae. The zooplankton now graze more actively during the winter than ever before. How might this affect the annual late-winter "bloom" of algae?

BIOLOGY.com Search Lesson 3.3 GO • Lesson Assessment

Active Reading

INTERPRET VISUALS Point out to students that the pyramid shows that both biomass and number of organisms decrease as the trophic level increases.

Ask Since a rabbit is much bigger than a blade of grass, how can more biomass be available at the level of grasses? *(There are many more blades of grass than rabbits. All together, the blades of grass make up more biomass than the rabbits.)*

KEY QUESTION Invite a volunteer to read aloud the question and answer. Remind students that the word *illustrate* means "show."

Ask At which trophic level is the most energy available, the most biomass, and the greatest number of individuals? *(first trophic level)*

Build Vocabulary

ACADEMIC WORDS Explain that the word *mass* has several meanings. With one exception (a religious service), all refer to an amount or size of something. Have students look up the word in a dictionary and use it correctly in a variety of sentences, such as these: *The mass of the iceberg was huge. A mass of people ran onto the field after the game. The mass of the cork was three grams.*

Speed Bump

Ask What is the difference between a pyramid of biomass and a pyramid of numbers? *(Biomass refers to the amount of living tissue, whereas numbers represent individual organisms.)*

Post Reading

MYSTERY CLUE Tell students that an algal bloom is a rapid increase in the size of an algae population. Zooplankton are tiny animal-like organisms that feed on algae much like cows graze on grass. Lead students to suggest how increased grazing by zooplankton would affect the algal bloom in Narragansett Bay.

Wrap-Up Activity

DINNER FOOD CHAIN Have students diagram a food chain for each food item they ate for dinner last night. Tell them to label the trophic level of each link in the food chains.

Check Understanding

1. food web
2. trophic level
3. Student answers may vary. More feeding by zooplankton decreases the biomass of the algae, so fewer algae reproduce, decreasing the size of the algal bloom.

LESSON 3.4

Pre-Reading

Objectives

3.4.1 Describe how matter cycles among the living and nonliving parts of an ecosystem.

3.4.2 Describe how water cycles through the biosphere.

3.4.3 Explain why nutrients are important in living systems.

3.4.4 Describe how the availability of nutrients affects the productivity of ecosystems.

Preview the Pages

1. Preview the visuals. Draw attention to the visual analogies on the first and last pages of the lesson. Remind students that analogies use familiar information to explain new information.
2. Have students look over the four cycles of matter diagrams. Point out that each diagram focuses on a different kind of matter. The arrows show matter's movement through the biosphere.
3. Refer students to the three paragraphs on this page and the one on the next page that are preceded by blue arrows. Tell them that the processes described are involved in the four cycles of matter.

Build Understanding

MAIN IDEAS AND DETAILS CHART Have students space the headings so they can fill in the details under the headings as they read. Students can also use the chart in Workbook B for guidance.

NATIONAL SCIENCE EDUCATION STANDARDS

UNIFYING CONCEPTS AND PROCESSES

I, II, III

CONTENT

B.2, C.4.a, C.4.d, C.4.e, C.5.e, C.5.f, D.2, F.3, F.4

For the corresponding lesson in *Miller & Levine Biology,* see pages 79–86.

3.4 Cycles of Matter

Key Questions

How does matter move through the biosphere?

How does water cycle through the biosphere?

What is the importance of the main nutrient cycles?

How does nutrient availability relate to the primary productivity of an ecosystem?

BUILD Understanding

Main Idea and Details Chart Make a main idea and details chart using the green and blue headings in this lesson. Fill in the details as you read to help you organize the information.

In Your Workbook Go to your workbook to learn more about making a main idea and details chart. Complete the chart for Lesson 3.4.

Recycling in the Biosphere

Energy passes in one direction from one trophic level to the next. Eventually it escapes into the environment as heat along the way. Energy in the form of sunlight is constantly entering the biosphere. But Earth doesn't receive a steady supply of new matter from space. Instead, matter is recycled. Elements, such as oxygen, carbon, and hydrogen, pass from one organism to another and from one ecosystem to another. They also pass among parts of the biosphere through closed loops called **biogeochemical cycles.** The flow of energy powers these cycles. Cycles of matter involve *bio*logical processes, *geo*logical processes, and *chemical* processes. Human activity can also play an important role. Matter is transformed as it moves through these cycles. It is never created or destroyed—just changed. The processes involved in biogeochemical cycles can be classified in the following ways:

▶ ***Biological Processes*** Biological processes consist of any and all activities performed by living organisms. These processes include eating, breathing, "burning" food, and eliminating waste products.

▶ ***Geological Processes*** Geological processes include volcanic eruptions and the formation and breakdown of rock. They also include major movements of matter within and below the surface of the earth.

▶ ***Chemical and Physical Processes*** Chemical and physical processes include the formation of clouds and precipitation. They also include the flow of running water and the action of lightning.

BUILD Connections

THE MATTER MILL

Nutrients are recycled through biogeochemical cycles. These cycles are powered by the one-way flow of energy through the biosphere.

3.4 Cycles of Matter

Lesson Objective

- Describe how matter cycles among the living and nonliving parts of an ecosystem.
- Describe how water cycles through the biosphere.
- Explain why nutrients are important in living systems.
- Describe how the availability of nutrients affects the productivity of ecosystems.

BUILD Vocabulary

A. The chart below shows key terms from the lesson with their definitions. Complete the chart by writing a strategy to help you remember the meaning of each term. One has been done for you.

Term	Definition	How I'm Going to Remember the Meaning
Biogeochemical cycle	Cycle of matter whereby elements pass from one organism to the next and among parts of the biosphere	*This cycle involves biological, geological, and chemical processes.*
Denitrification	The process by which nitrogen gas is released in the atmosphere after bacteria change nitrates into nitrogen	*In denitrification, nitrogen gas is dispelled into the atmosphere.*
Limiting nutrient	The nutrient that is short in supply and thus puts limits on the productivity of an ecosystem	*When you reach your limit, you have to stop.*
Nitrogen fixation	The process by which bacteria convert nitrogen gas into ammonia	*ammonia NH_4 = Nitrogen fixation*
Nutrient	A chemical substance that an organism needs to sustain life	*Eating a balanced diet provides good nutrition for the body.*

B. As you work through this lesson, you may find these terms in the activities. When you need to write a key term or a definition, **highlight** the term or the definition.

BUILD Understanding

Main Idea and Details Chart A main idea and details chart helps you organize information as you read. Copy the chart below in your notebook. Write details about each main idea in the right-hand column.

Main Idea	Details
1. Recycling in the biosphere	*Energy travels straight through an ecosystem, but matter is recycled.*
2. The water cycle	*Water continuously moves between the oceans, atmosphere, and land.*
3. Nutrient cycles	*Organisms need nutrients to survive. Nutrients pass through organisms and the environment in biogeochemical cycles. Nutrients include C, N, and P.*
4. Nutrient limitations	*If plenty of light and water are available, the primary productivity of an ecosystem may be limited by the availability of nutrients.*

The Water Cycle

The water cycle involves the movement of water between the ocean, the atmosphere, land, and living things. The diagram below shows the water cycle.

Use the words in the box to label the parts of the water cycle.

condensation runoff	root uptake transpiration	evaporation precipitation

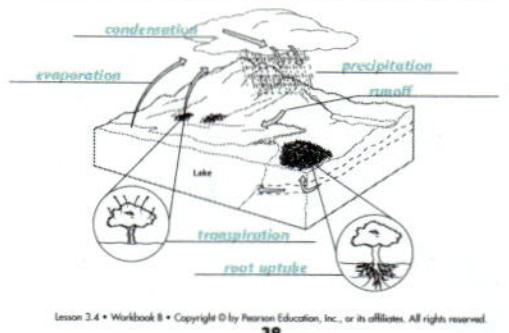

▶ ***Human Activity*** Mining and burning fossil fuels, and burning forests, affect the cycles of matter. So does clearing land for building and farming. Manufacturing and using fertilizers also influence cycles of matter.

These cycles pass the same atoms and molecules around again and again. Just think—carbon atoms in your body may once have been part of the tail of a dinosaur!

Key Question How does matter move through the biosphere? **Matter is recycled within and between ecosystems.**

The Water Cycle

Water constantly moves between the oceans, the atmosphere, and land. It can be inside organisms or outside them. Water molecules usually enter the atmosphere by evaporating from bodies of water or from plant leaves. When it evaporates, water becomes a gas called water vapor.

Winds can transport water vapor over great distances. In cooling air, the vapor condenses into tiny droplets that form clouds. When the droplets become large enough, they fall to Earth's surface as precipitation. This can be in the form of rain, snow, sleet, or hail. Precipitation can become runoff, flowing along the surface of land into a river or stream. It can also be absorbed into the ground, becoming groundwater. Groundwater enters plants through their roots or flows into bodies of water. If it penetrates deeply into the ground, it becomes part of underground reservoirs. Water that re-enters the atmosphere through transpiration or evaporation begins the cycle again.

Key Question How does water cycle through the biosphere? **Water continuously moves between the oceans, the atmosphere, and land. Sometimes the water is outside living organisms and sometimes it is inside them.**

BUILD Vocabulary

biogeochemical cycle
a process in which elements, chemical compounds, and other forms of matter are passed from one organism to another and from one part of the biosphere to another

ROOT WORDS
The term *biogeochemical* is formed by combining the words *biological, geological,* and *chemical.*

The Water Cycle This diagram shows the main processes in the water cycle. Scientists estimate that it can take as long as 4000 years to complete one cycle.

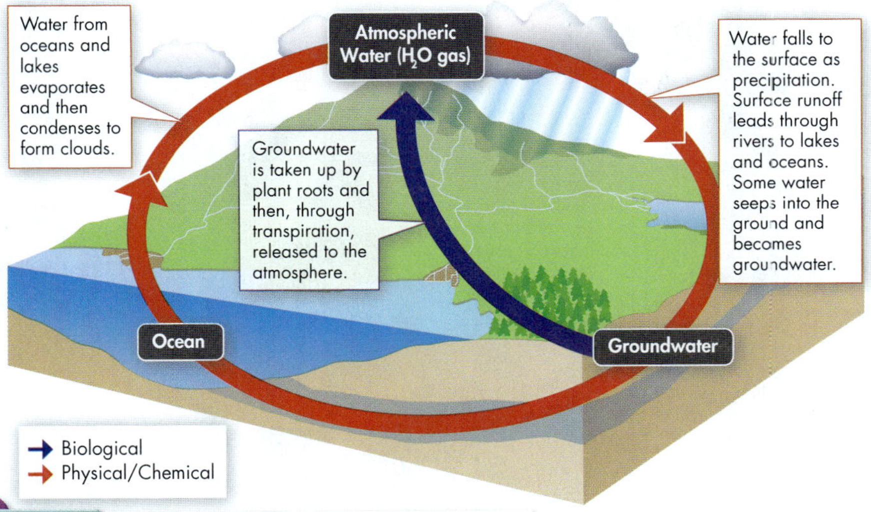

BIOLOGY.com Search Lesson 3.4 GO • InterActive Art

69

Focus on Academic Language

• reclassify • redefine • redraw • reevaluate • reorient • rephrase • rethink

Use the words *cycle* and *recycle* to introduce the prefix *re-* and how it might be used in test questions.

Ask What is a cycle? *(a process that repeats itself over and over)* What does *recycle* mean? *(to cycle again)*

Say Adding the prefix *re-* to a root word means to do the action of the root word again.

Work with the class to clarify what each of the above terms means and how it might be used in a test. For example, a test might ask students to redefine, or *define again,* a term from the chapter, using different words.

During Reading

Build Connections

VISUAL ANALOGY Refer students to the analogy on the previous page. Tell students that a water wheel has boards set at an angle to catch the water flowing over them. The water pushes on the boards, causing the wheel to turn.

Ask How is the flow of water over the water wheel similar to the flow of energy in the biosphere? *(The flow of water powers the turning of the wheel. The flow of energy in the biosphere powers the cycling of matter.)*

Active Reading

INTEPRET DIAGRAMS Point out to students that the arrows in The Water Cycle diagram are explained in the key.

Ask Which part of the water cycle involves biological processes? *(Water is taken up by plants from the ground and then released into the atmosphere.)*

Ask What kind of process is the evaporation of water from oceans? *(physical)*

KEY QUESTION Invite a volunteer to read aloud the first Key Question and answer.

Ask Why must matter on Earth by recycled? *(Earth doesn't receive a steady supply of new matter.)*

KEY QUESTION Read aloud the second Key Question and answer.

Ask How are you part of the water cycle? *(Sample answers: by drinking water and excreting water in urine)*

Build Vocabulary

ROOT WORDS Have students look up the words *biological, geological,* and *chemical* in a dictionary. Relate the meanings of the words to the term *biogeochemical cycle.*

Ask What makes up matter? *(chemicals)* What kind of matter is described by the word *biochemical*? *(chemicals in living things)* What kind of matter is described by the word *geochemical*? *(chemicals in Earth)*

Speed Bump

Ask In general, what does a biogeochemical cycle do? *(It moves matter from one part of the biosphere to another.)*

Ask What are the two main processes by which water enters the atmosphere? *(transpiration and evaporation)*

During Reading continued

Active Reading

USE CONTEXT CLUES Focus students' attention to the second sentence of this page.

Say This sentence is comparing difference cycles of matter.

Ask What word in the sentence signals the comparison? *(like)* What is being compared to water? *(nutrients)*

Ask How does comparing nutrients to water help you understand how nutrients move through the biosphere? *(Sample answer: I already know how water moves in a biogeochemical cycle. Now I know that nutrients cycle in a similar way.)*

Tell students to look for words, such as *like, similarly,* and *the same as* that signal comparisons.

INTERPERT DIAGRAMS Ensure students make the connection between the key and the colored arrows in the visual The Carbon Cycle. Have them locate the carbon reservoirs (gray boxes) and the processes that take in and release carbon (white boxes).

Have volunteers point out one process associated with each of the categories in the key.

Say Notice that two arrows change from blue to green.

Ask What does that mean? *(A biological process changes to a geological process.)*

Ask How does that happen in the carbon cycle? *(The geological process of fossil fuel formation begins with the biological process of decomposition.)*

Ask How are some of the processes part of the cycle even though their arrows aren't circular? *(Human processes (orange arrows) and geological processes (green arrows) release carbon into the atmosphere, making it available for other processes that continue the cycle.)*

Speed Bump

Ask Which processes within the carbon cycle involve humans? Explain. *(biological—releasing carbon through cellular respiration; human—releasing carbon through burning of forests and fossil fuels)*

Nutrient Cycles

Nutrients are chemical substances organisms need to build tissues and carry out life functions. Like water, they pass through organisms and the environment through biogeochemical cycles. The three most important cycles move carbon, nitrogen, and phosphorus through the biosphere.

The Carbon Cycle Carbon is a major component of all organic compounds, including carbohydrates, lipids, proteins, and nucleic acids. There are several major carbon reservoirs in the biosphere. For example, carbon is found in calcium carbonate ($CaCO_3$). This mineral is found in animal skeletons and several kinds of rocks. Carbon and oxygen form carbon dioxide gas (CO_2). CO_2 is an important component of the atmosphere and is also dissolved in oceans. Coal, oil, natural gas, and forests also are important carbon reservoirs.

The Carbon Cycle Carbon is found in several large reservoirs in the biosphere. In the atmosphere, it is found as carbon dioxide gas (CO_2); in the oceans, as dissolved carbon dioxide; on land, in organisms, rocks, and soil; and underground, as coal, petroleum, and calcium carbonate.

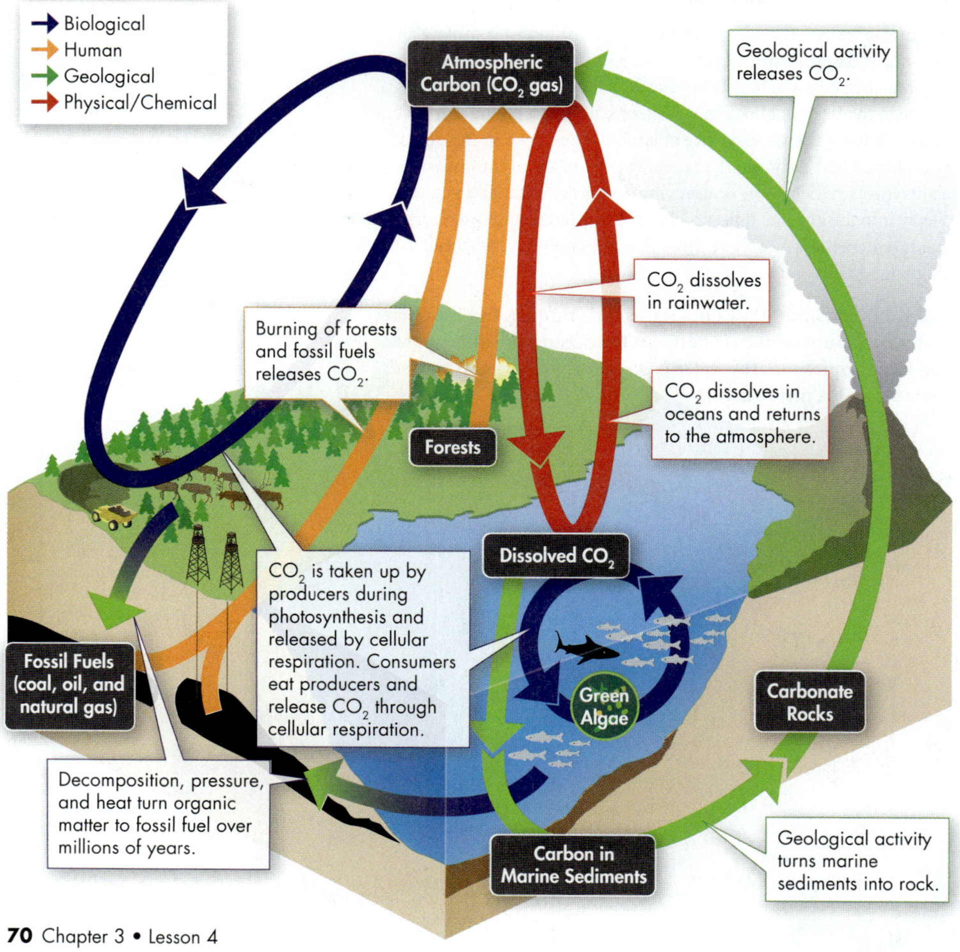

Hands-On Learning

CREATE A MÖBIUS STRIP Have students work in pairs to make a Möbius strip that describes one pathway in the carbon cycle. Write the following directions on the board for students to follow.

1. Cut two strips of paper 1-inch wide from the long edge of a sheet of notebook paper. Tape the strips together to form one long strip.
2. On one side of the strip, write: *Organisms die. Decomposers release carbon dioxide into the atmosphere.* Write one sentence on each half of the strip.
3. Flip the strip over. Write these sentences: *Plants take in carbon dioxide. Plants make carbohydrates. Consumers eat plants.* Space the sentences over the length of the strip.
4. Twist the strip one-half turn and tape the free ends together. Draw directional arrows between the sentences. Start at one sentence and follow the cycle.

Challenge students to create a Möbius strip for the nitrogen cycle.

The figure to the left shows the carbon cycle. Notice how carbon often moves between different parts of the cycle as carbon dioxide. Stored carbon is found in organic matter, fossil fuels, rocks, and the skeletons of organisms.

The Nitrogen Cycle All organisms need nitrogen to make amino acids and nucleic acids. Nitrogen exists as a gas (N_2) in the atmosphere. Nitrogen is found in soil in compounds such as ammonia (NH_3), and in nitrate (NO_3^-) and nitrite (NO_2^-) ions. These substances also exist in wastes and decomposing organic matter. Several forms of dissolved nitrogen can be found in the oceans and other large water bodies.

Nitrogen gas is the most common form of nitrogen on Earth. But only certain kinds of bacteria can use nitrogen gas directly. These bacteria use a process called **nitrogen fixation** to convert nitrogen gas to ammonia. Some of these bacteria live in the soil and on the roots of certain plants, such as peanuts and peas. Other kinds of bacteria convert ammonia into nitrates and nitrites. Primary producers use these compounds to make proteins and nucleic acids. Consumers eat producers and reuse the nitrogen to make their own proteins and nucleic acids. Decomposers release ammonia, nitrates, and nitrites from wastes and dead organisms. Producers can reuse these compounds. There are also some soil bacteria that convert nitrates to nitrogen gas. The process of converting nitrates to nitrogen gas is called **denitrification.** Denitrification provides the bacteria with energy. Humans also add nitrogen to the biosphere by manufacturing and using fertilizers. Runoff often carries excess fertilizer into surface water or groundwater.

BUILD Vocabulary

nitrogen fixation
the process of converting nitrogen gas into nitrogen compounds that plants can absorb and use

denitrification
the process by which soil bacteria convert nitrates into nitrogen gas

ACADEMIC WORDS
The verb *accumulate* means "to collect or gather." Carbon accumulates, or collects, in the soil and in the oceans where it cycles among organisms or is turned into fossil fuels.

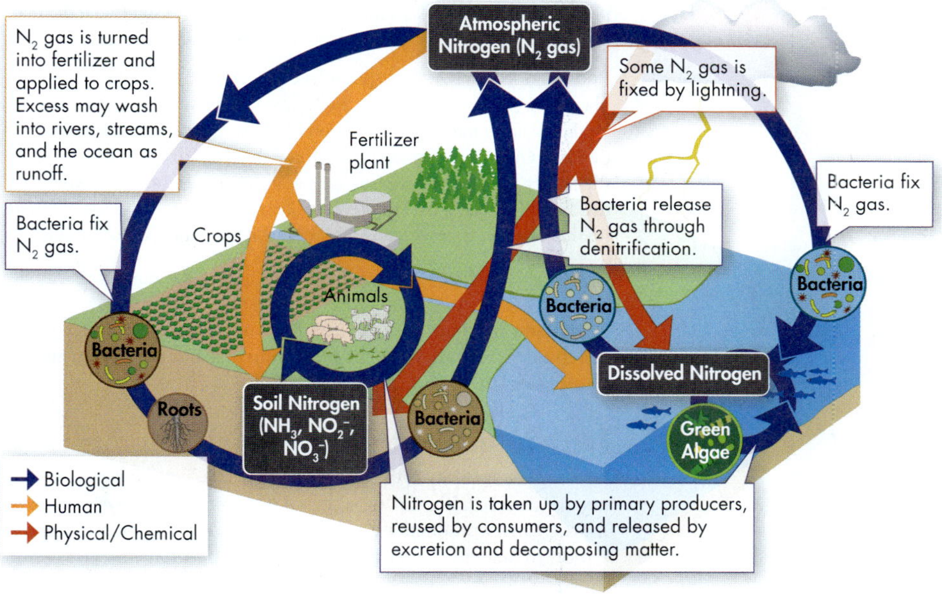

The Nitrogen Cycle The atmosphere is the largest reservoir of nitrogen in the biosphere. Nitrogen also cycles through the soil and through the tissues of living organisms.

Students Say

"To review the nitrogen cycle, which was really confusing because of all the vocabulary, we folded a sheet of paper into four squares, then wrote four terms. I wrote *nitrogen fixation, denitrification, nitrate ions,* and *nitrite ions,* but other people had different terms. Then we walked around the room getting other students to fill in our squares with a fact about one of the terms. When we had four different people write something they knew about one of our terms, we could sit down. After 10 minutes, our teacher called on people to share one of the boxes on their paper, and we could check to see if we had the same thing on our paper. Or we could add a fact if we wanted a better study guide."

Active Reading

INTERPRET DIAGRAMS Focus students' attention on the diagram The Nitrogen Cycle. Ask a volunteer to read the caption and labels.

Ask Which boxes indicate nitrogen reservoirs? *(the gray ones)* What do the white boxes indicate? *(processes that add nitrogen to the reservoirs or release nitrogen from the reservoirs)*

Have volunteers point out one process associated with each of the categories in the key.

Ask Is there a difference between the biological cycles that occur on the left side of the diagram and those on the right? Explain. *(The processes are the same, except that those on the left occur in land organisms and those on the right occur in organisms that live in water.)*

Ask Why do the blue arrows near the center of the diagram form a complete circle? *(The arrows represent nitrogen moving from the land to plants to animals and back to the land without moving into the atmosphere.)*

DRAW CONCLUSIONS Focus students' attention on the role of bacteria in the nitrogen cycle.

Ask What two processes in the nitrogen cycle involve bacteria? *(nitrogen fixation and denitrification)*

Ask Why could these processes be considered opposites? (In nitrogen fixation, bacteria convert nitrogen gas to nitrogen compounds. In denitrification, bacteria convert nitrogen compounds to nitrogen gas.)

Build Vocabulary

ACADEMIC WORDS Review the meaning of the word *accumulate*. Ask a volunteer to use the word in a sentence that relates to the nitrogen cycle. *(Sample answer: Nitrogen compounds, such as ammonia, accumulate in soil.)*

Speed Bump

Ask What must happen to nitrogen gas before it can be used by primary producers? *(It must be changed to nitrites or nitrates.)*

During Reading continued

Active Reading

COMPARE AND CONTRAST Ask a volunteer to read aloud the first paragraph about the phosphorus cycle. Clarify for students that phosphates are compounds containing phosphorus.

Ask How is the phosphorus cycle different from the other nutrient cycles? *(Phosphorus is not abundant in the biosphere. It does not enter the atmosphere in large amounts.)*

MAKE CONNECTIONS Draw a simple food chain on the board—primary producer, first level consumer, and decomposer.

Ask How does phosphorus cycle through this food chain? *(The primary producer takes in phosphorus from the soil and uses it to make DNA and RNA. When a consumer eats and digests the producer, the phosphorus is released and can be used to make the consumer's DNA and RNA. When the consumer dies, decomposers release the phosphorus compounds into the environment, which are taken in by other primary producers.)*

KEY QUESTION Invite a volunteer to read aloud the Key Question and answer.

Ask Through what four main parts of the biosphere do nutrients cycle? *(atmosphere, land, water, and organisms)*

Speed Bump

Ask Why is phosphorus essential to living organisms? *(It is an important part of molecules such as DNA and RNA.)*

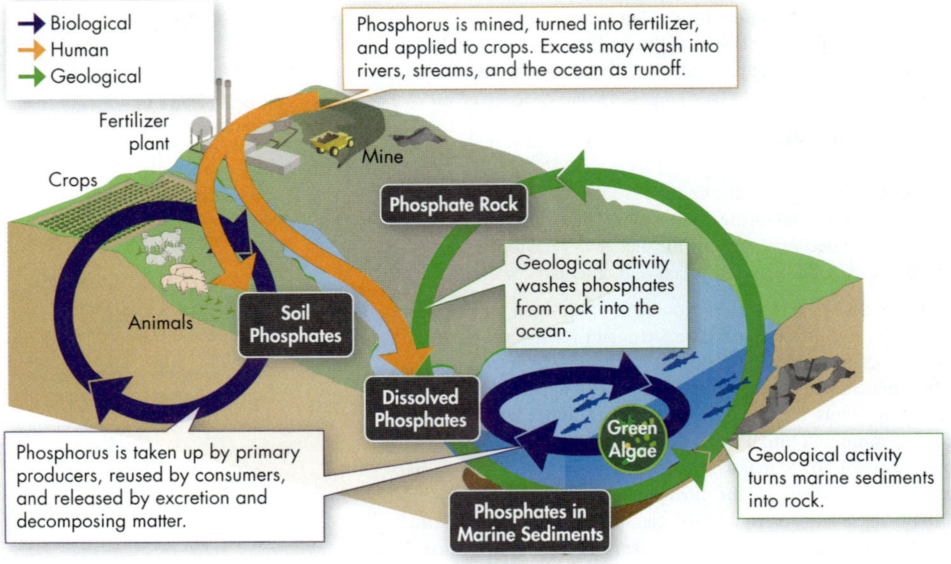

The Phosphorus Cycle Phosphorus in the biosphere cycles among the land, ocean, sediments, and living organisms. Unlike other nutrients, phosphorus is not found in significant quantities in the atmosphere.

The Phosphorus Cycle Phosphorus is essential to living organisms. It is an important part of molecules such as DNA and RNA. Although phosphorus is important biologically, it is not abundant in the biosphere. Phosphorus does not enter the atmosphere in large amounts. Phosphorus remains mostly on land and in the ocean as inorganic phosphate. On land it is found in phosphate rock and soil minerals. In the ocean, it may be dissolved or found in phosphate sediments.

Rocks and sediments release phosphate as they wear down. Some phosphate stays on land and cycles between organisms and soil. Plants bind phosphate into organic compounds when they absorb it from soil or water. Organic phosphate moves through the food web to the rest of the ecosystem. Other phosphate washes into rivers and streams, where it dissolves. This phosphate may eventually make its way to the ocean, where marine organisms process and incorporate it into biological compounds.

Key Question What is the importance of the main nutrient cycles?
Organisms need nutrients to build tissues and carry out life functions. The main nutrient cycles move carbon, nitrogen, and phosphorus through the biosphere.

Nutrient Limitation

Primary productivity is the rate at which primary producers create organic material. If even one essential nutrient in an ecosystem runs short, primary productivity will be limited. A nutrient whose supply limits productivity is called a *limiting nutrient.*

72 Chapter 3 • Lesson 4

ELL Focus on ELL: Build Background

ALL SPEAKERS The term *primary productivity* introduced under the heading Nutrient Limitation may be easily confused with *primary producers*. To help students differentiate, compare and contrast the terms.

Have students identify the terms' common word—*primary*—and the root of the second word in each term—*product* and *produce,* respectively. Then, have them look up the definitions of all three words.

Ask How does *produce* relate to *producer*? *(The producer is one that produces, or makes.)* How does *productivity* relate to *product*? *(Productivity is a measure of the amount of products made, or produced.)*

Nutrient Limitation in Soil On many farms, growth of crop plants is limited by a lack of one or more nutrients in the soil. That's why farmers use fertilizers! Most fertilizers contain large amounts of nitrogen, phosphorus, and potassium. These nutrients help plants grow better in poor soil. Micronutrients such as calcium, magnesium, sulfur, iron, and manganese are necessary in relatively small amounts. These elements are sometimes included in specialty fertilizers. Chemical fertilizers do not contain carbon. Plants get carbon from carbon dioxide during photosynthesis. All nutrient cycles work together like the gears in the figure to the right. If a nutrient runs short, the whole system slows down or even stops.

Nutrient Limitation in Aquatic Ecosystems The open oceans of the world are nutrient-poor compared to many land areas. In saltwater environments, nitrogen is typically the limiting nutrient. Phosphorus is typically the limiting nutrient in freshwater environments.

Sometimes an aquatic ecosystem receives a large input of a limiting nutrient. For instance, a heavy rain may cause runoff from heavily fertilized fields. The extra nutrients can cause a dramatic increase in the amount of algae and other primary producers. If there are not enough consumers to eat the algae, an algal bloom can occur. During an algal bloom, algae can cover the water's surface and disrupt the functioning of an ecosystem.

BUILD Connections

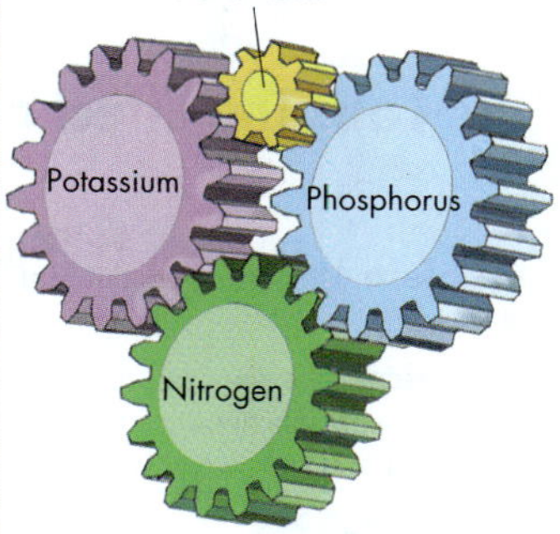

INTERLOCKING NUTRIENTS
The movement of each nutrient through ecosystems depends on the movement of all the others, because all are needed for living systems to function.

Key Question How does nutrient availability relate to the primary productivity of an ecosystem?
The availability of nutrients may limit primary productivity, even when sunlight and water are plentiful.

CHECK Understanding

Apply Vocabulary

Use the highlighted words from the lesson to complete each sentence correctly.

1. Bacteria convert nitrogen gas into ammonia using a process called _______.
2. Nutrients move between organisms, the atmosphere, the land, and the water in a series of processes called _______.

Critical Thinking

3. **Explain** How does the way that matter flows through an ecosystem differ from the way that energy flows?
4. **Relate Cause and Effect** A farmer plants a field of corn and fertilizes it with a low-nitrogen fertilizer. The corn receives plenty of sunlight and water, but does not grow well. What might be preventing the corn from growing?
5. **Write to Learn** Answer the fourth clue to the mystery. Think about how nitrogen enters the water. Is the increase in nitrogen in the water caused by nitrogen fixation or denitrification?

MYSTERY CLUE

Recently, researchers discovered that levels of dissolved nitrogen in the bay have increased. Since human activity hasn't changed much, which organisms in the bay might be responsible? (Hint: See p. 71.)

BIOLOGY.com Search Lesson 3.4 GO • Lesson Assessment • Visual Analogy

Check Understanding

1. nitrogen fixation
2. biogeochemical cycles
3. Matter is recycled through an ecosystem, while energy flows in one direction from primary producers through the decomposers and detritivores.
4. When plants do not grow well even with ample sunlight and water, it is a sign that they may not be receiving enough of one or more nutrients. Because the farmer used a low-nitrogen fertilizer, the nitrogen might be a limiting nutrient in the field.
5. Sample answer: With little change in human activity, nitrogen runoff has not likely increased. Researchers had noticed changes in the activity of the mud-dwelling bacteria. Bacteria are involved in both nitrogen fixing and denitrification. Because denitrification releases nitrogen gas into the atmosphere, nitrogen-fixing bacteria must be the cause.

Active Reading

MAKE AN ANALOGY Help students understand the term *limiting nutrients* by imagining they are making five dozen brownies. They have plenty of flour, sugar, chocolate, and butter, but they have only a few eggs.

Ask What would keep you from making all five dozen brownies? *(the number of eggs available)* What is the "limiting nutrient" in your plans? *(the eggs)*

KEY QUESTION Read aloud the Key Question and answer.

Ask What does *primary productivity* mean? *(Primary productivity is the rate at which primary producers make organic material.)*

USE VISUALS Read aloud the caption to the visual Interlocking Nutrients. Point out to students that in this gear system, moving one gear moves the others.

Ask Would the gears turn more slowly or more quickly after a heavy rain washed phosphorus from fertilized fields into a nearby pond? Why? *(The gears would turn more quickly because the added phosphorus would increase primary productivity in the pond.)*

Post Reading

ASSESSMENT TIPS For Question 3, point out that the question is asking to contrast the flow of matter with the flow of energy in an ecosystem.

Tell students that Question 4 describes the effect (poor corn growth), and they must identify the cause.

MYSTERY CLUE Refer students to The Nitrogen Cycle diagram to identify organisms that contribute to the level of nitrogen dissolved in bodies of water.

Wrap-Up Activity

BIOGEOCHEMICAL CYCLES Have students work in pairs. Assign one cycle of matter to each pair, ensuring that all four cycles discussed in this lesson are assigned. Have pairs choose an ecosystem, such as a forest, desert, or pond, and draw a labeled diagram that traces the path of a water molecule or a nutrient atom (carbon, nitrogen, or phosphorus) through that ecosystem. Tell students that the path should have at least three steps and include both living and nonliving things. Have pairs display and explain their diagram to the class.

EXTENSION Have students write three or more sentences that describe the cycle they drew.

Pre-Lab

Invite a volunteer to read the Problem and the Big Idea paragraph.

Ask Which nutrient cycles is this lab related to? *(nitrogen cycle and phosphorus cycle)* What is the primary producer in this lab? *(algae)*

Be sure students understand the meaning of the word *intentionally.*

Ask If a person is intentionally increasing the amount of nutrients, is that person aware that he or she is increasing it? *(yes)*

Review the Background Questions and Pre-Lab Questions as a class.

Lab

Tell students they will perform the chapter lab *The Effect of Fertilizer on Algae* described in **Lab Manual B.**

SAFETY Students should wear protective safety goggles and gloves during this lab. Remind students to handle glassware with care and to wash their hands when they have completed the lab.

Post-Lab

EXTEND THE EXPERIMENT Have students repeat the procedure using different amounts of the plant food to see what amount maintains optimal algae growth for the size of the ecosystem being tested.

NATIONAL SCIENCE EDUCATION STANDARDS

UCP II

CONTENT C.4.a, C.4.d, C.4.e, C.5.b, C.5.e, F.4, F.5

INQUIRY A.1.b, A.2.a

Real-World Lab

GUIDED INQUIRY

Pre-Lab: The Effect of Fertilizer on Algae

Problem How do excess nutrients affect the growth of algae?

Materials test tubes, test-tube rack, glass-marking pencil, dropper pipettes, algae culture, 25-mL graduated cylinder, spring water, plant food, cotton balls, grow light

Lab Manual Chapter 3 Lab

Skills Focus Predict, Compare and Contrast, Infer

Connect to the Big idea In a healthy ecosystem, nutrients cycle among primary producers, consumers, and decomposers. The growth of primary producers is limited by the availability of nutrients. Humans can intentionally increase the amount of nutrients in an ecosystem. For example, farmers may add fertilizer to the soil in which they grow crops. But the addition of nutrients to an ecosystem is not always planned. For example, runoff from soil that contains fertilizer may flow into coastal waters or freshwater ponds. In this lab, you will observe what happens when algae that live in those waters are provided with excess nutrients.

Background Questions

a. Review What is a limiting nutrient?

b. Explain Why do farmers use fertilizers?

c. Classify What role do algae play in freshwater ecosystems?

Pre-Lab Questions

Preview the procedure in the lab manual.

1. Design an Experiment What is the independent variable in this experiment?

2. Predict After four days, how will you be able to tell which test tube has more algae?

3. Control Variables Why will you grow *Chlorella* in spring water instead of pond water?

BIOLOGY.com Search Chapter 3 GO

Visit Chapter 3 online to test yourself on chapter content and to find activities to help you learn.

Untamed Science Video Help the Untamed Science crew explore food relationships as they turn the ecological pyramid upside down.

Art in Motion View a short animation showing the different levels of organization.

Art Review Review your understanding of which organisms are producers and which are consumers with this drag-and-drop activity.

InterActive Art Build your understanding of the water cycle with this animation.

Tutor Tube Get some clarification on producers and consumers and learn how the flow of matter and energy is not what you may think!

Visual Analogies Compare a recycling center to decomposers in this activity. Compare nutrient limitation to a series of cogs in this activity.

Pre-Lab Answers

BACKGROUND QUESTIONS

a. A limiting nutrient is a chemical substance that limits primary productivity of an ecosystem.

b. Most fertilizers contain large amounts of nitrogen, phosphorus, and potassium, which are essential nutrients for plants.

c. Algae are the main photosynthetic producers in freshwater ecosystems. They produce energy-rich compounds that are used by other organisms.

PRE-LAB QUESTIONS

1. The independent variable is the presence of fertilizer.

2. Sample answer: In the test tube with more algae, the liquid will have a darker green color and be less transparent to light.

3. Sample answer: Pond water is likely to contain many species of algae. Using spring water ensures that both test tubes have *Chlorella,* and that neither test tube has a different algal species.

3 CHAPTER Summary

3.1 What Is Ecology?

- Ecology is the scientific study of interactions among organisms and between organisms and their physical environment.
- A biotic factor is any living part of the environment.
- An abiotic factor is any nonliving part of the environment.
- Modern ecologists use three methods in their work: observation, experimentation, and modeling.

biosphere (p. 56)
ecology (p. 56)
population (p. 56)
community (p. 56)
biotic factor (p. 57)
abiotic factor (p. 58)

3.2 Energy, Producers, and Consumers

- Primary producers are the first producers of energy-rich compounds that are later used by other organisms.
- Organisms that rely on the other organisms for energy and nutrients are called consumers.

autotroph (p. 60)
primary producer (p. 60)
heterotroph (p. 61)
consumer (p. 61)

3.3 Energy Flow in Ecosystems

- Energy flows through an ecosystem in a one-way stream, from primary producers to various consumers.
- Pyramids of energy show the relative amount of energy available at each trophic level of an ecosystem. A pyramid of biomass illustrates the relative amount of living organic matter available at each trophic level. A pyramid of numbers shows the relative number of individual organisms at each trophic level.

food chain (p. 63)
food web (p. 63)
trophic level (p. 66)
ecological pyramid (p. 66)
biomass (p. 67)

3.4 Cycles of Matter

- Matter is recycled within and between ecosystems.
- Water continuously moves between the oceans, the atmosphere, and land. Sometimes the water is outside living organisms and sometimes it is inside them.
- Organisms need nutrients to build tissues and carry out life functions. The main nutrient cycles move carbon, nitrogen, and phosphorus through the biosphere.
- The availability of nutrients may limit primary productivity, even when sunlight and water are plentiful.

biogeochemical cycle (p. 68)
nitrogen fixation (p. 71)
denitrification (p. 71)

BIOLOGY.com Search Chapter 3 GO • Match It • Chapter Assessment

UbD Transfer the Big Idea

TRANSFER Have students write one or two paragraphs from the point of view of a producer in an ecosystem. The paragraphs should be written in the first person and should include the following points:

- What interactions occur between the producer and the other organisms in the ecosystem?
- What interactions occur between the producer and the nonliving things in the ecosystem?
- What role does the producer play in the movement of matter and energy in the ecosystem?

Chapter Summary

Think Visually

USE GRAPHICS Have students complete the following concept map about cycles of matter by adding kinds of organisms to the numbered circles.

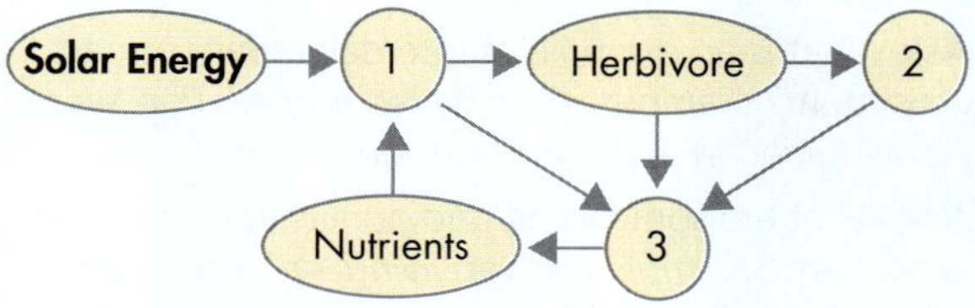

Answers

1. autotroph or primary producer
2. carnivore, omnivore, or consumer
3. decomposer or detritivore

BIOLOGY.com Study Online

REVIEW RESOURCES

Art in Motion Students can zoom in or out to see the levels of organization of living things on Earth.

Data Analysis Students simulate a classic ecological data collection method to learn how data are used to monitor a site.

Tutor Tube This tutorial clarifies how matter flows between producers and consumers.

Visual Analogy Students compare the reuse of building materials to the recycling of nutrients.

InterActive Art This animation lets students interact with the water cycle.

Visual Analogy Students can trace the movement of nutrients through the environment.

Vocabulary Review The *Flash Cards* and *Match It* are an interactive vocabulary review.

Editable Worksheets Editable pages of Study Workbook B, Lab Manual B, and the Assessment Resources Book are available online.

Lesson Overviews Lesson Overviews are based on Miller & Levine Biology. However, they are available on CD-ROM and can be edited for your class.

ASSESSMENT RESOURCES

Chapter Assessment Have students take an online version of the Chapter 3 Check Understanding.

Standardized Test Prep Students can take an online version of the Standardized Test Prep. You will receive their scores along with ideas for remediation.

Diagnostic and Benchmark Tests Use these tests to monitor your students' progress and supply remediation.

Assess

Big idea Invite a volunteer to read the Big Ideas and the question.

Ask What are examples of matter and energy in an ecosystem? *(Sample answers: matter—water, carbon, nitrogen; energy—light energy, chemical energy)*

Ask What is an example of interdependence in an ecosystem? *(Sample answers: feeding relationships, plants taking in water and carbon dioxide)*

Ask What are some living and nonliving parts of an ecosystem? *(Sample answer: living—plants, animals, bacteria; nonliving—nutrients, water, energy)*

Ask How do living things depend on nonliving things for survival? *(Sample answer: All organisms need the right amounts of nutrients to survive.)*

Have students write their answers in their notebooks.

CONSTRUCTED RESPONSE Paragraphs will vary. Suggest that students begin their paragraphs with the main idea as a topic sentence and add sentences that reflect supporting details. Remind them that a new main idea starts a new paragraph. Be sure students understand that the *Hints* should be used as a guide to help them answer the questions.

Write Question 1 on the board and read it aloud.

Ask What are the key words in the question? *(energy, trophic level, organisms)*

Underline these words.

Ask What energy does the question ask about? *(the amount of energy available at each trophic level)* What is being asked about this energy? *(why it limits the number of organisms that can be supported at each trophic level)*

Read the first hint.

Ask What does the word *transferred* mean? *(passed along)* If 100 units of energy are available at the second trophic level, how many will be transferred to the third trophic level? *(10 units)*

Read the second hint.

Ask What are some examples of life processes? *(Sample answers: growth, breathing, reproduction)*

Model writing the first one or two sentences, using the hints on the board and in the text. *(Sample: A limited amount of energy is available to each trophic level. The organisms that make up each trophic level use the energy for life processes, such as growth, breathing, and reproduction.*

3 CHECK Understanding

Assess the Big idea: Matter and Energy, Interdependence in Nature

Write an answer to the question below.

Q: How do Earth's living and nonliving parts interact and affect the survival of organisms?

Constructed Response

Write an answer to each of the numbered questions below. The answer to each question should be one or two paragraphs. To help you begin, read the **Hints** below the questions.

1. **Why does the amount of energy available at each trophic level limit the number of organisms it can support?**

   Hint Only 10 percent of the energy from a given trophic level is transferred to the next trophic level.

   Hint Most of the energy transferred to a trophic level is used for life processes.

2. **How does a wildfire affect the carbon cycle?**

   Hint Trees are a part of an ecosystem's biomass.

3. **Why is it important for farmers to fertilize land used for crops?**

   Hint Primary productivity is limited by the availability of nutrients.

   Hint Which nutrients are found in high concentrations in plant material?

Foundations for Learning Wrap-Up

The layered book you made as you read the chapter is a tool you can use to arrange your thoughts about the biosphere.

Activity 1 Working with a partner, take turns explaining the important concepts in each lesson using your layered book.

Activity 2 Working in small groups, create two fishbone maps like the one shown below. One map will represent Matter and Energy. The second map will focus on Interdependence in Nature. Fill in the maps with details from your layered book. Use the maps to answer the following questions:

- How do plants and animals interact to move matter and energy through the biosphere?
- How do the biogeochemical cycles interact to move matter and energy between land, water, and air?

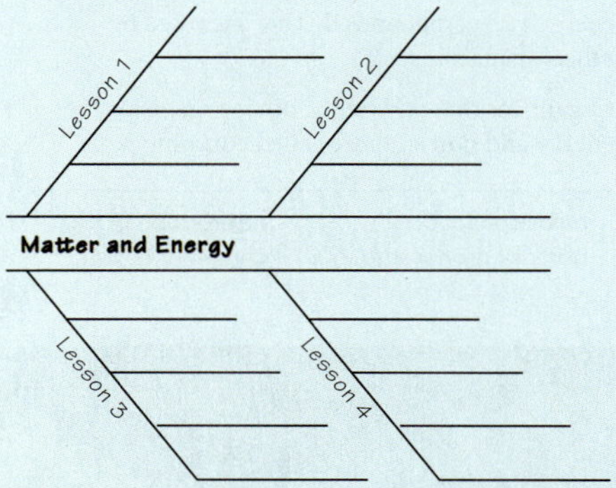

For Question 2, pair students and have them use the hints and your modeling to answer the question. Have students share their paragraphs with the class.

For Question 3, have students write the paragraphs independently for homework or as an anchor activity.

Foundations for Learning Wrap-Up

Spot check summaries and have students complete at least one summary statement for each heading before pairs work together in Activity 1.

Divide the class into groups of 3 or 4 for Activity 2. Groups should use details from all group members' summaries to create the two maps. Remind students to write the answers to the questions in their notebooks.

3.1 What Is Ecology?

Understand Key Concepts

1. The term that ecologists use to indicate life on a global scale is
   - a. ecosystem.
   - b. biome.
   - c. biosphere.
   - d. ecology.
2. Which term describes a group of different species that live together in a defined area?
   - a. a population
   - b. a community
   - c. an ecosystem
   - d. a biosphere

Test-Taking Tip

Anticipate the Answer Think of your answer before looking at the answer choices. Then select the answer choice that comes closest to your own. In question 2, answer **b** is correct because a community is the only term that includes different species in a specific area.

Think Critically

3. **Pose Questions** You live near a pond that you have observed for years. One year you notice the water contains a massive overgrowth of green algae. What is a question you might have about this unusual growth?

3.2 Energy, Producers, and Consumers

Understand Key Concepts

4. Primary producers are organisms that
   - a. rely on other organisms for their energy and food supply.
   - b. consume plant and animal remains and other dead matter.
   - c. use energy from the environment to make complex organic molecules from inorganic molecules.
   - d. obtain energy by eating only plants.
5. Which of the following organisms is a decomposer?

a. 

c. 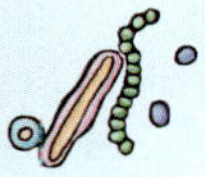

b. 

d. 

6. Which of the following describes how ALL consumers get their energy?
   - a. directly from the sun
   - b. from eating primary producers
   - c. from inorganic chemicals like hydrogen sulfide
   - d. from eating organisms that are living or were once living

Think Critically

7. **Classify** Classify each of the following as an herbivore, a carnivore, an omnivore, or a detritivore: earthworm, bear, cow, snail, snake, human.

3.3 Energy Flow in Ecosystems

Understand Key Concepts

8. The series of steps in which a large fish eats a small fish that has eaten algae is a
   - a. food web.
   - b. food chain.
   - c. pyramid of numbers.
   - d. pyramid of biomass.
9. The total amount of living tissue at each trophic level in an ecosystem can be shown in a(n)
   - a. energy pyramid.
   - b. pyramid of numbers.
   - c. biomass pyramid.
   - d. biogeochemical cycle.
10. What happens to energy in a trophic level that is not stored or used for life processes?

Think Critically

11. **Sequence** Describe a food chain of which you are a member. You may draw or use words to describe the chain.

Answers

Lesson 3.1

UNDERSTAND KEY CONCEPTS

1. c

2. b

THINK CRITICALLY

3. Sample answer: What are the biotic factors that affect this pond?

Lesson 3.2

UNDERSTAND KEY CONCEPTS

4. c

5. c

6. d

THINK CRITICALLY

7. earthworm—detritivore; bear—omnivore; cow—herbivore; snail—herbivore; snake—carnivore; human—omnivore

Lesson 3.3

UNDERSTAND KEY CONCEPTS

8. b

9. c

10. The remaining energy is usually lost as heat.

THINK CRITICALLY

11. Sample answer: grass→cow→me

Assessment Tips

Pair students and have them work together to answer the Check Understanding questions on this page and the following page.

For Question 3, suggest students think of the massive overgrowth as an algal bloom, as explained in Lesson 3.4.

For Question 7, ensure students know that to classify means to put into groups.

For Question 15, suggest that students review the Nitrogen Cycle diagram in Lesson 3.4.

ASSESSMENT

Lesson 3.4

UNDERSTAND KEY CONCEPTS

12. a

13. Water enters the atmosphere through evaporation from a body of water or from the leaves of plants (transpiration).

14. Nutrient limitation occurs when a nutrient in short supply limits the primary productivity of an ecosystem.

THINK CRITICALLY

15. Student answers may vary. Answers should indicate that students understand that the large increase in nitrogen in the water caused a disturbance in the food web that affected the ability of the trout to survive.

Connecting Concepts

16. Productivity increases rapidly at first as rainfall increases. The rate of increase eventually slows, but productivity continues to increase.

17. Primary productivity is affected by the supply of nutrients in the soil and by the amount of sunlight the plants receive.

CHAPTER MYSTERY After students read the Chapter Mystery, discuss interactions between organisms and the environment.

Ask In which abiotic factor did a change occur that started the series of changes described in the Chapter Mystery? *(water temperature)*

Ask What is an example of an interaction between two living things in the bay that changed because of the increase in water temperature? *(Sample answer: Warm-water shrimp now stay in the bay all winter, feeding on baby flounder.)*

Ask What is an example of an interaction between a living thing and a nonliving thing that has happened because of the increase in water temperature? *(Bacteria in the bay now fix nitrogen instead of denitrifying the water.)*

3 CHECK Understanding

3.4 Cycles of Matter

Understand Key Concepts

12. Nutrients move through an ecosystem in
- **a.** biogeochemical cycles.
- **b.** water cycles.
- **c.** energy pyramids.
- **d.** ecological pyramids.

13. List two ways in which water enters the atmosphere in the water cycle.

14. What is meant by "nutrient limitation"?

Think Critically

15. **Infer** Ecologists discovered that trout were dying in a stream that ran through some farmland. Nitrogen fertilizer had been used on the crops growing on this farmland. What do you think happened in the stream that caused the fish to die?

Connecting Concepts

Use Science Graphics

The graph below shows the effect of annual rainfall on the rate of primary productivity in an ecosystem. Use the graph to answer questions 16 and 17.

The Effect of Rainfall on Plant Productivity

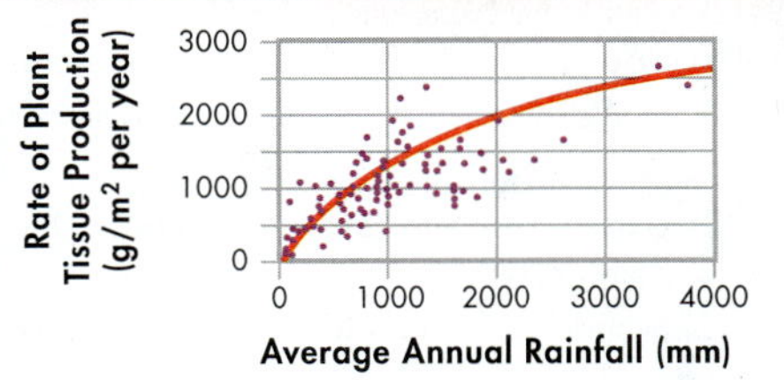

16. **Interpret Graphs** What happens to productivity as rainfall increases?

17. **Apply Concepts** What factors other than water might affect primary productivity?

solve the CHAPTER MYSTERY

CHANGES IN THE BAY

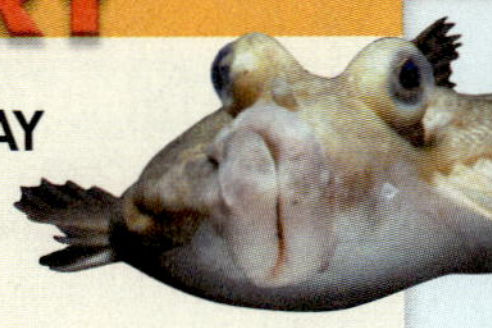

One hypothesis suggests that rising water temperatures have caused most of the changes in Narragansett Bay. The bay's temperature has risen more than 1.5°C (3°F) since 1960. This warmth encourages bluefish to stay in the bay later in the fall. It also allows predatory warm-water shrimp to remain in the bay all winter. These shrimp feed on baby flounder. Warmer water also enables zooplankton to graze heavily on marine algae. This stops the late-winter algal bloom. The algal bloom used to provide organic carbon to the entire food web.

The result of these changes seems to be a shift in the activities of bacteria. The bacteria are responsible for transforming nitrogen. When the spring bloom provided organic carbon, bacteria denitrified the water. This released nitrogen into the atmosphere. Now, the bacterial community has changed and actually fixes nitrogen. This brings more of it into the water. No one knows how this will affect the long-term health of the bay and coastal waters.

1. **Compare and Contrast** In one paragraph, compare the way the bay used to be with the way the bay is now. Be sure to include changes in the food web and nitrogen cycle.

2. **Infer** Narragansett Bay harbors sea jellies that prefer warm water and have previously been present only in summer and early fall. These sea jellies eat fish eggs, fish larvae, and zooplankton. If the bay continues to warm, what do you think might happen to the population of sea jellies in the bay?

UntamedScience Finding out about Narragansett Bay is only the beginning. Take a video field trip with the ecogeeks of Untamed Science to see where this mystery leads.

BIOLOGY.com Search Chapter 3 GO • Untamed Science Video • Chapter Mystery

CHAPTER MYSTERY ANSWERS

1. Sample answer: In the past, the bay was cooler. Bluefish and predatory shrimp migrated away from the bay during winter. There was a late-winter algal bloom. Now, the water temperature is higher. Bluefish and shrimp no longer migrate. The shrimp prey on baby flounder, decreasing that population. Zooplankton eat the algae, eliminating the algal bloom. The bloom provided nutrients, including carbon. Bacteria used to denitrify. Now, the bacteria fix nitrogen.

2. Student answers may vary. The population of sea jellies should increase and be present later into the fall.

Untamed Science In the video, **Nature's Tangled Web,** the crew of Untamed Science explores feeding relationships and discovers how some energy pyramids do not follow the usual pattern.

Standardized Test Prep

Multiple Choice

1. A group of individuals that belong to a single species and that live together in a defined area is termed a(n)
   A population.
   B ecosystem.
   C community.
   D biome.

2. Which of the following is NOT true about matter in the biosphere?
   A Matter is recycled in the biosphere.
   B Biogeochemical cycles transform and reuse molecules.
   C The total amount of matter decreases over time.
   D Water and nutrients pass between organisms and the environment.

3. Which is a source of energy for Earth's living things?
   A wind energy only
   B sunlight only
   C wind energy and sunlight
   D sunlight and chemical energy

4. Which of the following is a primary producer?
   A a producer, like algae
   B a carnivore, like a lion
   C an omnivore, like a human
   D a detritivore, like an earthworm

5. Human activities, such as the burning of fossil fuels, move carbon through the carbon cycle. Which other processes also participate in the carbon cycle?
   A biological processes only
   B geochemical processes only
   C chemical processes only
   D a combination of biological, geological, and chemical processes

6. What are the physical, or nonliving components of an ecosystem called?
   A abiotic factors
   B temperate conditions
   C biotic factors
   D antibiotic factors

Questions 7–8

The diagrams below represent the amount of biomass and the numbers of organisms in an ecosystem.

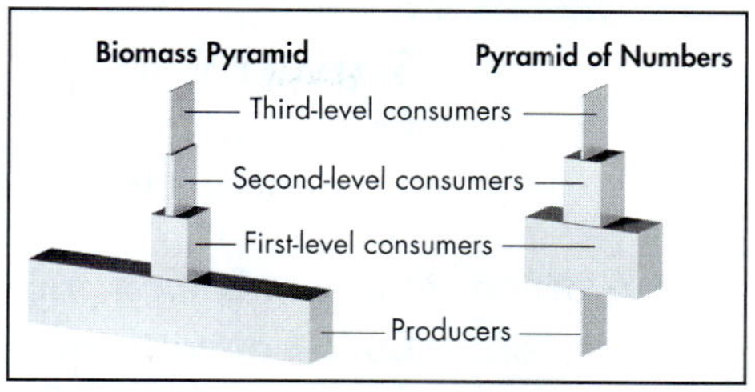

7. What can you conclude about the ecosystem from the pyramid of numbers shown?
   A There are more first-level consumers than producers.
   B There are more third-level consumers than second-level consumers.
   C There are more producers than first-level consumers.
   D There are more second-level consumers than first-level consumers.

8. What can you conclude about the producers in the ecosystem based on the two pyramids shown?
   A The producers in the ecosystem are probably very small organisms.
   B There are no producers in the ecosystem.
   C The producers in the ecosystem are probably large organisms.
   D Decomposers in the ecosystem outnumber the producers in the ecosystem.

Open-Ended Response

9. What ultimately happens to the bulk of matter in any trophic level of a biomass pyramid—that is, the matter that does not get passed to the trophic level above?

If You Have Trouble With . . .

Question	1	2	3	4	5	6	7	8	9
See Lesson	3.1	3.4	3.2	3.2	3.4	3.1	3.3	3.3	3.3

Test-Taking Tip

USE TIME WISELY

Tell students to scan the questions to identify those that are about unfamiliar topics and might require pure guesswork on their part. Explain that those questions should be saved for last. After they have completed questions about familiar topics, students should then return to the questions about unfamiliar topics. Explain that using reasoning to eliminate one or more obviously incorrect choices is a technique that can be useful, even if the question is about an unfamiliar topic.

Before the Test

Review with students the Test-Taking Tip.

Point out the word *NOT* in Question 2.

Point out the word *ONLY* in the answer choices for Questions 3 and 5. Remind students to read the question carefully and to decide whether each choice is the only correct choice.

Be sure students know that the word *conclude* in Questions 7 and 8 means "to tell a fact they figured out."

Remind students that the paragraph they write to answer the open-ended question should contain the same facts as another student's paragraph, but the order of the facts, examples used, and style of writing probably will not be the same.

Answers

1. A
2. C
3. D
4. A
5. D
6. A
7. A
8. C
9. Student answers may vary. The bulk of matter in a trophic level ultimately dies and is broken down by decomposers to inorganic molecules, which re-enter the biogeochemical cycles. These molecules either are reincorporated into more organic matter or become part of geological or atmospheric compounds.

CHAPTER PLANNER 4 Ecosystems and Communities

Chapter Contents	NSES	Time	Core Resources
Chapter Preview			**Student Edition,** pp. 80–81 **Chapter Mystery,** p. 81 **Foundations for Learning,** p. 81
4.1 Climate Weather and Climate • Factors That Affect Climate	I, II, III, B.5, B.6, D.1	½ period ¼ block	**Student Edition,** pp. 82–84 **Build Connections,** p. 83 **Study Workbook B** 4.1 Worksheets **Biology.com** *Visual Analogy:* The Greenhouse Effect • 4.1 Lesson Assessment
4.2 Niches and Community Interactions The Niche • Competition • Predation, Herbivory, and Keystone Species • Symbioses	I, II, III, A.2.d, C.4.b, C.4.c, C.4.d, C.5.e	1 period ½ block	**Student Edition,** pp. 85–87 **Study Workbook B** 4.2 Worksheets **Biology.com** 4.2 Lesson Assessment
4.3 Succession Primary and Secondary Succession • Climax Communities	I, II, III, IV, C.4.c, C.4.e, F.5	1 period ½ block	**Student Edition,** pp. 88–90 **Study Workbook B** 4.3 Worksheets **Biology.com** *Art in Motion:* Primary and Secondary Succession • 4.3 Lesson Assessment
4.4 Biomes The Major Biomes • Other Land Areas	I, III, V, A.1.c, A.2.d, C.4.c, C.6.b, D.1	1 period ½ block	**Student Edition,** pp. 91–95 **Inquiry Into Scientific Thinking,** p. 95 **Study Workbook B** 4.4 Worksheets **Biology.com** 4.4 Lesson Assessment
4.5 Aquatic Ecosystems Conditions Underwater • Freshwater Ecosystems • Estuaries • Marine Ecosystems	I, V, C.4.b, C.4.c, C.5.b, C.5.e	1 period ½ block	**Student Edition,** pp. 96–99 **Study Workbook B** 4.5 Worksheets **Biology.com** *Art Review:* Ocean Zones • 4.5 Lesson Assessment
Chapter Pre-Lab	F.3, F.4	1 period ½ block	**Student Edition,** p. 100 **Lab Manual B** *Abiotic Factors and Plant Selection*

Tools for ELL Instruction

ELL Handbook describes teaching strategies to make *Miller & Levine Biology: Foundation Edition* more accessible to ELL students.

Spanish Study Workbook is a Spanish translation of Study Workbook A.

Multilingual Glossary is the *Miller & Levine Biology* glossary translated into ten languages.

Tools to Challenge Students

Study Workbook A includes more challenging worksheets and activities such as Chapter Mystery activities that reinforce 21st Century Skills.

Lab Manual A labs are geared toward on-level students. The Lab Manual includes fifteen labs in addition to the Chapter Labs.

Biology.com Data Analysis activities offer students a way to practice data gathering and analyzing skills.

Visual Quizzes in the Assessment Resources Book are geared toward on-level students and use visuals from *Miller & Levine Biology.*

Additional Resources

Biology.com Untamed Science Video • Flash Cards

Study Workbook A 4.1 Worksheets
Spanish Study Workbook 4.1 Worksheets

Study Workbook A 4.2 Worksheets
Lab Manual B *Data Analysis:* Predator-Prey Dynamics • *Hands-On Activity:* Modeling Predator-Prey Interactions
Spanish Study Workbook 4.2 Worksheets
Biology.com *Data Analysis:* The Intertidal Zone
Assessment Resources Book Visual Quizzes

Study Workbook A 4.3 Worksheets
Spanish Study Workbook 4.3 Worksheets
Assessment Resources Book Visual Quizzes

Study Workbook A 4.4 Worksheets
Lab Manual B *Data Analysis:* Which Biome?
Spanish Study Workbook 4.4 Worksheets

Study Workbook A 4.5 Worksheets
Spanish Study Workbook 4.5 Worksheets

Lab Manual A *Abiotic Factors and Plant Selection*

Chapter Review

Student Edition Chapter Summary, p. 102
Study Workbook B Chapter 4 Review
Transparencies, pp. 44–54
Biology.com Untamed Science Video • Editable Worksheets of Study Workbooks A and B and Lab Manuals A and B • Chapter 4 Flash Cards and Crossword

Untamed Science DVD • Classroom Resources CD (includes lesson presentations and editable worksheets)

Chapter Assessment

Student Edition Check Understanding, pp. 101–105
Study Workbook B Chapter 4 Taking a Standardized Test
Assessment Resources Book Chapter 4 Test B
Biology.com Chapter 4 Assessment • Editable Worksheets of Chapter 4 Visual Quizzes* and Chapter 4 Tests A and B

Exam*View* *Assessment Suite* • Classroom Resources CD (includes lesson presentations and editable worksheets)

Time: 1 period, 1/2 block

* Visual Quizzes are based on visuals from *Miller & Levine Biology*. Review carefully to make sure they are appropriate for your class.

Pressed for Time?

Preview the Chapter Discuss the Key Questions for Lesson 4.2 and introduce the vocabulary for Lessons 4.2 and 4.3.

Cover the Chapter Quickly Have students read all of Lesson 4.2, *Primary and Secondary Succession* in Lesson 4.3, and go over the Primary Succession and Secondary Succession visuals. For Lesson 4.4, have students read Defining Biomes in *The Major Biomes*, and instruct students to use pp. 92–94 as a reference while doing the Inquiry Into Scientific Thinking activity. Assign *Conditions Underwater* in Lesson 4.5 and briefly discuss the Ocean Zones visual.

Assess Assign the 4.2 Check Understanding, Questions 1–3 and 6 in the 4.3 Check Understanding, Questions 1–4 in the 4.4 Check Understanding, Questions 2 and 3 in the 4.5 Check Understanding, and Questions 4–13 and 16 in the Chapter 4 Check Understanding.

CHAPTER 4

Get the Picture

Have a volunteer read the photo caption.

Say Cheetahs are carnivores. They hunt other animals for food. A cheetah's prey includes zebras and other herbivores. Herbivores are plant eaters that depend on the grasses and shrubs of the savannah for their food.

Ask What organisms do you see in the photograph besides the cheetah? *(plants)*

Ask What can you infer would happen if all the plants disappeared from the savanna? *(Zebras and other plant eaters would have nothing to eat. They might die or move away, leaving the cheetah with nothing to eat.)*

Activate Prior Knowledge

Ask What are the most important things that people need to stay alive? *(air to breathe, water to drink, food to eat, shelter for warmth and protection)*

Say Most living organisms need oxygen, water, food, and some sort of protection.

Ask What are some examples of ways organisms get the things they need? *(Answers will vary. Examples should relate to interdependence in nature, such as a bird that depends on insects for food and the branches of a tree for a safe place to build a nest.)*

4 Ecosystems and Communities

Big idea **Interdependence in Nature**

Q: How do abiotic and biotic factors shape ecosystems?

Cheetah looking out across the savanna at the Masai Mara National Reserve in Kenya

BIOLOGY.com Search Chapter 4 GO • Flash Cards

80

NATIONAL SCIENCE EDUCATION STANDARDS

UNIFYING CONCEPTS AND PROCESSES

I, II, III, IV, V

CONTENT

B.5, B.6, C.4.b, C.4.c, C.4.d, C.4.e, C.5.b, C.5.e, C.6.b, D.1, F.3, F.4, F.5

INQUIRY

A.1.c, A.1.d

UbD Connect to the Big Idea

Invite a volunteer to read aloud the chapter number and the chapter title. Read the Big Idea and question. Emphasize the question, **How do abiotic and biotic factors shape ecosystems?**

Set the Purpose

PREVIEW AND PREDICT Draw the graphic organizer on the board. Point out that *interdependence* means that living things depend on each other and the environment.

Ask What are the abiotic factors in an ecosystem? *(nonliving things)* What are biotic factors? *(living things)*

Guide students to the INSIDE feature on the next page, and invite a volunteer to read each lesson number and title. Discuss how the lesson titles relate to the questions in the graphic organizer. Remind students to search for answers to the questions as they read.

CHAPTER 4

INSIDE:

- 4.1 Climate
- 4.2 Niches and Community Interactions
- 4.3 Succession
- 4.4 Biomes
- 4.5 Aquatic Ecosystems

CHAPTER MYSTERY

THE WOLF EFFECT

During the 1920s, hunting and trapping eliminated wolves from Yellowstone National Park. For decades, ecologists hypothesized that the loss of wolves had changed the park ecosystem. Wolves are important predators of elk and other large grazing animals. But there were no before-and-after data, so it was impossible to test that hypothesis directly.

Then, in the mid-1990s, wolves were reintroduced to Yellowstone. Researchers watched park ecosystems carefully. Soon, the number of elk in parts of the park began to fall just as predicted. But, unpredictably, forest and stream communities have changed, too. Could a "wolf effect" be affecting organisms in the park's woods and streams?

Read for Mystery Clues As you read this chapter, look for connections among Yellowstone's organisms and their environment. Then, solve the mystery.

FOUNDATIONS for Learning

Before you read the chapter, write each vocabulary word on an index card. As you read the chapter, write down the definition of each term on the back of the card in your own words. You can put the cards in categories to help you remember the definitions. Make a second set of cards containing the different types of ecosystems. At the end of the chapter are two activities that use the cards to answer the question: How do abiotic and biotic factors shape ecosystems?

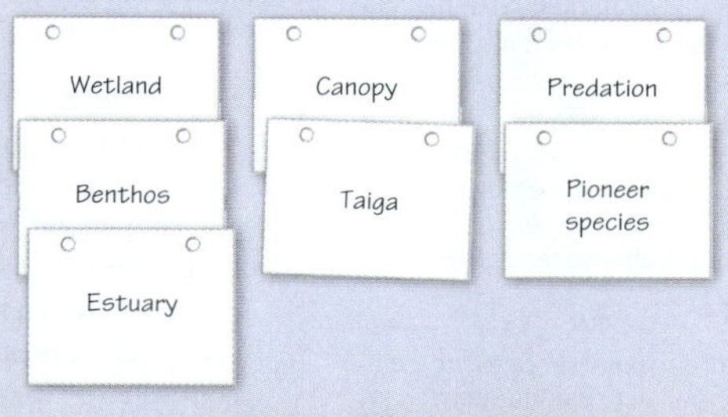

- Untamed Science Video
- Chapter Mystery

Chapter 4 Big Idea: Interdependence in Nature

Chapter 4: How do abiotic and biotic factors shape ecosystems?

- **4.1:** What factors affect global climate?
- **4.2:** How do organisms interact with one another?
- **4.3:** How do ecosystems change over time?
- **4.4:** What are the characteristics of the major biomes?
- **4.5:** What are the characteristics of aquatic ecosystems?

Foundations for Learning

Pass out the index cards. Help students understand why learning new words is important.

Ask Why do you think vocabulary is important? *(Knowing the definitions of words makes it possible to understand text that uses those words.)*

Ask Why do you think sorting vocabulary terms into categories can help you remember the definitions? *(Words that fall into the same category are related in some way. Knowing the meaning of one word can help you understand other words in the same category, even if you don't know their definitions.)*

Ask For example, if the word *benthos* is put in a category with *wetland* and *estuary,* what can you infer about *benthos*? *(that it is associated with aquatic ecosystems)*

Remind students that they will use these cards as a review and study aid. Familiarize students with the activity by modeling one term and its definition on an index card.

Untamed Science™ A volcanic eruption can quickly destroy an existing ecosystem, but it also makes room for a new one. In this short movie, **From Lava to Life,** the Untamed Science crew explores ecological succession.

CHAPTER MYSTERY **Ask** What do you know about wolves? *(Answers may refer to pack behavior, howling, hunting for prey, or wolves' close relation to domestic dogs.)*

Invite a volunteer to read aloud the Chapter Mystery. Have students predict how the reintroduction of wolves into Yellowstone might be affecting organisms living in the park's woods and streams.

Tell students that they will find clues in the chapter that will help them solve this mystery.

Digital Support

BIOLOGY.com Extend your reach by using the digital assets offered at **Biology.com.**

LESSON 4.1

Pre-Reading

Objectives

4.1.1 Differentiate between weather and climate.

4.1.2 Identify the factors that influence climate.

Preview the Pages

1. Preview the vocabulary terms in this lesson by having students find them in the text. Remind students that, as they read, they will be writing down the definition of each term in their own words, as directed in the Foundations for Learning activity on the previous page.
2. Preview the diagram The Greenhouse Effect. Encourage students to share what they know about greenhouses. Explain that the straight, thick arrows in the diagram represent light, and the squiggly, thin arrows represent heat.
3. Preview the visual Winds and Currents. Explain that these visuals show how heat from the sun is transported, or carried, from one area of Earth to other areas by winds and ocean currents. Review how to use keys to interpret the map and diagram.

Activate Prior Knowledge

Ask What are some words that you would use to describe the weather? *(Sample answers: warm, cold, rainy, windy, humid, cloudy)*

Ask How do you think Alaska's climate is different from Florida's climate? *(Sample answers: Alaska's climate is much colder than Florida's. Alaska gets more snow, but Florida gets more rain.)*

NATIONAL SCIENCE EDUCATION STANDARDS

UNIFYING CONCEPTS AND PROCESSES

I, II, III

CONTENT

B.5, B.6, D.1

For the corresponding lesson in *Miller & Levine Biology,* see pages 96–98.

4.1 Climate

Key Questions

- *What is climate?*
- *What factors determine global climate?*

BUILD Understanding

Preview Visuals Before you read, look at the Climate Zones diagram. What does it tell you about how the heat from the sun affects Earth's climates?

In Your Workbook Go to your workbook for more help with previewing visuals.

BUILD Vocabulary

climate
the average year-to-year conditions of temperature and precipitation in an area over a long period of time

microclimate
environmental conditions within a small area that differ significantly from the climate of the surrounding area

greenhouse effect
the process in which certain gases (carbon dioxide, methane, and water vapor) trap sunlight energy in Earth's atmosphere as heat

PREFIXES

The prefix *hemi-* in *hemisphere* means "half." The Northern Hemisphere is the northern half of Earth.

Weather and Climate

When you think of climate, you might think of major storms like Hurricane Katrina or a drought in the southeastern states. But, big storms and seasonal droughts are better described as weather instead of climate. So, what is climate and how is it different from weather? How do climate and weather affect organisms and ecosystems?

Weather and climate both involve variations in temperature, precipitation, and other environmental factors. Weather is the day-to-day condition of Earth's atmosphere. Weather where you live may be clear and sunny one day but rainy and cold the next. **Climate** refers to average conditions over long periods of time. A region's climate is defined by year-after-year patterns of temperature and precipitation.

However, environmental conditions can vary over small distances, creating **microclimates.** For example, south-facing sides of trees and buildings in the Northern Hemisphere receive more sunlight. They are often warmer and drier than north-facing sides. These differences can be very important to many organisms.

Key Question What is climate?
A region's climate is defined by year-after-year patterns of temperature and precipitation.

Factors That Affect Climate

What causes differences in climate? Global climate is shaped by many factors, including solar energy trapped in the biosphere. Latitude and the transport of heat by winds and ocean currents also help shape global climate.

Solar Energy and the Greenhouse Effect The main force that shapes our climate is solar energy that arrives as sunlight and strikes Earth's surface. Some solar energy is reflected back into space, and some is absorbed and converted into heat. Some of that heat radiates back into space, and some is trapped in the biosphere. The balance between heat that stays in the biosphere and heat lost to space determines Earth's average temperature. This balance is largely controlled by the concentrations of carbon dioxide, methane, and water vapor in the atmosphere. These gases are called greenhouse gases.

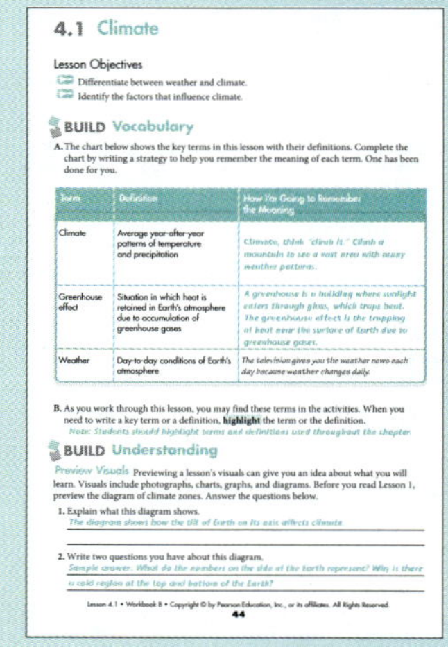

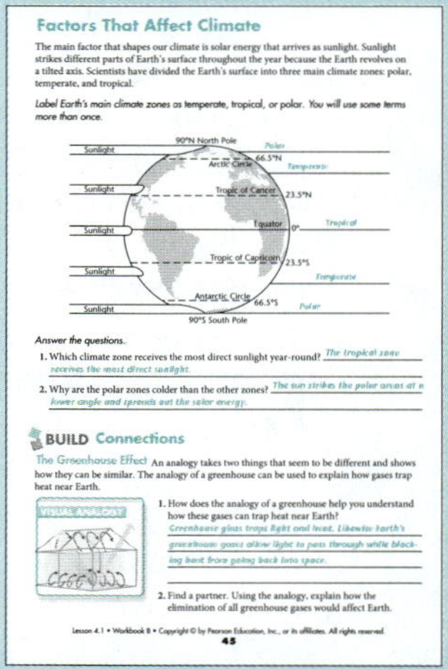

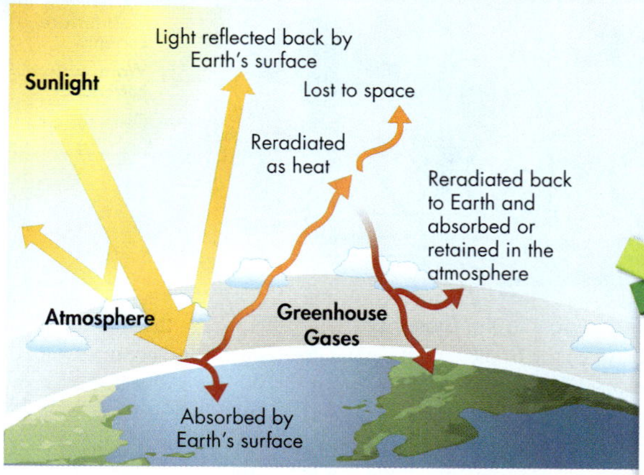

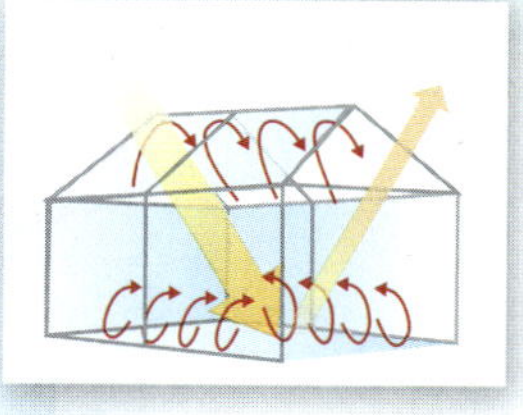

BUILD Connections

THE GREENHOUSE EFFECT

Greenhouse gases in the atmosphere allow solar radiation to enter the biosphere. But they slow down the radiation of Earth's heat back to space.

In Your Workbook Go to your workbook to explore the greenhouse analogy further.

Greenhouse gases act like the glass in a greenhouse. The glass allows visible light to enter but traps heat. This effect is called the **greenhouse effect.** If greenhouse gas concentrations rise, they trap more heat, warming Earth. If their concentrations fall, more heat escapes, cooling Earth. Without the greenhouse effect, Earth would be about 30° Celsius cooler than it is today.

Latitude and Solar Energy Near the equator, solar energy is intense because the sun is almost directly overhead at noon all year. But this same amount of solar energy is spread out over a much larger area near the poles than near the equator. This spreading is caused by the curvature of the Earth. So Earth's polar areas annually receive less intense solar energy, and therefore heat, from the sun. This difference in heat distribution creates three different climate zones: tropical, temperate, and polar.

The tropical zone (tropics) includes the areas near the equator. The tropical zone receives nearly direct sunlight all year. It is located between 23.5° north and 23.5° south latitudes. The two temperate zones are on either side of the tropical zone between 23.5° and 66.5° north and south latitudes. Beyond the temperate zones are the polar zones, between 66.5° and 90° north and south latitudes. Temperate and polar zones receive very different amounts of solar energy at different times of the year. This variation is because Earth's axis is tilted. The tilt causes solar radiation to strike the Earth at an angle. The angle varies from summer to winter as the Earth revolves around the sun. During winter in the temperate and polar zones, the sun is much lower in the sky. The days are also shorter, and solar energy is less intense.

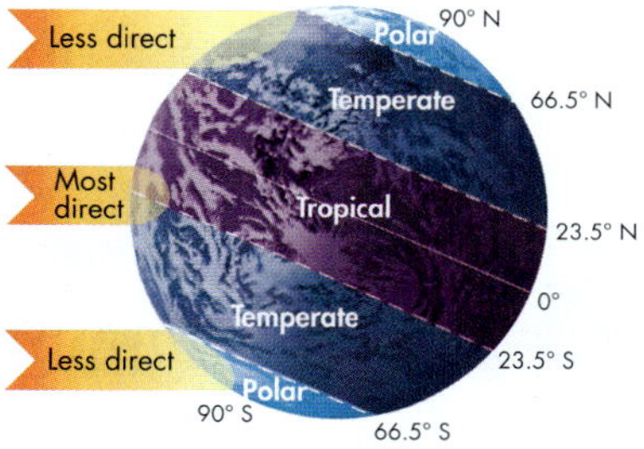

Climate Zones Earth's climate zones are produced by unequal distribution of the sun's heat on Earth's surface. Polar regions receive less solar energy per unit area than tropical regions. This means they receive less heat, too. The tilt of Earth's axis causes the distribution of sunlight to change over the course of the year.

BIOLOGY.com Search Lesson 4.1 GO • Visual Analogy

Hands-On Learning

TILT OF THE EARTH Provide student pairs with a flashlight, a 12-inch ruler, and a sheet of graph paper.

1. Have teams lay their graph paper on a flat surface. One student holds the ruler so that one end rests on the paper. Another turns on the flashlight and holds it 12 inches from the paper at a 90-degree angle. The other student counts the number of squares inside the beam of light and records the data.
2. Have each team repeat Step 1 holding the light at a 45-degree angle.
3. **Ask** What does the number of squares represent? *(amount of surface area over which a given amount of light energy is spread)* Discuss how this activity models how Earth's tilt affects the amount of solar energy received in temperate and polar regions.

Build Understanding

PREVIEW VISUALS Have students preview the Climate Zones diagram. Then, direct students to Workbook B and have them answer the questions about the diagram.

Ask Which climate zone is closest to Earth's equator? *(tropical)* Farthest? *(polar)* Which climate zone do you live in? *(Answers may vary. Most of the United States falls within the temperate zone.)*

During Reading

Build Vocabulary

PREFIXES Point out that the prefix *micro-* means "small."

Ask What is it about a microclimate that is small? *(the area it applies to)*

Active Reading

KEY QUESTION Invite a volunteer to read aloud the question and answer from the previous page.

Say Last month it rained a lot in Seattle, Washington.

Ask Is this a description of Seattle's weather or Seattle's climate? *(weather)*

Build Connections

VISUAL ANALOGY Explain that this visual analogy compares the glass in a greenhouse to greenhouse gases in the atmosphere.

Ask What traps heat in Earth's biosphere? *(greenhouse gases in the atmosphere)*

Ask What traps heat in a greenhouse? *(the glass walls and ceiling of the greenhouse)*

USE VISUALS Help students understand the Climate Zones diagram.

Ask What do the yellow bars represent? *(solar energy streaming toward Earth)*

Explain that each bar represents an equal amount of solar energy. Help students see, however, that the same amount of energy gets spread out over a larger surface area near the poles than near the equator.

Speed Bump

Ask What causes Earth's three types of climate zones? *(uneven distribution of heat caused by the curvature of the Earth)*

Ask Why are there cooler and warmer seasons in temperate regions? *(Earth's axis is tilted, which causes solar radiation to hit Earth at an angle. The angle varies from summer to winter.)*

LESSON 4.1

During Reading continued

Active Reading

USE VISUALS Help students use the Winds and Currents visuals to determine the directions winds and currents move.

Ask Do southeast trade winds move toward the west or toward the east? *(west)*

Ask In what direction do cold surface currents in the Northern Hemisphere generally move? *(southward toward the equator)* In what direction do warm surface currents in the Northern Hemisphere generally move? *(northward toward the polar region)*

Explain that these currents redistribute heat in the Northern Hemisphere, moving warm water from the tropical region north toward the polar region.

KEY QUESTION Invite a volunteer to read aloud the Key Question and answer. Then, have students find the approximate location of New York City on the Ocean Currents map.

Ask What are two factors that affect the climate in New York City? *(Sample answer: New York's latitude, and the warm ocean current that flows along the east coast of the United States)*

Speed Bump

Ask How does solar energy cause wind? *(It heats an area of Earth's surface, which then heats the air above it, causing the air to rise. Cool air moves in to take the place of the rising warm air. The movement of the air is wind.)*

Ask What causes most ocean currents? *(wind)*

Post Reading

ASSESSMENT TIPS Advise students to review the highlighted vocabulary terms and their definitions before they begin the assessment.

For Question 4, remind students how to set up *x* and *y* axes for graphing data.

Wrap-Up Activity

USE DIAGRAMS Have students use colored pencils to draw a diagram showing how the sun's energy causes wind. Instruct them to use yellow arrows to show sunlight reaching Earth's surface, red to show heated air rising, and blue to show cooler air sinking.

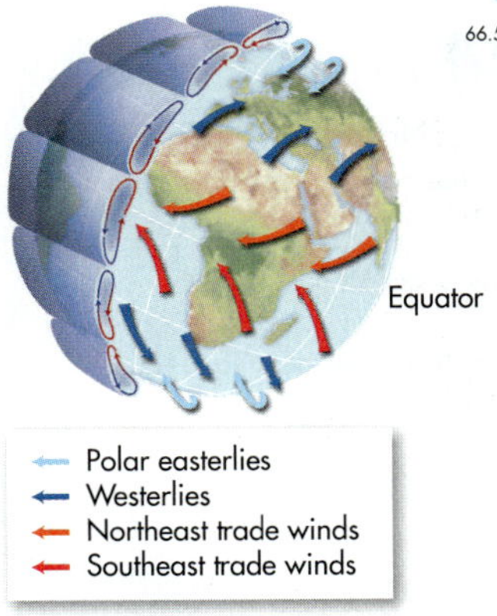

Winds and Currents Earth's winds (above left) and ocean currents (above right) interact to help produce climate patterns. The paths of winds and currents are the result of heating and cooling, Earth's rotation, and geographic features.

Heat Transport in the Biosphere The difference in the amount of heat between the tropics and the poles creates wind and ocean currents. These currents transport heat and moisture. When air is heated in warm areas such as the tropics, it expands. This warm air is less dense, so it rises. After it rises, it spreads north and south. As the air spreads, it cools and sinks. At the same time, chilled air at the poles sinks toward Earth's surface. It pushes air at the surface outward. This air warms as it travels over the surface, so it rises. These upward and downward movements of air create winds. Winds transport heat from regions of rising warmer air to regions of sinking cooler air. Earth's rotation causes winds to blow generally from west to east over the temperate zones. Winds blow from east to west over the tropics and the poles.

Similar patterns of heating and cooling occur in the oceans. Winds push surface water, causing ocean currents. These currents transport enormous amounts of heat. Air that passes over warm currents picks up moisture and heat. Air that passes over cool currents is cooled. In this way, surface currents affect the weather and climate of nearby landmasses. Deep ocean currents are caused by cold water near the poles sinking and flowing along the ocean floor. This water rises in warmer regions through a process called upwelling.

Key Question What factors determine global climate?
Global climate is shaped by many factors, including solar energy trapped in the biosphere. Latitude and the transport of heat by winds and ocean currents also shape global climate.

CHECK Understanding

Apply Vocabulary

Use the highlighted words from the lesson to complete each sentence correctly.

1. The _______ is caused by gases in the atmosphere trapping heat, causing the Earth to warm.
2. A _______ develops when environmental conditions vary over small distances.

Critical Thinking

3. **Explain** How is climate different from weather?
4. **Quick Write** Research the average monthly high temperature (in °C) for Quito, Ecuador. Quito is a city on the equator. Plot the temperature data in a line graph. Describe the pattern shown.

BIOLOGY.com Search Lesson 4.1 GO • Lesson Assessment

Check Understanding

1. greenhouse effect
2. microclimate
3. Climate refers to the average conditions of temperature and precipitation in an area over a long time. Weather refers to day-to-day variations.
4. Graphs should show that average monthly temperatures are very similar from month to month for the entire year, with an average temperature of about 15°C.

4.2 Niches and Community Interactions

The Niche

Different organisms live in different places. For example, one organism may live on a coral reef and another in the desert. What determines where an organism can live? Each species has a range of conditions under which it can grow and reproduce. These conditions help define where and how an organism lives.

Tolerance Every species has its own range of tolerance. Tolerance is the ability to survive and reproduce under a range of environmental conditions. When an environmental condition is outside of an organism's optimum range, the organism becomes stressed. It uses more energy to maintain homeostasis, and less for growth and reproduction. A species has an upper and lower limit of tolerance for every environmental factor. It cannot survive beyond those limits. A species' tolerance for environmental conditions helps determine its **habitat.** A habitat is the general place where an organism lives.

Defining the Niche Describing a species' habitat is important. But ecologists also study a species' ecological "job"—where and how it "makes a living." An organism's job is part of its **niche** (nitch). A niche describes what an organism does and how it interacts with the environment. A niche includes the way a species obtains **resources** to survive and reproduce. Resources include any necessity of life, such as water, nutrients, light, food, and space.

Key Question What is a niche?
A niche is the range of physical and biological conditions in which a species lives. It includes the way the species obtains what it needs to survive and reproduce.

Competition

Organisms often try to use a limited resource in the same place at the same time as other organisms. This process is called competition. Plant roots compete for water and nutrients in the soil. Animals compete for food, mates, and places to live and raise their young. Competition can occur among members of the same species (intraspecific competition) and between members of different species (interspecific competition).

Key Questions

- *What is a niche?*
- *How does competition shape communities?*
- *How do predation and herbivory shape communities?*
- *What are the three primary ways that organisms depend on each other?*

BUILD Understanding

Concept Map Use the highlighted vocabulary words to create a concept map that organizes the information in this lesson.

In Your Workbook Go to your workbook for help completing a concept map. Complete the concept map in Lesson 4.2.

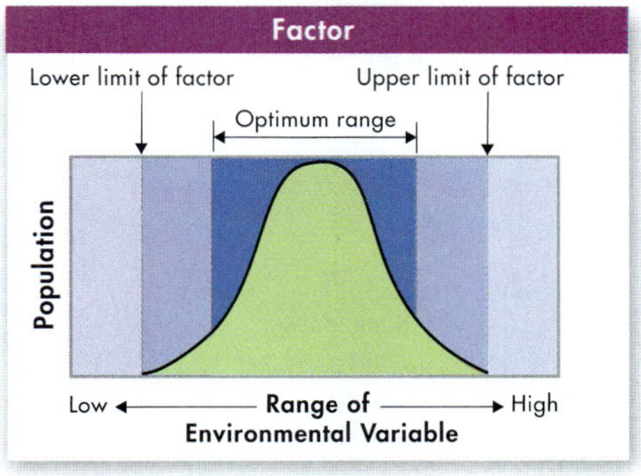

Tolerance A single environmental factor, such as sunlight or temperature, can affect population size. Organisms become more rare in zones of physiological stress (medium blue) and are absent from zones of intolerance (light blue).

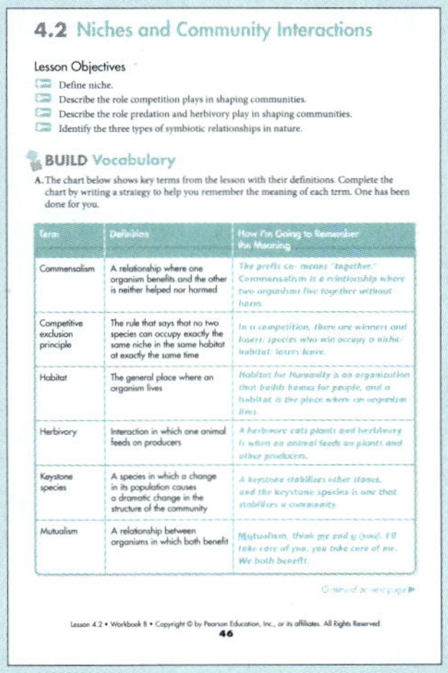

4.2 Niches and Community Interactions

Lesson Objectives
- Define niche.
- Describe the role competition plays in shaping communities.
- Describe the role predation and herbivory play in shaping communities.
- Identify the three types of symbiotic relationships in nature.

BUILD Vocabulary

A. The chart below shows key terms from the lesson with their definitions. Complete the chart by writing a strategy to help you remember the meaning of each term. One has been done for you.

Term	Definition	How I'm Going to Remember the Meaning
Commensalism	A relationship where one organism benefits and the other is neither helped nor harmed	*The prefix co- means "together." Commensalism is a relationship where two organisms live together without harm.*
Competitive exclusion principle	The rule that says that no two species can occupy exactly the same niche in the same habitat at exactly the same time	*In a competition, there are winners and losers; species who win occupy a niche/habitat; losers leave.*
Habitat	The general place where an organism lives	*Habitat for Humanity is an organization that builds homes for people, and a habitat is the place where an organism lives.*
Herbivory	Interaction in which one animal feeds on producers	*A herbivore eats plants and herbivory is when an animal feeds on plants and other producers.*
Keystone species	A species in which a change in its population causes a dramatic change in the structure of the community	*A keystone stabilizes other stones, and the keystone species is one that stabilizes a community.*
Mutualism	A relationship between organisms in which both benefit	*Mutualism, think me and u (you). I'll take care of you, you take care of me. We both benefit.*

Continued on next page ▶

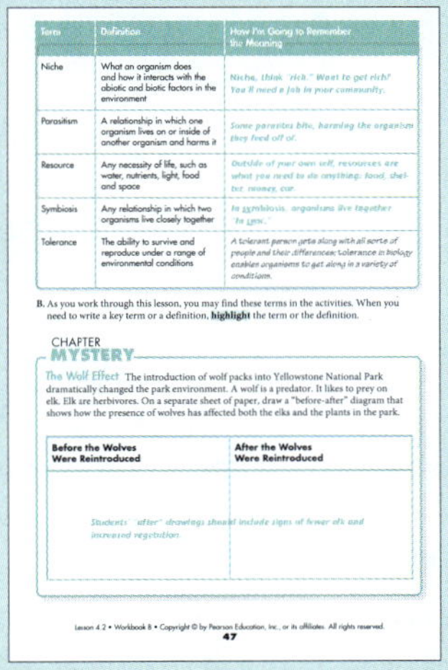

Term	Definition	How I'm Going to Remember the Meaning
Niche	What an organism does and how it interacts with the abiotic and biotic factors in the environment	*Niche, think "rich." Want to get rich? You'll need a job in your community.*
Parasitism	A relationship in which one organism lives on or inside of another organism and harms it	*Some parasites bite, harming the organism they feed off of.*
Resource	Any necessity of life, such as water, nutrients, light, food and space	*Outside of your own self, resources are what you need to do anything: food, shelter, money, car.*
Symbiosis	Any relationship in which two organisms live closely together	*In symbiosis, organisms live together "in sync."*
Tolerance	The ability to survive and reproduce under a range of environmental conditions	*A tolerant person gets along with all sorts of people and their differences; tolerance in biology enables organisms to get along in a variety of conditions.*

B. As you work through this lesson, you may find these terms in the activities. When you need to write a key term or a definition, **highlight** the term or the definition.

CHAPTER MYSTERY

The Wolf Effect The introduction of wolf packs into Yellowstone National Park dramatically changed the park environment. A wolf is a predator. It likes to prey on elk. Elk are herbivores. On a separate sheet of paper, draw a "before-after" diagram that shows how the presence of wolves has affected both the elks and the plants in the park.

Before the Wolves Were Reintroduced	After the Wolves Were Reintroduced
Students' "after" drawings should include signs of fewer elk and increased vegetation.	

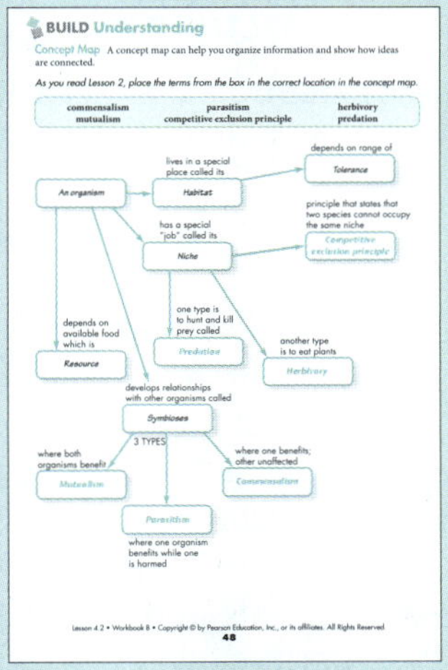

BUILD Understanding

Concept Map A concept map can help you organize information and show how ideas are connected.

As you read Lesson 2, place the terms from the box in the correct location in the concept map.

commensalism	parasitism	herbivory
mutualism	competitive exclusion principle	predation

LESSON 4.2

Pre-Reading

Objectives

4.2.1 Define *niche.*

4.2.2 Describe the role competition plays in shaping communities.

4.2.3 Describe the role predation and herbivory play in shaping communities.

4.2.4 Identify the three types of symbiotic relationships in nature.

Preview the Pages

1. Preview the vocabulary terms in this lesson by having students locate the words in the text and say them aloud.
2. Preview the Tolerance graph. Use an example, such as amount of sunlight falling on a population of plants, to explain the concept of optimum range. Below the optimum range there is too little light for most plants to grow. Above the optimum there is too much sunlight, and plants may die. The plant population is largest when the amount of sunlight is in the optimum range.

Build Understanding

CONCEPT MAP Direct students to Workbook B. Encourage students to complete the concept map as they read the lesson.

KEY QUESTION Read aloud the Key Question and answer. Then choose a familiar organism, such as a cactus, and work as a class to define its niche. Prompt with questions such as: What type of soil does a cactus live in? *(sandy soil)* Does a cactus serve as a food source for other organisms? *(yes)*

NATIONAL SCIENCE EDUCATION STANDARDS

UNIFYING CONCEPTS AND PROCESSES

I, II, III

CONTENT

C.4.b, C.4.c, C.4.d, C.5.e

For the corresponding lesson in *Miller & Levine Biology,* see pages 99–104.

During Reading

Active Reading

DRAW A DIAGRAM Have a volunteer read aloud the Dividing Resources section. Then, have students draw a simple diagram of a tree that shows the three different areas inhabited by the three bird species discussed.

KEY QUESTION Read aloud the first Key Question and answer.

Ask Why can't two species share the same niche in the same habitat? *(If two species try to share the same niche, competing for the same resources, one species will always win.)*

INTERPRET GRAPHS Guide students as they interpret the Competitive Exclusion graph. Review that a paramecium is a single-celled organism common in freshwater ecosystems. Explain that the graph shows changes in the populations of two paramecium species being grown in cultures.

Ask Which line represents the *P. caudatum* population when *P. aurelia* is not present? *(blue dotted line)* When the two species are grown in the same culture, which one excludes, or outcompetes, the other? *(P. aurelia outcompetes P. caudatum.)*

Ask Can you infer why the population of *P. caudatum* grows for the first few days before starting to drop? *(For the first few days, there are enough resources in the culture for both species to survive.)*

Build Vocabulary

ROOT WORDS Challenge students to name other words that have the root *herb*. *(Sample answers: herbal, herbicide, herbaceous)* Have students define each word with the help of a dictionary.

KEY QUESTION Read aloud the second Key Question and answer.

Ask Which vocabulary term on this page describes the relationship between sea otters and sea urchins? *(predation)* Which term describes the relationship between sea urchins and kelp? *(herbivory)*

Ask How does a large sea urchin population affect a sea otter population? *(A large sea urchin population supports a large sea otter population.)*

Speed Bump

Ask What are some examples of predator–prey relationships? *(Sample answers: cat–mouse, hawk–snake, spider–insect)* What are some examples of herbivore–plant relationships? *(Sample answers: mice–seeds, butterflies–flower nectar, rabbits–weeds)*

BUILD Vocabulary

habitat an area where an organism lives, including the biotic and abiotic factors that affect it

niche the full range of physical and biological conditions in which an organism lives and the way in which the organism uses those conditions

resource any necessity of life, such as water, nutrients, light, food, or space

predation an interaction in which one animal (the predator) captures and feeds on another animal (the prey)

herbivory an interaction in which one animal (the herbivore) feeds on producers (such as plants)

ROOT WORDS

The word *herb* means "a flowering plant without woody stems." When used as a root word, it usually refers to the leaves and flowers of plants. Animals that practice herbivory often prefer to eat leaves and flowers, but they may eat other plant parts as well.

Competitive Exclusion

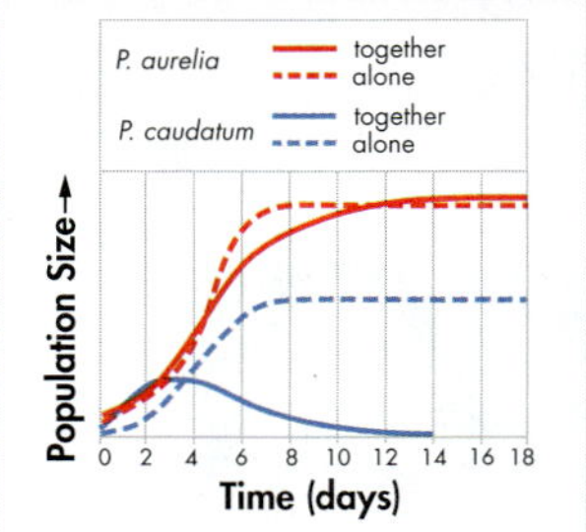

Competitive Exclusion This graph shows what happens when two species of paramecia compete for the same resources. In separate cultures, the two species easily survive. But in the same culture, one species outcompetes the other. The losing species dies off.

The Competitive Exclusion Principle No two species can occupy the same niche in the same habitat at the same time. This fact is an ecological rule called the competitive exclusion principle. Direct competition between different species for a limited resource almost always produces a winner and a loser. The losing species dies out.

Dividing Resources Species usually divide similar resources instead of competing for them. For example, three species of warblers can all live in the same trees and feed on insects. One species feeds on high branches. Another feeds on low branches. The third feeds in the middle. In this way, each species has its own niche.

Key Question How does competition shape communities? **Competition causes species to divide resources. It helps determine the number and kinds of species in a community. It also determines the niche each species occupies.**

Predation, Herbivory, and Keystone Species

What happens if a group of animals eats all the available food in the area? They will no longer have anything to eat! That's why predator-prey and herbivore-plant interactions are very important.

Predator-Prey Relationships **Predation** (pree DAY shun) is an interaction in which one animal (predator) captures and feeds on another animal (prey). Predators can affect the size of prey populations in a community. For example, owls are predators that help regulate the populations of mice and other small mammals.

Herbivore-Plant Relationships **Herbivory** is an interaction in which one animal (herbivore) feeds on producers (such as plants). Herbivores can affect the size and distribution of plant populations in a community. For example, dense populations of white-tailed deer are eliminating some plants from many places in the United States.

Keystone Species Sometimes a population change in a single species can cause dramatic changes in its community. This species is called a keystone species. Pacific sea otters are a keystone species. These otters eat large numbers of sea urchins. Sea urchins are herbivores. Their favorite food is kelp, giant algae that grow in the ocean. A century ago, sea otters were nearly eliminated by hunting. Sea urchin numbers then skyrocketed, and the urchins devoured the kelp. Once the sea otters were protected as an endangered species, the otter population recovered. The sea urchin populations dropped, and the kelp thrived again.

Key Question How do predation and herbivory shape communities? **Predators can affect where prey populations can survive. Herbivores can help determine where populations of certain plants can survive.**

ELL Focus on ELL: Extend Language

INTERMEDIATE SPEAKERS Review how the suffix *–ion* or *–ition* is used to change verbs into nouns. Write the words *compete* and *competition* on the board. Ask a volunteer to identify which is a verb and which is a noun and to define the two words. Explain that the suffix *–ion* (or *–ition*) means "the act of," "the process of," or "result of."

Say Competition is the act of competing.

Have students look for other words on this page that use the suffix *–ion* or *–ition.* Make a list of these words on the board *(exclusion, predation, interaction, population, distribution),* along with the verb form of each word *(exclude, predate, interact, populate, distribute).* Point out that the noun forms of these words are formed in different ways. For example, the *d* in *exclude* is changed to an *s* before adding the suffix.

Have students work in pairs to define each of the words on the board. Then have them practice using suffixes to form nouns from the verbs *exhibit, protect,* and *divide.* *(exhibition, protection, division)*

LESSON 4.2

Symbioses

Any relationship in which two species live closely together is called **symbiosis** (sim by OH sis). Biologists recognize three main classes of symbiotic relationships in nature: mutualism, parasitism, and commensalism.

Mutualism The sea anemone uses its sting to capture prey and to protect itself from predators. But clownfish are immune to anemone stings. They hide from predators in the anemone's deadly tentacles. When an anemone is attacked, the clownfish fiercely chase away the much larger fish. This relationship is an example of **mutualism,** because both animals benefit from the relationship.

Parasitism Ticks live on the bodies of mammals, feeding on their blood and skin. This is an example of **parasitism** (PAR uh sit iz um). In this relationship, an organism lives inside or on another organism and harms it. The parasite obtains all or part of its nutrients from the host organism. Parasites weaken but usually do not kill their host.

Commensalism Small marine animals called barnacles often attach themselves to a whale's skin. The barnacles have no effect on the whale. They benefit from the constant movement of water and food particles past the swimming whale. This is an example of **commensalism** (kuh MEN sul iz um). In this relationship, one organism benefits and the other is neither helped nor harmed.

Key Question What are the three primary ways that organisms depend on each other? **The three main classes of symbiotic relationships are mutualism, parasitism, and commensalism.**

BUILD Vocabulary

symbiosis
a relationship in which two species live closely together

mutualism
the symbiotic relationship in which both species benefit from the relationship

parasitism
the symbiotic relationship in which one organism lives on or inside another organism and harms it

commensalism
the symbiotic relationship in which one organisms benefits and the other is neither helped nor harmed

ROOT WORDS
The word *commensal* means "eating together at the same table." So organisms that practice commensalism could be said to be dinner companions.

CHECK Understanding

Apply Vocabulary

Use the highlighted words from the lesson to complete each sentence correctly.

1. A ________ is something that is essential for an organism's life.
2. A relationship in which the interaction between organisms is helpful to both is called ________.

Critical Thinking

3. **Compare and Contrast** What is the difference between a predator and a parasite?
4. **Write to Learn** Answer the first clue of the mystery.

MYSTERY CLUE

One of the favorite prey species of the wolves in Yellowstone is elk. How do you think this relationship could affect the ability of certain *plants* to grow in Yellowstone? (Hint: See p. 86.)

BIOLOGY.com Search Lesson 4.2 GO • Lesson Assessment

Check Understanding

1. resource
2. mutualism
3. A predator kills its prey, but a parasite can obtain nutrition by weakening rather than killing its host. Parasites are small and usually live on or inside much larger hosts. Predators live independently from their prey and may be larger or smaller in size.
4. Wolves help keep the elk population in check. A decrease in the wolf population would result in an increase in the elk population and could cause some plant populations to be stressed by overgrazing, which would limit plant growth.

Active Reading

MAKE INFERENCES Help students think about why a parasite does not usually kill its host.

Ask What does the parasite get from the host? *(The parasite gets all or most of its food from the host. Some students may also recognize that the parasite is sheltered by the host.)*

Ask What would happen to the parasite if the host were killed? *(If the parasite were not able to find another host quickly, the parasite would also die.)*

Build Vocabulary

ROOT WORDS Tell students that *symbios,* the root word of *symbiosis*, means "living with." So symbiotic organisms could be said to be roommates.

KEY QUESTION Read aloud the question and the answer.

Ask What does *symbiotic* mean? *(living in symbiosis)*

Speed Bump

Ask What kind of symbiotic relationship does a visually impaired person have with a seeing-eye dog? *(mutualism)*

Ask What is the difference between parasitism and commensalism? *(The host is harmed in parasitism, but not in commensalism.)*

Post Reading

ASSESSMENT TIPS For Question 3, suggest that students think of specific examples of predators *(lions, sharks)* and parasites *(mosquitoes, ticks)* as they write their answers.

MYSTERY CLUE After students read the clue, remind them that elk are herbivores. Have them reread the paragraph on Herbivore-Plant Relationships before answering. Students should consider how a change in the population of herbivores can affect the population of plants.

Wrap-Up Activity

MAKE A CHART Have students work in pairs to create a chart comparing the different types of relationships between organisms discussed in this lesson. The chart should include the following relationships: predator–prey, herbivore–plant, mutualism, parasitism, and commensalism. Encourage students to include drawings in the chart that illustrate each type of relationship.

LESSON 4.3

Pre-Reading

Objectives

4.3.1 Describe how ecosystems recover from a disturbance.

4.3.2 Compare succession after a natural disturbance with succession after a human-caused disturbance.

Preview the Pages

1. Have students find each vocabulary term in the text and read it aloud. Ask students where they may have heard any of these words before.
2. Preview the drawings showing primary and secondary succession. Point out the timelines, and make sure students understand that both show changes in an environment over time. Discuss the differences in the timelines and in the beginning stages of each drawing.

Activate Prior Knowledge

Ask What kinds of natural disasters have you heard about? What happened to plants and animals as a result of those disasters? *(Sample answers: tornado knocked down trees in a woodland, which destroyed nesting habitat for birds; hurricane ripped up trees in wetlands, which caused flooding in other areas.)*

Build Understanding

COMPARE/CONTRAST TABLE Read aloud the Build Understanding box. Direct students to Workbook B, where they will be able to work on their compare/contrast tables. Point out that the table will help students organize the information in the lesson.

NATIONAL SCIENCE EDUCATION STANDARDS

UNIFYING CONCEPTS AND PROCESSES

I, II, III, IV

CONTENT

C.4.c, C.4.e, F.5

For the corresponding lesson in *Miller & Levine Biology,* see pages 106–109.

4.3 Succession

Key Questions

How do communities change over time?

Do ecosystems return to "normal" following a disturbance?

BUILD Understanding

Compare/Contrast Table As you read, create a table comparing primary and secondary succession.

In Your Workbook Go to your workbook to learn more about creating a compare/contrast table.

Primary and Secondary Succession

In 1883, an eruption blew the volcanic island of Krakatau in the Indian Ocean to pieces. A tiny barren island remained. By 1929, a forest with over 300 species covered the island. Today it is a mature rain forest. Krakatau is an example of **ecological succession.** Succession is a series of more-or-less predictable changes that occur in a community over time. As succession proceeds, the number of different species present typically increases.

Primary Succession Volcanic explosions and retreating glaciers can create new land or sterilize existing areas. Both events leave only exposed bare rock—no soil or plant life. Succession that begins in an area that has no existing community is called **primary succession.**

The first species to colonize barren areas are called **pioneer species.** One pioneer that grows on bare rock is lichen. Lichen is an example of mutualism between a fungus and an alga. Lichens fix atmospheric nitrogen into useful forms for other organisms. Over time, they break down rock and add organic material to form soil. Some grasses are also pioneer species.

Secondary Succession Existing communities are not always completely destroyed by disturbances. In these cases, **secondary succession** occurs. Secondary succession proceeds faster than primary succession. The soil remains, so new and surviving vegetation regrows rapidly.

Primary Succession Primary succession occurs on newly exposed surfaces. In Glacier Bay, Alaska, a retreating glacier exposed barren rock. Over the course of more than 100 years, a series of changes has led to the hemlock and spruce forest currently found in the area. Changes in this community will continue for centuries.

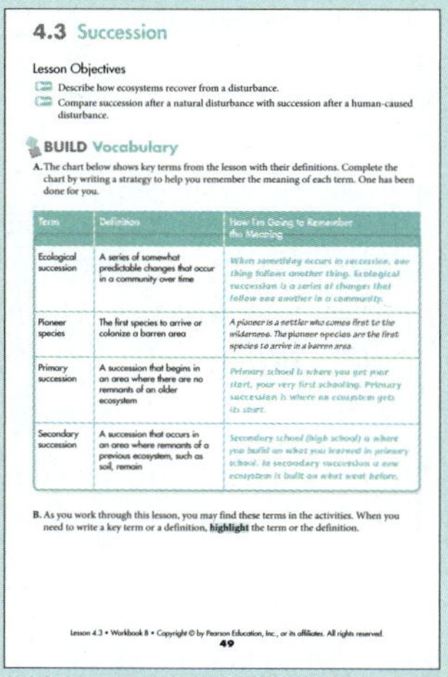

4.3 Succession

Lesson Objectives

- Describe how ecosystems recover from a disturbance.
- Compare succession after a natural disturbance with succession after a human-caused disturbance.

BUILD Vocabulary

A. The chart below shows key terms from the lesson with their definitions. Complete the chart by writing a strategy to help you remember the meaning of each term. One has been done for you.

Term	Definition	How I'm Going to Remember the Meaning
Ecological succession	A series of somewhat predictable changes that occur in a community over time	*When something occurs in succession, one thing follows another thing. Ecological succession is a series of changes that follow one another in a community.*
Pioneer species	The first species to arrive or colonize a barren area	*A pioneer is a settler who comes first to the wilderness. The pioneer species are the first species to arrive in a barren area.*
Primary succession	A succession that begins in an area where there are no remnants of an older ecosystem	*Primary school is where you get your start, your very first schooling. Primary succession is where an ecosystem gets its start.*
Secondary succession	A succession that occurs in an area where remnants of a previous ecosystem, such as soil, remain	*Secondary school (high school) is where you build on what you learned in primary school. In secondary succession a new ecosystem is built on what went before.*

B. As you work through this lesson, you may find these terms in the activities. When you need to write a key term or a definition, **highlight** the term or the definition.

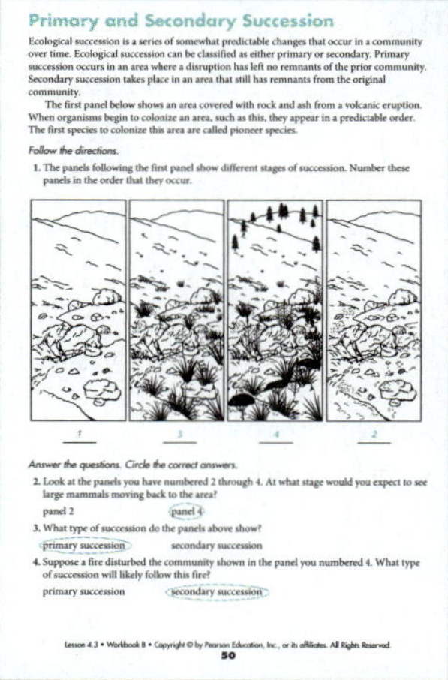

Primary and Secondary Succession

Ecological succession is a series of somewhat predictable changes that occur in a community over time. Ecological succession can be classified as either primary or secondary. Primary succession occurs in an area where a disruption has left no remnants of the prior community. Secondary succession takes place in an area that still has remnants from the original community.

The first panel below shows an area covered with rock and ash from a volcanic eruption. When organisms begin to colonize an area, such as this, they appear in a predictable order. The first species to colonize this area are called pioneer species.

Follow the directions.

1. The panels following the first panel show different stages of succession. Number these panels in the order that they occur.

Answer the questions. Circle the correct answers.

2. Look at the panels you have numbered 2 through 4. At what stage would you expect to see large mammals moving back to the area?
   panel 2 — panel 4
3. What type of succession do the panels above show?
   primary succession — secondary succession
4. Suppose a fire disturbed the community shown in the panel you numbered 4. What type of succession will likely follow this fire?
   primary succession — secondary succession

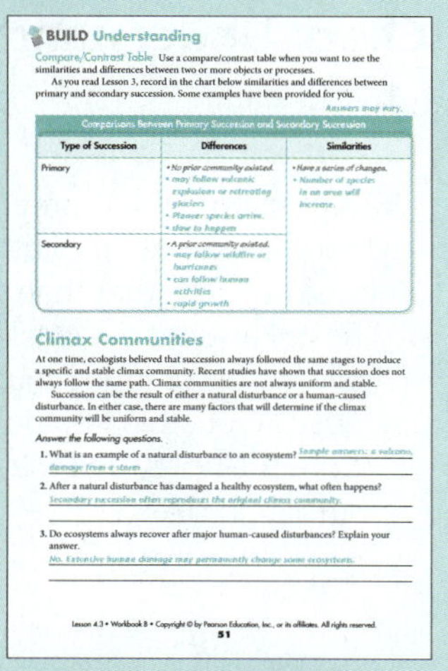

BUILD Understanding

Compare/Contrast Table Use a compare/contrast table when you want to see the similarities and differences between two or more objects or processes.

As you read Lesson 3, record in the chart below similarities and differences between primary and secondary succession. Some examples have been provided for you.

Answers may vary.

Comparisons Between Primary Succession and Secondary Succession

Type of Succession	Differences	Similarities
Primary	• No prior community existed. • may follow volcanic explosions or retreating glaciers • Pioneer species arrive. • slow to happen	• Have a series of changes. • Number of species in an area will increase.
Secondary	• A prior community existed. • may follow wildfire or hurricanes • can follow human activities • rapid growth	

Climax Communities

At one time, ecologists believed that succession always followed the same stages to produce a specific and stable climax community. Recent studies have shown that succession does not always follow the same path. Climax communities are not always uniform and stable.

Succession can be the result of either a natural disturbance or a human-caused disturbance. In either case, there are many factors that will determine if the climax community will be uniform and stable.

Answer the following questions.

1. What is an example of a natural disturbance to an ecosystem? *Sample answers: a volcano, damage from a storm*
2. After a natural disturbance has damaged a healthy ecosystem, what often happens? *Secondary succession often reproduces the original climax community.*
3. Do ecosystems always recover after major human-caused disturbances? Explain your answer. *No. Extensive human damage may permanently change some ecosystems.*

Secondary succession often follows a wildfire, hurricane, or other natural disturbance. To us, these events are disasters, but many species are adapted to them. Forest fires kill some trees, but spare others—and may stimulate their seeds to germinate. Secondary succession can also follow human activities like logging and farming.

Why Succession Occurs Every organism changes the environment it lives in. In one model of succession, as one species alters its environment, other species find it easier to compete for resources and survive. Lichens add organic matter and form soil. Then mosses and other plants are able to colonize and grow. Organic matter accumulates and soon other species move in. Tree branches produce shade and cooler temperatures nearer the ground. Over time, more and more species can find suitable niches and survive.

Key Question How do communities change over time? **Ecosystems change over time, especially after disturbances. As some species die out, new species move in.**

Climax Communities

Ecologists used to think that succession in a given area always proceeded through the same stages. The end result was a specific and stable community called a climax community. But recent studies have shown that succession doesn't always follow the same path. Climax communities are not always uniform and stable.

Succession After Natural Disturbances Natural disturbances are common in many communities. Healthy coral reefs and tropical rain forests recover from storms. Healthy temperate forests and grasslands recover from wildfires. Sometimes succession in a healthy ecosystem reproduces the original climax community after a natural disturbance. But multiple disturbances take place at different times in some ecosystems. This causes climax communities that are not uniform. They look like patchwork quilts because different areas are in varying stages of succession. Some climax communities are disturbed so often that they can't really be called stable.

BUILD Vocabulary

ecological succession
a series of gradual changes that occur in a community following a disturbance

primary succession
succession that occurs in an area in which no trace of a previous community is present

pioneer species
the first species to populate an area during succession

secondary succession
a type of succession that occurs in an area that was only partially destroyed by disturbances

WORD ORIGINS
The origin of the word *succession* is the Latin word *succedere*, meaning "to come after." Ecological succession involves changes that occur one after the other as species move into and out of a community.

Secondary Succession Secondary succession occurs in disturbed areas where parts of previous ecosystems still remain. This series shows changes taking place in abandoned fields in the Carolinas. Over the last century, the fields have passed through several stages and matured into oak forests. Changes will continue for years to come.

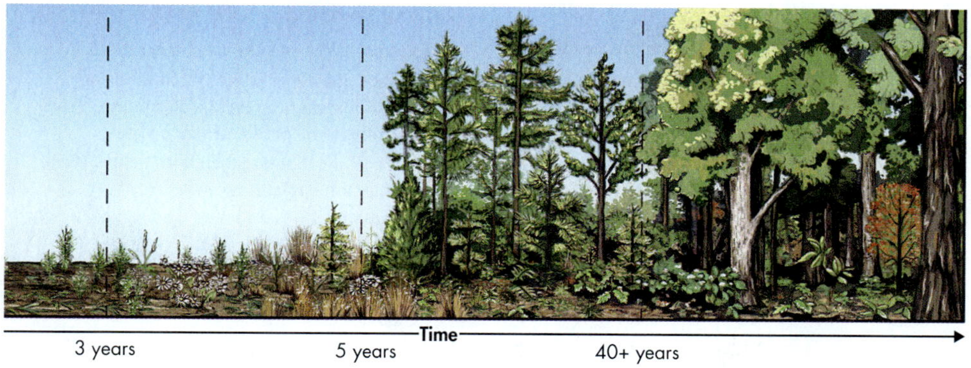

ELL Focus on ELL: Build Background

ALL SPEAKERS Many students may be familiar with the concept of being a pioneer. Help students use their understanding of this concept to better understand the role that pioneer species play in primary succession.

Invite volunteers to share what they know about the pioneers who settled the western United States.

Ask What did these pioneers do? *(They traveled west in covered wagons, bringing few possessions. They built homes in areas where no person from their culture had settled before. They cultivated the land. Gradually, more people followed them, and towns and cities grew up around them.)*

Ask How is this similar to the role that pioneer species play in primary succession? *(Pioneer species are the first living things to colonize new land. They change the environment, making it more suitable for other living things. Gradually, other species colonize the area as well.)*

During Reading

Build Vocabulary

WORD ORIGINS Explain to students that the origin of the word *primary* is the Latin word *primus*, meaning "first."

Say Primary succession occurs in areas where there is no trace of a former community. Primary succession involves changes that occur when the first organisms begin to live in an area.

Ask What other terms can you think of that use the word *primary*? *(Sample answers: primary school, primary elections, primary colors)*

Active Reading

USE DIAGRAMS Direct students' attention to the Primary Succession and Secondary Succession visuals. Have them compare the visuals' timelines.

Ask Did the primary or the secondary succession proceed faster? *(secondary)* Why? *(because parts of the previous ecosystem still remained in the disturbed area)*

Ask In the example of primary succession, how long did it take for young trees to appear? *(between 35 and 80 years)*

KEY QUESTION Read aloud the Key Question and answer.

Ask Do all of the same species remain in a community that is going through ecological succession? Why or why not? *(No; because conditions keep changing, the habitat keeps changing. With each change, it becomes suitable for different species.)*

Speed Bump

Ask What often happens to the number of species in a community as succession progresses? *(The number of species increases.)*

Ask What causes succession? *(As organisms change the environment they live in, other organisms are able to compete and survive.)*

During Reading continued

Active Reading

MAKE CONNECTIONS Help students realize that they might have firsthand experience with examples of secondary succession after human-caused disturbances.

Say Perhaps you've seen small plants growing in cracked pavement or on land where a building has been torn down.

Ask What kinds of plants and animals did you notice? *(Sample answers: weeds, grasses, insects, birds)*

FIND THE MAIN IDEA Help students find the main idea in the Studying Patterns of Succession section.

Ask How did researchers learn about primary succession on Mount Saint Helens and Krakatau? *(by comparing initial conditions and the stages of succession over time)*

Ask What is the main idea of this section? *(Scientists study succession by observing and comparing stages of succession in different communities.)*

KEY QUESTION Read aloud the Key Question and answer.

Ask What would be needed for an ecosystem to return to its original climax community after a disturbance? *(the same conditions that originally produced that community)*

Speed Bump

Ask What makes it difficult for a tropical rain forest to grow back once it has been cleared for farming? *(Soil and microclimates change in a way that prevents the re-growth of rain forest trees.)*

Post Reading

ASSESSMENT TIPS Review the highlighted vocabulary words with students before they begin the assessment. For Question 6, remind students to think about the starting conditions for primary and secondary succession.

Wrap-Up Activity

PERFORM A QUIZ SHOW Have teams write questions about the lesson content on one side of a card and answers on the other side. Organize a quiz show to see which team gets the most correct answers.

Studying Succession These Forest Service rangers are surveying plants and animals that have returned to Mount Saint Helens. The volcano erupted in 1980, leaving only barren land for miles.

Succession After Human-Caused Disturbances In North America, land cleared for farming and then abandoned often passes through succession. In some cases, the original climax community is restored. But restoration does not always happen. Ecosystems may or may not recover from extensive human-caused disturbances. For example, when tropical rain-forest land is cleared and farmed, the microclimate and soil often change. These changes prevent regrowth of the original community.

Studying Patterns of Succession Ecologists study succession by comparing different cases and looking for similarities and differences. For example, researchers swarmed over Mount Saint Helens as soon as it was safe. These researchers might also have studied Krakatau. In both places, primary succession proceeded through predictable stages. Seeds, spores, or adult stages of the first plants and animals traveled over long distances. Hardy pioneer species helped stabilize loose volcanic debris, which allowed later plant species to take root.

Studies of Krakatau and Mount Saint Helens prove that early stages of primary succession are slow. Opportunity can play a large role in determining which species colonize at different times.

Key Question Do ecosystems return to "normal" following a disturbance?
In healthy ecosystems, secondary succession often reproduces the original climax community after a natural disturbance. But ecosystems may or may not recover from widespread disturbances caused by humans.

CHECK Understanding

Apply Vocabulary

Use the highlighted words from the lesson to complete each sentence correctly.

1. The first species to colonize an area during ecological succession is called a _______.
2. Succession that begins after a wildfire is usually _______, because soil is already present.
3. Succession that begins after a volcanic eruption is usually _______, because it starts with bare rock.

Critical Thinking

4. **Explain** What is a climax community?
5. **Relate Cause and Effect** What kinds of conditions might prevent a community from returning to its predisturbance state?
6. **Quick Write** Look at the photo below. Write a paragraph explaining what sort of succession this is, and why.

BIOLOGY.com Search Lesson 4.3 GO • Lesson Assessment

Check Understanding

1. pioneer species
2. secondary succession
3. primary succession
4. a community that is the end result of ecological succession
5. Possible answers: A community that experiences frequent wildfires or volcanic eruptions may not be able to return to a climax community before the next disturbance occurs.

Human-caused disturbances such as the burning and cultivation of rain forests can also prevent a community from returning to its predisturbance state—in this case, the soil is so nutrient-poor that vegetation of the climax community cannot grow back.

6. This represents primary succession because the dune started out with no soil or vegetation.

4.4 Biomes

The Major Biomes

Why are the characteristics of biological communities different from one place to another? You learned that latitude affects global climate. So does the heat transported by winds. But regions with similar latitudes and prevailing winds may have different climates and biological communities. Why? Because other factors can also influence climate. These factors include an area's nearness to an ocean or a mountain range.

Regional Climates Oregon, for example, borders the Pacific Ocean, which has cold currents flowing from north to south. Cold currents make the region cooler in summer than other places at the same latitude. Similarly, prevailing winds traveling west to east push moist air upward against the Cascade Mountains. The air expands and cools, causing the moisture to condense and form clouds. The clouds then drop rain or snow on western Oregon. But the air descends on the eastern side of the mountains, becoming warmer and drier. So, much less rain falls in east Oregon. West and east Oregon have very different regional climates. Therefore, the areas also have different plant and animal communities.

Key Questions

What abiotic and biotic factors characterize biomes?

What areas are not easily classified into a major biome?

BUILD Understanding

Preview Visuals Before you read, preview the biome map. Study the names of the biomes carefully.

In Your Workbook As you read, examine the photographs of the biomes. Complete the table in Lesson 4.4 by filling in the missing information.

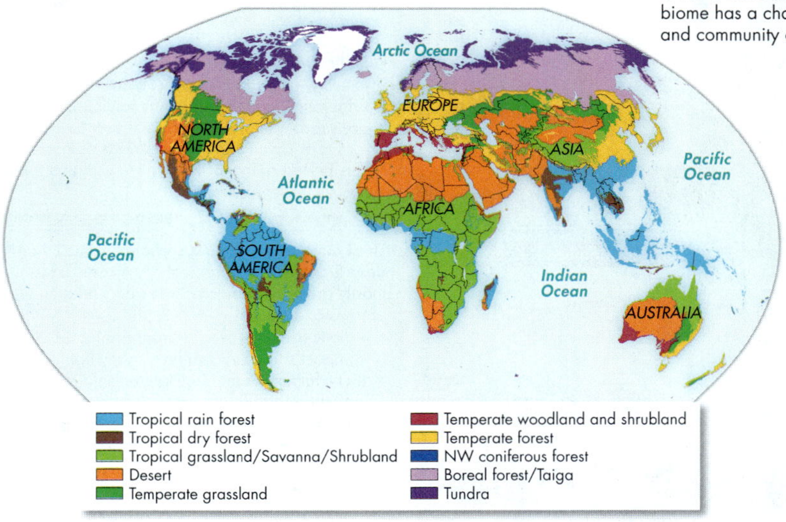

Biomes This map shows the locations of the world's major biomes. Each biome has a characteristic climate and community of organisms.

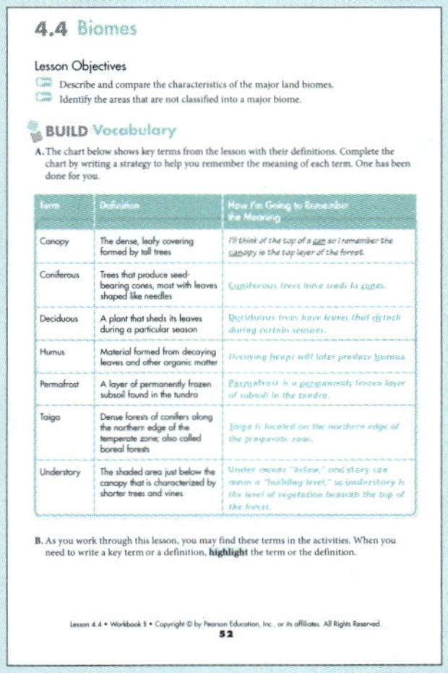

4.4 Biomes

Lesson Objectives

- Describe and compare the characteristics of the major land biomes.
- Identify the areas that are not classified into a major biome.

BUILD Vocabulary

A. The chart below shows key terms from the lesson with their definitions. Complete the chart by writing a strategy to help you remember the meaning of each term. One has been done for you.

Term	Definition	How I'm Going to Remember the Meaning
Canopy	The dense, leafy covering formed by tall trees	I'll think of the top of a can so I remember the canopy is the top layer of the forest.
Coniferous	Trees that produce seed-bearing cones, most with leaves shaped like needles	Coniferous trees have seeds in cones.
Deciduous	A plant that sheds its leaves during a particular season	Deciduous trees have leaves that detach during certain seasons.
Humus	Material formed from decaying leaves and other organic matter	Decaying heaps will later produce humus.
Permafrost	A layer of permanently frozen subsoil found in the tundra	Permafrost is a permanently frozen layer of subsoil in the tundra.
Taiga	Dense forests of conifers along the northern edge of the temperate zone; also called boreal forests	Taiga is located on the northern edge of the temperate zone.
Understory	The shaded area just below the canopy that is characterized by shorter trees and vines	Under means "below," and story can mean a "building level," so understory is the level of vegetation beneath the top of the forest.

B. As you work through this lesson, you may find these terms in the activities. When you need to write a key term or a definition, **highlight** the term or the definition.

Lesson 4.4 • Workbook B • Copyright © by Pearson Education, Inc., or its affiliates. All Rights Reserved.
52

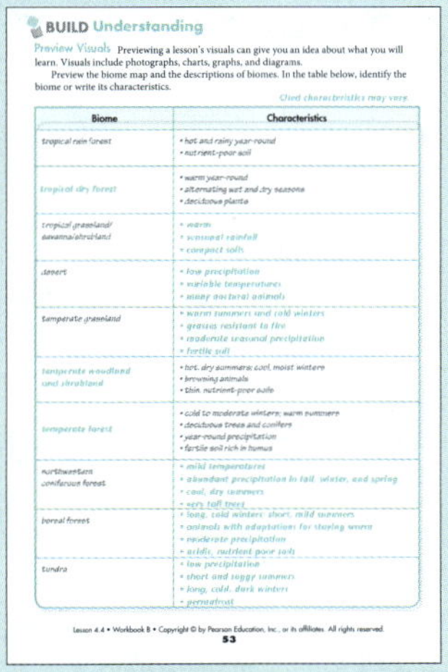

BUILD Understanding

Preview Visuals Previewing a lesson's visuals can give you an idea about what you will learn. Visuals include photographs, charts, graphs, and diagrams.

Preview the biome map and the descriptions of biomes. In the table below, identify the biome or write its characteristics.

Cited characteristics may vary.

Biome	Characteristics
tropical rain forest	• hot and rainy year-round • nutrient-poor soil
tropical dry forest	• warm year-round • alternating wet and dry seasons • deciduous plants
tropical grassland/savanna/shrubland	• warm • seasonal rainfall • compact soils
desert	• low precipitation • variable temperatures • many nocturnal animals
temperate grassland	• warm summers and cold winters • grasses resistant to fire • moderate seasonal precipitation • fertile soil
temperate woodland and shrubland	• hot, dry summers; cool, moist winters • browsing animals • thin, nutrient-poor soils
temperate forest	• cold to moderate winters; warm summers • deciduous trees and conifers • year-round precipitation • fertile soil rich in humus
northwestern coniferous forest	• mild temperatures • abundant precipitation in fall, winter, and spring • cool, dry summers • very tall trees
boreal forest	• long, cold winters; short, mild summers • animals with adaptations for staying warm • moderate precipitation • acidic, nutrient-poor soils
tundra	• low precipitation • short and soggy summers • long, cold, dark winters • permafrost

Lesson 4.4 • Workbook B • Copyright © by Pearson Education, Inc., or its affiliates. All rights reserved.
53

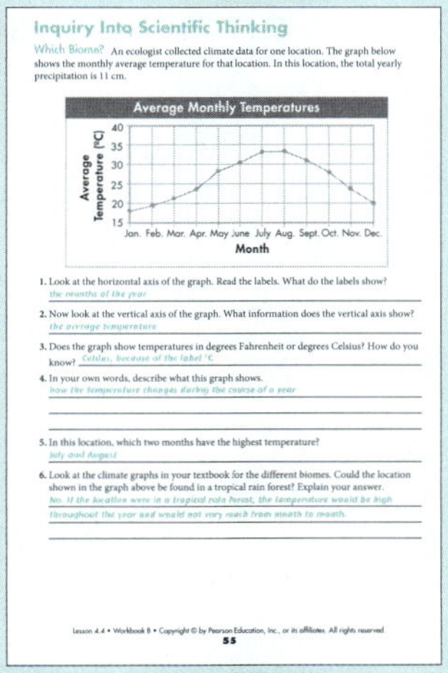

Inquiry Into Scientific Thinking

Which Biome? An ecologist collected climate data for one location. The graph below shows the monthly average temperature for that location. In this location, the total yearly precipitation is 11 cm.

1. Look at the horizontal axis of the graph. Read the labels. What do the labels show? the months of the year
2. Now look at the vertical axis of the graph. What information does the vertical axis show? the average temperature
3. Does the graph show temperatures in degrees Fahrenheit or degrees Celsius? How do you know? Celsius, because of the label °C
4. In your own words, describe what this graph shows. how the temperature changes during the course of a year
5. In this location, which two months have the highest temperature? July and August
6. Look at the climate graphs in your textbook for the different biomes. Could the location shown in the graph above be found in a tropical rain forest? Explain your answer. No. If the location were in a tropical rain forest, the temperature would be high throughout the year and would not vary much from month to month.

Lesson 4.4 • Workbook B • Copyright © by Pearson Education, Inc., or its affiliates. All rights reserved.
55

LESSON 4.4

Pre-Reading

Objectives

4.4.1 Describe and compare the characteristics of the major land biomes.

4.4.2 Identify the areas that are not classified into a major biome.

Preview the Pages

1. Preview the vocabulary terms by having students find them in the text. Invite students to discuss any terms with which they are familiar.
2. Preview the images and climate graphs for the major biomes. Point out that the *y* axis for each graph has two separate scales—one for temperature and one for precipitation. Make sure students understand that the higher the red line, the higher the temperature, and the taller the bar, the greater the rainfall amount. Point out also that the *x* and *y* axes for all the graphs are the same, so that data can be easily compared.

Build Understanding

PREVIEW VISUALS Direct students to Workbook B, where they'll find a partially completed table. Have students complete the table as they read.

NATIONAL SCIENCE EDUCATION STANDARDS

UNIFYING CONCEPTS AND PROCESSES

I, III, V

CONTENT

C.4.c, C.6.b, D.1

INQUIRY

A.1.c, A.2.d

For the corresponding lesson in ***Miller & Levine Biology,*** see pages 110–116.

Ecosystems and Communities 91

LESSON 4.4

During Reading

Active Reading

FIND THE MAIN IDEA Have a volunteer read aloud the first paragraph of this lesson on the previous page. Question the students to elicit the main idea of the paragraph.

Ask What is the single most important factor that influences the kind of biological community present in an area? *(climate)*

KEY QUESTION Read aloud the Key Question and answer. Then, ask students to define *abiotic* and *biotic*. *(Abiotic factors are physical, nonliving factors that shape an ecosystem. Biotic factors are all of the living parts of the ecosystem.)*

USE DIAGRAMS Direct students' attention to the map of biomes.

Ask Which kind of tropical forest covers the most area? *(rain forest)* Does the desert biome appear on all of Earth's continents? *(yes)*

Have students find their own location on the map. Then challenge them to find two other places on the map that are in the same biome.

Ask In what ways would you expect ecosystems in those places to be similar to our ecosystem? *(Answers may vary but should address similarities of climate and similarities of plant and animal life.)*

MAKE CONNECTIONS Ask students who have lived in or visited other parts of the country or the world to point to those areas on the World Biomes map and describe the climate and vegetation.

COMPARE AND CONTRAST Have students compare the descriptions of the tropical rain forest and tropical dry forest biomes.

Ask Which of these two biomes has rain all year round? *(tropical rain forest)*

Ask Why are animals in the tropical rain forest unlikely to practice estivation? *(There's plenty of water year round.)*

SCIENCE SUPPORT Extend students' knowledge by telling them that tall trees growing in the poor, shallow soil of tropical rain forests often have buttress roots for support. In order to get enough sunlight, epiphytic plants grow on the branches of tall plants rather than in the soil.

Speed Bump

Ask Why aren't all plant and animal communities in the same biome exactly alike? *(Differences in elevation, soil conditions, and exposure to wind and other weather factors can cause different habitats within the same biome.)*

Defining Biomes Earth's terrestrial ecosystems are classified into ten groups of regional climate communities called biomes. A biome is described by its abiotic factors, such as climate and soil type. Each biome has a seasonal pattern of temperature and precipitation. A biome is also described by its biotic factors, such as plant and animal life. Organisms in each biome have adaptations so they can live and reproduce in the environment. Plant and animal communities often vary even if they are in the same biome. Variations can be caused by differences in exposure, elevation, or local soil conditions. Human activity or community interactions may also change local conditions.

Key Question What abiotic and biotic factors characterize biomes? **Biomes are described in terms of abiotic factors like climate and soil type. They are also described by biotic factors like plant and animal life.**

Tropical rain forests have more species than all the other biomes combined. They get at least 2 meters of rain a year! Tall trees form a dense, leafy covering called a **canopy.** The canopy shades a layer of shorter trees and vines called the **understory.**
- **Abiotic factors** hot and wet year-round; thin, nutrient-poor soils subject to erosion
- **Biotic factors–Plant life:** Understory plants use large leaves to compete for limited light. **Animal life:** Animals are active all year. Many use camouflage to hide from predators.

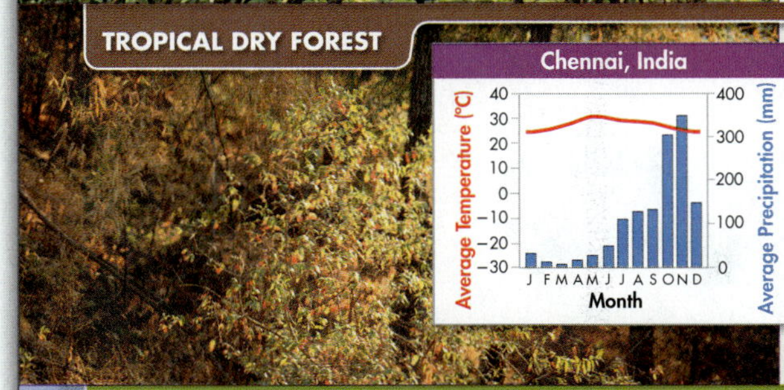

Tropical dry forests grow in areas where rainy seasons alternate with dry seasons.
- **Abiotic factors** warm year-round; usually, a period of rain is followed by a long period of drought
- **Biotic factors–Plant life:** Some plants lose leaves to survive the dry season. A plant with this adaptation is *deciduous.* **Animal life:** During dry season, many animals reduce their need for water by entering long periods of inactivity. This period of inactivity is called estivation, and it is similar to hibernation.

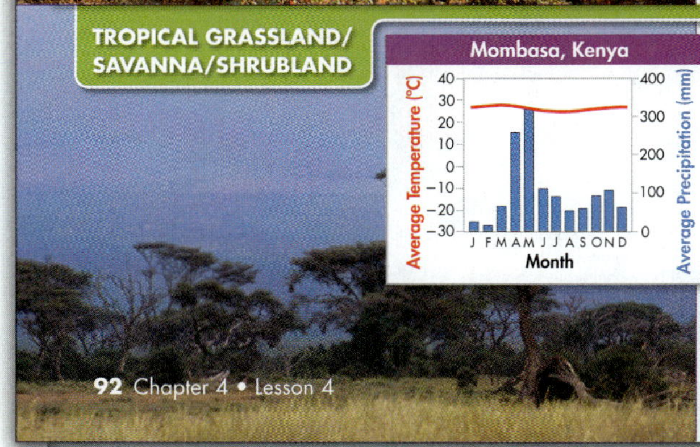

This biome receives more seasonal rainfall than deserts but less than tropical dry forests. It is mostly grass, with isolated trees and small groves of trees and shrubs.
- **Abiotic factors** warm; seasonal rainfall; compact soils; frequent fires set by lightning
- **Biotic factors–Plant life:** Plant adaptations include waxy leaf coverings and seasonal leaf loss. **Animal life:** Many animals migrate during the dry season to find water. Some smaller animals burrow and remain inactive during the dry season.

Focus on Academic Language

• canopy • understory

Help familiarize students with the multiple uses of *canopy* and the meanings of the words that combine to make *understory.*

Say The word *canopy* can be used to describe something other than treetops. For example, *canopy* can be used to describe: (1) a cloth covering over a bed, (2) the part of a parachute that opens and fills with air, and (3) the transparent cover of an aircraft's cockpit.

Write these other uses of *canopy* on the board. Have students write a sentence that correlates with each usage.

Ask What are the definitions of the two words that combine to form *understory*? *(Under means "below something." Story means "a floor or level in a building.")* What does *understory* mean in the text about tropical rain forests? *(a layer of shorter trees and vines below the level of the taller trees in a forest)*

DESERT

Deserts have less than 25 centimeters of precipitation annually. Deserts vary greatly, depending on elevation and latitude. Many deserts undergo extreme daily temperature changes, alternating between hot and cold.

- **Abiotic factors** low precipitation; variable temperatures
- **Biotic factors–Plant life:** Many plants store water in their tissues. They also minimize leaf surface area to cut down on water loss. **Animal life:** Many desert animals get water from the food they eat. Many are active at night to avoid the hottest parts of the day.

TEMPERATE GRASSLAND

Plant communities are mostly grasses, which have been maintained by periodic fires and heavy grazing. Their soils are fertile and ideal for growing crops. For this reason, most have been converted for agriculture.

- **Abiotic factors** warm to hot summers; cold winters; moderate seasonal precipitation; occasional fires
- **Biotic factors–Plant life:** Grassland plants are resistant to grazing and fire. **Animal life:** Predation is a threat for smaller animals because of the open environment. Camouflage and burrowing are common adaptations.

TEMPERATE WOODLAND AND SHRUBLAND

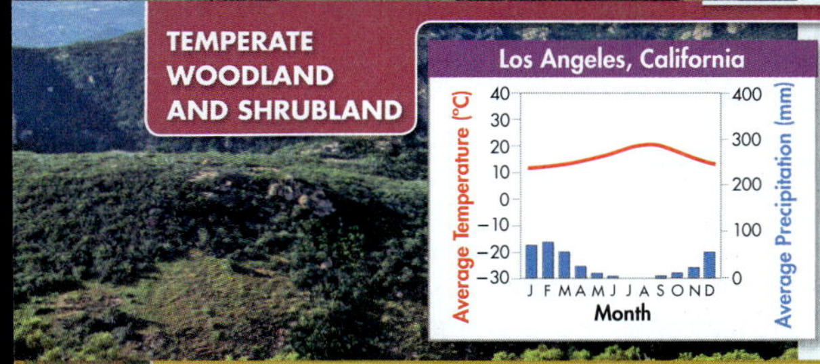

Open woodlands have large areas of grasses and wildflowers mixed with oak and other trees. Fire is a constant threat because of dense, low, oily plants.

- **Abiotic factors** hot dry summers; cool moist winters; thin, nutrient-poor soils; periodic fires
- **Biotic factors–Plant life:** Plants in this biome have adapted to drought. They have tough waxy leaves that resist water loss. **Animal life:** Animals tend to be browsers that eat varied diets of grasses, leaves, shrubs, and other vegetation.

TEMPERATE FOREST

Temperate forests are mostly made up of deciduous and evergreen coniferous (koh NIF ur us) trees. Soils are fertile, and are often rich in **humus.** Humus is a material formed from decaying leaves and other organic matter.

- **Abiotic factors** cold to moderate winters; warm summers; year-round precipitation; fertile soils
- **Biotic factors–Plant life:** Deciduous trees drop their leaves in autumn and go dormant in winter. **Animal life:** Animals must cope with changing weather. Bare trees leave animals exposed in winter.

NORTHWESTERN CONIFEROUS FOREST

The forest has a variety of conifers, flowering trees, and shrubs. Moss often covers tree trunks and the forest floor. This biome is sometimes called a "temperate rain forest" because of its lush vegetation.

- **Abiotic factors** mild temperatures; abundant precipitation in fall, winter, and spring; cool dry summers
- **Biotic factors–Plant life:** This biome is less diverse than a tropical rain forest. Ample water and nutrients support lush, dense plant growth. Trees here are among the world's tallest. **Animal life:** Camouflage helps insects and ground-dwelling mammals avoid predation.

ELL Focus on ELL: Access Content

BEGINNING SPEAKERS Use the following questions and answers to lead a conversation about how the biomes on these pages are similar and different. Have students use the sentence frames to help structure their answers, using details from the text.

Q: How does the amount of precipitation vary between biomes?

A: The _____ biome gets more rainfall than the _____ biome.

Q: How do plants in different biomes get the things they need to survive?

A: Plants in the _____ biome get _____ by _____.

Q: What adaptations help animals survive in the different biomes?

A: Animals survive by _____.

Active Reading

SCIENCE SUPPORT Provide students with additional details about the adaptations of plants in grasslands and deserts. Explain that some grasses in tropical grasslands have a high silica content that makes them less appetizing to grazing herbivores. Also, unlike most plants, grasses grow from their bases, not their tips. This means that they can continue to grow after being grazed. Some desert plants, such as cacti, have an adaptation that allows them to conserve moisture during photosynthesis, when plants open their pores to take in carbon dioxide and release oxygen. These plants, called C4 plants, open their pores only at night, preventing loss of moisture on hot, dry days.

DRAW CONCLUSIONS Guide students to draw conclusions about the tropical grassland, desert, and temperate grassland biomes, using the descriptions in the text and other details that you provide.

Ask What abiotic factors affect the ability of organisms to survive in deserts? *(Sample answers: low precipitation, extreme temperature changes)* What are some ways that plants have adapted to these abiotic factors? *(Sample answers: Plants store water in tissues. Some plants open their pores only at night to prevent loss of moisture during hot daytime hours.)*

Ask What is a key biotic factor that affects the ability of plants to survive in grasslands? *(heavy grazing by animals)* What are some adaptations that allow grassland plants to survive heavy grazing? *(Sample answers: Grasses grow from bases, not tips. Some grasses have a high silica content that makes them unappetizing to herbivores.)*

MAKE INFERENCES Tell students that coniferous trees have needlelike leaves that are coated in a waxy substance. Ask them how these adaptations might help conifers survive in temperate forests. *(These adaptations minimize water loss.)*

Speed Bump

Ask What adaptations help plants survive with little water in the desert? *(small leaf surface area, ability to store water in tissues)*

Ask Why is it an advantage for plants growing in limited light to have large leaves? *(The large leaves capture more sunlight than smaller leaves would.)*

Ask Why is predation a constant threat for animals in temperate grasslands? *(The animals are not able to quickly find a place to hide from predators.)*

LESSON 4.4

During Reading continued

Active Reading

USE VISUALS As students read the description of each biome, have them return to the biomes map and locate the areas on Earth where that biome is found.

DRAW CONCLUSIONS Have students use the descriptions of the temperate woodland and temperate forest biomes to draw conclusions.

Ask Why do oily plants increase the risk of fire? *(Plant oils fuel a fire once it gets started.)*

Ask Why does forest soil contain more rich humus than woodland soil does? *(A forest has more trees, and therefore more leaves that drop to the ground and decay.)*

COMPARE AND CONTRAST Have students compare the descriptions of boreal forest and tundra biomes.

Ask Which biome has more precipitation? *(boreal forest)*

Ask In which biome would you be more likely to find tall trees with large roots systems? *(boreal forest)* Why are such trees not well suited for the tundra biome? *(The tundra has strong winds, permafrost, and cycles of thawing and freezing in the top layer of soil.)*

Build Vocabulary

WORD ORIGINS The word *humus* is from Latin and Greek roots that mean "on the ground" or "of the earth." These root words also gave rise to the word *humble*. Discuss how the meaning of *humble* relates to these roots.

KEY QUESTION Read aloud the Key Question and answer.

Ask Which land biome has environmental conditions similar to a mountainside covered with coniferous forest and blanketed with snow? *(boreal)*

Speed Bump

Ask How does the temperature in a mountain environment change as you move to higher and higher elevations? *(It gets colder.)*

Ask Which of Earth's poles is covered with sea ice and inhabited by polar bears? *(North Pole)*

Ask What kinds of organisms can live on ice? *(algae, marine mammals, insects, mites)*

BOREAL FOREST

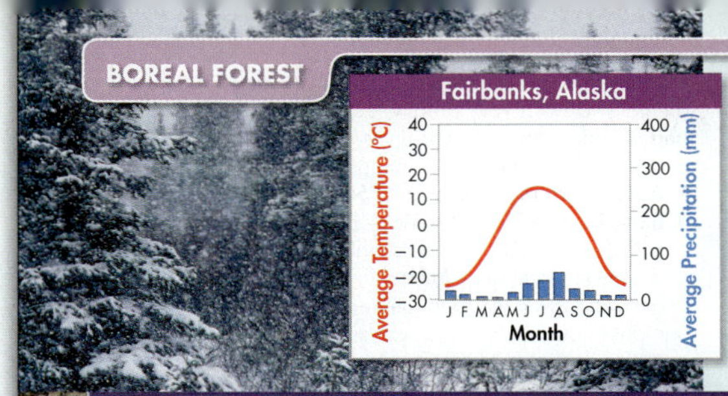

Boreal forests, or **taiga** (TY guh), are dense forests of coniferous evergreens. They are found along the northern edge of the temperate zone.

- **Abiotic factors** long cold winters; short mild summers; moderate precipitation; high humidity; acidic, nutrient-poor soils
- **Biotic factors–Plant life:** Conifers are well suited to this biome. Their shape sheds snow. Wax-covered, needlelike leaves reduce water loss. **Animal life:** To stay warm most animals have small limbs and ears, and fat or downy feathers for insulation.

TUNDRA

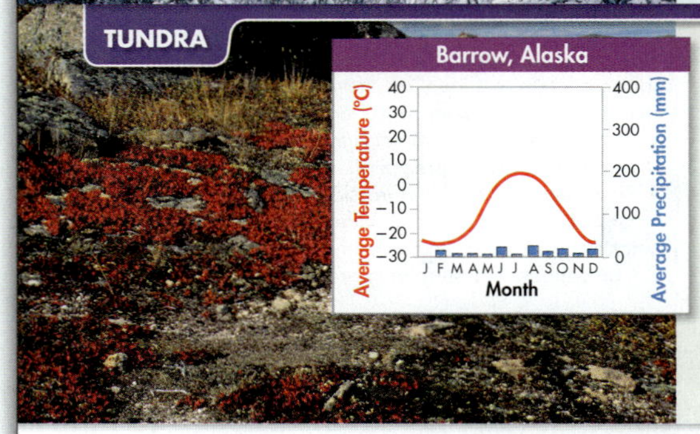

The tundra is identified by **permafrost,** a layer of permanently frozen subsoil. In summer, the ground thaws to a depth of a few centimeters and becomes soggy. In winter, the top layer of soil freezes again. This cycle of thawing and freezing rips and crushes plant roots. Cold temperatures, high winds, a short growing season, and humus-poor soils also limit plant height.

- **Abiotic factors** strong winds; low precipitation; short and soggy summers; long, cold, dark winters; permafrost
- **Biotic factors–Plant life:** Mosses and other plants are low-growing to avoid damage from frequent strong winds. **Animal life:** Many animals migrate to avoid long harsh winters. Animals that live in the tundra year-round have adaptations such as natural antifreeze to limit heat loss.

BUILD Vocabulary

canopy a dense covering formed by the leafy tops of tall rain forest trees

understory the layer in a rain forest found underneath the canopy formed by shorter trees and vines

humus material formed from decaying leaves and other organic matter

taiga a biome with long cold winters and a few months of warm weather; dominated by coniferous evergreens; also called boreal forest

permafrost a layer of permanently frozen subsoil found in the tundra

WORD ORIGINS

The word *taiga* is Russian for "dense evergreen forest." Gradually, it became the name of the biome where these Russian forests are found.

Other Land Areas

Some land areas are not easily defined in terms of a typical community of plants and animals. For this reason, mountain ranges and polar ice caps are not usually classified into biomes.

Mountain Ranges Mountain ranges exist on all continents and in many biomes. Conditions such as temperature and precipitation vary with elevation. Exposure to wind increases and soil types and organisms change as you move up from the valley to the mountain summit. For example, the base of the Rocky Mountains in Colorado is grassland. As you climb upward, you enter pine woodland, then a forest of spruce and other conifers. Aspen and willow trees grow in thickets along protected valley streambeds. Near the top, winds are strong and batter open fields of wildflowers. The stunted vegetation resembles tundra. Glaciers are often found at the peaks.

Polar Ice Caps Polar regions border the tundra and are cold year-round. There are few plants, but some algae grow on snow and ice. Animal life includes marine mammals, insects, and mites. In the north, the Arctic Ocean is covered with sea ice and inhabited by polar bears. In the south, the continent of Antarctica is inhabited by many species of penguins. The ice there is nearly 5 kilometers thick in places.

Key Question What areas are not easily classified into a major biome? **Mountain ranges and polar ice caps are not usually classified into biomes. They are not easily defined in terms of a typical community of plants and animals.**

Analyze and Conclude

1. *x* axis: month; *y* axis: average temperature
2. The temperature remains steady in Location A and rises during the summer months in Location B.
3. You would expect to find Location A in a rain forest biome because the total precipitation is more than 2 m and the average temperature varies little from month to month. You would expect to find Location B in a desert biome because precipitation is very low and the temperature varies seasonally from warm to hot.

INQUIRY into Scientific Thinking

Which Biome?

An ecologist collected climate data from two locations. The graph shows the monthly average temperatures in the two locations. The total yearly precipitation in Location A is 273 cm. In Location B, the total yearly precipitation is 11 cm.

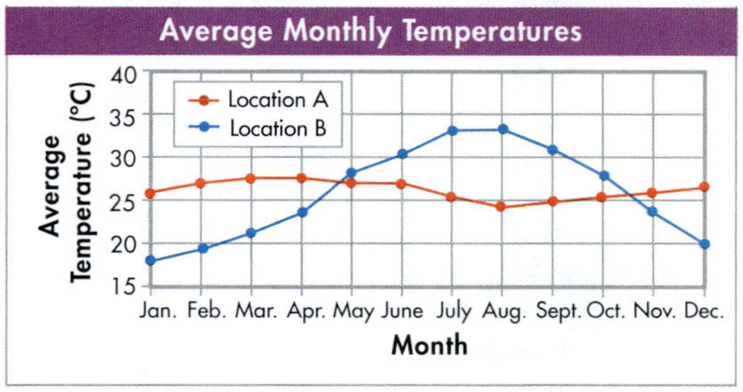

Analyze and Conclude

1. Interpret Graphs What is the variable plotted on the horizontal (X) axis? What is the variable plotted on the vertical (Y) axis?

2. Interpret Graphs How would you describe the temperature over the course of the year in the two locations?

3. Draw Conclusions In which biome would you expect to find each location, given the precipitation and temperature data? Explain your answer.

In Your Workbook Go to your workbook for more help wth this activity.

CHECK Understanding

Apply Vocabulary

Use the highlighted words from the lesson to complete each sentence correctly.

1. The ________ in a tropical rain forest is made up of shorter trees and vines that live in the shade of the tallest trees.

2. The soils in temperate forests are often rich in ________, which is made up of decaying leaves and organic matter.

3. Dense coniferous forests and animals with small ears and limbs are characteristics of the ________ biome.

Critical Thinking

4. Explain How are biomes classified?

5. Write to Learn Answer the second clue of the mystery. Be sure to include the effect of microclimates in your answer.

MYSTERY CLUE

Yellowstone has high mountain slopes and valleys with streams. Moose and elk prefer to graze in the valleys instead of on the mountain slopes. How do you think their preference might affect Yellowstone's plant communities? (Hint: See p. 94.)

BIOLOGY.com Search Lesson 4.4 GO • Lesson Assessment

Check Understanding

1. understory

2. humus

3. boreal forest

4. Biomes are classified by their abiotic factors, such as climate and soil type, and biotic factors, such as plant and animal life.

5. Answers may vary. Moose and elk are herbivores. Students should understand that these species prefer the valleys with streams because of the types of plants available and the easy access to water. Also, the microclimate of the valleys would influence these species' preference, because protected valleys have a warmer climate than the slopes of the mountains, making it easier for the two species to survive harsh winters. If elk and moose populations rise, the plant populations in the valleys would decrease from the additional grazing.

Inquiry Into Scientific Thinking

ANALYZING DATA Explain to students that they will analyze temperature and precipitation data for two ecosystems to determine which biome each ecosystem belongs to. Draw students' attention to the graph of temperature data. Make sure they understand that the graph shows data for two different ecosystems.

Ask At first glance, do you think these two ecosystems belong to the same biome? Why or why not? *(No, because one has much higher summer temperatures than the other.)*

Lead a Discussion

Discuss why it is important, when comparing biomes, to gather data on average temperature for each month in the year, rather than just calculating the average temperature for an entire year. Explain that average temperature for the whole year could be similar for two very different biomes. For example, a temperate forest and a northwestern coniferous forest could have very similar annual temperatures, but the temperate forest has hotter summers. That difference leads to different kinds of environmental conditions, which would in turn influence which species could survive there.

Post Reading

ASSESSMENT TIPS For Question 4, remind students that classifying involves sorting things into groups or categories. Explain that the question is asking which characteristics are used to sort ecosystems into biomes.

MYSTERY CLUE Remind students to think about predator-prey and herbivore-plant relationships as they read the clue. Discuss how moose and elk might affect the plant communities they feed on.

Wrap-Up Activity

PLAY 20 QUESTIONS Have a student secretly select one biome. Then, have the student answer questions from other students whose goal is to identify the biome. Repeat with other students and biomes as time permits.

LESSON 4.5

Pre-Reading

Objectives

4.5.1 Discuss the factors that affect aquatic ecosystems.

4.5.2 Identify the major categories of freshwater ecosystems.

4.5.3 Describe the importance of estuaries.

4.5.4 Describe and compare the distinct ocean zones that make up marine ecosystems.

Preview the Pages

1. Preview with students all of the vocabulary terms, pronouncing each term aloud. Have students use context clues to define the terms.
2. Preview the photographs of freshwater ecosystems. Remind students that fresh water contains very little salt. Point out that estuaries are saltwater wetlands.
3. Preview the Ocean Zones diagram. Show that the area to the left of the intertidal zone is the part of the beach that is rarely underwater.

Activate Prior Knowledge

Ask What are some bodies of water in our area? How would you describe each one? *(Accept all answers that reflect an understanding of the characteristics of streams, lakes, wetlands, estuaries, the ocean, or other water bodies.)*

Build Understanding

COMPARE/CONTRAST TABLE Direct students to the table in Workbook B. Point out that it will help them organize the information in the lesson.

NATIONAL SCIENCE EDUCATION STANDARDS

UNIFYING CONCEPTS AND PROCESSES

I, V

CONTENT

C.4.b, C.4.c, C.5.b, C.5.e

For the corresponding lesson in ***Miller & Levine Biology,*** see pages 117–121.

4.5 Aquatic Ecosystems

Key Questions

What factors affect life in aquatic ecosystems?

What are the major categories of freshwater ecosystems?

Why are estuaries so important?

How do ecologists usually classify marine ecosystems?

BUILD Understanding

Compare/Contrast Table As you read, note the similarities and differences between freshwater and marine ecosystems in a compare/contrast table.

In Your Workbook Go to Lesson 4.5 in your workbook to learn more about making a compare/contrast table.

The Photic Zone Sunlight penetrates only a limited distance into aquatic ecosystems. Whatever the depth, the photic zone is the only area in which photosynthesis can occur.

Conditions Underwater

We call our planet "Earth." But almost three fourths of Earth's surface is covered with water. What is life like underwater?

Underwater organisms are affected by a variety of environmental factors. These factors include the water depth, temperature, flow, and amount of dissolved nutrients. Runoff from land can affect some factors, so distance from shore also shapes marine communities.

Water Depth Sunlight only penetrates a short distance through water. This region is called the photic zone. It ranges from 200 meters deep (tropical seas) to less than a few meters (swamps). Photosynthetic algae called phytoplankton live here. They are eaten by tiny free-floating animals called zooplankton. Below the photic zone is the dark aphotic zone, where photosynthesis cannot occur.

Many aquatic organisms live on, or in, rocks and sediments on the bottoms of lakes, streams, and oceans. These organisms are called the **benthos,** and their habitat is the benthic zone. In shallow waters the benthos are in the photic zone. Algae and rooted aquatic plants grow here. In the aphotic zone, chemosynthetic autotrophs are the only primary producers.

Temperature and Currents Aquatic habitats are warmer near the equator and colder near the poles. Temperature in aquatic habitats also varies with depth. The deepest parts of lakes and oceans are often colder than surface waters. Currents can dramatically affect water temperature, too. They can carry water that is noticeably warmer or cooler than normal into a given area.

Nutrient Availability Aquatic organisms need certain substances to live, including oxygen, nitrogen, potassium, and phosphorus. The type and availability of these dissolved substances varies within and between bodies of water. This difference greatly affects the types of organisms that can survive there.

Key Question What factors affect life in aquatic ecosystems? **Aquatic organisms are affected primarily by the water's depth, temperature, flow, and amount of dissolved nutrients.**

4.5 Aquatic Ecosystems

Lesson Objectives

- Discuss the factors that affect aquatic ecosystems.
- Identify the major categories of freshwater ecosystems.
- Describe the importance of estuaries.
- Describe and compare the distinct ocean zones that make up marine ecosystems.

BUILD Vocabulary

A. The chart below shows key terms from the lesson with their definitions. Complete the chart by writing a strategy to help you remember the meaning of each term. One has been done for you.

Term	Definition	How I'm Going to Remember the Meaning
Aphotic zone	Dark region in a body of water where no photosynthesis occurs; below the photic zone	The prefix a- means "not," and a photo is associated with light, so the aphotic zone is the dark region of the ocean where light cannot reach.
Benthos	Organisms living on, or in, rocks and sediments at the bottom of lakes, streams, or oceans	Benthos are the organisms living at the bottom of a body of water.
Estuary	A special kind of wetland, formed where a river meets the sea	Remove the letters "tua" from the middle of estuary. The S and R will meet. An estuary is a place where the river meets the sea.
Photic zone	Sunlit region near the surface of a body of water in which photosynthesis can occur	When you take a "photo," you need light, and the photic zone is a region of water that has light for photosynthesis.
Plankton	Phytoplankton (photosynthetic algae) and zooplankton (free floating small animals)	When you "walk the plank," you are jumping off a ship into the ocean, and plankton are small organisms that are found in the ocean.
Wetland	Ecosystem in which water either covers the soil or is present at or near the surface for at least part of the year	A wetland is an area that is wet.

B. As you work through this lesson, you may find these terms in the activities. When you need to write a key term or a definition, **highlight** the term or the definition.

BUILD Understanding

Compare/Contrast Table Use a compare/contrast table when you want to see the similarities and differences between two or more objects or processes.

As you read Lesson 5, record the similarities and differences of freshwater and marine ecosystems in the chart below. An example has been provided.

Comparisons between Freshwater and Marine Ecosystems

Ecosystem	Differences	Similarities
Freshwater	• shallow photic zone • organisms not under the influence of tides	• Both have zones. • temperature of water changes with water depth • have autotrophs and heterotrophs • organisms affected by environmental factors
Marine	• Photic zone may be up to 200 m deep. • chemosynthetic autotrophs present • organisms affected by the tides	

Marine Ecosystems

Marine ecosystems are divided into regions or zones. These zones are based on depth of water and distance from shore. The open ocean is divided into two main zones according to light penetration: the photic and the aphotic zone.

Follow the directions.

1. Find the photic zone in the diagram. Color it yellow.
2. Use these terms to label the marine zones: open ocean, intertidal zone, coastal ocean.

Answer the questions.

3. Which best describes the aphotic zone?
   A. the layer of the ocean where photosynthesis occurs
   (B.) the deep, dark area of the ocean
   C. the shore of the ocean
   D. the area from water surface down to 100 meters
4. Which ocean location would contain phytoplankton?
   A. aphotic zone
   B. benthos
   C. ocean trench
   (D.) photic zone

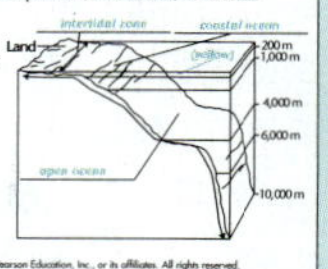

Freshwater Ecosystems

Only 3 percent of Earth's surface water is fresh water. Freshwater ecosystems can be divided into three main categories: rivers and streams, lakes and ponds, and freshwater wetlands.

Rivers and Streams Rivers, streams, creeks, and brooks often originate from underground water sources in mountains or hills. Water has little plant life here. Downstream, sediments build up and plants establish themselves. Farther downstream, water may wander slowly through flat areas. Animals in many rivers and streams depend for food on plants and other animals that live along the banks.

Lakes and Ponds Food webs in lakes and ponds often are based on plankton and attached algae and plants. **Plankton** is a general term that includes both phytoplankton and zooplankton. Water typically flows in and out of lakes and ponds. It also circulates between the surface and the benthos during at least some seasons. This movement distributes heat, oxygen, and nutrients.

Freshwater Wetlands A **wetland** is an ecosystem in which water either covers the soil or is present at or near the surface for at least part of the year. Water may flow through wetlands or remain still. Wetlands are often nutrient-rich and highly productive, and they serve as breeding grounds for many organisms. Freshwater wetlands purify water by filtering pollutants. Wetlands help to prevent flooding by absorbing and slowly releasing water. Three main types of freshwater wetlands are freshwater bogs, freshwater marshes, and freshwater swamps. Saltwater wetlands are called estuaries.

Key Question What are the major categories of freshwater ecosystems?

Freshwater ecosystems can be divided into three main categories: rivers and streams, lakes and ponds, and freshwater wetlands.

BUILD Vocabulary

benthos
organisms that live attached to or near the bottom of lakes, streams, or oceans

plankton
typically small organisms that drift in, and may swim through, aquatic environments; includes both phytoplankton and zooplankton

wetland
an ecosystem in which water either covers the soil or is present at or near the surface for at least part of the year

estuary
a kind of wetland formed where a river meets the ocean

WORD ORIGINS

The word *benthos* is Greek for "depth of the sea." Many of the most unusual benthic organisms live in the deepest parts of the open ocean.

Freshwater Ecosystems Freshwater ecosystems include streams, lakes, and freshwater wetlands (bogs, marshes, and swamps).

ELL Focus on ELL: Extend Language

ALL SPEAKERS Write the word *plankton* on the board. Explain that this word comes from a Greek word that means "to drift." *Drift* means "to be carried by a current of water, wind, or air."

Ask Why is *plankton* an appropriate name for microscopic organisms that float in water? *(These organisms cannot control where they go. They are carried by water currents.)*

Write the prefixes *phyto-* and *zoo-* on the board.

Ask What does *phyto-* mean? *(plant)*

Ask What does *zoo-* mean? *(animal)*

Ask What is the difference between phytoplankton and zooplankton? *(Phytoplankton are microscopic organisms capable of photosynthesis. They form the base of many aquatic ecosystems. Zooplankton are consumers that are not capable of photosynthesis.)*

During Reading

Build Vocabulary

WORD ORIGINS Tell students that the vocabulary term *estuary* comes from the Latin word *aestuarium*, which means "heat, surge, tide." Explain that estuaries contain a mix of fresh and salt water and are affected by ocean tides.

Active Reading

MAKE INFERENCES Direct students' attention to the photograph of the photic zone.

Ask What could affect how clear the water is in the photic zone? *(Sample answers: amount of sediment in runoff from shore, amount of phytoplankton growing in the water)*

Ask Why do phytoplankton grow only in the photic zone? *(Phytoplankton need sunlight for photosynthesis. Sunlight does not penetrate below the photic zone.)*

KEY QUESTION Read aloud the Key Question and answer on the previous page.

Ask How does the presence of dissolved nutrients affect aquatic organisms? *(Different species need different kinds and amounts of nutrients. Nutrient levels in the water help determine which species can live there.)*

FIND THE MAIN IDEA Instruct students to draw a two-column chart. Have them label the columns Main Idea and Details. Then, have a volunteer read aloud the paragraph Lakes and Ponds.

Ask What is the main idea of the paragraph? *(Food webs in lakes and ponds are often based on plankton, algae, and plants.)*

Say Read through the paragraph again. Write details about the main idea in the right column.

KEY QUESTION Read aloud the Key Question and answer on this page.

Ask What are three helpful things that freshwater wetlands do? *(purify water, prevent flooding, serve as a breeding ground for many organisms)*

Speed Bump

Ask What are two conditions that are necessary for plants to grow in the benthic zone? *(Sample answer: the presence of sunlight and sediments)*

Ask Are rivers and streams the only freshwater ecosystems that contain moving water? Explain. *(No, water in lakes, ponds, and wetlands can also flow.)*

During Reading continued

Active Reading

USE CONTEXT CLUES Have a volunteer read aloud the first paragraph in Estuaries to help them determine what *commercially valuable* means.

Say *Commercially* comes from the word *commerce*.

Ask What does the word *commerce* make you think of? *(Sample answer: buying and selling)* What does *commercially valuable* mean? *(able to be sold at a profit)*

MAKE INFERENCES Have students look at the photograph of the fish, and read the caption.

Ask What kinds of adaptations do you think the hatchetfish has that enables it to live in the aphotic zone? *(adaptations that allow it to live with high pressure, frigid temperatures, and total darkness)*

KEY QUESTION Read aloud the first Key Question and answer on this page.

Ask What might make a fish species ecologically important? *(t may be a species that plays a key role in a food web. It may be a keystone species.)*

Speed Bump

Ask What is an example of an estuary that is located in a temperate climate? *(salt marsh)* A tropical climate? *(mangrove swamp)*

Active Reading

KEY QUESTION Read aloud the second Key Question and answer on this page.

Ask What is the relationship between ocean depth and distance from shore? *(Depth typically increases with distance from shore.)*

USE VISUALS Refer students to the diagram on the next page as they read about each marine ecosystem. Be sure students understand that the diagram is divided horizontally as well as vertically. Check understanding by pointing to one of the organisms and asking students which zones it occupies. *(Sample answer: squid—aphotic zone in the open ocean)*

Ask Do you think the organisms shown in the diagram are drawn to scale? Explain. *(Sample answers: No, because the blue whale is much larger and the phytoplankton much smaller compared to other organisms.*

Ocean Zone diagram: *Photic zone* sea lion, herring, blue whale, great white shark, swordfish, flying fish; *200–1000 m* krill-like shrimp, ocean sunfish, bigeye tuna, cod, giant squid; *1000–4000 m* viperfish, dragonfish, bathypelagic anglerfish, snipe eel; *4000–10,000 m* rattail, gulper eel, tripod fish

Estuaries Salt marshes and mangrove swamps are estuaries. These are areas where fresh water from rivers meets salt water.

Estuaries

Estuaries (ES tyoo ehr ee) are wetlands that form where a river meets the sea. They contain a mixture of fresh water and salt water and are affected by ocean tides. Many are shallow, so enough sunlight reaches the benthos to power photosynthesis. Estuaries support an astonishing amount of biomass. They serve as breeding and nursery grounds for many important fish and shellfish species.

Salt marshes are temperate estuaries. Salt marshes have salt-tolerant grasses above the low-tide line and seagrasses below water. The Chesapeake Bay in Maryland is a salt marsh. Mangrove swamps, such as the Everglades, are tropical estuaries. They are identified by several species of salt-tolerant trees, called mangroves.

Key Question Why are estuaries so important? **Estuaries serve as breeding and nursery grounds for many ecologically and commercially important fish and shellfish species.**

Marine Ecosystems

Marine ecosystems usually occupy specific zones within the ocean, based on depth and distance from shore.

Intertidal Zone Organisms in the intertidal zones are subjected to regular and extreme changes in temperature. At high tide, they are submerged in seawater and often battered by waves and currents. At low tide, they are exposed to air and sunlight.

Coastal Ocean The coastal ocean extends from the low-tide mark to the outer edge of the continental shelf. The continental shelf is the relatively shallow border that surrounds the continents. Water here is brightly lit and often supplied with nutrients by freshwater runoff from land. This makes coastal oceans highly productive. Kelp forests and coral reefs are two exceptionally important coastal communities.

Open Ocean The open ocean begins at the edge of the continental shelf. It is divided into the photic zone and the aphotic zone.

▶ ***The Open Ocean Photic Zone*** The open ocean typically has low nutrient levels and supports only the smallest species of phytoplankton. Still, because of its enormous area, most photosynthesis on Earth occurs in the sunlit top 100 meters of the open ocean.

▶ ***The Open Ocean Aphotic Zone*** The permanently dark aphotic zone includes the deepest parts of the ocean. Its food webs are usually based on organisms that fall from the photic zone. Some food webs here are based on chemosynthetic primary producers.

Creature From the Deep This slender hatchetfish lives in the aphotic zone of the Pacific Ocean. It can be found off the coast of California.

Key Question How do ecologists usually classify marine ecosystems? **Ecologists divide the ocean into zones based on depth and distance from shore.**

Students Say

"Our teacher gives us a list of topics from the chapter, and with a partner, we write information on index cards in riddle form, like 'Because of me, you have brown eyes and your ear lobes are attached. What am I?' Your answer goes on the back of the card. Whenever another pair of students knows the answer, they get a point, and if you and your partner stump them, you get a point."

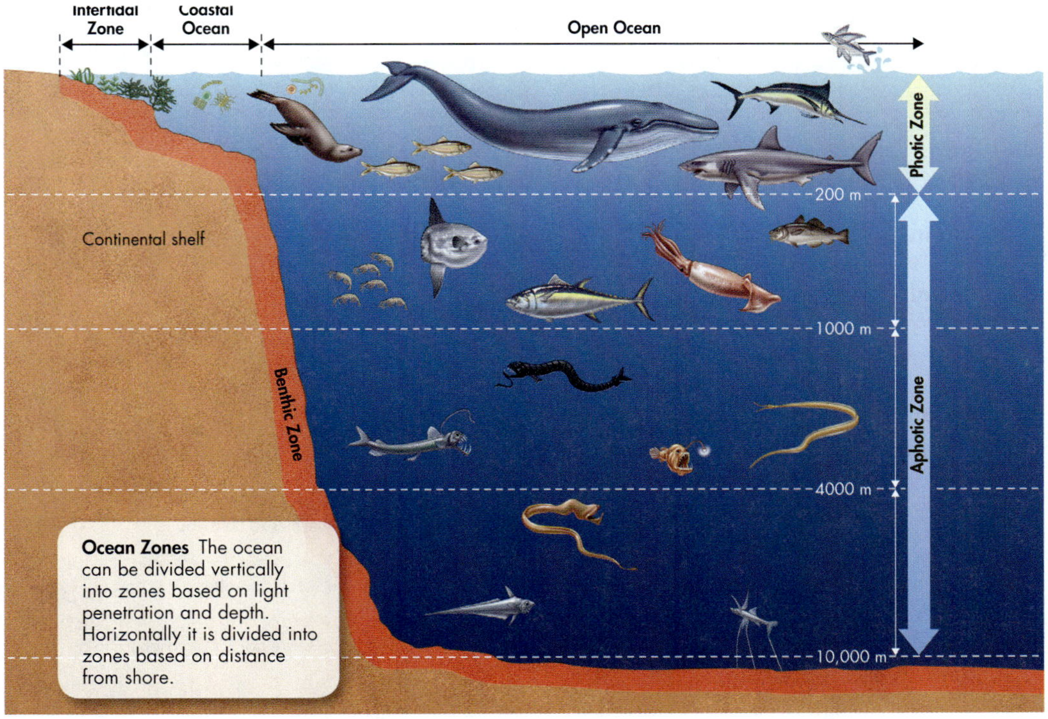

Ocean Zones The ocean can be divided vertically into zones based on light penetration and depth. Horizontally it is divided into zones based on distance from shore.

CHECK Understanding

Apply Vocabulary

Use the highlighted words from the lesson to complete each sentence correctly.

1. _______ is the group of organisms that is often at the base of food webs in open-water aquatic ecosystems.

2. Aquatic organisms that grow in the sediments at the bottom of a pond or lake are called the _______.

Critical Thinking

3. Explain What are the primary abiotic factors that affect life underwater?

4. Apply Concepts Why is it important to protect estuaries?

5. Write to Learn Answer the third clue to the mystery. Where do you think the streams in Yellowstone most likely originate?

MYSTERY CLUE

How might the presence or absence of plants along stream banks affect life in Yellowstone's streams? (**Hint:** See p. 97.)

BIOLOGY.com Search Lesson 4.5 GO • Art Review • Lesson Assessment

Check Understanding

1. Plankton

2. benthos

3. water depth, temperature, current (flow)

4. They are breeding and nursery grounds for ecologically and economically important aquatic species.

5. Answers may vary. Animals in the water depend on plants along the stream banks for food. If the plants disappear, the animal species that depend on them for food might die out.

Active Reading

USE VISUALS Call students' attention to the continental shelf in the Ocean Zones diagram.

Ask Which marine ecosystems have benthos within the photic zone? *(intertidal zone, coastal ocean)*

Ask How does food availability in the photic zone compare with food availability in the aphotic zone? *(Algae and phytoplankton, which are an important food source, can only live in the photic zone because light is available for photosynthesis. Students may also point out that freshwater runoff from land increases nutrient levels in intertidal and coastal zones.)*

Speed Bump

Ask What is life like for organisms that live in the intertidal zone? *(Because of ocean tides, they are alternately submerged in water and exposed to air. They are also subject to waves crashing onto shore.)*

Ask Why is the coastal ocean highly productive? *(It is in the photic zone, and freshwater runoff from land brings in plenty of nutrients for living organisms.)*

Post Reading

ASSESSMENT TIPS Review the highlighted vocabulary words with students before they begin the assessment. If students have trouble with Question 4, suggest that they reread the section about estuaries and look for the answer there.

MYSTERY CLUE Suggest that students reread the section on rivers and streams. Draw their attention to the following sentence: **Animals in many rivers and streams depend on plants and animals that live along their banks for food.** Ask students to provide examples of animals in rivers or streams that might depend on plants for food.

Wrap-Up Activity

DRAW PICTURES Invite students to draw and label a diagram that shows a stream, a lake or pond, a wetland or estuary, and the ocean shore. Students could show a stream flowing down from a mountainside into a lake, and water flowing from the lake into a river that in turn flows into an estuary before reaching the intertidal zone and coastal ocean. Have students draw at least two animals in the scene, showing how each depends on biotic or abiotic parts of the ecosystem. Have students explain to a partner what their drawing shows.

Pre-Lab

Review the meaning of the word *tolerance* (the ability to survive and reproduce under a range of environmental conditions). Then, discuss how organisms' adaptations help them to tolerate the abiotic factors in their environment. Invite volunteers to share examples from the chapter. *(Sample answer: By storing water in their stems, cacti can tolerate the lack of precipitation in a desert environment.)*

Review the Background Questions and Pre-Lab Questions as a class.

Lab

Tell students they will perform the chapter lab Abiotic Factors and Plant Selection described in **Lab Manual B.**

Post-Lab

GARDENS IN DIFFERENT BIOMES Challenge student pairs to plan a garden for a location in a different biome. Have each pair select a biome whose climate is very different from their own. Then, have students research the abiotic conditions in their chosen biome and plan a garden. Conclude with a class discussion about how gardens in the various biomes would differ.

NATIONAL SCIENCE EDUCATION STANDARDS

CONTENT F.3, F.4

Real-World Lab

OPEN-ENDED INQUIRY

Pre-Lab: Abiotic Factors and Plant Selection

Problem How can you decide which plants will thrive in a garden?

Materials plant hardiness zone map, plant catalogs, graph paper, tape measure or metersticks

Lab Manual Chapter 4 Lab

Skills Focus Classify, Analyze Data, Use Models

Connect to the Big idea Why are white birch trees abundant in Minnesota, but not in the Florida Keys? Why do coconut palms grow in the Florida Keys, but not in Minnesota? Simply put, white birch trees could not tolerate the hot summers in the Keys and coconut palms could not tolerate the cold winters in Minnesota. A plant's habitat is determined by its range of tolerance for temperature and other abiotic factors. In other words, abiotic factors limit where a given plant can live.

In this lab, you will plan a garden for a specific location. You will select plants for the garden that can tolerate the abiotic factors in this location.

Background Questions

a. **Review** What is an abiotic factor? List three examples other than temperature.

b. **Review** What kinds of resources do plants need?

c. **Relate Cause and Effect** Give an example of an adaptation that helps a plant survive in a biome with low precipitation.

Pre-Lab Questions

Preview the procedure in the lab manual.

1. **Predict** How will knowing the plant hardiness zone for your area help you plan a garden?

2. **Relate Cause and Effect** What is the relationship between the last frost and the length of the growing season?

3. **Infer** A plant species grows well in one location in a small garden but does not grow as well in another location. Suggest one possible reason for this difference.

BIOLOGY.com Search Chapter 4 GO

Visit Chapter 4 online to test yourself on chapter content and to find activities to help you learn.

Untamed Science Video Join the Untamed Science crew as they explore succession after a volcanic eruption on Hawaii.

Visual Analogy Compare Earth's atmosphere to a greenhouse.

Art in Motion View a short animation that brings succession to life.

Art Review Review your understanding of ocean zones with this drag-and-drop activity.

Pre-Lab Answers

BACKGROUND QUESTIONS

a. Abiotic factors are non-living factors that influence the growth of plants. Possible abiotic factors are sunlight, precipitation, wind, humidity, and soil type.

b. Sample answer: Plants need water, light, nutrients, and space to grow.

c. Possible answers include leaves with limited surface area or leaf pores that open only at night.

PRE-LAB QUESTIONS

1. Sample answer: Knowing the plant hardiness zone will provide information about some abiotic factors. (Some students may know that suppliers of seeds and plants usually provide a range of hardiness zones.)

2. Sample answer: The earlier the last frost occurs, the longer the growing season.

3. Sample answer: One location may have direct sunlight all day and the other may be in the shade for most of the day.

4 CHAPTER Summary

4.1 Climate

- A region's climate is defined by year-after-year patterns and averages of temperature and precipitation.
- Global climate is shaped by many factors, including the amount of solar energy that is trapped in the biosphere. The transport of heat by winds and ocean currents also shapes global climate.

climate (p. 82)
microclimate (p. 82)
greenhouse effect (p. 83)

4.2 Niches and Community Interactions

- A niche is the range of physical and biological conditions in which a species lives. It includes the way that a species obtains what it needs to survive and reproduce.
- Competition causes species to divide resources. It helps determine the number and kinds of species in a community. It also helps to shape the niche each species occupies.
- Predators and herbivores can affect the size of other populations in a community. Predators can affect where prey populations can survive. Herbivores can help determine where populations of certain plants can survive and grow.
- The three main classes of symbiotic relationships in nature are mutualism, parasitism, and commensalism.

habitat (p. 85)
niche (p. 85)
resource (p. 85)
predation (p. 86)
herbivory (p. 86)
symbiosis (p. 87)
mutualism (p. 87)
parasitism (p. 87)
commensalism (p. 87)

4.3 Succession

- Ecosystems change over time, especially after disturbances. As some species die out, new species move in.
- In healthy ecosystems, secondary succession often reproduces the original climax community after a natural disturbance. But ecosystems may or may not recover from widespread disturbances caused by humans.

ecological succession (p. 88)
primary succession (p. 88)
pioneer species (p. 88)
secondary succession (p. 88)

4.4 Biomes

- Biomes are described in terms of abiotic factors like climate and soil type. They are also described by biotic factors like plant and animal life.
- Mountain ranges and polar ice caps are not usually classified into biomes. They are not easily defined in terms of a typical community of plants and animals.

canopy (p. 92)
understory (p. 92)
humus (p. 93)
taiga (p. 94)
permafrost (p. 94)

4.5 Aquatic Ecosystems

- Aquatic organisms are affected primarily by the water's depth, temperature, flow, and amount of dissolved nutrients.
- Freshwater ecosystems can be divided into three main categories: rivers and streams, lakes and ponds, and freshwater wetlands.
- Estuaries serve as spawning and nursery grounds for many ecologically and commercially important fish and shellfish species.
- Ecologists divide the ocean into zones based on depth and distance from shore.

benthos (p. 96)
plankton (p. 97)
wetland (p. 97)
estuary (p. 98)

BIOLOGY.com Search Chapter 4 GO • Crossword • Chapter Assessment

UbD Transfer the Big Idea

TRANSFER Have groups of students select a land or aquatic ecosystem that has been disturbed by human activity, is threatened by human actions, or has been restored to health after a human-caused disturbance. Have each group make a scrapbook showing the biotic and abiotic factors important to the ecosystem and how human activities have affected them or could affect them. Encourage student groups to include potential solutions to the problems they describe, if possible. Tell students they can use photographs, drawings, maps, graphs, and short passages of text to convey the information.

Chapter Summary

Think Visually

USE GRAPHICS Have students create a concept map that includes the following terms: *abiotic factors, biotic factors, community interactions, predation, competition, symbiosis, nutrients, ecosystems, light,* and *oxygen*.

ANSWERS
Student concept maps should show that ecosystems are determined by abiotic factors, including nutrients, light, and oxygen; and by biotic factors, including the community interactions of predation, competition, and symbiosis.

BIOLOGY.com Study Online

REVIEW RESOURCES

Visual Analogy Find out how Earth's atmosphere acts like the glass in a greenhouse.

Data Analysis Students analyze tolerance data to explain the zonation pattern of intertidal species.

Art in Motion An animation of primary and secondary succession helps students understand the processes.

Art Review Students can show an understanding of the ocean's zones with this drag-and-drop activity.

Vocabulary Review The *Flash Cards* and *Crossword* provide an interactive way to review chapter vocabulary.

Editable Worksheets Pages of Study Workbook B, Lab Manual B, and the Assessment Resources Book are available online. These documents can be easily edited using a word-processing program.

Lesson Overviews Lesson Overviews are based on Miller & Levine Biology. However, they are available on CD-ROM and can be edited for your class.

ASSESSMENT RESOURCES

Chapter Assessment Have students take an online version of the Chapter 4 Check Understanding.

Standardized Test Prep Students can take an online version of the Standardized Test Prep. You will receive their scores along with ideas for remediation.

Diagnostic and Benchmark Tests Use these tests to monitor your students' progress and supply remediation.

ASSESSMENT

Assess

Big idea Invite a volunteer to read the Big Idea and the question. Write the words *biotic* and *abiotic* on the board.

Ask What are the important abiotic factors in any ecosystem? *(Sample answers: latitude, amount of light and water, temperature, soil type, prevailing winds)*

Ask What are biotic factors in an ecosystem? *(the plants, animals, and other organisms that live in the ecosystem)*

Ask How do abiotic and biotic factors shape an ecosystem? *(Abiotic factors such as climate and soil type determine what kinds of organisms can survive in an ecosystem. Interactions between organisms influence population sizes and the niche that each organism occupies. Examples of these interactions include competition, symbiotic relationships, and predator-prey and herbivore-plant relationships.)*

CONSTRUCTED RESPONSE Remind students that a paragraph often begins with a main idea followed by supporting details. The most important details should come before less important details.

Write Question 1 on the board and read it aloud. Underline *abiotic factors* and *primary succession*.

Say This question is about how a new community begins in a barren area that starts out with no trace of its former community.

Read the first hint.

Ask Are volcanic rock and ash abiotic or biotic factors? *(abiotic)* What has to happen before plants can grow here? *(Soil has to be formed by pioneer species such as lichens and moss.)*

Read the next hint.

Ask How would primary succession differ in different biomes? *(Initial stages would not be very different, but types of plants that colonize once soil develops would differ according to abiotic factors characteristic of that biome.)*

Model writing the first one or two sentences using the hints on the board and in the text. *(Sample: The first organisms to appear after the eruption would be lichens that help break down rock. Most plant species would be unable to grow in the area until soil forms.)*

4 CHECK Understanding

Assess the Big idea **Interdependence in Nature**

Write an answer to the question below.

Q: How do abiotic and biotic factors shape an ecosystem?

Constructed Response

Write an answer to each of the numbered questions below. The answer to each question should be one or two paragraphs. To help you begin, read the **Hints** below the questions.

1. **How do abiotic factors determine which organisms are involved in primary succession after a volcanic eruption?**

   Hint Volcanic eruptions deposit large amounts of volcanic rock and ash.

   Hint Volcanoes can be found in many different biomes.

2. **How does introducing an invasive species into an ecosystem demonstrate the competitive exclusion principle?**

   Hint Introduced species often have no predators or parasites in their new environments. How might this affect their competition with native species?

3. **What type of symbiosis occurs between a cow and the bacteria in its stomach?**

   Hint The bacteria can digest cellulose, but the cow cannot.

Foundations for Learning Wrap-Up

Use the index cards you prepared when you started the chapter as tools to organize your thoughts about ecosystems and communities.

Activity 1 Working with a partner, take turns picking vocabulary index cards at random. Identify each term as a biotic factor or an abiotic factor in an ecosystem and explain why it is biotic or abiotic.

Activity 2 Working in a small group, divide all the vocabulary and ecosystem cards into groups by lesson. Construct concept maps by arranging the cards on a table and explain how the terms are related.

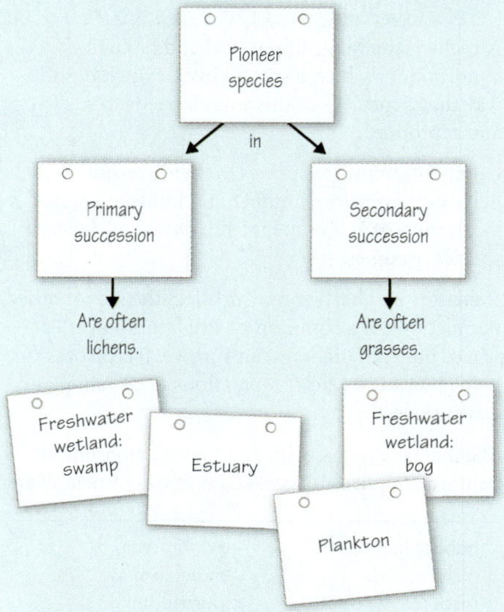

For Question 2, pair students and have them use the hint and your modeling to answer the question. Have students share their paragraphs with the class.

For Question 3, have students write the paragraphs independently for homework or as an anchor activity.

Foundations for Learning Wrap-Up

Have students take out their index cards. Be sure that students have written all the vocabulary and definitions. Either assign groups or pair students, and explain the activities.

4.1 Climate

Understand Key Concepts

1. The average temperature in a certain valley is usually higher than that of the surrounding countryside. This means that the valley has its own
   a. weather.
   b. climate.
   c. rainfall.
   d. microclimate.

Test-Taking Tip

Choose Among Similar Answers Sometimes questions have answer choices that are similar to each other. It may help if you first define each of the answer choices. In question 2, answer choices **a**, **b**, and **d** are similar. The correct answer is **d** because microclimate refers to the conditions in a small area. A valley is a small area.

2. Distinguish between *weather* and *climate*.

Think Critically

3. **Apply Concepts** Based on the relative positions of the sun and Earth, explain why Earth has climate zones and seasons.

4.2 Niches and Community Interactions

Understand Key Concepts

4. The relationship between a tick and its host is an example of
   a. mutualism.
   b. parasitism.
   c. commensalism.
   d. succession.

5. What is the competitive exclusion principle?

Think Critically

6. **Compare and Contrast** How are predation and parasitism different?

4.3 Succession

Understand Key Concepts

7. Fires, hurricanes, and other natural disturbances can result in
   a. commensalism.
   b. competition.
   c. parasitism.
   d. succession.
8. The first organisms to repopulate an area affected by a volcanic eruption are called
   a. keystone species.
   b. climax species.
   c. primary producers.
   d. pioneer species.
9. Describe two major causes of ecological succession.

Think Critically

10. **Relate Cause and Effect** Why does secondary succession usually proceed faster than primary succession?

4.4 Biomes

Understanding Key Concepts

11. Permafrost characterizes the biome called
    a. taiga.
    b. boreal forest.
    c. savanna.
    d. tundra.
12. Why are there generally few plants in a desert?

Think Critically

13. **Apply Concepts** Although the amount of precipitation is low, most parts of the tundra are very wet during the summer. How would you explain this apparent contradiction?

Assessment Tips

Encourage students to work in pairs or in small groups to answer the questions.

For Questions 4 and 6, students need to understand the three main classes of symbiotic relationships. Tell them to reread the Symbioses section in Lesson 4.2.

For Questions 14 and 16 students should review the Ocean Zones diagram in Lesson 4.5.

Have students review the Productivity of Aquatic Land Ecosystems table before answering Questions 17 and 18. Explain that biomes that support many plants and animals will be the most productive.

Answers

Lesson 4.1

UNDERSTAND KEY CONCEPTS

1. d

2. weather—day-to-day variations in environmental conditions; climate—year-to-year variations in environmental conditions

THINK CRITICALLY

3. Earth's axis is tilted, causing the intensity of sunlight reaching the surface in temperate and polar regions to change as the Earth orbits the sun every year. Sunlight always hits the equator with the same intensity, which makes seasonal changes in tropical areas minimal.

Lesson 4.2

UNDERSTAND KEY CONCEPTS

4. b

5. No two species can occupy exactly the same niche in the same habitat at the same time.

THINK CRITICALLY

6. Predation: Predators hunt and kill prey for food. Parasitism: Parasites get nutrients from the host, usually without killing the host.

Lesson 4.3

UNDERSTAND KEY CONCEPTS

7. d

8. d

9. Sample answers: natural disturbances, such as wildfires, hurricanes, volcanic eruptions; human-caused disturbances such as clearing grasslands or forests for cultivation or for roads and buildings

THINK CRITICALLY

10. Secondary succession usually proceeds faster because soil and a few plants are already in place when secondary succession starts.

Lesson 4.4

UNDERSTAND KEY CONCEPTS

11. d

12. The low levels of precipitation in a desert can support only a few plants.

THINK CRITICALLY

13. Because of permafrost, only the top few centimeters of the ground thaw in the summer. What little precipitation that does fall cannot soak into the permanently frozen subsoil, and instead, makes the ground soggy.

ASSESSMENT

Lesson 4.5

UNDERSTAND KEY CONCEPTS

14. c

15. Both are estuaries—mixtures of salt and fresh water that form where freshwater rivers and streams empty into oceans.

THINK CRITICALLY

16. Sample answer: Some animals may have large eyes that gather as much light as possible. Other animals may have adaptations that allow them to survive in very cold conditions.

Connecting Concepts

17. coral reef

18. Three-fourths of Earth's surface is covered with water. Even though the average productivity per square meter is low, there is so much more area that the accumulated amount of photosynthesis taking place is much larger than in other ecosystems.

CHAPTER MYSTERY After students have read through the Chapter Mystery, discuss the results of removing wolves from Yellowstone National Park, and the results of reintroducing them to the park.

Ask What happened when wolves were removed from the park? *(The elk population increased. Grazing by larger numbers of elk reduced the size of plant communities, especially along stream banks. Aquatic food webs were changed.)*

Ask What happened when wolves were reintroduced into the park? *(The elk population decreased. Plants grew back along the banks of streams, and populations of smaller herbivores increased)*

4 CHECK Understanding

4.5 Aquatic Ecosystems

Understanding Key Concepts

14. Kelp forests and coral reefs are types of communities found in which marine ecosystem?

a. intertidal zone

b. open ocean photic zone

c. coastal ocean

d. open ocean aphotic zone

15. How are salt marshes and mangrove swamps alike?

Think Critically

16. **Infer** The deep ocean lies within the aphotic zone and is very cold. What are some unique characteristics that might enable animals to live in the deep ocean?

Connecting Concepts

Use Science Graphics

The following table presents primary productivity (measured in grams of organic matter produced per year per square meter) for several ecosystems. Use the table below to answer questions 17 and 18.

Productivity of Aquatic and Land Ecosystems

Ecosystem	Average Primary Productivity
Aquatic Ecosystems	
Coral reef	2500
Estuary	1800
Open ocean	125
Land Ecosystems	
Tropical rain forest	2200
Tropical savanna	900
Tundra	90

17. **Interpret Tables** According to the table, which ecosystem is most productive?

18. **Infer** The open ocean is among the least productive ecosystems per square meter. But most of Earth's photosynthesis takes place in its photic zone. How might the amount of water that covers Earth explain why these two facts can both be true?

solve the CHAPTER MYSTERY

THE WOLF EFFECT

Removing wolves from Yellowstone National Park contributed to an increase in the number of elk. These elk grazed so heavily that the seedlings and shoots of aspens, willows, and other trees could not grow. This was especially a problem along streams. Fewer trees meant beavers built fewer dams, which led to an increase in runoff and erosion. Aquatic food webs broke down, affecting birds, fish, and other animals. The recent reintroduction of wolves has caused a decrease in the overall elk population. The presence of wolves seems to have reduced elk grazing along certain streams. That may be in part because wolves are killing more elk. It may also be in part because elk have learned to stay away from places like stream banks and valleys. Wolves can attack them most easily in these areas.

In recent years, researchers have shown that streamside vegetation is exhibiting secondary succession. The aspen and willow trees are starting to grow back. There have been numerous other changes as well. Fewer elk mean more food for smaller animals. The increase in small prey, in turn, has brought diverse predators into the community. Carcasses abandoned by the wolves provide food for scavengers. In short, organisms from every trophic level have been affected by the Yellowstone wolves.

1. **Predict** The Yellowstone wolf and elk are linked through a predator-prey relationship. If a disease were to strike the elk population, how would this affect the wolves?

2. **Apply Concepts** Why might you consider the wolves to be a keystone species?

Untamed Science **Never Stop Exploring Your World.** The mystery of the Yellowstone wolves is just the beginning. Take a video field trip with the ecogeeks of Untamed Science to see where this mystery leads.

CHAPTER MYSTERY ANSWERS

1. In order to survive, the wolves would have to consume more of other prey species. If that were not possible, the wolf population would drop.

2. Accept reasonable answers. A population change in a keystone species can dramatically change the size of other populations in the community. Student answers should point to the many species that were affected by removal and then reintroduction of wolves as evidence that wolves are a keystone species.

Untamed Science Students can learn more about succession by watching **From Lava to Life.**

Standardized Test Prep

Multiple Choice

1. The factor that generally has the greatest effect on determining a region's climate is its
   A longitude.
   B abundant plant species.
   C distance from the equator.
   D closeness to a river.

2. All of the following are abiotic factors that affect global climate EXCEPT
   A latitude. C solar energy.
   B longitude. D ocean currents.

3. The way an organism makes its living, including its interactions with biotic and abiotic factors of its environment, is called the organism's
   A habitat. C lifestyle.
   B niche. D biome.

4. If a newly introduced species fills a niche that is normally occupied by a native species, the two species compete. One of the species may die out as a result of
   A competitive exclusion.
   B predation.
   C commensalism.
   D mutualism.

5. Photosynthetic algae are MOST likely to be found in
   A the open-ocean benthic zone.
   B the aphotic zone.
   C the photic zone.
   D ocean trenches.

6. The water in an estuary is
   A salt water only.
   B poor in nutrients.
   C fresh water only.
   D a mixture of fresh water and salt water.

7. In which biome do organisms have the greatest tolerance to dry conditions?
   A tundra C tropical savanna
   B desert D boreal forest

Questions 8–9

Month-by-month climate data for the city of Lillehammer, Norway, is shown in the table below.

Climate Data for Lillehammer, Norway

Month	Average Temperature (°C)	Average Precipitation (mm)
Jan.	−8.1	38.1
Feb.	−6.2	27.9
Mar.	−3.9	30.5
Apr.	3.3	35.6
May	8.9	45.7
June	13.9	63.5
July	16.4	81.3
Aug.	14.2	88.9
Sept.	9.5	58.4
Oct.	3.9	63.5
Nov.	−3.8	50.8
Dec.	−6.1	48.3

8. Which type of graph would be BEST suited to showing the precipitation data from the table?
   A bar graph C pie chart
   B pictograph D scatter plot

9. For a given set of data, the range is the difference between highest and lowest points. The average annual temperature range, in °C, for Lillehammer is approximately
   A −8.
   B 8.5.
   C 16.5.
   D 24.5.

Open-Ended Response

10. Why are lichens especially well adapted to play the role of pioneer organisms in an ecological succession?

If You Have Trouble With . . .

Question	1	2	3	4	5	6	7	8	9	10
See Lesson	4.1	4.1	4.2	4.2	4.5	4.5	4.4	4.1	4.1	4.3

Test-Taking Tip

INTERPRET GRAPHS

Tell students to carefully review a graph before they answer any questions that relate to it. Make sure students read the title, axis labels, and any other information on the graph. Suggest they mentally summarize what the graph shows before reading any questions. Once they are sure they fully understand the graph, then suggest they read and answer the questions.

Before the Test

Review the Test-Taking Tip with students.

Point out the word *EXCEPT* in Question 2 and the word *MOST* in Question 5.

Remind students that the paragraphs they write to answer an open-ended question should contain the same facts as everyone else's paragraphs, but the order of the facts, examples used, and style of writing probably will not be the same.

Answers

1. C
2. B
3. B
4. A
5. C
6. D
7. B
8. A
9. D
10. Lichens can grow on bare rock. They fix nitrogen from the atmosphere by changing it into a form plants and other organisms can use, break down rock, and help build soil.

CHAPTER PLANNER 5 Populations

Chapter Contents	NSES	Time	Core Resources
Chapter Preview			**Student Edition,** pp. 106–107 **Chapter Mystery,** p. 107 **Foundations for Learning,** p. 107
5.1 How Populations Grow Describing Populations • Population Growth • Exponential Growth • Logistic Growth	II, III, IV, A.2.d, C.4.d, C.4.e, F.2	1 period 1/2 block	**Student Edition,** pp. 108–111 **Study Workbook B** 5.1 Worksheets **Biology.com** 5.1 Lesson Assessment
5.2 Limits to Growth Limiting Factors • Density-Dependent Limiting Factors • Density-Independent Limiting Factors	I, III, IV, A.1.b, A.2.a, A.2.d, C.4.d, C.4.e, F.2	1 period 1/2 block	**Student Edition,** pp. 112–116 **Inquiry Into Scientific Thinking,** p. 116 **Study Workbook B** 5.2 Worksheets **Biology.com** *Art Review:* Limiting Factors • *InterActive Art:* Moose-Wolf Populations on Isle Royale • 5.2 Lesson Assessment
5.3 Human Population Growth Historical Overview • Patterns of Human Population Growth	II, III, A.2.a, A.2.d, C.4.d, C.4.e, F.2, F.3, F.6, G.2	1/2 period 1/4 block	**Student Edition,** pp. 117–119 **Study Workbook B** 5.3 Worksheets **Biology.com** *Art in Motion:* Age Structure of World Population • 5.3 Lesson Assessment
Chapter Pre-Lab	A.1.b, A.1.c, A.2.a, F.2	1 period 1/2 block	**Student Edition,** p. 120 **Lab Manual B** *The Growth Cycle of Yeast*

Tools for ELL Instruction

ELL Handbook describes teaching strategies to make *Miller & Levine Biology: Foundation Edition* more accessible to ELL students.

Spanish Study Workbook is a Spanish translation of Study Workbook A.

Multilingual Glossary is the *Miller & Levine Biology* glossary translated into ten languages.

Tools to Challenge Students

Study Workbook A includes more challenging worksheets and activities such as Chapter Mystery activities that reinforce 21st Century Skills.

Lab Manual A labs are geared toward on-level students. The Lab Manual includes fifteen labs in addition to the Chapter Labs.

Biology.com Data Analysis activities offer students a way to practice data gathering and analyzing skills.

Visual Quizzes in the Assessment Resources Book are geared toward on-level students and use visuals from *Miller & Levine Biology.*

Additional Resources

Biology.com Untamed Science Video • Flash Cards

Study Workbook A 5.1 Worksheets
Lab Manual B *Data Analysis:* Multiplying Rabbits
Spanish Study Workbook 5.1 Worksheets
Biology.com *Data Analysis:* Invasion of the Zebra Mussels
Assessment Resources Book Visual Quizzes

Study Workbook A 5.2 Worksheets
Lab Manual B *Hands-On Activity:* How Competition Affects Growth
Spanish Study Workbook 5.2 Worksheets

Study Workbook A 5.3 Worksheets
Spanish Study Workbook 5.3 Worksheets
Assessment Resources Book Visual Quiz

Lab Manual A *The Growth Cycle of Yeast*

Chapter Review

Student Edition Chapter Summary, p. 121
Study Workbook B Chapter 5 Review
Transparencies, pp. 55–63
Biology.com Untamed Science Video • Editable Worksheets of Study Workbooks A and B and Lab Manuals A and B • Chapter 5 Flash Cards and Match It

Untamed Science DVD • Classroom Resources CD (includes lesson presentations and editable worksheets)

Chapter Assessment

Student Edition Check Understanding, pp. 122–125
Study Workbook B Chapter 5 Taking a Standardized Test
Assessment Resources Book Chapter 5 Test B
Biology.com Chapter 5 Assessment • Editable Worksheets of Chapter 5 Visual Quizzes* and Chapter 5 Tests A and B

Exam*View* *Assessment Suite* • Classroom Resources CD (includes lesson presentations and editable worksheets)

Time: 1 period, 1/2 block

* Visual Quizzes are based on visuals from *Miller & Levine Biology*. Review carefully to make sure they are appropriate for your class.

Pressed for Time?

Preview the Chapter Introduce the Key Questions for Lessons 5.1 and 5.2. Preview the Exponential Growth and Logistic Growth graphs and the Limiting Factors visual.

Cover the Chapter Quickly In Lesson 5.1, have students read *Describing Populations, Population Growth,* and the introductions to the *Exponential Growth* and *Logistic Growth* sections. Discuss the Exponential Growth and Logistic Growth graphs. Have students read all of Lesson 5.2, focusing on the Moose and Wolf Populations on Isle Royale graph.

Assess Assign Questions 2, 4, and 5 in the 5.1 Check Understanding, the 5.2 Check Understanding, and Questions 1, 2, 4, and 6–10 in the Chapter 5 Check Understanding.

CHAPTER 5

Get the Picture

Tell students that the picture shows a population of red crabs, and remind them that a population consists of organisms belonging to the same species living in the same area. Guide students to see that organisms in two animal populations are visible, red crabs and humans, as well as several plant populations.

Ask Why are the red crabs on the move? *(They are leaving the forest to reach the sea, where they breed.)* Why do all the adult crabs make this trip at the same time? *(They have to make the trip to participate in breeding.)*

Explain that for a population to survive, the organisms that make up the population must reproduce.

Activate Prior Knowledge

Help students make the connection that the factors that organisms require to live also determine how a population changes.

Ask What do both you and a red crab need to live? *(Sample answers: food, water, oxygen, shelter)*

Ask What types of abiotic factors and community interactions might affect the number of organisms living in a place? *(Sample answers: abiotic factors such as rainfall and temperatures of an area, and community interactions such as competition, predation, and parasitism)*

Remind students that the abiotic factors described in Chapter 3 determine where particular organisms live, and explain that changes in abiotic conditions affect populations. Explain also that the community interactions discussed in Chapter 4, such as those between prey and predators, help determine whether a population grows, shrinks, or stays the same.

NATIONAL SCIENCE EDUCATION STANDARDS

UNIFYING CONCEPTS AND PROCESSES

I, II, III, IV

CONTENT

C.4.d, C.4.e, F.2, F.3, F.6, G.2

INQUIRY

A.1.b, A.1.c, A.2.a, A.2.d

5 Populations

Big idea Interdependence in Nature

Q: What factors contribute to changes in populations?

BIOLOGY.com Search Chapter 5 GO • Flash Cards

106

UbD Connect to the Big Idea

Invite a volunteer to read aloud the chapter number and title. Read the Big Idea and question. Emphasize the question, **What factors contribute to changes in populations?**

Set the Purpose

PREVIEW AND PREDICT Draw the graphic organizer on the board. Refer students to the Big Idea. Invite students to infer what the prefix *inter-* means. *(between, among)*

Ask What do you think the word *interdependence* means? *(dependence between)* How might interdependence in nature relate to the questions in the graphic organizer? *(Changes in a population depend on many factors in nature. In turn, changes in populations can affect these factors.)*

Remind students to search for answers as they read.

INSIDE:
- 5.1 How Populations Grow
- 5.2 Limits to Growth
- 5.3 Human Population Growth

Millions of red crabs live on Christmas Island in the Indian Ocean. Each year, all of the adult crabs move from forest to sea to breed.

CHAPTER
MYSTERY

A PLAGUE OF RABBITS

In 1859, an Australian farmer released 24 English rabbits on his ranch. In ten years, more than 2 million rabbits were hunted on that ranch alone! Soon, millions of rabbits spread across the country and caused lots of trouble. They ate native plants that used to feed native animals. The loss of those food plants pushed native animals close to extinction. The rabbits also made life miserable for sheep and cattle ranchers. Yet those rabbits weren't a problem in England. Why were they such a problem in Australia? Could they be stopped? How?

Read for Mystery Clues As you read this chapter, look for clues to help you predict why this rabbit population grew so much. Think about factors that affect population growth. Then, solve the mystery at the end of the chapter.

FOUNDATIONS for Learning

Before you read this chapter, read the lesson titles and headings. Then, fold a piece of paper into three columns. In the first column, list what you know about populations. In the second column, write questions about what you want to learn. In the third column, you will write what you have learned. There are two activities at the end of the chapter that will require you to use your paper to answer the question: What factors contribute to changes in populations?

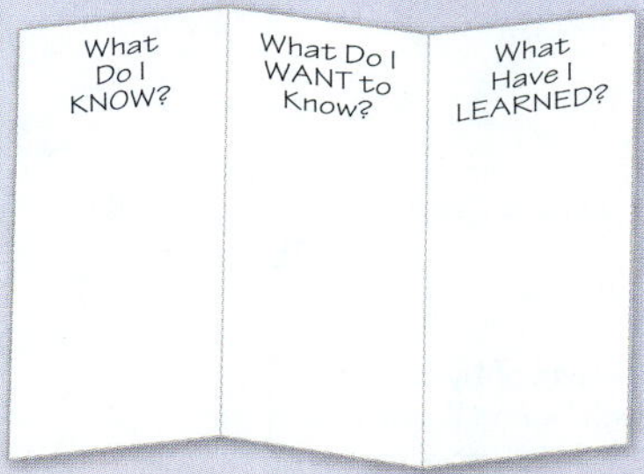

- Untamed Science Video
- Chapter Mystery

Chapter 5 Big Idea: Interdependence in Nature

Chapter 5: What factors contribute to changes in populations?

- **5.1:** How do populations grow?
- **5.2:** What factors limit a population's growth?
- **5.3:** How is the human population growing?

Foundations for Learning

Have students take out one sheet of paper, and demonstrate how to fold the paper. Help students understand the importance of this activity.

Ask How do you think it might help to identify what you already know about populations? *(Identifying what we know helps us build on that knowledge.)* How might writing the questions help you understand this chapter? *(Writing questions before reading helps us focus on what we want to learn more about.)*

Remind students that they will use this paper as a study aid as they read this chapter and will complete it at the end of the chapter. Familiarize students with the activity by modeling with an example. (first column: A population consists of a single species.; second column: What affects population growth?)

Untamed Science™

In **Census Consensus,** the ecogeeks of Untamed Science investigate how scientists count individuals in a population—including the ones that try to hop away.

CHAPTER MYSTERY

Invite a volunteer to read aloud the Chapter Mystery.

Ask What growing populations have you observed? *(Sample answers: fish in a tank, weeds in the yard, people in a town)*

Have students predict how the chapter question, **What factors contribute to changes in populations?**, will help reveal the solution to this mystery.

Tell students that they will find clues in the chapter to answer the question, **Why did the rabbit population grow so much in Australia?**

Digital Support

BIOLOGY.com Extend your reach by using the digital assets offered at **Biology.com.**

LESSON 5.1

Pre-Reading

Objectives

5.1.1 List the characteristics used to describe a population.

5.1.2 Identify factors that affect population growth.

5.1.3 Describe exponential growth.

5.1.4 Describe logistic growth.

Preview the Pages

1. Preview the visuals and read the captions and labels. When previewing the Exponential Growth graph, point out the units of measure on the axes of the graph. Indicate how the specific labels on this graph differ from the general ones on the Logistic Growth graph.
2. Review the vocabulary words by having students find the words in the text and read each aloud. Ask students to identify any words that are familiar as they relate to human populations.
3. Review the use of exponents and the way in which they are written.

Build Understanding

CONCEPT MAP Direct students to Workbook B, where they will find a partially filled-in concept map that they should complete as they read. Remind students that a concept map will help them understand the relationships between key vocabulary terms in the lesson.

NATIONAL SCIENCE EDUCATION STANDARDS

UNIFYING CONCEPTS AND PROCESSES

II, III, IV

CONTENT

C.4.d, C.4.e, F.2

INQUIRY

A.2.d

For the corresponding lesson in *Miller & Levine Biology,* **see pages 130–135.**

5.1 How Populations Grow

Key Questions

How do ecologists study populations?

What factors affect population growth?

What happens during exponential growth?

What is logistic growth?

BUILD Understanding

Concept Map As you read, make a concept map that organizes the information in this lesson.

In Your Workbook Go to your workbook for help in completing the concept map. Complete the concept map for Lesson 5.1.

Describing Populations

About 60 years ago, a fish farmer in Florida tossed some hydrilla plants into a canal. These few plants, which were brought here from Asia, reproduced quickly. Today, hydrilla can be found across Florida and many other states. Because hydrilla grow so thickly, they kill other plants and animals. Why do hydrilla grow so quickly?

Ecologists try to answer such questions. Ecologists study a population's range, density, growth rate, and age structure.

Geographic Range A population's geographic range describes the spaces where it lives. Different populations can have very different ranges. For example, the range of a bacterial population may be tiny. But the population of cod in the western Atlantic Ocean ranges from Greenland to North Carolina.

Density and Distribution Population density is the number of individuals of a species per unit area. In a meadow, for example, you might find a dense population of grasses but a less dense population of spiders. Population distribution describes how the organisms are spread across their range. The distribution of a population may be random, uniform, or clumped.

Growth Rate A population's growth rate tells you if the population size is getting bigger, smaller, or staying the same. The population of hydrilla in Florida has a very high growth rate. Populations can also decrease in size. The cod population, which is decreasing, has a negative growth rate.

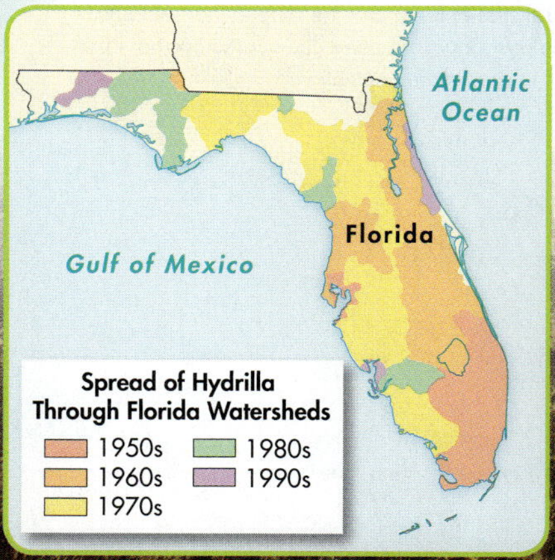

Invasive Hydrilla Hydrilla has spread throughout most of Florida in just a few decades. Efforts to control the waterweed cost millions of dollars a year.

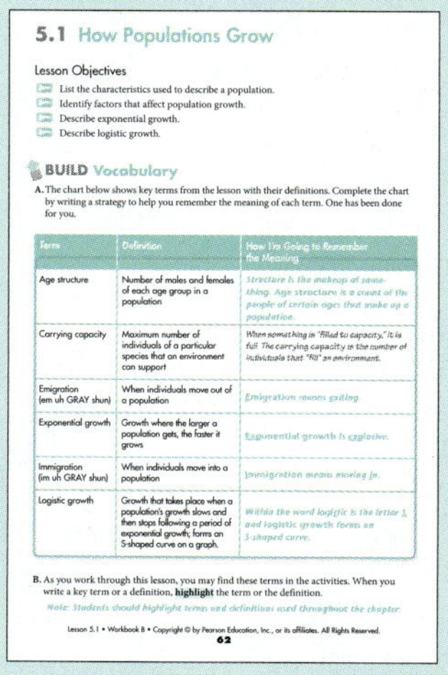

5.1 How Populations Grow

Lesson Objectives

- List the characteristics used to describe a population.
- Identify factors that affect population growth.
- Describe exponential growth.
- Describe logistic growth.

BUILD Vocabulary

A. The chart below shows key terms from the lesson with their definitions. Complete the chart by writing a strategy to help you remember the meaning of each term. One has been done for you.

Term	Definition	How I'm Going to Remember the Meaning
Age structure	Number of males and females of each age group in a population	*Structure is the makeup of something. Age structure is a count of the people of certain ages that make up a population.*
Carrying capacity	Maximum number of individuals of a particular species that an environment can support	*When something is "filled to capacity," it is full. The carrying capacity is the number of individuals that "fill" an environment.*
Emigration (em uh GRAY shun)	When individuals move out of a population	*Emigration means exiting.*
Exponential growth	Growth where the larger a population gets, the faster it grows	*Exponential growth is explosive.*
Immigration (im uh GRAY shun)	When individuals move into a population	*Immigration means moving in.*
Logistic growth	Growth that takes place when a population's growth slows and then stops following a period of exponential growth; forms an S-shaped curve on a graph.	*Within the word logistic is the letter S, and logistic growth forms an S-shaped curve.*

B. As you work through this lesson, you may find these terms in the activities. When you write a key term or a definition, **highlight** the term or the definition.

Note: Students should highlight terms and definitions used throughout the chapter.

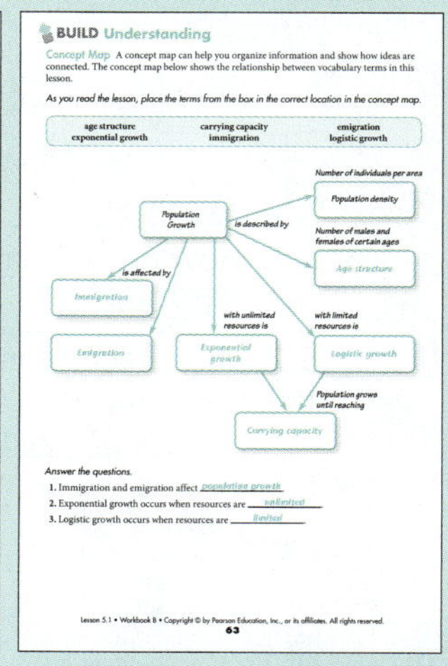

BUILD Understanding

Concept Map A concept map can help you organize information and show how ideas are connected. The concept map below shows the relationship between vocabulary terms in this lesson.

As you read the lesson, place the terms from the box in the correct location in the concept map.

age structure	carrying capacity	emigration
exponential growth	immigration	logistic growth

Answer the questions.

1. Immigration and emigration affect *population growth*.
2. Exponential growth occurs when resources are *unlimited*.
3. Logistic growth occurs when resources are *limited*.

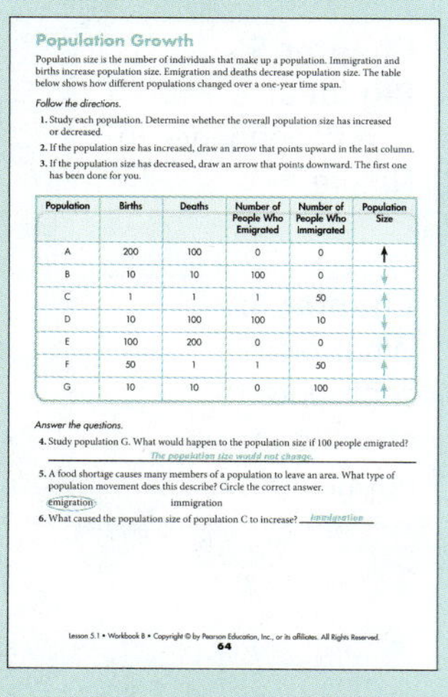

Population Growth

Population size is the number of individuals that make up a population. Immigration and births increase population size. Emigration and deaths decrease population size. The table below shows how different populations changed over a one-year time span.

Follow the directions.

1. Study each population. Determine whether the overall population size has increased or decreased.
2. If the population size has increased, draw an arrow that points upward in the last column.
3. If the population size has decreased, draw an arrow that points downward. The first one has been done for you.

Population	Births	Deaths	Number of People Who Emigrated	Number of People Who Immigrated	Population Size
A	200	100	0	0	↑
B	10	10	100	0	↓
C	1	1	1	50	↑
D	10	100	100	10	↓
E	100	200	0	0	↓
F	50	1	1	50	↑
G	10	10	0	100	↑

Answer the questions.

4. Study population G. What would happen to the population size if 100 people emigrated? *The population size would not change.*
5. A food shortage causes many members of a population to leave an area. What type of population movement does this describe? Circle the correct answer.
   (emigration) immigration
6. What caused the population size of population C to increase? *Immigration*

Age Structure The **age structure** of a population describes the number of males and females of each age in the population. Age structure is important because in animals only the females between certain ages can reproduce.

Key Question How do ecologists study populations?
Ecologists study a population's range, density, growth rate, and age structure.

Population Growth

The size of a population changes based on how many individuals are added to it or removed from it. How are individuals added or removed?

Birthrate and Death Rate A population may grow if more individuals are born than die in any period of time. In this case, the birthrate is higher than the death rate. If the death rate is higher, the population may shrink. If these rates are about the same, the population may stay the same size.

Immigration and Emigration When individuals move into a population's range, the process is called **immigration** (im uh GRAY shun). When individuals move out of the range, it is called **emigration** (em uh GRAY shun). If immigration rates are higher than emigration rates, the population will grow.

Key Question What factors affect population growth?
Population growth is affected by birthrate and death rate. It can also be affected by immigration and emigration.

Exponential Growth

Imagine that a population has all the food and space it needs. It also has protection from predators and disease. Imagine also that its waste products are all removed. Under these conditions, the population will grow.

Organisms That Reproduce Rapidly Imagine a single bacterium that divides to form two cells every 20 minutes. Those two cells divide to form four cells. Those four cells divide again. So, after three 20-minute periods, there are $2 \times 2 \times 2$, or 8 cells, which can also be written using an exponent: 2^3 cells.

This situation is called exponential (eks poh NEN shul) growth. In **exponential growth,** the larger a population gets, the more quickly it grows. When you graph the growth, the graph is in the shape of a J.

BUILD Vocabulary

age structure
the numbers of males and females in different age groups of a population

immigration
the movement of individuals into an area occupied by an existing population

emigration
the movement of individuals out of an area

exponential growth
a growth pattern in which the individuals in a population reproduce at a constant rate

RELATED WORD FORMS
An exponent refers to the number of times a number is multiplied by itself. The adjective *exponential* describes something that can be expressed using exponents.

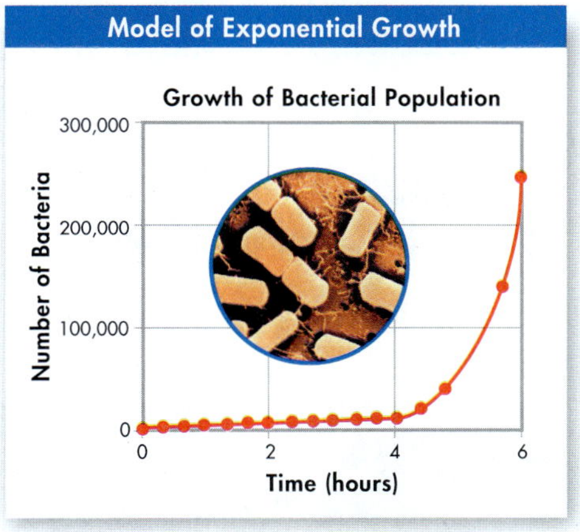

Exponential Growth When there are plenty of resources and few predators or diseases, populations will grow exponentially. This graph shows the characteristic J-shaped curve of exponential growth.

During Reading

Active Reading

KEY QUESTION Read aloud the first Key Question. Refer students to the subheads under Describing Populations to answer the question.

SELF-QUESTION Demonstrate how asking questions and reviewing the text for answers can help clarify new information. Read the third paragraph on this page aloud.

Ask Does this make sense? What do we call the process by which organisms leave their population? *(emigration)* How does this compare to immigration? *(It is the opposite. Through immigration, organisms join a population.)*

Say Sometimes it helps to reread information to find the answers to questions.

Have students reread the paragraph.

Ask Which factor causes a population to shrink: emigration or immigration? *(emigration)*

KEY QUESTION Invite a student to read aloud the second Key Question and answer.

Ask Will a population grow or shrink if its birthrate is higher than its death rate? *(grow)* Will the population grow or shrink if its emigration rate is higher than its immigration rate? *(shrink)* When asking this last question, emphasize the *e* in *emigration* and *i* in *immigration* to distinguish the two words.

USE VISUALS Refer students to the Model of Exponential Growth graph.

Ask How fast does the population grow during the first four hours? *(The population grows slowly.)* How does the growth rate during the last two hours compare to the growth rate of the first four hours? *(The population grows much more quickly during the last two hours.)*

Say The growth rate of a population increases as its size increases.

Build Vocabulary

RELATED WORD FORMS Invite a volunteer to read aloud the definition of *exponential growth* and the paragraphs in the section Exponential Growth. Write the following on the board: *Population A: 2, 4, 8, 16* and *Population B: 2, 5, 7, 8.*

Say These are the sizes of two different populations each year over four years.

Ask Which population is growing? *(Both are.)* Which population shows exponential growth? *(Population A)* How do you know? *(Population A grows more quickly each year. Its growth can be expressed using an exponent: 2^4.)*

Hands-On Learning

RECORD CHANGE IN A POPULATION Pair students. Give 15 small squares each of red and blue paper to each pair. Explain that each red square represents one female (F) in a population, and each blue square represents one male (M) in that population.

1. Have students record the following information in their notebooks: Original population age structure: 0–20 years old—2 M, 1 F; 20–40 years old—2 M, 3 F; 50–80 years old—1 M, 3 F. Population changes during one month: birthrate—1 M, 3 F; death rate—2 M, 3 F; emigration rate—2 M, 1 F; immigration rate—4 M, 5 F.
2. Tell students these steps. **1:** "Construct" the original population using the proper number of red and blue squares. Determine and record the original population size by gender. *(5 males, 7 females)* **2:** Add or subtract individuals according to the changes given. **3:** Determine and record the final population size by gender. *(6 males, 11 females)* Note whether the population is growing, shrinking, or staying the same size. *(growing)*

During Reading continued

Active Reading

INFER Have students reread the paragraph Organisms in New Environments.

Ask How many hydrilla plants did the farmer originally place into the Florida canal? *(a few)* What happened to the hydrilla population over 60 years? *(The population spread across Florida and many other states.)* What does this suggest about the growth rate of hydrilla? *(It suggests that hydrilla grew exponentially.)*

KEY QUESTION Read aloud the Key Question and answer.

Ask If a population is showing exponential growth, does its growth rate level off at a particular population size? Explain. *(No, in exponential growth, the population continues to grow more and more quickly.)*

USE VISUALS Refer students to the Logistic Growth graph.

Say A logistic growth graph is shaped like an S. Each part of the S shows the growth rate of a population during a different phase of growth.

Ask Where on the graph is population growth most rapid? *(where the line of the S is steepest)* Where on the graph is population size not changing much? *(where the line of the S is least steep or somewhat level)*

Speed Bump

Ask Do all populations that show exponential growth reproduce at the same rate? Explain. *(No, some reproduce very slowly and some very rapidly.)*

Ask How many growth phases do populations go through during logistic growth? *(three)*

Ask What growth phase is a population likely to go through when the organisms have plenty of food and space and are free of disease? *(exponential growth)*

Organisms in New Environments Does exponential growth ever happen in nature? Yes! That's what happened with hydrilla, and with many other species that humans have introduced into new habitats. When a species is moved into a new habitat, its population often grows exponentially for a time. But think about this situation for a minute. Can this growth continue without stopping?

Key Question What happens during exponential growth? **In exponential growth, the larger a population gets, the more quickly it grows.**

Logistic Growth

Exponential growth presents us with a puzzle. Obviously, bacteria and hydrilla don't cover the planet. This means that populations in the real world don't grow exponentially for very long. Sooner or later, population growth slows down. But why?

Phases of Growth To begin to answer this question, let's see what happens when a few individuals are brought into a new area. Usually the population will go through three phases. You have read a bit about the first phase—exponential growth. In the next phases, the growth slows and eventually stops.

Logistic Growth Population growth slows down as resources become limited. This graph shows the characteristic S-shaped curve of logistic growth.

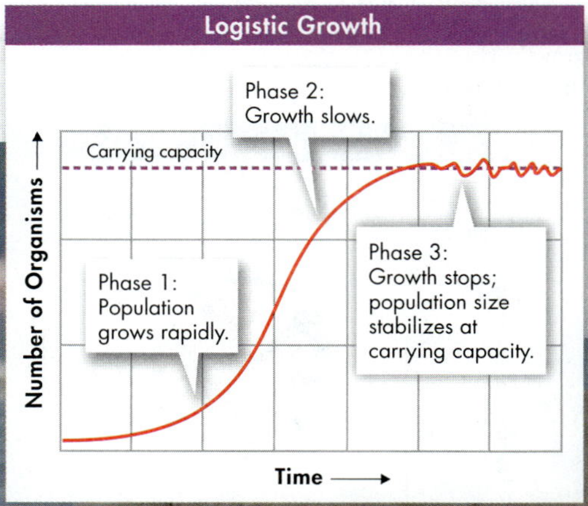

- ***Phase 1: Exponential Growth*** After a short time, the population begins to grow exponentially. There is plenty of food. There is plenty of space. The death rate is low. The birthrate is high. The rate of growth increases more and more quickly.
- ***Phase 2: Growth Slows Down*** In most populations, exponential growth does not last very long. At some point, the growth must slow down. The population keeps growing, but it does not grow as quickly.
- ***Phase 3: Growth Stops*** At some point, the growth rate drops to zero. The population size stays about the same.

ELL Focus on ELL: Build Background

ALL SPEAKERS Read aloud the caption for the Logistic Growth graph. Make sure students understand what it means that the S-shaped curve is "characteristic" of logistic growth. Have students draw a T-chart with the headings Exponential Growth and Logistic Growth. Direct students to include in the correct columns the following features that are characteristic of either type of growth or of both types: S-shaped curve, J-shaped curve, growth that slows, rapid growth, three phases, growth that stops, growth affected by death rate.

The Logistic Growth Curve When a population's growth slows down and stops, it is called **logistic growth.** A logistic growth curve is shaped like an S.

What kinds of changes in a population cause logistic growth? A population grows when the number of organisms added to the population is more than the number of organisms that leave the population. Population growth may slow for several reasons. Growth may slow because the birthrate slows. Growth may slow because the death rate increases. Immigration and emigration can also affect growth rates. If immigration decreases or emigration increases, growth will slow.

Carrying Capacity In the third phase of logistic growth, the population stops growing. Population growth stops when the birthrate and death rate are the same and when immigration equals emigration. Look again at the logistic growth curve. There is a broken, horizontal line in the graph where the population size stays about the same. This population size is the largest number of individuals an area can support. The largest number of individuals that an area can support is called the **carrying capacity.** The population size may go up or down somewhat. However, the size stays very near the carrying capacity.

Key Question What is logistic growth?
Logistic growth happens when the growth of a population begins as exponential growth and then slows and stops.

BUILD Vocabulary

logistic growth
a growth pattern in which a population's growth rate slows or stops, following a period of exponential growth

carrying capacity
the largest number of individuals of a population that a given environment can support

RELATED WORD FORMS
The word *logistics* means the way someone handles the resources of a project or event. The adjective *logistic* is used in this sense to describe growth that can be supported by the available resources.

CHECK Understanding

Apply Understanding

Use the highlighted words from the lesson to complete each sentence correctly.

1. ______ sometimes happens when a new kind of organism is introduced in an area.

2. The number of males and females of each age group in a certain population is called the ______.

3. The number of organisms an area can support over a long period of time is called the ______.

Critical Thinking

4. Relate Cause and Effect More dandelion seedlings develop in a lawn than dandelion plants are pulled. What is likely to happen to the lawn's dandelion population?

5. Explain Describe logistic growth.

6. Apply Concepts A few European gypsy moths were accidentally released from a laboratory near Boston. Describe what might have happened to the population over the next several years.

7. Write to Learn Answer the first clue of the mystery. Think about the different phases of population growth.

MYSTERY CLUE

What kind of growth is the rabbit population in Australia demonstrating? Why does that cause problems? (**Hint:** See p. 110.)

BIOLOGY.com Search Lesson 5.1 GO • Lesson Assessment

Check Understanding

1. Exponential growth

2. age structure

3. carrying capacity

4. The dandelion population is likely to grow larger.

5. Logistic growth is population growth that slows down as resources become limited. The population grows rapidly at first before slowing and stopping at carrying capacity.

6. Because the gypsy moths were released into a new environment, the population might have grown exponentially over the next few years.

7. The rabbit population is demonstrating exponential growth. That causes problems as the rabbits use up resources needed by other organisms.

Active Reading

DRAW CONCLUSIONS Have students reread the paragraphs describing the logistic growth curve.

Ask Which of these populations is most likely to show logistic growth: one with a high birthrate, one in which diseases are becoming common, or one with a high immigration rate? *(one in which diseases are becoming common)* Why? *(Diseases are likely to increase the death rate in a population.)*

KEY QUESTION Invite a volunteer to read aloud the Key Question and answer.

Ask Do populations showing logistic growth always grow slowly? Explain. *(No, during the first phase of logistic growth, the population increases rapidly, undergoing exponential growth.)*

Build Vocabulary

RELATED WORD FORMS Invite students to use the word *logistics* in a sentence. Ask them to give an example of being in charge of the logistics of an event or situation. *(Sample answers: planning the logistics of a party, coordinating logistics for getting to an after-school job, arranging the logistics of participating in sports and completing homework)*

Post Reading

ASSESSMENT TIPS Review the four factors that affect population growth: birthrate, death rate, immigration, and emigration.

For Question 4, have students compare the birthrate and death rate of the dandelion population.

For Question 5, have students look at the Logistic Growth graph to help them write their answers.

MYSTERY CLUE Refer students to the graphs showing exponential growth and logistic growth. Have them consider the rate at which the rabbit population is growing and whether that rate is slowing or stopping.

Wrap-Up Activity

PREDICT Divide students into small groups. Tell students that each group represents a population. Give each group eight file cards. On each card, have students write one of the following statements: *Food is in short supply. Birthrate is steady. Many organisms immigrate. Death rate is high. Many organisms emigrate. There is plenty of space. Reproduction rate is high. Wastes cannot be removed.* Ask students to shuffle the cards before selecting three. Have them decide from the cards whether their population is most likely to grow, shrink, or remain the same size.

LESSON 5.2

Pre-Reading

Objectives

5.2.1 Identify factors that determine carrying capacity.

5.2.2 Identify the limiting factors that depend on population density.

5.2.3 Identify the limiting factors that do not depend on population density.

Preview the Pages

1. Preview all the visuals and read the captions and labels.
2. Preview the vocabulary with students. Have them note that each term includes the words *limiting* and *factor.*
3. Introduce the concept of a limiting factor. Relate the discussion to raising fish in an aquarium or growing plants in a window box or garden. Ask students to name factors that could limit the growth of the fish or plant populations.

Build Understanding

LESSON PREVIEW Direct students to Workbook B, where they will find a series of questions that will guide them as they preview the lesson headings.

NATIONAL SCIENCE EDUCATION STANDARDS

UNIFYING CONCEPTS AND PROCESSES

I, III, IV

CONTENT

C.4.d, C.4.e, F.2

INQUIRY

A.1.b, A.2.a, A.2.d

For the corresponding lesson in ***Miller & Levine Biology,*** see pages 137–141.

5.2 Limits to Growth

Key Questions

What factors determine carrying capacity?

What limiting factors depend on population density?

What limiting factors do not usually depend on population density?

BUILD Understanding

Preview Before you read this lesson, write down the headings and look at the figures. As you read, write the main idea under each heading.

In Your Workbook Go to your workbook for help previewing the lesson.

Limiting Factors

Now that you have seen *how* populations tend to grow in nature, you can explore *why* they grow as they do. What factors cause the growth to slow down? What determines the carrying capacity of an environment for a species?

Think about the hydrilla discussed in Lesson 5.1. In Asia, the population reached its carrying capacity. Then, it stopped growing. In the United States, hydrilla just keeps growing. The same thing happened when a few nonnative gypsy moths were accidentally released near Boston. Within a few years, these plant-eating pests had spread across the northeastern United States. Gypsy moths and hydrilla do not seem to have a carrying capacity. What is happening?

A **limiting factor** is any factor that controls the size of a population. The number of predators in an area may be a limiting factor. How much food is available may be a limiting factor. Limiting factors keep most natural populations from growing out of control. Limiting factors determine the carrying capacity of an environment for a species.

Some factors depend on the density of the population. They depend on how many individuals live in one place. Other factors affect a population no matter how many individuals live in an area.

Key Question What factors determine carrying capacity? **Limiting factors determine the carrying capacity of a population.**

Density-Dependent Limiting Factors

Some limiting factors affect populations most when the individuals live close together. These limiting factors depend on the population density and are called **density-dependent limiting factors.** There are several different kinds of density-dependent limiting factors.

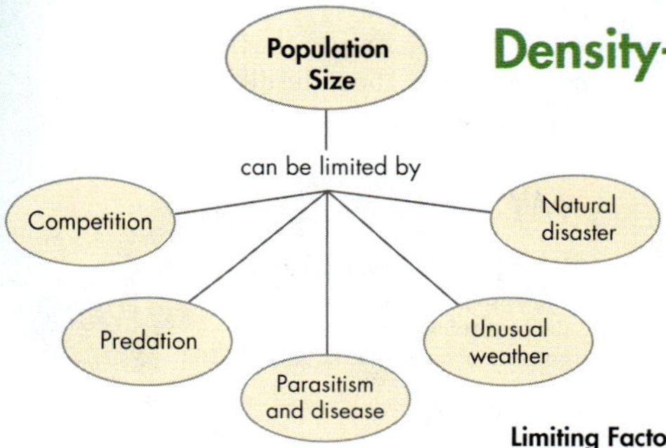

Limiting Factors Many factors can limit population growth. Some of these factors depend on population density. Others do not.

BIOLOGY.com Search Lesson 5.2 GO • Art Review

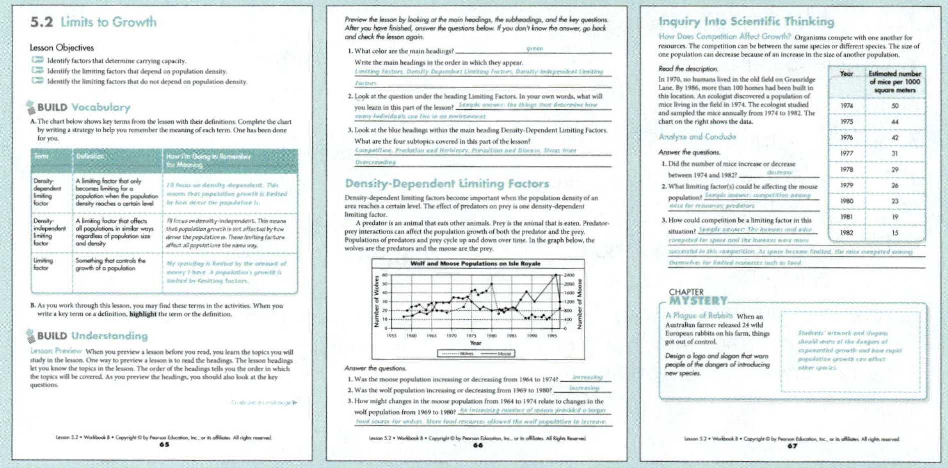
5.2 Limits to Growth

Inquiry Into Scientific Thinking

Density-Dependent Limiting Factors

CHAPTER MYSTERY

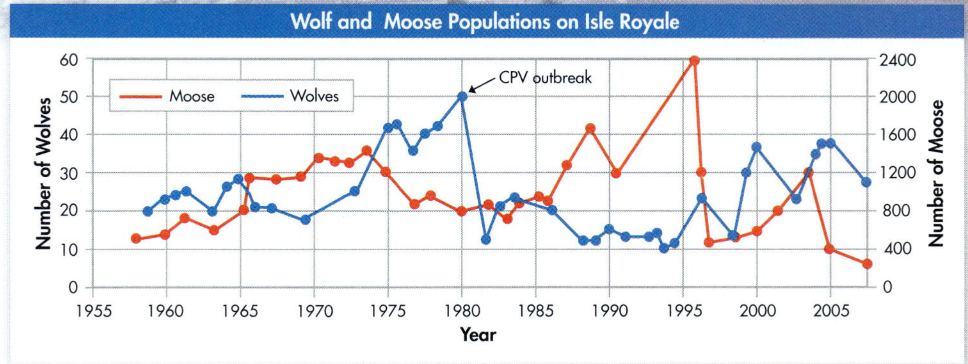

Moose and Wolf Populations on Isle Royale Moose and wolves on Isle Royale affect each other's population growth. In this case, the moose population was also affected by changes in food supply. The wolf population was also impacted by a virus (CPV) in 1980.

Predators One important density-dependent factor is the relationship between predators and prey. The populations of moose and wolves on Isle Royale in Lake Superior are one example. When the moose population is high, wolves catch them easily. So the wolf population then grows. More wolves mean that more moose are killed than are born. The moose population falls. With fewer moose, the wolves starve. The wolf population falls. Eventually, the moose population rises again, and the cycle repeats.

People can also limit population sizes. In New England, people are predators of codfish. People catch more and more fish each year. The cod population has gotten much smaller. If people killed fewer cod, the population could rise again. Scientists are studying the cod population. They want to learn how many fish can be caught each year without risking the entire population.

BUILD Vocabulary

limiting factor
a factor that causes the growth of a population to decrease

density-dependent limiting factor
a limiting factor that depends on population size

ACADEMIC WORDS
You learned that the word *factor* is a feature that contributes to an event or result. A limiting factor, however, is a feature that contributes to a change in population growth.

Focus on Academic Language

• parasite • cycle

Write the words on the board and review their meanings. Explain that the word *parasite* comes from the Greek *parasitos:* the prefix *para-* means "beside or alongside of," and *sitos* means "food."

Say Parasites live on or in other organisms to get their food. Unlike predators, they usually do not kill their host.

Draw a circle on the board, with arrows pointing clockwise, to represent a cycle. Label the top of the circle with an X.

Say The word *cycle* comes from the Greek *kyklos,* which means "circle." A cycle is a series of events repeated over and over.

Ask How would you move through a cycle on this circle? *(from point X all the way around the circle back to point X)*

During Reading

Active Reading

KEY QUESTION Read aloud the Key Question and answer on the previous page.

Ask If food is the most important limiting factor in a population, what is likely to happen to the population's carrying capacity if a drought decreases food supplies? *(The carrying capacity is likely to decrease.)*

GENERATE QUESTIONS Have students read the section Density-Dependent Limiting Factors. Instruct them to think of questions as they read and to write the questions in their notebooks. Invite students to share and discuss their questions aloud.

Refer students to the Limiting Factors concept map on the previous page.

Ask Which of the factors shown in the concept map do you think are density-dependent? *(Answers may vary.)*

USE VISUALS Invite a volunteer to read aloud the description of the interactions between wolves and moose on Isle Royale. Refer students to the Moose and Wolf Populations on Isle Royale graph.

Ask What happens to the moose population when the wolf population decreases? *(It increases.)*

Ask As the moose population increases, what eventually happens to the wolf population? *(It increases.)* Why? *(There is an abundance of food, so more wolves can survive.)*

Ask How does the decreasing moose population affect the wolf population? *(The wolf population decreases also.)*

Ask According to what you've read, what eventually will happen to both populations? *(They will cycle.)*

Build Vocabulary

ACADEMIC WORDS Have students say each vocabulary word aloud and read its definition. Point out the word *limiting.*

Say Think of other ways in which we use the word *limiting* or *limits.*

Ask If you take care of young children, what are some limits that you place on them? *(Sample answers: limits on junk food, television time, or staying up past bedtime)* In this situation, what are the limiting factors in keeping a child healthy and happy? *(type of food, amount of television viewing, amount of sleep)*

Speed Bump

Ask What does a limiting factor control? *(the size of a population)*

Ask What are three examples of density-dependent limiting factors? *(Sample answers: competition, predators, parasites, disease, stress)*

LESSON 5.2

During Reading continued

Active Reading

SUMMARIZE Have students summarize each paragraph in one or two sentences in their notebooks.

Ask What density-dependent limiting factors have affected the population sizes of wolves and moose on Isle Royale? *(predation, competition for food, disease)*

Ask How does overcrowding affect populations besides increasing competition for resources? *(It causes some organisms to fight, leading to stress.)*

KEY QUESTION Read aloud the Key Question and answer. Have students close their books.

Ask What do we call limiting factors that have a greater effect when populations are crowded? *(density-dependent limiting factors)*

Build Vocabulary

PREFIXES Review with students that the prefix *in-* in *independent* means "not."

Ask What are some other words that begin with the prefix *in-* meaning "not"? *(Sample answers: inactive, incapable, inconvenient, inefficient)*

Say But sometimes, *in-* at the beginning of a word does not mean "not."

Ask What are some examples of words that begin with *in-* when it isn't a prefix meaning "not"? *(Sample answers: insult, insurance, individual, industry, infection)*

Speed Bump

Ask What factors that determine population growth does competition affect? *(birthrates, death rates)*

Ask What resources do organisms often compete for in crowded areas? *(food, water, space, sunlight)*

Active Reading

READ, COVER, REMEMBER, RETELL Have students read the last paragraph on this page silently. Then instruct them to cover the text with their hands and think about what they read.

Ask In your own words, what was the main idea of this paragraph? *(Density-independent limiting factors affect populations the same, regardless of the populations' sizes.)*

Ask Why don't the effects of density-independent limiting factors vary with population size? *(Answers may vary but should center on the idea that density-independent limiting factors are often large-scale events that drastically change an area.)*

Parasites The ticks on this hedgehog are parasites. They can spread diseases to the hedgehog.

Herbivores From the plant's point of view, herbivores are predators. So it makes sense that the populations of herbivores and plants go up and down over time. Moose on Isle Royale eat balsam fir. When there are a lot of moose living in parts of Isle Royale, the fir population falls. Then, the moose begin to starve. As the moose population drops, the fir population goes back up.

Parasites and Disease Parasites and organisms that cause disease feed on their hosts. This feeding weakens the host and can cause disease or death. When hosts live close together, the hosts come into contact with each other more often. Therefore, diseases and parasites spread more easily from one host to the next. The wolf population on Isle Royale dropped quickly around 1980. This decline was due to a viral disease that spread quickly through the large, dense wolf population. Only three females and ten males survived.

Competition An area can support only a certain number of organisms. When an area is crowded, organisms compete for food, water, space, sunlight, and other essentials. Some individuals may not get enough to live. Others may not get enough to raise offspring. Competition can lower birthrates, increase death rates, or both.

Competition is a density-dependent limiting factor because resources are used up faster if there are more individuals that need them. Space and food are often related to each other. Many grazing animals compete for territories. They use the territories to graze and raise offspring. If they do not have a territory, then they cannot find a mate and produce offspring.

Overcrowding In some populations, overcrowding causes fighting. Too much fighting may lead to stress. Stress can make it harder for the body to fight sickness. In some cases, stress can even cause females to kill their offspring. So, stress from overcrowding can lower birthrates, raise death rates, and increase rates of emigration.

Key Question What limiting factors depend on population density?
Competition depends on population density. Predators, herbivores, parasites, and diseases affect crowded populations more. Stress can also affect a dense population.

BUILD Vocabulary

density-independent limiting factor
a limiting factor that affects all populations in similar ways, regardless of population size

PREFIXES
The prefix *in-* in *independent* means "not." Density-independent limiting factors do not depend on the density of the population.

Density-Independent Limiting Factors

Some limiting factors do not depend on how many individuals live close together. These factors affect all populations in almost the same ways and are called **density-independent limiting factors.** Bad weather, such as a hurricane or drought, is a density-independent limiting factor. Natural disasters, such as wildfires, do not depend on density either. Large storms can sometimes kill most of a population. For example, insects may be washed away by a heavy rain.

ELL Focus on ELL: Access Content

BEGINNING SPEAKERS Students may be confused by the difference between density-dependent limiting factors and density-independent limiting factors. Divide students into pairs. Instruct each pair to sketch out a cluster diagram that is centered on the topic "Limiting Factors." Tell students that their cluster diagram should branch into density-dependent limiting factors and density-independent limiting factors, and from there, into details of how these factors affect populations. Tell pairs that one student should read aloud the appropriate sections of the text while the other student constructs the diagram. Students should reverse roles so that each reads and contributes to the diagram.

True Density Independence? It is sometimes difficult to say that a factor is truly density independent. On Isle Royale, the moose population grew quickly after the wolf population dropped. Then, a very cold winter came. Snow covered the plants the moose ate. Many moose died. If there had been fewer moose, they may have found enough food. In this case, two factors worked together to limit the population.

Controlling Introduced Species People have tried many things to control hydrilla in the United States. Density-independent limiting factors have not solved the problem. Using chemicals and machines to kill or remove the hydrilla have worked only temporarily. So far, the best plan of control seems to be the introduction of sterilized grass carp fish to eat the plants.

Key Question What limiting factors do not usually depend on population density?
Unusual weather and natural disasters can act as density-independent limiting factors.

Effects of a Severe Drought on a Population During a drought, dead fish lie rotting on the banks of the Paraná de Manaquiri River in Brazil.

CHECK Understanding

Use the highlighted words from the lesson to complete each sentence correctly.

1. A flood is an example of a ______ because it usually affects all populations in the same way.
2. Limiting factors that affect large, dense populations more than small, scattered populations are called ______.

Critical Thinking

3. **Apply Concepts** How do limiting factors affect the growth of populations?
4. **Relate Cause and Effect** What is the relationship between competition and population size?
5. **Apply Concepts** Arctic foxes eat lemmings. What is likely to happen to the lemming population if the arctic fox population were to increase quickly?
6. **Write to Learn** Answer the second clue of the mystery. Think about factors that might be different in England than in Australia.

MYSTERY CLUE

What factors do you think could limit the size of a rabbit population? (**Hint:** See p. 114.)

BIOLOGY.com Search Lesson 5.2 GO • Lesson Assessment

Active Reading

MAKE CONNECTIONS Review the definition of *density-independent limiting factors* on the previous page. Then read aloud the first paragraph on this page—True Density Independence?

Say Picture a bird species whose population used to exist across a large area but is now rare and limited to one forest. A wildfire destroys most of the forest and the population vanishes.

Ask Is a wildfire a density-dependent or density-independent limiting factor? *(density-independent)* But would the bird population likely have survived if it were larger? *(Yes, if it covered a geographic range outside of the destroyed forest.)*

Use students' answers to reinforce the idea that it is sometimes difficult to determine whether a factor is truly density independent.

KEY QUESTION Read aloud the Key Question and answer.

Have students give examples of unusual weather conditions and natural disasters that can act as density-independent limiting factors.

Post Reading

ASSESSMENT TIPS For Question 4, review the meanings of *cause* and *effect.* Point out that in the question, population size is the cause, and competition is the effect.

For Question 5, remind students of a similar example.

Ask What other organisms in this lesson have the same relationship as the Arctic fox and the lemmings? *(wolves and moose, moose and balsam fir)* What do we call an animal that eats other organisms to survive? *(predator)*

MYSTERY CLUE Review with students the two different types of limiting factors and examples of each. *(density-dependent limiting factors: competition, predation, disease, overcrowding; density-independent limiting factors: bad weather, natural disasters)* Instruct students to consider which of those factors are most likely to affect a rabbit population.

Wrap-Up Activity

USING VISUALS Display photos of different communities of organisms. Have students identify the populations that make up each community and list factors that might limit the size of each population.

EXTENSION: Have students identify both density-dependent and density-independent limiting factors that might affect the populations.

Check Understanding

1. density-independent limiting factor
2. density-dependent limiting factors
3. Limiting factors control the carrying capacity, the size of the population that an area can support. If a population is near its carrying capacity, limiting factors will keep it from growing larger.
4. When an area becomes crowded, organisms compete for resources. Because some organisms will not obtain enough resources to survive or to reproduce, competition lowers birthrates, increases death rates, or both.
5. The Arctic fox is a predator and the lemmings are its prey. If the Arctic fox population were to increase quickly, the lemming population would likely decrease.
6. Possible limiting factors include availability of food, predators, competition, disease, parasites, or weather events and natural disasters.

LESSON 5.2

Inquiry Into Scientific Thinking

MAKE CONNECTIONS Explain to students that competition for resources within a population can be particularly great when the organisms in that population cannot move.

Ask If lizards need to bask on sun-warmed rocks, and the brightest sunlight moves from one rock to another throughout the day, what do lizards do in response? *(They move from one rock to another.)* If short plants are in a crowded forest, can they move to get out from under the shade of the tree branches? *(no)* Which trees in a forest get the most direct sunlight? *(the tallest ones)*

PREDICT Ask volunteers to predict what might happen if they changed the activity.

Ask If you added two more cups with seeds—one with 7 seeds, and one with 11 seeds—what do you think would happen with these additional cups? *(Answers may vary but will likely suggest some intermediate result: Plants in the additional cups will probably grow taller than those in the cup containing 15 seeds, but shorter than those in the cup containing 3 seeds.)*

Lead a Discussion

Talk about competition between organisms for available resources.

Say When gardeners plant flower beds, they make sure to space the plants apart from each other.

Ask What is the reason for spacing the plants? *(to minimize competition between the plants for resources, such as sunlight, water, and minerals in the soil)*

Ask Why do you think gardeners space some types of plants farther apart from each other than other types? *(Different types of plants have different needs for resources. For example, some plants might need more water than others. Spacing those plants farther apart would make more water available to each plant.)*

INQUIRY into Scientific Thinking

How Does Competition Affect Growth?

Individuals who live near each other compete for what they need. There is usually only so much food, water, space, and sunlight to go around. The individuals who get enough of these resources live longer, healthier lives. They tend to produce more offspring. Individuals who do not get enough may not survive. Some individuals may get enough food or water to survive but not enough to raise offspring. In this way, competition can affect the size of a population.

Competition is a density-dependent limiting factor. Resources are used up more quickly by large, dense populations. Small, scattered populations do not use as many resources. In these populations, competition is not usually a limiting factor.

In Lesson 5.2, you learned about competition and other limiting factors. In this activity, you will explore how competition affects the growth of bean sprouts.

1. Label two paper cups 3 and 15. Prepare the two cups by completing the following steps:
   - Make five or six small holes in the bottom of each cup.
   - Fill each cup two-thirds full with potting soil.
   - Plant 3 bean seeds in cup 3.
   - Plant 15 bean seeds in cup 15.
2. Water both cups so that the soil is moist but not wet.
3. Put the cups in a location that gets bright, indirect light. Water the cups equally as needed.
4. Count the seedlings every other day for two weeks. Draw a chart like the one below to record your data. (You planted your seeds on Day 0.)

Analyze and Conclude

1. Observe What differences did you observe between the two cups?

In Your Workbook Get more help for this activity in your workbook.

	Day 2	Day 4	Day 6	Day 8	Day 10	Day 12	Day 14
Date							
Cup 3							
Cup 15							

Materials

bean seeds, 2 paper cups, potting soil, waterproof marker

Analyze and Conclude

1. The seedlings in cup 15 will be smaller and less robust than the seedlings in cup 3. Some seedlings in cup 15 may die.

5.3 Human Population Growth

Historical Overview

For most of human history, the human population grew slowly. Food was hard to find. The death rate was high because of predators and disease. Today, four human babies are born every second. The human population could reach 9 billion in your lifetime. What changes have made it possible for the human population to grow so quickly?

Exponential Human Population Growth As life became easier, the human population began to grow more rapidly. During the Industrial Revolution in the 1800s, people learned how to ship goods around the world. Food supplies became more reliable. Healthcare and medicine improved. People learned more about sanitation and nutrition. Because of all these improvements, death rates dropped. But the birthrate stayed very high. This combination of lower death rates and high birthrates led to exponential growth of the human population.

World Population Growth Slows The human population growth rate was highest around 1962–1963. The population is still growing very rapidly. However, the rate of growth is slowing.

It took 123 years for the population to go from 1 billion in 1804 to 2 billion in 1927. It took only 33 years to reach 3 billion. The time it took to add another billion people continued to decrease until 1999. Since then, the growth rate has slowed. What has been happening to slow human growth?

Key Question How has the size of the human population changed over time? **For a long time, the population grew slowly. In the 1800s, the population began to grow exponentially.**

Key Questions

How has the size of the human population changed over time?

Why do population growth rates differ among countries?

BUILD Understanding

Preview Visuals Before you read, preview the graphs throughout the lesson. Make a list of questions about the graphs. Then, as you read, write down the answers to your questions.

In Your Workbook Go to your workbook to review the graphs.

Human Population Growth Over Time For a long time, the human population grew very slowly. Technological advances lowered death rates. Then, the global population began to increase much more quickly.

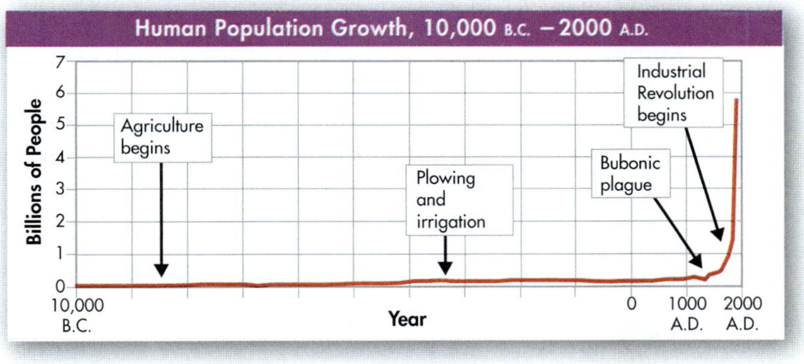

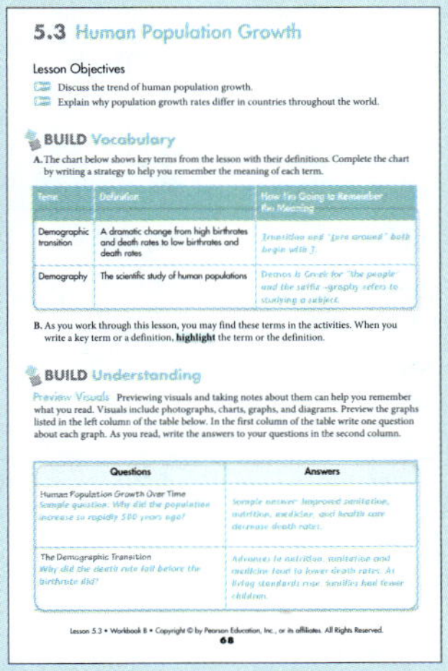

5.3 Human Population Growth

Lesson Objectives

- Discuss the trend of human population growth.
- Explain why population growth rates differ in countries throughout the world.

BUILD Vocabulary

A. The chart below shows key terms from the lesson with their definitions. Complete the chart by writing a strategy to help you remember the meaning of each term.

Term	Definition	How I'm Going to Remember the Meaning
Demographic transition	A dramatic change from high birthrates and death rates to low birthrates and death rates	Transition and "turn around" both begin with T.
Demography	The scientific study of human populations	Demos is Greek for "the people" and the suffix -graphy refers to studying a subject.

B. As you work through this lesson, you may find these terms in the activities. When you write a key term or a definition, **highlight** the term or the definition.

BUILD Understanding

Preview Visuals Previewing visuals and taking notes about them can help you remember what you read. Visuals include photographs, charts, graphs, and diagrams. Preview the graphs listed in the left column of the table below. In the first column of the table write one question about each graph. As you read, write the answers to your questions in the second column.

Questions	Answers
Human Population Growth Over Time Sample question: Why did the population increase so rapidly 500 years ago?	Sample answer: Improved sanitation, nutrition, medicine, and health care decrease death rates.
The Demographic Transition Why did the death rate fall before the birthrate did?	Advances in nutrition, sanitation and medicine lead to lower death rates. As living standards rise, families had fewer children.

68

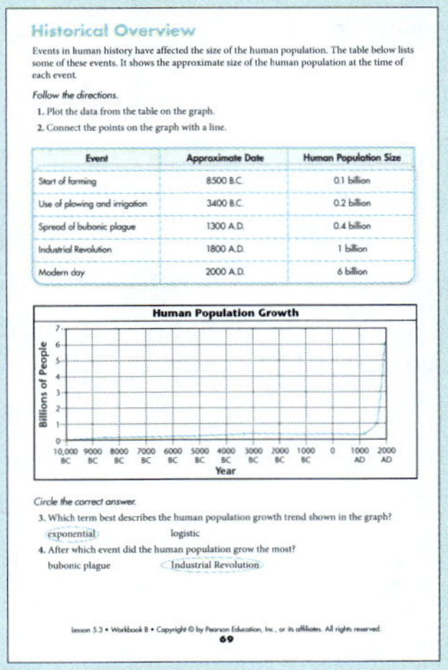

Historical Overview

Events in human history have affected the size of the human population. The table below lists some of these events. It shows the approximate size of the human population at the time of each event.

Follow the directions.

1. Plot the data from the table on the graph.
2. Connect the points on the graph with a line.

Event	Approximate Date	Human Population Size
Start of farming	8500 B.C.	0.1 billion
Use of plowing and irrigation	3400 B.C.	0.2 billion
Spread of bubonic plague	1300 A.D.	0.4 billion
Industrial Revolution	1800 A.D.	1 billion
Modern day	2000 A.D.	6 billion

Circle the correct answer.

3. Which term best describes the human population growth trend shown in the graph?
   exponential logistic
4. After which event did the human population grow the most?
   bubonic plague Industrial Revolution

69

LESSON 5.3

Pre-Reading

Objectives

5.3.1 Discuss the trend of human population growth.

5.3.2 Explain why population growth rates differ in countries throughout the world.

Preview the Pages

1. Preview the vocabulary with students. Have students find the words in the text and in the graphs.
2. Preview the graphs. Have students read all graph titles and labels on axes.
3. Make the connection that exponential growth as described in Lesson 5.1 is the type of growth exhibited by humans.

Activate Prior Knowledge

Talk about whether students have observed the human population size in their town or neighborhood change over time. Ask what changes they have noticed in the age structure in their community.

Build Understanding

PREVIEW VISUALS Direct students to the partially filled-in chart of questions and answers in Workbook B. Explain that the major ideas in this lesson are best represented visually and that generating questions before they study the graphs will help them discover the main ideas of the graphs. Remind students to complete the chart as they finish studying each graph.

NATIONAL SCIENCE EDUCATION STANDARDS

UNIFYING CONCEPTS AND PROCESSES

II, III

CONTENT

C.4.d, C.4.e, F.2, F.3, F.6, G.2

INQUIRY

A.2.a, A.2.d

For the corresponding lesson in *Miller & Levine Biology,* see pages 142–145.

During Reading

Active Reading

KEY QUESTION Read aloud the Key Question and answer from the previous page. Have students locate on the graph Human Population Growth Over Time the time when the human population began to grow exponentially.

INTERPRET GRAPHS Guide students to the Human Population Growth graph on the previous page.

Ask How fast did the human population grow over most of human history? *(The human population did not grow or grew only very slowly.)*

Now direct students to the graph The Demographic Transition.

Ask In what phase of the demographic transition was worldwide population growth before the Industrial Revolution? *(Stage I)* Birthrates are high during Stage I, so why don't populations grow very fast during that stage? *(Death rates also are high.)*

Ask What happened to the world population from 10,000 B.C. to the 1800s? *(It stayed the same or grew slowly.)* Why? *(Both birthrates and death rates were high.)* What happened to the U.S. and world population from the 1800s until 1963? *(It grew rapidly.)* Why? *(Birthrates remained high but death rates dropped.)* What has happened to the growth rate of the U.S. population since 1963? *(It has slowed.)* Why? *(Birthrates have dropped and are closer to death rates.)*

Build Vocabulary

PREFIXES Have students find the words *demography* and *demographic transition* in the text. As a group, pronounce each term.

Ask Which part of the words *demography* and *demographic* tells you that they involve people? *(demo-)*

Say The Greek word *kratia* means "rule."

Ask What do you think the word *democracy* means? *(rule by the people)*

Speed Bump

Ask Are population growth rates slowing down everywhere in the world? *(no)* Which countries have the highest population growth rates? *(China and India)*

Ask Would a country with a fast-growing population have more adults than children, or more children than adults? *(more children than adults)*

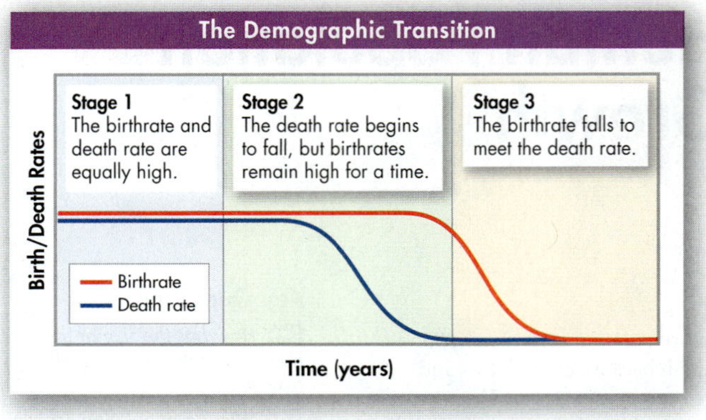

The Demographic Transition
Stage 1: Birthrates and death rates are high for most of history.
Stage 2: Advances in nutrition, sanitation, and medicine lower death rates. Birthrates remain high. There are many more births than deaths. The population increases exponentially.
Stage 3: As living standards rise, families have fewer children and the birthrate falls. Population growth slows. The demographic transition is complete when the birthrate meets the death rate.

Patterns of Human Population Growth

The world population growth rate has slowed a lot recently. In some countries, however, the growth rate is still high. In fact, most of the world's population growth is happening in only ten countries. India and China are leading the growth.

Scientists have identified several things that affect population growth. The scientific study of human populations is called **demography** (duh MAH gruh fee). Demography looks at characteristics of human populations. It tries to explain how those populations will change over time. Scientists study birthrates, death rates, and the age structure of a population. They use this information to explain why some countries have high growth rates and some countries grow more slowly.

The Demographic Transition For a long time, human societies had very high birthrates and death rates. Then, death rates dropped and birthrates remained high. Over the past century, birthrates in the United States, Japan, and much of Europe also fell. The growth rates in these countries slowed dramatically. These countries have completed the demographic transition. In the **demographic transition,** a population changes from having high birthrates and death rates to having low birthrates and death rates. There are three stages to the transition, as shown above.

Age Structure and Population Growth The age structure in a country helps predict how quickly a population will grow. In the United States, there are nearly equal numbers of people in each age group. This structure suggests that the population of the United States will grow slowly. In countries such as Guatemala, there are many more young children than teenagers. There are many more teenagers than adults. The population of Guatemala may double in the next 30 years.

BUILD Vocabulary

demography
the scientific study of human populations

demographic transition
a change in a population from high birthrates and death rates to low birthrates and death rates

PREFIXES
The prefix *demo-* in *demography* and *demographic* comes from the Greek word *demos,* which means "people." This prefix is also used in *democracy.*

ELL Focus on ELL: Extend Language

INTERMEDIATE AND ADVANCED SPEAKERS Pair up students. Draw a simple word map on the board, and distribute copies of the Vocabulary Word Map graphic organizer. (See page A8 in Workbook B or GO17 in the Transparencies book.) Write the following list of words on the board: *demography, geography, photography, oceanography.* Remind students that the vocabulary word *demography* is defined as "the study of human populations," with the word part *demo-* meaning "people." The word part *-graphy,* derived from the Greek *graphia,* meaning "writing," is defined as "a process or method of writing, recording, or representing." Instruct each pair of students to fill in the word map for these similar words, listing for each their own definition, the dictionary definition, a word clue that focuses on the parts of each word, a picture, and an association with the word. Invite volunteers to share the components of their word maps.

Future Population Growth People who study demography look at many factors to predict how the human population will grow. Age structure is one important factor. They also study how diseases affect death rates. Right now, AIDS is causing very high death rates in Africa and parts of Asia.

The world population may reach 9 billion people by the year 2050. Will the human population level out to a logistic growth curve? Countries growing rapidly right now would need to complete the demographic transition.

It is expected that the world population will grow more slowly over the next 50 years than it did over the last 50 years. The growth rate will probably still be higher than zero in 2050. If so, our population will continue to grow. In the next chapter, you will learn about the effect of human population growth on the biosphere.

Key Question Why do population growth rates differ among countries?
Birthrates, death rates, and age structure are different in different countries. These differences cause some countries to have high growth rates and others to have low growth rates.

Age Structure of World Population

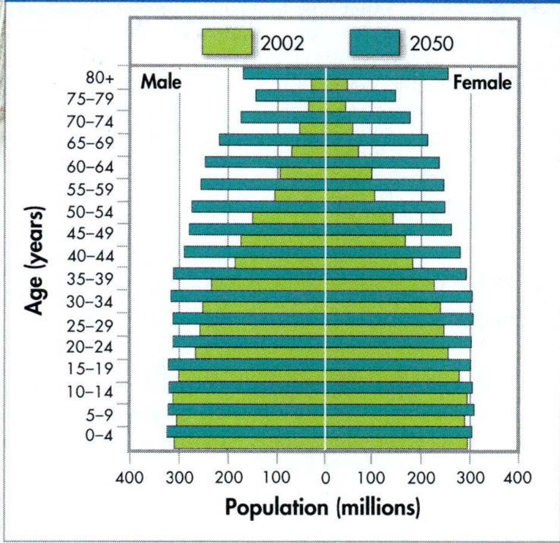

A Growing Population The information for this graph was taken from the U.S. Census Bureau, International Database. The graph shows the projected age structure of the world population in 2050. The photo above shows a housing complex in Hong Kong, China. Each apartment building is home to thousands of people.

CHECK Understanding

Apply Vocabulary

Use the highlighted terms from the lesson to complete each sentence correctly.

1. People who study ________ look at characteristics, such as age structure, death rate, and birthrate to predict how human populations will change.

2. In the third stage of the ________, the birthrate and death rate are equal.

Critical Thinking

3. Apply Concepts If every married couple in a country were to decide to have only one child, what would happen to the country's population growth?

4. Relate Cause and Effect The human population grew very slowly for most of human history. In about 1800, the population began growing exponentially. What factors contributed to this growth pattern?

5. Write to Learn In one or two paragraphs, describe the three stages of demographic transition. Explain how the population growth rate changes at each stage.

BIOLOGY.com Search Lesson 5.3 GO • Lesson Assessment • Art in Motion

Active Reading

USE GRAPHS Have students reread the caption and labels for Age Structure of World Population. Instruct them to focus on the differences between the age structure for 2002 and the age structure predicted for 2050.

Ask When do children most outnumber adults, in 2002 or 2050? *(2002)* What does this difference say about future population growth? *(It should slow down.)* Which age group will increase more between 2002 and 2050: 30–34 year-olds or 80+ year-olds? *(80+ year-olds)*

KEY QUESTION Read aloud the Key Question and answer.

Ask Why might one country have a much higher death rate than another country? *(The country with the lower death rate might have better medical care, less poverty, and better nutrition and sanitation than the country with the higher death rate.)*

Post Reading

ASSESSMENT TIPS Review the highlighted vocabulary words with students before they begin the assessment.

For Question 5, refer students to The Demographic Transition graph to help them write their answers.

Wrap-Up Activity

MAKE CONNECTIONS Divide students into small groups. Explain that for this activity they are all going to be demographers, researchers who study the makeup of populations.

Say The students in your group and all of their immediate family will make up a population that you are going to describe with an age structure graph.

Explain that students should tally the ages and sex (male or female) of people in the combined families of all group members, including parents, brothers, sisters, and grandparents. Instruct students to use these numbers to construct an age structure graph like that in the lesson, picturing individuals from 0 to 80+ years old.

Ask Would your population likely grow rapidly, slowly, or not at all? Why? *(Answers may vary with the group but should depend on the proportion of children and teenagers (compared to adults) in each population. The greater the proportion of children and teenagers, the faster the population will grow.)*

Check Understanding

1. demography
2. demographic transition
3. The population growth rate would decrease, and the population size would decrease as the birthrate dropped below the death rate.
4. Exponential growth was caused by the greater availability of food, through better means of shipping, and improvements in healthcare, medicine, sanitation, and nutrition.
5. The paragraphs should reflect the content of The Demographic Transition graph: Stage I features high birthrates and death rates (little population growth); Stage II features dropping death rates but still-high birthrates (rapid population growth); and Stage III features birthrates dropping to equal the low death rates (slow population growth).

CHAPTER LAB

Pre-Lab

Review the definition and examples of *density-dependent limiting factors (limiting factors that depend on population size: food availability, predation, disease, competition, parasites, and stress).*

Ask What does it mean that a factor is "dependent" on population size? *(The effect of the factor varies directly with the population's size; the factor affects the population more when the population is larger than when it is smaller.)* What happens to a population's growth when resources become limited? *(It slows or stops.)*

Review the Background Questions and the Pre-Lab Questions as a class.

Lab

Tell students they will perform the chapter lab *The Growth Cycle of Yeast* described in **Lab Manual B**.

SAFETY Remind students to handle glass items with care and to follow your directions about disposal of lab materials. Have them wash their hands with soap and warm water when they complete the lab.

Post-Lab

DESIGN A GRAPH Have pairs of students draw a logistic growth curve for a population of their choice and label the three phases. Tell students to "invent" numbers for their population's size as the population goes through the different phases of growth. Instruct them to label the population size at two points in each phase. Point out that students' numbers should only reflect population growth as it increases, decreases, or holds steady during logistic growth.

Skills Lab

GUIDED INQUIRY

Pre-Lab: The Growth Cycle of Yeast

Problem What type of population growth occurs in a yeast culture?

Materials yeast culture, stirring rod, dropper pipettes, microscope slides, coverslips, microscope, 10-mL graduated cylinder, test tubes, test-tube rack, graph paper

Lab Manual Chapter 5 Lab

Skills Measure, Calculate, Interpret Graphs

Connect to the Big idea Populations depend on, and are limited by, their environments. A population can grow when its members have the resources they need to survive and reproduce. Factors that can limit those resources include natural disasters, such as forest fires, and competition from other species. Predation and disease are also limiting factors for populations.

In nature, populations often experience cycles of growth and decline. In this lab, you will investigate whether such a cycle occurs in yeast populations.

Background Questions

a. **Review** What is the carrying capacity of a population?

b. **Sequence** Briefly describe the three phases of logistic growth.

c. **Relate Cause and Effect** Describe two different ways that a population might achieve a growth rate of zero.

d. **Classify** After two weeks of hot and sunny days with very little rain, the blades of grass in a backyard began to wither and die. Were any of the factors that caused the decline of the grass population dependent on density? Explain.

Pre-Lab Questions

Preview the procedure in the lab manual.

1. **Infer** Why was grape juice used to prepare the yeast cultures instead of plain water?

2. **Calculate** Suppose you have to do one dilution of your culture before you are able to count the yeast cells. If you count 21 yeast cells in the diluted sample, how many yeast cells were in the same area of the undiluted sample? MATH

3. **Predict** What do you think will happen to a yeast population between Day 3 and Day 7? Give reasons for your answer.

BIOLOGY.com Search Chapter 5 GO

Visit Chapter 5 online to test yourself on chapter content and to find activities to help you learn.

Untamed Science Join the Untamed Science crew as they learn the latest techniques for counting populations.

Art in Motion View a short animation that brings age-structure diagrams to life.

Art Review Review your understanding of limiting factors with this drag-and-drop activity.

InterActive Art Manipulate factors such as starting population size, birthrate, and death rate to see how they would impact moose and wolf populations over time.

Pre-Lab Answers

BACKGROUND QUESTIONS

a. the maximum number of a species that a particular environment can support

b. In Phase I, resources are unlimited and the population grows rapidly. In Phase II, the rate of growth slows. In Phase III, the rate of growth drops to zero and the population levels off.

c. Sample answer: The number of births is equal to the number of deaths, or as many people moved into the population as moved out.

d. Sample answer: The small amount of rain is a density-dependent factor because the blades had to compete for this resource.

PRE-LAB QUESTIONS

1. The grape juice contains nutrients that the yeast need to grow and reproduce.

2. 21 × 10 or 210 yeast cells

3. Sample answers: The nutrients will be sufficient to sustain rapid growth through Day 7. A lack of nutrients will begin to limit and slow down growth by Day 7.

NATIONAL SCIENCE EDUCATION STANDARDS

CONTENT F.2

INQUIRY A.1.b, A.1.c, A.2.a

5 CHAPTER Summary

5.1 How Populations Grow

- Ecologists study a population's range, density, growth rate, and age structure.
- Population growth is affected by birthrate and death rate. It can also be affected by immigration and emigration.
- In exponential growth, the population grows more and more quickly.
- Logistic growth happens when the growth of a population slows and then stops.

age structure (p. 109)
immigration (p. 109)
emigration (p. 109)
exponential growth (p. 109)
logistic growth (p. 111)
carrying capacity (p. 111)

5.2 Limits to Growth

- Limiting factors determine the carrying capacity of a population.
- Competition depends on population density. Predators, herbivores, parasites, and diseases affect crowded populations more than small, scattered populations. Stress can also affect a dense population.
- Unusual weather and natural disasters can act as density-independent limiting factors.

limiting factor (p. 112)
density-dependent limiting factor (p. 112)
density-independent limiting factor (p. 114)

5.3 Human Population Growth

- For a long time, the population grew slowly. In the 1800s, the population began to grow exponentially.

- Birthrates and death rates are different in different countries. The age structures are different. Countries with high birthrates and low death rates are growing quickly. The United States, Japan, and much of Europe have completed the demographic transition, and their growth rates have slowed.

demography (p. 118)
demographic transition (p. 118)

BIOLOGY.com Search Chapter 5 GO • Crossword • Chapter Assessment

121

UbD Transfer the Big Idea

TRANSFER Have students use what they have learned about population growth, limiting factors, carrying capacity, and human population growth patterns to write a short science-fiction story. Have students set the story in the future, at the time when the human population reaches Earth's carrying capacity. Stories should be creative but scientifically accurate. Before students write, lead a discussion about which resources might be limited in this situation and how the stress of overcrowding might affect humans. Remind students to include details about limiting factors acting on the human population and to include an accurate definition of carrying capacity. Instruct students also to include accurate information from the text about how human populations grow. Invite students to share their stories in small groups or as a class.

Chapter Summary

Think Visually

USE GRAPHICS Have students create a table in which they describe the phases of logistic growth. Students' tables should include a description of Phase I, during which the population grows exponentially; Phase II, during which population growth slows; and Phase III, during which population growth stops and the population size remains fairly constant.

BIOLOGY.com Study Online

REVIEW RESOURCES

Data Analysis Students analyze data to make predictions about zebra mussel population growth.

Art Review Reinforce limiting factors using this drag-and-drop activity.

InterActive Art Students manipulate factors to see how predator and prey relationships would change over several generations.

Art in Motion This animated time series shows change in the world population age structure.

Vocabulary Review The *Flash Cards* and *Crossword* provide an interactive way to review chapter vocabulary.

Editable Worksheets Pages of Study Workbook B, Lab Manual B, and the Assessment Resources Book are available online. These documents can be easily edited using a word-processing program.

Lesson Overviews Lesson Overviews are based on Miller & Levine Biology. However, they are available on CD-ROM and can be edited for your class.

ASSESSMENT RESOURCES

Chapter Assessment Have students take an online version of the Chapter 5 Check Understanding.

Standardized Test Prep Students can take an online version of the Standardized Test Prep. You will receive their scores along with ideas for remediation.

Diagnostic and Benchmark Tests Use these tests to monitor your students' progress and supply remediation.

ASSESSMENT

Assess

Big idea Invite a volunteer to read the question. Write the words *birthrate* and *death rate* at the top of the board. Then write the words *density-dependent limiting factors* and *density-independent limiting factors* underneath.

Ask What must happen, in terms of the birthrate and death rate, for a population to grow? *(The birthrate must be higher than the death rate.)*

Ask How do limiting factors affect population growth? *(They determine a population's carrying capacity, the number of organisms of the same species that an area can support.)* Can you name one example each of a density-dependent limiting factor and a density-independent limiting factor? *(Sample answers: density-dependent limiting factors—food availability, space, predators, competition, disease; density-independent limiting factors—hurricanes, drought, wildfires)*

Ask What is the difference between the two types of limiting factors? *(Density-dependent limiting factors depend on population size; density-independent factors affect populations the same regardless of their size.)* Have students write their answers in their notebooks.

CONSTRUCTED RESPONSE Paragraphs will vary. Explain the activity carefully. Be sure students understand that the *Hints* should be used as a guide to help them answer the questions. Review how to write a paragraph.

Write Question 1 on the board and read it aloud. Instruct students to underline *introducing organisms* and *new environment*.

Say This question is asking what effect an introduced organism has on its new environment.

Read the first hint.

Ask If an organism whose population has been controlled in part by predators is introduced into a new environment where it does not have predators, what is likely to happen to the introduced organism's population? *(It will grow.)*

Read the second hint.

Instruct students to underline in their list of resources any that are associated with limiting factors.

Ask How can one population act as a limiting factor to another population? *(One population can contribute to density-dependent limiting factors, such as competition, predation, parasites, and disease.)*

Model writing the first one or two sentences, using the hints from the text and suggestions volunteered from students' lists. *(Sample: It is hard to know how an organism introduced into a new environment is going to affect the other organisms there. The introduced population might grow rapidly, compete with other populations for food and space, and damage the environment.)*

For Question 2, pair students to answer the question. Have students share their paragraphs with the class.

Foundations for Learning Wrap-Up

Have students take out their three-columned paper chart. Be sure they have filled in the first two columns as previously instructed. Either assign groups or pair students, and explain the activities.

5 CHECK Understanding

Assess the Big idea Interdependence in Nature

Write an answer to the question below.

Q: What factors contribute to changes in populations?

Constructed Response

Write an answer to each of the questions below. The answer to each question should be one or two paragraphs long. To help you begin, read the **Hints** below each of the questions.

1. **Why should people be cautious about introducing organisms into new environments?**

   Hint When organisms are introduced into new environments, they may not face natural predators.

   Hint A limiting factor is a factor that controls the growth of a population.

2. **Natural populations do not grow exponentially for very long. Why not?**

   Hint In exponential growth, the larger a population gets, the faster it grows.

   Hint Carrying capacity is the maximum number of organisms of a species an area can support.

Foundations for Learning Wrap-Up

Use the folded paper chart you made before reading the chapter as a tool to help you organize your thoughts about populations.

Activity 1 On your own, reread the questions you asked at the beginning of the chapter. Write the answers to these questions in the third section of your page, under the heading, What Have I Learned?

Activity 2 Trade charts with a partner. Compare your questions and answers. If one of you had any difficulty answering the questions, find the answers together. Also, double-check the information you each wrote in the first and third columns. Is everything correct? If not, work together to correct your charts.

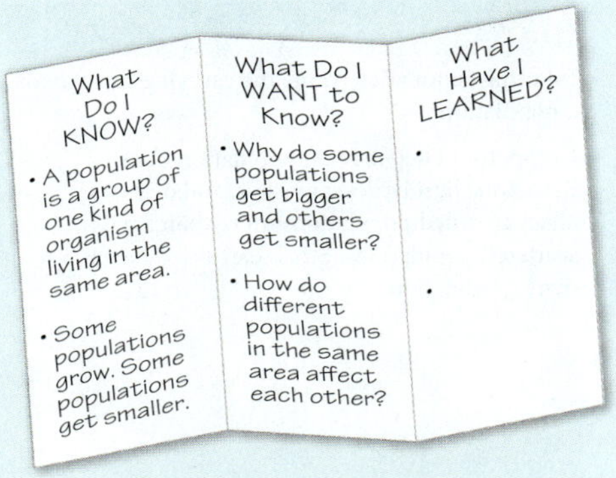

122 Check Understanding

5.1 How Populations Grow

Understanding Key Concepts

1. The space taken up by a population is known as its
   a. growth rate. **c.** age structure.
   b. geographic range. **d.** population density.
2. The graph below represents
   a. carrying capacity. **c.** logistic growth.
   b. exponential growth. **d.** age structure.

One Population Over Time

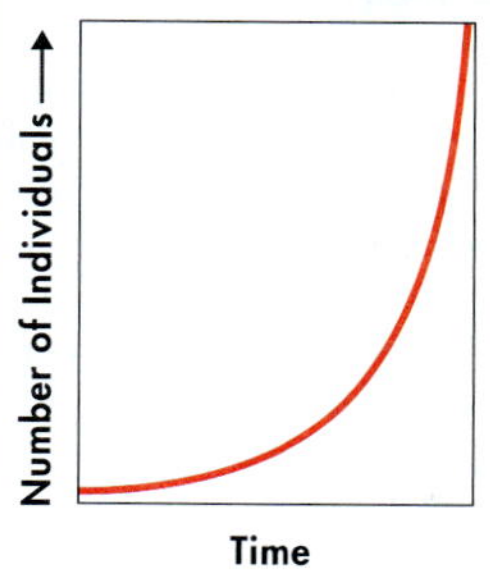

Test-Taking Tip

Interpret Visuals Look carefully at visuals. In the graph in question 2, the number of individuals is increasing very quickly over time. The growth rate is increasing over time as well. The graph of logistic growth is shaped like the letter S. The graph of exponential growth is shaped like the letter J. So, answer **c** is not correct. Answer **b** is the correct answer.

3. The largest population size that can survive in a given area is called
   a. logistic growth. **c.** exponential growth.
   b. carrying capacity. **d.** population density.
4. What is the difference between immigration and emigration?

Think Critically

5. **Use Analogies** How is the carrying capacity of a road like the carrying capacity of an ecosystem?

5.2 Limits to Growth

6. A limiting factor that depends on the population size is called a
   a. density-dependent limiting factor.
   b. density-independent limiting factor.
   c. predator-prey relationship.
   d. parasitic relationship.
7. Which of the following is a density-independent limiting factor?
   a. predators **c.** competition
   b. unusual weather **d.** overcrowding
8. Describe how the populations of moose and wolves on Isle Royale affect each other. How does an increase in the population of one affect the population of the other?
9. Why are parasites considered a density-dependent limiting factor?

Think Critically

10. **Predict** What would happen to a population of predators if there were a sudden increase in food for the prey? Explain your answer.

5.3 Human Population Growth

11. What is the scientific study of human populations called?
    a. immigration
    b. emigration
    c. demographic transition
    d. demography
12. What is the last stage of the demographic transition?
    a. The birthrate and death rate are both very high.
    b. The death rate is low, and the birthrate is very high.
    c. The birthrate and death rate are both low.
    d. The birthrate is low, and the death rate is very high.

Assessment Tips

Pair students and have them work together to answer the questions. Remind students that they will need to recall the differences between exponential growth and logistic growth in populations.

For Question 10 remind students that when they are asked to predict, they need to apply information they have learned to a new situation. For Question 14 have students consider what the age structure diagrams for these small towns would look like.

Answers

Lesson 5.1

UNDERSTAND KEY CONCEPTS

1. b

2. b

3. b

4. Immigration occurs when individuals move into a population's range. This adds to the population. Emigration is when individuals move out of the population's range. Emigration removes individuals from the population.

THINK CRITICALLY

5. The carrying capacity of an ecosystem is the number of individuals an ecosystem can support. If the population grows beyond the carrying capacity, there are not enough resources to support the population. If there are more cars than the carrying capacity of the road, the traffic slows to a crawl. There is not enough space (a limiting resource) to support the cars.

Lesson 5.2

UNDERSTAND KEY CONCEPTS

6. a

7. b

8. Wolves are predators of moose. If the moose population drops dramatically, the wolves will begin to starve and die. As the wolf population falls, the moose death rate will also fall. The moose population will begin to rise again. More moose means more food for the wolves. The wolf population will then begin to rise. The wolves will prey on the moose, dropping the moose population again. The cycle repeats.

9. Parasites cannot usually live for very long outside of the host's body. If the hosts live close together, the parasites spread more quickly. Therefore, parasites do not affect small, scattered populations as much.

THINK CRITICALLY

10. The increase in food supply for the prey would probably cause the prey population to grow. A larger prey population represents an increased food supply for the predator. Initially, the predator population size would increase. Eventually, the predators would kill enough prey that food became scarce. The predator population would begin to decrease.

Lesson 5.3

UNDERSTAND KEY CONCEPTS

11. d

12. c

13. For a long time, the human population grew very slowly. Improvements in health care, medicine, sanitation, and nutrition around the 1800s lowered death rates. The population grew exponentially. The global population has continued to grow very quickly, but the rate has begun to slow down.

THINK CRITICALLY

14. The population of the first small town is likely to decrease. The birthrate will be very small, and the death rate may be high. On the other hand, the birthrate in the second small town may be very high, and the death rate is likely to be small. The population of the second small town is likely to increase.

Connecting Concepts

15. 1 billion people in 1804; 6 billion people in 1999

16. The population grew exponentially until 1999. From 1804 until 1999, the population added a billion people more and more quickly. Growth slowed after 1999. The data suggests that the growth will slow more between now and 2050.

CHAPTER MYSTERY After students have read through the Chapter Mystery, discuss reasons why the rabbit population expanded so rapidly in Australia, as well as approaches that have been taken to manage the rabbit population.

Ask What are two reasons the rabbit population grew so quickly in Australia? *(There were no rabbits in Australia before these were released, so there were no natural predators or parasites, and rabbits reproduce rapidly.)* Did the rabbit population reach carrying capacity? *(no)* How do you know? *(Its growth did not level off on its own.)*

Ask How have the rabbits damaged the environment? *(They have completely consumed native trees and shrubs, causing them to appear locally extinct.)*

Ask Which factor did Australians focus on to reduce the rabbit population: the birthrate or the death rate of the population? *(They focused on the death rate, by killing off existing rabbits.)* How might Australians focus on the birthrate to reduce the size of the rabbit population? *(They could use a technique that would prevent the rabbits from reproducing.)*

5 CHECK Understanding

13. Describe the general trend of human population growth over time.

Think Critically

14. Compare and Contrast One small town is made up mainly of senior citizens. How would you expect this town's population to change over the next several years? Compare this growth to that of a small town that is made up mostly of newly married couples in their twenties.

Connecting Concepts

Use Science Graphics

The following data come from the United Nations Department of Economic and Social Affairs, Population Division. The table shows when the world population reached 1 billion and when it reached or will reach each additional billion. Use the data table to answer questions 15 and 16.

World Population Milestones

Population (billion)	Year	Time Interval (years)
1	1804	—
2	1927	123
3	1960	33
4	1974	14
5	1987	13
6	1999	12
7	2012	13
8	2027	15
8.9	2050	23

15. Observe When did the world population reach 1 billion people? When did it reach 6 billion?

16. Interpret Tables Describe the trend in population growth since the 1-billion-people mark.

solve the CHAPTER MYSTERY

A PLAGUE OF RABBITS

Before the farmer let those 24 rabbits go in 1859, Australia had no wild rabbits. So the rabbits had very few predators, parasites, and diseases. Rabbits reproduce quickly. The population jumped from 24 to millions in less than ten years. Large rabbit populations cause a lot of damage.

Australians have tried many things to manage the population. They have tried fencing and poisoning. They have tried to ruin burrows. People have used parasites and disease. In the 1950s, a rabbit virus was deliberately introduced. It killed many rabbits. But eventually, the rabbit populations rose again. Later, a new virus that causes rabbit hemorrhagic disease (RHD) was introduced. Rabbit populations dropped again. In several places, the results were dramatic. Native trees and shrubs returned. It had been thought that these plants were locally extinct! Native animals also recovered. But the RHD virus and rabbits appear to have reached a new balance. Rabbit populations are rising again.

1. Predict Wildcats and foxes were also introduced to Australia. These predators have come to depend on rabbits as prey. How do you think wildcats and foxes would be affected by a large drop in the rabbit population?

Untamed Science™ **Never Stop Exploring Your World.** Finding the solution to the rabbit population mystery is only the beginning. Take a video field trip with the ecogeeks of Untamed Science to see where this mystery leads.

CHAPTER MYSTERY ANSWERS

1. If wildcats and foxes depend on rabbits as prey, then a large drop in the rabbit population would likely result in a drop in populations of both wildcats and foxes. With fewer rabbits, these predators probably would not have enough to eat to maintain their numbers.

Untamed Science™ Students can discover more about how scientists estimate population numbers by watching **Census Consensus.**

Standardized Test Prep

Multiple Choice

1. The movement of individuals into an area is called
   A immigration.
   B emigration.
   C population growth rate.
   D population density.

2. All other things being equal, the size of a population will decrease if
   A birthrate exceeds the death rate.
   B immigration rate exceeds emigration rate.
   C death rate exceeds birthrate.
   D birthrate equals death rate.

3. Which of the following is NOT an example of a density-dependent limiting factor?
   A natural disaster
   B predator
   C competition
   D disease

4. A population like that of the United States with an age structure of roughly equal numbers in each of the age groups can be predicted to
   A grow rapidly over a 30-year-period and then stabilize.
   B grow little for a generation and then grow rapidly.
   C fall slowly and steadily over many decades.
   D show slow and steady growth for some time into the future.

5. In the presence of unlimited resources and in the absence of disease and predation, what would probably happen to a bacterial population?
   A logistic growth
   B exponential growth
   C endangerment
   D extinction

6. Which of the following statements best describes human population growth?
   A The growth rate has remained constant over time.
   B Growth continues to increase at the same rate.
   C Growth has been exponential in the last few hundred years.
   D Birthrate equals death rate.

7. Which of the following refers to when a population's birthrate equals its death rate?
   A limiting factor
   B carrying capacity
   C exponential growth
   D population density

Questions 8–9

Use the graph below to answer the following questions.

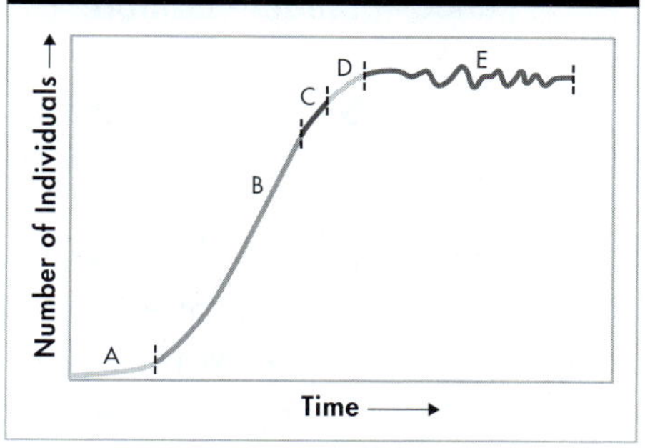

8. Which time interval(s) in the graph shows exponential growth?
   A D and E
   B A and B
   C C and D
   D E only

9. Which time interval(s) in the graph depicts the effects of limiting factors on the population?
   A A only
   B A and B
   C C, D, and E
   D C and D

Open-Ended Response

10. When a nonnative species is imported into a new ecosystem, the population sometimes runs wild. Explain why this might be the case.

If You Have Trouble With . . .

Question	1	2	3	4	5	6	7	8	9	10
See Lesson	5.1	5.1	5.2	5.3	5.1	5.3	5.1	5.1	5.2	5.2

Test-Taking Tip

REPHRASE THE QUESTION

Advise students to rewrite difficult test questions in their own words. This will help them better understand what the questions are asking so they can write more appropriate answers. However, caution them to avoid changing the meaning of questions when they rephrase them. Suggest that they ask for clarification, if possible, of any question they still do not understand after rephrasing.

Before the Test

Review with students the Test-Taking Tip.

Point out the word *NOT* in Question 3.

Remind students that the open-ended response question means that they should write a paragraph to answer the question. Have them paraphrase the question in their notebooks, and remind them about the steps you took when you modeled a paragraph. Specifically point out that you identified the meaning of the key words before starting to write.

Answers

1. A
2. C
3. A
4. D
5. B
6. C
7. B
8. B
9. C
10. Sample answer: The new ecosystem may lack the limiting factors present in the species' native ecosystem, allowing the population to grow exponentially. Introduced organisms may destroy populations of native organisms, and in doing so may further reduce their competition. It can be very difficult, or impossible, to remove a species once it has been introduced without upsetting the rest of the ecosystem.

CHAPTER PLANNER 6 Humans in the Biosphere

Chapter Contents	NSES	Time	Core Resources
Chapter Preview			**Student Edition,** pp. 126–127 **Chapter Mystery,** p. 127 **Foundations for Learning,** p. 127
6.1 A Changing Landscape The Effect of Human Activity • Sustainable Development	I, III, A.1.b, A.2.a, C.4.d, C.4.e, C.5.e, F.3, F.4, F.5, F.6	1/2 period 1/4 block	**Student Edition,** pp. 128–131 **Inquiry Into Scientific Thinking,** p. 130 **Study Workbook B** 6.1 Worksheets **Biology.com** 6.1 Lesson Assessment
6.2 Using Resources Wisely Soil Resources • Freshwater Resources • Atmospheric Resources	I, II, III, A.2.d, C.4.b, C.4.d, C.4.e, C.5.e, F.3, F.4, F.5, F.6	1 period 1/2 block	**Student Edition,** pp. 132–137 **Build Connections,** p. 134 **Study Workbook B** 6.2 Worksheets **Biology.com** *Art in Motion:* Biological Magnification • 6.2 Lesson Assessment
6.3 Biodiversity The Value of Biodiversity • Threats to Biodiversity • Conserving Biodiversity	I, II, III, IV, A.2.a, C.4.c, C.4.d, C.4.e, C.5.e, F.2, F.3, F.4, F.5, F.6	1 period 1/2 block	**Student Edition,** pp. 138–142 **Study Workbook B** 6.3 Worksheets **Biology.com** *Art Review:* Threats to Biodiversity • 6.3 Lesson Assessment
6.4 Meeting Ecological Challenges Ecological Footprints • Ecology in Action	I, II, III, IV, A.2.a, C.4.a, C.4.c, C.4.d, C.4.e, C.5.e, C.6.b, F.2, F.3, F.4, F.5, F.6, G.1	1 period 1/2 block	**Student Edition,** pp. 143–149 **Build Connections,** p. 143 **Study Workbook B** 6.4 Worksheets **Biology.com** *Visual Analogy:* Ecological Footprints • 6.4 Lesson Assessment
Chapter Pre-Lab	II, A.1.b, A.1.c, A.2.a, A.2.d, F.4, F.5	1 period 1/2 block	**Student Edition,** p. 150 **Lab Manual B** *Acid Rain and Seeds*

Tools for ELL Instruction

ELL Handbook describes teaching strategies to make *Miller & Levine Biology: Foundation Edition* more accessible to ELL students.

Spanish Study Workbook is a Spanish translation of Study Workbook A.

Multilingual Glossary is the *Miller & Levine Biology* glossary translated into ten languages.

Tools to Challenge Students

Study Workbook A includes more challenging worksheets and activities such as Chapter Mystery activities that reinforce 21st Century Skills.

Lab Manual A labs are geared toward on-level students. The Lab Manual includes fifteen labs in addition to the Chapter Labs.

Biology.com Data Analysis activities offer students a way to practice data gathering and analyzing skills.

Visual Quizzes in the Assessment Resources Book are geared toward on-level students and use visuals from *Miller & Levine Biology.*

Additional Resources

Biology.com Untamed Science Video • Flash Cards

Study Workbook A 6.1 Worksheets
Spanish Study Workbook 6.1 Worksheets

Study Workbook A 6.2 Worksheets
Lab Manual B *Data Analysis:* Vehicle Emissions
Spanish Study Workbook 6.2 Worksheets

Study Workbook A 6.3 Worksheets
Lab Manual B *Data Analysis:* Saving the Golden Lion Tamarin
Spanish Study Workbook 6.3 Worksheets
Biology.com *Data Analysis:* Measuring Biodiversity

Study Workbook A 6.4 Worksheets
Lab Manual B *Hands-On Activity:* Overpackaging
Spanish Study Workbook 6.4 Worksheets
Assessment Resources Book Visual Quiz

Lab Manual A *Acid Rain and Seeds* • *Oil-Eating Bacteria*

Chapter Review

Student Edition Chapter Summary, p. 151 • Unit Project, p. 156
Study Workbook B Chapter 6 Review
Transparencies, pp. 64–77
Biology.com Untamed Science Video • You're the Director • Editable Worksheets of Study Workbooks A and B and Lab Manuals A and B • Chapter 6 Flash Cards and Crossword

Untamed Science DVD • Classroom Resources CD (includes lesson presentations and editable worksheets)

Chapter Assessment

Student Edition Check Understanding, pp. 152–155
Study Workbook B Chapter 6 Taking a Standardized Test
Assessment Resources Book Chapter 6 Test B • Unit 2 Test B
Biology.com Chapter 6 Assessment • Editable Worksheets of Chapter 6 Visual Quiz,* Chapter 6 Tests A and B, Unit 2 Tests A and B

Exam*View* *Assessment Suite* • Classroom Resources CD (includes lesson presentations and editable worksheets)

Time: 1 period, 1/2 block

* Visual Quizzes are based on visuals from *Miller & Levine Biology*. Review carefully to make sure they are appropriate for your class.

Pressed for Time?

Preview the Chapter Introduce the vocabulary for Lessons 6.2 and 6.3, and have students preview the major headings in these lessons.

Cover the Chapter Quickly Assign *Sustainable Development* in Lesson 6.1, Lesson 6.2, and the Biological Magnification visual. Assign *The Value of Biodiversity* and *Threats to Biodiversity* in Lesson 6.3, and the Habitat Fragmentation visual. Assign *Ecological Footprints* in Lesson 6.4.

Assess Assign Questions 2 and 4 in the 6.1 Check Understanding, the entire 6.2 Check Understanding, Questions 1–4 in the 6.3 Check Understanding, and Questions 1 and 3 in the 6.4 Check Understanding. Assign the Think Visually activity on p. 151, and Questions 1, 2, 5–11, 13, and 14 in the Chapter 6 Check Understanding.

Get the Picture

Have students examine the photo of Earth, and ask a volunteer to read the caption aloud. Point out each continent and other landmasses, such as Greenland, and ask a volunteer to name it. Do the same for recognizable countries such as India and China.

Ask On which continent do you see more lights, North America or South America? *(North America)*

Ask Do you think North America uses more energy than South America does? *(yes)* Why do you think so? *(Sample answer: Lights require energy to stay lit, and many more lights are used in North America.)*

Ask Where else in the world does the photo suggest a high use of energy? *(Europe, India, eastern China, and Japan)*

Activate Prior Knowledge

Encourage students to discuss the pros and cons of high energy use.

Ask Besides powering our lights, what are some other things that we use energy for in our daily lives? *(Sample answer: powering computers and televisions, transportation, heating and cooling our homes and schools)*

Say The high use of energy has enabled millions of people living in North America to have a high standard of living. Standard of living is the level of comfort enjoyed by a society.

Ask What are some disadvantages of high energy use? *(Sample answer: It can cause air, water, and land pollution. It uses up Earth's resources.)*

Say In this chapter, we will learn how resources are used on Earth, and we will explore ways that people can meet their needs without harming the environment.

NATIONAL SCIENCE EDUCATION STANDARDS

UNIFYING CONCEPTS AND PROCESSES

I, II, III, IV

CONTENT

C.4.a, C.4.b, C.4.c, C.4.d, C.4.e, C.5.e, C.6.b, F.2, F.3, F.4, F.5, F.6, G.1

INQUIRY

A.1.b, A.1.c, A.2.a, A.2.d

6 Humans in the Biosphere

Big idea Interdependence in Nature

Q: How have human activities shaped local and global ecology?

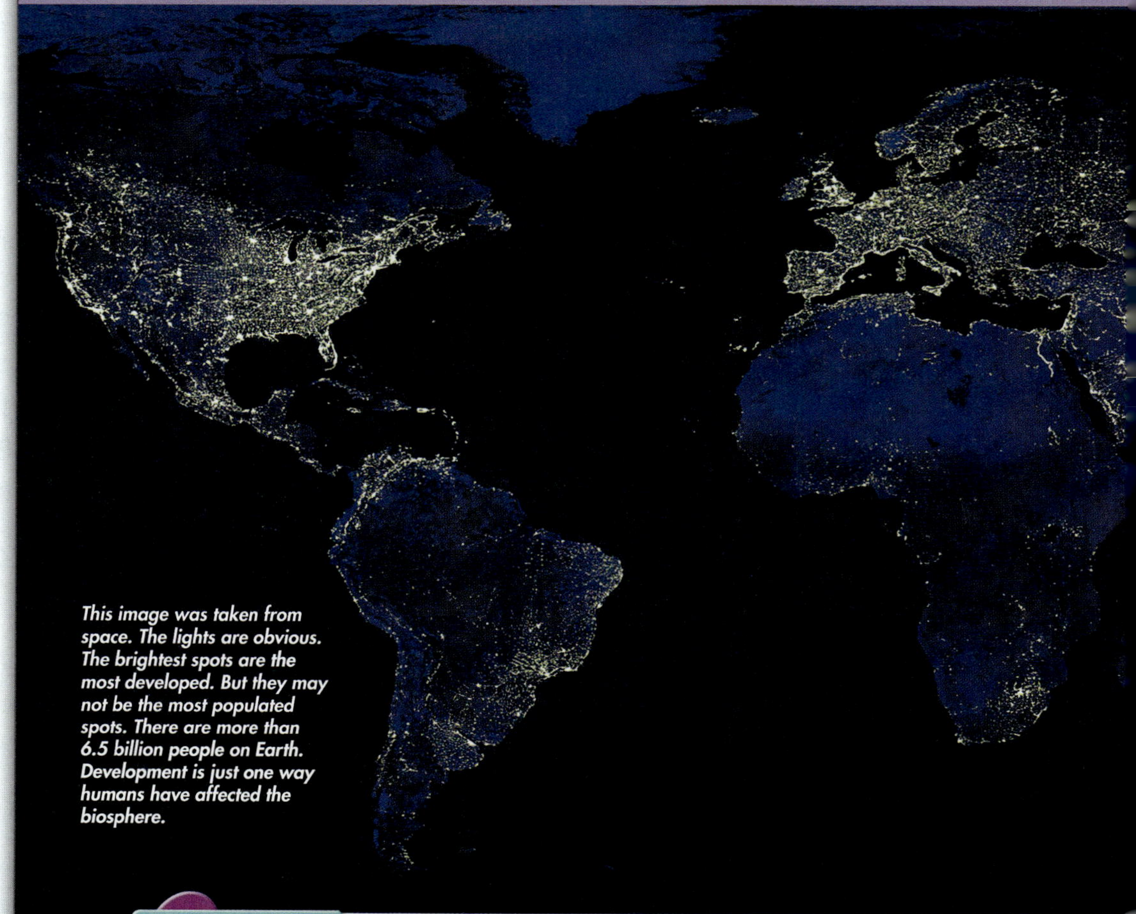

This image was taken from space. The lights are obvious. The brightest spots are the most developed. But they may not be the most populated spots. There are more than 6.5 billion people on Earth. Development is just one way humans have affected the biosphere.

BIOLOGY.com Search Chapter 6 GO • Flash Cards

126

UbD Connect to the Big Idea

Invite a volunteer to read aloud the chapter number and the chapter title. Read the Big Idea and question. Emphasize the question, **How have human activities shaped local and global ecology?**

Set the Purpose

PREVIEW AND PREDICT Draw the graphic organizer on the board. Guide students to the INSIDE feature on the next page, and invite a volunteer to read each lesson number and title.

Ask How do these titles relate to the questions in the graphic organizer? *(Sample answer: The title and question for Lesson 6.1 are related because human activity changes the environment.)*

Remind students to search for answers to the questions in the graphic organizer as they read the chapter.

INSIDE:

CHAPTER MYSTERY

MOVING THE *MOAI*

Easter Island is a tiny island off the coast of Chile. The original islanders came from Polynesia. They called themselves Rapa Nui. They carved hundreds of huge stone statues called *moai* (MOH eye). Each statue weighs between 10 and 14 tons. About 800 years ago, the Rapa Nui moved the huge stones from quarries to places all around the island. Nearly all theories about this process suggest that strong, large logs were needed to move the *moai*. Yet, by the time Europeans landed on the island in 1722, there was no sign of any trees large enough to provide such logs. What had happened?

Read for Mystery Clues As you read this chapter, look for clues about the interactions of the Rapa Nui with their island environment. Then solve the mystery at the end of the chapter.

FOUNDATIONS for Learning

Before you read this chapter, make a mini book. Fold two pieces of notebook paper in half and then in half again in the other direction. Staple along the longer folded edge. This edge will be the spine of your book. Cut the folds along the folded edge. You should now have sixteen pages. Write the title of the chapter on the front cover of your mini book and each of the key questions within the chapter at the top of its own page. As you read the chapter, answer each key question in your own words. Use the remaining pages for additional notes.

- Untamed Science Video
- Chapter Mystery

Foundations for Learning

Demonstrate how to make a mini book. Help students understand the value of this activity as a study tool.

Ask How does your mini book help you organize information? *(Each page is devoted to one key question and summarizes information from a lesson.)*

Ask How can answering the key questions in your own words be helpful? *(Sample answer: Writing an answer in my own words will help me understand and remember the answer better.)*

Ask What's the purpose of having all the key questions in one mini book? *(It makes it easy to review the key questions, and that's a helpful way to remember the main ideas in each of the lessons.)*

Remind students that they will answer the Key Questions as they read the lesson.

Untamed Science The ecogeeks of Untamed Science go behind the scenes to uncover critical research and conservation programs in progress at the Houston Zoo in **What Do Zoos . . . Do?**

CHAPTER MYSTERY Have students read over the introduction to the Chapter Mystery and predict what happened to the trees on Easter Island. Use their predictions to help students start connecting the Chapter Mystery to the Big Idea of interdependence in nature.

Digital Support

BIOLOGY.com Extend your reach by using the digital assets offered at **Biology.com.**

Chapter 6 Big Idea: Interdependence in Nature

Chapter 6: How have human activities shaped local and global ecology?

- **6.1:** How does human activity affect the environment?
- **6.2:** How can we use our natural resources wisely?
- **6.3:** Why is it important to protect and conserve biodiversity?
- **6.4:** How can we change our behaviors to help protect our planet?

LESSON 6.1

Pre-Reading

Objectives

6.1.1 Describe human activities that can affect the biosphere.

6.1.2 Describe the relationship between resource use and sustainable development.

Preview the Pages

1. Preview the visuals. Have students examine the photographs of two different areas of the Hawaiian Islands. Point out that the photographs show an undeveloped and a developed area.
2. Have volunteers find and read the vocabulary words in the text. For each word, have students cover the definition and attempt to define the word from its context. Then, have them compare their definitions to those in the margin.
3. Direct students' attention to the two photographs of Hennepin and Hopper Lakes. Tell them that the inset shows how the area appeared before restoration.

Activate Prior Knowledge

Show students pictures of undeveloped areas.

Say Suppose these areas were scheduled to have housing developments, highways, or shopping malls built on them.

Ask What might be some of the positive results of such developments? *(increased employment, new housing for residents, convenient shopping)* What might be some of the negative results of such developments? *(loss of trees and other natural vegetation, loss of habitat for some species, loss of natural recreation areas and beauty)*

NATIONAL SCIENCE EDUCATION STANDARDS

UNIFYING CONCEPTS AND PROCESSES

I, III

CONTENT

C.4.d, C.4.e, C.5.e, F.3, F.4, F.5, F.6

INQUIRY

A.1.b, A.2.a

For the corresponding lesson in *Miller & Levine Biology,* see pages 154–157.

6.1 A Changing Landscape

Key Questions

How do our daily activities affect the environment?

What is the relationship between resource use and sustainable development?

BUILD Understanding

Lesson Preview Before you read this lesson, write down the headings and look at the figures. Write one or two sentences about what topics you think will be covered. As you read, write the main idea under each heading.

In Your Workbook Go to your workbook to identify the main ideas of this lesson.

The Effect of Human Activity

People from Polynesia began to settle in Hawaii around 1600 years ago. These island people had customs that protected the natural resources. They did not catch fish during spawning season. If they cut down a coconut tree, they planted two trees to replace it. These settlers changed their environment. But they managed their islands carefully. Local ecosystems provided enough fresh water, fertile soil, fish, and other needs to support the population.

Europeans began moving to Hawaii in the late 1700s. These new settlers didn't understand the limits of island ecosystems. They imported many plants and animals. They cleared forests to plant crops. They covered land with housing and tourist developments. What happened?

Living on Island Earth Humans, like all living things, depend on Earth's life-support systems. We also change the environment when we obtain food, get rid of wastes, and build homes. These changes are most obvious on islands, such as Hawaii. Because islands are small, they have limited resources. If local ecosystems cannot supply food and other basic needs, they must be brought in from far away. That's the situation in Hawaii today. Fresh water is getting scarce. Food must be brought in from thousands of miles away.

On large continents, there used to be plenty of space, food, and water. So most of us don't think of land, food, or water as limited resources. But human activity has now used or changed half of all land not covered by ice and snow. As our population reaches 7 billion, we may be getting close to Earth's carrying capacity. Earth begins to feel like a very large island. Humans affect the biosphere of Earth through agriculture, development, and industry.

The Lesson of Hawaii The picture on the left is of Kalalau Valley on Kauai. It looks almost untouched by humans. The picture on the right is of Waikiki Beach on Oahu. Human development is very visible.

6.1 A Changing Landscape

Lesson Objectives

- Describe human activities that can affect the biosphere.
- Describe the relationship between resource use and sustainable development.

BUILD Vocabulary

A. The chart below shows key terms from the lesson with their definitions. Complete the chart by writing a strategy to help you remember the meaning of each term. One has been done for you.

Term	Definition	How I'm Going to Remember the Meaning
Monoculture	The practice of planting large fields with a single variety of crop year after year	*The prefix mono- means "one." Monoculture is the practice of planting one type of crop year after year.*
Nonrenewable resource	A resource that cannot be remade by natural processes	*I know that the prefix non- means "not." I also know that I need to renew a magazine subscription in order to keep getting it. So a nonrenewable resource is a resource that cannot start again by itself.*
Renewable resource	Resource that can be remade quickly and that is replaceable	*Both replaceable and renewable begin with r, and a renewable resource is one that is replaceable.*

B. As you work through this lesson, you may find these terms in the activities. When you need to write a key term or a definition, **highlight** the term or the definition.
Note: Students should highlight terms and definitions used throughout the chapter.

BUILD Understanding

Main Ideas The green headings let you know that this lesson has two main topics.

As you read, describe the main idea of each section.

Sample answers:

Main Heading	Main Idea
The Effect of Human Activity	*We affect the environment by our daily activities, such as agriculture, development, and industry.*
Sustainable Development	*Sustainable development provides for human needs while respecting ecosystem goods and services, such as renewable and nonrenewable resources.*

The Effect of Human Activity

Humans rely on Earth's life-support systems. We affect the environment when we obtain food, eliminate waste, and prepare places to live. Hawaii serves as a good example of this effect because of its small size. Island living also highlights an area's ability to care for humans.

Decide how each of the following activities by Hawaiian settlers affected the environment. In the column labeled "Effect," write a plus sign (+) if the effect was positive and a negative sign (–) if the effect was negative. The first one has been done for you.

Human Activity	Example in Hawaii	Effect
Agriculture	Farmed monoculture crop of pineapples, straining water supply	–
	Imported invasive plant and animal species	–
Development	Cut down forest for housing and tourism	–
	Planted two coconut trees for every one coconut tree cut down	+
Industry	Prohibited fishing industry from fishing during spawning	+
	Allowed factory to spill waste into air, water, and soil	–

Answer the questions.

1. How do humans affect the environment? *Sample answer: We affect the environment when we obtain food, eliminate waste, and prepare places to live.*
2. How is Earth like an island? *The Earth has limited resources to care for humans just like an island.*
3. How can the tourist industry in Hawaii help protect the island? *Sample answer: The tourist industry can give tourists information that will help them avoid damaging the environment.*

Sustainable Development

Sustainable development provides a way for humans to meet their needs without causing long-term harm to the environment. Humans working in this way use renewable and nonrenewable resources wisely. Remember: Renewable resources can be produced again, while nonrenewable resources cannot be produced within a reasonable amount of time.

Study the resources listed in the box below. Decide whether or not each resource is renewable. Then write the resource under its proper heading in the chart.

coal, marsh grass, natural gas, oil, pine tree, wind

Renewable Resources	Nonrenewable Resources
pine tree	*coal*
wind	*natural gas*
marsh grass	*oil*

Answer the questions. Circle the correct answer.

1. Sustainable development is the wise use of ________
   renewable resources / nonrenewable resources / both renewable and nonrenewable resources
2. When fossil fuels are depleted, they are essentially gone ________
   for a little while / forever
3. A ________ cannot be produced within a reasonable amount of time.
   renewable resource / nonrenewable resource
4. What is an example of a good or service provided by an ecosystem?
   Sample answer: a wetland purifying water by filtering it
5. What is sustainable development?
   Sustainable development is development that provides for human needs while preserving the ecosystems that produce natural resources.
6. Which of the following is an example of sustainable development? Circle the right answer.
   When building a new house, the builders leave many of the trees in place.
   After they paint a building, the painters dump the leftover paint in a nearby stream.
   A new highway goes through a wetland.

Agriculture Agriculture has been vital to human history. Once people could grow a dependable supply of food, they could gather together and live in towns and cities. Living in these large groups encouraged the growth of civilization.

Farmers have been able to double the world's food production over the last 50 years. One way they did this is through monoculture. **Monoculture** is the practice of clearing a lot of land to plant a single crop year after year. Farmers use machines to efficiently sow, tend, and harvest the crops. This is how farmers have supplied food for nearly 7 billion people. But monoculture can pollute soil and water. Also, running farm machinery and producing fertilizer both use a lot of fossil fuels.

Development In modern times, small settlements grew into crowded cities. Suburbs then grew out from those cities. All those dense human communities produce lots of wastes. If wastes are not treated properly, they can pollute the land, air, and water resources. Also, spreading development uses up farmland and divides natural habitats into fragments.

Industrial Growth The Industrial Revolution of the 1800s changed human society. Industry and science have led to many modern conveniences. We have comfortable homes and clothing and many electrical devices for work and play. It takes a lot of energy to produce and power these modern conveniences. We get most of this energy by burning fossil fuels—coal, oil, and natural gas. Also, industries have often gotten rid of wastes by dumping them into the air, water, and soil. All these activities can harm the environment.

Key Question How do our daily activities affect the environment?
People affect the environment through agriculture, development, and industrial growth. These activities affect the quality of Earth's natural resources, including soil, water, and the atmosphere.

BUILD Vocabulary

monoculture
a farming strategy of planting a single, highly productive crop, year after year

PREFIXES
The prefix *mono-* in monoculture means "one, alone, single." Monoculture is the practice of planting a single productive crop, year after year.

Monoculture This farmer is using a tractor to plow a large field of soybeans.

Hands-On Learning

MAKE FLYERS Have students make flyers to alert schoolmates about some of the environmental issues they are learning about in this lesson. Follow these steps:

1. Have groups of two or three students review the lesson and identify an environmental issue they would like to address in their flyer.
2. Have each group design its flyer. Remind students to address these questions: How much text will be needed to summarize the environmental issue? What kinds of visuals will be needed to support the text? Provide various art supplies. Tell students that their flyer should attract attention so that schoolmates will be likely to read it.
3. Have groups write the text for their flyer. The text should briefly describe the issue and explain what can be done about it.
4. Suggest students post their finished flyer in a place where all can see it, or pass it out at busy locations around school. Be sure to follow school policy for passing out flyers.

Build Understanding

LESSON PREVIEW Refer students to Workbook B, where they'll find a two-column chart to use in previewing the lesson. Tell students that previewing a lesson will help them prepare for what they will learn and increase their understanding.

During Reading

Active Reading

USE GRAPHIC ORGANIZERS Draw a word web on the board with a large central circle joined by lines to smaller circles. In the central circle, write: *Human activities that affect the environment.* Instruct students to copy the web on a sheet of paper. As students read the paragraphs with blue heads on the previous page and this one, tell them to complete the word web by writing examples of human activities that affect the environment. Students can add circles as needed.

Build Vocabulary

PREFIXES Have pairs of students use a dictionary to find the meanings of the following words in which the prefix *mono-* means "one, alone, or single": *monochromatic, monocle, monocracy,* and *monogamy.* (monochromatic—*one color;* monocle—*eyeglass for one eye;* monocracy—*rule by one person;* monogamy—*married to one person)*

KEY QUESTION Read aloud the Key Question and answer.

Ask How has population growth contributed to changes in the environment? *(As the human population grows, more crops must be grown, more land developed, and more energy used.)*

Speed Bump

Ask Why might introducing new plants and animals to a Hawaiian island cause native plants and animals to die out? *(If the new organisms have no natural predators, they will spread quickly and overtake the native organisms.)*

Ask How do we get most of the energy used for heating our homes and powering electrical devices? *(by burning fossil fuels)*

During Reading continued

Inquiry Into Scientific Thinking

COMPARE AND CONTRAST Make sure students understand what they are to compare in this activity. Have them read the introduction and procedure.

Ask What is your goal in this activity? *(to find out how much garbage I throw away in a day)*

Ask What will you be comparing in this activity? *(the amounts of recyclable garbage, reusable garbage, and trash that I produce)*

PREDICT Have students predict which kind of garbage they will produce in the largest amount.

Ask How can you determine if your prediction is correct? *(First, I will find the total amount of garbage I produce. Then, I will figure out what percentage of the total I put in each container.)*

Lead a Discussion

Have students discuss the value of using durable, reusable products instead of disposable versions.

Ask How does using reusable lunch bags or lunch boxes instead of paper bags help the environment? *(Sample answers: Less garbage will be placed in landfills. Fewer resources will be used to make the single container than the many paper bags I would use in a school year.)*

Active Reading

USE VISUALS Have students discuss the inset of farmland and the photo of the same area as a wetland. Point out that this area was wetland first, then farmland, and then wetland again.

Ask What is the value of this area as a wetland? *(It provides wildlife habitat, natural flood control, and a natural filter for the area's water.)*

SCIENCE SUPPORT Explain to students that a wetland may not be wet all year. In a wetland, ground water is at or near the surface for at least part of the year. Wetlands are often nutrient-rich and highly productive, and they serve as breeding grounds for many organisms.

Speed Bump

Ask What resources do people depend on from healthy ecosystems? *(clean air and water)*

Ask How can water in polluted water sources be made safe to drink? *(The water must be treated to make it drinkable.)*

INQUIRY into Scientific Thinking GUIDED INQUIRY

Reduce, Reuse, Recycle

Each person on Earth affects the biosphere. The garbage you throw out each day is one way you affect the biosphere. In this activity, you will find out how much you throw away in one day.

1. Get three containers to hold your trash for one day. Label the containers Reuse, Recycle, and Trash.
2. Throughout the day, collect your dry garbage.
3. Sort the garbage into the three containers. Items that cannot be reused or recycled will go into the Trash container. Be sure to wash your hands when you are done.

Analyze and Conclude

1. Analyze Data Compare the amount of garbage in each of your containers. About what percentage of the total did you put into each container?

2. Predict What do you think happens to the things you throw out? Think of at least three ways that garbage can impact living things.

3. Evaluate List three ways you can reduce the amount of trash you throw out each day.

In Your Workbook Get more help for this activity in your workbook.

Sustainable Development

We depend on healthy ecosystems to provide natural resources such as clean air and water. We usually take clean air and water for granted. But if the environment cannot provide them, our cities and towns must spend money to provide them. For example, drinking water is often filtered by wetlands. Water then flows through streams and rivers and is stored in lakes. But if the water sources or wetlands are polluted or damaged, water quality may fall. When this happens, cities and towns must build and run expensive water treatment plants to provide safe drinking water.

Ecosystem Services Hennepin and Hopper Lakes is a wetland in Illinois. It was drained for farming in 1900. The smaller image shows how the area looked when it was farmland. In 2003, the area was restored. The larger image shows how it looks now.

130 Chapter 6 • Lesson 1

Materials

gloves, large trash bags, boxes, or bins

Analyze and Conclude

1. Answers may vary.
2. Sample answer: Most of the trash goes to a landfill. Some items are recycled. Landfills take away from natural habitats for living things. Chemicals that leak out from landfills can contaminate water. Animals mistaking trash for food may become ill.
3. Sample answer: I can recycle more items, reuse some materials instead of throwing them away, and stop buying items that are heavily packaged.

Renewable and Nonrenewable Resources There are two main types of natural resources. A **renewable resource** can be made or replaced by a healthy ecosystem. A single southern white pine is a renewable resource. A new tree can grow in place of an old tree that dies or is cut down. But some resources are nonrenewable. A **nonrenewable resource** cannot be replaced in a reasonable amount of time. Fossil fuels are nonrenewable resources. They take millions of years to form. When we use up the fossil fuels that are already here, they are basically gone forever.

Sustainable Resource Use We can learn to use natural resources to meet our needs without causing long-term environmental damage. Using resources this way is called sustainable development. Sustainable development should cause no long-term harm to the soil, water, and climate. It should use as little energy and materials as possible. It must be able to withstand droughts, floods, and heat waves or cold snaps. Sustainable development must also work well enough to help people improve their standard of living.

Key Question What is the relationship between resource use and sustainable development?
Sustainable development uses natural resources to meet human needs without causing long-term damage to the environment.

BUILD Vocabulary

renewable resource
a resource that can be produced or replaced by healthy ecosystem functions

nonrenewable resource
a resource that cannot be replenished by natural processes within a reasonable amount of time

RELATED WORD FORMS
The word *renew* is used to mean "to extend the checkout time for a library book" and "to begin again." These uses are related to renewable resources and nonrenewable resources. Renewable resources can be used over and over again. Nonrenewable resources cannot.

CHECK Understanding

Apply Vocabulary
Use the highlighted words from the lesson to complete each sentence correctly.

1. When farmers plant the same crop in a field year after year, they are practicing ________.
2. Coal, oil, and natural gas are examples of ______ because they cannot be replaced in a reasonable amount of time.

Critical Thinking

3. **Relate Cause and Effect** How might practices that allow farmers to grow more crops affect a developing nation's population? How might the practices affect the nation's environmental health?
4. **Compare and Contrast** Explain why energy from the sun is a renewable resource but energy from oil is a nonrenewable resource.
5. **Apply Concepts** Imagine that a large shopping center is being built near your school. Until now, the land was basically untouched. There was a clump of trees and a large grassy field. How might the shopping center affect your local ecosystem?
6. **Write to Learn** Answer the first clue of the mystery. Keep in mind what can happen when an organism is introduced to a new environment.

MYSTERY CLUE
Easter Island's first colonists brought with them banana trees, taro root, and chickens. They may also have brought small mammals, such as rats. What impact might these new organisms have had on the island's ecosystems? (Hint: See p. 128.)

BIOLOGY.com Search Lesson 6.1 GO • Lesson Assessment

Check Understanding

1. monoculture
2. nonrenewable resources
3. More crops would feed more people. The nation's population would probably increase. But the farming could also pollute the nation's soil and water and would use a lot of fossil fuels.
4. The sun is always providing energy. It takes millions of years to produce oil. If we use all of the available oil, it would take millions of years to replace it. Therefore, oil is nonrenewable.
5. Sample answer: The construction will use a lot of resources, including fossil fuels. The new shopping center will likely use a lot of water and energy. Also, the organisms that lived on the unused land may be displaced or die.
6. When these new organisms arrived on Easter Island, they had an unlimited amount of food and space. Their populations probably grew exponentially. They may have competed with native plants and animals. The native plants and animals may have died off.

Active Reading

USE GRAPHIC ORGANIZERS Have students make a T-chart to organize information about renewable and nonrenewable resources. Tell them to add details about each kind of resource in the appropriate column, along with examples.

SELF-QUESTION Tell students that turning statements into questions can help them find and remember important information.

Read aloud the first sentence of the paragraph Sustainable Resource Use.

Say I can turn this sentence into a question by asking *How can we learn to use natural resources to meet our needs without causing long-term environmental damage?* Now I can look for the answer as I read. Read the rest of the paragraph aloud.

Say I see that at least part of the answer is to use sustainable development. Discuss what sustainable development is.

KEY QUESTION Read the Key Question and answer aloud.

Ask Does *sustainable* mean that we should not use resources? *(no)* Explain. *(Resources can be used, but in a way that avoids environmental damage.)*

Build Vocabulary

RELATED WORD FORMS Explain that to sustain something means to make it continue to exist or happen. Have students relate this definition to *sustenance* and *sustainable*.

Ask What things do you need for sustenance? *(food, air, water, etc.)*

Post Reading

ASSESSMENT TIPS For Question 3, have students discuss what makes populations decrease or increase. For Question 4, make a list of resources on the board. Ask volunteers to explain why each is renewable or nonrenewable.

MYSTERY CLUE If students struggle to answer the mystery clue, have them reread the first paragraph of the lesson, which describes changes that occurred on Hawaii after settlers introduced new plants and animals.

Wrap-Up Activity

DESIGN A COMMUNITY PROJECT Have groups of students brainstorm actions that can help meet a local environmental need. Ideas might include starting or expanding a recycling program, reducing consumption of disposable products, or organizing community clean-up days. Have each group write a plan and share it with the class.

LESSON 6.2

Pre-Reading

Objectives

6.2.1 Describe how human activities affect soil and land.

6.2.2 Describe how human activities affect water resources.

6.2.3 Describe how human activities affect air resources.

Preview the Pages

1. Before reading, have students preview the green and blue section headings and discuss how the lesson is organized. Explain that understanding the structure of the lesson will help students understand the text as they read.
2. Preview with students all of the vocabulary terms, pronouncing each term aloud. Have students use context clues to define the terms.
3. Have students examine the map on the next page. Discuss what the colors in the key indicate.
4. Direct students' attention to the Biological Magnification visual. Point out that the visual shows how the concentration of pollutants increases as you go up the food chain.

Build Understanding

CONCEPT MAP Refer students to Workbook B where they will find a partially completed concept map with information from this lesson. When they finish reading a section, students should go back to the workbook to add additional details to the map.

NATIONAL SCIENCE EDUCATION STANDARDS

UNIFYING CONCEPTS AND PROCESSES

I, II, III

CONTENT

C.4.b, C.4.d, C.4.e, C.5.e, F.3, F.4, F.5, F.6

INQUIRY

A.2.d

For the corresponding lesson in *Miller & Levine Biology,* see pages 158–165.

6.2 Using Resources Wisely

Key Questions

Why is soil important, and how do we protect it?

What are the primary sources of water pollution?

What are the major forms of air pollution?

BUILD Understanding

Concept Map As you read, make a concept map to organize the information in this lesson.

In Your Workbook Go to your workbook for help completing the concept map.

Soil Resources

Our economy is built on the use of natural resources. We cannot just stop using land for farming. Yet, we need to protect the health of ecosystems that supply us with renewable resources. How can we get what we need from local and global environments in a sustainable way?

You may not think of soil as a resource. Yet, many things you need depend on healthy soil. Food production depends on crops. Paper comes from trees. All these plants we rely on need healthy soil.

The mineral- and nutrient-rich portion of the soil is called topsoil. Good topsoil absorbs and holds the water that plants need but also allows some water to drain. It is rich in nutrients but low in salts. Topsoil can be renewable if it is managed well. Good topsoil is produced by many years of interactions between soil and plants. These processes take a long time to form good topsoil. But topsoil can be damaged or lost quickly. For example, large areas across the Great Plains in the United States were poorly farmed for years. Then, a very bad drought hit in the 1930s. The combination of bad farming and drought eroded topsoil. Crops failed. The area became like a desert. People began to call the Plains the "dust bowl." Thousands of people lost their homes and their jobs.

Soil Erosion The dust bowl was caused, in part, by the way the prairie was used as farmland. Soil erosion happens when wind or water take soil away. Soil erosion is often worse if land is plowed and left bare between plantings. Plant roots hold soil in place. If plants are taken out of the soil, then the soil is easily washed away. Sometimes, farming, overgrazing, and very little rain can turn farmland into desert. This is what happened in the Great Plains in the 1930s. The process is called **desertification.** About 40 percent of Earth's land is thought to be at risk for desertification.

The Dust Bowl A ranch in Boise City, Idaho, is about to be hit by a cloud of dry soil on April 15, 1935.

6.2 Using Resources Wisely

Lesson Objectives

- Describe how human activities affect soil and land.
- Describe how human activities affect water resources.
- Describe how human activities affect air resources.

BUILD Vocabulary

A. The chart below shows key terms from the lesson with their definitions. Complete the chart by writing a strategy to help you remember the meaning of each term. One has been done for you.

Term	Definition	How I'm Going to Remember the Meaning
Biological magnification	Process by which the strength of a pollutant increases as it moves up the food chain	*Magnify* means "to increase." Biological *magnification* is the increase in the strength of a pollutant as it moves up the food chain.
Deforestation	The loss of forests	The prefix de- means "to remove." Deforestation is the removal of forests.
Desertification	Process in which farmland is turned into desert	To desert also means "to leave." In desertification farmers must leave because their farmland has turned to desert.

B. As you work through this lesson, you may find these terms in the activities. When you need to write a key term or a definition, **highlight** the term or the definition.

CHAPTER MYSTERY

Easter Island Easter Island was once covered with palm trees. Islanders cut down many of these trees to eat their seeds. In the box on the right, draw a picture of Easter Island after deforestation. Surround your picture with words or phrases that describe the effect deforestation had on Easter Island.

Students' drawings should show a landscape without trees. Sample phrases: "soil erosion," "no wood," and "harm water supply."

BUILD Understanding

Concept Map A concept map can help you organize information and show you how ideas are connected.

acid rain, automobile emission standards, crop rotation, deforestation, desertification, drip irrigation, greenhouse gases, industrial chemicals, nonpoint sources, residential sewage, smog

As you read the lesson, place the terms in the correct location in the concept maps.

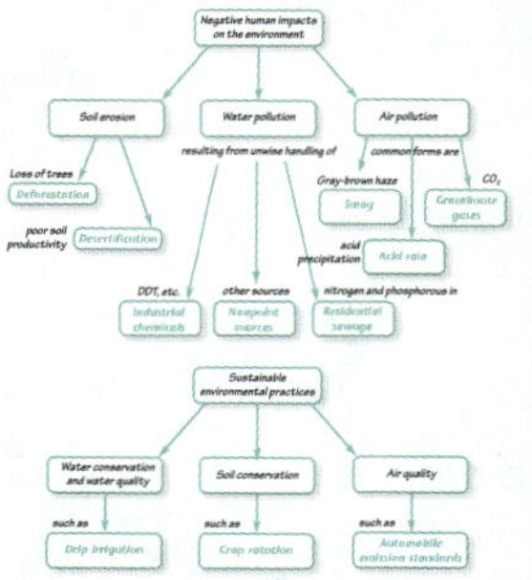

Using Resources Wisely

Resources like soil, water, and air are not likely to run out. However, they can become too polluted too quickly to help support life on Earth. Humans can slow down negative effects on soil, water, and air resources.

Study the chart below. The first column lists three resources: soil, water, and air. The second column describes how humans have affected each resource. In the third column, you will find ways in which humans can slow down the effects. But read carefully! One way in each row is incorrect. Circle the incorrect choice.

Resource	Human Effect	Ways to Slow Down the Effect
Soil	Deforestation and continuous plowing lead to soil erosion.	• Plow following Earth's contours. • Rotate crops. • Clear-cut forests. • Plant and harvest tree farms in sustainable ways.
Water	Pesticides and too much sewage lead to limited supplies of fresh water.	• Dump sewage directly into rivers and streams. • Treat sewage. • Protect the watershed. • Conserve water.
Air	The burning of fossil fuels leads to air pollution.	• Use technology that reduces chemicals in the air. • Increase the burning of natural gas. • Walk more, drive less. • Plant more trees to absorb carbon dioxide.

Answer the questions.

1. Why is it important to protect soil, water, and air? *If these resources become polluted, they cannot help support life on Earth.*
2. What is one way humans can protect water? *Sample answer: treat sewage and reduce the use of pesticides*
3. Why is smog a problem? *Smog can cause respiratory problems in people.*

The loss of forests, called **deforestation,** can also harm soils. Unfortunately, more than half of the world's old-growth forests have already been lost. Old-growth forests are forests that have never been cut. Deforestation can lead to severe soil erosion. Soils and microclimates can change so much that trees cannot grow there again. For example, topsoil in tropical rain forests is usually thin and not very fertile. The soil is often useful for only a few years after the forest is cut down.

Soil Use and Sustainability Soil is most likely to erode when it is completely bare. Leaving stems and roots in the soil between plantings can help hold soil in place. Also, different plants use different nutrients from the soil. When farmers plant different crops at different times of the year or in different years, it is called crop rotation. Crop rotation can help stop erosion and the loss of nutrients.

Another way to limit soil erosion is by contour plowing. Contour plowing is the practice of planting crops across the slope of the land instead of down the slope. This method reduces water runoff and erosion.

What are options for sustainable forestry? Cutting down only some of the mature trees within a forest can allow growth of younger trees without damaging the whole forest or its soil. Tree farms can also protect soil and make the trees themselves a renewable resource.

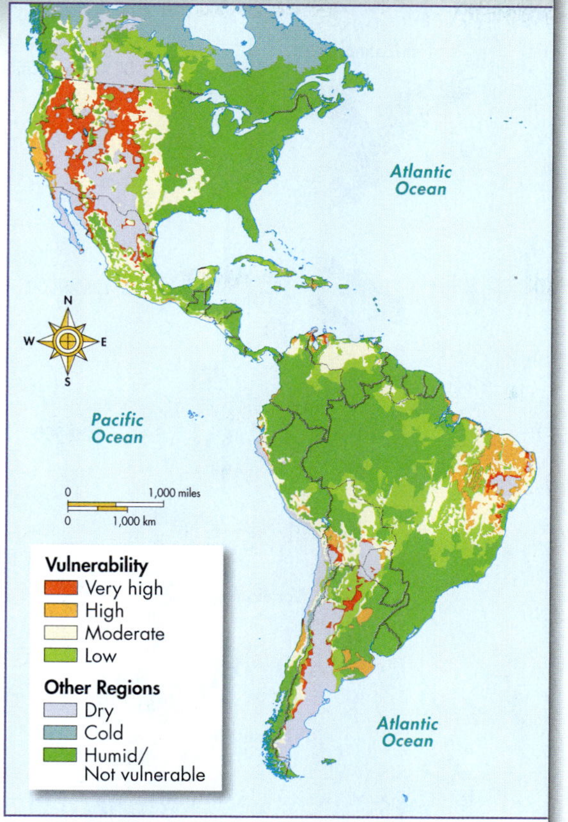

Desertification Risk The U.S. Department of Agriculture identifies the risk of desertification. The risk is based on soil type and climate.

Key Question Why is soil important, and how do we protect it?
Soil allows plants, including crops, to grow. Farming practices such as crop rotation and leaving stems and roots in the soil between plantings can help protect soil from erosion and a loss of nutrients. Cutting only some mature trees and using tree farms can also protect soil.

Freshwater Resources

Fresh water is usually considered a renewable resource. However, some sources of fresh water are not renewable. The Ogallala aquifer runs under eight states from South Dakota to Texas. It took more than a million years to fill. Today, more water is being pumped out than is returning through rain. The aquifer is likely to run dry in 20 to 40 years.

In many parts of the world, the amount of fresh water is limited. Only 3 percent of the water on Earth is fresh water. We must be careful to protect the ecosystems that collect and purify fresh water.

BUILD Vocabulary

desertification
lower land productivity caused by overfarming, overgrazing, seasonal drought, and climate change

deforestation
the destruction of forests

PREFIXES
The prefix *de-* in deforestation means "remove" or "away from." Deforestation is the removal of forests.

ELL Focus on ELL: Extend Language

BEGINNING AND INTERMEDIATE SPEAKERS Have students make vocabulary word maps to help them learn lesson vocabulary terms and any other difficult terms they encounter. Have students call out difficult words while you write them on the board. Choose one word and demonstrate how to draw a simple word map. On their word maps, students should include the following features: the term's definition in their own words; a dictionary definition; a pronunciation guide for the term; and translations for the term. Encourage students to add drawings, diagrams, or other clues to help them remember each term's meaning.

During Reading

Active Reading

CAUSE AND EFFECT Have students draw a two-column chart on lined paper. They should label one column Causes and the other Effects.

Ask What happens in a cause-and-effect relationship? *(One thing makes something else happen.)*

Say I am going to begin reading the paragraphs under Soil Resources. When I get to the first cause-and-effect relationship in the text, raise your hand. *(Good topsoil absorbs and holds the water that plants need but also allows some water to drain.)*

Ask What should you write in the Causes column in your chart? *(good topsoil)* What should you write under Effects? *(absorbs and holds the water that plants need but also allows some water to drain)*

Pair students to find more cause-and-effect relationships as they read through the remaining paragraphs under Soil Resources.

MAKE CONNECTIONS Have students examine the Desertification Risk map and study its key. Display a United States map that identifies cities and states.

Ask Where are we located on the Desertification Risk map? *(Answers may vary.)*

Ask What is our risk of desertification? *(Answers may vary according to the map.)*

Build Vocabulary

PREFIXES Tell students that the prefix *de-* can also mean "to reverse the action of or undo."

Say *Dehumidify* and *defrost* are examples of words that have the *de-* prefix.

Ask What do each of these words mean? (dehumidify—*reverse humidity or take humidity away;* defrost—*remove frost)*

KEY QUESTION Read the Key Question and answer.

Ask Plants are the base of nearly all land food webs. Why does that make it important to protect soil? *(If soil were unhealthy and could not grow plants, food webs would collapse.)*

Speed Bump

Ask How can deforestation affect soil? *(It can lead to soil erosion.)*

Ask How does crop rotation help stop the loss of nutrients in soil? *(Different plants use different nutrients from the soil. By planting different crops at different times, farmers avoid depleting nutrients.)*

During Reading continued

Active Reading

FIND THE MAIN IDEA Have students determine the main idea in each paragraph under Freshwater Resources, beginning on the previous page.

Say An *explicit* main idea is one or two sentences in the paragraph that state the main idea of the paragraph. An explicit main idea might be found anywhere in the paragraph—beginning, middle, or end.

Say An *implicit* main idea is a main idea that is not given in one or two sentences in the text and must be put into the reader's own words.

Read aloud the first paragraph under the heading.

Ask Is the main idea of this paragraph explicit or implicit? *(explicit)* Which sentence or sentences state the main idea? *(Fresh water is usually considered a renewable resource. However, some sources of fresh water are not renewable.)*

Have pairs of students read each remaining paragraph and write its main idea on a sheet of paper. Have students label the main ideas explicit or implicit. Then have students use the main ideas to summarize information from the topic.

Build Connections

VISUAL SUMMARY Point out that the pyramid visual represents a food chain. Tell students that the orange dots show the concentration of DDT at each trophic level. Have students discuss the changes in the number of dots from level to level.

Ask Does the concentration of the pollutant increase or decrease as it goes up the food chain? *(increases)* Does the number of consumers increase or decrease as the pollutant becomes more concentrated? *(decreases)*

Point out that the pyramid is a model for how the concentration of the pollutant is magnified from one trophic level to the next higher level.

Say The numbers are not actual concentrations. The numbers represent how the concentration of the pollutant increases by a factor of 10 at each trophic level, starting with producers.

Ask How much more concentrated is DDT in the large fish than in open water? *(1 million times more)*

Speed Bump

Ask How can fresh water be both a renewable and a nonrenewable resource? *(Most fresh water is renewable, but an aquifer can be considered a nonrenewable resource if water is being pumped out faster than rainwater can fill it.)*

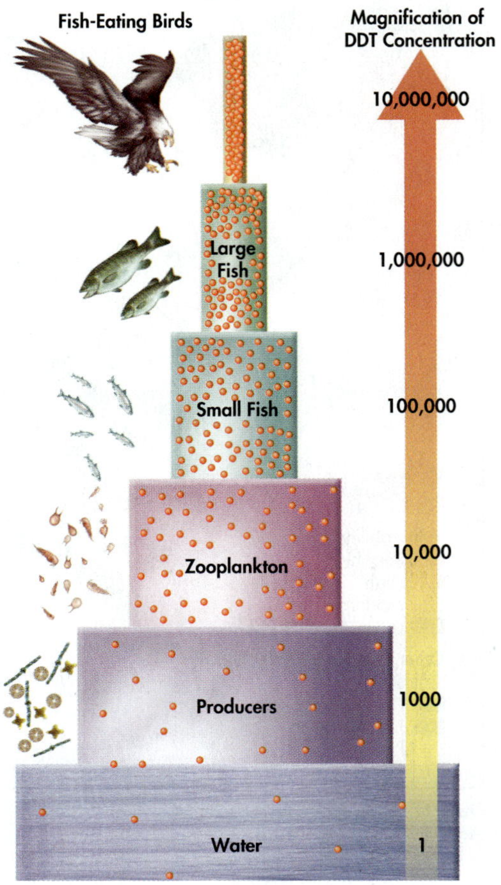

BUILD Connections

BIOLOGICAL MAGNIFICATION

In the process of biological magnification, a pollutant like DDT—represented by the orange dots—becomes more concentrated as it passes up the food chain from producers to consumers.

In Your Workbook Do you understand how the concentration of a pollutant changes in the food web? Go to your workbook to find out.

Water Pollution Freshwater sources can be affected by different kinds of pollution. A pollutant is any harmful material that can enter land, water, or air. When a pollutant enters water supplies from one spot, the source is called a point source. A factory or an oil spill are examples of point sources. When pollutants enter from many smaller sources, the sources are called nonpoint sources. Grease and oil washed off of roads are from nonpoint sources. Chemicals released into the air by factories and cars are also nonpoint sources.

▶ ***Industrial and Agricultural Chemicals*** Sometimes, a pollutant remains in the ecosystem for a long time. PCBs are a group of chemicals that were used in industry until the 1970s. These chemicals can still be found in parts of the Great Lakes and some coastal areas. Heavy metals like lead and mercury are also industrial pollutants.

Monoculture has increased the use of pesticides and insecticides. These chemicals can enter the water supply. Pesticides can be very dangerous pollutants. For example, DDT is a cheap, long-lasting type of pesticide that controls agricultural pests and mosquitoes. But when DDT gets into rivers and lakes, it can have serious effects.

These effects are caused by a process called **biological magnification.** Biological magnification begins when a pollutant is picked up by a primary producer, such as plants or algae. If the pollutant is not broken down, it stays in the plant tissues. Then, herbivores eat the plants. If the herbivores do not break down the pollutant, it gets concentrated in their tissues. When carnivores eat the herbivores, the compound is concentrated even more. Pollutant concentration keeps getting higher as the pollutant gets passed along the food chain. In the highest trophic levels, a pollutant may reach 10 million times its original concentration in the water.

High concentrations of pollutants can cause serious problems for wildlife and humans. High DDT concentrations in pelicans, falcons, and eagles caused these birds to lay eggs with thin, fragile shells. These eggs often didn't hatch, so the populations of these birds dropped.

Focus on Academic Language

- point source
- nonpoint source
- pesticides
- insecticides

Ask What is a clue or memory device that can help you understand the difference between point source pollution and nonpoint source pollution? *(If you can point to the pollution source, it is point source pollution. If there is no way to point to a single source, then it is nonpoint source pollution.)*

Point out to students that the text uses the terms *pesticides* and *insecticides.* Explain that the word *pesticide* is a broad term that includes insecticides, fungicides, and other substances that get rid of a broad range of "pests" that destroy crops. An insecticide is classified as a type of pesticide because some insects are considered pests.

Ask Is a fungicide an insecticide? *(no)* Is a fungicide a pesticide? *(yes)*

Drip Irrigation One way to protect water supplies is to use less water. These cabbages are supplied water directly through drip irrigation. Tiny holes in water hoses allow farmers to deliver water only where it is needed.

▶ ***Sewage*** Have you ever stopped to think what happens after you flush your toilet? After you flush, the wastes become sewage. Sewage carries a lot of nitrogen and phosphorus. Healthy ecosystems can deal with some amount of these nutrients. However, large amounts of sewage can cause many bacteria and algae to grow. These organisms use up the water's oxygen. This depletion of the oxygen causes places called "dead zones." Other organisms are not likely to survive in dead zones. Sewage also carries tiny organisms that can spread disease.

Water Quality and Sustainability One way to make sure we have enough water is to protect ecosystems involved in the water cycle. For example, plants and forests around a water supply help to absorb excess nutrients and filter out pollutants. We must protect the land containing these plants. The land whose groundwater, streams, and rivers drain into the same body of water is called a watershed. In order to clean up a body of water, you must protect the whole watershed.

Cutting pollution rates can improve the water quality in a watershed. Sewage treatment can help prevent dead zones. In place of chemicals, farmers can sometimes use predators and parasites to keep insect populations small. Farmers may also use less-poisonous sprays and crop rotation.

Using less water is also important. One way to use less water in agriculture is to water the crops by drip irrigation. In drip irrigation, water is given directly to the roots of the plants that need it.

Key Question What are the primary sources of water pollution? **The primary sources of water pollution are industrial and agricultural chemicals, residential sewage, and nonpoint sources.**

BUILD Vocabulary

biological magnification
the increasing concentration of a harmful substance in organisms at higher trophic levels in a food chain or food web

RELATED WORD FORMS

To magnify means "to increase the apparent or actual size of an object." In this context, the noun *magnification* refers to the increased concentration of a pollutant in organisms that are at higher levels of the food chain.

Hands-On Learning

WATERSHED CONSERVATION Do the following activity to help students understand that cleaning up pollution in a local area does only limited good if nothing is done about water pollution elsewhere in the watershed.

Show students a map of their state or their region. Have them identify a river that runs through or close to their community. Then, have them try to identify the river's watershed. What body of water does the river run into? What rivers or streams feed into the river? Suggest various places where the watershed might be polluted, and have students identify communities downstream from that source that would be affected by the pollution.

Active Reading

SCIENCE SUPPORT Share with students these facts about the Ogallala aquifer. It underlies parts of eight states: South Dakota, Wyoming, Nebraska, Kansas, Colorado, New Mexico, Oklahoma, and Texas. It is the main source of water in the region for home and industrial water use, although most of the water from the aquifer is used for crop irrigation. The aquifer is recharged by rainfall and snowmelt, but more water is being pumped out than is recharged. The water level has declined by as much as 100 feet in some areas since the 1940s.

USE VISUALS Direct students' attention to the photo showing drip irrigation. Have a volunteer read its caption.

Ask What is another way that crops can be irrigated? *(by spraying them with water)*

Ask Why is spraying more wasteful than drip irrigation? *(Sample answer: Water is sprayed where it's not needed. Additionally, more evaporation might occur when water is sprayed.)*

Build Vocabulary

RELATED WORD FORMS Tell students that the root of the word *magnification* is *magni,* which comes from the Latin *magnus* or *magnificare* and means "large" or "great." Explain that the verb *magnify* comes from the same root.

Ask What does the suffix *-ation* mean? *(the act of)* What does *magnification* mean literally? *(the act of making something larger)*

KEY QUESTION Have a volunteer read the Key Question and answer.

Ask What manufacturing and farming practices do you know about that might be sources of water pollution? *(Answers may vary but will likely include the use of chemicals.)*

Speed Bump

Ask What is an area in a body of water called that has little oxygen due to bacterial and algal growth from sewage? *(a dead zone)*

Ask What is a watershed? *(an area of land whose groundwater, streams, and rivers drain into the same body of water)*

Ask How does drip irrigation conserve water? *(It gives water directly to the roots of the plants that need it.)*

During Reading continued

Active Reading

USE GRAPHIC ORGANIZERS Draw a graphic organizer on the board for recording main ideas and details. Have students copy the graphic organizer into their notebooks and then use it to analyze how the first paragraph of the Atmospheric Resources section is organized. For information on Main Ideas and Details Charts see page A7 in Workbook B.

Ask What is the main idea of this paragraph? *(The atmosphere has a direct effect on health.)* Is the main idea explicit or implicit? *(explicit)*

Have students record the main idea in the graphic organizer. Then ask them to record supporting details from the paragraph.

Say The first sentence of the paragraph presents the main idea, and each of the other sentences provides an example that supports the main idea.

MAKE CONNECTIONS Help students see how the function of a greenhouse is similar to that of Earth's atmosphere.

Ask What is the purpose of a greenhouse? *(A greenhouse lets us grow plants inside even when the weather outside is too cold.)*

Ask What holds heat inside a greenhouse? *(Heat from the sun is held inside by the glass panes of the greenhouse.)* How is Earth's atmosphere like a greenhouse? *(The atmosphere also holds in heat.)*

Speed Bump

Ask What are three greenhouse gases in Earth's atmosphere? *(carbon dioxide, methane, and water vapor)*

Ask How can greenhouse gases be both helpful and harmful? *(Greenhouse gases regulate Earth's temperature, making life possible. Too high levels of greenhouse gases, however, can cause Earth's atmosphere to hold too much heat, which causes global warming and climate change.)*

Ask What are the solid components of air pollution called? *(particulates)*

Smog The 2008 Summer Olympics were held in Beijing, China. For many weeks before the games, factories were closed and fewer cars were allowed on the streets. Yet Beijing remained under a blanket of dense smog.

Atmospheric Resources

The atmosphere has a direct effect on health. The atmosphere gives us the oxygen we need to breathe. The upper atmosphere contains a form of oxygen called ozone. Ozone protects us from the sun's ultraviolet radiation. The atmosphere also contains other gases such as carbon dioxide, methane, and water vapor. These gases are called greenhouse gases. Greenhouse gases help keep the Earth's temperature stable.

The atmosphere cannot be "used up." So, it is not important to call it renewable or nonrenewable. Instead, it is important to understand what human activities do to it.

Air Pollution Many things, including the burning of fossil fuels, can cause air pollution. Some forms of air pollution are smog, acid rain, greenhouse gases, and particulates. When air is not clean, it can be hard to breathe. People who suffer from asthma and allergies will notice this effect the most. Certain types of air pollution can also cause global climate change.

▶ ***Smog*** If you have been to a large city, then you have probably seen smog. Smog is a gray-brown haze in the air. It is formed by chemical reactions among air pollutants. These reactions form ozone. High in the atmosphere, ozone protects life on Earth. Near ground level, ozone is dangerous, especially to people with difficulty breathing. In the 2008 Summer Olympics, many athletes were concerned about the level of smog in the air.

▶ ***Acid Rain*** Burning fossil fuels releases chemicals that combine with water vapor in the air to form acids. When the acids fall as rain, it is called acid rain. In some areas, acid rain damages leaves and changes the chemistry of soil and surface water. It can release toxic elements from soil, allowing them to move through the biosphere.

▶ ***Greenhouse Gases*** When we burn forests and fossil fuels, we release carbon dioxide into the atmosphere. Many kinds of farming release methane into the air. Natural concentrations of these greenhouse gases in the atmosphere control Earth's temperature. But by raising greenhouse gas concentrations, human activity plays a role in causing global warming and climate change.

▶ ***Particulates*** Particulates are tiny pieces of ash and dust. These are released by some industrial processes and by some diesel engines. Very small particulates can cause serious health problems when they enter the lungs.

ELL Focus on ELL: Extend Language

ALL SPEAKERS Point out to students that the term *smog* may remind them of another English word, *fog.* Explain that the word *smog* first appeared in writing in 1905, when a newspaper reported that "there was something produced in great cities which was not found in the country, and that was smoky fog, or what was known as *smog.*"

Ask What do you think caused this smoky fog? *(Sample answer: air pollution from burning coal)*

Tell students that the smog we see today is somewhat different than the smog that occurred during the Industrial Revolution. The average amount of smoke in the air has decreased, and the amount of ground-level ozone has increased. Nevertheless, we still use the word *smog,* put together from the words *smoke* and *fog.*

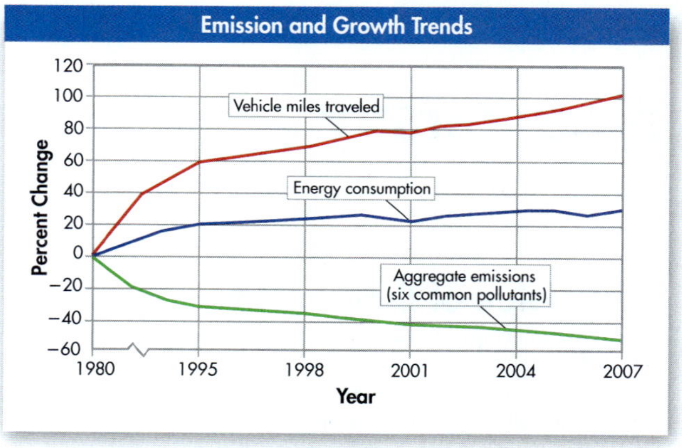

Air Pollution Trends This graph summarizes findings of the United States Environmental Protection Agency (EPA). It shows the total percent change from 1980–2007 for energy consumption, vehicle mileage, and the combined concentration of six common pollutants—carbon monoxide, lead, nitrogen oxides, organic compounds, particulates, and sulfur dioxide.

Air Quality and Sustainability It is hard to clean up the air. Air doesn't stay in one place and doesn't "belong" to anyone. But many countries have made rules about car and truck emissions. These efforts are improving the air quality around the world. At one time, all gasoline had the metal lead in it. Car and truck exhaust carried lead into the air. Lead from the air washed onto land and into rivers and streams. The United States banned lead in gasoline in 1996. Now the lead levels in soils, rivers, and streams are much lower than they used to be.

Key Question What are the major forms of air pollution?
Common forms of air pollution include smog, acid rain, greenhouse gases, and particulates.

CHECK Understanding

Apply Vocabulary

Use the highlighted words from the lesson to complete each sentence correctly.

1. When a forest is cleared so that farmers can plant crops, the process is called ______.
2. Pollutant concentrations may be much higher in consumers than in producers because of ______.

Critical Thinking

3. **Explain** Why is it important to protect soil resources?
4. **Apply Concepts** How can sewage affect a water source?
5. **Interpret Graphs** Look at the "Emission and Growth Trends" graph above. Describe the trends in emissions since 1980. Is this trend expected based on the rest of the data shown in the graph? Explain your answer.
6. **Write to Learn** Answer the second clue of the mystery. Before you answer, reread the introduction to the chapter mystery.

MYSTERY CLUE

Forests of palm trees with strong, tall trunks and edible seeds once covered most of Easter Island. Why would the islanders have cut down these forests? What effect would deforestation have had? (Hint: See page 133.)

BIOLOGY.com Search Lesson 6.2 GO • Lesson Assessment

Check Understanding

1. deforestation
2. biological magnification
3. We depend on many things that need healthy soil. We use plants as food, for clothing, and to build shelters. If soil is polluted or erodes away, it can take a very long time to re-form.
4. Sewage contains large amounts of nitrogen and phosphorus. When a lot of nitrogen and phosphorus enter a body of water, too many bacteria and algae can grow. These organisms use up the oxygen in the water and cause a dead zone in which other organisms can't survive. Sewage can also carry diseases.
5. Sample answer: The overall trend is a decrease in emissions. This is unexpected, because the trend for vehicle miles traveled and the trend for energy consumption have risen.
6. The islanders probably cut down these trees to move the large moai statues. They may also have cut down the trees for firewood, to build their homes, or to make room for farmland.

LESSON 6.2

Active Reading

INTERPRET GRAPHS Direct students' attention to the graph on this page. Have a volunteer read the caption. Then have another volunteer identify the trend line for each statistic.

Ask Why were these trend lines put together on a single graph? *(They are easy to compare and contrast when they are next to each other.)*

Ask How did the number of vehicle miles traveled in the United States change between 1980 and 2007? *(It increased by 100 percent.)* How did the total amount of emissions change over the same period? *(It decreased by about 50 percent.)*

Have students write two or three sentences in their notebooks that summarize what the graph shows about air pollution trends. *(Sample answer: People in the United States drive more than ever and use more energy than ever, but some pollutant emissions are decreasing. Emissions rules seem to be working.)*

KEY QUESTION Read the question and answer. **Ask** In what ways could you try to lessen your impact on our atmosphere? *(Sample answers: Ride a bike or walk more often instead of riding in a car. Take a bus. Use less energy at home.)*

Speed Bump

Ask How might global warming be slowed and eventually reversed? *(Sample answer: We could reduce the amount of greenhouse gases like carbon dioxide that we release into the atmosphere.)*

Post Reading

ASSESSMENT TIPS If students struggle with Question 3, have them discuss the many ways we depend on plants besides as food. For Question 4, have students review the components of sewage described in the text. Have them infer the consequences of letting sewage go into a pond.

MYSTERY CLUE Discuss the Mystery Clue as a class. Ask students to think about how we use trees today.

Wrap-Up Activity

MAKE A CHART Have students work in small groups to make a two-column chart that summarizes the information from the lesson. Have each group choose one of the resources featured in the lesson *(soil, fresh water, or atmospheric resources)*. In the first column, students should summarize human activities that harm that resource. In the second column, students should describe sustainable practices that lessen the negative effects of each harmful activity.

LESSON 6.3

Pre-Reading

Objectives

6.3.1 Define biodiversity and explain its value.

6.3.2 Identify current threats to biodiversity.

6.3.3 Describe how biodiversity can be preserved.

Preview the Pages

1. Preview the vocabulary words throughout the lesson. Have a volunteer find each word and read the definition aloud.
2. Preview the visuals and have students read captions silently. Students may express surprise at the small size of the infant panda. Tell students that pandas are very small and helpless at birth, much like kangaroos.

Activate Prior Knowledge

Ask students to name an animal species that has gone extinct or that has been pushed to the brink of extinction in the past 100 years. *(Sample answers: bald eagle, passenger pigeon, giant panda)*

Ask What caused the animal's population to decline? *(Sample answers: hunting, habitat loss, pesticide use)*

Ask What steps can be taken to save endangered animals? *(Sample answers: create parks and other protected areas, regulate hunting and pesticide use)*

Build Understanding

T-CHART Tell students that this activity will help them practice organizing information and identifying main ideas.

NATIONAL SCIENCE EDUCATION STANDARDS

UNIFYING CONCEPTS AND PROCESSES

I, II, III, IV

CONTENT

C.4.c, C.4.d, C.4.e, C.5.e, F.2, F.3, F.4, F.5, F.6

INQUIRY

A.2.a

For the corresponding lesson in *Miller & Levine Biology,* see pages 166–172.

6.3 Biodiversity

Key Questions

Why is biodiversity important?

What are the most significant threats to biodiversity?

How do we preserve biodiversity?

BUILD Understanding

T-Chart As you read the lesson, make a T-chart. In the left column, write the headings in the lesson. In the right column, write the main ideas from each heading.

In Your Workbook Go to your workbook to learn more about T-charts.

BUILD Vocabulary

biodiversity
the total of the variety of organisms in the biosphere; also called biological diversity

ACADEMIC WORDS

Variety and *variation* are nouns that refer to the differences among a group of things. Biodiversity is the total of all the different kinds of organisms, ecosystems, and genetic information in the biosphere.

The Value of Biodiversity

There are so many different kinds of living things on Earth. From deserts to pine forests, variety is "the spice of life." But variety gives us more than just interesting things to see. Our well-being is closely tied to the well-being of many other organisms. **Biodiversity** is the total of all the genetically based variation in all organisms in the biosphere. What are the different types of biodiversity? What is the value of biodiversity to society?

Types of Biodiversity There are three levels of biodiversity: ecosystem diversity, species diversity, and genetic diversity. Ecosystem diversity means the many different kinds of ecosystems that exist in the biosphere. Many different habitats, communities, and ecological processes are important to biodiversity. Species diversity is the number of different species in the biosphere or in a particular place. Biologists have studied and named more than 1.8 million species. They think that there are at least 30 million that no one has discovered yet. Single-celled organisms make up most of the diversity. But new species of multicellular organisms, like the snake below, are also being found.

Genetic diversity can mean all the different forms of genetic information carried by one species or by all organisms on Earth. In many ways, genetic diversity is the most basic kind of biodiversity. Genetic diversity is also the hardest kind to see and appreciate. If a population loses genetic diversity, it is less likely to be able to survive changes in its environment.

A New Species This tiny snake is native to the island of Barbados. It is one of many species that has been discovered recently. Photos of the snake were released in 2008.

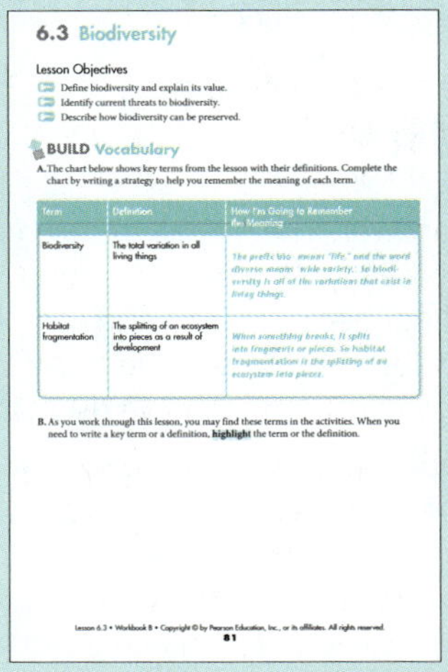

6.3 Biodiversity

Lesson Objectives

Define biodiversity and explain its value.
Identify current threats to biodiversity.
Describe how biodiversity can be preserved.

BUILD Vocabulary

A. The chart below shows key terms from the lesson with their definitions. Complete the chart by writing a strategy to help you remember the meaning of each term.

Term	Definition	How I'm Going to Remember the Meaning
Biodiversity	The total variation in all living things	*The prefix bio- means "life," and the word diverse means "wide variety." So biodiversity is all of the variations that exist in living things.*
Habitat fragmentation	The splitting of an ecosystem into pieces as a result of development	*When something breaks, it splits into fragments or pieces. So habitat fragmentation is the splitting of an ecosystem into pieces.*

B. As you work through this lesson, you may find these terms in the activities. When you need to write a key term or a definition, **highlight** the term or the definition.

Lesson 6.3 • Workbook B • Copyright © by Pearson Education, Inc., or its affiliates. All rights reserved.
81

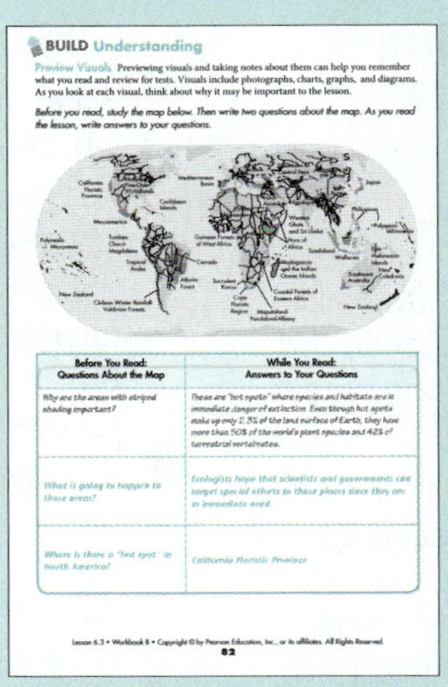

BUILD Understanding

Preview Visuals Previewing visuals and taking notes about them can help you remember what you read and review for tests. Visuals include photographs, charts, graphs, and diagrams. As you look at each visual, think about why it may be important to the lesson.

Before you read, study the map below. Then write two questions about the map. As you read the lesson, write answers to your questions.

Before You Read: Questions About the Map	While You Read: Answers to Your Questions
Why are the areas with striped shading important?	*These are "hot spots" where species and habitats are in immediate danger of extinction. Even though hot spots make up only 2.3% of the land surface of Earth, they have more than 50% of the world's plant species and 42% of terrestrial vertebrates.*
What is going to happen to these areas?	*Ecologists hope that scientists and governments can target special efforts to those places since they are in immediate need.*
Where is there a "hot spot" in North America?	*California Floristic Province*

Lesson 6.3 • Workbook B • Copyright © by Pearson Education, Inc., or its affiliates. All Rights Reserved.
82

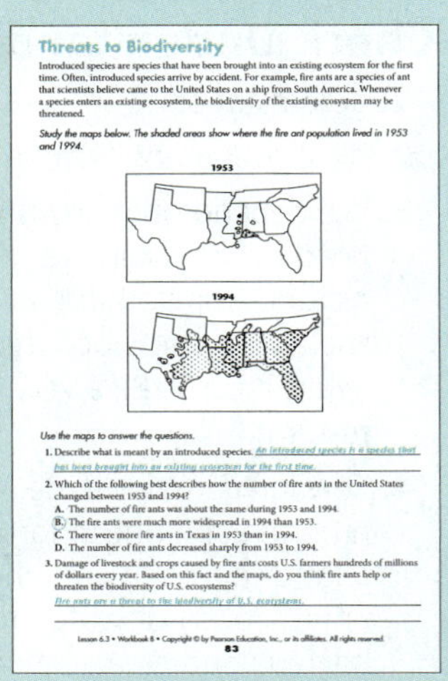

Threats to Biodiversity

Introduced species are species that have been brought into an existing ecosystem for the first time. Often, introduced species arrive by accident. For example, fire ants are a species of ant that scientists believe came to the United States on a ship from South America. Whenever a species enters an existing ecosystem, the biodiversity of the existing ecosystem may be threatened.

Study the maps below. The shaded areas show where the fire ant population lived in 1953 and 1994.

1953

1994

Use the maps to answer the questions.

1. Describe what is meant by an introduced species. *An introduced species is a species that has been brought into an existing ecosystem for the first time.*
2. Which of the following best describes how the number of fire ants in the United States changed between 1953 and 1994?
   A. The number of fire ants was about the same during 1953 and 1994.
   B. The fire ants were much more widespread in 1994 than 1953.
   C. There were more fire ants in Texas in 1953 than in 1994.
   D. The number of fire ants decreased sharply from 1953 to 1994.
3. Damage of livestock and crops caused by fire ants costs U.S. farmers hundreds of millions of dollars every year. Based on this fact and the maps, do you think fire ants help or threaten the biodiversity of U.S. ecosystems?
   Fire ants are a threat to the biodiversity of U.S. ecosystems.

Lesson 6.3 • Workbook B • Copyright © by Pearson Education, Inc., or its affiliates. All rights reserved.
83

Valuing Biodiversity Biodiversity is one of Earth's greatest natural resources. Biodiversity is very important in medicine and agriculture. Healthy and diverse ecosystems are important for soil, water, and air quality. Because you can't touch, taste, or smell biodiversity, it is easy to forget its value. But when biodiversity is lost, the biosphere and humanity may face serious consequences.

▶ ***Biodiversity and Medicine*** Many medicines are based on substances in wild species. Aspirin and penicillin were first made using wild species. The chemicals in wild species are used to treat depression and cancer. These chemicals are put together based on directions found in genes. When we lose biodiversity, we lose genetic information that may have been useful in medicine.

▶ ***Biodiversity and Agriculture*** Genetic diversity is also important in agriculture. Most crop plants have wild relatives. These wild plants may carry genes for such important features as disease resistance and pest resistance. If we lose the wild plants, we may lose those genes.

▶ ***Biodiversity and Healthy Ecosystems*** The number and variety of species in an ecosystem can affect the ecosystem's health. The loss of species may make an ecosystem less valuable to people. Sometimes changes in the population of a single species can change the whole ecosystem. Such important species are often called keystone species. Also, healthy and diverse ecosystems are important for soil, water, and air quality.

Key Question Why is biodiversity important?
Biodiversity is important in medicine and agriculture. Also, healthy and diverse ecosystems are important for soil, water, and air quality.

Medicinal Plants Compounds in this foxglove plant are used to make medicine to treat heart disease.

Potato Diversity The genetic diversity of wild potatoes in South America can be seen in the colorful varieties shown here. The International Potato Center in Peru houses a "library" of more than 4500 tuber varieties.

Keystone Species The sea otter is a keystone species. Sea otters eat sea urchins. Sea urchins eat kelp. When the otter population falls, the sea urchin population goes up. When the sea urchin population goes up, the kelp population drops.

Focus on Academic Language

• genetic • penicillin • keystone species

Remind students that *genetic* has the root word *gene*. **Ask** What is a gene? *(A gene is a molecule in a cell that holds inheritable information about an organism's characteristics.)*

Tell students that penicillin is derived from a mold. **Ask** What does penicillin do? *(It is a medicine that helps fight bacterial infections.)*

Tell students that a keystone is the top center stone in an arch, upon which the sides lean for support. If the keystone is removed, the arch will collapse. **Ask** What happens when a keystone species is removed from an ecosystem? *(The ecosystem fails or collapses.)*

During Reading

Build Vocabulary

ACADEMIC WORDS Tell students that a particular term may be a shortened form of another term.

Say *Bio-* means "biological." *Diversity* means "a variety or a collection of several different things."

Ask For what term is *biodiversity* a shortened form? *(biological diversity)*

Ask Can you think of other terms in ecology that are shortened forms of another term? *(biogeochemical, geothermal)*

Active Reading

USE MATH Have a volunteer read aloud the first paragraph under Types of Biodiversity on the opposite page.

Say Biologists have identified about 1.8 million species. They think about 30 million more species have yet to be discovered.

Ask What percentage of Earth's species have been discovered? *(1.8 million ÷ 31.8 million = 0.06, or about 6 percent)* Does this number surprise you? *(Students will likely be surprised that so many species are yet to be discovered.)*

USE VISUALS Direct students' attention to the Potato Diversity photograph. Have a volunteer read the caption.

Ask Why don't we see this variety of potatoes in our grocery stores? *(Grocery stores sell cultivated potatoes, not wild ones.)*

Ask Why would someone want to keep a "library" of potato types? *(to preserve genetic diversity; Wild potatoes may carry genes for important features such as disease and pest resistance.)*

KEY QUESTION Have a volunteer read the Key Question and answer.

Ask How can you see interdependence in biodiversity? *(Loss of biodiversity can affect the entire ecosystem's health. The loss of one species can cause the increase or decrease in size of other populations.)*

Speed Bump

Ask Which has more biodiversity, a desert or a field of corn? Why? *(A desert has more biodiversity because it has a greater variety of organisms. A corn field has only one species of plant, even though it may have more individual plants.)*

During Reading continued

Active Reading

INTERPRET ANALOGIES Have a student read aloud the first paragraph under Threats to Biodiversity. Then discuss the analogy that compares the loss of biodiversity to destroying a library before its books are read.

Ask Why, in this analogy, does it make sense to say that the books haven't been read before they are destroyed? *(Scientists think that only about 6 percent of Earth's species have been discovered. Some species go extinct before we can discover them and study them.)*

USE VISUALS Have a volunteer read the caption of the Habitat Fragmentation photograph. Discuss what might happen to an animal population, such as a deer population, that was living in this area before the area was developed.

Ask What happened to the deer's habitat as the houses and roads were built? *(The habitat was fragmented, or broken into smaller pieces.)*

Have students point to some of the habitat islands where deer might still live.

Ask What might happen to the deer if they try to cross the roads from one piece of habitat to another? *(They might be killed by cars.)* And what might happen if they have to stay in one of the small pieces of habitat? *(They might not have enough food and space. Their population might go down.)*

Ask Would it help if the habitat islands could be connected by pathways, allowing deer to move safely between islands? *(Sample answer: That might help deer get more food, but their habitat is still reduced by the area covered by development.)*

Build Vocabulary

ACADEMIC WORDS Discuss examples of fragmentation.

Say Fragmentation is the process of shattering or breaking up into fragments, or pieces. Habitat fragmentation is one kind of fragmentation.

Ask What other kinds of fragmentation have you heard about? *(Sample answers: fragmentation of rocks, computer-memory, forests, societies)*

Speed Bump

Ask How does habitat fragmentation affect species competition? *(Competition increases for resources both within species and among species.)*

BUILD Vocabulary

habitat fragmentation the splitting of ecosystems into pieces

ACADEMIC WORD
The noun *fragment* means "a broken or incomplete part." Habitat fragmentation forms broken and incomplete parts of an ecosystem.

Threats to Biodiversity

Species have been changing and dying out since life began. In fact, scientists think that over 99 percent of the species that have ever lived are now extinct. So extinction is not new. But human activity today is causing the greatest wave of extinctions since the dinosaurs died off. Scientists compare this loss of biodiversity to destroying a library before its books are read.

People reduce biodiversity by changing habitats, hunting, and introducing species into new places. Pollution and climate change also lower biodiversity.

Altered Habitats Often, people change a natural habitat into farmland or land for housing. When this happens, the number of species living nearby will go down. Some species may even become extinct. A species can be at risk even if a habitat is not completely destroyed. Development often breaks a habitat into smaller pieces. This process is called **habitat fragmentation.** This fragmentation leaves little habitat "islands." Competition for food, space, and other resources is very high. Some organisms are forced to leave. Others do not survive.

Hunting and the Demand for Wildlife Products People can push species to extinction by hunting them. In the 1800s, hunting wiped out the Carolina parakeet and the passenger pigeon. Endangered species in the United States are now protected from hunting. But hunting still threatens rare animals in Africa, South America, and Southeast Asia. Some animals are hunted for meat. Others are hunted for their hides or skins. Sometimes, animals are hunted to be sold as pets. Habitat fragmentation is very hard on hunted animals. Fragmentation leaves the hunted animals with fewer hiding places.

Habitat Fragmentation Housing developments in Florida have formed the pattern of forest "islands" shown here. Habitat fragmentation limits biodiversity and the potential size of populations.

ELL Focus on ELL: Build Background

ALL SPEAKERS Students may have some prior knowledge of the value of biodiversity, but language barriers may stand in the way of expressing that knowledge. Encourage nonnative speakers to share information about their own cultures and histories.

Ask What are some ideas that people have from other cultures about biodiversity and its value? *(Answers may vary, but a sample answer may be similar to the following: Costa Rica leads the world in forming wildlife preserves aimed at preserving biodiversity. Ecotourism is an important economic factor for Costa Rica.)*

Have students write responses in their native language, and then prepare a translation in English. Have proficient English speakers assist, if necessary.

Introduced Species In Chapter 5, you read about rabbits in Australia and about hydrilla plants in Florida. When a species is put into a new environment, its population can grow very quickly. The new species may out-compete native plants and animals. When this happens, the new species is called invasive.

Pollution Many pollutants can also lower biodiversity. For example, DDT prevents birds from laying healthy eggs. In the United States, many bird populations decreased because of DDT. Acid rain can hurt plants and animals on land and in the water. The increased carbon dioxide in the air is dissolving in oceans. The dissolved carbon dioxide makes the oceans more acidic. This increased acidity threatens the biodiversity on coral reefs and in other ocean ecosystems.

Climate Change A change in climate can be a major threat to biodiversity. Species are adapted to their environment. A species may survive only within a certain temperature range. Species also need a certain amount of rain. If temperatures or rainfall change, organisms may need to move to survive. Species in fragmented habitats are even more affected by climate change. If global temperatures rise as much as many researchers predict, many species will become extinct.

Key Question What are the most significant threats to biodiversity?
Biodiversity is threatened by changing habitats, hunting, and introduced species. Pollution and climate change also threaten biodiversity.

Conserving Biodiversity

What can we do to protect biodiversity? Should we focus on one species? Should we try to save an entire ecosystem? We need to do both. We also need to consider human needs and interests.

Protecting Individual Species The Association of Zoos and Aquariums (AZA) tries to protect organisms, one species at a time. A key part of their plans is a captive breeding program. Members of the AZA choose and manage mating pairs. They are trying to increase genetic diversity. The goal is to return the organisms to the wild. The AZA is currently trying to protect more than 180 species. The giant panda is one of these species.

Preserving Habitats and Ecosystems Scientists are not only trying to save individual species. Now, they have begun to focus on protecting entire ecosystems. This can help many species at once. Governments and conservation groups work to set aside land for parks and reserves. The United States has national parks, forests, and other protected areas.

Saving an Individual Species A number of efforts are being made to save the giant panda. These efforts include captive breeding and reintroduction programs. Here, a researcher examines an infant panda at China's Wolong Nature Preserve.

Active Reading

USE GRAPHIC ORGANIZERS Draw a word web on the board with a large central circle joined by lines to smaller circles. In the central circle, write *Threats to Biodiversity*. Instruct students to copy the web on a sheet of paper. As students read the paragraphs under Threats to Biodiversity on the opposite page and this one, tell them to list conditions that threaten biodiversity in the smaller circles. Students can add circles as needed to include details about each condition.

RECOGNIZE ABBREVIATIONS Tell students that the full name of the pollutant DDT is very long. Write its name on the board and pronounce it: dichloro-diphenyl-trichloroethane (dy-KLAW-ruh–dy-FEH-nul–try-KLAW-roh-EH-thayn).

Ask Can you see how this name is abbreviated as DDT?

Ask a volunteer to circle or underline the letters that make up the abbreviation.

SCIENCE SUPPORT Remind students that hydrilla is an aquatic plant native to Africa, Australia, and parts of Asia. Decades ago, it was brought to Florida and sold as an aquarium plant. It is now well established throughout bodies of water in the southern states, where it costs millions of dollars each year to control. Hydrilla grows in thick mats, mostly submerged. It poses a potentially deadly threat to divers and interferes with recreation. It also absorbs nutrients more efficiently than native plants. Early eradication efforts included pulling out the plants, grinding them, and dumping them back into the water. It was not known then that each small piece of stem could grow to a mature plant.

KEY QUESTION Read the Key Question and answer aloud.

Ask Can you identify a threat to biodiversity in our community? *(Answers may vary.)*

Speed Bump

Say Explain what a threat to biodiversity is. *(A threat to biodiversity is any disturbance that causes a species to no longer be able to compete or exist normally in the natural environment.)*

Ask What is captive breeding? *(Captive breeding is the management of mating pairs in a zoo environment with the goal of returning a population to the wild.)*

Hands-On Learning

CARTOON STRIPS Have student pairs create cartoon strips that illustrate ways of conserving biodiversity.

1. Provide colored pencils or markers and chart paper. As an alternative to chart paper, students can tape together four blank sheets of plain white paper.
2. Have the class brainstorm several ways to conserve biodiversity, using the text as a resource. Write their ideas on the board.
3. Assign each idea to a group. Some ideas may be assigned to more than one group.
4. Have each group make a six- to eight-panel cartoon strip that explains one way to conserve biodiversity.
5. Share completed strips by posting them on a bulletin board. Discuss the message in each strip.

During Reading continued

Active Reading

MAKE CONNECTIONS Encourage students to describe conservation efforts in their area that they are familiar with.

Ask What conservation practices are taking place in your home? neighborhood? school? city parks? city industries? *(Answers may vary.)*

KEY QUESTION Have a volunteer read aloud the Key Question and answer.

Ask Why must human interests be considered in conservation efforts? *(Sample answer: People are more likely to support conservation efforts if they think it is in their own best interests.)*

Speed Bump

Ask How can governments help people and businesses adopt new habits to protect biodiversity? *(Offer incentives or rewards to people and businesses who make efforts to change their ways.)*

Post Reading

ASSESSMENT TIPS Advise students to review the highlighted vocabulary words before they begin the assessment. For Question 3, encourage students to make a graphic organizer before they answer the question. Then have them answer the question in complete sentences.

MYSTERY CLUE Have students infer where the nonnative rats came from. *(The ships probably brought them to the island as "stowaways.")* Then, have students speculate about how the rats may have affected the coconut palm population. If students struggle, have them reread the Introduced Species section.

Wrap-Up Activity

DEBATE Have students form teams and debate the following scenario. Ask them to consider possible effects on biodiversity, and encourage them to use concepts and vocabulary from the lesson.

A company has proposed building several hundred wind turbines in your state. The wind turbines will provide clean power to thousands of homes, reducing the amount of greenhouse gases and other pollutants. However, the wind turbines will have some environmental effects. They will kill some birds, though not enough to cause population declines. They will be located in an area that is home to numerous species of plants and animals. Should people in your state support building the turbines?

Ecotourism A tourist gets an elephant-size kiss from one of the rescued elephants at Thailand's Elephant Nature Park. This kind of park helps protect biodiversity and brings tourist dollars to the area.

Biologists face the challenge of finding the best areas to protect. They have identified ecological hot spots to make sure conservation efforts happen in places where they are most needed. An ecological hot spot is a place where many species and habitats are in immediate danger of extinction. Biologists hope that identifying these spots will help them focus on the right places.

Considering Local Interests In order to protect biodiversity, people need to change their habits. Sometimes, they even need to change their jobs. In these cases, it is helpful to give money or other rewards in support for the changes. In the United States, the government has given tax breaks to people who buy hybrid cars or who put solar panels on their houses. In Africa, Central America, and Southeast Asia, countries have set aside land for parks and nature reserves.

Industries are also encouraged to change their habits. For example, companies may be rewarded for releasing less carbon into the environment. These examples show that conservation efforts work best when they are based on scientific information and when they benefit the affected community.

Key Question How do we preserve biodiversity?
To preserve biodiversity, we need to protect individual species as well as whole ecosystems. We also need to keep human interests and needs in mind.

CHECK Understanding

Apply Vocabulary

Use the highlighted words from the lesson to complete each sentence correctly.

1. The number of different species in the biosphere or in a particular place is one level of ______.
2. The process of ______ often leaves small "islands" of space for animals and plants to live between more developed land.

Critical Thinking

3. **Apply Concepts** How are people helped by biodiversity?
4. **Relate Cause and Effect** What usually happens to the size and diversity of populations after development causes habitat fragmentation?
5. **Explain** Why is it important to consider local interests when finding ways to protect local ecosystems?
6. **Write to Learn** Answer the third clue in the mystery. In your answer, use the terms *introduced species* and *compete.*

MYSTERY CLUE

Almost all of the coconut shells on Easter Island show signs of having been gnawed on by rats, which are not native to the Island. Coconuts hold the seeds of the coconut palm tree. How do you think the rats affected the coconut palm population? (**Hint:** See p. 141.)

BIOLOGY.com Search Lesson 6.3 GO • Lesson Assessment

Check Understanding

1. biodiversity
2. habitat fragmentation
3. Answers may vary. Students should list benefits in medicine, agriculture, and ecosystem services.
4. Habitat fragmentation usually causes the size and diversity of populations to decrease. Competition for food, space, and water causes some organisms to leave or makes it impossible for them to survive.
5. People may need to change habits and some may lose their jobs. Thinking of local interests will help gather support and will help people see the benefits of protecting the ecosystem.
6. When a species is introduced into a new environment, its population may grow quickly. An introduced species may compete with native species for food, space, and resources, or eat native species. The introduced rats may have destroyed most of the coconuts and their seeds, which prevented the growth of new palm trees.

6.4 Meeting Ecological Challenges

Ecological Footprints

Every year, the United States Environmental Protection Agency (EPA) gives out up to ten Environmental Youth Awards. These awards are given to students with ideas that protect the environment in ways that meet the needs of local people. One winner encouraged people who fish to stop using lead weights. These weights put lead in the water. The lead can make the plants and animals that live in or drink the water sick. Another group of winners found ways to reduce waste at their school. Their program saved the school more than a million dollars. This kind of leadership will help us find a new direction for the future.

What is our impact on the biosphere? Think about the kind and amount of resources each of us uses. Ecologists use a concept called the ecological footprint. The **ecological footprint** describes the total area of the land and water ecosystems that provide the resources that each person uses. Ecological footprints include resources such as energy, food, water, and shelter. They also include the resources needed to absorb wastes such as sewage and greenhouse gases. An ecological footprint can be used to determine the carrying capacity for humans.

Footprint Limitations Ecologists talk about the ecological footprint of individuals, of countries, and of the world's population. It is difficult, if not impossible, to calculate an exact footprint. The idea is very new. There is not yet a standard way of measuring footprint size. Also, footprints give only a "snapshot" of the resources used at a particular point in time.

Key Questions

How does the average ecological footprint in America compare to the world's average?

How can ecology guide us toward a sustainable future?

BUILD Understanding

Compare/Contrast Table As you read, make a table comparing the challenges associated with the ozone layer, fisheries, and global climate. Note the problem observed, the causes identified, and the solutions implemented.

In Your Workbook Go to your workbook for help completing the compare/contrast table for Lesson 6.4.

BUILD Connections

ECOLOGICAL FOOTPRINTS

The food you eat, the miles you travel, and the electricity you use all contribute to your—and the population's—ecological footprint.

In Your Workbook Do you understand the analogy of the ecological footprint? Go to your workbook for extra practice.

BIOLOGY.com Search Lesson 6.4 GO • Visual Analogy

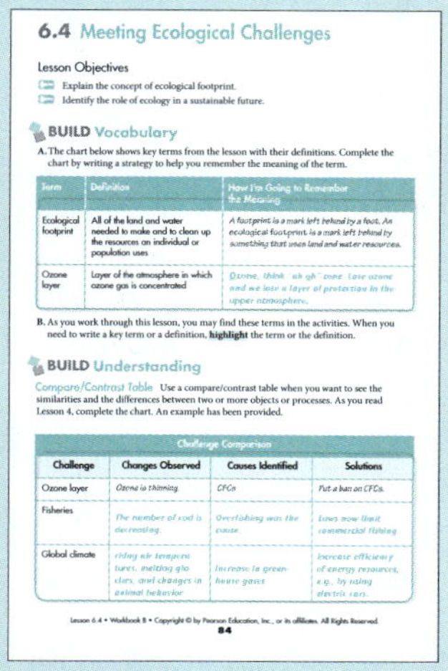

6.4 Meeting Ecological Challenges

Lesson Objectives

- Explain the concept of ecological footprint.
- Identify the role of ecology in a sustainable future.

BUILD Vocabulary

A. The chart below shows key terms from the lesson with their definitions. Complete the chart by writing a strategy to help you remember the meaning of the term.

Term	Definition	How I'm Going to Remember the Meaning
Ecological footprint	All of the land and water needed to make and to clean up the resources an individual or population uses	A footprint is a mark left behind by a foot. An ecological footprint is a mark left behind by something that uses land and water resources.
Ozone layer	Layer of the atmosphere in which ozone gas is concentrated	Ozone, think "uh oh" zone. Lose ozone and we lose a layer of protection in the upper atmosphere.

B. As you work through this lesson, you may find these terms in the activities. When you need to write a key term or a definition, **highlight** the term or the definition.

BUILD Understanding

Compare/Contrast Table Use a compare/contrast table when you want to see the similarities and the differences between two or more objects or processes. As you read Lesson 4, complete the chart. An example has been provided.

Challenge Comparison

Challenge	Changes Observed	Causes Identified	Solutions
Ozone layer	Ozone is thinning.	CFCs	Put a ban on CFCs.
Fisheries	The number of cod is decreasing.	Overfishing was the cause.	Laws now limit commercial fishing
Global climate	rising air temperatures, melting glaciers, and changes in animal behavior	Increase in greenhouse gases	Increase efficiency of energy resources, e.g., by using electric cars.

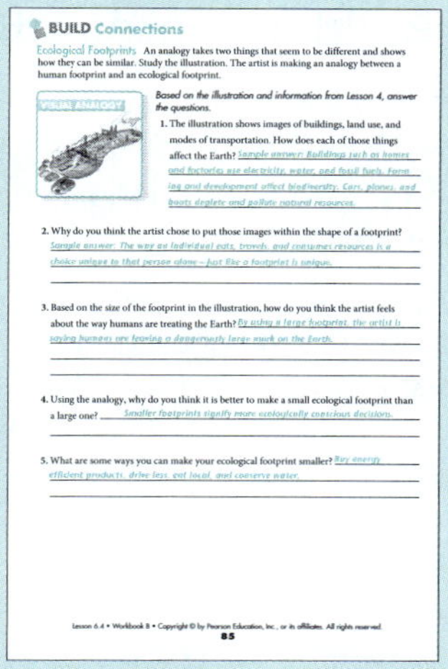

BUILD Connections

Ecological Footprints An analogy takes two things that seem to be different and shows how they can be similar. Study the illustration. The artist is making an analogy between a human footprint and an ecological footprint.

VISUAL ANALOGY

Based on the illustration and information from Lesson 4, answer the questions.

1. The illustration shows images of buildings, land use, and modes of transportation. How does each of those things affect the Earth? Sample answer: Buildings such as homes and factories use electricity, water, and fossil fuels. Farming and development affect biodiversity. Cars, planes, and boats deplete and pollute natural resources.

2. Why do you think the artist chose to put those images within the shape of a footprint? Sample answer: The way an individual eats, travels, and consumes resources is a choice unique to that person alone—just like a footprint is unique.

3. Based on the size of the footprint in the illustration, how do you think the artist feels about the way humans are treating the Earth? By using a large footprint, the artist is saying humans are leaving a dangerously large mark on the Earth.

4. Using the analogy, why do you think it is better to make a small ecological footprint than a large one? Smaller footprints signify more ecologically conscious decisions.

5. What are some ways you can make your ecological footprint smaller? Buy energy efficient products, drive less, eat local, and conserve water.

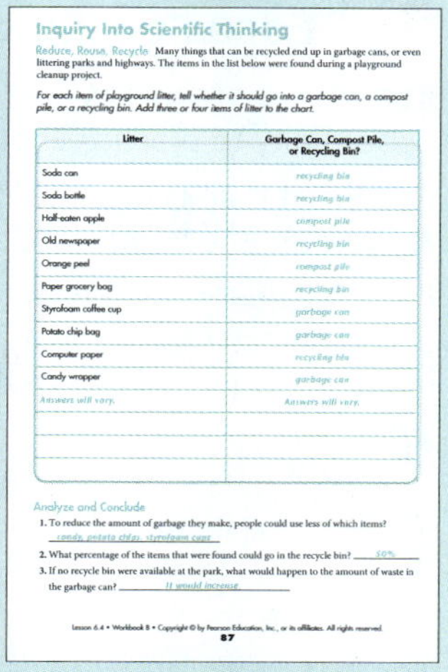

Inquiry Into Scientific Thinking

Reduce, Reuse, Recycle Many things that can be recycled end up in garbage cans, or even littering parks and highways. The items in the list below were found during a playground cleanup project.

For each item of playground litter, tell whether it should go into a garbage can, a compost pile, or a recycling bin. Add three or four items of litter to the chart.

Litter	Garbage Can, Compost Pile, or Recycling Bin?
Soda can	recycling bin
Soda bottle	recycling bin
Half-eaten apple	compost pile
Old newspaper	recycling bin
Orange peel	compost pile
Paper grocery bag	recycling bin
Styrofoam coffee cup	garbage can
Potato chip bag	garbage can
Computer paper	recycling bin
Candy wrapper	garbage can
Answers will vary.	Answers will vary.

Analyze and Conclude

1. To reduce the amount of garbage they make, people could use less of which items? candy, potato chips, styrofoam cups

2. What percentage of the items that were found could go in the recycle bin? 50%

3. If no recycle bin were available at the park, what would happen to the amount of waste in the garbage can? It would increase.

LESSON 6.4

Pre-Reading

Objectives

6.4.1 Explain the concept of ecological footprint.

6.4.2 Identify the role of ecology in a sustainable future.

Preview the Pages

1. Preview the vocabulary words. Have students locate each word and read its definition aloud.
2. Tell students to expect a "case study" format in this lesson. Have students scan the studies. Explain that in these studies they will learn about different issues that present ecological challenges and how the challenges are being met.

Build Understanding

COMPARE/CONTRAST TABLE Read aloud the Build Understanding box. Clarify that the ozone layer, fisheries, and global climate are the topics of the three case studies in this lesson. Refer students to Workbook B, where they will find a partially completed compare/contrast table to fill in.

Build Connections

VISUAL ANALOGY Discuss the analogy of the ecological footprint.

Ask What does a human footprint measure? *(the amount of space a human foot takes up on a surface)*

Ask What does an ecological footprint measure? *(how much land and water area a human population needs to produce its resources and to absorb wastes.)*

NATIONAL SCIENCE EDUCATION STANDARDS

UNIFYING CONCEPTS AND PROCESSES

I, II, III, IV

CONTENT

C.4.a, C.4.c, C.4.d, C.4.e, C.5.e, C.6.b, F.2, F.3, F.4, F.5, F.6, G.1

INQUIRY

A.2.a

For the corresponding lesson in *Miller & Levine Biology,* see pages 173–179.

During Reading

Active Reading

USE VISUALS Ask a volunteer to read aloud the caption to the Relative Footprint map.

Ask What does the range of colors in the color key, from green to red, show? *(least to greatest ecological deficit, or least to greatest ecological footprint)*

Say The inset map shows the true size and shape of each country and continent.

Ask On the large map, why are some continents shown larger than they actually are? Why are some shown smaller? *(Areas with large ecological footprints are shown to be bigger than they actually are. Areas with small ecological footprints are shown to be smaller than they actually are.)*

Say Look back at the image of Earth from space on the first pages of this chapter. Compare that image with the Relative Footprints map.

Ask What do you notice about the areas with the densest light? *(These areas are also the areas with the largest ecological deficits.)*

KEY QUESTION Read aloud the first Key Question and answer on this page.

Ask What does the large ecological footprint of the United States show about how we live? *(We consume much more than people in many other parts of the world, and we produce more wastes.)*

KEY QUESTION Read aloud the second Key Question and answer.

Ask Which of the three steps may be the most difficult to do? *(changing our behavior)* Explain. *(People have to be willing to make sacrifices in the way they live.)*

Build Vocabulary

WORD ORIGINS Have a volunteer look up the word *ecology* in the dictionary.

Ask What is the word origin of *ecology*? *(from the Greek* oikos*, meaning "house")*

Say The association between a word's origin and its use often makes sense. But sometimes it doesn't. The suffix *-ology* means "the study of," so you might think *ecology* means "the study of a house."

Ask Does this meaning make sense? Explain. *(Sample answer: Yes, our "house" is where we live, just as our environment is where we live.)*

Speed Bump

Ask Which country has the largest ecological footprint? *(the United States)*

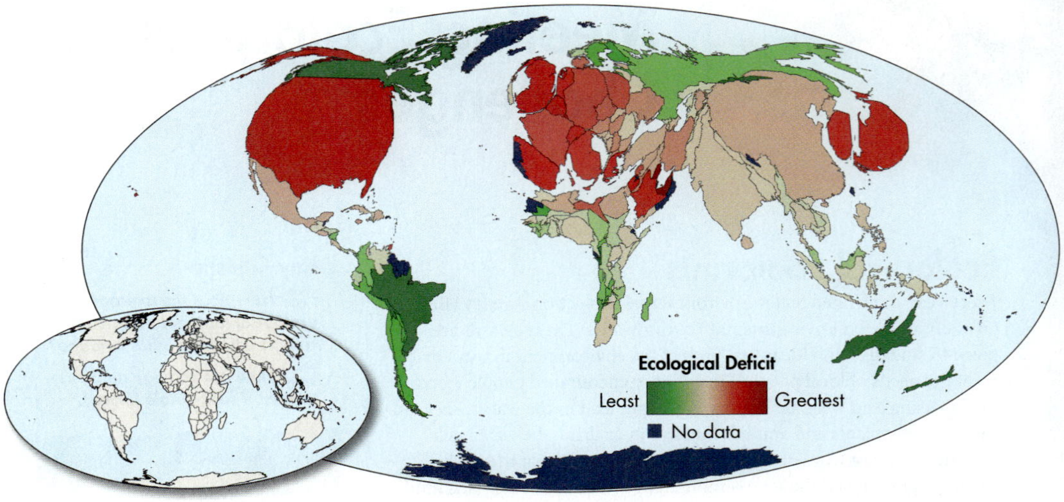

Relative Footprint This world map shows each country in proportion to its ecological footprint. The average American has an ecological footprint more than four times the world's average. By contrast, the average person in the African nation of Zambia has a footprint a little over one fourth the global average. Compare each country's footprint size on the color map to its actual size on the smaller map.

Comparing Footprints Calculating an exact footprint is almost impossible. However, this idea can be very useful when comparing different populations. For example, each person in the United States uses more than four times as many resources as the worldwide average. On average, each person in the United States uses almost twice as many resources as someone in England. To calculate the ecological footprint of an entire country, researchers first calculate the footprint of an average citizen. Then they multiply that footprint by the population of the country.

Key Question How does the average ecological footprint in America compare to the world's average?
The average American has an ecological footprint of more than four times the global average.

BUILD Vocabulary

ecological footprint
the total amount of functioning ecosystem needed both to provide the resources a human population uses and to absorb the wastes that population generates

ozone layer
an atmospheric layer in which ozone gas is relatively concentrated; protects life on Earth from harmful ultraviolet rays in sunlight

WORD ORIGINS
The word *ozone* comes from the Greek word that means "to smell." Ozone gas smells a little like chlorine gas.

Ecology in Action

Our ecological footprints are very important to the future of the biosphere. Global population growth and technology are also important. We often hear more stories about ecological challenges than ecological successes. But good ecological research can make a difference. Ecologists follow three basic steps: (1) recognize the problem; (2) find the cause; (3) change behavior. Following these steps can lead us to a sustainable future.

Key Question How can ecology guide us toward a sustainable future?
By (1) recognizing the problem, (2) determining the cause, and (3) changing our behavior, we can make a difference.

144 Chapter 6 • Lesson 4

ELL Focus on ELL: Build Background

BEGINNING SPEAKERS Use a think/pair/share activity to give students the opportunity to practice English by explaining science concepts.

Ask What is an ecological footprint? *(the amount of ecosystem needed to provide the resources a human population uses)*

Provide students with time to think about and write a response to the question. Give additional time for students to discuss the question with a partner. When pairs are ready, allow them to share their answer with the class.

Case Study #1: Atmospheric Ozone

Near the ground, ozone is a pollutant. However, ozone forms an important layer high in the atmosphere. This layer is called the **ozone layer.** The ozone layer absorbs the ultraviolet (UV) rays from the sun. UV can cause sunburn and skin cancer. It also damages eyes and lowers resistance to disease. In extreme cases, UV can damage plants and algae. By absorbing UV light, the ozone layer is like a global suncreen.

The following is an ecological success story. Over four decades, society has recognized a problem, identified its cause, and cooperated internationally to address the global problem.

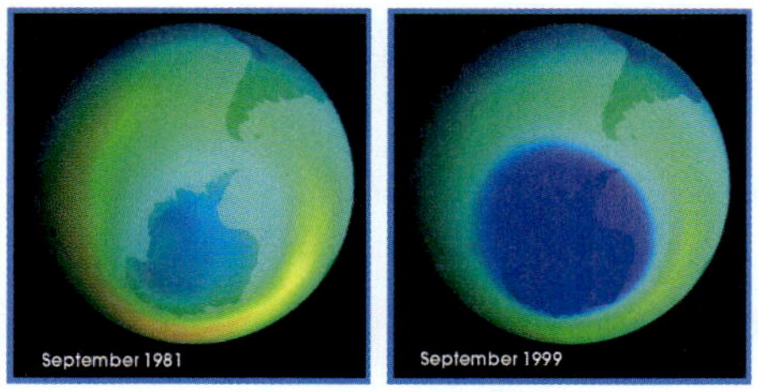

The Disappearing Ozone

CFC-Containing Refrigerators

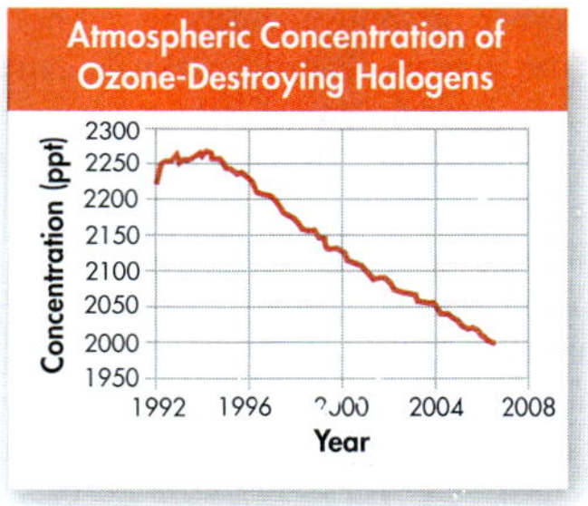

The Decline of CFCs

1 Recognizing a Problem: "Hole" in the Ozone Layer In the 1970s, scientists realized that the ozone concentration over Antarctica was dropping during the winter. The area was called an ozone hole. Of course, there was not a real hole in the atmosphere. There was less ozone. For several years, the ozone hole was getting bigger each year.

2 Researching the Cause: CFCs In 1974, research showed that gases called chlorofluorocarbons (CFCs) could hurt the ozone layer. CFCs had been used in spray cans, refrigerators, and air conditioners. They were also used to make plastic foams.

3 Changing Behavior: Regulation of CFCs Once the research was accepted by scientists, the rest was up to governments. There was a big response. Following the recommendations of ozone researchers, 191 countries agreed to ban most uses of CFCs. CFCs can stay in the atmosphere for a hundred years. So, the effects of CFCs on the ozone layer are still visible. But current data suggest that the ozone hole will "close" in the next fifty years.

Active Reading

RECOGNIZE PATTERNS Have students read again the answer to the second Key Question on the opposite page. Then ask them to scan the three case studies in this lesson.

Ask Do you recognize a pattern? *(yes)* How are the paragraphs organized in each case study? *(Each case study is organized according to the three steps used in ecological research: 1. recognize the problem, 2. find the cause, 3. change behavior.)*

USE VISUALIZATION Have a volunteer read the last sentence in the first paragraph of Case Study #1.

Ask Describe what you visualize as the effect of ozone in the upper atmosphere. *(Sample answer: It's like a giant tube of sunscreen smoothed all over the atmosphere.)* What is this kind of comparison called? *(an analogy)*

INTERPRET GRAPHS Direct students' attention to the graph.

Ask Did the regulation of CFC use appear to have an effect on CFC levels in the atmosphere? *(yes)* What lead you to that conclusion? *(Once the regulations were in place, the levels of CFCs in the atmosphere started to drop.)*

Speed Bump

Ask How does ozone in the upper atmosphere protect life on Earth? *(by absorbing ultraviolet rays from the sun that can cause sunburn, cancer, and other health problems)*

Ask What caused the thinning of the ozone layer? *(the use of CFCs in spray cans and as refrigerants)*

Focus on Academic Language

- ultraviolet
- chlorofluorocarbons

Review with students where ultraviolet radiation is found on the electromagnetic spectrum. Provide an Earth science or physical science textbook, and have students examine a diagram of the spectrum.

Ask Where is ultraviolet on the spectrum? *(next to visible light)* Why do you think ultraviolet might cause damage to cells that leads to cancer? *(Ultraviolet light has a lot of energy.)*

Help students break up the term *chlorofluorocarbons* into meaningful parts that reveal its chemical makeup. *(chloro-, chlorine; fluoro-, fluorine; and carbon)* Tell students that CFCs also contain hydrogen. Compare the abbreviation of CFCs to that of DDT. Explain that, like other chemical names, both names are based on the kinds of atoms in the chemicals, and that such names are often long, requiring the use of abbreviations to talk about the chemicals more easily.

During Reading continued

Active Reading

APPLY ACADEMIC VOCABULARY Have students examine The Decline of Cod graph, in particular the key. Point out that the blue line on the graph indicates "catch" and the red line indicates "biomass."

Say *Biomass* means "the total mass of living things in an area." In this case, it is the total mass of cod in the northwestern Atlantic Ocean.

Ask What does the red line show? *(the cod biomass in the northwestern Atlantic)*

Say *Catch* means "the total mass of the cod caught."

Ask What does the blue line show? *(the cod mass actually caught)*

USE CONTEXT CLUES Have students use text in the paragraphs on this page to help them understand the content in the visuals.

Have a volunteer read aloud paragraph 1. Then discuss the fishing trends shown in the graph.

Say The graph shows a large increase in catch during the late 1970s and early 1980s. Which sentences in the paragraph explain what led to this increase? *(fourth and fifth)*

Say The graph shows a large drop in the cod population happening during the 1990s. Which sentences in the paragraph confirm this information? *(sixth and seventh)*

Have a volunteer read aloud paragraph 2.

Ask Which sentence in the paragraph helps explain the meaning of the term *overfishing*? *(It became clear that more fish were being caught than were being born.)*

Have a volunteer read aloud paragraph 3.

Ask How is aquaculture part of the solution to this problem? *(Aquaculture, the farming of aquatic animals, is an alternative to fishing for wild fish.)*

Speed Bump

Ask What do you think needs to be done to stop the decline in commercial fish populations? *(Sample answer: All countries need to agree on and follow rules to protect commercial fish populations. Lawmakers need to design the rules to have as little negative impact as possible on the fishing industry.)*

Case Study #2: North Atlantic Fisheries

In 1950, fishermen around the world caught about 19 million tons of seafood. In 1997, fishermen caught more than 90 million tons. People thought that the fish supply was an endless, renewable resource. However, commercial fish populations are now dropping dramatically. This problem is one that society is still working on.

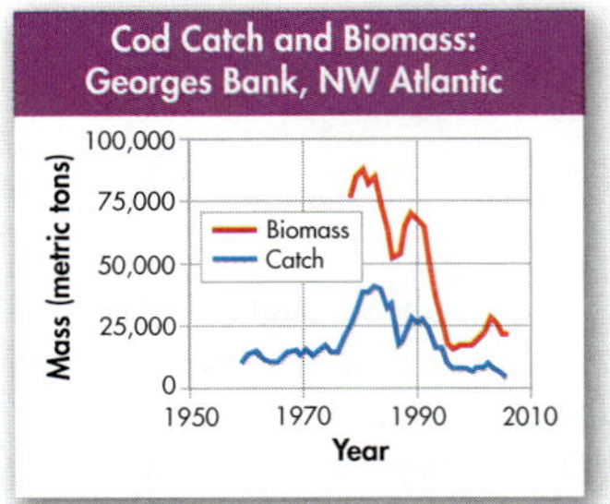

The Decline of Cod As the mass of the cod populations dropped, the catch also decreased.

Overfishing

Aquaculture

1 Recognizing a Problem: More Work, Fewer Fish The amount of cod caught rises and falls each year. Sometimes, these differences are because of natural differences in ocean ecosystems. Often, however, low fish catches happen when boats start taking too many fish. From the 1950s to the 1970s, fishermen began using larger boats and more high-tech equipment to find fish. At first, the fishermen caught more fish. Then the catches began falling again. The mass of cod in the ocean has decreased significantly since the 1980s. Fishermen cannot catch what isn't there.

2 Researching the Cause: Overfishing Ecologists gathered data including age structure and growth rates. Fish populations were getting smaller. By the 1990s, the cod populations had dropped so low that researchers were afraid the fish might disappear forever. It became clear that more fish were being caught than were being born.

3 Changing Behavior: Regulation of Fisheries The United States has set up rules for commercial fishing. These rules state how many fish of what size can be caught. Certain places have been closed to fishing until populations rise again. Other places are closed during breeding seasons. These rules are helping some populations but not all. Aquaculture, the farming of aquatic animals, can also help. It offers an option for commercial fishing.

Progress has been slow. Some countries are not setting up their own rules. Some countries are afraid that these rules take away jobs and income for fishermen. Scientists and lawmakers are still trying to find a solution. They are trying to find ways to protect the fish populations with as little negative impact on the industry as possible.

Students Say

"I liked VIP. That means 'Very Important Point.' The teacher would give us a sticky note and we'd cut it into four strips. While reading, we'd indicate a point in the book that interested us with one of the four slim strips. After reading, we'd discuss our VIPs in small groups."

Case Study #3: Climate Change

Global climate is affected by the cycles of matter and everything that humans do. This includes cutting and burning forests, manufacturing, driving cars, and making electricity. The Intergovernmental Panel On Climate Change (IPCC) began in 1988. Its goal is to provide the best scientific information on climate change. The 2007 report from the IPCC gives the most reliable current information on climate change.

1 Recognizing the Problem: Global Warming The IPCC report states that global temperatures are rising. This rise in average temperature is called global warming. The report also covers climate change. Temperature, rain, and other factors can change because of global warming. Scientists studied physical and biological evidence of climate change.

- **Physical Evidence** Earth's temperatures are getting warmer, sea ice is melting, and sea levels are rising. Eleven of the twelve years between 1995 and 2006 were among the warmest years ever recorded. People began recording temperatures in 1850. Earth's average global temperature rose by 0.74°C between 1906 and 2005. The largest changes are taking place in and near the Arctic Circle. Alaska's average temperature rose 2.4°C in fifty years. Sea level has risen 1.8 mm each year since 1961. This rise is being caused by warmer water expanding and by melting glaciers, ice caps and polar sheets.

A Warming Earth

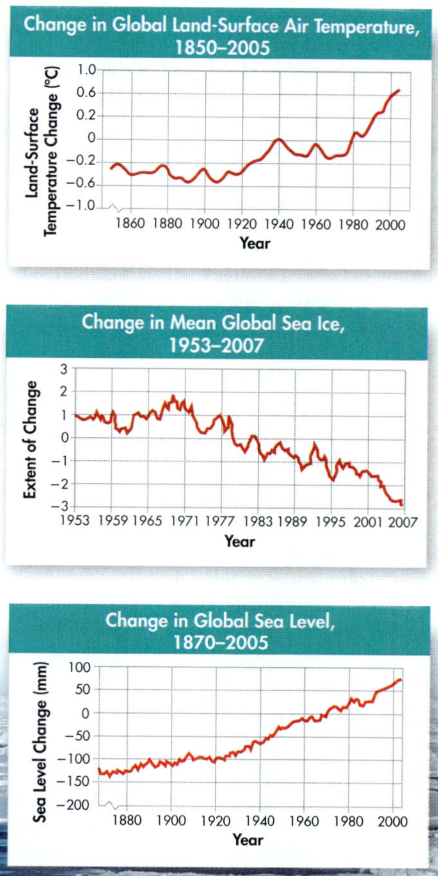

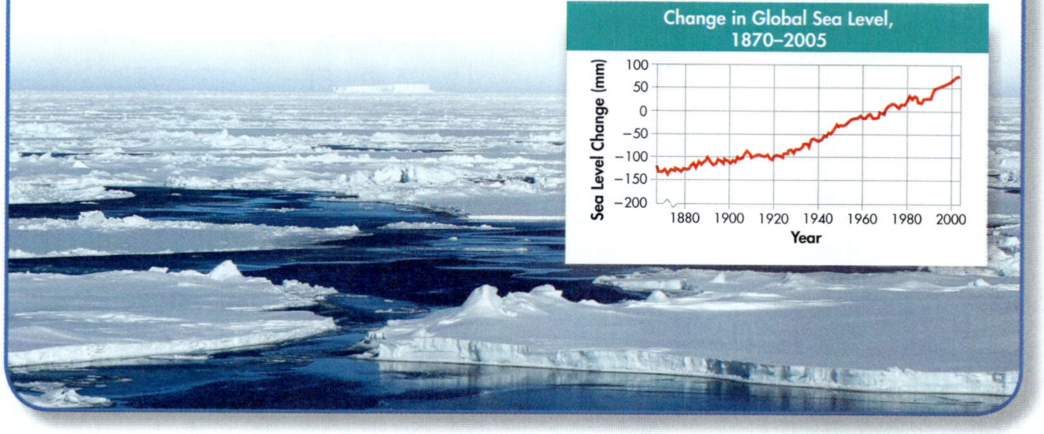

ELL Focus on ELL: Access Content

BEGINNING AND INTERMEDIATE SPEAKERS Tell students that Earth is often referred to as the "globe."

Ask What does *global,* the adjective form of *globe,* mean? *("all around the planet," or "worldwide")*

Say The phrase *global warming* suggests an average warmer temperature around the globe. Climate change is an effect of global warming.

Ask Does climate change necessarily mean a warmer climate? *(no)* Why or why not? *(Climates may become wetter or drier, and some may actually become cooler, as a result of climate change.)*

Write *cause and effect* on the board.

Ask What is the relationship between global warming and climate change? Think about cause and effect to help you answer. *(Global warming is the cause, and climate change is the effect.)*

LESSON 6.4

Active Reading

SET A PURPOSE FOR READING Have students read the title for Case Study #3 and then skim the overall text of the study.

Ask What do you predict this study is about? *(global warming and climate change)*

Point out that the text in this case study is much longer than the text in the two previous studies. Additional paragraphs have been added to give evidence of the problem—global warming. The study continues for the next two pages.

MAKE INFERENCES Have a volunteer read the introductory paragraph of Case Study #3 aloud.

Ask Who is the authority that gathered the data in graphs on this page? *(the IPCC, the Intergovernmental Panel On Climate Change)*

Say The prefix *inter-* means "between or among."

Ask What can you infer about the panel from the word *intergovernmental*? *(that it is a panel made up of representatives from several governments or countries)*

INTERPRET GRAPHS Have students examine the first graph, Change in Global Land-Surface Air Temperature, 1850–2005.

Ask What is the interval between grid lines for years? *(20 years)* Does this mean measurements were taken every 20 years? *(no)* Then what does it mean? *(The graph was drawn that way to make the data easy to read.)*

Ask What year was land-surface air temperature first recorded according to this graph? *(about 1850)*

Have students examine the next two graphs.

Ask What are the trends for change in sea ice and change in sea level? *(Sea ice is decreasing, and sea level is increasing.)*

SCIENCE SUPPORT Tell students that the U.S. Department of the Interior announced in 2008 that it is listing the polar bear as a threatened species under the Endangered Species Act. Polar bears use ice floes as a platform for hunting and eating their prey, mostly seals. Explain that, due to global warming, ice floes are melting so fast during the Arctic summers that bears are having difficulty finding floes large enough to support them and have to swim farther between floes. Some scientists predict that, if warming trends continue in the Arctic, two-thirds of the world's polar bears could disappear by 2050.

Speed Bump

Ask What is the term for the rise in worldwide average temperatures over time? *(global warming)*

Ask What is happening to sea ice that gives evidence of global warming? *(Sea ice is melting.)*

LESSON 6.4

During Reading continued

Active Reading

INTERPRET GRAPHS Direct students' attention to the Greenhouse Gas Emissions circle graph.

Say Data about percentages are often shown on a circle graph.

Ask What does the word *sector* mean in the title of this graph? *(a specific source of greenhouse gas emissions)*

Ask Which sector gives off the most greenhouse gases? *(energy supply)* What is energy supply? *(energy-producing plants)*

MAKE INFERENCES Direct students' attention to the Greenhouse Gas Concentration graph.

Ask What do the colored lines on the graph represent? *(different greenhouse gases)*

Ask What is the trend you see with the three greenhouse gases? *(Their concentrations have increased sharply over the last 200 years.)*

Say You can use the data on this graph to make inferences about concentrations of greenhouse gases in the future.

Ask What can you infer about the concentrations of these gases in the year 2100? *(If the present trend continues, concentrations of these greenhouse gases will continue to increase in our atmosphere.)*

SCIENCE SUPPORT Students may wonder how scientists can learn about atmospheric concentrations of greenhouse gases in the distant past. Explain that one important method is the use of ice cores. Ice cores are cylinders of ice drilled out of glaciers and polar ice sheets. Trapped within these cores are molecules that were in the air at the time the ice was deposited. Each layer of ice in a core corresponds to a single year. By analyzing the molecules in a layer of ice, scientists can get a clear picture of the climate at that time. An ice core drilled through the thick ice cap in the middle of Antarctica shows changes in Earth's climate up to 800,000 years ago.

Speed Bump

Ask Why is a small change in average temperature so important in nature? *(Each organism is adapted to the conditions in its environment. Some species may survive only within a narrow temperature range.)*

Case Study #3: Climate Change *(continued)*

Biological Evidence This yellow-bellied marmot is coming out of hibernation over a month earlier than it used to.

• **Biological Evidence** Small changes in climate that people barely notice can be important to other organisms. Temperature, humidity, and rain affect where an organism can live. If temperatures rise, organisms need to move to cooler places. They may move away from the equator or higher up a mountainside. Plant flowering and animal breeding are also affected by rising temperatures. These organisms may respond as though spring were beginning earlier.

The IPCC report is based on 75 different studies. These studies covered 1700 species of plants and animals. Many species and communities are reacting to the rising temperatures.

❷ **Researching the Cause: Models and Questions** Earth's climate has changed often during its history. Concentrations of carbon dioxide and several other greenhouse gases have risen over the last 200 years. People and their activities are adding more carbon dioxide to the atmosphere than the carbon cycle can remove. Most scientists around the world agree that this added carbon dioxide is increasing the greenhouse effect. In turn, the biosphere is holding more heat.

• **How Much Change?** How much are global temperatures expected to rise? Researchers use computer models based on climate data. The models are complex, and they make many assumptions. So, there is some debate about the predictions. Several models suggest that, by the end of this century, average temperatures may rise by as much as 6.4°C over the average in the year 2000.

• **Possible Effects of Climate Change** What does all of this mean? Some changes may damage ecosystems from tundra to the Amazon rain forest. The western United States is likely to get drier. But the Sahara Desert may become greener. Many places may flood because of rising sea levels. Other places in North America may face more droughts.

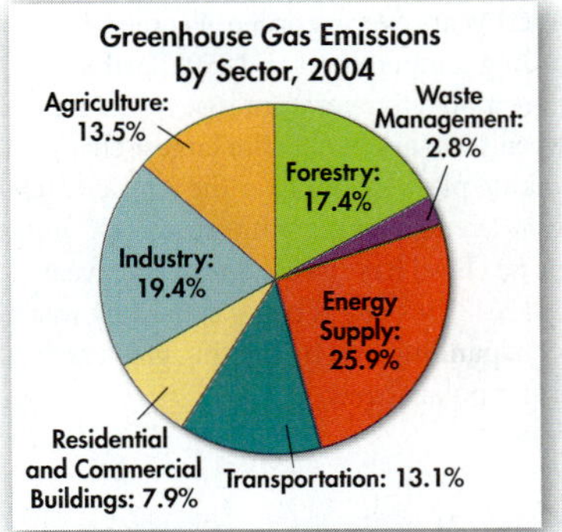

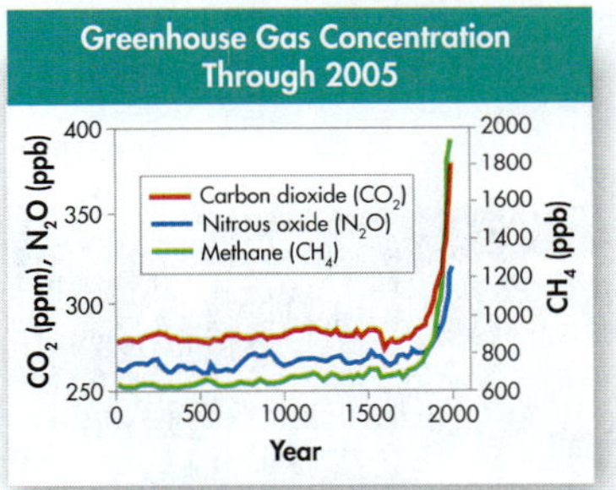

Hands-On Learning

RISING WATERS Explain that in the United States, generally speaking, the land of the East and Gulf coasts slopes more gently than the land of the West coast. Demonstrate what this means regarding rising sea levels.

1. Make a rectangular "island" out of wet sand, soil, or modeling clay in the middle of a cake pan or other tray. Make one side of the island steep, and the other side a gentle slope. Place small, plastic objects, such as dice, on both ends of the island.
2. Explain that the steep side represents the West coast and the gentle side represents the East and Gulf coasts. The objects represent communities. The bottom of the pan represents sea level.
3. Slowly pour water into the pan. Add enough water to "flood" the beach area on the West coast and several of the communities on the East coast. Let students see how the gentle slope allows the floodwaters to reach much farther inland than the steep slope does.

3 Changing Behavior: The Challenges Ahead You have read how research has led to changes that are protecting the ozone layer and helping to bring back fish populations. Global climate change presents many challenges. The changes in behavior will need to be major. The changes will affect economies and many other fields besides biology. Some changes will depend on new technology for renewable energy and more efficient energy use.

Nations around the world have begun to work out agreements to protect the atmosphere and the climate. As the world and our nation work through these challenges, remember that the world is our island of life. Hopefully, there will come a day when we reach the common goal of preserving the quality of life on Earth.

Little Changes, Big Results We have begun to see electric cars, recycled products, and green buildings in response to the need for more efficient energy use.

CHECK Understanding

Apply Vocabulary

Use the highlighted words from the lesson to complete each sentence correctly.

1. All of the energy you use, food you eat, and garbage you throw out contribute to your ______.
2. Beginning in the 1970s, scientists began to realize that the ______ over Antarctica was getting thinner.

Critical Thinking

3. **Apply Concepts** How is the concept of the ecological footprint limited? When is the concept most useful?
4. **Sequence** Describe the three steps ecologists can use to lead us toward a sustainable future.
5. **Write to Learn** Look back at the Warming Earth graphs on page 147. Describe what these graphs tell us about global climate change.

BIOLOGY.com Search Lesson 6.4 GO • Lesson Assessment

Check Understanding

1. ecological footprint
2. ozone layer
3. It is currently impossible to measure an exact ecological footprint. The footprint concept also gives an idea of the resources used only at a particular moment in time. The concept can be very useful, however, when comparing the footprints of different groups.
4. 1) Recognize an environmental problem. 2) Research the problem to find out the cause. 3) Change behaviors to solve the problem.
5. Sample answer: The first graph shows that Earth's temperature is rising. It shows that the average temperature in 2005 was about 1°C warmer than the average temperature in 1850. The second graph shows that the sea ice around the world is melting. The third graph shows that sea levels are rising. Sea levels in 2000 were about 200 mm higher than in 1880. These all provide physical evidence of global climate change.

Active Reading

USE VISUALS Discuss how the practices shown in the pictures could affect a person's ecological footprint.

Focus on the photo of the electric car.

Ask How does this car differ from most cars you see? *(It is very small and is plugged into an electrical outlet.)*

Ask Would driving this car make your ecological footprint smaller or larger? *(smaller)* Explain. *(The car requires fewer nonrenewable resources to build and operate than a larger, gasoline-powered car does.)*

Focus on the photo of the building with grass on its roof.

Ask How do you think having a grass roof on your home might reduce your ecological footprint? *(Sample answers: A grass roof might insulate a building, reducing the amount of energy used for heating and cooling. A grass roof might absorb pollutants in rainwater and carbon dioxide from the air.)*

Speed Bump

Ask Of the problems described in the three case studies, which problem will require the greatest number of people to change their behavior? *(climate change)* Explain. *(Sample answer: The greenhouse gases causing global warming are released when we travel in cars, when we heat our homes, and when power plants generate electricity. These actions affect almost everyone in our society.)*

Post Reading

ASSESSMENT TIPS Have students review all of the vocabulary words before beginning the assessment. If students have difficulty with Question 3, have them reread the Footprint Limitations section with a partner. Suggest they read, cover, remember, and retell.

Wrap-Up Activity

DEVELOP A CASE STUDY Have students use the case study format to analyze another ecological challenge. The challenge could be something related to their community or a resource management problem described earlier in this chapter (such as soil erosion or water pollution). Have students present their case study in an illustrated brochure. Remind them that their case study should: 1) describe a problem, 2) identify the cause, and 3) tell about behavior changes that contribute to solving the problem.

Pre-Lab

Activate students' prior knowledge about acid rain.

Say Describe an acid. *(An acid is a substance with a pH value below 7. Strong acids can be corrosive, capable of eating away other materials.)*

Ask What is acid rain? *(rain with a pH below 5)*

Ask What causes acid rain? *(burning fossil fuels releases chemicals that combine with water vapor and form acids)*

Ask What things on Earth might acid rain harm? *(plants, animals, structures)*

Review the Background Questions and the Pre-Lab Questions as a class.

Lab

Tell students they will perform the chapter lab *Acid Rain and Seeds* described in **Lab Manual B.**

SAFETY Make sure students wear goggles and a lab apron. Caution them to be careful in handling glassware and other breakable materials and to avoid getting the vinegar and food coloring on their skin. Tell students to wash their hands at the end of the lab.

Post-Lab

COMPARING DATA: Post copies of each group's graph. Ask students to identify how the graphs are similar and how they are different. The overall pattern should be the same—the rate of germination increases as the pH increases, but eventually begins to decrease again. There should be variation in the number of seeds germinating at each pH.

Design Your Own Lab

OPEN-ENDED INQUIRY

Pre-Lab: Acid Rain and Seeds

Problem How does acid rain affect seed germination?

Materials large test tubes, test-tube rack, glass-marking pencil, 25-mL graduated cylinder, solutions of vinegar and water, pH paper, dried beans, paper towels, zip-close plastic bags, stick-on labels, hand lens, graph paper

Lab Manual Chapter 6 Lab

Skills Focus Design an Experiment, Organize Data, Measure, Graph

Connect to the Big idea Every organism alters its environment in some way. Elephants uproot trees, prairie dogs dig tunnels, and corals build reefs. But no other organism has as much impact on the global environment as humans. One of the ways that humans affect global ecology is by burning fossil fuels. The burning produces carbon dioxide, which can accumulate in the atmosphere and cause climate change. Other products react with water to form nitric and sulfuric acids. Rain that contains these acids can damage many things, including stone statues and growing plants. In this lab, you will investigate the effect of acid rain on seeds.

Background Questions

a. Review What does a pH scale measure?

b. Review Which solution is more acidic, one with a pH of 4.0 or one with a pH of 5.0, and why?

c. Explain Use the water cycle to trace the path from acids in water vapor to plants.

Pre-Lab Questions

Preview the procedure in the lab manual.

1. Compare and Contrast How are the solutions you will use in this experiment similar? How are they different?

2. Using Models What do the solutions represent?

3. Infer How will you know that a seed has germinated?

BIOLOGY.com Search Chapter 6 GO

Visit Chapter 6 online to test yourself on chapter content and to find activities to help you learn.

Untamed Science Video The Untamed Science crew visits a zoo to learn about the important work that goes on behind the scenes.

Art in Motion View a short animation of biological magnification.

Art Review Review your understanding of the various threats to biodiversity with this activity.

Visual Analogy Compare human impact on the biosphere to a footprint in this activity.

Pre-Lab Answers

BACKGROUND QUESTIONS

a. A pH scale measures the concentration of H^+ ions in a solution.

b. The solution with a pH of 4.0 is more acidic because it has a greater concentration of H^+ ions (ten times as many per unit volume).

c. Sample answer: Water vapor in the atmosphere can condense and fall to Earth's surface as rain, which can contain dissolved acids. Some rain is absorbed into the soil where it can enter plants through their roots.

PRE-LAB QUESTIONS

1. Sample answer: As the vinegar is diluted with water, the intensity of the color will decrease. The food coloring provides a visual indication for the decreased concentration of acid (H^+ ions) in the solutions.

2. The solutions are all a mixture of vinegar and water. Each solution has a different concentration of vinegar (and a different pH).

3. Sample answer: The seed coat will crack and a root will be visible.

NATIONAL SCIENCE EDUCATION STANDARDS

UCP II

CONTENT F.4, F.5, C.4.e

INQUIRY A.1.b, A.1.c, A.2.a, A.2.d

6 CHAPTER Summary

6.1 A Changing Landscape

- People affect the environment through agriculture, development, and industrial growth.
- Sustainable development uses natural resources to meet human needs without causing long-term damage to the environment.

monoculture (p. 129)
renewable resource (p. 131)
nonrenewable resource (p. 131)

6.2 Using Resources Wisely

- Soil allows plants, including crops, to grow. Farming practices such as crop rotation and leaving stems and roots in the soil between plantings can help protect soil from erosion and a loss of nutrients. Cutting only some mature trees and using tree farms can also protect soil.
- The primary sources of water pollution are industrial and agricultural chemicals, residential sewage, and nonpoint sources.
- Common forms of air pollution include smog, acid rain, greenhouse gases, and particulates.

desertification (p. 132)
deforestation (p. 133)
biological magnification (p. 134)

6.3 Biodiversity

- Biodiversity is important in medicine and agriculture. Also, healthy and diverse ecosystems are important for soil, water, and air quality.
- Biodiversity is threatened by changing habitats, hunting, and introduced species. Pollution and climate change also threaten biodiversity.
- To preserve biodiversity, we need to protect whole ecosystems as well as individual species. We also need to figure out how to preserve the environment while helping people improve their standard of living.

biodiversity (p. 138)
habitat fragmentation (p. 140)

6.4 Meeting Ecological Challenges

- The average American has an ecological footprint of more than four times the global average.
- By (1) recognizing a problem in the environment, (2) determining the cause, and (3) changing our behavior, we can move toward a sustainable future.

ecological footprint (p. 143)
ozone layer (p. 145)

BIOLOGY.com Search Chapter 6 GO • Crossword • Chapter Assessment

UbD Transfer the Big Idea

TRANSFER Have each student write a short magazine article describing a success story of a fictional environmental activist. The article should describe an environmentalist who influenced a community to change a development plan in order to conserve biodiversity or use resources wisely. The development could be an apartment complex, a shopping mall, or even a park. Encourage students to be creative. Their article could include portions of an interview with the activist.

Explain to students that their description of human impact on the environment should be plausible and accurate according to what they have learned in this chapter, even though their story can be fictional. Inform students that the success story does not necessarily mean there was no development, and to consider a story in which all parties regard it as a success.

Chapter Summary

Think Visually

USE GRAPHICS Have students create a flowchart that shows the steps in the biological magnification of DDT. Their flowchart should show how DDT enters the food web and its effects on organisms.

SAMPLE ANSWERS

1. Primary producers in an aquatic food web pick up DDT in the water.
2. Zooplankton consume primary producers. Students should indicate in steps 2–6 that the DDT concentration is magnified by 10.
3. Small fish consume zooplankton.
4. Large fish consume small fish.
5. Fish-eating birds consume large fish.
6. The concentration of DDT in the fish-eating birds is high enough to cause thin, fragile egg shells.
7. The hatching rate falls, and the population of the birds decreases.

BIOLOGY.com Study Online

REVIEW RESOURCES

Art in Motion Students watch an animation of biological magnification.

Data Analysis Students collect virtual biodiversity data from two sites and calculate a biodiversity index.

Art Review Students review threats to biodiversity in this drag-and-drop activity.

Visual Analogy Explore ecological footprints through this visual analogy.

Vocabulary Review The *Flash Cards* and *Crossword* provide an interactive way to review chapter vocabulary.

Editable Worksheets Pages of Study Workbook B, Lab Manual B, and the Assessment Resources Book are available online. These documents can be easily edited using a word-processing program.

Lesson Overviews Lesson Overviews are based on Miller & Levine Biology. However, they are available on CD-ROM and can be edited for your class.

ASSESSMENT RESOURCES

Chapter Assessment Have students take an online version of the Chapter 6 Check Understanding.

Standardized Test Prep Students can take an online version of the Standardized Test Prep. You will receive their scores along with ideas for remediation.

Diagnostic and Benchmark Tests Use these tests to monitor your students' progress and supply remediation.

Assess

Big idea Invite a volunteer to read the question. Discuss the kinds of human activities that affect global ecology.

Say Remember that most human activities are not designed to harm the ecology of an area.

Ask How do human activities sometimes harm the environment? *(Sample answer: All living things change the environment. People affect the environment when we use resources for things such as getting food, building homes, and manufacturing products. We also affect the environment when we produce and eliminate wastes. Our impact on the environment increases as the human population grows.)*

Ask Why are other organisms affected when humans change the environment? *(All living things are interdependent, and all living things rely on Earth's resources. When humans harm Earth's resources by causing pollution, habitat fragmentation, and other problems, it affects the ability of other living things to get what they need.)*

CONSTRUCTED RESPONSE Paragraphs will vary. Be sure that students understand that the *Hints* should be used as a guide to help them answer the questions. Review that students' paragraphs should have a main idea supported by details. Remind students that their main ideas can be explicit or implicit.

Write Question 1 on the board and read it aloud.

Ask What is the term for the increasing concentration of pollutants in higher levels of a food chain? *(biological magnification)*

Ask a volunteer to read the first hint.

Ask What happens to the DDT when it is consumed by an organism? *(It collects in the organism's tissues.)*

Ask What happens if another organism eats the first organism? *(The DDT enters the body tissues of the second organism.)*

Read aloud the second hint.

Ask Does a primary consumer eat only one primary producer? *(No, a primary consumer will eat many primary producers.)*

Help students get started by writing the first one or two sentences of a response on the board. *(Example: When a primary producer picks up a pollutant from the environment, the pollutant collects in the producer's tissues. A consumer may eat many producers, consuming a lot of the pollutant in the process.)*

6 CHECK Understanding

Assess the Big idea Interdependence in Nature

Write an answer to the question below.

Q: How have human activities shaped local and global ecology?

Constructed Response

Write an answer to each of the questions below. The answer to each question should be one or two paragraphs long. To help you begin, read the **Hints** below each of the questions.

1. **Why does the concentration of pollutants, such as DDT, increase in higher levels of a food chain?**

   Hint When DDT is consumed by an organism, the body does not break it down.

   Hint Primary producers can pick up a pollutant directly from the environment.

2. **Describe the relationship between agriculture and soil quality.**

   Hint Soil erosion happens when wind or water takes soil away.

   Hint Topsoil is the mineral- and nutrient-rich portion of soil.

   Hint Topsoil can be a renewable resource, but it can be lost or damaged quickly.

Foundations for Learning Wrap-Up

Use the mini book you created before reading the chapter as a tool to help you organize your thoughts about humans in the biosphere.

Activity 1 Working with a partner, review your answers to the key questions. Did you both answer the questions correctly and completely? If not, work together to make any corrections.

Activity 2 Working in a small group, use your mini books to prepare a lecture on how human activities have shaped ecosystems. In your lecture, mention how your own activities shape local and global ecology. Present one idea about what you could do as a school to reduce your ecological footprint.

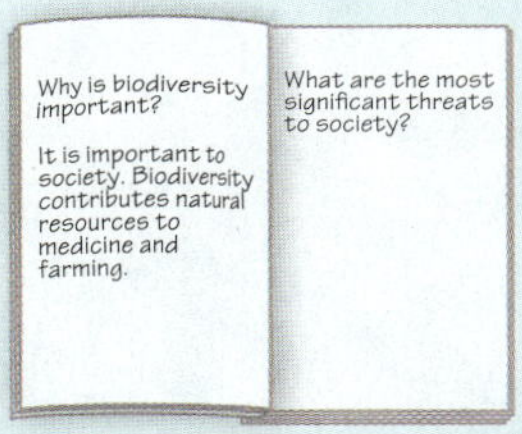

152 Check Understanding

For Question 2, pair students and have them use the hints and your modeling to answer the question. Have students share their paragraphs with the class.

Foundations for Learning Wrap-Up

Have students take out their mini books. Check to see that students have written all of the Key Questions in their mini books and have recorded answers as they read the lessons.

Either assign groups or pair students, and explain the activities.

6.1 A Changing Landscape

Understand Key Concepts

1. A resource that can be replenished or replaced is called
   a. common.
   b. renewable.
   c. nonrenewable.
   d. recycled.
2. The concept of using natural resources to meet the needs of humans without causing long-term environmental damage is called
   a. monoculture.
   b. sustainable development.
   c. recycling.
   d. resource renewal.

Test-Taking Tip.

Read All of the Answer Choices Take the time to read all of the answer choices carefully. Question 2 asks for the word or phrase that is defined in the question. Decide whether you know the definitions of the answer choices. Once you define all of the answer choices, you'll find that **b** is the correct answer. Sustainable development is the use of natural resources in an environmentally conscious way.

3. Describe how Hawaiian settlers negatively affected the islands after the 1700s.
4. Explain how Earth is like an island.

Think Critically

5. **Compare and Contrast** How are renewable resources and nonrenewable resources alike? How are they different?

6.2 Using Resources Wisely

6. The loss of topsoil through the action of water or wind is called
   a. acid rain.
   b. erosion.
   c. desertification.
   d. monoculture.
7. Look closely at the image of the food web shown below. In which organism would the concentration of a pesticide probably be highest?
   a. hawk
   b. rabbit
   c. frog
   d. grasses

8. How can forests be managed in a sustainable way?
9. What are some of the common sources of water pollution?

Think Critically

10. **Relate Cause and Effect** How does the use of fossil fuels negatively impact Earth's atmosphere?

6.3 Biodiversity

11. Which of the following is NOT a term for one of the three levels of biodiversity?
    a. species diversity
    b. population diversity
    c. genetic diversity
    d. ecosystem diversity
12. A place where many species and habitats are in immediate danger of extinction is called a(n)
    a. altered habitat.
    b. habitat fragment.
    c. ecological hot spot.
    d. dead zone.
13. What are the major threats to biodiversity?

Think Critically

14. **Compare and Contrast** Explain the difference between species diversity and ecosystem diversity.

Assessment Tips

Pair students and have them work together to answer the questions. Encourage students to look back and review vocabulary and important graphics before beginning the assessment.

For Question 5, encourage students to make a Venn diagram before writing their answer. Question 7 requires that students examine a diagram. Tell them to recall what they know about biological magnification. Point out the word *NOT* in Question 11.

Answers

Lesson 6.1

UNDERSTAND KEY CONCEPTS

1. b

2. b

3. Hawaiian settlers cut trees to plant farms and introduced nonnative plants, pigs, chickens, dogs, and rats.

4. Sample answer: Earth is like an island because the resources here are limited—we can't get water or soil from anywhere else. It's also like an island because humans can't leave to live anywhere else.

THINK CRITICALLY

5. Sample answer: Both renewable and nonrenewable resources are made by natural ecosystems. However, renewable resources can be produced or replaced by a healthy ecosystem while nonrenewable resources cannot be replaced within a reasonable length of time.

Lesson 6.2

UNDERSTAND KEY CONCEPTS

6. b

7. a

8. Tree farms can protect soil and make the trees themselves a renewable resource if only some of the mature trees within a forest are cut down. This allows growth of younger trees without damaging the whole forest or its soil. Also, farmers can plant new trees to replace the trees they do cut down each year.

9. industrial and agricultural chemicals, residential sewage, and nonpoint sources such as grease and oil washed off streets by rain or the chemicals released into the air by factories and automobiles

THINK CRITICALLY

10. The burning of fossil fuels releases acids that cause acid rain; ozone that causes smog; greenhouse gases; and particulates.

Lesson 6.3

UNDERSTAND KEY CONCEPTS

11. b

12. c

13. altering habitats, hunting, introducing new species, releasing pollution into food webs, and climate change

THINK CRITICALLY

14. Species diversity is the number of different species in the biosphere or in a particular area. Ecosystem diversity refers to the variety of habitats, communities, and ecological processes in the biosphere.

Lesson 6.4

UNDERSTAND KEY CONCEPTS

15. a

16. Sample answer: Organisms move toward cooler places away from the equator and from warm lowlands to cooler, higher altitudes. Plants flower and animals breed earlier as though spring begins earlier.

THINK CRITICALLY

17. The ozone layer protects living things from the harmful effects of the sun's ultraviolet rays.

18. The steps taken include making rules about how many fish could be caught in U.S. waters, closing certain areas to fishing until stocks recover, closing some areas seasonally to allow fishes to breed, and using aquaculture as an alternative to fishing.

Connecting Concepts

19. I would expect the trend to continue—the size of the catch of Atlantic bluefin tuna would continue to drop.

20. Sample answer: Governments should make rules about the fishing of bluefin tuna. The catch for at least the next decade should be strictly limited. This would give the population a chance to grow again to previous numbers.

CHAPTER MYSTERY After students have read through the Chapter Mystery, discuss how the actions of the Rapa Nui resulted in the destruction of the forests on Easter Island.

Ask What activities of the Rapa Nui resulted in the destruction of Easter Island's forests? *(They cleared trees for agriculture, cut trees to make canoes, and used logs in rolling the moai.)*

Ask How did the rats reach Easter Island? *(They came as stowaways on the Rapa Nui's boats.)*

Ask How did this invasive species contribute to the deforestation of the island? *(The rats ate coconuts and their seeds before the seeds could grow.)*

6 CHECK Understanding

6.4 Meeting Ecological Challenges

Understand Key Concepts

15. The increase in the average temperature on Earth is called

a. global warming. **c.** atmospheric change.

b. climate change. **d.** Earth warming.

16. What are some of the biological effects of climate change?

Think Critically

17. Explain How is the ozone layer important to living things?

18. Apply Concepts Describe some of the steps taken to solve the problem of overfishing cod in the North Atlantic.

Connecting Concepts

Use Science Graphics

The graph shows the amount of bluefin tuna caught by the United States in the Atlantic Ocean between 2002 and 2006. Use the graphics to answer questions 19 and 20.

U.S. Catch of Atlantic Bluefin Tuna

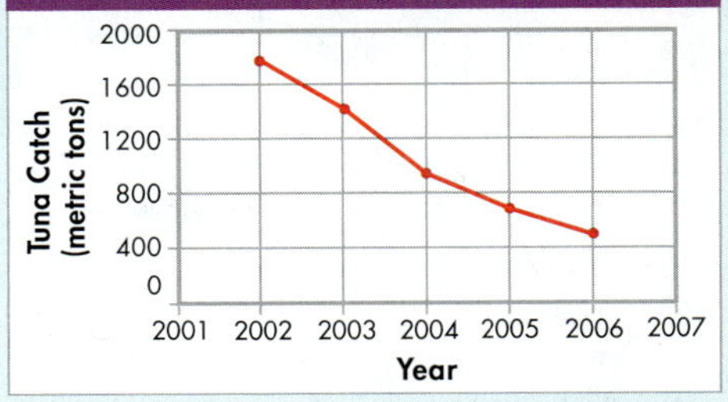

19. Predict What trend would you expect to see in the annual catch from 2006 to 2007?

20. Propose a Solution What recommendations would you make to help the bluefin tuna population recover in the next decade or two?

solve the CHAPTER MYSTERY

MOVING THE *MOAI*

Easter Island did not have as much biodiversity as the Hawaiian Islands. This made it less resistant to ecological damage. The Rapa Nui cut palm trees for agriculture and for wood to make fishing canoes. They also used large logs from palm trees to move *moai*. They cleared fields and did not protect the soil from erosion. Healthy topsoil washed away.

The Rapa Nui brought rats to the island. Rat populations grew exponentially. Hordes of hungry rats killed young palm trees. They ate coconuts and their seeds.

The combination of human activities and introduced species destroyed almost all of Easter Island's forests. Because of deforestation and the difficult climate, the island has been able to house very few people ever since.

1. Relate Cause and Effect Easter Island is about half the size of Long Island, New York. How did its small size affect the outcome of deforestation and the introduction of rats?

2. Compare and Contrast Gather information about the geography, climate, and biological diversity on the Hawaiian Islands and on Easter Island. How do you think the differences made the islands respond differently to human settlement?

Untamed Science **Never Stop Exploring Your World.** The mystery of Easter Island is just the beginning. Take a video field trip with the ecogeeks of Untamed Science to see where this mystery leads.

CHAPTER MYSTERY ANSWERS

1. Sample answer: Because of the small size of the island, relatively few species lived there, and the populations of those species were relatively small. The lack of species diversity and genetic diversity made the native species less able to survive the changes in their environment.

2. Sample answer: Easter Island is smaller, the climate is harsher, and the biological diversity is more limited than in the Hawaiian Islands. Those differences made the ecosystem on Easter Island much more vulnerable to the disturbances of human activities than were the ecosystems on the Hawaiian Islands.

Untamed Science Have students watch the video **What Do Zoos . . . Do?** in which the Untamed Science crew explores a zoo's research and education initiatives to protect biodiversity.

Standardized Test Prep

Multiple Choice

1. Which of the following statements about renewable resources is TRUE?
   A They are found only in tropical climates.
   B They can never be depleted.
   C They are replaceable by natural means.
   D They never regenerate.

2. Which of the following is a nonrenewable resource?
   A wind
   B fresh water
   C coal
   D topsoil

3. Which of the following is NOT a direct effect of deforestation?
   A decreased productivity of the ecosystem
   B soil erosion
   C biological magnification
   D habitat destruction

4. The total variation in all organisms in the biosphere is called
   A biodiversity.
   B species diversity.
   C ecosystem diversity.
   D genetic diversity.

5. Ozone is made up of
   A hydrogen.
   B oxygen.
   C nitrogen.
   D chlorine.

6. Ozone depletion in the atmosphere has been caused by
   A monoculture.
   B CFCs.
   C suburban sprawl.
   D soil erosion.

7. In a food chain, concentrations of harmful substances may increase in higher levels in the food chain in a process known as
   A biological magnification.
   B genetic drift.
   C biological succession.
   D pesticide resistance.

Questions 8–9

Fire ants first arrived in the United States in 1918, probably on a ship traveling from South America to Alabama. The maps below show the geographic location of the U.S. fire ant population in 1953 and 2001.

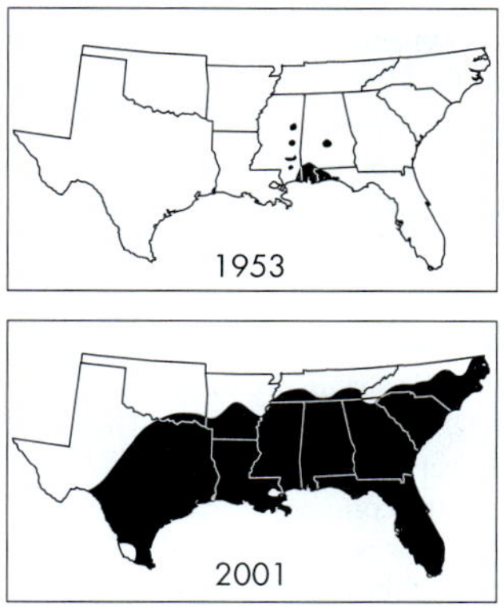

8. Which of the following statements about fire ants in the United States is TRUE?
   A They reproduce slowly.
   B They are a native species of the United States.
   C They are an introduced species.
   D They do not compete with other ant species.

9. By 2010, fire ants are MOST likely to
   A have spread to a larger area.
   B have reached their carrying capacity.
   C die out.
   D return to South America.

Open-Ended Response

10. Describe how ecologists use the ecological footprint concept.

If You Have Trouble With . . .

Question	1	2	3	4	5	6	7	8	9	10
See Lesson	6.1	6.2	6.2	6.3	6.2	6.4	6.2	6.3	6.3	6.4

Test-Taking Tip

USE TIME WISELY

Advise students that if they are taking a long time to answer a question, they should move on to other questions and come back to the difficult question later. While answering other questions, they may remember the information needed to answer the skipped question.

ASSESSMENT

Before the Test

Review with students the Test-Taking Tip.

Call attention to words that are emphasized in the questions, such as the word *TRUE* in Questions 1 and 8, the word *NOT* in Question 3, and the word *MOST* in Question 9.

Remind students that an open-ended response means that they'll have to write a paragraph to answer the question. Have them paraphrase the question in their notebooks, and remind them about modeling a paragraph.

Answers

1. C
2. C
3. C
4. A
5. B
6. B
7. A
8. C
9. A
10. Sample answer: They use it for making comparisons among different populations. For example, according to one data set, the average American has an ecological footprint over four times larger than the global average.

UNIT 2

Plan Ahead

Have students preview the Unit 2 Project a few days before the day of debate. Suggest they review ways in which human activities can affect the environment. Then, divide the class into groups of four. Assign each member of the groups one of the four roles for the debate. Tell students they might search online to find evidence to support the position assigned. Encourage students to talk to classmates with the same role about evidence and arguments that could be persuasive in a debate.

Materials Internet access for research

Monitor the Project

Suggest students make lists of evidence and arguments supporting the position of the role they have taken. Ask individual students how they are preparing for the debate and what evidence they have found that supports their position. On debate day, each group could be given about 10 minutes to debate the issue, or students with the same role could collaborate in a debate that includes the entire class.

Project Assessment

Make sure students use the rubric and reflection questions to assess their work. Then, use the rubric to assign a final score. If desired, talk with students about any differences between their self-assessment scores and your assigned score.

Unit Project

Development Debate

A large company wants to build a new factory on your town's wetlands. Many people in the town are opposed to the idea, claiming it will disturb the local ecosystem and cause problems for residents. Others support the development, arguing that the new factory will bring jobs and money into the town. Representatives have been called in to debate the issue before the town council.

Your Task Take on one of the stakeholder roles listed below. Find evidence to support that point of view and debate the issue in class. The roles are

- Conservation ecologist
- CEO of the company
- Town mayor who supports the development
- Resident of the town who lives next to the wetlands

Be sure to

- justify your arguments with credible information.
- present your arguments in a clear and convincing manner.

Reflection Questions

1. **Score your performance using the rubric below. What score did you give yourself?**
2. **What did you do well in this project?**
3. **What about your performance needs improvement?**
4. **After hearing various sides of the argument, meet with a partner and discuss which side you agree with the most. Justify your opinion.**

Assessment Rubric

Score	Evidence Provided	Quality of Performance
4	Student justifies his/her argument with sophisticated and highly credible information.	Ideas are presented in a highly convincing and clear manner. Student shows a deep understanding of the issues involved.
3	Student justifies his/her argument with logical and credible information.	Ideas are presented in an effective and clear manner. Student shows a solid understanding of the issues involved.
2	Student provides some credible information, but other points are weak or inaccurate.	Some ideas are presented in an unclear manner. Student shows a limited understanding of the issues involved.
1	Student provides mostly illogical and invalid evidence to support his/her argument.	Most ideas are presented in an unclear manner. Student shows a very limited understanding of the issues involved.

NATIONAL SCIENCE EDUCATION STANDARDS

UCP I, II, III

CONTENT C.4.a, C.4.c, C.4.e, F.3, F.4, F.5, F.6

INQUIRY A.1.f, A.2.e, A.2.f

21st Century Skills

To be successful in the 21st century, students need skills and learning experiences that extend beyond subject area mastery. The Unit 2 Project helps students build the following 21st Century Skills: *Information and Media Literacy; Communication Skills; Critical Thinking and Systems Thinking; Problem Identification, Formulation, and Solution; Interpersonal and Collaborative Skills; Self-Direction; Accountability and Adaptability;* and *Social Responsibility.*

FOCUS ON CREATIVITY AND INTELLECTUAL CURIOSITY Extend this Unit Project by asking small groups of students to use what they have learned from the debate to write a screenplay about an environmental controversy within a community.

For more practice building 21st Century Skills, see The Chapter Mystery pages in **Study Workbook A.**

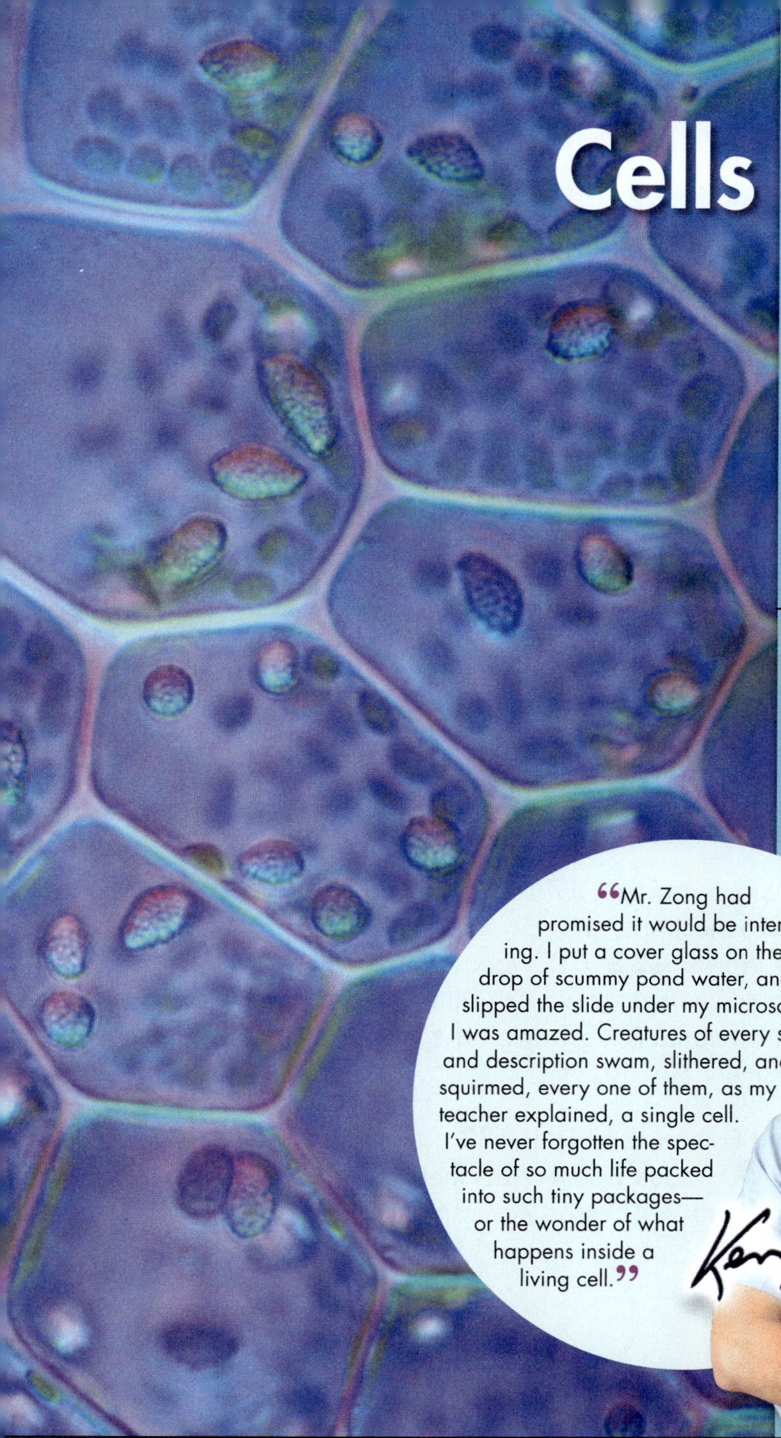

Cells

UNIT 3

INTRODUCE the Big ideas

- **Cellular Basis of Life**
- **Homeostasis**
- **Growth, Development, and Reproduction**

"Mr. Zong had promised it would be interesting. I put a cover glass on the drop of scummy pond water, and slipped the slide under my microscope. I was amazed. Creatures of every shape and description swam, slithered, and squirmed, every one of them, as my teacher explained, a single cell. I've never forgotten the spectacle of so much life packed into such tiny packages—or the wonder of what happens inside a living cell."

Ken Miller

157

Dear Colleague,

I can still remember the first time I looked through a microscope and saw a living cell. It was in Paul Zong's ninth grade biology class in my hometown high school in New Jersey. After carefully instructing us in the proper use of the microscope, our teacher placed a drop of water on every student's slide and told us to have a look. I couldn't believe my eyes. Glistening creatures swam across the field of view. They twisted and turned, I thought, almost as if they were alive. I think I said that out loud, because I can remember Mr. Zong's deep, gentle laugh and a pat on my shoulder. "They are alive, Kenny! They're alive just like you and me."

Later that year, I transformed a corner of the small room I shared with my brother into a miniature laboratory. A tiny desk lamp glowed day and night, providing energy for nearly a dozen test-tube colonies of Euglena*. At the end of the year, those cells would become a science project, the very first research I would ever do on my own. What stuck with me from that first experience was the realization that the cell is life itself. Everything that we associate with life, from growth and reproduction to digestion and movement, happens at the level of the cell.*

I won a ribbon that year for my study of light's effect on the growth of Euglena*. Although I have long since misplaced the ribbon, I hope I never lose the greater gift that came from a year of study in that biology classroom—a sense of amazement that returns every time I sit down at a microscope in my laboratory.*

I hope that you and your students will find some of that amazement written into the pages of this unit. As a cell biologist, I especially hope to give students an appreciation of the roles that cells play in every aspect of life. In these four chapters, we have done our best to explain how cells live and grow, how they transform energy, and how they pass information along from one generation to the next.

Ken Miller

Cell Structure and Function

Chapter Contents	NSES	Time	Core Resources
Chapter Preview			**Student Edition,** pp. 158–159 **Chapter Mystery,** p. 159 **Foundations for Learning,** p. 159
7.1 Life Is Cellular The Discovery of the Cell • Exploring the Cell • Prokaryotes and Eukaryotes	I, III, V, A.1.b, A.1.c, A.2.a, A.2.c, C.1.a	1 period 1/2 block	**Student Edition,** pp. 160–163 **Inquiry Into Scientific Thinking,** p. 162 **Study Workbook B** 7.1 Worksheets **Biology.com** 7.1 Lesson Assessment
7.2 Cell Structure Cell Organization • Organelles That Store, Clean Up, and Support • Organelles That Build Proteins • Organelles That Capture and Release Energy • Cellular Boundaries	I, V, C.1.a, C.1.b, C.1.c, C.1.d, C.1.e, C.2.a	2 periods 1 block	**Student Edition,** pp. 164–175 **Build Connections,** pp. 164, 168, and 174 **Study Workbook B** 7.2 Worksheets **Biology.com** *Art Review:* Plant and Animal Cells • *Visual Analogy:* A Cell as a Living Factory • *Tutor Tube:* Plants Have Mitochondria Too • 7.2 Lesson Assessment
7.3 Cell Transport Passive Transport • Active Transport	III, IV, V, A.2.a, C.1.a, C.1.b	1 period 1/2 block	**Student Edition,** pp. 176–180 **Build Connections,** p. 179 **Study Workbook B** 7.3 Worksheets **Biology.com** *InterActive Art:* Diffusion and Osmosis • *Art in Motion:* Active Transport • 7.3 Lesson Assessment
7.4 Homeostasis and Cells The Cell as an Organism • Multicellular Life	I, III, IV, V, C.1.a, C.1.d, C.1.f, C.5.d	1/2 period 1/4 block	**Student Edition,** pp. 181–183 **Study Workbook B** 7.4 Worksheets **Biology.com** 7.4 Lesson Assessment
Chapter Pre-Lab	II, V, A.1.b, A.1.c, A.2.a, C.1.a, C.1.d	1 period 1/2 block	**Student Edition,** p. 184 **Lab Manual B** *Detecting Diffusion*

Tools for ELL Instruction

ELL Handbook describes teaching strategies to make *Miller & Levine Biology: Foundation Edition* more accessible to ELL students.

Spanish Study Workbook is a Spanish translation of Study Workbook A.

Multilingual Glossary is the *Miller & Levine Biology* glossary translated into ten languages.

Tools to Challenge Students

Study Workbook A includes more challenging worksheets and activities such as Chapter Mystery activities that reinforce 21st Century Skills.

Lab Manual A labs are geared toward on-level students. The Lab Manual includes fifteen labs in addition to the Chapter Labs.

Biology.com Data Analysis activities offer students a way to practice data gathering and analyzing skills.

Visual Quizzes in the Assessment Resources Book are geared toward on-level students and use visuals from *Miller & Levine Biology.*

Additional Resources

Biology.com Untamed Science Video • Flash Cards

Study Workbook A 7.1 Worksheets
Spanish Study Workbook 7.1 Worksheets
Assessment Resources Book Visual Quiz

Study Workbook A 7.2 Worksheets
Lab Manual B *Hands-On Activity:* Making a Model of a Cell
Spanish Study Workbook 7.2 Worksheets
Assessment Resources Book Visual Quiz

Study Workbook A 7.3 Worksheets
Spanish Study Workbook 7.3 Worksheets

Study Workbook A 7.4 Worksheets
Lab Manual B *Data Analysis:* Mitochondria Distribution in the Mouse
Spanish Study Workbook 7.4 Worksheets
Biology.com *Data Analysis:* Maximizing Mitochondria

Lab Manual A *Detecting Diffusion* • *Osmosis*

Chapter Review

Student Edition Chapter Summary, p. 185
Study Workbook B Chapter 7 Review
Transparencies, pp. 78–98
Biology.com Untamed Science Video • Editable Worksheets of Study Workbooks A and B and Lab Manuals A and B • Chapter 7 Flash Cards and Match It

Untamed Science DVD • Classroom Resources CD (includes lesson presentations and editable worksheets)

Chapter Assessment

Student Edition Check Understanding, pp. 186–189
Study Workbook B Chapter 7 Taking a Standardized Test
Assessment Resources Book Chapter 7 Test B
Biology.com Chapter 7 Assessment • Editable Worksheets of Chapter 7 Visual Quizzes* and Chapter 7 Tests A and B

Exam*View* *Assessment Suite* • Classroom Resources CD (includes lesson presentations and editable worksheets)

Time: 1 period, 1/2 block

* Visual Quizzes are based on visuals from *Miller & Levine Biology.* Review carefully to make sure they are appropriate for your class.

Pressed for Time?

Preview the Chapter Introduce the Key Questions for Lesson 7.2 and preview the Lesson 7.2 vocabulary.

Cover the Chapter Quickly Have students read *The Discovery of the Cell* and *Prokaryotes and Eukaryotes* in Lesson 7.1. Assign all of Lesson 7.2 and go over the Typical Cells visual. Assign all of Lesson 7.3, focusing on the Osmotic Pressure and Active Transport visuals.

Assess Assign Questions 1–4 and 6 in the 7.1 Check Understanding, the 7.2 Check Understanding, the 7.3 Check Understanding, and Questions 1–3 and 5–13 in the Chapter 7 Check Understanding.

Get the Picture

Tell students that the picture shows several diatoms as viewed through a microscope. Make sure students note the explanation that diatroms are single-celled algae. Point out the different shapes and sizes of the diatoms. Explain that although the diatoms differ in shape and size, they are all one-celled organisms.

Ask Are all cells of all organisms alike? *(no)*

Activate Prior Knowledge

Help students understand that cells are the basic unit of life.

Ask What are some things that are made up of many parts? *(cars, watches, bikes)*

Ask How would they be different if they were made of one big part instead of many? *(Answers will vary.)*

Ask Is there any value to having things made of many small, similar parts? *(Answers will vary.)*

7 Cell Structure and Function

Big ideas Cellular Basis of Life, Homeostasis

Q: How are cell structures adapted to their functions?

NATIONAL SCIENCE EDUCATION STANDARDS

UNIFYING CONCEPTS AND PROCESSES

I, II, III, IV, V

CONTENT

C.1.a, C.1.b, C.1.c, C.1.d, C.1.e, C.1.f, C.2.a, C.5.d

INQUIRY

A.1.b, A.1.c, A.2.a, A.2.c

UbD Connect to the Big Idea

Invite a volunteer to read aloud the chapter number and the chapter title. Read the Big Ideas and the question. Emphasize the question, **How are cell structures adapted to their functions?**

Set the Purpose

PREVIEW AND PREDICT Draw the graphic organizer on the board.

Ask How do the words *structure* and *function* in this title relate to the question in the graphic organizer? *(The question asks about how cell structures are suited for their functions. The title of the chapter in general suggests that the chapter will be about cell structure and function.)* As you review the questions, explain that structure has to do with organelles; homeostasis has to do with how organelles function.

CHAPTER 7

INSIDE:

- 7.1 Life Is Cellular
- 7.2 Cell Structure
- 7.3 Cell Transport
- 7.4 Homeostasis and Cells

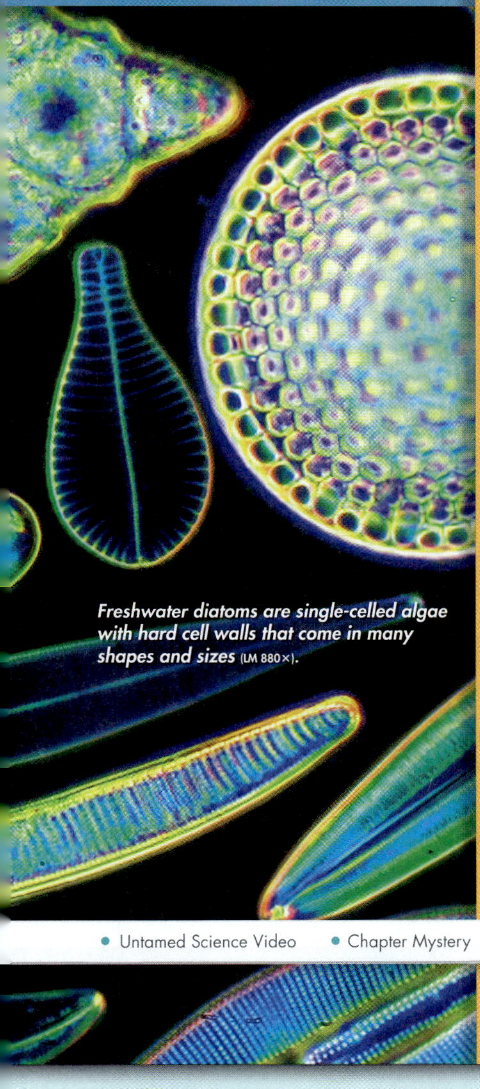

Freshwater diatoms are single-celled algae with hard cell walls that come in many shapes and sizes (LM 880×).

- Untamed Science Video
- Chapter Mystery

CHAPTER MYSTERY

DEATH BY . . . WATER?

Michelle was healthy and 25 years old when she ran in her first marathon. The hot and humid weather made everyone sweat a lot, so Michelle drank water whenever she could. Slowly, she began to feel weak and confused. At the end of the run, Michelle staggered into a medical tent. Her head hurt and she felt nauseous. After telling a volunteer how she felt, she collapsed onto the floor. Thinking she was dehydrated, the volunteer gave Michelle even more water. But, Michelle felt worse and worse. They rushed her to the hospital, where she had a seizure and fell into a coma. Why did giving Michelle water make her condition worse?

Read for Mystery Clues As you read this chapter, look for clues to help you predict how water made Michelle sick. Then, solve the mystery at the end of the chapter.

FOUNDATIONS for Learning

Just as you have special body structures that let you do different things, cells have special parts that help keep them alive. Before you read the chapter, get 15 index cards and punch holes in them as shown below. As you read each lesson, write on the cards the name of each cell part. Draw a picture of the cell part and briefly describe what it does. At the end of the chapter are two activities that use the cards. They will help you answer the question: How are cell structures adapted to their functions?

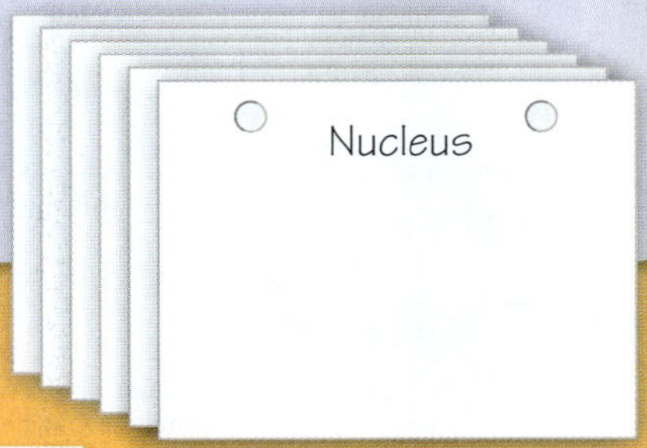

Chapter 7
Big Ideas: Cellular Basis of Life, Homeostasis

Chapter 7: How are cell structures adapted to their functions?

- **7.1:** Why is it important to study cells?
- **7.2:** How do cell structures enable a cell to carry out basic life processes?
- **7.3:** How does a cell transport materials across the cell membrane?
- **7.3:** How does a cell maintain homeostasis both within itself and as part of a multicellular organism?

Foundations for Learning

Pass out the index cards. To save time, you may wish to punch holes in the cards before class. Help students understand the value of this activity.

Remind them that they will use these cards to make a flip chart that can be used as a study aid. Familiarize students with the activity by demonstrating one card. Point out the name of the cell part, the diagram of the cell part, and the description of its function.

Ask Why do you think some structures, such as DNA, RNA, and a cell membrane, are found in the cells of all organisms? Why do you think other structures, such as a nucleus, are found only in cells of certain kinds of organisms? *(Answers may vary. Students should recognize that not every life process is carried out by every kind of cell or organism.)*

Untamed Science™ The Untamed Science crew explores how saltwater and freshwater fish maintain water balance and homeostasis in **Fishing for the Right Balance.**

CHAPTER MYSTERY Invite a volunteer to read aloud the Chapter Mystery.

Ask Have you ever been urged to drink plenty of water while exercising or participating in a sport? Why do you think drinking water is important? *(Sample answer: to prevent dehydration)*

Ask Can too much water be bad for you? *(yes)* What do you think happened to Michelle's cells? *(At this point, accept all reasonable answers.)*

Tell students that they will find clues in the chapter to help them answer these questions.

Digital Support

BIOLOGY.com Extend your reach by using the digital assets offered at **Biology.com.**

LESSON 7.1

Pre-Reading

Objectives

7.1.1 State the cell theory.

7.1.2 Describe how the different types of microscopes work.

7.1.3 Distinguish between prokaryotes and eukaryotes.

Preview the Pages

1. Preview the visuals and read the captions and labels. Be sure students recognize that the colors in the cells were added to make the cell structures more visible.
2. Review the vocabulary words. Be sure students pronounce the words *eukaryote* and *prokaryote* correctly.
3. Challenge students to infer the functions of some structures, such as the cilia and flagellum on the prokaryotic cell. Then ask them to compare the plant and animal cells.

Build Understanding

CONCEPT MAP Direct students to Workbook B, where they will find more information about making concept maps.

7.1 Life Is Cellular

Key Questions

What is the cell theory?

How do microscopes work?

How are prokaryotic and eukaryotic cells different?

BUILD Understanding

Concept Map Make a concept map to organize information about cells. As you read, fill in the circles to show the most important parts of the concept and how they are connected.

In Your Workbook Go to your workbook to learn more about how to use a concept map. Complete the concept map activity for Lesson 7.1.

The Discovery of the Cell

There is an old saying that goes "Seeing is believing." A great example of this concept is the discovery of the cell. You cannot see cells with your eyes alone. Without tools to make cells visible, for centuries, scientists did not even know that there were cells. The microscope changed all of this.

Early Microscopes In the late 1500s, eyeglass makers in Europe found that they could use several glass lenses together to make very small things easy to see. Before long, they had built the first true microscopes from these lenses.

In 1665, Robert Hooke used a microscope to look at thin slices of cork from plants. The cork seemed to be made of many tiny, empty boxes. Hooke called these boxes "cells." Today we know that **cells,** which are the basic units of life, are not empty boxes. In fact, they are full of working parts, each with its own job. Cells are the smallest living unit of any organism. Around the same time that Hooke looked at cork, Anton van Leeuwenhoek looked at water. To his amazement, the microscope allowed him to see a world of tiny living things that seemed to be everywhere. He even found living bacterial cells in his own mouth.

The Cell Theory Scientists made many discoveries about cells in the next 200 years. These discoveries are summarized in the **cell theory,** a fundamental concept of biology. The cell theory states:

- All living things are made up of cells.
- Cells are the basic units of structure and function in living things.
- New cells come from existing cells.

Key Question What is the cell theory?
The cell theory states that all living things are made up of cells, that cells are the basic units of structure and function in living things, and that new cells come from existing cells.

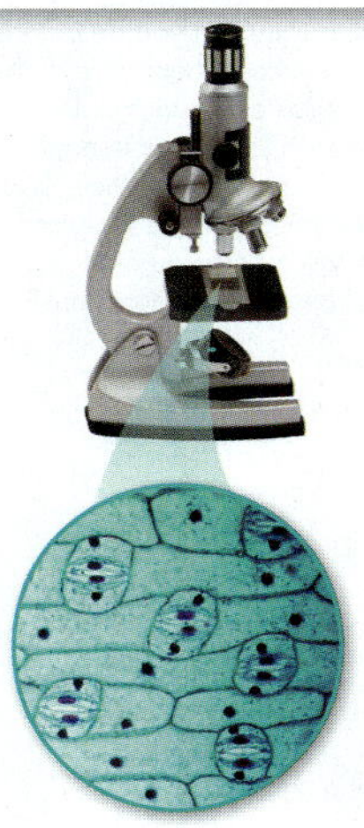

LM 35×

Cell Stains This onion leaf skin has been stained to show the shapes of the cells and some of their parts.

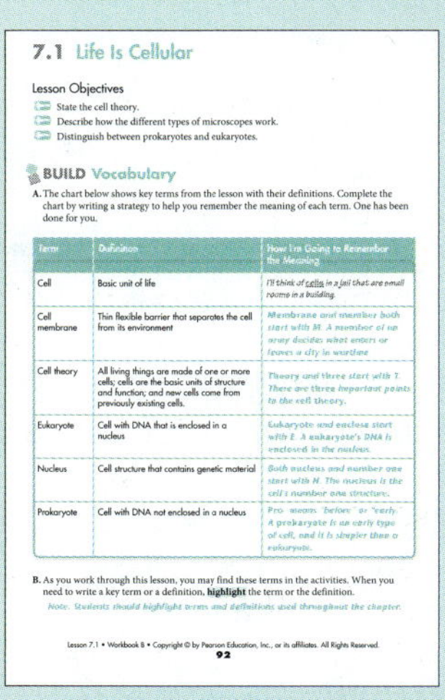

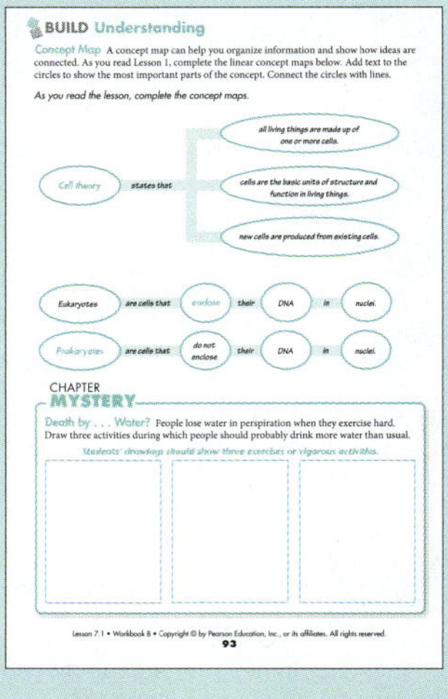

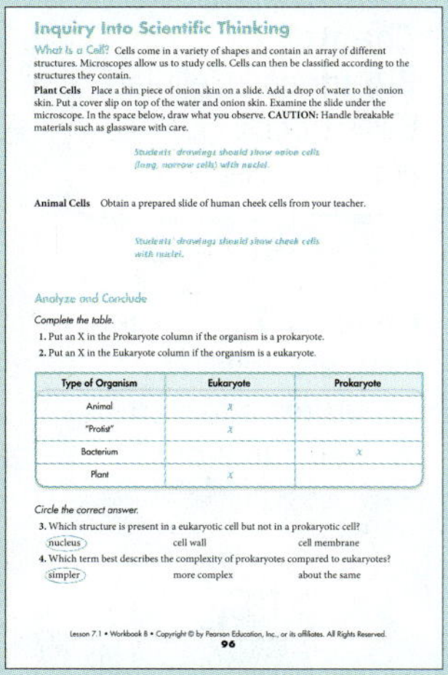

NATIONAL SCIENCE EDUCATION STANDARDS

UNIFYING CONCEPTS AND PROCESSES

I, III, V

CONTENT

C.1.a

INQUIRY

A.1.b, A.1.c, A.2.a, A.2.c

For the corresponding lesson in *Miller & Levine Biology,* see pages 190–194.

Exploring the Cell

A microscope gives you a larger view of something very small, such as a cell. Light microscopes use glass lenses to focus light and magnify the object. Electron microscopes produce even higher magnifications by using electrons instead of light.

Light Microscopes and Cell Stains The type of microscope that you are probably most familiar with is a compound light microscope. A light microscope lets light pass through an object. Its two lenses focus the light to form an image. Some of the light scatters as it passes through the lenses. Because of this scattering of light, light microscopes can make clear images of objects only to a magnification of about 1000 times. Most living cells are nearly clear like glass. This makes it very hard to see what is inside them. So scientists use stains or dyes to show the different structures inside cells.

Electron Microscopes To study very small things, such as a DNA molecule, scientists use electron microscopes. Instead of focusing light, electron microscopes focus beams of electrons using magnetic fields. Electron microscopes can show things that are 1 billionth of a meter in size.

There are two major types of electron microscopes: transmission and scanning. Transmission electron microscopes send beams of electrons through thin slices of cells and tissues. In scanning electron microscopes, a beam of electrons is scanned over the surface of an object to give a three-dimensional image of the object's surface. Electron microscopes can be used only to look at cells and tissues that are no longer living. They make pictures that are black and white. Scientists often use computers to add "false colors" that make some structures stand out. False colors can be added to images from electron microscopes and light microscopes.

Key Question How do microscopes work?
Most microscopes use lenses to magnify the image of an object by focusing light or electrons.

BUILD Vocabulary

cell
the basic unit of all forms of life

cell theory
a fundamental concept of biology that states that all living things are composed of cells, that cells are the basic units of structure and function in living things, and that new cells are produced from existing cells

MULTIPLE MEANINGS
The word *theory* means different things in everyday speech and in science. In everyday speech it means "a guess." In science, however, a theory is a well-tested explanation that brings together many observations and hypotheses.

Images From Microscopes Images of yeast cells from a light microscope (LM 500×), transmission electron microscope (TEM 4375×), and a scanning electron microscope (SEM 3750×)

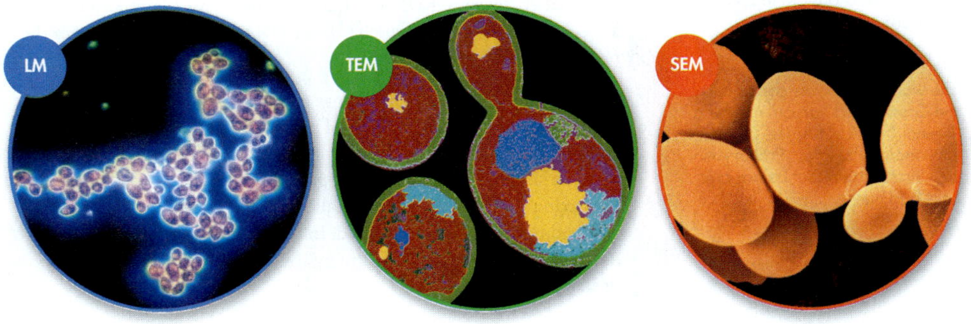

ELL Focus on ELL: Build Background

INTERMEDIATE AND ADVANCED SPEAKERS Write the words *focus* and *scatter* on the board. Explain that both words are used commonly in physical science when discussing light. Have students read this page to find the words. Using three wavy lines to represent a beam of light, draw a diagram showing how light behaves when it is focused and when it is scattered. Ask students to use different words to describe how objects appear in scattered light *(blurry, fuzzy, dull)* versus focused light *(clear, sharp, defined).*

Write the word *magnify* on the board. Ask students to find the word in the text and figure out what the word means. Explain that the lenses in eyeglasses and microscopes are basically the same—they both magnify objects by focusing light.

During Reading

Active Reading

KEY QUESTION Invite a volunteer to read aloud the Key Question and answer on the previous page.

Ask How many parts make up the cell theory? *(three)*

Point out the three bulleted items in the text that list the parts of the cell theory.

COMPARE AND CONTRAST Have students reread the page. Help them draw a Venn diagram to compare and contrast light microscopes and electron microscopes.

Ask What characteristic do light microscopes and electron microscopes share? *(They both magnify objects.)*

Ask What is different about how light microscopes and electron microscopes magnify things? *(Light microscopes focus light; electron microscopes focus a beam of electrons.)*

Ask What can you look at with a light microscope that you cannot look at with an electron microscope? *(living organisms)*

Ask What is different about how the cells are colored depending on the type of microscope used? *(light microscope—colored with stain before they are viewed; electron microscope—computer adds color after image is taken)*

Build Vocabulary

MULTIPLE MEANINGS Give students several examples of theories, both scientific and common. Have students decide whether each statement is a common theory or a scientific theory.

Ask On this page, can you find other words with more than one meaning? *(Sample answers: cell, work, light, pass, form, beams, fields, tissues, stand)*

KEY QUESTION Invite a volunteer to read aloud the question and answer.

Ask What part of a light microscope focuses the light? *(the lenses)*

Speed Bump

Ask What kind of microscope would be best for comparing two colors of human hair? *(a light microscope)*

Ask What kind of electron microscope would be best for examining tiny openings on the surface of a cell? *(scanning)* The DNA inside the nucleus of a cell? *(transmission)*

Inquiry Into Scientific Thinking

MAKE CONNECTIONS Remind students that their skeletons are what support their bodies. Explain that because plants do not have skeletons, they need something else to provide them with support.

Ask What did you notice about the plant cell you examined that might help support the plant? *(The plant cell had a thick wall around the outside.)*

MAKE INFERENCES Ask students if they noticed a tiny dot in some of the cells they examined.

Ask What do you think the dot is? *(a nucleus)* Which cell did not have a dot? *(the bacterial cell)*

Ask What features helped you to identify a plant cell? *(thick walls, green color, square shape)*

Lead a Discussion

Talk about the four kinds of cells students observed.

Ask What do you think the cells of your own bodies look like? Are they green? Do they each have a nucleus? Are they surrounded by a thick wall? Have students justify their responses. *(Guide students' responses to conclude that their cells are not green and have nuclei but no cell walls.)*

Build Vocabulary

WORD ORIGINS Point out the word *protists* on the following page. Explain that *protist* comes from a Greek word *(protistos)* that means "very first."

Say If *prokaryote* means "before nucleus," and *protist* means "very first," you might infer that protists are prokaryotic cells. But protists are thought to be the very first eukaryotes.

Ask What would you expect the very first eukaryotes to be like? *(unicellular with a nucleus)*

Speed Bump

Ask What is the function of DNA? *(It carries the genetic information of the cell.)*

Ask Where is the DNA in a eukaryotic cell? *(inside the nucleus)*

INQUIRY into Scientific Thinking

What Is a Cell?

All living things are made up of cells. Cells are not usually visible without a microscope. In this activity, you will use a light microscope to observe several cells.

1. Obtain a slide of a plant leaf or a slice of a plant stem.
   - Look at the slide using a microscope.
   - Sketch one or more cells.
   - Describe the shape of the cells and how the different parts of a cell look.
2. Repeat step 1 with slides of nerve cells, bacteria, and paramecia.
3. Compare the cells.
   - List what they have in common.
   - List some of their differences.

Analyze and Conclude

Classify Classify the cells you studied into two or more groups. Explain what characteristics you used to put each cell in its group.

In Your Workbook Go to your workbook for more help with this activity.

BUILD Vocabulary

nucleus
a structure that contains the cell's genetic material in the form of DNA

eukaryote
an organism whose cells contain a nucleus

prokaryote
a unicellular organism that lacks a nucleus

WORD ORIGINS
The word *prokaryote* comes from the Greek word *karyon*, meaning "kernel," or nucleus. The prefix *pro-* means "before." Prokaryotic cells first came about before any cells had nuclei.

Prokaryotes and Eukaryotes

Cells come in many different shapes and sizes. Even though they are different, at some point in their lives all cells have DNA. DNA is the molecule that carries genetic information. Also, all cells are surrounded by a thin, flexible barrier called a cell membrane.

There are two main kinds of cells. One kind has a nucleus, and the other does not. The **nucleus** (plural: nuclei) holds the cell's DNA. There are two membranes around the nucleus. The nucleus also controls many of the cell's activities. **Eukaryotes** (yoo KAR ee ohts) are cells that enclose their DNA in nuclei. **Prokaryotes** (pro KAR ee ohts) are cells that do not have nuclei. Their DNA is not separated from the rest of the cell.

Prokaryotes Prokaryotic cells are generally smaller and simpler than eukaryotic cells. The DNA of prokaryotic cells is not held inside a nucleus. Instead, the DNA floats freely in the cell. Even though they are simpler than eukaryotes, prokaryotes do all the activities that living things must do to be called "alive." They grow, reproduce, respond to the environment, and, in some cases, glide along surfaces or swim through liquids. The living things that we call bacteria are prokaryotes.

Materials

compound microscope; slides of plant leaves or stems, nerve cells, bacteria, and paramecia

Analyze and Conclude

Sample answer: Cells that were square in shape and green in color were plant cells. Cells that were very small and didn't have a large dot inside them were bacterial cells. I classified all other cells as animal cells.

Eukaryotes Eukaryotic cells are generally larger and more complex than prokaryotic cells. Most eukaryotic cells have dozens of structures and membranes inside them. Many of these structures have special jobs. In eukaryotic cells, the nucleus separates the DNA from the rest of the cell. Eukaryotes can be very different from one another. Some, like the ones commonly called "protists," live as single cells. Others make up large organisms with many cells—plants, animals, and fungi.

Key Question How are prokaryotic and eukaryotic cells different? **Prokaryotic cells do not have a nucleus. In eukaryotic cells, the nucleus separates the DNA from the rest of the cell.**

Cell Types Eukaryotic cells (including plant and animal cells) are usually more complex than prokaryotic cells.

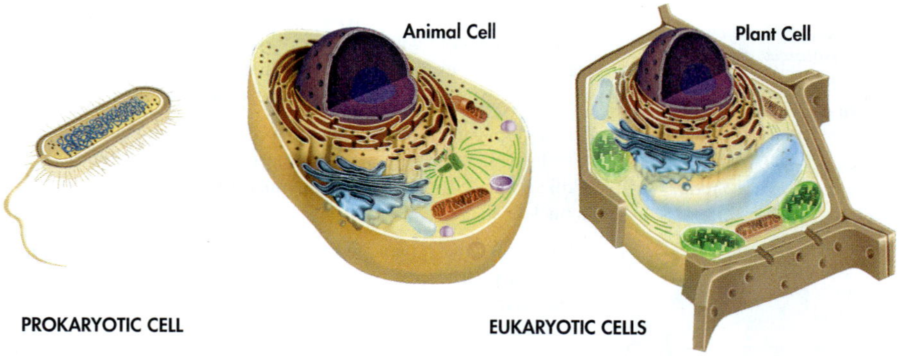

CHECK Understanding

Apply Vocabulary

Use the highlighted words from the lesson to complete each sentence correctly.

1. ______ are the basic units of all living things.
2. According to the _____, all living things are made of cells.
3. Cells whose DNA is held in a nucleus are _____.

Critical Thinking

4. **Apply Concepts** Bacteria cells are living in a test tube. What does the cell theory tell you about where these cells came from?
5. **Infer** A picture from a microscope is black and white and shows only the surface features of the cell. What type of microscope most likely made this picture?
6. **Compare and Contrast** How are prokaryotes and eukaryotes alike? What is the main difference between them?
7. **Write to Learn** Answer the first clue of the mystery.

MYSTERY CLUE

At the hospital, a nurse took some of Michelle's blood and looked at it under a microscope. The red blood cells looked swollen. What kind of microscope was most likely used? (**Hint:** See p. 161.)

BIOLOGY.com Search Lesson 7.1 GO • Lesson Assessment

Check Understanding

1. Cells
2. cell theory
3. eukaryotes
4. The cell theory says that they must have come from an existing cell.
5. The picture is most likely from a scanning electron microscope.
6. Both types of organisms are made up of cells that are enclosed in a cell membrane and have DNA. The main difference between them is that eukaryotic cells have nuclei, while prokaryotic cells do not.
7. A light microscope could magnify a red blood cell enough to tell that it was swollen.

Active Reading

USE VISUALS Review the diagram Cell Types. Explain that each cell in the diagram has been cut in half, or bisected. The nuclei have not been completely bisected, so some organelles stick out above the surface plane in three dimensions.

Ask Why do you think the artist drew the nucleus this way? *(Students may suggest that it was to emphasize that the cell is three-dimensional and show what the outer edges of the nuclei look like.)*

Ask What can you say about the sizes of bacteria, plant, and animal cells? *(Bacterial cells are much smaller than either plant or animal cells.)*

Ask How else do the cells differ? *(Sample answers: The plant cell has large green organelles (chloroplasts) that the other two cells do not have; the plant cell has a very large blue structure (vacuole); the eukaryotic cells are much larger than the prokaryotic cell)*

Point out the cell wall on the plant cell. Relate it to what students observed in the Inquiry Into Scientific Thinking lab.

KEY QUESTION Invite a student to read the question and answer aloud.

Ask Name two ways in which prokaryotic and eukaryotic cells are similar. *(They both contain DNA and are surrounded by a cell membrane.)*

Post Reading

ASSESSMENT TIPS Have students return to the Cell Types diagram to review the meanings of the vocabulary words.

For Question 5, remind students that *infer* means the answer is not stated directly in the book. Also, be sure students understand the term *surface features* refers to how an object appears on the outside.

For Question 6, tell students to make sure they answer both questions.

MYSTERY CLUE Have students reread the section Exploring the Cell. Refer them again to the three photomicrographs of yeast cells. Ask students how they would expect red blood cells to look when viewed under each kind of microscope.

Wrap-Up Activity

CREATE A TIMELINE Have pairs of students make a timeline showing the discovery of cells, the invention of early microscopes, and some of the people and events that led to the formulation of the cell theory. Students will have to research dates for some of these events.

LESSON 7.2

Pre-Reading

Objectives

7.2.1 Describe the structure and function of the cell nucleus.

7.2.2 Describe the role of vacuoles, lysosomes, and the cytoskeleton.

7.2.3 Identify the role of ribosomes, endoplasmic reticulum, and Golgi apparatus in making proteins.

7.2.4 Describe the function of the chloroplasts and mitochondria in the cell.

7.2.5 Describe the function of the cell membrane.

Preview the Pages

1. Preview the lesson's vocabulary words by having students find the words in the text and read each one aloud. *Mitochondria, endoplasmic reticulum, vacuoles, and lysosomes* may be especially difficult to pronounce.
2. Say aloud the names of the organelles in each micrograph. Show where the names occur in the text.
3. Walk students through the Making Proteins diagram. Compare the cell structures with cell structures shown in the micrographs and diagrams.

Build Understanding

VENN DIAGRAM Have students review the characteristics of prokaryotes and eukaryotes from Lesson 7.1 before they make their Venn diagrams. Invite volunteers to draw their diagrams on the board. Correct and redirect students as necessary.

NATIONAL SCIENCE EDUCATION STANDARDS

UNIFYING CONCEPTS AND PROCESSES

I, V

CONTENT

C.1.a, C.1.b, C.1.c, C.1.d, C.1.e, C.2.a

For the corresponding lesson in *Miller & Levine Biology,* see pages 196–207.

7.2 Cell Structure

Key Questions

What is the role of the cell nucleus?

What are the functions of vacuoles, lysosomes, and the cytoskeleton?

What organelles help make and transport proteins?

What are the functions of chloroplasts and mitochondria?

What is the function of the cell membrane?

BUILD Understanding

Venn Diagram Create a Venn diagram that shows how prokaryotes and eukaryotes are the same and different.

In Your Workbook Go to your workbook to learn more about making a Venn diagram. Complete the Venn diagram for Lesson 7.2.

BUILD Connections

THE CELL AS A FACTORY

Each different machine and worker helps make a factory run. In much the same way, the different parts in a cell help the cell to survive.

In Your Workbook Go to your workbook for more help understanding how a cell is like a factory.

Cell Organization

Like a factory, a eukaryotic cell is a busy place with many different parts. At first glance, things may seem chaotic and disorganized. But if you look closely, you can classify those many parts into simpler categories. For example, it's easy to divide each eukaryotic cell into two major parts: the nucleus and the cytoplasm.

The **cytoplasm** is the portion of the cell outside the nucleus. The nucleus and cytoplasm work together to keep the cell alive. Prokaryotic cells have cytoplasm, too, even though they do not have a nucleus.

Many structures in plant and animal cells act like specialized organs. Thus, these structures are known as **organelles** or "little organs." Understanding what each organelle does for the cell can help you understand the functions of cells.

The Cell as a Factory In some ways, a eukaryotic cell is much like a factory. The different organelles of the cell are like the specialized machines, trained workers, and assembly lines of the factory. And, like workers at factories, cells follow instructions and make products. In this chapter, you will read about many ways that the parts of a cell are like the parts of a factory.

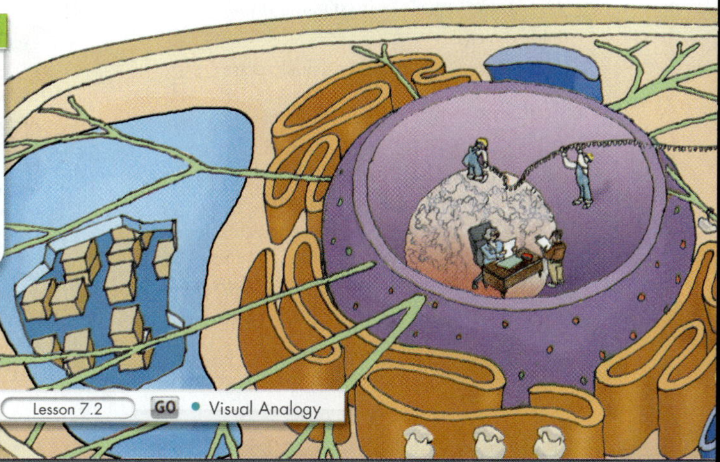

164 BIOLOGY.com Search Lesson 7.2 GO • Visual Analogy

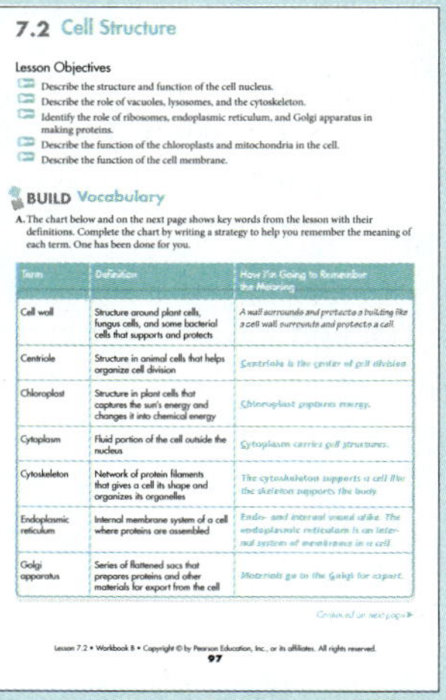

7.2 Cell Structure

Lesson Objectives

- Describe the structure and function of the cell nucleus.
- Describe the role of vacuoles, lysosomes, and the cytoskeleton.
- Identify the role of ribosomes, endoplasmic reticulum, and Golgi apparatus in making proteins.
- Describe the function of the chloroplasts and mitochondria in the cell.
- Describe the function of the cell membrane.

BUILD Vocabulary

A. The chart below and on the next page shows key words from the lesson with their definitions. Complete the chart by writing a strategy to help you remember the meaning of each term. One has been done for you.

Term	Definition	How I'm Going to Remember the Meaning
Cell wall	Structure around plant cells, fungus cells, and some bacterial cells that supports and protects	*A wall surrounds and protects a building like a cell wall surrounds and protects a cell.*
Centriole	Structure in animal cells that helps organize cell division	*Centriole is the center of cell division.*
Chloroplast	Structure in plant cells that captures the sun's energy and changes it into chemical energy	*Chloroplast captures energy.*
Cytoplasm	Fluid portion of the cell outside the nucleus	*Cytoplasm carries cell structures.*
Cytoskeleton	Network of protein filaments that gives a cell its shape and organizes its organelles	*The cytoskeleton supports a cell like the skeleton supports the body.*
Endoplasmic reticulum	Internal membrane system of a cell where proteins are assembled	*Endo- and internal sound alike. The endoplasmic reticulum is an internal system of membranes in a cell.*
Golgi apparatus	Series of flattened sacs that prepares proteins and other materials for export from the cell	*Materials go to the Golgi for export.*

Continued on next page ▶

Lesson 7.2 • Workbook B • Copyright © by Pearson Education, Inc., or its affiliates. All rights reserved.
97

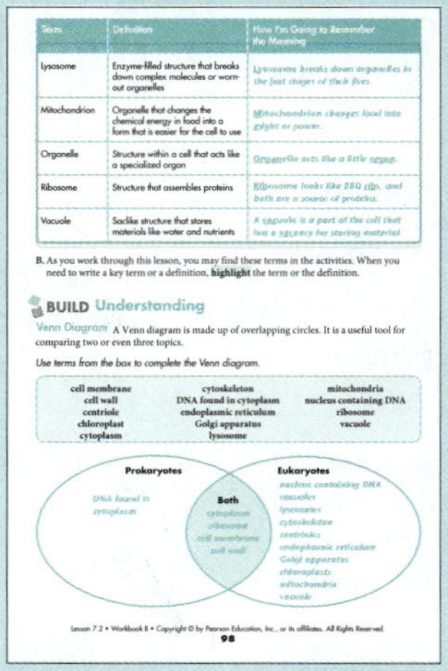

Term	Definition	How I'm Going to Remember the Meaning
Lysosome	Enzyme-filled structure that breaks down complex molecules or worn-out organelles	*Lysosome breaks down organelles in the last stages of their lives.*
Mitochondrion	Organelle that changes the chemical energy in food into a form that is easier for the cell to use	*Mitochondrion changes food into light or power.*
Organelle	Structure within a cell that acts like a specialized organ	*Organelle acts like a little organ.*
Ribosome	Structure that assembles proteins	*Ribosome looks like BBQ ribs, and both are a source of proteins.*
Vacuole	Saclike structure that stores materials like water and nutrients	*A vacuole is a part of the cell that has a vacancy for storing material.*

B. As you work through this lesson, you may find these terms in the activities. When you need to write a key term or a definition, **highlight** the term or the definition.

BUILD Understanding

Venn Diagram A Venn diagram is made up of overlapping circles. It is a useful tool for comparing two or even three topics.

Use terms from the box to complete the Venn diagram.

cell membrane, cell wall, centriole, chloroplast, cytoplasm, cytoskeleton, DNA found in cytoplasm, endoplasmic reticulum, Golgi apparatus, lysosome, mitochondria, nucleus containing DNA, ribosome, vacuole

Lesson 7.2 • Workbook B • Copyright © by Pearson Education, Inc., or its affiliates. All Rights Reserved.
98

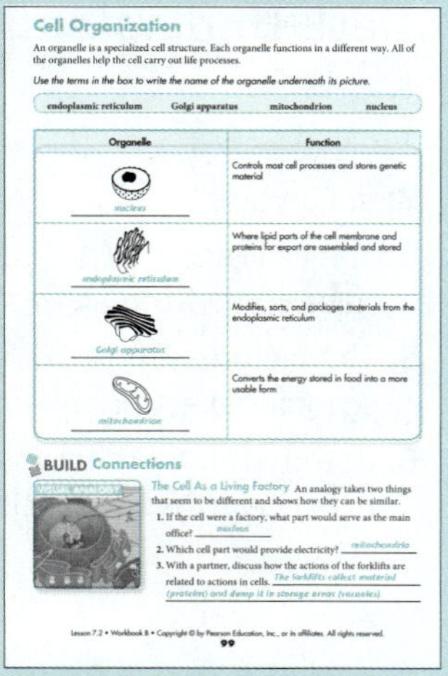

Cell Organization

An organelle is a specialized cell structure. Each organelle functions in a different way. All of the organelles help the cell carry out life processes.

Use the terms in the box to write the name of the organelle underneath its picture.

endoplasmic reticulum, Golgi apparatus, mitochondrion, nucleus

Organelle	Function
nucleus	Controls most cell processes and stores genetic material
endoplasmic reticulum	Where lipid parts of the cell membrane and proteins for export are assembled and stored
Golgi apparatus	Modifies, sorts, and packages materials from the endoplasmic reticulum
mitochondrion	Converts the energy stored in food into a more usable form

BUILD Connections

The Cell As a Living Factory An analogy takes two things that seem to be different and shows how they can be similar.

1. If the cell were a factory, what part would serve as the main office? *nucleus*
2. Which cell part would provide electricity? *mitochondria*
3. With a partner, discuss how the actions of the forklifts are related to actions in cells. *The forklifts collect material (proteins) and dump it in storage areas (vacuoles).*

Lesson 7.2 • Workbook B • Copyright © by Pearson Education, Inc., or its affiliates. All rights reserved.
99

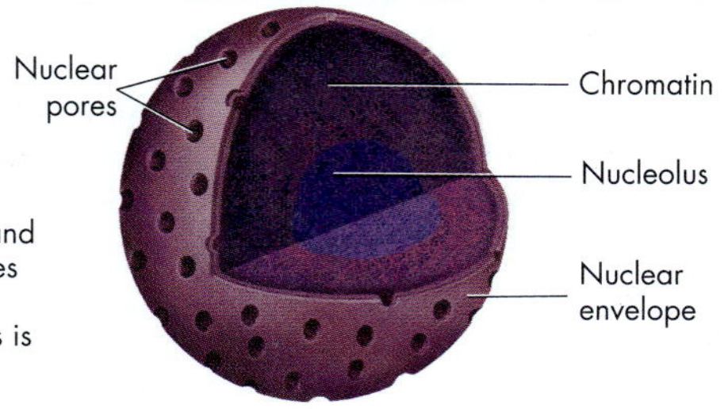

The Nucleus
The nucleus holds DNA and controls most of what goes on in the cell. The small, dense area in the nucleus is the nucleolus.

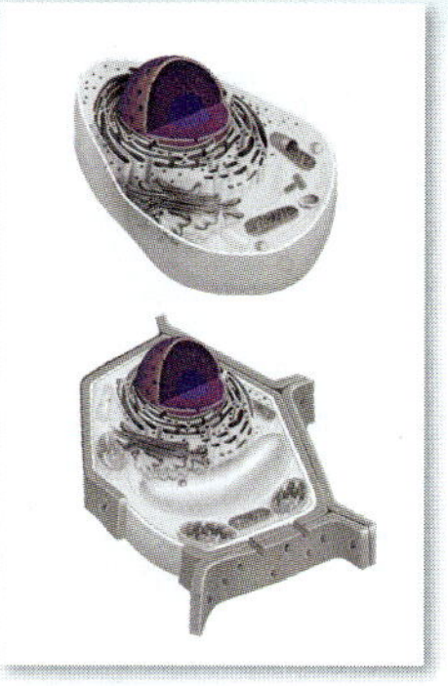

The Nucleus In the same way that the main office controls a large factory, the nucleus is the control center of the cell. The nucleus holds nearly all of the cell's DNA. DNA contains the coded directions for making proteins and other important molecules. Prokaryotic cells do not have a nucleus, but they have DNA that codes the same kinds of directions.

The nucleus is surrounded by a nuclear envelope made up of two membranes. The nuclear envelope is dotted with thousands of holes. These holes allow material to move into and out of the nucleus. Messages, directions, and blueprints move in and out of a factory's main office. In a similar way, proteins, RNA, and other molecules move through the holes in the nuclear envelope to and from the rest of the cell.

Most nuclei also have a small dense spot known as the nucleolus (noo KLEE uh lus). The nucleolus is where ribosomes are first put together. Ribosomes are used by the cell to build proteins.

Key Question What is the role of the cell nucleus?
The nucleus holds nearly all of the cell's DNA, which has the coded directions for making proteins and other important molecules.

BUILD Vocabulary

cytoplasm
the fluid portion of the cell outside the nucleus

organelle
a specialized structure that performs important cellular functions within a eukaryotic cell

WORD ORIGINS
The prefix *cyto-* comes from the Greek word *kytos*, which refers to "cell." The Greek word *plasma* means "something molded."

ELL Focus on ELL: Access Content

ALL SPEAKERS To help students visualize the nucleus and better understand the cutaway drawings in the lesson, make a model of the nucleus. Cut away a quarter portion of a tennis ball or large hollow rubber ball. Shred a cotton ball into tiny pieces. Place a smaller solid ball and the tiny pieces of cotton inside. Explain that the solid ball represents the nucleolus and the cotton represents DNA. Have students compare the structure of the model with the structure of the nucleus in the diagram on this page.

During Reading

Build Connections

VISUAL ANALOGY Tell students that making analogies can help them learn and remember new information. Emphasize the visual analogy of the cell as a factory.

Ask If a cell were a factory, where would the main office most likely be? *(in the nucleus)* Why? *(because the nucleus is the control center of the cell, as the main office controls the activities in the factory)* How does the nucleus direct the cell's activities? *(Thin strands of chromatin in the nucleus have genetic material, which has the instructions for directing the cell's functions.)*

Active Reading

USE VISUALS Point out the small cutaway cell figures shown throughout the lesson. Have students use these cutaway drawings to focus on where cell parts are located in plant and animal cells.

Ask Look at the diagram of the nucleus. What kinds of cells contain a nucleus? *(eukaryotic cells)*

KEY QUESTION Invite a volunteer to read aloud the question and answer.

Ask Why does the nuclear envelope have thousands of holes in it? *(proteins, RNA, and other molecules move in and out of the nucleus through the holes.)*

Build Vocabulary

WORD ORIGINS Have students look through the lesson and find another cell part that contains the prefix *cyto-*. *(cytoskeleton)*

Ask What does this word mean? *(cell skeleton, the structure that gives the cell its shape)*

Speed Bump

Ask How many membranes make up the nuclear envelope? *(two)*

Ask Why is it called an "envelope"? *(It surrounds, or covers, the nucleus.)*

Ask In the analogy of a cell and a factory, what role does the nucleus play? *(the main office)*

During Reading continued

Active Reading

SUMMARIZE Have students summarize each paragraph in one or two sentences in their notebooks.

Say Vacuoles and vesicles store things.

Ask How do vacuoles and vesicles compare in size? *(Vacuoles are larger.)*

Ask What is the function of a contractile vacuole in a paramecium? *(It pumps extra water out of the cell.)*

DRAW CONCLUSIONS Read the first paragraph of Vacuoles and Vesicles. Pause after the last sentence.

Ask What do you think has happened inside the plant when it wilts because it doesn't have enough water? *(The vacuoles have lost water and, therefore, pressure.)*

USE CONTEXT CLUES Have a volunteer read the two sentences describing the contractile vacuole in a paramecium.

Say The word *contractile* comes from the verb *contract*, which means to "draw together or squeeze."

Ask Why does the vacuole contract? *(to squeeze excess water out of the cell)*

MAKE CONNECTIONS Remind students about the analogy of a cell as a factory.

Say Eukaryotic cells are much like factories.

Ask How is a custodian's work in a factory similar to the work of a lysosome in a cell? *(They both clean.)*

Ask What part of a factory serves the same function as a vacuole? *(a storage room)*

Speed Bump

Ask What materials do vacuoles store? *(water, salts, proteins, and sugars)*

Ask What is the function of lysosomes? *(Lysosomes break down materials into smaller molecules and get rid of "junk" from the cell.)*

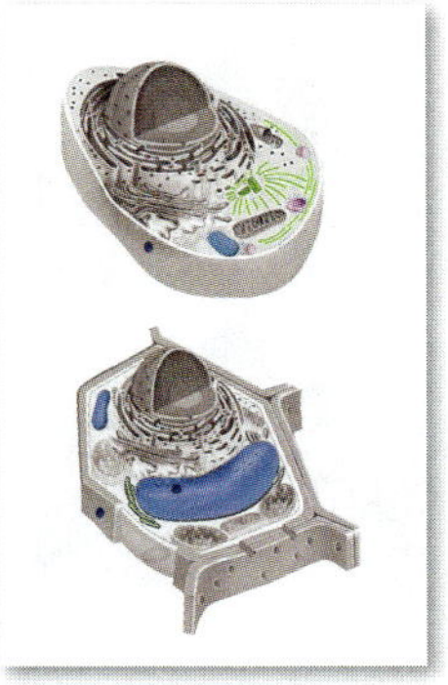

Organelles That Store, Clean Up, and Support

Many of the organelles outside the nucleus of a eukaryotic cell have specific functions. They include structures called vacuoles, lysosomes, and the cytoskeleton. These organelles are the cell's storage space, cleanup crew, and support structure.

Vacuoles and Vesicles Every factory needs a place to store things. So does every cell. Many cells have large, membrane sacs called **vacuoles** (VAK yoo ohlz). Vacuoles store materials like water, salts, proteins, and sugars. In many plant cells, there is a single, large central vacuole filled with liquid. The pressure of this large vacuole makes the cell firm and lets the plant hold up heavy parts, such as leaves and flowers.

Vacuoles are also found in some single-celled organisms and in some animals. A paramecium has an organelle called a contractile vacuole. By contracting over and over, this vacuole pumps extra water out of the cell.

Nearly all eukaryotic cells have much smaller membrane sacs called vesicles. Vesicles store and move materials between organelles as well as to and from the outside of the cell.

Lysosomes Even the neatest, cleanest factory needs a cleanup crew. That's where lysosomes (LY suh sohmz) come in. **Lysosomes** are small organelles filled with enzymes that break down lipids, sugars, and proteins into smaller molecules. These smaller molecules can be used by the rest of the cell. Lysosomes also help break down organelles that are no longer useful. They carry out the important job of removing "junk" that might otherwise pile up in the cell. Many human diseases are linked to lysosomes that stop working correctly. Animal cells have lysosomes, and a few special kinds of plant cells also have them.

Vacuoles The central vacuole of plant cells stores salts, proteins, and sugars. A paramecium's contractile vacuole keeps the water levels just right by pumping water out.

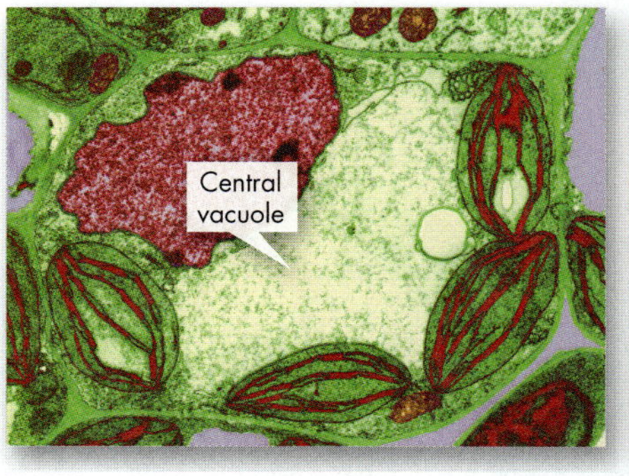

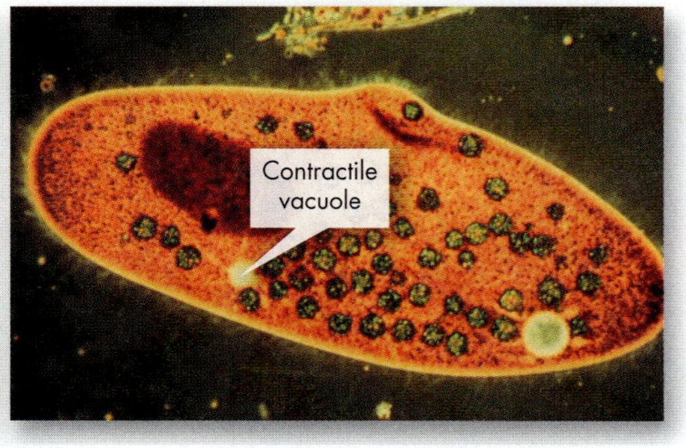

Hands-On Learning

DIAGRAM ANIMAL AND PLANT CELLS When you begin this lesson, give each student an outline of a plant cell and an animal cell. As students learn about each cell part, have them draw the cell part on the outline of the appropriate cell or cells. The drawings should show the cell parts in correct locations within the cell. Check students' completed diagrams, and then encourage them to keep the diagrams for review and as a study guide.

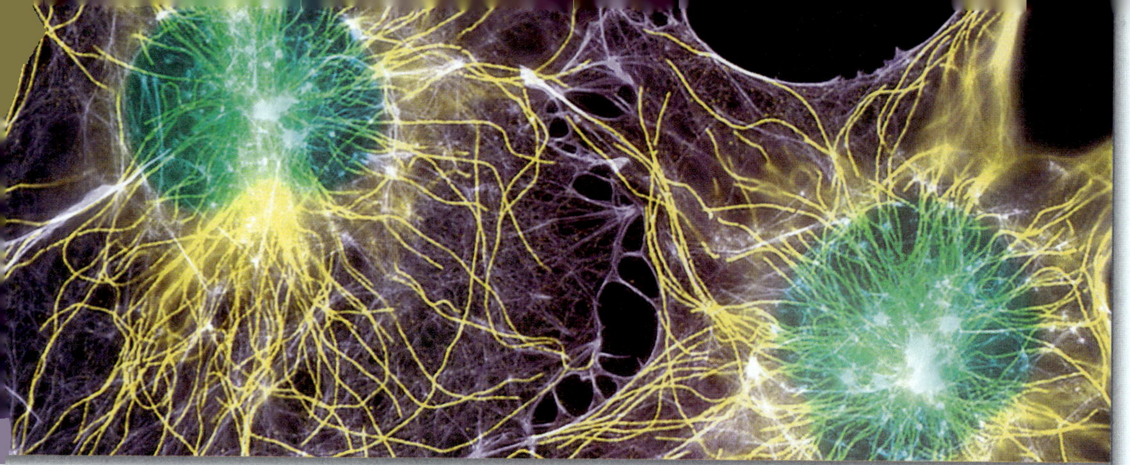

LM 1175×

Cytoskeleton The cytoskeleton supports and gives shape to the cell. Dyes have been used to color different parts of the cell in this picture. Microfilaments are pale purple, microtubules are yellow, and nuclei are green.

The Cytoskeleton A factory building has strong beams and columns that hold up its walls and roof. Eukaryotic cells get their shape from a web of proteins known as the **cytoskeleton.** Some parts of the cytoskeleton help move materials between different parts of the cell. These parts are much like conveyor belts in a factory. Some cells use parts of the cytoskeleton to help them move. The cytoskeleton includes two types of protein filaments called microfilaments and microtubules.

▶ ***Microfilaments*** Microfilaments are threadlike structures made up of a protein called actin. In some cells they form an internal flexible web. This web supports the cell. Microfilaments also help some cells move. Microfilaments are built, taken apart, and then built again in different parts of the cell, allowing amoebas and other cells to crawl along surfaces.

▶ ***Microtubules*** Microtubules are like thin, hollow pipes. They are made up of proteins known as tubulins. Microtubules help the cell keep its shape. They are also found in hairlike organelles called cilia and flagella that help some cells swim. Cilia and flagella whip back and forth, pushing cells along or moving fluids along the surfaces of cells.

Microtubules are also important in cell division. They form a structure known as the mitotic spindle. This structure helps to separate the different sets of DNA that each daughter cell will get. In animal cells, organelles called centrioles (SEN tree ohlz) are also formed from tubulins. Centrioles are found near the nucleus and help organize cell division. Plant cells do not have centrioles.

Key Question What are the functions of vacuoles, lysosomes, and the cytoskeleton?
Vacuoles store materials. Lysosomes break down large molecules and old organelles. The cytoskeleton helps the cell keep its shape and move.

BUILD Vocabulary

vacuole
the cell organelle that stores materials such as water, salts, proteins, and carbohydrates

lysosome
the cell organelle that breaks down lipids, carbohydrates, and proteins into small molecules that can be used by the rest of the cell

cytoskeleton
the network of protein filaments in a eukaryotic cell that gives the cell its shape and internal organization and is involved in movement

WORD ORIGINS
The prefix *lys-* comes from the Greek word *lysis*, which refers to "loosening" or "dissolving." The suffix *-some* comes from the Greek word *soma*, which means "body." A lysosome is a small body in the cell that dissolves, or breaks up, wastes in the cell.

Focus on Academic Language

- spindle

Write the word on the board.

Say The rods on the back of some chairs are called spindles.

The spindles on a chair separate parts of the chair. Likewise, a mitotic spindle separates sets of DNA.

Active Reading

READ, COVER, REMEMBER, RETELL Have students read silently the two paragraphs about microtubules. After they have read, have them cover the text with their notebooks and think about what they read. Then have students make a list of three functions of microtubules.

Say Microtubules are important parts of a cell.

Ask What are microtubules made of? *(proteins called tubulins)*

Ask Name two cell life processes that depend on microtubules. *(cell division and movement)*

USE CONTEXT CLUES Have students describe in their own words what the phrase *internal flexible web* means as it relates to microfilaments. *(Sample answer: a mass of filaments that bend and change shape, found inside a cell)* Point out that in the Cytoskeleton picture the microfilaments are purple as indicated in the caption.

KEY QUESTION Invite a volunteer to read aloud the question and answer.

Ask How do microfilaments help an amoeba move? *(They build up and break down at different points inside the cell. This changes the cell's shape so that the amoeba "crawls" along a surface.)*

Build Vocabulary

WORD ORIGINS Tell students that *vacuole* comes from a Latin word *(vacuus)* that means "hollow" or "empty," like a storage container before it is filled. Have students think of a common word that sounds similar to *vacuole* and comes from the same Latin root. *(vacuum)*

Speed Bump

Ask What are microfilaments and microtubules? *(the protein structures that make up the cytoskeleton)*

Ask What do microfilaments look like? *(threads)* What do microtubules look like? *(hollow pipes, tubes)*

Ask What are some organelles and structures that are made out of tubulin? *(cilia, flagella, centrioles, mitotic spindles)*

During Reading continued

Active Reading

FIND THE MAIN IDEA Read the paragraph Ribosomes aloud as students follow along.

Ask What does this tell you about ribosomes? *(Sample answers: what ribosomes are, where they are located, how they function.)*

Turn the answers into questions and write them on the board. Have students write answers to the *what, where,* and *how* of this paragraph in their notebooks.

Ask What important molecules are made on ribosomes? *(proteins)*

USE ANALOGIES Invite volunteers to take turns reading the numbered captions on the Making Proteins diagram. After each step is read, stop the read-aloud, refer to the diagram, and emphasize the factory analogy. For example, after the first step is read:

Ask If this were a factory, what are the proteins? *(the product being made or assembled)* Who are the workers? *(the ribosomes)* Who supervises the ribosomes? *(DNA)*

USE VISUALS Tell students to look at the five numbered steps in the Making Proteins diagram. Have them write the numbers 1–5 on sticky notes. Tell them to place the appropriate sticky note where each step is discussed in the text.

Speed Bump

Ask Why is the rough ER bumpy? *(It has ribosomes on its surface.)*

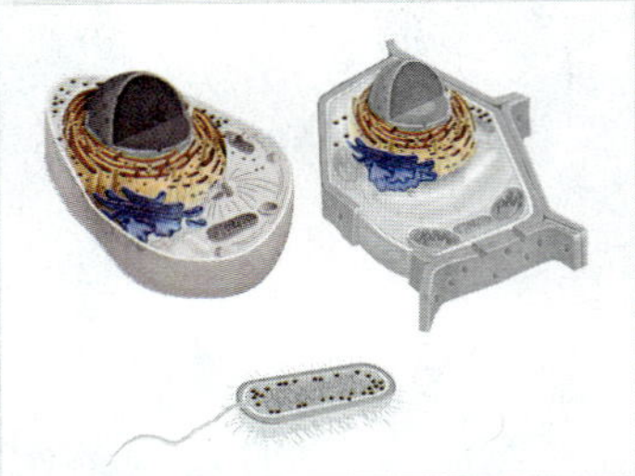

Organelles That Build Proteins

Living things are always working, building new molecules all the time, especially proteins. Proteins help carry out chemical reactions and make up important parts of the cell. Because proteins are so important, a big part of the cell is used for making them and moving them around.

Ribosomes Proteins are made on ribosomes. **Ribosomes** are small units of RNA and protein found in the cytoplasm of all cells. Ribosomes make proteins by following coded directions that come from DNA. Each ribosome is like a factory worker that makes proteins on orders that come from its DNA "boss." Many ribosomes are free in the cytoplasm. Others are attached to the endoplasmic reticulum.

Endoplasmic Reticulum The **endoplasmic reticulum** (en doh PLAZ mik rih TIK yuh lum), or ER, is a membrane system where lipid parts of the cell membrane are put together. Proteins and other materials that are shipped from the cell are also put together on the ER. There are two types of ER: rough and smooth.

▶ ***Rough ER*** Proteins are made on the rough endoplasmic reticulum, or rough ER. It is called "rough" because the ribosomes on its surface make it bumpy. Newly made proteins leave these ribosomes and enter the rough ER. Enzymes inside the rough ER make changes to some of the proteins. Some proteins made on the rough ER will leave the cell. Others are membrane proteins and proteins that go to lysosomes and other specialized locations inside the cell.

BUILD Connections

MAKING PROTEINS

Together, ribosomes, the endoplasmic reticulum, and the Golgi apparatus make, package, and ship proteins.

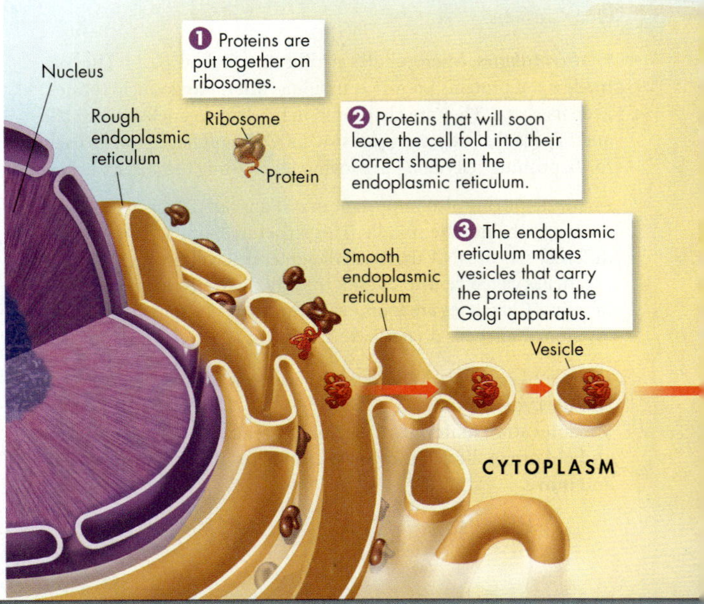

168 Chapter 7 • Lesson 2

Students Say

"We made a cell model out of candy! We used a jawbreaker for the nucleus, jelly beans for mitochondria, toothpicks for the cytoskeleton, Skittles for the lysosomes, Gummi bears for the ribosomes, and licorice for the ER. We put them all in a plastic sandwich bag, which was the cell membrane. We had to make a key telling what candy did what function in the cell."

▶ ***Smooth ER*** The other part of the ER is known as smooth endoplasmic reticulum, or smooth ER. It is called "smooth" because there are no ribosomes on its surface. In many cells, the smooth ER has groups of enzymes that do special tasks. They make membrane lipids and get rid of toxins, such as drugs. Liver cells, which play a key part in getting rid of drugs, often have large amounts of smooth ER.

Golgi Apparatus In eukaryotic cells, proteins made on the rough ER move next into an organelle called the **Golgi apparatus.** This organelle looks like a stack of flat membrane sacs. It is like the "shipping station" of the cell. Proteins have molecular "address tags" that point out where they need to go. The ER "reads" these tags and bundles the proteins into tiny vesicles. The vesicles bud from the ER and carry the proteins to the Golgi apparatus. The Golgi apparatus changes, sorts, and packages proteins and other materials from the ER. Some of them are stored in the cell, while others are sent outside the cell.

Key Question What organelles help make and transport proteins?

Proteins are put together on ribosomes in the cytoplasm and on the rough ER. The Golgi apparatus changes, sorts, and packages proteins and other materials for "shipment."

BUILD Vocabulary

ribosome
a cell organelle consisting of RNA and protein found throughout the cytoplasm in a cell; the site of protein synthesis

endoplasmic reticulum
the internal membrane system found in eukaryotic cells; the place where lipid components of the cell membrane are assembled

Golgi apparatus
an organelle in cells that modifies, sorts, and packages proteins and other materials from the endoplasmic reticulum for storage in the cell or release outside the cell

ACADEMIC WORDS
In everyday speech, the word *apparatus* means "equipment" or "machine." The Golgi apparatus is named for the man who first discovered it, an Italian doctor named Camillo Golgi who won a Nobel Prize in 1906.

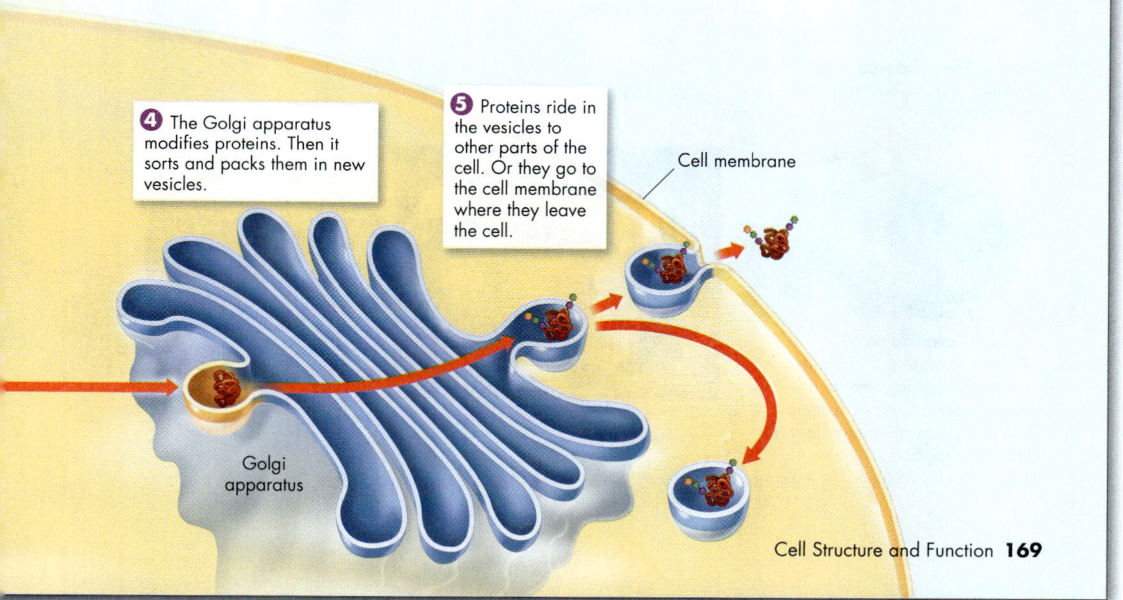

Cell Structure and Function **169**

Active Reading

DRAW A DIAGRAM Making sure each group includes one proficient reader, pair students to construct a concept map that shows the relationship between proteins, ribosomes, and the endoplasmic reticulum.

Ask Where are proteins made? *(on ribosomes)*

Ask What happens to proteins after they are shipped to the ER? *(The ER chemically changes some proteins and then ships them to where they have to go.)*

SELF-QUESTION Demonstrate how asking questions and reviewing the text for answers can help clarify complex information. Read aloud the paragraph under the heading Golgi Apparatus.

Ask What does it mean that the Golgi apparatus is like a cell's "shipping station"? In other words, what does the Golgi apparatus do that makes it similar to a shipping station in a factory?

Say I know the answer to my question is in this paragraph. Let's reread it to find out.

Have students reread the paragraph. Focus on the last two sentences.

Ask What do these sentences tell us about how the Golgi apparatus is like a shipping station? *(It changes, sorts, and packages proteins and other materials from the ER. Some of these are shipped to other places outside the cell.)*

KEY QUESTION Invite a volunteer to read aloud the question and answer.

Ask Why is it important to say rough ER instead of just ER? *(There are two types of the ER, the rough ER and the smooth ER and the two perform different tasks.)*

Build Vocabulary

ACADEMIC WORDS Ask students for examples of something that could be called an apparatus. (exercise or medical equipment such as a treadmill, weight-lifting machine, or x-ray machine).

Speed Bump

Ask What is the function of smooth ER? *(It makes membrane lipids and gets rid of toxins in the cell.)*

Ask Is it rough ER or smooth ER that sends proteins to the Golgi apparatus? *(rough ER)*

Hands-On Learning

CELLULAR ROLE PLAY Have students demonstrate the organization of a cell by taking on the roles of its different parts. One student can represent the nucleus while other students assume the roles of the other organelles such as the ribosomes, rough and smooth ER, vesicles, and so on. Give students time to research and practice their roles. Suggest they use props to represent materials such as DNA, proteins, water, salts, and sugars. Then invite the "actors" to role-play the jobs of the organelles in the cell. Suggest one group member act as narrator to describe the actions and functions of the different organelles.

LESSON 7.2

During Reading continued

Active Reading

DRAW CONCLUSIONS Read aloud the text on the page and have students follow along. When you come to important information, change your voice to emphasize its importance.

Ask How do plants get food? *(They make their own food through photosynthesis.)*

Ask How do animals get energy if they have no chloroplasts for making food? *(They must eat food.)*

Ask How are chloroplasts and mitochondria related? *(Chloroplasts capture energy from the sun and change it into food that contains chemical energy. Mitochondria change the chemical energy in food into compounds the cell can use.)*

KEY QUESTION Invite a volunteer to read aloud the question and answer.

Ask Why do plant cells need mitochondria if they make their own food? *(Without mitochondria, plant cells would not be able to use the chemical energy that is stored in the food.)*

Speed Bump

Ask What kinds of cells contain mitochondria? *(nearly all eukaryotic cells)*

Ask Why are mitochondria called the "power plants" of cells? *(They convert energy into a form that cells can use.)*

Ask How do animals get energy if they have no chloroplasts for making food? *(They must eat food.)*

Ask What is the energy source that powers photosynthesis? *(the sun)*

Organelles That Capture and Release Energy

All living things need a source of energy. Factories are connected to the local power company, but how do cells get energy? Most cells are powered by food molecules that are built using energy from the sun.

Chloroplasts Plants and some other living things have chloroplasts (KLAWR uh plasts). **Chloroplasts** are like solar power plants. They take the energy from sunlight and change it into energy stored in food. This process is called photosynthesis (foh toh SIN thuh sis). Two membranes surround chloroplasts. Inside the chloroplasts are large stacks of other membranes. These hold the green pigment chlorophyll.

Mitochondria Nearly all eukaryotic cells, including plant cells, have mitochondria (myt oh KAHN dree uh; singular: mitochondrion). **Mitochondria** are the power plants of the cell. Mitochondria change the chemical energy stored in food into compounds that are easier for the cell to use. Like chloroplasts, mitochondria are surrounded by two membranes—an outer membrane and an inner membrane. The inner membrane is much larger than the outer one. It is folded up inside the organelle.

Chloroplasts and mitochondria have their own genetic information in the form of small DNA molecules.

Key Question What are the functions of chloroplasts and mitochondria?
Chloroplasts take the energy from sunlight and change it into food in a process called photosynthesis. Mitochondria change the chemical energy stored in food into compounds that are easier for the cell to use.

Cellular Powerhouses Chloroplasts and mitochondria both help get energy for the cell.

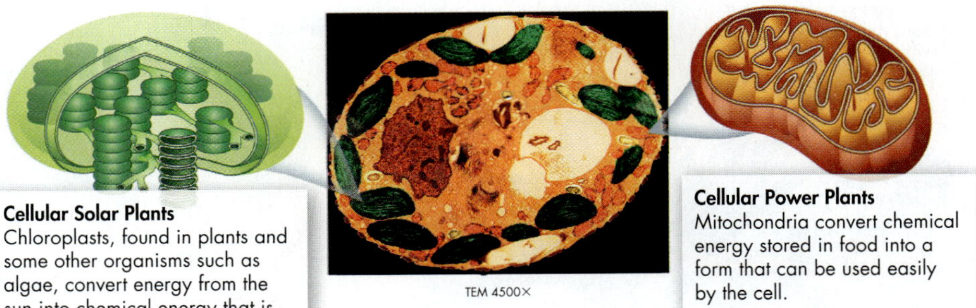

Cellular Solar Plants
Chloroplasts, found in plants and some other organisms such as algae, convert energy from the sun into chemical energy that is stored as food.

Cellular Power Plants
Mitochondria convert chemical energy stored in food into a form that can be used easily by the cell.

Hands-On Learning

MODEL MITOCHONDRION MEMBRANES Point out that the inner membrane of mitochondria is much larger than the outer membrane. **Say** We are going to do an activity to explore the most efficient way to pack a large membrane into a small space.

Divide students into groups. Give each group a 30-cm piece of yarn, a 100-cm piece of yarn, four pieces of masking tape, and a ruler. Tell students that the yarn represents membranes. Have students use the shorter piece of yarn to make a bean-shaped outline of a mitochondrion outer membrane and tape it to the desk. Challenge them to find a way to pack the most yarn into the outer membrane with all yarn touching the desk. When students have finished, explain that the inner membrane is highly folded to provide the most surface area in the smallest space.

Ask What is an example of an organ in our body that is highly folded? *(Sample answer: small intestine)*

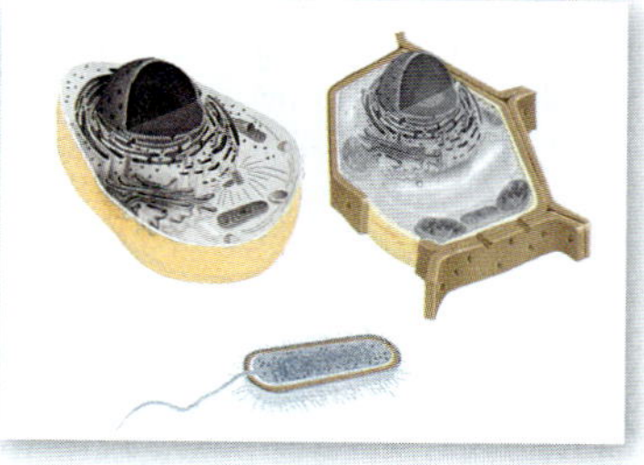

Cellular Boundaries

A working factory needs walls and a roof to protect it from the weather outside. The roof and walls also keep the factory's products safe until they are ready to be shipped out. Cells have similar needs, and they meet them in a similar way. All cells are surrounded by a barrier known as the cell membrane. Many cells also have a stiff layer around the membrane known as a cell wall.

Cell Walls The main job of the cell wall is to support, shape, and protect the cell. Most prokaryotes and many eukaryotes, such as plant cells, have cell walls. Animal cells do not have cell walls. Cell walls lie outside the cell membrane. Most cell walls can let water, oxygen, carbon dioxide, and other materials pass through easily. Cell walls give plants the strength they need to stand up straight. In trees and other large plants, nearly all of the tissue we call wood is made up of cell walls. In fact, the cellulose fiber used to make paper, including the pages of this book, comes from these cell walls.

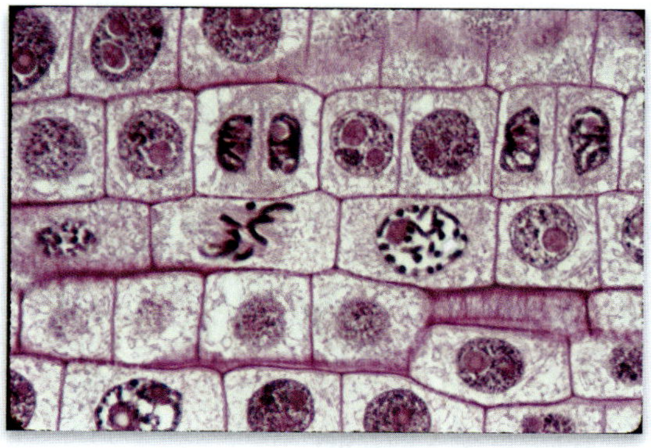

Cell Walls Rigid cell walls give these onion cells their shape (LM 700×).

Cell Membranes All cells have cell membranes. They are made up of a double-layered sheet called a **lipid bilayer.** The lipid bilayer makes membranes flexible and lets them form a strong barrier between the cell and its surroundings. The cell membrane controls what enters and leaves the cell. It also protects and supports the cell.

BUILD Vocabulary

chloroplast
an organelle found in cells of plants and some other organisms that captures the energy from sunlight and converts it into chemical energy

mitochondrion
a cell organelle that converts the chemical energy stored in food into compounds that are more convenient for the cell to use

lipid bilayer
a flexible double-layered sheet that makes up the cell membrane and forms a barrier between the cell and its surroundings

WORD ORIGINS
The Greek root *chloro-* means "green." The root *-plast* means "shape." Chloroplasts are the green-colored, round shapes in the cells of leaves.

Hands-On Learning

MODEL A CELL WALL AND MEMBRANE To help students visualize how a cell wall supports and protects the cell, have them inflate a balloon and put it inside a small cardboard box.

Ask What does the balloon represent? *(the cell membrane)*

Ask What does the cardboard box represent? *(the cell wall)*

Build Vocabulary

WORD ORIGINS Remind students that word parts are often combined to form new words. Ask students what other science word they know with the word part *chloro-*. *(chlorophyll)*

Ask Why do you think this word has *chloro-* in its name? *(Chlorophyll is a green pigment.)*

Active Reading

USE ANALOGIES To help students understand cell barriers, have them visualize a walled city surrounded by a moat. Tell students that anyone with a boat can cross the moat, but only citizens can pass through the gates.

Ask Which functions more like a cell wall, the moat or the gate? *(the moat)* Why? *(The moat is less selective, because anyone with a boat can cross it.)*

COMPARE AND CONTRAST Read the paragraphs about cell walls and cell membranes.

Ask How are cell walls and cell membranes alike? *(Both surround and protect cells.)*

Ask How are they different? *(The cell wall is rigid but the membrane is flexible.)*

PREDICT Ask students to predict what would happen if a cell wall did not let most things pass through. *(The cell would die.)*

MAKE CONNECTIONS
Explain that a lobster has an exoskeleton, which means that its skeleton is on the outside of its body.

Ask How is a lobster's shell like a cell wall? *(The lobster's shell supports, shapes, and protects it.)*

Ask What other animals have an exoskeleton that functions like a cell wall? *(Sample answers: crabs, shrimp, insects)*

Speed Bump

Ask What kind of cell does not have a cell wall? *(an animal cell)*

LESSON 7.2

During Reading continued

Active Reading

INTERPRET DIAGRAMS Direct students to the Cell Membrane diagram and read the caption.

Ask Where are the water-loving heads of the lipids found? *(on the outsides of the bilayer)*

Ask Where are the oil-loving tails of the lipids found? *(inside the bilayer)*

Ask If a cell membrane were like a peanut butter sandwich, what would the bread represent? *(the lipid heads)*

USE CONTEXT CLUES Direct students to the Cell Membrane diagram.

Ask What does the word *channel* mean in this diagram? *(a tube or duct)* What do you think the protein channels are for? *(Sample answer: passages for certain materials)*

USE CONTEXT CLUES Demonstrate how to use context clues to define unfamiliar words. Begin by having students study the labels in the diagram as you read aloud the second paragraph. Write the words *hydrophilic* and *hydrophobic* on the board.

Ask What does the text say about lipids that can help you figure out what the words *hydrophilic* and *hydrophobic* mean? *(that heads are "water-loving" and tails are "oil-loving")*

Say *Hydro* means "water."

Say What else do we know? We know that both words begin with *hydro.* We also know that heads are "water-loving."

Ask What does *hydrophilic* probably mean? *("water loving")*

Ask If you have a phobia about touching earthworms, do you most likely love to touch them or hate to touch them? *(hate to touch them)*

Ask What does the word hydrophobic probably mean? *(water-hating)*

Speed Bump

Ask What are lipids? *(large molecules made mostly of carbon and hydrogen atoms)*

Ask If water molecules tried to get through a cell membrane, what part of a membrane lipid would help keep them out of the cell? *(the tails)*

▶ ***Lipids*** As you can tell from the name, each layer of the lipid bilayer is made up of lipids. Lipids are large molecules made mostly from carbon and hydrogen atoms. The special lipids in the cell membrane have two parts: a head and a tail. The head is a chemical group that mixes well with water. The head is called hydrophilic, or water-loving. The tail is made up of fatty acid chains that mix well with oil. The tail is called hydrophobic, or water-hating. The difference between these heads and tails is the secret to why a cell membrane is good at its job.

When these lipids are in water, they form an ordered layer. Their oil-loving fatty acid tails stick together, while their water-loving heads mix with the water. When two of these layers come together, they are like a sandwich. The water-loving parts of the lipids form the outside layer. The oily parts of the lipids stick together to form the inner layer. A lipid bilayer is the result. As you can see in the picture below, the head groups of lipids in a bilayer are on the outside and inside of the cell. The fatty acid tails form an oily layer inside the membrane that keeps water out.

Cell Membrane The cell membrane keeps track of the movement of materials in and out of the cell. Cell membranes are made up of a lipid bilayer that has proteins and sugars set into it.

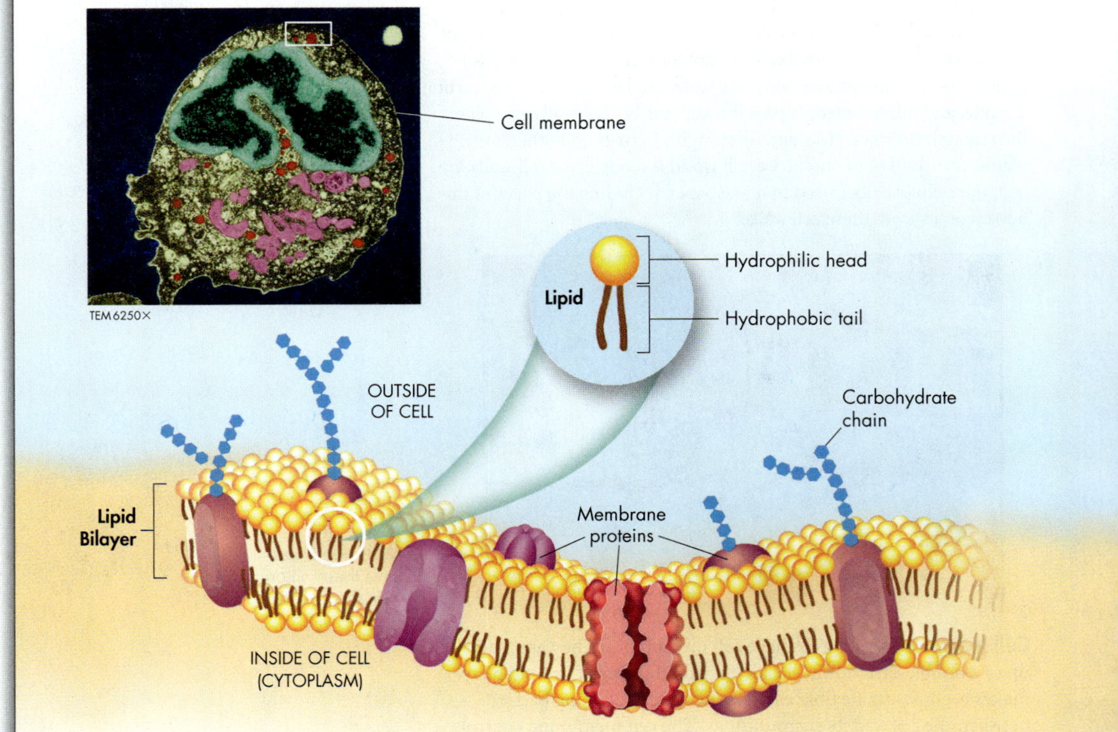

Hands-On Learning

SHOW CELL MEMBRANE MOVEMENT Help students understand that the cell membrane is not a solid structure and that the parts of the membrane move around within it. Have students model the lipid bilayer with a tray of yellow tennis balls (lipids) and a few red balls (protein channels) scattered among the yellow ones. Show that all the balls can move, and as they do, the positions of the red balls (the proteins) also change.

Ask What is it about the structure of a cell membrane that allows proteins to move about as you observed in this activity? *(The lipid bilayer gives cells a flexible structure and proteins can "float" in the lipids.)*

▶ *How Membranes Work* The lipid bilayers of most cell membranes contain many different proteins. Some of the proteins form channels and pumps that help to move material across the cell membrane. Sugar molecules are attached to many of these proteins. They act like name cards that help cells tell each other what they are doing. Other proteins attach directly to the cytoskeleton. They help the cell react to its surroundings by using the membranes to help the cell move or change shape.

As you know, some things are allowed to enter and leave a factory, and some are not. The same is true for living cells. Although many substances can cross cell membranes, some are too large to cross the lipid bilayer. Others have too much charge to pass through it. If a substance is able to cross a membrane, the membrane is said to be permeable to it. A membrane is not permeable to substances that cannot pass across it. Most cell membranes are selectively permeable, which means that some substances can pass across them and others cannot.

Key Question What is the function of the cell membrane? **The cell membrane keeps track of what enters and leaves the cell and also protects and supports the cell.**

CHECK Understanding

Apply Vocabulary

Use the highlighted words from the lesson to complete each sentence correctly.

1. A large sac that stores water, salts, and other materials for the cell is called a ______.
2. ______ are filled with enzymes that can break down substances in cells to be used later.
3. Microtubules and microfilaments make up the ______, which helps the cell hold its shape.
4. The ______ makes up the cell membrane, which acts as a flexible protective boundary for the cell.

Critical Thinking

5. **Review** What are the two major parts of a eukaryotic cell?
6. **Use Analogies** How is the role of the nucleus in a cell similar to the role of the captain on a sports team?
7. **Apply Concepts** How do contractile vacuoles help maintain water balance in a paramecium?
8. **Sequence** Describe the steps in making, packaging, and exporting a protein from a cell.
9. **Review** What is the function of mitochondria?
10. **Write to Learn** In your own words, explain how lipid molecules are arranged in the cell membrane. In your answer, use the words *bilayer, hydrophilic,* and *hydrophobic.*

BIOLOGY.com Search Lesson 7.2 GO • Lesson Assessment

Check Understanding

1. vacuole **2.** Lysosomes **3.** cytoskeleton **4.** lipid bilayer **5.** The two major parts of a eukaryotic cell are the nucleus and the cytoplasm. **6.** The captain is the leader on the team and may give out directions to the other players and control the game. The nucleus contains almost all of the DNA, the coded directions for making proteins and other molecules. **7.** A paramecium has a contractile vacuole. It contracts over and over to pump water out of the cell. **8.** Proteins are put together on ribosomes. Proteins that are to be released from the cell are made on ribosomes that are attached to the rough ER. Next, the proteins move to the Golgi apparatus, where they are modified, sorted, and packaged. From the Golgi apparatus, proteins are "shipped" to the cell membrane in vesicles. **9.** Mitochondria change the chemical energy stored in food into compounds that are easier for the cell to use. **10.** The cell membrane lets some materials in and blocks other materials. It is selectively permeable. The cell membrane must be permeable to water. If Michelle drank too much water, some of it might have gone into her cells and caused them to swell.

LESSON 7.2

Active Reading

SUMMARIZE Have students read both paragraphs and summarize each in one or two sentences.

Ask What is the main idea of the first paragraph? *(Sample answer: Proteins floating in the lipid bilayer do several things that are necessary for membranes to function.)*

Ask What kinds of materials may be prevented from passing through a cell membrane? *(Materials with too much charge and those that are too large.)*

Ask What term is used to describe a membrane that allows only certain materials to pass through? *(selectively permeable)*

KEY QUESTION Invite a volunteer to read aloud the question and answer.

Ask What are some of the functions of membrane proteins? *(They act as channels and pumps to move materials across the membrane, and they help the cell react to its environment.)*

Post Reading

ASSESSMENT TIPS For Question 6, direct students to the diagram of a nucleus.

For Question 8, make sure students realize that the word sequence means to put things in order. Have students make a graphic organizer showing the sequence of steps in making, packaging, and exporting a protein. Refer students to the Making Proteins diagram. Have them use complete sentences in their graphic organizers so that the organizers can serve as their answers.

MYSTERY CLUE Have students reread the section How Membranes Work. Ask them to discuss with a partner how the structure of the cell membrane may have contributed to too much water entering Michelle's cells. Students should conclude that the cell membranes may have allowed too much water into the cells because there was too much water outside the cells.

Wrap-Up Activity

MAKING MODELS Have students make a three-dimensional model of a cell or organelle. Have available a variety of craft items, such as construction paper, yarn, beads, balloons, plastic bubble wrap, tape, cardboard tubes, flexible straws, and glue. Allow students to use other items as desired. Encourage creativity in materials.

Build Connections

VISUAL SUMMARY Use the Comparing Typical Cells diagram to review the parts of bacterial, plant, and animal cells. Divide the class into small groups, and have each group submit a question about a cell function to be answered by the class. Tell students that they are to answer questions using specific vocabulary terms shown in the diagram.

Ask What cell part allows a cell to release energy? *(mitochondrion)*

Ask What two cell parts store materials for the cell? *(vacuoles and vesicles)*

Ask What two cell parts are made of microtubules? *(cytoskeleton and centrioles)*

Ask Which of the following cell parts are found in prokaryotic cells: nucleus, ribosomes, mitochondria, rough ER? *(ribosomes)*

Ask Which of the following cell parts are found in plant cells but not in animal cells: ribosomes, nucleus, chloroplasts, Golgi apparatus? *(chloroplasts)*

Ask What cell part packages and ships proteins? *(Golgi apparatus)*

Active Reading

SELF-QUIZ Have students turn to the Cell Types diagram at the end of Lesson 7.1. Explain that it is the same as the diagram on this page, except that the cell parts are not labeled. Tell students they can use the Cell Types diagram to quiz themselves and the Comparing Typical Cells diagram on this page to check their answers.

BUILD Connections

COMPARING TYPICAL CELLS

Eukaryotic cells have many different kinds of organelles. Some of these organelles are also found in prokaryotic cells. The table on the facing page compares prokaryotic cells, animal cells, and plant cells.

In Your Workbook Do you know the differences between animal cells, plant cells, and prokaryotic cells? Go to your workbook to test yourself.

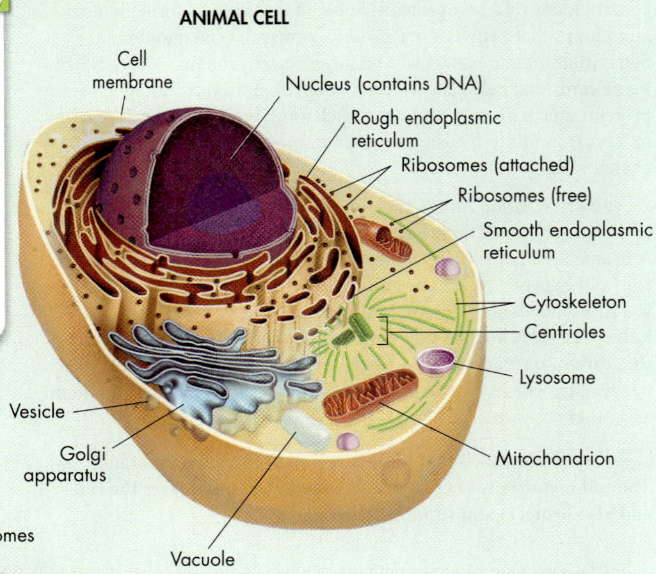

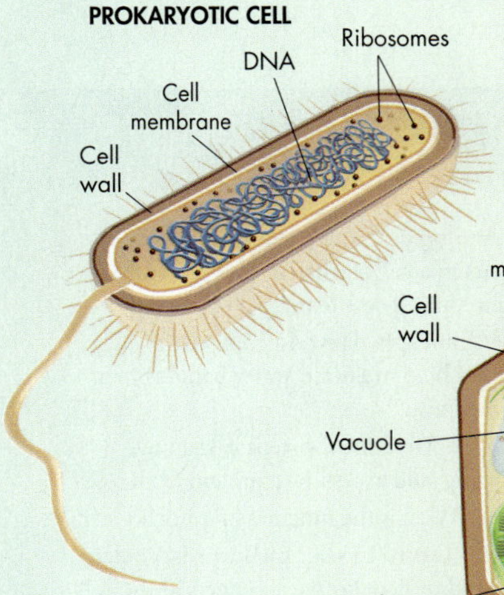

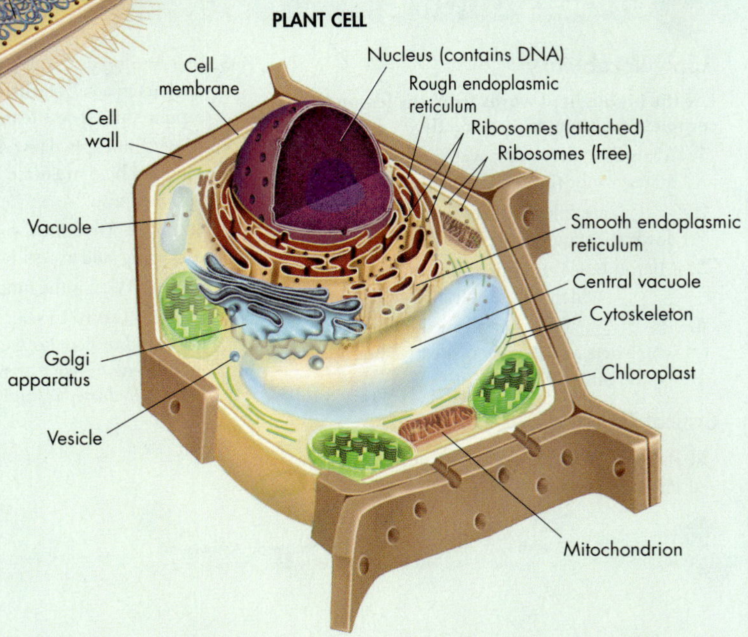

174 BIOLOGY.com Search Lesson 7.2 GO • Art Review • Tutor Tube

Focus on Academic Language

• assemble • convert • maintain • modify • regulate • synthesize

Write the words on the board. Review with students the meanings of the words and the cell parts associated with them.

Ask Which of these words means to change or alter the structure of something? *(modify, convert)*

Ask Which organelle *modifies* things? *(the Golgi apparatus)* What does the Golgi apparatus modify? *(proteins and other materials)*

Point out that these words are all common English words that are used to explain science-related situations. Have students write two sentences for each word: one using the word in a non-science way and the other as it is used in this lesson.

	Structure	Function	Prokaryote	Eukaryote: Animal	Plant
Cellular Control Center	Nucleus	Contains DNA	*Prokaryote DNA is found in cytoplasm.*	✓	✓
Organelles That Store, Clean-Up, and Support	Vacuoles and vesicles	Store materials		✓	✓
	Lysosomes	Break down and recycle macromolecules		✓	✓ (rare)
	Cytoskeleton	Maintains cell shape; moves cell parts; helps cells move	*Prokaryotic cells have protein filaments similar to actin and tubulin.*	✓	✓
	Centrioles	Organize cell division		✓	
Organelles That Build Proteins	Ribosomes	Synthesize proteins	✓	✓	✓
	Endoplasmic reticulum	Assembles proteins and lipids		✓	✓
	Golgi apparatus	Modifies, sorts, and packages proteins and lipids for storage or transport out of the cell		✓	✓
Organelles That Capture and Release Energy	Chloroplasts	Convert solar energy to chemical energy stored in food	*In some prokaryotic cells, photosynthesis occurs in association with internal photosynthetic membranes.*		✓
	Mitochondria	Convert chemical energy in food to usable compounds	*Prokaryotes carry out these reactions in the cytoplasm rather than in specialized organelles.*	✓	✓
Cellular Boundaries	Cell wall	Shapes, supports, and protects the cell	✓		✓
	Cell membrane	Regulates materials entering and leaving cell; protects and supports cell	✓	✓	✓

Hands-On Learning

CELL FUNCTIONS Have students write cell functions on individual index cards. Then, have students select a card and name the part or organelle with that function and the kind of cell in which it is found. For example: "Traps sunlight"; *chloroplast, plant cell.*

Active Reading

INTERPRET CHARTS AND TABLES Call students' attention to the table, which summarizes the structures and functions of cells and can help students distinguish between prokaryotes and eukaryotes. Discuss how the table is set up into columns, rows, and cells. Then give students practice interpreting a table.

Ask What does a checkmark in the Animal or Plant column of the table mean? *(The structure is present in a cell.)*

Ask What does an empty box mean? *(The structure is not present in the cell.)*

Tell students to use the table to answer the following questions.

Ask What is a structure that is not found in a plant cell? *(centrioles)*

Ask What are three structures that are found in all prokaryotes? *(ribosomes, cell walls, cell membranes)*

Ask What does a cell wall do? *(shapes, supports, and protects the cell)*

MAKE CONNECTIONS Write individual names of cell parts on each of several index cards. Give a card to each student. Place students in pairs and ask each pair to write a sentence or two describing how the two cell parts are related to each other. For example, chloroplasts and mitochondria are related because a chloroplast traps energy in food molecules and the mitochondria release that energy. If students have trouble making connections, work with them to discuss how the terms might be related, and then have the students write new sentences.

Cell Structure and Function **175**

LESSON 7.3

Pre-Reading

Objectives

7.3.1 Describe passive transport.

7.3.2 Describe active transport.

Preview the Pages

1. Preview the vocabulary words by having students find the words in the text and read them aloud.
2. Preview the visuals and read the captions and labels. Be sure students recognize that the curved structures on this page represent cell membranes, and each drawing illustrates different concentrations of solutions inside and outside the cell membrane.
3. Explain that the two U-shaped tubes in the Osmosis diagram represent *before* osmosis takes place and *after* osmosis takes place.

Activate Prior Knowledge

Ask students to list the different ways items might enter or leave a country, such as by air or water, as packages, in people's luggage, and so on. Then ask students to predict how each different method might relate to how materials enter and leave a cell.

Build Understanding

COMPARE/CONTRAST TABLE Direct students to Workbook B, where they will find a compare/contrast table. Remind them that putting information into a table format like this one will help organize the similarities and differences between two related ideas.

NATIONAL SCIENCE EDUCATION STANDARDS

UNIFYING CONCEPTS AND PROCESSES

III, IV, V

CONTENT

C.1.a, C.1.b

INQUIRY

A.2.a

For the corresponding lesson in *Miller & Levine Biology,* see pages 208–213.

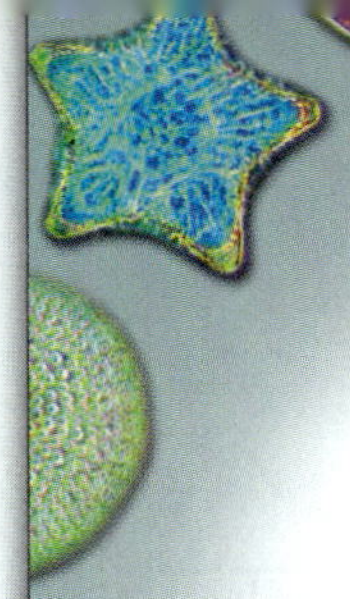

7.3 Cell Transport

Key Questions

What is passive transport?

What is active transport?

BUILD Understanding

Compare/Contrast Table As you read, create a compare/contrast table for passive and active transport.

In Your Workbook Go to your workbook for help making the compare/contrast table.

Passive Transport

Nearly every nation tries to control the things that move across its borders. A cell has a border, too. And the cell needs to control what crosses its border. The cell's border is the cell membrane.

One of the most important jobs of the cell membrane is to keep up the right balance between the liquid in the cell and the liquid around the cell. It does so by keeping track of the movement of molecules across the membrane.

Diffusion Cytoplasm contains many different substances dissolved in water. These substances move in this solution, run into each other, and spread out randomly. They tend to move from an area where there are more of them to an area where there are fewer of them. Think about when you add sugar to tea. The sugar molecules will move away from their original positions and move throughout the hot liquid. The process by which particles move from an area of high concentration (close together) to an area of lower concentration (spread apart) is known as **diffusion** (dih FYOO zhun).

Suppose there is more of a substance on one side of a cell membrane than on the other. If the substance can cross the membrane, its particles will diffuse toward the area of lower concentration. The particles will keep moving to the other side until the sides are even. Once there are equal numbers on both sides, the particles will move from both sides at equal rates.

Diffusion does not need energy from the cell. The movement of materials across the cell membrane without using cellular energy is called passive transport.

Diffusion Diffusion is the process by which molecules of a substance move from an area of higher concentration (where they are closer together) to an area of lower concentration (where they are more spread out).

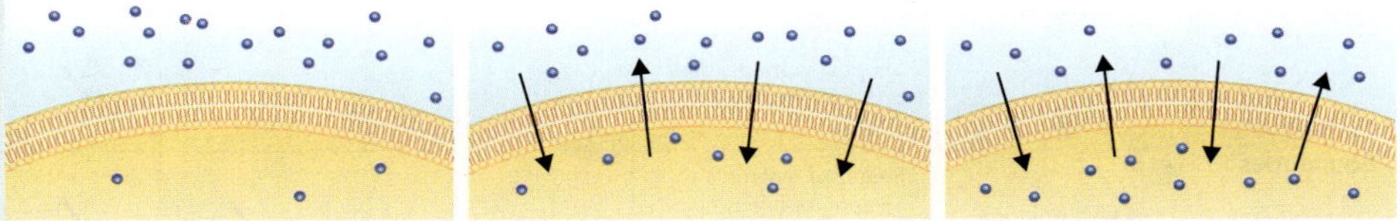

There are more particles on one side of the membrane than on the other. Particles are always moving across the membrane in both directions.

Diffusion causes particles to move in one direction at a faster rate. They move from the side of the membrane with the higher concentration to the side with the lower concentration.

Once both sides are even, particles keep moving across the membrane in both directions at the same rate. There is no net change in concentration.

 BIOLOGY.com Search Lesson 7.3 GO • InterActive Art

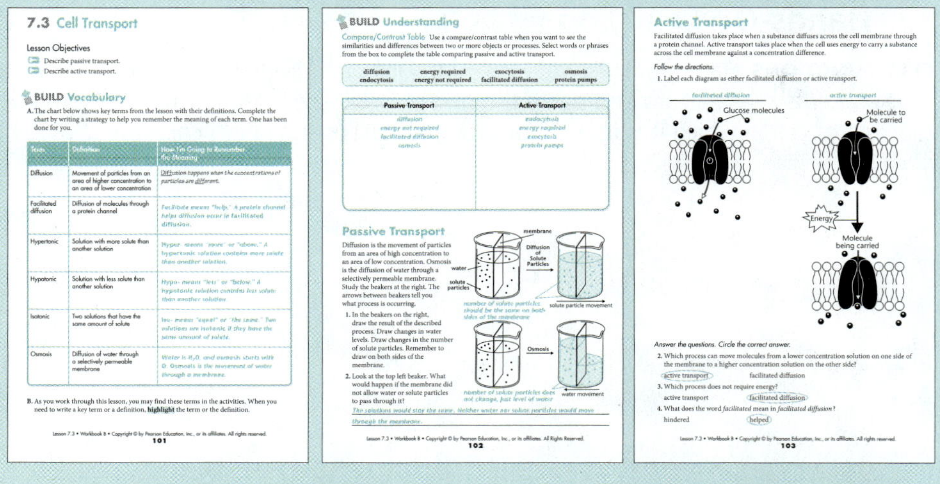

Facilitated Diffusion Remember that cell membranes are built around lipid bilayers. The lipids let small, uncharged molecules pass most easily. But many charged ions, such as Cl^-, and large molecules, such as the sugar glucose, seem to pass right through cell membranes much more quickly than they should. It's almost as if they have a shortcut.

In fact, they do. Proteins in the cell membrane act as carriers, or channels. They make it easy for certain molecules to cross. For example, red blood cells have proteins that let glucose pass through them in either direction. This process, in which molecules that cannot diffuse across the membrane alone instead pass through protein channels, is known as **facilitated diffusion.** This action does not use energy. Facilitated diffusion is a form of passive transport.

Osmosis Most cell membranes are selectively permeable. That term means they allow some things to cross, but not others. Water is a good example. The movement of water across a selectively permeable membrane is called **osmosis.** Water molecules do not mix well with lipid bilayers. So water has a hard time passing across some cell membranes. However, many cells have water channel proteins, known as aquaporins (ak wuh PAWR inz) in their membranes. These channels let water pass right through. In osmosis, as in all diffusion, molecules move from higher concentrations to lower concentrations—lower concentrations of water.

▶ ***How Osmosis Works*** Look at the picture of the U-shaped tube below. The barrier at the bottom of the tube will let water through in both directions. It will not let sugar pass. To start, there are more sugar molecules on the right side of the barrier than on the left. Even though water molecules move in both directions, they move at a greater rate toward the side with the concentrated sugar solution. Water will tend to move across the barrier until the sugar molecules on both sides are evenly spaced.

BUILD Vocabulary

diffusion
the process by which particles tend to move from an area where they are more concentrated to an area where they are less concentrated

facilitated diffusion
the process of diffusion in which molecules pass across the membrane through cell membrane channels

osmosis
the diffusion of water through a selectively permeable membrane

ACADEMIC WORDS
The word *facilitate* means "to help." In facilitated diffusion, protein channels facilitate the diffusion of large molecules across the membrane.

A Water Channel Protein
Water can pass into and out of the cell through aquaporins in the cell membrane.

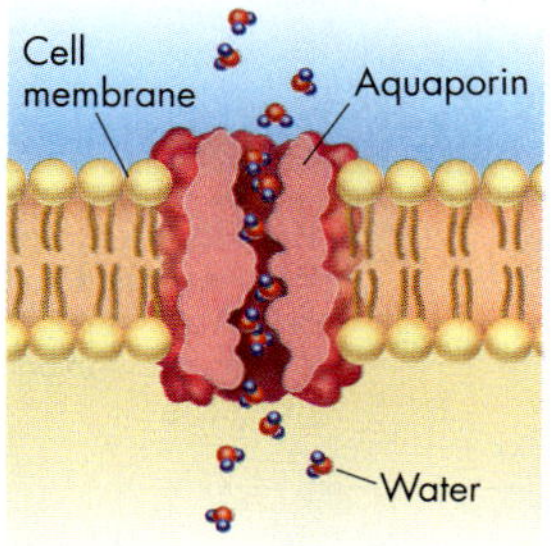

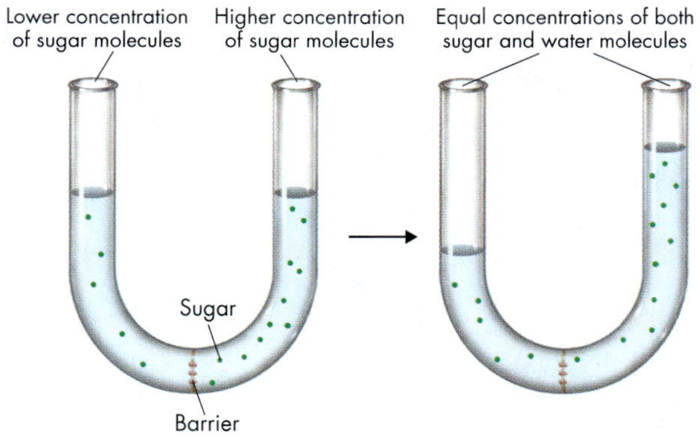

Osmosis In a laboratory experiment, water can move through a barrier that will not let sugar pass through it. Water molecules move from one side to the other until the sugar is evenly spaced. In the end, there will be more solution on one side of the barrier than on the other.

ELL Focus on ELL: Extend Language

ALL SPEAKERS Help students understand that the term *concentration* has multiple meanings. Students may be more familiar with *concentration* when it means the act of thinking, or focusing attention on a particular activity. Explain that when talking about solutions, concentration refers to the amount of solute in a given amount of another material.

Have students compare a snack-size clear bag of cereal with five marshmallows in it to another bag of equal size of the same kind of cereal with fifteen marhsmallows in it. Emphasize that everything is equal except the numbers of marshmallows. Ask students which bag has the greater concentration of marshmallows. *(the bag with fifteen marshmallows)*

During Reading

Active Reading

MAKE INFERENCES Point out the lesson title to students.

Ask What does *transport* mean? *(to move things from place to place)* What is cell transport? *(moving materials in and out of cells)* What materials might a cell need to transport? *(Nutrients and raw materials to build cell parts need to be transported into a cell; wastes need to be transported out of a cell.)*

USE VISUALS Have students examine the Diffusion diagram.

Ask How do the first and second pictures differ? *(The first shows more particles outside than inside the cell. The second shows a greater number of particles moving in than moving out, as indicated by more arrows pointed in than out.)*

Ask How do you know that both sides are even in the third picture? *(There are equal numbers of particles on both sides of the membrane and the same number of arrows point in and out.)*

PREDICT Before students read the paragraph *How Osmosis Works* have them look at the U-shaped tube on the left of the Osmosis diagram. Point out the barrier at the bottom of the tube. Have students predict how water molecules will move in this osmosis experiment.

Then have students read the paragraph. Review students' predictions and discuss how they compare to what they've just learned.

Build Vocabulary

ACADEMIC WORDS Point out to students that academic words often build on each other. Write the word *diffusion* on the board.

Ask What does diffusion mean? *(to move from an area of higher concentration to an area of lower concentration)*

Ask What does facilitate mean? *(to help)*

Speed Bump

Ask What is osmosis? *(the movement of water through membranes by facilitated diffusion)*

Ask In what direction do water molecules move during osmosis? *(from a higher concentration of water molecules to a lower concentration of water molecules)*

During Reading continued

Active Reading

USE VISUALS Some students may be confused by the two-headed arrow in the first drawing of the Osmotic Pressure diagram, which shows an isotonic condition. Explain that this arrow means water is moving both into and out of the cell at the same time.

Ask Why is water moving in both directions across the cell membrane? *(The concentration of the solutions is equal on both sides of the cell membrane.)*

MAKE CONNECTIONS Relate the illustrations in the bottom row of the Osmotic Pressure diagram to what happens when a lawn care company sprays a mixture of fertilizer and water onto grass.

Say If too much fertilizer and too little water are sprayed on grass, the grass may die and the lawn may turn brown.

Ask Is the fertilizer and water mixture hypotonic or hypertonic compared to the inside solution of the grass cells? *(hypertonic)* After putting the fertilizer mixture on the grass, would more water enter or leave the grass cells? *(More water would leave the cells.)* Why? *(Osmotic pressure would cause the water to move out of the cell, into the hypertonic solution.)*

KEY QUESTION Invite a volunteer to read aloud the question and answer.

Ask What does the word *passive* mean in terms of transport? *(No energy is needed.)*

Ask What three kinds of passive transport have you learned about? *(diffusion, facilitated diffusion, and osmosis)*

Build Vocabulary

ACADEMIC WORDS Discuss the meaning of pressure. Point out that a force is a push or pull.

Ask How does a force differ from pressure? *(Both involve a push or a pull, but pressure is the push or pull applied over an area.)*

BUILD Vocabulary

osmotic pressure
the pressure that must be applied to prevent osmotic movement across a selectively permeable membrane

ACADEMIC WORDS
The word *pressure* means "force applied over an area." Water pressure is caused by the force of water molecules hitting the sides of their container. Osmotic pressure is caused by the difference in the forces of water molecules hitting either side of the cell membrane.

▶ ***Osmotic Pressure*** Sometimes cell membranes have solutions that are the same on both sides. Sometimes the solutions are different. When the solutions inside and outside of the cell are the same, the solutions are said to be isotonic. Isotonic means "same strength" and refers to the concentration of the dissolved particles, not the water. When the solution outside of the cell has a higher concentration than the inside solution, the outside solution is hypertonic. This word means "above strength." When the solution outside of the cell has a lower concentration, it is hypotonic, or "below strength."

Sometimes there are differences in concentration of salts, sugars, proteins, and other dissolved molecules on one side of the cell membrane. These differences produce a force known as **osmotic pressure,** which can cause the movement of water out of or into a cell through osmosis. Osmotic pressure can cause an animal cell in a hypertonic solution to shrink as water moves out. An animal cell in a hypotonic solution will swell as water moves into it. Eventually, the cell may burst like an overinflated balloon. In plant cells, osmotic pressure can cause changes in the size of the central vacuole. A shrunken central vacuole does not give the same support to the cell as a full one. This loss of support is why plants wilt when they do not get enough water.

Osmotic Pressure Osmotic pressure inside an animal cell causes the cell to swell and burst. In plant cells, osmotic pressure inside the central vacuole pushes the cell's insides against the cell wall.

Key Question What is passive transport?
The movement of materials across the cell membrane without using energy is called passive transport.

The Effects of Osmosis on Cells

Solution	**Isotonic:** The concentration is the same inside and outside the cell. Water molecules move at the same rate in both directions.	**Hypertonic:** The outside solution has a higher concentration than the solution inside the cell. Movement of water molecules out of the cell causes it to shrink.	**Hypotonic:** The outside solution has a lower concentration than the solution inside the cell. Movement of water molecules into the cell causes it to swell.
Animal Cell	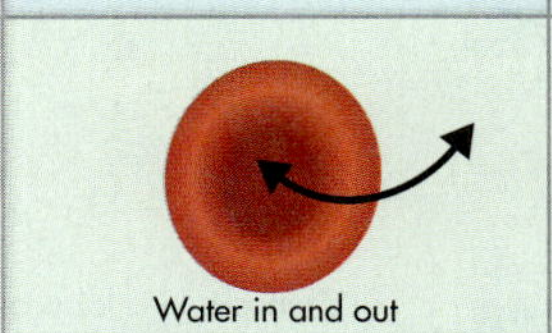 Water in and out	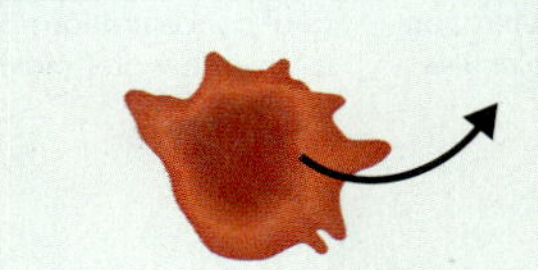 Water out	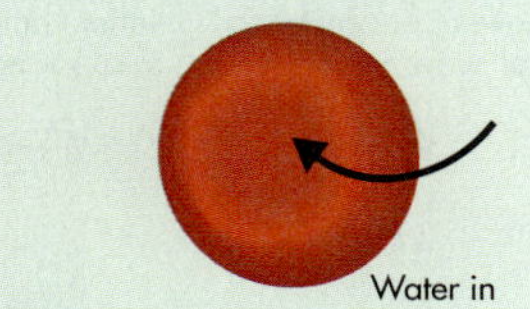 Water in
Plant Cell	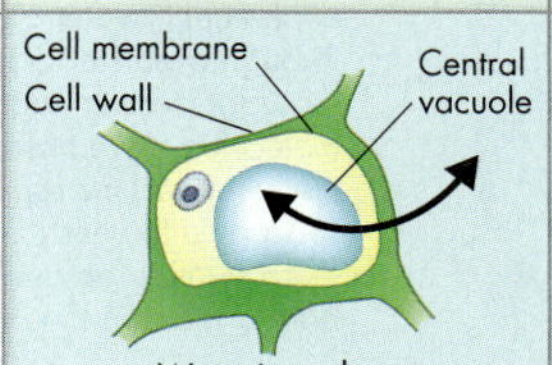 Cell membrane; Cell wall; Central vacuole; Water in and out	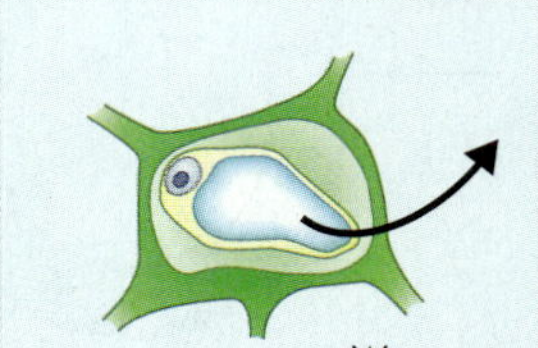 Water out	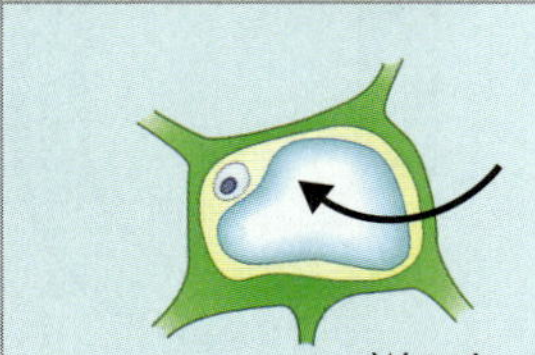 Water in

Focus on Academic Language

• isotonic • hypertonic • hypotonic

Explain that there are two kinds of particles in a solution: the particles of the substance that is dissolved (the solute) and the particles of the liquid that is doing the dissolving (the solvent). Emphasize that the three terms refer *only* to the concentration of the dissolved substance, or solute, not the solvent (in most cases, water.)

Write the prefixes *hypo-* and *hyper-* on the board. Explain that *hypo-* means "under" or "less than usual" and *hyper-* means "over" or "more than usual."

Say The word *isotonic* means the concentration of solute is the same inside and outside a cell. The words *hypertonic* and *hypotonic* also refer to the levels of solute inside and outside of a cell.

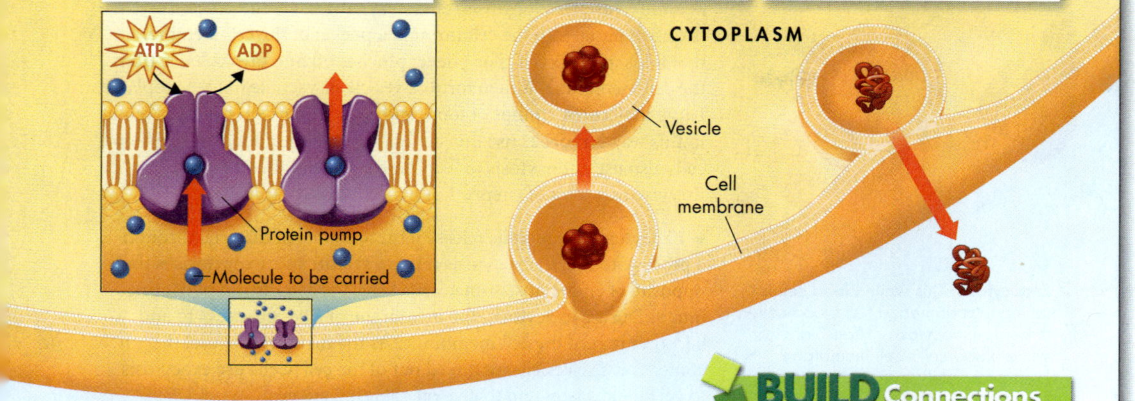

Protein Pumps
Chemical energy is used to pump ions across the cell membrane. Protein pumps grab ions on one side of the membrane, change shape, and let go of them on the other.

Endocytosis
The membrane forms a pocket around a particle outside the cell. The pocket pinches closed on the inside of the cell, forming a vesicle. The vesicle breaks loose, bringing the particle into the cell.

Exocytosis
A vesicle inside the cell carries a particle to the cell membrane. The membrane of the vesicle sticks to and becomes part of the cell membrane. The particle is forced out of the cell.

Active Transport

Diffusion allows particles to move from the side of the membrane where they are crowded to the side where they are not so crowded. But what happens when the cell needs to move even more particles to the side that is already more concentrated? This takes energy. Active transport is the movement of particles across the cell membrane using energy. The major kinds of active transport are shown in the picture above. Small molecules or ions move across a cell membrane through the work of protein pumps found in the membrane. Larger molecules and big clumps of material can also be moved across the cell membrane by processes known as endocytosis and exocytosis.

Moving Molecules Small molecules and ions are carried across membranes by proteins that act like pumps. Many cells use protein pumps to move calcium, potassium, and sodium ions across cell membranes. Changes in protein shape seem to be an important part of the pumping process. Chemical energy causes a change in the shape of the protein, so the protein binds the substance on one side of the membrane, then lets it go on the other side. A lot of the energy used by cells in their daily activities goes toward keeping active transport working. Using energy lets cells move substances to concentrated areas, even when diffusion might tend to move these substances in the opposite direction.

BUILD Connections

ACTIVE TRANSPORT

Energy from the cell is needed to move particles from an area of lower concentration to an area of higher concentration.

In Your Workbook Do you understand the three kinds of active transport? Go to your workbook for more practice.

BIOLOGY.com Search Lesson 7.3 GO • Art in Motion

Active Reading

CONTRAST Discuss the differences between active and passive transport. Have students write down these differences in the compare/contrast table in their workbooks.

Ask How is active transport different from passive transport, in terms of energy used? *(Active transport uses energy; passive transport does not.)*

Build Connections

VISUAL SUMMARY Make a copy of the diagram and corresponding text on active transport (this page and the next page). Mark the three labels in the diagram as A, B, and C. Also mark descriptions in the text with the corresponding letters. Make copies of these pages for each student.

Have students tape their pages together and then draw a line from each lettered part of the diagram to the sentence describing that part. Then have students read the sentence aloud.

Ask Why does a cell need to move materials by active transport? *(The cell often needs to move materials from a less concentrated area to a more concentrated area.)*

Ask How is facilitated transport similar to active transport using a protein pump? *(Both use membrane proteins to move materials across the cell membrane.)*

Ask What organelle plays a role in active transport of large molecules? *(vesicles)*

Speed Bump

Ask How does a protein pump release the substance it's transporting to the other side of the membrane? What does it use in the process? *(It uses chemical energy to change the shape of the protein and let go of the substance on the other side.)*

Focus on Academic Language

- endocytosis
- exocytosis

Say The prefix *endo-* means "within" and *exo-* means "outside."

Ask How do the meanings of the prefixes relate to the words? *(Endocytosis is when a particle is taken within the cell, and exocytosis is when a particle is forced outside the cell.)*

Have students work in groups to come up with mnemonic devices to help them remember the difference between endocytosis and exocytosis.

During Reading continued

Active Reading

SUMMARIZE Have students imagine they are amoebas. Tell them that amoebas take in bacteria for food. Have students summarize, without using the word *endocytosis,* how they would take in a bacterial cell.

KEY QUESTION Invite a volunteer to read aloud the question and answer.

Ask Why must large particles be moved into a cell by endocytosis rather than by protein pumps? *(The particles are too big to fit through the protein pumps.)*

Speed Bump

Ask Why is endocytosis an example of active transport and not facilitated diffusion? *(Endocytosis requires energy to engulf a particle and form a vesicle around it. Facilitated diffusion does not require energy.)*

Post Reading

ASSESSMENT TIPS For Question 3 suggest that students make a Venn diagram and then answer the question.

MYSTERY CLUE Suggest that students review the first three pages of this lesson.

Ask What cellular process is involved with Michelle's problems during and after the race? *(osmosis)* What happens to water during this process? *(It moves into an area with a higher concentration of solutes.)* When the concentration of salts and minerals in Michelle's blood got low, in what direction would osmosis move water across the cell membrane? *(from a lower concentration of solutes outside the cell, to a higher concentration of solutes inside the cell)*

Wrap-Up Activity

Give student groups a teabag and a cup containing warm water. Have them put the teabag into the water.

Ask How do you know that diffusion takes place in the cup? *(The water changed color.)* Were there more tea particles in the water near the bag or away from the bag? *(near the bag)* Did the tea particles spread out in the cup? How do you know? *(Yes, over time, the color spread out.)* Is the teabag permeable to tea particles? *(yes)*

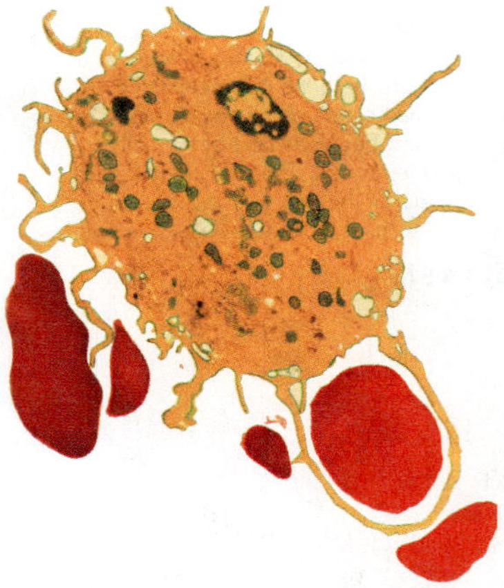

Endocytosis This white blood cell is taking in a damaged red blood cell through endocytosis. "Arms" of the white blood cell's cell membrane have completely surrounded the red blood cell (TEM 5300×).

Moving Larger Particles Larger molecules and even solid clumps of material cannot pass through protein pumps. They must be moved from one side of the membrane to the other through vesicles. Two types of active transport that use vesicles to move larger particles are endocytosis and exocytosis.

▶ ***Endocytosis*** Endocytosis (en doh sy TOH sis) is the process of taking material into the cell through a pocket of the cell membrane that folds into the cell. The pocket pinches off and breaks loose from the cell membrane. It then forms a vesicle or vacuole in the cytoplasm. Large molecules, clumps of food, and even whole cells can be taken up in this way. Amoebas use this method for taking in food. White blood cells also use endocytosis to "eat" damaged cells. Taking in material in this way uses a lot of energy. It is a form of active transport.

▶ ***Exocytosis*** Many cells release large amounts of material through a process known as exocytosis (ek soh sy TOH sis). Exocytosis is the opposite of endocytosis—materials leave the cell instead of coming into it. During exocytosis, the membrane of a vesicle sticks to and becomes part of the cell membrane. As the membranes fuse, the contents of the vesicle are forced out of the cell. Cells use exocytosis to get rid of wastes and to give off chemical signals.

Key Question What is active transport?
The movement of materials from an area of lower concentration to an area of higher concentration is known as active transport. Active transport requires energy.

CHECK Understanding

Apply Vocabulary

Use the highlighted words from the lesson to complete each sentence correctly.

1. Diffusion of particles through protein channels is called ______.
2. Osmosis produces a force known as ______.

Critical Thinking

3. **Compare and Contrast** How is endocytosis similar to exocytosis? How are they different?
4. **Apply Concepts** What would happen to the cells of a saltwater plant if the plant were placed in fresh water?
5. **Write to Learn** Answer the second clue of the mystery.

MYSTERY CLUE

As Michelle ran, she sweated. Sweating made her lose salts from her bloodstream. As she drank more water, the concentration of salts and minerals in her bloodstream got lower. How do you think these things made Michelle's problem worse? (Hint: See p. 176.)

BIOLOGY.com Search Lesson 7.3 GO • Lesson Assessment

Check Understanding

1. facilitated diffusion
2. osmotic pressure
3. Endocytosis and exocytosis use vesicles to move large particles from one side of the cell membrane to the other. Endocytosis brings materials into the cell. Exocytosis moves materials out of the cell.
4. The central vacuole would swell with water, pushing the cell's organelles against the cell membrane.
5. Sweating made her lose salts, and drinking water further diluted her blood. The solution outside of Michelle's cells was not very concentrated compared to that inside the cells. Water flowed into the cells to balance this difference in concentration. The cells began to swell from taking in too much water.

7.4 Homeostasis and Cells

The Cell as an Organism

It is sometimes hard to remember that all living things have certain characteristics in common. All living things are made of cells. They all have the same kinds of organelles. But, clearly, living things are not all the same.

Sometimes, a single cell is the whole organism. In fact, there are many more single-celled organisms on Earth than there are organisms with many cells. Just like other living things, single-celled organisms must maintain **homeostasis**—relatively constant internal physical and chemical conditions. To stay in this state single-celled organisms grow, respond to the environment, change food or sunlight into useful energy, and reproduce.

Single-celled organisms include both prokaryotes and eukaryotes. Prokaryotes, especially bacteria, are very adaptable. They live almost everywhere—in the soil, on leaves, in the ocean, in the air, and even within the human body. Many eukaryotes also spend their lives as single cells. Some types of algae are single cells found in oceans, lakes, and streams around the world. Yeasts, or unicellular fungi, have an important job breaking down nutrients, making them available for other living things.

Key Question How do individual cells maintain homeostasis? **To maintain homeostasis, unicellular organisms grow, respond to the environment, change food or sunlight into useful energy, and reproduce.**

Single-Celled Life Single-celled organisms, like this freshwater protozoan, must be able to carry out all of the functions necessary for life (SEM 600×).

Key Questions

How do individual cells maintain homeostasis?

How do the cells of multicellular organisms work together to maintain homeostasis?

BUILD Understanding

Preview Visuals Before you read, look at the Levels of Organization image at the top of the next page. Then write two questions you have about the figure. As you read, write answers to your questions.

In Your Workbook Go to your workbook for help answering your questions. Complete the preview visuals activity for Lesson 7.4.

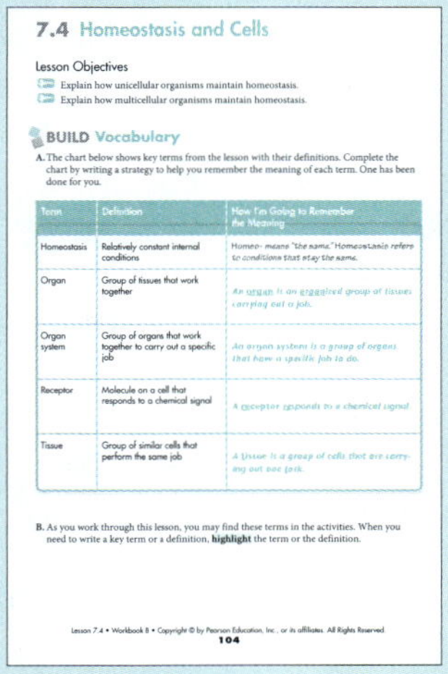

7.4 Homeostasis and Cells

Lesson Objectives

Explain how unicellular organisms maintain homeostasis.

Explain how multicellular organisms maintain homeostasis.

BUILD Vocabulary

A. The chart below shows key terms from the lesson with their definitions. Complete the chart by writing a strategy to help you remember the meaning of each term. One has been done for you.

Term	Definition	How I'm Going to Remember the Meaning
Homeostasis	Relatively constant internal conditions	Homeo- means "the same." Homeostasis refers to conditions that stay the same.
Organ	Group of tissues that work together	An organ is an organized group of tissues carrying out a job.
Organ system	Group of organs that work together to carry out a specific job	An organ system is a group of organs that have a specific job to do.
Receptor	Molecule on a cell that responds to a chemical signal	A receptor responds to a chemical signal.
Tissue	Group of similar cells that perform the same job	A tissue is a group of cells that are carrying out one task.

B. As you work through this lesson, you may find these terms in the activities. When you need to write a key term or a definition, **highlight** the term or the definition.

Lesson 7.4 • Workbook B • Copyright © by Pearson Education, Inc., or its affiliates. All Rights Reserved.
104

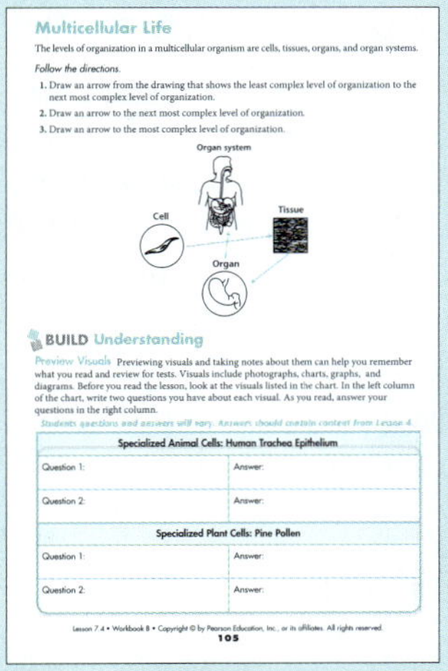

Multicellular Life

The levels of organization in a multicellular organism are cells, tissues, organs, and organ systems.

Follow the directions.

1. Draw an arrow from the drawing that shows the least complex level of organization to the next most complex level of organization.
2. Draw an arrow to the next most complex level of organization.
3. Draw an arrow to the most complex level of organization.

Organ system
Cell
Tissue
Organ

BUILD Understanding

Preview Visuals Previewing visuals and taking notes about them can help you remember what you read and review for tests. Visuals include photographs, charts, graphs, and diagrams. Before you read the lesson, look at the visuals listed in the chart. In the left column of the chart, write two questions you have about each visual. As you read, answer your questions in the right column.

Students' questions and answers will vary. Answers should contain content from Lesson 4.

Specialized Animal Cells: Human Trachea Epithelium	
Question 1:	Answer:
Question 2:	Answer:
Specialized Plant Cells: Pine Pollen	
Question 1:	Answer:
Question 2:	Answer:

Lesson 7.4 • Workbook B • Copyright © by Pearson Education, Inc., or its affiliates. All rights reserved.
105

LESSON 7.4

Pre-Reading

Objectives

7.4.1 Explain how unicellular organisms maintain homeostasis.

7.4.2 Explain how multicellular organisms maintain homeostasis.

Preview the Pages

1. Discuss specialization. Compare specialization to a sports team with players who play different positions. Discuss the benefits of having players on a team who specialize in certain positions as opposed to players who can play all positions.
2. Review with students all the vocabulary, pronouncing each word aloud. Have students find the words in the text.
3. Have students view the Levels of Organization picture. Make sure they understand that each image is part of the image that follows.

Build Understanding

PREVIEW VISUALS Direct students to Workbook B, where they'll find a two-column chart. Invite them to look at each of the visuals in the chapter, and write down two questions they have about each one.

KEY QUESTION Invite a volunteer to read aloud the question and answer.

Ask How does being able to respond to the environment help a unicellular organism maintain homeostasis? *(The organism can sense harmful things in the environment and move away from them. It can sense a food source and get energy from it.)*

NATIONAL SCIENCE EDUCATION STANDARDS

UNIFYING CONCEPTS AND PROCESSES

I, III, IV, V

CONTENT

C.1.a, C.1.d, C.1.f, C.5.d

For the corresponding lesson in *Miller & Levine Biology*, see pages 214–217.

During Reading

Active Reading

COMPARE/CONTRAST Explain that single-celled organisms and multicellular organisms maintain homeostasis differently.

Ask What activities do single-celled organisms perform to maintain homeostasis? *(grow, respond to environment, change the energy in food or sunlight into useful energy, reproduce)*

Ask How are the cells in a multicellular organism different from single-celled organisms? *(The cells are specialized.)*

Ask How does this specialization help a multicellular organism maintain homeostasis? *(Different kinds of cells perform different functions for the organism.)*

USE VISUALS Review the Levels of Organization diagram with students. Explain how the diagram is an example of how cells are specialized in humans, and how they are organized. Point out the muscle cell in the diagram, and note how it and many other muscle cells form smooth muscle tissue.

Ask In the diagram, what does smooth muscle tissue along with other types of tissue form? *(stomach)* What system is the stomach a part of? *(digestive system)* What function does this system perform? *(breaks down food so the body can use it)*

Invite students to think of some other kinds of cells in the body that are organized into organs and systems. *(Sample answer: bone cell → bone tissue → bone → skeletal system)*

Build Vocabulary

PREFIXES Have students say the word *homeostasis,* find it in the text, and read the definition.

Ask How does the word *homeostasis* relate to the meaning "the same"? *(Homeostasis describes the state in which conditions within a cell are kept relatively the same.)*

Speed Bump

Ask Why can't a cell taken from a multicellular organism live on its own? *(Cells of multicellular organisms are specialized. They depend on the roles of other specialized cells to fulfill their needs.)*

Ask List the levels of organization from cell to organism in order from simplest to most complex. *(cell, tissue, organ, organ system, organism)*

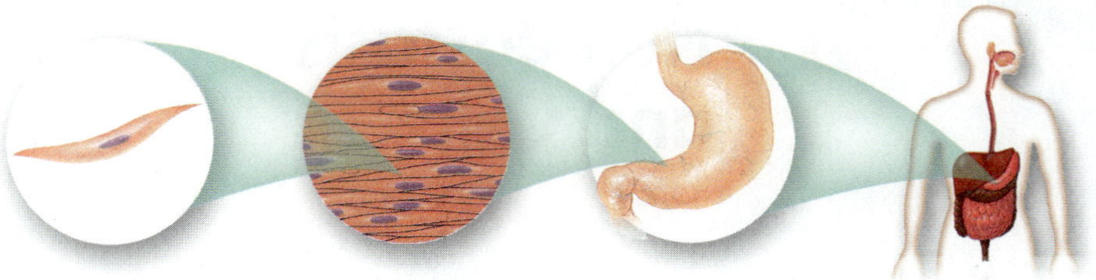

Levels of Organization From simplest to most complex, the levels of organization in a many-celled organism are cells, tissues, organs, and organ systems.

Multicellular Life

The cells of human beings and other many-celled organisms do not live on their own. They need other cells to survive. Like the members of a winning baseball team, they must work together. In baseball, each player plays a different position. To play the game well, players communicate with one another, sending and receiving signals. Cells in a many-celled organism work the same way. They become specialized for certain jobs. They communicate with one another to maintain homeostasis.

Cell Specialization The cells of a many-celled organism are specialized. Different kinds of cells have different jobs. Some cells are specialized for movement. Others must react to the environment. Still others must make substances that the organism needs. No matter what its job, each specialized cell helps the organism maintain homeostasis. For example, cells that line the air passages in your lungs have cilia (SIL ee uh). These hairlike parts sweep dust away to keep your lungs clean. These specialized cells are packed with mitochondria that make the chemical energy that powers their cilia.

Levels of Organization The specialized cells of many-celled organisms are organized into tissues. A **tissue** is a group of similar cells that that work together to do a particular job. Many tasks in the body are too complicated to be carried out by just one kind of tissue. In these cases, groups of tissues work together as an **organ.** For example, each muscle in your body is an organ. Within a muscle, however, there are nervous tissues and connective tissues. Each kind of tissue has a special job that helps the organ do its job. A group of organs that work together to do a job is called an **organ system.** For example, the stomach, pancreas, and intestines work together as the digestive system to break down food and take up nutrients.

BUILD Vocabulary

homeostasis
the relatively constant internal physical and chemical conditions that organisms maintain

tissue
a group of similar cells that perform a particular function

organ
a group of tissues that work together to perform closely related functions

organ system
a group of organs that work together to perform a specific function

PREFIXES
The prefix *homeo-* means "the same." Living things are always trying to maintain homeostasis. They must keep their internal physical and chemical conditions about the same at all times, in spite of changes in their environments.

182 Chapter 7 • Lesson 4

ELL Focus on ELL: Build Background

BEGINNING SPEAKERS To help students understand the concept of levels of organization, prepare a labeled four-level diagram using something familiar. For example, you might show a CD by a rock band, a group of CDs by the same band, the rock-music section of a store, and the entire store. Ask student to think of other examples of levels of organization. Invite them to create a diagram like the one they see in their text. *(Answers will vary. Sample answer: student → class → grade → school)*

Help students correlate the diagram in the text to their diagrams. Point to each level in the Levels of Organization diagram, then have students point to the corresponding level in their own diagrams. As you point, read the labels aloud.

Cellular Communication Cells in a large organism use chemical signals to communicate. These chemical signals are passed from one cell to another. They can speed up or slow down the activities of the cells that receive them. They can even cause a cell to change what it is doing in an instant.

In order to communicate, certain cells form connections, or cellular junctions, to neighboring cells. Some of these junctions hold cells together firmly. Others allow small molecules carrying chemical messages or signals to pass from one cell to the next. To "understand" one of these chemical signals, a cell must have a receptor that sticks to the chemical signal. Some receptors are on the cell membrane. Other kinds of receptors are inside the cytoplasm.

The chemical signals sent by different kinds of cells can cause important changes in cells and tissues. For example, the signal that causes heart muscle cells to contract begins in a place known as the pacemaker. Ions carry the signal from cell to cell through a special connection known as a gap junction. This junction can cause the millions of heart muscle cells to contract as one in a single heartbeat.

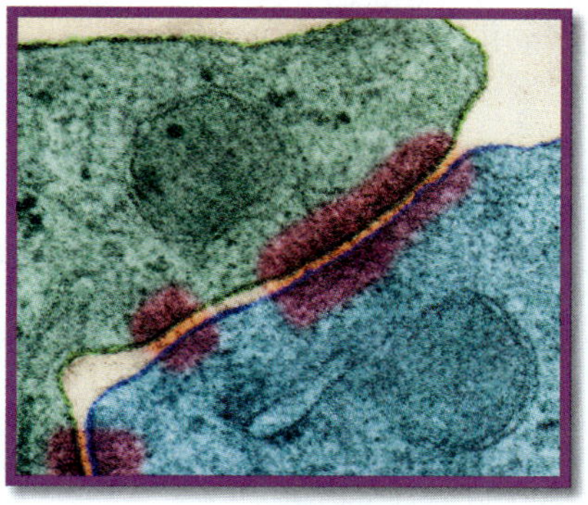

Cellular Junctions Some junctions hold cells together in tight formations. This kind of junction can be seen in dark pink in this picture of cells lining the central nervous system (blue and green) (TEM 21,600×).

Key Question How do the cells of multicellular organisms work together to maintain homeostasis?
The cells of many-celled organisms have specialized jobs and communicate with one another to maintain homeostasis.

CHECK Understanding

Apply Vocabulary

Use the highlighted words from the lesson to complete each sentence correctly.

1. _______ is the relatively constant internal physical and chemical conditions of a cell or organism.
2. Many similar specialized cells working together make up a(n) _______.
3. Many organs make up a(n) ______.

Critical Thinking

4. **Review** What is homeostasis?
5. **Explain** What do unicellular organisms do to maintain homeostasis?
6. **Predict** Using what you know about the ways muscles move, predict which organelles would be most common in muscle cells.
7. **Apply Concepts** The contractile vacuole is an organelle found in some single-celled organisms. Contractile vacuoles pump out fresh water that builds up in the cell from osmosis. How is this activity an example of the way the cell maintains homeostasis?
8. **Sequence** List the different levels of organization in many-celled organisms from most complex to least complex.
9. **Write to Learn** Write a paragraph that explains how cells in a many-celled organism are like teammates on a sports team.

BIOLOGY.com Search Lesson 7.4 GO • Lesson Assessment

LESSON 7.4

Active Reading

APPLY ACADEMIC VOCABULARY Discuss the relationship between *receptor* and *receive*.

Say Receptors are receivers.

Ask What do receptors in cells receive? *(chemical signals from another cell)*

KEY QUESTION Invite a volunteer to read aloud the question and answer.

Ask How do the cells of a multicellular organism communicate with each other? *(They form tight connections with other cells, send chemical signals to cells in the form of small molecules, and have receptors to receive messages from other cells.)*

Post Reading

ASSESSMENT TIPS Review vocabulary with students. If students have trouble with Question 7, have them review in Lesson 7.3 what happens to cells in hypotonic solutions. For Question 8, ask students what it means to sequence something. Have students sequence what they do in the morning to get ready for school.

Wrap-Up Activity

INTERVIEW Pair students. Have one student play a reporter who is interviewing a cell, played by the other student. The "cell" has just received a signal from another cell warning that environmental conditions have changed and the cell is in danger of getting too much water. Have the reporter ask questions to find out how the cell will respond

Check Understanding

1. Homeostasis
2. tissue
3. organ system
4. Homeostasis is the relatively stable internal physical and chemical conditions of an organism.
5. They maintain homeostasis by growing, responding to changes in their environment, transforming energy, and reproducing.
6. Answers may vary. Predictions should note that muscle cells need energy to move and, therefore, probably contain a large number of mitochondria.
7. The contractile vacuole helps maintain water balance in paramecia by expelling excess water that would otherwise accumulate and burst the cell.
8. organ system, organ, tissue, cell
9. Sample answer: Cells in a many-celled organism are like teammates on a football team. They all have different jobs to do, but they must communicate and work together to do a bigger job. The different jobs of the football players are their different positions. The different jobs of the cells are the roles they play in the body. The bigger job for the football team is to win the game. The bigger job for the organism is to maintain homeostasis.

Pre-Lab

Review with students the relationship between osmosis and the concentration of solutes in a solution.

Ask Say you have areas with a lot of dissolved salt and areas with little dissolved salt. In what direction do water particles move between the two areas? *(from the area with little dissolved salt, or low salt concentration, to the area with a lot of dissolved salt, or high salt concentration)*

Review the Background Questions and the Pre-Lab Questions as a class.

Lab

Tell students they will perform the chapter lab *Detecting Diffusion* described in **Lab Manual B.**

SAFETY Students should wear safety goggles, gloves, and a lab apron. Caution students that iodine solution can irritate the eyes and skin and can stain clothing. Have them rinse off any solution that spills on skin or clothing.

Post-Lab

OBSERVE OSMOSIS Form students into groups. Give each group two jars. Have students half-fill each jar with water and label one jar "Salt water" and the other "Water." Have students dissolve 2 teaspoons of table salt in the saltwater jar. Then have students place a carrot stick in each jar. After 30 minutes, have students remove the carrot sticks, and ask them to explain what happened.

Ask Was the salt concentration higher in the salt water or in the carrot cells? *(in the salt water)*

Skills Lab

GUIDED INQUIRY

Pre-Lab: Detecting Diffusion

Problem How can you determine whether solutes are diffusing across a membrane?

Materials 400-mL beaker, 25-mL graduated cylinder, 1% starch solution, plastic sandwich bag, twist tie, iodine solution, forceps

Lab Manual Chapter 7 Lab

Skills Focus Use Models, Analyze Data, Infer

Connect to the Big idea The cell membrane forms a thin flexible barrier between a cell and its surroundings. The cell membrane controls what enters the cell and what leaves the cell. Diffusion is the process responsible for much of the movement across a cell membrane. During diffusion, solutes move from an area of high concentration to an area of lower concentration. When water is the molecule that is diffusing, the process is called osmosis. Proteins embedded in the membrane can facilitate the diffusion of many particles, including water. In this lab, you will model the diffusion of small molecules.

Background Questions

a. **Review** What does it mean to say that a membrane is selectively permeable?

b. **Explain** Does the movement of molecules stop when the concentration of a solute is equal on both sides of a membrane? Explain.

c. **Compare and Contrast** What is the main difference between passive transport and active transport?

Pre-Lab Questions

Preview the procedure in the lab manual.

1. **Draw Conclusions** How will you know whether starch has diffused across the membrane?
2. **Draw Conclusions** How will you know whether iodine has diffused across the membrane?
3. **Use Analogies** How is a window screen similar to a cell membrane?

BIOLOGY.com Search Chapter 7 GO

Visit Chapter 7 online to test yourself on chapter content and to find activities to help you learn.

Untamed Science Video Travel to the ocean's depths with the Untamed Science crew to explore how fish maintain water homeostasis.

Art in Motion View a short animation that explains the different types of active transport.

Art Review Review your understanding of plant and animal cell structures with this activity.

InterActive Art Build your understanding of osmosis and diffusion with these animations.

Visual Analogy Compare the structures of the cell to the parts of a factory.

Tutor Tube Hear suggestions from the tutor for help in remembering cell structures.

Pre-Lab Answers

BACKGROUND QUESTIONS

a. Some substances can pass through the membrane while other substances cannot pass through the membrane.

b. The movement does not stop, but an equal number of molecules move in each direction to maintain equilibrium.

c. The movement of substances through passive transport does not require energy; the movement of substances through active transport requires energy.

PRE-LAB QUESTIONS

1. If the solution outside the plastic bag turns blue black then starch diffused out of the bag.
2. If the solution inside the plastic bag turns blue black, then iodine diffused into the bag.
3. Sample answer: Like a cell membrane, a window screen prevents the passage of some things, such as insects, while allowing other things, such as air, to flow in and out of a house.

NATIONAL SCIENCE EDUCATION STANDARDS

UCP II, V

CONTENT C.1.a, C.1.d

INQUIRY A.1.b, A.1.c, A.2.a

7 CHAPTER Summary

7.1 Life Is Cellular

- The cell theory states that (1) all living things are made up of cells, (2) cells are the basic units of structure and function in living things, and (3) new cells come from existing cells.
- Most microscopes use lenses to magnify the image of an object by focusing light or electrons.
- Prokaryotic cells do not have a nucleus. In eukaryotic cells, the nucleus holds the cell's genetic material.

cell (p. 160)
cell theory (p. 160)
nucleus (p. 162)
eukaryote (p. 162)
prokaryote (p. 162)

7.2 Cell Structure

- The nucleus holds nearly all of the cell's DNA.
- Vacuoles store water, salts, proteins, and sugars. Lysosomes break down old organelles and large molecules so that they can be used by the cell. The cytoskeleton helps the cell move and keep its shape.
- Proteins are put together on ribosomes.
- Proteins that will leave the cell and many kinds of membrane proteins are made on the rough ER. The Golgi apparatus then changes, sorts, and packages proteins for storage in the cell or for leaving the cell.
- Chloroplasts take the energy from sunlight and change it into chemical energy stored in food in a process called photosynthesis. Mitochondria change the chemical energy of food into compounds that are easier for the cell to use.
- The cell membrane keeps track of what enters and leaves the cell. It also protects and supports the cell.

cytoplasm (p. 164)
organelle (p. 164)
vacuole (p. 166)
lysosome (p. 166)
cytoskeleton (p. 167)
ribosome (p. 168)
endoplasmic reticulum (p. 168)
Golgi apparatus (p. 169)
chloroplast (p. 170)
mitochondrion (p. 170)
lipid bilayer (p. 171)

7.3 Cell Transport

- Passive transport is the movement of materials across the cell membrane without using energy. Diffusion, facilitated diffusion, and osmosis are kinds of passive transport.
- Active transport uses energy. It is the movement of materials from an area of lower concentration to an area of higher concentration. Protein pumps, endocytosis, and exocytosis are kinds of active transport.

diffusion (p. 176)
facilitated diffusion (p. 177)
osmosis (p. 177)
osmotic pressure (p. 178)

7.4 Homeostasis and Cells

- To maintain homeostasis, single-celled organisms grow, respond to the environment, get and use energy, and reproduce.
- The cells of living things that have many cells become specialized for particular jobs. These cells must communicate with one another to maintain homeostasis.

homeostasis (p. 181)
tissue (p. 182)
organ (p. 182)
organ system (p. 182)

BIOLOGY.com Search Chapter 7 GO • Match It • Chapter Assessment

UbD Transfer the Big Idea

TRANSFER Form students into groups. Tell them that each group will design a single-celled eukaryotic organism that will live in a special environment that they will design. Have students first make a list of the features of their environment, including amount of water and light. Then, students should decide what life processes their organism will perform, including the kind of food the organism will eat and how it will get that food, or if the organism will make its own food. Finally, have students make a large diagram of their cell. It should contain the organelles needed for its life processes, with labels that tell why the organism needs each organelle.

Chapter Summary

Think Visually

USE GRAPHICS Have students use the terms *diffusion, facilitated diffusion, osmosis, active transport, endocytosis,* and *exocytosis* to create a concept map about the ways substances can move into and out of cells.

ANSWERS
Answers may vary. Check accuracy of concept maps. To make certain that students clearly understand the differences among these concepts, suggest they add a caption or a small drawing by each term.

BIOLOGY.com Study Online

REVIEW RESOURCES

Visual Analogy Cell parts are compared with factory parts and their functions in this online activity.

Art Review Students label structures in typical plant cells and animal cells.

Tutor Tube Students are reminded that plants have mitochondria in addition to chloroplasts.

InterActive Art Explore diffusion and osmosis with an animation and an activity.

Art in Motion The different types of active transport are reviewed in this short animation.

Data Analysis Students analyze how mitochondrial function is related to human health and longevity.

Vocabulary Review The *Flash Cards* and *Match It* provide an interactive way to review chapter vocabulary.

Editable Worksheets Pages of Study Workbook B, Lab Manual B, and the Assessment Resources Book are available online. These documents can be easily edited using a word-processing program.

Lesson Overviews Lesson Overviews are based on Miller & Levine Biology. However, they are available on CD-ROM and can be edited for your class.

ASSESSMENT RESOURCES

Chapter Assessment Have students take an online version of the Chapter 7 Check Understanding.

Standardized Test Prep Students can take an online version of the Standardized Test Prep. You will receive their scores along with ideas for remediation.

Diagnostic and Benchmark Tests Use these tests to monitor your students' progress and supply remediation.

ASSESSMENT

Assess

Big idea Invite a volunteer to read the question. Review the meaning of *homeostasis.* Write the words *cell membrane, nucleus, ribosome,* and *mitochondrion* on the board.

Ask What are four things organisms must do to maintain homeostasis? *(grow, respond to environment, change the energy in food or sunlight into usable energy, reproduce)*

Ask How does the cell membrane help the cell maintain homeostasis? *(It prevents needed materials from leaving a cell and lets needed materials in.)* How does the nucleus help? *(It tells the cell when to make needed proteins)* What about ribosomes? *(They are involved in protein production.)* The mitochondria? *(They release energy for these processes.)*

Ask Would most cells be able to survive if they were missing any of these cell parts? Why or why not? *(No, these processes are needed for life.)*

CONSTRUCTED RESPONSE Paragraphs will vary. Explain the activity carefully. Be sure students understand that the *Hints* should be used as a guide to help them answer the questions. Review how to write a paragraph.

For Question 1, write it on the board and read it aloud. Underline the words *prokaryotic, plant cells,* and *animal cells.*

Say This question asks you to compare three kinds of cells: prokaryote, plant, and animal.

Read the first hint.

Ask Do plant cells or animal cells share any of these features with prokaryotic cells? If so, which features? *(Plant cells share both cell walls and photosynthesis with prokaryotes. Animal cells share neither feature.)*

Read the next hint.

Say Organelles are important cell parts. Both plant and animal cells have organelles that are enclosed in membranes.

Ask Why don't prokaryotes have organelles that are enclosed in membranes? *(They are much simpler cells than eukaryotes and first came about before any cells had nuclei.)*

Model writing the first one or two sentences using the hints on the board and the text. (Sample: Prokaryotic cells are very simple and have no nuclei, although they do have cell walls. Plant and animal cells are both eukaryotic cells. They both contain many of the same kinds of organelles.)

For Question 2, pair students and have them use the hints and your modeling to answer the question. Have students share their paragraphs with the class. Or have students write the paragraphs independently for homework or as an anchor activity.

7 CHECK Understanding

Assess the Big idea **Cellular Basis of Life, Homeostasis**

Write an answer to the question below.

Q: How are cell structures adapted to their functions?

Constructed Response

Write an answer to each of the questions below. The answer to each question should be one or two paragraphs long. To help you begin, read the **Hints** below each of the questions.

1. **Infer** Prokaryotic cells have some structures in common with plant cells and some structures in common with animal cells. Which two types of cells are MOST likely more closely related to one another? Explain your reasoning.

   Hint Prokaryotic cells have cell walls, and some can carry out photosynthesis like plant cells can. However, they do not have a nucleus.

   Hint Plant and animal cells have organelles that are enclosed by membranes, which prokaryotic cells do not have.

2. **Analyze Concepts** What happens to lipids when they are dropped into water? How does this action explain the structure of cell membranes?

   Hint Lipids have a head that mixes well with water and a tail that mixes well with oil.

   Hint Cell membranes are made up of lipid bilayers.

Foundations for Learning Wrap-Up

Use the index cards you prepared when you started the chapter as a tool to help you organize what you learned about the structure and function of cell parts.

Activity 1 Recall that different kinds of cells can have different cell parts. Sort the cards into four piles. The first pile should be cell parts that all cells have. The second pile should be cell parts that only animal cells have. The third pile should be cell parts that only plant cells and animal cells have. The fourth pile should be cell parts that only plant cells and bacteria cells (prokaryotes) have. Label the bottom right-hand corner of each card with the letters *A*, *P*, and *B* (animal, plant, and bacteria) accordingly to show which type of cell has each cell part.

Activity 2 Using metal or plastic rings and a piece of cardboard, combine your cards into a flip chart as shown below. Work with a partner to write a lesson on cell parts that uses the flip chart to explain important concepts from the chapter. Then use the flip chart to present your lesson to a classmate.

Foundations for Learning Wrap-Up

Have students take out their index cards and sort them into piles. Be sure students have written the proper function for each organelle. Either assign groups or pair students, and explain the activities.

7.1 Life Is Cellular

Understand Key Concepts

1. In many cells, the structure that controls the cell's activities is the
   a. cell membrane. c. nucleolus.
   b. organelle. d. nucleus.
2. Cells can be very different in size, shape, and the types of organelles they have. However, at some point all cells have DNA and
   a. a cell wall. c. mitochondria
   b. a cell membrane. d. a nucleus.

Test-Taking Tip

Watch for Qualifiers When you are answering a question, watch out for words such as *all*, *none*, *always*, and *not*. In question 2, notice the word *all*. This question is asking what part ALL cells have in common. You know animal cells do not have a cell wall and prokaryotes do not have a nucleus or organelles. Mitochondria are organelles. All cells must have a cell membrane. The correct answer is b.

3. What three statements make up the cell theory?

Think Critically

4. **Infer** How did the invention of the microscope help scientists come up with the cell theory?

7.2 Cell Structure

Understand Key Concepts

5. In eukaryotic cells, the cell's genetic information is found in the
   a. ribosomes. c. nucleus.
   b. lysosomes. d. cell membrane.
6. The organelles that break down lipids, sugars, and proteins into small molecules that can be used by the cell are called
   a. vacuoles. c. ribosomes.
   b. lysosomes. d. microfilaments.
7. Cell membranes are made up of
   a. lipid bilayers. c. carbohydrates.
   b. mitochondria. d. ribosomes.
8. What is the function of a ribosome?

Think Critically

9. **Infer** The pancreas is an organ found in many animals. It makes enzymes that are used by other parts of the digestive system. Enzymes are proteins that help speed up reactions in the body. Which type of cell structure(s) might produce those enzymes? Explain your answer.
10. **Classify** For each of the following, tell whether the structure is found only in eukaryotes or in both eukaryotes and prokaryotes: cell membrane, mitochondria, ribosome, Golgi apparatus, nucleus, cytoplasm, and DNA.

7.3 Cell Transport

Understand Key Concepts

11. The movement of water molecules across a membrane that does not let all materials pass through it is known as
    a. exocytosis. c. endocytosis.
    b. pumping. d. osmosis.
12. Can other substances besides water undergo osmosis? Explain.

Think Critically

13. **Apply Concepts** What would happen to a sample of your red blood cells if they were placed in a hypotonic solution? Explain.

Assessment Tips

Pair students and have them work together to answer the questions.

Remind them that *infer* means that the answers to Questions 4, 9, 15, 16, and 18 are not found directly in the text.

Point out that Questions 2, 5, 6, and 7 require students to know the function of certain organelles. Have students review these functions in the table located after Lesson 7.2.

For Questions 10, 17, and 18, review as a group the similarities and differences between prokaryotes and eukaryotes. Have students share their answers with the class and explain their reasoning for each answer.

Read the Test Taking Tip as a group and ask volunteers to identify assessment questions that include qualifiers. *(14)*

Answers

Lesson 7.1

UNDERSTAND KEY CONCEPTS

1. d
2. b
3. All living things are made up of cells. Cells are the basic units of structure and function in living things. New cells come from existing cells.

THINK CRITICALLY

4. Cells cannot be seen with the naked eye. Microscopes helped scientists finally see that living things were made of cells.

Lesson 7.2

UNDERSTAND KEY CONCEPTS

5. c
6. b
7. a
8. Ribosomes are the site where proteins are made.

THINK CRITICALLY

9. Enzymes are proteins, which are put together on ribosomes.
10. cell membrane: both; mitochondria: eukaryotes; ribosome: both; Golgi apparatus: eukaryote; nucleus: eukaryote; cytoplasm: both; DNA: both

Lesson 7.3

UNDERSTAND KEY CONCEPTS

11. d
12. Osmosis is the diffusion of water through a selectively permeable membrane. Only water can move by osmosis.

THINK CRITICALLY

13. The blood cells would swell and burst, because the concentration of solute is higher inside the cells than in the solution outside. Therefore, the solution outside the cells would tend to diffuse into the cells.

ASSESSMENT

Lesson 7.4

UNDERSTANDING KEY CONCEPTS

14. c

THINK CRITICALLY

15. I would expect skin cells to have fewer mitochondria than muscle cells because muscle cells need a lot of energy to contract. Mitochondria help get the cell energy that it can use.

16. The area in the heart tissue known as the pacemaker must not be communicating properly with the other heart cells. This person's pacemaker cells might be unable to send signals, or this person's heart cells might be unable to receive signals from the natural pacemaker.

Connecting Concepts

17. *Escherichia coli* is prokaryotic, a human erythrocyte is eukaryotic, a human ovum is eukaryotic, *Saccharomyces cerevisiae* is eukaryotic, and *Streptococcus pneumoniae* is prokaryotic.

18. It is more similar in size to the eukaryotic cells given here, so it is probably a eukaryote.

CHAPTER MYSTERY Have students read Solve the Chapter Mystery and reread the clues. Ask students to use the diagram Osmotic Pressure to review conditions that can affect the flow of water into and out of cells and the osmotic pressure in the cells. Explain that the lack of salts in the water Michelle was drinking upset the homeostasis of her body.

Ask Besides water, which substances did Michelle lose from her body in sweat? *(sodium and other salts)*

Ask How did the loss of those substances affect homeostasis in her body? Why? *(Because the loss of salts drastically upset the equilibrium between water in her cells and water in her bloodstream, cells in her body were damaged.)*

Ask Which drawing in the diagram Osmotic Pressure shows what happened to Michelle's red blood cells? *(the drawing of the animal cell in the column on the right)* What type of environment existed in her blood? *(a hypotonic environment)*

7 CHECK Understanding

7.4 Homeostasis and Cells

Understand Key Concepts

14. Which of the following is true of ALL single-celled organisms?
- **a.** They are all prokaryotes.
- **b.** They are all bacteria.
- **c.** They all reproduce.
- **d.** They all have a nucleus.

Think Critically

15. Infer Would you expect skin cells to contain more or less mitochondria than muscle cells? Explain your answer.

16. Infer Artificial pacemakers are devices that help keep heart muscle cells contracting at a steady rate. If a person needs an artificial pacemaker, what does that mean about his or her heart cells' ability to send and receive signals?

Connecting Concepts

Use Science Graphics

Use the data table to answer questions 17 and 18.

Cell Sizes

Cell	Approximate Diameter
Escherichia coli (bacterium)	0.5–0.8 μm
Human erythrocyte (red blood cell)	6–8 μm
Human ovum (egg cell)	100 μm
Saccharomyces cerevisiae (yeast)	5–10 μm
Streptococcus pneumoniae (bacterium)	0.5–1.3 μm

17. Classify Tell whether each of the cells listed is prokaryotic or eukaryotic.

18. Infer *Chlamydomonas reinhardtii* is a single-celled organism with an approximate diameter of 10 μm. Based on the information given here, do you think it is a prokaryote or a eukaryote? Explain your answer.

solve the CHAPTER MYSTERY

DEATH BY…WATER?

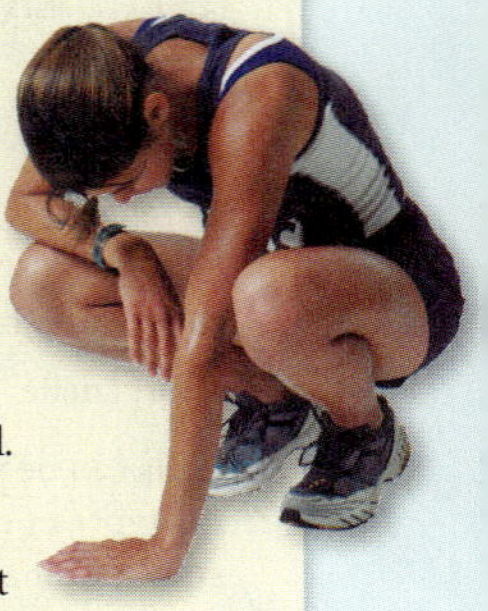

During the race, Michelle drank plenty of water. However, she didn't replace the salts she lost in her sweat. As a result, her blood became hypotonic. Osmotic pressure led the cells in her brain (and throughout her body) to swell.

As Michelle's blood became less concentrated, cells in her brain sent chemical signals to her kidneys to stop removing salts from her bloodstream. However, she continued to sweat. Sweating caused her to continue to lose salt through her skin.

By the end of the race, Michelle had lost a lot of salt and minerals. She had taken in so much water that homeostasis had broken down. Her cells were damaged by osmotic pressure that was unbalanced.

When Michelle went to the hospital, the doctors found that she had hyponatremia, or water intoxication. Left untreated, this condition can lead to death.

1. Relate Cause and Effect When a person sweats, water and salts are lost from the body. Michelle drank water but did not replace salts. How did this affect her cells?

2. Infer Imagine that Michelle drank both water and sports drinks containing the salts she lost. Do you think she would have felt better or worse? Explain.

3. Infer Do you think that hyponatremia happens because of osmosis or active transport? Explain your reasoning.

Untamed Science **Never Stop Exploring Your World.** Michelle's mysterious illness is just the beginning. Take a video field trip with the ecogeeks of Untamed Science to see where the mystery leads.

CHAPTER MYSTERY ANSWERS

1. When her body lost electrolytes and gained water, her body fluids became less concentrated (hypotonic) than the inside of her cells. The cells took in more water through osmosis and began to swell.

2. She would have replaced electrolytes with those in the sports drinks, so her body fluids would have a concentration of electrolytes that would have been more similar to the inside of her cells. She probably would have felt better, because her cells would not have swelled as much.

3. Hyponatremia happens because of osmosis. Water moves into the cells because the salt concentration inside the cells is greater than the salt concentration outside the cells.

Untamed Science Students can watch **Fishing for the Right Balance** to find out how saltwater and freshwater fishes maintain water balance.

Standardized Test Prep

Multiple Choice

1. Animal cells have all of the following EXCEPT
A mitochondria.
B chloroplasts.
C a nucleus.
D a cell membrane.

2. The nucleus includes all of the following structures EXCEPT
A cytoplasm. **C** DNA.
B a nuclear envelope. **D** a nucleolus.

3. The human brain is an example of a(n)
A cell.
B tissue.
C organ.
D organ system.

4. Which cell structures are sometimes found attached to the endoplasmic reticulum?
A chloroplasts
B nuclei
C mitochondria
D ribosomes

5. Which process always involves the movement of materials from inside the cell to outside the cell?
A diffusion
B exocytosis
C endocytosis
D osmosis

6. Which of the following is an example of active transport?
A facilitated diffusion
B osmosis
C diffusion
D endocytosis

7. The difference between prokaryotic and eukaryotic cells involves the presence of
A a nucleus.
B genetic material in the form of DNA.
C chloroplasts.
D a cell membrane.

Questions 8 and 9

In an experiment, scientists placed plant cells in solutions that had different concentrations of the sugar sucrose. Then the scientists measured the rate at which the cells absorbed sucrose from the solution. The results are shown in the graph below.

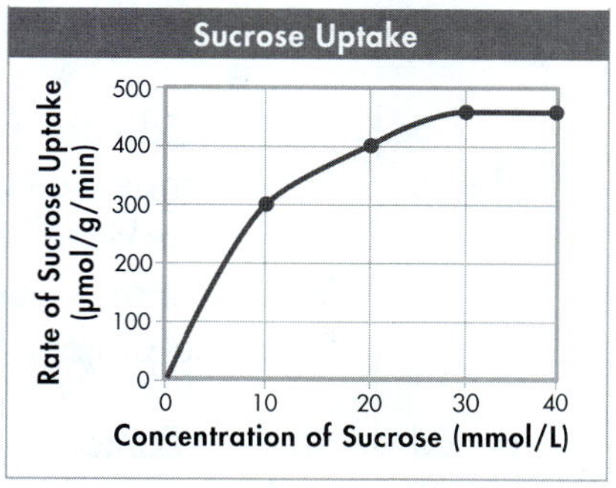

8. The graph shows that overall, sucrose
A cannot be absorbed by plant cells.
B uptake increased as concentration increased.
C uptake decreased as concentration increased.
D uptake stayed the same as concentration increased.

9. Based on the graph, the rate of sucrose uptake
A increased at a constant rate from 0 to 30 mmol/L.
B decreased at varying rates from 0 to 30 mmol/L.
C was less at 25 mmol/L than at 5 mmol/L.
D was constant between 30 and 40 mmol/L.

Open-Ended Response

10. What would you expect to happen if you placed a typical cell in fresh water?

If You Have Trouble With . . .

Question	1	2	3	4	5	6	7	8	9	10
See Lesson	7.2	7.2	7.4	7.2	7.3	7.3	7.1	7.3	7.3	7.3

Before the Test

Review with students the Test-Taking Tip.

Point out the word *EXCEPT* in Questions 1 and 2.

Remind students that the open-ended response question means that they should write a paragraph to answer the question. Have them paraphrase the question in their notebooks, and remind them about the steps you took when you modeled a paragraph. Specifically point out that you identified the meaning of the key words before starting to write.

Answers

1. B
2. A
3. C
4. D
5. B
6. D
7. A
8. B
9. D
10. Sample answer: A typical cell has dissolved salt particles in its cytoplasm, while fresh water does not have very much, if any, dissolved salt. The cell would thus be in a hypotonic solution. This means that the cell would take in water through osmosis from the outside solution. This would cause the cell to swell.

Test-Taking Tip

INTERPRET GRAPHS

Tell students that when answering a question based on experimental data, they should read the description of the experiment carefully to determine the steps followed. Explain that they should examine the description of the experiment and the labels on the graph axes to determine the independent and dependent variables. Suggest they look for any trends in the data. For example, on a graph, "if *x* increases, what happens to *y*?"

Cell Structure and Function **189**

Photosynthesis

Chapter Contents	NSES	Time	Core Resources
Chapter Preview			**Student Edition,** pp. 190–191 **Chapter Mystery,** p. 191 **Foundations for Learning,** p. 191
8.1 Energy and Life Chemical Energy • ATP Comes From Food	I, V, C.1.b, C.4.b, C.5.a, C.5.b, C.5.c, C.5.d	1/2 period 1/4 block	**Student Edition,** pp. 192–194 **Build Connections,** p. 193 **Study Workbook B** 8.1 Worksheets **Biology.com** *Visual Analogy:* ATP as a Charged Battery • 8.1 Lesson Assessment
8.2 Photosynthesis: An Overview Chlorophyll and Chloroplasts • High-Energy Electrons • A Summary of Photosynthesis	II, III, A.1.b, A.2.a, B.6, C.1.b, C.1.e, C.5.a, C.5.b	1 period 1/2 block	**Student Edition,** pp. 195–198 **Build Connections,** p. 196 **Inquiry Into Scientific Thinking,** p. 198 **Study Workbook B** 8.2 Worksheets **Biology.com** *InterActive Art:* Photosynthesis • *Visual Analogy:* Carrying Electrons • *Tutor Tube:* Sorting out Light-Independent and Light-Dependent Reactions • 8.2 Lesson Assessment
8.3 The Process of Photosynthesis The Light-Dependent Reactions: Making ATP and NADPH • Light-Independent Reactions: Making Sugars • What Affects Photosynthesis	III, V, B.3, B.6, C.1.a, C.1.b, C.1.e, C.5.a, C.5.b	1 1/2 periods 3/4 block	**Student Edition,** pp. 199–203 **Study Workbook B** 8.3 Worksheets **Biology.com** *Art in Motion:* Light-Dependent Reactions • *Art Review:* Light-Independent Reactions • 8.3 Lesson Assessment
Chapter Pre-Lab	II, A.1.b, A.2.a, C.1.e, C.5.b	1 period 1/2 block	**Student Edition,** p. 204 **Lab Manual B** *Plant Pigments and Photosynthesis*

Tools for ELL Instruction

ELL Handbook describes teaching strategies to make *Miller & Levine Biology: Foundation Edition* more accessible to ELL students.

Spanish Study Workbook is a Spanish translation of Study Workbook A.

Multilingual Glossary is the *Miller & Levine Biology* glossary translated into ten languages.

Tools to Challenge Students

Study Workbook A includes more challenging worksheets and activities such as Chapter Mystery activities that reinforce 21st Century Skills.

Lab Manual A labs are geared toward on-level students. The Lab Manual includes fifteen labs in addition to the Chapter Labs.

Biology.com Data Analysis activities offer students a way to practice data gathering and analyzing skills.

Visual Quizzes in the Assessment Resources Book are geared toward on-level students and use visuals from *Miller & Levine Biology.*

Additional Resources

Biology.com Untamed Science Video • Flash Cards

Study Workbook A 8.1 Worksheets
Spanish Study Workbook 8.1 Worksheets
Assessment Resources Book Visual Quiz

Study Workbook A 8.2 Worksheets
Spanish Study Workbook 8.2 Worksheets
Biology.com *Data Analysis:* Shedding Light on Marine Algae
Assessment Resources Book Visual Quiz

Study Workbook A 8.3 Worksheets
Lab Manual B *Data Analysis:* Rates of Photosynthesis
Spanish Study Workbook 8.3 Worksheets
Assessment Resources Book Visual Quiz

Lab Manual A *Plant Pigments and Photosynthesis*

Chapter Review

Student Edition Chapter Summary, p. 205
Study Workbook B Chapter 8 Review
Transparencies, pp. 99–110
Biology.com Untamed Science Video • Editable Worksheets of Study Workbooks A and B and Lab Manuals A and B • Chapter 8 Flash Cards and Crossword

Untamed Science DVD • Classroom Resources CD (includes lesson presentations and editable worksheets)

Chapter Assessment

Student Edition Check Understanding, pp. 206–209
Study Workbook B Chapter 8 Taking a Standardized Test
Assessment Resources Book Chapter 8 Test B
Biology.com Chapter 8 Assessment • Editable Worksheets of Chapter 8 Visual Quizzes* and Chapter 8 Tests A and B

Exam*View* *Assessment Suite* • Classroom Resources CD (includes lesson presentations and editable worksheets)

Time: 1 period, 1/2 block

* Visual Quizzes are based on visuals from *Miller & Levine Biology.* Review carefully to make sure they are appropriate for your class.

Pressed for Time?

Preview the Chapter Introduce students to the vocabulary terms for Lesson 8.1 and preview the visual The Stages of Photosynthesis.

Cover the Chapter Quickly Have students read all of Lesson 8.1 and all of Lesson 8.2, focusing on the visual The Stages of Photosynthesis.

Assess Assign the 8.1 Check Understanding, the 8.2 Check Understanding, and Questions 1–8 in the Chapter 8 Check Understanding.

Get the Picture

Tell students that the picture is a plant leaf under a microscope. Guide students to recognize that the rectangular, box-like structures are individual plant cells.

Ask What parts of this drawing are green? *(The tiny, round dots around the edges of each cell.)*

Point out that each tiny, round dot is an organelle or cell structure, called a chloroplast.

Ask Do you know what the function of a chloroplast is?

Explain that chloroplasts convert light energy into chemical energy through photosynthesis.

Invite students to guess what the prefix *chloro-* means. *(from the Greek* khloros *which means "green")*

Ask What do you think *chlorophyll* might mean? (Students should say something about *green.*)

Activate Prior Knowledge

Help students make the connection between energy from the sun and the food we eat.

Ask What did you have for breakfast (or lunch) today? *(Answers may vary.)* Why did you need to eat? *(Students may say that they were hungry.)*

Guide students to recognize that their bodies need food for energy.

Ask What do plants need to survive? *(Answers may include soil, minerals, water, and light.)* Remind students that, like us, plants use food for energy.

Emphasize the importance of light to plant survival. Explain that plants use photosynthesis to make food. Ask students to name plant cell organelles they can recall from Chapter 7.

NATIONAL SCIENCE EDUCATION STANDARDS

UNIFYING CONCEPTS AND PROCESSES

I, II, III, V

CONTENT

B.3, B.6, C.1.a, C.1.b, C.1.e, C.4.b, C.5.a, C.5.b, C.5.c, C.5.d

INQUIRY

A.1.b, A.2.a

8 Photosynthesis

Big idea Cellular Basis of Life

Q: How do plants and other organisms capture energy from the sun?

Leaf cells from Canadian pondweed (Elodea canadensis) (LM 2430×)

UbD Connect to the Big Idea

Invite a volunteer to read aloud the chapter number and the chapter title. Read the Big Idea and question. Emphasize the question, **How do plants and other organisms capture energy from the sun?**

Set the Purpose

PREVIEW AND PREDICT Draw the graphic organizer on the board. Guide students to the INSIDE feature and invite a volunteer to read each lesson number and title.

Ask What does *cellular basis of life* mean? *(Life is based on cells.)*

Ask How is this related to the ability of some organisms to capture and use energy from the sun? *(Photosynthesis happens at the cellular level.)*

CHAPTER
MYSTERY

INSIDE:

- 8.1 Energy and Life
- 8.2 Photosynthesis: An Overview
- 8.3 The Process of Photosynthesis

OUT OF THIN AIR?

When a tiny seed grows into a huge tree, where does all the extra mass come from? More than 300 years ago, a Flemish doctor named Jan van Helmont decided to find out. He chose a young willow tree that weighed just 2 kilograms. He planted it in a pot with 90 kilograms of dry soil and placed it in bright sunlight. He watered the plant as needed. Five years later, he took the tree from the pot and weighed it. It weighed about 77 kilograms. Where did the extra 75 kilograms come from? Did it come from the soil, the water—or right out of thin air?

Read for Mystery Clues As you read this chapter, look for clues to help you discover where the willow tree's extra mass came from. Then, solve the mystery at the end of the chapter.

FOUNDATIONS for Learning

Plant cells use light energy to make useful chemicals. These reactions take place in special parts of the cell. Before you read the chapter, label index cards with the words *water, carbon dioxide, oxygen, sugars,* and *light.* Label one sheet of paper as shown. At the end of the chapter are two activities that use the cards and sheets of paper to help answer the question: How do plants and other organisms capture energy from the sun?

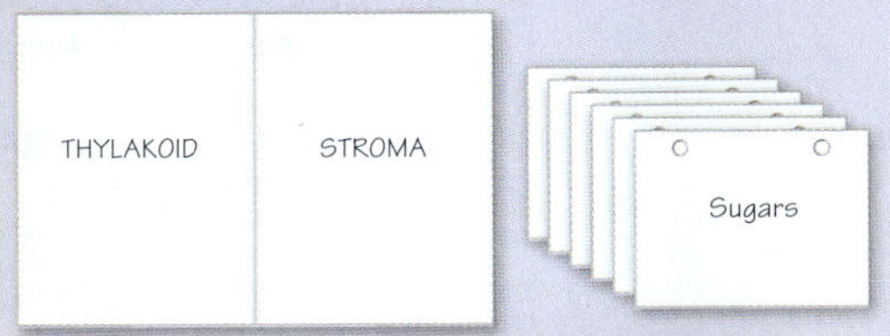

- Untamed Science Video
- Flash Cards

Chapter 8
Big Idea: Cellular Basis of Life

Chapter 8
How do plants and other organisms capture energy from the sun?

- **8.1:** How do organisms store energy?
- **8.2:** What cellular structures and molecules are involved in photosynthesis?
- **8.3:** How do photosynthetic organisms convert the sun's energy into chemical energy?

CHAPTER 8

Foundations for Learning

Pass out the index cards. Ask students to get out two sheets of paper. Help students understand the value of this activity.

Ask Why do you think energy is an important idea in this chapter? *(Answers may vary. Students should recognize that living things need energy.)*

Ask How will putting these words in groups help you understand this chapter? *(Putting related words in groups helps us understand how they are related, and helps us remember them.)*

Remind students that they will use these cards and sheets of paper as a study aid as they read this chapter. Familiarize students with the activity by modeling with the card for *light.* Demonstrate placing the *light* card on the sheet labeled *thylakoid.*

Untamed Science — The Untamed Science crew travels to Panama to learn about ongoing investigations regarding how plants respond to increased levels of CO_2 in the video **Soaking up the CO_2.**

CHAPTER MYSTERY — Invite a volunteer to read aloud the Chapter Mystery.

Ask Where are houseplants typically placed in a home? *(near a window)*

Encourage students to guess how plant care might relate to the Chapter Mystery. Have students predict how the chapter question, **How do plants and other organims capture energy from the sun?**, will help reveal the solution to this mystery.

Tell students that they will find clues in the chapter to answer the question, **Where did the extra 75 kilograms come from—the soil, the water, or right out of thin air?**

Digital Support

BIOLOGY.com — Extend your reach by using the digital assets offered at **Biology.com.**

LESSON 8.1

Pre-Reading

Objectives

8.1.1 Describe the role of ATP in cellular activities.

8.1.2 Explain where plants get the energy they need to produce food.

Preview the Pages

1. Preview the visuals and read the captions and labels. Point out that the energy-storage molecule ATP shown on the bottom of this page is shown again in the diagram on the next page.
2. Review the vocabulary words by having students find the words in the text and read them aloud.
3. Review the formulas for water and carbon dioxide. *(H_2O and CO_2)* Remind students that chemical bonds join atoms to form compounds.

Build Understanding

COMPARE/CONTRAST TABLE Direct students to Workbook B, where they will find a partially filled-in compare/contrast table. Remind students that a compare/contrast table will help them to see the similarities and differences between two things.

Speed Bump

Ask What is another name for *adenosine triphosphate? (ATP)*

Ask What are the three parts of ATP? *(adenine, ribose, and phosphate)*

Ask Why are the three phosphate groups important? *(ATP uses the groups to store and release energy.)*

NATIONAL SCIENCE EDUCATION STANDARDS

UNIFYING CONCEPTS AND PROCESSES

I, V

CONTENT

C.1.b, C.4.b, C.5.a, C.5.b, C.5.c, C.5.d

For the corresponding lesson in *Miller & Levine Biology,* see pages 226–228.

8.1 Energy and Life

Key Questions

Why is ATP useful to cells?

What happens during photosynthesis?

BUILD Understanding

Compare/Contrast Table As you read, create a table that compares autotrophs and heterotrophs. Think about how they get energy. Include in your table a few examples of each type of living thing.

In Your Workbook Go to your workbook to learn more about making a compare/contrast table. Complete the compare/contrast table for Lesson 8.1.

Chemical Energy

Energy is the ability to do work. Nearly everything you do takes energy. Lights, radios, and computers all use electrical energy to run. When you toast marshmallows on a campfire, you are using heat energy to warm the food. A car runs on fuel—the chemical energy in gasoline. All these things need energy to work.

Living things need energy, too. It takes plenty of energy to play soccer or other sports. Sometimes it is not as easy to tell that something uses energy. Did you know that even when you are sleeping, your cells are busy using energy? They are using energy to build new molecules and get rid of wastes. Without the ability to get and use energy, cells would die.

Energy Changes Form Light, heat, and electricity are just three forms of energy. Chemical energy is another form of energy. Chemical energy is the energy stored in chemical bonds. Energy can change from one form to another. When you light a candle, the wax burns. Chemical bonds between carbon and hydrogen atoms in the wax are broken. New bonds form between these atoms and oxygen from the air. What you get are new molecules of carbon dioxide and water. The bonds in the new molecules store less energy than before. The extra chemical energy that came from the old bonds changes form. The extra energy becomes the heat and light given off by the candle's flame.

ATP Living things use chemical energy. One of the most important chemicals that cells use to store and give off energy is called **adenosine triphosphate** (uh DEN uh seen try FAHS fayt). It's known as **ATP** for short. ATP is made up of adenine, a sugar called ribose, and three phosphate groups. The phosphate groups are the key to why ATP can store and release energy.

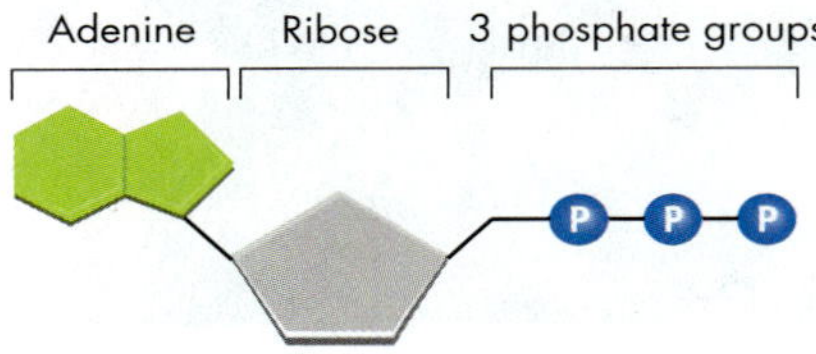

ATP ATP is the basic energy source used by all types of cells.

8.1 Energy and Life

Lesson Objectives

Describe the role of ATP in cellular activities.

Explain where plants get the energy they need to produce food.

BUILD Vocabulary

A. The chart below shows key terms from the lesson with their definitions. Complete the chart by writing a strategy to help you remember the meaning of each term. One has been done for you.

Term	Definition	How I'm Going to Remember the Meaning
Adenosine triphosphate (ATP)	Compound that cells use to store and release energy	ATP makes a cell able to do work.
Autotroph	A living thing that makes its own food	Auto- means "self-," so an autotroph is an organism that feeds itself.
Heterotroph	A living thing that gets its food by consuming other living things	Hetero- means "other." A heterotroph relies on other organisms for food.
Photosynthesis	Process by which autotrophs use sunlight, water, and carbon dioxide to produce carbohydrates and oxygen	Photo- means "light." Photosynthesis uses light to make its products.

Note: Students should highlight terms and definitions used throughout the chapter.

B. As you work through this lesson, you may find these terms in the activities. When you need to write a key term or a definition, **highlight** the term or the definition.

BUILD Connections

ATP as a Charged Battery An analogy takes two things that seem to be different and shows how they can be similar.

1. How is a partially charged battery like ADP? *ADP can be recharged and so can a weak battery.*
2. Why do cells contain only a small amount of ATP?
   - (A.) ATP cannot store large amounts of energy for a long time.
   - B. ATP releases energy too quickly.
   - C. ATP cannot carry out active transport.
   - D. ATP has only two phosphate groups.

BUILD Understanding

Compare/Contrast Table Use a compare/contrast table when you want to see the similarities and differences between two or more objects or processes. As you read, place an "X" in the box next to characteristics that are true of an autotroph or a heterotroph. One has been done for you.

	Autotroph	Heterotroph
Directly uses sunlight for energy	X	
An oak tree is an example.	X	
A mushroom is an example.		X
Makes its own food	X	
Consumes other living things to get energy		X
Is capable of photosynthesis	X	

Chemical Energy and ATP

Chemical Energy Living things use chemical fuels. One of the most important compounds that cells use to store and release energy is ATP.

Follow the directions. Then answer the questions.

1. Trace the outlines of the adenosine molecule and the three phosphate groups.
2. Cut out each phosphate group.
3. Arrange the phosphate groups on the adenosine molecule to form a molecule of ATP.

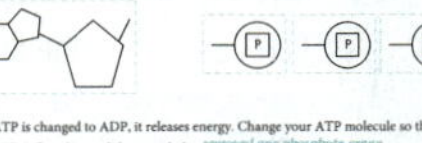

4. When ATP is changed to ADP, it releases energy. Change your ATP molecule so that it forms ADP. What change did you make? *removed one phosphate group*
5. ADP can be converted into ATP. Change your ADP molecule into ATP. What change did you make? *added one phosphate group*
6. Fill in the blanks with *two* or *three*. The primary difference in the structures of ATP and ADP is that ATP has *three* phosphate groups and ADP has *two* phosphate groups.

Storing Energy Adenosine diphosphate (ADP) is similar to ATP. However, ADP has only two phosphate groups instead of three. This difference is the key to how living things store energy. When a cell has extra energy, it can store small amounts of it by adding a phosphate group to ADP molecules, which turns ADP to ATP.

Releasing Energy Cells can give off the energy stored in ATP by breaking the chemical bonds between the second and third phosphate groups. A cell can add or subtract these phosphate groups whenever it needs to store or use energy.

Key Question Why is ATP useful to cells?
ATP can easily release and store energy by breaking and re-forming the bonds between its phosphate groups.

Using Chemical Energy Cells use ATP in many ways. For example, cells use it to power active transport. ATP provides the energy needed to move material into and out of many cells. ATP also helps proteins in muscles to slide closer together. This motion makes muscles flex. Cells that crawl, twist, and swim depend on ATP as well. Energy from ATP powers many other important events, such as the making of proteins and responses to chemical signals outside of the cell.

Short-Term Storage ATP is so useful that you might think cells would be packed with it. In fact, most cells have only a small amount—enough to last for a few seconds of activity. The reason is that ATP is great for giving off a small amount of energy very quickly. But, it is not good for storing large amounts for a long time. The sugar glucose is much better for energy storage. A single glucose molecule stores more than 90 times the energy of a molecule of ATP. So, it makes sense for cells to store energy in the form of glucose. Cells can then use the stored glucose to make ATP and ADP as needed.

ATP Comes From Food

Once cells use up their supply of ATP, they must somehow make more of it. So, where do living things get the energy they use to make ATP? The simple answer is that it comes from the chemical compounds that we call food. But, different living things get their food from different sources.

BUILD Vocabulary

adenosine triphosphate (ATP)
one of the principal chemical compounds that living things use to store and release energy

PREFIXES
The prefix *di-* means "two" and the prefix *tri-* means "three." A molecule of adenosine *tri*phosphate (ATP) has three phosphate groups, while adenosine *di*phosphate (ADP) has only two.

BUILD Connections

MOLECULAR BATTERY

When a phosphate group is added to an ADP molecule, it makes ATP. ADP has some energy, but not as much as ATP. ADP is like a partially charged battery. It can be fully charged by adding a phosphate group.

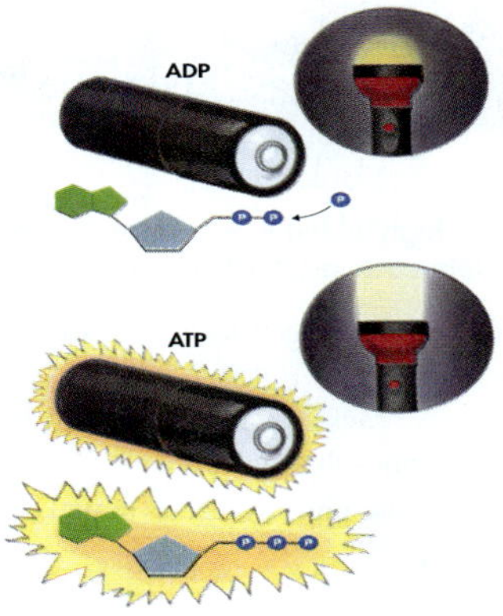

In Your Workbook Go to your workbook to learn more about how an ATP molecule stores energy.

ELL Focus on ELL: Extend Language

BEGINNING SPEAKERS Have students fill in a cluster diagram for Chemical Energy. Ask them to write the terms in the blue heads in the boxes surrounding the circle. Then, ask them to write one or two words in each box that describe the term. Review the diagrams with students.

During Reading

Active Reading

FIND THE MAIN IDEA Invite a volunteer to read the four blue headings on this page. Then, ask the volunteer to read aloud the paragraph Storing Energy.

Ask Which idea is the most important idea in this paragraph? *(A cell can store extra energy in ATP.)*

INFER Have students read the Short-Term Storage paragraph.

Ask Why is ATP not used for long-term energy storage? *(ATP is great for giving off small amounts of energy quickly, but glucose is much better for storing energy.)*

KEY QUESTION Read aloud the Key Question and answer.

Ask What do you think would happen to a cell if it did not have any ATP? *(Answers may vary.)* Lead students to conclude that a cell without ATP will die.

Build Connections

VISUAL ANALOGY An analogy is a kind of comparison: It takes two things that are different and shows how they can be similar.

Help students understand how the diagram compares a type of energy-storage molecule to a battery. Draw students' attention to the diagram of a molecule that goes with each battery and flashlight beam. Make sure students can identify the molecule with two phosphate groups as ADP and the one with three phosphate groups as ATP.

Ask What's the difference in the flashlight beams in ADP and ATP? *(ATP beam is much brighter than ADP beam because ATP has more stored energy.)*

Ask How does the energy stored in ADP compare to a battery? What about ATP? *(ADP doesn't store as much energy, so it's like a partially charged battery. ATP is like a fully charged battery.)*

Build Vocabulary

PREFIXES Have students find the words *adenosine diphosphate* and *adenosine triphosphate* in their text. As a group, pronounce each term.

Ask Which part of the name *adenosine triphosphate* tells you how many phosphate groups ATP contains? *(The prefix* tri- *which means "three.")* How many phosphate groups does ADP contain? *(two)* How do you know? *(The prefix* di- *means "two.")*

Ask What are some other words that begin with *di-* or *tri-*? *(divide, dialogue, diameter, tricycle, triangle)*

During Reading continued

Active Reading

COMPARE AND CONTRAST Have students reread the paragraphs about heterotrophs and autotrophs.

Ask How are heterotrophs and autotrophs alike? *(They both need food to survive.)* How are they different? *(Heterotrophs get their food by eating other living things, but autotrophs make their own food.)*

USE VISUALS Refer to the picture of the cheetah.

Ask How does sunlight provide food for the cheetah? *(The hare ate grass which used energy from sunlight to make its own food.)*

KEY QUESTION Read aloud the Key Question and answer.

Build Vocabulary

WORD ROOTS Explain the meaning of the roots: *troph*—feeder; *hetero*—other; and *auto*—self.

Ask What root tells how autotrophs get their food? (auto; *They make their own food.)* What other words do you know that begin with *auto*-? *(autobiography, autograph)*

Speed Bump

Ask Why are autotrophs so important? *(All living things would die without autotrophs.)*

Ask Autotrophs are also known as producers. Why? *(They produce their own food and food for other organisms.)*

Post Reading

ASSESSMENT TIPS For Question 4, ask students to think about how heterotrophs and autotrophs get energy.

MYSTERY CLUE Remind students that the food the tree makes gives it energy. Discuss how making food helps the willow tree gain mass. *(The tree uses energy to make proteins and carbohydrates, which have mass.)*

BUILD Vocabulary

heterotroph
a living thing that gets food by consuming, or eating, other living things

autotroph
a living thing that can capture energy from sunlight or chemicals to make its own food

photosynthesis
the process by which plants and other autotrophs use light energy to convert water and carbon dioxide into oxygen and high-energy carbohydrates such as sugars and starches

WORD ROOTS
Photosynthesis comes from the Greek words *photo*, meaning "light," and *synthesis*, meaning "putting together." Thus, *photosynthesis* means "using light to put something together."

Heterotrophs Eat Food Living things that get food by eating other living things are known as **heterotrophs.** Some heterotrophs get their food by eating plants, such as grasses. Others, such as the cheetah shown at right, get food by eating other animals. Still other heterotrophs—mushrooms, for example—get food by breaking down the tissues of dead things.

Autotrophs Make Food The energy in nearly all food molecules first came from the sun. Plants, algae, and some bacteria can use light energy from the sun to make their own food. Living things that make their own food are called **autotrophs.** All life on Earth depends on autotrophs to get energy from sunlight and store it in the molecules that make up food. The process by which autotrophs use the energy of sunlight to produce high-energy carbohydrates—sugars and starches—that can be used as food is known as **photosynthesis.**

Key Question What happens during photosynthesis?
During photosynthesis, plants change the energy of sunlight into chemical energy stored in the bonds of carbohydrates.

Autotrophs and Heterotrophs
Grass uses sunlight to make food. African hares get energy by eating grass. Cheetahs, in turn, get their energy by eating other animals, like the hare.

CHECK Understanding

Apply Vocabulary

Use the highlighted words from the lesson to complete each sentence correctly.

1. An animal that gets its food by eating other animals is a(n) ______.
2. ______ provides a cell with energy when it loses a phosphate group.
3. ______ use photosynthesis to make sugars.

Critical Thinking

4. **Explain** Why are decomposers, such as mushrooms, heterotrophs and not autotrophs?

5. **Write to Learn** Answer the first clue of the mystery. In your answer explain how autotrophs get food and how that might lead to an increase in mass.

MYSTERY CLUE

Like all plants, the willow tree van Helmont planted was an autotroph. How might its ability to use the sun's energy to make food help it gain mass? (Hint: See p. 193.)

BIOLOGY.com Search Lesson 8.1 GO • Lesson Assessment

Check Understanding

1. heterotroph
2. Adenosine triphosphate (ATP)
3. Autotrophs
4. Mushrooms cannot make their own food. They must get it from the remains of other living things.
5. The tree is an autotroph, which means it uses photosynthesis to make its own food. The food gives it energy, which it can use to make material for growth. These compounds have mass, so the tree increases in mass as it makes and stores the compounds.

8.2 Photosynthesis: An Overview

Chlorophyll and Chloroplasts

Our lives, and the lives of nearly every living thing on Earth, depend on the sun and photosynthesis. For photosynthesis to take place, autotrophs must capture light energy from the sun.

Light Energy from the sun travels to Earth in the form of light. Sunlight, which our eyes see as "white" light, is actually a mixture of different wavelengths. Many of these wavelengths are visible to our eyes. Our eyes see the different wavelengths of visible light as different colors: shades of red, orange, yellow, green, blue, indigo, and violet.

Pigments Plants gather the sun's energy with light-absorbing molecules called pigments. The most important pigment in plants is chlorophyll (KLAWR uh fil). Out of all the different colors of light, the chlorophyll found in plants absorbs blue-violet light and red light the best. Chlorophyll does not absorb green light very well. Leaves look green because they reflect green light.

Key Question What role do pigments play in the process of photosynthesis?
Living things that carry out photosynthesis use pigments to get energy from sunlight.

Chloroplasts In plants, photosynthesis takes place inside organelles called chloroplasts. These organelles hold many flat, bag-shaped membranes called **thylakoids** (THY luh koydz). Thylakoids are connected to one another and arranged in stacks. These stacks are called grana. Chlorophyll and other pigments are found in these thylakoid membranes. The liquid-filled space around the thylakoids is known as the **stroma.** The picture below shows what makes up a chloroplast.

The Chloroplast In plants, photosynthesis takes place inside chloroplasts.

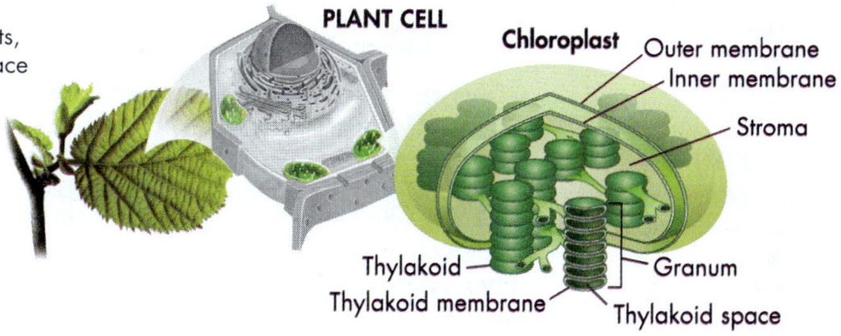

Key Questions

- *What role do pigments play in the process of photosynthesis?*
- *What are electron carrier molecules?*
- *What are the reactants and products of photosynthesis?*

BUILD Understanding

KWL Chart Make a chart with three columns labeled What I **K**now, What I **W**ant to Know, and What I **L**earned. Before you read the lesson, fill in the first two columns. Complete the last column as you read the lesson.

In Your Workbook Go to your workbook to learn more about making a KWL chart. Complete the KWL chart for Lesson 8.2.

BIOLOGY.com Search Lesson 8.2 GO • Visual Analogy

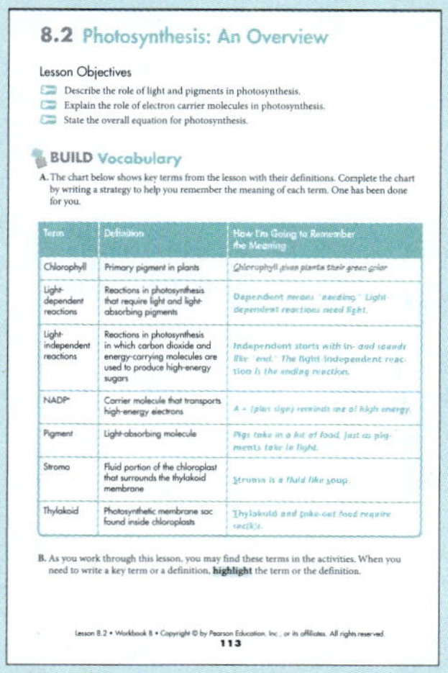

8.2 Photosynthesis: An Overview

Lesson Objectives
- Describe the role of light and pigments in photosynthesis.
- Explain the role of electron carrier molecules in photosynthesis.
- State the overall equation for photosynthesis.

BUILD Vocabulary

A. The chart below shows key terms from the lesson with their definitions. Complete the chart by writing a strategy to help you remember the meaning of each term. One has been done for you.

Term	Definition	How I'm Going to Remember the Meaning
Chlorophyll	Primary pigment in plants	*Chlorophyll gives plants their green color*
Light-dependent reactions	Reactions in photosynthesis that require light and light-absorbing pigments	*Dependent means "needing." Light-dependent reactions need light.*
Light-independent reactions	Reactions in photosynthesis in which carbon dioxide and energy-carrying molecules are used to produce high-energy sugars	*Independent starts with in- and sounds like "end." The light independent reaction is the ending reaction.*
$NADP^+$	Carrier molecule that transports high-energy electrons	*A + (plus sign) reminds me of high energy.*
Pigment	Light-absorbing molecule	*Pigs take in a lot of food, just as pigments take in light.*
Stroma	Fluid portion of the chloroplast that surrounds the thylakoid membrane	*Stroma is a fluid like soup.*
Thylakoid	Photosynthetic membrane sac found inside chloroplasts	*Thylakoid and take-out food require sacks.*

B. As you work through this lesson, you may find these terms in the activities. When you need to write a key term or a definition, **highlight** the term or the definition.

113

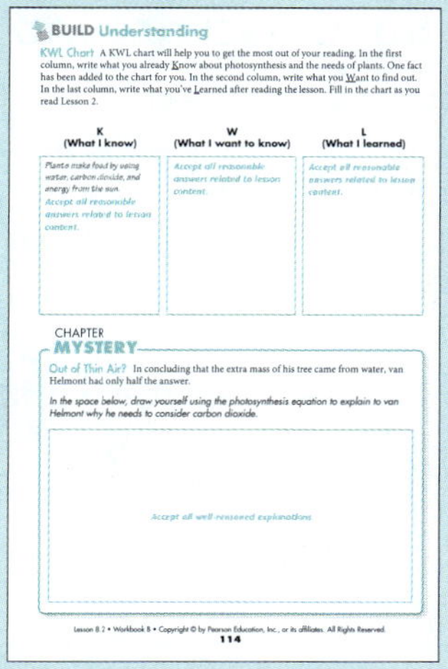

BUILD Understanding

KWL Chart A KWL chart will help you to get the most out of your reading. In the first column, write what you already Know about photosynthesis and the needs of plants. One fact has been added to the chart for you. In the second column, write what you Want to find out. In the last column, write what you've Learned after reading the lesson. Fill in the chart as you read Lesson 2.

K (What I know)	W (What I want to know)	L (What I learned)
Plants make food by using water, carbon dioxide, and energy from the sun. *Accept all reasonable answers related to lesson content.*	*Accept all reasonable answers related to lesson content.*	*Accept all reasonable answers related to lesson content.*

CHAPTER MYSTERY

Out of Thin Air? In concluding that the extra mass of his tree came from water, van Helmont had only half the answer.

In the space below, draw yourself using the photosynthesis equation to explain to van Helmont why he needs to consider carbon dioxide.

Accept all well-reasoned explanations.

114

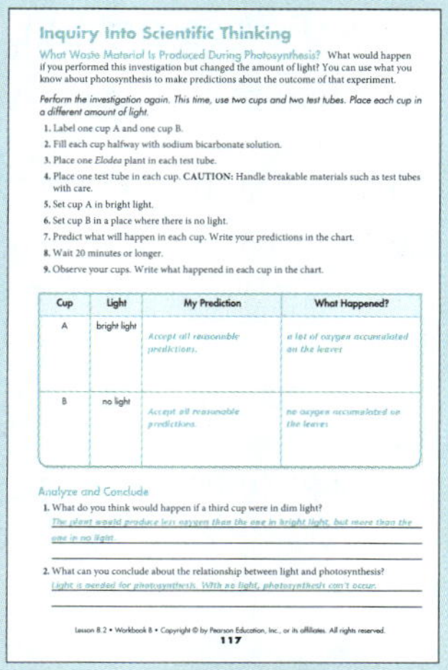

Inquiry Into Scientific Thinking

What Waste Material Is Produced During Photosynthesis? What would happen if you performed this investigation but changed the amount of light? You can use what you know about photosynthesis to make predictions about the outcome of that experiment.

Perform the investigation again. This time, use two cups and two test tubes. Place each cup in a different amount of light.

1. Label one cup A and one cup B.
2. Fill each cup halfway with sodium bicarbonate solution.
3. Place one *Elodea* plant in each test tube.
4. Place one test tube in each cup. **CAUTION:** Handle breakable materials such as test tubes with care.
5. Set cup A in bright light.
6. Set cup B in a place where there is no light.
7. Predict what will happen in each cup. Write your predictions in the chart.
8. Wait 20 minutes or longer.
9. Observe your cups. Write what happened in each cup in the chart.

Cup	Light	My Prediction	What Happened?
A	bright light	*Accept all reasonable predictions.*	*a lot of oxygen accumulated on the leaves*
B	no light	*Accept all reasonable predictions.*	*no oxygen accumulated on the leaves*

Analyze and Conclude

1. What do you think would happen if a third cup were in dim light?
The plant would produce less oxygen than the one in bright light, but more than the one in no light.

2. What can you conclude about the relationship between light and photosynthesis?
Light is needed for photosynthesis. With no light, photosynthesis can't occur.

117

Pre-Reading

Objectives

8.2.1 Explain the role of light and pigments in photosynthesis.

8.2.2 Explain the role of electron carrier molecules in photosynthesis.

8.2.3 State the overall equation for photosynthesis.

Preview the Pages

1. Review with students the vocabulary words, pronouncing each word aloud.
2. Preview all the visuals and their captions/labels.
3. Explain that the Stages of Photosynthesis diagram shows the events taking place inside a chloroplast. Make sure students see the relationship between the chloroplast in this diagram and the diagram of a chloroplast on this page.

Activate Prior Knowledge

Ask students what they know about the differences between animal and plant cells. Guide students to discuss that only plant cells have chloroplasts.

Build Understanding

KWL CHART Direct students to Workbook B where they'll find a partially filled-in KWL chart. Ask them to fill in the first two boxes with what they know and want to find out about photosynthesis.

NATIONAL SCIENCE EDUCATION STANDARDS

UNIFYING CONCEPTS AND PROCESSES

II, III

CONTENT

B.6, C.1.b, C.1.e, C.5.a, C.5.b

INQUIRY

A.1.b, A.2.a

For the corresponding lesson in *Miller & Levine Biology,* see pages 230–234.

During Reading

Active Reading

KEY QUESTION Read aloud the Key Question and answer on the previous page.

Ask What is a pigment? *(a light-absorbing molecule)*

DRAW DIAGRAMS Have students draw and label a chloroplast in their notebooks. Beneath their drawings have them draw two-column charts with the headings *Structure* and *Function*. Have students write the names of the structures in the first column and the function descriptions in the second column.

SELF-QUESTION Demonstrate how asking questions and reviewing the text for answers can help clarify new information. Read the first paragraph on this page aloud.

Ask Does this make sense? What does it mean that the energy levels of electrons go up?

Say Sometimes it helps to reread information to find the answers to questions. Let's reread the first four sentences of the paragraph.

Ask When chlorophyll takes in light from the sun, where in the chlorophyll does much of the energy go? *(to the electrons in the chlorophyll)*

Ask When energy is added to the chlorophyll's electrons, what happens? *(The energy levels of the electrons go up.)*

KEY QUESTION Read aloud the first Key Question and answer.

Ask How does $NADP^+$ transfer energy to other parts of the cell? *(It transfers electrons to other molecules.)*

KEY QUESTION Read aloud the second Key Question and answer.

Ask What is needed for photosynthesis? *(light, water, and carbon dioxide)* What is produced during photosynthesis? *(sugars and oxygen)*

Build Connections

VISUAL ANALOGY Discuss the visual analogy of an oven mitt and $NADP^+$.

Ask What does the oven mitt represent? *(the electron carrier $NADP^+$)* What does the potato stand for? *(a pair of high-energy electrons and an H^+ ion)*

Build Vocabulary

CHEMICAL NAMES Tell students that ATP and $NADP^+$ are actually initialisms of chemical names. Show students how initialisms are formed.

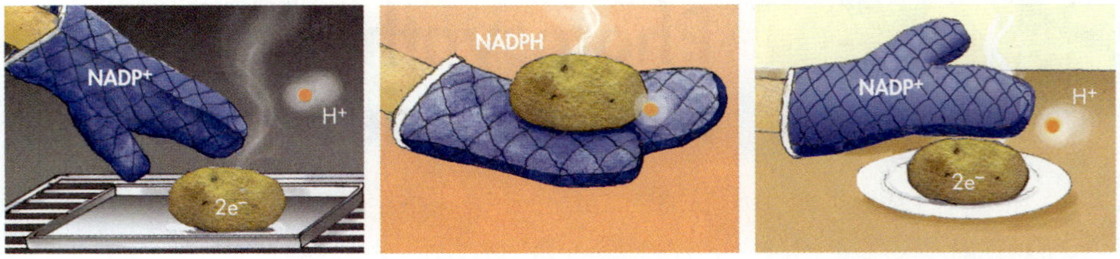

BUILD Connections

CARRYING ELECTRONS

$NADP^+$ is a carrier molecule that transports pairs of high-energy electrons (and an H^+ ion). Compare its role to that of an oven mitt used to transport a hot baked potato.

BUILD Vocabulary

thylakoid
a baglike photosynthetic membrane found in chloroplasts

stroma
the liquid-filled space surrounding the thylakoids in chloroplasts

$NADP^+$
one of the molecules that accepts high-energy electrons from chlorophyll and carries them to other molecules during photosynthesis

light-dependent reaction
a reaction of photosynthesis that uses light energy to make ATP and NADPH

light-independent reaction
a reaction of photosynthesis that uses ATP, NADPH, and carbon dioxide to make high-energy sugars

CHEMICAL NAMES
When a plus sign appears in a chemical name, it means the chemical has a positive charge and can accept electrons.

High-Energy Electrons

Light is a form of energy. Anything that absorbs light takes in energy. When chlorophyll takes in light, much of that energy is moved directly to electrons in the chlorophyll. The energy levels of the electrons go up. A stream of these high-energy electrons is what makes photosynthesis work.

These high-energy electrons cannot keep their energy for long on their own. They need a special carrier. Plant cells use molecules to carry high-energy electrons from chlorophyll to other places. One of these carrier molecules is **$NADP^+$** (nicotinamide adenine dinucleotide phosphate). It accepts and holds two high-energy electrons, along with a hydrogen ion (H^+). This process changes $NADP^+$ into NADPH. The NADPH carries the electrons to other parts of the cell. There, the electrons and their energy can be used to help build useful molecules, such as sugars.

Key Question What are electron carrier molecules?
Electron carrier molecules are compounds that can transfer a pair of high-energy electrons, along with most of their energy, to another molecule.

A Summary of Photosynthesis

Photosynthesis has many steps. It usually has 6-carbon sugars ($C_6H_{12}O_6$) as the final product. All of the steps of photosynthesis can be summed up in the following equation:

In Symbols:

$$6CO_2 + 6H_2O \xrightarrow{\text{light}} C_6H_{12}O_6 + 6O_2$$

In Words:

$$\text{Carbon dioxide} + \text{Water} \xrightarrow{\text{light}} \text{Sugars} + \text{Oxygen}$$

Plants use the sugars from photosynthesis to make other carbohydrates, such as starches. The sugars are also used to make other compounds, such as proteins and lipids.

Key Question What are the reactants and products of photosynthesis?
Photosynthesis uses light energy to change water and carbon dioxide (reactants) into sugars and oxygen (products).

ELL Focus on ELL: Build Background

ALL SPEAKERS Write the words *equation, reactants,* and *products* on the board. Make sure students understand the meaning of the three words. Have students find the equation for photosynthesis in their texts.

Invite students to work in pairs. Ask them to copy the equation for photosynthesis onto a sheet of paper. Then, have partners draw a circle around each reactant and a square around each product.

Discuss students' responses. Then, review the chemical formulas for carbon dioxide (CO_2), water (H_2O), and six-carbon sugars ($C_6H_{12}O_6$) as they appear in the equation for photosynthesis.

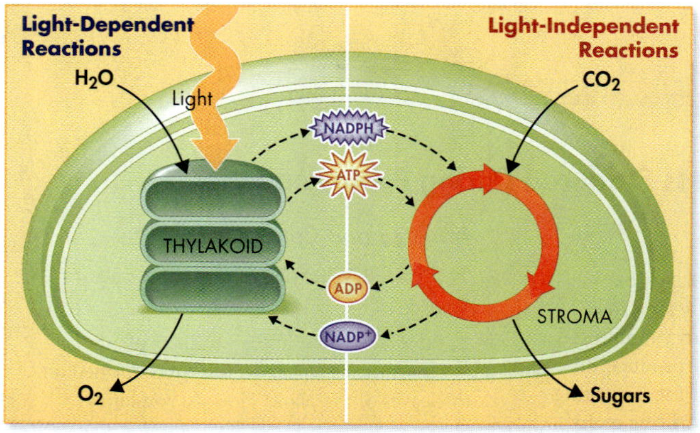

The Stages of Photosynthesis There are two stages of photosynthesis: light-dependent reactions and light-independent reactions.

Light-Dependent Reactions The equation for photosynthesis looks simple, but that is because it does not show all of the smaller steps. In fact, there are two sets of reactions. The first set is the **light-dependent reactions.** They take place in the thylakoids. These reactions need light and light-absorbing pigments to take place. They use energy from sunlight and electrons and hydrogen ions from water to make high-energy ATP and NADPH. Oxygen gas, the oxygen we breathe every day, is also released.

Light-Independent Reactions The second set of reactions in photosynthesis is called the **light-independent reactions.** In this set, plants use the ATP and NADPH made in the light-dependent reactions. They make high-energy sugars out of carbon dioxide from the air. Light-independent reactions do not need light. They take place in the stroma.

CHECK Understanding

Apply Vocabulary

Use the highlighted words from the lesson to complete each sentence correctly.

1. The set of reactions that takes place in the stroma are called the ______.

2. The light-dependent reactions need ______, a carrier molecule that is able to accept high-energy electrons and a hydrogen ion.

Critical Thinking

3. Describe the overall process of photosynthesis. Include the reactants and the products.

4. Write to Learn Answer the second clue of the mystery. Think about the things that plants need for photosynthesis. Which of these things might cause the plant to gain mass?

MYSTERY CLUE

Van Helmont concluded that water must have supplied the extra mass gained by the tree. In fact, he had only half of the answer. What photosynthesis reactant was he missing in his conclusion? (**Hint:** See p. 196.)

BIOLOGY.com Search Lesson 8.2 GO • Lesson Assessment

Check Understanding

1. light independent reactions
2. $NADP^+$
3. In photosynthesis, light energy, water and carbon dioxide are used to make sugars and release oxygen as a waste product.
4. Photosynthesis requires both water and carbon dioxide. The carbon dioxide provides much of the mass that eventually becomes sugars. When Van Helmont concluded that water supplied the extra mass, he was not accounting for the carbon dioxide, which plants get from air.

Active Reading

USE DIAGRAMS Remind students that the diagram shows a chloroplast. Point out that the diagram has a left-hand half and a right-hand half.

Say One half shows the light-dependent reactions. The other half shows the light-independent reactions.

Ask Where do light-dependent reactions take place? *(thylakoid)* What two things does a chloroplast need in light-dependent reactions? *(light and H_2O)* What's the output? *(O_2)*

Invite a volunteer to explain light-independent reactions. *(Be sure students see that the reaction takes place in the stroma, that CO_2 is the input, and sugars are the output.)*

Say NADPH and $NADP^+$, and ATP and ADP, are interdependent and recycled between the two reactions.

Ask What happens to the ATP and NADPH produced by the light-dependent reactions? *(They are used by the light-independent reactions.)*

Speed Bump

Ask How many stages are in photosynthesis? *(two)* What are they? *(light independent and light dependent)*

Ask The prefix *in-* means "without." What is a light-independent reaction doing without? *(sunlight)*

Post Reading

ASSESSMENT TIPS Have students revisit the diagram of a chloroplast and the Stages of Photosynthesis diagram.

For Question 3, review the meanings of *products* and *reactants.*

MYSTERY CLUE Remind students that a reactant is an element or compound that enters into a chemical reaction, and that water is one of the reactants in photosynthesis. Tell students to use the Stages of Photosynthesis diagram to find the reactants and products of photosynthesis.

Wrap-Up Activity

INTERVIEWS Divide students into pairs. Have one student represent the light-dependent reactions and the other the light-independent reactions. Have partners interview one another to find out what they need, where they will get it, what they will produce, and where their products will go.

Inquiry Into Scientific Thinking

MAKE CONNECTIONS Remind students that all green plants perform photosynthesis, even those that live in water. Explain that this experiment captures a product of photosynthesis that usually cannot be seen.

Ask Why is it a good idea to use an aquatic plant instead of a land-dwelling plant in this activity? *(Oxygen bubbles can be seen as they form.)*

PREDICT Ask volunteers to predict what might happen if you changed the activity in different ways.

Ask What do you think would happen if you used plain water instead of sodium bicarbonate solution? *(Fewer bubbles would form.)*

Ask How would your observations change if you placed the cup in ordinary room light instead of bright light? *(Fewer bubbles would form.)* Why? *(Less light would decrease photosynthesis.)* What do you think would happen if you then moved the cup to a sunny windowsill? *(Bubbles would form more quickly.)* Why? *(Photosynthesis would increase.)*

Ask What do you think will happen in your cup during the next 24 hours? *Hint:* Think about how conditions in the room will change. *(Bubbles will stop forming at night as the light-dependent reactions stop.)*

USE DIAGRAMS Look at the art showing the *Elodea* in the test tube.

Ask Why do you think this Inquiry asked you to place the test tube upside down? *(If the test tube wasn't upside down, the bubbles would escape through the solution and into the air. There would be no way to measure how much oxygen collected.)*

Lead a Discussion

Talk about how photosynthesis on a global level provides the oxygen needed by living things. Like the *Elodea* in the test tube, many of the world's photosynthetic organisms live in aquatic environments. Phytoplankton, for example, produce a large percentage of Earth's oxygen.

Ask How could ocean pollution affect the supply of oxygen? *(Widespread pollution could kill large numbers of phytoplankton, decreasing the amount of oxygen in the air.)*

INQUIRY into Scientific Thinking

What Waste Product Is Produced During Photosynthesis?

What is a waste product? It is like garbage in your kitchen. You may eat the food that comes out of a jar, but your body cannot use the jar. The empty jar is a waste product. But you might be able to use the jar for something else, like storing leftover soup. During many reactions, cells make waste products as well as needed products. Sometimes these waste products cannot be used by the cell at all. At other times, the waste product is used in another part of the cell or by other cells. In this activity, you will study photosynthesis in a water plant called *Elodea*. You will see one product of photosynthesis. You will decide if it is a waste product for this plant.

1. Fill a clear, plastic cup halfway with sodium bicarbonate solution. This solution has carbon dioxide in it.
2. **a.** Place a piece of plant in a test tube. Make sure that the cut part of the plant is at the bottom.
   b. Add sodum bicarbonate solution. **CAUTION:** *Be careful with glass.*
3. **a.** Hold your finger over the open end of the test tube.
   b. Turn the test tube over. Lower it to the bottom of the cup. Make sure no air is inside the test tube.
4. Place the cup in bright light.
5. Wait 20 minutes.
6. Look closely at the leaves. Write down what you see.

Analyze and Conclude

1. Observe What did you see on the plant's leaves?

2. Infer Think about the products of photosynthesis. What do you think formed on the leaves of the plant? How do you know?

3. Draw Conclusions Do you think that the product that formed on the leaves is a waste product? Explain.

4. Infer Think about the reactants of photosynthesis. What do you think happened to the carbon dioxide in the sodium bicarbonate solution?

5. Apply Concepts Which plant organelle carries out photosynthesis? How can you tell that this plant's cells have that organelle?

In Your Workbook Get more help for this activity in your workbook.

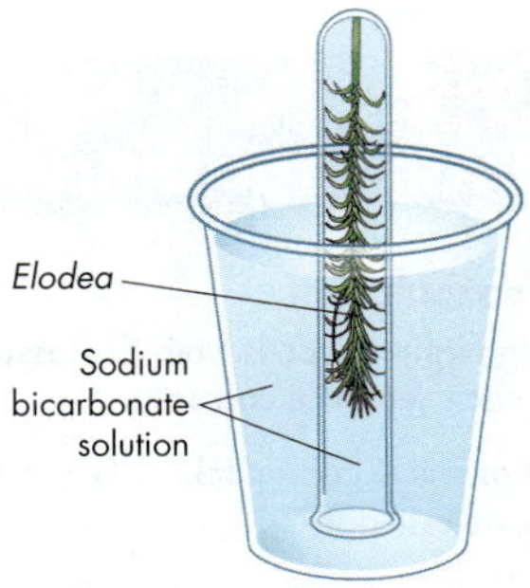

Materials

large clear plastic cup, sodium bicarbonate solution, elodea, large test tube

Analyze and Conclude

1. gas bubbles
2. The products of photosynthesis are sugars and oxygen. The bubbles that formed are a gas, so it must be oxygen.
3. The oxygen that was removed from the plant is excess oxygen not used for cellular respiration. It is a waste product of photosynthesis.
4. The reactants of photosynthesis are carbon dioxide and water. The plant probably used the carbon dioxide during photosynthesis.
5. The plant organelle that carries out photosynthesis is the chloroplast. Chloroplasts contain chlorophyll, which is green. This plant is green and produces gas when placed in light, so it must have chloroplasts.

8.3 The Process of Photosynthesis

The Light-Dependent Reactions: Making ATP and NADPH

Remember that the steps of photosynthesis are divided into two sets. They are the light-dependent reactions and the light-independent reactions. The light-dependent reactions cannot take place in the dark. They are the reason that plants need light.

The light-dependent reactions take place in the thylakoids of chloroplasts. These baglike membranes hold groups of chlorophyll and proteins known as **photosystems.** Photosystems take in sunlight and use it to add energy to electrons. These electrons are passed to a set of carriers in the thylakoid membrane. There are two photosystems, and they are named in order of their discovery—not in the order that they do their work.

Photosystem II The pigments of photosystem II absorb light energy and release high-energy electrons. These electrons then get passed down the **electron transport chain.** This chain is a group of carrier proteins. They use energy from the electrons to pump H^+ ions inside the thylakoid.

As light shines on chlorophyll, more and more electrons enter the electron transport chain. Chlorophyll gets new electrons when enzymes break up water molecules (H_2O). As plants take electrons from water, oxygen and hydrogen are left behind. The oxygen goes into the air that we breathe. The hydrogen becomes ions that are left inside the thylakoid.

ATP Formation In photosystem II, hydrogen ions build up inside the thylakoid. Some of these ions come from water breaking up. Others come from the electron transport chain. The building up of H^+ ions inside the thylakoid makes the outside of the thylakoid more negatively charged than the inside. This difference in charge and number of H^+ ions on either side of the membrane is what powers the making of ATP.

H^+ ions cannot cross the membrane directly. Instead, they pass through a protein called ATP synthase that is like a revolving door in the membrane. The buildup of H^+ ions on one side causes them to pass through ATP synthase and force it to turn. As it turns, it adds a phosphate group to ADP to make ATP.

Key Questions

What happens during the light-dependent reactions?

What happens during the light-independent reactions?

What factors affect photosynthesis?

BUILD Understanding

Flowchart As you read, create a flowchart that clearly shows the steps of the light-dependent reactions.

In Your Workbook Go to your workbook to learn more about making a flowchart. Complete the flowchart for Lesson 8.3.

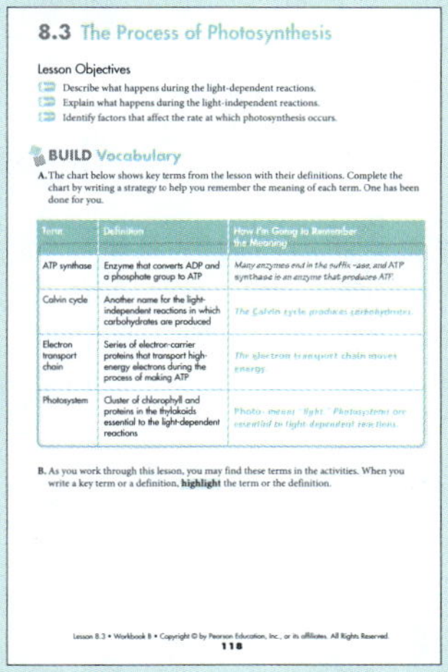

8.3 The Process of Photosynthesis

Lesson Objectives

Describe what happens during the light-dependent reactions.

Explain what happens during the light-independent reactions.

Identify factors that affect the rate at which photosynthesis occurs.

BUILD Vocabulary

A. The chart below shows key terms from the lesson with their definitions. Complete the chart by writing a strategy to help you remember the meaning of each term. One has been done for you.

Term	Definition	How I'm Going to Remember the Meaning
ATP synthase	Enzyme that converts ADP and a phosphate group to ATP	*Many enzymes end in the suffix -ase, and ATP synthase is an enzyme that produces ATP.*
Calvin cycle	Another name for the light-independent reactions in which carbohydrates are produced	
Electron transport chain	Series of electron-carrier proteins that transport high-energy electrons during the process of making ATP	
Photosystem	Cluster of chlorophyll and proteins in the thylakoids essential to the light-dependent reactions	

B. As you work through this lesson, you may find these terms in the activities. When you write a key term or a definition, **highlight** the term or the definition.

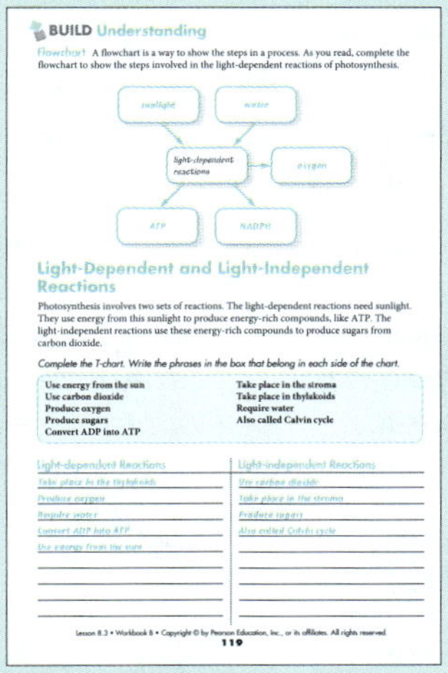

BUILD Understanding

Flowchart A flowchart is a way to show the steps in a process. As you read, complete the flowchart to show the steps involved in the light-dependent reactions of photosynthesis.

Light-Dependent and Light-Independent Reactions

Photosynthesis involves two sets of reactions. The light-dependent reactions need sunlight. They use energy from this sunlight to produce energy-rich compounds, like ATP. The light-independent reactions use these energy-rich compounds to produce sugars from carbon dioxide.

Complete the T-chart. Write the phrases in the box that belong in each side of the chart.

Use energy from the sun	**Take place in the stroma**
Use carbon dioxide	**Take place in thylakoids**
Produce oxygen	**Require water**
Produce sugars	**Also called Calvin cycle**
Convert ADP into ATP	

Light-dependent Reactions	Light-independent Reactions

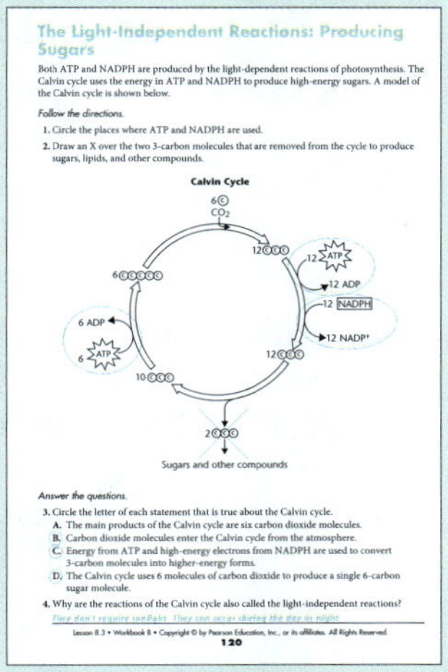

The Light-Independent Reactions: Producing Sugars

Both ATP and NADPH are produced by the light-dependent reactions of photosynthesis. The Calvin cycle uses the energy in ATP and NADPH to produce high-energy sugars. A model of the Calvin cycle is shown below.

Follow the directions.

1. Circle the places where ATP and NADPH are used.
2. Draw an X over the two 3-carbon molecules that are removed from the cycle to produce sugars, lipids, and other compounds.

Calvin Cycle

Answer the questions.

3. Circle the letter of each statement that is true about the Calvin cycle.
   A. The main products of the Calvin cycle are six carbon dioxide molecules.
   B. Carbon dioxide molecules enter the Calvin cycle from the atmosphere.
   C. Energy from ATP and high-energy electrons from NADPH are used to convert 3-carbon molecules into higher-energy forms.
   D. The Calvin cycle uses 6 molecules of carbon dioxide to produce a single 6-carbon sugar molecule.
4. Why are the reactions of the Calvin cycle also called the light-independent reactions?

LESSON 8.3

Pre-Reading

Objectives

8.3.1 Describe what happens during the light-dependent reactions.

8.3.2 Describe what happens during the light-independent reactions.

8.3.3 Identify factors that affect the rate at which photosynthesis occurs.

Preview the Pages

1. Review with students all the vocabulary, pronouncing each word aloud. Have students find the words in the text and in the diagrams.
2. Preview all the visuals, and read the captions and labels.
3. Be sure that students know that the Light-Dependent Reactions diagram and the Light-Independent Reactions diagram represent the two stages of photosynthesis. To get a complete understanding of photosynthesis, they must view both diagrams. Have students compare each of these diagrams with the Stages of Photosynthesis diagram in the previous lesson.

Build Understanding

FLOWCHART Direct students to Workbook B, where they will find a flowchart for this lesson. Remind them that a flowchart will break down a process into a series of steps. Suggest that students make two different flowcharts, one for each photosystem.

NATIONAL SCIENCE EDUCATION STANDARDS

UNIFYING CONCEPTS AND PROCESSES

III, V

CONTENT

B.3, B.6, C.1.a, C.1.b, C.1.e, C.5.a, C.5.b

For the corresponding lesson in *Miller & Levine Biology,* see pages 235–241.

LESSON 8.3

During Reading

Active Reading

USE VISUALS Review the diagram, pointing out that students should follow the yellow arrow from left to right. Read the captions in the five boxes aloud, pausing after each box to trace the activity in the diagram.

Ask Does this diagram show the light-dependent reactions or the light-independent reactions? *(light-dependent reactions)* How can you tell, without reading the diagram label? *(These reactions use sunlight. That means they must be the light-dependent reactions.)*

Ask What does the bright yellow arrow show? *(the path followed by high-energy electrons)*

MAKE CONNECTIONS Ask students if they have ever passed through a turnstile when entering a sports arena, an amusement park, or a train station. Explain that a turnstile turns, or rotates, each time a person goes through it.

Ask Which part of this diagram is acting like a turnstile? *(ATP synthase)* What does it do? *(ATP synthase rotates, producing energy that is used to convert ADP to ATP.)*

Ask What causes ATP synthase to rotate? *(H^+ ions passing through the membrane)*

USE GRAPHIC ORGANIZERS On the board, write the names of the steps for light-dependent reactions. Have students arrange the steps in a flowchart. Then, have students describe each step in simplified language and discuss where in the text each step is mentioned.

KEY QUESTION Read aloud the Key Question and answer.

Ask Where does the oxygen gas released during photosynthesis come from? *(water)*

Speed Bump

Again, direct students' attention to the diagram.

Ask Why does Photosystem II come before Photosystem I? *(The photosystems were named in the order they were discovered, not in the order they do their work. Photosystem I was discovered before Photosystem II.)*

Photosystem I The electrons from photosystem II lost energy when they pumped ions across the membrane. In photosystem I, light gives them energy again. Then, they go through another electron transport chain. Here, they are used to make $NADP^+$ into NADPH, which goes on to the next set of reactions.

Summary of Light-Dependent Reactions The light-dependent reactions give off oxygen gas. They also make ATP and NADPH. These compounds are important for the cell. They provide the energy needed to build sugars in the light-independent reactions.

Key Question What happens during the light-dependent reactions?
The light-dependent reactions use water and energy from sunlight to make oxygen and change ADP and $NADP^+$ into the energy carriers ATP and NADPH.

Light-Dependent Reactions
The light-dependent reactions of photosynthesis take place in the thylakoids of the chloroplast. They use energy from sunlight to produce ATP, NADPH, and oxygen.

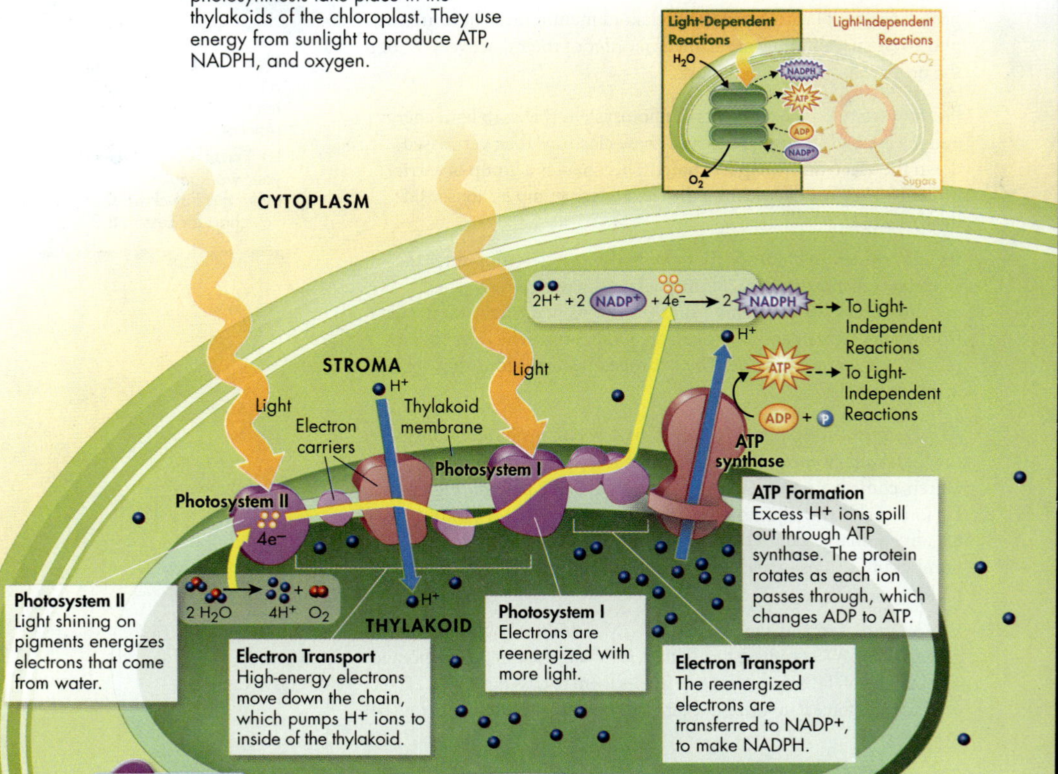

200 BIOLOGY.com Search Lesson 8.3 GO • Art in Motion

ELL Focus on ELL: Access Content

ALL SPEAKERS Students may have trouble keeping track of the physical locations inside the cell where photosynthesis takes place. Students can lose their orientation as they follow events occurring inside a thylakoid.

Have students work in small groups. On a sheet of paper, have groups draw a large square, which will serve as a plant cell. Then, have students work together to label the cytoplasm and sketch a chloroplast, a thylakoid, and the stroma. Have each group read aloud the text on the previous page about ATP formation, pausing each time they read the words *inside, outside, side, pass through,* and *turn,* while pointing to the corresponding location in their drawing. As a wrap-up, have a class discussion summarizing ATP formation.

Seeing Green The color green of most plants is caused by the reflection of green light by the pigment chlorophyll.

Light-Independent Reactions: Making Sugars

The ATP and NADPH made by the light-dependent reactions store a lot of chemical energy. However, they are not very stable. They last only a few minutes. During the light-independent reactions—also called the Calvin cycle—plants build high-energy sugars. These sugars are stable, so they can store energy for a long time. The Calvin cycle is named after the American scientist Melvin Calvin.

Carbon Dioxide Enters the Cycle One of the reactants of the Calvin cycle is carbon dioxide. It comes from the air around a plant. An enzyme joins carbon dioxide molecules with molecules that are already in the cell. Then, one chemical step at a time, the cycle produces energy-rich carbohydrates. These steps use energy from ATP and NADPH.

Sugar Production With each turn of the cycle, a few energy-rich carbohydrate molecules leave the cycle. This is the step that makes the Calvin cycle important. These molecules become the building blocks that the plant cell uses to make sugars, lipids, amino acids, and other compounds. The plant will use these compounds as the food and building materials it needs to grow.

The rest of the higher-energy molecules stay in the cycle. Enzymes use ATP to change them to molecules that will join with new carbon dioxide molecules. Then, the cycle begins again.

Summary of Light-Independent Reactions The Calvin cycle uses compounds made in the light-dependent reactions and carbon dioxide from the air to make sugars. The plant uses the sugars to meet its energy needs and to build molecules needed for growth. When other living things eat plants, they, too, get the energy they need.

Key Question What happens during the light-independent reactions?
During the light-independent reactions, ATP and NADPH are used to make high-energy sugars.

BUILD Vocabulary

photosystem
the light-collecting units of the chloroplast

electron transport chain
a series of proteins in which high-energy electrons are used to change ADP to ATP

WORD ROOTS
The Greek word *photo* means "light." A photosystem is a system of pigment molecules that collects energy from light.

Focus on Academic Language

- chain • cycle • stable

Write the words on the board and review their meanings. Remind students that in *electron transport chain*, *chain* means "a series of connected events that happen one after the other." For example, a row of dominoes falling shows a *chain* reaction.

Say A chain is different from a cycle. A *cycle* is a sequence of events that is repeated again and again. Think of a bicycle. Each full turn of the wheels is a *cycle.*

Tell students that the word *stable* has multiple meanings. Here it means "steady and not likely to change."

Ask What does it mean when I say, Food prices are stable?

Ask Which is more stable, ATP and NADPH or sugars? *(sugars)* Which will store energy longer? *(sugars)*

Build Vocabulary

WORD ROOTS Remind students that learning word roots can help them understand new words.

Ask What words do you know that contain the root word *photo-*? *(Sample answers: photosynthesis, photograph, photocopy.)*

Active Reading

SUMMARIZE Have students summarize the Summary of Light-Independent Reactions paragraph in one or two sentences in their notebooks. Invite students to read aloud their summaries.

Say Plants store energy as sugars, such as $C_6H_{12}O_6$.

Ask Where does a plant get the carbon atoms needed to make these sugars? *(from CO_2 in the air)*

DRAW CONCLUSIONS Make sure students understand the relationship between the light independent and light dependent reactions.

Ask Since the light-independent reactions do not require sunlight, could these reactions keep going if sunlight weren't available? *(no)* Why not? *(These reactions need ATP and NADPH, the products of the light-dependent reactions. Without sunlight, plants could not make ATP and NADPH.)*

Ask What happens to ATP and NADPH after they give up their energy during the light-independent reactions? *(They become ADP and $NADP^+$, which will be reused during the light-dependent reactions.)*

USE CONTEXT CLUES Read the Sugar Production paragraph aloud.

Say Cover your book while I read the first paragraph. Raise your hand when you hear the word *building*.

Ask What molecules serve as building blocks? *(higher-energy molecules that leave the Calvin cycle)*

Ask What does the plant do with these building blocks? *(uses them as food and building materials so it can grow)*

INFER Invite a volunteer to read the second paragraph. Allow for a pause after the last sentence.

Ask What happens to the higher-energy molecules that are not used to make sugars and other compounds? *(They stay in the Calvin cycle and pick up new carbon dioxide molecules as the cycle goes around again.)*

KEY QUESTION Read aloud the Key Question and answer.

Ask Which is left over after the Calvin cycle—$NADP^+$ or NADPH? *($NADP^+$)* ADP or ATP? *(ADP)*

Ask Which is more stable—$NADP^+$ or NADPH? *($NADP^+$)* ADP or ATP? *(ADP)*

During Reading continued

Active Reading

INTERPRET DIAGRAMS Guide students to the diagram of the Calvin cycle. Read each label as you walk through the diagram.

Ask Why is the Calvin cycle called a *cycle*? *(because molecules in the last step are used in the first step)*

DRAW CONCLUSIONS Have students consider the effects of temperature change on photosynthesis.

Ask What happens to photosynthesis if the temperature drops below 0°? *(Photosynthesis slows down and is inefficient.)*

KEY QUESTION Read aloud the Key Question and answer.

Ask When you purchase a new plant, it often comes with information about its care. What three pieces of information do you think are normally included? *(how much light and water the plant needs and what temperature it grows best in)*

Speed Bump

Ask What does the Calvin cycle get from the air? *(carbon dioxide)*

Ask What do plants use some of the 3-carbon-atom molecules for in the Calvin cycle? *(to make other molecules that are used to make compounds such as sugars)*

The End Results The two parts of photosynthesis work together. The light-dependent reactions change the energy of sunlight into chemical energy. The light-independent reactions use that energy to make sugars from carbon dioxide and water. Both plants and animals use the sugars and oxygen from photosynthesis to live and grow.

Factors Affecting Photosynthesis

Temperature, Light, and Water Photosynthesis can speed up or slow down depending on several factors. Temperature is one factor that can affect the rate of photosynthesis. Photosynthesis enzymes work best between 0°C and 35°C. Light intensity is another factor. Very bright light speeds up photosynthesis. Water levels also affect the rate of photosynthesis.

Key Question What factors affect photosynthesis?
Three important factors that affect photosynthesis are temperature, light intensity, and the availability of water.

Light-Independent Reactions The light-independent reactions of photosynthesis take place in the stroma. They use ATP and NADPH from the light-dependent reactions to make high-energy sugars.

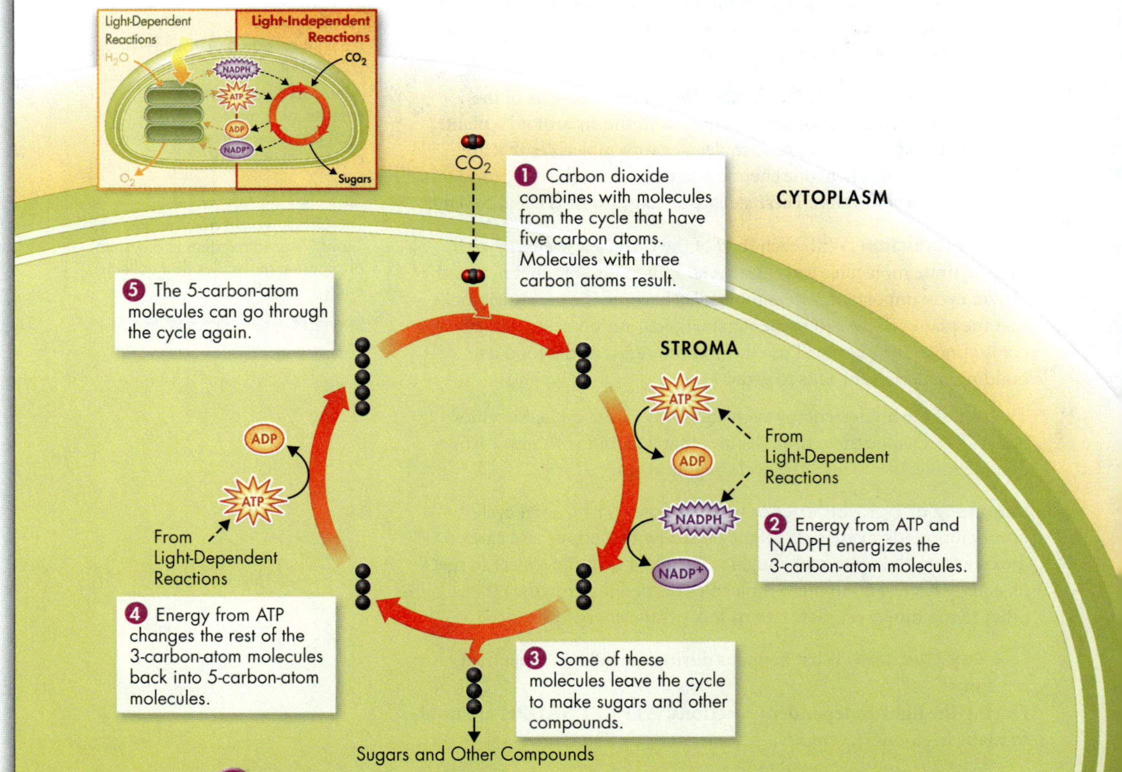

202 BIOLOGY.com Search Lesson 8.3 GO • Art Review

Hands-On Learning

DEMONSTRATE THE CALVIN CYCLE Select five students to form a circle to demonstrate the Calvin cycle. Choose one person to begin. Hand that student three tennis balls. Have the student hand the three balls to the person on the left, who continues passing them around the circle, keeping all three balls together. Intervene just before the balls return to the first person; replace one ball with a new ball. Have the students repeat the cycle.

Say I took away one ball. What does this ball represent? *(sugars)* I replaced it with a new ball. What did the new ball stand for? *(carbon dioxide)* I left two balls in before the CO_2 was added. What did they represent? *(other high-energy molecules)*

Ask Why did I leave two balls in the circle? *(so the cycle could keep going)*

Ask What's missing from this demonstration? How did all these molecules have the energy to keep going? *(They got energy from ATP and NADPH.)*

Photosynthesis in Extreme Conditions Plants lose water through the tiny openings in their leaves that let in carbon dioxide. When it is hot, most plants close these openings to keep from drying out. But, doing so means less carbon dioxide enters, which slows photosynthesis. Some plants that live in dry, sunny areas have special ways to save water and still carry out photosynthesis.

C4 Plants C4 plants have a special chemical pathway that gets carbon into the Calvin cycle even when there is not much carbon dioxide available. The pathway uses extra ATP but lets the plants carry out photosynthesis when it is hot. Corn and sugar cane are examples of C4 plants.

CAM Plants CAM plants save water by taking air into their leaves only at night. In the dark, carbon dioxide is used to make acids. During the day, these acids are turned back into carbon dioxide for photosynthesis. Some examples of CAM plants are pineapple trees, many desert cacti, and ice plants.

CAM Plants Plants like this ice plant can survive in dry places. The leaves let in carbon dioxide only at night, minimizing water loss.

CHECK Understanding

Apply Vocabulary

Use the highlighted words from the lesson to complete each sentence correctly.

1. Energy produced by the _______ is used to pump hydrogen ions from the stroma to the inside of the thylakoid.

2. A ______ is a cluster of pigments and proteins that allows plants to absorb light energy and transfer it to electrons.

Critical Thinking

3. **Sequence** Put the following events of the light-dependent reactions in the order in which they occur: photosystem I, photosystem II, making NADPH, and pumping hydrogen ions into the inside of the thylakoid.

4. **Compare and Contrast** List at least three differences between the light-dependent and light-independent reactions of photosynthesis.

5. **Write to Learn** Answer the third clue of the mystery. How would labeling carbon atoms in carbon dioxide tell apart the products of photosynthesis?

MYSTERY CLUE

Melvin Calvin was able to label carbon atoms in carbon dioxide. He used these labeled atoms to show what happens to the carbon that is used during the light-independent reactions. Where does this carbon end up? (**Hint:** See p. 201.)

BIOLOGY.com Search Lesson 8.3 GO • Lesson Assessment

Active Reading

USE CONTEXT CLUES Tell students to read the first paragraph.

Ask In this paragraph, what does *extreme conditions* refer to? *(very hot, dry conditions)* How did you know that? *(The last sentence starts with the words* some plants that live in dry, sunny areas.*)*

DRAW A DIAGRAM Have students refer to the flowchart they began earlier in this lesson. Invite volunteers to draw their flowcharts on the board. Have students work together as a group to finish the flowchart. Correct and redirect students as necessary.

Ask How can a flowchart help you understand this lesson? *(It breaks down a process into a series of steps, which makes it easier to understand.)*

Post Reading

ASSESSMENT TIPS Have students review the diagrams for light-independent reactions and light-dependent reactions before they begin the assessment. Invite students to work in pairs to summarize the events in the diagrams to one another.

MYSTERY CLUE Ask students to return to the Calvin Cycle diagram. Discuss with students where they think Melvin Calvin's labeled carbon atoms could possibly have wound up.

Wrap-Up Activity

KWL CHART Have students finish the KWL chart they made during the previous lesson. Ask students to work in groups to write down what they learned about photosynthesis.

Check Understanding

1. electron transport chain

2. photosystem

3. 1) photosystem II, 2) pumping hydrogen ions into the inside of the thylakoid, 3) photosystem I, 4) making NADPH

4. Sample answer: The light-dependent reactions cannot take place in the dark, while the light-independent reactions can. The light-dependent reactions produce ATP and NADPH, while the light-independent reactions use them up. The light-dependent reactions require water, while the light-independent reactions require carbon dioxide.

5. Sample answer: The final products of the light-independent reactions are oxygen and sugars ($C_6H_{12}O_6$). Oxygen does not contain any carbon atoms, but sugars do. The labeled carbon from the carbon dioxide ends up being used during photosynthesis to make the sugars.

CHAPTER LAB

Pre-Lab

Review the meaning and importance of pigments. Remind students that pigments are molecules that absorb light. The most important pigment in plants is chlorophyll. This green pigment absorbs blue-violet and red light.

Ask How does chlorophyll make plants look green? *(It reflects green light.)*

Tell students that plants also contain red and orange pigments. They reflect red and orange light and absorb other colors of light.

Review the Background Questions and the Pre-Lab Questions as a class.

Lab

Tell students they will perform the chapter lab *Plant Pigments and Photosynthesis* described in **Lab Manual B.**

SAFETY Make sure students work in a well-ventilated area. Caution them to avoid breathing in the fumes and to handle glassware with care. Students should wash their hands thoroughly after the lab.

Post-Lab

VIEW A RAINBOW Pass light through a prism. Students should see that the light separates into the colors of the rainbow. Explain that white light is made up of different colors of light, each with a different wavelength. The separation occurs because each of them refracts, or bends, a different amount. Draw a model on the board. Show that photosynthetic pigments absorb some colors well but not others.

Skills Lab

GUIDED INQUIRY

Pre-Lab: Plant Pigments and Photosynthesis

Problem Do red leaves have the same pigments as green leaves?

Materials paper clips, one-hole rubber stoppers, chromatography paper strips, metric ruler, green and red leaves, coin, sheet of paper, large test tubes, test-tube rack, glass-marking pencil, 10-mL graduated cylinder, isopropyl alcohol, colored pencils

Lab Manual Chapter 8 Lab

Skills Focus Predict, Analyze Data, Draw Conclusions

Connect to the Big idea Almost all life on Earth depends, directly or indirectly, on energy from sunlight. Photosynthesis is the process in which light energy is captured and converted to chemical energy. Many reactions are required for this conversion, which takes place in the chloroplasts of plant cells. Some of the reactions depend on light and some do not. Plant pigments play a major role in the light-dependent reactions. In this lab, you will use chromatography to compare the pigments in red leaves with those in green leaves.

Background Questions

a. **Compare and Contrast** What do all plant pigments have in common? How are they different?

b. **Review** Why do most leaves appear green?

c. **Review** What property makes chlorophyll so important for photosynthesis?

Pre-Lab Questions

Preview the procedure in the lab manual.

1. **Design an Experiment** What is the purpose of this lab?
2. **Control Variables** What is the control in this lab?
3. **Design an Experiment** Why must the pigment line be at least 2 cm from the bottom of the paper?
4. **Predict** Will red leaves contain the same amount of chlorophyll as green leaves? Why or why not?

BIOLOGY.com Search Chapter 8 GO

Visit Chapter 8 online to test yourself on chapter content and to find activities to help you learn.

Untamed Science Video Journey to Panama with the Untamed Science crew to discover how CO_2 affects plant growth.

Data Analysis Look at pigment color data in the ocean to find out how marine algae photosynthesize in the blue light available underwater.

Tutor Tube Learn how to sort out the products and reactants in both the light-dependent and light-independent reactions.

Art Review Focus on the thylakoid membrane to review your knowledge of the light-dependent reactions.

InterActive Art Bring the components of photosynthesis together to run an animation.

Art in Motion Watch the steps of the light-dependent reactions in motion at the molecular level.

Visual Analogies Compare ATP production to a charged battery. See how the electron transport chain is like passing a hot potato.

Pre-Lab Answers

BACKGROUND QUESTIONS

a. Sample answer: All the pigments are light-absorbing molecules. They differ in how well they absorb different wavelengths of light.

b. Leaves reflect green light because chlorophyll does not absorb light well in the green spectrum.

c. Sample answer: Much of the energy absorbed by chlorophyll molecules is transferred directly to electrons in the chlorophyll molecule.

PRE-LAB QUESTIONS

1. The purpose is to find out whether red leaves have the same pigments as green leaves.
2. Students may say that the green leaf is the control.
3. Sample answer: When the strips are placed in the test tube, the pigment line must be above the alcohol.
4. Sample answer: Red leaves will have less chlorophyll than green leaves because they are not green.

NATIONAL SCIENCE EDUCATION STANDARDS

UCP II

CONTENT C.1.e, C.5.b

INQUIRY A.1.b, A.2.a

8 CHAPTER Summary

8.1 Energy and Life

- ATP can easily release and store energy by breaking and re-forming the bonds between its phosphate groups.
- This ability makes ATP very useful as a basic energy source for all cells.
- During photosynthesis, plants change the energy of sunlight into chemical energy stored in the bonds of carbohydrates.

adenosine triphosphate (ATP) (p. 192)
heterotroph (p. 194)
autotroph (p. 194)
photosynthesis (p. 194)

8.2 Photosynthesis: An Overview

- Living things that carry out photosynthesis use pigments to get energy from sunlight.
- An electron carrier is a compound that can transfer a pair of high-energy electrons, along with most of their energy, to another molecule.
- Photosynthesis uses light energy to change water and carbon dioxide (reactants) into sugars and oxygen (products).

thylakoid (p. 195)
stroma (p. 195)
NADP⁺ (p. 196)
light-dependent reactions (p. 197)
light-independent reactions (p. 197)

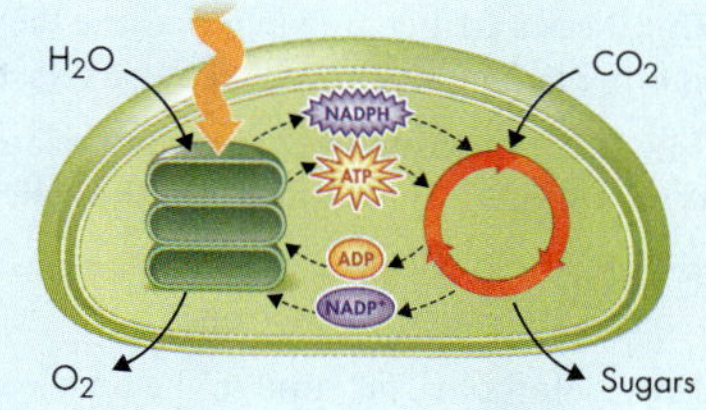

8.3 The Process of Photosynthesis

- The light-dependent reactions use energy from sunlight to make oxygen and change ADP and NADP⁺ into the energy carriers ATP and NADPH.
- During the light-independent reactions, ATP and NADPH are used to make high-energy sugars.
- Three important factors that affect photosynthesis are temperature, light intensity, and the availability of water.

photosystem (p. 199)
electron transport chain (p. 199)

UbD Transfer the Big Idea

TRANSFER In a class discussion, have students review the difference between heterotrophs and autotrophs. Point out that now that they have learned about photosynthesis, they should have a much better understanding of how autotrophs make their own food, as well as how autotrophs provide heterotrophs with the food they need to produce the ATP used by their cells for energy. Then, ask each student to create a series of scenes that "runs the movie backward"—beginning with a human activity and ending with sunlight being absorbed by chlorophyll. Give students a choice of how they create their scenes. For example, they could draw a series of cartoons, make a flowchart, or create a numbered list.

Chapter Summary

Think Visually

USE GRAPHICS Have students complete the following flowchart about photosynthesis.

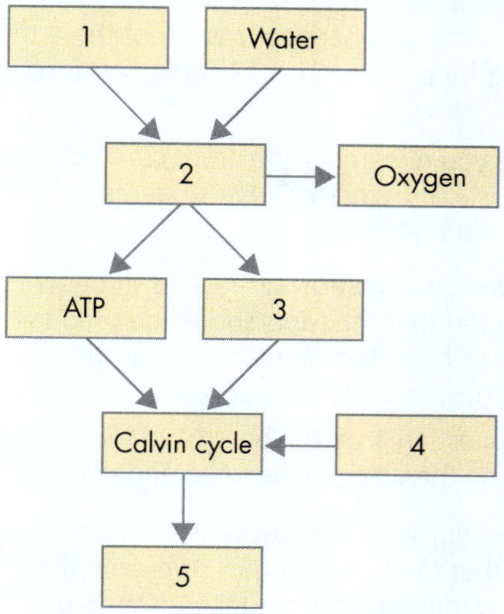

Answers

1. light **2.** light-dependent reactions
3. NADPH **4.** carbon dioxide **5.** sugars

BIOLOGY.com Study Online

REVIEW AND ASSESSMENT RESOURCES

Visual Analogies Students use analogies to understand how cells release stored energy from ATP, and how they transport high-energy electrons.

Tutor Tube These tutorials help students identify the reactants and products of photosynthesis.

Data Analysis Students gather and analyze data about photosynthesis in marine algae.

Vocabulary Review The *Flash Cards* and *Crossword* are an interactive vocabulary review.

Editable Worksheets Editable pages of Study Workbook B, Lab Manual B, and the Assessment Resources Book are available online.

Lesson Overviews Lesson Overviews are based on Miller & Levine Biology. However, they are available on CD-ROM and can be edited for your class.

Chapter Assessment Have students take an online version of the Chapter 8 Check Understanding.

Standardized Test Prep Students can take an online version of the Standardized Test Prep. You will receive their scores along with ideas for remediation.

Diagnostic and Benchmark Tests Use these tests to monitor progress and supply remediation.

ASSESSMENT

Assess

Big idea Invite a volunteer to read the question. Write the word *photosynthesis* at the top of the board. Then, write the words *oxygen, sugars, water, light,* and *carbon dioxide* underneath.

Ask Do you remember what things are needed for photosynthesis to happen? Which of these things are required for photosynthesis? *(light, carbon dioxide, water)*

Ask Do you remember the products of photosynthesis? *(sugars, oxygen)* Have students write their answers in their notebooks.

CONSTRUCTED RESPONSE Be sure students understand that the *Hints* should be used as a guide to help them answer the questions. Review how to write a paragraph.

Write Question 1 on the board and read it aloud. Underline the words *depend on* and *environment.*

Say This question is about yeasts, which are living things that feed off of sugars. We know from the question that their food is taken from the environment. This means that yeasts must depend on other living things for food.

Read the first hint.

Ask Where do many yeasts get their food? *(from plant products in their environment)* Do you remember what we call living things that get food by eating other organisms? *(heterotrophs)*

Read the second hint.

Write *heterotrophs* on the board under the question.

Say Yeasts do not make their own food, so they are heterotrophs. From what kind of living things do yeasts get food? *(autotrophs)*

Say In Lesson 8.1 we learned that the energy needed to make ATP comes from food and that different living things get their food from different sources. Write *heterotrophs eat food* and *autotrophs make food* on the board under the question.

Model writing the first one or two sentences using the hints on the board and the text. *(Sample: Some living things, called autotrophs, make their own food. Others, called heterotrophs, get their food by eating other organisms.)*

8 CHECK Understanding

Assess the Big idea: Cellular Basis of Life

Write an answer to the question below.

Q: How do plants and other organisms capture energy from the sun?

Constructed Response

Write an answer to each of the questions below. The answer to each question should be one or two paragraphs. To help you begin, read the **Hints** below the questions.

1. **Yeasts are single-celled fungi that feed off of sugars in their environment. How do yeasts depend on other living things for survival?**

   Hint Many yeasts get their food from plants that produce sugars in fruits and saps.

   Hint Yeasts are heterotrophs because they rely on other living things to make sugars for them.

2. **How is the function of chlorophyll related to its very specific location within the cell?**

   Hint Chlorophyll absorbs sunlight and transfers its energy to electrons.

   Hint Chlorophyll is found in the thylakoid, where many steps of photosynthesis take place.

3. **The light-independent reactions are sometimes called the "dark reactions" to tell them apart from the light-dependent reactions. Do the steps of the light-independent reactions ever depend on light? Explain.**

   Hint The photosystems need light to energize electrons. These electrons are used to make ATP and NADPH.

   Hint NADPH and ATP are needed to carry out the light-independent reactions.

Foundations for Learning Wrap-Up

Use the index cards and notebook page you prepared when you started the chapter as a tool to help you organize your thoughts about photosynthesis.

Activity 1 Working with a partner, use the index cards and notebook page to create a diagram to show the two stages of photosynthesis. Then, use the index cards to create a chemical equation that summarizes photosynthesis.

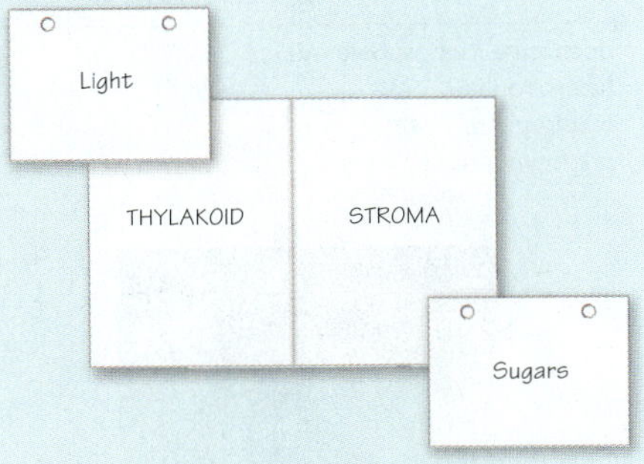

Activity 2 With your partner, label four more index cards with NADPH, ATP, ADP, and $NADP^+$. Using your diagram from Activity 1, add the cards to the diagram in the correct positions to complete the stages of photosynthesis. Draw arrows to show the movement of the molecules between the stages.

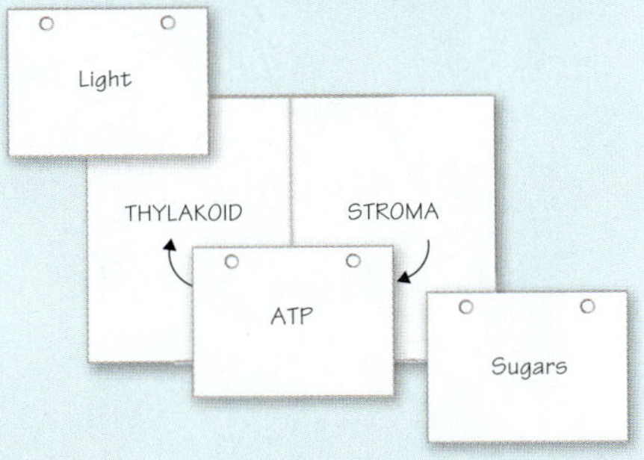

206 Check Understanding

For Question 2, pair students and have them use the hints and your modeling to answer the question. Have students share their paragraphs with the class.

For Question 3, have students write the paragraphs independently for homework or as an anchor activity.

Foundations for Learning Wrap-Up

Have students take out their six index cards and two sheets of paper. Be sure that students have written all the correct terms on the cards and on the sheets of paper. Either assign groups or pair students, and explain the activities.

8.1 Energy and Life

Understand Key Concepts

1. Which of the following are autotrophs?
   a. deer **c.** leopards
   b. grasses **d.** mushrooms

Test-Taking Tip

Eliminate Incorrect Answers Read all of the answer choices carefully. Even if you are not sure of the correct answer, you may be able to cross out one or two of the answer choices that are incorrect. In question 1, answers **a** and **c** are incorrect because they are animals, and animals are heterotrophs. So, you can eliminate these answer choices.

2. Where do plants get the energy they need to carry out photosynthesis?
   a. air **c.** sunlight
   b. soil **d.** water
3. How does a molecule of ATP compare to a molecule of glucose in terms of the amount of energy stored?

Think Critically

4. **Relate Cause and Effect** How might the disappearance of all autotrophs on Earth affect other living things? Explain.

8.2 Photosynthesis: An Overview

Understand Key Concepts

5. In addition to light and chlorophyll, photosynthesis requires
   a. carbon dioxide and sugars.
   b. carbon dioxide and water.
   c. oxygen and sugars.
   d. oxygen and water.
6. What is the function of $NADP^+$?
   a. It absorbs light.
   b. It is an electron carrier.
   c. It is a light-reflecting pigment.
   d. It acts as long-term energy storage.
7. Some plants have leaves that are bright red. How do red leaves differ from green leaves in the types of light they absorb and reflect?

Think Critically

8. **Predict** Why would a plant grow better in white light than in green light? Explain your answer.

8.3 The Process of Photosynthesis

Understand Key Concepts

9. What drives the formation of ATP by ATP synthase?
   a. buildup of H^+ ions
   b. burning oxygen
   c. light absorption
   d. splitting carbon dioxide
10. Which substance from the light-dependent reactions of photosynthesis is a source of energy for the light-independent reactions?
    a. ADP **c.** NADPH
    b. H_2O **d.** pyruvic acid
11. The light-independent reactions of photosynthesis are also known as the
    a. ATP cycle. **c.** carbon cycle.
    b. Calvin cycle. **d.** sugar cycle.
12. Identify the chloroplast parts labeled A, B, and C. In which part(s) do the light-dependent reactions take place? In which part(s) do the light-independent reactions take place?

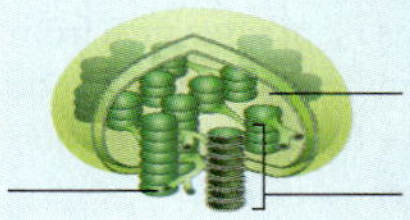

Assessment Tips

Pair students and have them work together to answer the questions. For Questions 1, 4, and 15, have students recall the differences between autotrophs and heterotrophs by reviewing the table they made in Lesson 8.1.

For Questions 2, 5, 6, 10, and 11, have students review the Stages of Photosynthesis diagram in Lesson 8.2. Have students share and discuss their answers with the class.

Answers

Lesson 8.1

UNDERSTAND KEY CONCEPTS

1. b
2. c
3. Glucose stores much more energy than ATP stores. However, ATP is more easily used as an immediate source of energy for the cell.

THINK CRITICALLY

4. Only autotrophs can make food from light energy. Heterotrophs must eat other living things for food. Without autotrophs, eventually heterotrophs would die out because they would eat up all the living things. There would be nothing to renew food sources on Earth.

Lesson 8.2

UNDERSTAND KEY CONCEPTS

5. b
6. b
7. Green leaves have green pigments, such as chlorophyll, that reflect green light and absorb other colors of light. Red leaves have red pigments that reflect red light and absorb other colors of light.

THINK CRITICALLY

8. Sample answer: A plant would not grow as well in green light as it would under white light, because chlorophyll does not absorb green light very well. White light is a mixture of different wavelengths, most of which are absorbed by chloroplasts.

Lesson 8.3

UNDERSTAND KEY CONCEPTS

9. a
10. c
11. b
12. Part A is the stroma, part B is a stack of thylakoids called a granum, and part C is a thylakoid. The light-dependent reactions take place in the membrane of the thylakoid. The light-independent reactions take place in the stroma (the liquid-filled space around thylaloids).

ASSESSMENT

THINK CRITICALLY

13. Photosynthesis is affected by temperature, light intensity, and the availability of water. Temperature affects the rate that enzymes work, both light-dependent and light-independent reactions rely on enzymes. Light intensity and water availability affect mostly the light-dependent reactions.

14. Over the next few days it would probably continue to carry out photosynthesis and make the sugars it needs to live and grow. Eventually, the plant would die because it would run out of the water and carbon dioxide that it needs to continue carrying out photosynthesis.

15. Plants would have difficulty carrying out the light-dependent reactions of photosynthesis. Because the light-independent reactions require products of the light-dependent reactions, these too would slow down. Plants would not be able to make the sugars they need to survive. Other animals that eat plants would have trouble getting the energy they need if there were fewer plants in an area.

Connecting Concepts

16. Line graphs should show bubbles produced per minute as a function of distance from light. The data should show a downward trend, a decrease in bubbles produced as distance increases.

17. Sample answer: Fewer bubbles are produced the farther away the light gets. I predict that there would be only one bubble produced per minute when the light is moved to 50 cm away.

18. As distance increases, the number of bubbles produced per minute decreases. The light intensity decreases the farther away the light source is from the plant. Photosynthesis decreases as light intensity decreases.

CHAPTER MYSTERY After students have read the Chapter Mystery, discuss what van Helmont concluded as well as what he could not know about photosynthesis.

Ask Why did van Helmont decide that the tree's added mass came from water and not the soil? *(The mass of the soil was about the same. Water was the only thing he added.)*

Ask Van Helmont was right that plants use water to build mass. How do plants use water to increase mass? *(Sample answer: Plants use water in the light-dependent reactions of photosynthesis. Enzymes break up water molecules. The electrons and hydrogen ions are picked up by electron carriers and then used in the light-independent reactions to produce high-energy sugars.)*

8 CHECK Understanding

13. Discuss three factors that affect the rate at which photosynthesis takes place. Which steps of photosynthesis are affected by these factors?

Think Critically

14. **Predict** Suppose you water a potted plant and place it by a window in a clear, airtight jar. Predict how the rate of photosynthesis might be affected over the next few days. What might happen if the plant were left there for several weeks?

15. **Form a Hypothesis** Many of the sun's rays may be blocked by dust or clouds formed by volcanic eruptions or pollution. What are some possible short-term and long-term effects of this on plants in an area? On other forms of life?

Connecting Concepts

Use Science Graphics

A water plant placed under bright light gives off bubbles of oxygen. The table below contains the results of an experiment in which the distance from the light to the plant was varied. Use the data table to answer questions 16–18.

Oxygen Production

Distance From Light (cm)	Bubbles Produced per Minute
10	39
20	22
30	8
40	5

16. **Graph** Use the data in the table to make a line graph.

17. **Interpret Graphics** Describe the trend in the data. How many bubbles would you predict if the light was moved to 50 cm away? Explain.

18. **Draw Conclusions** What relationship exists between the plant's distance from the light and the number of bubbles produced? What is taking place to cause this relationship? Explain your answer.

solve the CHAPTER MYSTERY

OUT OF THIN AIR?

Most plants grow out of the soil, so you might think that soil adds to plant mass. But, at the end of Jan van Helmont's experiment with the willow tree, he discovered that the mass of the soil was about the same—even though the tree had gained nearly 75 kilograms. Van Helmont decided that the mass must have come from water, because water was the only thing he had added. The tree grew parts that did contain atoms like those found in water, but what van Helmont didn't know was that the new parts also contained carbon. We now know that most of that carbon comes from carbon dioxide in air. Thus, mass comes from two sources: carbon dioxide and water. What form does the added mass take? Think about the origin of the word carbohydrate, from *carbo-*, meaning "carbon," and *hydrate*, meaning "to combine with water," and you have your answer.

1. **Infer** Although soil does not add much to plant mass, how might it help plants grow?
2. **Infer** Imagine that a scientist measures the exact mass of carbon dioxide and water that entered a plant and the exact mass of the sugars produced. Would these masses be identical? Why or why not?
3. **Apply Concepts** What do plants do with all of the sugars they make through photosynthesis? (*Hint:* Plant cells have mitochondria in addition to chloroplasts. What do mitochondria do?)

UntamedScience™ Finding the solution to the mystery is only the beginning. Take a video field trip with the ecogeeks of Untamed Science to see where the mystery leads.

CHAPTER MYSTERY ANSWERS

1. Soil has nutrients that help plants build many important compounds that contain nitrogen and phosphorus, for example.

2. The masses would not be identical. The plant gives off oxygen, which causes it to lose mass. It also gets some nutrients from soil, which increases its mass.

3. Plants use the sugars they make through photosynthesis to build cell parts and as fuel for cell processes. Mitochondria are organelles that take sugars and oxygen and use them to produce energy, carbon dioxide, and water.

Untamed Science™ Students can explore the connection between plant growth, carbon, and photosynthesis deep in the jungles of Panama by watching **Soaking Up the CO_2.**

Standardized Test Prep

Multiple Choice

1. Autotrophs differ from heterotrophs because they
A use oxygen to burn food.
B do not require oxygen to live.
C make carbon dioxide as a product of using food.
D make their own food from carbon dioxide and water.

2. The principal pigment in plants is
A chlorophyll.
B oxygen.
C ATP.
D NADPH.

3. Which of the following is NOT produced in the light-dependent reactions of photosynthesis?
A NADPH
B sugars
C hydrogen ions
D ATP

4. Which of the following correctly summarizes photosynthesis?
A $H_2O + CO_2 \xrightarrow{light} sugars + O_2$
B $sugars + O_2 \xrightarrow{light} H_2O + CO_2$
C $H_2O + O_2 \xrightarrow{light} sugars + CO_2$
D $sugars + CO_2 \xrightarrow{light} H_2O + O_2$

5. The color of light that is LEAST useful to a plant during photosynthesis is
A red.
B blue.
C green.
D violet.

6. The first step in photosynthesis is the
A synthesis of water.
B production of sugars.
C breakdown of carbon dioxide.
D absorption of light energy.

7. In a typical plant, all of the following factors are necessary for photosynthesis EXCEPT
A chlorophyll.
B light.
C oxygen.
D water.

Questions 8–10

A scientist mashes up spinach leaves to make a liquid, or extract, that has lots of pigment in it. She places a drop of the liquid at one end of a strip of paper towel. After the liquid dries, she hangs the paper in a test tube containing alcohol so that only the tip of the paper is in the alcohol. As the alcohol is absorbed and moves up the paper, the different pigments separate as shown below.

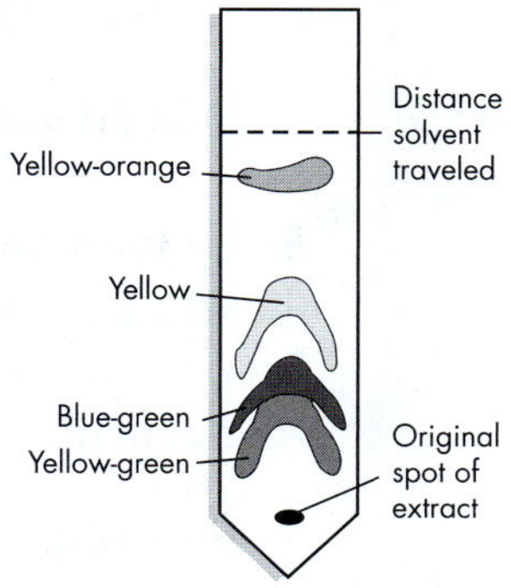

8. Which pigment traveled the shortest distance?
A yellow-orange
B yellow
C blue-green
D yellow-green

9. Which of the following is a valid conclusion that can be drawn from this information?
A Spinach leaves use only chlorophyll during photosynthesis.
B Spinach leaves contain several pigments.
C Spinach leaves contain more orange pigment than yellow pigment.
D Spinach leaves are yellow-orange rather than green.

10. In which organelle would MOST of these pigments be found?
A vacuoles
B centrioles
C mitochondria
D chloroplasts

Open-Ended Response

11. Describe how high-energy electrons are ultimately responsible for driving the reactions of photosynthesis.

If You Have Trouble With . . .

Question	1	2	3	4	5	6	7	8	9	10	11
See Lesson	8.1	8.2	8.2	8.2	8.2	8.3	8.3	8.2	8.2	8.2	8.3

Test-Taking Tip

INTERPRET VISUALS

When a paragraph and related questions accompany a visual, tell students to first carefully read the paragraph and all the labels on the visual. The accompanying paragraph often will put the visual in some context, as well as provide a description or explanation of the visual itself. Reading the labels carefully lets the reader know what specific structures, events, or measurements are shown on the visual. After reading the paragraph and labels, look for any trends or comparisons that the visual presents. After taking these steps to interpret the visual, read and answer the question.

Before the Test

Review with students the Test-Taking Tip.

Point out that the word *principal* in Question 2 means "main."

Point out the word *NOT* in Question 3.

Point out the word *LEAST* in Question 5.

Point out the word *first* in Question 6.

Point out the word *EXCEPT* in Question 7. Remind students that they're looking for the answer that isn't true.

Remind students that this open-ended response means that they'll have to write a paragraph to answer the question. Have them paraphrase the question in their notebooks, and remind them about the steps you took when you modeled a paragraph. Specifically point out that you identified the meaning of the key words before starting to write.

Answers

1. D
2. A
3. B
4. A
5. C
6. D
7. C
8. D
9. B
10. D
11. Light is absorbed by electrons in pigments, increasing the electrons' energy level. These high-energy electrons are used in the light-dependent reactions to convert ADP and $NADP^+$ into the molecules ATP and NADPH. In the light-independent reactions, ATP and NADPH are then used to produce high-energy sugars.

CHAPTER PLANNER 9 Cellular Respiration and Fermentation

Chapter Contents	NSES	Time	Core Resources
Chapter Preview			**Student Edition,** pp. 210–211 **Chapter Mystery,** p. 211 **Foundations for Learning,** p. 211
9.1 Cellular Respiration: An Overview Chemical Energy and Food • Overview of Cellular Respiration • Comparing Photosynthesis and Cellular Respiration	I, V, A.1.c, A.2.a, A.2.d, B.3, C.1.a, C.1.b, C.5.a, C.5.c, C.5.d	1 period 1/2 block	**Student Edition,** pp. 212–215 Inquiry Into Scientific Thinking, p. 213 **Study Workbook B** 9.1 Worksheets Biology.com *Art in Motion:* Opposite Processes: Respiration and Photosynthesis • 9.1 Lesson Assessment
9.2 The Process of Cellular Respiration Glycolysis • The Krebs Cycle • Electron Transport and ATP Synthesis • The Totals	I, V, B.3, C.1.a, C.1.b, C.5.c, C.5.d	1 1/2 periods 3/4 block	**Student Edition,** pp. 216–222 **Build Connections,** pp. 217, 219, and 221 **Study Workbook B** 9.2 Worksheets Biology.com *InterActive Art:* Cellular Respiration and Fermentation • *Art Review:* Electron Transport and ATP Synthesis • *Tutor Tube:* Oxygen as the Acceptor of Cellular Respiration Waste • 9.2 Lesson Assessment
9.3 Fermentation Fermentation • Energy and Exercise	I, B.3, C.1.a, C.1.b, C.5.c, C.5.d	1 period 1/2 block	**Student Edition,** pp. 223–225 **Study Workbook B** 9.3 Worksheets Biology.com 9.3 Lesson Assessment
Chapter Pre-Lab	A.1.b, A.1.c, A.2.a, A.2.c	1 period 1/2 block	**Student Edition,** p. 226 **Lab Manual B** *Comparing Fermentation Rates of Sugars*

Tools for ELL Instruction

ELL Handbook describes teaching strategies to make *Miller & Levine Biology: Foundation Edition* more accessible to ELL students.

Spanish Study Workbook is a Spanish translation of Study Workbook A.

Multilingual Glossary is the *Miller & Levine Biology* glossary translated into ten languages.

Tools to Challenge Students

Study Workbook A includes more challenging worksheets and activities such as Chapter Mystery activities that reinforce 21st Century Skills.

Lab Manual A labs are geared toward on-level students. The Lab Manual includes fifteen labs in addition to the Chapter Labs.

Biology.com Data Analysis activities offer students a way to practice data gathering and analyzing skills.

Visual Quizzes in the Assessment Resources Book are geared toward on-level students and use visuals from *Miller & Levine Biology.*

Additional Resources

Biology.com Untamed Science Video • Flash Cards

Study Workbook A 9.1 Worksheets
Lab Manual B *Data Analysis:* You Are What You Eat
Spanish Study Workbook 9.1 Worksheets

Study Workbook A 9.2 Worksheets
Spanish Study Workbook 9.2 Worksheets
Assessment Resources Book Visual Quiz

Study Workbook A 9.3 Worksheets
Lab Manual B *Hands-On Activity:* CO_2 and You
Spanish Study Workbook 9.3 Worksheets
Biology.com *Data Analysis:* Lactic Acid and Athletes

Lab Manual A *Comparing Fermentation Rates of Sugars* • *Photosynthesis and Respiration*

Chapter Review

Student Edition Chapter Summary, p. 227
Study Workbook B Chapter 9 Review
Transparencies, pp. 111–121
Biology.com Untamed Science Video • Editable Worksheets of Study Workbooks A and B and Lab Manuals A and B • Chapter 9 Flash Cards and Match It

Untamed Science DVD • Classroom Resources CD (includes lesson presentations and editable worksheets)

Chapter Assessment

Student Edition Check Understanding, pp. 228–231
Study Workbook B Chapter 9 Taking a Standardized Test
Assessment Resources Book Chapter 9 Test B
Biology.com Chapter 9 Assessment • Editable Worksheets of Chapter 9 Visual Quiz* and Chapter 9 Tests A and B

Exam*View* *Assessment Suite* • Classroom Resources CD (includes lesson presentations and editable worksheets)

Time: 1 period, 1/2 block

* Visual Quizzes are based on visuals from *Miller & Levine Biology*. Review carefully to make sure they are appropriate for your class.

Pressed for Time?

Preview the Chapter Have students read the Key Questions for Lesson 9.1 and preview the Stages of Cellular Respiration visual.

Cover the Chapter Quickly Have students read all of Lesson 9.1 focusing on the Stages of Cellular Respiration visual. Assign *The Totals* in Lesson 9.2 and *Fermentation* in Lesson 9.3.

Assess Assign the 9.1 Check Understanding, Question 7 in the 9.2 Check Understanding, Questions 1 and 2 in the 9.3 Check Understanding, and the Chapter 9 Standardized Test Prep except Questions 2, 3, and 11.

Get the Picture

Tell students that the picture shows several cellular structures, or organelles. Point out that the red organelles are called mitochondria.

Ask Do you know what the function of mitochondria is? *(Mitochondria change the energy in food to energy that cells can use.)*

Guide students to observe the threadlike shapes within the mitochondria. Connect this observation to the word itself. Explain that the prefix *mito-* comes from the Greek work *mitos,* which means "thread." The second part of the word, *chondrian,* comes from the Greek *khondrion,* which means "granule," or "tiny grain."

Ask What feature of the mitochondria in the picture do you think *granule* describes? *(their small size)*

Activate Prior Knowledge

Remind students of the function of mitochondria.

Ask Why do cells need energy? *(to carry out cell activities)* Guide students to think about the different amounts of energy needed by different types of cells.

Ask Which of your cells would you expect to contain more mitochondria—a muscle cell or a skin cell? *(muscle cell)* Why? *(Muscle cells need more energy than skin cells.)*

9 Cellular Respiration and Fermentation

Big idea **Cellular Basis of Life**

Q: How do organisms obtain energy?

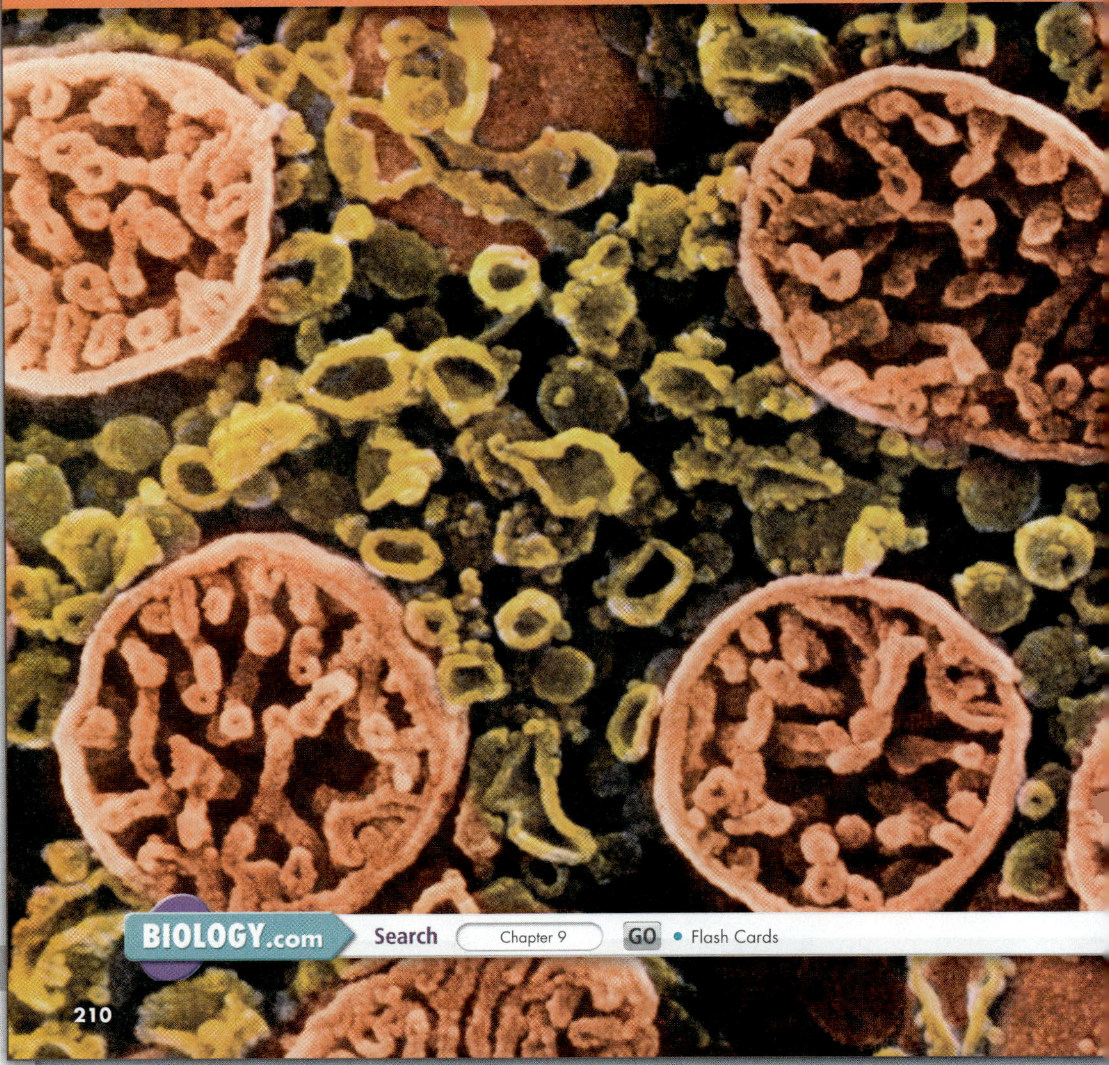

BIOLOGY.com Search Chapter 9 GO • Flash Cards

210

NATIONAL SCIENCE EDUCATION STANDARDS

UNIFYING CONCEPTS AND PROCESSES

I, V

CONTENT

B.3, C.1.a, C.1.b, C.5.a, C.5.c, C.5.d

INQUIRY

A.1.b, A.1.c, A.2.a, A.2.c, A.2.d

UbD Connect to the Big Idea

Invite a volunteer to read aloud the chapter number and title. Read the Big Idea and question. Emphasize the question, **How do organisms obtain energy?**

Set the Purpose

PREVIEW AND PREDICT Draw the graphic organizer. Point to the INSIDE feature on the next page, and have a student read the lesson numbers and titles.

Ask How do you think the chapter title relates to the questions in the graphic organizer? (*Cellular respiration* and *fermentation* will likely be new and difficult terms for students. Make sure they understand that they will learn about these two terms in the chapter.) Remind students as they read to search for the meanings of these two words, and how the chapter title relates to how organisms get energy.

INSIDE:
- 9.1 Cellular Respiration: An Overview
- 9.2 The Process of Cellular Respiration
- 9.3 Fermentation

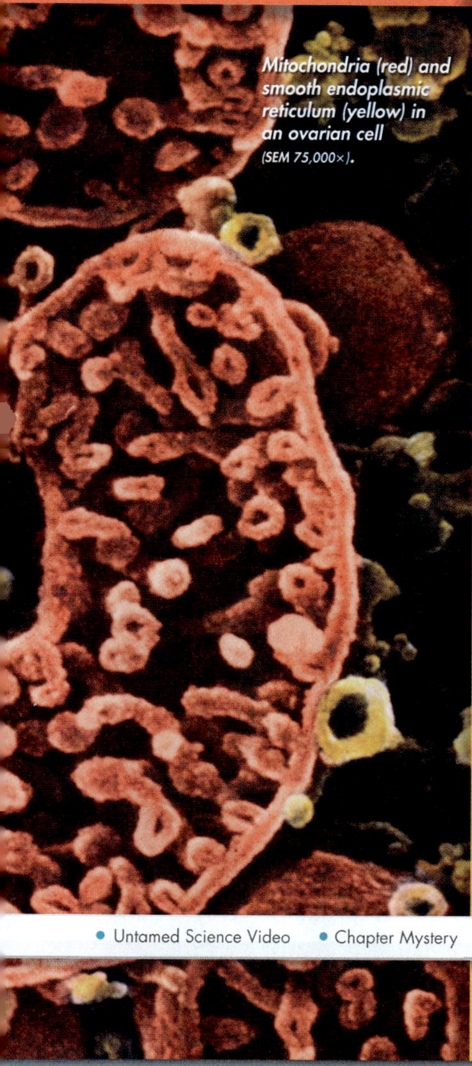

Mitochondria (red) and smooth endoplasmic reticulum (yellow) in an ovarian cell (SEM 75,000×).

• Untamed Science Video • Chapter Mystery

CHAPTER **MYSTERY**

DIVING WITHOUT A BREATH

You know what it is like to feel out of breath. Just a few minutes of hard exercise can have some people huffing and puffing for air. But what if you couldn't get air? What if you were asked to hold your breath and exercise? Before too long, you'd run out of oxygen and pass out. You would not be able to get enough oxygen. It may seem like a bad idea to hold your breath while exercising, but there are animals that do it all the time—whales. Because whales are air-breathing mammals, they still need to come to the surface for air. But after a breath, some whales can dive underwater for as long as 45 minutes! How is that possible? Diving takes a lot of energy. How do whales stay active for so long on only one breath?

Read for Mystery Clues As you read this chapter, look for clues to help you discover how whales can stay underwater so long. Then, solve the mystery at the end of the chapter.

FOUNDATIONS for Learning

Cells need chemical energy to stay alive. Special parts of the cell help them get this energy. Before you read the chapter, label six index cards with the words *water*, *pyruvic acid*, *carbon dioxide*, *oxygen*, *energy*, and *glucose*. Label one sheet of paper with the words *cytoplasm* and *mitochondrion*. At the end of the chapter are two activities that use the cards and sheet of paper. They will help you answer the question: How do organisms obtain energy?

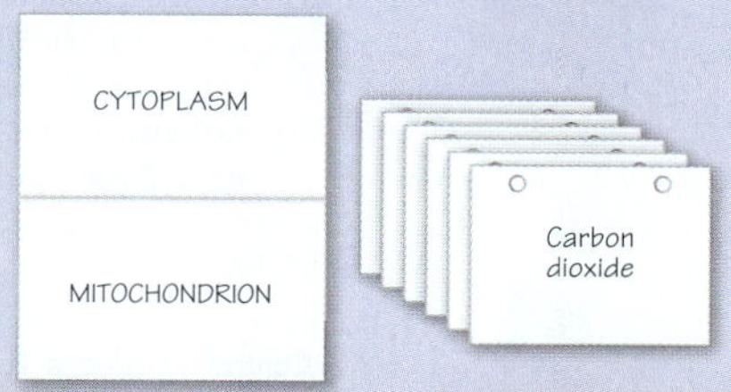

Chapter 9
Big Idea: Cellular Basis of Life

Chapter 9: How do organisms obtain energy?

- **9.1:** Why do most organisms undergo the process of cellular respiration?
- **9.2:** How do cells release energy from food in the presence of oxygen?
- **9.3:** How do cells release energy from food without oxygen?

Foundations for Learning

Pass out six index cards and two sheets of unlined paper to each student.

Ask Why is sorting terms important? *(Sorting terms helps you see how they are alike and how they are different.)*

Ask What do you have to know about the different categories before you can sort the terms? *(You have to know what a chemical is, what energy is, and some cell parts.)*

Familiarize students with the activity by modeling with the card for *oxygen*. Demonstrate placing the *oxygen* card on the word *mitochondria*. Explain that the *oxygen* card goes on *mitochondria* because oxygen is part of the cellular respiration process, which takes place in the mitochondria.

Point out that they might not be able to sort all of the terms until they have read the whole chapter.

Untamed Science™ Dive below the ocean surface to explore how marine mammals can survive on a single breath for as long as they do in the short video **Take a Deep Breath.**

CHAPTER MYSTERY **Ask** How do you feel during hard exercise, such as bicycling up a steep hill? *(out of breath and tired)*

Point out that their body cells need more oxygen during hard exercise. Explain that heart rate and breathing increase to provide the cells with more oxygen.

Ask a volunteer to read aloud the Chapter Mystery. Have students predict how this mystery relates to the chapter title and which of the terms on their index cards might play an important role in why whales can stay underwater for so long.

Digital Support

BIOLOGY.com Extend your reach by using the digital assets offered at **Biology.com.**

LESSON 9.1

Pre-Reading

Objectives

9.1.1 Explain where organisms get the energy they need for life processes.

9.1.2 Define cellular respiration.

9.1.3 Compare photosynthesis and cellular respiration.

Preview the Pages

1. Have students find the vocabulary words in the text and make a list of them. As a group, discuss the meaning of each word.
2. Read the captions and labels in the Stages of Cellular Respiration diagram and be sure students can follow the multiple arrows. Then, guide students to the Opposite Processes diagram. Identify the heterotroph *(rabbit)* and autotroph *(plant)*. Point out that the equations for photosynthesis and cellular respiration are reversed.

Build Understanding

PREVIEW VISUALS Refer students to Workbook B for guidance on building a T-chart. They can use their T-charts to write down questions they have about the diagram The Stages of Cellular Respiration. Then they can record the answers as they read.

NATIONAL SCIENCE EDUCATION STANDARDS

UNIFYING CONCEPTS AND PROCESSES

I, V

CONTENT

B.3, C.1.a, C.1.b, C.5.a, C.5.c, C.5.d

INQUIRY

A.1.c, A.2.a, A.2.d

For the corresponding lesson in *Miller & Levine Biology,* see pages 250–253.

9.1 Cellular Respiration: An Overview

Key Questions

Where do organisms get energy?

What is cellular respiration?

What is the relationship between photosynthesis and cellular respiration?

BUILD Understanding

Preview Visuals Before you read, study The Stages of Cellular Respiration figure. Make a list of questions that you have about the figure. As you read, write down the answers to the questions.

In Your Workbook Go to your workbook to learn more about Previewing Visuals. Complete the Previewing Visuals activity for Lesson 9.1.

Chemical Energy and Food

How do you feel when you are hungry? Do you feel sluggish, dizzy, or a little weak? That weakness comes from a lack of the food you need for energy. But what does food have to do with energy?

Food gives living things the energy they need to grow. Some living things, such as plants, are autotrophs (AW toh trohfs). They make their own food through photosynthesis. Other living things are heterotrophs (HET ur oh trohfs). They have to eat other living things for food. For all living things, food molecules store chemical energy. Living things release that energy when they break those food molecules down.

Energy in food can be measured in units called calories. A calorie is the amount of energy needed to raise the temperature of 1 gram of water 1 degree Celsius. The Calorie you see on food labels is actually a kilocalorie, or 1000 calories. The amount of energy in different kinds of food varies. This is because of differences in the way the atoms in food molecules are bonded together. For example, 1 gram of the sugar glucose gives off 3811 calories of energy when it is burned. But, 1 gram of the fat found in beef gives off 8893 calories of energy when it is burned.

Cells break down food molecules over time, getting a little bit of chemical energy at key steps. This chemical energy helps cells use the energy in foods to make compounds such as ATP. ATP and other useful energy compounds directly power the activities of the cell.

Key Question Where do organisms get energy?
Organisms get the energy they need from food.

A Controlled Release The energy stored in food is released slowly. If the energy were given off all at once, most of it would be lost as light and heat. That is what happens when a marshmallow catches fire.

9.1 Cellular Respiration: An Overview

Lesson Objectives

- Explain where organisms get the energy they need for life processes.
- Define cellular respiration.
- Compare photosynthesis and cellular respiration.

BUILD Vocabulary

The chart below shows key terms from the lesson with their definitions. Complete the chart by writing a strategy to help you remember the meaning of each term. One has been done for you.

Term	Definition	How I'm Going to Remember the Meaning
Aerobic	Process that requires oxygen	Aero- means "air"; oxygen is part of air.
Anaerobic	Process that does not require oxygen	An- means "not," so the word anaerobic means "no oxygen."
calorie	Amount of energy needed to raise the temperature of 1 gram of water 1 degree Celsius	calorie = energy
Cellular respiration	Process that uses oxygen to release energy from food	The process occurs in cells. It requires oxygen; and so does respiration in the lungs.

BUILD Understanding

Preview Visuals Previewing visuals and taking notes about them can help you remember what you read and review for tests. As you look at each visual, think about why it may be important to the lesson.

Draw a T-Chart in your notebook. Make a list of questions you have about the diagram The Stages of Cellular Respiration. List those questions in the left column. Read about the stages of cellular respiration in your textbook. As your questions are answered, record these answers in the right column. You can see a sample below.

Check students' answers for accuracy.

Your Question Before Reading	Your Answer After Reading
What do the six circles drawn below the word Glucose mean?	The six circles represent the six carbon atoms found in glucose.

Overview of Cellular Respiration

Cellular respiration is the process that uses oxygen to release energy from food.

Use the terms in the box below to label the diagram of cellular respiration.

electron transport	Krebs cycle
glycolysis	mitochondrion

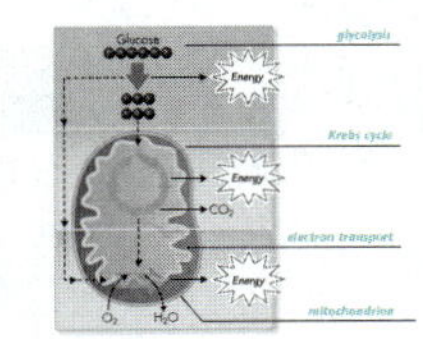

Answer the questions.

1. Where does glycolysis take place?
   A. mitochondrion C. cell membrane
   (B.) cytoplasm D. nucleus
2. Where do the Krebs cycle and the electron transport chain take place?
   A. ribosome (C.) mitochondrion
   B. nucleus D. cytoplasm
3. Which reactant is used in the electron transport chain?
   A. water (C.) oxygen
   B. carbon dioxide D. sugar
4. Which stage(s) of cellular respiration are aerobic?
   A. glycolysis C. electron transport
   B. Krebs cycle (D.) both B and C

Inquiry Into Scientific Thinking

You Are What You Eat Read the label on a bag of potato chips and you'll see the number of Calories in one serving. Calories are units of energy. You add Calories to your body when you eat. You burn those Calories to get energy for survival. When you are active, you increase the amount of Calories you burn.

Look at the two charts that follow. The chart on the left shows some common foods and the number of Calories per serving in these foods. The chart on the right shows some typical exercises and the approximate number of Calories burned by that activity.

Food item	Calories in 1 serving
Pepperoni pizza (1 slice)	198
Baked potato (1 plain)	145
Broccoli (1 cup)	44
Pear (1)	65
BLT sandwich (1)	437

Exercise	Calories Burned (50-kilogram woman)
Walking 3.2 km/hr (1 hr)	144
Running 13 km/hr (1 hr)	625
Bicycling (60 min)	330
Swimming, leisurely (45 min)	175

Study the food and exercise charts above. Use the information in the charts and your math skills to answer the questions about Mary, a young woman who weighs 50 kilograms.

1. Mary ate five slices of pepperoni pizza. What combination of activities would allow her to burn all of the Calories she consumed?
   A. 1 hour of running
   B. bicycling for 1 hour and walking for 2 hours
   (C.) Swimming for 1.5 hours, bicycling for 2 hours
   D. bicycling for 2 hours and swimming for 45 minutes
2. Mary is eating 1500 Calories per day. She has no time to exercise today. What combination of foods would NOT keep her within that guideline?
   A. 2 BLT sandwiches, 2 slices of pepperoni pizza, and 2 pears
   B. 2 baked potatoes with 3 cups of broccoli, 1 BLT sandwich, 1 pear
   (C.) 4 slices of pepperoni pizza and 2 BLT sandwiches
   D. 3 pears, 2 cups of broccoli, 2 slices of pizza, 1 BLT sandwich
3. Which of the following food combinations provides fewer than 200 Calories?
   A. a slice of pepperoni pizza and a BLT sandwich
   (B.) a baked potato and a cup of broccoli
   C. a BLT sandwich and a pear
   D. a slice of pepperoni pizza and a pear

INQUIRY into Scientific Thinking

You Are What You Eat

Different types of food can store very different amounts of energy. Most foods have a mix of proteins, carbohydrates, and fats. One gram of protein or a carbohydrate such as glucose has about 4 Calories. One gram of fat, though, has about 9 Calories. The table to the right shows what is found in one serving of some common foods.

Composition of Some Common Foods

Food	Protein (g)	Carbohydrate (g)	Fat (g)
Apple, 1 medium	0	22	0
Bacon, 2 slices	5	0	6
Chocolate, 1 bar	3	23	13
Eggs, 2 whole	12	0	9
2% milk, 1 cup	8	12	5
Potato chips, 15 chips	2	14	10
Skinless roasted turkey, 3 slices	11	3	1

1. Interpret Data The food from the table that has the most protein is ______. The food that has the most carbohydrates is ______. The food that has the most fat is ______.

2. Calculate

a. How many Calories total are there in 2 slices of bacon?

b. How many Calories total are there in 3 slices of roasted turkey?

c. What is the difference in total Calories between these two foods? Why is there a difference?

3. Calculate Walking uses around 5 Calories per minute. At that rate, how many minutes would you have to walk to burn the Calories in one chocolate bar? Use the equations below to help you with your calculations.

____ Calories from protein
\+ ____ Calories from carbohydrates
\+ ____ Calories from fat
= ____ Total Calories in a chocolate bar

$$\frac{\text{\_\_\_\_ Total Calories}}{\text{5 Calories/minute}} = \text{\_\_\_\_ minutes}$$

In Your Workbook Get more help for this activity in your workbook.

Overview of Cellular Respiration

Living things can get energy from food through cellular respiration. **Cellular respiration** is the process that releases energy from food, such as the simple sugar glucose, when there is oxygen present. Cellular respiration gives off carbon dioxide, water, and energy. The process can be summarized like this:

$$6O_2 + C_6H_{12}O_6 \longrightarrow 6CO_2 + 6H_2O + \text{Energy}$$

But cellular respiration does not take place in just one step. If it did, all of the energy from sugar would be given off at once. Most of it would be lost in the form of light and heat. A living cell has to release the chemical energy in food molecules a little bit at a time. The cell uses that energy to make ATP.

Key Question What is cellular respiration?
Cellular respiration is the process that releases energy from food when oxygen is present.

BUILD Vocabulary

cellular respiration
the process that releases energy by breaking down glucose and other food molecules in the presence of oxygen

MULTIPLE MEANINGS
Respiration is another word for breathing, the exchange of oxygen and carbon dioxide within the lungs. Cellular respiration requires the same exchange of gases, only within the living cell.

Analyze and Conclude

1. two whole eggs; one chocolate bar; one chocolate bar
2. a. Two slices of bacon have 20 Calories from protein, 54 Calories from fat, and 74 Calories total.

   b. Three slices of roasted turkey have 44 Calories from protein, 12 Calories from carbohydrates, 9 Calories from fat, and 65 Calories total.

   c. The bacon has 9 more Calories than the turkey. This difference is due mostly to the greater amount of fat in bacon.
3. A chocolate bar has 12 Calories from protein, 92 Calories from carbohydrates, 117 Calories from fat, and 221 Calories total. Walking uses 5 Calories per minute. It would take about 45 minutes to burn the Calories in one chocolate bar.

Inquiry Into Scientific Thinking

INTERPRET CHARTS & TABLES If students have difficulty interpreting the information in the table, have them align a sheet of paper with the column heads and write on the paper the number of calories per gram for protein, carbohydrate, and fat. The numbers should align vertically with the column heads. Students can then move this paper below any row to see the number of calories by which to multiply.

Lead a Discussion

Discuss with students how small differences in words can result in large differences in meanings. Show students the label of a small chocolate bar and have them locate the number of Calories listed.

Ask How does the number of Calories on the label compare to the number of calories for a chocolate bar from this activity? *(The number on the label is much less.)*

Point out that a chocolate bar contains about 200 Calories, which is about 200,000 calories. Use this information to reinforce the relationship between a Calorie and a calorie as explained in the text.

During Reading

Active Reading

INTERPRET SYMBOLS Review with students the meaning of the symbols in the equation. Write the equation on the board and have students copy it in their notebooks. Have students write the equation in words. *(Oxygen + Glucose → Carbon dioxide + Water + Energy)*

KEY QUESTION Have one student read aloud the Key Question and another student read the answer.

Ask What two materials must be present for cellular respiration to occur? *(oxygen and food)*

Ask How would your energy level change if you did not eat for a day? *(It would decrease.)*

Build Vocabulary

MULTIPLE MEANINGS Point out the word *present* in the Overview of Cellular Respiration paragraph.

Ask What does this word mean in this sentence? *(exists at this place and time)*

Ask Can you think of another meaning for this word? *(a gift)*

During Reading continued

Active Reading

USE VISUALS Be sure that students understand the diagram: cytoplasm (yellow) surrounds the (orange) mitochondria. As they read this page, have students refer to the visual that shows the stages of cellular respiration. Be sure that students realize that the black circles represent carbon atoms.

Ask How many carbon atoms does the glucose molecule have? *(six)* What happens to the glucose molecule during glycolysis? *(It is split into two molecules of three carbons each.)* What are the new molecules called? *(pyruvic acid)* What happens to the molecules of pyruvic acid? *(They move to the next stage, called the Krebs cycle.)*

Ask Where does the Krebs cycle take place? *(mitochrondria)*

Ask What is given off at each stage of cellular respiration? *(energy)*

Ask Look at the electron transport part of the diagram. What do the arrows tell you? *(oxygen is taken in and water and energy are produced)*

SUMMARIZE Have students use the text and the visual to make a numbered list that summarizes the steps of cellular respiration. Be sure their summaries include the relative differences in energy given off at each stage.

Build Vocabulary

PREFIXES For each vocabulary word, have students write the word and underline the prefix. Explain that a prefix is a letter or group of letters added to the root word to change its meaning.

Ask What are some other words that begin with the prefix *an-* or *aero-*? *(anemia (without enough red blood cells), anesthetic (without feeling), aerodynamic (air resistant structure), aeronautics (having to do with air navigation))*

Ask What is something that is anaerobic doing "without"? *(air or oxygen)*

Speed Bump

Ask Which stages of cellular respiration are aerobic? *(Krebs cycle and electron transport chain)*

Ask Which stage happens in a cell's cytoplasm? *(glycolysis)* Is it aerobic or anaerobic? *(anaerobic)*

Ask Does the Krebs cycle depend on oxygen directly or indirectly? *(indirectly, by way of the electron transport chain)*

The Stages of Cellular Respiration
There are three stages to cellular respiration: glycolysis, the Krebs cycle, and the electron transport chain.

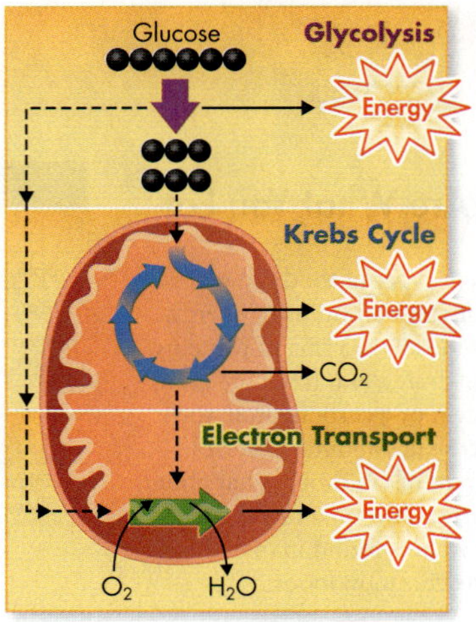

BUILD Vocabulary

aerobic
a process that requires oxygen

anaerobic
a process that does not require oxygen

PREFIXES
The prefix *an-* means "without" and the prefix *aero-* means "air." *Aerobic* means "with air" and *anaerobic* means "without air."

Three Main Stages Cellular respiration gets energy from food in three stages. These stages are glycolysis, the Krebs cycle, and the electron transport chain. When looking at these stages, it is helpful to use an example. We will use the sugar glucose. Glucose enters the first stage, known as glycolysis (gly KAHL ih sis). Only a small amount of energy is used to make ATP during this stage. The rest is still locked in the bonds of a molecule called pyruvic (py ROO vik) acid. Pyruvic acid enters the second stage, the Krebs cycle, where a little more energy is given off. Most of the energy from cellular respiration comes from the last stage, the electron transport chain. This stage uses oxygen and reactants from the other two stages to finish the job.

Oxygen and Energy You get oxygen from the air you breathe. Oxygen is used at the end of the electron transport chain. Any time a cell needs more energy, it needs more oxygen, too. That is why you breathe hard when you exercise. You use up energy and need more oxygen, which means taking more breaths.

Chemical pathways that need oxygen are called **aerobic,** which means "in air." The Krebs cycle and electron transport chain are both aerobic. Even though the Krebs cycle does not directly need oxygen, it cannot run without the electron transport chain, which does need oxygen. Glycolysis, however, does not need oxygen. It does not depend on an aerobic process, either. Glycolysis is **anaerobic,** which means "without air."

Mitochondria (myt oh KAHN dree uh) are the organelles most important in cellular respiration. Glycolysis actually takes place in the cytoplasm of a cell, but the Krebs cycle and electron transport chain take place inside the mitochondria. If oxygen is not present, another anaerobic path, known as fermentation, keeps glycolysis running. You will learn more about fermentation later in this chapter.

ELL Focus on ELL: Extend Language

INTERMEDIATE AND ADVANCED SPEAKERS Prepare a list of words in the reading that have multiple meanings or uses that might be confusing to English learners. Examples include *stage, bond, chain,* and *process*. Help students distinguish the different meanings and usages of these words by using each term in a sentence that reflects its meaning. *(Sample answers: A stage is a step in a process. A stage is where actors perform plays.)*

214 Chapter 9 • Lesson 1

Comparing Photosynthesis and Cellular Respiration

If nearly all living things break down food through cellular respiration, why doesn't Earth run out of oxygen? Where does all of the carbon dioxide go? Where do the cells get food molecules from? Cellular respiration is balanced by photosynthesis. The energy from each process flows in opposite directions. The equations for photosynthesis and cellular respiration are the reverse of each other. Photosynthesis removes carbon dioxide from the air. Cellular respiration puts it back. Photosynthesis gives off oxygen and makes sugars. Cellular respiration uses that oxygen to get energy from sugars. Cellular respiration takes place in nearly all life: plants, animals, fungi, protists, and most bacteria. Photosynthesis, on the other hand, takes place only in plants, algae, and some bacteria.

Key Question What is the relationship between photosynthesis and cellular respiration? **Photosynthesis produces food molecules and removes carbon dioxide from the air, and cellular respiration puts it back. Photosynthesis gives off oxygen, and cellular respiration uses that oxygen to release energy from food.**

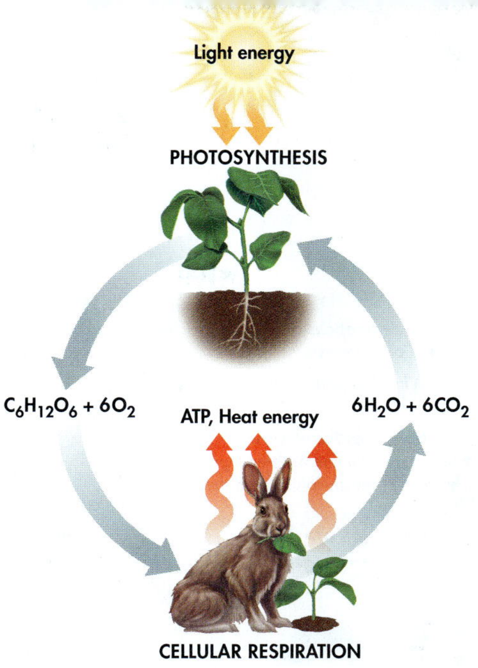

Opposite Processes Photosynthesis and cellular respiration can be thought of as opposite processes.

CHECK Understanding

Apply Vocabulary

Use the highlighted words from the lesson to complete each sentence correctly.

1. A process that needs oxygen to take place is called _____.
2. Glycolysis is _____ because it can happen when there is no oxygen.

Critical Thinking

3. **Infer** Do plant cells carry out cellular respiration? What organelle do they have that helps you know the answer?
4. **Compare and Contrast** How are the equations for cellular respiration and photosynthesis similar? How are they different?
5. **Apply Concepts** Aerobics is a kind of exercise that involves rhythmic movement and stretching, usually to music. What does the name *aerobics* tell you about this kind of exercise?
6. **Write to Learn** Answer the first clue of the mystery. Think about the difference between aerobic and anaerobic processes.

MYSTERY CLUE

If whales stay underwater for 45 minutes or more, do you think they mostly use aerobic or anaerobic processes? (**Hint:** See p. 214.)

BIOLOGY.com Search Lesson 9.1 GO • Lesson Assessment • Art in Motion

Check Understanding

1. aerobic
2. anaerobic
3. Plants carry out cellular respiration in mitochondria.
4. The two equations have the same chemicals. These chemicals are on opposite sides of the reaction arrow. Also, energy flows in opposite directions in the two equations.
5. The name *aerobics* tells you that this kind of exercise requires oxygen. Doing aerobics makes you breathe harder because it is an aerobic process.
6. Whales must use mostly anaerobic processes to get the energy they need. Anaerobic processes do not need oxygen. The whales cannot get more oxygen until they go back up for more air, but they still need energy while they are underwater.

Active Reading

SET PURPOSE FOR READING Have a volunteer read aloud the three questions that begin the paragraph.

Ask What do you think you will learn in this paragraph? *(the answers to these questions)* Point out that questions at the beginning of a passage set a purpose for reading that passage—to find answers to those questions. After students read this paragraph, revisit these opening questions and discuss the answers.

USE VISUALS Use the diagram to help students compare photosynthesis and cellular respiration.

Ask How does this diagram show that plants and animals depend on one another? *(Sample answer: Plants provide glucose and much of the oxygen that animals need, and animals provide much of the carbon dioxide and some water that plants need.)*

KEY QUESTION Have students create a compare/contrast table in their notebooks to summarize the differences between photosynthesis and cellular respiration.

Post Reading

ASSESSMENT TIPS For Question 4, review the diagram on this page, focusing on the chemical reactants and products.

MYSTERY CLUE For Question 6, discuss what students already know about whales, especially that whales need oxygen to survive, but cannot breathe underwater.

Ask Considering what you know about whales, are all the energy-producing processes in a whale anaerobic? *(No, whales need to breathe, which shows that some of the processes are aerobic.)*

Wrap-Up Activity

PLAY A VOCABULARY GAME Divide students into groups of four. In each group, one student chooses an important term from the lesson without telling anyone what it is. The other three students take turns asking this student a question about the term. The student can answer only *yes* or *no*. After one round of questions, students guess what the secret term is. Repeat until all students have been questioned about their secret terms.

LESSON 9.2

Pre-Reading

Objectives

9.2.1 Describe what happens during glycolysis.

9.2.2 Describe what happens during the Krebs cycle.

9.2.3 Explain how high-energy electrons are used by the electron transport chain.

9.2.4 Identify how much ATP cellular respiration generates.

Preview the Pages

1. Provide students with an example of an initialism, such as ADP (unlike an acronym, which is pronounced as a word). Then, have students skim the lesson and make a list of initialisms used and what they stand for. Tell them they will use this as a reference when reading the lesson. For initialisms that have been used before (ADP, ATP, $NADP^+$), direct students to the appropriate chapter or the glossary.
2. Read the captions and labels of each diagram. Encourage students to think of questions about the diagrams. Tell students to create a preview visuals chart in their notebooks to organize their questions. Point out that looking at a visual before, during, and after reading the text can help clarify content.

Build Understanding

COMPARE/CONTRAST TABLE Refer students to Workbook B where they will find a partially completed compare/contrast table that will help them organize information about the three stages of cellular respiration.

NATIONAL SCIENCE EDUCATION STANDARDS

UNIFYING CONCEPTS AND PROCESSES

I, V

CONTENT

B.3, C.1.a, C.1.b, C.5.c, C.5.d

For the corresponding lesson in *Miller & Levine Biology,* see pages 254–260.

9.2 The Process of Cellular Respiration

Key Questions

What happens during the process of glycolysis?

What happens during the Krebs cycle?

How does the electron transport chain use high-energy electrons from glycolysis and the Krebs cycle?

How much ATP does cellular respiration generate?

BUILD Understanding

Compare/Contrast Table As you read, make a compare/contrast table for glycolysis, the Krebs cycle, and the electron transport chain. In your table, show where each step takes place, what reactants it uses, what products it makes, and how many molecules of ATP are made.

In Your Workbook Go to your workbook to learn more about making a compare/contrast table. Complete the compare/contrast table for Lesson 9.2.

Glycolysis

Food burns. In fact, flour is so flammable that is has caused many explosions. Because of this, you should never store flour above a stove. But how do living things get the energy they need from food without starting a fire?

Living things use a series of reactions and steps to get energy from food molecules. The first set of reactions in cellular respiration is known as **glycolysis.** In one of the reactions of glycolysis, a 6-carbon sugar made from glucose is split into two. The final step produces two molecules of pyruvic acid. Pyruvic acid has 3 carbons. Energy is given off as the bonds in glucose are broken and formed again between different atoms.

Making ATP Glycolysis gives off energy. But the cell needs to put in a little energy to get things going. At the beginning, 2 ATP molecules are used up. Glycolysis makes 4 ATP molecules. This gives the cell a net gain of 2 ATP molecules for each molecule of glucose that enters glycolysis.

Making NADH One of the steps of glycolysis passes 4 electrons to an electron carrier. This carrier is called **NAD^+,** or nicotinamide adenine dinucleotide. Like $NADP^+$ in photosynthesis, each NAD^+ molecule can pick up a hydrogen ion and carry a pair of high-energy electrons. The new molecule is known as NADH. It holds the electrons until they can be moved to other molecules. Later on, when there is oxygen present, these high-energy electrons can be used to make even more ATP.

Breaking Down Sugars Animals get sugars from food. Sugars such as glucose are then broken down by cells through the process of glycolysis.

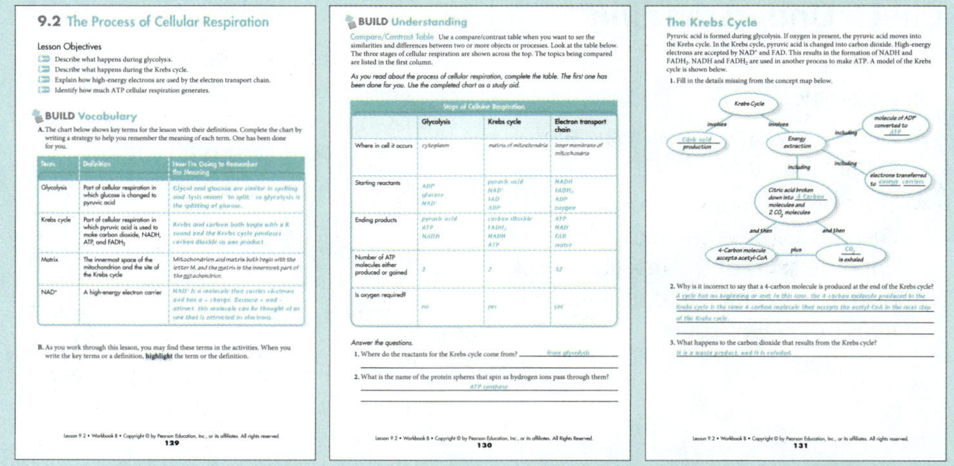

9.2 The Process of Cellular Respiration

Lesson Objectives

- Describe what happens during glycolysis.
- Describe what happens during the Krebs cycle.
- Explain how high-energy electrons are used by the electron transport chain.
- Identify how much ATP cellular respiration generates.

BUILD Vocabulary

A. The chart below shows key terms for the lesson with their definitions. Complete the chart by writing a strategy to help you remember the meaning of each term. One has been done for you.

Term	Definition	How I'm Going to Remember the Meaning
Glycolysis	Part of cellular respiration in which glucose is changed to pyruvic acid	
Krebs cycle	Part of cellular respiration in which pyruvic acid is used to make carbon dioxide, NADH, ATP, and $FADH_2$	
Matrix	The innermost space of the mitochondrion and the site of the Krebs cycle	Mitochondrion and matrix both begin with the letter M, and the matrix is the innermost part of the mitochondrion.
NAD^+	A high-energy electron carrier	

B. As you work through this lesson, you may find these terms in the activities. When you write the key terms or a definition, **highlight** the term or the definition.

129

BUILD Understanding

Compare/Contrast Table Use a compare/contrast table when you want to see the similarities and differences between two or more objects or processes. Look at the table below. The three stages of cellular respiration are shown across the top. The topics being compared are listed in the first column.

As you read about the process of cellular respiration, complete the table. The first one has been done for you. Use the completed chart as a study aid.

	Glycolysis	Krebs cycle	Electron transport chain
Where in cell it occurs	cytoplasm	matrix of mitochondria	inner membrane of mitochondria
Starting reactants			
Ending products			
Number of ATP molecules either produced or gained	2	2	32
Is oxygen required?	no	yes	yes

Answer the questions.

1. Where do the reactants for the Krebs cycle come from?
2. What is the name of the protein spheres that spin as hydrogen ions pass through them? ATP synthase

130

The Krebs Cycle

Pyruvic acid is formed during glycolysis. If oxygen is present, the pyruvic acid moves into the Krebs cycle. In the Krebs cycle, pyruvic acid is changed into carbon dioxide. High-energy electrons are accepted by NAD^+ and FAD. This results in the formation of NADH and $FADH_2$. NADH and $FADH_2$ are used in another process to make ATP. A model of the Krebs cycle is shown below.

1. Fill in the details missing from the concept map below.

2. Why is it incorrect to say that a 4-carbon molecule is produced at the end of the Krebs cycle?
3. What happens to the carbon dioxide that results from the Krebs cycle?

131

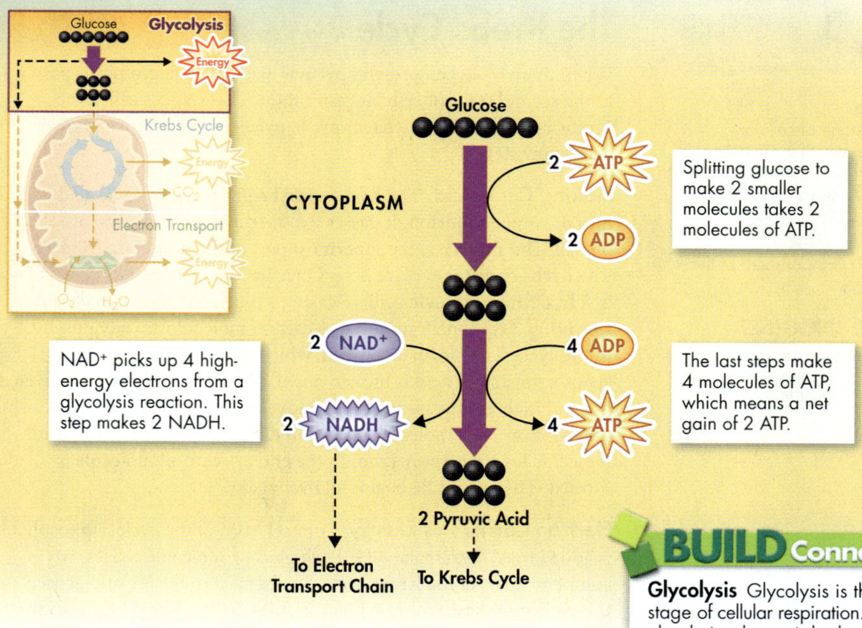

BUILD Connections

Glycolysis Glycolysis is the first stage of cellular respiration. During glycolysis, glucose is broken down into 2 molecules of pyruvic acid. ATP and NADH are also made.

Why Glycolysis Is Useful During glycolysis, 4 ATP molecules are made from 4 ADP molecules. Given that 2 ATP molecules are used to start the process, there is a net gain of just 2 ATP molecules. Even though the energy given off by glycolysis is small, the process is so fast that cells can make thousands of ATP molecules in just a few milliseconds. The speed of glycolysis can be a big plus when the cell suddenly needs a lot of energy.

Besides speed, another useful thing about glycolysis is that it does not need oxygen. This means that glycolysis can quickly get chemical energy to cells when there is not any oxygen present. When oxygen is present, however, the pyruvic acid and NADH made during glycolysis become the materials needed for the other stages of cellular respiration.

Key Question What happens during the process of glycolysis? **During glycolysis, 1 molecule of glucose, which has 6 carbon atoms, is changed into 2 molecules of pyruvic acid, which each have 3 carbon atoms.**

BUILD Vocabulary

glycolysis
The first set of reactions in cellular respiration in which a molecule of glucose is broken into two molecules of pyruvic acid

NAD^+
nicotinamide adenine dinucleotide: an electron carrier involved in glycolysis

WORD ORIGINS
The Greek word *glukus* means "sweet." The Latin word *lysis* means to "loosen." The term *glycolysis* thus means "loosening" or "breaking glucose."

Hands-On Learning

MODEL MOLECULES Provide students with six white paper circles, 12 blue paper circles, and six pink paper circles. Have them each place a C, H, or O on each of their circles to identify the circles as six carbon atoms in a glucose ($C_6H_{12}O_6$) molecule (white circles), 12 hydrogen atoms in a glucose molecule (blue circles), and six oxygen atoms in a glucose molecule (pink circles). Have students use the circles to model the formation of two pyruvic acid ($C_3H_4O_3$) molecules from one molecule of glucose.

Ask What atoms are left over after the two molecules of pyruvic acid form? *(four hydrogen atoms)*

Point out that the hydrogen atoms are used in other parts of the cycle.

During Reading

Build Connections

VISUAL SUMMARY Tell students that the Glycolysis diagram summarizes the process described in the text. Remind them that the curved arrows represent inputs and outputs of the process. Then draw students' attention to the inset visual on the left, showing the stages of cellular respiration.

Ask Where have you seen this visual before? *(in the previous lesson)* What does this visual show? *(the three stages of cellular respiration)* Why is the top section highlighted? *(because glycolysis is the focus of this diagram)*

Active Reading

READ, COVER, REMEMBER, RETELL Read the text on this page aloud. Then have students reread it silently. Tell students to cover the words. Then, have partners use what they remember from the text to explain the Glycolysis diagram to each other.

KEY QUESTION Have students read the question and answer. Tell students that they can remember the difference between glycolysis and glucose by understanding the suffix *-ose*.

Say In biochemistry, if a word ends in *-ose*, then it is always a sugar or other carbohydrate.

Build Vocabulary

WORD ORIGINS Point out that the suffix *-lysis* is also added to other root words to indicate a process that breaks down substances or processes. For example, *electrolysis* is the process that breaks down water into hydrogen and oxygen. Other words include *analysis* and *paralysis*.

Ask What word other than *glycolysis* likely comes from the Greek word *glukus*, which means "sweet"? *(glucose)*

Ask How do you know that glucose is a sugar? *(Glukus means "sweet.")* How else do you know? *(It has the suffix -ose.)*

Speed Bump

Ask What do the six dark balls at the top of the diagram represent? *(the six carbon atoms in a molecule of glucose)*

Ask What are the products of glycolysis? *(two NADH molecules, two pyruvic acid molecules, and a net gain of two ATP molecules)*

Ask What happens to each product? *(The NADH is used in the electron transport chain, the pyruvic acid is used in the Krebs cycle, and the ATP is used by cells as a source of energy.)*

During Reading continued

Active Reading

USE VISUALS Have students compare the Mitochondrion diagram with the picture of mitochondria at the beginning of the chapter. Challenge them to identify the membranes, inner membranes, and intermembrane spaces of the mitochondria in the picture.

USE CONTEXT CLUES As students read the lesson, have them list terms they do not know, such as *FAD* and *FADH₂*. For each term, have them work in pairs to use context clues and the diagram on the next page to help them understand the meanings.

USE GRAPHIC ORGANIZERS Create a flowchart on the board to illustrate the exchange of carbon atoms during the Krebs cycle.

Ask What does the text mean when it says that one molecule of glucose results in two complete turns of the Krebs cycle? *(One molecule of glucose makes two molecules of pyruvic acid. The Krebs cycle goes around once for every molecule of pyruvic acid.)*

KEY QUESTION Refer students to the diagram of the Krebs cycle on the next page. Have them compare the answer to the Key Question with the process shown in the diagram.

Build Vocabulary

WORD ORIGINS Remind students that knowing a term's origin can help them remember what it means. In science, it is common to name something after the person who discovered it.

Say Think back to the last chapter, which was about photosynthesis.

Ask What is the process called in photosynthesis that gets its name from the scientist who discovered it? If needed, give students the name *Melvin Calvin* as a hint. *(the Calvin cycle, named after Melvin Calvin)*

Speed Bump

Ask What is the difference in the carbon atoms in pyruvic acid and citric acid? *(Pyruvic acid contains three carbon atoms, and citric acid contains six carbon atoms.)*

Ask Look at the Krebs Cycle diagram. Why is the process shown called a cycle? *(The four-carbon molecule that forms citric acid is also a final product of the process and can be used again.)*

BUILD Vocabulary

Krebs cycle
The second stage of cellular respiration in which pyruvic acid is broken down into carbon dioxide in a series of energy-extracting reactions.

matrix
the innermost compartment of the mitochondrion

WORD ORIGINS
The Krebs cycle was named after the British biologist Hans Krebs, who won the Nobel Prize in 1953 for its discovery.

The Krebs Cycle

When there is oxygen present, pyruvic acid made in glycolysis passes to the second stage of cellular respiration. This stage is called the **Krebs cycle.** During the Krebs cycle, pyruvic acid is broken down over many steps.

Making Citric Acid The Krebs cycle begins when pyruvic acid goes into the mitochondrion. It passes two membranes and goes into the matrix. The **matrix** is the innermost part of the mitochondrion. The Krebs cycle takes place here. Once inside the matrix, 1 carbon is split off from pyruvic acid, releasing a molecule of carbon dioxide. The other 2 carbon atoms join a 4-carbon molecule already present in the cycle. They form citric acid, which has 6 carbons. As the cycle keeps going, citric acid is broken down. More carbon dioxide is given off, and high-energy electrons are moved to energy carriers, such as NAD^+. As each molecule of carbon dioxide leaves, the citric acid molecule loses a carbon atom. In the end a 4-carbon molecule is formed. Then the cycle begins all over again.

Electron Carriers as Energy For each turn of the cycle, a phosphate is added to ADP to make ATP. Each glucose molecule causes two complete turns of the Krebs cycle. So, for each molecule of glucose, the cycle makes two more ATP molecules. When we add these to the two made in glycolysis, so far we have gained just 4 ATP molecules. That doesn't sound like much. However, most of the chemical energy released in the cycle goes to produce high-energy electrons. These are passed to the electron carriers NAD^+ and FAD. This changes them into NADH and $FADH_2$.

What happens to the carbon dioxide, ATP, and electron carriers produced in the Krebs cycle? Carbon dioxide is not useful to the cell. It is expelled every time you breathe out. The ATP molecules are very useful. They are immediately available as energy for cellular activities. When oxygen is present, the electron carrier molecules are used make huge amounts of ATP.

Key Question What happens during the Krebs cycle?
During the Krebs cycle, pyruvic acid is broken down into carbon dioxide in a series of steps that releases chemical energy.

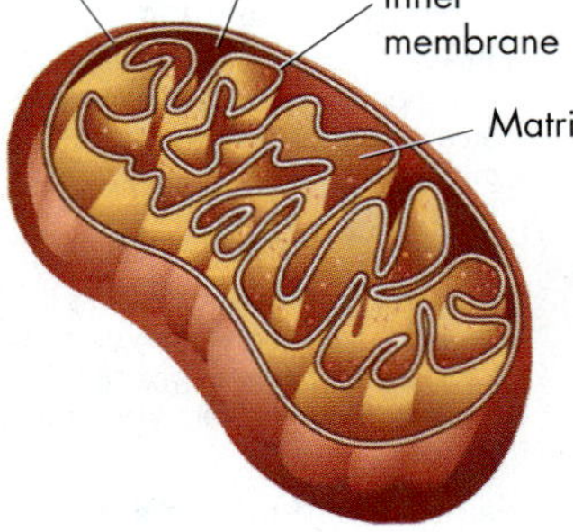

Mitochondrion

Mitochondrion The Krebs cycle takes place within the matrix, or the innermost space, of the mitochondrion.

ELL Focus on ELL: Access Content

ALL SPEAKERS Assign students to groups of two or three. Have them take turns reading a paragraph of text. After each paragraph is read, have the group summarize in their own words what was read. Encourage students to use the headings and visuals to help them summarize the content. Then, have them share their summaries with the group.

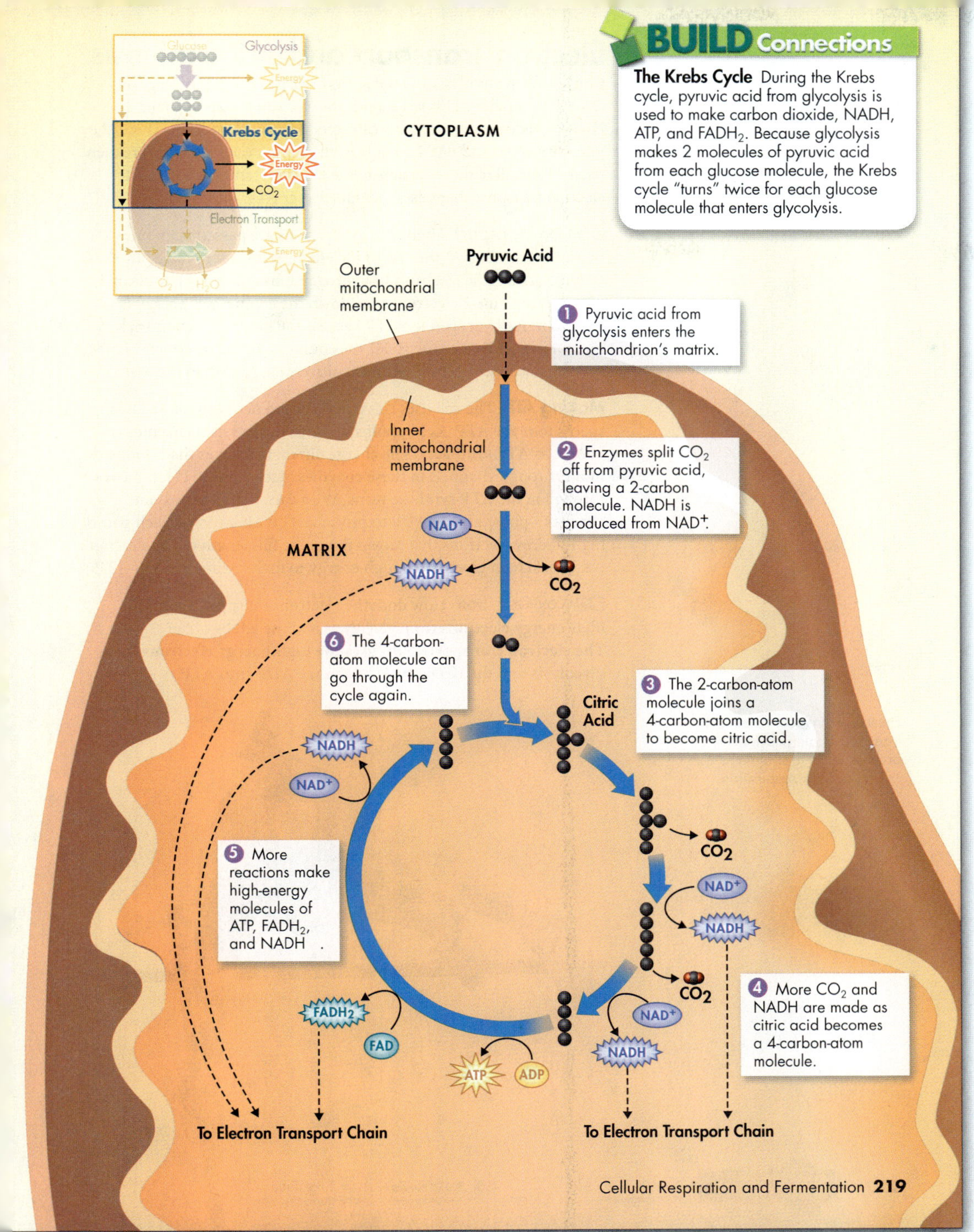

Focus on Academic Language

- membrane
- turn
- extracting
- reaction

Have students find each of these terms on this or the previous page. Ask them to use context clues to define each term.

The word *turn* can have several meanings. Point out to students that *turn* as used here refers to one Krebs cycle ending and another one beginning.

To clarify the meanings of these terms, have students use each one in a sentence. Remind students to use them with the appropriate meanings and in their own words.

Build Connections

VISUAL SUMMARY Give each student five index cards. Using the diagram on this page and the accompanying text, have each student write a different question about the Krebs cycle on each of the cards. Have students then trade cards and answer each other's questions. Then, have students discuss the questions and answers in small groups. If they cannot agree on the answer to a question, discuss it as a class.

Ask Did any question you answered relate the Krebs cycle to mitochondria? *(Answers may vary.)* Why might this be important? *(The Krebs cycle occurs in mitochondria.)*

Be sure students recognize that the CO_2 shown in the diagram is not an atom. The formula indicates that it is a molecule made up of one carbon atom and two oxygen atoms.

Active Reading

SCIENCE SUPPORT Explain that most of the steps that occur during cellular respiration happen because of the unique bonding ability of carbon atoms. Carbon can bond easily with itself as well as with the atoms of other elements. As a result, a tremendous number of carbon compounds exist in living things. When a bond to a carbon atom is broken, that carbon atom is available to bond to numerous other atoms.

Speed Bump

Ask How does energy change form during the Krebs cycle? *(Sample answer: When bonds in molecules are broken, some of the chemical energy stored in the bonds becomes chemical energy stored in new bonds. Also, some of the energy is released as heat.)*

Ask Why does glycolysis have to occur before the Krebs cycle? *(The Krebs cycle requires pyruvic acid, which is produced during glycolysis.)*

Ask According to the diagram, what happens to the NADH and $FADH_2$ molecules produced in the Krebs cycle? *(They go to the last stage of cellular respiration, the electron transport chain.)*

During Reading continued

Active Reading

DRAW A DIAGRAM Have students help you construct a flowchart of electron transport and ATP synthesis on the board. Begin with this step: *NADH and $FADH_2$ enter the electron transport chain.* As volunteers read paragraphs aloud, have students suggest the next steps in the process. Guide students to understand how each stage is related.

Ask Where do the NADH and $FADH_2$ come from? *(glycolysis and the Krebs cycle)*

SUMMARIZE Have students summarize each paragraph in one or two sentences in their notebooks.

Ask Why is the stage of electron transport called a chain? *(because high-energy electrons are passed from one carrier to the next)*

KEY QUESTION Have students cover the Key Question and read only the answer. From the answer, have them state a question that could be used to obtain that answer. Write suggested questions on the board. Then, have students compare their questions to the Key Question.

Speed Bump

Ask What materials produced by glycolysis and the Krebs cycle carry electrons into the electron-transport stage of cellular respiration? *(NADH from glycolysis and NADH and $FADH_2$ from the Krebs cycle)*

Ask What two reactions take place in the step where electrons are added to a molecule? *(ADP changes to ATP; hydrogen ions and oxygen form water.)*

Ask What happens each time electrons are passed? *(Hydrogen ions move across the membrane.)*

Ask What happens in the intermembrane space? *(Hydrogen ions build up and the space becomes more positively charged than the matrix.)*

Ask What makes the H^+ ions move through the ATP synthases? *(the difference in the electrical charge between the matrix and the intermembrane space)*

Ask What does the ATP synthase do? *(As it turns, it changes ADP into ATP.)*

Electron Transport and ATP Synthesis

Carriers from glycolysis and the Krebs cycle go into the last stage of cellular respiration. This last stage is the electron transport chain. The NADH made during glycolysis can enter the mitochondrion. There, they join the NADH and $FADH_2$ made by the Krebs cycle. Electrons are then passed from all of those carriers to the electron transport chain. The electron transport chain uses them to change ADP into ATP.

Electron Transport High-energy electrons are passed from one carrier to the next. At the end of the electron transport chain, an enzyme adds these electrons to hydrogen ions and oxygen to form water. Every time 2 electrons pass down the chain, their energy is used to move hydrogen ions (H^+) across the inner mitochondrial membrane. As this happens, H^+ ions build up in the intermembrane space. That space becomes more positively charged than the matrix.

Making ATP The cell uses this charge difference between the two spaces to make ATP. As in photosynthesis, the cell uses enzymes known as ATP synthases. The charge difference makes the H^+ ions move through channels in these enzymes. Each time a H^+ ion moves through it, the ATP synthase turns. With each turn, the enzyme attaches a phosphate to ADP to produce ATP. On average, each pair of high-energy electrons that moves down the full length of the electron transport chain gives off enough energy to make 3 molecules of ATP.

Key Question How does the electron transport chain use high-energy electrons from glycolysis and the Krebs cycle?
The electron transport chain uses the high-energy electrons from glycolysis and the Krebs cycle to change ADP into ATP.

Electron Transport The electron transport chain provides most of the energy that athletes use during cellular respiration.

Students Say

"A game called "Stray or Stay" helped me know if I really understood the Krebs cycle because I had to explain it to someone else. If I forgot something, other people in the group could help explain it. If I couldn't explain it well, I'd choose to "stay" until I could.

Here's how it works. The teacher gives each group of six students three red and three green cards. Each group has a chart depicting a different stage of the cycle. Three students in each group who can explain the events during this stage do so. They then take the green cards and "stray" to the next group, clockwise. The students with the red cards "stay." Three more (new) students with green cards "stray" to the group. All students put their cards in a pile. Three of the students who can explain the stage pick up green cards and explain to those with red cards. Someone might keep a red card and "stay" more than once. If by the third round, someone isn't "straying," the teacher knows that more intervention might be needed."

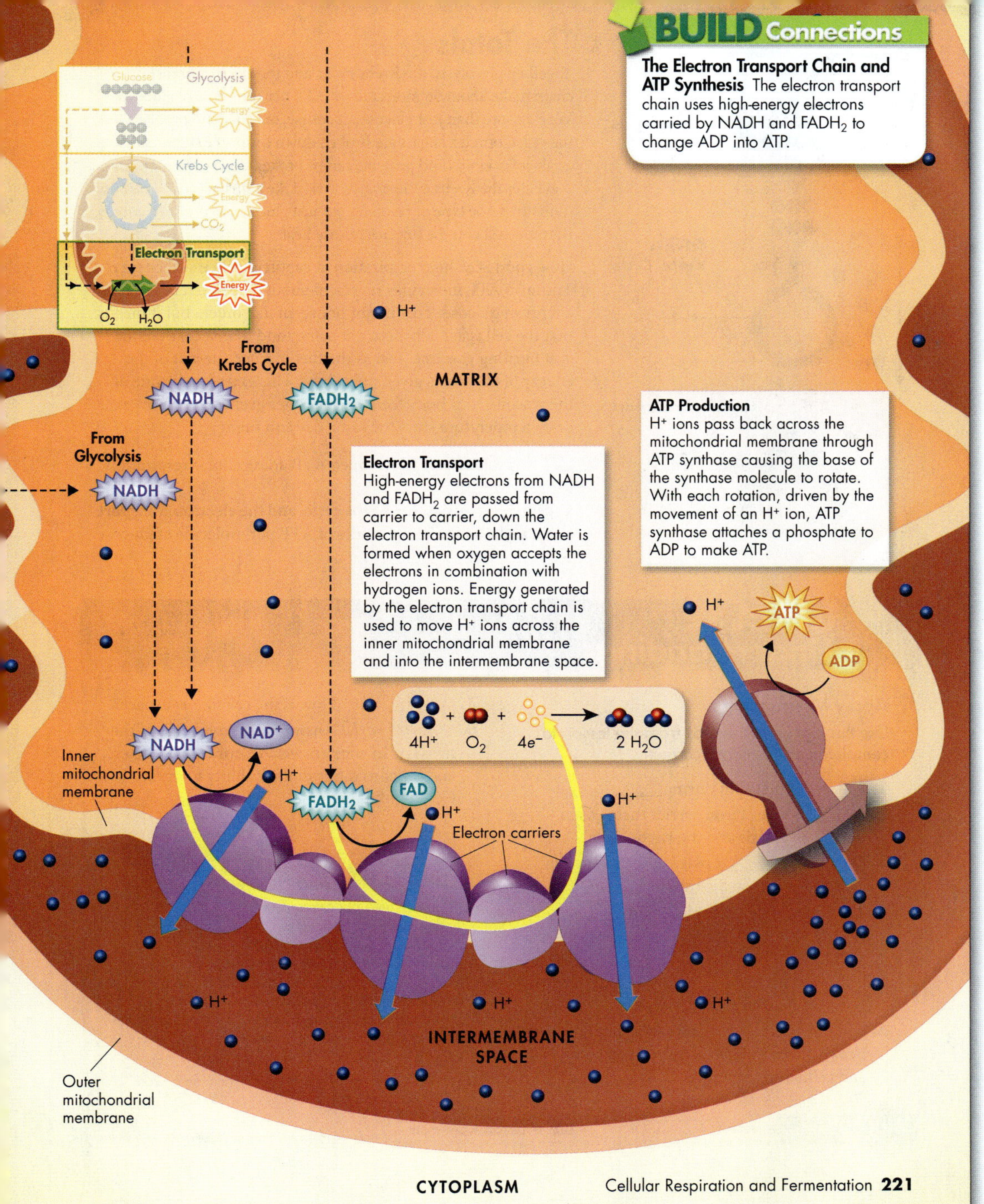

Build Connections

VISUAL SUMMARY Ask students questions about the diagram that will help clarify the process of electron transport.

Ask Which contains more hydrogen ions than it did before the process started—the space between membranes or the matrix? *(the space between membranes)*

Ask What causes the formation of ADP from ATP? *(Hydrogen ions pass through the ATP synthase. Every time a hydrogen ion passes through, the synthase rotates, and ADP changes to ATP.)*

Ask When do hydrogen ions move across the inner membrane? *(when electrons are passed down the chain)*

Active Reading

DRAW A DIAGRAM Redraw the diagram on the board in a simplified form. Start by sketching the electron transport chain. Draw a line to represent the membrane and five circles to represent the electron carrier proteins in the membrane. Then, draw an NADH molecule approaching the chain. Have students direct you as you draw the electrons from NADH entering the chain. Show how electrons move down the chain, continuing until they reach oxygen. Also show how hydrogen ions move through the carrier proteins from the matrix into the intermembrane space.

SELF-QUESTION Have students reread individually or in pairs Lessons 9.1 and 9.2. As they read, have them use index cards to write questions they still have about cellular respiration. Have them place their questions in a box. Pull the questions out one at a time. Have volunteers answer the questions, and then discuss the answers as a group.

Focus on Academic Language

- electron
- hydrogen ion
- particle
- proton

Review with students the definitions of a molecule, an atom, and the particles that make up an atom. Emphasize that an electron is a negatively charged particle, and a proton is a positively charged particle. Remind them that both particles are found in atoms. Explain that an ion is an atom that has either gained or lost electrons. Clarify that the ion is positively charged if the atom has lost electrons, and it is negatively charged if it has gained electrons. Point out that a hydrogen ion has the same makeup as a single proton.

During Reading continued

Active Reading

KEY QUESTION After reading the Key Question and answer, have students reread the last paragraph. Review the meaning of *generate*.

Ask What percent of the energy in one glucose molecule is contained in one ATP molecule? *(one percent of the energy)* Why? *(One glucose molecule makes 36 ATP molecules and those 36 molecules represent 36 percent of the total amount of energy in glucose.)*

Remind students that energy can change forms, but it is not lost or gained during chemical reactions. This is called the law of conservation of energy.

Speed Bump

Ask Why can your cells get energy from any foods? *(All foods contain macromolecules that can be broken down into molecules that can enter the Krebs cycle.)*

Ask During the process of cellular respiration, what happens to the 64 percent of energy in a molecule of glucose that is not converted to ATP? *(The remainder of the energy in glucose is given off as heat.)*

Post Reading

ASSESSMENT TIPS Advise students to read questions carefully and make their answers specific.

For Question 7, review the definition of *analogy*. *(a comparison that shows how two different things are similar in some way)*

MYSTERY CLUE For Question 8, make sure students notice that they are to compare whales and humans. So, their answers must mention both whales and humans.

Ask A whale breathes to bring in oxygen, but what does it breathe to get rid of? *(to rid its body of CO_2)* Suggest that students look back at the Krebs cycle to refresh their memories about CO_2 being given off during the process.

Wrap-Up Activity

MODEL A PROCESS On an index card, write the name or formula for each type of molecule or atom involved in cellular respiration. Give one or two cards to each student. Tell students that the contents of the card or cards they are given determine their "identity." Starting with glycolysis, have students move and arrange themselves to model what takes place during each stage of cellular respiration. Be sure students display their cards so that other students can see what they represent.

Energy Totals Breaking down glucose through cellular respiration makes 36 molecules of ATP.

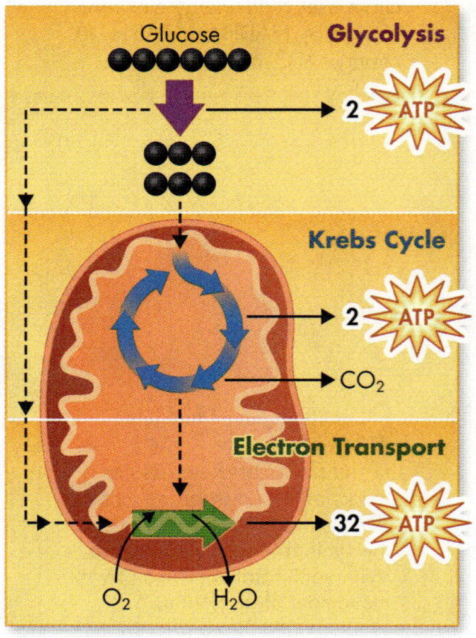

The Totals

Together, glycolysis, the Krebs cycle, and the electron transport chain make about 36 molecules of ATP per molecule of glucose. We eat more than just glucose, of course, but that's no problem for the cell. Complex sugars are broken down to simple sugars such as glucose. Lipids and proteins can be broken down into molecules that enter the Krebs cycle or glycolysis. Like a furnace, a cell can use more than one type of reactant, not just glucose, to produce energy. Both the cell and the furnace release heat.

How good is cellular respiration at capturing chemical energy? Those 36 ATP molecules represent about 36 percent of the total energy of glucose. That might not seem like much, but it means that the cell is actually better at using food than a car's engine is at burning gasoline! Just as the fuel in a car, the rest of the energy of glucose is given off as heat. This extra heat is one of the reasons your body feels warmer after exercise, and why your body temperature stays the same day and night.

Key Question How much ATP does cellular respiration generate?
Together, glycolysis, the Krebs cycle, and the electron transport chain make about 36 molecules of ATP per molecule of glucose.

CHECK Understanding

Apply Vocabulary

Use the highlighted words from the lesson to complete each sentence correctly.

1. Glucose is first split during ______.
2. ______ is an electron carrier that is used during glycolysis and the Krebs cycle.
3. During the ______, pyruvic acid is broken down into carbon dioxide.
4. The Krebs cycle takes place in the ______.

Critical Thinking

5. **Compare and Contrast** How is the function of NAD^+ similar to that of $NADP^+$?
6. **Infer** Why do you often breathe faster during heavy exercise? How is this related to oxygen and cellular respiration?
7. **Use Analogies** How is the cell like a furnace?
8. **Write to Learn** Answer the second clue of the mystery. In your answer, describe how whales and humans differ in their ability to stand CO_2 buildup in blood.

MYSTERY CLUE

When we are underwater, our bodies react to a buildup of CO_2 in the blood. This buildup makes us want to go to the surface and gasp for breath. The average human can hold his or her breath for only about a minute. Whales stay underwater for much longer. What does this suggest about a whale's ability to stand CO_2? (**Hint:** See p. 218.)

BIOLOGY.com Search Lesson 9.2 GO • Lesson Assessment

Check Understanding

1. glycolysis
2. NAD^+
3. Krebs cycle
4. matrix
5. Both are electron carriers that accept a hydrogen ion and electrons and take them to other parts of the cell to power other processes.
6. When you exercise, your cells need energy and must carry out more cellular respiration. This uses up oxygen, so you breathe faster and more deeply in order to get more.
7. Both a furnace and a cell can use more than one type of reactant to produce energy.
8. Some whales can stay underwater for at least 45 times longer than humans. This suggests that they must be able to withstand a greater buildup of CO_2 in the blood.

9.3 Fermentation

Fermentation

How do your cells get energy when you use up oxygen so quickly that you cannot get more oxygen fast enough? And what about organisms that live in places where there isn't oxygen? How do their cells get energy from food?

Glycolysis can make ATP quickly and it does not need oxygen. However, in just a few seconds, glycolysis fills all of the cell's NAD^+ molecules with electrons. Without oxygen, the electron transport chain does not run, so there is nowhere for the NADH molecules to leave their electrons. So, NADH doesn't get converted back to NAD^+. Without NAD^+, the cell cannot keep glycolysis going. That's where fermentation comes in. When oxygen is not present, glycolysis is followed by **fermentation.** During fermentation, cells change NADH to NAD^+ by passing the electrons back to pyruvic acid. The NAD^+ can then fuel glycolysis again. The cells can make ATP again. Fermentation is an aerobic process and takes place in the cytoplasm. Sometimes glycolysis and fermentation are called anaerobic respiration. There are two different kinds—alcoholic fermentation and lactic acid fermentation.

Alcoholic Fermentation Yeasts and a few other microorganisms use alcoholic fermentation. A summary of alcoholic fermentation after glycolysis is as follows:

$$\text{Pyruvic acid} + \text{NADH} \longrightarrow \text{Alcohol} + CO_2 + NAD^+$$

Humans use alcoholic fermentation to make alcoholic beverages. It is also what causes bread dough to rise. When yeast cells in the dough run out of oxygen, the dough begins to give off tiny bubbles of carbon dioxide. These bubbles form the air spaces you see in a slice of bread. The small amount of alcohol made in the dough evaporates when the bread is baked.

Key Questions

How do organisms generate energy when oxygen is not available?

How does the body produce ATP during different stages of exercise?

BUILD Understanding

Venn Diagram As you read, create a Venn diagram that compares alcoholic fermentation and lactic acid fermentation.

In Your Workbook Go to your workbook to learn more about making a Venn diagram. Complete the Venn diagram for Lesson 9.3.

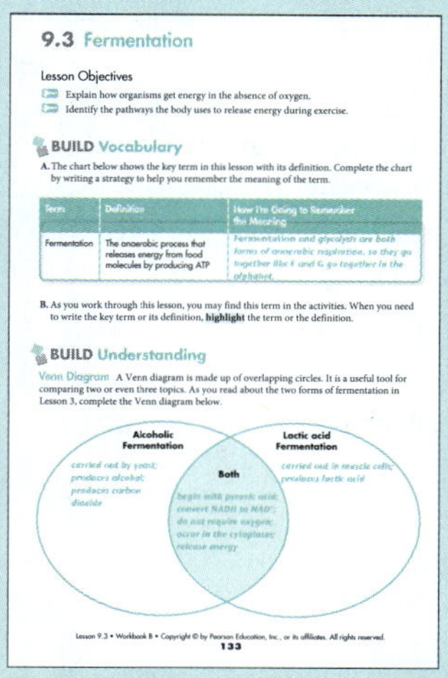

9.3 Fermentation

Lesson Objectives

Explain how organisms get energy in the absence of oxygen.

Identify the pathways the body uses to release energy during exercise.

BUILD Vocabulary

A. The chart below shows the key term in this lesson with its definition. Complete the chart by writing a strategy to help you remember the meaning of the term.

Term	Definition	How I'm Going to Remember the Meaning
Fermentation	The anaerobic process that releases energy from food molecules by producing ATP	Fermentation and glycolysis are both forms of anaerobic respiration, so they go together like F and G go together in the alphabet.

B. As you work through this lesson, you may find this term in the activities. When you need to write the key term or its definition, **highlight** the term or the definition.

BUILD Understanding

Venn Diagram A Venn diagram is made up of overlapping circles. It is a useful tool for comparing two or even three topics. As you read about the two forms of fermentation in Lesson 3, complete the Venn diagram below.

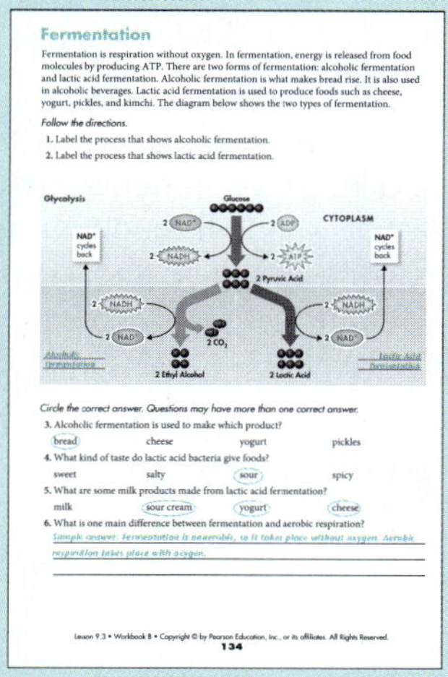

Fermentation

Fermentation is respiration without oxygen. In fermentation, energy is released from food molecules by producing ATP. There are two forms of fermentation: alcoholic fermentation and lactic acid fermentation. Alcoholic fermentation is what makes bread rise. It is also used in alcoholic beverages. Lactic acid fermentation is used to produce foods such as cheese, yogurt, pickles, and kimchi. The diagram below shows the two types of fermentation.

Follow the directions.

1. Label the process that shows alcoholic fermentation.
2. Label the process that shows lactic acid fermentation.

Circle the correct answer. Questions may have more than one correct answer.

3. Alcoholic fermentation is used to make which product?
   bread cheese yogurt pickles
4. What kind of taste do lactic acid bacteria give foods?
   sweet salty sour spicy
5. What are some milk products made from lactic acid fermentation?
   milk sour cream yogurt cheese
6. What is one main difference between fermentation and aerobic respiration?
   Sample answer: Fermentation is anaerobic, so it takes place without oxygen. Aerobic respiration takes place with oxygen.

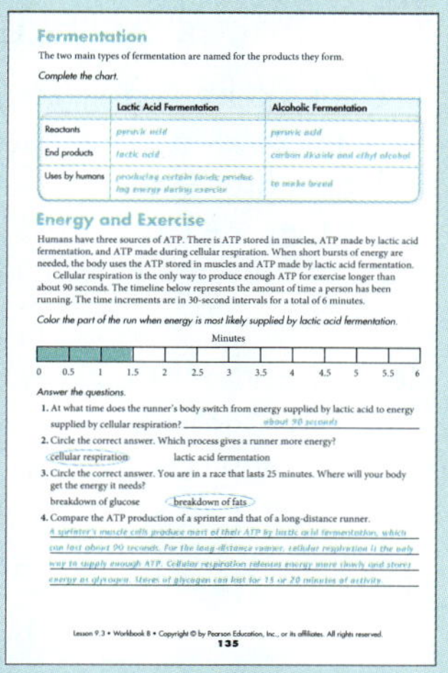

Fermentation

The two main types of fermentation are named for the products they form.

Complete the chart.

	Lactic Acid Fermentation	Alcoholic Fermentation
Reactants	pyruvic acid	pyruvic acid
End products	lactic acid	carbon dioxide and ethyl alcohol
Uses by humans	producing certain foods; producing energy during exercise	to make bread

Energy and Exercise

Humans have three sources of ATP. There is ATP stored in muscles, ATP made by lactic acid fermentation, and ATP made during cellular respiration. When short bursts of energy are needed, the body uses the ATP stored in muscles and ATP made by lactic acid fermentation.

Cellular respiration is the only way to produce enough ATP for exercise longer than about 90 seconds. The timeline below represents the amount of time a person has been running. The time increments are in 30-second intervals for a total of 6 minutes.

Color the part of the run when energy is most likely supplied by lactic acid fermentation.

Answer the questions.

1. At what time does the runner's body switch from energy supplied by lactic acid to energy supplied by cellular respiration? about 90 seconds
2. Circle the correct answer. Which process gives a runner more energy?
   cellular respiration lactic acid fermentation
3. Circle the correct answer. You are in a race that lasts 25 minutes. Where will your body get the energy it needs?
   breakdown of glucose breakdown of fats
4. Compare the ATP production of a sprinter and that of a long-distance runner.
   A sprinter's muscle cells produce most of their ATP by lactic acid fermentation, which can last about 90 seconds. For the long-distance runner, cellular respiration is the only way to supply enough ATP. Cellular respiration releases energy more slowly and stores energy as glycogen. Stores of glycogen can last for 15 or 20 minutes of activity.

Pre-Reading

Objectives

9.3.1 Explain how organisms get energy in the absence of oxygen.

9.3.2 Identify the pathways the body uses to release energy during exercise.

Preview the Pages

1. Ask volunteers to read the two Key Questions. As a group, discuss what the questions are asking.
2. Preview the visuals, giving special attention to the Fermentation diagram. Invite students to review what they have learned about glycolysis, and relate that to the diagram.

Activate Prior Knowledge

Help students make the connection between glycolysis and fermentation.

Ask What kind of molecules are broken down during glycolysis? *(glycogen)* What is produced by the process of glycolysis? *(pyruvic acid, NADH)*

Point out that what happens to the pyruvic acid molecules depends on whether oxygen is present. If oxygen is not present, the pyruvic acid undergoes the process of fermentation.

Build Understanding

VENN DIAGRAM Have students refer to the Venn diagram in Workbook B. Tell students to make a list of characteristics that apply to either alcoholic or lactic acid fermentation.

NATIONAL SCIENCE EDUCATION STANDARDS

UNIFYING CONCEPTS AND PROCESSES

I

CONTENT

B.3, C.1.a, C.1.b, C.5.c, C.5.d

For the corresponding lesson in ***Miller & Levine Biology,*** see pages 262–265.

LESSON 9.3

During Reading

Build Vocabulary

RELATED WORD FORMS Have students name the related noun forms of the following verbs from the student text: *transport, summarize, bubble, evaporate, store. (transportation, summary, bubble, evaporation, store)* Point out that sometimes the noun and verb forms are spelled the same way *(bubble, store)*, but the words are used differently.

Active Reading

COMPARE AND CONTRAST Have students compare and contrast the equation for alcoholic fermentation and the equation for lactic acid fermentation.

Ask How are the products of lactic acid fermentation and alcoholic fermentation the same? *(NAD^+ is produced in both equations.)* How are they different? *(Lactic acid fermentation produces lactic acid, and alcoholic fermentation produces alcohol and carbon dioxide.)*

SCIENCE SUPPORT Point out to students that although the Fermentation diagram shows both alcoholic and lactic acid fermentation, these processes usually do not occur simultaneously. Most organisms carry out either one or the other. For example, lactic acid fermentation is carried out in muscle tissues. Microorganisms such as yeast carry out alcoholic fermentation.

Ask Which form of fermentation is carried out in the human body? *(lactic acid fermentation)*

INTERPRET VISUALS Have students compare the glycolysis part of the diagram to the more expanded diagram of glycolysis in Lesson 9.2.

Say Four molecules of ATP are produced by glycolysis. However, according to the diagram of glycolysis on this page, only two molecules of ATP are formed by glycolysis.

Ask What explanation can you give for this? *(Four molecules are produced during glycolysis, but two are used to provide energy to start the process.)*

Say This diagram only shows the number of ATP molecules *gained* through glycolysis. Point out that the NADH formed during glycolysis is changed back to NAD^+ during both types of fermentation. The NAD^+ molecules then cycle back to form more NADH during glycolysis.

Ask What is missing from the diagram of fermentation that makes the process anaerobic? *(oxygen)*

KEY QUESTION Have students individually read the Key Question and answer. Point out that both alcoholic and lactic acid fermentation change NADH back into NAD^+.

Ask How does the cell use the NAD^+ to produce more ATP? *(through glycolysis)*

BUILD Vocabulary

fermentation
the process by which cells release energy in the absence of oxygen

RELATED WORD FORMS
The noun *fermentation* and the verb *ferment* are related word forms. Dough that is beginning to ferment is just starting to undergo the process of fermentation.

Lactic Acid Fermentation Most living things carry out fermentation by changing pyruvic acid into lactic acid. Unlike alcoholic fermentation, lactic acid fermentation does not give off carbon dioxide. Like alcoholic fermentation, lactic acid fermentation makes NAD^+ so that glycolysis can continue. Lactic acid fermentation after glycolysis can be written as:

$$\text{Pyruvic acid} + \text{NADH} \longrightarrow \text{Lactic acid} + \text{NAD}^+$$

Bacteria that use lactic acid fermentation are used to make many kinds of foods. Examples are cheese, yogurt, buttermilk, and sour cream. The acid is part of why these foods have a sour taste. Many kinds of pickles, sauerkraut, and kimchi are also made using lactic acid fermentation.

People also use lactic acid fermentation. Many of the cells in our body are able to make ATP by lactic acid fermentation when they are without oxygen for a few seconds. Muscle cells use this kind of fermentation when they need a lot of ATP for quick bursts of activity.

Lactic acid builds up in muscles after a burst of activity. The only way to get rid of lactic acid is in a chemical pathway that uses extra oxygen. This is why exercise can leave you huffing and puffing.

Key Question How do organisms generate energy when oxygen is not available?
In the absence of oxygen, fermentation releases energy from food molecules by making ATP.

Fermentation In alcoholic fermentation, pyruvic acid from glycolysis is changed into alcohol and carbon dioxide. In lactic acid fermentation, pyruvic acid from glycolysis is changed into lactic acid.

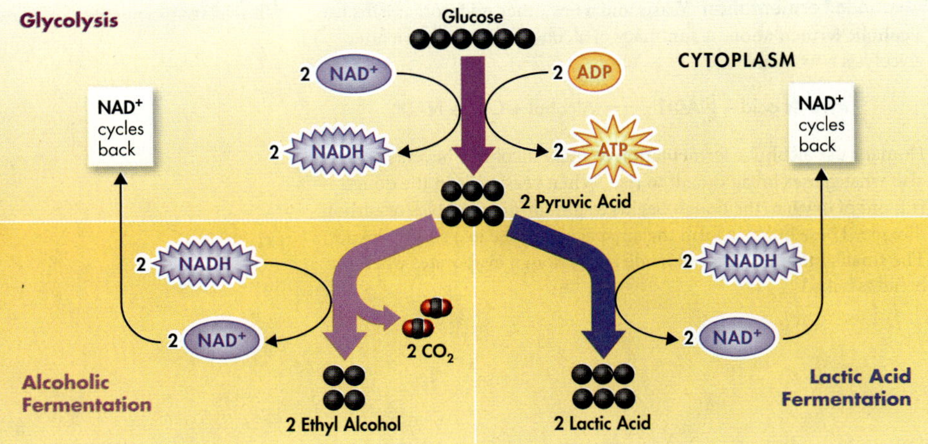

ELL Focus on ELL: Build Background

ALL SPEAKERS English language learners will likely be familiar with some foods that have undergone fermentation. Provide students with a sample list of such foods, including bread, cheese, and pickles. Have students work in groups to create a list of foods they are familiar with that are made through fermentation. Encourage them to think of foods that are unique to their own culture, such as kimchi. Review lists with students.

Remind students that lactic acid fermentation produces lactic acid, which makes food sour. Have students review their lists and decide which foods are produced by lactic acid fermentation and which are produced by alcoholic fermentation.

Energy and Exercise

Think about the last time you ran a race. At each stage of the race your body used different paths to get energy. You have three main sources of ATP: ATP already in muscles, ATP made by lactic acid fermentation, and ATP made by cellular respiration. At the beginning of a race, the body uses all three sources. However, stored ATP and lactic acid fermentation can give energy only for a short time.

Quick Energy Cells usually have only a little ATP stored up from cellular respiration. At the very beginning of a race, the muscles have only enough for a few seconds of activity. Usually by the 50-meter mark, that store of ATP is nearly gone. Then, the muscle cells make their ATP by lactic acid fermentation, which gives enough ATP to last about 90 seconds, or about 200 or 300 meters.

Long-Term Energy For exercise longer than about 90 seconds, cellular respiration is the only way to keep making ATP. Your body stores energy in muscle in the form of the carbohydrate glycogen. These stores are usually enough to last for 15 or 20 minutes of activity. After that, your body begins to break down other stored molecules, including fats, for energy.

Key Question How does the body produce ATP during different stages of exercise?
For quick bursts of energy, the body uses ATP already in muscles and ATP from lactic acid fermentation. For exercise longer than 90 seconds, the body uses cellular respiration.

Energy and Exercise During a race, runners rely on the energy supplied by ATP to make it to the finish line.

CHECK Understanding

Apply Vocabulary

Use the highlighted words from the lesson to complete the following sentence correctly.

1. When there is no oxygen present, the cell uses _____ to make ATP.

Critical Thinking

2. **Compare and Contrast** What reactants and products do the two types of fermentation have in common?
3. **Apply** Why do runners breathe heavily after a sprint race?
4. **Sequence** List the body's sources of energy in the order in which they are used during a long-distance race.
5. **Infer** Why might aerobic exercise help someone lose weight?
6. **Write to Learn** Answer the third clue of the mystery. In your answer explain what whales do with lactic acid from fermentation.

MYSTERY CLUE

Whales use lactic acid fermentation to get much of their energy during a deep dive. If they can't breathe in to get oxygen, what do they do with all of the lactic acid made by fermentation? (**Hint:** See p. 224.)

BIOLOGY.com Search Lesson 9.3 GO • Lesson Assessment

Check Understanding

1. fermentation
2. They both use pyruvic acid and NADH (reactants) to make NAD^+ (product).
3. When the race is over, the only way for the runner to get rid of lactic acid is through a chemical pathway that requires extra oxygen.
4. First, the body uses ATP that was already in the muscles. Next, the body gets ATP through lactic acid fermentation. Finally, the body makes ATP through cellular respiration.
5. During long periods of aerobic exercise, the body uses cellular respiration. After 15 or 20 minutes of aerobic activity, the body uses up all the glycogen stores in muscles. After that, the body begins to burn stored fat.
6. The lactic acid builds up in the whales' muscles until the whales take in enough oxygen to break down the lactic acid.

Active Reading

DRAW CONCLUSIONS Tell students that this first paragraph tells about three main sources of ATP. Remind them that, when running a race, the ATP stored in muscles and lactic acid fermentation can only give energy to the body for a short time.

Ask After the human body can no longer get energy from stored ATP and lactic acid fermentation, where will it get the energy it needs to continue in the race? *(cellular respiration)*

KEY QUESTION After reading the Key Question and answer, have students think about what activities they have done today. Ask them to list the activities according to whether most of their energy to do the activity came from fermentation or from cellular respiration. They can list the activities vertically in a continuum with *Fermentation* at the top and *Cellular Respiration* at the bottom.

SCIENCE SUPPORT Energy is stored in muscles and in the liver in the form of glycogen. Glycogen is formed by bonding several glucose molecules together. When glucose is needed for glycolysis, each glycogen molecule can be broken down into several glucose molecules. These glucose molecules then produce the pyruvic acid that is used in both cellular respiration and fermentation.

Ask As stored glycogen is used, what does the body use for another energy source? *(other stored molecules such as fat)*

Post Reading

ASSESSMENT TIPS To answer Question 3, ask students to think about what is produced during lactic acid fermentation *(lactic acid)* and where it is stored *(muscles)*.

When answering questions involving steps, such as Question 4, encourage students to draw a diagram of the steps before answering the question.

To answer Question 5, ask students to think about what the body uses for energy after glycogen stores are depleted. *(fat)*

MYSTERY CLUE Have students reread the last paragraph on the previous page to obtain information about what happens to lactic acid when oxygen is not available.

Wrap-Up Activity

OBSERVE ALCOHOLIC FERMENTATION Provide students with a bread recipe and have them identify the ingredients that are necessary for fermentation. *(sugar, yeast)* Display bread dough that hasn't risen and bread dough that has risen. Allow students to feel the differences in the two balls of dough. Ask what caused the dough to rise. *(carbon dioxide)*

CHAPTER LAB

Pre-Lab

Review the relationship between glycolysis and fermentation. Make sure students understand that glucose and energy from a cell are needed to start glycolysis.

Ask What is the purpose of fermentation? *(to release to the cell energy that was stored in glucose)*

Ask What are some organisms that carry out fermentation? *(Sample answers: humans, yeast, bacteria)*

Remind students that each type of sugar has a slightly different chemical makeup. In this lab, they will explore which type of sugar yeast can ferment the fastest.

Review the Background Questions and the Pre-Lab Questions as a class.

Lab

Tell students they will perform the chapter lab *Comparing Fermentation Rates of Sugars* described in **Lab Manual B.**

SAFETY Students should wear goggles and an apron when using a water bath. Remind them to handle glass objects with care and to alert you immediately if a glass object breaks. Tell students to avoid contact with the surface of the hot plate. To avoid electric shocks, they should make sure cords, connections, and their hands are dry. Remind students to wash their hands thoroughly at the end of the lab.

Post-Lab

CREATE A MENU Have students create a menu that includes as many items as possible made by processes that include fermentation. Menu items can include ingredients that are made from fermentation, even if the entire item is not, such as a salad that contains pickles.

NATIONAL SCIENCE EDUCATION STANDARDS

INQUIRY A.1.b, A.1.c, A.2.a, A.2.c

Real-World Lab

GUIDED INQUIRY

Pre-Lab: Comparing Fermentation Rates of Sugars

Problem How does the type of sugar affect the rate of fermentation?

Materials probe interface, gas pressure probe, hot plate, 400-mL beaker, thermometer, ring stand, test-tube clamp, medium test tube, test-tube rack, sugar solution, yeast suspension, pipettes, vegetable oil, 1-hole rubber stopper, plastic tubing with lock fitting

Lab Manual Chapter 9 Lab

Skills Focus Predict, Measure, Analyze Data, Infer

Connect to the Big idea In most cells, the pathways that release energy from food start with the conversion of glucose to pyruvic acid. This process does not require oxygen. When oxygen is present, however, pyruvic acid can react to form a compound that is used in the second stage of cellular respiration. When oxygen is not present, the pyruvic acid can be used in an anaerobic pathway. This alternate pathway from glucose to ATP is called fermentation. In this lab, you will use yeast to ferment sugars and compare the rates of fermentation.

Background Questions

a. **Review** What is the importance of the NAD^+ that is produced during the fermentation of pyruvic acid?

b. **Review** What other products are produced besides NAD^+ when yeast ferment sugar?

c. **Compare and Contrast** How are simple sugars different from disaccharides? (If needed, review Lesson 2.3 in your textbook.)

d. **Use Analogies** What do fermentation and a detour that drivers must use when roads are closed have in common?

Pre-Lab Questions

Preview the procedure in the lab manual.

1. **Infer** What is the purpose of adding a layer of vegetable oil above the sugar and yeast mixture?
2. **Relate Cause and Effect** How will the rate of fermentation affect the gas pressure in a test tube?
3. **Predict** Which of the sugars do you think will have the highest rate of fermentation, and why?

BIOLOGY.com Search Chapter 9 GO

Visit Chapter 9 online to test yourself on chapter content as well as find activities to help you learn.

Untamed Science Video Go underwater with the Untamed Science crew to discover why marine mammals can stay submerged for such a long time.

Tutor Tube Improve your understanding of respiration by working "backward" from a breath of oxygen.

Art Review Review the components of electron transport and ATP synthesis.

InterActive Art See glycolysis and the Krebs cycle in action.

Art in Motion See how matter and energy cycle between photosynthesis and respiration.

Pre-Lab Answers

BACKGROUND QUESTIONS

a. Without NAD^+, glycolysis will stop and the production of ATP will cease.

b. Alcohol and carbon dioxide are produced.

c. Simple sugars are the smallest carbohydrate molecules. Disaccharides are formed when two simple sugar molecules are joined together.

d. Fermentation and a detour are both less efficient alternate routes, in one case to the release of energy and in the other case to a destination.

PRE-LAB QUESTIONS

1. The layer of vegetable oil will prevent oxygen from reaching the sugar and yeast mixture. In the presence of oxygen, fermentation will not take place.
2. One of the products of fermentation is carbon dioxide gas. The faster the reaction, the greater the amount of gas in the tube and the greater the pressure will be.
3. Students are likely to choose glucose because they know that the first step in fermentation is the conversion of glucose to pyruvic acid.

9 CHAPTER Summary

9.1 Cellular Respiration: An Overview

- Living things get the energy they need from food.
- Cellular respiration is the process that releases energy from food when oxygen is present.
- Photosynthesis produces food molecules. Photosynthesis takes carbon dioxide from the air. Cellular respiration puts it back. Photosynthesis puts oxygen into the air. Cellular respiration uses that oxygen to get energy from food.

cellular respiration (p. 213)
aerobic (p. 214)
anaerobic (p. 214)

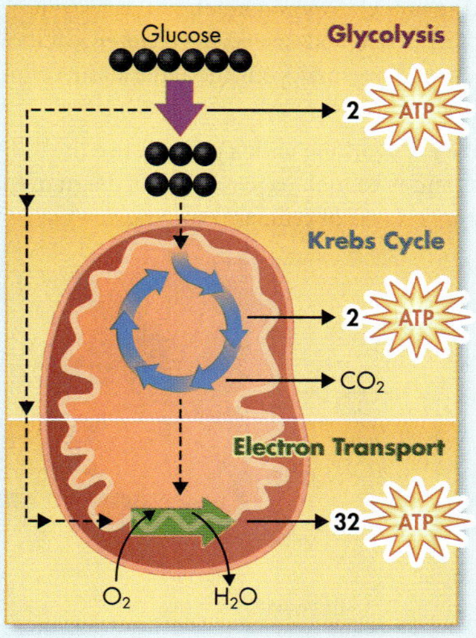

9.2 The Process of Cellular Respiration

- During glycolysis, 1 molecule of glucose, a 6-carbon compound, is changed into 2 molecules of pyruvic acid, a 3-carbon compound.
- During the Krebs cycle, pyruvic acid is broken down into carbon dioxide.
- The electron transport chain uses the high-energy electrons from glycolysis and the Krebs cycle to change ADP into ATP.
- Together, glycolysis, the Krebs cycle, and the electron transport chain make about 36 molecules of ATP per molecule of glucose.

glycolysis (p. 216)
NAD$^+$ (p. 216)
Krebs cycle (p. 218)
matrix (p. 223)

9.3 Fermentation

- When there is no oxygen, fermentation releases energy from food molecules by making ATP.
- For short, quick bursts of energy, the body uses ATP already in muscles and ATP made by lactic acid fermentation.
- For exercise longer than about 90 seconds, cellular respiration is the only way to continue making ATP.

fermentation (p. 223)

BIOLOGY.com Search Chapter 9 GO • Match It • Chapter Assessment

UbD Transfer the Big Idea

TRANSFER Have students keep an exercise journal for a week. Point out that exercise is not limited to formal exercise sessions. Students can include informal exercise, such as walking to school or raking leaves. Each time they exercise, have them write down the following:

- what type of exercise they performed
- how long they exercised
- how they felt after they exercised

Also have them note any changes in appetite on days when they exercised more than usual. Once their journals are complete, have students write a short essay describing how their journal entries relate to the content of this chapter. Make sure they discuss how their cells likely obtained the energy they used when exercising.

Chapter Summary

Think Visually

USE GRAPHICS Have students complete the following compare/contrast table.

Comparing Cellular Respiration and Fermentation		
Characteristic	**Cellular Respiration**	**Fermentation**
Starting reactants	1	2
Pathways involved	3	4
End products	5	6
Number of ATP molecules produced	7	8

Answers

1. glucose and oxygen
2. glucose
3. glycolysis, Krebs cycle, electron transport chain
4. glycolysis and either lactic acid or alcoholic fermentation
5. carbon dioxide, water, and energy
6. either carbon dioxide, alcohol, and NAD$^+$ or lactic acid and NAD$^+$
7. 36
8. 2

BIOLOGY.com Study Online

REVIEW AND ASSESSMENT RESOURCES

Art in Motion This animation highlights the relationship between photosynthesis and cellular respiration.

InterActive Art This short animation demonstrates the processes involved in cellular respiration.

Tutor Tube This online tutorial provides extra help on the electron transport chain and ATP production.

Data Analysis Students analyze data to learn more about lactic acid and exercise.

Vocabulary Review The *Flash Cards* and *Match It* are an interactive vocabulary review.

Editable Worksheets Editable pages of Study Workbook B, Lab Manual B, and the Assessment Resources Book are available online.

Lesson Overviews Lesson Overviews are based on Miller & Levine Biology. However, they are available on CD-ROM and can be edited for your class.

Chapter Assessment Have students take an online version of the Chapter 9 Check Understanding.

Standardized Test Prep Students can take an online version of the Standardized Test Prep. You will receive their scores along with ideas for remediation.

Diagnostic and Benchmark Tests Use these tests to monitor your students' progress and supply remediation.

ASSESSMENT

Assess

Big idea Invite a volunteer to read the Big Idea and the question.

Ask What raw materials does your body have to take in to obtain energy? *(food and oxygen)*

Ask What process do these raw materials undergo to provide energy for the body? *(cellular respiration)* What are the different stages of this process? *(glycolysis, the Krebs cycle, and electron transport.)*

Ask If no oxygen is available, what process is used? *(fermentation)*

Ask Where do these processes take place? *(Cellular respiration occurs in the mitochondria. Fermentation occurs in the cytoplasm.)*

CONSTRUCTED RESPONSE Remind students that paragraphs usually contain one main idea and several supporting details. Guide students to write each paragraph by stating the main idea and then adding supporting details.

For Question 1, write the question on the board and read it aloud.

Say This question is about reactants and products of photosynthesis and respiration. Remind students of the chemical equations for these two reactions.

Read the first hint.

Ask What are the reactants of photosynthesis? *(CO_2, H_2O, energy)* What are the products? *(glucose, O_2)* What are the reactants and products of cellular respiration? *(reactants: glucose, O_2; products: CO_2, H_2O, energy)* What does it mean that photosynthesis and cellular respiration are opposite processes? *(The reactants of one are the products of the other.)*

Read the second hint.

Ask What does carbon dioxide have to do with plant processes? *(It is a product of cellular respiration and a reactant for photosynthesis.)*

Model using the hints to write the main ideas for two paragraphs that answer Question 1. From what they learned in the chapter readings, have students fill in details that describe these main ideas. *(The first paragraph should show that the reactants of cellular respiration are the products of photosynthesis, and vice versa, making them opposite processes. The second paragraph should explain that carbon dioxide given off by cellular respiration in a plant is used as a reactant in photosynthesis. It is used and is not a waste product.)*

9 CHECK Understanding

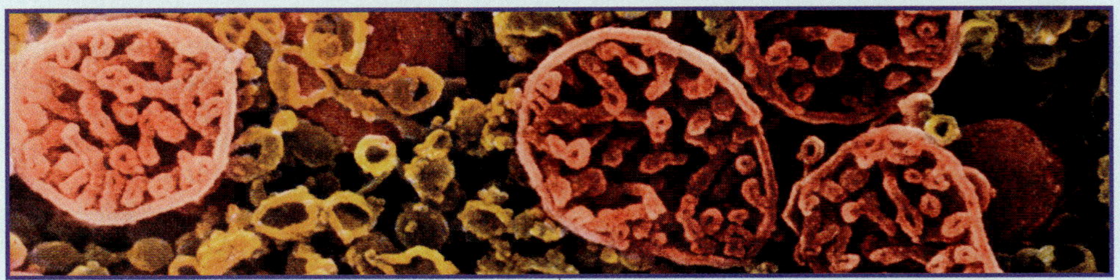

Assess the Big idea: Cellular Basis of Life

Write an answer to the question below.
Q: How do organisms obtain energy?

Constructed Response

Write an answer to each of the questions below. The answer to each question should be one or two paragraphs. To help you begin, read the **Hints** below the questions.

1. **Carbon dioxide is a waste product of cellular respiration. Plants carry out both photosynthesis and cellular respiration. Why is it wrong to say that carbon dioxide is a waste product in plants?**

   Hint Photosynthesis and cellular respiration are opposite processes.

   Hint The reactants for photosynthesis are carbon dioxide and water.

2. **When you run a race, you usually have a burst of energy at the beginning. But, that burst of energy soon fades, and you eventually settle down to a steady pace. How do the different pathways that the body takes to get energy explain how different parts of a race feel?**

   Hint Muscle cells get energy from three different sources during a race.

   Hint Stored ATP and lactic acid fermentation can give energy only for a short time.

3. **Some kinds of bread are lighter and more airy than others. How might rising and baking temperature affect how dense bread is?**

   Hint Enzymes in cellular process often work faster at higher temperatures.

   Hint The air bubbles in bread are caused by fermentation.

Foundations for Learning Wrap-Up

Use the index cards and notebook pages you prepared when you started the chapter as a tool to help you organize your thoughts about cellular respiration.

Activity 1 Working with a partner, use the index cards and notebook page to create a diagram to show the three stages of cellular respiration.

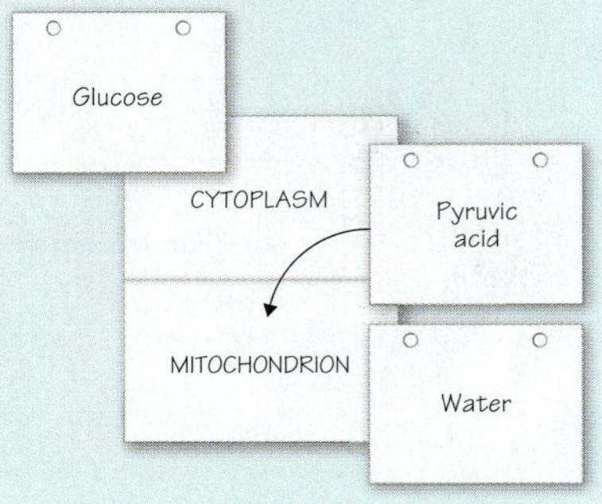

Activity 2 Working with a partner, label six more index cards with the chemical symbols for each word on the index cards from Activity 1. Then label three more index cards with +, +, and →. Now use the index cards to create a chemical equation that summarizes cellular respiration.

For Question 2, pair students and have them use the hints and your modeling to answer the question. Have students share their paragraphs with the class.

For Question 3, have students write the paragraphs independently for homework or as an anchor activity.

Foundations for Learning Wrap-Up

Have students take out the six cards and two sheets of paper they labeled at the beginning of the chapter. Be sure students have written all the correct terms on the cards and paper, and explain the activities. Have students work with partners to complete the activities.

9.1 Cellular Respiration: An Overview

Understand Key Concepts

1. Cells use the energy stored in food to make
   a. ADP.
   b. ATP.
   c. glucose.
   d. water.
2. The process that gets energy from food when oxygen is present is
   a. ATP synthase.
   b. cellular respiration.
   c. photosynthesis.
   d. synthesis.
3. The first step in getting energy in the cell by breaking down glucose is known as
   a. the Krebs cycle.
   b. electron transport.
   c. fermentation.
   d. glycolysis.

Test-Taking Tip

Watch For Key Words Read questions carefully for key words such as *first, most,* or *last*. Question 3 asks for the "*first* step in getting energy in the cell by breaking down glucose." Answers **a**, **b**, and **c** are all steps in getting energy. But, **d** is the first step. So **d** glycolysis, is the correct answer.

4. What does it mean if a process is *anaerobic?* Which part of cellular respiration is anaerobic?

Think Critically

5. **Compare and Contrast** Why are cellular respiration and photosynthesis considered opposite reactions?

9.2 The Process of Cellular Respiration

Understand Key Concepts

6. The net gain of energy from glycolysis is
   a. 8 ADP molecules.
   b. 4 ATP molecules.
   c. 3 pyruvic acid molecules.
   d. 2 ATP molecules.
7. The Krebs cycle takes place within the
   a. chloroplast.
   b. cytoplasm.
   c. mitochondrion.
   d. nucleus.
8. The electron transport chain uses the high-energy electrons from the Krebs cycle to
   a. make glucose.
   b. convert glucose to pyruvic acid.
   c. convert pyruvic acid to citric acid.
   d. move H^+ ions across the inner mitochondrial membrane.
9. What powers the making of ATP by ATP synthase?
   a. the splitting of glycogen
   b. the spinning of the mitochondrion
   c. the movement of ions across the membrane
   d. the movement of electrons through membrane channels
10. What happens to high-energy electrons made during the Krebs cycle?

Think Critically

11. **Compare and Contrast** How is the function of NAD^+ in cellular respiration similar to that of $NADP^+$ in photosynthesis?
12. **Use Models** Draw and label a mitochondrion surrounded by cytoplasm. Show where glycolysis, the Krebs cycle, and the electron transport chain take place in a eukaryotic cell.

9.3 Fermentation

Understand Key Concepts

13. Because fermentation takes place without oxygen, it is said to be
    a. aerobic.
    b. anaerobic.
    c. cyclic.
    d. oxygen-rich.

Assessment Tips

Have students work individually or in pairs to answer the questions. In Questions 6–12, encourage students to refer to the diagrams in Lesson 9.2 as needed.

For Question 6, remind students that the term *net gain* means the amount gained after additions and losses are added and removed.

Point out that in answering most of the questions students need to know what goes into each stage of cellular respiration, what happens during the stage, and the final products. Suggest that they create a question help sheet with this information before they begin answering questions. Refer students to the text and diagrams in Lesson 9.2 for cellular respiration and Lesson 9.3 for fermentation. Encourage students to help each other, if necessary.

Answers

Lesson 9.1

UNDERSTAND KEY CONCEPTS

1. b

2. b

3. d

4. To be anaerobic means to take place when oxygen is not present. Glycolysis is the part of cellular respiration that is anaerobic.

THINK CRITICALLY

5. The products of photosynthesis are the reactants of cellular respiration, and vice versa.

Lesson 9.2

UNDERSTAND KEY CONCEPTS

6. d

7. c

8. d

9. c

10. They are transported by the electron carriers NAD and $FADH_2$ to the electron transport chain to drive ATP synthesis.

THINK CRITICALLY

11. Both act as electron carriers by accepting a hydrogen ion and a pair of high-energy electrons, which they transport to other parts of the cell for use in another part of the process.

12. Drawings should show that glycolysis takes place in the cytoplasm, the Krebs cycle takes place in the matrix of the mitochondria, and the electron transport chain takes place in the inner mitochondrial membrane.

Lesson 9.3

UNDERSTAND KEY CONCEPTS

13. b

LESSON 9.3

14. a

15. a

THINK CRITICALLY

16. Cellular respiration takes place in the mitochondria. If someone's muscle cells have more mitochondria, they will be better able to make a lot of needed ATP through increased rates of cellular respiration.

Connecting Concepts

17. There are about 9 Calories in 1 gram of a lipid, 4 Calories in 1 gram of a carbohydrate, and 4 Calories in 1 gram of a protein. The different chemical structures of the different types of molecules allow them to store different amounts of energy in their bonds.

18. There are about 4 Calories in 1 gram of a protein. There must be about 4.5 grams of protein in this food to account for the number of Calories in a single serving.

CHAPTER MYSTERY After students have read through the Chapter Mystery, discuss different ways whales can stay underwater for so long without oxygen.

Ask When respiration occurs, carbon dioxide is produced. How do whales stay underwater without coming to the surface to get rid of excess carbon dioxide? *(Their blood can handle large amounts of carbon dioxide.)*

Ask How do whales get energy once the oxygen is used? *(through lactic acid fermentation)* Where do whales store lactic acid? *(in their muscles)* Why doesn't the extra acid cause acidosis? *(Whale muscles can handle a lot of lactic acid.)* What happens to the lactic acid when whales come to the surface? *(The whale breathes in oxygen and clears away lactic acid buildup.)*

9 CHECK Understanding

14. The process carried out by yeast that causes bread dough to rise is

a. alcoholic fermentation.

b. cellular respiration.

c. lactic acid fermentation.

d. mitosis.

15. Which of the following is a product of alcoholic fermentation?

a. carbon dioxide **c.** oxygen

b. lactic acid **d.** pyruvic acid

Think Critically

16. Predict Often, regular exercise causes the number of mitochondria in muscle cells to go up. How might that situation help someone do better at activities that use up a lot of energy?

Connecting the Concepts

Use Science Graphics

Use the nutritional information to answer questions 17 and 18.

17. Apply Concepts On average, how many Calories are there in 1 gram of a lipid, carbohydrate, and protein? Why the differences?

18. Calculate How many grams of protein must there be in this food in order to account for the number of Calories per serving shown in the label?

solve the CHAPTER MYSTERY

DIVING WITHOUT A BREATH

To be able to stay underwater for 45-minutes at a time, whales use many different tricks. For example, whale blood can handle the large amount of CO_2 that is made during the Krebs cycle. This lets whales stay underwater for a long time without making them feel a need to go to the surface. The Krebs cycle and electron transport need oxygen. And once the oxygen is used—and it's used quickly!—whale muscles use lactic acid fermentation to get energy. In humans, lactic acid causes the pH of the blood to drop. If the blood gets too acidic, a dangerous state called acidosis can take place. Whale muscles can handle a lot of lactic acid. The lactic acid stays in the muscles without causing acidosis. When whales come up after a long dive, they breathe in oxygen, which clears away the lactic acid buildup.

1. Relate Cause and Effect Why must whales have blood that can hold a lot of CO_2?

2. Predict Myoglobin is a molecule that stores oxygen in muscles. Would you expect to find more or less myoglobin on average in the muscle cells of whales than in the muscle cells of humans? Explain.

3. Infer How might being able to dive into very deep water be an advantage for whales?

UntamedScience **Never Stop Exploring Your World.** Finding the solution to the mystery is only the beginning. Take a video field trip with the ecogeeks of Untamed Science to see where the mystery leads.

CHAPTER MYSTERY ANSWERS

1. They have to be able to stay underwater for a long time and they cannot have high levels of CO_2 trigger the need to come to the surface, as it does in humans.

2. You would expect to find much more myoglobin in the muscle cells of whales because whales need to store a lot of oxygen during their long dives.

3. Sample answer: They can hunt for more food and avoid predators in the larger area of deep water.

Untamed Science For more on marine mammals and cellular respiration, suggest students watch **Take a Deep Breath.**

Standardized Test Prep

Multiple Choice

1. What raw materials are needed for cellular respiration?
   - A glucose and carbon dioxide
   - B glucose and oxygen
   - C carbon dioxide and oxygen
   - D oxygen and lactic acid

2. During the Krebs cycle
   - A hydrogen ions and oxygen form water.
   - B the cell gives off a small amount of energy through fermentation.
   - C each glucose molecule is broken down into 2 molecules of pyruvic acid.
   - D pyruvic acid is broken down into carbon dioxide over many steps.

3. Which substance is needed to begin glycolysis?
   - A ATP
   - B NADP
   - C pyruvic acid
   - D carbon dioxide

4. In eukaryotic cells, MOST of cellular respiration takes place in the
   - A nuclei.
   - B cytoplasm.
   - C mitochondria.
   - D cell walls.

5. Which substance is broken down during glycolysis?
   - A carbon
   - B NAD^+
   - C glucose
   - D pyruvic acid

6. The human body can use all of the following as energy sources EXCEPT
   - A ATP in muscles.
   - B glycolysis.
   - C lactic acid fermentation.
   - D alcoholic fermentation.

7. During cellular respiration, which of the following are given off?
   - A CO_2 and O_2
   - B H_2O and O_2
   - C O_2 and H_2O
   - D CO_2 and H_2O

8. Which of the following is an aerobic process?
   - A the Krebs cycle
   - B glycolysis
   - C alcoholic fermentation
   - D lactic acid fermentation

Questions 9 and 10

The graph below shows the rate of alcoholic fermentation for yeast at different temperatures.

Rate of Fermentation Versus Temperature

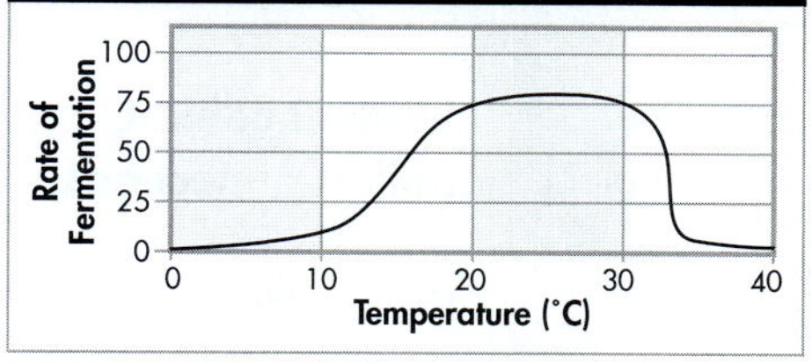

9. According to the graph, what is the relationship between the rate of fermentation and temperature?
   - A The rate of fermentation continually increases as temperature increases.
   - B The rate of fermentation continually decreases as temperature increases.
   - C The rate of fermentation increases with temperature at first, then it rapidly decreases.
   - D The rate of fermentation decreases with temperature at first, then it rapidly increases.

10. Which statement could explain the data shown in the graph?
    - A The molecules that carry out fermentation perform best at temperatures above 30°C.
    - B The yeast begins releasing carbon dioxide at 30°C.
    - C The yeast cannot survive above 30°C.
    - D The molecules that carry out fermentation perform best at temperatures below 10°C.

Open-Ended Response

11. Explain how a sprinter gets energy during a 30-second race. Is the process aerobic or anaerobic? Compare it to a long-distance runner getting energy during a 5-kilometer race.

If You Have Trouble With . . .

Question	1	2	3	4	5	6	7	8	9	10	11
See Lesson	9.1	9.2	9.2	9.1	9.2	9.3	9.1	9.1	9.3	9.3	9.3

Before the Test

Review with students the Test-Taking Tip.

Point out the word *MOST* in Question 4 and the word *EXCEPT* in Question 6.

Remind students that the paragraphs they write to answer an open-ended question should contain the same facts as another student's paragraph, but the order of the facts, examples used, and style of writing probably will not be the same.

Answers

1. B
2. D
3. A
4. C
5. C
6. D
7. D
8. A
9. C
10. C
11. Sample answer: During a 30-second race, a runner gets energy from the small amounts of ATP already in the muscles and from ATP made by lactic acid fermentation. This process does not depend on oxygen and therefore is anaerobic. During a 5-kilometer race, the only way for the runner to continue making ATP is through cellular respiration. Because cellular respiration takes place only when oxygen is present, this process is aerobic.

Test-Taking Tip

READ ALL THE ANSWER CHOICES When answering multiple-choice questions, remind students to first try to eliminate any choices they know to be wrong before choosing the correct answer. If the choices are numeric, estimating the correct answer can help eliminate choices that are not near the estimated value. If the choices are not numeric, reading the question with each choice substituted into the place for the answer can help eliminate incorrect choices.

Cellular Respiration and Fermentation **231**

CHAPTER PLANNER 10 Cell Growth and Division

Chapter Contents	NSES	Time	Core Resources
Chapter Preview			**Student Edition,** pp. 232–233 **Chapter Mystery,** p. 233 **Foundations for Learning,** p. 233
10.1 Cell Growth, Division, and Reproduction Limits to Cell Size • Cell Division and Reproduction	I, II, IV, A.1.c, A.2.a, C.1.c, C.2.a	1 period 1/2 block	**Student Edition,** pp. 234–238 **Build Connections,** pp. 234, 235 **Inquiry Into Scientific Thinking,** p. 238 **Study Workbook B** 10.1 Worksheets **Biology.com** *Visual Analogy:* Growing Pains • 10.1 Lesson Assessment
10.2 The Process of Cell Division Chromosomes • The Cell Cycle • Mitosis • Cytokinesis	I, II, C.1.c, C.2.b	2 periods 1 block	**Student Edition,** pp. 239–244 **Build Connections,** p. 244 **Study Workbook B** 10.2 Worksheets **Biology.com** *Art Review:* Eukaryotic Chromosome • *InterActive Art:* Mitosis • *Tutor Tube:* Unraveling Chromosome Vocabulary • 10.2 Lesson Assessment
10.3 Regulating the Cell Cycle Controls on Cell Division • Cancer: Uncontrolled Cell Growth	I, II, C.1.d	1 period 1/2 block	**Student Edition,** pp. 245–247 **Study Workbook B** 10.3 Worksheets **Biology.com** *Art in Motion:* Growth of Cancer Cells • 10.3 Lesson Assessment
10.4 Cell Differentiation From One Cell to Many • Stem Cells and Development • Frontiers in Stem Cell Research	I, II, A.2.a, C.1.d, C.1.f	1 period 1/2 block	**Student Edition,** pp. 248–251 **Study Workbook B** 10.4 Worksheets **Biology.com** 10.4 Lesson Assessment
Chapter Pre-Lab	II, A.1.b, A.2.a, C.1.f	1 period 1/2 block	**Student Edition,** p. 252 **Lab Manual B** *Regeneration in Planaria*

Tools for ELL Instruction

ELL Handbook describes teaching strategies to make *Miller & Levine Biology: Foundation Edition* more accessible to ELL students.

Spanish Study Workbook is a Spanish translation of Study Workbook A.

Multilingual Glossary is the *Miller & Levine Biology* glossary translated into ten languages.

Tools to Challenge Students

Study Workbook A includes more challenging worksheets and activities such as Chapter Mystery activities that reinforce 21st Century Skills.

Lab Manual A labs are geared toward on-level students. The Lab Manual includes fifteen labs in addition to the Chapter Labs.

Biology.com Data Analysis activities offer students a way to practice data gathering and analyzing skills.

Visual Quizzes in the Assessment Resources Book are geared toward on-level students and use visuals from *Miller & Levine Biology.*

Additional Resources

Biology.com Untamed Science Video • Flash Cards

Study Workbook A 10.1 Worksheets
Lab Manual B *Hands-On Activities:* Comparing Surface Area to Volume
Spanish Study Workbook 10.1 Worksheets

Study Workbook A 10.2 Worksheets
Spanish Study Workbook 10.2 Worksheets
Biology.com *Data Analysis:* Timing the Cell Cycle
Assessment Resources Book Visual Quiz

Study Workbook A 10.3 Worksheets
Lab Manual B *Data Analysis:* The Rise and Fall of Cyclins
Spanish Study Workbook 10.3 Worksheets
Assessment Resources Book Visual Quiz

Study Workbook A 10.4 Worksheets
Lab Manual B *Data Analysis:* Cellular Differentiation of *C. elegans*
Spanish Study Workbook 10.4 Worksheets

Lab Manual A *Regeneration in Planaria*

Chapter Review

Student Edition Chapter Summary, p. 253 • Unit Project, p. 258
Study Workbook B Chapter 10 Review
Transparencies, pp. 122–136
Biology.com Untamed Science Video • You're the Director • Editable Worksheets of Study Workbooks A and B and Lab Manuals A and B • Chapter 10 Flash Cards and Crossword

Untamed Science DVD • Classroom Resources CD (includes lesson presentations and editable worksheets)

Chapter Assessment

Student Edition Check Understanding, pp. 254–257
Study Workbook B Chapter 10 Taking a Standardized Test
Assessment Resources Book Chapter 10 Test B • Unit 3 Test B
Biology.com Chapter 10 Assessment • Editable Worksheets of Chapter 10 Visual Quizzes,* Chapter 10 Tests A and B, and Unit 3 Tests A and B

Exam*View* *Assessment Suite* • Classroom Resources CD (includes lesson presentations and editable worksheets)

Time: 1 period, 1/2 block

* Visual Quizzes are based on visuals from *Miller & Levine Biology*. Review carefully to make sure they are appropriate for your class.

Pressed for Time?

Preview the Chapter Have students read the Key Questions for Lesson 10.2 and preview the visual The Cell Cycle.

Cover the Chapter Quickly Have students read *Limits to Cell Size* in Lesson 10.1 and go over the Growing Pains visual. Have them read all of Lesson 10.2 and review the Eukaryotic Cell Cycle and The Cell Cycle visuals. Then, have them read *Controls on Cell Division* in Lesson 10.3 and *From One Cell to Many* in Lesson 10.4.

Assess Assign Question 1 in the 10.1 Check Understanding, the 10.2 Check Understanding, Questions 1, 3, 5, and 6 in the 10.3 Check Understanding, Question 2 in the 10.4 Check Understanding, and the Chapter 10 Standardized Test Prep except Questions 2 and 6.

CHAPTER 10

Get the Picture

Tell students the picture is called a micrograph, which is a picture that is magnified. Guide students to see the prefix *micro-* and ask what other words they know with the prefix *micro-*. *(microscope)*

Ask How would the real picture of cells compare to this micrograph? Be sure students understand that cells are actually much smaller and that these colors are artificial. Have them describe what they see, and then draw the cell in metaphase in their notebooks.

Activate Prior Knowledge

Ask What are the stages people go through during their lives? *(Students are likely to name or describe the stages such as baby, childhood, teenager, etc.)* Point out that cells, like people, undergo a life cycle, called a cell cycle.

Explain that during a cell cycle, cells grow and mature but unlike human life, a cell starts over when it divides.

10 Cell Growth and Division

Big idea Growth, Development, and Reproduction

Q: How does a cell produce a new cell?

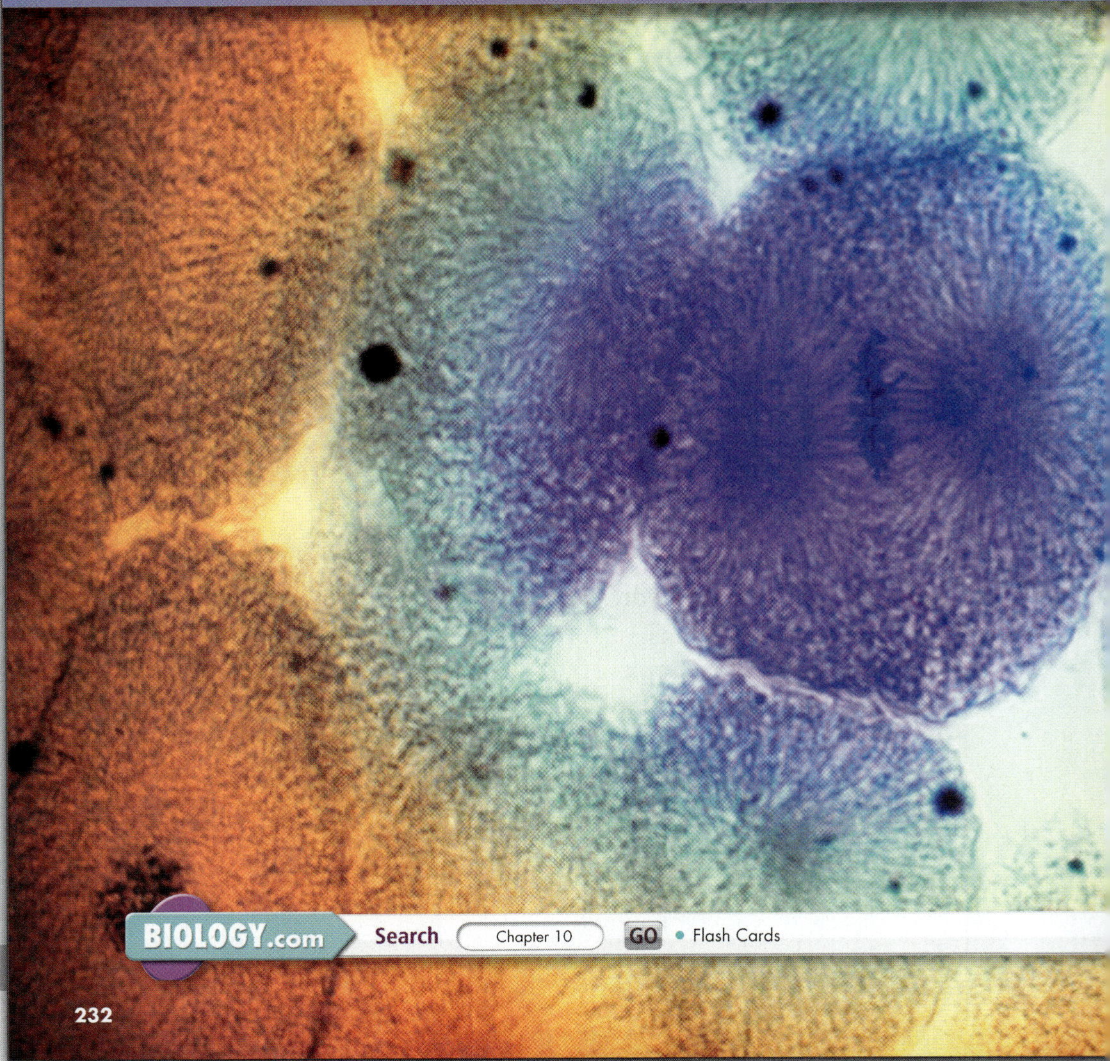

BIOLOGY.com Search Chapter 10 GO • Flash Cards

232

NATIONAL SCIENCE EDUCATION STANDARDS

UNIFYING CONCEPTS AND PROCESSES

I, II, IV

CONTENT

C.1.c, C.1.d, C.1.f, C.2.a, C.2.b

INQUIRY

A.1.b, A.1.c, A.2.a

UbD Connect to the Big Idea

Invite a volunteer to read aloud the chapter number and the chapter title. Read the Big Idea and question. Emphasize the question, **How does a cell produce a new cell?**

Set the Purpose

PREVIEW AND PREDICT Draw the graphic on the board. Guide students to the INSIDE feature on the opposite page. Invite a volunteer to read each lesson number and title.

Ask How do you think this title relates to the questions in the graphic organizer? Remind students to search for the answer as they read and to also ask themselves why cells divide.

INSIDE:

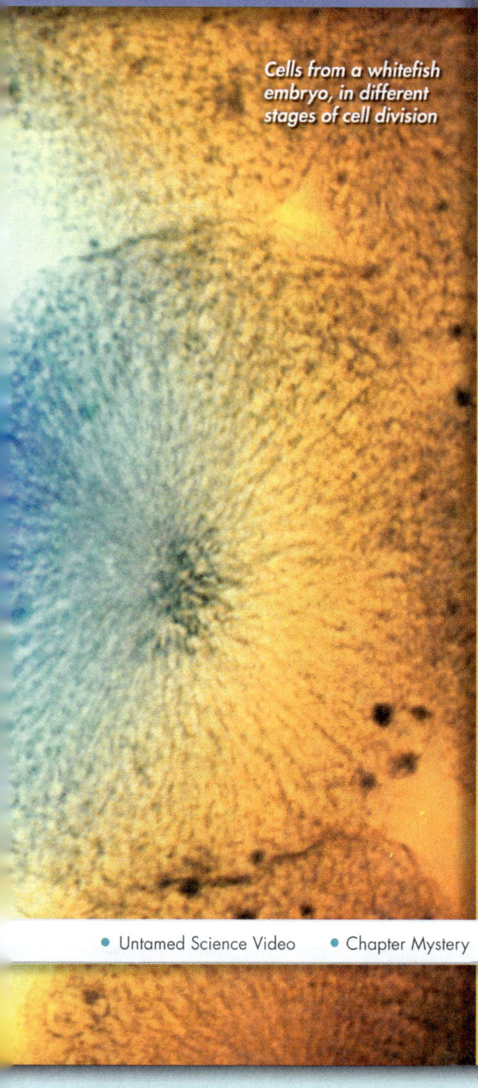

Cells from a whitefish embryo, in different stages of cell division

- Untamed Science Video
- Chapter Mystery

CHAPTER **MYSTERY**

PET SHOP ACCIDENT

Julia stared into the salamander tank in horror. As an assistant in a pet shop, Julia had mistakenly put a small salamander in a tank with a large one. As she realized her error, the large salamander bit off one of the small salamander's legs.

Julia put the injured salamander in its own tank. She was sure it would die before her shift ended. She was wrong! Days and weeks passed. Every time Julia checked on the salamander, she was amazed. How did the salamander's body react to losing a limb?

Read for Mystery Clues As you read this chapter, look for clues to help you predict what will happen to the salamander. Think about the cell processes that might be involved. Then, solve the mystery at the end of the chapter.

FOUNDATIONS for Learning

Think of the key vocabulary as building blocks for the Big Idea. Before your read the chapter, write each vocabulary term on an index card. Then write a definition in your own words as you read each lesson. As shown below, you can put the cards into categories to help you remember their meaning. At the end of the chapter are two activities that use the cards to help answer the question: How does a cell produce a new cell?

Process	Cell structures	Not sure
mitosis	chromosome	cancer
cytokinesis	chromatid	centromere
cell division		

Chapter 10
Big Idea: Growth, Development, and Reproduction

Chapter 10: How does a cell produce a new cell?

- **10.1:** Why do cells divide?
- **10.2:** How do cells divide?
- **10.3:** How does a cell control the process of cell division?
- **10.4:** How does a single undifferentiated cell lead to a complex multicellular organism?

CHAPTER 10

Foundations for Learning

Pass out the index cards. Help students understand the importance of this activity.

Ask Why do you think vocabulary is so important? *(Vocabulary is essential to reading comprehension.)*

Ask Why do you think putting vocabulary terms into categories helps you remember the definitions? *(Putting words into categories tells you that the words are related in some way, and once you understand one word in that category, you have some preunderstanding about another word in that category that you may not know.)*

What do words in the cell structure category, for example, tell you? *(The words all relate to parts of a cell.)*

Remind students that they will use these cards as a review and study aid. Familiarize students with the activity by modeling one term and its definition on an index card.

Untamed Science™ The short movie **Tails of Cell Division** will take students on a field trip that starts with a hunt for salamanders in an American woodland and ends with scientists studying echinoderm regeneration in Sweden.

CHAPTER MYSTERY Ask students if they know what a salamander is. Invite a volunteer to read aloud the Chapter Mystery. Have students predict how this mystery relates to the chapter title and to predict what will happen to the salamander.

Tell students that they will find clues in the chapter to answer the question, **How did the salamander's body react to losing a limb?**

Digital Support

BIOLOGY.com Extend your reach by using the digital assets offered at **Biology.com.**

LESSON 10.1

Pre-Reading

Objectives

10.1.1 Explain the problems that growth causes for cells.

10.1.2 Compare asexual and sexual reproduction.

Preview the Pages

1. Preview the vocabulary words by having students find the words in the text. Ask students where they may have heard the words before. Have students read the text and write a definition of sexual reproduction in their own words. Ask students to use the words *two cells* and *join* in the definition.
2. As you preview the visuals, be sure students see that the hydra is part of Asexual Reproduction.
3. Review the formula for finding surface area and volume of a cube.

   SA = length × width × number of sides
   V = length × width × height.

Build Understanding

T-CHART Direct students to Workbook B where they'll find a partially filled-in T-chart. Remind students that a T-chart will help them organize information.

NATIONAL SCIENCE EDUCATION STANDARDS

UNIFYING CONCEPTS AND PROCESSES

I, II, IV

CONTENT

C.1.c, C.2.a

INQUIRY

A.1.c, A.2.a

For the corresponding lesson in *Miller & Levine Biology,* see pages 274–278.

10.1 Cell Growth, Division, and Reproduction

Key Questions

What are some of the difficulties a cell faces as it increases in size?

How do asexual and sexual reproduction compare?

BUILD Understanding

T-Chart As you read the lesson, make a T-chart. Start with a two-column chart. In the left column, write the headings in the lesson. In the right column, rewrite the headings as questions. Then, write an answer to each question in the space beneath the question.

In Your Workbook Go to your workbook to learn more about making a T-chart. Complete the T-chart for Lesson 10.1.

Limits to Cell Size

Cells grow bigger, just as you do. However, most cells will grow only so big, then divide. There are two main reasons why cells divide instead of continuing to grow. First, the larger a cell becomes, the more demands the cell has on its DNA. Second, a large cell has more trouble moving enough food in and enough waste out.

Information "Overload" The information that cells need to live and grow is stored in a molecule called deoxyribonucleic acid, or DNA. When a cell is small, its DNA can meet all of its needs. However, as a cell grows, its DNA does not. If a cell grew too large, the result would be "information overload." A cell's DNA could not keep up with the cell's need for information.

You can compare the cell to a growing town. Suppose a small town has a library with a few thousand books. If more people move into the town, more people will borrow books. Sometimes people may have to wait to borrow popular books. Just like the town, a larger cell would need more from its genetic "library" (DNA). After a while, the DNA would not be able to meet the needs of the growing cell—it might be time to build a new "library."

Exchanging Materials Cell size affects how well a cell can move materials in and out. Food, oxygen, and water enter a cell through its cell membrane. Waste products leave a cell in the same way. The larger a cell becomes, the more nutrients it needs and the more waste it produces. But as a cell grows, the relationship between the surface area of its cell membrane and the size of its volume changes. This change is another reason why cell size is limited.

BUILD Connections

READING DNA

Like a book, DNA contains information. A cell cannot survive without the information stored in its DNA.

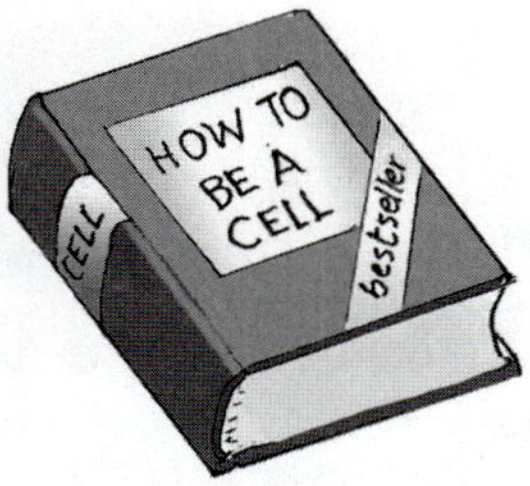

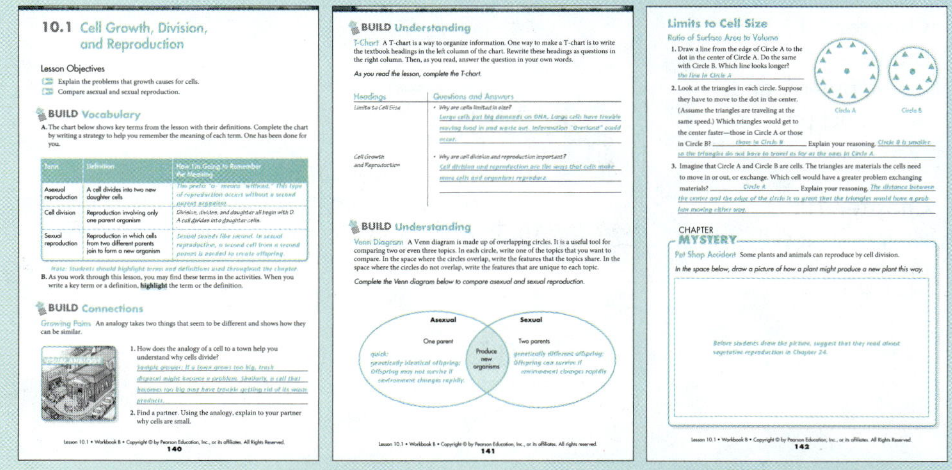

10.1 Cell Growth, Division, and Reproduction

Lesson Objectives

Explain the problems that growth causes for cells.

Compare asexual and sexual reproduction.

BUILD Vocabulary

A. The chart below shows key terms from the lesson with their definitions. Complete the chart by writing a strategy to help you remember the meaning of each term. One has been done for you.

Term	Definition	How I'm Going to Remember the Meaning
Asexual reproduction	A cell divides into two new daughter cells	The prefix "a-" means "without." This type of reproduction occurs without a second parent organism.
Cell division	Reproduction involving only one parent organism	Division, divides, and daughter all begin with D. A cell divides into daughter cells.
Sexual reproduction	Reproduction in which cells from two different parents join to form a new organism	Sexual sounds like second. In sexual reproduction, a second cell from a second parent is needed to create offspring.

Note: Students should highlight terms and definitions used throughout the chapter.

B. As you work through this lesson, you may find these terms in the activities. When you write a key term or a definition, **highlight** the term or the definition.

BUILD Connections

Growing Pains An analogy takes two things that seem to be different and shows how they can be similar.

1. How does the analogy of a cell to a town help you understand why cells divide?
   Sample answer: If a town grows too big, trash disposal might become a problem. Similarly, a cell that becomes too big may have trouble getting rid of its waste products.
2. Find a partner. Using the analogy, explain to your partner why cells are small.

Lesson 10.1 • Workbook B • Copyright © by Pearson Education, Inc., or its affiliates. All Rights Reserved.
140

BUILD Understanding

T-Chart A T-chart is a way to organize information. One way to make a T-chart is to write the textbook headings in the left column of the chart. Rewrite these headings as questions in the right column. Then, as you read, answer the question in your own words.

As you read the lesson, complete the T-chart.

Headings	Questions and Answers
Limits to Cell Size	• Why are cells limited in size? Large cells put big demands on DNA. Large cells have trouble moving food in and waste out. Information "Overload" could occur.
Cell Growth and Reproduction	• Why are cell division and reproduction important? Cell division and reproduction are the ways that cells make more cells and organisms reproduce.

BUILD Understanding

Venn Diagram A Venn diagram is made up of overlapping circles. It is a useful tool for comparing two or even three topics. In each circle, write one of the topics that you want to compare. In the space where the circles overlap, write the features that the topics share. In the space where the circles do not overlap, write the features that are unique to each topic.

Complete the Venn diagram below to compare asexual and sexual reproduction.

Asexual: One parent; quick; genetically identical offspring; Offspring may not survive if environment changes rapidly.

Both: Produce new organisms

Sexual: Two parents; genetically different offspring; Offspring can survive if environment changes rapidly.

Lesson 10.1 • Workbook B • Copyright © by Pearson Education, Inc., or its affiliates. All rights reserved.
141

Limits to Cell Size

Ratio of Surface Area to Volume

1. Draw a line from the edge of Circle A to the dot in the center of Circle A. Do the same with Circle B. Which line looks longer? the line in Circle A
2. Look at the triangles in each circle. Suppose they have to move to the dot in the center. (Assume the triangles are traveling at the same speed.) Which triangles would get to the center faster—those in Circle A or those in Circle B? those in Circle B Explain your reasoning. Circle B is smaller, so the triangles do not have to travel as far as the ones in Circle A.
3. Imagine that Circle A and Circle B are cells. The triangles are materials the cells need to move in or out, or exchange. Which cell would have a greater problem exchanging materials? Circle A Explain your reasoning. The distance between the center and the edge of the circle is so great that the triangles would have a problem moving either way.

Circle A   Circle B

CHAPTER MYSTERY

Pet Shop Accident Some plants and animals can reproduce by cell division.

In the space below, draw a picture of how a plant might produce a new plant this way.

Before students draw the picture, suggest that they read about vegetative reproduction in Chapter 24.

Lesson 10.1 • Workbook B • Copyright © by Pearson Education, Inc., or its affiliates. All Rights Reserved.
142

BUILD Connections

GROWING PAINS

Lots of growth can mean lots of trouble—both in a town and in a cell.

▶ ***Ratio of Surface Area to Volume*** As a cell grows, the size of its cell membrane does not grow as fast as its volume. To see how, picture a cell as a cube. In a cube with sides that measure 1 cm in length, the ratio of surface area to volume is 6/1 or 6 to 1. Increase the length to 3 cm. The ratio becomes 54/27 or 2 to 1. The ratio of surface area to volume gets smaller. For a growing cell, this difference means the cell has more trouble moving enough food in and all its waste out.

▶ ***Traffic Problems*** Think of the example of the town again. Suppose the town only has a two-lane main street. As the town grows, more people use this street. The increase in traffic means it takes longer to move in and out of the town. A cell that continues to grow would experience similar problems.

Key Question What are some of the difficulties a cell faces as it increases in size?
A cell's DNA may not be able to meet all of its information needs. Also, a large cell has trouble moving enough nutrients in and wastes out.

BIOLOGY.com Search Lesson 10.1 GO • Visual Analogy

Hands-On Learning

CALCULATE SURFACE AREA Pair students and hand out the materials. Materials: two identical rectangular blocks of wood or small cardboard boxes, masking tape, and metric ruler

1. On the board: What do you think happens to the surface area when the volume doubles? Have students record the question and their answer in their notebooks.
   $a = l \times w = \_\_ cm^2$
   $v = l \times w \times h = \_\_ cm^3$
2. Explain that area units are centimeters squared; for volume, the units are centimeters cubed. Remind students to find the area of each of the six surfaces, then add them together for the total surface area.
3. Tell students the steps. Step 1: Calculate and record the volume and surface area of one block or box. Step 2: Write a ratio comparing surface area to volume. Step 3: Tape the two objects together and repeat Steps 1 and 2. Step 4: Evaluate your findings in your notebook.

During Reading

Build Connections

VISUAL ANALOGY Discuss the visual analogy about a book and a cell. Tell students that an analogy is a kind of comparison: It takes two things that seem to be different and shows how they can be similar. Tell students that making analogies can help them learn and remember new information.

Ask Why can't a cell survive without the information stored in its DNA? *(The DNA has all the information a cell needs to live and grow.)*

Active Reading

SELF-QUESTION Start by pointing out that the surface area is the outside of the cell; it's the membrane. The volume is everything inside the cell. Demonstrate how asking questions helps to clarify new information. Read aloud the first paragraph.

Ask Does this make sense? What does it mean that the cell membrane is not growing as fast as the volume?

Say Sometimes I have to re-read information, and when I do, I can find the answer. Have students reread the last sentence on the previous page.

Ask What does this tell me about the ideal cell size? *(The smaller the volume, the more efficient the rate of exchange of materials.)*

KEY QUESTION Invite a volunteer to read aloud the question and answer.

Ask What are a cell's information needs? *(what it needs to know to live and grow)* What's the difference between nutrients and waste? *(nutrients are what cells need to grow; waste is what cells need to remove)*

Build Connections

VISUAL ANALOGY Point out that this visual analogy is comparing a cell to a town. Have students explain how a cell is like a town. If students are having trouble, encourage them to come up with other analogies to help them make sense of what happens when a cell grows too large.

Ask What happens in a movie theater filled with people if only one door is open and everyone wants to leave at once?

During Reading continued

Active Reading

VISUALIZE/DIAGRAM Point out how helpful it is to sketch ideas so that they make sense. Draw and label the following diagram on the board. Have students copy it into their notebooks.

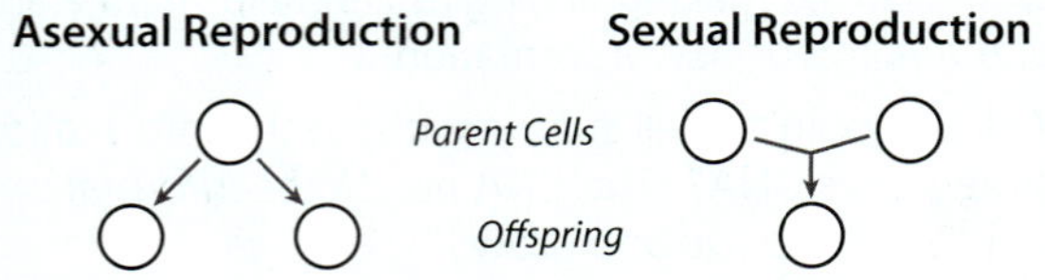

Point to the diagrams as you explain why asexual reproduction results in genetically identical offspring while sexual reproduction produces offspring with a combination of genetic information from both parents.

Build Vocabulary

PREFIXES Have students pronounce each vocabulary word and read the definition. Then, have them find the word and the definition in the text.

Point out the prefix *a-* means "without" as in *asexual reproduction.*

Ask What is asexual reproduction doing "without"? *(without a second cell)*

Speed Bump

Ask Can a cell keep growing forever? *(no)*

Ask Why do cells divide? What are the two main reasons? *(1. When a cell grows larger, its DNA does not, so the cell ends up making more demands on its DNA. 2. The larger the cell, the harder it is for the cell to exchange food and waste.)*

Ask In this lesson, what do the ratios compare? *(The ratios compare the cell's surface area to its volume.)*

Ask Is it better for a cell to be smaller in size or larger? Why? *(Smaller size is better; a smaller cell can more efficiently exchange materials across the cell membrane.)*

BUILD Vocabulary

cell division
a process by which a cell divides into two new daughter cells

asexual reproduction
a process by which a single parent reproduces by itself

sexual reproduction
a process by which two cells from different parents fuse, or join together, to produce the first cell of a new organism

PREFIXES
The prefix *a-* in *asexual* means "without." *Asexual reproduction* is reproduction without the fusion of cells.

Cell Division and Reproduction

Before it becomes too large, a growing cell divides. The two new cells are referred to as "daughter" cells. The process by which a cell divides into two cells is called **cell division.**

During the process of cell division, a cell makes a copy of its DNA. Each daughter cell gets its own copy. This copying solves the problem of information overload. Cell division also decreases a cell's volume, which allows for a better exchange of materials in and out of the cell. Cell division can also result in reproduction. Reproduction is the process by which organisms produce offspring—new organisms.

Asexual Reproduction For an organism made up of just one cell, cell division may be the only form of reproduction it needs. All the organism has to do is to copy its DNA and then divide. Reproduction by a single parent is called **asexual reproduction.** Offspring of asexual reproduction have the same genetic information as their parent.

Many multicellular organisms reproduce asexually. Hydras are small animals that live in ponds. They reproduce asexually by budding. As cells divide, the bud grows. The bud eventually separates from the parent. A budding hydra is shown below.

Sexual Reproduction Another way for an organism to reproduce is by **sexual reproduction.** Sexual reproduction results from the joining of two cells. Each cell comes from a different parent. Offspring of sexual reproduction have genetic information from both parents. Most animals and plants reproduce sexually. Some single-celled organisms also reproduce sexually. This type of reproduction happens when two cells come together and share DNA.

Asexual Reproduction Cell division leads to reproduction in single-celled organisms and some multicellular organisms.

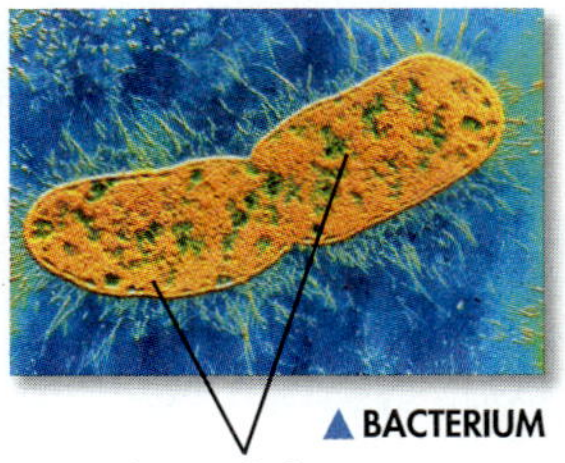

▲ BACTERIUM

Daughter cells form and then separate into two bacteria.

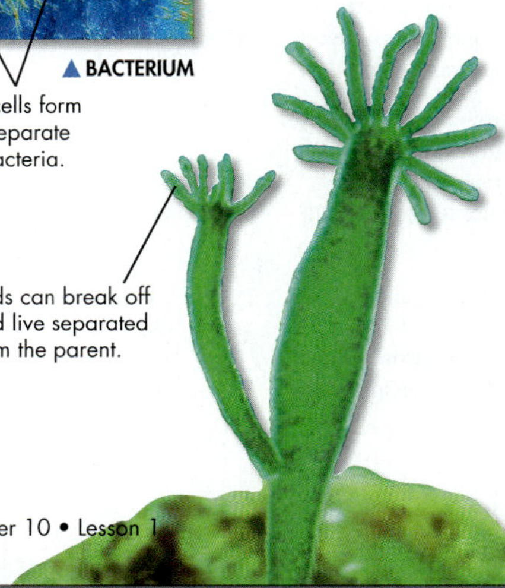

HYDRA

Buds can break off and live separated from the parent.

Sexual Reproduction Penguins reproduce sexually. A baby penguin grows from a single cell that contains DNA from both parents. ▶

Focus on Academic Language

• multicellular • advantage • well suited for • ideal conditions • adjust

Review with students the meaning of *multicellular* ("having many cells"). Explain that the prefix *multi-* means "many."

Ask Can you guess what these words mean? *multiplex* (theater with many screens); *multitask* (do many tasks at the same time); *multitalented* (having many talents); *multiple* (many parts).

Write *advantage, well suited for, ideal conditions,* and *adjust* on the board. Pair students and have them find the words in the Advantages section. Have them use context clues to predict the meanings and explain their reasoning in their notebooks.

Advantages Each method of reproduction has advantages. The main advantages of asexual reproduction are that it is quick and that it produces genetically identical offspring. An organism that is well suited for its environment can reproduce very quickly. The result is a large number of equally well-suited offspring. Asexual reproduction is one reason bacteria are able to grow so quickly when they find themselves in ideal conditions.

The main advantage of sexual reproduction is that offspring are genetically different from their parents. Sexual reproduction lets species "try out" new combinations of genetic information from one generation to the next. If the environment changes rapidly, some members of a species may be able to adjust to those changes.

Disadvantages Each method of reproduction also has disadvantages. In asexual reproduction, the lack of genetic diversity can be a disadvantage. Asexually reproducing organisms may not have needed characteristics if their environment changes rapidly. Sexual reproduction is generally slower than asexual reproduction, since it takes two parents instead of one to produce offspring.

Key Question How do asexual and sexual reproduction compare?
Asexual reproduction involves one parent—offspring have the same genetic material as the parent. Sexual reproduction involves two parents—offspring inherit genetic material from both parents.

BUILD Understanding

Venn Diagram Use a Venn diagram to compare and contrast asexual and sexual reproduction.

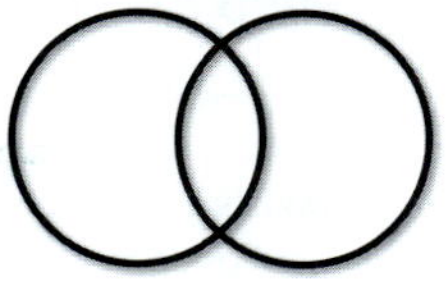

In Your Workbook Go to your workbook to learn more about making a Venn diagram. Complete the Venn diagram started for you.

CHECK Understanding

Apply Vocabulary

Use the highlighted words from the lesson to complete each sentence correctly.

1. For a growing cell, ______ solves the problems of overloading DNA and not being able to get enough materials in or out.
2. Offspring produced by ______ have a mix of DNA from two parents.
3. Cell division in a single-celled organism is a form of ______.

Critical Thinking

4. **Compare and Contrast** How does DNA in offspring produced by asexual reproduction compare to DNA in offspring produced by sexual reproduction?
5. **Apply Concepts** Describe advantages and disadvantages of asexual reproduction. Describe advantages and disadvantages of sexual reproduction.
6. **Write to Learn** Answer the first clue of the mystery. Write a paragraph that includes the terms *DNA* and *cell division.*

MYSTERY CLUE

As the salamander's wound heals, its body cells are dividing to repair the damage. How is this type of cell division similar to asexual reproduction? (Hint: See p. 236.)

BIOLOGY.com Search Lesson 10.2 GO • Lesson Assessment

Check Understanding

1. cell division
2. sexual reproduction
3. asexual reproduction
4. Offspring produced by asexual reproduction have the same DNA as their parent and are genetically the same. Offspring produced by sexual reproduction have the DNA of both parents; they are genetically different from their parents.
5. In sexual reproduction, offspring are genetically different, allowing species to adapt to changing environments. However, it needs two parents and is slower. Asexual reproduction is quick and needs only one parent, but lacks genetic diversity.
6. Sample answer: The cell division in the salamander is similar to asexual reproduction—the new cells will have the same DNA as the parent cells.

Build Understanding

Have students finish the Venn diagram in their notebooks.

MAKING INFERENCES Read aloud the first two paragraphs.

Ask Have you ever heard of strep throat? It's a bad sore throat caused by the bacteria *Streptococcus.* Strep throat is highly contagious — you can catch it easily. Strep throat can spread quickly through a school. Why? *(Bacteria are organisms that reproduce asexually; under the right conditions, they multiply quickly.)*

Active Reading

KEY QUESTION Read aloud the Key Question.

Ask What's the key word that can help you begin your answer? *(compare)* If you saw this question on a test, what should you do first? *(Answers may vary but should include a graphic organizer.)* Students could refer to the Venn diagrams they produced for the Build Understanding activity and then answer the question in complete sentences in their notebooks.

Post Reading

ASSESSMENT TIPS Review the highlighted vocabulary words with students before they begin the assessment.

For Question 5, pair students. Ask them to create a graphic organizer that will help them answer the question. Then, using the graphic organizer, have them answer the question in complete sentences.

MYSTERY CLUE Discuss with students what they think happens to the misssing limb of the salamander after a few days. *(As its wound heals, the salamander's body cells are dividing to repair the damage.)*

Wrap-Up Activity

COMPOSE A TEST Having students prepare a test is a valuable strategy in helping them identify the most relevant information in the lesson. Divide students into small groups. Each group is responsible for writing one test with five multiple-choice items including four options for each item or five fill-in-the-blanks, or five matching, as well as one constructed response. Each group passes its test until each group has another group's test to answer. Eventually, each test is given back to the group who originated the test to be corrected.

Inquiry Into Scientific Thinking

MAKE CONNECTIONS Explain that a ratio is a comparison.

Explain the ratio of vowels to consonants is 5 vowels to 21 consonants, or 5:21.

Ask students to calculate the ratio of the letters of their last name to the letters in the alphabet. (x:26) Ask students to suggest other ratios. (sun to planets, 1:8; dollars to cents, 1:100; pints to gallons, 8:1)

Say In this case, the ratio compares the amount of surface area a cell has to its volume. A ratio can be expressed as 2:1, or as a fraction, 2/1. In the table, the ratios are reduced to their simplest form to make it more obvious how the cube's surface area grows in relationship to its volume.

INTERPRET CHARTS AND TABLES Review the information in the table. Have a volunteer read the title. Remind students that rows go across; columns go up and down. Explain the first row: The cube at the head of each column represents a particular cell. All the information in that column is about that cell. The second row shows how to find the surface area; you multiply length times width times six.

Ask Why do you multiply by six? *(There are six surfaces on a cube.)* The third row shows how to find the volume; length × width × height.

Ask What does the fourth and last row show? *(the ratio of surface area to volume)*

USE DIAGRAMS Draw a cube on the board. Ask a volunteer to write *l, w, h* on the corresponding sides of the cube. Pair students. Have each student draw a cube in their notebooks and add a one-digit number to the *l, h,* and *w.* Then have students switch notebooks and calculate the SA of the cube. Then students should return the notebooks to their owner, who will check the answer.

Lead a Discussion

Help students realize that surface area is an important concept that applies to everyday life, not just to the efficiency of cells.

Ask Can you think of two products that you use every day in which surface area has an effect on how the product performs? *(Sample answers: towels, sponges, pots/pans)*

INQUIRY into Scientific Thinking

Modeling the Relationship Between Surface Area and Volume

As a cell grows, its volume increases faster than its surface area. In Lesson 10.1, you saw the relationship between surface area and volume described as a ratio. A ratio is simply a comparison of two numbers. A ratio can be written as a fraction. The table shows how the ratios for 1-cm and 3-cm cubes are calculated.

Look at the calculations for a 3-cm cube.

- First, you calculate the surface area. The formula for the surface area of a cube is

  (length × width) × 6 sides
  (3 cm × 3 cm) × 6 = 54 cm^2

- Next, calculate the volume. The formula for the volume of a cube is

  length × width × height
  3 cm × 3 cm × 3 cm = 27 cm^3

- Finally, determine the ratio of surface area to volume. Write the ratio as a fraction.

$$\frac{54}{27} = \frac{2}{1}$$

Notice that the fraction 54/27 has been simplified to 2/1. The ratio can also be written as 2 : 1, which is read as "2 to 1."

Analyze and Conclude

1. Predict How would the ratio for a 2-cm cube compare to cubes of 1 cm and 3 cm?

2. Calculate Use the formulas in the table to calculate the ratio of surface area to volume for a 2-cm cube. How accurate was your prediction?

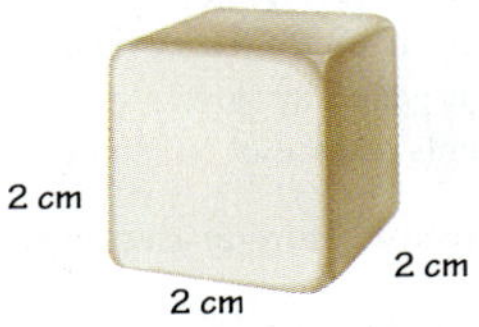

SA = 2 cm × 2 cm × 6 = _____ cm^2
V = 2 cm × 2 cm × 2 cm = _____ cm^3
Ratio = _____ : _____

In Your Workbook Get more help for this activity in your workbook.

Ratio of Surface Area to Volume in Cells		
Cell Size	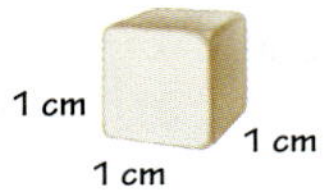 1 cm, 1 cm, 1 cm	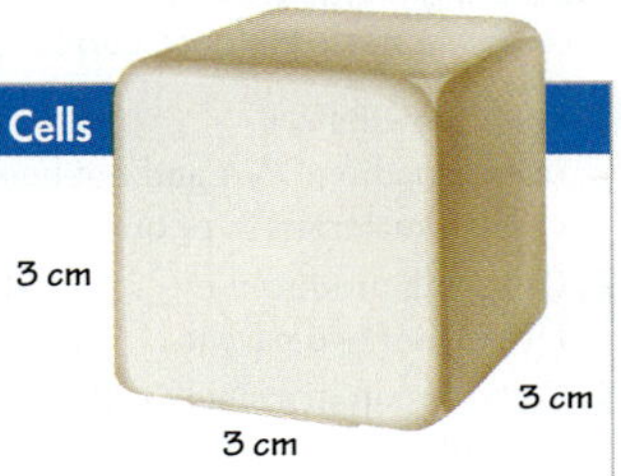 3 cm, 3 cm, 3 cm
Surface Area (length × width) × 6 sides	1 cm × 1 cm × 6 = 6 cm^2	3 cm × 3 cm × 6 = 54 cm^2
Volume length × width × height	1 cm × 1 cm × 1 cm = 1 cm^3	3 cm × 3 cm × 3 cm = 27 cm^3
Ratio of Surface Area to Volume	6/1 = 6 : 1	54/27 = 2 : 1

238

Analyze and Conclude

1. Encourage students to work in pairs to compare their calculations and check their answers. The ratio for a 2-cm cube is greater than the ratio for a 3-cm cube and less than the ratio for a 1-cm cube.
2. surface area, 24 cm^2; volume, 8 cm^3; ratio, 24/8 = 3:1

10.2 The Process of Cell Division

Chromosomes

Cells carry genetic information in DNA molecules. Even small cells like the bacterium *E. coli* are packed with DNA. In fact, the total length of the DNA molecule in *E. coli* is almost 1000 times the length of the cell. Clearly, that DNA has to be folded into the cell very carefully. Cells do this by packaging each molecule of DNA into a structure called a **chromosome.** Chromosomes make it possible to separate DNA precisely during cell division.

Prokaryotic Chromosomes Prokaryotic cells, such as *E. coli*, do not have a nucleus. The DNA molecules of prokaryotic cells are found in the cytoplasm. Usually, the genetic material is contained in a single, circle-shaped chromosome.

Eukaryotic Chromosomes The DNA molecules of eukaryotic cells are found in the cell nucleus. You are a eukaryote, as are all other animals and all plants. Eukaryotic cells have much more DNA than prokaryotes. And they package it into many chromosomes. The cells in your body have 46 chromosomes.

The chromosomes in eukaryotic cells form a close relationship with special proteins called histones. This combination of chromosomes and proteins is called **chromatin** (KROH muh tin). As a cell gets ready to divide, chromatin condenses, and individual chromosomes become visible inside the cell.

Key Question What is the job of chromosomes in cell division?
Chromosomes make it possible to separate DNA precisely during cell division.

Key Questions

What is the job of chromosomes in cell division?

What are the main events of the cell cycle?

What events occur during each phase of mitosis?

How do daughter cells split apart after mitosis?

BUILD Understanding

Two-Column Chart Use a two-column chart to take notes on the cell cycle. List the phases of the cell cycle in the first column. Describe the phases in the second column. Draw pictures to help you remember.

In Your Workbook Refer to your workbook for suggestions about how to use a two-column chart to organize your notes.

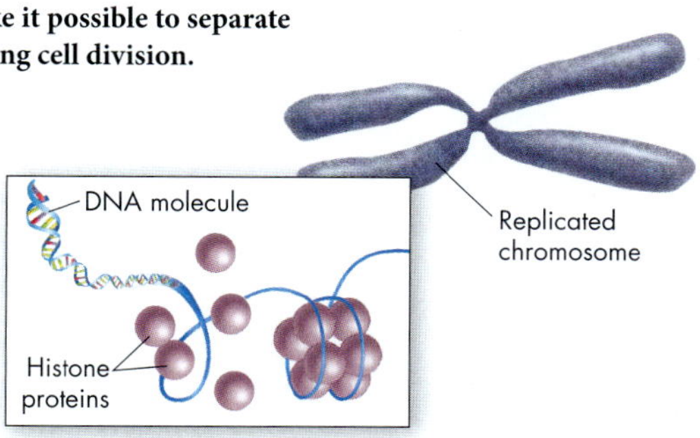

Eukaryotic Chromosome Histone proteins help to keep long molecules of DNA organized in eukaryotic cells. Before a cell divides, DNA molecules form into tighter and tighter coils that take the form of a chromosome.

BIOLOGY.com Search Lesson 10.2 GO • Art Review • Tutor Tube

Pre-Reading

Objectives

10.2.1 Describe the role of chromosomes in cell division.

10.2.2 Name the main events of the cell cycle.

10.2.3 Describe what happens during the four phases of mitosis.

10.2.4 Describe the process of cytokinesis.

Preview the Pages

1. Preview with students all the vocabulary, pronouncing each word aloud. Have students find the words in the text. Review word origins by having students name other words that include *cycle* and *cent*. Review the prefix *cyto-*.
2. Preview all the visuals, and read the captions and labels.
3. In The Cell Cycle diagram, be sure that students see that each phase of the cycle has an artist's drawing of the stage next to a micrograph.

Build Understanding

TWO-COLUMN CHART Read aloud the Build Understanding box. Clarify that prokaryotes and eukaryotes have different cell cycles. Suggest that students create two different charts, one for each cell cycle.

LESSON 10.2

NATIONAL SCIENCE EDUCATION STANDARDS

UNIFYING CONCEPTS AND PROCESSES

I, II

CONTENT

C.1.c, C.2.b

For the corresponding lesson in *Miller & Levine Biology,* see pages 279–285.

10.2 The Process of Cell Division

Lesson Objectives
- Describe the role of chromosomes in cell division.
- Name the main events of the cell cycle.
- Describe what happens during the four phases of mitosis.
- Describe the process of cytokinesis.

BUILD Vocabulary

A. The chart below shows key terms from the lesson with their definitions. Complete the chart by writing a strategy to help you remember the meaning of each term. One has been done for you.

Term	Definition	How I'm Going to Remember the Meaning
Cell cycle	Series of events a cell goes through as it grows and divides	I'll think of the my own life cycle—starting as a cell dividing and growing.
Centromere	Area where two sister chromatids are attached	Centromere is the spot in the center of the chromosome.
Chromatid	One of two identical "sister" parts of a replicated chromosome	The D at the end of chromatid will remind me that it is duplicated.
Chromatin	A combination of DNA and protein found in eukaryotic cells	The last two syllables of chromatin sound like mutton. Mutton is meat which is a good source of protein.
Chromosome	Made up of DNA; contains the genetic information needed to make new cells and carry out cell functions	Chromosomes contain cell information.
Cytokinesis	Part of the eukaryotic cell cycle during which the cytoplasm divides	Cytokinesis and cytoplasm both start with cyto-, which means cell.
Interphase	Long period of the cell cycle between one cell division and the next	Inter- means between. Interphase means between phases.
Mitosis	Part of the eukaryotic cell cycle during which the nucleus divides	Mitosis makes two

B. As you work through this lesson, you may find these terms in the activities. When you write a key term or a definition, **highlight** the term or the definition.

BUILD Understanding

Two-Column Chart A two-column chart is a way to take notes about what you read using words and pictures.

Look at the two-column chart below. The topic is at the top of the chart. In the left column, stages of the cell cycle are listed. In the right column, you will write a description and draw a picture of each stage.

As you read about the cell cycle in your textbook, complete the chart. Use the completed chart as a study aid.

The Cell Cycle	
Stage	**Description of Stage**
Interphase G_1	In G_1, the cell grows and produces new proteins and organelles. Check students' diagrams for accuracy.
S	In S phase, new DNA is synthesized when the chromosomes are replicated.
G_2	During the G_2 phase, many of the organelles and molecules required for cell division are produced.
M Phase: Cell Division Mitosis	The cell nucleus divides.
Cytokinesis	The cytoplasm divides.

The Cell Cycle

The Prokaryotic Cell Cycle The diagram on the left shows how a prokaryotic cell divides. The stages of cell division are shown in order they happen.

Use the flowchart on the right to describe the steps in prokaryotic cell division. In a flowchart, arrows connect one step to the next.

Follow the directions.

1. Write the stages in prokaryotic cell division in order in the flowchart.

Cell Division

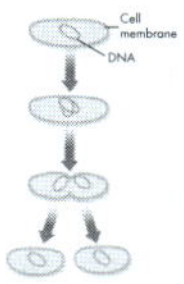

Flowchart

Before division, the DNA forms one circle. → DNA duplicates. → Membrane indents. → Two new cells are formed.

Answer the questions.

2. The process of cell division in prokaryotic cells is called binary fission.
3. Will the new cells have the same genetic material as the parent cell, or will each cell have different genetic material? same
4. Is this asexual or sexual reproduction? asexual
5. An example of an organism that reproduces with this kind of reproduction is a
   A. salamander.
   B. histone.
   C. red blood cell.
   D. bacterium.

During Reading

Active Reading

INFER Read the first paragraph about Eukaryotic Chromosomes on the previous page. Pause after the last sentence.

Ask Where are the chromosomes in your cells located? *(in the nucleus)*

USE CONTEXT CLUES Invite a volunteer to read the second paragraph and the caption for the Eukarytic Chromosomes diagram. Point out how students can use context clues to figure out the meaning of new or unfamiliar words.

Ask What does *condense* mean in relation to chromosomes? *(to form into tighter and tighter coils)*

KEY QUESTION Read aloud the Key Question from the previous page.

Ask What is a chromosome? *(structure that contains genetic information)*

INTERPRET DIAGRAMS Guide students to the diagram of Binary Fission after reading the last paragraph on this page.

Ask Is binary fission sexual or asexual reproduction? *(asexual)* How do you know? *(One cell divides to form two genetically identical cells.)*

Point out the DNA as a single, circle-shaped chromosome.

Redraw the diagram on the board as a four-stage cycle similar to the Eukaryotic Cell Cycle visual to show students that this diagram represents a cycle even though it is not a circle.

Build Vocabulary

WORD ORIGINS Invite students to think of other cycles referred to in science.

Ask How do you visualize the water cycle? What are other kinds of cycles? *(Accept all reasonable answers.)*

Speed Bump

Ask Imagine you're a cell: You're about to divide. What is the first thing you do? *(Make a complete copy of my DNA, genetic information.)* Why? *(to provide DNA for each daughter cell)*

Ask During cell division, fibers form between the DNA molecules. What would happen if these fibers did not tighten? *(The cytoplasm would not divide so there would be no way for the cell to separate into two cells.)*

BUILD Vocabulary

chromosome
a threadlike structure that contains the genetic information that is passed from one generation of cells to the next

chromatin
a substance found in eukaryotic chromosomes that consists of DNA tightly coiled around histones

cell cycle
a series of events a cell goes through as it grows and divides

WORD ORIGINS

The words *cycle* and *circle* share the same root, *kyklos*, a Greek word meaning "wheel." Cycles are often drawn as circles because they are a series of events that repeat—or go around again and again.

The Cell Cycle

Cells go through a series of events as they grow and then divide. Altogether, these events are called the **cell cycle.** During the cell cycle, a cell grows, prepares for division by making a copy of its DNA, and then divides to form two daughter cells. Each daughter cell then begins the cycle again.

Before any cell divides, it must copy, or replicate, its DNA. What this means, of course, is that the cell must make, or synthesize, new strands of DNA so that each cell receives a complete set of genetic information.

The Prokaryotic Cell Cycle The prokaryotic cell cycle includes cell growth, DNA replication, and cell division. Cell division can happen quickly under good conditions. Most prokaryotic cells replicate, or copy, their DNA once they have grown to a certain size. When DNA replication is complete, or almost complete, the cell begins to divide. The process of cell division in prokaryotes is a form of asexual reproduction called binary fission.

After the chromosome is replicated, the two DNA molecules attach to different parts of the cell membrane. A group of fibers form between the molecules. As the fibers tighten, the cell is pinched across its center. The tightening fibers divide the cytoplasm and separate the cell into two cells. Binary fission results in two cells with the same genetic material.

Binary Fission
Cell division in a single-celled organism produces two genetically identical organisms.

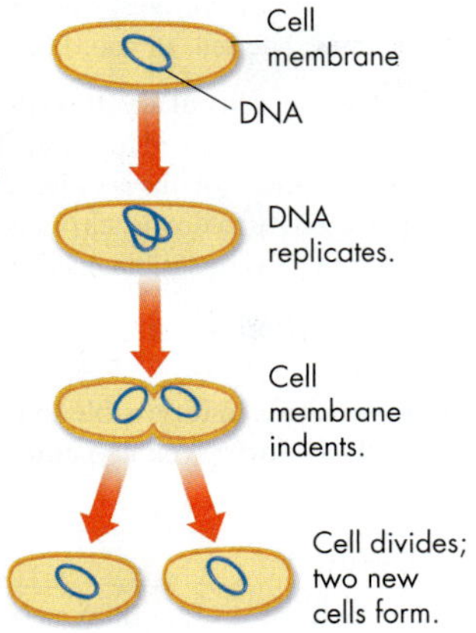

Hands-On Learning

THINKING ABOUT DIVISION Ask pairs of students to brainstorm various ways to equally divide a pile of several-sized rubber bands between themselves. Tell them to apply what they know about how a cell divides its genetic information to help them come up with the most efficient method for evenly distributing rubber bands. As a class, discuss and evaluate each pair's method.

The Eukaryotic Cell Cycle The eukaryotic cell cycle has four parts or phases: G_1, S, G_2, and M. The period of growth between cell divisions is referred to as **interphase.** It includes G_1, S, and G_2. A cell prepares to divide during the fourth phase M. At the end of M phase, the cell divides into two daughter cells. The cycle begins again. Refer to the figure below as you read the description of each phase of the eukaryotic cell cycle.

▶ ***G_1 Phase: Cell Growth*** A cell does most of its growing during the G_1 phase. It increases in size and produces new proteins and organelles. The G in G_1 and G_2 stands for "gap." However, both G phases are periods of growth and activity.

▶ ***S Phase: DNA Replication*** The G_1 phase is followed by the S phase. The S stands for "synthesis." During the S phase, new DNA is synthesized when chromosomes are replicated.

▶ ***G_2 Phase: Preparing for Cell Division*** When DNA replication is complete, a cell enters the G_2 phase. During this phase, many of the organelles and molecules needed for cell division are produced. At the end of the G_2 phase, a cell is ready to begin the process of cell division.

▶ ***M Phase: Cell Division*** The M in M phase stands for **mitosis** (my TOH sis). Mitosis is the first stage of the M phase, a series of events that lead to the division of the cell nucleus. **Cytokinesis** (sy toh kin NEE sis) is the second stage. It ends with the division of the cytoplasm. At the end of the M phase, two new daughter cells begin the cycle again.

Key Question What are the main events of the cell cycle?
During the cell cycle, a cell grows, prepares for division, and divides to form two daughter cells. The eukaryotic cell cycle has two parts: interphase and cell division.

BUILD Vocabulary

interphase
a long period of the cell cycle between one cell division and the next

mitosis
the part of the eukaryotic cell cycle during which the nucleus divides

cytokinesis
the part of the eukaryotic cell cycle during which the cytoplasm divides

PREFIXES
The prefix *cyto-* is a word part that has come to mean "cell." It comes from a Greek word *kytos* that means "hollow vessel." It is used in *cytokinesis* and *cytoplasm*. The first people to see cells using a microscope thought that cells were empty.

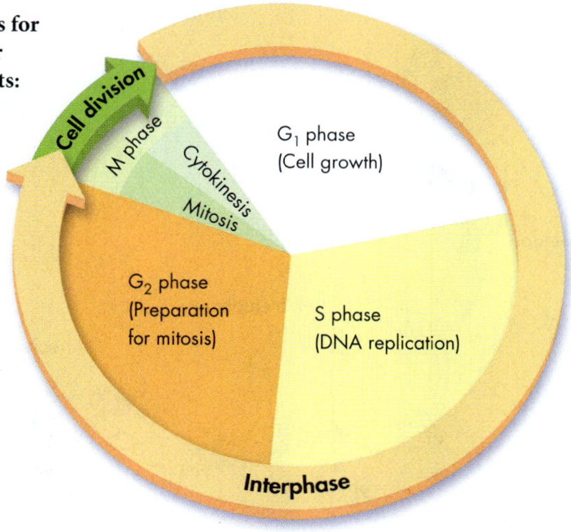

Eukaryotic Cell Cycle
The length of the yellow arrow indicates that eukaryotic cells spend most of their time in interphase. Cell division happens quickly.

ELL Focus on ELL: Access Content

INTERMEDIATE AND ADVANCED SPEAKERS Pair students and choose a selection for them to read, such as M Phase. Allow students to reread the selection. Have students refer to any visual support in the book. Have students write down what they have learned about the topic. Ask each pair to share their responses with the class. As a wrap-up, have a class discussion in which students evaluate the responses, highlighting important points.

Active Reading

FIND THE MAIN IDEA Invite volunteers to take turns reading each paragraph for the three phases of Interphase. After each phase is read, stop the read aloud, refer to the diagram, and stress the main idea for that phase as students write them in their notebooks.

G_1 phase: The cell increases in size and produces new proteins and organelles.

S phase: New DNA is synthesized and chromosomes are replicated, or copied.

G_2 phase: Organelles and molecules needed for cell division are produced.

INTERPRET DIAGRAMS Guide students to the diagram.

Ask What does the diagram remind you of? *(a pie chart)*

Ask What does each piece of pie represent? *(the amount of time, or period, each phase takes in the process of cell division)* Point out that the larger the wedge of pie, the more time that phase takes. Have students compare the green and yellow arrows.

Say If you look at just the arrows, this diagram tells you that there are really two stages or events in the eukaryotic cell cycle: interphase and cell division.

KEY QUESTION Read aloud the Key Question and answer.

Ask How many main events occur during cell division? *(2)* What are they? *(mitosis and cytokinesis)*

Build Vocabulary

PREFIXES Remind students that prefixes are added to root words to change their meanings. Tell students that the word *interphase* has the prefix *inter-*, which means "something between or among similar things." Ask students for the meanings of other words with *inter-*, such as *international, interference, intercept.*

Speed Bump

Ask What divides during mitosis? *(the cell nucleus)*

Ask What does the cell nucleus contain? *(DNA, genetic information)*

Ask Does mitosis happen in bacterial cells? Explain. *(No. Mitosis only happens in eukaryotic cells; bacteria are prokaryotes; prokaryotes do not have a cell nucleus.)*

LESSON 10.2

During Reading continued

Active Learning

DRAW A DIAGRAM Point out that the diagram can help students get a better idea of the process of mitosis. To help them process and retain what they are learning, remind them to copy and label diagrams in their notebooks. They should also write one sentence about what the diagram shows or explains. If there are several parts to a diagram, have them write a separate sentence for each part.

USE VISUALS Read aloud the paragraph describing Prophase. When you get to the part about the spindle forming, stop.

Ask Who knows what a spindle is? *(Expect shrugs.)* Refer students to the diagram showing Prophase. Do you have a better sense of the word now that you've seen a picture? *(Yes.)* Explain that when reading their science book, it's always a good idea to look over any pictures, diagrams, photographs, and charts.

KEY QUESTION Read the question aloud. Challenge students to draw and label in their notebooks a diagram of the Eukaryotic Cell Cycle that includes the phases of mitosis.

Build Vocabulary

WORD ORIGINS Point out that *chromatid* has the same beginning as *chromosome* and *chromatin* — *chrom-*.

Ask How can this help you to remember the meanings of the words? *(Guide students to see that they can use the endings of the words* -atid, -osome, *and* -matin. *For example,* osome *in* chromosome *is from* soma *which means* "body," so chromosome *is a body that contains genetic information.)*

Refer students back to the image of the Eukaryotic Chromosome. Invite a volunteer to draw it on the board. Point out the chromatids.

Say Each half of this X-shaped structure is a chromatid. (Draw a line through the centromere.) These sister chromatids (point to both sides of the X shape) have identical genetic information and are connected at the centromere (point to center of X shape).

Speed Bump

Ask During anaphase, what happens to the centromere when the sister chromatids separate? *(The centromere splits.)*

Ask During telophase, what do you find in the nuclear envelope? *(a full set of chromosomes)*

Ask During what phase does a chromatid become a chromosome? *(during anaphase, when the sister chromatids separate)*

BUILD Vocabulary

chromatid
one of two identical "sister" parts of a replicated chromosome

centromere
an area where two sister chromatids are attached

WORD ORIGINS
Both *centromere* and *centriole* have the word part *cent-*, which is similar to the word *center*. The Latin word *centrum* means "center."

Mitosis

Mitosis is divided into four phases: prophase, metaphase, anaphase, and telophase. Refer to the figure below as you read about the phases of mitosis.

Prophase Prophase is the first phase of mitosis. The replicated chromosomes become visible as DNA starts to condense. Each strand of DNA in a doubled chromosome is called a **chromatid.** The sister chromatids are joined at an area called the **centromere.** The nuclear envelope breaks down, and a spindle begins to form. A pair of structures called centrioles help to organize the spindle fibers.

Metaphase The second phase of mitosis is metaphase. The chromosomes line up at the center of the cell. Spindle fibers connect the centromere of each chromosome to the spindle.

Anaphase Anaphase is the third phase of mitosis. The centromeres split, and the sister chromatids separate. A complete set of chromosomes move to opposite sides of the cell.

Telophase The fourth and final phase is telophase. The chromosomes spread out into a tangle of chromatin. A nuclear envelope reforms around each group of chromosomes. The spindle begins to break apart. Mitosis is complete, but cell division still has one more step to go.

Key Question What events occur during each phase of mitosis? **During prophase, chromosomes become visible as DNA condenses and the nucleus breaks down. During metaphase, chromosomes line up across the center of the cell, attached to the spindle. During anaphase, sister chromatids separate and move apart. During telophase, a nucleus forms around each set of chromosomes.**

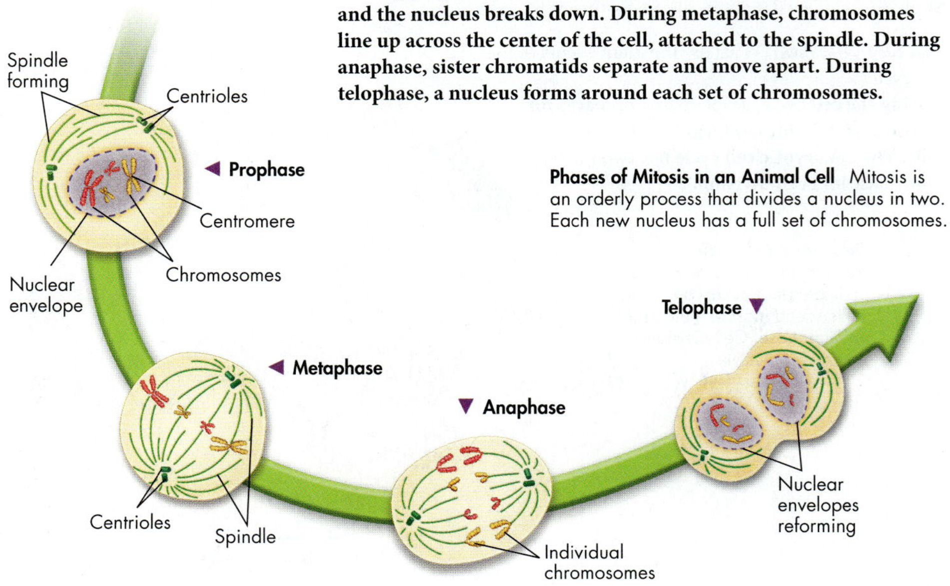

Phases of Mitosis in an Animal Cell Mitosis is an orderly process that divides a nucleus in two. Each new nucleus has a full set of chromosomes.

Focus on Academic Language

• spindle • centrioles

Students may need help understanding and pronouncing *spindle* (a fan-like microtubule structure that helps separate the chromosomes during mitosis) and *centrioles* (these two tiny structures in the cytoplasm take up position at the ends of the cell). Ask them to locate spindles and centrioles on the diagram of Prophase.

Cytokinesis

Mitosis produces two nuclei, each with a complete set of chromosomes. All that is left to complete the M phase is the division of the cytoplasm. Cytokinesis completes cell division by splitting one cell into two daughter cells. The process is different in animal and plant cells.

Cytokinesis in Animal Cells In animal cells, the cell membrane is pulled toward the center of the cell. Eventually the cytoplasm is divided into two nearly equal parts. Each part contains its own nucleus and organelles.

Cytokinesis in Plant Cells In plant cells the cell membrane is attached to a stiff cell wall. It is not flexible enough to pinch in. Instead a cell plate forms between the two nuclei. The cell plate develops into cell membranes that separate the two daughter cells. Next, a cell wall forms between the two new membranes. Each part contains its own nucleus and organelles.

Key Question How do daughter cells split apart after mitosis?
Cytokinesis divides the cytoplasm, splitting one cell into two. In animal cells, the cell membrane pinches in. In plant cells, a cell plate forms.

Cytokinesis
The division of the cytoplasm occurs differently in animal cells and plant cells.

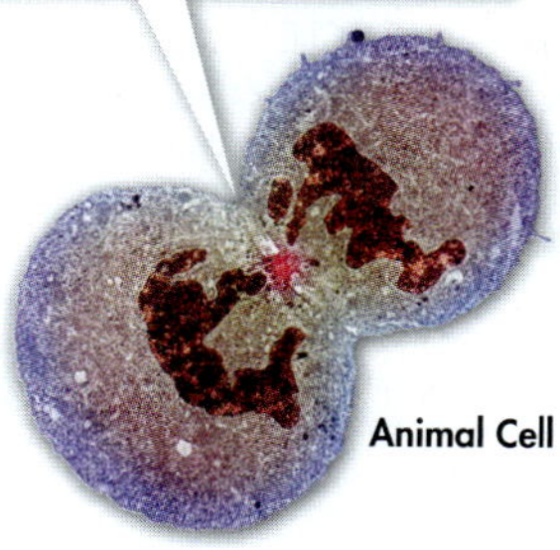

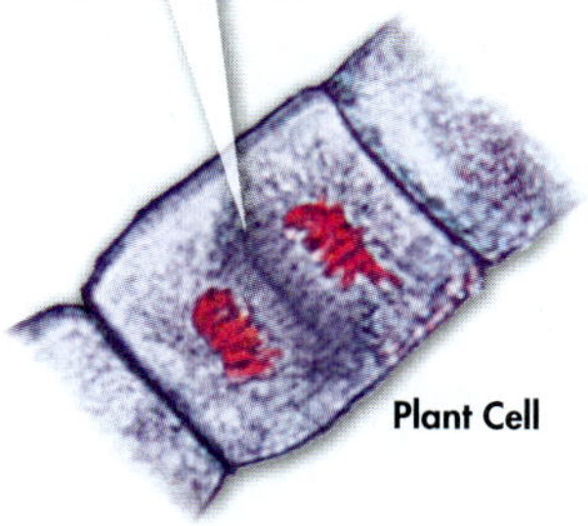

CHECK Understanding

Apply Vocabulary

Use the highlighted words from the lesson to complete each sentence correctly.

1. In prokaryotes, most cells have a single ______ that contains all the genetic material.
2. In eukaryotes, DNA and proteins combine to form ______.
3. The cell cycle of a eukaryote is divided into two stages: ______ and cell division.
4. In a eukaryotic cell, the chromosomes divide and separate during ______.

Critical Thinking

5. **Sequence** Draw a flowchart that follows a chromosome through a complete eukaryotic cell cycle.
6. **Compare and Contrast** How is cytokinesis in animals and plant cells alike and different?
7. **Write to Learn** Answer the second clue of the mystery. Think about what happens when you cut your finger. How quickly does your body repair the skin? What does that suggest?

MYSTERY CLUE

How would you expect the salamander's wound to affect the cell cycle of the cells that surround the wound? (**Hint:** See p. 241.)

BIOLOGY.com Search Lesson 10.2 GO • Lesson Assessment

Active Reading

DRAW CONCLUSIONS Invite a volunteer to read aloud the header and the first paragraph.

Ask If mitosis is about the division of the cell nucleus, what is cytokinesis about? *(the division of the cell cytoplasm)* What else, other than the cytoplasm, divides during cytokinesis in animal cells? *(the cell membrane, the two nuclei, organelles, chromosomes)*

INTERPRET DIAGRAMS What do you see that's different about these cells? What do you see that's the same? *(Answers may vary.)*

KEY QUESTION Read aloud the Key Question and answer. Repeat the answer aloud. Ask students to use a Venn diagram to show how cytokinesis is a different process in plant and animal cells. They can draw and complete a Venn diagram in their notebooks.

Speed Bump

Ask If a cell plate does not form in a plant cell during cytokinesis, is cell division still possible? Explain. *(No, the cell plate makes it possible for a plant cell to divide into two daughter cells.)*

Post Reading

ASSESSMENT TIPS For Question 6, encourage students to use the Venn diagram they created for the Key Question activity, or create one to help them organize their answer.

MYSTERY CLUE Discuss with students what they think might happen to the cell cycles of the cells surrounding the salamander's wound. Suggest students reread the information on eukaryotic cell cycles.

Wrap-Up Activity

PLAY THREE IN A ROW Divide students into groups of three and pass out 15 index cards to each group. Each student takes five cards. On separate cards, have one student draw each phase of mitosis; have another student print the names of each phase; have the third student write a brief description of each phase. Shuffle the cards. The object is to line up the three cards that correctly show, name, and describe each phase of mitosis. Students shuffle the cards after each turn. Students can refer to the drawing to check the line-up.

Check Understanding

1. chromosome
2. chromatin
3. interphase
4. mitosis
5. Students' drawings should contain the same stages as the Cell Cycle diagram on page 241.
6. Alike: Cytoplasm is divided to form two separate daughter cells with identical DNA. Different: Animal cell, membrane draws inward to separate the cytoplasm into two cells; plant cell, a cell plate forms developing a cell wall separating two daughter cells.
7. Sample answer: The cells surrounding the wound would move out of interphase and into cell division quickly to seal the wound and begin healing.

Build Connections

VISUAL SUMMARY Be sure that students understand that this art and the art Eukaryotic Cell Cycle both represent the cell cycle, but they present the information in different ways. Be sure that students see that the green color and orange color represent the same phases of the cycle.

Have students look at both pictures of the cell cycle as they answer.

Ask Look at both pictures of the cell cycle. Which color represents cell division? *(green)* Which color represents interphase? *(orange)*

Ask How are these two pictures the same? *(Both are of the cell cycle.)* What does the *M* in *M phase* stand for? *(mitosis)*

Ask What are the four phases of mitosis? *(prophase, metaphase, anaphase, telophase)* How will you remember them? Invite students to create a mnemonic device such as *Please Make A Taco, PMAT.*

Ask How are cytokinesis and mitosis related? *(cytokinesis is the second stage of mitosis)*

Ask Looking at the picture, which phases are a part of the M phase? *(prophase, metaphase, anaphase, telophase, cytokinesis)*

Ask What is important about interphase? *(It is the period when the cell grows before and after cell division.)*

BUILD Connections

THE CELL CYCLE

The phases of mitosis and cytokinesis are shown here as part of the the cell cycle. These stages are typical of eukaryotic cells.

In Your Workbook **Can you recognize all the phases of the cell cycle in a group of plant cells? Go to your workbook to find out.**

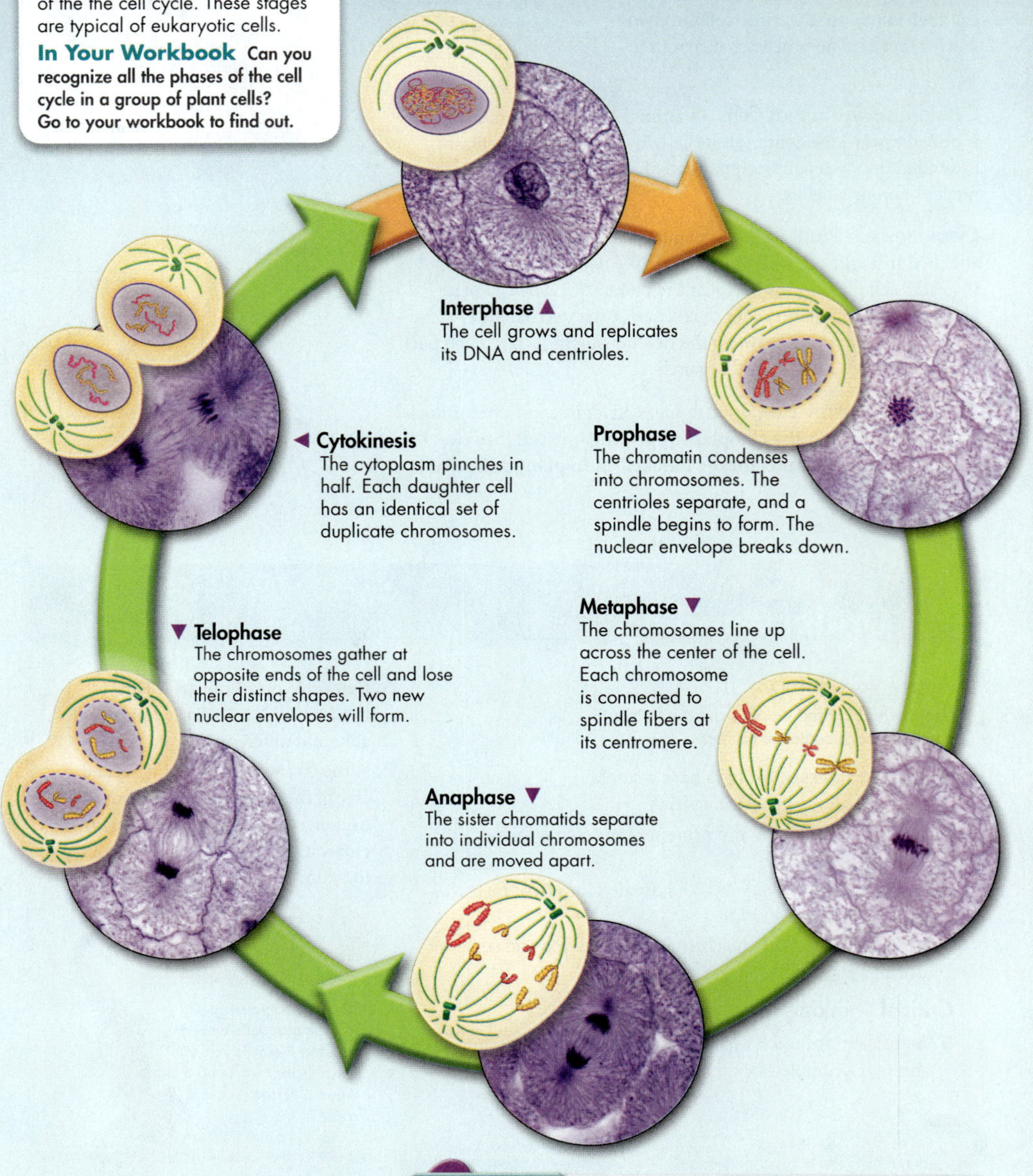

BIOLOGY.com Search Lesson 10.2 GO • InterActive Art

Focus on Academic Language

• condense • connected • separate • distinct • identical • duplicate • replicate

Write the words on the board and invite students to write them in their notebooks. Tell students that each of these words, or a form of the words, is in one of the descriptions of the phases of the cell cycle. Have students find the words and from the context, write a synonym or define it in their own words. Model the activity for students.

Say I found the word *replicates* in interphase. *Replicates* means "to reproduce." The cell grows and *reproduces* its DNA and centrioles.

After students are finished, as a class review their synonyms or definitions, *(condenses:* becomes more compact; *connected:* joined or linked; *separate:* break apart; *distinct:* clearly different; *identical:* exactly the same; *duplicate:* exact copy)

10.3 Regulating the Cell Cycle

Controls on Cell Division

The cell cycle is a process cells go through as they grow and divide. In multicellular organisms, the cell cycle is carefully controlled. In your body, for example, skin cells grow and divide quickly. However, most of the cells in your muscles and nerves do not grow and divide. How do cells know when to divide?

Some cells respond to outside signals. For example, cells grown in a small culture dish will stop growing and dividing when they touch other cells. What happens when cells are removed from the dish? The remaining cells begin to grow and divide again. They stop again when they come in contact with other cells.

Something like this happens in your body when you cut your skin or break a bone. The cells at the edges of the injury begin to divide quickly. These new cells allow the injury to heal. When the healing process is almost done, the rate of cell division slows down. The cell cycle returns to normal.

Key Questions

How is the cell cycle regulated?

How do cancer cells differ from other cells?

BUILD Understanding

Concept Map Make a concept map to organize the information in the lesson. Start with the heading "Regulating the Cell Cycle." Then add the smaller green headings.

In Your Workbook Go to your workbook for help in completing the concept map.

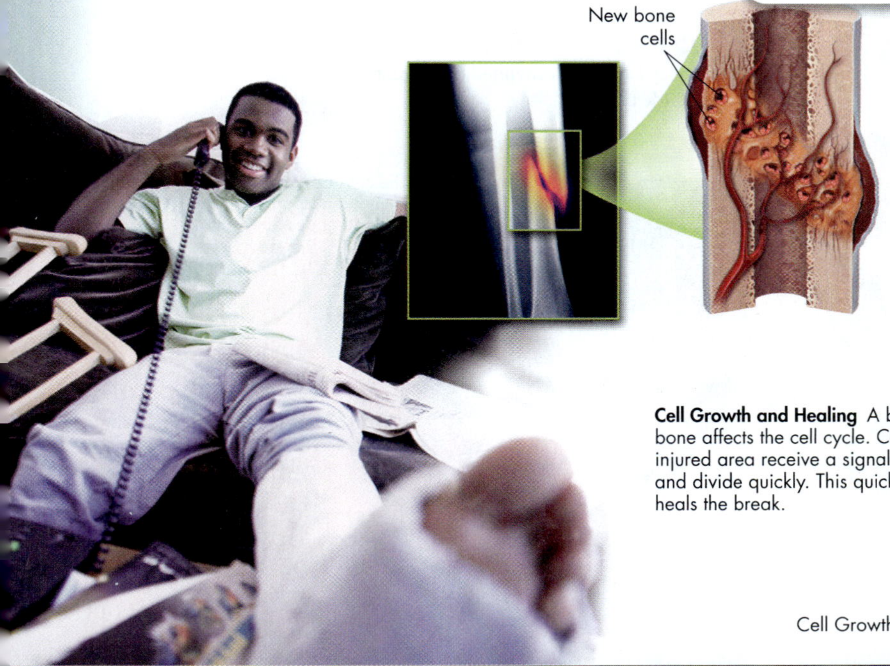

Cell Growth and Healing A broken bone affects the cell cycle. Cells in the injured area receive a signal to grow and divide quickly. This quick growth heals the break.

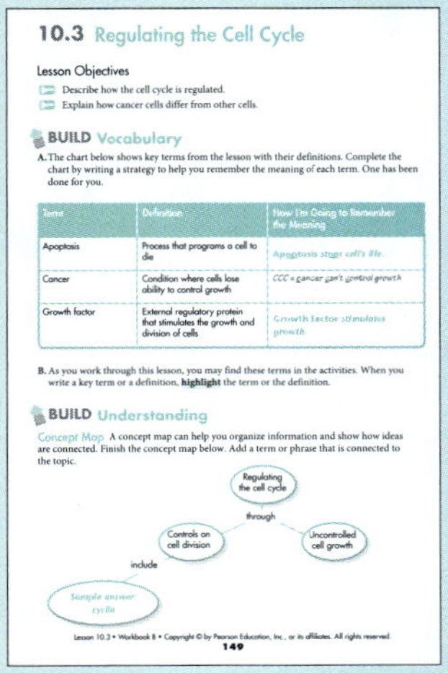

10.3 Regulating the Cell Cycle

Lesson Objectives

Describe how the cell cycle is regulated.

Explain how cancer cells differ from other cells.

BUILD Vocabulary

A. The chart below shows key terms from the lesson with their definitions. Complete the chart by writing a strategy to help you remember the meaning of each term. One has been done for you.

Term	Definition	How I'm Going to Remember the Meaning
Apoptosis	Process that programs a cell to die	Apoptosis stops cell's life.
Cancer	Condition where cells lose ability to control growth	CCC = cancer can't control growth
Growth factor	External regulatory protein that stimulates the growth and division of cells	Growth factor stimulates growth.

B. As you work through this lesson, you may find these terms in the activities. When you write a key term or a definition, **highlight** the term or the definition.

BUILD Understanding

Concept Map A concept map can help you organize information and show how ideas are connected. Finish the concept map below. Add a term or phrase that is connected to the topic.

149

Pre-Reading

Objectives

10.3.1 Describe how the cell cycle is regulated.

10.3.2 Explain how cancer cells are different from other cells.

Preview the Pages

1. Point out the relationship between the photo of the young man with the broken leg, the X-ray, and the drawing.
2. Be sure that students understand that apoptosis—programmed cell death—is important for the health of an organism and not just a sign of disease. For example, dead skin cells on the surface of the skin are constantly shed and must be replaced with "new" dead skin cells.

Activate Prior Knowledge

Ask students to identify some controls that they experience every day, such as a class schedule or traffic signals. Then ask them what would happen if these controls didn't exist.

Build Understanding

CONCEPT MAP To help students complete their concept maps in Workbook B, have them return to the book and point out how headings can help them organize the concepts and the relationships among them. Invite volunteers to draw their concept maps on the board.

NATIONAL SCIENCE EDUCATION STANDARDS

UNIFYING CONCEPTS AND PROCESSES

I, II

CONTENT

C.1.d

For the corresponding lesson in *Miller & Levine Biology*, see pages 286–290.

During Reading

Active Reading

USE VISUALIZATION To help students understand apoptosis, have them visualize a duck's foot after they look at the photograph and read the caption.

Ask What's the difference between a duck's foot and the mouse's? Did the cells between a duck's toes undergo apoptosis? *(no)* So what happens to the duck's foot? *(The cells do not die so the toes stay connected to form a webbed foot.)*

USE GRAPHIC ORGANIZERS Have students draw a Venn diagram in their notebooks to compare internal and external regulators.

Build Vocabulary

ACADEMIC WORDS Have students pronounce each vocabulary word and read the definition. Then, have them find the word and the definition in the text. Point out that the word *factor* is used with other words, too, such as *risk*.

Ask What do you think the term *risk factor* means? Can you give me any examples of risk factors? *(speeding, overeating)*

KEY QUESTION Read aloud the Key Question and answer. Ask students to answer the question in their own words.

Speed Bump

Ask What could go wrong if the cell cycle is not carefully regulated? *(Chromosomes may not be replicated before mitosis begins, or cells may not divide in response to an injury.)*

BUILD Vocabulary

growth factor
an external regulatory protein that stimulates the growth and division of cells

apoptosis
a process in which a cell is programmed to die

cancer
a disorder in which some of the body's own cells lose the ability to control growth

ACADEMIC WORDS
In the term *growth factor*, the word *factor* means "something that contributes to a result or a process." Growth factors contribute to many processes, including the healing of broken bones.

Regulatory Proteins For many years, biologists searched for the signals that controlled the cell cycle. They have found many different proteins that regulate, or control, the cell cycle.

▶ ***External Regulators*** Proteins that respond to events outside the cell are called external regulators. External regulators can cause the cell cycle to speed up or slow down. **Growth factors** are important external regulators. Growth-factor proteins control the cell cycle when you have an injury, such as a broken bone. They are also involved in controlling the cell cycle during the development of an embryo.

▶ ***Internal Regulators*** Proteins that respond to events inside the cell are called internal regulators. Internal regulators control the stages of the cell cycle. For example, internal regulators make sure that chromosomes have been replicated before mitosis starts. One important group of regulatory proteins are called cyclins. Cyclins regulate the timing of the cell cycle in eukaryotic cells.

Apoptosis Your body is producing new cells every day. Your body is also removing cells every day. Some cells die by accident because of damage or injury. Other cells, such as skin cells, are "programmed" to die. The process that brings an end to the cell cycle is called **apoptosis** (AYP up TOH sis). During apoptosis, a cell goes through a series of steps that lead to its death.

Apoptosis shapes the structure of tissues and organs in plants and animals. For example, look at the photos of a mouse foot below. Compare the shape of the embryonic foot to the adult foot. Cells between the mouse's toes are programmed to die during tissue development. The cells that remain form into toes.

Key Question How is the cell cycle regulated?
The cell cycle is controlled by regulatory proteins both inside and outside the cell.

Apoptosis The cells between a mouse's toes undergo apoptosis during a late stage of development.

◀ Adult foot

◀ Embryonic foot

ELL Focus on ELL: Build Background

ALL SPEAKERS Because cancer is a common disease, English language learners may be familiar with aspects of the disease. Have students make a KWL chart. Have students brainstorm what they already know about cancer, and write all responses on the board. Then, have students categorize the responses and record their individual responses.

Before they begin to read, encourage students to discuss what they want to learn and have them record their responses.

After reading, have students fill in the What Have I Learned? column.

Cancer: Uncontrolled Cell Growth

Cancer is a disorder that shows what happens when cell growth is not controlled. Cancer develops when the body loses the ability to control growth in some of its cells. Cancer cells divide without control because they do not respond to the signals that regulate the cell cycle.

Cancerous Tumors Cancer cells form into a mass of cells called a tumor. The cancer cells can spread to nearby tissues and to other parts of the body. As cancer cells grow and divide, they use more and more of the nutrients needed by healthy cells. They can also prevent organs and tissues from functioning normally. Not all tumors are cancerous.

What Causes Cancer? Cancers are caused by damage to a cell's DNA. The damage, or defect, occurs in that part of the genetic material that regulates cell growth and division. Some of these defects are caused by tobacco smoking, radiation exposure, and viral infection.

Treatments for Cancer If a tumor is found early, doctors can often remove it with surgery. Skin cancer can usually be treated this way. In addition, beams of radiation can kill cancer cells. Doctors can also treat cancer with chemical compounds that kill or slow the growth of cancer cells. This treatment is called chemotherapy.

Key Question How do cancer cells differ from other cells?
Cancer cells do not respond to the signals that regulate the growth of most cells. As a result, cells divide uncontrollably.

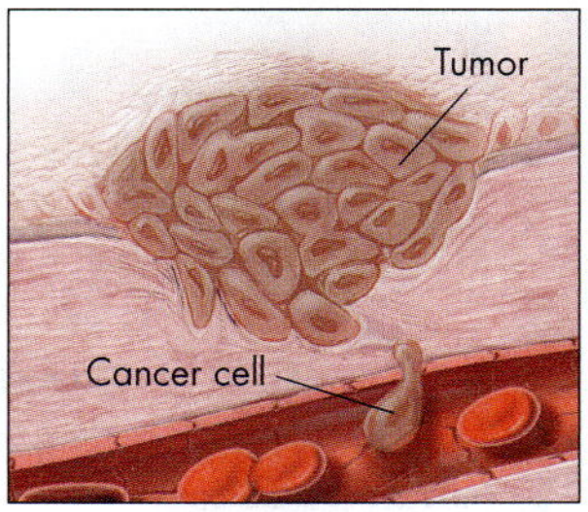

Cell Growth and Cancer Cancer starts when a cell begins to divide without control. A tumor forms. Cancer cells are dangerous because they can get into the bloodstream. From there, they can spread throughout the body.

CHECK Understanding

Use the highlighted words from the lesson to complete each sentence correctly.

1. ________ is a normal process that causes a cell to die.
2. Abnormal cells that do not die but continue to grow and divide cause ________.
3. External regulatory proteins that increase the rate of cell division during an organism's development are called ________.

Critical Thinking

4. **Infer** Why is cancer considered a disease of the cell cycle?
5. **Apply Concepts** How would the pattern of apoptosis differ in a duck's webbed foot compared to the pattern of apoptosis in a mouse's foot?
6. **Explain** What are cyclins? What do cyclins do?
7. **Write to Learn** Answer the third clue of the mystery. The image of the broken bone on the first page of this lesson will help you answer the question.

MYSTERY CLUE

How might regulatory proteins be involved in wound healing in the salamander? (**Hint:** See p. 246.)

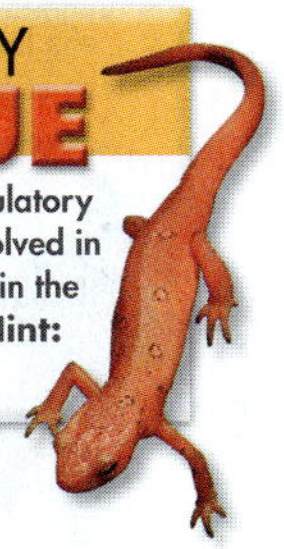

BIOLOGY.com Search Lesson 10.3 GO • Art in Motion • Lesson Assessment

Check Understanding

1. Apoptosis
2. cancer
3. growth factors
4. Cancer is considered a disease of the cell cycle because cancer occurs when changes in a cell prevent the cell from regulating how often it divides.
5. Not all of the cells between a duck's toes would undergo apoptosis. Some of the cells would remain, causing a duck to have webbed feet.
6. Cyclins are a group of internal regulatory proteins. Cyclins regulate the timing of the cell cycle in eukaryotic cells.
7. Regulatory proteins might signal that the process of cell division should speed up to promote quicker healing of the salamander.

Active Reading

USE GRAPHIC ORGANIZERS Draw the graphic organizer shown and have students draw it in their notebooks. Rephrase the blue heads in the book as questions. Have students write the questions and then answer them in the graphic organizer.

What is cancer?
(It's an uncontrolled cell growth.)

- What are cancerous tumors?
- What causes cancer?
- What are treatments for cancer?

KEY QUESTION Read aloud the Key Question and answer.

Ask Why is uncontrollable cell division a problem? *(Cancer cells use the nutrients that normal cells need to grow. Cancer cells can also prevent organs from functioning the way they are supposed to function.)*

Post Reading

ASSESSMENT TIPS Review the answers with students. If some students are having difficulty, suggest they refer to the lesson and find words in the text that are in the questions.

Ask For Question 4, what does it mean to infer? How does it clue you to the answer? *(Infer means "understanding information that is not in the text by reading between the lines." I have to read carefully and imagine what is happening or use the images the words give me.)"*

MYSTERY CLUE If students are having difficulty, refer them to the Cell Growth and Healing visual at the beginning of the lesson.

Wrap-Up Activity

CONDUCT AN INTERVIEW Pair students. One student is a reporter and the other is a cell, who just got a signal from a broken arm. The reporter should ask questions to find out how the break happened and what action the cell has to take. As the reporter, be sure the cell uses the words *signal, external regulators, growth factors*, and *cell cycle*. Have students switch roles but this time the cell's DNA is damaged. As the reporter, be sure the cell uses the words *cancer, tumor, uncontrolled growth*, and *chemotherapy*.

LESSON 10.4

Pre-Reading

Objectives

10.4.1 Describe the process of differentiation.

10.4.2 Define stem cells and explain their importance.

10.4.3 Identify the possible benefits and issues relating to stem cell research.

Preview the Pages

1. Discuss specialization with students. Encourage a discussion with the class about the benefits of having players on a sports team who specialize in certain positions as opposed to players who don't play a particular position.
2. Preview the visuals, with special attention to the Embryonic Stem Cells diagram. Invite students to share what they already know about stem cells.

Build Understanding

COMPARE/CONTRAST TABLE Direct students to Workbook B where they will find a partially completed compare/contrast table. Remind students that putting information into a table format will separate the most essential ideas and words from the nonessential and that they can use the table as a study guide later.

NATIONAL SCIENCE EDUCATION STANDARDS

UNIFYING CONCEPTS AND PROCESSES

I, II

CONTENT

C.1.d, C.1.f

INQUIRY

A.2.a

For the corresponding lesson in ***Miller & Levine Biology,*** see pages 292–297.

10.4 Cell Differentiation

Key Questions

How do cells become specialized for different functions?

What are stem cells?

What are some possible benefits and issues associated with stem cell research?

BUILD Understanding

Compare/Contrast Table Certain types of cells have the ability to differentiate. Use a table to write what you learn about these cells.

In Your Workbook Go to your workbook to see how to organize your compare/contrast table.

From One Cell to Many

Each of us started life as just one cell—a single fertilized egg. So did your pet dog or cat. The fertilized egg goes through an early stage of development called an **embryo.** As it grows, the embryo develops and changes. During the development process, the embryo's cells become different from one another. They differentiate.

What Is Differentiation? Cells become specialized through the process of **differentiation.** Your body has more than 200 different cell types. Notice that the word *differentiate* contains the word *different.* A differentiated cell has become different from the original cell that produced it.

What Is a Specialized Cell? Differentiated cells are specialized to perform certain jobs. In a plant, for example, some cells carry out photosynthesis. Other plant cells store sugars. Other plant cells create pathways for moving materials up and down. In your body, differentiated cells carry out certain jobs that keep you alive.

Key Question How do cells become specialized for different functions?
During the development of an organism, cells differentiate into many types of cells.

Specialized Plant Cells A buttercup plant is made up of cells that are specialized to carry out specific functions.

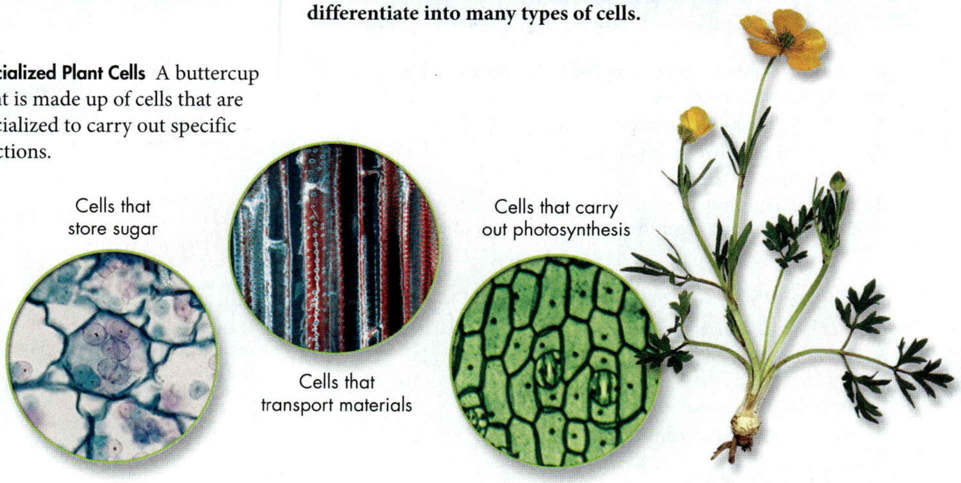

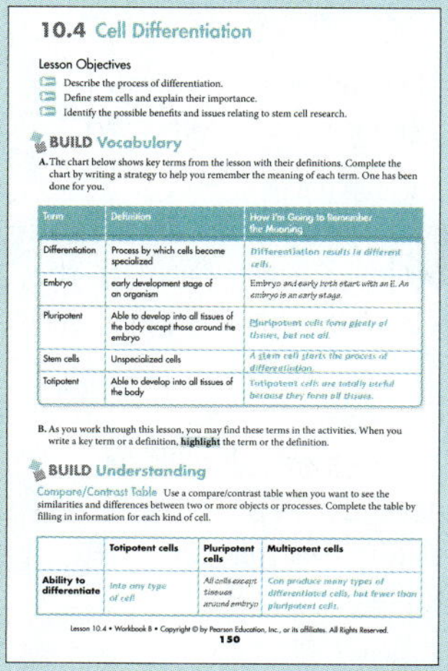

10.4 Cell Differentiation

Lesson Objectives

- Describe the process of differentiation.
- Define stem cells and explain their importance.
- Identify the possible benefits and issues relating to stem cell research.

BUILD Vocabulary

A. The chart below shows key terms from the lesson with their definitions. Complete the chart by writing a strategy to help you remember the meaning of each term. One has been done for you.

Term	Definition	How I'm Going to Remember the Meaning
Differentiation	Process by which cells become specialized	Differentiation results in different cells.
Embryo	early development stage of an organism	Embryo and early both start with an E. An embryo is an early stage.
Pluripotent	Able to develop into all tissues of the body except those around the embryo	Pluripotent cells form plenty of tissues, but not all.
Stem cells	Unspecialized cells	A stem cell starts the process of differentiation.
Totipotent	Able to develop into all tissues of the body	Totipotent cells are totally useful because they form all tissues.

B. As you work through this lesson, you may find these terms in the activities. When you write a key term or a definition, **highlight** the term or the definition.

BUILD Understanding

Compare/Contrast Table Use a compare/contrast table when you want to see the similarities and differences between two or more objects or processes. Complete the table by filling in information for each kind of cell.

	Totipotent cells	Pluripotent cells	Multipotent cells
Ability to differentiate	Into any type of cell	All cells except tissues around embryo	Can produce many types of differentiated cells, but fewer than pluripotent cells.

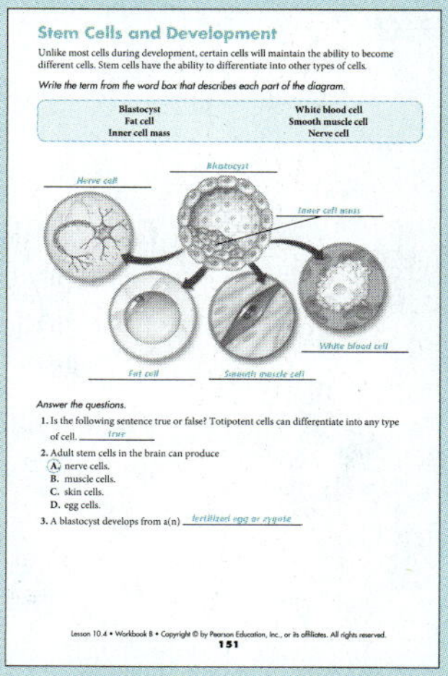

Stem Cells and Development

Unlike most cells during development, certain cells will maintain the ability to become different cells. Stem cells have the ability to differentiate into other types of cells.

Write the term from the word box that describes each part of the diagram.

Blastocyst	White blood cell
Fat cell	Smooth muscle cell
Inner cell mass	Nerve cell

Answer the questions.

1. Is the following sentence true or false? Totipotent cells can differentiate into any type of cell. true
2. Adult stem cells in the brain can produce
   A. nerve cells.
   B. muscle cells.
   C. skin cells.
   D. egg cells.
3. A blastocyst develops from a(n) fertilized egg or zygote

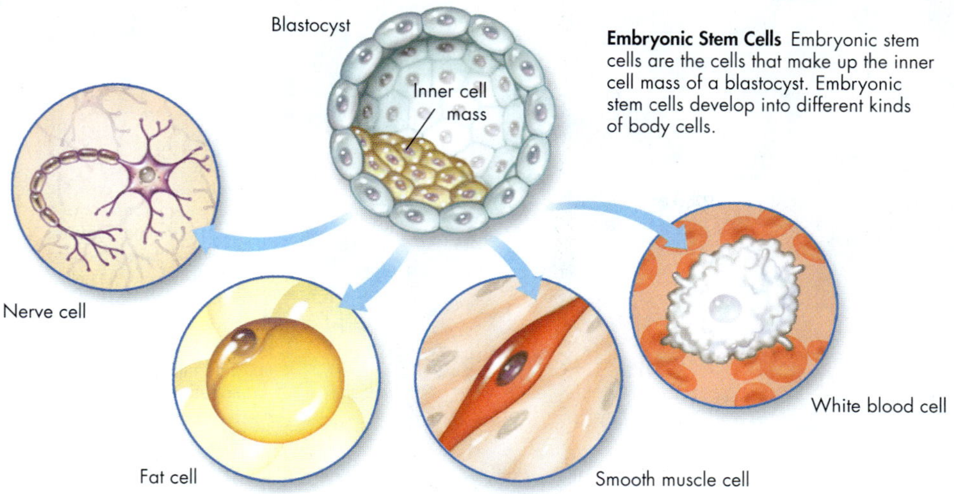

Embryonic Stem Cells Embryonic stem cells are the cells that make up the inner cell mass of a blastocyst. Embryonic stem cells develop into different kinds of body cells.

Stem Cells and Development

A fertilized egg divides over and over to form new cells. The fertilized egg and the cells produced by the first few divisions are totipotent. **Totipotent** (toh TIP uh tunt) cells can differentiate into any type of cell.

Human Development Early in development, a human embryo forms into a blastocyst. A blastocyst is a hollow ball of cells with a group of cells inside called the inner cell mass. The outer cells form tissues that attach the embryo to its mother. The inner cell mass becomes the embryo itself.

The cells of the inner cell mass are pluripotent. **Pluripotent** (plu RIP uh tunt) cells can develop into all the types of cells in the body. However, pluripotent cells cannot form the tissues that surround the embryo. Therefore, the cells of the inner cell mass are not totipotent.

Stem Cells Differentiated, specialized cells develop from unspecialized cells known as **stem cells.** Researchers are very interested in stem cells, because stem cells can develop into other types of cells.

▶ ***Embryonic Stem Cells*** The pluripotent cells that make up the early embryo are called embryonic stem cells. They will eventually produce all of the cells in the body. Scientists have experimented with stem cells grown in the laboratory. For example, scientists have been able to get mouse embryonic stem cells to develop into nerve cells, muscle cells, and even into sperm and egg cells.

BUILD Vocabulary

embryo
an organism in its early stage of development

differentiation
a process in which cells become specialized in structure and function

totipotent
having the ability to develop into all the tissues of the body

pluripotent
able to develop into all the body's cell types, but not capable of developing into the tissues surrounding the embryo

PREFIXES
The prefix *toti-* means "all." The word part *potent* means "powerful." A totipotent cell has the power to develop into all the tissues of the body.

Students Say

"The teacher would divide our class into four groups. The first group got two minutes to tell what they knew about a topic, such as plant cell structure. Each group member was required to say at least one fact, and the rest of the class gave a thumbs up or thumbs down, so the teacher could award a point for the fact. No point was given if most of the thumbs were down because the same fact was already given, or the fact was not correct. You had to be fast, and you had to pay attention."

During Reading

Active Reading

SUMMARIZE Have students read each paragraph on the previous page and summarize each in one or two sentences in their notebooks.

Ask What's an embryo? *(a fertilized egg in early stages of development)* What is a specialized cell? *(a cell that performs specific functions)* Have students read their summaries aloud to be sure they're accurate.

SCIENCE SUPPORT Guide students to the Specialized Plant Cells visual and the Embryonic Stem Cells visual. Point out that both plant cells and the human cells become specialized.

Ask Look at the buttercup cells. How are these cells alike? *(They're all plant cells.)* What's different about these cells? *(Each cell is differentiated and specialized.)*

Ask Why do you think plants could grow throughout their lives but humans cannot? *(Certain cells in the plants' roots and stems are not fixed in their development and can undergo rapid cell division and differentiation. Once human cells differentiate, they usually lose their ability to become other kinds of cells.)*

KEY QUESTION Read the Key Question on the previous page aloud. Then read the answer as a class. Have students close the book.

Ask In your own words, what is *differentiation*? *(process through which cells become specialized)* What would happen if cells didn't differentiate? *(If cells only divided without differentiating, all our cells would be identical.)*

Build Vocabulary

PREFIXES Model pronunciation of vocabulary words, have students repeat, and then read the definitions together. Write *toti-*, *pluri-* and *multi-* on the board and ask what each prefix means. Note that *multipotent* is defined on the next page. *(all, most, many)*

Ask How are you going to remember the meanings of *totipotent, pluripotent* and *multipotent cells*? Help students to infer that each cell's ability to differentiate is related to its prefix, from most flexible (totipotent) to most limited (multipotent).

During Reading continued

Active Reading

USE VISUALS Review the diagram, pointing out that stem cells are **1.** removed, then **2.** injected, then **3.** differentiate.

Ask Why are cells taken from bone marrow to inject into the heart? *(bone marrow contains some adult stem cells)*

Ask What happens to the stem cells when they are injected into the damaged heart muscle? *(The bone marrow cells differentiate into heart muscle cells.)*

KEY QUESTION Ask the question and read the answer as a class.

Ask What's the difference between adult stem cells and embryonic stem cells? *(Embryonic stem cells are cells in the early stages of the embryo. They will produce all the cells in the body. Adult stem cells are found in adults. They can produce a limited number of differentiated cells.)*

Speed Bump

Ask Some skin cells can also differentiate into several types of skin cells. Are these cells multipotent or pluripotent? *(multipotent)*

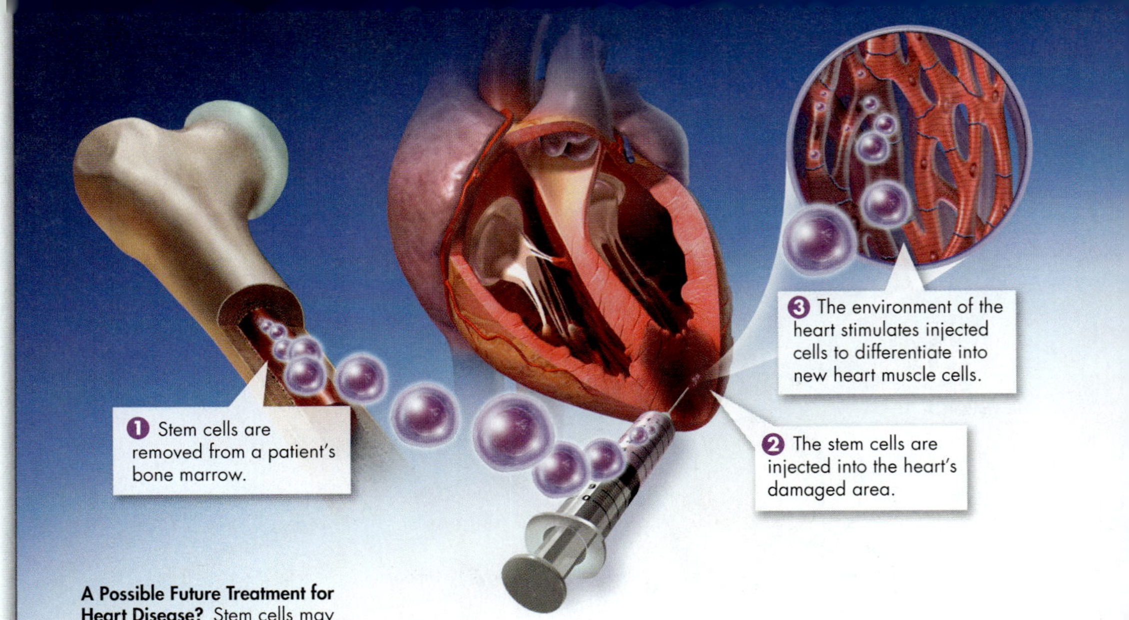

A Possible Future Treatment for Heart Disease? Stem cells may someday be used to reverse the damage caused by a heart attack. The diagram shows one method that scientists are investigating.

▶ ***Adult Stem Cells*** Adult organisms also contain some types of stem cells. For example, adult stem cells produce new skin and blood cells. Skin cells and blood cells only live a short time. Therefore, they need to be replaced all the time.

Adult stem cells are called **multipotent** (mul TIP uh tunt) cells. Multipotent means that they can produce many, but not all, types of differentiated cells. For example, adult stem cells in the bone marrow can develop into several types of blood cells.

Key Question What are stem cells?
Stem cells are the unspecialized cells from which differentiated cells develop.

Frontiers in Stem Cell Research

Scientists want to learn how stem cells keep the ability to differentiate into so many different types of cells. Scientists would also like to know why some cells in adults stay multipotent, but other cells lose their ability to differentiate.

Possible Benefits Research on stem cells is important. That is because scientists hope that stem cells might improve human health. Someday, stem cells might be used to fix cells that have been damaged. For example, stem cells might be used to repair heart-attack damage.

BUILD Vocabulary

stem cell
an unspecialized cell that can give rise to one or more types of specialized cells

multipotent
having the ability to produce many, but not all, kinds of differentiated cells

PREFIXES
The prefix *multi-* means "many." Multipotent cells can differentiate into many types of cells.

ELL Focus on ELL: Extend Language

BEGINNING SPEAKERS Prepare a Word Wall and post the key terms: *differentiation, specialization, blastocyst, frontier, ethical,* and any other words that students may have difficulty with along with the definition and any visuals. The title of this chapter can highlight how adding *-tion* changes the meaning of a word, so underline the *-tion* in *differentiation* and *specialization*. For more information on making a Word Wall, see page A4 in Workbook B.

Throughout the lesson, point out the words *differentiation, differentiate,* and *different.* Have students discuss how the meaning of each word changes with the different endings.

Ethical Issues Stem cell research may someday help humans. However, people disagree about whether certain kinds of stem cell research is right or wrong.

To obtain embryonic stem cells, scientists must usually destroy the embryo. Some people believe embryos should have the same protections as any person. These people object to research involving embryonic stem cells. Other people support research with embryonic stem cells. They believe it can improve and save human lives.

Scientists may be able to use technology to address some objections to embryonic stem cell research. For example, scientists hope to find a way to remove a few embryonic stem cells without damaging the embryo. In addition, scientists have been able to make adult stem cells act like pluripotent embryonic stem cells.

Key Question What are some possible benefits and issues associated with stem cell research?
Stem cells offer the possible benefit of using undifferentiated cells to repair or replace badly damaged cells and tissues. Human embryonic stem cell research is controversial because it involves issues of life and death.

CHECK Understanding

Apply Vocabulary

Use the highlighted words from the lesson to complete each sentence correctly.

1. A fertilized egg is a __________ cell, since all the different tissues of the body, including those that surround the embryo, develop from it.
2. The process in which cells become specialized to do different jobs is called __________.
3. Early in development, a human __________ forms into a hollow ball of cells called a blastocyst.
4. Specialized body cells develop from embryonic __________ cells.
5. Cells that are __________, such as those that produce skin and blood cells, can develop into some, but not all, types of specialized cells.

Critical Thinking

6. **Interpret Visuals** Look again at the picture called Embryonic Stem Cells. In your own words, describe what the picture shows.
7. **Compare and Contrast** How are totipotent and pluripotent cells alike? How are they different?
8. **Explain** Explain the potential benefits of stem cell research and the problems associated with this research.
9. **Write to Learn** Answer the fourth clue of the mystery. In your answer, explain how undifferentiated cells might be useful to the adult salamander.

MYSTERY CLUE

Some adult salamander cells never completely differentiate. What ability do these cells retain? (**Hint:** See p. 249.)

BIOLOGY.com Search Lesson 10.4 GO • Lesson Assessment

Check Understanding

1. totipotent
2. differentiation
3. embryo
4. stem
5. multipotent
6. Sample answer: Embryonic stem cells exist inside the blastocyst's inner cell mass. They can develop into the many different types of body cells.
7. Totipotent and pluripotent cells have the ability to develop into all the tissues of the body. Pluripotent cells cannot develop into tissues surrounding the embryo.
8. Sample answer: Stem cells may improve the quality of life for humans. To obtain stem cells, scientists usually have to destroy an embryo.
9. Sample answer: They retain the ability to differentiate. If an organism retains undifferentiated cells, it can grow new tissues.

Active Reading

DRAW A DIAGRAM Pair students. Have each pair create one graphic organizer to compare and contrast the benefits and issues surrounding stem cell research.

KEY QUESTION Ask the question and invite students to share their graphic organizers with the class.

Post Reading

ASSESSMENT TIPS For Question 7, have students refer to the compare/contrast table they worked on at the beginning of the lesson and rewrite the information in complete sentences.

MYSTERY CLUE Remind students that the salamander in the story eventually grew a new limb.

Ask What kinds of tissues are found in a salamander limb? *(bone, blood, and nerve tissue)* Help students to see that certain cells in the adult salamander must therefore retain the ability to differentiate into these types of tissues.

Wrap-Up Activity

PLAY CONCENTRATION Pair students. Pass out index cards. Have students write the prefixes or suffixes on index cards. Then, have them write the roots on separate index cards. Shuffle both piles together and arrange in rows and columns. Students take turns matching a prefix to a root. (For example: *embryo + nic; toti + potent; differentia + tion; ethic + al; un + specialized; special + ized; re + search, etc.)*
If time is limited, you may want to make the cards before class and distribute to students.

Some prefixes and suffixes: *nic, tion, al, ized, re, toto, pluri, multi, blasto*

Some root words: *embryo, ethic, special, search, potent, cyst*

CHAPTER LAB

Pre-Lab

Review the meaning of *hypothesis (a testable idea)*. Remind students about the relationship between dependent and independent variables by creating a chart like this one. Elicit examples from students to fill in the chart.

Hypothesis *(question)*	Independent Variable *(what I change)*	Dependent Variable *(what I see happen)*

Ask students for definitions for *regeneration* and *reproduction*.

Ask What other animal has the ability to regrow body parts? *(salamander)*

Review the Background Questions and Pre-Lab Questions as a class.

Lab

Tell students they will perform the chapter lab *Regeneration in Planaria* described in **Lab Manual B.**

SAFETY Students should handle the sharp scalpel with care as they cut the planarians. They should wear gloves while handling planaria and wash their hands after the lab.

Post-Lab

SILENT MOVIE Make a planarian flip book showing a planarian as it regenerates. Have students share their flip books.

Design Your Own Lab

OPEN-ENDED INQUIRY

Pre-Lab: Regeneration in Planaria

Problem How potent are the stem cells in planaria?

Materials fresh water or spring water, planarians, petri dishes, glass-marking pencil, forceps, scalpel, dissecting microscope, glass microscope slide, lens paper, pipette, small paintbrush, clear ruler

Lab Manual Chapter 10 Lab

Skills Focus Form a Hypothesis, Design an Experiment, Draw Conclusions

Connect to the Big idea All cells come from existing cells. When most cells in a multicellular organism divide, they produce cells just like themselves. However, some cells can differentiate to form different types of cells. These cells enable an organism to repair tissue after an injury or in some cases to regenerate body parts. In this lab, you will investigate the ability of planarians to regenerate body parts.

Background Questions

a. Compare and Contrast What is the difference between totipotent stem cells and multipotent stem cells?

b. Apply Concepts What type of stem cell enables your body to produce cells, such as skin and blood cells that are constantly replaced by the body?

c. Apply Concepts What type of stem cell enables a salamander to regenerate its tail?

d. Compare and Contrast In what way is regeneration of a body part similar to asexual reproduction? In what way is it different?

Pre-Lab Questions

Preview the procedure in the lab manual.

1. Draw Conclusions Suppose a cut planarian regenerates completely. What would you conclude about its stem cells? Suppose the regeneration is incomplete. What would you conclude?

2. Control Variables In this lab, you will use an uncut planarian as a control. Explain why you need this control.

3. Infer Two planarians are cut at different locations. Regeneration occurs in one planarian, but not in the other. Based on these results, what might you infer about stem cells in planarians?

BIOLOGY.com Search Chapter 10 GO

Visit Chapter 10 online to test yourself on chapter content and to find activities to help you learn.

Untamed Science Video Journey with the Untamed Science crew to a research facility in Sweden to learn why scientists are studying regeneration in brittle stars.

Visual Analogy Compare a growing cell to a growing city to understand limits on cell size.

Tutor Tube Confused about chromosome terminology? Tune in to Tutor Tube to unravel the vocabulary of chromatin.

Art Review Test your knowledge of the structure of a eukaryotic chromosome.

InterActive Art See the phases of mitosis in action.

Art in Motion See what happens when cancerous cells invade normal tissue.

Pre-Lab Answers

BACKGROUND QUESTIONS

a. Totipotent cells can form all body cells and the tissues that support an embryo. Multipotent cells can form many types of body cells.

b. multipotent cells

c. totipotent cells

d. Asexual reproduction and regeneration both depend on cell division and differentiation. However, asexual reproduction produces two organisms, but regeneration results in a single repaired organism.

PRE-LAB QUESTIONS

1. Sample answer: If a cut planarian regenerates completely, its stem cells are totipotent. If the regeneration is incomplete, the cells are multipotent.

2. Sample answer: I will be able to compare the growth that takes place when a planarian is not cut to the growth that takes place in the cut planarians.

3. Students might infer that stem cells are not found at every location in a planarian.

NATIONAL SCIENCE EDUCATION STANDARDS

UCP II

CONTENT C.1.f

INQUIRY A.1.b, A.2.a

10 CHAPTER Summary

10.1 Cell Growth, Division, and Reproduction

- Cells grow only so big before they divide. The larger a cell becomes, the more demands the cell places on its DNA. Also, a larger cell has more difficulty in moving enough nutrients in and wastes out.
- Organisms produced by asexual reproduction are genetically the same as the single cell that produced them.
- Organisms produced by sexual reproduction grow from a single cell that contains genetic information from two parents.

cell division (p. 236)
asexual reproduction (p. 236)
sexual reproduction (p. 236)

10.2 The Process of Cell Division

- Chromosomes are made up of DNA that carries a cell's genetic information.
- The eukaryotic cell cycle includes interphase, mitosis, and cytokinesis.
- Mitosis is divided into four phases in which replicated chromosomes and a cell's nucleus divide equally. In prophase, the chromosomes condense and the nucleus breaks down. In metaphase, the chromosomes line up at the center of the cell. In anaphase, sister chromatids separate. In telophase, a nucleus reforms around each set of separated chromosomes.
- Cytokinesis completes the process of cell division. It splits one cell into two.

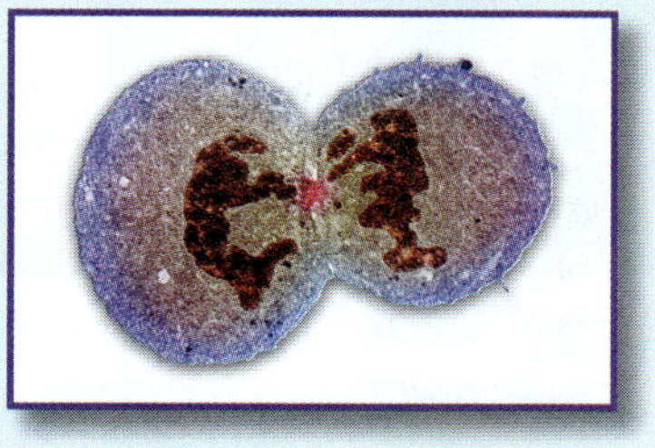

chromosome (p. 239)
chromatin (p. 239)
cell cycle (p. 240)
interphase (p. 241)
mitosis (p. 241)
cytokinesis (p. 241)
chromatid (p. 242)
centromere (p. 242)

10.3 Regulating the Cell Cycle

- Internal and external regulatory proteins help control a cell's growth and division.
- Apoptosis removes old or unneeded cells from an organism.
- Cancer cells do not respond to the signals that tell most cells to stop growing.

growth factors (p. 246)
apoptosis (p .246)
cancer (p. 247)

10.4 Cell Differentiation

- During the development of an organism, cells differentiate into many types of cells.
- Stem cells have the ability to differentiate into other types of cells.
- Stem cell research may offer many medical benefits, but it also raises ethical concerns.

embryo (p. 248)
differentiation (p. 248)
totipotent (p. 249)
pluripotent (p. 249)
stem cells (p. 249)
multipotent (p. 250)

BIOLOGY.com Search Chapter 10 GO • Match It • Chapter Assessment

UbD Transfer the Big Idea

TRANSFER Ask students to make a list of what types of cells and tissues humans can regenerate, such as hair and nails. Have partners compare lists and talk about how humans are different from salamanders and planarians in regards to tissue regeneration. Ask them to discuss why humans may be more limited in potential than the other two types of organisms.

Chapter Summary

Think Visually

USE GRAPHICS Have students fill in the following cycle diagram of the cell cycle.

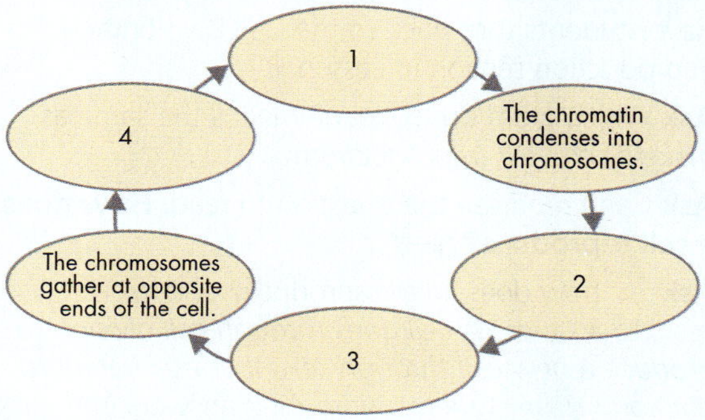

ANSWERS

1. The cell grows and replicates its DNA and centrioles. **2.** The chromosomes line up across the middle of the cell. **3.** The sister chromatids separate into individual chromosomes and move apart. **4.** The cell membrane pinches the cytoplasm in half.

BIOLOGY.com Study Online

REVIEW RESOURCES

Visual Analogy Virtual comparisons of cell parts with town structures provide insight into limits on cell size.

Tutor Tube Online tutorials provide extra help on chromosome structure and the cell cycle.

Data Analysis Students can gather data from a virtual sample of plant tissue to find the relative durations of the different phases of the cell cycle.

InterActive Art Students learn why mitosis occurs in phases, but cell division is a continuous process.

Vocabulary Review The *Flash Cards* and *Match It* provide an interactive vocabulary review.

Editable Worksheets Editable pages of Study Workbook B, Lab Manual B, and the Assessment Resources Book are available online.

Lesson Overviews Lesson Overviews are based on Miller & Levine Biology. However, they are available on CD-ROM and can be edited for your class.

ASSESSMENT RESOURCES

Chapter Assessment Have students take an online version of the Chapter 10 Check Understanding.

Standardized Test Prep Students can take an online version of the Standardized Test Prep. You will receive their scores along with ideas for remediation.

Diagnostic and Benchmark Tests Use these tests to monitor your students' progress and supply remediation.

ASSESSMENT

Assess

Big idea Invite a volunteer to read the question. Write the words *produce, reproduce,* and *reproduction* on the board.

Ask Do you remember what these words have in common? *(They all refer to making or creating.)* Have students turn back to the Cell Division and Reproduction section in Lesson 1.

Ask Which word on this page means the same as *make* or *create*? *(reproduction)*

Ask Can I rephrase the question to read, **How does a cell reproduce?** *(yes)*

Ask So, how does a cell reproduce? Ask for volunteers to answer. *(A cell goes through cell division to produce a new cell through sexual or asexual reproduction.)* Have students write their answers in their notebooks.

CONSTRUCTED RESPONSE Paragraphs may vary. Explain the activity carefully. Be sure that students understand that the *Hints* should be used as a guide to help them answer the questions. Review how to write a paragraph.

For Question 1, write it on the board and read it aloud. Underline *cell growth* and *exchange.*

Say This question is about cell growth and exchange of materials. Remind students of the town analogy in Lesson 10.1.

Read the first hint.

Ask *Comes into* and *go out* means "exchange." Do you remember what materials come into and go out of a cell? Write *waste leaves the cell, food, oxygen and water go into the cell* on the board under the question.

Read the next hint.

Say In Lesson 1 we learned that as the cell grows, the ratio of surface area to volume gets smaller. Write *SA:V ratio gets smaller* on the board under the question.

Model writing the first one or two sentences using the hints on the board and the text *(Sample: The size of a cell affects its ability to exchange materials. Food, oxygen, and water come into a cell; waste goes out. The bigger a cell grows, the more food it needs and the more waste it produces.)*

10 CHECK Understanding

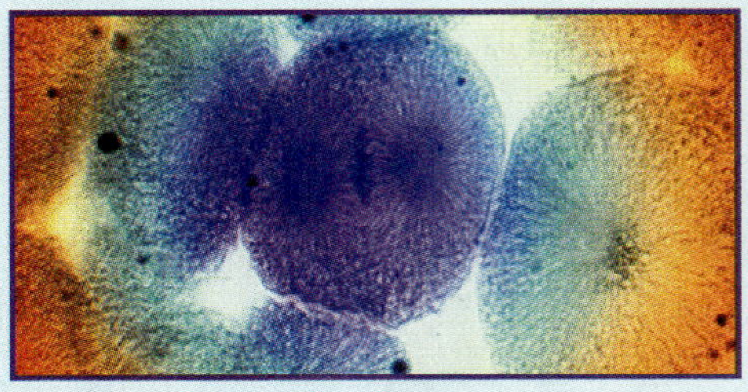

Assess the Big idea: Growth, Development, and Reproduction

Write an answer to the question below.
Q: How does a cell produce a new cell?

Constructed Response

Write an answer to each of the numbered questions below. The answer to each question should be one or two paragraphs. To help you begin, read the **Hints** below the questions.

1. **How does cell growth affect the ability of a cell to exchange materials with its environment?**

   Hint The exchange of materials is important to cells. Materials must come into a cell. Materials must go out of a cell.

   Hint As a cell grows, the relationship of its surface area to its volume changes.

2. **What are the main events of interphase?**

   Hint There are the three phases that make up interphase. List the phases in the order that they happen.

   Hint Look at the diagram called Eukaryotic Cell Cycle.

3. **In the future, how might stem cells be important for dealing with human health problems?**

   Hint Stem cells do things that most other cells cannot do.

Foundations for Learning Wrap-Up

Use the index cards you prepared when you started the chapter as a tool to help you organize your thoughts about cell growth and division.

Activity 1 Working with a partner, choose five index cards at random. Explain how each word is related to the other. If the words are not related, explain why. Take turn choosing cards.

Activity 2 Working in a small group, choose a group of cards whose terms are related. Construct a concept map by placing the cards on a table and describing how the words are connected. Make sure you use each of the cards at least once.

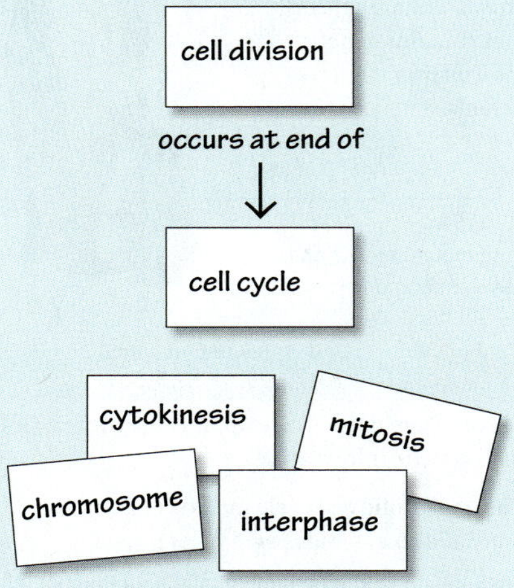

You may want to add a card for each phase of mitosis.

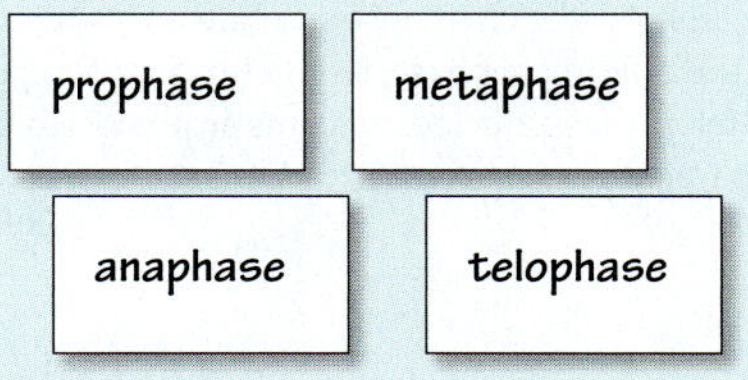

254 Check Understanding

For Question 2, pair students and have them use the hints and your modeling to answer the question. Have students share their paragraphs with the class.

For Question 3, have students write the paragraphs independently for homework or as an anchor activity.

Foundations for Learning Wrap-Up

Have students take out their index cards. Be sure that students have written all the vocabulary and definitions. Either assign groups or pair students, and explain the activities.

10.1 Cell Growth, Division, and Reproduction

Understand Key Concepts

1. A cell is limited in size by access to the genetic information contained in its
   a. cytoplasm.
   b. cell membrane.
   c. DNA.
   d. organelles.

Test-Taking Tip

Anticipate the Answer Think of your own answer before looking at the answer choices. Select the answer choice that comes closest to your own answer. In question 1, answer **c** is correct because genetic information is contained in DNA. All the other answers refer to cell parts. They are intended to distract you from the right answer.

2. Which material is it important for a cell to remove?
   a. water
   b. waste
   c. food
   d. oxygen
3. Identify one way in which asexual and sexual reproduction are alike and one way in which they are different.

Think Critically

4. **Relate Cause and Effect** In a rapidly changing environment, which organisms have an advantage—those that reproduce asexually or those that reproduce sexually? Explain.

10.2 The Process of Cell Division

Understand Key Concepts

5. For a cell to divide successfully, it must first
   a. form a new cell plate or cell membrane.
   b. divide its cytoplasm and organelles.
   c. increase its number of chromosomes.
   d. copy its genetic information.
6. The cell cycle in a eukaryotic cell is divided into
   a. prophase, anaphase, and telophase.
   b. interphase, metaphase, and cytokinesis
   c. interphase, mitosis, and cytokinesis.
   d. interphase, anaphase, and telophase.
7. What happens to a chromosome during cell division?
8. Identify the four phases of mitosis and give a brief description of what happens in each phase.

Think Critically

9. **Compare and Contrast** How does cell division in prokaryotes differ from cell division in eukaryotes?

10.3 Regulating the Cell Cycle

Understand Key Concepts

10. The timing in the cell cycle in eukaryotic cells is believed to be controlled by a group of closely related proteins known as
    a. chromatids.
    b. cyclins.
    c. centromeres.
    d. centrioles.
11. What type of regulatory proteins are activated when you break a bone?
    a. cyclins
    b. internal regulators
    c. bone cells
    d. growth factors
12. What effect do cancer cells have on the healthy cells that surround them?
13. What role does apoptosis play in a tadpole whose tail shrinks as it develops into a frog?

Think Critically

14. **Compare and Contrast** How are cancer cells different from other cells? How are they alike?

Assessment Tips

Answer Question 1 as a class. Then, read the tip. Point out that anticipating an answer can help you eliminate other choices. When you read *genetic information*, thinking of DNA helps you to get the correct answer.

Read Question 2. Ask what other words students know for *remove*. *(take out)* Review the tip, and ask students for the answer. Be sure that students understand that all the other answers are important for a cell to take in, not remove.

For Question 3, read the question aloud. Point out the words *alike* and *different*.

Ask What graphic organizer could you use to organize your ideas when you want to compare and contrast? *(Venn diagram)* Draw and fill in a Venn diagram. Then, model how to write two or three sentences as an answer.

For Question 4, review *asexual* and *sexual reproduction*. Have students write the answer in their notebooks.

ASSESSMENT

Answers

Lesson 10.1

UNDERSTAND KEY CONCEPTS

1. c **2.** b

3. Alike: Both produce offspring.
Different: Asexual involves one parent. Sexual involves two parents.

THINK CRITICALLY

4. those that reproduce sexually, because offspring are genetically different from their parents; The differences would allow some members of the species to adjust to the environmental changes.

Lesson 10.2

UNDERSTAND KEY CONCEPTS

5. d **6.** c

7. Every time a cell divides, each chromosome is replicated (copied). Each daughter cell gets a copy so that it has a complete and accurate set of DNA.

8. prophase, metaphase, anaphase, telophase
Prophase: chromosomes condense and the mitotic spindle starts to form. Metaphase: the duplicated chromosomes line up in the middle of the cell and the centromeres attach to the spindle. Anaphase: chromosomes, separate and move to opposite ends. Telophase: the chromosomes spread out and nuclear envelopes begin re-forming around genetic material.

THINK CRITICALLY

9. Prokaryotes divide by binary fission: the DNA replicates and each DNA molecule attaches to a different part of the cell membrane. The cell pinches and divides. In eukaryotes, cell division occurs in two stages: mitosis and cytokinesis.

Lesson 10.3

UNDERSTAND KEY CONCEPTS

10. b **11.** d

12. As cancer cells grow, they use the nutrients that healthy cells need. They can also prevent tissues and organs from functioning properly.

13. Apoptosis causes the cells in the tail to begin to go through steps that cause them to die.

THINK CRITICALLY

14. Cancer cells differ in that their growth is uncontrolled. But they use the same cell division process as normal cells.

ASSESSMENT

Lesson 10.4

UNDERSTAND KEY CONCEPTS

15. a

16. d

17. Differentiation is a process by which a cell becomes specialized. If cells didn't differentiate, all cells would be the same and no cell would have a specific function.

18. Scientists are trying to find ways to remove embryonic stem cells without damaging the embryo. Scientists are also trying to find ways to make an adult stem cell act like an embryonic stem cell.

THINK CRITICALLY

19. Harvesting skin cells would be much less difficult on a patient than harvesting bone marrow cells. Skin cells that are pluripotent could be injected into a damaged heart and forced to differentiate into heart muscle.

Connecting Concepts

20. In the bone. Because blood cells live only a short time, they need to be replaced. Bone marrow cells can develop into blood cells.

21. Because cancer cells have lost the ability to control growth and divide uncontrollably, they would be "long lived" and "can divide."

CHAPTER MYSTERY After students have read through the Chapter Mystery, discuss the topics of dedifferentiation and specialization.

Ask What is dedifferentiation? *(The process by which specialized cells lose their specialized characteristics.)*

Ask How did the process of dedifferentiation change the cells in the salamander's leg? *(The cells were no longer specialized.)*

Say Name some types of specialized cells that are in both your legs and the salamander's limb. *(Sample answers: muscle, nerve, skin, bone cells)*

Ask How does the regeneration of the salamander's limb differ from the healing of a broken human leg bone? *(For the salamander's limb, differentiation results in unspecialized cells that can then differentiate into whatever type of cells are needed to build a whole new limb. When a human bone heals, bone cells can repair bone cells, but they cannot dedifferentiate to form new types of tissue.)*

10 CHECK Understanding

10.4 Cell Differentiation

Understand Key Concepts

15. Which type of cell can differentiate into any kind of cell?

a. totipotent **c.** multipotent
b. pluripotent **d.** differentiated

16. An organism at an early stage of development is

a. a stem cell. **c.** a cytokinesis.
b. a chromosome. **d.** an embryo.

17. What is cell differentiation? How is it important to an organism's development?

18. Describe how scientists may help ease the ethical concerns about stem cell research.

Think Critically

19. **Relate Cause and Effect** Suppose researchers discover how to make skin stem cells pluripotent. How might this affect treatments for heart attack patients?

Connecting Concepts

Use Science Graphics

Use the data table to answer the questions 20 and 21.

Life Spans of Various Human Cells

Cell Type	Life Span	Cell Division
Red blood cells	<120 days	Cannot divide
Cardiac (heart) muscle	Long-lived	Cannot divide
Smooth muscle	Long-lived	Can divide
Neuron (nerve cell)	Long-lived	Most do not divide

20. **Compare and Contrast** Where would you expect to find active stem cells, in the bone where blood cells are produced or in the heart? Explain your thinking.

21. **Infer** Suppose cancer cells were added to the table. What would probably be written in the "Cell Division" column? Explain.

solve the CHAPTER MYSTERY

PET SHOP ACCIDENT

Julia checked the hurt salamander each day. About a month after the accident, Julia realized that a new leg was going to replace the lost one! Salamanders are one of a few vertebrates that can regenerate, or grow, a complete limb.

Look at the pictures that show how a new leg develops. Then answer the questions.

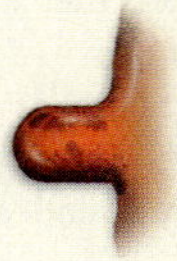

Week 1: Dedifferentiation Cells in the wounded area dedifferentiate. This means they are unable to differentiate. Cells can no longer perform their specialized jobs.

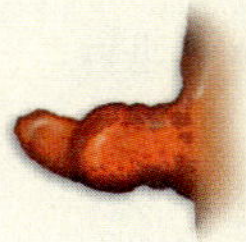

Week 3: Forming a web of cells The dedifferentiated cells move to the wounded area. There they form a tangled web of growing, undifferentiated cells.

Week 5: Redifferentiation The web of cells redifferentiate, again becoming differentiated. These new cells produce the new leg. The leg will continue to grow until it is full size.

1. **Relate Cause and Effect** Cells in the salamander's leg first had to dedifferentiate. Why was this a key part of growing a new leg?

2. **Classify** What type of stem cells do you think the tangled web of cells contain?

Untamed Science™ **Never Stop Exploring Your World.** Finding the solution to the pet shop mystery is only the beginning. Take a video field trip with the ecogeeks of Untamed Science to see where the mystery leads.

CHAPTER MYSTERY ANSWERS

1. Dedifferentiation needs to occur to provide the stem cells necessary to form a new limb.

2. The tangled web of cells contains pluripotent stem cells that could differentiate into many types of cells.

Untamed Science™ For more on cell differentiation, suggest students watch **Tails of Cell Division.**

Standardized Test Prep

Multiple Choice

1. Which statement is true regarding a cell's surface area-to-volume ratio?
   A As the size of a cell increases, its volume decreases.
   B As the size of a cell decreases, its volume increases.
   C Larger cells will have a greater surface area-to-volume ratio.
   D Smaller cells will have a greater surface area-to-volume ratio.

2. Which of the following is *not* an advantage of asexual reproduction?
   A simple and efficient
   B produces large number of offspring quickly
   C increases genetic diversity
   D requires one parent

3. At the beginning of cell division, a chromosome consists of two
   A centromeres. C chromatids.
   B centrioles. D spindles.

4. What regulates the timing of the cell cycle in eukaryotes?
   A chromosomes C nutrients
   B cyclins D DNA and RNA

5. The period between cell divisions is called
   A interphase. C G_3 phase.
   B prophase. D cytokinesis.

6. Which of the following is true about totipotent cells?
   A Totipotent cells develop in the inner cell mass.
   B Totipotent cells are differentiated cells.
   C Totipotent cells can differentiate into any type of cell and tissue.
   D Adult stem cells are totipotent cells.

7. A cell enters anaphase before all of its chromosomes have attached to the spindle. This may indicate that the cell is not responding to
   A internal regulators. C growth factors.
   B cyclins. D apoptosis.

Questions 8–10

The spindle fibers of a dividing cell were labeled with a fluorescent dye. At the beginning of anaphase, a laser beam was used to stop the dye from glowing on one side of the cell, thereby marking the fibers, as shown in the second diagram. The laser did not inhibit the normal function of the fibers.

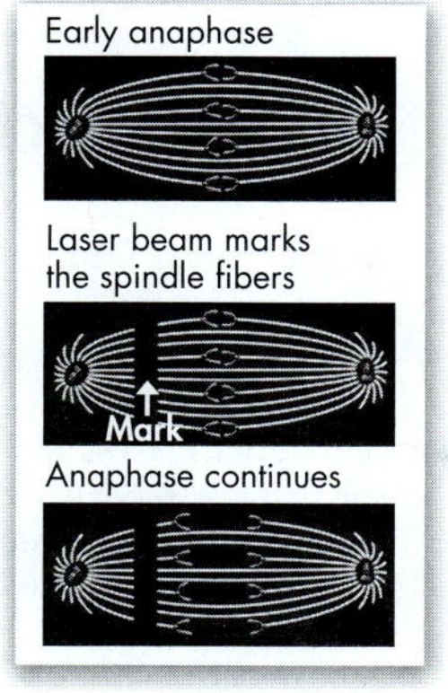

8. This experiment tests a hypothesis about
   A how chromosomes migrate during cell division.
   B how fluorescent dyes work in the cell.
   C the effect of lasers on cells.
   D why cells divide.

9. The diagram shows that the spindle fibers
   A shorten on the chromosome side of the mark.
   B lengthen on the chromosome side of the mark.
   C shorten on the centriole side of the mark.
   D lengthen on the centriole side of the mark.

10. A valid conclusion that can be drawn from this experiment is that the spindle fibers break down
    A at the centrioles.
    B in the presence of dye.
    C when marked by lasers.
    D when they are attached by chromosomes.

Open-Ended Response

11. Explain why careful regulation of the cell cycle is important to multicellular organisms.

If You Have Trouble With . . .

Question	1	2	3	4	5	6	7	8	9	10	11
See Lesson	10.1	10.1	10.2	10.3	10.2	10.4	10.3	10.2	10.2	10.2	10.3

ASSESSMENT

Before the Test

Review with students the Test-Taking Tip.

Point out the word *NOT* in Question 2.

Remind students that an open-ended response means that they'll have to write a paragraph to answer the question. Have them paraphrase the question in their notebooks, and remind them about the steps in writing a paragraph.

Answers

1. D
2. C
3. C
4. B
5. A
6. C
7. A
8. A
9. A
10. A
11. Careful regulation of the cell cycle ensures that each cell in a multicellular organism grows, divides, stops dividing, or dies appropriately. If cell cycle regulation goes awry, then diseases such as cancer can result.

Test-Taking Tip

INTERPRET VISUALS

When answering questions that contain a visual element, remind students to look carefully at the visual and read through any descriptive text that goes with it. Then, suggest they read and answer the test questions. For students who struggle to answer questions with visual prompts, suggest that they make a quick mental summary of what the visual is showing before reading the test questions.

Plan Ahead

Have students read through the Unit 3 Project a few days in advance of writing their comic books to give them time to think about the task. Suggest they review differences between prokaryotic and eukaryotic cells, as well as differences between plant and animal cells. You may want to have several "superhero" comic books on hand to give students examples of how they might portray their "superhero cell."

Materials colored pencils or pens, paper

Monitor the Project

Tell students the comic book should have a minimum of four pages, but may have as many as they can create in the given time period, as long as each page is well done. As they work on their comic books, walk around the room and ask students questions that will help them focus both on the cell being portrayed and the story line of the comic book.

Ask What kind of cell is your superhero, and what structures and organelles does that kind of cell include?

Ask How does your story line provide insight into the way your particular cell works?

Project Assessment

Make sure students use the rubric and reflection questions to assess their work. Then, use the rubric to assign a final score. Note that it is important to value the creativity of students' work as well as the content when you score their projects. If desired, talk with students about any differences between their self-assessment scores and your assigned score.

NATIONAL SCIENCE EDUCATION STANDARDS

UCP I, II, V

CONTENT C.1.a, C.1.b, C.1.c, C.1.d, C.1.e, C.1.f

INQUIRY A.2.a

Unit Project

Superhero Cell

Do you like reading comics? Have you ever designed a comic book of your own? Here's your chance! A high school teacher has contacted you asking for a comic book on cells and cell processes. She has told you that her students are just about to start studying cells and need a good introduction to the topic. You've been tasked with developing the story line and visuals that will provide the students with a basic understanding of cell structure and function. Remember that sometimes a picture can be worth a thousand words—so be creative!

Your Task Write a comic book about a "superhero cell" for an audience of high school students.

Be sure to

- incorporate important concepts and details about the structure and function of various organelles and cell processes.
- provide insight into the ways cells work and interact with their environment.
- be entertaining and creative.

Reflection Questions

1. **Score your project using the rubric below. What score did you give yourself?**
2. **What did you do well on this project?**
3. **What about your project needs improvement?**
4. **Exchange your comic book with a classmate and have him/her read it. What did your partner like about your comic book? What did he/she think could use improvement?**

Assessment Rubric

Score	Scientific Content	Quality of Comic Book
4	The comic book includes accurate details about the structures and functions of several organelles and cell processes. It provides exceptional insight into how a cell works and interacts with its environment.	The comic book is thoughtfully and creatively written and illustrated.
3	The comic book includes mostly accurate details about the structure and functions of organelles and cell processes. It provides good insight into how a cell works and interacts with its environment.	The comic book is well written and includes some creativity. Illustrations are clear.
2	The comic book includes a few details about the structure and functions of organelles and cell processes, with some inaccuracies. It provides some insight into how a cell works and interacts with its environment.	The comic book needs some edits and could use more creativity. Some parts of the story line and illustrations are difficult to follow.
1	The comic book includes vague and inaccurate information about the structure and functions of organelles and cell processes. It provides little insight into how a cell works and interacts with its environment.	The comic book needs significant edits and includes very little creativity. Story line and illustrations are unclear.

21st Century Skills

To be successful in the 21st century, students need skills and learning experiences that extend beyond subject area mastery. The Unit 3 Project helps students build the following 21st Century Skills: *Communication Skills; Creativity and Intellectual Curiosity; Interpersonal and Collaborative Skills; Self-Direction;* and *Accountability and Adaptability.*

FOCUS ON CRITICAL THINKING AND SYSTEMS THINKING Extend this Unit Project by having pairs of students work together to create a comic book that features a cell in a multicellular organism that is part of an organ or organ system, such as the digestive system or circulatory system. Explain that the story line should focus on how cells cooperate in a multicellular organism.

For more practice building 21st Century Skills, see The Chapter Mystery pages in **Study Workbook A.**

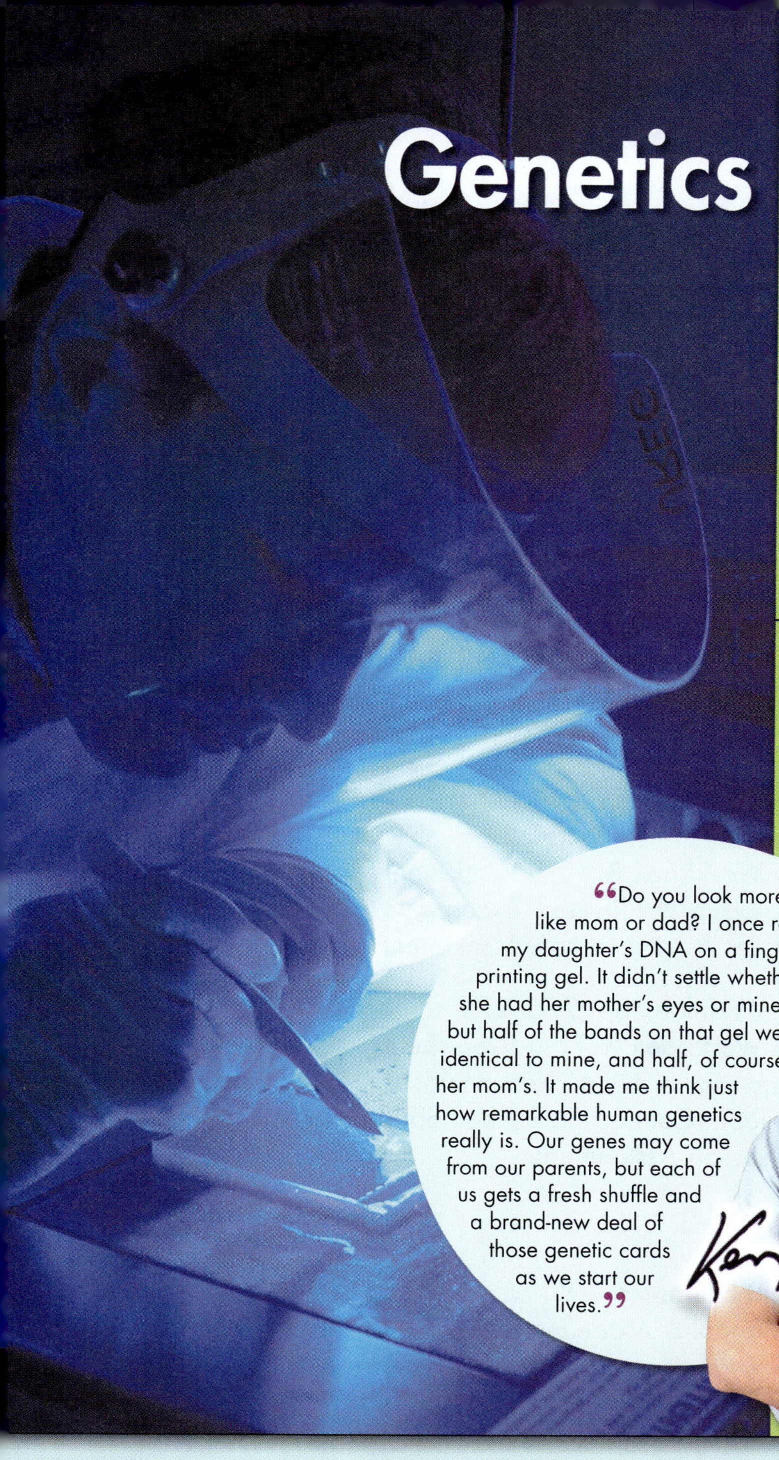

Genetics

UNIT 4

Chapters

INTRODUCE the **Big ideas**

- **Information and Heredity**
- **Cellular Basis of Life**
- **Science as a Way of Knowing**

"Do you look more like mom or dad? I once ran my daughter's DNA on a fingerprinting gel. It didn't settle whether she had her mother's eyes or mine, but half of the bands on that gel were identical to mine, and half, of course, to her mom's. It made me think just how remarkable human genetics really is. Our genes may come from our parents, but each of us gets a fresh shuffle and a brand-new deal of those genetic cards as we start our lives."

Ken Miller

259

UNIT 4

Dear Colleague,

A few months ago, I had the honor of speaking at a symposium in honor of Gregor Mendel. To prepare for my talk for the meeting, I read over Mendel's famous paper, a dry scientific report with the unassuming title of "Investigations of Plant Hybridization." As I turned the pages, I wondered if Mendel himself had any idea what he was starting. It would take decades, of course, but eventually his work led to a revolution in biology. Once, we naturalists were merely observers of life. Biologists today, of course, are active participants who study, shape, analyze, and even change living things.

What struck me most as I looked over Mendel's paper was not the way in which it laid out the basic principles of genetics. Mendel's great contribution, it seems to me, was his insistence that life itself could be studied, analyzed, and understood along systematic, rational lines. This attitude infuses every paragraph of his work, and it was picked up, almost unconsciously, by the scientists who rediscovered and extended that work at the beginning of the twentieth century.

Today, we have an opportunity to bring that sense of discovery to our students as never before. The rise of Mendelian genetics led to intense curiosity about the chemical nature of the gene. That, in turn, led to the identification of DNA as the genetic material, as well as the discovery of its double-helical structure. The result of all that curiosity, of course, is a new understanding of life that we can bring to all of our students. DNA carries the genetic code, and with it the fundamental instructions that operate our cells and interact with the environment to build our bodies. It's both our heritage and the legacy that we pass along to new generations. It's also something we share with every other living thing on this planet. That revolution in understanding may have begun with Gregor Mendel, but it hasn't stopped, even today. Indeed, the most important message we may be able to give our students is that the really interesting work is just beginning.

Ken Miller

CHAPTER PLANNER 11 Introduction to Genetics

Chapter Contents	NSES	Time	Core Resources
Chapter Preview			**Student Edition,** pp. 260–261 **Chapter Mystery,** p. 261 **Foundations for Learning,** p. 261
11.1 The Work of Gregor Mendel The Experiments of Gregor Mendel • Segregation	II, C.2.a, C.2.b, G.1, G.3	$^1/_2$ period $^1/_4$ block	**Student Edition,** pp. 262–265 **Study Workbook B** 11.1 Worksheets **Biology.com** 11.1 Lesson Assessment
11.2 Applying Mendel's Principles Probability and Punnett Squares • Independent Assortment • A Summary of Mendel's Principles	II, A.2.d, C.2.a, C.2.b, G.2	$1^1/_2$ periods $^3/_4$ block	**Student Edition,** pp. 266–270 **Build Connections,** p. 268 **Study Workbook B** 11.2 Worksheets **Biology.com** *InterActive Art:* Punnett Squares • 11.2 Lesson Assessment
11.3 Other Patterns of Inheritance Beyond Dominant and Recessive Alleles • Genes and the Environment	II, A.1.c, A.2.a, C.2.a, C.2.b, G.2	$^1/_2$ period $^1/_4$ block	**Student Edition,** pp. 271–274 **Inquiry Into Scientific Thinking,** p. 274 **Study Workbook B** 11.3 Worksheets **Biology.com** *Art Review:* Exceptions to Mendel's Principles • 11.3 Lesson Assessment
11.4 Meiosis Chromosome Number • Phases of Meiosis • Comparing Meiosis and Mitosis • Gene Linkage and Gene Maps	I, II, A.2.a, A.2.b, C.2.a, C.2.b	2 periods 1 block	**Student Edition,** pp. 275–279 **Study Workbook B** 11.4 Worksheets **Biology.com** *Art in Motion:* Meiosis • *Tutor Tube:* Connecting Punnett Squares to Meiosis • 11.4 Lesson Assessment
Chapter Pre-Lab	C.1.a, C.1.c, C.1.d	1 period $^1/_2$ block	**Student Edition,** p. 280 **Lab Manual B** *Modeling Meiosis*

Tools for ELL Instruction

ELL Handbook describes teaching strategies to make *Miller & Levine Biology: Foundation Edition* more accessible to ELL students.

Spanish Study Workbook is a Spanish translation of Study Workbook A.

Multilingual Glossary is the *Miller & Levine Biology* glossary translated into ten languages.

Tools to Challenge Students

Study Workbook A includes more challenging worksheets and activities such as Chapter Mystery activities that reinforce 21st Century Skills.

Lab Manual A labs are geared toward on-level students. The Lab Manual includes fifteen labs in addition to the Chapter Labs.

Biology.com Data Analysis activities offer students a way to practice data gathering and analyzing skills.

Visual Quizzes in the Assessment Resources Book are geared toward on-level students and use visuals from *Miller & Levine Biology.*

Additional Resources

Biology.com Untamed Science Video • Flash Cards

Study Workbook A 11.1 Worksheets
Spanish Study Workbook 11.1 Worksheets

Study Workbook A 11.2 Worksheets
Lab Manual B *Hands-On Activity:* How Are Dimples Inherited?
Spanish Study Workbook 11.2 Worksheets
Assessment Resources Book Visual Quiz

Study Workbook A 11.3 Worksheets
Lab Manual B *Data Analysis:* Human Blood Types
Spanish Study Workbook 11.3 Worksheets

Study Workbook A 11.4 Worksheets
Lab Manual B *Data Analysis:* Calculating Haploid and Diploid Numbers
Spanish Study Workbook 11.4 Worksheets
Biology.com *Data Analysis:* Gene Location and Crossing-Over
Assessment Resources Book Visual Quiz

Lab Manual A *Modeling Meiosis* • *Independent Assortment and Gene Linkage*

Chapter Review

Student Edition Chapter Summary, p. 281
Study Workbook B Chapter 11 Review
Transparencies, pp. 137–147
Biology.com Untamed Science Video • Editable Worksheets of Study Workbooks A and B and Lab Manuals A and B • Chapter 11 Flash Cards and Match It

Untamed Science DVD • Classroom Resources CD (includes lesson presentations and editable worksheets)

Chapter Assessment

Student Edition Check Understanding, pp. 282–285
Study Workbook B Chapter 11 Taking a Standardized Test
Assessment Resources Book Chapter 11 Test B
Biology.com Chapter 11 Assessment • Editable Worksheets of Chapter 11 Visual Quizzes* and Chapter 11 Tests A and B

Exam*View* *Assessment Suite* • Classroom Resources CD (includes lesson presentations and editable worksheets)

Time: 1 period, 1/2 block

* Visual Quizzes are based on visuals from *Miller & Levine Biology*. Review carefully to make sure they are appropriate for your class.

Pressed for Time?

Preview the Chapter Preview the table Mendel's Seven F_1 Crosses on Pea Plants and introduce students to the vocabulary in Lesson 11.2.

Cover the Chapter Quickly Have students read *The Experiments of Gregor Mendel* in Lesson 11.1, all of Lesson 11.2, and *Beyond Dominant and Recessive Alleles* in Lesson11.3. Assign *Chromosome Number* and the introductory paragraph of *Phases of Meiosis,* and go over the Meiosis visual in Lesson 11.4.

Assess Assign Questions 1–3 in the 11.1 Check Understanding, the entire 11.2 Check Understanding, Questions 1–4 in the 11.3 Check Understanding, and Questions 1 and 3 in the 11.4 Check Understanding. In the Chapter 11 Check Understanding, assign Questions 1, 3, 8, and 11.

Get the Picture

Have students look at the picture.

Say The puppies in a single litter do not all look alike.

Ask What can you infer about how the parents of this litter looked? *(Sample answers: The parents had the same physical features as the puppies. The coats of the parents may also have been different colors. One parent may have been black, and the other may have been yellow or brown.)*

Activate Prior Knowledge

Ask When you hear the word *trait*, what comes to mind? *(a physical characteristic)* Explain that a trait is a characteristic that an organism can pass on to its offspring through its genes.

Ask Can you give me an example of a trait that is passed from parent to offspring? *(Sample answers: hair color, height, freckles, dimples, shape of ears)*

Ask How do you think traits are passed from parent to offspring? *(Traits from each parent are passed to offspring during sexual reproduction. Some students may also know that each parent carries two copies of each gene.)*

Ask Why do you think siblings look alike but not the same? *(How siblings look depends on the combination of genes they get from their parents.)*

NATIONAL SCIENCE EDUCATION STANDARDS

UNIFYING CONCEPTS AND PROCESSES

I, II

CONTENT

C.1.a, C.1.c, C.1.d, C.2.a, C.2.b, G.1, G.2, G.3

INQUIRY

A.1.c, A.2.a, A.2.b, A.2.d

11 Introduction to Genetics

Big idea Information and Heredity

Q: How does biological information pass from one generation to another?

Labrador retrievers may be black, brown, or yellow. Each puppy in a litter can have a different color based on the genes it inherits.

BIOLOGY.com Search Chapter 11 GO • Flash Cards

260

UbD Connect to the Big Idea

Invite a volunteer to read aloud the chapter number and the chapter title. Read the Big Idea and question. Emphasize the question, **How does biological information pass from one generation to another?**

Set the Purpose

PREVIEW AND PREDICT Draw the graphic organizer on the board. Guide students to the INSIDE feature on the next page, and invite a volunteer to read each lesson number and title. Write the words *genetics, generation,* and *inheritance* on the board.

Ask Can you predict what the words *genetics* and *generation* have in common? What's the connection between *generation* and *inheritance*? Remind students to search for answers to the questions as they read.

CHAPTER 11

INSIDE:

- Untamed Science Video
- Chapter Mystery

CHAPTER MYSTERY

GREEN PARAKEETS

Susan's birthday was coming up. Parakeets make great pets, so Susan's parents wanted to give two birds to her as a birthday present. At the pet store, they chose two healthy green parakeets—one male and one female. They knew that green was Susan's favorite color.

Susan was happy about her birthday present. She fed the birds and kept their cage clean. A few weeks later, Susan found three small eggs in the birds' nest. When the eggs hatched, Susan was amazed. None of the chicks was green! One chick was white, one was blue, and one was yellow. Why weren't any of them green? What had happened to the green color of the birds' parents?

Read for Mystery Clues As you read this chapter, look for clues to help you figure out why the parakeet chicks were colored differently than their parents. Then solve the mystery at the end of the Chapter.

FOUNDATIONS for Learning

Before you read the chapter, write each vocabulary word on the open side of the Undercover Vocabulary Tool. As you read the chapter, fill in the definitions. You can use different color highlighters to show the different categories. At the end of the chapter are two activities that use the tool to help answer the question: How does information in cells pass from one generation to another?

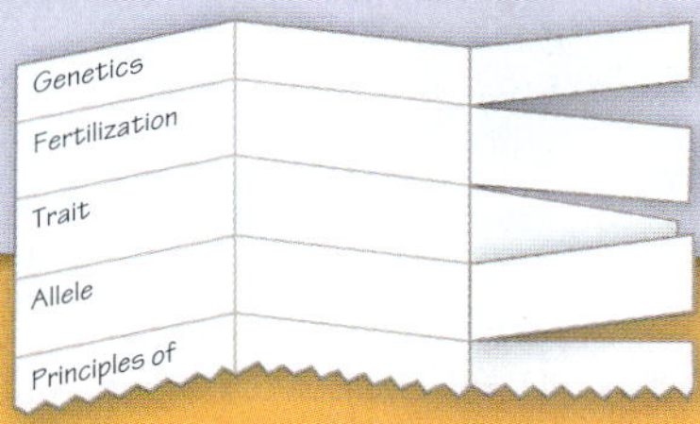

Chapter 11 Big Idea: Information and Heredity

Chapter 11: How does biological information pass from one generation to another?

- **11.1:** How does an organism pass its characteristics on to its offspring?
- **11.2:** How can you predict the outcome of a genetic cross?
- **11.3:** How can interactions between alleles, genes, and the environment affect an organism's traits?
- **11.4:** How does a cell divide to make cells with exactly half of the original cell's genetic information?

Foundations for Learning

Model how to make and use the Undercover Vocabulary Tool. Model one vocabulary word and its meaning so that students can see what they're going to do.

Ask Do you think having the vocabulary words and their definitions in one place is helpful? How do you think you'll use this tool? *(Answers may vary. Sample answers: I can review words and definitions anywhere. I can use it to quiz myself before a test.)*

Untamed Science Take a trip back in time with the ecogeeks in **Genetics Takes Root** to see Mendel in action.

CHAPTER MYSTERY Invite a volunteer to read aloud the Chapter Mystery. Have students brainstorm some reasons why none of the chicks were green, and list the reasons on the board. As a hint, suggest that students reread the Big Idea and the chapter question.

Have students tell which reason they think best explains why the parakeet chicks were colored differently than their parents. As they read through the chapter, have students gather information that shows that their prediction was correct or incorrect.

Digital Support

BIOLOGY.com Extend your reach by using the digital assets offered at **Biology.com.**

LESSON 11.1

Pre-Reading

Objectives

11.1.1 Describe Mendel's studies and conclusions about inheritance.

11.1.2 Describe what happens during segregation.

Preview the Pages

1. Preview the Build Vocabulary features throughout the lesson. Have students read each word and definition aloud, and then find the word in the text. Be sure students understand the difference between a gene and an allele.
2. Preview the visuals. Make sure students understand that each row of plants in the diagrams on the last two pages of this lesson represents a different generation.

Activate Prior Knowledge

Have students look at the diagram of the pea flower at the bottom of this page. Point out the male parts and female parts.

Ask What kind of reproduction is this plant going to undergo, sexual or asexual? *(sexual)* Why do you say *sexual*? *(pollen is transferred from the male parts to the female parts)* What's the name for when female and male reproductive cells join? *(fertilization)*

Build Understanding

TWO-COLUMN CHART Direct students to Workbook B to finish the two-column chart. Then pair students. Have partners take turns quizzing each other on the details and main ideas.

NATIONAL SCIENCE EDUCATION STANDARDS

UNIFYING CONCEPTS AND PROCESSES

II

CONTENT

C.2.a, C.2.b, G.1, G.3

For the corresponding lesson in *Miller & Levine Biology,* see pages 308–312.

11.1 The Work of Gregor Mendel

Key Questions

Where does an organism get its unique characteristics?

How are different forms of a gene passed to offspring?

BUILD Understanding

Two-Column Chart Before you read, draw a line down the center of a piece of paper. On the left side, write the main ideas in this lesson. On the right side, note the details and examples that support each of those ideas.

In Your Workbook Go to your workbook to learn more about making a two-column chart for Lesson 11.1.

The Experiments of Gregor Mendel

Every living thing has a set of characteristics inherited from its parent or parents. The scientific study of this biological inheritance is **genetics.** Genetics is the key to understanding what makes each organism unique.

The modern science of genetics was started by a monk named Gregor Mendel. Mendel was in charge of the garden in the monastery where he lived. In this garden, he did work that changed biology forever.

Mendel used ordinary garden peas in his studies. He used peas partly because peas are small and easy to grow. A single pea plant can make hundreds of offspring. Also, pea plants grow quickly and have many traits that are easy to see.

The Role of Fertilization When Mendel started his experiments, he knew that the male part of each flower makes pollen. Pollen contains the plant's male reproductive cells, called sperm. Mendel also knew that the female part of each flower makes reproductive cells called eggs. During sexual reproduction, male and female reproductive cells join in a process known as **fertilization** to make a new cell. In peas, this new cell develops into a tiny embryo wrapped inside a seed.

The male and female reproductive cells of pea plants are in every pea flower. Since pea flowers have both kinds of cells, a pea flower normally pollinates itself. A plant grown from a seed made by self-pollination has just one parent, because both reproductive cells came from the same plant.

True-Breeding Plants Mendel's monastery garden had several stocks of pea plants. These plants were "true-breeding." True-breeding plants are self-pollinating. They make offspring identical to themselves. The traits in each generation are the same. A **trait** is a specific characteristic, such as seed color or plant height, of an individual. A trait may vary from one individual to another. For example, one stock of Mendel's seeds made only tall plants. Another stock made only short ones. One stock made only green seeds. Another stock made only yellow seeds.

Pea Flower

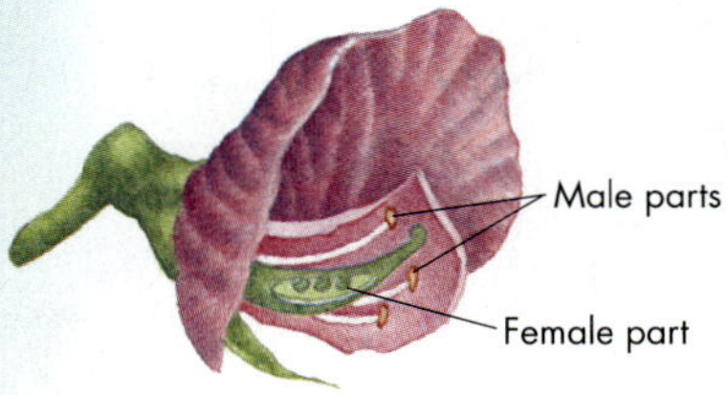

Flower Parts Flowers have male parts and female parts. During fertilization, pollen is transferred from the male parts to the female parts.

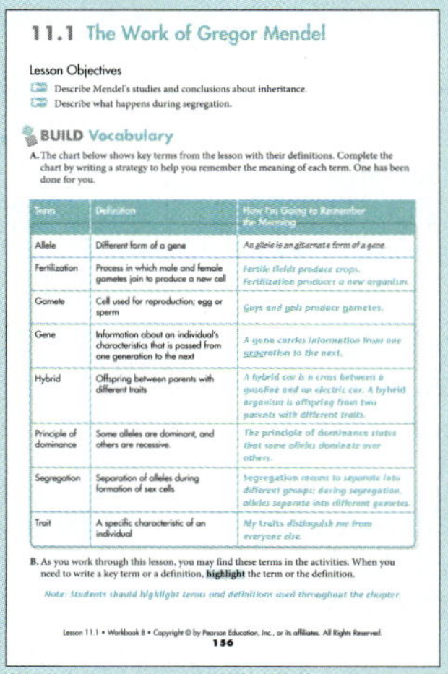

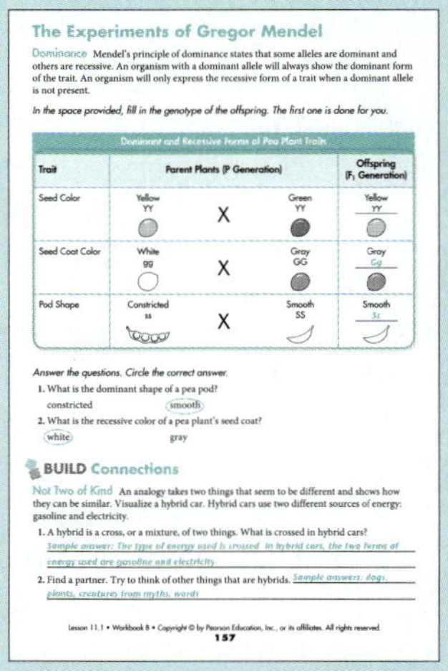

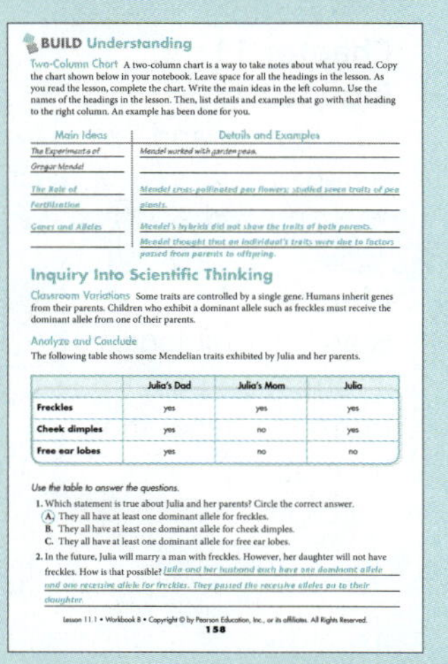

Crossbreeding Mendel wanted to learn more about how traits are passed from parent to offspring. So, he used pollen from one stock of plants to fertilize the female parts of flowers from other stocks of plants. This process is called cross pollination and produces a plant that has different parents.

Mendel studied seven different traits of pea plants. Each of the traits had two contrasting forms. Seed color is one example of a trait Mendel studied. The seeds were either green or yellow. Mendel wanted to see what would happen if he made a plant from two parents that had different traits. The offspring of crosses between parents with different traits are called hybrids.

Two Conclusions When a yellow-seed plant was crossed with a green-seed plant, all the offspring produced yellow seeds. It seemed that the green trait had disappeared. The same thing happened with each of the other traits. Mendel drew two conclusions from this.

▶ ***Passing Traits to Offspring*** First, Mendel found that an individual's characteristics are controlled by factors that are passed from parent to offspring. Today, scientists call these factors **genes** (jeenz). Each of the traits Mendel studied was controlled by a single gene that existed in two different forms. Each form of the gene controlled one form of a trait. For example, the gene for plant height came in one form that made tall plants and another form that made short plants. The different forms of the same gene are called **alleles** (uh LEELZ).

▶ ***Dominant and Recessive Alleles*** Mendel's second conclusion explains why some of the traits seemed to disappear in the offspring. This conclusion is called the **principle of dominance.** This principle states that some alleles are dominant and others are recessive (ree SESS iv). An organism with at least one dominant allele for a form of a trait will show that form of the trait. An organism with a recessive allele for a form of a trait will show that form only when the dominant allele for the trait is not there. In Mendel's experiments, the allele for tall plants was dominant, and the allele for short plants was recessive. Similarly, the allele for yellow seeds was dominant over the recessive allele for green seeds.

Key Question Where does an organism get its unique characteristics?
An individual's unique characteristics are determined by factors that are passed from parent to offspring.

BUILD Vocabulary

genetics
the scientific study of heredity

fertilization
the process in sexual reproduction in which male and female reproductive cells join to form a new cell

trait
a specific characteristic of an individual

gene
a factor that is passed from parent to offspring

allele
one of a number of different forms of a gene

principle of dominance
states that some alleles are dominant and others are recessive

WORD ORIGINS
The word *fertilization* is from the Latin word *fertilis,* which means "to be fruitful" or "to make offspring."

Mendel's F_1 Crosses This table shows three of the characteristics Mendel studied.

Mendel's F_1 Crosses on Pea Plants

	Seed Shape	Seed Color	Seed Coat
P	Round X Wrinkled	Yellow X Green	Gray X White
	↓	↓	↓
F_1	Round	Yellow	Gray

ELL Focus on ELL: Extend Language

INTERMEDIATE AND ADVANCED SPEAKERS Remind students that a synonym is a word that has the same or nearly the same meaning as that of another word. Have students read the definition of *allele* in the Build Vocabulary box and take special notice of the word *forms* in the definition. Then have them reread Passing Traits to Offspring.

Ask What words could be used in place of *forms* and still have the same meaning? *(varieties, kinds, sorts)*

During Reading

Active Reading

USE GRAPHIC ORGANIZERS Use a word web to help students learn the relationships among the many terms and the topic of genetics. Write the word *genetics* in the center circle, and *trait, allele,* and *gene* in the outer circles. Invite students to copy the circles and words in their notebooks.

Ask What's an allele? *(one form of a gene)* What's a gene? *(a gene determines a trait)* What's a trait? *(specific characteristic of an individual)*

Write the answers in the correct circles then draw arrows between the words to show the connections.

As you read the lesson, repeat the process by inviting students to suggest more words, give definitions, and show the relationships.

Build Vocabulary

WORD ORIGINS Point out that *fertilization* means "to be fruitful."

Ask Do you see any connection between *fertile* and *fruit*? *(Seeds are formed after fertilization. In flowering plants, the seeds are enclosed within a fruit.)*

USE CONTEXT CLUES Read aloud the first paragraph under the Crossbreeding heading.

Ask What are the two different stocks that resulted in cross breeding? *(pollen (males) from one plant and eggs (females) from another plant)*

SUMMARIZE Have students reread the Passing Traits to Offspring section. Ask them to write one or two sentences in their own words that summarize Mendel's first conclusion. *(Sample answer: An individual's traits are controlled by genes. Genes are passed from parent to offspring.)*

Ask What controls the appearance of a trait? *(a gene)* What is each form of a trait called? *(an allele)*

FIND THE MAIN IDEA Have students read Dominant and Recessive Alleles.

Ask What is the paragraph about? *(the principle of dominance)* What is the principle of dominance? *(Some alleles are dominant and others are recessive.)*

KEY QUESTION Invite a volunteer to read aloud the question and answer.

Ask What is a gene? *(factor that is passed from parent to offspring)*

Speed Bump

Ask How many parent plants take part in self-pollination? *(one)* How many parent plants are involved in cross breeding? *(two)*

LESSON 11.1

During Reading continued

Active Reading

INTERPRET DIAGRAMS Have students study the visual on this page.

Ask What do the labels *P*, F_1, and F_2 mean? *(P means the parental generation;* F_1 *means the first generation of offspring, and* F_2 *means the second generation of offspring.)*

Ask What do the X labels mean? *(The two plants were crossed with each other.)*

Point out that the plants below the arrows are the offspring of the two parent plants above the arrows.

MAKE CONNECTIONS Help students make connections between the text in the Recessive Traits Reappear section and the visual. Explain that the text focuses primarily on the bottom two parts of the diagram. Have volunteers read aloud the text one sentence at a time. After each sentence, discuss how the sentence relates to the diagram.

Ask Look at the F_2 row. Which plant shows the recessive allele and what percentage is it? *(short; 25%)*

Build Vocabulary

WORD ORIGINS Ask students how "a piece cut off" applies to the segregation of alleles in Mendel's crosses. *(The allele for shortness was cut off, or separated from, the allele for tallness.)*

Speed Bump

Ask What kinds of offspring are produced from a cross between two recessive short plants? Why? *(Only recessive short plants are produced because the parents had only alleles for shortness.)*

Ask What kinds of offspring are produced from a cross between a recessive short plant and a pure-breeding tall plant? Why? *(Only dominant tall plants are produced because each offspring had one dominant allele that prevents the recessive allele from being expressed.)*

Ask What kinds of offspring are produced from a cross between the offspring of the previous cross? (In other words, what kinds of offspring are found in the F_2 generation?) *(both tall and short)* Why? *(If two copies of the recessive allele end up in the same plant, that plant will be short. If the offspring plant ends up with even one copy of the dominant allele, the plant will be tall.)*

BUILD Vocabulary

segregation
the separation of alleles

gametes
sex cells

WORD ORIGINS
Like the word *segment*, the word *segregate* comes from the Latin root word *segmentum*, which means "a piece cut off."

Segregation

When doing genetic crosses, we call each original pair of plants the P, or parental, generation. Their offspring are called the F_1 generation. Mendel had a question about his F_1 generation of pea plants. Had the recessive alleles simply disappeared, or were they still in the new plants? To find out, he had all seven kinds of F_1 hybrids self-pollinate. The offspring of an F_1 cross are called the F_2 generation. So, Mendel crossed the F_1 generation with itself to make the F_2 offspring.

Recessive Traits Reappear Mendel found that the traits controlled by the recessive alleles reappeared in the second generation. About one fourth of the F_2 plants showed the trait controlled by the recessive allele. Why did the recessive alleles seem to disappear in the F_1 generation and then reappear in the F_2 generation?

Explaining the F_1 Cross Mendel figured out that only the dominant alleles determined the traits of the F_1 generation. However, the trait controlled by the recessive allele did show up in some of the F_2 second generation of plants. So, Mendel realized that the allele for shortness had separated from the allele for tallness. When did this separation, or **segregation,** of alleles happen? Mendel thought that the alleles for tallness and shortness in the F_1 plants must have segregated from each other during the formation of the gametes. **Gametes** (GAM eetz) are the sperm and egg cells that combine during fertilization.

Making Gametes Was Mendel right? Let's assume that the tall F_1 plants inherited two alleles controlling their height. They inherited an allele for tallness from their tall parent and an allele for shortness from their short parent. Because the allele for tallness is dominant, all the F_1 plants are tall. When each parent makes gametes, the alleles for each gene segregate from one another. Therefore, each gamete carries only one allele for each gene. So, each F_1 plant makes two kinds of gametes. Some have the tall allele, and some have the short allele.

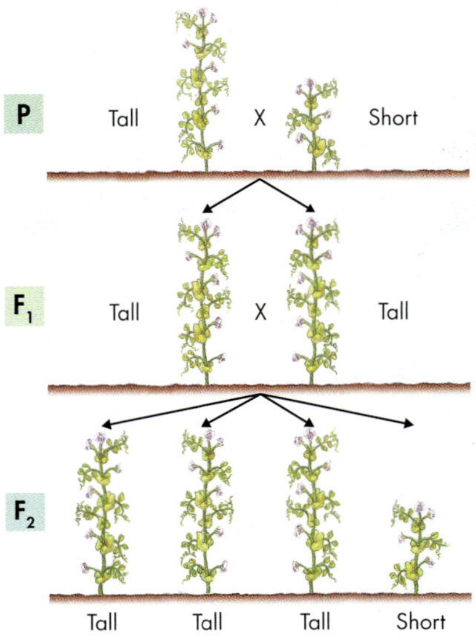

Results of the F_1 Cross Recessive traits reappeared in the second generation when the first generation of plants were allowed to self-pollinate.

Hands-On Learning

BEAN CROSSES Give each student a white bean *(r)* and a red bean *(R)* and tell them they'll be doing a Mendelian cross. Tell them that the letters represent alleles. The original beans are "true-breeding." The white bean represents a white-flowering plant. The red bean represents a red-flowering plant. The first generation will have the allele combination *Rr* and red flowers. Have students randomly "pollinate" with ten other students. To pollinate, one student hides one bean in each hand. The partner chooses one hand, which becomes the allele from one parent. Partners switch roles and the second bean is the allele from the second parent. Students record the color for each pollination. Point out that *rr* represents white-flowering plants and *Rr* and *RR* represent red-flowering plants. Have students tally the results and determine the ratio of white-flowering plants to red-flowering plants.

Look at the segregation diagram to see how alleles separate during gamete formation and then pair up again in the F_2 generation. A capital letter shows a dominant allele. A lowercase letter shows a recessive allele. Now we can see why the recessive trait for height, *t*, reappeared in Mendel's F_2 generation. Each F_1 plant in Mendel's cross made two kinds of gametes. Some had the allele for tallness, and some had the allele for shortness. Whenever a gamete that carried the *t* allele paired with another gamete that carried the *t* allele to make an F_2 plant, that F_2 plant was short. Every time a gamete carrying the *T* allele paired with another gamete carrying either the *T* or *t* allele, they made a tall plant. In other words, the F_2 generation had new combinations of alleles.

Key Question How are different forms of a gene passed to offspring?
When gametes are made, the alleles for each trait separate from each other. That way each gamete carries only one allele from each gene.

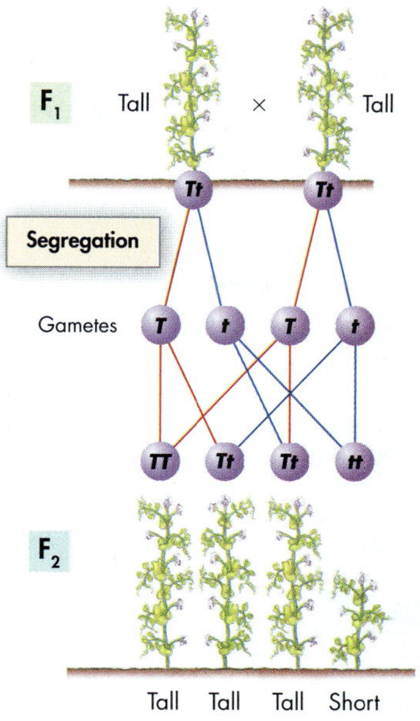

When gametes are made, the two alleles of each gene separate so that each gamete carries only a single copy of each gene. Each F_1 plant makes two types of gametes: those with an allele for tallness and those with an allele for shortness.

CHECK Understanding

Apply Vocabulary

Use the highlighted words from the lesson to complete each sentence.

1. The inherited factors that determine traits are called _______.
2. _______ is the process of male and female gametes joining.

Critical Thinking

3. **Compare and Contrast** How are dominant and recessive alleles alike and different?
4. **Write to Learn** Answer the first clue of the mystery. Write a paragraph that includes the terms *gene*, *allele*, and *trait*.

MYSTERY CLUE

Parakeets come in four colors: white, green, blue, and yellow. How many alleles might there be for feather color? (Hint: See p. 263.)

BIOLOGY.com Search Lesson 11.1 GO • Lesson Assessment

Check Understanding

1. genes
2. Fertilization
3. Both dominant and recessive alleles are genes that determine traits. However, the form of the trait determined by recessive alleles is seen only if no copy of the dominant allele is present.
4. Sample answer: The color of a feather is a trait. Traits are determined by genes. Each parakeet carries two alleles for each gene. Because four feather colors are possible, there are probably four different possible alleles.

LESSON 11.1

Active Reading

USE VISUALS Students may be confused by the many crossed lines in the visual on this page.

Ask What do the red lines represent? *(the dominant alleles)* What is the dominant allele? *(T, tall)* What do the blue lines represent? *(the recessive alleles)* What is the recessive allele? *(t, short)*

Say Look at the short plant in the F_2 generation and the *tt* directly above it. Now work backwards.

Ask How many lines and which color are attached to *tt*? *(2, blue)* Have students trace the lines back to the parent plants. Repeat for each F_2 plant, so that students see that one allele comes from each parent.

KEY QUESTION Read aloud the question and answer.

Ask What is a gamete? *(an egg cell or a sperm cell)*

Ask When do a parent's alleles separate from each other? *(when gametes are made)*

Post Reading

ASSESSMENT TIPS Have students review the highlighted vocabulary words before they begin the assessment.

For Question 3, point out that a graphic organizer can help them answer compare/contrast questions.

MYSTERY CLUE Have students reread the section Passing Traits to Offspring. Ask students to predict how many alleles there might be for feather color. Guide them to conclude that there are more than two alleles since there are four possible colors.

Wrap-Up Activity

TRAIT TRENDS On the board, draw a table with four columns. Title the first column *Trait 1*. The row entries are *Dimples, Free earlobes,* and *Can roll tongue*. Title the third column *Trait 2*. The row entries are *No dimples, Attached earlobes,* and *Can't roll tongue*. Then review what each trait looks like. Explain that Trait 1 traits are dominant, and Trait 2 traits are recessive. Have students copy the table in their notebooks. Have pairs of students enter data for each other by adding tic marks to the table in columns 2 or 4 as appropriate. Record the total number of students in the class. Challenge students to identify any trends in the class. *(Answers may vary.)*

Have students write a paragraph explaining why people look so different from one another. *(Students' paragraphs should indicate the large number of combinations that can result when many traits are considered.)*

LESSON 11.2

Pre-Reading

Objectives

11.2.1 Explain how geneticists use the principles of probability to make Punnett squares.

11.2.2 Explain the principle of independent assortment.

11.2.3 Explain how Mendel's principles apply to all organisms.

Preview the Pages

1. Review how to multiply fractions. Remind students of the difference between fractions and ratios. If someone has three pennies and one nickel, then 3/4 of the coins are pennies, but the ratio of pennies to nickels is 3:1.
2. Review the pronunciation of *homozygous*, *heterozygous*, *phenotype*, and *genotype*. Point out the prefixes. Have students guess at the meanings based on the prefix.
3. Review probability. Explain that probability is a way of estimating things that are not absolutely known. A "50% chance of rain" means there is a 1:1 probability of rain.
4. Preview the How to Make a Punnett Square visual and the visuals on page 269.

Build Understanding

PREVIEW VISUALS Have students preview the visual on p. 267. Then, direct students to Workbook B where they will find a sample KWL chart. Students should complete the chart when they have finished reading the lesson.

NATIONAL SCIENCE EDUCATION STANDARDS

UNIFYING CONCEPTS AND PROCESSES

II

CONTENT

C.2.a, C.2.b, G.2

INQUIRY

A.2.d

For the corresponding lesson in *Miller & Levine Biology*, see pages 313–318.

11.2 Applying Mendel's Principles

Key Questions

How can we use probability to predict traits?

How do alleles segregate when more than one gene is involved?

What did Mendel contribute to what we know about genetics?

BUILD Understanding

Previewing Visuals Before you read, look at one of the figures in the lesson. Try to understand the purpose of that figure. As you read, compare your original impression to what you learn about the figure in the text. After you read, revise your statement if it was incorrect. Write a new statement if needed.

In Your Workbook Go to your workbook to learn more about previewing visuals. Complete the preview statement for Lesson 11.2.

Probability and Punnett Squares

We cannot always know exactly how a genetic cross will turn out. However, by carefully studying what happens in a cross, Mendel began to figure out how genetic information is inherited. This information can be used to find the possible outcomes of a genetic cross.

Whenever Mendel performed a cross with pea plants, he carefully identified and counted the offspring. So he had plenty of data to analyze. He noticed trends. For example, whenever he crossed two plants that were hybrids for stem height (*Tt*), about three fourths of the offspring were tall and about one fourth were short.

Mendel figured out that the principles of probability could be used to explain these results of his genetic crosses. **Probability** (prah buh BILL uh tee) is the likelihood that a particular event will happen. For example, think about flipping a coin. There are only two possible outcomes of a coin flip. The coin can land either heads up or tails up. The chance, or probability, of either outcome is the same. So, the probability that a single coin flip will land heads up is 1 chance in 2. This can be described as a probability of 1/2, or 50 percent.

If you flip a coin three times in a row, what is the probability that it will land heads up every time? Each coin flip is a separate event, with a 1/2 probability of the coin landing heads up. So, the probability of flipping three heads in a row is:

$$1/2 \times 1/2 \times 1/2 = 1/8$$

As you can see, you have 1 chance in 8 of flipping heads three times in a row. Remember: Past coin flips do not affect future coin flips. You may have just flipped 3 heads in a row. Those flips do not change the probability of the next flip. The probability for that flip is still 1/2.

Using Segregation to Predict Outcomes The same type of probability used to predict coin flips can also be used to help understand the outcomes of genetic crosses.

Probability When you flip a coin, the chance that the coin will land heads up is 1/2, or 50 percent.

11.2 Applying Mendel's Principles

Lesson Objectives

- Explain how geneticists use the principles of probability to make Punnett squares.
- Explain the principle of independent assortment.
- Explain how Mendel's principles apply to all organisms.

BUILD Vocabulary

A. The chart below shows key terms from the lesson with their definitions. Complete the chart by writing a strategy to help you remember the meaning of each term. One has been done for you.

Term	Definition	How I'm Going to Remember the Meaning
Genotype	Genetic makeup	*Genotype is an organism's set of genes.*
Heterozygous	Organism that has different alleles for a gene	*Hetero- means "different." Heterozygous alleles are different.*
Homozygous	Organism that has two identical alleles for a gene	*Homo- means "same." Homozygous alleles are the same.*
Independent assortment	Genes for different traits segregate independently when gametes are formed	*Independent means "not controlled." When gametes form, the distribution of alleles for one gene is not controlled by the distribution of alleles for another gene.*
Phenotype	Physical traits	*Phenotype is an organism's physical traits.*
Probability	The likelihood that a particular event will occur	*Probability has to do with chance. Probability is the chance that an event will happen.*
Punnett square	Diagram that can be used to predict the offspring of a genetic cross	*Punnett square can be used to predict offspring.*

B. As you work through this lesson, you may find these terms in the activities. When you need to write a key term or a definition, **highlight** the term or the definition.

BUILD Understanding

Preview Visuals You can use a KWL chart to preview a visual. Study the figure of Segregation and Probability in your textbook. Write the information that you already know about it in the left-hand column. In the middle column, write something you want to know about the figure. After reading the section, write two things that you learned.

K (What I know)	W (What I want to know)	L (What I learned)
The two plants at the top of the figure are peas.	*Sample answer: What are the circles in the center of the picture?*	*Sample answer: The circles in the center of the picture represent alleles of traits.*

Probability and Punnett Squares

Follow the directions.

1. Table A lists some characteristic of pea plants. Write *Ph* in the right column if the characteristic describes a phenotype. Write *Ge* if the characteristic is a genotype.
2. Table B lists some genotypes of pea plants. Write *He* in the right column if the genotype is heterozygous. Write *Ho* if the genotype is homozygous.

Table A

Characteristic	Phenotype or Genotype?
tall	*Ph*
short	*Ph*
Tt	*Ge*
TT	*Ge*
yellow seed color	*Ph*
yy	*Ge*
Yy	*Ge*
rr	*Ge*

Table B

Genotype	Heterozygous or Homozygous?
RR	*Ho*
tt	*Ho*
Rr	*He*
Tt	*He*
YY	*Ho*
Yy	*He*
Gg	*He*
GG	*Ho*
gg	*Ho*

Probability and Punnett Squares (continued)

In the space provided, fill in the missing genotypes. Write Tt *or* tt. *In this example, T = tall and t = short.*

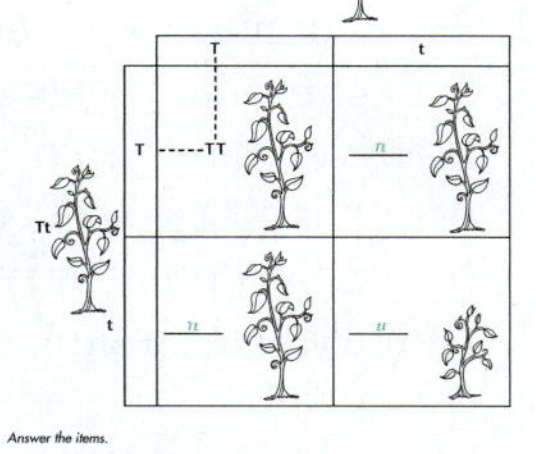

Answer the items.

1. Write the phenotypes of the three genotypes shown above.
   TT *tall*
   tt *short*
   Tt *tall*
2. If two heterozygous plants create four offspring, how many do you predict would be tall? How many do you predict would be short? *Sample answer: Three offspring will be tall and one offspring will be short.*

When alleles segregate to form gametes, there are only two possible outcomes, just like when you flip a coin. Since a pea plant has two alleles for every gene, the probability that a gamete will get either allele is 1/2. If two pea plants that have one of each kind of allele (*Tt*) are crossed, what is the probability that an offspring plant will be short? Remember that to be short, a plant needs two *t* alleles. The chance of a gamete getting either allele is 1/2. Two gametes are in each cross. So, the probability of both gametes creating a *tt* plant is $1/2 \times 1/2 = 1/4$.

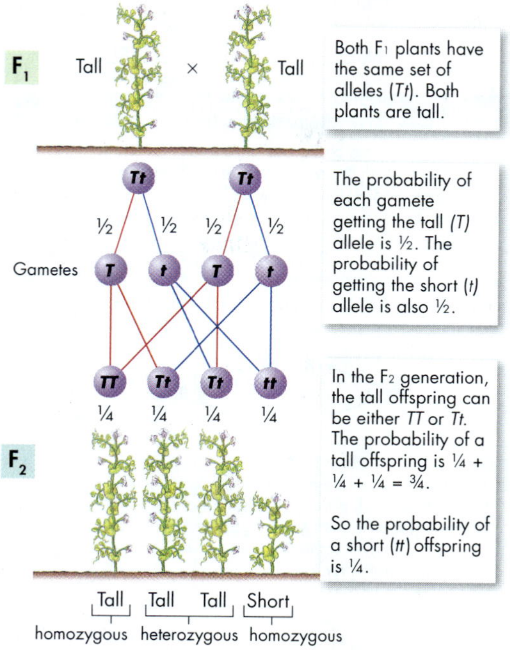

Organisms with the same characteristics can have different combinations of alleles. For example, the tall pea plants could be either *TT* or *Tt*. If an organism has two identical alleles of a gene, then the organism is **homozygous** (hoh moh ZY gus). The tall (*TT*) plants are homozygous because they have two identical alleles (*TT*) for a gene. The short plants are also homozygous because they have two identical alleles (*tt*) for a gene. If an organism has two different alleles for the same gene, then the organism is **heterozygous** (het ur oh ZY gus). The parents in this cross are heterozygous because they have two different alleles (*Tt*) for that gene.

Genotype and Phenotype How many different combinations of the alleles *t* and *T* are possible? Three. The options are *Tt*, *TT*, and *tt*. How many different forms of the plant are possible? Two. The options are tall and short. Plants with the combinations *TT* and *Tt* are tall. Plants with the combination *tt* are short. Mendel noticed that all of the tall pea plants had the same **phenotype** (FEE nuh typ), or physical traits. They did not, however, have the same **genotype** (JEE nuh typ), or genetic makeup.

Using Punnett Squares One of the best ways to predict the outcome of a genetic cross is by drawing a simple diagram known as a Punnett square. Making a Punnett square is easy. You begin with a square. Then write all of the alleles that could be in the gametes from one parent along the top of the square. Then, write all possible alleles in the gametes from the other parent along the left side. Next, write every possible combination of alleles into the boxes inside the square. The next page shows instructions for making a Punnett square.

Key Question How can we use probability to predict traits?
Punnett squares use probability to predict combinations of alleles in a genetic cross.

BUILD Vocabulary

probability
the likelihood that a particular event will occur

homozygous
having two identical alleles for a particular gene

heterozygous
having two different alleles for a particular gene

phenotype
the physical characteristics of an organism

genotype
the genetic makeup of an organism

PREFIXES
The prefix *hetero-* means "different." The two alleles in a heterozygous organism are different from one another.

Hands-On Learning

COIN CROSSES Students can use coins to predict the color of fur for the offspring of two cats. Both parents have brown fur *(Bb)*. The allele for brown fur *B* is dominant. The allele for white fur *b* is recessive. **Say** Put a small piece of masking tape on each side of two quarters (or large buttons). Then write *B* on one side and *b* on the other. Toss both coins ten times. Write the letter combinations that you get from each toss.

Ask

1. How many times did you get the *bb* combination? *(Students should get* bb *about 1:4 or 25% of the time.)*
2. What's the probability that the next toss will show *bb*? *(1:4 or 25%)*
3. What's the probability that the cats' offspring will have white fur? *(1:4; brown fur is dominant so has a genotype of* Bb *or* BB*, and white fur will show up in a genotype* bb*.)*

LESSON 11.2

During Reading

Active Reading

USE DIAGRAMS Make sure students understand the difference between *phenotype* and *genotype*. Help students remember the difference by pointing out that *physical* and *phenotype* begin with *ph*. Explain that *genotype* shares the root *gen* with *genetic*. Have students look at the visual on this page and note that tall plants have one phenotype but two possible genotypes.

Ask What are the two possible genotypes of a tall plant? *(TT and Tt)*

Ask What is the phenotype of a plant that has two alleles for shortness? *(short)*

USE CONTEXT CLUES Have students silently reread the Using Segregation to Predict Outcomes section. Have them locate and read aloud the phrases that identify a homozygous organism *("two identical alleles of a gene")* and a heterozygous organism *("two different alleles for the same gene")*. Then, ask students to read the paragraph under the heading Genotype and Phenotype. Have them explain which phenotype is presented by a heterozygous genotype. *(An organism with a heterozygous genotype will present the dominant phenotype.)*

Build Vocabulary

PREFIXES Tell students that the prefix *hetero-* means "different." Have them locate the bold-faced vocabulary word *heterozygous* and its definition in the text.

Ask What is different in a heterozygous individual? *(The individual has two different alleles for a trait.)*

Ask What prefix means the opposite of *hetero-*? What does it mean? *(The prefix* homo- *means "the same.")*

KEY QUESTION Invite a volunteer to read aloud the question and answer.

Ask Does a Punnett square determine the genotype of each offspring from a cross? Explain. *(No, it just predicts the probability of offspring having a certain genotype.)*

Speed Bump

Ask Is it possible for two individuals to have the same phenotype but different genotypes? *(Yes, both* TT *and* Tt *individuals have a tall phenotype.)*

Ask A snapdragon plant with white flowers has an *rr* genotype. What kind of gametes can this plant make? *(only* r *gametes)*

Build Connections

VISUAL SUMMARY Walk students through the one-factor-cross column first. Make sure they understand how the alleles combine. Point out that combinations are simply pairings of alleles from a particular row and column. Have students answer the following questions about the one-factor cross.

Ask What alleles can each parent pass on to offspring? *(Each parent has one* B *allele and one* b *allele.)*

Ask Why does the Punnett square in step 3 have four squares? *(Four pairings of alleles from the parents are possible.)*

Next, call on volunteers to explain each step in the two-factor cross. Make sure students understand that each parent's genotype includes two genes and that gametes get only one allele for each gene. Tell students that one way to double-check their work in step 4 is to make sure the letter above each column appears in the cells below it. Similarly, the letter to the left of the rows must appear in each cell in that row. Have students answer the following questions about the two-factor cross.

Ask How many alleles for height does each parent have? *(two;* T, *and* t*)*

Ask How many alleles for pod color does each parent have? *(two;* G, *and* g*)*

Ask How many different kinds of gametes can each parent produce? *(four;* TG, Tg, tG, *and* tg*)*

Ask Does the choice of allele for height that goes into a gamete affect the choice of allele for color that goes into that gamete? *(No, no matter which allele for height goes into a gamete, there are still two choices for color:* G *or* g*.)*

Give the class practice problems in making and completing both types of Punnett squares.

BUILD Connections

HOW TO MAKE A PUNNETT SQUARE

One-Factor Cross

1 Start With the Parents

Write the genotypes of the two organisms that will serve as parents in a cross. In this example we will cross a male and female osprey, or fish hawk, that are heterozygous for large beaks. They each have genotypes of *Bb*.

Bb* and *Bb

2 Figure Out the Gametes

Determine what alleles would be found in all of the possible gametes that each parent could produce.

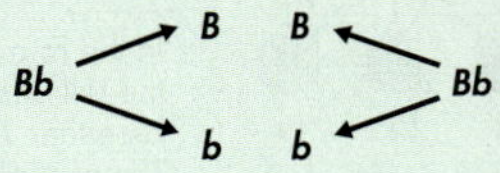

3 Line Them Up

Draw a table with enough squares for each pair of gametes from each parent. In this case, each parent can make two different types of gametes, *B* and *b*. Enter the genotypes of the gametes produced by both parents on the top and left sides of the table.

	B	b
B		
b		

4 Write Out the New Genotypes

Fill in the table by combining the gametes' genotypes.

	B	b
B		Bb
b	Bb	

	B	b
B	BB	Bb
b	Bb	bb

5 Figure Out the Results

Determine the genotype and phenotype of each offspring. Calculate the percentage of each. In this example, ¾ of the chicks will have large beaks, but only ½ will be heterozygous for this trait (*Bb*).

	B	b
B	BB	Bb
b	Bb	bb

Two-Factor Cross

1 Start With the Parents

In this example we will cross two pea plants that are heterozygous for size (tall and short alleles) and pod color (green and yellow alleles). The genotypes of the two parents are *TtGg* and *TtGg*.

TtGg* and *TtGg

2 Figure Out the Gametes

Determine what alleles would be found in all of the possible gametes that each parent could produce.

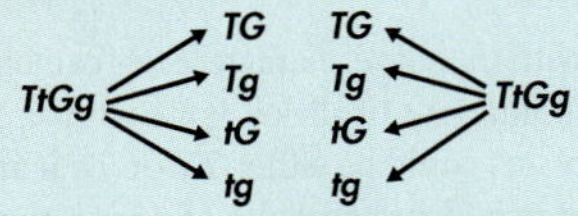

3 Line Them Up

In this case, each parent can make 4 different types of gametes, so the table needs to be 4 rows by 4 columns, or 16 squares.

	TG	tG	Tg	tg
TG				
tG				
Tg				
tg				

4 Write Out the New Genotypes

Fill in the table by combining the gametes' genotypes.

	TG	tG	Tg	tg
TG				
tG			TtGg	
Tg				
tg		ttGg		

	TG	tG	Tg	tg
TG	TTGG	TtGG	TTGg	TtGg
tG	TtGG	ttGG	TtGg	ttGg
Tg	TTGg	TtGg	TTgg	Ttgg
tg	TtGg	ttGg	Ttgg	ttgg

5 Figure Out the Results

In this example, the color of the squares represents pod color. Alleles written in black indicate short plants, while alleles written in red indicate tall plants.

	TG	tG	Tg	tg
TG	TTGG	TtGG	TTGg	TtGg
tG	TtGG	ttGG	TtGg	ttGg
Tg	TTGg	TtGg	TTgg	Ttgg
tg	TtGg	ttGg	Ttgg	ttgg

ELL Focus on ELL: Build Background

BEGINNING SPEAKERS As you describe each step in making a Punnett square, use vocabulary words as often as possible. Pair the vocabulary words *homozygous* or *heterozygous* with the vocabulary words *dominant* or *recessive* to describe the genotype of each individual in the Punnett square. For example, in step 4 of the one-factor cross, point out that *BB* is homozygous dominant, *bb* is homozygous recessive, and *Bb* is heterozygous. Remind students that the words *dominant* or *recessive* are not used when describing a heterozygous individual. Phrase questions so that students use vocabulary words in their answers.

Independent Assortment

Mendel wondered if alleles for one trait affected the alleles for another trait. For example, does the gene that determines the shape of a seed have anything to do with the gene for seed color? To find out, Mendel set up a cross that enabled him to study two different genes at the same time.

Getting the Seeds To get the seeds he needed, Mendel crossed true-breeding plants that made only round yellow peas with plants that produced only wrinkled green peas. The round yellow peas had the genotype *RRYY*. The wrinkled green peas had the genotype *rryy*. All of the F_1 offspring from this cross had round yellow peas with the genotype *RrYy*. This result showed that the alleles for yellow and round peas are dominant.

Showing Independent Assortment In the second part of this experiment, Mendel crossed two of these new *RrYy* plants from the F_1 generation to make the F_2 generation offspring. Remember that each plant was formed by the fusion of a gamete carrying the dominant *RY* alleles with another gamete carrying the recessive *ry* alleles. In Mendel's experiment, the F_2 plants made 556 seeds. Mendel sorted the seeds by trait. He noticed that 315 of the seeds were round and yellow, just like one parent. He saw that 32 seeds were wrinkled and green, just like the other parent. However, he also saw that 209 seeds had other combinations of phenotypes. Those included round and green peas and wrinkled and yellow peas. Those peas had to have allele combinations that were not found in either parent. This outcome showed that the alleles for seed shape segregated independently of those for seed color. Or to put it another way, genes that segregate independently do not influence each other's inheritance.

Mendel had discovered the principle of independent assortment. The principle of **independent assortment** states that genes for different traits can segregate independently when gametes are made. Independent assortment helps explain the many genetic variations we see in all living things.

Key Question How do alleles segregate when more than one gene is involved?

Genes for different traits segregate independently.

BUILD Vocabulary

independent assortment
one of Mendel's principles that states that genes for different traits can segregate independently during the formation of gametes

ACADEMIC WORDS
The term *assort* means "to sort out." When alleles undergo independent assortment, they sort out, or separate from each other. Genes for one trait act separately from genes for another trait.

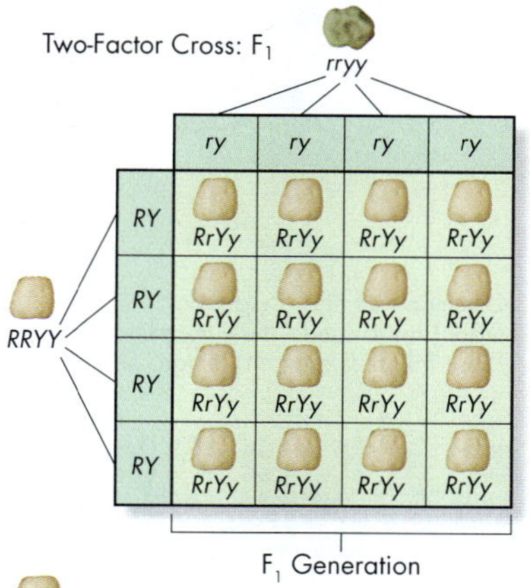

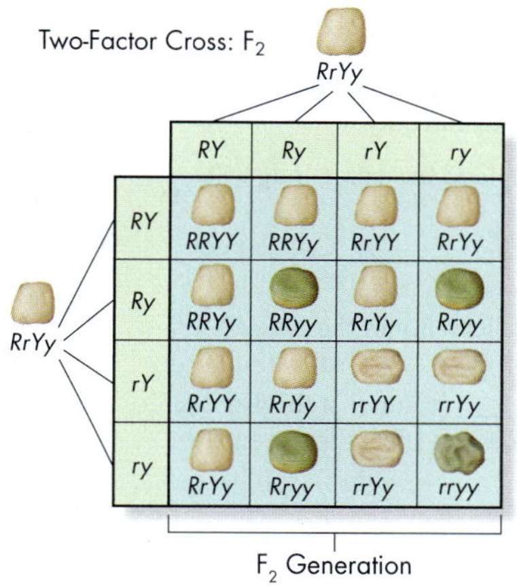

BIOLOGY.com Search Lesson 11.2 GO • InterActive Art

Students Say

"I found it helpful when our class was told to adjust our reading speed when reading content. Some of the class read so fast, and when we all slowed down, everything was easier to understand. I think we just needed permission, since all our lives we've been trying to read fast."

LESSON 11.2

Active Reading

USE SYMBOLS Have students practice writing the symbols for various genotypes. Model one example.

Ask What is the genotype symbol for a plant that is heterozygous for height and homozygous recessive for seed color? Write *Ttyy* on the board. Ask students to write the symbols for other genotypes as you name them.

Ask What words describe a plant with the genotype *ttYY*? *(homozygous recessive for height and homozygous dominant for seed color)*

Ask What is the phenotype of a plant with the genotype *ttYY*? *(a short plant that produces yellow seeds)*

SELF-QUESTION Read aloud the two paragraphs under Showing Independent Assortment.

Say As I read, I ask myself, "Does this make sense?" and "What does it mean that the alleles for seed shape segregated independently of those for seed color?"

Have students reread the first paragraph.

Ask Which phrase tells me the answer? *(209 seeds had other combinations of phenotypes)*

Ask How does this phrase answer my question? *(It says that the alleles for the two traits in each parent didn't stay together when gametes were formed.)*

Build Vocabulary

ACADEMIC WORDS Point out that *assort* means "to sort out." Have students brainstorm items in their lives that they sort out and the criteria they use for sorting. *(clothing by color and type, homework papers by course, and sporting equipment by sport)*

Say The suffix *-ment* added to *assort* changes the verb to a noun, *assortment*, "the result of assorting."

KEY QUESTION Read aloud the question and answer.

Ask What phenotypes would Mendel have observed if the alleles did not segregate independently? *(The* RY *would always stay together. The* ry *would always stay together. Only round, yellow seeds and wrinkled, green seeds would be observed.)*

Speed Bump

Ask Why didn't Mendel know from the first part of his experiment whether genes segregated independently? *(All the offspring had dominant alleles, so they showed the dominant phenotypes.)*

Ask What evidence that alleles segregated independently did Mendel have in his second cross? *(All combinations of phenotypes resulted.)*

During Reading continued

Active Reading

FIND THE MAIN IDEA Have a volunteer read aloud the first paragraph, including the bulleted list, under A Summary of Mendel's Principles.

Ask Which principle defines a gene? *(the first principle)* Which principle defines an allele? *(the second principle)*

Ask If an adult organism has two copies of each gene, how does the organism reproduce? *(sexually)*

Ask What does it mean that the alleles for different genes segregate independently during gamete formation? *(A parent's alleles do not stay together. Instead, they split up into different gametes.)*

Have students write the paragraph's main idea in their notebooks. *(There are four basic principles of modern genetics based on Mendel's principles.)*

KEY QUESTION Invite a volunteer to read aloud the question and answer. Point out that *basis* means "starting point."

Ask Why is Mendel's work considered the starting point of genetics? *(Mendel's work was the first significant work on genetics and other scientists continue to build upon what Mendel contributed.)*

Speed Bump

Ask Do Mendel's principles work only for pea plants? Explain. *(No, they work for all organisms that reproduce sexually.)*

Post Reading

ASSESSMENT TIPS To help students with Questions 5 and 7, review the steps for making a Punnett square.

For Question 5, have students make a Punnett square and fill in what they know about each parent's alleles. Tell them to leave a blank if they can't identify an allele. Point out that they can work backward to infer the parent's alleles from the offspring's phenotypes.

Wrap-Up Activity

MAKING GENETIC CROSSES Bring some lilies or tulips into class and have students carefully examine the male and female parts of the flower. Have students use scissors to cut off the male stamens of one flower. Have them use a small paintbrush to transfer pollen from the stamens of another flower to the top of the female pistil with no stamens. Explain that this is how Mendel performed a cross between two plants.

Common fruit flies are good organisms to use to study heredity. These fruit flies are on a lemon.

A Summary of Mendel's Principles

Today, Mendel's principles are the foundation of the modern science of genetics. These principles are below.

- Biological characteristics are inherited through units called genes. Genes are passed from parents to offspring.
- Sometimes there are two or more forms (alleles) of a gene for a single trait. Some forms of a gene may be dominant and others may be recessive.
- In most sexually reproducing organisms, each adult has two copies of each gene. Organisms get one copy from each parent. These alleles segregate from each other when gametes are made.
- Alleles for different genes usually segregate independently of each other.

Pea plants are not the only organisms used to study genetics. At the beginning of the 1900s, the American geneticist Thomas Hunt Morgan began to test Mendel's theories using fruit flies. The fruit fly is a good subject for genetic studies because it quickly produces hundreds of offspring. Using the fruit fly, Morgan and others tested all of Mendel's principles. The principles worked as well in fruit flies as they did in peas.

Key Question What did Mendel contribute to what we know about genetics? **Mendel's principles of heredity form the basis of modern genetics.**

CHECK Understanding

Apply Vocabulary

Use the highlighted words from the text to complete each sentence correctly.

1. ______ is the likelihood that something will happen.
2. If a pea plant has the ______ Tt, the plant will be tall.
3. If a rabbit has the alleles BB for the gene for fur color, the rabbit is ______ for that trait.
4. The principle of ______ states that the way one allele segregates does not affect the segregation of other alleles.

Critical Thinking

5. **Infer** Two pea plants with smooth seed pods are crossed. Some of the offspring have wrinkled seed pods. What can you infer about the genotypes of both parents?
6. **Review** What did Mendel conclude determines biological inheritance?
7. **Write to Learn** Make a Punnett square showing a cross between two pea plants. One plant is tall Tt, and the other is tall TT.

BIOLOGY.com Search Lesson 11.2 GO • Lesson Assessment

270 Chapter 11 • Lesson 2

Check Understanding

1. Probability
2. genotype
3. homozygous
4. independent assortment
5. Both parents are heterozygous for that trait, and smooth is the dominant trait.
6. Biological inheritance is determined by factors (genes) that are passed from one generation to the next.
7. Check students' Punnett squares. 50% are TT; 50% are Tt.

11.3 Other Patterns of Inheritance

Beyond Dominant and Recessive Alleles

We now know that heredity is a bit more complicated than Mendel thought. For example, most genes have more than two alleles. Also, many traits are controlled by more than one gene. Geneticists need to understand these other patterns of inheritance. That way, they can predict how complex traits will be inherited.

Codominance Sometimes neither allele is dominant over the other. If alleles have **codominance,** the phenotypes made by both alleles show up at the same time. For example, in certain kinds of chickens, the allele for black feathers is codominant with the allele for white feathers. Heterozygous chickens are speckled with black and white feathers. Because of codominant alleles, the black and white colors both appear. Many human genes also show codominance.

Incomplete Dominance The cross between two four o'clock plants shows a common exception to Mendel's principles. Some alleles are neither dominant or recessive. Cases where one allele is not completely dominant are called **incomplete dominance.** In incomplete dominance, the heterozygous phenotype lies somewhere between the two homozygous phenotypes. Four o'clock plants are an example of this. A cross between red-flowered (*RR*) and white-flowered (*WW*) plants results in offspring with pink flowers (*RW*).

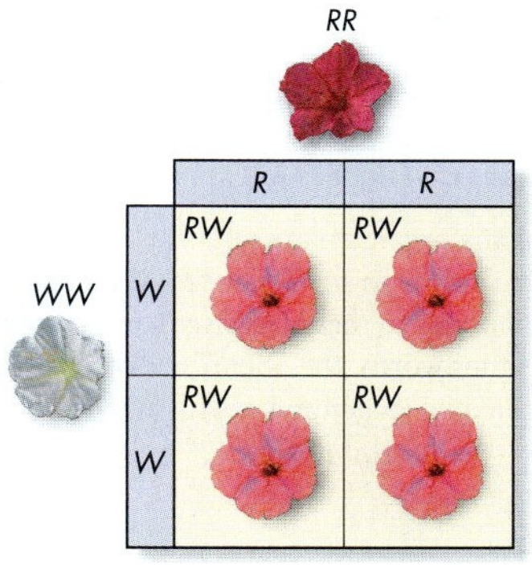

Incomplete Dominance
In four o'clock plants, the alleles for red and white flowers show incomplete dominance. Heterozygous (*RW*) plants have pink flowers.

Key Questions

What are some exceptions to Mendel's principles?

Does the environment have a role in how genes determine traits?

BUILD Understanding

Finding Main Ideas Before you read the lesson, create a list of all the lesson's green and blue headings in order. Leave space under each heading for notes. As you read, write the main idea of each paragraph under the heading above it.

In Your Workbook Your workbook has a table for recording main ideas and details.

BIOLOGY.com Search Lesson 11.3 GO • Art Review

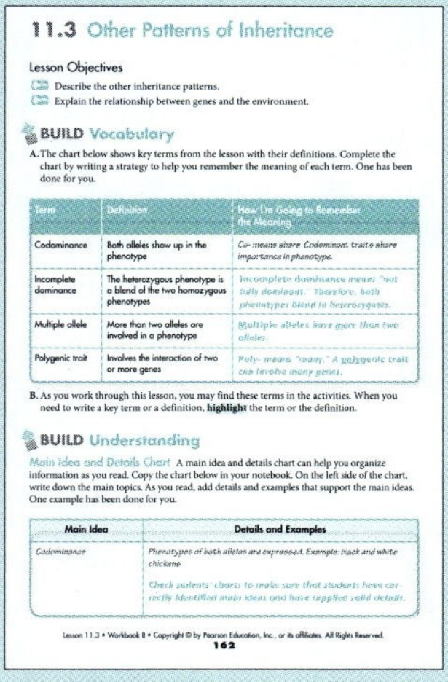

11.3 Other Patterns of Inheritance

Lesson Objectives

Describe the other inheritance patterns.

Explain the relationship between genes and the environment.

BUILD Vocabulary

A. The chart below shows key terms from the lesson with their definitions. Complete the chart by writing a strategy to help you remember the meaning of each term. One has been done for you.

Term	Definition	How I'm Going to Remember the Meaning
Codominance	Both alleles show up in the phenotype	Co- means share. Codominant traits share importance in phenotype.
Incomplete dominance	The heterozygous phenotype is a blend of the two homozygous phenotypes	Incomplete dominance means "not fully dominant." Therefore, both phenotypes blend in heterozygotes.
Multiple allele	More than two alleles are involved in a phenotype	Multiple alleles have more than two alleles.
Polygenic trait	Involves the interaction of two or more genes	Poly- means "many." A polygenic trait can involve many genes.

B. As you work through this lesson, you may find these terms in the activities. When you need to write a key term or a definition, **highlight** the term or the definition.

BUILD Understanding

Main Idea and Details Chart A main idea and details chart can help you organize information as you read. Copy the chart below in your notebook. On the left side of the chart, write down the main topics. As you read, add details and examples that support the main ideas. One example has been done for you.

Main Idea	Details and Examples
Codominance	Phenotypes of both alleles are expressed. Example: black and white chickens Check students' charts to make sure that students have correctly identified main ideas and have supplied valid details.

162

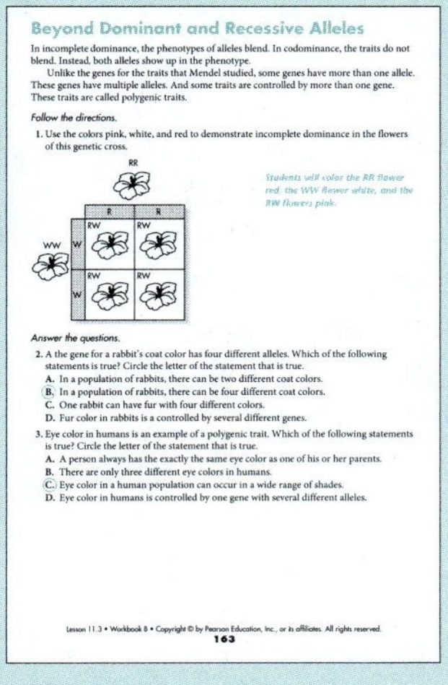

Beyond Dominant and Recessive Alleles

In incomplete dominance, the phenotypes of alleles blend. In codominance, the traits do not blend. Instead, both alleles show up in the phenotype.

Unlike the genes for the traits that Mendel studied, some genes have more than one allele. These genes have multiple alleles. And some traits are controlled by more than one gene. These traits are called polygenic traits.

Follow the directions.

1. Use the colors pink, white, and red to demonstrate incomplete dominance in the flowers of this genetic cross.

RR; R R; WW; W W; RW RW RW RW

Students will color the RR flower red, the WW flower white, and the RW flowers pink.

Answer the questions.

2. A the gene for a rabbit's coat color has four different alleles. Which of the following statements is true? Circle the letter of the statement that is true.
A. In a population of rabbits, there can be two different coat colors.
B. In a population of rabbits, there can be four different coat colors.
C. One rabbit can have fur with four different colors.
D. Fur color in rabbits is a controlled by several different genes.

3. Eye color in humans is an example of a polygenic trait. Which of the following statements is true? Circle the letter of the statement that is true.
A. A person always has the exactly the same eye color as one of his or her parents.
B. There are only three different eye colors in humans.
C. Eye color in a human population can occur in a wide range of shades.
D. Eye color in humans is controlled by one gene with several different alleles.

163

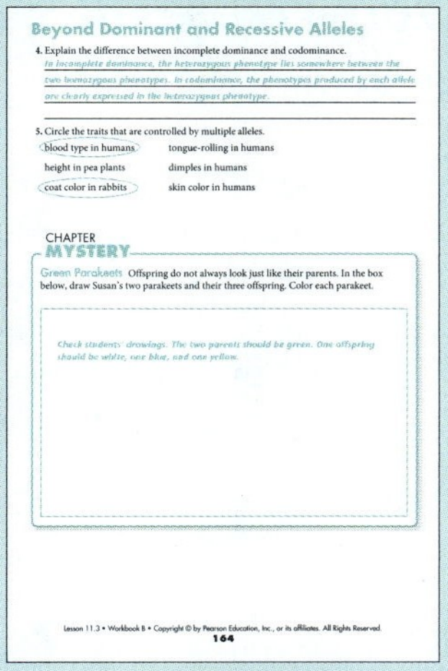

Beyond Dominant and Recessive Alleles

4. Explain the difference between incomplete dominance and codominance.
In incomplete dominance, the heterozygous phenotype lies somewhere between the two homozygous phenotypes. In codominance, the phenotypes produced by each allele are clearly expressed in the heterozygous phenotype.

5. Circle the traits that are controlled by multiple alleles.

blood type in humans	tongue-rolling in humans
height in pea plants	dimples in humans
coat color in rabbits	skin color in humans

CHAPTER MYSTERY

Green Parakeets Offspring do not always look just like their parents. In the box below, draw Susan's two parakeets and their three offspring. Color each parakeet.

Check students' drawings. The two parents should be green. One offspring should be white, one blue, and one yellow.

164

LESSON 11.3

Pre-Reading

Objectives

11.3.1 Describe the other inheritance patterns.

11.3.2 Explain the relationship between genes and the environment.

Preview the Pages

1. Have volunteers read aloud the vocabulary terms and definitions. Make sure students understand the difference between multiple alleles and multiple genes.
2. Preview the Inquiry Into Scientific Thinking activity, *Human Blood Types*. Review the circle graph and labels. Have students add up each slice on the graph to total 100%.

Activate Prior Knowledge

Tell students to think about all the eye colors that humans have and about the many shades of each color. Ask students if they think eye color is controlled by just one gene. Lead students to conclude that more than one gene is likely responsible for the color of human eyes and/or more than two alleles per gene.

Build Understanding

MAIN IDEAS AND DETAILS CHART Direct students to Workbook B, where they will find a main idea and details chart. Remind students that putting information into a table format like this one will help them summarize the four kinds of inheritance.

NATIONAL SCIENCE EDUCATION STANDARDS

UNIFYING CONCEPTS AND PROCESSES

II

CONTENT

C.2.a, C.2.b, G.2

INQUIRY

A.1.c, A.2.a

For the corresponding lesson in *Miller & Levine Biology,* see pages 319–321.

LESSON 11.3

During Reading

Active Reading

SUMMARIZE Form teams of four students each. Assign each team member a number from 1 to 4. Have students regroup by numbers, such as all ones together. Have each group create a summary of one of the exceptions to Mendel's principles: ones: Codominance; twos: Incomplete Dominance; threes: Multiple Alleles; fours: Polygenic Traits. Then, instruct students to re-form their original teams, and share their summaries.

USE VISUALS Have students examine the Punnett square on the previous page for incomplete dominance.

Ask Why isn't a lower case letter used to represent an allele? *(Neither allele is recessive so a lower case letter isn't used.)*

Ask What color flowers would you expect to see if two plants with pink flowers were crossed? Have students make a Punnett square to find the answer. *(one red, two pink, and one white)*

KEY QUESTION Read aloud the question and answer. Point out that the blue heads under Beyond Dominant and Recessive Alleles each identify a type of exception.

Ask Why are multiple alleles an exception to Mendel's principles? *(Mendel said that two factors control the inheritance of a trait.)*

Repeat the question with codominance *(Mendel identified only dominant and recessive alleles.)*, incomplete dominance *(Mendel identified only dominant and recessive alleles.)*, and polygenic traits *(Mendel did not identify single traits controlled by multiple genes.)*.

Build Vocabulary

PREFIXES Read aloud the meaning of the prefix co-.

Ask What does it mean if you cooperate? *(work together)*

Discuss with students other words they know with the prefix co-. Have them explain the meaning of each. *(Sample answers: couple, connect, collect)*

Speed Bump

Ask How are codominance and incomplete dominance different? *(In codominance, the phenotypes made by two alleles show up at the same time. In incomplete dominance, the phenotypes for two alleles blend.)*

Ask What kind of traits often show a wide range of phenotypes? *(polygenic traits)*

BUILD Vocabulary

codominance
a situation in which the phenotypes produced by both alleles are completely expressed

incomplete dominance
a situation in which one allele is not completely dominant over another allele

multiple alleles
a gene that has more than two alleles

polygenic traits
a trait controlled by two or more genes

PREFIXES
The prefix *co-* means *jointly* or *together*. Books written by coauthors are written by two people who share credit equally. Codominant alleles contribute equally to the phenotype of a trait.

Multiple Alleles So far, our examples have described genes for which there are only two alleles. In nature, such genes are not common. Many genes exist in several different forms. A gene with more than two alleles is said to have **multiple alleles.** An individual, of course, usually has only two copies of each gene. Within a population, however, there are many different alleles. Two individuals having two different alleles each could together have four different alleles for the same gene.

One of the best examples of multiple alleles is coat color in rabbits. A rabbit's coat color is controlled by one gene that has four possible alleles. The four known alleles display a pattern of simple dominance that can produce four coat colors. Many other genes have multiple alleles, including the human genes for blood type.

Polygenic Traits Many traits are controlled by several genes at work at the same time. Traits controlled by two or more genes are said to be **polygenic traits** (pahl ih JEN ik traytz). Polygenic means "many genes." For example, at least three genes are involved in making the reddish-brown pigment in the eyes of fruit flies. Different combinations of alleles for these three genes produce very different eye colors. Polygenic traits often show a wide range of phenotypes. The wide range of skin color in humans comes about partly because more than four different genes are likely to control this trait.

Be sure that you don't confuse polygenic traits with traits determined by one gene with multiple alleles. In multiple alleles, many different alleles exist for a gene in a population. However, an individual organism only has a single pair of these alleles. In polygenic traits, each individual organism has many different gene pairs that all work together on one trait.

Key Question What are some exceptions to Mendel's principles?
Traits can be controlled by incomplete dominant alleles. They can be controlled by codominant alleles, or by genes with more than two possible alleles. Traits can also be controlled by several genes.

Genes and the Environment

The characteristics of any organism are not only controlled by the genes that organism inherits. Genes provide a plan for development. What happens to that plan also depends heavily on the environment. In other words, the phenotype of an organism is only partly determined by its genotype. This is true no matter what the organism is—plant, fruit fly, or human being.

Focus on Academic Language

- pigment

Ask a volunteer to read the first paragraph under Polygenic Traits aloud. Then reread the following sentence to the class: *For example, at least three genes are involved in making the reddish-brown pigment in the eyes of fruit flies.* Point out that a *pigment* is a substance that affects color.

Ask students to rewrite that sentence in their own words. They may use more than one sentence. *(For example, at least three genes are involved in making the substance that controls eye color in fruit flies.)*

Consider the western white butterfly, *Pontia occidentalis*. It is found throughout western North America. People had noted for years that western whites hatching in the summer had different color patterns on their wings than those hatching in the spring. Scientific studies showed the reason. Butterflies hatching in the shorter days of springtime had greater levels of pigment in their wings. More pigment makes the markings of the springtime butterflies appear darker. In other words, the environment in which the butterflies develop affects how genes make pigments for the wings.

Key Question Does the environment have a role in how genes determine traits? **Environmental conditions can change gene expression and influence genetically controlled traits.**

Temperature and Wing Color Western white butterflies that hatch in the spring have darker wing patterns than those that hatch in the summer.

CHECK Understanding

Use the highlighted words from the text to complete each sentence correctly.

1. When neither allele is dominant over the other, and both phenotypes are fully expressed, the alleles are said to be __________.

2. In __________, both alleles of a gene are expressed and the phenotype is a blend of the phenotypes of both alleles.

3. In __________, several genes control the phenotype.

Critical Thinking

4. Review Describe two inheritance patterns besides simple dominance.

5. Apply How could you test to see if soil temperature was responsible for a variation in flower color?

6. Write to Learn Answer the third clue by writing a paragraph about how polygenic traits are different from the kinds of traits Mendel observed.

MYSTERY CLUE

Green feathers don't actually have green pigments. Rather, they contain a mixture of blue and yellow pigments. Could feather color be controlled by more than one gene? (**Hint:** See p. 272.)

BIOLOGY.com Search Lesson 11.3 GO • Lesson Assessment

Check Understanding

1. codominant

2. incomplete dominance

3. polygenic traits

4. Answers should describe two of the following patterns of inheritance: incomplete dominance, codominance, multiple alleles, or polygenic traits.

5. Sample answer: Conduct an experiment in which genetically identical plants are grown in different soil temperatures. All other factors, including amount of sunlight, air temperature, and amount of water, must be identical. If these plants develop flowers with different colors, the soil temperature affected the flower color.

6. Sample answer: If the color of the feathers is controlled by just one gene with two alleles, it would be impossible to have four colors of feathers. If two genes controlled the color, there could be four alleles. That way, each allele could be responsible for one color.

Active Reading

MAKE INFERENCES Have students reread the section Genes and the Environment.

Ask How do you know that the environment had a role in determining wing color? *(Wing color was different in different seasons.)*

Ask What environmental factor was responsible? *(day length)*

Ask How might this factor influence wing color? *(Sample answer: Sunlight may turn off pigment production.)*

KEY QUESTION Invite a volunteer to read aloud the question and answer.

Ask What environmental factor might cause an arctic hare to be white in winter and gray in summer? *(Sample answer: temperature or day length)*

Post Reading

ASSESSMENT TIPS Before students start the assessment, review the highlighted words in the text.

For Question 4, point out that the term *simple dominance* refers to Mendel's dominant and recessive gene theory.

If students have trouble with Question 5, have them work in pairs or small groups to brainstorm possible experiments. Then, have groups exchange lists and discuss which experiments might work and why.

To help with Question 6, remind students that Mendel's peas had one gene per trait and have students reread the Polygenic Traits section for comparison.

MYSTERY CLUE Have students discuss whether they think feather color is polygenic. Tell them to reread the Polygenic Traits section. Lead students to conclude that there are likely two genes controlling feather color, one for each pigment.

Wrap-Up Activity

MODEL DOMINANCE AND INCOMPLETE DOMINANCE Give students sheets of blue, green, and yellow acetate or cellophane. Have them hold yellow and green together up to a light source, then repeat with yellow and blue.

Ask What color do you see when you combine green and yellow? *(green)*

Ask Does this model dominance or incomplete dominance? Explain. *(dominance; Green is dominant over yellow because it masks yellow.)*

Ask What color do you see when you combine blue and yellow? *(green)*

Ask Does this model dominance or incomplete dominance? Explain. *(incomplete dominance; Neither color is dominant. They blend to form green.)*

Inquiry Into Scientific Thinking

VISUALIZE/DIAGRAM Draw four circles on the board to represent red blood cells. Draw small triangles on the surface of the first circle and label the circle Type A. Draw small squares on the surface of the second circle and label the circle Type B. Draw both triangles and squares on the third circle and label it Type AB. Label the fourth circle Type O. Have students copy the diagrams into their science notebooks and explain how the diagram provides a model for protein markers and blood types.

USE GRAPHIC ORGANIZERS In their notebooks, have students make a four-column chart with four rows to categorize blood donors and receivers. Have them label the columns Blood Type, Markers Present, Give Blood Safely To, and Receive Blood Safely From. Rows should be labeled with the four blood types. Have students reread the Inquiry to fill in the chart.

SUMMARIZE After students have made their four-column charts, have them answer the following questions.

Ask Can a person with type A blood safely receive type B blood? Explain. *(No, type B blood has a different antigen than type A.)*

Ask Can a person with type AB blood safely receive type B blood? Explain. *(Yes, type AB blood contains both antigens A and B. Type B blood contains B antigens.)*

Lead a Discussion

Talk about safe blood transfusions. Explain that all donated blood is tested for blood type and for disease-causing agents. Point out that transfusing blood of the wrong blood type can be very serious and might cause a person to die.

Ask Why is a person with type O blood called a universal donor? *(This person can give blood to people with all other types.)*

Ask Why is a person with type AB blood called a universal acceptor? *(This person can get blood of all other types.)*

Ask What type of blood do you think a blood center wants the most? *(Type O^-, because it has no antigens and can be safely given to any person.)*

INQUIRY into Scientific Thinking

Human Blood Types

Red blood cells have protein markers on them. These antigens can cause a reaction if a person receives the wrong blood type during a blood transfusion. Human blood type A carries an A antigen, type B has a B antigen, and type AB has both antigens. Type O carries neither antigen. The gene for these antigens has three alleles: A, B, and O. A and B are dominant over O. When A and B appear together, they are codominant.

For a transfusion to succeed, it must not put a new antigen into the body of the person getting the blood. So, a person with type A blood may get type O. However, a person with type O may not get type A blood.

Another gene controls a second type of antigen, known as Rh factor. Rh^+ individuals carry this protein. People with Rh^- do not carry it. The pie graph of the U.S. population shows the percentage of each blood type.

Blood Groups in the U.S. Population

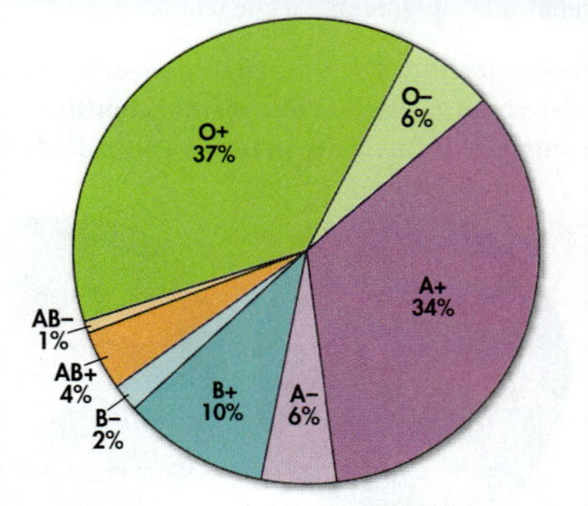

Analyze and Conclude

1. **Interpret Graphs** Which blood type makes up the greatest percentage of the U.S. population?

2. **Calculate** What percentage of the U.S. population has Rh^+ blood?

3. **Predict** Alleles are either dominant or recessive. The allele for type O is recessive to the alleles for types A and B. The allele for Rh^+ is dominant over the allele for Rh^-.

- Could a person with O^+ blood have two parents with O^- blood? Explain.
- Could a person with O^+ blood have a daughter with AB^+ blood? Explain your answers.

4. **Infer** A person with type A, B, or AB blood may receive a transfusion of type O. However, a person with type O may not get types A or AB blood. A person with Rh^+ blood can get Rh^- blood, but a person with Rh^- blood cannot get Rh^+ blood.

Use the graph to answer the questions that follow.

- Which blood type can be donated to the largest percentage of individuals? Explain your answer.
- Which type can be donated to the smallest percentage of people? Explain your answer.

274 Chapter 11 • Lesson 3

Analyze and Conclude

1. O^+

2. 85% are Rh^+, 15% are Rh^-

3. No, a person with O^+ blood could not have two parents with O^- blood because both parents would be homozygous recessive for the Rh factor. They do not have any Rh^+ alleles to pass on. A person with O^+ blood could not have an AB^+ daughter because a person with O^+ blood has only O alleles to pass on.

4. O^- can be donated to the most people because it contains no antigens. AB^+ blood can be given only to those people who have all three antigens. This is only 4% of the U.S. population.

11.4 Meiosis

Chromosome Number

What happens in cells to make sure they pass on the correct number of genes? To understand how the number of genes is controlled, we need to learn what happens to chromosomes when gametes are made. Chromosomes are strands of DNA and protein inside the cell nucleus, and they are the carriers of genes. Genes are located in specific places on chromosomes.

Diploid Cells Think about the fruit flies that Morgan used. A body cell in an adult fruit fly has eight chromosomes. Four of the chromosomes come from its male parent, and four come from its female parent. These two sets of chromosomes are **homologous** (hoh MAHL uh gus), because each of the four chromosomes from the male parent has a similar chromosome from the female parent. A cell that has both sets of homologous chromosomes is said to be **diploid,** meaning "two sets." The diploid cells of most adult organisms have two complete sets of chromosomes and so two complete sets of genes. The diploid number of chromosomes can be shown using the symbol 2N. So, for fruit flies, the diploid number is 8, which can be written as 2N = 8.

Key Question How many sets of genes are found in most adult organisms?
The diploid cells of most adult organisms contain two complete sets of inherited chromosomes and so two complete sets of genes.

Haploid Cells Some cells have only a single set of genes on a single set of chromosomes. These cells are **haploid,** meaning "one set." The gametes of sexually reproducing organisms, such as fruit flies and peas, are haploid. For fruit fly gametes, the haploid number is 4. This amount can be written as N = 4.

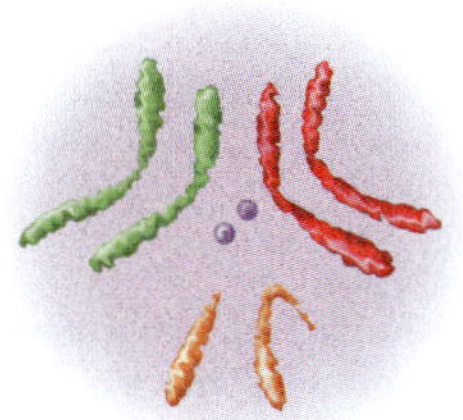

Fruit Fly Chromosomes These chromosomes are from a fruit fly. Each of the fruit fly's body cells is diploid, containing eight chromosomes.

Key Questions

How many sets of genes are found in most adult organisms?

What happens during each phase of meiosis?

How is meiosis different from mitosis?

How can two alleles from different genes be inherited together?

BUILD Understanding

Compare/Contrast Table Before you read, make a compare/contrast table to show the differences between mitosis and meiosis. As you read, complete the table.

In Your Workbook Refer to your workbook for suggestions about how to use a compare/contrast table.

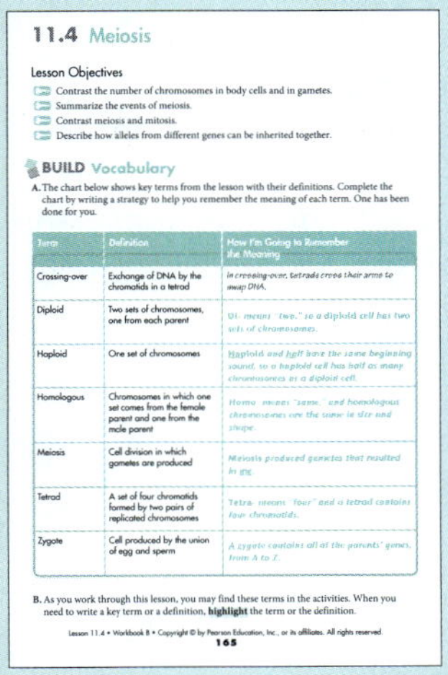

11.4 Meiosis

Lesson Objectives
- Contrast the number of chromosomes in body cells and in gametes.
- Summarize the events of meiosis.
- Contrast meiosis and mitosis.
- Describe how alleles from different genes can be inherited together.

BUILD Vocabulary

A. The chart below shows key terms from the lesson with their definitions. Complete the chart by writing a strategy to help you remember the meaning of each term. One has been done for you.

Term	Definition	How I'm Going to Remember the Meaning
Crossing-over	Exchange of DNA by the chromatids in a tetrad	In crossing-over, tetrads cross their arms to swap DNA.
Diploid	Two sets of chromosomes, one from each parent	Di- means "two," so a diploid cell has two sets of chromosomes.
Haploid	One set of chromosomes	Haploid and half have the same beginning sound, so a haploid cell has half as many chromosomes as a diploid cell.
Homologous	Chromosomes in which one set comes from the female parent and one from the male parent	Homo- means "same," and homologous chromosomes are the same in size and shape.
Meiosis	Cell division in which gametes are produced	Meiosis produced gametes that resulted in me.
Tetrad	A set of four chromatids formed by two pairs of replicated chromosomes	Tetra- means "four" and a tetrad contains four chromatids.
Zygote	Cell produced by the union of egg and sperm	A zygote contains all of the parents' genes, from A to Z.

B. As you work through this lesson, you may find these terms in the activities. When you need to write a key term or a definition, **highlight** the term or the definition.

Lesson 11.4 • Workbook B • Copyright © by Pearson Education, Inc., or its affiliates. All rights reserved.
165

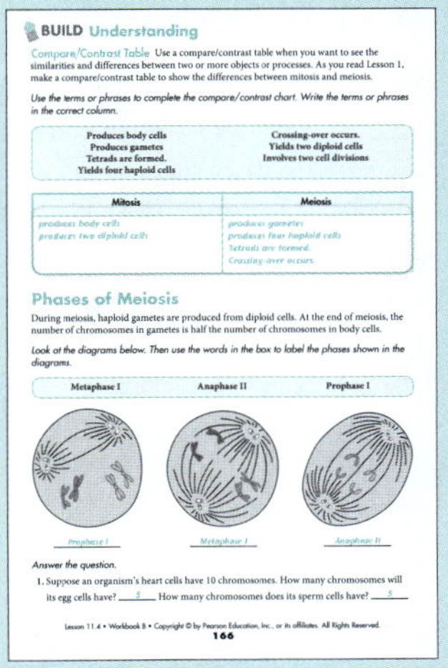

BUILD Understanding

Compare/Contrast Table Use a compare/contrast table when you want to see the similarities and differences between two or more objects or processes. As you read Lesson 1, make a compare/contrast table to show the differences between mitosis and meiosis.

Use the terms or phrases to complete the compare/contrast chart. Write the terms or phrases in the correct column.

Produces body cells / Produces gametes / Tetrads are formed. / Yields four haploid cells / Crossing-over occurs. / Yields two diploid cells / Involves two cell divisions

Mitosis	Meiosis
produces body cells produces two diploid cells	produces gametes produces four haploid cells Tetrads are formed. Crossing-over occurs.

Phases of Meiosis

During meiosis, haploid gametes are produced from diploid cells. At the end of meiosis, the number of chromosomes in gametes is half the number of chromosomes in body cells.

Look at the diagrams below. Then use the words in the box to label the phases shown in the diagrams.

Metaphase I / Anaphase II / Prophase I

Prophase I — Metaphase I — Anaphase II

Answer the question.

1. Suppose an organism's heart cells have 10 chromosomes. How many chromosomes will its egg cells have? 5 How many chromosomes does its sperm cells have? 5

Lesson 11.4 • Workbook B • Copyright © by Pearson Education, Inc., or its affiliates. All Rights Reserved.
166

LESSON 11.4

Pre-Reading

Objectives

11.4.1 Contrast the number of chromosomes in body cells and in gametes.

11.4.2 Summarize the events of meiosis.

11.4.3 Contrast meiosis and mitosis.

11.4.4 Describe how alleles from different genes can be inherited together.

Preview the Pages

1. Preview the visuals with students. Remind students to focus on the number and position of chromosomes in each diagram.
2. Have students pay special attention to the colors of the chromosomes to visualize crossing-over.

Activate Prior Knowledge

Have students recall what they know about cell structures. Challenge them to predict the location of Mendel's hereditary factors within the cell.

Build Understanding

COMPARE/CONTRAST TABLE Direct students to the compare/contrast table in Workbook B. After students have completed their table, have them think of phrases they could write in a third column headed Both Mitosis and Meiosis. *(division of nucleus, involves chromosomes, etc.)*

KEY QUESTION Read aloud the question and answer. Ask students where the chromosomes in diploid cells come from. *(both male and female parents)*

NATIONAL SCIENCE EDUCATION STANDARDS

UNIFYING CONCEPTS AND PROCESSES

I, II

CONTENT

C.2.a, C.2.b

INQUIRY

A.2.a, A.2.b

For the corresponding lesson in *Miller & Levine Biology,* see pages 323–329.

During Reading

Active Reading

Build Vocabulary

WORD ORIGINS Point out that *chromosome* means "colored bodies."

Ask What is the connection between *chromosome* and *colored bodies*? *(When cells are dyed so that scientists can study them, the chromosomes absorb much of the dye.)*

USE VISUALS Have students identify the following in the visual of meiosis I: chromosomes, chromatids, a tetrad, and spindles. Review the close-up of crossing-over. Have students create a flowchart showing the steps of meiosis using index cards, two different colors, and Xs for the chromosomes. Then, have students write a caption for each step.

KEY QUESTION Invite a volunteer to read aloud the question and answer. Refer students to the captions they wrote in Use Visuals. Have volunteers use their captions to explain what happens at each phase.

Speed Bump

Ask In what part of meiosis I do tetrads appear? *(in prophase I)*

Ask What crosses over in meiosis I? *(the chromatids of homologous chromosomes)*

Ask What separates in anaphase I? *(homologous chromosome)*

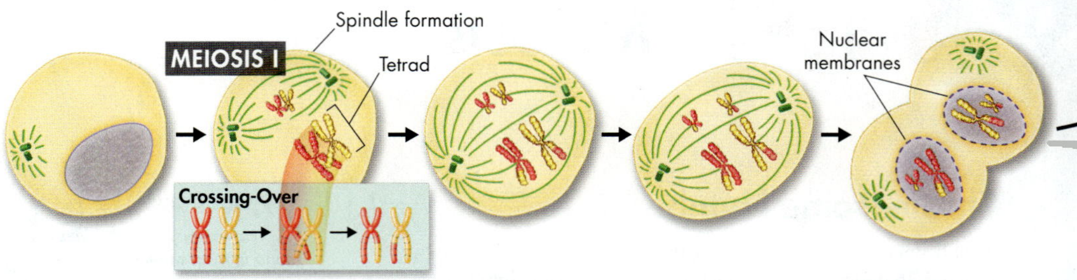

Interphase Cells undergo a round of DNA replication, forming duplicate chromosomes.

Prophase I Each chromosome pairs with its corresponding homologous chromosome to form a tetrad.

Metaphase I Spindle fibers attach to the chromosomes.

Anaphase I The fibers pull the homologous chromosomes to opposite ends of the cell.

Telophase I and Cytokinesis Nuclear membranes form. The cell separates into two cells.

Meiosis During meiosis, the number of chromosomes per cell is cut in half through the separation of the homologous chromosomes. The result of meiosis is 4 haploid cells that are genetically different from one another and from the original cell.

Phases of Meiosis

How are haploid cells made? Haploid cell are made by meiosis. **Meiosis** (my OH sis) is a process in which the number of chromosomes per cell is cut in half through the separation of homologous chromosomes in a diploid cell. Meiosis involves two rounds of cell division, called meiosis I and meiosis II. Let's see how meiosis takes place in a cell that has a diploid number of 4 (2N = 4).

Meiosis I Just before meiosis I, each chromosome is copied, or replicated. As in mitosis, each replicated chromosome is made of two identical chromatids joined at the center.

▶ ***Prophase I*** In prophase of meiosis I, each copied chromosome matches up with its homologous chromosome. This pairing makes a structure called a tetrad, which has four chromatids. As the tetrads form, they undergo a process called **crossing-over.** Chromatids from homologous chromosomes cross over one another, as shown in the art. Then, alleles from the crossed sections are exchanged. Crossing-over produces new combinations of alleles on each chromatid.

▶ ***Metaphase I and Anaphase I*** A spindle forms and attaches to each chromosome in the tetrad. The homologous chromosomes separate in anaphase I.

▶ ***Telophase I and Cytokinesis*** A nuclear membrane forms around each cluster of chromosomes in telophase I. During cytokinesis, the cell splits and two new haploid cells are made. The set of alleles in each cell is different from the set in the other cell.

Key Question What happens during each phase of meiosis I? **During prophase I, each copied chromosome matches up with its homologous chromosome. Crossing-over increases the allele combinations on chromatids.**

BUILD Vocabulary

homologous
the term used to refer to chromosomes in which one set comes from the male parent and one set comes from the female parent

diploid
the term used to refer to a cell that contains two sets of homologous chromosomes

haploid
the term used to refer to a cell that contains only a single set of genes

WORD ORIGINS
The word *diploid* comes from the Greek root *di-*, which means "double" or "twice."

ELL Focus on ELL: Access Content

ALL SPEAKERS Point out to learners that it might be easier to learn science vocabulary in English if they haven't learned the vocabulary in their native language. Have students make flashcards, writing the word on one side and the definition on the other: *chromosome, homologous, diploid, haploid, meiosis, tetrad, replicate,* and *zygote*. You may also want to pronounce the words aloud and have students repeat in order to help them with concepts as well as vocabulary.

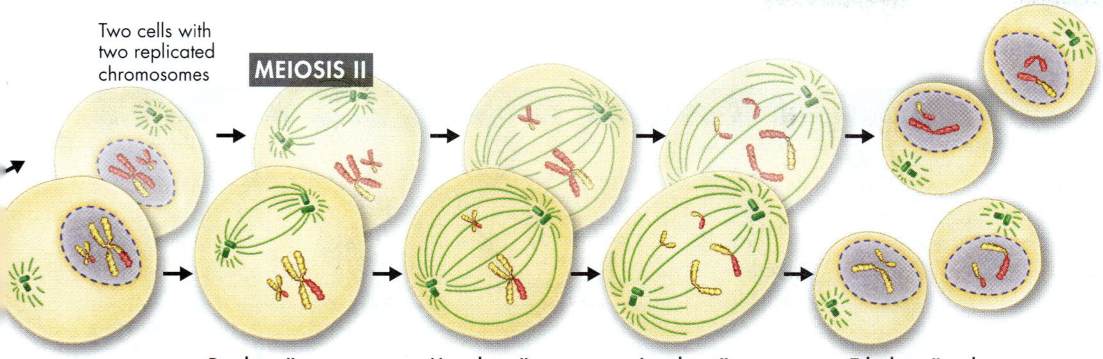

Prophase II
Meiosis I results in two haploid (N) daughter cells, each with half the number of chromosomes as the original cell.

Metaphase II
The chromosomes line up in a similar way to the metaphase stage of mitosis.

Anaphase II
The sister chromatids separate and move toward opposite ends of the cell.

Telophase II and Cytokinesis
Meiosis II results in four haploid daughter cells.

In metaphase I, the spindle fibers form. In anaphase I, the chromosomes separate. Then, in telophase, a nuclear membrane forms around the chromosomes. During cytokinesis, the cell splits, and two new cells are made.

Meiosis II The two cells get ready to divide once more. The chromosomes do not replicate before this division.

▶ ***Prophase II*** As the cells enter prophase II, their chromosomes become visible. Each prophase chromosome consists of two identical sister chromatids.

▶ ***Metaphase II, Anaphase II, Telophase II, and Cytokinesis*** During metaphase II, the chromosomes line up in the middle of the cell. As the cell enters anaphase II, the paired chromatids separate. When anaphase II is complete, the separated chromosomes cluster at opposite ends of the cell. In telophase II, the nuclear membrane forms around each cluster of chromosomes. In the example shown here, each of the four daughter cells produced in meiosis II receives two chromosomes. The four daughter cells now contain the haploid number (N)—two chromosomes each.

Key Question What happens during each phase of meiosis II? **As the cells enter prophase II, their chromosomes become visible. During metaphase II, the chromosomes line up in the middle of the cell. In the second anaphase, the chromatids separate. In the second telophase and cytokinesis, four new haploid cells are formed.**

Gametes to Zygotes The haploid cells made by meiosis II are gametes, which are important to heredity. Gametes, such as eggs and sperm, join in fertilization to form a **zygote** (ZY goht), which has new combinations of alleles. The zygote undergoes cell division by mitosis and grows into an organism.

BUILD Vocabulary

meiosis
the process in which the number of chromosomes per cell is cut in half through the separation of homologous chromosomes in a diploid cell

crossing-over
the process in which homologous chromosomes exchange portions of their chromatids during meiosis

zygote
a fertilized egg

WORD ORIGINS
The word *meiosis* comes from a Greek word that means "to lessen or reduce." Because of meiosis, gametes have a reduced number of chromosomes.

BIOLOGY.com Search Lesson 11.4 GO • Art in Motion

Build Vocabulary

WORD ORIGINS Like *meiosis, haploid* and *diploid* have Greek origins.

Say The word *haplous* in Greek means "single."

Ask What does *haploid* mean? *(refers to cells that have a single set of chromosomes)* Can you predict what the prefix *diplo-* means? *(double)* What does *diploid* mean? *(refers to cells that have two sets of chromosomes)*

Active Reading

COMPARE AND CONTRAST Have students compare the diagrams of meiosis I and meiosis II. Be sure that students see that in meiosis I, the chromosome pairs separate into two different cells. In meiosis II, the centromeres split, and the chromosome copies separate.

Ask How many sex cells are formed at the end of meiosis? *(four)* Are the resulting cells haploid or diploid? *(haploid)*

MAKE INFERENCES Have students think about why meiosis occurs. Remind students that sex cells have half as many chromosomes as body cells.

Ask Why must meiosis happen before fertilization? *(Meiosis must first happen to form sperm and egg cells, otherwise sex cells would have a full set of 46 chromosomes. When the two sex cells combined, the offspring would have 92 chromosomes.)*

KEY QUESTION Invite a volunteer to read aloud the question and answer.

Ask How does prophase II differ from prophase I? *(Tetrad formation and crossing-over do not occur in prophase II.)* Where in meiosis II do the two alleles for a trait separate? *(The alleles separate during anaphase II.)*

Speed Bump

Ask What are some differences between meiosis I and meiosis II? *(Sample answer: Homologous chromosomes separate during meiosis I but not during meiosis II. The chromatids of each chromosome separate during meiosis II.)*

Ask Do diploid cells become haploid cells in meiosis I or meiosis II? *(in meiosis I)*

ELL Focus on ELL: Access Content

INTERMEDIATE SPEAKERS Have students use information from the Meiosis visual to prepare a three-column chart.

The first column should explain each step in the student's first language. The second column should be a sketch of the step with labels. The third column should have a corresponding explanation in English. Students can make a copy of the table to share with a beginner.

During Reading continued

Active Reading

USE VISUALS Be sure that students understand that this table compares mitosis and meiosis, not meiosis I and meiosis II.

Ask How is prophase of mitosis different from prophase of meiosis I? *(Tetrads form and crossing-over occurs during meiosis but not in mitosis.)*

Ask How do chromosomes line up differently in metaphase of mitosis and meiosis I? *(In mitosis, chromosomes line up at random. In meiosis I, homologous chromosomes pair up.)*

Ask How is separation of sister chromatids different in anaphase of mitosis and meiosis I? *(In mitosis, the chromatids separate from each other in anaphase. In meiosis I, the chromatids stay together.)* When do they separate in meiosis? *(in anaphase of meiosis II)*

Ask What is the result of mitosis? *(Two diploid daughter cells are formed that are identical to the parent cell. The chromosomes are not doubled.)*

Ask What is the result of meiosis I? Use the words *haploid, alleles,* and *doubled* in your answer. *(Two haploid daughter cells with different combinations of alleles from the parent cell are formed. Each chromosome is still doubled.)*

Ask What is the result of meiosis II? *(Four haploid daughter cells are formed. The chromosomes are no longer doubled.)*

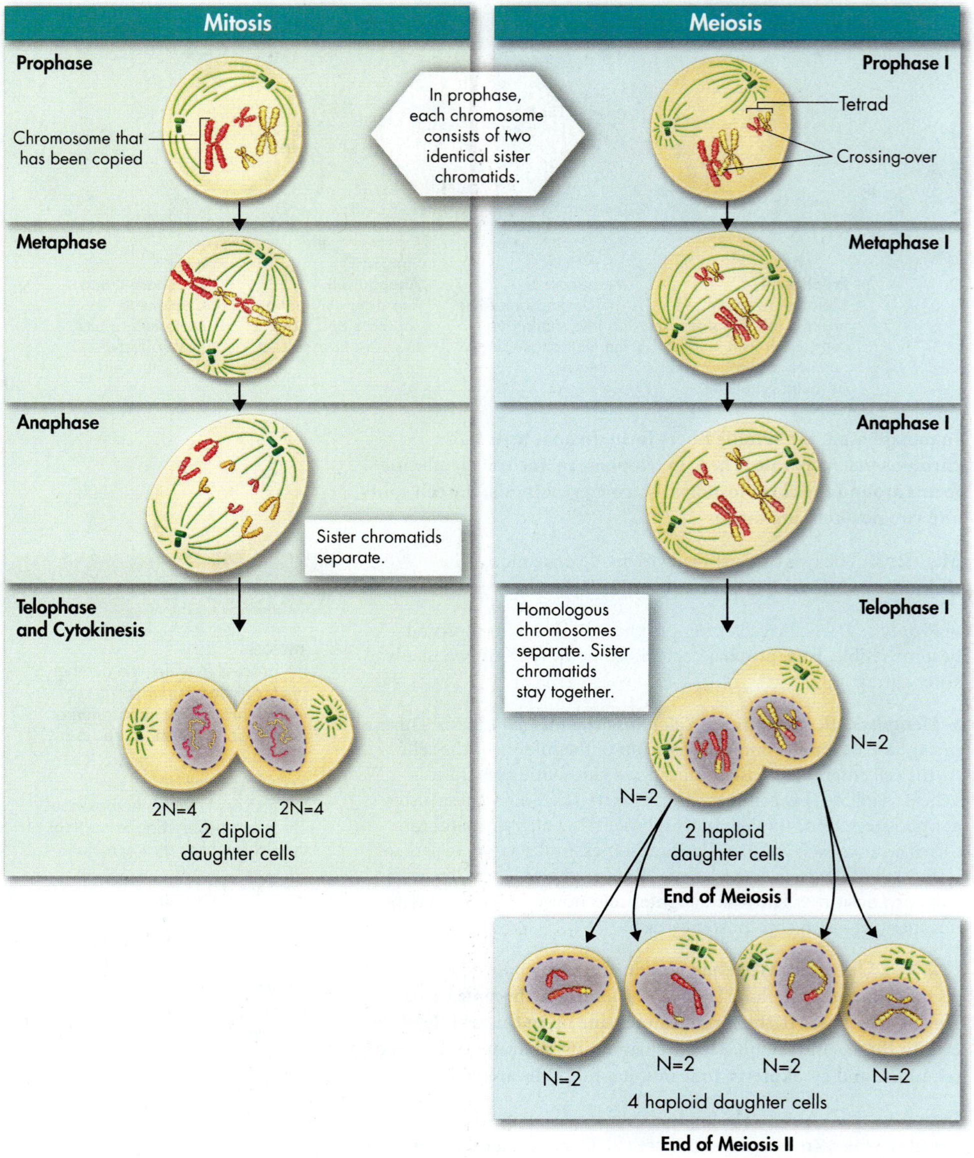

278 Chapter 11 • Lesson 4

Hands-On Learning

TETRAD MODELS Invite students to demonstrate how a tetrad forms from two homologous chromosomes using chenille stems and beads. Follow these steps:

1. Pair students and pass out different color stems and a bead to each pair. Tell students the stems represent sister chromatids and the beads represent centromeres.
2. Tell students to overlap the stems. **Ask** What does this represent? *(crossing-over)*
3. Tell students to cut and tape the stems back together to simulate the separation and recombination of chromosomes. **Ask** Will crossing-over cause a different phenotype in the offspring of true-breeding parents? *(No, the homologous chromosomes are homozygous for the particular trait because they have the same allele for the gene that encodes the trait. Crossing-over will cause portions of the chromosome to be mixed, but is not detected if the alleles are identical.)*

Comparing Meiosis and Mitosis

Even though mitosis and meiosis both involve chromosomes and cell division, they are very different. Mitosis can be a form of asexual reproduction. Meiosis is an early step in sexual reproduction. Review the table to compare the two processes.

Key Question How is meiosis different from mitosis? **Mitosis can be a form of asexual reproduction. It does not change the number of chromosomes. It makes two identical diploid cells. Meiosis is a step in sexual reproduction. It cuts the chromosome number in half. It makes four genetically different haploid cells.**

Gene Linkage and Gene Maps

Genes that are on the same chromosome are usually inherited together because they are linked and do not assort independently during meiosis. Thomas Hunt Morgan's research on fruit flies helped show gene linkage. Morgan used a fly with reddish-orange eyes and miniature wings in a series of test crosses. The genes for those two traits were almost always inherited together.

Linked genes are not always inherited together. When crossing-over happens during meiosis, linked genes can be separated. The farther apart genes are on a chromosome, the more likely crossing-over will separate them. This trend means that the rate of crossing-over can be used to locate and even to map genes on a chromosome.

Key Question How can two alleles from different genes be inherited together? **Alleles of different genes tend to be inherited together when those genes are on the same chromosome.**

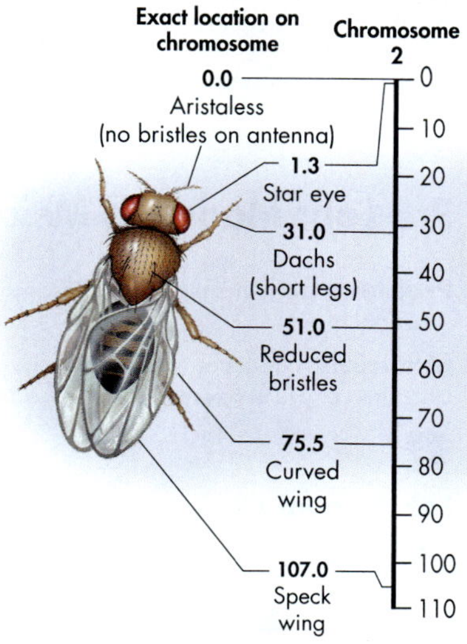

Gene Map Chromosomes are made of many genes linked together. Gene maps show how far apart these genes are from each other.

CHECK Understanding

Apply Vocabulary

Use the highlighted words from the text to complete each sentence correctly.

1. A cell that contains two sets of homologous chromosomes is ______.
2. A ______ is formed when an egg is fertilized by sperm.

Critical Thinking

3. **Summarize** What happens in meiosis I and II?
4. **Write to Learn** Answer the mystery clue.

MYSTERY CLUE

Traits can be controlled by more than one gene. White is the least common color found in parakeets. What does this fact suggest about the genotypes of both green parents? (Hint: See p. 271.)

BIOLOGY.com Search Lesson 11.4 GO • Lesson Assessment

Check Understanding

1. diploid
2. zygote
3. Sample answer: Meiosis I creates two haploid daughter cells that have half the number of chromosomes as the original cell. Meiosis II uses these new cells to create four haploid daughter cells. Each of these cells can become a gamete.
4. Sample answer: Both parents carry recessive alleles that account for the white appearance. White parakeets are likely homozygous recessive for both blue and white pigment genes. Both parents must be heterozygous for both pigments.

Active Reading

KEY QUESTION Read aloud the first question. Ask students to suggest and then make a graphic organizer to put the answers given in an organized format. Invite students to draw their organizers on the board. Invite other students to add information to the organizers.

KEY QUESTION Have a volunteer read aloud the second question and answer. Remind students that genes are located on specific places on chromosomes.

Ask What would Mendel have seen if the two genes were linked on the same chromosome? *(They would have been inherited together. If two daughter plants had the same seed shape, they would also have had the same seed color.)*

Speed Bump

Ask What does it mean for two genes to be linked? *(The genes are located on the same chromosome and are usually inherited together, unless they are separated during crossing-over.)*

Post Reading

ASSESSMENT TIPS Advise students to review the highlighted vocabulary words before they begin the assessment.

MYSTERY CLUE Have students reread the material about multiple alleles and polygenic traits in Lesson 11.3. Also, remind students that when a phenotype is coded for by recessive alleles, it is often less common in a population, especially if the dominant alleles occur frequently in the population. Draw some crosses on the board, and label the genotypes. Lead students to conclude that white parakeets are likely homozygous recessive for both blue and white pigment genes. Both parents must be heterozygous for both pigments.

Wrap-Up Activity

MODELING GENE LINKAGE Have students revisit the close-up image of crossing-over in the meiosis I diagram. Point out that chunks of the chromosomes, not individual genes, are exchanged between chromosomes. Have students model gene linkage in crossing-over by using different colors of modeling clay to represent each chromosome. Students can pull apart chunks of one-color clay and attach them to the other color.

LESSON 11.4

Pre-Lab

Explain to students that they will model meiosis.

Ask What is the purpose of making a meiosis model? *(A model helps to visualize a cell process that cannot easily be seen. Making a model helps me to learn about the details of meiosis.)*

Discuss with students how gametes are formed.

Ask Why is it important for gametes to have only half the number of chromosomes that body cells have? *(When gametes combine, the zygote will have the correct number of chromosomes.)*

Go through the mathematics needed to answer Question 3. If students are not on level mathematically, answer the questions as a class on the board.

Review the Background Questions and the Pre-Lab Questions as a class.

Lab

Tell students they will perform the chapter lab *Modeling Meiosis* described in **Lab Manual B.**

Post-Lab

FLIP BOOK Have students make a meiosis flip book. Have them draw a diagram of each stage of meiosis on a separate index card. The diagrams should all be about the same size and in the same position on the card. When the cards are finished, staple them together. As the cards are flipped, the process of meiosis will appear to be animated.

NATIONAL SCIENCE EDUCATION STANDARDS

CONTENT C.1.a, C.1.c, C.1.d

Skills Lab

GUIDED INQUIRY

Pre-Lab: Modeling Meiosis

Problem How does meiosis increase genetic variation?

Materials pop-it beads, magnetic centromeres, large sheet of paper, colored pencils, scissors

Lab Manual Chapter 11 Lab

Skills Use Models, Sequence, Draw Conclusions

Connect to the Big idea Inherited traits are passed from parents to offspring in the form of genes. Offspring produced by sexual reproduction receive one set of genes from each parent when the reproductive cells, or gametes, combine. Meiosis is the process by which gametes are produced. During meiosis, new combinations of genes form when genes cross over from one homologous chromosome to the other. Also, the sorting of chromatids among gametes is random. Both crossing-over and sorting lead to greater diversity in the genes of a population.

In this lab, you will model the steps of meiosis and track what happens to alleles as they move from diploid cells to haploid gametes.

Background Questions

a. **Review** What are alleles?

b. **Sequence** What happens during prophase I of meiosis? What happens during metaphase I? What happens during anaphase I?

c. **Compare and Contrast** In what ways does meiosis differ from mitosis?

Pre-Lab Questions

Preview the procedure in the lab manual.

1. **Control Variables** Why must you use the same number of beads when you construct the second chromosome in Step 1?
2. **Infer** Why is the longer chromosome pair used to model crossing-over?
3. **Calculate** A diploid cell has two pairs of homologous chromosomes. How many different combinations of chromosomes could there be in the gametes? MATH

BIOLOGY.com Search Chapter 11 GO

Visit Chapter 11 online to test yourself on chapter content and to find activities to help you learn.

Untamed Science Video Travel back in time with the Untamed Science explorers as they prove Mendel was no pea brain!

Art in Motion View a short animation that brings the process of meiosis to life.

Art Review Review your understanding of multiple alleles, incomplete dominance, and other exceptions to Mendel's principles.

InterActive Art Build your understanding of Punnett squares with this animation.

Data Analysis Investigate the connection between crossing-over and gene location.

Tutor Tube Hear suggestions from the tutor for help remembering what happens to chromosomes during meiosis.

Pre-Lab Answers

BACKGROUND QUESTIONS

a. Alleles are different forms of the same gene.

b. During prophase I, homologous chromosomes form tetrads, the tetrads line up across the center of the cell during metaphase I and are pulled to opposite ends of the cells during anaphase I.

c. During meiosis, homologous chromosomes separate, two cell divisions occur, and daughter cells have half as many chromosomes. During mitosis, homologous chromosomes are not separated, only one cell division occurs, and the number of chromosomes per cell does not change.

PRE-LAB QUESTIONS

1. Each chromosome in the homologous pair needs an allele for each gene.
2. The chances of crossing-over are greater on the longer chromosome.
3. There could be four different combinations (ignoring any variation due to crossing-over).

11 CHAPTER Summary

11.1 The Work of Gregor Mendel

- An individual's unique characteristics are determined by factors that are passed from parent to offspring.
- When gametes are made, the alleles for each trait separate from each other. That way each gamete carries only one allele from each gene.

genetics (p. 262)
fertilization (p. 262)
trait (p. 262)
gene (p. 263)
allele (p. 263)
principle of dominance (p. 263)
segregation (p. 264)
gamete (p. 264)

11.2 Applying Mendel's Principles

- Punnett squares use probability to predict combinations of alleles in a genetic cross.
- Genes for different traits segregate independently.
- Mendel's principles of heredity form the basis of modern genetics.

probability (p. 266)
homozygous (p. 267)
heterozygous (p. 267)
phenotype (p. 267)
genotype (p. 267)
independent assortment (p. 269)

11.3 Other Patterns of Inheritance

- Traits can be controlled by incomplete dominant alleles, codominant alleles, more than two possible alleles, and several genes.
- Environmental conditions can affect gene expression and influence genetically determined traits.

codominance (p. 271)
incomplete dominance (p. 271)
multiple alleles (p. 272)
polygenic trait (p. 272)

11.4 Meiosis

- The diploid cells of most adult organisms contain two complete sets of inherited chromosomes and so two complete sets of genes.
- Meiosis is a process of cell division that results in gametes that have half the number of chromosomes that other body cells have. In prophase I, copied chromosomes pair with their matching homologous chromosomes. At metaphase I, paired chromosomes line up across the center of the cell. In anaphase I, chromosome pairs move toward opposite ends of the cell. In telophase I, a nuclear membrane forms around each cluster of chromosomes. Cytokinesis then forms two new cells. As the cells enter prophase II, their chromosomes become visible. The final four phases of meiosis II result in four haploid daughter cells.
- Meiosis results in four genetically different haploid cells. Mitosis results in two genetically identical diploid cells.
- Alleles of different genes tend to be inherited together when those genes are on the same chromosome.

homologous (p. 275)
diploid (p. 275)
haploid (p. 275)
meiosis (p. 276)
crossing-over (p. 276)
zygote (p. 277)

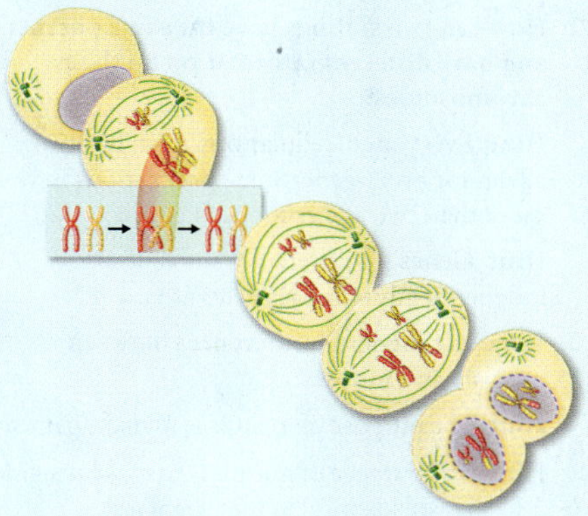

UbD Transfer the Big Idea

TRANSFER Have students think like animal breeders. Tell them they're beginning with true-breeding animal parents that have different genetic traits. Students will assign two or three traits to each parent such as blonde or black coat, sheds or doesn't shed, and long or short ears. Because certain traits in the animals are more desirable than others, they want to determine all phenotypes and genotypes as best they can before they start breeding. After deciding on the kind of animal to breed and the two or three traits, have students label each allele as dominant or recessive. Then have them use Punnett squares to determine the possible genotypes or phenotypes in the first and second generations of offspring.

Chapter Summary

Think Visually

USE GRAPHICS Have students fill in the following terms to create a concept map: *alleles, genes, homologous chromosomes, dominant, heredity, recessive.*

ANSWERS

Answers may vary. Students' concept maps should include that alleles are versions of genes, genes are located on chromosomes, genes help determine traits, and traits can be dominant or recessive.

BIOLOGY.com Study Online

REVIEW RESOURCES

InterActive Art Students can use this interactive activity to learn more about Punnett squares.

Art Review Students can use this drag-and-drop activity to review inheritance patterns using Punnett squares.

Art in Motion This short animation shows students how genetic material is separated during meiosis.

Tutor Tube To help students better understand the process of meiosis, the tutor makes connections between meiosis and Punnett squares.

Data Analysis Students analyze the connection between crossing-over and gene location.

Vocabulary Review The *Flash Cards* and *Match It* provide an interactive way to review chapter vocabulary.

Editable Worksheets Pages of Study Workbook B, Lab Manual B, and the Assessment Resources Book are available online. These documents can be easily edited using a word-processing program.

Lesson Overviews Lesson Overviews are based on Miller & Levine Biology. However, they are available on CD-ROM and can be edited for your class.

ASSESSMENT RESOURCES

Chapter Assessment Have students take an online version of the Chapter 11 Check Understanding.

Standardized Test Prep Students can take an online version of the Standardized Test Prep. You will receive their scores along with ideas for remediation.

Diagnostic and Benchmark Tests Use these tests to monitor your students' progress and supply remediation.

ASSESSMENT

Assess

Big idea Invite a volunteer to read the question. Write the words *trait, gene, alleles, meiosis,* and *gamete* on the board.

Ask How can these words be used in a short paragraph to answer the question? *(Sample: Traits are physical characteristics that are inherited as cellular information. Each trait is controlled by a gene, which can have different forms called alleles. The alleles are distributed to gametes during the process of meiosis.)*

Ask Does the process you just described involve sexual reproduction or asexual reproduction? *(sexual reproduction)* How do you know? *(Gametes from two parents are involved.)*

Have students write their answers in their notebooks.

CONSTRUCTED RESPONSE Remind students that the first sentence in a paragraph is often the topic sentence. This topic sentence should relay the main idea of the paragraph.

Write the first question on the board and read it aloud. Underline the word *characteristics*.

Say This question is about traits. Mendel's principles explain how traits are inherited.

Read the first hint.

Ask How did Mendel cross his pea plants? *(He put pollen from the male part of one flower onto the female part of another flower.)*

Read the next hint.

Say The traits of parents and offspring are often similar.

Ask What does Mendel's principle of dominance say? *(Some alleles are dominant and others are recessive. A dominant allele prevents a recessive allele from being expressed.)*

Model writing the first one or two sentences of the answer using the hints on the board and the text. *(Sample: Mendel studied the inheritance of traits in peas to find out how traits passed from one generation to another. He learned that each trait is controlled by two factors.)*

For Question 2, pair students and have them use the hints and your modeling to answer the question. Have students share their paragraphs with the class.

For Question 3, have students write the paragraphs independently for homework or as an anchor activity.

11 CHECK Understanding

Assess the Big idea **Information and Heredity**

Write an answer to the question below.

Q: How does biological information pass from one generation to another?

Constructed Response

Write an answer to each of the numbered questions below. The answer to each numbered question should be one or two paragraphs long. To help you begin, read the **Hints** below each of the questions.

1. **What did Mendel discover about biological characteristics?**

   Hint Mendel crossed pea plants and studied offspring.

   Hint Mendel learned that the traits of the parents affected the traits of the offspring.

2. **How can two siblings have the same parents and have different information on their chromosomes?**

   Hint Every multicellular organism has two alleles for every gene. Some populations have more than two alleles for each gene.

   Hint Genes for different traits can sort independently when gametes are made.

3. **What are the main differences between meiosis and mitosis?**

   Hint The purpose of meiosis is to make gametes.

   Hint The purpose of mitosis is to make daughter cells that are identical to the parent cell.

Foundations for Learning Wrap-Up

Use the undercover vocabulary tool you made as you read the chapter to help you review the material on heredity and meiosis.

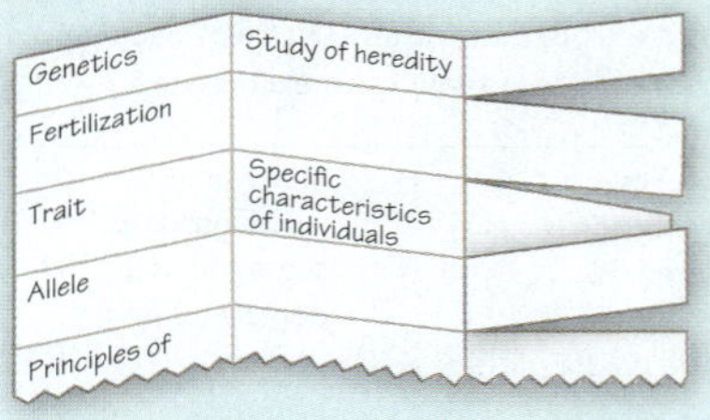

Activity 1 Form groups of three. Have one person read a definition from the terms on the undercover vocabulary tool. Have the other two students try to identify the terms associated with the definitions. Take turns reading the definitions.

Activity 2 Use scissors to carefully cut the definitions and terms from your undercover vocabulary tool. When all the terms and definitions are on separate pieces of paper, mix them up. Try to match the terms and the definitions.

Foundations for Learning Wrap-Up

Have students take out their Undercover Vocabulary Tools. Explain the activities. Monitor student responses in Activity 1 to make sure students have written the proper definition for each vocabulary word. For Activity 2, either pair students or have them work individually.

11.1 The Work of Gregor Mendel

Understand Key Concepts

1. Different forms of a gene are called
   a. hybrids.
   b. dominant factors.
   c. alleles.
   d. recessive factors.

Test-Taking Tip

Use Time Wisely Before you start your test, look at the kinds of questions being asked. Usually, multiple-choice questions take less time to answer than critical-thinking questions. Be sure not to spend so much time on multiple-choice questions that you won't have time to answer the essay questions.

Think Critically

2. **Infer** Suppose Mendel crossed two pea plants and got both tall and short offspring. What could have been the genotypes of the two original plants? What genotype could *not* have been present?

11.2 Applying Mendel's Principles

Understand Key Concepts

3. Organisms that have two identical alleles for a particular trait are said to be
   a. hybrid.
   b. heterozygous.
   c. homozygous.
   d. dominant.
4. A Punnett square is used to determine the
   a. probable outcome of a cross.
   b. actual outcome of a cross.
   c. result of incomplete dominance.
   d. result of meiosis.
5. The physical characteristics of an organism are called its
   a. genetics.
   b. heredity.
   c. phenotype.
   d. genotype.
6. List the four basic principles of genetics that Mendel discovered in his experiments. Briefly describe each of these principles.

Think Critically

7. **Apply Concepts** In guinea pigs, the allele for a rough coat (*R*) is dominant over the allele for a smooth coat (*r*). A heterozygous guinea pig (*Rr*) and a homozygous recessive guinea pig (*rr*) have a total of nine offspring. Is it possible for all the offspring to have smooth coats? Explain your answer.

11.3 Other Patterns of Inheritance

Understand Key Concepts

8. A situation in which a gene has more than two alleles is known as
   a. complete dominance.
   b. codominance.
   c. polygenic dominance.
   d. multiple alleles.
9. Are an organism's characteristics determined only by its genes? Explain your answer.

Think Critically

10. **Interpret Visuals** Genes that control hair or feather color in some animals are expressed differently in the winter than in the summer. How might such a difference be beneficial to the ptarmigan shown here?

Assessment Tips

Have students work individually or in pairs to answer the questions.

Questions 3, 6, and 7 require students to understand Mendel's principles. Have them review these principles at the end of Lesson 11.2.

For Questions 4, 7, and 13, have students review how to make a Punnett square and recognize what information it provides.

Have students review the steps of meiosis before they answer Questions 11 and 12.

Have students share their answers with the class and explain their reasoning for each answer.

Explain that sometimes a single assessment question will ask more than one question, such as in Question 2. Tell students that after they write the answer to a question, they should look back at the question to make sure they have answered all parts.

Answers

ASSESSMENT

Lesson 11.1

UNDERSTAND KEY CONCEPTS

1. c

THINK CRITICALLY

2. The original genotypes and the crosses could have been *Tt* × *tt* or *Tt* × *Tt*. The genotype *TT* could not have been present; if it were, all the offspring would be tall.

Lesson 11.2

UNDERSTAND KEY CONCEPTS

3. c
4. a
5. c
6. (1) The inheritance of biological characteristics is determined by genes. (2) Where there are two or more forms (alleles) of the gene for a single trait, some forms of the gene may be dominant and others recessive. (3) In most sexually reproducing organisms, each adult has two copies of each gene, one from each parent. These genes are segregated when gametes form. (4) The alleles for different genes (actually, the chromosomes) usually segregate independently.

THINK CRITICALLY

7. The result of each fertilization is independent of any previous fertilizations. It is possible for all offspring to have smooth coats. Each offspring could receive a recessive allele from both its parents.

Lesson 11.3

UNDERSTAND KEY CONCEPTS

8. d
9. No, genes provide a plan for development, but how the plan unfolds depends on the environment.

THINK CRITICALLY

10. The color helps the ptarmigan hide from predators. In winter, its white coat color blends in with its snowy surroundings.

ASSESSMENT

Lesson 11.4

UNDERSTAND KEY CONCEPTS

11. d

THINK CRITICALLY

12. Students' tables should show that both processes involve cell division. For mitosis, students should include that it occurs in most body cells, that it forms two diploid daughter cells, and that chromosome structure is not usually affected. For meiosis, students should include that it occurs in sex cells, that it forms four haploid cells, and that crossing-over occurs.

Connecting Concepts

13. 66 smooth and 66 wrinkled

14. Yes, the observed numbers are close to the expected values. No other cross would predict a ratio close to 50 percent for each trait.

CHAPTER MYSTERY Have students read Solve the Chapter Mystery and reread the clues.

Ask Can parakeet color be inherited by simple Mendelian dominance? *(No, simple dominance would not produce four colors in offspring.)*

Ask Which two pigments blend to form green? *(blue and yellow)*

Have students recall from Lesson 11.3 the cross between red-flowered and white-flowered plants that resulted in pink-colored flowers.

Ask What was the inheritance pattern in that example? *(incomplete dominance)* Could feather color of parakeets be inherited by incomplete dominance? *(No, incomplete dominance could not explain the appearance of white offspring.)*

Have students answer the questions, which will help them solve the mystery.

11 CHECK Understanding

11.4 Meiosis

11. Unlike mitosis, meiosis in male mammals results in the formation of

a. one haploid gamete.

b. three diploid gametes.

c. four diploid gametes.

d. four haploid gametes.

Think Critically

12. Compare and Contrast Make a table that compares and contrasts meiosis and mitosis.

Connecting Concepts

Use Science Graphics

In pea plants, the coat, or covering, of the seed is either smooth or wrinkled. Suppose a researcher has two plants—one that makes smooth seeds and another that makes wrinkled seeds. The researcher crosses the wrinkled-seed plants and the smooth-seed plants, obtaining the following data. Use the data to answer questions 13 and 14.

13. Infer Mendel knew that the allele for smooth (*R*) seeds was dominant over the allele for wrinkled (*r*) seeds. If this cross was $Rr \times rr$, what numbers would fill the middle column?

Results of Seed Experiment

Phenotype	Number of Plants in the F_1 Generation	
	Expected	Observed
Smooth seeds		60
Wrinkled seeds		72

14. Analyze Data Are the observed numbers consistent with the hypothesis that the cross is $Rr \times rr$? Explain your answer.

solve the CHAPTER MYSTERY

GREEN PARAKEETS

After talking to the owner of the pet store, Susan realized she had a rare gift. White parakeets are very uncommon. The pet shop owner told Susan that two genes control feather color. A dominant *Y* allele results in a yellow pigment. The dominant *B* allele controls melanin production.

- If the genotype contains a capital *Y* (either *YY* or *Yy*) and a capital *B*, the offspring will be green.
- If the genotype contains a capital *Y* (either *YY* or *Yy*) and two lowercase *b*'s, the offspring will be yellow.
- If the genotype contains two lowercase *y*'s, and a capital *B*, the offspring will be blue.
- If the genotype contains two lowercase *y*'s and two lowercase *b*'s, the offspring will be white. The Punnett square below shows the possible genotypes for the cross that produced Susan's baby parakeets. Copy this Punnett square onto a piece of paper. Use it to answer the questions that follow.

	?	?	?	?
?	BBYY	BBYy	BbYY	BbYy
?	BBYy	BByy	BbYy	Bbyy
?	BbYY	BbYy	bbYY	bbYy
?	BbYy	Bbyy	bbYy	bbyy

1. Use Models Write "blue" in the squares that represent a blue parakeet. Write "yellow" in the squares that represent a yellow parakeet. Write "green" in the squares that represent a green parakeet.

2. Use Models Write "white" in the square that represents a white parakeet.

3. Apply Concepts What are the genotypes of the parents in this cross?

BIOLOGY.com Search Chapter 11 GO • Untamed Science Video • Chapter Mystery

CHAPTER MYSTERY ANSWERS

1. Students' Punnett squares should show a two-factor cross between green parents that are heterozygous for both yellow and blue pigments. The three offspring with *yyBb* or *yyBB* genotypes would be blue. The three offspring with *YYbb* or *Yybb* would be yellow. The nine offspring with *YYBB, YyBB, YYBb,* or *YyBb* genotypes would be green.

2. The one offspring with a *yybb* genotype would be white.

3. The parents are both *BbYy.*

Untamed Science Follow the crew of Untamed Science in **Genetics Takes Root** as they describe the famous pea plant experiments that germinated modern genetics.

Standardized Test Prep

Multiple Choice

1. What happens to the chromosome number during meiosis?
A It doubles.
B It stays the same.
C It halves.
D It becomes diploid.

2. Which ratio did Mendel find in his F_2 generation?
A 3 : 1
B 1 : 3 : 1
C 1 : 2
D 3 : 4

3. During which phase of meiosis is the chromosome number reduced?
A anaphase I
B metaphase I
C telophase I
D telophase II

4. Two pink-flowering plants are crossed. The offspring flower as follows: 25% red, 25% white, and 50% pink. What pattern of inheritance does flower color in these flowers follow?
A dominance
B multiple alleles
C incomplete dominance
D polygenic traits

5. The physical characteristics of an organism are its
A heredity.
B genotype.
C genetics.
D phenotype.

6. Alleles for the same trait are separated from each other during the process of
A cytokinesis.
B meiosis I.
C meiosis II.
D metaphase II.

7. Which of the following is NOT one of Gregor Mendel's principles?
A The alleles for different genes usually segregate independently.
B Some forms of a gene may be dominant.
C The inheritance of characteristics is determined by factors (genes).
D Crossing-over occurs during meiosis.

Questions 8–10

The Punnett square below shows a cross between two pea plants each with round seeds.

	?	?
R	*RR*	*Rr*
R	*RR*	*Rr*

8. The unknown genotype is
A Rr.
B rR.
C rr.
D RR.

9. Which statement is true about the cross in the Punnett square?
A Both parents are heterozygous for the trait.
B Both parents are homozygous for the trait.
C One parent is heterozygous and the other is homozygous for the trait.
D The trait is controlled by codominant alleles.

10. What percentage of the offspring of this cross will produce round seeds?
A 0%
B 25%
C 50%
D 100%

If You Have Trouble With . . .

Question	1	2	3	4	5	6	7	8	9	10
See Lesson	11.4	11.1	11.4	11.3	11.4	11.4	11.2	11.2	11.2	11.2

ASSESSMENT

Before the Test

Review the Test-Taking Tip with students.

Point out the word *NOT* in Question 7.

Make sure students know to review the Punnett square before Questions 8–10 carefully before trying to answer those questions.

Answers

1. C
2. A
3. D
4. C
5. D
6. B
7. D
8. A
9. C
10. D

Test-Taking Tip

ANTICIPATE THE ANSWER

Tell students to read the question stem and try to anticipate what the answer will be. Then, have them read all of the answer choices carefully to see which choice best matches the answer they anticipated. Reinforce that they should read all the answer choices before selecting one.

CHAPTER PLANNER 12 DNA

Chapter Contents	NSES	Time	Core Resources
Chapter Preview			**Student Edition,** pp. 286–287 **Chapter Mystery,** p. 287 **Foundations for Learning,** p. 287
12.1 Identifying the Substance of Genes Bacterial Transformation • Bacterial Viruses • The Role of DNA	II, V, A.2.a, A.2.b, A.2.c, A.2.e, A.2.f, C.1.c, C.2.a, G.1, G.2, G.3	1 period 1/2 block	**Student Edition,** pp. 288–291 **Build Connections,** p. 290 **Study Workbook B** 12.1 Worksheets **Biology.com** *Art in Motion:* Hershey-Chase Experiment • *Visual Analogy:* The Main Functions of DNA and Books • 12.1 Lesson Assessment
12.2 The Structure of DNA The Components of DNA • Solving the Structure of DNA • The Double-Helix Model	II, V, A.2.a, A.2.b, A.2.c, A.2.e, A.2.f, C.1.c, C.2.a, G.1, G.3	1 period 1/2 block	**Student Edition,** pp. 292–295 **Study Workbook B** 12.2 Worksheets **Biology.com** *Tutor Tube:* Memory Tricks for Base Pairing • 12.2 Lesson Assessment
12.3 DNA Replication Copying the Code • Replication in Living Cells	I, V, A.1.b, A.1.d, C.2.a, C.2.c	1 period 1/2 block	**Student Edition,** pp. 296–299 **Inquiry Into Scientific Thinking,** p. 299 **Study Workbook B** 12.3 Worksheets **Biology.com** *Art Review:* Eukaryotic and Prokaryotic DNA Replication • *InterActive Art:* DNA Replication • 12.3 Lesson Assessment
Chapter Pre-Lab	A.1.b, A.2.a, C.2.a	1 period 1/2 block	**Student Edition,** p. 300 **Lab Manual B** *Extracting DNA*

Tools for ELL Instruction

ELL Handbook describes teaching strategies to make *Miller & Levine Biology: Foundation Edition* more accessible to ELL students.

Spanish Study Workbook is a Spanish translation of Study Workbook A.

Multilingual Glossary is the *Miller & Levine Biology* glossary translated into ten languages.

Tools to Challenge Students

Study Workbook A includes more challenging worksheets and activities such as Chapter Mystery activities that reinforce 21st Century Skills.

Lab Manual A labs are geared toward on-level students. The Lab Manual includes fifteen labs in addition to the Chapter Labs.

Biology.com Data Analysis activities offer students a way to practice data gathering and analyzing skills.

Visual Quizzes in the Assessment Resources Book are geared toward on-level students and use visuals from *Miller & Levine Biology.*

Additional Resources

Biology.com Untamed Science Video • Flash Cards

Study Workbook A 12.1 Worksheets
Spanish Study Workbook 12.1 Worksheets

Study Workbook A 12.2 Worksheets
Lab Manual B *Data Analysis:* Base Percentages
Spanish Study Workbook 12.2 Worksheets
Biology.com *Data Analysis:* Tracking Illegal Whaling
Assessment Resources Book Visual Quiz

Study Workbook A 12.3 Worksheets
Lab Manual B *Hands-On Activity:* Modeling DNA Replication
Spanish Study Workbook 12.3 Worksheets
Assessment Resources Book Visual Quiz

Lab Manual A *Extracting DNA*

Chapter Review

Student Edition Chapter Summary, p. 301
Study Workbook B Chapter 12 Review
Transparencies, pp. 148–156
Biology.com Untamed Science Video • Editable Worksheets of Study Workbooks A and B and Lab Manuals A and B • Chapter 12 Flash Cards and Match It

Untamed Science DVD • Classroom Resources CD (includes lesson presentations and editable worksheets)

Chapter Assessment

Student Edition Check Understanding, pp. 302–305
Study Workbook B Chapter 12 Taking a Standardized Test
Assessment Resources Book Chapter 12 Test B
Biology.com Chapter 12 Assessment • Editable Worksheets of Chapter 12 Visual Quizzes* and Chapter 12 Tests A and B

Exam*View* *Assessment Suite* • Classroom Resources CD (includes lesson presentations and editable worksheets)

Time: 1 period, 1/2 block

* Visual Quizzes are based on visuals from *Miller & Levine Biology.* Review carefully to make sure they are appropriate for your class.

Pressed for Time?

Preview the Chapter Preview the DNA Nucleotides visual and read the Key Questions for Lesson 12.2.

Cover the Chapter Quickly Have students read *The Role of DNA* in Lesson 12.1, all of Lesson 12.2, and *Copying the Code* in Lesson 12.3, focusing on the DNA Replication visual.

Assess Assign Question 5 in the 12.1 Check Understanding, the entire 12.2 Check Understanding, Questions 1–4 in the 12.3 Check Understanding, and Questions 3 and 6–14 in the Chapter 12 Check Understanding.

Get the Picture

Have students look at the photograph. Read aloud the caption.

Ask How would you describe the shape of this sculpture? *(Sample answer: The overall shape is a spiral. The rungs are two parts joined at the middle.)*

Ask What do you notice about the rungs that would suggest a pattern? *(Sample answer: The rungs are four different colors. A green rung is attached to a blue rung. A yellow rung is attached to a red rung.)* If you pulled apart the rungs in this sculpture, could you match them up by color? *(yes)*

Explain that the overall shape and the matching rungs model important parts of the structure of a DNA molecule, and that many scientists worked for years to discover DNA's structure. Then tell students to think about the structure of a regular ladder.

Ask Does the structure of a ladder give you a clue as to what it is used for? *(Yes, it allows you to climb to reach something.)* Explain that in the same way, the structure of a DNA molecule reflects its function.

Activate Prior Knowledge

Invite students to recall what they have learned about DNA up to this point as you write the responses on the board. Suggest they think about the role of DNA in cell division and in inheritance. You might jump-start students' thinking by writing on the board *When a cell divides, it makes a copy of its DNA*.

12 DNA

Big ideas Information and Heredity, Cellular Basis of Life

Q: What is the structure of DNA, and how does it function in genetic inheritance?

BIOLOGY.com Search Chapter 12 GO • Flash Cards

286

NATIONAL SCIENCE EDUCATION STANDARDS

UNIFYING CONCEPTS AND PROCESSES

I, II, V

CONTENT

C.1.c, C.2.a, C.2.c, G.1, G.2, G.3

INQUIRY

A.1.b, A.1.d, A.2.a, A.2.b, A.2.c, A.2.e, A.2.f

UbD Connect to the Big Idea

Invite a volunteer to read aloud the chapter number and the chapter title. Read the Big Idea and question. Emphasize the question, **What is the structure of DNA, and how does it function in genetic inheritance?**

Set the Purpose

PREVIEW AND PREDICT Guide students to the INSIDE feature on the next page, and invite a volunteer to read each lesson number and title. Ask students why "cellular basis of life" is part of the Big Idea when we talk about DNA. *(DNA is found in all cells.)*

Point out that the chapter discusses the discovery of the function of DNA first and structure second. Remind students to relate DNA's structure and function as they read the chapter.

CHAPTER 12

INSIDE:

- 12.1 Identifying the Substance of Genes
- 12.2 The Structure of DNA
- 12.3 DNA Replication

Genes are made of DNA. This sculpture at the University of California at Berkeley models the structure of DNA.

- Untamed Science Video
- Chapter Mystery

CHAPTER **MYSTERY**

UV LIGHT

"Put on your sunscreen!" You can hear people say this at beaches on a sunny day. Why do people say it? Sunlight can harm the skin. The most dangerous part of sunlight is the part we can't see: the ultraviolet (UV) rays of the electromagnetic spectrum. Overexposure to UV light harms skin cells. It can also cause a deadly form of skin cancer that kills nearly 10,000 Americans each year. Why is UV light so dangerous? How can these particular wavelengths of light harm our cells to the point of causing cell death and cancer?

Read for Mystery Clues As you read this chapter, look for clues to help you solve the question of why UV light is so harmful to skin cells. Then, solve the mystery.

FOUNDATIONS for Learning

The main ideas from the chapter can be found as the answers to the Key Questions. They are the building blocks of the Big Idea. Before you read the chapter, go through and use the answers to the Key Questions to make Connected Ideas cards. Write only one idea per card. Use several cards to cover all the ideas in long answers. As you read through the chapter, make additional cards that describe important discoveries and the steps of processes you learn. At the end of the chapter are two activities that use the cards to help answer the question: What is the structure of DNA, and how does it function in genetic inheritance?

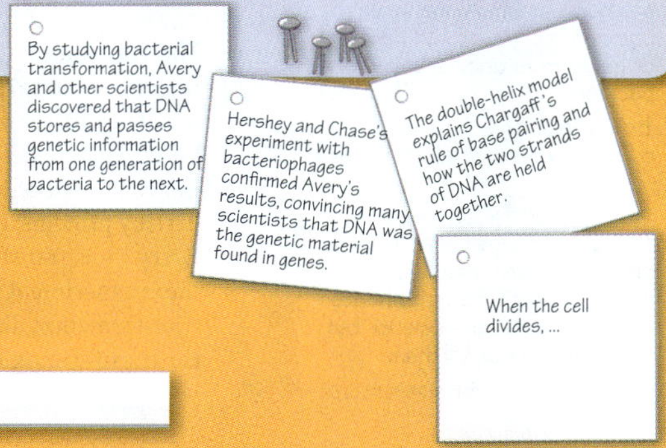

DNA **287**

Chapter 12 Big Idea: Information and Heredity

Chapter 12: What is the structure of DNA, and how does it function in genetic inheritance?

12.1: How did scientists determine that DNA is responsible for storing, copying, and transmitting genetic information?

12.2: How was the basic structure of DNA discovered?

12.3: How do cells copy their DNA?

Foundations for Learning

Pass out index cards. Provide time for students to write the answers to the Key Questions. Have them organize the cards by lesson to help them when they study. Suggest they write the lesson and page numbers on the cards. Then they can add cards with important discoveries and the steps of processes as they read.

Ask Besides Key Questions, what other parts of the text are clues to important ideas to include on your cards? *(Sample answer: headings and subheadings)*

Using a hole punch and paper brads, show students how to connect certain cards to string together related ideas. Remind them that they will use these connected ideas cards as a review and study aid.

Untamed Science

In **Super Sleuth DNA** the Untamed Science crew unlocks the hidden information that can be found in DNA evidence from a crime scene.

CHAPTER MYSTERY

Have a volunteer read the Chapter Mystery. Invite students to brainstorm about how the Chapter Mystery relates to DNA. Then ask them to predict why UV light is so dangerous. Tell students that they will find clues in the chapter to answer the question, **Why is UV light so harmful to skin cells?**

Digital Support

BIOLOGY.com Extend your reach by using the digital assets offered at **Biology.com.**

LESSON 12.1

Pre-Reading

Objectives

12.1.1 Summarize the process of bacterial transformation.

12.1.2 Describe the role of bacteriophages in identifying genetic material.

12.1.3 Identify the role of DNA in heredity.

Preview the Pages

1. Have students locate the vocabulary words in the text. Point out that both words relate to bacteria, which contain DNA. Explain that bacteria are often used in experiments because they are small and grow quickly.
2. Explain that the diagram of a T4 bacteriophage is based on the actual structure of this virus as seen through an electron microscope.

Activate Prior Knowledge

Ask students how they know they have a cold if they can't see cold viruses. *(Sample answer: They have runny noses, sneeze, and are congested.)* Point out that they have experienced the effects of disease without ever seeing the microorganisms. In the same way, scientists knew what DNA did before they were able to see it.

Build Understanding

FLOWCHART Tell students to look at the flowchart in Workbook B. Explain that it is a time line about the discovery of the structure and function of DNA. Remind students to complete the flowchart as they read.

NATIONAL SCIENCE EDUCATION STANDARDS

UNIFYING CONCEPTS AND PROCESSES

II, V

CONTENT

C.1.c, C.2.a, G.1, G.2, G.3

INQUIRY

A.2.a, A.2.b, A.2.c, A.2.e, A.2.f

For the corresponding lesson in *Miller & Levine Biology,* see pages 338–343.

12.1 Identifying the Substance of Genes

Key Questions

What clues did bacterial transformation give about the gene?

What role did bacterial viruses play in identifying genetic material?

What is the role of DNA in heredity?

BUILD Understanding

Flowchart As you read this section, make a flowchart that shows how scientists came to understand the molecule known as DNA.

In Your Workbook Go to your workbook to learn more about making a flowchart. Complete the flowchart for Lesson 12.1.

BUILD Vocabulary

transformation
a process in which one strain of bacteria is changed by a gene or genes from another strain of bacteria

bacteriophage
a kind of virus that infects bacteria

SUFFIXES
The suffix *-phage* comes from the Greek word *phagien,* which means "to eat." Bacteriophages do not actually eat bacteria, but they can harm or kill them.

Bacterial Transformation

How do genes work? You need to know what genes are made of to answer that question. The first scientist to help figure this out was Frederick Griffith. He found two similar types (or strains) of bacteria in mice. One of the strains (S strain) caused pneumonia. Pneumonia is a deadly lung disease. The other strain (R strain) was harmless. Griffith wondered why the S strain was deadly but the R strain was harmless.

Griffith's Experiments When Griffith injected mice with disease-causing bacteria, the mice developed pneumonia and died. When he injected mice with harmless bacteria, the mice stayed healthy. Griffith thought the S strain might make a toxin that made mice sick. So, he heated some of the S cells to kill them. Then he injected the heat-killed S cells into mice. The mice survived. This suggested that the pneumonia was not caused by a toxin.

Next Griffith mixed heat-killed S cells with R cells. He injected the mixture into mice. Griffith was surprised because the mice got sick with pneumonia. He found live disease-causing bacteria in some of the dead mice. How did this happen if he had injected dead S cells?

Transformation The heat-killed bacteria had made the harmless bacteria deadly. Griffith wondered if a particular molecule from the deadly strain had changed the harmless R cells into S cells. He called this process **transformation.** If he was right, this molecule was passed along to the transformed bacteria. Identifying the molecule could be the key to understanding heredity.

The Molecular Cause of Transformation What molecule causes transformation? In 1944, a team led by Oswald Avery tried to find out. They thought knowing this molecule would allow them to know what made up genes. Avery and his team took a mixture of molecules out of heat-killed bacteria. They treated this mixture to destroy proteins, lipids, carbohydrates, and RNA, one at a time. Every time, transformation still occurred. But then they tried one more experiment in which they destroyed the DNA. This time, transformation did not happen. They concluded that DNA was the transforming molecule.

288

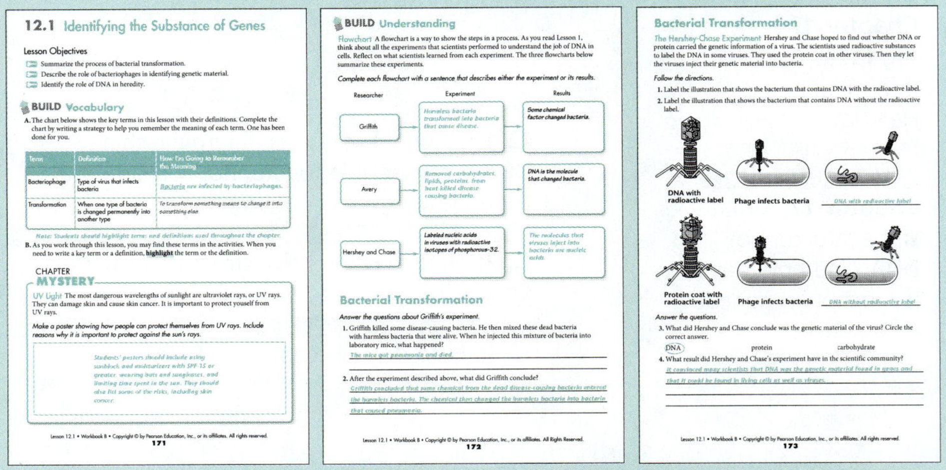

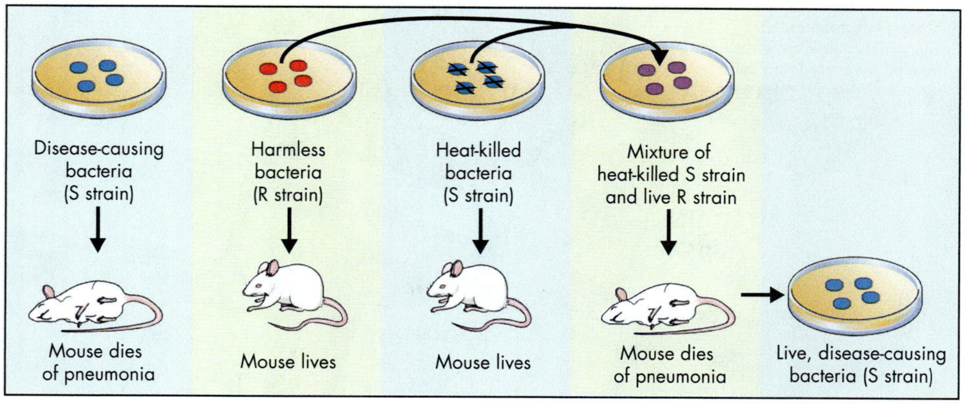

Griffith's Experiments Griffith injected mice with four different samples of bacteria. Disease-causing bacteria that had been heat-killed did not kill the mice. Harmless bacteria did not kill the mice. But when the two strains were mixed together, the mice died. Griffith concluded that genetic information could be passed from one bacterial strain to another.

Key Question What clues did bacterial transformation give about the gene? **Avery and other scientists discovered that DNA stores and passes genetic information from one generation of bacteria to the next.**

Bacterial Viruses

Other scientists tried to confirm Avery's discovery. In 1952, Alfred Hershey and Martha Chase used viruses to study DNA. Viruses are tiny, nonliving particles that can infect living cells.

Bacteriophage A **bacteriophage** is a kind of virus that infects bacterial cells. A bacteriophage sticks to the surface of the cell and injects its genetic information into it. The viral genes make many new bacteriophages, which destroy the bacterium. When the cell splits open, hundreds of new viruses burst out.

The Hershey-Chase Experiment Hershey and Chase used a bacteriophage that had a DNA core and a protein coat. They wanted to find out which part of the virus—the protein coat or the DNA core—entered bacterial cells. They made different parts of these viruses radioactive using two different tags. They used radioactive ^{35}S to tag the protein coats. They used radioactive ^{32}P to tag the DNA. Then, they infected bacterial cells with the viruses. After infection, Hershey and Chase separated the bacteria from the viruses. What they found was that the bacteria had received ^{32}P from the DNA but not ^{35}S from the coat. This finding confirmed that DNA is the molecule of heredity.

T4 Bacteriophage

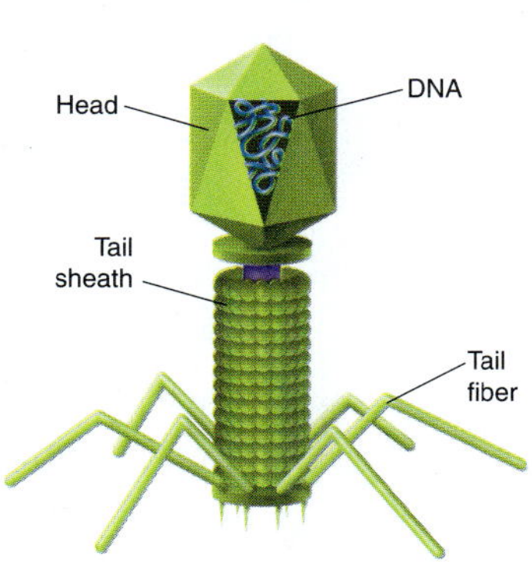

Key Question What role did bacterial viruses play in identifying genetic material?
Hershey and Chase's experiment with bacteriophages confirmed Avery's results, convincing many scientists that DNA was the genetic material found in genes.

BIOLOGY.com Search Lesson 12.1 GO • Art in Motion

During Reading

Active Reading

INTERPRET DIAGRAMS Have students study the figure of Griffith's experiment as you read aloud the first three paragraphs of Bacterial Transformation on the previous page. Reinforce that the S strain bacteria were dead when they were injected along with living R strain.

Ask What did Griffith expect to happen when he injected living R strain and dead S strain together? *(The mice would live.)* What did Griffith conclude caused the R strain to become deadly? *(A gene from the S strain moved into the R strain.)*

KEY QUESTION Read aloud the first Key Question and answer. Review the definition of *transformation*.

DRAW CONCLUSIONS Make sure students understand that the methods were different in the Avery and Hershey-Chase experiments, but the goal and results were the same.

Ask If Avery had already discovered that DNA passes genetic information, why did Hershey and Chase do their own experiment? *(Sample answer: to confirm Avery's results)*

Say When different scientists conduct similar experiments and get the same results, then the results are more likely to be valid.

KEY QUESTION Have a volunteer read aloud the second Key Question and answer.

Ask How did the structure of the bacteriophage make it easier to identify DNA as the genetic material in genes? *(Hershey and Chase had only two substances—protein and DNA—to tag.)*

Build Vocabulary

SUFFIXES Read aloud the text for Suffixes on the previous page. Tell students that when they see *-phage* as a word part, they know the word relates to eating. Write these prefixes and their meanings on the board. *(dys—bad; phyto—plant; hema—blood)* Then write these words and invite students to offer a definition: *dysphagia* (difficulty eating, specifically swallowing), *phytophage* (plant eater), hematophage (blood eater).

Speed Bump

Ask How did Avery show that DNA, and no other substance, was responsible for transformation? *(Transformation occurred every time he did the experiment except when DNA was destroyed.)*

Hands-On Learning

MODEL HERSHEY-CHASE Hand out the following materials to small groups: several rubber bands, two paper clips of different colors, construction paper, small plastic bag, scissors, and masking tape.

1. Explain that a virus has two basic structural components: an outer protein coat and inner genetic material. Show an unshelled walnut (avoid peanuts) as a model.
2. Have students model the Hershey-Chase Experiment. *(Sample model: Attach one colored paper clip to rubber bands to represent tagged DNA and the other colored paper clip to a piece of construction paper to represent the tagged coat. Wrap and tape the paper around the rubber bands to represent a virus. Open the coat enough to pull out the DNA and place it in the plastic bag, or cell.)*
3. Ask students to use their model to write in their own words how the Hershey-Chase experiment helped scientists know that DNA is in genes.

LESSON 12.1

During Reading continued

Build Connections

VISUAL ANALOGY Direct students' attention to the art on these two pages. Read aloud all the captions.

Ask How is DNA like a how-to book? *(Both contain instructions on how to do something.)* Point out that both DNA and a book store information.

Say Suppose a group of people had only one book on how to build a house, and they had never built one before.

Ask What might be some disadvantages of having only one book? *(Sample answer: It would take a long time to build the house because everyone would need time to read the book. As people worked, they might make mistakes because they couldn't refer to the book when they needed to.)* Point out that in the same way, every cell needs its own copy of DNA so the cell has the information it needs at all times to function normally.

Ask What problems might occur if everyone did have a copy of the book, but several pages were missing from each one? *(Sample answer: The text where the missing pages start and end would not go together. They would not make sense. The people would not be able to build the house correctly.)* Point out that in the same way, cells must get a complete copy of DNA to function normally.

Active Reading

SUMMARIZE Model how to summarize information. Read aloud the Storing Information paragraph.

Say Summarizing information helps me remember it. When I summarize, I restate the main ideas in a sentence or two. First, I identify the main ideas. The first sentence is a main idea. The next two sentences are supporting details—they give examples or details about the main idea. How can I put these sentences together? I see that the details are talking about traits. So I can write *DNA stores information that determines traits.*

Have students read the rest of The Role of DNA and then work with a partner to identify the main ideas and supporting details. Then have the pair discuss how to summarize what they have read by restating the main ideas in a sentence or two. Invite volunteers to share their statements.

Speed Bump

Ask What might happen if mitosis occurred before DNA was copied? *(One of the daughter cells would not get a full copy of the genetic information it needs to function.)*

Storing Information
The genetic material stores information needed by every living cell.

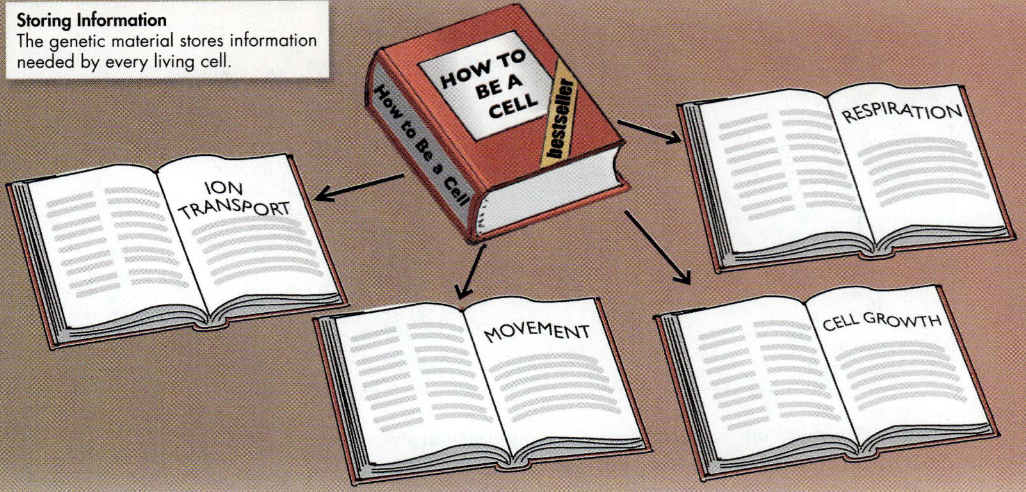

BUILD Connections

THE MAIN FUNCTIONS OF DNA

Like DNA, the book in this diagram contains coded instructions for a cell to carry out important biological processes. The book, like DNA, can also be copied and passed along to the next generation. These three tasks—storing, copying, and passing on information—are also the three main functions of DNA.

The Role of DNA

After scientists learned that genes were made of DNA, they had another question. How could DNA do all the things that genes were known to do? The DNA that makes up genes must be capable of storing, copying, and passing on the genetic information in a cell. These three functions are similar to the way you might share an important book.

Storing Information The main job of DNA is to store information. The genes that make a flower purple must carry the information needed to make purple pigment. Genes for blood type and eye color must have the information needed for their jobs, as well. Other genes have to do even more. Genes control patterns of development. The instructions that cause a single cell to develop into an oak tree, a goldfish, or a dog must be written into the DNA of each of these organisms.

Copying Information A cell must make a complete copy of every one of its genes before the cell divides. After the experiments of Avery, Hershey, and Chase, scientists wanted to know how DNA is copied. The answer to this puzzle had to wait until the scientists could figure out the structure of DNA. Within a few weeks of learning DNA's structure, scientists suggested a way DNA could be copied. You will learn about this mechanism later in the chapter.

BIOLOGY.com Search Lesson 12.1 GO • Visual Analogy

ELL Focus on ELL: Extend Language

INTERMEDIATE AND ADVANCED SPEAKERS Help students relate English syntax to meaning. Write these sentences on the board: **The flower is purple. He held a purple flower.**

Ask Which part of speech is *purple*? *(adjective)* Point out that *purple* follows the verb in the first sentence and comes before the noun in the second sentence. Explain that these positions for adjectives are common in English sentences.

Then write on the board and read aloud the following phrase from the text: *the genes that make a flower purple*. Explain that the adjective follows the noun for emphasis. The important idea is that genes cause the flower's color. Read aloud the phrase but say *purple* before *flower*. Discuss how this changes the emphasis.

Speakers of French and Spanish will recognize this unusual construction in those languages—only in reverse (adjective before noun for emphasis).

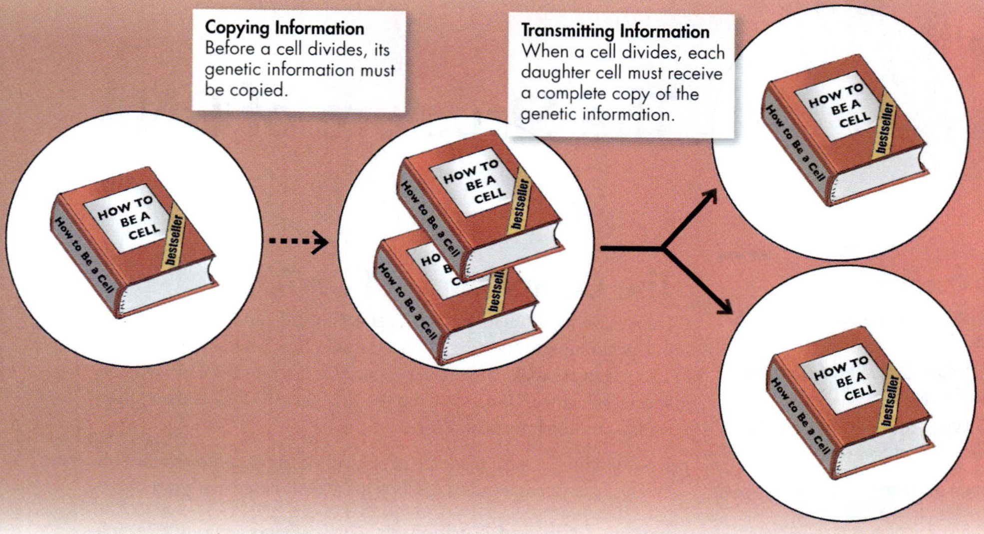

Transmitting Information Mendel's work showed that genes are passed down from parents to offspring. Therefore, DNA molecules must be carefully sorted and passed along when cells divide. Such careful sorting is especially important when reproductive cells are made in meiosis. Remember, the chromosomes of eukaryotic cells contain genes made of DNA. The loss of any DNA during meiosis might mean a loss of valuable genetic information that offspring might need to survive.

Key Question What is the role of DNA in heredity? **The DNA that makes up genes must be capable of storing, copying, and passing on the genetic information in a cell.**

CHECK Understanding

Apply Vocabulary

Use the highlighted words from the lesson to complete each sentence correctly.

1. A _______ is a virus that infects bacteria.

2. _______ is a process that changes the nature of cells.

Critical Thinking

3. Review Define *transformation.*

4. Review Briefly describe how Avery and his group determined which molecule was most important for transformation.

5. Review What are the three key roles of DNA?

6. Quick Write Create a time line that shows the sequence of the scientific discoveries in this lesson.

BIOLOGY.com Search Lesson 12.1 GO • Lesson Assessment

Check Understanding

1. bacteriophage

2. Transformation

3. the process in which one strain of bacteria is changed by a gene from another strain

4. Sample answer: They destroyed the proteins, lipids, carbohydrates, and RNA in killed, harmful bacteria, and injected the bacteria into mice. Transformation occurred. Then they repeated the experiment but destroyed only DNA. Transformation did not occur.

5. storing, copying, and transmitting genetic information

6. The time line should show that Griffith discovered that a chemical changed harmless cells into harmful ones. Then Avery discovered that this chemical is DNA. Hershey and Chase confirmed that genes were made of DNA.

Active Reading

DRAW CONCLUSIONS Have students recall the difference between meiosis and mitosis.

Say If a mistake was made in passing along DNA molecules during mitosis of skin cells, that error might continue to be passed to the next generations of skin cells.

Ask If a mistake was made in passing along DNA molecules during meiosis, which cells in an offspring might inherit that error? *(all of them)*

KEY QUESTION Read aloud the Key Question and answer.

Ask How do the headings under The Role of DNA help answer the question? *(The answer is given in the subheadings.)*

Speed Bump

Have students make their own analogy of the role of DNA. Remind them to include all three major roles. *(Sample answer: A hard drive on a computer is like DNA because it stores information. You can copy that information into a document and transmit it through email over the Internet.)*

Post Reading

ASSESSMENT TIPS For Question 4, have students write out the steps Avery used in his experiment.

For Question 6, suggest students use the connected ideas cards they started at the beginning of the lesson. They should have added cards on important discoveries.

Wrap-Up Activity

SHUFFLE, SWAP, SEQUENCE Divide students into six small groups, and give each group a stack of index cards. Assign two groups the Griffith experiment, two groups the Avery experiment, and two groups the Hershey-Chase experiment. Tell them to work as a team to sequence the steps in the experiment, writing each step on a separate index card. Then have them shuffle the cards and swap them with a neighboring group. Each group should use the cards to sequence the experiment. Use this activity to reinforce the importance of complete and accurate genetic information being passed on when cells divide. Discuss the difficulty of completing the activity if a step were unclear or missing.

LESSON 12.2

Pre-Reading

Objectives

12.2.1 Identify the chemical components of DNA.

12.2.2 Discuss the experiments leading to the identification of DNA as the molecule that carries the genetic code.

12.2.3 Describe the steps leading to the development of the double-helix model of DNA.

Preview the Pages

1. Ask students to locate these words on the first page: *deoxyribonucleic, nitrogenous, adenine, guanine, cytosine, and thymine.* Pronounce each word and have students repeat after you.
2. Preview the art of DNA molecules. Explain that they look different because the art on the first and last pages is two-dimensional. The art of the double helix shows what the molecule looks like in three dimensions.
3. Draw attention to the photographs. Point out that scientists built on one another's work to discover the structure of DNA. Rosalind Franklin built on the work by Chargaff. Watson and Crick built on Franklin's work.

Build Understanding

T-CHART Direct students to Workbook B, where they will find a partially completed T-chart. Remind students that a T-chart is a way of organizing information from the lesson and can be used as a review and study aid.

NATIONAL SCIENCE EDUCATION STANDARDS

UNIFYING CONCEPTS AND PROCESSES

II, V

CONTENT

C.1.c, C.2.a, G.1, G.3

INQUIRY

A.2.a, A.2.b, A.2.c, A.2.e, A.2.f

For the corresponding lesson in *Miller & Levine Biology,* see pages 344–348.

12.2 The Structure of DNA

Key Questions

What are the chemical parts of DNA?

What clues helped scientists solve the structure of DNA?

What does the double-helix model tell us about DNA?

BUILD Understanding

T-Chart Make a T-chart with all the green and blue headings in the left column. As you read, find the key ideas for each heading. Write down a few key words from each main idea in the right column.

In Your Workbook Go to your workbook to learn more about T-charts. Finish the chart for Lesson 12.2.

The Components of DNA

Deoxyribonucleic acid, or DNA, can be copied and passed from one generation to the next. It also specifies how proteins are made. These features make DNA a very special molecule. DNA also has a special structure. Understanding the structure of DNA is the key to understanding how genes work.

DNA is a nucleic acid made up of nucleotides joined into long strands or chains by covalent bonds. Let's look at each of these parts more closely.

Nucleic Acids and Nucleotides Nucleic acids are long molecules found in cell nuclei. Like other big molecules, nucleic acids are made up of smaller subunits, or parts. These parts are linked together to form long chains. Nucleotides are the building blocks of nucleic acids. These nucleotides are made up of three basic parts: a 5-carbon sugar called deoxyribose, a phosphate group, and a nitrogenous base.

Nitrogenous Bases and Covalent Bonds Nitrogenous bases are bases that have nitrogen in them. DNA has four kinds of nitrogenous bases: adenine (AD uh neen) guanine (GWAH neen), cytosine (SY tuh zeen), and thymine (THY meen). Biologists often refer to the nucleotides in DNA by the first letters of their base names: A, G, C, and T. The nucleotides in a strand of DNA are joined by covalent bonds formed between the sugar of one nucleotide and the phosphate group of the next. The nitrogenous bases stick out sideways from the nucleotide chain. The nucleotides can be joined together in any order. So, any order of bases is possible. The structure of DNA makes it very good at absorbing ultraviolet (UV) light. This property is one way that scientists can measure the amount of DNA in a solution.

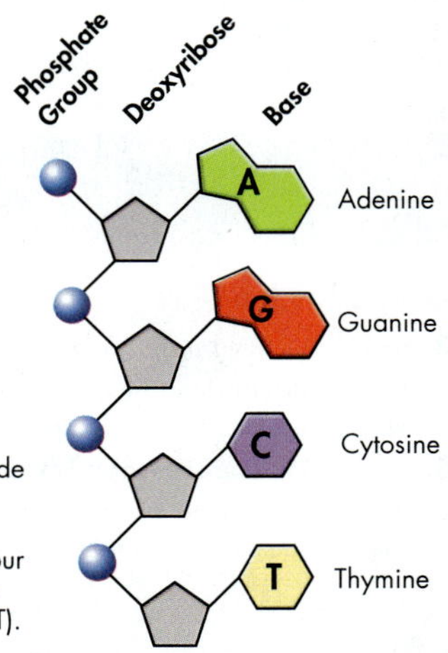

DNA Nucleotides DNA is made up of nucleotides. Each nucleotide has a deoxyribose molecule, a phosphate group, and a nitrogen-containing base. The four bases are adenine (A), guanine (G), cytosine (C), and thymine (T).

Key Question What are the chemical parts of DNA?
DNA is a nucleic acid made up of nucleotides joined into long strands or chains by covalent bonds.

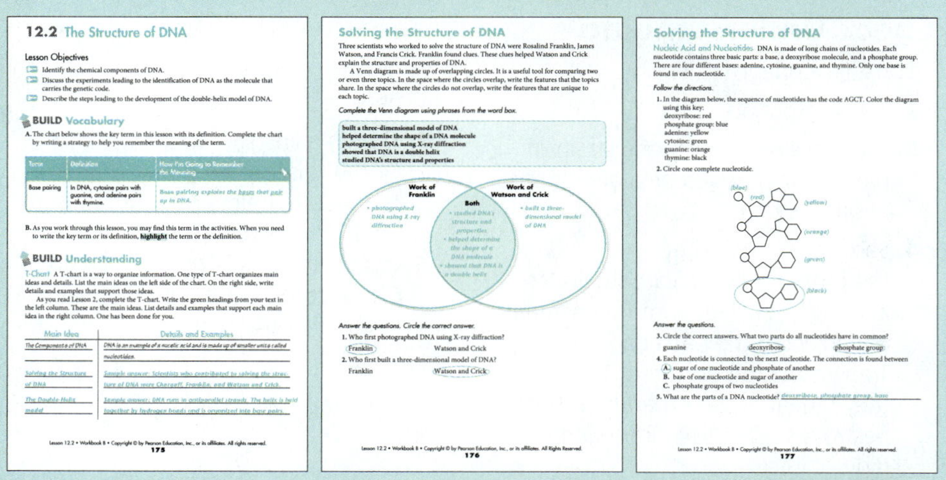

Rosalind Franklin made X-ray diffraction pictures that helped determine the structure of DNA.

Erwin Chargaff

Franklin's X-ray diffraction photograph, May 1952

Solving the Structure of DNA

Knowing that DNA is made from long chains of nucleotides was only the beginning of understanding the structure of this molecule. The next step was to figure out how those chains are arranged.

Chargaff's Rule Erwin Chargaff had discovered years earlier that the percentages of adenine [A] and thymine [T] bases are almost equal in any sample of DNA. The same thing is true for guanine [G] and cytosine [C]. The observation that [A] = [T] and [G] = [C] became known as "Chargaff's rule." DNA samples from organisms as different as bacteria and humans obeyed this rule. However, no one knew why they did.

Franklin's X-Rays In the early 1950s, Rosalind Franklin began to study DNA. She used a technique called X-ray diffraction to get information about the structure of the DNA molecule. Her X-ray pictures showed that the strands in DNA are twisted around each other. This shape is known as a helix. She also showed that DNA is made of two strands. Other clues suggested that the nitrogenous bases are near the center of the DNA molecule.

The Work of Watson and Crick Meanwhile, James Watson and Francis Crick were also trying to understand the structure of DNA. They tried to build models of DNA out of cardboard and wire. Nothing worked. The pieces did not fit. Then, in 1953, Watson saw a copy of Franklin's X-ray picture. Immediately, he realized that it was the missing piece of the puzzle. Watson rushed back to tell Crick about the picture, and a few weeks later they had solved the structure of DNA.

Key Question What clues helped scientists solve the structure of DNA? **The clues in Franklin's X-ray pattern allowed Watson and Crick to build a model that explained the specific structure and properties of DNA.**

Clues to the Structure of DNA Erwin Chargaff and Rosalind Franklin both helped solve the puzzle of the structure of DNA.

During Reading

Active Reading

KEY QUESTION Read aloud the Key Question and answer on the previous page.

Ask Can you find this answer in more than one place? *(yes, in the text or by looking at the diagram)*

USE DIAGRAMS Make an overhead transparency of the diagram showing the components of DNA. Draw a circle around a nucleotide.

Ask What is this subunit, or part, called? *(a nucleotide)* What is it made of? *(a phosphate group, a deoxyribose molecule, and a nitrogenous base)* Point to each component as students name it.

Ask What is formed when the nucleotides are put together in chains? *(a nucleic acid, or DNA)*

PREDICT Have volunteers read aloud Solving the Structure of DNA. Refer them to the photo of the DNA model at the beginning of the chapter. Then ask them how they think the model relates to Chargaff's discovery of the percentage of bases being equal. *(The bases match up.)*

Tell students they will check their prediction after reading the rest of this lesson.

KEY QUESTION Have a volunteer read aloud the Key Question and answer.

Ask What did Franklin's X-ray pictures show? *(DNA has two strands that are twisted around each other.)*

Speed Bump

Ask Which chemical parts are the same in every nucleotide? *(the phosphate group and deoxyribose molecule)* Which parts can be different? *(The nitrogenous bases can be one of four different kinds.)*

ELL Focus on ELL: Build Background

INTERMEDIATE AND ADVANCED SPEAKERS Help students use what they already know about chemical bonds and nucleic acids to learn about the components of DNA.

Write these terms on the board: *nucleic acid, molecule, covalent bond*. Ask students to brainstorm what they already know about each term. Encourage them to look at the visuals in Chapter 2 to help with recall. Summarize their responses on the board.

Draw a sketch of the visual DNA Nucleotides on the board and add further details. Describe aloud everything you draw. Draw a phosphate group showing one phosphorus atom and four oxygen atoms. Draw a large pentagon and write *C* in each of the five corners to represent a deoxyribose molecule with five carbon atoms. Then draw an arrow to the line between a deoxyribose molecule and a phosphate group. Point out that in this covalent bond, carbon atoms in deoxyribose share electrons with oxygen atoms in the phosphate group. Have students work together to add their own diagrams of bases.

During Reading continued

Active Reading

IDENTIFY TEXT PATTERNS Read aloud the first paragraph under The Double-Helix Model. Then give students a moment to scan the rest of the passage.

Ask Does the first paragraph tell you what base pairing is or how the strands are held together? *(no)* How do you find that information? *(by reading the rest of the passage)*

Explain that this paragraph is an introduction to the text under the green heading; it gives readers a sense of what they will read about. An introduction sets a purpose for reading—to find out more. Point out that this pattern is used throughout the book.

CLARIFY WORD USAGE Draw students' attention to the words *double* and *helix* that appear in the first two sentences under The Double-Helix Model.

Ask Which part of speech is *helix* in the first sentence? *(noun)* Which part of speech is *double*? *(adjective)* In the second sentence, which part of speech is *model*? *(noun)* Which part of speech is *double-helix*? *(adjective)*

Point out that when *helix* is used as a noun and modified by *double*, the words are not hyphenated. When both words together modify a noun, they are hyphenated. Remind students that this is called a *compound adjective*. The two words together modify *model*.

INTERPRET DIAGRAMS Ask a volunteer to read the text under Antiparallel Strands. Then have students observe the computer model of DNA. Make sure they understand that the model is three-dimensional.

Ask If the bases meet in the center, where might you find the phosphate and deoxyribose groups? *(the edges or the spiral itself)* What do you think the yellow balls represent? *(deoxyribose)* Why do you think this? *(because they are large and make up the edges, not the center)*

MAKE CONNECTIONS Help students see why DNA is described as a twisted ladder. Pass out chenille stems and two different colors of twist ties. Have students work with partners to make a "ladder" by attaching twist ties to parallel stems and to other twist ties. Then one student can hold the ends of the paired stems while the partner twists them into a spiral shape. Alternatively, you can pass around a flexible model of DNA and have students bend and unbend it.

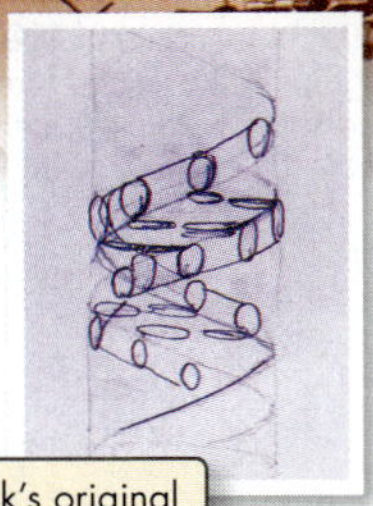

A Double Helix Watson and Crick determined that DNA has the structure of a double helix.

The Double-Helix Model

DNA is a double helix. A double helix looks like a twisted ladder. In the double-helix model of DNA, the two strands twist around each other like spiral staircases. Watson and Crick realized that the double helix explained Franklin's X-ray pattern. The shape also explains many of the most important properties of DNA. The double-helix model explains the reasons behind Chargaff's rule. It also explains how the two strands of DNA are held together. This model can even tell us how DNA carries genetic information.

Antiparallel Strands One of the surprising discoveries of the double-helix model is that the two strands of DNA run in opposite directions. As biochemists say, these strands are "antiparallel." Because of this arrangement, the nitrogenous bases on both strands meet at the center of the molecule. The structure also allows each strand of the double helix to carry a sequence of nucleotides. These bases are arranged almost like letters in a four-letter alphabet.

Hydrogen Bonding At first, Watson and Crick could not explain what forces held the two strands of DNA's double helix together. Then, they discovered that hydrogen bonds could form between certain nitrogenous bases. Hydrogen bonds are fairly weak forces. These bonds have just enough force to hold the two strands of DNA together.

Why would a molecule as important as DNA be held together by weak bonds? Well, if the two strands of the helix were held together by strong bonds, it might not be possible to pull them apart. As we will see, DNA's strands have to be able to separate easily to function.

294 Chapter 12 • Lesson 2

Focus on Academic Language

- antiparallel

Help students understand why DNA strands are called *antiparallel* and not *opposite*. Tell them to cross their arms in front of them on the desk. They should place the right forearm palm down on top of the left forearm, which is also palm down.

Ask What does *parallel* mean? *(side by side)* What does *anti* mean? *(opposite)* Are your forearms parallel to each other? *(yes)* Are they opposite, or exact mirror images of each other? *(no)*

Point out that their forearms are opposite in the direction in which their fingertips are pointing. Have students compare this orientation with what they observe in the diagram of base pairing. Explain that the top left strand starts with a phosphate molecule and ends with a deoxyribose. The right strand begins and ends in the opposite way. This is the meaning of *antiparallel*.

Base Pairing Watson and Crick's DNA model showed that hydrogen bonds could hold the bases together along the center of the molecule. However, these bonds would form only between certain base pairs. Adenine paired with thymine, and guanine paired with cytosine. This nearly perfect fit between A–T and G–C nucleotides is known as **base pairing.**

Once they discovered the fit between bases, Watson and Crick realized that base pairing explained Chargaff's rule. It gave a reason why [A] = [T] and [G] = [C]. For every adenine in a double-stranded DNA molecule, there has to be one thymine. For each guanine, there is one cytosine. Their model explained Chargaff's observations. It also accounted for Rosalind Franklin's findings. At last, Watson and Crick were confident that they had come to the right conclusion.

Key Question What does the double-helix model tell us about DNA?
The double-helix model explains Chargaff's rule of base pairing and how the two strands of DNA are held together.

BUILD Vocabulary

base pairing
the principle that bonds in DNA can form only between adenine and thymine and between guanine and cytosine

WORD ORIGINS
The word *pair* is from the Latin word *paria,* meaning "equal" or "alike."

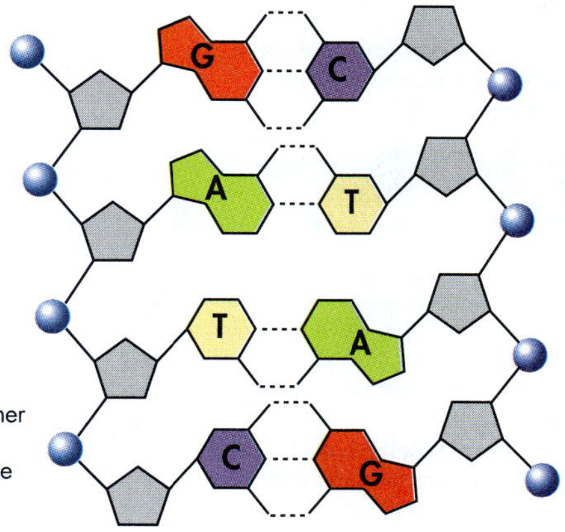

Base Pairing The two strands of DNA are held together by hydrogen bonds. The bonds here are represented by the dashes between the nitrogenous bases adenine and thymine, and between guanine and cytosine.

CHECK Understanding

Apply Vocabulary

1. The principle of _______ states that bonds in DNA can form only between adenine and thymine and between guanine and cytosine.

Critical Thinking

2. Review List the chemical components of DNA.

3. Relate Cause and Effect Why are hydrogen bonds so important to the structure of DNA?

4. Review Describe the discoveries that led to the modeling of DNA.

5. Review Describe Watson and Crick's model of the DNA molecule.

6. Write to Learn Answer the first mystery clue.

MYSTERY CLUE

The energy from UV light can excite electrons in a substance. The excited electrons can cause chemical changes. What chemical changes might occur in the nitrogenous bases of DNA? (**Hint:** See p. 294.)

BIOLOGY.com Search Lesson 12.2 GO • Lesson Assessment

Check Understanding

1. base pairing

2. 5-carbon sugar molecules, phosphate groups, and nitrogenous bases

3. Hydrogen bonds hold together the paired bases. Because these bonds are weak, the two strands of DNA are easily separated.

4. Chargaff determined that in a DNA sample, adenine and thymine are present in equal proportions, and the same is true for guanine and cytosine. Franklin's X-ray pictures of DNA revealed a spiral structure. Watson and Crick used these clues to model a double helix with matching base pairs.

5. DNA consists of two antiparallel strands of nucleotides wound around each other. The A and T bases and C and G bases pair between the strands, joined by hydrogen bonding.

6. The bonds between the bases are fragile. UV light might change the way the bases fit together and the way they bond in the DNA molecule.

LESSON 12.2

Active Reading

PREDICT After students read the text under Base Pairing, have them discuss whether the prediction they made earlier about Chargaff's discovery was correct. Then tell them to look at the left side of the DNA molecule in the Base Pairing diagram.

Ask If the bases on the left side continue as A A C T, what will be the matching bases? *(T T G A)*

INFER Ask students what the effect on the structure and function of DNA would be if the sugars and bases were joined by hydrogen bonds? *(The bonds would break and the molecule might not function normally.)*

KEY QUESTION Read aloud the Key Question and answer.

Ask How does the arrangement of base pairs relate to Chargaff's rules? *(Adenine and thymine are present in equal percentages because they always pair; the same is true for guanine and cytosine.)*

Build Vocabulary

WORD ORIGINS Ask students how the base pairs in DNA are alike. *(They have a similar structure and function.)*

Speed Bump

Ask If the percentage of adenine in human DNA is 30%, what is the percentage of thymine? *(30%)* Why? *(For every adenine in a DNA molecule, there has to be one thymine; adenine and thymine are base pairs.)*

Post Reading

ASSESSMENT TIPS Tell students that Question 3 is asking for the effects of hydrogen bonds in the DNA molecule.

For Question 5, suggest students refer to the text and the picture of the three-dimensional model.

MYSTERY CLUE Remind students that chemical changes occur when bonds form or break. Ask them to infer which bonds are most likely to be broken in a DNA molecule.

Wrap-Up Activity

WRITE RIDDLES Ask students to brainstorm about key people and terms in this lesson as you list them on the board. Divide the class into small groups and assign each group an equal number of the words, for which they make up short riddles. The rest of the class guesses the answer. *(Example: "My partner is thymine." [adenine])*

Pre-Reading

Objectives

12.3.1 Summarize the events of DNA replication.

12.3.2 Compare DNA replication in prokaryotes with that of eukaryotes.

Preview the Pages

1. Direct students to the micrograph of a DNA molecule on this page. Explain that a micrograph is a photo of an object as seen through a microscope. Tell students a double strand of DNA is 2 nanometers in diameter. A nanometer is a billionth of a meter. By comparison, the diameter of a human hair is about 90,000 nanometers.
2. Compare the colors of the bases and the phosphates in the DNA replication diagram with those in the Base Pairing diagram in Lesson 12.2. *(They are the same.)*
3. Point out the diagrams of prokaryotic DNA and eukaryotic DNA. Have students recall that eukaryotic cells have a nucleus. The cells of prokaryotes lack a nucleus.

Build Understanding

PREVIEW VISUALS Direct students to Workbook B, where they will write their questions. Suggest they read the labels and think about what they already know about a DNA molecule before they write their questions.

NATIONAL SCIENCE EDUCATION STANDARDS

UNIFYING CONCEPTS AND PROCESSES

I, V

CONTENT

C.2.a, C.2.c

INQUIRY

A.1.b, A.1.d

For the corresponding lesson in *Miller & Levine Biology,* see pages 350–353.

12.3 DNA Replication

Key Questions

What role does DNA polymerase play in copying DNA?

How does DNA replication differ in prokaryotic cells and eukaryotic cells?

BUILD Understanding

Preview Visuals Before you read, study the DNA Replication figure on the next page. Make a list of questions about the figure. As you read, write down the answers to your questions.

In Your Workbook Go to your workbook to learn more about previewing visuals.

BUILD Vocabulary

replication
the process of copying DNA prior to cell division

USE SUFFIXES
The suffix *-tion* means "process of." A replica is a copy. So, replication is the process of making a copy.

Copying the Code

When Watson and Crick discovered the structure of DNA, they immediately recognized something important. Each strand of the double helix has all the information needed to make the other strand. Because each strand can be used to make the other strand, the strands are said to be complementary.

The Replication Process Before a cell divides, it makes a copy of its DNA in a process called **replication.** Replication makes sure that each daughter cell has the same complete set of DNA molecules. During replication, the DNA molecule first separates into two strands. Then, the process makes two new strands following the rules of base pairing. Each strand of the double helix of DNA is a template, or model, for making the new strand.

The two strands of the double helix separate, the same way two sides of a zipper come apart. This separation makes two replication forks. As each new strand forms, new bases are added following the rules of base pairing. Adenine (A) is always paired with thymine (T). Guanine (G) is always paired to cytosine (C). For example, a strand that has the base sequence TACGTT makes a strand with the base sequence ATGCAA. The end result is two DNA molecules. Each is identical to the other and to the original DNA molecule.

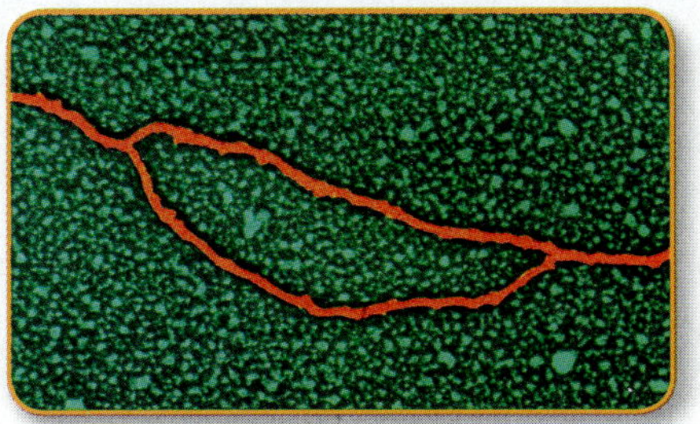

TEM 60,000×

Replication Forks The DNA molecule comes apart the way the two sides of a zipper come apart. The result is two areas called replication forks, where the DNA molecule can be copied. This micrograph shows a pair of replication forks in human DNA.

 BIOLOGY.com Search Lesson 12.3 GO • InterActive Art

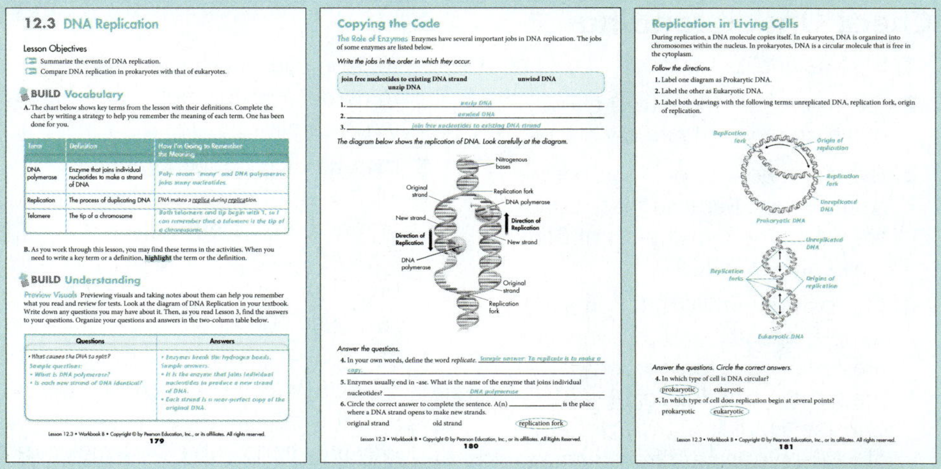

The Role of Enzymes DNA replication is carried out by special proteins called enzymes. These enzymes pull apart a molecule of DNA. They break the hydrogen bonds between base pairs. Then, they unwind the two strands. Each strand then serves as a template for making a new strand. The main enzyme involved in DNA replication is called **DNA polymerase** (PAHL ih mur ayz). DNA polymerase is an enzyme that joins individual nucleotides to make a new strand of DNA. DNA polymerase produces the sugar-phosphate bonds that join nucleotides together to form the new strands. DNA polymerase also checks each new DNA strand, so that each molecule is a close copy of the original.

Key Question What role does DNA polymerase play in copying DNA?
DNA polymerase is an enzyme that joins individual nucleotides to produce a new strand of DNA.

Telomeres The tips of chromosomes are known as **telomeres.** DNA in these regions is hard to replicate. Cells use a special enzyme called telomerase to fix this problem. Telomerase makes it less likely that genes will be damaged or lost during replication of rapidly dividing cells. Telomerase is often switched off in normal adult cells. But, in cancer cells, telomerase may be switched on. This may be one of the reasons why cancer cells are able to grow and divide rapidly.

BUILD Vocabulary

DNA polymerase
the principal enzyme involved in DNA replication

telomere
repetitive DNA at the end of a eukaryotic chromosome

PREFIXES
The prefix *telo-* means "end." The root *-mere* means "part." The telomere is the end part of the chromosome.

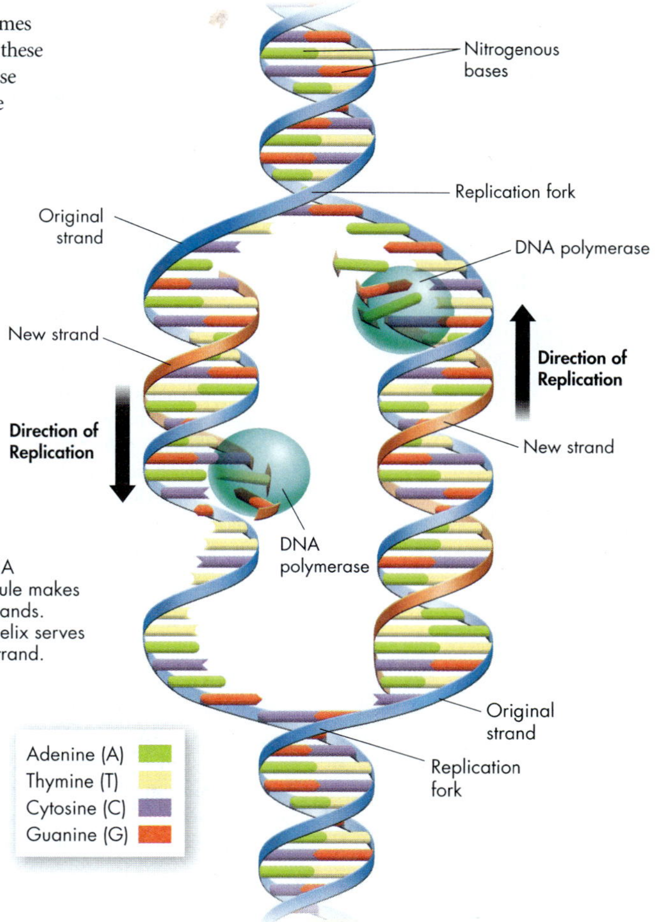

DNA Replication During DNA replication, the DNA molecule makes two new complementary strands. Each strand of the double helix serves as a template for the new strand.

ELL Focus on ELL: Access Content

ALL SPEAKERS Provide cloze prompts to reinforce key ideas and science terms.

For beginning speakers, distribute simplified paragraphs about the replication process and the role of enzymes. Leave key words blank. Provide a word bank. An example paragraph follows:

During DNA _____ *(replication)*, DNA makes a copy of itself. First, _____ *(enzymes)* separate the DNA molecule into two strands. This forms two replication _____ *(forks)*. Each strand is a _____ *(template)* for making the other strand. Then the enzyme _____ *(DNA polymerase)* copies each strand. One DNA molecule makes _____ *(two)* new DNA molecules.

For intermediate speakers, add sentences and terms, and omit the word bank.

For advanced speakers, have them make up their own cloze prompts and then switch with a partner to fill in the blanks.

During Reading

Build Vocabulary

USE SUFFIXES Read aloud the text under Use Suffixes. Have students locate *replication* in the text. Then read aloud the first paragraph under The Replication Process. Ask students what is being copied in replication. *(each strand of DNA)*

Active Reading

CLARIFY WORD USAGE Point out the difference between *complementary* (making something complete) and *complimentary* (free or giving praise).

USE DIAGRAMS Ask students to study the diagram of DNA replication as you read aloud The Role of Enzymes. Explain that a template is a pattern.

Ask Which chemical components of DNA pull apart when the strands separate? *(the nitrogenous bases)* Does the new strand look exactly like the original strand? Explain. *(No, the bases in the new strand are complementary.)*

Pass out strands of yarn in two different colors to student pairs. Tell them one color represents the original DNA strands and the other color represents the new DNA strands. Give students a few moments to wind and separate the strands to show replication.

Ask When replication is complete, what color are the strands? *(Each set has two different colored strands.)*

SEQUENCE Have student pairs construct a flowchart that shows the sequence of steps in replication. Suggest they draw the steps and write a caption for each. Tell them the structures in their drawings can be short and simple, but enough of the structures should be included to illustrate the captions clearly.

KEY QUESTION Ask a volunteer to read the Key Question and answer.

Ask How is the function of DNA polymerase like putting together puzzle pieces? *(Nucleotides fit together only in a certain way.)*

Build Vocabulary

PREFIXES Make sure students understand that *telo-* is different from *tele-*, which means "distant."

Ask A telomere is the end part of what? *(a DNA molecule)*

Speed Bump

Ask Why does a DNA molecule contain one strand that is older than the other? *(Each molecule has one original strand and one new strand.)*

Ask What does the enzyme telomerase do? *(adds DNA at the ends of chromosomes to replace DNA that is lost during replication)*

During Reading continued

Active Reading

COMPARE/CONTRAST Have students look at the diagrams on this page and identify differences between them. Read aloud the first three paragraphs. Ask students to raise their hands when you read something that is different between prokaryotic and eukaryotic cells.

Then have them work with a partner to complete a compare/contrast table of replication in living cells. Ask students to share how they constructed their charts as you draw one and write their responses on the board.

USE MODELS Show students a set of two yarn strands that have been twisted and tied into a circle to represent prokaryotic DNA. Demonstrate how, during prokaryotic DNA replication, the two strands are separated at a single starting point and then unwound in two directions as replication takes place. Point out the two replication forks so that students can observe them. Then provide student pairs with strands of yarn in two different colors and have them make their own models of prokaryotic DNA. Have students use the models to do their own demonstration of prokaryotic DNA replication.

KEY QUESTION Read aloud the Key Question and answer. Clarify that "goes in two directions" for prokaryotes and "goes in both directions" for eukaryotes is a difference, not a similarity. In prokaryotes, once replication starts, it proceeds in two directions only at two places. In eukaryotes, replication proceeds in two directions in many different places.

Speed Bump

Ask How might differences in the amount of DNA help explain why prokaryotic cells usually divide faster than eukaryotic cells? *(Sample answer: Because eukaryotic cells have more DNA than prokaryotic cells, it takes more time to copy it.)*

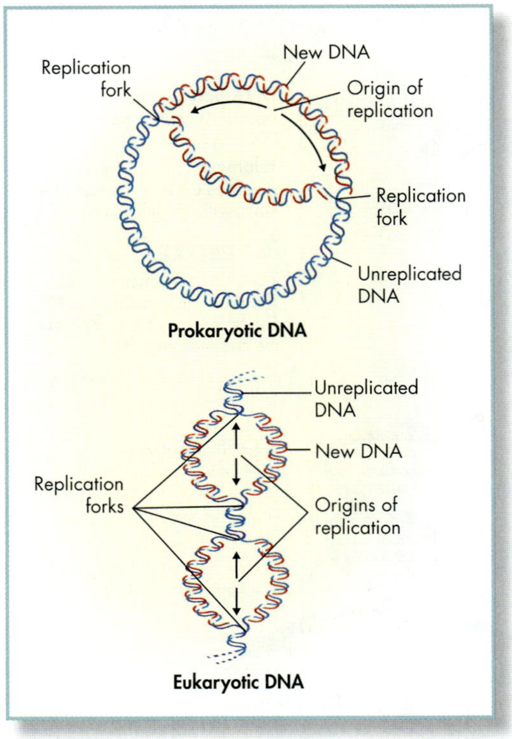

Differences in DNA Replication Replication in most prokaryotic cells (top) begins at one point and goes in two directions until the entire DNA ring is copied. In eukaryotic cells (bottom), replication happens at many starting points on chromosomes. Replication ends when all the chromosomes are copied.

Replication in Living Cells

Where is DNA found inside a cell? The cells of most prokaryotes have a single, ring-shaped DNA molecule in the body of the cell. This ring contains almost all of the cell's genetic information. Eukaryotic cells have much more DNA. Their DNA is found in chromosomes in the nucleus.

Prokaryotic DNA Replication In most prokaryotes, DNA replication starts when special proteins bind to one starting point on the chromosome. Replication usually goes in two directions until the whole chromosome is copied. Often, the two chromosomes made by replication are attached to different points inside the cell membrane. They are separated when the cell splits to make two new cells.

Eukaryotic DNA Replication Eukaryotic chromosomes are bigger than those of prokaryotes. In eukaryotic cells, replication may begin at many places on the DNA. Replication moves in both directions. Proteins check to make sure the copies are accurate. However, mistakes happen, and then the order of the bases in DNA changes. When the base order changes, the information on a gene may change. Sometimes this can have serious consequences.

As you learned in Chapter 10, the chromosomes stay together until anaphase of mitosis. Then, the chromosomes separate, and each new cell has a complete set of genes coded in DNA.

Key Question How does DNA replication differ in prokaryotic cells and eukaryotic cells?
In most prokaryotic cells, replication starts from a single point, and it continues in two directions until the whole chromosome is copied. In eukaryotic cells, replication may begin in hundreds of places on the DNA molecule. Replication then occurs in both directions until each chromosome is completely copied.

Materials

colored paper, tape, scissors

Analyze and Conclude

1. DNA polymerase
2. A is always matched with T. C is always matched with G.
3. Answers will vary, but may include observations about the size, scale, and materials. Also, DNA is twisted into a double helix while the student model is flat.

INQUIRY into Scientific Thinking

Modeling DNA Replication

1. Cut out small squares of blue and gray paper to represent phosphate and sugar molecules.
2. Cut out small strips of purple, green, red, and yellow paper to represent the four nitrogenous bases.
3. Build a set of five nucleotides using your paper strips and tape.
4. Using your nucleotides, tape together a single strand of DNA. For example you could make a pattern of bases that reads, "ATTGC."
5. Exchange your paper strand with a partner's strand.
6. Model DNA replication by creating a strand that is complementary to your partner's original strand.

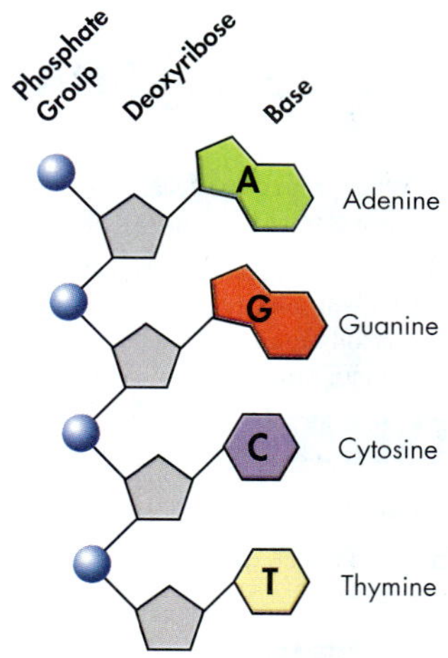

In Your Workbook Get more help for this activity in your workbook.

Analyze and Conclude

1. Use Models When you taped the nucleotides together, what enzyme were you modeling?

2. Analyze How does your model follow Chargaff's rule?

3. Evaluate How does your model differ from real DNA?

CHECK Understanding

Apply Vocabulary

Use the highlighted word from the lesson to complete the sentence correctly.

1. _______ are the tips of chromosomes.

2. _______ is the primary enzyme used in DNA replication.

Critical Thinking

3. Review How is DNA replicated?

4. Apply Concepts What is the role of DNA polymerase in DNA replication?

5. Compare and Contrast How is replication in eukaryotic cells different from replication in prokaryotic cells?

6. Write to Learn Answer the mystery clue.

MYSTERY CLUE

UV light can cause chemical changes in bases. How might these chemical changes in bases affect the process of DNA replication? (**Hint:** See p. 297.)

BIOLOGY.com Search Lesson 12.3 GO • Lesson Assessment • Art Review

Check Understanding

1. Telomeres
2. DNA polymerase
3. The DNA molecule separates into two strands at the replication fork. Each strand serves as a template for the attachment of complementary bases.
4. DNA polymerase joins individual nucleotides to produce a new strand of DNA and checks each newly generated DNA strand.
5. Replication in eukaryotic cells proceeds in both directions from multiple starting points; replication in prokaryotic cells begins at a single starting point and proceeds in two directions.
6. Answers should include the idea that if a base is changed, then it might bond with a different base from what it did in the original pattern. The information in the newly replicated DNA molecule will be different from the information in the original strand. The change will continue to be replicated as cells continue to divide.

Inquiry Into Scientific Thinking

USE DIAGRAMS Read aloud steps 1–3 in the activity. Then refer students to the Base Pairing diagram in Lesson 12.2.

Ask How does this diagram differ from the one you are going to make? *(The one we are going to make represents only one strand of DNA.)*

Review each subunit of a nucleotide and have students describe how they will represent them according to the steps in the activity.

Read aloud Steps 4–6.

Ask Is it necessary to write the letters on the bases? Why? *(No, the bases are color coded.)* What does it mean to make a complementary strand? *(The bases you make should join with the bases of your partner's molecule.)* How will you match the bases if they don't have letters on them? *(The color representing adenine goes with the color for thymine, and the same is true for cytosine to match with guanine.)*

Lead a Discussion

Talk about how knowing the base pairing rules allows you to predict the base sequence of a second strand of DNA if you know the sequence of the first strand. To reinforce this concept, have students make connections between base pairing and everyday objects.

Ask What kinds of everyday objects fit together only in a certain way? *(Sample answer: jigsaw puzzle pieces, zipper teeth, lock and key, gloves and hands, shoes and feet, toy building pieces that snap together)*

Post Reading

ASSESSMENT TIPS Point out to students that although the title of Question 5 asks them to compare and contrast, the actual question asks them only to describe how replication differs.

MYSTERY CLUE Have students read the Mystery Clue. Remind them that the strands in a DNA molecule serve as templates for making new strands.

Wrap-Up Activity

GO FOUR! Have students copy to individual index cards each caption they wrote when they sequenced replication. Divide the class into groups of four. Have each group combine and shuffle the cards of all its members. When you give the signal, each group tries to be the first to arrange in order all four different sets of cards.

CHAPTER LAB

Pre-Lab

Ask What does *extract* mean? *(take out or remove)*

Say Scientists use the word *extract* to refer to removing DNA because *extract* has the added meaning of removing with effort or force. You would not say, "I extracted the book from the table" if you were just picking up the book.

Ask Why do you think scientists use care when they extract DNA from cells to study it? *(to avoid damaging the molecules)*

To help students understand how it's possible that DNA can be 1 meter long, compare it to the human small intestine, which is about 7 meters long when stretched out.

Review the Background Questions and the Pre-Lab Questions as a class.

Lab

Tell students they will perform the chapter lab *Extracting DNA* described in **Lab Manual B**.

SAFETY Students should wear safety goggles and gloves while performing this lab. Have students wash their hands when they complete the lab. Students should use caution with glassware and follow the lab instructions exactly.

Post-Lab

CLOSE-UP OBSERVATION Have students remove some of the DNA precipitate with a dissecting probe or paper clip shaped into a hook and place it on a microscope slide. They can add a drop of methylene blue to the DNA, cover it with a coverslip, and observe it under a microscope. Ask them to describe what they see and to explain why they cannot distinguish the double-helix structure.

Skills Lab

GUIDED INQUIRY

Pre-Lab: Extracting DNA

Problem What properties of DNA can you observe when you extract DNA from cells?

Materials self-sealing plastic freezer bag, ripe strawberry, detergent solution, 25-mL graduated cylinder, cheesecloth, funnel, test tube, test tube rack, chilled ethanol, stirring rod

Lab Manual Chapter 12 Lab

Skills Focus Predict, Observe, Draw Conclusions

Connect to the Big idea Not surprisingly, the molecules that store genetic information are long molecules. If the DNA from a human cell were unfolded, the double helix structure would be about one meter long. Yet, most of a cell's DNA can be folded and tightly packed inside the cell's tiny nucleus. How can scientists remove DNA from the nucleus so that it can be studied and analyzed? In this lab, you will learn that extracting DNA from living tissue is not as difficult as you might think.

Background Questions

a. **Review** Describe the structure of a DNA molecule.

b. **Review** What type of bond holds the strands of DNA together?

c. **Apply Concepts** How does the strength of those bonds affect how DNA functions?

Pre-Lab Questions

Preview the procedure in the lab manual.

1. **Apply Concepts** Why do strawberry cells need DNA?

2. **Form a Hypothesis** If you observe a cell nucleus under a compound microscope, you will not see a molecule of DNA. Why will you be able to see the DNA you extract?

3. **Predict** Use what you know about DNA to predict some of the physical properties of DNA.

BIOLOGY.com Search Chapter 12 GO

Visit Chapter 12 online to test yourself on chapter content and to find activities to help you learn.

Untamed Science Video The Untamed Science CSI crew unravels the secrets of DNA left at the scene of a crime.

Art in Motion View an animation that re-creates the Hershey-Chase experiments.

Art Review Review your understanding of both prokaryotic and eukaryotic DNA replication.

InterActive Art Drag-and-drop base pairs to build your own strand of DNA while you practice the process of DNA replication.

Tutor Tube Tune into the tutor to find out hints for remembering which bases pair together.

Visual Analogy Compare transcription and translation with the process of publishing a book.

Pre-Lab Answers

BACKGROUND QUESTIONS

a. In DNA, nucleotides are joined together in long strands by covalent bonds. A double helix structure forms when two anti-parallel strands twist around each other.

b. hydrogen bonds

c. Because hydrogen bonds are relatively weak, the strands of DNA can separate and be replicated.

PRE-LAB QUESTIONS

1. Sample answer: Strawberry cells need DNA to produce the proteins that control reactions within the cells.

2. The clump of DNA will contain DNA from many cells.

3. Sample answer: The solid DNA will be made up of thin, long threads. The solid will be flexible rather than rigid (because the DNA must be able to fold up).

NATIONAL SCIENCE EDUCATION STANDARDS

CONTENT C.2.a

INQUIRY A.1.b, A.2.a

12 CHAPTER Summary

12.1 Identifying the Substance of Genes

- By studying bacterial transformation, Avery and other scientists discovered that DNA stores and passes genetic information from one generation of bacteria to the next.
- Hershey and Chase's experiment with bacteriophages confirmed Avery's results, convincing many scientists that DNA was the genetic material found in genes.
- The DNA that makes up genes must be capable of storing, copying, and transmitting the genetic information in a cell.

transformation (p. 288)
bacteriophage (p. 289)

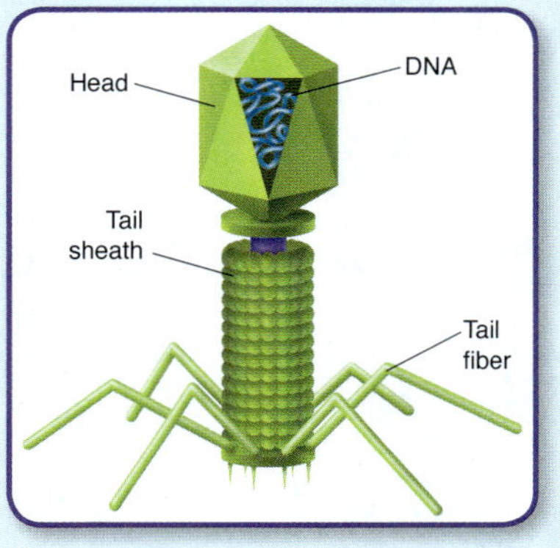

12.2 The Structure of DNA

- DNA is a nucleic acid made up of nucleotides joined into long strands or chains by covalent bonds.
- The clues in Franklin's X-ray pattern allowed Watson and Crick to build a model that explained the specific structure and properties of DNA.
- The double-helix model explains the reasons behind Chargaff's rule and how the two strands of DNA are held together.

base pairing (p. 295)

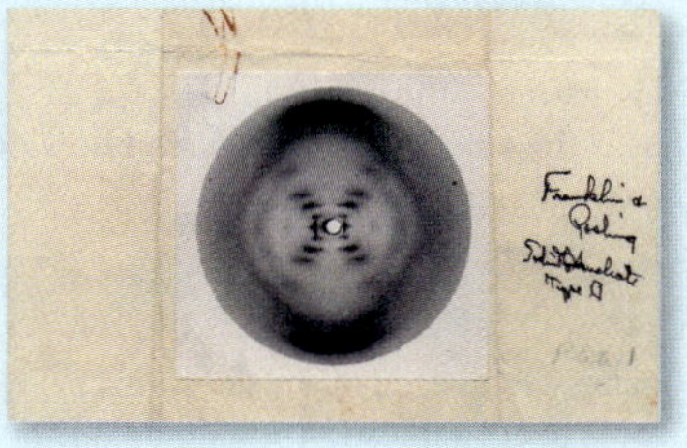

12.3 DNA Replication

- DNA polymerase is an enzyme that joins individual nucleotides to produce a new strand of DNA.
- In most prokaryotic cells, replication starts from a single point, and it continues in two directions until the entire chromosome is copied.
- In eukaryotic cells, replication may begin in hundreds of places on the DNA molecule. Replication then moves in both directions until each chromosome is completely copied.

replication (p. 296)
DNA polymerase (p. 297)
telomere (p. 297)

UbD Transfer the Big Idea

TRANSFER Divide the class into groups of four. Have them work together to compose a script for a television panel interview lasting 5–10 minutes. All members should work together to write questions and answers posed to three "experts" by one interviewer. Each expert will answer questions about one lesson in the chapter. Students should use the Key Questions as guides, but the questions and answers should be in the students' own words. Then have each group present their TV interview for the class. Students may read from their scripts during their presentations. Encourage students to be creative, for example, an expert on Lesson 12.3 might introduce herself as Dr. Double and repeat her answers.

Chapter Summary

Think Visually

USE GRAPHICS Have students complete the following flowchart about DNA replication.

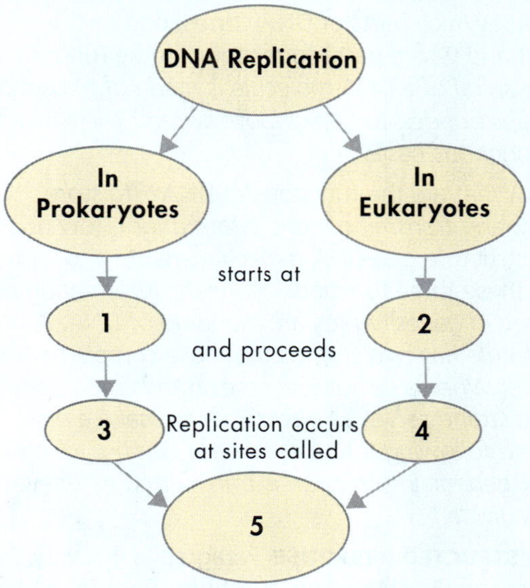

ANSWERS

1. a single point
2. up to hundreds of places **3.** in two directions
4. in both directions **5.** replication forks

BIOLOGY.com Study Online

REVIEW AND ASSESSMENT RESOURCES

Visual Analogy Students will watch the ways in which DNA can be compared to a how-to book.

Data Analysis Students can analyze DNA data to assess the relatedness of different species.

Tutor Tube This online tutorial offers some strategies for remembering Chargaff's rule.

Vocabulary Review The *Flash Cards* and *Match It* are an interactive vocabulary review.

Editable Worksheets Editable pages of Study Workbook B, Lab Manual B, and the Assessment Resources Book are available online.

Lesson Overviews Lesson Overviews are available on CD-ROM and can be edited for your class.

ASSESSMENT RESOURCES

Chapter Assessment Have students take an online version of the Chapter 12 Check Understanding.

Standardized Test Prep Students can take an online version of the Standardized Test Prep. You will receive their scores along with ideas for remediation.

Diagnostic and Benchmark Tests Use these tests to monitor your students' progress and supply remediation.

Assess

Big idea Have students look at the photo. Read aloud the Big Idea and the question.

Ask How does the shape of the structure in the photo model a DNA molecule? *(It looks like a twisted ladder, which is the shape of a DNA molecule.)* Which parts of DNA correspond to the spiral itself and which parts correspond to the rungs? *(The spiral of a DNA molecule is made of phosphate groups bonded to deoxyribose sugars. The rungs are nitrogenous bases.)*

Ask What are the functions of DNA? *(to store, copy, and transmit genetic information)* How does the structure of a DNA molecule enable it to carry out these three functions? *(Genetic information is stored in genes, which are made up of DNA. A DNA molecule has two strands that have complementary bases. When a cell divides, the strands separate and each strand serves as a template to make a new strand following the rules of base pairing. In this way, genetic information is transmitted whenever cells divide.)*

CONSTRUCTED RESPONSE Paragraphs will vary. Read aloud the directions and questions. Be sure students understand that the *Hints* should be used as a guide to help them answer the questions.

Remind students that the writing process starts with prewriting.

Say When you prewrite, you brainstorm what you are going to write. You gather and organize facts and details. You don't worry at this point about topic sentences, punctuation, or grammar. Once your ideas are on paper, you can work on putting them together and polishing your writing, or you can change or omit them.

Pair students and read aloud the first hint under Question 1.

Say The hint is there because it relates to the question and should be included in your writing.

Give students two minutes to discuss with their partner how they think the hint is related to the question. When time is up, tell them to write their ideas as quickly as possible. Then, tell students to read what they have written and decide if they need more facts or details. They can look these up in the book.

Repeat this exercise with each hint before students work on editing their paragraphs.

12 CHECK Understanding

Assess the Big idea Information and Heredity, Cellular Basis of Life

Write an answer to the question below.

Q: What is the structure of DNA, and how does it function in genetic inheritance?

Constructed Response

Write an answer to each of the questions below. The answer to each question should be one or two paragraphs. To help you begin, read the **Hints** below the questions.

1. **How did Hershey and Chase confirm that genetic material was a nucleic acid?**

   Hint Bacteriophages are viruses that infect bacteria.

   Hint Viruses have a protein coat around a nucleic acid core.

2. **How did Chargaff's rule help Watson and Crick determine the structure of DNA?**

   Hint Adenine always pairs with thymine, and cytosine always pairs with guanine.

3. **If a change happens in the genetic code in the gamete of a eukaryotic parent, will offspring have that change, too?**

   Hint Changes in the genetic code can sometimes be copied when the DNA is copied.

Foundations for Learning Wrap-Up

Use the Connected Ideas cards you made as you read the chapter to help you organize your thoughts about DNA structure and replication.

Activity 1 Combine your cards with the cards of a partner. Using as many cards you can, make a time line of the history of the discoveries that led to understanding the structure of DNA.

Activity 2 With a partner, make cards to complete the two chains of connected ideas shown below. One chain shows the process of DNA replication in eukaryotes. The other chain shows the process of DNA replication in prokaryotes.

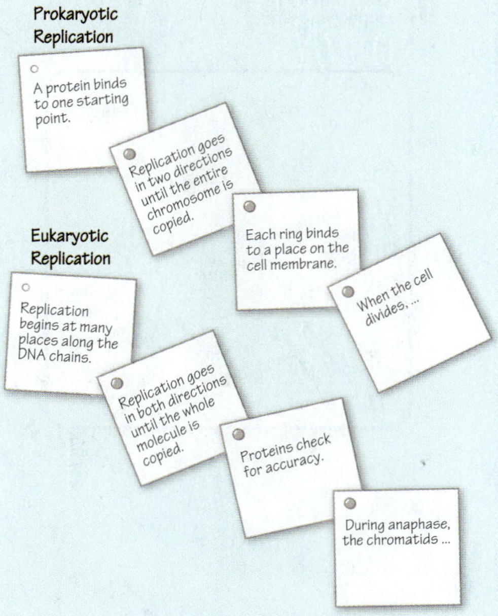

302 Check Understanding

For Question 2, pair students and have them use the hints and your modeling to answer the question. Have students share their paragraphs with the class.

For Question 3, have students write the paragraphs independently for homework or as an anchor activity.

Foundations for Learning Wrap-Up

To help students with Activity 1, explain that Chargaff made his discovery in the late 1940s. Remind students that DNA replication in both groups of organisms has similarities as well as differences and to make additional cards if needed.

12.1 Identifying the Substance of Genes

Understand Key Concepts

1. The process by which one strain of bacterium is apparently changed into another strain is called
   a. transcription.
   b. transformation.
   c. duplication.
   d. replication.

Test-Taking Tip

Use Root Words As you read the answers, look for root words to help you choose the correct answer. The root words for the answers in question 1 are *transcrip*, *transform*, *duplica*, and *replica*. Remember that *transform* means "to change." Now read the question again. The question asks which process changed one strain of bacterium. You know that transformation means change, so **b** is the correct answer.

2. Bacteriophages are
   a. a form of bacteria.
   b. enzymes.
   c. coils of DNA.
   d. viruses.
3. Before DNA could be shown to be the genetic material in cells, scientists had to show that it could
   a. tolerate high temperatures.
   b. carry and make copies of information.
   c. be modified in response to environmental conditions.
   d. be broken down into small subunits.
4. Briefly describe the conclusion that could be drawn from the experiments of Frederick Griffith.

Think Critically

5. **Evaluate** Avery and his team identified DNA as the molecule that caused transformation. How did they control variables in their experiment to make sure that only DNA caused the change?

12.2 The Structure of DNA

Understand Key Concepts

6. A nucleotide does NOT contain a
   a. 5-carbon sugar.
   b. protein.
   c. nitrogen base.
   d. phosphate group.
7. According to Chargaff's rule of base pairing, which of the following is true about DNA?
   a. A = T, and C = G
   b. A = C, and T = G
   c. A = G, and T = C
   d. A = T = C = G
8. The bonds that hold the two strands of DNA together come from
   a. the attraction of phosphate groups for each other.
   b. strong bonds between nitrogenous bases and the sugar-phosphate backbone.
   c. hydrogen bonds between nitrogenous bases.
   d. carbon-to-carbon bonds in the sugar portion of the nucleotides.
9. Describe the parts and structure of a DNA nucleotide.

Think Critically

10. **Infer** Adenine and guanine are longer than thymine and cytosine. What would happen to the parallel strands if adenine were paired with guanine?

12.3 DNA Replication

Understand Key Concepts

11. The diagram below shows the process of DNA
    a. replication.
    b. digestion.
    c. transformation.
    d. transpiration.

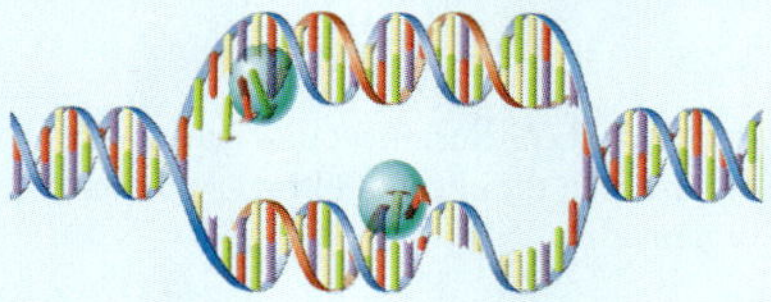

Answers

Lesson 12.1

UNDERSTAND KEY CONCEPTS

1. b
2. d
3. b
4. Chemical factors from dead bacteria can move into living bacteria and change the genetic characteristics of the living bacteria.

THINK CRITICALLY

5. Avery and his team used enzymes to destroy various biological molecules. Then, they showed that when only DNA was destroyed, genetic information could not be transferred. However, when they destroyed other biological molecules (and not DNA), transformation did happen.

Lesson 12.2

UNDERSTAND KEY CONCEPTS

6. b
7. a
8. c
9. A nucleotide has three parts: a five-carbon sugar called deoxyribose, a phosphate group, and a nitrogenous base.

THINK CRITICALLY

10. The two long bases would cause a bulge in the DNA, and the backbones would no longer be parallel.

Lesson 12.3

UNDERSTAND KEY CONCEPTS

11. a

ASSESSMENT

Assessment Tips

Read Question 1 and the options. Ask students for the correct answer. Then read and discuss the tip.

Read aloud Question 3. Point out that option d is true, but it does not answer the question. Emphasize that just because an option might be a true statement does not necessarily mean it is the correct answer to the question. The key phrase in the question is *genetic material.* Students must decide which option is a function of genetic material. Tell students that when they are deciding between two options, to go back and carefully read the question again.

To help them answer Question 10, suggest students sketch a diagram of what such a DNA molecule might look like.

For Question 13, discuss what is being compared. Tell students that if they rush through reading the question, they might answer how new strands compare, not how new molecules compare. Remind students that a DNA molecule has two strands.

ASSESSMENT

12. Base pairing is the principle that hydrogen bonds form only between certain base pairs: adenine to thymine and cytosine to guanine.

13. DNA separates into two strands, the two new complementary strands are generated following the rules of base pairing. Each new DNA molecule has one strand from the original molecule and one newly synthesized strand, making each new DNA molecule an exact duplicate of the original.

THINK CRITICALLY

14. Alike: You start with one copy and end up with two identical copies. Different: In photocopying, you end up with one original copy and one completely new copy. In DNA replication, you end up with two sets of strands, each containing an original strand and a new strand.

Connecting Concepts

15. b

16. Ultraviolet light, particularly between 250 and 270 nm wavelengths, is harmful to living organisms.

CHAPTER MYSTERY After students have read through the chapter mystery, discuss the bonds that hold together nitrogenous bases.

Ask How would you describe the bonds between bases? *(They are hydrogen bonds, which means they are relatively weak.)* Why might the bonds between bases be more susceptible to damage from UV light than covalent bonds in other parts of the DNA molecule? *(The hydrogen bonds between bases are easier to break.)*

Ask How might damage to the bases cause a change in the genetic information that is transmitted? *(The information that is copied during replication might be incorrect. If passed during mitosis, it would affect only one type of cell. If passed during meiosis, all cells of the offspring might be affected.)*

Take this opportunity to tell students that sunscreens contain chemicals that interact with the skin to protect it from UV rays. However, according to the CDC, sunscreen is not meant to allow you to spend more time in the sun than you would otherwise. Expert dermatologists say that no sunscreen offers 100% protection and that wearing a high SPF gives some people a false sense of security. That's why it is important to use sunscreen with other measures, such as staying out of the sun at peak hours.

12 CHECK Understanding

12. What is meant by the term *base pairing*?

13. Explain the process of replication. When a DNA molecule is replicated, how do the new molecules compare to the original molecule?

Think Critically

14. Use Analogies How is photocopying a page similar to DNA replication? Think of the original materials, the copying process, and the final products. Explain how the two processes are alike. Identify major differences.

Connecting Concepts

Use Science Graphics

A scientist studied the effect of exposing DNA to various wavelengths of ultraviolet light. The scientist determined the number of copying errors made after exposure to ultraviolet rays. The graph shows the results. Use the graph to answer questions 15 and 16.

DNA Replication Errors

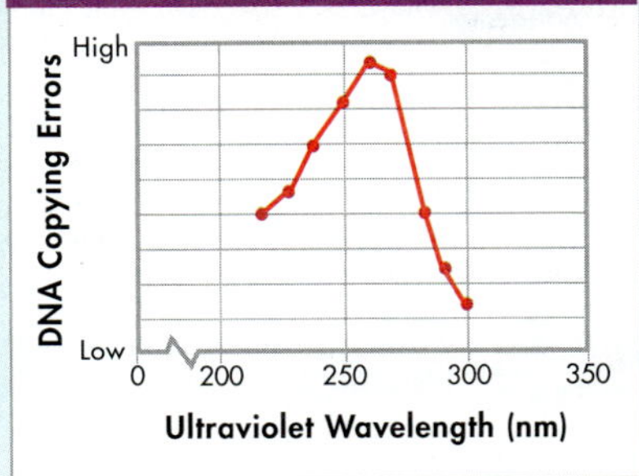

15. Interpret Graphs The most damaging effects of ultraviolet light on DNA replication occur closest to which wavelength?

a. 200 **c.** 300

b. 250 **d.** 350

16. Infer What conclusion would you draw from the graph about the effect of ultraviolet light on living organisms?

solve the CHAPTER MYSTERY

UV LIGHT

Nucleotides are small parts that are linked together to make the larger DNA molecule. The three parts of each nucleotide include a nitrogenous base—adenine, cytosine, guanine, or thymine (A, C, G, or T). The energy of ultraviolet light can cause chemical changes between the nitrogenous bases in DNA.

1. Predict Think about the structure of DNA. What sorts of problems could too much UV light cause in the DNA molecule? How might these changes affect the way DNA works?

2. Relate Cause and Effect Analyze the effects that UV light might have on skin cells. Why is UV light so dangerous? Why is the skin particularly vulnerable to it?

Untamed Science **Never Stop Exploring Your World.** Finding the connection between UV light and DNA is only the beginning. Take a video field trip with the ecogeeks of Untamed Science to see where the mystery leads.

CHAPTER MYSTERY ANSWERS

1. The energy from UV light can excite electrons in the nitrogenous bases, which absorb light, to the point at which they cause chemical changes. They might cause the formation of new bonds or the breaking of old ones.

2. The energy absorbed from UV light can excite electrons in the nitrogenous bases, which can cause existing bonds to break and new bonds to form.

Untamed Science Have students watch the short video **Super Sleuth DNA** to see how scientists use DNA to solve crimes.

Standardized Test Prep

Multiple Choice

1. During replication, which sequence of nucleotides would bond with the DNA sequence TATGA?
A TATGA
B ATACT
C CACTA
D AGTAT

2. The scientist(s) responsible for the discovery of bacterial transformation is (are)
A Watson and Crick.
B Avery.
C Griffith.
D Franklin.

3. Which of the following does NOT describe the structure of DNA?
A double helix
B long chain of nucleotides
C contains adenine-guanine pairs
D sugar-phosphate backbone

4. What did Hershey and Chase's work show?
A Genes are probably made of DNA.
B Genes are probably made of protein.
C Viruses contain DNA but not protein.
D Bacteria contain DNA but not protein.

5. The two "backbones" of the DNA molecule consist of
A adenines and sugars.
B phosphates and sugars.
C adenines and thymines.
D thymines and sugars.

6. When prokaryotic cells copy their DNA, replication begins at
A one point on the DNA molecule.
B two points on opposite ends of the DNA molecule.
C dozens to hundreds of points along the molecule.
D opposite ends of the molecule.

7. Compared to eukaryotic cells, prokaryotic cells contain
A much more DNA.
B much less DNA.
C twice as much DNA.
D the same amount of DNA.

Questions 8 and 9

Under ideal conditions, a single bacterial cell can reproduce every 20 minutes. The graph shows how the total number of cells under ideal conditions can change over time.

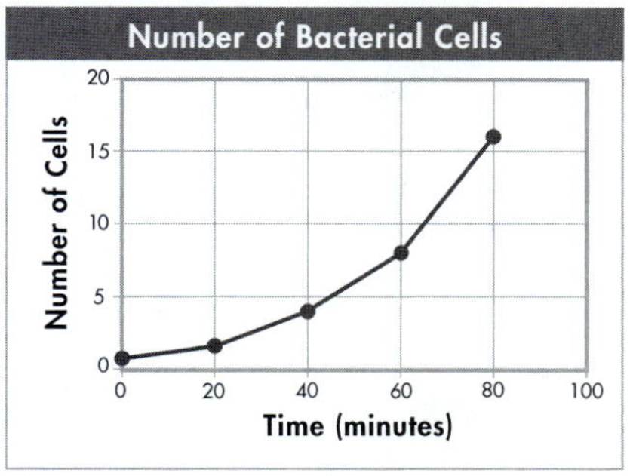

8. How many cells are present after 80 minutes?
A 1
B 2
C 16
D 32

9. If the DNA of this bacterium is 4 million base pairs in length, how many total molecules of A, T, C, and G are required for replication to be successful?
A 2 million
B 4 million
C 8 million
D 32 million

Open-Ended Response

10. Describe how eukaryotic cells are able to keep such large amounts of DNA in the small volume of the cell nucleus. (**Hint** Review the structure of a chromosome in Chapter 10.)

If You Have Trouble With . . .

Question	1	2	3	4	5	6	7	8	9	10
See Lesson	12.3	12.1	12.2	12.1	12.2	12.3	12.3	12.3	12.3	12.3

Test-Taking Tip

USE SCRATCH PAPER

Tell students that when they are asked to find the solution to a problem, such as the complementary sequence of DNA or RNA, they should first solve the problem on scratch paper. They should then compare their answer with the options provided. This method will help them avoid answer choices that are very similar, but not identical to the correct answer choice.

Before the Test

Review with students the Test-Taking Tip.

Point out the word *NOT* in Question 3.

Tell students that thinking of analogies, such as packing clothes in a suitcase, might help them answer Question 10. Remind them that analogies help readers visualize abstract concepts. Suggest they include an analogy to help explain their answer.

Answers

1. B
2. C
3. C
4. A
5. B
6. A
7. B
8. C
9. C
10. Sample answer: In eukaryotic cells, DNA is tightly wound and coiled around proteins and coiled again and again to form chromosomes. Like placing clothes in a suitcase, a very long DNA strand can be folded to fit within the shape of a cell.

CHAPTER PLANNER 13 RNA and Protein Synthesis

Chapter Contents	NSES	Time	Core Resources
Chapter Preview			**Student Edition,** pp. 306–307 **Chapter Mystery,** p. 307 **Foundations for Learning,** p. 307
13.1 RNA The Role of RNA • RNA Synthesis	II, V, B.3, C.2.a	1 period 1/2 block	**Student Edition,** pp. 308–310 **Build Connections,** p. 308 **Study Workbook B** 13.1 Worksheets **Biology.com** *Visual Analogy:* Master Plans and Blueprints • *Art in Motion:* RNA Processing • 13.1 Lesson Assessment
13.2 Ribosomes and Protein Synthesis The Genetic Code • Translation • The Molecular Basis for Heredity	II, V, A.2.a, B.2, B.3, C.1.c, C.2.a, G.3	1 1/2 periods 3/4 block	**Student Edition,** pp. 311–315 **Build Connections,** p. 315 **Study Workbook B** 13.2 Worksheets **Biology.com** *InterActive Art:* Transcription and Translation • *Tutor Tube:* Why Are Proteins So Important? • 13.2 Lesson Assessment
13.3 Mutations Types of Mutations • Effects of Mutations	II, IV, V, C.2.c, C.3.a, F.1, F.4	1 period 1/2 block	**Student Edition,** pp. 316–319 **Study Workbook B** 13.3 Worksheets **Biology.com** *Art Review:* Types of Mutations • 13.3 Lesson Assessment
13.4 Gene Regulation and Expression Prokaryotic Gene Regulation • Eukaryotic Gene Regulation • Genetic Control of Development • Environmental Influences	I, II, IV, V, A.2.a, A.2.b, A.2.c, B.3, C.1.d, C.1.f, C.3.c, G.1	1 1/2 periods 3/4 block	**Student Edition,** pp. 320–325 **Inquiry Into Scientific Thinking,** p. 325 Study Workbook B 13.4 Worksheets **Biology.com** 13.4 Lesson Assessment
Chapter Pre-Lab	C.1.d	1 period 1/2 block	**Student Edition,** p. 326 **Lab Manual B** *From DNA to Protein Synthesis*

Tools for ELL Instruction

ELL Handbook describes teaching strategies to make *Miller & Levine Biology: Foundation Edition* more accessible to ELL students.

Spanish Study Workbook is a Spanish translation of Study Workbook A.

Multilingual Glossary is the *Miller & Levine Biology* glossary translated into ten languages.

Tools to Challenge Students

Study Workbook A includes more challenging worksheets and activities such as Chapter Mystery activities that reinforce 21st Century Skills.

Lab Manual A labs are geared toward on-level students. The Lab Manual includes fifteen labs in addition to the Chapter Labs.

Biology.com Data Analysis activities offer students a way to practice data gathering and analyzing skills.

Visual Quizzes in the Assessment Resources Book are geared toward on-level students and use visuals from *Miller & Levine Biology.*

Additional Resources

Biology.com Untamed Science Video • Flash Cards

Study Workbook A 13.1 Worksheets
Spanish Study Workbook 13.1 Worksheets

Study Workbook A 13.2 Worksheets
Spanish Study Workbook 13.2 Worksheets
Biology.com *Data Analysis:* A Complicated Operon
Assessment Resources Book Visual Quiz

Study Workbook A 13.3 Worksheets
Spanish Study Workbook 13.3 Worksheets
Assessment Resources Book Visual Quiz

Study Workbook A 13.4 Worksheets
Lab Manual B *Data Analysis:* The Discovery of RNA Interference
Spanish Study Workbook 13.4 Worksheets
Biology.com *Data Analysis:* Identifying *Lac* Operon Mutants

Lab Manual A *From DNA to Protein Synthesis*

Chapter Review

Student Edition Chapter Summary, p. 327
Study Workbook B Chapter 13 Review
Transparencies, pp. 157–171
Biology.com Untamed Science Video • Editable Worksheets of Study Workbooks A and B and Lab Manuals A and B • Chapter 13 Flash Cards and Match It

Untamed Science DVD • Classroom Resources CD (includes lesson presentations and editable worksheets)

Chapter Assessment

Student Edition Check Understanding, pp. 328–331
Study Workbook B Chapter 13 Taking a Standardized Test
Assessment Resources Book Chapter 13 Test B
Biology.com Chapter 13 Assessment • Editable Worksheets of Chapter 13 Visual Quizzes* and Chapter 13 Tests A and B

Exam*View Assessment Suite* • Classroom Resources CD (includes lesson presentations and editable worksheets)

Time: 1 period, 1/2 block

* Visual Quizzes are based on visuals from *Miller & Levine Biology.* Review carefully to make sure they are appropriate for your class.

Pressed for Time?

Preview the Chapter Introduce the Lesson 13.1 vocabulary and discuss the visuals Transcribing DNA Into RNA and Translation.

Cover the Chapter Quickly Have students read *The Role of RNA* and the Transcription section of *RNA Synthesis* in Lesson 13.1. Assign *The Genetic Code* and *Translation* in Lesson 13.2, and go over the Translation visual. Assign *Types of Mutations* in Lesson 13.3.

Assess Assign Questions 1–4 in the 13.1 Check Understanding, Questions 1, 2, and 4 in the 13.2 Check Understanding, Questions 1, 2, 4, and 6 in the 13.3 Check Understanding, and the Chapter 13 Standardized Test Prep except for Questions 5, 6, and 9.

13 RNA and Protein Synthesis

Big idea **Information and Heredity**

Q: How does information flow from DNA to RNA to direct the synthesis of proteins?

BIOLOGY.com Search Chapter 13 GO • Flash Cards

306

CHAPTER 13

Get the Picture

Have students look at the picture. Then, invite a volunteer to read aloud the caption. Guide students to recognize that the two tigers are the same species. In fact, the tigers could be in the same family.

Ask How do these tigers differ? *(One has orange and black fur, and the other has white and brown fur.)*

Ask What caused one tiger to have white fur? *(a mutation)*

Tell students they will learn about mutations in this chapter.

Activate Prior Knowledge

Say Genetic information is passed from generation to generation.

Ask Which material in a cell transmits genetic information from one generation to the next? *(DNA)*

Ask What are the other functions of DNA? *(storing and copying genetic information)*

Have students recall that the genetic information in an organism's cells determines the organism's characteristics.

NATIONAL SCIENCE EDUCATION STANDARDS

UNIFYING CONCEPTS AND PROCESSES

I, II, IV, V

CONTENT

B.2, B.3, C.1.c, C.1.f, C.2.a, C.2.c, C.3.a, C.3.c, F.1, F.4, G.1, G.3

INQUIRY

A.2.a, A.2.b, A.2.c

UbD Connect to the Big Idea

Invite a volunteer to read aloud the chapter number and the chapter title. Read the Big Idea and question. Emphasize the question, **How does information flow from DNA to RNA to direct the synthesis of proteins?**

Set the Purpose

PREVIEW AND PREDICT Draw the graphic organizer on the board. Guide students to the INSIDE feature on the next page, and invite a volunteer to read each lesson number and title.

Ask Where is DNA located in a cell? *(in the nucleus)* What must happen to the information that is stored in DNA before it can be used by the cell? *(The information must move out of the nucleus into the cytoplasm.)*

Say When the information reaches the cytoplasm, it is used to make proteins. Encourage students to speculate how this happens, using the lesson titles as hints.

CHAPTER 13

INSIDE:

- 13.1 RNA
- 13.2 Ribosomes and Protein Synthesis
- 13.3 Mutations
- 13.4 Gene Regulation and Expression

One of these Bengal tigers has normal coloring. The other tiger has a genetic mutation that affects its coloring.

- Untamed Science Video
- Chapter Mystery

CHAPTER MYSTERY

MOUSE-EYED FLY

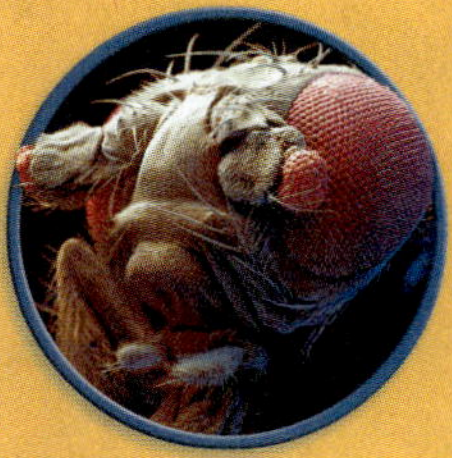

It was not a science fiction movie. The animal in the lab was real. It had two forward-looking eyes. It also had eyes on its knees and eyes on its hind legs. It even had eyes in the back of its head! Yet, as strange as it looked, this animal was not a monster. It was simply a fruit fly with eyes in very strange places. These eyes looked like the fly's normal compound eyes. However, a mouse gene transplanted into the fly's DNA had made them. How could a mouse gene make extra eyes in a fly?

Read for Mystery Clues This gene normally controls the growth of eyes in mice. As you read, look for clues to explain how it could cause a fly to grow eyes in unusual places. Then, solve the mystery.

FOUNDATIONS for Learning

Create an Important Facts Envelope by drawing and cutting out the shape below. Fold the flaps in order, taping flap 3 to flaps 1 and 2. As you read the chapter, record important facts on slips of paper and put them in the envelope. Facts can include key terms and drawings of processes. The facts and terms can be organized into themes, such as "Mutations." At the end of the chapter are two activities that use the facts to help answer the question: How does information flow from DNA to RNA to direct the synthesis of proteins?

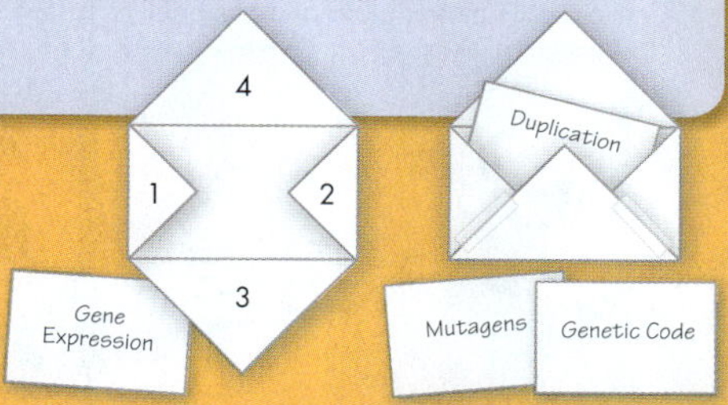

Chapter 13 Big Idea: Information and Heredity

Chapter 13: How does information flow from DNA to RNA to direct the synthesis of proteins?

- **13.1:** What is RNA?
- **13.2:** How do cells make proteins?
- **13.3:** What happens when a cell's DNA changes?
- **13.4:** How do cells regulate gene expression?

Foundations for Learning

Have students construct their Important Facts Envelope. Assist those who are having trouble folding and taping the envelope correctly. Tell students to cut from their notebooks blank slips of paper that will fit inside the envelope. Then, help students understand the value of this activity.

Ask What kind of information might you include on a slip of paper for a vocabulary word? *(Sample answer: the vocabulary word, the definition in your own words, an illustration of what the word means)*

Ask What type of information might you include on a slip of paper for a process? *(Sample answer: a description of the process, an explanation of why it is important, an illustration of the process)*

Remind students that they will use the contents of this envelope as a study aid as they read the chapter.

Untamed Science™ To further explore the mystery of the mouse-eyed fly, students can take a video field trip with the ecogeeks of Untamed Science in **Tales of a Mutant Leopard.**

CHAPTER MYSTERY Remind students that DNA determines an organism's inherited characteristics. Invite a volunteer to read aloud the Chapter Mystery. Have students predict how this mystery relates to the chapter title and how the fruit fly was able to produce extra eyes.

Tell students that they will find clues in the chapter that will help them solve this mystery.

Digital Support

BIOLOGY.com Extend your reach by using the digital assets offered at **Biology.com.**

Pre-Reading

Objectives

13.1.1 Contrast RNA and DNA.

13.1.2 Explain the process of transcription.

Preview the Pages

1. Preview the lesson vocabulary by having volunteers read aloud each term and its definition. Point out that each term includes *RNA*.
2. Preview the visuals. Direct students' attention to the diagram Transcribing DNA Into RNA. Have students look over the information in the key and locate DNA and RNA in the diagram.

Activate Prior Knowledge

Have students recall that a DNA molecule is made of nucleotides arranged in a double helix.

Ask What are the three parts of a DNA nucleotide? *(a deoxyribose molecule, a phosphate group, and a nitrogen-containing base)* What are the four bases in DNA? *(adenine (A), guanine (G), cytosine, (C), and thymine (T))*

Say In this lesson, you will learn about another nucleic acid, RNA.

Build Understanding

PREVIEW VISUALS Direct students to Workbook B, where they will find a prediction chart for recording their notes. Remind students that previewing visuals will help them read for understanding as they read the lesson.

NATIONAL SCIENCE EDUCATION STANDARDS

UNIFYING CONCEPTS AND PROCESSES

II, V

CONTENT

B.3, C.2.a

For the corresponding lesson in *Miller & Levine Biology,* see pages 362–365.

13.1 RNA

Key Questions

How does RNA differ from DNA?

How does the cell make RNA?

BUILD Understanding

Preview Visuals Before you read, preview the Transcribing DNA Into RNA figure on the last page of this lesson. Write a prediction of how you think a cell makes RNA. Then, as you read, take notes on how a cell makes RNA. After you read, compare your notes to your prediction.

In Your Workbook Go to your workbook and complete the prediction chart in Lesson 13.1.

The Role of RNA

We know that DNA is the genetic material. We also know that its sequence of nucleotides carries our genetic code. But DNA cannot put genes into action by itself. Ribonucleic acid, or **RNA,** is also needed. RNA, like DNA, is a nucleic acid made of a long chain of nucleotides.

Genes are made of instructions coded into DNA that tell cells how to build proteins. The first step in decoding these instructions is to copy part of the base sequence from DNA into RNA. RNA then uses these instructions to direct the making of proteins. Proteins help to determine an organism's characteristics.

Comparing DNA and RNA There are three important differences between RNA and DNA:

- The sugar in RNA is ribose, not deoxyribose.
- RNA is usually single-stranded, not double-stranded.
- RNA has uracil in place of thymine.

These chemical differences make it easy for enzymes in the cell to tell DNA and RNA apart. You can compare the different roles played by DNA and RNA molecules in the making of proteins to the two types of plans builders use. A master plan has all the information needed to make a building. That plan is valuable. So, builders use only copies of the plan, called blueprints, on the building site. The cell's "master plan" is its DNA, which it uses to prepare RNA "blueprints" for protein synthesis.

BUILD Connections

MASTER PLANS AND BLUEPRINTS

The roles of DNA and RNA molecules are like the two types of plans used by builders. DNA is like the master plan used by builders. RNA is like a cheap, disposable copy of that plan.

In Your Workbook Explore how DNA and RNA act like a cell's master plans and blueprints in Lesson 13.1.

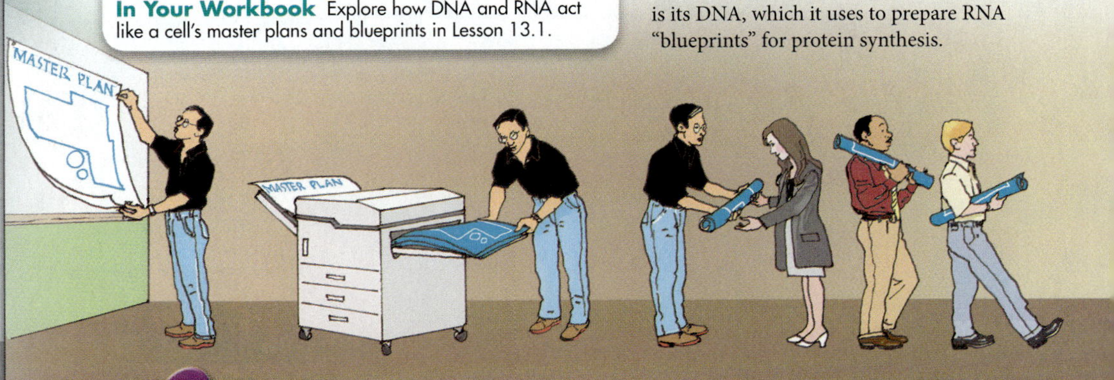

BIOLOGY.com Search Lesson 13.1 GO • Visual Analogy

13.1 RNA

Lesson Objectives

- Contrast RNA and DNA.
- Explain the process of transcription.

BUILD Vocabulary

A. The chart below shows key terms from the lesson with their definitions. Complete the chart by writing a strategy to help you remember the meaning of each term.

Term	Definition	How I'm Going to Remember the Meaning
Exon	Coding region of a gene that is expressed; exons are kept and put together to form mRNA	*If a puzzle piece is an exact fit, you keep it in place. Exons are pieces that are kept.*
Intron	Noncoding sequence of a gene that is cut out and discarded	*An intron is the part of RNA that is in between exons.*
Messenger RNA	mRNA; RNA molecules that carry instructions for assembling proteins	*Messenger RNA carries the message of how to make proteins.*
Ribosomal RNA	rRNA; RNA molecules that make up ribosomes, on which proteins are assembled	*Ribosomal RNA makes up ribosomes.*
RNA	Ribonucleic acid; nucleic acid that helps put the genetic code in DNA into action	*RNA molecules are the carriers of the code for proteins.*
RNA polymerase	An enzyme that separates DNA strands and uses one strand to make RNA	*RNA is a polymer. RNA polymerase is the enzyme that builds the polymer.*
Transcription	Process in which RNA molecules are made using DNA as a template	*Transcribe means writing down what someone says. In transcription, RNA molecules are written from DNA.*
Transfer RNA	tRNA; RNA molecule that transfers amino acids to the ribosomes	*Transfer RNA transfers amino acids.*

Note: Students should highlight terms and definitions used throughout the chapter.

B. As you work through this lesson, you may find these terms in the activities. When you need to write a key term or a definition, **highlight** the term or the definition.

BUILD Understanding

Preview Visuals Previewing the visuals in a lesson can give you an idea about what you will learn. Visuals include photographs, charts, graphs, and diagrams.

Before you read the lesson, preview the Transcribing DNA into RNA diagram. Then use the chart below to predict how you think a cell makes RNA. As you read the lesson, add notes to your chart about how RNA is made. After you read, compare your prediction to your notes.

Prediction	Notes
Students' predictions should explain how they think a cell makes RNA.	*Students' notes should explain how a cell makes RNA.*
How my notes compare to my prediction:	

BUILD Connections

Master Plans and Blueprints An analogy takes two things that seem to be different and shows how they can be similar. In this visual analogy, DNA is compared with a master plan, and RNA is compared with blueprints.

Answer the questions. Use the terms DNA, RNA, *and* nucleus *in your answers.*

1. The master plan stays in the office. Why is DNA like the master plan? *DNA stays in the nucleus.*
2. Blueprints go out of the office to the building site. Why is RNA like a set of blueprints? *RNA goes outside the nucleus to the building site.*
3. Find a partner. Using the analogy, explain the different functions of RNA and DNA to your partner.

RNA Synthesis

In transcription, RNA polymerase separates the two DNA strands. RNA then uses one strand as a template to make a complementary strand of RNA. RNA contains the nucleotide uracil instead of the nucleotide thymine.

Follow the directions.

1. Label the RNA
2. Label the DNA.
3. Use the key to label the missing nucleotides marked on the diagram.

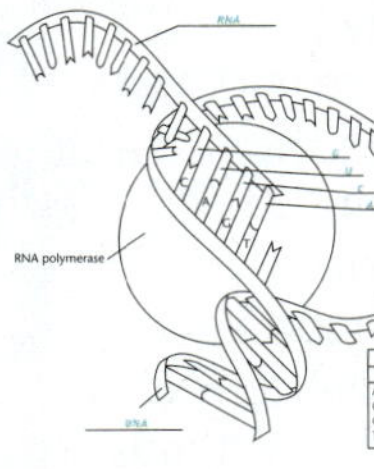

Key	
DNA	**RNA**
A = Adenine	A = Adenine
C = Cytosine	C = Cytosine
G = Guanine	G = Guanine
T = Thymine	U = Uracil

Answer the questions. Circle the correct answer.

4. In DNA, ________ is always paired with cytosine.
   adenine — (guanine) — uracil
5. In RNA, ________ replaces thymine.
   (uracil) — cytosine — adenine
6. In DNA, ________ is paired with adenine.
   uracil — (thymine) — guanine

Key Question How does RNA differ from DNA?
There are three main differences between RNA and DNA. The sugar in RNA is ribose instead of deoxyribose. RNA is usually single-stranded and not double-stranded. RNA contains uracil in place of thymine.

Functions of RNA Think of an RNA molecule as a disposable copy of a piece of DNA. It is a working copy of a single gene. RNA has many jobs. But most RNA molecules just make proteins. This process is called protein synthesis. RNA controls how amino acids are made into proteins. There are three main kinds of RNA: messenger RNA, ribosomal RNA, and transfer RNA.

▶ ***Messenger RNA*** Most genes have instructions for assembling amino acids into proteins. **Messenger RNA** (mRNA) molecules carry copies of this information to ribosomes within the cell.

▶ ***Ribosomal RNA*** Proteins are put together on ribosomes. Ribosomes are small organelles made of two parts, or subunits. These subunits have several **ribosomal RNA** (rRNA) molecules and up to 80 different proteins.

▶ ***Transfer RNA*** A third kind of RNA molecule carries amino acids to the ribosome. This kind of RNA molecule is known as **transfer RNA** (tRNA).

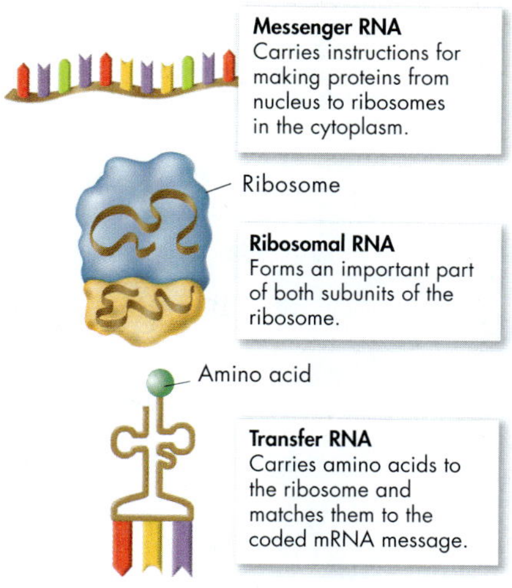

Types of RNA The three main types of RNA are messenger RNA, ribosomal RNA, and transfer RNA.

RNA Synthesis

To understand how genes work, we need to learn how cells make RNA using the information found in DNA.

Transcription Most of the work of making RNA takes place during **transcription.** In transcription, RNA molecules are produced that are complementary to the DNA sequences in genes. The order of the RNA bases complements the base sequences of the DNA. In eukaryotes, RNA is made in the cell's nucleus. Then it moves to the cytoplasm to help make proteins.

Transcription requires an enzyme known as *RNA polymerase*. This enzyme is like DNA polymerase and binds to DNA during transcription. It pulls the DNA strands apart. One strand is used as a pattern to make a complementary strand of RNA. Hundreds or even thousands of RNA copies can be made from a single gene.

Key Question How does the cell make RNA?
In transcription, segments of DNA act as templates, or patterns, to make complementary RNA molecules.

BUILD Vocabulary

ribonucleic acid (RNA)
a single-stranded nucleic acid that contains the sugar ribose

messenger RNA
a type of RNA that carries copies of instructions for the assembly of amino acids into proteins from DNA to the rest of the cell

ribosomal RNA
a type of RNA that combines with proteins to form ribosomes

transfer RNA
a type of RNA that carries each amino acid to a ribosome during protein synthesis

transcription
the synthesis of an RNA molecule from a DNA template, or pattern

BIOLOGY.com Search Lesson 13.1 GO • InterActive Art

ELL Focus on ELL: Access Content

ALL SPEAKERS Pair beginning and intermediate speakers with advanced speakers to extend their understanding of the process of transcription. Have partners divide a sheet of paper into four equal squares. In the upper left square, have them write the word *transcription*. In the upper right square, ask students to sketch the transcription process, using the diagram on the next page as a guide. In the lower left square, ask them to write a definition of *transcription* in their own words. Then, tell students to write in the lower right square an original sentence about transcription. Have pairs share their work with the class.

During Reading

Build Connections

VISUAL ANALOGY Remind students that an analogy is a kind of comparison: it takes two things that are different and shows how they can be similar. Emphasize the analogy of builders' plans and DNA and RNA.

Ask How is the blueprint in the illustration made? *(It is copied from the master plan.)*

Ask How is RNA like the blueprint and DNA like the master plan? *(RNA is copied from DNA.)*

Active Reading

KEY QUESTION Invite a volunteer to read aloud the first Key Question and answer. Draw a Venn diagram on the board. Label one side DNA and the other side RNA. Use the Venn diagram to compare and contrast DNA and RNA.

COMPARE AND CONTRAST Refer students to the Types of RNA figure.

Ask How are the three types of RNA the same? *(All are single-stranded and made of the same kinds of nucleotides.)*

Ask How are the three types of RNA different? *(They have different functions in making proteins.*

Have students find in the text the abbreviations used for the three types of RNA, and compare the abbreviations to the actual terms.

KEY QUESTION Read aloud the second Key Question and answer. Have students locate on the Transcribing DNA Into RNA diagram where RNA is being made. *(in the cell nucleus, in the area where RNA polymerase is shown)*

Speed Bump

Ask What is the main function of RNA? *(to direct the making of proteins)*

Ask What do proteins help determine? *(an organism's characteristics)*

Ask What are the three main types of RNA? *(messenger RNA, ribosomal RNA, and transfer RNA)*

During Reading continued

Active Reading

INTERPRET VISUALS Refer students to the Transcribing DNA Into RNA diagram.

Ask What has happened to the DNA molecule that allows it to be used as a pattern for making RNA? *(The two strands of the DNA molecule have been pulled apart by RNA polymerase.)*

Ask To what DNA base does the uracil in RNA bind? *(adenine)*

USE GRAPHIC ORGANIZERS Have students reread the section on RNA synthesis. As a class, make a flowchart on the board that shows the sequence of steps in RNA synthesis. The flowchart should begin with RNA polymerase binding to a DNA strand. It should end with RNA editing. Tell students to copy the flowchart in their notebooks.

Speed Bump

Ask What happens during transcription? *(A part of a DNA molecule is used as a pattern to make messenger RNA.)*

Post Reading

ASSESSMENT TIPS Refer students to the Types of RNA visual to help them answer Questions 1–3.

For Question 4, refer students to the bulleted list on the first page of the lesson.

For Question 6, have students review the flowchart they copied from the board into their notebooks.

Wrap-Up Activity

MATCH UP Give each student two index cards, one labeled *DNA* and the other labeled *RNA*. Say aloud the following words in random order: *deoxyribose, ribose, double-stranded, single-stranded, master plan, blueprint, cytosine, guanine, thymine, adenine,* and *uracil.* As you say each word, students should raise the card or cards that correctly identify the nucleic acid each word describes.

Transcribing DNA Into RNA
During transcription, the enzyme RNA polymerase uses one strand of DNA as a template to put together nucleotides to make a strand of RNA.

Promoters How does RNA polymerase know where to start and stop making a strand of RNA? The enzyme binds only to promoters, regions of DNA that have specific base sequences. Promoters are signals in the DNA that show RNA polymerase exactly where to begin making RNA. Other signals in DNA cause transcription to stop when a new RNA molecule is finished.

RNA Editing Many RNA molecules are edited before they are ready to be used. Bits and pieces called introns are cut out and discarded from these RNAs. The remaining pieces, called exons, are then spliced back together.

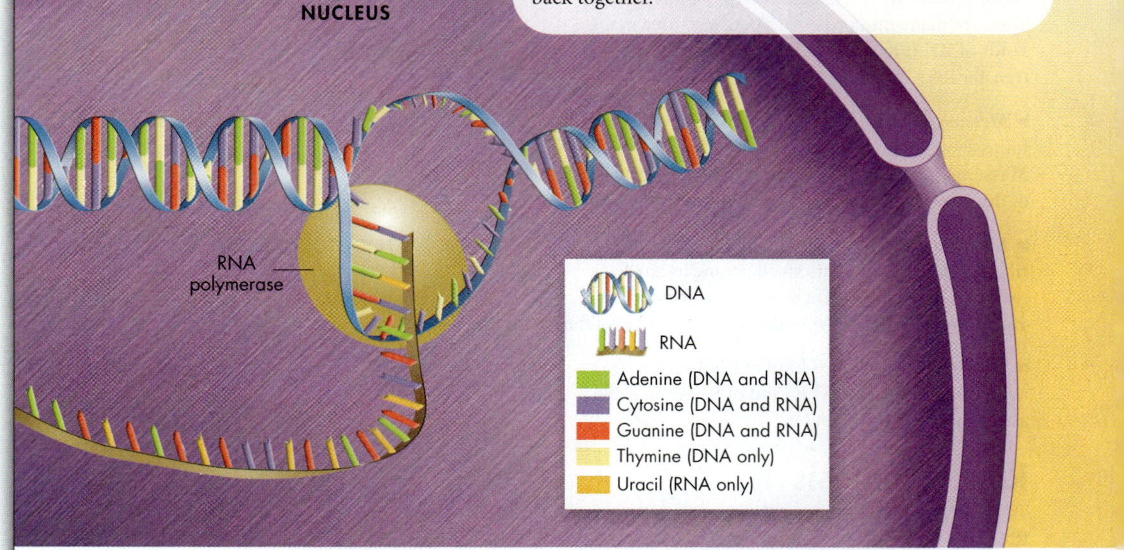

CHECK Understanding

Apply Vocabulary

Use the highlighted words from the lesson to complete each sentence correctly.

1. _______ carries amino acids.
2. _______ combines with up to 80 proteins to form an important organelle in the cytoplasm.
3. The information from DNA is transcribed into _______ and then carried out of the nucleus.

Critical Thinking

4. **Explain** Describe three main differences between RNA and DNA.
5. **Predict** What do you think would happen if introns were not removed from RNA?
6. **Quick Write** Using the terms *RNA polymerase* and *transcription*, describe how RNA gets made.

BIOLOGY.com Search Lesson 13.1 GO • Lesson Assessment • Art in Motion

Check Understanding

1. Transfer RNA
2. Ribosomal RNA
3. messenger RNA
4. RNA contains the sugar ribose instead of deoxyribose, is generally single-stranded rather than double-stranded, and contains uracil instead of thymine.
5. Sample answer: If introns were not removed, the instructions carried by mRNA for assembling amino acids into a protein might be incorrect, and the resulting protein might not function properly.
6. During transcription, the enzyme RNA polymerase binds to DNA and separates the DNA strands. It then uses one strand of DNA as a template to join nucleotides and make a complementary strand of RNA.

13.2 Ribosomes and Protein Synthesis

The Genetic Code

The DNA bases in genes are like a code. You have to know the code to understand what the bases mean.

Remember that the first step in decoding genetic messages is using transcription to make RNA from a sequence of DNA bases. This RNA holds a code for making proteins. Proteins are made of long chains of amino acids called **polypeptides.** Up to 20 different amino acids are found in polypeptides.

The shape and function of a protein are determined by its amino acids and their sequence. How is the order of RNA bases related to the sequence of amino acids? RNA contains four different bases: adenine (A), cytosine (C), guanine (G), and uracil (U). These bases are like the letters of a language. We call this language the **genetic code.** Each word in the genetic code is three "letters," or three bases. Each three-base set is called a **codon.** A codon specifies one amino acid.

How to Read Codons There are four different bases in RNA. These four bases mean there are 64 possible three-base codons in the genetic code. Most amino acids can be specified by more than one codon. For example, six different codons—UUA, UUG, CUU, CUC, CUA, and CUG—specify the amino acid leucine.

Start and Stop Codons Special codons tell the cell where to start and stop translating RNA. The codon AUG acts as the "start" codon for protein synthesis. After the start codon, mRNA is read three bases at a time.

Key Questions

What is the genetic code, and how is it read?

What role does the ribosome play in assembling proteins?

What is the "central dogma" of molecular biology?

BUILD Understanding

Two-Column Chart In the left column of a two-column chart, write the section headings. In the right column, list the main ideas.

In Your Workbook Go to your workbook and complete the chart for Lesson 13.2.

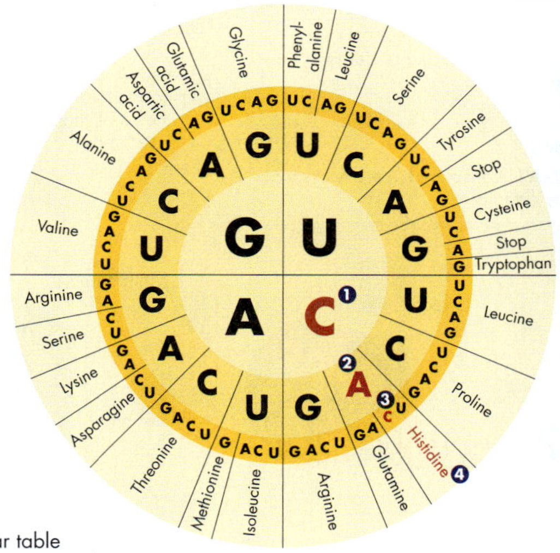

1. To decode the codon CAC, find the first letter in the set of bases at the center of the circle.
2. Find the second letter of the codon A, in the "C" quarter of the next ring.
3. Find the third letter, C, in the next ring, in the "C-A" grouping.
4. Read the name of the amino acid in that sector—in this case histidine.

Reading Codons The circular table shows the animo acid that corresponds to each of the 64 codons.

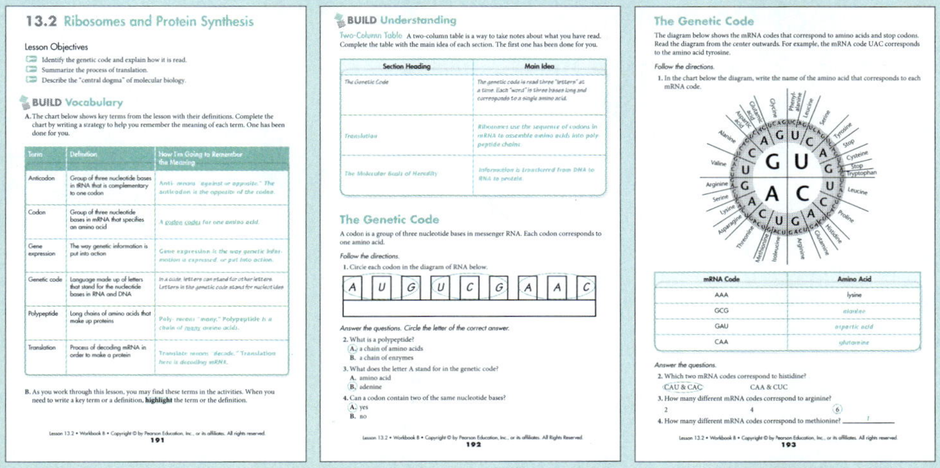

LESSON 13.2

Pre-Reading

Objectives

13.2.1 Identify the genetic code and explain how it is read.

13.2.2 Summarize the process of translation.

13.2.3 Describe the "central dogma" of molecular biology.

Preview the Pages

1. Have students find the vocabulary words in the text. Help them pronounce each word aloud. Point out similarities among the words *genetic code, codon,* and *anticodon.*
2. Preview all the visuals. Point out that the Translation visual extends over two pages. Help students connect the lettered captions in the visual to the lettered headings in the text.
3. Point out that the Gene Expression visual on the last page of the lesson provides a summary of the processes described in Lessons 13.1 and 13.2. The visual illustrates how genetic information is transferred from DNA to RNA to protein.

Build Understanding

TWO-COLUMN CHART Direct students to Workbook B, where they will find a two-column chart for this lesson. Ask students to list more main ideas from the first section. Have students complete their chart by listing the main ideas for Translation and The Molecular Basis for Heredity as they read the lesson.

NATIONAL SCIENCE EDUCATION STANDARDS

UNIFYING CONCEPTS AND PROCESSES

II, V

CONTENT

B.2, B.3, C.1.c, C.2.a, G.3

INQUIRY

A.2.a

For the corresponding lesson in *Miller & Levine Biology,* see pages 366–371.

LESSON 13.2

During Reading

Active Reading

KEY QUESTION Read aloud the Key Question and answer.

Ask What molecule holds copies of the genetic code? *(messenger RNA)*

USE ANALOGIES Use the English language as an analogy to help students understand the process of translation.

Say In the English language, the letters of the alphabet are used to form words, and sequences of words form sentences.

Ask What do messenger RNA bases form? *(codons)* What do sequences of codons form? *(polypeptides)*

Say A capital letter signals the beginning of a sentence, and a period signals the end of the sentence.

Ask What signals the beginning and end of a polypeptide? *(A start codon signals the beginning of a polypeptide, and a stop codon signals the end.)*

Build Vocabulary

ROOT WORDS Challenge students to find other words on this page that include the root word *code*. *(decoding, anticodon)* Discuss each word's meanings in relation to that of the root word.

Speed Bump

Ask In which part of the cell does translation take place? *(cytoplasm)*

Ask What role does mRNA have in the translation process? *(It carries the code for making proteins.)*

Ask What role does tRNA have in the translation process? *(tRNAs bring the correct amino acids into the ribosome. Each tRNA carries only one kind of amino acid and has an anticodon that is complementary to a codon on the mRNA.)*

BUILD Vocabulary

polypeptide
a long chain of amino acids that makes proteins

genetic code
a collection of codons of mRNA, each of which directs the incorporation of a particular amino acid into a protein during protein synthesis

codon
a group of three nucleotide bases in mRNA that specify a particular amino acid to be incorporated into a protein

ROOT WORDS
The word *codon* was created by adding the suffix *-on* to the word *code*. Other scientific terms, such as *electron*, were formed in a similar way.

Translation continues until one of three different "stop" codons is reached. Then, translation stops and the polypeptide is complete.

Key Question What is the genetic code, and how is it read?
The genetic code is a code for making proteins. The genetic code is read three "letters" at a time. Each "word" is three bases long and corresponds to a single amino acid.

Translation

The sequence of nucleotide bases in each mRNA strand is a set of instructions. The instructions give the order in which amino acids should be joined to make a polypeptide. Once a polypeptide is made, it folds into its final shape or joins with other polypeptides. Once it folds or joins with other polypeptides, it becomes a working protein.

To put together a complex toy, you need instructions. Once you read them, you can put the parts together. In the cell, the ribosome reads the instructions and puts together proteins. Ribosomes use the sequence of codons in mRNA to assemble amino acids into polypeptide chains. The decoding of an mRNA message into a protein is a process called **translation.**

Steps in Translation Transcription produces the mRNA molecules that are used in protein synthesis, or translation. In eukaroytes, transcription occurs in the cell's nucleus. After transcription, mRNA leaves the nucleus, and translation takes place in the cytoplasm. The figure below shows this process.

Translation The cell uses information from messenger RNA to make proteins during translation.

Messenger RNA
Messenger RNA is transcribed in the nucleus and then enters the cytoplasm.

A Transfer RNA
Translation begins at AUG, the start codon. Each transfer RNA has an anticodon whose bases are complementary to the bases of a codon on the mRNA strand. The ribosome positions the start codon to attract its anticodon, which is part of the tRNA that binds methionine. The ribosome also binds the next codon and its anticodon.

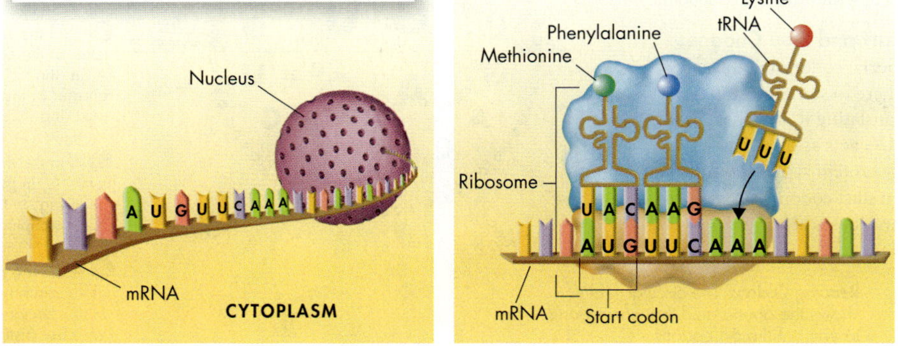

Hands-On Learning

DECODING A CODED MESSAGE Introduce the idea of a code by challenging the class to decode a coded message. On the board, write the following sequence of numbers:

9 3-1-14 18-5-1-4 20-8-9-19 3-15-4-5

Tell students that each number represents a letter of the alphabet: 1 = a, 2 = b, 3 = c, and so on. After students have correctly decoded the message *(I can read this code)*, explain that messenger RNA also contains a code and, in this lesson, they will learn how it is decoded.

A ***Transfer RNA*** Translation begins when a ribosome attaches to an mRNA molecule in the cytoplasm. As each codon passes through the ribosome, tRNAs bring the correct amino acids into the ribosome. One at a time, the ribosome then attaches these amino acids to the growing polypeptide chain.

Each tRNA molecule carries just one kind of amino acid. Each tRNA molecule also has three unpaired bases. This set of bases is called the **anticodon.** Each tRNA anticodon is complementary to one mRNA codon.

The anticodon of the tRNA molecule for methionine is UAC. It will pair with the methionine codon, AUG. The ribosome has a second binding site for a tRNA molecule for the next codon. Notice that the second codon in the figure is UUC. A tRNA molecule with an AAG anticodon will bind to this codon. This tRNA molecule brings the amino acid phenylalanine into the ribosome.

B ***Joining Amino Acids*** Next, the ribosome helps make a peptide bond between the first and second amino acids. At the same time, the bond holding the first tRNA molecule to its amino acid is broken. That tRNA moves into an exit site, from which it leaves the ribosome. The ribosome moves to the next codon. Another tRNA then carries in the amino acid specified by that codon.

BUILD Vocabulary

translation
a process by which the sequence of bases of an mRNA is converted into the sequence of amino acids of a protein

anticodon
a group of three bases on a tRNA molecule that are complementary to the three bases of a codon of mRNA

PREFIXES
The prefix *anti-* is Latin for "against." When added to science terms, it can mean "opposite." An anticodon, then, has the opposite, or complementary, base sequence of its matching codon.

B The Polypeptide "Assembly Line"
The ribosome joins the two amino acids—methionine and phenylalanine—and breaks the bond between methionine and its tRNA. The tRNA floats away from the ribosome, allowing the ribosome to bind another tRNA. The ribosome moves along the mRNA, from right to left, binding new tRNA molecules and amino acids.

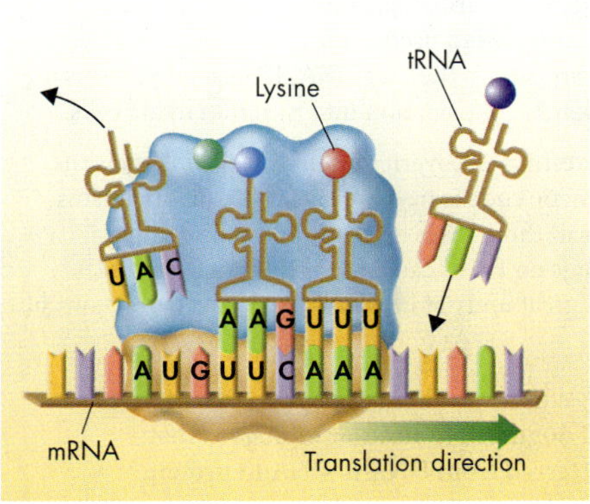

C Completing the Polypeptide
The process continues until the ribosome reaches one of the three stop codons. Once the polypeptide is complete, it and the mRNA are released from the ribosome.

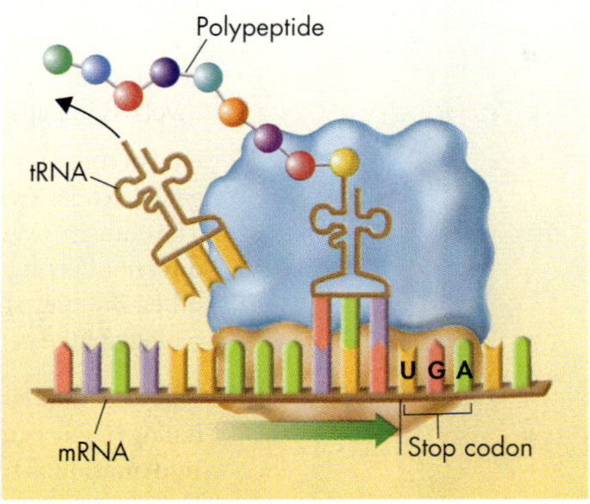

LESSON 13.2

Active Reading

MAKE CONNECTIONS Students may confuse the words *transcription* and *translation* because of their similar spellings and their close connection. Remind students that transcription is the process of making messenger RNA from DNA.

Ask How is *translation* connected to *transcription*? *(During translation, the messenger RNA that was made during transcription is used to make proteins.)*

USE VISUALS Ask a volunteer to read aloud the lettered paragraphs in the text and the corresponding lettered captions in the visual, pausing between each letter.

Ask What is mRNA? *(abbreviation for* messenger RNA*)* What is tRNA? *(abbreviation for* transfer RNA*)*

Ask In Picture A, what is the anticodon of the first tRNA molecule that binds with the mRNA shown? *(UAC)* What is the complementary codon on the mRNA? *(AUG)* What amino acid is attached to the first tRNA? *(methionine)*

Point out the bond between the first two amino acids shown in Picture B.

Say The ribosome moves along the mRNA strand like a bead sliding along a string, translating the mRNA as it moves and producing a growing peptide chain.

Draw students' attention to the stop codon shown in Picture C.

Ask Why do you think the stop codon causes translation to stop? *(There is no tRNA that has an anticodon that is complementary to the stop codon.)*

Build Vocabulary

PREFIXES Have students think of other words that begin with the prefix *anti-* (for example, *antiaging, antibacterial, antiviral, antifreeze*) and use each word in a sentence.

Speed Bump

Ask What is the final product of the translation process? *(a protein, or polypeptide)*

ELL Focus on ELL: Build Background

BEGINNING SPEAKERS English language learners are probably familiar with translating words from English to their native language.

Say If you find an unfamiliar word in a book, you might use a dictionary to learn the meaning of the word in your native language. The dictionary translates the word from one language to another. Just as unfamiliar words need to be translated, so does the genetic code carried by messenger RNA in a cell.

On the board, write this sentence for students to complete,

During translation in a cell, the sequence of __________ in messenger RNA is translated into a sequence of __________.

Ask a volunteer to complete the sentence. *(codons, amino acids)*

During Reading continued

Active Reading

KEY QUESTION Read aloud the first Key Question and answer.

Ask If you saw this question on a test, which illustration in the lesson could you visualize to help you answer the question? *(the Translation illustration across the bottom of the two previous pages)* Have a volunteer refer to the illustration and summarize the role of the ribosome in assembling proteins.

SUMMARIZE Have students make a three-column table on a sheet of paper with the headings Messenger RNA, Transfer RNA, and Ribosomal RNA. Tell them to summarize in each column the role each type of RNA plays during translation.

KEY QUESTION Read aloud the second Key Question and answer.

Ask What two processes are involved in the "central dogma" of molecular biology? *(transcription and translation)*

Speed Bump

Ask In what two ways do proteins produce an organism's traits? *(as enzymes and as building parts in cells)*

Ask What do molecular biologists study? *(organisms at the molecular level)*

C ***The Chain Grows*** The polypeptide chain grows until the ribosome reaches a "stop" codon on the mRNA. Then, the ribosome releases both the newly formed polypeptide and the mRNA molecule. The process of translation is complete.

Key Question What role does the ribosome play in assembling proteins?
Ribosomes use the sequence of codons in mRNA to assemble amino acids into polypeptide chains.

The Roles of tRNA and rRNA in Translation All three kinds of RNA are put to work in the ribosome during translation. The mRNA molecule carries the coded message that directs the process. The tRNA molecules bring the correct amino acid for each codon on the mRNA. The rRNA and many proteins make up the ribosomes.

The Molecular Basis for Heredity

Gregor Mendel might have been surprised to learn that genes have instructions for putting proteins together. He might have asked what proteins have to do with traits of pea plants.

The answer is that proteins have everything to do with such traits. Remember that many proteins are enzymes. Enzymes help speed up and regulate chemical reactions. A gene that codes for an enzyme to make pigment controls flower color. Another gene makes proteins that control tissue growth in a leaf. Basically, proteins are tiny tools. Each one is designed to build or run a part of a living cell.

Once scientists learned that genes were made of DNA, other discoveries soon followed. Soon, a new scientific field called molecular biology had formed. Molecular biology tries to explain living organisms by studying them at the molecular level. It uses molecules like DNA and RNA as tools to understand living things. One of the earliest findings is now called the field's "central dogma," or basic rule. The central dogma states that information is transferred from DNA to RNA to protein. In real life, there are many exceptions to this "dogma." For example, some viruses pass information from RNA to DNA. But the central dogma is a useful rule that helps to explain how genes work. Gene expression is the way DNA, RNA, and proteins are involved in putting genetic information into action in living cells.

One of the most interesting discoveries of molecular biology is the genetic code. The genetic code is nearly universal to all organisms. The code is always read three bases at a time and in the same direction. Living things on Earth can be very different from each other. But they show great unity at life's most basic level, the molecular biology of the gene.

Key Question What is the "central dogma" of molecular biology? **The central dogma of molecular biology is that information is transferred from DNA to RNA to protein.**

314 Chapter 13 • Lesson 2

Focus on Academic Language

• field • element • central • level • role

Write the words on the board. Have students find in a dictionary the multiple meanings of each of the first four words and write the definitions in their notebooks. Then, tell students to find the words on this page and circle in their notebooks the meanings of the words that are used on the page.

Point out that the word *role* is a homophone. Explain that a homophone is a word that is pronounced the same as another word but has a different spelling and meaning. For example, *role* is a homophone of *roll.* Have students look up the definitions of these words and write them in their notebooks. Then, have them find the word *role* on this page and circle in their notebooks the definition of the word used on the page.

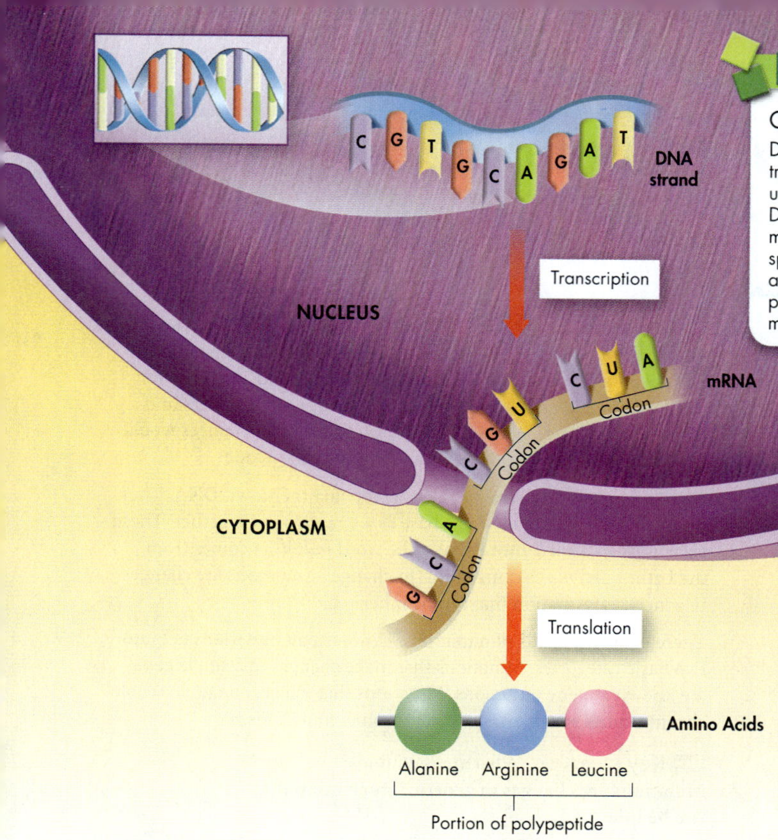

BUILD Connections

GENE EXPRESSION

DNA carries information for the traits of an organism. The cell uses the sequence of bases in DNA as a template for making mRNA. The codons of mRNA spell out the sequence of amino acids in a protein. Those proteins play a key role in making an organism's traits.

CHECK Understanding

Apply Vocabulary

Use the highlighted words from the lesson to complete the sentence correctly.

1. A(n) _______ is a group of three bases on a tRNA molecule that are complementary to the three bases of a codon of mRNA.

Critical Thinking

2. Apply Concepts Using the Reading Codons circular table, identify the amino acids specified by codons UGG, AGG, and UGC.

3. Explain What does the term *gene expression* mean?

4. Explain How does a cell interpret the genetic code?

5. Write to Learn Answer the first clue for the chapter mystery. Include the term *codon* in your answer.

MYSTERY CLUE

What features of the genetic code make it possible for a mouse's gene to work inside the cells of a fly? (**Hint:** See p. 314.)

BIOLOGY.com Search Lesson 13.2 GO • Lesson Assessment

Check Understanding

1. anticodon

2. tryptophan, arginine, cysteine

3. It refers to the way in which DNA, RNA, and proteins are involved in putting genetic information into action in living cells.

4. The genetic code is read one codon, or three bases, at a time. Each codon calls for one amino acid.

5. A mouse's gene works inside the cells of a fruit fly because almost every organism uses the same molecular code. In almost all organisms, the same amino acids are assigned to particular codons, and the code is always read three bases at a time and in the same direction.

Build Connections

VISUAL SUMMARY Point out to students that the diagram on this page is a combination of the Transcribing DNA Into RNA diagram in Lesson 13.1 and the Translation diagram in this lesson.

Ask What is the role of DNA in gene expression? *(DNA carries information for the traits of an organism.)*

Ask In what form is this information? *(the sequence of bases in DNA)*

Ask How do mRNA codons spell out the sequence of amino acids in a protein? *(The codons are complementary to anticodons on tRNA, which carry amino acids to the ribosomes.)*

Ask What basic rule of molecular biology does this diagram show? *(the "central dogma," that information is transferred from DNA to RNA to protein)*

Post Reading

ASSESSMENT TIPS Review the highlighted vocabulary words with students before they begin the assessment.

For Question 2, remind students to begin with the letter in the center of the Reading Codons circle, and work out toward the edge of the circle.

For Question 3, suggest students reread the second paragraph of The Molecular Basis for Heredity before answering the question.

For Questions 4 and 5, review the meanings of the terms *genetic code* and *codon*.

CHAPTER MYSTERY Have students read the last paragraph on page 314 before writing their answer.

Ask What does the text mean when it says that living things "show great unity" in the molecular biology of their genes? *(that the genetic code in genes is consistent in nearly all organisms)*

Wrap-Up Activity

COMPLEMENTING Pair students and give each pair a strip of paper with a sequence of 12 DNA bases written across the bottom. Do not include bases that are complementary to the mRNA stop codons: UAA, UAG, and UGA. Tell students the strip represents a portion of a gene. Have them write in the middle of the strip, above the DNA bases, the mRNA codons that are complementary to the DNA bases. Then, tell pairs to exchange strips and write above the codons the complementary anticodons. Have pairs return the strips and discuss any inaccuracies they found.

LESSON 13.3

Pre-Reading

Objectives

13.3.1 Define mutations and describe the different types of mutations.

13.3.2 Describe the effects mutations can have on genes.

Preview the Pages

1. Preview with students all of the vocabulary, pronouncing each word. Point out the recurrence of the word *mutation* in the vocabulary and the similarity of the words *mutagen* and *mutation.*
2. Preview all of the visuals, and read their captions. Point out the unusual features of the organisms shown on this page. Tell students that these unusual structures were caused by mutations.
3. Preview the Point Mutations and Chromosomal Mutations diagrams. Explain that a point mutation involves a single nucleotide in DNA, whereas a chromosomal mutation involves a large number of nucleotides.

Activate Prior Knowledge

Ask What determines the sequence of amino acids in a particular protein? *(the sequence of bases in the gene that codes for that protein)*

Ask What do you think might happen if the sequence of bases in the gene changed? *(Sample answer: The sequence of amino acids in the protein might change.)*

NATIONAL SCIENCE EDUCATION STANDARDS

UNIFYING CONCEPTS AND PROCESSES

II, IV, V

CONTENT

C.2.c, C.3.a, C.3.c, F.1, F.4

For the corresponding lesson in *Miller & Levine Biology,* see pages 372–376.

13.3 Mutations

Key Questions

What are mutations?

How do mutations affect genes?

BUILD Understanding

Preview Visuals Before you read the lesson, look at the Point Mutations and Chromosomal Mutations figures. As you read, note the changes produced by various gene and chromosomal mutations.

In Your Workbook Complete the chart in Lesson 13.3 of your workbook.

Types of Mutations

The order of bases in DNA is like the letters of a coded message. What would happen if a few of those letters changed and, so, accidentally changed the message? Could the cell still understand the code's meaning? What effects do you think those changes would have on genes and the polypeptides for which they code?

Sometimes cells do make mistakes in copying their own DNA. They insert the wrong base or skip a base as a strand is put together. These variations are called **mutations.** The word *mutation* comes from the Latin word *mutare*, meaning "to change." Mutations are changes in genetic information that can be inherited.

There are many kinds of mutations. However, all mutations fall into two basic categories. Mutations that make changes in a single gene are known as gene mutations. Mutations that make changes in whole chromosomes are known as chromosomal mutations.

Key Question What are mutations?

Mutations are changes in genetic information that can be inherited.

The shape of this flower is caused by a mutation that affects the growing areas of the flower tissue.

A mutation in a gene has led to an extra set of wings on this fruit fly. (SEM 20×)

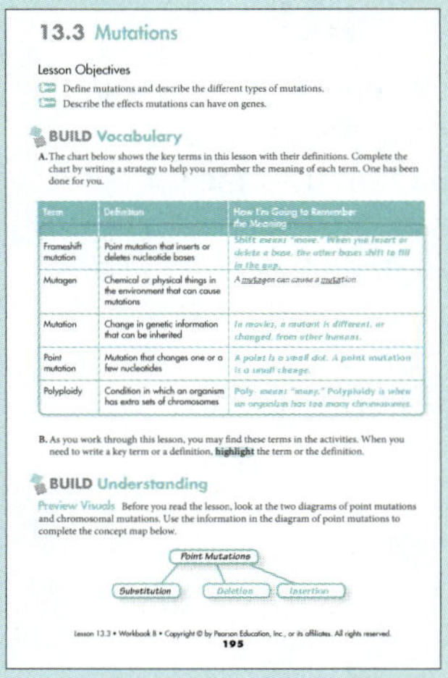

13.3 Mutations

Lesson Objectives

Define mutations and describe the different types of mutations.

Describe the effects mutations can have on genes.

BUILD Vocabulary

A. The chart below shows the key terms in this lesson with their definitions. Complete the chart by writing a strategy to help you remember the meaning of each term. One has been done for you.

Term	Definition	How I'm Going to Remember the Meaning
Frameshift mutation	Point mutation that inserts or deletes nucleotide bases	Shift means "move." When you insert or delete a base, the other bases shift to fill in the gap.
Mutagen	Chemical or physical things in the environment that can cause mutations	A mutagen can cause a mutation
Mutation	Change in genetic information that can be inherited	In movies, a mutant is different, or changed, from other humans.
Point mutation	Mutation that changes one or a few nucleotides	A point is a small dot. A point mutation is a small change.
Polyploidy	Condition in which an organism has extra sets of chromosomes	Poly- means "many." Polyploidy is when an organism has too many chromosomes.

B. As you work through this lesson, you may find these terms in the activities. When you need to write a key term or a definition, **highlight** the term or the definition.

BUILD Understanding

Preview Visuals Before you read the lesson, look at the two diagrams of point mutations and chromosomal mutations. Use the information in the diagram of point mutations to complete the concept map below.

Point Mutations: Substitution, Deletion, Insertion

Lesson 13.3 • Workbook B • Copyright © by Pearson Education, Inc., or its affiliates. All rights reserved.
195

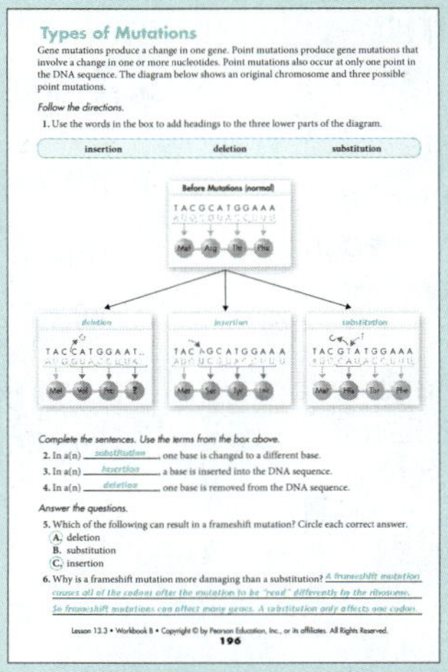

Types of Mutations

Gene mutations produce a change in one gene. Point mutations produce gene mutations that involve a change in one or more nucleotides. Point mutations also occur at only one point in the DNA sequence. The diagram below shows an original chromosome and three possible point mutations.

Follow the directions.

1. Use the words in the box to add headings to the three lower parts of the diagram.

insertion deletion substitution

Complete the sentences. Use the terms from the box above.

2. In a(n) substitution, one base is changed to a different base.
3. In a(n) insertion a base is inserted into the DNA sequence.
4. In a(n) deletion one base is removed from the DNA sequence.

Answer the questions.

5. Which of the following can result in a frameshift mutation? Circle each correct answer.
A. deletion
B. substitution
C. insertion
6. Why is a frameshift mutation more damaging than a substitution? A frameshift mutation causes all of the codons after the mutation to be "read" differently by the ribosome. So frameshift mutations can affect many genes. A substitution only affects one codon.

Lesson 13.3 • Workbook B • Copyright © by Pearson Education, Inc., or its affiliates. All Rights Reserved.
196

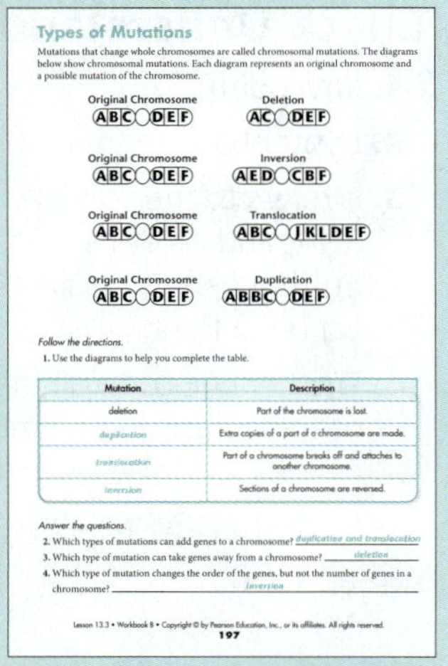

Types of Mutations

Mutations that change whole chromosomes are called chromosomal mutations. The diagrams below show chromosomal mutations. Each diagram represents an original chromosome and a possible mutation of the chromosome.

Follow the directions.

1. Use the diagrams to help you complete the table.

Mutation	Description
deletion	Part of the chromosome is lost.
duplication	Extra copies of a part of a chromosome are made.
translocation	Part of a chromosome breaks off and attaches to another chromosome.
inversion	Sections of a chromosome are reversed.

Answer the questions.

2. Which types of mutations can add genes to a chromosome? duplication and translocation
3. Which type of mutation can take genes away from a chromosome? deletion
4. Which type of mutation changes the order of the genes, but not the number of genes in a chromosome? inversion

Lesson 13.3 • Workbook B • Copyright © by Pearson Education, Inc., or its affiliates. All rights reserved.
197

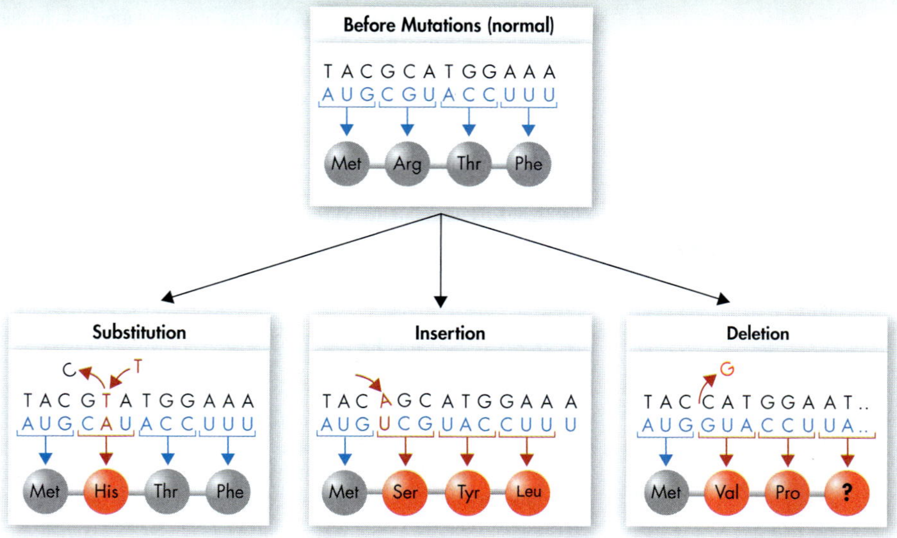

Point Mutations These diagrams show how changes in just one nucleotide can change the order of amino acids in proteins.

Gene Mutations Gene mutations that involve changes in one or a few nucleotides are known as **point mutations.** They happen at a single point in the DNA sequence. Point mutations include substitutions, insertions, and deletions, and they usually happen during replication. If a gene in one cell is changed, the change can be passed on during cell division. Every daughter cell will have the mutation. The figure above shows how point mutations occur.

▶ ***Substitutions*** In a substitution, one base is changed to a different base. Substitutions usually affect a single amino acid. Sometimes they have no effect at all. For example, a mutation might change one codon of mRNA from CCC to CCA. The codon would still call for the amino acid proline. But changing CCC to ACC would replace proline with the amino acid threonine.

▶ ***Insertions and Deletions*** Insertions and deletions are point mutations. An insertion adds a new base to the DNA sequence. A deletion removes a base from the DNA sequence. The effects of these changes can be dramatic. Remember that the genetic code is read three bases at a time. Reading the bases three at a time happens even if a nucleotide is added or taken away. After a change, the sets shift in every codon that comes after the mutation.

Insertions and deletions are also called **frameshift mutations** because they shift the "reading frame" of the genetic code. A shift in the reading frame can change every amino acid that follows the mutation. This change can alter a protein so much that it cannot do its job.

BUILD Vocabulary

mutation
a change in the genetic material of a cell

point mutation
a gene mutation in which a single base pair in DNA has been changed

frameshift mutation
a mutation that shifts the "reading frame" of the genetic message by inserting or deleting a nucleotide

Hands-On Learning

POINT MUTATIONS Provide pairs of students with sticky notes and five different colored markers. Using a different color for each letter, have them write *A* on eight sticky notes, *T* on three notes, *U* on six notes, *G* on five notes, and *C* on five notes.

Say You are going to model a normal gene and three kinds of point mutations in that gene, using your sticky notes.

Instruct students to use their sticky notes to recreate the DNA and mRNA shown in the Before Mutations (normal) portion of the Point Mutations figure. Tell students to position their sticky notes in a way that leaves a space between each group of three bases, or codon. Then, have them recreate each of the point mutations by changing the sticky notes as needed.

Ask Which of the mutations required that you reposition the codons? *(insertion and deletion)* Which of the mutations were frameshift mutations? *(insertion and deletion)*

Build Understanding

PREVIEW VISUALS Explain that the letters in the Point Mutations diagram represent bases, while the letters in the Chromosomal Mutations diagram represent segments of a chromosome. Direct students to the partially completed chart in Workbook B. Have them complete the chart with notes as they read. Point out that they can use the notes as a study aid later.

During Reading

Active Reading

KEY QUESTION Read aloud the Key Question and answer from the previous page.

Ask What are the two basic kinds of mutations? *(gene mutations and chromosomal mutations)*

INTERPRET DIAGRAMS Have students study the Point Mutations diagrams. Read the caption aloud. Remind students that point mutations usually happen during replication, when a cell makes a copy of its DNA prior to cell division.

Ask What is the top strand of bases in the diagrams? *(DNA)* What is the bottom strand? *(messenger RNA)* What do the circles represent? *(amino acids)*

Explain that the three letters in each circle are the first three letters in the name of a particular amino acid. For example, in the first circle, the letters *Met* stand for the amino acid methionine.

Draw students' attention to the arrows showing the movement of bases into and out of the DNA strand. Call on volunteers to explain in their own words what is occurring during each type of point mutation.

Ask How many codons are affected in a substitution? *(one)* How many codons are affected in an insertion or deletion? Why? *(many; The change affects every codon that comes after the mutation.)*

Speed Bump

Ask What are the three types of point mutations? *(substitution, insertion, and deletion)*

Ask What is a frameshift mutation? *(A frameshift mutation is a mutation that changes the reading frame of a gene.)*

During Reading continued

Active Reading

MAKE CONNECTIONS On the board, write the verbs *delete, duplicate,* and *invert* and their definitions: *Delete* means "to erase." *Duplicate* means "to copy." *Invert* means "to reverse."

Have students look at the deletion, duplication, and inversion mutations in the Chromosomal Mutations diagram. Invite volunteers to explain how the meanings of the verbs relate to the manner in which the mutations occur.

SCIENCE SUPPORT Elaborate on the topic of sickle cell disease. Tell students that the allele for sickle cell disease is recessive. A person must be homozygous for the recessive allele to have the disease. Then, tell them that an often-fatal disease called malaria is common in parts of Africa. Mosquitoes spread this disease. People who carry one or two alleles for sickle cell disease are resistant to malaria. In areas where malaria is prevalent, heterozygotes are more fit than people who are homozygous for the normal allele. This explains why sickle-cell disease is fairly common in some populations, and why people in these populations have a relatively high risk of inheriting two copies of the mutant gene.

Speed Bump

Ask How do gene mutations and chromosomal mutations differ? *(A gene mutation is a change in a single gene. A chromosomal mutation is a change in the number or structure of chromosomes.)*

Ask What are the four types of chromosomal mutations? *(deletion, duplication, inversion, and translocation)*

Ask What is a mutagen? *(A mutagen is a chemical or physical agent in the environment that causes a mutation.)*

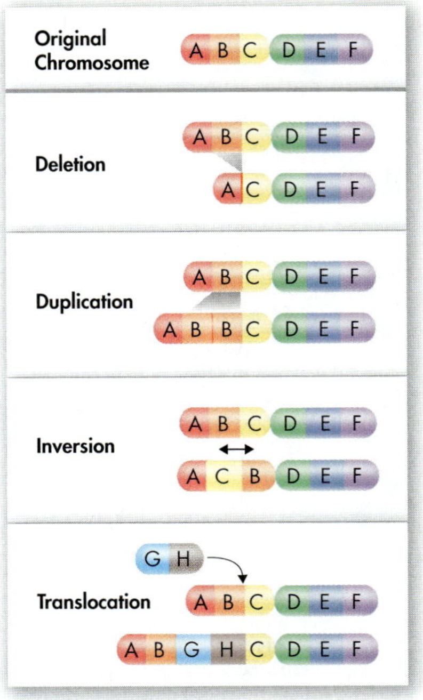

Chromosomal Mutations Four types of mutations cause changes in whole chromosomes. The four types of chromosomal mutations are deletion, duplication, inversion, and translocation.

Chromosomal Mutations A chromosomal mutation is a change in the number or structure of chromosomes. These mutations can change the location of genes on chromosomes. They can also change the number of copies of some genes.

There are four types of chromosomal mutations: deletion, duplication, inversion, and translocation. Deletion happens when part or all of a chromosome is lost. Duplication happens when an extra copy of all or part of a chromosome is made. Inversion happens when parts of a chromosome change direction. Translocation happens when part of one chromosome breaks off and attaches to another one.

Effects of Mutations

Genetic material can be altered by natural events or by artificial means. The resulting mutations may or may not affect an organism. Many mutations are made by mistakes in genetic processes. For example, some point mutations are caused by mistakes in DNA replication. An incorrect base is inserted into DNA about once in every 10 million bases. However, small changes in genes can build up over time.

Mutagens Some mutations are caused by mutagens. **Mutagens** are chemical or physical agents in the environment. Chemical mutagens include some pesticides, tobacco smoke, and pollutants. Physical mutagens include X-rays and ultraviolet light. These mutagens can cause mutations at high rates. Sometimes the cell can repair the DNA. But when the cell cannot fix the DNA, the sequence changes become permanent.

Harmful and Helpful Mutations Mutations can help or harm organisms. However, most mutations have little or no effect on genes. So, there is little or no effect on the function of the proteins they make. In any case, mutations are important. Without mutations, there would be no genetic variation. Without genetic variation, species could not evolve.

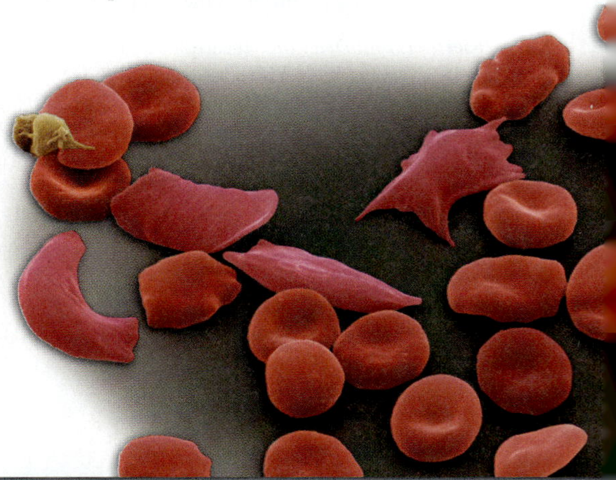

Effects of a Point Mutation Sickle cell disease affects the shape of red blood cells. The round cells in the image are normal red blood cells. The crescent and star-shaped cells are sickled cells. (SEM 1700×)

ELL Focus on ELL: Extend Language

INTERMEDIATE AND ADVANCED SPEAKERS Write the word *translocation* on the board. Tell students that the prefix *trans-* means "across, beyond, or change."

Ask What do you think the word *translocation* means? *(a change in location or position)*

Invite students to think of other words that begin with the prefix *trans-* and write them on the board with their meanings. *(Sample words: transatlantic—across the Atlantic; transform—change form; transcontinental—across a continent; transport—to carry across a distance)*

If students cannot think of words with the prefix *trans-*, have them look for words in a dictionary. Use a history book for the example of *transcontinental railroad.*

▶ ***Harmful Effects*** Some of the most harmful mutations make big changes in protein shape or gene activity. The proteins made by these mutations can get in the way of biological activities. For example, some cancers are caused by mutations. Sickle cell anemia is caused by a point mutation in one of the polypeptides of hemoglobin. Hemoglobin is the oxygen-carrying protein in red blood cells. The mutant protein causes red blood cells to change shape, giving them a sickle-like appearance. These sickled cells interfere with blood flow and cause severe problems in the bloodstream.

▶ ***Helpful Effects*** Some of the changes made by mutations can help an organism or species. These mutations make genes with functions that are useful to organisms in different environments. Mutations have made many African mosquitoes resistant to the chemicals once used to kill them. This resistance may be bad news for people. However, it is very helpful to the insects. Beneficial mutations happen in people, too. For example, one mutation makes bones stronger. Another mutation increases resistance to HIV, the virus that causes AIDS.

Plant and animal breeders often make use of "good" mutations. For example, sometimes a full set of chromosomes does not separate during meiosis. The gametes made may produce triploid (3N) or tetraploid (4N) organisms. The condition in which an organism has extra sets of chromosomes is called **polyploidy.** Polyploid plants are often larger and stronger than diploid plants.

Key Question How do mutations affect genes?
The effects of mutations on genes vary widely. Some have little or no effect, and some produce beneficial variations. Some negatively disrupt gene function.

Polyploid Plants The fruit of the Tahiti lime is seedless because of polyploidy. Changes to the ploidy number of citrus plants can affect the size and strength of the trees, the quality of the fruit, and the number of seeds in their fruit.

BUILD Vocabulary

mutagen
a chemical or physical agent in the environment that interacts with DNA and may cause a mutation

polyploidy
a condition in which an organism has extra sets of chromosomes

WORD ORIGINS
The word *mutagen* is a Latin word that means "producer of change." Mutagens change an organism's genetic information.

CHECK Understanding

Apply Vocabulary

Use the highlighted words from the lesson to complete each sentence correctly.

1. A gene mutation in which a single base pair in DNA has been changed is a ______.
2. A mutation that shifts the "reading frame" of the genetic message by inserting or deleting a nucleotide is a ______.
3. A chemical that causes a mutation is called a ______.

Critical Thinking

4. **Explain** Describe the two main types of mutations.
5. **Apply Concepts** How do mutations affect living things?
6. **Write to Learn** Write the following DNA sequence GCTAATCGGCTA. Below this sequence, show how the sequence might look after a substitution mutation. Repeat this for a deletion mutation and an insertion mutation.

BIOLOGY.com Search Lesson 13.3 GO • Lesson Assessment

Build Vocabulary

WORD ORIGINS Point out to students that the words *mutation* and *mutagen* have a common Latin root that means "to change."

Ask How is the word *mutagen* related to the word *mutation? (A mutagen may cause a mutation in DNA.)*

Active Reading

KEY QUESTION Read aloud the Key Question and answer.

Ask What is an example of a harmful effect of a genetic mutation? *(Sample answers: some cancers, sickle cell disease)* What is an example of a beneficial effect? *(Sample answers: resistance to HIV, polyploidy in plants)*

Post Reading

ASSESSMENT TIPS Refer students to the Point Mutations and the Chromosomal Mutations diagrams to help them answer Question 4.

Refer students having difficulty with Question 5 back to the Point Mutations diagram.

Wrap-Up Activity

MUTATION MATCH-UP Have students form pairs, and give each pair seven index cards, one for each type of point mutation and chromosomal mutation shown in the lesson diagrams. Tell students to cut each card in half. On one half, have them write the name of one type of mutation, such as substitution. On the other half, have them draw a diagram that illustrates that mutation. (As an alternative, copies of the diagrams can be given to students, and they can paste the diagrams onto their cards.) Have students shuffle their cards and then match the name of each type of mutation with the correct diagram.

Check Understanding

1. point mutation
2. frameshift mutation
3. mutagen
4. Gene mutations involve changes in one or a few nucleotides. Chromosomal mutations involve changes in the number or structure of chromosomes.
5. Sometimes, mutations have no effect on the organism; and sometimes they have negative effects. However, mutations are a source of genetic variation for living things. Sometimes variation can help organisms adapt to different or changing environments. Mutations are also necessary for species to evolve.
6. Student answers should show a clear understanding of the differences between substitution, deletion, and insertion mutations.

LESSON 13.4

Pre-Reading

Objectives

13.4.1 Describe gene regulation in prokaryotes.

13.4.2 Explain how most eukaryotic genes are regulated.

13.4.3 Relate gene regulation to development in multicellular organisms.

Preview the Pages

1. Have students preview the vocabulary words and their definitions. Students might find the words *operon* and *operator* to be confusing. Tell them the operator controls the operon.
2. Preview the visuals. Tell students the picture on this page is a microscopic view of a bacterium. Have them look for similarities and differences between the figure Gene Expression in Prokaryotes and the figure The TATA Box and Transcription.
3. Students might find the Blocking Gene Expression diagram surprising because of the unusual kinds of RNAs shown, including double-stranded RNA. Tell students that mRNA, tRNA, and ribosomal RNA are the main types of RNA in cells, but not the only types.

Activate Prior Knowledge

Have students review prokaryotes and eukaryotes.

Ask How do prokaryotic cells differ from eukaryotic cells? *(Prokaryotic cells have a single DNA molecule and no nucleus or organelles.)*

NATIONAL SCIENCE EDUCATION STANDARDS

UNIFYING CONCEPTS AND PROCESSES

I, II, IV, V

CONTENT

B.3, C.1.d, C.1.f, C.3.c, G.1

INQUIRY

A.2.a, A.2.b, A.2.c

For the corresponding lesson in *Miller & Levine Biology,* see pages 377–383.

13.4 Gene Regulation and Expression

Key Questions

How are prokaryotic genes regulated?

How are genes regulated in eukaryotic cells?

What controls the development of cells and tissues in multicellular organisms?

BUILD Understanding

Concept Map As you read, create a concept map from the major headings of the lesson. Then identify the important ideas in each section to complete your map.

In Your Workbook
Complete the concept map in Lesson 13.4.

Prokaryotic Gene Regulation

Think of a library filled with how-to books. Would you ever need to use all of those books at the same time? Of course not. If you wanted to know how to fix a leaky faucet, you'd find a book on plumbing. But, you would ignore the one on carpentry. Now picture a tiny bacterium like *E. coli,* which contains more than 4000 genes. These genes code for proteins that do everything from building cell walls to breaking down food. Do you think that *E. coli* uses every gene in its genetic library at the same time?

Most bacteria transcribe only the genes they need at any one time. For example, some genes produce enzymes used to digest certain types of food molecules. If these food molecules are not present, there is no need for these enzymes. As you might expect, bacteria turn these genes off when they are not needed. By controlling gene expression in this way, bacteria respond to change.

How do bacteria control genes? DNA-binding proteins in prokaryotes regulate genes by controlling transcription. Some of these proteins help switch genes on. Others turn genes off. How does an organism know when to turn a gene on or off? One way bacteria control making proteins is through operons. An **operon** is a group of genes that are regulated together. The genes in an operon have related jobs. *E. coli* provides a clear example of how this works. *E. coli* has 4288 genes that code for proteins. Three of these genes must be turned on together before the bacterium can use the sugar lactose as a food. Since these genes are operated together, they are called the *lac* operon.

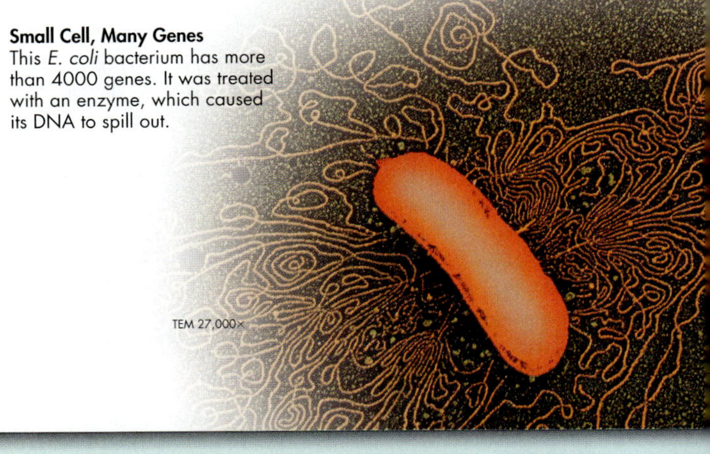

Small Cell, Many Genes
This *E. coli* bacterium has more than 4000 genes. It was treated with an enzyme, which caused its DNA to spill out.

13.4 Gene Regulation and Expression

Lesson Objectives

- Describe gene regulation in prokaryotes.
- Explains how most eukaryotic genes are regulated.
- Relate gene regulation to development of multicellular organisms.

BUILD Vocabulary

A. The chart below shows key terms from the lesson with their definitions. Complete the chart by writing a strategy to help you remember the meaning of each term.

Term	Definition	How I'm Going to Remember the Meaning
Homeotic gene	Master control gene that regulates how cells develop into specific parts of the body	Homeotic genes control the development of different organs.
Operon	Group of genes that are regulated together	An operon is a group of genes that operate together.
Operator	Region where the lac repressor can bind to DNA	When the lac repressor binds to the operator, it turns off.
RNA interference	Blocking gene expression with small RNA molecules	Interfere means "block." RNA interference blocks gene expression.

B. As you work through this lesson, you may find these terms in the activities. When you need to write a key term or a definition, **highlight** the term or the definition.

BUILD Understanding

Concept Map Complete the concept map by filling in the main ideas of Lesson 4. Use the following terms: transcription factors, DNA-binding proteins, master control genes.

Prokaryotic Gene Regulation	involves	DNA-binding proteins
Eukaryotic Gene Regulation	often involves	Transcription factors
Genetic Control of Development	involves	Master control genes

Prokaryotic Gene Regulation

DNA-binding proteins called repressors control transcription. They bind to areas of DNA called operators. They stop RNA polymerase from transcribing the genes on the DNA strand.

Answer the following questions. For each question, circle the letter of the correct answer.

1. What is the function of the *lac* operon in *E. coli*?
   A. It regulates reproduction.
   B. It regulates cell division.
   C. It enables the bacterium to use lactose for food.
   D. It uses three proteins to make lactose.
2. What does the *lac* repressor do?
   A. It prevents the *lac* operon from working by preventing transcription.
   B. It makes RNA polymerase, which prevents the *lac* operon from working.
   C. It makes lactose.
   D. It makes glucose.
3. What happens when lactose is *E. coli*'s only food source?
   A. The presence of lactose keeps RNA polymerase from functioning.
   B. The presence of lactose enables the *lac* operon genes to function.
   C. The presence of lactose allows DNA to be synthesized.
   D. The presence of lactose makes the lac repressor go into action.
4. What does the promoter do?
   A. It prevents the synthesis of lactose.
   B. It is the place where RNA polymerase can bind to DNA and therefore start the synthesis of mRNA.
   C. It is the place where RNA polymerase can bind to DNA and enable the *lac* repressor gene to work.
   D. It lets the *lac* repressor bind to DNA.

In the space below, draw what happens to the repressor when lactose is present.

Students' drawings should resemble the illustration in the textbook.

Eukaryotic Gene Regulation

In many eukaryotes, a short region of DNA containing the base sequence TATATA or TATAAA is known as the TATA box. The TATA box marks the beginning of a gene. It also helps position the RNA polymerase. When RNA polymerase binds in the correct position, transcription can occur.

Follow the directions.

1. Label the *TATA box,* the *gene,* and the *RNA polymerase* in the diagram below.

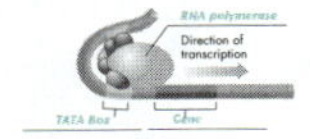

Complete the sentences.

2. The TATA box helps RNA polymerase bind in the correct place.
3. The TATA box marks the beginning of the gene.

CHAPTER MYSTERY

Mouse-Eyed Fly The chapter mystery involves an experiment in which scientists transplanted a mouse gene into a fruit fly. The mouse gene is involved in the formation of eyes. The fruit fly then grew eyes in odd places, such as its legs.

Imagine you are writing a Web article about this experiment. To learn more about the experiment, you will interview one of the scientists. In the space below, write some questions that you might ask the scientist. One question has been done for you.

- When you planned the experiment, what was your hypothesis?

Sample questions: Why did you choose a mouse gene to transplant, instead of the gene of another insect? Were the eyes that developed more like mouse eyes, or more like normal fruit-fly eyes? How did you insert the mouse gene into the fruit fly's genome?

Encourage students to try to find answers to the questions they write. You might find a geneticist to talk to the class about the experiment.

The *Lac* Operon Why must *E. coli* be able to turn the *lac* genes on and off? Lactose is made up of two simple sugars, galactose and glucose. To use lactose for food, the bacterium must first move lactose across its cell membrane. Then it has to break the bond between glucose and galactose. These jobs are done by proteins coded for by the genes of the *lac* operon. The bacterium must transcribe these genes and make the proteins if lactose is its only food source. If grown on another food source, the bacterium does not need these proteins.

Remarkably, the bacterium almost seems to "know" when the proteins from these genes are needed. When lactose is not around, the *lac* genes are turned off by proteins. These proteins bind to DNA, blocking transcription.

Promoters and Operators On one side of the operon's three genes are two control regions. The first is a promoter (P). This is a site where RNA polymerase can bind to begin transcription. The other region is called the **operator** (O). The O site is where a DNA-binding protein known as the *lac* repressor can bind to DNA.

▶ ***The* Lac *Repressor Blocks Transcription*** When the *lac* repressor binds to the O region, RNA polymerase cannot reach the *lac* genes. RNA polymerase must bind to P to start transcription. So, the binding of the repressor protein switches the operon "off." It prevents the transcription of genes.

▶ ***Lactose Turns the Operon "On"*** If the repressor protein is always around, how can the *lac* genes ever be switched on? Besides its DNA binding site, the *lac* repressor protein has a binding site for lactose, too. When lactose is in the growth medium, it moves into the cell. It attaches to the *lac* repressor, changing the shape of the repressor protein. The change in shape causes the protein to fall off the operator. Now the repressor is no longer bound to the O site. RNA polymerase binds to the promoter and transcribes the genes of the operon. This transcription of the operon genes means that whenever lactose is present, the operon is automatically switched on.

Key Question How are prokaryotic genes regulated?
DNA-binding proteins in prokaryotes regulate genes by controlling transcription.

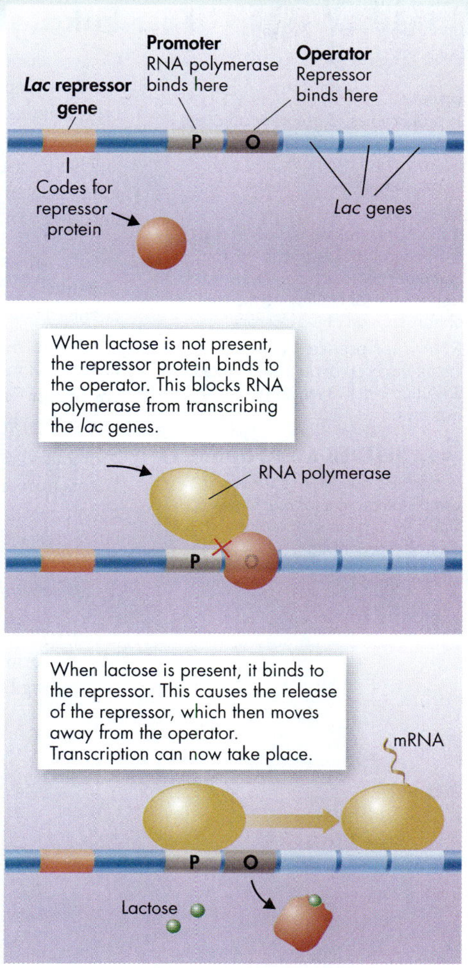

Gene Expression in Prokaryotes The *lac* genes in *E. coli* are turned off by repressors. They are turned on in the presence of lactose.

Focus on Academic Language

• lactose • galactose • glucose

Review with students the relationship among lactose, galactose, and glucose. Lactose is a sugar that is commonly found in the milk of mammals. Lactose is a double-ring sugar composed of glucose and galactose. Find a structural drawing of lactose in a textbook, reference book, or on the Internet. Draw the structure on the board or project it on a screen. Have a volunteer label or point out each sugar.

Ask Do the names of the sugars have a common suffix? *(yes; -ose)*

Say The suffix *-ose* denotes a carbohydrate, particularly a sugar.

Build Understanding

CONCEPT MAP Ask a volunteer to read aloud the green headings in this lesson. Tell students they will use the headings and details under the headings to complete a concept map in Workbook B. Remind students that a concept map is a good tool for organizing information and studying for tests.

During Reading

Active Reading

INTERPRET DIAGRAMS Have students study the Gene Expression in Prokaryotes diagrams.

Ask Why is mRNA being transcribed in the bottom diagram but not in the middle diagram? *(In the middle diagram, the lac repressor protein is bound to the operator and is blocking the RNA polymerase from transcribing.)*

Ask Why did the lac repressor protein in the bottom diagram fall off the operator? *(Lactose attached to the lac repressor protein, which changed its shape, causing it to fall off.)*

MAKE A GRAPHIC ORGANIZER Help students use the diagrams to make two flowcharts in their notebooks. The first flowchart should summarize the steps that occur at a bacterium's lac operon when lactose is not present. The other flowchart should show the steps that occur when lactose is present.

KEY QUESTION Read aloud the Key Question and answer.

Ask How do the DNA-binding proteins control transcription? *(They block RNA polymerase from transcribing genes.)*

Speed Bump

Ask What is an operon? *(a group of genes that have related jobs and are regulated together)*

Ask What are an operon's two control regions? *(the promoter and the operator)*

Ask What does each control region do? *(The promoter binds with RNA polymerase, which allows the operon to be transcribed. The operator binds with a DNA-binding protein, which blocks RNA polymerase and prevents transcription.)*

LESSON 13.4

During Reading continued

Active Reading

COMPARE AND CONTRAST Have students make a chart in their notebooks that compares and contrasts gene regulation in prokaryotes and eukaryotes. The chart should include information about the following: number of genes controlled, RNA polymerase, DNA binding site for the RNA polymerase, presence or absence of an operator, and transcription factors. Suggest pairs of students use their charts to ask and answer questions.

KEY QUESTION Have a student read aloud the Key Question and answer.

Ask What is the name for the DNA sequence that is referred to in the answer? *(TATA box)*

Build Vocabulary

MULTIPLE MEANINGS Note that a person who is licensed to drive a motor vehicle is often referred to as the operator, especially on insurance forms and accident reports.

Ask How are a motor vehicle operator and a genetic operator similar? *(They both control something.)* When is a motor vehicle operator most like the operator in an operon? *(when he or she stops the vehicle and shuts off the engine)*

Speed Bump

Ask In general, how is eukaryotic gene regulation different from prokaryotic gene regulation? *(Most eukaryotic genes are controlled individually, and the systems for controlling transcription are more complicated.)*

Ask What is the function of the TATA box in eukaryotic gene regulation? *(It binds to a transcription factor to form a binding site for RNA polymerase.)*

BUILD Vocabulary

operon
in prokaryotes, a group of adjacent genes that share a common operator and promoter and are transcribed into a single mRNA

operator
a short DNA region, adjacent to the promoter of a prokaryotic operon, that binds repressor proteins responsible for controlling the rate of transcription of the operon

RNA interference
the introduction of double-stranded RNA into a cell to inhibit gene expression

MULTIPLE MEANINGS
A machine operator controls a complicated piece of machinery. Similarly, a genetic operator controls transcription.

Eukaryotic Gene Regulation

Gene regulation in eukaryotes is much like gene regulation in prokaryotes, but there are differences. Most genes are controlled individually and use more complicated systems to control transcription. One important control is a piece of DNA called the TATA box. It is about 25 or 30 base pairs long and has the sequence TATATA or TATAAA. This region is usually found just in front of a gene. It binds a protein that helps move RNA polymerase into place to begin its job.

Transcription Factors Gene expression in eukaryotic cells is regulated in many ways. One way is to control transcription by using transcription factors. These proteins control gene expression by binding to regulatory DNA sequences, such as the TATA box. For example, some transcription factors can make transcription easier by opening up tightly packed chromatin. Several transcription factors usually have to bind before RNA polymerase binds to the promoter region.

Promoters have many binding sites for transcription factors. Certain factors turn on many genes at once. Factors may be controlled by chemical signals such as steroid hormones. These chemical messengers enter cells and bind to receptor proteins. These "receptor complexes" then act as transcription factors that bind to DNA. This binding allows a single chemical signal to turn on many genes.

Key Question How are genes regulated in eukaryotic cells?
Transcription factors control the expression of eukaryotic genes by binding DNA sequences in regulatory regions.

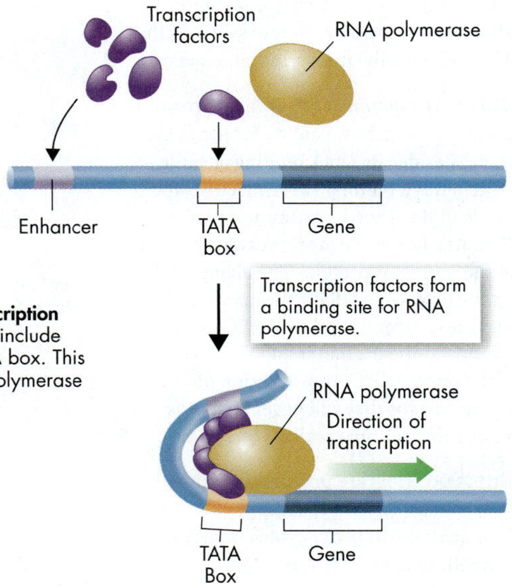

The TATA Box and Transcription
Many eukaryotic genes include a region called the TATA box. This region helps put RNA polymerase in place to do its job.

ELL Focus on ELL: Access Content

INTERMEDIATE AND ADVANCED SPEAKERS Before students read this page, distribute copies of a KWL chart. Have students recall what they know about gene regulation in prokaryotes and record this information in the K column (What I Know). Then, have them record in the W column (What I Want to Know) any questions they have about gene regulation in eukaryotes. As students read the section Eukaryotic Gene Regulation, have them write the answers to their questions in the L column (What I Learned).

Cell Specialization Why is gene regulation in eukaryotes more complicated than in prokaryotes? Every eukaryotic cell nucleus has all the genes of the body. But not all cells use every gene. This ability to selectively use genes means that nerve cells will not make enzymes that are only needed in the liver. Complex gene regulation is what makes specialization possible.

RNA Interference Other kinds of RNA are in the cell besides the major groups of RNA. Very small RNA molecules play a large role in regulating gene expression. They do so by interfering with mRNA.

After they are made by transcription, these small RNA molecules fold into double-stranded loops. An enzyme called the "Dicer" enzyme cuts, or dices, these double-stranded loops into microRNA (miRNA). Each miRNA molecule is about 20 base pairs in length. The two strands of the loops then separate. Next, one of the miRNA pieces attaches to a group of proteins. This forms what is known as a silencing complex. This complex binds to and destroys any mRNA containing the complementary sequence to the miRNA. The miRNA sticks to mRNA molecules and stops them from passing on their protein-making instructions. Using a silencing complex to block gene expression is called **RNA interference** (RNAi).

The Promise of RNAi Technology The discovery of RNAi has made it possible for researchers to switch genes on and off. They do this by inserting double-stranded RNA into cells. This technology is a powerful way to study gene expression in the laboratory. RNAi technology may also provide a way for medical scientists to turn off genes from viruses and cancer cells. So, RNAi may provide new ways to treat, and maybe even cure, diseases.

Genetic Control of Development

Controlling gene expression helps shape the way a multicellular organism develops. Each kind of specialized cell in an adult begins from the same fertilized egg cell. Different sets of genes are turned on and off as the embryo develops. Gene regulation helps cells change so that every cell can do a job. This kind of cell change and development is called cell differentiation.

Blocking Gene Expression MicroRNAs attach to mRNA molecules and stop them from passing on their protein-making instructions.

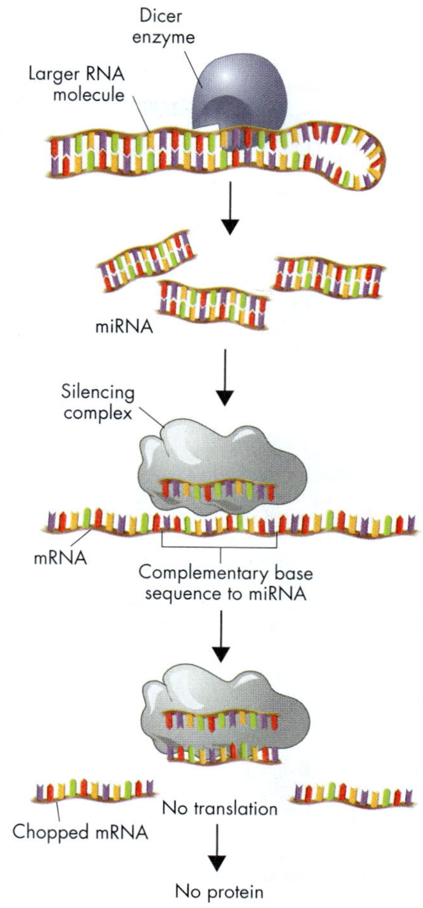

Active Reading

MAKE INFERENCES Have a student read aloud the Cell Specialization paragraph.

Ask Why can't a nerve cell in a eukaryote make liver enzymes? *(The genes that code for liver enzymes aren't transcribed in a nerve cell.)*

Say In a multicellular organism, every cell nucleus contains all the genes of the body, but not all cells use every gene.

Ask Do you think a single-celled prokaryote, such as a bacterium, uses every gene? *(Sample answer: I think a single-celled prokaryote is probably able to turn all of its genes on and off, and use them to produce proteins as needed.)*

USE DIAGRAMS Have students find three types of RNA in the Blocking Gene Expression diagram: "looped" RNA, microRNA, and messenger RNA. Emphasize that RNA is always single-stranded when it is transcribed.

Ask How did the double-stranded looped RNA form? *(A single-stranded RNA molecule folded back on itself.)*

Draw students' attention to the dicer enzyme and the silencing complex. Discuss the meanings of the words *dice* and *silence*. Tell students that the words *dicer* and *silencer* in the text are adjectives. They each describe the noun that follows it.

Ask What is produced when the dicer enzyme dices the looped RNA? *(microRNA)*

Point out that microRNA attaches to a group of proteins to form a silencing complex.

Ask What does the silencing complex silence? *(It silences the protein-making instructions in messenger RNA.)* How does it do that? *(It binds to portions of messenger RNA that are complementary to the microRNA within the silencing complex and destroys the mRNA.)*

DRAW CONCLUSIONS Discuss the meanings of the words *interfere* and *interference* with students.

Ask Why is *RNA interference* a good name for the technology that uses silencing complexes to block gene expression? *(The silencing complexes interfere with the translation of messenger RNA.)*

Speed Bump

Ask How does a cell become specialized? *(by having only certain genes expressed)*

Ask What is the role of microRNA in blocking gene expression? *(MicroRNA within a silencing complex binds with complementary bases in mRNA, causing the mRNA to be destroyed.)*

Focus on Academic Language

- differentiation

Write the word *differentiation* on the board, dividing it into two parts *(different* and *-iation).*

Say The suffix *-iation* means "process of."

Ask What do you think *differentiation* means? *(the process of making something different)*

Ask What does *cell differentiation* mean? *(the process of making a cell different)*

Point out that cell differentiation occurs when an unspecialized cell becomes a blood cell, skin cell, or some other type of specialized cell. Controlling the expression of genes leads to cell differentiation.

During Reading continued

Build Vocabulary

ROOT WORDS Invite a student to read the explanation of the meaning of the root *homeo.*

Ask How are homeotic genes similar to each other? *(All homeotic genes share a similar DNA sequence.)*

Active Reading

INTERPRET VISUALS Draw students' attention to the colored Hox genes in the fruit fly chromosome and the similarly colored fruit fly embryo segments and body parts.

Ask What do the green-colored genes control? *(the development of the fly's head)*

Point out that the order of the genes in the fruit fly chromosome is the same as the order of the segments and body parts they control.

FIND THE MAIN IDEA After students read the Environmental Influences section, help them make an outline of the main idea and supporting details.

Ask What is the main idea? *(Environmental factors can change gene expression.)*

Have students write the main idea in their notebooks.

Ask What is one supporting detail? *(Sample answer: The lac operon in* E. coli *is switched on only when lactose is the only food in the bacteria's environment.)* Have students add this detail as a bullet point under the main idea.

Ask students to identify other supporting details and have them add more bullet points under the main idea.

KEY QUESTION Read aloud the Key Question and answer.

Ask What are homeotic genes? *(master control genes that determine body parts)*

Speed Bump

Ask What is metamorphosis? *(a series of changes from one life stage to another)*

Ask What role does gene expression play in metamorphosis? *(As organisms move through their life stages, their cells differentiate to make new organs.)*

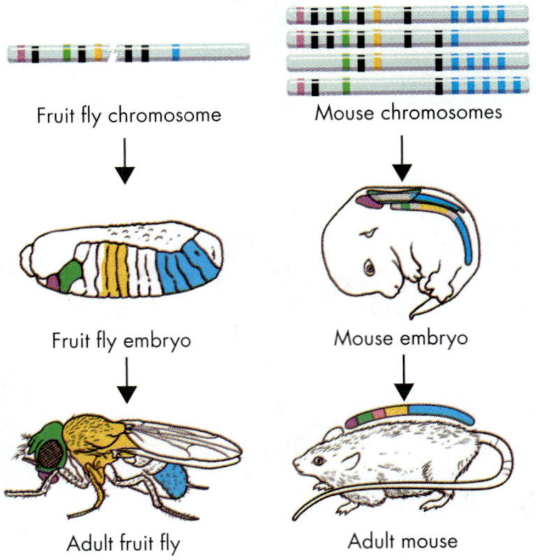

Hox Genes and Body Development
The basic body plan of a fruit fly is determined by a series of hox genes. The genes are arranged side by side on a single chromosome. Mammals, including mice, have several clusters of similar hox genes, arranged in nearly the same order. The colored areas on the fly and mouse show the approximate areas affected by genes of the corresponding colors.

Homeotic Genes In many animals, a set of master control genes regulates organs that develop in specific parts of the body. These genes are known as **homeotic genes,** and they control the identity of different body parts.

Homeobox and Hox genes All homeotic genes share a similar DNA sequence, called the homeobox sequence. Homeobox genes code for transcription factors that turn on other genes. These genes are important in cell differentiation. They control features such as the presence of wings or legs.

Flies have a group of homeobox genes known as Hox genes. These Hox genes are located side by side in a single group. They determine the identities of each part of a fly's body. A Hox gene mutation can change which organs develop in certain parts of the body. Other animals, including humans, also have Hox genes. The genes always tell cells how to differentiate. So, nearly all animals share the same basic tools for building the different parts of the body.

Environmental Influences

An organism's environment plays a role in cell differentiation, too. In all kinds of organisms, environmental factors like temperature can change gene expression. One example is the *lac* operon in *E. coli.* It is switched on only when lactose is the only food source in the bacteria's environment.

Metamorphosis is another example of how organisms can alter gene expression in response to environmental changes. Metamorphosis involves a series of changes from one life stage to another. It is usually regulated by factors inside and outside of the body. As organisms move through their life stages, their cells differentiate to make new organs. Tadpoles may speed up their metamorphosis based on the conditions around them. This more rapid metamorphosis may happen if a pond is drying up quickly or if food is becoming scarce. The environmental changes are translated into hormonal changes. The hormones act to regulate gene expression, which controls the speed of metamorphosis. Temperature and population size can also affect the speed of metamorphosis.

BUILD Vocabulary

homeotic gene
a class of regulatory genes that determine the identity of body parts and regions in an animal embryo. Mutations in these genes can transform one body part into another.

ROOT WORDS
The root *homeo* means "similar." Homeotic genes are similar to each other.

Key Question What controls the development of cells and tissues in multicellular organisms?
Master control genes are like switches that trigger particular patterns of development and differentiation in cells and tissues.

Analyze and Conclude

1. c
2. Sample answer: In worms with the twitching response, the unc-22 protein that controls muscle contractions was not made.
3. Sample answer: Introns are small pieces that are removed and discarded from pre-mRNA before it leaves the nucleus. Active mRNA is made of exons. Since introns are not normally part of active mRNA, silencing complexes made with intron-based miRNA probably would have no effect on muscle because there most likely would not be any complementary base sequences on muscle mRNA.

INQUIRY into Scientific Thinking

The Discovery of RNA Interference

In 1998, Andrew Fire and Craig Mello performed an experiment that helped explain how RNA interference happens. They used RNA from a large gene called unc-22, which codes for a protein found in muscle cells. They made short pieces of mRNA that matched with two parts of the gene. Then they injected the mRNA into egg cells of the worm *C. elegans*. Some of their results are shown in the table.

Injections of mRNA into *C. elegans* Eggs

Portion of Gene Used to Produce mRNA	Strand Injected	Result in Adult Worm
Unc-22 (exon 21–22)	Sense	Normal
	Antisense	Normal
	Sense + Antisense	Twitching
Unc-22 (exon 27)	Sense	Normal
	Antisense	Normal
	Sense + Antisense	Twitching

1. Interpret Data Which strand caused adult worms to twitch?

a. single-stranded mRNA (the "sense" strand)

b. its complementary strand ("antisense")

c. double-stranded RNA ("sense + antisense")

2. Infer Twitching happens when muscle cells cannot control their contractions. What does this suggest about the unc-22 protein in some of the worms?

3. Predict The scientists made their mRNA fragments from exons. Do you think their experiment would have worked if the fragments were made from introns? Explain your answer.

In Your Workbook Get more help for this activity in your workbook.

CHECK Understanding

Apply Vocabulary

Use the highlighted word from the lesson to complete the sentence correctly.

1. ______ are a class of regulatory genes that determine the identity of body parts and regions in an animal embryo.

Critical Thinking

2. Review How is the *lac* operon regulated?

3. Explain What is a promoter?

4. Review Describe how most eukaryotic genes are controlled.

5. Compare and Contrast How is gene regulation in prokaryotes and eukaryotes similar? How is it different?

6. Write to Learn Answer the third mystery clue.

MYSTERY CLUE

What do you think controls the growth and development of eyes in flies and mice? (**Hint:** See p. 324.)

BIOLOGY.com Search Lesson 13.4 GO • Lesson Assessment

Check Understanding

1. Homeotic genes
2. The lac operon is regulated by a repressor protein that prevents RNA polymerase from binding to the lac promoter site when lactose is absent.
3. a site on prokaryotic DNA where RNA polymerase can begin transcription
4. Most eukaryotic genes are controlled by a TATA box, which helps position RNA polymerase. Transcription factors also bind to regulatory regions of DNA and control gene expression.
5. Gene regulation in both prokaryotes and eukaryotes involves a repressor protein that prevents transcription. However, most eukaryotic genes are controlled individually and have more complex regulatory sequences.
6. homeobox genes

Inquiry Into Scientific Thinking

SELF-QUESTION Demonstrate how asking questions can help clarify new information. Point out that Question 3 mentions exons and introns.

Say To answer the question, I ask myself, *Do I know what exons and introns are?* Sometimes I have to reread information to find the answer to a question.

Reread RNA Editing in Lesson 13.1 aloud as students follow along. Then discuss the difference between exons and introns.

Lead a Discussion

Talk about how controlling gene expression might treat human genetic diseases.

Say Huntington's disease is caused by a single mutant gene. The gene produces a protein that causes brain abnormalities, which in turn interfere with coordination, speech, and mental abilities.

Discuss how RNA interference technology might be used to treat Huntington's disease. *(A microRNA complementary to the messenger RNA transcribed from the mutant gene could be injected into a person with the disease. The microRNA would prevent the messenger RNA from being translated, so the protein that causes brain abnormalities would not be made.)*

Post Reading

ASSESSMENT TIPS For Questions 2 and 3, refer students to the Gene Expression in Prokaryotes diagrams and the flowcharts in their notebooks.

Refer students to the TATA Box and Transcription visual to help them answer Question 4.

Students can use the compare and contrast chart they made to answer Question 5.

MYSTERY CLUE Refer students to the Hox Genes and Body Development visual. Have them locate the gene on the fruit fly chromosome that controls the development of fruit fly's eyes.

Wrap-Up Activity

WHO AM I? Assign to each student one of the following "characters" from the lesson: lac repressor gene, promoter, operator, operon, RNA polymerase, transcription factor, TATA box, dicer enzyme, microRNA, silencing complex, and homeotic gene. In front of the class, have students take turns describing who they are and what they do, without ever identifying themselves. Invite volunteers to guess the identity of each character.

CHAPTER LAB

Pre-Lab

Guide students to the Transcribing DNA Into RNA diagram in Lesson 13.1 and the Translation diagram in Lesson 13.2. Have them define *transcription* and *translation* in their own words.

Ask How is modeling transcription and translation helpful in understanding the processes? *(Modeling helps us visualize processes that cannot be seen.)*

Review the Background Questions and the Pre-Lab Questions as a class.

Lab

Tell students they will perform the chapter lab *From DNA to Protein Synthesis* described in **Lab Manual B.**

Post-Lab

CREATE A CODE Divide students into small groups. Have each group create a three-symbol code for each letter of the alphabet and then use their code to write a message about protein synthesis. Have groups pass their code and coded message to another group to decode.

Skills Lab

GUIDED INQUIRY

Pre-Lab: From DNA to Protein Synthesis

Problem What are the steps involved in making a protein?

Lab Manual Chapter 13 Lab

Skills Focus Use Models, Sequence

Connect to the Big idea One of the most important tasks in a cell is the assembly of proteins from amino acids. This task always begins on ribosomes that are located throughout a cell's cytoplasm. The directions for the assembly of proteins are stored in DNA molecules. The information is carried to the ribosomes by a form of RNA called messenger RNA, or mRNA. In this lab, you will model the transcription of DNA and the translation of mRNA.

Background Questions

a. **Review** Is the following sequence from a DNA or mRNA molecule? How can you tell?

CUAAUGCCCUAGGGCACU

b. **Compare and Contrast** How are transcription and translation similar? How are they different?

c. **Sequence** List the following molecules in the order in which they take part in protein synthesis: amino acid, DNA, mRNA, tRNA.

Pre-Lab Questions

Preview the procedure in the lab manual.

1. **Sequence** Describe briefly the process you will use to decode the messages.
2. **Compare and Contrast** What role do stop codons play in protein synthesis? What are they used for in the coded messages?
3. **Predict** Which six letters will not appear in the coded messages. Give a reason for your answer.

BIOLOGY.com Search Chapter 13 GO

Visit Chapter 13 online to test yourself on chapter content and to find activities to help you learn.

Untamed Science Video Watch the Untamed Science explorers as they search for examples of how mutations have benefitted a species.

Art in Motion Watch how RNA is processed to make mRNA.

Art Review Review your understanding of different types of mutations with this drag-and-drop activity.

InterActive Art Build your understanding of transcription and translation with these animations.

Visual Analogy Compare DNA and RNA to the master plans and blueprints of a builder.

Tutor Tube Tune into the tutor to find out why proteins are so important!

Pre-Lab Answers

BACKGROUND QUESTIONS

a. mRNA, because the sequence contains uracil instead of thymine

b. Sample answer: During both transcription and translation, large molecules are synthesized from smaller units. During transcription, nucleotides are assembled into complementary mRNA molecules. During translation, amino acids are assembled into proteins.

c. DNA, mRNA, tRNA, amino acid

PRE-LAB QUESTIONS

1. Transcribe the DNA to mRNA; translate the mRNA to amino acids; find the single-letter abbreviation for each amino acid.
2. In protein synthesis, a stop codon is used to mark the end of a protein's synthesis. In the coded messages, stop codons are used to represent spaces between words.
3. The letters *B, J, O, U, X,* and *Z* will not appear in the messages because these letters are not used as single-letter abbreviations for amino acids.

NATIONAL SCIENCE EDUCATION STANDARDS

CONTENT C.1.d

13 CHAPTER Summary

13.1 RNA

- There are three main differences between RNA and DNA. The sugar in RNA is ribose instead of deoxyribose. RNA is generally single-stranded and not double-stranded. RNA contains uracil in place of thymine.
- RNA is made through a process called transcription. In transcription, stretches of DNA serve as templates to make complementary RNA molecules.

RNA (p. 308)
messenger RNA (p. 309)
ribosomal RNA (p. 309)
transfer RNA (p. 309)
transcription (p. 309)

13.2 Ribosomes and Protein Synthesis

- The genetic code is a code for making proteins. The genetic code is read three "letters" at a time. Each "word" is three bases long and corresponds to a single amino acid.
- Ribosomes use the sequence of codons in mRNA to put together amino acids into polypeptide chains. Long, functional polypeptide chains are proteins.
- The central dogma of molecular biology is that information is passed from DNA to RNA to protein. This means that the code used to make DNA, RNA, and proteins is almost universal among organisms.

polypeptide (p. 311)
genetic code (p. 311)
codon (p. 311)
translation (p. 312)
anticodon (p. 313)

13.3 Mutations

- Mutations are changes in genetic information that can be inherited. These changes in the genetic code can affect individuals. If the change is passed to offspring, it can affect whole species.
- The effects of mutations on genes vary widely. Some have little or no effect. Some make beneficial variations that are passed on to offspring. Some negatively disrupt gene function and harm the organism.

mutation (p. 316)
point mutation (p. 317)
frameshift mutation (p. 317)
mutagen (p. 318)
polyploidy (p. 319)

13.4 Gene Regulation and Expression

- DNA-binding proteins in prokaryotes regulate genes by controlling transcription.
- Transcription factors control the expression of eukaryotic genes by binding DNA sequences in regulatory regions.
- Master control genes are like switches that turn on and off development and differentiation in cells. The environment also can turn gene expression on or off.

operon (p. 320)
operator (p. 321)
RNA interference (p. 323)
homeotic gene (p. 324)

Chapter Summary

Think Visually

USE GRAPHICS Have students complete the following flowchart about protein synthesis.

1
↓
Translation begins at the start codon.
↓
2
↓
The polypeptide is complete.

ANSWERS

1. mRNA is transcribed and edited in the nucleus and then goes to a ribosome in the cytoplasm.
2. As the ribosome reads each codon, tRNA brings the correct amino acid to the ribosome until a "stop" codon is reached.

BIOLOGY.com Study Online

REVIEW RESOURCES AND ASSESSMENT RESOURCES

Visual Analogy Using master plans and blueprints as an analogy for DNA and RNA helps students comprehend the different roles of the two molecules.

Tutor Tube Students will hear about the importance of proteins in determining phenotype.

Data Analysis Students can analyze and interpret data on mutations in the *lac* operaon of *E. coli.*

Vocabulary Review The *Flash Cards* and *Crossword* are an interactive way to review vocabulary.

Lesson Overviews Lesson Overviews are based on Miller & Levine Biology. However, they are available on CD-ROM and can be edited for your class.

Chapter Assessment Have students take an online version of the Chapter 13 Check Understanding.

Standardized Test Prep Students can take an online version of the Standardized Test Prep.

Diagnostic and Benchmark Tests Use these tests to monitor your students' progress and supply remediation.

UbD Transfer the Big Idea

TRANSFER Have students pretend that the process of gene regulation in eukaryotes has just been discovered. Ask them to write a newspaper story reporting on the discovery. Tell students the story should describe what happens during the process of gene regulation. All the factors involved in eukaryotic gene regulation (genes, transcription factors, RNA polymerase, TATA box) should be included in the story. Encourage students to make illustrations for their news story that help clarify the process. Also invite students to include quotations from scientists explaining why gene regulation is important. You may want to read excerpts from a couple of science stories in newspapers or online news organizations to provide models for a journalistic account of events.

ASSESSMENT

Assess

Big idea Invite a volunteer to read the question. Write *mRNA, tRNA,* and *rRNA* on the board.

Ask For what purpose do RNA molecules work together in a cell? *(They work together to use the genetic code in the nucleus to synthesize proteins in the cytoplasm.)*

Ask Which RNA molecule carries the genetic code from the nucleus into the cytoplasm? *(mRNA)*

CONSTRUCTED RESPONSE Paragraphs will vary. Explain the activity carefully. Be sure students understand that the *Hints* should be used as a guide to help them answer the questions. Review how to write a paragraph.

Write Question 1 on the board and read it aloud. Underline the words *proteins* and *characteristics.*

Ask What are some examples of an organism's characteristics? *(Sample answers: flower color, eye color, blood type, number of legs)*

Read the first hint.

Ask What kind of proteins speed up and control chemical processes in cells? *(enzymes)*

Read the second hint.

Ask What are some body parts that are made of protein? *(Sample answers: muscles, skin, bones)*

Ask If body parts are made of proteins, and proteins help body processes function, how can an organism's characteristics be controlled? *(by controlling proteins)*

On the board, model writing the first one or two sentences using the hints. *(Sample: An organism's characteristics include its body parts and the chemical processes that occur inside its cells. Both kinds of characteristics involve proteins.)*

For Question 2, pair students and have them use the hints and your modeling to answer the question. Have students share their paragraphs with the class.

For Question 3, have students write the paragraphs independently for homework or as an anchor activity.

13 CHECK Understanding

Assess the Big idea — Information and Heredity

Write and answer the question below.
Q: How does information flow from DNA to RNA to direct the synthesis of proteins?

Constructed Response

Write an answer to each of the questions below. The answer to each question should be one or two paragraphs. To help you begin, read the **Hints** below the questions.

1. **Why does controlling an organism's proteins control its characteristics?**

   Hint Proteins help your body processes function.

   Hint Many of the body's parts are made of protein.

2. **How might a point mutation in a gene affect proteins?**

   Hint Genes are transcribed into mRNA.

   Hint mRNA is translated into protein using tRNA and rRNA.

3. **Why is a TATA box necessary for a eukaryotic gene to be expressed properly?**

   Hint A TATA box is usually found just before a gene.

   Hint A TATA box binds transcription factor proteins that help position RNA polymerase.

Foundations for Learning Wrap-Up

Activity 1 Working with a partner, figure out which important facts you both chose to record. Examine which facts one partner recorded that the other did not record. Each partner can explain the importance of his or her unmatched facts.

Activity 2 In groups of three, use the material in your Important Facts Envelopes to create information collages. (Be sure each person's initials are on the back of each fact.) Then, spread the facts out on a table, and put the facts in groups, such as "Translation," "Transcription," "or "Prokaryotes." Explain why each fact belongs in a particular group.

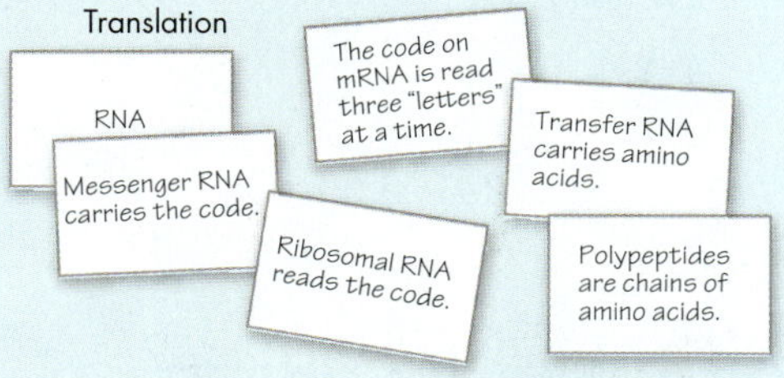

Foundations for Learning Wrap-Up

Have students take out their Important Facts Envelope. Be sure students have written on the slips all of the important facts in the chapter and placed them in the envelope. Assign groups, and explain the activities.

13.1 RNA

Understand Key Concepts

1. Which of the following describes RNA?
   a. RNA is usually double-stranded and contains the base thymine.
   b. RNA is usually single-stranded and contains the base uracil.
   c. RNA is longer than DNA and uses five bases to encode information.
   d. RNA is made in the nucleus of eukaryotic cells and stays there to carry out its functions.
2. The process by which the genetic code of DNA is copied into a strand of RNA is called
   a. translation. c. transformation.
   b. transcription. d. replication.
3. How does the enzyme that makes RNA know where to start transcribing the DNA?

Think Critically

4. **Apply Concepts** Suppose you start with the DNA strand ACCGTCAC. What would a messenger RNA strand transcribed from this DNA strand look like?

Test-Taking Tip

Use Scratch Paper Question 4 asks you to write the messenger RNA sequence that would be transcribed from the DNA sequence ACCGTCAC. Write the DNA sequence on scratch paper. Write the letters that represent the RNA bases directly underneath the DNA letters:

ACCGTCAC

UGGCAGUG

Once you have the right sequence, transfer it to your answer sheet.

13.2 Ribosomes and Protein Synthesis

Understand Key Concepts

5. In messenger RNA, each codon calls for a particular
   a. nucleotide. c. amino acid.
   b. enzyme. d. promoter.
6. The process of making proteins on the ribosome based on instructions from messenger RNA is called
   a. transcription. c. translation.
   b. transformation. d. molecular biology.
7. What is a codon?
8. How do anticodons work?

Think Critically

9. **Applying Concepts** A code on a DNA molecule for a specific amino acid is CTA. What would the messenger RNA codon be? What would the transfer RNA anticodon be?

13.3 Mutations

Understand Key Concepts

10. Changes in DNA sequences that affect genetic information are known as
    a. replications. c. transformations.
    b. mutations. d. translations.
11. A substance that can cause a change in the DNA code of an organism is called a
    a. toxin. c. nitrogenous base.
    b. mutagen. d. nucleotide.
12. Name and give examples of two major types of mutations.
13. How does a deletion mutation differ from a substitution mutation?

Think Critically

14. **Analyze** Can mutations have a positive effect? Explain your answer.

Assessment Tips

Pair students and have them work together to answer the questions. For Question 4, review the test-taking tip. Also, remind students that RNA uses uracil in place of thymine.

For Questions 4 and 9, tell students they won't find the answer to the question in the lesson. Rather, they have to use what they learned to figure out the answer.

For Questions 5–8, have students review the Translation visual in Lesson 13.2.

Suggest that students use the Gene Expression in Prokaryotes visual in Lesson 13.4 and the concept map they made in Workbook B to help them summarize gene expression in prokaryotes for Question 17.

Point out to students that they will need to use the information in the table to answer Questions 20 and 21.

Answers

Lesson 13.1

UNDERSTAND KEY CONCEPTS

1. b

2. b

3. The enzyme knows to start transcribing DNA at a promoter, which is a region of DNA that has specific base sequences.

THINK CRITICALLY

4. UGGCAGUG

Lesson 13.2

UNDERSTAND KEY CONCEPTS

5. c

6. c

7. a three-base code "word" in the genetic code that specifies a particular amino acid, start, or stop

8. Each tRNA brings to the ribosome the amino acid specified by its complementary codon. At the ribosome, anticodons in tRNA form bonds with the complementary codons in mRNA, and tRNA adds its amino acid to the polypeptide chain.

THINK CRITICALLY

9. GAU; CUA

Lesson 13.3

UNDERSTAND KEY CONCEPTS

10. b

11. b

12. Sample answer: Two major types of mutations are gene mutations and chromosomal mutations. An example of a gene mutation is an insertion mutation, in which a single extra base is inserted into a codon. An example of a chromosomal mutation is an inversion, in which part of a chromosome is reversed.

13. A deletion mutation in a gene occurs when a base is lost from a codon. This shifts the "reading frame," so all the codons after the point of deletion are affected. A substitution mutation occurs when a single base in a codon is replaced by a different base. This does not shift the "reading frame."

THINK CRITICALLY

14. Mutations can have a positive effect. For example, a mutation could produce a protein with a new or altered function that might be useful to an organism in a changing environment.

ASSESSMENT

Lesson 13.4

15. b

16. c

17. In prokaryotes, DNA-binding proteins regulate genes by helping switch genes on or off before transcription.

18. A homeobox gene is a gene that codes for a transcription factor that activates other genes important to cell development and differentiation. Homeobox genes are expressed in certain regions of the body, and they determine factors like the presence of wings or legs.

19. Differentiation is controlled by the regulation of gene expression.

Connecting Concepts

20. The mutation would have no effect because the codons would still code for alanine.

21. With different amino acids, a protein might have a different structure and function.

CHAPTER MYSTERY Have students read the Chapter Mystery. Discuss the role of *Pax6* as a master control gene.

Ask What does the *Pax6* gene control in both mice and fruit flies? *(development of eyes)*

Ask What type of gene must *Pax6* be? Why? *(It must be a homeobox gene because it controls other genes that are responsible for development of major body structures.)*

Point out that *Pax6* codes for a transcription factor.

Ask How does *Pax6* control other genes? *(The transcription factor that* Pax6 *codes for forms a binding site for RNA polymerase, thereby allowing the genes to be transcribed.)*

13 CHECK Understanding

13.4 Gene Regulation and Expression

Understand Key Concepts

15. An expressed gene
- **a.** functions as a promoter.
- **b.** is transcribed into RNA.
- **c.** codes for just one amino acid.
- **d.** is made of mRNA.

16. Blocking gene expression in eukaryotes with microRNA strands is called RNA
- **a.** transcription.
- **b.** translation.
- **c.** interference.
- **d.** digestion.

17. How is gene expression controlled in prokaryotes?

18. What is a homeobox gene?

Think Critically

19. **Explain** How is cell differentiation controlled?

Connecting Concepts

Use Science Graphics

Use the data table to answer questions 20 and 21.

Codon Translation

Amino Acid	mRNA Codons
Alanine (Ala)	GCA, GCG, GCU, GCC
Valine (Val)	GUA, GUG, GUU, GUC
Leucine (Leu)	CUA, CUG, CUU, CUC, UUA, UUG

20. **Relate Cause and Effect** The table shows mRNA codons for three amino acids. Suppose a substitution mutation occurred at the third nucleotide position of the codons for alanine. What would happen to the resulting protein?

21. **Infer** The three amino acids shown in the table have very similar—but not identical—properties. What substitution mutations could cause one of these amino acids to be switched for another? What might be the result?

solve the CHAPTER MYSTERY

MOUSE-EYED FLY

Years ago scientists discovered a fly gene they called eyeless. Mutations that turn off this gene cause flies to develop without eyes. Scientists later discovered a mouse gene, called *Pax6*, that was similar to eyeless. Transplanting an activated *Pax6* gene into a fruit fly causes eyes to grow in odd places. These eye growths happen even though mouse eyes and fly eyes are very different.

How can the *Pax6* gene perform the same role in animals that are so different? It probably began very early in the history of life. Then eyes were just patches of light-sensitive cells. They were found on the skin of the common ancestors of all animals. Master control genes like *Pax6* kept working as those organisms evolved and diversified. But their functions changed. All animals, including insects, worms, sea urchins, and humans, share many genes like *Pax6*.

1. **Explain** How are fly eyes and mouse eyes similar? How are they different?

2. **Infer** The *Pax6* and eyeless genes do not code for parts of the actual eye. These genes code for transcription factors. Think about the effect of *Pax6* when it is inserted into a fly. Why does it make sense that *Pax6* is a transcription factor?

Untamed Science™ **Never Stop Exploring Your World.** Finding the solution to the mouse-eyed fly is only the beginning. Take a video field trip with the ecogeeks of Untamed Science to see where this mystery leads.

CHAPTER MYSTERY ANSWERS

1. Both types of eyes have the same function. They sense light and make vision possible. However, fly eyes and mouse eyes look very different and have different structures.

2. If *Pax6* and eyeless coded for parts of the eye, then the fly would have developed mouse eyes in strange places instead of extra fly eyes.

Untamed Science™ In the **Tales of a Mutant Leopard** the ecogeeks lead students on a field trip to find out more about mutations in animals.

Standardized Test Prep

Multiple Choice

1. How does RNA differ from DNA?
   A RNA contains uracil and deoxyribose.
   B RNA contains ribose and thymine.
   C RNA contains uracil and ribose.
   D RNA contains adenine and ribose.

2. How would the DNA sequence GCTATA be transcribed to mRNA?
   A GCUAUA
   B CGATAT
   C CGAUAU
   D GCUTUT

Questions 3–4

Use the chart below to answer the questions.

First Base in Code Word	Third Base in Code Word	Second Base in Code Word: A	G	U	C
A	A	Lys	Arg	Ile	Thr
A	G	Lys	Arg	Met	Thr
A	U	Asn	Ser	Ile	Thr
A	C	Asn	Ser	Ile	Thr
G	A	Glu	Gly	Val	Ala
G	G	Glu	Gly	Val	Ala
G	U	Asp	Gly	Val	Ala
G	C	Asp	Gly	Val	Ala
U	A	"Stop"	"Stop"	Leu	Ser
U	G	"Stop"	Trp	Leu	Ser
U	U	Tyr	Cys	Phe	Ser
U	C	Tyr	Cys	Phe	Ser
C	A	Gln	Arg	Leu	Pro
C	G	Gln	Arg	Leu	Pro
C	U	His	Arg	Leu	Pro
C	C	His	Arg	Leu	Pro

3. Which of the following codons signifies the end of translation?
   A CAA
   B UGA
   C AUC
   D CCA

4. Which of the chains of amino acids corresponds to the nucleotide sequence UCAAGCGUA?
   A glu-cys-pro
   B glu-asp-"stop"
   C thr-arg-met
   D ser-ser-val

5. In eukaryotes, functional messenger RNA molecules are made from
   A exons spliced together after introns are removed.
   B introns spliced together after exons are removed.
   C exons spliced together with introns.
   D long pieces of RNA shortened by the Dicer enzyme.

6. Promoters are
   A genes that code for individual proteins.
   B proteins that bind with DNA and prevent transcription.
   C DNA sequences near operons that regulate transcription.
   D small molecules that bind with repressor proteins.

Questions 7–8

Use the diagrams below to answer the questions.

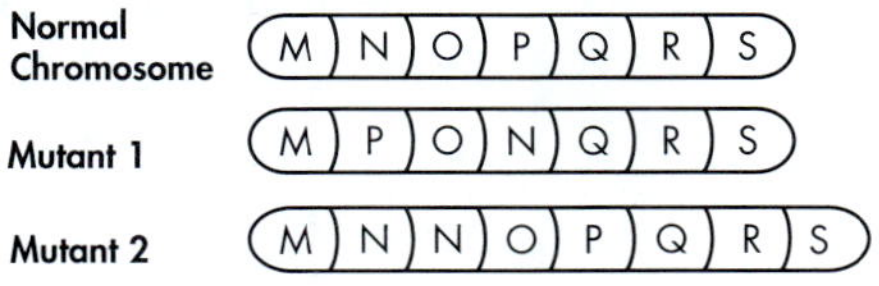

7. Mutant 1 is a(n)
   A deletion.
   B translocation.
   C inversion.
   D duplication.

8. Mutant 2 is a(n)
   A deletion.
   B translocation.
   C inversion.
   D duplication.

Open-Ended Response

9. What is the function of the *lac* repressor system in *E. coli*?

If You Have Trouble With . . .

Question	1	2	3	4	5	6	7	8	9
See Lesson	13.1	13.1	13.2	13.2	13.1	13.1	13.3	13.3	13.4

Test-Taking Tip

INTERPRET DATA TABLES

Tell students that when they are answering questions that refer to a data table, they should read the table carefully and determine what information the table contains.

Say Read all headings carefully. Look at the top of the table and on each side of the table to see if extra headings are included. Read all the data included in the table, and carefully read the question that goes with the table to determine exactly what the question is asking. Then, use the data to answer the question.

Before the Test

Review with students the Test-Taking Tip.

Point out that the word *signifies* in Question 3 means "signals."

Remind students that an open-ended response means that they'll have to write a paragraph to answer the question. Have them paraphrase the question in their notebooks, and remind them about the steps you took when you modeled a paragraph for the Constructed Response activity.

Answers

1. C
2. C
3. B
4. D
5. A
6. C
7. C
8. D
9. The *lac* repressor system controls the production of enzymes needed to digest lactose. When lactose is absent and the enzymes are not needed, repressor proteins turn off the genes, so the enzymes are not produced. When lactose is present and the enzymes are needed, lactose prevents the repressor proteins from turning off the genes, so the enzymes are produced.

CHAPTER PLANNER 14 Human Heredity

Chapter Contents	NSES	Time	Core Resources
Chapter Preview			**Student Edition,** pp. 332–333 **Chapter Mystery,** p. 333 **Foundations for Learning,** p. 333
14.1 Human Chromosomes Karyotypes • Transmission of Human Traits • Human Pedigrees	I, II, A.1.b, A.1.c, A.1.d, A.2.d, C.1.d, C.2.a, C.2.b	1 period 1/2 block	**Student Edition,** pp. 334–338 **Inquiry Into Scientific Learning,** p. 338 **Study Workbook B** 14.1 Worksheets **Biology.com** *Art Review:* A Human Karyotype • *InterActive Art:* Pedigrees • *Tutor Tube:* Skipping a Generation • 14.1 Lesson Assessment
14.2 Human Genetic Disorders From Molecule to Phenotype • Chromosomal Disorders	II, IV, V, B.2, C.1.d, C.2.a, C.2.b, C.2.c	1/2 period 1/4 block	**Student Edition,** pp. 339–341 **Study Workbook B** 14.2 Worksheets **Biology.com** *Art in Motion:* Nondisjunction Disorders • 14.2 Lesson Assessment
14.3 Studying the Human Genome Manipulating DNA • The Human Genome Project	I, II, III, A.2.b, A.2.c, A.2.f, B.2, C.1.d, C.2.a, C.2.b, C.2.c, F.6, G.1	1 period 1/2 block	**Student Edition,** pp. 342–345 **Build Connections,** p. 343 **Study Workbook B** 14.3 Worksheets **Biology.com** 14.3 Lesson Assessment
Chapter Pre-Lab	A.1.b, A.2.a, A.2.b, A.2.c, A.2.f, C.2.a, G.1	1 period 1/2 block	**Student Edition,** p. 346 **Lab Manual B** *Using DNA to Identify Human Remains*

Tools for ELL Instruction

ELL Handbook describes teaching strategies to make *Miller & Levine Biology: Foundation Edition* more accessible to ELL students.

Spanish Study Workbook is a Spanish translation of Study Workbook A.

Multilingual Glossary is the *Miller & Levine Biology* glossary translated into ten languages.

Tools to Challenge Students

Study Workbook A includes more challenging worksheets and activities such as Chapter Mystery activities that reinforce 21st Century Skills.

Lab Manual A labs are geared toward on-level students. The Lab Manual includes fifteen labs in addition to the Chapter Labs.

Biology.com Data Analysis activities offer students a way to practice data gathering and analyzing skills.

Visual Quizzes in the Assessment Resources Book are geared toward on-level students and use visuals from *Miller & Levine Biology.*

Additional Resources

Biology.com Untamed Science Video • Flash Cards

Study Workbook A 14.1 Worksheets
Spanish Study Workbook 14.1 Worksheets
Biology.com *Data Analysis:* Blood Types and Cholera
Assessment Resources Book Visual Quiz

Study Workbook A 14.2 Worksheets
Lab Manual B *Data Analysis:* The Geography of Malaria
Spanish Study Workbook 14.2 Worksheets

Study Workbook A 14.3 Worksheets
Lab Manual B *Hands-On Activity:* Modeling Restriction Enzymes
Spanish Study Workbook 14.3 Worksheets
Assessment Resources Book Visual Quiz

Lab Manual A *Using DNA to Identify Human Remains*

Chapter Review

Student Edition Chapter Summary, p. 347
Study Workbook B Chapter 14 Review
Transparencies, pp. 172–181
Biology.com Untamed Science Video • Editable Worksheets of Study Workbooks A and B and Lab Manuals A and B • Chapter 14 Flash Cards and Crossword

Untamed Science DVD • Classroom Resources CD (includes lesson presentations and editable worksheets)

Chapter Assessment

Student Edition Check Understanding, pp. 348–351
Study Workbook B Chapter 14 Taking a Standardized Test
Assessment Resources Book Chapter 14 Test B
Biology.com Chapter 14 Assessment • Editable Worksheets of Chapter 14 Visual Quizzes* and Chapter 14 Tests A and B

Exam*View* *Assessment Suite* • Classroom Resources CD (includes lesson presentations and editable worksheets)

Time: 1 period, 1/2 block

* Visual Quizzes are based on visuals from *Miller & Levine Biology.* Review carefully to make sure they are appropriate for your class.

Pressed for Time?

Preview the Chapter Preview the visuals A Human Karyotype and Pedigree Example and introduce the Lesson 14.1 vocabulary.

Cover the Chapter Quickly Have students read all of Lesson 14.1 and go over the Pedigree Example visual. Have students read *Manipulating DNA* in Lesson 14.3.

Assess Assign the 14.1 Check Understanding, Questions 1 and 2 from the 14.3 Check Understanding, and Questions 1–5, 13, 14, 16, and 17 from the Chapter 14 Check Understanding.

Get the Picture

Focus attention on the picture. Explain that these students show a small sample of the variation that exists among humans.

Ask What are some differences in inherited traits you can see among these students? *(Sample answers: sex, height, skin color, hair color, eye color)* What are some differences that are not inherited? *(Sample answers: hair length, use of makeup, clothing)* Tell students that many human traits, such as weight and even skin color, have a genetic basis but are also influenced by the person's environment.

Activate Prior Knowledge

Ask What have you learned so far about heredity? *(Answers may vary.)*

Write some of the students' responses on the board. Ask students to begin a KWL chart for human heredity in their notebooks. Encourage them to use the responses on the board to fill in the K (What I Know) column of their chart. They should list any questions they have in the W (What I Want to Know) column.

Then, have students review key vocabulary terms from Chapters 11, 12, and 13. Be sure they can distinguish the meanings of related terms such as *gene* and *allele*, *homozygous* and *heterozygous*, and *dominant* and *recessive*.

At the end of the chapter, have students revisit their KWL chart to fill in the L (What I Learned) column.

14 Human Heredity

Big idea Information and Heredity

Q: How can we use genetics to study human inheritance?

Notice that none of these students look alike. The diversity of traits in the human race comes from DNA.

BIOLOGY.com Search Chapter 14 GO • Flash Cards

332

NATIONAL SCIENCE EDUCATION STANDARDS

UNIFYING CONCEPTS AND PROCESSES

I, II, III, IV, V

CONTENT

B.2, C.1.d, C.2.a, C.2.b, C.2.c, F.6, G.1

INQUIRY

A.1.b, A.1.c, A.1.d, A.2.b, A.2.c, A.2.d, A.2.f

UbD Connect to the Big Idea

Invite a volunteer to read aloud the chapter number and the chapter title. Read the Big Idea and question. Emphasize the question, **How can we use genetics to study human inheritance?**

Set the Purpose

PREVIEW AND PREDICT Draw the graphic organizer on the board. Guide students to the INSIDE feature on the next page. Invite a volunteer to read each lesson number and title.

Say All variations of the genes in all of the people make up the human genome.

Ask How do the Big Idea and the questions in the graphic organizer relate to this fact? *(They tell you what is involved in learning about the human genome.)*

Remind students to search for answers to the graphic organizer questions as they read.

INSIDE:

CHAPTER
MYSTERY

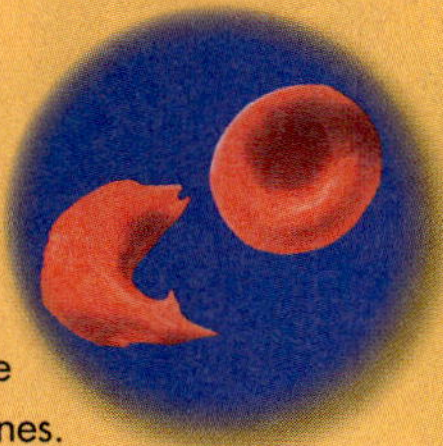

THE CROOKED CELL

When Ava visited her Uncle Eli in the hospital, he seemed tired and pale. He told Ava he had sharp pains in his bones. "I've got sickle cell disease," Uncle Eli said, short of breath. "I just hope it doesn't run in your side of the family." That evening, Ava checked the Internet for information about her uncle's disease. She saw photos of red blood cells shaped like the letter C. They were not like the healthy, round blood cells of a normal individual. Ava learned that these sickle-shaped cells are rigid and sticky. In blood vessels, they form clumps that can block blood flow and cause organ damage. "Am I at risk for carrying the sickle cell trait?" Ava wondered. To find out, she would need to investigate her family history—and her own cells.

Read for Mystery Clues As you read this chapter, look for clues that would help Ava discover whether she might carry the sickle cell trait. Then, solve the mystery.

FOUNDATIONS for Learning

Every main idea in the chapter is made clear by a series of smaller, supporting ideas. As you read through the chapter, make layered books for the main ideas you find. Write the main idea on the first page of the book. Then, write the supporting ideas on the pages that follow. Each page should include a supporting idea and notes about that idea. At the end of the chapter are two activities that help you answer the question: How can we use genetics to study human inheritance?

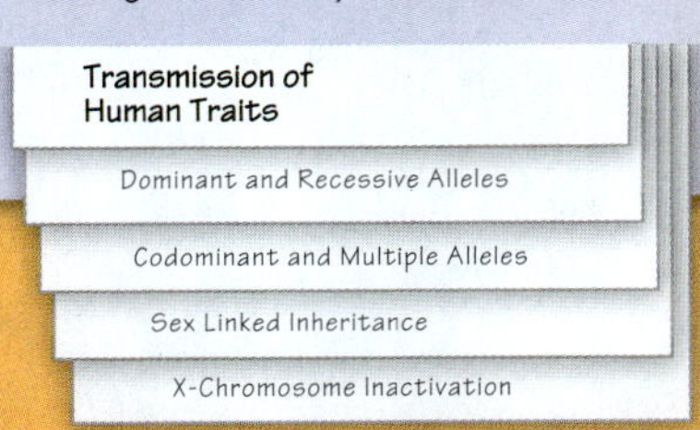

- Untamed Science Video
- Chapter Mystery

Chapter 14 Big Idea: Information and Heredity

Chapter 14: How can we use genetics to study human inheritance?

- **14.1:** How does studying the human genome help us draw conclusions about the inheritance of traits?
- **14.2:** What causes some human genetic disorders?
- **14.3:** How do we study the human genome, and what have we learned so far?

Foundations for Learning

Demonstrate one way to make a layered book using two blank sheets of notebook paper. Provide the following instructions.

Say Mark the length of one sheet into thirds and then fold the top third down so that it covers the middle third. Then, fold the second sheet so that its top edge is about 3 cm from its bottom edge. Now, tuck this second folded sheet inside the first one. Staple the pages three times along the top.

Ask How many books will you need to make to address all the main ideas in this chapter? *(one for each main idea)*

Explain that students will use their layered books in review activities at the end of the chapter.

Untamed Science™ Take a trip with the crew of Untamed Science into the microscopic world of chromosomes in **What Color Are My Genes?** to find out how colorblindness is inherited.

CHAPTER MYSTERY Invite a volunteer to read aloud the Chapter Mystery.

Ask How can Ava begin investigating her family history? *(Sample answer: She can ask her grandparents, aunts, uncles, cousins, and other family members what they know about the health of family members and whether any other family member has sickle cell disease.)* What are some questions Ava should try to answer about human red blood cells? *(Sample answers: What is their function in the body? What gives them their shape? Do genes play a role in the shape and function of red blood cells?)*

Tell students that they will find clues in the chapter to answer the questions surrounding this mystery.

Digital Support

BIOLOGY.com Extend your reach by using the digital assets offered at **Biology.com.**

LESSON 14.1

Pre-Reading

Objectives

14.1.1 Identify the types of human chromosomes in a karyotype.

14.1.2 Describe the patterns of the inheritance of human traits.

14.1.3 Explain how pedigrees are used to study human traits.

Preview the Pages

1. Have volunteers read aloud the green headings. Encourage students to use them as guides for selecting main ideas for their layered books.
2. Preview the vocabulary by having volunteers read aloud the words and definitions. As they read the definition of *karyotype,* explain that a micrograph is a picture taken with a microscope.
3. Preview the visuals by having volunteers read aloud each caption and label. Tell students that the chromosomes in the karyotype were removed from a human cell and stained before being photographed through a microscope.

Build Understanding

SPIDER MAP Direct students to Workbook B, where they will find a partially filled in spider map for organizing main ideas and details about human chromosomes. Encourage students to complete the spider map as they read the lesson.

NATIONAL SCIENCE EDUCATION STANDARDS

UNIFYING CONCEPTS AND PROCESSES

I, II

CONTENT

C.1.d, C.2.a, C.2.b

INQUIRY

A.1.b, A.1.c, A.1.d, A.2.d

For the corresponding lesson in *Miller & Levine Biology,* see pages 392–397.

14.1 Human Chromosomes

Key Questions

What is a karyotype?

What patterns of inheritance do human traits follow?

How can pedigrees be used to analyze human inheritance?

BUILD Understanding

Spider Map Before you read, create a spider map of the major headings in the lesson. As you read, look for the main ideas and supporting details in each lesson.

In Your Workbook Go to your workbook for help creating spider maps. Complete the spider map in Lesson 14.1.

Karyotypes

What makes us human? We could try looking inside the cell with a microscope for the answer. But human cells look similar to cells from any other animal. To find what makes us human, we have to look deeper into the cell. That means we must look at the human **genome.** A genome is the full set of genetic information that an organism carries in its DNA. Remember that genes are on chromosomes. Chromosomes are bundles of DNA and protein found in the nuclei of eukaryotic cells.

To see human chromosomes clearly, cell biologists take pictures of cells during mitosis. The chromosomes are condensed and can be seen clearly at this time. Scientists then cut out the chromosomes from the photographs. They arrange the chromosome images into a picture known as a **karyotype** (KAR ee uh typ). A karyotype shows the complete diploid set of chromosomes. They are lined up together in pairs and arranged from largest to smallest. A typical human cell has 46 chromosomes, arranged in 23 pairs.

Key Question What is a karyotype?
A karyotype shows the complete diploid set of chromosomes. They are grouped in pairs and arranged from largest to smallest.

Sex Chromosomes Two of the 46 chromosomes in the human genome are known as **sex chromosomes** because they determine an individual's sex. Females have two copies of the X chromosome. Males have one X chromosome and one Y chromosome. All egg cells carry a single X chromosome. Half of all sperm cells carry an X chromosome and half carry a Y chromosome. The X chromosome has more than 1200 genes. The Y chromosome is smaller, and has about 140 genes.

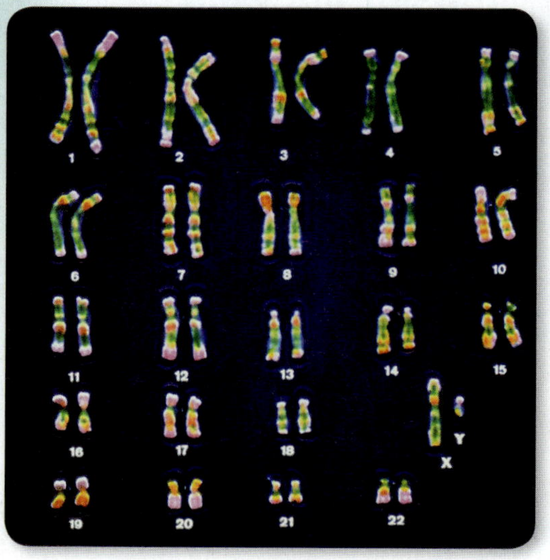

A Human Karyotype The nucleus of a human body cell has 23 pairs of chromosomes. These chromosomes have been cut out of a photograph and arranged to make a karyotype.

BIOLOGY.com Search Lesson 14.1 GO • Art Review

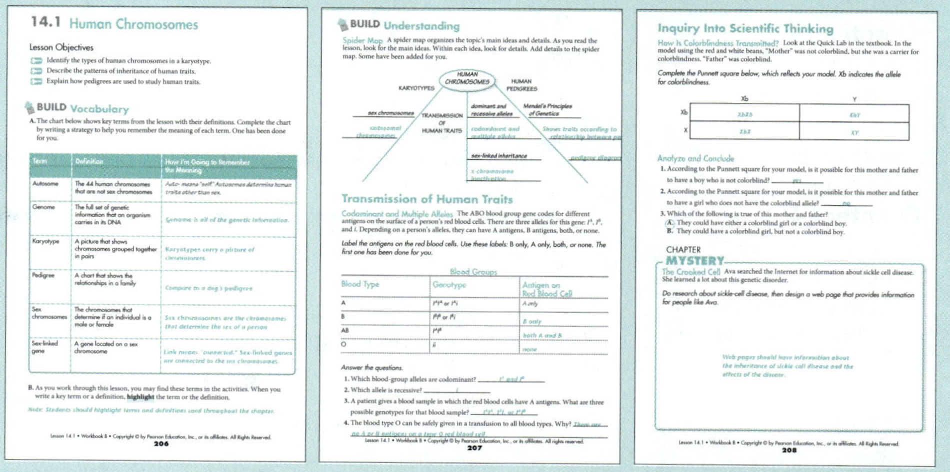

Autosomal Chromosomes The whole human genome has 46 chromosomes. There are 2 sex chromosomes. The other 44 are autosomal chromosomes, or *autosomes*. Most of the genetic information in the cell is on the autosomes.

Transmission of Human Traits

Scientists have learned many things about human genetics in recent years by studying DNA. Human genes, like the genes of other organisms, follow Mendelian patterns of inheritance.

Dominant and Recessive Alleles Many human traits follow a pattern of simple dominance. For example, a gene known as MC1R helps determine skin and hair color. A person with red hair usually has two recessive alleles for this gene. Dominant alleles for the MC1R gene help produce darker hair colors.

Codominant and Multiple Alleles Some alleles for human genes are codominant. One example is the ABO blood group, the set of alleles that determines your blood type. There are three alleles: I^A, I^B, and *i*. Alleles I^A and I^B are codominant. They make protein markers called antigens on the surface of red blood cells. People with alleles I^A and I^B make both A and B antigens, so they are blood type AB. The *i* allele is recessive. People with alleles I^AI^A or I^Ai make only the A antigen. They have blood type A. Those with I^BI^B or I^Bi alleles are type B. Those with two *i* alleles (*ii*) make no antigen. They have blood type O.

Key Question What patterns of inheritance do human traits follow?
Many human traits follow a pattern of simple dominance. The alleles for many human genes are codominant.

Blood Groups

Phenotype (Blood Type)	Genotype	Antigen on Red Blood Cell	Safe Transfusions	
			To	From
A	I^AI^A or I^Ai	A	A, AB	A, O
B	I^BI^B or I^Bi	B	B, AB	B, O
AB	I^AI^B	A and B	AB	A, B, AB, O
O	*ii*	None	A, B, AB, O	O

Human Blood Groups This table shows the relationship between genotype and phenotype for the ABO blood groups. It also shows which blood types can safely be transfused into people with other blood types.

Female XX

Male XY	X	X
X	XX	XX
Y	XY	XY

Sex Ratios Human egg cells have one X chromosome. Sperm cells have either one X chromosome or one Y chromosome. Half the offspring will be male. Half will be female.

BUILD Vocabulary

genome
the entire set of genetic information that an organism carries in its DNA

karyotype
a photograph of the complete diploid set of chromosomes grouped together in pairs, arranged in order

sex chromosome
one of two chromosomes that determines an individual's sex

sex-linked gene
a gene located on a sex chromosome

pedigree
a chart that shows the presence or absence of a trait according to the relationships within a family across several generations

ROOT WORDS
The prefix *karyo-* comes from the Greek word *karyon*, which means "nut or kernel." In biology, this prefix refers to the nucleus of a cell. The word *type* means "characteristic." So a karyotype shows the chromosomes that are characteristic of a cell's nucleus.

Focus on Academic Language

• dominant • codominant • recessive • allele • I^A • I^B • *i*

Write the words *dominant, codominant,* and *recessive* on the board. Point out that each word is an adjective used in genetics to modify the word *allele*.

Ask How do you refer to the form of a gene that is always expressed? *(dominant allele)* The form that can be hidden by another allele? *(recessive allele)*

Point out that *codominant* is similar to *dominant,* except for the prefix *co-*, which means "equally."

Ask What is the meaning of *codominant? (equally dominant)*

Write I^A, I^B, and *i* on the board. Explain that *I* and *i* stand for the first letter of the word *immunoprotein,* which is a type of chemical that can be an antigen.

Ask Which alleles for ABO blood type are dominant? *(I^A, I^B)* If you inherited the I^A and I^B alleles, which would be dominant? *(The alleles would be codominant.)*

During Reading

Active Reading

KEY QUESTION Read the Key Question and answer on the previous page.

Ask In which paired set of chromosomes do the chromosomes look different? *(the XY pair)*

MAKE CONNECTIONS Remind students that Punnett squares are tools used to predict the outcomes of crosses, or matings.

Say The data in a Punnett square can be thought of in two ways. In one sense, the data indicate what percent of all offspring from a cross will have a certain genotype or phenotype. This is useful for studying species that have large numbers of offspring.

Ask In the Punnett square on this page, what percentage of the offspring will be male? *(50%)*

Say The data in a Punnett square are also used to find the likelihood, or probability, that an individual offspring will inherit a particular phenotype or genotype. Like the outcome of a single toss of a coin, each offspring is an independent event. The probability of a genotype or phenotype indicated by a Punnett square is the same for each offspring.

Ask What is the likelihood that any one offspring will be male? *(50%)*

INTERPRET DIAGRAMS Continue to refer students to the Punnett square for the inheritance of sex in humans.

Ask Does the fact that a couple's first offspring is a girl change the second offspring's chance of being a girl? *(no)*

KEY QUESTION Invite a volunteer to read aloud the Key Question and answer.

Ask How many different alleles are involved in the inheritance of human ABO blood type? *(three)*

Build Vocabulary

ROOT WORDS Have students think about other words in biology that contain the root word *karyo*. *(eukaryote, prokaryote)* Ask students to explain why these terms would contain a word root that means "nucleus." *(One difference between the two types of cells is that one has a nucleus and one does not.)*

Speed Bump

Ask How many X chromosomes and Y chromosomes do human females have? *(two X, no Y)* Human males? *(one X, one Y)*

Ask Can two parents with type O blood have a child with type A, B, or AB blood? *(no)*

LESSON 14.1

During Reading continued

Active Reading

FIND THE MAIN IDEA Remind students that the main idea of a paragraph might not be stated in one sentence. Have a volunteer read aloud the X-Chromosome Inactivation section.

Ask What is the main idea of the first paragraph? *(Sample answer: In the cells of females, one of the two X chromosomes becomes inactive and becomes a Barr body.)* Explain that a summary of the paragraph's last two sentences best communicates the paragraph's main idea.

INTERPRET SYMBOLS To prepare students for analyzing the pedigree on the next page, have volunteers read aloud the text boxes in the diagram Pedigree Example.

VISUALIZE Emphasize that each family member represented in a pedigree, male or female, has two copies of the gene for the trait being studied. (Note that for a sex-linked trait, males would have only one copy of a gene.) Explain that the copies may be the same allele (homozygous) or they may be different alleles (heterozygous).

Ask How could you show that a person is heterozygous for a trait? *(Sample answer: Only one half of a circle or square could be shaded.)*

KEY QUESTION Invite a volunteer to read aloud the Key Question and answer.

Ask From which parent can a male inherit colorblindness? *(his mother)*

Speed Bump

Ask Why are recessive phenotypes of sex-linked traits less common among females than among males? *(Females must have two copies of the recessive allele to have the recessive phenotype.)*

Ask Why doesn't X-chromosome inactivation occur in males? *(Males have only one X chromosome in their cells.)*

X-Chromosome Inactivation Female calico cats are tricolored, meaning they have three colors. The spot color on their fur is controlled by a gene on the X chromosome. Spots can be either orange or black. Their color depends on which X chromosome is turned off in that patch of skin.

Sex-Linked Inheritance The genes located on X and Y chromosomes show a pattern of inheritance called sex-linkage. A **sex-linked gene** is a gene on a sex chromosome. Only males have the genes on the Y chromosome. Since only males have a Y chromosome, boys inherit it from their father. The X chromosome comes from their mother. The fact that males have only one X chromosome can cause genetic problems.

For example, humans have three genes for color vision on their X chromosome. A defective allele for any of these genes results in colorblindness. A male's only copy of each allele is on his X chromosome. If he has one of these defective alleles, he will be unable to tell certain colors apart. Females have two X chromosomes, so they have two alleles for each of the color-vision genes. Since the defective alleles are recessive, in a female both copies of the recessive allele must be present to produce colorblindness. Because of this, colorblindness and other sex-linked recessive disorders show up much more often in males than in females.

Key Question What patterns of inheritance do human traits follow? **Because the X and Y chromosomes determine sex, the genes located on them show a pattern of inheritance called sex-linkage.**

X-Chromosome Inactivation One X chromosome is enough for cells in males. But what about cells in females? A cell only needs one X chromosome to function normally. Female cells adjust to having two X chromosomes by randomly inactivating genes on one of them. The inactive chromosome makes a dense area in the nucleus called a Barr body.

X-chromosome inactivation also occurs in other female mammals. For example, cats have a gene that controls the color of coat spots on their X chromosome. One X chromosome may have an allele for orange spots. The other may have an allele for black spots. Different X chromosomes are inactivated in various parts of a female's skin. This causes a mixture of orange and black spots in her fur. Male cats have just one X chromosome, so their spots are only one color.

Human Pedigrees

You can use Mendel's basic laws of genetics to find out if a trait is dominant or recessive. These same laws help you find out if the gene for that trait is autosomal or sex-linked. You can use a chart called a **pedigree** to show the pattern of inheritance for a trait. A pedigree shows the presence or absence of a trait according to the relationships between parents, siblings, and offspring.

The pedigree on the next page shows how one human trait passes through three generations of a family. At the top of the chart is a grandfather who had the white forelock trait. Two of his three children inherited the trait. Three grandchildren have the trait, but two do not.

BIOLOGY.com Search Lesson 14.1 GO • InterActive Art • Tutor Tube

ELL Focus on ELL: Extend Language

BEGINNING SPEAKERS Tell students that, in English, many compound adjectives are made by using a hyphen to join two words. Explain that hyphenated compound adjectives usually modify a noun that follows them.

Say Find sentences on this page that include the term *sex-linked* before a noun. Write these sentences in your notebooks.

Ask What part of speech is the hyphenated term in these sentences? *(an adjective)*

Write other examples of hyphenated compound adjectives on the board, such as *red-headed, big-hearted,* and *eight-year-old.* Have students practice using these words in sentences in their notebooks. Remind them that the compound adjective should immediately precede the word it modifies.

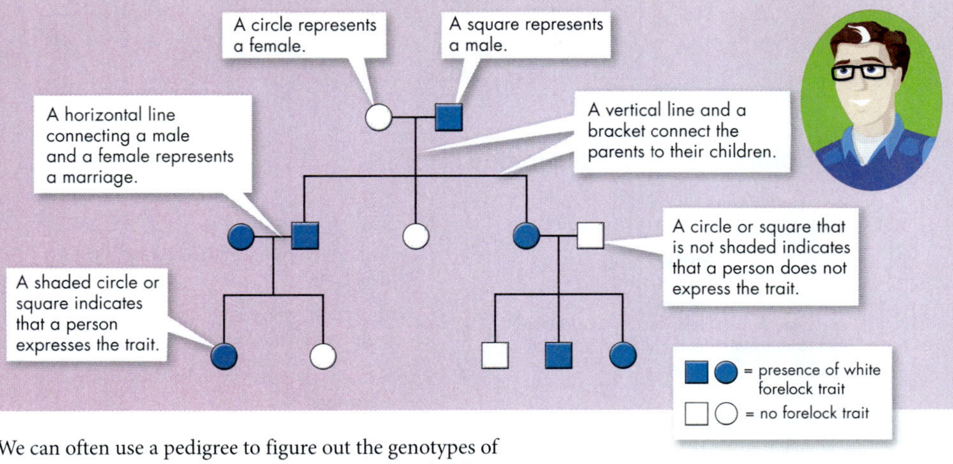

Pedigree Example This diagram shows what the symbols in a pedigree represent.

We can often use a pedigree to figure out the genotypes of family members. In the pedigree above, the allele for the white forelock trait is dominant. So, someone without the trait must have two recessive alleles. The grandfather at the top of the chart has this trait. So do two of his children. But the third child does not have a white forelock. So, the grandfather must be heterozygous for this trait.

Key Question How can pedigrees be used to analyze human inheritance? **The information learned from pedigrees helps determine the nature of genes and alleles associated with inherited human traits.**

CHECK Understanding

Apply Vocabulary

Use the highlighted words from the lesson to complete each sentence correctly.

1. A _______ contains all the instructions that are needed to build an organism.

2. An organism's paired chromosomes arranged in order from largest to smallest is called a _______.

Critical Thinking

3. Explain What does a pedigree show?

4. Predict Suppose a woman with type O blood and a man with type AB blood have children. What are the children's possible genotypes?

5. Write to Learn Answer the first clue of the mystery.

MYSTERY CLUE

Two sickle cell alleles are needed to produce sickle cell disease. Males and females develop sickle cell disease in equal frequencies. What do these statements suggest about the location of the gene responsible for the disorder? (**Hint:** See p. 336.)

BIOLOGY.com Search Lesson 14.1 GO • Lesson Assessment

Check Understanding

1. genome

2. karyotype

3. A pedigree shows the presence or absence of a trait over several generations within a family. It can be used to determine inheritance patterns or the possibility of inheritance for a future generation.

4. I^Ai and I^Bi

5. Because males and females develop sickle cell disease in equal frequencies, this suggests that the allele is located on an autosomal chromosome.

Active Reading

INTERPRET DIAGRAMS Remind students that some human traits are sex-linked but most are not. Ask questions to help students interpret the pedigree on this page. Explain that a forelock is a section, or lock, of hair on the front of the head.

Ask Does the white forelock trait appear to be sex-linked? *(no)* Why not? *(because males usually inherit sex-linked traits from their mothers, but the male in the second row of this pedigree inherited the trait from his father; also because men and women seem to be equally likely to inherit the trait)*

KEY QUESTION Read the Key Question and answer.

Ask What kind of information about a human trait can be inferred from a pedigree? *(Sample answers: whether the trait is dominant or recessive, whether the gene for the trait is autosomal or sex-linked)*

Speed Bump

Ask How many generations of the family are shown in the pedigree on this page? *(three)*

Post Reading

ASSESSMENT TIPS For Question 3, remind students that pedigrees show many things. Encourage them to write more than a one-sentence answer.

For Question 4, refer students to the Human Blood Groups table. Remind them that to predict means to use facts to say what you think will happen. Explain that the answer to this question cannot be found in the text.

MYSTERY CLUE After students have read the clue, remind them that if two sickle cell alleles are needed for a person to have sickle cell disease, then the sickle cell allele must be recessive to the allele for blood cells with a normal shape. Ask students if recessive sex-linked traits tend to appear equally in both males and females.

Wrap-Up Activity

MAKE PEDIGREES Have students make two of the four possible pedigrees for the calico cat shown on the previous page. *(calico female × orange-spotted male, calico female × black-spotted male, orange-spotted female × black-spotted male, black-spotted female × orange-spotted male)* Students' pedigrees should show the possible parents of the calico cat, the sex and colors of spots for each cat, and a key.

Inquiry Into Scientific Thinking

MAKE CONNECTIONS Explain that in this activity, students will use models to explore the inheritance of colorblindness in humans. Point out that there are many kinds of models.

Ask What is an example of a model? *(Sample answers: someone who poses for pictures to advertise clothing or other products; a small-scale object, such as a model airplane; a style of car or truck)*

Guide students to recognize that in science, models may be objects, pictures, or ideas that are used to explain other objects or processes. A model shows the relationships between different parts, or the way the parts work together.

USE VISUALS Refer students to the picture of the beans.

Ask What are some advantages of using the beans to model human sex chromosomes? *(Sample answers: The beans have two colors, which can be used to represent the two sex chromosomes. The beans are easy to manipulate.)*

Ask What allele for color vision does a red bean have? *(It has no allele for color vision.)*

USE CHARTS AND TABLES Demonstrate how students should record data in the data table by holding up a red bean and an unmarked white bean.

Say These beans represent the sex chromosomes for an individual.

Ask What should be written in the table under *Colors? (red and white)* under *Sex of Individual? (male)* under *Number of X-linked Alleles for Colorblindness? (none)* under *Colorblind? (no)*

Lead a Discussion

Have groups compare their results.

Ask Why was it important to calculate class totals? *(to increase the sample size for making calculations)*

Say Suppose a colorblind mother and a father with normal color vision have three daughters with normal color vision.

Ask If the parents have another daughter, what is the probability that she also will have normal color vision? *(100 percent, unless the daughter inherits from her father an X chromosome with a mutation for colorblindness)*

INQUIRY into Scientific Thinking

How Is Colorblindness Transmitted?

1. Copy the data table below into your notebook.

Trial	Colors	Sex of Individual	Number of X-linked Alleles for Colorblindness	Colorblind? (Yes/No)
1				
2				
3				
4				
5				
6				
7				
8				
9				
10				

2. Label one plastic cup Mother and a second plastic cup Father.
3. The white beans represent X chromosomes.
   - Use a black marker to make a dot on each of 2 white beans. The dot represents the X-linked allele for colorblindness.
   - Place 1 marked bean into each plastic cup.

4. Place 1 more white bean into the cup labeled Mother.
5. Red beans represent Y chromosomes. Place 1 red bean into the cup labeled Father.
6. Close your eyes and pick one bean from each cup. This represents how each parent contributes a sex chromosome to a fertilized egg.
7. Record the data about the beans you picked.
   - Record the colors of the 2 beans in your data table.
   - Record the sex of an individual who would carry this pair of sex chromosomes.
   - Record how many X-linked alleles the individual has.
   - Put the beans back in the cups they came from.
   - Determine whether the individual would have colorblindness.
8. Repeat steps 6 and 7 for a total of 10 pairs of beans.

Analyze and Conclude

1. Relate Cause and Effect How do human sex chromosomes keep the numbers of males and females roughly equal?

2. Calculate Calculate the class totals for each data column.

a. How many females were colorblind?

Number of colorblind females = ________

Percentage of females who were colorblind = ________

b. How many males were colorblind?

Number of colorblind males = ________

Percentage of males who were colorblind = ________

3. Use Models Does your model accurately represent how colorblindness is transmitted in a population? Explain. (*Hint:* See p. 336.)

In Your Workbook Get more help for this activity in your workbook.

Materials

2 plastic cups, black marker, 3 white beans, 1 red bean

Analyze and Conclude

1. Only the father's chromosome will determine sex. The chance is 50 percent that a child will receive an X chromosome from the father and 50 percent that a child will receive a Y chromosome from the father.
2. a. About 50 percent of the females will be colorblind.
   b. About 50 percent of the males will be colorblind.
3. No, because in actual human populations, the percentage of colorblind people is lower, and more males are affected than females.

14.2 Human Genetic Disorders

From Molecule to Phenotype

Have you ever heard the expression "It runs in the family"? Family and friends may have said that about your smile or the shape of your ears. But what could this mean when they talk about genetic diseases and disorders? What is a genetic disorder?

We know that genes are made of DNA. We also know that they interact with the environment to make an individual's characteristics, or phenotype. The link between molecule and trait is often that simple and direct. Genotype and phenotype are directly connected, since changes in a gene's DNA sequence can change proteins. Changes in the proteins cells make may directly affect one's phenotype. In other words, there is a molecular basis for genetic disorders.

Disorders Caused by Individual Genes Thousands of genetic disorders are caused by changes in individual genes. These changes often affect specific proteins that do important jobs.

▶ ***Sickle-Cell Disease*** This disorder is caused by a flawed allele for a polypeptide in hemoglobin. Hemoglobin is the oxygen-carrying protein in red blood cells. The polypeptide made from the flawed allele makes hemoglobin molecules stick together at times. The attached molecules make long fibers that force the cells to be shaped like a sickle. This shape gives the disorder its name. Normal red blood cells are flexible. They can squeeze through tiny capillaries—the narrowest blood vessels in the body. Sickle-shaped cells are more rigid and tend to get stuck in the capillaries. If the blood stops moving through the capillaries, then damage to cells, tissues, and even organs can happen.

▶ ***Huntington's Disease*** Huntington's disease is caused by a dominant allele for a protein found in brain cells. The allele for this disease has a long string of bases. In the string, the codon CAG repeats over and over again, more than 40 times. CAG codes for the amino acid glutamine. No one knows for sure why this long string of glutamine causes the disease. People who have Huntington's disease suffer from decreasing mental abilities and uncontrollable movements. The symptoms usually do not appear until middle age. The greater the number of CAG codon repeats, the earlier the disease appears, and the more severe its symptoms are.

Key Questions

How do small changes in DNA molecules affect human traits?

What are the effects of errors in meiosis?

BUILD Understanding

Two-Column Chart Before you read, make a two-column chart. In the first column, write three questions you have about genetic disorders. As you read, fill in answers to your questions in the second column.

In Your Workbook Go to your workbook to learn more about making a two-column chart. Complete the chart in Lesson 14.2.

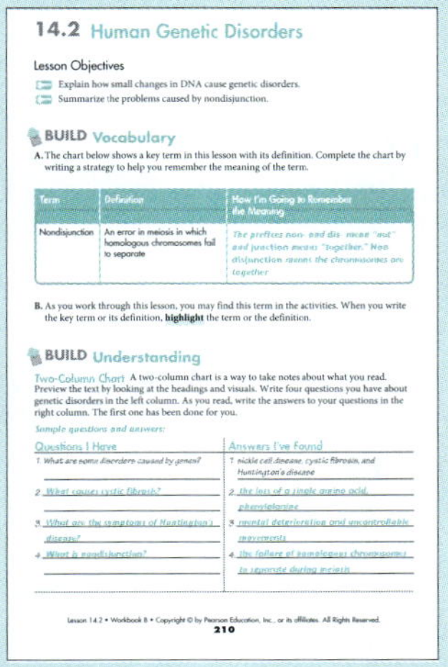

14.2 Human Genetic Disorders

Lesson Objectives

Explain how small changes in DNA cause genetic disorders.

Summarize the problems caused by nondisjunction.

BUILD Vocabulary

A. The chart below shows a key term in this lesson with its definition. Complete the chart by writing a strategy to help you remember the meaning of the term.

Term	Definition	How I'm Going to Remember the Meaning
Nondisjunction	An error in meiosis in which homologous chromosomes fail to separate	The prefixes non- and dis- mean "not" and junction means "together." Non disjunction means the chromosomes are together

B. As you work through this lesson, you may find this term in the activities. When you write the key term or its definition, **highlight** the term or the definition.

BUILD Understanding

Two-Column Chart A two-column chart is a way to take notes about what you read. Preview the text by looking at the headings and visuals. Write four questions you have about genetic disorders in the left column. As you read, write the answers to your questions in the right column. The first one has been done for you.

Sample questions and answers:

Questions I Have	Answers I've Found
1. What are some disorders caused by genes?	1. sickle cell disease, cystic fibrosis, and Huntington's disease
2. What causes cystic fibrosis?	2. the loss of a single amino acid, phenylalanine
3. What are the symptoms of Huntington's disease?	3. mental deterioration and uncontrollable movements
4. What is nondisjunction?	4. the failure of homologous chromosomes to separate during meiosis

Lesson 14.2 • Workbook B • Copyright © by Pearson Education, Inc., or its affiliates. All Rights Reserved.
210

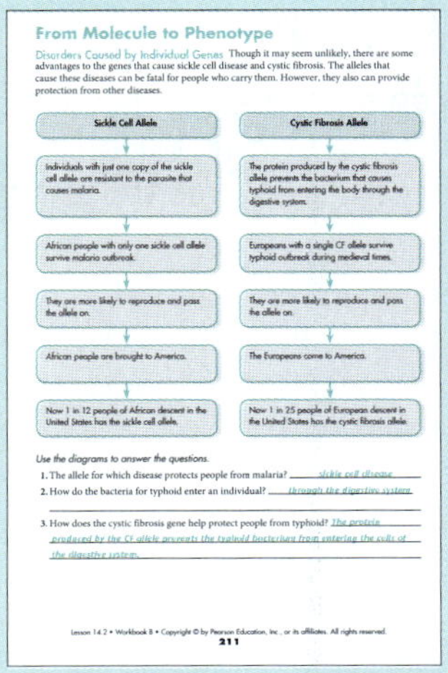

From Molecule to Phenotype

Disorders Caused by Individual Genes Though it may seem unlikely, there are some advantages to the genes that cause sickle cell disease and cystic fibrosis. The alleles that cause these diseases can be fatal for people who carry them. However, they also can provide protection from other diseases.

Sickle Cell Allele
↓ Individuals with just one copy of the sickle cell allele are resistant to the parasite that causes malaria.
↓ African people with only one sickle cell allele survive malaria outbreak.
↓ They are more likely to reproduce and pass the allele on.
↓ African people are brought to America.
↓ Now 1 in 12 people of African descent in the United States has the sickle cell allele.

Cystic Fibrosis Allele
↓ The protein produced by the cystic fibrosis allele prevents the bacterium that causes typhoid from entering the body through the digestive system.
↓ Europeans with a single CF allele survive typhoid outbreak during medieval times.
↓ They are more likely to reproduce and pass the allele on.
↓ The Europeans come to America.
↓ Now 1 in 25 people of European descent in the United States has the cystic fibrosis allele.

Use the diagrams to answer the questions.

1. The allele for which disease protects people from malaria? sickle cell disease
2. How do the bacteria for typhoid enter an individual? through the digestive system
3. How does the cystic fibrosis gene help protect people from typhoid? The protein produced by the CF allele prevents the typhoid bacterium from entering the cells of the digestive system.

Lesson 14.2 • Workbook B • Copyright © by Pearson Education, Inc., or its affiliates. All rights reserved.
211

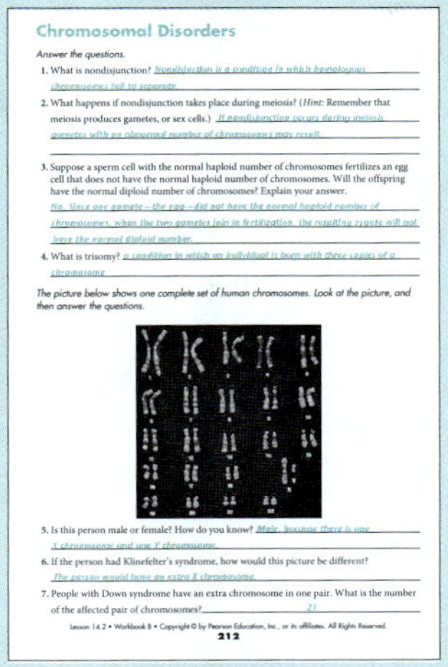

Chromosomal Disorders

Answer the questions.

1. What is nondisjunction? Nondisjunction is a condition in which homologous chromosomes fail to separate.
2. What happens if nondisjunction takes place during meiosis? (*Hint:* Remember that meiosis produces gametes, or sex cells.) If nondisjunction occurs during meiosis, gametes with an abnormal number of chromosomes may result.
3. Suppose a sperm cell with the normal haploid number of chromosomes fertilizes an egg cell that does not have the normal haploid number of chromosomes. Will the offspring have the normal diploid number of chromosomes? Explain your answer. No. Since one gamete—the egg—did not have the normal haploid number of chromosomes, when the two gametes join in fertilization, the resulting zygote will not have the normal diploid number.
4. What is trisomy? a condition in which an individual is born with three copies of a chromosome

The picture below shows one complete set of human chromosomes. Look at the picture, and then answer the questions.

5. Is this person male or female? How do you know? Male, because there is one X chromosome and one Y chromosome.
6. If the person had Klinefelter's syndrome, how would this picture be different? The person would have an extra X chromosome.
7. People with Down syndrome have an extra chromosome in one pair. What is the number of the affected pair of chromosomes? 21

Lesson 14.2 • Workbook B • Copyright © by Pearson Education, Inc., or its affiliates. All Rights Reserved.
212

LESSON 14.2

Pre-Reading

Objectives

14.2.1 Explain how small changes in DNA cause genetic disorders.

14.2.2 Summarize the problems caused by nondisjunction.

Preview the Pages

1. Have students read the definition of *nondisjunction.* Refer them to the Nondisjunction visual to illustrate the meaning of the word.
2. Preview the visual Mutations Cause Cystic Fibrosis. Have students follow the numbered events that lead to cystic fibrosis.

Build Understanding

TWO-COLUMN CHART Direct students to Workbook B, where they will find a partially completed two-column chart. Have students record additional questions about human genetic disorders and then fill in the answers as they read. After students are finished reading, they can research any questions that weren't answered.

Speed Bump

Ask What is the phenotype of sickle-cell disease? *(red blood cells shaped like a sickle)*

Ask What type of allele causes Huntington's disease, dominant or recessive? *(dominant)*

NATIONAL SCIENCE EDUCATION STANDARDS

UNIFYING CONCEPTS AND PROCESSES

II, IV, V

CONTENT

B.2, C.1.d, C.2.a, C.2.b, C.2.c

For the corresponding lesson in *Miller & Levine Biology,* see pages 398–401.

During Reading

Active Reading

INTERPRET VISUALS Direct students' attention to the visual Mutations Cause Cystic Fibrosis. Have a volunteer read aloud the caption and then the steps, pausing after each one.

Ask In the first picture, what do the letters in the colored boxes represent? *(the names of amino acids)* Why is one of the amino acids crossed out? *(Three bases were deleted in the DNA. As a result, one amino acid is missing from the CFTR protein.)*

Ask What is crossed out in the second picture, and why? *(The CFTR protein, a chloride ion channel, is crossed out. Students might infer that it is destroyed in the cell because it is abnormal.)*

Ask What organs are shown in the third picture? *(the lungs)* How does cystic fibrosis affect the lungs? *(They become blocked with thick mucus.)*

MAKE CONNECTIONS Write *trisomy* on the board. Have students identify the prefix in this word *(tri-)* and the root *(-some),* which refers to "chromosomes."

Ask What does *tri-* mean? *(three)* What does *trisomy* mean? *(three chromosomes)* Can you think of other examples of words that use this prefix? *(Sample answers: tricycle, triangle, tripod)*

Remind students of the use of the word *tri-colored* in the discussion of calico cats in the previous section.

Have students compare the use of the prefix *tri-* in a noun such as *trisomy* and in a hyphenated compound adjective such as *tri-colored.*

KEY QUESTION Invite a volunteer to read aloud the Key Question and answer.

Ask What is the smallest change in a protein that can result in a genetic disorder? *(a change in one of the protein's amino acids)*

Build Vocabulary

PREFIXES Review the meaning of the prefix *non-*.

Ask What are some other words that begin with the prefix *non-*? *(Sample answers: nonsense, nondairy, nonprofit)* Discuss the meanings of the words.

Speed Bump

Ask What kind of change in DNA causes cystic fibrosis? *(a deletion of three bases)*

Ask How can inheriting one allele for sickle-cell disease be an advantage? *(People with one copy of the sickle-cell allele are resistant to the parasite that causes malaria.)*

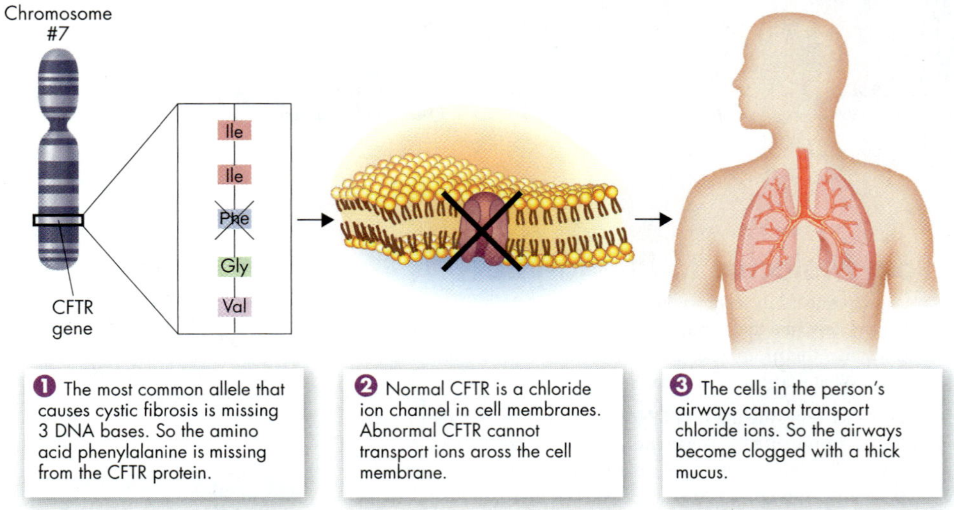

Mutations Cause Cystic Fibrosis
CF is usually caused by the deletion of three bases in the DNA of a single gene. So, the body does not make normal CFTR. CTFR is a protein needed to transport chloride ions.

BUILD Vocabulary

nondisjunction
an error in meiosis in which the homologous chromosomes fail to separate properly

PREFIXES
The prefix *non-* is a negative prefix that means "not." Adding a negative prefix gives a new word that means the opposite of the root word. The word *disjunction* means "coming apart." So *nondisjunction* means "not coming apart."

▶ ***Cystic Fibrosis*** Known as CF for short, cystic fibrosis is usually caused by the deletion of three bases in one gene. This gene codes for a protein called CFTR. The loss of these bases removes one amino acid from CFTR. This causes the protein to fold incorrectly, so the protein cannot do its job.

The CF allele is recessive, so two copies of the defective allele are needed to cause the disorder. People with one normal copy of the CF allele are unaffected by CF. Children with CF have trouble digesting food. They also make thick, heavy mucus that blocks their lungs and airways.

Key Question How do small changes in DNA molecules affect human traits?
Changes in a gene's DNA sequence can change proteins by altering their amino acid sequences. This may directly affect a person's phenotype.

Genetic Advantages Diseases such as sickle cell disease and CF are still common in human populations. Why are the alleles for these diseases still around if they can kill those who have them? Sometimes having just one recessive allele can be an advantage.

A parasite that lives in red blood cells causes a disease called malaria. But the parasite cannot live in sickled blood cells. People with one copy of the sickle cell allele are resistant to the parasite. This resistance gives them a big advantage against malaria.

More than 1000 years ago, terrible epidemics of typhoid fever spread through cities in Europe. Typhoid is caused by a bacterium that enters the body through cells in the digestive system. The protein made by the CF allele helps block this bacterium from entering the cells. People with a single CF allele were less likely to die from typhoid.

ELL Focus on ELL: Build Background

INTERMEDIATE AND ADVANCED SPEAKERS Write the words *disease, disorder, syndrome,* and *epidemic* on the board. Invite volunteers to explain the meaning of these words. Explain that all of these words are used in reference to illnesses or disabilities.

Have students find each of the words in this lesson and write a definition based on its use in the lesson. Then, have them look up the words in a dictionary. Ask students to make a vocabulary word map for each of the four words. Suggest either the Frayer model or a cluster diagram for making the word maps. For more information on the Frayer model and cluster diagrams see pages A6 and A4 in Workbook B respectively.

340 Chapter 14 • Lesson 2

Chromosomal Disorders

Meiosis usually works perfectly. Each human gamete, or sex cell, gets exactly 23 chromosomes. But occasionally something goes wrong. The most common error in meiosis occurs when homologous chromosomes do not separate. This mistake is called nondisjunction. **Nondisjunction** means "not coming apart."

If nondisjunction happens during meiosis, gametes can end up with the wrong number of chromosomes. This can lead to disorders. For example, if two copies of an autosomal chromosome fail to separate during meiosis, an individual may be born with three copies of that chromosome. This condition is called a trisomy. Down syndrome is trisomy of chromosome number 21. People with Down syndrome often have mental retardation and other birth defects.

Nondisjunction of sex chromosomes also causes problems. A female who inherits only one X chromosome usually has Turner's syndrome. Her sex organs do not develop at puberty, so she cannot have offspring. A male with an extra X chromosome has Klinefelter's syndrome. The extra chromosome interferes with meiosis, so the male usually cannot reproduce. There is no evidence of someone being born without any X chromosomes. So the genes on this chromosome are necessary for an embryo to survive and develop.

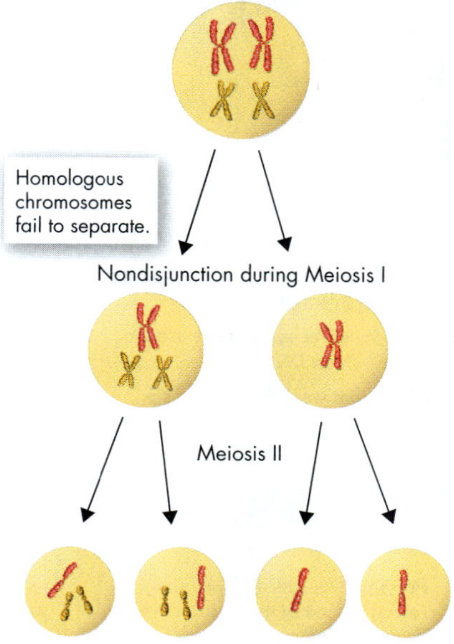

Nondisjunction This failure of meiosis causes gametes to have an abnormal number of chromosomes.

Key Question What are the effects of errors in meiosis? **If nondisjunction occurs during meiosis, gametes may receive an abnormal number of chromosomes. This change can lead to a disorder of chromosome numbers.**

CHECK Understanding

Apply Vocabulary

Use the highlighted words from the lesson to complete the sentence correctly.

1. When homologous chromosomes do not separate properly during meiosis, it is called _______.

Critical Thinking

2. **Explain** How can a small change in a person's DNA cause a genetic disorder?
3. **Apply Concepts** How does nondisjunction cause chromosomal disorders?
4. **Write to Learn** Answer the second clue in the mystery. Think about how proteins are made.

MYSTERY CLUE

Sickle cell disease occurs because a single amino acid is different. This difference causes a change in one of the hemoglobin proteins. What could produce this change? (**Hint:** See p. 340.)

BIOLOGY.com Search Lesson 14.2 GO • Lesson Assessment • Art in Motion

Check Understanding

1. nondisjunction
2. It can affect the structure and function of a protein, which may cause a disorder.
3. Chromosomes fail to separate, causing gametes to have abnormal numbers of chromosomes.
4. Sample answer: A substitution mutation that changes one or more of the nucleotides in a codon that codes for an amino acid can cause that codon to code for a different amino acid. The visible effect of the mutation is the change in the shape of the red blood cell to the characteristic "sickle," or "crescent."

Active Reading

USE VISUALIZATION Direct students' attention to the diagram showing nondisjunction. Have a volunteer identify the structures shown and explain what is happening. Then, ask students to visualize the normal process of meiosis.

Ask In the diagram, how many chromosomes would the final four daughter cells have if meiosis had happened normally? *(two, one of each kind)*

KEY QUESTION Read the question and answer.

Ask What is nondisjunction? *(homologous chromosomes not separating properly during meiosis)*

Speed Bump

Ask What is the result of having three number 21 chromosomes? *(Down syndrome)*

Ask How many chromosomes would the body cells of a woman born with Turner's syndrome have? *(45)*

Ask What technique can be used to confirm that a person has an extra chromosome? *(making a karyotype of the person's chromosomes)*

Post Reading

ASSESSMENT TIPS

For Question 3, refer students to the Nondisjunction diagram. Explain that *apply* means "use your knowledge to explain different situations."

MYSTERY CLUE Before students read the clue, review the three kinds of DNA point mutations: substitution, insertion, and deletion.

Wrap-Up Activity

CLUSTER DIAGRAM Have students work in small groups to create a cluster diagram that summarizes human genetic disorders. The diagram should differentiate disorders caused by gene mutations from those caused by chromosomal abnormalities and should give examples of disorders for each category. Provide chart paper and other materials such as markers, yarn, and construction paper for students to use in making their diagrams visually appealing. Invite groups to display their diagrams in the classroom.

LESSON 14.3

Pre-Reading

Objectives

14.3.1 Summarize the methods of DNA analysis.

14.3.2 State the goals of the Human Genome Project and explain what we have learned so far.

Preview the Pages

1. Preview the vocabulary by having students find the highlighted words. Help students with the pronunciation of more difficult words such as *electrophoresis* and *bioinformatics*.
2. Preview the visuals. Point out that the caption headings in the visual How Scientists Manipulate DNA are the same as the paragraph headings on the previous page. The information in the visual should be read in the order shown in the text.

Activate Prior Knowledge

Help students recall the cause of cystic fibrosis. *(a deletion of three DNA bases in the gene that codes for CFTR protein)*

Say To identify the defective gene, scientists first had to sequence the DNA bases in the normal gene. In this lesson, you'll learn about the techniques scientists use to sequence DNA.

Build Understanding

PREVIEW VISUALS Direct students to Workbook B, where they will find a partially filled-in Preview Visuals chart. Encourage students to complete the chart before they read the lesson.

NATIONAL SCIENCE EDUCATION STANDARDS

UNIFYING CONCEPTS AND PROCESSES

I, II, III

CONTENT

B.2, C.1.d, C.2.a, C.2.b, C.2.c, F.6, G.1

INQUIRY

A.2.b, A.2.c, A.2.f

For the corresponding lesson in *Miller & Levine Biology*, see pages 403–409.

14.3 Studying the Human Genome

Key Questions

What techniques are used to study human DNA?

What were the goals of the Human Genome Project, and what have we learned so far?

BUILD Understanding

Preview Visuals Before you read Lesson 3, look at the illustrations on the following page. Think of three questions you have about the illustrations.

In Your Workbook Write your questions in the chart in Lesson 14.3. As you read, write the answers to your questions.

BUILD Vocabulary

restriction enzyme
an enzyme that cuts DNA at a sequence of nucleotides

gel electrophoresis
a procedure used to analyze DNA fragments by placing a mixture of DNA fragments at one end of a porous gel and applying an electrical voltage to the gel

WORD ORIGINS

The root word *phoresis* is from the Greek *phorein*, which means "to carry." In electrophoresis, an electric current carries suspended particles through a medium.

Manipulating DNA

Only a few decades ago, computers were big machines found only in laboratories or universities. Today, many people carry small, powerful computers to school and work. Decades ago, the human genome was unknown. Now, we can see our entire genome on the Internet. How long will it be before having a copy of your own genome is as common as carrying a cell phone?

Since discovering the genetic code, biologists have dreamed of reading the human genome. For a long time, it seemed impossible because working with molecules as large as DNA is difficult. Then scientists discovered tools to cut, separate, and replicate DNA. Now they can read the base sequences in DNA from any cell.

Cutting DNA To study DNA, scientists first cut it into smaller pieces using bacterial enzymes called **restriction enzymes.** Restriction enzymes cut a DNA molecule into exact pieces called restriction fragments. Different enzymes cut DNA in different places, and into pieces of different sizes.

Separating DNA After DNA is cut, scientists use a technique called **gel electrophoresis** to separate and analyze the fragments of DNA. A mixture of DNA fragments is put at one end of a gel. An electric current makes the DNA move toward one end of the gel. Smaller DNA fragments move faster and farther than large ones. The result is a pattern of bands on the gel. The pattern is based on the size of the DNA fragments.

Reading DNA After the DNA fragments have been separated, researchers read, or sequence, the DNA. Single-stranded DNA fragments are put in a test tube with DNA polymerase. Then the bases, A, T, G, and C, are added. DNA polymerase uses the bases to make many new DNA strands. Some of the bases are labeled with a chemical dye. When a dye-labeled base is added to a new DNA strand, the synthesis of that strand stops. When all the synthesis is done, the result is a series of color-coded DNA strands. The strands are different lengths. Researchers can then separate these strands by size on a gel. The order of colored bands on the gel tells the exact sequence of bases in the DNA.

Key Question What techniques are used to study human DNA? **Scientists use tools that cut, separate, and then replicate DNA base by base. Now they can read the base sequences in DNA from any cell.**

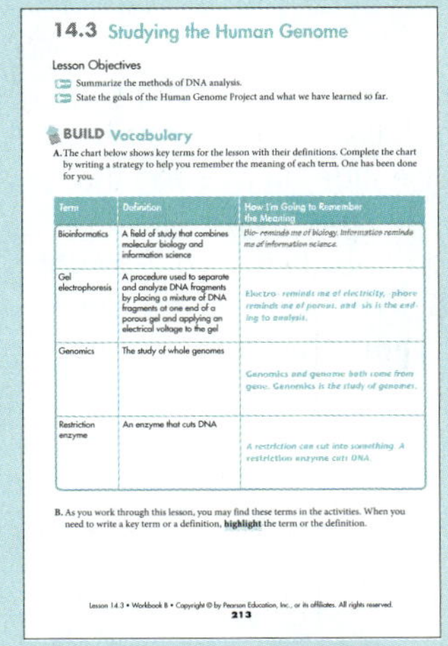

14.3 Studying the Human Genome

Lesson Objectives

Summarize the methods of DNA analysis.

State the goals of the Human Genome Project and what we have learned so far.

BUILD Vocabulary

A. The chart below shows key terms for the lesson with their definitions. Complete the chart by writing a strategy to help you remember the meaning of each term. One has been done for you.

Term	Definition	How I'm Going to Remember the Meaning
Bioinformatics	A field of study that combines molecular biology and information science	Bio- reminds me of biology. Information reminds me of information science.
Gel electrophoresis	A procedure used to separate and analyze DNA fragments by placing a mixture of DNA fragments at one end of a porous gel and applying an electrical voltage to the gel	Electro- reminds me of electricity, phore reminds me of porous, and sis is the ending to analysis.
Genomics	The study of whole genomes	Genomics and genome both come from gene. Genomics is the study of genomes.
Restriction enzyme	An enzyme that cuts DNA	A restriction can cut into something. A restriction enzyme cuts DNA.

B. As you work through this lesson, you may find these terms in the activities. When you need to write a key term or a definition, **highlight** the term or the definition.

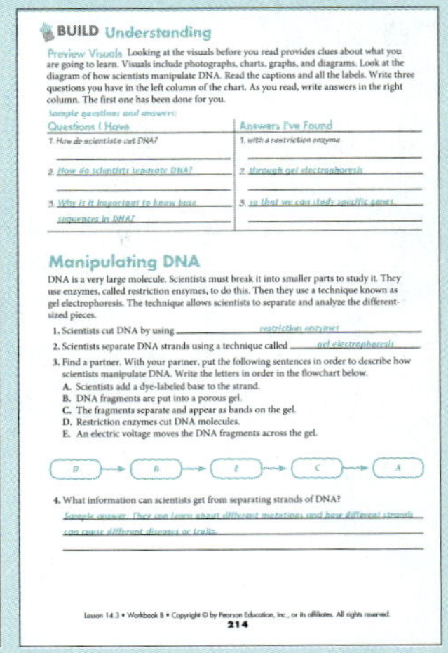

BUILD Understanding

Preview Visuals Looking at the visuals before you read provides clues about what you are going to learn. Visuals include photographs, charts, graphs, and diagrams. Look at the diagram of how scientists manipulate DNA. Read the captions and all the labels. Write three questions you have in the left column of the chart. As you read, write answers in the right column. The first one has been done for you.

Sample questions and answers:

Questions I Have	Answers I've Found
1. How do scientists cut DNA?	1. with a restriction enzyme
2. How do scientists separate DNA?	2. through gel electrophoresis
3. Why is it important to know base sequences in DNA?	3. so that we can study specific genes.

Manipulating DNA

DNA is a very large molecule. Scientists must break it into smaller parts to study it. They use enzymes, called restriction enzymes, to do this. Then they use a technique known as gel electrophoresis. The technique allows scientists to separate and analyze the different-sized pieces.

1. Scientists cut DNA by using restriction enzymes
2. Scientists separate DNA strands using a technique called gel electrophoresis
3. Find a partner. With your partner, put the following sentences in order to describe how scientists manipulate DNA. Write the letters in order in the flowchart below.
   A. Scientists add a dye-labeled base to the strand.
   B. DNA fragments are put into a porous gel.
   C. The fragments separate and appear as bands on the gel.
   D. Restriction enzymes cut DNA molecules.
   E. An electric voltage moves the DNA fragments across the gel.

D → B → E → C → A

4. What information can scientists get from separating strands of DNA?
   Sample answer: They can learn about different mutations and how different strands can cause different diseases or traits.

BUILD Connections

HOW SCIENTISTS MANIPULATE DNA

Scientists use tools to cut, separate, replicate, and sequence DNA. Knowing the sequence of DNA allows us to study specific genes.

Cutting DNA
A restriction enzyme is like a key that fits only one lock. The *Eco*RI restriction enzyme can only recognize the base sequence GAATTC. It cuts each strand of DNA between the G and A bases. This cut leaves single-stranded overhangs with the sequence AATT. The overhangs are called "sticky ends." The ends can bond, or "stick," to a DNA fragment with the complementary base sequence.

Separating DNA
Gel electrophoresis is used to separate DNA fragments. The fragments are first cut by restriction enzymes. Then, the fragments are put into wells on a gel. The gel is like a slice of gelatin. An electric voltage moves the shorter fragments faster than the longer fragments. Within an hour or two, the fragments all separate. Fragments of the same size show up as a band on the gel.

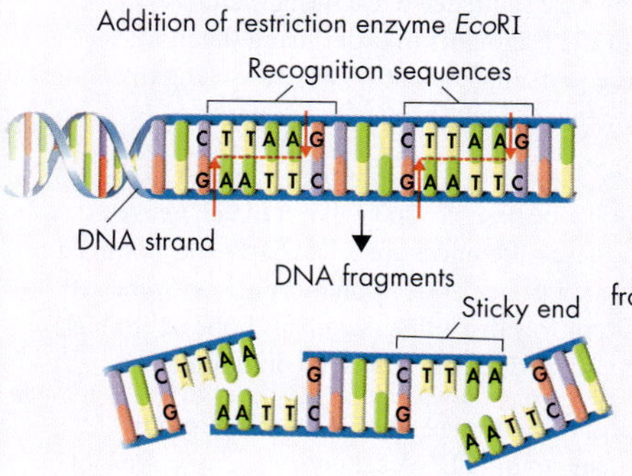

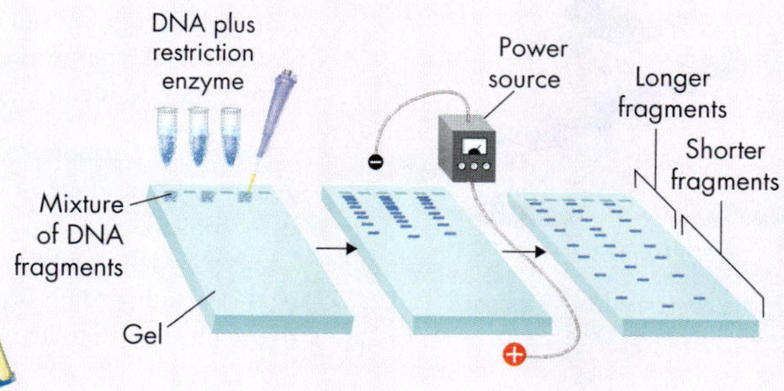

Reading DNA
Some dye-labeled bases are mixed in with the regular bases added to make the new DNA. Each time a labeled base is added to the strand, DNA replication for that strand stops. Each kind of base is labeled with a different color. This labeling results in color-coded DNA fragments of different lengths. Gel electrophoresis is used to separate the fragments. Then scientists can "read" the DNA sequence from the gel.

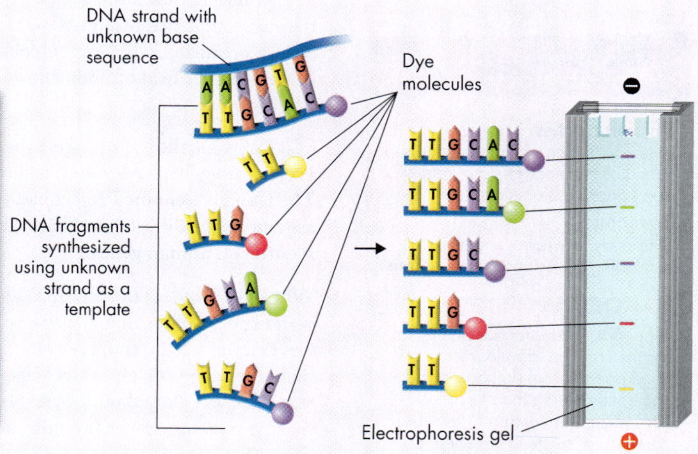

Base sequence as "read" from the order of the bands on the gel from bottom to top: **T G C A C**

Hands-On Learning

MANIPULATING DNA Have students work in pairs and use sticky notes of four different colors to model the techniques shown on this page.

1. Have pairs label 12 green sticky notes with the letter *A*, 12 yellow notes *T*, 7 red notes *G*, and 7 purple notes *C*.
2. Tell students to use their notes to model the nucleotides that make up the unwound portion of the DNA molecule in the diagram.
3. Have students model the cutting of the DNA into fragments, as the red arrows and dotted lines show. They can use a pencil to represent *Eco*RI cutting the DNA. Encourage students to model how the fragments could join at their "sticky ends."
4. Tell students to model the single-stranded DNA with dye molecules attached. They can use unlabeled notes of the correct color to represent the dyes. They can stack the fragments as they would be in a gel.

Ask If the next longest DNA strand ended with the red dye, what would be the next base in the DNA? *(G)*

During Reading

Build Connections

VISUAL SUMMARY Have students study each technique as they read about it in the text. Note that DNA molecules are studied by cutting them into smaller pieces.

Ask What do you think the word *restriction* means in the term *restriction enzyme*? *(Sample answer: It probably has something to do with making shorter pieces of DNA.)*

Ask What is the name of the restriction enzyme used in this diagram? *(EcoRI)*

Say *Eco* refers to the bacterium *E. coli.* The *R* stands for *restriction*. The last symbol is the Roman numeral for *one*. The name is pronounced "Eco R one."

Point out the way *Eco*RI cuts a strand of DNA.

Ask How do you think the single-stranded ends could be used? *(Sample answer: They could be used to attach marked pieces of DNA together.)*

Write the word *electrophoresis* on the board and underline the root *electro*. Explain that separating DNA fragments by gel electrophoresis is another basic technique used to study DNA.

Ask What does gel electrophoresis use to separate the DNA fragments? *(electric current)* What do you think *electro* means? *(Sample answer: electricity)*

Ask Why do some DNA fragments move farther through the gel than other fragments? *(Smaller fragments move faster.)* Why do scientists separate the fragments? *(Sample answer: in order to study particular segments of DNA molecules)*

Say Explain that you read words and sentences to find out the information in a book. Letters give a word its particular meaning.

Ask What are the "letters in the alphabet" that are used to make "words" in DNA "sentences"? *(the bases A, G, C, and T)* What else do scientists need to know about DNA in order to understand the instructions it contains? *(the number and arrangement of its bases)*

KEY QUESTION Read the Key Question and answer from the previous page aloud.

Ask What is the overall purpose of the techniques used to study human DNA? *(to determine the sequence of bases in the DNA)*

Speed Bump

Ask How do scientists cut DNA into fragments? *(They use restriction enzymes)* How do scientists separate and read DNA fragments? *(They use gel electrophoresis.)*

During Reading continued

Active Reading

MAKE AN ANALOGY Ask students to imagine a city street that many other streets intersect.

Ask How are the streets marked? *(with street signs)* What do people use the signs for? *(for locating places)*

Explain that scientists use DNA markers, or signs, to locate places in DNA. The signs in DNA are regions of known DNA sequences.

APPLY ACADEMIC VOCABULARY On the board, write the following pairs of words: *cat / fat, flow / flaw,* and *distances / distanced*. For each pair, stack the words and circle the letters that are different.

Say Each pair of words represents base sequences in the same DNA region in two different people. The circled letters represent SNPs. Compare the meanings of each pair of words.

Ask What effect do the SNPs have in the first two pairs of words? *(They completely change the meanings of the words.)* What about in the third pair of words? *(The meanings are only slightly different.)*

Say Some SNPs result in changes in the proteins that are coded for by the DNA. Other SNPs have no effect.

KEY QUESTION Read aloud the question and answer.

Ask How many base pairs make up the human genome? *(3 billion)*

Build Vocabulary

ROOT WORDS Point out that the terms *bioinformatics* and *genomics* have the same suffix, *-ics,* which means "related to" or "derived from." Explain that this suffix is often used to form words that mean "the study of."

Ask What are some other words in science that have the suffix *-ics?* *(Sample answers: genetics, mathematics, statistics, physics)*

Speed Bump

Ask What was the importance of bioinformatics to the Human Genome Project? *(Sample answer: It allowed researchers to analyze and share large amounts of data.)*

Shotgun Sequencing
This method rapidly sorts DNA fragments by using overlapping base sequences.

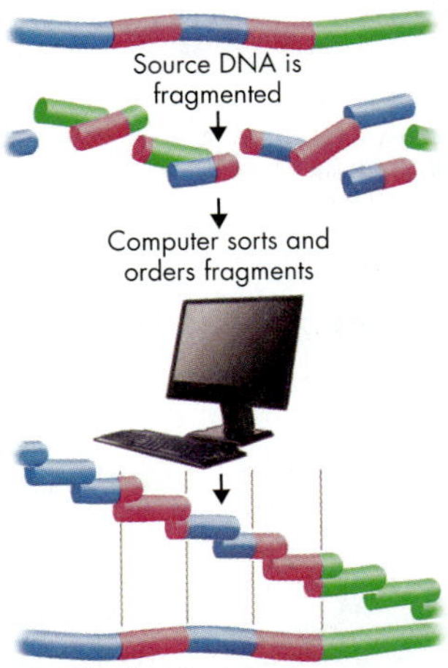

Overlapping sequences are matched and aligned to determine the complete DNA sequence.

The Human Genome Project

In 1990, the United States and other countries began the 13-year Human Genome Project. The project had two main goals. One goal was to sequence all 3 billion base pairs of human DNA. The other was to identify all human genes. DNA sequencing was at the center of the Human Genome Project. How do scientists handle a sequencing project that big? First, they break up the whole genome into smaller pieces. Next, they determine the base sequences in regions of a DNA strand that are far apart. These regions act as markers, like mile markers on a long highway. Using the markers, researchers can find and return to specific places in the DNA.

Sequencing and Identifying Genes Once researchers have marked the DNA strands, they cut them into random fragments. Then they determine the base sequence of each fragment. Computer programs put the fragments in order using the markers. This is called "shotgun sequencing." Scientists are now using this information to identify genes.

Comparing Sequences Most people have almost identical genomes. On average, only one base in 1200 will not match between two people. These single base differences are called SNPs (pronounced "snips"), which stands for single nucleotide polymorphisms. Some sets of closely linked SNPs occur together often. Scientists hope to be able to use these SNPs to identify various diseases and conditions.

Sharing Data The Human Genome Project was completed in 2003. Copies of the human genome DNA sequence are now available on the Internet. One of the key research areas of the project was a new field called **bioinformatics.** By 2003, scientist had determined the entire sequence. Scientists in this field use computer tools to collect, organize, and interpret biological data. Bioinformatics also began a similar field of study known as **genomics**—the study of whole genomes.

Key Question What were the goals of the Human Genome Project, and what have we learned so far?
The Human Genome Project had two main goals. One was to sequence all 3 billion base pairs of human DNA. The other was to identify all human genes.

What We Have Learned In 2000, scientists announced that a working copy of the human genome was complete. By 2003, scientists had determined the entire sequence. Scientists had learned that the human genome contains three billion nucleotide bases. They found surprises, too. Only about 2 percent of our genome has codes for making proteins. Many chromosomes have large areas with very few genes. The scientists discovered that much of our genome is made up of genetic code from viruses. The project also identified genes linked to many diseases and disorders. Finally, the project found three million locations where single base pairs are different in humans.

BUILD Vocabulary

bioinformatics
the application of mathematics and computer science to store, retrieve, and analyze biological data

genomics
the study of whole genomes, including genes and their functions

ROOT WORDS
The root word *informatics* refers to the creation, development, and operation of databases and other computing tools to collect, organize, and interpret data. The prefix *bio-* refers to life sciences. In bioinformatics, life science data are collected in databases.

ELL Focus on ELL: Access Content

ALL SPEAKERS Have students work in groups of two or three to make flowcharts that summarize the way scientists manipulate DNA. Give students the following terms to use in constructing their flowcharts: *reading DNA, gel electrophoresis, cutting DNA, restriction enzyme, shotgun sequencing,* and *separating DNA.* Have each group compare their flowchart with that of another group, discuss any differences, and decide what corrections, if any, should be made.

Key Question What were the goals of the Human Genome Project, and what have we learned so far?

The Human Genome Project pinpointed genes. It matched some sequences in those genes with many diseases and disorders. It also identified about three million places where single-base DNA differences occur in humans.

New Questions The scientists in this project knew that their data could cause ethical, legal, and social problems. The problem areas include privacy, fairness in the use of and access to the data, and medical issues. In May 2008, the Genetic Information Nondiscrimination Act was passed into law. The act makes sure that U.S. companies cannot discriminate on the basis of information they learn from genetic tests.

What's Next There is much more to learn about the human genome. The 1000 Genomes Project began in 2008. It studies the genomes of 1000 people and builds a careful list of human variation. Data from the project will be used in studies of development and disease. The information may hold the key to successful research on new drugs. It may also lead to therapies to save human lives and keep people healthy.

Announcements
The first details of the human genome appeared in two well-known scientific journals in February 2001.

CHECK Understanding

Apply Vocabulary

Use the highlighted terms from the lesson to complete each term correctly.

1. A scientist can use a _______ to cut a strand of DNA at a specific place.
2. _______ is the process that uses electrical current to separate DNA fragments by size.
3. _______ is a field of study that grew out of bioinformatics and includes the study of genes and their functions.

Critical Thinking

4. **Explain** What is the Human Genome Project?
5. **Apply Concepts** How might the Human Genome Project be used to benefit humankind?
6. **Write to Learn** Answer the third clue to the chapter mystery. How many bases are involved in the mutation?

MYSTERY CLUE

Scientists used SNPs to locate the sickle cell allele in genes that code for hemoglobin proteins. What does this tell you about the sickle cell mutation? (**Hint:** See p. 344.)

BIOLOGY.com Search Lesson 14.3 GO • Lesson Assessment

Check Understanding

1. restriction enzyme
2. Gel electrophoresis
3. Genomics
4. The Human Genome Project was an international effort to sequence all 3 billion base pairs of human DNA and to identify all human genes.
5. The project is benefitting humankind by helping scientists learn about the causes of genetic disorders.
6. Sample answer: The sickle cell mutation affects the formation of the protein hemoglobin. Scientists used tests for SNPs to find the allele for sickle-cell disease. Since SNPs are single bases, this means that only one base is involved in the mutation.

Active Reading

MAKE INFERENCES Have students identify characteristics scientists discovered about the human genome as you list them on the board.

Say Scientists also completed the genomes of several other organisms. They found that more than 40 percent of the proteins coded for by our genome are very similar to proteins in many of those organisms, including fruit flies, worms, and yeast.

Ask What inference can be made about this similarity between coded proteins in humans and other organisms? *(Humans and other organisms have a common evolutionary origin.)*

KEY QUESTION Read the question and answer.

Ask What information did scientists find surprising about the human genome? *(Many chromosomes have large areas with very few genes.)*

Speed Bump

Ask What percent of the human genome codes for proteins? *(about 2 percent)*

Ask What is the 1000 Genomes Project, and what are its goals? *(The 1000 Genomes Project studies the genomes of 1000 people. Its goal is to learn more about human diseases and possible treatments for them.)*

Post Reading

ASSESSMENT TIPS For Question 5, explain that *apply* means "use your knowledge to predict outcomes." Encourage students to draw on all they have learned about DNA and human heredity to think of ways in which knowing the DNA sequence of the human genome could benefit humans.

MYSTERY CLUE After students read the clue, remind them that the information in DNA is read in sequences of three bases called codons. Most codons code for a particular amino acid in proteins.

Wrap-Up Activity

GENERATE QUESTIONS Have students work in pairs to develop two open-ended questions about the topics in this lesson and a detailed answer for each one. Then, have each pair exchange only their questions with another pair and answer the exchanged questions. Have the question writers compare their own answers with those written by the pair of students they exchanged with and assess whether they wrote good questions.

Pre-Lab

Invite a volunteer to read aloud the lab title and problem. Discuss meanings of the word *remains.* Explain that in the lab, *remains* refers to the parts, such as bones and teeth, of a person who has died.

Review the skills students will use for this lab. Remind them that analyzing data involves looking carefully for patterns and that identifying patterns allows scientists to draw conclusions. Before assigning the reading and questions, review the processes involved in inheriting traits.

Ask How do traits pass from one generation of a family to the next? *(through sexual reproduction)* What cells are involved in sexual reproduction? *(egg and sperm)* By what two processes are eggs and sperm produced? *(meiosis and cell division)* How does the number of chromosomes in eggs and sperm compare to that in other cells of the body? *(Eggs and sperm have half the number of chromosomes found in other cells.)*

Review the Background Questions and Pre-Lab Questions as a class.

Lab

Tell students they will perform the chapter lab *Using DNA to Identify Human Remains* described in **Lab Manual B.**

Post-Lab

DIAGRAMMING Ask students to draw a diagram that contrasts the inheritance of nuclear DNA and mitochondrial DNA. Diagrams should show meiosis and fertilization. For simplicity, have students limit the nuclear DNA to one pair of homologous chromosomes and the mtDNA to one loop.

NATIONAL SCIENCE EDUCATION STANDARDS

CONTENT C.2.a, G.1

INQUIRY A.1.b, A.2.a, A.2.b, A.2.c, A.2.f

Forensics Lab

GUIDED INQUIRY

Pre-Lab: Using DNA to Identify Human Remains

Problem How can pedigrees help scientists identify human remains?

Lab Manual Chapter 14 Lab

Skills Focus Analyze Data, Draw Conclusions

Connect to the Big idea The nucleus is not the only location in a cell where DNA can be found. DNA is also found in the mitochondria of cells. This mitochondrial DNA, or mtDNA, exists as small loops, rather than long strands. Unlike nuclear DNA, mtDNA is inherited only from the mother. Thus, except for mutations, the sequence of nucleotides in mtDNA remains constant over many generations.

Less than one percent of a cell's DNA is mtDNA, but in that percentage are many copies of the small mtDNA molecules. So when forensic scientists cannot collect a suitable sample of nuclear DNA, they look for mtDNA. Usable mtDNA can often be found even after a body decays or is burned. In this lab, you will explore how mtDNA was used to help confirm the identity of bones that scientists thought belonged to members of the Romanov family.

Background Questions

a. Review What is a pedigree?

b. Explain In a pedigree, what does a circle represent? What does a square represent?

c. Infer How do you know that mtDNA isn't sorted and recombined during meiosis?

Pre-Lab Questions

Preview the procedure in the lab manual.

1. Infer The tsar and tsarina had five children. Did all seven family members have the same mtDNA? Give a reason for your answer.

2. Predict To confirm that bones belonged to the tsar's children, which living relative would be more useful—a relative of the tsar or a relative of the tsarina? Why?

The Romanovs ruled Russia for 300 years until the Bolshevik Revolution of 1918 resulted in the execution of Tsar Nicholas II and his family.

3. Infer If two people have the same mtDNA, what can you infer about their biological relationship?

BIOLOGY.com Search Chapter 14 GO

Visit Chapter 14 online to test yourself on chapter content and to find activities to help you learn.

Untamed Science Video The Untamed Science crew identifies the chromosomes that carry genes for colorblindness.

Art in Motion View a short animation that explains nondisjunction.

Art Review Review your understanding of karyotypes with this drag-and-drop activity.

InterActive Art Learn all about pedigrees and how to make them with this animation.

Tutor Tube Why do traits sometimes "skip a generation"? Tune in to the tutor to find out.

Pre-Lab Answers

BACKGROUND QUESTIONS

a. A pedigree is a chart that is used to analyze the pattern of inheritance of a trait across several generations.

b. In a pedigree, a circle represents a female and a square represents a male.

c. Sample answer: The sequence of nucleotides in mtDNA remains constant over many generations. (All the daughter cells must have identical mtDNA.)

PRE-LAB QUESTIONS

1. The tsarina and the children had the same mtDNA because the children inherited their mtDNA from the tsarina. The tsar had different mtDNA, which he inherited from his mother.

2. A living relative of the tsarina would be more useful because that relative would have the same mtDNA as the children.

3. Sample answer: The two people have a common ancestor on the maternal side of their family trees.

14 CHAPTER Summary

14.1 Human Chromosomes

- A karyotype shows the complete diploid set of chromosomes. They are grouped in pairs and arranged from largest to smallest.
- Many human traits follow a pattern of simple dominance. The alleles for other human genes are codominant.
- Because the X and Y chromosomes determine sex, the genes located on them show a pattern of inheritance called sex-linkage.
- The information learned from pedigrees helps determine the nature of genes and alleles associated with inherited human traits.

genome (p. 334)
karyotype (p. 334)
sex chromosome (p. 334)
sex-linked gene (p. 336)
pedigree (p. 336)

14.2 Human Genetic Disorders

- Changes in a gene's DNA sequence can change proteins by altering their amino acid sequences. This change may directly affect a person's phenotype.
- If nondisjunction happens during meiosis, gametes may receive an abnormal number of chromosomes. This change can lead to a disorder of chromosome numbers.

nondisjunction (p. 341)

14.3 Studying the Human Genome

- Scientists use tools that cut, separate, and then replicate DNA base by base. Now they can read the base sequences in DNA from any cell.
- The Human Genome Project had two main goals. One was to sequence all 3 billion base pairs of human DNA. The other was to identify all human genes.
- The Human Genome Project pinpointed genes. It matched some sequences in those genes with many diseases and disorders. It also identified about 3 million locations where single-base DNA differences occur in humans.

restriction enzyme (p. 342)
gel electrophoresis (p. 342)
bioinformatics (p. 344)
genomics (p. 344)

BIOLOGY.com Search Chapter 14 GO • Chapter Assessment

UbD Transfer the Big Idea

TRANSFER Tell students that genetic counselors and genetic tests are available for identifying a person's risk for certain health problems, such as heart disease and cancer, and for passing on genetic disorders to offspring. Have students work in small groups to discuss the questions that follow.

1. Would you want to know if you are at risk for developing serious diseases? If so, which ones? If not, why not?
2. If you found out that you are likely to develop a serious disease, what, if anything, would you do about it? Explain.
3. Would you share the information you got from a genetic counselor or a genetic test with your doctor, family members, or the person you plan to marry? Explain.

After the discussion, have students write a private journal entry that describes what they would do in the event of finding out they will likely develop a serious disease or pass on a disease to their children.

Chapter Summary

Think Visually

USE GRAPHICS Have students create a concept map that uses the following terms: *nondisjunction, autosomes, sex chromosomes, Down syndrome, Turner's syndrome,* and *Klinefelter's syndrome.*

ANSWERS
Students' concept maps should show that inheritable disorders can be caused by nondisjunction of autosomes (in the case of Down syndrome) or sex chromosomes (in the case of Turner's syndrome and Klinefelter's syndrome).

BIOLOGY.com Study Online

REVIEW RESOURCES

Art Review Drag-and-drop items allow students to construct a karyotype.

Data Analysis Students analyze the connection between blood type O and a higher susceptibility to cholera.

InterActive Art Students can interact with pedigrees online.

Tutor Tube Short, online tutorial shows how pedigrees help students understand how a trait can "skip a generation."

Art in Motion Students watch an animation that shows how nondisjunction occurs during meiosis.

Vocabulary Review The *Flash Cards* and *Crossword* provide an interactive way to review chapter vocabulary.

Editable Worksheets Pages of Study Workbook B, Lab Manual B, and the Assessment Resources Book are available online. These documents can be easily edited using a word-processing program.

Lesson Overviews Lesson Overviews are based on Miller & Levine Biology. However, they are available on CD-ROM and can be edited for your class.

ASSESSMENT RESOURCES

Chapter Assessment Have students take an online version of the Chapter 14 Check Understanding.

Standardized Test Prep Prep Students can take an online version of the Standardized Test Prep. You will receive their scores along with ideas for remediation.

Diagnostic and Benchmark Tests Use these tests to monitor your students' progress and supply remediation.

ASSESSMENT

Assess

Big idea Invite a volunteer to read the Big Idea and the question. Write the words *karyotyping, pedigrees, gene sequencing,* and *bioinformatics* on the board.

Ask What do these words have in common? *(All are techniques for studying heredity.)*

Ask Which of these techniques requires the use of a microscope and a camera? *(karyotyping)*

Ask Which of these techniques require the use of a computer? *(gene sequencing and bioinformatics)*

Ask Which technique does not require the use of any special equipment? *(making pedigrees)*

Ask How do scientists collect and analyze information about human heredity? *(They use a variety of tools, such as microscopes and computers, and a variety of techniques, such as karyotyping, analyzing pedigrees, gene sequencing, and bioinformatics.)*

CONSTRUCTED RESPONSE Paragraphs will vary. Explain the activity carefully. Be sure students understand that the *Hints* should be used as a guide to help them answer the questions. Review how to organize the topics to include in a paragraph by making an outline.

For Question 1, write it on the board and read it aloud. Underline *sex-linked traits* and *inherited.*

Say This question asks you to explain how traits that appear more often in one sex than in the other are passed on to offspring.

Read the first hint.

Ask How many copies of the X chromosome does a male have? *(one)*

Read the next hint.

Say The Y chromosome, which causes an offspring to be male, has few genes and does not carry the same genes as the X chromosome carries.

Ask How many alleles for each sex-linked trait does a male carry? *(one)*

Model writing the first one or two sentences of a response using the hints on the board and in the text. *(Sample: Sex-linked traits are controlled by alleles on the X chromosome. Males have one X chromosome, and females have two.)*

14 CHECK Understanding

Assess the Big idea — Information and Heredity

Write an answer to the question below.

Q: How can we use genetics to study human inheritance?

Constructed Response

Write an answer to each of the questions below. The answer to each question should be one or two paragraphs. To help you begin, read the **Hints** below the questions.

1. **How are sex-linked traits inherited?**

   Hint Females have two copies of the alleles on the X chromosome.

   Hint Sex-linked traits are controlled by alleles on the X chromosome.

2. **How can a substitution mutation in a gene affect the ability of a cell to function properly?**

   Hint A substitution mutation replaces one DNA base with another base in a codon.

   Hint A change in a DNA base may change the amino acid coded for by the codon.

3. **Describe the tools and processes scientists use to manipulate DNA.**

   Hint Restriction enzymes cut DNA into short fragments at known locations.

   Hint Gel electrophoresis separates DNA fragments by their size.

Foundations for Learning Wrap-Up

Use the layered books you made as you read this chapter as tools to study and review the concepts of the chapter.

Activity 1 Exchange a layered book you made with a partner who made a similar book. Read the books aloud to each other and compare them. Add notes to each book, and correct any errors you find.

Activity 2 Look through your layered books and find one about the Human Genome Project or genetic disorders. With a partner, make notes on the book about either social or ethical problems that came up because of the Human Genome Project. Or make notes about a genetic advantage of having one of the genetic disorders.

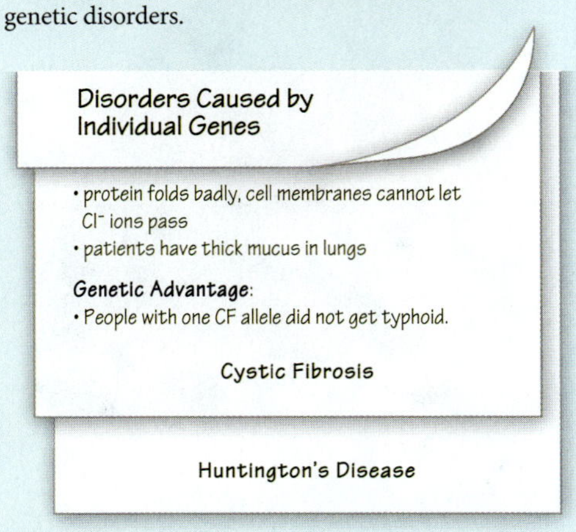

348 Check Understanding

For Question 2, pair students and have them use the hints and your modeling to answer the question. Have students share their paragraphs with the class.

For Question 3, have students write the paragraphs independently for homework or as an anchor activity.

Foundations for Learning Wrap-Up

Have students take out the layered books they prepared while reading the chapter. Make sure they have written supporting ideas for the main idea of each layered book. Explain the activities, and have students complete them.

14.1 Human Chromosomes

Understand Key Concepts

1. A normal human diploid zygote contains
   a. 23 chromosomes.
   b. 46 chromosomes.
   c. 44 chromosomes.
   d. XXY chromosomes.
2. A chart that traces the inheritance of a trait in a family is called a(n)
   a. pedigree.
   b. karyotype.
   c. genome.
   d. autosome.
3. An example of a trait that is determined by multiple alleles is
   a. cystic fibrosis.
   b. ABO blood groups.
   c. Down syndrome.
   d. colorblindness.
4. Could a person with blood type alleles I^A and I^B have blood type A? Explain your answer (refer to the table on page 335).

Think Critically

5. **Predict** If a male is born colorblind, what do we know about his mother?

14.2 Human Genetic Disorders

Understand Key Concepts

6. A mutation involving a change in a single DNA base pair
   a. will definitely result in a genetic disease.
   b. will have no effect on the organism's phenotype.
   c. will produce a positive change.
   d. may have an effect on the organism's phenotype.
7. Cystic fibrosis is caused by
   a. nondisjunction of an autosome.
   b. a small change of three base pairs in DNA.
   c. nondisjunction of a sex chromosome.
   d. deletion of an entire gene from a chromosome.
8. What is a chromosomal disorder?
9. Analyze the human karyotype below. Which chromosomal disorder does it show?

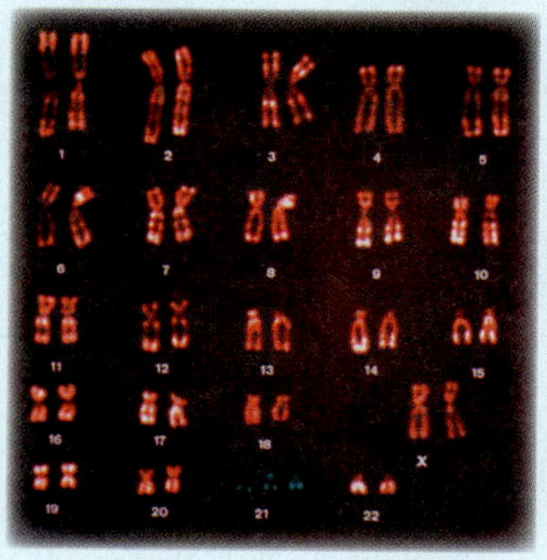

Think Critically

10. **Infer** Can a genetic counselor use a karyotype to identify a carrier of cystic fibrosis? Explain your answer.

14.3 Studying the Human Genome

Understand Key Concepts

11. The human genome consists of approximately how many DNA base pairs?
    a. 30,000
    b. 3,000,000
    c. 300,000,000
    d. 3,000,000,000
12. If you sequence short pieces of DNA and then use a computer to find overlapping sequences that map to a much longer DNA fragment, you are using
    a. genomics.
    b. hapmaps.
    c. shotgun sequencing.
    d. "sticky ends."

Answers

Lesson 14.1

UNDERSTAND KEY CONCEPTS

1. b

2. a

3. b

4. No, the I^A and I^B alleles are codominant. When both alleles are present in an individual, that person has blood type AB.

THINK CRITICALLY

5. Because the disorder is the result of a recessive allele on the X chromosome, the boy's mother is either a carrier of the gene for the disorder or she is colorblind.

Lesson 14.2

UNDERSTAND KEY CONCEPTS

6. d

7. b

8. A chromosomal disorder is a disorder caused by nondisjunction during meiosis, resulting in gametes with an abnormal number of chromosomes.

9. The karyotype shows trisomy of chromosome 21. The disorder is Down syndrome.

THINK CRITICALLY

10. No, cystic fibrosis is caused by a gene mutation. Karyotypes can detect abnormalities only in chromosome number.

Lesson 14.3

UNDERSTAND KEY CONCEPTS

11. d

12. c

Assessment Tips

Answer the first two questions as a class. Discuss why the answers are correct. Then ask volunteers to find information in the text that verifies the answer. Pair students and have them work together to answer the rest of the questions on this page and the next.

For Questions 6 and 7, suggest students read each answer choice along with the sentence stem to help them determine the correct answer.

Question 10 requires students to draw a conclusion based on information they have learned. Be sure students provide a reason for their answer to this question—not just yes or no.

Read aloud Question 13. Then have a volunteer read aloud the Test-Taking Tip. Be sure students understand the logic behind the elimination of each choice.

13. a

14. Answers should include descriptions of restriction enzymes, gel electrophoresis, and the use of colored labels in sequencing DNA.

THINK CRITICALLY

15. Scientists in the field of bioinformatics use computers to collect, organize, and interpret biological data.

Connecting Concepts

16. Humans require a Y chromosome for maleness.

17. no, because fruit fly males and females and human females do not have Y chromosomes

CHAPTER MYSTERY After students read the clue, build a pedigree on the board as you lead a class discussion of the inheritance of sickle cell disease.

Ask How many generations should the pedigree include? Explain. *(three—Ava's grandparents, their children, Uncle Eli and Ava's father, Uncle Eli's children, and the children of Ava's father and mother)*

Ask What symbols should be used to represent Ava's grandparents? *(a shaded circle for Ava's grandmother and an unshaded square for Ava's grandfather)*

Ask What symbols should be used to represent Uncle Eli and Ava's father? *(a shaded square for Uncle Eli and an unshaded square for Ava's father)*

Ask What can you conclude about the allele for hemoglobin that Ava's mother, Ava, and Ava's siblings inherited? *(All inherited at least one allele for normal hemoglobin but could also have inherited an allele for sickle cell disease.)*

14 CHECK Understanding

13. Cutting DNA into small pieces that can be sequenced is done by

a. restriction enzymes. **c.** gel electrophoresis.

b. DNA polymerase. **d.** RNA polymerase.

Test-Taking Tip

Choose Among Similar Answers Sometimes eliminating a known incorrect answer only helps a little. In question 13, we can eliminate **c** because it does not involve a cutting process. We also know that cutting is done with enzymes. But we still have three enzymes as possible answers. The names of the enzymes give us clues. We eliminate **d** because it refers to RNA. Since we know that polymerases build strands of DNA, this means that the only possible correct answer is **a.**

14. Describe the tools and processes that scientists use to manipulate human DNA.

Think Critically

15. Explain What is bioinformatics?

Connecting Concepts

Use Science Graphics

Use the data table to answer questions 16–17.

Chromosomes and Phenotypes

Sex Chromosomes	Fruit Fly Phenotype	Human Phenotype
XX	Female	Female
XY	Male	Male
X	Male	Female
XXY	Female	Male

16. Interpret Tables Which organism requires a Y chromosome for maleness?

17. Predict Do you think the genes on the Y chromosome are necessary for survival? Explain your answer.

solve the CHAPTER MYSTERY

THE CROOKED CELL

Ava looked into her family's medical history. She found out that Uncle Eli's mother (Ava's grandmother) also had sickle cell disease. However, Uncle Eli's father did not. One of Ava's uncle's four children also had the disease. But Ava's father, who is Eli's only sibling, does not have sickle cell disease. Ava's mother does not have sickle cell disease, either. Ava's two siblings also show no signs of the disease.

1. Apply Concepts Does sickle cell disease follow a simple dominance pattern of heredity? Give evidence from the chapter and its clues to support your answer.

2. Draw Conclusions Think about your answer to question 1. What can you conclude about the inheritance of sickle cell disease in Ava's family? What might be Ava's chances of being a carrier of the sickle cell trait?

3. Classify What kind of medical test could Ava request that would help determine whether or not she has the sickle cell trait? Explain your answer.

UntamedScience **Never Stop Exploring Your World.** Finding out about Ava's risk of sickle cell disease is only the beginning. Take a video field trip with the ecogeeks of Untamed Science to see where the mystery leads.

CHAPTER MYSTERY ANSWERS

1. Yes, it does. Students should recognize that the sickle cell alleles are inherited in a dominant-recessive pattern. Sickle cell disease is autosomal (not sex-linked) recessive. A person must inherit two recessive alleles for sickle cell disease to develop.

2. Ava's grandmother must be homozygous recessive for the trait. In order for Eli to have received two recessive alleles, he must have received one from his father. Therefore, Ava's grandfather must be heterozygous for the trait. Ava's father must be heterozygous for the trait as well. (He received a normal allele from his father and a recessive allele from his mother.) Assuming that Ava's mother is not a carrier, Ava has a 50% chance of being a carrier of the sickle cell trait.

3. Sample answer: Protein amino-acid sequencing of hemoglobin (specifically beta-globin) or DNA analysis of the gene that codes for that protein would show whether Ava is a carrier because the allele that causes sickle cell disease is known.

Untamed Science Suggest students follow the ecogeeks of Untamed Science in **What Color Are My Genes?** to find out how colorblindness is inherited.

Standardized Test Prep

Multiple Choice

1. Which of the following disorders can be observed in a human karyotype?
   A colorblindness
   B trisomy 21
   C cystic fibrosis
   D sickle cell disease

2. Which of the following disorders is a direct result of nondisjunction?
   A sickle cell disease
   B Turner's syndrome
   C Huntington's disease
   D cystic fibrosis

3. A woman is homozygous for A blood type. A man has AB blood type. What is the probability that the couple's child will have type B blood?
   A 0%
   B 50%
   C 75%
   D 100%

4. Cystic fibrosis is a genetic disorder caused by a
   A single base substitution in the gene for hemoglobin.
   B deletion of an amino acid from a chloride channel protein.
   C defective gene found on the X chromosome.
   D trisomy of chromosome 21.

5. The technique used to separate DNA strands of different lengths is
   A gel electrophoresis.
   B shotgun sequencing.
   C restriction enzyme digestion.
   D bioinformatics.

6. The study of whole genomes, including genes and their functions, is called
   A bioinformatics.
   B information science.
   C life science.
   D genomics.

7. DNA can be cut into shorter sequences by proteins known as
   A haplotypes.
   B polymerases.
   C restriction enzymes.
   D restriction fragments.

Questions 8–9

A student traced the recurrence of a widow's peak hairline in her family. Based on her interviews and observations, she drew the pedigree shown below.

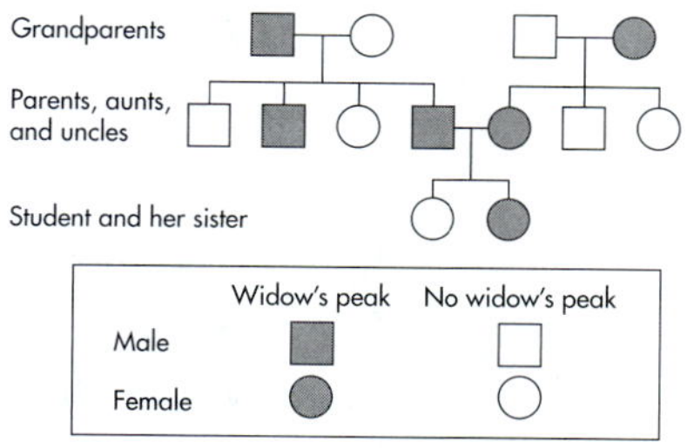

8. Which pattern of inheritance is consistent with the pedigree?
   A sex-linked inheritance
   B complete dominance
   C codominance
   D multiple alleles

9. What are the probable genotypes of the student's parents?
   A Mother—Ww; Father—ww
   B Mother—ww; Father—ww
   C Mother—WW; Father—Ww
   D Mother—Ww; Father—Ww

Open-Ended Response

10. Explain how the allele for sickle cell disease, which is a harmful allele when a person is homozygous, can be beneficial when a person is heterozygous.

If You Have Trouble With . . .

Question	1	2	3	4	5	6	7	8	9	10
See Lesson	14.1	14.2	14.1	14.2	14.3	14.3	14.3	14.1	14.1	14.2

Test-Taking Tip

READ ALL THE ANSWER CHOICES

Tell students to take the time to read each question completely, including all of the answer choices. They should consider each possible choice before determining which answer is the best.

Before the Test

Review with students the Test-Taking Tip.

Point out that the answers to Questions 8 and 9 can be found by studying the pedigree.

Remind students that *open-ended response* means that they'll have to write a paragraph to answer the question. Have them paraphrase the question in their notebooks, and remind them about the steps in writing a paragraph.

Answers

1. B
2. B
3. B
4. B
5. A
6. D
7. C
8. B
9. D
10. It is harmful in homozygous individuals because all of their hemoglobin has the defective form of the protein and the resulting defective red blood cells do not function normally, causing clots that can damage tissues. It is beneficial in heterozygous individuals because they are generally healthy and being carriers of the trait makes them resistant to the malaria parasite.

CHAPTER PLANNER 15 Genetic Engineering

Chapter Contents	NSES	Time	Core Resources
Chapter Preview			**Student Edition,** pp. 352–353 **Chapter Mystery,** p. 353 **Foundations for Learning,** p. 353
15.1 Selective Breeding Selective Breeding • Increasing Variation	III, IV, V, A.2.b, C.2.b, C.2.c, C.3.a, E.1, E.2, G.1, G.3	1/2 period 1/4 block	**Student Edition,** pp. 354–356 **Study Workbook B** 15.1 Worksheets **Biology.com** 15.1 Lesson Assessment
15.2 Recombinant DNA Copying DNA • Changing DNA • Transgenic Organisms	III, V, A.2.c, C.1.a, C.1.c, C.1.f, C.2.a, E.1, E.2, G.1, G.3	1 period 1/2 block	**Student Edition,** pp. 357–361 **Study Workbook B** 15.2 Worksheets **Biology.com** *Tutor Tube:* The Whys and Hows of Gel Electrophoresis • *Art in Motion:* Plasmid DNA Transformation • 15.2 Lesson Assessment
15.3 Applications of Genetic Engineering Agriculture and Industry • Health and Medicine • Personal Identification	III, V, A.1.c, C.1.a, C.1.c, C.1.d, C.2.a, E.1, E.2, G.1, G.3	1/2 period 1/4 block	**Student Edition,** pp. 362–366 **Inquiry Into Scientific Thinking,** p. 363 **Study Workbook B** 15.3 Worksheets **Biology.com** *Art Review:* Identifying Individuals • 15.3 Lesson Assessment
15.4 Ethics and Impacts of Biotechnology Profits and Privacy • Safety of Transgenics • Ethics of the New Biology	II, C.1.c, C.2.a, E.2, F.4, F.6	1/2 period 1/4 block	**Student Edition,** pp. 367–369 **Study Workbook B** 15.4 Worksheets **Biology.com** 15.4 Lesson Assessment
Chapter Pre-Lab	A.1.b, A.1.c, A.2.a, A.2.b, A.2.e, E.2	1 period 1/2 block	**Student Edition,** p. 370 **Lab Manual B** *Using DNA to Solve Crimes*

Tools for ELL Instruction

ELL Handbook describes teaching strategies to make *Miller & Levine Biology: Foundation Edition* more accessible to ELL students.

Spanish Study Workbook is a Spanish translation of Study Workbook A.

Multilingual Glossary is the *Miller & Levine Biology* glossary translated into ten languages.

Tools to Challenge Students

Study Workbook A includes more challenging worksheets and activities such as Chapter Mystery activities that reinforce 21st Century Skills.

Lab Manual A labs are geared toward on-level students. The Lab Manual includes fifteen labs in addition to the Chapter Labs.

Biology.com Data Analysis activities offer students a way to practice data gathering and analyzing skills.

Visual Quizzes in the Assessment Resources Book are geared toward on-level students and use visuals from *Miller & Levine Biology.*

Additional Resources

Biology.com Untamed Science Video • Flash Cards

Study Workbook A 15.1 Worksheets
Spanish Study Workbook 15.1 Worksheets

Study Workbook A 15.2 Worksheets
Spanish Study Workbook 15.2 Worksheets
Assessment Resources Book Visual Quiz

Study Workbook A 15.3 Worksheets
Lab Manual B *Data Analysis:* Genetically Modified Crops in the United States
Spanish Study Workbook 15.3 Worksheets
Assessment Resources Book Visual Quiz

Study Workbook A 15.4 Worksheets
Spanish Study Workbook 15.4 Worksheets
Biology.com *Data Analysis:* Who Sees Your DNA? Privacy and Genetic Data

Lab Manual A *Using DNA to Solve Crimes*

Chapter Review

Student Edition Chapter Summary, p. 371 • Unit Project, p. 376
Study Workbook B Chapter 15 Review
Transparencies, pp. 182–192
Biology.com Untamed Science Video • You're the Director • Editable Worksheets of Study Workbooks A and B and Lab Manuals A and B • Chapter 15 Flash Cards and Crossword

Untamed Science DVD • Classroom Resources CD (includes lesson presentations and editable worksheets)

Chapter Assessment

Student Edition Check Understanding, pp. 372–375
Study Workbook B Chapter 15 Taking a Standardized Test
Assessment Resources Book Chapter 15 Test B • Unit 4 Test B
Biology.com Chapter 15 Assessment • Editable Worksheets of Chapter 15 Visual Quizzes,* Chapter 15 Tests A and B, and Unit 4 Tests A and B

Exam*View* *Assessment Suite* • Classroom Resources CD (includes lesson presentations and editable worksheets)

Time: 1 period, 1/2 block

* Visual Quizzes are based on visuals from *Miller & Levine Biology.* Review carefully to make sure they are appropriate for your class.

Pressed for Time?

Preview the Chapter Introduce students to the Lesson 15.1 and 15.3 vocabulary.

Cover the Chapter Quickly Assign students to read *Increasing Variation* in Lesson 15.1 and all of Lesson 15.3.

Assess Assign Questions 2 and 3 in the 15.1 Check Understanding, the 15.3 Check Understanding, and Questions 4, 10, and 11 in the Chapter 15 Check Understanding.

Get the Picture

Have students compare the cats in the photograph and read the caption. Explain that the researchers who developed the cat that glows in the dark hope experiments like this will help scientists better understand genetic diseases in animals and humans. Use the photo as a discussion prompt to access students' prior knowledge about genetic engineering. Allow students to express their opinions about genetic engineering as well.

Ask How do you think the researchers were able to modify the cat's normal DNA to make the cat glow red in the dark? *(Sample answer: The researchers placed a "glowing" gene into the cat's DNA.)*

Ask How could this technology be used to help people? *(Answers may vary. Students might suggest this could be used to cure diseases or grow crops.)*

Activate Prior Knowledge

Review what students have learned about the connection between a *genotype* and *phenotype*.

Say You know that the order of bases in an organism's DNA forms a genetic code.

Ask How does the genetic code in DNA influence an organism's traits? *(The genetic code specifies what proteins are produced in the organism's cells. Proteins control the chemical reactions in cells that determine the organism's traits.)*

Ask Why would scientists want to make changes to an organism's DNA? *(to change one or more of the organism's traits)*

Ask What are some situations where it might be useful to change an organism's traits? *(Sample answer: inventing new kinds of crops)*

NATIONAL SCIENCE EDUCATION STANDARDS

UNIFYING CONCEPTS AND PROCESSES

II, III, IV, V

CONTENT

C.1.a, C.1.c, C.1.d, C.1.f, C.2.a, C.2.b, C.2.c, C.3.a, E.1, E.2, F.4, F.6, G.1, G.3

INQUIRY

A.1.b, A.1.c, A.2.a, A.2.b, A.2.c

15 Genetic Engineering

Big idea Science as a Way of Knowing

Q: How and why do scientists manipulate DNA in living cells?

By cloning cells and changing genes, scientists in Korea have given house cats the ability to glow in the dark. The cloned cat on the left has a fluorescent protein that turns red when exposed to UV light in a dark room. The cloned cat on the right lacks the red fluorescent protein, so it does not give off a red glow.

BIOLOGY.com Search Chapter 15 GO • Flash Cards

UbD Connect to the Big Idea

Invite a volunteer to read aloud the chapter number and the chapter title. Read the Big Idea and question. Emphasize the question, **How and why do scientists manipulate DNA in living cells?**

Set the Purpose

PREVIEW AND PREDICT Draw the graphic organizer on the board. Guide students to the INSIDE feature on the next page, and invite a volunteer to read each lesson number and title.

Ask How do you think the lesson titles relate to the questions in the graphic organizer?

Remind students to search for answers to the questions as they read and to also ask themselves why scientists manipulate the DNA in living cells.

INSIDE:

- 15.1 Selective Breeding
- 15.2 Recombinant DNA
- 15.3 Applications of Genetic Engineering
- 15.4 Ethics and Impacts of Biotechnology

CHAPTER
MYSTERY

A CASE OF MISTAKEN IDENTITY

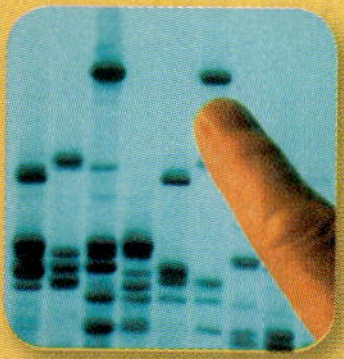

In the summer of 1998, an elderly Indiana woman was brutally attacked. It was dark, and she couldn't see who hurt her. Soon afterward, police found a man only a few blocks from the victim's house. He was unconscious. His clothing was stained with blood. There were also scratches on his forearms. The man said that he had passed out following a drunken fight. He couldn't remember what had happened afterward. The blood type of the stains on his clothing matched the victim's blood type. The police thought they had their man.

Hours later, the police knew they had the wrong suspect. They began to search again for the real attacker, who was later caught, tried, and convicted.

Read for Mystery Clues As you read this chapter, look for clues to help you determine how the police knew they had the wrong suspect. Then, solve the mystery.

FOUNDATIONS for Learning

Before you read the chapter, make two six-sided cubes. First, draw a flat cube in the shape below. Then, cut the paper and fold it into a cube. As you read the chapter, write the vocabulary words on the sides of the cubes. Write one vocabulary word per side. At the end of the chapter are two activities that use the word cubes to help you answer the question: How and why do scientists manipulate DNA in living cells?

- Untamed Science Video
- Chapter Mystery

Chapter 15
Big Idea: Science as a Way of Knowing

Chapter 15: How and why do scientists manipulate DNA in living cells?

- **15.1:** How do humans take advantage of naturally occurring variation among organisms?
- **15.2:** How do scientists study and work with specific genes?
- **15.3:** How do humans use genetic engineering?
- **15.4:** What are some of the ethical issues raised by genetic engineering?

Foundations for Learning

Pass out paper and scissors for the word cubes. Help students understand the value of this activity.

Ask Why do you think understanding vocabulary words is important? *(Understanding vocabulary is essential to reading comprehension.)*

Ask How can writing the words on the paper cubes help you review the definitions? *(Sample answers: Writing the words helps me remember them. When I roll the cubes, I can review the words randomly.)*

Ask How could you use the two cubes to review the relationship between vocabulary words? *(You could roll two cubes and then explain how the two face-up terms are related to each other.)*

Remind students that they will use these cubes as a review and study aid.

Untamed Science™ Watch and see how pigeon breeding helps the Untamed Science crew unravel the mysteries of genetic engineering in the video **Designer Pigeons.**

CHAPTER MYSTERY Invite a volunteer to read aloud the Chapter Mystery. Have students guess how the mystery relates to the chapter title and predict how police determined the first suspect was not the real attacker.

Tell students they will find clues in the chapter to answer the question, **How did the police know they had the wrong suspect?**

Digital Support

BIOLOGY.com Extend your reach by using the digital assets offered at **Biology.com.**

LESSON 15.1

Pre-Reading

Objectives

15.1.1 Explain the purpose of selective breeding.

15.1.2 Explain how people increase genetic variation.

Preview the Pages

1. Use the photographs of different dog breeds as a springboard for a discussion to find out what students already know about genetic variation and selective breeding.
2. Review the vocabulary words by having students find the words and their definitions in the text. Ask students where they may have heard the words before.
3. Preview the Ploidy Numbers table at the end of the lesson. Remind students that the letter *N* represents one set of chromosomes. Review that human gametes are haploid, or N, while human body cells are diploid, or 2N. Point out that some organisms have more than two complete sets of chromosomes, as shown in the table.

Build Understanding

KWL CHART Direct students to Workbook B where they'll find a partially completed KWL chart. Remind students that after reading the text, they should check the K column for errors. Have them rewrite any incorrect statements.

NATIONAL SCIENCE EDUCATION STANDARDS

UNIFYING CONCEPTS AND PROCESSES

III, IV, V

CONTENT

C.2.b, C.2.c, C.3.a, E.1, E.2, G.1, G.3

INQUIRY

A.2.b

For the corresponding lesson in *Miller & Levine Biology,* see pages 418–420.

15.1 Selective Breeding

Key Questions

What is selective breeding used for?

How do people increase genetic variation?

BUILD Understanding

KWL Chart Use a KWL chart to organize your notes for this lesson. Before you begin reading, write down what you already know (K) about selective breeding. Write what you want to know (W) in the chart. Then after you read the lesson, go back and add notes about what you have learned (L).

In Your Workbook Go to your workbook to see an example of a KWL chart. Complete the KWL chart for Lesson 15.1.

Selective Breeding

You've had popcorn at the movies. You've probably made it at home, too. Where does it come from? Believe it or not, popcorn was one of the first organisms humans improved for our own benefit. Corn as we know it was first grown as a crop at least 6000 years ago by Native Americans living in what is now Mexico. A tiny kernel of popped corn found in a cave in New Mexico is more than 5000 years old!

Humans have bred many organisms for our own benefit. Visit a dog show, and what do you see? You see dogs of all shapes and sizes. There are tiny Chihuahuas and huge Great Danes. There are Labrador retrievers with short coats and poodles with curly hair. You can see wolfhounds with long noses and bulldogs with short noses. The differences among breeds of dogs are so great that someone might think they are different species. They're not, of course, but where did these big differences come from?

The answer is selective breeding. Humans have kept and bred dogs for thousands of years. Humans are always looking to make animals that are better hunters, better retrievers, or better companions. We've done so by **selective breeding,** letting only those animals with desired characteristics produce the next generation.

Dog Breeds There are more than 150 dog breeds. New breeds are still being developed.

354 Chapter 15 • Lesson 1

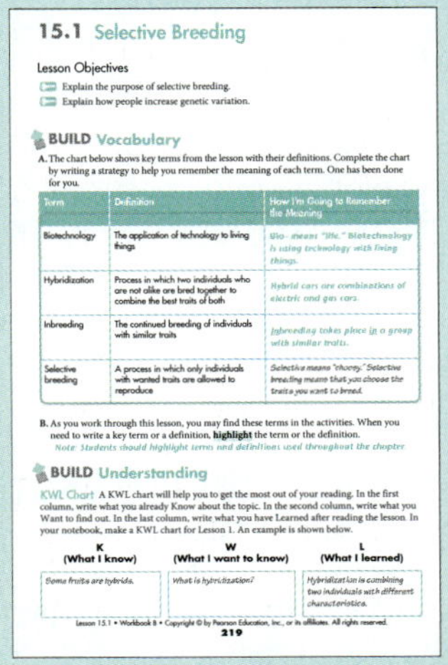

15.1 Selective Breeding

Lesson Objectives

Explain the purpose of selective breeding.

Explain how people increase genetic variation.

BUILD Vocabulary

A. The chart below shows key terms from the lesson with their definitions. Complete the chart by writing a strategy to help you remember the meaning of each term. One has been done for you.

Term	Definition	How I'm Going to Remember the Meaning
Biotechnology	The application of technology to living things	*Bio- means "life." Biotechnology is using technology with living things.*
Hybridization	Process in which two individuals who are not alike are bred together to combine the best traits of both	*Hybrid cars are combinations of electric and gas cars.*
Inbreeding	The continued breeding of individuals with similar traits	*Inbreeding takes place in a group with similar traits.*
Selective breeding	A process in which only individuals with wanted traits are allowed to reproduce	*Selective means "choosy." Selective breeding means that you choose the traits you want to breed.*

B. As you work through this lesson, you may find these terms in the activities. When you need to write a key term or a definition, **highlight** the term or the definition.
Note: Students should highlight terms and definitions used throughout the chapter.

BUILD Understanding

KWL Chart A KWL chart will help you to get the most out of your reading. In the first column, write what you already Know about the topic. In the second column, write what you Want to find out. In the last column, write what you have Learned after reading the lesson. In your notebook, make a KWL chart for Lesson 1. An example is shown below.

K (What I know)	W (What I want to know)	L (What I learned)
Some fruits are hybrids.	*What is hybridization?*	*Hybridization is combining two individuals with different characteristics.*

Lesson 15.1 • Workbook B • Copyright © by Pearson Education, Inc., or its affiliates. All rights reserved.
219

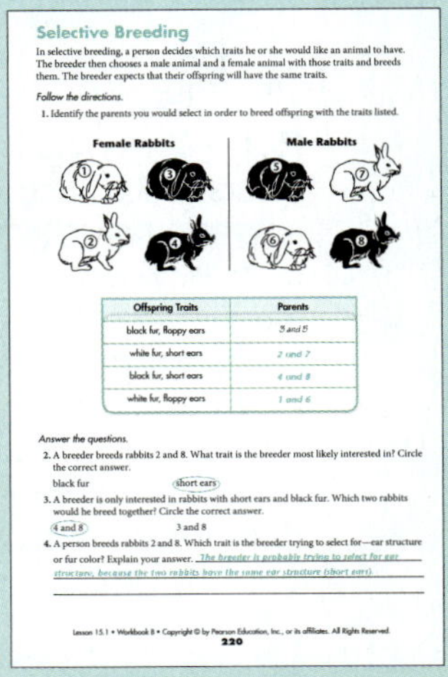

Selective Breeding

In selective breeding, a person decides which traits he or she would like an animal to have. The breeder then chooses a male animal and a female animal with those traits and breeds them. The breeder expects that their offspring will have the same traits.

Follow the directions.

1. Identify the parents you would select in order to breed offspring with the traits listed.

Female Rabbits — Male Rabbits

Offspring Traits	Parents
black fur, floppy ears	*3 and 5*
white fur, short ears	*2 and 7*
black fur, short ears	*4 and 8*
white fur, floppy ears	*1 and 6*

Answer the questions.

2. A breeder breeds rabbits 2 and 8. What trait is the breeder most likely interested in? Circle the correct answer.
black fur — short ears (circled)
3. A breeder is only interested in rabbits with short ears and black fur. Which two rabbits would he breed together? Circle the correct answer.
4 and 8 (circled) — 3 and 8
4. A person breeds rabbits 2 and 8. Which trait is the breeder trying to select for—ear structure or fur color? Explain your answer. *The breeder is probably trying to select for ear structure, because the two rabbits have the same ear structure (short ears).*

Lesson 15.1 • Workbook B • Copyright © by Pearson Education, Inc., or its affiliates. All Rights Reserved.
220

For thousands of years, we have produced new varieties of plants and animals. We did this by selectively breeding for particular traits. Long before Europeans came to the New World, Native Americans had selectively bred teosinte (tee oh SIN tee). Teosinte is a wild grass native to central Mexico. Teosinte was selectively bred to make corn. Corn is a far more productive and nutritious plant than teosinte. Corn is also now one of the world's most important crops. The two most common methods of selective breeding are hybridization and inbreeding.

Hybridization American botanist Luther Burbank may have been the greatest selective breeder of all time. He developed more than 800 kinds of plants. As one of his tools, Burbank used hybridization. **Hybridization** is crossing organisms that have different traits to bring together the best of both organisms. The offspring made by such crosses are called hybrids. Hybrids are often hardier than either parent. Many of Burbank's hybrid crosses mixed the disease resistance of one plant with the food-making ability of another. The results were new lines of plants that had the traits farmers needed to make more food.

Inbreeding Breeders often use a method called inbreeding to keep desired characteristics in a line of organisms. **Inbreeding** is the continued breeding of individuals with similar characteristics. Many breeds of dogs are inbred. Inbreeding helps keep the characteristics that make each breed unique and helps make sure that those characteristics are preserved. Although inbreeding may seem like a good way to keep certain characteristics, it's risky business. Inbred organisms are genetically very similar. Therefore, inbreeding increases the probability that organisms may inherit alleles that lead to genetic disorders. For example, inherited hip problems are common to many breeds of dogs.

Key Question What is selective breeding used for?
Selective breeding takes advantage of naturally occurring genetic variation. Humans use selective breeding to pass wanted traits on to the next generation of organisms.

Increasing Variation

The great variety found in wild plants and animals makes selective breeding possible. But sometimes breeders want more variation than they find in the wild. Breeders can add to the genetic variation in a population by introducing mutations.

When scientists change the genetic makeup of an organism, they are using biotechnology. **Biotechnology** is the use of a technological process, invention, or method on living organisms. Selective breeding is one form of biotechnology important in farming and medicine, but there are many others.

Corn From Teosinte Modern corn was selectively bred from teosinte at least 6000 years ago. Over time, the hard case around the kernel disappeared, which left the rows of soft kernels that we enjoy today.

BUILD Vocabulary

selective breeding
a method of breeding that allows only those organisms with desired characteristics to produce the next generation

hybridization
a breeding technique that involves crossing dissimilar individuals to bring together the best traits of both organisms

inbreeding
the continued breeding of individuals with similar characteristics to maintain the derived characteristics of a kind of organism

biotechnology
the process of manipulating the organisms, cells, or molecules to produce specific products

PREFIXES
Remember that the prefix *bio-* refers to living things. Also remember that *technology* refers to scientific tools. Biotechnology uses the tools of science to study and alter living things.

ELL Focus on ELL: Extend Language

ALL SPEAKERS Write the words *quickly, selectively, genetically,* and *recently* on the board.

Say All these words are adverbs. An adverb tells how, when, or where something happens. Adverbs modify, or tell about, verbs; they may appear before or after a verb. Many, but not all, adverbs end in *-ly.*

Present the following sentences to students. Have them identify the adverb in each sentence and explain whether it tells how, when, or where.

Inbred organisms are genetically *alike. (how)*

Teosinte was selectively *bred to make corn. (how)*

Until recently, *plant and animal breeders worked with variations. (when)*

Remind students that some adjectives, such as the word *friendly,* can also end in *-ly,* so they must pay attention to the word's role in the sentence.

During Reading

Active Reading

USE VISUALS Direct students to the pictures of teosinte and modern corn, and read the caption.

Ask How is modern corn different from teosinte? *(Sample answers: Corn plants produce more food. Corn kernels are softer and bigger, and yellow instead of green. Corn is more nutritious.)*

Ask Teosinte belonged to the grass family. To what family does corn belong? *(grass)*

USE GRAPHIC ORGANIZERS Tell students that there are positive and negative consequences to selective breeding, hybridization, and inbreeding. Model for students how to build a T-chart listing the advantages and disadvantages of inbreeding.

advantages	disadvantages
keeps unique traits	*increases chance of disorder*

Have students copy the chart in their notebooks and make similar charts for selective breeding and hybridization.

Build Vocabulary

PREFIXES Remind students that they can use prefixes they already know to figure out the meaning of new words.

Ask What do you think the word *biofuel* means? *(fuel made out of living things, such as plants)*

Ask What do you think the word *biohazard* means? *(living things, such as microorganisms, that pose a hazard to human health)*

KEY QUESTION Have a volunteer read the Key Question and answer.

Ask What is a trait? *(A trait is a feature determined by the genes of an organism.)*

Ask What is selective breeding? *(a way of breeding that uses only organisms with desired traits to produce the next generation)*

Speed Bump

Ask Why do humans use selective breeding? *(to produce plants or animals that have a desired set of traits)*

Ask What are some advantages of hybrid plants? *(They may be hardier, disease resistant, and more productive.)*

During Reading continued

Active Reading

USE VISUALS Help students decode the information in the Ploidy Numbers table.

Say We know that gametes are haploid, meaning that they have a single set of unpaired chromosomes. The haploid number, *N*, is the number of chromosomes in one chromosome set. Domestic oats have 7 chromosomes per set.

Say We also know that most cells are diploid, meaning that they have two sets of chromosomes, or 2N.

Ask What is the diploid number, 2N, for domestic oats? How did you get it? *(14; 2 × 7 = 14)*

Say Some organisms have more than two sets of chromosomes in their cells, a condition known as polyploidy. In the table, the chromosome number is the total number of chromosomes in a cell. For domestic oats that number is 42.

Ask How many sets of chromosomes does a polyploid oat cell have? *(6; 42 ÷ 7 = 6)*

Ask How many total chromosomes does a 3N banana plant cell have? *(33; 3 × 11 = 33)*

KEY QUESTION Invite a student to read the Key Question. Ask a second student to read the answer.

Ask What is genetic variation? *(having different genetic characteristics among members of a species)*

Ask What are mutations? *(changes in DNA)*

Speed Bump

Ask A mule is produced by crossing a female horse and a male donkey. What type of selective breeding is this? *(hybridization)*

Ask Why would farmers wish to grow polyploidy plants? *(Polyploids are often larger and stronger than normal diploid plants.)*

Post Reading

ASSESSMENT TIPS Review the highlighted vocabulary words with students before they begin the assessment.

Wrap-Up Activity

MAKE A VOCABULARY WORD MAP Direct students to make a vocabulary word map in their notebooks using the term *Selective Breeding* in the center oval. Students should include the terms *hybridization* and *inbreeding*. They can complete the map by listing details of each process.

Bacterial Mutations Mutations are changes in DNA that parents can pass to offspring. Mutations often happen on their own. However, breeders can also make mutations happen using radiation and chemicals. Many mutations hurt or kill the organism. Some do not. With luck and patience, breeders can grow mutants that have new, helpful characteristics. This way of making mutants has been very useful with bacteria. Because they are small, millions of bacteria can be treated with radiation or chemicals at the same time. Since so many bacteria are treated, the chances of making a useful mutant are good. This technique has let scientists develop hundreds of useful kinds of bacteria. For example, scientists are working to make bacteria that can clean up radioactive substances and metal pollution.

Polyploid Crops

Plant	Probable Ancestral Haploid Number	Chromosome Number	Ploidy Level
Domestic oat	7	42	6N
Peanut	10	40	4N
Sugar cane	10	80	8N
Banana	11	22, 33	2N, 3N
Cotton	13	52	4N

Ploidy Numbers Polyploid plants are often larger than other plants. So, farmers grow many polyploid varieties of crops on purpose. This table shows some polyploid crops.

Polyploid Plants Drugs that keep chromosomes from separating during meiosis have been very useful in plant breeding. These drugs can make cells that have many times the normal number of chromosomes. Plants grown from these cells are called polyploid because they have many sets of chromosomes. Polyploidy is rare in animals. Plants, however, are much more likely than animals to survive with extra sets of chromosomes. Polyploidy can quickly make new kinds of plants that are larger and stronger than their diploid relatives. Scientists have made some important crop plants, including bananas and many varieties of citrus fruits, in this way.

Key Question How do people increase genetic variation?

Breeders can increase the genetic variation in a population by introducing mutations, which are the ultimate source of biological diversity.

CHECK Understanding

Apply Vocabulary

Use the highlighted words from the lesson to complete each sentence correctly.

1. ________ is the process of crossing two individuals that are not genetically alike.

Critical Thinking

2. **Explain** How can breeders introduce mutations?
3. **Review** What is the relationship between genetic variation and mutations?
4. **Apply Concepts** Suppose you are a geneticist trying to develop a sunflower with red petals and a short stem. As you compare the sunflowers you have on hand, what genetic variations would you look for?
5. **Write to Learn** Answer the following question in two or three sentences: How is selective breeding a kind of biotechnology?

BIOLOGY.com Search Lesson 15.1 GO • Lesson Assessment

Check Understanding

1. Hybridization
2. Breeders can use radiation and chemicals to introduce mutations.
3. Mutations are the ultimate source of genetic variation. The more mutations there are within a population, the greater the genetic variation.
4. You would look for reddish petals and shorter stems.
5. Sample answer: Biotechnology is the use of a technological process or tool on living organisms. Because selective breeding is a technological process that helps humans breed organisms with desired traits, it is a kind of biotechnology.

15.2 Recombinant DNA

Copying DNA

Suppose you have an electronic game you want to change. If the game depends on a coded program in a computer microchip, you'd need to get the existing program out of the microchip. Then, you'd have to read the program and change it to your new program code. Finally, you would put the changed code back into the microchip. Believe it or not, this is very similar to genetic engineering.

How do you change an organism? Until recently, plant and animal breeders had to work with variations that already existed in nature. Now, scientists can transfer genes for particular traits from one organism to another. How does this gene transfer happen? Scientists first take DNA from cells. The DNA is then cut into pieces using restriction enzymes. Restriction enzymes cut DNA at specific base sequences. From a large DNA molecule, these enzymes may produce hundreds or even thousands of pieces. These pieces can then be separated by size.

If we were to cut DNA from the human genome, we might end up with millions of pieces of DNA. How do we find the DNA of a single gene among millions of pieces? It is kind of like finding a needle in a haystack. There is a huge pile of hay and one needle. Actually, you can find a needle in a haystack. You can toss the hay in front of a powerful magnet until something sticks. The hay won't stick, but a needle made of iron or steel will. Believe it or not, this method is similar to the way scientists find specific genes.

Finding Genes Douglas Prasher, a biologist, wanted to find a specific gene in a jellyfish. The gene codes for a molecule called green fluorescent protein, or GFP. This natural protein in the jellyfish absorbs energy from light and makes parts of the jellyfish glow.

To find the gene, Prasher studied the amino acid sequence of part of the GFP protein. Then, he figured out which mRNA sequence would make that protein. That mRNA sequence told him the DNA sequence of the gene. By making a piece of RNA that would match that DNA sequence, Prasher made the DNA "magnet" to find the GFP gene.

Key Questions

How do scientists copy the DNA of living organisms?

How is recombinant DNA used?

How can genes from one organism be inserted into another organism?

BUILD Understanding

Preview Visuals Before you read, preview the PCR Method figure. Write down any questions you have about the figure. As you read, find answers to your questions.

In Your Workbook Go to your workbook for help with previewing visuals.

Fluorescent Gene The Pacific Ocean jellyfish gives off a bluish glow. A protein in the jellyfish absorbs the blue light and makes green fluorescence. This protein, called GFP, is now used in genetic engineering.

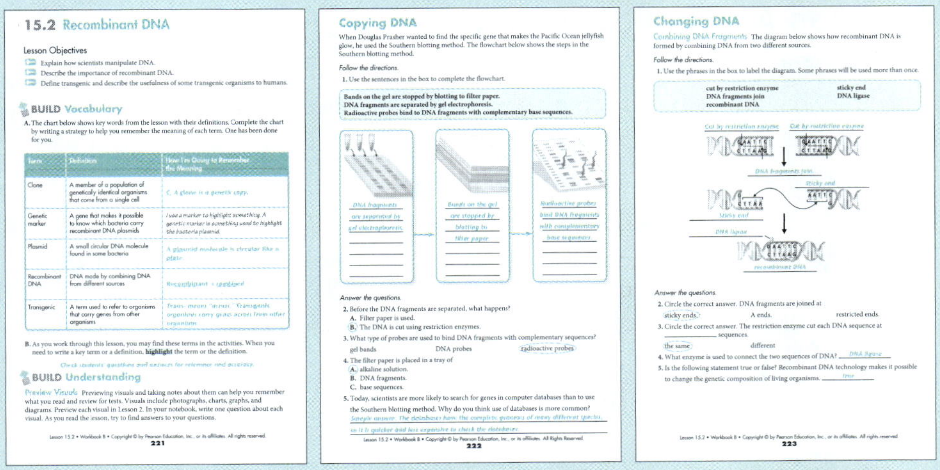

LESSON 15.2

Pre-Reading

Objectives

15.2.1 Explain how scientists manipulate DNA.

15.2.2 Describe the importance of recombinant DNA.

15.2.3 Define *transgenic* and describe the usefulness of some transgenic organisms to humans.

Preview the Pages

1. Preview the visuals. Point out that several of them include arrows between images. Explain that each of these visuals shows a step-by-step procedure. Remind students that it is important to keep the events in the correct sequence.
2. Preview the highlighted vocabulary words. Encourage students to check the definitions of these words as they encounter them in the text.
3. Remind students of procedures used to obtain fragments of DNA for study. Guide a discussion about using restriction enzymes to cut DNA and gel electrophoresis to separate DNA fragments.

Build Understanding

PREVIEW VISUALS Direct students to Workbook B where they will find instructions on how to preview visuals in this lesson.

NATIONAL SCIENCE EDUCATION STANDARDS

UNIFYING CONCEPTS AND PROCESSES

III, V

CONTENT

C.1.a, C.1.c, C.1.f, C.2.a, E.1, E.2, G.1, G.3

INQUIRY

A.2.c

For the corresponding lesson in *Miller & Levine Biology*, see pages 421–427.

LESSON 15.2

During Reading

Active Reading

MAKE CONNECTIONS Review what an analogy is. *(a comparison that shows similarities between two different things)* Then, read aloud the first three paragraphs under Copying DNA on the previous page and have students search for analogies.

Ask What analogy does the text make in the first paragraph? *(The text compares genetic engineering to reprogramming a computer microchip.)* How does this analogy help you understand genetic engineering? *(Sample answer: It makes me realize that genetic engineering is really about changing the genetic information that controls the characteristics of a living thing.)*

Lead a similar discussion about the "needle in a haystack" analogy in the third paragraph.

INTERPRET DIAGRAMS Direct students' attention to the diagram of the PCR method on this page.

Ask What do the arrows and the numbers in this diagram tell you? *(Sample answers: PCR is a process. The sequence of steps is important.)*

Ask What do the colors red and blue indicate? *(Blue indicates an original strand of DNA; red indicates a primer or a newly copied DNA strand.)*

MAKE INFERENCES Point out that the number of DNA copies multiplies with each cycle of PCR. Write the following information on the board.

Number of PCR cycles	1	2	3	4	5
Number of DNA copies	2	4	8		

Ask How many copies of the DNA will exist after five PCR cycles? *(32)* after six cycles? *(64)*

KEY QUESTION Invite a volunteer to read aloud the Key Question and answer.

Say The answer to this Key Question has four steps.

Ask How does each sentence in the answer compare to each of the four steps in the PCR diagram? *(They match each other. The diagram provides the answer.)*

Speed Bump

Ask How did Prasher make a "tool" to identify the GFP gene? *(He looked at the amino acid sequence of the protein and then made a piece of RNA that would match the gene's DNA sequence.)*

Ask For what purpose do researchers use PCR? *(to make copies of a desired gene)*

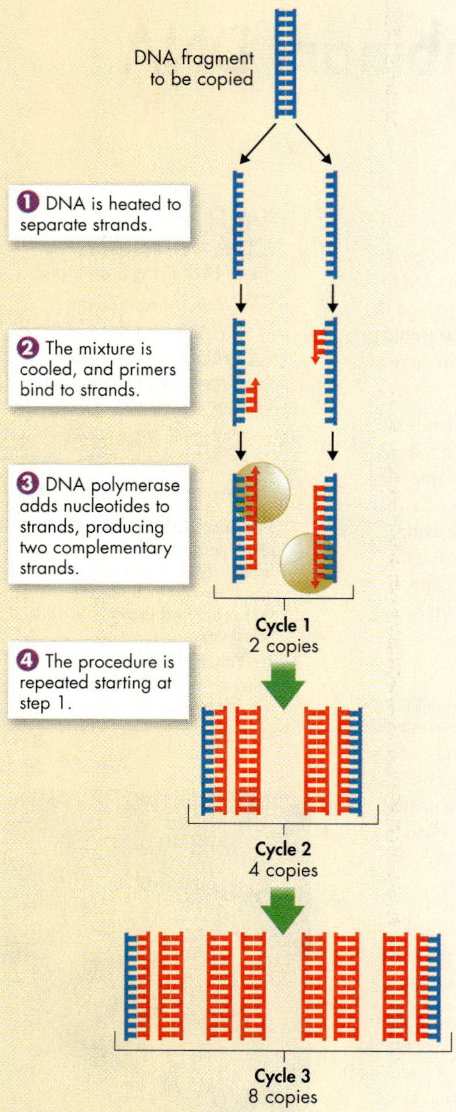

The PCR Method Polymerase chain reaction is used to make multiple copies of a gene. This method is useful even when only tiny amounts of DNA are available.

Polymerase Chain Reaction Once they find a gene, biologists often need to make many copies of it. To do this, biologists use a technique known as **polymerase chain reaction (PCR).** At one end of the original piece of DNA, a biologist adds a short piece of DNA that complements a portion of the sequence. At the other end, the biologist adds another short piece of complementary DNA. These short pieces are known as primers because they prepare, or prime, a place for DNA polymerase to start working.

The first step in using the PCR method to copy a gene is to use heat to separate the strands of a piece of DNA. Then, as the DNA cools, primers bind to the single strands. Next, DNA polymerase starts copying the DNA between the primers. These copies become templates used to make even more copies of the gene. In this way, just a few dozen cycles of copying the DNA can make billions of copies of a gene.

Key Question How do scientists copy the DNA of living organisms?
The first step in using the polymerase chain reaction method to copy a gene is to heat a piece of DNA, which separates its two strands. Then, as the DNA cools, primers bind to the single strands. Next, DNA polymerase starts copying the region between the primers. These copies can serve as templates to make more copies.

Changing DNA

Do you remember reading about Griffith's experiments on bacterial transformation? During transformation, a cell takes in DNA from outside the cell. Then that added DNA becomes part of the cell's own genome. Today, biologists know that Griffith's mixture of heat-killed bacteria had pieces of DNA in it. When he mixed those pieces of DNA with live bacteria, a few bacteria took up the DNA. Those bacteria cells changed. Griffith could do this only with DNA taken from other bacteria.

Combining DNA Fragments Today, scientists can make the DNA for any gene whose sequence they know. They can then put those genes into living cells. Changing genes for a practical purpose is called genetic engineering. How do scientists do this?

Focus on Academic Language

• recombinant

Write the word on the board and pronounce it. Then circle *combin* and write *combine*.

Say *Recombinant* comes from *combin,* which means "join together."

Ask What is being joined together in recombinant DNA? *(pieces of DNA)*

Explain that to recombine means to take one or more things that were part of something else and combine them in a new way. Be sure students understand that the DNA pieces come from different sources and are recombined to make a new complete DNA molecule.

First, they build a DNA sequence with the gene or genes they want to put into a cell. Machines known as DNA synthesizers can make short pieces of DNA. These pieces made in a lab can then be joined to natural pieces of DNA using enzymes such as DNA ligase (LY gays). These enzymes join DNA molecules together. These same enzymes make it possible to take a gene from one organism and attach it to the DNA of another organism. DNA molecules made this way are called **recombinant DNA.** This technology works because any pair of complementary DNA sequences is likely to bond. They will bond even if each sequence comes from a different organism. Recombinant-DNA technology makes it possible to change the genetic makeup of living organisms. By working with DNA in this way, scientists can study how genes are put together and how they work.

Key Question How is recombinant DNA used?
Recombinant-DNA technology—joining together DNA from two or more sources—makes it possible to change the genetic composition of living organisms.

Plasmids and Genetic Markers Scientists working with recombinant DNA soon learned a few useful tricks. In addition to their own chromosomes, some bacteria have small circular DNA molecules called **plasmids.** One reason plasmids are useful is that they have their own replication "start" signals. That way, when the cell copies its own DNA, it copies the plasmid, too.

Plasmids are often used in recombinant DNA studies. Scientists insert a desired piece of DNA into a plasmid. Then they use transformation to add the recombinant plasmid to bacteria. The bacteria will then copy the new DNA along with the rest of the cell's genome.

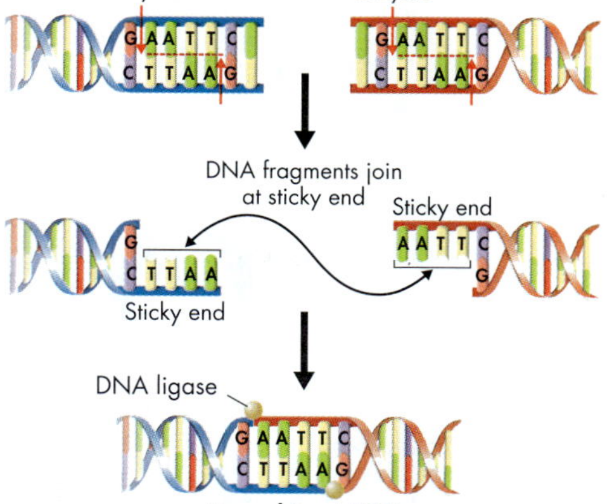

Joining DNA Pieces Together Recombinant DNA molecules are made up of DNA from different sources. Restriction enzymes cut DNA at specific sequences, making "sticky ends." Sticky ends are single-stranded bases of DNA at the end of a piece of DNA. DNA ligase allows two single-stranded pieces that are complementary to each other to stick together.

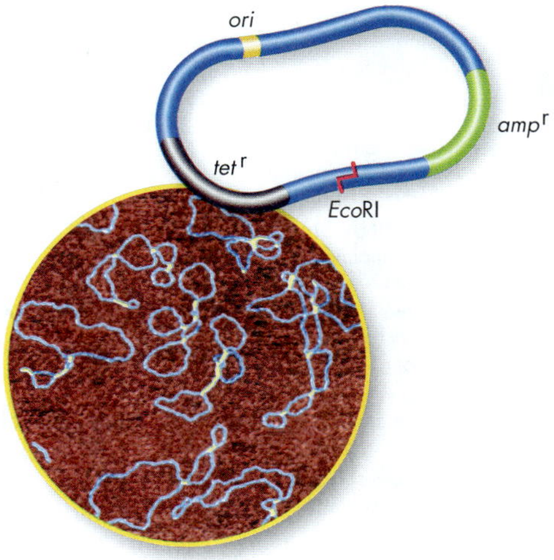

A Plasmid Map Plasmids used for genetic engineering have a start signal, called the origin of replication (*ori*). They have a restriction enzyme cutting site, such as *Eco*R1. They also have genetic markers, like the antibiotic resistance genes *tet*^{r} and *amp*^{r} shown here.

Hands-On Learning

MODELING RECOMBINANT DNA Direct students' attention to the visuals on this page and the next showing how DNA pieces join together and how plasmid DNA is transformed. Point out that in both cases, restriction enzymes are used to produce "sticky ends."

Provide scissors and construction paper in two different colors and ask pairs of students to use the materials to model the making of recombinant DNA. *(A model might show a circular recombinant DNA molecule like the one depicted on the opposite page, formed from two pieces of construction paper cut with complementary ends.)*

Ask How did you get the pieces of DNA from different sources to join together? *(The ends were cut with complementary overhangs.)* How do scientists make sure that two sticky ends will fit together? *(They use the same restriction enzyme.)*

Ask students to manipulate their models to show the steps as they read through the steps and captions for each visual.

Active Reading

SELF-QUESTION Demonstrate how asking questions and reviewing the text for answers can clarify information. Read aloud the text with the heading Combining DNA Fragments.

Ask Does this make sense? How can scientists make DNA for a certain gene and put it into living cells?

Ask Where in the paragraph can I find an answer to my question? *(in the sentences about DNA ligase)*

Ask What do these sentences tell me about how enzymes such as DNA ligase help build DNA and put it into living cells? *(DNA pieces made in a lab are attached to natural pieces of DNA using enzymes. The enzymes make it possible to take a gene from one organism and attach it to the DNA of another organism.)*

USE VISUALS Direct students to the diagram showing how DNA pieces join together.

Ask Which two kinds of enzymes are shown in the diagram? *(restriction enzymes and DNA ligase)*

Ask What does the squiggly arrow show? *(It points out the "sticky ends" on the DNA molecules. Because the sticky ends have complementary bases, they will bond together.)*

Ask What are sticky ends? *(single-stranded bases of DNA at the end of a piece of DNA)*

KEY QUESTION Ask a volunteer to read the Key Question and its answer.

Ask What is recombinant DNA? *(DNA from different sources put together)*

Speed Bump

Ask How can the DNA of a cell be changed? *(Enzymes can join a gene from one organism to the DNA of another organism.)*

Ask What's the purpose of a replication "start" signal? *(It signals the cell to copy the recombinant DNA along with its own DNA.)*

Ask What are plasmids? *(small circular DNA molecules found in the cytoplasm of some bacteria)*

During Reading continued

Active Reading

USE VISUALS Direct students to the Plasmid DNA Transformation diagram and read the caption.

Ask What is Eco*RI*? *(the restriction enzyme used to cut the human gene and the plasmid)*

Ask What color is the human gene in the diagram? *(purple)* What color is the plasmid? *(blue)* Why is the recombinant DNA two colors? *(because it contains DNA from the human and the bacterium)*

Ask Why is Eco*RI* used to cut the human gene and the plasmid? *(so they will have complementary "sticky ends" and attach to one another)*

SUMMARIZE Direct students to read and then summarize the paragraph on transforming bacteria.

Ask To transform bacteria, what must be true about the plasmid? *(It must have a signal that starts replication.)*

Ask What is a genetic marker? *(a gene that makes it possible to distinguish bacteria that carry the recombinant DNA from those that don't)*

Build Vocabulary

PREFIXES Invite students to locate other words in the lesson that contain the prefix *trans-*. *(transform, transplant)* Discuss how the prefix relates to the meaning of these words. Then, write the word *antibiotic* on the board and circle the prefix *anti-*.

Ask What does the prefix *anti-* mean? *(opposed to or against)*

Say Predict the meanings of these words: *antibiotic, antifungal, anticancer, antiviral.* Have students use a dictionary to check their predictions.

Speed Bump

Ask What is the role of restriction enzymes in transforming DNA? *(They cut DNA into pieces. They produce complementary "sticky ends" on pieces of DNA, allowing DNA from different sources to join.)*

Ask Explain how scientists can use recombinant DNA to learn what a specific gene does. *(They put recombinant DNA containing the specific gene into the genome of a host. They check the genetic marker to be sure the gene was copied. Then, they study the traits of the genetically changed organism. Any different or unusual traits are likely the result of the transferred gene.)*

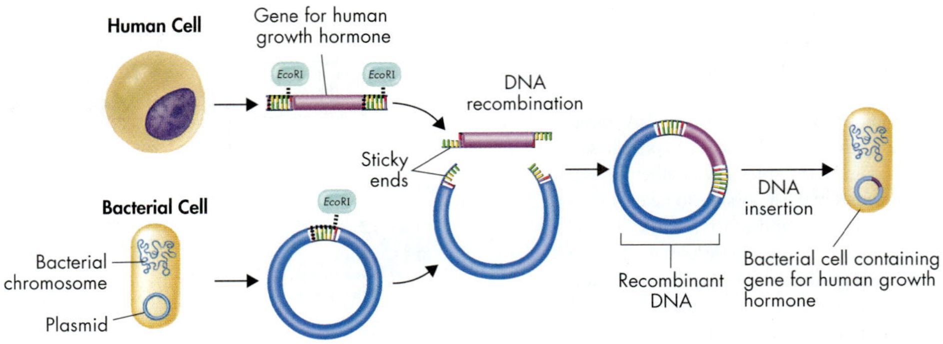

Plasmid DNA Transformation Scientists can insert a piece of DNA into a plasmid if both the plasmid and piece of DNA have been cut by the same restriction enzyme. In this example, the gene for human growth hormone is added to a bacterial cell. First a human gene is inserted into bacterial DNA. Then the new combination of genes is put back into a bacterial cell. The cell then copies the recombinant DNA over and over again.

Transforming Bacteria How do scientists transform bacteria? First, a piece of DNA is joined to a plasmid. The plasmid DNA has a signal that starts replication. The plasmid also has one or more genetic markers. A genetic marker allows a scientist to identify which bacteria carry the plasmid. An example of a genetic marker is a gene for resistance to a specific antibiotic. This gene keeps the bacteria alive in the presence of that antibiotic. Researchers mix recombinant plasmids with bacteria. Only about one cell in a million may take up the plasmids, but that's more than enough. After transformation, the bacteria are treated with the antibiotic. Only those few cells that have been transformed survive, because only they carry the antibiotic-resistance gene.

BUILD Vocabulary

polymerase chain reaction a technique that allows biologists to make many copies of a gene

recombinant DNA DNA produced by combining DNA from different sources

plasmid a small, circular piece of DNA located in the cytoplasm of many bacteria

transgenic a term used to refer to an organism that contains genes from other organisms

clone a member of a population of genetically identical cells produced from a single cell

PREFIXES

The prefix *trans-* means "cross." Genes in transgenic organisms have crossed from one organism to another.

Transgenic Organisms

Most living things share the same genetic code. So, scientists can make organisms that are transgenic. **Transgenic** organisms have genes from other organisms. They can be made by putting recombinant DNA into the genome of a host. The DNA molecules used for transformation of plant and animal cells have genetic markers so scientists can tell which cells have been transformed. Genetic engineers can make transgenic plants, animals, and microorganisms. By studying the traits of a genetically changed organism, scientists can learn about the function of the transferred gene. These studies helped us learn much more about genes.

Transgenic Plants Many plant cells can be transformed using *Agrobacterium*. This bacterium inserts a small DNA plasmid that makes tumors in a plant's cells. Scientists can turn off the plasmid's tumor-making gene and replace it with a piece of recombinant DNA. The recombinant plasmid can then be used to transform plant cells.

Transgenic Animals Scientists can transform animal cells, too. The egg cells of many animals are large enough that DNA can be injected into the nucleus. Once the DNA is in the nucleus, enzymes help insert the new DNA into the chromosomes of the host cell.

ELL Focus on ELL: Access Content

INTERMEDIATE SPEAKERS Some students may be overwhelmed by the number of processes involving DNA in this lesson. Tell students that one way to organize the processes is to focus on what each is designed to do. Offer the following questions and partial answers. Have students write complete answers in their notebooks.

What is PCR used for? PCR is used to make _____ of a gene. (copies)

What is recombinant DNA used for? Recombinant DNA is used to _____ the genetic makeup of living organisms. (change)

What are plasmids used for? Plasmids are used to _____ recombinant DNA into a bacterium and to _____ the bacterium to copy the recombinant DNA. (carry, signal)

What is a genetic marker used for? A genetic marker is used to _____ that the recombinant DNA is being copied. (show)

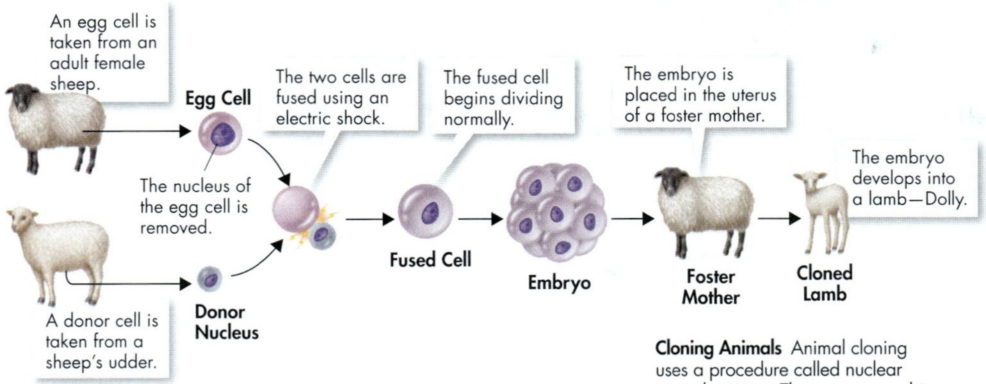

Cloning Animals Animal cloning uses a procedure called nuclear transplantation. The process combines an egg cell with a donor nucleus to make an embryo.

DNA molecules can also be made with two ends that can recombine with specific sequences in the host chromosome. Once they do, the host gene found between those two sequences is replaced with a new gene. This kind of gene replacement has helped scientists figure out the roles of genes in many organisms.

Key Question How can genes from one organism be inserted into another organism? **Transgenic organisms can be produced by the insertion of recombinant DNA into the genome of a host organism.**

Cloning A **clone** is a member of a group of genetically identical cells made from a single cell. Cloning uses a single cell from an adult organism to grow a new individual. That individual, then, is genetically identical to the organism from which the cell was taken.

Cloned colonies of bacteria and other microorganisms are easy to grow. Multicellular organisms, such as animals, are harder to grow. In 1997, Scottish scientist Ian Wilmut announced that he had made a sheep, called Dolly, by cloning. The figure above shows the process.

CHECK Understanding

Apply Vocabulary

Use the highlighted words from the lesson to complete each sentence correctly.

1. ______ are small rings of DNA found in bacteria and yeast.

Critical Thinking

2. Infer Why would a scientist want to know the sequence of a DNA molecule?

3. Explain How do scientists use recombinant DNA?

4. Write to Learn Answer the first mystery clue.

MYSTERY CLUE

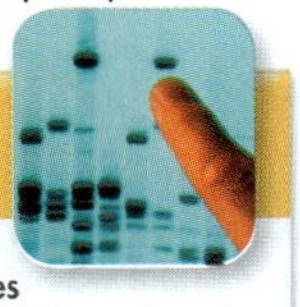

How could restriction enzymes be used to analyze the DNA evidence found on the suspect? (**Hint:** See page 357.)

BIOLOGY.com Search Lesson 15.2 GO • Lesson Assessment

Check Understanding

1. Plasmids
2. Knowing the sequence allows scientists to find individual genes.
3. Scientists use recombinant DNA to change the genetic composition of living organisms and to investigate the structure and function of genes.
4. Restriction enzymes would be used to help analyze the DNA found at the scene. For example, if investigators wanted to make more copies of the DNA so that they could sequence it, they could use restriction enzymes to cut up the DNA before they use PCR to make copies.

Active Reading

SEQUENCING Work with students to establish the correct order of the following steps. Have them write the correct order in their notebooks.

(4) The embryo develops until birth.

(2) The egg cell is joined with a donor cell that has a nucleus.

(1) The nucleus is removed from an unfertilized egg cell.

(3) The egg develops into an embryo.

USE VISUALS Help students distinguish between the egg cell and the donor nucleus.

Ask Why is the nucleus removed from the egg cell? *(so that DNA from an adult can be used; the nucleus of the adult cell has a full set of chromosomes)*

Ask Why can't an egg cell, with its own nucleus in place, be placed in the uterus of a foster mother and allowed to grow? *(An egg cell has only half the chromosomes needed for an individual to grow.)*

Ask Why does the cloned lamb look different from its foster mother? *(The nuclei of the cloned lamb's cells carry the genetic information from the donor nucleus. The donor individual had a white face.)*

KEY QUESTION Read aloud the Key Question and its answer.

Ask What is a transgenic organism? *(An organism with genes from other organisms.)*

Speed Bump

Ask What is cloning? *(a process in which a single adult cell is used to produce a new organism)*

Ask What is unusual about a cloned individual? *(It is genetically the same as the organism from which the nucleus was taken.)*

Post Reading

ASSESSMENT TIPS Review vocabulary words with students before they begin the assessment.

Ask For Question 2, what does it mean to infer? *(*Infer *means "to use prior knowledge or observations to draw conclusions that are not stated in the text.")*

MYSTERY CLUE Review with students the role of restriction enzymes in copying DNA.

Wrap-Up Activity

WRITE INSTRUCTIONS Direct student pairs to write step-by-step instructions in the style of a recipe for one of the processes in this lesson, such as the PCR method. Pairs can exchange instructions and determine if all the necessary steps are included.

LESSON 15.3

Pre-Reading

Objectives

15.3.1 Describe the benefits of genetic engineering as they relate to agriculture and industry.

15.3.2 Explain how recombinant DNA technology can improve human health.

15.3.3 Summarize the process of DNA fingerprinting and explain its uses.

Preview the Pages

1. Preview the vocabulary words with students. Have them write each word in their notebooks along with their prediction of the word's meaning. Encourage them to check, and correct if necessary, the meanings as they read this lesson.
2. Preview the Genetically Engineered Crops line graph in the Inquiry activity on the next page. Make sure that students know how to read the graph's key and that they understand the significance of the colored lines.
3. Remind students that mRNA is messenger RNA. mRNA carries genetic information from the DNA to other parts of the cell.

Build Understanding

MAIN IDEA AND DETAILS CHART Direct students to Workbook B, where they will find a partially completed main idea and details chart. Remind students that their charts can be used later as study guides about the uses of genetic engineering.

NATIONAL SCIENCE EDUCATION STANDARDS

UNIFYING CONCEPTS AND PROCESSES

III, V

CONTENT

C.1.a, C.1.c, C.1.d, C.2.a, E.1, E.2, G.1, G.3

INQUIRY

A.1.c

For the corresponding lesson in *Miller & Levine Biology,* see pages 428–434.

15.3 Applications of Genetic Engineering

Key Questions

How can genetic engineering benefit agriculture and industry?

How can recombinant-DNA technology improve human health?

How is DNA used to identify individuals?

BUILD Understanding

Main Idea and Details Chart Taking notes is a way to organize the information you read. One way to take notes is to make a main idea and details chart while you read. Include details or evidence that supports each important idea.

In Your Workbook Go to your workbook for more about using a main idea and details chart to organize information.

Agriculture and Industry

Have you eaten any corn, potatoes, or soy products recently? If so, then you have probably eaten something that was modified by genetic engineering. Everything we eat and much of what we wear come from living organisms. Researchers have used genetic engineering to try to improve the products we get from plants and animals.

GM Crops Since 1996, genetically modified (GM) plants have become an important part of our food supply. One type of modification has already been very useful to agriculture. It uses bacterial genes that make a protein called Bt toxin. This toxin is harmless to humans and most other animals. However, it kills the insects that eat it. So, plants with the Bt gene do not have to be sprayed with pesticides. They make more food per acre than unmodified plants, too. Scientists are also adding other traits to crops. Some are making plants that are resistant to viruses. Others are making crops that are slower to rot and spoil.

GM Animals Transgenic animals are also becoming more important to our food supply. For example, many dairy farms now raise cows that have been injected with hormones made by recombinant-DNA techniques. The hormones help the cows make more milk. Pigs can be genetically modified so they make leaner meat and cleaner wastes. Using growth-hormone genes, scientists have made transgenic salmon that grow much more quickly than wild salmon. This makes the fish easier to farm.

Key Question How can genetic engineering benefit agriculture and industry? **Ideally, genetic modification could lead to better, less expensive, and more nutritious food as well as less harmful manufacturing processes.**

Antibacterial Goat Milk
Scientists are working to add a gene for a natural antibiotic to the DNA of goats. Milk from these goats may help prevent infections in young children who drink it.

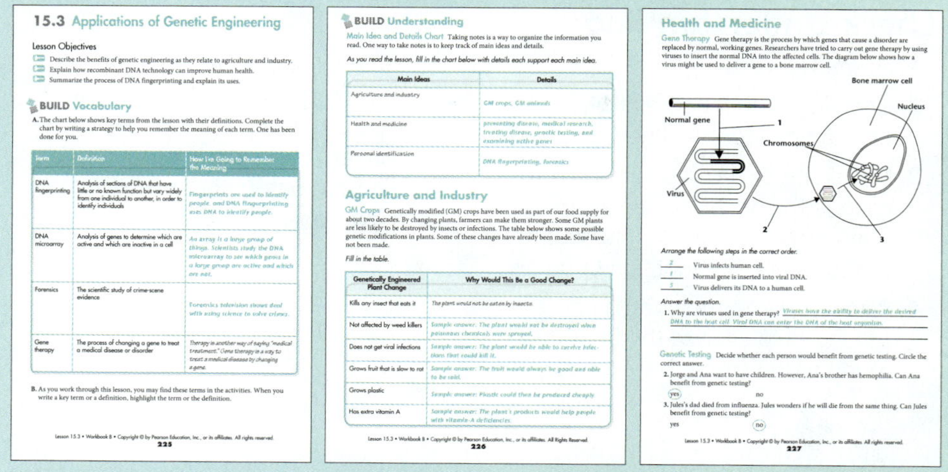

15.3 Applications of Genetic Engineering

Lesson Objectives

- Describe the benefits of genetic engineering as they relate to agriculture and industry.
- Explain how recombinant DNA technology can improve human health.
- Summarize the process of DNA fingerprinting and explain its uses.

BUILD Vocabulary

A. The chart below shows key terms from the lesson with their definitions. Complete the chart by writing a strategy to help you remember the meaning of each term. One has been done for you.

Term	Definition	How I'm Going to Remember the Meaning
DNA fingerprinting	Analysis of sections of DNA that have little or no known function but vary widely from one individual to another, in order to identify individuals	Fingerprints are used to identify people, and DNA fingerprinting uses DNA to identify people.
DNA microarray	Analysis of genes to determine which are active and which are inactive in a cell	An array is a large group of things. Scientists study the DNA microarray to see which genes in a large group are active and which are not.
Forensics	The scientific study of crime-scene evidence	Forensics television shows deal with using science to solve crimes.
Gene therapy	The process of changing a gene to treat a medical disease or disorder	Therapy is another way of saying "medical treatment." Gene therapy is a way to treat a medical disease by changing a gene.

B. As you work through this lesson, you may find these terms in the activities. When you write a key term or a definition, highlight the term or the definition.

Lesson 15.3 • Workbook B • Copyright © by Pearson Education, Inc., or its affiliates. All rights reserved.
225

BUILD Understanding

Main Idea and Details Chart Taking notes is a way to organize the information you read. One way to take notes is to keep track of main ideas and details.

As you read the lesson, fill in the chart below with details each support each main idea.

Main Ideas	Details
Agriculture and industry	GM crops, GM animals
Health and medicine	preventing disease, medical research, treating disease, genetic testing, and examining active genes
Personal identification	DNA fingerprinting, forensics

Agriculture and Industry

GM Crops Genetically modified (GM) crops have been used as part of our food supply for about two decades. By changing plants, farmers can make them stronger. Some GM plants are less likely to be destroyed by insects or infections. The table below shows some possible genetic modifications in plants. Some of these changes have already been made. Some have not been made.

Fill in the table.

Genetically Engineered Plant Change	Why Would This Be a Good Change?
Kills any insect that eats it	The plant would not be eaten by insects.
Not affected by weed killers	Sample answer: The plant would not be destroyed when poisonous chemicals were sprayed.
Does not get viral infections	Sample answer: The plant would be able to survive infections that could kill it.
Grows fruit that is slow to rot	Sample answer: The fruit would always be good and able to be sold.
Grows plastic	Sample answer: Plastic could then be produced cheaply.
Has extra vitamin A	Sample answer: The plant's products would help people with vitamin-A deficiencies.

Lesson 15.3 • Workbook B • Copyright © by Pearson Education, Inc., or its affiliates. All Rights Reserved.
226

Health and Medicine

Gene Therapy Gene therapy is the process by which genes that cause a disorder are replaced by normal, working genes. Researchers have tried to carry out gene therapy by using viruses to insert the normal DNA into the affected cells. The diagram below shows how a virus might be used to deliver a gene to a bone marrow cell.

Arrange the following steps in the correct order.

2 Virus infects human cell.
1 Normal gene is inserted into viral DNA.
3 Virus delivers its DNA to a human cell.

Answer the question.

1. Why are viruses used in gene therapy? Viruses have the ability to deliver the desired DNA to the host cell. Viral DNA can enter the DNA of the host organism.

Genetic Testing Decide whether each person would benefit from genetic testing. Circle the correct answer.

2. Jorge and Ana want to have children. However, Ana's brother has hemophilia. Can Ana benefit from genetic testing?
(yes) no
3. Jules's dad died from influenza. Jules wonders if he will die from the same thing. Can Jules benefit from genetic testing?
yes (no)

Lesson 15.3 • Workbook B • Copyright © by Pearson Education, Inc., or its affiliates. All rights reserved.
227

INQUIRY into Scientific Thinking

Genetically Modified Crops in the United States

U.S. farmers have started using many GM crops. Soybeans, cotton, and corn have been modified to tolerate herbicides and resist insect damage. The graph at the right shows the extent to which these crops were adopted between 1996 and 2007. The modified traits shown here include herbicide tolerance (HT) and insect resistance (Bt).

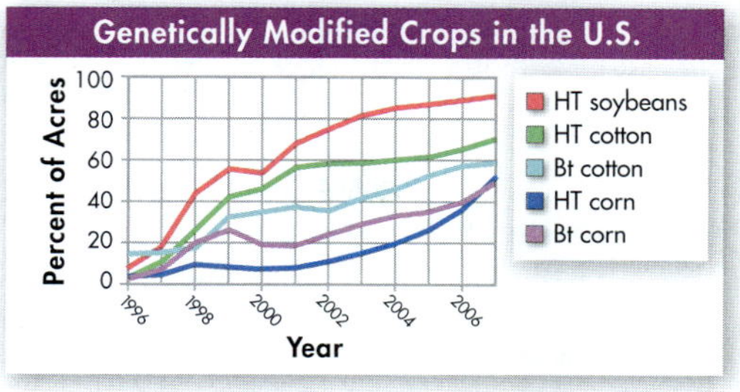

1 Analyze Data Which two crops were most widely and rapidly adopted?

2 Draw Conclusions Why do you think the levels of adoption fell at certain points over the period?

3 Explain Why can't the line showing the adoption of modified soybeans continue to grow past the 100 percent mark of the graph?

4 Infer What would happen to herbicide-tolerant plants when farmers sprayed herbicide on the crops to kill weeds growing near them?

In Your Workbook Get more help for this activity in your workbook.

Health and Medicine

In a way, biotechnology has always been a part of medicine. Early physicians used substances from plants and animals to cure their patients. In the twentieth century, the use of vaccinations saved countless lives. Today, recombinant-DNA technology is the source of some of the most important and exciting advances in the prevention and treatment of disease.

Preventing Disease One interesting development in transgenic technology is golden rice. This rice contains increased amounts of provitamin A. A lack of provitamin A causes infant blindness and other problems. There is hope that provitamin A-rich golden rice will help prevent these problems. Some scientists are making transgenic plants and animals that make human antibodies to fight disease. Several laboratories have made transgenic sheep and pigs that make important human proteins in their milk. Many of these proteins can be taken from the milk and used as medicines in people.

Vitamin-Rich Rice Golden rice is a GM plant that has high levels of provitamin A. Two genes engineered into the rice genome help the grains produce and accumulate provitamin A.

Medical Research Transgenic animals are often used as test subjects in medical research. They can sometimes model human diseases caused by defective genes. Scientists can study a disease and then use the models to test new drugs that may help treat the disease.

During Reading

Active Reading

KEY QUESTION Read aloud the Key Question and answer on the previous page.

Ask What is genetic engineering? *(changing genes in living things for a practical purpose)*

Inquiry Into Scientific Thinking

INTERPRET GRAPHS Review the graph title, axis labels, and key.

Ask What do *HT* and *Bt* mean? *(HT is a gene that makes plants herbicide-tolerant. Bt is an insect-resistance gene.)*

Ask What does the green line represent? *(the percentage of cotton acres planted with HT cotton)*

Ask In 2007, which was planted over more acres, HT or Bt cotton? Explain. *(HT cotton—70% of acres; Bt cotton—roughly 60% of acres)*

Ask In 2007, were some cotton crops both herbicide-tolerant and insect resistant? Explain. *(Yes, 70% of cotton acres were HT cotton, and 60% were Bt cotton. That adds up to 130% of the planted acres. So, some must have been both HT and Bt.)*

Lead a Discussion

In 2006, 22 countries planted a total of 252 million acres of GM crops. Of these GM crops, 53% were grown in the United Sates. Lead a discussion about the pros and cons of widespread planting of GM crops. GM crops can be more convenient to grow and can produce foods that are easier to transport and sell. Critics are concerned about food allergens, transfer of antibiotic resistance markers, and overall unknown effects of GM crops on the food supply. GM crops also pose a risk of harming naturally grown plants. Encourage interested students to research additional pros and cons.

Speed Bump

Ask What are three examples of the way plants have been genetically modified? *(Sample answers: adding genes to plants to make them resistant to insects, viruses, and herbicides)*

Ask What are two advantages of soybeans that resist insects? *(Fewer soybeans are lost to insects and less pesticide is needed.)*

Ask Why do scientists use genetically modified animals to model some human diseases? *(Doing so allows scientists to test new drugs without having to test them on human patients.)*

Analyze and Conclude

1. HT soybeans and HT cotton
2. Sample answer: Perhaps the price of seed for these GM crops was higher than the expected increase in profit from growing the seed.
3. Sample answer: When the line for any crop reaches 100% on this graph, all the plants of that type grown in the U.S. would be genetically modified plants.
4. Sample answer: Those crops aren't affected by the herbicides farmers use to kill weeds growing on their fields. So, the weeds would die, but the crops would survive.

During Reading continued

Active Reading

USE VISUALS Read aloud the two paragraphs under Treating Disease. Then, discuss what is happening in the diagram showing how gene therapy can be used.

Say Scientists want to add a normal, healthy gene to the patient's cells. First, they put the normal gene into a virus. Then, they infect the patient's cells with the genetically engineered virus.

Ask What happens after the virus moves inside a host cell? *(It should insert the healthy gene into the host cell's DNA.)*

Point out the blue (healthy) gene among the chromosomes in the nucleus of the host cell. Emphasize that the virus has been used to deliver the healthy gene to the patient's cells.

INTERPRET DIAGRAMS Ask students to compare the Gene Therapy diagram with the diagram in the previous lesson showing the transformation of plasmid DNA.

Ask How are these two processes alike? *(Both show a human gene being inserted into another source.)* How are they different? *(One process uses a bacterium, the other a virus.)*

Build Vocabulary

USING PRIOR KNOWLEDGE Point out to students that they can use what they already know to determine the meaning of new phrases.

Ask What do you know about a person's fingerprints? *(Everyone has unique fingerprints.)* How can you apply that knowledge to DNA fingerprints? *(Everyone has unique DNA.)*

KEY QUESTION Read aloud the Key Question and answer.

Ask What are some diseases recombinant-DNA technology helps prevent or treat? *(Sample answers: It is used to make proteins to treat diabetes, hemophilia, and growth disorders. It is also used to treat and test for different genetic disorders.)*

Speed Bump

Ask What part does a virus play in gene therapy? *(A virus carries the corrective gene into the patient.)*

Ask How do genetic tests work? *(Some genetic tests use a piece of complementary DNA to find a defective allele. Some search for changes in cutting sites of restriction enzymes.)*

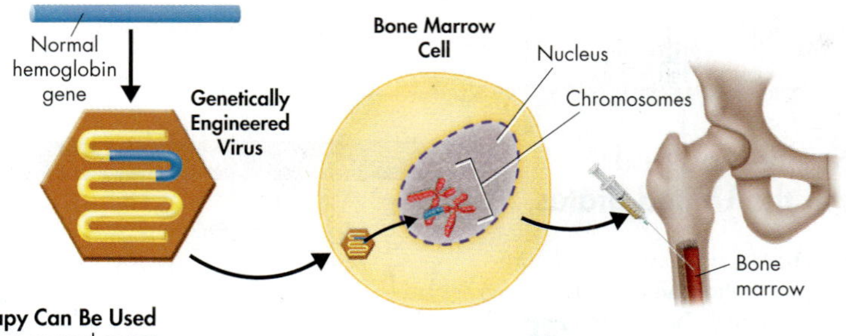

How Gene Therapy Can Be Used
Gene therapy uses normal genes to add to or replace defective genes or to boost a normal function like immunity.

Treating Disease Recombinant DNA can be used to make important proteins that can save human lives. Products now made in genetically engineered bacteria include insulin to treat diabetes, blood-clotting factors for hemophiliacs, and human growth hormone.

Scientists are now trying to learn how to fix faulty genes in people. **Gene therapy** is the process of changing a gene to treat a disease or disorder. In gene therapy, a missing or faulty gene is replaced with a normal, working gene. This process lets the body make the protein or enzyme it needs. How are genes added to human cells? First, scientists make a virus that cannot reproduce and hurt people. Then they put DNA that has the working gene into their modified virus. Next, they infect the patient's cells with the virus carrying the gene. With luck, the virus will insert the healthy gene into the cells that need it and correct the problem.

Genetic Testing Suppose a couple wants to have children. Could they find out if they are carrying an allele for a genetic disorder? Genetic tests are now available for diagnosing hundreds of different disorders. For example, the CF allele has slightly different DNA sequences than the normal allele. So, genetic tests use a piece of DNA that is complementary to a defective allele to find that allele. Other genetic tests search for changes in cutting sites of restriction enzymes.

Examining Active Genes By studying which genes are active in cells, scientists can learn how cells work. Active genes make mRNA. So, scientists use a test to see what kinds of mRNA are in different kinds of cells. DNA microarray technology lets scientists learn the activity levels of thousands of genes at once. A **DNA microarray** is a glass slide to which spots of sequenced, single-stranded DNA have been tightly attached. Scientists then get mRNA from a cell they want to study. They make and label single-stranded DNA from the mRNA. Then, they add their labeled DNA to the wells of the array. They can tell which of the genes of the cell are active because the wells with those genes will also have labeled DNA.

Key Question How can recombinant-DNA technology improve human health?
Today recombinant-DNA technology is the source of some of the most important and exciting advances in the prevention and treatment of disease.

BUILD Vocabulary

gene therapy
a process of changing a gene to treat a medical disease or disorder. An absent or faulty gene is replaced by a normal working gene.

DNA microarray
the glass slide or silicon chip that carries thousands of different kinds of single-stranded DNA fragments arranged in a grid

DNA fingerprinting
a tool used by biologists to determine whether two samples of genetic material are from the same person

forensics
the scientific study of crime scene evidence

USING PRIOR KNOWLEDGE
You may have learned that physical therapy is a way to help people recover from physical problems. Gene therapy helps people recover from genetic disorders.

Focus on Academic Language

- hemophiliac

Write the word *hemophiliac* on the board. Circle the prefix *hemo-* and point out that it means "blood." Explain that blood does not clot properly in people who have hemophilia.

Ask What might happen to a person if his or her blood didn't clot? *(Small cuts and bruises could result in extensive bleeding and become life-threatening.)*

Encourage students to use a dictionary to look up words with the prefix *hemo-* and to relate the meanings of each word to blood.

1 Chromosomes contain many regions with repeated DNA sequences that do not code for proteins. These vary from person to person. Here, one sample has 12 repeats between genes A and B, while the second sample has 9 repeats between the same genes.

2 Restriction enzymes are used to cut the DNA into fragments containing genes and repeats. Note that the repeat fragments from these two samples are of different lengths.

3 The restriction fragments are separated according to size using gel electrophoresis. The DNA fragments containing repeats are then labeled using radioactive probes. This labeling produces a series of bands—the DNA fingerprint.

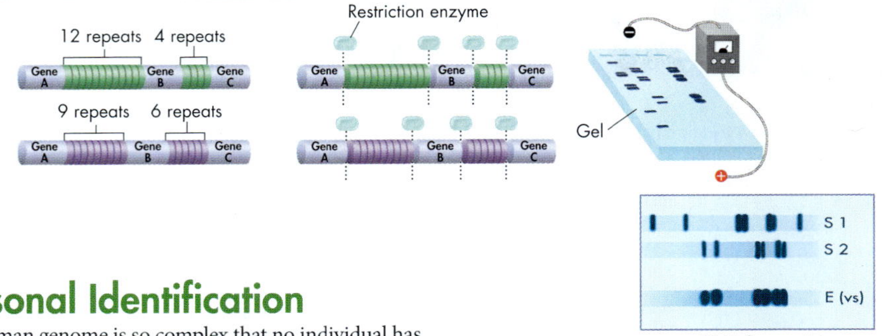

DNA fingerprint

Identifying Individuals
DNA fingerprinting is useful in solving crimes. The diagram shows how scientists match DNA evidence from a crime scene with two possible suspects.

Personal Identification

The human genome is so complex that no individual has DNA exactly like another individual. Identical twins are the only exception. Molecular biology has used this fact to make a powerful tool called DNA fingerprinting that is used to identify individuals. **DNA fingerprinting** analyzes pieces of DNA that may have little or no function in a cell. However, these pieces, called restriction fragments, vary widely from one person to another. How do scientists make a DNA fingerprint? DNA samples can be taken from blood or any other tissue. Then, restriction enzymes cut the DNA into pieces. Next, gel electrophoresis separates the pieces by size. A DNA probe finds the fragments that have highly variable regions. This is shown as a series of DNA bands on the gel. If enough enzymes and probes are used, the pattern of bands from one person is unlike anyone else's pattern of bands.

Forensic Science DNA fingerprinting has been used in the United States since the late 1980s. It has been of great help in **forensics**—the scientific study of crime scene evidence. DNA fingerprinting has helped solve crimes, convict criminals, and even overturn guilty verdicts. To date DNA evidence has saved more than 110 persons who were sentenced to death for crimes they didn't commit.

DNA forensics is used in wildlife conservation as well. African elephants are a highly vulnerable species. Poachers kill the animals illegally for their valuable tusks. To stop the ivory trade, African officials now use DNA fingerprinting to identify the herds from which the ivory has been taken.

BIOLOGY.com Search Lesson 15.3 GO • Art Review

Active Reading

USE VISUALS Help students understand the comparison being made in the Identifying Individuals diagram.

Say Step 1 shows samples of DNA from two people who are suspects in a crime. Look closely at the samples.

Ask What do the green and purple areas on these DNA samples represent? *(repeated DNA sequences that do not code for proteins)* How are the two samples different? *(The number of repeats is different on each sample.)*

Ask In step 2, what is the function of the restriction enzymes? *(They cut the DNA into fragments, separating the repeats from the genes.)*

Say In step 3, the fragments are sorted by size and then used to create the DNA fingerprint. On the DNA fingerprint shown here, the top two sets of DNA bands are from suspect 1 and suspect 2. The set of bands at the bottom are bands of DNA from the evidence found at the crime scene.

Ask Which suspect's DNA does the evidence DNA match? Explain. *(suspect 2, because that person's DNA has the same number of bands as the evidence DNA, and the bands are in the same position)*

USE GRAPHIC ORGANIZERS Remind students that using a flowchart can help them understand and remember the steps in a process. Ask students to draw a flowchart depicting the three steps shown in the Identifying Individuals diagram. Remind students to use their own words and sketches.

Ask How can DNA fingerprints be used to identify individuals? *(If enough enzymes and probes are used, the pattern of bands can distinguish one individual from others.)*

MAKE CONNECTIONS Help students relate the information about DNA fingerprints to news or television shows they are familiar with.

Ask What kind of samples are useful for DNA fingerprinting? *(Sample answer: samples from blood, body fluids, or hair)*

Ask What part does DNA fingerprinting play in criminal investigations? *(It can help solve crimes, convict criminals, and clear people accused or convicted of crimes they didn't commit.)*

Speed Bump

Ask What does *forensics* refer to? *(the scientific study of crime scene evidence)*

Ask What are restriction fragments? *(fragments of DNA cut by restriction enzymes)*

ELL Focus on ELL: Build Background

ALL SPEAKERS Many English language learners are likely to be familiar with television shows that showcase DNA forensics. Help students tie their new knowledge of DNA fingerprinting to what they see in these shows.

Say Often a television program shows a crime scene investigator (CSI) placing a small tissue sample in a test tube and adding some substance to it.

Ask What might the CSI be adding to the tissue? *(restriction enzymes, preservatives)*

Say Shortly afterward, the CSI holds up a film showing DNA bands.

Ask What step was not shown? *(The step in which gel electrophoresis is used to separate the restriction fragments by size.)*

Say In the next scene, the CSI compares two sets of DNA bands.

Ask What is the CSI doing? *(trying to see if the pattern of bands in the suspect's DNA fingerprint matches the pattern of bands of the evidence DNA)*

During Reading continued

Active Reading

SUMMARIZE Have students read each paragraph and summarize it in one or two sentences in their notebooks.

Ask How can DNA help determine who a child's biological father is? *(Some of the child's alleles will not match the mother's. These must come from the father.)*

Ask How can Y chromosomes be used to trace ancestors? *(The Y chromosome is passed from father to son with very few changes.)*

KEY QUESTION Read aloud the Key Question and answer.

Ask Why might researchers use several restriction enzymes when comparing DNA fingerprints? *(Increasing the number of restriction enzymes increases the number of DNA fragments. This gives more fragments to separate and analyze.)*

Speed Bump

Ask Where in the cell is mtDNA? *(in mitochondria)*

Ask Can researchers use mtDNA to determine the identity of a person's grandfather? Explain. *(No, mtDNA passes from mother to child. It could be used to identify a maternal grandmother, but not a grandfather.)*

Post Reading

ASSESSMENT TIPS Encourage students to review the visual of DNA fingerprinting before answering Question 4.

MYSTERY CLUE Discuss the kinds of evidence investigators collect at crime scenes. Suggest students review the process of DNA fingerprinting before answering Question 5.

Wrap-Up Activity

COMPOSE A TEST Have students form groups of four. Ask each group to compose a test based on the lesson content, with five fill-in-the-blank items, five matching items, and one constructed response question. Direct groups to exchange tests with one another. Tests should be returned to test writers for scoring. Discuss results with the entire class.

Identifying Individuals
The pattern on this DNA fingerprint can belong to only one person.

Establishing Relationships How can we figure out who is the biological father of a child? DNA fingerprinting makes it easy to find alleles carried by the child that do not match those of the mother. Any such alleles must come from the child's biological father. The alleles will show up in the father's DNA fingerprint.

When genes are passed from parent to child, genetic recombination mixes up the molecular markers used for DNA fingerprinting. So, ancestry can be difficult to trace. There are two ways to solve this problem. The Y chromosome never undergoes crossing over, and only males carry it. So, Y chromosomes pass directly from father to son with few changes. The small DNA molecules found in mitochondria pass from mother to child with few changes. Mitochondria are organelles that release energy from sugar to power cells. They also carry DNA. It rarely changes. The DNA in mitochondria is passed from mother to child in the cytoplasm of the egg cell. These genetic markers can be used to trace ancestry over many generations.

Because mitochondrial DNA (mtDNA) is passed directly from mother to child, your mtDNA is the same as your mother's mtDNA, which is the same as her mother's mtDNA. This means that if two people have an exact match in their mtDNA, then there is a very good chance that they share a common maternal ancestor. Y-chromosome analysis has been used in the same way and has helped researchers settle longstanding historical questions.

Key Question How is DNA used to identify individuals?
DNA fingerprinting analyzes sections of DNA that may have little or no function but that vary widely from one individual to another.

CHECK Understanding

Apply Vocabulary

Use the highlighted words from the lesson to complete each sentence correctly.

1. An absent or faulty gene is replaced by a normal working gene using ______.
2. The study of blood or other crime scene evidence is ______.

Critical Thinking

3. **Apply Concepts** What is golden rice, and how has it helped people?
4. **Explain** What are the steps in DNA fingerprinting?
5. **Write to Learn** Write a paragraph that answers the chapter mystery clue.

MYSTERY CLUE

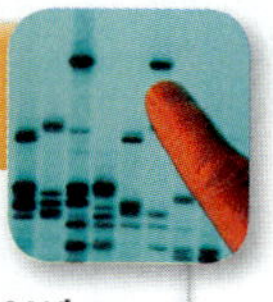

What kind of evidence do you think investigators collected at the crime scene? What kinds of tests would they have run on this evidence? What would the tests have to show before the suspect was released? (Hint: See p. 365.)

BIOLOGY.com Search Lesson 15.3 GO • Lesson Assessment

Check Understanding

1. gene therapy
2. forensics
3. Golden rice is a GM rice that has increased amounts of provitamin A. It may help prevent infant blindness.
4. Restriction enzymes cut a small sample of human DNA. The restriction segments are separated by size. A DNA probe detects a series of variously sized DNA bands that can be matched to another sample.
5. Sample answer: Investigators would gather samples of blood, saliva, hair, or sweat in order to make a DNA fingerprint. For the man to be released, his DNA fingerprint would have to be different from the DNA fingerprint that was produced from evidence at the crime scene.

15.4 Ethics and Impacts of Biotechnology

Profits and Privacy

Who owns your genes? Could someone else use the information in your genes? When biotechnology and drug companies develop GM plants and animals, they protect their discoveries and innovations with patents. A patent is a legal tool that gives an individual or company the exclusive right to profit from its innovations for a number of years.

Patenting Life When you think about patents, you might think about an inventor protecting a new machine. However, molecules and DNA sequences can be patented, too. One fifth of the known genes in the human genome have been patented. Even laboratory techniques like PCR have been patented.

The ability to patent is meant to encourage new discovery. After all, patent holders have a good chance of making money. Sometimes, though, patent holders demand high fees that block other scientists from doing their research.

Now think about the information held in your own genome. Do you have exclusive rights to your DNA? Should you, like patent holders, be able to keep your genetic information confidential? When it comes to your own DNA, how much privacy are you entitled to?

Key Question What privacy issues does biotechnology raise? **Do you have exclusive rights to your DNA? Should you, like patent holders, be able to keep your genetic information confidential?**

Genetic Ownership People who serve in the U.S. armed forces give a DNA sample when they begin serving. Those DNA samples are used, if needed, to identify the remains of people who die in the line of duty. But what if the government wants to use a person's DNA sample for other tests? What if health-insurance companies want to refuse coverage to people who have a gene that makes them more likely to get an illness? The U.S. Congress debated these issues. The Genetic Information Nondiscrimination Act was signed into law in 2008. This act protects Americans against discrimination based on their genetic information.

Key Questions

What privacy issues does biotechnology raise?

Are GM foods safe?

Should genetic modifications to humans and other organisms be closely regulated?

BUILD Understanding

Two-Column Table As you read the lesson, think about the ethical issues involved in genetic engineering. Write the positives in a column titled Pros. Write the negatives in a column titled Cons.

In Your Workbook Go to your workbook to learn more about using a two-column table.

Unknown Identities The Tomb of the Unknowns in Arlington National Cemetery holds the remains of unknown American soldiers. DNA testing now helps identify the remains of all U.S. military personnel who die in the line of duty.

LESSON 15.4

Pre-Reading

Objectives

15.4.1 Describe some of the issues that relate to biotechnology.

15.4.2 Identify some of the pros and cons of genetically modified food.

15.4.3 Describe some of the ethical issues relating to biotechnology.

Preview the Pages

1. Encourage students to scan text headings and then discuss with a partner what topics they think will be covered. Suggest they list any questions they have about the topics.
2. Review these terms before beginning the lesson: *transgenic, GM organism, ethics,* and *privacy.*

Build Understanding

TWO-COLUMN TABLE Tell students that another word for *pro* is *advantage*; another word for *con* is *disadvantage*.

Ask What other synonyms can you give for *pro* and *con*? *(Sample answers: benefit, drawback; good thing, bad thing)*

Refer students to Workbook B. Impress upon them that the pros and cons they record in the two-column table should be facts from the lesson or their own outside research.

NATIONAL SCIENCE EDUCATION STANDARDS

UNIFYING CONCEPTS AND PROCESSES

II

CONTENT

C.1.c, C.2.a, E.2, F.4, F.6

For the corresponding lesson in *Miller & Levine Biology,* see pages 436–439.

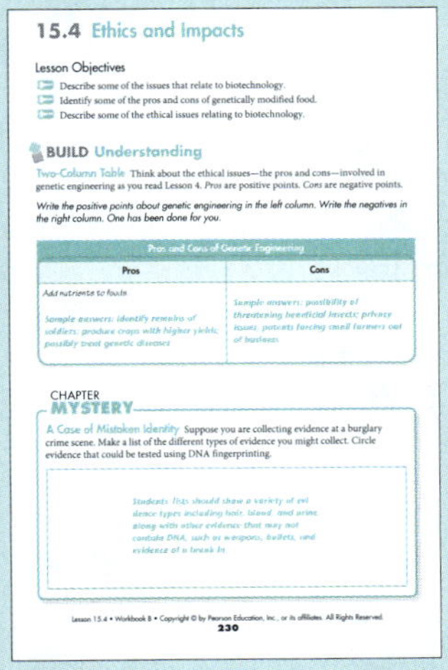

15.4 Ethics and Impacts

Lesson Objectives

Describe some of the issues that relate to biotechnology.
Identify some of the pros and cons of genetically modified food.
Describe some of the ethical issues relating to biotechnology.

BUILD Understanding

Two-Column Table Think about the ethical issues—the pros and cons—involved in genetic engineering as you read Lesson 4. *Pros* are positive points. *Cons* are negative points.

Write the positive points about genetic engineering in the left column. Write the negatives in the right column. One has been done for you.

Pros and Cons of Genetic Engineering	
Pros	Cons
Add nutrients to foods Sample answers: identify remains of soldiers; produce crops with higher yields; possibly treat genetic diseases	Sample answers: possibility of threatening beneficial insects; privacy issues; patents forcing small farmers out of business

CHAPTER MYSTERY

A Case of Mistaken Identity Suppose you are collecting evidence at a burglary crime scene. Make a list of the different types of evidence you might collect. Circle evidence that could be tested using DNA fingerprinting.

Students' lists should show a variety of evidence types including hair, blood, and urine along with other evidence that may not contain DNA, such as weapons, bullets, and evidence of a break in.

Lesson 15.4 • Workbook B • Copyright © by Pearson Education, Inc., or its affiliates. All Rights Reserved.
230

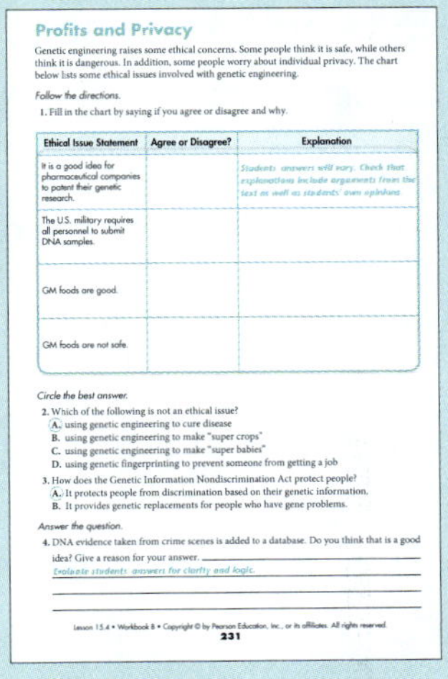

Profits and Privacy

Genetic engineering raises some ethical concerns. Some people think it is safe, while others think it is dangerous. In addition, some people worry about individual privacy. The chart below lists some ethical issues involved with genetic engineering.

Follow the directions.

1. Fill in the chart by saying if you agree or disagree and why.

Ethical Issue Statement	Agree or Disagree?	Explanation
It is a good idea for pharmaceutical companies to patent their genetic research.		Students' answers will vary. Check that explanations include arguments from the text as well as students' own opinions.
The U.S. military requires all personnel to submit DNA samples.		
GM foods are good.		
GM foods are not safe.		

Circle the best answer.

2. Which of the following is not an ethical issue?
A. using genetic engineering to cure disease
B. using genetic engineering to make "super crops"
C. using genetic engineering to make "super babies"
D. using genetic fingerprinting to prevent someone from getting a job

3. How does the Genetic Information Nondiscrimination Act protect people?
A. It protects people from discrimination based on their genetic information.
B. It provides genetic replacements for people who have gene problems.

Answer the question.

4. DNA evidence taken from crime scenes is added to a database. Do you think that is a good idea? Give a reason for your answer.
Evaluate students' answers for clarity and logic.

Lesson 15.4 • Workbook B • Copyright © by Pearson Education, Inc., or its affiliates. All rights reserved.
231

During Reading

Active Reading

KEY QUESTION Ask a volunteer to read aloud the Key Question and answer on the previous page.

Ask What is a patent? *(a legal tool that gives someone the exclusive right to profit from the patented item or process)* What is an exclusive right? *(when only one party has the right to use something; others must get permission from the owner to use it)*

USE GRAPHIC ORGANIZERS Point out to students that a single cause can sometimes have many effects. Draw a chart on the board showing the following cause and effects: GM crops → give higher yields → use less land → use less energy

Ask Why do GM crops use less land? *(They produce higher yields, meaning that the same amount of food can be grown on a smaller area of land.)*

Start a second diagram on the board beginning with *GM crops → insect resistant →*. Ask students to help you fill in the rest of the diagram. *(need fewer chemicals → help reduce pollution)*

MAKE INFERENCES Point out that GM seeds are more expensive than conventional seeds.

Ask How do you think the high prices of GM seeds might affect small farms? *(Sample answer: Small farms may not be able to buy GM seeds. This could mean that small farms will have lower crop yields than industrial farms. It could also mean that small farms would have to charge more for their crops.)*

KEY QUESTION Ask a volunteer to read the Key Question and answer. Note that an unintended consequence is an effect of an action that was not expected.

Ask GM soy and GM corn are used in many commercially processed foods. How could the use of GM products in these foods lead to unintended consequences? *(Sample answer: People might consume large amounts of GM products without knowing it. If long-term use of these products turns out to be dangerous, people could suffer health effects.)*

Speed Bump

Ask What is a pro of growing GM foods? *(Sample answer: They have higher yields and are resistant to certain insects and chemicals.)*

Ask What is a con of growing GM foods? *(Sample answers: They have not been sufficiently tested for safety. They could put small farmers out of business.)*

Safety of Transgenics

GM crops are controversial. People argue about the safety of using them. Are the foods from GM crops the same as those prepared from traditionally bred crops?

Pros of GM Foods The companies making seeds for GM crops would say that GM plants are better and safer than other crops. Farmers choose them because they have higher yields. They use less land and energy and help lower costs. They are insect resistant. So, they need fewer chemicals and help reduce pollution. Careful studies have not supported concerns about their safety. These studies seem to show that GM foods are safe to eat.

Cons of GM Foods Critics point out that no studies have tested whether long-term use of these foods might cause danger. Some worry that the insect resistance in GM plants may also hurt helpful insects. Herbicide-resistant plants allow farmers to use more weed killers, increasing pollution. Company patents on costly GM seeds may put small or poor farmers out of business.

Labeling Foods
Some people want GM foods labeled so consumers can decide whether or not to use GM products.

In the United States, federal regulations treat GM foods and non-GM foods the same. So, GM foods do not have to face extra safety tests before entering the U.S. market. No additional labeling is required to identify a product as genetically modified unless its ingredients are very different from its non-GM counterpart. Some states, however, have introduced state legislation to require the labeling of GM foods. This would give consumers a choice.

Key Question Are GM foods safe? **Careful studies of such foods have provided no scientific support for concerns about their safety, and it does seem that foods made from GM plants are safe to eat. Even if GM food itself presents no hazards, there are many serious concerns about the unintended consequences that a shift to GM farming and ranching may have on agriculture.**

Ethics of the New Biology

You've seen how easy it is to move genes from one organism to another. For example, the GFP gene can be taken from a jellyfish and spliced onto genes coding for important cellular proteins. This ability has led to significant new discoveries about how cells function.

ELL Focus on ELL: Extend Language

INTERMEDIATE AND ADVANCED SPEAKERS Pair students and distribute copies of a word map graphic organizer. Write the following list of words on the board: *herbicide, homicide, genocide, pesticide,* and *insecticide.* Remind students that the word *herbicide* is defined as "an agent used to destroy or restrain plant growth." Its root, *herb,* means "plant." The suffix *-cide,* derived from the Latin *cida,* means "cut" or "kill." Instruct each pair of students to fill in the word map for these similar words, listing for each their own definition, the dictionary definition, a word clue that focuses on the parts of each word, and an association with the word. Invite volunteers to share the components of their word maps.

The same GFP technology was used to make fluorescent zebra fish. Experiments moving the GFP gene have taught scientists much about cells and proteins. But, just because we have the technology to change an organism's characteristics, should we?

It would be great if genetic engineering led to cures for genetic diseases. However, should biologists try to make people taller or change their sex or looks? What will happen to the human species when we can design our bodies or those of our children? What will happen if biologists want to clone human beings? In a democratic nation, everyone is responsible for making sure that the tools science has given us are used wisely. This means that you need to be ready to help decide what should and should not be done with the human genome.

Gaining More Understanding
These fluorescent zebra fish were originally bred to help scientists find environmental pollutants. Today, the techniques used to produce such fish are widely used in research. Fluorescent labels are helping researchers to understand cancer and other diseases.

Key Question Should genetic modifications to humans and other organisms be closely regulated?
The issue: Just because we have the technology to modify an organism's characteristics, are we justified in doing so?

CHECK Understanding

Critical Thinking

1. **Apply Concepts** How could biotechnology affect your privacy?
2. **Explain** What are genetically modified foods?
3. **Form an Opinion** Should a vegetarian be concerned about eating a GM plant that contains DNA from a pig gene? Support your answer with details from the text.
4. **Explain** What are the main concerns about genetic engineering discussed in this chapter?
5. **Write to Learn** Answer the question in the mystery clue below.

MYSTERY CLUE

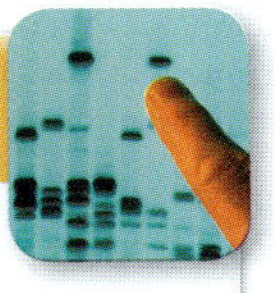

What privacy considerations, if any, should investigators have taken into account when obtaining the DNA evidence? (**Hint:** See p. 367.)

BIOLOGY.com Search Lesson 15.4 GO • Lesson Assessment

Active Reading

MAKE CONNECTIONS Draw students' attention to the photograph of the fluorescent fish. Point out that scientists made this type of fish by transferring the GFP (green fluorescent protein) gene to the fish's DNA. Open a discussion on ethics by asking the following questions.

Ask How do you feel about scientists transferring the GFP gene so that they can study cellular proteins and learn how cells function? *(Answers may vary.)* What is your opinion of transferring the GFP gene so that your pet fish can glow in the dark? *(Answers may vary.)*

Ask Would you feel differently if the gene was transferred to dogs or horses, so that they could glow in the dark? Explain. *(Answers may vary.)*

KEY QUESTION Ask a volunteer to read the Key Question to the class.

Ask How is this question different from other Key Questions in this chapter? *(It is not a factual question; it requires judgment.)*

Have a second volunteer read the answer aloud.

Ask What is different about this answer? *(It is another question. It is not a clear answer.)*

Ask Who should be in charge of deciding which genetic modifications are allowed? *(Answers may vary. Students should justify their responses and be respectful of differing opinions.)*

Post Reading

ASSESSMENT TIPS Remind students that they can refer to the two-column chart they began at the start of the lesson to answer Question 4.

MYSTERY CLUE Lead a discussion about the constitutional rights of citizens of the United States, including privacy rights. Read aloud Amendment IV of the Bill of Rights to the U.S. Constitution, and discuss whether students think collecting DNA from a person would violate their right against "unreasonable searches and seizures."

Wrap-Up Activity

NEWSCAST EDITORIALS Have students work in pairs to prepare a mock newscast editorial. One student can present the pros of genetic modification; the partner can present the cons. Each student should cover GM in plants, animals, and humans. Each should address privacy issues.

Check Understanding

1. Sample answer: It could be used to perform tests I didn't order. Health insurance companies could decide not to insure me if I have an illness-related allele.
2. Sample answer: Genetically modified foods are foods produced from GM crops.
3. Answers may vary but should show an understanding of how plants are genetically modified with animal genes.
4. Sample answer: whether genetic engineering will be used wisely and ethically
5. Answers may vary.

CHAPTER LAB

Pre-Lab

Review the section in Lesson 15.3 on the use of DNA in forensics.

Ask What kinds of samples are needed for forensic testing? *(Sample answers: blood, hair, skin)*

Ask What do technicians look for when they compare DNA evidence with a suspect's DNA fingerprint? *(They compare the pattern of dark bands formed by the DNA. The pattern of bands from one person is unlike anyone else's pattern of bands.)*

Review the Background Questions and the Pre-Lab Questions as a class.

Lab

Tell students they will perform the chapter lab *Using DNA to Solve Crimes* described in **Lab Manual B.**

SAFETY Students should wear aprons, protective eyewear, and gloves. Remind students to watch out for electric cords and batteries and to keep them away from water. They should wash their hands after the procedure is completed.

Post-Lab

COURT TESTIMONY Have students prepare their conclusions for use as testimony in a court case. Each conclusion should be supported by laboratory observations.

NATIONAL SCIENCE EDUCATION STANDARDS

CONTENT

E.2

INQUIRY

A.1.b, A.1.c, A.2.a, A.2.b, A.2.e

Forensics Lab

GUIDED INQUIRY

Pre-Lab: Using DNA to Solve Crimes

Problem How can DNA samples be used to connect a suspect to a crime scene?

Materials gel block, electrophoresis chamber, buffer solution, 250-mL beaker, metric ruler, DNA samples, micropipettes with metal plungers, 9-volt batteries, electric cords, staining tray, DNA stain, 100-mL graduated cylinder, clock or timer

Lab Manual Chapter 15 Lab

Skills Focus Measure, Compare and Contrast, Draw Conclusions

Connect to the Big idea Scientists who worked on the Human Genome Project had to develop methods for sequencing and identifying genes. Those methods have since been used for many other applications. For example, genetically altered bacteria are used to produce large amounts of life-saving drugs. Another example is the use of DNA evidence to solve crimes. In this lab, you will prepare and compare DNA "fingerprints," or profiles.

Background Questions

a. Review What characteristic of the human genome makes DNA a powerful tool for solving crimes?

b. Review What do the segments of DNA that are used to make DNA profiles have in common?

c. Apply Concepts When forensic scientists want to determine whether two DNA samples come from the same person, they analyze more than one section of DNA. Why would the results be less reliable if the scientists compared only one section of DNA?

Pre-Lab Questions

Preview the procedure in the lab manual.

1. **Control Variables** Why must you use a new pipette to load each DNA sample?
2. **Relate Cause and Effect** Why will the DNA samples separate into bands as they move through the gel?
3. **Infer** Why is purple tracking dye added to the DNA samples?

BIOLOGY.com Search Chapter 15 GO

Visit Chapter 15 online to test yourself on chapter content and to find activities to help you learn.

Untamed Science Video Pigeon breeding helps the Untamed Science crew unravel the mysteries of genetic engineering.

Art in Motion View a short animation that brings bacterial transformation to life.

Art Review Review your understanding of DNA fingerprinting with this drag-and-drop activity.

InterActive Art Build your understanding of key chapter concepts with these animations.

Pre-Lab Answers

BACKGROUND QUESTIONS

a. Except for identical twins, each individual has unique DNA.

b. The segments are regions with repeated DNA sequences that do not code for proteins.

c. Sample answer: The results would be less reliable because the probability of two people having an identical sequence in only one DNA section is much greater than the probability of two people having identical sequences in multiple sections.

PRE-LAB QUESTIONS

1. Using the same pipette would cause all samples to be contaminated, except the first sample.
2. The DNA samples contain fragments of different lengths. Shorter fragments will travel through the gel faster than longer fragments.
3. Because the DNA will not be visible until it is stained, the dye provides a visual indicator of when to stop the electrophoresis.

15 CHAPTER Summary

15.1 Selective Breeding

- Humans use selective breeding, which takes advantage of naturally occurring genetic variation, to pass wanted traits on to the next generation of organisms.
- Breeders can increase the genetic variation in a population by introducing mutations, which are the ultimate source of biological diversity.

selective breeding (p. 354) inbreeding (p. 355)
hybridization (p. 355) biotechnology (p. 355)

15.2 Recombinant DNA

- The first step in using the polymerase chain reaction method to copy a gene is to heat a piece of DNA, which separates its two strands. Then, as the DNA cools, primers bind to the single strands. Next, DNA polymerase starts copying the region between the primers. These copies can serve as templates to make still more copies.
- Recombinant-DNA technology is the joining together of DNA from two or more sources. This technology makes it possible to change the genetic makeup of living organisms.
- Transgenic organisms can be produced by the insertion of recombinant DNA into the genome of a host organism.

polymerase chain reaction (p. 358)
recombinant DNA (p. 359)
plasmid (p. 359)
transgenic (p. 360)
clone (p. 361)

15.3 Applications of Genetic Engineering

- Ideally, genetic modification could lead to better, less expensive, and more nutritious food as well as to less harmful manufacturing processes.
- Recombinant-DNA technology is advancing the prevention and treatment of disease.
- DNA fingerprinting analyzes sections of DNA that may have little or no function within the cell. However, these sections vary widely from one individual to another.

gene therapy (p. 364)
DNA microarray (p. 364)
DNA fingerprinting (p. 365)
forensics (p. 365)

15.4 Ethics and Impacts of Biotechnology

- Should you, like patent holders, be able to keep your genetic information confidential?
- Careful studies of GM foods have provided no scientific support for concerns about their safety.
- There are many concerns about unintended consequences that a shift to GM farming and ranching may have on agriculture.
- Just because we have the technology to modify an organism's characteristics, are we justified in doing so?

UbD Transfer the Big Idea

TRANSFER Have students work in pairs to write a scene of a television show in which a detective talks with a police lab technician about the process of DNA fingerprinting. Students can use characters from a favorite TV show or make up their own characters. In this scene, the detective has come to the lab technician with a blood sample from a crime scene. The detective wants to understand how DNA fingerprinting works but has a weak science background. So, the lab technician explains in simple, non-technical language what DNA and genes are, which samples can be used, how restriction enzymes and gel electrophoresis work, and what the results of DNA fingerprinting mean. Invite students to act out one or two of the scripts for the class.

Chapter Summary

Think Visually

USE GRAPHICS Have students fill in the following concept map.

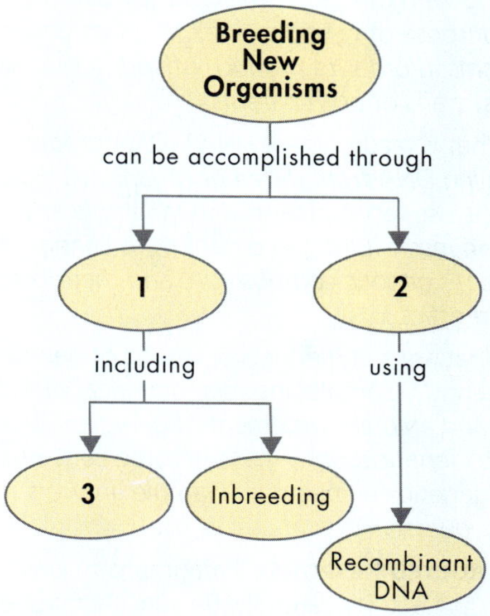

ANSWERS

1. Selective breeding **2.** Genetic engineering
3. Hybridization

BIOLOGY.com Study Online

REVIEW RESOURCES

Tutor Tube The online tutor shares analogies to help students understand gel electrophoresis.

Art in Motion In this animation, students can observe the steps of DNA transformation.

Data Analysis Students explore how genetically modified crop plants can help prevent nutrient deficiencies.

Vocabulary Review The *Flash Cards* and *Crossword* provide an interactive way to review chapter vocabulary.

Editable Worksheets Editable pages of Study Workbook B, Lab Manual B, and the Assessment Resources Book are available online.

Lesson Overviews Lesson Overviews are based on Miller & Levine Biology. However, they are available on CD-ROM and can be edited for your class.

ASSESSMENT RESOURCES

Chapter Assessment Have students take an online version of the Chapter 15 Check Understanding.

Standardized Test Prep Students can take an online version of the Standardized Test Prep. You will receive their scores along with ideas for remediation.

Diagnostic and Benchmark Tests Use these tests to monitor your students' progress and supply remediation.

ASSESSMENT

Assess

Big idea Invite a volunteer to read the question. Write the words *manipulate DNA* at the top of the board.

Ask What is the purpose of inbreeding? *(to produce offspring with the same traits as their parents)* What is the purpose of hybridization? *(to cross dissimilar organisms in order to produce offspring that have the best traits of both parents)*

Ask What is recombinant DNA? *(DNA produced by combining DNA from different organisms)* How do scientists use recombinant DNA? *(Sample answers: They can insert it into an organism to change the organism's genetic composition, and therefore its characteristics.)*

Ask What are some effects or results of genetic engineering? *(Sample answers: genetically modified plants and animals, treatment of genetic diseases, genetic identification)* What are some concerns about genetic engineering? *(Sample answers: privacy, ownership, and ethical use)*

CONSTRUCTED RESPONSE Paragraphs may vary. Explain the activity carefully. Be sure that students understand that the *Hints* should be used as a guide to help them answer the questions. Review how to write a paragraph.

Write Question 1 on the board and read it aloud. Underline *breeders* and *genetic variations.*

Say This question is about breeding organisms and genetic variations. We learned about selective breeding in Lesson 15.1.

Read the first hint.

Ask What are mutations? *(Mutations are changes in DNA that parents can pass to offspring.)* What are ways that people can cause mutations in organisms? *(Breeders can make mutations happen using radiation and chemicals.)*

Read the next hint.

Ask What is one way to change the number of chromosomes in an organism? *(Some drugs stop the chromosomes in an organism's cells from splitting during meiosis, so the cells have more than the normal number of chromosomes.)* What do we call plants with extra sets of chromosomes? *(polyploid plants)*

Model writing the first one or two sentences using the hints on the board and the text. *(Sample answer: To produce organisms with new traits, breeders can cause mutations using radiation or chemicals. This is usually done using bacteria. Breeders also can produce polyploid plants.)*

15 CHECK Understanding

Assess the Big idea Science as a Way of Knowing

Write an answer to the question below.

Q: How and why do scientists manipulate DNA in living cells?

Constructed Response

Write an answer to each of the questions below. The answer to each question should be one or two paragraphs long. To help you begin, read the **Hints** below each of the questions.

1. **How do breeders produce genetic variations that are not found in nature?**

   Hint Some treatments can cause mutations.

   Hint Some treatments can change the numbers of chromosomes in an organism.

2. **Describe what happens during a polymerase chain reaction.**

   Hint A polymerase chain reaction makes many copies of a gene.

   Hint Each new copy made can serve as a template to make more genes.

3. **Your friend suggests that genetic engineering makes it possible for biologists to produce an organism with any combination of characteristics—an animal with the body of a frog and the wings of a bat, for example. Do you think this is a reasonable statement? Explain your answer.**

   Hint All organisms share the same genetic code.

   Hint Organisms have millions of genes.

Foundations for Learning Wrap-Up

Use the cubes you made as you read the chapter. The cubes will help you review what you have learned about genetic engineering.

Activity 1 Working with a partner, take turns rolling one of the cubes. Read the word on top of the cube. Say the definition of the word. Your partner will tell you if you are correct. If you don't know the definition, you lose your turn. Continue until you have defined all the words at least once. The first player to get all the definitions correct is the winner.

Activity 2 Working with a partner, use both cubes. Take turns rolling both cubes. Tell how the words from the tops of the cubes are related. For example, for the cubes below, you could say, "Scientists use *plasmids* to make *transgenic* plants."

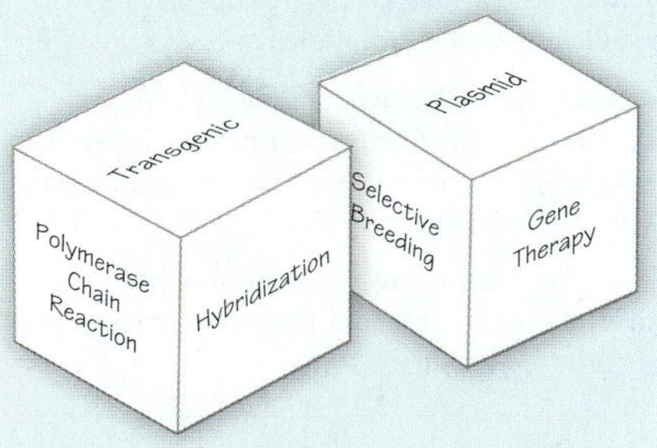

For Question 2, pair students and have them use the hints and your modeling to answer the question. Have students share their paragraphs with the class.

For Question 3, have students write the paragraphs independently for homework or as an anchor activity.

Foundations for Learning Wrap-Up

Have students take out their word cubes. Check that students have recorded all vocabulary words on the sides of the cubes—one word per side—as they read the chapter. Pair students and explain the activities.

15.1 Selective Breeding

Understand Key Concepts

1. Crossing individuals with different traits to bring together their best characteristics is called
   a. domestication.
   b. inbreeding.
   c. hybridization.
   d. polyploidy.
2. Crossing individuals with similar characteristics so that those characteristics will appear in their offspring is called
   a. inbreeding.
   b. hybridization.
   c. recombination.
   d. polyploidy.
3. Taking advantage of naturally occurring variations in organisms to pass wanted traits on to future generations is called
   a. selective breeding.
   b. forensics.
   c. gene therapy.
   d. mutation.
4. What is polyploidy? When is this condition useful?

Think Critically

5. **Compare and Contrast** Hybridization and inbreeding are important methods used in selective breeding. How are the methods similar? How are they different?

Test-Taking Tip

Use Graphic Organizers Sometimes a question, such as question 5, asks you to compare and contrast two processes. Before you begin writing your answer, organize your ideas with a Venn diagram. Label one circle Hybridization. Label the other Inbreeding. Where the circles overlap, write Both.

15.2 Recombinant DNA

Understand Key Concepts

6. Organisms that have genes from other organisms are called
   a. transgenic. c. donors.
   b. mutagenic. d. clones.
7. A member of a population of genetically identical cells produced from a single cell is a
   a. clone. c. mutant.
   b. plasmid. d. sequence.
8. How does a transgenic plant differ from a hybrid plant?

Think Critically

9. **Apply Concepts** Describe one or more advantages of producing insulin and other proteins through genetic engineering.

15.3 Applications of Genetic Engineering

Understand Key Concepts

10. Which of the following characteristics is often genetically engineered into crop plants?
    a. improved flavor
    b. resistance to herbicides
    c. different colors
    d. thicker stems

Think Critically

11. **Infer** Suppose a human's bone marrow was removed, changed genetically, and returned to his body. Would the change be passed on to the patient's children? Explain your answer.

Assessment Tips

Pair students and have them work together to answer the questions on this page and the next.

Remind students that the answer to Question 5 must address both parts of the question: how the methods are alike and how they are different.

Make sure students realize that the answers to Questions 11 and 15 are not directly in the textbook. Students will have to use concepts they have learned, and apply them to the situations described in the questions. For example, in Question 15, students must apply what they know about complementary DNA and restriction enzymes.

Have students share their answers with the class and explain their reasoning for each answer.

Answers

Lesson 15.1

UNDERSTAND KEY CONCEPTS

1. c

2. a

3. a

4. Polyploidy describes an organism with more than the typical number of chromosomes. It can be useful to produce new species of plants that are larger and stronger than their diploid relatives.

THINK CRITICALLY

5. They are similar in that both involve selective breeding that takes advantage of naturally occurring genetic variation to produce wanted characteristics in the next generation. They are different in that hybridization crosses dissimilar individuals to bring together the best of both organisms, while inbreeding involves breeding individuals of similar characteristics.

Lesson 15.2

UNDERSTAND KEY CONCEPTS

6. a

7. a

8. A transgenic plant is made by inserting DNA from another individual, while a hybrid plant contains genes from only the parent plants.

THINK CRITICALLY

9. Sample answer: One advantage of producing insulin or other proteins through genetic engineering is that huge cultures of bacteria can be cheaply grown, so huge supplies of insulin can be made much more inexpensively than taking insulin from mammals. Also, recombinant DNA can be used to make human insulin, rather than having to rely on insulin made by other animals, which may be different and cause reactions in some people.

Lesson 15.3

UNDERSTAND KEY CONCEPTS

10. b

THINK CRITICALLY

11. The change would not be passed on because genetically altering bone marrow would not affect the DNA in gametes involved in reproduction.

ASSESSMENT

Lesson 15.4

UNDERSTAND KEY CONCEPTS

12. c

13. b

THINK CRITICALLY

14. Sample answer: Expensive GM seeds may force small farmers out of business, especially in the developing world.

Connecting Concepts

15. Complementary strand: TACTCTAGATGCCTTAAGAGTTCAAGCTTAGC On the original strand, *Bgl*III will cut after the fourth nucleotide and *EcoRI* will cut after the thirteenth nucleotide. *Hind*III will not cut the strand.

CHAPTER MYSTERY After students have read through the Chapter Mystery, review with them the evidence that suggested the first suspect was guilty of the crime. *(His clothing was stained with blood that matched the victim's blood type, there were scratches on his arms, and he couldn't remember what he had done.)* Then, discuss the process investigators used to exonerate this suspect.

Ask How would investigators have tested the blood on his clothing for DNA? *(They would have used DNA fingerprinting.)*

Call on students to describe details of the DNA fingerprinting process.

Ask In reaching the conclusion that the suspect was not the attacker, what would investigators have compared the results of that DNA fingerprint to? *(They would have compared the DNA fingerprint of the blood on his clothing to the DNA fingerprint of the victim.)*

15 CHECK Understanding

15.4 Ethics and Impacts of Biotechnology

Understand Key Concepts

12. The right to profit from a new genetic technology is protected by

a. getting a copyright for the method.

b. discovering a new gene.

c. obtaining a patent.

d. publishing its description in a journal.

13. Which of the following is most likely to be used in a court case to determine who the father of a particular child is?

a. microarray analysis **c.** gene therapy

b. DNA fingerprinting **d.** genetic engineering

Think Critically

14. Explain What is one argument used by critics of genetically modified foods?

Connecting Concepts

Use Science Graphics

Use the table below to answer question 15.

DNA Restriction Enzymes

Enzyme	Recognition Sequence
*Bgl*III	A↓G A T C T T C T A G↑A
*Eco*RI	G↓A A T T C C T T A A↑G
*Hind*III	A↓A G C T T T T C G A↑A

15. Apply Concepts Copy the following DNA sequence and write its complementary strand ATGAGATCTACGGAATTCTCAAGCTTCGA ATCG. Where will each restriction enzyme in the table cut the DNA strand?

solve the CHAPTER MYSTERY

A CASE OF MISTAKEN IDENTITY

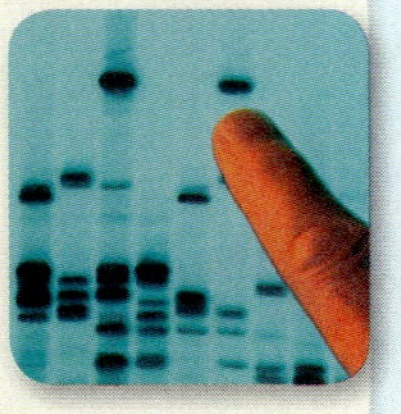

The first suspect was lucky: Twenty years earlier, it would have been an open-and-shut case. But by 1998, DNA fingerprinting was widely available. After the police took the suspect into custody, forensic scientists tested the DNA in the bloodstains on his shirt. Within a few hours, they knew they had the wrong suspect. Before long, the police caught the real attacker, who was then tried in a court and convicted of the crime.

1. Infer How did the investigators determine that the person they took into custody was not a suspect?

2. Apply Concepts Red blood cells don't have a nucleus; white blood cells do. Did the DNA evidence from the bloodstains come from the red blood cells, the white blood cells, or both? Explain your answer.

3. Predict What if the initial suspect was related to the victim? Would that have changed the result? Why or why not?

4. Predict What if this crime happened before DNA fingerprinting was discovered? What do you think the police would have done after they took in the first suspect?

Untamed Science **Never Stop Exploring Your World.** Finding the solution to the case of mistaken identity is only the beginning. Take a video field trip with the ecogeeks of Untamed Science to see where the mystery leads.

BIOLOGY.com Search Chapter 15 GO • Untamed Science Video • Chapter Mystery

CHAPTER MYSTERY ANSWERS

1. DNA fingerprinting showed investigators that the blood on his clothing didn't match the victim's.

2. The DNA evidence from the bloodstains came from the white blood cells because red blood cells don't have nuclei, so they don't have DNA. When enough probes are used, DNA fingerprinting can distinguish samples of even the closest relatives.

3. It probably wouldn't have changed the result, because members of the same family have similar but not identical DNA.

4. Sample answer: The man might have been arrested and tried. Because the blood type of the blood on his clothing matched the victim's blood type and because he didn't remember where he'd been, jurors may have convicted him in a trial.

Untamed Science Students can see the Untamed Science crew unravel the mysteries of genetic engineering in animals by watching **Designer Pigeons.**

Standardized Test Prep

Multiple Choice

1. Polyploidy may produce new types of organisms that are larger and stronger than their diploid relatives in
- **A** animals.
- **B** plants.
- **C** bacteria.
- **D** fungi.

2. Which of the following characteristics does NOT apply to a plasmid?
- **A** made of DNA
- **B** found in bacterial cells
- **C** has circular loops
- **D** found in animal cells

3. To separate DNA fragments from one another, scientists use
- **A** polymerase chain reaction.
- **B** DNA microarrays.
- **C** gel electrophoresis.
- **D** restriction enzymes.

4. Recombinant DNA technology involves
- **A** cutting DNA into nucleotides.
- **B** selective breeding.
- **C** joining DNA from two or more sources.
- **D** producing a cloned organism.

5. The activity of thousands of genes at one time can be followed using
- **A** polymerase chain reaction.
- **B** plasmid transformation.
- **C** restriction enzymes.
- **D** DNA microarrays.

6. Genetically engineered crop plants can benefit farmers by
- **A** increasing crop yield.
- **B** introducing chemicals into the environment.
- **C** increasing an animal's resistance to antibiotics.
- **D** changing the genomes of other crop plants.

7. Genetic markers allow scientists to
- **A** clone animals.
- **B** separate strands of DNA.
- **C** make antibiotics.
- **D** identify transformed cells.

Questions 8–9

The graph below shows the number of accurate copies of DNA produced by polymerase chain reaction.

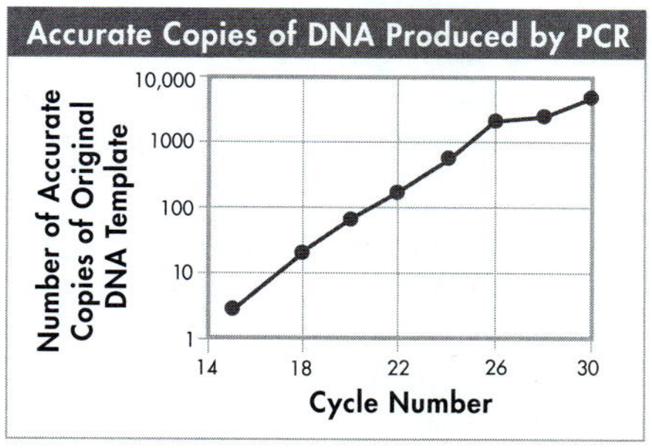

8. What can you conclude about cycles 18 through 26?
- **A** PCR produced accurate copies of template DNA at an exponential rate.
- **B** The amount of DNA produced by PCR doubled with each cycle.
- **C** The DNA copies produced by PCR were not accurate copies of the original DNA template.
- **D** The rate at which PCR produced accurate copies of template DNA fell in later cycles.

9. Based on the graph, which of the following might have happened between cycles 26 and 28?
- **A** PCR stopped producing accurate copies of the template.
- **B** The rate of reaction slowed.
- **C** All of the template DNA was used up.
- **D** A mutation occurred.

Open-Ended Response

10. Why are bacteria able to make human proteins when a human gene is inserted in them with a plasmid?

If You Have Trouble With . . .										
Question	1	2	3	4	5	6	7	8	9	10
See Lesson	15.1	15.2	15.2	15.2	15.3	15.4	15.2	15.2	15.2	15.3

Test-Taking Tip

ELIMINATE INCORRECT ANSWERS

If students are confused by a question, suggest they start by eliminating each answer choice that they know is not correct. After eliminating incorrect choices, they can read over the remaining choices and choose the best answer.

Before the Test

Review the Test-Taking Tip on eliminating incorrect answers. Point out that if students eliminate the answer choices they know are incorrect, they will increase their chances of choosing the correct answer from the remaining choices.

Point out the word *NOT* in Question 2.

Remind students that an open-ended response means that they'll need to write a paragraph to answer the question. Have them paraphrase the question in their notebooks, and remind them about the steps in writing a paragraph. Encourage them to leave enough time to read over their finished paragraph to make certain that it answers the question.

Answers

1. B

2. D

3. C

4. C

5. D

6. A

7. D

8. A

9. B

10. Bacteria are able to make human proteins when a human gene is inserted into them because the genetic code is universal. Proteins are made up of amino acids, and DNA codes for the amino acids of a protein. A DNA sequence, such as ATG, specifies a certain amino acid. It doesn't matter if the cell is a bacterial cell or a human cell, it will interpret the code the same way.

Plan Ahead

Have students preview the Unit 4 Project a few days before making the collage. Suggest that they start collecting magazine, newspaper, and Internet images and articles that they want to use. You may also want to gather additional source materials that students can use the day they create their collages.

Materials magazines and newspapers, posterboard, glue or tape, scissors, colored markers or pens, additional art materials

Monitor the Project

Suggest students start by laying out the basic images and text in their collages before pasting or taping them down. As they work, walk around the classroom and ask students questions that will help them organize and design their collages.

Ask How does this image show why DNA is important to a cell?

Ask How can you show in pictures and words how DNA and genetics influence society?

Remind students to address all three central questions in their collages.

Project Assessment

Make sure students use the rubric and reflection questions to assess their work. Then, use the rubric to assign a final score. Note that it is important to value the creativity of students' work as well as the content when you score their projects. If desired, talk with students about any differences between their self-assessment scores and your assigned score.

Unit Project

Genetics Collage

Genetics is a fascinating field of study and is becoming increasingly important to society. A local genetics laboratory in your town wants to increase public awareness of the importance of genetics. To do so, it has decided to hold a scholarship competition. The scholarship will go to the student(s) who create the best educational collage related to topics in genetics.

Your Task Use magazine and newspaper clippings, Internet sources, and art materials to make a colorful collage. The images should relate to three central questions.

1) Why is DNA important to a cell?
2) Why is DNA important to you, as a human being?
3) Why is DNA important to society as a whole?

Be sure to

- communicate answers to the above questions in the images, words, and phrases you choose.
- carefully design your collage so that it is clear and organized.

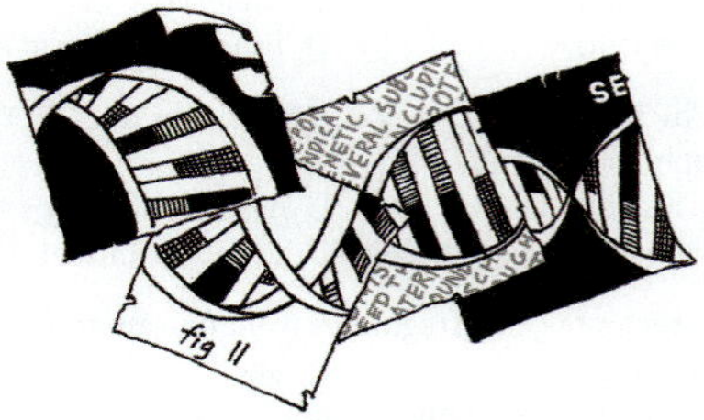

Reflection Questions

1. **Score your collage using the rubric below. What score did you give yourself?**
2. **What did you do well in this project?**
3. **What about your collage needs improvement?**
4. **What could a person who didn't know much about DNA learn from your collage?**

Assessment Rubric

Score	Scientific Content	Quality of Collage
4	Collage includes many important and thoughtful images related to the three central questions. Student demonstrates a deep understanding of genetics topics.	The collage is clear, organized, and creative.
3	Collage includes important images related to the three central questions. Student demonstrates an adequate understanding of genetics topics.	The collage is well designed and organized.
2	Collage is missing some important ideas and/or includes several insignificant ideas. Student demonstrates a limited level of understanding of genetics topics.	The collage could be better designed and organized.
1	Collage is missing several important ideas. Student demonstrates significant misunderstandings.	The collage is unclear and lacks a solid design.

NATIONAL SCIENCE EDUCATION STANDARDS

UCP II

CONTENT C.2.a, F.6, G.1

21st Century Skills

To be successful in the 21st century, students need skills and learning experiences that extend beyond subject area mastery. The Unit 4 Project helps students build the following 21st Century Skills: *Information and Media Literacy; Creativity and Intellectual Curiosity; Self-Direction;* and *Accountability and Adaptability.*

FOCUS ON INFORMATION AND MEDIA LITERACY Extend this Unit Project by having small groups of students evaluate the scientific accuracy of the media sources they used when making their collages. Then, ask them to consider the articles and images they chose (as well as the ones they did not) and discuss the intended audience of this information. Suggest students also discuss how the source's point of view may have influenced the information they presented.

For more practice building 21st Century Skills, see The Chapter Mystery pages in **Study Workbook A.**

376 Unit 4 Project

Evolution

UNIT 5

Chapters

INTRODUCE the **Big ideas**

- **Evolution**
- **Unity and Diversity of Life**

"Darwin's theory of evolution by natural selection is often called 'the most important scientific idea that anyone has ever had.' Evolutionary theory provides the best scientific explanation for the unity and diversity of life. It unites all living things in a single tree of life and reminds us that humans are part of nature. As researchers explore evolutionary mysteries, they continue to marvel at Darwin's genius and his grand vision of the natural world."

Joe Levine

377

UNIT 5

Dear Colleague,

In 1992, a doctor at Tufts Medical School published a book entitled The Antibiotic Paradox, *explaining how bacteria are becoming resistant to nearly all known antibiotics. The author of the book, Stuart Levy, has argued that medicine and public health would be much better off if medical students were taught as much about Darwin as they are about Pasteur.*

Huh? Why should doctors learn so much about Darwin? Because bacteria evolve under pressure from natural selection—just as Darwin proposed. Once humans made antibiotics part of the environment, bacteria evolved resistance to those drugs.

Darwin's theory of evolution has been described as "the most important scientific idea that anyone has ever had." Ever since Darwin, new branches of science have appeared and matured, gathering information about the living world beyond Darwin's wildest dreams. Any of that evidence—from biochemistry, molecular genetics, geology, and physics—could have confirmed or negated Darwin's work. Astonishingly, all those new data supported, strengthened, and expanded Darwin's insights. Evolutionary theory now informs every aspect of biological thought, from global ecology to medicine.

But, evolutionary theory does more than just describe these phenomena; it enables us to predict how organisms will respond to events around them. Evolutionary theory informs new treatments for AIDS, new approaches to the production and use of antibiotics, and new strategies for using insecticides. These and other applications explain why understanding evolution is vital to making informed judgments about many issues in the modern world.

The goal of this unit is to help students understand the evolutionary worldview. As scientists and teachers, we believe very strongly that the purpose of education is to promote understanding, not to compel belief. That applies to evolution, too, which, if properly taught, should never threaten the beliefs of students. As biologists, we genuinely feel, as Darwin wrote, "There is grandeur in this view of life . . ." We hope you agree.

Joe Levine

CHAPTER PLANNER 16 Darwin's Theory of Evolution

Chapter Contents	NSES	Time	Core Resources
Chapter Preview			**Student Edition,** pp. 378–379 **Chapter Mystery,** p. 379 **Foundations for Learning,** p. 379
16.1 Darwin's Voyage of Discovery Darwin's Big Journey • Observations Aboard the *Beagle*	II, IV, A.2.a, A.2.b, C.3.a, G.1, G.2, G.3	1/2 period 1/4 block	**Student Edition,** pp. 380–383 **Study Workbook B** 16.1 Worksheets **Biology.com** 16.1 Lesson Assessment
16.2 Ideas That Shaped Darwin's Thinking An Ancient, Changing Earth • Lamarck's Evolutionary Hypotheses • Population Growth • Artificial Selection	I, II, IV, V, A.1.b, A.1.c, A.1.e, A.2.a, A.2.e, A.2.f, C.3.a, C.4.d, D.3, F.2, F.3, G.1, G.2, G.3	1 period 1/2 block	**Student Edition,** pp. 384–387 **Inquiry Into Scientific Thinking,** p. 386 **Study Workbook B** 16.2 Worksheets **Biology.com** *Art in Motion:* The Ladder of Life • 16.2 Lesson Assessment
16.3 Darwin Presents His Case Evolution by Natural Selection • Common Descent	I, II, IV, V, A.2.a, A.2.f, C.3.a, C.3.b, C.3.c, C.3.d, C.3.e, C.4.d, C.6.c, G.1, G.2, G.3	1 period 1/2 block	**Student Edition,** pp. 388–391 **Build Connections,** pp. 389 and 390 **Study Workbook B** 16.3 Worksheets **Biology.com** 16.3 Lesson Assessment
16.4 Evidence of Evolution Biogeography • Comparing Body Structure and Embryos • The Age of Earth and Fossils • Genetics and Molecular Biology • Testing Natural Selection	I, II, III, IV, V, A.2.a, A.2.b, A.2.c, A.2.e, A.2.f, C.1.f, C.2.a, C.3.a, C.3.c, C.3.d, D.3, E.2, G.1, G.2, G.3	1 1/2 periods 3/4 block	**Student Edition,** pp. 392–397 **Build Connections,** pp. 394 and 396 **Study Workbook B** 16.4 Worksheets **Biology.com** *Art Review:* Homologous and Analogous • *Visual Analogy:* Finch Beak Tools • 16.4 Lesson Assessment
Chapter Pre-Lab	IV, C.2.a, C.2.c, C.3.d, C.3.e	1 period 1/2 block	**Student Edition,** p. 398 **Lab Manual B** *Amino Acid Sequences: Indicators of Evolution*

Tools for ELL Instruction

ELL Handbook describes teaching strategies to make *Miller & Levine Biology: Foundation Edition* more accessible to ELL students.

Spanish Study Workbook is a Spanish translation of Study Workbook A.

Multilingual Glossary is the *Miller & Levine Biology* glossary translated into ten languages.

Tools to Challenge Students

Study Workbook A includes more challenging worksheets and activities such as Chapter Mystery activities that reinforce 21st Century Skills.

Lab Manual A labs are geared toward on-level students. The Lab Manual includes fifteen labs in addition to the Chapter Labs.

Biology.com Data Analysis activities offer students a way to practice data gathering and analyzing skills.

Visual Quizzes in the Assessment Resources Book are geared toward on-level students and use visuals from *Miller & Levine Biology.*

Additional Resources

Biology.com Untamed Science Video • Flash Cards

Study Workbook A 16.1 Worksheets
Spanish Study Workbook 16.1 Worksheets

Study Workbook A 16.2 Worksheets
Spanish Study Workbook 16.2 Worksheets

Study Workbook A 16.3 Worksheets
Spanish Study Workbook 16.3 Worksheets
Biology.com *Data Analysis:* Natural Selection
Assessment Resources Book Visual Quiz

Study Workbook A 16.4 Worksheets
Lab Manual B *Hands-On Activity:* Comparing Bones • *Data Analysis:* Molecular Homology in *Hoxc8*
Spanish Study Workbook 16.4 Worksheets

Lab Manual A *Amino Acid Sequences: Indicators of Evolution*

Chapter Review

Student Edition Chapter Summary, p. 399
Study Workbook B Chapter 16 Review
Transparencies, pp. 193–202
Biology.com Untamed Science Video • Editable Worksheets of Study Workbooks A and B and Lab Manuals A and B • Chapter 16 Flash Cards and Crossword

Untamed Science DVD • Classroom Resources CD (includes lesson presentations and editable worksheets)

Chapter Assessment

Student Edition Check Understanding, pp. 400–403
Study Workbook B Chapter 16 Taking a Standardized Test
Assessment Resources Book Chapter 16 Test B
Biology.com Chapter 16 Assessment • Editable Worksheets of Chapter 16 Visual Quizzes* and Chapter 16 Tests A and B

Exam*View* *Assessment Suite* • Classroom Resources CD (includes lesson presentations and editable worksheets)

Time: 1 period, 1/2 block

* Visual Quizzes are based on visuals from *Miller & Levine Biology*. Review carefully to make sure they are appropriate for your class.

Pressed for Time?

Preview the Chapter Preview the Adaptations and Natural Selections visuals, and introduce the Key Questions for Lessons 16.3 and 16.4.

Cover the Chapter Quickly Have students read *Darwin's Big Journey* in Lesson 16.1 and all of Lesson 16.3. Discuss the Natural Selection visual. Assign *Biogeography, Comparing Body Structure and Embryos, The Age of Earth and Fossils,* and *Genetics and Molecular Biology* in Lesson 16.4 and discuss the visuals Homologous Limb Bones and Evidence From Fossils.

Assess Assign Questions 2 and 3 in the 16.1 Check Understanding, the 16.3 Check Understanding, Questions 1–7 in the 16.4 Check Understanding, and the Chapter 16 Standardized Test Prep, except Questions 2–4.

CHAPTER 16

Get the Picture

Tell students that the picture shows many individual organisms of the same species, the Cuban tree snail.

Say The snails are very similar, but they also show differences, or variations.

Ask In what ways do the snails appear similar? *(shell shape and size, striped pattern, multicolored)* How do the snails appear different? *(size of shell, sizes and numbers of stripes, exact colors)*

Ask If the yellow snails were better able to reproduce than the other snails, which color of snail would probably be most common in the species? *(yellow)*

Activate Prior Knowledge

Remind students that organisms pass their characteristics, or traits, to their offspring.

Ask What controls an organism's traits? *(genes)*

Say There can be different varieties of a trait.

Ask What is an example of a human trait that has different varieties? *(Sample answers: eye color, hair color, blood type)*

Ask What do scientists call the different forms of a gene that control the different varieties of a trait? *(alleles)*

16 Darwin's Theory of Evolution

Big idea Evolution

Q: What is natural selection?

BIOLOGY.com Search Chapter 16 GO • Flash Cards

NATIONAL SCIENCE EDUCATION STANDARDS

UNIFYING CONCEPTS AND PROCESSES

I, II, III, IV, V

CONTENT

C.1.f, C.2.a, C.2.c, C.3.a, C.3.b, C.3.c, C.3.d, C.3.e, C.4.d, C.6.c, D.3, E.2, F.2, F.3, G.1, G.2, G.3

INQUIRY

A.1.b, A.1.c, A.1.e, A.2.a, A.2.b, A.2.c, A.2.e, A.2.f

UbD Connect to the Big Idea

Invite a volunteer to read aloud the chapter number and the chapter title. Read the Big Idea and question. Emphasize the question, **What is natural selection?**

Set the Purpose

PREVIEW AND PREDICT Draw the graphic organizer on the board. Guide students to the INSIDE feature on the next page, and invite a volunteer to read each lesson number and title.

Say Charles Darwin was the scientist who developed the theory of evolution by natural selection. **Ask** What do you think people mean when they say that something evolves? *(It changes over time.)* When you make a selection, what are you doing? *(You are choosing one thing over another.)* What does the word *natural* mean in the term *natural selection*? *(taking place in nature)*

INSIDE:

- 16.1 Darwin's Voyage of Discovery
- 16.2 Ideas That Shaped Darwin's Thinking
- 16.3 Darwin Presents His Case
- 16.4 Evidence of Evolution

The shells of these Cuban tree snails show variation within a species. Variation provides the raw material for evolution.

- Untamed Science Video
- Chapter Mystery

CHAPTER **MYSTERY**

SUCH DIFFERENT HONEYCREEPERS

Imagine walking in the misty rain forest on an island in Hawaii. This is home for birds found nowhere else on Earth. A red and black bird flies to a nearby branch. The bird is an 'i'iwi. It uses its long, curved beak to get the nectar from the flowers of a tree.

The 'i'iwi is one of a number of species of Hawaiian honeycreepers. The different species eat different foods. Some eat nectar. Others eat insects, seeds, or fruit. Many honeycreepers eat only the seeds or nectar of plants that are unique to Hawaii.

How did all these birds get to Hawaii? How did they come to eat such specialized foods?

Read for Mystery Clues As you read this chapter, look for clues that help explain the number and diversity of Hawaiian honeycreepers. Then, solve the mystery at the end of the chapter.

FOUNDATIONS for Learning

Cut a stiff piece of paper so it is square. Fold the square diagonally to make a triangle. Fold that triangle again to make a smaller triangle. Open the paper. You will see an X. Color in one quarter; you will not write on this part of the paper. Three triangular sections now remain. Label each with one of these topics: Struggle for Existence, Variation and Adaptation, and Survival of the Fittest. As you read, take notes about each of these topics on the back of the correct triangle. At the end of the chapter are two activities that will use your notes to help answer the question: What is natural selection?

Chapter 16
Big Idea: Evolution

Chapter 16:
What is natural selection?

16.1: What patterns of biodiversity did Darwin observe while traveling aboard the *Beagle*?

16.2: How did other scientists' work help Darwin develop his theory of natural selection?

16.3: What is Darwin's theory of evolution by natural selection?

16.4: What are the main lines of scientific evidence that support Darwin's theory of evolution by natural selection?

Foundations for Learning

Model how to label and use the paper shape. Have a volunteer read aloud the topics to be labeled in the three triangles.

Ask How might focusing on the phrases on the triangles help you answer the question, **What is natural selection?** *(Sample answer: The phrases are important ideas that are connected to natural selection.)*

Ask On these two opening pages, what did you read about that is related to any of these phrases? *(Sample answers: variation in snails, variation in honeycreepers, the chapter question)*

Untamed Science™ Take a trip to Hawaii in the video **It Happened in Hawaii** with the ecogeeks of Untamed Science to find out why these tiny islands hold big potential for evolution.

CHAPTER MYSTERY Invite a volunteer to read aloud the Chapter Mystery.

Have students predict how this mystery relates to the chapter title. Guide students to discuss how the honeycreepers might have evolved, or changed over time. Ask students to identify the variation in honeycreeper species described in the Chapter Mystery. *(what the honeycreepers eat)* Have students brainstorm whether this variation makes the birds more or less likely to compete with one another. *(probably less likely)* Remind students to focus on the chapter question, **What is natural selection?,** as they look for clues to help them solve the Chapter Mystery.

Digital Support

BIOLOGY.com Extend your reach by using the digital assets offered at **Biology.com.**

LESSON 16.1

Pre-Reading

Objectives

16.1.1 State Charles Darwin's contribution to science.

16.1.2 Describe the three patterns of biodiversity noted by Darwin.

Preview the Pages

1. Have students read aloud all lesson headings and subheadings. Point out the three patterns of biological diversity noted by Darwin, listed in the section Observations Aboard the *Beagle*.
2. Preview the visuals. Tell students the portrait on this page is of Charles Darwin. Ask a volunteer to read aloud the names of the continents and oceans shown on the map of Darwin's voyage. Have students compare the traits of the two tortoises and of the armadillo and glyptodant.

Build Understanding

PREVIEW VISUALS Direct students to the preview visuals activity in Workbook B. Have them complete the activity before reading the lesson.

Build Vocabulary

RELATED WORD FORMS Remind students that the word *evolve* is a verb and *evolution* is a noun.

Ask What other nouns on this page end in *–tion*? *(observation, contribution)* To what verbs are these nouns related? *(observe, contribute)*

NATIONAL SCIENCE EDUCATION STANDARDS

UNIFYING CONCEPTS AND PROCESSES

II, IV

CONTENT

C.3.a, G.1, G.2, G.3

INQUIRY

A.2.a, A.2.b

For the corresponding lesson in *Miller & Levine Biology,* see pages 450–453.

16.1 Darwin's Voyage of Discovery

Key Questions

What was Charles Darwin's contribution to science?

What three patterns of biodiversity did Darwin note?

BUILD Understanding

Preview Visuals Before you read, look at the map of Darwin's trip on the next page. Describe the path that the ship took around the world.

In Your Workbook Go to your workbook to learn about previewing visuals.

BUILD Vocabulary

evolution
change over time; the process by which modern organisms have descended from ancient organisms

RELATED WORD FORMS
The noun *evolution* is related to the verb *evolve,* which means "to change over time."

Darwin's Big Journey

If you had met young Charles Darwin, you would not have guessed that he would grow up to be a famous scientist. As a boy, Darwin was not a top student. He would rather watch birds or read for pleasure than study. Yet he would go on to develop one of the most important scientific theories of all time.

Darwin's Time Charles Darwin was a naturalist and scientist from England. He grew up in the early 1800s, a time when the scientific view of the world was changing. Geologists were learning that the world was very, very old. They also suggested that the world we see today is constantly changing. Biologists were saying that life on Earth had also changed. The process of change over time is called **evolution.**

Sailing Around the World While still a young man, Darwin was invited to sail around the world on a ship called the *Beagle*. The captain and his crew would be mapping the coast of South America during the five-year journey. Darwin planned to collect plant and animal specimens along the way. At the time, no one knew how important the trip would be. Darwin's observations during the journey would lead to his theory of evolution.

Darwin's Work Darwin's theory explains how modern organisms evolved over millions of years. It also tells how modern organisms came from shared, or *common*, ancestors. But Darwin's work does more than explain life's history. Evolution reminds us that life is always changing. An understanding of evolution helps us to explain and find ways to overcome many challenges we face today. These challenges include drug-resistant bacteria and deadly new viruses.

Key Question What was Charles Darwin's contribution to science? **Darwin developed a theory of evolution that explains how modern organisms evolved over long periods of time from common ancestors.**

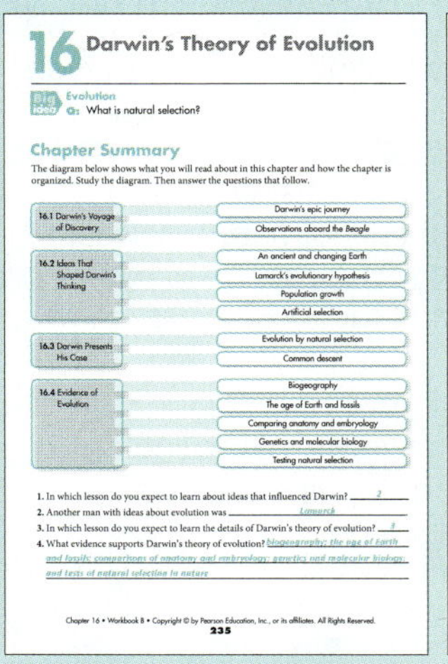
16 Darwin's Theory of Evolution

Evolution
Q: What is natural selection?

Chapter Summary

The diagram below shows what you will read about in this chapter and how the chapter is organized. Study the diagram. Then answer the questions that follow.

16.1 Darwin's Voyage of Discovery: Darwin's epic journey; Observations aboard the *Beagle*
16.2 Ideas That Shaped Darwin's Thinking: An ancient and changing Earth; Lamarck's evolutionary hypothesis; Population growth; Artificial selection
16.3 Darwin Presents His Case: Evolution by natural selection; Common descent
16.4 Evidence of Evolution: Biogeography; The age of Earth and fossils; Comparing anatomy and embryology; Genetics and molecular biology; Testing natural selection

1. In which lesson do you expect to learn about ideas that influenced Darwin? *2*
2. Another man with ideas about evolution was *Lamarck*
3. In which lesson do you expect to learn the details of Darwin's theory of evolution? *3*
4. What evidence supports Darwin's theory of evolution? *biogeography; the age of Earth and fossils; comparisons of anatomy and embryology; genetics and molecular biology; and tests of natural selection in nature*

235

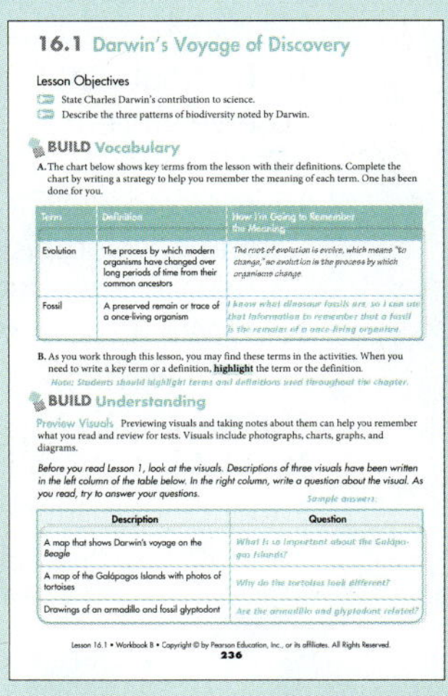
16.1 Darwin's Voyage of Discovery

Lesson Objectives
State Charles Darwin's contribution to science.
Describe the three patterns of biodiversity noted by Darwin.

BUILD Vocabulary

A. The chart below shows key terms from the lesson with their definitions. Complete the chart by writing a strategy to help you remember the meaning of each term. One has been done for you.

Term	Definition	How I'm Going to Remember the Meaning
Evolution	The process by which modern organisms have changed over long periods of time from their common ancestors	*The root of evolution is evolve, which means "to change," so evolution is the process by which organisms change*
Fossil	A preserved remain or trace of a once-living organism	*I know what dinosaur fossils are, so I can use that information to remember that a fossil is the remains of a once-living organism*

B. As you work through this lesson, you may find these terms in the activities. When you need to write a key term or a definition, **highlight** the term or the definition.
Note: Students should highlight terms and definitions used throughout the chapter.

BUILD Understanding

Preview Visuals Previewing visuals and taking notes about them can help you remember what you read and review for tests. Visuals include photographs, charts, graphs, and diagrams.

Before you read Lesson 1, look at the visuals. Descriptions of three visuals have been written in the left column of the table below. In the right column, write a question about the visual. As you read, try to answer your questions.

Sample answers:

Description	Question
A map that shows Darwin's voyage on the *Beagle*	*What is so important about the Galápagos Islands?*
A map of the Galápagos Islands with photos of tortoises	*Why do the tortoises look different?*
Drawings of an armadillo and fossil glyptodont	*Are the armadillo and glyptodont related?*

236

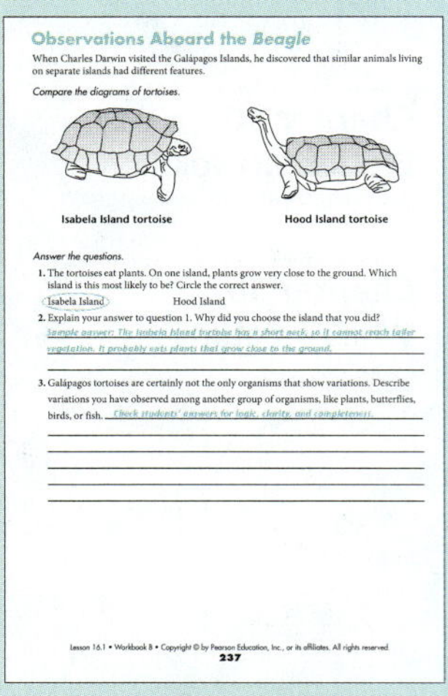
Observations Aboard the *Beagle*

When Charles Darwin visited the Galápagos Islands, he discovered that similar animals living on separate islands had different features.

Compare the diagrams of tortoises.

Isabela Island tortoise

Hood Island tortoise

Answer the questions.

1. The tortoises eat plants. On one island, plants grow very close to the ground. Which island is this most likely to be? Circle the correct answer.
Isabela Island   Hood Island
2. Explain your answer to question 1. Why did you choose the island that you did? *Sample answer: The Isabela Island tortoise has a short neck, so it cannot reach taller vegetation. It probably eats plants that grow close to the ground.*
3. Galápagos tortoises are certainly not the only organisms that show variations. Describe variations you have observed among another group of organisms, like plants, butterflies, birds, or fish. *Check students' answers for logic, clarity, and completeness.*

237

Observations Aboard the *Beagle*

The time Darwin spent studying birds and collecting bugs prepared him for his long trip around the world. On his trip, he studied and took notes about the amazing new plants and animals he saw. Darwin was surprised by the many different kinds of organisms he found. In just one day, he collected 68 species of beetles in the South American forest.

Darwin wanted to do more than just collect bugs and birds. He wanted to make sense of what he found. As he traveled, he noticed three patterns of biological diversity:

1. Species vary around the world.
2. Species vary locally.
3. Species vary over time.

Species Vary Around the World Darwin visited habitats in South America, Australia, and Africa. In the grasslands of South America, he found large, flightless birds called rheas. Rheas look and act a lot like ostriches. Yet ostriches live only in Africa. Rheas live only in South America. In Australia's grasslands, Darwin found another flightless bird called the emu. He wondered why there were no large, flightless birds living in the grasslands of the northern parts of the world.

Darwin also noticed habitats in Australia that were ideal for rabbits. Why did no rabbits live there? And why were there no kangaroos in England? What did these patterns of distribution mean?

Key Question What did Darwin notice about biodiversity around the world? **Places around the world that had similar habitats often had different animals that were ecologically similar.**

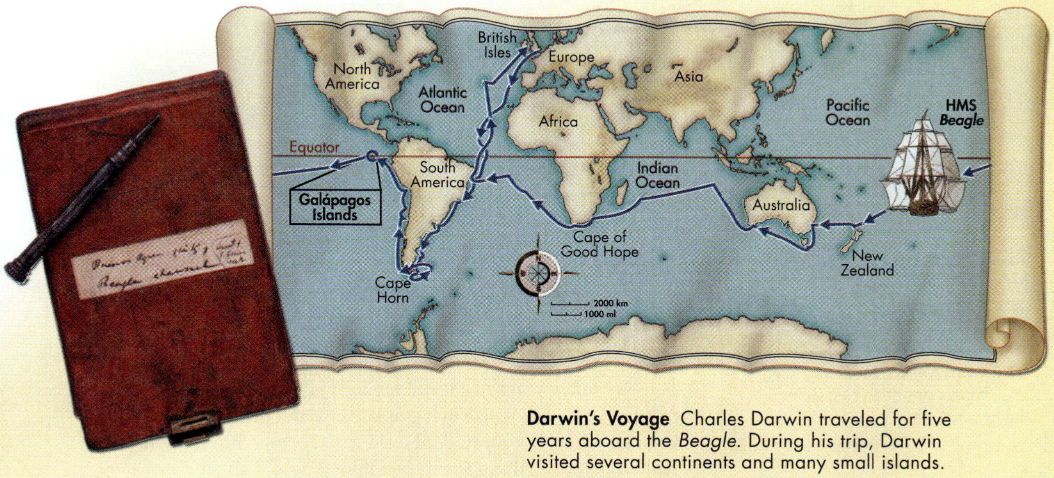

Darwin's Voyage Charles Darwin traveled for five years aboard the *Beagle*. During his trip, Darwin visited several continents and many small islands.

During Reading

Active Reading

KEY QUESTION Read aloud the Key Question and answer from the previous page.

Ask What does the word *evolution* mean? *(change over time)*

FIND THE MAIN IDEA Have students reread the first two paragraphs under Observations Aboard the *Beagle*.

Ask What is this section about? *(the great variety of species Darwin observed during his trip around the world)* What three patterns of species variation did Darwin observe? *(species vary around the world, locally, and over time)*

INTERPRET DIAGRAMS Invite volunteers to describe aloud the route of the *Beagle* illustrated in the visual Darwin's Voyage at the bottom of this page.

Ask Why didn't Darwin compare biodiversity in North America or Asia? *(The* Beagle *did not travel near North America or Asia.)*

KEY QUESTION Have one student read aloud the Key Question and another student read aloud the answer.

Ask According to Darwin's observations, how likely would it be that similar mouse-like species would be found in similar habitats far apart? *(very likely)*

Speed Bump

Ask What theory did Charles Darwin develop? *(a theory of evolution)*

Ask What did Darwin do as he traveled around the world on the *Beagle* that led to his theory of evolution? *(He collected and studied many different kinds of plant and animal species.)*

ELL Focus on ELL: Access Content

BEGINNING SPEAKERS Write on the board these three terms: *continents, habitats,* and *bird species.* Ask volunteers to look up and read aloud the definitions of *continent* and *habitat.* Then, divide the class into pairs. Instruct each pair to make a three-column table and label the columns with the terms listed on the board. Direct students to reread this page and look for names of continents, types of habitats, and different bird species. Have pairs list these words under the correct column heading.

Ask Which bird species observed by Darwin live in the same kind of habitat but on different continents? *(rhea, ostrich, and emu)* What is their habitat? *(grassland)* Where do they live? *(rhea: South America; ostrich: Africa; emu: Australia)*

During Reading continued

Active Reading

PREDICT Have students reread Species Vary Locally and the captions for the visuals.

Say Imagine that another Galápagos Island, "Island X," contains a lot of vegetation found only at a medium height, just above a tortoise's head.

Ask How might the shells and necks of tortoises on Island X compare to those of tortoises on Isabela Island and Hood Island? *(They would probably have longer necks and more open shells than tortoises on Isabela Island, but shorter necks and less open shells than tortoises on Hood Island.)* Why would you predict these differences? *(The vegetation height on Island X is between that on Isabela and Hood Islands.)*

KEY QUESTION Read aloud the Key Question and answer.

Ask Is biodiversity likely to be higher or lower in an area where environments are more similar, compared to an area where environments are different? *(lower)*

Build Vocabulary

WORD ORIGINS Have students read the definition of *fossil*.

Ask What is the meaning of *ancient*? *(very old)*

Say The word *ancient* comes from the Latin root *ante,* which means "before." Ancient organisms existed a very long time before now.

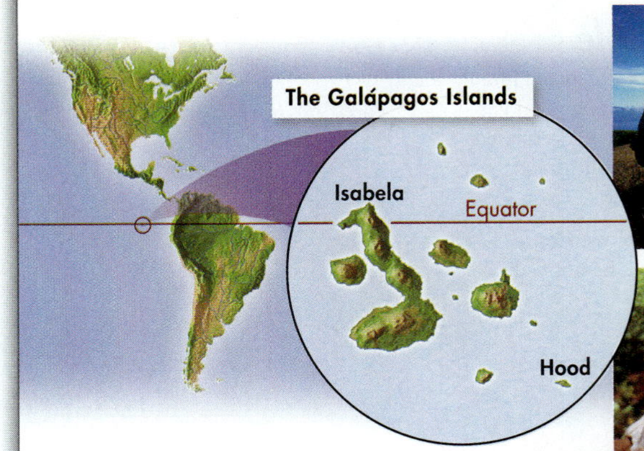

Isabela Island Tortoise Tortoises from Isabela Island have dome-shaped shells and short necks. Vegetation on this island is abundant and close to the ground.

Hood Island Tortoise The shells of Hood Island tortoises are curved and open around their long necks and legs. This helps them to reach the island's sparse, high vegetation.

Tortoise Diversity The tortoises of the Galápagos Islands have shells with different shapes. The shapes are related to the environment on the island where the tortoise lives.

Species Vary Locally Darwin was puzzled over patterns he saw on a local level, too. For example, there were two kinds of rheas living in South America. One lived in the grasslands of Argentina. A different-looking rhea lived in the colder environment further south. Could their differences be related to their different environments?

The Galápagos Islands off the west coast of South America were also full of puzzles. The islands were small and close to each other, but often had different environments. Darwin noticed differences in the large land turtles called tortoises that lived on the islands. At first he did not think much about them. Then someone showed him that each island had its own type of tortoise with a unique shell. He wondered about the connection between environment and the features of an organism.

Darwin also noticed several kinds of small brown birds on the different Galápagos islands. The birds' beaks looked so different from each other that Darwin thought they were unrelated. Darwin sent samples of the birds to experts in England. Later he would learn how closely related—and important—these little birds were.

Key Question What did Darwin notice about biodiversity within a local area? **Darwin noticed that different, related species often lived in different habitats within a local area.**

Species Vary Over Time Darwin also collected fossils. **Fossils** are the preserved remains of ancient organisms. Some fossils Darwin collected did not look like living organisms. But others did. For example, the extinct glyptodont lived where armadillos live today. The glyptodont looked like a giant armadillo. Why did glyptodonts and armadillos look alike? Why did glyptodonts disappear?

BUILD Vocabulary

fossil the preserved remains or traces of ancient organisms

WORD ORIGINS

The word *fossil* comes from the Latin word *fossilis,* meaning "dug up." Fossils are often dug out of the ground.

Hands-On Learning

TRACE FOSSILS Explain to students that there are different types of fossils.

Say Some fossils are actual parts of organisms, preserved in rock.

Ask What are some examples of this type of fossil? *(Sample answers: dinosaur bones, invertebrate shells, leaves of plants)*

Say Other fossils, called trace fossils, are records of the movement or behaviors of living things.

Ask What are some examples of trace fossils? *(Sample answers: tracks, burrows, tooth marks)*

Provide students with modeling clay and have them model the formation of trace fossils by pressing their hands into the clay.

Related Organisms? Ancient glyptodonts were similar to armadillos that live today, but far bigger. The art shows the size of an armadillo and a glyptodont. Darwin wondered if the ancient glyptodont was related to the modern armadillo.

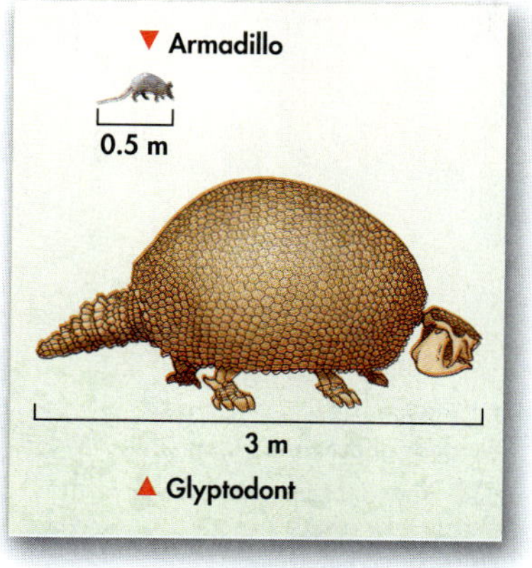

Key Question What did Darwin notice when he compared some fossils to living species? **Darwin noticed that some fossils of extinct animals were similar to living species.**

Solving the Puzzle On the long trip home, Darwin thought about the patterns he had seen. The little brown birds he had collected were already causing excitement back home. Experts had found that they were all finches! These finches were similar to a species of finch that lives in South America. Darwin began to wonder if different Galápagos species might have evolved from ancestors in South America. The evidence he gathered on his trip suggested that species change over time by some natural process.

CHECK Understanding

Apply Vocabulary

Use the highlighted words from the lesson to complete each sentence correctly.

1. On his journey, Darwin collected ______, the preserved remains of ancient organisms.
2. The process of change over time is called ______.

Critical Thinking

3. **Review** What did Darwin's theory of biological evolution explain?
4. **Relate Cause and Effect** What ideas were changing in the scientific community at the time that Darwin traveled on the *Beagle*? How might these new ideas have influenced Darwin?
5. **Review** What three kinds of variations among organisms did Darwin observe during the voyage of the *Beagle*?
6. **Infer** Darwin found fossils of many organisms that did not resemble any living species. How might this finding have affected his understanding of life's diversity?
7. **Write to Learn** Answer the question in the mystery clue below. Remember that honeycreepers eat different kinds of foods. How might the sources of food differ from island to island?

MYSTERY CLUE

Like the small brown birds on the Galápagos, Hawaiian honeycreepers live on islands with different environments. How might these differences have affected honeycreeper evolution? (**Hint:** See p. 382.)

BIOLOGY.com Search Lesson 16.1 GO • Lesson Assessment

Check Understanding

1. fossils
2. evolution
3. The theory explained how modern organisms changed slowly, over a very long time, and described how modern organisms came from shared ancestors.
4. Geologists were suggesting that Earth was very old and had changed over time. Biologists were suggesting the same for life on Earth. These new ideas may have influenced Darwin as he formed his theory of evolution.
5. Darwin observed that (1) species can be different around the world, (2) species can be different on a local level, and (3) species can change over time.
6. He might have guessed that some species become extinct and some species survive with changes that have evolved over time.
7. The different environments would favor different traits on each island. Over time, evolution could lead to very different species.

Active Reading

COMPARE AND CONTRAST Have students study the photograph and art picturing the glyptodont fossil and the armadillo.

Ask How are the glyptodont fossil and the armadillo similar? *(They have similar shells, tails, legs, feet, and body shapes.)* How are they different? *(The armadillo is much smaller than the glyptodont.)*

Ask Why did Darwin think the ancient glyptodont might be related to the modern armadillo? *(Fossil glyptodonts are similar to modern armadillos. Also, glyptodonts lived where armadillos live today.)*

KEY QUESTION Have a student read aloud the question and its answer.

Ask In what ways besides their appearance might extinct species have been similar to living species? *(Sample answers: what they ate, where they lived, how they moved, how they reproduced)*

Speed Bump

Ask When Darwin observed different-looking tortoises living on the Galápagos Islands, what did he think the difference might be related to? *(differences in the tortoises' environments)*

Ask What evidence did Darwin have to suggest that extinct species were similar to living species? *(fossils)*

Post Reading

ASSESSMENT TIPS Suggest that students having difficulty answering Questions 3 and 4 reread the first page of this lesson.

For Question 6, remind students that to infer means to use given information and apply it to different situations.

MYSTERY CLUE Before students compare honeycreepers to the Galápagos brown birds, encourage them to review the differences in Galápagos tortoises that are related to their different environments.

Wrap-Up Activity

PREPARE A QUIZ Divide students into pairs, and instruct them to prepare a ten-item true or false "quiz" based on the lesson's text and visuals. Then, group pairs into sets of two pairs (four students). Have each pair listen to the other pair's quiz and state whether the statements are true or false. Declare as winners the student pairs who identify the most true or false statements correctly. If desired, continue grouping winning pairs for a class tournament.

LESSON 16.2

Pre-Reading

Objectives

16.2.1 Identify the conclusions drawn by Hutton and Lyell about Earth's history.

16.2.2 Describe Lamarck's hypothesis of evolution.

16.2.3 Describe Malthus's view of population growth.

16.2.4 Explain the role of inherited variation in artificial selection.

Preview the Pages

1. Preview the pages for the names of scientists. Encourage students to make flash cards as they read the lesson with the name of a scientist on one side and the scientist's ideas on the other. Point out to students that they can use the flash cards as study aids.
2. Preview the visuals by reading the captions aloud. Discuss ideas that might be familiar, such as how the Grand Canyon formed. Identify unfamiliar ideas, such as how Lamarck would explain the development of long-legged birds.

Activate Prior Knowledge

Explain that learning about evolution requires applying knowledge of population growth and genetics.

Ask What two factors cause populations to grow? *(births and immigration)* What two factors cause populations to decline? *(deaths and emigration)*

Ask Can you look at two organisms and tell exactly what traits their offspring will inherit? *(no)* Why not? *(Inheritance depends on how alleles sort and combine.)*

NATIONAL SCIENCE EDUCATION STANDARDS

UNIFYING CONCEPTS AND PROCESSES

I, II, IV, V

CONTENT

C.3.a, C.4.d, D.3, F.2, F.3, G.1, G.2, G.3

INQUIRY

A.1.b, A.1.c, A.1.e, A.2.a, A.2.e, A.2.f

For the corresponding lesson in *Miller & Levine Biology,* see pages 454–458.

16.2 Ideas That Shaped Darwin's Thinking

Key Questions

What did Hutton and Lyell conclude about Earth's history?

How did Lamarck propose that species evolve?

What was Malthus's view of population growth?

How is inherited variation used in artificial selection?

BUILD Understanding

Main Idea and Details Chart As you read a lesson, you can organize the information in a main idea and details chart. Write all the green headings in the Main Idea column. Then write details about each main idea in the Supporting Details column.

In Your Workbook Go to your workbook to learn more about main idea and details charts.

An Ancient, Changing Earth

Like all scientists, Darwin developed his ideas by building on the work of others. Darwin worked in an exciting time of discovery. Naturalists were finding connections between animals and their environments.

In Darwin's time, many people believed that Earth was only a few thousand years old. They also believed that Earth had not changed much in that time. But scientists were beginning to find evidence that supported different ideas about Earth's history. These ideas were part of the new science of geology. Two of the best-known geologists were James Hutton and Charles Lyell. The ideas of both these geologists influenced Darwin's thinking.

Hutton and Geological Change Hutton described how geological processes shape the land. He suggested that great forces under Earth's surface push mountains upward. These same mountains are then worn down by wind and rain. Most of the processes Hutton described are very, very slow. To make the formations he observed, Earth would have to be much older than a few thousand years. Hutton introduced a concept called *deep time* to explain his ideas. Deep time is the idea that Earth's history is so long that it is difficult to imagine.

Ancient Rocks These rock layers in the Grand Canyon were laid down over millions of years. The river then slowly carved out the canyon over many more years.

16.2 Ideas That Shaped Darwin's Thinking

Lesson Objectives

- Identify the conclusions drawn by Hutton and Lyell about Earth's History.
- Describe Lamarck's hypothesis of evolution.
- Describe Malthus's view of population growth.
- Explain the role of inherited variation in artificial selection.

BUILD Vocabulary

A. The chart below shows a key term from the lesson with its definition. Complete the chart by writing a strategy to help you remember the meaning of the term.

Term	Definition	How I'm Going to Remember the Meaning
Artificial selection	Selection in which nature provides variation of traits, but humans decide which are the most useful and breed organisms to have those traits	*Artificial means "not in nature," so I know that artificial selection is made by humans*

B. As you work through this lesson, you may find this term in the activities. When you need to write the key term or its definition, **highlight** the term or the definition.

BUILD Understanding

Main Idea and Details Chart As you read a lesson, use all the green headings as the main ideas. Write them in the Main Idea column. Write details about each main idea in the Supporting Details column. One has been done for you.

Main Idea	Supporting Details
An Ancient, Changing Earth	James Hutton said that Earth was extremely old.
Lamarck's Evolutionary Hypothesis	*Lamarck came up with a hypothesis about evolution before Darwin, but it was incorrect.*
Population Growth	*When populations grow unchecked, resources become scarce.*
Artificial Selection	*Breeders and farmers select useful traits among natural variations in their crops or livestock.*

Lamarck's Evolutionary Hypotheses/ Population Growth/Artificial Selection

Like all scientists before and after him, Darwin was inspired and influenced by the ideas of others.

humans, nature, Malthus, geology, artificial, Lamarck, Hutton

Use the clues below to complete the crossword puzzle with words from the box.

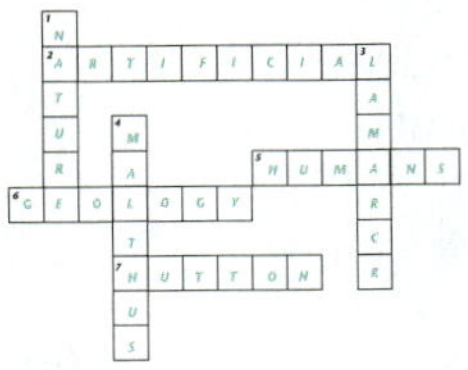

ACROSS

2. ______ selection
5. In artificial selection, these select the useful variations.
6. Darwin was influenced by Lyell's book *Principles of* ______
7. One of two scientists who recognized that the Earth was very old.

DOWN

1. In natural selection, this determines the variations.
3. His theory of evolution was based on acquired traits.
4. Darwin was inspired by his ideas on population growth.

Lyell's *Principles of Geology* Lyell argued that the laws of nature do not change over time. He concluded that geological processes in the past worked like the geological processes of today. Ancient volcanoes released lava, just as volcanoes do now. Rivers in the past cut deep canyons, just as rivers do today. Over time, these processes formed the modern landscape. Such changes would take millions of years.

Key Question What did Hutton and Lyell conclude about Earth's history? **Hutton and Lyell concluded that Earth is extremely old. Also, the processes that changed Earth in the past are the same processes that operate in the present.**

Lamarck's Evolutionary Hypotheses

Darwin was not the first scientist to say that species evolve over time. But earlier scientists disagreed on *how* evolution happened. The French scientist Jean-Baptiste Lamarck was one of the first scientists to hypothesize how evolution works. Lamarck suggested two ideas.

Lamarck's Ideas Lamarck proposed that all organisms are born with a desire to become better. He hypothesized that organisms could change their bodies so that they worked better in their environment. Lamarck also thought that organisms could change their bodies by using them in new ways. For example, consider a bird that walks in water looking for food. Long legs would be very useful to that bird. According to Lamarck's ideas, the bird's legs could become longer by stretching each day. Traits that change during the life of an organism are called *acquired characteristics*.

Lamarck also thought that acquired characteristics could be passed on to offspring. This idea is called the *inheritance of acquired traits*. Thus, over a few generations, the legs of a wading bird would get longer and longer.

Acquired Characteristics? This long-legged water bird is a stilt. According to Lamarck, its long legs were the result of the bird's effort to stretch and use its legs in new ways.

Evaluating Lamarck's Hypotheses Today scientists know that most of Lamarck's ideas were wrong. Animals do not have a desire to become better. Evolution is not a process that makes animals more "perfect" over time. Scientists also know that there is no inheritance of acquired traits. Acquired characteristics cannot be passed on to offspring.

But Lamarck's work did help Darwin in several ways. Lamarck made a connection between an organism's body and its environment. He was one of the first scientists to suggest that organisms can change over time. He also tried to explain evolution using natural processes. Even though he was incorrect, Lamarck helped other biologists, including Darwin, form their ideas.

Key Question How did Lamarck propose that species evolve? **Lamarck suggested that individual organisms could change during their lifetimes by using or not using parts of their bodies. He also suggested that individuals could pass these changes on to their offspring.**

BIOLOGY.com Search Lesson 16.2 GO • Art in Motion

ELL Focus on ELL: Build Background

ALL SPEAKERS Write the words *inherited* and *acquired* on the board. Have a volunteer look up the words in a dictionary and read aloud their definitions. Reinforce the difference between *acquired* and *inherited* by asking students to explain the difference between a person with acquired money and another with inherited money. *(The person with acquired money gained it on his or her own. The person with inherited money received it from ancestors or other relatives.)* Invite students to identify other examples of items a person may acquire and items a person may inherit.

During Reading

Build Understanding

MAIN IDEA AND DETAILS CHART Direct students to Workbook B, where they will find a main idea and details chart. Tell students that putting information into this type of format will help them summarize the ideas that shaped Darwin's thinking.

Active Reading

READ, COVER, REMEMBER, RETELL Have students read Lyell's *Principles of Geology* silently. Then, instruct them to cover the text with their hands and think about what they read.

Ask In your own words, what was the subject or main idea of this paragraph? *(Sample answer: Today's geological processes are the same that Earth experienced in the distant past.)*

Ask What are some examples of these geological processes? *(volcanoes erupting, rivers forming canyons)*

KEY QUESTION Read aloud the first Key Question and its answer.

Ask What led Hutton and Lyell to conclude that Earth is extremely old? *(They reasoned that it would take a very long time for geological processes to produce the formations they observed on Earth.)*

USE GRAPHIC ORGANIZERS As a class, make a flowchart on the board that illustrates Lamarck's idea of inheritance of acquired traits. The flowchart might include the following five steps, using the water-bird species from the text as an example:

1. A bird uses its legs to walk in water.
2. Long legs are more useful to the bird than short legs, so the bird stretches its legs.
3. Over time, the bird's legs become longer.
4. The bird passes its acquired trait for longer legs to its offspring.
5. Over time, the legs of the water-bird species would get longer and longer.

KEY QUESTION Have a student read aloud the second Key Question and its answer.

Ask If scientists know that Lamarck was wrong, why do they find his work worthwhile? *(He was one of the first scientists to explore the idea of evolution.)*

Speed Bump

Ask What did Hutton and Lyell think about geological processes on Earth? *(that geological processes shaped Earth's landscape over a long time)*

Ask How did Lamarck think organisms evolved? *(by passing their acquired traits to their offspring)*

During Reading continued

Active Reading

MAKE CONNECTIONS Remind students that in Chapter 5, they learned about density-dependent limiting factors that affect populations.

Ask Why are disease and starvation likely to be serious problems when human populations are overcrowded? *(Sample answers: Contagious diseases spread more easily when people are in close contact with one another. Food supplies may not be large enough.)*

KEY QUESTION Read the Key Question and answer aloud.

Ask Based on Malthus's ideas about population growth, what would be the benefits of a population that is not overcrowded? *(Sample answers: enough food and space for all individuals; less disease)*

Inquiry Into Scientific Thinking

MAKE CONNECTIONS Help students relate the lab to Darwin's ideas about the survival of offspring.

Ask In peppers, what represents one offspring? *(one seed)*

Ask Do you think that, in nature, all pepper seeds end up in good environments, with the right soil and enough sunlight and rain? *(probably not)* In terms of seed production, which peppers probably have an advantage of higher offspring survival and greater trait variety? *(the peppers that produce the most seeds)*

PREDICT Remind students that reproduction requires energy.

Ask If it costs a plant energy to produce seeds, what would be more advantageous for the plant than producing a huge amount of seeds? *(The plant might produce a "medium" amount of seeds so that enough offspring will survive but at less of an energy cost to the plant.)*

Lead a Discussion

Have students apply what they learned to their everyday lives.

Say Picture yourself either planting a vegetable in a garden or preparing a flowerpot for a window.

Ask Once your soil is ready, do you plant one seed or several? *(several)* Why? *(Not all the seeds will survive to become adult plants.)*

Overcrowding in London This drawing from the 1800s shows the crowded conditions in London when Darwin lived.

Population Growth

The ideas of the Thomas Malthus also helped Darwin. Malthus was an English economist who noticed that people were being born faster than people were dying. This led to overcrowding. Malthus thought that overcrowding led to conditions that would slow down population growth. These conditions included war, disease, and starvation.

Darwin saw that Malthus's ideas were true for all organisms. A maple tree can produce thousands of seeds every year. An oyster can lay millions of eggs. If all the offspring of a single species survived, that species would soon cover the world. Of course, this doesn't happen. Most offspring die. Only a few survive. Darwin thought it was important that only a few individuals survive to reproduce. It could be a key part of evolution. He wondered which individuals survive. Why did they survive?

Key Question What was Malthus's view of population growth? **Malthus reasoned that if the human population were to grow without control, then there would not be enough space and food for everyone to live.**

INQUIRY into Scientific Thinking

Variation in Peppers

As Darwin noted, most organisms produce more offspring than can survive in the environment. This may seem like a waste, but it does make evolutionary sense. Having many offspring makes it more likely that some will survive to reproduce. And more offspring means a wider variety of traits. This makes it is more likely that some traits will be helpful if the environment changes.

In this lab you will examine variations in the colors of peppers and the number of seeds produced.

1. Get a green, yellow, red, or purple bell pepper.
2. Slice open the pepper and count the number of seeds.
3. Compare your data with the data of other students who have peppers of a different color.

Pepper Color	Number of Seeds

Analyze and Conclude

1. **Calculate** Use the equation below to find the average number of seeds for all of the peppers.

$$\text{Average number of seeds} = \frac{\text{total number of seeds}}{\text{total number of peppers}}$$

2. **Calculate** Determine the difference from the average for each of the peppers in your group.
3. **Infer** What other traits could you measure that might affect the success of the pepper plants?

In Your Workbook Get more help for this activity in your workbook.

386

Materials

bell peppers of different colors, knife

Analyze and Conclude

1. Answers will depend on the total number of seeds students count for the peppers.
2. Answers will depend on the number of seeds students count for each type of pepper.
3. Sample answers: fruit size, fruit wall thickness, fruit sweetness, seed size, seed thickness

Artificial Selection Darwin used artificial selection in breeding fancy pigeons at his home outside London.

Artificial Selection

Darwin also studied the work of farmers who bred plants and animals. Breeders knew that some trees produced larger fruits than other trees. Some cows produced more milk than others. Breeders selected only the plants and animals with the best traits for breeding. These traits were passed on to their offspring. Over time, selective breeding could form trees with even bigger fruit, or cows that gave even more milk. Nature provided the variation. Humans selected the traits they found useful. Darwin called this process **artificial selection.**

Darwin had no idea how heredity worked. He did know that variation occurs in wild species as well as farm species. Darwin saw how important this variation was to the process of evolution. He now had all the information he needed to explain evolution. When published, his theory would change the way people understood the living world.

Key Question How is variation used in artificial selection?
In artificial selection, nature provides the variation, and humans select the traits they find useful.

BUILD Vocabulary

artificial selection
the selective breeding of plants and animals to promote desirable traits in offspring

ACADEMIC WORDS
The noun *process* means "a series of actions or changes that take place in a definite manner." Artificial selection is a process. So is evolution.

CHECK Understanding

Apply Vocabulary

Use the highlighted words from the lesson to complete each sentence correctly.

1. Breeders use ______ to produce better crops and livestock.

Critical Thinking

2. Explain How did Hutton and Lyell help to shape Darwin's thinking?

3. Apply Concepts What parts of Lamarck's theory of evolution were incorrect?

4. Explain How does artificial selection work?

5. Write to Learn Imagine that you are Darwin and you have just read Malthus's ideas about population. Write down Malthus's key ideas. Next to these ideas, write how they might apply to evolution.

BIOLOGY.com Search Lesson 16.2 GO • Lesson Assessment

Check Understanding

1. artificial selection

2. Hutton and Lyell were geologists who introduced the concept of "deep time"—stretches of time long enough for evolution to occur—and the idea of slow, steady changes leading to big changes over time.

3. Lamarck thought incorrectly that animals could change their bodies through effort and that the changes in an organism's body could be passed on to the next generation.

4. Domestic organisms have a variety of traits. Farmers choose only the organisms with the most desired traits for breeding. These are passed on and become more common.

5. Sample answer: Malthus thought that overcrowding would lead to war, disease, and starvation. Darwin realized that this problem applies to populations of all organisms. If every offspring survived, then all organisms would run out of space and food.

Active Reading

KEY QUESTION Have a volunteer read aloud the question and answer. Then, have students study the visuals of the fancy pigeons.

Ask What traits did breeders select for in fancy pigeons? *(feather color, breast shape, beak shape, tail length)*

Build Vocabulary

ACADEMIC WORDS Tell students that they have already explored many processes in science, such as photosynthesis, cellular respiration, and cell division. Encourage them to name processes that they perform in their everyday lives. *(Sample answers: opening a locker, applying makeup, making lunch, getting dressed)*

Speed Bump

Ask What observation did Darwin make about the offspring of a species? *(Only some of the offspring survive and reproduce.)*

Ask What do humans select in the process of artificial selection? *(plant or animal traits that are desirable or useful to humans)* How do humans select for those traits? *(by breeding plants or animals that have those traits)*

Post Reading

ASSESSMENT TIPS Have students use their main idea and details chart to help them answer the questions.

Wrap-Up Activity

MODEL ARTIFICIAL SELECTION Divide students into small groups. Assign each group one of the following organisms: dairy cow, tomato plant, show dog, racehorse, wheat, rose variety, or fish raised in a fish farm. Instruct each group to list traits that they would select if they were to breed several generations of their organism. Tell students that the trait must be one that would appear in nature for that species. Have each group illustrate their organism and present their list and illustration to the class.

LESSON 16.3

Pre-Reading

Objectives

16.3.1 Describe the conditions under which natural selection occurs.

16.3.2 Explain the principle of common descent.

Preview the Pages

1. Have students find the vocabulary terms in the text and in the visuals. Compare and contrast the vocabulary terms *natural selection* and *artificial selection* (from the previous lesson).
2. Preview the visuals by reading aloud the captions. Connect the captions in the Adaptations visual to the appropriate photos. Point out that the numbered labels in the visual Natural Selection correspond to the subheadings on the previous pages.

Activate Prior Knowledge

Remind students that the interactions between predators and prey are an important part of a community.

Ask What types of traits would be helpful to a prey species? *(Sample answers: ability to hide from, scare off, or run quickly away from predators)* What types of traits would be helpful to a predator species? *(Sample answers: ability to detect or capture prey)*

Build Understanding

PREVIEW VISUALS Direct students to Workbook B, where they should follow the activity provided for this lesson. As students look over the images of grasshoppers, encourage them to read the headings of the accompanying labels.

NATIONAL SCIENCE EDUCATION STANDARDS

UNIFYING CONCEPTS AND PROCESSES

I, II, IV, V

CONTENT

C.3.a, C.3.b, C.3.c, C.3.d, C.3.e, C.4.d, C.6.c, G.1, G.2, G.3

INQUIRY

A.2.a, A.2.f

For the corresponding lesson in *Miller & Levine Biology,* see pages 460–464.

16.3 Darwin Presents His Case

Key Questions

Under what conditions does natural selection occur?

How does Darwin's theory explain the connection between living and extinct species?

BUILD Understanding

Preview Visuals Before you read this lesson, look at the pictures of grasshoppers on page 390. Read the information in the figure. Then write three questions you have about it. As you read, answer your questions.

In Your Workbook Go to your workbook to learn more about previewing visuals.

Evolution by Natural Selection

Soon after reading Malthus and thinking about artificial selection, Darwin worked out his theory of natural selection. Most of his scientific friends thought his ideas were brilliant. But Darwin did not feel ready to publish. He did not want to be attacked for his ideas, as Lamarck had been. Darwin did not share his theory with the public for 20 years. He spent this time collecting more information to support his theory.

In 1858, Darwin read the unpublished work of another scientist, Alfred Russel Wallace. Darwin was shocked to find that Wallace's thoughts on evolution were almost the same as his own! If Darwin did not act quickly, Wallace would be given credit for describing natural selection. The next year, Darwin published his first book on evolution. It was called *On the Origin of Species*.

The Struggle for Existence In his book, Darwin combined his own thoughts with ideas from Malthus and Lamarck. Like Malthus, Darwin saw that organisms produce more offspring than can survive. Members of a population must compete for food, water, living space, and other things needed to live. Those that do not get enough will not survive to reproduce. Darwin called this *the struggle for existence*.

Variation and Adaptation Darwin knew that members of a population have variation in traits. Many of these traits, called *heritable traits*, can be passed on to offspring. Darwin hypothesized that some heritable traits make survival easier. Any heritable trait that helps an organism survive in its environment is called an **adaptation.** Adaptations can be body parts, such as sharp claws or strong wings. Adaptations can also be behaviors, such as avoiding predators.

Survival of the Fittest Like Lamarck, Darwin saw how the body of each organism works well in its environment. Darwin knew that differences in adaptations affect fitness. **Fitness** describes how well an organism can survive and reproduce in its environment.

Some individuals have adaptations that make survival and reproduction easier. These individuals have high fitness. Individuals without these adaptations are more likely to die, and have less chance to produce offspring. These individuals have low fitness. This difference in success is called *survival of the fittest*.

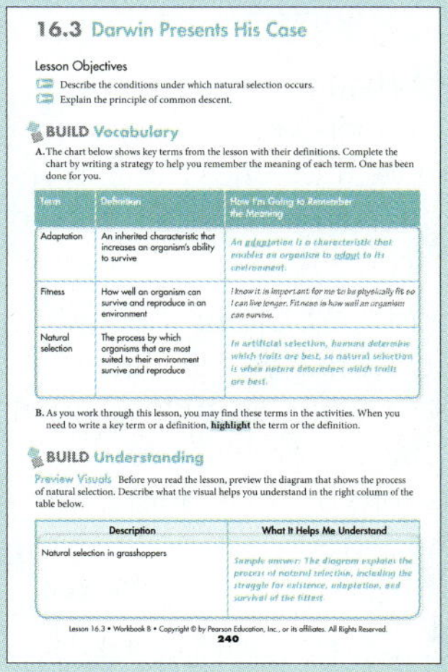

16.3 Darwin Presents His Case

Lesson Objectives

Describe the conditions under which natural selection occurs.

Explain the principle of common descent.

BUILD Vocabulary

A. The chart below shows key terms from the lesson with their definitions. Complete the chart by writing a strategy to help you remember the meaning of each term. One has been done for you.

Term	Definition	How I'm Going to Remember the Meaning
Adaptation	An inherited characteristic that increases an organism's ability to survive	An adaptation is a characteristic that enables an organism to adapt to its environment.
Fitness	How well an organism can survive and reproduce in an environment	I know it is important for me to be physically fit so I can live longer. Fitness is how well an organism can survive.
Natural selection	The process by which organisms that are most suited to their environment survive and reproduce	In artificial selection, humans determine which traits are best, so natural selection is when nature determines which traits are best.

B. As you work through this lesson, you may find these terms in the activities. When you need to write a key term or a definition, **highlight** the term or the definition.

BUILD Understanding

Preview Visuals Before you read the lesson, preview the diagram that shows the process of natural selection. Describe what the visual helps you understand in the right column of the table below.

Description	What It Helps Me Understand
Natural selection in grasshoppers	Sample answer: The diagram explains the process of natural selection, including the struggle for existence, adaptation, and survival of the fittest

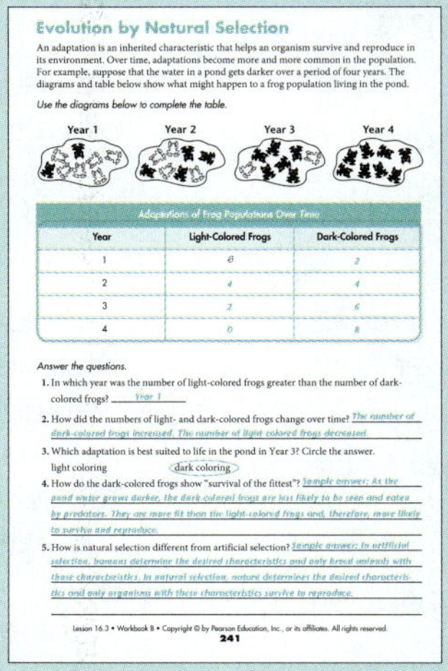

Evolution by Natural Selection

An adaptation is an inherited characteristic that helps an organism survive and reproduce in its environment. Over time, adaptations become more and more common in the population. For example, suppose that the water in a pond gets darker over a period of four years. The diagrams and table below show what might happen to a frog population living in the pond.

Use the diagrams below to complete the table.

Adaptations of Frog Populations Over Time		
Year	Light-Colored Frogs	Dark-Colored Frogs
1	6	2
2	4	4
3	2	6
4	0	8

Answer the questions.

1. In which year was the number of light-colored frogs greater than the number of dark-colored frogs? Year 1
2. How did the numbers of light- and dark-colored frogs change over time? The number of dark-colored frogs increased. The number of light colored frogs decreased.
3. Which adaptation is best suited to life in the pond in Year 3? Circle the answer.
   light coloring   (dark coloring)
4. How do the dark-colored frogs show "survival of the fittest"? Sample answer: As the pond water grows darker, the dark-colored frogs are less likely to be seen and eaten by predators. They are more fit than the light-colored frogs and, therefore, more likely to survive and reproduce.
5. How is natural selection different from artificial selection? Sample answer: In artificial selection, humans determine the desired characteristics and only breed animals with those characteristics. In natural selection, nature determines the desired characteristics and only organisms with these characteristics survive to reproduce.

BUILD Connections

ADAPTATIONS

Adaptations take many forms.

A scorpionfish's coloring is an example of camouflage—an adaptation that allows an organism to blend into its background and avoid predation.

Adaptations can be forms of behavior. Here, a crane is displaying defensive behavior in an effort to scare off the fox.

Scarlet king snake

Eastern coral snake

Mimicry is an adaptation in which an organism copies a more dangerous organism. The scarlet king snake is harmless, but it looks like, or mimics, the poisonous eastern coral snake, so predators avoid it, too.

Natural Selection Darwin named his method of evolution *natural selection*. In **natural selection,** organisms with traits that are the best match to their environment survive and leave more offspring. In some ways, natural selection is like artificial selection. In both forms of selection, only some individuals reproduce. In artificial selection, the breeder selects which individuals reproduce. In natural selection, the environment determines which individuals survive and reproduce.

Natural selection happens whenever certain conditions exist. First, more individuals are born than can survive. Second, there is heritable variation in the population. Third, some individuals have traits that make them more likely to survive and reproduce.

Populations continue to evolve as they become better adapted to their environment. Populations are also likely to change if the environment changes. Traits that were well suited to the old environment may not work as well in the new environment. New adaptations may evolve.

Key Question When does natural selection occur?
Natural selection occurs when more individuals are born than can survive (the struggle for existence), there is heritable variation (variation and adaptation), and individuals have different fitness rates (survival of the fittest).

BUILD Vocabulary

adaptation a heritable characteristic that increases an organism's ability to survive and reproduce in an environment

fitness how well an organism can survive and reproduce in its environment

natural selection the process by which organisms that are most suited to their environment survive and reproduce most successfully; also called survival of the fittest

RELATED WORD FORMS

The verb *inherit* and the adjective *heritable* are related word forms. A trait that can be *inherited*, or passed on to offspring, is said to be *heritable*.

Focus on Academic Language

• behavior • defensive • conditions

Write the words on the board. Discuss as a class each word's part of speech and its meaning as suggested by context clues in sentences on this page.

Say Behaviors are any actions or responses of animals, including people.

Ask What do we mean when we say a child shows bad behavior? *(Sample answer: The child might be mean, difficult, or out of control.)*

Ask What does it mean to show defensive behavior? *(to act in a way that protects you)*

Say Conditions are a set of requirements for a situation to take place.

Ask What are the conditions you must meet to graduate from high school? *(Sample answer: passing courses, attending classes, not getting into trouble)*

During Reading

Active Reading

USE GRAPHIC ORGANIZERS Draw on the board a Venn diagram for students to copy in their notebooks. Label the two circles *Natural Selection* and *Artificial Selection,* respectively. Instruct students to reread the Natural Selection section on this page and the Artificial Selection section in Lesson 16.2 before filling in the Venn diagram.

Build Connections

VISUAL SUMMARY Explain to students that the photos illustrate different types of adaptations in animals. Have students review the definition of *adaptation*. Remind them that all adaptations are heritable, including behaviors.

Ask What three types of adaptations are shown in the photos? *(camouflage, mimicry, and defensive behavior)* How do animals benefit from all of these adaptations? *(The adaptations provide protection from predators.)*

Say Match the following examples of adaptations to the three types in the figure: a harmless butterfly resembles a butterfly that is poisonous to predators *(mimicry);* a grasshopper looks like a leaf *(camouflage);* a cat flashes its fangs at a dog *(defensive behavior).*

Build Vocabulary

RELATED WORD FORMS Explain that just as the verb *inherit* has a related adjective form, *heritable,* the adjective *variable* has a related verb form, *vary*. A variable trait is one that varies, or is not always the same.

Ask What are some things in your daily life that are variable? *(Sample answers: weather, school menu, friends' moods)* Do you vary your hairstyle? Do you vary your style of clothes? *(Answers may vary.)*

KEY QUESTION Invite a student to read aloud the question and answer.

Ask By what process did Darwin think evolution occurred? *(by natural selection)*

Speed Bump

Ask What does the phrase *survival of the fittest* mean? *(an individual in a population has a trait that makes it more likely to survive and reproduce)*

Ask Why isn't it considered natural selection when a farmer breeds cows that produce more milk? *(The environment isn't selecting the trait, the farmer is.)*

Build Connections

VISUAL SUMMARY Invite volunteers to read aloud the caption and the numbered labels. Discuss as a class the events pictured by the illustrations.

Ask What three conditions for natural selection are shown on this page? *(struggle for existence, variation and adaptation, survival of the fittest)*

Ask How would you describe each condition? *(More organisms are born than can survive; there is heritable variation in the population; some individuals have traits that make them better able to survive.)*

Say Remember that adaptations help an organism survive in its environment.

Ask What grasshopper adaptation is being illustrated? *(green body color)*

Ask How might natural selection have worked differently if these grasshoppers lived in a rocky area where both green and yellow colors were visible? *(Green grasshoppers would no longer have a higher fitness than yellow grasshoppers. Yellow and green grasshoppers would be equally likely to survive.)*

Ask What do scientists mean when they say that individuals, like the green grasshoppers in this example, have a higher fitness than other individuals in a population? *(The individuals with higher fitness are more likely to survive and reproduce.)*

Ask If green grasshoppers were sterile—unable to reproduce—would they still have a higher fitness than yellow grasshoppers? *(no)* Why not? *(The green grasshoppers wouldn't produce offspring.)*

Speed Bump

Ask In what way is the grasshopper's camouflage an adaptation? *(It is a heritable variation that increases the grasshopper's ability to survive and reproduce, by protecting it from predators.)*

Ask If all the individuals in a population were identical, could natural selection act on that population? *(no)* Why not? *(There has to be heritable variation for natural selection to occur.)*

BUILD Connections

NATURAL SELECTION

This population of grasshoppers changes over time because of natural selection.

1 The Struggle for Existence

Organisms produce more offspring than can survive. Grasshoppers can lay more than 200 eggs at a time. Only a few of these offspring survive to reproduce.

2 Variation and Adaptation

There is variation in nature. Some variations that are inherited are adaptations. Adaptations increase the chance that an individual will survive and reproduce. In this population, body color is inherited—some grasshoppers are green and some are yellow. Green coloring is an adaptation that allows grasshoppers to blend into their environment.

3 Survival of the Fittest

The color green serves as a form of camouflage. It is harder for predators to see green grasshoppers than yellow grasshoppers. Therefore, green grasshoppers have higher fitness than yellow grasshoppers. In this environment, more green grasshoppers are able to survive and reproduce than yellow grasshoppers.

4 Natural Selection

Over time, green grasshoppers become more common than yellow grasshoppers. This is because
(1) more grasshoppers are born than can survive,
(2) individuals vary in color, and color is a heritable trait,
(3) green individuals have a higher fitness in this environment.

ELL Focus on ELL: Extend Language

INTERMEDIATE AND ADVANCED SPEAKERS Tell students that even simple words can have more than one meaning. Explain, for example, that the word *time* is used in two ways on this page. Have students find a location where it is used to mean "an interval" in time *(over time)* and a location where it is used to mean "a point" in time *(at a time)*. Use a timeline or a graph with time plotted on the *x*-axis to help students distinguish the two meanings. Have students look up and write down two other definitions of *time.* Then ask them to write sentences using these meanings. Invite students to read their sentences aloud and describe the meanings they used.

Common Descent

According to Darwin's theory, individuals that are well adapted to their environment survive and reproduce. Every organism alive today descended from parents who survived and reproduced. Those parents descended from their parents. This line of organisms continues back through time to the very earliest life forms.

In a similar way, Darwin proposed that species living today are descended from older species that survived and reproduced. Over time, those older species evolved into new species in a process he called *descent with modification*. Modification means "change." It can take thousands or even millions of years for new species to evolve. This length of time—deep time—was an idea Darwin borrowed from Hutton and Lyell. Deep time gave enough time for natural selection to work. Darwin pointed to the fossil record as evidence of descent with modification.

Darwin used the idea that species change over time to explain the great variety of life on Earth. Darwin drew the first evolutionary tree to show how organisms share ancient ancestors. This idea is called *common descent*. A single "tree of life" links all living things. If you look back in time far enough, all organisms are related.

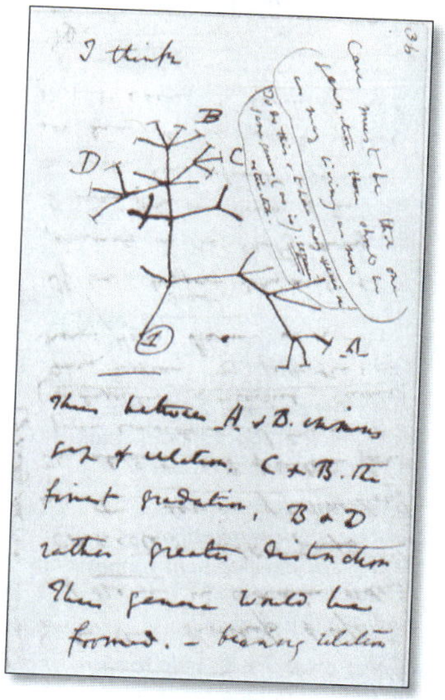

Descent With Modification This is a page from one of Darwin's notebooks. It shows the first evolutionary tree ever drawn. Note that Darwin wrote "I think" just above the tree.

Key Question How does common descent explain the connection between living and extinct species? **According to the principle of common descent, all species—living and extinct—are descended from ancient common ancestors.**

CHECK Understanding

Apply Vocabulary

Use the highlighted words from the lesson to complete each sentence correctly.

1. Organisms with high ______ are more likely to survive and reproduce.

2. Sharp teeth and claws are ______ that help tigers capture their prey.

3. The traits that are most helpful for survival in an environment become more common through the process of ______.

Critical Thinking

4. Apply Concepts What do evolutionary trees show? What does a tree of life imply about all species living and extinct?

5. Relate Cause and Effect Briefly describe the process that causes a population to change by natural selection. Use the words *overproduction*, *variation*, *adaptation*, and *survival of the fittest*.

6. Write to Learn Answer the question in the mystery clue below.

MYSTERY CLUE

How might natural selection explain the history of the Hawaiian honeycreepers? (**Hint:** See p. 389.)

BIOLOGY.com Search Lesson 16.3 GO • Lesson Assessment

Check Understanding

1. fitness

2. adaptations

3. natural selection

4. Evolutionary trees show how living species are descended from ancient extinct species. The tree of life implies that all species on Earth are descended from common ancestors.

5. Overproduction of offspring leads to competition within a species. Variation in traits within a species includes adaptations, traits that make survival easier. Individuals with adaptations are more likely to survive and reproduce, a result of natural selection known as *survival of the fittest*.

6. In the past, there may have been a single species of honeycreeper. Overproduction of offspring and environmental conditions forced competition. Competing birds showed variation in traits. Some honeycreepers had adaptations, traits that helped them survive and reproduce. These traits became more common over time.

During Reading *continued*

Active Reading

SET PURPOSE FOR READING Have a volunteer read aloud the heading, the caption for the visual, and the Key Question (not the answer).

Ask What topics do you think will be covered in the paragraphs on this page? *(common descent, evolutionary trees, the connection between living and extinct species)*

Point out that the headings, questions, and visuals within a section help set the purpose for reading that section.

WRITE IN YOUR OWN WORDS Instruct students to read the paragraphs under the heading *Common Descent.* Have them write notes about the ideas they read.

Say You can often remember a new idea better if you put it in your own words.

Direct students to write in their own words a short paragraph that states the ideas they have just learned. Invite students to share and discuss their paragraphs with the class.

KEY QUESTION Invite a volunteer to read aloud the question and answer.

Ask What does an evolutionary tree show? *(common descent, or how organisms share ancient ancestors)*

Post Reading

ASSESSMENT TIPS For Question 4, have students review the Common Descent section before answering.

For Questions 5 and 6, have students review the Natural Selection visual on the previous page. Remind them that what happens at one step in a process affects the next step.

MYSTERY CLUE Remind students that natural selection in honeycreepers will follow the conditions described in the Natural Selection visual.

Wrap-Up Activity

PRACTICE RECIPROCAL TEACHING Instruct students to review the lesson and write five questions about ideas in the lesson. Explain that each question should ask in some way **What is it?, What does it do?,** or **What does it mean?** Also, tell students to prepare answers to their questions that include examples or supporting details. Pair students and have them take turns quizzing each other with their prepared questions and answers. If time permits, have students share their questions with the class.

LESSON 16.4

Pre-Reading

Objectives

16.4.1 Explain how geologic distribution of species relates to their evolutionary history.

16.4.2 Explain how fossils and the fossil record document the descent of modern species from ancient ancestors. (Note: Text objectives are listed in the same order as in *Miller & Levine Biology.* However, the content in this lesson is presented in a slightly different order. Objective 16.4.3 is presented before objective 16.4.2.)

16.4.3 Describe what homologous structures and embryology suggest about the process of evolutionary change.

16.4.4 Explain how molecular evidence can be used to trace the process of evolution.

16.4.5 Explain the results of the Grants' investigation of adaptation in Galápagos finches.

Preview the Pages

1. Preview the visuals by having students read the captions. Point out the color-coding in the Homologous Limb Bones diagram. Explain that the circled drawings in the Evidence from Fossils visual and Finch Beak Tools visual provide additional information about the animals shown.
2. Draw students' attention to two types of physical features used as evidence for evolution: limb structure in the Comparing Body Structure and Embryos section and The Age of Earth and Fossils section, and beak shape in the Testing Natural Selection section.

NATIONAL SCIENCE EDUCATION STANDARDS

UNIFYING CONCEPTS AND PROCESSES

I, II, III, IV, V

CONTENT

C.1.f, C.2.a, C.3.a, C.3.c, C.3.d, D.3, E.2, G.1, G.2, G.3

INQUIRY

A.2.a, A.2.b, A.2.c, A.2.e, A.2.f

For the corresponding lesson in *Miller & Levine Biology,* see pages 465–473.

16.4 Evidence of Evolution

Key Questions

How does the geographic distribution of species today relate to their evolutionary history?

What do homologous structures and similarities in embryonic development suggest about the process of evolutionary change?

How do fossils help to show the descent of modern species from ancient ancestors?

How can molecular biology be used to trace the process of evolution?

What does recent research on the Galápagos finches show about natural selection?

BUILD Understanding

Concept Map Make a concept map that shows the kinds of evidence that support the theory of evolution.

In Your Workbook Go to your workbook to learn more about making a concept map. Complete the concept map for Lesson 16.4.

Biogeography

When Darwin published *On the Origin of Species* in 1859, scientists were not able to test all of his ideas. But since then, scientists have discovered a great deal of evidence. This evidence comes from many different sciences—biogeography, anatomy, geology, chemistry, genetics, and even molecular biology. Amazingly, evidence from all of these areas supports Darwin's basic ideas about evolution.

Darwin used the science of biogeography to support his theory. **Biogeography** is the study of where organisms live and where their ancestors lived in the past. Darwin noted two important patterns: (1) Closely related organisms that live in different environments often show great differences. (2) Distantly related organisms that live in similar environments are often similar.

Closely Related but Different Darwin used biogeography to explain his observations in the Galápagos. He hypothesized that birds on the islands were descended from birds in South America. Natural selection in different environments had led to differences among the closely related populations.

Distantly Related but Similar Darwin compared plants and animals living in similar environments around the world. For example, he found ground-dwelling birds in the grasslands of Europe, Australia, and Africa. These organisms looked alike, but were not closely related. Because they lived in similar habitats, natural selection had led to similar adaptations.

Key Question How does the geographic distribution of species today relate to their evolutionary history?
Patterns linking adaptations to environments around the world provide evidence for natural selection.

Comparing Body Structure and Embryos

In Darwin's time, researchers had discovered that many different animals have similar structures. For example, the front legs of amphibians, birds, and reptiles all have the same basic bones! Darwin suggested that animals with similar structures are related. Their structures are inherited from a common ancestor.

392 Chapter 16 • Lesson 4

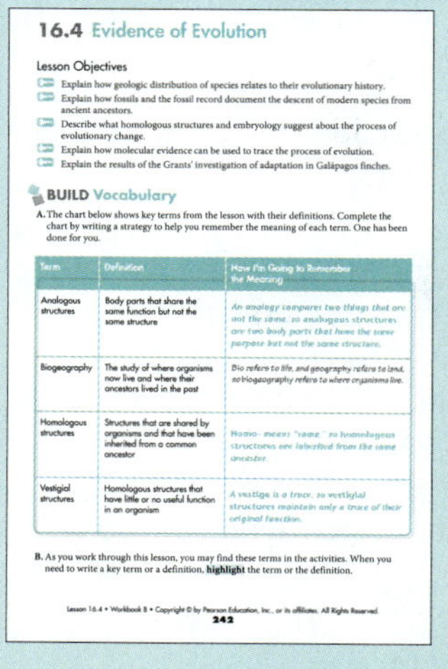

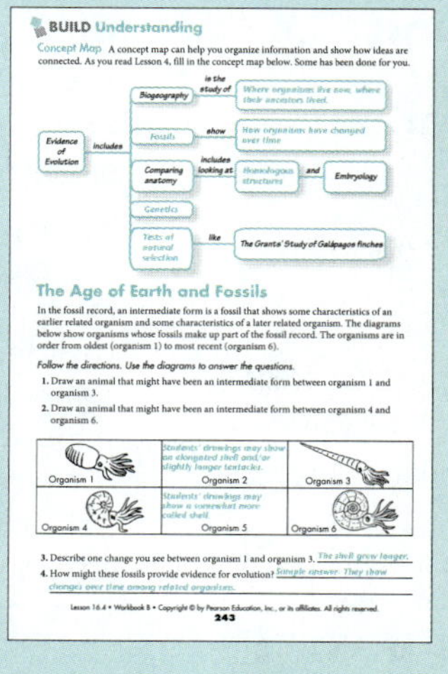

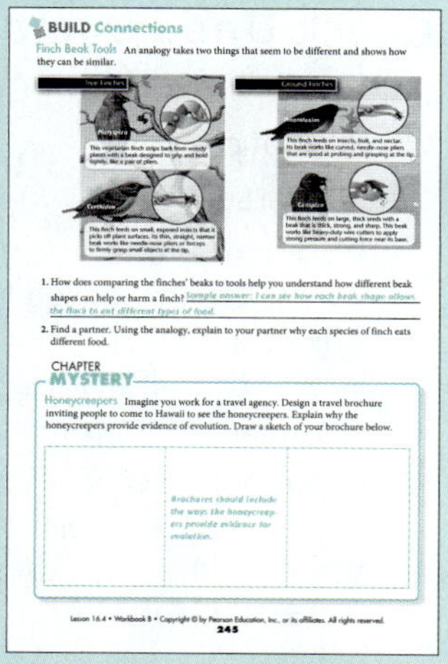

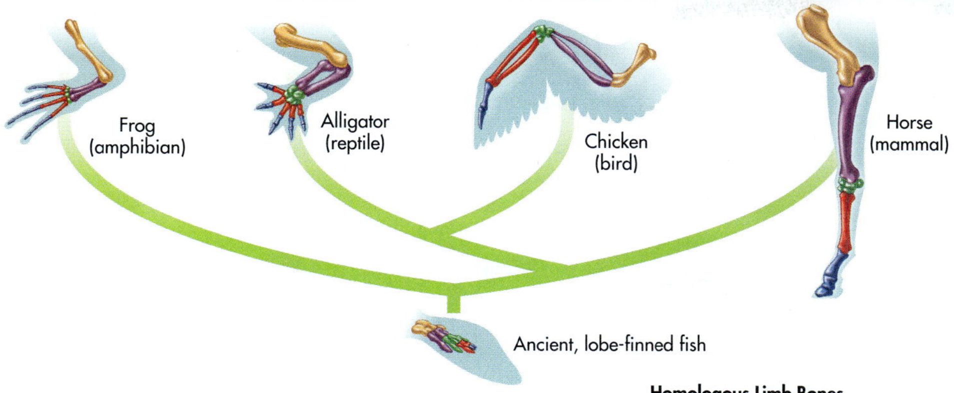

Homologous Limb Bones
The limb bones of seemingly unrelated animals have the same basic structures. Matching colors allow you to compare homologous structures. The limbs evolved from an ancient fish, their common ancestor.

Homologous Structures Structures that are shared by related species and that have been inherited from a common ancestor are called **homologous structures.** These structures often show modifications, or differences. The more alike homologous structures are, the more closely related the organisms. Bird limbs are more like reptile limbs than mammal limbs. This means that birds and reptiles are more closely related than birds and mammals. Homologous structures do not always work the same way. A bird's wing and horse's leg are homologous, even though they work differently.

Key Question What do homologous structures suggest about the process of evolutionary change? **Evolutionary theory explains that homologous structures adapted to different purposes are the result of descent with modification from a common ancestor.**

Analogous Structures Sometimes body parts share a common function but are not structurally related. These structures are called **analogous structures.** The wings of bees, bats, and birds are analogous. Analogous structures do not show common descent.

Vestigial Structures Some homologous structures do not have important functions. **Vestigial structures** are inherited structures that have lost much of their original function. Vestigial structures may show evolutionary relationships. For example, the hipbones of dolphins are vestigial structures. They link dolphins to ancestors that used hipbones as they walked on land.

Embryology The early stages of many animals with backbones look very similar. You may look nothing like a chicken now. But your embryonic cells grew in similar patterns. These patterns of growth produced homologous tissues and organs. These similarities are evidence of common descent.

Key Question What do similarities in embryonic development suggest about the process of evolutionary change?
Similar patterns of embryo development provide evidence that organisms have descended from a common ancestor.

BUILD Vocabulary

biogeography
the study of past and present distribution of organisms

homologous structures
structures that are similar in different species of common ancestry

analogous structures
structures that are similar in function but not structure; they do not suggest common ancestry

vestigial structure
a structure that is reduced in size and has little or no function

WORD ORIGINS
The word *homologous* comes from the Greek word *homos*, which means "same." Homologous structures may not look exactly the same, but they share certain characteristics and a common ancestor.

ELL Focus on ELL: Access Content

ALL SPEAKERS Assist students in sorting out the meanings of *homologous, analogous,* and *vestigial,* and in comparing these different kinds of structures. Draw on the board a three-column compare/contrast table for students to copy in their notebooks. Guide students to use Homologous, Analogous, and Vestigial as column headings. Have students use Definition of the Word, Type of Structure, and Example of Structure as row headings. Work together as a class to write the definitions. Instruct students to work in pairs to reread this page and complete the chart.

During Reading

Build Understanding

CONCEPT MAP Direct students to Workbook B, where they will find an example of a concept map that they should complete as they read. Help students start the concept map by sketching it on the board and asking volunteers to suggest main ideas or vocabulary words from the lesson.

KEY QUESTION Read aloud the Key Question and answer from the previous page.

Ask What did Darwin observe about distantly related organisms that live in similar environments in different areas? *(They often have similar adaptations.)*

Active Reading

INTERPRET DIAGRAMS Refer students to the Homologous Limb Bones diagram. Remind them that homologous structures are shared by very different, but distantly related, organisms. Have students find the five different color-coded bones in each limb.

Ask How are the limbs similar in the different organisms? *(They are made up of five bones and used for movement.)* What structures evolved from the bones colored blue? *(fish: tip of fin; frog: flipper; alligator: webbed foot; chicken: wing tip; horse: hoof)*

KEY QUESTION Read aloud the first Key Question and answer on this page.

Ask How did animal limb bones evolve to be adapted to different purposes? *(As animals moved to different environments, natural selection favored adaptations that were helpful in the new environments.)*

Build Vocabulary

WORD ORIGINS Explain that other words that indicate similarity come from the Greek *homos*.

Ask What does *homozygous* mean? *(having two of the same alleles for a gene)*

KEY QUESTION Ask a volunteer to read aloud the second Key Question and answer on this page.

Ask If you viewed pictures of two embryos of different species developing similarly, could you tell whether the adults would look alike? *(no)* Why? *(Different organisms share similar patterns of embryo development.)*

Speed Bump

Ask How are a bird's wing and a horse's leg homologous even though they don't work in the same way? *(They have the same basic structure.)*

During Reading continued

Active Reading

SELF-QUESTION Demonstrate how asking questions and reviewing the text for answers can help make new information easier to understand. Read the page aloud.

Ask Why is it necessary that Earth be very old for Darwin's theory of evolution to be true? *(Natural selection takes place very slowly.)* Why couldn't Darwin use the fossil record to completely support his theory? *(He didn't have enough examples of fossils across time.)*

Say Sometimes it helps to reread information to find answers to questions. Reread the fourth paragraph.

Ask What does it tell you about how the fossil record has changed since Darwin's time? *(More gaps in the fossil record have been filled.)*

KEY QUESTION Ask a volunteer to read aloud the Key Question and its answer.

Ask What does the fossil series on this page and the next show? *(the evolution of modern whales)*

Build Connections

VISUAL SUMMARY Point out that the Evidence from Fossils visual extends to the next page. Make sure students understand that each animal shown is based on fossil evidence and represents a branch in the evolutionary history of whales.

Have students read the labels, and help them pronounce the animals' names. (artiodactyl—AHRD-ee-uh-DAK-tuhl; *Pakicetus*—pak-uh-SEE-tuhs; *Ambulocetus*—am-byoo-loh-SEE-tuhs; *Rodhocetus*—roh-duh-SEE-tuhs; *Basilosaurus*—bas-uh-loh-SAW-ruhs; *Dorudon*—dawr-oo-dahn; *Mysticetes*—mihs-tuh-SEE-teez; *Odontocetes*—oh-dahn-tuh-SEE-teez)

Direct students' attention to the traits shown in the circles. **Say** The evolution of each trait shows how the animal lived in its environment.

Ask How might being able to live in water and on land have been an advantage for *Ambulocetus*? *(Sample answers: It could find food in both places. It could escape predators in one environment by moving to the other.)*

Ask What benefit might *Basilosaurus* have gained with its streamlined body that *Rodhocetus* did not have? *(Sample answer:* Basilosaurus *might have been able to swim faster to catch food or avoid predators.)*

Ask What is an obvious difference between *Rodhocetus* and modern whales? *(Modern whales don't have hind limbs.)*

The Age of Earth and Fossils

Darwin knew that natural selection would take a very long time. For his theory to be true, Earth would have to be incredibly old.

The Age of Earth. Today, geologists use radioactivity to find the age of rocks and fossils. Their work supports Darwin's theory. Radioactive dating suggests that Earth is about 4.5 billion years old. That's plenty of time for evolution to take place!

Recent Fossil Finds When Darwin lived, enough fossils had been found to convince Darwin that life had changed over time. But there were not enough fossils to show clearly how modern species had descended from their ancestors. The fossil record was incomplete.

Today, hundreds of new fossil discoveries show clearly how many modern species have evolved from older species. Scientists have found fossils of animals that were "in between" dinosaurs and birds. Other fossils link fishes to land animals. New finds even connect land animals to whales. The fossil record is still incomplete. But new discoveries continue to support Darwin's theory.

Key Question How do fossils help to show the descent of modern species from ancient ancestors? **Many recently discovered fossils form series that trace the evolution of modern species from extinct ancestors.**

BUILD Connections

EVIDENCE FROM FOSSILS

Recently, scientists have discovered more than 20 fossils that show how whales evolved from ancestors that walked on land. The drawings of the ancient animals are based on fossil evidence.

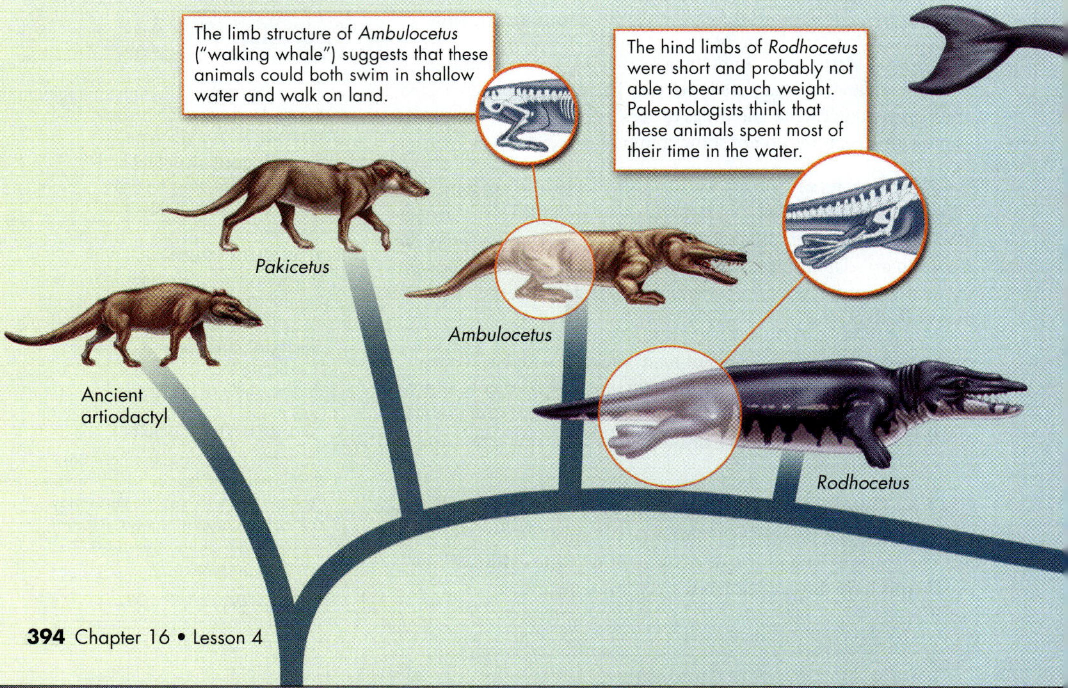

Hands-On Learning

FICTIONAL ANCESTOR Provide students with colored pencils and drawing paper. Ask them to make a drawing of a fictional ancestor of the modern whale that evolved after *Rodhocetus* but before *Basilosaurus,* showing details of the animal's hind limbs. Invite students to share their drawings with the class. Have them describe the traits of their animal and explain why they chose those traits.

Genetics and Molecular Biology

Darwin did not understand how heredity worked. Since his time, biologists have learned a great deal about heredity. Darwin would be glad to know that modern genetics strongly supports evolution.

Common Genetic Code All living cells use DNA and RNA to make proteins and to pass on genetic information. The code is nearly the same for all living things—bacteria, yeasts, plants, fungi, and animals. The shared genetic code is powerful evidence that all organisms evolved from a common ancestor that used this code.

Homologous Molecules Biologists have found many homologous molecules. For example, the protein cytochrome *c* is important in cellular respiration. Versions of this protein are found in almost all living things. Homologous proteins and genes are evidence of common descent.

Key Question How can molecular biology be used to trace the process of evolution? **The universal genetic code and homologous molecules provide molecular evidence of evolution.**

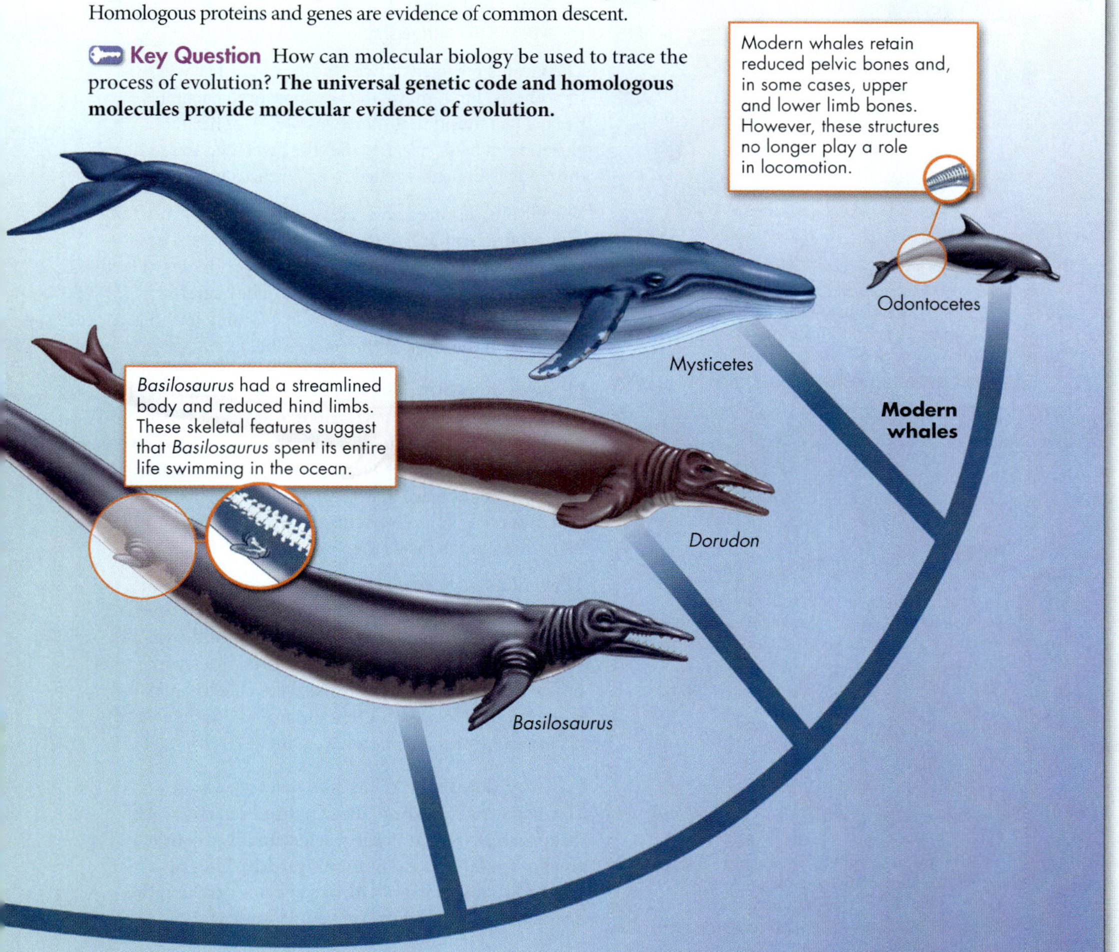

Active Reading

DRAW CONCLUSIONS Reinforce the link between modern organisms and their ancestors.

Say Suppose you learned that nearly all modern mammals have a molecule X that controls an important cell process. A second cell process is controlled by different molecules—Q, Y, or Z—in different groups of mammals.

Ask Without additional information, what molecule or molecules can you conclude were found in the common ancestor of these mammal groups? *(molecule X)*

APPLY ACADEMIC VOCABULARY Have students look back at the definition of *homologous structures*.

Ask How would you define the term *homologous molecules*? *(Sample answer: molecules that are similar in different species of common ancestry)*

KEY QUESTION Read aloud the Key Question and the answer.

Ask How might scientists' view of evolution be different if different groups of organisms had different genetic codes? *(Scientists might not think all organisms evolved from a common ancestor.)*

Speed Bump

Ask How have recent fossil finds supported Darwin's theory of evolution? *(They have shown series of changes in organisms over time.)*

Ask Why is cytochrome c considered a homologous molecule? *(Versions of this protein are found in almost all living things.)*

Focus on Academic Language

- common
- support

Have students find the words *common* and *support* on this page. Call on volunteers to define the words using context clues. Then, instruct students to look up definitions of the adjective form of *common* and verb form of *support* to share with the class.

Ask Which definition of *common* is used on this page? *(belonging to or shared by two or more things)*

Encourage students to use a different meaning of *common* in a sentence. *(Students will likely use the word as meaning "occurring frequently," as in "Black shoes are common in a shoe store.")*

Ask Which definition of *support* is used on this page? *(to uphold or defend an idea)*

Have students use a different meaning of *support* in a sentence. *(Students will likely use the word as meaning "to carry the weight of an object," as in "The posts support the sign.")*

During Reading continued

Active Reading

SUMMARIZE Have students work in groups of three to write in their notebooks a three-part summary of the studies performed by the Grants. The first part should summarize the tests the Grants performed. The second part should summarize the results of the tests, and the third part should summarize the conclusions drawn from the results. Follow with a class discussion about the Grants' work, allowing students to refer to their summaries during the discussion.

KEY QUESTION Invite a student volunteer to read aloud the Key Question and answer.

Ask How did the finch-beak variation that the Grants studied help certain finches survive when the environment changed? *(In dry years, when fewer small seeds were available, finches with large beaks were able to open larger seeds.)*

Build Connections

VISUAL ANALOGY Help students make the connection that variations in finch-beak shape and size resulted from natural selection.

Ask Do all tree finches have similar beaks and all ground finches have similar beaks? *(No, there are variations in beaks within each group.)*

Focus students' attention on the drawings in the circles, which compare the birds' beaks to different kinds of tools, to clarify how the beaks are used. Explain that this figure is making an analogy, taking a concept we're unfamiliar with (how variation in birds' beaks allows them to feed on different foods) and comparing it to a concept we are familiar with (how different tools work best for different jobs).

Ask How is *Certhidea's* beak similar to forceps? *(The beak is thin, straight, and narrow.)* How does this finch use its beak to obtain food? *(It uses the tip of the beak to firmly grasp small insects.)*

Ask How are the beaks of tree finches and those of ground finches adapted to the birds' different environments? *(Tree finch beaks can strip bark or pick off insects from surfaces or narrow openings. Ground finch beaks can crush seeds or get food from soft plants.)*

Ask In a drought period when few insects or seeds were produced, which of the four birds pictured would be most likely to survive? *(*Platyspiza*)* Why? *(Its beak is adapted to strip bark.)*

BUILD Connections

FINCH BEAK TOOLS

Finch beaks can be compared to different tools. The different beaks allow each species of finch to pick up and eat a different kind of food.

Testing Natural Selection

The best way to gather evidence about evolution is to observe natural selection in action. Some scientists have designed laboratory experiments using bacteria that let them test ideas about natural selection. Others have carried out experiments with guppies. These studies support Darwin's ideas. But the best way to test ideas about natural selection is to observe it in the wild. Peter and Rosemary Grant did just that. They studied finches on the Galápagos Islands.

A Testable Hypothesis Darwin hypothesized that all the Galápagos finches had descended from a common ancestor. He thought that natural selection had caused their beaks to change shape.

The Grants performed two studies to test Darwin's hypothesis. First, they carefully measured each bird's beak. They found that there was plenty of heritable variation in beak size. Second, they watched to see if differences in beak size led to differences in fitness.

The Grants captured and released nearly every medium ground finch on one of the Galápagos Islands. They recorded which birds were alive and reproducing. They also recorded which birds had died. They carefully measured the beak of each bird. The Grants found variation in the sizes of the finches' beaks.

Natural Selection The Grants studied the finches for many years, during normal and dry seasons. When food was hard to find, the finches with the largest beaks were most likely to survive. Over a period of just decades, they observed an increase in average beak size. The Grants had observed natural selection in the wild. And it was happening much faster than expected!

The Grants also showed that variations help a species survive when the environment changes. In dry years, there were not enough small seeds for all of the finches. Competition among the finches favored birds with larger beaks. Finches with larger beaks were able to break open larger seeds. Thus, a trait that did not matter in a wet year became a helpful adaptation in a dry year.

Key Question What does research on the Galápagos finches show about natural selection? **The Grants showed that natural selection occurs in wild finch populations, sometimes quickly. They also showed that variation within a species can help the species adapt to and survive environmental changes.**

Hands-On Learning

TOOLS FOR THE JOB Emphasize to students that finch-beak variation illustrates how natural selection favors traits that are "the right tool for the job."

1. On a central table, gather several tools like those shown on this page (for example, pliers, forceps, pruning shears, needle-nose pliers, and tweezers). Tell students that the tools represent bird beaks.
2. Also place several shallow bowls containing different "foods" eaten by birds, such as acorns, sunflower seeds, gummy bears (to represent large insects), grains of rice (to represent small insects), and small woody twigs.
3. Have students take turns using the tools to perform tasks: cracking the nuts and seeds, gathering the insects, and stripping bark from the twigs. Caution students to handle the instruments with care.
4. Have a class discussion about which tool was best for each task. Ask volunteers to relate their results to the Grants' findings about variation in bird beaks.

Survival of the Fittest and Beak Size
The Grants studied finches on the Galápagos Islands. This graph shows the survival of a population of ground finches. It shows that birds with larger beaks survived at a higher rate than birds with smaller beaks.

Evaluating Evolutionary Theory Like any scientific theory, evolutionary theory is constantly reviewed with each new bit of information. New studies, tools, and techniques have continued to support most of Darwin's hypotheses. Today, evolution is important in all branches of biology, from ecology to medicine.

Many important questions about evolution are still being studied: Exactly how did life begin? How do new species arise? Why do some species become extinct? Biologists often disagree about answers to these questions. But biologists agree that evolution *is* happening. To biologists, evolution is the key to understanding the natural world.

CHECK Understanding

Apply Vocabulary

Use the highlighted words from the lesson to complete each sentence correctly.

1. Body parts of unrelated organisms that serve the same function are examples of ______.
2. Body parts in related organisms that are structurally similar are examples of ______.
3. Inherited structures that have lost much of their original function are called ______.

Critical Thinking

4. **Explain** What is biogeography?
5. **Compare and Contrast** What is the difference between homologous and analogous structures? Which are more important to evolutionary biologists? Why?
6. **Apply Concepts** How do recent fossil finds help to support evolution?
7. **Review** How do modern genetics and molecular biology support the theory of evolution?
8. **Draw Conclusions** How do the data collected by the Grants show that variation is important to the survival of a species?
9. **Write to Learn** Answer the question in the mystery clue below. The Hawaiian islands have different environments. Some islands are very wet. Others are much drier. How might these environments affect the natural selection of honeycreepers?

MYSTERY CLUE

How can biogeography be used to explain why some species of honeycreepers are found only on the Hawaiian Islands? (Hint: See p. 392.)

BIOLOGY.com Search Lesson 16.4 GO • Lesson Assessment

Check Understanding

1. analogous structures
2. homologous structures
3. vestigial structures
4. Biogeography is the study of where organisms live now and in the past.
5. Homologous structures share a common ancestry but not always a common function. Analogous structures share a common function but do not share an ancestor. Homologous structures are more important because they suggest evolutionary relationships.
6. Recent fossil finds help fill the gaps in the fossil record.
7. They revealed that the genetic code and homologous molecules evolved from ancient common ancestors.
8. Variation in beak size made it possible for finches to survive dry periods.
9. Related species evolve differences depending on their environments. The unique environments on the Hawaiian Islands lead to unique birds.

Active Reading

USE VISUALS Remind students that the Grants found evidence of natural selection in the wild. Have students review the Survival of the Fittest and Beak Size graph and caption.

Ask Birds with which beak size showed the greatest percentage of survival? *(about 11.5 mm)*

Ask What can you conclude about birds with beaks that are 6 to 8 mm in size? *(Those birds die because their beaks are too small to obtain food.)*

Speed Bump

Ask How did natural selection lead to an increase in finch-beak size over time? *(Finches with large beaks were able to break open larger seeds, making more food available to them. Thus, these birds were more likely to survive and reproduce.)*

Ask Are all scientists' questions about evolutionary theory likely to be answered in your lifetime? *(no)* Why not? *(Evolutionary theory is constantly reviewed as new information is gained.)*

Post Reading

ASSESSMENT TIPS Remind students to review the highlighted vocabulary words before they answer the questions.

Have students reread Homologous Structures and Analogous Structures before answering Question 5.

Have students review the Evidence from Fossils visual before answering Question 6.

MYSTERY CLUE Prompt students to remember that variation in traits must exist for natural selection to occur. Remind them of the role the environment plays in how organisms' traits evolve, as illustrated by the Galápagos finches.

Wrap-Up Activity

FILL IN THE BLANK Divide the class into two teams. Assign each team sections within this lesson to reread and review. Instruct all students to write, in their own words, two fill-in-the-blank sentences that describe a concept in their assigned sections. *(Sample: Organisms with high ______ are more likely to survive and reproduce.)* Once students have completed their sentences, direct a quiz game by taking turns writing one team's sentence on the board and having the other team complete the sentence. A team is declared the winner by completing the most sentences correctly.

LESSON 16.4

Pre-Lab

Read aloud Glossary definitions of *amino acid* and *protein*.

Ask Strings of which molecules make up proteins? *(amino acids)* Do all organisms have the same amino acid sequence for a particular protein? *(no)* What role do genes play in protein formation? *(The base sequences of genes determine the amino acid sequences of proteins.)*

Review the Background Questions and the Pre-Lab Questions as a class.

Lab

Tell students they will perform the chapter lab *Amino Acid Sequences: Indicators of Evolution* described in **Lab Manual B.**

Post-Lab

FINDING RELATIVES Write on the board the abbreviations for the 20 amino acids: *Gly, Ala, Val, Leu, Ile, Met, Phe, Trp, Pro, Ser, Thr, Cys, Tyr, Asn, Gln, Asp, Glu, Lys, Arg,* and *His.* Divide students into groups of five or six. Pass out to each student a piece of construction paper. Tell students to cut out 20 small squares from the paper and label each square with a different amino acid abbreviation. Direct all students in a group to combine their "amino acids" into a pile. Have each student randomly pick out 10 squares from the group pile and then assemble a "protein" by gluing or taping the squares, in the order picked, onto a piece of notebook paper. Tell students to compare how closely the proteins in their group are related to each other, based on similarities and differences in their amino acid sequences. Discuss group results as a class.

Skills Lab

GUIDED INQUIRY

Pre Lab: Amino Acid Sequences: Indicators of Evolution

Problem How can you use proteins to determine how closely organisms are related?

Materials light-colored highlighting pen, graph paper

Lab Manual Chapter 16 Lab

Skills Focus Analyze Data, Graph, Draw Conclusions

Connect to the Big idea For years, scientists who studied evolution had to rely on only visible differences among organisms. Then a new source of evidence emerged. Biochemists were able to unravel the sequences of bases in DNA and amino acids in proteins. Scientists are able to use this data to confirm relationships based on anatomy. They also use the data to show that some species that appear very different are in fact more closely related than had been thought.

Biologists can compare the sequences of amino acids in a protein for two species. In general, when the total number of differences is small, the species are closely related. When the total number of differences is large, the species are more distantly related.

In this lab, you will compare amino acid sequences for one protein and analyze the results of a similar comparison for another protein. You will use both sets of data to predict relatedness among organisms.

Background Questions

a. Review What are homologous molecules?

b. Explain Why might scientists use molecules instead of anatomy to figure out how closely rabbits and fruit flies are related?

c. Relate Cause and Effect Amino acid sequences in the proteins of two species are similar. What can you conclude about the DNA in those species, and why?

Pre-Lab Questions

Preview the procedure in the lab manual.

1. **Predict** Based only on their anatomy, rank gorillas, bears, chimpanzees, and mice from most recent common ancestor with humans to least recent.
2. **Use Analogies** You tell a story to a second person who tells it to a third person, and so on. As the story is retold, changes are introduced. Over time, the number of changes increases. How is this process an analogy for what happens to DNA over time?
3. **Infer** Hemoglobin from two species is compared. On the long protein chains, there are three locations where the amino acids are different. Where would you place the common ancestor of the two species on the "tree of life," and why?

BIOLOGY.com Search Chapter 16 GO

Visit Chapter 16 online to test yourself on chapter content and to find activities to help you learn.

Untamed Science Video Islands are rich environments for evolution, as you will find out with the Untamed Science crew.

Art in Motion This animation shows how fossil layers accumulate and are later exposed.

Art Review Review homologus and analogous structures in vertebrates.

Visual Analogy See how different types of finch beaks function like tools.

Data Analysis Collect population data for several generations of grasshoppers and then analyze how the population changed due to natural selection.

Pre-Lab Answers

BACKGROUND QUESTIONS

a. Molecules that are similar in related organisms.

b. Because rabbits and fruit flies are distantly related, they don't have many visible structures that can be used to establish their relatedness.

c. DNA directly influences the sequence of amino acids in a protein. So if two species have similar sequences of amino acids in their proteins, the order of bases in that segment of their DNA must also be similar.

PRE-LAB QUESTIONS

1. Students are likely to say that chimpanzees and gorillas have a more recent common ancestor with humans than do bears and mice.
2. Students can compare retelling the story to the replication of DNA and the changes in the story to mutations. The number of changes will increase in both over time.
3. The common ancestor should be placed relatively recently along the tree, based on the limited differences between the proteins and the time required for mutations to occur.

NATIONAL SCIENCE EDUCATION STANDARDS

UCP IV

CONTENT C.2.a, C.2.c, C.3.d, C.3.e

16 CHAPTER Summary

16.1 Darwin's Voyage of Discovery

- Darwin developed a theory of evolution that explains how organisms evolved over long periods of time from common ancestors.
- Darwin noticed that (1) different places around the world that had similar habitats often had different species that were ecologically similar; (2) different, related species often lived in different habitats within a local area; and (3) some fossils of extinct animals were similar to living species.

evolution (p. 380)
fossil (p. 382)

16.2 Ideas That Shaped Darwin's Thinking

- Hutton and Lyell concluded that Earth is extremely old. Also, the processes that changed Earth in the past are the same processes that operate in the present.
- Lamarck suggested that individual organisms could change during their lifetime by using or not using different parts of their bodies. He also suggested that individuals could pass these acquired traits on to their offspring. This would cause species to change over time.
- Malthus reasoned that if the human population were to grow without control, there would not be enough space and food for everyone to live.
- In artificial selection, nature provides the variation of traits. Humans select the traits they find useful.

artificial selection (p. 387)

16.3 Darwin Presents His Case

- Natural selection occurs whenever more individuals are born than can survive, there is heritable variation, and some individuals have higher fitness than others.
- The principle of common descent says that all species are descended from ancient common ancestors. This is true of living and extinct species.

adaptation (p. 388)
fitness (p. 388)
natural selection (p. 389)

16.4 Evidence of Evolution

- Patterns linking adaptations to environments around the world provide evidence for natural selection.
- Many recently discovered fossils form series that trace the evolution of modern species from extinct ancestors.
- Homologous structures and patterns of embryo development provide evidence that species have descended, with modification, from a common ancestor.
- Many recently discovered fossils form series that trace the evolution of modern species from extinct ancestors.
- The universal genetic code and homologous molecules provide molecular evidence of common descent.
- The Grants observed natural selection occurring in a wild finch population. They also showed that variation within a species can help the species adapt to and survive environmental changes.

biogeography (p. 392)
homologous structure (p. 393)
analogous structure (p. 393)
vestigial structure (p. 393)

UbD Transfer the Big Idea

TRANSFER Tell students to suppose that a newspaper has published an article questioning whether evolution occurs. Ask students to write a short letter to the editor of the newspaper in which they argue that evolution does occur, using evidence from the chapter to support their argument. Instruct students to describe in their letters two or more examples of evidence, using one or two sentences for each example. You might wish to have a class discussion about the evidence before students write their letters..

Chapter Summary

Think Visually

USE GRAPHICS Using the information in this chapter, have students create a concept map that links the following terms: *adaptation, artificial selection, biogeography, camouflage, Charles Darwin, Charles Lyell, evolution, fitness, fossil, homology, James Hutton, Jean-Baptiste Lamarck, mimicry, natural selection,* and *Thomas Malthus.*

ANSWERS
Students' concept maps should show the following relationships: Hutton, Lamarck, and Malthus influenced Darwin's theory of evolution by natural selection; natural selection occurs when organisms in a population differ in fitness and it leads to adaptations such as camouflage and mimicry; evidence for evolution includes fossils, artificial selection, homologies, and biogeography.

BIOLOGY.com Study Online

REVIEW RESOURCES

Art in Motion An animation of canyon formation shows how fossil layers accumulate and are later exposed.

Data Analysis Students collect population data to see how variation in a population of grasshoppers changes over time due to natural selection.

Art Review Students identify homologous and analogous structures in this drag-and-drop activity.

Visual Analogy Animations show the functions of specialized pliers and finch beaks.

Vocabulary Review The *Flash Cards* and *Crossword* provide an interactive way to review chapter vocabulary.

Editable Worksheets Pages of Study Workbook B, Lab Manual B, and the Assessment Resources Book are available online. These documents can be easily edited using a word-processing program.

Lesson Overviews Lesson Overviews are based on Miller & Levine Biology. However, they are available on CD-ROM and can be edited for your class.

ASSESSMENT RESOURCES

Chapter Assessment Have students take an online version of the Chapter 16 Check Understanding.

Standardized Test Prep Students can take an online version of the Standardized Test Prep. You will receive their scores along with ideas for remediation.

Diagnostic and Benchmark Tests Use these tests to monitor your students' progress and supply remediation.

ASSESSMENT

Assess

Big idea Invite a volunteer to read the question. Write on the board the words *evolution, natural selection,* and *Charles Darwin.*

Ask What is the connection between these words? *(Charles Darwin developed the theory of evolution by natural selection.)*

Have students review the Evolution by Natural Selection section in Lesson 16.3.

Ask What is the word for a heritable trait that improves an individual's ability to survive and reproduce? *(adaptation)*

Ask What is the relationship between adaptations and natural selection? *(Organisms that are better adapted to their environment are more likely to survive and reproduce.)*

Ask What is the result of natural selection? *(evolution of species)*

CONSTRUCTED RESPONSE Paragraphs may vary. Explain the activity carefully. Be sure students understand that the *Hints* should be used as a guide to help them answer the questions. Review how to write a paragraph. Tell them that writing notes about topics you intend to cover in a paragraph can help you organize your thoughts before you write. Have students copy Question 1 into their notebooks. Ask a volunteer to read it aloud. Encourage students to take notes during your discussion.

Say To answer this question, you will need to remember ideas developed by scientists who influenced Darwin.

Read the first hint.

Ask What did Hutton and Lyell mean by *deep time*? *(Earth is extremely old.)* How did Darwin use the idea of deep time in his theory of evolution? *(Darwin thought that it can take millions of years for species to evolve.)*

Read the second hint.

Ask What did Malthus observe about the differences between human birth and death rates? *(More humans were being born than were dying.)* What conditions did these differences cause? *(problems related to overcrowding, such as starvation and disease)*

Model writing the first one or two sentences using the hints and details offered by students. *(Sample: Darwin's theory of evolution was influenced by the work of other scientists. The idea that Earth is extremely old, which was proposed by the geologists Hutton and Lyell, influenced Darwin's idea that species may evolve over millions of years.)*

16 CHECK Understanding

Assess the Big idea Evolution

Write an answer to the question below.
Q: What is natural selection?

Constructed Response

Write an answer to each of the numbered questions below. The answer to each numbered question should be one or two paragraphs. To help you begin, read **Hints** below the questions.

1. **How was Darwin's theory of evolution influenced by the work of other scientists?**

   Hint Geologists such as Hutton and Lyell developed the concept of deep time.

   Hint Thomas Malthus described conditions that affect human populations. These conditions can also affect other organisms.

2. **How would Darwin explain the evolution of the long legs of the bird in the picture on page 385? How would his explanation differ from that of Lamarck?**

   Hint Read the caption to review some of Lamarck's ideas.

3. **Look back at the art of the natural selection of grasshoppers on page 390. Explain how conditions could change so that yellow grasshoppers are better adapted to their environment than green grasshoppers. What would happen to the relative numbers of green and yellow grasshoppers in the population?**

   Hint The word *camouflage* refers to a way of hiding in an environment. In your answer, explain how camouflage can be an adaptation.

Foundations for Learning Wrap-Up

You can use the notes you took about the steps of natural selection as a tool for review. First, you will make your sheet of paper into a pyramid. To do this, cut to the center along a fold next to the colored triangle. (Do not cut all the way through.) Now you have two flaps. Tape the flap with writing over the colored flap, so the two flaps make one side. You now have a three-sided pyramid.

Activity 1 Turn each side of your pyramid into a question. For example, "What is the struggle for existence?" Write the answer on a separate sheet of paper. Then look inside your pyramid to check your answer. Make any corrections or additions that are needed on your answer sheet. Then move on to the next question.

Activity 2 Work with a partner. Quiz each other on the stages of natural selection. Begin with a population of green grasshoppers that live in a habitat with green grass. Use the sides of your pyramid as clues. For example, you could ask, "How does the struggle for existence affect the grasshoppers?" If you cannot answer a question, check inside the pyramid.

Then take turns suggesting other kinds organisms that live in other environments. For example, you might describe the process of natural selection on a population of orange and pink fish that live on a coral reef. Be creative!

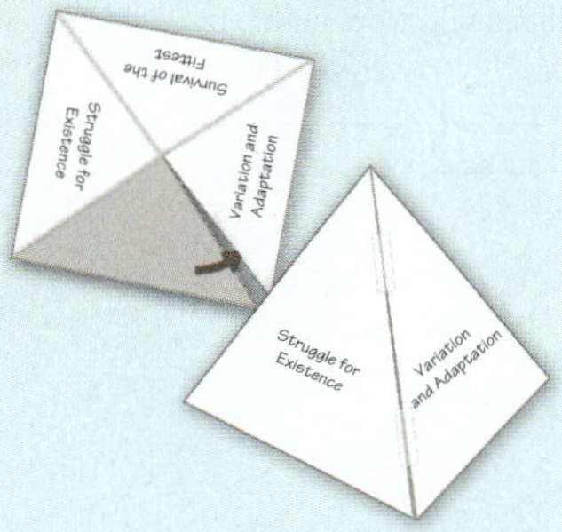

For Question 2, pair students and have them use the hints and your modeling to answer the question. Have students share their paragraphs with the class.

For Question 3, have students write the paragraphs independently for homework or as an anchor activity.

Foundations for Learning Wrap-Up

Have students take out their review paper about the steps of natural selection. Ask volunteers to share the notes they have written. Be sure that students assemble their pyramids correctly, and then explain the activities. Have students complete the first activity individually and the second activity in pairs.

16.1 Darwin's Voyage of Discovery

Understand Key Concepts

1. Who observed variations in the characteristics of plants and animals on the Galápagos Islands?
   a. James Hutton
   b. Thomas Malthus
   c. Charles Lyell
   d. Charles Darwin

Test-Taking Tip

Read Carefully! Sometimes the difference between the correct answer and an incorrect answer is a single word. In question 1, you might be tempted to answer **c**, because Charles Lyell is similar to Charles Darwin. Try to think of the answer before you read the choices, then carefully read each choice.

2. In different parts of the world, Darwin found unrelated species that looked alike because
   a. the species lived in different environments.
   b. the species lived in similar environments.
   c. the species were closely related.
   d. the species once lived in the same place.
3. What connection did Darwin make between the Galápagos tortoises and their environments?

Think Critically

4. **Relate Cause and Effect** Why was Darwin's trip on the *Beagle* so important to his development of the theory of natural selection?

16.2 Ideas That Shaped Darwin's Thinking

Understand Key Concepts

5. Which of the following ideas proposed by Lamarck was later found to be incorrect?
   a. Acquired characteristics can be inherited.
   b. All species are descended from other species.
   c. Living things change over time.
   d. There is a relationship between an organism and its environment.
6. Which of the following would an animal breeder use to increase the amount of milk given by his herd of cows?
   a. overproduction
   b. genetic isolation
   c. acquired characteristics
   d. artificial selection
7. According to Malthus, what factors limit human population growth?

Think Critically

8. **Relate Cause and Effect** Lamarck made a very significant contribution to science, even though his explanation of evolution was wrong. Explain how Lamarck helped other scientists.
9. **Infer** Could artificial selection happen without inherited variation? Explain your answer.

16.3 Darwin Presents His Case

Understand Key Concepts

10. An inherited characteristic that increases an organism's ability to survive and reproduce in its environment is called a(n)
    a. vestigial structure.
    b. adaptation.
    c. homologous structure.
    d. variation.
11. What do evolutionary trees show?
    a. all life depends on trees
    b. all life is changing due to natural selection
    c. all living species are descended from earlier ancestors.
    d. all life evolves to the point of perfection
12. What is fitness, in evolutionary terms?

Think Critically

13. **Infer** Many species of birds build nests in which they lay eggs and care for their young. How does this behavior relate to reproductive fitness?

Assessment Tips

Have students work in pairs to answer the questions on this page and the following page. For Question 2, have students review the three patterns of biological diversity Darwin observed. For Question 8, have students reread Evaluating Lamarck's Hypotheses. Tell students to review the differences between homologous, analogous, and vestigial structures and to use this knowledge to answer Question 15. For Question 17, have students discuss results presented in Testing Natural Selection.

Answers

Lesson 16.1

UNDERSTAND KEY CONCEPTS

1. d

2. b

3. Closely related Galápagos tortoises exhibited different traits depending on the environment of the island where they lived.

THINK CRITICALLY

4. The trip exposed Darwin to a much greater diversity of life than did his home in England. He was able to see patterns of biogeography that could be seen only during world travel.

Lesson 16.2

UNDERSTAND KEY CONCEPTS

5. a

6. d

7. According to Malthus, population growth is limited by overcrowding and lack of food, which in turn lead to war, famine, and disease.

THINK CRITICALLY

8. Lamarck was one of the first naturalists to suggest that species are not fixed, and he tried to explain evolution scientifically using natural processes. He also recognized that there is a link between an organism's environment and its body structures. Although Lamarck's explanation of how evolutionary change occurs was wrong, these other ideas paved the way for later biologists, including Darwin.

9. Artificial selection could not happen without inherited variation, because breeders select individuals carrying a desired version of a trait over those with other variations of that trait.

Lesson 16.3

UNDERSTAND KEY CONCEPTS

10. b

11. c

12. Individuals that have high fitness are more likely to survive and have more offspring than low-fitness individuals. Their inherited characteristics are more likely to be passed on to future generations.

THINK CRITICALLY

13. Birds that build better nests have a better chance of hatching all of their eggs. Well-constructed and well-hidden nests are safer from accidents and from predators. More young will survive to reproduce from well-made nests.

Lesson 16.4

UNDERSTAND KEY CONCEPTS

14. a

15. c

16. DNA is common to all life on Earth, and the code is almost identical in all organisms. This suggests that all organisms alive today are related and that DNA was inherited from an ancient common ancestor.

THINK CRITICALLY

17. The Grants measured all the beaks of the medium ground finch. They recorded the survival of different birds and continued the study for many years. The Grants showed that natural selection is occurring often and more rapidly than many thought possible.

Connecting Concepts

18. The environment of the mice looks brown, so the brown mice are better adapted. Brown mice will blend in better to their environment and be harder for predators to see.

CHAPTER MYSTERY After students have read the Chapter Mystery, discuss similarities between the Galápagos finches and the Hawaiian honeycreepers.

Ask What do Galápagos finches and Hawaiian honeycreepers have in common, in terms of where they are found? *(Both are found on a specific group of islands and nowhere else.)* How are their special adaptations similar? *(Both types of birds have adaptations that allow them to eat different types of food.)* How do scientists think the honeycreeper species are related to each other? *(The species are closely related and descended from a common ancestor.)* How does this compare to relationships between the Galápagos finch species? *(The finch species are also closely related to each other and descended from a common ancestor.)* Why do you think these similar examples of natural selection in birds took place on islands instead of on a large continent? *(The birds were separated from each other and evolved in different environments.)*

16 CHECK Understanding

16.4 Evidence of Evolution

Understand Key Concepts

14. Series of related fossils are important evidence of evolution because they show
- **a.** how organisms changed over time.
- **b.** how animals behaved in their environments.
- **c.** how the embryos of organisms develop.
- **d.** molecular homologies.

15. The wing of a bird that cannot fly is an example of a(n)
- **a.** analogous structure.
- **b.** homologous structure.
- **c.** vestigial structure.
- **d.** molecular structure.

16. How does DNA provide evidence for common descent?

Think Critically

17. Evaluate Darwin hypothesized that natural selection led to the different beak shapes in the Galapagos finches. Describe how the Grants tested this hypothesis. Did their data support Darwin's hypothesis?

Connecting Concepts

Use Science Graphics

Use the illustration below to answer question 18.

18. Infer Based on what you can see, are the brown mice or white mice better adapted to their environment? Explain why.

solve the CHAPTER MYSTERY

SUCH DIFFERENT HONEYCREEPERS

Hawaiian honeycreepers like the i'iwi have a lot in common with the Galápagos finches. Like the finches, the honeycreepers are small birds found nowhere else on Earth. They live on islands far from the mainland. And like the finches, the 20 known species of honeycreepers are closely related to each other.

This suggests that the honeycreepers are all descended, with modification, from a recent common ancestor. Honeycreepers are recent in evolutionary terms. Biologists think the ancestor first arrived on the islands between 3 and 4 million years ago.

Many of the honeycreepers have specialized diets. Their adaptations allow them to use different food sources on the different islands. Today, habitat loss is making it harder for the honeycreepers to survive. Some species have already become extinct since humans settled on the islands.

1. Infer Imagine a small group of birds landed on one of the Hawaiian islands millions of years ago. This small population then reproduced. Do you think all of the descendants would have stayed on that one island? Explain your answer.

2. Infer Do you think that environmental conditions are the same everywhere on all of the Hawaiian Islands? How might the environment have affected the evolution of honeycreepers?

3. Form a Hypothesis Explain how the different species of honeycreepers in Hawaii today might have evolved from one ancestral species.

UntamedScience™ **Never Stop Exploring Your World.** Finding the solution to the honeycreeper mystery is only the beginning. Take a video field trip to Hawaii with the ecogeeks of Untamed Science to see where the mystery leads.

CHAPTER MYSTERY ANSWERS

1. No, because most species tend to overproduce, the island would soon become overcrowded. Some birds would leave to escape competition.

2. Environmental conditions vary from island to island. On a single island, the environment is different in the lowlands than in the mountains. Birds might evolve adaptations to thrive in one of the different environments.

3. A population of migrating birds might have been blown to the islands during a storm. The birds would naturally have had heritable variations. Competition and different environments would make it easier for birds with certain traits to survive. These birds would have more offspring, and these traits would become more common. Eventually birds flew to other islands, where the process of evolution by natural selection continued.

Untamed Science™ Students can learn more about the Hawaiian honeycreeper mystery by watching the video **It Happened in Hawaii.**

Standardized Test Prep

Multiple Choice

1. Which scientist formulated the theory of evolution through natural selection?
   A Charles Darwin
   B Thomas Malthus
   C James Hutton
   D Jean-Baptiste Lamarck

2. Lamarck's ideas about evolution were wrong because he proposed that
   A species change over time.
   B species descended from other species.
   C acquired characteristics can be inherited.
   D species are adapted to their environments.

3. Lyell's *Principles of Geology* influenced Darwin because it explained how
   A organisms change over time.
   B adaptations occur.
   C Earth must be very old.
   D the Galápagos Islands formed.

4. A farmer's use of the best livestock for breeding is an example of
   A natural selection.
   B artificial selection.
   C extinction.
   D adaptation.

5. The ability of an individual organism to survive and reproduce in its natural environment is called
   A natural selection.
   B evolution.
   C descent with modification.
   D fitness.

6. Which of the following is an important concept in Darwin's theory of evolution by natural selection?
   A descent with modification
   B homologous molecules
   C processes that change the surface of Earth
   D the tendency toward perfection

7. Which of the following provides evidence for evolution?
   A fossil record
   B homologous structures of living organisms
   C geographical distribution of living things
   D all of the above

8. DNA and RNA provide evidence of evolution because
   A all organisms have nearly identical DNA and RNA.
   B no two organisms have exactly the same DNA.
   C each RNA codon specifies just one amino acid.
   D in most organisms, the same codons specify the same amino acids.

9. A bird's wings are homologous to a(n)
   A fish's tailfin.
   B alligator's claws.
   C dog's front legs.
   D mosquito's wings.

Questions 10 and 11

The birds shown below are two of the species of finches Darwin found on the Galápagos Islands.

Woodpecker Finch

Large Ground Finch

10. What process produced the two different types of beaks shown?
    A artificial selection
    B natural selection
    C geographical distribution
    D disuse of the beak

11. The large ground finch obtains food by cracking seeds. Its short, strong beak is an example of
    A the struggle for existence.
    B the tendency toward perfection.
    C an adaptation.
    D a vestigial organ.

Open-Ended Response

12. Compare and contrast the processes of artificial selection and natural selection.

If You Have Trouble With . . .

Question	1	2	3	4	5	6	7	8	9	10	11	12
See Lesson	16.1	16.2	16.2	16.2	16.3	16.3	16.4	16.4	16.4	16.3	16.3	16.3

Test-Taking Tip

USE TIME WISELY

Tell students that when they take a test, they need to pay attention to the time that they are spending on sections of the test and on individual questions. If they do not know the answer to a multiple-choice question, they should skip it and come back to it later, if time permits. Explain to students that if the test contains a constructed response question that requires a long answer, they need to allocate time to plan their responses as well as to write them.

Before the Test

Review with students the Test-Taking Tip.

Remind students that an open-ended response means that they'll have to write a paragraph to answer the question. Have them paraphrase the question in their notebooks, and remind them about the steps in writing a paragraph.

Answers

1. A
2. C
3. C
4. B
5. D
6. A
7. D
8. D
9. C
10. B
11. C
12. Both artificial and natural selection change the heritable traits of a population or species over time. In both cases, organisms with certain traits are more likely to survive and reproduce than organisms with other traits. In artificial selection, a breeder or farmer decides which organisms reproduce. In natural selection, environmental conditions determine which organisms reproduce.

Evolution of Populations

Chapter Contents	NSES	Time	Core Resources
Chapter Preview			**Student Edition,** pp. 404–405 **Chapter Mystery,** p. 405 **Foundations for Learning,** p. 405
17.1 Genes and Variation Genetics Joins Evolutionary Theory • Sources of Genetic Variation • Single-Gene and Polygenic Traits	I, II, III, IV, A.2.d, C.2.a, C.2.b, C.2.c, C.3.a	1 period 1/2 block	**Student Edition,** pp. 406–408 **Study Workbook B** 17.1 Worksheets **Biology.com** *Art Review:* Frequency and Dominance • 17.1 Lesson Assessment
17.2 Evolution as Genetic Change in Populations How Natural Selection Works • Genetic Drift • Evolution Versus Genetic Equilibrium	I, II, III, IV, A.1.b, A.1.c, A.2.a, A.2.d, C.2.c, C.3.a, C.3.b, C.3.c	1 1/2 periods 3/4 block	**Student Edition,** pp. 409–413 **Inquiry Into Scientific Thinking,** p. 413 **Study Workbook B** 17.2 Worksheets **Biology.com** *Art in Motion:* Natural Selection • 17.2 Lesson Assessment
17.3 The Process of Speciation Isolating Mechanisms • Speciation in Darwin's Finches	II, III, IV, V, A.2.a, A.2.b, A.2.e, A.2.f, C.2.c, C.3.a, C.3.b, C.3.d, C.6.c, G.1	1/2 period 1/4 block	**Student Edition,** pp. 414–416 **Study Workbook B** 17.3 Worksheets **Biology.com** *Tutor Tube:* Organizing the Vocabulary of Speciation • 17.3 Lesson Assessment
17.4 Molecular Evolution Molecular Clocks • Gene Duplication • Developmental Genes and Body Plans	II, III, IV, V, C.2.a, C.2.c, C.3.a, C.3.b, C.3.c, C.3.d, C.3.e	1 period 1/2 block	**Student Edition,** pp. 417–419 **Study Workbook B** 17.4 Worksheets **Biology.com** 17.4 Lesson Assessment
Chapter Pre-Lab	IV, A.2.a, A.2.b, C.3.a, C.3.d, C.4.c, C.6.c, G.3	1 period 1/2 block	**Student Edition,** p. 420 **Lab Manual B** *Competing for Resources*

Tools for ELL Instruction

ELL Handbook describes teaching strategies to make *Miller & Levine Biology: Foundation Edition* more accessible to ELL students.

Spanish Study Workbook is a Spanish translation of Study Workbook A.

Multilingual Glossary is the *Miller & Levine Biology* glossary translated into ten languages.

Tools to Challenge Students

Study Workbook A includes more challenging worksheets and activities such as Chapter Mystery activities that reinforce 21st Century Skills.

Lab Manual A labs are geared toward on-level students. The Lab Manual includes fifteen labs in addition to the Chapter Labs.

Biology.com Data Analysis activities offer students a way to practice data gathering and analyzing skills.

Visual Quizzes in the Assessment Resources Book are geared toward on-level students and use visuals from *Miller & Levine Biology.*

Additional Resources

Biology.com Untamed Science Video • Flash Cards

Study Workbook A 17.1 Worksheets
Spanish Study Workbook 17.1 Worksheets
Assessment Resources Book Visual Quiz

Study Workbook A 17.2 Worksheets
Lab Manual B *Hands-On Activity:* Modeling Natural Selection • *Data Analysis:* Allele Frequency
Spanish Study Workbook 17.2 Worksheets
Assessment Resources Book Visual Quiz

Study Workbook A 17.3 Worksheets
Spanish Study Workbook 17.3 Worksheets
Biology.com *Data Analysis:* Galápagos Finches: Evolution in Action

Study Workbook A 17.4 Worksheets
Spanish Study Workbook 17.4 Worksheets

Lab Manual A *Competing for Resources* • *Ecosystems and Speciation*

Chapter Review

Student Edition Chapter Summary, p. 421
Study Workbook B Chapter 17 Review
Transparencies, pp. 203–214
Biology.com Untamed Science Video • Editable Worksheets of Study Workbooks A and B and Lab Manuals A and B • Chapter 17 Flash Cards and Match It

Untamed Science DVD • Classroom Resources CD (includes lesson presentations and editable worksheets)

Chapter Assessment

Student Edition Check Understanding, pp. 422–425
Study Workbook B Chapter 17 Taking a Standardized Test
Assessment Resources Book Chapter 17 Test B
Biology.com Chapter 17 Assessment • Editable Worksheets of Chapter 17 Visual Quizzes* and Chapter 17 Tests A and B

Exam*View* *Assessment Suite* • Classroom Resources CD (includes lesson presentations and editable worksheets)

Time: 1 period, 1/2 block

* Visual Quizzes are based on visuals from *Miller & Levine Biology.* Review carefully to make sure they are appropriate for your class.

Pressed for Time?

Preview the Chapter Introduce students to the first Key Question for each lesson and preview the Alleles in a Population and Selection on a Single-Gene Trait visuals.

Cover the Chapter Quickly Have students read Lesson 17.1 and go over the Alleles in a Population visual. For Lesson 17.2, have them read *How Natural Selection Works* and *Genetic Drift* focusing on the Selection on a Single-Gene Trait and Selection on Polygenic Traits visuals. Then, have students read *Isolating Mechanisms* in Lesson 17.3 and *Molecular Clocks* in Lesson 17.4. Discuss the Molecular Clock diagram.

Assess Assign the 17.1 Check Understanding, Questions 1, 2, 4, and 5 from the 17.2 Check Understanding, Questions 1–4 from the 17.3 Check Understanding, Questions 1 and 6 from the 17.4 Check Understanding, and Questions 1–7, 11, 12, 14, and 16–18 in the Chapter 17 Check Understanding.

Get the Picture

Draw students' attention to the photograph of the two butterflies.

Ask Do you think these butterflies are of the same species? *(Answers may vary, but they are the same species.)*

Ask Are the butterflies identical? *(no)*

Have a volunteer read the caption aloud. Discuss the different patterns on the wings: the black spots are slightly larger on the left butterfly and the black spots make a slightly different pattern on each butterfly; the D-shaped pattern of spots on the right butterfly is more rounded compared to the angular shape of the backwards D shape on the left butterfly. You might want to sketch the patterns on the board to help students see the distinction.

Ask How might a different wing pattern on a butterfly help it survive? *(Sample answer: Some patterns might be more attractive to or more easily noticed by a predator. Some patterns might better camouflage the butterfly.)*

Activate Prior Knowledge

Remind students of the role that genes play in passing traits from one generation to another.

Ask Would the genes of the two butterflies be identical? *(no)* Why not? *(The genes would have slight differences that control the wing patterns.)*

Have students sketch a wing pattern that they think would help a butterfly survive.

Ask Would populations of butterflies that have this trait be likely to pass the trait on to the next generation? *(Sample answer: yes)*

NATIONAL SCIENCE EDUCATION STANDARDS

UNIFYING CONCEPTS AND PROCESSES

I, II, III, IV, V

CONTENT

C.2.a, C.2.b, C.2.c, C.3.a, C.3.b, C.3.c, C.3.d, C.3.e, C.6.c

INQUIRY

A.1.b, A.1.c, A.2.a, A.2.b, A.2.d, A.2.e, A.2.f

17 Evolution of Populations

Big idea **Evolution**

Q: How can populations evolve to form new species?

At first glance, these two butterflies look exactly the same. But if you look closely, you can see that the patterns on their wings are slightly different. Variations among members of a population make evolution possible. Sometimes evolution leads to new species.

BIOLOGY.com Search Chapter 17 GO • Flash Cards

404

UbD Connect to the Big Idea

Invite a volunteer to read aloud the chapter number and the chapter title. Read the Big Idea and question. Emphasize the question, **How can populations evolve to form new species?**

Set the Purpose

PREVIEW AND PREDICT Draw the graphic organizer on the board. Guide students to the INSIDE feature on the next page and invite a volunteer to read each lesson number and title.

Ask How do you think the lesson titles relate to the questions in the graphic organizer? Remind students to search for answers to the questions as they read and to ask themselves how populations evolve.

CHAPTER 17

INSIDE:

- Untamed Science Video
- Chapter Mystery

CHAPTER MYSTERY

KILLER DISEASE

In 1918, a new disease began to kill millions of people. Around the world, more than 40 million people died. What caused this terrible disease? It was an influenza virus. This virus was a strain of "the flu" that you can catch again and again. Every year different strains of these viruses appear and cause the flu.

How did this one virus become so deadly? Could such a deadly flu ever happen again? Scientists are searching for the answers to these questions. They are still very worried about the influenza virus, and the chance of another deadly disease. That is why they are watching the "bird flu" so closely.

Read for Mystery Clues As you read this chapter, look for clues that explain how new types of influenza virus appear. How is evolution involved? Then, solve the mystery at the end of the chapter.

FOUNDATIONS for Learning

Fold a piece of paper into three equal sections. Then lay the paper out flat. At the top of the page, write "Sources of Genetic Variation," At the top of the middle section, write "Natural Selection." At the top of the bottom section, write "Isolation." As you read, make notes or drawings that relate to each heading. These notes will help you answer the question: How can populations evolve to form new species?

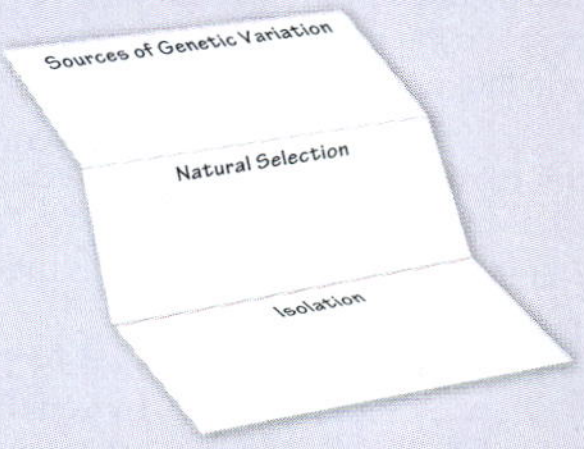

Chapter 17
Big Idea: Evolution

Chapter 17: How can populations evolve to form new species?

- **17.1:** How do genes make evolution possible?
- **17.2:** What causes a population's gene pool to change?
- **17.3:** How do new species form?
- **17.4:** What can genes tell us about an organism's evolutionary history?

Foundations for Learning

Students can use their own papers for this activity. Tell them to fold their papers horizontally.

Ask Does this activity show how these three concepts relate to each other? *(no)*

Ask What relationship does it show? *(It shows how each of the three topics relates to the Big Idea question.)*

Point out that when students make notes or drawings about one of the topics, they should ask themselves whether what they wrote also relates to another of the topics. Tell students that when they decide what to write, they should keep in mind that each note or drawing should help them find out how populations evolve and form new species.

Untamed Science

In **Evolution at a Snail's Pace**, students explore speciation mechanisms in Hawaii, where geographic isolation contributes to evolutionary patterns similar to those in Darwin's finches.

CHAPTER MYSTERY

Ask students if they ever had the flu.

Ask How did you feel? Have you always felt the same way when you had the flu, or did you feel worse sometimes? *(Answers may vary.)*

Have a student volunteer read the Chapter Mystery and predict how evolution is involved in the appearance of new strains of viruses.

Digital Support

BIOLOGY.com Extend your reach by using the digital assets offered at **Biology.com.**

LESSON 17.1

Pre-Reading

Objectives

17.1.1 Define *evolution* in genetic terms.

17.1.2 Identify the main sources of genetic variation in a population.

17.1.3 State what determines the number of phenotypes for a trait.

Preview the Pages

1. Preview the vocabulary words and their definitions. Discuss with students the meaning of the word *pool* in *gene pool.* Emphasize that a pool is a group of things that have a common purpose. Elicit other examples of *pool* used this way, such as car pool.
2. As you preview the visuals, ask students to identify how the individuals in each group are alike and different.
3. Refer students to the Single-Gene Trait and Polygenic Trait graphs. Make sure students understand that the scales used on the graphs differ. Point out that the scale of the *y*-axis on the Single-Gene Trait graph is labeled in percentage units. The *y*-axis of the Polygenic Trait graph is not labeled in units. Ask students to identify what is shown on the *x*-axis of each graph.

Build Understanding

CONCEPT MAP Direct students to Workbook B, where they will find a partially filled-in concept map. Remind students that concept maps help them take notes and identify main ideas and supporting details.

NATIONAL SCIENCE EDUCATION STANDARDS

UNIFYING CONCEPTS AND PROCESSES

I, II, III, IV

CONTENT

C.2.a, C.2.b, C.2.c, C.3.a

INQUIRY

A.2.d

For the corresponding lesson in *Miller & Levine Biology,* see pages 482–486.

17.1 Genes and Variation

Key Questions

How is evolution defined in genetic terms?

What are the sources of genetic variation?

What determines the number of phenotypes for a given trait?

BUILD Understanding

Concept Map As you read, construct a concept map to describe the sources of genetic variation.

In Your Workbook Go to your workbook to learn more about making a concept map.

BUILD Vocabulary

gene pool all the genes, including all the different alleles for each gene, that are present in a population at any one time

allele frequency the number of times that an allele occurs in a gene pool, compared to the total number of alleles in that pool for the same gene

single-gene trait a trait controlled by one gene that has two alleles

polygenic trait a trait controlled by two or more genes

PREFIXES

The prefix *poly-* means "several or many." *Polygenic* means "many genes."

Genetics Joins Evolutionary Theory

When Darwin developed his theory of evolution, he did not know how heredity worked. He did not know the source of variation in a population. Mendel's studies were published while Darwin was alive. But no one understood the importance of Mendel's work. Mendel's work was rediscovered around 1900. When scientists combined Mendel's work with Darwin's theory, they could understand how traits were inherited. Today, evolution can be described in terms of genetics.

Genotype and Phenotype in Evolution Most plants and animals have two sets of genes—one set from each parent. These genes come in different forms, called alleles. The set of alleles found in an organism is called its genotype. An organism's genotype and its environment determine its phenotype. Phenotype is an organism's appearance and other characteristics, or traits.

Natural selection acts on phenotypes. It does not act directly on genes. In any population, some individuals have phenotypes (or traits) that are a better fit for their environment. Those individuals have higher fitness. In genetic terms, higher fitness means that they have more offspring and pass on more of their genes to the next generation.

Gene Pools A population is a group of individuals that mate and produce offspring. A **gene pool** contains all the alleles of all the genes in a population. In a gene pool, some alleles are common. Others are rare. **Allele frequency** is the number of times an allele occurs in a gene pool, compared to the total number of alleles for the same gene in that pool. Allele frequency is a percentage. The diagram on the next page shows the allele frequencies for fur color in a population of mice. In this population, the frequency of the dominant allele *B* (black) is 40 percent. The frequency of the recessive allele *b* (brown) is 60 percent.

If the frequency of an allele changes over time, the population is evolving. Only populations evolve, not individuals.

Key Question How is evolution defined in genetic terms? **In genetic terms, evolution involves a change in the frequency of alleles in a population over time.**

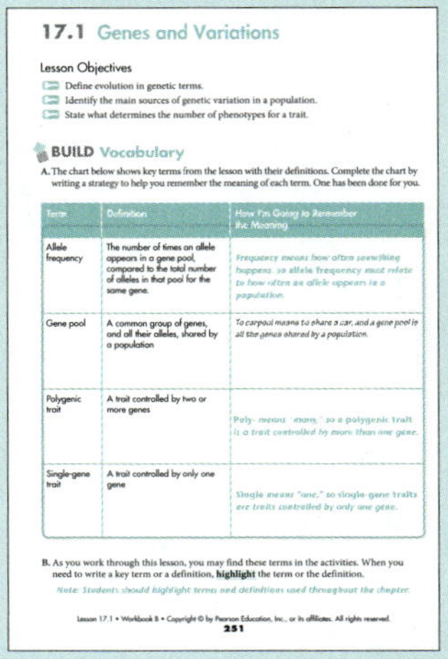

17.1 Genes and Variations

Lesson Objectives

Define evolution in genetic terms.

Identify the main sources of genetic variation in a population.

State what determines the number of phenotypes for a trait.

BUILD Vocabulary

A. The chart below shows key terms from the lesson with their definitions. Complete the chart by writing a strategy to help you remember the meaning of each term. One has been done for you.

Term	Definition	How I'm Going to Remember the Meaning
Allele frequency	The number of times an allele appears in a gene pool, compared to the total number of alleles in that pool for the same gene.	Frequency means how often something happens, so allele frequency must relate to how often an allele appears in a population.
Gene pool	A common group of genes, and all their alleles, shared by a population	To carpool means to share a car, and a gene pool is all the genes shared by a population.
Polygenic trait	A trait controlled by two or more genes	Poly- means "many," so a polygenic trait is a trait controlled by more than one gene.
Single-gene trait	A trait controlled by only one gene	Single means "one," so single-gene traits are traits controlled by only one gene.

B. As you work through this lesson, you may find these terms in the activities. When you need to write a key term or a definition, **highlight** the term or the definition.

Note: Students should highlight terms and definitions used throughout the chapter.

Lesson 17.1 • Workbook B • Copyright © by Pearson Education, Inc., or its affiliates. All rights reserved.
251

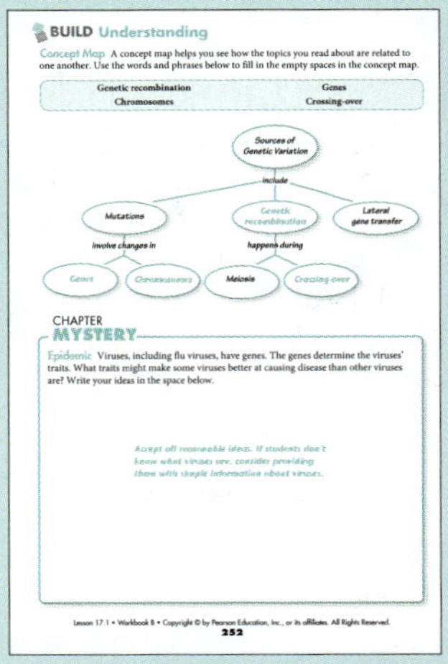

BUILD Understanding

Concept Map A concept map helps you see how the topics you read about are related to one another. Use the words and phrases below to fill in the empty spaces in the concept map.

Genetic recombination Genes
Chromosomes Crossing-over

CHAPTER MYSTERY

Epidemic Viruses, including flu viruses, have genes. The genes determine the viruses' traits. What traits might make some viruses better at causing disease than other viruses are? Write your ideas in the space below.

Accept all reasonable ideas. If students don't know what viruses are, consider providing them with simple information about viruses.

Lesson 17.1 • Workbook B • Copyright © by Pearson Education, Inc., or its affiliates. All Rights Reserved.
252

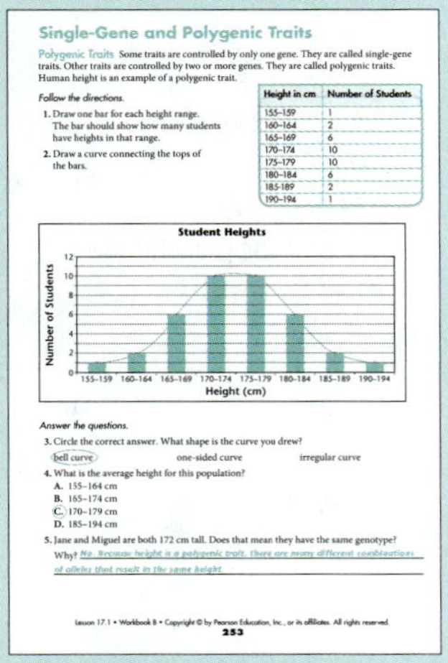

Single-Gene and Polygenic Traits

Polygenic Traits Some traits are controlled by only one gene. They are called single-gene traits. Other traits are controlled by two or more genes. They are called polygenic traits. Human height is an example of a polygenic trait.

Follow the directions.

1. Draw one bar for each height range. The bar should show how many students have heights in that range.
2. Draw a curve connecting the tops of the bars.

Height in cm	Number of Students
155–159	1
160–164	2
165–169	6
170–174	10
175–179	10
180–184	6
185-189	2
190–194	1

Answer the questions.

3. Circle the correct answer. What shape is the curve you drew?

bell curve one-sided curve irregular curve

4. What is the average height for this population?
A. 155–164 cm
B. 165–174 cm
C. 170–179 cm
D. 185–194 cm

5. Jane and Miguel are both 172 cm tall. Does that mean they have the same genotype? Why? No. Because height is a polygenic trait, there are many different combinations of alleles that result in the same height.

Lesson 17.1 • Workbook B • Copyright © by Pearson Education, Inc., or its affiliates. All rights reserved.
253

Sources of Genetic Variation

The members of a population differ from each other. Genetics explains the source of these variations.

Mutations A mutation is a genetic change. Most mutations are neutral and have little effect on fitness. Some mutations lower fitness by making survival and reproduction more difficult. Others increase fitness by making survival and reproduction easier. To play a role in evolution, a mutation must be passed from one generation to the next. The only mutations that can be passed on are those that are carried in egg or sperm cells.

Sexual Reproduction During sexual reproduction, the genes from two parents combine in new ways. This process can produce millions of different gene combinations. Crossing-over also mixes genes. Crossing-over happens during meiosis when chromosome pairs trade pieces of DNA.

Sexual reproduction creates new genotypes. But it does not change the frequency of alleles in the whole population. Sexual reproduction is like shuffling a deck of cards. You get new combinations of cards. But the deck always has the same number of kings and queens.

Lateral Gene Transfer Some single-celled organisms can pass genes from one individual to another. For example, many bacteria trade genes on plasmids. The passing of genes to an organism that is not an offspring is called lateral gene transfer. Lateral gene transfer is important in the evolution of single-celled organisms.

Key Question What are the sources of genetic variation?
Three sources of genetic variation are mutations, genetic recombination during sexual reproduction, and lateral gene transfer.

Genetic Variation Genetic variation can produce differences in appearance, such as the different colors of the kernels on these ears of corn. Other kinds of genetic variation, such as resistance to disease, cannot be seen.

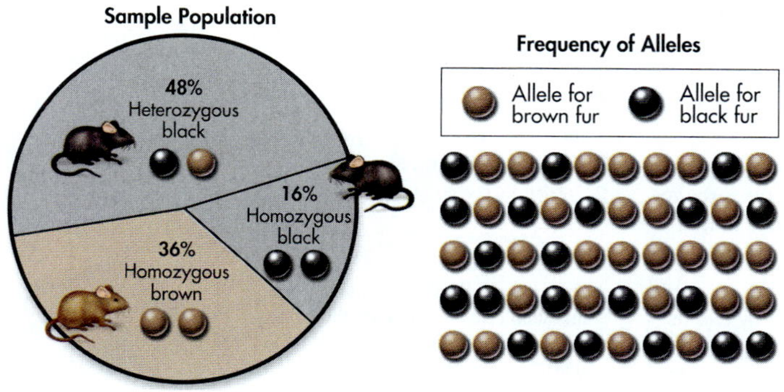

Alleles in a Population To determine whether a population is evolving, scientists study its allele frequencies. This diagram shows allele frequencies for fur color in a mouse population.

During Reading

Build Vocabulary

PREFIXES Remind students that a prefix is added to the beginning of a root word to change its meaning. Point out that the prefix *poly-* means "several or many."

Ask What other prefix also means "several or many"? *(multi-, as in* multicellular*)*

Active Reading

KEY QUESTION Have a volunteer read the question and answer on the previous page.

Say In a field of wildflowers, one plant is twice as tall as all the other plants of the same species.

Ask Is this an example of evolution? *(no)* Why? *(Only the tall plant has a change in genes. Evolution is a change over time in genes of a population, not of an individual.)*

DRAW CONCLUSIONS Emphasize that allele frequency is the number of times a particular allele occurs in a gene pool compared with the number of times other alleles occur. The information is stated as a percentage for each allele. The total percentage of all alleles must equal 100.

Say Suppose a gene pool has three forms of an allele: A, B, and C.

Ask If the frequency of allele A decreases in the gene pool, what would happen to the frequency of the other two alleles? *(One or both would increase in frequency because the total of the three allele frequencies must equal 100 percent.)*

KEY QUESTION After students read the question and answer on this page, have them find the blue run-in heads under Sources of Genetic Variation. Point out that these heads are a list of sources. Have students explain each source in their own words.

ELL Focus on ELL: Build Background

BEGINNING SPEAKERS Write the term *transfer* on the board and say it aloud. Then transfer a book from one hand to another.

Say I transfer the book from hand to hand.

Demonstrate transferring other objects from one place to another. Then draw two bacteria on the board. Place an arrow from each bacterium to the other. Label the arrow *Genes.*

Say Bacteria transfer genes to each other.

Label the diagram on the board *Lateral gene transfer,* and have students find the term in the text on this page. Read aloud the text, using the diagram to facilitate comprehension.

Speed Bump

Ask Does sexual reproduction affect genotypes? *(yes)*

Ask Does it affect the resulting phenotypes? *(yes)*

Ask What effect does sexual reproduction have on the overall gene pool? *(Sexual reproduction leads to new genotypes and phenotypes but does not affect allele frequency in the gene pool.)*

Ask Does an organism show only single-gene traits, only polygenic traits, or a combination of these types of traits? *(In general, an individual will show a combination of single-gene and polygenic traits.)*

During Reading continued

Active Reading

READ, COVER, REMEMBER, RETELL Ask students to reread the Single-Gene Traits section. Have them then cover the section and write what a single-gene trait is and how many genotypes and phenotypes are possible for the trait. Have students share their notes with a partner. Then have them repeat this process for the Polygenic Traits section.

KEY QUESTION Have students silently read the question and the answer.

Ask What is a phenotype? *(an organism's physical appearance)*

Ask A trait has many phenotypes. Is it most likely a single-gene trait or a polygenic trait? *(polygenic)*

Speed Bump

Ask The thickness of corn stalks can vary. Is the thickness of a corn stalk a single-gene trait or a polygenic trait? *(It is a polygenic trait.)*

Ask How do you know? *(because corn stalks can be of many different thicknesses)*

Post Reading

ASSESSMENT TIPS Before answering Questions 1–3, have students review the highlighted vocabulary terms and their definitions.

For Question 6, review with students that *range* refers to the set of all possible values. If many possible values exist, the range is large.

MYSTERY CLUE Discuss the connection between mutations and variation.

Ask Would a high mutation rate result in more or fewer changes in the genes of individuals in a population? *(more)*

Ask How would that affect the gene pool of the population? *(The allele frequencies would change.)*

Wrap-Up Activity

MAKE A GRAPH Have students measure in centimeters the height of each student in the class. Attach a growth chart to a wall. Have each student stand against it while a partner holds a ruler on his or her head and calls out a measurement. Have students record the data in a table and then use the data to create a line graph similar to the Polygenic Trait graph. The *x*-axis of the graph should show height; the *y*-axis should show the number of individuals. Discuss how their graph compares to the Polygenic Trait graph.

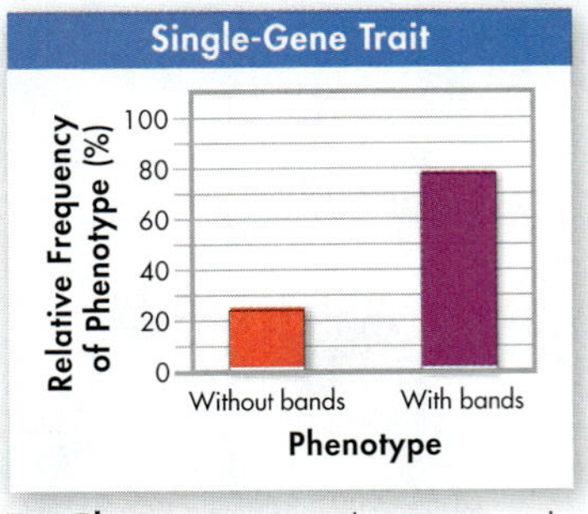

Two Phenotypes A single gene controls whether or not a snail's shell has bands.

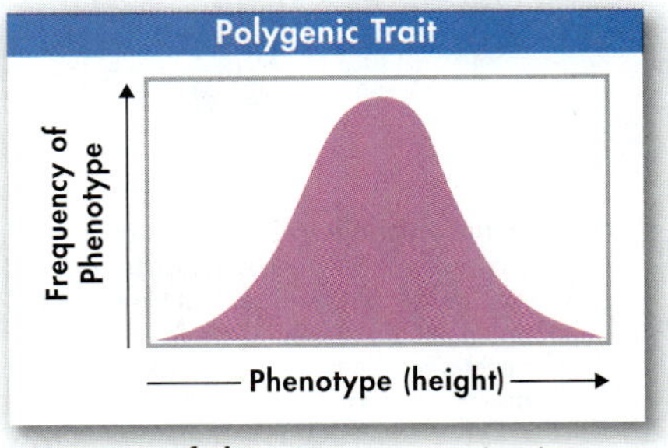

A Range of Phenotypes Human height is controlled by several genes. Most individuals are of average height, so the graph's curve is highest in the middle.

Single-Gene and Polygenic Traits

Genes control phenotype in different ways. In some cases, a single gene controls a trait. In other cases, several genes work together to control a trait.

Single Gene Traits A trait controlled by just one gene is called a **single-gene trait.** Single-gene traits may have just two or three phenotypes. For example, in one type of snail, a single gene controls stripes on the shell. The gene has two alleles, one for striped shells and one for plain shells. There are only two possible phenotypes: with or without stripes.

Polygenic Traits A trait that is controlled by two or more genes is a **polygenic trait.** Polygenic traits can have many possible genotypes and phenotypes. In other words, there is a wide range of phenotypes for polygenic traits.

One example of a polygenic trait in humans is height. Individuals can be very tall, very short, or any height in between. You can measure the height of the members of your own class. Most students will be of average height. A graph of your data might be similar to the one shown at left. The shape of this graph is like the curve of a bell. A bell curve is typical for polygenic traits.

Key Question What determines the number of phenotypes for a given trait?
The number of phenotypes for a trait depends on how many genes control the trait.

CHECK Understanding

Apply Vocabulary

Use the highlighted words from the lesson to complete each sentence correctly.

1. A ______ is the complete set of genes in a population.
2. The number of times that an allele occurs in a gene pool, compared to the total number of alleles in that pool for the same gene is called ______.
3. A trait that is controlled by two or more genes is called a ______ trait.

Critical Thinking

4. **Explain** What indicates that evolution is taking place in a population? Use the word *alleles* in your answer.
5. **Review** What are three sources of genetic variation?
6. **Compare and Contrast** How does the range of phenotypes for single-gene traits differ from the range for polygenic traits?
7. **Write to Learn** Answer the mystery clue below.

MYSTERY CLUE

The genes of flu viruses have very high mutation rates. How might this affect the amount of variation in the virus gene pool? (**Hint:** See p. 407.)

BIOLOGY.com Search Lesson 17.1 GO • Lesson Assessment

Check Understanding

1. gene pool
2. allele frequency
3. polygenic
4. Evolution takes place where there is a change in the frequency of alleles in the gene pool of a population.
5. mutations, sexual reproduction, and lateral gene transfer
6. Single-gene traits have only two or three phenotypes; polygenic traits have a range of phenotypes, often in a bell curve.
7. A high rate of mutation would increase the amount of variation.

17.2 Evolution as Genetic Change in Populations

How Natural Selection Works

The word *fitness* can describe how healthy you are. But in evolution, *fitness* has a different meaning. *Fitness* describes individuals that have traits that help them survive and reproduce. Their traits are favored by natural selection. Individuals with high fitness have more offspring and pass on more of their genes. The alleles that produce these traits become more common in the population. This is the process of evolution.

Natural Selection on Single-Gene Traits In single-gene traits, natural selection can lead to changes in allele frequencies. When natural selection favors one trait over another, the allele for the favored trait becomes more common over time.

Imagine a population of brown lizards. A single gene that controls color mutates to form two new alleles: red and black. What happens to lizards with these new alleles? Red lizards are easier to see. They often get eaten before they can reproduce. For that reason, the red allele will probably not become common. Black lizards can warm up in the sun more quickly than brown lizards. The warmer black lizards can run from predators more quickly than brown lizards. Therefore, the frequency of the black allele is likely to increase. Over time, black lizards will become more common.

Key Question How does natural selection affect single-gene traits? **Natural selection on single-gene traits can lead to changes in allele frequencies. This leads to changes in phenotype frequencies.**

Key Questions

How does natural selection affect single-gene and polygenic traits?

What is genetic drift?

What five conditions can disturb genetic equilibrium?

BUILD Understanding

Preview Visuals Before you read, look at the chart below. What trend does the chart seem to show? How is it related to evolution?

In Your Workbook Go to your workbook to learn more about previewing visuals. Complete the previewing visuals chart for Lesson 17.2.

Selection on a Single-Gene Trait
Natural selection on a single-gene trait can cause changes in allele frequencies. Populations evolve as the frequency of alleles change, as seen in the lizards here.

Effect of Color Mutations on Lizard Survival

Initial Population	Generation 10	Generation 20	Generation 30
80%	80%	70%	40%
10%	0%	0%	0%
10%	20%	30%	60%

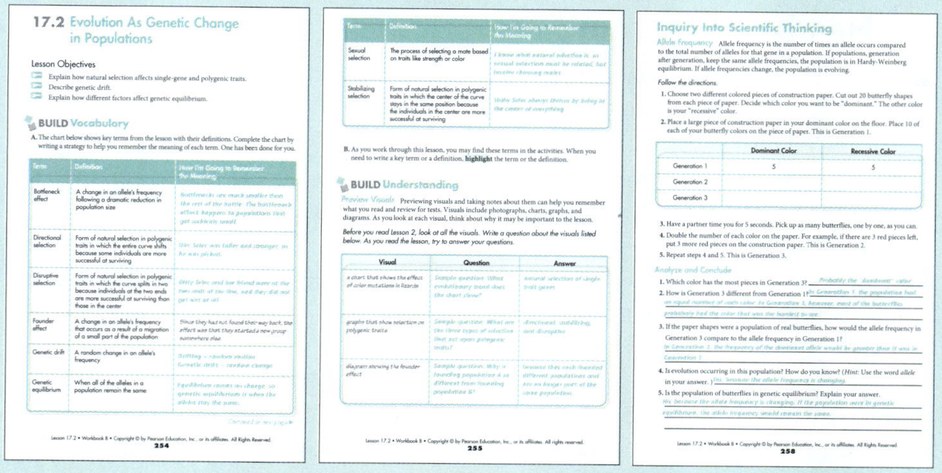

LESSON 17.2

Pre-Reading

Objectives

17.2.1 Explain how natural selection affects single-gene and polygenic traits.

17.2.2 Describe genetic drift.

17.2.3 Explain how different factors affect genetic equilibrium.

Preview the Pages

1. Preview the lesson vocabulary. Say aloud each word and read the definition. Ask volunteers to restate the definitions in their own words.
2. Address misconceptions about the meaning of *selection* in evolution. Students might consider *selection* to be a conscious choice on the part of an individual rather than a mechanism acting on a population.
3. Preview all visuals, and read the captions and labels. As you examine the three graphs that show natural selection on polygenic traits, point out the difference between the dotted line and the solid line. Make sure students understand that the dotted line should be compared to the solid line on the same graph to show the effect of natural selection on the population.

Build Understanding

PREVIEW VISUALS Direct students to Workbook B, where they will find a partially filled-in chart that will help them preview the lesson visuals. Remind students that when they preview a visual, they should describe it, identify important information from it, and ask questions about it.

NATIONAL SCIENCE EDUCATION STANDARDS

UNIFYING CONCEPTS AND PROCESSES

I, II, III, IV

CONTENT

C.2.c, C.3.a, C.3.b, C.3.c

INQUIRY

A.1.b, A.1.c, A.2.a, A.2.d

For the corresponding lesson in *Miller & Levine Biology,* see pages 487–492.

During Reading

Active Reading

CAUSE/EFFECT After students read Natural Selection on Single-Gene Traits on the previous page, help them identify cause-and-effect relationships.

Ask What causes more red lizards to be eaten than brown or black lizards to be eaten? *(They are easier to see.)*

Ask How does this affect the frequency for red coloring in the population? *(This allele becomes less common.)*

Have students identify other cause-and-effect relationships in the information.

KEY QUESTION Read aloud the Key Question and answer from the previous page. Then have students look again at the chart. Ask a volunteer to summarize how the allele frequency for black lizards changed from the initial population to Generation 30. *(The frequency changed from 10 percent to 60 percent.)*

USE VISUALS Point out that in each graph on this page, the original distribution of phenotypes has a bell-shaped curve. As students read each section of text, discuss how the distribution on the corresponding graph changes.

Ask If large seeds became scarce and more small seeds were available, how would the curve for directional selection be affected? *(The curve would probably shift to the left.)*

Ask If medical advances led to large babies having a better chance of surviving, how might the curve for stabilizing selection change? *(The curve would shift to the right.)*

Ask If more medium-size seeds became available and small and large seeds become scarce, how would the curve for disruptive selection change? *(The curve would shift back to a bell shape.)*

KEY QUESTION Invite a volunteer to read aloud the question and the answer. Then write the following cause-and-effect graphics on the board. Have students identify the cause for each.

[ ] → directional selection
Cause

[ ] → stabilizing selection
Cause

[ ] → disruptive selection
Cause

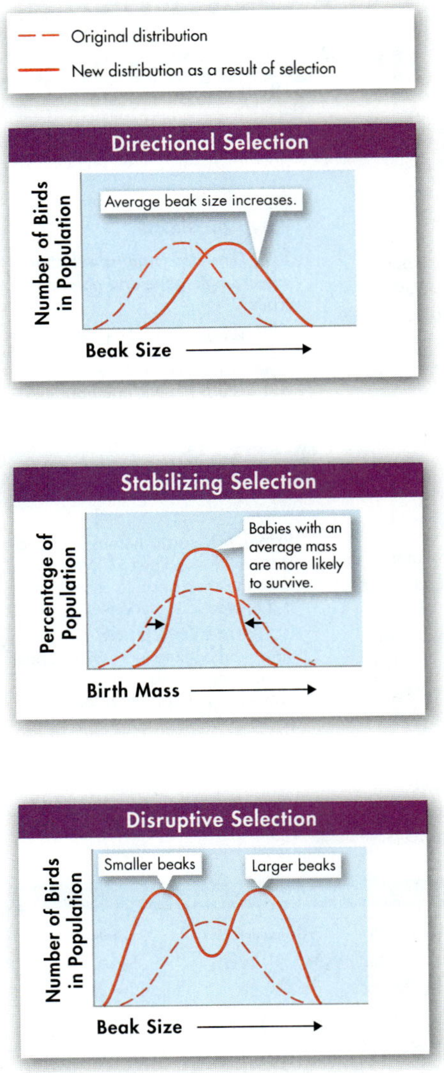

Selection on Polygenic Traits Natural selection on polygenic traits has one of three patterns. These patterns are directional selection, stabilizing selection, or disruptive selection.

Natural Selection on Polygenic Traits

Traits that are controlled by more than one gene produce a range of phenotypes. This range can often be shown by a bell curve. Natural selection can act on polygenic traits in one of three ways: directional selection, stabilizing selection, or disruptive selection.

▶ ***Directional Selection*** Sometimes natural selection favors organisms at one end of the bell curve. When individuals at one end of the curve have higher fitness than the others, **directional selection** takes place.

Imagine a population of birds that live on an island. The island has many large seeds with thick shells, but few small seeds. Birds with bigger beaks would have more to eat than birds with smaller beaks. Birds with bigger beaks would live longer and have more offspring. Over time, the average size of beaks would get bigger. Natural selection would move in one direction—toward larger beaks.

▶ ***Stabilizing Selection*** Sometimes natural selection favors the average individuals. When individuals near the center of the bell curve have the highest fitness, **stabilizing selection** takes place. Stabilizing selection shifts the ends of the curve closer to the middle.

Human babies are affected by stabilizing selection. Very small babies are usually less healthy and less likely to survive. Very large babies may have trouble being born. In this example, the fitness of the extremes is lower than that of the average.

▶ ***Disruptive Selection*** Sometimes the most extreme traits are most likely to survive and reproduce. The average types have a harder time surviving and reproducing. This situation is called **disruptive selection.** Disruptive selection can eventually create two distinct phenotypes. These phenotypes are shown by a curve with two peaks.

Suppose a population of birds lives on an island where medium-size seeds are rare. Most seeds are either large or small. Birds with very small or very large beaks would be more likely to survive. The population might split into two groups: one with smaller beaks and one with larger beaks.

Key Question How does natural selection affect polygenic traits? **Natural selection on polygenic traits can affect the relative fitness of phenotypes. Natural selection produces one of three types of selection: directional selection, stabilizing selection, or disruptive selection.**

Hands-On Learning

MAKE A VISUAL MODEL Have students work in small groups to use common art materials to prepare a visual model of one type of selection described below. The model should include labels that explain the type of selection and what is happening.

- An anteater's source of food is ants, some of which are deep in the soil or deep in an ant hill. Anteaters with longer tongues have more access to the food source. *(directional selection)*
- Large spiders are easily seen by predators. Small spiders have difficulty eating large insects they catch. Spiders of average size are not easily seen and can easily eat insects they catch. *(stabilizing selection)*
- A type of crab ranges in color from almost white to dark brown. Light-colored crabs are not easily seen on sand. Dark brown crabs are not easily seen on dark rocks. Tan crabs can be seen on either dark or light surfaces. *(disruptive selection)*

Genetic Drift

Natural selection is not the only cause of evolutionary change. In small populations, chance events can cause evolution. A random change in the frequency of the alleles in a gene pool is called **genetic drift.** Genetic drift may be caused by the bottleneck effect or the founder effect.

Genetic Bottlenecks Sometimes, a disaster can kill off most of a population. For example, a flood or disease may leave only a few individuals alive. The small group of survivors might have a set of alleles that is very different from the lost population. When this small population grows, its gene pool will be different from the original population. A change in allele frequency following a dramatic loss of population is called the **bottleneck effect.** The bottleneck effect can greatly reduce the genetic diversity in a population.

The Founder Effect Genetic drift can also happen when a few individuals move into a new habitat. The small group may have a set of alleles that is different from the main population. The **founder effect** is a change in allele frequency that results from a small group starting a new population.

The hundreds of species of fruit flies on the different Hawaiian islands demonstrate the founder effect. All of these species came from a single species of fruit fly in South America. Each Hawaiian species has allele frequencies that differ from the original species.

Key Question What is genetic drift? **Genetic drift is the change in allele frequency that happens in small populations due to random events.**

BUILD Vocabulary

directional selection a form of natural selection in which individuals at one end of a distribution curve have higher fitness than individuals in the middle or at the other end of the curve

stabilizing selection a form of natural selection in which individuals near the center of a distribution curve have higher fitness than individuals at either end of the curve

disruptive selection a form of natural selection in which individuals at the upper and lower ends of a distribution curve have higher fitness than individuals near the middle of the curve

genetic drift a random change in allele frequency caused by a series of chance occurrences that cause an allele to become more or less common

bottleneck effect a change in allele frequency following a dramatic reduction in the size of a population

founder effect a change in allele frequency as a result of the migration of a small group of a population

ACADEMIC WORDS

The adjective *random* means "lacking a pattern" or "happening by chance." A random change happens by chance.

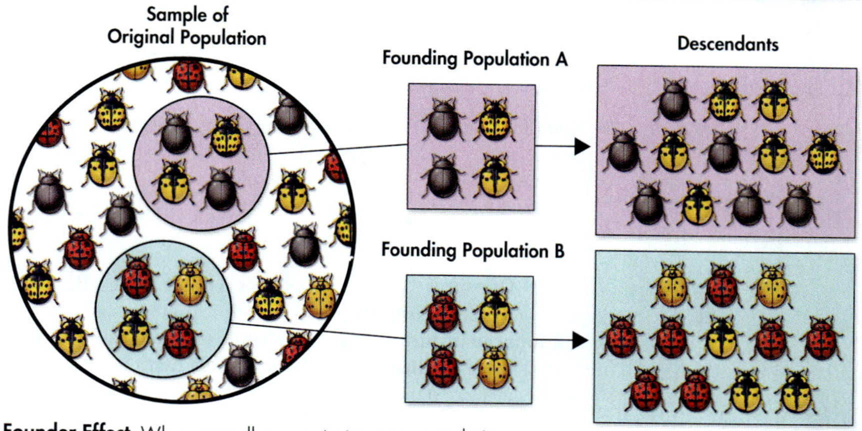

Founder Effect When a small group starts a new population, the descendants can differ from the original population. In the art, a diverse population gives rise to two very different populations.

Focus on Academic Language

• founder • dramatic • descendants

Pair students and have them locate the words in their reading of the student page. Have them write in their notebooks what they think each word means and their reasoning. Tell them also to tell how each word is related to genetic drift.

Active Reading

USE VISUALS Guide students to the Founder Effect diagram.

Ask What does the large circle on the left represent? *(the original population)*

Ask Individuals of how many colors are represented in that population? *(three—red, yellow, and gray)*

Ask What do the small center boxes represent? *(groups of individuals that moved to new habitats)*

Ask How many different habitats did they populate? *(two)*

Discuss how those two populations differ from the original population. *(Each founding population is composed of only two colors.)*

Ask What do the boxes on the right represent? *(the descendants of the individuals that moved to the two new habitats)*

Discuss how the descendant populations compare with the founding populations. *(They are composed of individuals with the same two colors.)*

KEY QUESTION Read aloud the question and the answer. Model genetic drift by showing students a bowl with ten beans each of five different types, such as pinto, kidney, navy, white, and lima beans. Tell students that the beans represent a small population. Point out that each type of bean (allele) in the population represents 20 percent of the total. Have a volunteer randomly select only some of the beans to represent alleles in the next generation. Discuss how the allele frequency changes in the resulting generation as a result of a random event. Also discuss whether genetic diversity is affected.

Build Vocabulary

ACADEMIC WORDS Invite students to suggest examples of random changes.

Ask What other words can you think of that have a meaning similar to *random*? *(Sample answers: chance, unplanned)*

Speed Bump

Ask What two processes can result in changes in allele frequencies, and thus evolution? *(natural selection and genetic drift)*

During Reading continued

Active Reading

USE GRAPHIC ORGANIZERS Have students create a spider map to help them understand and remember the conditions of the Hardy-Weinberg principle. Have them write *Hardy-Weinberg principle* in the center and label the lines from the center with the conditions. Then have them place details or descriptions of the conditions on lines off of the condition lines. See page 56 in the ELL Handbook for information on spider maps.

KEY QUESTION Divide students into groups of six. Have one student in each group read the question. Then have the other students take turns identifying conditions that can disturb genetic equilibrium and cause evolution to occur. Have each student name a condition and then explain it. For example, a student might say, "One condition that can disturb genetic equilibrium is non-random mating, which means that individuals choose mates with certain traits. This causes the frequency of those traits to increase."

Build Vocabulary

WORD ORIGINS Read aloud the information about the word *equilibrium*. Point out the words *balanced* and *stable* in the definition. If necessary, discuss their meanings.

Ask What is balanced and stable when a population is in genetic equilibrium? *(the allele frequencies)*

Ask What does it mean that the allele frequencies are stable? *(The relative amounts of alleles in the population do not change.)*

Speed Bump

Ask Which population is more likely to be in genetic equilibrium, a very large population of birds living on the continent of South America or a small population of the same species living on an island off the coast of South America? *(the large population)*

Ask Why must no mutations occur if a population is to remain in genetic equilibrium? *(Mutation can produce new alleles.)*

BUILD Vocabulary

genetic equilibrium
a situation in which the allele frequencies in a population remain the same

Hardy-Weinberg principle
the principle that states that allele frequencies in a population remain constant unless one or more factors cause those frequencies to change

sexual selection
the process by which individuals select mates on the basis of heritable traits

WORD ORIGINS
The word *equilibrium* comes from two Latin words: *aequus*, which means "level," and *libra*, which means"a balance." Equilibrium is a state in which all the forces are balanced by each other, so the condition is stable.

Choosing a Mate In many species, mating is not random. For example, female peacocks prefer mates with big, colorful tail feathers. This is an example of sexual selection.

Evolution Versus Genetic Equilibrium

To understand how populations evolve, it helps to imagine a population that does *not* evolve. If a population is not evolving, the allele frequencies in the gene pool do not change. This condition is called **genetic equilibrium.**

Sexual Reproduction and Allele Frequency In sexual reproduction, genes are shuffled. This produces many new gene combinations. But meiosis and fertilization do not change the allele frequencies in the total population. Therefore, sexual reproduction alone does not affect genetic equilibrium.

The Hardy-Weinberg Principle The **Hardy-Weinberg principle** states that allele frequencies in a population will stay in genetic equilibrium unless something causes them to change. This principle predicts that five conditions can upset genetic equilibrium and cause a population to evolve.

▶ ***Nonrandom Mating*** In genetic equilibrium, individuals must mate with other individuals at random. But in many species, individuals choose mates that have particular traits, such as size or color. When individuals pick mates with certain heritable traits, the process is called **sexual selection.** If mates are selected for a particular trait, the frequency of that trait will increase.

▶ ***Small Population*** Genetic drift usually does not affect a large population. But it can affect small populations greatly. Therefore, evolution takes place more easily in a small population.

▶ ***Movement Into or Out of the Population*** Individuals that join a population may add new alleles to the gene pool. Individuals that leave can lower the frequency of certain alleles. Either direction of movement will affect the genetic equilibrium.

▶ ***Mutations*** Mutations can create new alleles. These new alleles can change the allele frequencies of the population.

▶ ***Natural Selection*** If different genotypes have different fitness, individuals with those genotypes will be more likely to survive. Genetic equilibrium will be upset. The population will evolve.

In real populations, one or more of these conditions is usually true. Therefore, most of the time, most species are evolving.

Key Question What five conditions can disturb genetic equilibrium? **The Hardy-Weinberg principle predicts that five conditions can disturb genetic equilibrium and cause evolution to occur. These conditions are (1) nonrandom mating, (2) small population size, (3) movement into or out of a population, (4) mutations, and (5) natural selection.**

Practice Problems

1. 94%
2. *SS*= 88.36%; *Ss*= 11.28%; *ss*= 0.36%
3. 36

INQUIRY into Scientific Thinking

Allele Frequency

The Hardy-Weinberg principle can be used to predict the frequencies of genotypes in a population that is not evolving. This equation can be used to predict the percentages of the possible genotypes in a population:

$$p^2 + 2pq + q^2 = 1, \text{ where}$$

p and q stand for two alleles of the same gene;
p^2 and q^2 stand for the homozygous genotypes;
pq stands for the heterozygous genotype.

Sample Problem

A population has two alleles for a gene: ***A*** (dominant) and ***a*** (recessive). In one generation, the frequency of allele ***A*** is 60% and the frequency of allele ***a*** is 40%. What will be the frequencies of each genotype in the next generation?

1. From genetics, you know that a cross of these two alleles can produce three genotypes: ***AA, Aa,*** and ***aa***. In the equation, allele ***A*** will be ***p***, and allele ***a*** will be ***q***.

$$p^2 + 2pq + q^2 = 1$$
$$AA + 2Aa + aa = 1, \text{ or } 100\%$$

2. Substitute the known frequencies for alleles *A* and *a*.

A = 60%, or 0.60 $\quad$ a = 40%, or 0.40
A^2 = 36%, or 0.36 $\quad$ a^2 = 16%, or 0.16

$$2Aa = 2\ (0.60 \times 0.40) = 2\ (0.24) = 0.48, \text{ or } 48\%$$

The next generation will have these frequencies: 36% ***AA;*** 48% ***Aa;*** 16% ***aa***. If the population does not show these frequencies, evolution is taking place.

Practice Problems

Some members of a human population have a genetic condition. The condition is controlled by two alleles, *S* (dominant) and *s* (recessive). The condition affects only individuals that are homozygous for the recessive allele (*ss*). Heterozygous individuals (*Ss*) do not have the condition. The *s* allele is found in 6% of the population.

1. Calculate What percentage of the population carries the *S* allele? (**Hint:** $p + q$ = 100%)

2. Calculate What are the frequencies of the *SS*, *Ss*, and *ss* genotypes?

3. Calculate In a population of 10,000 people, how many people would have the condition?

CHECK Understanding

Apply Vocabulary

Use the highlighted words from the lesson to complete each sentence correctly.

1. The change in allele frequency after a dramatic drop in the size of a population is called the ________.

2. The change in allele frequency caused by the migration of a small group is called the ________.

3. The ________ identifies conditions necessary for gene frequencies in a population to remain constant.

Critical Thinking

4. Relate Cause and Effect How does natural selection affect a single-gene trait?

5. Compare and Contrast Compare directional selection and disruptive selection.

6. Relate Cause and Effect What is genetic equilibrium? What five conditions are necessary to maintain genetic equilibrium?

7. Write to Learn Answer the mystery clue below.

MYSTERY CLUE

Normally, your immune system can kill invading viruses. What would happen if a flu virus had a mutation that your immune system could not fight? (Hint: See p. 409.)

BIOLOGY.com Search Lesson 17.2 GO • Lesson Assessment

413

Check Understanding

1. bottleneck effect

2. founder effect

3. Hardy-Weinberg principle

4. Natural selection on single-gene traits can favor one phenotype over the other, which leads to changes in allele frequencies.

5. Directional selection favors one of the extreme phenotypes for a trait. Disruptive selection favors phenotypes at both ends of the curve.

6. Genetic equilibrium is a condition in which allele frequencies in a population remain constant. Five conditions necessary for genetic equilibrium are random mating, a large population, no movement in or out of the population, no mutations, and no natural selection.

7. Viruses with the mutation would be able to reproduce more rapidly than those without it. The virus population would evolve, because natural selection would occur. The frequency of the new allele would increase in the population.

Inquiry into Scientific Thinking

MAKE CONNECTIONS Review what students learned in Lesson 11.2 about using probability to predict traits.

Ask What tool can we use to predict the probability that the offspring of two parents will have a particular genotype? *(a Punnett square)*

Say A Punnett square allows us to predict the frequencies of genotypes among a group offspring. In this activity, we'll learn about an equation that allows scientists to predict the frequencies of genotypes among a whole population.

Discuss the sample problem as a class. Make sure students understand how to use the equation before allowing them to move on to the practice problems.

Lead a Discussion

Discuss how scientists might use the Hardy-Weinberg equation.

Say The equation predicts the genotype frequencies we'd expect to see in a population that is not evolving. However, as we've learned, most species are evolving.

Ask Why do you think it is useful to know the genotype frequencies that would occur if a population was not evolving? *(Sample answer: This information provides a baseline that can be used to measure genetic change that is actually occurring.)*

Post Reading

ASSESSMENT TIPS Review with students the highlighted vocabulary words for this lesson.

For Question 2, point out that migration refers to the movement of the group from one habitat to another.

For Question 5, suggest that students use a Venn diagram or T-chart to help with the comparison.

MYSTERY CLUE Discuss with students how the inability of the immune system to fight the virus would affect the size of the virus population.

Ask Which would reproduce more rapidly, viruses with the new allele or those without it? *(viruses with the new allele)*

Wrap-Up Activity

MODEL DISRUPTIVE SELECTION Have students work in small groups to make up a scenario that is an example of disruptive selection. Then have them make a disruptive selection graph using pictures to show the two extreme traits and the average trait for the scenario.

LESSON 17.3

Pre-Reading

Objectives

17.3.1 Identify the types of isolation that can lead to the formation of new species.

17.3.2 Describe the current hypothesis about Galápagos finch speciation.

Preview the Pages

1. Preview the vocabulary terms by having students read each term and its definition. Discuss each definition until students are comfortable with the meaning.
2. As you preview the visuals that refer to Darwin's finches, emphasize that each visual does not stand alone. Starting with the first image, the visuals show the sequence of the events that resulted in different finch species.

Build Understanding

COMPARE/CONTRAST TABLE Direct students to Workbook B, where they will find a partially filled-in compare/contrast table. Remind students that comparing helps them recognize how things are alike, and contrasting shows how things are different.

NATIONAL SCIENCE EDUCATION STANDARDS

UNIFYING CONCEPTS AND PROCESSES

II, III, IV, V

CONTENT

C.2.c, C.3.a, C.3.b, C.3.d, C.6.c, G.1

INQUIRY

A.2.a, A.2.b, A.2.e, A.2.f

For the corresponding lesson in *Miller & Levine Biology,* see pages 494–497.

17.3 The Process of Speciation

Key Questions

What types of isolation lead to the formation of new species?

What is the current hypothesis about Galápagos finch speciation?

BUILD Understanding

Compare/Contrast Table In a compare/contrast table, describe the three mechanisms of reproductive isolation.

In Your Workbook Go to your workbook to learn more about using compare/contrast tables. Complete the compare/contrast table for Lesson 17.3.

BUILD Vocabulary

species
a population whose members can breed and produce fertile offspring

speciation
the formation of a new species

reproductive isolation
the separation of a species or population so that members can no longer interbreed

geographic isolation
a form of reproductive isolation in which two populations are separated by geographic barriers such as rivers, mountains, or bodies of water

Isolating Mechanisms

Biologists define a **species** as a population whose members can breed and produce fertile offspring. How does one species become two? If members of a population can no longer mate, a new species may evolve. The formation of a new species is called **speciation.**

Breeding connects the gene pool of a species. If a species is split into two parts, genetic changes in one part of the gene pool cannot spread to the other. When two populations can no longer mate and produce offspring, **reproductive isolation** has occurred. Geography, behavior, and time can separate populations from each other.

Geographic Isolation When populations are separated by a barrier, such as a river, mountain, or ocean, **geographic isolation** can occur. For example, the Abert's and Kaibab squirrels once belonged to the same population. About 10,000 years ago, a small population became isolated north of the Grand Canyon. Since then, genetic drift and natural selection have led to two different-looking populations.

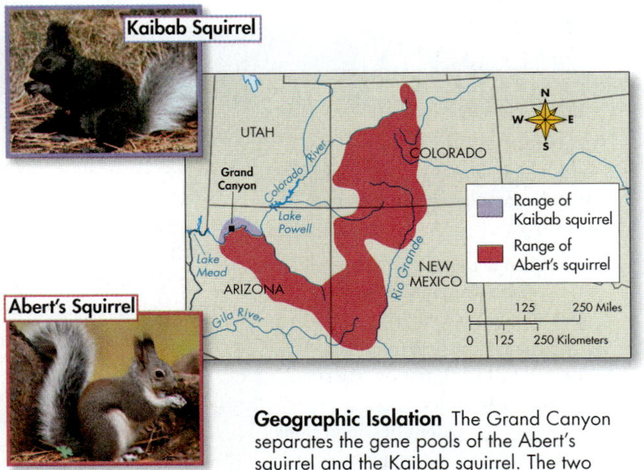

Geographic Isolation The Grand Canyon separates the gene pools of the Abert's squirrel and the Kaibab squirrel. The two populations are still the same species, but they have evolved striking differences.

17.3 The Process of Speciation

Lesson Objectives

Identify the types of isolation that lead to the formation of new species.

Describe the current hypothesis about Galápagos finch speciation.

BUILD Vocabulary

A. The chart below shows key terms from the lesson with their definitions. Complete the chart by writing a strategy to help you remember the meaning of each term. One has been done for you.

Term	Definition	How I'm Going to Remember the Meaning
Behavioral isolation	When two populations have different behaviors that keep them from reproducing	*It's just like it sounds—different behaviors keep groups isolated, or apart.*
Geographic isolation	When two populations are separated by geographic features, such as rivers or mountains, that keep them from reproducing	*The geography/mountain is so big that we are isolated/kept apart.*
Reproductive isolation	Separation of populations so that they cannot reproduce	*Reproduce = have kids Isolation = alone Have kids alone? Can't.*
Species	A population or group of populations whose members can interbreed and produce fertile offspring	*Since there are all kinds of crossbred dogs, all dogs must belong to the same species.*
Speciation	The forming of a new species	*Species + evolution = speciation*
Temporal isolation	When two populations reproduce at different times, so that they cannot reproduce together	*Temporal means time, so temporal isolation = kept apart by a timing issue.*

B. As you work through this lesson, you may find these terms in the activities. When you need to write a key term or a definition, **highlight** the term or the definition.

BUILD Understanding

Compare/Contrast Table Use a compare/contrast table when you want to see the similarities and differences between two or more processes.

As you read about the different types of reproductive isolation, fill in the table below.

What Leads to Reproductive Isolation	
Type of Isolation	Description
Behavioral	*when populations cannot mate because they have different courtship rituals*
Geographic	*when populations cannot mate because they are separated by geographical features*
Temporal	*when populations cannot mate because they mate at different times*

Populations Become Isolated

Speciation occurs when a new species evolves. Many things contribute to speciation.

Read the chart below. Then, use phrases in the word box to identify each process that leads to speciation. The first row has been completed for you.

changes in the gene pool **competition** **geographic isolation** **behavioral isolation** **founders arrive**

Process	How It Leads to Speciation
founders arrive	A population arrives in a new place.
geographic isolation	Populations are separated by a geographic barrier and do not share a gene pool.
changes in the gene pool	Populations evolve new traits in response to natural selection in their environments.
competition	Species evolve in a way that reduces competition between them.
behavioral isolation	Groups within a population are separated by different courtship rituals.

Answer the questions.

1. Finches prefer to find mates with the same size beaks. Finches in Group A have larger beaks than those in Group B, so finches from the two groups will not mate. What is this an example of? *behavioral isolation*
2. Do populations that are geographically isolated from one another share a gene pool? Why or why not? *No. They are separated by a geographic barrier so they cannot mate with one another.*

Speciation in Darwin's Finches

Charles Darwin proposed that the different species of finches on the Galapagos Islands once had a common ancestor. Over time, he said, natural selection led to each species of finch. A current hypothesis says that speciation in the Galapagos finches occurred through the founding of a new population, geographic isolation, changes in the new population's gene pool, behavioral isolation, and ecological competition.

Use the phrases in the box to complete the chart.

insects that live inside dead wood **small seeds** **large, thick-shelled seeds**

Galápagos Islands Finches				
Shape of head and beak				
Main food	fruits	*insects that live inside dead wood*	*small seeds*	*large, thick-shelled seeds*
Feeding adaptation	parrotlike beak	uses cactus spines	pointed crushing beak	large crushing beak
Habitat	trees	trees	ground	ground

Answer the questions.

1. How does the large crushing beak help the fourth finch survive? *It allows the finch to crush and eat large seeds.*
2. Circle the finch that would be least likely to survive if the insect population decreased.
3. Both the first and second finch live in trees. Suppose one season there is less fruit in the trees. Which finch is more likely to survive and reproduce?
   A. the first finch
   (B.) the second finch

Behavioral Isolation If two interbreeding populations develop different behaviors, such as different mating dances, then **behavioral isolation** can happen. Two nearly identical species of birds—eastern meadowlarks and western meadowlarks—share the same habitat. These two species do not interbreed because their mating songs are different.

Temporal Isolation A third isolating mechanism is temporal isolation. **Temporal isolation** happens when populations that live in the same habitat reproduce at different times.

Key Question What types of isolation lead to new species? **New species can arise from behavioral isolation, geographic isolation, and temporal isolation.**

Speciation in Darwin's Finches

Since Darwin, many scientists have studied the Galápagos finches. The Grants showed that under the right conditions, natural selection can happen quickly. This knowledge makes the Galápagos finches a good choice for describing how speciation might work in real life.

Speciation probably began when a small number of finches founded a new population. This population was geographically isolated from the main population. Over time, the gene pool changed. According to the currently accepted hypothesis, isolation, competition, and natural selection formed new species that could no longer interbreed.

Founders Arrive Many years ago, a few finches from South America arrived on one of the islands. Maybe they were blown there by a storm. The small group survived and reproduced. Because of the founder effect, the allele frequencies of this new population were different from the original population.

Geographic Isolation These finches do not usually fly over open water, so the population was geographically isolated. The founder effect, geographic isolation, and natural selection led to the evolution of a new species of finch. Eventually some of these finches moved to a new island, where the process was repeated.

Changes in Gene Pools The environment on the second island may have had different plants with larger seeds. Directional selection would have favored birds with larger beaks. Over time, a population with larger beaks evolved.

BUILD Vocabulary

behavioral isolation a form of reproductive isolation in which two populations develop differences in courtship rituals or other behaviors that prevent them from breeding

temporal isolation a form of reproductive isolation in which two or more species reproduce at different times

WORD ORIGINS

Temporal comes from the Latin *tempus*, meaning "time." In temporal isolation, the time of reproduction separates populations.

ELL Focus on ELL: Access Content

BEGINNING SPEAKERS Guide students to summarize the information in Speciation in Darwin's Finches. List these terms on the board: *Galápagos Islands, new island, reproduce, South America, species, survive.* Provide the frames below and have students use word list to complete the sentences.

1. Finches come to the ______________. *(Galápagos Islands)*
2. The finches survive and ______________. *(reproduce)*
3. Finches on the island do not mix with finches of ______________. *(South America)*
4. Over time, a new ____________ of finch forms. *(species)*
5. Some of the finches fly to a ______________. *(new island)*
6. There, more finches with larger beaks ____________. *(survive)*
7. Over time, another ____________ of finch forms. *(species)*

During Reading

Build Vocabulary

WORD ORIGINS Have students pronounce each vocabulary term in the Build Vocabulary boxes on the previous page and this page. Then have them use dictionaries to find the word origins for two terms of their choice.

Ask How do the meanings of the word origins compare to the definitions on the student page? *(Answers may vary.)*

Active Reading

DRAW CONCLUSIONS Refer students to the last two diagrams on this page.

Ask In the speciation of Darwin's finches, why did geographic isolation have to occur before changes happened in the two gene pools? *(If geographic isolation had not occurred first, there would not have been two different populations, and the gene pools would not have become separated.)*

USE GRAPHIC ORGANIZERS Have students use two different types of graphic organizers to summarize and clarify lesson concepts. They can use a cluster diagram to show how geographic isolation, behavioral isolation, and temporal isolation relate to reproductive isolation. Have them use a flowchart to show the steps in speciation of Darwin's finches.

KEY QUESTION Read aloud the question and answer. Challenge students to give a specific example for each type of isolation.

Speed Bump

Ask What is the relationship among reproductive isolation, geographic isolation, behavioral isolation, and temporal isolation? *(Geographic isolation, behavioral isolation, and temporal isolation are all causes of reproductive isolation.)*

Ask What must be different about the reproductive behavior of species that are isolated by temporal isolation? *(They reproduce at different times of the day or year.)*

During Reading continued

Active Reading

DRAW A DIAGRAM Have small groups choose an animal and then brainstorm a scenario in which behavioral isolation causes the animal to evolve into a separate species.

Ask Why did the population evolve into two species? *(Answers may vary, but should include that over time the gene pool in the two groups changed in different ways, resulting in two different species.)*

Have students draw diagrams similar to those on this page to illustrate their scenarios.

KEY QUESTION Have a student volunteer read aloud the question and answer. Then have other volunteers explain the answer using the diagrams on this page and the previous page.

Speed Bump

Have students suggest captions for the diagrams on this and the previous page.

Post Reading

ASSESSMENT TIPS Review the vocabulary terms and definitions with students before they answer Questions 1–3.

For Question 4, point out the label *Explain* and the word *How.* Explain that these words signal that students should answer the question by telling about a process or sequence of events.

For Question 5, refer students to the diagrams of Darwin's finches.

MYSTERY CLUE Discuss what students know about geographic isolation.

Ask Why might the isolation of viruses in an animal be considered geographic isolation? *(The individuals in the isolated group cannot interact with individuals in other populations.)*

Ask How does geographic isolation contribute to speciation? *(When populations are isolated, genetic changes in one population cannot spread to the other. Over time, the two populations become different enough that they can no longer mate.)*

Wrap-Up Activity

SEQUENCING Students can work in pairs to write on separate index cards a brief summary of each step in the speciation of Darwin's finches. Then have them shuffle the cards and put them back in the correct order.

Behavioral Isolation Imagine that a few birds from the second island cross back to the first island. Will the two populations interbreed? Probably not. Finches choose mates carefully. They like mates that look like themselves. A small-beaked bird will not breed with a large-beaked bird. Even though they share the same habitat, the two populations are now isolated by behavior. They have become two different species.

Competition and More Evolution The two new species live together and compete for seeds. During the dry season, it is difficult to find enough seeds to eat. Individuals with very small beaks have greater success than birds with average-size beaks. Small-beaked birds can specialize in eating very small seeds that birds with average-size beaks cannot eat. Over time the differences lead to reproductive isolation. A third species evolves.

Over many years, this process could have repeated itself again and again on the different islands. The combination of geographic isolation, behavioral isolation, and natural selection could have produced the 13 different finch species found in the Galápagos today.

Key Question What is the current hypothesis about Galápagos finch speciation?
According to this hypothesis, speciation in Galápagos finches happened by the founding of a new population, geographic isolation, changes in the new population's gene pool, behavioral isolation, and ecological competition.

CHECK Understanding

Apply Vocabulary
Use the highlighted words from the lesson to complete each sentence correctly.

1. The formation of a new species is called _____.
2. The form of reproductive isolation in which populations are separated by rivers, mountains, or oceans is called ________.
3. The form of reproductive isolation in which species reproduce at different times is called ________.

Critical Thinking

4. **Explain** How can differences in behavior lead to reproductive isolation?
5. **Sequence** Scientists hypothesize that a series of events lead to the evolution of different species of finches on the Galápagos islands. List the following events in the correct order: behaviors prevent the birds from interbreeding; changes in gene pool; geographic isolation; a small group of birds arrives from South America.
6. **Write to Learn** Answer the question in the mystery clue below. Use the word *gene pool* in your answer.

MYSTERY CLUE

A population of viruses inside an animal is isolated from other virus populations. How might this isolation affect virus evolution? (**Hint:** See p. 414.)

BIOLOGY.com Search Lesson 17.3 GO • Lesson Assessment

Check Understanding

1. speciation
2. geographic isolation
3. temporal isolation
4. Behaviors such as mating rituals might become different in two separate populations. Individuals may not recognize or respond to the behaviors of the other population.
5. birds arrive from South America; geographic isolation; changes in gene pool; behaviors prevent birds from interbreeding
6. Isolation would prevent the exchange of genetic material. The gene pool of the viruses in the animal would be separated from the rest of the gene pool. This small population could evolve in a new direction.

17.4 Molecular Evolution

Molecular Clocks

A genome is the complete set of genes found in an organism. Each genome has many genes. For example, your body has about 25,000 working genes! Scientists often study genomes to learn how organisms have evolved. By comparing the genes of different organisms, scientists can test different hypotheses. Some hypotheses suggest how closely organisms are related. Other hypotheses suggest how long ago evolutionary lines split apart. One way to test these hypotheses is by using DNA as a molecular clock. A **molecular clock** uses mutation rates in DNA to estimate how long ago two organisms shared a common ancestor.

Ticking Mutations Some kinds of mutations happen at a steady rate, like seconds ticking on a clock. As time passes, more and more mutations happen. Some mutations help or hurt the survival of an organism. Because these mutations are affected by natural selection, they cannot be used as "molecular ticks." But many mutations do not affect an organisms's fitness. Such mutations are said to be *neutral*. Neutral mutations change DNA without affecting an organism's fitness.

Neutral mutations can collect in the DNA of different species at about the same rate. Scientists can compare the number of neutral mutations in specific sequences of DNA from different organisms. The more differences there are, the more mutations have happened. By counting mutations, researchers can estimate how long it has been since the two organisms shared a common ancestor.

Key Questions

- *What are molecular clocks?*
- *Where do new genes come from?*
- *How may Hox genes be involved in evolutionary change?*

BUILD Understanding

Flowchart A flowchart is a way to show the steps in a process. As you read about gene duplication, make a flowchart that shows how duplicate genes evolve.

In Your Workbook Go to your workbook to learn more about making a flowchart. Complete the flowchart for Lesson 17.4.

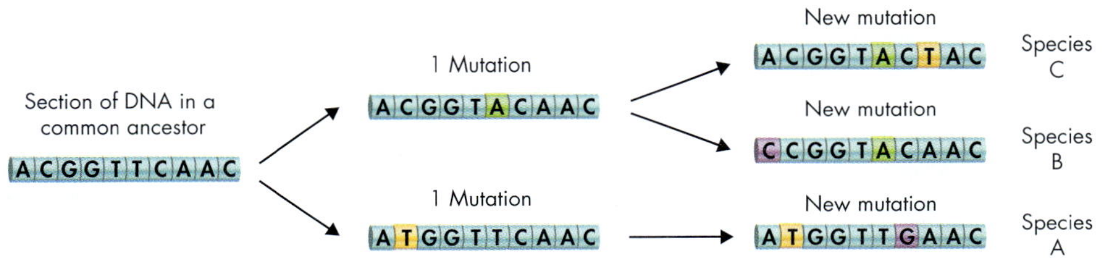

Molecular Clock By comparing DNA sequences, biologists can estimate how closely related species are. Two species that have only a few different mutations shared an ancestor recently. Species that have many different mutations shared an ancestor long ago.

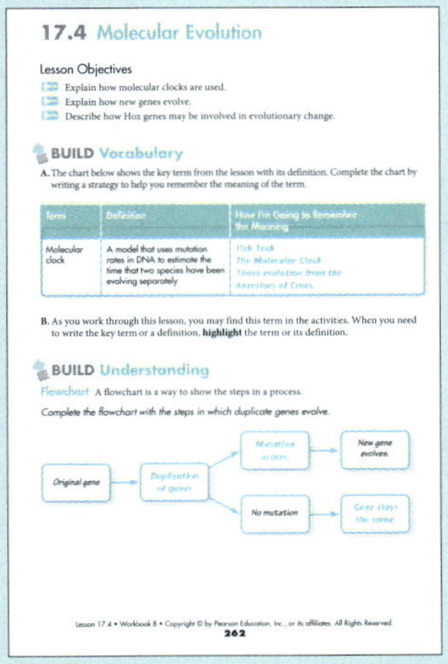
17.4 Molecular Evolution

Lesson Objectives

- Explain how molecular clocks are used.
- Explain how new genes evolve.
- Describe how Hox genes may be involved in evolutionary change.

BUILD Vocabulary

A. The chart below shows the key term from the lesson with its definition. Complete the chart by writing a strategy to help you remember the meaning of the term.

Term	Definition	How I'm Going to Remember the Meaning
Molecular clock	A model that uses mutation rates in DNA to estimate the time that two species have been evolving separately	

B. As you work through this lesson, you may find this term in the activities. When you need to write the key term or a definition, **highlight** the term or its definition.

BUILD Understanding

Flowchart A flowchart is a way to show the steps in a process.

Complete the flowchart with the steps in which duplicate genes evolve.

Original gene → Duplication of genes → Mutation occurs. → New gene evolves.
No mutation → Gene stays the same.

Lesson 17.4 • Workbook B • Copyright © by Pearson Education, Inc., or its affiliates. All Rights Reserved.
262

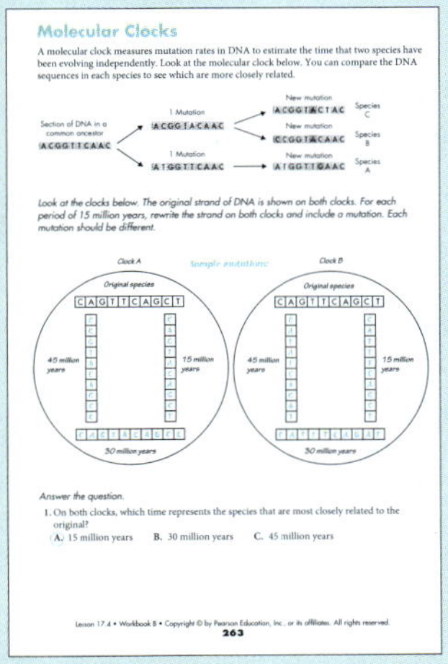
Molecular Clocks

A molecular clock measures mutation rates in DNA to estimate the time that two species have been evolving independently. Look at the molecular clock below. You can compare the DNA sequences in each species to see which are more closely related.

Section of DNA in a common ancestor ACGGTTCAAC; 1 Mutation ACGGTACAAC; 1 Mutation ATGGTTCAAC; New mutation ACGGTACTAC Species C; New mutation CCGGTACAAC Species B; New mutation ATGGTTGAAC Species A

Look at the clocks below. The original strand of DNA is shown on both clocks. For each period of 15 million years, rewrite the strand on both clocks and include a mutation. Each mutation should be different.

Clock A — Original species CAGTTCAGCT — 15 million years — 30 million years — 45 million years
Clock B — Original species CAGTTCAGCT — 15 million years — 30 million years — 45 million years

Answer the question.

1. On both clocks, which time represents the species that are most closely related to the original?
A. 15 million years B. 30 million years C. 45 million years

Lesson 17.4 • Workbook B • Copyright © by Pearson Education, Inc., or its affiliates. All rights reserved.
263

LESSON 17.4

Pre-Reading

Objectives

17.4.1 Explain how molecular clocks are used.

17.4.2 Explain how new genes evolve.

17.4.3 Describe how Hox genes may be involved in evolutionary change.

Preview the Pages

1. Preview the lesson by guiding students to the Key Questions. Read the questions aloud. Point out that the questions give clues to what students will learn in this lesson. Remind them that after reading the lesson, they should be able to answer each question.
2. When previewing the Molecular Clock diagram, have students recall that DNA sequences are represented by using the first letter of each base contained in the sequence. Point out that A stands for adenine, G for guanine, C for cytosine, and T for thymine.
3. Have students skim the lesson and list terms they do not know, such as *Hox genes*. Tell them as they read the lesson, they should use context clues to find the meanings.

Build Understanding

FLOWCHART Direct students to Workbook B, where they will find a partially filled-in flowchart. Remind students that flowcharts are helpful in sequencing the steps of a process, such as how duplicate genes evolve. Remind students to read the chart from left to right.

NATIONAL SCIENCE EDUCATION STANDARDS

UNIFYING CONCEPTS AND PROCESSES

II, III, IV, V

CONTENT

C.2.a, C.2.c, C.3.a, C.3.b, C.3.c, C.3.d, C.3.e

For the corresponding lesson in *Miller & Levine Biology*, see pages 498–501.

During Reading

Active Reading

INTERPRET DIAGRAMS Have students examine the Molecular Clock diagram on the previous page.

Ask What base change occurs in the first mutation at the top? *(The second thymine (T) is replaced by an adenine (A).)*

Have students describe the other four mutations shown in the diagram. Discuss the effect these mutations might have on survival if they are passed on to offspring. *(Depending on the changes, the mutation might have little or no effect. A few mutations make survival easier or more difficult.)*

KEY QUESTION Have students read the first Key Question and the answer.

Ask How are a classroom clock and a molecular clock alike and different? *(Both can be used to tell how much time has passed. The classroom clock displays exact times and is used for small time spans. A molecular clock is used to estimate much longer time spans.)*

KEY QUESTION Read aloud the second Key Question and answer. Refer students to the diagram on this page. Point out that the cylindrical shapes represent genes.

Ask What part of the diagram shows the duplication of an original gene? *(the part that shows the single gene duplicating to become two genes)*

Ask What does the change on the left from a single blue gene to a blue and purple gene represent? *(a mutation)*

Ask How do the two final genes differ? *(The gene on the left has a mutation. It is different from the original gene; the one on the right does not have a mutation. It is like the original gene.)*

Build Vocabulary

RELATED WORD FORMS Have a student read the term and its definition. Point out that the suffix *-ar* is added to *molecule* to form the adjective *molecular.*

Ask What are some other words that add *-ar* to form an adjective? *(Sample answers:* linear, columnar, circular*)*

Have students tell the meaning of each adjective they name and identify the noun from which it is formed.

Speed Bump

Ask Why are neutral mutations more likely to be passed down over time than are negative mutations? *(An organism with a negative mutation is more likely to be eliminated by natural selection.)*

BUILD Vocabulary

molecular clock
a method that uses mutation rates in DNA to estimate the length of time that two species have been evolving independently

RELATED WORDS FORMS
The adjective *molecular* comes from the noun *molecule.* A molecular clock uses DNA molecules to measure the speed of evolution.

Gene Duplication
In this diagram, a gene is first duplicated. Then one of the two genes mutates.

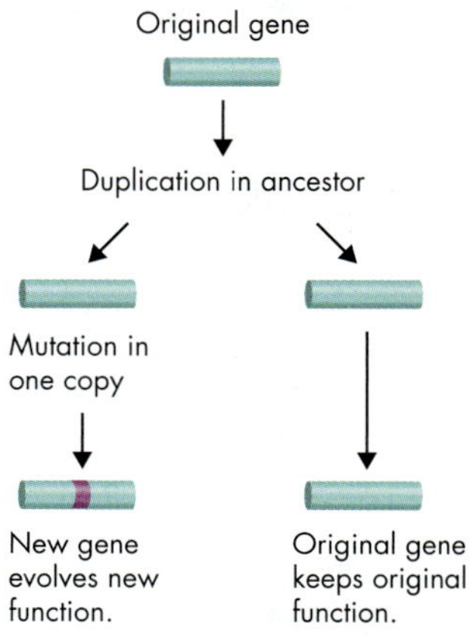

Rate of Mutation Not all genes mutate at the same rate. You can think of a gene that mutates quickly as the second hand on a clock. These genes are useful for measuring evolution that happens over a short time. Other genes mutate more slowly. Such genes are used to measure evolution that takes place over a much longer time.

Key Question What are molecular clocks?
Molecular clocks are pieces of DNA that mutate at a known rate. Scientists use molecular clocks to estimate how much time has passed since two species shared a common ancestor.

Gene Duplication

Humans have about 25,000 working genes. Where did all these genes come from? Modern genes probably evolved from a much smaller number of genes in the earliest forms of life.

Copying Genes One way that the number of genes grew so large was by the duplication, or copying, of existing genes. Duplication can happen during meiosis as chromosome pairs exchange DNA. Sometimes one chromosome ends up with extra DNA during the exchange.

Duplicate Genes Evolve Sometimes a duplicated gene mutates. The mutation changes the way the gene works. The mutated copy of the gene performs a new role. The original gene continues to perform its old role.

Gene Families Several copies of a gene can form a group of related genes that is called a gene family. Genes in a family usually make slightly different proteins that carry out similar roles. For example, one gene family makes several kinds of oxygen-carrying proteins. This gene family evolved from just one original gene.

Key Question Where do new genes come from?
One way that new genes evolve is through the duplication, and then modification, of existing genes.

Developmental Genes and Body Plans

An exciting new way of studying evolution looks at gene activity during development. Small changes in the way genes work in an embryo can lead to big changes in the adult body.

Hox Genes and Evolution Hox genes are a group of genes that play an important role in animal development. Hox genes determine which parts of the embryo develop arms, legs, or wings. Hox genes also control the size and shape of the adult body parts.

ELL Focus on ELL: Extend Language

ALL SPEAKERS Write the word *develop* on the board. Ask students to find two other words within the text on these two pages that contain *develop. (developmental, development)* Write those words on the board. Have students discuss how the meaning of each word changes with different endings. Ask students to write or say each word in a sentence.

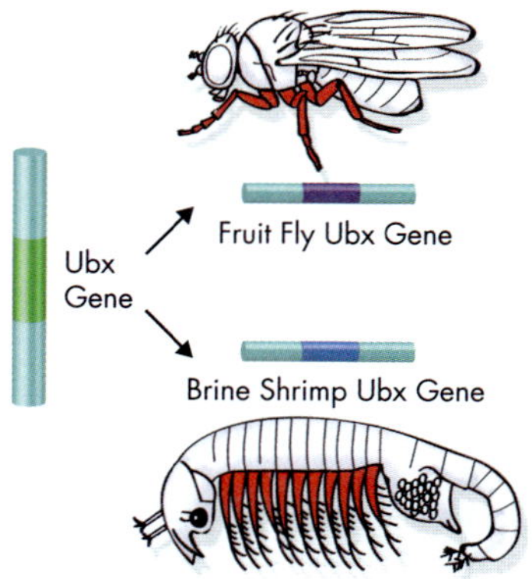

Changes in a Hox Gene
Fruits flies and brine shrimp share a common ancestor. The Hox gene called Ubx directs the development of legs in both animals. Millions of years ago, mutations on Ubx caused fruit flies to have six legs and brine shrimp to have many legs.

Shrimp and insects share an ancient common ancestor. This ancestor had dozens of legs. Shrimp still have many legs, but insects have only six. What happened to those other legs? A single mutation in a Hox gene turns off the growth of some pairs of legs. Thus, a change in one Hox gene can cause a major difference between two groups of animals.

Timing Is Everything The growth of an embryo is carefully controlled by genes. Growth starts and stops at exact times. Small changes in starting and stopping times can make a big difference in the final organism. For example, small changes in timing can make the difference between long, slender fingers and short, stubby toes.

Key Question How may Hox genes be involved in evolutionary change? **Even small changes in Hox genes can produce large changes in adult animals.**

CHECK Understanding

Apply Vocabulary

Use the highlighted words from the lesson to complete each sentence correctly.

1. When scientists use the rates of mutation in DNA to estimate how long two species have been evolving separately, they are using a _______.

Critical Thinking

2. **Relate Cause and Effect** How can crossing-over result in gene duplication?
3. **Relate Cause and Effect** How can a duplicate gene become a new gene with a different function?
4. **Review** What is the function of Hox genes?
5. **Relate Cause and Effect** Why have small changes in Hox genes had such a great effect on the evolution of animal diversity?
6. **Write to Learn** How does time affect the number of differences in the genes of two species with a common ancestor?

BIOLOGY.com Search Lesson 17.4 GO • Lesson Assessment

Check Understanding

1. molecular clock
2. During crossing-over, there may be unequal swapping of DNA. One of the homologous chromosomes may receive extra DNA that carries part or all of a gene.
3. A mutation in the duplicate gene would change the structure of the protein it produces, which in turn would change the way it functions.
4. Hox genes control the shape of an animal's body, including the development of its wings, legs, or arms.
5. Small changes in the development of an embryo can cause large changes in the body of the adult, such as the number of legs or the size of limbs.
6. The more differences there are between the DNA sequences of the two species, the more time has passed since the two species shared a common ancestor.

Active Reading

COMPARE/CONTRAST Refer students to the visual on this page.

Ask What characteristics do these two organisms share? *(Sample answer: Both have hard outer coverings and segmented bodies.)*

Discuss their differences. For example, the fruit fly has wings, and the brine shrimp appears to have a "tail."

Ask Why do scientists think that the organisms came from a common ancestor? *(They have similar DNA that contains genes that have mutated.)*

KEY QUESTION Have the class read the question and answer aloud.

Ask What are Hox genes? *(Hox genes are a group of genes that determine which parts of the embryo develop arms, legs, or wings. They also control the size and shape of adult body parts.)*

Have a volunteer use the visual on this page to explain to the class how Hox genes can produce large changes.

Post Reading

ASSESSMENT TIPS Remind students that they can use key words in a question to find information in the text to help answer the question.

For Question 2, have students recall what *crossing over* is. The section Copying Genes refers to chromosome pairs exchanging DNA during meiosis, but it does not call the process *crossing over.*

Point out to students that Questions 3 and 5 have presented the effects. In each case, students are expected to explain the cause.

Wrap-Up Activity

MODEL MUTATIONS Have students work individually or in small groups to cut 16 identical squares of paper. Tell them to label four squares with letters for each of the bases of DNA—A, G, C, and T. Have them use these squares to model the mutations shown in the first visual in this lesson. Then have them use the squares to model additional mutations for each of the species shown.

Pre-Lab

Review the meanings of *competition* and *speciation.* Then read the Problem aloud and make sure students understand it. Ask a volunteer to restate the Problem using different words.

Have students examine the pictures of the birds. Explain that each bird is a different species of finch on the Galápagos Islands.

Ask How does each species differ? *(They have different sizes and shapes of heads and beaks.)* What does this tell you about their eating habits? *(They each eat different foods and get those foods in different ways.)*

Review the Background Questions and Pre-Lab Questions as a class.

Lab

Tell students they will perform the chapter lab *Competing for Resources* described in **Lab Manual B.**

SAFETY Before using any of the seeds, ask students about any possible seed or nut allergies. Double-check with the school nurse or other school authority since nut allergies in particular can be extremely dangerous. If allergies exist, substitute similar-sized objects.

Have students wash their hands and their work surfaces thoroughly after handling the seeds and nuts.

Advise students to be cautious when using sharp objects.

Post-Lab

DESIGN YOUR OWN LAB Encourage students to design a similar lab with different tools and different objects to pick up. Have students explain why they chose each tool and each object to pick up.

Skills Lab

GUIDED INQUIRY

Pre-Lab: Competing for Resources

Problem How can competition lead to speciation?

Materials assorted tools, large and small seeds, large and small paper plates, timer or clock with second hand

Lab Manual Chapter 17 Lab

Skills Focus Use Models, Predict, Apply Concepts

Connect to the Big idea Speciation is not easy to see in nature. Usually, new phenotypes take years to emerge or become common enough to be noticed. Also, new phenotypes can be difficult to track in a complex environment. For scientists who want to study speciation, islands can provide an ideal environment.

Peter and Rosemary Grant spent years studying finches on the Galápagos Islands. They measured and recorded the traits and diets of hundreds of birds. During a year with a severe drought, the Grants were able to observe natural selection in action as food became scarce. In this lab you will model variation in bird beaks and diet to demonstrate the impact of competition on survival and speciation.

Background Questions

a. **Review** What is speciation?

b. **Relate Cause and Effect** How did geographic isolation lead to speciation among the Galápagos finches?

c. **Compare and Contrast** How does an adaptation differ from other inherited traits?

Pre-Lab Questions

Preview the procedure in the lab manual.

1. **Use Models** In this lab, what do the different types of tools represent?

2. **Predict** Which tools do you think will work best for picking up small seeds? Which will work best for picking up large seeds?

3. **Design an Experiment** Why will the time you have to collect seeds be limited?

BIOLOGY.com Search Chapter 17 GO

Visit Chapter 17 online to test yourself on chapter content and to find activities to help you learn.

Untamed Science Video Climb the cliffs of Hawaii with the Untamed Science crew to discover how geographic isolation can result in a new species.

Art Review Review your understanding of alleles and allele frequencies in a population.

Art in Motion Watch how different types of selection change the types of individuals that comprise a population.

Tutor Tube Learn more about the mechanisms of speciation from the tutor.

Pre-Lab Answers

BACKGROUND QUESTIONS

a. Speciation is the formation of a new species.

b. Sample answer: Over time, populations of finches on different islands no longer shared the same gene pool.

c. An adaptation is a trait that increases an organism's ability to survive and reproduce in its environment.

PRE-LAB QUESTIONS

1. The tools represent different types of beaks.

2. Answers will vary with the types of tools. Tweezers or small spoons might work w with small seeds. Pliers or ice tongs migh work well with large seeds.

3. Sample answer: Limiting the time gives advantage to those with the tools that a better adapted to pick up a certain type seed.

NATIONAL SCIENCE EDUCATION STANDARDS

UCP IV

CONTENT C.3.a, C.3.d, C.4.c, C.6.c

INQUIRY A.2.a, A.2.b

17 CHAPTER Summary

17.1 Genes and Variation

- Evolution is change in the frequency of alleles in a population over time.
- Three sources of genetic variation are mutations, genetic recombination during sexual reproduction, and lateral gene transfer.
- The number of phenotypes produced for a trait depends on how many genes control the trait.

gene pool (p. 406)
allele frequency (p. 406)
single-gene trait (p. 408)
polygenic trait (p. 408)

17.2 Evolution as Genetic Change in Populations

- Natural selection on single-gene traits can lead to changes in allele frequencies. This can lead to changes in phenotype frequencies.
- Natural selection on polygenic traits can affect the relative fitness of phenotypes. This selective pressure can cause one of three types of selection: directional selection, stabilizing selection, and disruptive selection.
- Genetic drift may occur in small populations. By chance, individuals that carry one allele may leave more offspring than other individuals leave. Over time, a series of chance events can cause an allele to become more or less common in a population.
- The Hardy-Weinberg principle predicts that five conditions can cause evolution to take place: (1) nonrandom mating, (2) small population size, (3) movement into or out of the population, (4) mutations, and (5) natural selection.

directional selection (p. 410)
stabilizing selection (p. 410)
disruptive selection (p. 410)
genetic drift (p. 411)
bottleneck effect (p. 411)
founder effect (p. 411)
genetic equilibrium (p. 412)
Hardy-Weinberg principle (p. 412)
sexual selection (p. 412)

17.3 The Process of Speciation

- When populations become reproductively isolated, they can evolve into separate species. Reproductive isolation can develop in a variety of ways. These include behavioral isolation, geographic isolation, and temporal isolation.
- Speciation in Galápagos finches most likely occurred by: founding of a new population, geographic isolation, changes in the population's gene pool, behavioral isolation, and ecological competition.

species (p. 414)
speciation (p. 414)
reproductive isolation (p. 414)
geographic isolation (p. 414)
behavioral isolation (p. 415)
temporal isolation (p. 415)

17.4 Molecular Evolution

- A molecular clock uses mutation rates in DNA to estimate the time that two species have been evolving separately.
- One way that new genes evolve is through the duplication, and then modification, of existing genes.
- Small changes in Hox gene activity during development can produce large changes in adult animals.

molecular clock (p. 417)

UbD Transfer the Big Idea

TRANSFER Ask each student to write a short science fiction story about a population of imaginary organisms on a distant planet. In their story, students should describe how the organisms evolve and come to have different phenotypes after several generations. The story should include details about genetic bottlenecks or severe environmental stresses, such as geographic isolation, that cause the population to undergo genetic change.

Chapter Summary

Think Visually

USE GRAPHICS Have students fill in the following concept map.

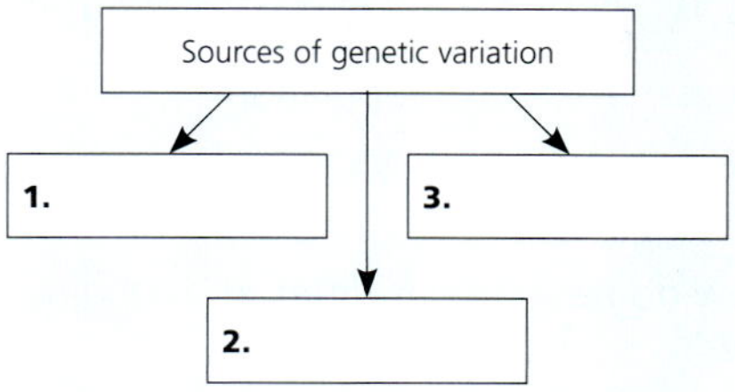

ANSWERS (can be in any order)

1. mutations
2. genetic recombination during sexual reproduction
3. lateral gene transfer

BIOLOGY.com Study Online

REVIEW RESOURCES

Art in Motion Students explore how the composition of a population changes under different types of natural selection pressure.

Data Analysis Students can analyze data from the Grants' study of Darwin's finches.

Tutor Tube The tutorial shows students how to use a concept map to distinguish the concepts of selection, adaptation, speciation, and evolution.

Vocabulary Review The *Flash Cards* and *Match It* provide an interactive way to review chapter vocabulary.

Editable Worksheets Pages of Study Workbook B, Lab Manual B, and the Assessment Resources Book are available online. These documents can be easily edited using a word-processing program.

Lesson Overviews Lesson Overviews are based on Miller & Levine Biology. However, they are available on CD-ROM and can be edited for your class.

ASSESSMENT RESOURCES

Chapter Assessment Have students take an online version of the Chapter 17 Check Understanding.

Standardized Test Prep Students can take an online version of the Standardized Test Prep. You will receive their scores along with ideas for remediation.

Diagnostic and Benchmark Tests Use these tests to monitor your students' progress and supply remediation.

ASSESSMENT

Assess

Big idea Write the Big Idea question on the board in one line. Underline the terms *populations, evolve,* and *new species.* Have students list details about each term, and write those details above and below the terms, creating a concept map. If students need help volunteering information, ask them leading questions about the terms.

Ask What makes up a population?

Ask What is evolution? What causes populations to evolve?

Ask How do new species differ from original species?

CONSTRUCTED RESPONSE Review how to write a paragraph. Remind students that paragraphs consist of a main idea and details to support the main idea. Point out that usually the main idea is stated first, followed by the details.

Explain that for the following questions, the hints will guide students as they write their paragraphs.

Write Question 1 on the board. Be sure students know the meaning of each term in the question. Then have them read the hints below the question.

Ask How does a population's environment affect genetic variation?

Have students write a sentence containing the main idea from the question itself. Then have them use the hints and the words *fitness* and *phenotype* to explain and describe the main idea.

For Question 2, pair students and have them use the hints and your modeling to answer the question. Have students share their paragraphs with the class.

For Question 3, have students write the paragraphs independently for homework or as an anchor activity.

17 CHECK Understanding

Assess the Big idea Evolution

Write an answer to the question below.

Q: How can populations evolve to form new species?

Constructed Response

Write an answer to each of the numbered questions below. The answer to each numbered question should be one or two paragraphs long. To help get ideas, read the **Hints** below the questions.

1. **Why is genetic variation in a population necessary for natural selection to take place? Use the words *fitness* and *phenotype* in your answer.**

   Hint The fitness of an organism depends upon the conditions of its environment.

2. **How does population size affect evolution?**

   Hint The Hardy-Weinberg principle describes the conditions under which a population will *not* evolve.

3. **On the Galápagos Islands, there is a species of finch that eats only fruit. Describe a process by which this species may have evolved from a species that eats many kinds of foods.**

   Hint Refer to the art on pages 415 and 416.

Foundations for Learning Wrap-Up

When you have finished taking notes, fold the paper into a triangle with the notes on the outside. Tape the ends together to make a table tent, as shown below.

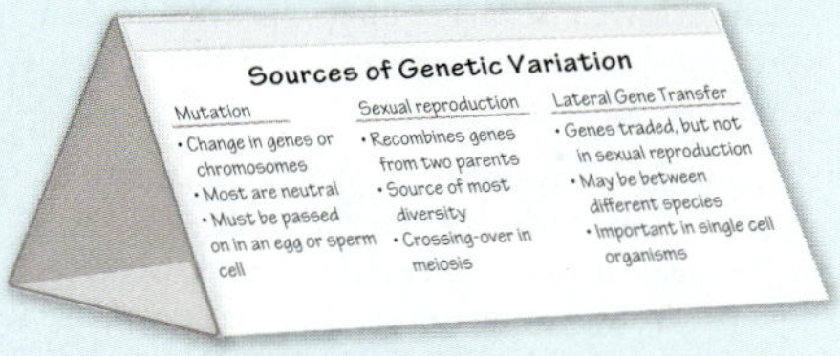

Activity 1 Work with a partner to review each of the main topics. Ask a partner to pick a side of your table tent—Sources of Genetic Variation, Natural Selection, or Isolation. Have your partner form a question or two about each topic, such as "What are three sources of genetic diversity?" While you answer, your partner will watch your table tent to be sure you have answered correctly and completely. When you have answered all three sides, change roles.

Activity 2 Use your table tent to explain how a new species of birds may have arisen on an island. Begin with the side labeled Sources of Genetic Variation. You may also refer to the art about the speciation in Darwin's finches in Lesson 17.3.

422 Check Understanding

Foundations for Learning Wrap-Up

Check each student's table tent to make sure the information on it is valid and complete. Explain the activities. Have students perform the first activity in pairs. Tell the students asking the questions that they are not limited to the information on the table tent. They can ask any valid question about any of the three topics on the table tent, but they have to first know the correct answers. Ask students to find new partners for Activity 2.

17.1 Genes and Variation

Understand Key Concepts

1. The combined genetic information of a population forms a
   a. genotype.
   b. gene pool.
   c. phenotype.
   d. bottleneck.

Test-Taking Tip

Read Before You Choose Read all the choices before you pick your answer. Sometimes a choice can be partly correct, but there is a better choice later in the list. In question 1, you might be tempted to answer **a**, because a genotype is a set of genes. But genotypes refer to individuals, not populations. So **b**, gene pool, is the best choice.

2. Mutations that improve an individual's ability to survive and reproduce are
   a. harmful.
   b. neutral.
   c. helpful.
   d. chromosomal.
3. Identify two ways that sexual reproduction is a source of genetic variation.

Think Critically

4. **Compare and Contrast** Compare the effects of two mutations: a mutation in a body cell and a mutation in an egg cell. Which mutation could affect evolution? Why?

17.2 Evolution as Genetic Change in Populations

Understand Key Concepts

5. In a population, short individuals have a greater fitness than average-size or tall individuals. This difference in fitness could lead to
   a. directional selection.
   b. stabilizing selection.
   c. disruptive selection.
   d. artificial selection.
6. A change in the allele frequency of a small population that is caused by chance is known as
   a. a gene pool.
   b. the Hardy-Weinberg principle.
   c. variation.
   d. genetic drift.
7. What is *fitness* in genetic terms?
8. How does meiosis in sexual reproduction affect the frequency of alleles in a population?

Think Critically

9. **Apply Concepts** Why is genetic drift most common in small populations?
10. **Infer** Genetic equilibrium is rare in most populations. Why do you think this is the case?

17.3 The Process of Speciation

Understand Key Concepts

11. When two populations no longer interbreed, what is the result?
    a. genetic equilibrium
    b. reproductive isolation
    c. stabilizing selection
    d. artificial selection
12. Two populations of turtles nest on an island. One of the species nests in June. The other species nests in August. These two species are separated by
    a. temporal isolation.
    b. geographical isolation.
    c. genetic isolation.
    d. behavioral isolation.
13. Why is variation important to the survival of a species when the environment changes?

Think Critically

14. **Relate Cause and Effect** Before a population splits into two different species, there must usually be reproductive isolation. Explain why this is so.

Assessment Tips

Have students work individually or in pairs to answer the questions. Encourage students to read the vocabulary terms and heads in each lesson as a review before answering the questions for that lesson.

For Question 5, refer students to the Selection on Polygenic Traits graphs in Lesson 17.2.

For Question 20, review what the information in each column of the data table shows.

Point out that there are several possible answers to Question 21.

Answers

Lesson 17.1

UNDERSTAND KEY CONCEPTS

1. b
2. c
3. Sexual reproduction leads to genetic variation by recombination and crossing-over.

THINK CRITICALLY

4. Only the mutation in the egg cell can affect evolution. Mutations in egg cells can be passed on. Mutations in body cells are not passed on to future generations.

Lesson 17.2

UNDERSTAND KEY CONCEPTS

5. a
6. d
7. Fitness describes an individual's ability to survive and pass its genes on to the next generation.
8. Meiosis does not affect the frequencies of alleles in a population.

THINK CRITICALLY

9. Sample answer: Other factors that can disrupt genetic equilibrium, such as migration or mutations, could affect allele frequencies more when the population size is small.
10. In an actual population, one or more of the conditions that can disrupt equilibrium is usually present.

Lesson 17.3

UNDERSTAND KEY CONCEPTS

11. b
12. a
13. With variation, it is likely that some individuals will have traits to help them survive in the new environment.

THINK CRITICALLY

14. If there is no reproductive isolation, the populations will still be interbreeding, their genes will be mixing, and there will be only one gene pool. Since they are interbreeding, they are still one species.

Lesson 17.4

UNDERSTAND KEY CONCEPTS

15. d

16. a

17. mutations that do not have a positive or negative impact on the fitness of an organism

18. Mutations that have no effect on phenotype are neutral, so they will occur at a steady rate. They will not be affected by natural selection.

THINK CRITICALLY

19. The original gene duplicates. Then one of the two resulting genes undergoes a mutation. This gene evolves a new function.

Connecting Concepts

20. Over time, the recessive allele *b* has become more common than dominant allele *B*.

21. Sample answer: The environment may have changed from a time when natural selection favored allele *B*. Now natural selection favors allele *b*.

CHAPTER MYSTERY After students have read through the Chapter Mystery, discuss the characteristics of evolution in flu viruses that allow them to quickly evolve new strains that are not easily affected by the host's immune system.

Ask Besides a high rate of mutations, what increases the genetic diversity in viruses? *(Different strains can swap genes if they infect the same host at the same time.)*

Ask What is the major stress in a flu virus's environment? *(the host's immune system)*

Ask How do new surface proteins increase a virus's fitness? *(They allow the virus to multiply repeatedly before being detected by the host's immune system.)*

Ask If a bird flu virus infected a human host, why would it be likely to cause a serious, even fatal, disease? *(The virus would probably have surface proteins the human immune system had never been exposed to and could not detect.)*

17 CHECK Understanding

17.4 Molecular Evolution

Understand Key Concepts

15. Hox genes are important to evolution because
a. they mutate often.
b. they are only found in egg and sperm cells.
c. they are immune to natural selection.
d. they have a large effect on the traits of an organism.

16. What are scientists counting when using a molecular clock?
a. mutations
b. genes
c. gene families
d. chromosomes

17. What are neutral mutations?

18. Why do molecular clocks use mutations that have no effect on phenotype?

Think Critically

19. Sequence Describe how duplicate genes can lead to the creation of new genes.

Connecting Concepts

Use Science Graphics

Use the data table to answer questions 20 and 21.

Frequency of Alleles

Year	Frequency of Allele *B*	Frequency of Allele *b*
1910	0.81	0.19
1930	0.49	0.51
1950	0.25	0.75
1970	0.10	0.90

20. Interpret Tables Describe the trend shown by the data in the table.

21. Form a Hypothesis What might account for the trend shown by the data?

solve the CHAPTER MYSTERY

KILLER DISEASE

The genes of flu viruses mutate often. Also, different strains of flu can swap genes. These mutations and the ability to gene swap gives the viruses plenty of genetic diversity.

Natural selection acts on viruses in the environment of your own body! Viruses have proteins on their surfaces that make them targets for your immune system. Viruses with proteins that your body recognizes will be destroyed. These viruses have low fitness. Viruses that your body can't recognize have a higher fitness.

Viruses evolve new proteins often. This makes them harder for your immune system to recognize. That's when you get sick. That's also why you can have the flu one winter, and then get another strain the next winter.

Influenza Virus

Rarely, flu virus evolution produces a very new protein. Your immune system is unable to recognize or fight these new proteins. Such flu viruses can be deadly, as was the influenza virus of 1918. If a strain like that appeared today, it would kill many people.

1. Infer Vaccines help prepare your body for an infection. Doctors recommend that people receive a new vaccine for the flu every year. People do not need to receive a new vaccine for measles every year. What does this suggest about a difference between the measles virus and flu viruses?

2. Apply Concepts Can you think of other issues in public health that relate directly to evolution?

UntamedScience **Never Stop Exploring Your World.** Finding the solution to the epidemic mystery is only the beginning. Take a video field trip with the ecogeeks of Untamed Science to see where the mystery leads.

424 BIOLOGY.com Search Chapter 17 GO • Untamed Science Video • Chapter Mystery

CHAPTER MYSTERY ANSWERS

1. The measles virus does not evolve as quickly as flu viruses, so your body's defenses do not need to be updated every year.

2. Sample answer: Improper use of antibiotics has helped bacteria evolve into forms that resist current medical treatments. Other quickly evolving viruses, such as HIV, also challenge health workers.

Untamed Science™ Students can explore speciation mechanisms at work in Hawaii by watching **Evolution at a Snail's Pace.**

Standardized Test Prep

Multiple Choice

1. Which of the following conditions is MOST likely to result in changes in allele frequencies in a population?
   A random mating
   B small population size
   C no migrations into or out of a population
   D absence of natural selection

2. Mutations and the genetic recombination that occurs during sexual reproduction are both sources of
   A genetic variation. C genetic equilibrium.
   B stabilizing selection. D genetic drift.

3. In a population of lizards, the smallest and largest lizards are more easily preyed upon than medium-size lizards. What kind of natural selection is MOST likely to occur in this situation?
   A genetic drift
   B sexual selection
   C stabilizing selection
   D directional selection

4. Populations of antibiotic-resistant bacteria are the result of the process of
   A natural selection. C genetic drift.
   B temporal isolation. D artificial selection.

5. If species A and B have very similar genes and proteins, what is probably true?
   A Species A and B share a relatively recent common ancestor.
   B Species A evolved independently of species B for a long period.
   C Species A is younger than species B.
   D Species A is older than species B.

6. When two species reproduce at different times, the situation is called
   A genetic drift.
   B temporal selection.
   C temporal isolation.
   D lateral gene transfer.

7. The length of time that two species have been evolving separately can be estimated using
   A genetic drift. C a molecular clock.
   B gene duplication. D Hox genes.

Questions 8–9

The graphs below show the changes in crab color at one beach.

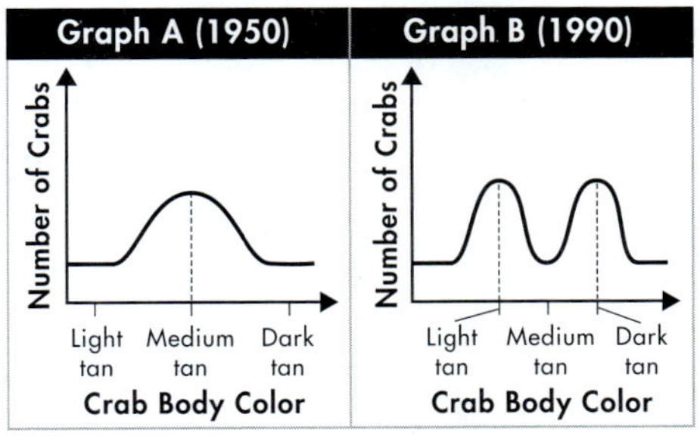

8. What process occurred over the 40-year period?
   A artificial selection
   B directional selection
   C stabilizing selection
   D disruptive selection

9. Which of the following is MOST likely to have caused the change in the distribution?
   A A new predator arrived that preferred dark-tan crabs.
   B A new predator arrived that preferred light-tan crabs.
   C A change in beach color made medium-tan crabs the least visible to predators.
   D A change in beach color made medium-tan crabs the most visible to predators.

Open-Ended Response

10. How does evolution change the relative frequency of alleles in a gene pool? Why does this happen?

If You Have Trouble With . . .

Question	1	2	3	4	5	6	7	8	9	10
See Lesson	17.3	17.1	17.2	17.2	17.4	17.3	17.4	17.2	17.2	17.1

Test-Taking Tip

INTERPRET GRAPHS

Remind students that different types of graphs are used for different purposes. A circle graph (pie chart) shows the parts of a whole. A bar graph shows the number of items in different categories. A line graph shows how the variable on the *y*-axis changes relative to the variable on the *x*-axis.

Help students interpret the graphs for Questions 8 and 9. They are line graphs that show how the number of crabs changes in relation to the color of crabs. Some students might be confused by the fact that there are no numbers on the axes. Point out that in some graphs, the important part is the overall trend shown, not the absolute quantities.

Before the Test

Review with students the Test-Taking Tip. Point out that the tip will help them answer Questions 8 and 9.

Point out the word *MOST* in Questions 1, 3 and 9.

Remind students that an open-ended response requires several sentences. Encourage them to write a draft copy of their answer before finalizing it.

Answers

1. B
2. A
3. C
4. A
5. A
6. C
7. C
8. D
9. D
10. Evolution causes some alleles to become more common in a gene pool and other alleles to become less common. This changes the relative frequency of the alleles in the gene pool. This might happen due to natural selection or genetic drift.

Classification

Chapter Contents	NSES	Time	Core Resources
Chapter Preview			**Student Edition,** pp. 426–427 **Chapter Mystery,** p. 427 **Foundations for Learning,** p. 427
18.1 Finding Order in Diversity Scientific Names • The Linnaean Classification System	I, IV, A.2.b, A.2.e, C.3.b, C.3.e, G.1, G.2, G.3	1 period 1/2 block	**Student Edition,** pp. 428–432 **Build Connections,** p. 429 **Study Workbook B** 18.1 Worksheets **Biology.com** *Art in Motion:* Using a Dichotomous Key • 18.1 Lesson Assessment
18.2 Modern Evolutionary Classification Evolutionary Classification • Cladograms • DNA in Classification	I, II, IV, A.2.e, C.3.a, C.3.b, C.3.d, C.3.e	1 period 1/2 block	**Student Edition,** pp. 433–437 **Inquiry Into Scientific Thinking,** p. 435 **Study Workbook B** 18.2 Worksheets **Biology.com** *InterActive Art:* Cladograms • 18.2 Lesson Assessment
18.3 Building the Tree of Life Changing Ideas About Kingdoms • The Tree of All Life	I, II, IV, A.2.e, A.2.f, C.3.a, C.3.b, C.3.d, C.3.e	1 period 1/2 block	**Student Edition,** pp. 438–441 **Study Workbook B** 18.3 Worksheets **Biology.com** *Art Review:* Three Domains • 18.3 Lesson Assessment
Chapter Pre-Lab	A.2.b, C.3.e	1 period 1/2 block	**Student Edition,** p. 442 **Lab Manual B** *Dichotomous Keys*

Tools for ELL Instruction

ELL Handbook describes teaching strategies to make *Miller & Levine Biology: Foundation Edition* more accessible to ELL students.

Spanish Study Workbook is a Spanish translation of Study Workbook A.

Multilingual Glossary is the *Miller & Levine Biology* glossary translated into ten languages.

Tools to Challenge Students

Study Workbook A includes more challenging worksheets and activities such as Chapter Mystery activities that reinforce 21st Century Skills.

Lab Manual A labs are geared toward on-level students. The Lab Manual includes fifteen labs in addition to the Chapter Labs.

Biology.com Data Analysis activities offer students a way to practice data gathering and analyzing skills.

Visual Quizzes in the Assessment Resources Book are geared toward on-level students and use visuals from *Miller & Levine Biology.*

Additional Resources

Biology.com Untamed Science Video • Flash Cards

Study Workbook A 18.1 Worksheets
Lab Manual B *Hands-On Activity:* Classifying Cerealites
Spanish Study Workbook 18.1 Worksheets

Study Workbook A 18.2 Worksheets
Spanish Study Workbook 18.2 Worksheets
Biology.com *Data Analysis:* A Friend for Lonesome George
Assessment Resources Book Visual Quiz

Study Workbook A 18.3 Worksheets
Lab Manual B *Data Analysis:* Comparing the Domains
Spanish Study Workbook 18.3 Worksheets
Biology.com *Data Analysis:* Bar-Coding Life

Lab Manual A *Dichotomous Keys*

Chapter Review

Student Edition Chapter Summary, p. 443
Study Workbook B Chapter 18 Review
Transparencies, pp. 215–225
Biology.com Untamed Science Video • Editable Worksheets of Study Workbooks A and B and Lab Manuals A and B • Chapter 18 Flash Cards and Crossword

Untamed Science DVD • Classroom Resources CD (includes lesson presentations and editable worksheets)

Chapter Assessment

Student Edition Check Understanding, pp. 444–447
Study Workbook B Chapter 18 Taking a Standardized Test
Assessment Resources Book Chapter 18 Test B
Biology.com Chapter 18 Assessment • Editable Worksheets of Chapter 18 Visual Quiz* and Chapter 18 Tests A and B

Exam*View* *Assessment Suite* • Classroom Resources CD (includes lesson presentations and editable worksheets)

Time: 1 period, 1/2 block

* Visual Quizzes are based on visuals from *Miller & Levine Biology*. Review carefully to make sure they are appropriate for your class.

Pressed for Time?

Preview the Chapter Preview the From Species to Kingdom and Interpreting a Cladogram visuals.

Cover the Chapter Quickly Have students read Lesson 18.1. Assign *Evolutionary Classification* and *Cladograms* in Lesson 18.2, and go over the Building a Cladogram visual. Have students read *Changing Ideas About Kingdoms* in Lesson 18.3 and discuss the Three Domains visual.

Assess Assign the 18.1 Check Understanding, Questions 1–5 in the 18.2 Check Understanding, and Questions 1, 2, and 4–6 in the 18.3 Check Understanding. In the Chapter 18 Check Understanding, assign Questions 1–11 and 14–16.

Get the Picture

Refer students to the photograph and read aloud the caption. Point out the large variety of birds in the collection.

Ask Why do you think the museum workers grouped the bird collection into different types of birds? *(Sample answer: to make it easier to compare the birds and study them)*

Ask What are some characteristics the museum workers might have used to group the birds into different types? *(Sample answers: body size, wing shape, beak shape, feather color)*

Activate Prior Knowledge

Remind students of the great diversity of organisms Darwin observed during his voyage.

Say Darwin collected many samples of finches and other birds. When Darwin returned from his voyage, a bird specialist placed the birds into groups, or classified them.

Ask How do you think classifying the birds helped Darwin? *(Sample answers: by allowing him to make comparisons more easily; by showing him how the birds were related to each other; by helping him understand how the birds evolved)*

NATIONAL SCIENCE EDUCATION STANDARDS

UNIFYING CONCEPTS AND PROCESSES

I, II, IV

CONTENT

C.3.a, C.3.b, C.3.d, C.3.e, G.1, G.2, G.3

INQUIRY

A.2.b, A.2.e, A.2.f

18 Classification

Big idea Unity and Diversity of Life

Q: What is the goal of biologists who classify living things?

The National Museum of Natural History has one of the largest collections of bird species in the world. Nearly 80 percent of all the world's species are found here.

426

UbD Connect to the Big Idea

Invite a volunteer to read aloud the chapter number and the chapter title. Read the Big Idea and question. Emphasize the question, **What is the goal of biologists who classify living things?**

Set the Purpose

PREVIEW AND PREDICT Draw the graphic organizer on the board. Guide students to the INSIDE feature on the next page. Invite a volunteer to read each lesson number and title.

Ask How might you use the words *unity* and *diversity* to describe the bird collection in the photograph? *(Sample answer: There is unity in that all the organisms are birds and have the same basic characteristics, such as wings, beaks, and feathers. There is diversity in that the characteristics have variations, such as shape, size, and color.)*

INSIDE:

- 18.1 Finding Order in Diversity
- 18.2 Modern Evolutionary Classification
- 18.3 Building the Tree of Life

- Untamed Science Video
- Chapter Mystery

CHAPTER
MYSTERY

GRIN AND BEAR IT

If you just looked at polar bears and brown bears, you would probably think that they were different species. Polar bears are much larger than brown bears. Their paws let them swim and walk on snow and ice. Their white fur hides them in the snow. In contrast, brown bears are brown. And their paws are not as big.

Polar bears and brown bears look very different. They also act differently. Do those differences mean that they are different species? Remember the definition of a species: "a group of similar organisms that can breed and produce fertile offspring." Polar bears and brown bears *can* interbreed and produce fertile offspring. This should make them the same species. Are they?

Read for Mystery Clues As you read this chapter, look for clues to help you decide if polar bears and brown bears are separate species. Then, solve the mystery at the end of the chapter.

FOUNDATIONS for Learning

To understand biological classification, you will need to learn a few new words. Write the following words on index cards: *Domain, Kingdom, Phylum, Class, Order, Family, Genus,* and *Species.*

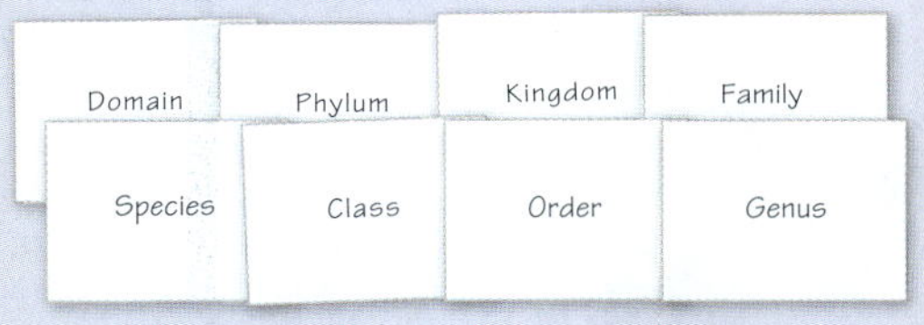

As you read the chapter, write the definition for each word on the back of the card. Also, write down examples of the terms. You will use these cards for two activities at the end of the chapter. They will help you answer the question: What is the goal of biologists who classify living things?

Chapter 18
Big Idea: Unity and Diversity of Life

Chapter 18: What is the goal of biologists who classify living things?

- **18.1:** Why do scientists classify organisms?
- **18.2:** How do evolutionary relationships affect the way scientists classify organisms?
- **18.3:** What are the major groups within which all organisms are currently classified?

Foundations for Learning

Give each student eight index cards. Have students write each word on a separate card.

Ask How might writing the definitions of the words be helpful to you? *(Sample answer: Writing the definitions can help me remember them.)*

Ask How might writing examples be helpful to you? *(Sample answer: Writing examples can help me understand the definitions of the words.)*

Tell students they will use their vocabulary cards for review activities at the end of the chapter.

Untamed Science™ Join the crew of Untamed Science as they investigate the classification of a fish species in **Finned Kin**.

CHAPTER MYSTERY Invite a volunteer to read aloud the Chapter Mystery.

Ask How does the mystery about polar bears and brown bears relate to the chapter title? *(The mystery is whether the two kinds of bears should be classified as the same species.)*

List on the board the characteristics of polar bears and brown bears described in the Chapter Mystery. Based on the characteristics, have students predict whether polar bears and brown bears are the same species. Tell students that as they read the chapter, they will find clues that will reveal if their predictions were correct.

Digital Support

BIOLOGY.com Extend your reach by using the digital assets offered at **Biology.com.**

LESSON 18.1

Pre-Reading

Objectives

18.1.1 Describe the goals of binomial nomenclature and systematics.

18.1.2 Identify the taxa in the classification system devised by Linnaeus.

Preview the Pages

1. As students preview the headings and scan the text, ask them to find the highlighted vocabulary terms. They should write the words on their index cards. As they read the lesson, they should write the definitions in their own words.
2. Ask volunteers to read aloud the visuals' captions and tell how they think the captions connect to the headings.
3. Point out the Build Connections feature on the next page. Explain that a dichotomous key is a tool that is used to identify organisms.

Activate Prior Knowledge

Have volunteers define *species* in their own words and tell how organisms are grouped into species. Write students' definitions on the board. Revisit them during the lesson to address misconceptions.

Build Understanding

PREVIEW VISUALS Direct students to Workbook B, where they can learn more about previewing visuals. As students look over the From Species to Kingdom diagram, have a volunteer read aloud the taxonomic groups and the names of the animals shown.

NATIONAL SCIENCE EDUCATION STANDARDS

UNIFYING CONCEPTS AND PROCESSES

I, IV

CONTENT

C.3.b, C.3.e, G.1, G.2, G.3

INQUIRY

A.2.b, A.2.e

For the corresponding lesson in *Miller & Levine Biology,* see pages 510–515.

18.1 Finding Order in Diversity

Key Questions

What are the goals of binomial nomenclature and systematics?

How did Linnaeus group species into larger taxa?

BUILD Understanding

Preview Visuals Before you read, look at the diagram From Species to Kingdom on p. 431. Notice all the levels of classification. As you read about each level, look at the diagram again.

In Your Workbook Go to your workbook to learn more about previewing visuals.

Binomial Nomenclature The scientific name of the polar bear is *Ursus maritimus,* which means "marine bear."

Assigning Scientific Names

The first step in studying the diversity of life is to describe and name each species. For names to be useful, everyone must use the same name for each species.

People in different places often use different words to describe the same organism. Using common names can cause problems. For example, the words *cougar, puma, panther,* and *mountain lion* are all common names for the same animal. Sometimes, people use the same name for different organisms. For example, in the United Kingdom, a *buzzard* is a kind of hawk. In the United States, a *buzzard* is a vulture. And of course, a person who does not speak English would not use any of these names!

By the 1700s, scientists knew that it was too confusing to call organisms by their common names. They agreed to use a Latin or Greek name for each species. At first, this did not help much. The early names were very long and complicated. The name for a tree might be "Oak with deeply divided leaves that have no hairs on their undersides and no teeth around their edges." Also, scientists used different characteristics to describe species.

Binomial Nomenclature In the 1730s, a Swedish scientist named Carolus Linnaeus developed a two-word naming system called **binomial nomenclature.** In binomial nomenclature, each species is assigned a scientific name that has two words. Scientific names are written in italics. The first word begins with a capital letter. The second word is lowercase.

The scientific name for the polar bear is *Ursus maritimus.* The first part of the name—*Ursus*—is the genus to which the organism belongs. A **genus** (plural: genera, JEHN er uh) is a group of closely related species. The genus *Ursus* contains five other species of bears, including *Ursus arctos,* the brown bear.

The second part of a scientific name is unique to each species. This part of the name often describes an important trait or the organism's habitat. The word *maritimus* comes from the Latin word for "sea." Polar bears live on ice that floats in the sea.

Key Question What is the goal of binomial nomenclature?
In binomial nomenclature, each species is given a two-part scientific name.

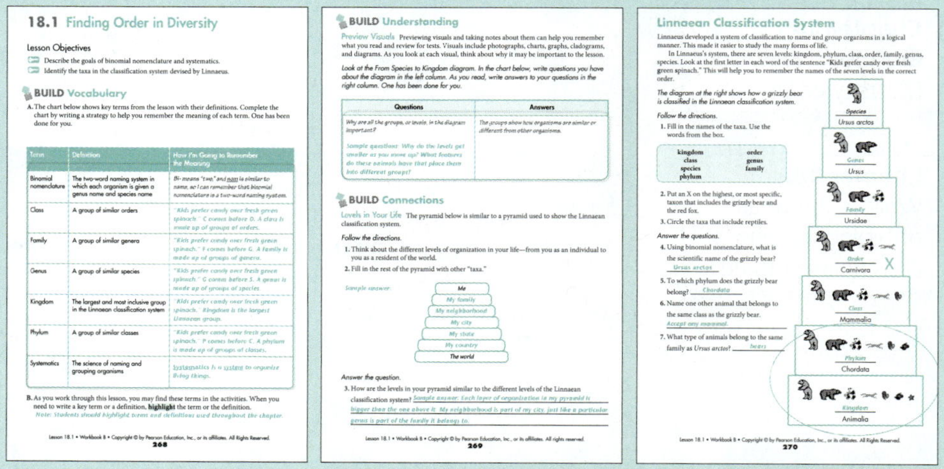

18.1 Finding Order in Diversity

BUILD Understanding

Linnaean Classification System

BUILD Connections

USING A DICHOTOMOUS KEY

A dichotomous key is used to identify organisms. The key is made up of a series of statements about visible traits. Each statement can be considered a question that can be answered yes or no. The answers lead you to new statements until you have identified your organism. Use the key to identify the leaf shown here. Note the clues to help you at each step.

Step	Leaf Characteristics	Tree
1a	Compound leaf (leaves divided into leaflets) . . . go to Step 2	
1b	Simple leaf (leaf not divided into leaflets) . . . go to Step 4	
2a	Leaflets all attached at a central point	Buckeye 
2b	Leaflets attached at several points . . . go to Step 3	
3a	Leaflets tapered with pointed tips	Pecan 
3b	Leaflets oval with rounded tips	Locust 
4a	Veins branched out from one central point . . . go to Step 5	
4b	Veins branched off main vein in middle of the leaf . . . go to Step 6	
5a	Heart-shaped leaf	Redbud 
5b	Star-shaped leaf	Sweet gum 
6a	Leaf with jagged edges	Birch
6b	Leaf with smooth edges	Magnolia 

Because your leaf is a simple leaf, you skip ahead to Step 4.

Continue reading the statements until you determine the identity of your leaf.

Because your leaf has jagged edges, you determine that it's from a birch tree.

Hands-On Learning

MAKE A DICHOTOMOUS KEY Give small groups of students a collection of similar objects, such as different kinds of pasta or dried legumes. Challenge groups to write a dichotomous key that can be used to identify each object in their collection. Then, have groups exchange their objects and dichotomous keys and use the keys to identify each object.

During Reading

Active Reading

SUMMARIZE Direct students to reread the section Binomial Nomenclature. Have them list in their notebooks the rules for writing the scientific names of species and give an example.

KEY QUESTION Invite a volunteer to read aloud the question on the previous page. Ask a second volunteer to read aloud the answer.

Ask What are the two parts of the scientific name for a species? *(The first part is the genus name. The second part is a descriptive name that is unique to the species.)*

Speed Bump

Ask What is the disadvantage of using common names for organisms? *(Common names can be confusing because people sometimes use different common names for the same organism.)*

Write on the board the scientific name for the polar bear, *Ursus maritimus.*

Ask To what genus does the polar bear belong? *(Ursus)* What part of the name describes the species? *(maritimus)*

Build Connections

VISUAL SUMMARY Remind students that a dichotomous key is a tool. Write the words *dichotomous* and *key* on the board. Underline the prefix *di-* and the word *key.*

Ask What does the prefix *di-* mean? *(two)* For what is a key used? *(to unlock something)*

Say A dichotomous key is a tool that uses two statements at a time to "unlock" the identity of an organism or other object.

As a class, use the key to identify the leaf at the top of the page. Read aloud the steps and the clues.

After you read step 4, ask, Does the appearance of the unknown leaf lead you to step 5 or step 6? *(step 6)*

After identifying the leaf as a birch leaf, ask, Can you use this particular key to identify any leaf you might find? *(no)* Why not? *(The key's statements describe only the collection of leaves shown.)*

Ask What would need to be done to make a key that identifies a larger collection of leaves? *(The characteristics of all the leaves in the collection would need to be evaluated so that contrasting statements that relate to all the leaves could be written for the key.)*

During Reading continued

Active Reading

IDENTIFY WORD FORMS Point out the irregular plural forms of the words *genus (genera)* and *taxon (taxa)*. Have students find sentences on this page that use the singular and plural forms of the word *genus* and ask a volunteer to read the sentences aloud.

CLARIFY WORD USAGE Point out the multiple meanings of the words *family, order, class,* and *kingdom.*

Say In biology, a *family* is a group of genera. In everyday language, a family is a group of people who are related to one another.

Have students use a dictionary to find other meanings of *order, class,* and *kingdom* and write sentences that convey their multiple meanings.

KEY QUESTION Read aloud the question and answer.

Ask What is systematics? *(Systematics is the science of naming and grouping organisms.)*

Build Vocabulary

PREFIXES Encourage students to think of other words that begin with the prefix *bi-* and give their meanings. *(Sample answers: bicycle — two wheels; biannual — two years; bidirectional — two directions)*

Say *Nomenclature* is a set or system of names or terms that describe something, such as the nomenclature of sports, music, or texting.

Ask What is the nomenclature for species? *(a two-part name consisting of a genus name and a name that describes the species)*

Carolus Linnaeus

Classifying Species Into Larger Groups In addition to naming organisms, biologists organize, or classify, species into larger groups. The science of naming and grouping organisms is called **systematics** (sis tuh MAT iks). In systematics, biologists try to place organisms into groups that have biological meaning. Organisms in a group should be more similar to one another than they are to organisms in other groups. Biologists call these groups *taxa* (singular: taxon).

You may not know it, but you use classification systems all the time. For example, you may talk about "teachers" or "pets." You also refer to smaller groups, such as "biology teachers" or "dogs." When you do this, you are using names for groups that many people understand.

Key Question What is the goal of systematics?
The goal of systematics is to organize living things into groups that have biological meaning.

BUILD Vocabulary

binomial nomenclature a classification system in which each species is assigned a two-part scientific name

genus a group of closely related species; the first part of the scientific name in binomial nomenclature

systematics the science of naming and grouping organisms

family in classification, a group of similar genera

order in classification, a group of closely related families

class in classification, a group of closely related orders

phylum in classification, a group of closely related classes

kingdom the largest and most inclusive group in the Linnaean classification system

PREFIXES
The prefix *bi-* means "two." In binomial nomenclature, the names of species have two parts.

The Linnaean Classification System

Linnaeus developed the system of binomial nomenclature. He also developed a classification system for organizing species into larger groups. Linnaeus classified species according to physical similarities and differences.

Linnaeus's original classification system had four levels, or ranks. Over time, his system grew to include seven levels: species, genus, family, order, class, phylum, and kingdom. As you read about these levels, look at the diagram of the classification of the Bactrian camel shown on the next page. The scientific name of the camel that has two humps is *Camelus bactrianus*. (Bactria was an ancient country in Asia.)

▶ ***Species*** A species is a group of individuals that can interbreed and produce fertile offspring.

▶ ***Genus*** As its name tells you, the Bactrian camel belongs to the genus *Camelus*. The genus *Camelus* also includes the dromedary camel, *Camelus dromedaries*, which has only one hump.

▶ ***Family*** Llamas belong to the genus *Lama*. Because camels and llamas share many traits, they are classified in the same family. A **family** is a group that includes genera that share similar traits. The family that includes *Camelus* and *Lama* is Camelidae.

▶ ***Order*** Closely related families are grouped together in the next larger rank—an **order.** Camels and llamas (Camelidae) are grouped with several other families, including the deer family and the cattle family. Together they form the order Artiodactyla. This order includes hoofed animals that have an even number of toes.

▶ ***Class*** Similar orders are grouped together into the next larger rank, called a **class.** The order Artiodactyla is placed with other mammals in the class Mammalia.

430 Chapter 18 • Lesson 1

Hands-On Learning

SCIENTIFIC NAMES Challenge students to create "scientific names" for pens.

1. Give small groups several index cards and pens of different colors and brands.
2. Tell students to make up scientific names for the pens that would enable others to tell which pens are related. Remind them that the names must have two parts. Have students write the names on separate cards.
3. Direct groups to exchange pens and cards.
4. Have groups match the pens they receive with the scientific names on the cards. Have groups confer to determine if their matches are correct.

Ask Which names were the most useful in identifying specific pens? *(Students will probably say those that used descriptive language.)*

Ask How could you tell by the scientific names that some pens were related? *(They had the same "genus" name.)*

▶ ***Phylum*** Classes are grouped into a **phylum.** The class Mammalia is grouped with birds, reptiles, amphibians, and fishes into the phylum Chordata. All of the members of this phylum have a nerve cord along their back.

▶ ***Kingdom*** The largest of Linnaeus's taxonomic groups is called the **kingdom.** The Bactrian camel is placed in the kingdom Animalia. Kingdom Animalia includes all animals.

Key Question How did Linnaeus group species into larger taxa? **Over time, Linnaeus's original classification system grew to include seven levels: species, genus, family, order, class, phylum, and kingdom.**

From Species to Kingdom The diagram shows how a Bactrian camel, *Camelus bactrianus*, is grouped within each level of classification. Only a few organisms are shown for the levels that are larger than genus.

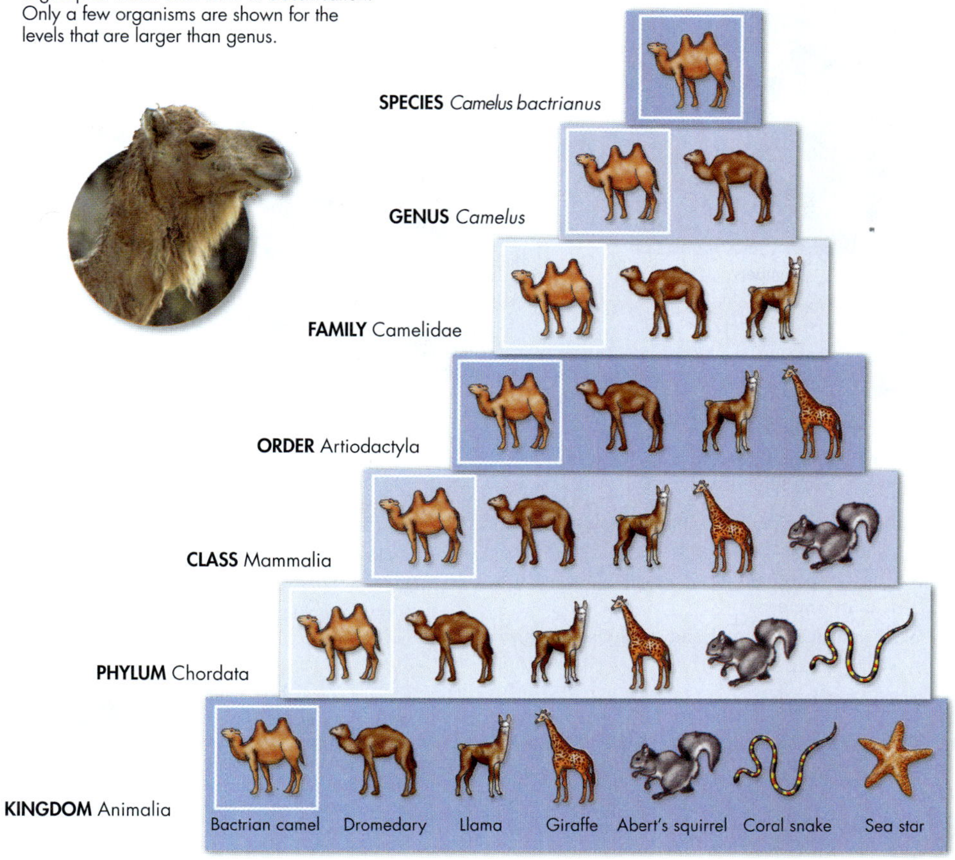

LESSON 18.1

Active Reading

INTERPRET DIAGRAMS Ask a volunteer to read aloud the names of the seven taxonomic groups in the From Species to Kingdom diagram, beginning with species. Then, have students compare the relative sizes of the bands in the diagram, which represent the different taxa.

Ask Which classification group is the largest? *(kingdom)* The smallest? *(species)*

Explain that animals in the smaller groups are more closely related to each other than animals in the larger groups are. Have students look at the animals shown in each taxon.

Ask Are camels more closely related to llamas or to giraffes? *(llamas)* How do you know? *(Camels, llamas, and giraffes all belong to the same order, but camels and llamas also belong to the same family, which is a smaller group.)*

KEY QUESTION Read aloud the question and answer. Ask volunteers to identify the name of each taxonomic group to which the Bactrian camel belongs, such as the genus *Camelus.* Help students pronounce difficult names, such as *Camelus bactrianus* (kuh-MEE-luhs bak-tree-AN-uhs), *Camelidae* (kuh-MEHL-uh-dee), and *Artiodactyla* (ahr-tee-oh-DAK-tuh-luh).

Speed Bump

Ask What is the order of Linnaeus's classification system from largest to smallest? *(kingdom, phylum, class, order, family, genus, species)*

Ask Similar orders are grouped together to form which taxonomic group? *(class)*

ELL Focus on ELL: Build Background

ALL SPEAKERS Discuss supermarkets with students as an example of a familiar hierarchical classification system. Draw on the board a diagram showing how the products in a supermarket are arranged in hierarchical order. For example, the diagram might show that the entire supermarket is divided into the produce department and other departments, that fruits are a subdivision within the produce department, that melons are a subdivision within fruits, and that cantaloupes are a subdivision within melons. Write the word *supermarket* at the bottom of the diagram, *produce department* above that, *fruits* above *produce department,* and so on. Have students compare the classification diagram on the board with the classification diagram on this page.

Ask Which classification group on the board is the second largest? *(produce department)* Which taxonomic group is the second largest? *(phylum)*

During Reading continued

Active Reading

USE CONTEXT CLUES Ensure students understand the meaning of *traditional classification.*

Say A tradition is an idea or custom that is handed down from person to person over time.

Ask What are some examples of traditions? *(Sample answers: Fourth of July fireworks, birthday cakes, graduation ceremonies)*

Ask Why might Linnaeus's system be considered traditional? *(It was developed a few hundred years ago and has been used since.)*

USE VISUALS Use the photographs on this page to highlight the problems with traditional classification. Read the caption aloud.

Ask Why might someone classify barnacles and limpets in the same group but barnacles and crabs in different groups? *(Barnacles and limpets look very similar; barnacles and crabs do not.)*

Ask Why should barnacles and crabs be classified in the same group? *(They are closely related.)*

Speed Bump

Ask On what did Linnaeus base his system of classification? *(similarities and differences of organisms that could be seen)*

Ask On what do modern scientists base their classification of organisms? *(evolutionary relationships of organisms)*

Post Reading

ASSESSMENT TIPS

Advise students to evaluate their answer to Question 6 before answering Question 7.

MYSTERY CLUE Have students consider the definition of *species* in light of the fact that brown bears and polar bears interbreed only under certain conditions.

Wrap-Up Activity

USING MNEMONICS Explain that a mnemonic technique or device helps a person remember something, such as a sequence of terms. For example, a mnemonic device to help remember the sequence of taxa, from largest to smallest, is: **K**ids **P**refer **C**heese **O**ver **F**resh **G**reen **S**pinach. Ask small groups to create their own mnemonics to help them remember the sequence of taxa. Have each group present its best mnemonic to the class.

Looks Can Be Deceiving Classifying organisms based on traits that are easy to see can lead to problems. Barnacles and limpets look very similar, yet they are not closely related. Barnacles and crabs do not look alike, yet they *are* closely related.

Barnacles

Limpets

Problems With Traditional Classification Individual species tend to define themselves by choosing with whom to mate. Thus, there is a "natural" definition of a species. But higher ranks, such as genus and family, are defined by biologists.

Linnaeus grouped organisms into larger taxa based on similarities and differences that could be seen. Linneaus chose his characteristics carefully. For this reason, many of his groups are still used in modern classification. But it is not always obvious which traits should be used to define a group. For example, would you call a dolphin a fish, because it looks like a fish? Or would you call it a mammal, because it is warmblooded and produces milk? Scientists have changed the way they name and group organisms to solve such problems.

When modern scientists classify organisms, they are interested in more than similarities and differences. They apply Darwin's ideas, and look for evolutionary relationships. Scientists today try to place species into groups that show how closely members are related to each other.

Crab

CHECK Understanding

Apply Vocabulary

Use the highlighted words from the lesson to complete each sentence correctly.

1. The largest group in Linnaeus's seven-level classification system is the ______.
2. The two-part naming system that gives each species a unique name is ______.
3. The science of naming and classifying organisms is called ______.

Critical Thinking

4. **Explain** Why do biologists classify organisms?
5. **Apply Concepts** The scientific name of a red maple is *Acer rubrum.* Which part of this name identifies the genus? Which part describes the species?
6. **Sequence** List the seven ranks in the Linnaean classification system, beginning with the smallest.
7. **Compare and Contrast** In which level of classification are the members more closely related—a kingdom or an order?
8. **Write to Learn** Answer the question in the mystery clue below. Use the word *species* in your answer.

MYSTERY CLUE

Polar bears and brown bears can interbreed and produce fertile offspring in zoos. In nature this rarely happens. What do you think this means about the relationship between them? (Hint: See p. 430.)

BIOLOGY.com Search Lesson 18.1 GO • Lesson Assessment

Check Understanding

1. kingdom
2. binomial nomenclature
3. systematics
4. to give each organism a unique and agreed-upon name; to organize living things into groups that have biological meaning or show their degree of relatedness
5. The genus is *Acer.* The species-descriptive name is *rubrum.*
6. species, genus, family, order, class, phylum, kingdom
7. Members of an order are more closely related than members of a kingdom.
8. The two bears are very closely related. They may belong to the same species, but behavior and geography probably prevent them from mating in nature.

18.2 Modern Evolutionary Classification

Evolutionary Classification

Linnaeus began classifying organisms more than one hundred years before Darwin published his ideas about a "tree of life." Darwin's ideas suggested a new way to classify organisms. Biologists today use evolutionary relationships to place species into larger groups, or taxons.

Common Ancestors Evolutionary classification groups species into larger taxa, in which all members are descended from a common ancestor. All members of any group should be more closely related to each other than to members of any other group. The larger the taxon, the farther back in time all of its members shared a common ancestor. The study of how living and extinct organisms are related is called phylogeny (fy LAHJ uh nee).

Clades Evolutionary classification puts organisms into groups called clades. A **clade** is a group of species that includes a single common ancestor and all of its descendants. A clade includes both living and extinct organisms. Clades must be monophyletic (mahn oh fy LET ik). A **monophyletic group** includes a single common ancestor and *all* of its descendants. Monophyletic groups cannot include any organisms that are *not* descended from the common ancestor.

Some taxa that were defined before evolutionary classification are monophyletic. But others are not. These groups have a common ancestor but do not include all of its descendants. Class Reptilia (reptiles) is not monophyletic. So in evolutionary classification, reptiles are not a true group.

Key Question What is the goal of evolutionary classification? **The goal of evolutionary classification is to group species into larger categories that show lines of evolutionary descent, rather than overall similarities and differences.**

Cladograms

In evolutionary classification, biologists compare carefully selected traits. They use these traits to determine the order in which groups of organisms branched off from a common ancestor. This information is used to link clades together in a cladogram. A **cladogram** is a drawing that shows relationships among species and larger taxa by showing how evolutionary lines branched off a common ancestor.

Key Questions

What is the goal of evolutionary classification?

What is a cladogram?

How are DNA sequences used in classification?

BUILD Understanding

T-Chart As you read a lesson, you can organize the information in a T-chart. Write all the key questions on the left side of the chart. Answer each question on the right side of the chart.

In Your Workbook Go to your workbook to learn more about T-charts.

BUILD Vocabulary

clade an evolutionary branch of a cladogram that includes a single ancestor and all its descendants

monophyletic group a group that consists of a single ancestral species and all its descendants and excludes any organisms that are not descended from that common ancestor

cladogram a diagram showing patterns of shared characteristics among species

derived character a trait that arose in the most recent common ancestor of an evolutionary line and was passed on to its descendants

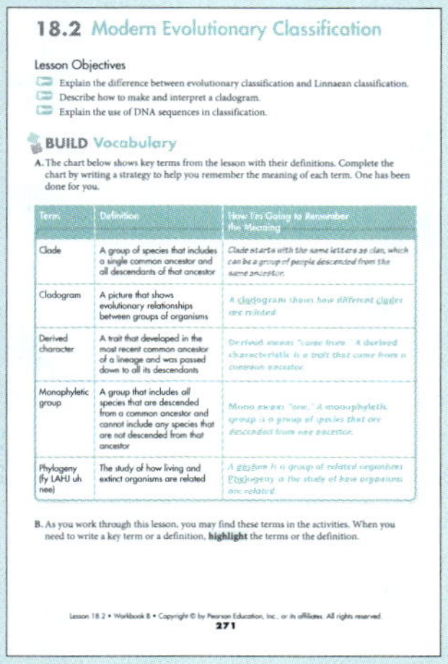

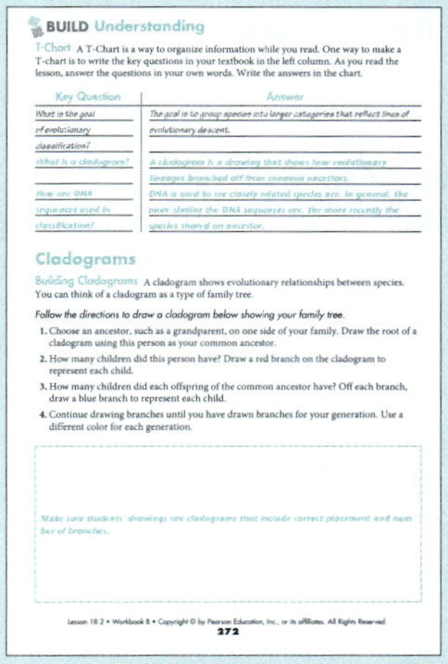

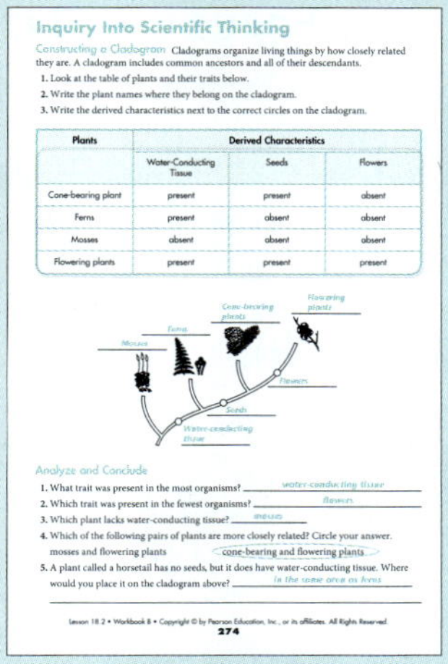

LESSON 18.2

Pre-Reading

Objectives

18.2.1 Explain the difference between evolutionary classification and Linnean classification.

18.2.2 Describe how to make and interpret a cladogram.

18.2.3 Explain the use of DNA sequences in classification.

Preview the Pages

1. Read aloud the lesson title. Ask students to hypothesize how the methods of classification presented in this lesson are different from those in the previous lesson.
2. Ask volunteers to read aloud the definitions of the vocabulary terms and point out where they can be found in the text.
3. Have students look at the visuals and read the captions and labels. Be sure students pay close attention to the cladograms, as they will encounter more cladograms in the next lesson and subsequent chapters. Point out to students that the Inquiry Into Scientific Thinking activity involves making a cladogram.

Build Understanding

T-CHART Remind students that making a T-chart starts with drawing a large *T* on a sheet of paper. Direct students to Workbook B to see a T-chart that has been started for them.

NATIONAL SCIENCE EDUCATION STANDARDS

UNIFYING CONCEPTS AND PROCESSES

I, II, IV

CONTENT

C.3.a, C.3.b, C.3.d, C.3.e

INQUIRY

A.2.e

For the corresponding lesson in *Miller & Levine Biology,* see pages 516–522.

During Reading

Active Reading

KEY QUESTION Invite a volunteer to read aloud the question and answer on the previous page.

Ask How is evolutionary classification different from Linnaean classification? *(Evolutionary classification focuses on how closely organisms are related to each other. Linnaean classification focuses on similarities and differences among organisms.)*

INTERPRET DIAGRAMS Make sure students understand what the colors and lines of the cladograms represent.

Ask What do the blue color and pink color represent? *(two different clades)*

Ask What color do you make when you blend blue and pink? *(purple)* What does the purple color represent? *(the common ancestors of the descendants belonging to the blue clade and pink clade)*

Point out the numbered lines in the second cladogram.

Ask Which numbered line is least closely related to the others? *(line 1)* How do you know? *(Line 1 branched off further back in time.)*

Direct students' attention to the third cladogram.

Ask Are birds more closely related to turtles or lizards? *(lizards)*

KEY QUESTION Read aloud the question and answer on this page. Refer students to the completed cladogram in the visual Building a Cladogram.

Ask Did amphibians or ray-finned fishes branch off first from the evolutionary line leading to mammals? *(ray-finned fishes)*

Speed Bump

Ask What is the relationship between a clade and a cladogram? *(A clade is one of the branches of a cladogram that includes a single ancestor and all its descendants.)*

Refer students to the completed cladogram on this page.

Ask Are lizards more closely related to crocodiles or to snakes? *(snakes)*

Building a Cladogram A cladogram shows the relationships among different evolutionary lines.

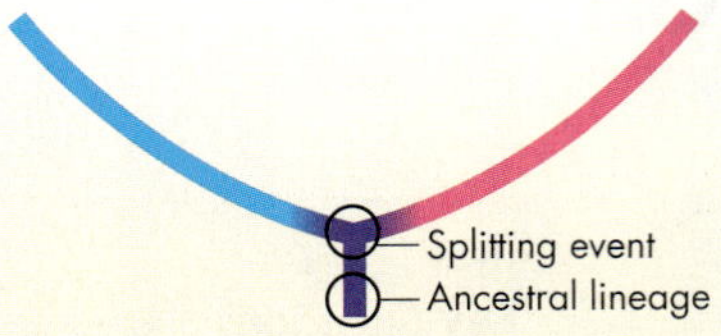

1 Cladograms are diagrams showing how evolutionary lines, or lineages, split from each other over time. This diagram shows a single ancestral lineage splitting into two. The point of splitting is called a "node" in the cladogram.

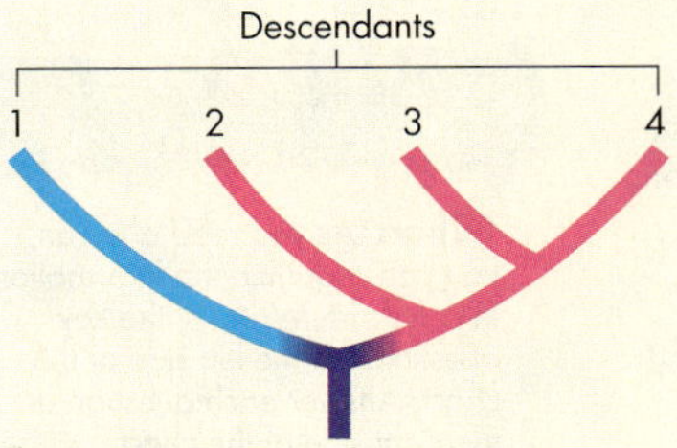

2 How recently lineages share a common ancestor reflects how closely the lineages are related to one another. Here, lineages 3 and 4 are each more closely related to each other than either of them is to any other lineage.

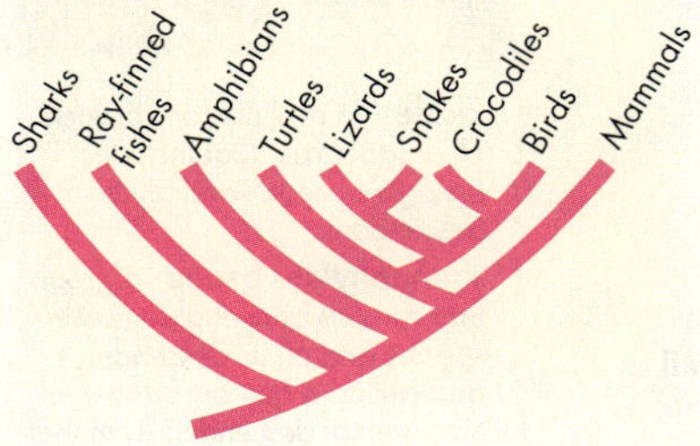

3 This cladogram shows the evolutionary relationships among vertebrates, animals with backbones.

Building Cladograms Look at the cladograms on the left. Each cladogram looks like a tree with branches. The bottom, or "root," shows the common ancestor of all the organisms in the cladogram. The point where each branch splits off is called a *node*. Each node represents a speciation event. In speciation, one species splits into two new species. A node shows the last point at which the two new lines shared a common ancestor.

In a cladogram, the pattern of the branches shows how closely organisms are related. Look at cladogram 2. Notice that lines 3 and 4 share a common ancestor more recently than either does with line 2. Therefore, lines 3 and 4 are more closely related to each other than either is to line 2.

Key Question What is a cladogram?
A cladogram links groups of organisms by showing how evolutionary lines branched off from common ancestors.

Derived Characters Biologists focus on particular characteristics, or traits, to put organisms into clades. A **derived character** is a trait that arose in the most recent common ancestor of a group and was passed on to its descendants. For example, having four limbs is a derived character in clade Tetrapoda. The common ancestor of all tetrapods had four limbs. Hair is a derived character for the clade Mammalia. All mammals have hair. But having four limbs is not a derived character for mammals. If it were, only mammals would have that trait.

Lost Traits Sometimes descendents lose a derived character. For example, snakes are reptiles, members of the clade Tetrapoda. But snakes do not have four limbs. During the process of evolution, snakes lost their legs.

Biologists must be careful about using the *absence* of a trait in classification. Distantly related groups can sometimes lose the same trait. For example, both whales and snakes have lost their legs. But whales and snakes are not very closely related. Snakes are more closely related to other reptiles than to whales.

Interpreting Cladograms The cladogram on the next page shows the evolutionary line of the cat family. At the lower left is the common ancestor of all animals with four legs. The position of the nodes shows the order in which different groups branched off. First, amphibians branched off from the line leading to clade Amniota. (Amniota includes reptiles and mammals.) Next, reptiles branched off from the line leading to the clade Mammalia, and so on. Notice that smaller clades fit within larger clades. For example, clade Amniota is part of clade Tetrapoda.

ELL Focus on ELL: Extend Language

INTERMEDIATE AND ADVANCED SPEAKERS Ask volunteers to point out words on this page that are often used to describe plants. *(tree, root, node, branch)* Tell students that "family trees" are diagrams used to show the relationships among people who belong to the same family. Have students write a few sentences that describe a large tree, such as an oak. Then, have them use their sentences as a base to write a few sentences that describe a cladogram. Discuss with students how writing about the tree helped them write about a cladogram.

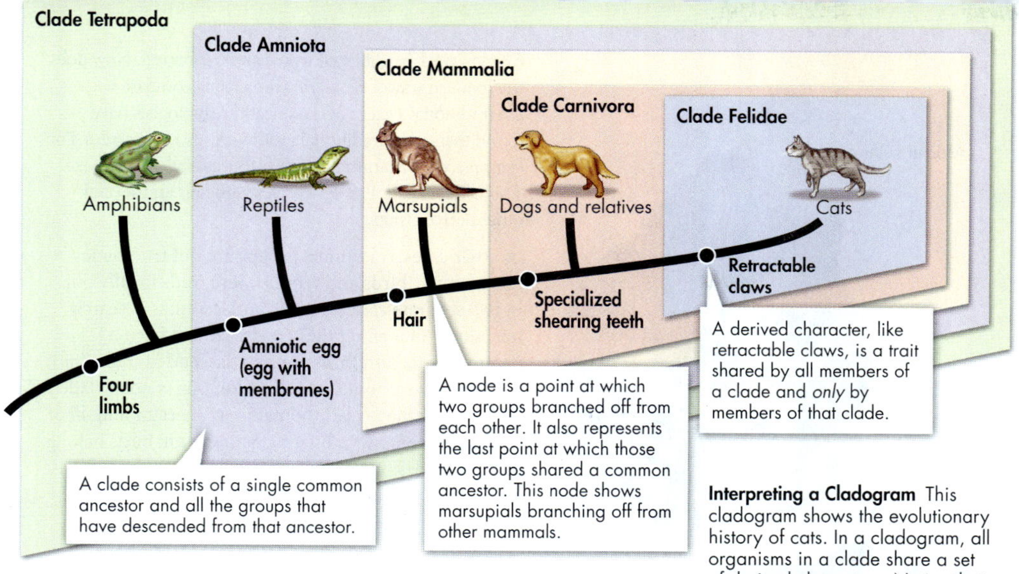

Interpreting a Cladogram This cladogram shows the evolutionary history of cats. In a cladogram, all organisms in a clade share a set of derived characters. Notice that smaller clades are nested within larger clades.

Derived characters are listed along the main trunk. Each trait is a derived character for all the organisms to the right. For example, marsupials, dogs, and cats all have hair. On the cladogram, they are to the right of the common ancestor that had hair. The last derived character on the cladogram is retractable claws. Members of the cat family can retract, or pull back, their claws. Members of the dog family do not have this trait.

INQUIRY into Scientific Thinking

Constructing a Cladogram

❶ Find the list of organisms in the table at right. Then look at the list of derived characters. The more characters that two organisms share, the more closely related they are.

❷ Find the organism in the table that is *least* closely related to the others. That organism will be the first organism on your cladogram.

❸ Use the information in the table to complete your cladogram. Put each organism on a separate branch.

❹ Write the derived characters along the main trunk.

In Your Workbook Go to your workbook for more help with cladograms.

Derived Characters in Organisms

Organism	Derived Character		
	Backbone	Legs	Hair
Earthworm	Absent	Absent	Absent
Trout	Present	Absent	Absent
Lizard	Present	Present	Absent
Human	Present	Present	Present

Analyze and Conclude

1. **Interpret Tables** What trait separates the earthworm from the other animals?

2. **Classify** Do you have enough information to decide where to put a frog on your cladogram? Explain your answer.

3. **Draw Conclusions** Does your cladogram indicate that lizards and humans share a more recent common ancestor than either does with an earthworm? Explain your answer.

Active Reading

USE DIAGRAMS Help students identify the information provided in the cladogram on this page.

Ask What is the oldest derived character in this cladogram? *(four limbs)*

Ask Which of the animals shown in the cladogram produce eggs with membranes? *(reptiles, marsupials, dogs, and cats)*

Ask To what clade or clades do dogs belong? *(clade Tetrapoda, clade Amniota, clade Mammalia, and clade Carnivora)*

Inquiry Into Scientific Thinking

INTERPRET TABLES Be sure students understand the information presented in the table.

Ask What two types of information are listed in the table? *(different kinds of organisms and derived characters)*

Ask What are the organisms? *(earthworm, trout, lizard, and human)*

Ask What are the derived characters? *(backbone, legs, and hair)*

Ask Which organisms listed in the table have legs? *(lizard and human)* Which organisms do not have a backbone? *(earthworm)*

Lead a Discussion

Help students use the information in the table to make a cladogram that shows the relationships among the organisms listed. A model of the cladogram is shown below.

Ask What descendants should the cladogram include? *(earthworm, trout, lizard, and human)*

Ask How many branches should the cladogram have? *(four)*

Ask Where should hair be labeled on the cladogram? *(between the branches leading to lizard and human)*

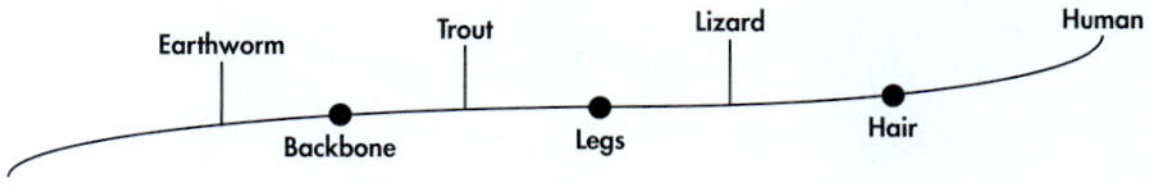

Analyze and Conclude

1. backbone

2. No, a frog has the same set of derived characters as the lizard—backbone and legs, but no hair. You cannot tell whether to put it before or after the lizard. You would need another derived character to tell where to insert the frog.

3. yes, because lizards and humans split at the last node on the cladogram

During Reading continued

Active Reading

DRAW CONCLUSIONS Have students reread the section Clades and Traditional Taxonomic Groups. Then, revisit the third cladogram in the Building a Cladogram visual.

Ask In which clade on the cladogram would you expect dinosaurs to be included? *(the clade that includes turtles, lizards, snakes, crocodiles, and birds)*

SCIENCE SUPPORT Tell students that researchers noticed a peculiar behavior in American vultures: When they become overheated, they urinate on their legs, relying on evaporation to cool them down. African vultures don't do this. The only other bird group this behavior has been observed in is storks.

Say Genes influence animal behavior.

Ask What does the absence of urine-cooling behavior in African vultures indicate? *(African vultures may not have the genes that control this behavior.)*

Speed Bump

Ask How does the DNA of organisms that share a recent common ancestor compare with the DNA of organisms that share a common ancestor farther back in time? *(The DNA of organisms that share a recent common ancestor is more similar.)*

African vulture

American vulture

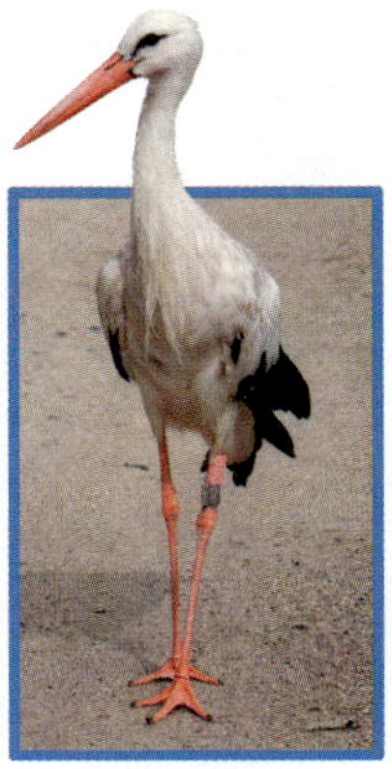

Stork

DNA and Classification Scientists today compare DNA to help classify organisms. African vultures and American vultures were once grouped together in the falcon family. DNA analysis has shown that American vultures are more closely related to storks than to African vultures.

Clades and Traditional Taxonomic Groups How does the modern way of classifying organisms compare with the traditional system of classification? Biologists have found that many traditional groups are also true clades. For example, class Mammalia is the same as clade Mammalia. This clade includes all vertebrates with hair and several other derived traits.

In other cases, traditional groups are not true clades. For example, birds and reptiles were traditionally put in two separate classes—Aves and Reptilia. Research has shown that all reptiles are descended from a common ancestor. Birds are also descended from this ancestor. So without birds, class Reptilia is not a clade. (Remember that a clade contains an ancestor and *all* of its descendants.) In modern classification, birds belong to three clades: Aves (birds themselves), Dinosauria, and Reptilia. Is it correct to call birds reptiles? An evolutionary biologist would say yes!

DNA in Classification

Most of the examples of evolutionary classification you have read about are based on physical traits such as fur, bones, and teeth. The goal of modern systematics is to understand the relationships of *all* life on Earth. If two organisms do not share any obvious features, how can you tell if they are related?

Genes as Derived Characters Remember that all species inherit genes on strands of DNA. Very different organisms often share a number of homologous genes. The similarities and differences in these homologous genes can tell how closely related organisms are. Because genes mutate at a steady rate, they can be used as molecular clocks. The more similar the genes in two organisms are, the more recently they shared a common ancestor. In these comparisons, genes and mutations are used as derived characters.

New Techniques Redraw Old Trees Information from DNA is helping to make evolutionary trees more accurate. For example, look at the three birds on the left. The African vulture looks like the American vulture in the middle. Both vultures were once classified in the same clade. But by studying their DNA, researchers have learned that American vultures are more closely related to storks than to African vultures.

Focus on Academic Language

• descendant • descended • ancestor

As students read silently or aloud, they may not see or hear the difference in the words *descendant* and *descended*. Write these two words on the board and underline the root word *descend* in each. Then, have students look up both words in a dictionary to determine that *descendant* is a noun and *descended* is a verb. Ask students to write sentences in which they use both words correctly with the word *ancestor*.

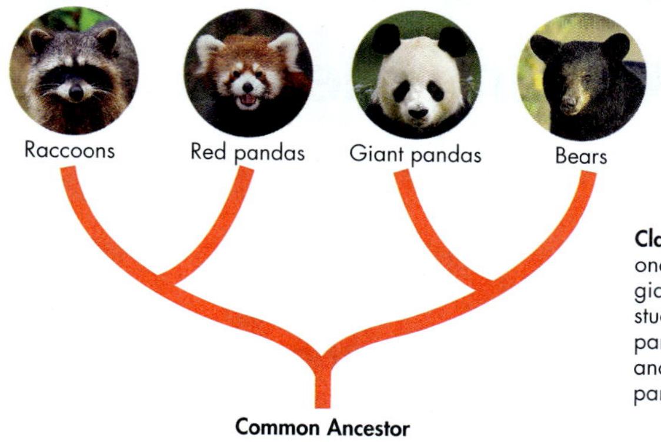

Classification of Pandas Biologists once classified the red panda and giant panda together. Recently, DNA studies have shown that the giant panda shares a more recent common ancestor with bears than with red pandas or raccoons.

Evidence from DNA has also helped biologists classify pandas. Giant pandas and red pandas share many traits. For example, both pandas have bones in the wrist that work like a human thumb. But studies of DNA show that giant pandas share more recent ancestors with bears than with raccoons. So giant pandas are now classified with bears in the family Ursidae. Red pandas are placed in a clade that also includes raccoons, seals, and weasels.

Key Question How are DNA sequences used in classification? **In general, the more derived genetic characters that two species share, the more recently they shared a common ancestor. This information also tells how closely they are related.**

CHECK Understanding

Apply Vocabulary

Use the highlighted words from the lesson to complete each sentence correctly.

1. _______ are groups of species that include a single common ancestor and all descendants of that ancestor.

2. A _______ is a diagram that shows the evolutionary relationships among clades.

3. A trait that evolved in a recent common ancestor and was passed on to the descendants of that ancestor is called a _______.

Critical Thinking

4. Explain How does evolutionary classification differ from traditional classification?

5. Apply Concepts What do the locations of derived characters on a cladogram show?

6. Relate Cause and Effect Explain how DNA analysis changed the classification of giant pandas.

7. Write to Learn Answer the question in the mystery clue below.

MYSTERY CLUE

When DNA is compared, some brown bears are more closely related to polar bears than they are to other brown bears. What do you think this means for the classification of polar bears? (**Hint:** See p. 436.)

BIOLOGY.com Search Lesson 18.2 GO • Lesson Assessment

Check Understanding

1. Clades

2. cladogram

3. derived character

4. Traditional classification used physical traits and similarities to group organisms. In evolutionary classification, organisms are grouped based on their evolutionary relationships.

5. Derived characters on a cladogram show the point at which an important trait evolved in an ancestor. Organisms to the right of the derived character are descended from the common ancestor and have inherited the important trait.

6. When DNA was compared, researchers found that giant pandas are more closely related to bears than to raccoons and red pandas, and the animals were regrouped.

7. Polar bears and brown bears may be members of the same species that have become isolated from each other. Over time, isolated populations may develop different characteristics as a result of natural selection or genetic drift.

Active Reading

KEY QUESTION Invite a volunteer to read aloud the question and answer.

Ask Why do two species that have a recent common ancestor have more similar DNA than two species that have a common ancestor further back in time? *(Fewer mutations occurred in their DNA because less time passed.)*

Speed Bump

Refer students to the cladogram on this page.

Ask Is the DNA of red pandas more similar to the DNA of bears or to the DNA of raccoons? *(raccoons)* Why? *(Red pandas and raccoons have a more recent common ancestor.)*

Post Reading

ASSESSMENT TIPS Tell students that in Question 4, the term *traditional classification* refers to the Linnaean classification system.

To help students answer Question 5, have them look at the Interpreting a Cladogram visual.

MYSTERY CLUE Refer students back to the discussion of genes as derived characters. Guide them to suggest that because some brown bears share a more recent common ancestor with polar bears than with other brown bears, brown bears and polar bears could be classified as one species.

Wrap-Up Activity

ANALYZE A CLADOGRAM Have pairs use the Interpreting a Cladogram visual in this lesson to make a table similar to the Derived Characters in Organisms table in the Inquiry Into Scientific Thinking activity. (For the Organism column, students should list *amphibians, reptiles, marsupials, dogs,* and *cats.* For the Derived Character columns, students should list *four limbs, amniotic egg, hair, shearing teeth,* and *retractable claws.*) Tell students to indicate within the table the presence or absence of each derived character in each organism.

LESSON 18.3

Pre-Reading

Objectives

18.3.1 Name the six kingdoms of life as they are currently identified.

18.3.2 Explain what the tree of life represents.

Preview the Pages

1. Point out the Classification of Living Things table. Read aloud the caption. Then, read the title of the table and the headings for the columns and rows.
2. Preview the visuals. Tell students the photograph on the next page shows a magnified view of bacteria.
3. Draw attention to the Tree of Life cladogram. Explain that it is a larger, more complex cladogram than those in the previous lesson. Point out that the color-coded key identifies the kingdoms of organisms included in the cladogram.

Build Understanding

CONCEPT MAP Begin the concept map as a class. Ask a volunteer to suggest what the major heading might be. Then, ask how many branches should be drawn from the main heading and what their headings would be. Students might also use the partially completed concept map in Workbook B.

NATIONAL SCIENCE EDUCATION STANDARDS

UNIFYING CONCEPTS AND PROCESSES

I, II, IV

CONTENT

C.3.a, C.3.b, C.3.d, C.3.e

INQUIRY

A.2.e, A.2.f

For the corresponding lesson in *Miller & Levine Biology,* see pages 523–528.

18.3 Building the Tree of Life

Key Questions

What are the six kingdoms of life as they are now identified?

What does the tree of life show?

BUILD Understanding

Concept Map As you read, construct a concept map describing the characteristics of the three domains.

In Your Workbook Go to your workbook to learn about making a concept map.

BUILD Vocabulary

domain a larger, more inclusive taxonomic category than a kingdom

Bacteria the domain of unicellular prokaryotes that have cell walls containing peptidoglycan

Archaea the domain of unicellular prokaryotes that have cell walls that do not contain peptidoglycan

WORD ORIGINS

The domain Archaea and kingdom Archaebacteria come from the Greek word *arkhaios*, which means "ancient or primitive." Archaea live in harsh environments that may resemble early, or primitive, conditions on Earth.

Changing Ideas About Kingdoms

Since the time of Linnaeus, biologists have worked to improve the system of classification. Classification today is based on the most recent discoveries and reflects evolutionary relationships. As scientists learn more and more about living and extinct organisms, the classification system will continue to change.

Linnaeus's original system had only two kingdoms: plants and animals. Animals were organisms that moved and used food for energy. Plants included everything that was not an animal. Over the years, classification at the level of the kingdom has changed greatly.

Five Kingdoms In the late 1800s, researchers discovered that single-celled organisms were very different from plants and animals. A third kingdom, called Protista, was created to contain all microorganisms. Later, biologists gave yeasts, molds, and mushrooms their own kingdom: Fungi. When biologists realized how different bacteria were from other organisms, they added a fifth kingdom: Monera. (Bacteria are prokaryotes, or organisms that lack a nucleus and other organelles.) The research and discoveries that expanded the classification system from two to five kingdoms took more than 200 years.

Six Kingdoms More recently, biologists studying bacteria reached a surprising conclusion. Kingdom Monera contained two very different kinds of bacteria. So the bacteria were divided into two new kingdoms: Eubacteria and Archaebacteria. These two new kingdoms give us the six kingdoms used in modern classification.

Key Question What are the six kingdoms of life as they are now identified?

The six-kingdom classification system includes the kingdoms Eubacteria, Archaebacteria, Protista, Fungi, Plantae, and Animalia.

Three Domains Genetic studies show that the two groups of bacteria are very different from each other. The bacteria are also very different from all the other organisms, or eukaryotes. To reflect these differences, biologists have added a new level of classification—the domain. A **domain** is larger and includes more than a kingdom.

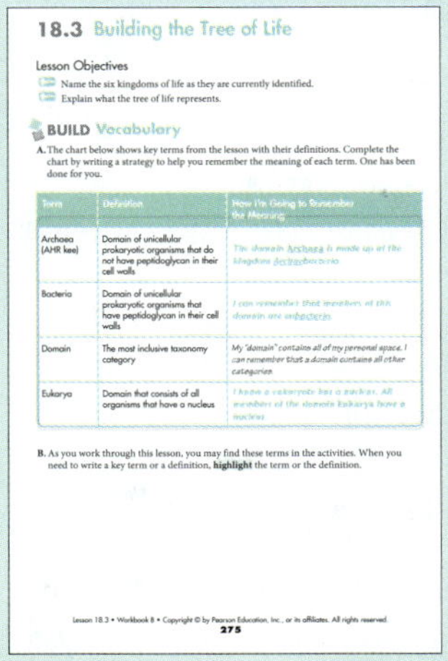

18.3 Building the Tree of Life

Lesson Objectives

Name the six kingdoms of life as they are currently identified.

Explain what the tree of life represents.

BUILD Vocabulary

A. The chart below shows key terms from the lesson with their definitions. Complete the chart by writing a strategy to help you remember the meaning of each term. One has been done for you.

Term	Definition	How I'm Going to Remember the Meaning
Archaea (AHR kee)	Domain of unicellular prokaryotic organisms that do not have peptidoglycan in their cell walls	[illegible]
Bacteria	Domain of unicellular prokaryotic organisms that have peptidoglycan in their cell walls	[illegible]
Domain	The most inclusive taxonomy category	My "domain" contains all of my personal space. I can remember that a domain contains all other categories
Eukarya	Domain that consists of all organisms that have a nucleus	I know a eukaryote has a nucleus. All members of the domain Eukarya have a nucleus

B. As you work through this lesson, you may find these terms in the activities. When you need to write a key term or a definition, **highlight** the term or the definition.

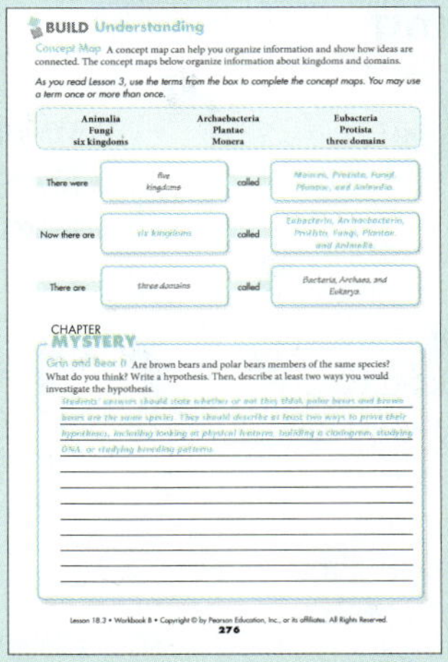

BUILD Understanding

Concept Map A concept map can help you organize information and show how ideas are connected. The concept maps below organize information about kingdoms and domains.

As you read Lesson 3, use the terms from the box to complete the concept maps. You may use a term once or more than once.

Animalia	Archaebacteria	Eubacteria
Fungi	Plantae	Protista
six kingdoms	Monera	three domains

There were | five kingdoms | called | Monera, Protista, Fungi, Plantae, and Animalia

Now there are | six kingdoms | called | Eubacteria, Archaebacteria, Protista, Fungi, Plantae, and Animalia

There are | three domains | called | Bacteria, Archaea, and Eukarya

CHAPTER MYSTERY

Grin and Bear It Are brown bears and polar bears members of the same species? What do you think? Write a hypothesis. Then, describe at least two ways you would investigate the hypothesis.

Students' answers should state whether or not they think polar bears and brown bears are the same species. They should describe at least two ways to prove their hypotheses, including looking at physical features, building a cladogram, studying DNA, or studying breeding patterns.

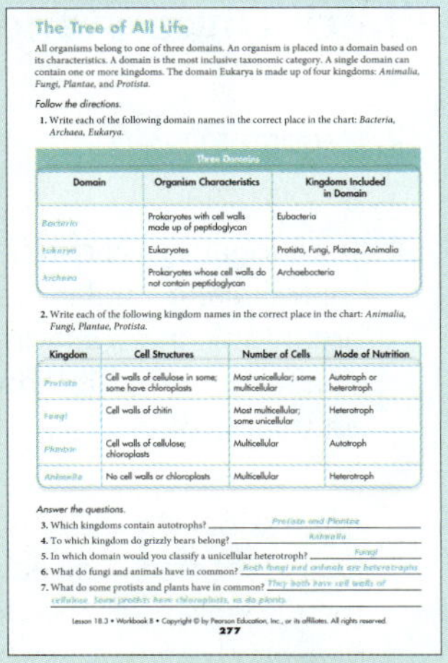

The Tree of All Life

All organisms belong to one of three domains. An organism is placed into a domain based on its characteristics. A domain is the most inclusive taxonomic category. A single domain can contain one or more kingdoms. The domain Eukarya is made up of four kingdoms: *Animalia, Fungi, Plantae,* and *Protista.*

Follow the directions.

1. Write each of the following domain names in the correct place in the chart: *Bacteria, Archaea, Eukarya.*

Three Domains		
Domain	Organism Characteristics	Kingdoms Included in Domain
Bacteria	Prokaryotes with cell walls made up of peptidoglycan	Eubacteria
Eukarya	Eukaryotes	Protista, Fungi, Plantae, Animalia
Archaea	Prokaryotes whose cell walls do not contain peptidoglycan	Archaebacteria

2. Write each of the following kingdom names in the correct place in the chart: *Animalia, Fungi, Plantae, Protista.*

Kingdom	Cell Structures	Number of Cells	Mode of Nutrition
Protista	Cell walls of cellulose in some; some have chloroplasts	Most unicellular; some multicellular	Autotroph or heterotroph
Fungi	Cell walls of chitin	Most multicellular; some unicellular	Heterotroph
Plantae	Cell walls of cellulose; chloroplasts	Multicellular	Autotroph
Animalia	No cell walls or chloroplasts	Multicellular	Heterotroph

Answer the questions.

3. Which kingdoms contain autotrophs? Protista and Plantae
4. To which kingdom do grizzly bears belong? Animalia
5. In which domain would you classify a unicellular heterotroph? Fungi
6. What do fungi and animals have in common? Both fungi and animals are heterotrophs
7. What do some protists and plants have in common? They both have cell walls of cellulose. Some protists have chloroplasts, as do plants.

Classification of Living Things

DOMAIN	Bacteria	Archaea	Eukarya			
KINGDOM	Eubacteria	Archaebacteria	"Protista"	Fungi	Plantae	Animalia
CELL TYPE	Prokaryote	Prokaryote	Eukaryote	Eukaryote	Eukaryote	Eukaryote
CELL STRUCTURES	Cell walls with peptidoglycan	Cell walls without peptidoglycan	Cell walls of cellulose in some; some have chloroplasts	Cell walls of chitin	Cell walls of cellulose; chloroplasts	No cell walls or chloroplasts
NUMBER OF CELLS	Unicellular	Unicellular	Most unicellular; some colonial; some multicellular	Most multicellular; some unicellular	Most multicellular: some green algae unicellular	Multicellular
MODE OF NUTRITION	Autotroph or heterotroph	Autotroph or heterotroph	Autotroph or heterotroph	Heterotroph	Autotroph	Heterotroph
EXAMPLES	*Streptococcus, Escherichia coli*	Methanogens, halophiles	*Amoeba, Paramecium,* slime molds, giant kelp	Mushrooms, yeasts	Mosses, ferns, flowering plants	Sponges, worms, insects, fishes, mammals

Three Domains Today organisms are grouped into three domains and six kingdoms. This table shows the key characteristics used to classify organisms into these major taxonomic groups.

There are three domains. Domain **Bacteria** contains the organisms in kingdom Eubacteria. Domain **Archaea** contains the organisms in kingdom Archaebacteria. Domain Eukarya contains the kingdoms Fungi, Plantae, and Animalia, and "Protista." Why are there quotation marks around the word "Protista"? Scientists have found that the members of the old kingdom Protista do not form a monophyletic group. The many kinds of unicellular eukaryotes do not have a single common ancestor. Therefore, "Protista" is not a true clade.

SEM 13,000×

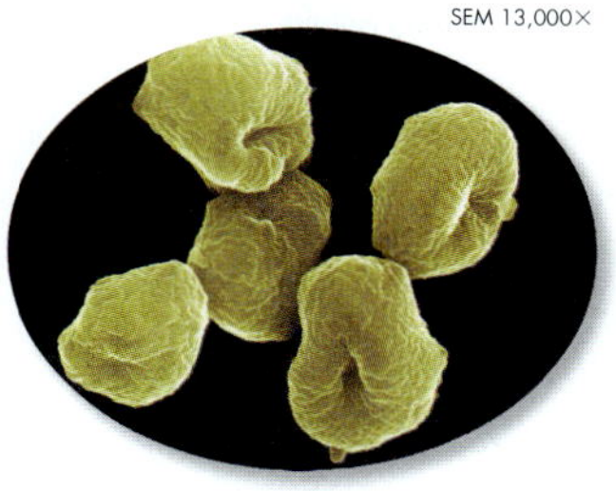

Sulfolobus This member of the domain Archaea lives in hot springs. It thrives in acidic environments that have large amounts of sulfur.

The Tree of All Life

Evolutionary classification is a rapidly changing science. It has a difficult goal—to show all life on a single evolutionary tree. Evolutionary biologists often change the way that organisms are grouped. They also change the names of groups. Cladograms show hypotheses about the relationships among groups. These hypotheses change as more evidence becomes available.

Domain Bacteria Members of the domain Bacteria are single-celled prokaryotes. They do not have a nucleus. They have thick cell walls that contain a substance called peptidoglycan. Bacteria are a very diverse group. Some live in the soil. Some make their own food by photosynthesis. And some are parasites that cause deadly diseases.

BIOLOGY.com Search Lesson 18.3 GO • Art Review

ELL Focus on ELL: Access Content

BEGINNING SPEAKERS As a class, draw on the board a timeline that summarizes changes in the way scientists classify organisms and when these changes took place.

Ask When should we start the timeline? *(in the 1730s, the time when Linnaeus developed his classification system)*

Ask What date should we put at the other end of the timeline? *(present day)*

Ask What is the first classification system we should place on the timeline? *(two-kingdom system: plants and animals)*

As you proceed through this lesson, have volunteers add other classification systems to the timeline.

During Reading

Build Vocabulary

WORD ORIGINS Tell students that the word *archaic* is also derived from the Greek *arkhaios.* Help students understand how *archaic* is used in everyday language.

Say A horse and carriage is an archaic form of transportation. The idea that Earth is flat is an archaic idea.

Ask What do you think the word *archaic* means? *(Sample answers: from earlier times, out-of-date, old-fashioned)*

Active Reading

SET PURPOSE FOR READING Read aloud the lesson title.

Ask Based on what you learned in the last lesson, what do you think the title of this lesson means? *(Sample answer: making a cladogram that shows how all organisms are related)*

Ask What do you think you will learn about in this lesson? *(Sample answer: the evolutionary relationships among all the organisms on Earth)*

KEY QUESTION Read aloud the question and answer on the previous page.

Ask Why was the kingdom Monera separated into the kingdoms Eubacteria and Archaebacteria? *(Kingdom Monera contained two very different kinds of bacteria.)*

USE CHARTS AND TABLES As students read about each domain and kingdom, have them use the Three Domains table to help them remember the characteristics of the organisms that belong to each group. Review with students the meanings of the words *prokaryote, eukaryote, unicellular, multicellular, autotroph,* and *heterotroph.*

Speed Bump

Ask How many kingdoms of organisms are used in modern classification? *(six)*

Ask Is a domain a larger or smaller group than a kingdom? *(larger)*

During Reading continued

Active Reading

INTERPRET DIAGRAMS Help students interpret the cladogram.

Ask What do the blue, purple, and peach background colors in the cladogram represent? *(the three domains)*

Point out the key.

Ask What do the different colored lines represent? *(the different kingdoms)*

Instruct students to trace with their fingers the origin of each kingdom.

Ask Which kingdom is not included in a single clade? *(kingdom Protista)*

Ask Which organisms evolved first, fungi or animals? *(fungi)*

COMPARE AND CONTRAST Have students make a Venn diagram in their notebooks that compares and contrasts the domain Bacteria and the domain Archaea.

Ask How are the organisms in these two domains similar? *(They are single-celled and prokaryotic.)* How are they different? *(Organisms in the domain Bacteria have peptidoglycan in their cell walls; organisms in the domain Archaea do not.)*

Speed Bump

Ask Why are protists found on several different branches of the cladogram? *(Some kinds of protists are not closely related.)*

Ask Which kingdom contains only multicellular organisms? *(kingdom Animalia)*

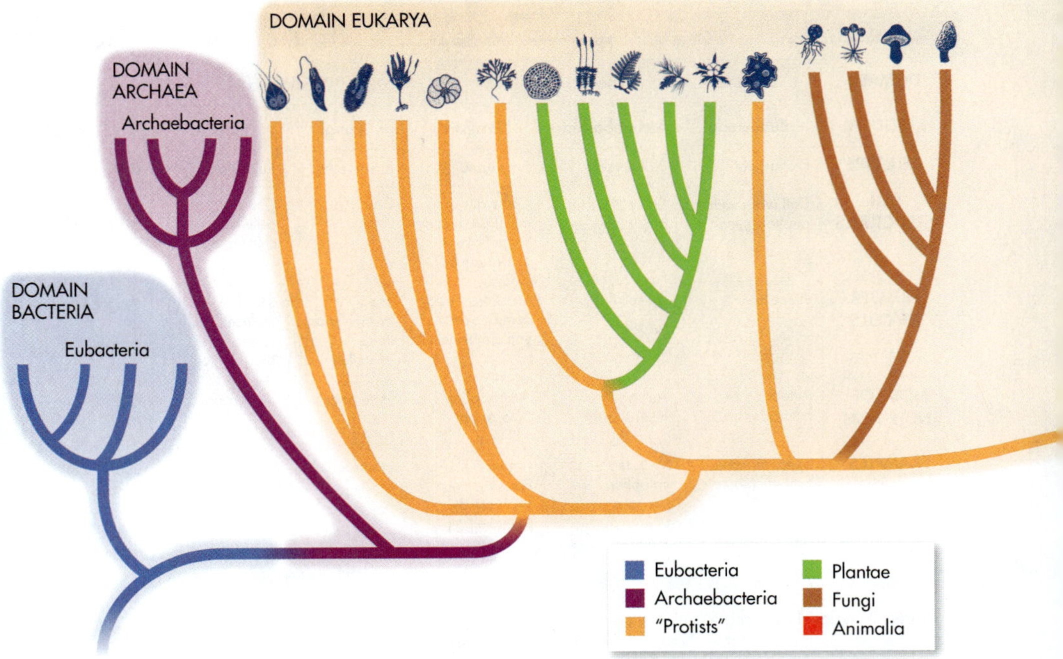

Tree of Life The tree of life shows the most recent hypothesis about how the major groups of organisms are related to one another.

Domain Archaea Members of the domain Archaea are also single-celled prokaryotes. They have unusual cell membranes. Their cell walls do not have peptidoglycan. These bacteria live in some of the most extreme environments on Earth. Some live in hot springs. Others live in very salty water. Still others live only in places that have no oxygen.

Domain Eukarya The domain **Eukarya** contains all of the organisms that have a nucleus. It includes four major groups from the six-kingdom system: "Protista," Fungi, Plantae, and Animalia.

▶ ***"Protists"*** In the cladogram above, notice that kingdom "Protista" is spread out over several different branches. This kingdom is not monophyletic, so many members are not closely related. Recent studies of these organisms have divided them into at least five clades. Each of these clades is more closely related to other groups than to other "protists." Most "protists" are single-celled organisms, but the brown algae are multicellular. Some "protists" are photosynthetic. Others are heterotrophic.

▶ ***Fungi*** Members of the kingdom Fungi are heterotrophs that have cell walls containing chitin. Most fungi feed on dead organisms. Fungi release enzymes that digest food outside of their body. The body of the fungus then absorbs the digested food. Most fungi, such as mushrooms, are multicellular. Some, such as yeasts, are single-celled.

Students Say

"Pick a Card, any card . . . Our teacher fans out index cards like playing cards. Lab partners choose a card, find a fact in the chapter, then write the fact on a small sticky note and attach it to the card. We hand the card to the teacher and choose another card. You might get a card with a sticky note already on it. So you have to add a different fact. This continues until each card has three sticky notes with three facts. Then the pair reads their card with the three facts to the class."

▶ ***Plantae*** Members of the kingdom Plantae make their own food by photosynthesis. They also have cell walls that contain cellulose. Most are multicellular. Green algae, once considered to be protists, are now classified as plants. Many green algae are single-celled organisms.

▶ ***Animalia*** Members of kingdom Animalia are multicellular and do not have cell walls. Animals are heterotrophic. To get energy, they feed on other organisms or the remains of other organisms. Most animals can move from place to place, at least in part of their life cycle.

Key Question What does the tree of life show?
The tree of life shows current hypotheses about the evolutionary relationships among the groups within the three domains of life.

BUILD Vocabulary

Eukarya the domain consisting of all organisms that have a nucleus

WORD ORIGINS
Members of domain Eukarya are called eukaryotes. *Eukaryote* comes from two Greek roots. *Eu* means "good" and *kary* means "nucleus." So, a eukaryote is an organism with a good, or true, nucleus.

CHECK Understanding

Apply Vocabulary

Use the highlighted words from the lesson to complete each sentence correctly.

1. The three _______ are very large groups that help to organize the six kingdoms.
2. The domain _______ includes protists, fungi, plants, and animals.
3. The domain _______ includes bacteria that live in very harsh environments.

Critical Thinking

4. **Review** What are the six kingdoms of life as they are now identified?
5. **Classify** Which kingdoms include prokaryotes? Which kingdoms include eukaryotes?
6. **Relate Cause and Effect** Why did biologists add the level of "domain"?
7. **Write to Learn** Do you think the tree of life cladogram will always stay the same as it appears in your book? Explain your answer.

BIOLOGY.com Search Lesson 18.3 GO • Lesson Assessment

Check Understanding

1. domains
2. Eukarya
3. Archaea
4. The six kingdoms are Eubacteria, Archaebacteria, Protista, Fungi, Plantae, and Animalia.
5. Prokaryotes are found in the kingdoms Eubacteria and Archaebacteria. Eukaryotes are found in the kingdoms Protista, Fungi, Plantae, and Animalia.
6. Researchers learned that the two kingdoms of bacteria were at least as different from each other as they were from the eukaryotes.
7. The tree of life has gone through many changes since Darwin's time. New information and new technology have helped researchers correct, improve, and enlarge the tree. It will probably change again as new information and technology become available.

Active Reading

USE GRAPHIC ORGANIZERS Have students expand their three-domain concept map by adding to the map the characteristics of the four kingdoms belonging to the domain Eukarya, described on these two pages.

KEY QUESTION Invite a volunteer to read aloud the question and answer.

Ask What are the three domains of life? *(domain Bacteria, domain Archaea, and domain Eukarya)*

Build Vocabulary

WORD ORIGINS Refer students to the Tree of Life cladogram.

Ask What is the name of another classification group that begins with the prefix *eu-*? *(kingdom Eubacteria)*

Ask What does the prefix tell you about the organisms in this kingdom? *(Sample answer: They are "good" or "true" bacteria.)*

Post Reading

ASSESSMENT TIPS Suggest that students look at the Three Domains table to find the answer to Question 5.

Explain to students that Question 6 describes an effect and that they are to identify the cause of the effect.

For Question 7, students might review the first page of this lesson.

Wrap-Up Activity

FLASH CARDS Divide the class into six groups, and assign one of the kingdoms to each group. Have each group make a set of at least five flash cards with characteristics and examples of organisms in their kingdom. Collect the flash cards and use them to review the lesson with the class.

CHAPTER LAB

Pre-Lab

Read aloud the Problem, and then review what a dichotomous key is by having a volunteer read aloud the Connect to the Big Idea paragraph. You might wish to refer students to the dichotomous key in Lesson 18.1.

Focus students' attention on the words *expert* and *amateur* in the Connect to the Big Idea paragraph. Discuss the meanings of the words.

Ask What kinds of experts might use a dichotomous key to identify organisms? *(Sample answers: scientists, zookeepers, museum workers)*

Ask What kinds of amateurs might use a dichotomous key to identify organisms? *(Sample answers: hikers, plant collectors, bird watchers)*

Review the Background Questions and Pre-Lab Questions as a class.

Lab

Tell students they will perform the chapter lab *Dichotomous Keys* described in **Lab Manual B.**

Post-Lab

USE A CLADOGRAM Divide the class into pairs. Refer them to the cladogram in Lesson 18.2 that shows the evolutionary history of cats. Tell students to use the derived characters and animals shown in the cladogram to make a dichotomous key that can be used to identify a cat.

Design Your Own Lab

OPEN-ENDED INQUIRY

Pre-Lab: Dichotomous Keys

Problem Can you construct a dichotomous key that can be used to identify organisms?

Materials reference materials

Lab Manual Chapter 18 Lab

Skills Focus Observe, Classify, Compare and Contrast, Sequence

Connect to the Big idea Given the enormous variety of life on Earth, not even experts can identify every organism they observe. Experts and amateurs use dichotomous keys to identify organisms. These keys are based on the appearance of organisms. A key is a series of paired statements. Readers select the statement that best describes an organism at each step until the organism is identified and named. In this lab, you will practice using a dichotomous key. Then you will construct your own key for a group of organisms.

Background Questions

a. Review Why do biologists prefer to identify an organism by its scientific name?

b. Compare and Contrast How is the way modern biologists group species into larger categories different from the system that Linnaeus used?

c. Review How many choices does a dichotomous key provide at each step?

Pre-Lab Questions

Preview the procedure in the lab manual.

1. **Observe** Name three different physical traits that are used in the shark dichotomous key.
2. **Apply Concepts** After you make a list of physical traits that you can use in your dichotomous key, how will you decide which trait to pick for the first step?

BIOLOGY.com Search Chapter 18 GO

Visit Chapter 18 online to test yourself on chapter content and to find activities to help you learn.

Untamed Science Video Hop on board with the Untamed Science crew to find out how organisms are classified.

Art in Motion View a short animation that explains how to use a dichotomous key.

Art Review How well do you know the characteristics of the three domains? Test yourself in this activity.

InterActive Art Build your understanding of cladograms with this animation.

Pre-Lab Answers

BACKGROUND QUESTIONS

a. An organism can have many different common names. By using a scientific name, biologists can be sure they are discussing the same organism.

b. Sample answer: Linnaeus based his groupings on easily observed traits. Modern biologists base their groupings on how recently organisms shared a common ancestor.

c. There are two choices at each step.

PRE-LAB QUESTIONS

1. Answers may include number of fins, body shape, mouth placement, and presence or absence of spines.
2. Sample answer: I will pick a trait that divides all the species into two smaller groups.

NATIONAL SCIENCE EDUCATION STANDARDS

UCP I

CONTENT C.3.e

INQUIRY A.2.b

442 Chapter 18 • Pre-Lab

18 CHAPTER Summary

18.1 Finding Order in Diversity

- In binomial nomenclature, each species is given a two-part scientific name.
- The goal of systematics is to organize living things into groups that have biological meaning.
- Over time, Linnaeus's original classification system was expanded to include seven levels: species, genus, family, order, class, phylum, and kingdom.

binomial nomenclature (p. 428)
genus (p. 428)
systematics (p. 430)
family (p. 430)
order (p. 430)
class (p. 430)
phylum (p. 431)
kingdom (p. 431)

18.2 Modern Evolutionary Classification

- The goal of evolutionary classification is to group species into larger categories that reflect lines of evolutionary descent, rather than overall similarities and differences.
- A cladogram links groups of organisms by showing how evolutionary lines branched off from common ancestors over time.
- In general, the more derived genetic characters that two species share, the more recently they shared a common ancestor. This information also tells how closely two species are related.

clade (p. 433)
monophyletic group (p. 433)
cladogram (p. 433)
derived character (p. 434)

18.3 Building the Tree of Life

- The six-kingdom system of classification includes the kingdoms Eubacteria, Archaebacteria, Protista, Fungi, Plantae, and Animalia.
- The tree of life shows current hypotheses about evolutionary relationships among the groups within the three domains of life.

domain (p. 438)
Bacteria (p. 439)
Archaea (p. 439)
Eukarya (p. 440)

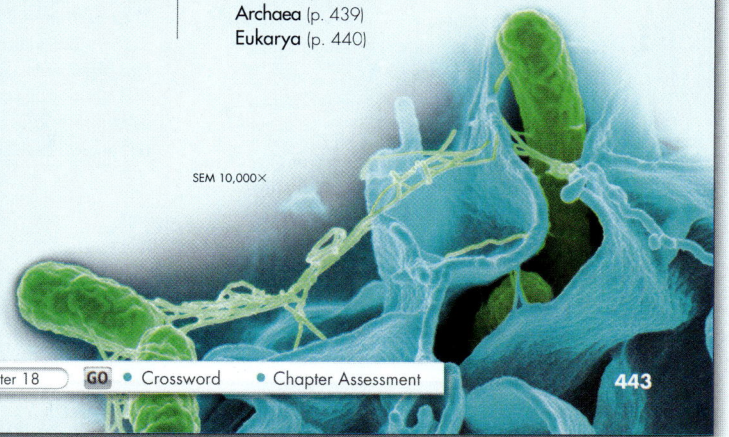

SEM 10,000×

UbD Transfer the Big Idea

TRANSFER Have students work in small groups to produce a field guide for five organisms found in a local park, vacant lot, garden, or lawn. If these settings are not available, show students a variety of pictures from a specific place such as a national park. Students should classify the organisms they choose by domain, kingdom, and any other taxa they can identify. Tell students to write descriptions of the organisms' characteristics, including those that support their classifications of the organisms.

Chapter Summary

Think Visually

USE GRAPHICS Have students use information from this chapter to complete the following Venn diagram.

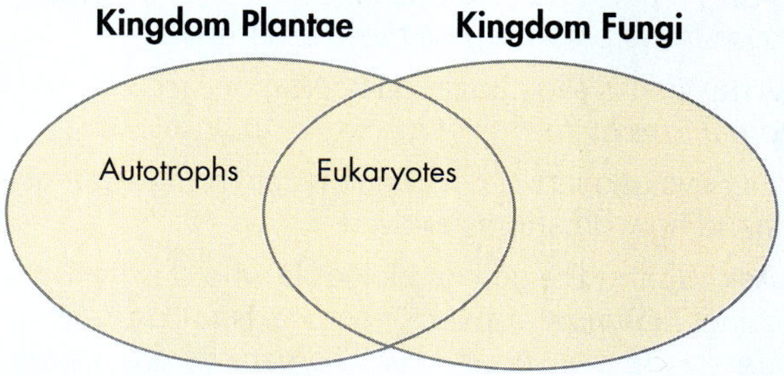

ANSWERS
Both: cell walls. Kingdom Plantae only: walls made of cellulose, chloroplasts. Fungi only: heterotrophs, cell walls made of chitin

BIOLOGY.com Study Online

REVIEW RESOURCES

Art in Motion This short animation can help students understand how to use a dichotomous key.

InterActive Art Students can practice using cladograms.

Art Review Use this drag-and-drop activity to help students review the characteristics of the three domains.

Data Analysis Students compare geographic relationships to genetic relationships to try to find a mate for Lonesome George, a Galápagos tortoise.

Vocabulary Review The *Flash Cards* and *Crossword* provide an interactive way to review chapter vocabulary.

Editable Worksheets Editable pages of Study Workbook B, Lab Manual B, and the Assessment Resources Book are available online.

Lesson Overviews Lesson Overviews are based on Miller & Levine Biology. However, they are available on CD-ROM and can be edited for your class.

ASSESSMENT RESOURCES

Chapter Assessment Have students take an online version of the Chapter 18 Check Understanding.

Standardized Test Prep Students can take an online version of the Standardized Test Prep. You will receive their scores along with ideas for remediation.

Diagnostic and Benchmark Tests Use these tests to monitor your students' progress and supply remediation.

ASSESSMENT

Assess

Big idea Invite a volunteer to read aloud the Big Idea and the question.

Ask What is classification? *(grouping objects by their similarities)*

Ask What is traditional classification of organisms? *(classifying organisms by their similar traits)*

Ask What is evolutionary classification? *(classifying organisms by their evolutionary relationships)*

Say Evolutionary classification helps biologists understand how organisms evolved.

Ask What is the goal of biologists who classify living things? *(Sample answer: Their goal is to construct a tree of life that shows how all organisms are related to one another.)*

CONSTRUCTED RESPONSE Paragraphs will vary. Suggest that students begin their paragraphs with the main idea as a topic sentence and add sentences that reflect supporting details. Remind them that a new main idea starts a new paragraph. Be sure students understand that the *Hints* should be used as a guide to help them answer the questions.

Write Question 1 on the board and read it aloud. Underline *species* and *two-part* name.

Say This question asks why giving each species its own scientific name is important.

Ask Why do common names for organisms cause confusion? *(Sample answers: There often is more than one name for the same organism. People in different places use different words to describe the same organism.)*

Read the hint.

Ask How is a species' two-part name unique? *(No other species has the same combination of genus name and species-descriptive name.)*

Model writing the first one or two sentences on the board. *(Sample answer: Biologists around the world have to communicate with each other about organisms. With so many different languages, this can be difficult.)*

18 CHECK Understanding

Assess the Big idea **Unity and Diversity of Life**

Write an answer to the question below.

Q: What is the goal of biologists who classify living things?

Constructed Response

Write an answer to each of the numbered questions below. The answer to each numbered question should be one or two paragraphs long. To help you begin, read the **Hints** below the questions.

1. **Why do biologists give each species its own two-part name?**

   Hint The two-part naming system is called binomial nomenclature. The first part of the name is the genus name. The second part is unique to each species.

2. **How are derived characters used to show evolutionary relationships?**

   Hint A cladogram is a visual way of organizing and presenting evolutionary relationships among a group of organisms.

3. **Is the kingdom Protista a true clade? Explain your answer.**

   Hint Look at the tree of life. Where are the protists located?

Foundations for Learning Wrap-Up

Use the index cards that you made at the beginning of the chapter as a tool for review.

Activity 1 Shuffle the cards until they are no longer in order. Then lay the cards on your desk and arrange them in order, from the largest group to the smallest. Repeat this activity until it becomes easy.

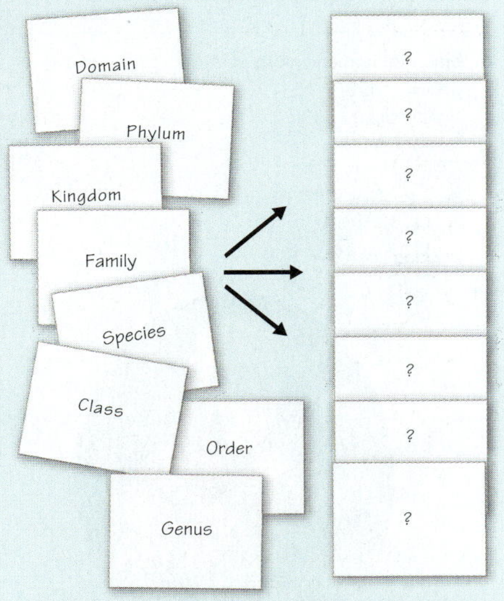

Activity 2 Work with a partner. Lay your cards on your desk to help you with the following activity. Your partner will use examples in the textbook, such as the camel chart on page 431. He or she will name two organisms, such as *camel* and *squirrel*. Name the smallest taxon that both animals belong to, such as *family* or *kingdom*. Check your answers in the textbook. Then, switch roles with your partner.

444 Check Understanding

For Question 2, pair students and have them use the hints and your modeling to answer the question. Have students share their paragraphs with the class.

For Question 3, have students write the paragraphs independently for homework or as an anchor activity.

Foundations for Learning Wrap-Up

Have students take out their index cards. Spot-check students' cards to see that they have listed all eight levels of classification. Explain the activities. Have students complete the first activity individually and the second activity in pairs.

18.1 Finding Order in Diversity

Understand Key Concepts

1. By looking at its name, you know that *Rhizopus nigricans* must be
   a. a plant.
   b. an animal.
   c. in the genus *nigricans.*
   d. in the genus *Rhizopus.*
2. A useful classification system does NOT
   a. show relationships.
   b. reveal evolutionary trends.
   c. use different names for the same organism.
   d. change the taxon of an organism based on new data.

Test-Taking Tip

Watch for Qualifiers When reading a question, watch out for the words NOT, EXCEPT, ALL, and ONLY. These words can change a correct answer into an incorrect answer. For example, the word NOT appears in question 2 above. Without this word, choices **a, b,** and **d** could all be correct, since they all describe a useful classification system. The word NOT is a signal to look for the one exception. The correct answer is **c,** because a useful classification system does NOT use different names for the same organism.

3. In Linnaeus's system of classification, orders are grouped together into
   a. genera. **c.** families.
   b. species. **d.** classes.
4. The largest and most inclusive rank in the Linnaean classification system is the
   a. kingdom. **c.** phylum.
   b. order. **d.** domain.
5. What features of binomial nomenclature make it useful for scientists around the world?

Think Critically

6. **Classify** Venn diagrams can be used to show groups within groups. The Venn diagram below shows four groups as circles—A, B, C, and D. Each shape is a taxonomic level. Shapes that overlap share common members. Shapes that do not overlap do not share members. Use these levels to label the shapes in the diagram: *kingdom Animalia, phylum Chordata, class Insecta,* and *class Mammalia.* Give reasons for your labels. (**Hint:** See the diagram on p. 431.)

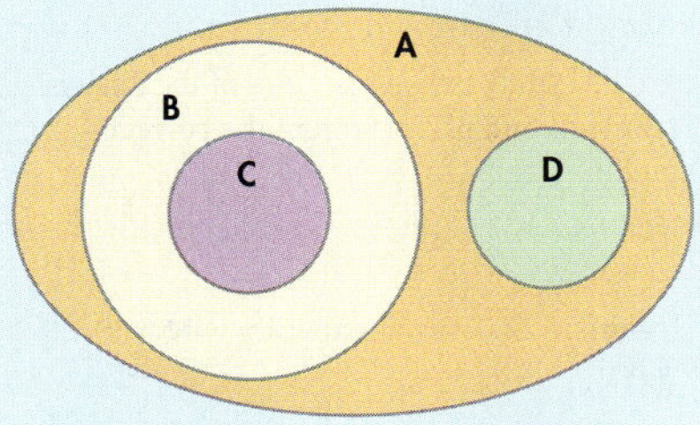

18.2 Modern Evolutionary Classification

Understand Key Concepts

7. A specific trait that is used to construct a cladogram is called a
   a. taxon. **c.** clade.
   b. structural feature. **d.** derived character.
8. A branch of a cladogram that consists of a single common ancestor and all its descendants is called
   a. cladistics. **c.** a clade.
   b. a kingdom. **d.** a class.
9. What does each node in a cladogram represent?

Think Critically

10. **Apply Concepts** Hair is a derived character for clade Mammalia. Having four limbs is not a derived character for this clade. Explain why this is so. For which clade is four limbs a derived character?

Answers

Lesson 18.1

UNDERSTAND KEY CONCEPTS

1. d
2. c
3. d
4. a
5. There is only one scientific name for each organism. The name is agreed upon by everyone and is in Latin or Greek instead of many different languages.

THINK CRITICALLY

6. Shape A is the *kingdom Animalia* because it is the largest, most inclusive group. Shape B is the *phylum Chordata* because it is the next biggest group. Shape C is *class Mammalia* because it is a smaller group within Chordata and Animalia. Shape D is *class Insecta* because it is not part of Chordata, but insects are animals.

Lesson 18.2

UNDERSTAND KEY CONCEPTS

7. d
8. c
9. a speciation event

THINK CRITICALLY

10. All mammals have hair, and only mammals have hair. Mammals have four legs, but so do many other vertebrates. Four legs is a derived character for tetrapods.

Assessment Tips

Have pairs work together to answer the questions on this page and the following.

Point out the qualifiers in Questions 2 and 11.

For Question 6, students might revisit the From Species to Kingdom diagram in Lesson 18.1.

The Three Domains table in Lesson 18.3 can help students answer Questions 11–13.

Be sure students are aware that Questions 14–16 refer to the cladogram shown.

ASSESSMENT

Lesson 18.3

UNDERSTAND KEY CONCEPTS

11. b

12. The organism must be single-celled, without a nucleus, and have thick cell walls that contain peptidoglycan.

THINK CRITICALLY

13. Organism A—kingdom Animalia; Organism B—kingdom Fungi

Connecting Concepts

14. Groups B and C share derived character 1.

15. The node represents the point at which groups B and C branched off from each other. This also shows the last time B and C shared a common ancestor.

16. Group A split off first.

CHAPTER MYSTERY After students read the Chapter Mystery, discuss the classification of polar bears and brown bears.

Ask Why have polar bears and brown bears been classified into different species in the Linnaean system? *(Linnaean classification groups organisms based on their observable similarities and differences. The differences between polar bears and brown bears—including color, teeth, body shape, metabolism, and behavior—determined their classification into different species.)*

Ask Why don't brown bears by themselves form a single clade? *(A clade is a group of organisms that includes a single common ancestor and all descendants of that ancestor. DNA analysis shows that some brown bears are more closely related to polar bears than to other brown bears. Therefore, for all brown bears to be in a single clade, that clade would have to include polar bears.)*

18 CHECK Understanding

18.3 Building the Tree of Life

Understand Key Concepts

11. Which of the following kingdoms contains ONLY heterotrophs?

a. Protista
b. Fungi
c. Plantae
d. Eubacteria

12. What characteristics are used to place an organism in the domain Bacteria?

Think Critically

13. **Classify** Study the descriptions of the following organisms and place them in the correct kingdoms.

Organism A: Multicellular eukaryote without cell walls.

Organism B: Unicellular eukaryote with cell walls of chitin.

Connecting Concepts

Use Science Graphs

The cladogram below shows the relationships among three imaginary groups of organisms—groups A, B, and C. Use the cladogram to answer questions 14–16.

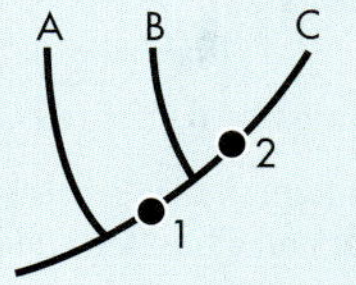

14. **Interpret Visuals** Which groups share derived character 1?

15. **Apply Concepts** What does the node, or fork, between characters 1 and 2 represent?

16. **Infer** Which group split off from the other groups first?

solve the CHAPTER MYSTERY

GRIN AND BEAR IT

Most biologists classify the polar bear, *Ursus maritimus*, as a separate species from the brown bear, *Ursus arctos*. The teeth, body shape, metabolism, and behavior of polar bears are very different from those of brown bears. But some researchers now wonder if the two are the same species.

Polar bears and brown bears can mate and produce fertile offspring. That ability is the definition of a species. But, in the wild, polar bears and brown bears almost never mate.

DNA analysis makes the question harder to answer, not easier. Different populations of brown bears have different genetic makeups. Some populations of brown bears have DNA that is different from that of polar bears. Other brown bears have DNA that is more similar to polar bears than to other brown bears. It may be that polar bears are not a separate species after all. It may also be that brown bears by themselves do not form a single clade.

1. **Classify** List the evidence that supports classifying polar bears and brown bears as two separate species. Then list the evidence that shows that polar bears and brown bears belong to the same species.

2. **Infer** What evidence exists that different populations of brown bears belong to different clades?

UntamedScience **Never Stop Exploring Your World.** Solving the mystery of scientific classification is only the beginning. Take a video field trip with the ecogeeks of Untamed Science to see where the mystery leads.

CHAPTER MYSTERY ANSWERS

1. Evidence for separate species: Polar bears and brown bears have different feet, metabolism, body shape, behavior, teeth, and color; they almost never mate in the wild. Evidence for the same species: Polar bears and brown bears can mate and produce fertile offspring; some populations of brown bears have DNA that is more similar to polar bears than to other brown bears.

2. Some populations of brown bears have DNA that is more similar to polar bears than to other brown bears. Other brown bears have DNA that is more similar to other brown bears than to polar bears. Some brown bears are clearly more closely related to polar bears than others.

Untamed Science Students can learn more about the classification of organisms by watching **Finned Kin.**

Standardized Test Prep

Multiple Choice

1. Which of the following is NOT a characteristic of Linnaeus's system for naming organisms?
A two-part name
B multipart name describing several traits
C name that identifies the organism's genus
D name that includes the organism's species identifier

2. In which of the following are the taxa in correct order?
A phylum, kingdom, species
B genus, order, family
C kingdom, phylum, class
D order, class, family

3. In the six-kingdom system of classification, which kingdoms contain unicellular organisms?
A Eubacteria only
B Eubacteria and "Protista" only
C Archaebacteria only
D Eubacteria, Archaebacteria, Plantae, and "Protista"

4. If species A and B have very similar genes, which of the following statements is probably true?
A Species A and B shared a relatively recent common ancestor.
B Species A evolved independently of species B for a long period.
C Species B is older than species A.
D Species A is older than species B.

5. The taxon called Eukarya is a(n)
A order.
B phylum.
C kingdom.
D domain.

6. Bacteria are classified into
A two domains.
B three domains.
C three species.
D three kingdoms.

Questions 7–9

The cladogram below shows the evolutionary relationships among four groups of plants. Use the cladogram to answer questions 7–9.

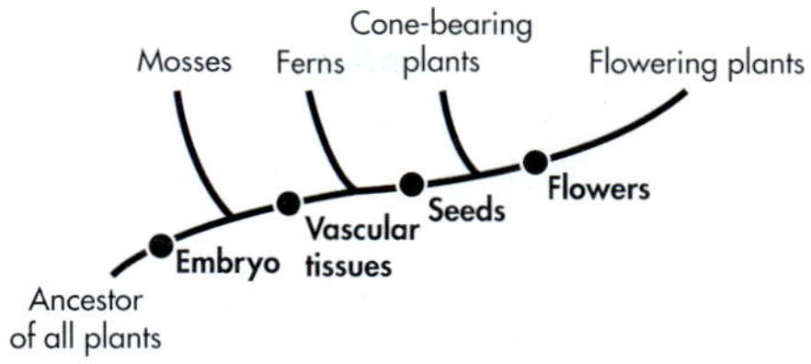

7. Which of the following groups, by themselves, do NOT form a clade?
A cone-bearing plants and flowering plants
B ferns, cone-bearing plants, and flowering plants
C mosses and ferns
D mosses, ferns, cone-bearing plants, and flowering plants

8. Which of the following groups share the most recent common ancestor?
A cone-bearing plants and flowering plants
B mosses and ferns
C mosses and cone-bearing plants
D ferns and flowering plants

9. Which derived character appeared first during the course of the plants' evolution?
A seeds
B flowers
C embryo
D vascular tissues

Open-Ended Response

10. Why have biologists changed many of Linnaeus's original classifications of organisms?

If You Have Trouble With . . .

Question	1	2	3	4	5	6	7	8	9	10
See Lesson	18.1	18.1	18.3	18.2	18.3	18.3	18.2	18.2	18.2	18.2

Test-Taking Tip

READ ALL THE ANSWER CHOICES

Tell students that it is a good idea to read all of the answer choices for multiple choice questions before choosing the correct answer. Explain that wrong choices are often intentionally written to seem as though they could be correct. Therefore, without reading all the choices, it is easy to make a mistake by selecting the first choice that appears to be correct.

Before the Test

Review with students the Test-Taking Tip.

Point out the word *NOT* in Questions 1 and 7.

Be sure students are aware that Questions 7–9 refer to the cladogram.

Remind students that the paragraphs they write to answer an open-ended question should contain the same facts as another student's paragraph, but the order of the facts, examples used, and style of writing probably will not be the same.

Answers

1. B
2. C
3. D
4. A
5. D
6. A
7. C
8. A
9. C
10. Sample answer: Linnaeus based his classification on observable similarities and differences among organisms. Modern evolutionary classification groups species into larger categories that reflect lines of evolutionary descent. Also, modern evolutionary classification uses similarities in DNA.

CHAPTER PLANNER 19 History of Life

Chapter Contents	NSES	Time	Core Resources
Chapter Preview			**Student Edition,** pp. 448–449 **Chapter Mystery,** p. 449 **Foundations for Learning,** p. 449
19.1 The Fossil Record Fossils and Ancient Life • Dating Earth's History • Geologic Time Scale • Processes Affecting Life's History	I, II, III, IV, V, A.1.b, A.1.c, A.2.a, A.2.d, B.1, C.3.a, C.3.c, C.3.d, D.3	1 period $^1/_2$ block	**Student Edition,** pp. 450–455 **Inquiry Into Scientific Thinking,** p. 452 **Build Connections,** p. 454 **Study Workbook B** 19.1 Worksheets **Biology.com** *Art in Motion:* Fossil Formation • *Visual Analogy:* Geologic Time as a Clock • 19.1 Lesson Assessment
19.2 Patterns and Processes of Evolution Speciation and Extinction • Rate of Evolution • Adaptive Radiation and Convergent Evolution • Coevolution	I, II, III, IV, V, A.1.e, A.2.a, A.2.e, C.3.a, C.3.b, C.3.c, C.3.d, C.3.e	1 period $^1/_2$ block	**Student Edition,** pp. 456–461 **Study Workbook B** 19.2 Worksheets **Biology.com** 19.2 Lesson Assessment
19.3 Earth's Early History The Mysteries of Life's Origins • Origin of Eukaryotic Cells • Sexual Reproduction and Multicellularity	I, II, III, IV, V, A.1.c, A.2.a, A.2.b, A.2.e, A.2.f, C.1.a, C.1.e, C.3.a, C.3.b, C.3.d, C.3.e, D.3, D.4, G.3	1 period $^1/_2$ block	**Student Edition,** pp. 462–469 **Study Workbook B** 19.3 Worksheets **Biology.com** *Art Review:* Conditions on the Early Earth • 19.3 Lesson Assessment
Chapter Pre-Lab	II, IV, C.3.b, C.3.c, C.3.d	1 period $^1/_2$ block	**Student Edition,** p. 470 **Lab Manual B** *Using Index Fossils*

Tools for ELL Instruction

ELL Handbook describes teaching strategies to make *Miller & Levine Biology: Foundation Edition* more accessible to ELL students.

Spanish Study Workbook is a Spanish translation of Study Workbook A.

Multilingual Glossary is the *Miller & Levine Biology* glossary translated into ten languages.

Tools to Challenge Students

Study Workbook A includes more challenging worksheets and activities such as Chapter Mystery activities that reinforce 21st Century Skills.

Lab Manual A labs are geared toward on-level students. The Lab Manual includes fifteen labs in addition to the Chapter Labs.

Biology.com Data Analysis activities offer students a way to practice data gathering and analyzing skills.

Visual Quizzes in the Assessment Resources Book are geared toward on-level students and use visuals from *Miller & Levine Biology.*

Additional Resources

Biology.com Untamed Science Video • Flash Cards

Study Workbook A 19.1 Worksheets
Spanish Study Workbook 19.1 Worksheets
Assessment Resources Book Visual Quiz

Study Workbook A 19.2 Worksheets
Lab Manual B *Data Analysis:* Extinctions Through Time
Spanish Study Workbook 19.2 Worksheets
Biology.com *Data Analysis:* Explaining Extinctions

Study Workbook A 19.3 Worksheets
Lab Manual B *Data Analysis:* Comparing Atmospheres
Spanish Study Workbook 19.3 Worksheets
Assessment Resources Book Visual Quiz

Lab Manual A *Using Index Fossils*

Chapter Review

Student Edition Chapter Summary, p. 471 • Unit Project, p. 476
Study Workbook B Chapter 19 Review
Transparencies, pp. 226–239
Biology.com Untamed Science Video • You're the Director • Editable Worksheets of Study Workbooks A and B and Lab Manuals A and B • Chapter 19 Flash Cards and Crossword

Untamed Science DVD • Classroom Resources CD (includes lesson presentations and editable worksheets)

Chapter Assessment

Student Edition Check Understanding, pp. 472–475
Study Workbook B Chapter 19 Taking a Standardized Test
Assessment Resources Book Chapter 19 Test B • Unit 5 Test B
Biology.com Chapter 19 Assessment • Editable Worksheets of Chapter 19 Visual Quizzes,* Chapter 19 Tests A and B, and Unit 5 Tests A and B

Exam*View* *Assessment Suite* • Classroom Resources CD (includes lesson presentations and editable worksheets)

Time: 1 period, 1/2 block

* Visual Quizzes are based on visuals from *Miller & Levine Biology.* Review carefully to make sure they are appropriate for your class.

Pressed for Time?

Preview the Chapter Introduce students to the chapter with the first two Key Questions for Lessons 19.1 and 19.2. Preview the pictures on pp. 466–469 and discuss how living things have changed over time.

Cover the Chapter Quickly Assign students to read *Fossils and Ancient Life* and *Dating Earth's History* in Lesson 19.1. Go over the Index Fossils visual. Assign all of Lesson 19.2 and *The Mysteries of Life's Origins* in Lesson19.3.

Assess Assign Questions 1–5 in the 19.1 Check Understanding, the 19.2 Check Understanding, and Questions 3 and 6 in the 19.3 Check Understanding. In the Chapter 19 Check Understanding, assign Questions 1, 3, 5–16, and 18–20.

CHAPTER 19

Get the Picture

Tell students that the photograph shows the remains of an ichthyosaur, which lived more than 100 million years ago. Explain that scientists learned a lot about ichthyosaurs by studying their fossils.

Ask How would you describe an ichthyosaur by looking at the fossil? *(Sample answer: It had limbs that looked like flippers, a tail, and a pointed beak.)*

Say The ichthyosaur was a reptile. Its name is a combination of *ichthyo-*, which comes from the Greek word for "fish," and *-saur,* from the Greek *sauros*, meaning "lizard."

Ask Which of the fossil's features remind you of a fish? *(flippers, the shape of its body)* Which features remind you of a lizard? *(tail)*

Ask What might fossils show you about how an organism lived? *(Sample answers: how it moved, how and what it ate, how it reproduced)*

Ask The ichthyosaur lived in the ocean. Many fossils of marine reptiles are found in Wyoming. What does this tell us about the ancient environment? *(It was underwater at one time.)*

Activate Prior Knowledge

Remind students that offspring inherit adaptations.

Ask What is an adaptation? *(an inherited trait that improves an organism's ability to survive and reproduce in an environment)*

Ask If the fossil record shows that the descendants of the ichthyosaur had legs instead of flippers, how might you explain that the descendants' legs were an adaptation? *(Sample answer: If the environment became drier, the legs would have given the descendants a better chance to survive and reproduce.)*

NATIONAL SCIENCE EDUCATION STANDARDS

UNIFYING CONCEPTS AND PROCESSES

I, II, III, IV, V

CONTENT

C.1.a, C.1.e, C.3.a, C.3.b, C.3.c, C.3.d, C.3.e, B.1, D.3, D.4, G.3

INQUIRY

A.1.b, A.1.c, A.1.e, A.2.a, A.2.b, A.2.d, A.2.e, A.2.f

19 History of Life

Big idea Evolution

Q: How do fossils help biologists understand the history of life on Earth?

Ichthyosaurs were reptiles that looked like dolphins. They lived in the seas during the Mesozoic Era. This ichthyosaur died around the time of giving birth.

BIOLOGY.com Search Chapter 19 GO • Flash Cards

448

UbD Connect to the Big Idea

Invite a volunteer to read aloud the chapter number and the chapter title. Read the Big Idea and question. Emphasize the question, **How do fossils help biologists understand the history of life on Earth?**

Set the Purpose

PREVIEW AND PREDICT Draw the graphic organizer on the board. Guide students to the INSIDE feature on the next page, and invite a volunteer to read each lesson number and title.

Ask What is evolution? *(change over time)* How would you describe the speed at which evolution occurs? *(very slowly)* How might fossils of species that lived over a long period of time provide evidence that the species evolved? *(The fossils might show gradual changes in the species over time.)*

INSIDE:

- 19.1 The Fossil Record
- 19.2 Patterns and Processes of Evolution
- 19.3 Earth's Early History

- Untamed Science Video
- Chapter Mystery

CHAPTER
MYSTERY

MURDER IN THE PERMIAN

About 250 million years ago, life on Earth nearly came to an end. This time is known as the Permian extinction, because it happened at the end of the Permian Period. The Permian extinction may be the greatest murder mystery in the history of the world. Whatever happened back then killed about 96 percent of marine species and 70 percent of terrestrial vertebrate species.

Scientists once thought that this extinction happened over a long period. But new fossils suggest that it took no more than 200,000 years. In geology, that is a short amount of time.

Read for Mystery Clues As you read this chapter, look for clues to help you decide what could have killed so much of life on Earth. Then, solve the mystery at the end of the chapter.

FOUNDATIONS for Learning

As you read the chapter, you will learn about key events in the history of life. To help remember these events, you will make folded tabs for 12 key events. Start with two sheets of notebook paper. Fold the sheets in half lengthwise. Cut each sheet into six pieces to make 12 folded tabs, as shown below. As you read, make notes about the key events on the inside of each tab. Be sure to include the time when each event took place. At the end of the chapter are two activities that use the folded tabs to help answer the question: How do fossils help biologists understand the history of life on Earth?

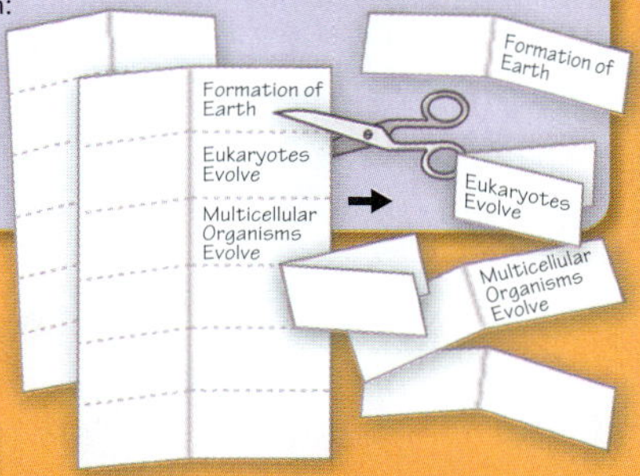

Chapter 19 Big Idea: Evolution

Chapter 19: How do fossils help biologists understand the history of life on Earth?

- **19.1:** How do scientists use fossils to study Earth's history?
- **19.2:** What are some patterns in which evolution has occurred?
- **19.3:** What happened during Earth's early history?

CHAPTER 19

Foundations for Learning

Ask students to take out two sheets of notebook paper and scissors. Then, demonstrate how to fold and cut the paper and label the tabs. Have a volunteer read aloud the events in the visual Geologic Time as a Clock in Lesson 19.1. Model an example of adding a picture to a tab.

Ask Why is it useful to add pictures? *(Pictures give you something to remember that connects to an idea.)*

Ask Why is it important to remember key events in the history of life? *(These events shaped life on Earth today and help us understand how life evolved.)*

Untamed Science What do fossils tell us? Students can answer this question by taking a trip back in time with the Untamed Science crew in **Stories in Stone**.

CHAPTER MYSTERY Invite a volunteer to read aloud the Chapter Mystery. Emphasize to students the extent and duration of the Permian extinction.

Ask How did scientists learn about the Permian extinction? *(by studying fossils)*

Ask How does the Permian extinction relate to the chapter title? *(The Permian extinction is part of the history of life on Earth.)*

Remind students to focus on the chapter question, **How do fossils help biologists understand the history of life on Earth?**, as they look for clues to help them solve the chapter mystery.

Digital Support

BIOLOGY.com Extend your reach by using the digital assets offered at **Biology.com.**

LESSON 19.1

Pre-Reading

Objectives

19.1.1 Explain what information fossils can reveal about ancient life.

19.1.2 Differentiate between relative dating and radiometric dating.

19.1.3 Identify the divisions of the geologic time scale.

19.1.4 Describe how environmental processes and living things have shaped life on Earth.

Preview the Pages

1. Preview the vocabulary words by having students find the words in the text and read each word aloud. Discuss with students any words that are familiar to them.
2. Preview the visuals by reading the captions aloud. Point out the different ways passage of time is represented in the Geologic Time Scale chart and the Geologic Time as a Clock diagram.
3. Preview the Radioactive Decay graph. Have a volunteer read aloud the caption and axes labels and identify the periods of time represented.

Build Understanding

MAIN IDEAS AND DETAILS CHART Direct students to Workbook B, where they will find a main ideas and details chart. Tell students that putting information into a chart will make it easier for them to answer the lesson's Key Questions. The chart can be used for a review and study guide later.

NATIONAL SCIENCE EDUCATION STANDARDS

UNIFYING CONCEPTS AND PROCESSES

I, II, III, IV, V

CONTENT

B.1, C.3.a, C.3.c, C.3.d, D.3

INQUIRY

A.1.b, A.1.c, A.2.a, A.2.d

For the corresponding lesson in *Miller & Levine Biology,* see pages 538–545.

19.1 The Fossil Record

Key Questions

What do fossils reveal about ancient life?

How do we date events in Earth's history?

How was the geologic time scale established, and what are its major divisions?

How have our planet's environment and living things interacted to shape the history of life on Earth?

BUILD Understanding

Main Ideas and Details Chart One way to take notes is to make a main ideas and details chart. In the left column, write down the main idea. In the right column, write down details that support the main idea.

In Your Workbook Go to your workbook to learn more about making a main idea and details chart. Complete the chart for Lesson 19.1.

Fossils and Ancient Life

Fossils are one of the most important sources of information about ancient life. All fossils together make up the history of life on Earth called the *fossil record.* The scientists who study fossils are called **paleontologists** (pay lee un TAHL uh jists).

Types of Fossils Many fossils are the preserved remains of organisms. Usually, only hard parts of organisms, such as shell, teeth, bone, or wood, are preserved. (Soft body parts usually rot or are eaten before they can be preserved.) Fossils can be as large as a whole animal. Others are as tiny as bacteria or pollen grains. Many fossils are just pieces of an organism—a few teeth or bits of bone. Some fossils are called *trace* fossils. Trace fossils are signs of activity, such as footprints, burrows, or even droppings.

What Fossils Show Only a tiny percentage of ancient organisms became fossils. Although incomplete, the fossil record holds enough information to teach us a great deal about ancient life. Without fossils, we would know almost nothing about extinct species. An **extinct** species has no living members.

Fossils can show how an organism lived or what it ate. By comparing fossils to living organisms, paleontologists can trace the evolution of a body part or a whole species. Bone structure and footprints can show how an animal moved. Fossils from plants, such as leaves and pollen, can tell whether an area was a swamp or a forest. When plant and animal fossils are found together, paleontologists can learn about ancient ecosystems.

Key Question What do fossils reveal about ancient life?
The fossil record shows the structure of ancient organisms, their environment, and the way in which they lived.

Fossil Fish Like this 50-million-year-old fossil of a fish, most fossils form from the hard parts of organisms.

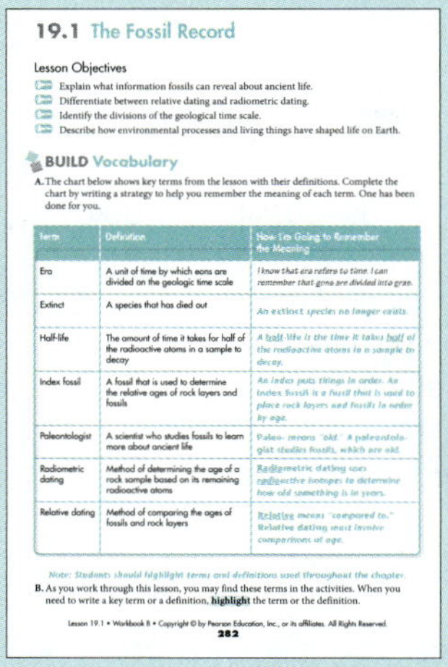

19.1 The Fossil Record

Lesson Objectives

- Explain what information fossils can reveal about ancient life.
- Differentiate between relative dating and radiometric dating.
- Identify the divisions of the geological time scale.
- Describe how environmental processes and living things have shaped life on Earth.

BUILD Vocabulary

A. The chart below shows key terms from the lesson with their definitions. Complete the chart by writing a strategy to help you remember the meaning of each term. One has been done for you.

Term	Definition	How I'm Going to Remember the Meaning
Era	A unit of time by which eons are divided on the geologic time scale	I know that era refers to time. I can remember that eons are divided into eras.
Extinct	A species that has died out	An extinct species no longer exists.
Half-life	The amount of time it takes for half of the radioactive atoms in a sample to decay	A half-life is the time it takes half of the radioactive atoms in a sample to decay.
Index fossil	A fossil that is used to determine the relative ages of rock layers and fossils	An index puts things in order. An index fossil is a fossil that is used to place rock layers and fossils in order by age.
Paleontologist	A scientist who studies fossils to learn more about ancient life	Paleo- means "old." A paleontologist studies fossils, which are old.
Radiometric dating	Method of determining the age of a rock sample based on its remaining radioactive atoms	Radiometric dating uses radioactive isotopes to determine how old something is in years.
Relative dating	Method of comparing the ages of fossils and rock layers	Relative means "compared to." Relative dating must involve comparisons of age.

Note: Students should highlight terms and definitions used throughout the chapter.

B. As you work through this lesson, you may find these terms in the activities. When you need to write a key term or a definition, **highlight** the term or the definition.

Lesson 19.1 • Workbook B • Copyright © by Pearson Education, Inc., or its affiliates. All Rights Reserved.
282

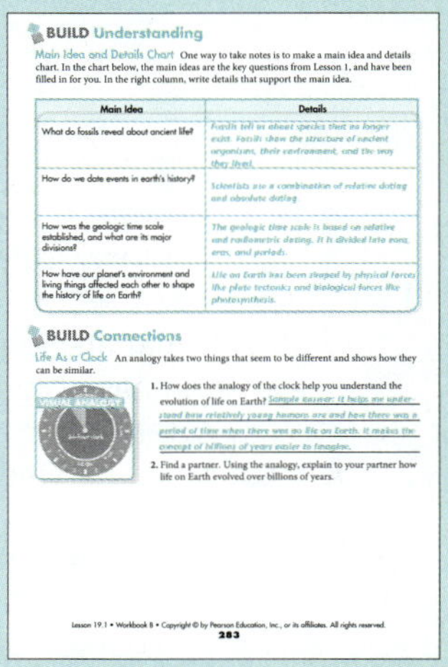

BUILD Understanding

Main Idea and Details Chart One way to take notes is to make a main idea and details chart. In the chart below, the main ideas are the key questions from Lesson 1, and have been filled in for you. In the right column, write details that support the main idea.

Main Idea	Details
What do fossils reveal about ancient life?	Fossils tell us about species that no longer exist. Fossils show the structure of ancient organisms, their environment, and the way they lived.
How do we date events in earth's history?	Scientists use a combination of relative dating and absolute dating.
How was the geologic time scale established, and what are its major divisions?	The geologic time scale is based on relative and radiometric dating. It is divided into eons, eras, and periods.
How have our planet's environment and living things affected each other to shape the history of life on Earth?	Life on Earth has been shaped by physical forces like plate tectonics and biological forces like photosynthesis.

BUILD Connections

Life As a Clock An analogy takes two things that seem to be different and shows how they can be similar.

1. How does the analogy of the clock help you understand the evolution of life on Earth? Sample answer: It helps me understand how relatively young humans are and how there was a period of time when there was no life on Earth. It makes the concept of billions of years easier to imagine.
2. Find a partner. Using the analogy, explain to your partner how life on Earth evolved over billions of years.

Lesson 19.1 • Workbook B • Copyright © by Pearson Education, Inc., or its affiliates. All rights reserved.
283

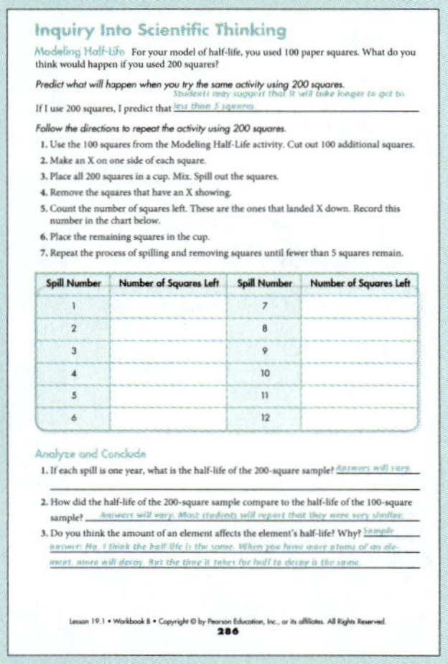

Inquiry Into Scientific Thinking

Modeling Half-Life For your model of half-life, you used 100 paper squares. What do you think would happen if you used 200 squares?

Predict what will happen when you try the same activity using 200 squares.

If I use 200 squares, I predict that Students may suggest that it will take longer to get to less than 5 squares.

Follow the directions to repeat the activity using 200 squares.

1. Use the 100 squares from the Modeling Half-Life activity. Cut out 100 additional squares.
2. Make an X on one side of each square.
3. Place all 200 squares in a cup. Mix. Spill out the squares.
4. Remove the squares that have an X showing.
5. Count the number of squares left. These are the ones that landed X down. Record this number in the chart below.
6. Place the remaining squares in the cup.
7. Repeat the process of spilling and removing squares until fewer than 5 squares remain.

Spill Number	Number of Squares Left	Spill Number	Number of Squares Left
1		7	
2		8	
3		9	
4		10	
5		11	
6		12	

Analyze and Conclude

1. If each spill is one year, what is the half-life of the 200-square sample? Answers will vary.
2. How did the half-life of the 200-square sample compare to the half-life of the 100-square sample? Answers will vary. Most students will report that they were very similar.
3. Do you think the amount of an element affects the element's half-life? Why? Sample answer: No. I think the half-life is the same. When you have more atoms of an element, more will decay. But the time it takes for half to decay is the same.

Lesson 19.1 • Workbook B • Copyright © by Pearson Education, Inc., or its affiliates. All Rights Reserved.
286

Dating Earth's History

To understand the fossil record, we need a time scale to tell what happened when. To find the age of rocks and fossils, scientists use relative dating and radiometric dating.

Sedimentary Rocks Fossils are usually found in sedimentary rock. Most sedimentary rocks form when sediments settle to the bottom of a body of water. Sediments are tiny particles, such as sand and clay. Over time, more and more layers of sediment build up. The weight of the many layers and chemical reactions turn the sediments into rock. As sediments build up, they bury dead organisms that have settled to the bottom. The remains of these organisms may become fossils.

Relative Dating When sedimentary rocks form, new layers are laid on top of older rocks. The layers on the bottom are usually older than the layers above them. Therefore, fossils in lower layers are usually older than fossils in upper layers. Scientists use the position of the layers to find the relative age of fossils. **Relative dating** places rock layers and their fossils in order according to time. Relative dating does not tell the age of a fossil in years.

To find the relative age of rock layers and their fossils, scientists use **index fossils.** Index fossils help scientists identify rocks and fossils of the same age that are found in different places. An index fossil must be easy to recognize. It should be found in only a few rock layers. (That would mean that it lived for only a short time.) But the layers that hold the index fossil must be found in many places.

BUILD Vocabulary

paleontologist a scientist who studies fossils

extinct the word used to describe a species that has died out and has no living members

relative dating a method of determining the age of a fossil by comparing its placement with that of fossils in other rock layers

index fossil a distinctive fossil that is used to establish and compare the relative ages of rock layers and the fossils they contain

WORD ORIGINS

The word *paleontologist* comes from the Greek word *palaios*, meaning "ancient." A paleontologist studies the remains of ancient life.

Index Fossils Each of these fossils is an index fossil. If the same index fossil is found in two different rock layers, the rock layers are probably about the same age. Use the index fossils to find which layers are missing from each location. Layers may be missing because they were never laid down or because they were worn away.

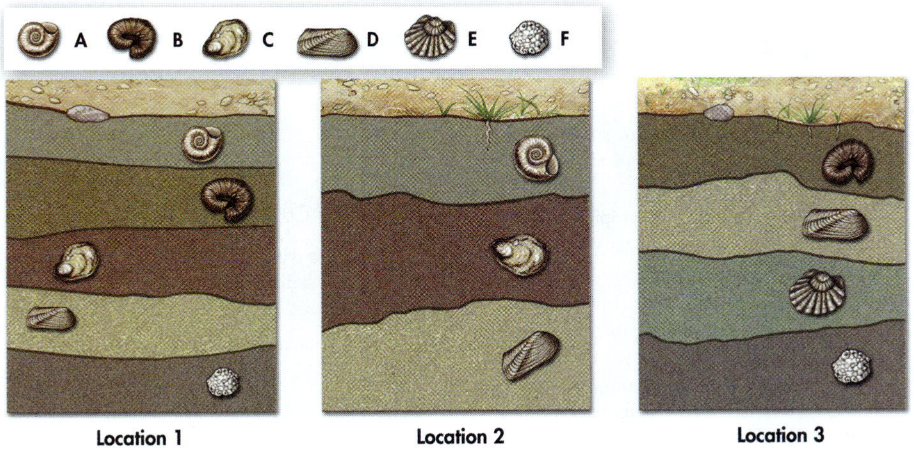

LESSON 19.1

During Reading

Active Reading

KEY QUESTION Read aloud the Key Question and answer on the previous page.

Ask In a swampy area, would you expect to find fossils of plants and animals that only live in swamps? *(no)* Why not? *(because the ancient environment in the same area may have been very different from the environment today)*

SUMMARIZE Read aloud the main heading and subheadings on this page. Then, have students read the text and summarize each paragraph in one or two sentences in their notebooks.

Ask How do sediments turn into rock? *(The weight of the sediment layers and chemical reactions change sediment into rock.)*

Ask Which of two rock layers would be deeper in the ground, the one that is older or the one that is younger? *(the one that is older)*

INTERPRET DIAGRAMS Direct students to the Index Fossils diagram. Have a volunteer read aloud the caption.

Ask Are all of the index fossils in the diagram found in each of the three locations? *(no)*

Ask How do the rock layers at Location 1 differ from those at Location 2? *(Only three of the layers at Location 1 are present in Location 2.)* What likely happened to the missing rock layers at Location 2? *(They were probably worn away.)*

Ask Which rock layer at Location 3 is the same age as the third layer from the top at Location 2? *(the second layer from the top at Location 3)*

Build Vocabulary

WORD ORIGINS Write the word *paleontology* on the board and underline *-logy*.

Say The suffix *-logy* comes from the Latin *-logia* and Greek *-logos*, meaning "the study of" or "science of."

Ask What do you think the word *paleontology* means? *(Sample answer: the scientific study of the remains of ancient life)*

Speed Bump

Ask In what types of environments are sediment layers usually deposited? *(in bodies of water)*

Ask To determine the relative ages of fossils, fossils in a layer are compared to what? *(fossils in other layers)*

Hands-On Learning

RELATIVE DATING Use an analogy to help students understand the concept of relative dating.

1. Tell students to imagine that on a student's desk at home, there is an empty juice bottle on top of a half-written English essay, which is on top of an open Spanish textbook, which is on top of a birthday card from a friend. Finally, at the very bottom of the pile, is a graded math test.
2. Sketch these "layers" on the board.
3. Have students write these activities in their notebooks in the order in which they took place: student drank juice, student studied Spanish, student took math test, student began English essay. *(student took math test, student studied Spanish, student began English essay, student drank juice)*
4. Invite a volunteer to infer the timing of the student's birthday. *(The birthday took place sometime after the student took the math test but before he/she studied Spanish.)*

During Reading continued

Active Reading

COMPARE AND CONTRAST Have students reread the Radiometric Dating section silently.

Ask How does radiometric dating of fossils differ from relative dating? *(Radiometric dating uses radioactive isotopes to determine the ages of fossils in years. Relative dating compares fossils in different rock layers to determine which fossils are younger or older than others.)*

INTERPRET GRAPHS Read aloud the caption for the Radioactive Decay graph. Point out the three half-lives on the graph.

Ask How far apart, in years, are Half-life 2 and Half-life 3 on the graph? *(1.26 billion years)*

Ask After how many years does one-eighth of an original sample of potassium-40 remain? *(3.78 billion years)*

KEY QUESTION Invite a student to read aloud the Key Question and answer.

Ask If a fossil in a museum is labeled as being 72.5 million years old, which method of dating was likely used to determine its age? *(radiometric dating)*

Inquiry Into Scientific Thinking

SUMMARIZE Help students understand how the lab models radioactive decay.

Ask In this lab, what do the squares with the hidden X represent? *(radioactive isotopes that haven't yet decayed)*

Ask What measurement represents the amount of time that passes as the isotopes decay? *(the number of spills)*

USE CHARTS AND TABLES Have students look over the data table.

Ask What will you record in the Number of Squares Left column? *(the number of squares after each spill that do not have the X showing)*

Point out to students that they will use the data in the table to make the graph described in Step 6.

Lead a Discussion

Tell students more about radioactive isotopes.

Say An isotope is any of several forms of an element that have the same number of protons but different numbers of neutrons. Radioactive decay occurs when the nucleus of an unstable isotope releases energy and particles.

Ask Why aren't stable isotopes used in radiometric dating? *(Stable isotopes don't decay over time.)*

Radiometric Dating In **radiometric dating,** scientists use radioactive isotopes to find the age of rocks and fossils. These isotopes decay, or break down, into nonradioactive isotopes at a steady rate. A **half-life** is the time it takes for half of the radioactive atoms in a sample to decay. After one half-life, half of the radioactive atoms have decayed. Radiometric dating gives the *absolute age*, or age in years.

Different radioactive isotopes decay at different rates, so they have different half-lives. Carbon-14 has a half-life of 5730 years. It is useful for finding the age of organisms that lived in the recent past—the last 60,000 years. Because carbon-14 is found in living organisms, it can be used to date fossils directly.

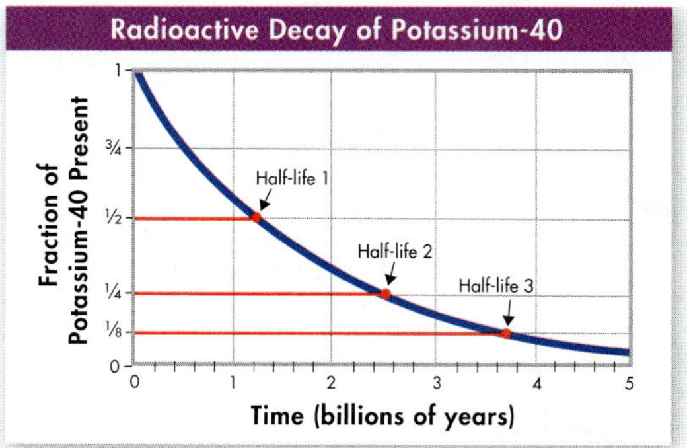

Radioactive Decay A half-life is the time it takes for half of the radioactive atoms in a sample to decay. The half-life of potassium-40 is 1.26 billion years.

Isotopes with longer half-lives are used to date older fossils. Isotopes used for dating very old fossils include potassium-40 (half-life: 1.26 billion years), uranium-238 (half-life: 4.5 billion years), and rubidium-87 (half-life: 48.8 billion years). These isotopes are found in rocks, but not in fossils. To date an older fossil, researchers first find the age of the rocks in which the fossil is found. They then use this information to infer the age of the fossil itself.

Key Question How do we date events in Earth's history? **Relative dating can be used to determine whether a fossil is older or younger than other fossils. Radiometric dating uses the proportion of radioactive to nonradioactive isotopes to find the absolute age of a sample.**

INQUIRY into Scientific Thinking

Modeling Half-Life

1. Make a data table like the one to the right.
2. Take a sheet of paper, and cut out 100 1-cm squares. Write an *X* on one side of each square. Put all the squares in a cup.
3. Mix the squares in the cup. Then spill them out.
4. Remove all the squares that have the *X* showing. These represent radioactive isotopes that have decayed into different isotopes.
5. Repeat steps 3 and 4 until there are five or fewer squares left.
6. Make a graph of your results. Put the number of spills on the *x*-axis. Put the number of squares left after each spill on the *y*-axis.

Spill Number	Number of Squares Left
1	
2	
3	
4	
5	

Analyze and Conclude

1. **Analyze Data** How many spills did you need to remove half the squares? To remove three fourths?
2. **Calculate** If each spill represents 100 years, what is the half-life of the squares?

In Your Workbook Get more help for this activity in your workbook.

Materials

1-cm grid paper, scissors, plastic or paper cup

Analyze and Conclude

1. one spill; two spills
2. 100 years

Geologic Time Scale

	Eon	Era	Period	Time (millions of years ago)
	Phanerozoic	Cenozoic	Quaternary	2–present
			Neogene	23–2
			Paleogene	66–23
		Mesozoic	Cretaceous	146–66
			Jurassic	200–146
			Triassic	251–200
		Paleozoic	Permian	299–251
			Carboniferous	359–299
			Devonian	416–359
			Silurian	444–416
			Ordovician	488–444
			Cambrian	542–488
Pre-cambrian Time	Protero-zoic			2500–542
	Archean			4000–2500
	Hadean			About 4600–4000

Geologic Time Scale The basic divisions of the geologic time scale are eons, eras, and periods. About 90 percent of Earth's history took place before the Cambrian Period.

Geologic Time Scale

Geologists and paleontologists have made a time line of Earth's history called the **geologic time scale.** The most recent version is shown above.

Establishing the Time Scale At first, scientists studied rock layers and index fossils to place them in order according to their relative ages. Then scientists noticed major changes in the fossil record between certain rock layers. Geologists used the boundaries between these layers to mark the end of one time division and the beginning of another. Years later, radiometric dating was used to find the ages of each of the layers. The geologic time scale is always being tested and corrected.

Key Question How was the geologic time scale established, and what are its major divisions? **The geologic time scale is based on both relative and absolute dating. The major divisions of the time scale are eons, eras, and periods.**

BUILD Vocabulary

radiometric dating a method for determining the age of a sample by comparing the amount of radioactive atoms to the nonradioactive atoms of the same element

half-life the length of time required for half of the radioactive atoms in a sample to decay

geologic time scale the time line used to represent Earth's history

PREFIXES

The prefix *radio-* means "radiation" or "radiant energy." In radiometric dating, scientists measure the amount of radiation given off by a sample.

Active Reading

INTERPRET CHARTS AND TABLES Have students look over the Geologic Time Scale chart.

Ask What are the names of the basic divisions of the geologic time scale? *(eons, eras, and periods)* Which include the most amount of time? *(eons)* Which include the least amount of time? *(periods)* Do all periods represent the same number of years? *(no)*

Ask What era is Earth in now? *(Cenozoic)* What period is Earth in now? *(Quaternary)* How long has Earth been in the Quaternary Period? *(1.8 million years)*

KEY QUESTION Read aloud the question and answer. Refer students to the Geologic Time Scale chart.

Ask What eon and era would a fossil belong to if it formed during the Silurian Period? *(the Phanerozoic Eon and Paleozoic Era)*

Build Vocabulary

PREFIXES Tell students that the prefix *geo-* means "Earth," "soil," or "ground." Have students locate words on this page that contain the prefix *geo- (geologists, geologic)* and read the sentences that include those words.

Ask Based on what you know about other similar words, how would you define the word *geologists*? *(Geologists are scientists who study Earth.)*

Speed Bump

Ask What dating method do scientists use to find the ages of fossils in years? *(radiometric dating)*

Ask What is the geologic time scale? *(the time line of Earth's history)*

ELL Focus on ELL: Extend Language

INTERMEDIATE AND ADVANCED SPEAKERS As a class, discuss the meaning of each component word of the vocabulary terms *radiometric dating, half-life,* and *geologic time scale.* Help students sort out any confusion over multiple meanings of a word. Discuss how the separate word definitions change when each component word is combined into the vocabulary term.

Have students work in pairs or small groups to complete a Frayer Model for each of the three vocabulary terms discussed. Students should start with the term's definition and then add a drawing, examples of the term, and a short sentence using the term. Choose an example to model on the board to get students started. At the conclusion of the lesson, ask each pair or group to choose one completed Frayer Model to share with the class. For more information on the Frayer Model, see page A6 in Workbook B, or G09 and G10 in the Transparencies book.

During Reading continued

Active Reading

USE GRAPHIC ORGANIZERS Have students read the first two paragraphs of the section Life on a Changing Planet. Before students read the rest of the section, have them prepare in their notebooks a KWL chart. Tell students to start their chart by listing in the first box what they already know about how Earth's physical environment has changed and how these changes have affected life on Earth. Direct them to list in the second box what they want to learn. Ask volunteers to share with the class what they have written. *(Charts will vary but might include what the term* plate tectonics *means, how climate has changed, or how comets and asteroids have affected Earth.)* After students read the section, give them time to complete their KWL charts with descriptions of the concepts they learned.

Build Connections

VISUAL ANALOGY Invite a student to read aloud the caption and labels for the Geologic Time as a Clock diagram. Direct students to look at the parts of the clock face and the time intervals.

Say This clock can help you better understand geologic time.

Ask What are two ways in which this clock is different from a regular clock? *(A regular clock shows the time of day; this clock shows the history of Earth. A regular clock shows 12 hours; this clock shows 24 hours.)*

Ask How does this clock show geologic time? *(It compares the length of geologic time to a 24-hour day.)* In terms of a 24-hour day, how long have humans existed? *(less than two minutes)*

Refer students to the color key.

Ask Why is there so little yellow on the clock? *(because yellow represents the Cenozoic Era, which has taken up only a very small portion of Earth's history)*

Build Vocabulary

WORD ORIGINS Remind students that plate tectonics explains the movement of Earth's continents.

Say The word *continent* comes from the Latin word *continens*, which means "to hold together" or "contain."

Ask What do you think is held together or contained on a continent? *(Sample answers: the land, landforms such as mountains and valleys, bodies of water, living things)*

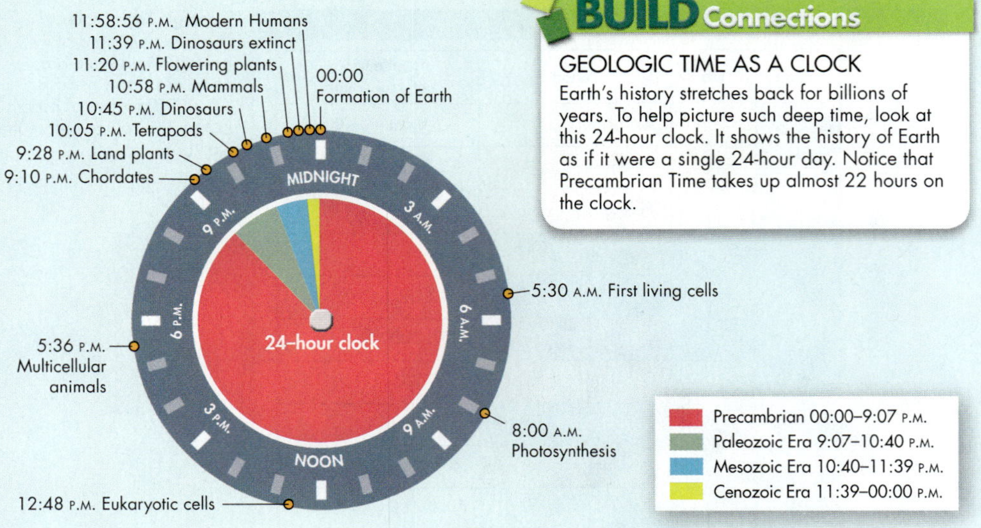

BUILD Connections

GEOLOGIC TIME AS A CLOCK

Earth's history stretches back for billions of years. To help picture such deep time, look at this 24-hour clock. It shows the history of Earth as if it were a single 24-hour day. Notice that Precambrian Time takes up almost 22 hours on the clock.

Divisions of the Geologic Time Scale Divisions of geologic time have different lengths. For example, the Cambrian Period lasted 54 million years. The Cretaceous Period lasted 80 million years. Eons are the largest divisions of time. Eons are divided into **eras.** Eras are further divided into **periods.** Periods range in length from hundreds of millions of years to just 2 million years.

Life on a Changing Planet

In our planet's long history, its physical environment has undergone many changes. These changes have greatly affected life on Earth.

The Physical Environment Some of the factors that have changed the physical environment are plate tectonics, changing climates, and collisions with objects from space.

▶ ***Geological Forces*** Geologic forces can form mountain ranges and move continents. The theory of **plate tectonics** states that Earth's surface is divided into huge plates. These plates move very slowly. Over long periods of time, their movement has pushed continents together, then pulled them apart again. The movement of the plates causes earthquakes and volcanoes.

The movement of the continents has affected the distribution of organisms. For example, Africa and South America are now separated by the Atlantic Ocean. But fossils of a reptile called *Mesosaurus* have been found on both Africa and South America. These fossils are evidence that these two continents were once joined together.

BUILD Vocabulary

era a major division of geologic time; usually divided into two or more periods

period a division of geologic time into which eras are subdivided

plate tectonics the theory that explains the slow movement of solid continental plates over Earth's mantle

WORD ORIGINS
The word *tectonics* from the Greek *tekton*, which means "builder." Plate tectonics is the theory that explains the movement of continents, formation of volcanoes, and mountain building.

 BIOLOGY.com Search Lesson 19.1 GO • Visual Analogy

Hands-On Learning

CONTINENTAL DRIFT Have students model the movement of continents over time.

1. Divide the class into small groups, and give each group seven small objects, such as erasers, paper clips, various pasta shapes, or pieces of index cards.
2. Refer groups to The Changing Face of Earth visual. Point out the seven continents on the Present Day globe. Then, have students locate the continents in the End of Permian Period globe. Tell them to arrange their objects on a desk to match the positions of the continents on that globe.
3. Have students move the objects together or apart as needed to model the movement of the continents in the following globes.
4. Ask each group to make a prediction about where similar fossils might be found on different continents. Encourage groups to share their predictions with the class and explain their reasoning.

▶ ***Changing Climate*** Earth's climate has changed greatly over the history of life. Many of these changes were caused by small shifts in temperature. For example, during the great ice ages, temperatures were only about 5°C cooler than they are now. These small changes had a big effect on living things.

▶ ***Comets and Asteroids*** Evidence shows that large objects from space have crashed into Earth several times. These collisions threw large amounts of dust into the air. This dust would have blocked enough sunlight to cause global cooling and other kinds of climate change. Such events could have caused organisms around the world to go extinct.

Biological Forces Living organisms also play a major role in shaping the environment. Earth's early oceans held very little oxygen. The first photosynthetic organisms began using carbon dioxide and giving off oxygen. Since then, our planet has never been the same! As levels of carbon dioxide dropped, the climate cooled.

Even today, organisms shape the landscape as they build soil from rock and sand. Plants, animals, fungi, and microorganisms are part of the cycles of carbon, nitrogen, and oxygen. Earth is indeed a living planet!

Key Question How have our planet's environment and living things interacted to shape the history of life on Earth? **Geological forces, changing climates, and collisions with comets and asteroids have changed the habitats of living organisms. The actions of living organisms have also changed conditions in the land, water, and atmosphere of Earth.**

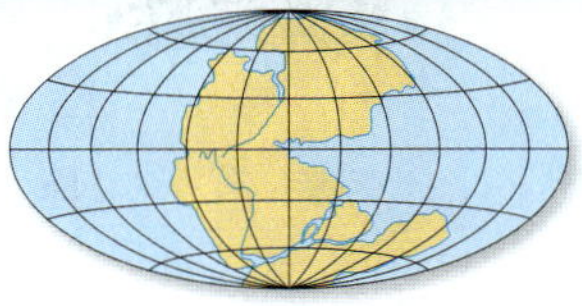

End of Permian Period At the end of the Permian Period, Earth's continents collided to form one giant landmass called Pangaea.

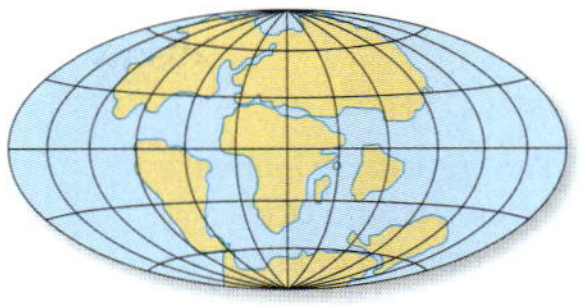

End of Cretaceous Period By the end of the Cretaceous Period, the continents as we know them began to drift apart.

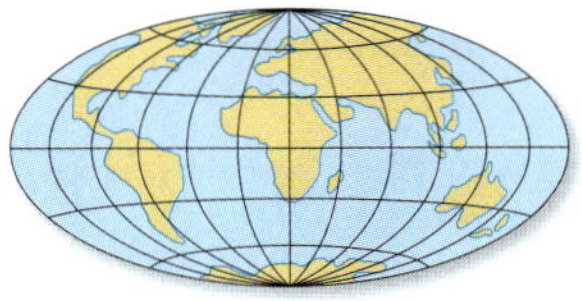

Present Day

The Changing Face of Earth Over the last 225 million years, the movement of the continents has changed Earth dramatically.

CHECK Understanding

Apply Vocabulary

Use the highlighted words from the lesson to complete each sentence correctly.

1. In _______, rock layers and their fossils are placed in order according to time.
2. Fossils that are used to determine the relative ages of rock layers are called _______.
3. The _______ of a radioactive element is the amount of time it takes for one half of the atoms in a sample to decay.

Critical Thinking

4. **Explain** What can a paleontologist learn from fossils?
5. **Review** What are the two ways in which geologists determine the age of fossils?
6. **Explain** How are eras and periods related?
7. **Relate Cause and Effect** Describe two ways that plate tectonics has affected life on Earth.
8. **Write to Learn** Answer the first clue in the mystery.

MYSTERY CLUE

Paleontologists discovered dramatic changes in the fossil record at the end of the Permian Period. What methods do you think they used to date that change at 251 million years ago? (Hint: See p. 452.)

BIOLOGY.com Search Lesson 19.1 GO • Lesson Assessment

Check Understanding

1. relative dating
2. index fossils
3. half-life
4. A paleontologist can learn about the structures of ancient organisms, infer evolutionary relationships with living organisms, track evolutionary changes over time, reconstruct ecosystems, and infer ancient environmental changes.
5. relative dating and radiometric dating
6. An era is divided into periods.
7. Plate tectonics has caused continental drift, which changes the distribution of organisms. It also causes volcanoes, which change the landscape and can provide habitats where new species can evolve.
8. The scientists probably used radiometric dating to date the end of the Permian Period, because this method gives an absolute age, or age in years. Relative dating does not give the absolute age.

Active Reading

SCIENCE SUPPORT Explain to students that Earth's surface is divided into large plates that float on partly melted rock and that the continents are part of the plates. As the plates move, so do the continents.

KEY QUESTION Have a student read aloud the Key Question and answer.

Ask How might dust in the air from a collision between Earth and an asteroid affect the way plants carry out photosynthesis? *(Dust could block sunlight and reduce photosynthesis.)* If plants died as a result, how might that change levels of oxygen on Earth? *(Oxygen levels might drop.)*

Speed Bump

Ask What are three nonliving factors that have affected life on Earth? *(movement of continents, climate change, collisions of comets and asteroids with Earth)*

Ask Organisms are involved in the cycles of which elements? *(carbon, nitrogen, oxygen)*

Post Reading

ASSESSMENT TIPS Refer students to the Geologic Time Scale chart to help them answer Question 6.

Point out that Questions 5 and 7 require two answers.

MYSTERY CLUE Discuss with students the two ways scientists date fossils and rock layers. Prompt them to note the difference between relative age and absolute age.

Wrap-Up Activity

MAKE AN INTERACTIVE QUIZ Divide the class into pairs, and ask each pair to use the information in the Geologic Time as a Clock diagram to prepare three quiz questions. Tell students the answers to their questions should be times during which events took place, the names of events that took place, or the names of eras. Sketch on the board the framework of the clock model. Label only the 24-hour time divisions. Then, have each pair take turns reading aloud their questions. Invite other pairs to respond when they know the correct answer. Direct those responding to come to the board to label the clock model with the answer to the question, or, if the answer is a time on the clock, to circle the correct time.

LESSON 19.2

Pre-Reading

Objectives

19.2.1 Identify the processes that influence survival or extinction of a species or clade.

19.2.2 Contrast gradualism and punctuated equilibrium.

19.2.3 Name two important patterns in microevolution.

19.2.4 Explain the evolutionary characteristics of coevolving organisms.

Preview the Pages

1. Help students with the pronunciations of the vocabulary terms. Point out the inclusion of the words *evolution* and *extinction* in the terms.
2. Preview the visuals. Point out that both The Evolution of Modern Birds cladogram and the Adaptive Radiation diagram show evolutionary relationships among animals. Note that both diagrams in Models of Evolution illustrate genetic change over time.

Activate Prior Knowledge

Guide students to recall that a cladogram is a diagram that shows patterns of shared characteristics among groups of related organisms.

Build Understanding

CONCEPT MAP Direct students to Workbook B, where they will find a concept map that they should complete as they read. Tell students that a concept map will help them understand the relationships between key terms and important concepts.

NATIONAL SCIENCE EDUCATION STANDARDS

UNIFYING CONCEPTS AND PROCESSES

I, II, III, IV, V

CONTENT

C.3.a, C.3.b, C.3.c, C.3.d, C.3.e

INQUIRY

A.1.e, A.2.a, A.2.e

For the corresponding lesson in *Miller & Levine Biology,* see pages 546–552.

19.2 Patterns and Processes of Evolution

Key Questions

What processes influence whether species and clades survive or become extinct?

How fast does evolution take place?

What are two patterns of macroevolution?

What is coevolution?

BUILD Understanding

Concept Map Make a concept map that includes the patterns of macroevolution shown in this lesson.

In Your Workbook Go to your workbook to learn more about making concept maps. Complete the concept map for Lesson 19.2.

Speciation and Extinction

The fossil record shows a parade of organisms that evolved, survived for a time, and then disappeared. More than 99 percent of all the species that ever lived are now extinct. How have so many different groups evolved? Why are so many extinct?

The study of Earth's history leaves no doubt that life has changed over time. Many changes took place within species. Others took place in larger groups and over longer periods of time. Major changes in structure, behavior, and ecology are **macroevolutionary patterns.** The ways that new species form and others become extinct are macroevolutionary patterns. The extinction of the dinosaurs is a macroevolutionary pattern. So is the increase in the number of flowering plants.

Macroevolution and Cladistics Paleontologists study fossils to learn about the history of life. Part of their work is classifying fossils. As with living species, cladistics is used to classify fossil species. Paleontologists use shared derived characteristics to place fossils in clades. Paleontologists put some fossils into clades that contain only extinct organisms. They classify others in clades with organisms that are alive today.

Cladograms often show that a fossil species is related to a living species. Usually this does not mean that the fossil was the direct ancestor of the living species. For example, the cladogram on the facing page shows many dinosaurs that descended from a line of common ancestors. These dinosaurs were not the direct ancestors of modern birds.

Adaptation and Extinction Throughout the history of life, organisms have lived in changing environments. When conditions change, the process of evolution allows some species to adapt. These species thrive in the new environment. Other species do not adapt. Eventually, those species become extinct.

The rate of speciation and extinction varies from clade to clade. Some clades contain many species that have survived for long periods of time. Other clades have only a few species, which soon become extinct.

Why are some clades more successful than others? Part of the answer to this question is species diversity. Variation among species in a clade is like the variation among individuals in a species. Both are "raw material" for evolution.

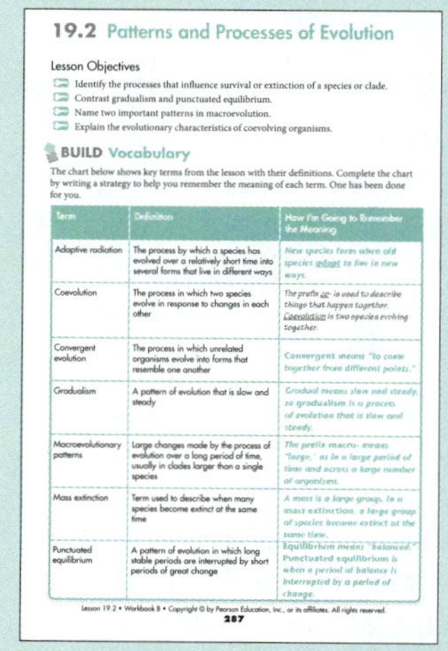

19.2 Patterns and Processes of Evolution

Lesson Objectives

- Identify the processes that influence survival or extinction of a species or clade.
- Contrast gradualism and punctuated equilibrium.
- Name two important patterns in macroevolution.
- Explain the evolutionary characteristics of coevolving organisms.

BUILD Vocabulary

The chart below shows key terms from the lesson with their definitions. Complete the chart by writing a strategy to help you remember the meaning of each term. One has been done for you.

Term	Definition	How I'm Going to Remember the Meaning
Adaptive radiation	The process by which a species has evolved over a relatively short time into several forms that live in different ways	New species form when old species adapt to live in new ways.
Coevolution	The process in which two species evolve in response to changes in each other	The prefix co- is used to describe things that happen together. Coevolution is two species evolving together.
Convergent evolution	The process in which unrelated organisms evolve into forms that resemble one another	Convergent means "to come together from different points."
Gradualism	A pattern of evolution that is slow and steady	Gradual means slow and steady, so gradualism is a process of evolution that is slow and steady.
Macroevolutionary patterns	Large changes made by the process of evolution over a long period of time, usually in clades larger than a single species	The prefix macro- means "large," as in a large period of time and across a large number of organisms.
Mass extinction	Term used to describe when many species become extinct at the same time	A mass is a large group. In a mass extinction, a large group of species become extinct at the same time.
Punctuated equilibrium	A pattern of evolution in which long stable periods are interrupted by short periods of great change	Equilibrium means "balanced." Punctuated equilibrium is when a period of balance is interrupted by a period of change.

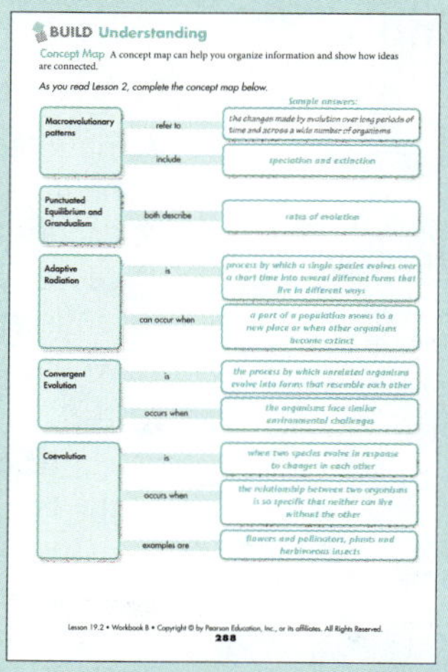

BUILD Understanding

Concept Map A concept map can help you organize information and show how ideas are connected.

As you read Lesson 2, complete the concept map below.

Sample answers:

Macroevolutionary patterns — refer to — the changes made by evolution over long periods of time and across a wide number of organisms
Macroevolutionary patterns — include — speciation and extinction
Punctuated Equilibrium and Grandualism — both describe — rates of evolution
Adaptive Radiation — is — process by which a single species evolves over a short time into several different forms that live in different ways
Adaptive Radiation — can occur when — a part of a population moves to a new place or when other organisms become extinct
Convergent Evolution — is — the process by which unrelated organisms evolve into forms that resemble each other
Convergent Evolution — occurs when — the organisms face similar environmental challenges
Coevolution — is — when two species evolve in response to changes in each other
Coevolution — occurs when — the relationship between two organisms is so specific that neither can live without the other
Coevolution — examples are — flowers and pollinators, plants and herbivorous insects

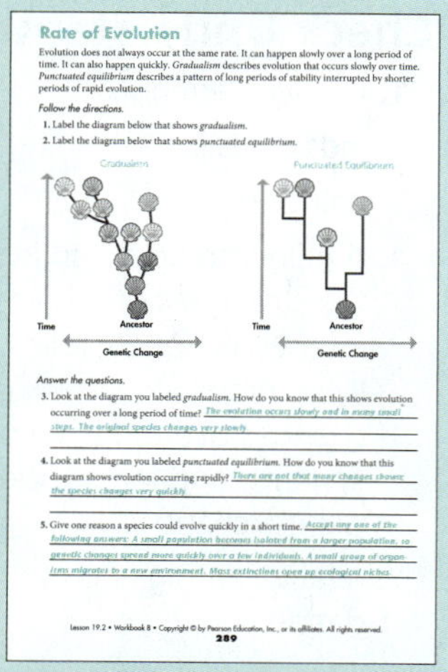

Rate of Evolution

Evolution does not always occur at the same rate. It can happen slowly over a long period of time. It can also happen quickly. *Gradualism* describes evolution that occurs slowly over time. *Punctuated equilibrium* describes a pattern of long periods of stability interrupted by shorter periods of rapid evolution.

Follow the directions.

1. Label the diagram below that shows *gradualism.*
2. Label the diagram below that shows *punctuated equilibrium.*

Answer the questions.

3. Look at the diagram you labeled *gradualism.* How do you know that this shows evolution occurring over a long period of time? The evolution occurs slowly and in many small steps. The original species changes very slowly.
4. Look at the diagram you labeled *punctuated equilibrium.* How do you know that this diagram shows evolution occurring rapidly? There are not that many changes shown; the species changes very quickly.
5. Give one reason a species could evolve quickly in a short time. Accept any one of the following answers: A small population becomes isolated from a larger population, so genetic changes spread more quickly over a few individuals. A small group of organisms migrates to a new environment. Mass extinctions open up ecological niches.

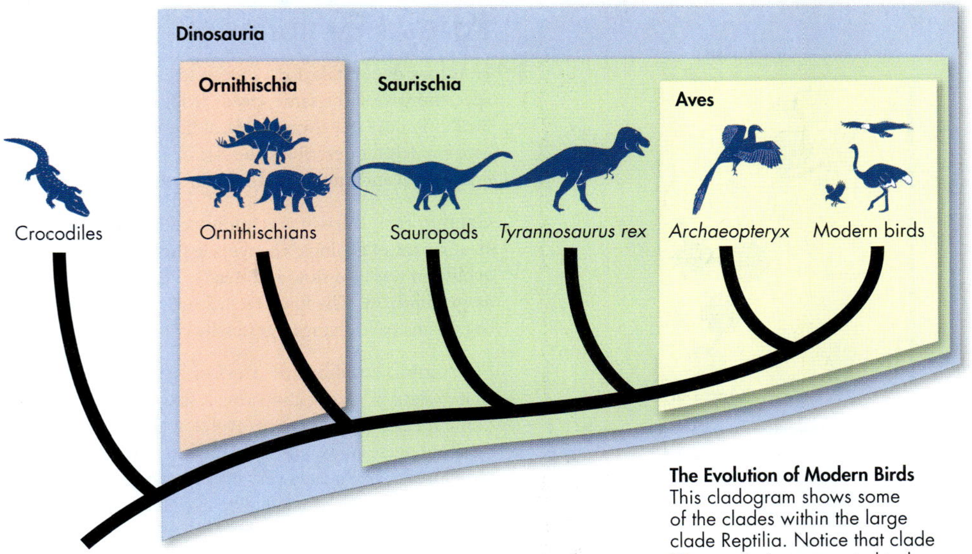

The Evolution of Modern Birds This cladogram shows some of the clades within the large clade Reptilia. Notice that clade Dinosauria is represented today by modern birds.

Genetic variation increases the chance that a species will survive in a changing environment. In the same way, the diversity of species in a clade increases the chance that a clade will survive. If new species evolve more quickly than species become extinct, a clade will survive. If extinctions happen faster than speciation, a clade will die out.

Clade Reptilia is a highly successful clade. It includes living organisms such as snakes, lizards, and turtles. It also includes dinosaurs, which lived for tens of millions of years. Most species in clade Dinosauria are now extinct. But clade Dinosauria also included some species that adapted to new conditions. Descendants of those species are still alive today. We call them birds.

Patterns of Extinction Some species become extinct because of the slow and steady process of natural selection. This "everyday" extinction is called **background extinction.** Other species go extinct in a relatively short period. This process is called **mass extinction.** In a mass extinction, entire ecosystems vanish. Some mass extinctions may have been caused by a single event, such as the large asteroid that hit Earth at the end of the Cretaceous period. Other mass extinctions were probably due to several causes working together, such as volcanic eruptions, changing climates, and changing sea levels.

Key Question What processes influence whether species and clades survive or become extinct? **If the rate of speciation in a clade is equal to or greater than the rate of extinction, the clade will survive. If the rate of extinction is greater than the rate of speciation, the clade will become extinct.**

BUILD Vocabulary

macroevolutionary patterns the changes in anatomy, phylogeny, behavior, and ecology, which take place in clades that are larger than a single species

background extinction extinction caused by the slow and steady process of natural selection

mass extinction an event during which many species become extinct during a relatively short time

PREFIXES
The prefix *macro-* comes from the Greek *makro-*, meaning "large." Macroevolutionary patterns are large patterns of evolution.

ELL Focus on ELL: Extend Language

ALL SPEAKERS Have students use the Think-Pair-Share strategy to help them understand the concepts of speciation and extinction. Pair beginning and intermediate speakers with advanced speakers. Have students take turns reading aloud each paragraph in the section Speciation and Extinction, one paragraph at a time. As a student finishes reading a paragraph, both students in the pair should write in their notebooks a question about the content of the paragraph. Direct students to focus on terms describing speciation or extinction or on steps and examples of those processes.

Have students in each pair take turns reading their questions and responding to each other's questions. Discuss as a class the questions and responses that each pair had discussed.

During Reading

Active Reading

USE DIAGRAMS Discuss the relationship between dinosaurs and birds, using The Evolution of Modern Birds cladogram.

Say The blue section of the cladogram shows clade Dinosauria.

Ask Which descendants of clade Dinosauria are alive today? *(modern birds)*

Ask Are modern birds direct descendants of *Tyrannosaurus rex*? *(no)* How are modern birds related to *Tyrannosaurus rex*? *(They share a common ancestor.)*

Build Vocabulary

PREFIXES Review with students the meaning of the prefix *macro-*.

Ask What do you think a macrofossil is? *(a large fossil)*

KEY QUESTION Read aloud the Key Question and answer. Discuss the meanings of *rate of speciation* and *rate of extinction*. Then, have students paraphrase the answer to include the meanings of the phrases.

Speed Bump

Ask How are dinosaurs, birds, snakes, and lizards related? *(All four types of organisms are members of the clade Reptilia. Dinosaurs and birds are also members of the clade Dinosauria.)*

Ask What type of extinction could occur if polar ice melted and organisms that rely on the ice could no longer survive? *(mass extinction)*

During Reading continued

Active Reading

INTERPRET DIAGRAMS Direct students to the Models of Evolution diagrams. Have them identify the ancestor and the four different-colored final descendants (light green, dark green, light blue, and pink) in each diagram.

Ask Do the shells show the same pattern of evolution in the two diagrams? *(no)*

Ask What kind of genetic change is represented by the curved paths in the top diagram? *(gradual change)*

Ask In the bottom diagram, what do the straight horizontal lines indicate? *(sudden genetic change)* What do the straight vertical lines indicate? *(periods of little genetic change)*

In the Gradualism diagram, point out the change in color that occurs at each stage of evolution between the ancestor and a final descendant.

Say I am looking at the Gradualism diagram. Between the ancestor and the dark green descendant, I see several different shades of blue and green. What do these colors represent? *(gradual changes that occurred in stages between an ancestor and a descendant)*

COMPARE AND CONTRAST Have students prepare a compare/contrast table in their notebooks. Tell them to label the two columns Gradualism and Punctuated Equilibrium, respectively. Give students time to reread this page and then fill in their charts.

Ask Which model of evolution matches Darwin's ideas of evolution? *(gradualism)*

Ask How does punctuated equilibrium differ from gradualism? *(Punctuated equilibrium describes evolution as having both periods of rapid change and periods of little change.)*

Ask What is the result of each type of evolution? *(genetic change over time)*

KEY QUESTION Invite a student to read aloud the Key Question and answer.

Ask What are the names of the two different patterns for describing the rate of evolution? *(gradualism and punctuated equilibrium)*

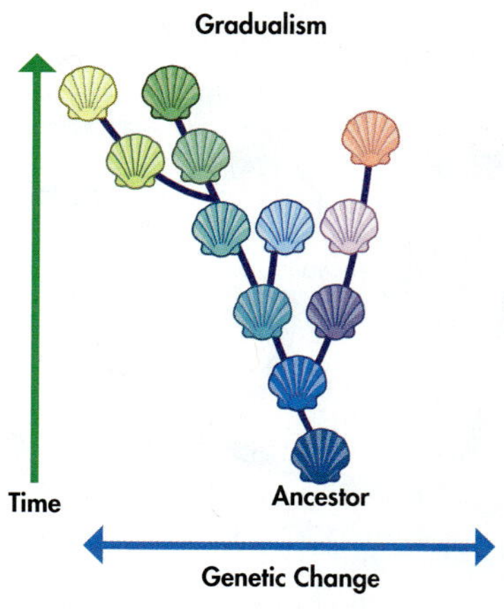

Gradualism involves a slow, steady change in a particular line of descent.

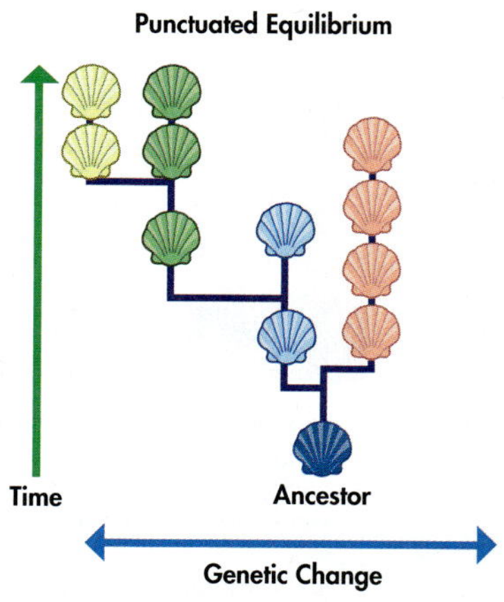

Punctuated equilibrium involves stable periods interrupted by rapid changes.

Models of Evolution Biologists have two different ways of looking at and thinking about the rate of evolution: gradualism and punctuated equilibrium. These simplified drawings compare the two patterns.

Rate of Evolution

How long does it take for a species to evolve? Do all species evolve at the same speed? Evidence shows that evolution does not always move at the same speed. Two models for describing the rate of evolution are gradualism and punctuated equilibrium.

Gradualism Darwin was impressed by the slow, steady pace of geologic change. He thought that evolution was also slow and steady. This idea is known as **gradualism.** The fossil record shows that many organisms have changed gradually over time.

Punctuated Equilibrium The fossil record also shows that evolution has not always been slow and steady. Some species do not seem to evolve at all. For example, fossils of horseshoe crabs show that their major body structures have changed very little over millions of years. Such species are said to be in *equilibrium.* They may be evolving genetically, but their structures do not change much.

Fossils of other organisms show periods of stability and rapid change. Equilibrium that is interrupted by periods of rapid change is **punctuated equilibrium.** Some biologists suggest that most new species are produced during these periods of rapid change. (But remember that evolution takes place according to the geologic time scale. To a geologist, "rapid change" may mean thousands of years!)

Rapid Change After Equilibrium Small, isolated populations can evolve rapidly. This rapid evolution is possible because genetic changes spread more quickly among fewer individuals. Rapid evolution may also happen when a small group moves to a new environment. That is what happened with the Galápagos finches.

Mass extinctions can lead to rapid evolution. After a mass extinction, there are fewer competing species. As a result, more "jobs" are available for new species. The fossil record shows that after a mass extinction, the groups that survive evolve rapidly. Biodiversity is eventually restored.

Key Question How fast does evolution take place? **Evidence shows that evolution has moved at different rates for different organisms at different times over the long history of life on Earth.**

Focus on Academic Language

- steady
- stable
- isolated

Have students find the sentences on this page that include the adjectives *steady, stable,* or *isolated.* Ask volunteers to define the words using context clues. Instruct other students to look up definitions of the three adjectives in a dictionary.

Ask How are the meanings of *steady* and *stable* similar? *(Both words can mean "not changing.")* What is a somewhat different meaning of *steady*? *(direct in movement)* What is an example of this definition of *steady*? *(Sample answers: steady car ride, steady drill, steady rainfall)*

Ask What is another way to describe a population that is isolated? *(Sample answer: a population that is by itself)* What other things can be described as isolated? *(Sample answers: a house far from others, the top of a distant mountain, a small island in the middle of the ocean)*

458 Chapter 19 • Lesson 2

Adaptive Radiation and Convergent Evolution

Two important patterns of macroevolution are adaptive radiation and convergent evolution. Darwin noticed both of these patterns as he traveled on the *Beagle*.

Adaptive Radiation Fossils show that a single species can evolve into a clade with many species. All of these species have variations on the original ancestor's body plan. The new species may fill different ecological niches. **Adaptive radiation** is the process by which a single species or small group of species evolves into several different forms that live in different ways. Sometimes, this process can happen relatively quickly. Adaptive radiation can happen when species move to a new environment or after a mass extinction. It can also happen when a new adaptation lets a species live in an environment that had not been used before.

▶ ***Adaptive Radiations in the Fossil Record*** Dinosaurs lived for about 150 million years during the Mesozoic. They arose from several adaptive radiations of reptiles. Fossils show that when dinosaurs were at their greatest diversity, mammals were small and not very diverse. After most of the dinosaurs became extinct, an adaptive radiation of mammals began. That radiation led to the great diversity of mammals of the Cenozoic Era.

▶ ***Modern Adaptive Radiation*** Two examples of adaptive radiation are the Galápagos finches and Hawaiian honeycreepers. Their stories are similar: A single species evolved into many different species. Both finches and honeycreepers evolved different beaks and behaviors that let them eat different kinds of food.

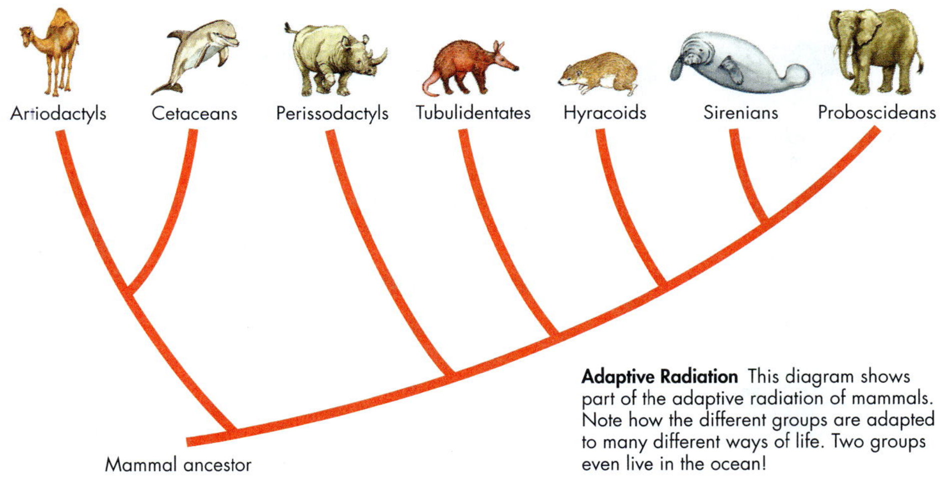

Adaptive Radiation This diagram shows part of the adaptive radiation of mammals. Note how the different groups are adapted to many different ways of life. Two groups even live in the ocean!

BUILD Vocabulary

gradualism
the evolution of a species by the gradual addition of small genetic changes over long periods of time

punctuated equilibrium
the pattern of evolution in which long stable periods are interrupted by brief periods of more rapid change

adaptive radiation
the process by which a single species or a small group of species evolves into several different forms that live in different ways

WORD ORIGINS
The word *radiation* means "to spread out in all directions." It comes from the Greek *radi-*, which means "ray," as in a ray of the sun. In adaptive radiation, species evolve in many directions.

ELL Focus on ELL: Access Content

BEGINNING SPEAKERS On the board, draw a vocabulary word map with a center circle attached to five surrounding boxes. Write *adaptive radiation* in the center circle. Have students work with a partner to copy and complete the word map in their notebooks. Explain that they should fill in the surrounding boxes with the following: dictionary or text definition of the term, definition in their own words, an idea that involves the term, a word clue, and a picture representing the term.

Circulate among the pairs as they work, asking questions about the construction of their word maps. Have pairs present their word maps to the class.

Active Reading

SUMMARIZE Have students read this page.

Say You can often discover problems you are having understanding a new idea when you try to write it in your own words.

Direct students to write in their own words a short paragraph that summarizes what they just read about adaptive radiation. Tell them to focus their paragraph on these questions: "What is being described?" "When does it take place?" "How does it work?" and "What are examples?" Invite students to share and discuss their paragraphs with the class. Discuss any problems students had understanding the concept of adaptive radiation.

INTERPRET DIAGRAMS Have a student read aloud the caption of the Adaptive Radiation diagram.

Ask What are some physical differences among these mammals? *(Sample answers: body size, size of ears, amount of hair or fur, legs vs. flippers, types of tails, presence of trunk or horn)*

Ask What are some different environments in which these mammals live? *(Sample answers: bodies of water, deserts, forests, grasslands)*

Ask According to the diagram, which mammal group is most closely related to elephants? *(Sirenians)*

Build Vocabulary

WORD ORIGINS Point out to students that *gradual* is part of the word *gradualism*.

Say The adjective *gradual* means "proceeding by steps or degrees." It comes from the Latin word *gradus*, which means "step."

Ask How can you use the word origin of *gradual* to reword the definition of *gradualism*? *(Sample answer: Gradualism is evolution that occurs in small, steady steps over a long period of time.)*

Speed Bump

Ask What is the difference between gradualism and punctuated equilibrium? *(Gradualism is a pattern of evolution that is slow and steady. Punctuated equilibrium is a pattern of evolution in which long stable periods are interrupted by brief periods of rapid change.)*

Ask How are fossils important in identifying adaptive radiations? *(Fossils show how species in a clade have evolved with variations on the original ancestor's body plan.)*

Ask Why were new ecological niches available for mammals to fill when their adaptive radiation began? *(Dinosaurs had become extinct.)*

During Reading continued

Active Reading

CAUSE AND EFFECT Discuss with students the cause of convergent evolution.

Ask What caused anteaters in different parts of the world to evolve similar features even though the animals aren't related? *(Natural selection favored features that allowed the animals to catch and eat insects.)*

Ask How do similar environments cause convergent evolution to occur among organisms? *(Organisms in similar environments tend to evolve similar structures because they face similar selection pressures.)*

Ask What do you think caused both dolphins and fishes, which are only very distantly related, to develop sleek shapes, muscular bodies, and smooth surfaces? *(In both groups, natural selection has favored adaptations that allow for strong swimming abilities.)*

Build Vocabulary

ACADEMIC WORDS Remind students that *convergent* is an adjective describing something that comes together from different directions.

Say Convergent lines in a painting or drawing are lines that come together.

Ask What are some other things that can be described as convergent? *(Sample answers: lines of people walking into a building, streets that meet at an intersection, subway lines that meet in the middle of a city)*

Active Reading

KEY QUESTION Invite a student to read aloud the Key Question and answer.

Ask What is macroevolution? *(major changes in groups of organisms over time)*

Speed Bump

Ask What kinds of structures might develop through convergent evolution in unrelated animals that catch prey? *(Sample answers: good eyesight, legs adapted for running, strong jaws, sharp teeth)*

Ask Why would it be difficult to find examples of convergent evolution by comparing organisms from the deserts to organisms from the deepest parts of the ocean? *(These two groups of organisms face very different natural selection pressures.)*

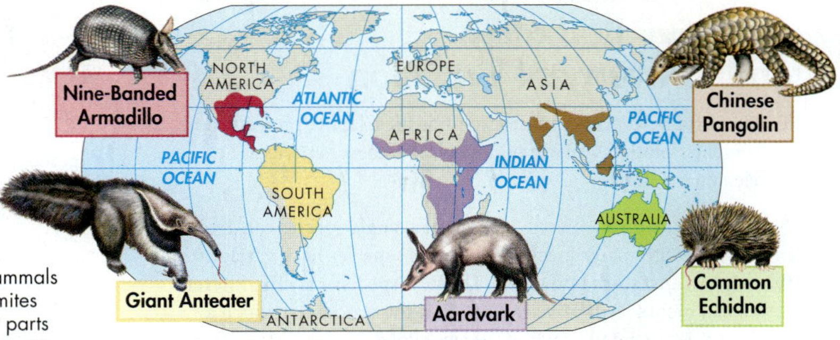

Convergent Evolution Mammals that feed on ants and termites are found in five different parts of the world. These species are unrelated. But they all have powerful front claws, a long hairless snout, and a tongue covered with sticky saliva. These adaptations are very useful for hunting and eating insects.

Convergent Evolution Sometimes, organisms in different places evolve in similar environments. These organisms start out with different structures. But they face similar selection pressures. In such cases, natural selection may cause different body structures to evolve in ways that let them carry out similar functions. Over time, the different organisms become more and more alike. This process is called convergent evolution. In **convergent evolution,** natural selection forms similar structures and characteristics in organisms that are not closely related.

During the voyage of the *Beagle*, Darwin noticed similarities among large grassland birds on different continents. Although they look alike, emus, rheas, and ostriches are not closely related. All three of these birds have long, strong legs let them run swiftly through tall grasses. These three flightless birds are an example of convergent evolution.

The different kinds of anteaters also show convergent evolution. Most continents have mammals that specialize in eating ants and termites. South America has giant anteaters, North America has armadillos, and Africa has aardvarks. All of these animals have long tongues. They also have strong front legs with claws that can pull apart termite mounds and ant nests. The similar adaptations of these distantly related animals let them eat the same kinds of food.

Key Question What are two forms of macroevolution?
Two important patterns of macroevolution are adaptive radiation and convergent evolution.

BUILD Vocabulary

convergent evolution the process by which unrelated organisms independently evolve similarities when adapting to similar environments

coevolution the process by which two species evolve in response to changes in each other over time

ACADEMIC WORDS
The word *convergent* is the adjective form of the verb *converge*, which means "to come together from different directions."

Coevolution

Sometimes the lives of two species are so closely connected that they evolve together. For example, many flowering plants can reproduce only if they are pollinated by a certain species of insect. The insect, in turn, depends on the nectar or pollen from those flowers for food. The process by which two species evolve in response to changes in each other is called **coevolution.**

Hands-On Learning

CONVERGENT EVOLUTION Have students imagine that space travel has advanced and people are being transported to distant planets A, B, C, and D to start civilizations there. The people do not have weapons and will have to adapt physically to their environment. Surprisingly, the planets all have very similar environments.

1. Write on the board the following characteristics of the planets: light levels are low, rain falls daily, many food sources are underwater, the only green plants are on high mountains, some animals are large and fast enough to catch and eat humans, and the landscape requires tall people to duck under rocks and trees.
2. Have students form four groups, and assign one planet to each group. Instruct each group to predict how the humans on their planet will evolve, by listing possible adaptations. *(Sample adaptations: good eyesight, short height, better swimming ability, tough skin, fast running)* Have groups compare their predicted adaptations to see if convergent evolution occurred on the different planets.

Flowers and Pollinators Coevolution of flowers and pollinators is common and can lead to unusual results. For example, Darwin discovered a flower with a long structure called a spur. The spur was 40 centimeters long. At the bottom of the spur was a supply of nectar. Nectar in flowers usually attracts pollinators. But what pollinator could reach down such a long tube? Darwin predicted that the plant must be pollinated by an insect with mouth parts that could reach to the bottom of the spur. About 40 years later, scientists discovered a moth with a 40-centimeter-long feeding tube. This moth matched Darwin's prediction!

Plants and Herbivorous Insects Coevolution also takes place between plants and the insects that eat them. Insects have been eating flowering plants ever since both groups evolved. In some cases, plants have evolved poisons to protect themselves from being eaten. But insects have evolved in response to the poisons. Natural selection has favored any individuals that are resistant to the poisons.

Milkweed plants are poisonous to many insects. But the caterpillars of monarch butterflies are able to eat their leaves. In fact, the monarch caterpillars store the milkweed poisons in their bodies. The poisons help protect the monarchs from predators!

Key Question What is coevolution? **The relationship between two organisms may become so specific that neither organism can survive without the other. Thus, an evolutionary change in one organism is usually followed by a change in the other organism.**

Coevolution Milkweed plants produce poisons that protect them from being eaten. But monarch butterflies have coevolved with the milkweed plants. Their caterpillars can eat milkweed leaves without harm. In fact, monarchs store the milkweed poisons in their own bodies. This poison protects the monarchs against *their* predators!

CHECK Understanding

Apply Vocabulary

Use the highlighted words from the lesson to complete each sentence correctly.

1. An event during which many species become extinct over a short period of time is a _______.
2. The slow, steady "everyday" rate of extinction that has happened throughout the history of life is called _______.
3. The process by which two species evolve in response to changes in each other over time is called _______.

Critical Thinking

4. **Review** How does variation within a clade affect the clade's chance of surviving environmental change?
5. **Compare and Contrast** Explain how punctuated equilibrium is different from gradualism.
6. **Review** What is adaptive radiation?
7. **Apply Concepts** Describe an example of coevolution.
8. **Write to Learn** Answer the second clue of the mystery. Use the words *cellular respiration* in your answer.

MYSTERY CLUE

Evidence indicates that before the Permian extinction, the oceans lost most of their oxygen. What effect do you think the loss of oxygen had on most organisms? (**Hint:** See p. 457.)

BIOLOGY.com Search Lesson 19.2 GO • Lesson Assessment

Check Understanding

1. mass extinction
2. background extinction
3. coevolution
4. More variation means some species are more likely to have features that will help them survive in the altered environment.
5. In punctuated equilibrium, species remain unchanged for long periods of time and then suddenly undergo rapid evolution. In gradualism, species evolve slowly and steadily over a long period of time.
6. Adaptive radiation is a rapid evolution of a single species into a number of different species that live in different ways.
7. Students might describe the milkweed and monarch butterfly or a flower and pollinator.
8. Without oxygen, organisms cannot carry out cellular respiration. Therefore, most eukaryotic organisms would die.

Active Reading

SCIENCE SUPPORT Explain to students that many examples of coevolution are well known in biology.

Say Biologists refer to the type of coevolution between plants and herbivorous insects as a "coevolutionary arms race." If the plant evolves a better "weapon," the insect also evolves a better "weapon." This triggers further evolutionary changes in the plant, and the "arms race" continues.

KEY QUESTION Read aloud the Key Question and answer.

Ask In terms of coevolution, why do fruit trees produce fewer offspring when bee populations in the same area decrease in size? *(The fruit trees have coevolved with bees to be pollinated by the bees when the bees get nectar from their flowers; the trees can't reproduce without the bees.)*

Post Reading

ASSESSMENT TIPS Before students answer Question 5, have them review the Models of Evolution diagrams and the compare/contrast table they made for gradualism and punctuated equilibrium.

Have students refer to the Adaptive Radiation diagram to help them answer Question 6.

For Question 7, direct students to reread the section Coevolution.

MYSTERY CLUE To help students evaluate the effects of oxygen loss in ancient seas, have them visualize what would happen to humans and other organisms if oxygen suddenly disappeared from Earth's atmosphere today.

Wrap-Up Activity

ACTING OUT EARTH'S HISTORY Write each vocabulary term on a separate index card. Shuffle the cards, and then call out the terms one at a time. Ask volunteers to read or say the definition of each term. Once all terms have been defined, divide the class into small groups. Put the vocabulary cards in a box, and have groups take turns taking a card from the box. Direct each group to act out silently, as a team, the definition of the term on their card. Name as winner the group that most often acted out the correct definitions of their terms.

LESSON 19.3

Pre-Reading

Objectives

19.3.1 Identify some of the hypotheses about early Earth and the origin of life.

19.3.2 Explain the endosymbiotic theory.

19.3.3 Explain the significance of sexual reproduction in evolution.

Preview the Pages

1. Read aloud the lesson's green headings. Ask students to predict what the text below each heading will discuss.
2. Preview the visuals. Have students read aloud the captions and labels. Point out that except for the Early Earth image on this page, the visuals in this lesson show step-by-step processes. Explain that the Miller-Urey Experiment visual shows an actual event. By contrast, the Origin of RNA and DNA visual and the Endosymbiotic Theory visual illustrate hypothetical processes.

Activate Prior Knowledge

Remind students that most of Earth's familiar life forms are eukaryotes. Ask a volunteer to define the term *eukaryotic cell. (a cell that contains a nucleus)*

Build Understanding

FLOWCHART Direct students to Workbook B, where they will find a partially completed flowchart. As students read Lesson 19.3, have them complete the flowchart to show steps that took place over time during Earth's early history.

NATIONAL SCIENCE EDUCATION STANDARDS

UNIFYING CONCEPTS AND PROCESSES

I, II, III, IV, V

CONTENT

C.1.a, C.1.e, C.3.a, C.3.b, C.3.d, C.3.e, D.3, D.4, G.3

INQUIRY

A.2.a, A.2.b, A.2.e, A.2.f

For the corresponding lesson in *Miller & Levine Biology,* see pages 533–558.

19.3 Earth's Early History

Key Questions

What do scientists hypothesize about early Earth and the origin of life?

What do scientists hypothesize about the origins of DNA and RNA?

What theory explains the origin of eukaryotic cells?

What is the evolutionary significance of sexual reproduction?

BUILD Understanding

Flowchart Construct a flowchart that shows what scientists hypothesize about early Earth. Show the major steps from the origin of Earth to the appearance of eukaryotic cells.

In Your Workbook Go to your workbook to learn about making a flowchart. Complete the flowchart for lesson 19.3.

Early Earth Violent volcanic eruptions helped shape Earth's early history.

The Mysteries of Life's Origins

How did life on Earth begin? What were the earliest forms of life? Scientists are working hard to answer these questions. As new evidence comes to light, hypotheses about early life are likely to change.

Evidence suggests that Earth formed from pieces of rock and dust. One or more huge objects crashed into the young Earth, causing it to melt. For millions of years, volcanoes erupted. Comets and asteroids crashed into the surface. About 4.2 billion years ago, Earth cooled enough for solid rocks to form. Rain fell, and oceans formed.

The Early Atmosphere The early atmosphere held little or no oxygen. It was mostly carbon dioxide, water vapor, and nitrogen. It also had smaller amounts of carbon monoxide, hydrogen sulfide, and hydrogen cyanide. A few deep breaths would have killed you! These gases probably made the sky orange. Dissolved iron probably made the oceans brown. This was the Earth on which life began.

Key Question What do scientists hypothesize about early Earth? **Earth's early atmosphere was mainly carbon dioxide, water vapor, and nitrogen, with lesser amounts of carbon monoxide, hydrogen sulfide, and hydrogen cyanide. It contained little or no oxygen.**

The First Organic Molecules Could organic molecules have formed on early Earth? In 1953, Stanley Miller and Harold Urey did an experiment to answer that question. They built a system of tubes and containers filled with water vapor, methane, ammonia, and hydrogen. They thought that this mixture of gases was like that of Earth's early atmosphere. To represent lightning, they hit the gases with sparks of electricity. After a week, they had made 12 amino acids. Amino acids are the building blocks of proteins!

We now know that Miller and Urey were wrong about the mixture of gases in the early atmosphere. But new experiments based on more current ideas have also made organic compounds. In 1995, Miller repeated the experiment with a more accurate mix of gases. This time, he made cytosine and uracil, two bases found in RNA.

Key Question What do scientists hypothesize about the origin of life? **Miller and Urey's experiment suggested how mixtures of the organic compounds necessary for life could have arisen from simpler compounds in the atmosphere of early Earth.**

 BIOLOGY.com Search Lesson 19.3 GO • Art Review

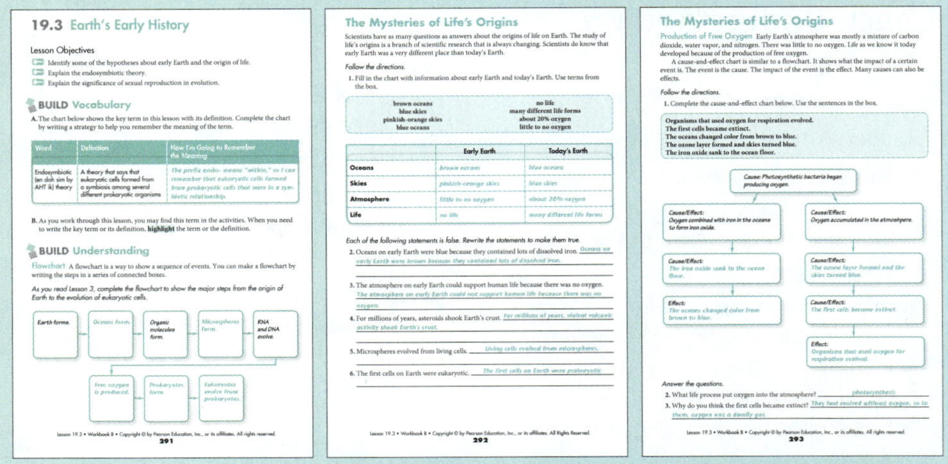

Formation of Microspheres The compounds made by the Miller-Urey experiments are a long way from living cells. How could living cells develop from this "stew" of organic compounds? The leap from nonlife to life is the greatest gap in scientific hypotheses about the history of life.

Large organic compounds sometimes form tiny bubbles called *microspheres*. Microspheres might have played a role in the evolution of cells. Like cells, they have selectively permeable membranes that allow water to pass through. Microspheres also have a simple means of storing and releasing energy. Several hypotheses suggest that microspheres may have developed the characteristics of living cells as early as 3.8 billion years ago.

Evolution of RNA and DNA Another unanswered question is the origin of RNA and DNA. Remember that cells are controlled by information stored in DNA. The information on DNA is transcribed onto RNA, which is used to form proteins. How could such a complex system evolve? The key to the answer may be RNA. Modern RNA is involved in many of life's reactions. The "RNA world" hypothesis suggests that RNA existed before DNA. The RNA-based system could then have evolved into the more complex DNA-based system.

Key Question What do scientists hypothesize about the origins of DNA and RNA?
The "RNA world" hypothesis proposes that RNA existed by itself before DNA. From this simple RNA-based system, several steps could have led to DNA-directed protein synthesis.

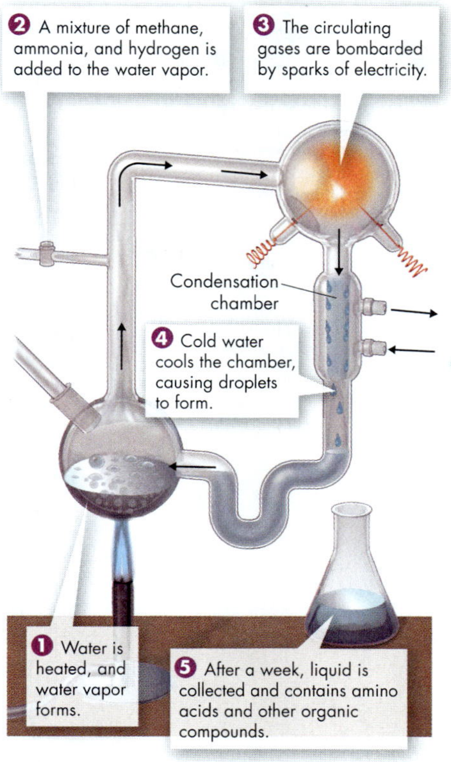

Miller-Urey Experiment Miller and Urey modeled conditions on ancient Earth. They produced amino acids by passing sparks through a mixture of gases.

Origin of RNA and DNA One hypothesis about the origin of life suggests that RNA evolved before DNA. Scientists have not yet demonstrated the later stages of this process in a laboratory setting.

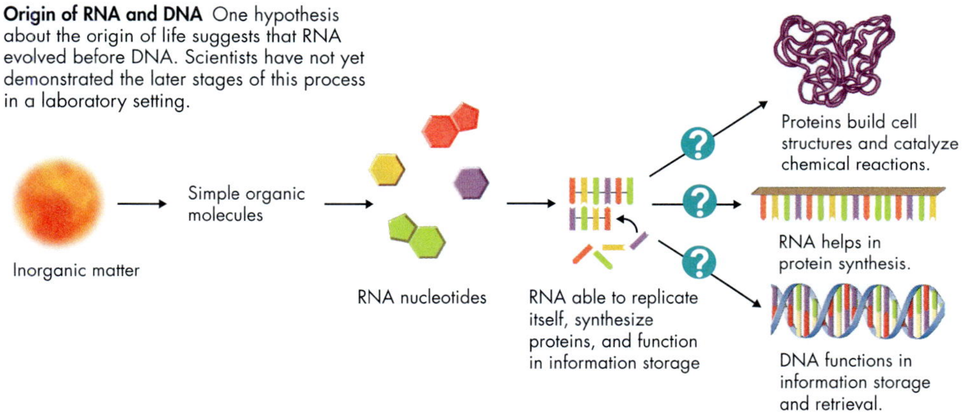

During Reading

Active Reading

KEY QUESTION Read aloud the first Key Question and answer on the previous page.

Ask What does *early Earth* mean? *(Sample answer: the time before there was life on Earth)*

USE VISUALS Explain that Miller and Urey performed an experiment to find out if the conditions on early Earth were right for organic molecules to have formed. Refer students to the visual showing the steps of the experiment.

Ask What mixture of gases did Miller and Urey use that modeled Earth's ancient atmosphere? *(water vapor, methane, ammonia, and hydrogen)*

Ask At what numbered step of the experiment did the scientists model lightning in the early atmosphere? *(step 3)* What did they use for lightning? *(sparks of electricity)*

Ask What did the experiment produce? *(amino acids and other organic compounds)*

KEY QUESTION Have a volunteer read aloud the second Key Question and answer on the previous page.

Ask What organic compounds necessary for life were produced in Miller and Urey's 1995 experiment? *(cytosine and uracil)*

KEY QUESTION Read aloud the question and answer on this page.

Ask In what three ways does DNA differ from RNA? *(DNA is double-stranded instead of single-stranded. DNA contains thymine instead of uracil. DNA contains deoxyribose instead of ribose.)*

Speed Bump

Ask What are microspheres? *(tiny bubbles made of large organic compounds that might have played a role in the evolution of cells)*

Ask How are microspheres similar to cells? *(They have selectively permeable membranes and a simple means of storing and releasing energy.)*

Hands-On Learning

LAB REPORT Have students work in pairs to write a lab report that summarizes the Miller-Urey experiment. Tell students the report should include a hypothesis, materials list, procedure, and description of results. You might use this activity as part of a discussion on the proper format for lab reports.

During Reading continued

Build Vocabulary

PREFIXES Have students use a dictionary to find the meanings of the word parts *sym-* and *-biotic.* *(sym- means "together;" -biotic means "related to living organisms")*

Ask Does the combined meanings of these three word parts—*endo-, sym-,* and *-biotic*—make sense, given what we know about endosymbiotic theory? Explain. *(Sample answer: Yes. A symbiotic relationship involves two organisms "living together." An endosymbiotic relationship involves one organism living within the body of another.)*

Active Reading

INTERPRET DIAGRAMS Ask a volunteer to read aloud the caption for the Endosymbiotic Theory diagram on the next page.

Ask Were the entering "guests" of the primitive eukaryotes also eukaryotes, or were they prokaryotes? *(prokaryotes)*

Ask Which of the organelles that evolved from this relationship are found in both plants and animals? *(mitochondria)* Which are found in plants but not animals? *(chloroplasts)*

MAKE CONNECTIONS Guide students to recall the definition of *symbiosis,* which they learned in Chapter 4. *(a relationship in which two species live close together)*

Ask What is the connection between symbiosis and the evolution of eukaryotes? *(Sample answer: Eukaryotes began to evolve when prokaryotes developed a symbiotic relationship with early eukaryotic cells by entering the cells and living inside them.)*

KEY QUESTION Read the Key Question and answer aloud.

Ask Which eukaryotic organelles are thought to have evolved from prokaryotes that lived inside early eukaryotic cells? *(mitochondria and chloroplasts)*

Speed Bump

Ask How did ancient photosynthetic bacteria change Earth's atmosphere? *(They released large amounts of oxygen.)*

Ask How does the DNA of eukaryotic organelles provide evidence for the endosymbiotic theory? *(The DNA of chloroplasts and mitochondria is like the DNA of bacteria.)*

BUILD Vocabulary

endosymbiotic theory
the theory that proposes that eukaryotic cells formed from a symbiotic relationship among several different prokaryotic cells

PREFIXES
The prefix *endo-* in *endosymbiotic theory* means "within." The endosymbiotic theory involves a symbiotic relationship between eukaryotic cells and the bacteria within them.

Production of Free Oxygen The oldest fossils are found in rocks that are more than 3.5 billion years old. These prokaryotic cells evolved without oxygen. (Remember that the early atmosphere had very little oxygen.) Later, photosynthetic bacteria became common. By 2.2 billion years ago, photosynthetic cells were giving off great amounts of oxygen. This oxygen changed the atmosphere. In a few hundred million years, the amount of oxygen reached today's levels. For early life, this new gas was a deadly poison! It caused many early life forms to die out. Other organisms evolved ways to protect themselves from oxygen. Many organisms evolved the ability to use oxygen for respiration.

Origin of Eukaryotic Cells

One of the most important events in the history of life was the evolution of eukaryotic cells. How did these complex cells evolve from their prokaryotic ancestors?

Endosymbiotic Theory About 2.2 billion years ago, some kinds of prokaryotes began to evolve features found in eukaryotes. According to the **endosymbiotic theory,** prokaryotes became part of those early eukaryotes. At first, prokaryotic "guests" lived inside the eukaryotes. Over time, the "guests" became a part of their hosts. The "guests" became the mitochondria and chloroplasts of modern eukaryotic cells. Mitochondria evolved from bacteria that helped their hosts use oxygen. Chloroplasts evolved from photosynthetic bacteria that made food for their host cells.

Modern Evidence In 1963, the American biologist Lynn Margulis found that the DNA of chloroplasts and mitochondria was like the DNA of bacteria. She also noted that the ribosomes of these organelles are very similar to those of bacteria. Finally, these organelles divide by fission, just like bacteria. The similarities between mitochondria, chloroplasts, and bacteria provide strong support for the endosymbiotic theory.

Key Question What theory explains the origin of eukaryotic cells? **The endosymbiotic theory proposes that a symbiotic relationship evolved over time between early eukaryotic cells and prokaryotic cells that lived inside them.**

Sexual Reproduction and Multicellularity

Sometime after eukaryotic cells first evolved, they began to reproduce sexually. This ability helped to increase the pace of evolution.

Sexual Reproduction Prokaryotes reproduce asexually. They copy their genes before dividing by fission into two daughter cells. The daughter cells have exactly the same genes as the parent cell. Genetic variation is possible only because of mutations in DNA.

ELL Focus on ELL: Build Background

ALL SPEAKERS Assign students to small groups that include a mix of different levels of speakers. Ask groups to locate the words *hosts, complex,* and *features* on this page. Have the groups use context clues to define each of the words. Discuss the definitions as a class. Then, present each group with a written copy of the following discussion topics:

- How did primitive eukaryotes benefit from their relationship with ancient bacteria?
- What characteristics of chloroplasts and mitochondria support the endosymbiotic theory?

Encourage students to practice using the words they defined as they carry out their discussions. Allow groups enough time to discuss the topics. Then, have one student from each group report the group's responses.

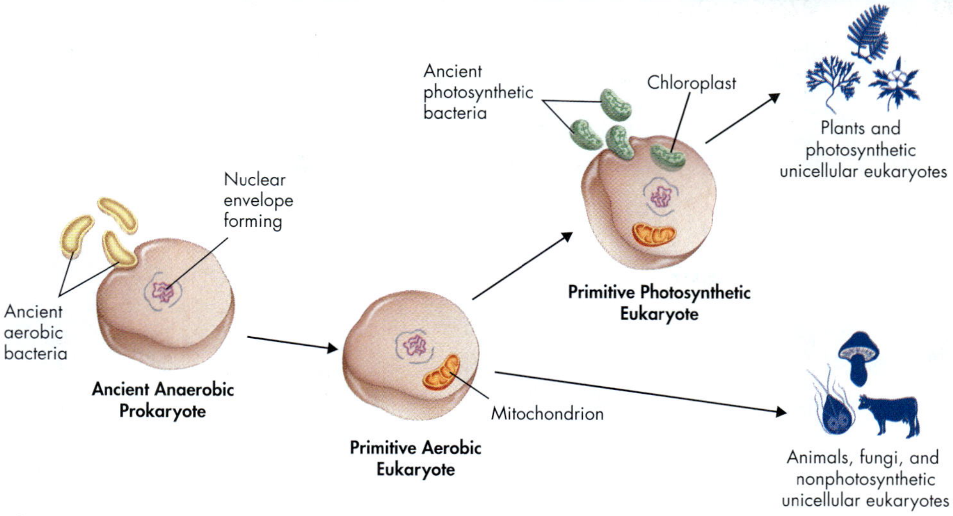

When eukaryotes reproduce sexually, the offspring receive genes from two parents. Meiosis and fertilization increase genetic variation through new gene combinations. Genetic variation is the raw material for natural selection. More variation means a population can evolve more quickly. Genetic variation helps populations adapt to new conditions.

Multicellularity Multicellular organisms evolved a few hundred million years after the evolution of sexual reproduction. Early multicellular organisms underwent a series of adaptive radiations. The result was a great diversity of organisms.

Key Question What is the evolutionary significance of sexual reproduction? **The development of sexual reproduction sped up evolutionary change because it increased genetic variation.**

Endosymbiotic Theory
The endosymbiotic theory states that eukaryotic cells evolved from a relationship formed with bacteria. Ancient bacteria may have entered primitive eukaryotic cells and remained there as organelles.

CHECK Understanding

Apply Vocabulary

Use the highlighted word from the lesson to complete the sentence correctly.

1. The _______ states that mitochondria and chloroplasts evolved from symbiotic bacteria living inside eukaryotes.

Critical Thinking

2. **Apply Concepts** What evidence supports the endosymbiotic theory?
3. **Review** Describe Earth's early atmosphere.
4. **Relate Cause and Effect** Why is the development of sexual reproduction so important in the history of life?
5. **Sequence** Put the following events in the order in which they occurred: *sexual reproduction, development of eukaryotic cells, free oxygen in atmosphere,* and *development of photosynthesis.*
6. **Write to Learn** Write a paragraph that describes the experiment conducted by Miller and Urey. What was their hypothesis? What was their procedure? What conclusion could they draw from their results?

BIOLOGY.com Search Lesson 19.3 GO • Lesson Assessment

Active Reading

READ, COVER, REMEMBER, RETELL Have students silently read the section Sexual Reproduction and Multicellularity. Then, instruct them to cover the text with their hands and think about what they read.

Ask In your own words, what was the subject or main idea of this section? *(Answers may vary but should address the idea that the evolution of sexual reproduction in eukaryotes helped increase the pace of evolution of life on Earth.)*

KEY QUESTION Have a volunteer read aloud the Key Question and answer.

Ask What two processes in sexual reproduction increase genetic variation? *(meiosis and fertilization)*

Post Reading

ASSESSMENT TIPS Explain that in Question 4, the development of sexual reproduction is the cause, and students are asked to describe the effect.

Students might refer to the flowcharts they started at the beginning of the lesson to help them answer Question 5.

Wrap-Up Activity

VISUALIZING EARLY EARTH Have students work in pairs to make a drawing, collage, or other visual that illustrates Earth's early history. The visual should convey scientists' hypotheses about the physical conditions on early Earth and the origin of life. Tell students their work should be creative but must also be scientifically correct. Encourage pairs to share their visuals with the class and explain them. Correct any misconceptions or inaccuracies.

Check Understanding

1. endosymbiotic theory
2. The DNA of chloroplasts and mitochondria is like the DNA of bacteria. The ribosomes of these organelles are like those of bacteria. Finally, these organelles divide by fission, just like bacteria.
3. no oxygen; mostly carbon dioxide, nitrogen, and water vapor; lesser amounts of carbon monoxide, hydrogen sulfide, and hydrogen cyanide
4. Sexual reproduction leads to greater genetic diversity, which speeds up evolutionary change.
5. 1) development of photosynthesis
2) free oxygen in the atmosphere
3) development of eukaryotic cells
4) sexual reproduction
6. Student paragraphs should clearly outline the steps illustrated in the Miller-Urey Experiment diagram in this lesson. Paragraphs should state that Miller and Urey hypothesized that the organic compounds necessary for life could have arisen from chemical reactions on early Earth and that the experiment supported their hypothesis.

During Reading

Active Reading

PREVIEW VISUALS Have students look over this four-page feature. Point out that the colors of the bands at the top of the pages match the colors used for the different divisions in the Geologic Time Scale chart in Lesson 19.1.

Ask What does the top band identify? *(the eon)* What do the other two bands identify? *(The middle band identifies the eras that make up the eon, and the bottom band identifies the periods within each era.)*

Ask What do the vertical lines on these pages connect? *(They connect each period with information about that period.)*

Ask What do the photos and drawings show? *(fossils and images of organisms from the various periods)*

SUMMARIZE Have students read about the Cambrian, Ordovician, and Silurian periods silently. Instruct them to write in their notebooks what they consider the main events of each period.

Ask What major evolution event took place during the Cambrian Period? *(the adaptive radiation of multicellular life, called the Cambrian Explosion)* What types of environments were formed then? *(shallow seas)* What occurred at the end of the Cambrian Period? *(a mass extinction)*

Ask What environments were most common during the Ordovician Period? *(shallow seas)* Which groups of animals were most common in the seas? *(invertebrates)*

Ask What happened during the Silurian Period that resulted in new environments? *(Land areas rose, which drained the seas and produced tropical habitats.)* What organisms appeared for the first time? *(fishes with true jaws, multicellular land plants, and arthropods)*

PHANEROZOIC EON

PALEOZOIC ERA

Cambrian Period | Ordovician Period | Silurian Period

Cambrian Period

▲ Elrathia

During the Cambrian Period, multicellular life had its greatest adaptive radiation. This adaptive radiation is called the Cambrian Explosion. Continents moved in ways that created shallow seas over much of Earth. Many organisms in the seas evolved hard shells and outer skeletons. The first jawless fishes appeared. The Cambrian ended with a mass extinction. Nearly one third of all animal groups died.

Silurian Period

During the Silurian Period, land areas rose, draining the shallow seas. Moist tropical habitats formed. Many different jawless fishes evolved. The first fishes with true jaws appeared. The first multicellular land plants evolved from green algae. Arthropods become the first animals to live on land.

▲ Sea Lily Fossil

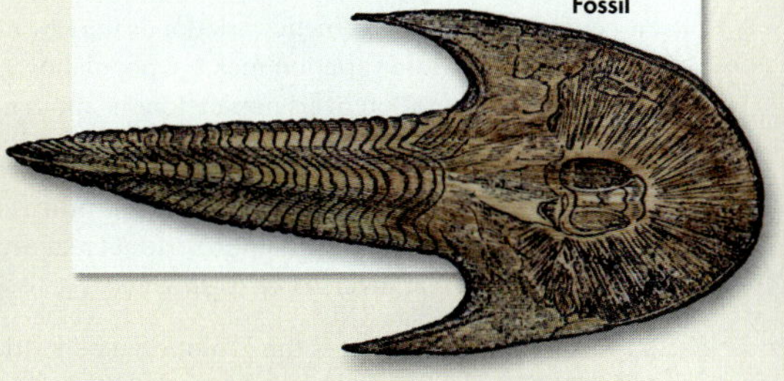

▲ Cephalaspis (raylike jawless fish)

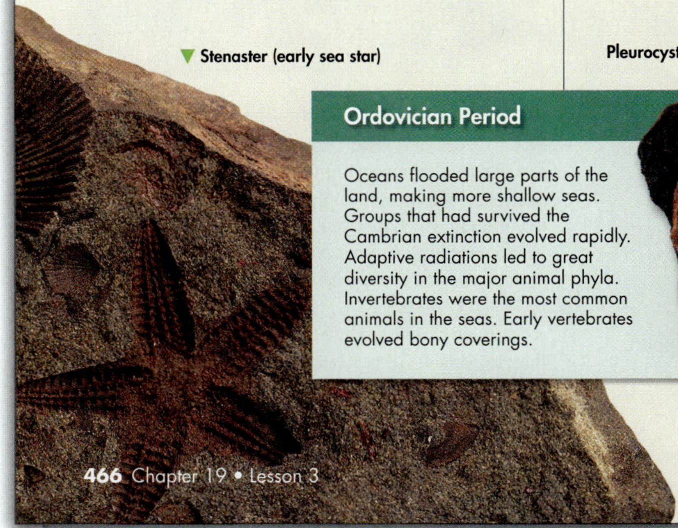

▼ Stenaster (early sea star)

Pleurocystes (early echinoderms) ▼

Ordovician Period

Oceans flooded large parts of the land, making more shallow seas. Groups that had survived the Cambrian extinction evolved rapidly. Adaptive radiations led to great diversity in the major animal phyla. Invertebrates were the most common animals in the seas. Early vertebrates evolved bony coverings.

Focus on Academic Language

• Cambrian Period • Ordovician Period • Silurian Period

Divide students into pairs, and assign each pair one of the geologic periods featured on this page. Have students develop a cluster diagram that describes their assigned period. Instruct them to write the name of the period in the center and surround it with characteristics of that period: the environment(s), organisms, and major events. Explain that the cluster diagram should extend to provide details of all these categories. You might also have students illustrate their cluster diagrams with simple pictures representing characteristics of the geologic period.

Draw an example of a cluster diagram on the board, and help students think of the first details to include in their own cluster diagrams. Discuss as a class each pair's cluster diagram. If time permits, repeat the activity for the geologic periods described on each of the next three pages.

Devonian Period | Carboniferous Period | Permian Period

Devonian Period

During the Devonian Period, invertebrates and vertebrates thrived in the seas. Fishes evolved jaws, bony skeletons, and scales. Sharks began their adaptive radiation. Certain groups of fishes evolved leglike fins; some of these evolved into the first amphibians. Some land plants, such as ferns, adapted to drier areas. Insects began their adaptive radiation on land.

Permian Period

▲ Crinoid

During the Permian Period, the diversity of invertebrates, vertebrates, and plants on land increased. Reptiles experienced their first adaptive radiation. This produced the ancestors of modern reptiles, dinosaurs, and mammals. The Permian Period ended with the biggest mass extinction of all time. More than 50 percent of animal families on land disappeared. More than 95 percent of species in the oceans became extinct.

Early Amphibian ▼

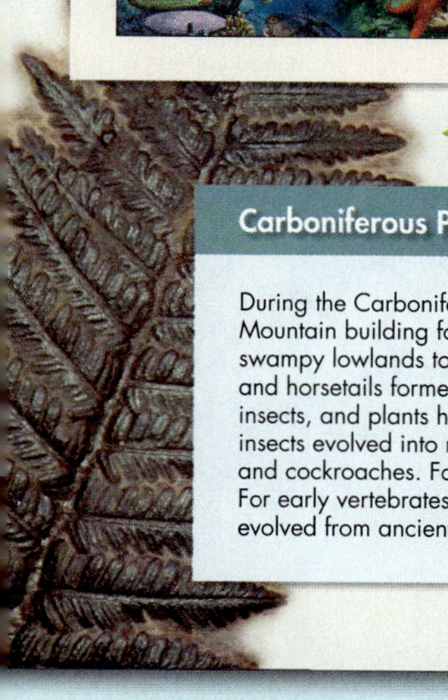

◄ Fossil Fern From Carboniferous Period

Carboniferous Period

During the Carboniferous Period, mountains pushed upward. Mountain building formed a wide range of habitats, from swampy lowlands to drier uplands. Giant ferns, club mosses, and horsetails formed huge swampy forests. Amphibians, insects, and plants had major adaptive radiations. Winged insects evolved into many forms, including huge dragonflies and cockroaches. For early plants, insects were predators. For early vertebrates, insects were food. The first reptiles evolved from ancient amphibians.

Hands-On Learning

NAME THAT PERIOD Have students identify characteristics of the different geologic periods.

1. Divide students into twelve groups, and give each student one index card.
2. Assign each group one of the twelve geologic periods described in this four-page feature.
3. Instruct students to record on their index card an event or fact related to their geologic period. Students should not include the name of the period on their card.
4. Collect the cards and shuffle them.
5. Have students take turns choosing a card.
6. After students read aloud their selected card, have them name the geologic period described by that event or fact.
7. If desired, the activity can be carried out as a competition among the twelve groups.

Active Reading

USE A GRAPHIC ORGANIZER Emphasize to students that each geologic period is characterized by the organisms that lived during that time. Have students prepare a three-column compare/contrast table for the periods described on this page. Instruct them to label the columns Devonian Period, Carboniferous Period, and Permian Period. Have them continue by labeling five rows Fishes, Amphibians, Reptiles, Insects, and Plants. Help students complete the first row together. First, read aloud the Devonian Period paragraph.

Ask What did you learn about the evolution of fishes during the Devonian Period? *(They evolved jaws, bony skeletons, and scales; some developed leglike fins and evolved into amphibians.)* Which group of fishes began adaptive radiation? *(sharks)*

Have students fill in their table entry for fishes in the Devonian Period. Next, read aloud the paragraphs for the Carboniferous and Permian periods.

Ask Were developments in fishes described for these periods? *(no)*

Tell students not to include in their table anything about fishes for those periods. Encourage students to complete their table by filling in details for the remaining groups of organisms.

Speed Bump

Ask Between the beginning of the Devonian Period and the end of the Permian Period, what types of environments were forming? *(land environments)*

Ask What land organisms were evolving during that time? *(amphibians, reptiles, plants, and insects)*

Ask What major event happened at the end of the Permian Period? *(the biggest mass extinction of all time)*

During Reading continued

Active Reading

USE VISUALS Refer students to the visuals on this page, and have volunteers read aloud the descriptions.

Ask What three periods make up the Mesozoic Era? *(Triassic, Jurassic, and Cretaceous periods)*

Point out the pterodactyl fossil.

Say Pterodactyls were flying reptiles, not dinosaurs. Unlike *Archaeopteryx,* pterodactyls did not have feathers.

Have a volunteer describe the Maiasaura nest. Tell students that Maiasaura were duck-billed herbivorous dinosaurs that laid eggs in nests and raised their young.

MAKE INFERENCES Draw students' attention to the information about dinosaurs on this page.

Say Dinosaurs were a successful group even though they became extinct.

Ask What information on this page supports this statement? *(Dinosaurs were the most diverse group of land animals during the Jurassic Period. Dinosaurs "ruled" Earth for about 150 million years.)*

Speed Bump

Ask During which period did the first dinosaurs and earliest mammals evolve? *(Triassic Period)*

Ask Which period included adaptive radiations of plants? *(Cretaceous Period)*

Ask Which period ended with a mass extinction that included the extinction of all dinosaurs except the ancestors of modern birds? *(Cretaceous Period)*

PHANEROZOIC EON

MESOZOIC ERA

Triassic Period | Jurassic Period | Cretaceous Period

Triassic Period

During the Triassic Period, fishes, insects, reptiles, and cone-bearing plants that survived the Permian extinction evolved rapidly. The first dinosaurs evolved. During the late Triassic, the earliest mammals evolved. These mammals were about the size of a mouse.

▲ Living Horsetails

▲ Horsetail Fossil

Cretaceous Period

▲ *T. rex* Skeleton

During the Cretaceous Period, dinosaurs like *Tyrannosaurus rex* roamed the land. Flying reptiles and birds flew in the sky. Turtles, crocodiles, and other reptiles swam with fishes and invertebrates in the seas. Leafy trees, shrubs, and flowering plants evolved and experienced adaptive radiations. The Cretaceous ended with another mass extinction. More than half of all plant and animal groups were wiped out. All dinosaurs except the ancestors of modern birds became extinct.

Jurassic Period

◀ Pterodactyl Fossil

During the Jurassic Period, dinosaurs became the most diverse group of animals on land. They "ruled" for about 150 million years. Different types of dinosaurs lived at different times. One line of dinosaurs evolved feathers and led to modern birds. *Archaeopteryx*, the first feathered fossil to be discovered, evolved during this time.

▼ *Maiasaura* Nest

Focus on Academic Language

• cone • roam • period

Write the words on the board, and have students locate each word as it appears on this page. (Note that *roamed* is the form of *roam* that appears.) Discuss as a class each word's part of speech and its meaning as suggested by context clues. Then, have students look up dictionary definitions of each word.

Ask Which definition of *cone* is used on this page? *(a mass of ovule-bearing or pollen-bearing scales in most conifers or cycads)* What other cones might you see? *(Sample answers: construction cones on the road, ice cream cone)*

Ask What are other examples of animals roaming? *(Sample answers: cows in a field, people at a mall)*

Ask What is meant by *period* as it is used here? *(a portion of time)* What other definition of *period* is important in writing? *(the mark used at the end of a statement)*

CENOZOIC ERA

Paleogene Period	Neogene Period	Quaternary Period

Paleogene Period

During the Paleogene Period, climates changed from warm and moist to cool and dry. Flowering plants, grasses, and insects thrived. In drier areas, open woods and grasslands replaced forests. After the extinction of the dinosaurs, mammals underwent a major adaptive radiation. Ancestors of cattle, deer, and sheep evolved and spread across the grasslands. In the oceans, the first whales evolved.

▲ Early Mammal

Neogene Period

During the Neogene Period, crashing continents pushed up modern mountain ranges. In North America, the Rockies, Cascades, and Sierra Nevada formed. Ice and snow built up at higher elevations and in the Arctic. Falling sea levels and moving continents created connections between North and South America, and between Africa, Europe, and Asia. These connections allowed land animals to move between continents. As climates became cooler and drier, grasslands increased. Grazing animals evolved digestive systems that allowed them to digest the grasses.

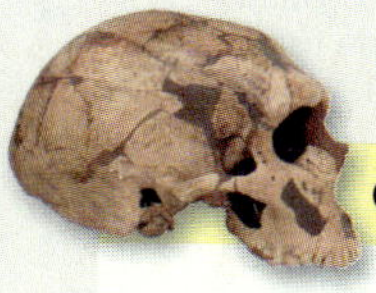

◄ Neanderthal Skull

Quaternary Period

During the Quaternary Period, Earth cooled. In a series of ice ages, glaciers moved over parts of Europe and North America. So much water was frozen in glaciers that sea levels fell by more than 100 meters. Then, about 20,000 years ago, the climate began to warm. Glaciers slowly melted, and sea levels rose. In the oceans, algae, coral, mollusks, fishes, and mammals thrived. Insects and birds shared the skies. On land, mammals, including bats, cats, dogs, cattle, and mammoths, became common. Between 6 and 7 million years ago, one group of mammals began an adaptive radiation that led to the ancestors and relatives of modern humans.

▲ Cave Paintings

ELL Focus on ELL: Extend Language

INTERMEDIATE AND ADVANCED SPEAKERS Help students develop a higher comfort level in participating in class discussions. Prepare several statements that can be used to start a conversation about events described on this page. Hand out a copy of these statements to students. Have them read the statements and then prepare their own written or spoken statements that logically follow the starter sentences. Examples of starter sentences to provide students include:

- Many organisms evolved after climates changed from warm and moist to cool and dry.
- Moving continents allowed animals to move to places they could not reach before.
- The end of the ice ages led to rising sea levels.

Have students take turns discussing their prepared responses.

Active Reading

CAUSE AND EFFECT Help students understand how changes in climate or landforms shaped ancient environments, which in turn affected the evolution of organisms. Have students complete a cause-and-effect diagram for an event that took place during one of the periods described on this page. Divide students into small groups, and assign each group one of the following events: climate change during the Paleogene Period, continental drift during the Neogene Period, ice ages during the Quaternary Period, or climate warming during the Quaternary. Instruct students to list in their cause-and-effect diagrams at least two effects that these events had on the environment and/or on organisms. Invite groups to share their diagrams with the class.

Speed Bump

Ask During the Paleogene Period, which animals underwent a major adaptive radiation? *(mammals)*

Ask During which period did connections form between continents, allowing land animals to move between them? *(Neogene Period)*

Ask In which period did the close relatives of modern humans evolve? *(Quaternary Period)*

CHAPTER LAB

Pre-Lab

Invite a volunteer to read aloud the Problem. Review the process of rock layer formation and the concept of relative age.

Ask When rock layers form in an area over time, which layer is the oldest? *(the bottommost layer)* Which layer is the youngest? *(the topmost layer)*

Ask What do scientists use to determine the relative ages of rock layers in different areas? *(index fossils)*

Review the Background Questions and the Pre-Lab Questions as a class.

Lab

Tell students they will perform the chapter lab *Using Index Fossils* described in **Lab Manual B.**

SAFETY Caution students to direct the pointed end of the scissors away from themselves and others.

Post-Lab

CHANGES IN ROCK LAYERS After students have sequenced the rock layers in their correct order, have them rearrange the layers to show possible ways rock layers can be altered, such as by being worn away or overturned as a result of land movements. Discuss as a class the consequences of these changes on determining the relative ages of rock layers.

NATIONAL SCIENCE EDUCATION STANDARDS

UCP II, IV

CONTENT C.3.b, C.3.c, C.3.d

Forensics Lab

GUIDED INQUIRY

Pre-Lab: Using Index Fossils

Problem How can fossils be used to determine the relative ages of rock layers?

Materials scissors

Lab Manual Chapter 19 Lab

Skills Focus Interpret Visuals, Sequence, Draw Conclusions

Connect to the Big idea When detectives work on a case, they may look for items with a time stamp, such as parking tickets and credit card slips. Such items can help detectives piece together a sequence of events. Events related to a crime usually occur within a relatively short period of time. In contrast, the events that paleontologists study will have occurred over millions of years. Placing these events in their proper order can be challenging. The clues that a paleontologist uses to sequence events in the history of life are fossils buried in rock layers. In this lab, you will use fossils to place rock layers in order from oldest to youngest.

Background Questions

a. **Review** What is a fossil? What are the characteristics of a good index fossil?

b. **Explain** What characteristic of radioactive decay allows scientists to assign specific ages to rock layers?

c. **Classify** How do fossils help geologists decide where one division of geologic time should end and another division begin?

Pre-Lab Questions

Preview the procedure in the lab manual.

1. **Organize Data** After you cut out the drawings of the rock layers, how will you begin the process of sorting the layers by age?

2. **Infer** *Desmatosuchus* was a crocodile relative that lived only during the Triassic Period. Horsetails are plants that first appeared in the Triassic Period and still exist. Which of these organisms would be more useful as an index fossil for the Triassic Period? Why?

3. **Use Analogies** Luke found a box of photos labeled 1970–1995. Each photo shows his entire extended family. No dates appear on the photos. Luke knows that his grandmother died in 1985 and his uncle was born in 1975. Luke's sister was born in 1990. How can Luke use this information to sort the photos into four batches? How are Luke's relatives similar to index fossils?

BIOLOGY.com Search Chapter 19 GO

Visit Chapter 19 online to test yourself on chapter content and to find activities to help you learn.

Untamed Science Video Go back in time with the Untamed Science crew to find out what fossils reveal.

Art in Motion View a short animation that shows how fossils form.

Art Review Review your understanding of the composition of Earth's early atmosphere as compared with the composition of Earth's current atmosphere.

Visual Analogy Compare geologic time to a 24-hour clock.

Pre-Lab Answers

BACKGROUND QUESTIONS

a. A fossil is the preserved remains of an ancient (usually extinct) organism. A good index fossil must be easy to identify. It must also come from a species that existed for a relatively short time over a wide geographic range.

b. Radioactive decay occurs at a steady rate over time.

c. Geologists noticed that major changes in the fossil record occurred at boundaries between certain rock layers. Those boundaries were used to divide geologic time into eons, eras, and periods.

PRE-LAB QUESTIONS

1. Sample answer: I will use the Key to Fossils to identify the fossils in each layer.

2. The crocodile relative would be more useful as an index fossil for the Triassic Period because it would not occur in rock layers from other periods.

3. Sample answer: Possible batches are photos with (1) grandmother only from 1970–1975, (2) grandmother and uncle from 1975–1985, (3) uncle only from 1985–1990, and (4) uncle and sister from 1990–1995. The people are like index fossils because they exist in specific time ranges relative to the overall range of the photos.

470 Chapter 19 • Pre-Lab

19 CHAPTER Summary

19.1 The Fossil Record

- From the fossil record, paleontologists learn about the structure of ancient organisms, their environment, and the way in which they lived.
- Relative dating allows paleontologists to determine whether a fossil is older or younger than other fossils. Radiometric dating uses the proportion of radioactive to nonradioactive isotopes to calculate the absolute age of a sample.
- The geologic time scale is based on both relative and absolute dating. The major divisions of the geologic time scale are eons, eras, and periods.
- Throughout Earth's history, geological forces, changing climates, and collisions with objects from space have changed the habitats of living organisms. Over time, the actions of living organisms have also changed conditions in the land, water, and atmosphere of Earth.

paleontologist (p. 450)
extinct (p. 450)
relative dating (p. 451)
index fossil (p. 451)
radiometric dating (p. 452)
half-life (p. 452)
geologic time scale (p. 453)
era (p. 454)
period (p. 454)
plate tectonics (p. 454)

19.2 Patterns and Processes of Evolution

- If the rate of speciation in a clade is equal to or greater than the rate of extinction, the clade will survive. If the rate of extinction in a clade is greater than the rate of speciation, the clade will become extinct.
- Evidence shows that evolution has moved at different rates for different organisms at different times over the long history of life on Earth.
- Two important patterns of macroevolution are adaptive radiation and convergent evolution.
- Adaptive radiation is the process by which a single species or a small group of species evolves quickly into several different species that live in different ways.
- Convergent evolution is the process by which unrelated organisms evolve into similar forms.
- The relationship between two organisms that are coevolving often becomes so specific that neither organism can survive without the other. Thus, an evolutionary change in one organism is usually followed by a change in the other organism.

macroevolutionary patterns (p. 456)
background extinction (p. 457)
mass extinction (p. 457)
gradualism (p. 458)
punctuated equilibrium (p. 458)
adaptive radiation (p. 459)
convergent evolution (p. 460)
coevolution (p. 460)

19.3 Earth's Early History

- Earth's early atmosphere contained little or no oxygen. It was mostly carbon dioxide, water vapor, and nitrogen, with lesser amounts of carbon monoxide, hydrogen sulfide, and hydrogen cyanide.
- Miller and Urey's experiment suggested how mixtures of the organic compounds necessary for life could have arisen from simpler compounds found in the atmosphere of early Earth.
- The "RNA world" hypothesis proposes that RNA existed by itself before DNA. From this simple RNA-based system, several steps could have led to DNA-directed protein synthesis.
- The endosymbiotic theory proposes that a symbiotic relationship evolved over time between early eukaryotic cells and the prokaryotic cells within them.
- The development of sexual reproduction sped up evolutionary change because sexual reproduction increases genetic variation.

endosymbiotic theory (p. 464)

UbD Transfer the Big Idea

TRANSFER Have students use what they learned about fossils, evolution, and geologic time to write about a fossil dig that takes place in the future. Tell students to imagine that paleontologists living millions of years from now uncover fossils of organisms living today. Direct students to write a short report that describes the paleontologists' findings. The report should include answers to these questions: Where did the paleontologists find fossils? What was preserved as fossils? What evidence do these fossils provide about changes in environments and the evolution of existing groups of organisms?

Chapter Summary

Think Visually

USE GRAPHICS Have students construct a table comparing the Paleozoic, Mesozoic, and Cenozoic eras. Direct them to include the approximate time periods for each era and identify the characteristic organisms.

ANSWERS
Tables will vary. Students can use the information in the Geologic Time Scale in Lesson 19.1 and in the time line pages at the end of Lesson 19.3 to complete their table.

BIOLOGY.com Study Online

REVIEW RESOURCES

Art in Motion Students can use this activity to see an animation of fossil formation.

Visual Analogy This activity expands on the analogy of geologic time as a clock.

Data Analysis Students can use data to infer likely causes of extinction events.

Art Review This drag-and-drop activity allows students to explore conditions on early Earth.

Vocabulary Review The *Flash Cards* and *Crossword* provide an interactive way to review chapter vocabulary.

Editable Worksheets Pages of Study Workbook B, Lab Manual B, and the Assessment Resources Book are available online. These documents can be easily edited using a word-processing program.

Lesson Overviews Lesson Overviews are based on Miller & Levine Biology. However, they are available on CD-ROM and can be edited for your class.

ASSESSMENT RESOURCES

Chapter Assessment Have students take an online version of the Chapter 19 Check Understanding.

Standardized Test Prep Students can take an online version of the Standardized Test Prep. You will receive their scores along with ideas for remediation.

Diagnostic and Benchmark Tests Use these tests to monitor your students' progress and supply remediation.

ASSESSMENT

Assess

Big idea Invite a volunteer to read the question aloud. Write on the board the terms *fossil, paleontologist, relative dating,* and *radiometric dating.*

Ask How are these four terms connected? *(Sample answer: Paleontologists use relative dating and radiometric dating to determine the ages of fossils.)*

Ask What is the geologic time scale? *(the time line used to represent Earth's history)* How does the dating of fossils relate to the geologic time scale? *(Paleontologists have constructed the geologic time scale by dating fossils.)*

Ask What types of changes on Earth are shown by fossils? *(changes in climates, changes in environments, changes in organisms)*

Have students write in their notebooks their answers to the question in the text.

CONSTRUCTED RESPONSE Paragraphs will vary. Review how to write a paragraph by building on main ideas. Help students use the *Hints* to develop the main ideas of their paragraphs. Ask a volunteer to read Question 1 aloud.

Say To answer this question, you will need to consider how fossils form and the rate at which evolution occurs.

Read the first hint.

Ask Where do scientists usually find fossils? *(in layers of sedimentary rock)* Where do layers of sedimentary rock usually form? *(at the bottom of bodies of water)*

Read the second hint.

Ask What do we mean when we say that fossils suggest that a species was in equilibrium at a certain time in history? *(The species did not seem to be evolving.)* What is punctuated equilibrium? *(a pattern of evolution in which equilibrium is interrupted by periods of rapid change)*

Model writing the first one or two sentences using the hints and details offered by students. *(Sample: The fossil record may not show the steady evolution of a particular species. One reason might be that during certain times, environmental conditions weren't right for the remains of those organisms to become fossils.)*

19 CHECK Understanding

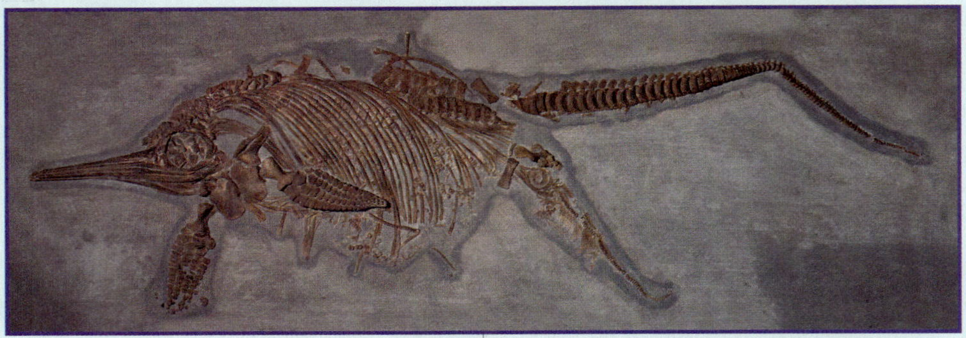

Assess the Big idea Evolution

Write an answer to the question below.

Q: How do fossils help biologists understand the history of life on Earth?

Constructed Response

Write an answer to each of the numbered questions below. The answer to each numbered question should be one or two paragraphs long. To help you begin, read the **Hints** below the questions.

1. **Why doesn't the fossil record always show a progression of changing fossils for the evolution of each species?**

   Hint Fossils can only form under the right conditions.

   Hint Use the terms *equilibrium* and *punctuated equilibrium* in your answer.

2. **What role have mass extinctions played in the history of life?**

   Hint Think of what happened to the mammals after most of the dinosaurs became extinct.

3. **How does the formation of sedimentary rock give paleontologists information about the sequence in which life forms appeared on Earth?**

   Hint To help plan your answer, review the art of index fossils on page 451.

Foundations for Learning Wrap-Up

Use the tabs that you made at the beginning of the chapter as a tool for review.

Activity 1 Lay your folded tabs on your desk, with the titles on top. Mix them so they are out of order. Then arrange them in order like the fossils in layers of rock, with the oldest events on the bottom. Repeat this activity until it becomes easy.

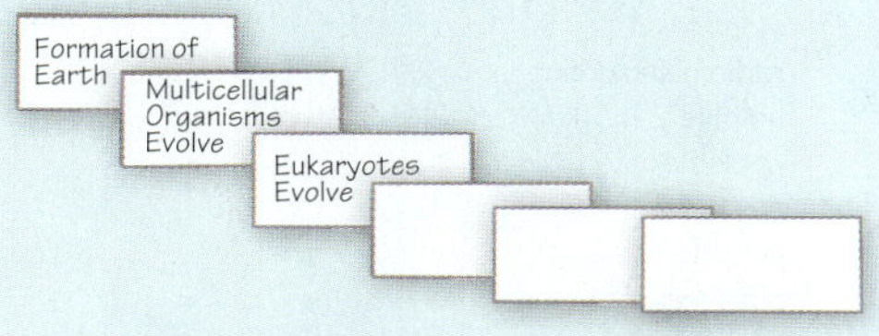

Activity 2 Work with a partner to make a time line out of your tabs. Cut a piece of string so it is 10 meters long. Fasten the string to a sturdy object at eye level. Attach the other end to an object or wall so that the string is level and tight. Hang the "Formation of Earth" tab at one end of the string. Then arrange all the tabs in order according to time.

Next, use the geologic time scale to place each event in the correct place on your string. Each meter of length stands for 400 million years. Each centimeter stands for 4 million years. Use the information from your textbook to help date each event. Your tabs will be very crowded at the end of the string!

For Question 2, pair students and have them use the hints and your modeling to answer the question. Have students share their paragraphs with the class.

For Question 3, have students write the paragraphs independently for homework or as an anchor activity.

Foundations for Learning Wrap-Up

Have students take out their prepared tabs. Be sure they have written notes and drawn pictures on their tabs. Ask volunteers to share the details and drawings they added.

Have students work individually to complete the first activity. Then assign groups, and explain the second activity. Provide each student pair with a 10-meter-long piece of string.

19.1 The Fossil Record

Understand Key Concepts

1. Scientists who specialize in the study of fossils are called
a. biologists.
b. zoologists.
c. geologists.
d. paleontologists.

Test-Taking Tip

Read Answers Carefully! In multiple choice questions, you might be tempted to answer the first response that seems correct. In question 1, you might answer **a** or **c.** But "**d.** paleontologists" is the best choice. Try to think of the answer before you read the choices. Then read each choice.

2. According to the theory of plate tectonics,
a. Earth's climate has changed many times.
b. Earth's continents move very slowly.
c. evolution occurs at different rates.
d. giant asteroids crashed into Earth in the past.

3. An isotope that can be used to determine the age of recent fossils is
a. carbon-14.
b. potassium-40.
c. rubidium-87.
d. uranium-238.

4. How have the activities of organisms affected Earth's environment?

Think Critically

5. **Relate Cause and Effect** Why have so few organisms become fossils?

6. **Calculate** The half-life of carbon-14 is 5730 years. What is the age of a fossil containing ¼ the amount of carbon-14 found in living organisms? Explain your calculation.

19.2 Patterns and Processes of Evolution

Understand Key Concepts

7. The process that produces similar structures in unrelated groups of organisms is
a. adaptive radiation.
b. coevolution.
c. convergent evolution.
d. macroevolution.

8. Cladograms that are based on the fossil record always show
a. only organisms that are direct ancestors of the others.
b. relationships based on shared derived characters.
c. that clades contain only extinct species.
d. relative ages of organisms in the clade.

9. Explain the process of adaptive radiation, and give an example.

Think Critically

10. **Compare and Contrast** How is mass extinction different from background extinction?

11. **Infer** What do you think would have happened to the evolution of flowering plants if there had been no pollinators present?

19.3 Earth's Early History

Understand Key Concepts

12. Earth's early atmosphere contained little or no
a. water vapor. **c.** nitrogen.
b. carbon dioxide. **d.** oxygen.

13. In their experiment that modeled conditions on early Earth, Miller and Urey used electric sparks to
a. simulate temperature.
b. simulate sunlight.
c. simulate lightning.
d. simulate atmospheric gases.

Assessment Tips

Have students work in pairs to answer the questions on this page and the following page.

For Question 6, have students review the Radioactive Decay graph in Lesson 19.1.

The visual in Lesson 19.3 that outlines the steps of the Miller-Urey experiment can help students answer Question 16.

Before students answer Questions 18–20, have them revisit the definitions and descriptions of *index fossils* and *relative dating* in Lesson 19.1.

Answers

Lesson 19.1

UNDERSTAND KEY CONCEPTS

1. d

2. b

3. a

4. Sample answer: The most dramatic example of life affecting Earth's environment is the production of oxygen by ancient photosynthetic organisms. Today, organisms continue to form soil and play a key role in the carbon, nitrogen, and oxygen cycles.

THINK CRITICALLY

5. Most organisms rot or decay before they are preserved. Only those that are buried in sediments are preserved.

6. If the fossil has ¼ the amount of carbon-14 as living organisms, then it has been dead for two half-lives. (½ x ½ = ¼) Two half-lives equal 11460 years. (5730 + 5730 = 11460) Therefore, the fossil is about 11460 years old.

Lesson 19.2

UNDERSTAND KEY CONCEPTS

7. c

8. b

9. Adaptive radiation occurs when a small population populates a new area or when an extinction removes most of the species from an area. In such cases, a single species or small group of species evolves rapidly into several species, each with different traits and ways of life. The diversity of Galápagos finches is the result of an adaptive radiation in modern times.

THINK CRITICALLY

10. Background extinction is the result of the slow and steady process of natural selection. It takes place over a very long period of time. In contrast, mass extinction takes place very quickly. Through mass extinction, entire ecosystems vanish because of events such as volcanic eruptions, changing sea levels, or an asteroid hitting Earth.

11. Without pollinators, there would not have been any pressure to coevolve characteristics to attract pollinators. Showy flowers, nectar, and scents would not have helped plants become pollinated and fertilized without pollinators, so these traits would not have evolved. There would be no flowering plants as we know them.

Lesson 19.3

UNDERSTAND KEY CONCEPTS

12. d

13. c

ASSESSMENT

14. Like cells, microspheres have a semi-permeable membrane that allows water and other materials to pass through. They also have a simple method for storing and releasing energy.

15. At first, the oxygen was deadly and killed off many organisms, but some organisms evolved ways to protect themselves from oxygen. Others evolved the ability to use oxygen for respiration.

THINK CRITICALLY

16. The water in the first step of the experiment simulated the oceans. The methane, ammonia, and hydrogen simulated what Miller and Urey thought had been the composition of the atmosphere. Rain is represented by the condensing water as it cools and drips down the tubes.

17. The eukaryotic cells were better able to use oxygen and to make food through photosynthesis when they were associated with the ancestors of mitochondria and chloroplasts.

Connecting Concepts

18. Fossils in layer C are probably older because they are deeper and would have formed before fossils in layer A.

19. Layer F is probably about the same age as layer B because they both contain the same index fossil.

20. A useful index fossil is easy to identify and is found only in a few layers of rock. The layers containing the index fossil must be found in many places for the index fossil to be useful.

CHAPTER MYSTERY After students have read the Chapter Mystery, have them review the clues given for the cause of the Permian extinction. Discuss as a class the various types of evidence for the major changes that took place during the Permian extinction.

Ask What are the three hypotheses discussed for why the Permian extinction occurred? *(Volcanic eruptions in Siberia released carbon dioxide, oxygen levels dropped in the oceans and on land, and dust released into the air after an asteroid hit Earth changed the climate.)*

Ask What do all three hypotheses have in common? *(All involve a major change in the environment.)*

19 CHECK Understanding

14. How are microspheres similar to living cells?

15. How did the addition of oxygen to Earth's atmosphere affect the evolution of life?

Think Critically

16. **Use Models** In the Miller-Urey experiment, what did the water and gases represent? What part of Miller and Urey's set-up represents rain?

17. **Relate Cause and Effect** How do you think that the eukaryotic cells that took in the ancestors of mitochondria and chloroplasts benefited from the relationship?

Connecting Concepts

Use Science Graphs

The diagram shows rock layers in two different places. Use the diagram to answer questions 18–20.

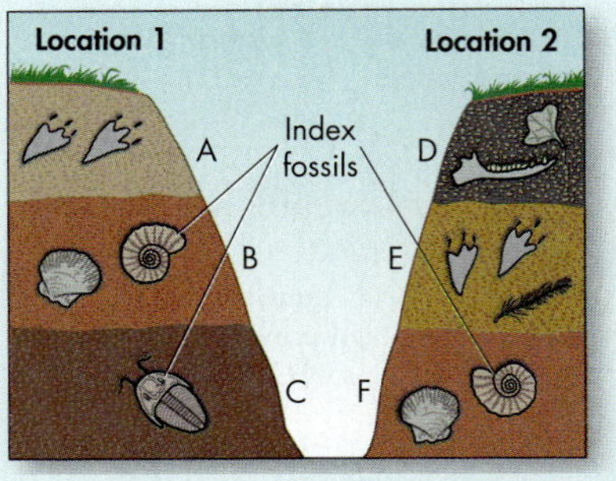

18. **Interpret Visuals** Which fossils are probably older, those in layer A or those in layer C? How do you know?

19. **Infer** Which rock layer in location 2 is probably about the same age as layer B in location 1? How do you know?

20. **Apply Concepts** What are the characteristics of a useful index fossil?

solve the CHAPTER MYSTERY

MURDER IN THE PERMIAN

Solving a mystery that is 250 million years old is not easy! Scientists have been working to solve this puzzle for years, and they are still working on it. Recently, scientists have studied the chemistry of Permian rocks and changes in the fossil record. These scientists have concluded that huge, long-lasting volcanic eruptions in Siberia released enormous amounts of carbon dioxide. This carbon dioxide caused massive changes in global climate. Was this the cause of the mass extinction?

Other researchers found that during the Permian, oxygen levels dropped to roughly half of what they are today. Huge parts of the oceans lost all oxygen. On land, there would have been very little oxygen. Mystery solved?

Finally, there is evidence that an asteroid hit Earth! Such a collision would throw so much dust into the air that global climates would have changed. To this day, paleontologists are testing different hypotheses to explain what caused the mass extinction. These hypotheses are constantly changing. In fact, they have probably changed again since this book was written!

1. **Compare and Contrast** How do current hypotheses about the Permian extinction compare with the main theory about the Cretaceous extinction?

2. **Form a Hypothesis** From the information in this book, suggest an explanation for the Permian mass extinction.

Untamed Science **Never Stop Exploring Your World.** Solving the mystery of the mass extinction is only the beginning. Take a video field trip with the ecogeeks of Untamed Science to see where the mystery leads.

CHAPTER MYSTERY ANSWERS

1. Current hypotheses propose that the Permian extinction had multiple causes, whereas the Cretaceous extinction may have had only one major cause—the effects of an asteroid impact on Earth's climate.

2. Accept all reasonable answers. Sample answer: Volcanic eruptions and the impact of an asteroid threw materials into the air that decreased sunlight and air quality, including a drop in oxygen in the atmosphere. Environmental conditions became so harsh that many species could not survive.

Untamed Science Students can explore fossils with the Untamed Science crew by watching **Stories in Stone.**

Standardized Test Prep

Multiple Choice

1. Index fossils are from species that lived
   A in a small area for a short time.
   B in a small area for a long time.
   C over a large area for a short time.
   D over a large area for a long time.

2. What happens if the rate of extinction in a clade is greater than the rate of speciation?
   A The clade will eventually become extinct.
   B The clade will continue to exist.
   C The species in the clade will become more varied.
   D The number of species in the clade will stay the same.

3. Which of the following is evidence for the endosymbiotic theory?
   A Mitochondria and chloroplasts contain DNA similar to bacterial DNA.
   B Mitochondria and chloroplasts have similar functions in the cell.
   C Mitochondria and chloroplasts have no DNA of their own.
   D Mitochondria and chloroplasts can live when removed from the eukaryotic cell.

4. Potassium-40 is useful for dating very old fossils because
   A it has a very long half-life.
   B it has a very short half-life.
   C most organisms contain more potassium than carbon.
   D it is found only in certain rock layers.

5. The movement of continents has played a significant role in evolution because
   A continents move rapidly and some organisms cannot adjust.
   B without the movement of continents, there would be no water on Earth.
   C the movement of continents has caused environments to change.
   D all mass extinctions are the result of continental drift.

Questions 6 and 7

The graph shows the decay of radioactive isotope atoms. Use the information in the graph to answer the questions that follow.

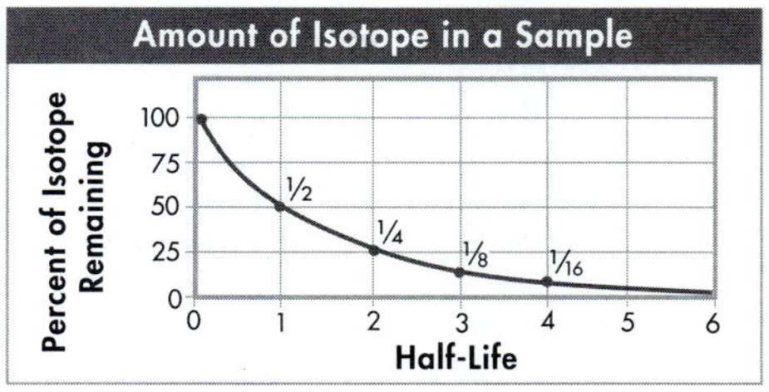

6. The half-life of thorium-230 is 75,000 years. How long will it take for $\frac{7}{8}$ of the original amount of thorium-230 in a sample to decay?
   A 75,000 years
   B 225,000 years
   C 25,000 years
   D 150,000 years

7. The half-life of potassium-40 is about 1.3 billion years. After four half-lives have passed, how much of the original sample will be left?
   A $\frac{1}{16}$
   B $\frac{1}{16} \times 1300$ million grams
   C $\frac{1}{4}$
   D $\frac{1}{4} \times 1300$ million grams

Open-Ended Response

8. How does the process by which sedimentary rock forms allow scientists to determine the relative ages of fossils?

If You Have Trouble With . . .

Question	1	2	3	4	5	6	7	8
See Lesson	19.1	19.2	19.3	19.1	19.1	19.1	19.1	19.1

Test-Taking Tip

REVIEW ANSWERS
Tell students to use any time remaining after they have answered all the questions on a test to review their answers. Point out that they should change only those responses they are sure are incorrect. If the answer booklet is separate from the test questions booklet, students should make sure the question numbers correctly correspond with their responses in the answer booklet.

Before the Test

Review with students the Test-Taking Tip.

Remind students that an open-ended response means that they'll have to write a paragraph to answer the question. Have them paraphrase the question in their notebooks, and remind them about the steps in writing a paragraph.

ANSWERS

1. C
2. A
3. A
4. A
5. C
6. B
7. A
8. Sedimentary rock usually forms when particles of sand, silt, clay, or lime settle in layers at the bottom of a body of water. These layers cover the bodies of dead organisms. Other layers with other organisms build up over the lower layers. Unless the layers of rock are disturbed, scientists can infer that fossil specimens in lower layers are older than specimens buried in more recent layers that were deposited on top.

History of Life 475

UNIT 5

Plan Ahead

Have students preview the Unit 5 Project a few days before writing to give them time to think through what they need to include in their scripts. If possible, provide students with an example of a script for a documentary. You may find one online or in a library. Point out that a script includes both text for a narrator to speak and descriptions of the images shown throughout the documentary. It might also include interviews with experts.

Monitor the Project

Suggest students begin by making a basic outline of how the documentary will proceed. As students write their scripts, circulate through the class, asking individual students how the script addresses the five misconceptions about evolution.

Ask What example are you including that demonstrates the process of natural selection?

Ask What are you including that addresses the misconception that gaps in the fossil record disprove evolution?

Project Assessment

Make sure students use the rubric and reflection questions to assess their work. Then, use the rubric to assign a final score. Note that it is important to value the creativity of students' work as well as the content when you score their projects. If desired, talk with students about any differences between their self-assessment scores and your assigned score.

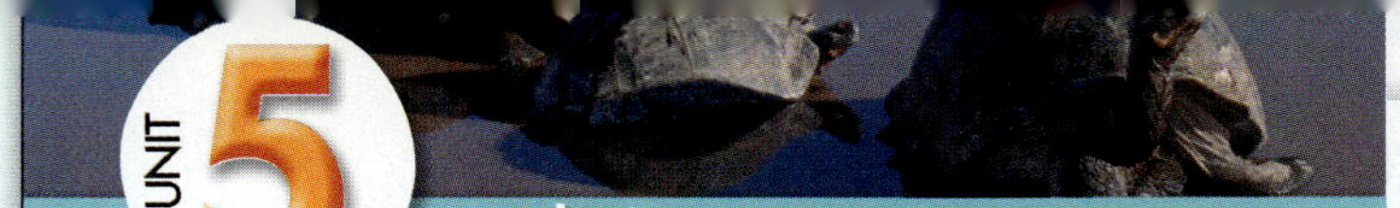

UNIT 5 Evolution

Unit Project

Evolution Documentary

Have you ever flipped through the channels and stopped on a documentary that caught your eye? And before you knew it an hour had passed? Documentaries can be a great way to learn about fascinating topics. Imagine you are a TV producer and have been hired to produce a documentary on evolution for a public television station. Your target audience is the general public.

Your Task Write a script for a 5–10 minute segment of an evolution documentary and present it to your class.

Be sure to

- discuss evidence for evolution by bringing in specific examples.
- present the information clearly and in an engaging manner.
- explain why the misconceptions listed below are *not* true:
  1) Evolution causes organisms to improve—life has gotten better over time.
  2) Evolution is not observable or testable.
  3) Gaps in the fossil record disprove evolution.
  4) Natural selection involves organisms "trying" to adapt.
  5) Natural selection is the only way that populations can change over time.

Reflection Questions

1. Score your documentary using the rubric below. What score did you give yourself?
2. What did you do well in this project?
3. What needs improvement?
4. What do you think a member of the general public would learn from your documentary?

Assessment Rubric

Score	Scientific Content	Quality of Documentary Script
4	Documentary provides accurate evidence for evolution and clearly corrects several misconceptions.	Information is presented in a clear, organized, and engaging manner.
3	Documentary provides some accurate evidence for evolution and attempts to correct misconceptions.	Information is presented in a clear and organized manner, but it could be more engaging.
2	Documentary provides little evidence for evolution and does not correct misconceptions well.	Information could be presented in a clearer manner. The script needs editing.
1	Documentary does not provide evidence for evolution and does not attempt to correct misconceptions.	Information is presented in a disorganized and confusing manner. The script needs a lot of editing.

21st Century Skills

To be successful in the 21st century, students need skills and learning experiences that extend beyond subject area mastery. The Unit 5 Project helps students build the following 21st Century Skills: *Information and Media Literacy; Communication Skills; Creativity and Intellectual Curiosity; Self-Direction;* and *Accountability and Adaptability.*

FOCUS ON PROBLEM IDENTIFICATION, FORMULATION, AND SOLUTION Extend this Unit Project by having students talk with other students, relatives, friends, and neighbors about their understanding of evolution. Then, have students discuss any misunderstandings of evolution they have encountered. Suggest they think of ways a documentary could address these misconceptions.

For more practice building 21st Century Skills, see The Chapter Mystery pages in **Study Workbook A.**

NATIONAL SCIENCE EDUCATION STANDARDS

UCP IV

CONTENT C.3.a, C.3.b, C.3.c, C.3.d

INQUIRY A.1.f, A.2.e, A.2.f

From Microorganisms to Plants

UNIT 6

Chapters

- **20** Viruses and Prokaryotes
- **21** Protists and Fungi
- **22** Introduction to Plants
- **23** Plant Structure and Function
- **24** Plant Reproduction and Response

INTRODUCE the Big ideas

- **Unity and Diversity of Life**
- **Structure and Function**
- **Growth, Development, and Reproduction**
- **Cellular Basis of Life**
- **Interdependence in Nature**

"In sports, a most valuable player, or MVP, is the person most responsible for a team's success. Which organisms are the MVPs of life? Certainly not us—we could disappear from this planet, and other forms of life would go on just fine. The real stars of life on Earth, its MVPs, are its microorganisms, fungi, and plants. We almost never notice these superstars, unless we look very closely, but without them our lives would be impossible."

Ken Miller

Dear Colleague,

It's only natural. Because we're animals ourselves, we tend to be more interested in the lives of other animals. We tend to think of evolution largely in terms of the origins of major groups of animals, and we often consider the well-being of ecosystems in terms of the animals that may or may not be present and part of the food chain. But, if we force ourselves to think in purely biological terms, we animals make up only a tiny part of the range and diversity of life on Earth. The real players, the ones that have changed this planet in fundamental ways, are microorganisms and plants. Students don't always appreciate this, and it's a special challenge to open their eyes to these remarkable forms of life.

Every time I take a walk through the woods near my house, I'm struck by the obvious ways in which plants dominate the landscape, and by the not-so-obvious ways in which decomposers such as bacteria, fungi, and slime molds shape the living world. Even more amazing are the protists, whose various members exemplify nearly every conceivable way of making a living on this little planet. Indeed, the cellular and biochemical diversity of these organisms is so great that, compared to them, all animals are pretty much the same. If that's a shock to your students, point out to them that biologists now suspect that huge numbers of bacteria live deep beneath the surface, thriving in darkness, often under some of the most extreme conditions imaginable. In fact, the abundance of underground bacteria is so great that their mass might dwarf all life found on Earth's surface.

Students may know that microorganisms and plants are essential to our lives, but many have a difficult time believing that these organisms are just as alive as you or I. Nonetheless, as teachers, one of the greatest gifts we can give our students is to extend their vision beyond the obvious—and I can think of no better way to do exactly that than to spend some time investigating these extraordinary organisms.

Ken Miller

Viruses and Prokaryotes

Chapter Contents	NSES	Time	Core Resources
Chapter Preview			**Student Edition,** pp. 478–479 **Chapter Mystery,** p. 479 **Foundations for Learning,** p. 479
20.1 Viruses The Discovery of Viruses • Viral Infections	V, A.2.a, F.1, G.3	1 period 1/2 block	**Student Edition,** pp. 480–484 **Build Connections,** p. 481 **Study Workbook B** 20.1 Worksheets **Biology.com** *Visual Analogy:* How a Lytic Virus Is Like an Outlaw • *InterActive Art:* Two Types of Virus Reproduction • 20.1 Lesson Assessment
20.2 Prokaryotes Classifying Prokaryotes • Structure and Function • The Importance of Prokaryotes	I, V, A.2.a, C.1.b, C.1.e, C.3.a, C.3.e, C.4.a, C.4.b	1 period 1/2 block	**Student Edition,** pp. 485–488 **Study Workbook B** 20.2 Worksheets **Biology.com** *Art Review:* Prokaryote Structure and Classification • 20.2 Lesson Assessment
20.3 Diseases Caused by Bacteria and Viruses Bacterial Diseases • Viral Diseases • Emerging Diseases	V, A.1.c, A.2.a, A.2.d, F.1,G.1	1/2 period 1/4 block	**Student Edition,** pp. 489–493 **Build Connections,** p. 491 **Inquiry Into Scientific Thinking,** p. 492 **Study Workbook B** 20.3 Worksheets **Biology.com** *Art in Motion:* Prion Infection Mechanism • 20.3 Lesson Assessment
Chapter Pre-Lab	A.1.b, A.2.a, F.1, F.5	1 period 1/2 block	**Student Edition,** p. 494 **Lab Manual B** *Controlling Bacterial Growth*

Tools for ELL Instruction

ELL Handbook describes teaching strategies to make *Miller & Levine Biology: Foundation Edition* more accessible to ELL students.

Spanish Study Workbook is a Spanish translation of Study Workbook A.

Multilingual Glossary is the *Miller & Levine Biology* glossary translated into ten languages.

Tools to Challenge Students

Study Workbook A includes more challenging worksheets and activities such as Chapter Mystery activities that reinforce 21st Century Skills.

Lab Manual A labs are geared toward on-level students. The Lab Manual includes fifteen labs in addition to the Chapter Labs.

Biology.com Data Analysis activities offer students a way to practice data gathering and analyzing skills.

Visual Quizzes in the Assessment Resources Book are geared toward on-level students and use visuals from *Miller & Levine Biology.*

Additional Resources

Biology.com Untamed Science Video • Flash Cards

Study Workbook A 20.1 Worksheets
Lab Manual B *Hands-On Activities:* How Do Viruses Differ?
Spanish Study Workbook 20.1 Worksheets
Assessment Resources Book Visual Quiz

Study Workbook A 20.2 Worksheets
Spanish Study Workbook 20.2 Worksheets
Assessment Resources Book Visual Quiz

Study Workbook A 20.3 Worksheets
Lab Manual B *Data Analysis:* MRSA on the Rise
Spanish Study Workbook 20.3 Worksheets
Biology.com *Data Analysis:* MRSA on the Rise

Lab Manual A *Controlling Bacterial Growth*

Chapter Review

Student Edition Chapter Summary, p. 495
Study Workbook B Chapter 20 Review
Transparencies, pp. 240–248
Biology.com Untamed Science Video • Editable Worksheets of Study Workbooks A and B and Lab Manuals A and B • Chapter 20 Flash Cards and Crossword

Untamed Science DVD • Classroom Resources CD (includes lesson presentations and editable worksheets)

Chapter Assessment

Student Edition Check Understanding, pp. 496–499
Study Workbook B Chapter 20 Taking a Standardized Test
Assessment Resources Book Chapter 20 Test B
Biology.com Chapter 20 Assessment • Editable Worksheets of Chapter 20 Visual Quizzes* and Chapter 20 Tests A and B

Exam*View* *Assessment Suite* • Classroom Resources CD (includes lesson presentations and editable worksheets)

Time: 1 period, 1/2 block

* Visual Quizzes are based on visuals from *Miller & Levine Biology*. Review carefully to make sure they are appropriate for your class.

Pressed for Time?

Preview the Chapter Introduce the vocabulary for Lesson 20.1. Preview the images on pp. DOL 6–DOL 9 in the Diversity of Life guide and discuss the variations that exist among different bacteria.

Cover the Chapter Quickly In 20.1, have students read the definition of a virus, the Viral Structure and Function section of *The Discovery of Viruses*, and *Viral Infections*. Discuss the visual Two Types of Bacteriophage Infection. Assign students to read all of Lesson 20.2 and discuss the Typical Bacterial Structure visual.

Assess Assign Questions 1–6 in the 20.1 Check Understanding, the 20.2 Check Understanding, and Questions 1–12 in the Chapter 20 Check Understanding.

CHAPTER 20

Get the Picture

Have a volunteer read aloud the caption of the picture.

Ask What do the bacteria on the petri dish look like? *(dots)* Is each dot a single bacterial cell or a group of bacteria? *(a group of bacteria, or a colony)* How do you know? *(Individual bacteria are microscopic.)* Why do you think scientists study bacteria? *(Sample answers: Bacteria can cause disease. Bacteria are found almost everywhere and play important roles in ecosystems.)*

Activate Prior Knowledge

Ask What experiences have you had with viruses and bacteria? *(Students are likely to describe being sick or cleaning surfaces or hands to prevent the spread of viruses or bacteria.)*

Ask What does the term *microbe* mean? *(The term* microbe *refers to germs or pathogens.)* The prefix *micro-* means "small." How does this prefix fit with the definition of *microbe*? *(Microbes are microscopic, or very small, in size.)*

Explain that viruses and bacteria are microbes. Tell students that the term *prokaryote* refers to single-celled living organisms, such as bacteria, that lack a nucleus. If necessary, have students review the terms *microbe* and *prokaryote* before beginning the chapter.

NATIONAL SCIENCE EDUCATION STANDARDS

UNIFYING CONCEPTS AND PROCESSES

I, V

CONTENT

C.1.b, C.1.e, C.3.a, C.3.e, C.4.a, C.4.b, F.1, F.5, G.1, G.3

INQUIRY

A.1.b, A.1.c, A.2.a, A.2.d

20 Viruses and Prokaryotes

Big idea Cellular Basis of Life

Q: Are all microbes that make us sick made of living cells?

A scientist looks at colonies of E. coli *bacteria on a petri dish.*

BIOLOGY.com Search Chapter 20 GO • Flash Cards

478

UbD Connect to the Big Idea

Invite a volunteer to read aloud the chapter number and the chapter title. Read the Big Idea and question. Emphasize the question, **Are all microbes that make us sick made of living cells?**

Set the Purpose

PREVIEW AND PREDICT Show students a sheet of paper you have folded into thirds and labeled as shown in the Foundations for Learning on the next page. Then, guide students to the INSIDE feature on the next page, and invite a volunteer to read each lesson number and title.

Ask How do you think these titles relate to the headings in the z-fold?

Help students link each column in your example z-fold to each lesson in this chapter.

INSIDE:
- 20.1 Viruses
- 20.2 Prokaryotes
- 20.3 Diseases Caused by Bacteria and Viruses

CHAPTER
MYSTERY

THE MAD COWS

In 1986, something strange began to happen to cattle in Great Britain. Without warning, the animals began acting strangely. They lost control of their movements and eventually died. Farmers watched helplessly as this "mad cow disease" spread through their cattle. In 1991, the disease affected more than 30,000 cattle.

Scientists studied the brains of cattle killed by mad cow disease. They found that large areas of the animals' brains had been destroyed. Under a microscope, holes in the diseased tissue made the brain look like a sponge. Because of this, the disease was given the name bovine spongiform encephalopathy, or BSE.

Read for Mystery Clues What caused BSE? As you read this chapter, look for clues that explain the cause of the disease. Then, solve the mystery.

FOUNDATIONS for Learning

Use a sheet of paper to make a z-fold similar to the one shown below. Use one column of the z-fold for each lesson. Write down what you learn as you read each lesson. At the end of the chapter are two activities that will help you answer the question: Are all microbes that make us sick made of living cells?

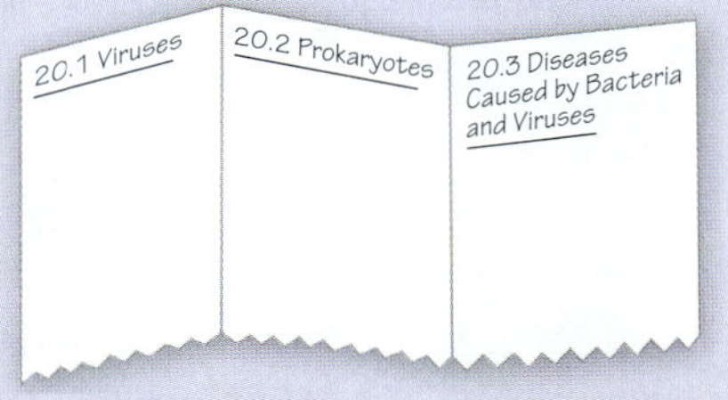

- Untamed Science Video
- Chapter Mystery

Chapter 20 Big Idea: Cellular Basis of Life

Chapter 20: Are all microbes that make us sick made of living cells?

- **20.1:** What is a virus?
- **20.2:** What are prokaryotes, and why are they important?
- **20.3:** How can we prevent bacterial and viral diseases from spreading?

CHAPTER 20

Foundations for Learning

Demonstrate how to make and label the z-fold.

Ask How does the z-fold graphic organizer help you learn about viruses and prokaryotes as you read? *(Sample answer: It gives me a place to write notes and helps me organize information.)* Encourage students to think about how they will use the z-fold to organize information as they read and to make comparisons between viruses and prokaryotes when it is completed.

Ask What is a living cell? *(Students should be able to describe the characteristics of a living cell in terms of structure and function.)*

Ask students to think about how the characteristics of a microbe might be different depending on whether the microbe is made of living cells.

Untamed Science — Have students take a video field trip to learn more about beneficial bacteria in **Bacteria, at Your Service**.

CHAPTER MYSTERY — Invite a volunteer to read aloud the Chapter Mystery. Have students predict how this mystery relates to the chapter title and what might have caused BSE. Tell students that they will find clues in the chapter that will help them solve the mystery.

Digital Support

BIOLOGY.com — Extend your reach by using the digital assets offered at **Biology.com.**

LESSON 20.1

Pre-Reading

Objectives

20.1.1 Explain how viruses reproduce.

20.1.2 Explain how viruses cause infection.

Preview the Pages

1. Review the characteristics of living things.
2. Preview the diagrams in this lesson. As you do, review the meaning of science terms used in the labels and captions, especially terms such as *DNA, RNA, nucleic acid, protein,* and *chromosome* that were introduced in previous chapters.
3. Have students find the definitions for *lytic infection* and *lysogenic infection* in the text. Then, point out that the diagram Two Types of Bacteriophage Infection illustrates both kinds of infection. Show students how to follow the arrow pathways to distinguish the infections.
4. Tell students that the table at the end of the lesson compares and contrasts viruses with living cells. **Ask** What characteristics does the table compare? *(structure, reproduction, genetic code, etc.)*

Build Understanding

VENN DIAGRAM Have students review the partially filled-in Venn diagram in Workbook B. Explain to students that there are distinct similarities and differences between viruses and living cells. Using the Venn diagram to compare and contrast will help them to gain a more complete understanding of both.

NATIONAL SCIENCE EDUCATION STANDARDS

UNIFYING CONCEPTS AND PROCESSES

V

CONTENT

F.1, G.3

INQUIRY

A.2.a

For the corresponding lesson in *Miller & Levine Biology,* see pages 574–579.

20.1 Viruses

Key Questions

How do viruses reproduce?

What happens after a virus infects a cell?

BUILD Understanding

Venn Diagram Make a Venn diagram to compare viruses and cells. List similarities and differences between viruses and cells. Fill in the Venn diagram as you read the lesson.

In Your Workbook Go to your workbook to learn more about Venn diagrams. Complete the Venn diagram for Lesson 20.1.

The Discovery of Viruses

Imagine that you have to solve a mystery. Farmers have begun to lose their tobacco crop to tobacco mosaic disease. First, the leaves turn yellow. Eventually, the leaves fall off, killing the plant. You use a light microscope to search for the pathogen, or disease-causing agent. But it is too small, even for a microscope. Although you cannot see the pathogen, you are sure that it is there.

Discovery of Viruses In 1892, the Russian biologist Dmitri Ivanovski studied tobacco mosaic disease. He showed that liquid taken from infected plants could cause the disease. But he could not find the pathogen itself. In 1897, the Dutch scientist Martinus Beijerinck suggested that tiny particles caused the disease. He named these particles *viruses*, after the Latin word for "poison."

Then, in 1935, the American biochemist Wendell Stanley produced crystals of tobacco mosaic virus. Living organisms do not crystallize. So, Stanley concluded that viruses were not truly alive. Most biologists still think this today. A **virus** is a nonliving particle. It is made of proteins, nucleic acids, and sometimes lipids. Viruses can reproduce only by infecting living cells.

Key Question How do viruses reproduce?
Viruses can reproduce only by infecting living cells.

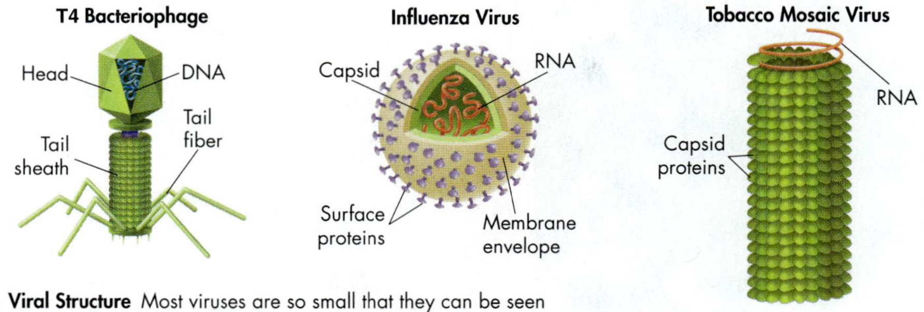

Viral Structure Most viruses are so small that they can be seen only with very powerful electron microscopes. The shapes of three kinds of viruses are shown here.

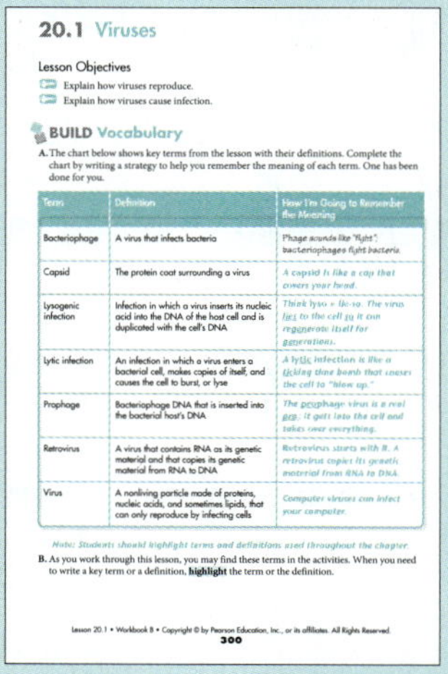

20.1 Viruses

Lesson Objectives

Explain how viruses reproduce.

Explain how viruses cause infection.

BUILD Vocabulary

A. The chart below shows key terms from the lesson with their definitions. Complete the chart by writing a strategy to help you remember the meaning of each term. One has been done for you.

Term	Definition	How I'm Going to Remember the Meaning
Bacteriophage	A virus that infects bacteria	Phage sounds like "fight"; bacteriophages fight bacteria.
Capsid	The protein coat surrounding a virus	A capsid is like a cap that covers your head.
Lysogenic infection	Infection in which a virus inserts its nucleic acid into the DNA of the host cell and is duplicated with the cell's DNA	Think lyso = lie-so. The virus lies to the cell so it can regenerate itself for generations.
Lytic infection	An infection in which a virus enters a bacterial cell, makes copies of itself, and causes the cell to burst, or lyse	A lytic infection is like a ticking time bomb that causes the cell to "blow up."
Prophage	Bacteriophage DNA that is inserted into the bacterial host's DNA	The prophage virus is a real pro: it gets into the cell and takes over everything.
Retrovirus	A virus that contains RNA as its genetic material and that copies its genetic material from RNA to DNA	Retrovirus starts with R. A retrovirus copies its genetic material from RNA to DNA.
Virus	A nonliving particle made of proteins, nucleic acids, and sometimes lipids, that can only reproduce by infecting cells	Computer viruses can infect your computer.

Note: Students should highlight terms and definitions used throughout the chapter.

B. As you work through this lesson, you may find these terms in the activities. When you need to write a key term or a definition, **highlight** the term or the definition.

Lesson 20.1 • Workbook B • Copyright © by Pearson Education, Inc., or its affiliates. All Rights Reserved.
300

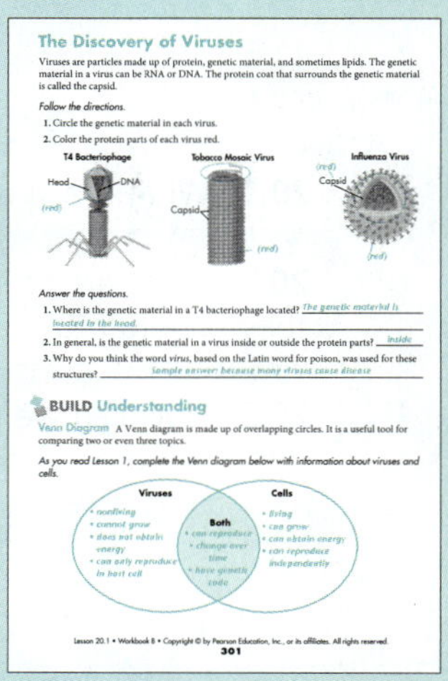

The Discovery of Viruses

Viruses are particles made up of protein, genetic material, and sometimes lipids. The genetic material in a virus can be RNA or DNA. The protein coat that surrounds the genetic material is called the capsid.

Follow the directions.

1. Circle the genetic material in each virus.
2. Color the protein parts of each virus red.

Answer the questions.

1. Where is the genetic material in a T4 bacteriophage located? The genetic material is located in the head.
2. In general, is the genetic material in a virus inside or outside the protein parts? inside
3. Why do you think the word *virus*, based on the Latin word for poison, was used for these structures? Sample answer: because many viruses cause disease

BUILD Understanding

Venn Diagram A Venn diagram is made up of overlapping circles. It is a useful tool for comparing two or even three topics.

As you read Lesson 1, complete the Venn diagram below with information about viruses and cells.

Lesson 20.1 • Workbook B • Copyright © by Pearson Education, Inc., or its affiliates. All rights reserved.
301

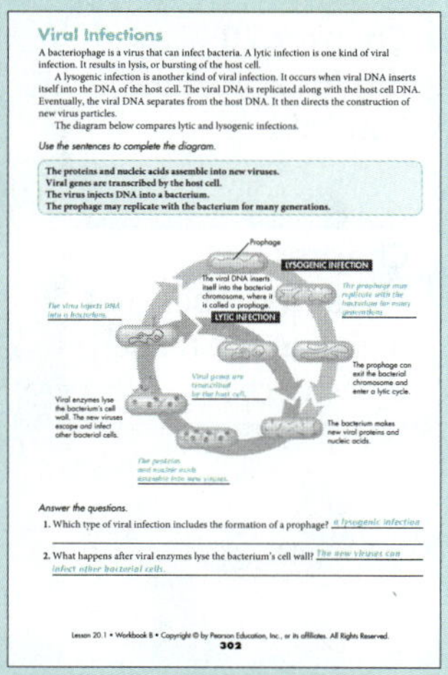

Viral Infections

A bacteriophage is a virus that can infect bacteria. A lytic infection is one kind of viral infection. It results in lysis, or bursting of the host cell.

A lysogenic infection is another kind of viral infection. It occurs when viral DNA inserts itself into the DNA of the host cell. The viral DNA is replicated along with the host cell DNA. Eventually, the viral DNA separates from the host DNA. It then directs the construction of new virus particles.

The diagram below compares lytic and lysogenic infections.

Use the sentences to complete the diagram.

The proteins and nucleic acids assemble into new viruses.
Viral genes are transcribed by the host cell.
The virus injects DNA into a bacterium.
The prophage may replicate with the bacterium for many generations.

Answer the questions.

1. Which type of viral infection includes the formation of a prophage? a lysogenic infection
2. What happens after viral enzymes lyse the bacterium's cell wall? The new viruses can infect other bacterial cells.

Lesson 20.1 • Workbook B • Copyright © by Pearson Education, Inc., or its affiliates. All Rights Reserved.
302

Viral Structure and Function Viruses come in many shapes and sizes. A protein coat called a **capsid** surrounds a virus particle. Some viruses also have a membrane around the capsid. The simplest viruses have only a few genes. Complex viruses may have hundreds of genes.

The structure of a virus allows it to infect living cells. Most viruses have proteins on their surface. These proteins bind to proteins on the surface of a host cell. Viral proteins "trick" the host cell into taking in the virus or its genetic material. Viruses must bind exactly to proteins on the host cell's surface. Therefore, most viruses infect only specific kinds of cells. Plant viruses infect plant cells. Most animal viruses infect only certain kinds of animals. Viruses that infect bacteria are called **bacteriophages** (bak TEER ee uh fayj uz). They infect only certain kinds of bacteria.

BUILD Vocabulary

virus a particle made of proteins, nucleic acids, and sometimes lipids that can replicate only by infecting living cells

capsid the protein coat surrounding a virus

bacteriophage a kind of virus that infects bacteria

WORD ORIGINS

The word *capsid* comes from the Latin word *capsa*, which means "box."

Viral Infections

What happens after a virus enters a host cell? The virus uses the host cell's genetic information to make copies of itself! Some viruses copy themselves immediately. This type of infection is called a lytic infection. Other viruses stay inactive for a period of time within the host cell. This type of infection is called a lysogenic infection.

BUILD Connections

HOW A LYTIC VIRUS IS LIKE AN OUTLAW

A virus that causes a lytic infection works much like an outlaw in the Wild West. Imagine that the virus is an outlaw. Then, imagine that the cell the virus takes over is a town.

First, the outlaw eliminates the town's existing authority.
Lytic Infection The host cell's DNA is chopped up.

Next, the outlaw demands to be outfitted with new equipment from the local townspeople.
Lytic Infection Viruses use the host cell to make viral DNA and viral proteins.

Finally, the outlaw forms a gang that leaves the town to attack new communities.
Lytic Infection The host cell bursts, releasing hundreds of virus particles.

BIOLOGY.com Search Lesson 20.1 GO • Visual Analogy

During Reading

Active Reading

KEY QUESTION Invite a volunteer to read the question on the previous page and its answer.

Ask How do living organisms use cells to reproduce? *(Living things use their own cells to reproduce, either sexually or asexually. The living cells have the ability and structures needed.)*

MAKE CONNECTIONS Show students samples of hook-and-loop fasteners to help them understand how proteins on the surface of a virus might bind to proteins on the host cell's surface. Have students compare the proteins on the surface of the influenza virus in the diagram with the hooks and loops.

Ask Will the hooks and loops stick to all types of materials? To plastic, for example? *(No, they stick only to certain materials.)*

Say In the same way, viruses can bind to and infect only specific kinds of cells.

Ask Why don't plant viruses infect animal cells? *(Plant viruses lack proteins that bind to animal cells.)*

Build Vocabulary

WORD ORIGINS Show students a box with some objects inside it.

Ask How is the capsid of a virus similar to a box? *(The capsid contains or surrounds the virus, much like a box contains or surrounds its contents.)*

Build Connections

VISUAL ANALOGY Remind students that an analogy can help them understand a concept by comparing it to something they know. Ask students if they have ever seen a western that features an outlaw. Relate the plot line of the film to the events in the visuals.

Ask How is a cell's DNA similar to a town's sheriff? *(The DNA contains the cell's instructions for making proteins and carrying out life processes. A town's sheriff maintains order. Both keep things running smoothly.)* What happens if an outlaw gets rid of the sheriff? *(The outlaw takes over the town.)*

Speed Bump

Ask How do viruses infect host cells? *(They first bind to proteins on the surface of a host cell, and then they trick the host cell into letting the virus or its genetic material into the cell.)*

Ask What happens after a virus enters a host cell? *(The virus uses the host cell's genetic information to make copies of itself.)*

ELL Focus on ELL: Build Background

ALL SPEAKERS Write the following words on the board: *host, surface,* and *bind.* Invite students to locate these words in the text and explain their meanings.

Point out that a host is often someone who entertains guests. For example, a party may take place at the host's home.

Ask How is a cell infected by a virus a host? *(The virus is inside the host cell.)*

Help students identify the surface of a virus and a host cell using pictures in the textbook or models.

Then, demonstrate how viral surface proteins can bind, or attach to a host cell's surface proteins using tabs of hook and loop fasteners or chenille stems twisted together. Make sure students understand that *bind* means "to attach or connect."

Encourage students to practice using these words in sentences of their own. Advanced speakers can investigate the multiple meanings of each of these words.

During Reading continued

Build Vocabulary

WORD ORIGINS Ask students to write out the terms *lytic* and *lysogenic* and to circle the part of each word that comes from the Greek *luein*.

Ask How are these two words alike in meanings? *(Both terms describe how a virus takes over and "breaks up" a host cell.)*

Active Reading

USE DIAGRAMS Have students trace the progression of the infections shown in the cycle diagram at the bottom of this page.

Ask What does a cycle diagram tell you about a process? *(A cycle indicates that the process continues through the steps over and over.)*

Ask At what place in the diagram does the cycle end and start over again? *(when the new viruses escape and infect other cells)*

KEY QUESTION Read the question aloud to the class and answer it together.

Ask Why do the lytic and lysogenic infections take place inside a living cell? *(A virus cannot reproduce on its own. It must use the structures and functions of living cells.)*

Speed Bump

Ask What is the difference between a lytic infection and a lysogenic infection in terms of time? *(In a lytic infection, the virus makes copies of itself and destroys the cell right away. In a lysogenic infection, the destruction of the host cell can take much longer to occur.)*

Ask What happens to a lysogenic infection when a prophage becomes active? *(It becomes a lytic infection.)*

BUILD Vocabulary

lytic infection
a type of infection in which a virus enters a cell, makes copies of itself, and causes the cell to burst

lysogenic infection
a type of infection in which a virus embeds its DNA into the DNA of the host cell and is replicated along with the host cell's DNA

retrovirus
an RNA virus that contains RNA as its genetic information

WORD ORIGINS
The adjective *lytic*, the verb *lyse*, and the prefix *lyso-* all come from the Greek word *luein*. This word means "to loosen or break up."

Lytic Infection In a **lytic infection,** a virus enters a cell, makes copies of itself, then causes the cell to burst. For example, a bacteriophage injects its DNA into a bacterial cell. Then the bacterial cell makes messenger RNA (mRNA) from the viral genes. This mRNA is translated into viral proteins that chop up the cell's DNA. The host cell makes new virus particles. Before long, the cell bursts open. Copies of the bacteriophage are released to infect other cells.

Lysogenic Infection In a **lysogenic infection,** the host cell is not taken over right away. Instead, the virus inserts its nucleic acid into the host cell's DNA. The viral DNA is copied along with the cell's DNA. The cell is not damaged. Daughter cells are infected by the virus.

Bacteriophage DNA that becomes part of the bacterial host's DNA is called a prophage. A prophage may be a part of the host cell's DNA for many generations. Changes in the environment, such as heat or chemicals, can cause the prophage to become active. Then the lysogenic infection becomes a lytic infection.

Key Question What happens after a virus infects a cell?
Inside living cells, viruses use their host's genetic information to make many copies of themselves. Some viruses replicate right away. Other viruses stay inactive for a long period of time within the host cell.

Two Types of Bacteriophage Infection Bacteriophages infect cells in two ways: lytic infection and lysogenic infection.

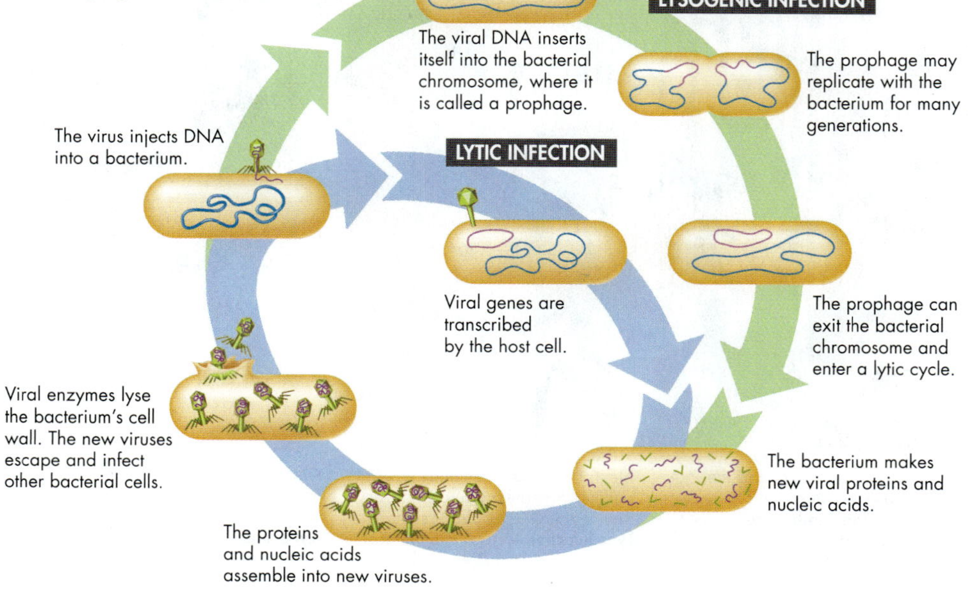

Focus on Academic Language

- replicate

Ask students to define the term *replicate* using their own words.

Then, have students observe the diagram of the lytic and lysogenic cycle. Point out that as part of the cycle, the viral nucleic acid is replicated with the host DNA.

Ask What does the word *replicate* tell you about the new viral nucleic acid? *(It indicates that it is an exact copy of the original nucleic acid.)*

Ask What makes something a good replica? *(Answers may vary. Students should express that a good replica is exactly like the original.)*

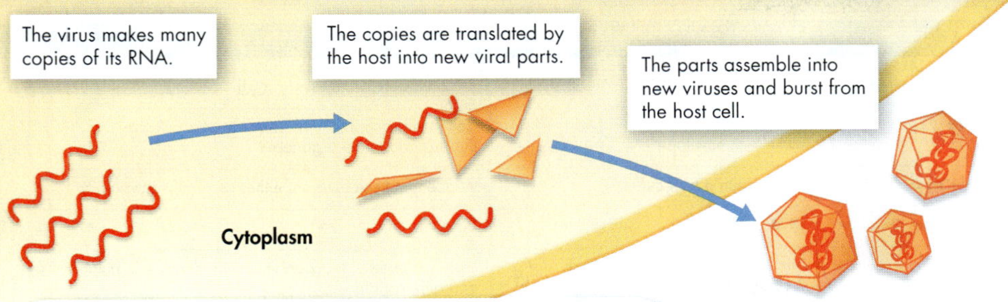

Common Cold Infection Once a cold virus has entered the host cell, it uses the host cell to copy itself.

A Closer Look at Two RNA Viruses About 70 percent of viruses have RNA rather than DNA. In humans, RNA viruses cause a wide range of infections, from mild colds to AIDS.

▶ ***The Common Cold*** What happens when you get a cold? A capsid lands on a cell, often in your nose, and is brought inside a host cell. Inside the host cell, a viral protein makes many new copies of the viral RNA. The host cell's ribosomes mistake the viral RNA for the host's own mRNA. Viral RNA is translated into capsids and other viral proteins. New virus particles form. Within 8 hours, the host cell bursts. Hundreds of new virus particles then infect other cells.

▶ ***HIV*** The disease called acquired immune deficiency syndrome (AIDS) is caused by the human immunodeficiency virus (HIV). HIV belongs to a group of RNA viruses called **retroviruses.** The genetic information of a retrovirus is copied from RNA to DNA instead of from DNA to RNA.

Retroviral infections are similar to lysogenic infections. When a retrovirus infects a cell, it makes a DNA copy of its RNA. This DNA copy is inserted into the DNA of the host cell. Like a prophage, the viral DNA may remain inactive for many cell cycles. Eventually, it makes new virus particles. These new particles attack the cells of the host's immune system—the system that normally fights infection.

HIV Infection A retrovirus such as HIV makes a DNA copy of itself and inserts it into the host cell's DNA. The viral DNA may stay inactive for many cell cycles.

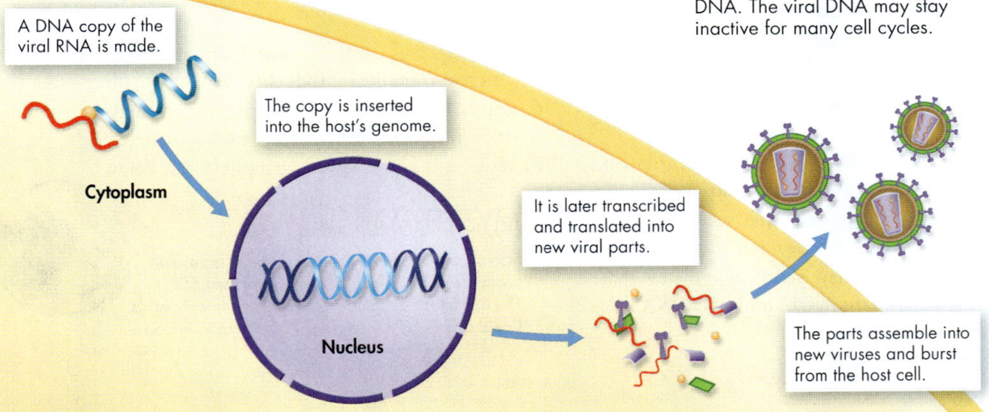

Active Reading

USE DIAGRAMS Have students use the diagrams to make a table that compares and contrasts an HIV infection with a common cold infection.

SUMMARIZE Have students write a summary statement for this page. Encourage them to use their own words but to limit the length to one sentence.

Ask What idea ties together all the information on this page? *(RNA viruses use host cells to make DNA copies of their RNA.)*

MAKE CONNECTIONS Point out to students the prefix *retro-* in the term *retrovirus.*

Ask What do you think of when you hear *retro-*? *(Sample answer: the past or moving backward)*

Draw a simple diagram showing how genetic information is copied from DNA to RNA in humans. Then point out that a retrovirus follows this concept in reverse, in essence "going backward."

Ask What other words contain the prefix *retro-*? *(Sample answers: retroactive, retrospect, retrofit)*

Speed Bump

Ask How are viruses linked to sickness in humans? *(Viral infections cause some illnesses, such as the common cold and AIDS.)*

Ask How is a retrovirus different from other viruses? *(A retrovirus copies genetic information from RNA to DNA instead of from DNA to RNA.)*

Hands-On Learning

DESIGN A BOARD GAME Place students into small groups. Give each group a piece of posterboard, some markers, and a pair of dice. Have students brainstorm and design a board game that models the activity of a viral infection. Tell students to use the dice to advance along a series of events and also to randomly determine if or when a lysogenic cycle is activated. When they have finished designing their board, have the groups test their product by playing their game.

During Reading continued

Active Reading

MAKE INFERENCES Have students form pairs. Within the pairs, have students take turns using the table to describe one aspect of either a living cell or a virus. Have the other person try to infer whether a living cell or a virus is being described without looking at the table or text.

Speed Bump

Ask Why are viruses considered parasites? *(Viruses rely on living host cells for nutrients and the machinery to reproduce. Viruses harm their host cell in this process.)*

Ask Why do scientists think that viruses evolved after other living things? *(Because viruses depend on other living cells, they probably did not evolve until after living things.)*

Post Reading

ASSESSMENT TIPS Review the highlighted vocabulary words with students before they begin the assessment.

For Question 6, pair students. Ask them to make a Venn diagram to help them answer the question. Then, have them answer the question in complete sentences, using the information in the diagram.

MYSTERY CLUE Discuss the Mystery Clue with students as a class. Ask students to think about how they would determine if a virus was responsible for BSE. Have students state how nucleic acids are connected to viral diseases. Students should infer that the infectious agent that causes BSE is not a virus, because viruses cause disease by infecting a host with nucleic acids, which are then replicated in the host.

Wrap-Up Activity

USE VOCABULARY TERMS Pair students and have them take turns choosing a vocabulary term from the lesson. The other person must then define the term and use it in a sentence that explains how the term fits in the lesson.

Viruses and Cells

Characteristic	Virus	Cell
Structure	DNA or RNA in capsid, some with envelope	Cell membrane, cytoplasm; eukaryotes also contain nucleus and many organelles
Reproduction	Only within a host cell	Independent cell division, either asexually or sexually
Genetic Code	DNA or RNA	DNA
Growth and Development	No	Yes; in multicellular organisms, cells increase in number and differentiate
Obtain and Use Energy	No	Yes
Response to Environment	No	Yes
Change Over Time	Yes	Yes

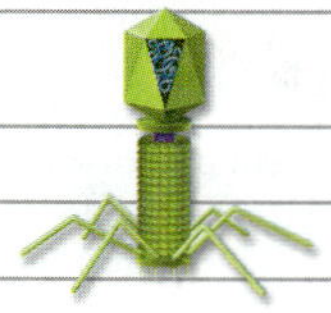

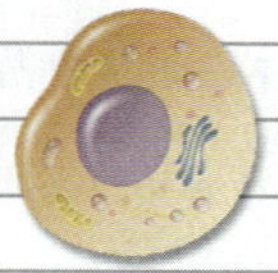

Comparing Viruses and Cells
The table above lists differences and similarities between viruses and living cells.

Viruses and Cells To grow and reproduce, viruses must infect living cells. Viruses rely on the nutrients and cellular machinery of their hosts. Therefore, viruses are parasites. Parasites rely on a host, harming it in the process.

Viruses are not alive. But they do have several characteristics of living things. After infecting living cells, viruses can reproduce. They have genetic material. They can evolve.

Although viruses are smaller and simpler than the smallest cells, it is unlikely that they appeared before living things. Because they depend on living organisms, viruses probably developed after living cells. Perhaps the first viruses evolved from the genetic material of living cells. Viruses, and the cells they infect, have been evolving for billions of years.

CHECK Understanding

Apply Vocabulary

Use the highlighted terms from the lesson to complete each sentence correctly.

1. A virus is surrounded by a _______ made up of proteins.
2. During a _______, viral DNA remains inactive as part of the host cell's DNA.
3. A _______ causes the host cell to burst open and release many virus particles.
4. _______ have genetic information that is copied from RNA instead of DNA.

Critical Thinking

5. **Compare and Contrast** How is viral reproduction different from the reproduction of a living cell?
6. **Compare and Contrast** How are the lytic and lysogenic cycles similar? How are they different?
7. **Write to Learn** Answer the mystery clue below.

MYSTERY CLUE

Scientists injected BSE-infected material from cow brain tissue into mice. The mice developed BSE. The BSE-infected material contained no nucleic acids. Could a virus be the cause of BSE? (**Hint:** See above.)

BIOLOGY.com Search Lesson 20.1 GO • Lesson Assessment

484

Check Understanding

1. capsid
2. lysogenic infection
3. lytic infection
4. Retroviruses
5. Sample answer: Viruses must reproduce within living cells. Living cells can reproduce on their own.
6. Sample answer: The lytic and lysogenic cycles are both ways in which viruses infect bacterial cells. The lytic cycle causes the cell to burst. During the lysogenic cycle, viruses remain inactive for a period of time within the cell. The genetic information replicates as the host cell multiplies.
7. Accept all reasonable answers. Students should recognize that because there are no nucleic acids in the infected material, the culprit is most likely not a virus.

20.2 Prokaryotes

Classifying Prokaryotes

When the microscope was first invented, we humans discovered something amazing: We share every corner of our world with microorganisms! The smallest and most abundant of these microorganisms are **prokaryotes** (pro KAR ee ohts). Prokaryotes are single-celled organisms that lack a nucleus. The DNA of prokaryotes is found in their cytoplasm. Biologists classify prokaryotes as either Bacteria or Archaea.

Key Question How are prokaryotes classified?
Prokaryotes are classified as Bacteria or Archaea, which are two of the three domains of life.

Bacteria Bacteria live almost everywhere—in fresh water, in salt water, and on land. They even live inside the bodies of humans and other eukaryotes. The art below shows the structure of a bacterium that lives in human intestines.

Most bacteria are surrounded by a cell wall that contains peptidoglycan. This molecule is made up of sugars and amino acids. Some bacteria are also protected by a membrane outside the cell wall. Bacteria may use flagella or pili (PY ly; singular: pilus) for movement. Pili also help them attach to surfaces or to other bacteria.

Key Questions

- *How are prokaryotes classified?*
- *How do prokaryotes vary in their structure and function?*
- *What roles do prokaryotes play in the living world?*

BUILD Understanding

Preview Visuals Look at the Prokaryotic Shapes figure on the next page. In your own words, describe the three shapes shown.

In Your Workbook Go to your workbook to complete the activity for Lesson 20.2.

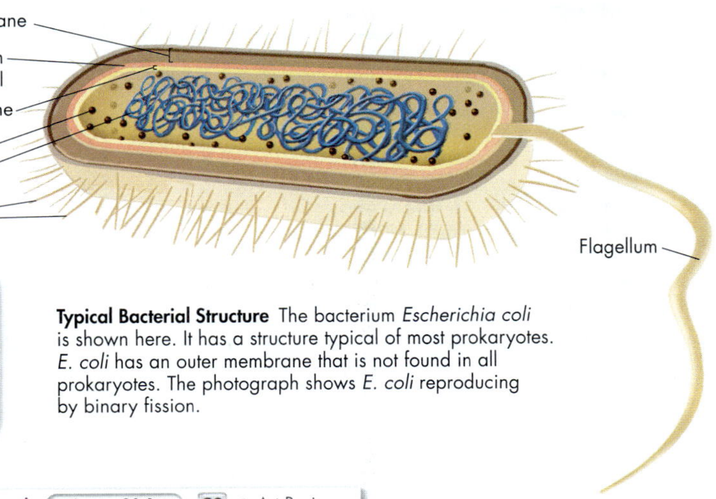

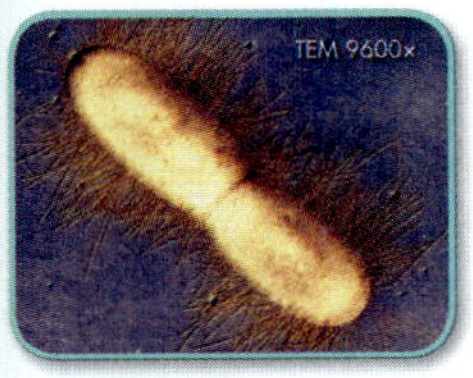

Typical Bacterial Structure The bacterium *Escherichia coli* is shown here. It has a structure typical of most prokaryotes. *E. coli* has an outer membrane that is not found in all prokaryotes. The photograph shows *E. coli* reproducing by binary fission.

BIOLOGY.com Search Lesson 20.2 GO • Art Review

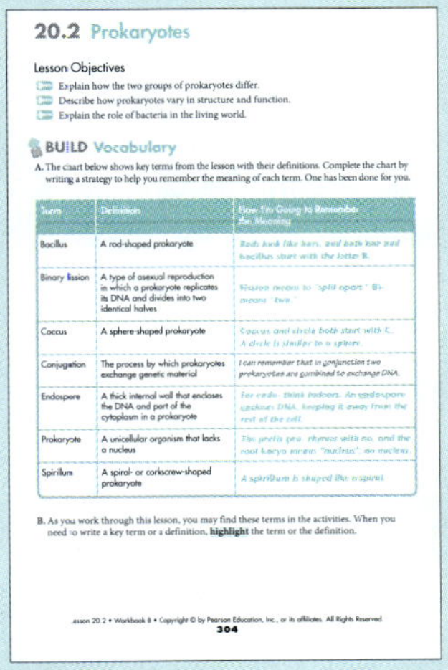

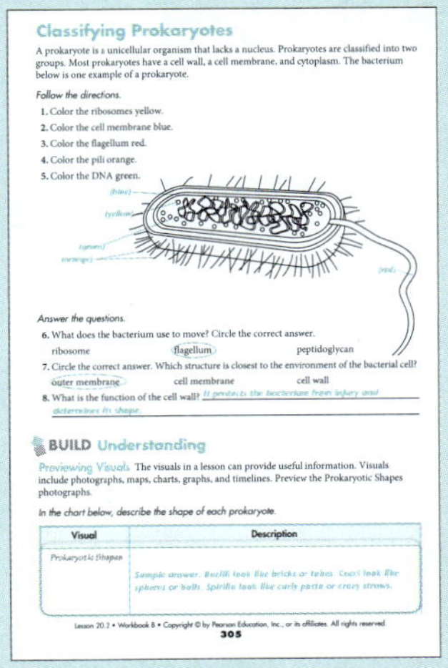

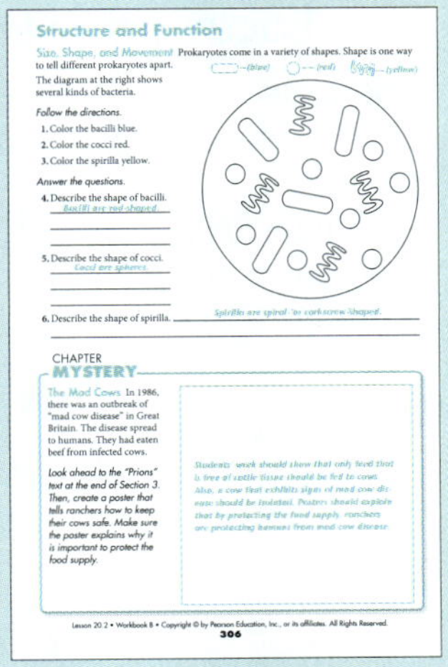

LESSON 20.2

Pre-Reading

Objectives

20.2.1 Explain how the two groups of prokaryotes differ.

20.2.2 Describe how prokaryotes vary in structure and function.

20.2.3 Explain the role of bacteria in the living world.

Preview the Pages

1. Review the vocabulary terms with students. Remind students that these terms refer to prokaryotes rather than viruses.
2. Preview the visuals. Remind students that the micrographs of prokaryotes show the organisms as seen through a powerful microscope.

Activate Prior Knowledge

Refer students to the Three Domains visual in Lesson 18.3.

Ask What are the three domains called? *(Bacteria, Archaea, and Eukarya)* Which domains include prokaryotes? *(Bacteria and Archaea)*

Build Understanding

PREVIEW VISUALS Direct students to Workbook B and have them complete the activity for the lesson. Remind them to focus on the shapes when looking at the visual and to ignore details and surface features.

NATIONAL SCIENCE EDUCATION STANDARDS

UNIFYING CONCEPTS AND PROCESSES
I, V

CONTENT
C.1.b, C.1.e, C.3.a, C.3.e, C.4.a, C.4.b

INQUIRY
A.2.a

For the corresponding lesson in *Miller & Levine Biology,* see pages 580–585.

During Reading

Active Reading

SUMMARIZE Have the class work together to compose a summary statement that explains why prokaryotes are classified in seperate domains from eukaryotes.

Ask What makes a prokaryote different from many eukaryotes? *(They are always single-celled, their cells lack a nucleus, and their DNA is found in the cytoplasm.)*

KEY QUESTION Read the Key Question on the previous page to the class and ask a volunteer to answer it.

Ask What is a prokaryote? *(a single-celled organism that lacks a nucleus)*

Ask How are eukaryotes classified? *(in Eukarya)*

USE GRAPHIC ORGANIZERS Have students construct a T-chart to compare and contrast archaea with bacteria as they read this section. Students can check their charts by referring to the Three Domains visual in Lesson 18.3.

INTERPRET CHARTS AND TABLES Ask students to look at the charts at the bottom of this page and the next page.

Ask How can you tell what kind of information the charts give? *(Look at the titles and headings.)*

Ask How are these two charts different? *(One includes information on how prokaryotes get food, and the other includes information on how prokaryotes release energy from food.)*

USE CONTEXT CLUES Ask students to look at the words that describe the modes of nutrition in the chart on this page.

Ask Which word part do all of these words have in common? *(-troph)*

Ask In what way are the meanings of these words all similar? *(They all refer to a kind of "feeder.")*

Ask What do you think *-troph* might mean? *(an eater or feeder)*

Speed Bump

Ask In what domains are prokaryotes classified? *(They are classified in Domain Bacteria and Domain Archaea.)*

Ask What are three common shapes of prokaryotes? *(rods, spheres, and spirals)*

For more on the diversity of Bacteria and Archaea, go to the Visual Guide. pp. DOL 6–9.

Archaea Archaea look very much like bacteria. Both are tiny, lack nuclei, and have cell walls. However, the cell walls of archaea do not have peptidoglycan. Also, their cell membranes contain different lipids. The DNA of archaea is more like the DNA of eukaryotes than the DNA of bacteria. Many archaea live in harsh environments. Some live in places with little or no oxygen, such as mud and animal digestive tracts. Others live in very salty water or hot springs.

Structure and Function

Prokaryotes come in many sizes and shapes. They also get and use energy in a number of different ways.

Prokaryotic Shapes Prokaryotes usually come in one of three shapes: rod-shaped bacilli (left), spherical cocci (middle), and corkscrew-shaped spirilla (right).

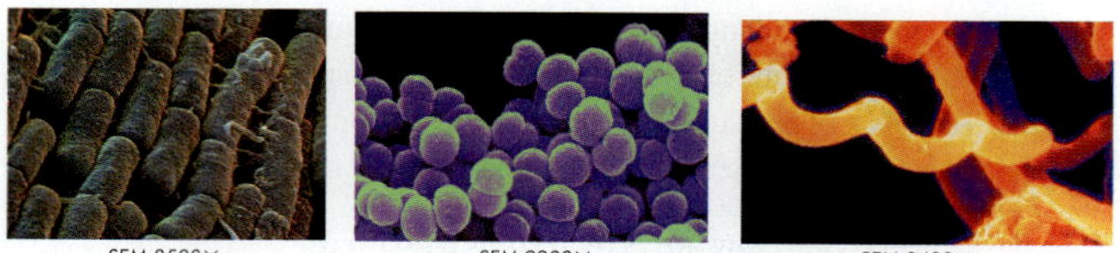

Size, Shape, and Movement Prokaryotes are much smaller than most eukaryotic cells. They range in size from 1 to 5 micrometers. Rod-shaped prokaryotes are called bacilli (buh SIL eye; singular: bacillus). Sphere-shaped prokaryotes are called cocci (KAHK sy; singular: coccus). Spiral and corkscrew-shaped prokaryotes are called spirilla (spy RIL uh; singular: spirillum). Some prokaryotes use flagella to move. Others put down a layer of slime and slowly glide along it. And some prokaryotes do not move at all.

Energy Capture and Release in Prokaryotes Prokaryotes have different ways of getting food and releasing energy from food. Some bacteria can get food and release energy in more than one way.

Energy Capture by Prokaryotes

Mode of Nutrition	How Energy Is Captured	Habitat	Example
Heterotroph "other feeder"	Take in organic molecules from environment or other organisms to use as both energy and carbon supply	Wide range of environments	*Clostridium*
Photoheterotroph "light and other feeder"	Like basic heterotrophs, but also use light energy	Where light is plentiful	*Rhodobacter, Chloroflexus*
Photoautotroph "light self-feeder"	Use light energy to convert CO_2 into carbon compounds	Where light is plentiful	*Anabaena*
Chemoautotroph "chemical self-feeder"	Use energy released by chemical reactions involving ammonia, hydrogen sulfide, etc.	In chemically harsh and/or dark environments: deep in the ocean, in thick mud, in digestive tracts of animals, in boiling hot springs	*Nitrosomonas*

ELL Focus on ELL: Access Content

BEGINNING SPEAKERS Direct students' attention to the photographs showing common shapes of prokaryotes. Then, provide pairs of students with modeling clay or home-made salt dough. Ask students to use the clay to make a model of each of the three shapes shown. Complete the activity by having students write the names *bacilli, cocci,* and *spirilla* on index cards next to each model.

Nutrition and Metabolism Like all organisms, prokaryotes need chemical energy, which they store in molecules such as sugars. Some prokaryotes make their own food. Others take in food for energy. Some can do both. Prokaryotes get energy from food by cellular respiration, fermentation, or both. Some can change the way they get or release energy depending on conditions in their environment.

Key Question How do prokaryotes vary in their structure and function? **Prokaryotes vary in their size and shape, in the way they move, and in the way they get and release energy.**

Growth, Reproduction, and Recombination When a prokaryote has nearly doubled in size, it can reproduce. First, it copies its DNA. Then, it divides in half, forming two identical cells. This kind of reproduction is known as **binary fission.** Binary fission is a form of asexual reproduction. When conditions are good, prokaryotes can divide quickly. Some can divide once every 20 minutes!

In organisms that reproduce sexually, genes are shuffled during meiosis. But prokaryotes reproduce asexually. How do their populations evolve? One way is through mutations. Mutations are random changes in DNA. These changes are passed on to daughter cells formed by binary fission.

Many prokaryotes exchange genetic information by **conjugation.** During conjugation, a hollow bridge forms between two bacterial cells. Genetic material moves from one cell to the other. This process makes populations more genetically diverse.

Formation of Endospores When conditions are not good, many prokaryotic cells can form an **endospore.** A thick internal wall forms around the DNA and part of the cytoplasm. Endospores can remain dormant for months or even years. By forming endospores, prokaryotes are able to live through very harsh conditions.

BUILD Vocabulary

prokaryote
a unicellular organism that lacks a nucleus

binary fission
a type of asexual reproduction in which an organism replicates its DNA and divides in half, producing two identical daughter cells

conjugation
the process in which paramecia and some prokaryotes exchange genetic information

endospore
a structure produced by prokaryotes in unfavorable conditions; a thick internal wall that encloses the DNA and a portion of the cytoplasm

PREFIXES
The prefix *endo-* comes from a Greek word that means "within." An endospore forms when a thick wall encloses DNA within a bacterial cell.

Energy Release by Prokaryotes

Mode of Metabolism	How Energy Is Released	Habitat	Example
Obligate aerobe "requiring oxygen"	Cellular respiration; must have ready supply of O_2 to release fuel energy	Oxygen-rich environments, such as near water surface or in animal lungs	*Mycobacterium tuberculosis*: Sometimes found in human lungs
Obligate anaerobe "requiring a lack of oxygen"	Fermentation; die in presence of oxygen	Environments lacking O_2, such as deep soil, animal intestines, or airtight containers	*Clostridium botulinum*: Sometimes found in improperly sterilized canned food, causing food poisoning
Facultative anaerobe "surviving without oxygen when necessary"	Can use either cellular respiration or fermentation as necessary	Oxygen-rich or oxygen-poor environments	*E. coli*: Lives aerobically in sewage and anaerobically in human large intestine

Hands-On Learning

MAKE BACTERIA MODELS Have students work in small groups to make models of bacteria that can be used to demonstrate binary fission and conjugation. Give students simple objects to use, such as plastic containers, yarn, and modeling clay.

Active Reading

MAKE CONNECTIONS Direct students to the chart on the previous page. Help them understand the different ways prokaryotes capture energy by discussing how more familiar organisms would fit into the chart.

Ask Which mode of nutrition in the chart best describes the way animals get food? *(heterotroph)* Which mode of nutrition best describes the way plants get food? *(photoautotroph)*

KEY QUESTION Read the question aloud and ask a volunteer to read the answer.

Ask What are some different ways bacteria move? *(Sample answers: using flagella or slime layers)*

USE MATH After students read the section on growth, reproduction, and recombination, discuss how binary fission can lead to rapid population growth.

Ask If a bacterial cell can divide once every 20 minutes when conditions are perfect, how many cells would there be after an hour from a single bacterial cell? *(eight)*

Draw a tree-type diagram on the board showing how a single organism will form eight organisms by dividing three times in an hour. Explain that the diagram shows that under perfect conditions, the population of bacteria can double every 20 minutes.

Ask If this is true, why isn't every inch of Earth covered with bacteria? *(Conditions for reproduction are rarely perfect.)*

Build Vocabulary

PREFIXES Ask students to use their own words to explain the meaning of the word *endospore.*

Ask Based on what you know about the meaning of the prefix *endo-*, what do you think an endoskeleton is? *(a skeleton that lies within the body)* The prefix *exo-* means "outside." Do humans and other vertebrates have an endoskeleton or an exoskeleton? *(endoskeleton)*

Speed Bump

Ask How do prokaryotes reproduce? *(asexually, through binary fission)*

Ask What is the difference between mutation and conjugation? *(Mutations are random changes in DNA. Conjugation is a way that some prokaryotes exchange genetic material.)*

During Reading continued

Active Reading

DRAW A DIAGRAM Have students draw a simple diagram, such as a circle graph, that illustrates the different ways that prokaryotes contribute to the living world. Encourage students to be creative and artistic with their depictions. Allow them to use photographs or objects to create a collage as part of their diagram.

Ask What characteristics of prokaryotes make them especially important to life on Earth? *(the ability to break down materials, the ability to serve as producers, the ability to fix nitrogen)*

KEY QUESTION Have a volunteer read the Key Question and answer aloud. Then, ask students to write a job description that describes the role that prokaryotes play as part of the living world. *(Answers should include the information listed under each of the subheadings on this page.)*

Speed Bump

Ask What might life on Earth be like if prokaryotes suddenly didn't exist? *(Students should describe scenarios in which dead organisms do not decompose, less food and energy is available, and the supply of nitrogen in soil is greatly reduced.)*

Post Reading

ASSESSMENT TIPS Review the highlighted vocabulary words with students before they begin the assessment.

For Question 5, ask students to consider how prokaryotes benefit plants.

For Question 6, remind students to review the T-chart they made at the beginning of this lesson to compare archaea and bacteria.

Wrap-Up Activity

MAKE AN OUTLINE Have students make a brief outline of the lesson. Tell them to use the headings from the text. Underneath each heading, have students summarize the main idea using their own words. Encourage them to use vocabulary words in their summaries.

Rhizobium The root nodules of this soybean plant contain *Rhizobium* bacteria. These bacteria are nitrogen fixers.

The Importance of Prokaryotes

Prokaryotes are necessary to the balance of the living world. How can such tiny organisms be so important?

Decomposers Prokaryotes break down, or decompose, dead organisms and wastes. This makes raw materials available to other organisms. If these materials were not recycled, life could not go on.

Producers Photosynthetic prokaryotes are some of the world's most important producers. They form the base of many food chains. For example, the tiny bacterium *Prochlorococcus* may be responsible for more than half of the primary production in the open ocean.

Nitrogen Fixers All organisms need nitrogen to make proteins, nucleic acids, and other molecules. Although nitrogen gas (N_2) makes up 80 percent of Earth's atmosphere, most organisms cannot use N_2 directly. Fortunately, a few prokaryotes can change N_2 into forms that can be used by other living things. This process is called nitrogen fixation. Nitrogen-fixing prokaryotes provide 90 percent of the nitrogen used by other organisms.

A few plants have symbiotic relationships with nitrogen-fixing bacteria. Nitrogen-fixing bacteria grow in nodules on the roots of plants such as clover and soybeans. The bacteria change nitrogen in the air into nitrogen that the plants need.

Human Use of Prokaryotes Bacteria are used to make many different foods and other products. For example, yogurt is made by the bacterium *Lactobacillus*. Some bacteria remove waste products from water. Others are used to make drugs in genetic engineering.

Key Question What roles do prokaryotes play in the living world? **Prokaryotes are decomposers, producers, and nitrogen fixers. In addition, some species are used in human industry.**

CHECK Understanding

Apply Vocabulary

Use the highlighted terms from the lesson to complete each sentence correctly.

1. To survive long periods of harsh conditions, bacteria may form ________.
2. A bacterium undergoes ________, forming two identical daughter cells.
3. ________ increases the genetic diversity of prokaryote populations.

Critical Thinking

4. **Classify** Which two domains of life contain only prokaryotes?
5. **Apply Concepts** Many farmers practice crop rotation. They may plant a field with corn one year and plant soybeans the next year. Why might they do this?
6. **Write to Learn** Write a paragraph comparing and contrasting archaea and bacteria.

BIOLOGY.com Search Lesson 20.2 GO • Lesson Assessment

488

Check Understanding

1. endospores
2. binary fission
3. Conjugation
4. The domains Archaea and Bacteria contain only prokaryotes.
5. Sample answer: Soybeans have nitrogen-fixing bacteria in nodules on their roots. These bacteria add nitrogen to the soil, which will fertilize the corn that is grown the following year.
6. Sample answer: Bacteria and archaea are both prokaryotes. They have cells walls, lack nuclei, and are very tiny. Unlike bacteria, archaea live in harsh environments, lack peptidoglycan in their cell walls, and have different lipids in their cell membranes. Archaea DNA is more like eukaryotic DNA than it is like bacterial DNA.

20.3 Diseases Caused by Bacteria and Viruses

Bacterial Diseases

The first person to clearly show that bacteria cause disease was the French chemist Louis Pasteur. Pasteur helped establish the *germ theory of disease.* The germ theory states that microorganisms cause many human and animal diseases. An organism that can cause sickness is a **pathogen.**

Disease Mechanisms Bacteria cause disease in two main ways. Some bacteria destroy living cells and tissues. Others release chemicals that upset the normal activities of the host. These chemicals are called toxins.

Key Question How do bacteria cause disease?
Bacteria cause disease by damaging host cells and tissues or by releasing chemicals that upset homeostasis.

Controlling Bacteria One of the most important things you can do to control bacteria is to wash your hands well and often. Also wash surfaces where bacteria might settle. Washing removes, but does not kill, bacteria. Disinfectants kill bacteria. Sterilization, or heating objects above 100°C, will also kill most bacteria. It is important to store and cook foods properly. The cold temperature in a refrigerator will slow the growth of bacteria.

Key Questions

- *How do bacteria cause disease?*
- *How do viruses cause disease?*
- *Why are emerging diseases a threat to human health?*

BUILD Understanding

Two-Column Chart Make a two-column chart. In the left column, write the key questions from the lesson. In the right column, write answers to the questions.

In Your Workbook Go to your workbook for help completing this two-column chart. Complete the chart for Lesson 20.3.

Human Bacterial Diseases Some common bacterial diseases are described in the table.

Some Human Bacterial Diseases

Disease	Effect on Body	Transmission
Lyme disease	"Bull's-eye" rash at site of tick bite, fever, fatigue, headache	Ticks transmit the bacterium *Borrelia burgdorferi.*
Tetanus	Lockjaw, stiffness in neck and abdomen, difficulty swallowing, fever, elevated blood pressure, severe muscle spasms	Bacteria enter the body through a break in the skin.
Tuberculosis	Fatigue, weight loss, fever, night sweats, chills, appetite loss, bloody sputum from lungs	Bacteria particles are inhaled.
Strep throat	Fever, sore throat, headache, fatigue, nausea	Direct contact with mucus from an infected person or direct contact with infected wounds or breaks in the skin

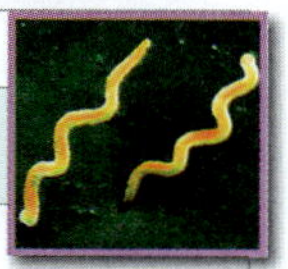
SEM 7300x

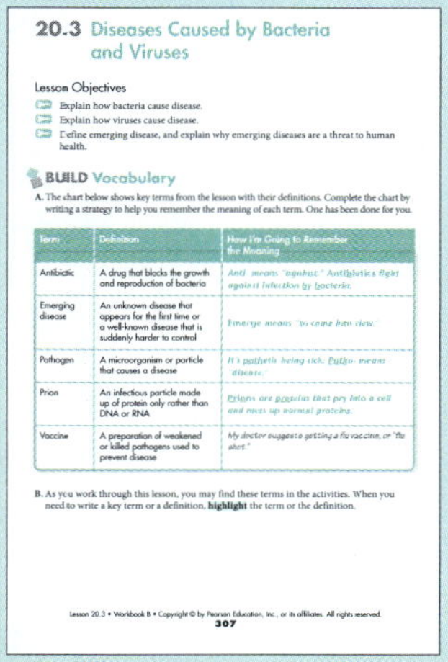
20.3 Diseases Caused by Bacteria and Viruses

Lesson Objectives
- Explain how bacteria cause disease.
- Explain how viruses cause disease.
- Define emerging disease, and explain why emerging diseases are a threat to human health.

BUILD Vocabulary

A. The chart below shows key terms from the lesson with their definitions. Complete the chart by writing a strategy to help you remember the meaning of each term. One has been done for you.

Term	Definition	How I'm Going to Remember the Meaning
Antibiotic	A drug that blocks the growth and reproduction of bacteria	Anti- means "against." Antibiotics fight against infection by bacteria.
Emerging disease	An unknown disease that appears for the first time or a well-known disease that is suddenly harder to control	Emerge means "to come into view."
Pathogen	A microorganism or particle that causes a disease	It's pathetic being sick. Patho- means "disease."
Prion	An infectious particle made up of protein only rather than DNA or RNA	Prions are proteins that pry into a cell and mess up normal proteins.
Vaccine	A preparation of weakened or killed pathogens used to prevent disease	My doctor suggests getting a flu vaccine, or "flu shot."

B. As you work through this lesson, you may find these terms in the activities. When you need to write a key term or a definition, **highlight** the term or the definition.

Lesson 20.3 • Workbook B • Copyright © by Pearson Education, Inc., or its affiliates. All rights reserved.
307

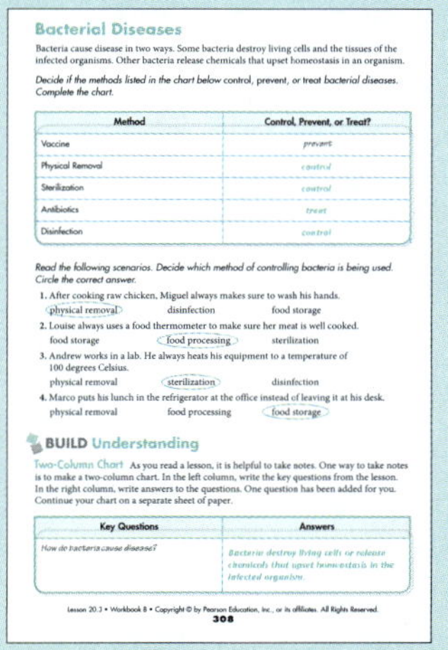
Bacterial Diseases

Bacteria cause disease in two ways. Some bacteria destroy living cells and the tissues of the infected organisms. Other bacteria release chemicals that upset homeostasis in an organism.

Decide if the methods listed in the chart below control, prevent, or treat bacterial diseases. Complete the chart.

Method	Control, Prevent, or Treat?
Vaccine	prevent
Physical Removal	control
Sterilization	control
Antibiotics	treat
Disinfection	control

Read the following scenarios. Decide which method of controlling bacteria is being used. Circle the correct answer.

1. After cooking raw chicken, Miguel always makes sure to wash his hands.
   physical removal   disinfection   food storage
2. Louise always uses a food thermometer to make sure her meat is well cooked.
   food storage   food processing   sterilization
3. Andrew works in a lab. He always heats his equipment to a temperature of 100 degrees Celsius.
   physical removal   sterilization   disinfection
4. Marco puts his lunch in the refrigerator at the office instead of leaving it at his desk.
   physical removal   food processing   food storage

BUILD Understanding

Two-Column Chart As you read a lesson, it is helpful to take notes. One way to take notes is to make a two-column chart. In the left column, write the key questions from the lesson. In the right column, write answers to the questions. One question has been added for you. Continue your chart on a separate sheet of paper.

Key Questions	Answers
How do bacteria cause disease?	Bacteria destroy living cells or release chemicals that upset homeostasis in the infected organism.

Lesson 20.3 • Workbook B • Copyright © by Pearson Education, Inc., or its affiliates. All Rights Reserved.
308

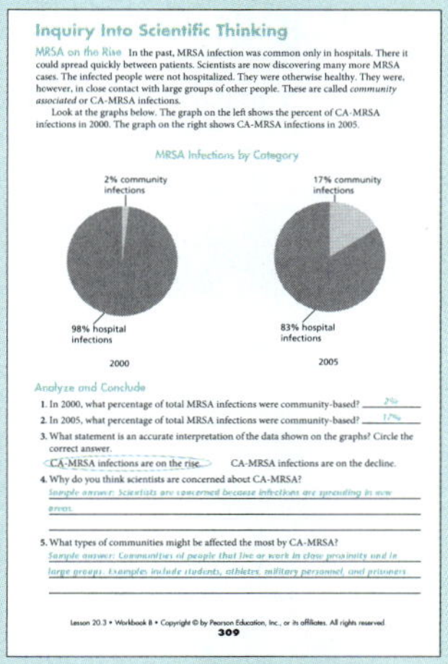
Inquiry Into Scientific Thinking

MRSA on the Rise In the past, MRSA infection was common only in hospitals. There it could spread quickly between patients. Scientists are now discovering many more MRSA cases. The infected people were not hospitalized. They were otherwise healthy. They were, however, in close contact with large groups of other people. These are called *community associated* or CA-MRSA infections.

Look at the graphs below. The graph on the left shows the percent of CA-MRSA infections in 2000. The graph on the right shows CA-MRSA infections in 2005.

Analyze and Conclude

1. In 2000, what percentage of total MRSA infections were community-based? 2%
2. In 2005, what percentage of total MRSA infections were community-based? 17%
3. What statement is an accurate interpretation of the data shown on the graphs? Circle the correct answer.
   CA-MRSA infections are on the rise.   CA-MRSA infections are on the decline.
4. Why do you think scientists are concerned about CA-MRSA?
   Sample answer: Scientists are concerned because infections are spreading in new areas.
5. What types of communities might be affected the most by CA-MRSA?
   Sample answer: Communities of people that live or work in close proximity and in large groups. Examples include students, athletes, military personnel, and prisoners.

Lesson 20.3 • Workbook B • Copyright © by Pearson Education, Inc., or its affiliates. All rights reserved.
309

LESSON 20.3

Pre-Reading

Objectives

20.3.1 Explain how bacteria cause disease.

20.3.2 Explain how viruses cause disease.

20.3.3 Define emerging disease and explain why emerging diseases are a threat to human health.

Preview the Pages

1. Direct students' attention to the charts listing human diseases caused by bacteria and viruses. Have a volunteer read the title of each chart and another volunteer read the example diseases. Discuss any disease students do not recognize.
2. Preview the Innovations in Vaccines visual. Points out that the visual provides a chronology of important discoveries in the development of vaccines. Explain that the term *chronology* refers to the order in which historical events occurred. Point out the numbered steps in the Prion Infections visual. Explain that the numbers tell the sequence of events.

Build Understanding

TWO-COLUMN CHART Have students find the two-column chart in Workbook B. Discuss the difference between controlling, preventing, and treating bacterial infection. Encourage students to use the example in Workbook B as a guide as they complete the rest of the chart.

NATIONAL SCIENCE EDUCATION STANDARDS

UNIFYING CONCEPTS AND PROCESSES

V

CONTENT

F.1, G.1

INQUIRY

A.1.c, A.2.a, A.2.d

For the corresponding lesson in *Miller & Levine Biology,* see pages 586–592.

LESSON 20.3

During Reading

Active Reading

KEY QUESTION Read the Key Question on the previous page and its answer aloud. Then, ask the class to find the paragraph in the text that contains the answer.

Ask How does damaging cells and tissues in an organism cause problems? *(Damaging cells and tissues can prevent organs or systems from functioning properly.)*

FIND MAIN IDEAS Ask students to list the blue subheadings on this page in their notebooks. Then, have students write the main idea of the paragraph under each subheading.

Say The main idea is the point of the paragraph. It is often the first sentence. The rest of the paragraph or section gives details and examples that support the main idea.

Ask What is the purpose of the blue subheadings? Why did the writer include them in the text? *(They make it easy for the reader to predict the main idea of each paragraph.)*

CLARIFYING WORD USAGE Make sure students can distinguish between preventing a disease and treating a disease.

Ask When a disease has been prevented, does the person become ill? *(no)*

KEY QUESTION Read the Key Question aloud and answer it together as a class.

Ask What is homeostasis? *(relatively constant internal physical and chemical conditions that organisms maintain)*

Build Vocabulary

WORD ORIGINS Review the definitions of the words *pathos* and *genes*. Read aloud the definition of *pathogen*.

Ask What does *pathology* mean? *(Sample answer: the study of diseases)* What do you think a pathologist is? *(Sample answer: someone who studies the damage caused by disease)*

Speed Bump

Ask Which treatment is useful in fighting a bacterial infection? *(antibiotics)*

Ask Are antibiotics a good way to treat colds? *(no)* Why or why not? *(Antibiotics treat bacterial infections. Colds are caused by viruses.)*

BUILD Vocabulary

pathogen
a disease-causing agent

vaccine
a preparation of weakened or killed pathogens used to produce immunity to a disease

antibiotic
a drug used to block the growth and reproduction of bacterial pathogens

WORD ORIGINS
Pathogen comes from the Greek words *pathos*, which means "suffering," and *genes*, meaning "produced." Infection with a pathogen produces suffering in the form of illness.

Preventing Bacterial Disease Infections can sometimes be prevented by vaccines. A **vaccine** is a preparation of weakened or killed pathogens or inactivated toxins. A vaccine helps the body become immune to a specific disease. Immunity is the body's ability to fight off pathogens or to deal with toxins.

Treating Bacterial Diseases Many drugs can be used to treat a bacterial infection. These drugs include **antibiotics,** such as penicillin and tetracycline. Antibiotics keep bacteria from growing and reproducing by interfering with bacterial proteins or cell processes. Antibiotics affect proteins or cell processes that are found only in bacteria. For this reason, they do not harm the host's cells.

Viral Diseases

Like bacteria, viruses cause disease by upsetting the body's normal homeostasis. Several common diseases caused by viruses are listed below. Viruses also cause serious diseases in plants and animals.

Disease Mechanisms Some viruses cause disease by destroying living cells. They may also change processes in cells and upset homeostasis. Often, viruses attack only certain kinds of cells. For example, poliovirus destroys cells in the nervous system, causing paralysis. Other viruses change the way cells grow and develop. Such changes can lead to cancer.

Key Question How do viruses cause disease?
Viruses cause disease by destroying living cells or by affecting processes in cells in ways that upset homeostasis in the host.

Human Viral Diseases
This table describes some common viral diseases.

Some Human Viral Diseases

Disease	Effect on Body	Transmission
Common cold	Sneezing, sore throat, fever, headache, muscle aches	Contact with contaminated objects; droplet inhalation
Influenza	Body aches, fever, sore throat, headache, dry cough, fatigue, nasal congestion	Flu viruses spread in respiratory droplets caused by coughing and sneezing.
AIDS (HIV)	Helper T cells, which are needed for normal immune-system function, are destroyed.	Sexual contact; contact with contaminated blood or body fluids; can be passed to babies during delivery or during breastfeeding.
Chicken pox	Skin rash of blisterlike lesions	Virus particles are spread in respiratory droplets caused by coughing and sneezing; highly contagious
Hepatitis B	Jaundice, fatigue, abdominal pain, nausea, vomiting, joint pain	Contact with contaminated blood or bodily fluids
West Nile Virus	Fever, headache, body ache	Bite from an infected mosquito ▶
Human papillomavirus (HPV)	Genital or anal warts, also cancer of the cervix, penis, and anus	Sexual contact

Hands-On Learning

RESEARCH VACCINES Have student groups research which vaccines are required to be given to children. Have each group make a list of the vaccines, the diseases they prevent, and the ages at which they are given.

Before the advent of the polio vaccine, hospitals were filled with polio-stricken children in machines, called *iron lungs*, that helped them breathe.

BUILD Connections

INNOVATIONS IN VACCINES

Many vaccines have been developed in the last 300 years. Today, there are vaccines against more than 20 diseases.

TEM 5000×

▲ Smallpox viruses

1769 Edward Jenner performs the first inoculation against smallpox, using the less harmful but similar cowpox virus.

1880s Louis Pasteur develops vaccines against anthrax and rabies.

1923 Albert Calmette and Camille Guerin develop a vaccine against tuberculosis.

1950s Jonas Salk develops a polio vaccine that uses killed viruses. Albert Sabin develops a polio vaccine that uses weakened viruses.

1981 A vaccine against hepatitis B that uses recombinant DNA gains government approval.

2006 A vaccine against human papillomavirus, a virus known to cause certain cancers, gains approval.

◀ Before vaccine development, the Red Cross made the public aware of the threat of tuberculosis using posters such as this one, circa 1919.

Preventing Viral Diseases In most cases, the best way to protect against viral diseases is prevention. Studies show that cold and flu viruses are passed on by hand-to-mouth contact. So it is important to wash your hands often. Cough or sneeze into a tissue rather than into your hands. Avoid contact with sick people.

Prevention of viral diseases often includes the use of vaccines. Some important events in the development of vaccines are shown above.

Treating Viral Diseases Viral diseases cannot be treated with antibiotics. Yet, there are some antiviral drugs. These drugs attack viral proteins that host cells do not have. Some antiviral drugs help people get over the flu more quickly. Other antiviral drugs help people who are infected with HIV live longer.

ELL Focus on ELL: Extend Language

INTERMEDIATE SPEAKERS Write the words *antibiotic* and *antiviral* on the board. Then, write out the word part *anti-* by itself. Tell students that *anti-* is a prefix that can be placed at the beginning of many words.

Ask What does the word part *anti-* indicate when it is used as a prefix? *(The prefix* anti- *means "against" or "opposed to.")*

Ask students to decipher terms such as *antifreeze* and *antidepressant.* Then, ask them to think of other words that include the prefix *anti-*. Have them use these words in sentences.

Active Reading

COMPARE AND CONTRAST Have students compare and contrast antibiotics with antiviral drugs. Have students write their comparisons side-by-side in a T-chart. Students should consider what the drugs do, how they work, and what effect (if any) they have on the host.

Build Connections

VISUAL SUMMARY Remind students that vaccines were not as common in the past as they are now. Point out that in the past many people died of diseases that are now prevented by vaccines.

Ask What are the large machines in the photograph? *(iron lungs)* Why are children inside these machines? *(These children have polio; the machines help them breathe.)* Why is polio uncommon today? *(Almost everyone receives a vaccine that prevents the disease.)*

Have students place the events listed on the photo in the form of a timeline. Tell students to use a ruler or graph paper to make the distances between the events proportional to the number of years between them.

Ask How does placing the events in this type of timeline help you to understand the events? *(Using a timeline with distances as a ratio of years makes it easy to see how much time passed between events.)* How would you characterize the progress that has been made in developing vaccines for diseases? Has progress happened quickly, or slowly over time? *(slowly, over hundreds of years)*

Speed Bump

Ask How do antiviral drugs work? *(They attack proteins that are on the viruses but not on the host cells.)*

Ask How does washing your hands help prevent viral diseases? *(Viruses are often passed on from hands to mouth. Washing hands keeps viruses off your hands.)*

During Reading continued

Build Vocabulary

WORD ORIGINS Tell students that the word *emerging* comes from a Latin word that means "to come out of" or "to arise from." Ask students how the origin of *emerging* relates to the meaning of the vocabulary term *emerging diseases. (Emerging diseases are diseases that are coming from new sources.)*

Active Reading

KEY QUESTION Read the question aloud and ask a volunteer to answer.

Ask Why do you think many travelers must get extra vaccinations when they visit a foreign country? *(They may encounter diseases in the foreign country that are not common where they live. To the travelers, these diseases are like emerging diseases because they do not have any resistance to them.)*

Speed Bump

Ask How do new viruses come about? *(Viruses may mutate and jump from one species to another.)*

Inquiry Into Scientific Thinking

MAKE GRAPHS Review how to use data to construct a line graph. You may wish to set up the graph on the board and start it as a class before having students finish it on their own.

Ask What should we plot on the *x*-axis? *(the year)* What should we plot on the *y*-axis? *(the number of hospital cases reported)*

Lead a Discussion

Ask students to think about how the rise of MRSA affects other parts of society. For example, how does it affect productivity if more people are sick? How does it affect hospitals and doctors?

BUILD Vocabulary

emerging disease
a disease that appears in the population for the first time, or an old disease that suddenly becomes harder to control

prion
a misfolded protein that causes disease

WORD ORIGINS
The word *prion* combines letters from the words *protein* and *infectious* with the ending *–on.* Scientific names for particles, such as *electron* and *neutron,* often have the *–on* ending.

Emerging Diseases

In the last thirty years, a number of new diseases have appeared. Other diseases that once seemed to be under control have become more dangerous. An unknown disease that appears in a population for the first time or a well-known disease that suddenly becomes harder to control is an **emerging disease.** Emerging diseases are dangerous because people have little or no resistance to them. Scientists have not developed medications or vaccinations to fight new diseases.

"Superbugs" Sometimes old diseases become harder to fight as the pathogens that cause them evolve. Pathogens often evolve in response to the medicines used against them. Penicillin was the first antibiotic used to treat infections. At first, it worked well. Diseases that had once killed people were cured. Yet, in just a few decades, penicillin did not work as well. Bacteria had evolved resistance to it. The drug no longer stopped some infections. Many bacteria are now resistant to many antibiotics. Doctors sometimes call these bacteria "superbugs."

New Viruses Sometimes, a virus evolves and jumps from one host species to another. Researchers have evidence that HIV did this. It likely moved from nonhuman primates into humans. Public health officials are worried about a new flu virus. A bird flu virus much like the most deadly human flu viruses has evolved. In a few cases, bird flu has infected humans. A major "jump" to humans may be possible.

Key Question Why are emerging diseases a threat to human health? **Humans have little or no resistance to emerging diseases. Scientists have not yet developed ways to control them.**

INQUIRY into Scientific Thinking

MRSA on the Rise

An especially dangerous bacterium has recently appeared. *Staphylococcus aureus* (MRSA) is resistant to many drugs. It causes skin infections that can be spread by close contact. Infection by MRSA can spread very quickly among people in hospitals who have weakened immune systems.

Incidence of MRSA

Year	Hospital Cases Reported
1993	1900
1995	38,100
1997	69,800
1999	108,600
2001	175,000
2003	248,300
2005	368,600

Analyze and Conclude

1. Graph The table in the right-hand corner shows the number of MRSA infections in hospitals in the United States from 1993 through 2005. Use the table to make a line graph.

2. Explain What trend does your line graph show?

3. Calculate By what percentage did MRSA infections increase between 1995 and 2005?

4. Relate Cause and Effect The average hospital stay in the United States lasts 4.6 days. That of the average MRSA-infected patient is 10.0 days. If the trend shown by the data above continues, how will MRSA infections affect hospital costs?

In Your Workbook Get more help for this activity in your workbook.

Analyze and Conclude

1. Student graphs should show a dramatic increase in MRSA infections over time.
2. Sample answer: MRSA cases increased between 1993 and 2005.
3. 368,600 – 38,100 = 330,500. 330,500 ÷ 38,100 x 100 = 867%
4. Accept all reasonable answers. Students should recognize that MRSA infections double the length of hospital stays, so increased MRSA infections will lead to much higher hospitalization costs.

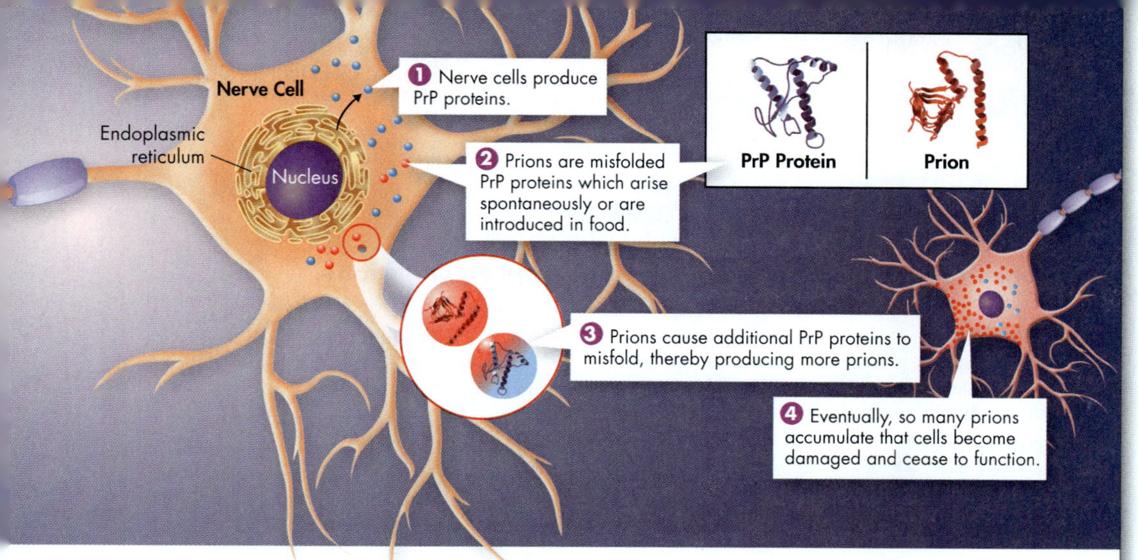

Prion Infections Prions are PrP proteins that are folded incorrectly. Contact with prions can cause other PrP proteins to misfold, too. Prions can damage nerve cells in the brain.

Prions In 1972, the American scientist Stanley Prusiner began to study a disease that was making sheep sick. At first, Prusiner thought a virus caused the disease. Then clumps of misfolded proteins were discovered in the brains of infected sheep. Prusiner called these proteins **prions.** A prion forms when a certain type of protein, called PrP, folds incorrectly. As prions build up, they damage brain cells. Prion infections are spread when animals eat other animals that are infected. Prions can infect many animals, including humans.

CHECK Understanding

Apply Vocabulary

Use the highlighted terms from the lesson to complete each sentence correctly.

1. A(n)________ may contain dead or weakened pathogens.

2. ________ are medications that can be used to treat bacterial infections.

Critical Thinking

3. Compare and Contrast How does the treatment of viral diseases differ from the treatment of bacterial diseases?

4. Explain Why are "superbugs" hard to control?

5. Write to Learn Answer the third clue of the chapter mystery. Describe the structure and effect of prion infections in your answer.

MYSTERY CLUE

BSE virtually disappeared when the British government banned the use of ground-up cattle protein in feed. What does this indicate about how the pathogen was passed on? Could prions be the cause of BSE? Why or why not? (Hint: See above.)

BIOLOGY.com Search Lesson 20.3 GO • Lesson Assessment • Art in Motion

Check Understanding

1. vaccine

2. Antibiotics

3. Sample answer: Antibiotics are used to treat bacterial diseases. However, they cannot be used to treat viral diseases. Sometimes an antiviral drug can be used to treat viral diseases.

4. Sample answer: Superbugs are resistant to many different antibiotics.

5. Accept all reasonable answers. Students should recognize that prions were likely the cause of BSE. Prion infection could have been spread if the ground up cattle protein in the feed was infected. When healthy cows ingested the infected feed, the misfolded prions infected their brain cells.

Active Reading

INTERPRET DIAGRAMS Have students read the captions on the diagram Prion Infections in order. Point out that the diagram uses color to differentiate the normal PrP proteins (blue) from the misfolded proteins (red).

Ask Where does this sequence of events take place in the body? *(inside a nerve cell in the brain)* How does this activity cause damage? *(Misfolded proteins increase in number until there are so many that the nerve cell is damaged and cannot function properly.)*

Speed Bump

Ask How do prions get in nerve cells? *(They can arise through mutations, or they can spread when animals eat other animals that are infected.)*

Post Reading

ASSESSMENT TIPS For Question 3, have students review the T-chart they made comparing and contrasting antibiotics with antiviral drugs.

MYSTERY CLUE Direct students to review how prion infections are spread from animal to animal. They should recognize that the transmission of prions through eating other infected animals fits the pattern of BSE.

Wrap-Up Activity

PERFORM A QUIZ SHOW Have students form small groups. Tell each small group to write ten quiz questions using content from the lesson. Then, collect all the questions and use a quiz-show format to see which group can answer the most questions correctly.

Pre-Lab

Review the meaning of *variable. (A variable is any factor that could affect the results of an experiment.)* Discuss the relationship between independent and dependent variables.

Ask What factor are you varying during this experiment? *(the type of antibiotic used)*

Say The type of antibiotic is the independent variable.

Ask What factor are you measuring during this experiment? *(the growth of bacteria)*

Say Bacterial growth is the dependent variable.

Ask What are some other variables that could affect the results of the experiment and need to be controlled? *(Sample answers: temperature, sunlight)*

Review the Background Questions and Pre-Lab Questions as a class.

Lab

Tell students they will perform the chapter lab *Controlling Bacterial Growth* described in **Lab Manual B.**

SAFETY Students should wear safety goggles and gloves while performing this lab. Be aware of students who may be allergic to penicillin or tetracycline. Remind them to follow your direction about disposal of lab materials. Have students wash their hands thoroughly with soap and water after completing the lab.

Post-Lab

ADVERTISEMENT Have students use their results to make a print advertisement for one of the antibiotic drugs they investigated. The advertisement should explain what the drug does, why it is useful, and why it is better than similar products.

NATIONAL SCIENCE EDUCATION STANDARDS

CONTENT F.1, F.5

INQUIRY A.1.b, A.2.a

Real-World Lab GUIDED INQUIRY

Pre-Lab: Controlling Bacterial Growth

Problem How can you determine the effectiveness of an antibiotic?

Materials agar plates, glass-marking pencil, bacterial cultures, glass beads, sterile micropipettes, disposal solution, forceps, paper disks with antibiotics and with distilled water, masking tape, metric ruler

Lab Manual Chapter 20 Lab

Skills Focus Observe, Measure, Draw Conclusions

Connect to the Big idea Bacteria can be found on and within the human body. Most of these bacteria are harmless and some are even beneficial to humans. But others can cause diseases. These pathogens need to be controlled. Physical removal of bacteria through hand washing is one of the most effective control methods. Proper food storage and preparation are also important, as are the vaccines that have been developed to help the body build up immunity to specific bacterial diseases.

What happens if a person does develop a bacterial infection? Then doctors use drugs, such as antibiotics, to fight the infection. These drugs are designed to kill bacteria but not to kill human cells. In this lab, you will compare the ability of two antibiotics to control the growth of two different types of bacteria.

Background Questions

a. Review What happens to a bacteria cell after it has grown to nearly double its size? What is this process called?

b. Explain How is genetic diversity increased in populations of bacteria?

c. Review What are the two general ways that bacteria can cause disease?

Pre-Lab Questions

Preview the procedure in the lab manual.

1. **Relate Cause and Effect** How will you know whether an antibiotic is able to control the growth of bacteria?
2. **Design An Experiment** Why is it important to leave space between the disks on the agar plates?
3. **Control Variables** Why must you avoid direct contact between your hands and the antibiotic disks?

BIOLOGY.com Search Chapter 20 GO

Visit Chapter 20 online to test yourself on chapter content and to find activities to help you learn.

Untamed Science Video Join the Untamed Science crew as they fire up the microscopes for a look at bacteria and all the ways they are good for us.

Art in Motion View a short animation of prion infection and see how misfolded proteins interact with normal proteins.

Art Review Review your understanding of the structure and classification of prokaryotes.

InterActive Art Build your understanding of lytic and lysogenic cycles.

Visual Analogy Compare an old west outlaw taking over a town to a lytic infection.

Pre-Lab Answers

BACKGROUND QUESTIONS

a. The cell's DNA is replicated and the cell divides in half, producing two identical cells. The process is called binary fission.

b. Genetic diversity is increased either through mutations or through conjugation, a process in which genetic material moves from one cell to another across a hollow bridge.

c. Bacteria can destroy living cells or release toxins that interfere with normal cell activity.

PRE-LAB QUESTIONS

1. There will be a clear space on the agar plate around the antibiotic disk.
2. Sample answer: If the disks are too close together, it will be difficult to determine which antibiotic was responsible for the results.
3. If my hands were to touch the disks, I might transfer bacteria from my hands, which could add an unwanted variable to the experiment.

20 CHAPTER Summary

20.1 Viruses

- Viruses can reproduce only by infecting living cells.
- Inside living cells, viruses use their host's genetic information to make many copies of themselves. Some viruses replicate right away. Other viruses stay inactive for a long period of time within the host cell.

virus (p. 480)
capsid (p. 481)
bacteriophage (p. 481)
lytic infection (p. 482)
lysogenic infection (p. 482)
retrovirus (p. 483)

20.2 Prokaryotes

- Prokaryotes are classified as Bacteria or Archaea, which are two of the three domains of life.
- Prokaryotes vary in their size and shape, in the way they move, and in the way they get and release energy.
- Prokaryotes are essential in maintaining the balance of the living world. They are decomposers, producers, and nitrogen fixers. In addition, some species are used in human industry.

prokaryote (p. 485)
binary fission (p. 487)
conjugation (p. 487)
endospore (p. 487)

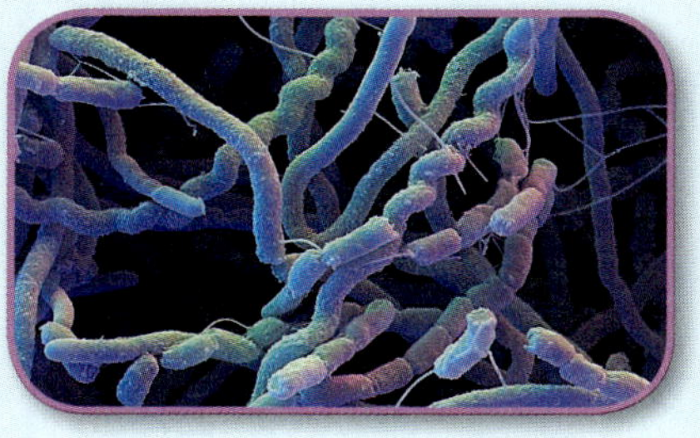

20.3 Diseases Caused by Bacteria and Viruses

- Bacteria cause disease by damaging host cells and tissues or by releasing chemicals that upset homeostasis.
- Viruses cause disease by destroying living cells or by affecting processes in cells in ways that upset homeostasis in the host.
- Humans have little or no resistance to emerging diseases. Scientists have not yet developed vaccines or medicines to control them.

pathogen (p. 489)
vaccine (p. 490)
antibiotic (p. 490)
emerging disease (p. 492)
prion (p. 493)

UbD Transfer the Big Idea

TRANSFER Have students use Internet or print resources to find current news stories on viral and bacterial diseases affecting human populations in different parts of the world. Tell students to choose a disease and make a pamphlet about it. The pamphlet should include a description of symptoms, the name of the pathogen, a classification of the pathogen, and information on preventing and treating the disease. If possible, students should include information on how the virus or bacteria infects the host.

Chapter Summary

Think Visually

USE GRAPHICS A flipbook consists of pages of sequential drawings that, when flipped, appear to move. Have students construct a flipbook movie of the steps in a lytic infection. Invite students to exchange flipbooks with a partner and write a review of their partner's flipbook.

ANSWERS
Students' flipbooks should include the steps of a lytic infection, including the virus injecting DNA into a bacterium, transcription of viral genes by the host cell, generation of viral proteins and nucleic acids by the infected bacterium, assembly of new viruses within the infected bacterium, lysis of the infected cell, and the release of new viruses.

BIOLOGY.com Study Online

REVIEW RESOURCES

Visual Analogy How is a virus like an outlaw? Students can go online to find out.

InterActive Art By working with a dynamic diagram, students will understand more about viral reproduction.

Art Review Activities help students compare archaea and bacteria, learn the structure of a typical prokaryotic cell, and match pictures of bacteria to the names of their shapes.

Art in Motion Encourage students to see what prions are like and how these tiny particles set off a chain reaction in the brain.

Data Analysis Students analyze data about MRSA to identify factors contributing to its increasing prevalence.

Vocabulary Review The *Flash Cards* and *Crossword* provide an interactive way to review chapter vocabulary.

Editable Worksheets Editable pages of Study Workbook B, Lab Manual B, and the Assessment Resources Book are available online.

Lesson Overviews Lesson Overviews are based on Miller & Levine Biology. However, they are available on CD-ROM and can be edited for your class.

ASSESSMENT RESOURCES

Chapter Assessment Have students take an online version of the Chapter 20 Check Understanding.

Standardized Test Prep Students can take an online version of the Standardized Test Prep. You will receive their scores along with ideas for remediation.

Diagnostic and Benchmark Tests Use these tests to monitor your students' progress and supply remediation.

ASSESSMENT

Assess

Big idea Read the Big Idea question aloud to the class. Draw a two-column table on board. Write *Microbe* at the top of one column and *Living Cells?* at the top of the second column.

Ask What is another name for *microbe? (Sample answers: pathogen, germ)*

Ask Which pathogens did you learn about in this chapter? *(bacteria, viruses, prions)*

Ask Which of these pathogens are made up of living cells? *(bacteria)*

Ask Which pathogens are not made up of living cells? *(viruses and prions)*

Complete the table as you answer these questions as a class.

CONSTRUCTED RESPONSE Tell students not to worry about the length of their paragraphs. A paragraph can be short or long. What matters is that the paragraph contains a clearly stated main idea followed by supporting information, or details. Tell students to reread their paragraphs after they write them. Students should evaluate whether the paragraphs would be understandable and convincing to someone who was reading them without having read the chapter. If not, students should add clarifying information to their paragraphs.
Read Question 1 aloud.

Ask What are the characteristics of a living organism? *(A living organism is made up of living cells, uses energy, and reproduces itself.)*

Read the first hint.

Ask What is significant about forming crystals? Is this a characteristic you associate with living organisms? *(Students should be able to describe the nature of crystals and explain that formation of crystals is not associated with living organisms.)*

Read the second hint.

Say Viruses can reproduce and have genetic material. These sound like characteristics of a living organism.

Ask How do viruses reproduce? *(Viruses reproduce inside living cells. Viruses cannot reproduce on their own.)*

Model writing the first two or three sentences using the hints on the board and in the text. *(Sample: Viruses share some characteristics with living organisms, but they are not living. They do not use energy as living things do. They also cannot reproduce on their own.)*

20 CHECK Understanding

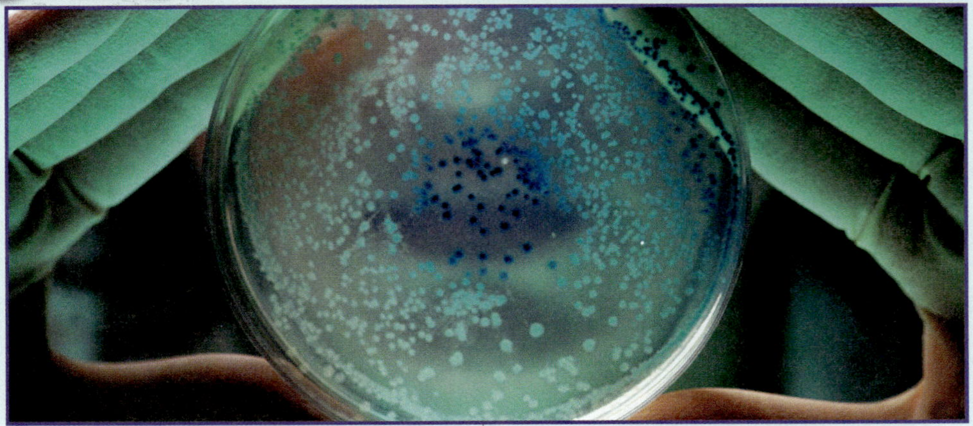

Assess the Big idea — Cellular Basis of Life

Write an answer to the question below.
Q: Are all microbes that make us sick made of living cells?

Constructed Response

Write an answer to each of the questions below. The answer to each question should be one or two paragraphs long. To help you begin, read the **Hints** below each of the questions.

1. **Are viruses living organisms?**

   Hint Viruses form crystals.

   Hint Viruses are able to reproduce, and they have genetic material.

2. **How are prokaryotes classified, and how are they important?**

   Hint Prokaryotes are single celled and lack nuclei.

   Hint Prokaryotes are organized into two groups.

3. **How is the reproduction of viruses similar to that of prokaryotes? How is it different?**

   Hint Refer to the art comparing a virus to an outlaw on page 481.

Foundations for Learning Wrap-Up

Use the z-fold you made as you read the chapter to help organize your thoughts about viruses, bacteria, and disease.

Activity 1 Working in a small group, write questions for a crossword puzzle. Base your questions on the notes you took on your z-fold. Draw your puzzle and trade it with another team.

Activity 2 Working with a partner, use your z-folds to make an informative brochure about viruses, bacteria, and the risks they pose to human health. Your notes on the z-fold will be the facts you present about viruses, bacteria, and disease. Be sure to make an inviting cover for your pamphlet. Add pictures to describe viruses and bacteria.

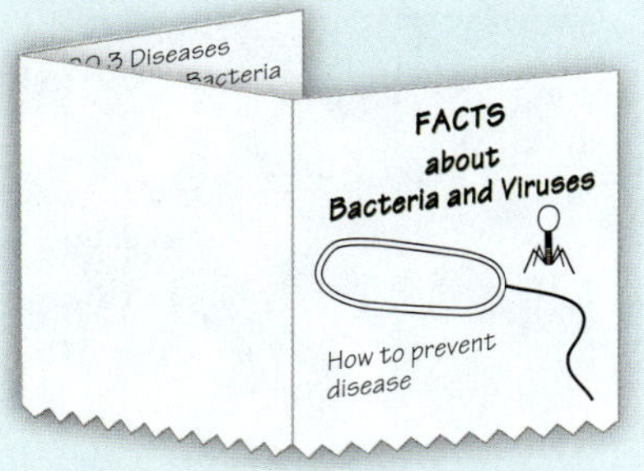

For Question 2, pair students and have them use the hints and your modeling to answer the question. Have students share their paragraphs with the class.

For Question 3, have students write the paragraphs independently for homework or as an anchor activity.

Foundations for Learning Wrap-Up

Have students take out their z-folds. Make sure students have completed each section with the appropriate information. Assign the groups or pairs for each activity and explain how to complete the activities before letting students begin.

20.1 Viruses

Understand Key Concepts

1. Particles that are made up of proteins, nucleic acids, and sometimes lipids and that can only reproduce by infecting living cells are called
   a. bacteria.
   b. capsids.
   c. prokaryotes.
   d. viruses.
2. One group of viruses has RNA as their genetic information. These viruses are called
   a. bacteriophages.
   b. capsids.
   c. prophages.
   d. retroviruses.

> **Test-Taking Tip**
>
> **Come Back to the Question** Sometimes, you might be stumped by a question. If you can't figure out the answer, skip the question and continue the test. Come back to the question after you have completed the other test questions. Doing so can help you organize your thoughts and come up with an answer.

3. Describe the events that take place during a lytic infection.
4. What is a prophage?

Think Critically

5. **Apply Concepts** Explain how a virus can spread in a bacterial population during a lysogenic infection.
6. **Apply Concepts** How could a mutation in a bacterial cell help the bacterial population become resistant to a bacteriophage?

20.2 Prokaryotes

Understand Key Concepts

7. Prokaryotes differ from all other organisms in that their cells
   a. lack nuclei.
   b. have cell walls.
   c. have ribosomes.
   d. lack nucleic acids.
8. Prokaryotes reproduce asexually by
   a. binary fission.
   b. conjugation.
   c. endospores.
   d. mutation.
9. List and describe the three main cell shapes of prokaryotes.
10. Describe two ways by which prokaryotes move.

Think Critically

11. **Apply Concepts** A scientist finds a new organism. The organism is single celled and has a cell wall containing peptidoglycan. The cell lacks a nucleus. To what domain does the organism belong? Explain your reasoning.
12. **Infer** Imagine that nitrogen-fixing bacteria disappeared. How would this affect other organisms in the world?

20.3 Diseases Caused by Bacteria and Viruses

Understand Key Concepts

13. Viruses cause disease by
    a. releasing toxins.
    b. undergoing conjugation.
    c. infecting then destroying cells.
    d. causing mutations in the host's DNA.

Assessment Tips

Answer questions to Lesson 20.1 as a class. After reading Question 2, read the Test-Taking Tip. Caution students that any time they skip a question on a test, they should make a note on their answer sheet to remind them which question was skipped. This will help them avoid putting the answers in the wrong spots after they skip a question.

Have students answer the questions for Lessons 20.2 and 20.3 on their own. Point out that many questions can quickly be answered by referring back to the figures and vocabulary in the chapter. Encourage students to be familiar with both elements of the chapter.

Answers

Lesson 20.1

UNDERSTAND KEY CONCEPTS

1. d

2. d

3. Sample answer: The virus injects its DNA into the host cell. Viral genes are transcribed by the host cell. The host cell makes new viral proteins and nucleic acids. New viruses are assembled. The cell bursts, and the new viruses leave to infect other cells.

4. Sample answer: A prophage is bacteriophage DNA that becomes part of the host cell's DNA during a lysogenic infection.

THINK CRITICALLY

5. Sample answer: During a lysogenic infection, the viral DNA inserts itself into the host's DNA. The viral DNA is replicated with the host's DNA. When the host reproduces, the daughter cells also have the viral DNA. The viral DNA is passed on to future generations and spread throughout the population until the virus becomes active and undergoes the lytic cycle.

6. Sample answer: Proteins on the membrane or capsid of a bacteriophage must bind to specific receptor proteins on a bacterium. If the mutation changes these proteins, the bacteriophage may no longer bind to the bacterium.

Lesson 20.2

UNDERSTAND KEY CONCEPTS

7. a

8. a

9. Sample answer: Bacilli are rod-shaped prokaryotes. Cocci are sphere-shaped prokaryotes. Spirilla are corkscrew-shaped prokaryotes.

10. Sample answer: Some prokaryotes have flagella for movement. Other prokaryotes secrete a slimy substance that they use to slide along a surface.

THINK CRITICALLY

11. The organism belongs to the domain Bacteria. Single-celled organisms that lack a nucleus, but contain peptidoglycan in their cell walls, are bacteria.

12. Sample answer: Nitrogen fixers are responsible for 90% of nitrogen fixation. If nitrogen gas is not converted, other organism cannot make proteins or nucleic acids. These organisms would die.

Lesson 20.3

UNDERSTAND KEY CONCEPTS

13. c

14. a

15. Sample answer: An emerging disease is a new disease or an existing disease that has suddenly become difficult to control.

16. Sample answer: Prions are PrP proteins that are folded incorrectly. They can cause other PrP proteins to misfold, also. As prions accumulate in the brain, they can damage brain cells. Prion infections spread when animals eat other animals that are infected.

THINK CRITICALLY

17. Antibiotics would not help because bird flu is caused by a virus and viruses cannot be treated with antibiotics.

Connecting Concepts

18. The bacteriophage infected and killed the bacterial cells in these areas.

19. The bacteria would become infected with the bacteriophage, resulting in clear areas on the fresh lawn of *E. coli.*

CHAPTER MYSTERY After students have read through the Chapter Mystery, discuss the concepts of prions and BSE.

Ask How was BSE similar to a viral disease? How was it different? *(BSE was similar to a viral disease in that it destroyed cells in the host. It was different from a viral disease in that it did not spread by contact, but instead was spread as prions were consumed from the body of an infected animal.)*

Ask How would you diagram the connection between BSE in cows and nVCJD in humans? *(Sample answer: Prions in meat and brain tissue of dead animal → cow eats infected material → cow is infected with prions → infected cow is slaughtered for meat → meat is consumed by humans → humans are infected by prions in meat)*

Ask How did understanding the cause of BSE allow officials to prevent the spread of the disease? *(After scientists understood that prions were the cause of BSE, government officials passed laws that prevented farmers from using ground-up cattle to feed cattle.)*

20 CHECK Understanding

14. Which of the following can be used to treat bacterial diseases but NOT viral diseases?

a. antibiotics

b. antiviral drugs

c. pathogens

d. vaccines

15. What is meant by the term *emerging disease*?

16. How do prions cause disease?

Think Critically

17. **Infer** Would antibiotics be effective in treating an outbreak of bird flu? Explain.

Connecting Concepts

Use Science Graphics

E. coli bacteria can be grown on agar in a petri dish. The bacteria cloud the agar surface, forming a bacterial "lawn." This photograph shows a lawn over which a solution containing bacteriophage particles has been poured. Use the photograph to answer questions 18–19.

18. **Interpret Visuals** What is the most reasonable explanation for the small, circular, clear areas on the bacterial lawn?

19. **Form a Hypothesis** Suppose you touched the tip of a glass rod to one of the clear areas and then touched it again to the surface of a petri dish with a fresh lawn of *E. coli*. What would happen to the new lawn of bacteria after several days?

solve the CHAPTER MYSTERY

THE MAD COWS

The mad cow disease that appeared in 1986 spread quickly among British cattle herds. People got a similar disease. This disease was called new variant Creutzfeld-Jacob disease (nvCJD). Many people died.

The British government made a law against feeding cattle using ground-up cattle. BSE and nvCJD almost disappeared. It now seems clear that "mad cow" and nvCJD were caused by prions. These prions were in the meat and brain tissue of infected cattle. When cattle ate the infected meat, the prions spread. People who ate meat from infected animals were also infected. Officials in Europe and the United States have made new laws about meat production to try to make sure there are no more outbreaks of BSE and nvCJD.

1. **Infer** The rapid rise of BSE between 1986 and 1991 ended when British authorities banned the use of meat and bone meal in cattle food. How does this support the hypothesis that BSE is caused by prions?

2. **Apply Concepts** Why did most scientists conclude that BSE was not caused by either viruses or bacteria?

UntamedScience **Never Stop Exploring Your World.** Finding what caused this disease is only the beginning. Take a video field trip with the ecogeeks of Untamed Science to explore the other side of the story. You'll see that not all microbes are "bad."

CHAPTER MYSTERY ANSWERS

1. Sample answer: The hypothesis was supported by the fact that the disease disappeared when ground-up cattle tissue was removed from feed.

2. The samples that infected mice did not contain nucleic acids, which are present in bacteria and viruses.

Untamed Science For a look at some beneficial bacteria, suggest students watch **Bacteria, at Your Service.**

Standardized Test Prep

Multiple Choice

1. A type of virus that infects bacterial cells is called a
   A capsid.
   B prion.
   C bacteriophage.
   D retrovirus.

2. Prokaryotic cells that have a spherical shape are called
   A cocci.
   B methanogens.
   C spirilli.
   D bacilli.

3. What is a capsid?
   A viral DNA that inserts into a host's DNA
   B a protein coat surrounding a virus
   C a type of plant virus
   D a rod-shaped bacterium

4. Which of the following is NOT used to identify specific prokaryotes?
   A type of nucleic acid
   B shape
   C movement
   D energy source

5. Which method is NOT used to protect food against microorganisms?
   A heating
   B freezing
   C sterilization
   D vaccination

6. Which illness is caused by a bacterium?
   A AIDS
   B polio
   C tuberculosis
   D common cold

7. Which process is used for the exchange of genetic information between two bacterial cells?
   A endospore formation
   B lysogenic cycle
   C conjugation
   D binary fission

8. All bacteria are classified as
   A eukaryotes.
   B protists.
   C archaea.
   D prokaryotes.

Questions 9–10

Use the graph below to answer the questions.

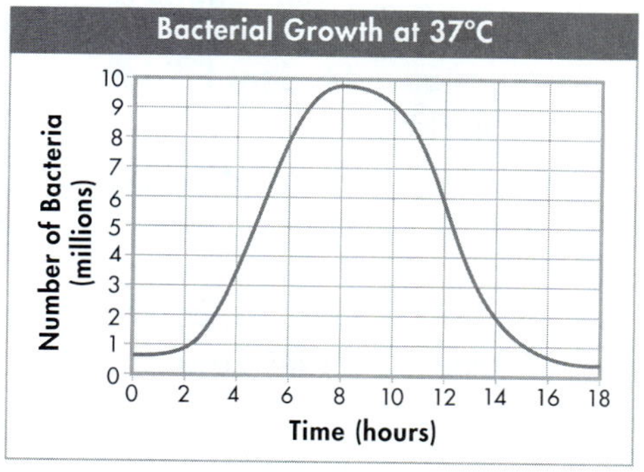

9. At which interval in the graph does the number of living bacteria increase at the greatest rate?
   A between hours 2 and 4
   B between hours 4 and 6
   C between hours 6 and 8
   D between hours 10 and 12

10. Which is the most likely reason for the decrease in bacteria shown?
   A The temperature of the bacterial culture was too high after 8 hours.
   B The bacteria stopped reproducing after 8 hours.
   C More nutrients were added to the culture at regular intervals.
   D Waste products from the bacteria accumulated in the nutrient solution.

Open-Ended Response

11. Explain why antibiotics can be useful in treating bacterial diseases but not in treating viral diseases.

If You Have Trouble With . . .

Question	1	2	3	4	5	6	7	8	9	10	11
See Lesson	20.1	20.2	20.1	20.2	20.3	20.3	20.2	20.2	20.2	20.2	20.3

Test-Taking Tip

WATCH FOR QUALIFIERS

Tell students to watch for questions containing the word *NOT* as a qualifier. For these questions, suggest that students begin by reading all the answer choices. They should eliminate any choice that is a characteristic of the question stem or is true of the question stem. Finally, students should check to see that their answer is correct by confirming that it does not fit the characteristic in question.

Before the Test

Review with students the Test-Taking Tip.

Point out the word *NOT* in Questions 4 and 5.

Remind students that an open-ended response means that they'll have to write a sentence or two to answer the question. Tell them that it is often helpful to write down notes about their answer first, and then put them together as a paragraph.

Answers

1. C
2. A
3. B
4. A
5. D
6. C
7. C
8. D
9. B
10. D
11. Antibiotics disrupt proteins or cell processes that are specific to bacteria; they do not affect viruses.

CHAPTER PLANNER 21

Protists and Fungi

Chapter Contents	NSES	Time	Core Resources
Chapter Preview			**Student Edition,** pp. 500–501 **Chapter Mystery,** p. 501 **Foundations for Learning,** p. 501
21.1 Protist Classification The First Eukaryotes • Ancestors and Descendants	II, V, A.1.b, A.1.c, A.2.a, A.2.b, A.2.e, C.3.a, C.3.d, C.3.e, G.3	1/2 period 1/4 block	**Student Edition,** pp. 502–504 **Inquiry Into Scientific Thinking,** p. 503 **Study Workbook B** 21.1 Worksheets **Biology.com** 21.1 Lesson Assessment
21.2 Protist Structure and Function How Protists Move • Protist Reproduction	I, IV, V, C.1.a	1/2 period 1/4 block	**Student Edition,** pp. 505–508 **Build Connections,** p. 506 **Study Workbook B** 21.2 Worksheets **Biology.com** *InterActive Art:* Amoeba and Paramecium • *Visual Analogy:* How Cells Move Like Boats • 21.2 Lesson Assessment
21.3 The Ecology of Protists Autotrophic Protists • Heterotrophic Protists • Symbiotic Protists	I, IV, V, C.4.b, C.4.c, C.5.d, F.1	1 period 1/2 block	**Student Edition,** pp. 509–513 **Study Workbook B** 21.3 Worksheets **Biology.com** *Art in Motion: Plasmodium* Life Cycle • 21.3 Lesson Assessment
21.4 Fungi What Are Fungi? • The Ecology of Fungi	I, IV, V, C.3.a, C.4.b, C.4.c, C.5.d	1 period 1/2 block	**Student Edition,** pp. 514–519 **Study Workbook B** 21.4 Worksheets **Biology.com** *Art Review:* Structure of a Mushroom • 21.4 Lesson Assessment
Chapter Pre-Lab	A.1.a, A.1.b, A.2.a	1 period 1/2 block	**Student Edition,** p. 520 **Lab Manual B** *Mushroom Farming*

Tools for ELL Instruction

ELL Handbook describes teaching strategies to make *Miller & Levine Biology: Foundation Edition* more accessible to ELL students.

Spanish Study Workbook is a Spanish translation of Study Workbook A.

Multilingual Glossary is the *Miller & Levine Biology* glossary translated into ten languages.

Tools to Challenge Students

Study Workbook A includes more challenging worksheets and activities such as Chapter Mystery activities that reinforce 21st Century Skills.

Lab Manual A labs are geared toward on-level students. The Lab Manual includes fifteen labs in addition to the Chapter Labs.

Biology.com Data Analysis activities offer students a way to practice data gathering and analyzing skills.

Visual Quizzes in the Assessment Resources Book are geared toward on-level students and use visuals from *Miller & Levine Biology.*

Additional Resources

Biology.com Untamed Science Video • Flash Cards

Study Workbook A 21.1 Worksheets
Spanish Study Workbook 21.1 Worksheets

Study Workbook A 21.2 Worksheets
Spanish Study Workbook 21.2 Worksheets

Study Workbook A 21.3 Worksheets
Spanish Study Workbook 21.3 Worksheets
Biology.com *Data Analysis:* Tiny Fossils and Ancient Oceans
Assessment Resources Book Visual Quiz

Study Workbook A 21.4 Worksheets
Lab Manual B *Data Analysis:* Mycorrhizae and Tree Height
Spanish Study Workbook 21.4 Worksheets
Assessment Resources Book Visual Quiz

Lab Manual A *Mushroom Farming*

Chapter Review

Student Edition Chapter Summary, p. 521
Study Workbook B Chapter 21 Review
Transparencies, pp. 249–258
Biology.com Untamed Science Video • Editable Worksheets of Study Workbooks A and B and Lab Manuals A and B • Chapter 21 Flash Cards and Crossword

Untamed Science DVD • Classroom Resources CD (includes lesson presentations and editable worksheets)

Chapter Assessment

Student Edition Check Understanding, pp. 522–525
Study Workbook B Chapter 21 Taking a Standardized Test
Assessment Resources Book Chapter 21 Test B
Biology.com Chapter 21 Assessment • Editable Worksheets of Chapter 21 Visual Quizzes* and Chapter 21 Tests A and B

Exam*View* *Assessment Suite* • Classroom Resources CD (includes lesson presentations and editable worksheets)

Time: 1 period, 1/2 block

* Visual Quizzes are based on visuals from *Miller & Levine Biology*. Review carefully to make sure they are appropriate for your class.

Pressed for Time?

Preview the Chapter Preview the images of protists and fungi in the Diversity of Life guide (pp. DOL 10–DOL 19).

Cover the Chapter Quickly Assign the first two paragraphs of *The First Eukaryotes* in Lesson 21.1. In Lesson 21.2, discuss the Amoeboid Movement and How Cells Move Like Boats visuals, along with their associated vocabulary terms, and have students read the introduction to *Protist Reproduction*. Assign students to read all of Lesson 21.3. Assign the introduction to *What Are Fungi?* and all of *The Ecology of Fungi* in Lesson 21.4.

Assess Assign Question 1 in the 21.1 Check Understanding, Questions 1, 2, and 4–6 in the 21.2 Check Understanding, the 21.3 Check Understanding, and Questions 3–8 in the 21.4 Check Understanding. In the Chapter 21 Check Understanding, assign Questions 1, 7, 11–17, and 20.

Get the Picture

Ask How many organisms do you see in this picture? *(Answers may vary. Students might list the frog, the tree, and the lichen.)*

Say One of the organisms—the lichen—is made of two organisms working together. One is a fungus. The other is a photosynthetic organism, like a green alga or a cyanobacterium.

Ask So, how many organisms can you see in this picture? *(four: the frog, the tree, the fungus, and the photosynthetic organism)*

Ask What does the word *interdependence* mean? *(Sample answer: Interdependence is when two or more things depend on each other.)* Invite students to recall what the prefix *inter-* means. *(between, among)*

Ask How are the organisms in this picture interdependent? *(Sample answer: The lichens hide the frog; animals that might want to eat the frog will be less likely to see it.)*

Activate Prior Knowledge

Ask What do you remember about the characteristics of protists and fungi? If necessary, refer students to Chapter 18 Classification.

Ask What are some ways that fungi are different from animals? *(Sample answers: They do not move. They have cell walls.)* How are they similar to animals? *(Sample answer: Their cells are eukaryotic. They are heterotrophs.)*

NATIONAL SCIENCE EDUCATION STANDARDS

UNIFYING CONCEPTS AND PROCESSES

I, II, IV, V

CONTENT

C.1.a, C.3.a, C.3.d, C.3.e, C.4.b, C.4.d, C.5.d, F.1, G.3

INQUIRY

A.1.a, A.1.b, A.1.c, A.2.a, A.2.b, A.2.e

21 Protists and Fungi

Big idea **Interdependence in Nature**

Q: How do protists and fungi affect the homeostasis of other organisms and ecosystems?

A tree frog hides among lichens. A lichen is a symbiotic relationship between a fungus and a photosynthetic organism, such as a green alga.

BIOLOGY.com Search Chapter 21 GO • Flash Cards

500

UbD Connect to the Big Idea

Invite a volunteer to read aloud the chapter number and the chapter title. Read the Big Idea and question. Emphasize the question, **How do protists and fungi affect the homeostasis of other organisms and ecosystems?**

Set the Purpose

PREVIEW AND PREDICT Draw the graphic organizer on the board. Guide students to the INSIDE feature on the next page, and invite a volunteer to read each lesson number and title.

Say Protists and fungi are very different organisms. That is why they are classified in different kingdoms.

Ask How do protists and fungi affect other living things in an ecosystem? *(Sample answer: Many protists are food for other organisms. Fungi act as decomposers.)*

Remind students to search for answers to the graphic organizer questions as they read.

INSIDE:

- Untamed Science Video
- Chapter Mystery

CHAPTER MYSTERY

"A BLIGHT OF UNUSUAL CHARACTER"

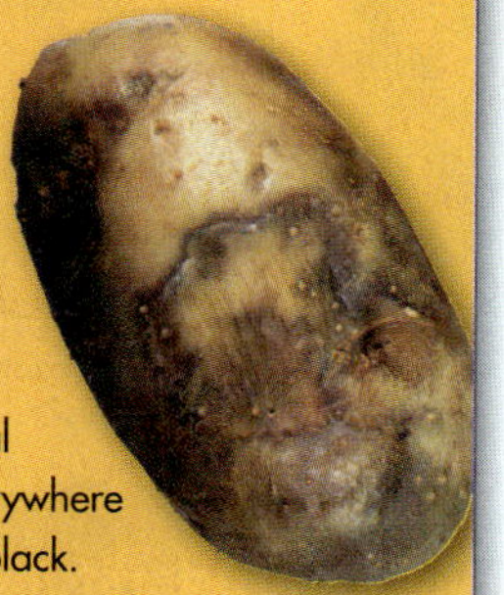

In the early 1800s, Irish farmers became heavily dependent on growing potatoes. Potatoes grew very well in the cool wet climate of Ireland. Soon potatoes became the main source of food for Irish families.

Then, during the summer of 1845, something strange began to happen. One magazine reported that "a blight of unusual character" was attacking the potatoes. Everywhere in Ireland, potatoes began to rot and turn black. Without their main food crop, many people starved. Others left Ireland to go to countries such as the United States. By the early 1900s, the population of Ireland was cut in half.

Read for Mystery Clues What caused this disaster? As you read this chapter, look for clues to help you find the cause of the potato blight. Then, solve the mystery.

FOUNDATIONS for Learning

If protists and fungi disappeared, our world would be a very different place. Make a double-door fold like the one shown below. Under Protists, write these headings: Autotrophs, Heterotrophs, Mutualists, Parasites. Under Fungi write Decomposers, Mutualists, Parasites. As you read the chapter, make notes under each of these headings. At the end of the chapter, you will find two activities that will help you answer the question: How do protists and fungi affect the homeostasis of other organisms and ecosystems?

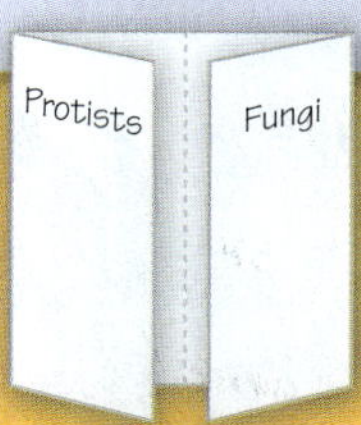

Chapter 21 Big Idea: Interdependence in Nature

Chapter 21: How do protists and fungi affect the homeostasis of other organisms and ecosystems?

- **21.1:** Why are "protists" difficult to classify?
- **21.2:** How do protists move and reproduce?
- **21.3:** What roles do protists play in the environment?
- **21.4:** What are fungi, and what roles do they play in the environment?

Foundations for Learning

Give each student a sheet of unlined paper. Holding a sheet of paper sideways, demonstrate how to fold the paper in a double-door fold as shown on the student page. Help students understand the importance of this activity.

Ask Why do you think you are using a double-door fold for this activity? *(The double-door design makes it easy to compare two different things.)*

Say Look at the headings you have written on each side.

Ask What is one way in which protists and fungi differ? *(Protists can be autotrophs or heterotrophs, but all fungi are heterotrophs.)*

Ask How are protists and fungi similar? *(Protists and fungi can be mutualists or parasites.)*

Remind students that they will use this sheet as a review and study aid.

Untamed Science™ The Untamed Science crew unearth some interesting facts about the structure and function of mushrooms in **Fungus Among Us**.

CHAPTER MYSTERY Invite a volunteer to read aloud the Chapter Mystery. Have students guess how the mystery relates to the chapter title and predict the cause of the potato blight.

Tell students they will find clues in the chapter to answer the question, **What caused this disaster?**

Digital Support

BIOLOGY.com Extend your reach by using the digital assets offered at **Biology.com.**

LESSON 21.1

Pre-Reading

Objectives

21.1.1 Explain what a protist is.

21.1.2 Describe how protists are related to other eukaryotes.

Preview the Pages

1. Review the terms *prokaryote* and *eukaryote* and their evolutionary relationship.
2. Preview the visuals and read the captions and labels.
3. Review the meaning of the term *cladogram,* an evolutionary diagram in which the end of each branch represents a different group.
4. Point out that the protist classification cladogram shows evolutionary relationships among the different groups of protists. Be sure students see that the cladogram also includes the plant, fungi, and animal kingdoms.

Build Understanding

PREVIEW VISUALS Have students complete the chart in Workbook B for previewing the visuals in this lesson. Remind students that a cladogram is a visual way to show evolutionary relationships among groups.

21.1 Protist Classification

Key Questions

What are protists?

How are protists related to other eukaryotes?

BUILD Understanding

Preview Visuals Look at the photos in the cladogram about protist classification. Do you recognize any of these protists? Write down two questions that you have about the cladogram. As you read the lesson, try to answer the questions.

In Your Workbook Go to your workbook to learn more about previewing visuals. Complete the activity for Lesson 21.1.

The First Eukaryotes

More than a billion years ago, a new kind of organism appeared. These organisms had one cell. Their fossils show they were the first eukaryotes. Eukaryotes have cells with nuclei. Single-celled eukaryotes are still with us today. They are often called *protists*, which means "first." Protists are eukaryotes that are not fungi, plants, or animals.

Most protists are single celled, but some have many cells. The largest protists—brown algae called kelp—have millions of cells. They also have tissues. Kelp is considered to be a protist because it is more closely related to certain single-celled protists than it is to members of any other kingdom.

Key Question What are protists? **Protists are eukaryotes that are not members of the fungi, plant, or animal kingdoms.**

Kingdom Protista Protists were once classified in kingdom Protista. But in recent years, biologists have discovered that kingdom Protista is far more diverse than any other eukaryotic kingdom. Many protists are more closely related to members of other kingdoms than to other protists. This causes a problem in classification. The members of a kingdom should be more like one another than they are like members of other kingdoms. This is not true of protists. As a result, biologists are still undecided as to the best way to classify protists.

Giant Kelp Otters wrap themselves in giant kelp, a multicellular protist, to keep from drifting out to sea while they sleep.

Diatoms The shells of diatoms have beautiful patterns. Diatoms are a group of tiny protists that live in the ocean.

SEM 960×

502

NATIONAL SCIENCE EDUCATION STANDARDS

UNIFYING CONCEPTS AND PROCESSES

II, V

CONTENT

C.3.a, C.3.d, C.3.e, G.3

INQUIRY

A.1.b, A.1.c, A.2.a, A.2.b, A.2.e

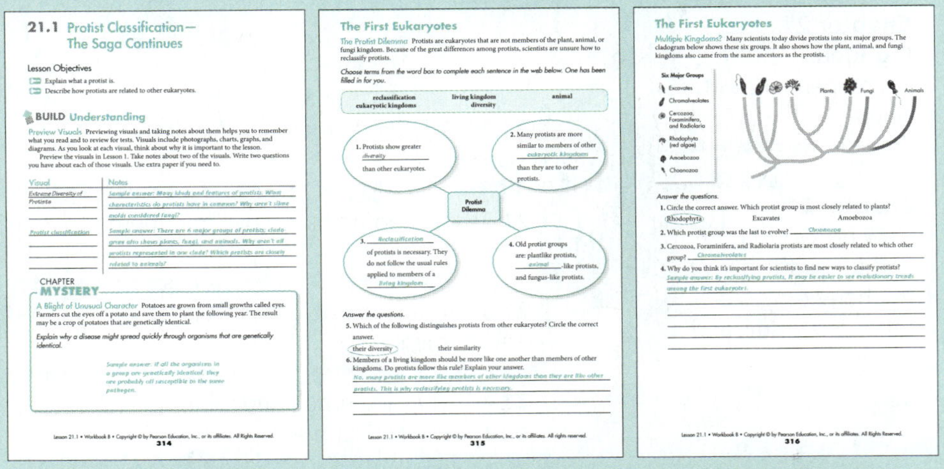

21.1 Protist Classification— The Saga Continues

Lesson Objectives

Explain what a protist is.

Describe how protists are related to other eukaryotes.

BUILD Understanding

Preview Visuals Previewing visuals and taking notes about them helps you to remember what you read and to review for tests. Visuals include photographs, charts, graphs, and diagrams. As you look at each visual, think about why it is important to the lesson.

Preview the visuals in Lesson 1. Take notes about two of the visuals. Write two questions you have about each of those visuals. Use extra paper if you need to.

Visual	Notes
Extreme Diversity of Protists	Sample answer: Many kinds and features of protists. What characteristics do protists have in common? Why aren't slime molds considered fungi?
Protist classification	Sample answer: There are 6 major groups of protists; cladogram also shows plants, fungi, and animals. Why aren't all protists represented in one clade? Which protists are closely related to animals?

CHAPTER MYSTERY

A Blight of Unusual Character Potatoes are grown from small growths called eyes. Farmers cut the eyes off a potato and save them to plant the following year. The result may be a crop of potatoes that are genetically identical.

Explain why a disease might spread quickly through organisms that are genetically identical.

Sample answer: If all the organisms in a group are genetically identical, they are probably all susceptible to the same pathogen.

The First Eukaryotes

The Protist Dilemma Protists are eukaryotes that are not members of the plant, animal, or fungi kingdom. Because of the great differences among protists, scientists are unsure how to reclassify protists.

Choose terms from the word box to complete each sentence in the web below. One has been filled in for you.

reclassification, living kingdom, animal, eukaryotic kingdoms, diversity

Protist Dilemma

1. Protists show greater diversity than other eukaryotes.
2. Many protists are more similar to members of other eukaryotic kingdoms than they are to other protists.
3. Reclassification of protists is necessary. They do not follow the usual rules applied to members of a living kingdom.
4. Old protist groups are: plantlike protists, animal-like protists, and fungus-like protists.

Answer the questions.

5. Which of the following distinguishes protists from other eukaryotes? Circle the correct answer.

their diversity   their similarity

6. Members of a living kingdom should be more like one another than members of other kingdoms. Do protists follow this rule? Explain your answer.

No, many protists are more like members of other kingdoms than they are like other protists. This is why reclassifying protists is necessary.

The First Eukaryotes

Multiple Kingdoms? Many scientists today divide protists into six major groups. The cladogram below shows these six groups. It also shows how the plant, animal, and fungi kingdoms also came from the same ancestors as the protists.

Answer the questions.

1. Circle the correct answer. Which protist group is most closely related to plants?

Rhodophyta   Excavates   Amoebozoa

2. Which protist group was the last to evolve? Choanozoa
3. Cercozoa, Foraminifera, and Radiolaria protists are most closely related to which other group? Chromalveolates
4. Why do you think it's important for scientists to find new ways to classify protists?

Sample answer: By reclassifying protists, it may be easier to see evolutionary trends among the first eukaryotes.

For the corresponding lesson in *Miller & Levine Biology,* see pages 602–605.

INQUIRY into Scientific Thinking

What Are Protists?

Protists can be found in oceans, rivers, streams, lakes, and ponds. Many are too small to see without a microscope.

Procedure

1. Place a drop of pond water on a microscope slide. Add a drop of methyl cellulose. Cover the sample with a coverslip. Look at the slide under the microscope at low and high magnifications.
2. Write down your observations. Draw each kind of organism that you see. Label any structures, such as the nucleus, that you can identify.

Analyze and Conclude

1. **Explain** For any of the organisms that move, describe their motion. Describe any structures involved in producing the motion.
2. **Evaluate** Did you see any structures that you think are used to gather food? That are used for reproduction? Explain.
3. **Apply Concepts** Are any of the organisms you saw bacteria, plants, or animals? Explain your answer.

In Your Workbook Get more help for this activity in your workbook.

Euglena Photosynthetic *Euglena* are common freshwater protists that can move (LM 250×).

Multiple Kingdoms? The most recent studies of protists divide them into six major groups, or clades. Each could be considered a kingdom. The fungi, plant, and animal kingdoms fit among these six protist groups. In fact, animals and fungi likely evolved from the same protist ancestor.

The Name "Protist" Most biologists today agree that the protists should not be lumped together into a single kingdom. Even so, they continue to call these organisms "protists." So, in this book, we will use the word "protist" to refer to these organisms.

Ancestors and Descendants

Fossils of eukaryotic cells have been found in rocks as old as 1.5 billion years. Evidence from fossils and DNA shows that eukaryotes evolved from prokaryotes. Eukaryotes are more closely related to present-day Archaea than to Bacteria.

Protists were the first eukaryotes. How are they related to other eukaryotes? It may be tempting to look among living protists to find the ancestors of the first plants or fungi. But it would be a mistake to do so. Protists have evolved over time. Those alive today are very different from their ancestors.

Slime Molds At a certain stage of their life cycle, protists called slime molds come together into colonies like this one (SEM 15×).

503

Materials

mixed protist culture, dropper, microscope slide, methyl cellulose, coverslip, compound microscope

Analyze and Conclude

1. Students should note that many organisms are moving along using small, hair-like structures such as cilia or long whip-like structures such as flagella. A few students may describe amoeboid movement.
2. Some students may observe gullets, as found in paramecia. A few students may observe protists undergoing conjugation or reproduction via mitosis.
3. Some students may note the presence of green algae, which are plants. Depending on the source of the sample, students may also see insect larvae or copepods, which are animals.

During Reading

Active Reading

KEY QUESTION Read aloud the Key Question and answer on the previous page.

Ask What are some examples of protists? *(Sample answers: diatom, brown algae, euglena)* How are protists similar to other eukaryotes? *(Sample answer: Their cells have nuclei.)*

Inquiry Into Scientific Thinking

DRAW CONCLUSIONS Ask students to examine their drawings. Then have students compare them to those of other students.

Ask Which organisms are plant-like? Which organisms are animal-like? *(Answers may vary.)*

Ask Why do you think these organisms are classified as protists? *(They cannot be classified as plants, animals, or fungi.)*

MAKE A TABLE Have students make a table listing each type of organism they observed and its characteristics.

Say Compare your table to that of another student.

Ask Are your tables the same? Why or why not? *(Each table is likely to be unique. Students might see different organisms or notice different characteristics.)*

Lead a Discussion

Say Think about items you classify in everyday life.

Ask What are some items you keep in separate cabinets in a kitchen? *(Sample answer: glassware, dishes, pots and pans, and bowls might be kept in four separate cabinets.)*

Ask What are some items you might find in a "junk drawer" in a kitchen? *(Sample answers: carryout menus, twist ties, matches, pencils, paperclips)* Point out that these items have little in common with each other but may be grouped together in a kitchen because they are unique from other kitchen items.

Ask How are protists like the items in a junk drawer? *(Protists don't fit into other kingdoms, so they form their own group.)* Help students develop an analogy between a junk drawer and how protists are classified.

Speed Bump

Ask Which two kingdoms contain prokaryotes? *(Archaea, Bacteria)*

Ask Which organisms were the first eukaryotes? *(protists)*

During Reading continued

Active Reading

INTERPRET DIAGRAMS Have students review the Protist Classification cladogram. Explain that the cladogram shows how protists and other eukaryotes are classified based on their evolutionary history.

Say Notice the four different colors in the cladogram—yellow, green, brown, and red.

Ask What does yellow represent? *(protists)* How many major groups of protists are shown? *(six)*

Ask What do green, brown, and red represent? *(plants, fungi, and animals, respectively)*

Ask What does the yellow "stem" at the bottom represent? *(an ancestor)*

Ask Why are protists, plants, fungi, and animals all shown in this cladogram? *(They are related and have a common evolutionary ancestor.)*

KEY QUESTION Invite a student to read aloud the question and answer.

Ask What do the histories of protists and other eukaryotes share? *(the common ancestor of all animals, plants, fungi, and protists)*

Speed Bump

Ask Why does the classification of protists frequently change? *(As scientists continue to learn more about protists, they change how protists are classified.)*

Post Reading

ASSESSMENT TIP For Question 3, tell students to think about a junk drawer and why some of its contents would be better put elsewhere.

For Question 4, ask students to think about which kingdoms are prokaryotic *(archaea, bacteria)* and which kingdoms are eukaryotic *(plants, fungi, animals).* Since prokaryotes appeared long before eukaryotes, the answer to Question 4 must be the three eukaryotic kingdoms.

Wrap-Up Activity

PERSUASIVE DEBATE Have pairs of students debate the question, **Should protists be classified in a single kingdom?** Have one student in each group argue the positive, while the other argues the negative. Each student must use facts from the lesson to support his or her point of view.

Euglena is classified as an excavate.

Brown algae, such as kelp (left), and diatoms (right) are examples of chromalveolates.

Slime molds are classified with the Amoebozoa.

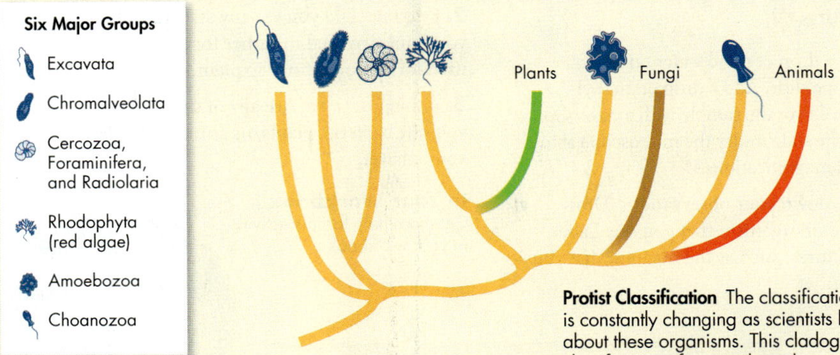

Protist Classification The classification of protists is constantly changing as scientists learn more about these organisms. This cladogram shows a classification of protists based on recent research.

For more on the diversity of protists, go to the Visual Guide. **pp. DOL 10–15**

Today there are as many as 300,000 species of protists. Most are single celled. Some live in colonies. Two groups of protists—the red algae and the brown algae—include true multicellular organisms. Fungi, plants, and animals evolved from the ancestors of these two groups. Biologists continue to study protists closely. They know that clues about the beginnings of diversity in plants, animals, and fungi can be found among the protists.

Key Question How are protists related to other eukaryotes?
Today's protists include groups whose ancestors were among the last to split from the ancestors of plants, animals, and fungi.

CHECK Understanding

Critical Thinking

1. **Explain** What is a protist?
2. **Compare and Contrast** Traditionally, protists were classified in kingdom Protista. How has the classification of protists changed?
3. **Relate Cause and Effect** What is a major reason why scientists need to reclassify protists?
4. **Apply Concepts** What three kingdoms arose from protist ancestors?
5. **Write to Learn** Scientists say that we should not look for the earliest ancestors of animals among modern-day protists. Write a short paragraph that explains why.

Check Understanding

1. A protist is a eukaryotic organism that does not belong to the fungi, plant, or animal kingdoms.
2. Protists are now divided into six groups, or clades. Each clade may represent a different kingdom.
3. Many protists are more like members of other kingdoms than they are like other protists. The members of a kingdom should be more like each other than like the members of other kingdoms.
4. The fungi, plant, and animal kingdoms arose from protist ancestors.
5. Students should recognize that protists have gone through evolution just like every other living thing. So, living protists differ from the ancient protists from which other eukaryotes evolved.

21.2 Protist Structure and Function

How Protists Move

Your body is packed with specialized systems. Organ systems help you move, break down food, and sense your environment. Imagine having only one cell. How would you carry out these functions? Single-celled protists don't have organs or organ systems. They have to do it all within a single cell. Despite these challenges, protists are very successful organisms. They successfully carry out all the activities necessary for life.

Protists move in several different ways. Some move by changing their shape. Others have specialized organelles for movement. Many protists do not move actively. These protists are carried by wind, water, or other organisms.

Amoeboid Movement Many single-celled protists move by changing their cell shape. First, a projection, or bulge, forms on the side of the cell. This projection is called a **pseudopod** (soo doh pahd). The pseudopod fills with cytoplasm. As cytoplasm flows into the pseudopod, the rest of the cell follows. This causes the cell to move. Amoebas are the best-known group of protists that move this way. As a result, this type of motion is called amoeboid movement.

Amoeboid movement is powered by a protein called actin. Actin is also found in the muscles of animals. In animals, actin plays an important role in muscle contraction.

Key Questions

How do protists move in the environment?

How do protists reproduce?

BUILD Understanding

Compare/Contrast Table As you read the lesson, make a table that compares and contrasts the ways in which protists move.

In Your Workbook Go to your workbook to learn more about comparing and contrasting. Complete the table for Lesson 21.2.

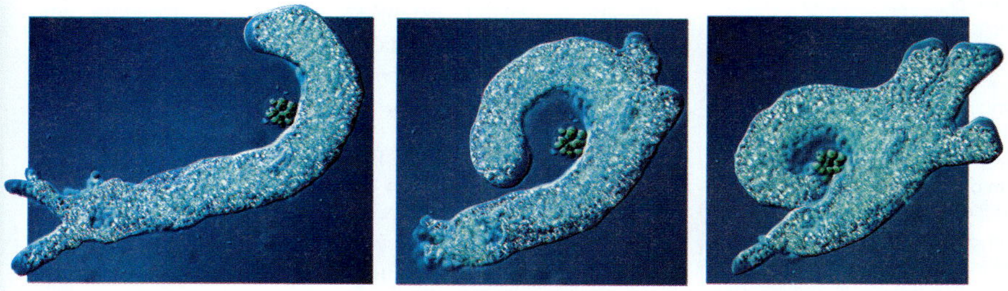

(LM 220×)

Amoeboid Movement An amoeba moves by extending a pseudopod away from its body. Cytoplasm streams into the pseudopod. Then the rest of the cell follows. Amoebas also use pseudopods to surround and take in food. In these photos, the amoeba is consuming algae.

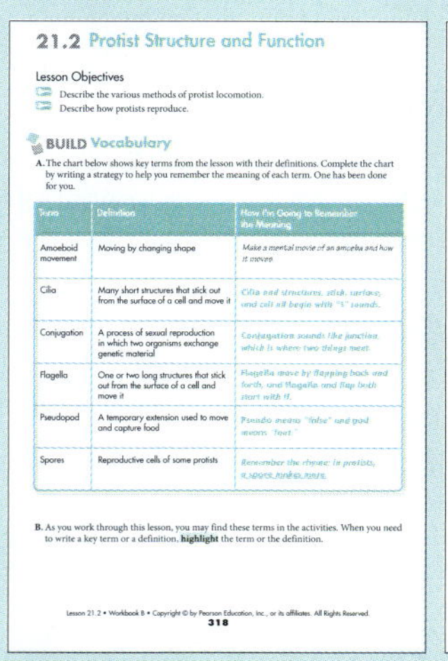

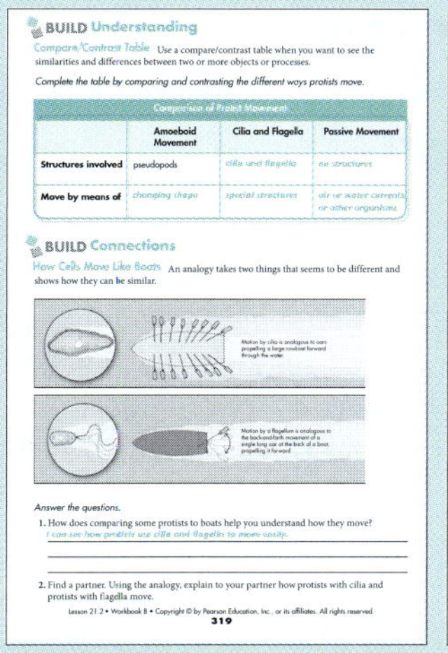

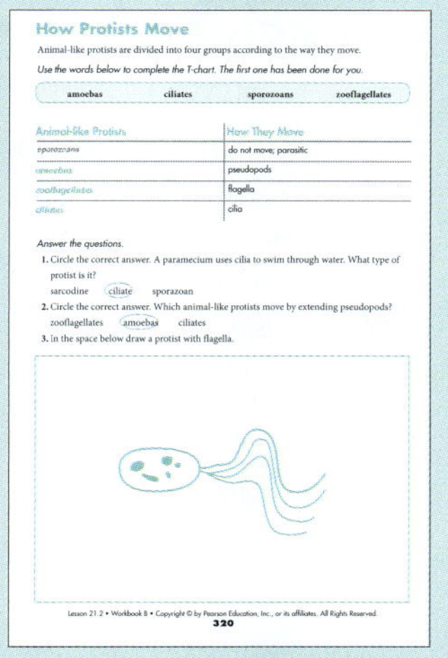

Pre-Reading

Objectives

21.2.1 Describe the various methods of protist locomotion.

21.2.2 Describe how protists reproduce.

Preview the Pages

1. Review with students all of the lesson vocabulary, pronouncing each word aloud. Have students find the words in the text.
2. Preview the visuals and read the captions and labels aloud. In the Water Mold Life Cycle diagram, be sure students understand that a life cycle diagram shows the sequential events that occur during an organism's lifetime. Point out that this includes two different pathways to show the events of asexual reproduction and sexual reproduction. Review these two types of reproduction, as needed.

Build Understanding

COMPARE/CONTRAST TABLE Tell students that the subheadings Amoeboid Movement, Cilia and Flagella, and Passive Movement describe some ways that protists move. Direct students to Workbook B, where they will find a compare/contrast table to complete for this lesson.

Speed Bump

Say The term *pseudopod* means "false foot." Think about amoeboid movement and explain why this term makes sense. *(Like a foot, a pseudopod enables an amoeba to move from place to place.)*

NATIONAL SCIENCE EDUCATION STANDARDS

UNIFYING CONCEPTS AND PROCESSES

I, IV, V

CONTENT

C.1.a

For the corresponding lesson in *Miller & Levine Biology,* see pages 606–609.

During Reading

Active Reading

COMPARE AND CONTRAST Point out that this page describes two general kinds of movement: active and passive.

Ask What does *passive* mean? *(not participating actively; affected by something else)*

Ask Which form of movement, active or passive, do ciliates and flagellates use? *(active)* Which form of movement is used by spore-producing protists? *(passive)*

Build Connections

VISUAL ANALOGY Discuss the visual analogy comparing movement by cilia or a flagellum with two ways in which a boat is moved.

Ask What do these protists have in common with boats? *(Both are found in wet environments.)*

Say Like the oars on a boat, cilia are evenly spaced and move in a regular pattern. A flagellum whips back and forth or spins in a pattern that propels a protist rapidly through water.

Ask Recall the Inquiry into Scientific Thinking activity in Lesson 21.1. What do you remember about the movements of the protists you observed? *(Students may refer to cilia, flagella, or pseudopods they observed.)*

Build Vocabulary

WORD ORIGINS Remind students that they may know other words that contain the root *pseudo*.

Ask What is a pseudonym? *(a name that is not someone's correct name)* Who might use a pseudonym? *(someone who did not want his or her identity known, such as an author who wishes to write anonymously)*

Ask Why is a pseudopod a "false" foot? *(It is a temporary projection, not an actual structure.)*

KEY QUESTION Read the Key Question and answer aloud. Challenge students to sketch in their notebooks an example of each type of protist movement.

Speed Bump

Ask How could you tell if a protist is a ciliate or a flagellate? *(Ciliates have many cilia over their surface; flagellates have only one or two flagella.)*

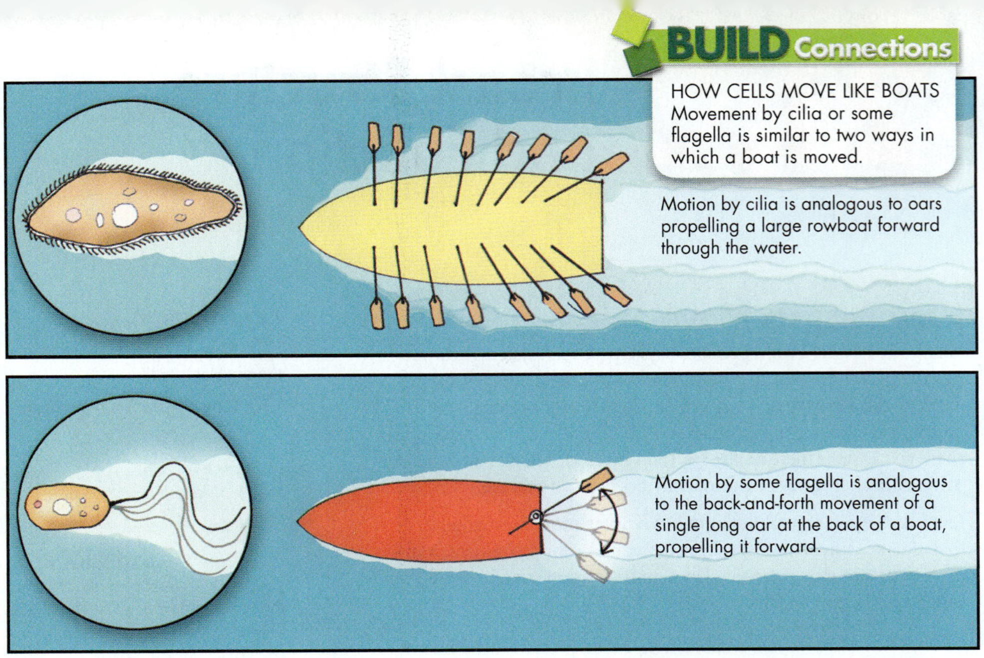

BUILD Vocabulary

pseudopod a temporary cytoplasmic projection used by some protists for movement

cilium a short, hairlike projection that produces movement

flagellum a structure used by protists for movement; produces movement in a wavelike motion

spore in prokaryotes, protists, and fungi, any of a variety of thick-walled life-cycle stages capable of surviving unfavorable conditions

conjugation a process in which paramecium and some prokaryotes exchange genetic information

alternation of generations a life cycle that has two alternating phases—a haploid (N) phase and a diploid (2N) phase

WORD ORIGINS

The word *pseudopod* comes from the Greek root *pseudo*, which means "false," and *pod*, which means "foot."

Cilia and Flagella Many protists move by means of cilia (SIL ee uh) and flagella (fluh JEL uh). These structures are supported by microtubules, which are part of a cell's cytoskeleton. Cilia and flagella have nearly identical internal structures. However, they produce motion in different ways.

Cilia (singular: cilium) are short and numerous. They move somewhat like oars on a boat. Protists that move with cilia are known as ciliates.

Flagella (singular: flagellum) are relatively long. Usually a cell has one or two. Some flagella spin like tiny propellers. But most flagella move like a wave from base to tip. Protists that move by flagella are called flagellates.

Passive Movement Many protists depend on air, water currents, or other organisms to carry them around. These protists form reproductive cells called **spores.** Some spores can enter other organisms and live as parasites. One spore-forming protist is *Plasmodium*. *Plasmodium* is carried by mosquitoes. It causes malaria. *Cryptosporidium* is spread through contaminated drinking water. It causes severe intestinal disease in humans.

Key Question How do protists move in the environment?
Some protists move by changing their cell shape. Some use cilia or flagella. Others do not move actively but are carried by wind, water, or other organisms.

Focus on Academic Language

• cilia • cilium • ciliates • flagella • flagellum • flagellates

Write these words on the board and review their meanings. *Cilia* is from *cilium,* the Latin word for "eyelid," later referring to eyelashes. *Flagella* is from *flagrum,* meaning "whip." Help students pronounce the terms *cilia* and *flagella.* Point out the difference between the singular and plural forms. Ask students to think of another term they have learned that forms the plural in the same way. *(Sample answer: bacterium/bacteria)* Then, have them practice using the singular and plural forms by making up sentences that contain the terms *cilium, cilia, flagellum,* and *flagella.* Remind students that protists that move with cilia are called *ciliates,* while protists that move by flagella are called *flagellates.*

Protist Reproduction

Some protists reproduce asexually. A few protists exchange genetic material during conjugation. Other protists have a life cycle that combines both asexual and sexual reproduction.

Cell Division Amoebas and other protists reproduce asexually by mitosis. First, the protist copies its genetic material. Then, the parent cell divides into two identical daughter cells. Mitosis lets protists reproduce quickly. However, mitosis produces cells that are genetically identical to the parent. Identical cells limit genetic diversity.

Conjugation Paramecia and most other ciliates reproduce asexually by mitosis. But when conditions are unfavorable, paramecia undergo conjugation. **Conjugation** is a process in which two organisms exchange genetic material. After conjugating, the cells reproduce by mitosis.

Paramecia have two types of nuclei: a macronucleus and one or more smaller micronuclei. The macronucleus controls the daily activities of the cell. Micronuclei contain the cell's chromosomes. During conjugation, the cells exchange haploid micronuclei. These micronuclei fuse to form a diploid micronucleus. Conjugation produces new combinations of genes, which can aid in evolution.

Sexual Reproduction Many protists have sexual life cycles that are complex. They alternate between diploid and haploid phases, in a process that is called **alternation of generations.** Water molds are one group of protists that have alternation of generations. Water molds reproduce asexually by producing diploid spores in a structure called a sporangium (spoh RAN jee um). Water molds also reproduce sexually. They undergo meiosis, forming male and female structures. These structures produce haploid nuclei. The nuclei fuse during fertilization, forming a diploid zygote.

Key Question How do protists reproduce? **Some protists reproduce asexually by mitosis. Others have life cycles that combine sexual and asexual forms of reproduction.**

Conjugation During conjugation, two paramecia attach to each other and exchange genetic information.

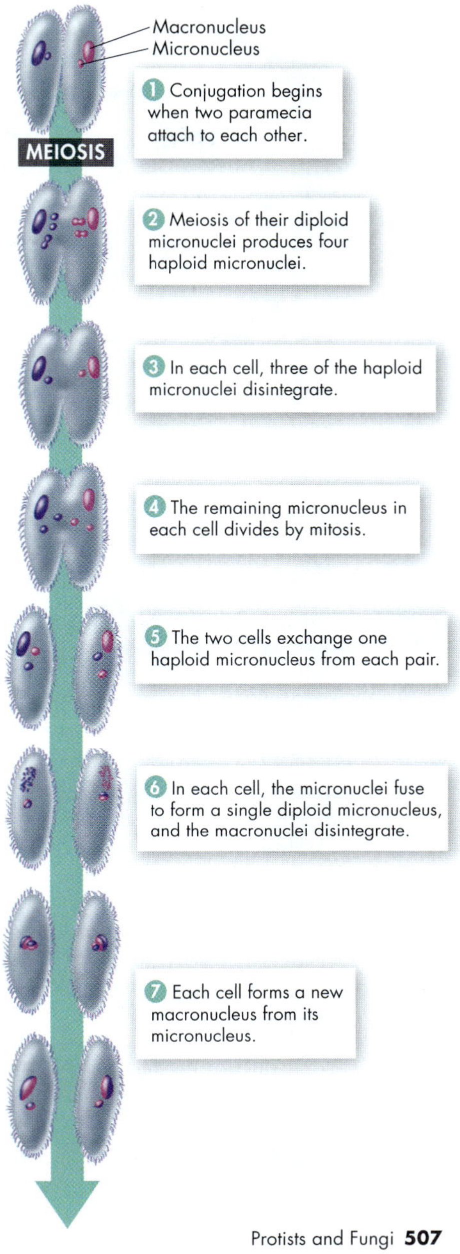

Active Reading

FIND THE MAIN IDEA Ask a student to read the first paragraph aloud.

Ask What is the main idea of this paragraph? *(Protists can reproduce asexually or sexually.)*

USE VISUALS Read aloud the paragraphs describing conjugation. Then have students refer to the diagram. Trace the events shown in the diagram by asking a different student volunteer to read each of the labels.

Ask What structure do cells exchange, or trade, during conjugation? *(micronuclei)*

MAKE INFERENCES Review the meaning of the terms *genetic material* and *genetic diversity.*

Ask What is genetic material? *(an organism's genes)*

Ask What does *genetic diversity* mean? *(a mix of different kinds of genes from different individuals)*

Say When change occurs, genetic diversity helps a population adapt. When a population has many kinds of genes, it is more likely to survive change.

Ask How does conjugation help paramecia? *(It produces new combinations of genes, which might be helpful if there is a change in the environment.)*

USE CONTEXT CLUES Read aloud the second paragraph under the heading Conjugation.

Say Does this make sense? What does it mean that micronuclei fuse to form a diploid micronucleus?

Say Sometimes if I don't know a word, I can figure out its meaning from the words around it.

Ask What does the word *fuse* mean? *(to join together to form a new whole)*

KEY QUESTION Read aloud the Key Question and answer.

Ask How does asexual reproduction differ from sexual reproduction? *(In asexual reproduction, new individuals are genetically identical to the parent. Sexual reproduction produces unique offspring and requires two parents.)*

Speed Bump

Ask Where are chromosomes found in paramecia? *(in micronuclei)*

ELL Focus on ELL: Access Content

ALL SPEAKERS Have students complete sentence frames to increase their comprehension of the water mold life cycle on the next page. First, have pairs of students review the events of the life cycle diagram. One member of each team should point out the following terms and explain them to the other team member: *asexual reproduction, sexual reproduction, haploid (N), diploid (2N), meiosis,* and *fertilization.* Then ask the members of each team to alternate. Ask students to define *alternate (to take turns).* When teams have finished, challenge students to use these terms to complete the sentence frames below. Terms may be used more than once.

The water mold life cycle has two kinds of reproduction: *(asexual reproduction)* and *(sexual reproduction)*. Water molds use *(asexual reproduction)* to form spores. In alternation of generations, a *(haploid (N))* phase alternates with a *(diploid (2N))* phase. A process called *(meiosis)* forms male and female structures. These structures produce *(haploid)* nuclei that fuse during *(fertilization)*.

During Reading continued

Active Reading

INTERPRET DIAGRAMS Guide students to the Water Mold Life Cycle diagram.

Ask Does this diagram remind you of any other diagram you've seen in this book? *(Sample answer: The diagram has a circle within a circle, similar to the lytic/lysogenic cycle diagram in Chapter 20.)*

Say This diagram shows a life cycle. It looks like a circle because it shows a sequence of events that are repeated again and again.

Ask What do the colors of the arrows represent? *(Orange shows the haploid part of the life cycle. Purple shows the diploid part of the life cycle.)*

Ask What do the colored arrows tell you about alternation of generations? *(They show that a haploid portion of the life cycle alternates with a diploid portion.)* If students have trouble answering this question, write the answer as above on the board, leaving blanks for key words such as *haploid* and *diploid*.

Speed Bump

Ask In water molds, which structures perform asexual reproduction? *(sporangia)* What do these structures produce? *(spores)*

Post Reading

ASSESSMENT TIPS Review the highlighted vocabulary words with students before they begin the assessment.

For Question 5, have students review the Build Connections diagram.

MYSTERY CLUE Discuss the method that the Irish farmers used to make new potato plants. Ask students whether they think this method was a form of asexual reproduction or sexual reproduction. *(asexual reproduction)* Suggest that students reread the third page of the lesson if they need to review the importance of genetic diversity.

Wrap-Up Activity

SEQUENCING Give students index cards. Have them write a key event of an alternation of generations life cycle on each card. Challenge students to place the events in the correct sequence. Then have students arrange the cards in a circle, to form a life cycle, on their desktops.

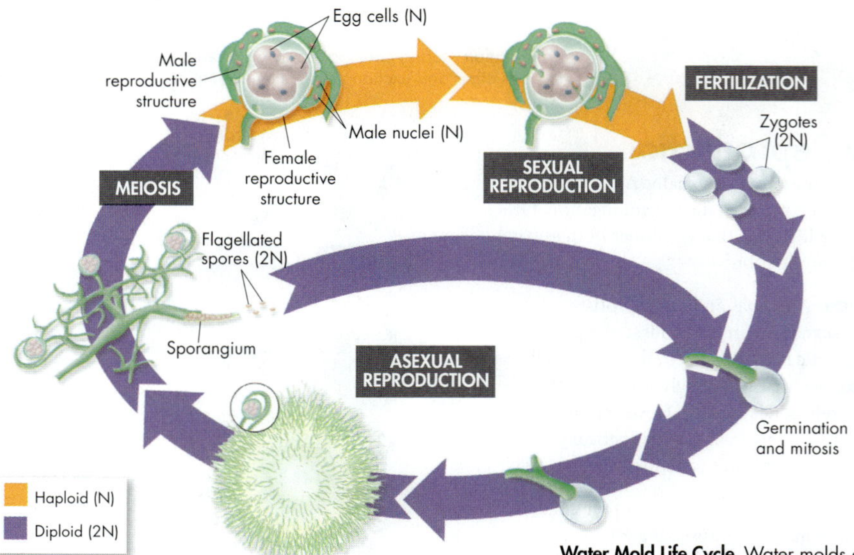

Water Mold Life Cycle Water molds grow into multicellular branching filaments that produce sporangia for asexual reproduction or male and female structures for sexual reproduction.

CHECK Understanding

Apply Vocabulary

Use the highlighted words from the lesson to complete each sentence correctly.

1. During amoeboid movement, a protist produces ________, or projections that fill with cytoplasm.
2. During ________, protists exchange genetic information.
3. Some protists form ________, reproductive cells that can enter the cells of other organisms and live as parasites.

Critical Thinking

4. **Explain** Summarize three ways in which protists move.
5. **Compare and Contrast** How does movement by means of cilia differ from movement by means of flagella?
6. **Review** Describe how protists reproduce.
7. **Write to Learn** Answer the first mystery clue. Use the words *asexual reproduction* and *sexual reproduction* in your answer.

MYSTERY CLUE

To produce more potatoes, Irish farmers cut the eyes, or buds, from a crop and saved them for the next year. As a result, the potatoes grown were genetically identical. How might this practice have made the potato blight worse? (**Hint:** See p. 507.)

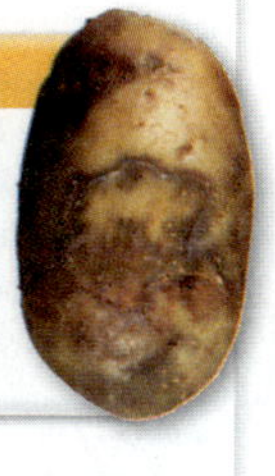

BIOLOGY.com Search Lesson 21.2 GO • Lesson Assessment

Check Understanding

1. pseudopods
2. conjugation
3. spores
4. Protists move using numerous cilia, a long flagellum, or by amoeboid movement.
5. Cilia work much like oars along the sides of a boat. Flagella work like a single oar at the back of the boat.
6. Protists reproduce asexually by mitosis and spore production and sexually by egg and sperm during fertilization.
7. Students should recognize that the potatoes were propagated asexually, which produces offspring that are identical to the parent. Sexual reproduction produces offspring that differ from the parents, increasing genetic diversity. The greater the genetic diversity, the better the chance that the potatoes can overcome a disease such as blight.

21.3 The Ecology of Protists

Autotrophic Protists

Seaweed in the ocean and the green scum on a pond are both commonly called *algae*. Many kinds of algae are protists.

Diversity of Algae Organisms called *algae* belong to several different groups. Some are prokaryotes. Others, such as green algae, are plants. Many are protists. All these organisms are autotrophs. They use energy from sunlight to make their own food by photosynthesis. Photosynthetic protists include many species of phytoplankton. Phytoplankton are small algae that float near the water's surface. Red and brown algae, euglenas, and dinoflagellates are also photosynthetic protists.

Ecological Roles Photosynthetic protists form the base of food chains in the ocean and in fresh water.

▶ ***Feeding Fish and Other Animals*** Phytoplankton carry out about half of the photosynthesis that takes place on Earth. Phytoplankton are eaten by organisms as diverse as shrimp and baleen whales. And when you eat fish such as tuna, you are indirectly getting energy from phytoplankton.

▶ ***Supporting Coral Reefs*** Algae play two key roles in coral reefs. They provide nutrients from photosynthesis to coral animals. They also provide the calcium carbonate that coral reefs need to grow.

Key Questions

How are photosynthetic protists important to ecosystems?

How do heterotrophic protists get food?

What types of symbiotic relationships involve protists?

BUILD Understanding

KWL Chart Before you read Lesson 21.3, make a chart with four columns. In the left column, write the titles and headings in the lesson. In the second column, write what you already know about each topic. In the third column, write what you expect to learn. After you read the lesson, write what you learned in the last column.

In Your Workbook Go to your workbook to learn more about making a KWL chart.

Baleen Whale

Coral Reef

Producers in the Food Chains Many ocean animals, including baleen whales, fishes, and reef animals, rely on photosynthetic protists for food.

21.3 The Ecology of Protists

Lesson Objectives

- Describe the ecological significance of photosynthetic protists.
- Describe how heterotrophic protists obtain food.
- Identify the symbiotic relationships that involve protists.

BUILD Vocabulary

A. The chart below shows key terms from the lesson with their definitions. Complete the chart by writing a strategy to help you remember the meaning of each term. One has been done for you.

Term	Definition	How I'm Going to Remember the Meaning
Algal bloom	Rapid increase in a protist population	*Bloom* rhymes with *boom*, which can mean a rapid growth.
Food vacuole	A small pocket in an amoeba's cytoplasm that stores food	*Vacuole rhymes with hole. Think of it as a hole for holding food.*
Gullet	A small opening in the side of a ciliate that traps food	*Gullet and gulp start with G, and I use my mouth to gulp food just like a paramecium uses a gullet to trap food.*
Plasmodium	A collection of amoeba-like cells that comes together during one stage in the slime mold life cycle	*The word plasmodium sounds a little slimy and podium and gold have the same starting sound.*

B. As you work through this lesson, you may find these terms in the activities. When you need to write a key term or a definition, **highlight** the term or the definition.

BUILD Understanding

KWL Chart A KWL chart is a good tool to use to preview a lesson before reading it. A KWL chart can be used to record what you know, what you want to know, and what you learned.

Before you read Lesson 3, read the lesson's title and headings. Think about what you already know about the lesson's topics. Complete the chart below by listing the title and each heading. Write what you already know about each topic. Finally, write what you expect to learn. As you read the lesson, you can add notes to this chart.

Title or Heading	I Already Know . . .	I Expect to Learn . . .	What I Learned . . .
The Ecology of Protists	*what protists are and the structure and function of protists*	*where different protists live and how they relate to other organisms*	*There are many different types of protists. Some make their own food and some eat other organisms. Some protists live in symbiotic relationships with other protists.*
Autotrophic Protists	*Autotrophs produce their own food.*	*ways autotrophic protists produce food; what their role is on Earth*	*Autotrophic protists are the base of aquatic food chains.*
Heterotrophic Protists	*Heterotrophs get food from other sources.*	*where amoebas, ciliates, slime molds, and protists that absorb get their food*	*Heterotrophic protists have specialized structures to obtain food.*
Symbiotic Protists	*A symbiotic relationship is when two organisms depend on one another.*	*what types of protists have symbiotic relationships*	*Some symbiotic relationships are beneficial to both organisms and some are harmful to one of the organisms in the relationship.*

Heterotrophic Protists

Ciliates Heterotrophic protists do not make their own food. Instead, they obtain food from other living organisms. A paramecium is an animal-like protist called a ciliate. It uses cilia for feeding and movement.

Follow the directions.

1. Label the cilia.
2. Use orange to color the structure that traps food.
3. Use blue to color the structures in the cytoplasm that temporarily store food.
4. Use green to color the structures that eliminate waste materials.

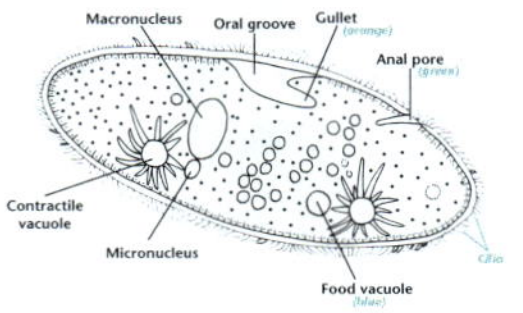

Answer the questions.

5. Circle the correct answer. Which of the following sweeps food particles into the gullet?
   (cilia) oral groove food vacuole
6. What is a gullet? *A gullet is an indentation that traps food.*
7. Waste materials leave the paramecium through the *anal pore*
8. Is the following statement true or false? Cilia have more than one function. They move the paramecium and they help it get food. *true*

LESSON 21.3

Pre-Reading

Objectives

21.3.1 Describe the ecological significance of photosynthetic protists.

21.3.2 Describe how heterotrophic protists obtain food.

21.3.3 Identify the symbiotic relationships that involve protists.

Preview the Pages

1. Explain that the protists in this lesson are grouped based on how they use energy. Point out that each green heading discusses a different group.
2. Have students find the vocabulary words in the text and read each word aloud.
3. Preview the visuals. Point out that the Slime Mold Life Cycle visual has the same color code as the Water Mold Life Cycle diagram in Lesson 21.2.

Activate Prior Knowledge

Have students brainstorm a simple food chain. *(Sample answer: grass, rabbit, fox)* Have them identify the base of the food chain and tell why it is important. Then, explain that some protists form the base of aquatic food chains.

Build Understanding

KWL CHART Direct students to the KWL chart in Workbook B. Remind students that they can use the KWL chart they fill in as a study guide later.

NATIONAL SCIENCE EDUCATION STANDARDS

UNIFYING CONCEPTS AND PROCESSES

I, IV, V

CONTENT

C.4.b, C.4.c, C.5.d, F.1

For the corresponding lesson in *Miller & Levine Biology,* see pages 610–616.

During Reading

Active Reading

COMPARE AND CONTRAST Remind students of the difference between autotrophs and heterotrophs.

Say Think about what you had for breakfast or lunch.

Ask Are humans autotrophs or heterotrophs? *(heterotrophs)* How do you know? *(We eat other organisms to get energy, so we are heterotrophs.)*

Say Autotrophs are "self-feeders" who make their own food; *auto-* means "self." Heterotrophs must get their food from other organisms; *hetero-* means "other."

Ask What is an example of an autotrophic protist? *(Sample answer: phytoplankton)*

Ask What is an example of a heterotrophic protist? *(Sample answer: amoeba or paramecium)*

KEY QUESTION Read the Key Question and answer aloud. Challenge students to draw a simple ocean or freshwater food chain that includes photosynthetic protists. *(Sample answer: phytoplankton, fish, shark)* Have students share their food chains and point out the importance of the photosynthetic protists.

READ, COVER, REMEMBER, RETELL Ask students to read silently the three paragraphs following the heading Heterotrophic Protists. Then tell students to cover the text with their hand and think about what was read.

Ask How does an amoeba feed? *(It surrounds a cell or particle, then takes it in.)*

Ask How does a ciliate feed? *(Its cilia sweep food particles into the gullet.)*

Build Vocabulary

MULTIPLE MEANINGS Remind students that a scientific name consists of two words: a genus name and a species name. A genus name is always capitalized and italicized. Write "amoeba" and *Amoeba* on the board. Ask students to identify which is the genus name.

Speed Bump

Ask What is kelp? *(a large multicellular brown alga)* Where would you find it? *(in the ocean)* How does it help other organisms? *(It provides shelter for many marine species.)*

BUILD Vocabulary

algal bloom
an increase in the amount of algae and other producers that results from a large input of a limiting nutrient

food vacuole
a small cavity in the cytoplasm of a protist that temporarily stores food

gullet
a dent in one side of a ciliate that allows food to enter the cell

plasmodium
the amoeboid feeding stage in the life cycle of a plasmodial slime mold

MULTIPLE MEANINGS
The name *Amoeba* refers to a genus of protists. Without italics or capitalization, the word *amoeba* can describe any protist that moves with pseudopods. In this sense, the adjective *amoeboid* describes the single-celled form of slime molds.

▶ ***Providing Shelter*** Giant kelp and *Sargassum* are multicellular brown algae. They provide shelter for marine species, such as sea otters and many kinds of fishes.

▶ ***Recycling Wastes*** Protists help break down and recycle nutrients in sewage and other waste materials. But if there is too much waste, populations of protists can quickly grow to huge numbers. This situation is called an **algal bloom.** An algal bloom can use up nutrients. As these protists die, their bodies decay. This decay removes oxygen from the water, killing fish and invertebrates. Blooms of protists called dinoflagellates cause red tides. Toxins from dinoflagellates can poison fish and shellfish.

Key Question How are photosynthetic protists important to ecosystems? **Photosynthetic protists form the base of food chains in the oceans and in freshwater environments.**

Heterotrophic Protists

Many protists are heterotrophs. They get food from other organisms. Some heterotrophic protists surround and digest their food. Others absorb nutrients from their surroundings.

Amoebas Amoebas feed by surrounding a cell or food particle. This food is held in a food vacuole. A **food vacuole** is a small cavity in the cytoplasm that temporarily stores food. The food is digested, and the nutrients are passed to the rest of the cell. The vacuole releases waste materials outside the cell.

Ciliates Ciliates sweep food into their **gullet,** a dent in one side of the organism. The particles are forced into food vacuoles at the base of the gullet. The food vacuoles fuse with lysosomes, which contain digestive enzymes. The vacuole releases waste materials through the anal pore.

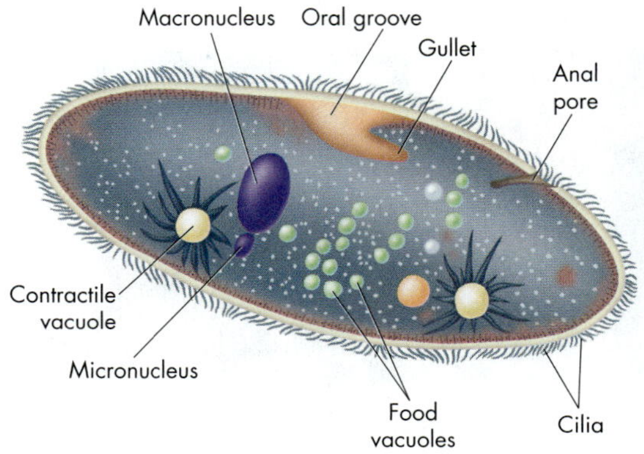

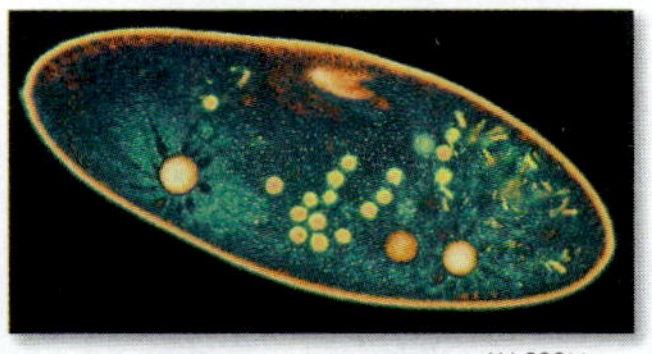

Feeding Structures of *Paramecium*
Cilia that line the oral groove and gullet move food into the cell. The food particles are surrounded, forming food vacuoles. Wastes are released through the anal pore.

Focus on Academic Language

• algae • algal • bloom • toxin

Write these words on the board. Help students understand how to pronounce and understand the words *algae* and *algal.* Explain that *algae* is a noun. *Algae* is the plural form of *alga.* The word *alga* is pronounced with a hard *g* sound, while *algae* is pronounced with a soft *g.*

Explain that *algal* (as in the phrase *algal bloom*) is an adjective. Like *alga,* it is pronounced with a hard *g.* Students may be unfamiliar with the use of the word *bloom* in this context. Use your hands to mimic the opening, or blooming, of a flower. Compare this spreading of petals to the spreading of algae during an algal bloom.

Help students distinguish between the noun *toxin* (meaning "poison") and the adjective *toxic.*

Ask student volunteers to find *algae, algal, bloom,* and *toxin* in the text. Then review additional ways to use these words in a sentence.

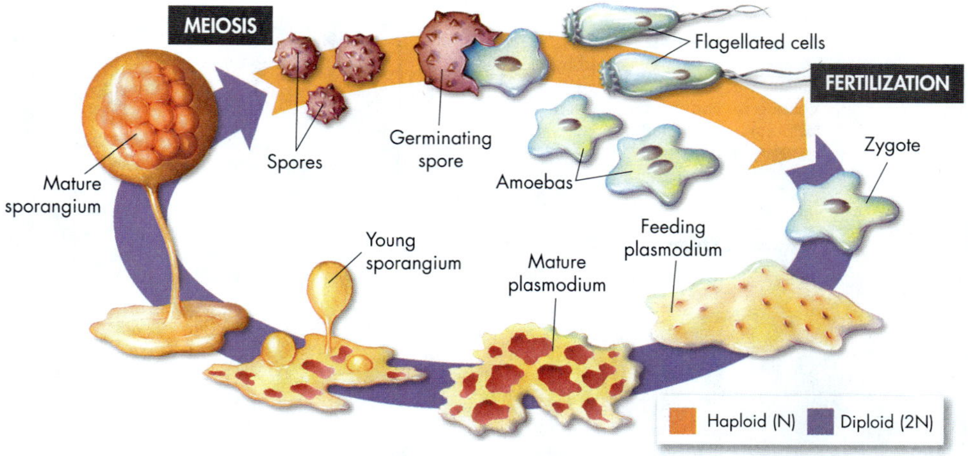

Slime Mold Life Cycle In the life cycle of a slime mold, the diploid feeding stage involves a collection of slime mold amoebas called a plasmodium. The plasmodium produces sporangia, which undergo meiosis and produce haploid spores. The spores grow into amoebalike or flagellated cells. Then the flagellated cells fuse to form diploid zygotes that repeat the cycle.

Slime Molds Slime molds live in damp places that are rich in dead organic matter, such as a forest floor or a compost pile. Because they are decomposers, slime molds play key roles in recycling nutrients.

Slime molds have complex life cycles. In the haploid phase, slime molds are single-celled amoebas or flagellated cells. In the diploid phase, the amoebas form a multicellular structure called a **plasmodium.** The plasmodium develops sporangia, which produce haploid spores.

Protists That Absorb Food Some protists absorb nutrients that other organisms have released into the environment. For example, water molds absorb molecules from decaying plants and animals through their cell walls and cell membranes.

Key Question How do heterotrophic protists get food?
Some protists surround and digest their food. Other protists absorb food molecules from their surroundings.

Symbiotic Protists

Many protists are involved in symbiotic relationships with other organisms. Recall that symbiosis is a relationship in which two species live closely together. Mutualism and parasitism are two kinds of symbiosis.

Mutualists Many protists are involved in mutualistic relationships. Both the protist and its host benefit. Termites and the protist *Trichonympha* are an example. Termites eat wood, but cannot break down the cellulose in wood. *Trichonympha* lives in the gut of the termite. It makes an enzyme called cellulase, which breaks down cellulose. This enzyme allows the termite to get energy from wood.

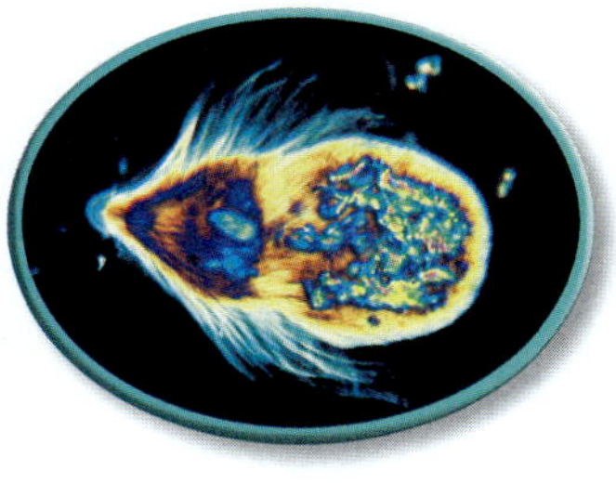

A Protist Mutualist *Trichonympha* lives in the gut of termites. This protist makes enzymes that break down the cellulose in wood (LM 250×).

BIOLOGY.com Search Lesson 21.3 GO • Art in Motion

Active Reading

USE VISUALS Use the Slime Mold Life Cycle diagram to explain the different stages that occur during the life cycle of a slime mold.

Say Slime molds take different forms at different times during their life cycle.

Ask Which stage consists of cells that cannot move? *(the sporangium and the spores)*

Ask When spores germinate, what cells are produced? *(individual flagellated cells or individual amoebas)*

Ask What happens when individual amoebas gather together? *(They form a plasmodium.)*

MAKE CONNECTIONS Students may not know what a compost pile is.

Say A compost pile is a collection of old organic materials, such as fruit or vegetable peels, lawn clippings, and dead leaves. As time passes, these materials decay, producing compost, which can be added to a garden.

Ask How do you think compost would help a garden? *(Sample answer: compost contains dead organic matter, which is rich in nutrients that plants need.)*

KEY QUESTION Read aloud the Key Question and answer.

Ask In which part of a ciliate is food digested? *(in food vacuoles)*

SUMMARIZE Ask students to summarize the relationship of termites and *Trichonympha*. *(Sample answer:* Trichonympha *helps termites break down cellulose in the wood the termites eat.)*

Ask Why is the relationship a good example of mutualism? *(Both species benefit. The termite can get nutrients from wood because* Trichonympha *breaks down the cellulose in wood so the termite can digest it.* Trichonympha *gets a safe home inside the termite and a steady supply of wood to eat.)*

Speed Bump

Ask How do slime molds recycle nutrients? *(Slime molds are decomposers, so they release nutrients from organic matter as it breaks down.)*

ELL Focus on ELL: Build Background

BEGINNING AND INTERMEDIATE SPEAKERS Students may have some background knowledge of mutualism and parasitism though not by those terms. Help students understand the difference between these two terms by discussing familiar examples.

For example, have a volunteer describe the relationship between bees and the flowering plants from which they gather nectar.

Ask How do the bees benefit from the relationship? *(The bees get food in the form of nectar.)* Are the plants helped or harmed by the bees? Explain. *(The bees spread pollen from plant to plant, helping plants to reproduce.)*

Ask Do both species benefit? *(yes)* So is this relationship an example of mutualism or parasitism? *(mutualism)*

Have volunteers share other examples of mutualism or parasitism. Explain that in an everyday context, *mutual* means "having shared feelings or a shared relationship." To have a mutual friend with someone means that you both share the same friend.

During Reading continued

Active Reading

USE VISUALS Ask students to look at the photos of water-borne protist parasites. Have volunteers read aloud the information next to each photo.

Say These protists live in the human intestine and are passed out of the body in feces. In places where sanitation is poor, the protists may contaminate water supplies and food. *Giardia* and *Cryptosporidium* are commonly reported in the United States. They can be found in lakes, swimming pools, and water parks.

Ask How could the water-borne spread of parasitic protists be prevented? *(Sample answer: treatment of water supplies to kill the parasites and asking people who have intestinal illnesses not to use these waters.)*

Ask What do you think might be some obstacles to these methods of prevention? *(Sample answer:* Cryptosporidium *is resistant to chlorine treatment. Using iodine or boiling water can be difficult or sometimes impossible if iodine or a means of boiling water is unavailable.)*

COMPARE AND CONTRAST Ask students to consider the similarities and differences among the diseases described on this page.

Ask What do African sleeping sickness, malaria, and the intestinal diseases described here have in common? *(Sample answer: All are caused by parasitic protists.)* How do they differ? *(Sample answer: African sleeping sickness and malaria are transmitted by insects. The intestinal diseases are spread by water.)*

KEY QUESTION Read aloud the Key Question and answer.

Ask What's the difference between a parasitic relationship and a mutualistic relationship? *(In a parasitic relationship, one organism is harmed. Both organisms benefit from a mutualistic relationship.)*

Speed Bump

Ask What do African sleeping sickness and malaria have in common? *(They are both caused by protists and spread by insects.)*

Water-Borne Protist Parasites
Water supplies contaminated by animal or human wastes can spread protist parasites, causing serious and sometimes deadly outbreaks of intestinal disease.

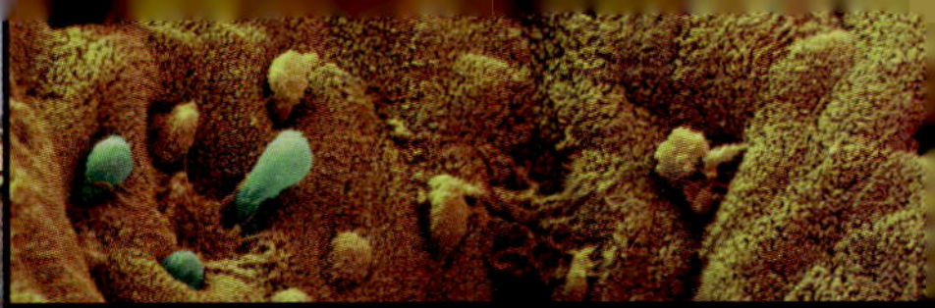

▲ *Entamoeba* causes a disease known as amebic dysentery. The amoebas live in the intestines, where they absorb food from the host. They also attack the wall of the intestine itself, shown above, destroying parts of it and causing severe bleeding. (SEM 2500×)

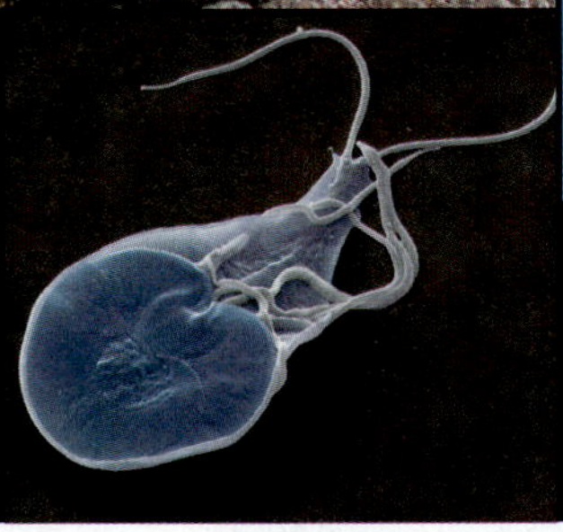

▶ The flagellated protist *Giardia* causes severe diarrhea and digestive system problems. Even crystal-clear streams may be contaminated with *Giardia*. It can be killed only by boiling water thoroughly or by adding iodine to the water. (SEM 1500×)

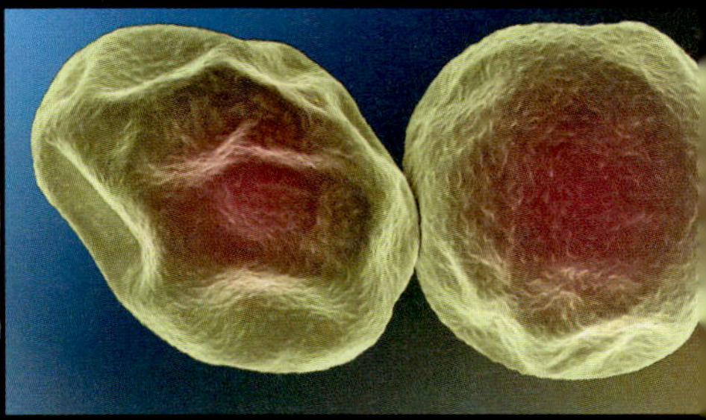

▲ *Cryptosporidium* is resistant to the chlorine compounds often used to sanitize drinking water. Therefore, it poses a special threat to public water systems. (SEM 16,000×)

Parasites and Disease Parasitic protists cause some of the world's deadliest diseases. In humans, they cause African sleeping sickness, malaria, and several intestinal diseases.

▶ ***African Sleeping Sickness*** Protists in the genus *Trypanosoma* cause African sleeping sickness. These protists destroy blood cells. They also infect other tissues, including nerve cells. Damage to the nervous system makes some people fall into a deep sleep, or even die. Trypanosomes are spread from person to person by the bite of the tsetse fly.

▶ ***Malaria*** Malaria is one of the world's most serious infectious diseases. More than 1 million people die from malaria every year. Malaria is caused by *Plasmodium*. This spore-forming protist is carried by the *Anopheles* mosquito. The art on the next page shows the life cycle of *Plasmodium*.

Key Question What types of symbiotic relationships involve protists? **Many protists are involved in mutualistic symbiosis, in which both they and their hosts benefit. Parasitic protists cause some of the world's deadliest diseases.**

512 Chapter 21 • Lesson 3

Focus on Academic Language

• infect • contaminated • threat • outbreak

Students may be unfamiliar with some or all of these words. Invite students to work in groups of four. Ask them to find these words in the text and to work together to tell what they mean. Discuss the meanings as a class, and ask students to use each word in a new sentence.

Point out that the words *infect* and *contaminate* are related.

Say Protists infect their host but contaminate a water supply.

Explain that, in this context, *infect* means "to invade," and *contaminate* means "to soil" or "to make unfit for use."

Ask What is the difference between a threat and an outbreak? *(A threat is something that might happen; an outbreak has already happened.)*

Ask What is the difference between the words *threat* and *threaten*? *(*Threat *is a noun;* to threaten *is a verb meaning "to make a threat.")*

Plasmodium Life Cycle *Plasmodium,* which causes malaria, needs two hosts to complete its life cycle: an *Anopheles* mosquito and a human.

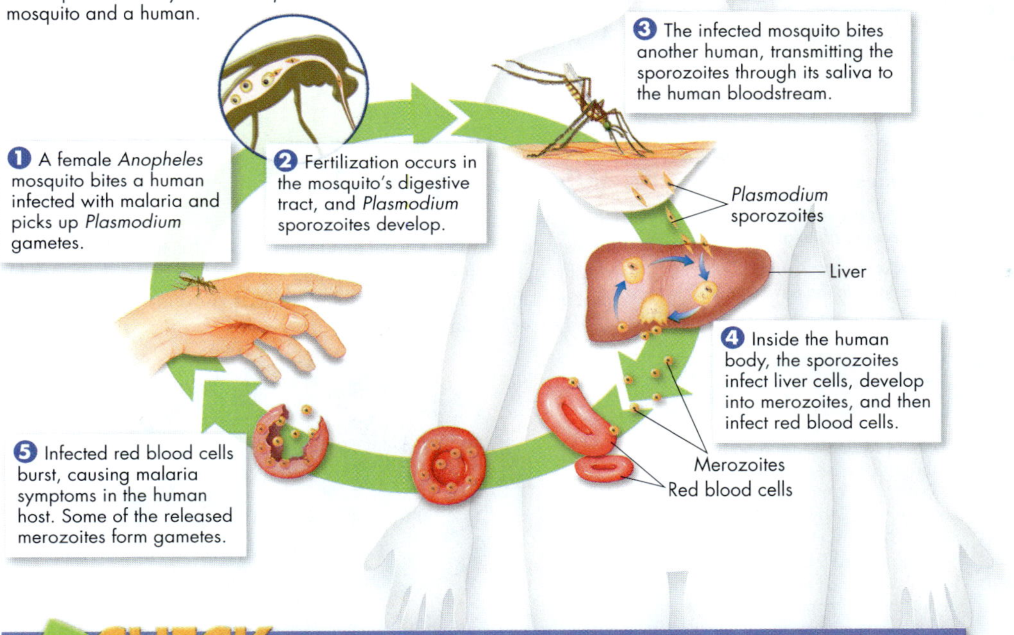

CHECK Understanding

Apply Vocabulary

Use the highlighted words from the lesson to complete each sentence correctly.

1. During the feeding stage, slime molds form a(n) ________.

2. During a(n) ________, the population of algae increases greatly.

3. A ciliate has a small dent on one side called a(n) ________, through which food enters the cell.

Critical Thinking

4. Explain What is the role of autotrophic protists in ocean ecosystems?

5. Compare and Contrast Compare how amoebas, ciliates, and water molds get food.

6. Explain Give one example of mutualism and one example of parasitism by protists.

7. Write to Learn Answer the second clue of the chapter mystery. Explain how the blight organism gets nutrients. Include the words *symbiosis* and *parasite.*

MYSTERY CLUE

The organism that caused the potato blight absorbs nutrients through its cell walls. What organism described in this lesson gets nutrients in a similar way? (**Hint:** See p. 511.)

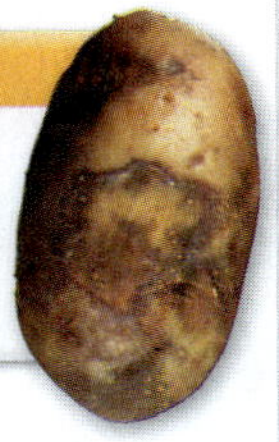

BIOLOGY.com Search Lesson 21.3 GO • Lesson Assessment

Check Understanding

1. plasmodium

2. algal bloom

3. gullet

4. They form the base of the food chains.

5. Amoebas surround food particles, forming food vacuoles in their cytoplasm. Ciliates take in food through their gullets. Water molds absorb nutrients from their environment.

6. mutualism: *Trichonympha* living in the guts of termites; parasitism: *Plasmodium,* the cause of malaria

7. Students should recognize that the organism functions similarly to a water mold. They should accurately describe the symbioses of mutualism and parasitism. Some students will accurately relate that the blight organism is a parasite.

Active Reading

USE VISUALS Have students review the life cycle of *Plasmodium.* Tell them that two hosts are required for this life cycle.

Ask What two hosts does *Plasmodium* need? *(humans and mosquitoes)*

Ask Why does *Plasmodium* need humans? *(in order to produce gametes, which form in the human bloodstream)*

Ask What would happen to *Plasmodium* if there were no mosquitoes? *(Their gametes would not be able to meet, so fertilization could not occur.)*

INFER Ask students to think about why so many people die from malaria each year.

Say In many parts of the world, it is difficult to reduce the number of cases of malaria. Why do you think this is so? *(Sample answers: Mosquitoes are very common. People do not know when they are infected. It is hard to avoid being bitten.)*

Ask How might officials slow the spread of malaria in an area? *(Sample answer: reducing the number of mosquitoes or instances of mosquito bites)*

Post Reading

ASSESSMENT TIPS Review the highlighted vocabulary words with students before they begin the assessment.

Encourage students who are having difficulty answering Question 5 to review the section Heterotrophic Protists. It may help them to complete a three-column chart as they review the material.

For Question 6, students should review the Symbiotic Protists section.

MYSTERY CLUE Discuss with students which organism could fit this description. If students are unsure, suggest that they reread the third page of this lesson.

Wrap-Up Activity

PROTIST INTERVIEW Tell students to work in pairs. Each pair will select two different protists (one autotrophic, one heterotrophic) from this lesson. Ask students to prepare a written profile containing their protist's name, "diet" (i.e., source of energy), location, and impact on the environment. When both partners are finished with their profile, have them interview each other to learn the details of each profile.

Pre-Reading

Objectives

21.4.1 Identify the defining characteristics of fungi.

21.4.2 Describe how fungi affect homeostasis.

Preview the Pages

1. Remind students that fungi are in their own kingdom and that the organisms in this lesson will be very different from the protists in the previous three lessons.
2. Review with students the vocabulary words, pronouncing each word aloud. Have students find the words in the text.
3. Preview the visuals and read the captions and labels. Point out that the color code in the Bread Mold Life Cycle is the same as the codes used in life cycle diagrams in the previous lessons.

Activate Prior Knowledge

Ask students to name all the different kinds of fungi they can think of. *(Sample answers: molds, yeasts, morels, mushrooms, mildew)* Ask them how fungi affect their lives. *(Sample answers: A fungus causes athlete's foot. Yeasts and mushrooms are used in foods. Molds can cause foods to spoil.)*

Build Understanding

CONCEPT MAP Direct students to Workbook B, where they will find a concept map to complete as they read through this lesson. Tell students that the concept map will help them understand and remember the information in this lesson.

NATIONAL SCIENCE EDUCATION STANDARDS

UNIFYING CONCEPTS AND PROCESSES

I, IV, V

CONTENT

C.3.a, C.4.b, C.4.c, C.5.d

For the corresponding lesson in *Miller & Levine Biology,* see pages 618–625.

21.4 Fungi

Key Questions

What are the basic characteristics of fungi?

How do fungi affect homeostasis in the environment?

BUILD Understanding

Concept Map As you read the lesson, use the new terms you learn to make a concept map. In your concept map, show how fungi affect other organisms in the environment.

In Your Workbook Go to your workbook to learn more about concept maps. Complete the concept map for Lesson 21.4.

What Are Fungi?

Imagine walking through the woods. Around a tree, you see a ring of many mushrooms. Which organism is bigger, the tree or the mushrooms? You may be surprised to learn that all the mushrooms are part of a single fungus. Most of this fungus is growing underground. It may be bigger than the tree!

Because many fungi grow from the ground, scientists once classified them as plants. But instead of carrying out photosynthesis, fungi are heterotrophs. They produce enzymes that digest food outside their bodies. Then their bodies absorb the resulting nutrients. Many fungi absorb nutrients from decomposing matter in soil. Others live as parasites, absorbing nutrients from their hosts.

Fungi are eukaryotes. They have cells walls that contain chitin (KY tun). **Chitin** is a complex carbohydrate that is also found in the skeletons of insects. The presence of chitin shows that fungi are more closely related to animals than to plants. (The cell walls of plants contain cellulose, not chitin.)

Key Question What are the basic characteristics of fungi?

Fungi are heterotrophic eukaryotes. They have chitin in their cell walls.

Scarlet Cup Fungus

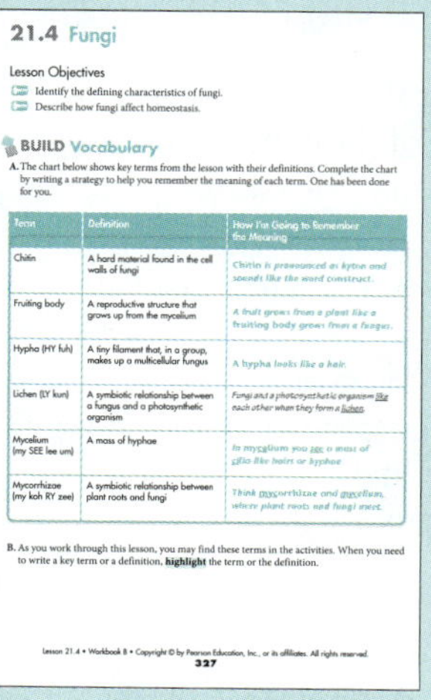

21.4 Fungi

Lesson Objectives

Identify the defining characteristics of fungi.

Describe how fungi affect homeostasis.

BUILD Vocabulary

A. The chart below shows key terms from the lesson with their definitions. Complete the chart by writing a strategy to help you remember the meaning of each term. One has been done for you.

Term	Definition	How I'm Going to Remember the Meaning
Chitin	A hard material found in the cell walls of fungi	Chitin is pronounced as kyton and sounds like the word construct.
Fruiting body	A reproductive structure that grows up from the mycelium	A fruit grows from a plant like a fruiting body grows from a fungus.
Hypha (HY fuh)	A tiny filament that, in a group, makes up a multicellular fungus	A hypha looks like a hair.
Lichen (LY kun)	A symbiotic relationship between a fungus and a photosynthetic organism	Fungi and a photosynthetic organism like each other when they form a lichen.
Mycelium (my SEE lee um)	A mass of hyphae	In mycelium you see a mass of cilia-like hairs or hyphae
Mycorrhizae (my koh RY zee)	A symbiotic relationship between plant roots and fungi	Think mycorrhizae and mycelium, where plant roots and fungi meet.

B. As you work through this lesson, you may find these terms in the activities. When you need to write a key term or a definition, **highlight** the term or the definition.

Lesson 21.4 • Workbook B • Copyright © by Pearson Education, Inc., or its affiliates. All rights reserved.

327

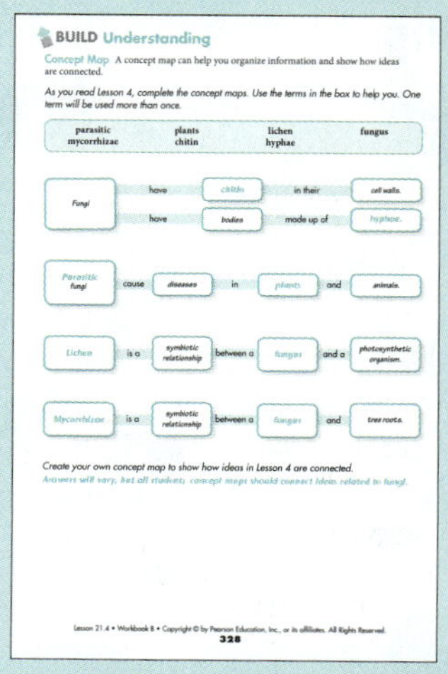

BUILD Understanding

Concept Map A concept map can help you organize information and show how ideas are connected.

As you read Lesson 4, complete the concept maps. Use the terms in the box to help you. One term will be used more than once.

parasitic mycorrhizae plants chitin lichen hyphae fungus

Create your own concept map to show how ideas in Lesson 4 are connected.

Answers will vary, but all students' concept maps should connect ideas related to fungi.

Lesson 21.4 • Workbook B • Copyright © by Pearson Education, Inc., or its affiliates. All Rights Reserved.

328

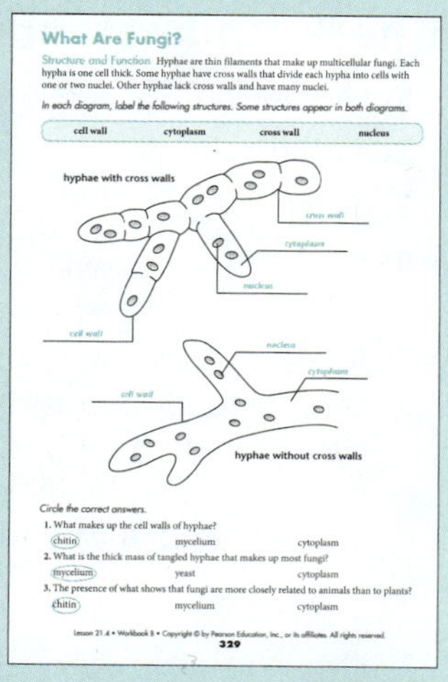

What Are Fungi?

Structure and Function Hyphae are thin filaments that make up multicellular fungi. Each hypha is one cell thick. Some hyphae have cross walls that divide each hypha into cells with one or two nuclei. Other hyphae lack cross walls and have many nuclei.

In each diagram, label the following structures. Some structures appear in both diagrams.

cell wall cytoplasm cross wall nucleus

Circle the correct answers.

1. What makes up the cell walls of hyphae?
   chitin mycelium cytoplasm
2. What is the thick mass of tangled hyphae that makes up most fungi?
   mycelium yeast cytoplasm
3. The presence of what shows that fungi are more closely related to animals than to plants?
   chitin mycelium cytoplasm

Lesson 21.4 • Workbook B • Copyright © by Pearson Education, Inc., or its affiliates. All rights reserved.

329

Structure and Function Fungi may have one cell or many. Yeasts are tiny fungi that live most of their lives as single cells.

Mushrooms and other fungi grow much larger. Their bodies are made up of cells that form long, slender branching filaments called **hyphae** (HY fee; singular: hypha). In most fungi, cross walls divide the hyphae into compartments resembling cells. Each compartment has one or two nuclei. Cytoplasm and organelles can move through openings in the cross walls.

What we call a mushroom is actually the **fruiting body,** or reproductive structure of the fungus. It grows from the **mycelium** (my SEE lee um; plural: mycelia), a mass of branching hyphae below the soil. The structure of a fungus is shown in the art below. The mycelia of some fungi live for many years and grow very large. Clusters and rings of mushrooms are often part of the same mycelium. That means they are all part of one organism.

BUILD Vocabulary

chitin
a complex carbohydrate that makes up the cell walls of fungi; also found in the exoskeletons of insects

hypha (pl. hyphae)
one of the many long, slender filaments that make up the body of a fungus

fruiting body
the reproductive structure of a fungus that grows from the mycelium

mycelium
a densely branched network of the hyphae of a fungus

WORD ORIGINS
The word *hypha* comes from a Greek word that means "web." If you look closely at the hyphae in a mushroom, they will look much like a tangled web.

Structure of a Mushroom The fruiting body of a mushroom is used for reproduction. The main part of the organism is the mycelium, which grows underground. The mycelium is made up of hyphae.

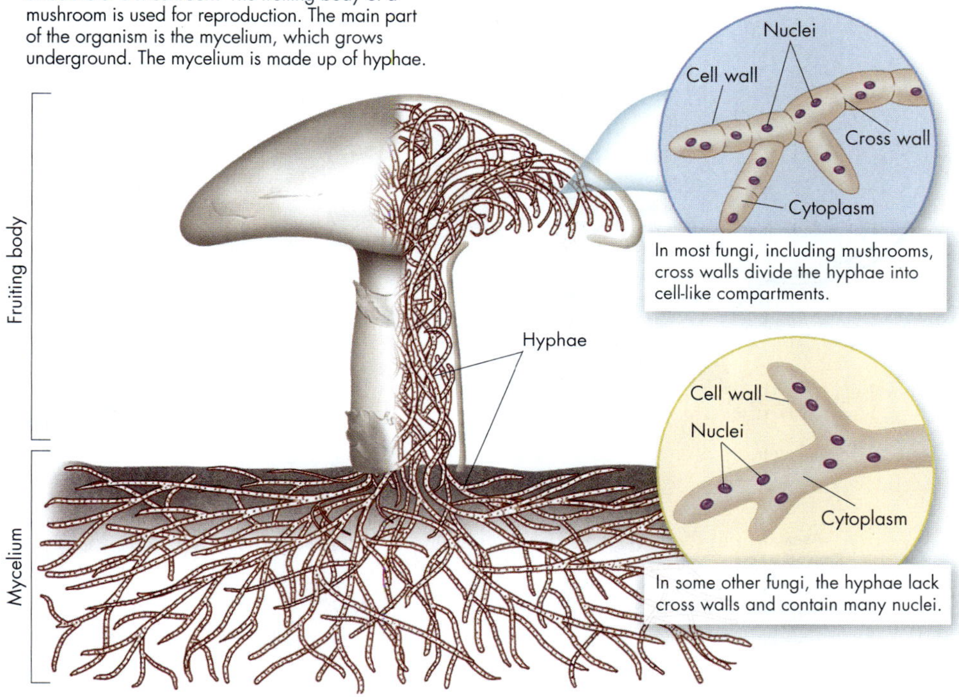

In most fungi, including mushrooms, cross walls divide the hyphae into cell-like compartments.

In some other fungi, the hyphae lack cross walls and contain many nuclei.

Hands-On Learning

COMPARE/CONTRAST DIGESTIVE SYSTEMS Many students have trouble understanding the idea that fungi digest foods *outside* their bodies. Ask each student to get out a sheet of notebook paper. Tell students that the outer surface of the paper represents an organism's digestive surface. Then have students roll the paper to form a tube. Explain that this tube represents a human's digestive system. Humans release digestive enzymes inside the digestive system, where nutrients are broken down and absorbed internally. Now have students unroll the paper. Tell them that the surface of the paper represents the surface of a fungus. Fungi produce enzymes that digest foods *outside* their bodies. Then, the nutrients from the food are absorbed.

EXTENSION Have students recall that as cell volume increases, the ratio of surface area to volume decreases. Point out that the structure of the mycelium provides a large ratio of surface area to volume for the organism's cells, or hyphae. This large ratio helps with the structure's function—to absorb nutrients from decaying matter.

During Reading

Active Reading

KEY QUESTION Read aloud the Key Question and answer on the previous page. Have students read the paragraph preceding the Key Question and pronounce and define *chitin.*

USE GRAPHIC ORGANIZERS Have students create a three-column KWL chart and complete the first two columns, listing what they ***know*** about fungi and what they ***want*** to know. Tell students that they will finish the chart at the end of the lesson.

INTREPRETING DIAGRAMS Refer students to the mushroom diagram. Have them examine the figure and read the caption and labels.

Ask Which part of the mushroom can be seen above the ground? *(the fruiting body, or reproductive structure)*

Ask What structures make up both the fruiting body and the mycelium? *(hyphae)*

Ask What do the circles on the right show? *(magnified views of the hyphae)*

Build Vocabulary

WORD ORIGINS Ask students to find examples of the singular and plural forms of *hypha* and *mycelium* in the text on this page. Have students become familiar with these terms by using each form in a new sentence.

Ask How are hyphae like a web? *(Sample answer: They are long and thin and tangled together like the threads of a spider web.)*

Speed Bump

Ask You are studying cells of an unknown organism. How could you find out if these cells are fungal cells? *(If the cell walls contain chitin, they are fungal cells.)*

Ask Which part of a mushroom is the reproductive structure? *(the fruiting body)* Which part of a mushroom absorbs nutrients? *(hyphae)*

During Reading continued

Active Reading

INTERPRET DIAGRAMS Refer students to the Bread Mold Life Cycle diagram. Be sure they understand that bread mold is a kind of fungus. Then have them compare this life cycle with that of the water mold shown in Lesson 21.2.

Say This diagram uses the same colors as the Water Mold Life Cycle diagram. Orange shows the haploid part of the life cycle. Purple shows the diploid part of the life cycle.

Ask Which part of the bread mold life cycle, haploid or diploid, happens most of the time? *(haploid)*

Say Most fungi reproduce both asexually and sexually. Notice that this diagram has two different circles, one for asexual reproduction and one for sexual reproduction.

Ask What is an advantage of asexual reproduction? *(It lets an organism reproduce quickly.)*

Ask How can sexual reproduction help a population of bread mold? *(It results in genetic variation, which can help a population adapt to changes in the environment.)*

SUMMARIZE Ask students to summarize fungal sexual reproduction. Pair each student with a partner.

Say Sometimes it is easier to understand information if you break it down into steps. Work with your partner to break down sexual reproduction in fungi into a series of steps.

Speed Bump

Ask What are three ways in which fungi can reproduce asexually? *(releasing spores, breaking off a hypha, or budding off a cell)*

Reproduction Fungi can reproduce asexually. They do so mainly by releasing spores that can travel through air and water. Other kinds of asexual reproduction include breaking off a hypha or budding off a cell.

Most fungi also reproduce sexually. Sexual reproduction often involves two mating types. They are not called male and female. Instead, they are called *plus* (+) and *minus* (−). When hyphae of opposite mating types meet, they fuse. The + and − nuclei are brought together in the same cell. The nuclei form pairs that divide together as the mycelium grows.

Many of the paired nuclei fuse to form diploid zygotes. Zygotes form within a structure called a zygospore. The zygospore germinates, producing a sporangium. The sporangium releases haploid spores that form by meiosis. Each spore has a different combination of parental genes. Each can grow into a new mycelium.

For more on the diversity of fungi, go to the Visual Guide. DOL 16–19

Diversity of Fungi More than 100,000 species of fungi have been identified. Biologists have used similarities and differences, along with DNA comparisons, to place the fungi into several groups. A major way in which these groups differ is their reproductive structures.

Bread Mold Life Cycle The bread mold *Rhizopus stolonifer* undergoes both asexual and sexual reproduction.

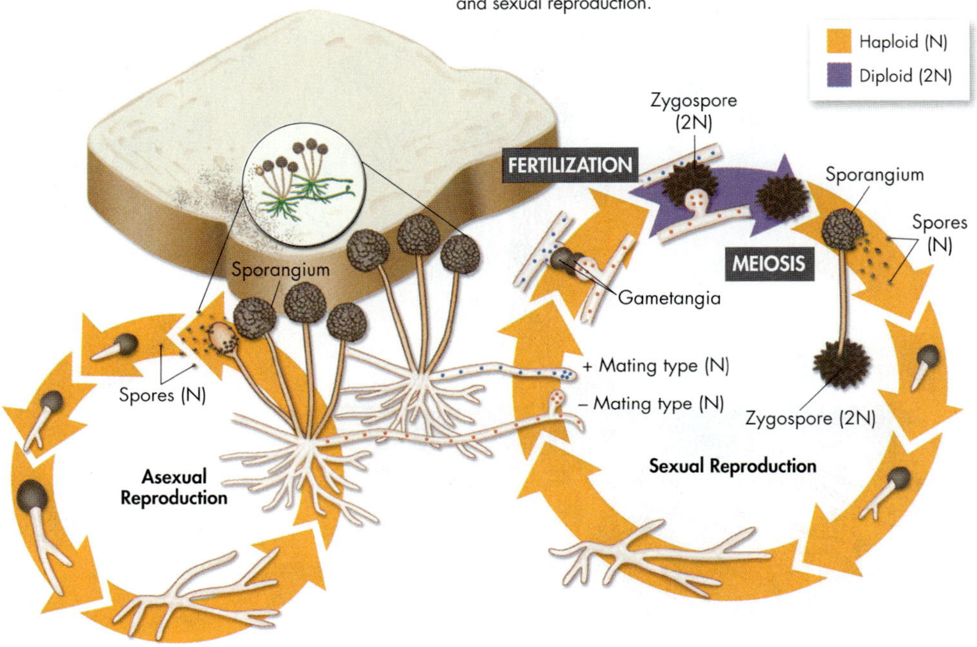

ELL Focus on ELL: Extend Language

BEGINNING AND INTERMEDIATE SPEAKERS Help students understand that the word *mold* has more than one meaning. Students have read about water molds and slime molds, which are types of protists. On this page, they are reading about bread mold. In each of these cases, *mold* refers to the form of an organism (protist or fungus) that grows in a moist environment.

Using clay and clay molds, demonstrate another meaning of the noun *mold*: a vessel or cavity used to shape soft substances. (Students may have used molds when studying about fossils, for example.) You can also shape clay with your hands into a ball or bowl, explaining that the movement of your fingers (e.g., the kneading) is called molding.

Challenge more advanced students to use a dictionary to find other meanings of *mold,* beyond those used here. Examples include a teacher or parent molding a child, that is, influencing them. Have them create a sentence for each meaning.

The Major Phyla of Fungi

Phylum		Distinguishing Features	Examples
Basidiomycota (club fungi)		Sexual spores found in club-shaped cell called a basidium	Mushrooms, puffballs, earthstars, shelf fungi, jelly fungi, rusts
Ascomycota (sac fungi)		Sexual spores found in saclike structure called an ascus	Morels, truffles, *Penicillium* species, yeasts
Zygomycota (common molds)		Tough zygospore produced during sexual reproduction that can stay dormant for long periods	*Rhizopus stolonifer* (black bread mold), molds found on rotting strawberries and other soft fruits, mycorrhizae associated with plant roots
Chytridomycota (chytrids)		Only fungi with flagellated spores	Many species are decomposers found in lakes and moist soil.

Phyla of Fungi This table summarizes the main differences among four phyla of fungi.

The Ecology of Fungi

Fungi play a major role in nearly every ecosystem. Some fungi that are parasites cause diseases in plants and animals.

Decomposition Many fungi feed by releasing digestive enzymes. These enzymes break down leaves, fruit, and other organic material into simple molecules. The molecules then diffuse into the fungus. Enzymes from mycelia of mushrooms speed up the decomposition of wastes and dead organisms.

Fungi are champions of decomposition. Fungi help recycle essential elements and nutrients. Many plants remove important elements and nutrients from the soil. If these materials were not returned to the soil, other organisms could not use them. Without these nutrients, organisms would die. Eventually, Earth would become barren and lifeless.

Key Question How do fungi affect homeostasis in the environment? **Many fungi are decomposers that help ecosystems maintain homeostasis by breaking down dead organisms and recycling essential elements and nutrients.**

Parasitism Many fungi are helpful to other organisms. But parasitic fungi cause many serious diseases.

▶ ***Plant Diseases*** Several parasitic fungi cause diseases that threaten food crops. Corn smut, for example, destroys corn kernels. Wheat rust affects North American wheat crops. Some mildews, which infect many different plants, are also fungi. In temperate areas, fungi destroy about 15 percent of crops. In tropical areas, a greater percentage is lost to fungal diseases.

Parasitic Fungi Some fungi are parasites.

Corn smut infests the kernels of a corn plant, reducing the farmer's crop yield.

A grasshopper is infected by a *Cordyceps* fungus.

Focus on Academic Language

• affect • infect • infection • temperate

Write the words on the board. Pair students and have them find the words on this page and the next. Have them predict each word's meaning using context clues and explain their reasoning in their notebooks.

Have students pronounce *affect* and *infect* several times. Be sure they enunciate so that they can hear the difference. Ask volunteers to use each word in a sentence.

Tell students that *temperate* means "mild, restrained, or moderate." Ask students if they can think of another word beginning with *temp-*. *(temperature)* Ask a volunteer to describe a temperate climate and to give an example.

Active Reading

INTERPRET TABLES Ask students to read the table.

Ask What is the main characteristic used to classify fungi into major phyla? *(how their reproductive spores differ)*

Ask Which phylum contains yeasts? *(Ascomycota)*

Ask What distinguishes Chytridomycota from the other three phyla? *(Its spores have flagella.)*

RELATING WORD FORMS Write the words *parasite, parasitic,* and *parasitism* on the board. Ask students to find examples of each word in the text.

Ask Which part of speech is each word? *(parasite—noun; parasitic—adjective; parasitism—noun)*

Ask What is a parasitic fungus? *(a fungus that is also a parasite)*

KEY QUESTION Read aloud the Key Question and answer.

Ask What is homeostasis? *(Homeostasis describes the state in which conditions within a cell are kept relatively constant.)*

Ask What do you think would happen to the environment if all fungi disappeared? *(Sample answer: Organic materials would no longer be broken down. Essential elements and nutrients would no longer be returned to the soil, so that other organisms could use them. Eventually, plants and other organisms would vanish from Earth.)*

READ, COVER, REMEMBER, RETELL Have students read silently the paragraph Plant Diseases. When they have finished reading, ask them to cover the text with their hand and think about what they read. Then have them retell the key information from the text without looking at their books.

Ask What are two food crops that are affected by diseases caused by parasitic fungi? *(Sample answer: corn and wheat)*

MAKE INFERENCES Tell students that corn smut is named for the sooty, black masses of spores on infected ears of corn.

Ask How do you think wheat rust got its name? *(It causes rust-colored spots on infected plants.)*

Challenge students to think about the relationship between plant diseases and climate.

Say Fungal diseases of plants are more common in tropical areas than in temperate areas.

Ask Why do you think this is so? *(Fungi grow best in moist areas.)*

Speed Bump

Ask How does the releasing of digestive enzymes help many fungi get food? *(The enzymes break down leaves, fruit, and other organic material into simple molecules that diffuse into the fungus.)*

During Reading continued

Active Reading

KEY QUESTION Read aloud the Key Question and answer.

Ask What are some harmful effects of parasitic fungi? *(Sample answer: Parasitic fungi destroy crops.)*

Build Vocabulary

RELATED WORD FORMS Explain that, like *symbiosis, mutualism* is a noun.

Ask How would you change the word *mutualism* into an adjective? *(mutualistic)*

DRAW CONCLUSIONS Invite a volunteer to read the first two paragraphs under Lichens. Then write the following sentence frames on the board: *In a lichen, the photosynthetic organism provides the fungus with* (food). *The fungus provides the photosynthetic organism with* (water and minerals).

Give students a minute to look at both sentences before asking for responses.

Say In a lichen, the two organisms can be described as partners.

Ask How are lichens an example of a mutualistic relationship? *(The photosynthetic organism and the fungus each benefit from the relationship; neither organism is harmed.)*

MAKE INFERENCES Explain the significance of lichens in ecological succession. Ask students to review the diagram Inside a Lichen.

Say Lichens are often the first living things to appear when new organisms begin to grow on bare rock. They break down rocks, helping to make soil. The fungus part of a lichen breaks down rock by sending its hyphae into cracks in a rock, eventually wedging the rock apart. The fungus chemically wears down the rock by producing acids.

Ask Why do you think lichens grow in places where few other organisms can survive? *(Even a barren environment provides what lichens need to survive. The fungus gets its energy from the algae or cyanobacteria, which in turn gets water and minerals from the fungus.)*

Speed Bump

Ask Why is the relationship between the *Cordyceps* fungus and grasshoppers considered parasitic? *(because the grasshoppers are harmed while the fungus benefits)*

Ask How do green algae and cyanobacteria get energy? *(through photosynthesis)*

BUILD Vocabulary

lichen
a symbiotic relationship between a fungus and a photosynthetic organism

mycorrhiza
a symbiotic association of plant roots and fungi

RELATED WORD FORMS

Symbiosis is a noun that describes two unlike organisms living closely together. *Symbiotic* is an adjective referring to a symbiosis.

▶ ***Animal Diseases*** Fungal diseases also affect animals, including insects, frogs, and mammals. One deadly example is caused by a fungus in the genus *Cordyceps*. This fungus infects grasshoppers in rain forests in Costa Rica. Microscopic spores become lodged in the grasshopper. The spores germinate and produce enzymes. These enzymes slowly eat through the insect's tough outer skeleton. The spores multiply in the insect's body, digesting cells and tissues. Eventually, the insect dies. Hyphae develop, covering the insect skeleton. Reproductive structures emerge from the grasshopper's remains. They produce more spores and spread the infection.

Key Question How do parasitic fungi affect other organisms? **Parasitic fungi can cause serious diseases in plants and animals.**

▶ ***Human Diseases*** Parasitic fungi also infect humans. For example, athlete's foot is caused by a fungus. Its mycelium grows in the outer layers of skin, causing sores. Spores from these sores can easily spread to other people.

The yeast *Candida albicans* also causes human diseases. It usually infects the mouth or vagina. *Candida* is usually kept in check by competition from bacteria and by the body's immune system. This balance can be upset by the use of antibiotics, which kill bacteria, or by damage to the immune system.

Lichens Some fungi form mutualistic relationships with photosynthetic organisms. Both partners benefit from these relationships. For example, a **lichen** (LY kun) is a mutualistic relationship between a fungus and a photosynthetic organism. The photosynthetic organism may be a green alga, or a photosynthetic bacterium called cyanobacteria. The green algae or cyanobacteria carry out photosynthesis. This provides the fungus with food. The fungus provides the green algae or cyanobacteria with water and minerals. The fungal hyphae protect the delicate green cells from bright sunlight.

Inside a Lichen The protective upper surface of a lichen is made up of densely packed fungal hyphae. Below this are layers of green algae or photosynthetic bacteria and loosely packed hyphae. The bottom layer attaches the lichen to a rock or tree.

Hands-On Learning

EXAMINE SPORES Explain that many fungi are able to reproduce quickly by producing large numbers of spores. Then help students observe the location and appearance of spores from a club fungus. Working in small groups, have students place a fresh mushroom cap, gill-side down, on a sheet of white paper and cover it with a plastic container. After one or two days, students will observe that spores have fallen to the paper in a characteristic pattern. Help students examine the spores under a microscope.

Lichens can grow in places where few other organisms can survive. For example, they grow on rocks in deserts and on the tops of mountains. They are often the first organisms to enter barren environments. They gradually break down the rocks on which they grow. In this way, lichens help in the early stages of soil formation.

Lichens are very sensitive to air pollution. They are among the first organisms affected when air quality gets worse.

Mycorrhizae Fungi also form mutualistic relationships with plant roots. These relationships are called **mycorrhizae** (my koh RY zee; singular: mycorrhiza). Hyphae of the fungi form a large web connecting the roots and the soil. The hyphae bring water and minerals to the roots. The fungi also make enzymes that free nutrients in the soil. The plants, in turn, give the fungi food from photosynthesis.

The partnership between plant and fungus does not end with a single plant. Mycorrhizae networks connect the roots of many plants. Some networks even connect plants of different species!

Many plants need mycorrhizae to grow. The seeds of orchids, for example, will not grow without mycorrhizal fungi. Many trees cannot live without mycorrhizae. Between 80 and 90 percent of all plant species may form mycorrhizae.

Mycorrhizae Research shows that mycorrhizae on Douglas fir roots help the trees get nutrients from other trees, including trees of other species.

Key Question How do fungi affect homeostasis in other organisms? **Some fungi form mutualistic relationships with photosynthetic organisms that help both partners.**

CHECK Understanding

Apply Vocabulary

Use the highlighted words from the lesson to complete each sentence correctly.

1. The ________ is the reproductive structure of a fungus.

2. Fungal cells that form long, slender filaments are called ________.

3. Fungal cell walls contain ________, a complex carbohydrate.

4. A ________ is the symbiotic association of plant roots and fungi.

Critical Thinking

5. Apply Concepts What characteristics do all fungi have in common?

6. Infer Tissue from several mushrooms found near the base of a tree were tested. They were found to be genetically identical. How would you explain this?

7. Explain Describe two kinds of symbiotic relationships that fungi form.

8. Apply Concepts Summarize two ways in which fungi maintain the homeostasis of a forest ecosystem.

9. Write to Learn Answer the third mystery clue. In your answer, use the words *chitin* and *hyphae*.

MYSTERY CLUE

Using a microscope, scientists found that the infected tissue from blighted potatoes had a network of hyphae. The hyphae cell walls did not contain chitin. Could the blight be caused by a fungus? (Hint: See p. 514.)

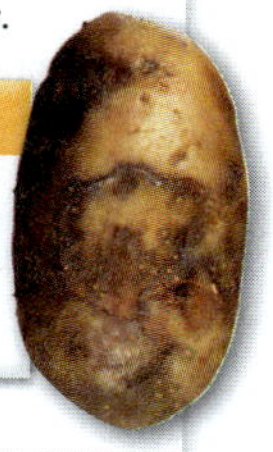

BIOLOGY.com Search Lesson 21.4 GO • Lesson Assessment

Check Understanding

1. fruiting body

2. hyphae

3. chitin

4. mycorrhiza

5. Fungi are heterotrophic eukaryotes that have chitin in their cell walls. They digest food outside their bodies and then absorb the nutrients. All can reproduce asexually.

6. The mushrooms are all growing from the same mycelium.

7. Parasitic fungi cause diseases such as athlete's foot in humans. Mutualistic fungi form relationships with photosynthetic organisms such as those found in lichens.

8. Fungi decompose dead organisms and wastes into products that can be used by other organisms. Fungi may also form mycorrhizae with the roots of plants. They provide the plants with nutrients and water.

9. Sample answer: The lack of chitin indicates that the organism is not a fungus. Possibly other types of organisms have structures like fungal hyphae.

Active Reading

COMPARE AND CONTRAST Explain that mycorrhizae are another example of a mutualistic relationship.

Say The word *mycorrhizae* is formed from two words. *Myco-* means "fungus"; *rhizae* means "roots."

Ask How do the plant roots in mycorrhizae help the fungi? *(They give food from photosynthesis to the fungi.)* How do the fungi help the plant roots? *(They help the plant take in water and minerals.)*

Ask How does the relationship between the fungus *Cordyceps* and grasshoppers differ from the relationship between mycorrhizae and a Douglas fir? *(The relationship between the* Cordyceps *and grasshoppers is a parasitic relationship. The relationship between mycorrhizae and a Douglas fir is a mutualistic relationship.)*

KEY QUESTION Read aloud the Key Question and answer. Help students understand how a mutualistic relationship can affect an organism's homeostasis.

USE GRAPHIC ORGANIZERS Have students return to the KWL chart they created at the beginning of the lesson. In the third column of the chart, ask students to add notes about what they have learned about fungi.

Ask Did you learn anything about fungi that contradicted something you thought you knew at the beginning of the lesson? *(Answers may vary.)*

Post Reading

ASSESSMENT TIPS If students have trouble answering Question 6, have them review the diagram of a mushroom's structure. Point out that most of the body of a fungus is underground. Make a sketch of a tree and mushrooms as described in the question. Draw the underground network of hyphae connecting the mushrooms.

For Question 8, suggest students create a graphic organizer. Then they can use the graphic organizer to answer the question in complete sentences.

MYSTERY CLUE Discuss with students the differences between protist cell structure and fungal cell structure. Suggest students reread the first page of the lesson.

Wrap-Up Activity

FUNGAL JOBS Have students write "job descriptions" for selected examples of fungi. Each student should then prepare a brief "resume" for each fungus, showing how that particular fungus is well suited for its job. Invite students to match fungal "candidates" with jobs.

CHAPTER LAB

Pre-Lab

Review the meaning of *hypothesis (a testable idea).* Remind students about the relationship between dependent and independent variables by creating a chart like this one. Elicit examples from students to fill in the chart.

Hypothesis (question)	Independent variable *(what I change)*	Dependent variable *(what I see happen)*

Ask What is a controlled experiment? *(an experiment in which only one variable is changed)* What is a control group? *(a group that is exposed to the same conditions as the experimental group except for one independent variable)*

Review the Background Questions and Pre-Lab Questions as a class.

Lab

Tell students they will perform the chapter lab *Mushroom Farming* described in **Lab Manual B.**

SAFETY Before beginning the lab, inquire if any students have an allergy to mushrooms. Tell students that organic media sometimes contain harmful microorganisms. Students should wear goggles and gloves while performing the experiment, and they should wash their hands thoroughly after completing the lab.

Post-Lab

DESIGN A FARM Have students develop plans for a large-scale mushroom farm.

Design Your Own Lab

GUIDED INQUIRY

Pre-Lab: Mushroom Farming

Problem How does the amount of available light affect mushroom growth?

Materials mushroom-growing kits, spray bottles with water, metric rulers

Lab Manual Chapter 21 Lab

Skills Focus Form a Hypothesis, Design an Experiment, Organize Data

Connect to the Big idea Fungi play an essential role in maintaining homeostasis in ecosystems. Many fungi speed up the decay of dead organisms and help recycle nutrients. Some fungi form symbiotic relationships with plants. These fungi deliver water and minerals to the plant roots. In turn, the plants supply the fungi with sugars. Do fungi grow better in some environments than in others? In this lab, you will investigate how the amount of light affects the growth and reproduction of a species of club fungi.

Background Questions

a. Review What are fungi?

b. Review Describe the general structure of a mushroom.

c. Explain How do fungi that are not in a symbiotic relationship obtain nutrients?

Pre-Lab Questions

Preview the procedure in the lab manual.

1. Infer Where will the mushrooms get the nutrients that they need to grow and develop?

2. Relate Cause and Effect Why will you have to wait about ten days to observe the mushrooms?

3. Apply Concepts The mushrooms you will grow are of a variety that is sold in food stores. Why do the instructions warn you not to eat the mushrooms?

BIOLOGY.com Search Chapter 21 GO

Visit Chapter 21 online to test yourself on chapter content and to find activities to help you learn.

Untamed Science Video Have a look at the fascinating world of mushrooms through the lenses of the Untamed Science crew.

Art in Motion View a short animation that shows the plasmodium life cycle and how malaria is transmitted by the *Anopheles* mosquito.

Art Review Review your understanding of the different structures of a mushroom.

InterActive Art Investigate the structures of an amoeba and paramecium.

Visual Analogy Compare the way boats move to the motion of flagella and cilia in a cell.

Pre-Lab Answers

BACKGROUND QUESTIONS

a. Fungi are heterotrophic eukaryotes with cell walls that contain chitin.

b. The mycelium, which is a mass of filaments called hyphae, is underground. The part of the mushroom that is above the ground is the fruiting body, or reproductive structure.

c. Fungi release enzymes that digest waste and dead organisms. The hyphae absorb the released nutrients.

PRE-LAB QUESTIONS

1. The growing medium contains food that the mushrooms can digest.

2. Sample answer: The mushroom spores must germinate and the mycelium form before the fruiting body can grow.

3. As a general rule, food should never be consumed in the lab. Plus, a lab is not an appropriate environment for growing food.

NATIONAL SCIENCE EDUCATION STANDARDS

INQUIRY A.1.a, A.1.b, A.2.a

21 CHAPTER Summary

21.1 Protist Classification

- Protists are eukaryotes that are not members of the fungi, plant, or animal kingdoms.
- Today's protists include groups whose ancestors were among the very last to split from the ancestors of plants, animals, and fungi.

21.2 Protist Structure and Function

- Some protists move by changing their cell shape. Some use cilia or flagella. Others do not move actively but are carried by wind, water, or other organisms.
- Some protists reproduce asexually by mitosis. Others have life cycles that combine asexual and sexual forms of reproduction.

pseudopod (p. 505)
cilium (p. 506)
flagellum (p. 506)
spore (p. 506)
conjugation (p. 507)
alternation of generations (p. 507)

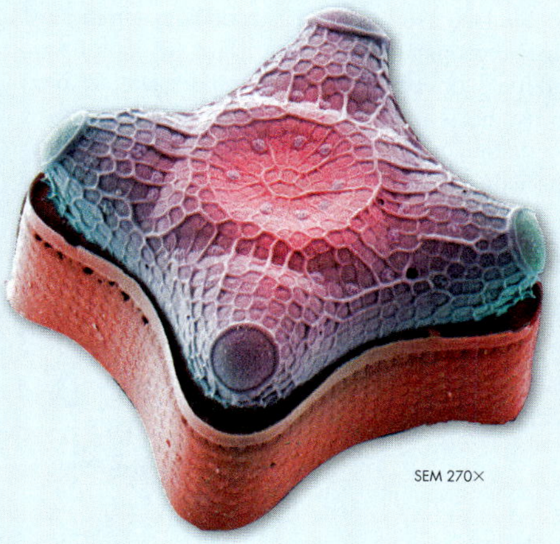
SEM 270×

21.3 The Ecology of Protists

- Photosynthetic protists form the base of food chains in the oceans and in freshwater environments.
- Some protists surround and digest their food. Other protists get food by absorbing molecules from their surroundings.
- Many protists are involved in mutualistic symbiosis, in which they and their hosts both benefit.
- These diseases include several kinds of intestinal disease, African sleeping sickness, and malaria. Parasitic protists cause some of the world's deadliest diseases.

algal bloom (p. 510)
food vacuole (p. 510)
gullet (p. 510)
plasmodium (p. 511)

21.4 Fungi

- Fungi are heterotrophic eukaryotes. They have chitin in their cell walls.
- Many fungi are decomposers that help ecosystems maintain homeostasis by breaking down dead organisms and recycling essential elements and nutrients.
- Parasitic fungi can cause serious diseases in plants and animals.
- Some fungi form mutualistic relationships with photosynthetic organisms that help both partners.

chitin (p. 514)
hypha (p. 515)
fruiting body (p. 515)
mycelium (p. 515)
lichen (p. 518)
mycorrhiza (p. 519)

BIOLOGY.com Search Chapter 21 GO • Crossword • Chapter Assessment

521

UbD Transfer the Big Idea

TRANSFER Have students work together to design a poster showing three groups of living things: autotrophic protists, heterotrophic protists, and fungi. The poster should explain how each group affects the homeostasis of other organisms and ecosystems. Students should identify at least one representative example for each group, and include descriptions of its environment and its source of nutrition. Students can draw their own illustrations or print out visuals of the organisms from the Internet.

Chapter Summary

Think Visually

USE GRAPHICS Have students make a concept map that illustrates the various ways that protists move, reproduce, and obtain food.

ANSWER
Students' concept maps should include passive movement and active movement by pseudopods, cilia, and flagella; sexual and asexual reproduction; and obtaining food through photosynthesis, capturing prey, and absorption.

BIOLOGY.com Study Online

REVIEW RESOURCES

Visual Analogy Students compare the motions of cilia and flagella to the motions of many paired oars and a single oar.

InterActive Art Students can interact with the art of an amoeba and a paramecium.

Data Analysis Students find out how protist microfossils are used to learn about conditions and changes in ancient oceans.

Art in Motion Students can watch an animated version of the *Plasmodium* life cycle.

Art Review Students can drag and drop structures to construct the features of a mushroom.

Vocabulary Review The *Flash Cards* and *Crossword* provide an interactive way to review chapter vocabulary.

Editable Worksheets Pages of Study Workbook B, Lab Manual B, and the Assessment Resources Book are available online. These documents can be easily edited using a word-processing program.

Lesson Overviews Lesson Overviews are based on Miller & Levine Biology. However, they are available on CD-ROM and can be edited for your class.

ASSESSMENT RESOURCES

Chapter Assessment Have students take an online version of the Chapter 21 Check Understanding.

Standardized Test Prep Students can take an online version of the Standardized Test Prep. You will receive their scores along with ideas for remediation.

Diagnostic and Benchmark Tests Use these tests to monitor your students' progress and supply remediation.

ASSESSMENT

Assess

Big idea Invite a volunteer to read the question. Write the word *homeostasis* on the board.

Ask Why is homeostasis important for living things? *(It enables an organism to keep a constant internal environment, even when things change outside the organism.)*

Ask How can a protist or a fungus affect an individual organism? *(Sample answer: It could produce food for the organism, or it could cause a disease that would weaken or kill the organism.)*

Ask How can a protist or a fungus affect an ecosystem? *(Sample answer: It could be an important part of the food chain for an ecosystem; it could cause a disease that might harm the ecosystem; or it could perform decomposition, recycling nutrients throughout an ecosystem.)*

CONSTRUCTED RESPONSE Paragraphs will vary. Explain the activity carefully. Be sure that students understand that the *Hints* should be used as a guide to help them answer the questions. Review how to write a paragraph.

For Question 1, write the question on the board and read it aloud. Underline *photosynthetic protists* and *aquatic environments*.

Ask What are photosynthetic protists? *(protists that use energy from sunlight to make food)*

Say This question is about photosynthetic protists that live in aquatic environments.

Ask What kind of aquatic environments is this question asking about? *(oceans)*

Ask How do other living things rely on photosynthetic organisms? *(Many organisms eat photosynthetic organisms, or they eat other organisms that eat photosynthetic organisms.)*

Read the hint aloud.

Ask What roles do photosynthetic protists play in these ecosystems? *(In coral reef ecosystems, algae provide calcium carbonate and food for coral animals. In kelp forests, giant kelp provide shelter for many species. In the open ocean, phytoplankton are eaten by fish and whales.)*

21 CHECK Understanding

Assess the Big idea: Interdependence in Nature

Write an answer to the question below.

Q: How do protists and fungi affect the homeostasis of other organisms and ecosystems?

Constructed Response

Write an answer to each of the numbered questions below. The answer to each numbered question should be one or two paragraphs long. To help get ideas, read the bulleted **Hints** below the question.

1. **Why are photosynthetic protists some of the most important organisms in aquatic environments? Give examples of three ocean ecosystems where protists play a key role.**

   Hint Ocean ecosystems include coral reefs, kelp forests, and the open ocean.

2. **How do protists and fungi recycle materials in an ecosystem?**

   Hint Some protists and many fungi are decomposers.

   Hint Fungi release digestive enzymes.

3. **How do parasites affect humans? Include specific details about at least one protist and one fungal parasite.**

   Hint Study the photos of water-borne parasites on page 512.

   Hint When parasites attack food crops, they also affect humans.

Foundations for Learning Wrap-Up

Use the double-door fold you made while reading the chapter to help review how protists and fungi affect the world.

Activity 1 Work with a partner to review the roles of protists and fungi in ecosystems. One of you will quiz the other. The person who has the double-door fold will ask the questions. For example, you could ask: How do fungi that are decomposers affect an ecosystem?

After four questions, switch sides. Continue reviewing the chapter concepts until you can easily answer the questions.

Activity 2 Cut your double-door fold in half to make two booklets. Add pages to each booklet. On these pages, draw pictures that show examples of protists and fungi for each ecological role. For example, for a protist that is an autotroph, you could draw a picture of a diatom or a giant kelp. Write a caption for each image that explains the concept shown.

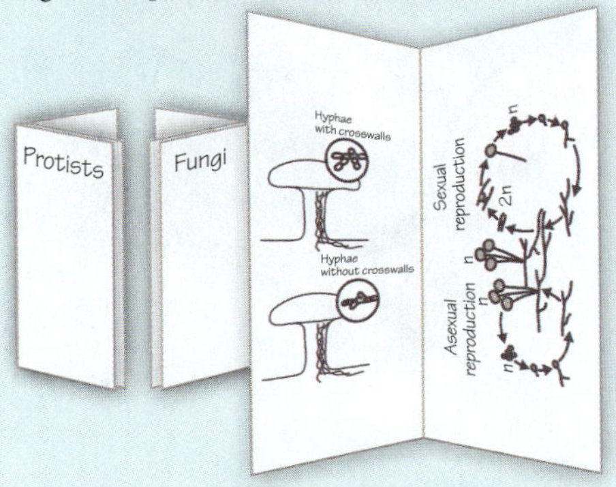

522 Check Understanding

For Question 2, pair students and have them use the hints and your modeling to answer the question. Have students share their paragraphs with the class.

For Question 3, have students write the paragraphs independently for homework or as an anchor activity.

Foundations for Learning Wrap-Up

Ask students to take out the double-door fold that they made earlier to record information about protists and fungi. Be sure that students have written logical notes below each of the headings. Assign students to work in pairs or small groups and explain the activities.

21.1 Protist Classification

Understand Key Concepts

1. Which of the following describes most protists?
   a. single-celled eukaryotes
   b. multicellular eukaryotes
   c. single-celled prokaryotes
   d. multicellular prokaryotes

Test-Taking Tip

Visualization Sometimes, it helps to picture a concept in your head or to draw it out on a scratch piece of paper. For example, in question 1, picture an organism for each answer. In this way, you can help eliminate incorrect answers and narrow your answer choices.

2. From what organisms did protists likely evolve?
3. Why is the word *protist* sometimes placed in quotation marks?

Think Critically

4. **Evaluate** Once, any organism that moved or that ate food was classified as an animal. Any organism that did not move or that made its own food was classified as a plant. Why can't this system be used to classify protists?

21.2 Protist Structure and Function

Understand Key Concepts

5. Which of the following is NOT true of amoebas?
   a. They have a rigid cell wall.
   b. They reproduce by mitosis.
   c. They move using pseudopods.
   d. The protein actin powers their movement.
6. Alternation of generations is the process of alternating between
   a. spores and eggs.
   b. diploid and haploid phases.
   c. male and female reproductive structures.
   d. positive and negative reproductive structures.
7. Compare the structure of a cilium and flagellum.
8. Is the process of conjugation in paramecia a form of reproduction? Why or why not?

Think Critically

9. **Apply Concepts** Some protists cannot move on their own. What generalization can you make about how these organisms survive?
10. **Infer** How do you think the ability to switch between asexual and sexual reproduction has aided the evolution of water molds and many other protists?

21.3 The Ecology of Protists

Understand Key Concepts

11. African sleeping sickness is caused by
   a. *Amoeba.*
   b. *Plasmodium.*
   c. *Trichonympha.*
   d. *Trypanosoma.*
12. Which of the following statements about photosynthetic protists is TRUE?
   a. Most photosynthetic protists are heterotrophs.
   b. All photosynthetic protists are closely related to plants.
   c. Giant kelp play an important role in the formation of coral reefs.
   d. Small photosynthetic organisms near the ocean's surface are called phytoplankton.
13. How do mosquitos spread malaria?
14. Describe the relationship between a termite and the protists that live in its gut.

Think Critically

15. **Apply Concepts** Holes in Earth's ozone layer may increase the amount of radiation that reaches the surface of the ocean. If this radiation affects the growth of phytoplankton, how do you think other organisms will be affected? Explain your answer.
16. **Infer** Study the diagram of the life cycle of *Plasmodium* on page 513. Do you think malaria could be transmitted by a blood transfusion? Why or why not?

Assessment Tips

Point out the word *NOT* in Question 5 and explain that students are looking for the answer choice that is false. They can begin by eliminating any answer choices that they know are true. Point out the word *TRUE* in Question 12. As a class, discuss what the question is asking.

Make sure students understand that Questions 21 and 22 both relate to the visual under Use Science Graphics.

Answers

Lesson 21.1

UNDERSTAND KEY CONCEPTS

1. a

2. Protists were the first eukaryotes. These eukaryotes evolved from prokaryotes.

3. Protists were once classified in one kingdom. However, many scientists now think protists should be classified in multiple kingdoms.

THINK CRITICALLY

4. Accept all reasonable answers. Students should recognize that some protists have a mix of these traits, so they can't be classified as either an animal or a plant. For example, some motile protists also have chloroplasts for photosynthesis.

Lesson 21.2

UNDERSTAND KEY CONCEPTS

5. a

6. b

7. Both cilia and flagella are made up of microtubules. A cilium is short and hairlike. A flagellum is long and whiplike.

8. No, because no offspring are produced.

THINK CRITICALLY

9. These organisms rely on wind, water, or other organisms to move. They may be parasites or mutualists if they rely on other organisms.

10. Protists can use asexual reproduction to make many copies of themselves very quickly when conditions are favorable. Sexual reproduction increases genetic diversity, which can aid survival when environmental conditions change.

Lesson 21.3

UNDERSTAND KEY CONCEPTS

11. d

12. d

13. A mosquito bites an infected human, picking up malaria gametes. Fertilization occurs in the mosquito's digestive tract and sporozoites form. The mosquito bites another human, spreading the sporozoites to the human. Inside the human, the sporozoites infect liver cells, develop into merozoites, and infect red blood cells. Red blood cells burst, causing disease.

14. The protists in a termite's gut break down cellulose. In turn, the protists are protected by the termite and get nutrients from the termite. This is a mutualistic relationship.

THINK CRITICALLY

15. Phytoplankton form the base of many food chains. If phytoplankton are affected positively, organisms will have more phytoplankton to eat and will produce more offspring. If phytoplankton are affected negatively, there will be fewer phytoplankton to eat.

16. Yes, if red blood cells are infected, they could be passed along by a transfusion.

ASSESSMENT

Lesson 21.4

UNDERSTAND KEY CONCEPTS

17. d

18. The mycelium is the underground structure of a fungus that is made up of hyphae.

19. Both contain a complex carbohydrate called chitin.

THINK CRITICALLY

20. Humans eat food, which is digested by enzymes in the stomach and other digestive organs. The resulting molecules are carried by the bloodstream to various parts of the body. Fungi release digestive enzymes into their environment and absorb the resulting molecules.

Connecting Concepts

21. Sample answer: The tree on the right, which had mycorrhizae, grew better.

22. Mycorrhizae provide nutrients that the lemon tree needs to grow. Without mycorrhizae, the tree does not get enough of these nutrients, so it grew much more slowly.

CHAPTER MYSTERY After students have read through the Chapter Mystery, discuss the topic of genetic diversity and the characteristics of protists and fungi.

Ask How was genetic diversity an important factor in the spread of potato blight? *(The new potato crop each year was grown from pieces of the previous year's potatoes. Thus, the potatoes grown each year were genetically identical, and the potato plants were not resistant to the blight.)*

Ask What do parasitic protists and parasitic fungi have in common? *(Sample answers: Both obtain food molecules from a host; both can cause diseases; both may have hyphae.)*

Ask How did knowledge of cell structure solve the mystery? *(The blight organism did not have cell walls containing chitin, which meant that it could not be a fungus.)*

21 CHECK Understanding

21.4 Fungi

Understand Key Concepts

17. Which of these is a symbiotic relationship between a fungus and the roots of a plant?
a. fruiting body **c.** mushroom
b. lichen **d.** mycorrhiza

18. Distinguish between the terms *hypha* and *mycelium*.

19. How are the cell walls of fungi similar to the exoskeleton of insects?

Think Critically

20. **Compare and Contrast** Both fungi and humans are heterotrophs. Compare the way fungi get food with the way humans do.

Connecting Concepts

Use Science Graphics

This photograph shows a comparison of lemon tree seedlings grown without mycorrhizae (left) to seedlings of the same age that were grown with mycorrhizae (right). Use the photograph to answer questions 21–22.

21. **Compare and Contrast** Which lemon tree shows more growth?

22. **Relate Cause and Effect** What is the most likely explanation for the results shown?

solve the CHAPTER MYSTERY

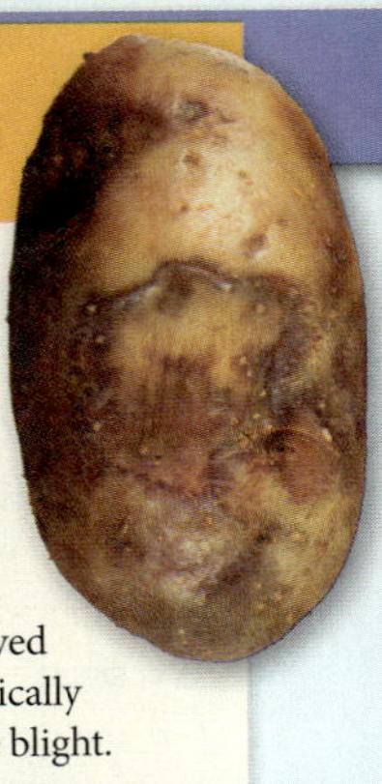

"A BLIGHT OF UNUSUAL CHARACTER"

In Ireland in the 1840s, conditions were just right for the rapid spread of potato blight. The disease lasted from year to year because the new crop was grown from potato eyes saved from the previous year's crop. Genetically identical crops could not survive the blight.

What clues in the chapter point to the blight's cause? The blight organism had hyphae. It got food by absorbing molecules through its cell walls. These traits are shared by water molds and fungi. But the organism that caused the blight did not have chitin in its cells. So it could not be a fungus. The organism that caused the blight is a water mold—a protist called *Phytophthora*. This name means "plant eater."

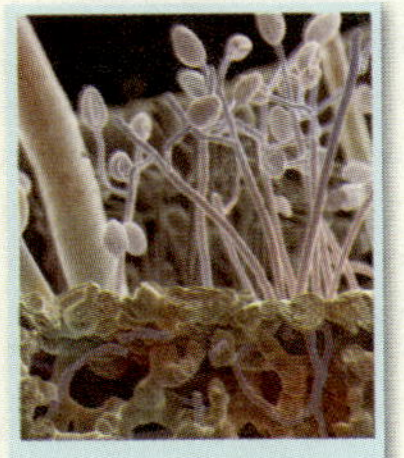

Phytophthora hyphae invading a potato (SEM 100×)

1. **Relate Cause and Effect** The weather in 1845 was unusually wet and cool. How might these weather conditions have favored the blight organism's life cycle?

2. **Infer** Scientists think that *Phytophthora* came to Ireland with some potatoes from South America. Why hadn't the same organism caused such widespread destruction there?

UntamedScience™ Finding the solution to this mystery is just the beginning. Take a video field trip with the ecogeeks of Untamed Science to see where this mystery leads.

 BIOLOGY.com Search Chapter 21 GO • Untamed Science Video

CHAPTER MYSTERY ANSWERS

1. Water molds grow well in wet conditions. Spores are flagellated, so the presence of water helps them spread easily.

2. Potatoes in South America likely were resistant to the blight organism. Also, they may have had more genetic diversity, which improved chances of survival.

Untamed Science™ Suggest students watch **Fungus Among Us** to find out more about the fascinating world of mushrooms.

Standardized Test Prep

Multiple Choice

1. All of the following are characteristics of some protists EXCEPT
   A peptidoglycan in the cell walls.
   B a membrane-bound nucleus.
   C flagella.
   D cilia.

2. The structures in *Amoeba* that help the organism move and feed are the
   A flagella.
   B cilia.
   C food vacuoles.
   D pseudopods.

3. In protists, the process of conjugation
   A is linked to photosynthesis.
   B results in an exchange of some genetic material.
   C produces offspring that are genetically identical to the parent.
   D decreases the genetic diversity of a population.

4. Which of the following statements about slime molds is FALSE?
   A Slime molds are eukaryotes.
   B Slime molds play an important part in recycling nutrients.
   C Slime molds are multicellular at some time during their life cycle.
   D Slime molds are photosynthetic protists.

5. Algal blooms can be made up of
   A paramecia.
   B lichens.
   C dinoflagellates.
   D *Trichonympha.*

6. Alternation of generations BEST describes sexual reproduction in
   A *Paramecium.*
   B water molds.
   C *Amoeba.*
   D yeast.

7. The primary carbohydrate found in the cell walls of fungi is
   A chitin.
   B actin.
   C cellulose.
   D starch.

Questions 8–10

Ripe grapes are covered with a grayish film, or bloom, that contains yeasts and other microorganisms. A group of students prepared three test tubes of fresh mashed grapes. They heated two of the test tubes to boiling and then cooled them. They inoculated one of those test tubes with live yeast, incubated all three test tubes at 30°C for 48 hours, and then examined the test tubes for signs of fermentation—an alcohol odor and bubbles. Their data are summarized in the table below.

Evidence of Fermentation

Test-Tube Contents	Alcohol Odor (yes or no)	Bubbles (yes or no)
Unheated grape mash	yes	yes
Boiled grape mash	no	no
Boiled grape mash inoculated with yeast	yes	yes

8. What is the independent variable in the students' investigation?
   A the presence of live yeast
   B an odor of alcohol
   C bubbles
   D time

9. What is the dependent variable in the students' investigation?
   A time
   B boiling
   C the presence of alcohol
   D the presence of yeast

10. What can you conclude from the students' results?
    A Uninoculated, boiled grape mash does not seem to ferment over a 48-hour period.
    B Boiled grape mash that contains live yeast undergoes fermentation.
    C Grape mash does not ferment unless live yeast is added.
    D Both A and B are correct.

Open-Ended Response

11. How does each of the partners in the lichen symbiosis benefit from the relationship?

If You Have Trouble With . . .

Question	1	2	3	4	5	6	7	8	9	10	11
See Lesson	21.2	21.2	21.2	21.3	21.3	21.2	21.4	21.4	21.4	21.4	21.4

Test-Taking Tip

USE TIME WISELY

Tell students that before they begin answering questions, they should determine the total number of questions on the test and how much time, on average, they have to answer each question. Then, they can try to allocate their time accordingly.

Before the Test

Review with students the Test-Taking Tip.

Point out the word *EXCEPT* in Question 1.

Point out the word *BEST* in Question 6.

Remind students that an open-ended response means that they'll have to write a paragraph to answer the question. Have them paraphrase the question in their notebooks, and remind them about the steps in writing a paragraph.

Answers

1. A
2. D
3. B
4. D
5. C
6. B
7. A
8. A
9. C
10. D
11. The photosynthetic partner supplies food to the fungus. The fungus provides water and minerals it absorbs from the surroundings and can protect the alga or cyanobacteria from too much sunlight.

CHAPTER PLANNER 22 Introduction to Plants

Chapter Contents	NSES	Time	Core Resources
Chapter Preview			**Student Edition,** pp. 526–527 **Chapter Mystery,** p. 527 **Foundations for Learning,** p. 527
22.1 What Is a Plant? Characteristics of Plants • What Plants Need • The Evolution of Plants • The Plant Life Cycle	I, IV, C.1.e, C.3.a, C.3.b, C.3.d, C.3.e, C.5.d	1 period 1/2 block	**Student Edition,** pp. 528–531 **Study Workbook B** 22.1 Worksheets **Biology.com** 22.1 Lesson Assessment
22.2 Seedless Plants Green Algae • Bryophytes • Vascular Plants	I, IV, V, C.3.b, C.3.d, C.3.e, C.5.d	1 period 1/2 block	**Student Edition,** pp. 532–535 **Study Workbook B** 22.2 Worksheets **Biology.com** *InterActive Art:* Plant Life Cycles • 22.2 Lesson Assessment
22.3 Seed Plants The Importance of Seeds • The Life Cycle of a Gymnosperm	I, IV, V, C.3.b, C.3.d, C.3.e, C.5.d	1 period 1/2 block	**Student Edition,** pp. 536–539 **Build Connections,** p. 537 **Study Workbook B** 22.3 Worksheets **Biology.com** *Art in Motion:* Gymnosperm Fertilization • 22.3 Lesson Assessment
22.4 Flowering Plants Flowers and Fruits • Angiosperm Diversity	I, IV, V, A.2.a, C.3.b, C.3.d, C.3.e	1 period 1/2 block	**Student Edition,** pp. 540–543 **Inquiry Into Scientific Thinking,** p. 541 **Study Workbook B** 22.4 Worksheets **Biology.com** *Tutor Tube:* Males, Females, and Sexual Reproduction in Plants—An Animal Perspective • *Art Review:* Comparing Monocots and Dicots • 22.4 Lesson Assessment
Chapter Pre-Lab	I, A.1.b, A.2.a, A.2.b, C.4.c, C.4.e, C.5.e, F.4	1 period 1/2 block	**Student Edition,** p. 544 **Lab Manual B** *Exploring Plant Diversity*

Tools for ELL Instruction

ELL Handbook describes teaching strategies to make *Miller & Levine Biology: Foundation Edition* more accessible to ELL students.

Spanish Study Workbook is a Spanish translation of Study Workbook A.

Multilingual Glossary is the *Miller & Levine Biology* glossary translated into ten languages.

Tools to Challenge Students

Study Workbook A includes more challenging worksheets and activities such as Chapter Mystery activities that reinforce 21st Century Skills.

Lab Manual A labs are geared toward on-level students. The Lab Manual includes fifteen labs in addition to the Chapter Labs.

Biology.com Data Analysis activities offer students a way to practice data gathering and analyzing skills.

Visual Quizzes in the Assessment Resources Book are geared toward on-level students and use visuals from *Miller & Levine Biology.*

Additional Resources

Biology.com Untamed Science Video • Flash Cards

Study Workbook A 22.1 Worksheets
Spanish Study Workbook 22.1 Worksheets
Assessment Resources Book Visual Quiz

Study Workbook A 22.2 Worksheets
Lab Manual B *Data Analysis:* Keeping Ferns in Check
Spanish Study Workbook 22.2 Worksheets
Biology.com *Data Analysis:* Bracken: Invasive and Deadly?

Study Workbook A 22.3 Worksheets
Spanish Study Workbook 22.3 Worksheets
Assessment Resources Book Visual Quiz

Study Workbook A 22.4 Worksheets
Lab Manual B *Hands-On Activity:* Are All Plants the Same?
Spanish Study Workbook 22.4 Worksheets

Lab Manual A *Exploring Plant Diversity* • *Using Pollen to Solve Crimes*

Chapter Review

Student Edition Chapter Summary, p. 545
Study Workbook B Chapter 22 Review
Transparencies, pp. 259–267
Biology.com Untamed Science Video • Editable Worksheets of Study Workbooks A and B and Lab Manuals A and B • Chapter 22 Flash Cards and Crossword

Untamed Science DVD • Classroom Resources CD (includes lesson presentations and editable worksheets)

Chapter Assessment

Student Edition Check Understanding, pp. 546–549
Study Workbook B Chapter 22 Taking a Standardized Test
Assessment Resources Book Chapter 22 Test B
Biology.com Chapter 22 Assessment • Editable Worksheets of Chapter 22 Visual Quizzes* and Chapter 22 Tests A and B

Exam*View* *Assessment Suite* • Classroom Resources CD (includes lesson presentations and editable worksheets)

Time: 1 period, 1/2 block

* Visual Quizzes are based on visuals from *Miller & Levine Biology.* Review carefully to make sure they are appropriate for your class.

Pressed for Time?

Preview the Chapter Introduce students to the Basic Needs of a Plant, Five Groups of Plants, and Reproduction in Seed Plants visuals.

Cover the Chapter Quickly Assign *Characteristics of Plants* in Lesson 22.1. In Lesson 22.2, have students read the Key Question and answer at the end of *Green Algae,* the introduction and Why Bryophytes Are Small in *Bryophytes,* and all of *Vascular Plants.* Assign *The Importance of Seeds* in Lesson 22.3, and *Flowers and Fruits* in Lesson 22.4.

Assess Assign Question 4 in the 22.1 Check Understanding, Questions 1–3 and 5 in the 22.2 Check Understanding, Questions 1–4 in the 22.3 Check Understanding, and Questions 2 and 4 in the 22.4 Check Understanding. In the Chapter 22 Check Understanding, assign Questions 1, 6, 7, 9–11, 14, 16, 19, and 20.

Get the Picture

Have students estimate how many different kinds of plants are in the foreground in the picture of the alpine meadow. Provide hand lenses if necessary. Students might count the different kinds of flowers or make a distinction among their leaves. Point out that the trees are plants too.

Ask What are the likely climate conditions in an alpine meadow? *(Sample answer: warm in the rainy spring and summer months; cold and snow-covered in the winter months)*

Activate Prior Knowledge

Have students imagine a picture of a park or other natural local area.

Ask What kinds of plants might you see? *(Students should give examples of local plants.)*

Ask Is our climate different from the climate in the alpine meadow? How? *(Answers may vary.)*

Ask Does the climate affect the kinds of plants that grow in a place? Explain your answer. *(Yes, plants are adapted for certain kinds of climates.)*

NATIONAL SCIENCE EDUCATION STANDARDS

UNIFYING CONCEPTS AND PROCESSES

I, IV, V

CONTENT

C.1.e, C.3.a, C.3.b, C.3.d, C.3.e, C.5.d, F.4

INQUIRY

A.1.b, A.2.a, A.2.b

22 Introduction to Plants

Big idea **Unity and Diversity of Life**

Q: What are the five main groups of plants, and how have four of these groups adapted to life on land?

An alpine meadow provides food for deer and other herbivores.

BIOLOGY.com Search Chapter 22 GO • Flash Cards

Connect to the Big Idea

Read aloud the chapter number and title. Read the Big Idea and question, **What are the five main groups of plants, and how have four of these groups adapted to life on land?**

Set the Purpose

PREVIEW AND PREDICT Draw the graphic organizer on the board. Turn to the INSIDE feature on the next page. Have a student read each lesson number and title.

Ask How do you think the word *unity* applies to the five different groups of plants? *(All plants have certain processes and needs in common.)* How do you think the word *diversity* applies to the five groups?

(Different groups of plants have different characteristics.) Remind students to search for answers to the questions as they read.

INSIDE:

- 22.1 What Is a Plant?
- 22.2 Seedless Plants
- 22.3 Seed Plants
- 22.4 Flowering Plants

CHAPTER
MYSTERY

STONE AGE STORYTELLERS

About 5300 years ago, a man died on a mountain pass in the Alps. His body was frozen in a glacier. In 1991, parts of the glacier melted, revealing his body. Scientists who studied the body called him Iceman. The body of Iceman and his possessions have provided an amazing amount of information about his life.

Some of the evidence found with Iceman comes from plants. He had a storage container made from bark and a bow and arrows made of wood. Scientists also found pollen from trees inside Iceman's digestive tract. What might these and other plant-related clues reveal about Iceman?

Read for Mystery Clues As you read the chapter, look for ways that plants can provide clues about the past. Use these clues to help solve the mystery of how Iceman lived and died.

FOUNDATIONS for Learning

You can use index cards to help you learn about the five groups of plants. You will need 20 cards, 4 for each group of plants. As you read, identify the main characteristics of each group. Write one characteristic on each card, as shown below. At the end of the chapter are two activities that use these cards to help you answer the question: What are the five main groups of plants?

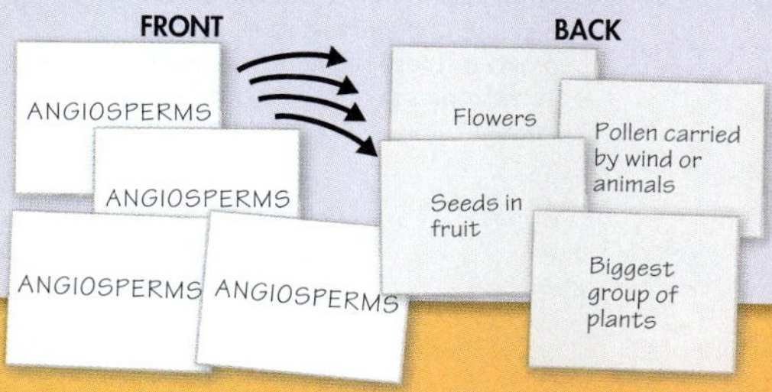

- Untamed Science Video
- Chapter Mystery

Chapter 22
Big Idea: Unity and Diversity of Life

Chapter 22: What are the five main groups of plants, and how have four of these groups adapted to life on land?

- **22.1:** What are the characteristics of plants?
- **22.2:** What are the characteristics of seedless plants?
- **22.3:** What are the characteristics of seed plants?
- **22.4:** What are the characteristics of flowering plants?

CHAPTER 22

Foundations for Learning

Pass out the index cards. Help students understand the importance of this activity.

Ask What are some characteristics that you think might describe the plant groups? *(Answers may vary. Students should mention some general characteristics of plants, such as leaf shape, size, or living on land or in water.)*

Ask How will writing down the main characteristics of each group show unity and diversity of plants? *(It will be easier to see characteristics that are alike—unity—and different—diversity.)*

Remind students that they will use these cards as a study aid as they read this chapter. Familiarize students with the activity by modeling with the cards for the first group of plants.

Untamed Science™ In **Paging Dr. Greenthumb**, students learn about some of the many chemicals plants make that humans use as medicines.

CHAPTER MYSTERY Make sure students know that a glacier is a moving body of ice. Then, invite a volunteer to read aloud the Chapter Mystery. Help students make the connection between preserving food in a freezer and how being frozen in a glacier preserved Iceman and his belongings.

Remind students that plants can provide clues about the past. For example, have students imagine that they put on a jacket they have not worn since last fall. They find a handful of dried leaves in a pocket.

Ask How could the leaves be a clue about something you did last fall? *(Sample answer: I was collecting leaves for a science assignment.)* Tell students to think of experiences like this as they read to solve the Chapter Mystery.

Digital Support

BIOLOGY.com Extend your reach by using the digital assets offered at **Biology.com.**

LESSON 22.1

Pre-Reading

Objectives

22.1.1 Describe what plants need to survive.

22.1.2 Describe how the first plants evolved.

22.1.3 Explain the process of alternation of generations.

Preview the Pages

1. Have students review the green headings and the associated blue headings to preview the main ideas and details of this lesson.
2. Preview the lesson visuals. Have volunteers read aloud the captions.
3. When previewing the Plant Life Cycle, review with students the difference between *haploid* and *diploid*.

Build Understanding

PREVIEW VISUALS Direct students to Workbook B, where they will find a partially filled-in table they can use to organize their notes. Remind students that previewing the visuals and taking notes will help them remember details and review for tests.

NATIONAL SCIENCE EDUCATION STANDARDS

UNIFYING CONCEPTS AND PROCESSES

I, IV

CONTENT

C.1.e, C.3.a, C.3.b, C.3.d, C.3.e, C.5.d

For the corresponding lesson in *Miller & Levine Biology,* see pages 634–638.

22.1 What Is a Plant?

Key Questions

What do plants need to survive?

How did plants adapt to life on land?

What feature defines most plant life cycles?

BUILD Understanding

Preview Visuals Before reading the lesson, look at the figure that lists the five main groups of plants. List anything that you already know about each group of plants.

In Your Workbook Go to your workbook to learn more about previewing visuals. Complete the chart for Lesson 22.1.

Characteristics of Plants

What color is life? Living things can be just about any color, of course. But imagine a place where living things are so plentiful that they block out the sun. Now, what color do you see?

If you've imagined a thick forest or a jungle, the color that you see is green. Green is the color of plants. Plants have adapted well to many environments. Therefore, plants dominate much of the surface of our planet. What are plants? Of course, trees, shrubs, flowers, and grasses are plants. Mosses and ferns are also plants. Biologists classify green algae as plants, too. What do all of these organisms have in common?

Cell Structure Plants are eukaryotes—their cells have nuclei and other organelles. The cell walls of all plants contain cellulose.

Getting Energy Most plants are autotrophs that make their own food by photosynthesis. Photosynthesis takes place in chloroplasts. During photosynthesis, the green pigments chlorophyll *a* and *b* capture energy from sunlight.

A few plants are heterotrophs. Instead of carrying out photosynthesis, these plants get their energy from other sources. Some plants are parasites. Others get their energy from the remains of dead organisms.

Plant Cells Plant leaves are green because of the pigments chlorophyll *a* and *b*. These pigments are found in chloroplasts. Plant cells also have cell walls containing cellulose.

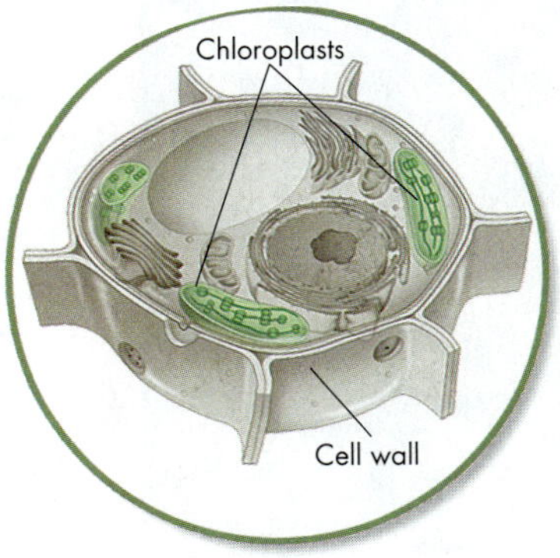

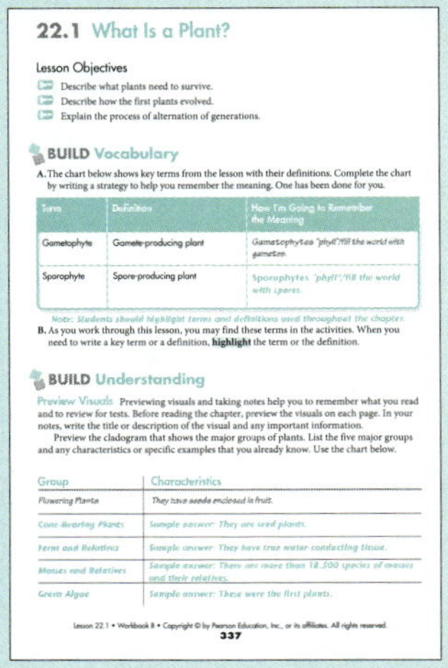

22.1 What Is a Plant?

Lesson Objectives

- Describe what plants need to survive.
- Describe how the first plants evolved.
- Explain the process of alternation of generations.

BUILD Vocabulary

A. The chart below shows key terms from the lesson with their definitions. Complete the chart by writing a strategy to help you remember the meaning. One has been done for you.

Term	Definition	How I'm Going to Remember the Meaning
Gametophyte	Gamete-producing plant	Gametophytes "phyll"/fill the world with gametes.
Sporophyte	Spore-producing plant	Sporophytes "phyll"/fill the world with spores.

Note: Students should highlight terms and definitions used throughout the chapter.

B. As you work through this lesson, you may find these terms in the activities. When you need to write a key term or a definition, **highlight** the term or the definition.

BUILD Understanding

Preview Visuals Previewing visuals and taking notes help you to remember what you read and to review for tests. Before reading the chapter, preview the visuals on each page. In your notes, write the title or description of the visual and any important information.

Preview the cladogram that shows the major groups of plants. List the five major groups and any characteristics or specific examples that you already know. Use the chart below.

Group	Characteristics
Flowering Plants	They have seeds enclosed in fruit.
Cone-Bearing Plants	Sample answer: They are seed plants.
Ferns and Relatives	Sample answer: They have true water-conducting tissue.
Mosses and Relatives	Sample answer: There are more than 18,500 species of mosses and their relatives.
Green Algae	Sample answer: These were the first plants.

Lesson 22.1 • Workbook B • Copyright © by Pearson Education, Inc., or its affiliates. All rights reserved.
337

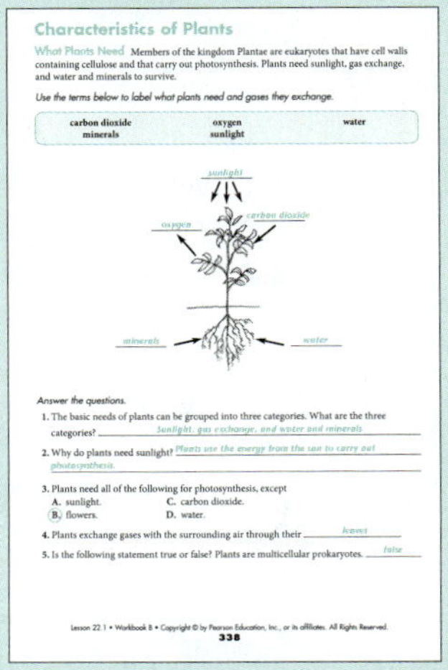

Characteristics of Plants

What Plants Need Members of the kingdom Plantae are eukaryotes that have cell walls containing cellulose and that carry out photosynthesis. Plants need sunlight, gas exchange, and water and minerals to survive.

Use the terms below to label what plants need and gases they exchange.

carbon dioxide, minerals, oxygen, sunlight, water

Answer the questions.

1. The basic needs of plants can be grouped into three categories. What are the three categories? Sunlight, gas exchange, and water and minerals
2. Why do plants need sunlight? Plants use the energy from the sun to carry out photosynthesis.
3. Plants need all of the following for photosynthesis, except
   A. sunlight. C. carbon dioxide.
   (B.) flowers. D. water.
4. Plants exchange gases with the surrounding air through their leaves
5. Is the following statement true or false? Plants are multicellular prokaryotes. false

Lesson 22.1 • Workbook B • Copyright © by Pearson Education, Inc., or its affiliates. All Rights Reserved.
338

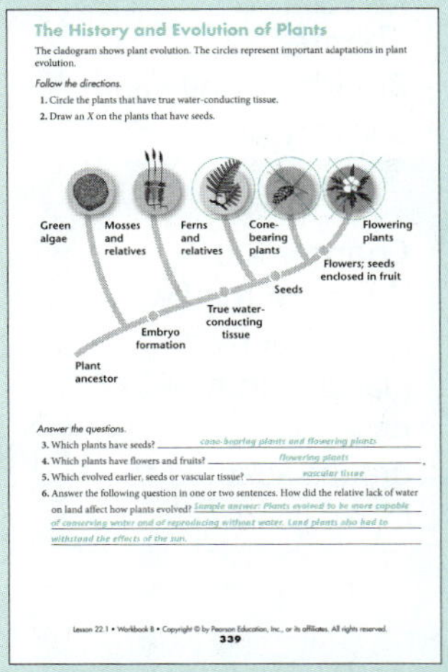

The History and Evolution of Plants

The cladogram shows plant evolution. The circles represent important adaptations in plant evolution.

Follow the directions.

1. Circle the plants that have true water-conducting tissue.
2. Draw an *X* on the plants that have seeds.

Answer the questions.

3. Which plants have seeds? cone-bearing plants and flowering plants
4. Which plants have flowers and fruits? flowering plants
5. Which evolved earlier, seeds or vascular tissue? vascular tissue
6. Answer the following question in one or two sentences. How did the relative lack of water on land affect how plants evolved? Sample answer: Plants evolved to be more capable of conserving water and of reproducing without water. Land plants also had to withstand the effects of the sun.

Lesson 22.1 • Workbook B • Copyright © by Pearson Education, Inc., or its affiliates. All rights reserved.
339

What Plants Need To survive, plants need to get energy from sunlight. They also need to take in carbon dioxide and release oxygen into the air. Plants also need water and minerals. Unlike animals, plants cannot move from place to place. But as you will see, plants have a number of adaptations that help them succeed without moving.

▶ ***Sunlight*** To carry out photosynthesis, plants need sunlight. Plants have many adaptations that help them gather sunlight. Leaves are often broad and flat, which lets them catch the sunlight. Stems hold leaves above the ground, exposing them to more light.

▶ ***Gas Exchange*** Plants need carbon dioxide for photosynthesis and oxygen for cellular respiration. Plants also need to release the extra oxygen produced during photosynthesis. They exchange these gases with the air and the soil.

▶ ***Water and Minerals*** Plants need water for photosynthesis. They also need water to help move nutrients throughout their bodies.

Land plants usually get water from the soil. As plants absorb water from the soil, they also absorb minerals. Minerals are nutrients needed for plant growth. Some plants rely on the process of diffusion to move water and nutrients throughout their bodies. Most plants have specialized tissues that carry water, minerals, and food.

Key Question What do plants need to survive?
Plants need sunlight, they need to exchange gases with the air, and they need water and minerals.

Basic Needs of a Plant All plants have the same basic needs: sunlight, water, minerals, and a way to exchange gases with the air.

The Evolution of Plants

For most of Earth's history, there were no plants on land. Fossils show that the ancestors of land plants lived in water. These early organisms were similar to the green algae of today. Most of these green algae were single-celled organisms. A few had many cells.

The First Land Plants Fossils of spores from land plants date back as far as 475 million years ago. The oldest fossils of plants themselves date back to about 425 million years ago. These early plants were small and did not have leaves or roots. To get water, they grew close to the ground in damp areas. Fossils also suggest that the first land plants needed water to reproduce.

ELL Focus on ELL: Access Content

ALL SPEAKERS Read aloud the paragraph about The First Land Plants, and emphasize the description of the early plants. Then, have students make drawings of early plants. Have them use simple labels to highlight the important information.

During Reading

Active Reading

USE CONTEXT CLUES As students read Characteristics of Plants, have them focus on the question, **What are plants?** Then, have them look for the phrase *plants are* in the remaining paragraphs.

Write the words *eukaryotes, autotrophs,* and *heterotroph* on the board.

Ask How would you define *eukaryotes, autotrophs,* and *heterotrophs? (Eukaroytes have cells with nuclei and other organelles. Autotrophs make their own food by photosynthesis. Heterotrophs get their energy from sources other than photosynthesis.)*

MAKE CONNECTIONS Have students compare and contrast themselves with plants.

Ask Are you a eukaryote? How do you know? *(Humans are eukaryotes because their cells contain nuclei and organelles.)*

Ask Are you an autotroph or a heterotroph? How do you know? *(Humans are heterotrophs because they need to consume food for energy.)*

USE VISUALS Help students make the connection between the text and the arrows in the Basic Needs of a Plant diagram. Have students read the first paragraph under What Plants Need.

Ask Which color arrows represent sunlight? *(yellow arrows)*

Have students read the Gas Exchange paragraph.

Ask Which arrows represent the first sentence in the Gas Exchange paragraph? *(the gray arrow and the red arrow)*

Ask Describe what the purple arrow and blue arrow show. *(Minerals and water move from the soil into plant roots.)*

KEY QUESTION Invite a volunteer to read aloud the question and answer. Discuss why all the parts of the answer are important to plants.

Ask What would happen to a plant that was kept in a dark place? *(It would not survive because it couldn't make food.)*

Speed Bump

Ask What do plants need to carry out photosynthesis? *(sunlight, carbon dioxide, and water)*

Ask Why do plants need minerals? *(Minerals are nutrients needed for plant growth.)*

LESSON 22.1

During Reading continued

Active Reading

USE DIAGRAMS Direct students to the cladogram. Remind them that in Chapter 18 they learned that cladograms show the relatedness of different groups of organisms. Be sure students understand that the colors of the circles on the cladogram refer to the same colors on the pie chart.

Ask Which group of plants has the most species? *(flowering plants)* The least? *(cone-bearing plants)*

Ask What does the cladogram show about the evolution of plants? *(All plants evolved from a common ancestor.)*

Explain that the green dots represent the relative time when the labeled characteristics appeared.

Ask Did the characteristic of seeds appear before or after the first ferns? *(after)*

Ask Which group evolved first? *(green algae)*

Ask What characteristic do both flowering and cone-bearing plants have, but ferns do not? *(seeds)*

Refer to the flowering plants in the diagram.

Ask Did the flowering plants evolve earlier or later than other plants? *(later)* What characteristic might help them be so successful? *(seeds enclosed in fruits)*

DRAW CONCLUSIONS Focus students' attention on the third paragraph *(The evolution of . . .)*.

Ask Why are plants the basis, or starting point, of an ecosystem? *(Sample answer: Plants capture the energy of sunlight and make that energy available to other organisms in an ecosystem.)*

Ask How did the movement of plants from watery environments to dry land make possible the evolution of other land species? *(Until plants inhabited land, most other species could not live there because they didn't have a source of energy.)*

KEY QUESTION Invite a volunteer to read aloud the question and answer.

Ask What is evolution? *(the process by which modern organisms have descended from ancient organisms)*

Ask Why would water loss be a problem for land plants? *(Plants cannot perform photosynthesis without water. If land plants lost a lot of water, they would have to replace it with water from the environment. Some environments would not be able to supply that much water.)*

Build Vocabulary

SUFFIXES Pronounce each vocabulary word. Review that the suffix *-phyte* means "plant." Then, have students look up the meanings of these words: *saprophyte, epiphyte, benthophyte,* and *zoophyte*. Ask students to tell how each word relates to plants.

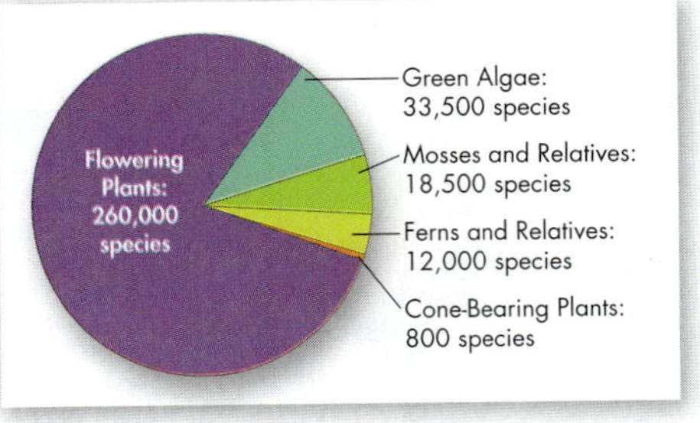

Five Groups of Plants Today, biologists classify plants into five major groups.

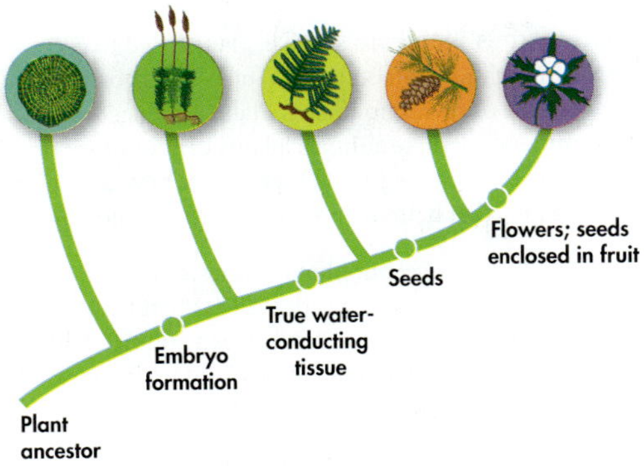

For more on the diversity of plants, go to the Visual Guide. pp. DOL 20–29.

Plants on Land Evolve Over time, adaptations arose that made plants more successful on land. These adaptations helped plants survive the drying rays of the sun. They also prevented the loss of water. Eventually, plants that did not need water to reproduce evolved. These plants could live in even the driest environments.

Several groups of plants evolved from the first land plants. One group developed into mosses. Another group gave rise to ferns, cone-bearing plants, and flowering plants. All of these groups are successful today. But they have very different adaptations.

The evolution of land plants changed the environments on Earth. As land plants grew, they formed the basis for new ecosystems. Plants also affected the evolution of other organisms, such as insects and birds.

An Overview of the Plant Kingdom Biologists divide the plant kingdom into five main groups, as shown in the pie graph above. This classification is based on four key features: the formation of embryos, specialized water-conducting tissues, seeds, and flowers. Plants that form embryos are called "land plants," even though some live in water.

Key Question How did plants adapt to life on land?
Evolution favored plants with adaptations that helped them survive the drying rays of the sun, prevented the loss of water, and allowed them to reproduce without water.

The Plant Life Cycle

Plants have a life cycle that sets them apart from most other organisms. This life cycle changes between two phases: a diploid (2N) phase and a haploid (N) phase. (Haploid organisms have one set of chromosomes in their cell nuclei. Diploid organisms have two sets of chromosomes.) The shift between phases is called **alternation of generations.**

Two Plants The multicellular diploid (2N) phase is known as the **sporophyte** (SPOH ruh fyt), or spore-producing plant. The multicellular haploid (N) phase is the **gametophyte** (guh MEET uh fyt), or gamete-producing plant.

BUILD Vocabulary

alternation of generations
a life cycle that has two alternating phases—a haploid (N) phase and a diploid (2N) phase

sporophyte
a spore-producing plant; the multicellular diploid phase of a plant life cycle

gametophyte
a gamete-producing plant; the multicellular haploid phase of a plant life cycle

SUFFIXES
The suffix *-phyte* comes from a Greek word that means "plant." Sporo*phytes* are plants that produce spores. Gameto*phytes* are plants that produce gametes.

ELL Focus on ELL: Extend Language

BEGINNING AND INTERMEDIATE SPEAKERS Remind students that compound adjectives are made up of two or more smaller words, each with its own meaning. When the smaller words are combined, the compound adjective has a new meaning.

Write the phrase *takeout food* on the board. Ask a volunteer to circle the two smaller words in *takeout* and to tell what the two smaller words mean. Ask another volunteer to tell what takeout food is. *(food you take out of a restaurant)*

Write the phrase *cone-bearing plant* on the board. Circle the two parts of the adjective *cone-bearing* and draw an arrow pointing at the hyphen. **Say** Sometimes the words in a compound adjective are joined by a hyphen.

Ask What is a cone? *(a reproductive structure on some types of plants)* What does *bearing* mean? *(producing)* What is a cone-bearing plant? *(a plant that produces cones)*

Have student find additional compound adjectives on this page and discuss their meanings.

Steps of the Life Cycle The plant life cycle has several basic steps. A sporophyte plant produces haploid spores. Spores are usually single-celled. These spores grow into gametophytes. Gametophytes have many cells. Each gametophyte produces reproductive cells called gametes—sperm cells and egg cells. During fertilization, a sperm and egg come together, producing a diploid zygote. The zygote develops into a new sporophyte, and the cycle begins again.

In many plants, the gametophyte is the largest stage of the life cycle. However, as vascular and seed plants evolved, their sporophyte stages became much larger. Modern seed plants have a tiny gametophyte that is contained within the sporophyte.

Key Question What feature defines most plant life cycles? **Plants have life cycles with two phases, or alternation of generations. These phases are a diploid (2N) sporophyte phase and a haploid (N) gametophyte phase.**

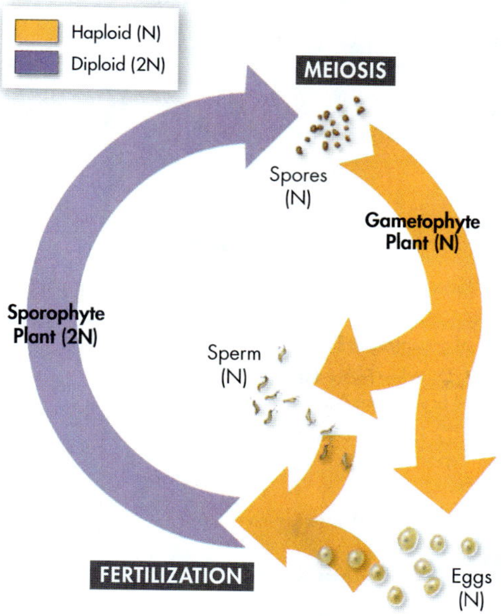

Plant Life Cycle Most plants have a two-phase life cycle: the haploid gametophyte phase and the diploid sporophyte phase.

CHECK Understanding

Apply Vocabulary

1. During the plant life cycle, the ______ phase produces egg and sperm cells.
2. During the ______ phase of the plant life cycle, the plant produces spores.
3. A life cycle that switches between haploid and diploid phases is called ______.

Critical Thinking

4. **Interpret Visuals** List four basic needs of a plant. Then, explain how the plant in the art on page 529 meets those basic needs.
5. **Explain** How did the dry environment on land affect the evolution of plants?
6. **Compare and Contrast** Compare the gametophyte and sporophyte stages of the plant life cycle. Which is haploid? Which is diploid?
7. **Write to Learn** Answer the first clue of the mystery. Use the words *chloroplast*, *chlorophyll*, and *photosynthesis* in your answer.

MYSTERY CLUE

Scientists extracted large amounts of chlorophyll from maple leaves found with Iceman. How could this information be used to determine the season when he died? (**Hint:** See p. 528.)

BIOLOGY.com Search Lesson 22.1 GO • Lesson Assessment

Check Understanding

1. gametophyte
2. sporophyte
3. alternation of generations
4. Plants need light, water, minerals, and a way to exchange gases. The plant's leaves catch sunlight and its roots take in water and minerals from soil. Its leaves take in carbon dioxide and give off oxygen.
5. Land plants evolved adaptations that made them resistant to drying out, capable of conserving water, and able to reproduce without it.
6. The gametophyte, which produces egg and sperm cells, is haploid. The sporophyte, which produces spores, is diploid.
7. Students should recognize that levels of chlorophyll change throughout the year. They should also recognize that chlorophyll is located in the chloroplasts and is needed for photosynthesis. Students may note that high levels of chlorophyll are common in maple leaves in spring and summer.

Active Reading

USING VISUALS Direct students' attention to the Plant Life Cycle diagram.

Ask Which color arrow or arrows show the reproductive cells that have only one set of chromosomes? *(orange)* What process results in reproductive cells having only one set of chromosomes? *(meiosis)*

Ask What process results in reproductive cells with two sets of chromosomes? *(fertilization)* Which color arrow or arrows show the reproductive cells with two sets of chromosomes? *(purple)*

KEY QUESTION Invite a volunteer to read the question and answer.

Ask How many sets of chromosomes are in diploid phase nuclei? *(two)* In the haploid phase? *(one)*

Ask What is a gametophyte? *(the gamete-producing plant)* A sporophyte? *(the spore-producing plant)*

Ask Why is the plant reproductive cycle known as *alternation of generations*? *(All plants cycle through the sporophyte stage and the gametophyte stage.)*

Post Reading

ASSESSMENT TIPS For Question 6, help students with a clue that *ha-* (in *haploid*) is similar to *half* or half the usual sets of chromosomes in cell nuclei.

MYSTERY CLUE Discuss with students the seasonal changes in plant life in your local area. Tell students that maple trees are common in many areas that have seasonal changes. Explain that in maples, leaf growth begins in the spring and continues throughout the summer, chlorophyll decreases and other color pigments appear in the fall, and dormancy occurs in the winter. Help students compare this cycle with changes in local trees.

Wrap-Up Activity

FIVE GROUPS OF PLANTS Divide students into groups and supply each group with five index cards and five plastic bags of varying sizes. Have the students label the bags from smallest to largest with plant group names in the order shown in the cladogram, beginning with green algae. On the first card, have students write *chlorophyll a and b* and *cell walls with cellulose,* and place that card inside the bag labeled *green algae*. Then have students write each plant characteristic shown on the cladogram on the remaining cards. Have them nest the cards and bags in the order shown in the cladogram. When finished, students should see that the type of plant in a certain bag has the characteristics of that bag and all the ones that are nested within it.

Introduction to Plants 531

LESSON 22.2

Pre-Reading

Objectives

22.2.1 Identify the characteristics of green algae.

22.2.2 Describe the adaptations of bryophytes.

22.2.3 Explain the importance of vascular tissue.

Preview the Pages

1. Refer students to the lesson title. Emphasize that this lesson explores plants that do not produce seeds. Have students scan the headings and visuals to identify some kinds of seedless plants.
2. Preview all the visuals. Point out the similarities between the Plant Life Cycle diagram in the previous lesson and the Moss and Fern Life Cycle diagrams in this lesson.

Activate Prior Knowledge

Remind students of the cladogram Five Groups of Plants in the previous lesson.

Ask What are the characteristics that green algae share with all other plants? *(presence of chlorophyll* a *and* b *and cell walls made of cellulose)*

Ask What kinds of typical plant structures do you think green algae do not have? *(Sample answer: roots, leaves, and stems)*

Build Understanding

VENN DIAGRAM Direct students to the Venn diagram in Workbook B. Emphasize that a Venn diagram is a tool for comparing and contrasting. Remind students that the characteristics that are comparative, or the same, are in the overlapping area.

NATIONAL SCIENCE EDUCATION STANDARDS

UNIFYING CONCEPTS AND PROCESSES

I, IV, V

CONTENT

C.3.b, C.3.d, C.3.e, C.5.d

For the corresponding lesson in ***Miller & Levine Biology,*** **see pages 639–645.**

22.2 Seedless Plants

Key Questions

What are the characteristics of green algae?

What factor limits the size of bryophytes?

How is vascular tissue important?

BUILD Understanding

Venn Diagram Make a Venn diagram to describe the similarities and differences among the three groups of seedless plants. Fill in your Venn diagram as you read the lesson.

In Your Workbook Go to your workbook to learn more about Venn diagrams. Complete the Venn diagram for Lesson 22.2.

Green Algae

Most algae are classified as protists. But green algae are classified as plants. Why? Because green algae share many characteristics with the land plants. Like land plants, the cells of green algae have cell walls with cellulose. Their chloroplasts have chlorophyll *a* and *b*. Finally, biologists have found that the genes of green algae and land plants are similar.

Most green algae live in fresh water or salt water. Some green algae live in moist areas on land. They absorb moisture and nutrients directly from their surroundings. Green algae lack the specialized tissues found in other plants.

Life Cycles of Green Algae Like other plants, many green algae show alternation of generations in their life cycles. Some live as haploid cells most of the time. They reproduce asexually by mitosis. Then, when living conditions become unfavorable, they reproduce sexually. To do this, they first produce haploid gametes. These gametes then fuse to form a diploid zygote. The zygote may remain inactive for many weeks or months. When conditions are favorable, the zygote enters meiosis and produces haploid cells. This begins the life cycle all over again.

Multicellular Green Algae Many kinds of green algae form colonies. Some form long chains of cells. Others, like the ball-shaped colonies of *Volvox*, are more complex. In a *Volvox* colony, some cells are specialized for reproduction. Colonial algae may provide a clue about how multicellular plants arose.

Key Question What are the characteristics of green algae? **Green algae live mostly in fresh water or salt water; some live in moist areas on land. Some green algae are multicellular.**

Spirogyra *(LM 140×)*

Volvox *(LM 50×)*

Multicellular Green Algae Some green algae live in colonies. *Spirogyra* colonies are made of long strands of cells. *Volvox* colonies form hollow balls, made of 500 to 50,000 cells.

22.2 Seedless Plants

Lesson Objectives

- Identify the characteristics of green algae.
- Describe the adaptations of bryophytes.
- Explain the importance of vascular tissue.

BUILD Vocabulary

A. The chart below shows key terms from the lesson with their definitions. Complete the chart by writing a strategy to help you remember the meaning of each term. One has been done for you.

Term	Definition	How I'm Going to Remember the Meaning
Phloem	Tissue that carries nutrients and carbohydrates produced by photosynthesis	*Phloem/flow-m: Phloem lets the materials made by photosynthesis flow throughout the plant.*
Vascular tissue	Tissue that carries water and nutrients	*Vascular/muscular. This tissue lets the plant grow strong and tall.*
Xylem	Tissue that carries water up from the roots of a plant	*The x of xylem has two lines crossing; think of water crossing from the soil up into the plant above ground.*

B. As you work through this lesson, you may find these terms in the activities. When you need to write a key term or a definition, **highlight** the term or the definition.

CHAPTER MYSTERY

The Iceman's Home Draw objects that the iceman may have had or made to survive.

Students might draw objects such as fur clothing, a bow and arrow, and food items.

BUILD Understanding

Venn Diagram A Venn diagram is made up of overlapping circles. It is a useful tool for comparing two or even three topics. In each circle, write one of the topics that you want to compare. In the space where the circles overlap, write the features that the topics share. In the space where the circles do not overlap, write the features that are unique to each topic.

As you read Lesson 2, make a Venn diagram showing the similarities and differences between the three major groups of seedless plants. Fill in the diagram below as you read.

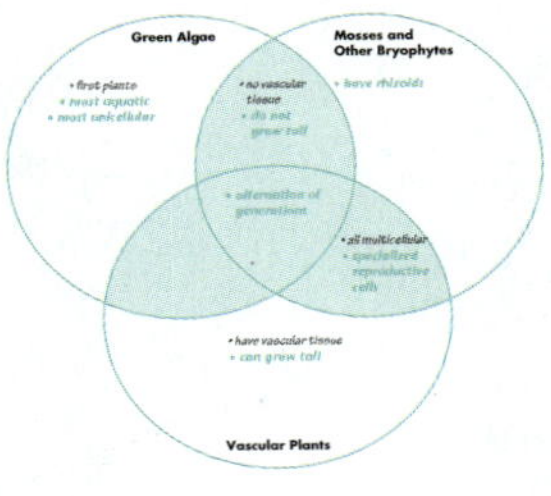

Mosses and Other Bryophytes

This diagram shows the gametophyte of a bryophyte plant. The sporophyte grows out of the gametophyte.

Follow the directions.

1. Use red to color the gametophyte portion of the plant.
2. Use blue to color the sporophyte portion of the plant.

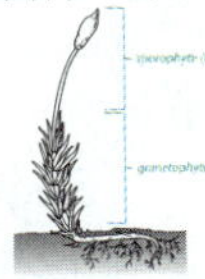

Answer the questions.

3. What is the role of the sporophyte? *It produces haploid spores.*
4. The thin, upright shoot of a moss plant is not considered a true stem because
   A. it is dependent on water.
   B. it contains lignin.
   (C.) it has no vascular tissue.
   D. it can tolerate low temperatures.
5. The bryophytes include all of the following plants except
   A. liverworts.
   (B.) ferns.
   C. hornworts.
   D. mosses.
6. Is the following statement true or false? The sporophyte is dependent on the gametophyte for water and nutrients. *true*

Complete the table.

Structure	Description	Structure Produces
Antheridia	*male reproductive structure*	Sperm
archegonia	Female reproductive structure	*egg cells*

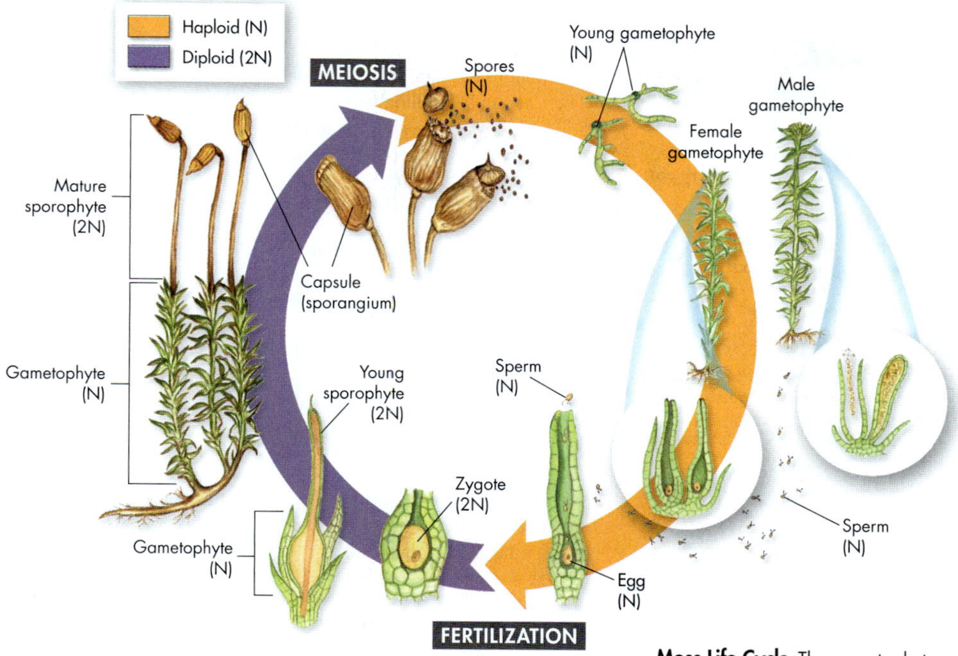

Moss Life Cycle The gametophyte is the dominant stage of the moss life cycle. Specialized structures in the male and female gametophytes produce sperm and eggs.

Bryophytes

Imagine a forest floor. The moist ground is covered with a soft green carpet made of mosses. Mosses are the most common members of a group of plants called **bryophytes** (BRY oh fyts). Unlike algae, bryophytes are multicellular plants that have specialized structures. Mosses lack true leaves, stems, and roots. Rootlike filaments called rhizoids (RY zoydz) hold mosses in place. Rhizoids also absorb water and minerals. Mosses have a thin, waxy coating that prevents them from drying out.

Why Bryophytes Are Small Most land plants carry water and nutrients in specialized tissue called **vascular tissue.** Vascular tissue contains long tubes with very hard cell walls. Bryophytes do not have vascular tissue. Instead, they absorb water from their surroundings by osmosis. Because they lack vascular tissue, bryophytes can draw up water no higher than a meter above the ground. The lack of vascular tissue limits the height of bryophytes. As a result, most bryophytes are small plants that grow only in damp soil.

Key Question What factor limits the size of bryophytes? **Bryophytes are small because they lack vascular tissue.**

Life Cycle The gametophyte is the dominant stage in the moss life cycle. The smaller sporophyte depends on the gametophyte for water and nutrients. Bryophytes produce sperm cells that swim using flagella. Sperm cells must swim through water to reach egg cells. Thus, bryophytes must live in damp habitats that have standing water for at least part of the year.

BUILD Vocabulary

bryophyte
a group of plants that have specialized reproductive organs but lack vascular tissue

vascular tissue
a specialized tissue in plants that carries water and nutrients

WORD ORIGINS
The word *vascular* comes from the Latin word *vas*, meaning "a vessel," or a container used to carry liquids. Vascular tissues carry liquids through a plant.

Focus on ELL: Build Background

ALL SPEAKERS To help students decipher the paragraph about the life cycles of green algae, focus on these pairs of terms: *sexually, asexually; favorable, unfavorable; mitosis, meiosis; haploid, diploid.*

Invite students to work in small groups to compare and contrast each pair of terms. Discuss their findings. Point out that the prefixes *a-* and *un-* mean "not."

Work with the class to develop a concept map that connects the terms. *(Favorable—mitosis—asexual—haploid; Unfavorable—meiosis—sexual—diploid)* Have students copy the concept map in their notebooks.

During Reading

Active Reading

USE VISUALS Point out the photographs of the water-dwelling organisms on the previous page.

Ask How are these two organisms similar? *(They are both types of multicellular green algae.)* How are they different? *(In Spirogyra, the colonies are made of long strands of cells. In Volvox, the cells form colonies shaped like hollow balls.)*

DRAW A DIAGRAM Help students recall the process of osmosis. Draw a cell in water on the board. Use Xs to represent water molecules. Show more water molecules outside the cell than inside it.

Ask Are there more water molecules inside or outside of the cell? *(outside)* What happens when more water molecules are outside of a cell than inside? *(The water molecules outside the cell move into the cell.)*

Now draw an arrow from a water molecule to the inside of the cell. Then draw an X inside. Erase the X outside. Continue until the Xs inside and outside are equal.

Ask Why do you think this is now called a state of equilibrium? *(because the water concentration inside and outside of the cell is now equal)*

Build Vocabulary

WORD ORIGINS Write the word *bryophyte* on the board.

Ask Can you remember from the last lesson what the suffix *-phyte* means? *(Students should recall that the suffix* -phyte *means "plant.")*

Explain that the prefix *bryo-* comes from the Greek word *bryon*, meaning "moss."

Ask Why is *bryophyte* a good name for this group of organisms? *(because mosses are members of the bryophyte group—a group of small plants without vascular tissue)*

KEY QUESTION Invite a volunteer to read the question and answer aloud.

Ask Why would having vascular tissue enable bryophytes to grow larger? *(Vascular tissue allows plants to draw up water from their surroundings. It also offers support.)*

Speed Bump

Ask How do green algae get moisture and nutrients? *(They absorb moisture and nutrients directly from their surroundings.)*

Ask What adaptations allow bryophytes to survive on land? *(rhizoids that hold them in place and a waxy coating that stops them from drying out)*

LESSON 22.2

During Reading continued

Active Reading

MAKE CONNECTIONS Tell students they also have a vascular system—a system of blood vessels.

Ask What does your vascular system do? *(It carries oxygen, nutrients, and wastes through the body.)*

Ask How is the human vascular system similar to a plant vascular system? *(In both humans and plants, the vascular system transports materials throughout the body.)*

USE GRAPHIC ORGANIZERS Have students create a two-column chart that compares and contrasts xylem and phloem. Tell them to refer to the chart periodically throughout the chapter and to use it as a study tool.

Build Vocabulary

WORD ORIGINS Have students read the vocabulary words and their definitions. Explain that although *phloem* is derived from Greek, it is actually a German word first used in 1858. Tell them that *xylem* is also a German word, first used in 1875. It also comes from a Greek word, *xylon,* which means "wood."

Ask Why might scientists have used this Greek word to describe the specialized cells of xylem? *(Sample answer: Xylem provides support to a plant, so it is like wood, which is supportive.)*

KEY QUESTION Invite a volunteer to read the question and answer.

Ask What tissues carry fluids against gravity? *(xylem and phloem)* What fluids do these tissues carry? *(xylem—water, phloem—solutions of nutrients and food)*

BUILD Vocabulary

xylem
a vascular tissue that carries water upward from the roots to every part of the plant

phloem
a vascular tissue that transports solutions of nutrients and carbohydrates produced by photosynthesis through the plant

WORD ORIGINS
The word *phloem* comes from the Greek word *phloios*, which means "bark." In many plants, such as trees, the inner layer of bark includes phloem.

Vascular Plants

For millions of years, only small plants, like mosses, grew on land. Then, about 420 million years ago, something remarkable happened. Suddenly, low-growing plants were joined by much taller plants and even trees. What happened? Fossils show that these new plants had vascular tissue. With the evolution of vascular tissues, plants were able to grow high above the ground.

Two Kinds of Vascular Tissue Vascular tissues allow fluids to travel through a plant body against the force of gravity. The vascular tissue that carries water from the roots to every part of a plant is **xylem** (ZY lum). Xylem is made up of hollow, tubelike cells called *tracheids* (TRAY kee idz). Tracheids are connected end to end. They have thick cell walls containing lignin. Lignin is a compound that toughens cell walls, making them strong enough to support the weight of the plant.

Vascular plants also have a second vascular tissue that is called phloem. **Phloem** (FLOH um) carries solutions of nutrients and food produced by photosynthesis. Like xylem, the cells of phloem are long and specialized to move fluids.

Key Question How is vascular tissue important? **Vascular tissue allows the movement of fluids throughout the body of a plant against the force of gravity.**

Seedless Vascular Plants Today there are three groups of vascular plants that do not have seeds: club mosses, horsetails, and ferns. Ferns are the most common seedless vascular plants. They have strong roots and underground stems called rhizomes (RY zohmz). Their large leaves are called fronds. Ferns can live in areas with little light. They commonly grow in habitats that are wet.

Fern Life Cycle The large plant you recognize as a fern is the sporophyte phase of the fern's life cycle. As the fern matures, haploid spores develop on the undersides of its fronds. The spores grow into tiny, heart-shaped gametophytes. The gametophyte stage lives independently of the sporophyte. The gametophyte produces sperm and eggs. For fertilization to take place, the sperm must have water in order to swim to the eggs. The diploid zygote grows into a sporophyte.

Vascular Tissue Horsetails are seedless vascular plants. The micrograph shows a magnified view of the vascular tissue, which is reinforced by rings of lignin. Lignin helps support the plant.

534 Chapter 22 • Lesson 2

Hands-On Learning

MODELING LIGNIN Have students model the function of lignin using the micrograph as a reference. Give pairs of students two plastic bags. Long, plastic newspaper bags work well. Ask students to cut thin strips of paper, about one-half inch wide. Have students use glue sticks or tape to attach the strips to the outside of one bag at intervals of about one-half inch. Tell students to try to make the two bags stand up.

Ask Which bag stands up the best? *(The one with the paper strips.)*

Ask How are the paper strips like rings of lignin in plant cells? *(They provide reinforcing strength.)*

Fern Life Cycle The dominant stage of the fern life cycle is the diploid sporophyte. In the diagram, the tiny, heart-shaped gametophyte has been enlarged to show the details.

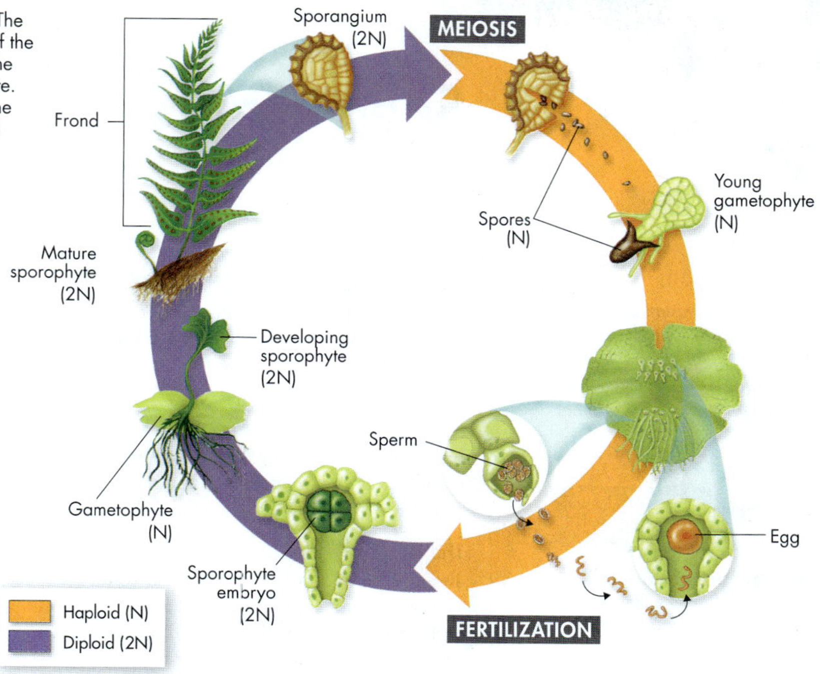

CHECK Understanding

Apply Vocabulary

Use the highlighted words from the lesson to complete each sentence correctly.

1. The evolution of _______ made it possible for land plants to grow larger.

2. _______ carries nutrients and food throughout the body of the plant.

3. _______ is made up of tracheids, specialized cells that form water-conducting tubes.

Critical Thinking

4. Compare and Contrast List four ways that green algae are similar to other plants. How are they different from land plants?

5. Relate Cause and Effect Why are ferns able to grow so much taller than bryophytes?

6. Interpret Visuals Study the art of the life cycles of the moss and the fern. Identify which phase is dominant in each of these life cycles.

7. Write to Learn Answer the next clue of the mystery. Include the words *bryophyte* and *vascular tissue* in your answer.

MYSTERY CLUE

When studying the stomach contents of Iceman, scientists found a piece of moss. They concluded that Iceman ate the moss by accident. What does the moss indicate about the area in which Iceman lived? (Hint: See p. 533.)

BIOLOGY.com Search Lesson 22.2 GO • Lesson Assessment • InterActive Art

Check Understanding

1. vascular tissue

2. Phloem

3. Xylem

4. Like other plants, green algae have cells walls with cellulose, carry out photosynthesis, have chlorophyll *a* and *b*, and have life cycles with alternation of generations. Unlike land plants, they mostly live in water. Many are single-celled and lack specialized tissues. They also lack embryos.

5. Ferns can grow taller than bryophytes because ferns have vascular tissue, which allows them to transport water upward.

6. Moss: gametophyte; fern: sporophyte

7. Sample answer: The moss indicates that Iceman probably lived in a moist area where bryophytes grow. Because they don't have vascular tissue, they must live in damp areas.

Active Reading

USING VISUALS Walk students through the steps of the Fern Life Cycle diagram, helping them to note similarities and differences between the fern life cycle and the moss life cycle presented earlier in the lesson. Then ask questions to check students' understanding of ferns.

Ask Is the fern sporophyte haploid or diploid? *(diploid)*

Ask Where do the spores develop? *(on the mature sporophyte, on the underside of the fronds)*

Ask What do the spores turn into as they grow? *(gametophytes)*

Ask Where are sperm and eggs produced? *(on the gametophytes)*

Speed Bump

Ensure that students understand that the presence of vascular tissue is necessary for plants to grow over a meter above the ground.

Ask Do bryophytes have vascular tissues? *(no)*

Ask How does vascular tissue enable plants to grow tall? *(Vascular tissues have thick cell walls strengthened by lignin, which makes the cells stronger. The stronger cells can pile up to greater heights without the plant collapsing or losing the ability to get water and nutrients.)*

Post Reading

ASSESSMENT TIPS For Question 4, point out that two questions are included—one asks for similarities and the other for differences.

For Question 6, elicit from volunteers the meaning of the words *phase (stage, part)* and *dominant (greater, stronger)* to clarify the question.

MYSTERY CLUE Suggest students review the Bryophytes section of this lesson. Tell them to look for a certain need that mosses have and that is emphasized in all three paragraphs. *(water)* Guide students to hypothesize that Iceman came from an area that was very moist, rainy, or had lakes, ponds, or streams.

Wrap-Up Activity

ARRANGE A LIFE CYCLE Have small groups of students put the following terms on individual index cards: *meiosis, spores, gametophyte, sperm and egg, fertilization, zygote,* and *sporophyte*. From memory, have students arrange the cards to show the life cycle of a fern or a moss. Instruct group members to check each other's index card arrangement against the diagrams in this lesson.

LESSON 22.3

Pre-Reading

Objectives

22.3.1 Describe the reproductive adaptations of seed plants.

22.3.2 Identify the reproductive structures of gymnosperms.

Preview the Pages

1. Preview the vocabulary words by having students find each word in the text and read aloud the sentence in which the word is boldfaced.
2. Direct students' attention to the green headings and the related blue subheadings for an overview of the main ideas and details presented in this lesson.
3. Preview all the visuals, and read the captions. Point out that the Pine Life Cycle diagram is based on the generalized life cycle diagram Plant Life Cycle from the first lesson in this chapter.

Build Understanding

PREVIEW VISUALS Have students work in pairs to complete the Preview Visuals table in Workbook B. Encourage them to add as much detail as possible.

NATIONAL SCIENCE EDUCATION STANDARDS

UNIFYING CONCEPTS AND PROCESSES

I, IV, V

CONTENT

C.3.b, C.3.d, C.3.e, C.5.d

For the corresponding lesson in *Miller & Levine Biology,* see pages 646–649.

22.3 Seed Plants

Key Questions

What adaptations allow seed plants to reproduce without standing water?

How does fertilization take place in gymnosperms without water?

BUILD Understanding

Preview Visuals Look at the figure describing the pine life cycle. Write down your first impressions about how it differs from the fern life cycle.

In Your Workbook Go to your workbook to learn more about previewing visuals. Complete the activity for Lesson 22.3.

BUILD Vocabulary

seed a plant embryo and a food supply inside a protective covering

gymnosperm a group of seed plants that bear their seeds directly on the scales of cones

angiosperm a group of seed plants that bear their seeds within a layer of tissue that protects the seed; also called flowering plants

pollen grain in seed plants, the structure that contains the entire male gametophyte

The Importance of Seeds

Acorns, pine nuts, dandelion seeds, corn kernels, beans—seeds can be found nearly everywhere. What are seeds? Are they gametes? Reproductive structures? Do they contain an egg or sperm? The answer is none of the above.

A **seed** is made up of a plant embryo and a food supply surrounded by a protective covering. Every seed holds a living plant, ready to sprout and grow when the conditions are right! Seeds allow plants to grow even in some of the driest places on Earth.

The First Seed Plants The earliest fossils of plants with seeds are about 360 million years old. These ancestors of today's seed plants had several adaptations that made them successful. Unlike mosses and ferns, seed plants do not need standing water for the fertilization of gametes. Adaptations that allow seed plants to reproduce without standing water include cones or flowers, the transfer of sperm by pollination, and the protection of embryos in seeds.

Cones and Flowers In seed plants, the dominant phase of the life cycle is the sporophyte. Male and female gametophytes grow and mature within the sporophyte. Gametophytes usually develop in reproductive structures known as cones or flowers. In fact, seed plants are divided into two groups based on which of these structures they have. Nearly all **gymnosperms** (JIM noh spurmz) bear their seeds directly on the scales of cones. Flowering plants, or **angiosperms** (AN jee oh spurmz), bear their seeds inside a layer of tissue that protects the seed.

Pollen In seed plants, the entire male gametophyte is contained in a tiny structure called a **pollen grain.** The pollen grain produces sperm. Sperm produced in a pollen grain do not swim through water to fertilize eggs. Instead, pollen grains are carried to the female reproductive structure in various ways. In some plant species, pollen grains are carried by wind. In other species, animals such as insects carry pollen grains to the female reproductive structure. The transfer of pollen from the male reproductive structure to the female reproductive structure of a plant is called **pollination.**

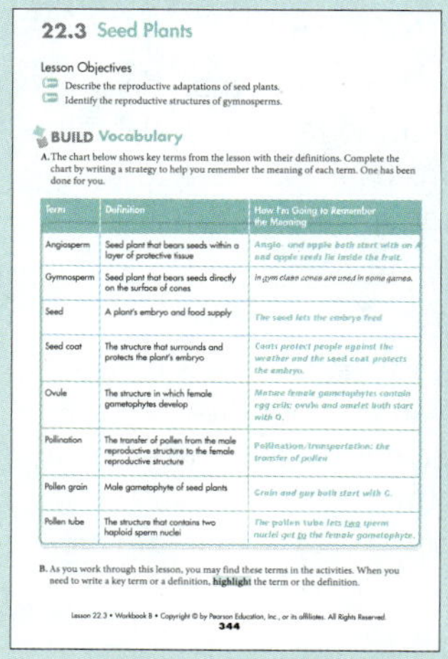

22.3 Seed Plants

Lesson Objectives

Describe the reproductive adaptations of seed plants.

Identify the reproductive structures of gymnosperms.

BUILD Vocabulary

A. The chart below shows key terms from the lesson with their definitions. Complete the chart by writing a strategy to help you remember the meaning of each term. One has been done for you.

Term	Definition	How I'm Going to Remember the Meaning
Angiosperm	Seed plant that bears seeds within a layer of protective tissue	*Angio- and apple both start with an A and apple seeds lie inside the fruit.*
Gymnosperm	Seed plant that bears seeds directly on the surface of cones	*In gym class cones are used in some games.*
Seed	A plant's embryo and food supply	*The seed lets the embryo feed*
Seed coat	The structure that surrounds and protects the plant's embryo	*Coats protect people against the weather and the seed coat protects the embryo.*
Ovule	The structure in which female gametophytes develop	*Mature female gametophytes contain egg cells; ovule and omelet both start with O.*
Pollination	The transfer of pollen from the male reproductive structure to the female reproductive structure	*Pollination/transportation: the transfer of pollen*
Pollen grain	Male gametophyte of seed plants	*Grain and guy both start with G.*
Pollen tube	The structure that contains two haploid sperm nuclei	*The pollen tube lets two sperm nuclei get to the female gametophyte.*

B. As you work through this lesson, you may find these terms in the activities. When you need to write a key term or a definition, **highlight** the term or the definition.

Lesson 22.3 • Workbook B • Copyright © by Pearson Education, Inc., or its affiliates. All Rights Reserved.
344

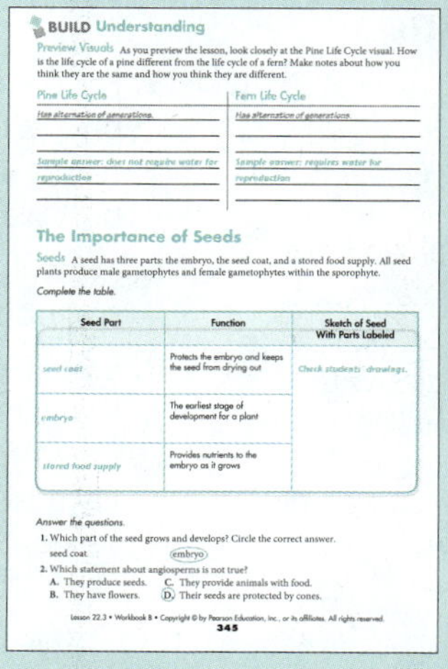

BUILD Understanding

Preview Visuals As you preview the lesson, look closely at the Pine Life Cycle visual. How is the life cycle of a pine different from the life cycle of a fern? Make notes about how you think they are the same and how you think they are different.

Pine Life Cycle	Fern Life Cycle
Has alternation of generations.	*Has alternation of generations*
Sample answer: does not require water for reproduction	*Sample answer: requires water for reproduction*

The Importance of Seeds

Seeds A seed has three parts: the embryo, the seed coat, and a stored food supply. All seed plants produce male gametophytes and female gametophytes within the sporophyte.

Complete the table.

Seed Part	Function	Sketch of Seed With Parts Labeled
seed coat	Protects the embryo and keeps the seed from drying out	*Check students' drawings.*
embryo	The earliest stage of development for a plant	
stored food supply	Provides nutrients to the embryo as it grows	

Answer the questions.

1. Which part of the seed grows and develops? Circle the correct answer.
seed coat (embryo)
2. Which statement about angiosperms is not true?
A. They produce seeds. C. They provide animals with food.
B. They have flowers. (D.) Their seeds are protected by cones.

Lesson 22.3 • Workbook B • Copyright © by Pearson Education, Inc., or its affiliates. All rights reserved.
345

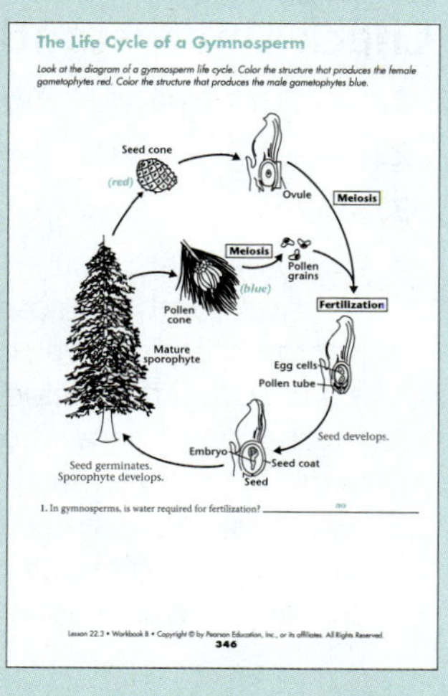

The Life Cycle of a Gymnosperm

Look at the diagram of a gymnosperm life cycle. Color the structure that produces the female gametophytes red. Color the structure that produces the male gametophytes blue.

1. In gymnosperms, is water required for fertilization? *no*

Lesson 22.3 • Workbook B • Copyright © by Pearson Education, Inc., or its affiliates. All Rights Reserved.
346

Seeds After fertilization, the zygote in a seed grows into a tiny plant—the embryo. The embryo is diploid. It is the early developmental stage of the plant's sporophyte. The embryo often stops growing while it is still small and contained in the seed. The embryo can remain in this state for weeks, months, or even years.

A tough **seed coat** surrounds and protects the embryo. The seed coat keeps the contents of the seed, including the embryo, from drying out. Seeds can survive long periods of cold, heat, or drought. When conditions are once again favorable, the embryo inside the seed begins to grow. The embryo uses nutrients from the food supply in the seed until it can carry out photosynthesis on its own.

Key Question What adaptations allow seed plants to reproduce without standing water?
Cones and flowers, the transfer of sperm by pollination, and the protection of embryos in seeds make it possible for seed plants to reproduce without standing water.

BUILD Vocabulary

pollination the transfer of pollen from the male reproductive structure to the female reproductive structure

seed coat the tough covering that surrounds and protects the plant embryo and keeps the contents of the seed from drying out

RELATED WORD FORMS
The verb *pollination* comes from the word *pollen*. Pollination occurs when pollen grains are carried to the female structures of a plant.

Students Say

"Remember the game "I'm going on vacation," where one person would say, "I'm going on vacation and I'm going to take an apple," and the next person would say, ". . . an apple and a basketball"? We do that to practice vocabulary in groups of three or four. The first person says a word and the definition. The second person repeats that and adds another vocab word and definition. We go around the circle a couple of times—repeating what's been said, and adding another word."

During Reading

Active Reading

FIND THE MAIN IDEA Help students dissect The First Seed Plants paragraph on the previous page. Remind students that the topic sentence of a paragraph reflects the main idea.

Ask What are some key words in this paragraph that might help you figure out what it is about? *(Sample answers:* seeds, adaptations, *and* water*)*

Ask Which sentence do you think is the topic sentence and which ones support the topic sentence? *(The second sentence is the topic sentence and the next two sentences support it.)*

SUMMARIZE In their notebooks, have students use one or two sentences to summarize the main idea and supporting details of The First Seed Plants paragraph.

KEY QUESTION Invite a volunteer to read the question and answer.

Ask How did these adaptations make possible the variety of plant life on Earth today? *(The adaptations allowed plants to live in many different environments. As plants adapted to different environments in different ways, new species developed.)*

Build Vocabulary

RELATED WORD FORMS Have students read the Related Word Forms note on the student page.

Ask What do you call an animal that carries pollen grains to a plant's female reproductive structures? *(a pollinator)*

Point out that the noun *pollinator* is formed by adding the suffix *-or*, meaning "one who," to the verb *pollinate*. Ask students to give examples of pollinators. *(Sample answers: butterfly, bee, hummingbird)*

Build Connections

VISUAL SUMMARY Direct students' attention to the visual Reproduction in Seed Plants. Help them connect the headings in the visual to the blue headings in the text on these two pages. Have a volunteer read each connecting paragraph.

Ask What part of the plant life cycle are cones and flowers? *(gametophytes)*

Ask How are pollen grains carried to the female cones or flowers? *(mainly by wind)*

Ask How do seeds differ between gymnosperms and angiosperms? *(gymnosperms—seeds directly on the scales; angiosperms—seeds enclosed in a layer of ovary tissue)*

LESSON 22.3

During Reading continued

Active Reading

SET PURPOSE FOR READING Help students change their focus from a general discussion of seeds on the previous pages of this lesson to one on gymnosperms.

Ask What kinds of plants are the focus of these two pages? *(gymnosperms: conifers, cycads, and gingkoes.)*

Ask With what kinds of conifers are you familiar? *(Sample answers: pines, firs, spruces, cedars, hemlocks, and yews.)*

SELF-QUESTION Have students reread this page individually or in pairs. As they read, have them use index cards to write questions they still have about the life cycle of a gymnosperm. Have them post their questions on a question board and search for answers while they read. After students are finished reading, read the questions, have volunteers attempt answers, and then discuss as a class.

USE VISUALS Direct students' attention to the photo on this page.

Ask Is this a male cone or a female cone? Explain your answer. *(It is a male cone because it is releasing pollen.)*

Ask How does the pollen reach the female cones? *(Wind carries the pollen.)*

Build Vocabulary

WORD ORIGINS Tell students that these words also come from the Latin word *ovum*: *ovoid*, *ovulate*. Give students time to look up these words in a dictionary.

Ask Which word is an adjective that means "egg-shaped"? *(ovoid)*

Ask Which word is a verb that means "to produce ovules"? *(ovulate)*

KEY QUESTION Invite a volunteer to read the question and answer.

Ask Have you heard of people having an allergy to pollen? *(probably)* How do people come in contact with pollen? *(It blows in the wind.)* Lead students to understand that the same process that leads to allergy symptoms in many people is a necessary stage in the life cycle of gymnosperms.

The Life Cycle of a Gymnosperm

The word *gymnosperm* means "naked seed." The name comes from the fact that the seeds of gymnosperms are exposed on the scales in their cones. The most common gymnosperms are conifers, such as pine, spruce, cedar, and fir trees. Gymnosperms also include rare plants such as cycads and ginkgoes. The life cycle of a pine tree is shown on the next page.

Pollen Cone The pollen cone of a pine tree releases large amounts of pollen. Wind blows the pollen grains to the seed cones.

Pollen Cones and Seed Cones Gymnosperms have two kinds of cones: pollen cones and seed cones. Pollen cones, also called male cones, produce pollen grains. Each tiny pollen grain is a complete male gametophyte. One nucleus in the pollen grain divides, forming two sperm nuclei.

Female cones, or seed cones, are usually much larger than pollen cones. Near the base of each scale of a seed cone are two ovules. **Ovules** (AHV yoolz) are the structures in which the female gametophytes develop. In the ovules, meiosis produces haploid cells. The haploid cells grow and divide to form female gametophytes. Female gametophytes may have thousands of cells. When fully grown, a female gametophyte has only a few large egg cells. Each of these egg cells is ready to be fertilized.

Pollination The life cycle of a conifer begins when male cones release large numbers of pollen grains. These pollen grains are carried away by the wind. Some of the pollen grains reach female cones. The scales of the female cone are sticky, and the pollen grains stick to them. These grains are pulled inside, toward the ovule.

Fertilization and Development If a pollen grain lands near an ovule, the grain splits open and begins to grow a **pollen tube.** The pollen tube contains two sperm nuclei. Once the pollen tube reaches the female gametophyte, one sperm nucleus fertilizes the egg. The other sperm nucleus breaks down.

Fertilization produces a zygote. The zygote grows into an embryo—the new sporophyte plant. The embryo is then surrounded by a seed coat, forming a seed. The seed then falls to the ground, or is carried away by wind, water, or animals. If it lands where conditions are right, the seed will grow into a new plant.

Key Question How does fertilization take place in gymnosperms without water?
In gymnosperms, wind carries pollen grains to the female cone, where fertilization takes place.

BUILD Vocabulary

ovule
in seed plants, the structure in which the female gametophytes develop

pollen tube
the structure in a seed plant that contains two haploid sperm nuclei

WORD ORIGINS
The word *ovule* comes from the Latin word *ovum*, which means "egg." Ovules are the structures inside which female gametophytes develop and produce haploid gametes, or eggs.

Focus on Academic Language

- fertilization
- development

Have students find the terms *fertilization* and *development* on this page. Ask them to use context clues to define each term. *(fertilization: the formation of a zygote by the joining of a sperm and egg; development: changes that a zygote goes through as it forms an embryo)*

Say Both words can have different meanings.

Ask When referring to lawns, crops, and gardens, what does *fertilization* mean? *(Students should describe a process of adding nutrients to the soil.)*

Ask Have you heard someone say, **"I live in a certain development"**? What does *development* mean in this case? *(Students should describe a grouping of homes or apartment buildings.)*

Have students use each word in a sentence for each of the two meanings described above.

Pine Life Cycle In the life cycle of pine trees and most other gymnosperms, the mature sporophytes produce male and female cones, which contain gametophytes.

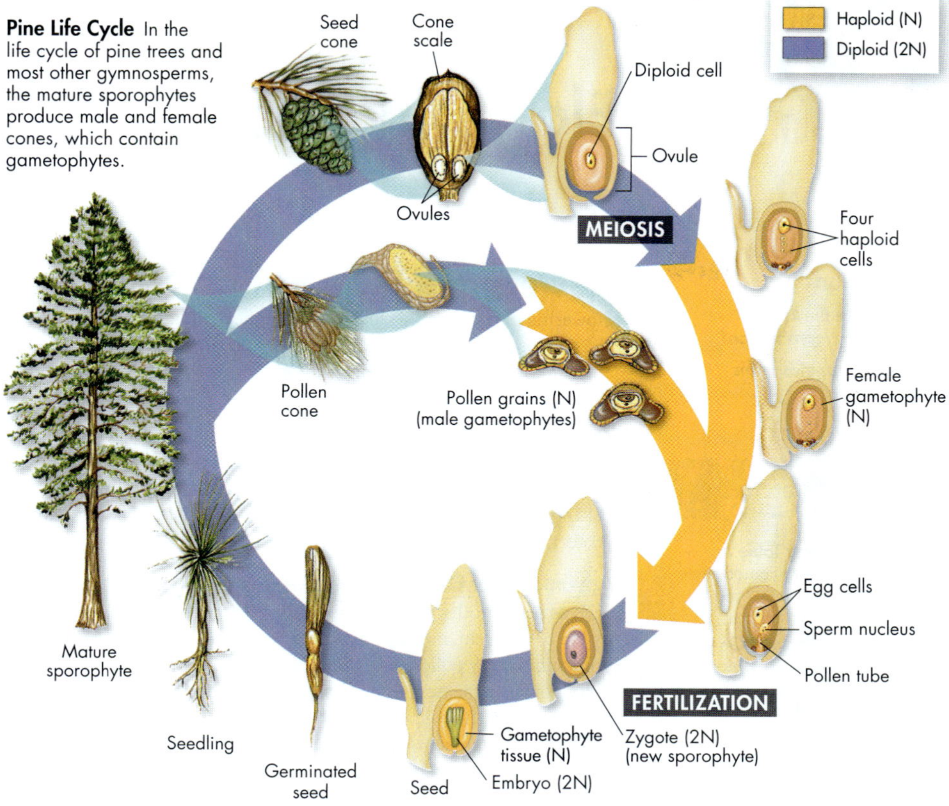

CHECK Understanding

Apply Vocabulary

Use the highlighted words from the lesson to complete each sentence correctly.

1. In seed plants, the male gametophyte is contained within a ________.

2. The young sporophyte embryo is protected within the seed by the ________.

3. ________ are seed plants that produce seeds in flowers.

Critical Thinking

4. **Review** List three adaptations of seed plants that allow them to reproduce without water.

5. **Apply Concepts** Classify the following structures of a pine tree as haploid or diploid: embryo, pollen grain, ovule, egg cell, sperm, zygote, seed cone, and seedling.

6. **Write to Learn** Answer the mystery clue. Use the word *conifer* in your answer.

MYSTERY CLUE

An unfinished bow made from conifer wood was found with Iceman. Why might Iceman carry a bow? What does the fact that the bow was not finished indicate about Iceman's journey?

BIOLOGY.com Search Lesson 22.3 GO • Lesson Assessment • Art in Motion

Check Understanding

1. pollen grain

2. seed coat

3. Angiosperms

4. cones and flowers, pollen grains or the transfer of sperm during pollination, and seeds

5. Pollen grains, egg cells, and sperm are haploid. Embryos, ovules, zygotes, seed cones, and seedlings are diploid.

6. Possible answers are that Iceman carried the bow for hunting or for protection. The unfinished bow could indicate that the trip was unexpected.

Active Reading

USE DIAGRAMS For each student, make an 11" x 17" copy of the two-page spread in the student book. Have students use highlighters to mark explanations in the text and connect them to matching parts of the Pine Life Cycle diagram.

Speed Bump

Students may think that a life cycle has a specific beginning or end. Place your finger at various places on the Pine Life Cycle diagram. Question the students at each place.

Ask Could this be the beginning of the life cycle? *(Answers may vary. Say* yes *at each place to help students understand that in a cyclic process, the cycle can begin anywhere.)*

Post Reading

ASSESSMENT TIPS For Question 4, point out to students that they are to list three adaptations.

For Question 5, suggest students review the meaning of *haploid* and *diploid* and then make a checkbox chart with those terms as the headings and the plant parts listed down the side. Tell students they can fill their charts in and use them for their answers for Question 5.

MYSTERY CLUE Be sure students understand that the bow referred to is from a bow and arrow. Discuss what the bow's material indicates about the Iceman's biome.

Wrap-Up Activity

20 QUESTIONS Divide students into groups of four. In each group, student A secretly chooses an important word from the lesson. The other three students take turns gathering clues by asking student A a question about the word. Student A can answer only *yes* or *no*. After one round of questions, students guess the identity of the secret word. Repeat until all students have had a chance to choose secret words and have been questioned about their secret words.

LESSON 22.4

Pre-Reading

Objectives

22.4.1 Identify the reproductive structures of angiosperms.

22.4.2 Identify some of the ways angiosperms can be categorized.

Preview the Pages

1. Preview the vocabulary words by having students find the words in the text and read each one aloud.
2. Direct students' attention to the visuals in this lesson. Ask students how they can tell what information each visual shows. *(by reading the title of the visual and the caption and labels as well as looking at the visual)*
3. Refer to the visual Angiosperm Clades. Remind students that a clade is a group of living things that includes a single common ancestor species and all its descendants.

Build Understanding

COMPARE/CONTRAST TABLE As a group, make a table that compares and contrasts what students know about gymnosperms with what they read about angiosperms in the second paragraph on this page. Then, have students complete the table for this lesson in Workbook B.

NATIONAL SCIENCE EDUCATION STANDARDS

UNIFYING CONCEPTS AND PROCESSES

I, IV, V

CONTENT

C.3.b, C.3.d, C.3.e

INQUIRY

A.2.a

For the corresponding lesson in *Miller & Levine Biology,* see pages 650–654.

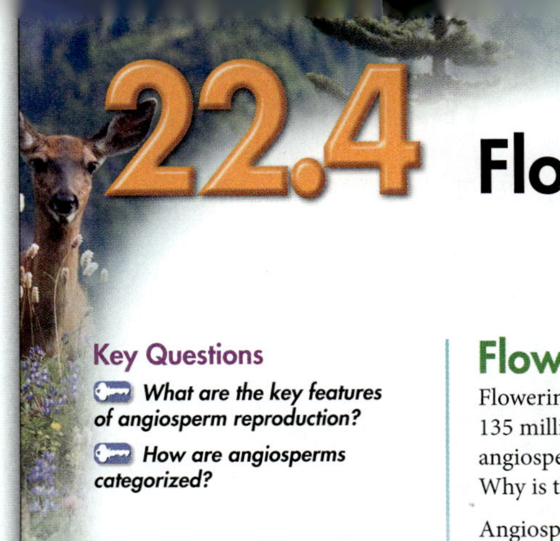

22.4 Flowering Plants

Key Questions

What are the key features of angiosperm reproduction?

How are angiosperms categorized?

BUILD Understanding

Compare/Contrast Table As you read, make a table to compare the three methods commonly used to put angiosperms into groups.

In Your Workbook Go to your workbook to learn more about using tables to compare and contrast characteristics. Complete the table for Lesson 22.4.

Flowers and Fruits

Flowering plants, or angiosperms, first appeared on land about 135 million years ago. They soon came to dominate the Earth. Today, angiosperms are the most abundant organisms in the plant kingdom. Why is this group of plants so successful?

Angiosperms have unique reproductive organs known as flowers. Flowers contain **ovaries,** which surround and protect seeds. The ovary gives angiosperms their name: *angiosperm* means "enclosed seed." After fertilization, ovaries develop into fruits. Fruits surround, protect, and help to spread the seeds.

Advantages of Flowers Most flowers attract animal pollinators, such as bees, moths, or hummingbirds. These animals are drawn by the color, scent, or shape of the flower. When they leave a flower, pollinators carry pollen with them. As they do this, the animals carry pollen to the next flower they visit. Pollination by animals is much more efficient than pollination by wind. Because they make pollination so efficient, flowers are a great evolutionary advantage to a plant.

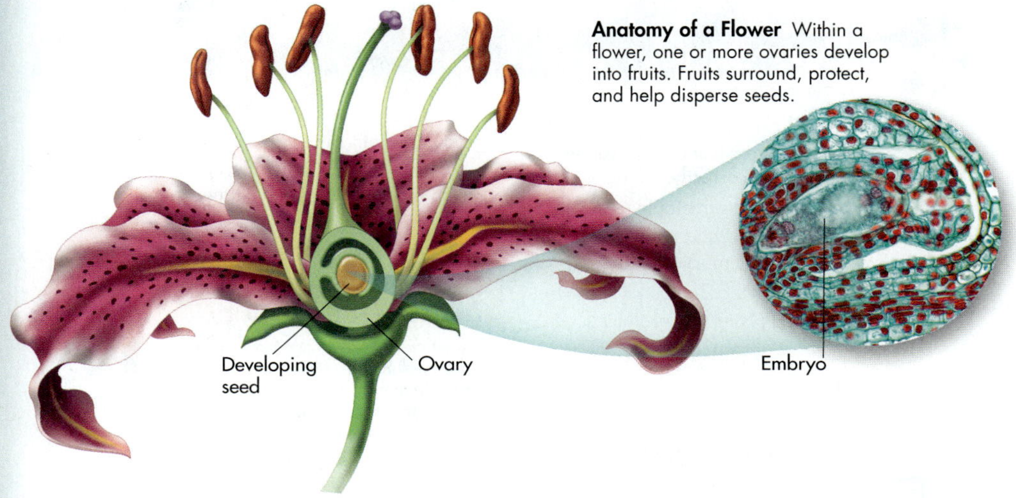

Anatomy of a Flower Within a flower, one or more ovaries develop into fruits. Fruits surround, protect, and help disperse seeds.

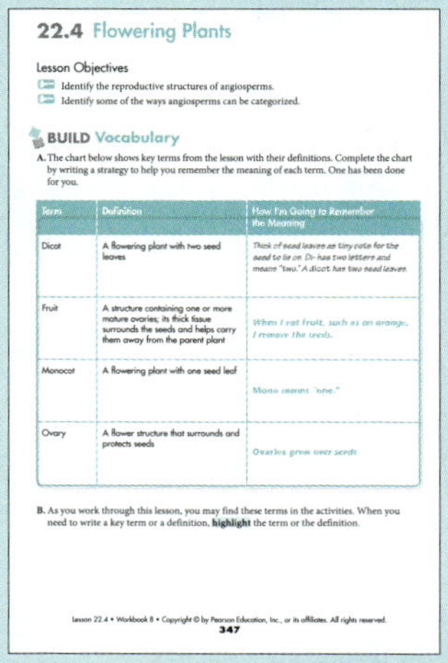

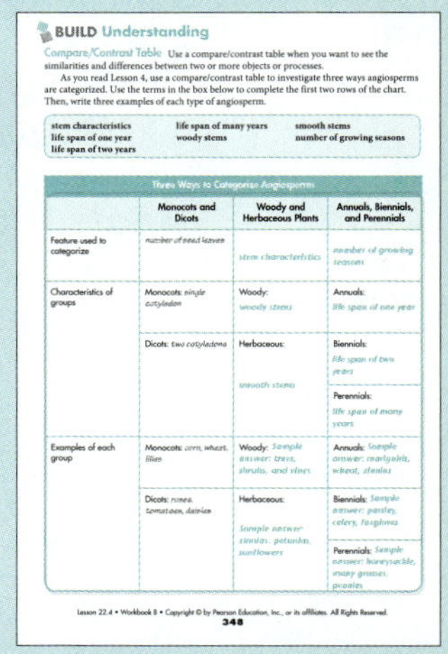

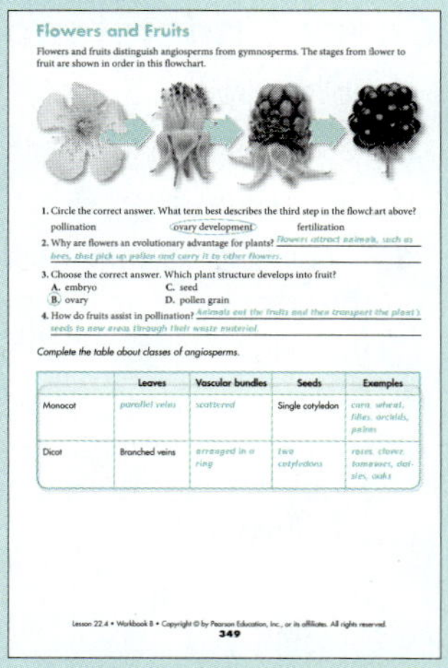

From Flower to Fruit After fertilization, the many ovaries of a blackberry flower develop into a cluster of individual fruits.

Advantages of Fruits After pollination, the ovary develops into a fruit. A **fruit** is a structure that contains one or more seeds. Fruits play an important role in spreading seeds. Consider what happens when a bird eats a fruit, such as a berry. Seeds from the berry enter the bird's digestive system. The bird may fly far from the parent plant before dropping the seeds. In this way, seeds can be spread over large areas of land. Thus, fruits are another reason why angiosperms are so successful.

Key Question What are the key features of angiosperm reproduction? **Angiosperms reproduce sexually by means of flowers. After fertilization, the ovaries develop into fruits. Fruits surround, protect, and help to spread the seeds.**

INQUIRY into Scientific Thinking

What Forms Do Fruits Take?

Fruits vary widely. Some fruits, such as apples and pears, are fleshy. Other fruits, such as dandelion and maple fruits, have structures specialized for dispersal by wind. Some fruits, such as those of cockleburs, have structures that ensure the fruit is caught in the fur of a passing animal.

Procedure

1. Use a hand lens to examine different fruits. Write down your observations and make sketches of the fruits.
2. Place each fruit in a petri dish and use a scalpel to dissect it. **CAUTION:** Use care with sharp instruments.
3. Locate the seeds within each fruit. Write down your observations and make sketches of the dissected fruits.

Analyze and Conclude

1. **Compare and Contrast** Make a table that compares the characteristics of the fruits you observed.
2. **Infer** Fleshy fruits are often spread by animals. Dry fruits are often spread by the wind. For each of the fruits you examined, identify how the fruit is likely spread. Explain your answers.

In Your Workbook Get more help for this activity in your workbook.

Materials

hand lens, variety of fruits, petri dishes, scalpel

Analyze and Conclude

1. Students should recognize how fruits are similar to each other and how they differ. Sample answer: For apples, pears, maple fruits, and acorns: All of the fruits contain seeds. Apples and pears are fleshy fruits that have multiple seeds. Maple fruits and acorns are dry fruits that have one seed. Acorns have a hard, protective shell that is not found in the other fruits.
2. Sample answer: The seeds of apples, pears, and acorns are likely spread by animals. Apples and pears have fleshy skin that animals can eat. Acorns are collected by squirrels and other animals. Maple seeds have "wings" that allow dispersal by wind.

During Reading

Active Reading

SUMMARIZE Have students summarize the Advantages of Flowers and Advantages of Fruits paragraphs with a focus on the role animals play.

Ask How do animals help with pollination? *(They carry pollen from flower to flower.)* How do they help disperse seeds? *(They eat fruits and drop seeds far from the parent plant.)*

Have students record their summary statements in their notebooks.

KEY QUESTION Invite a volunteer to read aloud the question and answer.

Ask What is the relationship between flowers and fruits? *(The ovary of a flower develops into a fruit.)*

Inquiry Into Scientific Thinking

USE VISUALIZATION Tell students to imagine shopping in the grocery store produce section and to call out the different types of produce they see. Write these examples on the board.

Ask Which of these are fruits? *(any that contain seeds, even if they are called vegetables)* Mark each fruit with a check as students reply.

Say Let's add to the list fruits that you might see in a lawn, on plants, or in pictures.

Ask What kinds of fruits do you see that are not in a grocery store? *(Answers may vary. Prompt students regarding plants in your local environment such as palm nuts, maple seeds, and cactus prickly pears.)*

APPLY ACADEMIC VOCABULARY Ensure students understand the meaning of the word *disperse*.

Ask What are some synonyms for the word *disperse*? *(Sample answers: scatter, spread, separate)*

Have students use *disperse*, or a form of the word, in sentences unrelated to plants or fruits. *(Sample answer: The crowd dispersed after the ball game.)*

Lead a Discussion

Guide students to understand that fruit structure is an adaptation for dispersal as well as protection. Remind students that there are about 260,000 species of flowering plants. Tell them that although the fruits they eat are likely produced on farms, thousands more flowering plants exist in the wild.

Ask Do you think fruits of wild plants are fleshy, dry, or both? *(both)* How do you think the seeds of fleshy fruits are dispersed? *(Animals eat the fleshy fruits and the seeds pass through their bodies and are dropped elsewhere.)* How do you think the seeds of dry fruits are dispersed? *(usually by the wind)*

Ask How does seed dispersal help the parent plant? *(There is less competition for resources for growth.)*

During Reading continued

Build Vocabulary

PREFIXES Point out to students that, like *di-*, the prefix *bi-* also means "two." Focus students' attention on the words *annual, biennial,* and *perennial* in the blue subheading.

Ask Why do you think these words are listed together? *(Students might guess that the second two words are a version of* annual *or that all of the words have something to do with time.)*

Say The word *annual* is from the Latin word for "year." The prefix *bi-* is Latin for "two." The prefix *per-* is Latin for "throughout."

Ask Now how do you think the words are related? *(Sample answer: They are all periods of time.* Annual *means "every year,"* biennial *means "every two years," and* perennial *means "continuing throughout many years.")*

Active Reading

INTERPRET CHARTS Be sure students understand how to read the Comparing Monocots and Dicots chart. Tell students to run their fingers across the top row of pictures as you explain that this row shows and describes the plant parts of monocots.

Ask How would you describe the leaves of monocots? *(They have parallel veins.)* Be sure students understand the meaning of *parallel*.

Have students run their fingers across the bottom row of pictures.

Ask How are dicot seeds different from monocot seeds? *(Dicot seeds have two cotyledons and monocot seeds have one cotyledon.)*

DRAW CONCLUSIONS Tell students to think of dandelions. Emphasize that dandelions are considered weeds.

Ask Why is it often difficult to completely pull dandelions out of a lawn? *(because they have long taproots)* So, what kind of plant is a dandelion? *(a dicot)*

KEY QUESTION Ask a volunteer to read aloud the question and answer.

Ask How is putting plants into categories helpful? *(Categories can help you group plants with similar characteristics so that you can identify the plants and know other characteristics to expect of them.)*

BUILD Vocabulary

ovary a structure in flowering plants that surrounds and protects seeds

fruit a structure that contains one or more matured ovaries

cotyledon the first leaf or first pair of leaves produced by the embryo of a seed plant

monocot an angiosperm with one seed leaf in its ovary

dicot an angiosperm with two seed leaves in its ovary

PREFIXES

Mono- means "one," and *di-* means "two." Monocots have one seed leaf. Dicots have two seed leaves.

Angiosperm Diversity

Flowering plants range in size from tiny duckweed—a plant smaller than your fingernail—to giant trees. Angiosperms can be described by the structure of their stems, their length of life, or the number of their seed leaves.

Woody and Herbaceous Plants One of the easiest ways to describe a plant is by the characteristics of its stem. Woody plants, such as trees and shrubs, have stems that are hard and live from year to year. Herbaceous (hur BAY shus) plants do not form wood as they grow. Their stems are softer than the stems of woody plants. Dandelions, petunias, and sunflowers are herbaceous plants.

Annuals, Biennials, and Perennials Plants can also be described by their length of life. Flowering plants called annuals grow, flower, form seeds, and die in a single year. Cucumbers, pansies, and marigolds are annuals. Many other plants continue to grow from year to year. Biennials, such as foxgloves, grow for two years. Perennials, such as asparagus, can live for many years.

Monocots and Dicots Until recently, scientists classified flowering plants according to the number of seed leaves, or **cotyledons** (kaht uh LEED uns) in their embryos. Plants with one seed leaf were called **monocots.** Monocots included wheat, lilies, and palms. Plants with two seed leaves were called **dicots.** Dicots included roses, tomatoes, and oak trees. Monocots and dicots also differed in the structure of their leaves, stems, roots, and flowers.

Key Question How are angiosperms categorized?
Angiosperms are often grouped according to the characteristics of their stems, their life span, or the number of their seed leaves.

Comparing Monocots and Dicots
This table compares the structure of monocots and dicots.

Characteristics of Monocots and Dicots

	Seeds	Leaves	Flowers	Stems	Roots
Monocots	Single cotyledon 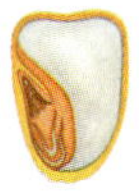	Parallel veins 	Floral parts often in multiples of 3 	Vascular bundles scattered throughout stem 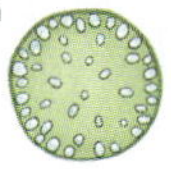	Fibrous roots 
Dicots	Two cotyledons 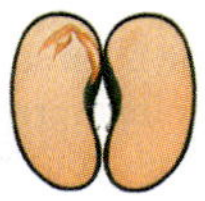	Branched veins 	Floral parts often in multiples of 4 or 5 	Vascular bundles arranged in a ring 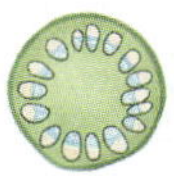	Taproot 

542 BIOLOGY.com Search Lesson 22.4 GO • Art in Review

Focus on Academic Language

• abundant • advantage • categorized • dominate • efficient • successful

Have students find each of the terms in this lesson and use context clues to define each one. Discuss and clarify meanings as necessary. Have students use each one in a sentence. Encourage students to form sentences that use the word in the context of plants as well as sentences that use the word in the context of other everyday experiences.

Amborella Clade Only one species still exists in this oldest branch of angiosperms. Its floral parts have a spiral arrangement.

Water Lily Clade The water lilies are another very old group. Early water lily flowers may have been no more than 1 cm across. Today water lilies are large and showy.

Magnoliids This clade contains a wide range of flowers, from species that have small, plain flowers to the dinner-plate sized *Magnolia* flower shown here.

Monocots This clade contains about 20 percent of all angiosperms. Monocots include important crops, such as rice, corn, and wheat. They also include orchids, lilies, and irises.

Eudicots About 75 percent of angiosperms are eudicots. This clade is nearly as old as the angiosperms themselves. Eudicots diversified greatly several times in their history.

Angiosperm Clades Five major clades of angiosperms are shown here. Scientists are still working out how angiosperm clades are related.

Modern Classification Recall that biologists classify organisms according to their evolutionary relationships. Recent studies of fossils and plant genetics show that the history of angiosperms is very complicated. These discoveries are changing how biologists classify flowering plants. The art shows one modern view of how angiosperms could be classifed.

Biologists still classify the monocots in a single group. Dicots, however, have been divided into several different groups. Although it is no longer used in scientific classification, the term *dicot* is useful for describing plant structures. In the following chapters, the term is used for that purpose.

CHECK Understanding

Apply Vocabulary

Use the highlighted words from the lesson to complete each sentence correctly.

1. A plant that has leaves with parallel veins and flowers with six petals is a ______.

2. After pollination and fertilization, the ______ of a flower develops into a fruit.

3. A ______ has two cotyledons.

Critical Thinking

4. Review What reproductive structures are found only in angiosperms? Briefly describe the function of each.

5. Compare and Contrast List three common ways of describing or grouping flowering plants. How are these three methods different from scientific classification?

6. Write to Learn Answer the next clue of the mystery. Use the words *monocot* and *seed leaf* in your answer.

MYSTERY CLUE

Scientists found fragments of wheat in Iceman's digestive tract. Scientists also discovered that the wheat was planted and grown by people. How did scientists determine that the seed fragments came from a monocot? (**Hint:** See p. 542.)

BIOLOGY.com Search Lesson 22.4 GO • Lesson Assessment

Check Understanding

1. monocot

2. ovary

3. dicot

4. Flowers and fruits are unique to angiosperms. Flowers make pollination more efficient by attracting pollinators; fruits protect and help disperse seeds.

5. Flowering plants can be described by the structure of their stems, the length of their life, or the number of their seed leaves. Unlike scientific classification, these methods of categorizing plants are not based on evolutionary relationships.

6. Accept all reasonable answers. Students should recognize that monocots have one seed leaf. Students may suggest that cultivated wheat differs from wild wheat in kernel size or genetics.

Active Reading

USE VISUALS Direct students' attention to the chart of Angiosperm Clades. Have volunteers read aloud the caption for each group.

Ask Which clade makes up 20 percent of all angiosperms and includes crop species? *(monocots)* Which clade has spiral parts in a floral arrangement? *(amborella)* Which clade has many different kinds of flowers? *(magnoliids)* Which clade makes up 75 percent of all angiosperms and is quite old? *(eudicots)* Which clade has flowers that are much larger now than they were in the past? *(water lily)*

Speed Bump

Ask How do biologists group flowering plants now? *(Biologists group flowering plants into monocots and several different groups of dicots.)*

Post Reading

ASSESSMENT TIPS For Question 4, have the class review the lesson and determine how many structures should be described in this question.

For Question 5, discuss with students the meaning of *scientific classification*.

MYSTERY CLUE Encourage students to review the chart comparing monocots and dicots and the visual summary of the different clades for help in determining how scientists identified the wheat.

Wrap-Up Activity

MONOCOT/DICOT INTERVIEWS Have pairs of students represent themselves as a monocot and dicot. Tell students to interview each other about their characteristics and to take notes during the interviews. After the second student is interviewed, tell each pair to compare notes to the text and add any additional important details.

CHAPTER LAB

Pre-Lab

Discuss the meaning of the word *survey* to ensure students understand the focus of the activity. Ask where students have heard the word before and what they think it means. *(Students may be familiar with the noun* survey, *meaning "a poll.")*

Have students find sentences on this page that include the word *survey.* Challenge students to use context clues to determine the part of speech and the meaning of *survey* in each instance. Then, have them look up the definition in a dictionary.

Ask What does the text mean when it says that you will survey a small ecosystem? *(We will do a detailed inspection of the ecosystem, gathering data on what plants are in the area.)*

Review the Background Questions and the Pre-Lab Questions as a class.

Lab

Tell students they will perform the chapter lab *Exploring Plant Diversity* described in **Lab Manual B.**

SAFETY Ask students about allergies to plants. Avoid areas with poison ivy, poison oak, or poison sumac. Students should wear gloves while handling plants and wash their hands after the lab.

Post Lab

CREATE A FIELD GUIDE Create a class field guide of the plants in your ecosystem. Have students combine information from their survey with other information that they find online or in published field guides. Create a separate page for each species, including a color drawing, a description, and notes on the plant's habitat.

NATIONAL SCIENCE EDUCATION STANDARDS

UCP I

CONTENT C.4.c, C.4.e, C.5.e, F.4

INQUIRY A.1.b, A.2.a, A.2.b

Real-World Lab

OPEN-ENDED INQUIRY

Pre-Lab: Exploring Plant Diversity

Problem How many different kinds of plants are in a small ecosystem?

Materials notebook, protective work gloves, measuring tape, tweezers, scissors, small plastic bags, labels, hand lens, field guides for plants, camera (optional)

Lab Manual Chapter 22 Lab

Skills Focus Observe, Measure, Classify, Infer

Connect to the Big idea There are more than 290,000 known species of plants that exist on Earth—from tiny green algae to large-leaved ferns to imposing redwood trees. What do these species have in common? They all need light, carbon dioxide, oxygen, water, and minerals to survive. But plant species vary in the way they obtain and retain resources.

Plants have adaptations that allow them to succeed in different habitats. Thus, you will not find 290,000 species of plants in your community. But you should be able to find a variety of plants. In this lab, you will survey a small ecosystem and identify as many plant species as possible.

Background Questions

a. Review What are the features that botanists use to divide the plant kingdom into five major groups?

b. Compare and Contrast Compare abiotic and biotic factors in an ecosystem.

c. Applying Concepts Suppose an ecosystem includes a stream or pond. Would you include that part of the ecosystem in your plant survey? Explain your answer.

Pre-Lab Questions

Preview the procedure in the lab manual.

1. **Design an Experiment** What are some ways that you can make sure that you survey all the plants in your ecosystem?
2. **Classify** What should you do if you are not sure that an organism is a plant?
3. **Infer** Why might you want to use a regional field guide rather than a national field guide when identifying plants?

BIOLOGY.com Search Chapter 22 GO

Visit Chapter 22 online to test yourself on chapter content and to find activities to help you learn.

Untamed Science Video The Untamed Science biologists interview plant experts to learn about healing chemicals manufactured by plants.

Tutor Tube Is this plant a girl? Compare animals and plants to understand plant reproduction.

Art Review See how well you can distinguish monocots and dicots.

InterActive Art Review and compare life cycles of vascular and nonvascular plants.

Art in Motion Follow the process of pollination and fertilization in a pine to see how a plant embryo is formed.

544 Chapter 22 • Pre-Lab

Pre-Lab Answers

BACKGROUND QUESTIONS

a. The features used to divide plants into five major groups are embryo formation, specialized water-conducting tissues, seeds, and flowers.

b. Biotic factors are all the living organisms in a ecosystem. Abiotic factors are the physical or nonliving parts of an ecosystem.

c. Sample answer: Yes, because some plants, such as green algae and water lilies, are aquatic.

PRE-LAB QUESTIONS

1. Answers may vary, but students might suggest dividing the area into smaller sections and investigating one section at a time. Or they may suggest doing the survey by type of plants—trees, shrubs, and so on.
2. Sample answer: I would record my observations of the organism and use reference materials to decide whether the organism is a plant.
3. Sample answer: The national field guide will include plants that don't grow nearby and exclude some plants that do grow nearby.

22 CHAPTER Summary

22.1 What Is a Plant?

- To survive, plants need sunlight, water, and minerals. They also need to exchange gases with the air.
- Over time, life on land favored the evolution of plants with adaptations that helped them survive the drying rays of the sun, prevented the loss of water, and allowed them to reproduce without water.
- Plants have alternation of generations, or life cycles with two phases. The two phases are a diploid (2N) sporophyte phase and a haploid (N) gametophyte phase.

alternation of generations (p. 530)
sporophyte (p. 530)
gametophyte (p. 530)

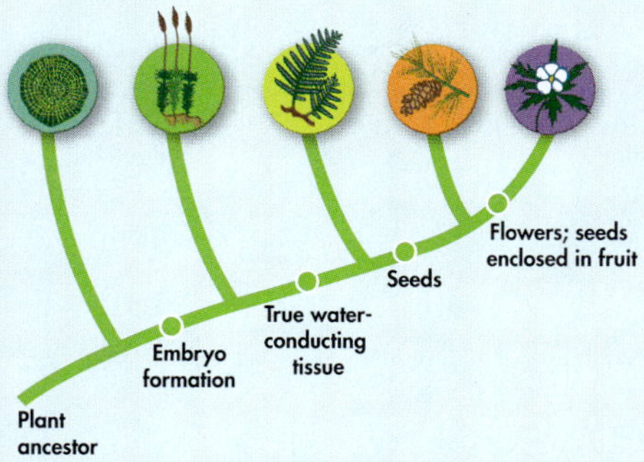

22.2 Seedless Plants

- Green algae live mostly in fresh water or salt water; some live in moist areas on land. Some green algae are multicellular.
- Bryophytes are small because they lack vascular tissue.
- Vascular tissues—xylem and phloem—allow the movement of fluids through the body of a plant against the force of gravity.

bryophyte (p. 533)
vascular tissue (p. 533)
xylem (p. 534)
phloem (p. 534)

22.3 Seed Plants

- Adaptations that allow seed plants to reproduce without standing water include cones or flowers, the transfer of sperm by pollination, and the protection of embryos in seeds.
- Gymnosperms do not need water for reproduction, because wind carries pollen grains to female cones, where fertilization takes place.

seed (p. 536)
gymnosperm (p. 536)
angiosperm (p. 536)
pollen grain (p. 536)
pollination (p. 536)
seed coat (p. 537)
ovule (p. 538)
pollen tube (p. 538)

22.4 Flowering Plants

- Angiosperms reproduce sexually by means of flowers. After fertilization, ovaries within flowers develop into fruits. Fruits surround, protect, and help to spread the seeds.
- Angiosperms are often grouped according to the characteristics of their stems, their life span, or the number of seed leaves.

ovary (p. 540)
fruit (p. 541)
cotyledon (p. 542)
monocot (p. 542)
dicot (p. 542)

UbD Transfer the Big Idea

TRANSFER Have students work in small groups to make a field guide for the five main groups of plants. The guide should be a booklet that high-school and middle school students could use to identify plants. Students should include drawings or photos from the local environment as examples of each kind of plant.

Chapter Summary

Think Visually

USE GRAPHICS Have students complete the following compare/contrast table about the major plant groups. Have students fill in the table with *yes* or *no* answers for each blank box.

Comparison of Major Plant Groups

	Vascular Tissue?	Dominant Sporophyte?	Seeds?
Green Algae	no	1. ______	2. ______
Bryophytes	3. ______	no	4. ______
Seedless Vascular Plants	5. ______	6. ______	no
Gymnosperms	7. ______	8. ______	9. ______
Angiosperms	yes	10. ______	yes

ANSWERS

1. no **2.** no **3.** no **4.** no **5.** yes **6.** yes **7.** yes **8.** yes **9.** yes **10.** yes

BIOLOGY.com Study Online

REVIEW RESOURCES

InterActive Art Students examine plant reproduction with these interactive life cycle diagrams.

Data Analysis Students will analyze data to learn more about bracken, a type of fern.

Art in Motion Students watch the fertilization and the development of a pine embryo.

Tutor Tube This short, online tutorial compares sexual reproduction in plants and animals.

Vocabulary Review The *Flash Cards* and *Crossword* are an interactive vocabulary review.

Editable Worksheets Editable pages of Study Workbook B, Lab Manual B, and the Assessment Resources Book are available online.

Lesson Overviews Lesson Overviews are based on Miller & Levine Biology. However, they are available on CD-ROM and can be edited for your class.

ASSESSMENT RESOURCES

Chapter Assessment Have students take an online version of the Chapter 22 Check Understanding.

Standardized Test Prep Students can take an online version of the Standardized Test Prep. You will receive their scores along with ideas for remediation.

Diagnostic and Benchmark Tests Use these tests to monitor your students' progress and supply remediation.

Assess

Big idea Invite a volunteer to read the Big Idea and the question.

Ask What are the five main groups of plants? *(green algae, bryophytes, seedless vascular plants, gymnosperms, and angiosperms)*

Ask What change in plants allowed them to grow very tall? *(evolution of vascular tissue)*

Ask How does vascular tissue help plants? *(It provides strength and structure as well as a way for water and minerals to travel from the roots to the rest of the plant and nutrients to travel from the leaves to the rest of the plant.)*

Ask Which two groups of plants can reproduce without water? *(gymnosperms and angiosperms)* Why is this important? *(It allows plants to grow away from moist and watery areas.)*

Ask Why do you think angiosperms are more successful than all other kinds of plants? *(Flowers enable pollination in varied ways, and fruits are adapted for widespread dispersal.)*

CONSTRUCTED RESPONSE Paragraphs may vary. Suggest that students begin their paragraphs with the main idea sentence and add sentences that provide supporting details. Remind them that a new main idea starts a new paragraph. Be sure students understand that the *Hints* should be used as a guide to help them answer the questions.

Write Question 1 on the board and read it aloud.

Ask What are the key words in the question? *(evolution, vascular tissue, structure)* Underline these words.

Say This question is about the evolution of vascular tissue. We know from the question that vascular tissue somehow is involved with the structure of a plant. This means that the evolution of vascular tissue enabled the evolution of different kinds of plants.

Read the first hint.

Ask What are the names of the vascular tissues? *(xylem and phloem)* Which tissue carries water and which carries nutrients? *(water: xylem, nutrients: phloem)*

Read the second hint.

Ask How might the lack of vascular tissue make these plants different from other land plants? *(Lack of vascular tissue might make them smaller than other land plants.)*

Model writing the first one or two sentences using the hints on the board and the text. *(Sample: The evolution of vascular tissue in land plants caused differences in structure. Land plants with vascular tissue had a way of getting water and nutrients to all parts of the plant without living near water.)*

For Question 2, pair students and have them use the hints and your modeling to answer the question. Have students share their paragraphs with the class.

For Question 3, have students write the paragraphs independently for homework or as an anchor activity.

22 CHECK Understanding

Assess the Big idea — Unity and Diversity of Life

Write an answer to the question below.

Q: What are the five main groups of plants, and how have four of these groups adapted to life on land?

Constructed Response

Write an answer to each of the numbered questions below. The answer to each question should be one or two paragraphs long. To help you begin, read the **Hints** below the questions.

1. **How did the evolution of vascular tissue affect the structure of land plants?**

   Hint Vascular tissue carries water and nutrients throughout the bodies of plants.

   Hint Mosses and other bryophytes do not have vascular tissue.

2. **Choose a group of seedless vascular plants and a group of seed plants. Then, compare the process of reproduction in these two groups.**

   Hint Ferns are seedless vascular plants. Conifers and flowering plants are seed plants.

   Hint Seeds have three parts: the embryo, the seed coat, and a supply of food. Each has an important role in protecting the young plant.

3. **Why are flowering plants the most numerous group of plants alive today?**

   Hint Angiosperms have two unique structures, flowers and fruits.

   Hint Animals play an important role in two parts of the plant life cycle.

Foundations for Learning Wrap-Up

Use the index cards you prepared when reading the chapter to help you organize your thoughts about the five types of plants.

Activity 1 Work with a partner to quiz each other on what you know about the five groups of plants. Shuffle your index cards and take turns sorting the cards based on whether each card describes green algae, bryophytes, seedless vascular plants, gymnosperms, or angiosperms.

Activity 2 Add one card to your deck and write Plants on it. Starting with the card labeled Plants, lay your cards on the table. Use pieces of yarn to create a cluster diagram that describes the five groups of plants.

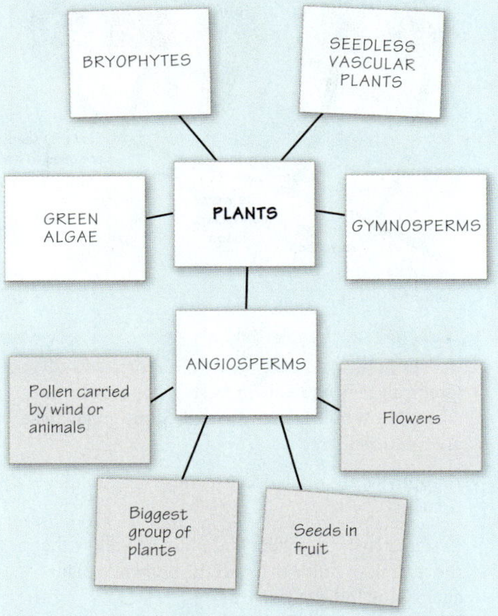

Foundations for Learning Wrap-Up

Have students take out the index cards on which they have been writing individual characteristics of plants throughout the chapter. Spot-check cards to be sure students understood that they were writing characteristics of the five groups of plants. Have students work with partners to complete the activities.

ASSESSMENT

22.1 What Is a Plant?

Understand Key Concepts

1. Which of the following is NOT a characteristic of plants?
   a. eukaryotic cell
   b. multicellular structure
   c. cell walls containing chitin
   d. chlorophyll

Test-Taking Tip

Process of Elimination When answering multiple-choice questions, identify answers that you know are wrong. In question 1, you know that all plants are eukaryotes and have chlorophyll. You know that plants have many cells. Therefore, the answer must be **c**, cell walls containing chitin.

2. Two gases that plants must exchange with the air are
   a. oxygen and nitrogen.
   b. oxygen and carbon dioxide.
   c. carbon dioxide and nitrogen.
   d. carbon dioxide and carbon monoxide.
3. Land plants likely evolved from organisms that are similar to which modern group of plants?
   a. green algae
   b. angiosperms
   c. gymnosperms
   d. seedless plants
4. List the two phases of the plant life cycle.

Think Critically

5. **Sequence** Four characteristics that biologists use to distinguish the major groups of plants are flowers, embryos, seeds, and vascular tissue. List these characteristics in the order they arose.

22.2 Seedless Plants

Understand Key Concepts

6. In vascular plants, water is carried from the roots to the body of the plant by
   a. xylem.
   b. phloem.
   c. cell walls.
   d. chloroplasts.
7. In ferns, underground stems are called
   a. fronds.
   b. rhizoids.
   c. rhizomes.
   d. vascular tissues.
8. Why are green algae classified as plants and not as protists?
9. What do seedless plants need for fertilization to occur?

Think Critically

10. **Explain** How does the structure of cells in vascular tissue play a role in the function of vascular tissue?

22.3 Seed Plants

Understand Key Concepts

11. The male gametophyte of seed plants is found in
    a. sperm.
    b. ovules.
    c. ovaries.
    d. pollen grains.
12. Which of the following is NOT part of a seed?
    a. ovule
    b. seed coat
    c. stored food
    d. young plant
13. Describe pollination in gymnosperms such as conifers.
14. What is the function of the seed coat?

Think Critically

15. **Compare and Contrast** Describe how gymnosperms and angiosperms are similar. Then, describe how they differ.

Assessment Tips

For Question 5, make sure students realize that *sequence* means "order" and encourage students to review the text, cladogram, and pie graph after the subhead The Evolution of Plants in Lesson 22.1.

For Question 10, have students review the text on vascular tissue as well as the photograph of the horsetail and its caption in Lesson 22.2. For Question 15, tell students they will need to draw conclusions about likenesses and differences after reviewing the first two pages of Lesson 22.3.

For Question 21, make sure students realize that the exact answer is not given in the textbook. Suggest students compare the plant life cycles in Lessons 22.1, 22.2, and 22.3 before deciding on an answer.

For Questions 22 and 23, have students review the visual table that compares monocots and dicots in Lesson 22.4.

Answers

Lesson 22.1

UNDERSTAND KEY CONCEPTS

1. c

2. b

3. a

4. haploid gametophyte phase and diploid sporophyte phase

THINK CRITICALLY

5. embryo, vascular tissue, seeds, flowers

Lesson 22.2

UNDERSTAND KEY CONCEPTS

6. a

7. c

8. Sample answer: Like land plants, green algae cells have cell walls with cellulose and chlorophyll *a* and *b*. Green algae also undergo alternation of generations.

9. Sample answer: Seedless plants need a watery environment so the sperm cells can swim to the eggs.

THINK CRITICALLY

10. Sample answer: Cells are hollow and tube-like. They are connected end to end, which allows fluids to flow from one cell to the next.

Lesson 22.3

UNDERSTAND KEY CONCEPTS

11. d

12. a

13. Pollen grains are transferred from male pollen cones to female seed cones by the wind.

14. A seed coat surrounds and protects the young plant embryo. The seed coat keeps the contents of the seed from drying out.

THINK CRITICALLY

15. Gymnosperms and angiosperms are both seed plants. Gymnosperms produce seeds in cones, but angiosperms reproduce using flowers and fruits.

ASSESSMENT

Lesson 22.4

UNDERSTAND KEY CONCEPTS

16. b

17. d

18. Woody plants have stems made primarily of cells with thick cell walls that support the plant body. Herbaceous plants have stems that are softer and may die down every year.

19. Animals eat fleshy fruits, and the seeds go through their digestive system. The seeds are carried away from the parent plant.

THINK CRITICALLY

20. Fruits that are spread by wind would be likely to have structures that catch the wind, such as broad leaf-like blades, or dandelion puffs. Fruits that are spread by animals may be fleshier, and have bright colors to attract animals.

21. In mosses, the gametophyte is the dominant phase. It carries out photosynthesis; the sporophyte is dependent on it for water and nutrients. In ferns, the gametophyte is much smaller than the sporophyte, but lives independently. In seed plants, the gametophyte is very tiny and dependent on the sporophyte for water and nutrients.

Connecting Concepts

22. a dicot, because it has five petals

23. Because it is a dicot, it should also have leaves with branched veins and may have a taproot.

CHAPTER MYSTERY After students read through the Chapter Mystery, discuss the clues and how scientists have interpreted those clues.

Ask What clues lead scientists to conclude that Iceman died in late spring? *(the leaves and pollen in his digestive tract)*

Ask Why might scientists think Iceman changed his elevation on his last day? *(Sample answer: The kind or amount of pollen in Iceman might indicate he was in places that had different kinds or numbers of trees.)*

Discuss how knowledge of plants is necessary for the interpretation of the clues found with Iceman.

Ask Why do archaeologists and other scientists often talk with other specialists about their findings? *(Sample answer: One scientist might find meaning in something that another scientist misses, or different scientists might interpret clues differently. By discussing other interpretations, scientists can come to better conclusions.)*

22 CHECK Understanding

22.4 Flowering Plants

Understand Key Concepts

16. In angiosperms, seeds are surrounded and protected by
- **a.** cones.
- **b.** fruits.
- **c.** flowers.
- **d.** seed leaves.

17. What kind of plants have life cycles that last several years?
- **a.** dicots
- **b.** annuals
- **c.** monocots
- **d.** perennials

18. How do woody plants differ from herbaceous plants?

19. How do fruits help spread seeds?

Think Critically

20. **Infer** Some fruits have adaptations that enable the seeds to be spread by wind. How might the structure of these fruits differ from the structure of fruits for seeds that are spread by animals?

21. **Compare and Contrast** Compare the size and function of the gametophyte in mosses, ferns, and seed plants.

Connecting Concepts

Use Science Graphics

Use the photograph to answer questions 22 and 23.

22. **Classify** Study the photograph of the flower above. Is this plant a monocot or dicot? Explain your answer.

23. **Infer** Would you expect the leaves of this plant to have parallel or branched veins? What you expect the root system to have a taproot? Explain your reasoning.

solve the CHAPTER MYSTERY

STONE AGE STORYTELLERS

Scientists learn more about Iceman as they study the plant materials found with him. The large amount of chlorophyll in the maple leaves is common in spring and summer. Pollen is usually produced in spring. The leaves and the pollen found in Iceman's digestive tract suggest that he died in late spring.

The unfinished bow suggests that Iceman's trip into the mountains was not planned. He may have been running away from an enemy. Scientists later learned that a stone arrowhead was lodged beneath Iceman's left shoulder blade. The arrowhead cut a major blood vessel.

How did Iceman and his people live? The wheat in his digestive tract and other grains found on his clothes show that Iceman's society practiced an early form of farming.

1. **Apply Concepts** A clump of moss was also found with Iceman's possessions. Scientists hypothesize that Iceman used the moss the way that we use paper towels today. What property of moss makes this possible?

2. **Infer** Pollen and seeds are the most reliable plant-related evidence at archeological sites and at modern-day crime scenes because they are long lasting. Relate this quality to their structure and function.

UntamedScience **Never Stop Exploring Your World.** Iceman used plants in ways that made sense for the time he lived in. Take a video field trip with the ecogeeks of Untamed Science to see how far humans have come in realizing what plants have to offer.

CHAPTER MYSTERY ANSWERS

1. Because mosses lack vascular tissue, they must be well adapted to absorb water through their cell walls. Iceman may have used clumps of moss to soak up water or blood from a cut.

2. Pollen has to be able to withstand a certain amount of harsh conditions so that the male gametophyte is protected as it travels. Seeds have a tough outer coat to protect the embryo inside. The seeds may need to last a long time to survive harsh conditions until the environment is favorable for growth.

Untamed Science Have students take a video field trip to explore the many drugs humans derive from plants by watching **Paging Dr. Greenthumb.**

Standardized Test Prep

Multiple Choice

1. Which of the following is a basic requirement of plants?
- **A** sunlight
- **B** carbon dioxide
- **C** water
- **D** all of the above

2. What stage in the alternation of generations is represented by fern fronds?
- **A** sporophyte
- **B** female gametophyte
- **C** male gametophyte
- **D** zygote

3. Which of the following is NOT a characteristic of dicots?
- **A** branched veins
- **B** taproot
- **C** parallel veins
- **D** two seed leaves

4. Which of the following is a structure associated with gymnosperms?
- **A** flower
- **B** cone
- **C** fruit
- **D** enclosed seed

5. In flowering plants, the mature plant ovary is also referred to as the
- **A** gymnosperm.
- **B** pollen grain.
- **C** fruit.
- **D** seed leaf.

6. Which is the largest group of plants?
- **A** gymnosperms
- **B** woody plants
- **C** angiosperms
- **D** bryophytes

Questions 7–9

A group of students placed a conifer sprig in a beaker of water. They measured the amount of oxygen given off during a set period of time to determine the rate of photosynthesis. They changed the temperature of the water in the beaker using an ice bucket and a hot plate. Their data are summarized in the graph below.

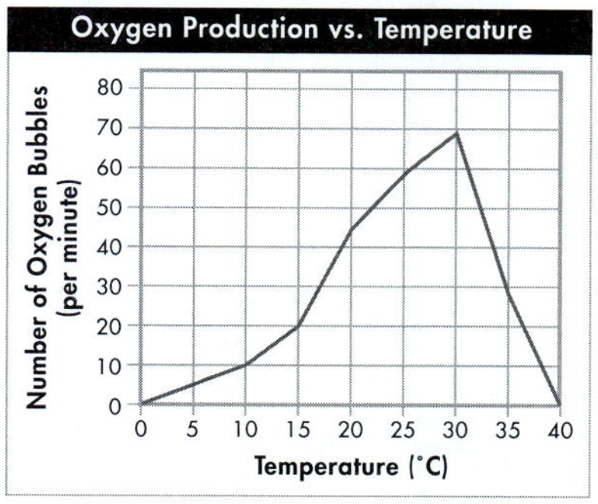

7. What is the independent variable?
- **A** light intensity
- **B** temperature
- **C** oxygen bubbles
- **D** photosynthesis rate

8. Which variable(s) should the students have held constant?
- **A** plant type
- **B** temperature
- **C** light intensity
- **D** plant type and light intensity

9. What can you conclude based on the data?
- **A** The higher the temperature, the more oxygen bubbles are released.
- **B** There is an optimum temperature for photosynthesis in this species of conifer.
- **C** All plants are most efficient at 30°C.
- **D** The lower the temperature, the more oxygen bubbles are released.

Open-Ended Response

10. Explain why seeds were an important adaptation for the success of plants on Earth.

If You Have Trouble With . . .

Question	1	2	3	4	5	6	7	8	9	10
See Lesson	22.1	22.2	22.4	22.3	22.4	22.4	22.3	22.3	22.3	22.3

Before the Test

Review with students the Test-Taking Tip.

In Question 1, be sure students know that the word *requirement* means "need."

Point out the word *NOT* in Question 3.

In Question 6, make sure students realize that *largest* refers to numbers of plants, not sizes of plants.

Remind students that the paragraphs they write to answer an open-ended question should contain many of the same facts as another student's paragraph, but the order of the facts, examples used, and style of writing probably will not be the same.

Answers

1. D
2. A
3. C
4. B
5. C
6. C
7. B
8. D
9. B
10. Seeds make it possible for plants to reproduce without being dependent on water. Seeds also enable plants to disperse their offspring over long distances and protect the embryos from harsh environmental conditions.

Test-Taking Tip

USE TIME WISELY

Explain to students that if they find a particular question difficult, they should put a light pencil mark beside the question and keep working. Tell them to not spend a lot of time trying to figure out the answer, because they might run out of time to finish the test. Point out that, as they answer later questions, they may find information that helps them answer the difficult question.

CHAPTER PLANNER 23 Plant Structure and Function

Chapter Contents	NSES	Time	Core Resources
Chapter Preview			**Student Edition,** pp. 550–551 **Chapter Mystery,** p. 551 **Foundations for Learning,** p. 551
23.1 Specialized Tissues in Plants Structure of Seed Plants • Tissue Systems in Plants • Plant Growth and Meristems	I, V, A.1.b, A.2.a, C.1.a, C.1.f, C.5.d	1/2 period 1/4 block	**Student Edition,** pp. 552–555 **Inquiry Into Scientific Thinking,** p. 553 **Study Workbook B** 23.1 Worksheets **Biology.com** 23.1 Lesson Assessment
23.2 Roots Root Structure and Growth • Root Functions	I, V, C.5.d	1 period 1/2 block	**Student Edition,** pp. 556–559 **Build Connections,** p. 558 **Study Workbook B** 23.2 Worksheets **Biology.com** *Art in Motion:* Water Passage Into a Root • 23.2 Lesson Assessment
23.3 Stems Structures and Function of Stems • Growth of Stems	I, V, C.5.d	1 period 1/2 block	**Student Edition,** pp. 560–563 **Study Workbook B** 23.3 Worksheets **Biology.com** *InterActive Art:* Primary and Secondary Growth • *Tutor Tube:* Understanding Plant Growth
23.4 Leaves Structure and Function of Leaves • Gas Exchange and Homeostasis	I, IV, V, C.1.e, C.5.d	1 period 1/2 block	**Student Edition,** pp. 564–566 **Study Workbook B** 23.4 Worksheets **Biology.com** *Art Review:* Anatomy of a Leaf • 23.4 Lesson Assessment
23.5 Transport in Plants Water Transport • Nutrient Transport	I, V, C.5.d	1/2 period 1/4 block	**Student Edition,** pp. 567–569 **Build Connections,** p. 567 **Study Workbook B** 23.5 Worksheets **Biology.com** *Visual Analogy:* Transpirational Pull •
Chapter Pre-Lab	II, III, A.1.a, A.1.b, A.1.f, A.2.a, C.1.f	1 period 1/2 block	**Student Edition,** p. 570 **Lab Manual B** Identifying Growth Zones in Roots

Tools for ELL Instruction

ELL Handbook describes teaching strategies to make *Miller & Levine Biology: Foundation Edition* more accessible to ELL students.

Spanish Study Workbook is a Spanish translation of Study Workbook A.

Multilingual Glossary is the *Miller & Levine Biology* glossary translated into ten languages.

Tools to Challenge Students

Study Workbook A includes more challenging worksheets and activities such as Chapter Mystery activities that reinforce 21st Century Skills.

Lab Manual A labs are geared toward on-level students. The Lab Manual includes fifteen labs in addition to the Chapter Labs.

Biology.com Data Analysis activities offer students a way to practice data gathering and analyzing skills.

Visual Quizzes in the Assessment Resources Book are geared toward on-level students and use visuals from *Miller & Levine Biology.*

Additional Resources

Biology.com Untamed Science Video • Flash Cards

Study Workbook A 23.1 Worksheets
Spanish Study Workbook 23.1 Worksheets

Study Workbook A 23.2 Worksheets
Spanish Study Workbook 23.2 Worksheets
Assessment Resources Book Visual Quiz

Study Workbook A 23.3 Worksheets
Lab Manual B *Data Analysis:* Reading a Tree's History
Spanish Study Workbook 23.3 Worksheets
Assessment Resources Book Visual Quiz

Study Workbook A 23.4 Worksheets
Spanish Study Workbook 23.4 Worksheets
Biology.com *Data Analysis:* Effects of Humidity and CO_2 on Plant Stomata

Study Workbook A 23.5 Worksheets
Spanish Study Workbook 23.5 Worksheets

Lab Manual A *Identifying Growth Zones in Roots*

Chapter Review

Student Edition Chapter Summary, p. 571
Study Workbook B Chapter 23 Review
Transparencies, pp. 268–279
Biology.com Untamed Science Video • Editable Worksheets of Study Workbooks A and B and Lab Manuals A and B • Chapter 23 Flash Cards and Match It

Untamed Science DVD • Classroom Resources CD (includes lesson presentations and editable worksheets)

Chapter Assessment

Student Edition Check Understanding, pp. 572–575
Study Workbook B Chapter 23 Taking a Standardized Test
Assessment Resources Book Chapter 23 Test B
Biology.com Chapter 23 Assessment • Editable Worksheets of Chapter 23 Visual Quizzes* and Chapter 23 Tests A and B

Exam*View* *Assessment Suite* • Classroom Resources CD (includes lesson presentations and editable worksheets)

Time: 1 period, 1/2 block

* Visual Quizzes are based on visuals from *Miller & Levine Biology.* Review carefully to make sure they are appropriate for your class.

Pressed for Time?

Cover the Chapter Quickly Have students read *Structure of Seed Plants* and *Tissue Systems in Plants* in Lesson 23.1, focusing on the Vascular Tissue visual. In Lesson 23.2, assign *Root Structure and Growth,* and read the Key Question and answer in *Root Functions.* Assign *Structures and Function of Stems* in Lesson 23.3, *Structure and Function of Leaves* in Lesson 23.4, and go over the Inside a Leaf visual. Assign *Water Transport* in Lesson 23.5.

Assess Assign Questions 1, 2, 4, and 5 in the 23.1 Check Understanding, Questions 1, 3, and 4 in the 23.2 Check Understanding, Questions 1–4 in the 23.3 Check Understanding, Questions 1, 2, and 4 in the 23.4 Check Understanding, and Questions 1, 2, and 4 in the 23.5 Check Understanding. In the Chapter 23 Check Understanding, assign Questions 1–4, 8, 9, 14, 16, 18–21, and 24.

CHAPTER 23

Get the Picture

Tell students the photograph shows an insect trapped on a leaf of a sundew, a carnivorous plant. Explain that carnivorous plants live in areas where the soil is nitrogen poor. They get nitrogen by digesting insects and other small animals.

Ask Why is nitrogen important to living things? *(Sample answers: Nitrogen is important for cell processes. Living things need nitrogen to grow.)*

Be sure students understand that proteins and nucleic acids contain nitrogen.

Tell students the sundew gets its name from the droplets that look like dew on its red leaf hairs. Explain that the droplets are a sticky fluid that contains digestive enzymes.

Ask How is the sundew able to capture and digest insects? *(Insects get stuck in the fluid, and the digestive enzymes break down the insects' tissues.)*

Activate Prior Knowledge

Help students recall that living things require food for energy and growth.

Ask How do you get energy and the materials your body needs to grow? *(by eating and digesting food)* How do most plants get energy and materials for growth? *(by undergoing photosynthesis, although some nutrients also come from the soil)* Emphasize that plants need carbon dioxide and water to perform photosynthesis. But they also need other nutrients, such as nitrogen, to make proteins and nucleic acids.

Tell students in this chapter they will study how the organization of plant cells, tissues, and organs helps them obtain materials they need for energy and growth.

NATIONAL SCIENCE EDUCATION STANDARDS

UNIFYING CONCEPTS AND PROCESSES

I, II, III, IV, V

CONTENT

C.1.a, C.1.e, C.1.f, C.5.d

INQUIRY

A.1.a, A.1.b, A.1.f, A.2.a

23 Plant Structure and Function

Big idea Structure and Function

Q: How are cells, tissues, and organs organized into systems that carry out the basic functions of a seed plant?

BIOLOGY.com Search Chapter 23 GO • Flash Cards

550

UbD Connect to the Big Idea

Invite a volunteer to read aloud the chapter number and the chapter title. Read the Big Idea and question. Emphasize the question, **How are cells, tissues, and organs organized into systems that carry out the basic functions of a seed plant?**

Set the Purpose

PREVIEW AND PREDICT Draw the graphic organizer on the board. Guide students to the INSIDE feature on the next page. Invite a volunteer to read each lesson title.

Say Plants have structures that are adapted to their functions. A structure is a part of an organism's body or the way the body or body part is put together, or organized. A function is what the body or body part does.

Ask What are some structures of a plant's body? *(Sample answers: leaves, roots, stems)* What are some functions of these structures? *(Sample answers: leaves—capture sunlight; roots—anchor the plant; stems—support the plant)*

INSIDE:

The leaves of this sundew plant are adapted to capture and digest live prey.

- Untamed Science Video
- Chapter Mystery

CHAPTER MYSTERY

THE HOLLOW TREE

Imagine that you are hiking through a Central American rain forest on a hot afternoon. You see many unusual plants and animals. A monkey calls from a far-off tree, and a dense fog covers the landscape. Then, you stumble on a root and look up. You are standing in front of a huge tree. Its trunk seems to be made up of many tangled branches. Edging closer, you stick your head through a large opening between the branches. You look straight up. Inside, you find that the tree is hollow! This tree, a kind of fig, is indeed unusual.

Read for Mystery Clues What happened to the inside of the fig tree? And how did the tree grow so big if it has no center? As you read this chapter, look for clues that explain the structure of this strange plant.

FOUNDATIONS for Learning

Like many living things, plants have cells, tissues, and organs. Fold a sheet of paper in half lengthwise. Then, use a pair of scissors to make five flaps as shown below. On the flaps, write Tissue, Stem, Root, Leaf, and Transport. As you read each lesson, write what you learn beneath the appropriate flap. At the end of the chapter are two activities that will help answer the question: How are cells, tissues, and organs organized into systems that carry out the basic functions of a seed plant?

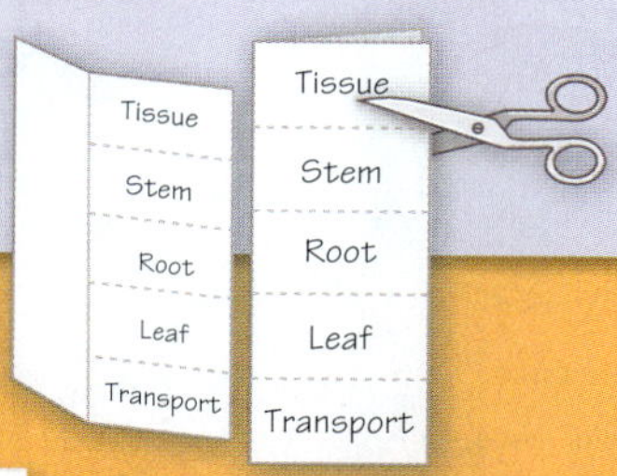

Chapter 23
Big Idea: Structure and Function

Chapter 23:
How are cells, tissues, and organs organized into systems that carry out the basic functions of a seed plant?

23.1: How are plant tissues organized?

23.2: How do the structure and function of roots help a plant carry out life processes?

23.3: How do the stucture and function of stems help a plant carry out life processes?

23.4: How do the structure and function of leaves help a plant carry out life processes?

23.5: How do plants move materials through their bodies?

Foundations for Learning

Have students use lined notebook paper. Help them understand how this organizer can be a helpful study tool. Emphasize that students are to record information about cells, tissues, and organs of plants as they read the chapter.

Ask What is the relationship between cells, tissues, and organs? *(Tissues are made up of many cells, and organs are made up of tissues.)*

Ask How can writing what you learn as you read help you study the chapter? *(Writing what you learn can help you remember it. The written notes can be used to review the chapter and prepare for tests.)*

Familiarize students with the activity by, as a group, writing the title of Lesson 23.1 beneath the first flap and recording one fact from that lesson under the title.

Untamed Science — Take a field trip around the world in **Plants Inside and Out** to see the amazing ways plants have adapted to their environments.

CHAPTER MYSTERY — Invite a volunteer to read aloud the Chapter Mystery. Ask students if they know what a fig is. Have students relate the mystery to the chapter title and hypothesize what happened to the inside of the fig tree.

Ask Do any other parts of a tree, besides the branches, look like the trunk? Which parts? *(The large roots of trees look like branches and are covered with bark just like the trunk.)*

Tell students that they will find clues in the chapter to answer the question, **What happened to the inside of the fig tree?**

Digital Support

BIOLOGY.com — Extend your reach by using the digital assets offered at **Biology.com.**

LESSON 23.1

Pre-Reading

Objectives

23.1.1 Identify the principal organs of seed plants.

23.1.2 Explain the primary functions of the main tissue systems of seed plants.

23.1.3 Contrast meristems with other plant tissues.

Preview the Pages

1. Preview the vocabulary words by having students find the words in the text and read each word aloud. Help students pronounce each word correctly.
2. Preview the visuals and read the captions and labels. Be sure students recognize that the blow-up pictures show details about how a section from each plant part would look under a microscope.
3. Call attention to some details in the blow-ups in the Vascular Tissue diagram. Point out the darkened spots in the cell walls of the tubes. Explain that they represent thin spots or holes in the cell walls.

Build Understanding

CONCEPT MAP Direct students to Workbook B, where they will find a partially filled-in concept map. Point out that the map summarizes the types of tissues that make up seed plants and the functions of the tissues. Encourage students to complete the concept map as they read about each type of tissue.

NATIONAL SCIENCE EDUCATION STANDARDS

UNIFYING CONCEPTS AND PROCESSES

I, V

CONTENT

C.1.a, C.1.f, C.5.d

INQUIRY

A.1.b, A.2.a

For the corresponding lesson in *Miller & Levine Biology,* see pages 664–668.

23.1 Specialized Tissues in Plants

Key Questions

What are the three main organs of seed plants?

What are the functions of the main tissue systems of seed plants?

How do meristems differ from other plant tissues?

BUILD Understanding

Concept Map As you read, make a concept map to organize the information in this lesson.

In Your Workbook Go to your workbook to learn more about concept maps. Complete the concept map for Lesson 23.1.

Structure of Seed Plants

Have you ever wondered if plants are really alive? Compared to animals, plants don't seem to do much. But if you could see inside a plant, this impression would disappear. Like animals, plants move materials, grow, repair themselves, and respond to their environment. Although their pace may seem slow, plants are, indeed, living organisms.

The cells of seed plants are organized into tissues, organs, and systems. The three main organs of seed plants are roots, stems, and leaves. Recall that an organ is a group of tissues that work together to carry out related functions.

Roots Roots hold plants in the ground and help hold the plant upright. Roots absorb water and nutrients from the soil. They then transport water and dissolved nutrients to the rest of the plant. Roots may also store food.

Stems Stems support the plant body. They contain vascular tissues that carry water and nutrients throughout the plant body. Stems help protect the plant against predators and disease. Stems also produce leaves and reproductive organs such as flowers.

Leaves Leaves are the plant's main photosynthetic organs. Oxygen and carbon dioxide move in and out of the leaf through tiny openings in the surface. Leaves also have adaptations that prevent water loss.

Key Question What are the three main organs of seed plants?

The three main organs of seed plants are roots, stems, and leaves.

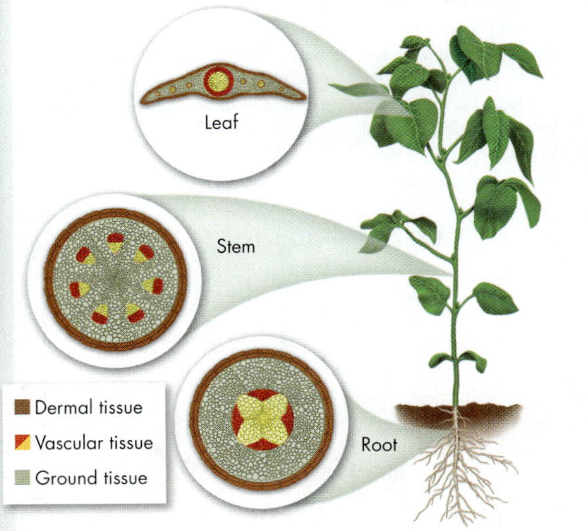

Main Organs of Plants These cross sections of the main organs of plants show that all three organs contain dermal tissue, vascular tissue, and ground tissue.

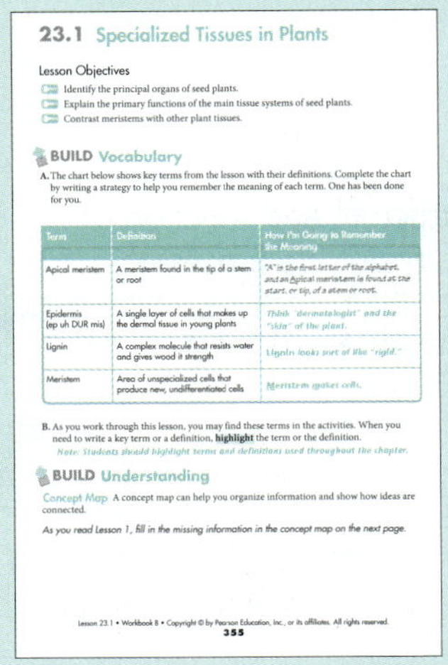

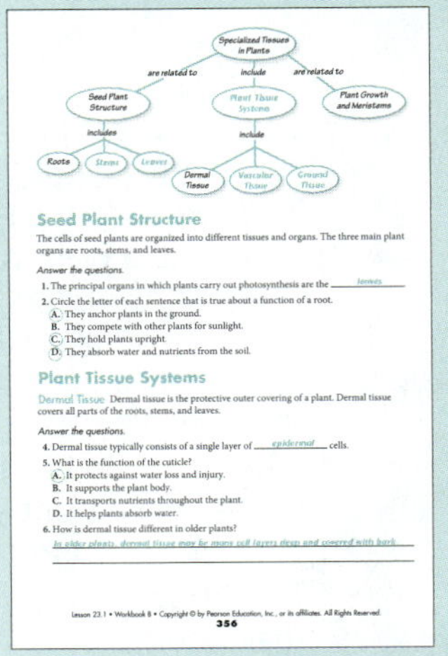

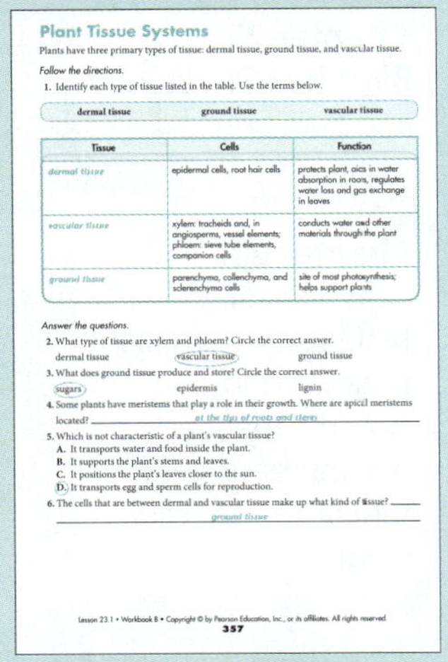

INQUIRY into Scientific Thinking

What Parts of Plants Do We Eat?

When you eat your vegetables, you may be eating the leaves, stems, roots, or even flowers of a plant!

Procedure

1. Examine the outside of an onion, a sweet potato, a cabbage, and an artichoke. Write down your observations of each vegetable. Include a sketch of each vegetable. Label the parts that you see. **CAUTION:** *Do not eat the vegetables.*
2. Use a sharp knife or scalpel to cut each vegetable in half. Draw what you see inside. **CAUTION:** *Use care with sharp instruments.*
3. Use your observations to classify each vegetable as a root, stem, leaf, or other plant part.

Analyze and Conclude

1. **Classify** How did you classify the onion? What characteristics did you use to make this decision?
2. **Infer** How did you classify the sweet potato? How is its structure related to its function?
3. **Infer** How did you classify the cabbage? What does its color tell you about its function?
4. **Infer** How did you classify the artichoke? What does its inner structure tell you about its function?

In Your Workbook Get more help for this activity in your workbook.

Tissue Systems in Plants

Roots, stems, and leaves are linked together by systems that run the length of the plant. These systems produce, store, and carry nutrients. They also provide support and protection for the plant. Plants have three main tissue systems: dermal, ground, and vascular.

Dermal Tissue Dermal tissue covers a plant the way that skin covers your body. In young plants, dermal tissue is made of a single layer of cells called the **epidermis** (ep uh DUR mis). The epidermis is often covered with a thick, waxy cuticle. The cuticle prevents water loss. In older plants, dermal tissue may have many layers of cells. It may be covered with bark.

Ground Tissue Ground tissue makes and stores food. Ground tissue also helps support the plant. The edible parts of plants such as potatoes and squash are mostly ground tissue.

Most ground tissue is made of **parenchyma** (puh RENG kih muh) cells. Parenchyma cells have a thin cell wall and a large central vacuole. In leaves, these cells contain many chloroplasts and carry out most of a plant's photosynthesis.

Ground tissue may also contain two types of cells with thicker cell walls. **Collenchyma** (kuh LENG kih muh) cells have strong cell walls that can bend. These flexible cells help support large plants. Chains of collenchyma make up the "strings" in a stalk of celery. **Sclerenchyma** (sklih RENG kih muh) cells have very thick, hard cell walls. These cells make some ground tissue, such as seed coats, tough and strong.

BUILD Vocabulary

epidermis
the single layer of cells that makes up dermal tissue of plants

parenchyma
the main type of ground tissue; contains cells with thin cell walls and large central vacuoles.

collenchyma
a type of ground tissue that has strong, flexible cell walls; helps support larger plants

sclerenchyma
a type of ground tissue with extremely thick and rigid cell walls that make ground tissue tough and strong

WORD ORIGINS
The terms *epidermis* and *dermal* have a common word root: *derm*. This root comes from the Greek word *derma*, which means "skin." Both terms are used to describe an outer, protective covering.

Materials

onion, potato, artichoke

Analyze and Conclude

1. The onion is a small stem surrounded by leaves, which are thin and flat and are attached to the stem at the base.
2. The sweet potato is a root. It may be swollen because it stores food.
3. The cabbage is a short stem to which many large leaves are attached. The green color indicates that cabbage leaves conduct photosynthesis.
4. The artichoke is a floral bud surrounded by modified leaves, which contain chloroplasts. The leaves are active in photosynthesis.

During Reading

Active Reading

KEY QUESTION Have volunteers read aloud the Key Question and answer on the previous page.

Ask What is an organ? *(a structure made of a group of tissues that work together to carry out related functions)*

INTERPRET DIAGRAMS Have students look at the Main Organs of Plants diagram on the previous page. Point out that the three main organs of plants—leaves, stems, and roots—all contain the same three types of tissue—dermal, ground, and vascular.

Ask How are the three types of tissue different in these organs? *(The tissues are arranged differently and carry out different functions in each organ.)*

Inquiry Into Scientific Thinking

SAFETY Have students wear safety goggles while performing this activity. Remind them to handle sharp instruments with care.

MAKE CONNECTIONS Remind students that most plants people eat are angiosperms, which are the only plants that produce fruits.

Ask What life process results in fruits? *(reproduction)* What is the function of fruits? *(protect and spread seeds)*

USE CONTEXT CLUES Have students read the introduction and pay special attention to what the word *vegetable* means.

Ask Which parts of plants are in the same sentence as the word *vegetables*? *(leaves, stems, roots, and flowers)* What do you think *vegetable* means? *(A vegetable is an edible leaf, stem, root, or flower.)*

Lead a Discussion

Invite students to name other types of vegetables. *(Sample answers: spinach, carrots, peas, broccoli)* Call on volunteers to classify each as a root, stem, leaf, or other plant part. *(spinach—consists of leaves; carrots—are roots; peas—are seeds; broccoli—consists of many tiny flower buds)* Then, challenge students to classify a tomato. Help them realize that a tomato, which contains seeds, is a fruit.

Build Vocabulary

WORD ORIGINS Invite students to think of other words that include the root *derm,* which means "skin." *(Sample answers: dermatologist, dermatitis)*

Say Elephants belong to a group of animals called pachyderms. *Pachy* means "thick."

Ask What kind of skin do elephants have? *(thick skin)*

During Reading continued

Active Reading

INTERPRET DIAGRAMS Have students look at the diagrams of phloem and xylem after reading the page. Encourage them to look for ways that the cells shown are similar and different.

Ask How would you describe the shape of the cells? *(They are shaped like tubes or pipes.)*

Ask What similarities do the walls of the vessel elements and sieve tube elements share? *(Both kinds of cells have many holes in their walls.)*

Ask Of the four types of cells found in vascular tissue, which ones are not shaped like tubes? *(companion cells)*

KEY QUESTION Invite a student to read aloud the question and answer.

Ask Which two types of tissue help support a plant's body? *(ground tissue and vascular tissue)*

Build Vocabulary

RELATED WORD FORMS Invite students to think of other examples of the adjective form of a noun.

Ask What is an adjective form of the noun *tube*? *(Sample answers: tube-like, tubular, tubed)*

Speed Bump

Ask What are the two types of vascular tissue in plants? *(xylem and phloem)*

Ask What does xylem transport? *(water)*

Ask What does phloem transport *(dissolved nutrients)*

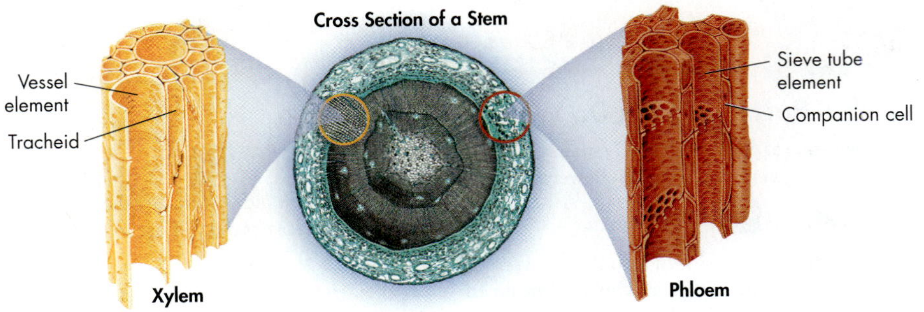

Vascular Tissue Xylem and phloem move water and nutrients throughout a plant.

Vascular Tissue Vascular tissue supports the plant body. It transports water and nutrients throughout the plant. One kind of vascular tissue is xylem, which carries water. The other kind of vascular tissue is phloem, which carries dissolved nutrients. Both xylem and phloem cells are long and thin, almost like pipes.

▶ ***Xylem*** Seed plants have xylem cells called *tracheids*. Tracheids have tough cell walls that help support the plant. These cell walls contain the compound **lignin,** which makes them stiff and strong. As they mature, tracheids die, leaving only their cell walls. Openings in tracheid cell walls let water flow from cell to cell. Thinner areas of tracheid cell walls, known as pits, allow water to move into ground tissue.

Angiosperms have a second kind of xylem made up of cells known as *vessel elements*. Vessel elements are wider than tracheids. Like tracheids, vessel elements die as they mature.

▶ ***Phloem*** Unlike xylem cells, phloem cells are alive at maturity. The main phloem cells are *sieve tube elements*. These cells are arranged end to end. They form tubes through which dissolved nutrients move. *Companion cells* surround sieve tube elements. Companion cells support phloem cells. They aid in the movement of substances in and out of the phloem.

Key Question What are the functions of the main tissue systems of seed plants? **Dermal tissue protects the plant. Ground tissue makes and stores food and supports the plant body. Vascular tissue supports the plant body and transports water and nutrients.**

BUILD Vocabulary

lignin
a substance in vascular plants that makes cell walls rigid

meristem
an area of unspecialized cells responsible for continuing growth throughout a plant's lifetime

apical meristem
a group of unspecialized cells that divide to increase the length of stems and roots

RELATED WORD FORMS
Apex and *apical* are related words. *Apex* is a noun that refers to a narrowed or pointed end. *Apical* is an adjective that describes something related to an apex.

Plant Growth and Meristems

When most animals become adults, they stop growing larger. This is not true for most plants. Even the oldest trees produce new leaves and new reproductive organs every year. How do plants do it? How do plants stay "forever young"? The secret of plant growth is found in meristems. **Meristems** are areas of unspecialized cells that produce new cells. These new cells are produced by mitosis. Meristems are found in places where plants grow rapidly, such as the tips of stems and roots.

Students Say

"I get more from the reading when the teacher hands us a web, and we fill in the web as we read. Then I know that there are five important things to know about protons, or there are six parts to this system."

Apical Meristems Meristems in the tips of stems and roots are called **apical meristems.** Unspecialized cells produced in apical meristems divide rapidly, increasing the length of stems and roots. At first, the new, thin-walled cells look very much alike. Gradually, they develop into cells with specialized structures and functions. This development process is called differentiation. As cells differentiate, they produce dermal, vascular, and ground tissue.

Meristems and Flower Development The specialized cells of flowers and cones are also produced by meristems. The development of flowers and cones begins when genes cause changes in the apical meristem. In flowering plants, apical meristems are changed into floral meristems. Floral meristems then produce all the specialized tissues found in flowers.

Key Question How do meristems differ from other plant tissues?
Meristems are areas of undifferentiated cells that produce new cells by mitosis. These new cells can then differentiate into specialized cells.

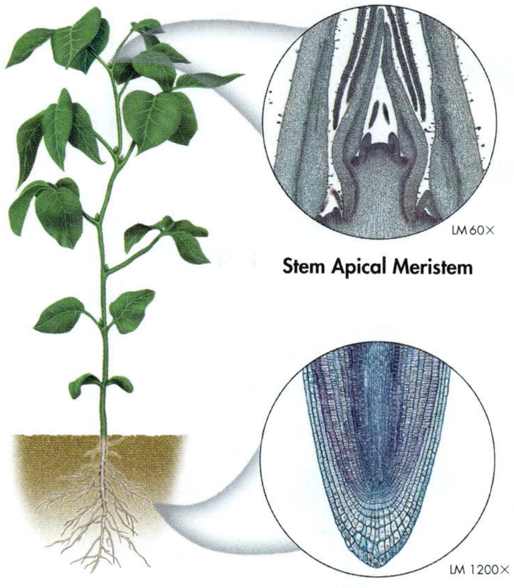

Apical Meristems Apical meristems are found in the growing tips of stems and roots. Within apical meristems, unspecialized cells are produced by mitosis.

CHECK Understanding

Apply Vocabulary

Use the highlighted words from the lesson to complete each sentence correctly.

1. The ________ is dermal tissue that is made up of a single layer of cells.

2. The compound that makes the cell walls of seed plants stronger is called ______.

3. The region that is made up of cells that divide to increase the length of stems and roots is called ______.

Critical Thinking

4. Review What are the three main organs of seed plants?

5. Interpret Diagrams Review the art of the main organs of a plant on page 552. How are the three main organs similar in structure?

6. Compare and Contrast Compare apical and floral meristems. How are they similar? How are they different?

7. Write to Learn Answer the first clue of the mystery.

MYSTERY CLUE

The tangled branches seen in the mature figs are not stems. The seeds of the fig species sprout high up in the branches of other forest trees. These trees are called hosts. What are the "branches" that make up the trunks of these figs? How do they grow? (**Hint:** See p. 552.)

BIOLOGY.com Search Lesson 23.1 GO • Lesson Assessment

Check Understanding

1. epidermis
2. lignin
3. meristem
4. roots, stems, and leaves
5. All contain dermal tissue, vascular tissue, and ground tissue.
6. Both apical and floral meristems are regions of undifferentiated cells that divide rapidly. Apical meristems are located at the tips of roots and stems; when they divide, they increase the length of those organs. Floral meristems produce the tissues that form flowers.
7. Students should conclude that the tangled "branches" are actually roots that grow downward from the top of host trees. Students should recognize that meristems, in particular apical meristems, in the tips of the roots make the roots longer.

Active Reading

DRAW CONCLUSIONS Call on a volunteer to read the main headings on this page.

Ask What do meristems do for a plant? *(Meristems cause a plant to grow.)* Then, have students read the page.

Ask What are some examples of plant parts that grow due to the unspecialized cells produced by meristems? *(stems, roots, and flowers)*

KEY QUESTION Read aloud the question and answer.

Ask How are the terms *unspecialized* and *undifferentiated* similar as they relate to cells? *(Cells produced by meristems have not gone through differentiation, the process that leads to cells with specialized structures and functions.)*

Speed Bump

Ask Where are apical meristems located? *(in the tips of stems and roots)*

Ask What do apical meristems produce? *(new undifferentiated cells)*

Post Reading

ASSESSMENT TIPS For Question 5, advise students to look carefully at the diagram's key and blow-ups to clarify the meanings of the different colors.

For Question 6, have students develop a two-column chart that lists similarities in one column and differences in the other. Then, have students use their charts to answer the questions.

MYSTERY CLUE After students read the clue, have them recall the three main organs of a plant. *(roots, stems, and leaves)* Help students understand that the branches of the fig tree are not stems. Encourage students to visualize in which direction these plant parts grew. Have them reread the Roots and Leaves sections if needed.

Ask Which other plant organ could form the branches of the fig—roots or leaves? *(roots)*

Wrap-Up Activity

MODELING Divide the class into small groups. Give each group three colors of modeling clay to represent the three main types of plant tissue: dermal, vascular, and ground. Have students build three-dimensional models showing the internal structure of a leaf, a stem, and a root. With some clay, model for them the difference between a two-dimensional tube in the Main Organs of Plants diagram and a three-dimensional clay tube shown in the Vascular Tissue diagram.

LESSON 23.2

Pre-Reading

Objectives

23.2.1 Describe the main tissues in a mature root.

23.2.2 Describe the different functions of roots.

Preview the Pages

1. Preview the vocabulary words by having students find the words in the text and read each word aloud. As a group, discuss the meaning of each word.
2. Preview the visuals. Read the captions and labels. Be sure students look carefully at where the line for each label points.
3. Make sure students realize that the Anatomy of a Root diagram is shown with the front part of the root sliced away.
4. Preview the Essential Plant Nutrients table. Clarify the meanings of unfamiliar words such as *deficiency, synthesis,* and *resistance.*

Activate Prior Knowledge

Show students a living plant, such as a dandelion, that was recently taken from the soil. Have volunteers point to its roots, stems, and leaves. Ask if roots have the same structure and function as stems and leaves. *(Students should infer they do not.)*

Build Understanding

VENN DIAGRAM Direct students to the Venn diagram in Workbook B. Explain that it shows the similarities and differences of taproot and fibrous root systems. Ask students to complete the diagram as they read.

NATIONAL SCIENCE EDUCATION STANDARDS

UNIFYING CONCEPTS AND PROCESSES

I, V

CONTENT

C.5.d

For the corresponding lesson in ***Miller & Levine Biology,*** see pages 669–673.

23.2 Roots

Key Questions

What are the main tissues in a mature root?

What are the functions of roots?

BUILD Understanding

Venn Diagram As you read, make a two-circle Venn diagram that compares the two types of root systems.

In Your Workbook Go to your workbook to learn more about Venn diagrams. Complete the Venn diagram for Lesson 23.2.

BUILD Vocabulary

root hairs small hairs on a root that produce a large surface area through which water and minerals can enter the root

cortex in plants, an area of ground tissue just inside the root through which water and minerals move

endodermis in plants, a layer of dermal tissue that completely surrounds the vascular cylinder

root cap the tough covering of the root tip that protects the meristem

PREFIXES

The prefix *endo-* means "inside." The endodermis is the inner layer of the cortex.

Root Structure and Growth

As soon as a seed begins to sprout, it puts out its first root. This root draws water and nutrients from the soil. Soon other roots branch out from the first root. These roots add length and surface area to the root system. Greater surface area lets the plant take in more water and nutrients. Rapid cell growth pushes the tips of the roots deeper into the soil. The new roots provide raw materials for the developing plant.

Types of Root Systems The two main types of root systems are taproots and fibrous roots. Taproot systems are found mainly in dicots. Fibrous root systems are found mainly in monocots. Recall from Chapter 22 that monocots and dicots are two groups of flowering plants.

▶ ***Taproot System*** In some plants, the main root grows long and thick. This root gives rise to much smaller branch roots. The main root is called a taproot. Taproots of oak and hickory trees grow so long that they can reach water several meters below the surface. Carrots, dandelions, radishes, and beets have short, thick taproots that store sugars and starches.

▶ ***Fibrous Root System*** In other plants, such as grasses, the root system begins with one root. This root is soon replaced by many other roots of about the same size. These roots grow separately from the base of the stem. They branch so much that no single root grows larger than the rest.

Two Root Systems Dandelions have a taproot system (left), while grasses have a fibrous root system (right).

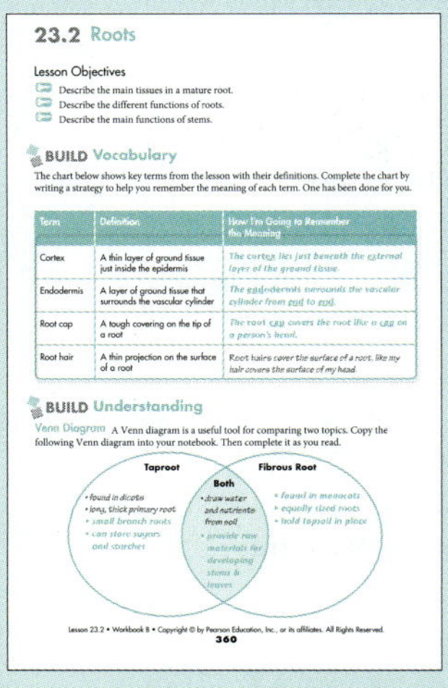

23.2 Roots

Lesson Objectives

- Describe the main tissues in a mature root.
- Describe the different functions of roots.
- Describe the main functions of stems.

BUILD Vocabulary

The chart below shows key terms from the lesson with their definitions. Complete the chart by writing a strategy to help you remember the meaning of each term. One has been done for you.

Term	Definition	How I'm Going to Remember the Meaning
Cortex	A thin layer of ground tissue just inside the epidermis	The cortex lies just beneath the external layer of the ground tissue.
Endodermis	A layer of ground tissue that surrounds the vascular cylinder	The endodermis surrounds the vascular cylinder from end to end.
Root cap	A tough covering on the tip of a root	The root cap covers the root like a cap on a person's head.
Root hair	A thin projection on the surface of a root	Root hairs cover the surface of a root, like my hair covers the surface of my head.

BUILD Understanding

Venn Diagram A Venn diagram is a useful tool for comparing two topics. Copy the following Venn diagram into your notebook. Then complete it as you read.

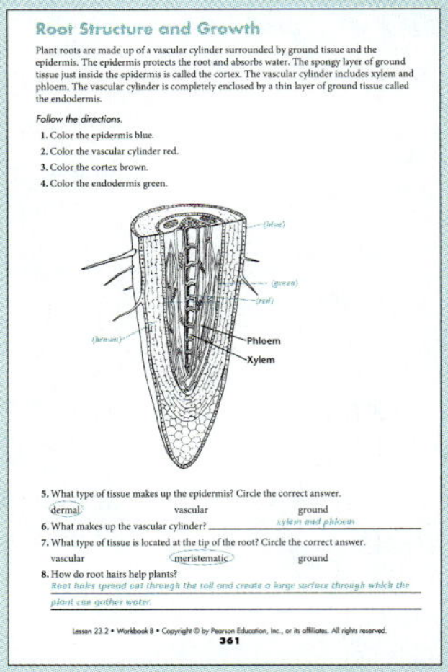

Root Structure and Growth

Plant roots are made up of a vascular cylinder surrounded by ground tissue and the epidermis. The epidermis protects the root and absorbs water. The spongy layer of ground tissue just inside the epidermis is called the cortex. The vascular cylinder includes xylem and phloem. The vascular cylinder is completely enclosed by a thin layer of ground tissue called the endodermis.

Follow the directions.

1. Color the epidermis blue.
2. Color the vascular cylinder red.
3. Color the cortex brown.
4. Color the endodermis green.

5. What type of tissue makes up the epidermis? Circle the correct answer.
   dermal vascular ground
6. What makes up the vascular cylinder? xylem and phloem
7. What type of tissue is located at the tip of the root? Circle the correct answer.
   vascular meristematic ground
8. How do root hairs help plants?
   Root hairs spread out through the soil and create a large surface through which the plant can gather water.

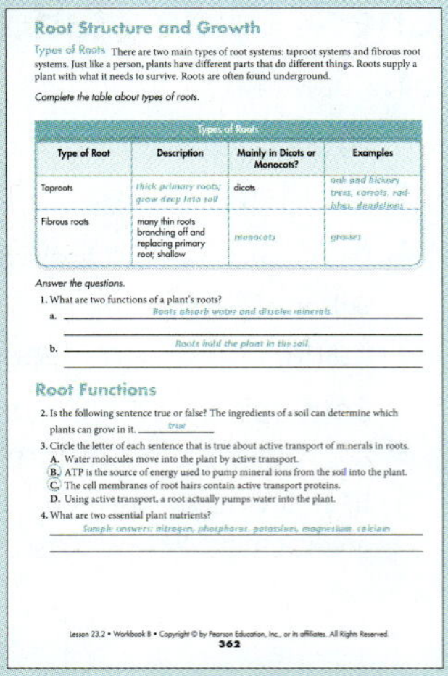

Root Structure and Growth

Types of Roots There are two main types of root systems: taproot systems and fibrous root systems. Just like a person, plants have different parts that do different things. Roots supply a plant with what it needs to survive. Roots are often found underground.

Complete the table about types of roots.

Types of Roots			
Type of Root	**Description**	**Mainly in Dicots or Monocots?**	**Examples**
Taproots	thick primary roots; grow deep into soil	dicots	oak and hickory trees, carrots, radishes, dandelions
Fibrous roots	many thin roots branching off and replacing primary root; shallow	monocots	grasses

Answer the questions.

1. What are two functions of a plant's roots?
   a. Roots absorb water and dissolve minerals.
   b. Roots hold the plant in the soil.

Root Functions

2. Is the following sentence true or false? The ingredients of a soil can determine which plants can grow in it. true
3. Circle the letter of each sentence that is true about active transport of minerals in roots.
   A. Water molecules move into the plant by active transport.
   B. ATP is the source of energy used to pump mineral ions from the soil into the plant.
   C. The cell membranes of root hairs contain active transport proteins.
   D. Using active transport, a root actually pumps water into the plant.
4. What are two essential plant nutrients?
   Sample answers: nitrogen, phosphorus, potassium, magnesium, calcium

Anatomy of a Root Roots contain the three tissue systems—dermal, ground, and vascular tissue. A mature root has an outside layer of dermal tissue, the epidermis. It also contains vascular tissue and a large area of ground tissue. The cells and tissues of a root are specialized to carry water and minerals.

▶ ***Dermal Tissue*** The epidermis protects the root. It also helps the root take in water and dissolved minerals. The cells of the epidermis project outward as long, thin **root hairs.** Root hairs reach into the spaces between soil particles. By increasing the surface area of the root that is in contact with the soil, root hairs help the root take in water and minerals.

▶ ***Ground Tissue*** Just inside the epidermis is an area of ground tissue called the **cortex.** After moving through the epidermis, water and minerals move through the cortex toward the center of the root. The cortex also stores the products of photosynthesis, such as starch.

A layer of ground tissue known as the **endodermis** completely surrounds the vascular tissue. The endodermis plays a key role in the movement of water and minerals into the center of the root.

▶ ***Vascular Tissue*** At the center of the root, xylem and phloem make up an area called the *vascular cylinder*. Dicot roots like the one shown at right have a central column of xylem.

▶ ***Apical Meristem*** Roots grow longer as apical meristems near the root tip produce new cells. The root tip is covered by a tough **root cap.** The root cap protects the meristem as the root tip forces its way through the soil. The root cap also releases a slippery substance that helps the root grow through the soil. Cells at the tip of the root cap are always being scraped away. New root cap cells are added by the meristem.

Key Question What are the main tissues in a mature root? **A mature root has an epidermis made up of dermal tissue. It contains ground tissue, which surrounds the vascular tissue.**

Anatomy of a Root A root has a central vascular cylinder that is surrounded by ground tissue and the epidermis.

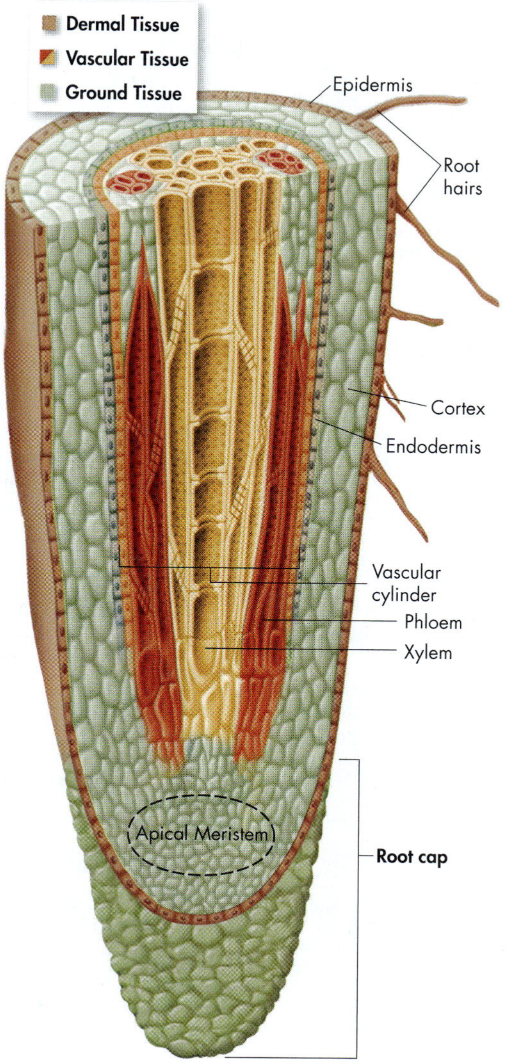

During Reading

Build Vocabulary

PREFIXES Tell students that the prefix *endo-* comes from the Greek word *endon,* meaning "inside." Explain that they can use this information to figure out the meaning of words with this prefix. For example, write the word *endoskeleton* on the board.

Ask If an animal has an endoskeleton, what do you think that means? *(Its skeleton is inside its body.)*

Invite volunteers to use a dictionary to find other words that begin with *endo-*. Examples include *endocrine, endoscope,* and *endothermic*. Help students pronounce the words. Then, have them read the definitions. Discuss how the meaning of each word relates to the meaning of the prefix.

Active Reading

READ, COVER, REMEMBER, RETELL After students read the page, form groups of four students, and have each student in a group silently reread one of the sections describing roots. Ask students to cover the page and retell to the group one fact they read.

INTERPRET DIAGRAMS Have students locate the different types of tissues in the Anatomy of a Root diagram. Encourage them to write a sentence that describes the cells in each type of tissue.

Ask What do you notice about the shapes of the cells in root tissues? *(The cells within the same tissue have similar shapes. The cells of different tissues have different shapes.)*

KEY QUESTION Invite a student to read the question and answer aloud.

Ask Between what root parts is the ground tissue inside a root located? *(between the epidermis and the vascular tissue)*

Speed Bump

Ask Which type of root system has many roots that are all about the same size? *(a fibrous root system)*

Ask From which layer of cells in a root do root hairs grow? *(the epidermis)*

Ask Which type of tissue in a root stores some products of photosynthesis? *(ground tissue)*

Hands-On Learning

EXAMINE A ROOT Carefully cut several tender fresh carrots in half lengthwise with a sharp knife. Let the halves wilt for about 24 hours. This will allow the tissues to shrink and somewhat separate from each other. Explain that a carrot is a root. Then pass around the carrot halves for students to examine. Have them compare the relative thickness of the different tissue layers *(epidermis—very thin; cortex—very thick and increasing in thickness away from the root tip; vascular cylinder—thick but not as thick as the cortex)* Encourage them to notice the smoothness of the inner layer of the cortex. Also, point out any small extensions coming from the vascular cylinder. Explain that these extensions are developing horizontal roots.

Ask What kind of tissue appears to make up most of the inside of a carrot? *(ground tissue)*

LESSON 23.2

During Reading continued

Active Reading

USE GRAPHIC ORGANIZERS Have students work in pairs to develop a fishbone map that summarizes the functions of roots. Encourage students to read the first paragraph on this page before they begin the map.

Ask What will be the main idea in your map? *(Roots have several important functions.)*

Say Notice the subheadings on this page. These describe how roots perform their functions.

Ask How many main branches should your fishbone map have? *(five)* Remind students to include details by adding several smaller branches to the main branches.

Speed Bump

Ask What are three main functions of roots? *(food storage, support, absorption of water and nutrients)*

Ask What kind of transport do plants use to take in nutrients? *(active transport)*

Ask By what process does water enter a root? *(osmosis)*

Essential Plant Nutrients

Nutrient	Some Roles in Plant	Result of Deficiency
Nitrogen (N)	• Proper leaf growth and color • Synthesis of amino acids, proteins, nucleic acids, and chlorophyll	• Stunted plant growth • Pale yellow leaves ▶
Phosphorus (P)	• Synthesis of DNA • Development of roots, stems, flowers, and seeds	• Poor flowering • Stunted growth
Potassium (K)	• Synthesis of proteins and carbohydrates • Development of roots, stems, and flowers • Resistance to cold and disease	• Weak stems • Stunted roots • Edges of leaves turn brown ▶
Magnesium (Mg)	• Synthesis of chlorophyll	• Thin stems • Mottled, pale leaves
Calcium (Ca)	• Cell growth and division • Cell wall structure • Cellular transport • Enzyme action	• Stunted growth • Curled leaves ▶

Plant Nutrients Plants get many essential nutrients from soil.

Root Functions

Roots have several important functions. Roots support a plant and hold it in the ground. They store food. Roots take in water and nutrients from the soil. But materials do not just "soak" into the root. A plant must use energy to take in water and nutrients.

Uptake of Nutrients To grow, plants need several nutrients from the soil. Those needed in largest amounts are nitrogen, phosphorus, potassium, magnesium, and calcium. Plants also need small amounts of other nutrients, called trace elements. Trace elements include sulfur, iron, zinc, boron, copper, manganese, and chlorine.

Active Transport of Nutrients In the membranes of cells in the epidermis are proteins that can move ions by active transport. (Active transport uses energy to move materials across cell membranes.) These proteins use energy to move nutrients from the soil into the root. This action causes a high concentration of nutrient ions in the root cells.

Water Movement by Osmosis Water moves into the root by osmosis. In osmosis, water moves across a membrane to a place where the concentration of ions is higher. Water "follows" the ions that have been pumped into the root by active transport.

Movement Into the Vascular Cylinder Water moves from the epidermis, through the cortex, and into the vascular cylinder. The endodermis surrounds the vascular cylinder. Where these tissues meet, their cell walls form a waterproof zone called a *Casparian strip.* The Casparian strip forces water and nutrients to move through the cells of the endodermis. Water cannot pass between the cells. The endodermis filters and controls water and nutrients that enter the vascular cylinder.

Root Pressure The Casparian strip allows water and nutrients to move in only one direction—from the cortex into the vascular cylinder. Water and nutrients cannot move from the vascular cylinder back to the cortex.

ELL Focus on ELL: Access Content

BEGINNING SPEAKERS Form groups of three students. Provide each group with 15 lined index cards. Tell students they will use the index cards to break up the information in the Essential Plant Nutrients table. Have each student in a group prepare five cards for one of the columns in the table. Tell students to prepare cards by writing the column head on the plain side of each card and then writing the information from one of the table cells shown under that head on the lined side of each card. Have each group combine and shuffle all cards and then sort the cards into five groups of three cards that represent one of the rows in the table. Encourage students to use the cards to quiz each other when reviewing the lesson.

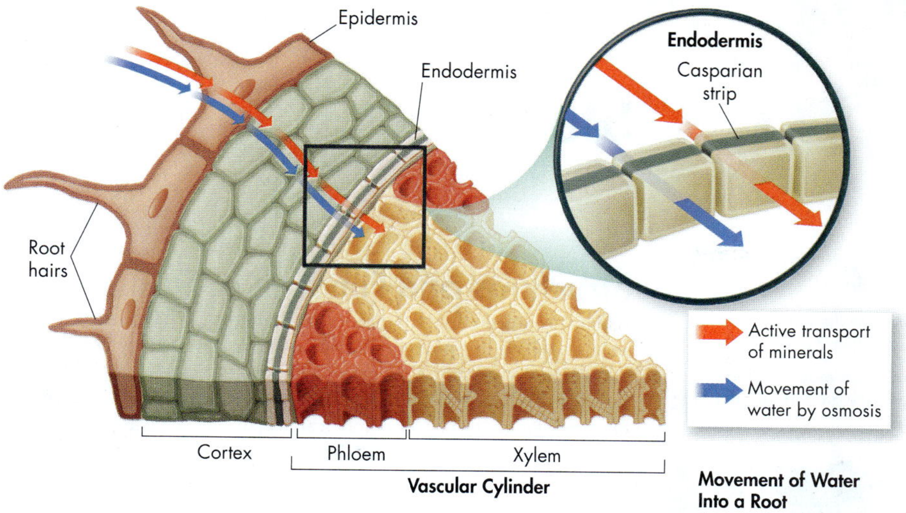

Movement of Water Into a Root
A root absorbs water and dissolved nutrients from soil.

Why is this one-way system important? As nutrients are pumped into the vascular cylinder, more water follows by osmosis. This movement of water causes strong pressure in the vascular cylinder. That water has just one place to go—up. Root pressure forces water through the vascular cylinder and into the xylem. As more water moves from the cortex into the vascular cylinder, water in the xylem is pushed upward. Root pressure starts the movement of water through the whole plant.

Key Question What are the functions of roots? **Roots support and anchor the plant, store food, and absorb water and nutrients.**

CHECK Understanding

Apply Vocabulary

Use the highlighted words from the lesson to complete each sentence correctly.

1. The tough cells of the ______ protect the apical meristem as the root grows through the soil.
2. The ______ surrounds the vascular cylinder.
3. Tiny projections from cells in the epidermis called ______ increase the surface area of a root.

Critical Thinking

4. **Classify** Roots contain dermal, vascular, and ground tissue. Name the tissue system to which each of these belongs: cortex, endodermis, epidermis, phloem, xylem.
5. **Compare and Contrast** Compare the role of active transport and the role of osmosis in taking in water and nutrients. How are these two processes connected?
6. **Write to Learn** Answer the mystery clue.

MYSTERY CLUE

The fig tree grows slowly until its roots reach the ground. Until its roots reach soil, how might the fig seedlings get nutrients?

BIOLOGY.com Search Lesson 23.2 GO • Lesson Assessment • Art in Motion

Check Understanding

1. root cap
2. endodermis
3. root hairs
4. cortex—ground tissue; endodermis—ground tissue; epidermis—dermal tissue; phloem and xylem—vascular tissue
5. The root uses active transport to move nutrients into the root. Water moves into the root by osmosis. When active transport increases the concentration of ions inside the root, water "follows" the concentration of ions.
6. Students should recognize that the fig tree cannot get nutrients from the soil. They will likely answer that the tree gets nutrients from rainwater. Some students may indicate that the fig tree gets nutrients from the leaf litter in the crevices of the host tree.

Active Reading

SUMMARIZE Have students summarize in their own words the role of the Casparian strip in controlling the movement of water and nutrients in a root. *(Sample answer: The Casparian strip prevents water and nutrients from moving out of the vascular cylinder.)*

KEY QUESTION Invite a volunteer to read aloud the question and answer. Then, have students look back at the photos of the two root systems.

Ask What feature of a root system helps roots anchor a plant? *(The branching of the root system.)*

Help students realize that branching increases the surface area of roots and that a greater surface area in contact with soil creates more friction that keeps the roots from easily being pulled out of the ground.

Speed Bump

Ask What is root pressure? *(Root pressure is the force that pushes water through a root's vascular cylinder and upward in the xylem.)*

Post Reading

ASSESSMENT TIPS For Question 4, have students look at the Anatomy of a Root diagram and carefully match colors in the key to the colors of the root structures.

For Question 5, encourage students to use a two-column chart to sort the roles of active transport and osmosis. Then, tell them to use their charts to answer the question.

MYSTERY CLUE For Question 6, have students read the clue and recall the functions of roots. *(anchor and support plant, store food, absorb water and nutrients)* Refer students to the section on root functions if they need help. Explain that above-ground roots also are able to perform all of these functions.

Ask What sources of water could above-ground roots absorb? *(rain, fog, dew)*

Wrap-Up Activity

OBSERVATION For each small group of students, provide a cup with several three- or four-day-old bean seedlings. Water the plants well before distributing them. Have students use scissors to cut off individual stems a few centimeters above the soil. **CAUTION:** *Students should handle scissors with care.* Students should observe the effects of root pressure as "beads" of liquid are pushed out of the tips of the cut stems.

LESSON 23.3

Pre-Reading

Objectives

23.3.1 Describe the main functions of stems.

23.3.2 Contrast the processes of primary growth and secondary growth in stems.

Preview the Pages

1. Preview the vocabulary by using each word in a sentence and having students write a possible definition of each term before reading the lesson.
2. Point out that the Primary and Secondary Growth diagram shows the changes that occur in a stem over a three-year period.
3. Explain that the Tree Rings photograph shows a cross section of a tree trunk including the wood and the bark. Ask students to describe the wood. *(It is made of rings of different widths and color.)* Encourage students to speculate as to the causes of the different widths and colors. Record their responses. Compare these responses with their new knowledge when they reach that point in the lesson.

Build Understanding

PREVIEW VISUALS Direct students to Workbook B, where they will find a partially filled-in preview visuals chart. Encourage students to write a sentence or two to describe what the visuals show.

NATIONAL SCIENCE EDUCATION STANDARDS

UNIFYING CONCEPTS AND PROCESSES

I, V

CONTENT

C.5.d

For the corresponding lesson in ***Miller & Levine Biology,*** see pages 674–679.

23.3 Stems

Key Questions

What are the three main functions of stems?

How do primary growth and secondary growth happen in stems?

BUILD Understanding

Preview Visuals Before you read, look at the figure that shows wood and bark formation. Define any familiar terms in your own words. List any terms you don't know. As you read, revise and add to your definitions.

In Your Workbook Go to your workbook to learn more about previewing visuals. Complete the table for Lesson 23.3.

Structures and Function of Stems

Asparagus, water chestnuts, onions, and potatoes. What do these vegetables have in common? They are all stems! Stems vary in size and shape. Some grow underground. Others reach high into the air. But all stems have several important functions. Stems produce leaves, branches, and flowers. Stems hold leaves up to the sun. Stems move water and nutrients throughout the plant. Some stems carry out photosynthesis or store materials.

Anatomy of a Stem Stems are surrounded by a layer of epidermal cells. These cells have thick cell walls and a waxy cuticle. The areas where leaves grow out of the stem are called **nodes.** Small buds are found where leaves attach to the nodes. **Buds** contain apical meristems that can produce new stems and leaves. In larger plants, stems develop woody tissue.

Primary and Secondary Growth
During primary growth, the apical meristem produces new cells. These cells cause the stem to get longer. During secondary growth, the stem grows wider.

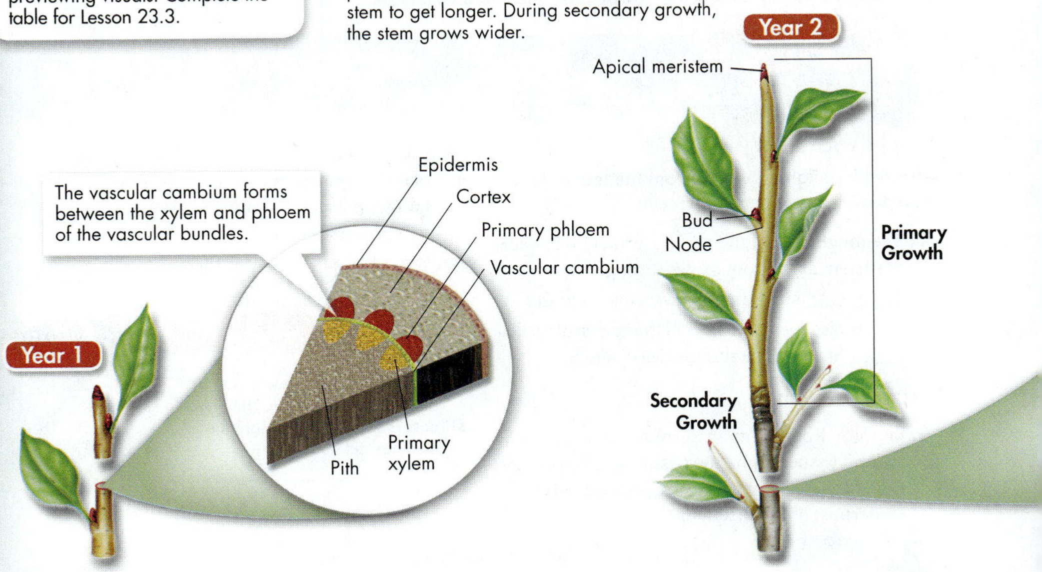

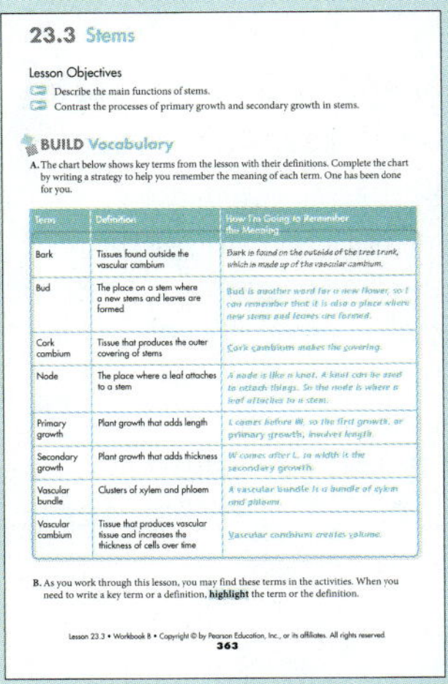

23.3 Stems

Lesson Objectives

Describe the main functions of stems.

Contrast the processes of primary growth and secondary growth in stems.

BUILD Vocabulary

A. The chart below shows key terms from the lesson with their definitions. Complete the chart by writing a strategy to help you remember the meaning of each term. One has been done for you.

Term	Definition	How I'm Going to Remember the Meaning
Bark	Tissues found outside the vascular cambium	Bark is found on the outside of the tree trunk, which is made up of the vascular cambium.
Bud	The place on a stem where a new stems and leaves are formed	Bud is another word for a new flower, so I can remember that it is also a place where new stems and leaves are formed.
Cork cambium	Tissue that produces the outer covering of stems	Cork cambium makes the covering.
Node	The place where a leaf attaches to a stem	A node is like a knot. A knot can be used to attach things. So the node is where a leaf attaches to a stem.
Primary growth	Plant growth that adds length	L comes before W, so the first growth, or primary growth, involves length.
Secondary growth	Plant growth that adds thickness	W comes after L, so width is the secondary growth.
Vascular bundle	Clusters of xylem and phloem	A vascular bundle is a bundle of xylem and phloem.
Vascular cambium	Tissue that produces vascular tissue and increases the thickness of cells over time	Vascular cambium creates volume.

B. As you work through this lesson, you may find these terms in the activities. When you need to write a key term or a definition, **highlight** the term or the definition.

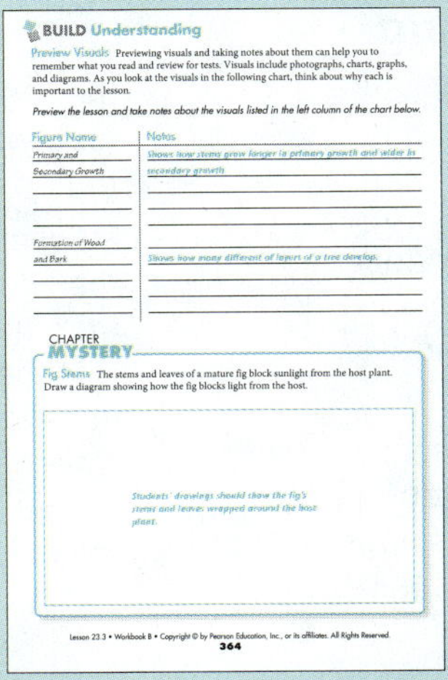

BUILD Understanding

Preview Visuals Previewing visuals and taking notes about them can help you to remember what you read and review for tests. Visuals include photographs, charts, graphs, and diagrams. As you look at the visuals in the following chart, think about why each is important to the lesson.

Preview the lesson and take notes about the visuals listed in the left column of the chart below.

Figure Name	Notes
Primary and Secondary Growth	Shows how stems grow longer in primary growth and wider in secondary growth
Formation of Wood and Bark	Shows how many different of layers of a tree develop.

CHAPTER MYSTERY

Fig Stems The stems and leaves of a mature fig block sunlight from the host plant. Draw a diagram showing how the fig blocks light from the host.

Students' drawings should show the fig's stems and leaves wrapped around the host plant.

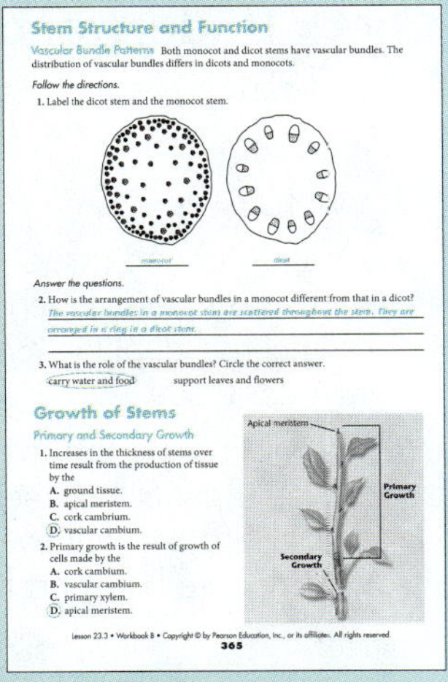

Stem Structure and Function

Vascular Bundle Patterns Both monocot and dicot stems have vascular bundles. The distribution of vascular bundles differs in dicots and monocots.

Follow the directions.

1. Label the dicot stem and the monocot stem.

monocot — dicot

Answer the questions.

2. How is the arrangement of vascular bundles in a monocot different from that in a dicot?
The vascular bundles in a monocot stem are scattered throughout the stem. They are arranged in a ring in a dicot stem.
3. What is the role of the vascular bundles? Circle the correct answer.
carry water and food — support leaves and flowers

Growth of Stems

Primary and Secondary Growth

1. Increases in the thickness of stems over time result from the production of tissue by the
A. ground tissue.
B. apical meristem.
C. cork cambrium.
D. vascular cambium.
2. Primary growth is the result of growth of cells made by the
A. cork cambium.
B. vascular cambium.
C. primary xylem.
D. apical meristem.

Apical meristem
Primary Growth
Secondary Growth

Vascular Bundle Patterns The arrangement of tissues in a stem differs among seed plants. In monocots, clusters of xylem and phloem tissue, called **vascular bundles,** are scattered throughout the stem. In most dicots and gymnosperms, vascular bundles form a ring.

Key Question What are three main functions of stems?
Stems produce leaves, branches, and flowers; they hold leaves up to the sun; and they carry water and nutrients throughout the plant.

Growth of Stems

Plants grow in ways that are very different from the ways that animals grow. When animals grow, they have set characteristics. For example, cows have four legs, ants have six legs, and spiders have eight. But roses, tomatoes, and other plants don't have a set number of branches, roots, or leaves. Even so, plant growth is carefully controlled. Each species has a typical size and shape.

Primary Growth Apical meristems produce new cells throughout the plant's lifetime. These cells add length to the roots and stems. This growth, which happens at the tips of shoots and roots, is called primary growth. **Primary growth** occurs when cells produced in the apical meristem become longer. Primary growth takes place in the stems of all seed plants.

BUILD Vocabulary

node the part on a growing stem where a leaf is attached

bud a plant structure containing apical meristem tissue that can produce new stems and leaves

vascular bundle clusters of xylem and phloem tissue in stems

primary growth a pattern of growth that takes place at the tips and shoots of a plant

secondary growth the type of growth in which stems increase in thickness

ACADEMIC WORDS
Adjectives are words that describe nouns. Some describe order. *Primary* identifies something that comes first. *Secondary* identifies what happens next. Primary growth happens first. Secondary growth follows primary growth.

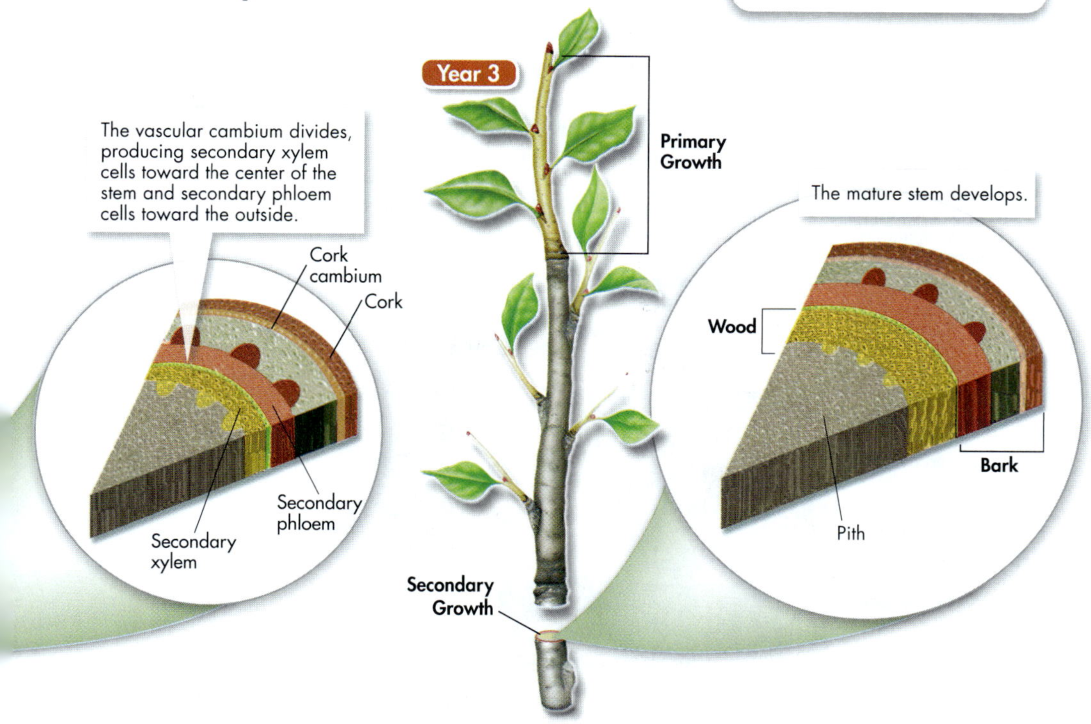

During Reading

Active Reading

KEY QUESTION Invite a volunteer to read aloud the question and answer. Then, direct students' attention to the diagram showing year 1 of a stem's growth.

Ask How is a stem able to carry water and nutrients throughout a plant? *(The stem contains vascular bundles, which consist of xylem and phloem.)*

SUMMARIZE After discussing the Primary and Secondary Growth diagram with the class, have students write a brief paragraph that describes each pie-shaped close-up of stem structure.

Build Vocabulary

ACADEMIC WORDS Discuss the meanings of the adjectives *primary* and *secondary.*

Ask What is an example of an everyday use of these adjectives? *(Sample answers: primary and secondary school, primary and secondary roads)*

Ask How do the words *primary* and *secondary* describe order in each example? *(Primary school happens first; secondary school follows. In a snowstorm, primary roads get plowed first; secondary roads get plowed next.)*

Ask In the Primary and Secondary Growth diagram, when does primary growth begin? *(during year 1)* When does secondary growth begin? *(during year 2)*

Speed Bump

Ask How is a node different from a bud? *(A node is a place on a stem where a leaf is attached, and a bud is a structure that can produce new stems and leaves.)*

Ask Which type of growth causes a stem to grow longer? *(primary)* Wider? *(secondary)*

Ask Why can't most monocots grow as big as many dicots? *(Secondary growth is rare in monocots, so less weight can be supported by their stems.)*

Hands-On Learning

EXAMINE A STEM Cut off several stalks of celery near their bases. Place the stalks in a container of water colored with red food coloring and allow them to sit for several hours or overnight. Then, have student groups examine the stalks. Help students cut a cross-section of their stalk with a knife, and have them observe it through a hand lens. Challenge students to find the location of the vascular bundles, which should turn red. Have students draw the cross-section of the stem in their notebooks. Also have them observe the celery leaves, which should develop a reddish fringe after the stalks have been in the water for several hours.

Ask Why did the leaves begin to turn red if they weren't submerged in the water? *(Xylem in the celery stem carried colored water to the leaves.)*

During Reading continued

Active Reading

COMPARE AND CONTRAST Invite a volunteer to read the third paragraph on this page.

Ask How are cambium and apical meristem similar? *(Both are types of meristem and both produce new cells.)* How are they different with regard to plant growth? *(Apical meristem causes primary growth. Cambium causes secondary growth.)*

Build Vocabulary

WORD ORIGIN Write the word *vascular* on the board. Tell students that *vascular* comes from the Latin word *vasculum,* which means "vessel."

Ask In biology, what is a vessel? *(a duct or tube)*

Ask How do the meanings of *vessel* and *vascular* relate to each other? *(*Vessel *is a noun that means "a duct or tube," and* vascular *is an adjective that means "having to do with ducts or tubes.")*

Speed Bump

Ask What do the vascular cambium and cork cambium produce in a stem? *(Vascular cambium produces vascular tissues. Cork cambium produces the outer covering of stems.)*

Ask What is wood made of? *(secondary xylem)*

Ask How is sapwood different from heartwood? *(Sapwood is lighter in color and conducts materials.)*

BUILD Vocabulary

vascular cambium a meristem that produces vascular tissues and increases the thickness of stems

cork cambium a meristem that produces the outer covering of stems during secondary growth

WORD ORIGIN

The word *cambium* comes from a Latin word that means "exchange." Both the vascular cambium and the cork cambium produce vascular tissue cells that will be involved in the exchange of materials within the plant.

Secondary Growth As a plant grows larger, the older stems and roots have more weight to support. They also need to move greater amounts of fluid through their vascular tissues. As a result, stems must increase in thickness as well as in length. The increase in thickness is known as **secondary growth.**

Secondary growth is common among dicots and conifers. These plants have meristems in their stems and roots that can produce secondary growth. Secondary growth lets shrubs and trees grow tall. The increase in stem width supports the extra weight. Secondary growth is rare in monocots. As a result, monocots are often limited in size.

In stems, secondary growth takes place in two tissues—vascular cambium and cork cambium. **Vascular cambium** is a meristem that produces vascular tissues. Another meristem, **cork cambium,** produces the outer covering of stems. Similar kinds of cambium tissue play a role in root growth. The addition of cells in the cambium layers increases the thickness of stems and roots.

Growth From the Vascular Cambium In the stem of a young dicot, vascular bundles are arranged in a ring. A thin layer of vascular cambium forms between the xylem and phloem of each vascular bundle. As cells in the vascular cambium divide, new layers of xylem and phloem form. This growth makes the stem get wider. Each year, the cambium makes new layers of vascular tissue. Over time, the stem grows thicker and thicker.

Formation of Wood Most wood is made up of layers of secondary xylem. The vascular cambium produces this xylem during secondary growth. The xylem builds up year after year. As woody stems grow thicker, the older xylem near the center of the stem no longer conducts water. Instead, it becomes what is known as *heartwood.* Heartwood usually darkens with age. Heartwood is surrounded by *sapwood,* which still conducts materials. Sapwood is usually a lighter color.

Formation of Wood and Bark This art shows the layers of wood and bark in a tree that has undergone several years of secondary growth.

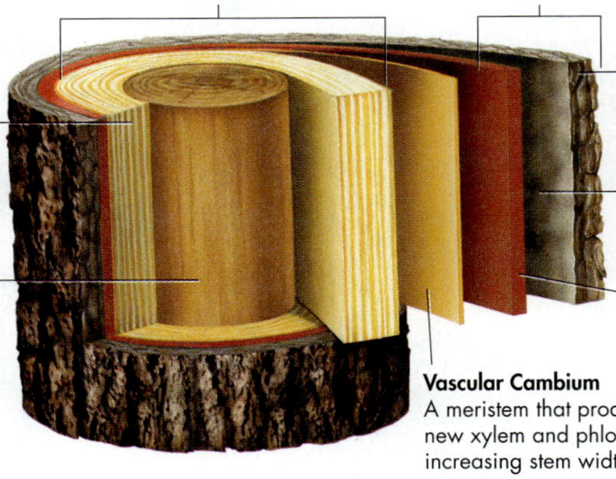

562 Chapter 23 • Lesson 3

Focus on Academic Language

• conifers • dicots • monocots • shrubs • trees

Have students copy these words into their notebooks and find them in the text.

Say Three of the words refer to plant groups that were discussed in the last chapter.

Ask Which words refer to the three plant groups? *(conifers, dicots, monocots)* Have students look up the words and write a definition for each. *(Conifers are gymnosperms that form seeds in cones. Dicots are flowering plants whose seeds have two cotyledons. Monocots are flowering plants whose seeds have one cotyledon.)*

Explain that the words *shrubs* and *trees* refer to types of plants within the three plant groups. Tell them that shrubs are relatively short woody plants with many stems growing from a base near the ground, and trees are relatively tall woody plants with a trunk from which branches grow above the ground. Invite volunteers to give examples of each.

Tree Rings In most of the temperate zone, tree growth is seasonal. When growth begins in the spring, the vascular cambium grows rapidly. Early growth produces large, light-colored xylem cells with thin cell walls. The result is a light-colored layer of wood. As the growing season continues, new cells become smaller and have thicker cell walls. As a result, a layer of darker wood forms. The alternation of light and dark wood produces tree rings. Little growth occurs during the winter or dry season, so there is usually a sharp boundary between rings.

Usually, each tree ring corresponds to a year of growth. By counting the rings in the cross section of a tree, you can estimate the tree's age. The size of the tree rings also provides information about weather conditions. For example, thick tree rings show that weather conditions were favorable for tree growth. There was plenty of rain. Thin rings show unfavorable conditions, such as a dry year.

Tree Rings A tree's age can be measured by counting its growth rings. The growing conditions for each year can be inferred by comparing the width and color of each ring.

Formation of Bark In a stem, all of the tissues found outside the vascular cambium make up the *bark*. Bark includes phloem, the cork cambium, and cork. During secondary growth, the vascular cambium adds secondary phloem to the bark.

Key Question How do primary growth and secondary growth happen in stems? **Primary growth is the result of the lengthening of cells produced in apical meristems. It takes place in all seed plants. Secondary growth increases the thickness of stems. It occurs in the vascular and cork cambiums of many dicots and conifers.**

CHECK Understanding

Apply Vocabulary

Use the highlighted words from the lesson to complete each sentence correctly.

1. The part of a stem where a leaf is attached is a ______.
2. A structure that holds apical meristem tissues that can produce new stems and leaves is a ______.
3. The clusters of xylem and phloem tissues found in stems are called ______.

Critical Thinking

4. **Apply Concepts** How do stems support the functions of leaves and roots?
5. **Explain** Which meristem is involved in primary growth? Which two meristems are involved in secondary growth? Briefly describe the role of each.
6. **Write to Learn** Answer the mystery clue question. Use the words *primary growth* and *secondary growth* in your answer.

MYSTERY CLUE

As the fig grows, the roots grow in both length and thickness. The roots wrap around the host tree's trunk. How might this growth affect the host? (**Hint:** See p. 561.)

BIOLOGY.com Search Lesson 23.3 GO • Lesson Assessment • Tutor Tube

Active Reading

KEY QUESTION After reading the Key Question and answer, have students look back at the Primary and Secondary Growth diagram on the previous pages. Invite volunteers to explain how the visual illustrates the facts in each sentence of the answer.

Speed Bump

Ask What are tree rings? *(alternating light- and dark-colored layers of wood)*

Ask What does the size of tree rings show about weather conditions? Give an example. *(They show how weather conditions changed in the past as the tree grew. Thick rings show favorable conditions for growth, such as plenty of rain.)*

Ask What causes the dark color of a tree ring? *(the small size and thick walls of new cells that form later in the growing season)*

Post Reading

ASSESSMENT TIPS For Question 4, have students think about how stems supply the materials needed by leaves and roots for their activities. Encourage students to record their thoughts in a T-chart.

For Question 5, have students refer to the diagrams of Primary and Secondary Growth and Formation of Wood and Bark to review the names of the meristems and their roles in plant growth. Also, make sure students notice that Question 5 includes three steps: two separate questions and a directive.

MYSTERY CLUE After students read the clue, discuss what might happen around the base of a large tree whose roots have undergone secondary growth. Have students refer to the Primary and Secondary Growth sections for details. *(Rocks and soil get pushed. Structures around the roots, such as sidewalks and roads, become warped or broken as the roots grow in width.)*

Wrap-Up Activity

MAKE FLIP CHARTS Have small groups make flip charts that illustrate how a tree trunk's growth rings might change over a period of 25 years. Provide each group with five index cards and an assortment of colored pencils. Direct students to create a separate card for each five-year period of tree growth. (Growth should be represented at years 5, 10, 15, 20, and 25.) Explain that each card should show a detailed cross-section of the tree truck, including growth rings with light and dark layers. Have students use the Tree Rings visual as a model.

Check Understanding

1. node **2.** bud
3. vascular bundles
4. Stems enable leaves to photosynthesize by transporting water from the roots to the leaves and by holding the leaves up to the sunlight. Stems carry sugars from the leaves to the roots so that root cells can carry out the active process of absorbing nutrients.
5. Apical meristems are involved in primary growth. They add new cells to roots and shoots. The vascular cambium and the cork cambium are involved in secondary growth. They increase a stem's width.
6. Students should recognize that primary growth increases length, and secondary growth increases width. Both types of growth increase the size and mass of the roots. These roots will put stress on the trunk of the host tree, and interfere with the flow of materials through the host tree's phloem.

LESSON 23.4

Pre-Reading

Objectives

23.4.1 Describe how the structure of a leaf enables it to carry out photosynthesis.

23.4.2 Explain how gas exchange in leaves relates to homeostasis.

Preview the Pages

1. Review the vocabulary words. Refer students to the pronunciation of *mesophyll,* and help them pronounce the word. Point out that *stomata* is the plural form of *stoma.*
2. Have students read the main headings in the lesson. Remind them that the word *homeostasis* means "stable internal conditions."
3. Preview the visuals. Help students link the captions in the Adaptations of Leaves visual with the correct photos.

Activate Prior Knowledge

Invite volunteers to share what they know about the sizes, shapes, and colors of leaves on familiar plants.

Ask In your mind, what does a typical leaf look like? *(Sample answer: Leaves are wide, thin, flat, and green.)*

Build Understanding

PREVIEW VISUALS Direct students to Workbook B, where they will find a preview visuals table. Discuss the example given, and have students complete the table before reading the section. Remind students that the table can be used as a study guide later.

NATIONAL SCIENCE EDUCATION STANDARDS

UNIFYING CONCEPTS AND PROCESSES

I, IV, V

CONTENT

C.1.e, C.5.d

23.4 Leaves

Key Questions

How is the structure of a leaf adapted to make photosynthesis more efficient?

What is the role of stomata in maintaining homeostasis?

BUILD Understanding

Preview Visuals Before you read, look at the figure of leaf anatomy. Find the three main tissue systems. After you read, identify which tissue system makes up leaf veins.

In Your Workbook Go to your workbook to learn more about previewing visuals.

Structure and Function of Leaves

Leaves are the world's most important factories. Using the energy captured in photosynthesis, plants make the sugars, starches, and oils that feed nearly all land animals. In photosynthesis, plants use carbon dioxide and water to produce sugars and oxygen.

Anatomy of a Leaf The structure of a leaf is ideal for carrying out photosynthesis. To collect as much light as possible, most leaves are thin and flat.

▶ ***Epidermis*** A tough epidermis covers the top and bottom of most leaves. The epidermis is coated with a waxy cuticle that limits the loss of water. Small openings in the epidermis called **stomata** (singular: stoma) let gases move in and out of the leaf.

▶ ***Mesophyll*** Photosynthesis takes place in specialized ground tissue called **mesophyll** (MES uh fil). Beneath the upper epidermis is *palisade mesophyll.* Palisade mesophyll is made of closely packed cells that absorb light. Beneath the palisade mesophyll is *spongy mesophyll,* which has many air spaces between cells.

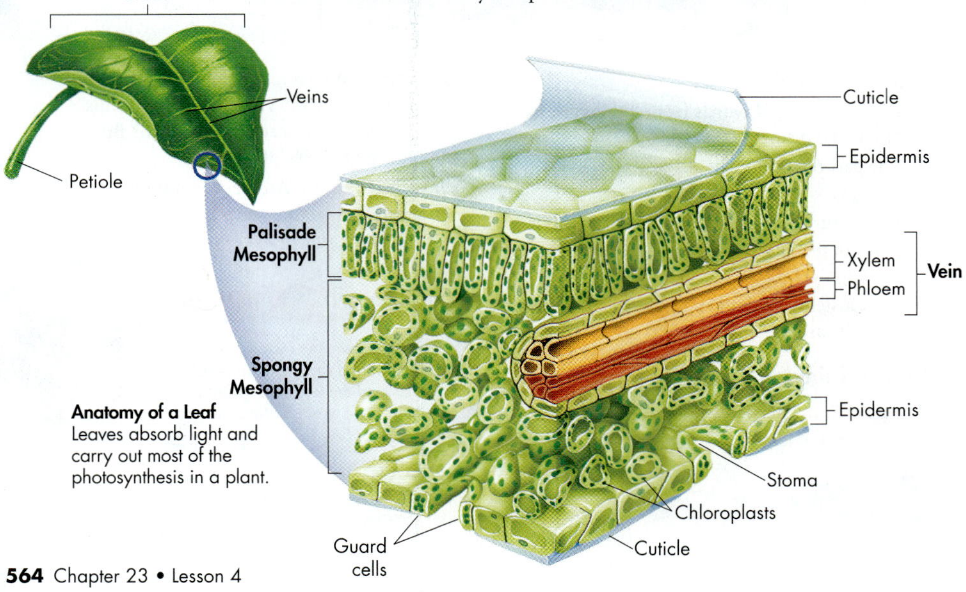

Anatomy of a Leaf Leaves absorb light and carry out most of the photosynthesis in a plant.

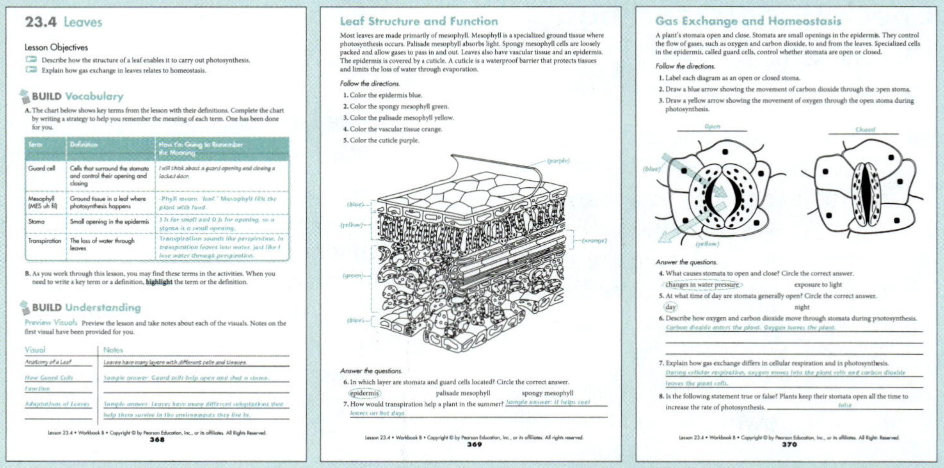

23.4 Leaves

Lesson Objectives

Describe how the structure of a leaf enables it to carry out photosynthesis.

Explain how gas exchange in leaves relates to homeostasis.

BUILD Vocabulary

A. The chart below shows key terms from the lesson with their definitions. Complete the chart by writing a strategy to help you remember the meaning of each term. One has been done for you.

Term	Definition	How I'm Going to Remember the Meaning
Guard cell	Cells that surround the stomata and control their opening and closing	I will think about a guard opening and closing a locked door.
Mesophyll (MES uh fil)	Ground tissue in a leaf where photosynthesis happens	-Phyll means "leaf." Mesophyll fills the plant with food.
Stoma	Small opening in the epidermis	S is for small and O is for opening, so a stoma is a small opening.
Transpiration	The loss of water through leaves	Transpiration sounds like perspiration. In transpiration leaves lose water, just like I lose water through perspiration.

B. As you work through this lesson, you may find these terms in the activities. When you need to write a key term or a definition, **highlight** the term or the definition.

BUILD Understanding

Preview Visuals Preview the lesson and take notes about each of the visuals. Notes on the first visual have been provided for you.

Visual	Notes
Anatomy of a Leaf	Leaves have many layers with different cells and tissues.
How Guard Cells Function	Sample answer: Guard cells help open and shut a stoma.
Adaptations of Leaves	Sample answer: Leaves have many different adaptations that help them survive in the environments they live in.

Lesson 23.4 • Workbook B • Copyright © by Pearson Education, Inc., or its affiliates. All Rights Reserved.
368

Leaf Structure and Function

Most leaves are made primarily of mesophyll. Mesophyll is a specialized ground tissue where photosynthesis occurs. Palisade mesophyll absorbs light. Spongy mesophyll cells are loosely packed and allow gases to pass in and out. Leaves also have vascular tissue and an epidermis. The epidermis is covered by a cuticle. A cuticle is a waterproof barrier that protects tissues and limits the loss of water through evaporation.

Follow the directions.

1. Color the epidermis blue.
2. Color the spongy mesophyll green.
3. Color the palisade mesophyll yellow.
4. Color the vascular tissue orange.
5. Color the cuticle purple.

(purple) (blue) (yellow) (orange) (green) (blue)

Answer the questions.

6. In which layer are stomata and guard cells located? Circle the correct answer.
   (epidermis) palisade mesophyll spongy mesophyll
7. How would transpiration help a plant in the summer? Sample answer: It helps cool leaves on hot days.

Lesson 23.4 • Workbook B • Copyright © by Pearson Education, Inc., or its affiliates. All rights reserved.
369

Gas Exchange and Homeostasis

A plant's stomata open and close. Stomata are small openings in the epidermis. They control the flow of gases, such as oxygen and carbon dioxide, to and from the leaves. Specialized cells in the epidermis, called guard cells, control whether stomata are open or closed.

Follow the directions.

1. Label each diagram as an open or closed stoma.
2. Draw a blue arrow showing the movement of carbon dioxide through the open stoma.
3. Draw a yellow arrow showing the movement of oxygen through the open stoma during photosynthesis.

Open Closed (blue) (yellow)

Answer the questions.

4. What causes stomata to open and close? Circle the correct answer.
   (changes in water pressure) exposure to light
5. At what time of day are stomata generally open? Circle the correct answer.
   (day) night
6. Describe how oxygen and carbon dioxide move through stomata during photosynthesis. Carbon dioxide enters the plant. Oxygen leaves the plant.
7. Explain how gas exchange differs in cellular respiration and in photosynthesis. During cellular respiration, oxygen moves into the plant cells and carbon dioxide leaves the plant cells.
8. Is the following statement true or false? Plants keep their stomata open all the time to increase the rate of photosynthesis. false

Lesson 23.4 • Workbook B • Copyright © by Pearson Education, Inc., or its affiliates. All Rights Reserved.
370

For the corresponding lesson in *Miller & Levine Biology,* see pages 680–684.

▶ ***Veins*** Veins of xylem and phloem run throughout the leaf and connect it to the stem. Xylem carries water from the roots into the leaf. Sugars produced in the leaf are moved into the veins. Then, phloem carries these sugars to the rest of the plant.

Key Question How is the structure of a leaf adapted to make photosynthesis more efficient? **The structure of a leaf allows the plant to collect light and carry out photosynthesis.**

Gas Exchange and Homeostasis

Plants need to exchange gases with the atmosphere. During photosynthesis, leaves take in carbon dioxide and give off oxygen. In respiration, plants take in oxygen and give off carbon dioxide.

The Role of Stomata Gases move in and out of the leaf through stomata. If stomata were open all the time, the plant would lose too much water. So, stomata are open long enough for photosynthesis to take place but not so long that the plant loses too much water. **Guard cells** are specialized cells that surround the stomata and control their opening and closing.

Changes in water pressure within the guard cells cause the stomata to open and close. When water is abundant, it flows into the leaf. This increases pressure in the guard cells. The stomata open. When water is scarce, water pressure within the guard cells decreases. Stomata close, reducing water loss by transpiration. Usually, stomata are open during the day, when photosynthesis takes place. They are closed at night. But if conditions are hot and dry, stomata may also close during the day.

Transpiration and Wilting The walls of mesophyll cells are moist, which allows gases to move in and out of the cells easily. Water evaporates from these surfaces and is lost to the atmosphere. **Transpiration** is the loss of water through leaves. Lost water may be replaced by water drawn into the leaf through xylem. If it is very hot or windy, water may be lost to transpiration faster than it is replaced. The plant may wilt.

Wilting is caused by the loss of water—and pressure—in a plant's cells. Normally, pressure in a plant's cells makes its leaves stiff. But if there is not enough water, the cell walls bend inward, and the plant wilts. When a leaf wilts, its stomata close. Transpiration slows down. Thus, wilting helps a plant save water.

Key Question What is the role of stomata in maintaining homeostasis? **Stomata are kept open long enough for photosynthesis to take place, but not so long that plants lose too much water.**

BUILD Vocabulary

stoma
(plural: stomata) a small opening in the epidermis of a plant that allows carbon dioxide, water, and oxygen to diffuse into and out of a leaf

mesophyll specialized ground tissue found in leaves; performs most of a plant's photosynthesis

guard cell
a specialized cell in the epidermis of plants that controls the opening and closing of stomata

transpiration
the loss of water from a plant through its leaves

PREFIXES
The prefix *meso-* means "middle." Mesophyll tissue is in the middle of a leaf, inside the epidermis and around the veins.

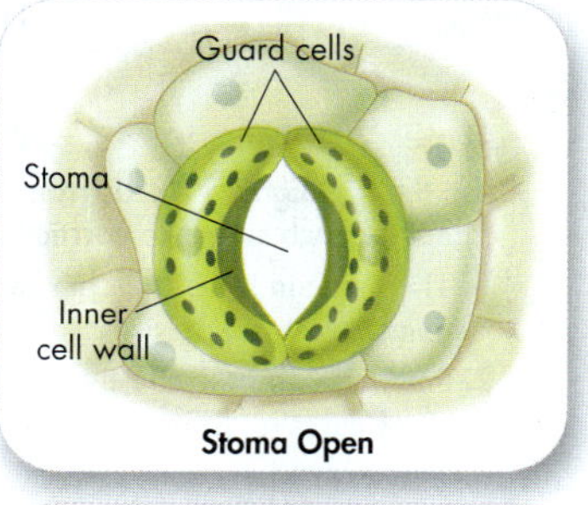

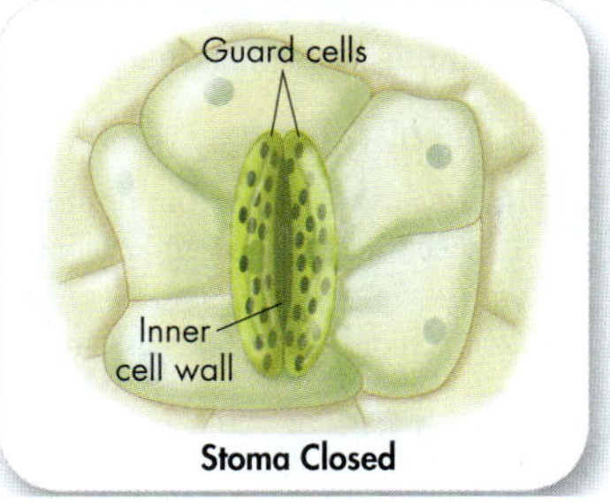

Guard Cell Function Stomata allow plants to balance the rate of photosynthesis and the loss of water.

ELL Focus on ELL: Build Background

ALL SPEAKERS Call students' attention to the vocabulary word *transpiration.* Explain that transpiration is the escape of water vapor from a plant into the environment.

Ask What is another word on this page that looks and sounds like *transpiration*? *(respiration)* What does this word mean? *(the process of taking in oxygen and giving off carbon dioxide)* Point out that both transpiration and respiration are processes, or actions.

Discuss the similarities and differences between transpiration and respiration. If necessary, have students research the terms. *(Both respiration and transpiration involve the movement of substances through the stomata. Respiration involves the exchange of oxygen and carbon dioxide. Transpiration does not involve an exchange—it is just the loss of water.)*

During Reading

Active Reading

MAKE CONNECTIONS Remind students that scientific terms often include descriptive words derived from Greek or Latin. Encourage students to look up any unfamiliar terms, such as *palisade mesophyll,* in a dictionary to find out what they mean. For example, *palisade* derives from the Latin word *palus,* meaning "stake." Tell students that a palisade is a barrier or fence made of a row of stakes.

Ask How does the palisade layer of a leaf's mesophyll resemble a fence made of a row of stakes? *(Like the stakes in a fence, the cells in the palisade layer are long and arranged side by side.)*

KEY QUESTION Read aloud the first question and answer.

Ask How does the shape of a leaf help it collect light? *(The leaf's thin, flat shape allows more of its cells to be exposed to light.)*

KEY QUESTION Invite a volunteer to read aloud the second question and answer.

Ask What are the three materials that pass through stomata? *(carbon dioxide, oxygen, and water)* Explain that water leaves a leaf as water vapor.

Build Vocabulary

PREFIXES Write the prefixes *epi-* and *trans-* on the board.

Say The prefix *epi-* means "on" or "over."

Ask Given that the prefix *epi-* means "on" or "over," where would you look for the epidermis of a leaf? *(on or over the leaf)*

Tell students that the prefix *trans-* means "across" or "through." Then, have a volunteer read the definition of *transpiration.* Explain that to get out of a leaf, water must travel across or pass through the epidermis.

Speed Bump

Ask What are the three main parts of a leaf? *(epidermis, mesophyll, and veins)*

Ask In what layer of cells are guard cells and stomata found? *(epidermis)*

Ask In what layers of cells are most chloroplasts found? *(palisade mesophyll and spongy mesophyll)*

Ask What type of tissue makes up the veins? *(vascular tissue: xylem and phloem)*

During Reading continued

Active Reading

COMPARE AND CONTRAST Have students read and discuss the Adaptations of Leaves visual.

Ask What do all of these plants have in common? *(All live in environments without an adequate supply of an important nutrient: pitcher plant—nitrogen; living stone, spruce, and cactus—adequate water.)* Which of these plants are named for their appearance? *(The pitcher plant resembles a water pitcher. The living stone looks like a rock.)*

SCIENCE SUPPORT Explain that the cactus and living stone have an adaptation that is not visible to the unaided eye. They have an internal structure that enables them to conduct CAM (crassulacean acid metabolism) photosynthesis, an adaptation for living in areas with too much light. The stomata of CAM plants open only at night to get in the carbon dioxide needed for photosynthesis.

Speed Bump

Ask How are the leaves of a spruce tree adapted to reduce water loss? *(Sample answer: The leaves are narrow, and they have a waxy epidermis.)*

Post Reading

ASSESSMENT TIPS For Question 4, suggest each student make a two-column, three-row chart with *light, water,* and *carbon dioxide* in the first column. In the second column, have them write where each material used in photosynthesis comes from.

For Question 5, have students review the Gas Exchange and Homeostasis section.

MYSTERY CLUE After students read the clue, tell them that the host tree would be in a situation similar to that of a plant growing on the ground in the shade. Have them reread the Anatomy of a Leaf section.

Ask How would the darkness in a heavily shaded area affect the process of photosynthesis for the host tree? *(Photosynthesis would be slowed or stopped because less light is reaching the leaves.)*

Wrap-Up Activity

WRITE A TEST Have students work in groups to write a test for this lesson with at least ten questions, including short answer and multiple choice. Groups can exchange tests, complete them, and then return them for grading by the test authors.

Adaptations of Leaves

The leaves of the plants shown here have adaptations that help the plants live in various areas and conditions.

◀ **Pitcher Plant** The leaf of a pitcher plant is modified to attract and then digest insects and other small prey. Such plants typically live in nutrient-poor soils and rely on animal prey as their source of nitrogen.

▼ **Living Stone** The two leaves of a living stone are adapted for hot, dry conditions. They are rounded, which minimizes the exposure of their surface to the air. They also have very few stomata.

Spruce The narrow leaves of a spruce tree contain a waxy epidermis as well as stomata that are sunken below the surface of the leaf. These adaptations reduce water loss from the leaves. ▶

Cactus Cactus leaves are actually nonphotosynthetic thorns that protect against herbivores. Most of the plant's photosynthesis is carried out in its stems. ▼

CHECK Understanding

Apply Vocabulary

Use the highlighted words from the lesson to complete each sentence correctly.

1. The tissue in leaves that performs most of a plant's photosynthesis is ______.
2. Small openings in the epidermis of a plant that allow gases to diffuse in and out of a leaf are called ______.
3. During the process of ______, water is lost through openings in a plant's epidermis.

Critical Thinking

4. **Explain** To carry out photosynthesis, leaves need light, water, and carbon dioxide. Identify the tissues or structures of a leaf that provide each of these materials.
5. **Apply Concepts** Imagine a field. The day is hot and it has not rained for a few weeks. Will the stomata of plants in the field be open or closed? Explain your reasoning.
6. **Write to Learn** Answer the mystery clue question. Write a paragraph that describes how leaf structure plays a role in photosynthesis and how the fig's growth affects this function in the host plant.

MYSTERY CLUE

The mature fig plant blocks sunlight from the host tree. How might this affect photosynthesis in the host? (**Hint:** See p. 564.)

BIOLOGY.com Search Lesson 23.4 GO • Lesson Assessment

Check Understanding

1. mesophyll
2. stomata
3. transpiration
4. Cells in the mesophyll capture sunlight; xylem in the veins brings water from the roots; carbon dioxide is allowed into the leaf by stomata.
5. The plants likely are not getting enough water and the day is hot. The stomata are likely closed, which will minimize the loss of water.
6. When light is blocked, rates of photosynthesis decrease. The host plant may not be able to produce enough food to meet its needs.

23.5 Transport in Plants

Water Transport

Look at a tall tree. Maybe there's a tree outside your school that is 15 meters tall or more. Imagine carrying a bucket of water to the top of that tree. You would have to do a lot of work! Now think of a giant redwood, 100 meters high. How does water reach the top?

Active transport and root pressure move water from the soil into a plant's roots. Water entering the vascular cylinder of the root creates pressure. This pressure pushes water upward into the stem. But root pressure does not produce nearly enough force to lift water up into trees. Other forces play a larger role.

Transpiration The major force in water transport is produced by the evaporation of water from leaves during transpiration. As water evaporates through open stomata, cell walls within the leaf begin to dry out. Plant cell walls contain cellulose, a complex carbohydrate. Paper is also made of cellulose. As you know, dry paper towels strongly attract water. Like a paper towel, the dry cell walls pull water from cells deeper inside the leaf. This pull is called transpirational pull. Transpirational pull also takes place in vascular tissue. It pulls water from the roots up through the xylem.

On a hot day, even a small tree may lose 100 liters of water to transpiration. The hotter and drier the air, the greater the amount of water lost. As a result of this loss, the plant draws up even more water from its roots.

Key Questions

What are the major forces that transport water in a plant?

What drives the movement of fluid through phloem tissue in a plant?

BUILD Understanding

Compare/Contrast Table As you read, make a table in which you compare and contrast water transport and nutrient transport.

In Your Workbook Go to your workbook to learn more about comparing and contrasting. Complete the table for Lesson 23.5.

BUILD Connections

TRANSPIRATIONAL PULL

Imagine a chain of circus clowns. The clowns are tied together and climbing a tall ladder. When the first clown reaches the top, he falls off. He pulls the clowns behind him up and over the top of the ladder. Water molecules move through a plant this way. The chain of water molecules in a plant extends from the leaves down to the roots. As molecules exit the leaves through transpiration, they pull up the molecules behind them.

BIOLOGY.com Search Lesson 23.5 GO • Visual Analogy

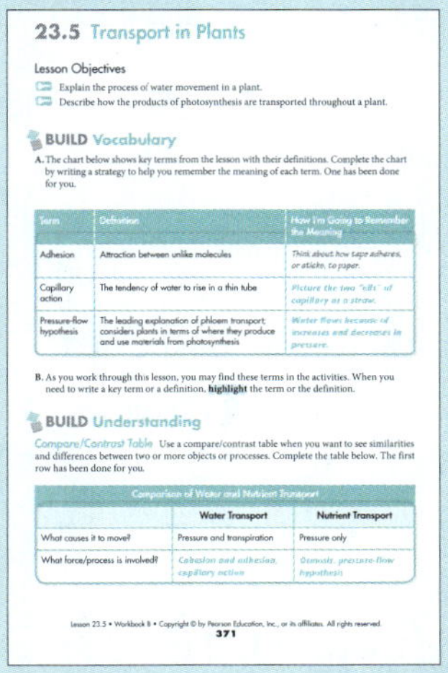

23.5 Transport in Plants

Lesson Objectives

Explain the process of water movement in a plant.

Describe how the products of photosynthesis are transported throughout a plant.

BUILD Vocabulary

A. The chart below shows key terms from the lesson with their definitions. Complete the chart by writing a strategy to help you remember the meaning of each term. One has been done for you.

Term	Definition	How I'm Going to Remember the Meaning
Adhesion	Attraction between unlike molecules	Think about how tape adheres, or sticks, to paper.
Capillary action	The tendency of water to rise in a thin tube	Picture the two "ells" of capillary as a straw.
Pressure-flow hypothesis	The leading explanation of phloem transport; considers plants in terms of where they produce and use materials from photosynthesis	Water flows because of increases and decreases in pressure.

B. As you work through this lesson, you may find these terms in the activities. When you need to write a key term or a definition, **highlight** the term or the definition.

BUILD Understanding

Compare/Contrast Table Use a compare/contrast table when you want to see similarities and differences between two or more objects or processes. Complete the table below. The first row has been done for you.

Comparison of Water and Nutrient Transport

	Water Transport	Nutrient Transport
What causes it to move?	Pressure and transpiration	Pressure only
What force/process is involved?	Cohesion and adhesion, capillary action	Osmosis, pressure-flow hypothesis

Lesson 23.5 • Workbook B • Copyright © by Pearson Education, Inc., or its affiliates. All rights reserved.

371

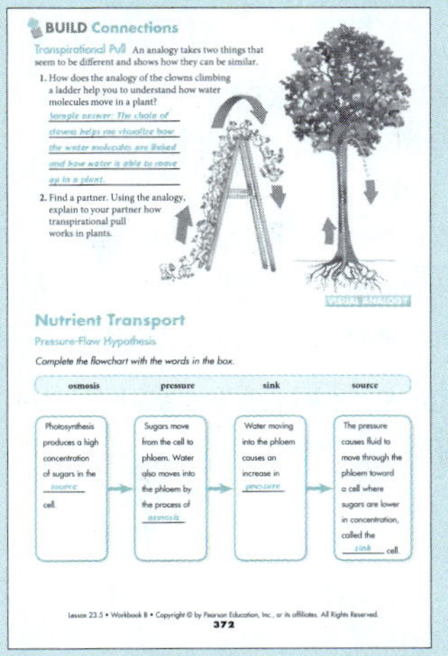

BUILD Connections

Transpirational Pull An analogy takes two things that seem to be different and shows how they can be similar.

1. How does the analogy of the clowns climbing a ladder help you to understand how water molecules move in a plant?
Sample answer: The chain of clowns helps me visualize how the water molecules are linked and how water is able to move up in a plant.

2. Find a partner. Using the analogy, explain to your partner how transpirational pull works in plants.

VISUAL ANALOGY

Nutrient Transport

Pressure-Flow Hypothesis

Complete the flowchart with the words in the box.

osmosis pressure sink source

Photosynthesis produces a high concentration of sugars in the source cell. → Sugars move from the cell to phloem. Water also moves into the phloem by the process of osmosis → Water moving into the phloem causes an increase in pressure → The pressure causes fluid to move through the phloem toward a cell where sugars are lower in concentration, called the sink cell.

Lesson 23.5 • Workbook B • Copyright © by Pearson Education, Inc., or its affiliates. All Rights Reserved.

372

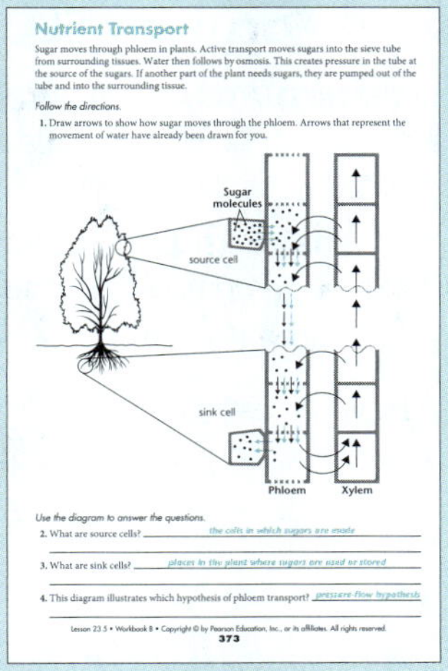

Nutrient Transport

Sugar moves through phloem in plants. Active transport moves sugars into the sieve tube from surrounding tissues. Water then follows by osmosis. This creates pressure in the tube at the source of the sugars. If another part of the plant needs sugars, they are pumped out of the tube and into the surrounding tissue.

Follow the directions.

1. Draw arrows to show how sugar moves through the phloem. Arrows that represent the movement of water have already been drawn for you.

Use the diagram to answer the questions.

2. What are source cells? the cells in which sugars are made

3. What are sink cells? places in the plant where sugars are used or stored

4. This diagram illustrates which hypothesis of phloem transport? pressure-flow hypothesis

Lesson 23.5 • Workbook B • Copyright © by Pearson Education, Inc., or its affiliates. All rights reserved.

373

Pre-Reading

Objectives

23.5.1 Explain the process of water movement in a plant.

23.5.2 Describe how the products of photosynthesis are transported throughout a plant.

Preview the Pages

1. Preview the vocabulary by having students find and read the sentences in the text that contain the highlighted vocabulary terms.
2. Have students read the caption and labels of each visual. Challenge them to match each vocabulary term with the visual that illustrates it.

Build Understanding

COMPARE/CONTRAST TABLE Direct students to the compare/contrast table in Workbook B. Identify the types of transport to be compared. Encourage students to complete the table by answering each question for each type of transport.

Build Connections

VISUAL ANALOGY Remind students that a visual analogy compares the similarities of two unlike things. Discuss the visual that compares transpirational pull with a chain of clowns on a ladder.

Ask What holds the clowns together? *(a rope)* Can you see what holds the chain of water molecules together? *(no)* What does the analogy mean? *(Like the clowns stick together with the rope, water molecules stick to each other.)* Have students look for this information as they read.

NATIONAL SCIENCE EDUCATION STANDARDS

UNIFYING CONCEPTS AND PROCESSES

I, V

CONTENT

C.5.d

For the corresponding lesson in *Miller & Levine Biology,* see pages 685–687.

During Reading

Active Reading

SET PURPOSE FOR READING Have a volunteer read aloud the two main headings in the lesson: Water Transport and Nutrient Transport.

Ask What do you think you will learn about in this lesson? *(how water and nutrients are transported in a plant)*

USE VISUALS Have students look at the three glass tubes in the Capillary Action visual.

Ask What do you notice about the tubes and the water? *(They have different widths and levels of water in them.)*

Ask What draws up the water molecules that are next to the inner surface of a tube? *(adhesion between the water molecules and the tube)*

Say Cohesion between water molecules draws up the rest of the column of water.

Ask What determines how high the water can rise in a tube? *(the width of the tube)*

KEY QUESTION Read aloud the question and answer.

Ask What forces make transpirational pull and capillary action possible? *(cohesion and adhesion)*

Help students understand that the clowns in the visual on the previous page represent water molecules and the rope represents the force of cohesion between the water molecules.

Build Vocabulary

ACADEMIC WORDS Read and discuss the meaning of *hypothesis* and the sample hypothesis in the definition for *pressure-flow hypothesis.* Have students work in pairs to write a hypothesis that answers the question, **What causes water to rise in the very thin tubes of a plant's xylem?** *(Sample answer: Water rises in xylem tubes by capillary action.)*

Speed Bump

Ask How do the forces of adhesion and cohesion differ? *(Adhesion is attraction between unlike molecules. Cohesion is attraction between like molecules.)*

Ask In which tissue does capillary action occur in a plant? *(xylem)*

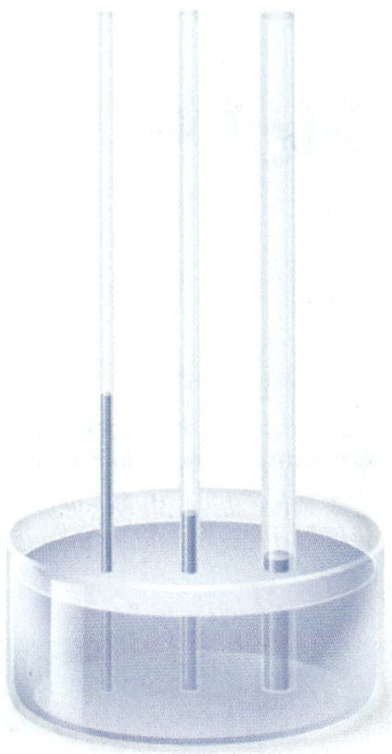

Capillary Action Capillary action causes water to move much higher in a narrow tube than in a wide tube.

Pulling Water Upward To pull water upward, plants make use of water's physical properties. One of these properties is cohesion. Cohesion is the attraction among molecules of the same substance. Water molecules are strongly attracted to one another. The cohesion of water is strong because water molecules form hydrogen bonds with each other. Water molecules also form hydrogen bonds with other substances. This results from the force of adhesion. **Adhesion** is the attraction between molecules that are unlike each other.

You can observe adhesion and cohesion at work. Here is how. Place a thin, empty glass tube in a dish of water. Water will move up into the tube. Why? Water is attracted to the walls of the tube. Water molecules are also attracted to one another. This tendency of water to rise in a thin tube is called **capillary action.**

Putting It All Together What do transpirational pull and capillary action have to do with the movement of water through xylem? Recall that xylem is made of tracheids and vessel elements. These cells connect to form long, hollow tubes. The walls of these hollow tubes are made of cellulose. Water adheres to cellulose. As water is lost by transpiration, adhesion pulls water from the wet inside of the leaf. This pull is so powerful that it reaches through the xylem down to the tips of the roots. It even reaches into the water in the soil. Together, transpiration and capillary action provide the major force that moves water upward through a plant.

Key Question What are the major forces that transport water in a plant?
Transpirational pull and capillary action are the major forces that move water from soil and through the xylem tissue to the rest of the plant.

BUILD Vocabulary

adhesion
a force of attraction between different kinds of molecules

capillary action
the tendency of water to rise in a thin tube

pressure-flow hypothesis
a hypothesis that explains how sap in phloem is transported through the plant from a sugar source to an area of lower sugar concentration

ACADEMIC WORDS
The word *hypothesis* means a possible answer to a scientific question. The pressure-flow hypothesis answers the question "What causes sap to flow within a plant?"

Nutrient Transport

How do sugars move in phloem? The leading explanation for transport in phloem is known as the **pressure-flow hypothesis.** Remember that unlike xylem cells, phloem cells are alive.

1. Active transport moves sugars into sieve tubes from the surrounding tissues.

2. Water follows the sugars into the sieve tubes by osmosis. This movement of water creates pressure in the sieve tube at the source of the sugars.

3. If sugars are needed in another area of the plant, they are pumped out of the sieve tube by active transport. Water then moves out of the sieve tube by osmosis. As the water leaves, it reduces the pressure in the tube at these areas.

ELL Focus on ELL: Extend Language

INTERMEDIATE AND ADVANCED SPEAKERS Start a word wall with the words *adhesion* and *cohesion.* Have students write the words in their notebooks and find them on this page.

Ask What are some similarities between the two words? *(*Adhesion *and* cohesion *have the same ending, or suffix,* -hesion.*)*

Explain that the suffix *-ion* indicates that the words *adhesion* and *cohesion* are nouns.

Ask What are some other examples of nouns that have the *-ion* ending? *(Sample answers: action, motion, corporation)*

Have students add their *-ion* nouns to the word wall.

568 Chapter 23 • Lesson 5

Thus, pressure drives the flow of nutrient-rich fluid, or sap. Fluid moves from places where sugars are made to places where sugars are used or stored. Differences in nutrient concentration cause fluids to move in directions that meet the needs of the plant. The pressure-flow system helps plants respond to the seasons. During the growing season, sugars from the leaves are moved into ripening fruits or into roots for storage. As the growing season ends, the plant drops its fruits. Then nutrients are stored in the roots. As the growing season again approaches, chemical signals stimulate phloem cells in the roots. They pump sugars back into the phloem. Then, the pressure-flow system carries these sugars to the stems and leaves. This nutrient-rich fluid supports rapid growth.

Key Question What drives the movement of fluid through phloem tissue in a plant?
Changes in nutrient concentration cause the movement of fluid through phloem to meet the needs of the plant.

Pressure-Flow Hypothesis This diagram shows the movement of sugars as explained by the pressure-flow hypothesis. Fluid starts at places where sugars are made (source cells). The fluid moves to places where sugars are used or stored (sink cells).

CHECK Understanding

Apply Vocabulary

Use the highlighted words from the lesson to complete each sentence correctly.

1. Water is attracted to cellulose by the force of ______.

2. ______ is the tendency of water to rise in a thin tube.

3. The ______ explains the movement of fluids through phloem.

Critical Thinking

4. Relate Cause and Effect On a hot, dry day, the stomata of a plant close. How could this affect the plant's rate of photosynthesis? Explain.

5. Compare and Contrast Compare the role of active transport with that of osmosis in the movement of materials through phloem.

6. Write to Learn Answer the mystery clue question. In your answer, use the words *xylem* and *cohesion*.

MYSTERY CLUE

Fig vines can grow to the top of very tall trees. How do the leaves at the top of the vine get the water they need for photosynthesis? (**Hint:** See p. 567.)

BIOLOGY.com Search Lesson 23.5 GO • Lesson Assessment

Check Understanding

1. adhesion

2. Capillary action

3. pressure-flow hypothesis

4. The rate would decrease because leaves would not be able to take in carbon dioxide.

5. Both active transport and osmosis move materials into and out of cells. Active transport is used to move sugars into and out of cells. Osmosis is the process by which water moves in and out of cells. Active transport requires energy, but osmosis does not require energy.

6. Transpiration and capillary action allow the xylem to transport water to the top of tall vines. Transpiration occurs when water evaporates from the leaves. As water evaporates, adhesion and cohesion work together to pull water up through the xylem. These forces can carry water to the tops of even the tallest trees.

Active Reading

INTERPRET DIAGRAMS Discuss the Pressure-Flow Hypothesis diagram. Point out the key, and have students locate the sugar molecules and the arrows that show the movements of water and sugar. Then, read aloud each numbered step on the previous page as the class finds the matching numbers in the diagram.

Clarify the meanings of the labels *Source cell (a cell that provides sugars to other parts of a plant)* and *Sink cell (a cell to which sugars move).*

KEY QUESTION Read aloud the question and answer.

Ask What is the general direction in which sugars move in phloem? *(Sugars move from where they are made to where they are used or stored.)*

Speed Bump

Ask How is sugar transported into and out of the phloem of a plant? *(by active transport)*

Ask How do the seasons relate to the fluid movement in plants? *(During the growing season, the fluids move sugars to ripening fruits or into roots for storage. During the non-growing seasons, the fluids move sugars to the roots to be stored.)*

Post Reading

ASSESSMENT TIPS For Question 5, help students identify in the Pressure-Flow Hypothesis diagram where the processes of active transport and osmosis are taking place. If students need more help, have them reread the numbered steps on the previous page and use a Venn diagram to help sort out the details.

MYSTERY CLUE Remind students that the vascular bundles in a stem extend all the way to the leaves. Help students understand that as long as the tubes in the vascular tissue remain unbroken, water will continue to move up in response to transpiration. For more information, direct students to the Pulling Water Upward section.

Wrap-Up Activity

THINK-PAIR-SHARE Have student pairs work together to develop a detailed answer to the question, **How does a water molecule from the soil move into and out of a plant?**

Give pairs time to construct a response, then have volunteers share their answers with the class.

CHAPTER LAB

Pre-Lab

Review the meaning of *primary growth. (growth that results from the elongation of cells produced by an apical meristem)* Also, review the events of cell differentiation as it applies to plants. *(During cell differentiation, unspecialized cells that are small and shaped like spheres elongate, or grow in length, and develop the specialized shapes and structures of the different types of cells that make up plant tissues.)*

Ask What does the word *apical* mean? *("at the top or tip")*

Ask What happens in a plant's apical meristem? *(rapid cell division that adds new cells to a shoot or a root)*

Ask What happens to the older cells in the apical meristem as new cells are made? *(The older cells are pushed out of the meristem.)*

Ask Do the cells made in an apical meristem differentiate while they are part of the meristem or after they have been pushed out? *(after)*

Review the Background Questions and Pre-Lab Questions as a class.

Lab

Tell students they will perform the chapter lab *Identifying Growth Zones in Roots* in **Lab Manual B.**

SAFETY Students should wash their hands before they leave the lab. Remind them to handle the glassware and plant material with care.

Post-Lab

OBSERVE ROOT STRUCTURE Give each group a microscope and a prepared slide of a longitudinal section of a root tip. Have students observe the very tip of the root under low power. Have students view different areas along the length of the root. Have them describe what they see. *(The cells in the root tissue away from the tip are longer and look different from the cells at the very tip as tissues become visible.)*

NATIONAL SCIENCE EDUCATION STANDARDS

UCP II, III

CONTENT C.1.f

INQUIRY A.1.a, A.1.b, A.1.f, A.2.a

Design Your Own Lab

GUIDED INQUIRY

Pre-Lab: Identifying Growth Zones in Roots

Problem Where does growth occur in plant roots?

Materials 150-mL beaker, paper towels, large bean seeds, petri dish, masking tape, metric ruler, fine-tip permanent marker

Lab Manual Chapter 23 Lab

Skills Focus Design an Experiment, Measure, Organize Data, Analyze Data

Connect to the Big idea A plant's root system absorbs nutrients, stores food, and provides support for the rest of the plant. As a plant grows, the root system must be able to absorb more nutrients, store more food, and provide more support. Thus, growth in the root system must keep pace with growth in the other parts of a plant. Where in a root does the growth occur that increases the root's length? In this lab, you will design an experiment to answer this question.

Background Questions

a. **Review** What are the three principal organs of seed plants?

b. **Compare and Contrast** What are the main functions of dermal tissue, vascular tissue, and ground tissue?

c. **Review** What are meristems?

Pre-Lab Questions

Preview the procedure in the lab manual.

1. **Predict** A root is marked at two points along its length. What will happen to the distance between these marks if the root grows longer only near the tip? What will happen if growth occurs evenly along the entire length of the root?

2. **Design an Experiment** The procedure in Part A asks you to use four seeds. Why not use two seeds instead?

3. **Design an Experiment** How will you keep track of which seedling is which?

BIOLOGY.com Search Chapter 23 GO

Visit Chapter 23 online to test yourself on chapter content and to find activities to help you learn.

Untamed Science Video The Untamed Science crew takes you to several exotic locations to see unique plant structures and adaptations.

Tutor Tube Tune in to Tutor Tube to see how new tissue growth makes plants taller.

Art Review Test your knowledge of leaf structures.

Art in Motion See how plant roots absorb nutrients and water molecules.

Visual Analogy Compare the motion of clowns climbing a ladder with water molecules being pulled up a tree.

Pre-Lab Answers

BACKGROUND QUESTIONS

a. roots, stems, and leaves

b. dermal tissue—protective outer layer; vascular tissue—support the plant body, transport water and nutrients; ground tissue—produce and store sugars, help support the plant

c. Meristems are regions of undifferentiated cells that produce new cells.

PRE-LAB QUESTIONS

1. The marks will stay the same distance apart if growth only occurs near the tip. If growth occurs evenly all along the root, then the distance between the marks will increase.

2. Sample answer: One seed might not germinate or one seedling might die before the experiment is done. A larger sample makes it more likely that the gathered data are not the result of abnormal growth.

3. Sample answer: Use a small piece of tape on the outside of the beaker to label the location of each seedling. Remove only one seedling at a time from the beaker when making measurements.

23 CHAPTER Summary

23.1 Specialized Tissues in Plants

- The three main organs of seed plants are roots, stems, and leaves.
- Dermal tissue is the protective outer covering of a plant. Ground tissue produces and stores sugars and supports the plant. Vascular tissue supports the plant body and transports water and nutrients throughout the plant.
- Meristems are areas of unspecialized cells in which mitosis produces new cells that are ready for differentiation.

epidermis (p. 553)
parenchyma (p. 553)
collenchyma (p. 553)
sclerenchyma (p. 553)
lignin (p. 554)
meristem (p. 554)
apical meristem (p. 554)

23.2 Roots

- A mature root has an outside layer called the epidermis. A mature root also contains vascular tissue and a large area of ground tissue.
- Roots support a plant, hold the plant in the ground, store food, and absorb water and dissolved nutrients from the soil.

root hair (p. 557)
cortex (p. 557)
endodermis (p. 557)
root cap (p. 557)

23.3 Stems

- Stems produce leaves, branches, and flowers. Stems hold leaves up to the sun. Stems transport substances throughout the plant.
- Primary growth of stems is the result of elongation of cells produced in the apical meristem. It takes place in all seed plants.
- In the stems of conifers and dicots, secondary growth takes place in meristems called the vascular cambium and cork cambium.

node (p. 560)
bud (p. 560)
vascular bundle (p. 561)
primary growth (p. 561)
secondary growth (p. 562)
vascular cambium (p. 562)
cork cambium (p. 562)

23.4 Leaves

- The structure of a leaf helps the plant collect light and carry out photosynthesis.
- Plants maintain homeostasis by keeping stomata open long enough for photosynthesis to take place, but not so long that plants lose too much water.

stoma (p. 564)
mesophyll (p. 564)
guard cell (p. 565)
transpiration (p. 565)

23.5 Transport in Plants

- Transpirational pull and capillary action provide most of the force needed to move water from soil through the xylem tissues of a plant.
- Changes in nutrient concentration drive the movement of fluid through phloem to meet the needs of the plant.

adhesion (p. 568)
capillary action (p. 568)
pressure-flow hypothesis (p. 568)

BIOLOGY.com Search Chapter 23 GO • Match It • Chapter Assessment

UbD Transfer the Big Idea

TRANSFER Have students imagine they are botanists who answer questions for a Web site about gardening. Pair students, and give them the following e-mail.

> Dear Green Thumb, I have a potted plant on my sunny patio. The plant wilts nearly every day. I have to water it constantly. Where does all that water go? Signed, Mystified

Have each pair write an e-mail that answers the question as completely as possible. Remind students that water passes through all parts of a plant and is used in several ways. Have two pairs exchange their e-mails and evaluate the answers. Tell students to judge whether the answer was complete and understandable, and to offer suggestions for improvement.

Chapter Summary

Think Visually

USE GRAPHICS Have students make a flowchart of the tissues through which water passes in a plant, beginning at the root hairs. Give students the following words to use in their flowcharts: *cortex, endodermis, epidermis, stomata, xylem, mesophyll.*

ANSWERS

Students' flowcharts should use the names of as many different tissues as possible.

BIOLOGY.com Study Online

REVIEW RESOURCES

Art in Motion Students watch water passing through the dermal tissue of a root and into xylem cells.

InterActive Art A short animation shows primary and secondary growth in plants.

Tutor Tube Short, online tutorial addresses misconceptions about how and where growth occurs in most plants.

Art Review Students can drag and drop labels to review the structure of a leaf.

Data Analysis This virtual data set shows how the activity of plant stomata changes in response to different atmospheric carbon dioxide levels.

Visual Analogy Transpirational pull is modeled by the animated movements of a chain of circus clowns.

Vocabulary Review The *Flash Cards* and *Match It* provide an interactive way to review chapter vocabulary.

Editable Worksheets Pages of Study Workbook B, Lab Manual B, and the Assessment Resources Book are available online. These documents can be easily edited using a word-processing program.

Lesson Overviews Lesson Overviews are based on Miller & Levine Biology. However, they are available on CD-ROM and can be edited for your class.

ASSESSMENT RESOURCES

Chapter Assessment Have students take an online version of the Chapter 23 Check Understanding.

Standardized Test Prep Students can take an online version of the Standardized Test Prep. You will receive their scores along with ideas for remediation.

Diagnostic and Benchmark Tests Use these tests to monitor your students' progress and supply remediation.

ASSESSMENT

Assess

Big idea Invite a volunteer to read the Big Idea and the question.

Ask What can the word *structure* mean? *(A structure can refer to a part of an organism's body or the way the body or a body part is put together.)*

Ask Which meaning of *structure* best describes the Big Idea in this chapter? *(the way a body or a body part is organized, or put together)*

Ask What are some functions of a seed plant? *(Sample answers: growth, transport of materials, reproduction, photosynthesis, and respiration)*

Have students write their answers in their notebooks.

CONSTRUCTED RESPONSE Paragraphs will vary. Explain the activity carefully. Be sure students understand that the *Hints* should be used as a guide to help them answer the questions. Review how to organize the topics to include in a paragraph by making an outline.

Write Question 1 on the board and read it aloud. Underline and point out in the question the key words *vascular tissue, specialized,* and *transport.*

Say This question asks for a summary of the structure and function of the cells in vascular tissue.

Read the first hint.

Ask Which lesson in this chapter introduces these terms? *(Lesson 23.1.)*

Ask What kind of vascular tissue has each of these kinds of cells? *(tracheids and vessel elements: xylem; sieve tube elements and companion cells: phloem)*

Read the next hint.

Say Xylem and phloem are responsible for water transport and nutrient transport.

Ask What about the structure of xylem allows capillary action to take place? *(Xylem cells form thin tubes that mostly lack divisions between individual cells.)*

Model writing the first one or two sentences using the hints on the board and the text. *(Sample: The xylem cells that make up a plant's vascular tissue are specialized to transport water and minerals. For example, the dead xylem cells resemble sections of pipe that carry water.)*

23 CHECK Understanding

Assess the Big idea Structure and Function

Write an answer to the question below.

Q. How are cells, tissues, and organs organized into systems that carry out the basic functions of a seed plant?

Constructed Response

Write an answer to each of the numbered questions below. The answer to each numbered question should be one or two paragraphs long. To help you begin, read the **Hints** below the question.

1. **Explain how the cells in the vascular tissue of a plant are specialized for transport of water and minerals.**

   Hint Use the words *tracheids, vessel elements, sieve tube elements,* and *companion cells* in your answer.

   Hint Review the information on Water Transport and Nutrient Transport in Lesson 23.5.

2. **How do the cells and tissues in the trunk of a tree help it support the weight of all of its branches and leaves?**

   Hint Trunks are stems that have undergone secondary growth.

   Hint Xylem contains tracheids and vessel elements.

3. **How do several different types of tissues in a leaf work together in the functioning of the leaf?**

   Hint The primary function of a leaf is to carry out photosynthesis.

   Hint Leaves contain epidermis, mesophyll, and vascular tissue.

Foundations for Learning Wrap-Up

Use the folded page you prepared when reading the chapter to help you organize your thoughts about how cells, tissues, organs, and organ systems work together to help a plant survive.

Activity 1 Working with a partner or small group, discuss how each part of a plant works with another part of a plant. Use each tab of your folded paper to guide your discussion. For example, for the tabs labeled Leaves and Stems, identify how leaves work with stems to help the plant survive.

Activity 2 Complete each horizontal cut for each of the five topics to create five separate booklets. On the inside of each booklet, add a labeled sketch of the major structures in the lesson. Use the booklets to review the concepts from each lesson.

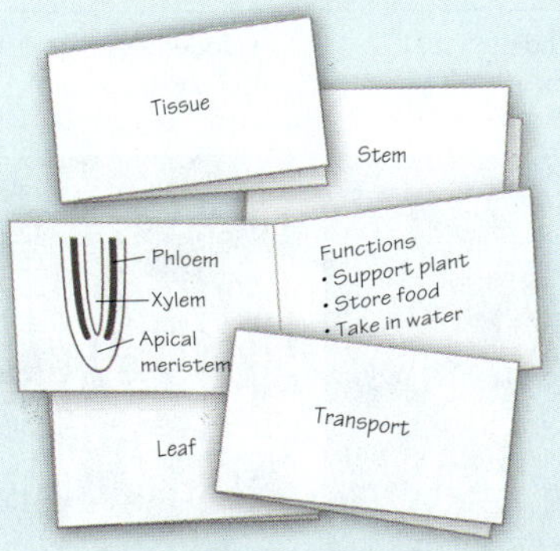

For Question 2, pair students and have them use the hints and your modeling to answer the question. Have students share their paragraphs with the class.

For Question 3, have students write the paragraphs independently for homework or as an anchor activity.

Foundations for Learning Wrap-Up

Have students take out the folded pages they prepared while reading the chapter. Be sure students have written information about how cells, tissues, and organs work together in plants. Explain the activities, and have pairs of students work together to complete them.

23.1 Specialized Tissues in Plants

Understand Key Concepts

1. The organ that supports the plant body and carries nutrients between different parts of the plant is the
   a. root.
   b. stem.
   c. leaf.
   d. flower.
2. Tracheids and vessel elements make up
   a. meristem.
   b. phloem.
   c. xylem.
   d. parenchyma.
3. What are the three main functions of ground tissue?
4. What is the main difference between mature xylem cells and mature phloem cells?

Think Critically

5. **Compare and Contrast** How are the epidermis of roots and leaves similar? How do they differ?

Test-Taking Tip

Using Lists It requires planning to answer an essay question. Before you start writing an answer, write a list of the points that you want to make. For a compare and contrast question, make two lists: one for similarities and one for differences.

23.2 Roots

Understand Key Concepts

6. Which of the following is a trace element that is absorbed by roots?
   a. nitrogen
   b. phosphorus
   c. zinc
   d. potassium
7. The waterproof barrier that separates the endodermis from the vascular cylinder is the
   a. guard cell.
   b. phloem.
   c. Casparian strip.
   d. cortex.
8. What are the two kinds of root systems?
9. How are root hairs important to plants?

Think Critically

10. **Explain** When you moved a plant to a large pot, you noticed that the plant's roots were very crowded in the old pot. After the plant was transplanted, it soon began to grow faster and looked healthier. Write a hypothesis that explains this observation.

23.3 Stems

Understand Key Concepts

11. A stem grows thicker when new tissue grows in
    a. vascular cambium.
    b. apical meristem.
    c. nodes.
    d. ground tissue.
12. Which of the following is part of the bark?
    a. heartwood
    b. sapwood
    c. xylem
    d. phloem
13. In stems, from what type of tissue does phloem develop?
14. What is the main difference between monocot and dicot stems?

Think Critically

15. **Infer** A plant undergoes only primary growth. Will it have a short life or a long life? Why?

23.4 Leaves

Understand Key Concepts

16. Most photosynthesis in a leaf takes place in the
    a. leaf vein.
    b. mesophyll.
    c. epidermis.
    d. cuticle.
17. Stomata open and close due to changes in the water pressure within
    a. phloem.
    b. xylem.
    c. guard cells.
    d. cell walls.
18. What is the function of the cuticle in a leaf?

Think Critically

19. **Infer** Cells in palisade mesophyll are closer together than cells in spongy mesophyll. How does this relate to the functions of these layers?

Assessment Tips

Pair students and have them work together to answer the questions.

In Questions 3, 4, and 14, make sure students note the word *main* and realize that there could be unimportant functions and differences that they should not include in their answers.

Questions 2, 5, 9, 11, 12, 13, and 19 require students to know the structures of plant tissues. Questions 1, 3, 7, 9, and 19 require students to know functions of plant tissues. Have students review these structures and functions.

For Questions 1, 8, 20, 21, 22, 23, and 24, have students review how materials are transported into and within plants.

Answers

Lesson 23.1

UNDERSTAND KEY CONCEPTS

1. b **2.** c

3. Ground tissue makes and stores sugars and supports the plant body.

4. Mature xylem cells are dead while mature phloem cells are alive.

THINK CRITICALLY

5. In both the root and the leaf, the epidermis is the outer layer that protects the inner tissues. In roots, the epidermis absorbs water and nutrients from the soil. It has long root hairs that increase its surface area. In leaves, the epidermis is covered by a waxy cuticle that prevents the loss of water. Stomata allow gases to enter the cell.

Lesson 23.2

UNDERSTAND KEY CONCEPTS

6. c **7.** c

8. taproots and fibrous roots

9. Root hairs increase the surface area for the absorption of water.

THINK CRITICALLY

10. Sample answer: The plant grew because its roots had more soil from which to obtain nutrients.

Lesson 23.3

UNDERSTAND KEY CONCEPTS

11. a **12.** d

13. vascular cambium

14. Monocot stems have scattered vascular bundles throughout the stem. In dicots, the vascular bundles are located in organized rings.

THINK CRITICALLY

15. A plant that undergoes only primary growth likely does not have a long life. As a plant gets older, it increases in size. Without the support of the tissues that grow during secondary growth, a plant cannot meet its needs. It will not survive.

Lesson 23.4

UNDERSTAND KEY CONCEPTS

16. b **17.** c

18. A cuticle limits the loss of water from a leaf.

THINK CRITICALLY

19. Sample answer: The palisade mesophyll captures light energy for photosynthesis. The spaces in spongy mesophyll allow carbon dioxide and oxygen to enter and exit the leaf. Plants need carbon dioxide for photosynthesis and oxygen to break down food.

Lesson 23.5

UNDERSTAND KEY CONCEPTS

20. a

21. d

22. pressure-flow hypothesis

23. Water enters the cells by osmosis.

THINK CRITICALLY

24. The rate of evaporation is greater on a hot, dry day than it is on a cold, rainy day. The rate of transpiration would be greater. So, transpirational pull would be stronger on a hot, dry day.

Connecting Concepts

25. between about 12:30 and 3:30 PM

26. As transpiration increases or decreases, water intake also increases or decreases, with changes in water intake lagging behind transpiration changes by a few hours.

CHAPTER MYSTERY Have students read Solve the Chapter Mystery and reread the clues. Then, discuss the functions of the three principal organs of plants.

Ask What is different about the growth of strangler fig roots from the growth of most other tree roots? *(The roots of the strangler fig start their growth high above the ground and grow down through the air into the ground. The roots of most trees begin and continue their growth underground.)*

Ask How do the organs of the strangler fig compete with those of the host tree? *(The leaves and stems of the fig compete with the leaves of the host for sunlight. The roots of the fig compete with the roots of the host for water and nutrients.)*

23 CHECK Understanding

23.5 Transport in Plants

Understand Key Concepts

20. Water molecules are attracted to each other by the force of

a. cohesion.
b. adhesion.
c. transpiration.
d. capillary action.

21. As water evaporates from a plant's leaves, it causes water to move through the plant by the force of

a. osmosis.
b. active transport.
c. capillary action.
d. transpirational pull.

22. What explains the movement of sugars and nutrients through phloem?

23. What happens to water when active transport moves sugars into phloem cells?

Think Critically

24. **Relate Cause and Effect** Would transpirational pull be stronger on a cold, rainy day or on a hot, dry day? Explain your answer.

Connecting Concepts

Use Science Graphics

Use the graph to answer questions 25–26.

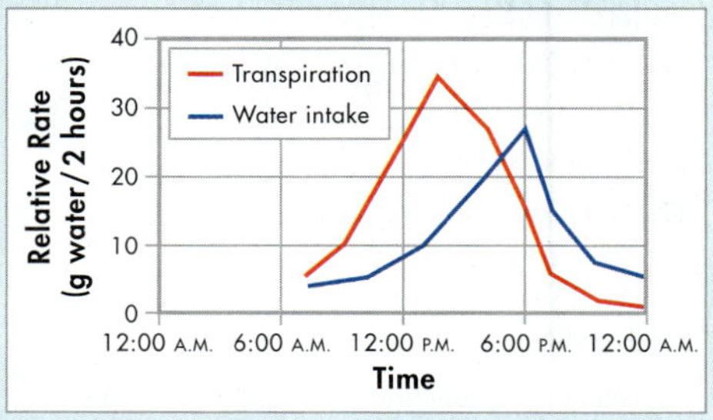

25. **Analyze Data** When is the greatest amount of water lost through transpiration?

26. **Draw Conclusions** What is the relationship between transpiration and water intake?

solve the CHAPTER MYSTERY

THE HOLLOW TREE

The life of a strangler fig starts with a sticky seed. This seed is dropped on a high tree branch by an animal such as a bird, bat, or monkey. At first, the plant grows slowly. This slow growth happens because the roots take in the few dissolved nutrients found in the rainwater and the leaf litter that collects in the host tree's branches. Eventually, the first roots grow down the host's trunk and enter the ground. Then, the fig's growth rate increases rapidly. The fig sends down many more roots. These roots become tangled together. They crush the host tree's bark, including the phloem. As a result, fewer nutrients can move through the phloem.

The fig's stems and leaves eventually grow taller than the host tree. This shades the host from the sun. The host performs less photosynthesis. The fig's roots compete with the host's roots for limited nutrients in the soil.

This triple punch—strangulation, competition for light, and competition for nutrients—usually kills the host. An impressive "hollow" fig tree is left behind.

1. **Use Analogies** One scientist has described the strangler fig as a "vegetable octopus." Explain how this analogy relates to the habits of the strangler fig.

2. **Infer** Plants that sprout and grow on top of other plants are called epiphytes. In what biome do you think epiphytes are most common? Explain your reasoning.

UntamedScience™ **Never Stop Exploring Your World.** Finding the solution to The Hollow Tree mystery is only the beginning. Take a video field trip with the ecogeeks of Untamed Science to see where the mystery leads.

CHAPTER MYSTERY ANSWERS

1. An octopus has tentacles that it uses to grasp prey. In a similar way, the strangler fig has many branching roots that wrap around the host.

2. Epiphytes are most common in tropical rainforests because the plant growth is so dense. Epiphytes grow on top of other plants to get more sunlight. Because the air is humid and rainfall is plentiful, epiphytes get water from the air or from rainwater that collects on their hosts.

Untamed Science™ Follow the Untamed Science crew in the video, **Plants Inside and Out,** and watch as they examine the unique plants that have found a way to survive in exotic locations.

Standardized Test Prep

Multiple Choice

1. Which of the following cell types is NOT found in a plant's vascular tissue?
   A tracheid
   B vessel element
   C guard cell
   D companion cell

2. Where in a plant does mitosis produce new cells?
   A meristems
   B chloroplasts
   C mesophyll
   D heartwood

3. Which tissues make up tree bark?
   A phloem
   B cork
   C cork cambium
   D all of the above

4. Which is NOT a factor in the movement of water through a plant's vascular tissues?
   A transpiration
   B capillary action
   C root pressure
   D meristems

5. All of the following conduct fluids in a plant EXCEPT
   A heartwood.
   B sapwood.
   C phloem.
   D xylem.

6. Where does most of the photosynthesis occur in a plant?
   A stomata
   B guard cells
   C vascular cambium
   D mesophyll tissue

7. Which of the following structures prevents the backflow of water into the root cortex?
   A palisade mesophyll
   B root cap
   C cambium
   D Casparian strip

8. Which of the following plants has a fibrous root system?
   A dandelion
   B beet
   C radish
   D grass

Questions 9–10

A student compared the average number of stomata on the top side and the underside of the leaves of different plants. Her data are summarized in the table below.

Average Number of Stomata (per square mm)

Plant	Top Surfaces of Leaves	Bottom Surfaces of Leaves
Pumpkin	29	275
Tomato	12	122
Bean	40	288

9. What generalization can be made based on the data?
   A All plants have more stomata on the top side of their leaves than on the bottom side.
   B Plants have fewer stomata on the top side of their leaves than on the bottom side.
   C Some plants have more stomata on the top side of their leaves than on the bottom side.
   D The number of stomata is the same from plant to plant.

10. Pumpkins, tomatoes, and beans all grow in direct sunlight. Assuming the plants receive plenty of water, stomata on the lower surface of their leaves
   A are always closed.
   B are usually clogged with dust.
   C are unlikely to close at night.
   D stay open during daylight hours.

Open-Ended Response

11. Contrast the functions of xylem and phloem.

If You Have Trouble With . . .

Question	1	2	3	4	5	6	7	8	9	10	11
See Lesson	23.1	23.1	23.3	23.5	23.3	23.4	23.2	23.2	23.4	23.4	23.5

Test-Taking Tip

USE TIME WISELY

Tell students to skip test questions they are having difficulty answering, keeping in mind the time remaining to complete the test. Point out that, sometimes, information in another question will jog their memory. Also, point out that they will feel less stressed if all other questions have been answered. Explain that when they are more relaxed, information will be easier to recall. Then, tell them that, if there is time remaining after they have gone through the entire test, they can go back to the questions they had trouble answering.

Before the Test

Review with students the Test-Taking Tip.

Point out the word *NOT* in Questions 1 and 4.

Point out the word *EXCEPT* in Question 5.

Remind students that an open-ended response means that they will have to write a paragraph to answer the question. Have them paraphrase the question in their notebooks, and remind them about the steps you took when you modeled writing a paragraph. Specifically point out that you identified the key words in the question before starting to write.

Answers

1. C
2. A
3. D
4. D
5. A
6. D
7. D
8. D
9. B
10. D
11. Xylem is dead tissue that conducts water and minerals from roots to leaves. It conducts in only one direction. Phloem is living tissue that conducts the products of photosynthesis from leaves to roots or in the reverse direction, depending on where sugars are needed in the plant.

Plant Reproduction and Response

Chapter Contents	NSES	Time	Core Resources
Chapter Preview			**Student Edition,** pp. *576–577* **Chapter Mystery,** p. *577* **Foundations for Learning,** p. *577*
24.1 Reproduction in Flowering Plants The Structure of Flowers • The Angiosperm Life Cycle • Vegetative Reproduction	V, A.1.b, A.1.c, A.2.a, C.3.b	1 1/2 periods 3/4 block	**Student Edition,** pp. 578–584 **Inquiry Into Scientific Thinking,** p. 579 **Study Workbook B** 24.1 Worksheets **Biology.com** *Art Review:* The Parts of a Flower • *Art in Motion:* The Development of Gametophytes • 24.1 Lesson Assessment
24.2 Fruits and Seeds Seed and Fruit Development • Seed Dispersal • Seed Dormancy and Germination	V, C.4.c	1/2 period 1/4 block	**Student Edition,** pp. 585–587 **Study Workbook B** 24.2 Worksheets **Biology.com** *Tutor Tube:* Flowers and Fruits • *Real-World Inquiry:* Fugitive Seeds? • 24.2 Lesson Assessment
24.3 Plant Hormones Hormones • Tropisms and Rapid Movements • Response to Seasons	IV, A.2.a, C.1.d, C.5.d, C.6.b	1 1/2 periods 3/4 block	**Student Edition,** pp. 588–592 **Build Connections,** p. 591 **Study Workbook B** 24.3 Worksheets **Biology.com** *InterActive Art:* Photoperiodism • 24.3 Lesson Assessment
24.4 Plants and Humans Agriculture • Fiber, Wood, and Medicine	I, C.4.e, F.3, F.4, F.5, F.6	1/2 period 1/4 block	**Student Edition,** pp. 593–595 **Study Workbook B** 24.4 Worksheets **Biology.com** 24.4 Lesson Assessment
Chapter Pre-Lab	A.1.b, A.2.a, C.1.d, C.1.e	1 period 1/2 block	**Student Edition,** p. 596 **Lab Manual B** *Plant Hormones and Leaves*

Tools for ELL Instruction

ELL Handbook describes teaching strategies to make *Miller & Levine Biology: Foundation Edition* more accessible to ELL students.

Spanish Study Workbook is a Spanish translation of Study Workbook A.

Multilingual Glossary is the *Miller & Levine Biology* glossary translated into ten languages.

Tools to Challenge Students

Study Workbook A includes more challenging worksheets and activities such as Chapter Mystery activities that reinforce 21st Century Skills.

Lab Manual A labs are geared toward on-level students. The Lab Manual includes fifteen labs in addition to the Chapter Labs.

Biology.com Data Analysis activities offer students a way to practice data gathering and analyzing skills.

Visual Quizzes in the Assessment Resources Book are geared toward on-level students and use visuals from *Miller & Levine Biology.*

Additional Resources

Biology.com Untamed Science Video • Flash Cards

Study Workbook A 24.1 Worksheets
Spanish Study Workbook 24.1 Worksheets
Assessment Resources Book Visual Quiz

Study Workbook A 24.2 Worksheets
Lab Manual B *Data Analysis:* Temperature and Seed Germination
Spanish Study Workbook 24.2 Worksheets

Study Workbook A 24.3 Worksheets
Lab Manual B *Hands-On Activity:* A-Mazing Plants
Spanish Study Workbook 24.3 Worksheets
Assessment Resources Book Visual Quiz

Study Workbook A 24.4 Worksheets
Lab Manual B *Data Analysis:* Increasing Crop Yields
Spanish Study Workbook 24.4 Worksheets
Biology.com *Data Analysis:* Plant Energy

Lab Manual A *Plant Hormones and Leaves* • *Using Pollen to Solve Crimes*

Chapter Review

Student Edition Chapter Summary, p. 597 • Unit Project, p. 602
Study Workbook B Chapter 24 Review
Transparencies, pp. 280–289
Biology.com Untamed Science Video • You're the Director • Editable Worksheets of Study Workbooks A and B and Lab Manuals A and B • Chapter 24 Flash Cards and Crossword

Untamed Science DVD • Classroom Resources CD (includes lesson presentations and editable worksheets)

Chapter Assessment

Student Edition Check Understanding, pp. 598–601
Study Workbook B Chapter 24 Taking a Standardized Test
Assessment Resources Book Chapter 24 Test B • Unit 6 Test B
Biology.com Chapter 24 Assessment • Editable Worksheets of Chapter 24 Visual Quizzes,* Chapter 24 Tests A and B, and Unit 6 Tests A and B

Exam*View Assessment Suite* • Classroom Resources CD (includes lesson presentations and editable worksheets)

Time: 1 period, 1/2 block

* Visual Quizzes are based on visuals from *Miller & Levine Biology.* Review carefully to make sure they are appropriate for your class.

Pressed for Time?

Preview the Chapter Introduce the first two Key Questions for Lessons 24.1 and 24.2, and preview the Angiosperm Life Cycle and Variety Among Fruits visuals.

Cover the Chapter Quickly In Lesson 24.1, have students read *The Structure of Flowers* and go over the Angiosperm Life Cycle visual. Assign *Seed and Fruit Development* and *Seed Dispersal* in Lesson 24.2. Have students read the introduction of *Hormones,* all of *Tropisms and Rapid Movements,* and the Photoperiod and Flowering section of *Response to Seasons* in Lesson 24.3.

Assess Assign Questions 2, 3, and 5 in the 24.1 Check Understanding, Questions 3–5 in the 24.2 Check Understanding, and Questions 1–5 in the 24.3 Check Understanding. In the Chapter 24 Check Understanding, assign Questions 1–3, 6, and 9–11.

Get the Picture

Have students cover the caption and see if they can guess what the picture shows. Tell them to use the chapter and lesson titles and the Big Idea to guide their guesses. Then, have students read the caption. Explain that *SEM* means that a scanning electron microscope was used to make the picture.

Ask How many more times larger do the pollen grains look than they really are? *(1375 times)* What does that tell you about the size of pollen grains? *(They are very tiny.)*

Ask What can you say about the quality of this image? *(It is very clear and looks three-dimensional.)* What do you think ragweed pollen grains would look like viewed through a regular optical microscope? *(Sample answer: They would be smaller and would not be as detailed.)*

Activate Prior Knowledge

Ask How many of you get "hay fever"? *(Answers may vary.)* What are the symptoms of hay fever? *(itching and running eyes and nose, fits of sneezing)*

Say Hay fever is an allergy caused by the pollen of ragweed and some other plants. Ragweed begins to bloom in August. It releases pollen during warm, dry, windy conditions.

Ask Why do you think ragweed doesn't release pollen in wet conditions? *(Pollen travels farther in dry, windy conditions. In wet conditions, the pollen would stick to the plant or be washed to the ground.)*

Say In this chapter, you will learn about many other effects of environmental conditions on the reproduction, development, and growth of plants.

NATIONAL SCIENCE EDUCATION STANDARDS

UNIFYING CONCEPTS AND PROCESSES

I, IV, V

CONTENT

C.1.d, C.1.e, C.3.b, C.4.c, C.4.e, C.5.d, C.6.b, F.3, F.4, F.5, F.6

INQUIRY

A.1.b, A.1.c, A.2.a

24 Plant Reproduction and Response

Big idea Growth, Development, and Reproduction

Q: How do changes in the environment affect the reproduction, development, and growth of plants?

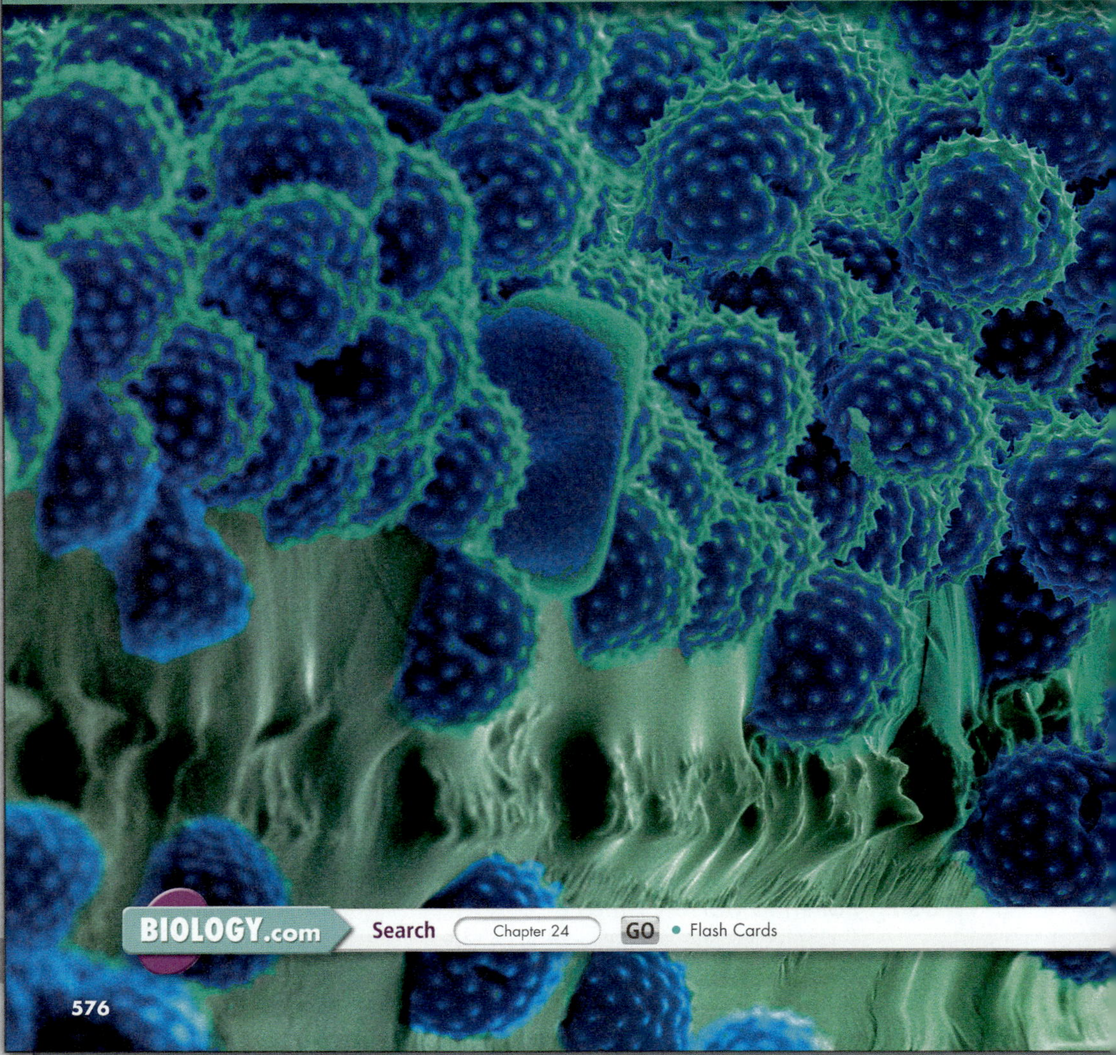

576

UbD Connect to the Big Idea

Invite a volunteer to read aloud the chapter number and the chapter title. Read the Big Idea and question. Emphasize the question, **How do changes in the environment affect the reproduction, development, and growth of plants?**

Set the Purpose

PREVIEW AND PREDICT Draw the graphic organizer on the board. Guide students to the INSIDE feature, and invite a volunteer to read each lesson number and title.

Say Most of the plants we grow are flowering plants. That's why this chapter is mostly about flowering plants. The first two lessons tell how flowering plants reproduce. The third lesson tells how flowering plants grow and respond. The fourth lesson tells how humans depend on plants.

Ask For what purposes do humans use plants? *(for food, clothing, building, heating)*

INSIDE:

- 24.1 Reproduction in Flowering Plants
- 24.2 Fruits and Seeds
- 24.3 Plant Hormones
- 24.4 Plants and Humans

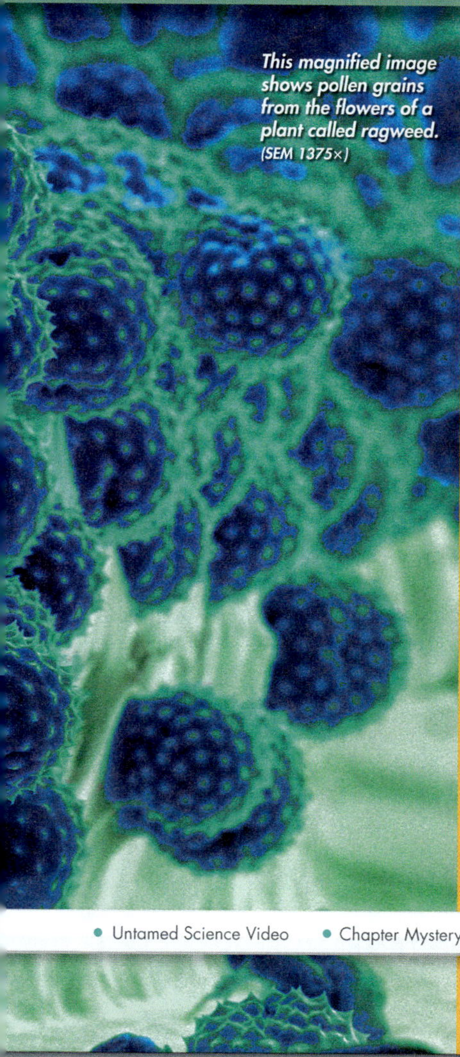

This magnified image shows pollen grains from the flowers of a plant called ragweed. (SEM 1375×)

• Untamed Science Video • Chapter Mystery

CHAPTER
MYSTERY

THE GREEN LEMONS

For years, a California warehouse stored freshly picked green lemons before they were shipped to market. The warehouse managers knew that the lemons would be a ripe yellow and ready to ship to market about five days after they arrived. Or so they thought. One year, for safety reasons, they replaced the warehouse's kerosene heaters with modern electric ones. Then, something surprising happened. When they began to pack their first shipment of five-day-old lemons, they had to call a halt. The lemons they expected to ship were still a bright, and very unripe, green.

Read for Mystery Clues What happened? Why didn't the lemons turn yellow? As you read this chapter, look for clues that explain why the lemons did not ripen.

FOUNDATIONS for Learning

As you read the chapter, make a layered book similar to the one shown below. On each flap, write down the name of the lesson. Then make notes about what you learn in each lesson. At the end of the chapter are two activities that use your layered book to help you to answer the question: How do changes in the environment affect the reproduction, development, and growth of plants?

Chapter 24
Big Idea: Growth, Development, and Reproduction

Chapter 24:
How do changes in the environment affect the reproduction, development, and growth of plants?

- **24.1:** How do flowering plants reproduce?
- **24.2:** How are fruits and seeds important adaptations for plants?
- **24.3:** How do plants respond to their environments?
- **24.4:** In what ways do humans depend on plants?

Foundations for Learning

Give the students the materials they will need to make a layered book.

Ask How many sections will you have in your layered book? *(four)* Why will you have this many sections? *(to have one section for each lesson)*

Ask How can using the Key Questions to make your book be helpful? *(Sample answer: Key Questions are clues to important information in the lessons.)*

Tell students that they will be able to use their layered books as a study aid when they prepare for the chapter test.

Untamed Science™ Plants move in mysterious ways . . . or so it seems. Follow the ecogeeks of Untamed Science as they get to the root of the problem, **How do plants move in response to their environment?** in the short video **Did That Plant Just Move?**

CHAPTER MYSTERY Ask students how they recognize when a fruit is ripe. *(Sample answer: color change, feels soft but not mushy)* Invite a volunteer to read aloud the Chapter Mystery. Have students guess how the mystery relates to the chapter title and to predict why the lemons remained green.

Tell students they will find clues in the chapter to answer the question, **Why didn't the lemons turn yellow?**

Digital Support

BIOLOGY.com Extend your reach by using the digital assets offered at **Biology.com.**

LESSON 24.1

Pre-Reading

Objectives

24.1.1 Identify the functions of various structures in a flower.

24.1.2 Explain how fertilization differs between angiosperms and other plants.

24.1.3 Describe vegetative reproduction.

Preview the Pages

1. Remind students that the green headings are the main topics of the lesson. The blue subheads explain what they will learn about each topic.
2. Preview the visuals. Point out that the flower diagrams are longitudinal sections, meaning they show what the eye would see if you sliced a flower lengthwise.

Activate Prior Knowledge

Ask What is an angiosperm? *(flowering plant whose seeds develop within a matured ovary)* What kinds of angiosperms are given as gifts on special occasions? *(Sample answers: roses, daisies, orchids, lilies)*

Ask Are trees, shrubs, and grasses angiosperms? *(yes)* Explain. (They have flowers that may be small and easy to miss during their life cycle.)

Build Understanding

TWO-COLUMN TABLE Direct students to the two-column table in Workbook B. Tell students to watch for the terms *male gametophyte* and *female gametophyte* as they read this lesson, and record details about each gametophyte in the table.

NATIONAL SCIENCE EDUCATION STANDARDS

UNIFYING CONCEPTS AND PROCESSES

V

CONTENT

C.3.b

INQUIRY

A.1.b, A.1.c, A.2.a

For the corresponding lesson in *Miller & Levine Biology,* see pages 696–703.

24.1 Reproduction in Flowering Plants

Key Questions

What are flowers?

How does fertilization in angiosperms differ from fertilization in other plants?

What is vegetative reproduction?

BUILD Understanding

Two-Column Table Make a two-column table. Write the heading Male Gametophyte at the top of one column. Write the heading Female Gametophyte at the top of the second column. As you read, take notes about each gametophyte.

In Your Workbook Complete the table for Lesson 24.1.

The Structure of Flowers

What makes a flower beautiful? Its color and shape? Of course, flowers vary greatly in color, size, and shape. But what is the point of a flower? To a plant, the whole point of a flower is sexual reproduction. Flowers help bring male and female gametes together and then protect the developing seeds.

Flowers are made up of four kinds of specialized leaves: sepals, petals, stamens, and carpels. The art below shows a typical flower with all of these structures.

Sepals and Petals The outermost flower parts are the sepals (SEE pulz). Often, sepals are green and look like leaves. Sepals surround and protect the flower bud as it develops. Inside the sepals are the petals. Petals, which are often brightly colored, attract pollinators to the flower.

Stamens and Carpels Within the petals are structures that produce gametophytes. Many flowering plants produce both male and female gametophytes. In other species, male and female gametophytes are produced on separate plants.

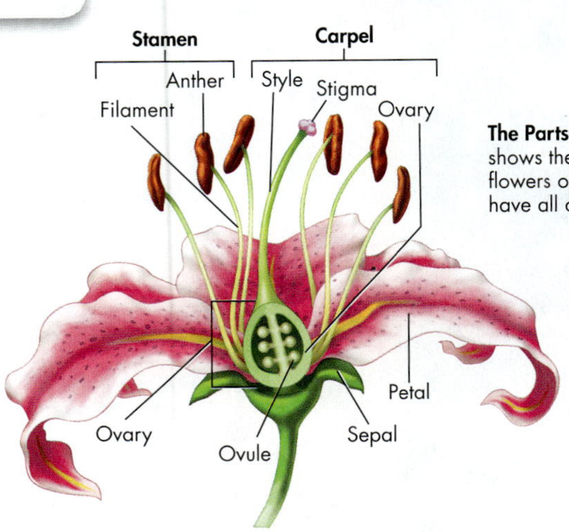

The Parts of a Flower This art shows the parts of a flower. The flowers of some species do not have all of the parts shown here.

 BIOLOGY.com Search Lesson 24.1 GO • Art Review

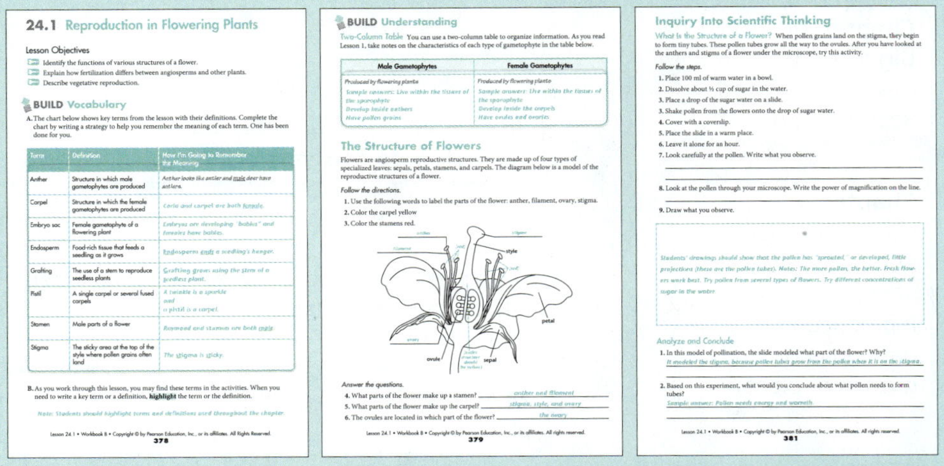

INQUIRY into Scientific Thinking

GUIDED INQUIRY

What Is the Structure of a Flower?

Procedure

1. Study a flower carefully. Use a hand lens if needed. Draw what you see, and label the parts of the flower.
2. Use forceps to remove an anther. Place the anther on a slide. Use a scalpel to cut one or more thin slices of the anther. **CAUTION:** *Be careful with sharp tools and slides. Place the slide on a flat surface before cutting.*
3. Lay the anther slices flat on the slide. Add a drop of water, and cover the anther slices with a coverslip. Use a microscope to look at the slices. Draw and label what you see.

Analyze and Conclude

1. Apply Concepts What structures did you find in the anthers? What is the function of these structures?

2. Apply Concepts What structures did you find in the ovaries? What is the function of these structures?

3. Infer Which parts of the flower will become seeds? Which parts will become fruit?

In Your Workbook Get more help for this activity in your workbook.

▶ ***Stamens*** **Stamens** are the male parts of a flower. Each stamen has a stalk called a filament. At the tips of filaments are **anthers.** Pollen grains, which are the male gametophytes, are produced in the anthers.

▶ ***Carpels*** The innermost parts of a flower are the carpels. **Carpels** produce and protect female gametophytes; later they produce seeds. Each carpel has a wide base called an ovary. The ovary contains one or more ovules. Female gametophytes grow inside the ovules. The carpel has a thin stalk called a style. At the top of the style is a sticky or feathery area known as the **stigma.** The stigma is specialized to capture pollen. Biologists sometimes call a single carpel or several carpels that are fused together a pistil.

Key Question What are flowers?
Flowers are reproductive organs that are made of four kinds of specialized leaves: sepals, petals, stamens, and carpels.

The Angiosperm Life Cycle

Like other plants, angiosperms have a life cycle with alternation of generations. The sporophyte phase is diploid (2N), and the gametophyte stage is haploid (1N). As with ferns and gymnosperms, the sporophyte of angiosperms is much larger than the gametophyte. In fact, the male and female gametophytes live within the tissues of the sporophyte.

BUILD Vocabulary

stamen the male part of the flower; includes an anther and a filament

anther a flower structure in which pollen grains are produced; part of the stamen

carpel the female part of the flower; the innermost part of a flower that produces and protects female gametophytes

stigma a sticky or feathery part of the carpel that is located at the top of a style; specialized to capture pollen

WORD ORIGINS
The word *carpel* comes from the Greek word *karpos*, which means "fruit." Carpels produce seeds and include the ovary, which develops into fruit.

Materials

flower, microscope slide, forceps, scalpel, water, dropper pipette, coverslips, microscope

Analyze and Conclude

1. Students may be able to observe pollen. Pollen grains form male gametophytes that can fertilize female gametophytes.
2. Students may be able to observe ovules. The ovules produce female gametophytes containing eggs that can be fertilized by male gametophytes.
3. The ovules will become the seeds. Generally, the ovary becomes the fruit.

During Reading

Inquiry Into Scientific Thinking

USE CONTEXT CLUES Help students correlate the steps in this Inquiry with the visual The Parts of a Flower and the text on these pages.

Ask For step 1, what could you look at to help you draw and label your flower parts correctly? *(labeled parts in the flower diagram)*

Ask For steps 2 and 3, what paragraph in the text should you reread to help you understand what you observe? *(the Stamens paragraph)*

USE MNEMONIC DEVICES Help students learn to use acronyms as memory aids. Write the sentence, **Seth peeled sticky carrots.**

Say Use this sentence to remember the parts of a flower: sepals, petals, stamens, and carpels.

Lead a Discussion

Have students consider floriculture, the cultivation of ornamental plants for aesthetic purposes.

Say Suppose you are making a flower plan for a window box or garden. You need to decide what kinds of flowers to plant for a pleasing effect.

Ask What characteristics of flowers do you need to think about? *(their structure, color, size, and shape)*

Encourage students to draw a flower plan for a window box or garden using colored pencils or markers.

Active Reading

KEY QUESTION Read the question and answer.

Ask What is a specialized leaf? *(a leaf adapted to a certain function)* Have students find each of these specialized leaves in the flower diagram.

Build Vocabulary

WORD ORIGINS Tell students that a carpel is also sometimes called a *pistil* because it resembles a pestle used to grind substances. Use an Internet photograph of a pestle and compare it to a pistil. Then, write the sentences below on the board and have students supply the underlined words.

Say A stigma, style and ovary make up a pistil. A pistil is shaped like a club.

Speed Bump

Ask Do the flowers of all flowering plants include stamens and carpels? *(No, the flowers of some plants have stamens or carpels, but not both. In some species, male and female gametophytes are produced on separate plants.)*

During Reading continued

Active Reading

INTERPRET DIAGRAMS Direct students' attention to The Development of Gametophytes diagram. Ask a volunteer to read the caption aloud.

Say Notice that this diagram has an upper panel and a lower panel.

Ask What process is shown in each panel? *(development of male gametophyte in upper panel, development of female gametophyte in lower panel)*

SEQUENCE Help students comprehend the sequence of events that occurs during the development of male and female gametophytes.

Ask Where in a flower do male gametophytes develop? *(inside anthers)* What happens first? *(A diploid cell in an anther divides by meiosis, producing four haploid cells. The haploid cells are pollen spores.)* What happens next? *(The nucleus of each pollen spore divides by mitosis to form two haploid nuclei surrounded by a thick cell wall.)* What is the result of this action? *(The nuclei and cell wall form a male gametophyte.)*

Ask Where in a flower do female gametophytes form? *(in ovules)* What happens first? *(A diploid cell in an ovule divides by meiosis to form four haploid cells.)* What happens to three of these cells? *(They disintegrate, or break down).* What happens to the fourth cell? *(Its nucleus divides by mitosis to form eight haploid nuclei.)* What is the result of this action? *(The eight nuclei and the membrane that surrounds them are the embryo sac, which is the female gametophyte.)*

Speed Bump

Ask What forms when a female gamete fuses with a male gamete? *(a zygote)*

Ask What does a zygote grow into? *(an embryo)*

Development of Male Gametophytes Male gametophytes develop inside anthers. First, meiosis produces four haploid cells called pollen spores. Then each pollen spore divides by mitosis. This division forms two haploid nuclei. A thick wall surrounds the two nuclei. Together, the nuclei and wall form the male gametophyte, or pollen grain. The art below shows this process.

Development of Female Gametophytes Female gametophytes form in ovules at the bottom of the carpel. Each ovule is a future seed. The ovary, which surrounds and protects the ovules, is the future fruit.

The process starts when a diploid cell in an ovule divides by meiosis. This division forms four haploid cells. Three of these cells break down. The fourth cell divides by mitosis, producing eight nuclei. These eight nuclei and the surrounding membrane make up the **embryo sac.** The embryo sac is the female gametophyte. The embryo sac is contained within the ovule.

Cell walls form around six of the eight nuclei. One of the eight nuclei is the nucleus of the egg, or female gamete. If fertilization takes place, the egg will fuse with a male gamete, forming a zygote. The zygote will grow into a new sporophyte.

The Development of Gametophytes The male gametophyte develops inside an anther. The female gametophyte develops inside a single ovule.

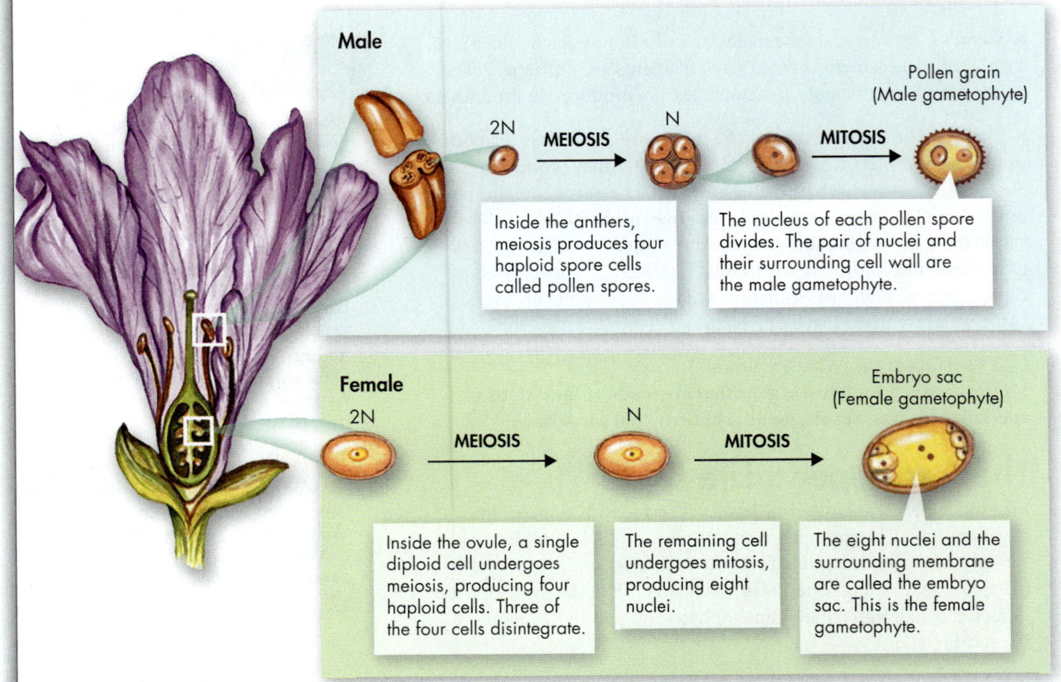

Hands-On Learning

MODEL GAMETOPHYTE DEVELOPMENT Have students make a model of the development of a male gametophyte, and a model of the development of a female gametophyte.

1. Pair students and pass out the following materials: construction paper, masking tape, glue, scissors, buttons or small plastic disks (game tokens), and string.
2. Tell students they can use buttons or disks to represent anthers, pollen spores, nuclei, ovules, and haploid cells. They can use string to represent cell walls and membranes. Arrows made of masking tape could indicate the processes of meiosis and mitosis.
3. Tell students to label the parts and processes that make up each model. Then, have the student pairs use their models to explain the development of gametophytes to each other. One student should explain the male gametophyte and the other student, the female gametophyte.

Pollination The flowers of an oak tree (left) are pollinated by wind. They are small and not very showy. However, they produce large amounts of pollen. In contrast, flowers that are pollinated by animals (right) are often large and brightly colored.

Pollination Pollination is the transfer of pollen to the stigma of the flower. Most angiosperms are pollinated by animals, but some are pollinated by wind. Wind-pollinated flowers are usually small and not very colorful. Wind-pollinated plants rely on the production of huge numbers of pollen grains to carry pollen from one plant to another.

Animal pollinators, such as insects, birds, and bats, carry pollen from flower to flower. Flowers that are pollinated by animals have many adaptations that attract pollinators. These adaptations include bright colors, strong scents, and sweet nectar. Animal pollinators, in turn, have adaptations that help them get nectar from the flowers. Hummingbirds, for example, have long thin beaks that can reach the nectar deep inside tube-shaped flowers.

Flowering plants and their pollinators have a mutualistic relationship. The animals are rewarded by a source of food—pollen and nectar. The plants are helped because animals take pollen directly from flower to flower. The efficiency of animal pollination may be one of the main reasons why angiosperms have become the most successful group of plants.

Fertilization If a pollen grain lands on the stigma of a flower, the process of fertilization begins. One of the male gametophyte's nuclei forms a pollen tube. The pollen tube grows into the style. Eventually it reaches the ovary and enters an ovule. The other nucleus divides, forming two sperm cells. These sperm cells reach the egg through the pollen tube.

Inside the embryo sac, the two sperm fertilize different nuclei. This process is called **double fertilization.** First, one of the sperm nuclei fuses with the egg nucleus. A diploid (2N) zygote forms. The zygote will grow into the embryo. Next, the second sperm nucleus fuses with two other nuclei in the embryo sac to form a triploid (3N) cell. This 3N cell will grow into tissue known as **endosperm.** Endosperm contains a rich supply of food that will nourish the seedling as it grows.

BUILD Vocabulary

embryo sac
the female gametophyte within the ovule of a flowering plant

double fertilization
the process of fertilization in angiosperms, in which the first event produces the zygote, and the second event produces the endosperm within the seed

endosperm
the food-rich tissue that nourishes a seedling as it grows

PREFIXES
The prefix *endo-* means "within." Endosperm is located within a seed.

Inside a Corn Seed The endosperm and embryo of a corn seed result from double fertilization.

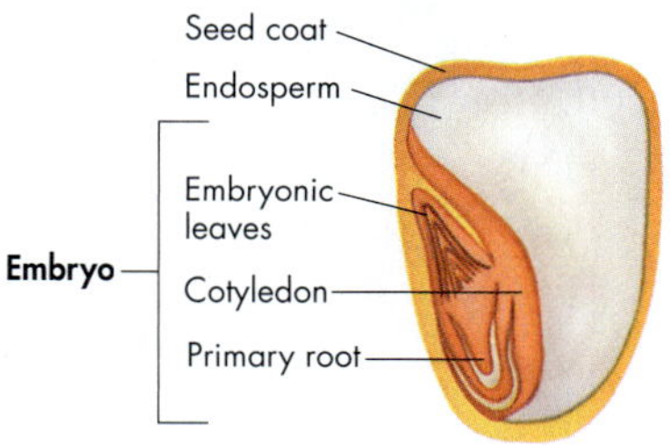

Focus on Academic Language

• pollen • pollinate • pollination • pollinator • efficiency

Write the words above on the board. Review the meaning of each term. Point out that a *pollen* grain is an actual male gametophyte. During pollination, pollen is transferred to the stigma of a flower. After pollination takes place, a flower is said to be *pollinated.* Tell students that a *pollinator* carries pollen from one plant to another. Wind, bees, butterflies, moths, birds, or bats can act as pollinators.

Say To be *efficient* means to be able to function well and to achieve a desired result by using as little time, money, materials, or energy as possible.

Ask Which has more pollen—an animal-pollinated flower or a wind-pollinated flower? *(wind-pollinated flower)* Why do you think more pollen is needed in such flowers? *(Sample answer: Most of the pollen blows to other places besides flowers.)*

Say When we talk about the *efficiency* of animal pollination, we mean it is more likely to cause pollination. Animal pollinators deliver pollen directly to the flower.

Active Reading

DRAW CONCLUSIONS Invite a volunteer to read aloud the caption for the Pollination photographs.

Ask How do wind-pollinated flowers differ from flowers that are pollinated by animals? *(Wind-pollinated flowers are usually small and do not have bright colors. Animal-pollinated flowers, however, usually have large, brightly colored petals, strong scents, and sweet nectar.)*

Say Pollen grains are very lightweight.

Ask What do you think would happen if pollen grains were heavy? *(They would be less likely to travel from plant to plant. It would be harder for fertilization to take place.)*

USE CONTEXT CLUES Remind students that flowers and their pollinators have evolved together. Ask a volunteer to read the third paragraph on this page about mutualistic relationships.

Ask Why do we say that flowering plants and their pollinators have a mutualistic relationship? *(Pollen and nectar provide a food source for animals, and animals take pollen from flower to flower. The plants and the pollinators both benefit.)*

DRAW A DIAGRAM Have a volunteer read aloud the paragraph about double fertilization at the bottom of the page. Then, have students make a flowchart that describes the two events involved in double fertilization.

Say First, list the steps as they occur during the first event. Then, list the steps for the second event.

Build Vocabulary

PREFIXES Tell students that *endo-* is a prefix they will encounter numerous times in biology.

Say For example, an endoskeleton, which you'll learn about in Chapter 28, is a skeleton found within an animal's body. The endocrine system, which you'll learn about in Chapter 34, is a system of glands that release hormones within the body.

Ask What do you think an endoparasite is? *(a parasite that lives within its host, in the host's tissues or organs)*

Speed Bump

Ask What is found inside a seed? *(the embryo and the endosperm)*

Ask What protects the outside of a seed? *(the seed coat)*

During Reading continued

Active Reading

INTERPRET DIAGRAMS Have students review the Angiosperm Life Cycle diagram. Point out the adult plant at the far left of the diagram. Draw students' attention to the use of the colored arrows (purple and orange) in the life cycle.

Say This angiosperm life cycle includes a sporophyte portion and a gametophyte portion.

Ask What does the purple color represent? *(the diploid or sporophyte portion of the life cycle)*

Ask What does the orange color represent? *(the haploid or gametophyte portion of the life cycle)*

Ask Which portion of the angiosperm life cycle has to do with the adult plant? *(sporophyte portion)*

Ask Where is the male gametophyte in this diagram? *(pollen grains)* Where is the female gametophyte? *(embryo sac)*

KEY QUESTION Read aloud the Key Question and answer. Challenge students to point to the products of double fertilization—the zygote (2N) and the endosperm (3N)—in the Angiosperm Life Cycle diagram.

Ask In this diagram, where can you first see the products of double fertilization? *(at the beginning of the purple arrow, near the* Fertilization *label)*

Ask How does double fertilization help angiosperms? *(Endosperm within the seed provides food for the embryo.)*

Speed Bump

Ask What does double fertilization help angiosperms conserve? *(energy)*

Ask What can angiosperms use the energy saved to make? *(many more seeds)*

Double fertilization may be another reason why angiosperms are so successful. By using endosperm to store food, a plant spends very little energy on producing seeds from ovules until double fertilization has actually taken place. The energy saved can be used to make many more seeds.

Key Question How does fertilization in angiosperms differ from fertilization in other plants?
Angiosperms have two fertilization events. One produces a zygote. The other produces a tissue called endosperm.

Angiosperm Life Cycle In the life cycle of an angiosperm, the developing seeds are protected inside the ovary.

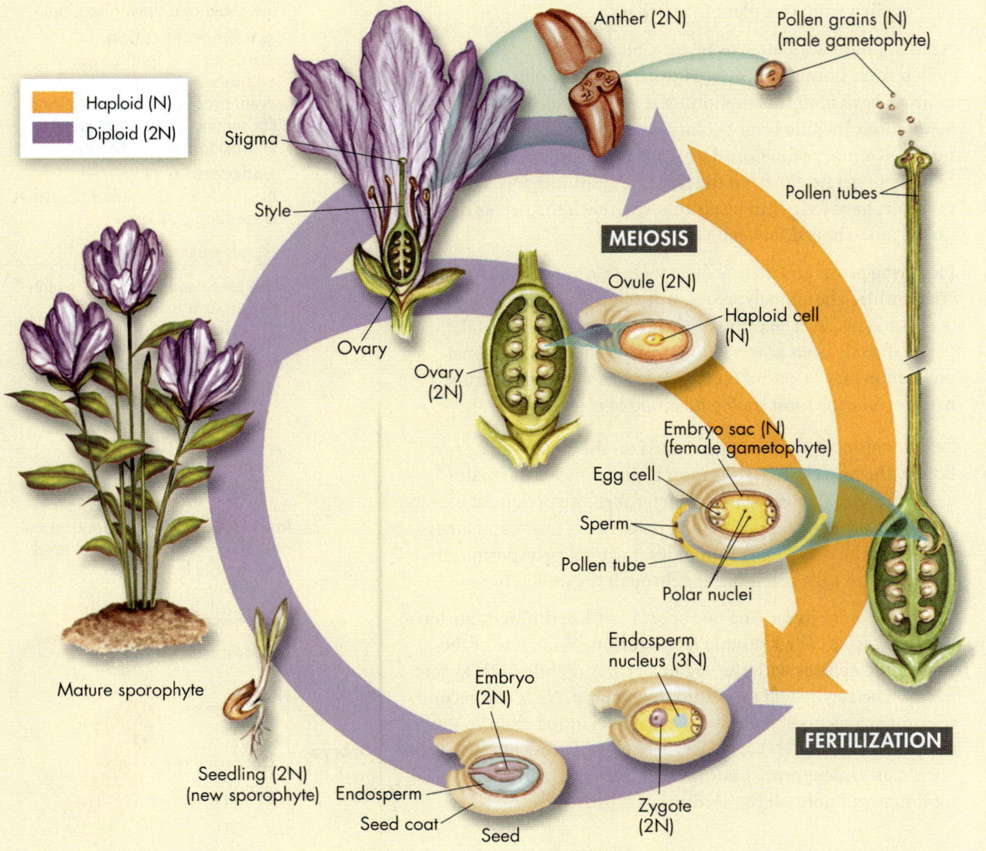

ELL Focus on ELL: Extend Language

ALL SPEAKERS English language learners may have trouble recalling the differences in meaning among terms ending in *-sperm*, which is from the Greek word for *seed*. Write the words *gymnosperm, angiosperm,* and *endosperm* on the board.

Say The word *gymnosperm* means "naked seed." The word *angiosperm* means "enclosed seed." The word *endosperm* means "inside the seed" and describes the food-rich tissue inside a seed.

Say The prefix *angio-* comes from the Greek word for *vessel,* which refers to the ripened ovary wall that contains angiosperm seeds.

Ask What do gymnosperms and angiosperms have in common? *(Sample answer: Both have seeds.)*

Challenge students to write a meaningful sentence containing at least two of these terms: *gymnosperm, angiosperm, endosperm.*

Vegetative Reproduction

Many flowering plants can reproduce asexually by **vegetative reproduction.** In vegetative reproduction, a single plant produces offspring that are genetically identical to itself. New plants are formed by mitosis. Vegetative reproduction does not require flowers, gametes, or fertilization.

Key Question What is vegetative reproduction?
Vegetative reproduction is the formation of new individuals by mitosis.

Types of Vegetative Reproduction Vegetative reproduction takes place naturally in many plants. New plants may grow from roots, leaves, or stems. Some species form plantlets, which are small plants that fall off the parent plant.

Vegetative reproduction does not involve pollination or seed formation. So, a single plant can reproduce quickly. If the parent plant is well adapted to an environment, its offspring can spread quickly. All of these offspring have exactly the same genes as the parent. One of the drawbacks of asexual reproduction is that it does not produce new combinations of genes. Genetic diversity helps a species survive if the environment changes.

BUILD Vocabulary

vegetative reproduction
a method of asexual reproduction in plants; the plant produces offspring that are genetically identical to itself

RELATED WORDS
The word *vegetative* comes from the Latin word *vegetus*, which means "to be lively, or alive." Words from this root usually refer to plants. For example, a *vegetable* is a plant that is grown for a part that can be eaten, such as its roots, stems, or leaves.

Examples of Vegetative Reproduction
Adaptations of the stem play a role in the vegetative reproduction of the plants shown here.

▲ A potato is an underground stem called a tuber that can grow whole new plants from buds called *eyes*.

▲ Strawberry plants send out long, trailing stems called stolons. Nodes that rest on the ground produce roots and upright stems and leaves.

Cholla and many other cactus species can reproduce by dropping sections of their stems. The small individuals growing at the base of the larger adults are, in fact, clones.

Active Reading

FIND THE MAIN IDEA Have a volunteer read aloud the first paragraph about vegetative reproduction.

Ask Which sentence in the paragraph states the main idea? *(In vegetative reproduction, a single plant produces offspring that are genetically identical to itself.)*

READ, COVER, REMEMBER, RETELL Ask students to read silently the text following the subhead *Types of Vegetative Reproduction*. When they have finished, tell students to cover the text with their hand and think about what they read.

Say Vegetative reproduction has advantages and disadvantages, or drawbacks.

Ask What is an advantage of vegetative reproduction? *(A single plant can reproduce quickly.)*

Ask What is a drawback of vegetative reproduction? *(There are no new combinations of genes.)*

Say Remember that genetic diversity is important to a plant species, not to an individual plant.

Ask When all of the plants in an area are genetically identical, how might this be harmful to the species? *(If a sudden environmental change, such as the appearance of a new disease or a change in climate, occurs, all of the members of a genetically identical population might die.)*

KEY QUESTION Read aloud the Key Question and answer.

Ask What is missing from vegetative reproduction? *(There are no flowers, gametes, or fertilization involved.)*

Build Vocabulary

RELATED WORDS Tell students that vegetative reproduction involves the growing, or vegetative, parts of the plant rather than the reproductive tissues and seeds. Ask students to think of other words that begin with *veg-*.

Ask What is a vegetarian? *(someone who eats vegetables but does not eat meat or fish)*

Speed Bump

Ask Describe an example of vegetative reproduction. *(Answers may vary. Students might describe using a plant root, leaf, or stem to grow a new plant.)*

Hands-On Learning

DEMONSTRATE VEGETATIVE REPRODUCTION Have students work in small groups to prepare a sweet potato for growth in the classroom. Or, to save materials, do this activity as a demonstration. You will need a sweet potato, a large jar, several toothpicks, and water. Insert three or four toothpicks around the center of the sweet potato, and suspend it in the jar so that the toothpicks rest atop the edge of the jar. Add water to the jar until the bottom half of the sweet potato is submerged. If possible, place the jar near a window. In one to three weeks, roots will emerge from the bottom of the potato, while leafy, green vines emerge from the top.

Ask Why is this growth considered to be the result of vegetative reproduction? *(A new plant has grown asexually. It grows from part of a single plant, without another plant being involved.)* What part of a plant is a sweet potato? *(Some students may know that a sweet potato is a root even though other kinds of potatoes are stems.)*

LESSON 24.1

During Reading continued

Active Reading

USE CONTEXT CLUES Have a volunteer read the first paragraph.

Ask What does it mean to propagate plants? *(to multiply plants by means of sexual or asexual reproduction)*

DRAW CONCLUSIONS Have volunteers read the second and third paragraphs. Tell students to think about why a grower might want to use the plant propagating methods mentioned.

Ask What would happen if you used a length of stem without buds for a cutting? *(The cutting would not produce a new plant.)* Why are buds needed? *(Buds contain meristems, which are tissues that can give rise to new cells.)*

Ask Have you eaten seedless grapes? Why do you think growers use cuttings to propagate seedless grapes? *(because they cannot grow them from seeds)*

Post Reading

ASSESSMENT TIPS If students have trouble with Question 6, ask them to reread the last paragraph under Fertilization. Have students summarize in two or three sentences how fertilization is unique in angiosperms and how it has helped them to become successful.

For Question 7, suggest students set up and fill out a 2 x 2 table, with *Advantages* and *Disadvantages* at the top and *Sexual reproduction* and *Asexual reproduction* along the side.

MYSTERY CLUE Have students identify the variable that changed after the lemons were harvested. Then ask them to brainstorm differences between heating with kerosene and heating with electricity.

Wrap-Up Activity

COMPARING REPRODUCTION TYPES Write the following terms on the board: *fertilization, flowers, gametes, grafting, meiosis, mitosis, pollination, roots, seeds,* and *stems.* Provide students with index cards. Tell students to write each term and its definition on a card. Ask students to sort the cards into two groups—terms that relate to sexual reproduction, and terms that relate to vegetative reproduction.

Grafting As it is just starting to bud, a branch from a lemon tree is grafted onto the branch of an established orange tree. Months later, the grafted branch bears lemons. This graft has formed a single plant bearing more than one species of fruit.

Plant Propagation People who grow, or propagate, plants often make use of vegetative reproduction. Two common ways of increasing the number of desirable plants are grafting and making cuttings.

To make a cutting, a grower cuts off a length of stem that has buds. (Remember that buds contain meristems.) The cut stem may then be partially buried in soil. Sometimes it is placed in a special mixture of nutrients that helps roots form.

Grafting is used to produce seedless plants and woody plants that cannot be grown from cuttings. In grafting, a bud or stem is cut from the parent plant. It is then attached to another plant. Grafting only works with plants that are closely related. A bud from a lemon tree can be grafted onto an orange tree, but not onto an oak tree.

CHECK Understanding

Apply Vocabulary

Use the highlighted words from the lesson to complete each sentence correctly.

1. In seeds, the _______ nourishes the seedling as it grows.
2. Pollen grains form inside the _______.
3. The _______ is the female part of a flower.
4. In the process of _______, a plant reproduces asexually by mitosis.

Critical Thinking

5. **Explain** List and describe the four kinds of specialized leaves that make up a flower.
6. **Apply Concepts** Describe a feature of fertilization that happens only in angiosperms.
7. **Compare and Contrast** Compare the advantages and disadvantages of sexual reproduction with those of asexual reproduction in flowering plants.
8. **Write to Learn** Answer the first clue of the mystery.

MYSTERY CLUE

Whether the lemons grew on a grafted branch did not affect their ripening. Something changed after the lemons were picked. This change kept the lemons from ripening. What was it? (Hint: See page 577.)

BIOLOGY.com Search Lesson 24.1 GO • Lesson Assessment

Check Understanding

1. endosperm 2. anther
3. carpel
4. vegetative reproduction
5. Sepals are the outermost leaves. They protect the flower bud. Petals are often brightly colored to attract pollinators. Stamens produce pollen. Carpels are the innermost part of the flower.
6. Double fertilization: A sperm fuses with the egg to form a zygote. Another sperm fuses with two nuclei in the embryo sac to form the seed's endosperm.
7. Sexual: produces new combinations of genetic traits that help species survive environmental changes, but takes time to produce offspring. Asexual: plants can multiply rapidly, but it does not result in new combinations of traits that may aid in species survival.
8. Many students will likely recognize that the change in heating systems caused the lack of ripening.

24.2 Fruits and Seeds

Seed and Fruit Development

What is a fruit? In everyday language, the word *fruit* describes sweet things, such as grapes or strawberries. But in biology, a fruit is a structure that surrounds and protects seeds. Therefore, many foods that we call vegetables, such as peas, corn, beans, cucumbers, and tomatoes, are also fruits.

Fruits develop from the ovary walls of a flower. After fertilization, nutrients from the plant flow into the flower tissues, feeding the developing seeds. As the seeds mature, the ovary walls around them thicken. A mature ovary wall may be fleshy, as in a tomato. Or it may be tough and dry, as in a peanut shell. (The peanuts themselves are seeds.)

Key Question How do fruits form?
As angiosperm seeds mature, the ovary walls thicken to form a fruit that surrounds the developing seeds.

Seed Dispersal

Fruits are rich in sugars and other nutrients. But fruits do not nourish seedlings. Seedlings get their food from endosperm. So why do angiosperms have seeds that are wrapped in another layer of nutrient-packed tissue? Think of wild blackberries. Each seed is enclosed in a sweet, juicy fruit. Blackberries are tasty treats for many animals that live in the forest. What good is a fruit if all it does is get eaten? Believe it or not, that's exactly the point! When an animal eats a fruit, it helps disperse, or scatter, the seeds.

Dispersal by Animals The seeds of many plants, especially those in fleshy fruits, are eaten by animals. These seeds have a tough coating that lets them pass through the animal's digestive system unharmed. Eventually, the seeds are deposited in the animal's droppings. Some dry fruits have burs that catch in an animal's fur. These fruits are then carried far away from the parent plant.

Key Questions

- *How do fruits form?*
- *How are seeds spread?*
- *What factors influence the dormancy and germination of seeds?*

BUILD Understanding

Flowchart Make a flowchart that shows the process of germination. Include the factors that affect germination. Indicate differences between monocots and dicots.

In Your Workbook Go to your workbook to learn more about flowcharts. Complete the flowchart for Lesson 24.2.

Variety Among Fruits Like the flowers from which they develop, fruits vary greatly in structure.

Strawberries ▼

Rose Hips ▼

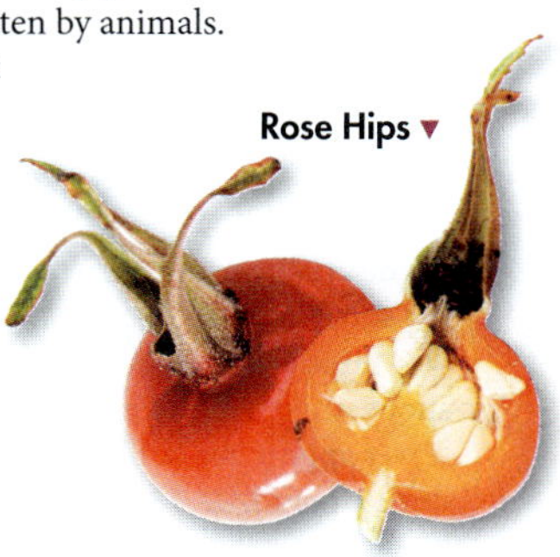

Peanut Shell ▼

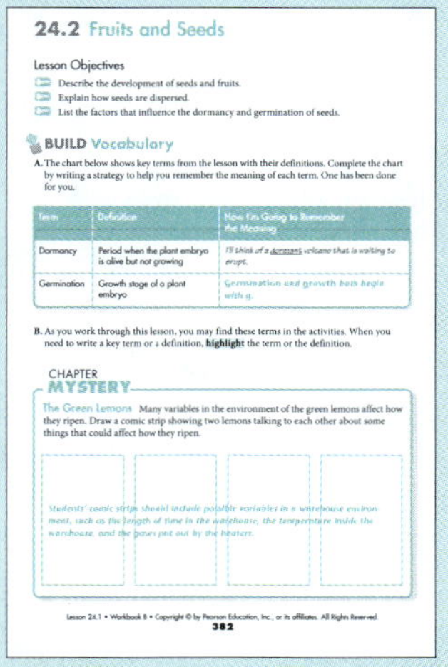

24.2 Fruits and Seeds

Lesson Objectives

- Describe the development of seeds and fruits.
- Explain how seeds are dispersed.
- List the factors that influence the dormancy and germination of seeds.

BUILD Vocabulary

A. The chart below shows key terms from the lesson with their definitions. Complete the chart by writing a strategy to help you remember the meaning of each term. One has been done for you.

Term	Definition	How I'm Going to Remember the Meaning
Dormancy	Period when the plant embryo is alive but not growing	I'll think of a dormant volcano that is waiting to erupt.
Germination	Growth stage of a plant embryo	Germination and growth both begin with g.

B. As you work through this lesson, you may find these terms in the activities. When you need to write a key term or a definition, **highlight** the term or the definition.

CHAPTER MYSTERY

The Green Lemons Many variables in the environment of the green lemons affect how they ripen. Draw a comic strip showing two lemons talking to each other about some things that could affect how they ripen.

Students' comic strips should include possible variables in a warehouse environment, such as the length of time in the warehouse, the temperature inside the warehouse, and the gases put out by the heaters.

Lesson 24.1 • Workbook B • Copyright © by Pearson Education, Inc., or its affiliates. All Rights Reserved.
382

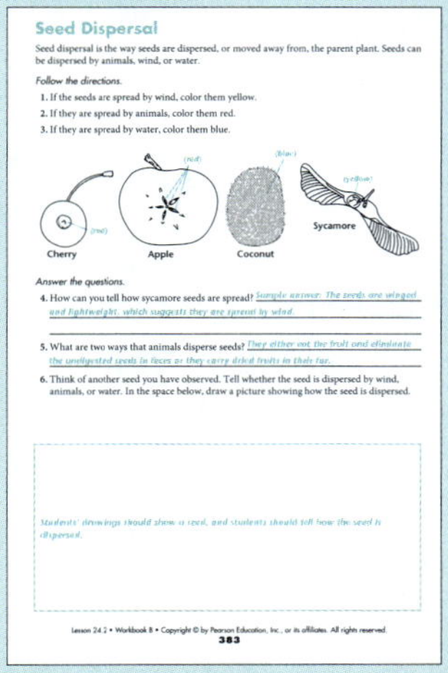

Seed Dispersal

Seed dispersal is the way seeds are dispersed, or moved away from, the parent plant. Seeds can be dispersed by animals, wind, or water.

Follow the directions.

1. If the seeds are spread by wind, color them yellow.
2. If they are spread by animals, color them red.
3. If they are spread by water, color them blue.

Answer the questions.

4. How can you tell how sycamore seeds are spread? *Sample answer: The seeds are winged and lightweight, which suggests they are spread by wind.*
5. What are two ways that animals disperse seeds? *They either eat the fruit and eliminate the undigested seeds in feces or they carry dried fruits in their fur.*
6. Think of another seed you have observed. Tell whether the seed is dispersed by wind, animals, or water. In the space below, draw a picture showing how the seed is dispersed.

Students' drawings should show a seed, and students should tell how the seed is dispersed.

Lesson 24.2 • Workbook B • Copyright © by Pearson Education, Inc., or its affiliates. All rights reserved.
383

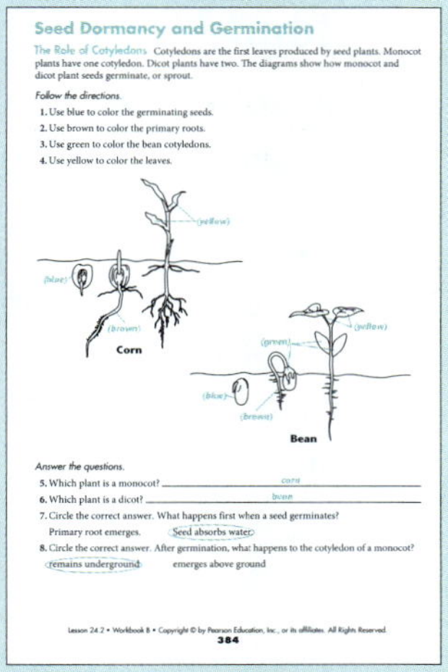

Seed Dormancy and Germination

The Role of Cotyledons Cotyledons are the first leaves produced by seed plants. Monocot plants have one cotyledon. Dicot plants have two. The diagrams show how monocot and dicot plant seeds germinate, or sprout.

Follow the directions.

1. Use blue to color the germinating seeds.
2. Use brown to color the primary roots.
3. Use green to color the bean cotyledons.
4. Use yellow to color the leaves.

Answer the questions.

5. Which plant is a monocot? *corn*
6. Which plant is a dicot? *bean*
7. Circle the correct answer. What happens first when a seed germinates?
   Primary root emerges. ~~~ Seed absorbs water. (circled)
8. Circle the correct answer. After germination, what happens to the cotyledon of a monocot?
   remains underground (circled) ~~~ emerges above ground

Lesson 24.2 • Workbook B • Copyright © by Pearson Education, Inc., or its affiliates. All Rights Reserved.
384

LESSON 24.2

Pre-Reading

Objectives

24.2.1 Describe the development of seeds and fruits.

24.2.2 Explain how seeds are dispersed.

24.2.3 List the factors that influence the dormancy and germination of seeds.

Preview the Pages

1. Ask a volunteer to read the green headings aloud. Explain that these headings list events of plant reproduction in the order in which they occur. Point out that *dispersal* means the "distribution or scattering of things over an area."
3. Preview the visuals and read the captions and labels. For the diagram on the last page of the lesson, review the distinguishing traits of monocots and dicots. Students having difficulty may need to review Lesson 22.4.

Build Understanding

FLOWCHART Direct students to Workbook B, where they will find a flowchart for this lesson. Explain that a flowchart shows the steps of a process in the order in which they occur.

NATIONAL SCIENCE EDUCATION STANDARDS

UNIFYING CONCEPTS AND PROCESSES

V

CONTENT

C.4.c

For the corresponding lesson in *Miller & Levine Biology,* see pages 704–707.

LESSON 24.2

During Reading

Active Reading

KEY QUESTION Read aloud the Key Question and its answer on the previous page.

Ask What do the strawberries, rose hips, and peanuts pictured on this page have in common? *(They are all fruits; they all contain seeds.)*

MAKE CONNECTIONS Have students reread the paragraph Dispersal by Animals on the previous page.

Ask What experiences have you had with seed dispersal described in this paragraph? *(Sample answers: Burs got all tangled in my dog's fur while she was running through a field. They stuck to my socks and jeans when I went running after her.)*

KEY QUESTION Read aloud the Key Question and answer.

Ask Why is it an advantage for seeds to be dispersed? *(Sample answer: Seeds that are spread away from the parent plant do not have to compete with the parent plant for resources such as nutrients, water, and sunlight.)*

DRAW CONCLUSIONS Bring in or show students pictures of the winged fruits of maple trees. Point out the winged structures and the seeds in the middle where the wings meet. If you have the fruits, let them drop and twirl to the ground. If using photos, ask students to visualize how the fruits move when they fall from the tree.

Ask How do you think these seeds get dispersed? *(by wind)* Why do you say that? *(The twirling motion and the wide, flat ends of the wings allow the wind to blow the seeds away.)*

Build Vocabulary

WORD ORIGINS The word *dormant* is the adjective form of *dormancy*.

Say Use the words *dormant* and *dormancy* to complete these sentences,

The seeds I planted in my garden are now in dormancy.

Warm temperatures will cause the dormant seeds to sprout.

Speed Bump

Ask What process marks the end of dormancy? *(germination)*

Ask What is the role of cotyledons? *(to store nutrients used by the growing plant)*

Dispersal of Seeds A bird called a Bohemian waxwing eats the berries of a mountain ash (left). Later, the bird will deposit the seeds in its droppings. Dandelion seeds (center) are spread by wind. The coconut fruit (right) can travel long distances by floating on water.

Dispersal by Wind and Water Some seeds are spread by wind. These seeds are usually contained in lightweight dry fruits. Consider a dandelion seed. It has a dry fruit that works like a parachute. Dandelion seeds can be blown far away from the parent plant. Other fruits travel by water. For example, coconut fruits can float long distances on seawater. That is why coconut trees can be found growing on islands that are far away from the mainland.

Key Question How are seeds spread?
Seeds in fleshy fruits are usually spread by animals. Seeds in dry fruits may be spread by animals, wind, or water.

BUILD Vocabulary

dormancy
a period of time during which a plant embryo is alive but not growing

germination
the growth of the plant embryo following dormancy

WORD ORIGINS
The word *dormancy* comes from the Latin word *dormire*, which means "to sleep." While seeds don't sleep, dormancy is a period of inactivity. The embryo in a dormant seed is alive, but it is not growing.

Seed Dormancy and Germination

When they reach a good location, some seeds begin to grow right away. But many other seeds enter a period of **dormancy.** During dormancy, the embryo is alive, but it is not growing. The length of dormancy varies among species.

Germination Dormancy ends with **germination,** which is when the plant embryo grows again. Environmental factors such as temperature and moisture can cause germination. Before germinating, seeds absorb water. The seed swells, breaking open the seed coat. The root begins to grow. Then the shoot—the part of the plant that grows above ground—begins to sprout. The art on the next page compares the germination of a monocot with that of a dicot.

The Role of Cotyledons Cotyledons, or seed leaves, store nutrients that are used by the growing plant. Monocots have one seed leaf, which usually stays underground. A sheath protects the monocot shoot as it pushes through the soil.

Dicots have two seed leaves. Dicots do not have a sheath to protect the shoot. Instead, the shoot bends to form a hook that forces its way through the soil. This protects the delicate tip of the plant. The shoot straightens as it grows out of the soil. In some dicots, the seed leaves appear above the ground. In other dicots, the seed leaves stay underground.

Hands-On Learning

BUILD A MODEL SEED Invite students to create a model of a seed.

1. Pair students and provide these materials in a central location: glue, tape, interlocking-fastener strips, cotton balls, pins, paper, scissors, clay, foil, and water in pans.
2. Have students build a model seed using materials of their choice.
3. Ask students to predict which method of seed dispersal will be most effective in moving their seed: wind, water, or an animal.
4. Have students write a paragraph describing their model seed and their prediction.

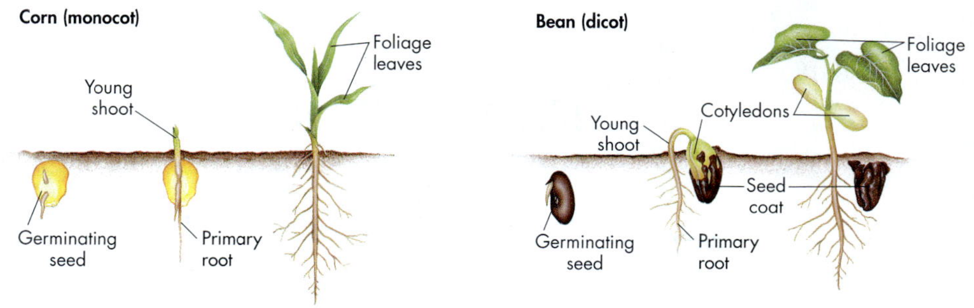

Comparing Germination in Monocots and Dicots Corn is a monocot. A corn seedling (left) grows directly upward. It is protected by a sheath of tissue that surrounds the developing leaves. A bean is a dicot. A bean seedling (right) forms a hook in its stem. This hook gently pulls the new plant through the soil.

Advantages of Dormancy Dormancy makes it possible for seeds to be dispersed over long distances. It also lets seeds wait to germinate until conditions for growth are ideal. Plants that live in cold climates often have seeds that will not germinate until after a long period of cold. This dormancy stops their seeds from germinating until winter is over. Their seeds can survive freezing, but their seedlings cannot.

In some species, only extreme conditions will end dormancy. For example, the cones of some pine trees do not release their seeds until there is a forest fire. High temperatures cause the cones to open. Once the seeds are released, they germinate. This adaptation allows pine trees to quickly grow again after a fire.

Key Question What factors influence the dormancy and germination of seeds? **Temperature and moisture can cause a seed to end dormancy and germinate.**

CHECK Understanding

Apply Vocabulary

Use the highlighted words from the lesson to complete each sentence correctly.

1. Some seeds undergo _______, so they do not grow right away.
2. The process in which the embryo in a seed starts to grow again is called _______.

Critical Thinking

3. **Explain** Describe how fruits form.
4. **Classify** Is a pumpkin a fruit? Why or why not?
5. **Explain** Describe two ways in which seeds are spread.
6. **Compare and Contrast** Compare the germination of a monocot with that of a dicot.
7. **Write to Learn** Answer the second clue of the mystery. Start a list of variables that affect seed development and fruit ripening. Add to your list as you continue reading the chapter.

MYSTERY CLUE

When the warehouse switched to electric heating, what might have changed in the lemons' environment? (**Hint:** See p. 586.)

BIOLOGY.com Search Lesson 24.2 GO • Lesson Assessment

Check Understanding

1. dormancy
2. germination
3. As seeds mature, the ovary walls thicken. Ovary walls may become fleshy fruits or tough, dry fruits.
4. Yes, it is a fruit. Pumpkins are fleshy fruits that surround many seeds.
5. Accept any two of the following: Seeds are spread by animals when they eat fruits or when the fruits catch in the animals' fur. Seeds can be spread by wind or by floating in water.
6. Monocots have one cotyledon, or seed leaf, that usually stays underground. A sheath protects the shoot as it grows through the soil. Dicots have two seed leaves. The dicot shoot forms a hook that pushes through the soil.
7. Kerosene heaters work by combustion, which means carbon dioxide was likely given off as kerosene burned. Electric heaters do not produce this gas. Students may infer that this was a factor.

Active Reading

INTERPRET DIAGRAMS Have students focus on the diagram Comparing Germination in Monocots and Dicots. Read the caption aloud.

Say A young plant that has just germinated is called a seedling.

Ask During germination, how do monocot and dicot seedlings protect the new shoot as it pushes through the soil? *(In monocots, a sheath protects the tip of the shoot. In dicots, the stem forms a hook that pushes through the soil before the shoot tip emerges.)*

MAKE INFERENCES Have students reread the paragraphs about the advantages of dormancy.

Ask What would happen if plants in cold climates had seeds that could not undergo dormancy? *(Instead of waiting until spring, the seeds would germinate in fall or winter and the seedlings would die.)*

KEY QUESTION Read aloud the Key Question and answer on this page.

Ask What does a seed do when it germinates? *(The seed swells, breaks open its seed coat. The root begins to grow and the shoot begins to sprout.)*

Post Reading

ASSESSMENT TIPS

If students have trouble with Question 4, help them think of the question in another way.

Ask How do you know that a plant part is a fruit? *(Students should recall that a fruit contains seeds.)*

Ask Does a pumpkin hold seeds?

MYSTERY CLUE If students are stumped, offer two suggestions: perhaps the kerosene heaters gave off a substance that the electric heaters did not, or the electric current somehow interfered with ripening.

Wrap-Up Activity

DESIGN AN ADVERTISEMENT Have pairs of students design an advertisement with visuals and text to "sell" a fruit to a potential "buyer." The buyer should be an animal that is adapted to disperse the seeds of the fruit. Information in the ad should include how the fruit forms, why it might appeal to the buyer, and how it would benefit the buyer.

LESSON 24.3

Pre-Reading

Objectives

24.3.1 Describe the effects of hormones on plant growth and development.

24.3.2 Identify three tropisms exhibited in plants.

24.3.3 Describe how plants respond to seasonal change.

Preview the Pages

1. Have students flip through the pages of this lesson, noting the three levels of headings: green, blue, and black italicized. Help students use these headings to create an outline shell for the lesson, leaving one or two lines of space after each heading. They can fill in the outline as they work through the lesson.
2. Tell students that the chart Effect of Photoperiod on Flowering explains why some plants flower only at certain times of the year.

Activate Prior Knowledge

Say Some of the most rapid periods of growth in the human body occur during adolescence. A teenager may grow three or four inches in just one year.

Ask What substances in your body control this growth? *(hormones)*

Tell students that plants also have hormones that control their growth.

Build Understanding

CONCEPT MAP Have students complete the Build Understanding activity in Workbook B.

NATIONAL SCIENCE EDUCATION STANDARDS

UNIFYING CONCEPTS AND PROCESSES

IV

CONTENT

C.1.d, C.5.d, C.6.b

INQUIRY

A.2.a

For the corresponding lesson in *Miller & Levine Biology,* see pages 708–714.

24.3 Plant Hormones

Key Questions

What roles do plant hormones play?

What are some examples of environmental stimuli to which plants respond?

How do plants respond to seasonal changes?

BUILD Understanding

Concept Map As you read, make a concept map that shows the effects of various hormones on plant growth.

In Your Workbook Go to your workbook to learn more about concept maps. Complete the concept map for Lesson 24.3.

Hormones

Plants grow in response to factors such as light, temperature, and moisture. But how do plant tissues "know" when to grow or reproduce? The actions of plant tissues and organs are controlled by hormones. **Hormones** are chemical signals that affect a plant's growth, activity, and development. Hormones also coordinate the way that a plant responds to its environment.

Cells in an organism affected by a particular hormone are called **target cells.** To respond to a hormone, a cell must have receptors. **Receptors** are usually proteins to which hormones bind. The way that a target cell responds depends on the kind of receptor. One kind of receptor might speed growth. Another might slow growth. As a result, a given hormone can affect roots differently from stems.

Auxins Hormones called auxins control cell elongation and the growth of new roots. Auxins are produced in apical meristems, then travel to the rest of the plant. Recall that apical meristems are found at the tips of stems.

Auxins control cell division in meristems. Apical meristems produce auxins that inhibit, or slow down, the growth of other buds on the same stem. The closer a bud is to the stem's tip, the more it is inhibited. This inhibition is called apical dominance. If you cut off the bud at the tip of a stem, the growth of the other buds on the stem is no longer inhibited. The other buds grow more quickly. The plant becomes bushier because the source of the auxins has been removed.

Auxins also explain why plants grow toward the light. When light hits one side of a shoot, auxins collect in the shaded areas. These cells grow longer. This causes the shoot to bend toward the light.

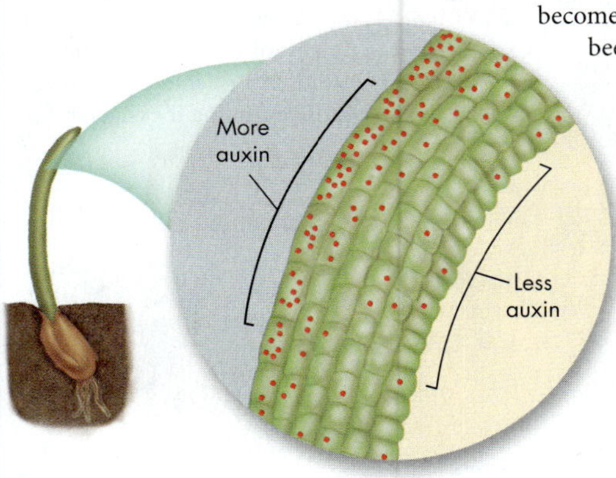

Auxins and Cell Elongation Cells grow longer on the shaded side of the shoot. This area has a higher concentration of auxins.

588

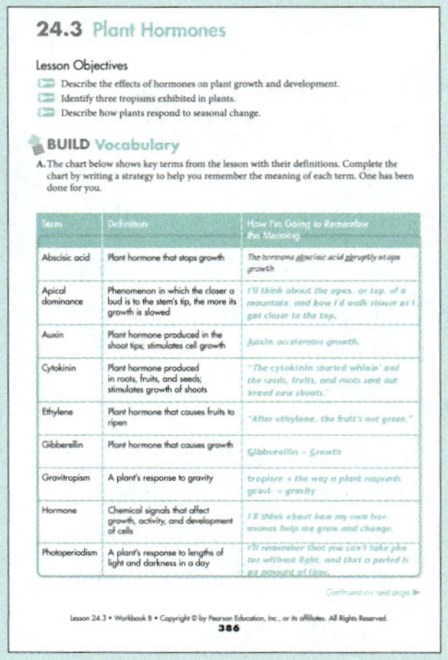

24.3 Plant Hormones

Lesson Objectives

- Describe the effects of hormones on plant growth and development.
- Identify three tropisms exhibited in plants.
- Describe how plants respond to seasonal change.

BUILD Vocabulary

A. The chart below shows key terms from the lesson with their definitions. Complete the chart by writing a strategy to help you remember the meaning of each term. One has been done for you.

Term	Definition	How I'm Going to Remember the Meaning
Abscisic acid	Plant hormone that stops growth	*The hormone abscisic acid abruptly stops growth.*
Apical dominance	Phenomenon in which the closer a bud is to the stem's tip, the more its growth is slowed	I'll think about the apex, or top, of a mountain, and how I'd walk slower as I got closer to the top.
Auxin	Plant hormone produced in the shoot tips; stimulates cell growth	Auxin accelerates growth.
Cytokinin	Plant hormone produced in roots, fruits, and seeds; stimulates growth of shoots	"The cytokinin started whirlin' and the seeds, fruits, and roots sent out brand new shoots."
Ethylene	Plant hormone that causes fruits to ripen	"After ethylene, the fruit's not green."
Gibberellin	Plant hormone that causes growth	Gibberellin = Growth
Gravitropism	A plant's response to gravity	tropism = the way a plant responds; gravi- = gravity
Hormone	Chemical signals that affect growth, activity, and development of cells	I'll think about how my own hormones help me grow and change.
Photoperiodism	A plant's response to lengths of light and darkness in a day	I'll remember that you can't take photos without light, and that a period is an amount of time.

Continued on next page ▶

386

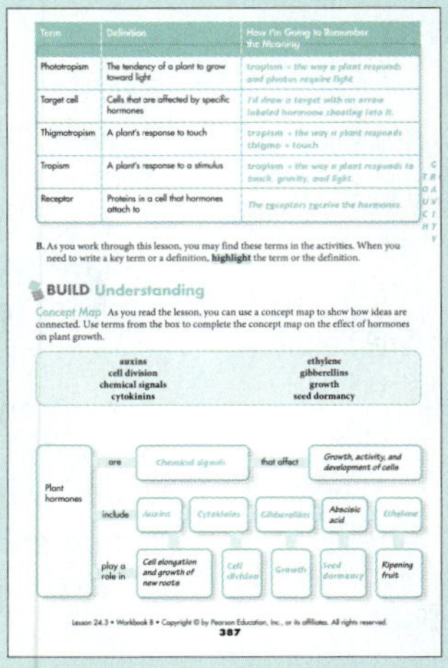

Term	Definition	How I'm Going to Remember the Meaning
Phototropism	The tendency of a plant to grow toward light	tropism = the way a plant responds and photos require light
Target cell	Cells that are affected by specific hormones	I'd draw a target with an arrow labeled hormone shooting into it.
Thigmotropism	A plant's response to touch	tropism = the way a plant responds; thigmo = touch
Tropism	A plant's response to a stimulus	tropism = the way a plant responds to touch, gravity, and light.
Receptor	Proteins in a cell that hormones attach to	The receptors receive the hormones.

B. As you work through this lesson, you may find these terms in the activities. When you need to write a key term or a definition, **highlight** the term or the definition.

BUILD Understanding

Concept Map As you read the lesson, you can use a concept map to show how ideas are connected. Use terms from the box to complete the concept map on the effect of hormones on plant growth.

auxins	ethylene
cell division	gibberellins
chemical signals	growth
cytokinins	seed dormancy

Plant hormones — are — Chemical signals — that affect — Growth, activity, and development of cells

include: Auxins | Cytokinins | Gibberellins | Abscisic acid | Ethylene

play a role in: Cell elongation and growth of new roots | Cell division | Growth | Seed dormancy | Ripening fruit

387

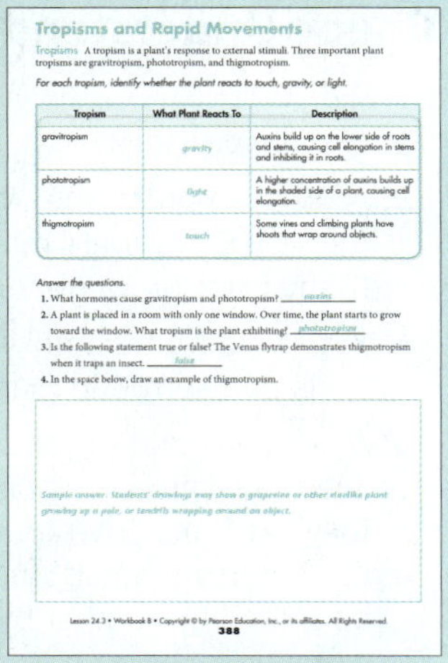

Tropisms and Rapid Movements

Tropisms A tropism is a plant's response to external stimuli. Three important plant tropisms are gravitropism, phototropism, and thigmotropism.

For each tropism, identify whether the plant reacts to touch, gravity, or light.

Tropism	What Plant Reacts To	Description
gravitropism	gravity	Auxins build up on the lower side of roots and stems, causing cell elongation in stems and inhibiting it in roots.
phototropism	light	A higher concentration of auxins builds up in the shaded side of a plant, causing cell elongation.
thigmotropism	touch	Some vines and climbing plants have shoots that wrap around objects.

Answer the questions.

1. What hormones cause gravitropism and phototropism? auxins
2. A plant is placed in a room with only one window. Over time, the plant starts to grow toward the window. What tropism is the plant exhibiting? phototropism
3. Is the following statement true or false? The Venus flytrap demonstrates thigmotropism when it traps an insect. false
4. In the space below, draw an example of thigmotropism.

Sample answer: Students' drawings may show a grapevine or other vinelike plant growing up a pole, or tendrils wrapping around an object.

388

Cytokinins Cytokinins are hormones that are produced in growing roots, fruits, and seeds. Cytokinins cause cell division. They interact with auxins to control the growth of new organs in injured plants and to balance root and shoot growth. The effects of cytokinins often oppose those of auxins. Auxins cause the growth of new roots and inhibit the growth of shoot tips. Cytokinins do the opposite.

Gibberellins Hormones called gibberellins promote growth. These hormones can produce great increases in size, particularly in stems and fruits. Gibberellins also stimulate germination.

Abscisic Acid The hormone abscisic acid keeps cells from dividing, thereby stopping growth. Abscisic acid and gibberellins work together to control seed dormancy. When a seed is fully grown, abscisic acid stops the growth of the embryo. The embryo becomes dormant. The embryo rests until environmental conditions change the balance of hormones. For example, a heavy spring rain may wash away the abscisic acid. (Gibberellins do not wash away as easily as abscisic acid.) Then, gibberellins in the seed can start germination.

Ethylene One plant hormone—ethylene—is a gas. Fruit tissues release small amounts of ethylene, which helps fruits ripen. Ethylene also causes plants to drop organs that are no longer needed. For example, leaves drop in autumn, and fruits drop after they ripen. In each case, ethylene signals cells at the base of the structure to seal it off from the rest of the plant.

Key Question What roles do plant hormones play? **Plant hormones serve as signals that control development of cells, tissues, and organs. They also control responses to the environment. The opposing effects of plant hormones balance each other, contributing to a plant's homeostasis.**

BUILD Vocabulary

hormone a chemical produced in one part of an organism that affects another part of the same organism

target cell a cell that has a receptor for a particular hormone

receptor a specific protein whose shape fits a specific hormone molecule

WORD ORIGINS

The word *hormone* comes from the Greek word *hormon*, which means "urge on." Hormones "urge" plants to grow or develop.

A Summary of Plant Hormones

Hormone	Some of the Effects	Where Found
Auxins	Promote cell elongation and apical dominance; stimulate growth of new roots	Produced in shoot apical meristem and transported elsewhere
Cytokinins	Stimulate cell division; affect root growth and differentiation; may work in opposition to auxins	Growing roots
Gibberellins	Stimulate growth; influence various developmental processes; promote germination	Meristems of shoot, root, and seed embryo
Abscisic acid	Inhibits cell division; promotes seed dormancy	Terminal buds; seeds
Ethylene	Stimulates fruits to ripen; causes plants to seal off and drop unnecessary organs, such as leaves in autumn	Fruit tissues; aging leaves and flowers

A Summary of Plant Hormones This table lists some of the effects of plant hormones and where the hormones can be found within the plant.

Focus on Academic Language

• oppose • inhibit • promote • stimulate • signal

Help students understand these words, which are often used to describe hormone actions. Write the words on the board.

Say *Oppose* means to be against something. When the effects of cytokinins *oppose* those of auxins, it means they work against the effects of auxins.

Say To *inhibit* something means to stop something from continuing. When auxins *inhibit* the growth of shoot tips, it means they stop shoot tips from developing.

Say To *promote* or *stimulate* something means to encourage its growth.

Say To *signal* something means to send a message.

Ask What is a word that means the opposite of *promote*? *(Sample answers: oppose, inhibit)*

Have students select a word from the list on the board and use it correctly in a sentence.

During Reading

Active Reading

COMPARE/CONTRAST Help students recognize how some plant hormones work in opposition to each other. Review the Auxins section on the previous page. Then, have a volunteer read the first paragraph on this page aloud.

Ask How do auxins affect root and shoot growth? *(They cause the growth of new roots and inhibit the growth of shoot tips.)* What hormones cause opposite effects? *(cytokinins)*

Have a volunteer read the third paragraph on this page aloud.

Ask How does abscisic acid affect a seed embryo? *(It stops the growth of the embryo causing it to become dormant.)* What opposite effect do gibberellins have on the dormant embryo? *(They stimulate germination.)*

Build Vocabulary

WORD ORIGINS Ask students to review the effects of the hormones shown in the Summary of Plant Hormones table.

Ask What does each hormone "urge" plants to do? *(Answers should match the information in the table.)*

KEY QUESTION Read the Key Question and answer aloud.

Ask What is homeostasis? *(a state of equilibrium or balance in a cell or organism)*

Speed Bump

Ask What are auxins and what do they do? *(Auxins are plant hormones that control the elongation of cells and the growth of new roots. Auxins make plants bend toward light.)*

Say Suppose you buy some pears that are not quite ripe enough to eat. You put them in a closed paper bag. You let the bag sit on your kitchen table for a day or two.

Ask Will your actions cause the pears to ripen faster? *(yes)* Why? *(The bag will trap the ethylene the pears give off inside it. The trapped gas will hasten the ripening of the pears.)*

LESSON 24.3

During Reading continued

Active Reading

READ, COVER, REMEMBER, RETELL Pair students. Ask them to read the text about tropisms silently. When they have finished, tell them to cover the text with their hand but leave the pictures of plant tropisms visible.

Say Without looking at your book, take turns describing to your partner what is happening to each plant. Use the pictures to help you remember.

Ask What is the role of auxins in tropisms? *(Auxins move toward shaded tissue, where they make cells grow longer, causing the plant to grow towards the light. Auxins also move to the lower side of horizontal stems, causing cells on this side to grow longer, thus making the stem grow upward.)*

Build Vocabulary

WORD ORIGINS Remind students that they can sometimes figure out the meaning of an unfamiliar word if they know the meaning of one of its word parts. Explain that the word *tropism* means "turning," so each of the pictured plants is turning toward or away from something.

Ask What word part of *phototropism* do you recognize? *(photo)*

Say The prefix *photo-* means "light." So *phototropism* means "turning toward light." *Gravity* is similar to the first word part in *gravitropism*. Gravity is a force that pulls things down. Gravitropism causes stems to bend against gravity and roots to bend with gravity.

Ask Which tropism comes from a Greek word that means "touch"? *(thigmotropism)* What do plants with this tropism do when they touch an object? *(wrap around it)*

KEY QUESTION Read the Key Question and answer aloud.

Ask What would happen to a plant that could not respond to environmental stimuli? *(Such a plant might not receive sufficient light, water, or nutrients. Its stems might not grow upward and its roots might not grow downward.)*

Speed Bump

Ask Many people who grow houseplants rotate their potted plants one-quarter turn each time they water them. Why? *(A plant will grow towards the light, so regular turning will help keep the plant symmetrical.)*

Three Plant Tropisms
Plants respond to light (left), gravity (middle), and touch (right).

Tropisms and Rapid Movements

Sometimes plants need to move in response to their environment. Many plant movements are slow. But some are so fast that even animals cannot keep up with them.

Tropisms Plants respond to environmental stimuli such as light, gravity, and touch. When plants sense these stimuli, they send signals to their roots and stems to change the direction of growth. These growth responses are called **tropisms.**

▶ ***Light*** The tendency of a plant to grow toward a light source is called **phototropism.** This response can be very quick. For example, young seedlings can bend toward a light source in just a few hours. Remember that changes in the concentration of auxin are responsible for phototropism. Experiments have shown that auxins move toward shaded tissue. This may be due to changes in membrane permeability caused by light.

▶ ***Gravity*** Auxins also affect **gravitropism,** the response of a plant to gravity. For reasons that are still not understood, auxins move to the lower sides of horizontal roots and stems. In stems, this movement causes the stem to bend upright. In roots, however, the movement of auxins causes roots to bend downward.

▶ ***Touch*** Some plants respond to touch, a process called **thigmotropism.** Vines and climbing plants show thigmotropism. When they encounter an object, they wrap around it. Other plants, such as grape vines, have tendrils. Tendrils grow from the base of the leaf and wrap tightly around any object they encounter.

Rapid Movements Some plant responses are surprisingly fast. One example of a plant that can respond rapidly is the carnivorous Venus' flytrap. When an insect lands on a flytrap's leaf, it triggers sensory cells on the leaf. Electrical signals are sent from cell to cell. Cell walls expand and osmotic pressure changes, causing the leaf to snap shut. The insect is trapped inside!

Key Question What are some examples of environmental stimuli to which plants respond?
Plants respond to environmental stimuli such as light, gravity, and touch.

BUILD Vocabulary

tropism
the movement of a plant toward or away from a stimulus

phototropism
the response of a plant to light

gravitropism
the response of a plant to the force of gravity

thigmotropism
the response of a plant to touch

photoperiodism
a plant's response to the relative lengths of light and darkness

WORD ORIGINS
The word *tropism* comes from a Greek word that means "turning." With a tropism, a plant will "turn" toward or away from a stimulus.

BIOLOGY.com Search Lesson 24.3 GO • InterActive Art

ELL Focus on ELL: Build Background

INTERMEDIATE AND ADVANCED SPEAKERS Students who have studied the actions of hormones in animals may initially find it hard to understand why plant hormones differ from animal hormones. Tell students that, unlike animals, plants cannot move to a new location when environmental conditions change. Since plants cannot get up and move to new places, they instead respond to changing conditions by growing either toward or away from a stimulus or by producing flowers and seeds.

Unlike animal hormones, plant hormones are not formed in specific organs and they are not made in tissues that are specialized for hormone production. Instead they are produced in plant parts—leaves, stems, and roots—that have other general functions.

Construct a two-column chart on the board. Use student input to compare characteristics of animals and plant hormones mentioned above.

Response to Seasons

Year after year, some plants flower in the spring. Other plants flower in summer. Still other plants flower in the fall. Plants such as chrysanthemums flower when nights are long and days are short. These plants are called short-day plants. Plants such as spinach and irises flower when nights are short and days are long. These are called long-day plants.

Photoperiod and Flowering In the early 1920s, scientists discovered that the flowering of tobacco plants depends on the amount of light they receive. Many plants respond to the relative lengths of darkness and light in a day, or the photoperiod. Their response is called **photoperiodism.**

Scientists have discovered that the pigment phytochrome (FYT oh krohm) plays a role in the way plants respond to photoperiod. Phytochrome absorbs red light. It also starts a number of signals within plant cells. These signals determine various plant responses, such as flowering and growth.

Key Question How do plants respond to seasonal changes?
Photoperiodism is a major factor in the timing of seasonal activities such as flowering and growth.

BUILD Connections

EFFECTS OF PHOTOPERIOD

How much light a plant receives and when it receives the light can affect when the plant flowers.

In Your Workbook Go to your workbook to learn more about the effect of photoperiod. Complete the table for Response to Seasons.

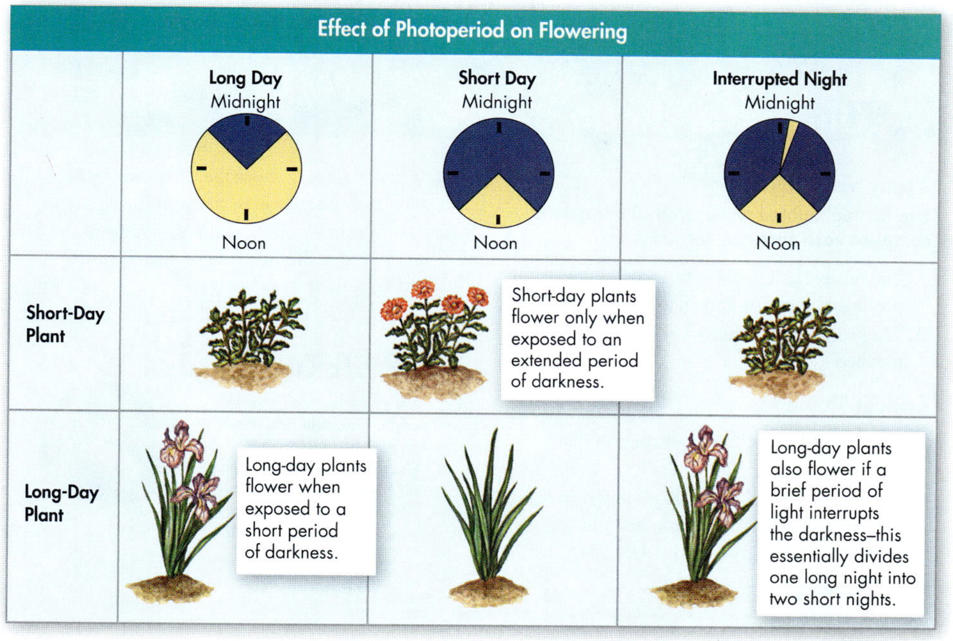

ELL Focus on ELL: Access Content

ALL SPEAKERS Some students might have difficulty relating the concept of photoperiods to seasonal patterns of flowering. Construct another way to present the information.

1. Ask a volunteer to draw a horizontal line on the board. Ask the student to write *Spring* underneath the line near the left end. Have the student write *Summer* in the middle and *Autumn* near the right end.
2. Ask students which seasons have short days versus long days. *(Spring and autumn have short days; summer has long days.)* Have volunteers write *short day* or *long day* under the appropriate season.
3. Help students determine where to write *short-day plant* and *long-day plant.* Have volunteers write *short-day plant* under spring and autumn; *long-day plant* under summer.

Active Reading

APPLY ACADEMIC VOCABULARY Call attention to the pigment *phytochrome* in the third paragraph on this page. Students may remember learning about pigments in Chapter 8.

Say Remember that a pigment is a light-absorbing molecule.

Ask What is phytochrome? *(a pigment that absorbs red light)* How does it affect the way plants respond to the length of daylight? *(It starts signals within plant cells that determine plant response, such as flowering and growth.)*

MAKE CONNECTIONS Tell students that commercial plant growers use photoperiod to raise plants that bloom at a certain time of year.

Ask What flowering plant can you think of that is sold mostly at a certain time of the year? *(Mention poinsettias if students do not.)* Explain that poinsettias are short-day plants. Starting in late September, they must have a day length of 12 hours or less in order to bloom for December holidays.

Build Connections

VISUAL SUMMARY Tell students that the chart represents different photoperiods. The rows show how short-day plants and long-day plants respond to each photoperiod.

Ask Which plant is more likely to flower in the summer? *(long-day plant)*

Ask Which plant needs long periods of darkness to flower? *(short-day plant)*

Ask How do short-day plants respond to interrupted night? *(The plants do not flower.)* Why? *(These plants need a long period of continuous darkness to flower.)*

KEY QUESTION Read the Key Question and answer aloud.

Ask How does photoperiodism benefit a plant? *(It enables a plant to bloom at the right time of year, when weather conditions are most likely to ensure successful flowering.)*

Speed Bump

Ask Are plants that flower in the winter likely to be short-day plants or long-day plants? *(short-day plants)* Why? *(because winter has short days)*

LESSON 24.3

During Reading continued

Active Reading

RECOGNIZE CAUSE AND EFFECT Help students understand the effects of autumn weather as plants prepare for winter dormancy. Ask students to distinguish between cause and effect for each change that occurs as winter approaches.

Say In the winter, plants face special challenges.

Ask What happens to plants in the winter? *(Sample answer: Some plants die; trees become dormant.)*

Ask What effects occur as trees get ready for winter? *(Leaves shut down. Materials move from leaves to roots. Meristems stop producing new leaves and instead produce thick scales that will protect buds. Xylem and phloem fill up with organic compounds and ions, making freezing less likely.)*

Ask What is the cause of these changes? In other words, how do plants "know" that winter is coming? *(In the fall, the days grow shorter, so phytochrome absorbs less light.)*

KEY QUESTION Have a volunteer read the Key Question and answer aloud.

Ask What would happen to a deciduous tree if it could not prepare for winter? *(Leaves and buds would freeze, making it difficult for the tree to survive in the springtime. Sap might also freeze. The tree might die.)*

Post Reading

ASSESSMENT TIPS For Questions 1 and 2, have students find a key word that should help them identify the answer. *(In both questions the key word is light.)*

For Question 4, ask students to recall where auxin is produced in the stem. *(at the tip, in the apical meristem)* Remind students that when a stem's tip is cut, lateral buds grow more quickly.

MYSTERY CLUE Have students use the information under the subhead Ethylene to help them answer the question.

Wrap-Up Activity

PLANT HORMONE DIARY Have students assume the role of a young deciduous tree. Ask them to create a one-year hormone diary that describes the effects of each plant hormone on that tree, from the moment of germination through winter dormancy. Encourage students to illustrate the diary to show the effects of each hormone.

Adaptations for Winter In autumn, the leaves of deciduous trees stop photosynthesis and fall from the trees. The meristems at the tips of the branches produce thick, waxy scales. These scales cover and protect the new stem and leaf buds throughout the winter.

Winter Dormancy In winter, many plants become dormant. Phytochrome controls the many changes that prepare plants for winter dormancy. As cold weather approaches, deciduous plants move materials from leaves to roots. They stop photosynthesis and seal off leaves from the rest of the plant.

▶ ***Leaf Loss*** Many plants lose their leaves during the colder months. Because days are shorter in the fall, the phytochrome in leaves absorbs less light. The production of auxin drops, while the production of ethylene increases. The change in the relative amounts of these two hormones causes leaves to shut down. The pigment chlorophyll breaks down. Other pigments in the leaves, such as the bright yellow and orange carotenoids, can now be seen. Bright reds come from freshly made anthocyanins.

▶ ***Changes to Meristems*** Hormones also produce important changes in apical meristems. Meristems stop producing new leaves. Instead, they produce thick, waxy scales that protect leaf buds from the cold. Xylem and phloem tissues fill up with ions and organic compounds. This solution acts like antifreeze in a car: It keeps the tree's sap from freezing.

Key Question How do plants respond to cold winters? **As cold weather approaches, deciduous plants stop photosynthesis. They move materials from leaves to roots and seal off leaves from the rest of the plant.**

CHECK Understanding

Apply Vocabulary

Use the highlighted words from the lesson to complete each sentence correctly.

1. The relative lengths of light and darkness produce a stimulus known as _______.

2. The growth of a plant toward a light source is called _______.

Critical Thinking

3. Explain What are the main roles of plant hormones?

4. Relate Cause and Effect Explain why a person who trims bushes should understand the effect of auxins on the growth of stems.

5. Apply Concepts Describe three examples of plant responses to external stimuli.

6. Write to Learn Answer the third clue of the mystery. Use the words *hormone*, *stimulus*, and *respond* in your answer.

MYSTERY CLUE

Kerosene heaters give off carbon dioxide and ethylene as they burn kerosene. Electric heaters do not give off these gases. Could this difference explain why the lemons didn't ripen? (**Hint:** See p. 589.)

BIOLOGY.com Search Lesson 24.3 GO • Lesson Assessment

Check Understanding

1. photoperiod

2. phototropism

3. Plant hormones control growth, development, and a plant's response to environmental stimuli.

4. Auxins are produced in the apical meristems. If the gardener trims off the apical meristems, the plant will become bushier.

5. Phototropism causes plants to grow toward light. Gravitropism causes plant shoots to grow away from gravity. Thigmotropism causes plant growth in response to touch.

6. Students should recognize that ethylene is a plant hormone that plays a role in ripening. The lemons responded to the stimulus ethylene by ripening.

24.4 Plants and Humans

Agriculture

Walk through a grocery store and you will see products made from hundreds of different plants. Which plants are most important? Are there plants that humans can't live without?

Origin of Agriculture Agriculture, or farming, is the foundation of human society. Evidence suggests that agriculture developed separately in many parts of the world about 10,000 to 12,000 years ago. Once people learned how to grow plants for food, they tended to stay in one place. This led to the development of governments and other social institutions. Even today, more humans focus on agriculture than on any other activity.

Worldwide Patterns Thousands of different plants are raised for food. But for the bulk of their food, most people in the world rely on the seeds of four angiosperms—rice, wheat, corn, and soybeans. Three of these crops are grasses. In the United States, about 80 percent of cropland is used to grow wheat, corn, soybeans, and hay. These four crops are used for both humans and livestock. Of these crops, wheat, corn, and hay are grasses.

Key Question Which crops are the major food supply for humans? **Worldwide, most people depend on a few crops, including rice, wheat, soybeans, and corn, for the major part of their food.**

Key Questions

Which crops are the major food supply for humans?

In addition to food, what are some benefits that humans get from plants?

BUILD Understanding

Preview Visuals Look at the last picture in the lesson. Identify the plants that provided the raw materials for the products shown. Then, list other products that come from plants.

In Your Workbook Go to your workbook to learn more about previewing visuals. Complete the activity for Lesson 24.4.

Plants and Agriculture Rice is a major crop in China and much of Southeast Asia.

593

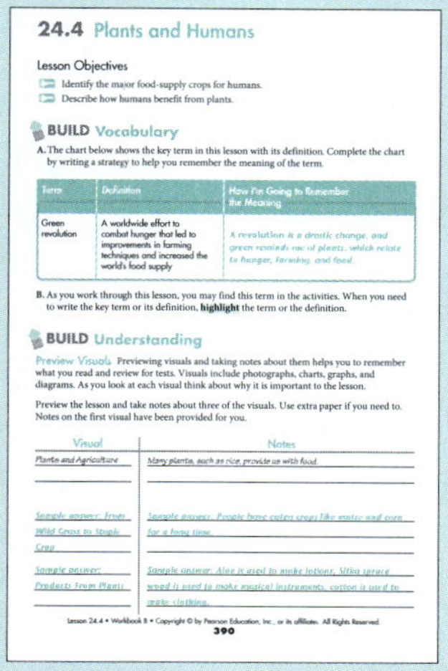

24.4 Plants and Humans

Lesson Objectives

- Identify the major food-supply crops for humans.
- Describe how humans benefit from plants.

BUILD Vocabulary

A. The chart below shows the key term in this lesson with its definition. Complete the chart by writing a strategy to help you remember the meaning of the term.

Term	Definition	How I'm Going to Remember the Meaning
Green revolution	A worldwide effort to combat hunger that led to improvements in farming techniques and increased the world's food supply	A revolution is a drastic change, and green reminds me of plants, which relate to hunger, farming, and food.

B. As you work through this lesson, you may find this term in the activities. When you need to write the key term or its definition, **highlight** the term or the definition.

BUILD Understanding

Preview Visuals Previewing visuals and taking notes about them helps you to remember what you read and review for tests. Visuals include photographs, charts, graphs, and diagrams. As you look at each visual think about why it is important to the lesson.

Preview the lesson and take notes about three of the visuals. Use extra paper if you need to. Notes on the first visual have been provided for you.

Visual	Notes
Plants and Agriculture	Many plants, such as rice, provide us with food.
Sample answer: From Wild Grass to Staple Crop	Sample answer: People have eaten crops like maize and corn for a long time.
Sample answer: Products From Plants	Sample answer: Aloe is used to make lotions, Sitka spruce wood is used to make musical instruments, cotton is used to make clothing.

Lesson 24.4 • Workbook B • Copyright © by Pearson Education, Inc., or its affiliates. All Rights Reserved.
390

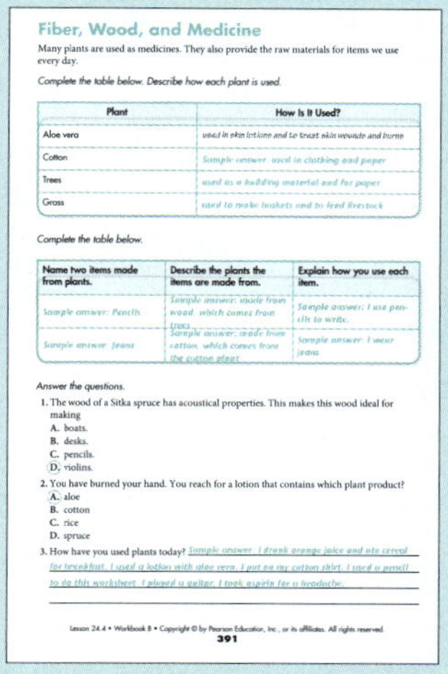

Fiber, Wood, and Medicine

Many plants are used as medicines. They also provide the raw materials for items we use every day.

Complete the table below. Describe how each plant is used.

Plant	How Is It Used?
Aloe vera	used in skin lotions and to treat skin wounds and burns
Cotton	Sample answer: used in clothing and paper
Trees	used as a building material and for paper
Grass	used to make baskets and to feed livestock

Complete the table below.

Name two items made from plants.	Describe the plants the items are made from.	Explain how you use each item.
Sample answer: Pencils	Sample answer: made from wood, which comes from trees	Sample answer: I use pencils to write.
Sample answer: Jeans	Sample answer: made from cotton, which comes from the cotton plant	Sample answer: I wear jeans.

Answer the questions.

1. The wood of a Sitka spruce has acoustical properties. This makes this wood ideal for making
   A. boats.
   B. desks.
   C. pencils.
   D. violins.
2. You have burned your hand. You reach for a lotion that contains which plant product?
   A. aloe
   B. cotton
   C. rice
   D. spruce
3. How have you used plants today? Sample answer: I drank orange juice and ate cereal for breakfast. I used a lotion with aloe vera. I put on my cotton shirt. I used a pencil to do this worksheet. I played a guitar. I took aspirin for a headache.

Lesson 24.4 • Workbook B • Copyright © by Pearson Education, Inc., or its affiliates. All rights reserved.
391

LESSON 24.4

Pre-Reading

Objectives

24.4.1 Identify the major food-supply crops for humans.

24.4.2 Describe how humans benefit from plants.

Preview the Pages

1. Preview all the visuals and read the captions and labels. Remind students that four angiosperms—rice, wheat, corn, and soybeans—provide most of the world's food.
2. Point out that the picture on this page shows rice farming. Explain that terracing creates flat land for farming in hilly areas and captures the large amounts of water needed for growing rice.

Build Understanding

PREVIEW VISUALS Ask students to complete the Build Understanding activity in Workbook B. Remind them that they can use their workbook activities to help them review the chapter material before a test.

NATIONAL SCIENCE EDUCATION STANDARDS

UNIFYING CONCEPTS AND PROCESSES

I

CONTENT

C.4.e, F.3, F.4, F.5, F.6

For the corresponding lesson in *Miller & Levine Biology,* see pages 715–718.

During Reading

Active Reading

KEY QUESTION Read the Key Question and answer on the previous page aloud.

Ask Which of these crops have you eaten today? *(Answers may vary.)* Point out that products of these crops are in many foods even if you don't realize it. Examples include cornstarch, corn oil, and various soy products.

COMPARE AND CONTRAST Point out the images of teosinte and modern corn on this page. Make sure students understand that a quarter is shown in both pictures.

Say You know what size a real quarter is.

Ask How can you use the quarter in each picture to determine the size of the ear of corn? *(compare the length of an ear to the length of a quarter)*

Ask Which plant's ears are the largest? *(modern corn)*

Ask A kernel is an individual grain on an ear. Which plant's kernels are the largest? *(modern corn)*

SCIENCE SUPPORT Explain that, while animal manure has long been an important natural fertilizer, it presents a food safety concern. Explain that pathogens such as *Salmonella, Campylobacter,* and *E. coli* are found in animal feces and can contaminate water and food. This contamination is of particular concern with foods that are eaten raw, such as fresh produce. Many outbreaks of disease linked to produce have been caused by manure contamination. Crops can be contaminated by direct application of untreated manure or by run-off water from grazing lands, feed lots, or dairies that flows onto nearby fields. Animal manure can be treated to destroy pathogens with technologies such as composting and drying.

Build Vocabulary

ACADEMIC WORDS Remind students that the word *revolution* is also used in an everyday context to mean a dramatic change in ideas or a new way of doing something. Have students write the word in a sentence. As volunteers read their sentences aloud, compare the different contexts of the word.

Speed Bump

Ask Which of these crops—rice, wheat, corn, or soybeans—is native to North America? *(corn)*

Ask Why did people start the green revolution? *(to increase the world's food supply to fight hunger)*

BUILD Vocabulary

green revolution
the development of highly productive crops and the use of modern agriculture to increase crop yields

ACADEMIC WORDS
The word *revolution* is used to describe a major change in society. During the Industrial Revolution, for example, improvements in technology changed the way goods were manufactured. The green revolution changed methods of farming in many parts of the world.

New Plants The introduction of new crop plants has often changed human history. Four hundred years ago, plants such as corn and peanuts were unknown in Europe. The introduction of these plants from the Americas changed European agriculture. We may think of potatoes as a German or Irish food. But, originally, potatoes came from South America.

Agriculture has been improved through selective breeding. Recall that humans use selective breeding to produce crops and animals that have desirable traits. The corn grown by Native Americans was selectively bred. It was developed thousands of years ago from teosinte (tee oh SIN tee), a grass from Mexico. Most crops grown today are the result of selective breeding. For example, cabbage, broccoli, and cauliflower were all developed from a single species of wild mustard.

Changes in Agriculture Between 1950 and 1970, people began a worldwide effort to fight hunger by improving methods of farming. This effort, called the **green revolution,** greatly increased the world's food supply. The green revolution used new varieties of seeds and fertilizers. Fertilizers provide plants with nutrients such as nitrogen, phosphorus, and potassium. These nutrients make plants grow larger and produce more food. For thousands of years, farmers have provided crops with nutrients from natural sources, such as animal remains and manure. Today, many farmers use artificial fertilizers.

Fertilizers and pesticides must be used with great care. Too much fertilizer can kill plants. Fertilizer that washes off fields can contaminate the water. Pesticides are chemicals that are used to kill pests such as insects. Because pesticides are poisons, they can harm wildlife and humans if not used correctly.

From Wild Grass to Crop The selective breeding of teosinte (left) began about 8000 years ago. It led to the development of modern corn (right). As you can see from the image of the quarter, modern corn has much larger kernels compared to its ancestor teosinte.

Focus on Academic Language

• selective • fertilizers • animal remains • manure • artificial fertilizers

Some students may be unfamiliar with some of the terms above. Write the words on the board.

Say The word *selective* means "to make careful choices." In selective breeding, a breeder chooses only those organisms with desired characteristics to produce the next generation. Selective breeding has produced modern corn, with characteristics such as large kernels and improved resistance to disease.

Say Fertilizers are organic materials added to the soil to support plant growth. Natural fertilizers may include animal remains (dead bodies) or manure (animal waste). Today many growers use artificial fertilizers, which are manufactured to contain the right amounts of chemicals required by a certain crop.

Ask What are some examples of natural fertilizers? *(animal remains, manure)*

Products From Plants Plants provide raw materials for many useful products.

◀ The succulent plant *Aloe vera* contains many chemicals that soothe and moisturize the skin. Extracts of this plant are used in many skin lotions as well as in burn and wound ointments.

Fiber, Wood, and Medicine

Some of the most important uses of plants have nothing to do with food. Plants produce materials that we use for homes and clothes. They also provide some of our most effective medicines. If you are reading this page in a printed book, you are looking at paper made from North American conifers. You may be sitting on a chair made from an oak tree. And you are probably wearing at least one piece of clothing made from the fibers of the cotton plant.

Key Question In addition to food, what are some benefits that humans get from plants? **Plants provide the raw materials for our homes and clothes, and some of our most effective medicines.**

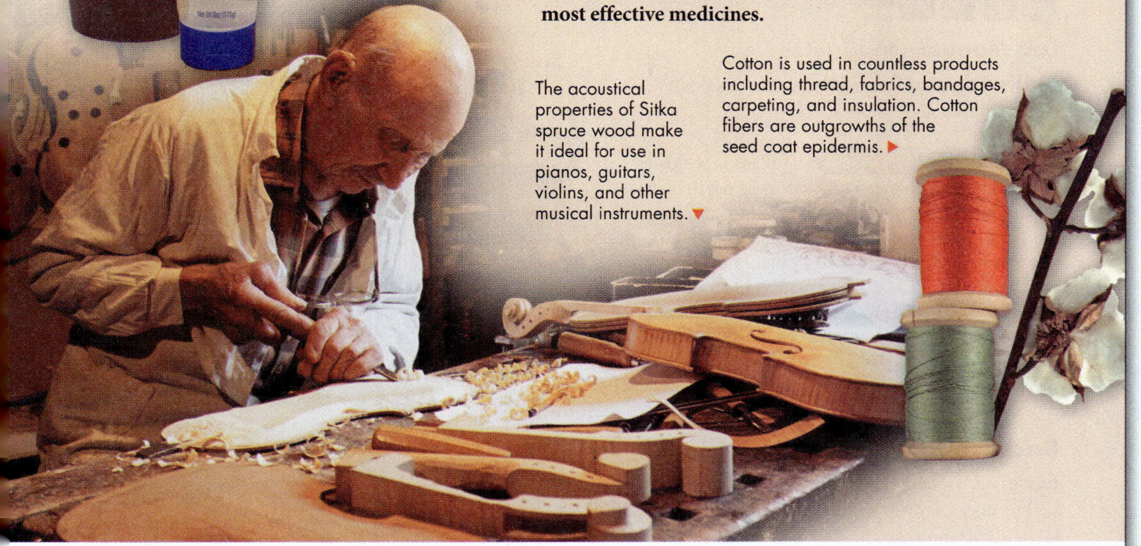

The acoustical properties of Sitka spruce wood make it ideal for use in pianos, guitars, violins, and other musical instruments. ▼

Cotton is used in countless products including thread, fabrics, bandages, carpeting, and insulation. Cotton fibers are outgrowths of the seed coat epidermis. ▶

CHECK Understanding

Apply Vocabulary

Use the highlighted words from the lesson to complete each sentence correctly.

1. The _______ led to the production of much larger amounts of food.

Critical Thinking

2. Review What four crops form the base of the world's food supply?

3. Explain In addition to providing food, why are plants important to human society? Give three examples.

4. Write to Learn What would life be like if humans had never developed agriculture? Write a short story that describes such a world.

BIOLOGY.com Search Lesson 24.4 GO • Lesson Assessment

Check Understanding

1. green revolution

2. Rice, wheat, soybeans, and corn form the base of the world's food supply.

3. Plants provide raw materials for our homes and clothing. They also provide powerful medicines.

4. Students should recognize that life would be very different without agriculture. People would have to rely on finding or hunting food. They would not have shelters and medicines as we have today. Some students may also note that there would be fewer people around in such a world. A few students may note that art, literature, and other social practices would be far less common because more time would be required to find food.

Active Reading

DRAW CONCLUSIONS Help students conclude that plants play important roles in their lives.

Ask What kinds of animal products have you used in the last 24 hours? *(Sample answers: leather belts, shoes, or other clothing from animals; meat, fish, milk, eggs, or other food from animals)*

Ask How would your list of animal products change if plants disappeared? *(Answers may vary, but students should recognize that animals depend on plants, too.)*

KEY QUESTION Have a volunteer read the Key Question and answer aloud.

Ask Look around the classroom. What are some examples of plant products that you see? *(Answers may vary.)*

Speed Bump

Ask How could plant extinctions affect the development of new medicines? *(If a plant species becomes extinct, scientists would never discover a particular medicine that the plant might have provided.)*

Post Reading

ASSESSMENT TIPS If students have trouble with Question 4, ask them to think about how they got the plant products they have eaten in the last seven days. Ask them to consider how they would get those foods or substitute foods if agriculture did not exist.

Wrap-Up Activity

AGRICULTURAL DISCOVERIES Invite students to imagine they are archaeologists excavating a community that is thousands of years old. They make discoveries that revolutionize our current thinking of how agriculture developed. What did they find? What conclusions can they draw? Suggest that students work in pairs. Have them make a report of their findings including maps, drawings, descriptions, and conclusions. Prompt students' thinking by suggesting such discoveries as dormant seeds found in a leather sack, agricultural tools, and ancient drawings showing the use of certain plants as medicine.

Pre-Lab

Review with students the purpose and parts of a controlled experiment.

Say In this lab, you will perform a controlled experiment about the effect of a hormone on leaf loss. You will need to recognize what the control is in the experiment.

Ask What is a controlled experiment? *(an experiment in which only one variable is changed)*

Ask What is an independent variable? *(what I change)*

Ask What is a dependent variable? *(what I see happen)*

Ask What is a control? *(a group or individual that is exposed to the same conditions as the experimental group except for one independent variable)*

Review the Background Questions and Pre-Lab Questions as a class.

Lab

Tell students they will perform the chapter lab *Plant Hormones and Leaves* described in **Lab Manual B.**

SAFETY Students should wear goggles and gloves and use care when handling scissors. They should wash their hands thoroughly with soap and warm water at the end of the lab. Check for students who might be allergic to the plants and fruits used.

Post-Lab

DESIGN AN EXPERIMENT Have students design an experiment to compare the effects of auxins on the leaves of different species of plants.

NATIONAL SCIENCE EDUCATION STANDARDS

CONTENT C.1.d, C.1.e

INQUIRY A.1.b, A.2.a

Real-World Lab

GUIDED INQUIRY

Pre-Lab: Plant Hormones and Leaves

Problem How does a plant hormone affect leaf loss?

Materials leafy plant, masking tape, permanent marker, scissors, string, toothpick, auxin paste, plastic container or tray

Lab Manual Chapter 24 Lab

Skills Focus Observe, Draw Conclusions, Apply Concepts

Connect to the Big idea A plant may flower in response to a change in the hours of daylight, or a plant may stop growing in response to colder temperatures. These responses to changes in the environment are coordinated by plant hormones, which regulate plant development. A hormone may stimulate roots to grow, seeds to germinate, or fruits to ripen. A hormone may inhibit cell division or promote dormancy of seeds or plants. In this lab, you will investigate the effect of a plant hormone on leaf loss.

Background Questions

a. Explain Do all plant cells respond to every plant hormone? Why or why not?

b. Explain What is a photoperiod? What is dormancy? How are they related?

c. Review In the fall, what happens to the production of auxins and ethylene in the leaves of flowering plants?

Pre-Lab Questions

Preview the procedure in the lab manual.

1. **Use Visuals** Draw a simple leaf. Label the blade and the petiole.
2. **Control Variables** What is the control in this experiment?
3. **Infer** How will auxins move from the paste to the base of the petiole?

BIOLOGY.com Search Chapter 24 GO

Visit Chapter 24 online to test yourself on chapter content and to find activities to help you learn.

Untamed Science Video Using time-lapse videography, the Untamed Science crew reveals how plants move in response to various stimuli.

Art in Motion Watch how meiosis and mitosis produce ova and pollen in an angiosperm.

Art Review Review the structures of a flower.

InterActive Art Change the length of day and see how it affects both short- and long-day plants.

Pre-Lab Answers

BACKGROUND QUESTIONS

a. No. To respond to a hormone, a cell must contain receptors to which a given hormone can bind.

b. Sample answer: Photoperiod is the relative length of daylight and darkness. Dormancy is the period during which an organism's growth and activity slow down or stop. Photoperiod is a stimulus for plants that respond to seasonal changes. Dormancy is one possible response to this stimulus.

c. The production of auxin decreases; the production of ethylene increases.

PRE-LAB QUESTIONS

1. Students should identify the flat area as the blade and the thin stalk as the petiole.
2. The control is the petiole that is not treated with auxin paste.
3. Sample answer: Auxins will move through the layer of vascular tissue (specifically phloem) in the petiole.

24 CHAPTER Summary

24.1 Reproduction in Flowering Plants

- Flowers are reproductive organs that are made up of four kinds of specialized leaves: sepals, petals, stamens, and carpels.
- The process of fertilization in angiosperms differs from fertilization in other plants. In angiosperms, there are two fertilization events. One produces the zygote. The other produces a tissue called endosperm within the seed.
- Vegetative reproduction is the formation of new individuals by mitosis. Vegetative reproduction does not include gametes, flowers, or fertilization.

stamen (p. 579)
anther (p. 579)
carpel (p. 579)
stigma (p. 579)
embryo sac (p. 580)
double fertilization (p. 581)
endosperm (p. 581)
vegetative reproduction (p. 583)

24.2 Fruits and Seeds

- As angiosperm seeds mature, the ovary walls thicken to form a fruit. The fruit surrounds and protects the developing seeds.
- Seeds are spread by animals, wind, and water. Seeds in fleshy fruits are usually spread by animals. Seeds that are spread by wind or water usually have lightweight fruits that can be carried by wind or float on water.
- Environmental factors such as temperature and moisture can cause a seed to end dormancy and germinate.

dormancy (p. 586)
germination (p. 586)

24.3 Plant Hormones

- Plant hormones serve as signals that control the development of cells, tissues, and organs. They also coordinate a plant's responses to the environment.
- The opposing effects of plant hormones balance each other, contributing to a plant's homeostasis.
- Plants respond to environmental stimuli such as light, gravity, and touch.
- The relative length of darkness and light, or photoperiod, affects seasonal activities such as flowering and growth.
- As cold weather approaches, deciduous plants stop photosynthesis. They move materials from leaves to roots and seal off leaves from the rest of the plant.

hormone (p. 588)
target cell (p. 588)
receptor (p. 588)
tropism (p. 590)
phototropism (p. 590)
gravitropism (p. 590)
thigmotropism (p. 590)
photoperiod (p. 591)

24.4 Plants and Humans

- Worldwide, most people depend on a few crops, including rice, wheat, soybeans, and corn, for the major part of their food.
- Plants produce the raw materials for our homes and clothes. They also provide some of our most powerful and effective medicines.

green revolution (p. 594)

BIOLOGY.com Search Chapter 24 GO • Crossword • Chapter Assessment

UbD Transfer the Big Idea

TRANSFER Ask students to imagine that they have been hired to plan a flower garden. The garden must contain flowering plants that have colorful blooms or attractive leaves. (Remind students that flowering plants include grasses.) Some of the plants must be blooming continuously from late spring until late fall. At least two of the plants must be able to reproduce asexually. The plants chosen for the garden should thrive in the local climate and rely on local pollinators. Students should include maps and/or sketches in their garden plan.

Chapter Summary

Think Visually

USE GRAPHICS Have students make a concept map showing the major concepts for reproduction in flowering plants.

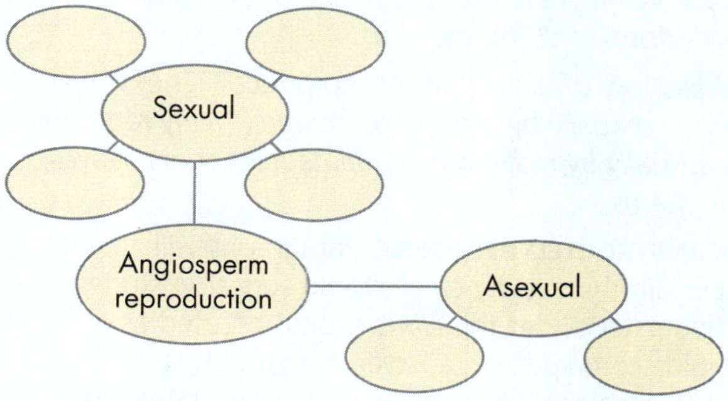

ANSWERS
Check students' concept maps for accuracy. Suggest they use vocabulary terms, headings, and subheadings as a guide for important concepts.

BIOLOGY.com Study Online

REVIEW RESOURCES

Art Review Students can drag and drop labels onto the diagram of a flower.

Art in Motion This animation shows meiosis and mitosis during angiosperm gamete production.

InterActive Art This activity allows students to examine how changing day length affects plants.

Data Analysis This activity allows students to analyze the potential fuel energy that can be obtained from plants.

Vocabulary Review The *Flash Cards* and *Crossword* are an interactive vocabulary review.

Editable Worksheets Editable pages of Study Workbook B, Lab Manual B, and the Assessment Resources Book are available online.

Lesson Overviews Lesson Overviews are based on Miller & Levine Biology. However, they are available on CD-ROM and can be edited for your class.

ASSESSMENT RESOURCES

Chapter Assessment Have students take an online version of the Chapter 24 Check Understanding.

Standardized Test Prep Students can take an online version of the Standardized Test Prep. You will receive their scores along with ideas for remediation.

Diagnostic and Benchmark Tests Use these tests to monitor your students' progress and supply remediation.

ASSESSMENT

Assess

Big idea Invite a volunteer to read the question. Write the words *grow, develop,* and *reproduce* on the board.

Ask Where does growth occur in a plant? *(in meristems at the shoot tips and root tips)*

Ask What chemicals in a plant control its growth and development? *(hormones)*

Ask How do angiosperms reproduce? *(They reproduce sexually by producing flowers. Many reproduce asexually by making new plants from roots, leaves, or stems.)*

CONSTRUCTED RESPONSE Paragraphs will vary. Explain the activity carefully. Be sure that students understand that the *Hints* should be used as a guide to help them answer the questions. Review that each paragraph in their answers should have a main idea and supporting details. Demonstrate how students can use a simple outline to plan their paragraphs before writing.

For Question 1, write it on the board and read it aloud. Underline *animals in an ecosystem* and *reproduction.*

Ask What is an ecosystem? *(a community of organisms and their environment)*

Say Name some animals in our ecosystem. *(Answers may vary.)*

Say The question is asking about the connection between animals and reproduction in flowering plants.

Read the first hint.

Ask What is the connection between animals and flowers? *(Sample answer: Flowers are required for reproduction. Many flowers are pollinated by animals. Pollination is an important step in plant reproduction.)*

Read the second hint.

Say We learned that animals can disperse seeds by eating fruits or by carrying them in their fur.

Model writing the first one or two sentences using the hints on the board and the text. *(Sample: Many plants rely on animals for pollination. For example, insects, birds, and bats carry pollen from one flower to another.)*

24 CHECK Understanding

Assess the Big idea **Growth, Development, and Reproduction**

Write an answer to the question below.

Q. How do changes in the environment affect the reproduction, development, and growth of plants?

Constructed Response

Write an answer to each of the numbered questions below. The answer to each numbered question should be one or two paragraphs long. To help get ideas, read the **Hints** below the question.

1. **How do animals in an ecosystem affect the reproduction of angiosperms?**

   Hint The key function of a flower is reproduction.

   Hint Fruits and seeds need to be dispersed.

2. **How do plants respond to light, gravity, and touch? Include the names of these three responses in your answer.**

   Hint Sometimes, a plant grows toward a stimulus. Sometimes, it grows away from a stimulus.

   Hint Sometimes, shoots and roots have different responses to the same stimulus.

3. **How does the interaction of hormones and the environment control the germination of seeds?**

   Hint Dormancy can ensure that a plant germinates under ideal conditions.

Foundations for Learning Wrap-Up

Use the layered book that you made while reading the chapter to help you review concepts related to how angiosperms reproduce and respond to the environment.

Activity 1 Work with a small group to make an illustrated book that describes how plants reproduce and how they respond to changes in their environment. You can add illustrations to your layered book, or you can make a new book.

Activity 2 The last page of your layered book is blank. Use this page to write review questions as shown below. Use these questions to help you learn the concepts in the chapter. When you are stumped, flip to the other pages in your layered book to find help.

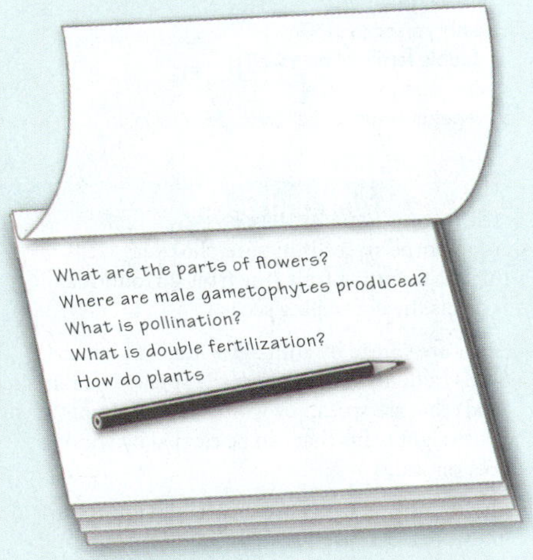

For Question 2, pair students and have them use the hints and your modeling to answer the question. Have students share their paragraphs with the class.

For Question 3, have students write the paragraphs independently for homework or as an anchor activity.

Foundations for Learning Wrap-Up

Ask students to take out the layered book they made earlier. Be sure they have written logical notes about the content of each lesson. Assign students to work in pairs or small groups to complete Activity 1. Students should complete Activity 2 independently. As an extension, they may find it helpful to share the questions they wrote in Activity 2 with a partner. Partners can quiz each other to review the important concepts in the chapter.

24.1 Reproduction in Flowering Plants

Understand Key Concepts

Use this art to answer questions 1–3.

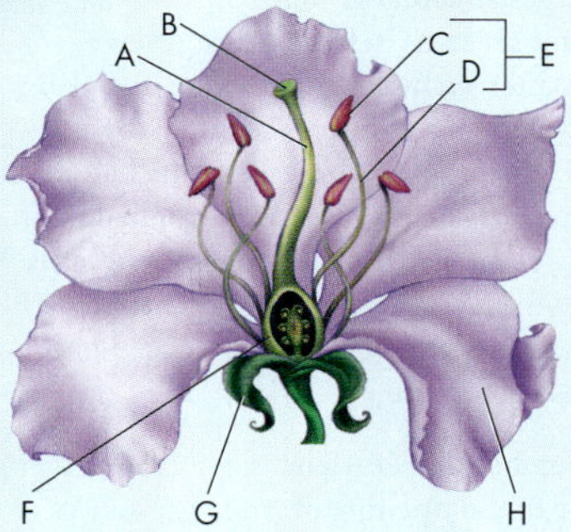

1. Which structure produces the male gametophytes?
   a. A **c.** C
   b. B **d.** D
2. Pollination takes place when a pollen grain lands on
   a. B. **c.** F.
   b. C. **d.** G.

> **Test-Taking Tip**
>
> **Read All Letters Carefully** Sometimes the answers are letters only. Don't get confused by the capital letter and the lowercase letter. As you read each answer choice, circle the corresponding letter in the diagram. Then write the name of the part next to the answer choice. So, in question 2, look at answer **a.** It is B, so circle B in the diagram. What is it? Write "stigma" next to B. This way, you can read the words, not just the letters.

3. What is the name for the entire female structure of a flower? Which three structures in the diagram make up the female part of a flower? Name each of these parts.

Think Critically

4. **Infer** Bees are important pollinators. Recently, scientists have discovered that bee populations are getting smaller, but they don't know why. What might happen to angiosperms if bees disappear?

24.2 Fruits and Seeds

Understand Key Concepts

5. A period during which an embryo is alive but not growing is called
   a. dormancy. **c.** germination.
   b. fertilization. **d.** pollination.
6. In the carpel, the thickened ovary wall will eventually become the
   a. cotyledon. **c.** female gametophyte.
   b. endosperm. **d.** fruit.
7. What is the function of seed dormancy? Give two examples of when it is helpful to a plant.

Think Critically

8. **Infer** Some flowers have stamens but lack carpels. Can these flowers produce fruits? Explain your reasoning.

24.3 Plant Hormones

Understand Key Concepts

9. The hormone that stimulates cell division and causes seeds to germinate is
   a. auxin. **c.** ethylene.
   b. cytokinin. **d.** gibberellin.

Think Critically

10. **Form a Hypothesis** Form a hypothesis that describes how plants such as vines benefit from thigmotropism.
11. **Evaluate** Spinach is a long-day plant. It grows best when night length is 10 hours or less. Will spinach grow well near the equator? Explain your reasoning.

Answers

Lesson 24.1

UNDERSTAND KEY CONCEPTS

1. c

2. a

3. The carpel is the female structure of a flower. In the diagram, these structures make up the carpel: A-style; B-stigma; F-ovary.

THINK CRITICALLY

4. Many flowering plants rely on bees for pollination. Without pollinators, these plants will be unable to reproduce. Eventually these plants may die out.

Lesson 24.2

UNDERSTAND KEY CONCEPTS

5. a

6. d

7. Dormancy helps seeds survive long cold periods. It also ensures that seeds sprout when sufficient moisture is available. For some seeds, special conditions, such as a forest fire, are needed to end dormancy.

THINK CRITICALLY

8. Sample answer: No, they cannot produce fruits. Seeds and fruits are produced by carpels. Without carpels, fruits cannot form.

Lesson 24.3

UNDERSTAND KEY CONCEPTS

9. d

THINK CRITICALLY

10. A vine shows thigmotropism when it comes in contact with an object such as a fence or another tree and wraps around it. This benefits the vine because the vine does not have to expend energy producing a strong trunk.

11. Spinach will not grow well near the equator. The period of darkness near the equator is 12 hours year round. The spinach would be exposed to too much darkness to ensure best growth.

Assessment Tips

Pair students and have them work together to answer the questions. Have students share their answers with the class and explain their reasoning for each answer.

Before students begin, make sure they understand that Questions 1, 2, and 3 refer to the diagram of the flower. Point out the word *entire* in Question 3. For Question 4, tell students to ask themselves what bees do for angiosperms.

For Question 8, tell students that *Explain your reasoning* means telling why you answered the question the way you did.

Make sure students understand that Questions 14–17 relate to the Use Science Graphics visual.

ASSESSMENT

Lesson 24.4

UNDERSTAND KEY CONCEPTS

12. a

THINK CRITICALLY

13. Sample answer: Seeds contain nutrient-rich endosperm. People can readily eat and use nutrients from endosperm. The rest of the plant may be difficult to eat, as is the case with woody plants.

Connecting Concepts

14. In the first setup, the plant was upright and grew upward. In the second setup, the plant was positioned on its side; the stem bent 90° and grew upward. In the third setup, the plant was positioned upside down; its stem bent 180° and grew upward

15. Even when a plant is placed in a horizontal or inverted position, the plant stem will grow upward due to negative gravitropism.

16. No, the fact that light sources were suspended above the plants means that these results could be due to phototropism, not gravitropism.

17. Sample answer: Set up the experiment so that light is coming from multiple directions at the same time. Light should be coming from all sides in each setup.

CHAPTER MYSTERY After students have read through the Chapter Mystery, discuss the topic of ethylene and fruit ripening. Ask student to discuss their experiences with fruit that looked good enough to eat but wasn't ready to eat.

Ask Why is ethylene considered a plant hormone? *(because it affects parts of growth and development such as fruit ripening and leaf drop)*

Ask How is ethylene different from all other plant hormones? *(It is a gas.)*

24 CHECK Understanding

24.4 Plants and Humans

Understand Key Concepts

12. Most human food comes from

a. angiosperms. **c.** gymnosperms.
b. conifers. **d.** trees.

Think Critically

13. **Infer** How is it possible that most of our food comes from seeds rather than the plant body?

Connecting Concepts

Use Science Graphics

Growth responses of plants to stimuli are called tropisms. A tropism is positive if the affected plant part grows toward the stimulus. The response is negative if the plant part grows away from the stimulus. The experiment shown below tested the effect of gravitropism on plant growth. The conclusion drawn from the experiment was that the plant stems grow upward due to negative gravitropism. Use the diagram to answer questions 14–17.

14. **Interpret Visual** Describe the three experimental setups and the result of each.

15. **Form a Hypothesis** What could be a probable hypothesis for this experiment?

16. **Evaluate** From the experimental setups shown, was the hypothesis successfully tested? Explain.

17. **Evaluate** What kinds of changes would you make to improve this experimental design?

solve the CHAPTER MYSTERY

THE GREEN LEMONS

The mystery of why the lemons didn't ripen was solved by a story from the nineteenth century. In those days, gas streetlights were commonly used in large cities. A few years after such streetlights were installed, people noticed a difference in nearby trees. The trees growing near the streetlights had developed short, thick stems. They also dropped their leaves much earlier than they should have. The hormone level in the trees had changed. This change happened because the streetlights released ethylene.

Remember that ethylene is a hormone that plays a role in the ripening of fruit. The source of the ethylene does not matter. The ethylene made by a plant is the same as the ethylene released by a kerosene heater. Because ethylene is a gas, it can diffuse through the air, cell walls, and membranes of a plant and its fruit. Replacing the kerosene heaters with electric heaters removed the source of ethylene in the warehouse. As a result, the lemons stayed green and did not ripen.

1. **Relate Cause and Effect** Tomatoes that are put in a paper bag with apples ripen much more quickly than those placed in open air. What would this suggest about the effects of ripe apples on unripe tomatoes?

2. **Propose a Solution** How could farmers, shippers, and grocers use the effects of ethylene to their advantage?

Untamed Science™ **Never Stop Exploring Your World.** Finding the solution to The Green Lemons mystery is only the beginning. Take a video field trip with the ecogeeks of Untamed Science to see where this mystery leads.

CHAPTER MYSTERY ANSWERS

1. The ripe apples produce ethylene, which is trapped in the paper bag and causes the tomatoes to ripen faster.

2. They could use ethylene to time the ripening of fruit to better fit their schedules. For example, farmers could pick produce before it is ripe and then shippers could apply ethylene just before the fruit is delivered. In this way, fruit wouldn't ripen and rot before it can be picked or delivered.

Untamed Science™ Students can further their understanding of plant tropisms by watching the short video **Did That Plant Just Move?**

Standardized Test Prep

Multiple Choice

1. Where in a flower are pollen grains produced?
A sepals
B carpels
C anthers
D ovary

2. Which part of the flower develops into a fruit?
A pollen tube
B sepals
C stigma
D ovary

3. Which flower structure includes all the others?
A style
B carpel
C stigma
D ovary

4. The trumpet honeysuckle has long, red, narrow tubular flowers. What is the most likely means of pollination?
A wind
B water
C bee
D hummingbird

5. All of the following are fruits EXCEPT
A tomato.
B corn.
C potato.
D cucumber.

6. Seeds that are contained in large, fleshy fruits are usually dispersed by
A animals.
B water.
C wind.
D rotting.

7. Which of the following causes fruit to ripen?
A auxin
B cytokinin
C ethylene
D gibberellin

8. Which is an example of thigmotropism?
A change in leaf color
B climbing vines
C blooming
D photoperiod

Questions 9–10

The results of an experiment are summarized in the art below.

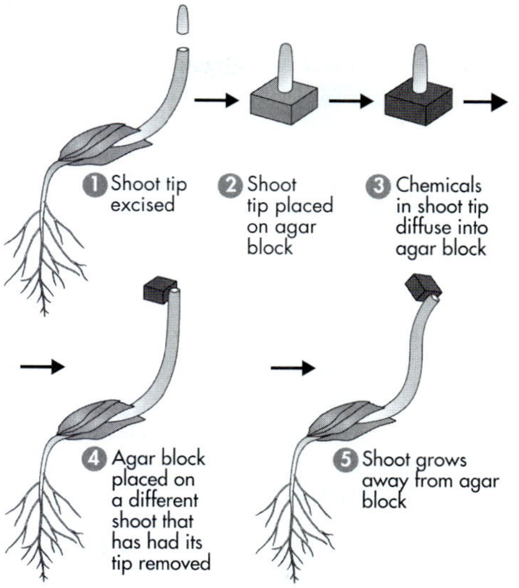

9. Which of the following can be concluded from the results of this experiment alone?
A Hormones are produced in the growing tips of plant roots.
B Plants grow toward the sun due to compounds produced in their stems.
C Agar blocks contain a variety of plant compounds.
D Compounds produced in shoot tips can cause stems to bend.

10. Applying your knowledge of specific plant hormones, explain the results.

Open-Ended Response

11. Describe why seed dormancy is a valuable adaptation that has helped explain the evolutionary success of seed plants.

If You Have Trouble With . . .

Question	1	2	3	4	5	6	7	8	9	10	11
See Lesson	24.1	24.2	24.1	24.1	24.2	24.2	24.3	24.3	24.3	24.3	24.2

Test-Taking Tip

INTERPRET EXPERIMENTAL DATA

Tell students that when answering questions about an experiment, they should read all of the questions and answers choices first. Then, they should read carefully about the experiment and/or examine any accompanying data, looking for the specific information needed to answer the questions.

Before the Test

Review with students the Test-Taking Tip.

Point out the word *EXCEPT* in Question 5.

Point out that Questions 9 and 10 refer to the diagram.

Remind students that an open-ended response means that they'll have to write a paragraph to answer the question. Have them paraphrase the question in their notebooks, and remind them about modeling a paragraph.

Answers

1. C
2. D
3. B
4. D
5. C
6. A
7. C
8. B
9. D
10. Auxins produced in the first shoot tip diffused from the agar block into the second shoot, and caused the cells in contact with the agar to increase in length. Cells on the side of the stem opposite the agar did not increase in length, and the shoot bent accordingly.
11. Dormancy allows seeds to survive winter or long periods of dry climate so that they do not germinate in unfavorable conditions. Seeds are much more likely to survive, germinate, and grow if they "wait" for the right conditions via dormancy.

Plan Ahead

Have students read through the description of the Unit 6 Project a few days before they begin designing their board game to get their creative ideas flowing. Tell them you will provide some materials for them to use, but they may want to bring other materials from home. Display a few popular board games in the classroom to give students ideas about how they might create a good game.

Materials posterboard, colored markers or pens, notecards, small objects used for board tokens, number cubes

Monitor the Project

Emphasize to students that the game they create should be one in which the winner is largely determined by who has the most knowledge, not the most luck. Moving to the finish line should not just depend on a throw of number cubes or some other chance action. Suggest a good way to start is by writing some basic rules of the game, to get a sense of how the game will proceed and how it will be won. After writing basic rules, students can design the game board and write questions to be used during the game. Suggest students "tweak" their rules after a trial run through the game.

Project Assessment

Make sure students use the rubric and reflection questions to assess their work. Then, use the rubric to assign a final score. Note that it is important to value the creativity of students' work as well as the content when you score their projects. If desired, talk with students about any differences between their self-assessment scores and your assigned score.

NATIONAL SCIENCE EDUCATION STANDARDS

UCP I

CONTENT C.3.d, C.3.e

Unit Project

Create a Board Game

Did you have any favorite board games as a kid? Remember how much fun they were to play? Now you get to relive that fun as an employee of an educational toy company! You have been asked to design a board game about the diversity of living things.

Your Task Create a board game based on the groups of living things discussed in this unit—bacteria, viruses, "protists," fungi, and plants. You will exchange games with classmates to test your knowledge and to evaluate the quality of their game.

Be sure to

- include a clever title and colorful, creative board.
- write clear "How to Play" instructions.
- write answerable game questions related to the groups of organisms you've learned about.
- design the game so that winning depends on mastery of the material.

Reflection Questions

1. **Score your game using the rubric below. What score did you give yourself?**
2. **What did you do well in this project?**
3. **What needs improvement?**
4. **What aspects of another group's game did you like? Why?**
5. **After playing the game, what topics did you do well with? What topics do you need to study more?**

Assessment Rubric

Score	Scientific Content	Quality of Game
4	Game includes challenging, but answerable, questions about all of the groups discussed in the unit. Winning the game depends on how well a player knows the material.	Game is very well designed and creative. "How to Play" instructions are clear and easy to follow.
3	Game includes answerable questions on all of the groups discussed in the unit. Winning the game depends on how well a player knows the material.	Game is designed effectively. "How to Play" instructions can be followed.
2	Game includes some answerable questions, but others are unclear or impossible to answer. Winning the game does not necessarily depend on how well a player knows the material.	Game design could use improvement. "How to Play" instructions are difficult to follow.
1	Many questions are vague and/or unanswerable. Winning the game does not depend on how well a player knows the material.	Game design shows little evidence of planning. "How to Play" instructions are impossible to follow.

21st Century Skills

To be successful in the 21st century, students need skills and learning experiences that extend beyond subject area mastery. The Unit 6 Project helps students build the following 21st Century Skills: *Creativity and Intellectual Curiosity; Interpersonal and Collaborative Skills; Self-Direction;* and *Accountability and Adaptability.*

FOCUS ON ACCOUNTABILITY AND ADAPTABILITY Extend this Unit Project by having small groups of students adapt elements of different students' games to a game that could be used by any biology student taking a more comprehensive biology test. Ask students to consider how material they have learned in other units could be incorporated into the game questions and how the board game could be both fun and instructive.

For more practice building 21st Century Skills, see The Chapter Mystery pages in **Study Workbook A.**

602 Unit 6 Project

Animals

UNIT 7

Chapters

INTRODUCE the **Big ideas**

- **Unity and Diversity of Life**
- **Evolution**
- **Structure and Function**

"Studies of animal structure and function show how evolutionary change has produced life's diversity. In each animal group, various 'body-part tool kits' have adapted and changed over time to form structures typical to each group. Those structures perform essential functions, most of which help—directly or indirectly—to maintain homeostasis. The body-part tool kits are shaped by genetics whose shared elements can be traced back over time and throughout the animal kingdom."

Dear Colleague,

Most biologists have a particular group of critters they are especially fond of, and I'm no exception. When I was a kid, I desperately wanted a pet, but I lived in an apartment building that prohibited dogs, cats, and birds. Then, I won a goldfish at a county fair. I named him Oscar. It was love at first sight, and the infatuation stuck. Years later, when professors steered me away from medical school to a marine biological station, fishes were on my mind.

Much of my graduate training focused on vertebrates, but an interest in rainforests and coral reefs also kept me informed about invertebrates—which was good, because of a sobering reality. If we think about "success" in terms of total numbers of species and individuals, invertebrates win hands down over the piddling few of us with backbones. Their diversity of body forms and ways of making a living are fascinating. A number of inverts, in fact, have life cycles bizarre enough to inspire sci-fi/horror films. (Compare the life cycle of the beast in "Alien" to those of certain solitary hymenopterans.) And, the more I learn about these beasts, the more I see the inspiration for Jonathan Swift, who wrote:

> *"So naturalists observe, a flea*
> *Has smaller fleas that on him prey;*
> *And these have smaller still to bite 'em;*
> *And so proceed ad infinitum."*

As fascinating as animals are though, I suspect that many of you look at this unit with mixed feelings. How could you cover all this stuff? The answer depends on your likes and dislikes (and state standards), but here's my take. We've rewritten and reorganized this material to focus on the evolutionary history of major animal groups and on the structures and functions that enable animals to carry out basic life processes and maintain homeostasis. The "march through the phyla" has been moved into our biodiversity appendix, which emphasizes taxonomic groups and their names. Hope you like the new conceptual organization of this unit!

CHAPTER PLANNER 25

Introduction to Animals

Chapter Contents	NSES	Time	Core Resources
Chapter Preview			**Student Edition,** pp. 604–605 **Chapter Mystery,** p. 605 **Foundations for Learning,** p. 605
25.1 What Is an Animal? Characteristics of Animals • Types of Animals • What Animals Do to Survive	I, II, IV, V, A.1.b, C.1.f, C.3.d, C.3.e, C.5.d, C.6.a, C.6.b	1 period 1/2 block	**Student Edition,** pp. 606–610 **Inquiry Into Scientific Thinking,** p. 608 **Build Connections,** p. 609 **Study Workbook B** 25.1 Worksheets **Biology.com** *InterActive Art:* Structure of a Sponge • 25.1 Lesson Assessment
25.2 Animal Body Plans and Evolution Features of Body Plans • The Cladogram of Animals	I, II, IV, V, C.1.f, C.3.a, C.3.b, C.3.d, C.3.e, C.5.d	1 1/2 periods 3/4 block	**Student Edition,** pp. 611–616 **Study Workbook B** 25.2 Worksheets **Biology.com** *Art Review:* Body Symmetry and Cavities • *Art in Motion:* Protostome and Deuterostome Early Development • 25.2 Lesson Assessment
Chapter Pre-Lab	I, V, A.1.b, A.2.a, C.1.f, C.5.d	1 period 1/2 block	**Student Edition,** p. 617 **Lab Manual B** *Comparing Invertebrate Body Plans*

Tools for ELL Instruction

ELL Handbook describes teaching strategies to make *Miller & Levine Biology: Foundation Edition* more accessible to ELL students.

Spanish Study Workbook is a Spanish translation of Study Workbook A.

Multilingual Glossary is the *Miller & Levine Biology* glossary translated into ten languages.

Tools to Challenge Students

Study Workbook A includes more challenging worksheets and activities such as Chapter Mystery activities that reinforce 21st Century Skills.

Lab Manual A labs are geared toward on-level students. The Lab Manual includes fifteen labs in addition to the Chapter Labs.

Biology.com Data Analysis activities offer students a way to practice data gathering and analyzing skills.

Visual Quizzes in the Assessment Resources Book are geared toward on-level students and use visuals from *Miller & Levine Biology.*

Additional Resources

Biology.com Untamed Science Video • Flash Cards

Study Workbook A 25.1 Worksheets
Spanish Study Workbook 25.1 Worksheets

Study Workbook A 25.2 Worksheets
Lab Manual B *Data Analysis:* Differences in Differentiation
Spanish Study Workbook 25.2 Worksheets
Biology.com *Data Analysis:* The Simplest Animal?
Assessment Resources Book Visual Quiz

Lab Manual A *Comparing Invertebrate Body Plans*

* Visual Quizzes are based on visuals from *Miller & Levine Biology.* Review carefully to make sure they are appropriate for your class.

Chapter Review

Student Edition Chapter Summary, p. 618
Study Workbook B Chapter 25 Review
Transparencies, pp. 290–296
Biology.com Untamed Science Video • Editable Worksheets of Study Workbooks A and B and Lab Manuals A and B • Chapter 25 Flash Cards and Match It

Untamed Science DVD • Classroom Resources CD (includes lesson presentations and editable worksheets)

Chapter Assessment

Student Edition Check Understanding, pp. 619–621
Study Workbook B Chapter 25 Taking a Standardized Test
Assessment Resources Book Chapter 25 Test B
Biology.com Chapter 25 Assessment • Editable Worksheets of Chapter 25 Visual Quiz* and Chapter 25 Tests A and B

Exam*View Assessment Suite* • Classroom Resources CD (includes lesson presentations and editable worksheets)

Time: 1 period, 1/2 block

Pressed for Time?

Preview the Chapter Introduce the Key Questions for each lesson and preview the Body Plans table in Lesson 25.2.

Cover the Chapter Quickly Have students read *Characteristics of Animals* and *Types of Animals* in Lesson 25.1. To cover Lesson 25.2, have students read the introduction to *Features of Body Plans,* and review the blue headings that name the eight main features of animal body plans. Go over the visuals Body Symmetry, Body Cavities, and Blastopore Formation, and assign the Segmentation: Repeating Parts and Cephalization: Getting a Head sections.

Assess Assign Questions 1–3 in the 25.1 Check Understanding, and Question 1 in the 25.2 Check Understanding. In the Chapter 25 Check Understanding, assign Questions 1, 3, 5, and 6.

Get the Picture

Tell students that the picture shows many coral reef organisms. Coral reefs are mounds or ridges, made up of living and nonliving things, that form on the ocean floor. Explain that most of the organisms in the picture are animals, even though some might not look like what students think are typical animals.

Ask Which of the organisms shown in the picture are animals? *(corals, anemones, sharks, and other fish)* Have students point to the locations of the various animals as they are named.

Ask Do these animals all have similar body plans? *(no)* What needs do all these animals share? *(Sample answers: They need oxygen and nutrients. They need to eliminate wastes and avoid predators.)*

Ask Do animals living on land have these same needs? *(yes)* Do you think land animals have the same body plans as coral reef animals? *(no)* Why are they different? *(Their bodies are adapted for life on land instead of life in water.)*

Activate Prior Knowledge

Remind students that Earth's animals, like all organisms, have evolved. Evolution has shaped animals' similarities and differences.

Ask Do you think the populations in a coral reef a million years ago looked the same as the populations in a coral reef today? *(no)* Why not? *(The populations have evolved.)*

Ask By what process of evolution have these populations become adapted to their environment? *(natural selection)*

25 Introduction to Animals

Big idea Unity and Diversity of Life

Q: What characteristics and traits define animals?

BIOLOGY.com Search Chapter 25 GO • Flash Cards

NATIONAL SCIENCE EDUCATION STANDARDS

UNIFYING CONCEPTS AND PROCESSES

I, II, IV, V

CONTENT

C.1.f, C.3.a, C.3.b, C.3.d, C.3.e, C.5.d, C.6.a, C.6.b

INQUIRY

A.1.b, A.2.a

UbD Connect to the Big Idea

Have a student read aloud the chapter number and title. Read the Big Idea and question. Emphasize the question, **What characteristics and traits define animals?**

Set the Purpose

PREVIEW AND PREDICT Draw the graphic organizer. Guide students to the INSIDE feature on the next page, and have a student read lesson numbers and titles.

Say Animals are alike in some ways and different in others.

Ask What do you think the words *unity* and *diversity* mean? (Unity *refers to how animals are alike.* Diversity *refers to how animals are different.)*

Ask Can you predict the answers to the questions in the graphic organizer?

Remind student to search for answers to these questions as they read.

CHAPTER 25

INSIDE:

- 25.1 What Is an Animal?
- 25.2 Animal Body Plans and Evolution

Though they look very different, the hundreds of animal species that make up or live near a coral reef share characteristics common to all animals.

- Untamed Science Video
- Chapter Mystery

CHAPTER MYSTERY

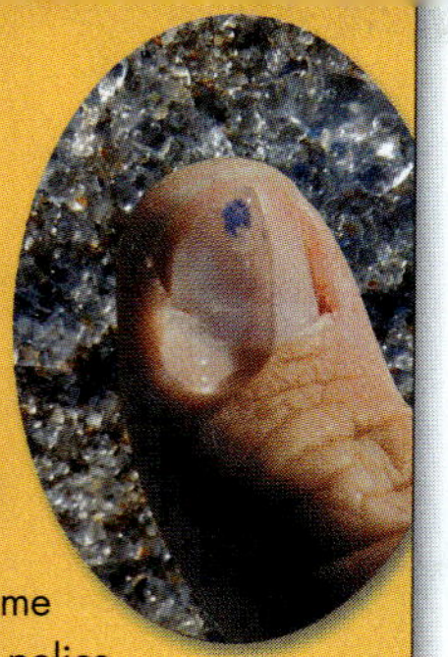

SLIME DAY AT THE BEACH

It was a warm October day in Boston when people started calling beach offices, aquariums, and even 9-1-1. Beaches near Boston were coated with a layer of jellylike ooze. People were confused and worried. Some thought there had been an oil spill. However, police and firefighters found that the ooze was not oil.

More slimy masses kept washing up on the beach. When people looked more closely, they saw that the slime was made of small, individual creatures. Each creature was transparent and the size of a fingernail. What were they?

Read for Mystery Clues As you read this chapter, look for clues to help you find out what the slime was.

FOUNDATIONS for Learning

You can use a layered book to review the main ideas in this chapter. Make a layered book by stapling pieces of paper together. Your book should have visible tabs, as in the diagram below. Write the name of the first lesson on the front page of your book. Write the main topics from the lesson on the tabs that are showing. Write notes for each main topic on the part of each paper that is not showing. Use both words and pictures in your notes.

Make another layered book for Lesson 2. Make a third layered book using the key questions in the chapter, but do not write notes in this third book yet. At the end of the chapter are two activities that use your layered books to help answer the question: What characteristics and traits define animals?

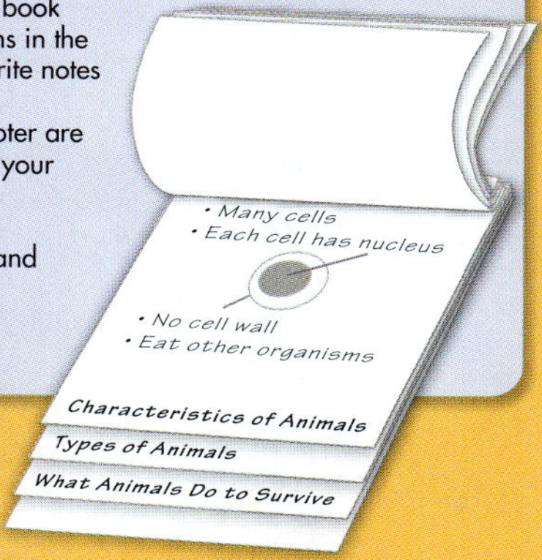

Chapter 25
Big Idea: Unity and Diversity of Life

Chapter 25: What characteristics and traits define animals?

- **25.1:** What is an animal?
- **25.2:** How have different body plans evolved?

Foundations for Learning

Using sheets of paper and a stapler, demonstrate how to assemble a layered book. Have students read aloud the titles of the lessons and main topics that will label the books' front pages and tabs. Model how to label and use the books.

Ask How might a layered book help you remember and review the lessons in this chapter? *(The labeled tabs will organize the ideas and make them available quickly.)*

Ask Why is it a good idea to include both words and pictures to review the main topics? *(The pictures can clarify ideas that might be confusing, and the words can focus on details that are complicated or difficult to remember.)*

Untamed Science Travel to Coconut Island with the Untamed Science crew to find out just what, exactly, an animal is.

CHAPTER MYSTERY Invite a volunteer to read aloud the Chapter Mystery. Have students predict how this mystery relates to the chapter title. Guide students to discuss how the chapter question, **What characteristics and traits define animals?** will help them understand more about the mystery slime creatures. Remind students that they will find clues in the chapter to help them identify these and other organisms.

Digital Support

BIOLOGY.com Extend your reach by using the digital assets offered at **Biology.com.**

LESSON 25.1

Pre-Reading

Objectives

25.1.1 List the characteristics that all animals share.

25.1.2 Differentiate between invertebrates and chordates

25.1.3 List and discuss the essential functions that animals perform in order to survive.

Preview the Pages

1. Preview the vocabulary words by having students find the words in the text and read each aloud. Ask students to consider how each word relates to differences or similarities between animals.
2. Point out in the text where invertebrates and chordates are compared. Direct students to use their layered books and the Venn diagram in Workbook B to help them compare the groups.
3. Preview the visuals. Ask volunteers to read aloud the captions and labels. Point out that the diagram Characteristics of Chordates does not show any particular animal. Rather, it is a generalized view to show the characteristics of this large group of animals.

Activate Prior Knowledge

Remind students that animals obtain energy as part of the flow of energy in ecosystems.

Say Animals are heterotrophs. This means they get their energy by eating other organisms. Autotrophs, such as plants, can produce their own food.

Ask What is the source of a plant's energy? *(the sun)*

NATIONAL SCIENCE EDUCATION STANDARDS

UNIFYING CONCEPTS AND PROCESSES

I, II, IV, V

CONTENT

C.1.f, C.3.d, C.3.e, C.5.d, C.6.a, C.6.b

INQUIRY

A.1.b

For the corresponding lesson in *Miller & Levine Biology,* see pages 730–735.

25.1 What Is an Animal?

Key Questions

What characteristics do all animals share?

What characteristics distinguish invertebrates and chordates?

What essential functions must animals perform to survive?

BUILD Understanding

Venn Diagram Use a Venn diagram to show the similarities and differences between invertebrates and chordates.

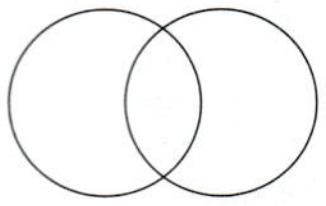

In Your Workbook Go to your workbook to learn more about making a Venn diagram. Complete the Venn diagram started for you.

For more on the diversity of animals, go to the Visual Guide. **pp. DOL 30–DOL 64**

Characteristics of Animals

A bald eagle circles a salt marsh looking for prey. Suddenly, it dives and captures a fish. At the bottom of the marsh, worms burrow beneath rocks. In the air above the marsh, mosquitoes buzz. The eagle, fish, worms, and mosquitoes are all animals. All members of the animal kingdom share certain characteristics. Animals are heterotrophs, organisms that get energy by eating other organisms. Animals are multicellular, or have bodies composed of many cells. Animal cells are eukaryotic—each cell has a nucleus. Animal cells do not have cell walls.

Key Question What characteristics do all animals share? **Animals are multicellular heterotrophs that get energy by eating other organisms. Animals have eukaryotic cells that do not have cell walls.**

Types of Animals

The animal kingdom is a large and diverse group. Animals can be classified into two main groups: invertebrates and chordates.

Invertebrates More than 95 percent of all animals can be lumped into a catch-all grouping called **invertebrates.** That term is used to describe animals that do not have a backbone. Invertebrates include sea stars, jellyfishes, worms, snails, and insects. They range in size from dust mites to giant squids that are more than 14 meters long. "Invertebrates" are not a proper clade according to evolutionary classification. Scientists classify invertebrates into 33 different phyla.

Chordates Fewer than 5 percent of animal species are **chordates** (KAWR dayts). Chordates are members of the phylum Chordata. All chordates have four characteristics that are present during at least one stage of life. Chordates have a hollow nerve cord that runs along the back (or dorsal) part of the body. They also have a **notochord** that is located below the nerve cord. A notochord is a long supporting rod that runs through the body just below the nerve cord. Most chordates only have a notochord when they are embryos.

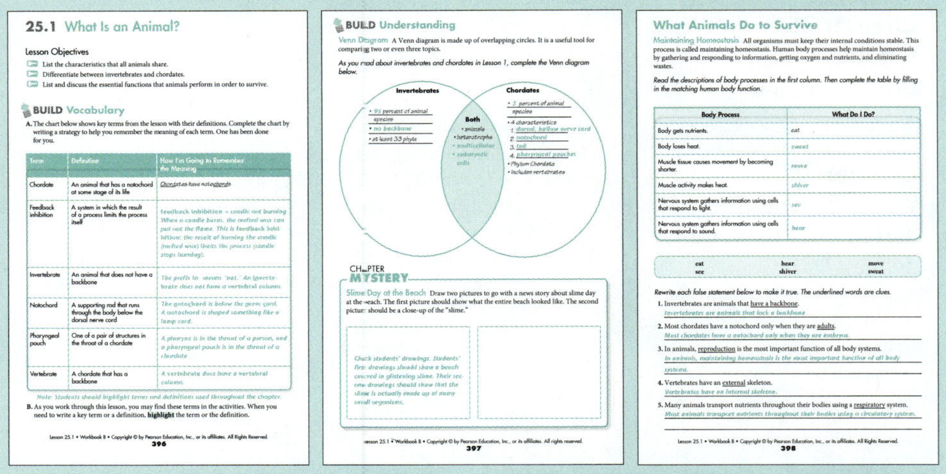

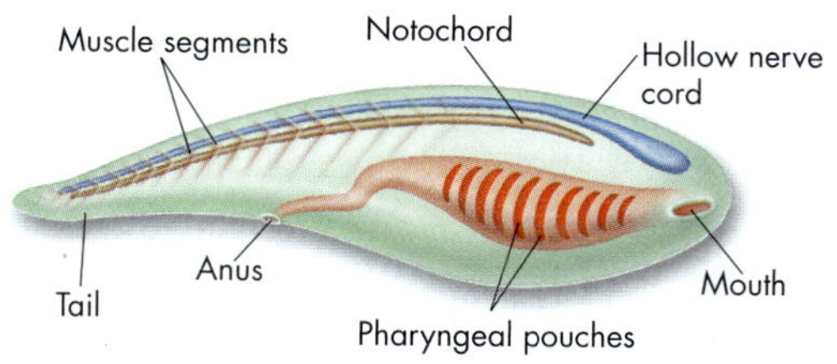

Characteristics of Chordates All chordates have a hollow nerve cord, a notochord, pharyngeal pouches, and a tail that extends past the anus. Some chordates have all these traits as adults. Other chordates only have some of these traits as embryos.

All chordates have a tail that extends past the anus. (The anus is the opening where wastes leave the body.) The tail disappears in some chordates as the embryo develops. All chordates also have **pharyngeal** (fuh RIN jee ul) **pouches.** Pharyngeal pouches are paired structures in the throat region. In some chordates, such as fishes, slits develop that connect pharyngeal pouches to the outside of the body. Pharyngeal pouches may develop into gills. Gills are structures that are used for gas exchange.

Phylum Chordata includes some odd aquatic animals known as nonvertebrate chordates. These chordates do not have a backbone. Most chordates develop a backbone, or vertebral column. Chordates with backbones are called **vertebrates.** Vertebrates include fishes, amphibians, reptiles, birds, and mammals.

Key Question What characteristics distinguish invertebrates and chordates?
Invertebrates lack a backbone. Chordates have all four of the following during at least one stage of life: a dorsal, hollow nerve cord; a notochord; a tail that extends past the anus; and pharyngeal pouches.

What Animals Do to Survive

Animals come in an amazing variety of body shapes, sizes, and colors. No matter what they look like, all animals must perform certain basic functions to stay alive. In addition, all animals must reproduce. Like all organisms, animals must keep their internal environments stable. They do this in three ways:

1. Animals gather and respond to information.
2. They take in and distribute oxygen and nutrients to cells.
3. They collect and get rid of carbon dioxide and other wastes.

Over time, members of different animal phyla have evolved very different body structures to perform the functions that keep them alive. You will study these structures in more detail in Chapters 27 and 28.

BUILD Vocabulary

invertebrate
an animal that lacks a backbone, or vertebral column

chordate
an animal that has, for at least one stage of its life, a dorsal, hollow nerve cord; a notochord; a tail that extends beyond the anus; and pharyngeal pouches

notochord
a long supporting rod that runs through a chordate's body just below the nerve cord

pharyngeal pouch
one of a pair of structures in the throat region of a chordate

vertebrate
an animal that has a backbone

PREFIXES
The prefix *in-* means "not." An invertebrate does *not* have a backbone. A vertebrate does have a backbone.

Build Understanding

VENN DIAGRAM Direct students to Workbook B, where they will find a partially completed Venn diagram that they should fill in as they read this lesson. Then pair students and have them quiz each other on the differences and similarities between chordates and invertebrates.

During Reading

Active Reading

KEY QUESTION Read aloud the Key Question on the previous page and its answer.

Ask Can animals carry out photosynthesis? *(no)* Do their cells have nuclei? *(yes)*

FIND THE MAIN IDEA Have students reread the section Types of Animals.

Ask What is this section about? *(the differences between invertebrates and chordates)*

Ask volunteers to list on the board some of the details included in the section.

KEY QUESTION Have a volunteer read aloud the Key Question on this page and its answer.

Ask Are humans chordates or invertebrates? *(chordates)* What structure can you feel on your own body that is a feature of most chordates but not invertebrates? *(backbone)*

USE DIAGRAMS Refer to each labeled structure in the Characteristics of Chordates diagram as students read about it in the text. Refer to the tail.

Ask If we're chordates, why don't we have a tail? *(A human has this trait only as an embryo.)*

Build Vocabulary

PREFIXES Review with students that the prefix *-in* in *invertebrate* means "not."

Ask What are some other words that begin with the prefix *in-*, meaning "not"? *(Sample answers: insensitive, independent, incomplete)* What other prefixes also mean "not"? *(non-, im-, un-)* What are some words that begin with these prefixes? *(Sample answers: nonfatal, imperfect, unnecessary)*

Speed Bump

Ask What do animals need to do in order to survive? *(gather and respond to information, breathe, take in nutrients, eliminate wastes, and reproduce)*

Ask How is reproduction important to survival? *(In order for the population to survive, many of the individuals need to reproduce.)*

Hands-On Learning

ANIMAL SURVIVAL Pair students and hand out these materials: six index cards, pen with blue ink, pen with red ink.

1. Have students read the text under What Animals Do to Survive.
2. Instruct each pair to write on separate cards three true statements and three false statements about animals and how they survive. True statements should be in blue and false statements in red. *(Sample true statement: Animals must keep their internal environments stable. Sample false statement: All animals are large.)*
3. Have pairs exchange cards and evaluate the statements for accuracy.
4. On the board, write *Animals and how they survive.* Under that, write *True* and *False*.
5. Have pairs read aloud the exchanged cards and identify them as true or false. Write each statement under the correct heading on the board, allowing time for class discussion.

LESSON 25.1

During Reading continued

Active Reading

MAKE CONNECTIONS Explain to students that maintaining homeostasis requires energy.

Say If your home gets too cold at night, a thermostat turns on the heat.

Ask What is needed to power the heating system? *(energy)* What is the source of that energy? *(Sample answers: electricity, natural gas, heating oil)*

Ask In your body, what powers your ability to sweat when you are hot or shiver when you are cold? *(energy)* What is the source of that energy? *(food)*

Build Vocabulary

WORD ORIGINS Discuss other forms of the word *inhibition*. Tell students that these words also mean "hold back."

Ask If lack of money inhibits you from buying gas for your car, what does that mean? *(You are held back from buying gas because you don't have enough money.)*

Ask What does it mean when you are inhibited from answering a teacher's question aloud because you don't know the answer? *(You are held back from speaking aloud.)*

Inquiry Into Scientific Thinking

MAKE INFERENCES Remind students that energy moves from one organism to another in a food chain.

Ask How does the hydra obtain energy? *(by eating other organisms, such as* Daphnia*)*

Ask If hydra are predators and *Daphnia* are prey, how do you think their bodies are adapted differently? *(Hydra are adapted to catch prey.* Daphnia *are adapted to escape predators.)*

Lead a Discussion

Discuss with students how the human body maintains homeostasis.

Ask How does a person's feeling of hunger help his or her body maintain homeostasis? *(Feeling hungry tells a person to eat food, which helps ensure that the body gets the nutrients it needs.)*

BUILD Vocabulary

feedback inhibition
a process in which a stimulus produces a response that opposes the original stimulus; also called negative feedback

WORD ORIGINS
The word *inhibition* comes from the Latin word *inhibitionem*, which means "a restraining." To restrain something means to hold it back.

Maintaining Homeostasis All organisms must keep their internal environments stable. This process is known as maintaining homeostasis. Maintaining homeostasis is the most important function of all animal body systems.

Homeostasis is often maintained by feedback inhibition. **Feedback inhibition,** or negative feedback, is a system in which the result of a process helps to limit the process itself. The heating system in a house uses feedback inhibition. If the rooms are too cold, the thermostat turns on the heat. When the rooms get warm enough, the thermostat turns off the heat. Your body's thermostat works in a similar way. If you get too cold, you shiver. Shivering causes your muscles to generate heat. If you get too hot, you sweat. Sweating helps you to lose heat.

Gathering and Responding to Information Animals use many body systems to gather and respond to information in their environment. In many animals, the nervous system gathers information. Specialized cells called receptors sense light, sound, chemicals, and touch. Information from the receptors travels to other nerve cells, which determine how to respond. Some invertebrates have a loose network of nerve cells with no real center. Other invertebrates and most chordates have large numbers of nerve cells concentrated in a brain.

INQUIRY into Scientific Thinking — GUIDED INQUIRY

How Hydra Feed

The hydra is a small aquatic invertebrate that is related to jellyfish and sea anemones. The hydra has a tubelike body with long, thin tentacles at the top. *Daphnia* are also small aquatic invertebrates. They are related to shrimp and crabs. Each *Daphnia* has a round body and an eyespot. In this lab, you will investigate how hydra feed on *Daphnia*.

1. Your teacher will provide you with hydra and *Daphnia*. Using a dropper pipette, gently place one hydra onto a well slide.
2. Let the hydra adjust to its surroundings for 5 to 10 minutes.
3. Using your dropper, add one *Daphnia* to the slide.
4. Observe the hydra under the microscope.

Analyze and Conclude

1. **Observe** What happens when the *Daphnia* is added to the same slide as the hydra?
2. **Draw Conclusions** How do the hydra's tentacles help it to maintain homeostasis?
3. **Quick Write** What else would you like to learn about how the hydra survives in its environment? Write down two new questions.

In Your Workbook Get more help for this activity in your workbook.

Materials

live hydra and *Daphnia* cultures, microscope, dropper pipette, depression slides

Analyze and Conclude

1. Sample answer: The hydra's tentacles begin to move toward the *Daphnia*. The *Daphnia* tries to swim away.
2. The hydra's tentacles allow it to capture food.
3. Sample answer: How does a hydra detect animals that it can eat? Does a hydra eat plants as well as animals?

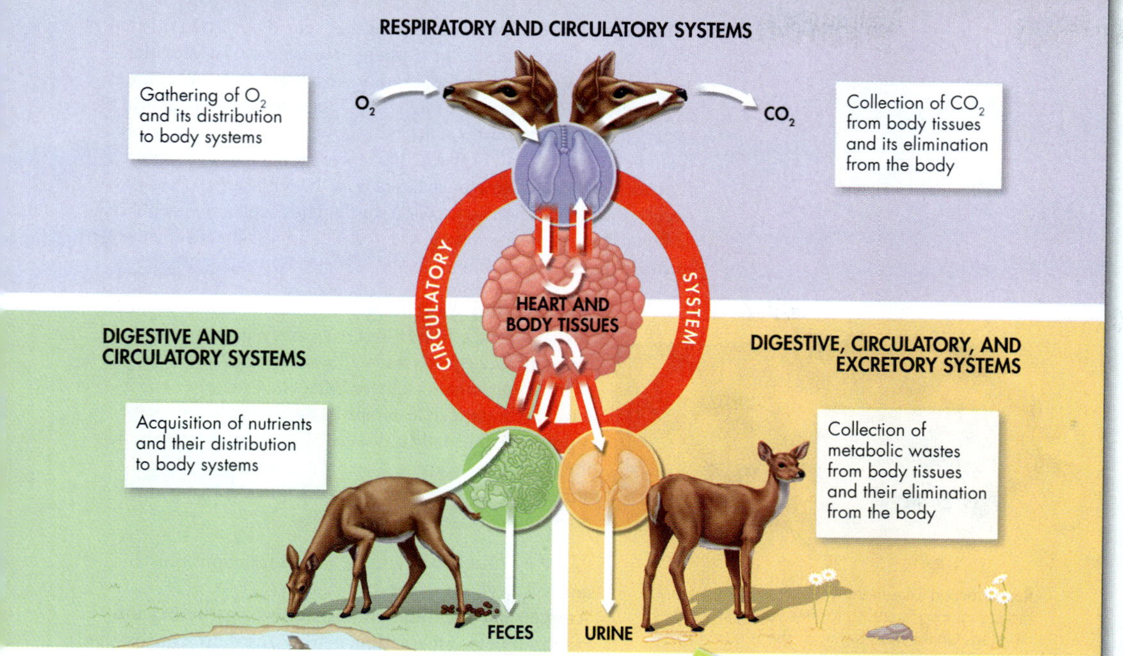

Animals often respond to the information they have processed by moving around. The nervous system causes muscles to become shorter. When muscles become shorter, they can move the animal's supporting structure, called a skeleton. Skeletons vary from phylum to phylum. Earthworms have fluid-filled, flexible skeletons. Insects and other invertebrates have skeletons outside their bodies. Vertebrates have skeletons inside their bodies that are made of bones.

Getting and Delivering Oxygen and Nutrients All animals must breathe to obtain oxygen. Small animals that live in wet places can "breathe" by allowing oxygen to diffuse across their skin. Larger animals use gills, lungs, or air passages to breathe.

All animals must eat to get nutrients. Many animals have digestive systems that break down food into nutrients. Then the body must deliver the oxygen and nutrients to its cells. Many animals use a circulatory system to carry oxygen and nutrients around the body. The circulatory system works together with the respiratory and digestive systems to keep an animal alive.

Collecting and Removing CO_2 and Other Wastes The activity of animal cells produces waste products. Some of these waste products contain nitrogen, often in the form of ammonia. Both carbon dioxide and ammonia are poisonous. In high concentrations, these substances can destroy cells. Therefore, the body must get rid of these wastes.

BUILD Connections

MOVING MATERIALS IN, AROUND, AND OUT OF THE BODY

The structures of an animal's respiratory, digestive, and excretory systems must work together with the structures of the animal's circulatory system.

ELL Focus on ELL: Access Content

ALL SPEAKERS Divide students into pairs. Have each pair make a three-column table with the headings from the diagram on this page: Respiratory and Circulatory Systems; Digestive and Circulatory Systems; and Digestive, Circulatory, and Excretory Systems. Write on the board the following words from the diagram: *collection, elimination, distribution,* and *acquisition*. Have students write the words in their tables under the appropriate headings as shown in the diagram. Tell them to look up the definition of each word in a dictionary, and write the definitions in their notebooks. Invite volunteers to define each word aloud.

Ask In the diagram, how are these four words used to describe the movement of materials into, within, and out of an animal's body? (Collection *refers to wastes being brought together before being moved out of the body;* elimination *refers to wastes being moved out of the body;* distribution *refers to oxygen and nutrients being moved to different parts of the body; and* acquisition *refers to taking in nutrients by eating food.)*

Build Connections

VISUAL SUMMARY Explain to students that this diagram shows how an animal's body systems work together to provide the animal's needs. Have volunteers read aloud the descriptions of each system and activity.

Say As you study the diagram, trace the arrows with your fingers. This will help you follow the movement of materials in the diagram.

Ask Which body systems are pictured? *(digestive, excretory, circulatory, and respiratory)* How are different colors used in the diagram? *(Sample answer: Different colors are used to highlight different processes.)* What do the arrows indicate? *(the direction in which materials move)*

Ask What do the arrows near the animal's head show? *(movement of oxygen into the lungs and carbon dioxide out of the lungs)*

Ask What body system carries oxygen from the lungs to body tissues, and carbon dioxide from body tissue to the lungs? *(circulatory system)*

Ask In what colored part of the diagram are nutrients entering the animal's digestive system and being carried by the circulatory system to the body's tissues? *(green part)* In what form do wastes from the digestive system leave the body? *(feces)*

Ask In what colored part of the diagram do metabolic wastes move into the excretory system? *(yellow part)* What system carries the wastes to the excretory system? *(circulatory system)* In what substance do these wastes leave the body? *(urine)*

Ask Which body system in the diagram works with all of the other three body systems? *(circulatory system)*

Speed Bump

Ask What does feedback inhibition help maintain in an animal? *(homeostasis)*

Ask Do all animals get oxygen directly from the air? *(no)* How else might animals get oxygen? *(Fish get oxygen from water as it flows across their gills.)*

During Reading continued

Active Reading

SUMMARIZE After students read the paragraph Reproducing, have them write in their notebooks two sentences that describe the advantage of each type of reproduction. *(Sample sentences: Sexual reproduction helps produce and maintain genetic diversity. Asexual reproduction allows animals to increase their numbers quickly.)*

KEY QUESTION Read aloud the Key Question and answer.

Ask What is the most important function of all animal body systems? *(maintaining homeostasis)*

Speed Bump

Ask Why is it that offspring produced through asexual reproduction do not increase that species' genetic diversity? *(The offspring are genetically identical to the parents.)*

Post Reading

ASSESSMENT TIPS For Questions 3 and 5, encourage students to refer to the Venn diagram they constructed at the beginning of the lesson.

MYSTERY CLUE Have students reread the section comparing chordates and invertebrates. Lead them to conclude that invertebrates would not have a stiff rod running along a tail. Therefore, the organisms must be a type of chordate.

Wrap-Up Activity

PLAY VOCABULARY CHARADES Write each vocabulary word from the lesson on a separate index card. Pair students and give each pair one or two index cards. Instruct students to work as partners to review the terms' definitions from the lesson and then to develop definitions or clues they can act out. Have class members guess the terms as pairs act out their definitions or clues.

Many animals use respiratory systems that get rid of carbon dioxide. Most animals have an excretory system that gets rid of wastes like ammonia. An excretory system is an organ system that concentrates or processes nitrogen-containing wastes. Some animals can get rid of nitrogen-containing wastes right away. Other animals have to store them before getting rid of them. Before waste products can be removed from the body, they have to be collected from the cells. Some sort of circulatory system is needed to collect wastes. The circulatory system brings carbon dioxide to the respiratory system. It also brings nitrogen-containing wastes to the excretory system.

Reproduction Like many vertebrates, this pygmy marsupial frog is caring for her young while they develop. Unlike most animals, she is carrying her eggs on her back!

Reproducing Most animals reproduce sexually. Sexual reproduction helps create and maintain genetic diversity. Genetic diversity allows a species to evolve as the environment changes. Many invertebrates and a few vertebrates can also reproduce asexually. Asexual reproduction usually produces offspring that are exactly the same as the parent. Asexual reproduction allows animals to increase their numbers quickly. However, it does not increase genetic diversity.

Key Question What essential functions must animals perform to survive?

Animals must maintain homeostasis by gathering and responding to information, getting and delivering oxygen and nutrients, and collecting and removing carbon dioxide and other wastes. They must also reproduce.

CHECK Understanding

Apply Vocabulary

Use the highlighted words from the lesson to complete each sentence correctly.

1. All _______ have a hollow nerve cord during at least one stage of life.
2. Fishes, amphibians, reptiles, birds, and mammals are all chordates that are also _______.

Critical Thinking

3. **Compare and Contrast** What is the defining characteristic of invertebrates? What are four characteristics of chordates?
4. **Relate** Name two waste products that are produced by an animal's cells. Why must these waste products be eliminated from an animal's body?
5. **Write to Learn** Answer the first clue of the mystery. Think about the characteristics of invertebrates and chordates. Which type of animal would have a stiff rod running along its tail?

MYSTERY CLUE

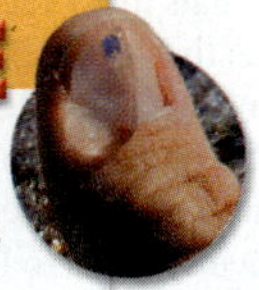

Scientists verified that the organisms were young animals that had a stiff rod running along the tail. What does this suggest about the slimy critters? (Hint: See p. 607.)

BIOLOGY.com Search Lesson 25.1 GO • Lesson Assessment

Check Understanding

1. chordates
2. vertebrates
3. Invertebrates lack a backbone. All vertebrates have a hollow nerve cord down their back, a notochord, a tail, and pharyngeal pouches during at least one stage of life.
4. Sample answer: Animal cells must get rid of carbon dioxide and ammonia. If organisms were unable to get rid of these wastes, the wastes would poison the body.
5. An animal with a stiff rod running along its tail must be a chordate.

25.2 Animal Body Plans and Evolution

Features of Body Plans

Why does a worm look so different from a fish? Why do butterflies have wings while dolphins have flippers? Body structures in animals that are alive today were shaped by millions of years of evolution. Each animal phylum has its own combination of body structures called a body plan. The eight main features of animal body plans are described below.

Levels of Organization As the first cells of most animals develop, they differentiate (or change) into specialized cells. These specialized cells are organized into tissues. Tissues combine during development to form organs. Organs work together to make up organ systems. Organ systems carry out complex functions for the body.

Body Symmetry The bodies of most animals show some kind of symmetry. Some animals have body parts that extend outward from the center. This type of symmetry is called **radial symmetry.** Radially symmetric animals can be sliced, like a pie, by imaginary planes beginning at the center. Some radially symmetric animals can be divided into an even number of "slices." Others, like sea stars, contain an odd number of radially symmetric parts.

Many animals have **bilateral symmetry.** In bilateral symmetry, a single imaginary plane divides the body into left and right sides that look the same. Animals with bilateral symmetry have a front (anterior) end and a back (posterior) end. Animals with bilateral symmetry also have an upper (dorsal) side and a lower (ventral) side. Your back is on your dorsal side. Your belly is on your ventral side.

Animals with radial symmetry have body parts that extend out from a central point. Animals with bilateral symmetry have distinct front (anterior) and back (posterior) ends and have right and left sides.

Key Questions

What are some features of animal body plans?

How are animal phyla defined?

BUILD Understanding

Concept Map You can use concept maps to show how ideas or concepts connect. As you read the lesson, draw a concept map showing the different features of animal body plans and the different types of each feature.

In Your Workbook Refer to your workbook to review the features of different body plans.

Radial Symmetry

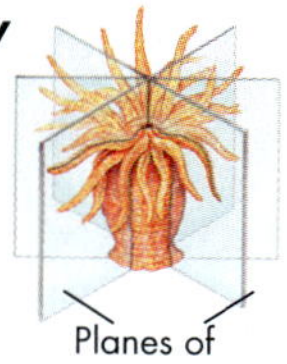

Bilateral Symmetry

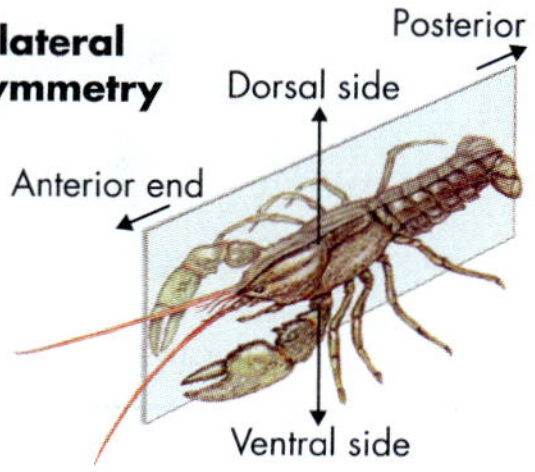

Body Symmetry Animals with radial symmetry have body parts that come out from a central point. Animals with bilateral symmetry have distinct front and back ends and left and right sides.

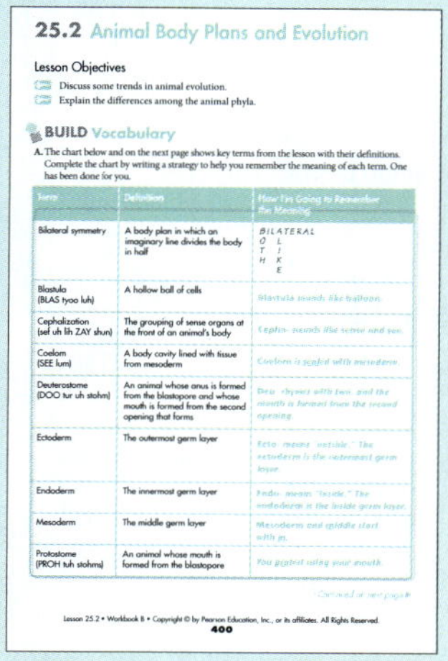

25.2 Animal Body Plans and Evolution

Lesson Objectives

Discuss some trends in animal evolution.

Explain the differences among the animal phyla.

BUILD Vocabulary

A. The chart below and on the next page shows key terms from the lesson with their definitions. Complete the chart by writing a strategy to help you remember the meaning of each term. One has been done for you.

Term	Definition	How I'm Going to Remember the Meaning
Bilateral symmetry	A body plan in which an imaginary line divides the body in half	BILATERAL / BOTH / LIKE
Blastula (BLAS tyoo luh)	A hollow ball of cells	Blastula sounds like balloon.
Cephalization (sef uh lih ZAY shun)	The grouping of sense organs at the front of an animal's body	Cepha- sounds like sense and see.
Coelom (SEE lum)	A body cavity lined with tissue from mesoderm	Coelom is sealed with mesoderm.
Deuterostome (DOO tur uh stohm)	An animal whose anus is formed from the blastopore and whose mouth is formed from the second opening that forms	Deu- rhymes with two, and the mouth is formed from the second opening.
Ectoderm	The outermost germ layer	Ecto- means "outside." The ectoderm is the outermost germ layer.
Endoderm	The innermost germ layer	Endo- means "inside." The endoderm is the inside germ layer.
Mesoderm	The middle germ layer	Mesoderm and middle start with m.
Protostome (PROH tuh stohms)	An animal whose mouth is formed from the blastopore	

Continued on next page ▶

Lesson 25.2 • Workbook B • Copyright © by Pearson Education, Inc., or its affiliates. All Rights Reserved.
400

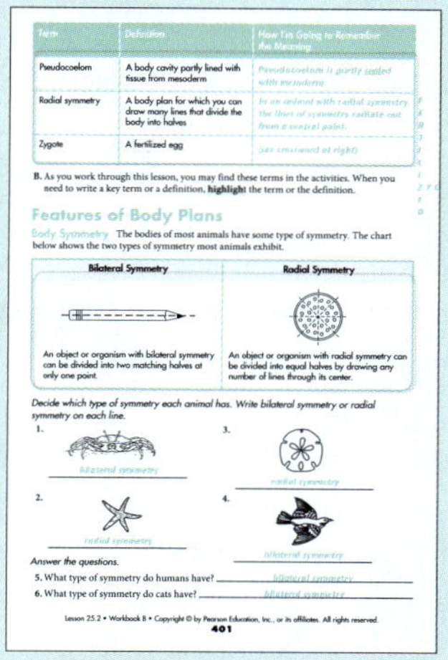

Term	Definition	How I'm Going to Remember the Meaning
Pseudocoelom	A body cavity partly lined with tissue from mesoderm	Pseudocoelom is partly sealed with mesoderm.
Radial symmetry	A body plan for which you can draw many lines that divide the body into halves	In an animal with radial symmetry the lines of symmetry radiate out from a central point.
Zygote	A fertilized egg	(see crossword at right)

B. As you work through this lesson, you may find these terms in the activities. When you need to write a key term or a definition, **highlight** the term or the definition.

Features of Body Plans

Body Symmetry The bodies of most animals have some type of symmetry. The chart below shows the two types of symmetry most animals exhibit.

Bilateral Symmetry	Radial Symmetry
An object or organism with bilateral symmetry can be divided into two matching halves at only one point.	An object or organism with radial symmetry can be divided into equal halves by drawing any number of lines through its center.

Decide which type of symmetry each animal has. Write bilateral symmetry or radial symmetry on each line.

1. bilateral symmetry
2. radial symmetry
3. radial symmetry
4. bilateral symmetry

Answer the questions.

5. What type of symmetry do humans have? bilateral symmetry
6. What type of symmetry do cats have? bilateral symmetry

Lesson 25.2 • Workbook B • Copyright © by Pearson Education, Inc., or its affiliates. All rights reserved.
401

LESSON 25.2

Pre-Reading

Objectives

25.2.1 Discuss some trends in animal evolution.

25.2.2 Explain the differences among the animal phyla.

Preview the Pages

1. Preview with students all the vocabulary, pronouncing each word aloud.
2. Preview the visuals. Point out the color-coding for the different germ layers shown in the Body Cavities, Blastopore Formation, and Body Plans visuals.
3. Point out that the headings in the first column of the Body Plans chart are similar to the blue headings in the lesson.

Activate Prior Knowledge

Remind students that in the previous lesson, they learned about two groups of animals: invertebrates and chordates. Have students recall the features that identify an animal as an invertebrate or a chordate. Explain that this lesson discusses the body features that scientists use to classify invertebrates and chordates into smaller groups.

Build Understanding

CONCEPT MAP Help students create a concept map for this lesson. Draw an oval on the board and write *Animal Body Plans* inside the oval. Then draw seven surrounding ovals and connect them to the central oval. Have students complete the map as described in the text. Students can go to Workbook B for a review activity on features of body plans.

NATIONAL SCIENCE EDUCATION STANDARDS

UNIFYING CONCEPTS AND PROCESSES

I, II, IV, V

CONTENT

C.1.f, C.3.a, C.3.b, C.3.d, C.3.e, C.5.d

For the corresponding lesson in *Miller & Levine Biology,* see pages 737–743.

During Reading

Active Reading

SELF-QUESTION Demonstrate how asking questions and reviewing the text for answers can help clarify new and complex information. Read aloud the paragraph Differentiation of Germ Layers.

Ask Does this make sense? When do germ layers appear in an animal? *(early in the development of the embryo)*

Say Germ layers are difficult to picture, but the paragraph describes their roles in development.

Ask What do the sentences tell me about the connection between germ layers and animal body systems? *(All body systems form from germ layers.)*

SCIENCE SUPPORT Explain that a zygote divides very rapidly early in animal development as it forms a blastula, but the zygote does not increase in size as it divides. Patterns of these cell divisions vary but are the same for all organisms within a species. Once the blastula is formed, its different regions respond to cell signals as development continues. Cells actually move from the outside of the blastula to the inside, through the blastopore, the opening. These cells then change to become the three germ layers: endoderm, mesoderm, and ectoderm.

Build Vocabulary

PREFIXES Read aloud the meanings of the prefixes *endo-* (within), *meso-* (middle), and *ecto-* (outside).

Say Parasites are organisms that survive by living on or in another organism.

Ask Do you think a parasite that lives in another organism is an endoparasite or an ectoparasite? *(endoparasite)* Is a parasite living on the skin of another organism an ectoparasite or an endoparasite? *(ectoparasite)*

Speed Bump

Ask What are the three types of germ layers? *(endoderm, mesoderm, ectoderm)*

Ask What do germ layers become? *(different tissues and organs of the body systems)*

Ask What is a coelom? *(a body cavity lined with mesoderm)*

Ask What is the name for invertebrates that do not have any body cavity surrounding their organs? *(acoelomates)*

BUILD Vocabulary

radial symmetry
a body plan in which body parts repeat around the center of the body

bilateral symmetry
a body plan in which a single imaginary line can divide the body into left and right sides that are mirror images of each other

endoderm
the innermost germ layer; develops into the linings of the digestive tract and much of the respiratory system

mesoderm
the middle germ layer; develops into muscles and much of the circulatory, reproductive, and excretory systems

ectoderm
the outermost germ layer; produces sense organs, nerves, and the outer layer of skin

coelom
a body cavity lined with mesoderm

pseudocoelom
a body cavity that is only partially lined with mesoderm

PREFIXES
You can use the prefixes for each word that describes a germ layer to remember where the germ layer is located in the embryo. The prefix *endo-* means "within." The prefix *meso-* means "middle." The prefix *ecto-* means "outside."

Differentiation of Germ Layers As fertilized eggs develop, the cells of most animals differentiate, or change, into three layers called germ layers. The **endoderm** is the inside germ layer. Its cells develop into the linings of the digestive system and most of the respiratory system. The **mesoderm** is the middle germ layer. Its cells develop into muscles and most of the circulatory, reproductive, and excretory systems. The **ectoderm** is the outer germ layer. Its cells develop into organs of the nervous system and the skin.

Formation of Body Cavity Most animals have some kind of body cavity. A body cavity is a space that is filled with fluid. A body cavity holds the organs and gives them space to grow. Your stomach and intestines are located in your body cavity. Many animals have a true **coelom** (SEE lum). A coelom is a body cavity that is completely lined with tissue that develops from the mesoderm. Some invertebrates have a **pseudocoelom** (soo doh SEE lum), which is only partly lined with tissue from the mesoderm. Other invertebrates, called acoelomates, do not have a body cavity at all.

Patterns of Embryo Development Every animal that reproduces sexually begins life as a fertilized egg, or **zygote** (ZY goht). As the zygote begins to develop, it develops into an embryo. As the embryo develops, it forms a hollow ball of cells called a **blastula** (BLAS tyoo luh). The blastula looks like an inflated balloon. As the blastula continues to develop, it folds in on itself. Imagine holding a partly inflated balloon in your hand and pushing your thumbs toward the center. This folding inward forms a tube down the middle of the blastula. This tube becomes a digestive tract. In a digestive tract, food enters through one opening, called the mouth. Wastes leave through another opening, called the anus.

The first opening that forms as the blastula folds is called a blastopore. This opening can become either the mouth or the anus, depending on the animal.

Body Cavities Acoelomates do not have a coelom between their body wall and digestive cavity. Pseudocoelomates have body cavities that are partly lined with tissue from the mesoderm.

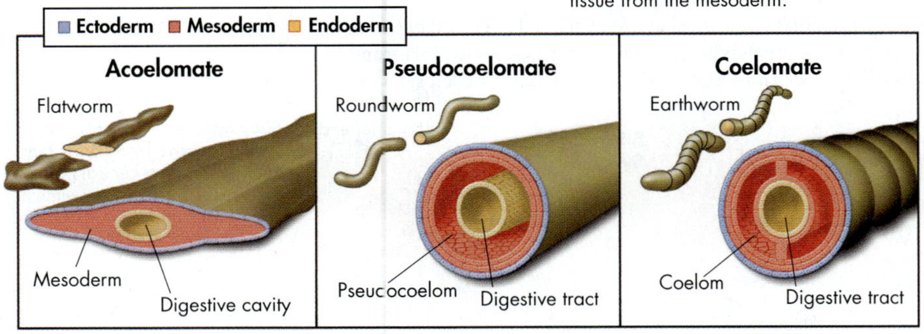

ELL Focus on ELL: Extend Language

INTERMEDIATE AND ADVANCED SPEAKERS Refer students to the labels *acoelomate, pseudocoelomate,* and *coelomate* in the diagram, and write the words on the board. Explain that the word *coelom* comes from a Greek word that means "cavity." Underline the prefix *a-* in *acoelomate*.

Ask Based on the meaning of the word acoelomate, what do you think the prefix *a-* means? *(not)* What are some other words that begin with the prefix *a-*? *(Sample answers: asexual, atypical, amoral)*

Underline the prefix *pseudo-*, and tell students that it means "false" or "apparently similar." Have students compare the pseudocoelom and coelom in the diagram.

Ask How is a pseudocoelom related to a coelom? *(A pseudocoelom is a partial coelom.)* What does a coelom have that a pseudocoelom does not have? *(an inner lining of mesoderm)*

Blastopore Formation A hollow ball of cells called a blastula forms during the early development of an animal embryo. An opening called a blastopore forms in this ball. In deuterostomes, such as fishes, the blastopore forms an anus. In protostomes, such as grasshoppers, the blastopore forms a mouth.

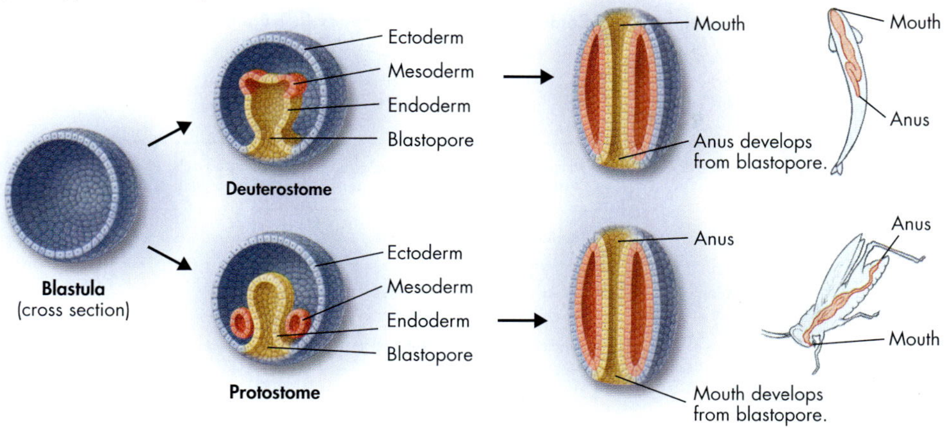

▶ ***Protostomes*** In phyla that are **protostomes** (PROH tuh stohms), the blastopore becomes the mouth. In protostomes, the anus forms from a second opening that develops at the opposite end of the tube. Most invertebrates are protostomes.

▶ ***Deuterostomes*** In **deuterostomes** (DOO tur uh stohms), the blastopore becomes the anus. The mouth is formed from the second opening that develops in the tube. Chordates and echinoderms are deuterostomes.

Segmentation: Repeating Parts As many animals that have bilateral symmetry develop, their bodies become divided into segments. Segments are body parts that are repeated. Worms, insects, and vertebrates usually have at least some parts that repeat on each side of the body. Bilateral symmetry and segmentation are often found together.

Cephalization: Getting a Head Animals with bilateral symmetry usually also have **cephalization** (sef uh lih ZAY shun). Cephalization is the concentration of sense organs and nerve cells at the front end of an animal. Cephalization often creates a head. Arthropods and vertebrates both show cephalization.

The heads of arthropods and vertebrates form when different body segments combine during embryo development. As the segments grow together, sense organs (such as eyes) and nerve cells become concentrated in the head. Animals with heads usually move in a "head-first" direction. Moving head first allows their sense organs to come into contact with new parts of their environment first.

BUILD Vocabulary

zygote a fertilized egg

blastula a hollow ball of cells that develops when a zygote undergoes a series of cell divisions

protostome an animal whose mouth is formed from the blastopore

deuterostome an animal in which the blastopore becomes an anus, and the mouth is formed from the second opening that develops

cephalization the concentration of sense organs and nerve cells at the anterior end of an animal

PREFIXES

The prefix *proto-* comes from a Greek word meaning "first." This can help you to remember that the blastopore becomes a mouth (the "first" opening of the digestive system) in a protostome.

Active Reading

USE VISUALIZATION Help students understand the patterns of embryo development described in the text by having them visualize a balloon partly filled with air.

Say Imagine that the balloon is a blastula and you are slowly poking your finger into the side of the balloon.

Ask What are you forming with your finger? *(a blastopore)* As you continue to push your finger to the other side of the balloon, what part of the embryo are you forming? *(the digestive tract)*

Ask If a mouth develops from the blastopore, what kind of animal embryo are you modeling—a protostome or a deuterostome? *(protostome)*

Build Vocabulary

PREFIXES Invite volunteers to tell the meanings of the prefixes *proto-* (first) and *deutero-* (second). Explain that knowing these meanings can help students understand similar words.

Say Often, engineers or scientists report that they are testing a prototype of some invention they are developing.

Ask Which version of the invention is a prototype, an early one or one nearly finished? *(an early one)*

Say The prefix *deutero-* comes from a Greek word. The word *deuce* has a similar meaning but comes from a Latin word.

Ask Which card in a deck of playing cards is called the deuce? *(the number two card)*

Speed Bump

Ask Does a jellyfish, which has radial symmetry, likely show segmentation? Explain. *(No, segmentation is common in animals with bilateral symmetry.)*

Ask Which group of invertebrates shows cephalization? *(arthropods)*

Focus on Academic Language

• segment • concentration

Explain that a segment is a repeated part of something.

Ask What are some examples of things that have repeated parts, or segments? *(Sample answers: centipede, sidewalk, orange)*

Have students find the word *concentration* in the section Cephalization: Getting a Head. Discuss different meanings of the word.

Ask What is the meaning of *concentration* in the sentence? *(the gathering of structures in one particular location)*

Tell students that *concentration* can mean the amount of a substance in a certain area or volume. It can also mean the act of focusing. Encourage volunteers to provide sentences for these other meanings of *concentration*. *(Sample sentences: This lemonade has a high concentration of sugar. Playing chess requires concentration.)*

During Reading continued

Active Learning

INTERPRET CHARTS Be sure students understand that the information presented in the Body Plans chart summarizes and builds on the descriptions of animal body plans presented in this lesson.

Say The colors in the chart link common features of body plans in different animal phyla.

Ask Which animal phylum lacks germ layers and symmetry? *(sponges)* Which two invertebrate phyla show radial symmetry? *(cnidarians and echinoderms)*

Say Notice that each of the features found in the chordate body plan is shared by at least one invertebrate phylum.

Ask Which invertebrate phylum shares the most features with chordates? *(arthropods)*

Say Imagine that you are told about a "mystery animal" that is a protostome and has segmentation, bilateral symmetry, and a true coelom.

Ask Is the mystery animal most likely a roundworm, arthropod, echinoderm, or chordate? *(arthropod)*

Body Plans The different features of body plans are shown for nine animal phyla below. The body plans of modern animals suggest that invertebrates and chordates evolved from a common ancestor.

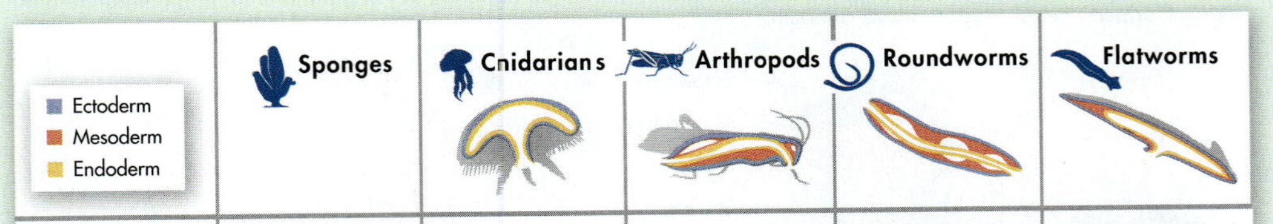

	Sponges	Cnidarians	Arthropods	Roundworms	Flatworms
Levels of Organization	Specialized cells	Specialized cells, tissues	Specialized cells, tissues, organs	Specialized cells, tissues, organs	Specialized cells, tissues, organs
Body Symmetry	Absent	Radial	Bilateral	Bilateral	Bilateral
Germ Layers	Absent	Two	Three	Three	Three
Body Cavity	–	Acoelom	True coelom	Pseudocoelom	Acoelom
Embryological Development	–	–	Protostome	Protostome	Protostome
Segmentation	Absent	Absent	Present	Absent	Absent
Cephalization	Absent	Absent	Present	Present	Present

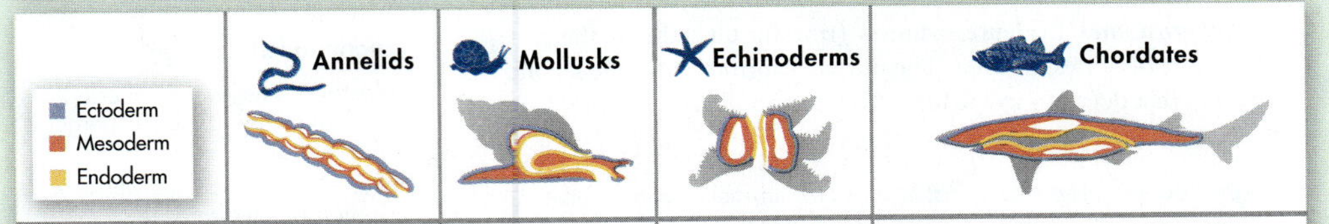

	Annelids	Mollusks	Echinoderms	Chordates
Levels of Organization	Specialized cells, tissues, organs	Specialized cells, tissues, organs	Specialized cells, tissues, organs	Specialized cells, tissues, organs
Body Symmetry	Bilateral	Bilateral	Radial (as adults)	Bilateral
Germ Layers	Three	Three	Three	Three
Body Cavity	True coelom	True coelom	True coelom	True coelom
Embryological Development	Protostome	Protostome	Deuterostome	Deuterostome
Segmentation	Present	Absent	Absent	Present
Cephalization	Present	Present	Absent (as adults)	Present

614 Chapter 25 • Lesson 2

Hands-On Learning

DESIGN AN ANIMAL Divide students into groups. Have each group design a "fantasy" animal by choosing variations of features from the Body Plans chart. Instruct each group to sketch its animal and present it to the class, describing its body plan. Encourage students to show both external and internal features.

Extension Have students make three-dimensional models of their fantasy animals, using clay, toothpicks, buttons, and other common materials.

Limb Formation: Legs, Flippers, and Wings Animals that have bilateral symmetry and segmentation usually have appendages on both sides of the body. Appendages can be bristles, legs, or other types of limbs. Dragonfly and bird wings, spider legs, dolphin flippers, and monkey arms are all appendages. Different kinds of appendages have evolved several times, and have been lost several times, in various animal groups.

Key Question What are some features of animal body plans? **Features of animal body plans include levels of organization, body symmetry, differentiation of germ layers, formation of body cavities, patterns of embryo development, segmentation, cephalization, and limb formation.**

The Cladogram of Animals

The features of animal body plans and other evidence provide information for building a cladogram, or phylogenetic tree, of animals. The cladogram below shows our current understanding of the relationships among animal phyla. Animal phyla are usually defined by their adult body plans and patterns of embryo development. For example, the phylum Arthropoda is defined by a body plan that includes bilateral symmetry, segmentation, cephalization, an external skeleton, and jointed legs. Every phylum has its own combination of ancient traits from ancestors and new traits that are found only in that phylum.

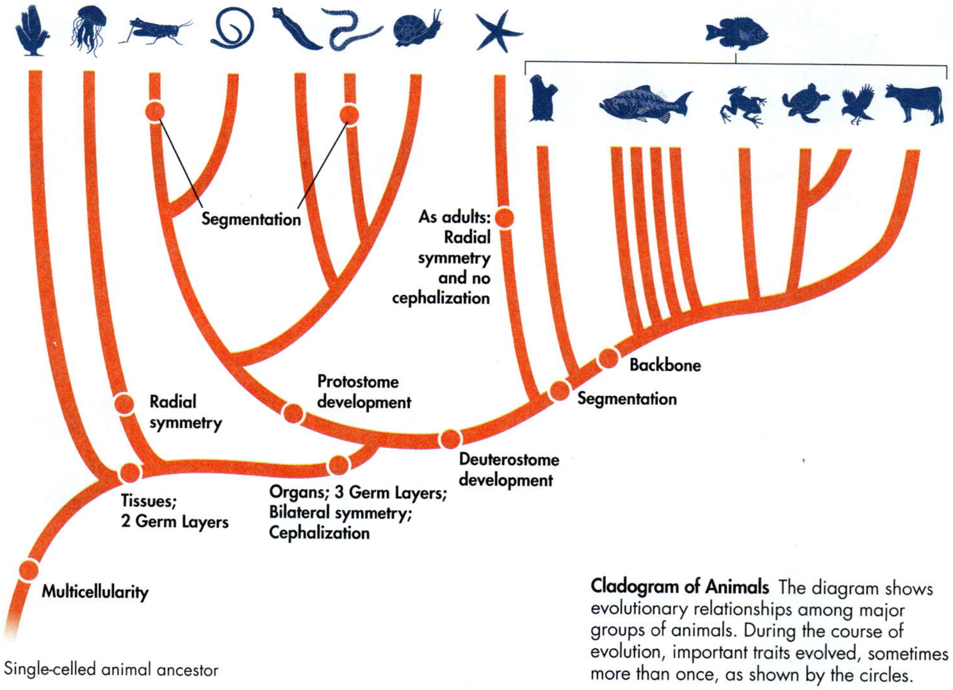

Cladogram of Animals The diagram shows evolutionary relationships among major groups of animals. During the course of evolution, important traits evolved, sometimes more than once, as shown by the circles.

Students Say

"We all got a sandwich baggie to keep our vocabulary note cards in. This helped because I could keep all of them from every chapter without losing them before the final exam. I used a quart freezer bag so I could write which class they were from on the white strip on the bag."

Active Reading

KEY QUESTION Invite a volunteer to read aloud the Key Question and answer.

Ask What does the term *animal body plan* mean? *(the combination of structures that make up an animal's body)*

INTERPRET DIAGRAMS Invite a volunteer to read aloud the caption to the cladogram and the traits labeled. Remind students that a cladogram shows the evolutionary relationships among species based on changes in traits over time.

Ask Which type of animal symmetry evolved twice? *(radial symmetry)* What evolutionary event separated chordates from most—but not all—invertebrates? *(evolution of deuterostome development)*

Ask Among chordates, which two animal groups are most closely related? *(birds and reptiles)* What is shown to be the common ancestor of all animals? *(a single-celled organism)*

Speed Bump

Ask Animals that have appendages on both sides of their body are likely to have which two other body plan features? *(bilateral symmetry and segmentation)*

Ask What information besides the traits of living organisms is used to construct a cladogram? *(genetic information and evidence from the fossil record)*

During Reading continued

Active Learning

USE A GRAPHIC ORGANIZER Have students draw a main ideas and details graphic organizer that has space for three main ideas (one for each paragraph on this page) and at least two details for every main idea. Help students with the first paragraph.

Ask What is the main idea of the first paragraph? *(A cladogram shows the sequence in which features evolved, not how body plans "improved" over time.)* Which details support this idea? *(Vertebrate body systems are not better than invertebrate body systems. Any animal body system is good enough.)*

KEY QUESTION Ask a volunteer to read aloud the Key Question and answer.

Ask What are the two patterns of embryo development? *(protostome and deuterostome)*

Speed Bump

Ask If an animal species' body plan is unsuccessful in its newly changed environment, what might happen to that species? *(It might go extinct.)*

Post Reading

ASSESSMENT TIPS If students are having trouble with Question 4, ask them what happens to individuals with body plans as described in the question. *(Individuals die without producing offspring.)* Then ask what would happen if all individuals in the phylum shared that body plan.

MYSTERY CLUE Have students review the cladogram. Lead them to conclude that the mystery creatures fall into the cladogram after echinoderms but before the development of a backbone.

Wrap-Up Activity

PLAY GO FISH Group students, and have each group make a deck of 29 playing cards, using paper or index cards. Tell students to label each card with information from the Body Plans chart: the name of each animal phylum *(nine total)* or an example of the seven body plan features *(specialized cells, symmetry absent, protostome, and so on; 20 total)*. Have students deal out four cards to each group member. Students should take turns removing a card from the top of the deck and then returning a card from their pile to the bottom of the deck. A student wins by matching up four cards that include the name of an animal phylum and three features that correctly describe that phylum's body plan.

Limb Variations Birds have evolved front limbs specialized as wings. Frogs have evolved four legs.

Differences Between Phyla The cladogram shows the sequence in which important body plan features evolved. You may think that a cladogram shows "improvements" from one phylum to the next. This notion is not correct. The body systems of vertebrates are not better than the body systems of invertebrates. Any system found in a living animal is good enough to allow the animal to survive and reproduce.

Changes Within Phyla Different groups within each phylum show different variations on basic body plans. For example, land vertebrates usually have four limbs. Many of these animals, such as squirrels and frogs, walk or hop on four limbs that we call legs. In birds, the front limbs have evolved into wings. In monkeys, the front limbs have evolved into arms. And through evolution, snakes have lost all of their limbs!

Evolutionary Experiments You can think of each phylum's body plan as an evolutionary "experiment." The first version of each major body plan appeared millions of years ago. Since that time, each body plan has changed as members of the phylum adapted to changing environments. If those changes enabled members of the phylum to survive and reproduce, the phylum still exists. If the body plan did not meet the changing survival challenge, groups in the phylum, or the entire phylum, have become extinct.

Key Question How are animal phyla defined?
Animal phyla are usually defined by their adult body plans and their patterns of embryo development.

CHECK Understanding

Apply Vocabulary

Use the highlighted words from the lesson to complete each sentence correctly.

1. A vertebrate has ______ symmetry, which means that it has right and left sides that are mirror images of each other.
2. The middle germ layer is called the ______.
3. A body cavity that is only partly lined with tissue from the mesoderm is a ______.

Critical Thinking

4. **Relate Cause and Effect** What will happen to a phylum over time if its body plan does not allow its members to survive and reproduce?
5. **Write to Learn** Answer the second clue of the mystery. Write your answer in paragraph form. Use the terms *segmentation*, *deuterostomes*, and *pharyngeal pouches* in your paragraph.

MYSTERY CLUE

The mystery creatures are deuterostomes. Their larvae have bilateral symmetry, a dorsal hollow nerve cord, and pharyngeal pouches—but no backbone. Where on the cladogram do they belong? (**Hint:** See p. 615.)

BIOLOGY.com Search Lesson 25.2 GO • Lesson Assessment

Check Understanding

1. bilateral
2. mesoderm
3. pseudocoelom
4. Members of the phylum will die, and the phylum or groups in the phylum will go extinct.
5. Because the animal has a dorsal hollow nerve cord and pharyngeal pouches, it must belong to the phylum Chordata. Because it does not have a backbone, it is not a vertebrate. The mystery animal belongs to the chordate group that has deuterostome development and segmentation, but does not have a backbone.

Skills Lab

GUIDED INQUIRY

Pre-Lab: Comparing Invertebrate Body Plans

Problem What characteristics can be used to classify invertebrates?

Materials compound microscope; prepared slides of cnidarian, roundworm, and earthworm cross sections; red, blue, and yellow colored pencils

Lab Manual Chapter 25 Lab

Skills Focus Observe, Classify, Compare and Contrast

Connect to the Big idea All members of Kingdom Animalia share a set of characteristics that define them as animals. However, the diversity within the kingdom is vast. For example, some animals have a backbone, but many do not. Some animals have radial symmetry, but many do not. In this lab, you will use preserved cross sections to compare the body plans of three invertebrates. You will pay particular attention to germ layers and body cavities.

Background Questions

a. Review Describe three characteristics that all animals share.

b. Review What are the three germ layers, and what structures do they give rise to?

c. Explain What is the function of a body cavity?

Pre-Lab Questions

Preview the procedure in the lab manual.

1. Compare and Contrast Which two features of animal body plans will you be comparing in this lab?

2. Apply Concepts Where will you look for tissue that formed from the ectoderm layer?

3. Infer Is a hydra smaller than, larger than, or about the same size as an earthworm? Base your answer on the procedure in this lab.

BIOLOGY.com Search Chapter 25 GO

Visit Chapter 25 online to test yourself on chapter content and to find activities to help you learn.

Untamed Science Video Learn how scientists determine that an organism is an animal as the Untamed Science crew visits a research facility on Coconut Island.

Art in Motion View an animation that shows the differences in early development in protostomes and deuterostomes.

Art Review Review your understanding of body symmetry with this drag-and-drop activity.

Pre-Lab Answers

BACKGROUND QUESTIONS

a. Sample answer: Animals are multicellular, obtain nutrients and energy by eating other organisms, and have eukaryotic cells.

b. The endoderm gives rise to linings of the digestive tract and the respiratory system. The mesoderm gives rise to muscles and much of the circulatory, reproductive, and excretory systems. The ectoderm gives rise to the sense organs, nerves, and outer layer of skin.

c. A body cavity provides a space in which organs can be suspended and room for those organs to grow.

PRE-LAB QUESTIONS

1. type of body cavity and number of germ layers

2. the outermost layer of each organism

3. Sample answer: A hydra is smaller than an earthworm. I can infer this because I will view the earthworm under low power, but I will need to use high power to view the hydra.

Pre-Lab

Review body symmetry, germ layers, and body cavities in animal body plans.

Ask Do all invertebrates have the same body symmetry? *(no)* How many germ layers are found in most developing animals? *(three)* What is the purpose of an animal's body cavity? *(It holds the animal's organs and gives them space to grow.)*

Review the Background Questions and the Pre-Lab Questions as a class.

Lab

Tell students they will perform the chapter lab *Comparing Invertebrate Body Plans* described in **Lab Manual B**.

SAFETY Caution students to keep water away from electrical equipment. Check for tangled cords, and disconnect microscopes when not in use. Tell students to be careful with sharp edges on microscope slides and to report broken glass to you.

Post-Lab

CLAY MODELS Have students make clay models of the cross sections they observed, using different colors of clay for the different layers. They can identify each part of a cross section with a label attached to a toothpick, inserting the toothpick into the appropriate part of the model.

NATIONAL SCIENCE EDUCATION STANDARDS

UCP I, V

CONTENT C.1.f, C.5.d

INQUIRY A.1.b, A.2.a

Chapter Summary

Think Visually

USE GRAPHICS Have students use information from this chapter to complete the following concept map.

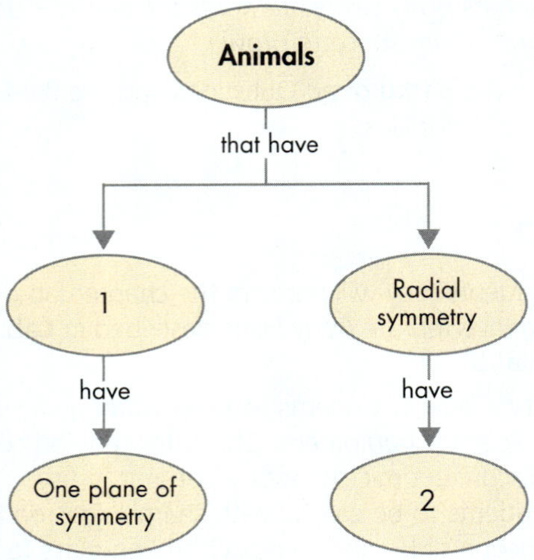

ANSWERS

1. bilateral symmetry

2. any number of planes of symmetry

BIOLOGY.com Study Online

REVIEW AND ASSESSMENT RESOURCES

InterActive Art Students can observe the structures and functions of a sponge.

Art Review Use this drag-and-drop labeling activity to help students review body symmetry and cavities.

Data Analysis Students analyze classification data to determine *Trichoplax's* place on the tree of life.

Vocabulary Review The *Flash Cards* and *Match It* provide an interactive way to review chapter vocabulary.

Editable Worksheets Editable pages of Study Workbook B, Lab Manual B, and the Assessment Resources Book are available online.

Lesson Overviews Lesson Overviews are based on Miller & Levine Biology. However, they are avail-able on CD-ROM and can be edited for your class.

ASSESSMENT RESOURCES

Chapter Assesment Have students take an online version of the Chapter 25 Check Understanding.

Standardized Test Prep Students can take an online version of the Standardized Test Prep. You will receive their scores along with ideas for remediation.

Diagnostic and Benchmark Tests Use these tests to monitor your students' progress and supply remediation.

25 CHAPTER Summary

25.1 What Is an Animal?

- Animals are members of the kingdom Animalia. Animals are heterotrophs that get nutrients and energy by eating other organisms. Animals have multicellular bodies made of eukaryotic cells that do not have cell walls.
- Invertebrates include all animals that lack a backbone.
- All chordates show four characteristics during at least one stage of life: a hollow nerve cord down the back, a notochord, a tail that extends past the anus, and pharyngeal pouches.
- Like all organisms, animals must maintain homeostasis. They do this by gathering and responding to information, obtaining and distributing oxygen and nutrients, and collecting and eliminating carbon dioxide and other wastes. Animals also reproduce.

invertebrate (p. 606)
chordate (p. 606)
notochord (p. 606)
pharyngeal pouch (p. 607)
vertebrate (p. 607)
feedback inhibition (p. 608)

25.2 Animal Body Plans and Evolution

- Features of animal body plans include levels of organization, body symmetry, differentiation of germ layers, formation of body cavities, patterns of embryo development, segmentation, cephalization, and limb formation.
- Animal phyla are usually defined according to adult body plans and patterns of embryo development.

radial symmetry (p. 611)
bilateral symmetry (p. 611)
endoderm (p. 612)
mesoderm (p. 612)
ectoderm (p. 612)
coelom (p. 612)
pseudocoelom (p. 612)
zygote (p. 612)
blastula (p. 612)
protostome (p. 613)
deuterostome (p. 613)
cephalization (p. 613)

Foundations for Learning Wrap-Up

Use the layered books you made when you started the chapter to help you to organize your thoughts about which characteristics define animals.

Activity 1 Get together in a group of three. Take out one of the layered books that summarizes the main topics in Lesson 1 or 2. Read aloud one of the main topics listed on the tabs of the book. Discuss the things that you should remember about that topic. Look inside the layered book to see if you have remembered all the important concepts.

Activity 2 Stay in your group of three. Take out all of the layered books that contain the key questions from the chapter. Work together to write notes and draw diagrams that answer each question. Do this without looking at your textbook.

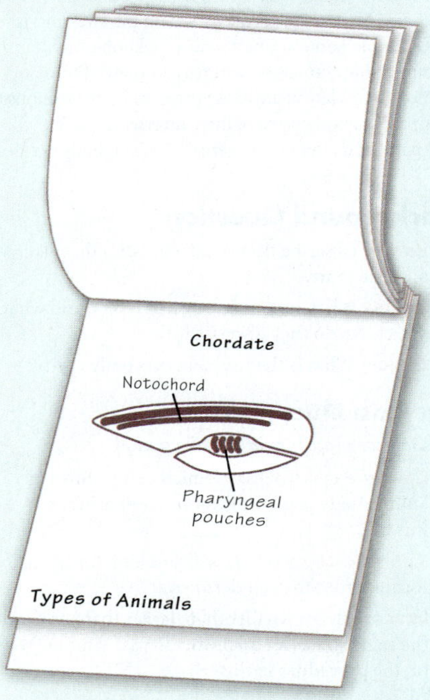

UbD Transfer the Big Idea

TRANSFER Ask students to imagine they have discovered a new animal. Then, divide the class into small groups, and ask each group to write a short article for a science journal announcing their discovery. In their article, students should describe the animal and tell how it shows the characteristics of all animals. The article should also use information about the animal's body plan to suggest and show where the animal belongs in the animal cladogram.

Foundations for Learning Wrap-Up

Have students take out their layered books. Be sure students have written notes for each main topic in Lessons 25.1 and 25.2. Assign groups, and explain the activities.

25 CHECK Understanding

Assess the **Big idea** Unity and Diversity of Life

Write an answer to the question below:

Q: What characteristics and traits define animals?

Constructed Response

Write an answer to each of the numbered questions below. The answer to each numbered question should be one or two paragraphs long. To help you begin, read the **Hints** below the questions.

1. **If you were presented with a small, living organism, how would you try to determine whether or not it was an animal?**

   Hint Remember the characteristics of organisms that belong to the animal kingdom.

2. **Animals need oxygen and nutrients in order to survive. How do animal systems work together to get oxygen and nutrients to the animal's cells?**

   Hint Review the Respiratory and Circulatory Systems diagram that shows how materials move in, around, and out of the body.

   Hint Think about which systems get oxygen and food into the body. Which system delivers oxygen and nutrients to cells?

3. **How have cephalization and segmentation led to the great diversity of animals?**

   Hint Remember how segmentation allowed the development of cephalization.

   Hint Refer to the chart of Body Plans and to the Cladogram of Animals to find examples of animals for your answer.

solve the CHAPTER MYSTERY

SLIME DAY AT THE BEACH

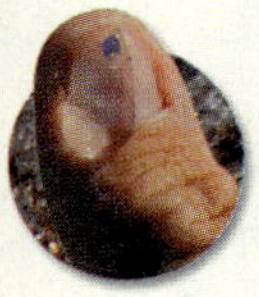

Although most people had never seen creatures like these before, biologists had no trouble identifying them. The slimy creatures were salps—descendants of the most ancient members of phylum Chordata. Salps belong to a group of chordates called tunicates. As adults, most tunicates live attached to rocks or the sea floor. Salps are different. The adults are free-swimming. They pump water in through their mouths and out the other end. This ability allows them to feed and move themselves through the water at the same time. Salps are usually found in tropical seas. They can be carried north by the Gulf Stream and are sometimes washed onto beaches by storms.

1. **Compare and Contrast** How are salps different from jellyfish?

Untamed Science™ **Never Stop Exploring Your World.**

Discovering what the slime was is only the beginning. Take a video field trip with the ecogeeks of Untamed Science to see where the mystery leads.

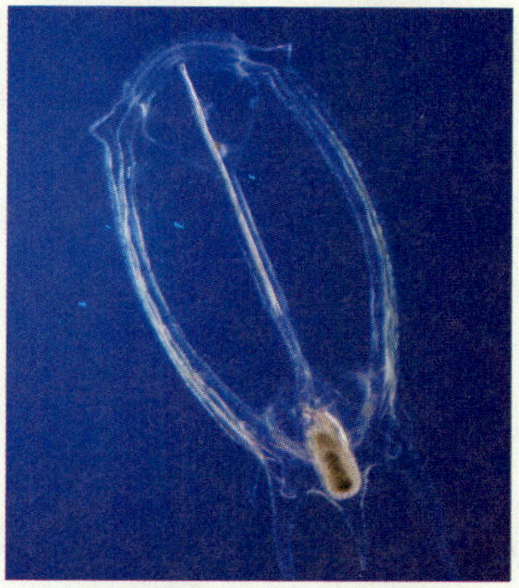

BIOLOGY.com Search Chapter 25 GO • Untamed Science Video • Chapter Mystery

CHAPTER MYSTERY ANSWER

1. Salps are chordates, whereas jellyfish are invertebrates. During development, salps have a notochord, a hollow nerve cord, a tail, and pharyngeal pouches. Jellyfish do not.

Untamed Science™ Suggest students watch **What Is an Animal?** to find out how scientists decide if an organism is an animal.

Assess

ASSESSMENT

Big idea Invite a volunteer to read the question.

Ask How do animals get the energy they need to live? *(They eat other organisms.)*

Ask What is the cellular makeup of all animals? *(Animals are multicellular. Their cells are eukaryotic and have no cell walls.)*

Ask What are the eight features of animal body plans? *(levels of organization, body symmetry, differentiation of germ layers, body cavity formation, patterns of embryological development, segmentation, cephalization, and limb formation)*

Have students write their answers in their notebooks.

CONSTRUCTED RESPONSE Paragraphs may vary. Remind students that paragraphs usually contain one main idea and several supporting details. Guide students to follow the *Hints* to help them answer the questions.

Write Question 1 on the board and read the first hint.

Ask What should be the main idea of your paragraph? *(how to determine whether or not an organism is an animal)*

Ask What are the characteristics of all animals? *(Animals eat other organisms, they are multicellular, and their cells have nuclei and lack cell walls.)*

Model writing the first one or two sentences, using the hint and any student suggestions.

Say The first sentence will express the main idea. *(Sample: I must observe the organism to see if it shows the characteristics that are found in all animals.)*

Say The next sentences should support the main idea by naming the animal characteristics to observe. *(Sample: Does the organism consist of multiple cells? Does it eat other organisms?)*

For Question 2, pair students and have them use the hints and your modeling to answer the question. Have students share their paragraphs with the class.

For Question 3, have students write the paragraphs independently for homework or as an anchor activity.

CHAPTER MYSTERY After students have read the Chapter Mystery, discuss the characteristics of salps.

Ask How are salps like all animals in the way they obtain nutrients? *(They eat other organisms.)*

Say Tunicates are nonvertebrate chordates.

Ask Knowing that salps are tunicates, how do salps differ from vertebrates? *(Salps lack a backbone.)*

ASSESSMENT

Answers

Lesson 25.1

UNDERSTAND KEY CONCEPTS

1. a

2. c

THINK CRITICALLY

3. Vertebrates have a backbone, and nonvertebrate chordates do not have a backbone.

4. An animal gets rid of carbon dioxide through its respiratory system. An animal gets rid of ammonia through its excretory system.

Lesson 25.2

UNDERSTAND KEY CONCEPTS

5. a

6. c

7. endoderm, mesoderm, and ectoderm

8. Answers must include two of the following: specialized cells, tissues and organs; bilateral symmetry; three germ layers; true coelom; segmentation; cephalization.

THINK CRITICALLY

9. multicellularity, tissues, deuterostome development, segmentation

10. Animals with bilateral symmetry usually have specialized anterior, posterior, dorsal, and ventral regions. Sense organs are clustered at the head end, encounter the environment first, and give the animal an advantage in orientation, navigation, feeding, and defense.

11. Sample answer: The more-complex body systems of vertebrates are not necessarily better than the "simpler" systems of invertebrates. Any system found in living animals functions well enough to allow these animals to survive and reproduce.

Connecting Concepts

12. at 3 P.M.

13. Body temperature remains stable even as outside temperature changes.

14. Sample answer: The human body maintains a constant internal temperature using feedback inhibition.

25 CHECK Understanding

25.1 What Is an Animal?

Understand Key Concepts

1. Which of the following is a characteristic of all chordates but is not found in invertebrates?
- **a.** a notochord
- **b.** four legs
- **c.** a circulatory system
- **d.** a skeleton

2. Most animals reproduce
- **a.** sexually by producing diploid gametes.
- **b.** asexually by cloning.
- **c.** sexually by producing haploid gametes.
- **d.** asexually by fusion.

Think Critically

3. **Classify** What characteristic distinguishes vertebrates from nonvertebrate chordates?

4. **Compare and Contrast** How does the way vertebrates gets rid of carbon dioxide differ from the way they get rid of ammonia?

25.2 Animal Body Plans and Evolution

Understand Key Concepts

5. Cnidarians and echinoderms have symmetry that extends from a central point. This type of symmetry is called
- **a.** radial.
- **b.** bilateral.
- **c.** circular.
- **d.** dorsal.

6. An animal whose mouth is formed from the blastopore is a(n)
- **a.** deuterostome.
- **b.** endoderm.
- **c.** protostome.
- **d.** mesoderm.

7. List the three germ layers.

8. Name two body plan characteristics that are shared by all arthropods and chordates.

Think Critically

9. **Sequence** List the following developments in the order of their appearance during evolution: tissues, deuterostome development, multicellularity, segmentation.

10. **Apply Concepts** Explain how bilateral symmetry was an important development in the evolution of animals.

11. **Infer** Why is it inaccurate to state that the cladogram of animals shows the improvements in body plans that have occurred over time?

Connecting Concepts

Use the graph to answer questions 12–14.

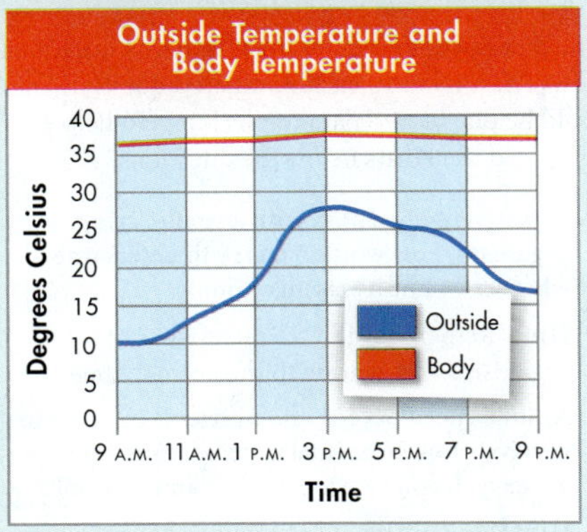

12. **Interpret Graphs** At what time of day is the body temperature closest to that of the outside environment?

13. **Draw Conclusions** What is the relationship between human body temperature and the temperature of the environment?

14. **Infer** How do you explain the shape of the graph for human body temperature?

Assessment Tips

Have students work individually or in pairs to answer the questions. For Question 1, have students refer to the Venn diagram they completed in Workbook B for Lesson 25.1. For Questions 7 and 8, have students review the Body Plans chart in Lesson 25.2. For Question 9, refer students to Cladogram of Animals in Lesson 25.2.

Standardized Test Prep

Multiple Choice

1. Which of the following is a type of tissue that arises in MOST animals during development?
   A endoderm
   B mesoderm
   C ectoderm
   D all of the above

2. Which of the following is NOT a characteristic of animals?
   A the ability to make their own food
   B the ability to move
   C eukaryotic cells
   D cells that lack cell walls

3. A hollow ball of cells formed after the zygote undergoes division is called a
   A coelom.
   B protostome.
   C deuterostome.
   D blastula.

4. Which trend did NOT occur during invertebrate evolution?
   A specialization of cells
   B development of a notochord
   C bilateral symmetry
   D cephalization

5. What is a function of the excretory system?
   A to supply cells with oxygen and nutrients
   B to rid the body of metabolic wastes
   C to gather information from the environment
   D to break down food

6. Animals often respond to information processed by their nervous system by moving around, using their
   A circulatory system.
   B excretory system.
   C musculoskeletal system.
   D digestive system.

7. The concentration of nerve tissue and organs in one end of the body is called
   A cephalization.
   B segmentation.
   C body symmetry.
   D nerve nets.

Questions 8 and 9

A biology student has two samples of earthworms in soil, as shown below. The student knows that, because the worms' body temperature changes with the environment, the worms in Sample A have a higher body temperature than those in Sample B. The student uses a stereomicroscope to count the number of heartbeats per minute for three worms from each sample.

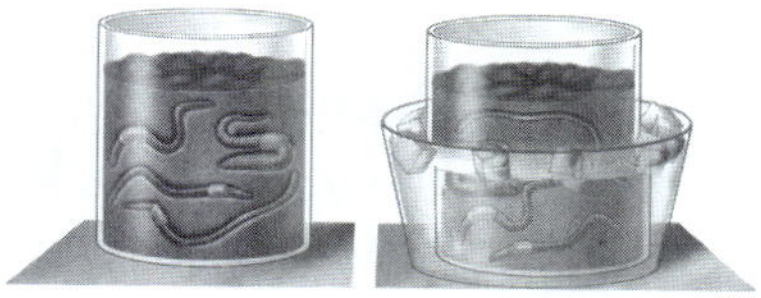

Sample A: At temperature of worms' soil environment

Sample B: In ice water

8. Look at the student's two samples. What can you conclude?
   A Sample A is the control.
   B Sample B is the control.
   C Either sample can serve as a control.
   D This is not a controlled experiment.

9. The student finds that the worms from Sample A have a faster heart rate than the worms from Sample B. What hypothesis might you form based on this observation?
   A The worms in Sample A are healthier than the worms in Sample B.
   B A decrease in body temperature corresponds to an increase in heart rate.
   C There is no relationship between body temperature and heart rate.
   D A decrease in body temperature corresponds to a decrease in heart rate.

Open-Ended Response

10. What characteristics distinguish invertebrates from nonvertebrate chordates?

If You Have Trouble With . . .

Question	1	2	3	4	5	6	7	8	9	10
See Lesson	25.2	25.1	25.2	25.2	25.1	25.1	25.2	25.1	25.1	25.1

Test-Taking Tip

READ ALL THE ANSWER CHOICES Caution students that when evaluating multiple-choice answers, they should read all the answer choices before making a selection. Even if the first choice seems correct, students should make certain the one they choose is the best answer.

Before the Test

Review with students the Test-Taking Tip.

Point out the word *MOST* in Question 1.

Point out the word *NOT* in Questions 2 and 4.

Remind students that an open-ended response means that they'll have to write a paragraph to answer the question. Have them paraphrase the question in their notebooks, and remind them about the steps in writing a paragraph.

Answers

1. D
2. A
3. D
4. B
5. B
6. C
7. A
8. A
9. D
10. Nonvertebrate chordates are animals that have a dorsal nerve cord, pharyngeal pouches, a tail that extends beyond the anus, and a notochord at some time during development. Invertebrates do not have all of the above characteristics.

CHAPTER PLANNER 26 Animal Evolution and Diversity

Chapter Contents	NSES	Time	Core Resources
Chapter Preview			**Student Edition,** pp. 622–623 **Chapter Mystery,** p. 623 **Foundations for Learning,** p. 623
26.1 Invertebrate Evolution and Diversity Origins of the Invertebrates • Cladogram of Invertebrates	I, II, III, IV, C.3.a, C.3.b, C.3.c, C.3.d, C.3.e	1/2 period 1/4 block	**Student Edition,** pp. 624–627 **Study Workbook B** 26.1 Worksheets **Biology.com** *InterActive Art:* Cladogram of Invertebrates • 26.1 Lesson Assessment
26.2 Chordate Evolution and Diversity Origins of the Chordates • Cladogram of Chordates	I, II, III, IV, V, C.3.a, C.3.b, C.3.c, C.3.d, C.3.e	1 period 1/2 block	**Student Edition,** pp. 628–633 **Build Connections,** p. 630 **Inquiry Into Scientific Thinking,** p. 632 **Study Workbook B** 26.2 Worksheets **Biology.com** *Art in Motion:* From Fins to Feet • 26.2 Lesson Assessment
26.3 Primate Evolution What Is a Primate? • Evolution of Primates • Hominine Evolution • The Road to Modern Humans	I, II, III, IV, V, C.3.a, C.3.b, C.3.c, C.3.d, C.3.e, C.6.a, C.6.c, C.6.d, G.2, G.3	1 period 1/2 block	**Student Edition,** pp. 634–637 **Study Workbook B** 26.3 Worksheets **Biology.com** *Art Review:* Comparison of Hominoids • 26.3 Lesson Assessment
Chapter Pre-Lab	III, IV, A.1.b, A.1.c, A.2.a, C.3.a, C.3.b, C.3.d, C.3.e	1 period 1/2 block	**Student Edition,** p. 638 **Lab Manual B** *Investigating Hominoid Fossils*

Tools for ELL Instruction

ELL Handbook describes teaching strategies to make *Miller & Levine Biology: Foundation Edition* more accessible to ELL students.

Spanish Study Workbook is a Spanish translation of Study Workbook A.

Multilingual Glossary is the *Miller & Levine Biology* glossary translated into ten languages.

Tools to Challenge Students

Study Workbook A includes more challenging worksheets and activities such as Chapter Mystery activities that reinforce 21st Century Skills.

Lab Manual A labs are geared toward on-level students. The Lab Manual includes fifteen labs in addition to the Chapter Labs.

Biology.com Data Analysis activities offer students a way to practice data gathering and analyzing skills.

Visual Quizzes in the Assessment Resources Book are geared toward on-level students and use visuals from *Miller & Levine Biology*.

Additional Resources

Biology.com Untamed Science Video • Flash Cards

Study Workbook A 26.1 Worksheets
Spanish Study Workbook 26.1 Worksheets
Assessment Resources Book Visual Quiz

Study Workbook A 26.2 Worksheets
Lab Manual B *Data Analysis:* Feather Evolution
Spanish Study Workbook 26.2 Worksheets

Study Workbook A 26.3 Worksheets
Lab Manual B *Hands-On Activity:* Binocular Vision
Spanish Study Workbook 26.3 Worksheets
Biology.com *Data Analysis:* Who is *H. floresiensis?*
Assessment Resources Book Visual Quiz

Lab Manual A *Investigating Hominoid Fossils*

Chapter Review

Student Edition Chapter Summary, p. 639
Study Workbook B Chapter 26 Review
Transparencies, pp. 297–305
Biology.com Untamed Science Video • Editable Worksheets of Study Workbooks A and B and Lab Manuals A and B • Chapter 26 Flash Cards and Crossword

Untamed Science DVD • Classroom Resources CD (includes lesson presentations and editable worksheets)

Chapter Assessment

Student Edition Check Understanding, pp. 640–643
Study Workbook B Chapter 26 Taking a Standardized Test
Assessment Resources Book Chapter 26 Test B
Biology.com Chapter 26 Assessment • Editable Worksheets of Chapter Visual Quizzes* and Chapter 26 Tests A and B

Exam*View* *Assessment Suite* • Classroom Resources CD (includes lesson presentations and editable worksheets)

Time: 1 period, 1/2 block

* Visual Quizzes are based on visuals from *Miller & Levine Biology*. Review carefully to make sure they are appropriate for your class.

Pressed for Time?

Preview the Chapter Preview the images of invertebrates on pp. 626–627 and briefly discuss the diversity of invertebrates. Read the Key Questions for Lessons 26.2 and 26.3.

Cover the Chapter Quickly Have students read *Origins of the Invertebrates* in Lesson 26.1 and go over the Cladogram of Invertebrates visual. In Lesson 26.2, assign the introduction to *Origins of the Chordates* and go over the Cladogram of Chordates visual. Have students read *What Is a Primate?,* the introduction to *Hominine Evolution,* and *The Road to Modern Humans* in Lesson 26.3.

Assess Assign Questions 2–6 in the 26.1 Check Understanding, Question 3 in the 26.2 Check Understanding, and the entire 26.3 Check Understanding.
In the Chapter 26 Check Understanding, assign Questions 1, 6, 9, and 13–19.

CHAPTER 26

Get the Picture

Ask students what the photograph shows. Students will most likely recognize the image as the skeleton of a snake.

Ask What features of the snake are easy to identify in this photograph? *(Sample answer: backbone, ribs, and head)*

Ask What other animals have a backbone? *(Students are likely to respond with* dog, cat, fish, *and other familiar animals.)* Shape the discussion so that students add humans to their list.

Ask Do all animals have a backbone? *(Allow for varied responses.)*

Lead a discussion to list some examples of animals that do not have backbones, such as insects, worms, and spiders.

Activate Prior Knowledge

Ask students what they remember about classifying organisms. Remind them that they classified plants earlier. Lead a discussion that includes the words *ancestor, descendant, adaptation,* and *evolution.*

Refresh students' memory of the cladogram as a classification tool. Draw a simple cladogram on the board, using the cladograms in Lesson 18.2 as a model.

Ask What part of the diagram shows an organism's ancestors? *(the trunklike horizontal line)*

Ask What part of the diagram represents an organism's descendants? *(the branchlike lines pointing upward from the trunk)*

Invite volunteers to label the appropriate parts of the diagram either *ancestor* or *descendant.*

NATIONAL SCIENCE EDUCATION STANDARDS

UNIFYING CONCEPTS AND PROCESSES

I, II, III, IV, V

CONTENT

C.3.a, C.3.b, C.3.c, C.3.d, C.3.e, C.6.a, C.6.c, C.6.d, G.2, G.3

INQUIRY

A.1.b, A.1.c, A.2.a

26 Animal Evolution and Diversity

Big idea **Evolution**

Q: How have animals descended from earlier forms through the process of evolution?

This skeleton of a gopher snake gives clues about how it is related to other vertebrates.

BIOLOGY.com Search Chapter 26 GO • Flash Cards

622

UbD Connect to the Big Idea

Invite a volunteer to read aloud the chapter title, the Big Idea, and the question. Emphasize the question, **How have animals descended from earlier forms through the process of evolution?**

Set the Purpose

PREVIEW AND PREDICT Draw the graphic organizer on the board. Guide students to the INSIDE feature on the next page, and invite a volunteer to read each lesson number and title.

Ask How do you think this title relates to the questions in the graphic organizer? Remind students to search for the answers as they read the chapter and to look for a link between evolution and diversity.

INSIDE:

- 26.1 Invertebrate Evolution and Diversity
- 26.2 Chordate Evolution and Diversity
- 26.3 Primate Evolution

CHAPTER
MYSTERY

FOSSIL QUEST

To Josh and Pedro, fossil hunting sounded like a great summer trip. But where should they go? Their parents said they must stay within the United States. Josh wanted to search for the very first animals. Pedro wanted to look for the ancestors of birds. Where could they find these fossils? Were any good sites close enough to keep their parents from worrying?

They realized that they needed to know when the animals they were searching for lived. Then, they needed to figure out where they could find rocks of the right ages. As you read this chapter, look for clues to the geologic time periods when the first animals and birds lived. Also look for clues about where the boys might find their fossils. Then, solve the mystery.

Read for Mystery Clues As you read this chapter, look for clues to help you discover where Josh and Pedro's "Fossil Quest" will take them. Then, solve the mystery at the end of the chapter.

FOUNDATIONS for Learning

Cladograms show the relationships among living things. Before you read the chapter, get three sheets of blank paper and three index cards. As you read each lesson, copy each cladogram from the book onto a sheet of paper. Then choose three animals mentioned in the chapter. Draw a picture of each animal and describe it. At the end of the chapter are two activities that use the cards and sheets of paper. They will help you answer the question: How have animals descended from earlier forms through the process of evolution?

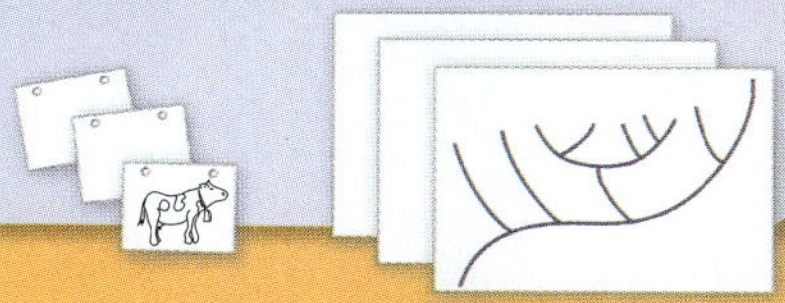

- Untamed Science Video
- Chapter Mystery

Chapter 26 Big Idea: Evolution

Chapter 26: How have animals descended from earlier forms through the process of evolution?

- **26.1:** How did invertebrates evolve?
- **26.2:** How did chordates evolve?
- **26.3:** How did primates evolve?

CHAPTER 26

Foundations for Learning

Pass out the index cards and sheets of paper. Help students grasp the importance of understanding the relationships among living things. Point out that evolution can explain the diversity of life on Earth. Remind students that cladograms serve as tools to help us recognize the ancestors and descendants of an organism.

Ask Why is it important to understand how different groups of animals evolved? *(Sample answer: Knowing about animal diversity and evolution can help us better understand and appreciate the world around us.)*

Tell students they can use the cladograms as a review and study aid. Familiarize students with the activity by referring them to the Cladogram of Nonchordate Invertebrates in Lesson 26.1. Suggest that they trace and label the cladogram.

Untamed Science Why does there always seem to be a new type of bug crawling out of the woodwork? Join the ecogeeks of Untamed Science in **Insects Rule!** as they investigate why there are so many different types of insects.

CHAPTER MYSTERY Open a discussion about fossil hunting. Ask students if they have been to any good places for finding fossils. Invite a student to read aloud the Chapter Mystery.

Ask Why might it be difficult to find fossils of the first animals and of the ancestors of birds at the same site? *(Sample answer: The first animals and the ancestors of birds lived at different times.)*

Tell students they will find clues in the chapter to answer the question, **Where can they find rocks of the right ages?**

Digital Support

BIOLOGY.com Extend your reach by using the digital assets offered at **Biology.com.**

LESSON 26.1

Pre-Reading

Objectives

26.1.1 Explain what fossil evidence indicates about the timing of the evolution of the first animals.

26.1.2 Interpret the cladogram of vertebrates.

Activate Prior Knowledge

Have students recall the six kingdoms of life: Eubacteria, Archaebacteria, Protista, Fungi, Plantae, and Animalia.

Ask What characteristics do all animals have that make them different from other organisms? *(multicellular, no cell walls)* How are animals different from each other? *(Answers may vary.)*

Say In this chapter, you will learn how animals are classified into groups.

Preview the Pages

1. Have students find the vocabulary words in the text and read each one aloud.
2. Point out the images and text relating to invertebrate classification on the last two pages of the lesson. Ask a volunteer to read aloud the headings. Explain that each heading is the name of an invertebrate group. Remind students that classification systems organize information, making it easier to understand.

Build Understanding

PREVIEW VISUALS Ask volunteers what questions they have about the cladogram. Direct students to use their questions to complete the Preview Visuals chart in Workbook B.

NATIONAL SCIENCE EDUCATION STANDARDS

UNIFYING CONCEPTS AND PROCESSES

I, II, III, IV

CONTENT

C.3.a, C.3.b, C.3.c, C.3.d, C.3.e

For the corresponding lesson in *Miller & Levine Biology,* see pages 752–756.

26.1 Invertebrate Evolution and Diversity

Key Questions

When did the first animals evolve?

What does the cladogram of invertebrates illustrate?

BUILD Understanding

Preview Visuals Before you read, preview the cladogram of invertebrates on p. 626. Write down any questions you have about it. As you read, write answers to your questions.

In Your Workbook Go to your workbook to learn more about how to preview visuals. Complete the Preview Visuals activity for Lesson 26.1.

For more on the diversity of animals, go to the Visual Guide. pp. DOL 30–64

Origins of the Invertebrates

The origins of the first animals are a great mystery. Fossils show that a huge number of animals first appeared between 530 and 515 million years ago. This period is called the "Cambrian Explosion." How did so many kinds of animals evolve so quickly? What simpler forms did they come from?

The first living things were single celled. For 3 billion years, all living things were single celled. Multicellular animals may have evolved from ancestors like choanoflagellates (koh AN uh FLAJ uh layts). These single-celled eukaryotes sometimes grow in groups. In many ways they are like sponges, the simplest animals.

Traces of Early Animals The first animals were tiny and had soft bodies. Soft bodies do not form fossils as well as hard body parts. Some early fossils come from sponges and animals like jellyfish. Scientists have also found trace fossils of tracks and burrows made by animals long ago. Such clues show that the first animals may have begun evolving 600 million years ago.

The Ediacaran Animals Some of the most exciting fossils of animals that lived before the Cambrian Period come from the Ediacara Hills of Australia. These strange fossils are about 565 to about 544 million years old. They show animals with body plans that are different from those of anything alive today. These animals had little cell or tissue specialization, and no organization into a front and back end. Some may have had algae living within their bodies. Others were segmented and had bilateral symmetry. Some seem to be related to jellyfishes and worms. Many were flat and lived on the bottom of shallow seas.

Fossil Clues Tiny fossils such as this 565-million-year-old embryo are rare and valuable finds (SEM 100×).

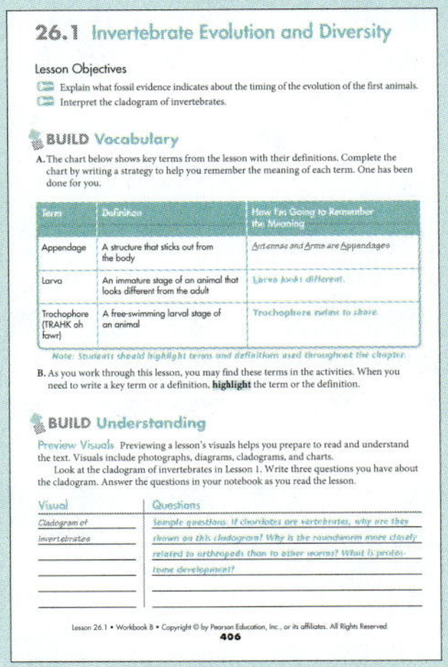

26.1 Invertebrate Evolution and Diversity

Lesson Objectives

Explain what fossil evidence indicates about the timing of the evolution of the first animals.

Interpret the cladogram of invertebrates.

BUILD Vocabulary

A. The chart below shows key terms from the lesson with their definitions. Complete the chart by writing a strategy to help you remember the meaning of each term. One has been done for you.

Term	Definition	How I'm Going to Remember the Meaning
Appendage	A structure that sticks out from the body	Antennas and Arms are Appendages
Larva	An immature stage of an animal that looks different from the adult	Larva looks different.
Trochophore (TRAHK oh fawr)	A free-swimming larval stage of an animal	Trochophore swims to share.

Note: Students should highlight terms and definitions used throughout the chapter.

B. As you work through this lesson, you may find these terms in the activities. When you need to write a key term or a definition, **highlight** the term or the definition.

BUILD Understanding

Preview Visuals Previewing a lesson's visuals helps you prepare to read and understand the text. Visuals include photographs, diagrams, cladograms, and charts.

Look at the cladogram of invertebrates in Lesson 1. Write three questions you have about the cladogram. Answer the questions in your notebook as you read the lesson.

Visual	Questions
Cladogram of invertebrates	Sample questions: If chordates are vertebrates, why are they shown on this cladogram? Why is the roundworm more closely related to arthropods than to other worms? What is protostome development?

Lesson 26.1 • Workbook B • Copyright © by Pearson Education, Inc., or its affiliates. All Rights Reserved.
406

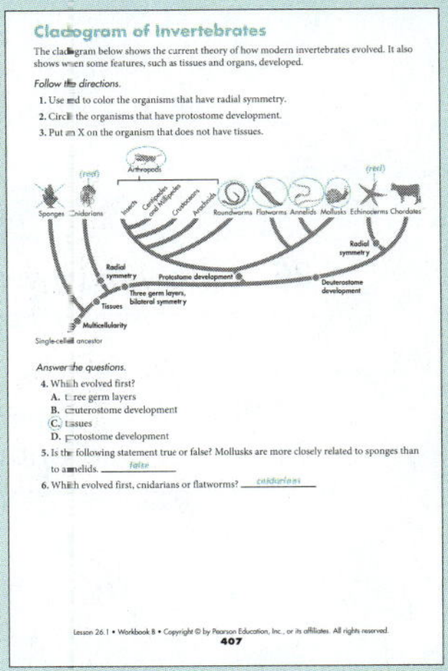

Cladogram of Invertebrates

The cladogram below shows the current theory of how modern invertebrates evolved. It also shows when some features, such as tissues and organs, developed.

Follow the directions.

1. Use red to color the organisms that have radial symmetry.
2. Circle the organisms that have protostome development.
3. Put an X on the organism that does not have tissues.

Answer the questions.

4. Which evolved first?
   A. three germ layers
   B. deuterostome development
   C. tissues
   D. protostome development
5. Is the following statement true or false? Mollusks are more closely related to sponges than to annelids. false
6. Which evolved first, cnidarians or flatworms? cnidarians

Lesson 26.1 • Workbook B • Copyright © by Pearson Education, Inc., or its affiliates. All rights reserved.
407

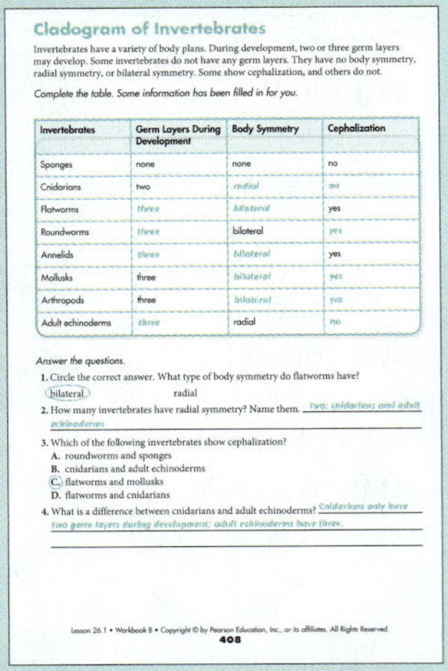

Cladogram of Invertebrates

Invertebrates have a variety of body plans. During development, two or three germ layers may develop. Some invertebrates do not have any germ layers. They have no body symmetry, radial symmetry, or bilateral symmetry. Some show cephalization, and others do not.

Complete the table. Some information has been filled in for you.

Invertebrates	Germ Layers During Development	Body Symmetry	Cephalization
Sponges	none	none	no
Cnidarians	two	radial	no
Flatworms	three	bilateral	yes
Roundworms	three	bilateral	yes
Annelids	three	bilateral	yes
Mollusks	three	bilateral	yes
Arthropods	three	bilateral	yes
Adult echinoderms	three	radial	no

Answer the questions.

1. Circle the correct answer. What type of body symmetry do flatworms have?
   bilateral   radial
2. How many invertebrates have radial symmetry? Name them. two; cnidarians and adult echinoderms
3. Which of the following invertebrates show cephalization?
   A. roundworms and sponges
   B. cnidarians and adult echinoderms
   C. flatworms and mollusks
   D. flatworms and cnidarians
4. What is a difference between cnidarians and adult echinoderms? Cnidarians only have two germ layers during development; adult echinoderms have three.

Lesson 26.1 • Workbook B • Copyright © by Pearson Education, Inc., or its affiliates. All Rights Reserved.
408

The Cambrian Explosion The Cambrian Period began about 542 million years ago. Many fossils of animals from this time can be found in China and Canada. These fossils show that over a period of 10 to 15 million years, animals evolved complex body plans. They had specialized cells, tissues, and organs. Many had body symmetry, segmentation, and a front end and a back end. Many had **appendages**, or parts such as legs or antennae that come out from the body. Some Cambrian animals had shells, skeletons, and other hard body parts. Animals with hard body parts are more likely to leave fossils than animals with soft bodies.

A number of Cambrian fossils are ancient members of modern invertebrate phyla. For example, *Marrella*, shown below, was an arthropod. Other Cambrian fossils come from strange groups that are now extinct. By the end of the Cambrian Period, all the basic animal body plans had evolved. Later evolutionary changes would form the structures of the animals we know today.

Modern Invertebrate Diversity Today, invertebrates are the most common animals on Earth. They live in nearly every ecosystem and are part of nearly every food web. There are far more invertebrates than so-called "higher animals," such as reptiles and mammals.

Key Question When did the first animals evolve?
Fossil evidence shows that the first animals began evolving long before the Cambrian Explosion.

Cladogram of Invertebrates

How are the groups of invertebrates related? On the top of the next page is a cladogram of invertebrates. It shows how scientists think that the major groups of living invertebrates are related to each other. It also shows the order in which some important features came about. Many of these features, such as tissues, first came about in the Cambrian animals.

BUILD Vocabulary

appendage
a structure, such as a leg or antenna, that extends from the body wall

larva
the immature stage of an organism

MULTIPLE MEANINGS
The word *append* means to "attach" or "add on." In everyday speech, an appendage is something that is added onto or attached to something larger. In biology, an appendage is a small part that is attached to a larger body.

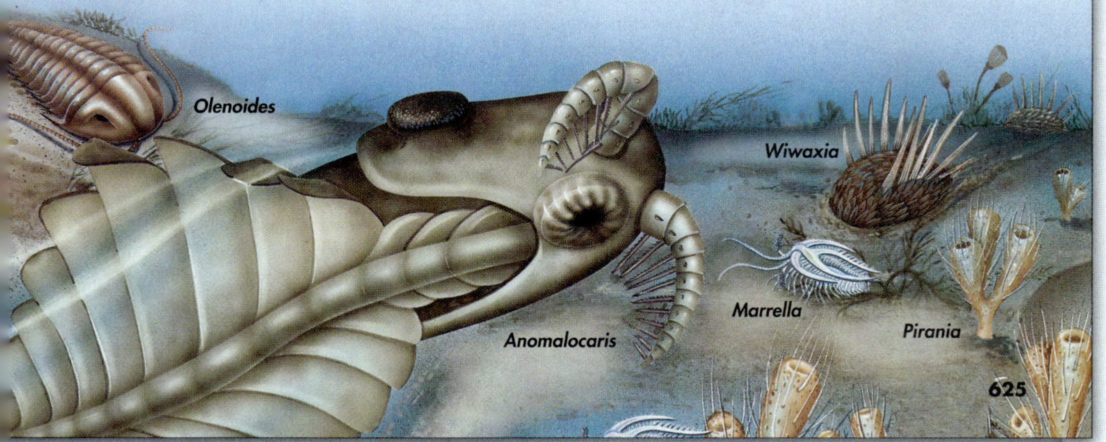

Cambrian Animals The Burgess Shale in Canada contains fossils of animals from the Cambrian Period. This picture shows how some of those animals may have looked when they were alive.

625

ELL Focus on ELL: Access Content

INTERMEDIATE AND ADVANCED SPEAKERS Pair students and tell them they will play the role of reporters. Have them read the section The Cambrian Explosion, looking for information for the 5Ws—who, what, where, when, why—and taking notes as they do. (Tell students they can assume scientists are the "who.") Each pair can record the appropriate information in their notebooks and then share their notes with the class. As a wrap-up, have a class discussion in which students evaluate the reporters' notes.

During Reading

Active Reading

MAKE INFERENCES Point out the Cambrian Animals visual to students. Make sure students realize that it is an artist's drawing, not a photograph.

Ask Why would it be impossible to show a photo of Cambrian animals? *(Only their fossils remain.)*

Ask How do scientists and artists know what Cambrian animals looked like? *(Scientists infer the appearances of the animals from the shapes and sizes of their fossils.)*

CLARIFY WORD USAGE Point out the word *trace* in the third paragraph on the previous page.

Say The word *trace* can be a noun, adjective, or verb.

Ask Where in the paragraph is *trace* used as a noun? *(in the heading, Traces of Early Animals)* Where is *trace* used as an adjective? *(trace fossils)*

Have a volunteer write a sentence on the board that uses *trace* as a verb. *(Sample sentence: Trace the picture that is in the book.)*

Build Vocabulary

MULTIPLE MEANINGS Direct students to find each vocabulary word and its definition in the text.

Ask Which of these body parts—arm, leg, tail, head, fin, trunk—are appendages? *(arm, leg, tail, fin)* Which are not? *(head, trunk)*

KEY QUESTION Read aloud the Key Question and its answer.

Say The word *first* is a clue to answering the question. You can find the answer to the question by scanning the text to find this word clue.

Ask What two pieces of information from the text can help you answer the Key Question? *("between 530 and 515 million years ago . . . is called the "Cambrian Explosion," and "trace fossils . . . show that the first animals may have begun evolving 600 million years ago")*

Speed Bump

Ask What happened during the Cambrian Explosion, and when did it occur? *(Many kinds of animals evolved. The Cambrian Explosion occurred between 530 and 515 million years ago.)*

Ask How did Cambrian animals differ from the first animals? *(Sample answers: hard body parts, appendages, complex body plans, more specialization)*

During Reading continued

Active Reading

INTERPRET DIAGRAMS Draw students' attention to the Cladogram of Nonchordate Invertebrates. Remind students that a cladogram shows the relationship between ancestors and descendents. Explain that the blue drawings at the top of the diagram show a representative organism of each invertebrate group.

Ask What is the primary feature of the ancestor of all invertebrates? *(single-celled)*

Ask Which invertebrate groups have radial symmetry? *(cnidarians and echinoderms)* How does the cladogram show this? *(A label on the cnidarians branch and on the echinoderms branch indicates radial symmetry.)*

KEY QUESTION Read aloud the Key Question and its answer.

Review the "trunk" and "branch" lines in the Cladogram of Nonchordate Invertebrates visual and the labels next to the lines.

USE GRAPHIC ORGANIZERS Draw an invertebrate compare/contrast table on the board. The first column should have the heading Nonchordate Invertebrate Groups and row entries of *Sponges, Cnidarians,* and *Arthropods*. Add six additional column headings to match the features identified in the cladogram, in the same order from left to right (do not repeat Radial symmetry). Have students draw the table in their notebooks. Suggest that they turn their notebooks sideways in order to fit seven columns. Have students refer to the Cladogram of Nonchordate Invertebrates to determine whether a particular invertebrate group has a particular feature. Tell them to write *yes* or *no* in the appropriate place in the table.

Ask Do you see any patterns in the table entries? *(Answers may vary. The pattern may not be easily recognized until students add the remaining invertebrate groups to the table.)*

Speed Bump

Ask What characteristic makes sponges the simplest animals? *(few specialized cells)*

Ask Do all worms have segmented bodies? Explain. *(No, nematodes and flatworms have unsegmented bodies. The bodies of annelids are segmented.)*

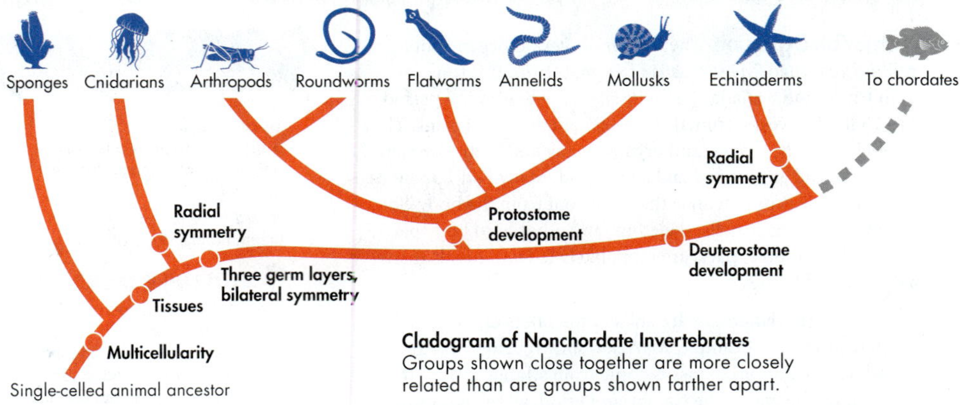

Cladogram of Nonchordate Invertebrates
Groups shown close together are more closely related than are groups shown farther apart.

Key Question What does the cladogram of invertebrates illustrate? **The cladogram shows current hypotheses about the evolutionary relationships of major groups of modern invertebrates. It also shows the order in which some important features evolved.**

▶ ***Sponges*** Sponges are members of the phylum Porifera (por IHF er uh). These animals have pores, or tiny openings, all over their bodies. With few specialized cells, sponges are some of the simplest animals. But like all animals, sponges are multicellular heterotrophs that do not have cell walls. p. DOL 31

▶ ***Cnidarians*** Jellyfishes, sea fans, sea anemones, hydras, and corals are in the phylum Cnidaria (ny DAYR ee uh). Cnidarians live in water. They have soft bodies, specialized tissues, and radial symmetry. They are carnivores that use rings of stinging tentacles around their mouths to capture prey. Some cnidarians live alone. Others live in groups called colonies. pp. DOL 32–33

▶ ***Arthropods*** Spiders, centipedes, insects, and crabs are members of the phylum Arthropoda (ahr THRAHP oh duh). Arthropods have bodies that are divided into segments and covered in a hard outer skeleton, or exoskeleton. Their appendages have joints. They also have cephalization. They live in oceans, fresh water, on land, and in the air. More than a million species have been identified. That is more than three times the number of all other animal species on Earth! pp. DOL 34–37

▶ ***Nematodes (Roundworms)*** Members of the phylum Nematoda are roundworms. They are unsegmented worms with bodies that have a pseudocoelom, a body space with tissue that loosely holds organs in place. Their digestive tracts have two openings—a mouth and an anus. Some live in soil or water. Others are parasites that live in plants or animals. p. DOL 38

Focus on Academic Language

- multicellular • exoskeleton • unsegmented

Review with students the meaning of the word *multicellular* ("having many cells"). Explain that the prefix *multi-* means "many."

Ask What other words do you know that begin with the prefix *multi-*? What do they mean? *(Sample answers: multicolored—many colored; multicultural—many cultures; multigenerational—many generations)*

Write *exoskeleton* and *unsegmented* on the board.

Ask What is the prefix in each word? (exo- *and* un-, *respectively)*

Pair students and have them find the words *exoskeleton* and *unsegmented* on the page. Ask them to determine the meanings of the words, using context clues and the meanings of the words' prefixes. Suggest that students write the words and their meanings in their notebooks.

▶ ***Flatworms*** Flatworms are in the phylum Platyhelminthes (plat ih hel MIN theez). These soft, unsegmented worms have tissues and organ systems. They have bilateral symmetry and cephalization. They do not have coeloms. Most of these worms are only a few millimeters thick. p. DOL 39

▶ ***Annelids*** Earthworms are members of the phylum Annelida (un NEL ih duh). So are leeches and some other worms that live in the ocean. Annelids have segmented bodies that look like stacks of little rings. They also have coeloms. pp. DOL 40–41

▶ ***Mollusks*** Snails, slugs, clams, and octopi belong to the phylum Mollusca. Most mollusks have soft bodies with an inner or outer shell. They have coeloms and complex organ systems. The **larva** (plural: larvae), or young stage, of a mollusk is called a trochophore (TRAHK oh fawr). Trochophores can swim through the water. (Many annelids also have trochophore larvae. These similar larvae show that annelids and mollusks are closely related.) pp. DOL 42–43

▶ ***Echinoderms*** Sea stars, sea urchins, and sand dollars are in the phylum Echinodermata (ee KY noh durm aht uh). Most adult echinoderms have five-part radial symmetry. Echinoderms have a spiny skin stretched over a hard inner skeleton. They also have a water vascular system—a group of water-filled tubes that end in tube feet. They use their tube feet for walking and for holding food. Like chordates, echinoderms are deuterostomes. pp. DOL 44–45

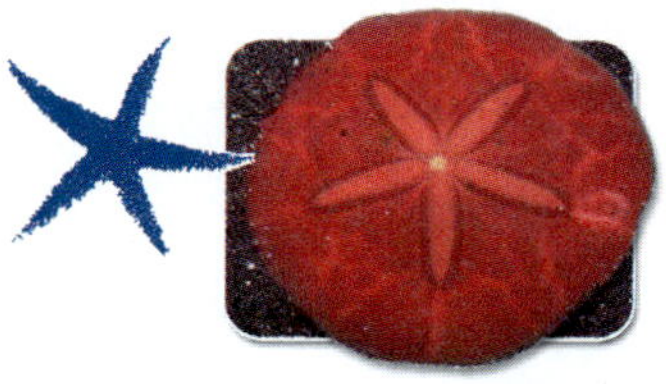

CHECK Understanding

Apply Vocabulary

Use the highlighted words from the lesson to complete each sentence correctly.

1. The young stage of some animals is called a ______.
2. Legs, arms, and antennae are kinds of ______.

Critical Thinking

3. **Describe** What was the Cambrian Explosion?
4. **Relate Cause and Effect** Why is it rare to find fossils of animals from before the Cambrian Period? What about these animals makes it rare for them to form fossils?
5. **Interpret Diagrams** Look at the cladogram of invertebrates. Which group is more closely related to echinoderms: mollusks or chordates? How can you tell?
6. **Sequence** Which feature of echinoderm ancestors evolved first—radial symmetry or deuterostome development? (**Hint:** Study the cladogram.)
7. **Write to Learn** Answer the first clue of the mystery. In your answer, tell when the earliest known animals are thought to have lived.

MYSTERY CLUE

How old would rocks need to be in order to hold fossils of the earliest known animals? (**Hint:** See p. 624.)

BIOLOGY.com Search Lesson 26.1 GO • Lesson Assessment

Check Understanding

1. larva
2. appendages
3. The Cambrian Explosion is a period of time between 530 and 515 million years ago when a huge number of animals evolved.
4. These animals had soft bodies, which don't form fossils as well as hard body parts.
5. Echinoderms are more closely related to chordates than to mollusks because echinoderms and chordates are deuterostomes. Mollusks are protostomes.
6. Deuterostome development evolved before radial symmetry.
7. The rocks would need to be at least 600 million years old, the age of fossils of early sponges and trace fossils of the first known early animals.

Active Reading

USE GRAPHIC ORGANIZERS Direct students to complete their invertebrate compare/contrast table by adding *Flatworms, Annelids, Mollusks,* and *Echinoderms* to the first column and recording in the table the features these invertebrates exhibit.

Ask What pattern do you see in the table now? *(Students should now recognize that the frequency of yes is greater on the left side of the table, where simpler features are listed. The frequency of no is greater on the right side of the table, where more-complex features are listed.)*

Speed Bump

Ask How do the skeletons of arthropods and echinoderms differ? *(Arthropods have an outer skeleton. Echinoderms have an inner skeleton.)*

Post Reading

ASSESSMENT TIPS For Question 3, prompt students to tell not only the time of the Cambrian Explosion but also its significance to the chapter Big Idea—evolution.

Point out to students that Question 4 describes an effect—it is rare to find fossils of animals from before the Cambrian period. The question is asking students to explain the cause.

MYSTERY CLUE Write the following time periods on the board: 530 to 515 million years ago; 565 to 544 million years ago; more than 600 million years ago. Have students look back in the lesson to find what occurred during each time period.

Wrap-Up Activity

FEATURE ADDITION Pair students and give each pair a copy of the Cladogram of Nonchordate Invertebrates visual. Tell students to write above the name of each invertebrate group at least two features of the group that are not provided on the cladogram. Have pairs exchange their cladograms with another student pair and add any features that are missing.

LESSON 26.2

Pre-Reading

Objectives

26.2.1 Describe the most ancient chordates.

26.2.2 Interpret the cladogram of chordates.

Preview the Pages

1. Review the vocabulary words by having students find the words in the text. As a class, discuss the meaning of each word.
2. Have students preview the visuals. Have them use the illustration titles and captions to predict what they will learn in this lesson.
3. Call attention to the blow-up pictures in the From Fins to Feet visual. Be sure students recognize that the blow-ups show the internal structures of the circled appendages.

Build Understanding

VENN DIAGRAM Draw a Venn diagram on the board. Use shading and crosshatching to emphasize that the diagram has three distinct areas. Tell students that a Venn diagram can be used to answer the questions, **How are these things alike? How are they different?** Indicate that one circle will contain features of nonvertebrate chordates, and the other circle will contain features of vertebrates.

Ask What question is answered by the features in the area where the circles overlap? *(How are nonvertebrate chordates and vertebrates alike?)*

Tell students they can learn more about Venn diagrams in Workbook B. Suggest that they complete the Venn diagram as they read Lesson 26.2.

NATIONAL SCIENCE EDUCATION STANDARDS

UNIFYING CONCEPTS AND PROCESSES

I, II, III, IV, V

CONTENT

C.3.a, C.3.b, C.3.c, C.3.d, C.3.e

For the corresponding lesson in *Miller & Levine Biology*, see pages 757–764.

26.2 Chordate Evolution and Diversity

Key Questions

What are the most ancient chordates?

What can we learn by studying the cladogram of chordates?

BUILD Understanding

Venn Diagram Create a Venn diagram that shows how nonvertebrate chordates and vertebrates are alike and different.

In Your Workbook Go to your workbook to learn more about making a Venn diagram. Complete the Venn diagram for Lesson 26.2.

Origins of the Chordates

Chordates are the animals we know best. Why? For one thing, they are large and easy to see. Some we keep as pets. Others we eat. Where did they all come from? What were the first chordates like?

The Earliest Chordates Studies of embryos suggest that the earliest chordates were related to the ancestors of echinoderms. The oldest known fossil of a chordate is *Pikaia* (pih KAY uh). When *Pikaia* was first found, scientists thought it was a worm. But it had a notochord and blocks of muscles that formed a pattern like that of simple modern chordates.

Another fossil called *Myllokunmingia* (MY loh kuhn min jee uh) is the earliest known vertebrate. It had muscles that form a pattern. It also had traces of fins and gills, and a head with sense organs. Its skull and skeleton were likely made of cartilage. **Cartilage** is a strong connective tissue that is softer than bone. It holds up many parts of a vertebrate's body. (Cartilage gives your ears their shape.)

Modern Chordate Diversity Today there are six groups of chordates. These are the nonvertebrate chordates and the five groups of vertebrates—fishes, amphibians, reptiles, birds, and mammals. Of these, fishes are the largest group. Today's chordate species are only a small part of the total number that have ever lived on Earth. pp. DOL 46–64

Key Question What are the most ancient chordates? **Studies of embryos show that the most ancient chordates were related to the ancestors of echinoderms.**

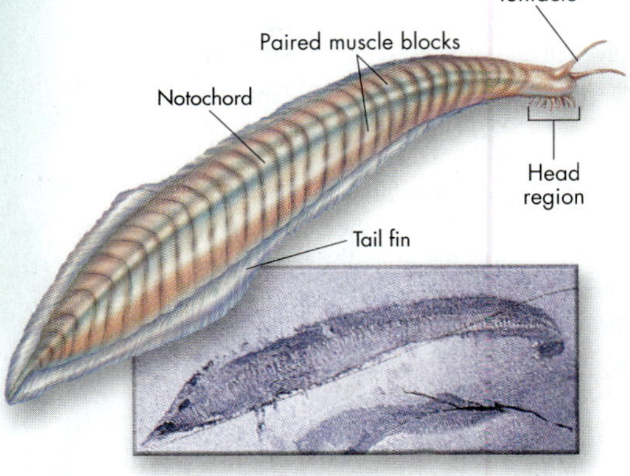

***Pikaia*, an Early Chordate**
Pikaia is the oldest known chordate.

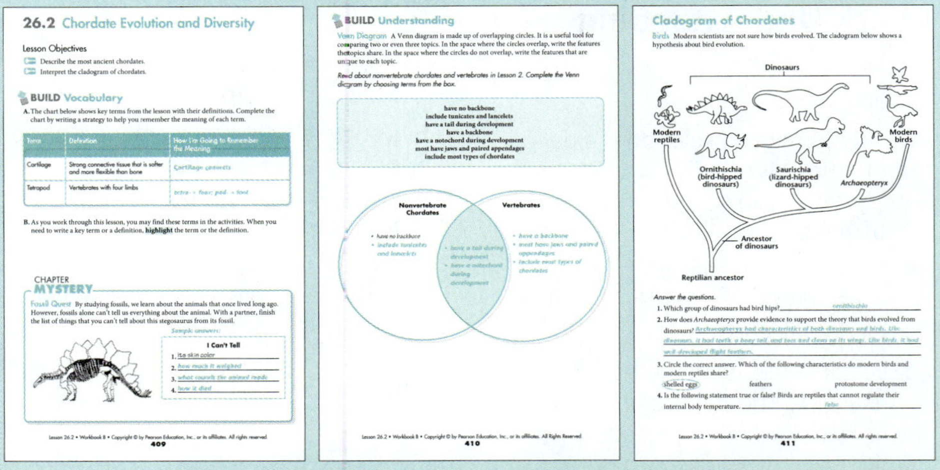
26.2 Chordate Evolution and Diversity

Lesson Objectives

- Describe the most ancient chordates.
- Interpret the cladogram of chordates.

BUILD Vocabulary

A. The chart below shows key terms from the lesson with their definitions. Complete the chart by writing a strategy to help you remember the meaning of each term.

Term	Definition	How I'm Going to Remember the Meaning
Cartilage	Strong connective tissue that is softer and more flexible than bone	Cartilage connects
Tetrapod	Vertebrates with four limbs	tetra- = four; pod- = foot

B. As you work through this lesson, you may find these terms in the activities. When you need to write a key term or a definition, **highlight** the term or the definition.

CHAPTER MYSTERY

Fossil Quest By studying fossils, we learn about the animals that once lived long ago. However, fossils alone can't tell us everything about the animal. With a partner, finish the list of things that you can't tell about this stegosaurus from its fossil.

Sample answers:

I Can't Tell

1. its skin color
2. how much it weighed
3. what sounds the animal made
4. how it died

Lesson 26.2 • Workbook B • Copyright © by Pearson Education, Inc., or its affiliates. All rights reserved.
409

BUILD Understanding

Venn Diagram A Venn diagram is made up of overlapping circles. It is a useful tool for comparing two or even three topics. In the space where the circles overlap, write the features the topics share. In the space where the circles do not overlap, write the features that are unique to each topic.

Read about nonvertebrate chordates and vertebrates in Lesson 2. Complete the Venn diagram by choosing terms from the box.

have no backbone
include tunicates and lancelets
have a tail during development
have a backbone
have a notochord during development
most have jaws and paired appendages
include most types of chordates

Nonvertebrate Chordates
- have no backbone
- include tunicates and lancelets

Both
- have a tail during development
- have a notochord during development

Vertebrates
- have a backbone
- most have jaws and paired appendages
- include most types of chordates

Lesson 26.2 • Workbook B • Copyright © by Pearson Education, Inc., or its affiliates. All Rights Reserved.
410

Cladogram of Chordates

Birds Modern scientists are not sure how birds evolved. The cladogram below shows a hypothesis about bird evolution.

Dinosaurs
Modern reptiles
Modern birds
Ornithischia (bird-hipped dinosaurs)
Saurischia (lizard-hipped dinosaurs)
Archaeopteryx
Ancestor of dinosaurs
Reptilian ancestor

Answer the questions.

1. Which group of dinosaurs had bird hips? ornithischia
2. How does *Archaeopteryx* provide evidence to support the theory that birds evolved from dinosaurs? Archaeopteryx had characteristics of both dinosaurs and birds. Like dinosaurs, it had teeth, a bony tail, and toes and claws on its wings. Like birds, it had well-developed flight feathers.
3. Circle the correct answer. Which of the following characteristics do modern birds and modern reptiles share?
   shelled eggs (circled) feathers protostome development
4. Is the following statement true or false? Birds are reptiles that cannot regulate their internal body temperature. false

Lesson 26.2 • Workbook B • Copyright © by Pearson Education, Inc., or its affiliates. All rights reserved.
411

Cladogram of Chordates Chordates include both vertebrates and nonvertebrates. All of the groups shown on the cladogram share a common invertebrate ancestor. The circles show the evolution of some important adaptations.

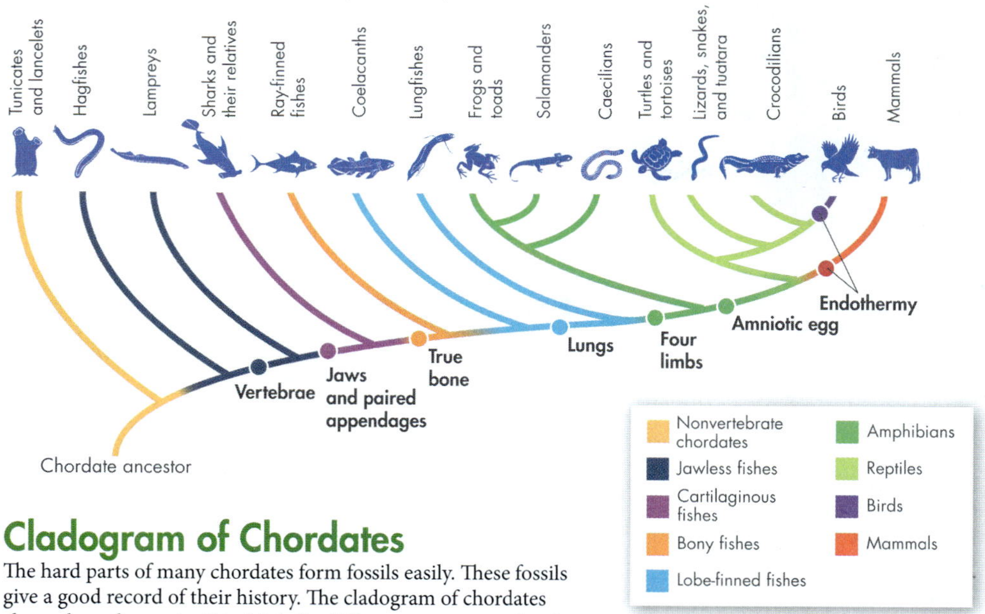

Cladogram of Chordates

The hard parts of many chordates form fossils easily. These fossils give a good record of their history. The cladogram of chordates above shows how scientists think the different groups are related. It also shows when features, such as jaws and limbs, came about.

Nonvertebrate Chordates Two groups of chordates do not have backbones. These nonvertebrate chordates are the tunicates and lancelets. Fossils show that their ancestors branched off from vertebrates more than 550 million years ago. Adult tunicates look more like sponges than humans. They do not have a notochord or a tail. But their larvae have all the key features of chordates. Lancelets look like small eels. They live on the sandy ocean bottom. pp. DOL 46–47

Jawless Fishes The earliest fishes evolved during the late Cambrian Period, about 510 million years ago. These odd-looking animals did not have true jaws or teeth. Their skeletons were made of cartilage. In the Devonian Period, jawless fishes ruled the seas. For this reason, this time is called the Age of Fishes. Some armored jawless fishes, such as those shown at right, became extinct about 360 million years ago. Other kinds of jawless fishes gave rise to modern jawless fishes: lampreys and hagfishes. p. DOL 49

Neither lampreys nor hagfishes have backbones. However, they do have notochords. As larvae, lampreys filter their food from water. As adults, they are parasites. Hagfishes look like big pink worms. They make slime, which helps them slip away from enemies. They can even tie their own bodies into knots!

Jawless Fishes of the Devonian These armored fishes did not have jaws. Their paired fins helped them swim.

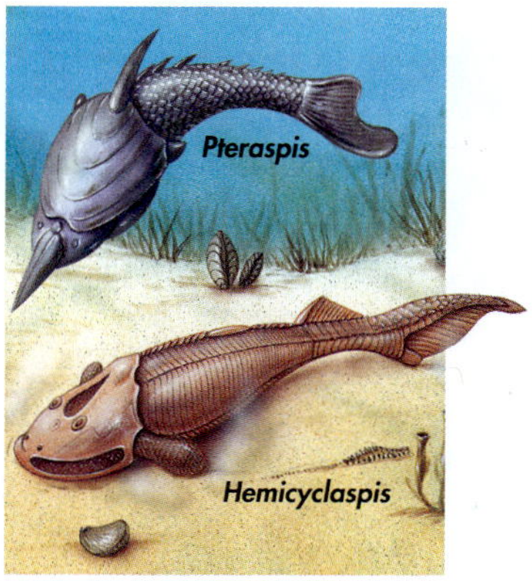

Focus on Academic Language

• nonvertebrate • invertebrate

Some students may read the adjective *nonvertebrate,* recognize the prefix *non-,* and assume the word has the same meaning as *invertebrate.* Point out that both words mean "not vertebrate." However, in the classification of organisms, the words *nonvertebrate* and *invertebrate* have different uses and are not interchangeable. Draw the following graphic organizer on the board to illustrate the difference.

INVERTEBRATES
no backbone and no notochord

CHORDATES

nonvertebrate
notochord but no backbone

vertebrate
backbone

During Reading

Active Reading

KEY QUESTION Read aloud the Key Question and its answer on the previous page.

Refer students to the Cladogram of Nonchordate Invertebrates in the previous lesson, and have students find the point where echinoderms and chordates evolved separately from a common ancestor.

INTERPRET DIAGRAMS Help students interpret the Cladogram of Chordates on this page.

Ask What do the different colors represent? *(different groups of chordates)* What do the labels along the trunk of the cladogram indicate? *(chordate features that evolved over time)*

Draw students' attention to the points where the lines branch.

Ask Are crocodiles more closely related to snakes or to turtles? *(snakes)* How do you know? *(The common ancestor of crocodiles and snakes is more recent than that of crocodiles and turtles.)*

READ, COVER, REMEMBER, RETELL Direct students to read silently the paragraphs about jawless fishes. After reading, have them cover the text with their notebooks and think about what they read.

Say Jawless fishes are some of the earliest vertebrate chordates. Although they do not have a backbone, they do have parts of what could be a skull, which is one of the reasons they are classified as vertebrates.

Ask What are their skeletons made of? *(cartilage)*

Ask What are the two types of jawless fishes living today? *(lampreys and hagfish)*

Ask How do lampreys eat without jaws? *(As larvae, they filter their food from the water. As adults, they are parasites.)*

Speed Bump

Ask What are the six groups of chordates? *(nonvertebrate chordates, fishes, amphibians, reptiles, birds, and mammals)*

Ask What is the main way in which nonvertebrate chordates differ from other chordates? *(Nonvertebrate chordates don't have a backbone.)*

During Reading continued

Active Reading

MAKE INFERENCES List the following four pieces of information on the board:

- Bony fishes have skeletons made of bone.
- Thin bones support the fins of bony fishes.
- Sharks have skeletons made of cartilage.
- Sharks' fins are not made of bone.

Ask What substance supports a shark's fins? *(cartilage)*

USE GRAPHIC ORGANIZERS A cluster diagram can help students organize the information about fishes. Draw a blank diagram on the board with four ovals as shown below, and label the center oval. Have students copy the diagram in their notebooks. Tell them to use the paragraph headings to fill in the three outer ovals.

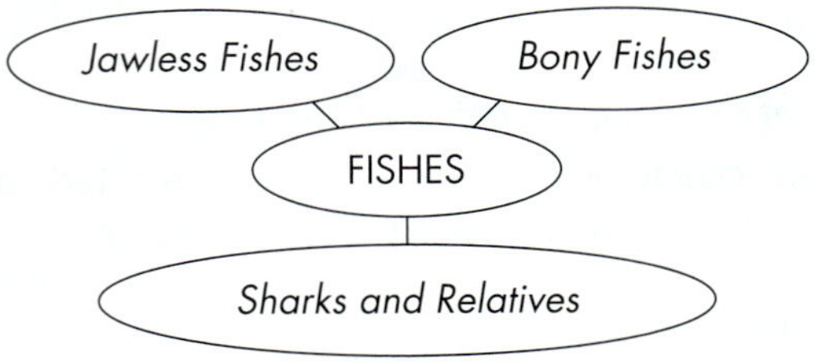

Then, have students add ovals, extending from the three outer ovals, that include examples and adaptations of the three groups of fish.

Build Connections

VISUAL SUMMARY Make sure students understand that the visual starts on this page and continues on the facing page. Help them focus on the details of the illustrations. Write the adaptations listed below on the board. Ask students to find them in the illustration and then number them in evolutionary sequence.

(3) stubby front fins with wrists

(1) front fins for steering

(2) larger, more mobile front fins

(4) front feet with toes

(6) true tetrapod

(5) hind feet with toes

Speed Bump

Ask How do sharks and their relatives differ from bony fishes? *(The skeletons of sharks and their relatives are made of cartilage, not bone.)*

Jaws *Dunkleosteus* lived 360 million years ago. Its jaws were so powerful that it could have bitten a modern shark in half!

Sharks and Their Relatives Other early fishes evolved a very useful adaptation for feeding: jaws. Jaws hold teeth and muscles. These parts make it possible to bite and chew.

Early fishes also evolved two pairs of fins, one pair toward the front of the body and one pair toward the back. Paired fins help fishes control their body movements. Tail fins and powerful muscles let them swim quickly.

These adaptations gave rise to the adaptive radiation of the Chondrichthyes (kahn DRIK theez): the sharks, rays, and skates. These animals have skeletons made of cartilage. Some, such as the great white shark, eat meat. Others, such as the manta ray, eat plankton. p. DOL 50

Bony Fishes Another group of early fishes evolved skeletons made of bone. This adaptation set in motion the adaptive radiation of the class Osteichthyes (ahs tee IK theez), which includes the bony fishes. p. DOL 51

▶ ***Ray-Finned Fishes*** Most fishes you know of, such as eels, catfish, and goldfish, are ray-finned fishes. Ray-finned fishes have skeletons made of bone. The name "ray-finned" describes the thin bones that hold up the skin in their fins. Most bony fishes also have fins, scales, and gills.

▶ ***Lobe-Finned Fishes*** Another group of bony fishes are the lobe-finned fishes. They evolved larger bones that support thick, fleshy fins. Modern lobe-finned fishes include lungfishes and coelacanths (SEE luh kanths). One ancient group of lobe-finned fishes gave rise to the four-limbed vertebrates, or **tetrapods.**

BUILD Connections

FROM FINS TO FEET

The feet of tetrapods evolved from the fins of ancient bony fish. This cladogram shows some animals in this line of evolution. All of the animals in the art are extinct.

Eusthenopteron was an early bony fish that used its muscular front fins for steering more than for swimming.

Panderichthys was a fish with sturdier, more mobile, and proportionately larger front fins than earlier fishes had.

Tiktaalik was not quite a fish and not quite a tetrapod. It had stout, stubby front fins with flexible wrists that likely enabled it to prop itself up on land, but it had no digits. It had gills and lungs.

630

Hands-On Learning

MODEL A FISH SKELETON Pair students and hand out these materials: photograph of a bony fish skeleton, such as a perch; modeling clay; chenille stems; scissors

Have students use the photograph to make a model of a bony fish skeleton. Tell them they can use the clay for the backbone, skull, and jaws and use the chenille stems for the smaller bones.

Ask students what advantages a skeleton offers a bony fish. Have them record their thoughts in their notebooks. *(Students should recognize that the bony skeleton provides structure [shape] and protection. The bony fins help in swimming.)*

Amphibians Frogs and toads are amphibians. The word *amphibian* means "double life." As larvae, these animals live in water. As adults, they can live on land. Most adults have lungs and can breathe air. But most amphibians need water to reproduce. Amphibians have moist skin. They do not have scales or claws. pp. DOL 52–53

▶ ***The "Fishapod"*** Adaptations for life on land evolved slowly, over millions of years. Some groups of lobe-finned fishes evolved sturdy fins. Over time, some of these appendages evolved features that looked like the limbs of tetrapods. Many fossils, discovered over many years, document these changes. Recently, scientists have discovered several fossils that show clearly how lobed fins evolved into front and hind limbs. One such fossil is *Tiktaalik*. This animal had a mix of fish and tetrapod features in many of its body parts. It had gills and could swim underwater. It also had lungs and could crawl on land. Many call it a "fishapod"—part fish, part tetrapod. It shows a clear middle stage between fishes and tetrapods.

▶ ***Adaptations to Life on Land*** To live on land, animals need more than legs. Early amphibians also evolved ways to breathe air and to protect themselves from drying out. With these adaptations, amphibians became the main vertebrates of the warm and wet Carboniferous Period, which lasted from 359 to 300 million years ago.

Their success did not last. Climate changes caused many swamps to dry up. Most amphibians became extinct by the end of the Permian Period, about 250 million years ago. Today, only three orders of amphibians survive—frogs and toads, salamanders, and caecilians (see SIL ee unz).

BUILD Vocabulary

cartilage
a type of connective tissue that supports the body and is softer and more flexible than bone

tetrapod
a vertebrate with four limbs

WORD ORIGINS
The word *tetrapod* comes from the Greek words *tetra*, meaning "four," and *pod*, meaning "foot."

***Tiktaalik*, the Fishapod** This 375-million-year-old fossil of *Tiktaalik* was found in Canada. Like a fish, it had gills. Like a tetrapod, it had lungs and fins with wrist bones.

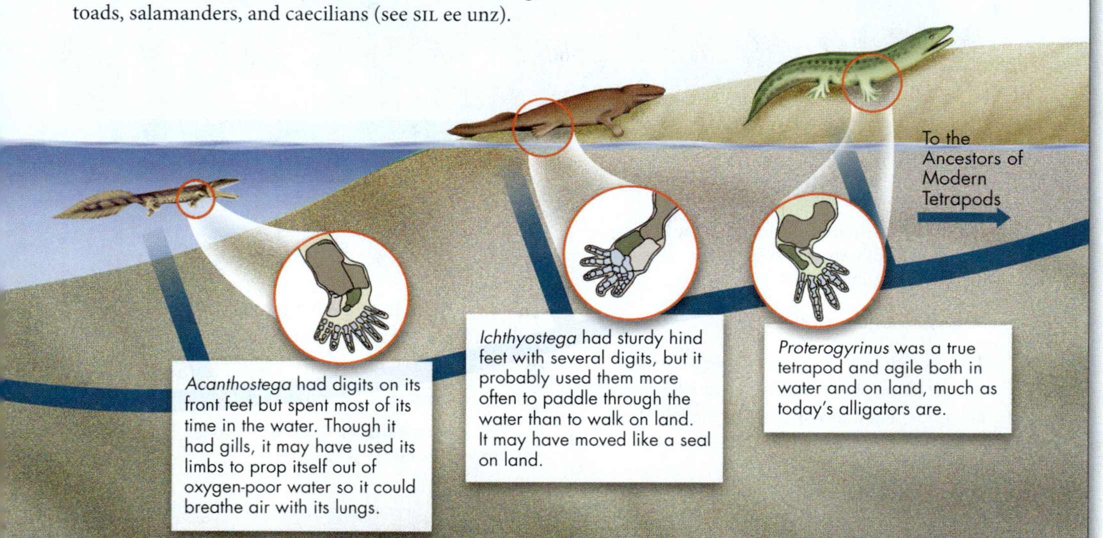

Focus on ELL: Build Background

BEGINNING SPEAKERS Students may be unfamiliar with some of the animals shown in the chordate cladogram, though most are familiar with frogs. Invite volunteers to tell what they know about the early life of a frog. *(A frog egg hatches as a tadpole. The tadpole lives in water and uses its tail to swim and its gills to breathe. As the tadpole matures, its gills are replaced with lungs, and it grows legs.)*

Ask What do the changes in the tadpole enable the adult frog to do? *(live on land)*

Ask How was *Tiktaalik* similar to both a tadpole and a frog? *(Like a tadpole,* Tiktaalik *had gills and could swim underwater. Like a frog,* Tiktaalik *had lungs and appendages for moving on land.)*

Active Reading

INTERPRET DIAGRAMS Encourage students to study both halves of the From Fins to Feet cladogram.

Ask Which animal, *Tiktaalik* or *Proterogyrinus,* is more closely related to the ancestors of modern tetrapods? How do you know? (Proterogyrinus; *the cladogram shows it is closer to the ancestors of modern tetrapods than* Tiktaalik *is.)*

DRAW CONCLUSIONS Discuss the characteristics of amphibians with students.

Ask If amphibians have adaptations for life on land, why did so many amphibians become extinct when their environment became drier? *(Amphibians need water to reproduce.)*

Build Vocabulary

WORD ORIGINS Tell students that a root word, such as *pod*, can be combined with different root words to produce words with different meanings. Have students look up in a dictionary the meanings of the following words from their Greek derivations: *tripod* (three footed), *arthropod* (jointed foot), *sauropod* (lizard foot), *pseudopod* (false foot).

Speed Bump

Ask What adaptations enabled vertebrates to make the transition to land? *(lungs, front and back appendages with fingers and toes, adaptations to protect themselves from drying out)*

During Reading continued

Active Reading

MAKE INFERENCES Ask a volunteer to read aloud the paragraph about birds on the next page and then repeat the second sentence. Have students use this sentence to infer what features birds have that are not mentioned in the paragraph. *(features of reptiles, such as well-formed lungs and eggs with shells)*

DRAW CONCLUSIONS Read aloud the third sentence in the paragraph about birds.

Ask Why is it important that birds have strong but light bones? *(The bones have to be strong for flying but not so heavy that they weigh down the bird.)*

Speed Bump

Ask What adaptations of reptiles help them live on dry land? *(scaly skin, well-formed lungs, strong legs, eggs with shells)*

Ask From what kind of reptile do scientists think birds evolved? *(feathered dinosaurs)*

Inquiry into Scientific Thinking

LEAD A DISCUSSION Review the information in the table. Point out the names of the different dinosaur groups in the first column.

Say When making a cladogram, it is often easier to work backward, from most recent to oldest.

Ask Which feathers listed in the table most likely evolved last? *(flight feathers)* How do you know? *(Modern birds have flight feathers.)*

Ask Which other feathers listed in the table most likely evolved last before flight feathers? *(true feathers)*

Tell students to continue to follow this sort of questioning pattern until they have placed the feather types in order from newest to oldest.

MAKE INFERENCES Help students use the meanings of the words *partial, true,* and *flight* to infer how the feathers differ. Lead them to understand that a partial feather has some characteristics of a feather. A true feather has all the characteristics of a feather. A flight feather is a specialized feather used for flying.

Ask Which feathers in the table are types of partial feathers? *(branched feathers and hairlike feathers)*

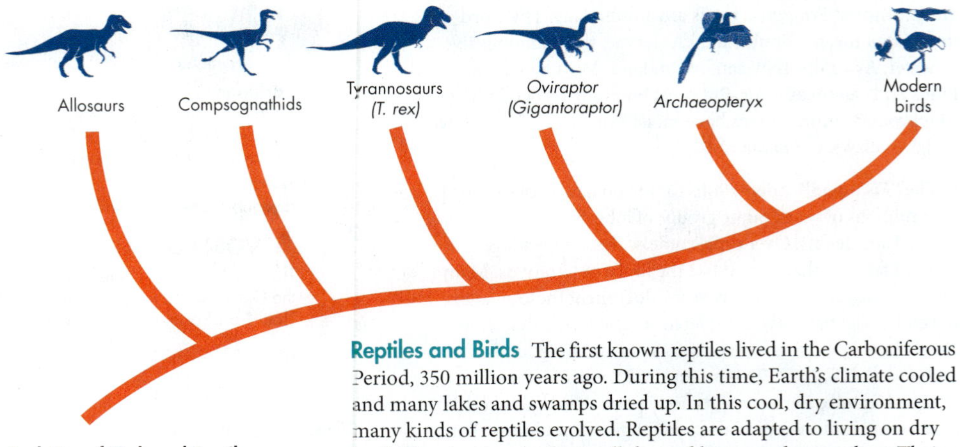

Evolution of Birds and Reptiles The cladogram shows current hypotheses about the relationship between living and extinct reptiles. None of the groups shown are direct ancestors of modern reptiles or modern birds.

Reptiles and Birds The first known reptiles lived in the Carboniferous Period, 350 million years ago. During this time, Earth's climate cooled and many lakes and swamps dried up. In this cool, dry environment, many kinds of reptiles evolved. Reptiles are adapted to living on dry land. They have scaly skin, well-formed lungs, and strong legs. Their eggs have shells and membranes that keep them from drying out. There are four groups of reptiles living today: lizards and snakes, crocodiles, turtles and tortoises, and the tuatara. pp. DOL 54–55

▶ ***Dinosaurs*** During the Triassic and Jurassic periods reptiles called dinosaurs lived all over the world. Some were small. Others were enormous. Some ate leafy plants. Others were carnivorous. Some dinosaurs even had feathers. The first birdlike fossils to be discovered came from the end of the Jurassic Period, about 150 million years ago. Fossils show that one group of feathered dinosaurs included the ancestors of modern birds.

INQUIRY into Scientific Thinking

Feather Evolution

The information in the table shows the kinds of feathers that were found in some groups of dinosaurs that lived before birds.

Group (listed alphabetically)	Feather Status
Allosaurs	None
Archaeopteryx	Flight feathers
Compsognathids	Hairlike feathers
Oviraptors	True feathers
Tyrannosaurs	Branched feathers

Analyze and Conclude

1. Organize Data Copy the cladogram of the Evolution of Reptiles and Birds into your notebook. Then review the information about drawing cladograms in Chapter 18.

2. Infer Draw circles on the cladogram to show when different kinds of feathers evolved.

3. Draw Conclusions Which type of feathers would you expect modern birds to have?

In Your Workbook Get more help for this activity in your workbook.

Analyze and Conclude

1. Check to see that students' cladograms are accurate.
2. Circles on the cladogram should be positioned to indicate first appearance of feather type. A "partial feathers" circle should precede Compsognathids, a "true feathers" circle should precede Oviraptors, and a "flight feathers" circle should precede *Archaeopteryx*.
3. Most modern birds can f y, so they probably have flight feathers.

▶ ***Mass Extinction*** At the end of the Cretaceous Period, about 66 million years ago, a mass extinction took place. Scientists hypothesize that it was caused by natural disasters, including volcanic eruptions and a fall in sea level. A huge asteroid smashing into Earth probably caused forest fires and dust clouds. After these events, dinosaurs and many other living things became extinct.

▶ ***Birds*** Modern birds make up the class Aves. Birds can keep their bodies warm when it is cold. Birds have feathers and strong, yet light bones. Their two legs are covered with scales and their front limbs are wings. pp. DOL 56–59

Mammals Members of the class Mammalia are mammals. They can be as small as mice or as large as whales. Mammals are covered in hair. They also breathe air, have four-chambered hearts, and are able to keep warm in cold weather. The mothers feed their young with milk from their mammary glands. Most mammals bear live young that have developed in the mother's body. pp. DOL 60–64

Mammals first evolved during the late Triassic Period, about 220 million years ago. They were very small and looked like tree shrews. While dinosaurs ruled, mammals were small. They were active mostly at night. After the great dinosaur extinction, about 65 million years ago, mammals began to get bigger and live in many different places. The Cenozoic Era is also called the Age of Mammals.

Key Question What can we learn by studying the cladogram of chordates? **The cladogram of chordates shows how chordate groups are related and the points at which different features evolved.**

***Archaeopteryx*, an Early Bird** This fossil comes from the late Jurassic Period, about 150 million years ago. *Archaeopteryx* (ahr kee AHP tur iks) was a small, running dinosaur that had feathers. The drawing shows how *Archaeopteryx* may have looked.

CHECK Understanding

Apply Vocabulary

Use the highlighted words from the lesson to complete each sentence correctly.

1. _______ is a strong, flexible tissue that supports chordate muscles and body parts.
2. Animals that have four appendages are called _______.

Critical Thinking

3. **Review** Name the group of animals whose ancestors were related to the earliest chordates.
4. **Compare and Contrast** What are two ways that bony fishes differ from ancient jawless fishes?
5. **Compare and Contrast** How do nonvertebrate chordates differ from other chordates?
6. **Relate Cause and Effect** How could the climate changes of the Permian Period have caused many kinds of amphibians to become extinct?
7. **Write to Learn** Answer the second clue of the mystery.

MYSTERY CLUE

How old would rocks need to be in order to contain fossils of the ancestors of birds? (**Hint:** See p. 632.)

BIOLOGY.com Search Lesson 26.2 GO • Lesson Assessment

Check Understanding

1. Cartilage
2. tetrapods
3. Early chordates were probably related to the ancestors of echinoderms.
4. Accept any two of the following differences: Jawless fishes do not have jaws, teeth, or hard bones in their skeletons; bony fishes do.
5. Nonvertebrate chordates do not have backbones, while all other chordates do.
6. Amphibians spend the early part of their life in water. When climate changes caused swamps and other wet areas to dry up, the amphibians had fewer places to grow and reproduce.
7. Fossils of birdlike ancestors, such as *Archaeopteryx*, date from the late Jurassic Period, about 150 million years ago.

Active Reading

SCIENCE CONNECTION Recognizing birds as descendants of dinosaurs should cause a change in their classification. Modern birds by themselves form a clade within the clade containing dinosaurs. Because the clade containing dinosaurs is part of a larger clade of reptiles, modern birds are also reptiles. The traditional class Reptilia includes living reptiles and dinosaurs but not birds. However, it is not a clade.

KEY QUESTION Ask a volunteer to read aloud the Key Question and its answer.

Refer students to the Cladogram of Chordates at the beginning of this lesson, and have them point out the different kinds of information that are shown on the diagram.

Speed Bump

Ask Why is the Cenozoic Era also called the Age of Mammals? *(Mammals began to get bigger and live in many different places during the Cenozoic Era.)*

Post Reading

ASSESSMENT TIPS Review with students the visuals Cladogram of Chordates and From Fins to Feet before they answer the assessment.

For Question 3, remind students that closely related organisms appear near one another on a cladogram.

For Question 6, review the meanings of *cause* and *effect*. Point out that in the question, climate change is the cause, and amphibian extinction is the effect.

MYSTERY CLUE Lead students to recall that fossil evidence suggests modern birds evolved from feathered dinosaurs.

Wrap-Up Activity

FLASH CARDS Have students form groups of three. Give each group six index cards. Tell them to write the name of one of the six chordate groups on the front of each card. On the back of the card, they should write one feature that distinguishes that chordate group from the others. Check the cards for accuracy. Then, allow groups to quiz one another using their feature cards as flash cards.

LESSON 26.3

Pre-Reading

Objectives

26.3.1 Identify the characteristics that all primates share.

26.3.2 Describe the major evolutionary groups of primates.

26.3.3 Describe the adaptations that enabled later hominine species to walk upright.

26.3.4 Describe the current scientific thinking about the genus *Homo*.

Preview the Pages

1. Preview the Cladogram of Primates. Point out that monkeys, chimpanzees, and gorillas belong to different groups of primates.
2. Have students find the vocabulary terms in the text. Refer them to the appropriate drawings of primates in the cladogram to illustrate the meaning of each term.

Build Understanding

T-CHART Direct students to Workbook B, where they will find a partially completed T-chart. Invite a volunteer to draw his or her chart on the board. Correct the chart as necessary. Students can revise their own charts as well.

NATIONAL SCIENCE EDUCATION STANDARDS

UNIFYING CONCEPTS AND PROCESSES

I, II, III, IV, V

CONTENT

C.3.a, C.3.b, C.3.c, C.3.d, C.3.e, C.6.a, C.6.c, C.6.d, G.2, G.3

For the corresponding lesson in ***Miller & Levine Biology*, see pages 765–772.**

26.3 Primate Evolution

Key Questions

What characteristics do all primates share?

What are the major evolutionary groups of primates?

What adaptations enabled later hominines to walk upright?

What is the current scientific thinking about the genus Homo?

BUILD Understanding

T-Chart Before you read Lesson 3, make a T-chart that lists the main topics (green headings) in the left column. As you read, write details for each topic in the right column.

In Your Workbook Go to your workbook to learn more about making a T-chart.

What Is a Primate?

Lemurs, monkeys, great apes, and humans are primates. All primates have evolved adaptations useful for life in trees. Primates have long fingers and toes that can curl around things and hold them tightly. Most primates have thumbs and big toes that can open and close against the other fingers like a hinge. All of us have nails instead of claws, and arms that can rotate at the shoulder. Primates also have a strong collar bone (clavicle) that helps tree-dwelling species to swing from branch to branch. Eyes on the front of our faces give us **binocular vision,** the ability to combine what we see with two eyes into one image that has depth. Finally, all primates have a well-developed cerebrum, or "thinking" part of the brain. p. DOL 64

Key Question What characteristics do all primates share?
Primates have long fingers and toes, nails, arms that rotate at the shoulder, strong clavicles, binocular vision, and large cerebrums.

Evolution of Primates

The ancestors of modern primates lived more than 65 million years ago. Today there are three major groups of primates. One group includes lemurs and lorises. They are small and have large eyes that help them see in the dark. A second group is the tarsiers. They have broad faces and widely separated nostrils. The third group, **anthropoids** (AN thruh poydz), includes monkeys, great apes, and humans.

Lemurs and lorises
Tarsiers
Anthropoids

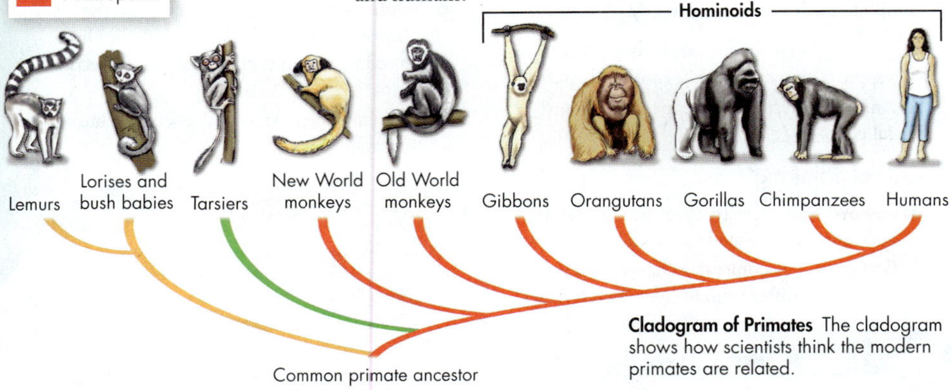

Cladogram of Primates The cladogram shows how scientists think the modern primates are related.

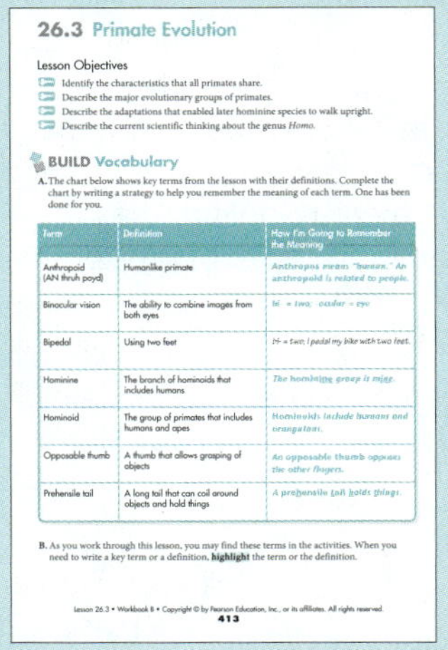

26.3 Primate Evolution

Lesson Objectives

- Identify the characteristics that all primates share.
- Describe the major evolutionary groups of primates.
- Describe the adaptations that enabled later hominine species to walk upright.
- Describe the current scientific thinking about the genus *Homo*.

BUILD Vocabulary

A. The chart below shows key terms from the lesson with their definitions. Complete the chart by writing a strategy to help you remember the meaning of each term. One has been done for you.

Term	Definition	How I'm Going to Remember the Meaning
Anthropoid (AN thruh poyd)	Humanlike primate	*Anthropos means "human." An anthropoid is related to people.*
Binocular vision	The ability to combine images from both eyes	*bi- = two; ocular = eye*
Bipedal	Using two feet	*bi- = two; I pedal my bike with two feet.*
Hominine	The branch of hominoids that includes humans	*The hominine group is mine.*
Hominoid	The group of primates that includes humans and apes	*Hominoids include humans and orangutans.*
Opposable thumb	A thumb that allows grasping of objects	*An opposable thumb opposes the other fingers.*
Prehensile tail	A long tail that can coil around objects and hold things	*A prehensile tail holds things.*

B. As you work through this lesson, you may find these terms in the activities. When you need to write a key term or a definition, **highlight** the term or the definition.

Lesson 26.3 • Workbook B • Copyright © by Pearson Education, Inc., or its affiliates. All rights reserved.
413

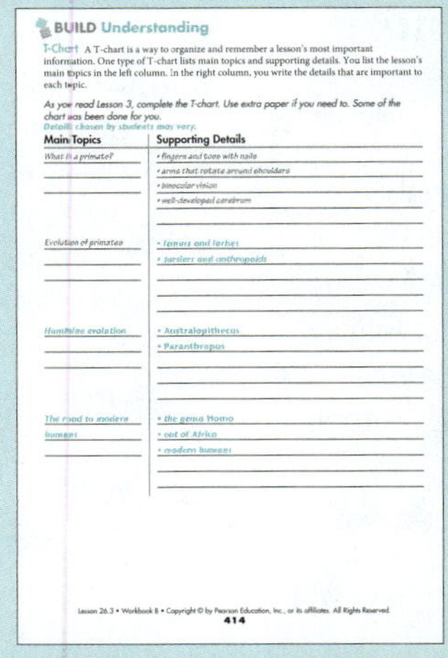

BUILD Understanding

T-Chart A T-chart is a way to organize and remember a lesson's most important information. One type of T-chart lists main topics and supporting details. You list the lesson's main topics in the left column. In the right column, you write the details that are important to each topic.

As you read Lesson 3, complete the T-chart. Use extra paper if you need to. Some of the chart has been done for you.
Details chosen by students may vary.

Main Topics	Supporting Details
What is a primate?	*• fingers and toes with nails*
	• arms that rotate around shoulders
	• binocular vision
	• well-developed cerebrum
Evolution of primates	*• lemurs and lorises*
	• tarsiers and anthropoids
Hominine evolution	*• Australopithecus*
	• Paranthropus
The road to modern humans	*• the genus Homo*
	• out of Africa
	• modern humans

Lesson 26.3 • Workbook B • Copyright © by Pearson Education, Inc., or its affiliates. All Rights Reserved.
414

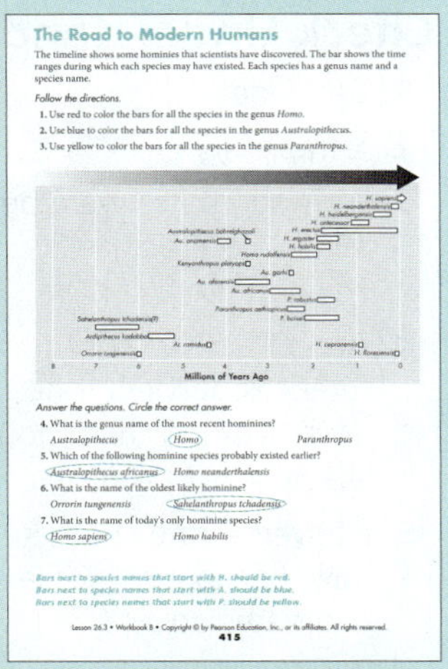

The Road to Modern Humans

The timeline shows some hominies that scientists have discovered. The bar shows the time ranges during which each species may have existed. Each species has a genus name and a species name.

Follow the directions.

1. Use red to color the bars for all the species in the genus *Homo*.
2. Use blue to color the bars for all the species in the genus *Australopithecus*.
3. Use yellow to color the bars for all the species in the genus *Paranthropus*.

Millions of Years Ago

Answer the questions. Circle the correct answer.

4. What is the genus name of the most recent hominines?
Australopithecus *Homo* *Paranthropus*
5. Which of the following hominine species probably existed earlier?
Australopithecus africanus *Homo neanderthalensis*
6. What is the name of the oldest likely hominine?
Orrorin tungenensis *Sahelanthropus tchadensis*
7. What is the name of today's only hominine species?
Homo sapiens *Homo habilis*

Bars next to species names that start with H. should be red.
Bars next to species names that start with A. should be blue.
Bars next to species names that start with P. should be yellow.

Lesson 26.3 • Workbook B • Copyright © by Pearson Education, Inc., or its affiliates. All rights reserved.
415

When the continents separated 45 million years ago, anthropoids were split into two groups. New World monkeys live in Central and South America and they have prehensile tails. A **prehensile tail** can wrap around a branch and hold on like an extra hand.

The other group of anthropoids are the Old World monkeys and great apes. They live in Africa and Asia. Old World monkeys climb trees. But they do not have prehensile tails. Great apes include gibbons, orangutans, gorillas, chimpanzees, and humans. Great apes are also called hominoids.

Key Question What are the major evolutionary groups of primates? **The three major groups of primates are the lemurs and lorises, the tarsiers, and the anthropoids.**

Hominine Evolution

The study of human ancestors and other primates shows that chimpanzees are our closest living relatives. Recent fossil discoveries show that the evolutionary lines that led to humans split from the lines that led to chimpanzees around 6 or 7 million years ago. DNA studies agree with fossil evidence about the time these lines split.

All hominoids in the line that led to humans are called hominines. The fossils of early hominines show changes in the slope of the skull, neck, spine, hip bones, and leg bones. These changes in shape let later species walk upright. Animals that walk on two feet are said to be bipedal. Standing and walking upright left the hands free for using tools. Each hand had an **opposable thumb** that could touch the tips of the fingers. This thumb made it easier to use tools. Hominines also evolved much larger brains than other apes.

BUILD Vocabulary

binocular vision the ability to merge visual images from both eyes, giving depth perception and a three-dimensional view of the world

anthropoid the primate group made up of monkeys, apes, and humans

prehensile tail a long tail that can coil tightly around a branch

opposable thumb a thumb that allows the hand to grasp objects and use tools

Time Line This time line shows hominine fossils and the time range during which each species probably lived. These time ranges may change as paleontologists gather new data. Right now, there are several different hypotheses about how these species are related. That's why we present these data as a time line, rather than as a cladogram.

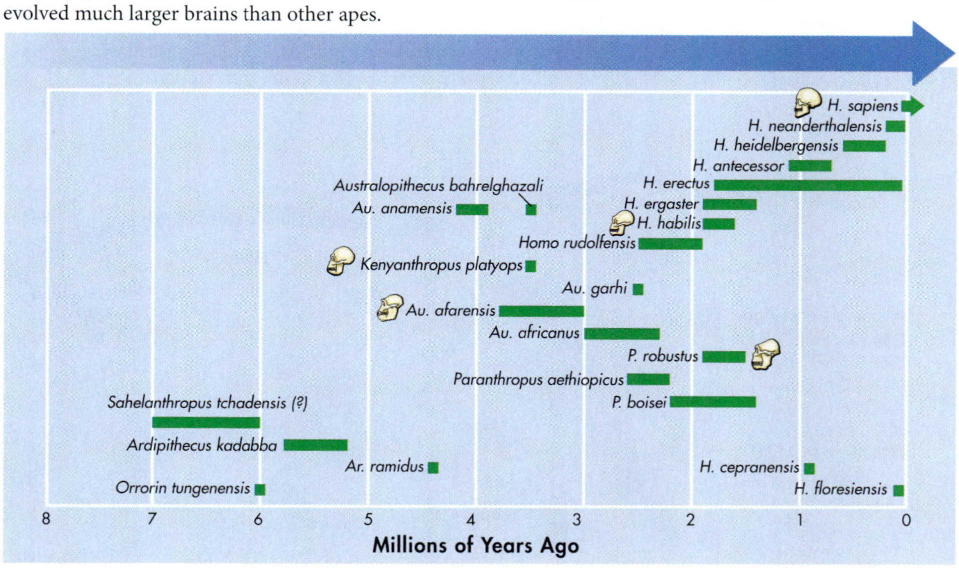

LESSON 26.3

During Reading

Active Reading

KEY QUESTION Invite a volunteer to read aloud the Key Question on the previous page and its answer.

Ask What are some examples of primates? *(Sample answer: lemurs, monkeys, apes, humans)*

USE VISUALS Have students read the section Evolution of Primates. Then, refer them to the Cladogram of Primates.

Ask If I drew a bracket on the cladogram to show which primate groups are anthropoids, where would the bracket begin and end? *(It would begin with New World monkeys and end with humans.)*

Direct students to read the rest of this page. Point out the hominoids in the cladogram.

Ask Which hominoid is a hominine? *(the human)*

MAKE AN ANALOGY Draw a simple family tree on the board. Discuss family trees and how hominine evolution can be compared to a shrub rather than to a tree.

Say The trunk of the tree represents the main ancestor and the branches represent the descendants.

Explain that a shrub has a number of trunks and more branches than a tree. Encourage students to read the rest of the Hominine Evolution section on the next page.

Ask Why would a family shrub be more appropriate than a family tree to show the relationship of humans to other hominines? *(There were many hominine species with many different ancestors. Not all the hominine ancestors are human ancestors.)*

KEY QUESTION Read aloud the Key Question and its answer on this page.

Encourage students to draw a three-column chart that lists the features of each of the three groups of primates.

Hands-On Learning

GOING THUMBLESS Have students form groups of four. Provide these materials: masking tape, scissors, paper clip, book, pencil and manual pencil sharpener, other items to manipulate.

1. One student from each group can act as the recorder. This student should tape the thumbs of each of the other group members to the sides of their hands.
2. Students with taped thumbs should attempt such tasks as picking up the paper clip, turning pages of the book, and sharpening the pencil.
3. The recorder should list the tasks attempted and record the group's answers to the following questions:

Which tasks were difficult to do without using an opposable thumb? *(Students will probably say that all tasks were difficult.)*

Why were the tasks difficult? *(It was difficult to grasp the objects.)*

What kind of task was the most difficult? *(Sample answer: the kind of task that required picking up a small object, such as a paper clip)*

Speed Bump

Ask Which of the following statements is supported by the primate cladogram?

Humans descended from monkeys. *(no)*

Humans and monkeys have a common primate ancestor. *(yes)*

During Reading continued

Active Reading

MAKE CONNECTIONS Have students think about a time when they have made footprints, such as when walking on wet sand or mud, and what their footprints looked like.

Ask How might a hominine's footprints show that it walked upright? *(Sample answer: The footprints have a deep heel mark, five toes in a line, and a space where the foot arches.)*

INTERPRET VISUALS Direct students' attention to the Out of Africa map. Make certain they understand the meanings of the colors and arrows.

Ask Why do you think the title of this map is Out of Africa? *(Sample answer: The earliest hominine fossils were found in Africa. The arrows show that some hominines left Africa and traveled to the Middle East.)* Where did hominines travel to from the Middle East? *(Europe and Asia)*

KEY QUESTION Ask a volunteer to read aloud the Key Question and its answer.

Ask Why was the ability to walk upright important to the evolution of hominines? *(Their hands were left free for using tools.)*

Speed Bump

Ask To what genus do modern humans and their closest relatives belong? *(Homo)*

Ask What evidence supports the hypothesis that *Homo sapiens* evolved in Africa and traveled outward from there? *(DNA from living humans)*

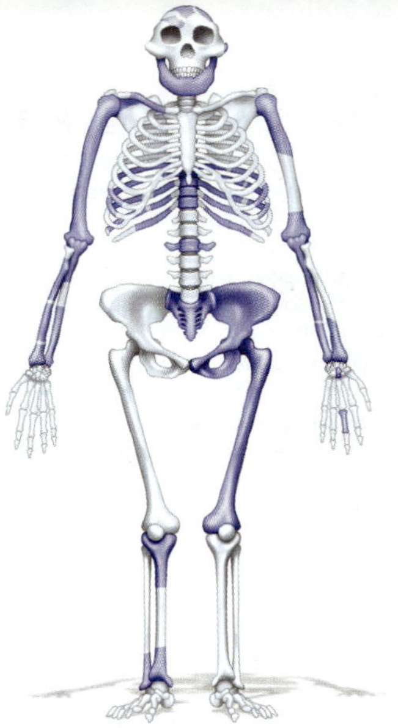

Lucy "Lucy" is the nickname of a fossil of *A. afarensis* that was discovered in Ethiopia. Lucy's skeleton was not complete. The bones that were found are dark blue.

Hominine Relationships Researchers have found fossils of at least 20 species of hominines. All these species are relatives of modern humans. But not all of them are human ancestors. What's the difference? Think of your family. Your relatives may include aunts, uncles, cousins, parents, grandparents, and great grandparents. But only your parents, grandparents, and great grandparents are your ancestors.

It isn't easy to figure out which hominines are human ancestors and which are just distant relatives. It is now clear that there have been several hominine adaptive radiations. Each of these radiations produced several species. These species are shown in the time line on p. 635. From fossils alone, it is hard to tell how some of these species are related to each other and to humans.

Fossils of Hominines Fossils tell how early hominines looked and lived. *Australopithecus afarensis* lived from about 4 million to 2.5 million years ago. A skeleton of *A. afarensis* called "Lucy" stood about 1 meter tall. One set of trace fossils is thought to be of *Australopithecus* footprints. These footprints show that the ability to walk upright evolved long before large brains.

Key Question What adaptations enabled later hominine species to walk upright? **The skull, neck, spine, hip bones, and leg bones of early hominines changed shape in ways that let later species walk upright.**

The Road to Modern Humans

About 2 million years ago in Africa, a new group of hominines arose. Scientists classify these species in the genus *Homo*. One set of fossils was found with tools made of stone and bone. It was named *Homo habilis* (HAB ih lus), which means "handy man." Another species, *Homo ergaster*, was larger than *H. habilis*. It had a bigger brain and a nose that was more like that of modern humans.

Out of Africa Fossils show that relatives and ancestors of modern humans left Africa in waves. (Skull symbols show where bones were found.)

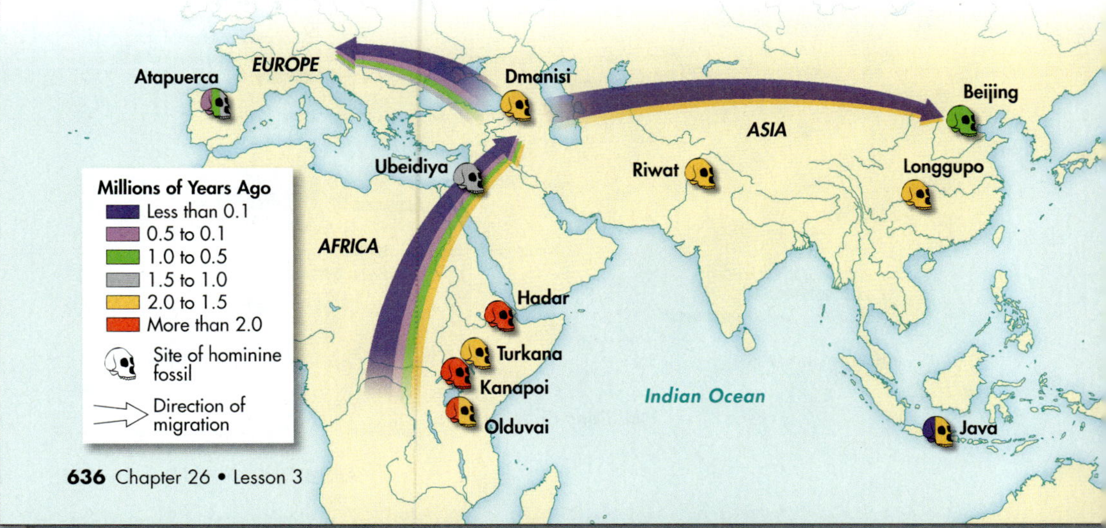

ELL Focus on ELL: Extend Language

ALL SPEAKERS Ask students to find the words *timeline, footprint,* and *upright* on this page. Tell them that each of these words is a compound word—a word made by joining two words. Discuss the meaning of each word in context.

Discuss the difference between closed compound words and those that take the open form (example: *post office*) or hyphenated form (example: *six-year-old*). Tell students that the only sure way to know if a compound word is closed, open, or hyphenated is to check a dictionary.

Write the following pairs of words on the board. Have students use a dictionary to determine which form closed compound words and which should be open or hyphenated.

hand puppet *(open)*	eye brow *(closed)*	night time *(closed)*
mass produced *(hyphenated)*	eye lid *(closed)*	night table *(open)*

Out of Africa—But When and Who? In time, some members of the genus *Homo* moved from Africa to other parts of the world. Where and when did our species, *Homo sapiens*, evolve? One hypothesis says that modern humans evolved in several different parts of the world at the same time. Another hypothesis says that modern humans evolved in Africa about 200,000 years ago. Then they moved out of Africa. Soon they took the place of other hominines that already lived elsewhere. DNA from living humans strongly supports the second hypothesis: *Homo sapiens* came out of Africa.

Modern Humans At least three other *Homo* species lived at the same time as early humans. The two most important were *Homo neanderthalensis* and *Homo sapiens*. Neanderthals began living in Europe and western Asia about 200,000 years ago. They lived in Europe until about 24,000 years ago. Neanderthals were very good hunters, made stone tools, and used fire. They lived in social groups and buried their dead with simple rituals.

Homo sapiens arrived in the Middle East from Africa about 100,000 years ago. About 50,000 years ago, they began making and using tools from stone and bones. They also painted on cave walls. They buried their dead with elaborate rituals.

Neanderthals and *Homo sapiens* lived side by side in the Middle East for about 50,000 years and in Europe for several thousand years. How and why did Neanderthals disappear? Did they inbreed with *Homo sapiens*? No one knows for sure. We do know this: For the last 24,000 years, *Homo sapiens* has been Earth's only hominine. Our species is the only surviving member of a clade that was once large and diverse.

Cro-Magnon Art This ancient cave painting from France shows that Cro-Magnons were skilled artists.

Key Question What is the current scientific thinking about the genus *Homo*? **Many species of *Homo* lived before *Homo sapiens* appeared. At least three other *Homo* species lived at the same time as early humans.**

CHECK Understanding

Apply Vocabulary

Use the highlighted words from the lesson to complete each sentence correctly.

1. _______ lets primates combine the images from two eyes into a single image.
2. A(n) _______ can wrap around a branch and hold on like a hand.

Critical Thinking

3. **Relate Cause and Effect** How was walking upright important to the evolution of hominines?
4. **Compare and Contrast** What is one way in which Neanderthals differed from *Homo sapiens*?
5. **Write to Learn** Describe the two hypotheses that explain how *Homo sapiens* spread throughout the world. How are these two hypotheses alike? How are they different?

BIOLOGY.com Search Lesson 26.3 GO • Lesson Assessment

Check Understanding

1. Binocular vision
2. prehensile tail
3. Walking upright freed arms and hands, allowing hominines to use tools.
4. Sample answer: *H. sapiens* had more elaborate burial rituals than *H. neanderthalensis*.
5. Sample answer: One hypothesis is that members of the *Homo* genus left Africa and that *Homo sapiens* evolved in several places around the world independently. The out-of-Africa model says that *Homo sapiens* evolved in Africa and then moved to other places around the world. The two models are similar in that they both say that the *Homo* ancestors originated in Africa. They are different in where they suggest the evolution to become *Homo sapiens* took place.

Active Reading

USE GRAPHIC ORGANIZERS Encourage students to make a T-chart that lists details about *Homo neanderthalensis* and *Homo sapiens.* Draw the chart on the board and work with students to record the first entries.

Homo neanderthalensis	*Homo sapiens*
in Europe and western Asia 200,000 years ago	in the Middle East 100,000 years ago

KEY QUESTION Read aloud the Key Question and its answer.

Ask How many living species are classified in the genus *Homo*? *(one)*

Speed Bump

Ask What happened to the species *Homo neanderthalensis*? *(The species became extinct, but no one knows how.)*

Post Reading

ASSESSMENT TIPS Review the highlighted vocabulary words with students before they begin the assessment.

For Question 4, tell students that the T-chart they made listing details about *Homo neanderthalensis* and *Homo sapiens* can help them answer the question.

Wrap-Up Activity

TIMELINE Have students work together in pairs to make a hominine timeline. Tell students the timeline should include the names of the hominine species discussed in the lesson and should extend from 4.5 million years ago to the present. Suggest that students divide the timeline into increments of 0.5 million years and label them. Check students' timelines for accuracy.

Extension: Next to the species names on the timelines, have students write details about the species described in the lesson.

Pre-Lab

Have students preview the text on this page and in the lab manual, skimming for words they don't understand. Make a list of these words on the board and then discuss their meanings. For words such as *relative, ancestor, hominoid,* and *hominine,* direct students to review Lesson 26.3, looking for context clues and definitions. Students should focus on the sections Evolution of Primates and Hominine Evolution.

Also review what it means to make inferences based on data. **Ask** What does *infer* mean? *(to draw a conclusion based on facts or clues)* How can you use data to infer? *(Sample answer: I can look for patterns in the data.)*

Review the Background Questions and Pre-Lab Questions as a class.

Lab

Tell students they will perform the chapter lab *Investigating Hominoid Fossils* in **Lab Manual B.**

Post-Lab

CEPHALIC INDEX Remind students that an index is a ratio that compares the value of two measurements. Tell them that a cephalic index is the ratio of head width to head length. This index is used to categorize pets, such as different breeds of dogs. This index is divided into three main categories. Have students use the Internet to define the three. Ask students who have dogs to measure their pet's head, calculate the cephalic index, and compare the index to the categories listed for that breed.

Forensics Lab

GUIDED INQUIRY

Pre-Lab: Investigating Hominoid Fossils

Problem What can a comparison of skulls and hands reveal about the evolution of humans?

Materials metric ruler, protractor

Lab Manual Chapter 26 Lab

Skills Focus Measure, Analyze Data, Compare and Contrast

Connect to the Big idea To learn about the evolution of humans, scientists study both close relatives and possible ancestors. Fossils of possible ancestors are rare, and complete skeletons are even rarer. Yet, scientists have gained valuable information from those fossils that have been found. In this lab, you will make measurements that a paleontologist might make after finding a fossil. Then, you will use your data to make inferences about human evolution.

Background Questions

a. **Review** What are hominoids, and what are hominines?

b. **Explain** Use the examples of chimpanzees and humans to explain the difference between evolutionary relatives and ancestors.

c. **Compare and Contrast** What is the difference between the locomotion of humans and the locomotion of chimpanzees?

Pre-Lab Questions

Preview the procedure in the lab manual.

1. **Use Models** What will you use instead of actual skulls and hands to make your measurements?
2. **Interpret Visuals** The bony cavities in a skull that protect the eyes are called orbits, or eye sockets. On the skulls, what does line AC measure? What does line BC measure?
3. **Use Analogies** Shoe sizes such as 9A and 11E (or 9 narrow and 11 extra-wide) are an example of an index. What two measurements are being compared in a shoe index?

BIOLOGY.com Search Chapter 26 GO

Visit Chapter 26 online to test yourself on chapter content and to find activities to help you learn.

Untamed Science Video Join the Untamed Science crew as they talk with insect experts to better understand why there are more than a million insects.

Art Review Review your understanding of different hominoids.

InterActive Art Build a cladogram of invertebrates.

Pre-Lab Answers

BACKGROUND QUESTIONS

a. Hominoids include gibbons, orangutans, gorillas, chimpanzees, and humans. Hominines include modern humans and all species more closely related to humans than chimpanzees.

b. Sample answer: Chimpanzees and humans are relatives who are descended from a common primate ancestor. Species that are not in the direct line of descent are not ancestors.

c. Chimpanzees walk on all four limbs while humans are bipedal, meaning that they walk upright on two limbs.

PRE-LAB QUESTIONS

1. I will be using images of skulls and hands
2. Line AC measures the distance from the bottom of the eye socket to the top of th skull. Line BC measures the distance from the top of the eye socket to the top of th skull.
3. The width of the shoe is being compared to the length of the shoe.

NATIONAL SCIENCE EDUCATION STANDARDS

UCP III, IV

CONTENT C.3.a, C.3.b, C.3.c, C.3.d, C.3.e

INQUIRY A.1.b, A.1.c, A.2.a

26 CHAPTER Summary

26.1 Invertebrate Evolution and Diversity

- Fossil evidence shows that the first animals began evolving long before the Cambrian Explosion.
- The cladogram of invertebrates shows current hypotheses about the evolutionary relationships of major groups of modern invertebrates. It also shows the order in which some important features evolved.

appendage (p. 625)
larva (p. 627)

26.2 Chordate Evolution and Diversity

- Studies of embryos show that the most ancient chordates were related to the ancestors of echinoderms.
- The cladogram of chordates shows how chordate groups are related and the points at which different features evolved.

cartilage (p. 628)
tetrapod (p. 630)

26.3 Primate Evolution

- Primates are mammals that have long fingers and toes, nails, arms that rotate at the shoulder, strong clavicles, binocular vision, and large cerebrums.
- The three major groups of primates are lemurs and lorises, tarsiers, and the anthropoids.
- The skull, neck, spine, hip bones, and leg bones of early hominines changed shape in ways that let later species walk upright.
- Many species of *Homo* lived before *Homo sapiens* appeared. At least three other *Homo* species lived at the same time as early humans.

binocular vision (p. 634)
anthropoid (p. 634)
prehensile tail (p. 635)
opposable thumb (p. 635)

UbD Transfer the Big Idea

TRANSFER Have students choose an animal with which they are familiar, such as a pet, an animal in a park, or a zoo animal. Direct students to write an informative pamphlet about the animal and include drawings or pictures. Tell students the pamphlet should identify the cladogram in this chapter that includes the animal and the name of the group in which this animal is classified. The pamphlet should also detail the animal's major adaptations, its ancestors, and the animal groups to which it is closely related.

Chapter Summary

Think Visually

USE GRAPHICS Have students copy a cladogram from the Animals section of the DOL and use information in the chapter and DOL to label the cladogram.

BIOLOGY.com Study Online

REVIEW RESOURCES

InterActive Art In this drag-and-drop activity, students will become more familiar with the cladogram of invertebrates.

Data Analysis Students investigate if fossils are of a species separate from *H. sapiens* or from a population of *H. sapiens.*

Vocabulary Review The *Flash Cards* and *Crossword* provide an interactive way to review chapter vocabulary.

Editable Worksheets Editable pages of Study Workbook B, Lab Manual B, and the Assessment Resources Book are available online.

Lesson Overviews Lesson Overviews are based on Miller & Levine Biology. However, they are available on CD-ROM and can be edited for your class.

ASSESSMENT RESOURCES

Chapter Assessment Have students take an online version of the Chapter 26 Check Understanding.

Standardized Test Prep Students can take an online version of the Standardized Test Prep. You will receive their scores along with ideas for remediation.

Diagnostic and Benchmark Tests Use these tests to monitor your students' progress and supply remediation.

ASSESSMENT

Assess

Big idea Invite a volunteer to read the Big Idea and the question. Then have students look back at the cladograms of invertebrates and chordates in Lessons 26.1 and 26.2.

Ask What do these cladograms show? *(how groups of invertebrates and chordates are related and the order in which different features evolved)*

Ask How would you summarize the relationship between invertebrates and chordates? *(Invertebrates and chordates share common structures that emphasize that all animals descended, with modifications, from common ancestors.)*

Ask What do we call the process through which organisms develop new adaptations and change over time? *(evolution)*

Have students write their answers in their notebooks.

CONSTRUCTED RESPONSE Paragraphs will vary. Explain the activity carefully. Be sure students understand that the *Hints* should be used as a guide to help them answer the questions. Review how to write a paragraph, stressing topic sentence/main idea and supporting details.

Write Question 1 on the board. Underline *ancestors, groups of animals,* and *cladogram.*

Read the first hint.

Ask If an animal group is in the center of a cladogram, what do you know about the groups to its left? *(They are related but they are missing at least one key characteristic of the group in the center.)* What do you know about the groups to its right? *(They are related but have key characteristics that are missing in the animal groups at the center.)*

Read the second hint.

Ask How does a cladogram show that several groups of animals have a common ancestor? *(The vertical lines from the animal groups lead to a common line that leads back to the ancestor.)*

Model writing the first one or two sentences, using the hints on the board and the text. *(Sample: A cladogram gives information about different animal groups and their ancestors. It shows which ancestors are common to different groups of animals.)*

26 CHECK Understanding

Assess the Big idea Evolution

Write an answer to the question below.

Q: How have animals descended from earlier forms through the process of evolution?

Constructed Response

Write an answer to each of the questions below. The answer to each question should be one or two paragraphs. To help you begin, read the **Hints** below the questions.

1. **Analyze Concepts** What can you tell about the ancestors of different groups of animals by looking at a cladogram?

   Hint Cladograms show evolutionary relationships.

   Hint Cladograms show when features evolved.

2. **Relate Cause and Effect** Life on Earth began in water. What are some of the ways that animals evolved to live on land?

   Hint Tetrapods can move on land, while fish cannot.

   Hint Breathing air requires different body parts than breathing underwater.

3. **Infer** When scientists find hominine fossils, they measure their size and decide their age. Why is it important for scientists to know the age as well as the size of these hominine bones?

   Hint Bone size helps scientists know how animals walked.

Foundations for Learning Wrap-Up

Use the three cladograms and the animal index cards you prepared while you read the chapter to help review what you learned. They will help you understand how different groups of animals are related.

Activity 1 Compare your three cladograms. Find the places where they overlap with one another. For example, the right end of the cladogram of invertebrates has a symbol that represents all chordates. The cladogram of chordates starts with an invertebrate ancestor. Attach these two cladograms where they overlap.

Next, attach the cladogram of primates where it overlaps with the cladogram of chordates. Then draw lines to connect the three cladograms.

You have now made a much larger cladogram. It shows details for only one group of chordates. Imagine how big your cladogram would be if it showed all the animals!

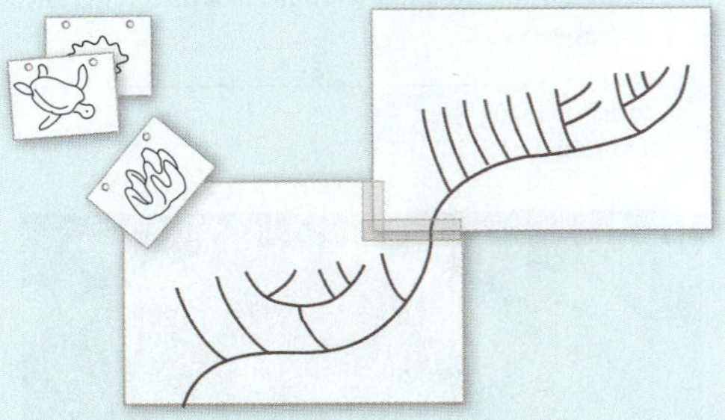

Activity 2 Study three index cards that each describe an animal in detail. Find where they belong in your large cladogram. Attach each one to the cladogram at the point where it belongs. Extend any lines that should connect with one another. Use the cladogram to find out how closely related your three animals are. Are they found within the same smaller cladogram? How closely related are they to humans? To a single-celled ancestor?

640 Check Understanding

For Question 2, pair students and have them use the hints and your modeling to answer the question. Have students share their paragraphs with the class.

For Question 3, have students write the paragraphs independently for homework or as an anchor activity.

Foundations for Learning Wrap-Up

Have students take out the cladograms and index cards they prepared. Be sure students have three cladograms and three index cards. Explain the activities and have students complete them.

26.1 Invertebrate Evolution and Diversity

Understand Key Concepts

1. The ancestors of many modern animal phyla first appeared during the
   a. Burgess Period.
   b. Cambrian Period.
   c. Precambrian Era.
   d. Ediacaran Period.

2. Which of the following groups has species that live in water, on land, and in the air?
   a. arthropods
   b. annelids
   c. mollusks
   d. echinoderms

Test-Taking Tip

Think of examples to help you eliminate answer choices. For question 2, think of what adaptations are needed to survive in the different places. Then think of examples of each of the different groups of animals. Earthworms and other kinds of worms are annelids. Worms do not have wings, so they cannot live in the air.

3. What evidence indicates that annelids and mollusks are closely related?

Think Critically

4. **Infer** Most cnidarians do not swim. To catch prey, they let water currents carry the prey to them. How is this behavior related to special structures they have?

5. **Compare and Contrast** How is the structure of echinoderms different from that of arthropods?

26.2 Chordate Evolution and Diversity

Understand Key Concepts

6. Which adaptation is NOT characteristic of reptiles?
   a. scaly skin
   b. shelled egg
   c. lungs
   d. gills

7. What feature separates birds from other groups of living animals?
   a. feathers
   b. two legs
   c. warm bodies
   d. wings

8. Dinosaurs became extinct at the end of the
   a. Triassic Period.
   b. Cretaceous Period.
   c. Carboniferous Period.
   d. Permian Period.

9. What adaptation lets birds live in places that are colder than those in which most reptiles live?

10. Which two major groups of fishes that still survive today evolved from the early jawed fishes?

Think Critically

11. **Infer** What structures of nonvertebrate chordates make these animals seem more closely related to vertebrates than to other groups of animals?

26.3 Primate Evolution

Understand Key Concepts

12. Anthropoids include monkeys and
    a. lemurs.
    b. lorises.
    c. tarsiers.
    d. humans.

13. Which of the following is characteristic of primates, but not of other mammals?
    a. body hair
    b. rotation at the shoulder joint
    c. notochord
    d. ability to control body temperature

14. The first hominines evolved about
    a. 30,000 years ago.
    b. 100,000 years ago.
    c. 6 to 7 million years ago.
    d. 120 million years ago.

15. What features do hominines have? Give an example of a hominine.

Assessment Tips

Direct pairs of students to work together to answer the questions. Tell students that to answer Questions 4 and 11 they have to infer, which means they will need to use information in the text, but that information doesn't directly answer the questions.

For Questions 17–19, encourage students to review the visual carefully before answering the questions.

Answers

Lesson 26.1

UNDERSTAND KEY CONCEPTS

1. b

2. a

3. Mollusks and annelids both have protostome development. Mollusks and many annelids also have trochophore larvae.

THINK CRITICALLY

4. Cnidarians have stinging tentacles that can catch prey that move through the water near them.

5. Echinoderms have an inner skeleton, while arthropods have an outer skeleton. Also, arthropods are segmented, while echinoderms are not.

Lesson 26.2

UNDERSTAND KEY CONCEPTS

6. d

7. a

8. b

9. Feathers help birds keep their body temperature warm even when it is cold.

10. sharks and their relatives and bony fishes

THINK CRITICALLY

11. Either as adults or as larvae, nonvertebrate chordates have the key chordate characteristics—a notochord, a nerve cord, pharyngeal pouches, and a tail that extends beyond the anus.

Lesson 26.3

UNDERSTAND KEY CONCEPTS

12. d

13. b

14. c

15. Hominines walk upright, can hold tools with an opposable thumb, and have large brains. *Homo erectus* is a hominine.

THINK CRITICALLY

16. Binocular vision lets a primate determine distance as it swings through trees. An opposable thumb lets the primate better grip things, such as branches while moving through trees.

Connecting Concepts

17. Devonian Period

18. Like a traditional cladogram, this kind of diagram shows the order in which certain groups may have evolved. This diagram also provides information about the geologic time scale and when each group may have evolved as well as the relative number of species in each group.

19. bony fishes

CHAPTER MYSTERY After students have read the Chapter Mystery, discuss the topics of fossil hunting and the way fossils provide evidence for the evolution of organisms.

Ask Why are there so few of the fossils Josh is interested in finding? *(Sample answer: The earliest animals were soft-bodied organisms, and fossils usually form from an organism's hard parts.)*

Ask Why can't fossils be found everywhere in all kinds of rock? *(Sample answer: Fossils can form in only some kinds of materials. Fossils are found mostly in sedimentary rocks.)*

Ask How do fossils provide evidence for the evolution of organisms? *(Sample answer: Fossils provide researchers with evidence of the structure of ancient organisms. By comparing different fossils, scientists can draw conclusions about how species changed over time.)*

26 CHECK Understanding

Think Critically

16. Apply Concepts How are binocular vision and opposable thumbs useful to a primate?

Connecting the Concepts

Use Science Graphics

The chart below shows the relative number of species in four groups of vertebrates over time. The shapes are wide at times when there were many species. The shapes are narrow at times when there were few species. Use the chart to answer questions 17–19.

17. Interpret Visuals During which period did amphibians first evolve?

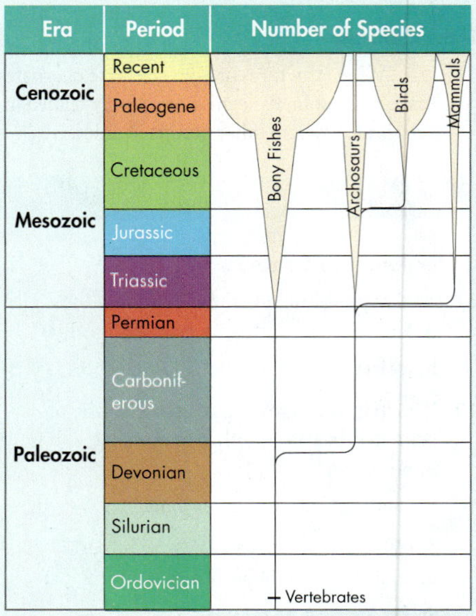

18. Compare and Contrast How is this kind of diagram similar to a traditional cladogram? What additional information can be learned from it?

19. Interpret Visuals Which of the groups shown has the greatest number of species today?

solve the CHAPTER MYSTERY

FOSSIL QUEST?

Josh would have a hard time finding rocks old enough to show traces of the first animals. Such fossils would be about 600 million years old. Most of these rocks have been destroyed by geological activity. Most sites are in China and Australia. None are in the United States.

Pedro found better news. Reptiles related to bird ancestors lived during the Cretaceous Period. There are a number of places where fossils of that age have been found. One such place is the Green River area in Utah. Because both boys like dinosaurs, they joined a field trip for teens to the Green River to search for bird ancestors.

1. Infer The earliest known animals lived during the Proterozoic Eon, while the ancestors of birds lived during the Cretaceous Period. Why is it so much harder to find fossils of the earliest known animals than it is to find fossils of the ancestors of birds? (*Hint:* See the graph below.)

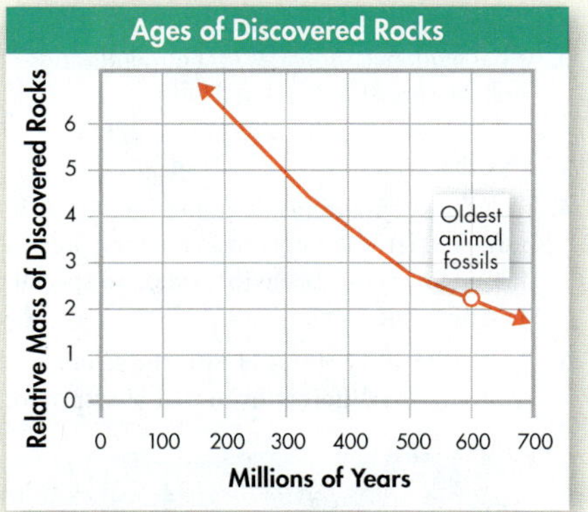

Untamed Science Finding the solution to the mystery is only the beginning. Take a video field trip with the ecogeeks of Untamed Science to see where the mystery leads.

CHAPTER MYSTERY ANSWERS

1. According to the graph, there are many more rocks of younger ages discovered than there are rocks of older ages discovered.

Untamed Science To help students understand why insects are such a diverse group, suggest they watch **Insects Rule!**

Standardized Test Prep

Multiple Choice

1. Which of the following is NOT a mollusk?
   A leech
   B squid
   C clam
   D snail

2. Which of the following invertebrates have segmented bodies?
   A flatworms
   B roundworms
   C cnidarians
   D annelids

3. All animals have some form of body symmetry EXCEPT
   A sponges.
   B jellyfishes.
   C worms.
   D arthropods.

4. Which of the following groups can be classified as nonvertebrate chordates?
   A sponges
   B tunicates
   C fishes
   D all of the above

5. Many scientists think that birds evolved from
   A mammal-like reptiles.
   B amphibians.
   C mammals.
   D dinosaurs.

6. Which of the following is NOT a characteristic of reptiles?
   A scaly skin
   B eggs with shells
   C lungs
   D mammary glands

7. Which of the following are hominoids?
   A all mammals
   B all primates
   C humans only
   D all great apes

8. When did the first true mammals appear?
   A Cretaceous Period
   B Triassic Period
   C Cenozoic Era
   D Carboniferous Period

Questions 9–11 Refer to the following cladogram.

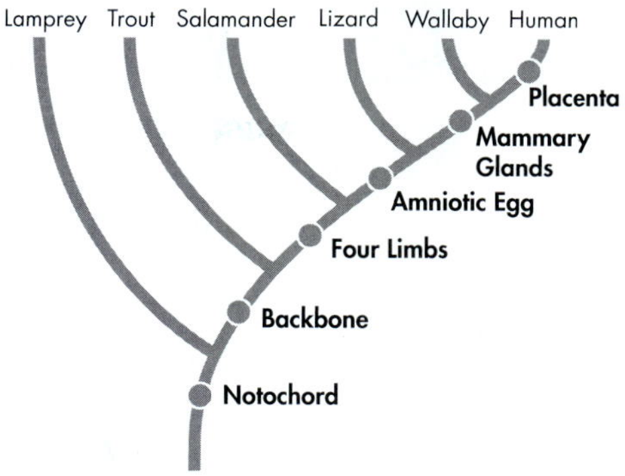

9. Which characteristic is shared by humans, wallabies, and trout?
   A placenta
   B notochord
   C four limbs
   D mammary glands

10. Which animals have the closest evolutionary relationship, as shown by the cladogram?
    A humans and wallabies
    B humans and lizards
    C humans and lampreys
    D humans and trout

11. A valid conclusion from this cladogram is that
    A salamanders, trout, and lampreys all have a backbone.
    B four limbs appeared in vertebrate evolution before the notochord appeared.
    C humans and lampreys share a common ancestor.
    D mammary glands appeared in vertebrate evolution after the placenta appeared.

Open-Ended Response

12. What is the difference between a hominoid and a hominine?

If You Have Trouble With . . .

Question	1	2	3	4	5	6	7	8	9	10	11	12
See Lesson	26.1	26.1	26.1	26.2	26.2	26.2	26.3	26.2	26.2	26.2	26.2	26.3

Test-Taking Tip

INTERPRET VISUALS

Suggest that before students attempt to answer questions about a diagram, they carefully study it to see the basic kind of information the diagram contains. In interpreting the diagram, students should look for any relationships or trends shown. For example, in the cladogram shown here, students should note the order in which the shared derived characters appeared. After thoroughly examining the diagram and interpreting the information it contains, students should read the questions that relate to the diagram. They should recheck all answers by reexamining the diagram.

Before the Test

Review with students the Test-Taking Tip.

Draw students' attention to the word *NOT* in Questions 1 and 6 and to the word *EXCEPT* in Question 3.

Remind students that an open-ended response means that they'll need to write a paragraph to answer the question in their notebooks. Have them paraphrase the question in their notebooks, and remind them about the steps in writing a paragraph.

Answers

1. A
2. D
3. A
4. B
5. D
6. D
7. D
8. B
9. B
10. A
11. C
12. Hominoids include all the great apes, including gibbons, orangutans, gorillas, chimpanzees, and humans. The hominoids in the lineage that led to humans are called hominines, which include modern humans and all other species more closely related to humans than to chimpanzees. Unlike other hominoids, hominines evolved large brains and the ability to walk upright.

Animal Systems I

Chapter Contents	NSES	Time	Core Resources
Chapter Preview			**Student Edition,** pp. 644–645 **Chapter Mystery,** p. 645 **Foundations for Learning,** p. 645
27.1 Feeding and Digestion Obtaining Food • Processing Food • Specialized Mouthparts	I, V, C.1.b, C.4.b, C.4.c, C.5.d, C.6.b	1 period 1/2 block	**Student Edition,** pp. 646–649 **Study Workbook B** 27.1 Worksheets **Biology.com** *Visual Analogy:* Specialized Teeth • 27.1 Lesson Assessment
27.2 Respiration Gas Exchange • Respiration in Aquatic Animals • Respiration in Land Animals	I, V, C.5.d	1 period 1/2 block	**Student Edition,** pp. 650–652 **Study Workbook B** 27.2 Worksheets **Biology.com** *Art Review:* Respiratory Systems • 27.2 Lesson Assessment
27.3 Circulation Open and Closed Circulatory Systems • Single- and Double-Loop Circulation	I, V, C.3.a, C.5.d	1/2 period 1/4 block	**Student Edition,** pp. 653–655 **Study Workbook B** 27.3 Worksheets **Biology.com** *InterActive Art:* Vertebrate Circulatory Systems • 27.3 Lesson Assessment
27.4 Excretion The Ammonia Problem • Excretion in Aquatic Animals • Excretion in Land Animals	I, V, A.1.b, A.2.a, C.1.a, C.5.d	1 period 1/2 block	**Student Edition,** pp. 656–659 **Build Connections,** p. 657 **Inquiry Into Scientific Thinking,** p. 658 **Study Workbook B** 27.4 Worksheets **Biology.com** *Art in Motion:* Excretion in Aquatic Animals • 27.4 Lesson Assessment
Chapter Pre-Lab	I, V, A.1.b, A.2.a, C.5.d	1 period 1/2 block	**Student Edition,** p. 660 **Lab Manual B** *Anatomy of a Squid*

Tools for ELL Instruction

ELL Handbook describes teaching strategies to make *Miller & Levine Biology: Foundation Edition* more accessible to ELL students.

Spanish Study Workbook is a Spanish translation of Study Workbook A.

Multilingual Glossary is the *Miller & Levine Biology* glossary translated into ten languages.

Tools to Challenge Students

Study Workbook A includes more challenging worksheets and activities such as Chapter Mystery activities that reinforce 21st Century Skills.

Lab Manual A labs are geared toward on-level students. The Lab Manual includes fifteen labs in addition to the Chapter Labs.

Biology.com Data Analysis activities offer students a way to practice data gathering and analyzing skills.

Visual Quizzes in the Assessment Resources Book are geared toward on-level students and use visuals from *Miller & Levine Biology.*

Additional Resources

Biology.com Untamed Science Video • Flash Cards

Study Workbook A 27.1 Worksheets
Lab Manual B *Hands-On Activity:* How a Bird "Chews" • *Data Analysis:* Protein Digestion
Spanish Study Workbook 27.1 Worksheets
Assessment Resources Book Visual Quiz

Study Workbook A 27.2 Worksheets
Spanish Study Workbook 27.2 Worksheets
Biology.com *Data Analysis:* Giant Insects of the Paleozoic

Study Workbook A 27.3 Worksheets
Spanish Study Workbook 27.3 Worksheets
Assessment Resources Book Visual Quiz

Study Workbook A 27.4 Worksheets
Spanish Study Workbook 27.4 Worksheets

Lab Manual A *Anatomy of a Squid* • *The Effect of Chemicals on Heart Rate*

Chapter Review

Student Edition Chapter Summary, p. 661
Study Workbook B Chapter 27 Review
Transparencies, pp. 306–315
Biology.com Untamed Science Video • Editable Worksheets of Study Workbooks A and B and Lab Manuals A and B • Chapter 27 Flash Cards and Match It

Untamed Science DVD • Classroom Resources CD (includes lesson presentations and editable worksheets)

Chapter Assessment

Student Edition Check Understanding, pp. 662–665
Study Workbook B Chapter 27 Taking a Standardized Test
Assessment Resources Book Chapter 27 Test B
Biology.com Chapter 27 Assessment • Editable Worksheets of Chapter 27 Visual Quizzes* and Chapter 27 Tests A and B

Exam*View* *Assessment Suite* • Classroom Resources CD (includes lesson presentations and editable worksheets)

Time: 1 period, 1/2 block

* Visual Quizzes are based on visuals from *Miller & Levine Biology.* Review carefully to make sure they are appropriate for your class.

Pressed for Time?

Preview the Chapter Have students read the first Key Question for each lesson and preview the visuals Digesting Food, Lungs, Open and Closed Circulatory Systems, and Excretion in Land Animals.

Cover the Chapter Quickly In Lesson 27.1, have students read *Processing Food* and discuss the Digesting Food visual. Assign *Gas Exchange* in Lesson 27.2 and *Open and Closed Circulatory Systems* in Lesson 27.3. Have students read *The Ammonia Problem* in Lesson 27.4 and discuss the Excretion in Land Animals visual.

Assess Assign Questions 1–3 and 6 in the 27.1 Check Understanding, Question 4 in the 27.2 Check Understanding, Questions 2, 4, and 5 in the 27.3 Check Understanding, and Question 4 in the 27.4 Check Understanding. In the Chapter 27 Check Understanding, assign Questions 2, 3, 6, 9–11, and 15–17.

Get the Picture

Have students examine the photograph of the bird and the zebra, and then read the caption.

Ask What is the bird doing on the zebra's head? *(The bird is eating ticks and insects from the zebra's head.)*

Ask Why is the relationship between the bird and zebra called mutualistic? *(The relationship benefits both organisms—the bird gets food and the zebra stays free of parasites.)*

Ask What physical characteristics does the bird have that helps it find and eat parasites? *(Sample answer: The bird has large eyes, so it probably can see the movement of insects beneath the zebra's fur more easily. The bird has a large beak that can capture squirming insects.)*

Say Think about how all animals have adaptations in body structures or behaviors that allow them to get essential materials and get rid of wastes.

Activate Prior Knowledge

Ask When a hummingbird hovers over a flower blossom, what can you infer about what is happening? *(The bird is feeding on the flower's nectar.)*

Ask What is one adaptation that the hummingbird has for feeding on nectar? *(Hummingbirds have long, specialized beaks. They also have the ability to hover.)*

Say Name other examples of feeding adaptations. *(Sample answers: suckers, stealth, claws, speed, pack behavior)*

NATIONAL SCIENCE EDUCATION STANDARDS

UNIFYING CONCEPTS AND PROCESSES

I, V

CONTENT

C.1.a, C.1.b, C.3.a, C.4.b, C.4.c, C.5.d, C.6.b

INQUIRY

A.1.b, A.2.a

27 Animal Systems I

Big idea **Structure and Function**

Q: How do the structures of animals allow them to take in essential materials and get rid of wastes?

BIOLOGY.com Search Chapter 27 GO • Flash Cards

644

UbD Connect to the Big Idea

Invite a volunteer to read aloud the chapter number and the chapter title. Read the Big Idea and question. Emphasize the question, **How do the structures of animals allow them to take in essential materials and get rid of wastes?**

Set the Purpose

PREVIEW AND PREDICT Draw the graphic organizer on the board. Move in sequence through the graphic organizer asking students to rephrase each item in their own words. Guide students to the INSIDE feature on the next page.

Ask How do you think each of these titles relates to the questions in the graphic organizer?

Have students discuss how structure relates to function in body systems. Have them predict how this will help them find answers to the questions in the graphic organizer as they read.

INSIDE:

- 27.1 Feeding and Digestion
- 27.2 Respiration
- 27.3 Circulation
- 27.4 Excretion

Red-billed oxpeckers are carnivores that have a mutualistic relationship with zebras. The birds eat ticks and insects that are parasites on the zebras. This action frees the zebras of these parasites.

CHAPTER **MYSTERY**

(NEAR) DEATH BY SALT WATER

It started as an adventure. Some college buddies tried their own version of a "survivor" experience. During summer vacation, the college friends were dropped on a small, tropical island. They had few supplies. They would be picked up in a few days.

The island was hot and dry. It had no fresh water. Some members of the group drank coconut milk when they were thirsty. One group member hated coconuts. He drank salt water to get his fluids. At first, he was fine. Then, he became sick and weak. He was dizzy, had headaches, and could not concentrate. His friends began to panic. What was happening? As you read the chapter, look for clues to explain why their friend became so ill. Then, solve the mystery.

FOUNDATIONS for Learning

Take a piece of paper and make a Z-fold, as shown below. Write the key questions for the first lesson of the chapter on the left panel of the fold. Predict the answers to these questions on the middle panel of the fold. Leave the right panel blank for now. Make three more Z-fold pages for the other three lessons of the chapter. Write the key questions and predict the answers for each of these lessons. At the end of the chapter are two activities that use the Z-fold charts to help answer the question: How do the structures of animals allow them to take in essential materials and get rid of wastes?

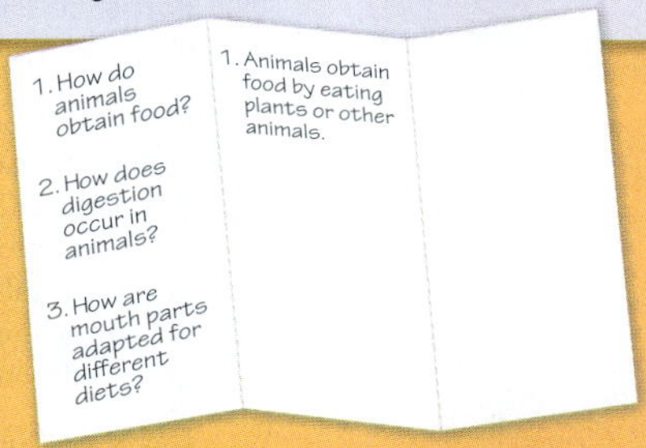

- Untamed Science Video
- Chapter Mystery

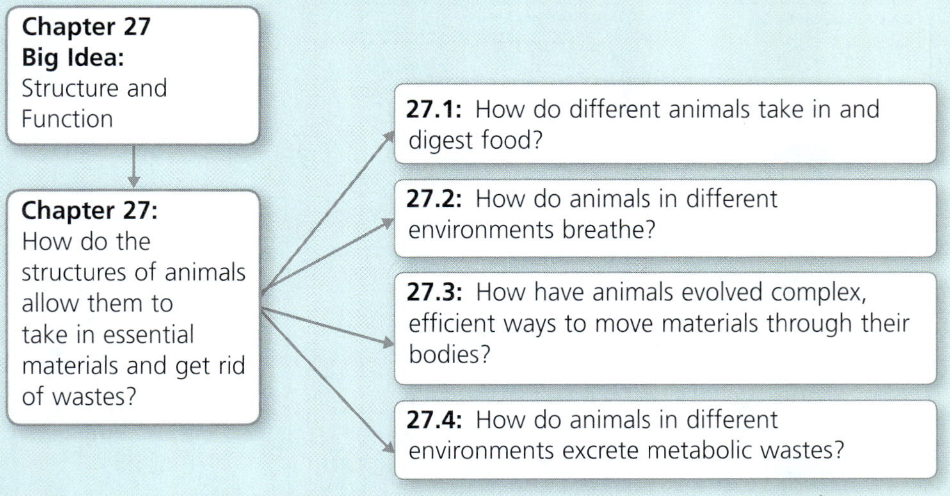

Foundations for Learning

Demonstrate to students how to make a Z-fold. Help students understand the importance of this activity.

Ask Why is it important to write the lesson's Key Questions? *(to focus attention on what information to look for as they read the lesson)*

Ask Why do you think predicting answers to questions before you read helps you understand the lesson? *(Sample answer: Predicting helps you find out what you already know and don't know about a topic. If you discover your prediction was wrong, you can recognize that and correct information as you read.)*

Remind students that they will find out if their answers are correct as they read the lessons. The activities at the end of the chapter will help them finalize the Z-fold charts. Encourage students to use their Z-folds as study aids.

Untamed Science — Follow the ecogeeks of Untamed Science as they explore the amazing adaptations of bears in **Bearly Asleep.**

CHAPTER MYSTERY — Have students read the Chapter Mystery and think about why one of the friends became ill and the others did not. Then, have students think about whether it was something that the other members of the group did (drinking the coconut milk) that prevented them from becoming ill, or if it was something they did not do (drinking salt water) that helped them survive.

Digital Support

BIOLOGY.com — Extend your reach by using the digital assets offered at Biology.com.

LESSON 27.1

Pre-Reading

Objectives

27.1.1 Describe the different ways animals get food.

27.1.2 Explain how digestion occurs in different animals.

27.1.3 Describe how mouthparts are adapted for an animal's diet.

Preview the Pages

1. Review the process of diffusion to help students understand intracellular digestion.
2. Preview the visuals and read the captions and labels. Have students examine the photograph of the bison. Ask what the bison is eating *(plants)* and link it to the heading Herbivores.
3. Have students find the vocabulary words in the text and read each word and its definition aloud.
4. Direct students' attention to the Digesting Food diagram of the sponge, cnidarian, and bird. Be sure students know what the gold-shaded areas indicate. *(The gold areas are the animals' specialized cells or organs for digestion.)*

Activate Prior Knowledge

Show students pictures of animals in a food web, such as a mouse, snake, hawk, deer, and cougar. Ask students to tell what each animal eats. Ask students to use the terms *carnivore* and *herbivore*. *(mouse, herbivore; snake, carnivore; hawk, carnivore; deer, herbivore; cougar, carnivore)*

NATIONAL SCIENCE EDUCATION STANDARDS

UNIFYING CONCEPTS AND PROCESSES

I, V

CONTENT

C.1.b, C.4.b, C.4.c, C.5.d, C.6.b

For the corresponding lesson in *Miller & Levine Biology*, see pages 782–786.

27.1 Feeding and Digestion

Key Questions

How do animals get food?

How does digestion occur in animals?

How are mouthparts adapted for different diets?

BUILD Understanding

Two-Column Table Before you read, look at the green headings in the lesson. As you read, make notes about the ideas developed in each heading.

In Your Workbook Refer to your workbook for suggestions about how to use a two-column chart to organize your notes.

Obtaining Food

All animals are heterotrophs that get their energy by eating food. Animals feed in many different ways. Tiny insects dine on blood. Bison feed on grasses. And giant blue whales feed on tiny plankton floating in the ocean. In fact, adaptations for feeding are a large part of what makes animals so interesting.

Filter Feeders Filter feeders eat by straining algae and tiny animals from water. Most filter feeders have gills or other structures that act like nets. Many invertebrate filter feeders, such as sponges, spend most of their lives in a single place. Vertebrate filter feeders, such as blue whales, usually feed while swimming.

Detritivores Detritivores eat bits of decaying plants and animals. They also eat bacteria and algae that live on decaying material. Many worms and crustaceans are detritivores.

Carnivores Carnivores eat other animals. Sharks and wolves are carnivores that use their sharp teeth to catch prey.

Herbivores Herbivores eat plants or parts of plants. Some herbivores can eat leaves. Others eat seeds or fruits.

Parasites Parasites live on or in another organism. They feed on the blood or tissues of their host. Many parasites cause disease.

Mutualists Some animals have mutualistic relationships that help both partners. For example, many coral animals have algae living in their tissues. The algae carry out photosynthesis, which is the corals' main source of energy. In addition, the algae recycle nutrients and help the corals make their calcium carbonate skeletons. The algae, in turn, get nutrients from the corals' wastes. The algae are also protected from algae eaters.

Key Question How do animals get food?
Some animals filter food out of water. Others feed on decaying material, other animals, or plants. Parasites feed on the tissues of living organisms.

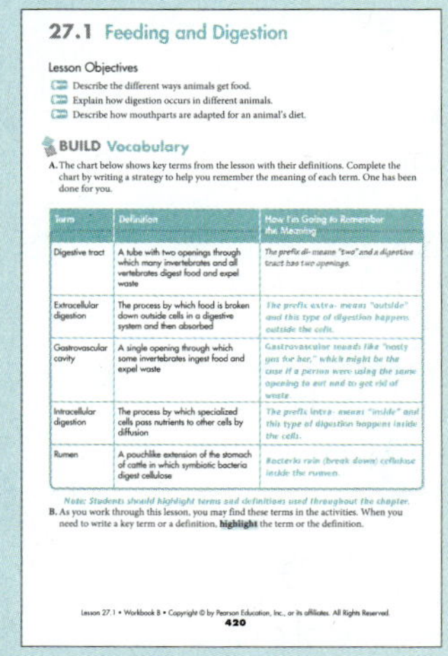

27.1 Feeding and Digestion

Lesson Objectives

Describe the different ways animals get food.
Explain how digestion occurs in different animals.
Describe how mouthparts are adapted for an animal's diet.

BUILD Vocabulary

A. The chart below shows key terms from the lesson with their definitions. Complete the chart by writing a strategy to help you remember the meaning of each term. One has been done for you.

Term	Definition	How I'm Going to Remember the Meaning
Digestive tract	A tube with two openings through which many invertebrates and all vertebrates digest food and expel waste	*The prefix di- means "two" and a digestive tract has two openings.*
Extracellular digestion	The process by which food is broken down outside cells in a digestive system and then absorbed	*The prefix extra- means "outside" and this type of digestion happens outside the cells.*
Gastrovascular cavity	A single opening through which some invertebrates ingest food and expel waste	*Gastrovascular sounds like "nasty gas for her," which might be the case if a person were using the same opening to eat and to get rid of waste.*
Intracellular digestion	The process by which specialized cells pass nutrients to other cells by diffusion	*The prefix intra- means "inside" and this type of digestion happens inside the cells.*
Rumen	A pouchlike extension of the stomach of cattle in which symbiotic bacteria digest cellulose	*Bacteria ruin (break down) cellulose inside the rumen.*

Note: Students should highlight terms and definitions used throughout the chapter.

B. As you work through this lesson, you may find these terms in the activities. When you need to write a key term or a definition, **highlight** the term or the definition.

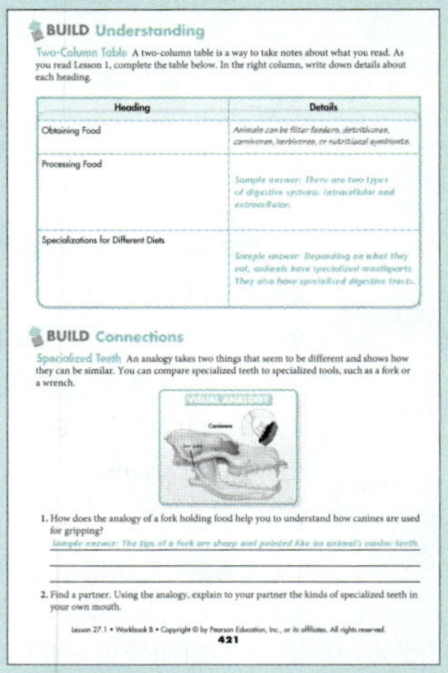

BUILD Understanding

Two-Column Table A two-column table is a way to take notes about what you read. As you read Lesson 1, complete the table below. In the right column, write down details about each heading.

Heading	Details
Obtaining Food	*Animals can be filter feeders, detritivores, carnivores, herbivores, or nutritional symbionts.*
Processing Food	*Sample answer: There are two types of digestive systems: intracellular and extracellular.*
Specializations for Different Diets	*Sample answer: Depending on what they eat, animals have specialized mouthparts. They also have specialized digestive tracts.*

BUILD Connections

Specialized Teeth An analogy takes two things that seem to be different and shows how they can be similar. You can compare specialized teeth to specialized tools, such as a fork or a wrench.

1. How does the analogy of a fork holding food help you to understand how canines are used for gripping?
Sample answer: The tips of a fork are sharp and pointed like an animal's canine teeth.

2. Find a partner. Using the analogy, explain to your partner the kinds of specialized teeth in your own mouth.

Processing Food

Getting food is just the first step. Next, food must be broken down, or digested. Then food can be absorbed, giving energy and nutrients to cells throughout the body.

Intracellular Digestion Sponges and many other simple animals digest food inside specialized cells. Nutrients move from the specialized cells to other body cells by diffusion. This digestive process is called **intracellular digestion.**

Extracellular Digestion Most complex animals rely on extracellular digestion. In **extracellular digestion,** food is broken down outside of cells in a digestive system.

▶ ***Gastrovascular Cavities*** Some invertebrates have a space in their bodies that is surrounded by tissues that carry out digestion and circulation. This space is called a **gastrovascular cavity.** Gastrovascular cavities have a single opening. Animals that have a gastrovascular cavity obtain food and get rid of wastes through the same opening. Some cells lining the cavity secrete enzymes that digest food. Other cells surround food and digest it in vacuoles. Nutrients are then absorbed and carried to the rest of the body.

BUILD Vocabulary

intracellular digestion
a type of digestion in which food is digested inside specialized cells that pass nutrients to other cells by diffusion

extracellular digestion
a type of digestion in which food is broken down outside the cells in a digestive system and is then absorbed

gastrovascular cavity
a digestive chamber with a single opening

PREFIXES
The prefix *intra-* means "on the inside, within." The prefix *extra-* means "outside of." Intracellular digestion takes place *inside* of cells and extracellular digestion takes place *outside* of cells.

Obtaining Food The orca, sea slug, barnacles, and cleaner shrimp get their food in different ways.

Carnivore – Orca

Herbivore – Sea Slug

Filter Feeders – Barnacles

Detritivore – Cleaner Shrimp

During Reading

Build Understanding

TWO-COLUMN TABLE Refer students to Workbook B for help in developing their two-column tables. Point out the green headings: Obtaining Food, Processing Food, and Specializations for Diets. Remind students that these headings highlight the main concepts addressed by the lesson.

Active Reading

KEY QUESTION Have one student read aloud the Key Question and answer from the previous page.

Ask How do orcas differ from sea slugs in the way they obtain food? *(An orca is a carnivore and eats animals; a sea slug is a herbivore and eats plants.)*

FIND THE MAIN IDEA As students read, have them determine and write down the main idea in each paragraph on the page. Have students use these main idea notes to write one summary sentence that states the main idea for the entire page.

Ask What is the main idea on this page? *(Digestion can occur in two main ways—intracellular and extracellular.)*

USE CONTEXT CLUES Have students silently reread the paragraph about intracellular digestion. Have them locate and read aloud the phrase that helps explain the word *diffusion.* *("move from the specialized cells to other body cells")*

Build Vocabulary

PREFIXES Write the prefixes *intra-* and *extra-* on index cards. Have students work with a partner to generate a list of words that can be paired with each index card. Students may use a dictionary or thesaurus. Have students discuss their words.

Ask What are some words that can be used with the prefixes *intra-* and *extra-*? *(Sample answers: extra-: terrestrial, ordinary, sensory; intra-: venous, mural, coastal.)*

Speed Bump

Ask Could a single organism have both intracellular and extracellular digestion? Explain. *(No, intracellular digestion is used by simple organisms. Extracellular digestion is used by more complex organisms.)*

Ask What type of digestion occurs in people? *(extracellular)*

Ask What process distributes nutrients among the cells during intracellular digestion? *(diffusion)*

Focus on Academic Language

• invertebrate • detritivore • mutualist • gastrovascular

Have students find each of these terms on these two pages. Ask them to use prefixes, suffixes, or context clues to define each term, and write pronunciation clues if needed. Begin by having students define *invertebrate* using a prefix and a root word. *(The prefix* in- *means "without." The root word* vertebrate *means "an animal that has a backbone." This word must mean "without a backbone.")*

Ask Is a detritivore the same as a decomposer? How do you know? *(No, a detritivore eats bits of decaying organisms. The material is already decaying.)*

Ask What is a context clue that helps you understand what a mutualist is? *("relationships that help both partners")*

Ask What is a good way to figure out how to pronounce a five-syllable word, such as gastrovascular? *(break it into separate syllables; use a dictionary.)*

During Reading continued

Active Reading

SELF-QUESTION As students read, have them ask themselves, "Does this make sense?" If the answer is *no,* then they should go back to the last part they understand. Have them read the next sentence and restate it in the form of a question, and then answer the question.

USE DIAGRAMS Have students review the visuals of the sponge, cnidarian, and bird. Then ask a volunteer to read the caption aloud. Tell students to cover the labels. Then have partners use what they remember from the text to describe the digestive systems of sponges, cnidarians, and birds.

SCIENCE SUPPORT Explain to students that there is a very large difference between elimination and excretion. Students will encounter excretion later on in this chapter, which is defined as the elimination of "metabolic" wastes. Feces are not metabolic wastes. They are simply indigestible materials and are eliminated, not excreted. Excreted wastes pass through a membrane before leaving the body.

Build Vocabulary

ACADEMIC WORDS Tell students that some words, called homonyms, can have entirely different meanings depending on how they are used. *Tract* is such a word. *Tract* could mean "a certain housing development," it could mean "a pamphlet," or it could mean "an expanse of land or water." Explain that in science, *tract* means "a system of organs or tissues working together."

Ask How do you know which meaning to apply to a homonym? *(by understanding the context)*

KEY QUESTION Read aloud the Key Question and ask a student to read the answer.

Ask What is a key difference between gastrovascular cavities and digestive tracts? *(gastrovascular cavities have one opening, digestive tracts have two openings)*

Speed Bump

Ask When food is digested in a digestive tract, is this intracellular digestion or extracellular digestion? *(extracellular)*

Ask What is the function of a bird's gizzard? *(The gizzard grinds food into little pieces.)*

Ask Which animals produce digestive enzymes that break down cellulose? *(none)*

BUILD Vocabulary

digestive tract
a tube that begins at the mouth and ends at the anus

rumen
a stomach chamber in cows and related animals in which bacteria digest cellulose

ACADEMIC WORDS
In biology, the word *tract* means "a system of organs and tissues that work together to perform one function." The *digestive tract* is a tube with organs such as a stomach and intestines that work together to break down food.

▶ ***Digestive Tracts*** Many invertebrates and all vertebrates digest food in a tube called a **digestive tract,** or gut. A digestive tract has two openings. Food moves through in one direction. Food enters through the mouth. Wastes leave through the anus.

Digestive tracts often have specialized structures that carry out different tasks as food passes through. Mouthparts such as teeth may cut or smash food into smaller pieces. Birds have a special organ called a gizzard that grinds food into smaller pieces. In many animals, the mouth secretes enzymes that start the process of chemical digestion. The stomach and intestines also secrete digestive enzymes. The intestines absorb nutrients that are released by digestion.

▶ ***Solid Waste Disposal*** Some materials in food cannot be digested. Organisms expel solid wastes called feces through either the single digestive opening or the anus.

Specialized Digestive Tracts Carnivorous invertebrates and vertebrates have short digestive tracts. These digestive tracts produce fast-acting enzymes. These enzymes are able to break down most animal tissues.

Plant tissues are more difficult to digest. No animal produces digestive enzymes that can break down the cellulose in plant tissue. Some herbivores have very long intestines that hold bacteria that are able to digest cellulose. Other herbivores have specialized pouches that hold mutualistic bacteria. For example, cows have a pouchlike extension of their esophagus called a **rumen.** Bacteria in the rumen digest cellulose. The food then moves back into the cow's mouth. The cow chews the food a second time, then swallows it again. This process is called "chewing the cud."

Key Question How does digestion occur in animals?
Some invertebrates, such as sponges, break down food using intracellular digestion. Most invertebrates and all vertebrates break down food using extracellular digestion.

Digesting Food Animals have different digestive structures with different functions.

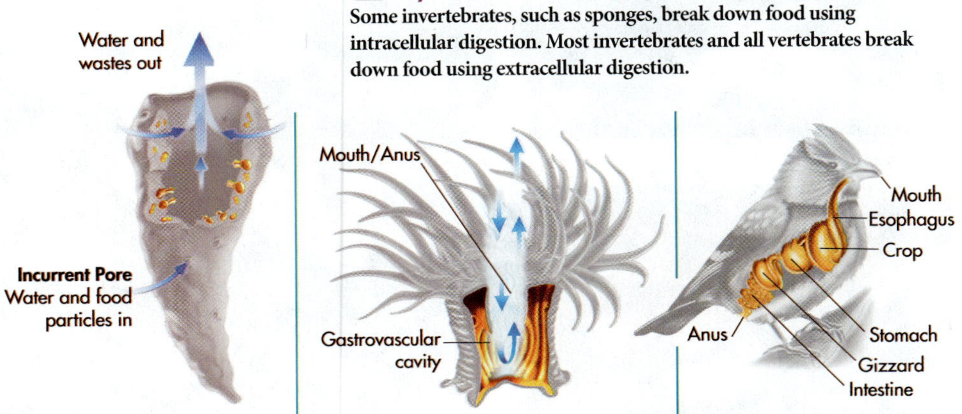

Sponge The sponge has one digestive opening and uses intracellular digestion.

Cnidarian The cnidarian has a gastrovascular cavity and uses extracellular digestion.

Bird The bird has a digestive tract and uses extracellular digestion.

ELL Focus on ELL: Build Background

BEGINNING AND INTERMEDIATE SPEAKERS To help English learners understand the differences between the teeth of mammalian herbivores and carnivores described on the next page, have students collect images (drawings, Internet printouts, clippings) of the teeth of mammalian carnivores and herbivores and sort the teeth into those two groups. Point out that mammals are animals that have fur and produce milk to feed their young. Then, instruct students to label each mammal in English using the word *herbivore* or *carnivore,* while explaining their labels either orally or in captions in their native language.

Specializations for Diets

Animals with specialized diets have evolved specialized mouthparts that help them to grab and digest food.

For example, mammals that are carnivores have sharp teeth that help them grab and hold prey. These teeth can also slice meat. The jaws of these carnivores move up and down to chop meat into small pieces.

Herbivores have mouthparts that grind leaves. The front teeth and lips of mammalian herbivores can grab and tear. The jaws of these herbivores move from side to side to grind leaves.

Key Question How are mouthparts adapted for different diets? **Carnivores usually have sharp mouthparts that can capture food, hold it, and cut it into pieces. Herbivores usually have mouthparts that can grind plants.**

BUILD Connections

SPECIALIZED TEETH

The jaws and teeth of mammals are well adapted to their diets. How are mammal teeth like objects that you have around the house?

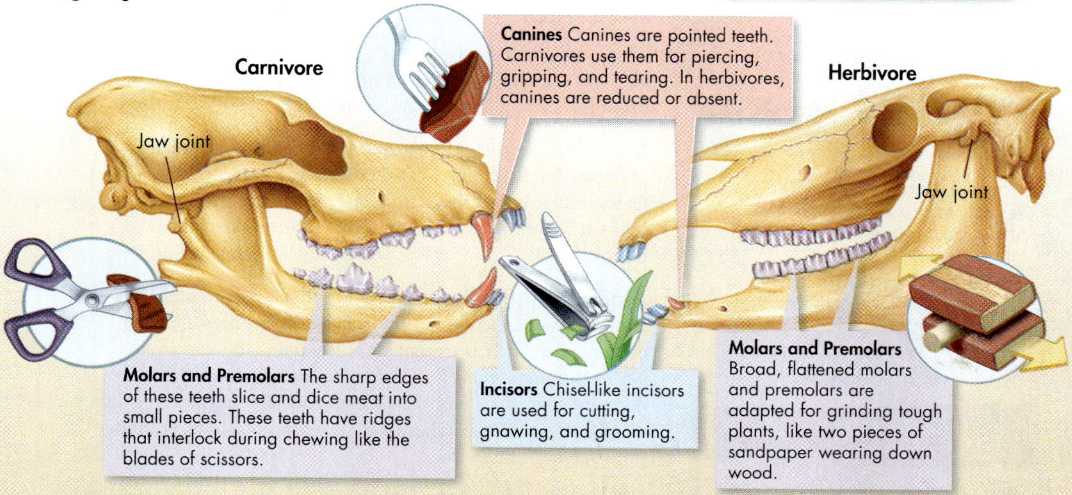

CHECK Understanding

Use the highlighted words from the lesson to complete each sentence correctly.

1. A digestive chamber that has a single opening is called a _______.
2. All vertebrates digest food in a tube called a _______.
3. The process by which food is broken down inside specialized cells is _______.

Critical Thinking

4. **Classify** Identify the following animals as a filter feeder, detritivore, parasite, or carnivore: a hawk that eats mice; a lobster that feeds on dead fish; a tick that feeds on a dog's blood; a blue whale that eats plankton.
5. **Compare and Contrast** How do the mouthparts of a leaf-eater differ from those of a meat-eater?
6. **Write to Learn** Describe how a cow digests food. Include the words *cellulose*, *rumen*, and *cud*.

BIOLOGY.com Search Lesson 27.1 GO • Lesson Assessment • Visual Analogy

Check Understanding

1. gastrovascular cavity
2. digestive tract
3. intracellular digestion
4. hawk, carnivore; lobster, detritivore; tick, parasite; blue whale, filter feeder
5. A leaf-eater has mouthparts that can grind plant material. A meat-eater has mouthparts that can capture food and cut it into pieces.
6. Sample answer: The cow grabs and tears the grass with its lips and front teeth. It grinds the grass with its teeth. Enzymes in the mouth help to break down the grass. The cow swallows the grass and it moves to the rumen. Bacteria in the rumen help to digest the cellulose in grass. The grass moves back into the cow's mouth. The cow chews and swallows the grass again. This process is called "chewing the cud."

Active Reading

SET A PURPOSE FOR READING Have a volunteer read aloud the question in the caption.

Ask Why do you think this question is asked in the caption? *(to help compare teeth to things we use every day, such as tools)*

SUMMARIZE Have students use the text and the visual to write a summary statement about how mouthparts are adapted to diet.

KEY QUESTION Ask a volunteer to read aloud the Key Question and answer.

Ask What do you think people would eat if they had suckers for mouthparts and no teeth? *(Sample answer: People would eat liquid food.)*

Build Connections

VISUAL ANALOGY As a class, discuss the different kinds of teeth animals have (premolars, molars, canines, incisors) and how these specialized mouthparts are adapted to an animal's diet. If you have any animals in the classroom—a snake, a lizard, gerbils, earthworms—have students observe their eating habits and make observations and drawings of the animal's mouthparts. Alternatively, use videos. Discuss the adaptations each animal has for its diet.

Post Reading

ASSESSMENT TIP Remind students to read each question carefully. Tell them not to assume they know what the question is until they have read it completely.

For Question 6, pair students to discuss what is being asked. Have them write a short outline of what information they would include to fully answer the question. For example, how does a cow find food, how does a cow eat food, and what structures does a cow have for eating and digesting?

Wrap-Up Activity

MAKE A LIVING VENN DIAGRAM Have groups of students write down similarities between the eating habits and digestive systems of herbivores and carnivores. Also, have them write down characteristics that are unique to each. They should write the entries in large letters on single pages of paper. Then set up a Venn diagram at the front of the class. Have volunteers carry their entry to the correct location—similarities in the center, differences on either end. Have another volunteer copy the living Venn diagram on the board.

LESSON 27.2

Pre-Reading

Objectives

27.2.1 Describe the characteristics of respiratory structures that all animals share.

27.2.2 Explain how aquatic animals breathe.

27.2.3 Identify the respiratory structures that enable land animals to breathe.

Preview the Pages

1. Write the vocabulary words on the board and read each word and its definition aloud.
2. Preview the visuals and read the captions and labels.
   Ask What are the orange body parts on the diagram of the fish? *(gills)* Make sure students understand that the gills are shown on both sides of the fish.
3. Direct students' attention to the silhouettes of the spider, insect, amphibian, reptile, and mammal.
   Ask How do you know which parts the captions refer to? *(The colored parts, arrows, lines, and blowouts call attention to the body parts and processes discussed in the caption.)*

Build Understanding

CONCEPT MAP Direct students to Workbook B, where they will find a partially completed concept map. Remind students that a concept map will help them quickly see the relationships between two different things as well as a sequence.

NATIONAL SCIENCE EDUCATION STANDARDS

UNIFYING CONCEPTS AND PROCESSES

I, V

CONTENT

C.5.d

For the corresponding lesson in *Miller & Levine Biology,* see pages 787–790.

27.2 Respiration

Key Questions

What characteristics do the respiratory structures of all animals share?

How do aquatic animals breathe?

What respiratory structures enable land animals to breathe?

BUILD Understanding

Concept Map As you read the lesson, create a concept map that shows the characteristics of the lungs of vertebrates.

In Your Workbook Refer to your workbook for suggestions about how to use a concept map to organize your notes.

BUILD Vocabulary

gill a feathery structure specialized for the exchange of gases with water

lung an organ used for respiration; a place where gases are exchanged between the blood and inhaled air

MULTIPLE MEANINGS

The word *respiration* has two different meanings. At the level of an organism, respiration is the exchange of oxygen and carbon dioxide with the air. At the level of a cell, respiration is the process by which food molecules are broken down to release energy. Because cellular respiration requires oxygen, the two processes are related.

Gas Exchange

In respiration, all animal cells take in oxygen and give off carbon dioxide. Animals must get this oxygen from their environment. In other words, all animals need to "breathe." Humans can drown because our lungs cannot take oxygen from water. Fishes have the opposite problem. Their gills cannot use oxygen from the air.

Living cells cannot actively pump oxygen and carbon dioxide across membranes. Instead, animals have respiratory structures that allow these gases to move in and out by passive diffusion.

Gas Diffusion and Membranes Gases diffuse from an area of higher concentration to an area of lower concentration. Gases diffuse best across a thin, moist membrane that is selectively permeable. The amount of diffusion that can take place increases as the surface area of a membrane increases.

Requirements for Respiration Respiratory structures contain a moist membrane with a large surface area. Only certain gases, such as oxygen and carbon dioxide, can move across this membrane. Respiratory structures also maintain different concentrations of carbon dioxide and oxygen on either side of the membrane. These different concentrations promote diffusion of those gases.

Key Question What characteristics do the respiratory structures of all animals share?

Respiratory structures provide a large surface area of moist, selectively permeable membrane. The membrane keeps a difference in the relative concentrations of oxygen and carbon dioxide on either side of it. This difference in concentration helps diffusion take place.

Respiration in Aquatic Animals

Animals such as flatworms and amphibians have thin-walled bodies that are wet. Oxygen and carbon dioxide can diffuse through the body coverings of these animals. Larger, active animals cannot rely on respiration through the skin alone. Many aquatic invertebrates and most aquatic chordates exchange gases through gills.

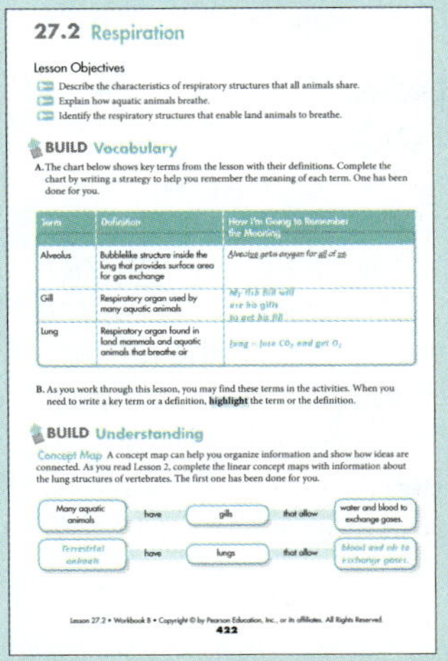

27.2 Respiration

Lesson Objectives

- Describe the characteristics of respiratory structures that all animals share.
- Explain how aquatic animals breathe.
- Identify the respiratory structures that enable land animals to breathe.

BUILD Vocabulary

A. The chart below shows key terms from the lesson with their definitions. Complete the chart by writing a strategy to help you remember the meaning of each term. One has been done for you.

Term	Definition	How I'm Going to Remember the Meaning
Alveolus	Bubblelike structure inside the lung that provides surface area for gas exchange	Alveolus gets oxygen for all of us
Gill	Respiratory organ used by many aquatic animals	My fish Bill will use his gills to get his fill
Lung	Respiratory organ found in land mammals and aquatic animals that breathe air	lung = lose CO_2 and get O_2

B. As you work through this lesson, you may find these terms in the activities. When you need to write a key term or a definition, **highlight** the term or the definition.

BUILD Understanding

Concept Map A concept map can help you organize information and show how ideas are connected. As you read Lesson 2, complete the linear concept maps with information about the lung structures of vertebrates. The first one has been done for you.

Many aquatic animals → have → gills → that allow → water and blood to exchange gases.

Terrestrial animals → have → lungs → that allow → blood and air to exchange gases.

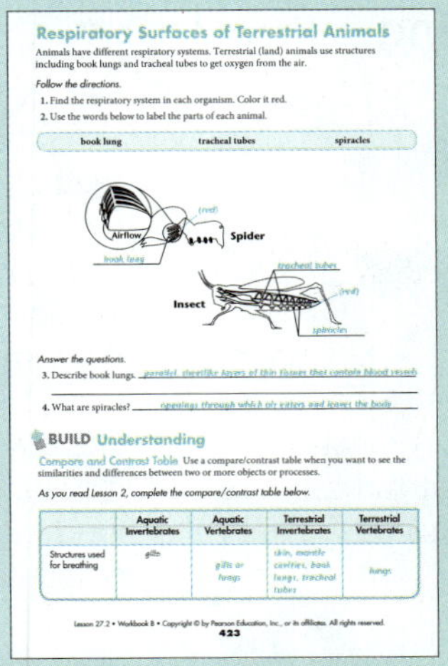

Respiratory Surfaces of Terrestrial Animals

Animals have different respiratory systems. Terrestrial (land) animals use structures including book lungs and tracheal tubes to get oxygen from the air.

Follow the directions.

1. Find the respiratory system in each organism. Color it red.
2. Use the words below to label the parts of each animal.

book lung tracheal tubes spiracles

Answer the questions.

3. Describe book lungs. parallel, sheetlike layers of thin tissues that contain blood vessels
4. What are spiracles? openings through which air enters and leaves the body

BUILD Understanding

Compare and Contrast Table Use a compare/contrast table when you want to see the similarities and differences between two or more objects or processes.

As you read Lesson 2, complete the compare/contrast table below.

	Aquatic Invertebrates	Aquatic Vertebrates	Terrestrial Invertebrates	Terrestrial Vertebrates
Structures used for breathing	gills	gills or lungs	skin, mantle cavities, book lungs, tracheal tubes	lungs

Respiration With Gills Fish respire with gills. As water passes over the gills, oxygen enters the capillaries and carbon dioxide exits.

Gills are feathery structures that expose a large surface area of membrane to water. Inside gills are many tiny blood vessels called capillaries. Animals pump water across their gills as blood flows inside. Pumping water over the gills helps to maintain differences between the oxygen and carbon dioxide concentrations in water and blood. When concentrations are different, diffusion can occur between the capillaries and water.

Reptiles that live in water and aquatic mammals such as whales, breathe with lungs. **Lungs** are organs that exchange oxygen and carbon dioxide between blood and air. Animals that breathe with lungs inhale air at the surface and hold their breath underwater.

Key Question How do aquatic animals breathe?
Most aquatic animals breathe through their skin or with gills. Aquatic reptiles and mammals breathe with lungs and must hold their breath underwater.

Respiration in Land Animals

Animals that live on land face a challenge that aquatic animals do not. Land animals must keep their respiratory membranes moist in dry conditions.

Respiration in Land Invertebrates Invertebrates that live on land have a wide variety of respiratory structures. Some land animals, such as earthworms, can respire across their skin as long as it stays moist. Land snails respire by using a mantle cavity, which is lined with moist tissue and blood vessels. Insects and spiders have very complex structures, as shown in the art.

Respiratory Structures of Land Invertebrates Invertebrates that live on land have a wide variety of respiratory structures that help their membranes stay moist, even in dry conditions.

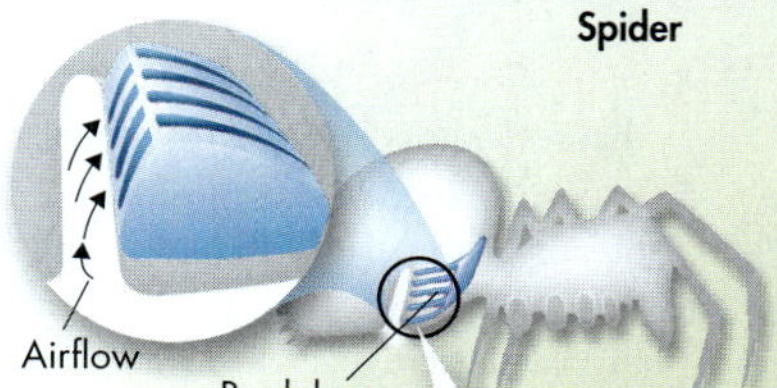

Spiders respire using organs called book lungs, which are made of parallel, sheetlike layers of thin tissues that contain blood vessels.

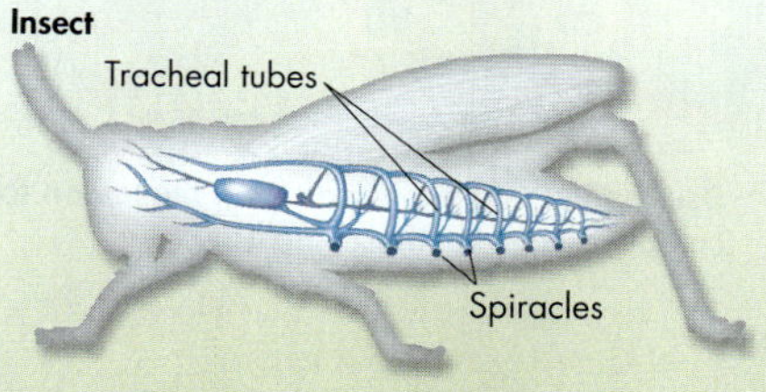

In most insects, a system of tracheal tubes extends throughout the body. Air enters and leaves the system through openings in the body surface called spiracles. In some insects, oxygen and carbon dioxide diffuse through the tracheal system, and in and out of body fluids. In other insects, body movements help pump air in and out of the tracheal system.

BIOLOGY.com Search Lesson 27.2 GO • Art Review

ELL Focus on ELL: Extend Language

INTERMEDIATE SPEAKERS Pair English language learners with native speakers. Have pairs work on the pronunciation of *concentration, permeable,* and *capillaries.* Ask English language learners to prepare a table of other troublesome words in this lesson, such as *alveolus,* and to add their own pronunciation guides and drawings, as needed.

Point out that the terms *concentration* and *capillaries* each have more than one meaning. In this context, *concentration* refers to the amount of oxygen or carbon dioxide in the blood; *capillaries* refers to small blood vessels. Students may be familiar with capillary tubes from performing labs.

During Reading

Build Vocabulary

MULTIPLE MEANINGS Tell students that often they will read a word that has two or more meanings. The meanings may or may not be related. Write the word *cell* on the board.

Ask You know that a cell is the basic unit of life, but what's another meaning of *cell*? *(a small room where prisoners are kept)*

Ask How are these two meanings of *cell* related? *(Sample answer: They are both relatively small structures that allow some, but not all, things to enter and leave.)*

Active Reading

KEY QUESTION Read the Key Question and answer on the previous page aloud.

Ask What are some respiratory structures? *(Sample answers: lungs, gills)*

USE VISUALS Have students examine the diagrams of the spider and the insect.

Say Sometimes we can remember words better when we relate them to something else.

Ask Why do you think a spider's lung structures are called *book lungs*? *(Because they are stacked in thin sheets that look like the pages of a book.)*

SUMMARIZE Pair students. Have one student read a paragraph aloud and the other student summarize the paragraph. Tell students to alternate until they have finished the assigned reading.

KEY QUESTION Read the Key Question and answer aloud.

Ask What sort of graphic tool might you use to compare the way different aquatic animals breathe? *(Answers will vary but should include a method of comparing, such as a T-chart or a compare/contrast table.)*

Speed Bump

Ask What happens in the lungs during respiration? *(Oxygen is inhaled and carbon dioxide is exhaled. These gases are also exchanged with the blood.)*

Ask By what process do gases move into and out of cells across a membrane? *(passive diffusion)*

Ask What is a special respiratory challenge faced by land animals? *(how to keep their respiratory membranes moist in dry conditions)*

During Reading continued

Active Reading

USE GRAPHIC ORGANIZERS Draw a spider map on the board with at least four legs. Write the name of an animal in the center. Ask students to name characteristics of the animal's breathing structures as you write them on the spider map legs.

Build Vocabulary

WORD ORIGINS Remind students that learning the origins of words can help them understand the meaning of new words. Make sure students understand that a cavity is an enclosed space.

Ask Which of the following uses of *alveolar* does not make sense: 1. sockets of the teeth; 2. number of toes; 3. cells of a honeycomb? *(2)*

KEY QUESTION Read the question and answer aloud. Then ask students to list the animals in the visuals in this lesson with their respiratory structures.

Post Reading

ASSESSMENT TIPS Review the lesson's vocabulary words with students before they begin the Check Understanding assessment.

For Question 3, have students prepare compare/contrast organizers.

For Question 7, have students work in pairs to make a sequence flowchart organizer.

Wrap-Up Activity

RESPIRATION AND EXERCISE Pair students. Have available a clock or watch with a second hand. One student will run in place and the other will be the recorder. Students with breathing difficulties or who are not feeling well should not be the runners. The recorder should count the breaths of the runner at rest for 15 seconds, then multiply by four and record the number of breaths per minute while resting. The runner should then run in place for two minutes. Afterward, the recorder should count and record the number of breaths per minute.

Ask What is the difference? *(There are more breaths per minute after running.)*

Ask Why is this so? *(because the body is getting rid of the waste gases that are produced by exercise, and because the tissues in the body need more oxygen during exercise)* Have teams prepare graphs and compare data.

■ Nostrils, mouth, and throat
■ Trachea
■ Lung

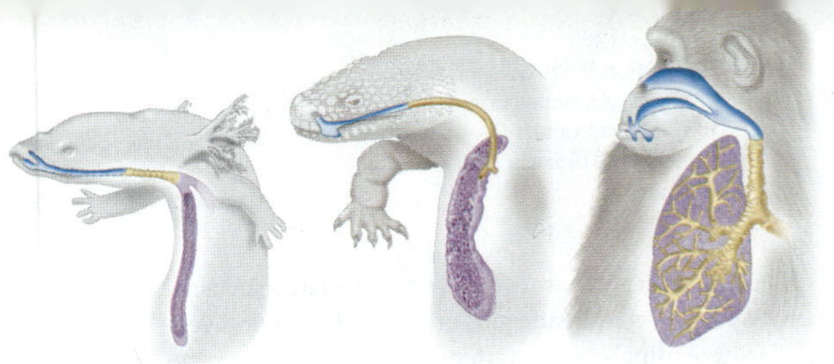

Amphibian | Reptile | Mammal

Lungs Vertebrates that live on land breathe with lungs. Lungs with a larger surface area can take in more oxygen and release more carbon dioxide.

Lung Structure in Vertebrates All land vertebrates breathe with lungs. Inhaling brings oxygen-rich air into the lungs. Oxygen diffuses into the blood through capillaries in the lungs. Carbon dioxide diffuses out of the blood and is exhaled.

▶ ***Amphibian, Reptilian, and Mammalian Lungs*** The internal surface area of lungs increases from amphibians to reptiles to mammals. The lung of an amphibian is like a sac with ridges. The lung of a reptile is small but often divided into chambers. A mammalian lung is large with many branches. Bubblelike structures called **alveoli** (al VEE uh ly; singular: alveolus) fill the lungs of mammals. These structures provide a large surface area for gas exchange. In the lungs of mammals and most other vertebrates, air moves in and out through the same passage. For this reason, some stale air is trapped in the lungs.

▶ ***Bird Lungs*** Air flows through bird lungs in one direction. One-way air flow is more efficient than the in-and-out air flow of mammals. Stale air does not get trapped. One-way flow allows birds to get enough oxygen to power their flight muscles.

Key Question What respiratory structures enable land animals to breathe? **Land animals breathe by using structures such as the skin, mantle cavities, book lungs, tracheal tubes, and lungs.**

BUILD Vocabulary

alveolus
a tiny air sac at the end of the bronchiole in the lungs that provides surface area for gas exchange to occur

WORD ORIGINS
The Latin word *alveolus* means "small cavity." A cavity is a hollow space. The *alveoli* in the lungs are air sacs that are small and hollow.

CHECK Understanding

Use the highlighted words from the lesson to complete each sentence correctly.

1. The feathery structures that allow fish to obtain oxygen from water are _______.
2. The tiny air sacs in lungs where gas exchange occurs are called _______.

Critical Thinking

3. **Explain** How are the respiratory structures of all animals similar?
4. **Apply Concepts** Explain why it is important that respiratory surfaces remain moist.
5. **Review** Which groups of aquatic animals breathe with gills? With lungs?
6. **Compare and Contrast** Compare the structures that land invertebrates and land vertebrates use to breathe.
7. **Write to Learn** Describe the events that occur when a mammal breathes in and out. Include the path of air through its lungs.

BIOLOGY.com Search Lesson 27.2 GO • Lesson Assessment

652

Check Understanding

1. gills
2. alveoli
3. They all have a large surface of moist, selectively permeable membrane for gas exchange. They all maintain differences in relative concentrations of oxygen and carbon dioxide on either side of their respiratory membrane so diffusion can happen.
4. Diffusion of gases happens best across moist surfaces.
5. Many aquatic invertebrates and most aquatic chordates breathe with gills. Aquatic reptiles and mammals breathe with lungs.
6. Land invertebrates breathe using their skin, mantle cavities, book lungs, or tracheal tubes. Land vertebrates breathe with lungs.
7. Sample answer: Oxygen enters the lungs when the mammal breathes in. Oxygen moves into the blood through lung capillaries. At the same time, carbon dioxide moves out of the blood and into the lungs. Carbon dioxide leaves the body when the mammal exhales.

27.3 Circulation

Open and Closed Circulatory Systems

How do animals get oxygen and nutrients to their cells? How do their cells get rid of carbon dioxide and other wastes? In small aquatic animals that are only a few cells thick, materials move by diffusion. But most animals move these materials in blood. Usually one or more hearts pump blood through a circulatory system. A **heart** is a hollow, muscular organ that pumps blood.

Open Circulatory Systems Arthropods and most mollusks have **open circulatory systems.** In an open circulatory system, blood is only partly contained in blood vessels. The vessels empty into a system of sinuses, or spongy holes. There, blood is in direct contact with body tissues. Blood collects in another set of sinuses. Eventually it flows back to the heart.

Closed Circulatory Systems Larger invertebrates and all vertebrates have **closed circulatory systems.** In a closed circulatory system, blood moves entirely within blood vessels. Nutrients and oxygen reach body tissues by diffusing across the walls of tiny blood vessels called capillaries. Blood that is completely contained in vessels can be pumped under high pressure. Thus, a closed circulatory system is more efficient than an open circulatory system.

Key Question How do open and closed circulatory systems compare? **In an open circulatory system, blood is only partly contained within blood vessels. In a closed circulatory system, blood is completely contained in blood vessels.**

Key Questions

How do open and closed circulatory systems compare?

How do the patterns of circulation in vertebrates compare?

BUILD Understanding

Cycle Diagram As you read, draw a cycle diagram showing how blood moves through a closed, two-loop circulatory system. Use five steps in your diagram.

In Your Workbook Refer to your workbook to see how to make a cycle diagram.

Open and Closed Circulatory Systems
A grasshopper has an open circulatory system. Blood leaves vessels and moves through sinuses before returning to the heart. Earthworms have closed circulatory systems. Blood stays in the vessels of a closed circulatory system.

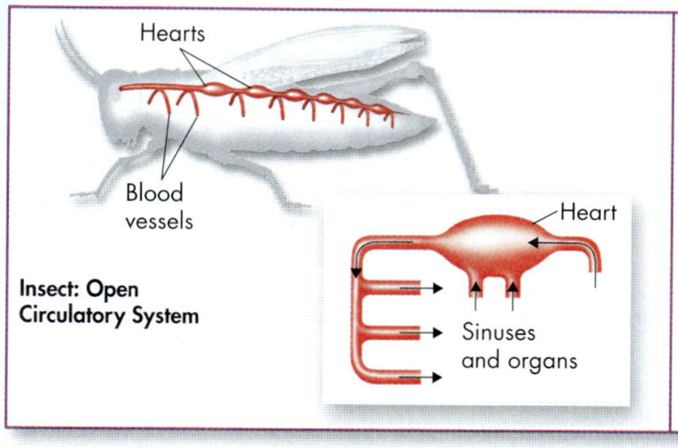

Insect: Open Circulatory System

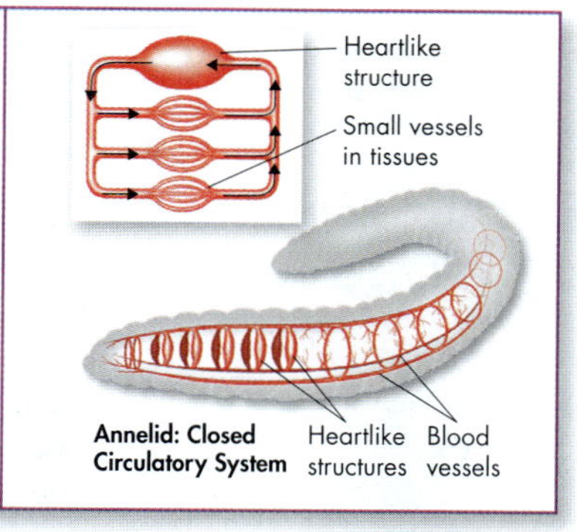

Annelid: Closed Circulatory System

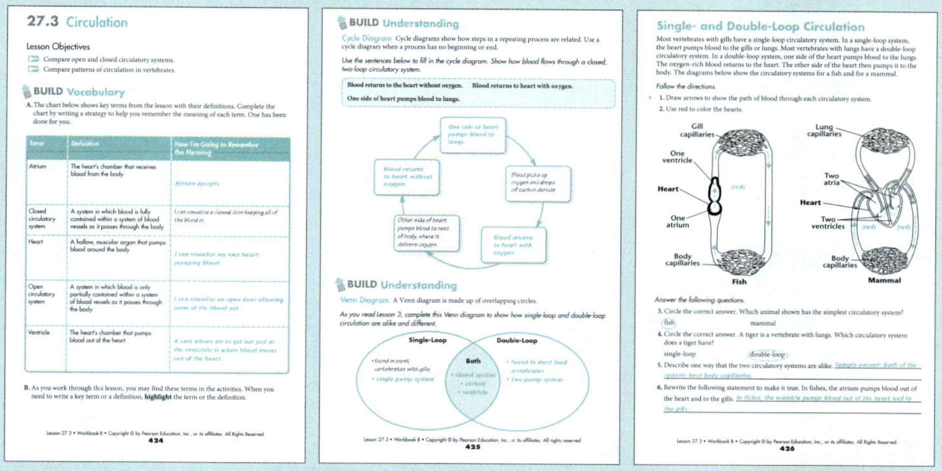

27.3 Circulation

Lesson Objectives
- Compare open and closed circulatory systems.
- Compare patterns of circulation in vertebrates.

BUILD Vocabulary

A. The chart below shows key terms from the lesson with their definitions. Complete the chart by writing a strategy to help you remember the meaning of each term. One has been done for you.

Term	Definition	How I'm Going to Remember the Meaning
Atrium	The heart's chamber that receives blood from the body	Atrium Accepts
Closed circulatory system	A system in which blood is fully contained within a system of blood vessels as it passes through the body	I can visualize a closed door keeping all of the blood in.
Heart	A hollow, muscular organ that pumps blood around the body	I can visualize my own heart pumping blood.
Open circulatory system	A system in which blood is only partially contained within a system of blood vessels as it passes through the body	I can visualize an open door allowing some of the blood out.
Ventricle	The heart's chamber that pumps blood out of the heart	A vent allows air to get out just as the ventricle is where blood moves out of the heart.

B. As you work through this lesson, you may find these terms in the activities. When you need to write a key term or a definition, **highlight** the term or the definition.

Lesson 27.3 • Workbook B • Copyright © by Pearson Education, Inc., or its affiliates. All Rights Reserved.
424

BUILD Understanding

Cycle Diagram Cycle diagrams show how steps in a repeating process are related. Use a cycle diagram when a process has no beginning or end.

Use the sentences below to fill in the cycle diagram. Show how blood flows through a closed, two-loop circulatory system.

Blood returns to the heart without oxygen. **Blood returns to heart with oxygen.**
One side of heart pumps blood to lungs.

BUILD Understanding

Venn Diagram A Venn diagram is made up of overlapping circles.

As you read Lesson 3, complete this Venn diagram to show how single-loop and double-loop circulation are alike and different.

Lesson 27.3 • Workbook B • Copyright © by Pearson Education, Inc., or its affiliates. All rights reserved.
425

Single- and Double-Loop Circulation

Most vertebrates with gills have a single-loop circulatory system. In a single-loop system, the heart pumps blood to the gills or lungs. Most vertebrates with lungs have a double-loop circulatory system. In a double-loop system, one side of the heart pumps blood to the lungs. The oxygen-rich blood returns to the heart. The other side of the heart then pumps it to the body. The diagrams below show the circulatory systems for a fish and for a mammal.

Follow the directions.

1. Draw arrows to show the path of blood through each circulatory system.
2. Use red to color the hearts.

Answer the following questions.

3. Circle the correct answer. Which animal shown has the simplest circulatory system?
fish mammal
4. Circle the correct answer. A tiger is a vertebrate with lungs. Which circulatory system does a tiger have?
single-loop double-loop
5. Describe one way that the two circulatory systems are alike. Sample answer: Both of the systems have body capillaries.
6. Rewrite the following statement to make it true. In fishes, the atrium pumps blood out of the heart and to the gills. In fishes, the ventricle pumps blood out of the heart and to the gills.

Lesson 27.3 • Workbook B • Copyright © by Pearson Education, Inc., or its affiliates. All Rights Reserved.
426

Pre-Reading

Objectives

27.3.1 Compare open and closed circulatory systems.

27.3.2 Compare patterns of circulation in vertebrates.

Preview the Pages

1. Preview the diagrams of the open and closed circulatory systems. Read the captions aloud and point out the difference between these two kinds of systems.
2. Have students find each vocabulary word in the text and read the sentence in which it appears.
3. Direct students' attention to the Single- and Double-Loop Circulation diagram. Have students trace the loop(s) in each visual. Link the number of loops to the terms *single* and *double.*

Activate Prior Knowledge

Say Just as organs work together as a system, body systems work together to accomplish body functions.

Ask How do the circulatory and respiratory systems work together? *(The circulatory system carries oxygen to cells and carbon dioxide from cells.)* The circulatory and digestive systems? *(The circulatory system carries nutrients to cells from the digested food.)*

Build Understanding

CYCLE DIAGRAM Refer students to Workbook B, where they will practice using a cycle diagram to show events in a repeating process and to show how the events are related.

NATIONAL SCIENCE EDUCATION STANDARDS

UNIFYING CONCEPTS AND PROCESSES

I, V

CONTENT

C.3.a, C.5.d

For the corresponding lesson in *Miller & Levine Biology,* see pages 791–793.

LESSON 27.3

During Reading

Active Reading

USE GRAPHIC ORGANIZERS Have students use a sequence flowchart or a cycle diagram to trace the path of blood through a closed circulatory system.

KEY QUESTION Read the Key Question on the previous page and have a student read the answer aloud.

Ask Which type of circulatory system do humans have? *(closed)*

INTERPRET DIAGRAMS Have students form small groups. Give groups about five minutes to compare the single-loop and double-loop circulatory system diagrams. Then have each group write a summary statement outlining two major differences and one similarity. Have a volunteer from each group read the summary statement.

Ask Which organ in the single-loop circulation is analogous to the lungs in a double-loop circulatory system? *(gills)*

KEY QUESTION Ask a volunteer to read the Key Question on this page and have another student read the answer.

Ask In a double-loop circulatory system, one pump sends blood to the body. Where does the other pump send blood to? *(the lungs)*

Build Vocabulary

WORD ORIGINS Tell students that most of the words in English are rooted in another language. The Latin word *circulare* is the root for many English words, such as *circulation* and *circular.* Explain that some very similar words have a somewhat different root: the Latin word *circum,* which means "around."

Ask Which word originates from the Latin *circum—circle, circumference,* or *circulate*? *(circumference)*

Speed Bump

Ask To where does blood flow after it moves through the lungs? *(back to the heart)*

Ask Which type of circulatory system does an invertebrate animal likely have if it has no gills or lungs? *(open)*

Single- and Double-Loop Circulation Most vertebrates that use gills for respiration have a single-loop circulatory system. Most vertebrates that use lungs have a double-loop circulatory system. (In diagrams of animals' circulatory systems, blood vessels carrying oxygen-rich blood are red. Blood vessels carrying oxygen-poor blood are blue.)

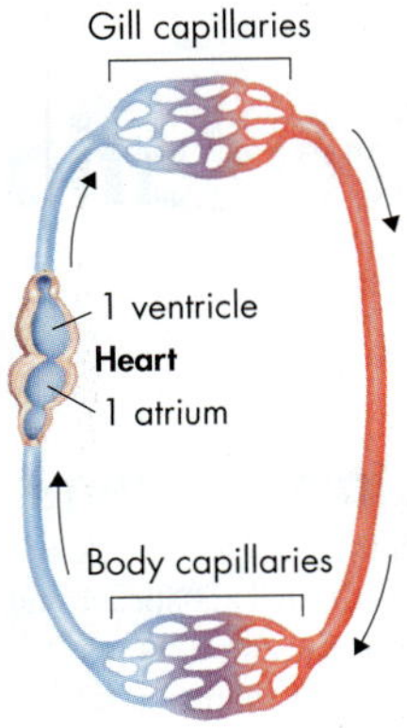

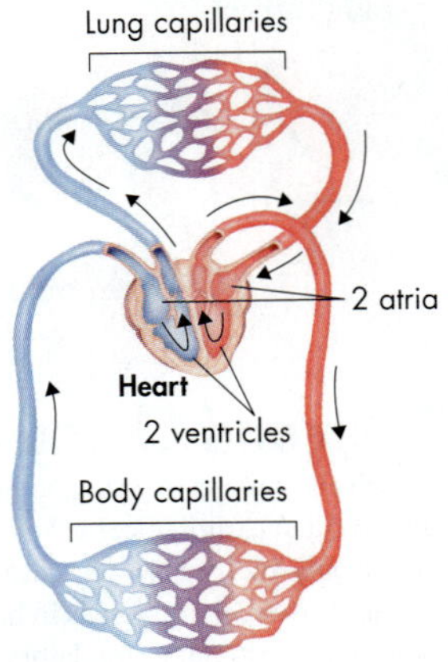

BUILD Vocabulary

heart a hollow muscular organ that pumps blood throughout the body

open circulatory system a type of circulatory system in which blood is only partly contained within a system of blood vessels as it travels through the body

closed circulatory system a type of circulatory system in which blood circulates entirely within blood vessels that extend throughout the body

atrium an upper chamber of the heart that receives blood that is about to enter the ventricle

ventricle a lower chamber of the heart that pumps blood out of the heart

WORD ORIGINS
The word *circulatory* comes from the Latin word *circulare,* which means "to form a circle." In a circulatory system, blood circles around the body.

Single- and Double-Loop Circulation

As vertebrates evolved, they developed more complex circulatory systems. These systems move materials to and from body tissues very efficiently.

Single-Loop Circulation Most vertebrates with gills have a single-loop circulatory system. A single pump forces blood around the body in one direction. In fishes, the heart has two chambers. The **atrium** (plural: atria) is the chamber that receives blood that is about to enter the ventricle. The **ventricle** is the chamber that pumps blood out of the heart and to the gills. In the gills, blood picks up oxygen and releases carbon dioxide. It then travels to the rest of the body, delivering oxygen to the body's cells. By the time blood returns to the atrium, it has become oxygen-poor.

Double-Loop Circulation As larger and more active land vertebrates evolved, the capillary networks in both their lungs and body tissues became larger. A single pump would not have been strong enough to pump blood through these larger systems. In reptiles, birds, and mammals, a double-loop, two-pump circulatory system evolved. Each loop is powered by one side of the heart.

In the first loop, one side of the heart forces oxygen-poor blood from the heart to the lungs. In the lungs, blood picks up oxygen and drops off carbon dioxide. Oxygen-rich blood then returns to the heart. In the second loop, the other side of the heart forces oxygen-rich blood to the rest of the body. Oxygen-poor blood returns to the heart. Then, the cycle begins again.

Key Question How do the patterns of circulation in vertebrates compare? **Most vertebrates that use gills for respiration have a single-loop, one-pump circulatory system. Most vertebrates that use lungs have a double-loop, two-pump circulatory system.**

BIOLOGY.com Search Lesson 27.3 GO • InterActive Art

Hands-On Learning

HEART RATE Have students investigate how their heart rate responds to various activities.

Have students make and record predictions. Will their heart rate increase or decrease if they stand, sit, lie down, or run? By how much?

First, model how to find your pulse in your wrist using two fingers (not the thumb). Then ask students to take and record their pulse after lying down for two minutes, sitting quietly for two minutes, standing for two minutes, and running in place for two minutes.

Ask Did your data support your predictions? Write a hypothesis that explains why your heart rate varied with different activity levels. *(Students should find that their heart rate increases with exercise because the circulatory system works with the respiratory system to deliver more oxygen to cells and to rid the body of waste gases produced by exertion.)*

Evolution of the Four-Chambered Heart A four-chambered heart is actually two separate pumps working next to each other. Where did the second pump come from? During the evolution of vertebrates, partitions arose that divided the original two chambers into four chambers. This separated oxygen-rich blood from oxygen-poor blood. You can get an idea of how the four chambers evolved by looking at the hearts of modern vertebrates.

Amphibian hearts usually have three chambers: two atria and one ventricle. The left atrium gets oxygen-rich blood from the lungs. The right atrium gets oxygen-poor blood from the body. Blood from both atria flows into the ventricle. The shape of the ventricle helps direct the movement of blood. Most oxygen-poor blood goes to the lungs. Most oxygen-rich blood goes to the rest of the body. But there is some mixing of oxygen-rich and oxygen-poor blood.

Reptilian hearts usually have three chambers. Their ventricles are usually divided a little more than those of typical amphibians. This partial division allows less mixing of oxygen-rich and oxygen-poor blood.

Mammals have a four-chambered heart. The right atrium gets oxygen-poor blood from the body. The right ventricle pushes that blood to the lungs. The left atrium gets oxygen-rich blood from the lungs. The left ventricle pushes oxygen-rich blood out to the rest of the body. Oxygen-rich and oxygen-poor blood cannot mix.

Reptilian Heart Under the armor-like hide of this crocodile lies a heart with two atria and one ventricle.

CHECK Understanding

Use the highlighted words from the lesson to complete each sentence correctly.

1. The upper chamber of the heart that receives blood that is about to enter the ventricle is called the ______.
2. In a(n) ______, blood empties into sinuses before returning to the heart.
3. The chamber of the heart that pumps blood out of the heart to the rest of the body is the ______.

Critical Thinking

4. **Compare and Contrast** Compare the structure of an open circulatory system to that of a closed circulatory system.
5. **Relate Cause and Effect** How does having a closed circulatory system benefit a large, active animal?
6. **Review** What are two different patterns of circulation found in vertebrates?
7. **Write to Learn** Answer the question in the mystery clue. What happens when you drink a thick liquid through a straw? Is it harder or easier than drinking water through a straw?

MYSTERY CLUE

Human blood is only about a third as salty as seawater. It needs to circulate through very small capillaries. What might happen if the water content of a person's blood were to drop too low? (Hint: See p. 654.)

BIOLOGY.com Search Lesson 27.3 GO • Lesson Assessment

Active Reading

MAKE INFERENCES Tell students that four-chambered hearts are found in all modern mammals.

Ask Do you think that large, active vertebrates, such as humans, would have been as successful had the heart not evolved? *(No, a four-chambered heart helps the circulatory system move blood more efficiently through an active animal's body.)*

Post Reading

ASSESSMENT TIPS Tell students to make flash cards of the vocabulary words and some of the main ideas in the lesson. Working in pairs, have them quiz each other. If students are having trouble with Question 4, have them return to the visuals of open and closed circulatory systems. Have students trace the circulatory pathways with their fingers.

MYSTERY CLUE Tell students that soup gets thicker if it is allowed to simmer because more and more of the water evaporates. The same happens to blood if there is less fluid, such as when our bodies become dehydrated by illness or overheating.

Wrap-Up Activity

INTERVIEW A BIOLOGIST Pair students. One will be a biologist and the other will be a reporter assigned to write an informative article about the circulatory system.

The writer will prepare interview questions about the function of the circulatory system in very small organisms, invertebrates without gills, vertebrates with gills, and vertebrates with lungs.

The biologist will answer the questions for the reporter. Both will then discuss the interview and make changes until they are satisfied with the answers.

Check Understanding

1. atrium
2. open circulatory system
3. ventricle
4. An open circulatory system has one or more hearts or heartlike organs that pump blood through vessels. The blood empties into sinuses and comes into direct contact with body tissue. It collects in another set of sinuses and returns to the heart. A closed circulatory system has a heart or heartlike organ that pumps blood entirely through a system of vessels.
5. A closed circulatory system can pump oxygen-rich blood to tissues at high pressure and can deliver oxygen to muscle tissues quickly to help an animal move.
6. Most vertebrates with gills have single-loop circulation. Most vertebrates with lungs have double-loop circulation.
7. Sample answer: If the water content were to drop too low, blood would get thicker. This would make it harder for blood to flow through tiny capillaries.

LESSON 27.4

Pre-Reading

Objectives

27.4.1 Describe the methods animals use to manage nitrogenous wastes.

27.4.2 Explain how aquatic animals eliminate wastes.

27.4.3 Explain how land animals eliminate wastes.

Preview the Pages

1. Have students preview all the visuals, read the captions, and prepare one question about each visual. Have them look for answers as they read.
2. Read the vocabulary words and write them on the board with their definitions. Scrutinize terms in the definitions to make sure they are understandable. For example, be sure students understand the meaning of *nitrogen-containing* and *eliminated* in the definition for *excretion*.
3. Direct students' attention to the visual Excretion in Land Animals.
   Ask What does the color indicate in these diagrams? *(the excretory system in each organism)*

Build Understanding

PREVIEW VISUALS Have students preview the Excretion in Aquatic Animals visual. Then, direct students to Workbook B where they will find information about previewing visuals.

27.4 Excretion

Key Questions

How do animals manage toxic nitrogen-containing waste?

How do aquatic animals eliminate wastes?

How do land animals remove wastes while conserving water?

BUILD Understanding

Preview Visuals Write a question you have about the figure called Excretion in Aquatic Animals. As you read the lesson, try to answer your question.

In Your Workbook Refer to your workbook to see how previewing visuals can help you to understand the ideas in this lesson.

BUILD Vocabulary

excretion the process by which nitrogen-containing wastes are eliminated from the body

kidney an organ of excretion that separates wastes and extra water from the blood

PREFIXES

The prefix *ex-* means "out of, from." *Excretion* helps to get wastes "out" of the body.

The Ammonia Problem

So far in this chapter, you have learned how animals get rid of carbon dioxide. But cells also produce other wastes. What are those wastes? And how do animals get rid of them?

When cells break down proteins, they create a nitrogen-containing waste called ammonia. Ammonia is a problem, because it is poisonous! Animals get rid of ammonia by the process of **excretion.** Some small animals that live in water get rid of ammonia by allowing it to diffuse out of their body. Big animals and smaller ones that live in dry places have excretory systems that process and remove ammonia.

Storing Nitrogen-Containing Wastes Animals that cannot get rid of ammonia right away have evolved ways to store nitrogen-containing wastes. Ammonia is too toxic, or poisonous, to store. Many animals convert ammonia into compounds that are less toxic. Insects, reptiles, and birds change ammonia into a sticky white compound called uric acid. Uric acid is much less toxic than ammonia and does not dissolve easily in water. Mammals and some amphibians convert ammonia to urea. Like uric acid, urea is less toxic than ammonia. Unlike uric acid, urea easily dissolves in water.

Maintaining Water Balance Excretory systems help keep the correct balance of water in blood and body tissues. Some systems get rid of water when they get rid of nitrogen-containing wastes. Other systems get rid of nitrogen-containing wastes but conserve water.

Many animals use **kidneys** to separate wastes and extra water from blood. Kidney cells cannot actively pump water across their membranes. Yet they need to separate water from wastes. Kidneys solve this problem by pumping ions to create areas with higher osmotic concentrations. Water moves passively into these areas by osmosis. So this process saves water, but it cannot get rid of extra salt.

Key Question How do animals manage toxic nitrogen-containing waste? **Animals either get rid of ammonia quickly or convert it to other nitrogen-containing compounds that are less toxic.**

NATIONAL SCIENCE EDUCATION STANDARDS

UNIFYING CONCEPTS AND PROCESSES

I, V

CONTENT

C.1.a, C.5.d

INQUIRY

A.1.b, A.2.a

For the corresponding lesson in *Miller & Levine Biology,* see pages 794–798.

27.4 Excretion

Lesson Objectives

- Describe the methods animals use to manage nitrogenous wastes.
- Explain how aquatic animals eliminate wastes.
- Explain how land animals eliminate wastes.

BUILD Vocabulary

A. The chart below shows key terms from the lesson with their definitions. Complete the chart by writing a strategy to help you remember the meaning of each term. One has been done for you.

Term	Definition	How I'm Going to Remember the Meaning
Excretion	The elimination of metabolic wastes	Excretion = elimination
Kidney	The organ that separates wastes and excess water from blood	I'll think of the Chapter Mystery and how the sick man's kidneys collected too much salt as they cleaned his blood.
Malpighian tubule	Structures in terrestrial invertebrates that move wastes from the blood to the gut	The prefix mal- means "bad" and that will remind me of bad waste being removed. I will also imagine a tube connecting the blood to the gut.
Nephridium	Structures in terrestrial invertebrates that filter body fluid	Nephridium filters fluid.

B. As you work through this lesson, you may find these terms in the activities. When you need to write a key term or a definition, **highlight** the term or the definition.

BUILD Understanding

Preview Visuals Previewing visuals and taking notes about them can help you remember what you read and review for tests.

Preview the diagram of excretion of aquatic animals. Write a question about the diagram. As you read, find the answer.

Visual	Question	Answer
Freshwater and saltwater fish showing excretion in aquatic animals	Sample question: How do saltwater and freshwater fish excrete waste differently?	Sample answer: Freshwater fish excrete water, while saltwater fish excrete salt.

BUILD Understanding

T-Chart A T-chart is a way to compare and contrast two things. Use each row to show how the two things relate to one idea. As you read Lesson 4, add details to the T-chart below to compare excretion in freshwater animals and saltwater animals.

Freshwater Animals	Saltwater Animals
Most freshwater invertebrates lose ammonia through simple diffusion.	Saltwater invertebrates lose ammonia through diffusion.
Sample answer: Some freshwater invertebrates and most fish excrete extra water through excretory tubules and pores in the skin.	Sample answer: Saltwater fish excrete salt across their gills.
Sample answer: Amphibians and some fish excrete water through diluted urine.	Sample answer: Saltwater fish have kidneys that produce concentrated urine.

Excretion in Terrestrial Animals

Terrestrial Invertebrates The excretory system in animals removes waste from the body. Nephridia are tubelike structures that filter body fluid. Malpighian tubules concentrate waste.

Follow the directions.

1. Find the excretory structures in each animal. Circle them in blue.
2. Label the nephridia and Malpighian tubules.

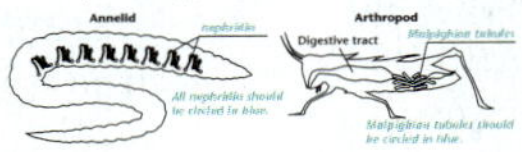

Answer the questions. Circle the correct answer.

3. Which animal shown uses nephridia to remove waste?
   (annelid) arthropod
4. Which animal shown eliminates waste through the digestive system?
   annelid (arthropod)
5. Which animal has Malpighian tubules?
   annelid (arthropod)

Excretion in Terrestrial Animals

Terrestrial Invertebrates and Vertebrates All land animals need water to survive. Their excretory processes use up some of this water. However, most land animals have adapted ways to minimize water loss.

The Venn diagram below compares the ways terrestrial invertebrates and terrestrial vertebrates eliminate waste and conserve water. Complete the diagram using the terms below.

insects, kidneys, Malpighian tubules, mammals, urea, uric acid, waste, water

Terrestrial Invertebrates
- Some convert ammonia into uric acid
- examples: insects and arachnids
- excretion carried out by Malpighian tubules

Both
- remove waste
- conserve water

Terrestrial Vertebrates
- Some convert ammonia into urea
- examples: mammals and amphibians
- excretion carried out by kidneys

CHAPTER MYSTERY

(Near) Death By Salt Water The sick survivor's friends tried to save his life. Since they were alone on the island, they had to use what was available to them.

Draw a picture showing one way the friends could help their sick friend.

Check students' drawings. Sample answer: Students' drawings could show the friends preventing the sick man from drinking salt water or helping him drink coconut milk instead. This would help decrease the salt concentration in his blood.

Excretion in Aquatic Animals

Most aquatic animals allow ammonia to diffuse out of their bodies. The surrounding water carries the ammonia away. But aquatic animals must still keep the right amount of water in their bodies. Animals have different challenges in fresh water and salt water.

Freshwater Animals Most freshwater animals have body fluids that are saltier than the water they live in. This higher concentration of salt causes water to move into their bodies and salts to diffuse out. Flatworms have cells called flame cells that remove extra water. Amphibians and freshwater fishes get rid of extra water by making lots of watery urine. Freshwater fishes also pump salt in through their gills.

Saltwater Animals Invertebrates that live is salt water usually have less of a problem with water balance than freshwater invertebrates. That is because the water concentration in their bodies is similar to that of the seawater around them. In contrast, the bodies of fishes are usually less salty than seawater. They tend to lose water to their surroundings. These fishes get rid of salt across their gills. Their kidneys conserve water by producing small amounts of concentrated urine.

BUILD Connections

EXCRETION IN AQUATIC ANIMALS

All animals must get rid of toxic ammonia and still keep the right amount of water in their bodies. Freshwater and saltwater fish do this in different ways.

Key Question How do aquatic animals eliminate wastes? **Aquatic animals generally allow ammonia to diffuse out of their bodies into the surrounding water.**

The bodies of freshwater animals, such as fishes, contain a higher concentration of salt than the water they live in.

So water moves into their bodies by osmosis, mostly across the gills. Salt diffuses out. If they didn't excrete water, they'd look like water balloons with eyes!

So they excrete water through kidneys that produce lots of watery urine. They don't drink, and they actively pump salt in across their gills.

The bodies of saltwater animals, such as fishes, contain a lower concentration of salt than the water they live in.

So they lose water through osmosis, and salt diffuses in. If they didn't conserve water and eliminate salt, they'd shrivel up like dead leaves.

So they conserve water by producing very little concentrated urine. They drink, and they actively pump salt out across their gills.

BIOLOGY.com Search Lesson 27.4 GO • Art in Motion

657

LESSON 27.4

During Reading

Build Vocabulary

PREFIXES Remind students that prefixes, such as *ex-,* may have more than one meaning. *Ex-* means "out of" as it is used in the word *excretory. Ex-* can also mean "completely" or "intensely," such as in the words *exaggerate* and *extreme,* and, with a hyphen, it can mean "former," as in *ex-president.*

Active Reading

KEY QUESTION Have a volunteer read the question on the previous page and its answer aloud.

Ask What is ammonia? *(a poisonous nitrogen-containing waste)*

USE VISUALS Call on volunteers to read aloud the text for a panel in the visual Excretion in Aquatic Animals. Then, call on another student to describe what the cartoon in that panel shows. Ask volunteers to summarize what the panel explains.

Ask Why don't freshwater fishes drink water? *(They need to get rid of water, not take in more.)*

Why do saltwater fishes lose water through osmosis? *(Water diffuses out of their bodies because their bodies contain a lower concentration of salt than the water they live in.)*

Build Connections

VISUAL ANALOGY As a class, discuss the two analogies in the visual, "like a water balloon with eyes" and "like dead leaves."

Ask How can an analogy help you understand a concept? *(It relates the concept to something you already understand.)*

Ask What word gives you a clue that an analogy follows? *(like)*

KEY QUESTION Have a student volunteer read the question and its answer aloud. Link the answer to the first two sentences under the heading Excretion in Aquatic Animals on this page.

Speed Bump

Ask Excretory systems have two functions. One is to excrete wastes. What is the other? *(to maintain water balance in blood and body tissues)*

Ask To maintain balance, does water in body tissues or in the environment move toward a high salt concentration or toward a low salt concentration? *(high)*

ELL Focus on ELL: Access Content

ALL SPEAKERS Students may have trouble distinguishing the movement of water and salt in saltwater and freshwater fishes. Have groups of students divide a sheet of paper in half and draw a large fish in the middle of each half. Ask students to label one side of the paper Fresh Water and the other side Salt Water. Then have students work together to provide arrows and labels showing the movement of salt and water in each fish. As a wrap-up, have students explain why freshwater fish produce watery urine while saltwater fish produce concentrated urine. *(Freshwater fish need to get rid of excess water while saltwater fish are conserving water.)*

During Reading continued

Inquiry Into Scientific Thinking

MAKE CONNECTIONS Write the word *ammonia* on the board, followed by two arrows, one pointing to *uric acid* and the other to *urea*.

Ask Why do animals convert ammonia to uric acid or urea? *(Ammonia is too toxic to store in the body and these animals are not able to get rid of ammonia right away.)* Which land animals convert ammonia to uric acid? *(insects, spiders, reptiles, birds)* Which land animals convert ammonia to urea? *(mammals, some amphibians)*

SCIENCE SUPPORT Solubility is a measure of the amount of a solute that can dissolve in a solvent. Help students link their observations of uric acid and urea mixed with water to the solubility of these substances. Point out that a substance that seems to dissolve, or disappear, in water is said to be soluble in water.

Lead a Discussion

Point out that birds and lizards have paste-like droppings, whereas mammals have liquid urine. It may be helpful to distinguish the excretion of uric acid and urea from feces, so that students do not confuse bird and lizard droppings with mammalian scat.

Ask How does the solubility of uric acid and urea relate to animal excretions? *(Animals that have less need to conserve water produce urine with urea dissolved in it. Animals living in a dry environment must conserve water and convert ammonia to uric acid, which is excreted as a paste-like substance.)*

Active Reading

SELF-QUESTION Tell students that if a statement or section does not make sense to them, they should go back to the point where they became confused. Tell them to read one sentence at a time, look away from the text, and silently repeat the sentence in their own words.

Build Vocabulary

WORD ORIGINS Tell students that a shortened form of *nephros* is often combined with suffixes to form words associated with *kidney.*

Say Infer the meaning of the following words. Use a dictionary if you don't know the suffix.

nephr + ologist *(a kidney specialist)*

nephr + itis *(inflammation of the kidney)*

nephr + ectomy *(surgical removal of the kidney)*

INQUIRY into Scientific Thinking

Water and Nitrogen Excretion

In this lab, you will investigate the differences between uric acid and urea.

1. Label one test tube Urea and the other Uric Acid. Place 2 grams of urea in the one labeled Urea. Place 2 grams of uric acid in the one labeled Uric Acid.
2. Add 15 mL of water to each test tube. Stopper and shake the test tubes for 3 minutes.
3. Observe each test tube. Record your observations.

Analyze and Conclude

1. **Observe** Which substance—urea or uric acid—is less soluble in water? Explain.
2. **Infer** Birds and reptiles excrete nitrogen-containing wastes in the form of uric acid. How does this adaptation help these animals to survive on land?

In Your Workbook Get more help for this activity in your workbook.

Excretion in Land Animals

Land animals also face challenges. In dry air, animals can lose large amounts of water through their respiratory membranes. And even though water may be scarce, they must use water to get rid of nitrogen-containing wastes. Land animals have evolved excretory systems that help conserve water.

Land Invertebrates Some invertebrates that live on land, such as earthworms and snails, produce urine in nephridia. **Nephridia** (singular: nephridium) are tubelike excretory structures that filter body fluid. Body fluid enters the nephridia through openings. The fluid becomes more concentrated as it moves along the tubes. Urine then leaves the body through excretory pores.

Insects and spiders convert ammonia into uric acid. Uric acid and other nitrogen-containing wastes are absorbed from body fluids by **Malpighian tubules.** These structures concentrate wastes and add them to digestive wastes moving through the gut. The uric acid and wastes form a thick paste that leaves the body through the anus. This paste does not contain much water, so this process reduces water loss.

Land Vertebrates In vertebrates, kidneys get rid of most nitrogen-containing wastes. Mammals and land-living amphibians convert ammonia to urea. Urea is excreted in urine. Most reptiles and birds convert ammonia to uric acid. This uric acid passes through ducts into a structure that also receives digestive wastes. The walls of this structure absorb most of the water from the wastes. The animal then excretes a thick, white paste that you would recognize as "bird droppings."

BUILD Vocabulary

nephridium
an excretory structure of earthworms and other annelids that filters body fluid

Malpighian tubule
a structure in most terrestrial arthropods that concentrates uric acid and adds it to digestive wastes

WORD ORIGINS
The word *nephridium* comes from the Greek word, *nephros*, which means "kidney." A nephridium is a structure in invertebrates that functions like a vertebrate kidney.

Materials

2 test tubes, 2 stoppers, test-tube rack, graduated cylinder, balance, urea, uric acid, glass-marking pencil

Analyze and Conclude

1. Uric acid is less soluble in water than urea is. The solution of urea appears to be clear, whereas the uric acid produced solid crystals in the water.
2. Land animals cannot get rid of ammonia by allowing it to diffuse across the skin. They cannot store ammonia because it is too toxic. The adaptation to convert ammonia to uric acid allows lizards and birds to store nitrogen-containing wastes safely until they are able to get rid of them.

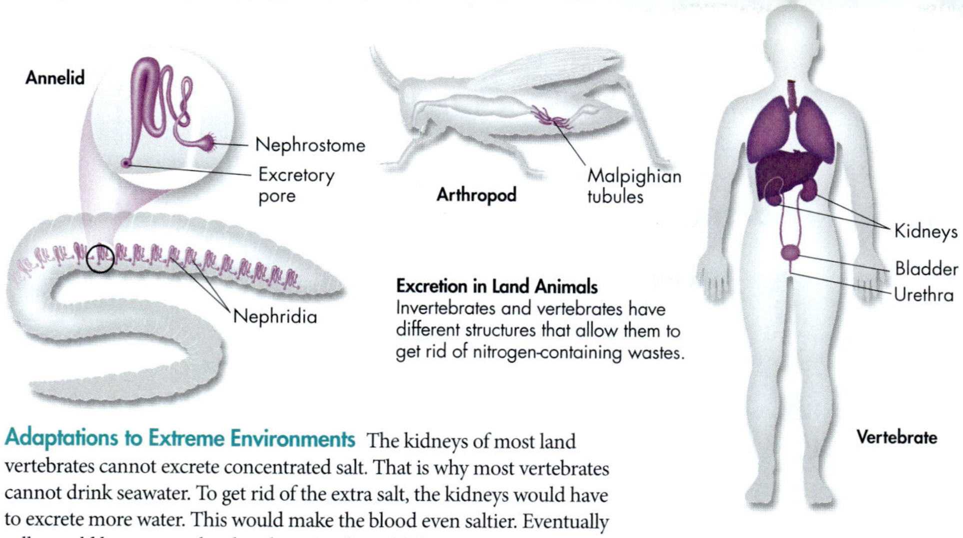

Excretion in Land Animals Invertebrates and vertebrates have different structures that allow them to get rid of nitrogen-containing wastes.

Adaptations to Extreme Environments The kidneys of most land vertebrates cannot excrete concentrated salt. That is why most vertebrates cannot drink seawater. To get rid of the extra salt, the kidneys would have to excrete more water. This would make the blood even saltier. Eventually cells would become so dry that the animal would die.

Many animals that live where fresh water is not available have adaptations that conserve water and get rid of extra salt. Some marine reptiles and birds have glands in their heads that excrete salt solutions. Kangaroo rats that live in the American southwest have remarkably efficient kidneys. They produce urine that is 25 times more concentrated than their blood!

Key Question How do land animals remove wastes while conserving water? **Land animals have different specialized structures that allow them to get rid of nitrogen-containing compounds while also conserving water.**

CHECK Understanding

Use the highlighted words from the lesson to complete each sentence correctly.

1. In a spider, waste fluid becomes concentrated inside _______.
2. Land vertebrates have organs called _______ that separate wastes and extra water from the blood.
3. Excretory structures in earthworms that create urine are called _______.

Critical Thinking

4. **Review** Why does ammonia create a problem for all animals?
5. **Compare and Contrast** How do the different water-balance needs of freshwater animals and saltwater animals explain the different ways that they excrete nitrogen-containing waste?
6. **Write to Learn** Answer the question in the mystery clue below. Think about what happens when there is extra salt in the blood.

MYSTERY CLUE

Humans have kidneys that have evolved to conserve salt, not to get rid of it. How could this have posed a problem for the sick "survivor"? (Hint: See above.)

BIOLOGY.com Search Lesson 27.4 GO • Lesson Assessment

Check Understanding

1. Malpighian tubules
2. kidneys
3. nephridia
4. Ammonia is a product of protein breakdown, but it is toxic to cells.
5. Water tends to move into the bodies of freshwater animals, so they have to produce a lot of watery urine as well as pump salt in across their gills. Water tends to move out of the tissues of saltwater animals, so they produce very concentrated urine to conserve water and pump salt out across their gills.
6. Human kidneys need to get rid of extra salt by making more urine. The process of making urine removes water from the blood. Water moves out of cells to replace the water that was removed from the blood. This causes cells and tissues to become dehydrated and can cause organs to fail.

Active Reading

USE VISUALS Point out that the three systems shown in the visual are representative examples, and that other land invertebrates and vertebrates have similar, if not identical, systems. In discussing each example, call on students to read aloud related material in the text.

INTERPRET DIAGRAMS Tell students that diagrams often label structures that are discussed in the text. Have students find the structure labeled *Excretory pore*. Ask students to find this term in the text on the previous page. Then call on a volunteer to explain its meaning. *(an opening to the outside of the body for excretion)*

KEY QUESTION Read the Key Question and answer aloud to the class.

Ask What tool could help you compare these specialized structures? *(Information can be organized in a table of land animals and their excretory organs.)*

Speed Bump

Ask Why can't land vertebrates drink seawater and remain healthy? *(Most land vertebrates cannot excrete concentrated salt.)*

Ask How do animals live in areas where fresh water is not available, such as in deserts? *(They have special adaptations that allow them to excrete salt and conserve water.)*

Post Reading

ASSESSMENT TIPS Have students review all the vocabulary in the lesson before they begin the Check Understanding assessment.

Students having difficulty with Question 4 should review the first page in this lesson.

For Question 5, have students work in pairs to construct compare/contrast tables.

MYSTERY CLUE Have students review why kidneys cannot excrete a large amount of excess salt. Then, ask them to think about the damage a buildup of salt could cause to the kidneys in terms of water balance in the cells.

Wrap-Up Activity

DEVELOP A QUIZ SHOW Play a game in which you give students the answers to questions about animal excretion and students respond with the right questions. Students can play the game in teams or as individuals. You might want to set up categories and point values.

Pre-Lab

Read aloud the Problem. Then review the animal body systems that students learned about in this chapter (digestive system, respiratory system, circulatory system, and excretory system).

Next, have a volunteer read aloud the Connect to the Big Idea paragraphs. Afterward, go back through the first paragraph one sentence at a time. Explain that this paragraph refers to several body systems without naming the systems by name. Have volunteers identify the systems being described in each sentence. Then ask students what body systems they think they'll be observing in this lab. *(digestive system, respiratory system, circulatory system, excretory system)*

Review the Background Questions and the Pre-Lab questions as a class.

Lab

Tell students that they will perform the chapter lab *Anatomy of a Squid* described in **Lab Manual B.**

SAFETY Students should wear goggles, a lab apron, and disposable plastic gloves when dissecting the animals. Caution students to handle the dissecting scissors and dissecting pins with care. Make sure they keep their hands away from their mouth, nose, and eyes during the lab and wash their hands thoroughly at the end of the lab.

Post-Lab

DRAW PICTURES Give students copies of the drawings from the lab manual and colored pencils. Ask them to use a different color to identify digestive, respiratory, circulatory, and excretory structures.

Skills Lab

GUIDED INQUIRY

Pre-Lab: Anatomy of a Squid

Problem What structures does a squid use to obtain nutrients and eliminate wastes?

Materials squid, dissecting tray, hand lens, forceps, dissecting scissors, dissecting pins, dissecting probe

Lab Manual Chapter 27 Lab

Skills Focus Observe, Infer, Sequence, Draw Conclusions

Connect to the Big idea All animals obtain their food by eating other organisms. All animals need a way to digest the food, and most animals need a way to circulate the absorbed nutrients to all the cells in the body. Animals also need to absorb oxygen from their environment for cellular respiration. Finally, animals need to rid their bodies of wastes.

The ways that animals meet these needs vary greatly. Often, different habitats require different structures. For example, an animal that must obtain its oxygen from air will not have the same respiratory structures as an animal that must obtain its oxygen from water. In this lab, you will dissect a squid and observe parts of several body systems.

Background Questions

a. **Compare and Contrast** How are a gastrovascular cavity and a digestive tract different?

b. **Review** What process takes place in all respiratory structures?

c. **Compare and Contrast** What is the difference between an open and a closed circulatory system?

Pre-Lab Questions

Preview the procedure in the lab manual.

1. **Interpret Visuals** What structure can you use to distinguish the ventral side of a squid from the dorsal side?
2. **Infer** Why is it important to lift the mantle while cutting it?
3. **Predict** What do you expect the gills to look like, and why?

BIOLOGY.com Search Chapter 27 GO

Visit Chapter 27 online to test yourself on chapter content and to find activities to help you learn.

Untamed Science Video Trek carefully with the Untamed Science crew as they get up close and personal with bears to learn about their adaptations.

Art in Motion What happens when fresh- and saltwater fishes excrete water or salt? Find out by watching this animation.

Art Review Review your knowledge of the different types of respiratory systems with this activity.

InterActive Art See how single- and double-loop circulation systems compare.

Visual Analogy Compare the structure and function of the types of teeth with common objects.

Pre-Lab Answers

BACKGROUND QUESTIONS

a. A gastrovascular cavity has a single opening through which food is ingested and wastes are expelled. A digestive tract has two openings, one for ingesting food and one for expelling wastes.

b. Gases diffuse across a selectively permeable membrane.

c. In an open circulatory system, blood is only partially contained within a system of blood vessels as it travels through the body. In a closed circulatory system, blood circulates entirely within blood vessels.

PRE-LAB QUESTIONS

1. The siphon is located on the ventral side.
2. Sample answer: If the mantle is not lifted, the scissors may cut through the organs that lie beneath it.
3. Sample answer: Because gas exchange takes place in the gills, I expect the gills to have a large surface area.

NATIONAL SCIENCE EDUCATION STANDARDS

UCP I, V

CONTENT C.5.d

INQUIRY A.1.b, A.2.a

660 Chapter 27 • Pre-Lab

27 CHAPTER Summary

27.1 Feeding and Digestion

- Most filter feeders have gills or other structures that strain algae and small animals from water. Detritivores eat decaying material. Carnivores eat other animals. Herbivores eat plants or parts of plants. Parasites feed on the blood or tissues of their host. Sometimes two types of animals have a relationship that helps both of them.
- Some invertebrates break down food through intracellular digestion. Most invertebrates and all vertebrates use extracellular digestion.
- Carnivores usually have sharp mouthparts that can capture food, hold it, and cut it into pieces. Herbivores usually have mouthparts that can grind plant material.

intracellular digestion (p. 647)
extracellular digestion (p. 647)
gastrovascular cavity (p. 647)
digestive tract (p. 648)
rumen (p. 648)

27.2 Respiration

- Respiratory structures provide a large surface area of moist, selectively permeable membrane. The membrane keeps a difference in the relative concentration of oxygen and carbon dioxide on either side of it. This difference in concentration allows diffusion to take place.
- Most aquatic animals breathe through their skin or with gills. Aquatic reptiles and mammals breathe with lungs and must hold their breath underwater.
- Invertebrates that live on land have a variety of different respiratory structures, including skin, mantle cavities, book lungs, and tracheal tubes. All vertebrates that live on land—reptiles, birds, mammals, and land-living amphibians—breathe with lungs.

gill (p. 651)
lung (p. 651)
alveolus (p. 652)

27.3 Circulation

- In an open circulatory system, blood is only partly contained within blood vessels. In a closed circulatory system, blood circulates entirely within blood vessels.
- Most vertebrates with gills have a single-loop circulatory system with a single pump. This pump forces blood through a single-loop circulatory system. Most vertebrates with lungs have a double-loop, two-pump circulatory system.

heart (p. 653)
open circulatory system (p. 653)
closed circulatory system (p. 653)
atrium (p. 654)
ventricle (p. 654)

27.4 Excretion

- Animals either get rid of ammonia quickly or convert it to other nitrogen-containing compounds that are less toxic.
- Aquatic animals generally allow ammonia to diffuse out of their bodies into the surrounding water.
- Some land invertebrates, including earthworms and snails, produce urine in nephridia. Insects and spiders convert ammonia to uric acid. Mammals and land amphibians convert ammonia to urea, which is excreted in urine. Most reptiles and birds convert ammonia into uric acid.

excretion (p. 656)
kidney (p. 656)
nephridium (p. 658)
Malpighian tubule (p. 658)

BIOLOGY.com Search Chapter 27 GO • Match It • Chapter Assessment

UbD Transfer the Big Idea

TRANSFER Have students work in small groups to make up an imaginary animal that might have been real if animals had evolved differently. Remind students that their animal should be well-adapted to its environment. Each group should prepare a "field report" about its animal, with descriptions and sketches, as if members of the group were researchers who had just discovered this new organism. The report should include information about the animal's methods of feeding, digestion, respiration, circulation, and excretion.

Chapter Summary

Think Visually

USE GRAPHICS Have student fill in the following table organizer.

Animal type	Aquatic invertebrates	2	Land invertebrates	4
Respiratory structures	1	Reptiles and mammals—lungs; Others—gills	3	Lungs

ANSWERS

1. Body covering; gills
2. Aquatic vertebrates
3. Skin; mantle cavities; book lungs; tracheal tubes
4. Land vertebrates

BIOLOGY.com Study Online

REVIEW RESOURCES

Visual Analogy This short animation compares the structure and function of teeth.

Data Analysis Students explore why insects were larger in the Paleozoic than they are today.

InterActive Art Students can watch this animation to help them compare single- and double-loop circulatory systems in vertebrates.

Art in Motion The challenges of maintaining water balance in freshwater and saltwater animals are the focus of this animation.

Vocabulary Review The *Flash Cards* and *Match It* provide an interactive way to review chapter vocabulary.

Editable Worksheets Editable pages of Study Workbook B, Lab Manual B, and the Assessment Resources Book are available online.

Lesson Overviews Lesson Overviews are based on Miller & Levine Biology. However, they are available on CD-ROM and can be edited for your class.

ASSESSMENT RESOURCES

Chapter Assessment Have students take an online version of the Chapter 27 Check Understanding assessment.

Standardized Test Prep Students can take an online version of the Standardized Test Prep. You will receive their scores along with ideas for remediation.

Diagnostic and Benchmark Tests Use these tests to monitor your students' progress and supply remediation.

ASSESSMENT

Assess

Big idea Invite a volunteer to read the question. Discuss the relationship between structure and function. Write the words *nephridium, Malpighian tubules,* and *kidney* on the board.

Ask What do these words have in common? *(They are all structures for excretion.)*

Ask Why do different animals have different excretory structures? *(Different animals in different environments have different needs. Each structure is adapted to a specific animal's needs.)*

Ask How do blood vessels relate to both the respiratory system and the circulatory system? *(Blood vessels are parts of the circulatory system that pick up oxygen in the lungs and deliver it to the cells. They also pick up carbon dioxide from cells and deliver it back to the lungs for exhalation.)*

CONSTRUCTED RESPONSE Paragraphs will vary. Explain the activity carefully. Be sure that students understand that the *Hints* should be used as a guide to help them answer the questions. Review how to organize important points into a paragraph. For Question 1, students may find it helpful to make a compare/contrast table.

Write Question 1 on the board and read it aloud. Underline *digestive tract* and *gastrovascular cavity.*

Read the first hint.

Say This question asks you to compare two structures: a digestive tract and a gastrovascular cavity.

Ask Which animals have a digestive tract? *(Sample answer: birds, reptiles, mammals)* Which animals have a gastrovascular cavity? *(Sample answer: cnidarians)*

Point out that both structures are efficient in the organisms they are found in. Ask students to think about how a digestive tract is more efficient for a vertebrate *(or other animal that has a digestive tract)* than a gastrovascular cavity would be. Encourage students to review the visuals that compare a gastrovascular cavity in a cnidarian with a digestive tract in a bird in Lesson 27.1.

Read the second hint.

Say Think about how a gastrovascular cavity differs from a digestive tract.

27 CHECK Understanding

Assess the Big idea Structure and Function

Write an answer to the question below:

Q: How do the structures of animals allow them to take in essential materials and get rid of wastes?

Constructed Response

Write an answer to each of the numbered questions below. The answer to each numbered question should be one or two paragraphs long. To help you begin, read the **Hints** below the questions.

1. **How is a digestive tract a more efficient structure for taking in and processing food than a gastrovascular cavity?**

   Hint A gastrovascular cavity has one opening. A digestive tract has two openings, and food moves in one direction.

   Hint A digestive tract may have specialized organs, such as a stomach and intestines.

2. **Do you think large, active vertebrates would have been likely to succeed without the evolution of closed circulatory systems? Explain your reasoning.**

   Hint Blood that is completely contained within blood vessels can be pumped under greater pressure. Therefore, it can travel a greater distance.

   Hint To move quickly, muscles need a large supply of oxygen.

3. **All animals need to control the amount of water within their bodies. They also need to get rid of nitrogen-containing wastes. Compare the way that earthworms and insects are able to control these functions.**

   Hint Excretion usually requires water. Earthworms and insects have different kinds of structures that help conserve water.

   Hint Review the art of excretory systems on page 659 as you plan your answer.

Foundations for Learning Wrap-Up

Use the Z-fold charts you made when you started the chapter to help you to organize your thoughts about how animals get food and get rid of wastes.

Activity 1 Get together in a group of three. Exchange the Z-fold charts for Lesson 1 with the members of your group. Look at the predicted answers for each key question. Discuss what is right about each answer and what might be missing. Do the same thing for the other lessons of the chapter.

Activity 2 Fold the Z-fold for Lesson 1 so that the predictions on the middle panel are hidden. Turn the Z-fold over. Write the answers for each of the key questions on the back of the Z-fold. Compare your predictions to the answers that you wrote after reading the chapter. Do the same thing for the other lessons of the chapter.

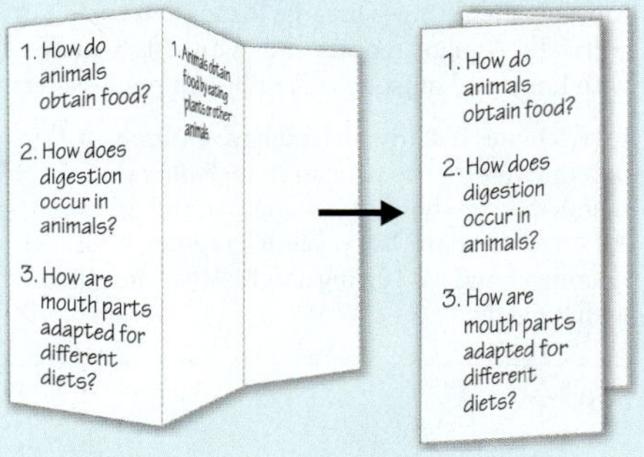

For Question 2, pair students and have them use the hints and your modeling to answer the question. Have students share their paragraphs with the class.

For Question 3, have students write the paragraphs independently for homework or as an anchor activity.

Foundations for Learning Wrap-Up

Have students take out their Z-folds. Check to see that students have written all the Key Questions and predictions and have revised their answers as they read the lessons. Either assign groups or pair students, and have them complete the activities.

27.1 Feeding and Digestion

Understand Key Concepts

1. Animals that get food by eating decaying bits of plant and animal material are called
   a. herbivores. c. detritivores.
   b. carnivores. d. filter feeders.

Test-Taking Tip

When you are answering multiple-choice questions, be sure to read all the choices before you choose an answer. One choice may seem right. However, when you read the other choices, you may find that another choice is more correct. For example, consider question 1. Herbivores eat plants, so at first you might think that the answer is **a**. But the question asks about decaying bits of plants, so the correct answer is **c**.

2. Which animal below relies primarily on intracellular digestion?
   a. sponge c. dragonfly
   b. clam d. earthworm
3. Compare the processes of intracellular and extracellular digestion.

Think Critically

4. **Infer** The skull of a mammal has no sharp canine teeth. The large teeth toward the back of the jaw are very flat. What did this mammal eat?

27.2 Respiration

Understand Key Concepts

5. Most terrestrial insects breathe using a network of structures called
   a. gills. c. book gills.
   b. tracheal tubes. d. book lungs.
6. For the exchange of carbon dioxide and oxygen, an animal's respiratory surfaces must be
   a. cold. c. hot.
   b. dry. d. moist.
7. Most fishes exchange gases by pumping water
   a. over their gills. c. over their atria.
   b. through their lungs. d. through their alveoli.
8. With what respiratory structures do aquatic reptiles and mammals breathe? Why are these structures inconvenient when these animals swim underwater?

Think Critically

9. **Infer** Snails that live on land have a respiratory structure called a mantle cavity. The mantle cavity is often covered with mucus. What might the function of the mucus be?

27.3 Circulation

Understand Key Concepts

10. Most arthropods have
    a. no circulatory system.
    b. an open circulatory system.
    c. a closed circulatory system.
    d. skin gills.
11. In a closed circulatory system, blood
    a. comes in direct contact with tissues.
    b. empties into sinuses.
    c. does not transport oxygen.
    d. remains within blood vessels.
12. Most vertebrates that have gills have a(n)
    a. double-loop circulatory system.
    b. accessory lung.
    c. single-loop circulatory system.
    d. four-chambered heart.
13. Describe the circulatory system of a mammal. Is it open or closed? State the number of loops and the number of heart chambers in the system.

Think Critically

14. **Compare and Contrast** What is the major structural difference between vertebrates that have single-loop circulatory systems and those that have double-loop systems?

Assessment Tips

Have students work together in small groups to answer the questions. Have each group share their answers with another group and compare their reasoning for each answer. Instruct the groups to resolve discrepancies.

Suggest that students make a graphic organizer to summarize critical information in each lesson before they begin answering questions.

Answers

Lesson 27.1

UNDERSTAND KEY CONCEPTS

1. c

2. a

3. Digestion happens inside specialized cells during intracellular digestion. Digestion happens outside of cells in a digestive tract or gastrovascular cavity during extracellular digestion.

THINK CRITICALLY

4. plants and plant parts

Lesson 27.2

UNDERSTAND KEY CONCEPTS

5. b

6. d

7. a

8. They breathe with lungs. Lungs are inconvenient when the animals are swimming underwater because they have to hold their breath and come up for air.

THINK CRITICALLY

9. The mucus could help to keep the respiratory surface moist.

Lesson 27.3

UNDERSTAND KEY CONCEPTS

10. b

11. d

12. c

13. A mammal has a closed, double-loop circulatory system with a four-chambered heart.

THINK CRITICALLY

14. Most vertebrates that have single-loop circulatory systems use gills for respiration. Most vertebrates that have double-loop circulatory systems use lungs for respiration.

ASSESSMENT

Lesson 27.4

UNDERSTAND KEY CONCEPTS

15. b

16. a

17. Ammonia is toxic to cells and cannot be stored in the body.

THINK CRITICALLY

18. The production of uric acid allows land animals to conserve water in a dry environment.

Connecting Concepts

19. Caffeine increases *Daphnia's* heart rate.

20. Five or more drops of coffee would make the heart rate increase to 190 or more beats per minute.

CHAPTER MYSTERY After students have read through the Chapter Mystery, discuss how land vertebrates with kidneys differ from saltwater organisms in their adaptations for conserving water or of getting rid of excess salt.

Ask How does water cross kidney membranes? *(Water crosses membranes by passive diffusion toward the higher concentration of salt. Salt is pumped across membranes and water follows.)*

Point out that if the kidney cells pumped too much salt across their membranes, too much water would follow, causing the body to dehydrate. That is one reason kidneys cannot excrete excess salt.

Ask In terms of salt concentration, what causes people to hold water in the cells to the point that hands or feet swell? *(The concentration of salt is higher inside the cells than outside, so water won't move out of the cells. Instead, water continues to move toward the inside of cells.)*

Ask How do saltwater fishes solve the problems of losing water to their salty environment and gaining salt from it? *(They produce very small amounts of concentrated urine, and they actively pump salt out across their gills.)*

Point out that humans are not adapted to a salty environment. They do not have adaptations for conserving water and pumping out salt.

27 CHECK Understanding

27.4 Excretion

Understand Key Concepts

15. The amount of salt and water in the body fluids of mammals is controlled by the
- **a.** lungs.
- **b.** kidneys.
- **c.** intestine.
- **d.** heart.

16. The elimination of nitrogen-containing wastes by the body is called
- **a.** excretion.
- **b.** circulation.
- **c.** respiration.
- **d.** digestion.

17. Why do most animals convert ammonia to urea or uric acid?

Think Critically

18. Apply Concepts Uric acid requires less water to excrete than urea does. How is the production of uric acid an advantage to animals that live on land?

Connecting Concepts

A student conducts an experiment to measure the effect of caffeine on the heart rate of a small pond-water animal called Daphnia. *The results are shown in the graph.*

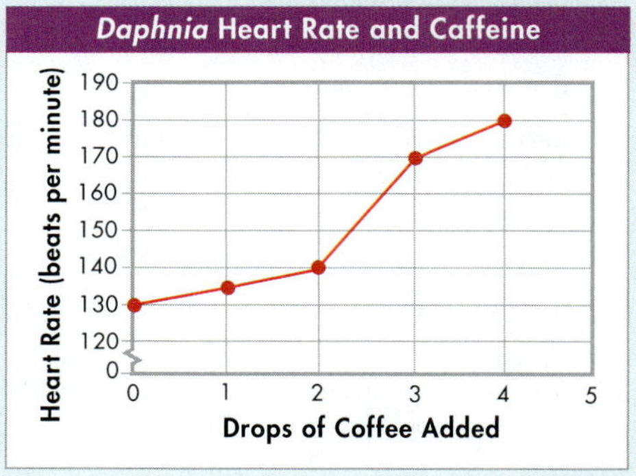

19. Interpret Graphs Describe the effect that caffeine has on the heart rate of *Daphnia*.

20. Predict What would be the effect of five or more drops of coffee on the heart rate of *Daphnia*?

solve the CHAPTER MYSTERY

(NEAR) DEATH BY SALT WATER

Luckily, the group that was supposed to pick up the friends came earlier than planned. They took the sick man to the hospital. He was diagnosed with severe dehydration. The doctors gave him water and fluids through an intravenous drip. If he had gone much longer without treatment, he would have died. What had happened? Why didn't his friends have the same problem?

As sailors have known for centuries, humans cannot drink salt water. But why?

Seawater is saltier than human blood and body fluids. Drinking seawater loads the body with extra salt. Human kidneys cannot make urine with salt concentrations high enough to get rid of the salt. The kidneys are then forced to excrete too much water in urine, which lowers the amount of water in blood. The blood becomes thick and cannot pass through capillaries. Cells and tissues begin to dry out. This dehydration can cause fatal kidney failure and heat stroke.

1. Compare and Contrast The other group members who did not drink seawater experienced water stress as well. What was going on in their circulatory and excretory systems? Why wasn't their water stress as serious as the water stress experienced by their friend?

2. Propose a Solution If you were marooned on an island that had no fresh water, what would be your plan for getting some?

Untamed Science Finding out what happened to the college student who drank salt water is only the beginning. Take a video field trip with the ecogeeks of Untamed Science to see where the mystery leads.

CHAPTER MYSTERY ANSWERS

1. Although the other group members took in some water by drinking the coconut "milk," they probably didn't get enough water to keep their circulatory and excretory systems functioning normally. The lack of fresh water caused some water stress in these group members.

2. Sample answer: You could look for fruit that contains a lot of water. You could try to capture rainwater and condensation.

Untamed Science To explore how bears have adapted to survive in extreme environments, have students watch the short video **Bearly Asleep**.

Standardized Test Prep

Multiple Choice

1. Animals that live on an animal and feed on its body tissues are called
   A parasites.
   B carnivores.
   C herbivores.
   D detritivores.

2. Examining the teeth of an animal can give information about whether it
   A practices intracellular or extracellular digestion.
   B is a filter feeder or a detritivore.
   C is a nutritional symbiont.
   D is a herbivore or a carnivore.

3. Movement of oxygen and carbon dioxide across a respiratory surface requires
   A that the respiratory surface be moist.
   B active transport by the cells of the respiratory surface.
   C alveoli.
   D an equal concentration of both gases on both sides of the membrane.

4. In an open circulatory system, blood
   A is confined to blood vessels at all times.
   B circulates around body tissues.
   C exchanges gases with lung alveoli.
   D is not required for exchanging gases with body cells.

5. In chordates with four-chambered hearts, there is
   A only one loop in the circulatory system.
   B mixing of oxygen-rich and oxygen-poor blood.
   C partial partition of the ventricle.
   D no mixing of oxygen-rich and oxygen-poor blood.

6. Most reptiles excrete wastes in the form of
   A urea.
   B ammonia.
   C uric acid.
   D toxins.

7. What is a function of the excretory system?
   A to supply cells with oxygen and nutrients
   B to rid the body of metabolic wastes
   C to exchange oxygen and carbon dioxide with the environment
   D to break down food

Questions 8–9

A biology student is investigating the relationship between cricket chirping and air temperature. She catches a cricket and places it in a jar. She leaves the jar outside, and each day she counts the number of chirps during a 15-second period. At the same time, she records the outside temperature near the cricket. Her data for a 5-day period are shown below.

Temperature and Cricket Chirping

Day	Number of Chirps in 15 Seconds	Outside Temperature (°C)
Monday	31	23
Tuesday	20	16
Wednesday	12	11
Thursday	29	21
Friday	25	19

8. At which of the following temperatures would a cricket be most likely to chirp 9 times in 15 seconds?
   A 10°C
   B 18°C
   C 0°C
   D 25°C

9. What can the student conclude from this experiment?
   A Crickets cannot chirp more than 31 times in 15 seconds.
   B The number of times a cricket chirps decreases when the temperature decreases.
   C The number of times a cricket chirps increases when the temperature decreases.
   D There is no relationship between the number of times a cricket chirps and temperature.

Open-Ended Response

10. Which types of vertebrates have double-loop circulation and which types have single-loop circulation?

If You Have Trouble With . . .

Question	1	2	3	4	5	6	7	8	9	10
See Lesson	27.1	27.1	27.2	27.3	27.3	27.4	27.4	27.3	27.3	27.3

Test-Taking Tip

USE TIME WISELY

Tell students that when they are taking a long time to answer a question on a test, they should consider moving on to the next question and coming back to the harder question later. When answering other questions, they may remember the information needed to answer the skipped question.

Before the Test

Review with students the Test-Taking Tip. Explain that if they take too long, they may run out of time to answer questions that they are sure about.

Remind students that an open-ended response means that they'll have to write a paragraph to answer the question. Have students paraphrase the question in their notebooks, and remind them about modeling a paragraph.

Answers

1. A
2. D
3. A
4. B
5. D
6. C
7. B
8. A
9. B
10. Most vertebrates that use lungs for respiration have double-loop circulation. These include mammals, reptiles, birds, and adult amphibians. Most vertebrates with gills have single-loop circulation. This includes fishes.

CHAPTER PLANNER 28 Animal Systems II

Chapter Contents	NSES	Time	Core Resources
Chapter Preview			**Student Edition,** pp. 666–667 **Chapter Mystery,** p. 667 **Foundations for Learning,** p. 667
28.1 Response How Animals Respond • Trends in Nervous System Evolution • Sensory Systems	I, V, C.1.f, C.3.a, C.5.d, C.6.a, C.6.b	1 period 1/2 block	**Student Edition,** pp. 668–673 **Build Connections,** p. 672 **Study Workbook B** 28.1 Worksheets **Biology.com** *Art Review:* Vertebrate Brains • 28.1 Lesson Assessment
28.2 Movement and Support Types of Skeletons • Muscles and Movement	I, V, C.5.d	1 period 1/2 block	**Student Edition,** pp. 674–677 **Study Workbook B** 28.2 Worksheets **Biology.com** *InterActive Art:* Sea Star Water Vascular System • *Art in Motion:* Muscles and Joints • 28.2 Lesson Assessment
28.3 Reproduction Asexual and Sexual Reproduction • Internal and External Fertilization • Development and Growth • Reproduction in Chordates	I, V, C.2.b, C.3.a, C.5.d, C.6.b, C.6.c	1 1/2 periods 3/4 block	**Student Edition,** pp. 678–683 **Build Connections,** p. 679 **Study Workbook B** 28.3 Worksheets **Biology.com** 28.3 Lesson Assessment
28.4 Homeostasis Interrelationships of Body Systems • Body Temperature Control	I, IV, V, A.1.c, A.2.a, C.3.a, C.5.d	1/2 period 1/4 block	**Student Edition,** pp. 684–687 **Inquiry Into Scientific Thinking,** p. 686 **Study Workbook B** 28.4 Worksheets **Biology.com** 28.4 Lesson Assessment
Chapter Pre-Lab	V, A.1.a, A.1.b, A.2.a	1 period 1/2 block	**Student Edition,** p. 688 **Lab Manual B** *Comparing Bird and Mammal Bones*

Tools for ELL Instruction

ELL Handbook describes teaching strategies to make *Miller & Levine Biology: Foundation Edition* more accessible to ELL students.

Spanish Study Workbook is a Spanish translation of Study Workbook A.

Multilingual Glossary is the *Miller & Levine Biology* glossary translated into ten languages.

Tools to Challenge Students

Study Workbook A includes more challenging worksheets and activities such as Chapter Mystery activities that reinforce 21st Century Skills.

Lab Manual A labs are geared toward on-level students. The Lab Manual includes fifteen labs in addition to the Chapter Labs.

Biology.com Data Analysis activities offer students a way to practice data gathering and analyzing skills.

Visual Quizzes in the Assessment Resources Book are geared toward on-level students and use visuals from *Miller & Levine Biology.*

Additional Resources

Biology.com Untamed Science Video • Flash Cards

Study Workbook A 28.1 Worksheets
Spanish Study Workbook 28.1 Worksheets
Assessment Resources Book Visual Quiz

Study Workbook A 28.2 Worksheets
Spanish Study Workbook 28.2 Worksheets

Study Workbook A 28.3 Worksheets
Spanish Study Workbook 28.3 Worksheets
Assessment Resources Book Visual Quiz

Study Workbook A 28.4 Worksheets
Lab Manual B *Data Analysis:* Soaking Up Sun
Spanish Study Workbook 28.4 Worksheets
Biology.com *Data Analysis:* Winter Survival

Lab Manual A *Comparing Bird and Mammal Bones* • *Observing Hydra*

Chapter Review

Student Edition Chapter Summary, p. 689
Study Workbook B Chapter 28 Review
Transparencies, pp. 316–325
Biology.com Untamed Science Video • Editable Worksheets of Study Workbooks A and B and Lab Manuals A and B • Chapter 28 Flash Cards and Match It

Untamed Science DVD • Classroom Resources CD (includes lesson presentations and editable worksheets)

Chapter Assessment

Student Edition Check Understanding, pp. 690–693
Study Workbook B Chapter 28 Taking a Standardized Test
Assessment Resources Book Chapter 28 Test B
Biology.com Chapter 28 Assessment • Editable Worksheets of Chapter 28 Visual Quizzes* and Chapter 28 Tests A and B

Exam*View* *Assessment Suite* • Classroom Resources CD (includes lesson presentations and editable worksheets)

Time: 1 period, 1/2 block

* Visual Quizzes are based on visuals from *Miller & Levine Biology.* Review carefully to make sure they are appropriate for your class.

Pressed for Time?

Preview the Chapter Introduce students to the first two Key Questions for Lessons 28.1 and 28.3. Discuss how each animal system in this chapter might be at work in the picture on pp. 666–667.

Cover the Chapter Quickly Read *How Animals Respond* in Lesson 28.1 and the Invertebrate Nervous Systems visual. Read the Key Questions and answers in Lesson 28.2 and use the Muscles and Joints visual to discuss animal movement and support. In Lesson 28.3, have students read *Asexual and Sexual Reproduction, Internal and External Fertilization,* and *Reproduction in Chordates.* Assign *Body Temperature Control* in Lesson 28.4.

Assess Assign Questions 1, 2, 5, and 7 in the 28.1 Check Understanding, Questions 3–5 and 7 in the 28.3 Check Understanding, and Questions 2, 3, 5, and 6 in the 28.4 Check Understanding. In the Chapter 28 Check Understanding, assign Questions 1, 3, 4, 12–15, 17, and 21–23.

Get the Picture

Have students look at the photograph.

Say This photograph shows a good example of how animals respond to their environment *appropriately*, or in a way that meets their needs.

Ask What are the penguins doing? *(huddling together to stay warm)*

Say The penguins' behavior is a response to something in their environment.

Ask What are they responding to? *(extreme cold)* Which body system tells the penguins that it is so cold out that they need to respond with this kind of behavior? *(nervous system)*

Activate Prior Knowledge

Ask How does your body react to cold temperatures? *(Sample answers: I shiver, I get goose bumps, my toes feel numb.)* What behavioral changes might you make in response to the cold? *(Sample answers: Stay inside, put on a coat, turn up the heat.)*

Point out that when people respond to a cold environment, they have some way of knowing it is cold. Explain that our body systems allow us to gather information about the environment so that we can respond and that this is true for all animals.

28 Animal Systems II

Big idea Structure and Function

Q: How do the body systems of animals allow them to collect information about their environments and respond appropriately?

BIOLOGY.com Search Chapter 28 GO • Chapter Mystery

666

NATIONAL SCIENCE EDUCATION STANDARDS

UNIFYING CONCEPTS AND PROCESSES

I, IV, V

CONTENT

C.1.f, C.2.b, C.3.a, C.5.d, C.6.a, C.6.b, C.6.c

INQUIRY

A.1.a, A.1.b, A.1.c, A.2.a

UbD Connect to the Big Idea

Invite a volunteer to read aloud the chapter number and the chapter title. Read the Big Idea and question. Emphasize the question, **How do the body systems of animals allow them to collect information about their environments and respond appropriately?**

Set the Purpose

PREVIEW AND PREDICT Draw the graphic organizer on the board. Guide students to the INSIDE feature on the next page, and invite a volunteer to read each lesson number and title.

Ask What is an example of how animals respond to the environment? How can you relate this example to structure or function? *(Sample answer: Bees have stingers (structure) they can use to protect themselves from enemies (function).)*

Have students think of examples for each lesson question. Remind them to keep structure and function in mind as they study each lesson.

CHAPTER MYSTERY

INSIDE:

Thick down feathers keep these young penguins warm. They share their warmth by huddling close together.

SHE'S JUST LIKE HER MOTHER!

In 2001, a surprising thing happened at the Henry Doorly Zoo in Omaha, Nebraska. A bonnethead shark gave birth to a female baby bonnethead. Workers at the zoo were shocked. For three years, there had been only three bonnethead sharks in the tank where the baby was born. All of these sharks were female.

Some female sharks are able to store sperm for later fertilization. Does this explain how the shark got pregnant?

Read for Mystery Clues As you read the chapter, look for clues that help explain how the baby bonnethead's mother got pregnant. Also, think about how sharks usually reproduce. Think about how that process affects genetic material in the offspring. Then, solve the mystery.

FOUNDATIONS for Learning

Science uses many vocabulary words. These words can help you to understand basic science concepts. Before you read the chapter, write down the vocabulary words on index cards. As you read, write the definitions on the back of the cards. At the end of the chapter are two activities that use these cards to help you to answer the question: How do the body systems of animals allow the animals to collect information about their environments and respond appropriately?

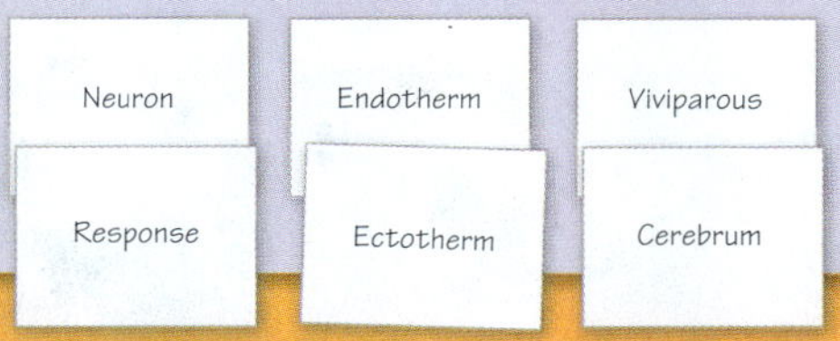

- Untamed Science Video
- Flash Card

Chapter 28
Big Idea: Structure and Function

Chapter 28: How do the body systems of animals allow them to collect information about their environments and respond appropriately?

- **28.1:** How do animals sense and respond to the environment?
- **28.2:** How are different animals adapted to move through their environments?
- **28.3:** What are the different strategies animals have evolved to help them produce offspring?
- **28.4:** How do animals maintain homeostasis?

CHAPTER 28

Foundations for Learning

Pass out the index cards. Provide time for students to write each vocabulary term from the chapter on a card. Have them identify each word they think they know and write a sentence in their notebooks that uses the word. Students will probably not be familiar with all of the words.

Ask Why do you think writing what you already know about a vocabulary word helps you better understand science concepts as you read about them? *(If you already know something about the vocabulary used to describe a science concept, learning about the concept will be easier.)*

Explain that as they study the lesson, they should write the definition on the backs of the cards.

Ask How might writing vocabulary words and definitions on separate cards help you understand the text? *(Sample answers: It helps you remember things when you write them. I can use the cards as flash-cards for studying.)*

Untamed Science™ Join the Untamed Science crew as they interview experts to learn more about how the sex of offspring is determined in some animals in **Boy? . . . Or Girl?**

CHAPTER MYSTERY Ask students whether they think sharks reproduce sexually or asexually. Have a volunteer read the Chapter Mystery. Invite students to brainstorm inferences about how a female shark got pregnant without a male present.

Tell students that they will find clues in the chapter to answer the question, **How did the baby bonnethead's mother get pregnant?**

Digital Support

BIOLOGY.com Extend your reach by using the digital assets offered at **Biology.com.**

LESSON 28.1

Pre-Reading

Objectives

28.1.1 Describe how animals respond to stimuli.

28.1.2 Summarize the trends in the evolution of nervous systems in animals.

28.1.3 Describe some of the different sensory systems in animals.

Preview the Pages

1. Have students locate the vocabulary words in the text. Point out that the definitions of some terms share basic elements but differ in important ways. For example, sensory neurons and interneurons are different types of neurons. Cerebrum and cerebellum are different parts of the brain.
2. As you preview the visuals, point out the examples of invertebrate nervous systems vs. vertebrate brains. Remind students that invertebrates do not have backbones or vertebral columns.
3. Tell students that the abbreviation *Hz* in the table of hearing ranges stands for *hertz,* which is a way of expressing a sound's frequency. The higher the frequency is, the higher the pitch, which is the lowness or highness of a sound.

Build Understanding

PREVIEW VISUALS Have students study the visual of neural circuits as a class. Tell them to use the T-chart in Workbook B to organize their questions. Then, have students work in pairs to complete the T-chart for all the visuals in this lesson.

NATIONAL SCIENCE EDUCATION STANDARDS

UNIFYING CONCEPTS AND PROCESSES

I, V

CONTENT

C.1.f, C.3.a, C.5.d, C.6.a, C.6.b

For the corresponding lesson in *Miller & Levine Biology,* see pages 808–813.

28.1 Response

Key Questions

How do animals respond to events around them?

What are the trends in nervous system evolution?

What are some types of sensory systems in animals?

BUILD Understanding

Preview Visuals Previewing visuals helps you prepare to read and understand the text. Before you read, preview the visual of neural circuits. Take note of any questions you have about it. As you read, try to answer your questions.

In Your Workbook Refer to your workbook for suggestions about how to use a T chart to organize your notes about the visuals in this lesson.

How Animals Respond

Imagine that you are at a favorite place, such as a beach. The sun feels warm on your face. You hear the waves crashing on the shore. You smell salt air. Now think about how your body experiences the place. Your senses gather information about how it looks, sounds, or smells. Your nervous system organizes this information. Your brain decides how to respond to it.

All animals experience and respond to their environments. Sometimes animals need to catch food. Other times, they need to escape from predators. Most animals have specialized nervous systems that allow them to respond to their environments. Nervous systems are made of specialized nerve cells, called **neurons.** Neurons work together to receive and organize information. Then, they "decide" how to respond.

Detecting Stimuli Information in the environment that causes an organism to react is called a **stimulus** (plural: stimuli). Chemicals, light, heat, and sounds can all be stimuli.

Specialized cells called **sensory neurons** are able to sense stimuli. Each type of sensory neuron responds to a particular stimulus. For example, sensory neurons in the ear sense sounds. Like humans, many other animals also respond to light, taste, sound, odor, temperature, gravity, and pressure. Some animals also have sensory cells that humans do not have. For example, some animals can sense weak electric currents or Earth's magnetic field.

Response Lions are predators that eat warthogs and other animals. When a warthog senses the presence of a lion, it begins to run. The lion responds by running after the warthog.

668 Chapter 28 • Lesson 1

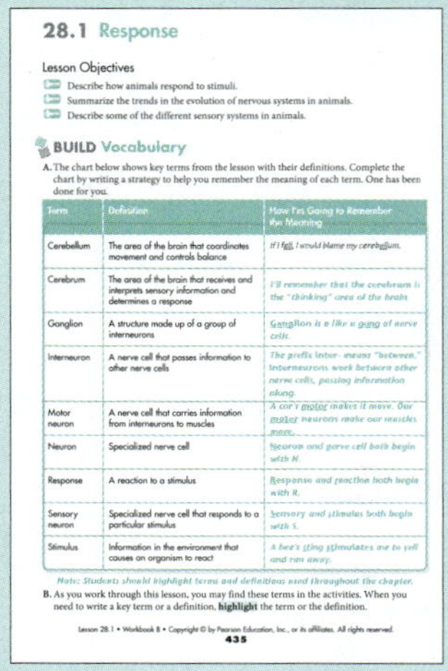
28.1 Response

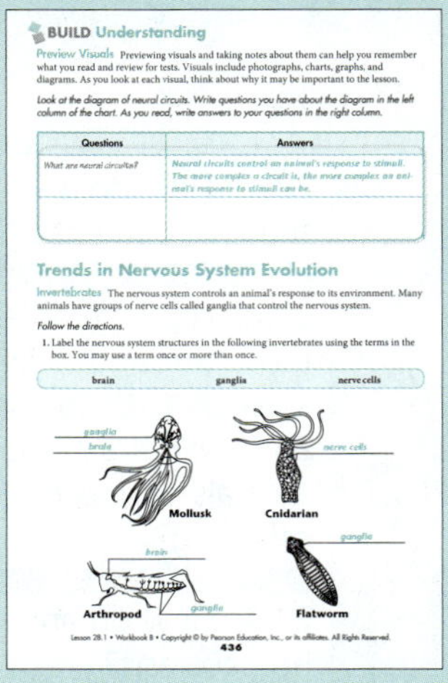
BUILD Understanding

Trends in Nervous System Evolution

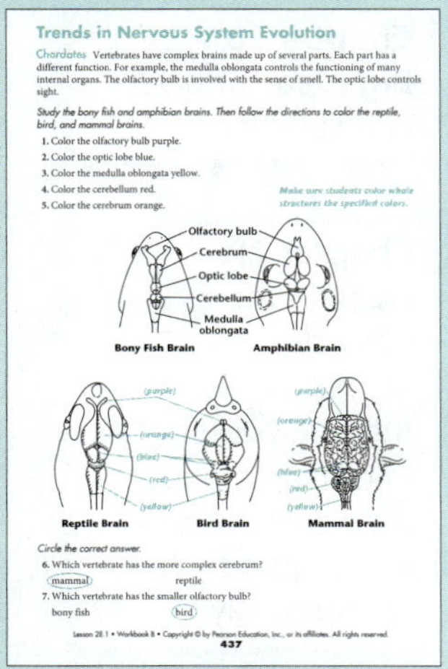
Trends in Nervous System Evolution

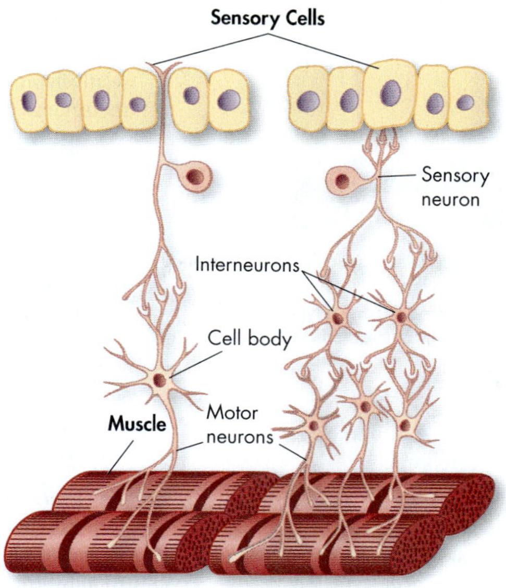

Neural Circuits In simple neural circuits, sensory neurons connect to motor neurons and allow fast but simple responses (left). In more complex neural circuits, interneurons and specialized sensory cells connect sensory neurons to motor neurons (right). Complex neural circuits allow a more complex response.

Processing Information When sensory neurons sense a stimulus, they pass information about the stimulus to other nerve cells. **Interneurons** receive information from sensory neurons and usually pass the information to other neurons. Interneurons process information and determine how an animal responds to stimuli.

The number of interneurons an animal has can determine how flexible and complex an animal's behavior can be. Some invertebrates, such as cnidarians and worms, have very few interneurons. These animals only have simple responses to stimuli. They may swim toward light or toward a chemical stimulus that signals food. They may move away from a stimulus that signals danger.

Vertebrates have more complex nervous systems with larger numbers of interneurons. The vertebrate brain is made of many interneurons, which is why the behaviors of vertebrates can be more complex than the behaviors of most invertebrates.

Responding A specific reaction to a stimulus is called a **response.** Waking up when you hear an alarm is a response. So is licking your lips when you smell good food.

Many body systems work together to allow an animal to respond to a stimulus. The nervous system directs the response to a stimulus. However, the response is usually carried out by cells and tissues that are not nerve cells. A lion's decision to chase prey is carried out by muscle cells that produce movement. Nerve cells called **motor neurons** carry "directions" from interneurons to muscle cells. Other responses to environmental conditions may be carried out by other body systems, such as the respiratory or circulatory systems.

Key Question How do animals respond to events around them? **An animal's body systems—including the nervous system and the muscular system—work together to respond to a stimulus.**

BUILD Vocabulary

neuron a nerve cell; specialized for carrying messages throughout the nervous system

stimulus a signal to which an organism responds

sensory neuron a type of nerve cell that receives information from sensory receptors and sends signals to the central nervous system

interneuron a type of neuron that processes information and may send information to motor neurons

PREFIXES

The prefix *inter-* means "between or among." Through the Internet, information is shared among all connected computers. Interneurons share information between neurons.

ELL Focus on ELL: Access Content

INTERMEDIATE SPEAKERS Help students distinguish between the different neurons.

Say The word *sensory* contains the root *sense*.

Ask What are the different senses in humans? *(smell, taste, touch, sight, hearing)* Reinforce that sensory neurons pick up environmental stimuli, such as light waves and sound waves.

Say The prefix *inter-* means "between."

Ask How does this meaning help you remember the function of interneurons? *(They pick up and pass information between sensory neurons and other neurons.)*

Ask What do you think of when you hear the word *motor*? *(Answers might include engine and noise, but should also relate to movement.)* What do motor neurons do? *(carry information to muscle cells)* What do muscles do? *(They make bones move.)*

Have students make a T-chart that lists the different types of neurons and a definition for each type in their own words.

During Reading

Active Reading

USE VISUALS Have a student read aloud the title and caption of the photograph of the lion and warthog.

Ask What two responses are shown in this photograph? *(A warthog runs in response to sensing a lion. The lion responds to the fleeing warthog by chasing it.)*

USE DIAGRAMS As you read these two pages aloud, pause at appropriate intervals to direct students' attention to the diagram Neural Circuits, and relate the text to appropriate parts of the diagram.

USE GRAPHIC ORGANIZERS Construct an outline of a flowchart on the board to show the main steps in a stimulus-response event. Tell students you will fill in the flowchart with the help of the class.

Say I look at the text and the diagram on this page to help me identify what steps to write in the flowchart. A flowchart shows action, so I look for words that tell me an action is taking place. The diagram helps me picture this action as it takes place in an animal's body.

Ask What happens first in an animal's body when a stimulus takes place? *(Sensory neurons sense the stimuli.)* What happens next? *(Sensory neurons pass information to other nerve cells.)*

Continue questioning and filling in the flowchart. Then, ask students to construct their own flowcharts that show how information flows through the nervous system of a lion that sees a warthog running.

Build Vocabulary

PREFIXES Have students think of other words that use the prefix *inter-*. *(interdependent, interval, international)* Discuss the meanings of these words as a group.

KEY QUESTION Read aloud the Key Question and answer. Point out that the events in an animal's response to a stimulus can be organized by the subheads on these pages.

Speed Bump

Ask What makes some animals have more complex behaviors than other animals? *(They may have more complex nervous systems with larger numbers of interneurons.)*

LESSON 28.1

During Reading continued

Active Reading

USE DIAGRAMS Direct students' attention to the diagram Invertebrate Nervous Systems showing invertebrate nervous systems and have a volunteer read its caption aloud.

Ask What do the nerve cells in the cnidarian make up? *(a nerve net)*

Ask Which animal or animals have radial symmetry? *(cnidarian)* How do you know? *(Its body parts repeat around an imaginary line drawn through the center of its body.)*

Ask Which animal or animals have bilateral symmetry? *(flatworm, arthropod, mollusk)* How do you know? *(They have left and right sides that are identical if you draw an imaginary line through their center.)* If necessary, use other images of an octopus or squid to help students see their bilateral symmetry.

Ask Which animal or animals show the most amount of cephalization? *(arthropod, mollusk)* How do you know? *(They have ganglia and a brain.)*

USE CONTEXT CLUES Have volunteers take turns reading aloud the paragraphs on this page.

Ask Which animal do you think has a more complex nervous system, a grasshopper or an octopus? *(octopus)* Which part of the text and diagram helped you answer this question? *(the last sentence under Heads and the diagram showing cerebral ganglia in the octopus)*

Build Vocabulary

WORD ORIGINS Point out that *cerebrum* and *cerebellum* share the word part *cere-*.

Ask What part of speech are *cerebrum* and *cerebellum?* *(nouns)*

Say *Cerebral* is another word that shares the word part *cere-*. It means "of the brain." *Cerebral* is an adjective.

Ask What suffix can I add to *cerebral* to make the adverb form of the word? *(-ly)*

Speed Bump

Ask Which nervous systems range from simple to complex—vertebrate or invertebrate? *(invertebrate)*

Ask Do invertebrates with brains have simple or complex nervous systems? *(complex)*

Ask What do brains allow invertebrates to do? *(complex behaviors and learning)*

Trends in Nervous System Evolution

Nervous systems vary greatly across the animal kingdom. Some are relatively simple. Others are complex. Nervous systems also differ in the amount of specialization and cephalization.

Invertebrates Invertebrate nervous systems range from simple collections of cells to complex systems with many interneurons.

▶ ***Nerve Nets, Nerve Cords, and Ganglia*** Cnidarians, such as jellyfishes, have simple nervous systems called nerve nets. In a nerve net, neurons are connected in a netlike arrangement. Neurons in a nerve net have few specializations. In radially symmetric invertebrates, such as sea stars, some interneurons are grouped together into nerves or nerve cords. These nerves form a ring around the animal's mouth and stretch out along its arms. In still other invertebrates, a number of neurons are grouped together into **ganglia** (singular: ganglion). Interneurons connect with each other in ganglia.

▶ ***"Heads"*** Animals that have bilateral symmetry often show cephalization. Cephalization (sef uh lih ZAY shun) is the concentration of sensory neurons and interneurons into a "head." Certain flatworms and roundworms have some cephalization. Cephalopod mollusks, such as octopi and squid, have more developed cephalization. So do many arthropods. In these animals, interneurons form ganglia throughout the body. The largest ganglia, called cerebral ganglia, are found in the head.

▶ ***Brains*** In some species of invertebrates, cerebral ganglia are organized into a brain. The brains of some cephalopods, such as octopi, allow complex behaviors and learning.

BUILD Vocabulary

response a specific reaction to a stimulus

motor neuron a type of nerve cell that carries directions from interneurons to either muscle cells or glands

ganglion a group of interneurons

cerebrum a part of the brain responsible for voluntary activities of the body; the "thinking" region of the brain

cerebellum a part of the brain that coordinates movement and controls balance

WORD ORIGINS

The Latin word for "brain" is *cerebrum*. The words *cerebrum* and *cerebellum* come from this Latin word. Both the cerebrum and cerebellum are parts of the brain.

Invertebrate Nervous Systems Invertebrate nervous systems have different amounts of cephalization and specialization. Flatworms have centralized nervous systems with small ganglia in their heads. Cnidarians have a nerve net which allows them to be very successful predators. Arthropods and cephalopod mollusks have a brain and specialized sensory organs.

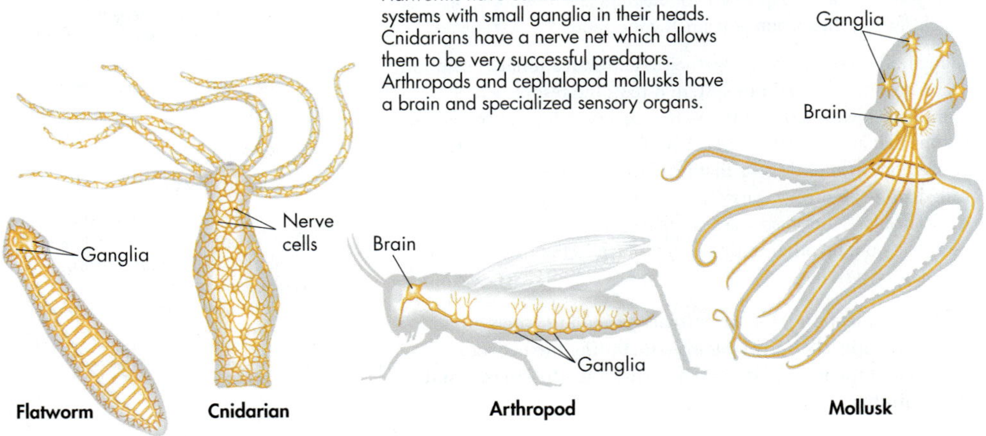

Focus on Academic Language

• specialization • cephalization

Write *specialization* on the board. Underline *-tion.* Remind students that this suffix indicates that a word is a noun. Write other *-tion* words on the board, such as *invention* and *concentration.*

Say *Specialization* is the noun form of the verb *specialize. Specialize* means to "focus on one area." Find the meanings of *specialization* in a dictionary. Which meaning is used in this lesson? *(structural adaptation of an organism for life in a particular environment)*

Say *Invention* is the noun form of the verb *invent,* and *concentration* is the noun form of the verb *concentrate.* Find the meanings of the nouns in a dictionary.

Say *Cephalization* ends in *-tion* but it is not the noun form of a verb. What is *cephalization* concerned with in this lesson? *(nervous systems of animals)*

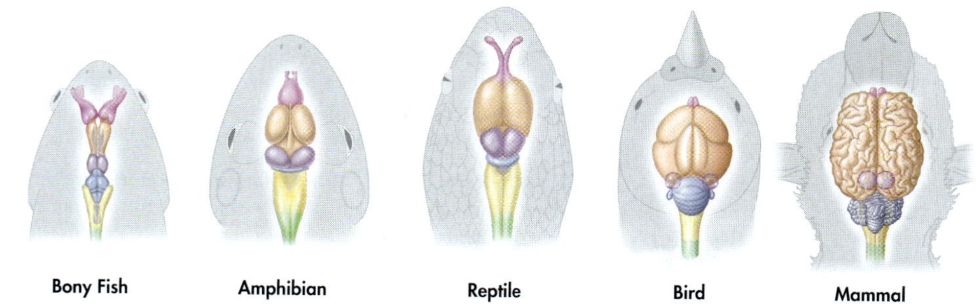

Olfactory bulb	Cerebellum
Cerebrum	Medulla oblongata
Optic lobe	Spinal cord

Vertebrate Brains The cerebrum and cerebellum increase in size from fishes to mammals. The cerebrum is the "thinking" region. In fishes, amphibians, and reptiles, the cerebrum is relatively small. In birds and mammals, the cerebrum is much larger and may contain folds that increase its surface area. The cerebrum is especially large in primates. The cerebellum is also highly developed in birds and mammals.

Chordates Chordates have complex nervous systems. Nonvertebrate chordates have no heads as adults. However, they still have cerebral ganglia. All vertebrates, on the other hand, have brains. Vertebrate brains are formed from many interneurons that are connected to each other. These interneurons are also connected with sensory neurons and with motor neurons in the head and the rest of the body.

▶ ***Parts of the Vertebrate Brain*** The cerebrum, cerebellum, medulla oblongata, optic lobes, and olfactory bulbs are all parts of the vertebrate brain. The **cerebrum** (SEHR uh brum) is the "thinking" part of the brain. The cerebrum receives sensory information and signals the body to respond. The cerebrum is also involved in learning, memory, and conscious thought. The **cerebellum** (sehr uh BEL um) controls balance and helps to coordinate body movement. The medulla oblongata (mih DUH luh ahb lahn GAHT uh) controls the way that many organs work. Optic lobes help an animal to see. Olfactory bulbs help an animal to smell. Vertebrate brains are connected to the rest of the body by a spinal cord. The spinal cord is a thick collection of nerves that is surrounded by the spine.

▶ ***Vertebrate Brain Evolution*** As vertebrates evolved, their brains became larger and more complex. Vertebrate brains increase in size and complexity from fishes, through amphibians and reptiles, to birds and mammals.

Key Question What are the trends in nervous system evolution? **Invertebrate nervous systems range from simple collections of nerve cells to complex systems that include ganglia and brains. All vertebrate nervous systems include brains. As vertebrates evolved, their brains became larger and more complex.**

Not Such a Bird Brain The brains of chickadees are very complex. The part responsible for remembering locations gets bigger when the bird stores food in the fall. When winter comes, the tiny bird is able to find hundreds of storage places. In spring, the bird's brain returns to normal size.

Hands-On Learning

GROUP STORIES Write each part of the brain (except the spinal cord) on a separate index card. Make enough so each student group of five will have one card of each part. Then write each type of vertebrate on five separate slips of paper and fold each.

Review the use of transition words and phrases, which help tie sentences together in a paragraph. Have the class brainstorm examples as you write them on the board. *(Examples: then, suddenly, next, because, but, however).*

Divide the class into groups of at least five students. Have one member of each group draw a slip with a type of vertebrate. (Provide extra for larger groups.) Provide each group with a stack of cards. Each member draws one card and writes a sentence showing the function of the brain part in the vertebrate. *(Example: Optic lobes helped the bird see its prey.)* Then, all group members work together to tie their sentences into a coherent story. Stories may be humorous or improbable, but should reflect correct use of the brain part. Have groups share their stories with the class.

Active Reading

DRAW A DIAGRAM The first paragraph on this page might pose some difficulty for struggling readers. Read aloud the paragraph. Then draw a blank Venn diagram on the board.

Ask What major group of animals does this paragraph discuss? *(chordates)* Into what two groups are chordates divided? *(nonvertebrate chordates and vertebrates)*

Say The paragraph compares and contrasts the nervous systems of the two subgroups of chordates. A Venn diagram helps us visualize, or see, this comparison.

Title the diagram Nervous Systems in Chordates. Label the left circle Nonvertebrate Chordates. Label the right circle Vertebrates. Ask volunteers to fill in the circles. *(left circle: cerebral ganglia in the brain; right circle: brain)*

Say In the overlapping area of the diagram, we write how the two things being compared are alike.

Ask What would you write here? *(hollow nerve cord running along the back of the body)*

INTERPRET DIAGRAMS Refer students to the diagrams of vertebrate brains. Point out that the drawings are not to scale, so actual sizes of the brains should be considered when making comparisons between types of vertebrates. Pronounce each structure listed in the key. Then have a volunteer find the function of each structure in Parts of the Vertebrate Brain and read the function aloud.

Ask How would you describe the size of the cerebrum as you go from fishes to mammals? *(It gets larger.)*

Remind students that some mammals, such as dogs, have a well-developed sense of smell.

KEY QUESTION Ask a volunteer to read the Key Question. Then read the answer. Explain that a trend refers to a general tendency to change in a certain way.

Ask How would you order the structures in the nervous systems of animals from simplest to most complex? *(nerve cells, nerve cords, ganglia, brain)*

Speed Bump

Ask Which type of vertebrate is able to learn and remember the most things? *(mammal)*

During Reading continued

Active Reading

COMPARE/CONTRAST Have volunteers read the text and the photograph captions on this page.

Ask What does the complex eye of the squid do that the simple eye of the scallop does not? *(form images)*

DRAW CONCLUSIONS Point out that the captions for the mosquito and the planarian do not mention whether the eye is simple or complex, but that readers can figure this out by putting together the information in the captions and the main text.

Ask Is the compound eye of a mosquito simple or complex? *(complex)* How do you know from what you read? *(The text states that complex eyes sense motion and color and form images. The caption states that compound eyes detect changes in movement and color and produce images.)*

Ask Is the eyespot of a planarian simple or complex? *(simple)* How do you know from what you read? *(The text states that flatworms have simple eyespots that sense light. The caption for the planarian states that it has an eyespot.)*

If students have difficulty answering these questions, guide them to look for the answers by connecting specific parts of the text and captions.

COMPARE/CONTRAST Have students compare the invertebrate eyes shown.

Ask Which type of invertebrate eye is most like a vertebrate eye? *(squid or octopus eye)*

Ask Can you answer the question just by looking at the photographs? *(yes)* Where could you find the answer without looking at the photographs? *(the caption for Squid)*

Say The caption does not mention the word *vertebrates.* How did you figure out the answer? *(Fishes and humans are vertebrates.)*

Point out that this is an example of how a reader can use prior knowledge to construct meaning beyond what is stated in the text. If readers did not know that fishes and humans are vertebrates, they would not know how to answer the question.

Speed Bump

Ask If you know that an animal has a complex nervous system, what can you infer about its sense organs? *(They are probably complex, too.)*

Sensory Systems

Animals that have complex nervous systems usually have well-developed sensory systems. Sensory systems in animals range from individual sensory neurons to sense organs.

Invertebrate Sense Organs Many invertebrates have sense organs that can sense light, sound, vibrations, movement, body position, and chemicals in air or water. Invertebrate sense organs range from simple to complex. Flatworms have simple eyespots that sense light. Invertebrates with more cephalization have specialized sensory tissues and complex sense organs. Some cephalopods and arthropods have complex eyes that sense motion and color. Complex eyes can also form images.

Invertebrate Eyes Invertebrate sense organs, such as the eyes shown in these photos, have many different structures. Some invertebrate eyes are simple. Some are complex.

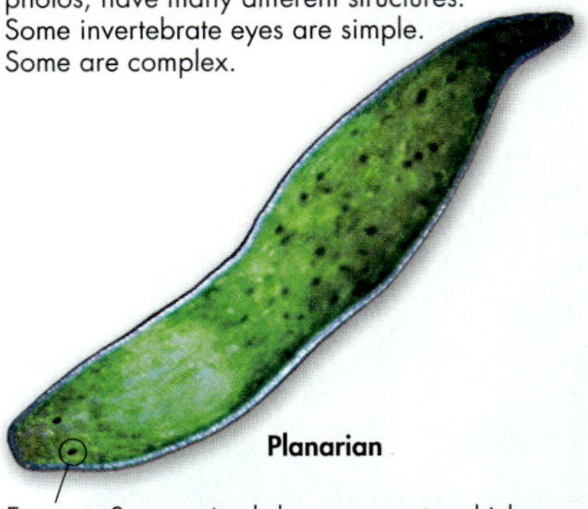

Eyespot: Some animals have eyespots, which are groups of cells that can detect changes in the amount of light (LM 50×).

Simple Eye: The 40–60 simple eyes of a scallop do not form images. They do, however, detect movement well enough to enable the scallop to escape its predators.

Compound Eye: The compound eyes of arthropods are made up of many lenses that detect minute changes in movement and color but produce less-detailed images than human eyes do.

Complex Eye: Octopi and squid have eyes as complex as fishes and humans, though their structures differ.

Focus on Academic Language

• enable • compound

Read aloud the second sentence under the photograph of the scallop. Have students use context clues to figure out the meaning of *enable. (to make able or make possible)* Tell students that the word is also used in everyday language. Provide an example, such as *The Internet enables us to search for information on many different topics.* Ask volunteers for several more sentences that use *enable.*

Read aloud the caption under Mosquito.

Ask What is a compound eye? *(It is made up of many lenses.)* What other phrases do you know that contain the word *compound? (a chemical compound, compound words, compound sentence)* What do all of these phrases have in common? *(They refer to something that has two or more parts.)*

Chordate Sense Organs Nonvertebrate chordates do not have many specialized sense organs. For example, lancelets have a cerebral ganglion with a pair of eyespots that can see light.

Most vertebrates have highly specialized sense organs. Many vertebrates have very sensitive organs of taste, hearing, and smell. Sense organs in some sharks are so sensitive that they can detect 1 drop of blood in 100 liters of water! Many fishes, amphibians, reptiles, birds, and mammals have color vision that is as good as, or better than, the color vision of humans.

Animal	Hearing Range (Hz)
Tree frog	50–4000
Canary	250–8000
Dog	67–45,000
Bat	2000–110,000
Human	30–23,000
Elephant	16–12,000
Bottlenose dolphin	75–150,000

All mammal ears have the same basic parts. However, there are big differences in each animal's hearing range. Bats and dolphins can hear sounds at very high frequencies. Elephants can hear sounds at very low frequencies.

Certain fishes can sense weak electric currents in water. Sharks use their "electric sense" to find their way around the oceans. Other species use electric currents to find prey in dark water. Some fishes use electric pulses to communicate with one another. Earth has a magnetic field that some animals can sense. Many birds use this magnetic field to find their way during migration.

Key Question What are some types of sensory systems in animals? **Sensory systems in animals range from individual sensory neurons to sense organs such as eyes or ears. Many vertebrates also have specialized sense organs that are sensitive to tastes, odors, vibrations, or electrical currents.**

Vertebrate Hearing Many vertebrates can hear sounds that humans cannot hear. Would you expect to be able to hear the highest pitch that a dog can hear? Explain.

CHECK Understanding

Apply Vocabulary

Use the highlighted words from the lesson to complete each sentence correctly.

1. Information in the environment that causes an animal to react is called a _______.
2. Specialized cells called _______ help an animal to detect stimuli.
3. In invertebrates, interneurons are grouped together into structures called _______.
4. The _______ is the "thinking" region of the brain.

Critical Thinking

5. **Explain** What is the job of a motor neuron?
6. **Compare and Contrast** Describe the degree of cephalization shown by cnidarians, flatworms, octopi, and vertebrates.
7. **Infer** What is the general relationship between how complex an animal's nervous system is and how complex its sensory system is?
8. **Write to Learn** The compound eyes of insects sense movement better than they see details. How might sensing movement be more important to an insect than seeing details?

BIOLOGY.com Search Lesson 28.1 GO • Lesson Assessment

Check Understanding

1. stimulus
2. sensory neurons
3. ganglia
4. cerebrum
5. A motor neuron carries messages from interneurons to muscles, so that an animal can move in response to a stimulus.
6. Cnidarians do not show cephalization. Flatworms have ganglia in their heads. Octopi have cerebral ganglia as well as brains. Vertebrates have brains.
7. Usually, animals that have complex nervous systems also have complex sensory systems. Animals with simpler nervous systems also have simpler sensory systems.
8. Sample answer: An insect predator may grab the insect very quickly. If the insect can sense movement from the predator, the insect may be able to get away before it is eaten. It is not important to the insect's survival to see every detail of the predator.

Active Reading

IDENTIFY CUE WORDS Remind students that writers often use special words or phrases that introduce major points or alert the reader to particular text structures. Examples of text structures are cause and effect, compare and contrast, and sequencing. Read aloud the third paragraph.

Ask Which word in the paragraph is a signal that you are about to read something opposite to what you just read? *(however)*

Point out that the cue word *however* signals a compare/contrast structure. In this paragraph, it is important for the reader to understand that the ears of different mammals have a similar structure but the mammals may hear differently.

INTERPRET TABLES Have students study the table of hearing ranges. Explain that the higher the number, the higher the pitch is.

KEY QUESTION Read aloud the Key Question.

Ask What are some functions of sensory systems in animals? *(Sample answer: to sense light, sound, vibrations, movement, and body position)*

Speed Bump

Ask Think of different habitats where you find vertebrates. How can you relate this to differences in sensory systems? *(Sample answer: Different sensory systems allow vertebrates to survive by responding appropriately in different environments.)*

Post Reading

ASSESSMENT TIPS For Question 6, clarify that *degree* means "amount or concentration."

For Question 7, have students use a T-chart to list descriptions of the nervous systems of several different animals in the left column and a description of their sensory systems in the right column. Guide students to identify a pattern.

Wrap-Up Activity

GUESS WHO Privately assign pairs of students an animal mentioned in this lesson, such as a planarian, fish, bird, or octopus. Give students a few minutes to write three sentences that describe its nervous and sensory systems. Have one member of each pair read aloud the sentences, and challenge the rest of the class to identify the animal.

LESSON 28.2

Pre-Reading

Objectives

28.2.1 Describe the three types of skeletons in animals.

28.2.2 Explain how muscles produce movement in animals.

Preview the Pages

1. Have students locate the vocabulary words as you pronounce them aloud. Ask them to write a few sentences about what they know about an organism's joints, ligaments, and tendons.
2. Preview the visuals. Point out that the circles accompanying the illustrations of the sea star and the grasshopper are close-ups of a part.

Activate Prior Knowledge

Ask What animals have skeletons similar to those of humans? *(Students will probably name mammals.)* Do invertebrates, such as insects, have skeletons? *(They may know they have exoskeletons.)*

Explain that all animals have skeletons that provide support. Tell students that, in this lesson, they will learn about skeletons and how animals move.

Build Understanding

COMPARE/CONTRAST TABLE Direct students to the compare/contrast table in Workbook B. Remind students to record information about the structure and functions.

NATIONAL SCIENCE EDUCATION STANDARDS

UNIFYING CONCEPTS AND PROCESSES

I, V

CONTENT

C.5.d

For the corresponding lesson in *Miller & Levine Biology,* see pages 814–818.

28.2 Movement and Support

Key Questions

What are the three types of skeletons?

How do muscles allow movement?

BUILD Understanding

Compare/Contrast Table As you read this lesson, create a table comparing and contrasting the three types of skeletons.

In Your Workbook Refer to your workbook to learn how to organize your ideas using a compare/contrast table.

Types of Skeletons

How does a salmon swim through the water and then leap into the air? How does a worm burrow into the ground? Animals move in many different ways. Yet, their body structures work in similar ways.

Skeletal Support To move efficiently, all animals must do two things. First, they must create a force. Then, they must use that force to push or pull themselves around. Stiff body parts help an animal move. Legs push against the ground. Wings push against the air. Body parts are made stiff and strong by skeletal systems.

▶ ***Hydrostatic Skeletons*** Some invertebrates, such as cnidarians and earthworms, have hydrostatic skeletons. **Hydrostatic skeletons** are fluid-filled body segments that work together with body cells that contract. A hydra has a hydrostatic skeleton. When a hydra closes its mouth, the cells that circle around its body tighten. This tightening allows the animal to get longer and reach out with its tentacles to grab prey. When the hydra opens its mouth, water flows out. Cells in its body wall contract, causing the hydra to become shorter.

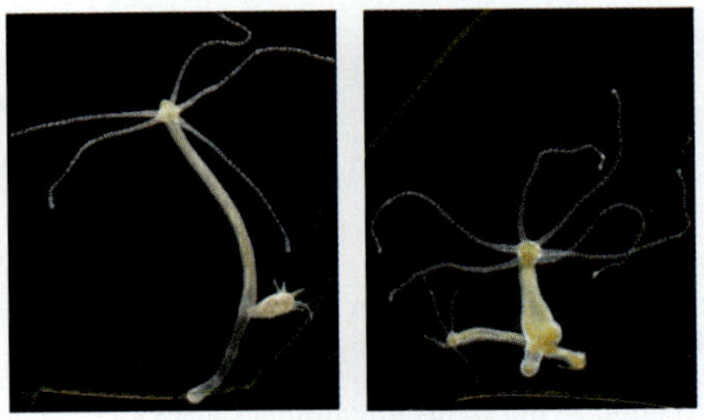

Hydrostatic Skeleton Hydras have hydrostatic skeletons. When a hydra closes its mouth, water that is trapped in its body causes it to get longer. When it opens its mouth, water is released and the hydra becomes shorter.

▶ ***Exoskeletons*** Many arthropods and mollusks have exoskeletons. An **exoskeleton** is a hard covering on the outside of an animal's body. Arthropods have exoskeletons made of a complex carbohydrate called chitin (KY tin). Most mollusks have exoskeletons that are shells made of calcium carbonate.

Exoskeletons have many advantages. Exoskeletons provide coverings that prevent the animals from drying out. They also provide protection from predators.

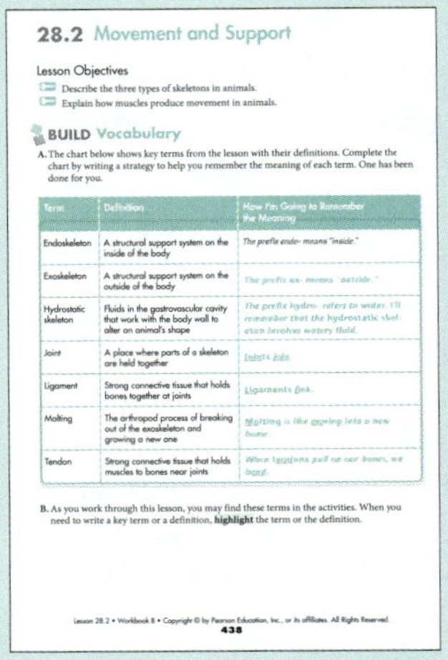

28.2 Movement and Support

Lesson Objectives

Describe the three types of skeletons in animals.

Explain how muscles produce movement in animals.

BUILD Vocabulary

A. The chart below shows key terms from the lesson with their definitions. Complete the chart by writing a strategy to help you remember the meaning of each term. One has been done for you.

Term	Definition	How I'm Going to Remember the Meaning
Endoskeleton	A structural support system on the inside of the body	*The prefix endo- means "inside."*
Exoskeleton	A structural support system on the outside of the body	*The prefix ex- means "outside."*
Hydrostatic skeleton	Fluids in the gastrovascular cavity that work with the body wall to alter an animal's shape	*The prefix hydro- refers to water. I'll remember that the hydrostatic skeleton involves watery fluid.*
Joint	A place where parts of a skeleton are held together	
Ligament	Strong connective tissue that holds bones together at joints	
Molting	The arthropod process of breaking out of the exoskeleton and growing a new one	
Tendon	Strong connective tissue that holds muscles to bones near joints	

B. As you work through this lesson, you may find these terms in the activities. When you need to write a key term or a definition, **highlight** the term or the definition.

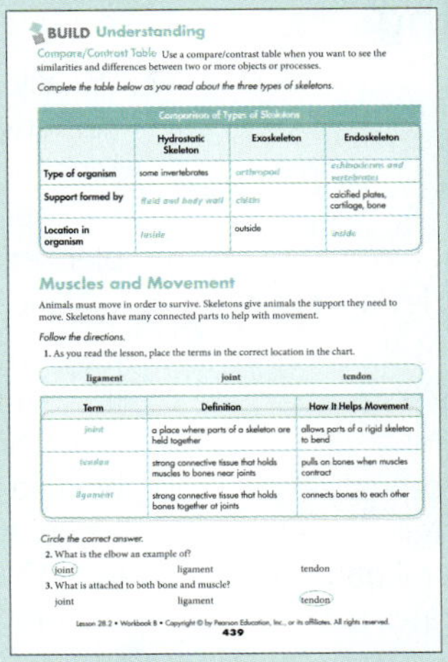

BUILD Understanding

Compare/Contrast Table Use a compare/contrast table when you want to see the similarities and differences between two or more objects or processes.

Complete the table below as you read about the three types of skeletons.

Comparison of Types of Skeletons

	Hydrostatic Skeleton	Exoskeleton	Endoskeleton
Type of organism	some invertebrates	*arthropod*	*echinoderms and vertebrates*
Support formed by	*fluid and body wall*	*chitin*	calcified plates, cartilage, bone
Location in organism	*inside*	outside	*inside*

Muscles and Movement

Animals must move in order to survive. Skeletons give animals the support they need to move. Skeletons have many connected parts to help with movement.

Follow the directions.

1. As you read the lesson, place the terms in the correct location in the chart.

ligament | joint | tendon

Term	Definition	How It Helps Movement
joint	a place where parts of a skeleton are held together	allows parts of a rigid skeleton to bend
tendon	strong connective tissue that holds muscles to bones near joints	pulls on bones when muscles contract
ligament	strong connective tissue that holds bones together at joints	connects bones to each other

Circle the correct answer.

2. What is the elbow an example of?
(joint) ligament tendon
3. What is attached to both bone and muscle?
joint ligament (tendon)

Exoskeletons also have disadvantages. When an arthropod needs to increase in size, it must break out of its old exoskeleton and grow a new one. This process is called **molting.** Exoskeletons are also relatively heavy. As an arthropod gets larger, its exoskeleton makes up a larger and larger portion of its body weight. This is one reason that some science fiction monsters could never exist. The legs of an elephant-size spider would break under the spider's weight!

▶ ***Endoskeletons*** Echinoderms and vertebrates have endoskeletons. An **endoskeleton** is a hard support system inside the body. Sea stars and other echinoderms have an endoskeleton composed of hard plates made of calcium. Sharks and some other fishes have endoskeletons made of cartilage. Other vertebrates have skeletons made of cartilage and bone.

Endoskeletons allow vertebrates to swim, fly, burrow, walk, crawl, or leap. All of these endoskeletons provide strong, lightweight support. Because endoskeletons are light in relation to the bodies that they support, vertebrates can grow very large.

Endoskeletons have some advantages and disadvantages over exoskeletons. Because an endoskeleton does not surround the body, it cannot protect the animal as an exoskeleton can. However, an endoskeleton can grow as an animal grows. Animals that have endoskeletons do not molt.

Joints Arthropods and vertebrates can move because their skeletons are divided into parts connected by joints. **Joints** are places where parts of a skeleton are held together in ways that allow the parts to move. In vertebrates, bones are connected to each other at joints by strong connective tissues called **ligaments.** Most joints are formed by a combination of ligaments, cartilage, and joint fluid. Joint fluid helps the bones to move without rubbing against each other.

Key Question What are the three types of skeletons? **The three main skeletal systems in animals are hydrostatic skeletons, exoskeletons, and endoskeletons.**

BUILD Vocabulary

hydrostatic skeleton
a skeleton made of fluid-filled body segments that work with cells that contract to allow the animal to move

exoskeleton
an external skeleton; a tough outer covering that protects and supports the bodies of many invertebrates

molting
the process of shedding an exoskeleton and growing a new one

endoskeleton
an internal skeleton; a structural support system within the body of an animal

PREFIXES
The prefix *exo-* means "outside." The prefix *endo-* means "within." And *exoskeleton* is outside of the body. An *endoskeleton* is inside, or within, the body.

Endoskeleton Endoskeletons are structural support systems that are inside of the body. Some invertebrates, such as the sea star, have endoskeletons composed of hard plates made of calcium.

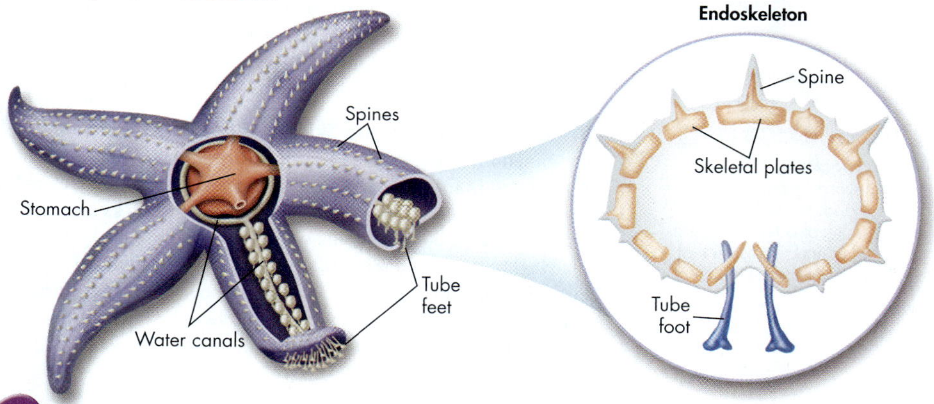

BIOLOGY.com Search Lesson 28.2 GO • InterActive Art

LESSON 28.2

During Reading

Active Reading

APPLY ACADEMIC VOCABULARY Read aloud the first two paragraphs on the previous page. Focus on the word *force*. Explain that a force is a push or a pull. Demonstrate how we use force to walk by pushing off the ground with each step. Use sentence frames to have students describe how other organisms use force to move. For example, Fish use force to move by ____. *(pushing their bodies against the water)*

MAKE INFERENCES Ask volunteers to read aloud the remaining text on these two pages. Have students list advantages and disadvantages of each type of skeleton. Note that this information is not given directly in the text for hydrostatic skeletons, but students should leave space on their paper for it.

Say You can read about the advantages and disadvantages of exoskeletons and endoskeletons and look at the photograph of the hydra to help you make inferences about hydrostatic skeletons.

Ask What environmental conditions does a hydra need in order to move? *(It must be surrounded by water.)* Why might having a hydrostatic skeleton instead of an exoskeleton be a disadvantage? *(Sample answer: A hydrostatic skeleton does not protect an animal from drying out.)* What might be an advantage of having a hydrostatic skeleton? *(Sample answer: It enables an animal to be very flexible.)*

KEY QUESTION Ask a volunteer to read aloud the Key Question and answer. Point out how the black subheads help to answer this question.

Build Vocabulary

PREFIXES Have students identify the words in the definitions of *exoskeleton* and *endoskeleton* that mean the same as *exo-* and *endo-*. *(exo-: external, outer; endo-: internal, within)*

Ask What other words can you think of with these prefixes? *(Sample answer:* exodus, exotic, endocrine, endospore*)*

Focus on Academic Language

- elephant-size

Students may have difficulty determining what compound adjectives modify. Read aloud the last sentence in the first paragraph on this page.

Say *Elephant-size* is a compound adjective; it is an adjective that is made up of two words. *Elephant* modifies *size,* not *spider.* What size? Elephant-size! The text is not talking about an elephant spider but about a spider the size of an elephant.

Ask What are some compound adjectives that describe animals? *(Sample answers: four-legged, fur-covered, smooth-skinned)*

Say Use one of the compound adjectives in a sentence, and tell what is being modified. *(Sample answer: A four-legged vertebrate walked through the woods.* Four *modifies* legged.*)*

Speed Bump

Ask How is an animal's size affected by its exoskeleton? by its endoskeleton? *(Exoskeletons are relatively heavy, so animals with these skeletons cannot grow very large. Endoskeletons are relatively light, so animals with these skeletons can grow very large.)*

Ask How do skeletons help animals move? *(Skeletons provide a rigid support that animals can apply force against.)*

During Reading continued

Active Reading

IDENTIFY CAUSE AND EFFECT Ask a volunteer to read the paragraphs under Opposite Muscle Groups.

Say Cause and effect is the relationship between two things when one thing makes something else happen. This paragraph clearly states a cause and effect relationship between muscles and joints.

Ask Which two sentences describe how muscles cause joints to move? *(the last two)* What signal word helped you identify cause and effect? *(when)* Explain that the word *then* is implied, that is,"When one muscle group contracts, then the joint bends."

Say Look at the visual Muscles and Joints.

Ask In the diagram of the human knee, what is the hamstring doing when it contracts? *(pulling on the knee joint)* What does it cause the joint to do? *(bend)* What causes the joint to straighten? *(The quadriceps pulls on the joint.)*

MAKE CHARTS Draw a Cause and Effect chart on the board with two columns, one for the person's leg and one for the grasshopper's leg in the diagram. Tell students to copy the chart on lined paper. Give them time to fill it in to show the causes and effects involved in muscle movements. Then, ask volunteers to explain how they completed the chart.

MAKE CONNECTIONS Help students make connections between the text in Opposite Muscle Groups and the muscles in their bodies. Tell students to extend their left arm and feel their biceps with their right hand. Demonstrate as you talk. Have them lift a book with their left arm and feel the biceps contract, then lower the book.

Ask Where is the muscle that contracted when you lowered the book? *(the back of the arm)*

Build Vocabulary

WORD ORIGINS Discuss the differences between the three terms.

Ask Why do you find ligaments near joints? *(That is where bones are connected to each other.)*

Speed Bump

Ask What would happen if an animal could not bend? *(It would not be able to move effectively.)*

Ask In what way do muscles work together? *(They work in pairs or groups that pull muscles in opposite directions.)*

Muscles and Joints These diagrams show how muscles work with both a vertebrate endoskeleton and an invertebrate exoskeleton to bend and straighten joints.

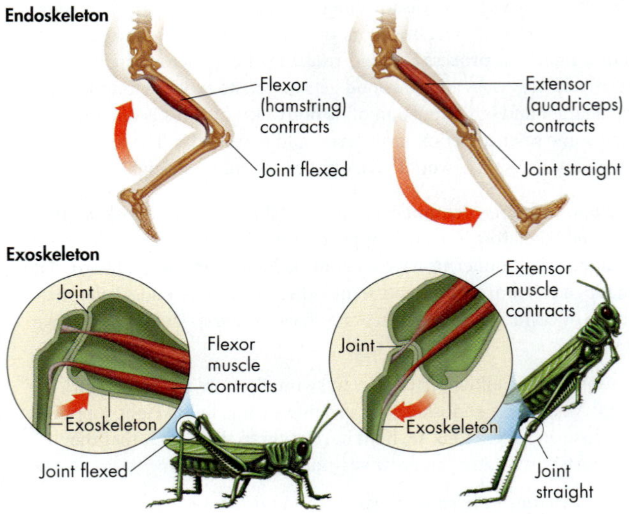

Muscles and Movement

Muscles are tissues that provide the force for movement. Muscles produce force when they contract, or get shorter. Muscles are attached to bone around the joints by tough connective tissue called **tendons.** Tendons pull on bones when muscles contract.

Opposite Muscle Groups Muscles do not produce force as they relax, or get longer. Muscles can only pull; they cannot push. If muscles only pull in one direction, how do they allow animal bodies to move in many different directions? Usually, muscles work together in pairs or groups that pull parts of the skeleton in opposite directions.

In arthropods, muscles are attached to the inside of the exoskeleton. In vertebrates, muscles are attached around the outside of bones. In both cases, muscle pairs or groups attach across joints. When one muscle group contracts, the joint bends. When the opposite group contracts, the joint straightens.

Vertebrate Muscular and Skeletal Systems Amazing combinations of bones, muscle groups, and joints have evolved in vertebrates. In many fishes and snakes, muscles are arranged in blocks on opposite sides of the backbone. These muscle blocks contract in waves down the body. This allows the body to move back and forth. In many amphibians and reptiles, the limbs stick out sideways from the body. Sideways motions of the backbone move these limbs forward and backward. In most mammals, the limbs extend straight down from their bodies. Opposite muscle groups allow these limbs to bend.

BUILD Vocabulary

joint
the place where one bone attaches to another bone

ligament
a tough connective tissue that holds bones together in a joint

tendon
a tough connective tissue that connects skeletal muscles to bones

WORD ORIGINS
The word *ligament* comes from the Latin word *ligamentum*, which means "band, tie, or ligature." You can think of a *ligament* as a "band" of connective tissue that "ties" muscles to bones.

Hands-On Learning

PURPOSE OF JOINTS Pair students and hand out masking tape and pieces of poster board, about 30 x 45 cm.

1. Have a partner in each pair roll a piece of poster board around the other partner's arm. Tell students to cover the elbow and secure the poster board with several pieces of tape to keep it closed.
2. Give the students wearing the poster board a few minutes to try to do simple tasks, such as scratch their heads, pick something up off the floor, and open or close a door or a cabinet.
3. Have the students remove the poster board and attach it to their partner's arm. The partner should attempt the same tasks.
4. When all students have completed the activity, ask them to write a paragraph that describes how restricting their elbow joint made it impossible for them to move their elbows. Then have them describe how joints aid movement.

Muscular and Skeletal Systems of Vertebrates A wide variety of bones, muscle groups, and joints have evolved in vertebrates. Differently shaped bones and muscles form limbs that are adapted for jumping (frog), flying through the air (hawk), and holding objects (raccoon).

The shapes and positions of bones, muscles, and joints in vertebrates relate to the functions that they perform. Limbs that allow animals to run or jump have very different shapes than limbs that allow animals to fly, swim, or hold objects. Scientists can understand how extinct vertebrates moved by studying the joints of fossil bones and the places where tendons and ligaments once attached to bones.

Key Question How do muscles allow movement? **Muscles work together in pairs or groups that are attached to different parts of a supporting skeleton.**

CHECK Understanding

Apply Vocabulary

Use the highlighted words from the lesson to complete each sentence correctly.

1. The hard covering on the outside of an arthropod is called a(n) ________.
2. The ________ of a sea star is composed of hard plates that are made of calcium.
3. Bands of connective tissue that connect bones at joints are called ________.
4. Bands of connective tissue that connect muscles to bones are called ________.

Critical Thinking

5. **Review** What is a hydrostatic skeleton?
6. **Compare and Contrast** How are exoskeletons and endoskeletons different from each other?
7. **Review** Which two body systems work together to allow animals to move?
8. **Review** What characteristics are common in the skeletons of all vertebrates?
9. **Write to Learn** Suppose that you found a vertebrate fossil and brought it to an expert. The expert showed you that the joint structure of the fossil was similar to the joint structure of a modern squirrel. Infer the kinds of movements for which the animal had been adapted. Explain your inferences in a paragraph.

BIOLOGY.com Search Lesson 28.2 GO • Lesson Assessment

LESSON 28.2

Active Reading

MAKE INFERENCES Have students read the paragraph and study the photographs on this page.

Ask How do bones and muscles help a raccoon hold objects? a frog jump? a hawk fly? *(The raccoon has many small bones and muscles in its paws that allow the paws to bend and move easily. The frog has relatively long bones and strong muscles in its legs to allow it to jump. The hawk has strong muscles where the wing meets the body to allow the hawk to move its wings.)*

KEY QUESTION Read aloud the Key Question and answer.

Ask How do hamstrings and quadriceps work together to help you bend and straighten your knee? *(Hamstrings pull on the knee joint to make it bend. Quadriceps pull on the knee joint to make it straighten.)*

Speed Bump

Ask What determines how a vertebrate moves? *(the shapes and positions of bones, muscles, and joints)*

Post Reading

ASSESSMENT TIPS Review the differences between joints, ligaments, and tendons.

For Question 6, tell students to refer to the compare/contrast table they started at the beginning of the lesson.

For Question 9, refer students to the last paragraph of the lesson and have them reread it if necessary.

Wrap-Up Activity

REVERSE QUIZ Divide the class into two teams. Tell students you are going to call out answers that can have more than one question. Teams will take turns calling out questions until they cannot think of any more. Call out vocabulary and other science terms. For example, you might call out, "Hydrostatic skeleton!" Teams might respond with "Which type of skeleton does an earthworm have?" "What allows a hydra to get longer?" Allow students to look at their books, but set a time limit on how long each team has to respond. The team with the greatest number of questions wins.

Check Understanding

1. exoskeleton
2. endoskeleton
3. ligaments
4. tendons
5. A hydrostatic skeleton is a skeleton made of fluid-filled body segments that work with muscles to allow the animal to move.
6. Exoskeletons are on the outside of the body and endoskeletons are on the inside. Animals need to molt exoskeletons as they grow. Endoskeletons grow as the animal grows. Exoskeletons are relatively heavy and endoskeletons are relatively light.
7. muscular system and skeletal system
8. Vertebrates have an endoskeleton that provides strong, lightweight support.
9. Sample answer: The fossil animal probably would have moved like a squirrel. It would have been able to climb trees, jump between trees, and run across the ground. It would also have been able to handle seeds or nuts with its front limbs.

LESSON 28.3

Pre-Reading

Objectives

28.3.1 Compare asexual and sexual reproduction.

28.3.2 Contrast internal and external fertilization.

28.3.3 Describe the different patterns of embryo development in animals.

28.3.4 Explain how terrestrial vertebrates are adapted to reproduction on land.

Preview the Pages

1. Preview the main heads. Note that a discussion of reproduction includes the fertilization event, whether fertilization is internal or external, and the development and growth of the offspring.
2. Have students volunteers locate and pronounce the vocabulary words in the text. Assist students with pronunciation as needed.
3. Preview the Alternating Reproductive Cycles visual. Point out that this is one complete cycle as opposed to separate cycles shown in Insect Metamorphosis later in the lesson.
4. Note that the egg in the diagram Amniotic Egg is fertilized. The eggs at a grocery store are unfertilized and do not contain features such as an embryo or network of blood vessels.

Build Understanding

CONCEPT MAP Direct students to Workbook B, where they will find a partially completed concept map. Suggest they use the heads and subheads in this lesson for the circles to indicate important ideas.

NATIONAL SCIENCE EDUCATION STANDARDS

UNIFYING CONCEPTS AND PROCESSES

I, V

CONTENT

C.2.b, C.3.a, C.5.d, C.6.b, C.6.c

For the corresponding lesson in *Miller & Levine Biology,* see pages 819–826.

28.3 Reproduction

Key Questions

How do asexual and sexual reproduction in animals compare?

How do internal and external fertilization differ?

Where do embryos develop?

How are terrestrial vertebrates adapted to reproduction on land?

BUILD Understanding

Concept Map As you read lesson 28.3, use concept maps to show how ideas or concepts connect. Use circles to show the most important parts of the concept. Connect the circles with lines.

In Your Workbook Refer to your workbook for suggestions about how to use a concept map to organize the main ideas in this lesson.

Asexual and Sexual Reproduction

Sexual reproduction can be dangerous. Just ask a male praying mantis, who may be eaten by his mate. Or ask a female deer, who carries young while trying to avoid predators. If sexual reproduction is dangerous, why do most animals engage in it? What are the advantages of sexual reproduction over asexual reproduction?

Asexual Reproduction Many invertebrates and a few chordates can reproduce asexually. Animals reproduce asexually in many ways. Some cnidarians divide in two. Some animals reproduce through budding, which produces new individuals from the parent's body wall. Females of some species produce eggs that develop without being fertilized, in a process called parthenogenesis (pahr thuh noh JEN uh sis). Parthenogenesis occurs in some invertebrates, but it is rare in vertebrates.

Asexual reproduction has advantages and disadvantages. Since it requires only one parent, individuals can reproduce quickly. However, since individuals carry DNA from only one parent, offspring have less genetic diversity than individuals that are produced sexually. A population with very little genetic diversity may not be able to adapt if the environment changes.

Sexual Reproduction Sexual reproduction involves meiosis, which produces reproductive cells called gametes. Gametes have half the number of chromosomes found in body cells. Male animals produce small gametes, called sperm, which swim. Females produce larger gametes, called eggs, which do not swim. When an egg and sperm join during fertilization, they produce a zygote that contains a full set of chromosomes.

Parthenogenesis Some whiptail lizard species reproduce only by parthenogenesis.

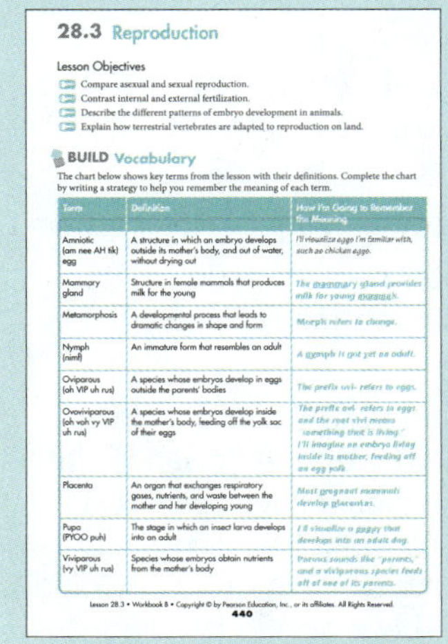

28.3 Reproduction

Lesson Objectives

- Compare asexual and sexual reproduction.
- Contrast internal and external fertilization.
- Describe the different patterns of embryo development in animals.
- Explain how terrestrial vertebrates are adapted to reproduction on land.

BUILD Vocabulary

The chart below shows key terms from the lesson with their definitions. Complete the chart by writing a strategy to help you remember the meaning of each term.

Term	Definition	How I'm Going to Remember the Meaning
Amniotic (am nee AH tik) egg	A structure in which an embryo develops outside its mother's body, and out of water, without drying out	I'll visualize eggs I'm familiar with, such as chicken eggs.
Mammary gland	Structure in female mammals that produces milk for the young	The mammary gland provides milk for young mammals.
Metamorphosis	A developmental process that leads to dramatic changes in shape and form	Morph refers to change.
Nymph (nimf)	An immature form that resembles an adult	A nymph is not yet an adult.
Oviparous (oh VIP uh rus)	A species whose embryos develop in eggs outside the parents' bodies	The prefix ovi- refers to eggs.
Ovoviviparous (oh voh vy VIP uh rus)	A species whose embryos develop inside the mother's body, feeding off the yolk sac of their eggs	The prefix ovi- refers to eggs and the root vivi means "something that is living." I'll imagine an embryo living inside its mother, feeding off an egg yolk.
Placenta	An organ that exchanges respiratory gases, nutrients, and waste between the mother and her developing young	Most pregnant mammals develop placentas.
Pupa (PYOO puh)	The stage in which an insect larva develops into an adult	I'll visualize a puppy that develops into an adult dog.
Viviparous (vy VIP uh rus)	Species whose embryos obtain nutrients from the mother's body	Parous sounds like "parents," and a viviparous species feeds off of one of its parents.

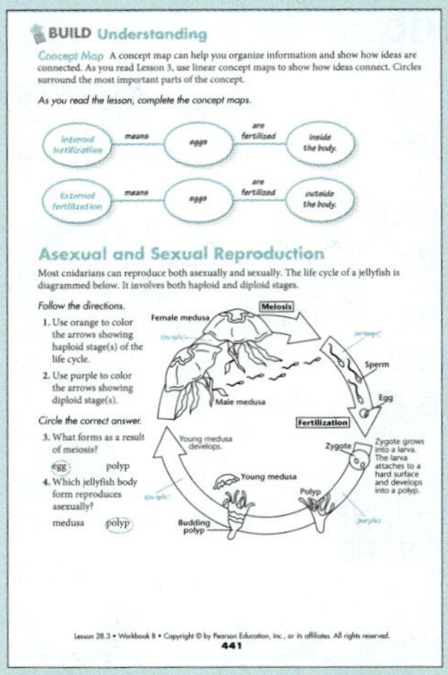

BUILD Understanding

Concept Map A concept map can help you organize information and show how ideas are connected. As you read Lesson 3, use linear concept maps to show how ideas connect. Circles surround the most important parts of the concept.

As you read the lesson, complete the concept maps.

Asexual and Sexual Reproduction

Most cnidarians can reproduce both asexually and sexually. The life cycle of a jellyfish is diagrammed below. It involves both haploid and diploid stages.

Follow the directions.

1. Use orange to color the arrows showing haploid stage(s) of the life cycle.
2. Use purple to color the arrows showing diploid stage(s).

Circle the correct answer.

3. What forms as a result of meiosis?
   egg polyp
4. Which jellyfish body form reproduces asexually?
   medusa polyp

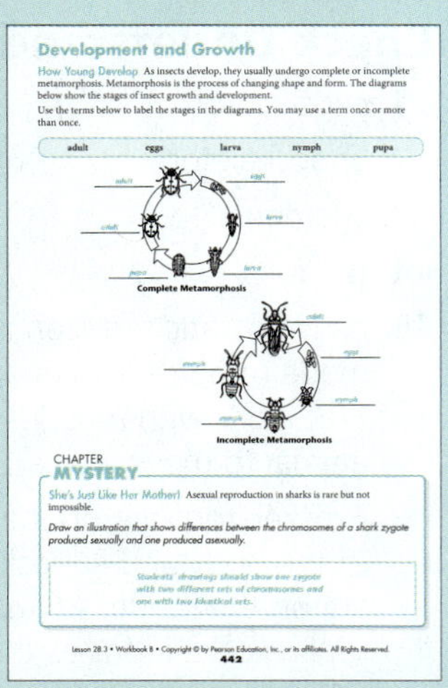

Development and Growth

How Young Develop As insects develop, they usually undergo complete or incomplete metamorphosis. Metamorphosis is the process of changing shape and form. The diagrams below show the stages of insect growth and development.

Use the terms below to label the stages in the diagrams. You may use a term once or more than once.

adult eggs larva nymph pupa

CHAPTER MYSTERY

She's Just Like Her Mother! Asexual reproduction in sharks is rare but not impossible.

Draw an illustration that shows differences between the chromosomes of a shark zygote produced sexually and one produced asexually.

Students' drawings should show one zygote with two different sets of chromosomes and one with two identical sets.

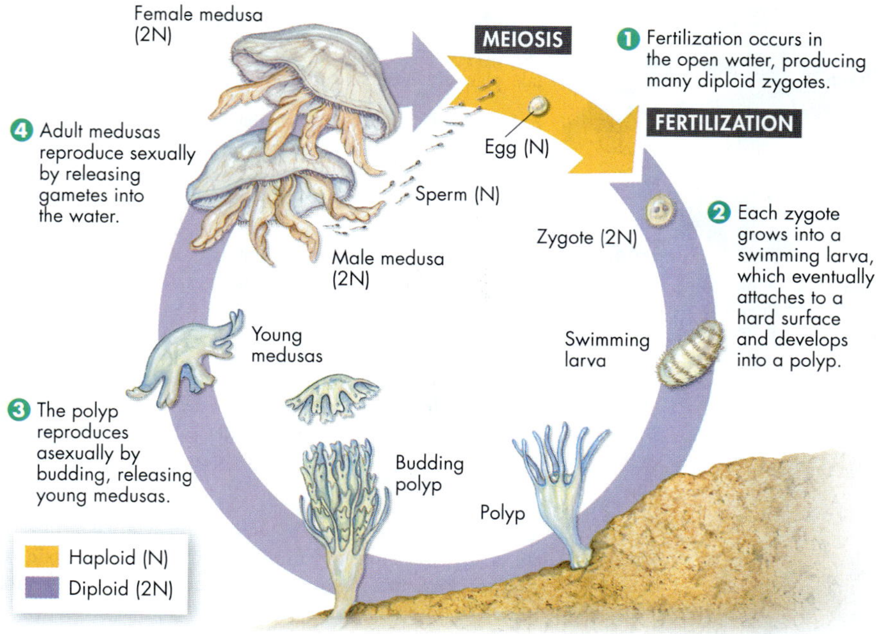

BUILD Connections

ALTERNATING REPRODUCTIVE CYCLES

The reproductive cycle of this jellyfish alternates between sexual and asexual reproduction. Medusas can reproduce sexually when a female releases eggs and a male releases sperm. Fertilization forms a zygote that grows into a larva. The larva develops into a polyp. The polyp reproduces asexually by budding. This budding forms a new medusa.

Sexual reproduction increases genetic diversity in a population by creating individuals with new combinations of genes. Populations with greater genetic diversity are better able to adapt to changing environmental conditions.

In most animals that reproduce sexually, individuals are either male or female. But some species of worms, mollusks, and fishes are hermaphrodites (hur MAF roh dytes). Hermaphrodites can be both male and female at the same time, or they can change from one sex to the other.

Key Question How do asexual and sexual reproduction in animals compare?

In asexual reproduction, one parent produces offspring. In sexual reproduction, gametes from two parents combine to produce offspring. Sexual reproduction increases genetic diversity in a population.

Reproductive Cycles Many invertebrates have life cycles that alternate between sexual and asexual reproduction. For example, many cnidarians alternate between two body forms: polyps and medusas (muh DOO suhs). In jellyfish, polyps produce medusas asexually by budding. The medusas reproduce sexually by releasing eggs and sperm into the water. A fertilized egg grows into a larva that swims freely. The larva may attach to a hard surface and develop into a polyp.

Hands-On Learning

COMPARE ASEXUAL AND SEXUAL REPRODUCTION Divide students into small groups and hand out the following materials: two different colors of chenille stems, cut into thirds for easier handling; light-colored heavy construction paper; glue.

1. Have students make models that show chromosomes being passed from parent to offspring in asexual and sexual reproduction. Students can glue the chenille stems to the paper and write a brief caption to explain what is happening. Tell students to label each step *haploid* or *diploid.*
2. For sexual reproduction, tell students to start with haploid cells from parents. You may wish to review the products of meiosis in Chapter 11. For asexual reproduction, have students recall that an organism makes a copy of its chromosomes before dividing.
3. Ask students to use their models to write a statement in their own words explaining why sexual reproduction increases genetic diversity.

During Reading

Active Reading

FIND THE MAIN IDEA Explain that the main idea of a paragraph is the most important concept. Other sentences explain or support the main idea.

Say The main idea of a paragraph is not always obvious. In the first paragraph under Asexual Reproduction, two broad ideas are expressed in the first two sentences. You can decide which is the more important idea, or the main idea, by thinking about which idea the rest of the sentences support.

Ask What is the main idea of this paragraph? *(Animals reproduce asexually in many ways.)*

Continue helping students find the main idea in each paragraph. Remind them it is not always the first sentence. For example, in the second paragraph on this page, the main idea is about hermaphrodites.

Build Connections

VISUAL SUMMARY Have students follow along as you read the caption and each numbered label in the Alternating Reproductive Cycles diagram. Make sure students understand that the short arrow is part of the cycle and represents the haploid phase. Review the meaning of *haploid* and *diploid.* Explain that a polyp is a vase-shaped form and a medusa is a bell-shaped form of a cnidarian.

Ask Which type of reproduction produced the polyp? *(sexual)* medusa? *(asexual)* How can both be diploid? *(The polyp grows from a zygote formed by the joining of a haploid egg and sperm. The medusa is produced through budding, in which the new organism receives a complete set of chromosomes.)*

KEY QUESTION Read aloud the Key Question and answer. Tell students that a comparison involves both similarities and differences.

Ask What is alike about both types of reproduction? *(Both produce offspring.)*

Ask What is different about the types of reproduction? *(Asexual reproduction requires only one parent. Sexual reproduction requires two.)*

Speed Bump

Ask What might be an advantage of parthenogenesis? *(Females can still reproduce when the number of males is scarce.)*

Write these words on the board and ask students to use them in sentences to contrast a medusa and a polyp: *formed by, asexually, sexually, reproduces. (Sample answer: A polyp is formed by sexual reproduction and reproduces asexually. A medusa is formed by asexual reproduction and reproduces sexually.)*

During Reading continued

Active Reading

SELF-QUESTION Demonstrate how asking questions helps to organize material. Read aloud the first sentence in the paragraph External Fertilization.

Say Notice the word *often* in this sentence. What does this mean? Some animals that live in water must not have external fertilization. Then how does fertilization take place for them?

Read aloud the first sentence in Internal Fertilization.

Say This sentence answers my question—it tells me that some animals that live in water have internal fertilization. Now I have a clear idea that some aquatic animals but very few land animals have external fertilization. Some aquatic animals and nearly all land animals have internal fertilization.

KEY QUESTION Read aloud the first Key Question and answer. Tell students that the words *internal* and *external* provide clues to help them remember the difference.

COMPARE/CONTRAST Have students look at the pictures in the visual Embryo Development. Tell students that the vocabulary in the paragraphs under Where Embryos Develop can be overwhelming, so they should focus first on the pictures and the ways embryos develop rather than on the text. Have volunteers describe each picture in terms of how embryo development is alike and different.

Ask Why do you think embryo development in the guppy is placed between that of the robin and the horse? *(Sample answer: The embryo development of guppies shares characteristics of both the robin and the horse. Embryos develop in eggs like robins' embryos, but inside the mother as in horses.)*

KEY QUESTION Ask a volunteer to read aloud the second Key Question and answer.

Ask Where can you find the answer in the text? *(in the sentence after Where Embryos Develop)*

External Fertilization One type of external fertilization is spawning. During spawning, females release eggs and males release sperm at the same time.

Embryo Development Embryos may develop inside or outside the body of a parent.

Robin - Oviparous

Guppy - Ovoviviparous

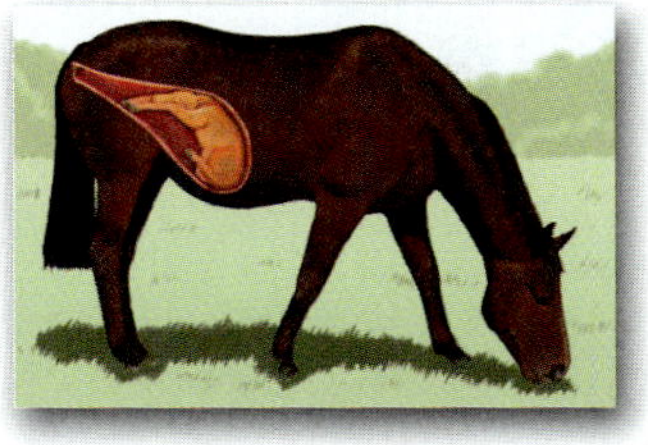

Horse - Viviparous

Internal and External Fertilization

Sperm fertilize eggs during sexual reproduction. Fertilization may take place either outside or inside the body of the female.

External Fertilization Animals that live in water often have external fertilization. External fertilization takes place outside the body of the female. Many invertebrates, such as corals, worms, and mollusks, have external fertilization. These animals release large numbers of eggs and sperm into the water at the same time. Chordates with external fertilization include the nonvertebrate chordates and many fishes and amphibians. They may spawn (release eggs and sperm) in pairs or in small groups.

Internal Fertilization Many aquatic animals and nearly all land animals reproduce by internal fertilization. Internal fertilization takes place inside the body of the female. In some animals, sperm is released into the environment and is later taken in by the female. In other animals, the male has an organ that deposits sperm inside the female.

Key Question How do internal and external fertilization differ? **In external fertilization, eggs are fertilized outside the body of the female. In internal fertilization, eggs are fertilized inside the body of the female.**

Development and Growth

When a fertilized egg starts to divide, it becomes an embryo. The care and protection given to developing embryos varies widely.

Where Embryos Develop Embryos develop either inside or outside the body of a parent.

▶ ***Oviparous Species*** In **oviparous** (oh VIP uh rus) species, embryos develop in eggs outside of the parents' bodies. Most invertebrates, many fishes and amphibians, most reptiles, all birds, and a few mammals are oviparous.

▶ ***Ovoviviparous Species*** In **ovoviviparous** (oh voh vy VIP uh rus) species, embryos develop in eggs inside their mother's body. The young rely on the yolk sac of their eggs for food. Young hatch inside the mother or outside right after the eggs leave her body. Some fishes, including some sharks, are ovoviviparous.

▶ ***Viviparous Species*** In **viviparous** (vy VIP uh rus) species, embryos get nutrients from their mother's body during development. Some insects, sharks, bony fishes, amphibians, and reptiles are viviparous. Most mammals are viviparous.

Key Question Where do embryos develop? **Embryos develop either inside or outside the mother's body.**

680 Chapter 28 • Lesson 3

Focus on Academic Vocabulary

• aquatic • embryo • young

Read aloud the first sentence after Internal Fertilization.

Ask What is an aquatic animal? *(one that lives in water)*

Explain that *aqua-* means "water." Have students brainstorm other words with this word part. *(Sample answer: aquarium, Aquarius, aquamarine)*

Remind students that an embryo is an organism formed from fertilization in its early stage of development. Some students may think of embryos only in terms of mammals. Make sure they understand that embryos develop from sexual reproduction of multicellular organisms, including plants.

Read aloud the second and third sentences in Ovoviviparous Species. Explain that we generally think of *young* as an adjective, but it also can serve as a noun, as in this sentence.

How Young Develop Most newborn mammals and newly hatched birds and reptiles look like small adults. In other groups of animals, young look very different from adults. As many animals develop, they go through metamorphosis. **Metamorphosis** is a process that causes many changes in the shape and form of a developing animal.

▶ ***Aquatic Invertebrates*** Many aquatic invertebrates have a larval stage that looks nothing like the adult animal. These larvae may float in open water before taking their adult form. Some groups, such as cnidarians, have a single larval stage. Other groups, such as crustaceans, go through several larval stages before becoming adults.

▶ ***Terrestrial Invertebrates*** Insects may go through incomplete or complete metamorphosis. Grasshoppers go through incomplete metamorphosis. In this process, immature forms called **nymphs** (nimfs) look like small adults. However, nymphs do not have adult structures such as wings or reproductive organs. These structures develop as the nymphs molt.

Butterflies go through complete metamorphosis. The larva eats, molts, and grows. After a final molt, it becomes a **pupa** (PYOO puh; plural: pupae). Body changes occur during the pupal stage. The adult that emerges is completely different from the larva.

Metamorphosis in arthropods is controlled by hormones. High levels of a juvenile hormone keep an insect in its larval stage during complete metamorphosis. When hormone levels decrease, the larva becomes a pupa. When hormone production stops, the pupa emerges as an adult.

▶ ***Amphibians*** Amphibians also go through metamorphosis that is controlled by hormones. This metamorphosis changes larvae that live in water into adults that live on land.

BUILD Vocabulary

oviparous
a species in which embryos develop in eggs outside the parent's body

ovoviviparous
a species in which the embryos develop within the mother's body but depend entirely on the yolk sac of their eggs

viviparous
animals that bear live young that are nourished directly by the mother as they develop

metamorphosis
a process of changes in shape and form of a larva into an adult

nymph
an immature form of an animal that looks like the adult form but does not have functional sexual organs

pupa
a stage of complete metamorphosis in which the larva develops into an adult

PREFIXES

The prefix *vivi-* comes from a Latin word meaning "alive." The prefix *ovi-* comes from a Latin word meaning "egg." A *viviparous* species is "born alive." An *oviparous* species hatches from an egg. An *ovoviviparous* species develops in an egg inside the mother but may hatch inside the mother to be "born alive."

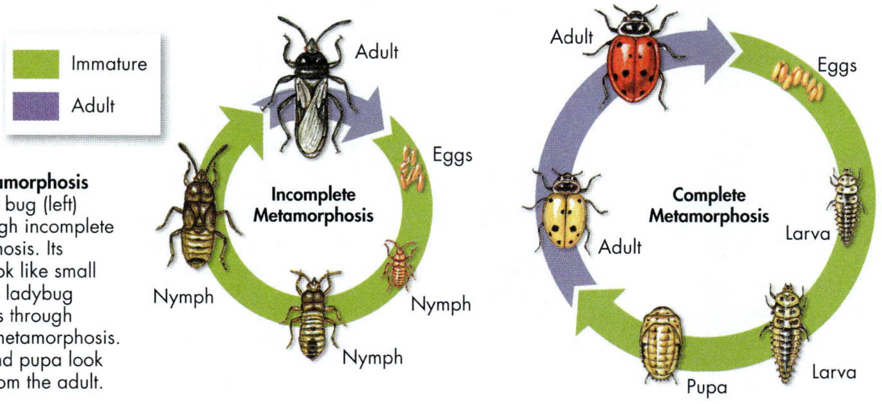

Insect Metamorphosis
The chinch bug (left) goes through incomplete metamorphosis. Its nymphs look like small adults. The ladybug (right) goes through complete metamorphosis. Its larva and pupa look different from the adult.

ELL Focus on ELL: Build Background

ALL SPEAKERS Show students pictures of different stages of metamorphosis in a butterfly and in a frog.

Ask What can you tell me about how the animals are changing? *(Students can relate personal experiences seeing animals in nature or share what they have learned in previous classes. Allow beginning students to speak in their native language and to have an advanced speaker translate their words into English.)*

After students have finished reading about metamorphosis in this lesson, show them the pictures again.

Ask What are the different stages of metamorphosis? How would you describe them? *(Guide students to name the different stages and to describe orally what it means for an animal to go through metamorphosis.)*

Assign pairs of students to write descriptions of metamorphosis.

Build Vocabulary

PREFIXES Write on the board the terms for the three kinds of embryo development. Draw vertical lines separating the word parts in each word. Explain that *-parous* means "producing offspring."

Ask Which word part is common to all three words? (-parous)

Strike through or erase *-parous* in each term. Explain that the words are easier to learn once readers understand that a big word part is the same in all. Read aloud the text under PREFIXES, pausing after each prefix to relate it to the term. Point out that *ovi-* and *ovo-* both come from the Latin word meaning "egg."

Active Reading

USE GRAPHIC ORGANIZERS Draw the outline of a compare/contrast table on the board with headings as shown below, but leave the cells blank. Add the title Metamophorsis to the table. Ask volunteers to read aloud the text on this page. Make sure students understand the meaning of the word *terrestrial. (of land, not water)*

Have the class tell you whether to put a √ (have) or an X (have not) in each cell. The table should be filled in as shown.

	Larva	Pupa	Nymph
Aquatic Invertebrates	√	×	×
Terrestrial Invertebrate Complete	√	√	×
Terrestrial Invertebrate Incomplete	×	×	√
Amphibians	√	×	×

Ask Which of these types of organisms do not have a larval stage? *(terrestrial incomplete)*

Ask How is the terrestrial complete different from the others? *(It has a pupal stage.)*

Ask Which two groups have similar stages? *(aquatic invertebrates and amphibians)*

Speed Bump

Ask If you saw a miniature grasshopper with no wings, which stage of metamorphosis could you infer it is in? *(nymph)*

Say The text states that insects molt as they go through metamorphosis. It does not state whether amphibians molt as they go through metamorphosis. Can you infer whether they do based on prior knowledge? Explain. *(Amphibians have endoskeletons and do not need to molt.)*

During Reading continued

Active Reading

DRAW CONCLUSIONS Explain that readers who draw conclusions identify information that is implied in addition to the information that is stated directly. Ask a volunteer to read aloud the first two paragraphs on this page.

Ask What general pattern can you see between the number of offspring and parental care? *(The more offspring produced, the less likely the parents are to take care of them.)*

Ask Why is an individual offspring of animals that produce few offspring more likely to survive than the individual offspring of animals that produce more? *(Parents that produce few young are more likely to protect them.)*

INTERPRET DIAGRAMS Have students study the Amniotic Egg diagram. Pronounce aloud the terms in the diagram. Remind students that a membrane is a thin tissue that covers a surface as a barrier or that surrounds a structure. Have volunteers read the labels for each membrane.

Ask How would you summarize the function of each membrane? *(amnion—protection from the environment, chorion—regulation of oxygen and carbon dioxide transport, yolk sac—food supply, and allantois—storage of wastes.)*

Build Vocabulary

WORD ORIGINS Read aloud the vocabulary terms and their definitions.

Ask Why is the word *mammal* a good word to distinguish mammals from other animals? *(Mammal comes from* mamma, *which means "breast." Mammals are the only animals that nourish their young with milk from mammary glands in a female's breast.)*

Speed Bump

Ask Why is a single young bear cub more likely to survive than any single tadpole? *(The bear cub receives protection and food from a parent.)*

Ask How did the amniotic egg help make it possible for animals to live on land? *(Animals could live away from water; they did not need water to reproduce.)*

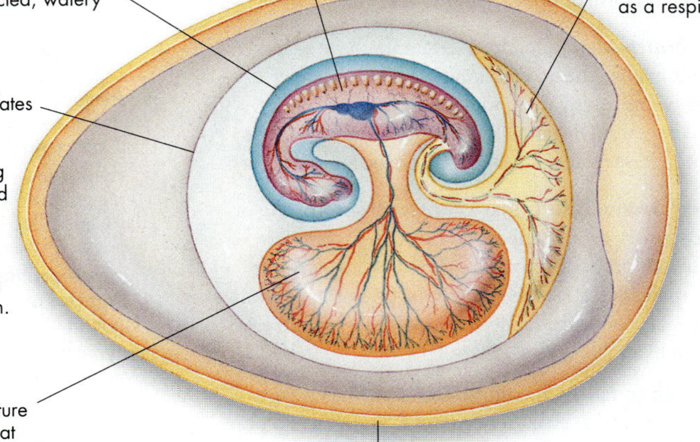

Amniotic Egg An amniotic egg contains four membranes and an outer shell. Although the egg is waterproof, it allows gases to pass through. The shell of a reptile egg is usually soft and leathery. The shell of a bird egg is usually hard and brittle.

Care of Offspring Some types of animals do not care for their young. Many aquatic invertebrates, fishes, and amphibians release many eggs and then leave. Some young are likely to survive, even though adults do not protect them.

Many animals do care for their young. Some amphibians carry developing young in their mouths, backs, or stomachs. Birds and mammals usually provide parental care. Species that provide long-term parental care usually give birth to fewer young than do species that give no parental care.

Reproduction in Chordates

Chordate reproduction was first adapted for water. Most fishes and amphibians lay eggs that must develop in water. Many terrestrial chordates have eggs that protect the developing embryo on land.

The Amniotic Egg Reptiles, birds, and a few mammals lay amniotic eggs. An **amniotic** (am nee AH tik) **egg** contains four membranes and an outer shell. The amniotic egg is one of the most important vertebrate adaptations to life on land. It provides a protected environment where the embryo can develop without drying out.

Reproduction in Mammals All mammals nourish their young with milk produced by the mother's **mammary glands.** There are three groups of mammals—monotremes, marsupials, and placentals. Each has different adaptations for reproduction.

▶ ***Monotremes*** Like reptiles, monotremes lay soft-shelled, amniotic eggs. Like mammals, young monotremes are fed with milk. Female monotremes secrete milk through pores on the abdomen.

BUILD Vocabulary

amniotic egg an egg composed of shell and membranes that create a protected environment in which the embryo can develop out of water

mammary gland a gland in female mammals that produces milk to nourish the young

placenta a specialized organ in placental mammals through which respiratory gases, nutrients, and wastes are exchanged between the mother and developing young

WORD ORIGINS

The word *mammary* comes from the Latin word *mamma*, which means "breast." Mammary glands are found in a female mammal's breast.

Hands-On Learning

EGG EXAM Divide the class into small groups and provide the following: uncooked and unfertilized chicken egg, bowl, hand lens, water. **Caution:** Tell students not to touch their mouths during the activity and to wash their hands afterwards.

1. Tell students to gently crack the egg into the bowl without breaking the yolk and use a hand lens to study the structures. Have them fill one part of the broken eggshell with water.
2. Point out these structures: 1) membrane attached to the inside of the shell (helps keep the egg from drying out); 2) clear, gel-like albumen—egg white (helps protect the embryo from drying out and cushions it); 3) chalazae—white rope-like structures on each end of the yolk (hold yolk in place; 4) blastodisc—small white spot on the yolk (develops into the embryo if fertilized).
3. Have students draw and label the structures they observe. Then have them summarize how an amniotic egg enables an embryo to survive on land.

▶ ***Marsupials*** Marsupials are born at such an early stage of development that they are almost embryos. Young marsupials crawl across their mother's fur and attach to a nipple in her pouch, or marsupium (mahr soo pee um). The young remain in the marsupium until they can survive on their own.

▶ ***Placentals*** Placental mammals are named for the placenta. A **placenta** is a specialized organ that allows the exchange of nutrients, gases, and wastes between the young and the mother. The placenta allows the embryo to develop for a long time inside the mother. Placental mammals are born at an advanced stage of development.

Key Question How are terrestrial vertebrates adapted to reproduction on land? **Reptiles, birds, and some mammals have amniotic eggs in which an embryo can develop outside the mother without drying out. Most mammals bear live young and feed their young with mother's milk.**

Mammals Mammal groups include monotremes such as the echidna (top), marsupials such as the wallaby (below), and placental mammals such as the harp seal (below left).

CHECK Understanding

Apply Vocabulary

Use the highlighted words from the lesson to complete each sentence correctly.

1. A species in which an embryo gets nutrients from the mother's body during development is called a(n) _______ species.
2. The process in which a developing animal changes in shape and form is _______.
3. Female mammals have _______ that produce milk to nourish their young.

Critical Thinking

4. **Review** Which type of reproduction results in the most genetic diversity?
5. **Review** Define the two types of fertilization.
6. **Compare and Contrast** What is the difference between a nymph and a larva?
7. **Interpret Visuals** In your own words, describe the functions of two of the membranes shown in the art of the amniotic egg.
8. **Write to Learn** Answer the first clue of the mystery. What kind of reproduction in vertebrates can produce offspring from only one parent?

MYSTERY CLUE

The females in the tank had not had contact with a male for three years. And female bonnetheads do not store sperm for longer than five months. So what happened? (**Hint:** See p. 680.)

BIOLOGY.com Search Lesson 28.3 GO • Lesson Assessment

Check Understanding

1. viviparous
2. metamorphosis
3. mammary glands
4. sexual reproduction
5. Internal fertilization takes place inside the body of the egg-producing individual. External fertilization takes place outside the body of the egg-producing individual.
6. A nymph is the stage of development of an animal that undergoes incomplete metamorphosis in which it resembles an adult but lacks functional sexual organs and some adult structures. A larva is a stage of development of an animal that undergoes complete metamorphosis. Larva look nothing like an adult and feed in different ways.
7. Student answers should reflect the functions of any two of the membranes.
8. In parthenogenesis, eggs of some species are able to develop without being fertilized.

Active Reading

SCIENCE SUPPORT Tell students that *monotreme* means "one hole." A monotreme has a single hole, or cloaca, like reptiles and birds, that is used for getting rid of wastes, laying eggs, and mating. The other groups of mammals have separate openings for wastes and reproduction. Explain that the placenta shares similarities with the amniotic egg. The allantois and yolk sac become the umbilical cord, and with the chorion, make up the placenta.

Have students use the text and new information to discuss the different reproductive adaptations in each group of mammals.

KEY QUESTION Read aloud the Key Question and answer. Point out that this question has an implied meaning. Readers must understand that being "adapted to reproduction on land" means animals changed from how they were reproducing previously, so the question is making a comparison.

Ask What is reproduction on land being compared to? *(reproduction in water)*

Post Reading

ASSESSMENT TIPS For Question 6, draw a T-chart on the board with *Nymph* and *Larva* as the column headings. Have volunteers state characteristics of each as you fill in the chart. Allow students to use the chart to answer the question.

For Question 7, refer students to the Amniotic Egg diagram.

MYSTERY CLUE Refer students to the first page of the lesson to review asexual reproduction.

Wrap-Up Activity

MATCH GAME Prepare a set of index cards with vocabulary and other science terms and their definitions from this lesson. On each card, write in the top half "I have (term)." Write in the bottom half "Who has (definition of another term)?" Shuffle the cards and hand out at least one card per student and one for yourself. Start the game by asking the question in the bottom half. The person who has the card with the matching term responds with the answer and then reads the question. The round ends when all cards have been read. When preparing the cards, make sure the answer on your card ends the round.

LESSON 28.4

Pre-Reading

Objectives

28.4.1 Explain how homeostasis is maintained in animals.

28.4.2 Describe the importance of body temperature control in animals.

Preview the Pages

1. Have students turn the lesson's headings and subheadings into simple questions that they write down. Tell them to write the answers as they study the lesson. They will use them in a quiz activity at the end of the lesson.
2. Ask students to look at the photographs of the ectotherm and endotherms and infer the meanings of these terms.

Activate Prior Knowledge

Ask what students recall about homeostasis. Write their ideas on the board.

Ask When you run, how does your body respond to help get more oxygen to muscles? *(breathes faster)*

Ask When your body needs energy, how does your body let you know you should eat? *(feelings of hunger, weakness, or lightheadedness)*

Build Understanding

VENN DIAGRAM Direct students to Workbook B. Suggest they write down similarities and differences on a separate sheet of paper before listing them in the diagram.

NATIONAL SCIENCE EDUCATION STANDARDS

UNIFYING CONCEPTS AND PROCESSES

I, IV, V

CONTENT

C.3.a, C.5.d

INQUIRY

A.1.c, A.2.a

For the corresponding lesson in *Miller & Levine Biology,* see pages 827–830.

28.4 Homeostasis

Key Questions

Why is the interdependence of body systems essential?

How do animals control their body temperature?

BUILD Understanding

Venn Diagram A Venn diagram is a useful tool for comparing two topics. Draw a Venn diagram to compare and contrast the ways that ectotherms and endotherms control body temperature.

In Your Workbook Refer to your workbook to learn how to fill in a Venn diagram.

Interrelationships of Body Systems

A herd of wildebeests moves across Africa's Serengeti Plain. The land is dusty and dry, so the herd is heading toward greener pastures. They move mechanically, so that their steps use as little energy as possible. Because they have not eaten in a long time, their bodies break down fat for energy. Their bodies also conserve water by producing very little urine. To survive the difficult journey, body systems work to keep internal conditions as constant as possible.

Recall that the word *homeostasis* is used to describe stable conditions inside the body. Homeostasis is necessary for an animal's survival. Animal cells must be supplied with glucose for energy. Cells must have enough water and be able to get rid of wastes. The cells of some animals must be kept at a constant temperature. Homeostasis must not fail, not even during dry conditions, extreme hunger, extreme heat, or extreme cold. If homeostasis fails, the animal could die.

All body systems work together to maintain homeostasis. Many body systems work closely with the circulatory system. After the respiratory system takes in oxygen, the circulatory system delivers the oxygen to cells. After the digestive system breaks food down into nutrients, the circulatory system delivers those nutrients to cells. The circulatory system collects wastes from body tissues. It then delivers these wastes to the lungs and kidneys of the excretory system.

The muscular and nervous systems work with other systems to maintain homeostasis. Muscles help to move food through the digestive system. The nervous system directs these muscle movements.

Interrelationships of Body Systems All body systems must work together to keep animals like these migrating wildebeests alive.

28.4 Homeostasis

Lesson Objectives

- Explain how homeostasis is maintained in animals.
- Describe the importance of body temperature control in animals.

BUILD Vocabulary

A. The chart below shows key terms from the lesson with their definitions. Complete the chart by writing a strategy to help you remember the meaning of each term. One has been done for you.

Term	Definition	How I'm Going to Remember the Meaning
Ectotherm	An animal that controls its body temperature by using heat from sources outside its body	The prefix ecto- means "outside." I know that therm refers to heat. (I think of a thermometer that measures heat.) So, an ectotherm gets heat from outside its body.
Endocrine gland	A structure that releases hormones into the blood	The prefix endo- means "inside." Endocrine glands make chemicals (hormones) inside the body.
Endotherm	An animal that controls its body's temperature by using heat made by (inside) its body	The prefix endo- means "inside." I know that therm refers to heat. So, an endotherm gets heat from inside its body.

B. As you work through this lesson, you may find these terms in the activities. When you need to write a key term or a definition, **highlight** the term or the definition.

BUILD Understanding

Venn Diagram A Venn diagram is made up of overlapping circles. It is a useful tool for comparing two or even three topics.

As you read Lesson 4 about the temperature-control strategies of ectotherms and endotherms, complete the Venn diagram below.

Ectotherm
- examples: lizard, fish, amphibian
- regulates body temperature by absorbing heat from, or losing heat to, its environment
- need less food than endotherms
- need warm climates

Both
- control body temperature
- have a source of heat, a way to conserve heat, and a way to get rid of excess heat

Endotherm
- examples: bird, mammal
- regulates body temperature by using heat generated by its own body
- need more food than ectotherms
- can live in cold climates

Interrelationship of Body Systems

The body systems of animals must work together to maintain homeostasis. The systems are all interconnected.

Complete each row with a description of how the two body systems work together. Use the terms below.

distribute oxygen | collect and release waste | distribute nutrients

- Circulatory system and respiratory system work together to distribute oxygen
- Circulatory system and excretory system work together to collect and release waste
- Circulatory system and digestive system work together to distribute nutrients

Inquiry Into Scientific Thinking

Comparing Ectotherms and Endotherms Think about areas of the world that are very hot and places that are very cold. Where in the world do ectotherms and endotherms live?

Several animals are listed in the chart, along with the area in which they live. Decide whether each animal is an ectotherm or an endotherm.

Animal	Where It Lives	Endotherm or Ectotherm?
Cardinal	northern United States and Canada	endotherm
Arctic fox	northern Canada and Alaska	endotherm
Armadillo lizard	southern Africa	ectotherm
Polar bear	polar regions	endotherm
King cobra	southeast Asian rain forest	ectotherm
Desert tortoise	Mojave Desert	ectotherm

Analyze and Conclude

1. What is one way that polar bears maintain their body temperature?
   Polar bears generate heat from their bodies, and they use their body fat and fur as insulation.
2. What conclusion can you draw about where ectotherms usually live?
   They usually live in areas that have warmer temperatures.
3. What other animal groups are ectotherms that are not listed in the chart above?
   fish and amphibians
4. List two endotherms that might be found in a rain forest.
   Sample answer: toucans and howler monkeys
5. Look at the animals in the chart above and think about how they compare in size. What might be the relationship between the size of an animal and its type of body heat regulation?
   The larger animals are usually endotherms.

Fighting Disease The controlled environment inside an animal's body is a comfortable place for both body cells and unwanted invaders. Pathogens can use oxygen and nutrients in the body to grow and reproduce. If pathogens are not destroyed, they may cause disease by harming the function of one or more body systems. Most animals have an immune system that can recognize "self" and "other." Once the immune system identifies the "others" in the body, it can attack the invaders and restore homeostasis.

Chemical Controls All vertebrates and many invertebrates have an endocrine system that controls many body processes. **Endocrine glands** control body activities by releasing chemicals called hormones into the blood. Hormones are carried by the blood to the organs that they control. Some hormones control growth and development.

Mammals have hormones that control the way that the body stores energy. Other hormones control the amount of water in the body. Still others control the amount of calcium in bones.

Key Question Why is the interdependence of body systems essential? **All body systems work together to maintain homeostasis.**

Body Temperature Control

Control of body temperature is important for homeostasis. Many body functions are influenced by temperature. Muscles cannot work well if they are too cold or too hot. Cold muscles contract slowly, making an animal slow to react. Muscles that are too hot may get tired easily.

An animal must do three things to control body temperature. First, the animal must find a source of heat. Second, it must find a way to conserve heat when temperatures in the environment are too cold. Third, it must find a way to get rid of extra heat when temperatures in the environment are too hot. An animal is called an ectotherm or an endotherm, based on the way that it can control body temperature.

Ectotherms An **ectotherm** is an animal that usually depends on sources of heat outside of its body to control its body temperature. Invertebrates, fishes, and amphibians are ectotherms. So are most reptiles.

Ectotherms have low metabolic rates when resting, so their bodies do not create much heat. When these animals are active, their muscles can create heat. However, most ectotherms do not have much body insulation. Therefore, they quickly lose heat to the environment.

Most ectotherms control body temperature by absorbing heat from the environment or losing heat to the environment. For example, a lizard may lie in the sun on a cool morning to absorb heat. In the afternoon, the lizard may raise its body off of a hot surface—in a process called "stilting"—in order to lower its body temperature. Ectotherms often use underground burrows to escape from extreme temperatures. On hot, sunny days, they can use the burrow to cool off. On cold nights, they can use the same burrow to conserve their body heat.

BUILD Vocabulary

endocrine gland
a gland that releases its secretions (hormones) directly into the blood, which carries these secretions to other areas of the body

ectotherm
an animal whose body temperature is determined mainly by the temperature of its environment

endotherm
an animal whose body temperature is regulated, at least in part, using heat created within its body

WORD ORIGINS

The prefix *ecto-* comes from the Greek word *ektos*, which means "outside." The suffix *–therm* comes from the Greek word *thermos*, which means "hot." An ectotherm gets its heat from the outside. In other words, it cannot make heat within its own body.

Ectotherm The shovel-snouted lizard is an ectotherm that lives in the Namib Desert in Africa. This desert is one of the hottest places on Earth. The lizard controls its body temperature by stilting—raising its body off of the hot sand by performing a kind of push-up.

During Reading

Active Reading

APPLY WORD ORIGINS Help students learn about homeostasis by applying knowledge of its word origins to the concept. Write the term on the board. Draw a line between *homeo-* and *-stasis,* explaining that the former means "the same" and the latter means "standing still."

Say Try to use the meanings of *homeo-* and *-stasis* to write a definition of *homeostasis*. *(Sample answer: Homeostasis describes the state in which conditions within a cell are kept relatively the same.)*

Have students identify examples of homeostasis from the text and from their own experiences. Discuss how they relate to the meaning of the word.

KEY QUESTION Ask a volunteer to read aloud the Key Question and answer.

Ask What does *interdependence* mean in the question? *(Different body systems depend on one another to function and keep the body alive.)*

IDENTIFY CAUSE AND EFFECT Have a volunteer read aloud the first paragraph under Body Temperature Control. Tell students to look for cue words in the paragraph, such as *because* and *if,* that can signal cause-and-effect relationships.

Ask What cue word do you see for a relationship in the third sentence of the paragraph? *(if)*

Say Sometimes cue words are not present and a relationship is implied. This is the case in the sentence, **Muscles that are too hot may get tired easily.** The reader must ask, "Did something cause something to happen?" *(cause: hot muscles; effect: tired muscles)*

Have students work in pairs to identify the cause-and-effect relationships in the rest of the text under Body Temperature Control.

Build Vocabulary

WORD ORIGINS Read aloud Word Origins.

Ask What do you think *endo-* means? *("inside")* Why do you say that? *(An endotherm uses heat generated inside its body.)*

Speed Bump

Ask What do you think would happen if an animal could not maintain homeostasis? *(It would not survive because body functions could not continue.)*

ELL Focus on ELL: Extend Language

INTERMEDIATE AND ADVANCED SPEAKERS Students may be confused by words that can act as different parts of speech. Write the word *control* on the board. Solicit responses as to the meaning of the word and confirm correct responses. Tell students that many words can be used as nouns, verbs, and adjectives. Read the first sentence on this page, the first sentence following Chemical Controls, and the first sentence in Body Temperature Control. Point out the words *controlled, controls,* and *control.*

Ask What part of speech is *controlled* in this context? *(adjective)* How do you know? *(Sample answer: It has the word form* -ed *and comes before the noun* environment.*)* Then discuss the use of *control* as a verb and as a noun in the other sentences.

Guide students to distinguish the meanings between *heat* and *contract* as verbs and as nouns, and *heated* and *contracted* as adjectives. Make sure students understand that *contract* can have more than one meaning.

Inquiry Into Scientific Thinking

INTERPRET GRAPHS Remind students that a line graph displays data that shows how one variable (the responding or dependent variable) changes in response to another variable (the manipulated or independent variable). The independent variable is changed by the experimenter.

Ask Which axis shows a condition changed by the experimenter? *(x, environmental temperature)* What does the *y* axis show? *(body temperature in response to environmental temperature)*

Students may have difficulty interpreting several lines in a single graph. Tell them to focus on one line at a time. Walk students through how to interpret the data at each point on the Pigeon line.

DRAW CONCLUSIONS Tell students that line graphs help to identify trends and may provide other information besides the relationship of two variables.

Ask Can you see a pattern between ectotherms and endotherms just by looking at the slopes? Explain. *(Yes, the slope of the endotherms' temperatures is more level than that of the ectotherms.)* What conclusion can you draw about the range of body temperatures of each group? *(Endotherms have a narrower range than ectotherms.)*

Explain that body temperatures must be kept to a certain range for an animal to survive. Homeostasis has limits. There is a point at which the body cannot compensate for heat loss or for overheating. Have students look at the data for 40° on the *x* axis.

Ask Why do you think the experimenters did not collect data at 50°? *(Animals cannot survive at that temperature.)*

Lead a Discussion

Discuss how a person's body temperature reacts to changing environmental temperatures.

Say Suppose you spend a short time outdoors in the southwestern U.S. desert in summer, then hike up a snow-covered mountain farther north.

Ask What would probably happen to your body temperature? *(Body temperature would remain relatively constant.)*

Ask What might your body do to keep you from getting too warm in the desert? *(sweat to get rid of extra heat)* What might it do to help you stay warm in the mountains? *(shiver to produce heat)*

INQUIRY into Scientific Thinking

GUIDED INQUIRY

Comparing Ectotherms and Endotherms

An ectotherm is an animal that controls its body temperature mainly by using heat from outside its body. An endotherm is an animal that controls its body temperature mostly by using heat that is produced within its body. The graph shows the internal body temperatures maintained by several ectotherms and endotherms at different environmental temperatures.

Temperature Control in Chordates

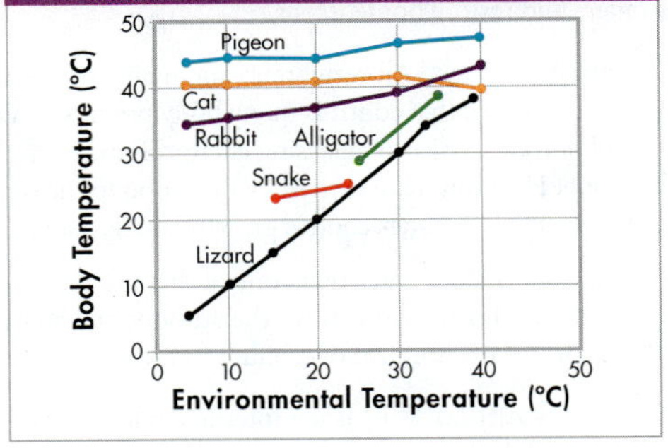

1. Interpret Graphs Which animal has the highest body temperature when the environmental temperature is between 0°C and 10°C? Which has the lowest body temperature under these conditions?

2. Infer Which animals represented in the graph are ectotherms? Which are endotherms? Explain your answers.

3. Predict If these animals lived in your area, would you expect all of them to be equally active year-round? Why or why not?

In Your Workbook Get more help for this activity in your workbook.

Endotherm Many endotherms, such as this dingo, pant when they are very warm. Panting allows moisture in the mouth and respiratory tract to evaporate. This process helps to cool the blood.

Endotherms An **endotherm** is an animal whose body temperature is controlled, at least in part, by using heat that is produced by its own body. Endotherms have high metabolic rates that create heat, even when the animals are resting. All birds and mammals are endotherms.

Endotherms have different ways of conserving heat. Birds conserve body heat by fluffing out their feathers. Mammals use a combination of body fat and hair as insulation.

Endotherms have different ways of getting rid of extra heat. Some birds and most mammals get rid of heat by panting. When an animal pants, it breathes with its mouth open and its tongue sticking out. Heat flows out of the body as water evaporates from the mouth and tongue. Humans sweat to reduce their body temperature. As sweat evaporates, it removes heat from the skin. It also removes heat from the capillaries just below the surface of the skin. As warm blood flows into these cooled capillaries, the blood loses heat.

Analyze and Conclude

1. highest: pigeon; lowest: lizard
2. The lizard, snake, and alligator are ectotherms. Their body temperatures fluctuate based on the temperature of the environment. The pigeon, cat, and rabbit are endotherms. Their body temperatures stay fairly level. Their body temperatures are regulated mostly from within.
3. Answers may vary. In all areas, the pigeon, cat, and rabbit would be expected to be active year-round. In colder areas, the alligator, snake, and lizard would probably not be as active during colder months.

Comparing Ectotherms and Endotherms There are advantages and disadvantages to being an endotherm. Endotherms can move around on cool nights or in cold weather, because they create and conserve body heat. But it takes a lot of fuel to feed the high metabolic rate that makes heat. The amount of food that it takes to keep a single cow alive would be enough to feed ten cow-size lizards!

There are also advantages and disadvantages to being an ectotherm. It is more energy efficient to be an ectotherm in a warm environment. However, large ectotherms can get into trouble if it gets very cold at night or if the environment stays cold for long periods of time. It takes a long time for an animal to warm up in the sun after a cold night. That is one reason why most large lizards and amphibians can only live in tropical or subtropical areas.

Evolution of Temperature Control The first animals were ectotherms. When did endothermy evolve? Current evidence suggests that endothermy evolved at least twice in vertebrates. It evolved once in the line of ancient reptiles that led to birds. It evolved again in the line of ancient reptiles that led to mammals.

Key Question How do animals control their body temperature? **An ectotherm controls body temperature by exchanging heat with the environment. An endotherm controls body temperature by using heat made by its own body.**

Endotherm Insulation Birds such as this pine grosbeak use feathers to stay warm. When a bird gets cold, its fluffy undercoat of down feathers stands up. Body heat becomes trapped in the air spaces between the feathers.

CHECK Understanding

Apply Vocabulary

Use the highlighted words from the lesson to complete each sentence correctly.

1. Structures that release hormones into the blood are called _______.
2. An animal that controls its body temperature mainly by absorbing heat from its surroundings is a(n) _______.
3. An animal that controls its body temperature mainly by producing its own heat is called a(n) _______.

Critical Thinking

4. **Explain** Give an example of how multiple body systems work together to maintain homeostasis.
5. **Compare and Contrast** Why must an endotherm eat more food than an ectotherm of the same size?
6. **Form a Hypothesis** Birds and mammals live in both warm and cold areas. Most reptiles and amphibians live in relatively warm areas. Form a hypothesis that would explain this difference.
7. **Write to Learn** Think about what you do during an average day. Write a paragraph describing how your body maintains homeostasis during some of your activities.

BIOLOGY.com Search Lesson 28.4 GO • Lesson Assessment

Check Understanding

1. endocrine glands
2. ectotherm
3. endotherm
4. Sample answer: The respiratory system works with the circulatory system to deliver oxygen to cells.
5. An endotherm needs a lot of fuel to feed the high metabolic rate that makes body heat.
6. Ectotherms tend to live in warm areas because they depend on the temperature of their environment to maintain body temperature.
7. Sample answer: I breathe all the time. My respiratory system brings oxygen into my body and removes carbon dioxide. My circulatory system brings oxygen to cells and takes away carbon dioxide. I also eat every day. My digestive system breaks down the food into nutrients. My circulatory system brings these nutrients to my body cells. My digestive system also moves solid wastes out of my body.

During Reading continued

Active Reading

COMPARE/CONTRAST Tell students to make sure they include advantages and disadvantages of being endotherms and ectotherms in the Venn diagram they started at the beginning of the lesson. Have students read the text in Comparing Ectotherms and Endotherms, and complete their diagrams.

KEY QUESTION Read aloud the Key Question and answer.

Remind students that though the question is written in the active voice, control of body temperature is not under conscious control.

Speed Bump

Ask Do you think it is wrong or misleading to say that an ectotherm is a cold-blooded animal? *(yes)* Explain. *(An ectotherm's body temperature is not always cold. It can get as high as that of an endotherm.)*

Invite students to share whether they have noticed outdoor pets or livestock needing more food in colder months than in warmer months.

Ask Why do you think outdoor animals need more food in winter than in summer? *(Outdoor animals need more energy to heat their bodies during colder months.)*

Post Reading

ASSESSMENT TIPS Review the vocabulary terms in this lesson and call on students to give definitions in their own words or tell something about the term.

For Question 6, remind students that a hypothesis is a statement that offers a possible explanation for an observation. Have them reread the second paragraph on this page if they need help.

For Question 7, suggest students think about body functions in addition to maintaining body temperature.

Wrap-Up Activity

PARTNER QUIZ Tell students to use the questions and answers they developed as directed in the Preview the Pages at the beginning of the lesson. Give them time to change their questions and answers if they would like. Then, have them work with a partner to quiz each other on the content. You may also wish to collect the questions and answers and make this a class activity.

Pre-Lab

Review that density is a measure of the amount of matter contained in a given volume. Write these formulas on the board, **Density = mass ÷ volume. Volume = length × width × height.**

If you think your students will need help with Part B of the lab, do the following. Draw this table, and ask students to help you complete it.

Purpose	Tool	
To Find Mass of an Object	To Find Volume of a Regular Object	To Find Volume of _an Irregular Object_
Balance	_Metric ruler_	Submerge in water

Ask What happens to the water level in a beaker when you drop an object in it? *(It rises because the object has taken up space and displaced the water.)* How can you determine the density of the object? *(The amount of water displaced is the object's volume. Measure the object's mass and then divide that by the volume to get the density.)*

Review the Background Questions and the Pre-Lab Questions as a class.

Lab

Tell students they will perform the chapter lab *Comparing Bird and Mammal Bones* described in **Lab Manual B.**

SAFETY Students should handle glassware with care and wash their hands after the activity. Students should wear gloves and an apron during this activity.

Post-Lab

COMPARE AGILITY Ask students to bend one of their wrists up and down as rapidly as they can, and then to rapidly bend a knee. Have them compare the agility (ease and quickness in moving and changing positions) of the movements with each other and with the size and density of the bones.

NATIONAL SCIENCE EDUCATION STANDARDS

UCP V

INQUIRY A.1.a, A.1.b, A.2.a

Real-World Lab

OPEN-ENDED INQUIRY

Pre-Lab: Comparing Bird and Mammal Bones

Problem Is the density of an animal's bones related to the way the animal moves?

Materials cross-sections of chicken, duck, and cow bones; hand lens; small chicken, duck, and cow bones; balance

Lab Manual Chapter 28 Lab

Skills Focus Form a Hypothesis, Design an Experiment, Measure

Connect to the Big idea In order to move, an animal must generate physical force and apply this force against the air, the water, or the ground. The force is generated by the contraction of muscles. In vertebrates, the muscles are attached to bones. The joints that connect bones bend or straighten when groups of muscles contract. There is a close link between the structure of an animal's skeletal and muscular systems and how the animal moves. In this lab, you will investigate whether there is a similar link between the density of bones and how an animal moves.

Background Questions

a. Review What type of skeleton do vertebrates have? List one advantage of this type of skeleton.

b. Explain Why are pairs of muscles or two different groups of muscles needed to bend and straighten a joint?

c. Apply Concepts Why do you think humans have only 4 bones in each arm and shoulder, but 27 bones in each wrist and hand? (**Hint:** Compare the movement of your arm and your hand when you button a shirt.)

Pre-Lab Questions

Preview the procedure in the lab manual.

1. **Compare and Contrast** Compare the type of data you will collect in Part A to the type of data you will collect in Part B.
2. **Predict** How might looking at cross sections of bones help you form a hypothesis about the relative density of the bones?
3. **Design an Experiment** Will you need to use samples with the same mass in Part B? Why or why not?

BIOLOGY.com Search Chapter 28 GO

Visit Chapter 28 online to test yourself on chapter content and to find activities to help you learn.

Untamed Science Video Join the Untamed Science crew as they interview experts to learn more about how the sex of offspring is determined in some animals.

Art in Motion Watch an animation that shows the motion of joints in both exoskeletons and endoskeletons.

Art Review Review your understanding of vertebrate brains.

InterActive Art Look at the structure and function of the water vascular system in a sea star.

Ask Why could you bend your wrist more rapidly than your knee? (The bones in my wrist are smaller and lighter than the bones in my knee.)

Pre-Lab Answers

BACKGROUND QUESTIONS

a. Sample answer: Vertebrates have an endoskeleton, which can grow as the vertebrate grows.

b. Sample answer: Muscle contraction can cause a tendon to pull on a bone but it cannot cause the tendon to push the bone back to its original position. So pairs of muscles are needed to generate a full range of motion around a joint.

c. Sample answer: Many small movements are required for a hand to grasp and manipulate objects.

PRE-LAB QUESTIONS

1. Sample answer: In Part A, the data will be qualitative. In Part B, the data will be quantitative.
2. Sample answer: I will be able to observe whether the bones are solid or contain air spaces.
3. Sample answer: No, because density is a ratio of mass to volume.

28 CHAPTER Summary

28.1 Response

- When an animal responds to a stimulus, its body systems—including its nervous system and muscular system—work together to create a response.
- Animal nervous systems show different degrees of cephalization and specialization. Invertebrate nervous systems range from simple collections of nerve cells to complex systems that include ganglia and brains. All vertebrate nervous systems include brains. As vertebrates evolved, their brains became larger and more complex.
- Sensory systems in animals range from individual sensory neurons to sense organs, such as eyes and ears. Many vertebrates also have sense organs that are sensitive to tastes, odors, vibrations, or electrical currents.

neuron (p. 668)
stimulus (p. 668)
sensory neuron (p. 668)
interneuron (p. 669)
response (p. 669)
motor neuron (p. 669)
ganglion (p. 670)
cerebrum (p. 671)
cerebellum (p. 671)

28.2 Movement and Support

- Animals have three main kinds of skeletal systems: hydrostatic skeletons, exoskeletons, and endoskeletons.
- In many animals, muscles work together in pairs or groups that are attached to different parts of a supporting skeleton.

hydrostatic skeleton (p. 674)
exoskeleton (p. 674)
molting (p. 675)
endoskeleton (p. 675)
joint (p. 675)
ligament (p. 675)
tendon (p. 676)

28.3 Reproduction

- Asexual reproduction requires only one parent, so individuals may reproduce quickly. However, offspring that are produced asexually have less genetic diversity than do offspring that are produced sexually. Sexual reproduction increases genetic diversity in a population by creating individuals with new combinations of genes.
- In internal fertilization, eggs are fertilized inside the body of the egg-producing individual. In external fertilization, eggs are fertilized outside the body of the egg-producing individual.
- Animals may be oviparous, ovoviviparous, or viviparous.
- Reptiles, birds, and a few mammals have evolved amniotic eggs in which an embryo can develop without drying out. Different mammals reproduce and develop in different ways. However, all mammals feed their young with mother's milk.

oviparous (p. 680)
ovoviviparous (p. 680)
viviparous (p. 680)
metamorphosis (p. 681)
nymph (p. 681)
pupa (p. 681)
amniotic egg (p. 682)
mammary gland (p. 682)
placenta (p. 683)

28.4 Homeostasis

- All body systems work together to maintain homeostasis.
- Most reptiles, invertebrates, fishes, and amphibians are ectotherms. Most ectotherms control body temperature mainly by getting heat from the environment or losing heat to the environment. Birds and mammals are endotherms. Endotherms have high metabolic rates that create heat even when they are resting.

endocrine gland (p. 685)
ectotherm (p. 685)
endotherm (p. 686)

UbD Transfer the Big Idea

TRANSFER TASK Ask students to choose an animal, other than humans, mentioned in this chapter and write a mini-autobiography, explaining the animal's nervous system, movement, reproduction, and body-temperature control. They should write four chapters, each reflecting a lesson. Encourage students to be creative but factual. Have them make a "book" by folding several sheets of letter size white paper in half and stapling them. They should give the book and each chapter a title, for example, "All About Me," and chapters "No Brainer," "On the Move," "All in the Family," and "Same Old, Same Old." Suggest students read all or parts of their book to the class without mentioning the animal, and have others guess the animal.

Chapter Summary

Think Visually

USE GRAPHICS Have students complete the following concept map about body temperature.

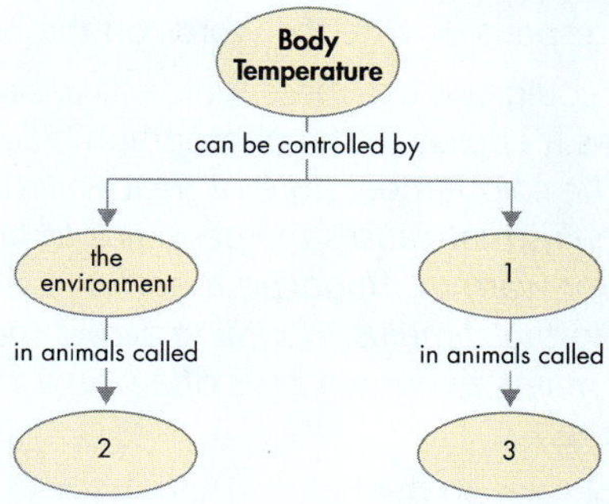

ANSWERS

1. heat generated by the body
2. ectotherms
3. endotherms

BIOLOGY.com Study Online

REVIEW RESOURCES

Art Review This drag-and-drop activity allows students to compare vertebrate brains.

InterActive Art Students can use this activity to learn more about the sea star water vascular system.

Art in Motion This activity animates the motion of joints, so students can compare joint function of exoskeletons and endoskeletons.

Data Analysis Students use data to learn more about strategies that mammals use to survive cold weather.

Vocabulary Review The *Flash Cards* and *Match It* provide an interactive way to review chapter vocabulary.

Editable Worksheets Pages of Study Workbook B, Lab Manual B, and the Assessment Resources Book are available online. These documents can be easily edited using a word-processing program.

Lesson Overviews Overviews are based on Miller & Levine Biology. However, they are available on CD-ROM and can be edited for your class.

ASSESSMENT RESOURCES

Chapter Assessment Have students take an online version of the Chapter 28 Check Understanding.

Standardized Test Prep Students can take an online version of the Standardized Test Prep. You will receive their scores along with ideas for remediation.

Diagnostic and Benchmark Tests Use these tests to monitor your students' progress and supply remediation.

Assess

Big idea Have students look at the photograph. Read aloud the Big Idea and the question.

Ask When you look at this photograph, which vocabulary terms from this chapter come to mind? *(Sample answer:* endotherm, homeostasis, stimulus, response*)* Write the words on the board.

Ask How could you use these words to answer the question as it applies to the photograph? *(Sample answer: The cold temperature of their environment is the stimulus that causes the penguins to huddle together for warmth. Huddling together is their response to the stimulus. Huddling allows their bodies to maintain homeostasis of a stable internal temperature.)*

Write responses on the board. Then, have students use them as a basis to write an answer to the Big Idea question that applies to animals in general.

CONSTRUCTED RESPONSE Paragraphs will vary. Read aloud the directions, questions, and *Hints*.

Say Readers use text structure to help them understand what they read. Text structure is the way the text is organized so that the writer clearly makes a point. One text structure is cause and effect.

Have students give other text structures as you write them on the board. *(Sample answers: sequence, compare/contrast, describe, explain)*

Read Question 1 aloud.

Ask What is the first hint in Question 1 asking you to do? *(identify sense organs in a cat and what they do)* What kind of text structure might you use to write this part of the paragraph? *(describe)* What types of words and information do you use when you describe something? *(Sample answer: use adjectives, provide details, define words)* What type of text structure might you use to write about the second hint? *(cause and effect)*

Tell students to list their ideas first and then think about how to make them into sentences and paragraphs. Model writing the first one or two sentences, using the hints and notes from your discussion. *(Sample: A cat can sense movement through its highly specialized eyes. It has very sensitive ears that can sense the sounds a mouse makes.)*

28 CHECK Understanding

Assess the Big idea — Structure and Function

Write an answer to the question below:

Q: How do the body systems of animals allow them to collect information about their environments and respond appropriately?

Constructed Response

Write an answer to each of the numbered questions below. The answer to each numbered question should be one or two paragraphs long. To help you begin, read the **Hints** below the questions.

1. **How does a cat's nervous system sense and respond to a mouse?**

   Hint What kinds of sense organs do cats have that allow them to sense stimuli such as movements, sounds, and vibrations?

   Hint Think about how the neurons in a cat's nervous system work together to sense and respond to stimuli in the environment.

2. **The cat runs after the mouse to try to catch it. Describe how a cat is able to run.**

   Hint List the body systems that work together to allow an animal to move.

   Hint Look at the diagram that shows how muscles and joints work together.

3. **A cat can spend time outside on a warm summer day and a cool summer night. How does the cat control its body temperature?**

   Hint Cats are endotherms.

Foundations for Learning Wrap-Up

Use the index cards that you have made to help you review the vocabulary in this chapter.

Activity 1 Take a stack of blank index cards. Write the main topics from this chapter on these cards. (Green headings are main topics.) Write Not Sure on one of the cards. Put the topic cards on a table. Arrange the vocabulary cards in groups under each topic. Put any cards that you are not sure about in the Not Sure category. Get together with a classmate and talk about how each word fits into its category. Help each other to put the Not Sure cards under the right topics.

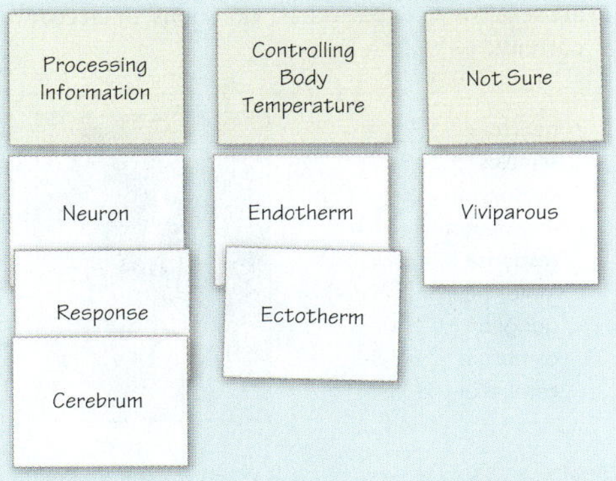

Activity 2 You can play a game to review the vocabulary words in this chapter. Get together with three of your classmates and divide yourselves into two teams. Shuffle all of the vocabulary cards and deal them out, face down, to both teams.

Set a timer for 2 minutes. When the timer starts, draw one of your vocabulary cards. Say words or sentences that relate to this vocabulary word to try to get your partner to say the word. If you accidentally say the vocabulary word, throw that card into a discard pile, so you won't get credit for it.

After 2 minutes, let the other team try. Play the game until one team is out of cards. At that point, the team with the most completed cards wins.

690 Check Understanding

For Question 2, pair students and have them use the hints and your modeling to answer the question. Have students share their paragraphs with the class.

For Question 3, have students write the paragraphs independently for homework or as an anchor activity.

Foundations for Learning Wrap-Up

Pair students and have them check each other's words for accuracy. You may wish to brainstorm as a class the topics for Activity 1. For Activity 2, monitor the noise level so that all teams can listen for the timer, or have one person in the team keep track of the time.

28.1 Response

Understand Key Concepts

1. An animal's reaction to a stimulus is called a
   a. response.
   b. ganglion.
   c. neuron.
   d. trigger.
2. The simplest nervous systems are called
   a. cephalopods.
   b. motor neurons.
   c. nerve nets.
   d. sensory neurons.
3. What kinds of environmental stimuli are some animals able to sense that humans cannot sense?

Test-Taking Tip

Rephrase the Question When you read a question that you do not understand, try to put the question in your own words. If it is written as a statement, make it into a question. Using simpler words and changing the way a question is written may help you to figure out how to answer a hard question.

Think Critically

4. **Compare and Contrast** List the three major types of neurons and compare their roles.
5. **Apply Concepts** Suppose a pet dog is having trouble coordinating its movements. Why might a veterinarian X-ray the dog's brain?

28.2 Movement and Support

Understand Key Concepts

6. Which of the following animals uses a hydrostatic skeleton to move?
   a. arthropod
   b. sponge
   c. fish
   d. earthworm
7. Muscles generate force
   a. only when they get longer.
   b. only when they get shorter.
   c. when they get longer or shorter.
   d. all the time.
8. How do a fish's muscles work when the fish swims?

Think Critically

9. **Compare and Contrast** List one advantage and one disadvantage of exoskeletons and endoskeletons.
10. **Infer** The largest land animals are vertebrates. Why are insects unable to grow as large as vertebrates?

28.3 Reproduction

Understanding Key Concepts

11. A species that has eggs that develop outside the mother's body is
    a. oviparous.
    b. viviparous.
    c. ovoviviparous.
    d. nonviparous.
12. What organ in many female mammals allows the exchange of gases, nutrients, and wastes between the mother and the developing young?
    a. marsupium
    b. amniotic egg
    c. mammary gland
    d. placenta
13. How did the evolution of the amniotic egg allow many animals to live on land?

Think Critically

14. **Compare and Contrast** Describe the differences between a newborn placental mammal and a newborn marsupial.
15. **Predict** Why would you expect most species that use external fertilization to reproduce in water?

Assessment Tips

Read aloud Question 5. Remind students that organs besides the brain are also involved with movement. Ask students which word ties movement to the brain. *(coordinating)*

For Question 7, explain that when an answer option contains a key word such as *all* or *only,* there can be no exceptions.

Point out to students that they must make an inference for Question 19; they must decide the body system to which blood belongs.

Answers

Lesson 28.1

UNDERSTAND KEY CONCEPTS

1. a

2. c

3. Sample answer: weak electric currents and Earth's magnetic field

THINK CRITICALLY

4. Sensory neurons sense stimuli, such as light or sound. Interneurons process information and determine how an animal responds to stimuli. Motor neurons carry "directions" from interneurons to muscles.

5. Sample answer: The veterinarian would be looking for anything wrong with the cerebellum, the part of the brain that controls how movements are coordinated.

Lesson 28.2

UNDERSTAND KEY CONCEPTS

6. d

7. b

8. Muscles are arranged in blocks on opposite sides of the backbone. These muscle blocks contract in waves down the body. This allows the body to move back and forth.

THINK CRITICALLY

9. Sample answer: Advantages: Exoskeletons protect the body from predators and from drying out. Endoskeletons provide lightweight support. Disadvantages: Exoskeletons must be molted as the animal grows. Endoskeletons do not provide a watertight covering.

10. An endoskeleton is light compared to an exoskeleton, so animals with endoskeletons, such as vertebrates, can get large. An insect cannot get large because its exoskeleton would become too heavy for its body.

Lesson 28.3

UNDERSTAND KEY CONCEPTS

11. a

12. d

13. The amniotic egg allows an embryo to develop on land without drying out. Species that produce an amniotic egg can live in very dry places and still reproduce.

THINK CRITICALLY

14. A placental animal is born at an advanced stage of development. A marsupial is born at a very early stage of development.

15. Sample answer: Water allows the gametes to move easily and come into contact with each other, so that eggs become fertilized.

ASSESSMENT

Lesson 28.4

UNDERSTAND KEY CONCEPTS

16. a

17. b

18. fights pathogens that have entered the body

THINK CRITICALLY

19. Endocrine glands in the endocrine system produce hormones, which are carried to the organs they affect by the blood in the circulatory system.

20. Sample answer: The circulatory system brings nutrients to the cells and removes wastes.

21. Birds conserve body heat by fluffing out their feathers. Mammals use a combination of body fat and hair as insulation.

Connecting Concepts

22. protects the developing embryo

23. the yolk sac; gives food to the developing embryo

CHAPTER MYSTERY After students have read through the Chapter Mystery, use the diagrams to clarify the difference between normal fertilization and automictic parthenogenesis.

Ask By what process do eggs and sperm form? *(meiosis)*

Guide students to recall that eggs and sperm are haploid, because they have one set of chromosomes.

Ask In normal fertilization, what two kinds of cells combine to form a zygote? *(egg and sperm)*

Point out that the zygote is diploid, having received one set of chromosomes from each parent. The two sets of chromosomes are homologous, but not identical.

Ask In automictic parthenogenesis, what happens to the single set of chromosomes in the mother's egg? *(It doubles, forming two identical sets of chromosomes.)* Then what happens to the egg? *(It divides into two cells, with each cell containing one set of chromosomes.)*

Point out that these two identical haploid cells combine to form a diploid zygote.

Ask What is the difference between the chromosomes of a shark zygote produced by normal fertilization and those of a shark zygote produced by automictic parthenogenesis? *(Sample answer: A zygote produced by normal fertilization has two homologous sets of chromosomes, one from each parent. A zygote produced by automictic parthenogenesis has two identical sets of chromosomes, both from the mother.)*

28 CHECK Understanding

28.4 Homeostasis

Understand Key Concepts

16. Stable conditions inside an animal's body are called
a. homeostasis.
b. ectothermy.
c. endothermy.
d. reactivity.

17. The main source of heat for an ectotherm is
a. its high rate of metabolism.
b. the environment.
c. its own body.
d. its food.

18. What does the immune system do?

Think Critically

19. Apply Concepts What two body systems work together to deliver hormones to the organs that they affect? Describe how this interaction takes place.

20. Apply Concepts Describe how the circulatory system helps maintain stable conditions inside the body.

21. Relate Cause and Effect Identify two ways in which endotherms conserve body heat.

Connecting Concepts

Use the diagram to answer questions 22 and 23.

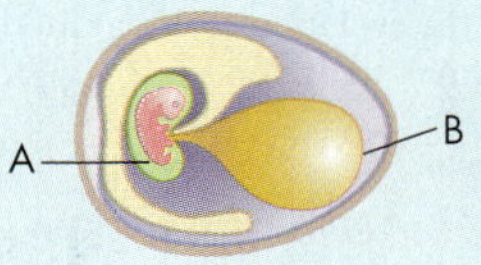

22. Intrepret Visuals What does the membrane labeled A do?

23. Intrepret Visuals What is membrane B? What does this membrane do?

solve the CHAPTER MYSTERY

SHE'S JUST LIKE HER MOTHER!

The baby bonnethead had been created by automictic parthenogenesis. In this process, the mother's unfertilized egg divided into two cells. These two cells came together to form a zygote with two sets of chromosomes that were exactly the same. (See below.) The identical nature of the chromosomes made the baby shark homozygous for every trait. She had two sets of alleles that were exactly the same.

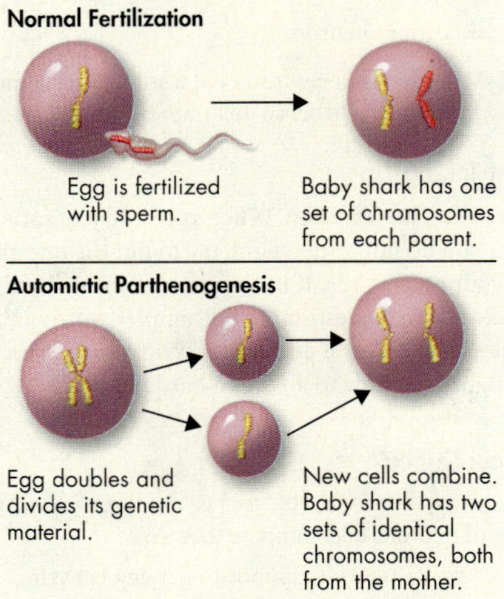

1. Apply Concepts Explain why the baby shark was not a clone, or exact genetic copy, of its mother. (**Hint:** The mother was conceived by sexual reproduction.)

Untamed Science™ **Never Stop Exploring Your World.** Finding a solution to what happened in the shark tank is just the beginning. Take a video field trip with the ecogeeks of Untamed Science to see where the mystery leads.

692 BIOLOGY.com Search Chapter 28 GO • Untamed Science Video • Chapter Mystery

CHAPTER MYSTERY ANSWER

1. The egg is produced through meiosis, and so is not an exact copy of the mother's genetic material. Every egg that the mother produces has different combinations of alleles that were inherited from the mother's mother or father. In automictic parthenogenesis, the haploid genetic material in an egg doubles and divides its genetic material. Two of the new cells (which contain the same genetic material) combine to form the diploid zygote.

Untamed Science™ Watch the Untamed Science crew in **Boy? . . . Or Girl?** as they interview experts to learn more about how the sex of offspring is determined in some animals.

Standardized Test Prep

Multiple Choice

1. What part of a vertebrate's brain is the "thinking" region?
A olfactory bulb
B cerebellum
C cerebrum
D medulla oblongata

2. Neurons that receive and send information from and to other neurons are called
A ganglia.
B motor neurons.
C sensory neurons.
D interneurons.

3. The skeleton of a shark is composed primarily of
A bone. C cartilage.
B vertebrae. D tendons.

4. Joints between bones of the human skeleton are held together mostly by
A tendons. C ligaments.
B muscles. D skin.

5. In oviparous species, embryos
A develop internally.
B obtain nutrients directly from the mother's body.
C obtain nutrients from the external environment.
D develop outside of the body.

6. Most animals reproduce sexually by producing
A buds.
B clones.
C haploid gametes.
D diploid gametes.

7. Maintaining homeostasis in multicellular organisms requires
A a properly functioning heart.
B a nervous system.
C hormones.
D all body systems working together.

8. Which of the following are endothermic?
A fish and amphibians
B mammals and birds
C reptiles and mammals
D all vertebrates

Questions 9–10

Study the illustration of a reptile brain and answer questions 9 and 10.

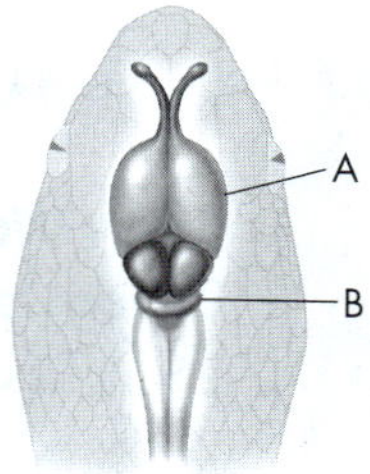

9. What is the name of the structure labeled A?
A cerebrum
B optic lobe
C cerebellum
D olfactory bulb

10. What are some functions of the structure labeled B?
A vision
B control of internal organ functions
C connection of the brain to the rest of the body
D coordination of movement and control of balance

Open-Ended Response

11. Why can't muscles function individually?

If You Have Trouble With . . .

Question	1	2	3	4	5	6	7	8	9	10	11
See Lesson	28.1	28.1	28.2	28.2	28.3	28.3	28.4	28.4	28.1	28.1	28.2

Test-Taking Tip

READ ALL ANSWER CHOICES CAREFULLY

Tell students that several answer choices for a multiple-choice question may be very similar. Students should read all of the answer choices carefully before choosing the best answer.

Before the Test

Review with students the Test-Taking Tip.

Point out in Question 8 that for A, B, or C to be correct, both groups of animals must be endothermic, not just one of the groups. For D, point out that *all* means "no exceptions."

Explain that rephrasing a question as part of an answer may help students identify exactly what a question is asking. For example, they could rephrase Question 11 as, **Muscles can't function individually because . . .** Remind students to include details to support their answer.

Answers

1. C
2. D
3. C
4. C
5. D
6. C
7. D
8. B
9. A
10. D
11. Sample answer: A muscle can exert force only when it contracts. The contraction of a single muscle can move a joint in only one way. Another muscle must contract in the opposite direction as the first muscle relaxes in order for the joint to move in the opposite way.

CHAPTER PLANNER 29 Animal Behavior

Chapter Contents	NSES	Time	Core Resources
Chapter Preview			**Student Edition,** pp. 694–695 **Chapter Mystery,** p. 695 **Foundations for Learning,** p. 695
29.1 Elements of Behavior Behavior and Evolution • Innate Behavior • Learned Behavior • Complex Behaviors	I, II, IV, A.1.c, A.2.b, C.3.a, C.6.b, C.6.c, C.6.d	1 period 1/2 block	**Student Edition,** pp. 696–700 **Inquiry Into Scientific Thinking,** p. 699 **Study Workbook B** 29.1 Worksheets **Biology.com** *Tutor Tube:* Sing . . . Sing a Song • 29.1 Lesson Assessment
29.2 Animals in Their Environments Behavioral Cycles • Social Behavior • Communication	I, II, IV, C.3.a, C.3.b, C.6.a, C.6.b, C.6.c	1/2 period 1/4 block	**Student Edition,** pp. 701–704 **Build Connections,** p. 703 **Study Workbook B** 29.2 Worksheets **Biology.com** *Art Review:* Animal Communication • *Art in Motion:* Social Behavior in Fiddler Crabs • 29.2 Lesson Assessment
Chapter Pre-Lab	II, A.1.a, A.1.b, A.1.d, A.1.f, A.2.a, C.6.b	1 period 1/2 block	**Student Edition,** p. 705 **Lab Manual B** *Termite Tracks*

Tools for ELL Instruction

ELL Handbook describes teaching strategies to make *Miller & Levine Biology Foundation Edition* more accessible to ELL students.

Spanish Study Workbook is a Spanish translation of Study Workbook A.

Multilingual Glossary is the *Miller & Levine Biology* glossary translated into ten languages.

Tools to Challenge Students

Study Workbook A includes more challenging worksheets and activities such as Chapter Mystery activities that reinforce 21st Century Skills.

Lab Manual A labs are geared toward on-level students. The Lab Manual includes fifteen labs in addition to the Chapter Labs.

Biology.com Data Analysis activities offer students a way to practice data gathering and analyzing skills.

Visual Quizzes in the Assessment Resources Book are geared toward on-level students and use visuals from *Miller & Levine Biology*.

Additional Resources

Biology.com Untamed Science Video • Flash Cards

Study Workbook A 29.1 Worksheets
Lab Manual B *Data Analysis:* Caring for Young
Spanish Study Workbook 29.1 Worksheets

Study Workbook A 29.2 Worksheets
Spanish Study Workbook 29.2 Worksheets
Biology.com *Data Analysis:* Tracking Shark Migrations
Assessment Resources Book Visual Quiz

Lab Manual A *Termite Tracks*

Chapter Review

Student Edition Chapter Summary, p. 706 • Unit Project, p. 710
Study Workbook B Chapter 29 Review
Transparencies, pp. 326–330
Biology.com Untamed Science Video • You're the Director • Editable Worksheets of Study Workbooks A and B and Lab Manuals A and B • Chapter 29 Flash Cards and Crossword

Untamed Science DVD • Classroom Resources CD (includes lesson presentations and editable worksheets)

Chapter Assessment

Student Edition Check Understanding, pp. 707–709
Study Workbook B Chapter 29 Taking a Standardized Test
Assessment Resources Book Chapter 29 Test B • Unit 7 Test B
Biology.com Chapter 29 Assessment • Editable Worksheets of Chapter 29 Visual Quiz,* Chapter 29 Tests A and B, and Unit 7 Tests A and B

Exam*View* *Assessment Suite* • Classroom Resources CD (includes lesson presentations and editable worksheets)

Time: 1 period, 1/2 block

* Visual Quizzes are based on visuals from *Miller & Levine Biology*. Review carefully to make sure they are appropriate for your class.

Pressed for Time?

Preview the Chapter Introduce the chapter with a discussion of the visual An Ant Society.

Cover the Chapter Quickly Have students read *Behavior and Evolution, Innate Behavior,* and the introduction to *Learned Behavior* in Lesson 29.1. In Lesson 29.2, have students read *Social Behavior.*

Assess Assign Questions 1 and 3 in the 29.1 Check Understanding, Questions 3, 4, and 5 in the 29.2 Check Understanding, and Questions 1, 2, 6–8, 10, and 11 in the Chapter 29 Check Understanding.

CHAPTER 29

Get the Picture

Invite a volunteer to read aloud the caption for the photograph of the wolves.

Say Wolves often cooperate with one another to kill prey.

Ask So, why are the wolves in the photograph fighting over the deer? *(They are competing to eat the deer.)*

Ask What do you think the wolf with its paw on the deer is trying to communicate to the other wolves? *(Sample answer: that the deer belongs to it)*

Ask Which wolf do you think will eat the deer? *(Sample answer: the wolf that wins the fight)*

Activate Prior Knowledge

Say Think about animals you have observed, such as pet dogs and cats, wild birds, or zoo animals.

Ask What are some ways these animals interact with one another? (*Sample answers: by playing, eating, fighting, making sounds, mating, raising young*)

Introduce the concepts that animals also interact with their environment and that their interactions with the environment and other animals affect their survival.

NATIONAL SCIENCE EDUCATION STANDARDS

UNIFYING CONCEPTS AND PROCESSES

I, II, IV

CONTENT

C.3.a, C.3.b, C.6.a, C.6.b, C.6.c, C.6.d

INQUIRY

A.1.a, A.1.b, A.1.c, A.1.d, A.1.f, A.2.a, A.2.b

29 Animal Behavior

Big idea **Evolution**

Q: How do animals interact with one another and their environments?

Timber wolves fighting over a meal

BIOLOGY.com Search Chapter 29 GO • Flash Cards

UbD Connect to the Big Idea

Invite a volunteer to read aloud the chapter number and the chapter title. Read the Big Idea and question. Emphasize the question, **How do animals interact with one another and their environments?**

Set the Purpose

PREVIEW AND PREDICT Draw the graphic organizer, then note the INSIDE feature on the next page. Have a student read each lesson number and title.

Ask What does *animal behavior* mean? *(Sample answer: the way an animal acts)*

Say Many animal behaviors are inherited and, thus, have resulted from evolution.

Ask What is biological evolution? *(change over time that leads to adaptation)*

Remind students to search for answers as they read.

INSIDE:
- 29.1 Elements of Behavior
- 29.2 Animals in Their Environments

- Untamed Science Video
- Chapter Mystery

CHAPTER
MYSTERY

ELEPHANT CALLER ID?

It is a hot, dusty afternoon in Africa's Etosha National Park. A group of elephants walks toward a watering hole. They begin to drink and splash around. Suddenly, they freeze. One elephant places her trunk flat on the ground. The tip of her trunk points toward her feet. Soon, the elephants clump together, pushing their young calves into the center of the group. Some of the elephants place weight on the front of their feet. Most of them keep their ears flattened against their heads. The group moves slowly away from the watering hole.

Read for Mystery Clues People who are watching the elephants cannot see or hear anything that could have scared the group. What are these elephants doing and why? As you read this chapter, look for clues that explain the behavior of this elephant group. Then, solve the mystery.

FOUNDATIONS for Learning

You can use table tents to help organize the ideas in this chapter. Fold a piece of construction paper into thirds. Color in the middle third of the paper. You will write notes on the other two thirds of the paper. Pick one of the main topics in Lesson 1. Use words and pictures to illustrate that topic, as shown in the diagram. Then, fold and tape the paper so that it makes a table tent. Make another table tent to illustrate one of the main topics from Lesson 2. At the end of the chapter are two activities that use table tents to help answer the question: How do animals interact with one another and their environments?

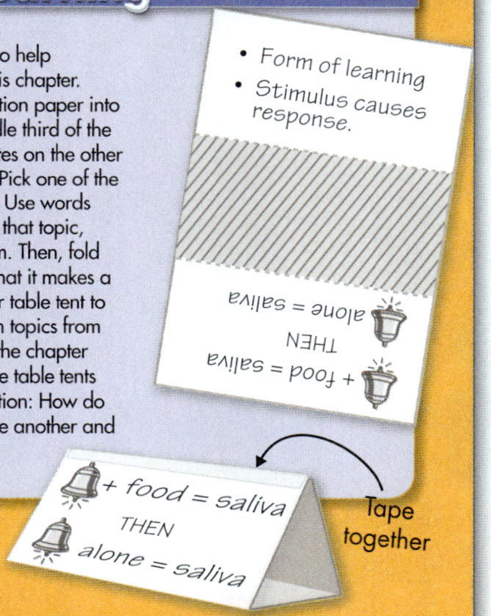

Chapter 29:
Big Idea: Evolution

Chapter 29:
How do animals interact with one another and their environments?

29.1: What are the elements of animal behavior?

29.2: How do the environment and other organisms affect an animal's behavior?

Foundations for Learning

Provide students with letter-size or legal-size construction paper or card stock to use in making table tents. Have them work individually to make a table tent for one major topic in each lesson. To be sure that all topics are covered, have the class count off from 1 to 4 for Lesson 29.1 and from 1 to 3 for Lesson 29.2. Explain that each 1 will make a tent for the lesson's first major topic, that the 2s will make tents for the second major topic, and so on.

Ask Why is it important to use pictures as well as words to make a table tent? *(Sample answers: Some people learn better from pictures. Pictures sometimes communicate more information than words do.)*

Encourage students to be brief but thorough in listing facts for the table tents. Remind them that table tents can be used as review and study aids.

Untamed Science™

The short movie **When Animals Go Digital** will take students on a field trip in which scientists use remote sensing to study animal behavior.

CHAPTER MYSTERY

Invite a volunteer to read aloud the Chapter Mystery.

Ask How did one of the elephants signal the group to move closer together? *(by placing her trunk flat on the ground and pointing the trunk toward her feet)* What might the large elephants be trying to do by placing young calves at the center of the group? *(Sample answer: protect the smallest and weakest elephants)*

Tell students that they will find clues in the chapter to answer the questions surrounding this mystery.

Digital Support

BIOLOGY.com Extend your reach by using the digital assets offered at **Biology.com.**

LESSON 29.1

Pre-Reading

Objectives

29.1.1 Identify the significance of behavior in the evolution of a species.

29.1.2 Explain what an innate behavior is.

29.1.3 Describe the major types of learning.

29.1.4 Explain what types of behaviors are usually considered complex.

Preview the Pages

1. Have students find each vocabulary word in the text and read it aloud, then copy it in their notebooks. Have them leave space for definitions.
2. Preview the visuals. Point out that many of the visuals in this lesson show a sequence of events. Explain that the events in each sequence do not happen one directly after another—some time has passed between each picture. For example, in the Habituation visual, many cars have passed between the blue car and the red car.

Activate Prior Knowledge

Ask students if they have ever taught a pet to follow a command or if they know someone who has. Encourage students to describe the method that was used to help the pet learn the new behavior.

Build Understanding

CONCEPT MAP Direct students to the concept map in Workbook B. Note that it summarizes the main types of animal behaviors in an organized manner. Encourage students to complete the concept map as they read about each type of behavior.

NATIONAL SCIENCE EDUCATION STANDARDS

UNIFYING CONCEPTS AND PROCESSES

I, II, IV

CONTENT

C.3.a, C.6.b, C.6.c, C.6.d

INQUIRY

A.1.c, A.2.b

For the corresponding lesson in *Miller & Levine Biology,* see pages 840–845.

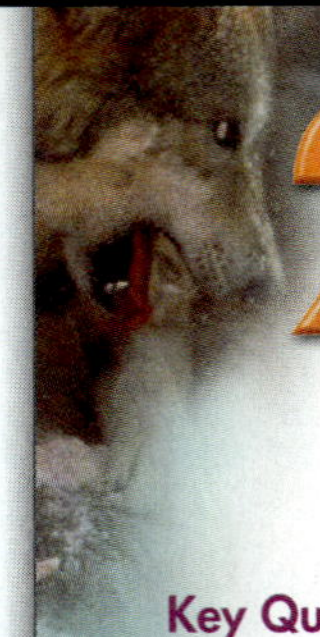

29.1 Elements of Behavior

Key Questions

How can behavior influence the evolution of animal species?

What is an innate behavior?

What are the major types of learning?

How do many complex behaviors arise?

BUILD Understanding

Concept Map As you read, create a concept map to organize the information in this lesson.

In Your Workbook Go to your workbook to learn more about making a concept map.

Behavior and Evolution

A young boy eats lunch at a table outside of a Caribbean restaurant. In the bushes nearby, an iguana watches closely. The dangerous-looking iguana crawls closer. When the boy sees the lizard, he jumps out of his chair. But the iguana is not interested in the boy's toes. This lizard is a vegetarian. The iguana quickly eats the boy's French fries that have fallen to the ground.

What's so interesting about this scene? This species of iguana usually lives in trees. These iguanas are normally shy and do not come close to humans. However, this iguana has learned that getting close to humans can mean an easy meal!

The activity of the iguana is an example of behavior. **Behavior** is the way that an organism reacts to a stimulus in its environment. Recall that a stimulus is any information that causes an organism to react.

Some behaviors are simple. A dog turns its head when it hears a noise. Other behaviors are complex. For example, some animals wash their food before eating it. The way that an animal behaves often depends on an internal condition, such as hunger. If the iguana had not been hungry, he probably would have stayed away from the boy.

Most behaviors are necessary for survival. For example, animals must be able to find and catch food, choose a habitat, avoid predators, and find mates. Many behaviors are influenced by genes. If these behaviors help an animal survive and reproduce, they will be passed on to the next generation. Eventually helpful behaviors will spread throughout the population, or even the species. Thus, certain behaviors can evolve by natural selection, just like physical traits.

Key Question How can behavior influence the evolution of animal species? **If an inherited behavior allows an animal to survive and reproduce, the behavior can spread throughout the population.**

696 Chapter 29 • Lesson 1

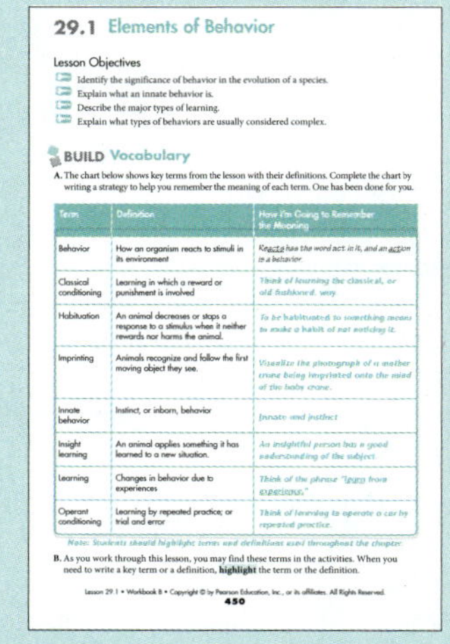

29.1 Elements of Behavior

Lesson Objectives

- Identify the significance of behavior in the evolution of a species.
- Explain what an innate behavior is.
- Describe the major types of learning.
- Explain what types of behaviors are usually considered complex.

BUILD Vocabulary

A. The chart below shows key terms from the lesson with their definitions. Complete the chart by writing a strategy to help you remember the meaning of each term. One has been done for you.

Term	Definition	How I'm Going to Remember the Meaning
Behavior	How an organism reacts to stimuli in its environment	Reacts has the word act in it, and an action is a behavior.
Classical conditioning	Learning in which a reward or punishment is involved	Think of learning the classical, or old fashioned, way.
Habituation	An animal decreases or stops a response to a stimulus when it neither rewards nor harms the animal.	To be habituated to something means to make a habit of not noticing it.
Imprinting	Animals recognize and follow the first moving object they see.	Visualize the photograph of a mother crane being imprinted onto the mind of the baby crane.
Innate behavior	Instinct, or inborn, behavior	Innate and instinct
Insight learning	An animal applies something it has learned to a new situation.	An insightful person has a good understanding of the subject.
Learning	Changes in behavior due to experiences	Think of the phrase "learn from experience."
Operant conditioning	Learning by repeated practice; or trial and error	Think of learning to operate a car by repeated practice.

Note: Students should highlight terms and definitions used throughout the chapter.

B. As you work through this lesson, you may find these terms in the activities. When you need to write a key term or a definition, **highlight** the term or the definition.

Lesson 29.1 • Workbook B • Copyright © by Pearson Education, Inc., or its affiliates. All Rights Reserved.
450

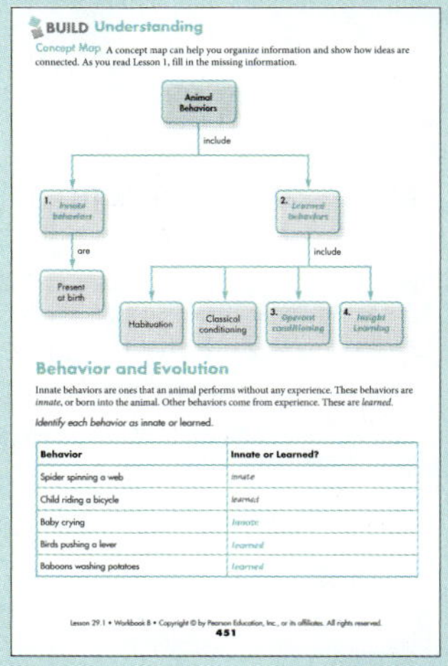

BUILD Understanding

Concept Map A concept map can help you organize information and show how ideas are connected. As you read Lesson 1, fill in the missing information.

Behavior and Evolution

Innate behaviors are ones that an animal performs without any experience. These behaviors are *innate*, or born into the animal. Other behaviors come from experience. These are *learned*.

Identify each behavior as innate or learned.

Behavior	Innate or Learned?
Spider spinning a web	innate
Child riding a bicycle	learned
Baby crying	innate
Birds pushing a lever	learned
Baboons washing potatoes	learned

Lesson 29.1 • Workbook B • Copyright © by Pearson Education, Inc., or its affiliates. All rights reserved.
451

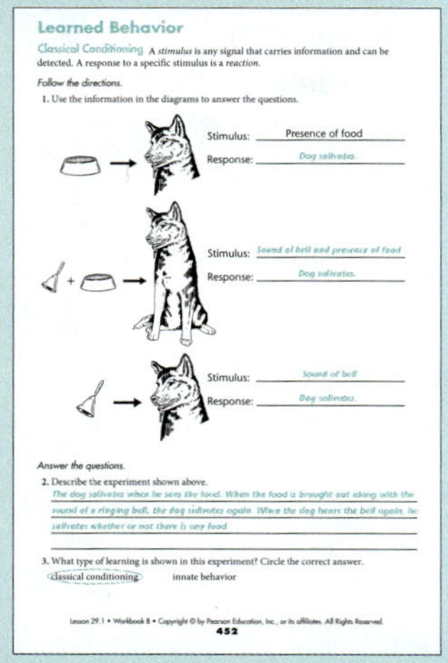

Learned Behavior

Classical Conditioning A *stimulus* is any signal that carries information and can be detected. A response to a specific stimulus is a *reaction*.

Follow the directions.

1. Use the information in the diagrams to answer the questions.

Answer the questions.

2. Describe the experiment shown above.
The dog salivates when he sees the food. When the food is brought out along with the sound of a ringing bell, the dog salivates again. When the dog hears the bell again, he salivates whether or not there is any food.

3. What type of learning is shown in this experiment? Circle the correct answer.
classical conditioning   innate behavior

Lesson 29.1 • Workbook B • Copyright © by Pearson Education, Inc., or its affiliates. All Rights Reserved.
452

Innate Behavior

How does a spider know how to spin a web? How do chicks know how to beg for food? These animals are showing **innate behaviors,** or instincts. An animal performs an innate behavior even though it has no previous experience with the stimulus to which it is responding. Innate behaviors allow animals to survive without the need for experience. An animal's genes and its environment interact to cause innate behaviors.

One example of an innate behavior is a newborn mammal's ability to suck milk from its mother's body. This innate behavior is simple. Other innate behaviors, such as making a spider web, are complex.

Key Question What is an innate behavior? **An innate behavior is a behavior that an animal performs even though it has no previous experience with the stimulus to which it is responding.**

Learned Behavior

If all behaviors were innate, the behaviors would be hard or impossible for an animal to change during its lifetime. If that were the case, individual animals would have a hard time responding to changes in their environment. But when many animals experience something new, they are able to respond by changing their behavior. Making changes to behaviors based on experience is called **learning.**

Many animals have the ability to learn. Organisms with simple nervous systems, such as sea stars and most other invertebrates, learn only rarely. In a few invertebrates and many chordates, learning is common. Scientists have identified four main types of learning: habituation, classical conditioning, operant conditioning, and insight learning.

Habituation The simplest type of learning is habituation. **Habituation** happens when an animal stops responding to a repeating stimulus that does not harm or reward the animal. Imagine crows eating near a road. At first, passing cars cause the crows to fly away. Over time, they learn that the cars will not harm them. They stop flying away. The crows have become habituated to the cars. In other words, a stimulus (the cars) no longer causes a response (flight). An animal can save energy by ignoring a stimulus that does not affect its survival.

BUILD Vocabulary

behavior the way an organism reacts to changes in its internal condition or to stimuli in its environment

innate behavior a behavior that an animal performs without any previous experience with the stimulus to which it is responding

learning a change in behavior as a result of experience

habituation a type of learning in which an animal decreases or stops its response to a repetitive stimulus that neither rewards nor harms the animal

WORD ORIGINS

The word *innate* is from the Latin word *innatus,* which means "inborn." An innate behavior is a behavior that is strongly shaped by genes an animal is born with.

Habituation Crows by the side of a road fly away when a car passes. After many cars have passed without harming the crows, the birds become habituated to the cars. From then on, the crows do not fly away when cars pass them.

ELL Focus on ELL: Build Background

ALL SPEAKERS Engage students by pointing out that they exhibit innate and learned behaviors in their daily lives. Provide examples of an innate behavior (such as jumping when surprised) and a learned behavior (such as brushing teeth before bed). Then, have students do a quick-write activity in which they list as many examples of each kind of behavior as possible in 5 minutes. Have students work in groups to compare and refine their lists. Point out that some behaviors are a combination. For example, the ability to speak is innate, but the language you speak is learned.

During Reading

Active Reading

KEY QUESTION Read aloud the Key Question on the previous page and its answer.

SCIENCE SUPPORT Tell students that innate behaviors result from the structure and function of the nervous system. Remind students that the nervous system develops along with all the body tissues, organs, and systems by a process that begins immediately after a sperm fertilizes an egg.

Ask What do eggs and sperm contribute to offspring? *(genes from the parents)* Emphasize that innate behaviors have a genetic basis but can be modified by the environment of a developing offspring.

Ask What determines inherited traits and enables them to be passed from parents to offspring? *(genes)*

Ask Which individuals are more likely to pass traits to offspring? *(individuals whose traits allow them to survive long enough to reproduce and those whose traits attract a mate)*

Build Vocabulary

WORD ORIGINS Point out that *innate* comes from a word that means "inborn."

Ask How are innate behavior and learned behavior different? *(Sample answer: Innate behavior is a behavior an animal has from birth. Learned behavior is behavior that an animal learns to do in response to a stimulus.)*

Explain to students that some innate behaviors develop over time. An animal may not be able to perform the behavior well at birth. For example, walking is an innate human behavior. However, infants are not able to walk. They walk once they are physically developed enough.

KEY QUESTION Discuss the Key Question and answer on this page.

Ask What is an example of an innate behavior, or instinct, in humans? *(Sample answers: infants suckling, babies crying, reflexes)*

Speed Bump

Ask How is learned behavior different from innate behavior? *(Learned behavior is changed behavior, and it is based on experience.)*

Ask How is habituation advantageous to an animal? *(The animal saves energy for use in activities that are important for survival.)*

LESSON 29.1

During Reading continued

Active Reading

FIND THE MAIN IDEA Students may have difficulty distinguishing classical conditioning and operant conditioning. Tell them that the two concepts are differentiated in the topic sentences of the corresponding two sections. Read and discuss the two sections as a class.

Ask What is the topic sentence for the section Classical Conditioning? *(the first sentence in the section)* Have a volunteer reread the sentence.

Repeat the exercise for the section Operant Conditioning. Point out that the topic sentence conveys the main idea of the section.

MAKE CONNECTIONS Tell students that in classical conditioning, the response is linked to a positive or negative experience and is often an innate behavior.

Ask What is an example of a positive experience you have had? *(Students should mention seeing or doing things that make them feel good or happy.)*

Explain that in operant conditioning, an animal learns how to behave in order to get a reward or avoid punishment.

Ask What is an example of a reward? *(Students should realize that a reward is something they want or like, and that they get a reward in return for something they have done.)*

KEY QUESTION Invite a volunteer to read aloud the question and answer.

Ask What is learning? *(making changes to behavior based on experience)*

Build Vocabulary

ACADEMIC VOCABULARY Have students pronounce the vocabulary words listed on this page and write the definitions in their notebooks. Point out that each term has the *-ing* ending, which is added to a verb and means "the act of." Explain that the root word for each term (*conditioning, learning,* and *imprinting*) is a verb (*condition, learn, imprint*).

Speed Bump

Ask By what type of conditioning does a wild bird learn to eat from a bird feeder? *(operant)*

Ask What is involved in conditioning but not in insight learning? *(repetition of a stimulus and response)*

Operant Conditioning

1. A dog randomly brushes its tail against a bell hanging on a doorknob.

2. The owner responds by opening the door. The ring-the-bell, open-the-door sequence happens a few times.

3. The dog then learns to ring the bell when it wants to go out.

Classical Conditioning In **classical conditioning,** an animal responds to a stimulus that has become linked to a positive or negative experience. The Russian scientist Ivan Pavlov first described classical conditioning around 1900. Pavlov noticed that dogs have an innate response to food—they salivate, or drool. Pavlov rang a bell every time he offered his dogs food. After a while, his dogs would salivate when they heard the bell, even when there was no food. Pavlov's experiment caused the dogs to salivate (a response) in reaction to a bell (stimulus) associated with food.

Operant Conditioning **Operant conditioning** happens when an animal learns to behave in a certain way to get a reward or to avoid punishment. The American scientist B.F. Skinner first described operant conditioning in the 1940s. Skinner conducted experiments using a box that became known as a "Skinner box." A Skinner box has a lever inside. An animal that is put inside the box will accidentally push the lever at some point. When it pushes the lever, the animal receives food. After it is rewarded several times, the animal learns that it can get food by pushing the lever.

Operant conditioning is often called trial-and-error learning. After a random behavior produces a reward, the animal tries to get the reward again (a trial). Most trials result in errors (no reward). However, sometimes a trial will produce a reward.

Insight Learning In **insight learning,** an animal applies something it has already learned to a new situation. The animal does not go through a period of trial and error. Insight learning is also called reasoning. Humans use insight learning to solve new problems. Insight learning is also common in some nonhuman primates. In one experiment, a hungry chimpanzee used insight learning to stack boxes in order to reach bananas overhead.

Key Question What are the major types of learning?
The four major types of learning are habituation, classical conditioning, operant conditioning, and insight learning.

BUILD Vocabulary

classical conditioning
a type of learning that occurs when an animal makes a mental connection between a stimulus and some kind of reward or punishment

operant conditioning
a type of learning in which an animal learns to behave in a certain way through repeated practice, to receive a reward or avoid punishment

insight learning
a type of behavior in which an animal applies something it has already learned to a new situation, without a period of trial and error; also called reasoning

imprinting
a type of behavior based on early experience; once imprinting has occurred, the behavior cannot be changed

ACADEMIC VOCABULARY
The term *insight* means "a clear or deep understanding of a complex situation." In insight learning, an animal understands a new situation and then responds to it.

Focus on Academic Language

• lever • reasoning

Students may be confused by the uses of the words *lever* and *reasoning* on this page. Tell them that in physics, a *lever* is a simple machine that consists of a bar used to lift or dislodge an object. Explain that in a Skinner box, the lever is connected to a device that releases food when the lever is pressed.

Tell students that the word *reason* has multiple meanings.

Ask What is one meaning of the word *reason*? *(Students will probably be familiar with* reason *meaning "an explanation" or "a cause.")*

Say *To reason* also means "to think and draw conclusions."

Tell students that *reasoning*, therefore, is "the act of thinking and drawing conclusions."

Imprinting in the Wild
These sandhill crane chicks have imprinted on their mother. Because they have imprinted on their mother, they will follow her during migration.

Complex Behaviors

Many complex behaviors combine innate behavior with learning. One example of complex behavior is imprinting. In **imprinting,** an animal has an innate urge to follow the first moving object that it sees. But the animal is not born knowing what that object will look like. It must learn from experience what to follow. Once imprinting has occurred, the behavior becomes fixed.

Usually young birds like the sandhill cranes shown above imprint on the sight of their mothers. Imprinting on their mothers improves their chance of survival. A young crane that stays close to its mother is protected from predators. It also learns where to get food.

INQUIRY into Scientific Thinking

Caring for Young

Can experience help animals be better parents? The data in the table are from studies of a seabird, the short-tailed shearwater. Each pair produces only one egg a year. If that egg breaks or if the chick dies, the female bird does not lay another egg. The graph shows the percentage of eggs that develop into young birds, in relation to the age or breeding experience of the parents. This ratio is called reproductive success.

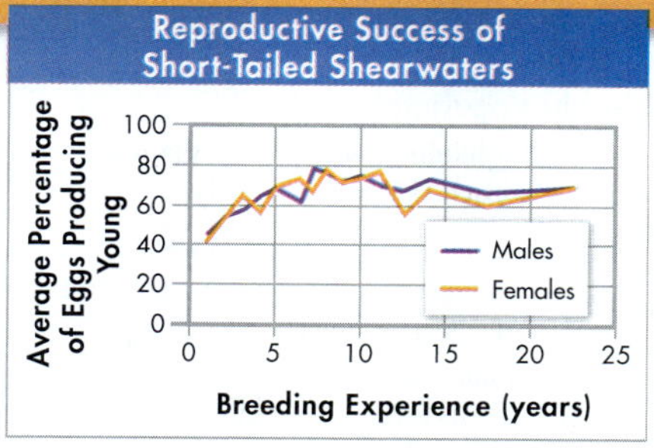

1. Interpret Tables What is the approximate success rate of a female shearwater with five years of breeding experience?

2. Compare and Contrast Are there obvious differences in reproductive success between male and female shearwaters?

3. Draw Conclusions Do older shearwaters have better reproductive success than younger birds have? Explain your answer.

4. Form a Hypothesis Do you think these birds learn to raise young more successfully over time?

In Your Workbook Get more help for this activity in your workbook.

Analyze and Conclude

1. about 70%
2. no
3. In general, yes; this is indicated by the rise in the graph until about 11 years. After 11 years, there is a slight decrease in reproductive success.
4. Most students will hypothesize that over time, shearwaters learn to raise young more successfully. Alternative hypothesis: Fertility of shearwaters is highest between 7 and 11 years of breeding experience.

Active Reading

USE CONTEXT CLUES Have students read the two paragraphs on this page. Then, have them return to the last sentence of the first paragraph.

Ask What behavior "becomes fixed"? *(the animal following the first object it saw after birth)*

MAKE INFERENCES Point out that sandhill cranes are large and have long legs for wading.

Say Sandhill cranes build nests on floating vegetation.

Ask What might be the advantage of this nesting behavior to young sandhill cranes? *(Sample answers: keeps them away from predators, reduces the chance that they will imprint on another species of bird)*

Speed Bump

Ask Why does a young bird usually follow its mother during migration? *(The young bird has imprinted on its mother.)*

Inquiry Into Scientific Thinking

INTERPRET GRAPHICS Have students examine the graph and its labels. Invite a volunteer to read the label for the y-axis (vertical axis). Explain that the graph is based on data from many individuals.

Ask Do you know from the information given how many birds were observed during these studies? *(no)*

Ask What does the word *percentage* refer to in the graph? *(the numbers 0 to 100, which indicate a ratio of eggs, specifically the number of young-producing eggs per 100 eggs produced)*

Ask What does the word *average* mean in the graph? *(The average is a sum of percentages divided by the number of percentages that make up that sum.)*

Lead a Discussion

Invite students to give examples of how their handwriting improved as they gained experience.

Ask What words would you use to describe your handwriting when you first began to write? *(Sample answers: shaky, sloppy, uneven, large)*

Ask How did your handwriting change as you gained experience? *(Sample answers: Neatness improved, lines were smoother, letters were smaller, and style became cursive.)*

During Reading continued

Active Reading

USE VISUALS Have students compare the hand puppet to an adult crane on the previous page.

Ask What similarities do you notice? *(In its shape and colors, the puppet resembles the head of an adult crane.)*

Then, have students compare the series of photographs on this page and on the previous page and read the captions.

Ask Why are the cranes following the plane? *(The puppet that the cranes had imprinted on is attached to the plane.)*

KEY QUESTION Read aloud the Key Question and answer.

Ask What are the two main types of animal behavior? *(innate behavior and learned behavior)*

Speed Bump

Ask What animal senses can imprinting involve? *(seeing, hearing, smelling)*

Post Reading

ASSESSMENT TIPS Advise students to review the highlighted vocabulary words before they begin the assessment.

MYSTERY CLUE After students read the clue, encourage them to think about the types of characteristics that tend to be inherited. Help students recall that characteristics that enable an individual to survive and reproduce in its environment are the ones most likely to be passed on to offspring.

Wrap-Up Activity

PLAY CHARADES Write the name of each type of behavior discussed in this lesson on a separate index card. Divide the class into two teams. Have one team select a card and work together to act out the behavior, while the other team guesses the behavior. If the second team guesses correctly, it's their turn. If not, the first team gets another turn. Give teams a chance to discuss any disagreements. Keep score to determine the winning team.

Imprinting in Captivity Recently hatched cranes that are raised in captivity imprint on a puppet (top left). Later, that puppet is used to help introduce these birds to the wild.

Researchers use imprinting to help raise endangered cranes in captivity. The researchers create hand puppets that look like adult cranes. When they show these puppets to baby cranes, the baby cranes imprint on the puppets. After the cranes have grown up, researchers use the puppets to train the cranes where to fly during migration.

Imprinting does not always involve vision. Animals can imprint on sights, sounds, odors, or other cues. Newly hatched salmon imprint on the odor of the stream in which they have hatched. When they are young, the salmon swim out to sea. Years later, adult salmon use that odor to help them return to their home stream to lay eggs.

Key Question How do many complex behaviors arise? **Many complex behaviors combine innate behavior with learning.**

CHECK Understanding

Apply Vocabulary

Use the highlighted words from the lesson to complete each sentence correctly.

1. A behavior that an animal performs even though it has no previous experience with the stimulus to which it is responding is called a(n)______.
2. An animal learns to behave in a certain way because it is given a reward. This type of learning is called ______.
3. An animal applies something that it has already learned to a new situation. This type of learning is called ______.

Critical Thinking

4. **Review** What is behavior?
5. **Review** Which aspect of imprinting is innate? Which aspect is learned?
6. **Write to Learn** Answer the first clue of the mystery.

MYSTERY CLUE

Elephants living elsewhere exhibit behaviors similar to those of the elephants at the watering hole. What does this suggest about the importance of these behaviors? (**Hint:** See p. 696.)

BIOLOGY.com Search Lesson 29.1 GO • Lesson Assessment

Check Understanding

1. innate behavior
2. operant conditioning
3. insight learning
4. Behavior is the way an organism reacts to stimuli in its environment.
5. The urge to follow the first moving object seen is innate. Animals learn what to follow.
6. Students should recognize that information about possible danger causes this behavior. Students might suggest that the behavior is complex and involves both innate and learned components.

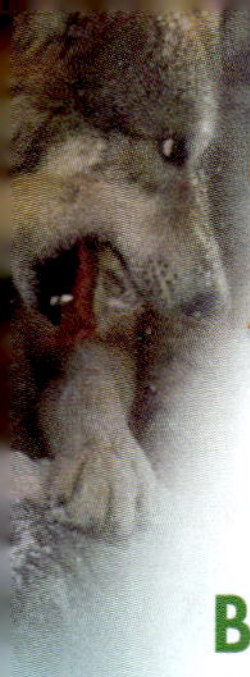

29.2 Animals in Their Environments

Behavioral Cycles

As night falls on a coral reef, animals there act like city commuters during rush hour. Some species have "day jobs" feeding in bright light. They form "traffic jams" as they move into reef caves to rest for the night. Species that work the "night shift" come out to feed in the dark. At dawn, the cycle reverses.

Daily Cycles The behavior of animals in the coral reef is an example of a circadian rhythm. **Circadian** (sur KAY dee un) **rhythms** are behavioral cycles that happen daily. The human behavior of going to school during the day and sleeping at night is an example of a circadian rhythm.

Seasonal Cycles Other behavioral cycles are seasonal. In temperate and polar areas, many species are active during spring, summer, and fall. These species enter a sleeplike state, or dormancy, during winter. Some mammals enter a kind of dormancy called hibernation. Dormancy allows animals to survive periods when food and other resources are not available.

Migration is the seasonal movement of an animal from one environment to another. Many species of animals migrate, and many animals migrate over huge distances. Migration allows animals to take advantage of good environmental conditions. Many songbirds live in tropical regions when it is winter in temperate areas. When these birds fly north in the spring, they find lots of food. They also find space to raise young.

Key Question How do environmental changes affect animal behavior?
Many animals respond to the cycles in the environment with daily or seasonal cycles of behavior.

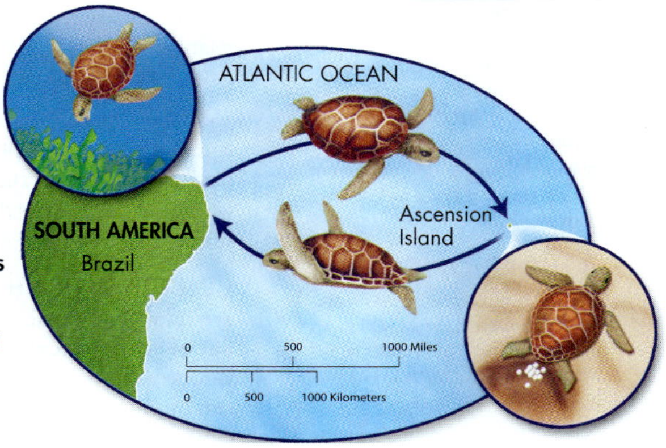

Seasonal Behavior of Green Sea Turtles
Each year, green sea turtles migrate back and forth between their feeding grounds on Brazil's coast and their nesting grounds on Ascension Island.

Key Questions

How do environmental changes affect animal behavior?

How do social behaviors increase an animal's evolutionary fitness?

How do animals communicate with others in their environments?

BUILD Understanding

KWL Chart Fill in a KWL chart as you read Lesson 29.2. In the first column, write down what you already know about the topic. In the second column, write what you want to find out. In the last column, write what you have learned after reading the lesson.

In Your Workbook Refer to your workbook for suggestions about how to use a KWL chart to organize your notes.

Pre-Reading

Objectives

29.2.1 Explain how environmental changes affect animal behavior.

29.2.2 Explain how social behaviors increase the evolutionary fitness of a species.

29.2.3 Summarize the ways that animals communicate.

Preview the Pages

1. Invite volunteers to read aloud the three Key Questions in this lesson. Encourage students to think about other questions they might have about these topics.
2. Preview the vocabulary words by having volunteers locate and pronounce the highlighted words in the lesson.
3. Preview the visuals and have volunteers read aloud all the captions and labels. Invite other volunteers to summarize in their own words what each visual shows. Correct any misconceptions.

Build Understanding

KWL CHART Direct students to Workbook B, where they will find a blank KWL chart. Remind them to fill in the K (What I Know) and W (What I Want to Know) columns of the chart before they read the lesson.

NATIONAL SCIENCE EDUCATION STANDARDS

UNIFYING CONCEPTS AND PROCESSES

I, II, IV

CONTENT

C.3.a, C.3.b, C.6.a, C.6.b, C.6.c

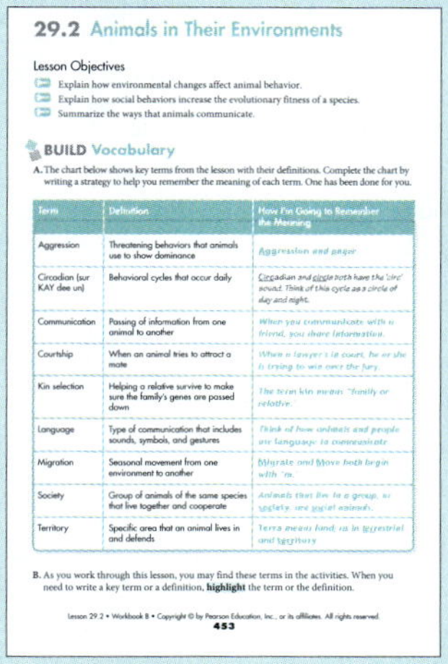

29.2 Animals in Their Environments

Lesson Objectives
- Explain how environmental changes affect animal behavior.
- Explain how social behaviors increase the evolutionary fitness of a species.
- Summarize the ways that animals communicate.

BUILD Vocabulary

A. The chart below shows key terms from the lesson with their definitions. Complete the chart by writing a strategy to help you remember the meaning of each term. One has been done for you.

Term	Definition	How I'm Going to Remember the Meaning
Aggression	Threatening behaviors that animals use to show dominance	Aggression and anger
Circadian (sur KAY dee un)	Behavioral cycles that occur daily	Circadian and circle both have the "circ" sound. Think of this cycle as a circle of day and night.
Communication	Passing of information from one animal to another	When you communicate with a friend, you share information.
Courtship	When an animal tries to attract a mate	When a lawyer is in court, he or she is trying to win over the jury.
Kin selection	Helping a relative survive to make sure the family's genes are passed down	The term kin means "family or relative."
Language	Type of communication that includes sounds, symbols, and gestures	Think of how animals and people use language to communicate.
Migration	Seasonal movement from one environment to another	Migrate and Move both begin with "m."
Society	Group of animals of the same species that live together and cooperate	Animals that live in a group, or society, are social animals.
Territory	Specific area that an animal lives in and defends	Terra means land, as in terrestrial and territory

B. As you work through this lesson, you may find these terms in the activities. When you need to write a key term or a definition, **highlight** the term or the definition.

Lesson 29.2 • Workbook B • Copyright © by Pearson Education, Inc., or its affiliates. All rights reserved.
453

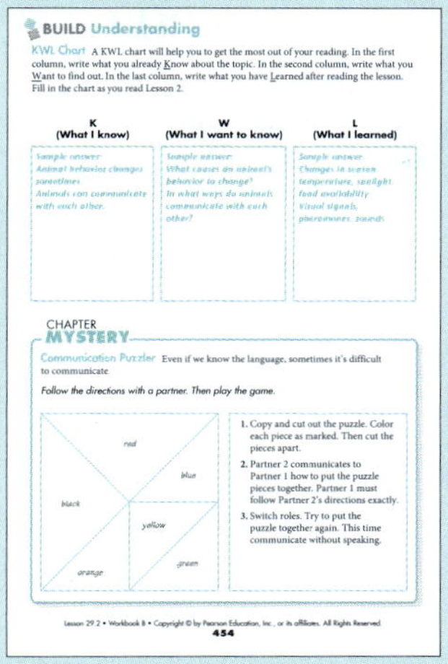

BUILD Understanding

KWL Chart A KWL chart will help you to get the most out of your reading. In the first column, write what you already Know about the topic. In the second column, write what you Want to find out. In the last column, write what you have Learned after reading the lesson. Fill in the chart as you read Lesson 2.

K (What I know)	W (What I want to know)	L (What I learned)
Sample answer: Animal behavior changes sometimes. Animals can communicate with each other.	Sample answer: What causes an animal's behavior to change? In what ways do animals communicate with each other?	Sample answer: Changes in season, temperature, sunlight, food availability. Visual signals, pheromones, sounds

CHAPTER MYSTERY

Communication Puzzler Even if we know the language, sometimes it's difficult to communicate.

Follow the directions with a partner. Then play the game.

1. Copy and cut out the puzzle. Color each piece as marked. Then cut the pieces apart.
2. Partner 2 communicates to Partner 1 how to put the puzzle pieces together. Partner 1 must follow Partner 2's directions exactly.
3. Switch roles. Try to put the puzzle together again. This time communicate without speaking.

Lesson 29.2 • Workbook B • Copyright © by Pearson Education, Inc., or its affiliates. All Rights Reserved.
454

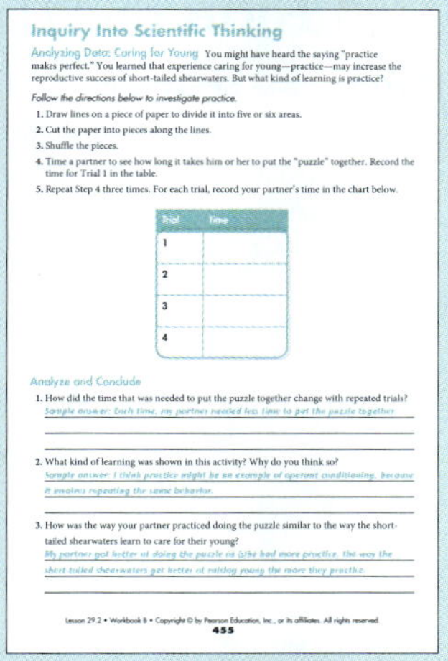

Inquiry Into Scientific Thinking

Analyzing Data: Caring for Young You might have heard the saying "practice makes perfect." You learned that experience caring for young—practice—may increase the reproductive success of short-tailed shearwaters. But what kind of learning is practice?

Follow the directions below to investigate practice.

1. Draw lines on a piece of paper to divide it into five or six areas.
2. Cut the paper into pieces along the lines.
3. Shuffle the pieces.
4. Time a partner to see how long it takes him or her to put the "puzzle" together. Record the time for Trial 1 in the table.
5. Repeat Step 4 three times. For each trial, record your partner's time in the chart below.

Trial	Time
1	
2	
3	
4	

Analyze and Conclude

1. How did the time that was needed to put the puzzle together change with repeated trials?
   Sample answer: Each time, my partner needed less time to put the puzzle together.
2. What kind of learning was shown in this activity? Why do you think so?
   Sample answer: I think practice might be an example of operant conditioning, because it involves repeating the same behavior.
3. How was the way your partner practiced doing the puzzle similar to the way the short-tailed shearwaters learn to care for their young?
   My partner got better at doing the puzzle as s/he had more practice, the way the short-tailed shearwaters get better at raising young the more they practice.

Lesson 29.2 • Workbook B • Copyright © by Pearson Education, Inc., or its affiliates. All rights reserved.
455

For the corresponding lesson in *Miller & Levine Biology,* see pages 847–851.

During Reading

Active Reading

READ, COVER, REMEMBER, RETELL Have students work with a partner to read the Behavioral Cycles section on the previous page and to record notes in their notebooks. Encourage partners to take turns reading each paragraph aloud. Then, have both partners discuss what they remember about the information in the paragraphs. Finally, have partners work together to prepare a summary of the paragraphs for their notebooks.

KEY QUESTION Read aloud the Key Question and answer on the previous page.

Ask What are some kinds of environmental changes? *(Sample answers: changes in temperature, amount of sunlight, amount of rainfall)*

COMPARE/CONTRAST Ask students to reread the Territoriality and Aggression section. Then, have them brainstorm ways that animals defend their territories with and without aggression. Tell students to record their examples in a two-column table. *(Sample answers: With Aggression—growling, charging, throwing things, attacking, biting; Without Aggression—scent marking, making scratches, vocalizing)*

Build Vocabulary

WORD ORIGINS Discuss the Latin derivations of the term *circadian rhythm*.

Say The word *society* also comes from a Latin word.

Ask What do you think that Latin word means? *(Students should predict that it refers to being with others.)* Tell students that the word *society* is from the Latin word *socius*, which means "companion."

Ask What does the phrase *an animal's evolutionary fitness* mean? *(the chance that an animal's genes will be passed on to offspring)*

KEY QUESTION Invite a volunteer to read aloud the Key Question and answer on this page.

Ask Why is territoriality considered a social behavior? *(It involves an interaction between animals.)*

Territorial Behavior A bear marks its territory with its fur and scent.

BUILD Vocabulary

circadian rhythm a behavioral cycle that happens daily

migration seasonal behavior in which animals travel from one environment to another

territory a specific area occupied and protected by an animal or group of animals

aggression threatening behavior that one individual uses to show dominance over another individual

courtship a type of behavior in which two animals communicate in a way that leads to mating

society a group of closely related individuals of the same species that work together for the benefit of the group

communication the passing of information from one individual to another

WORD ORIGINS

The word *circadian* comes from two Latin words: *circa*, which means "about," and *diem*, which means "day." Circadian rhythms have a daily cycle.

Social Behavior

When animals interact with each other, they are performing social behaviors. Social behaviors help animals get food and other resources, compete for mates, and protect their offspring. Social behaviors include territoriality, aggression, courtship, and forming of societies.

Territoriality and Aggression Many animals behave in ways that keep other animals from using limited resources. Some animals live in a specific area, or **territory,** that they defend against competitors. Territories usually contain resources such as food, water, nesting sites, shelter, and potential mates. Animals often mark their territory with their fur or scent. If another animal comes into their territory, the "owner" chases it away. Animals may show aggression when defending their territories. **Aggression** is threatening behavior that is used to show the dominance, or greater power, of one animal over another.

Courtship An animal that reproduces sexually must find and mate with another member of its species. During **courtship** behavior, members of one sex (usually males) show that they are ready to mate. Members of the opposite sex (usually females) choose which individual they will accept. Animals may produce sounds, visual displays, or chemical signals to attract mates.

In some species, courtship involves rituals. Rituals are a series of behaviors that are always performed in the same way. Many rituals include signals and responses that continue until mating happens.

Animal Societies An animal **society** is a group of animals of the same species that live close to each other and work together. Societies can protect individuals from predators. They can also improve an animal's ability to find food, protect a territory, guard young, or fight off rivals. For example, African wild dogs hunt in packs. The females in the pack may take turns guarding all the pups, while other adults hunt for prey.

Members of a society are often closely related to one another. The help of relatives may improve the evolutionary fitness of an individual, because closely related individuals share many genes. Helping a relative increases the chance that the genes of an individual will be passed on to offspring.

Social insects, such as ants, bees, and wasps, have some of the most complex animal societies. In colonies of social insects each individual has a specific job. In ant colonies, all females except the queen are workers. The queen is the only female that reproduces. Male ants function only to fertilize the queen.

Key Question How do social behaviors increase an animal's evolutionary fitness?
Social behaviors can help animals claim or defend territories or resources, choose mates, and form social groups. These behaviors can increase an animal's evolutionary fitness.

 BIOLOGY.com Search Lesson 29.2 GO • Art in Motion

ELL Focus on ELL: Extend Language

BEGINNING SPEAKERS Start a word wall by writing the vocabulary terms on separate sheets of paper and posting them in the classroom. As students read the lesson, have volunteers post definitions and pictures that explain each term. Also, encourage students to develop undercover vocabulary sheets for the chapter and use them to study and review the vocabulary terms before the chapter test.

Communication

Social behavior involves more than one individual, so it requires communication. **Communication** is the passing of information from one individual to another. Animals may communicate using chemical, visual, or sound signals. Some species also use their own kind of language. The kinds of signals a species uses depend on the sense organs it has and the environment in which it lives.

Chemical Signals Animals with good senses of smell, such as insects, fishes, and mammals, can communicate with chemicals. Some animals, such as bees, ants, and lampreys, release chemicals called pheromones (FEHR uh mohnz). Pheromones can mark a territory or signal that an animal is ready to mate. Pheromones affect the behavior of other individuals of the same species.

BUILD Connections

AN ANT SOCIETY

In a leaf-cutter ant society, only a single queen reproduces. Different groups of ants within the colony perform other tasks.

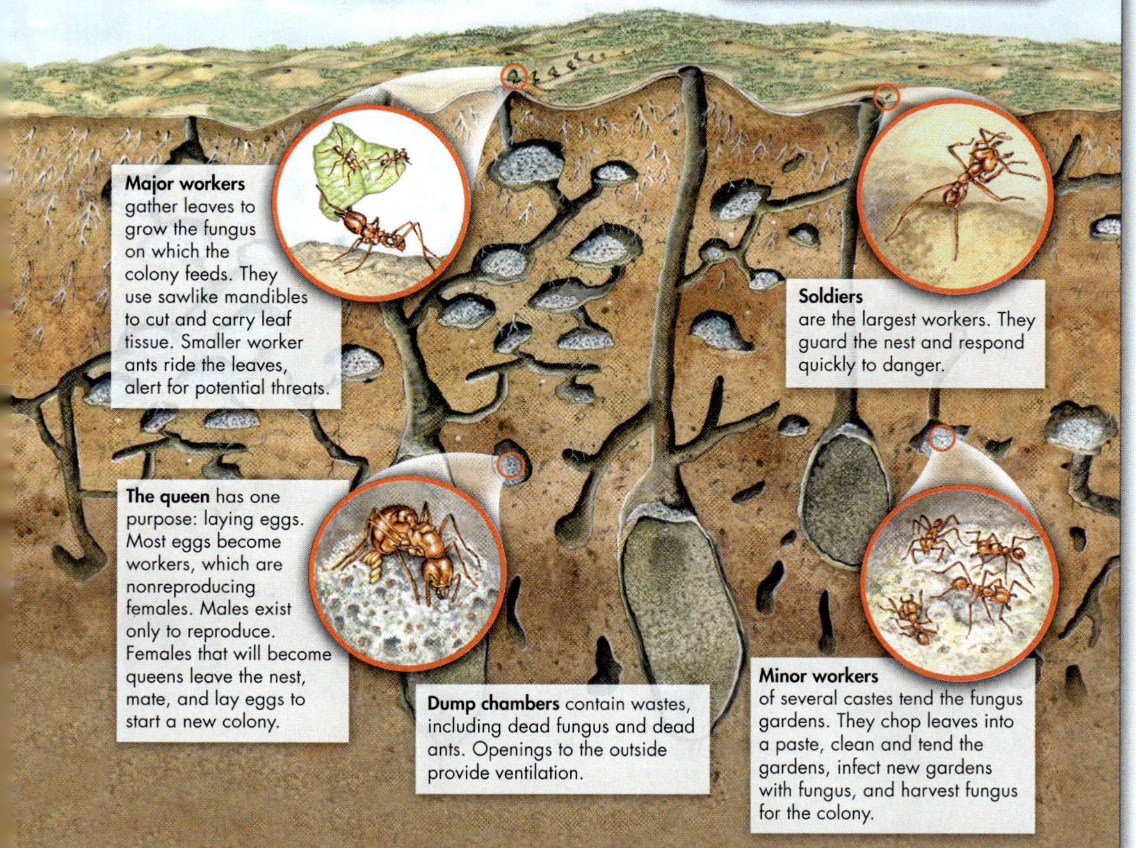

Focus on Academic Language

- pheromones

Point out that the word *pheromones* begins with the letters *ph*, which together are pronounced as an *f* sound. Have a volunteer pronounce the word.

Ask What are some other words that begin with *ph*? *(Sample answers: phone, photograph, physical, physics, pharmacy)*

Explain that the word *pheromone* comes from the Greek word *pherein*, meaning "to carry or bear," and the word *hormone*, which is a chemical substance that is produced in one part of the body but affects other parts of the body.

Ask What, then, is a pheromone? *(It is a chemical that is carried by an animal and affects other nearby animals.)*

Build Connections

VISUAL SUMMARY Have volunteers read the captions in the An Ant Society visual as you discuss the visual with the class.

Ask What is the light-colored material in the many small chambers? *(the fungus that grows on the leaves the ants bring into the colony)*

Ask How is the fungus important to the colony? *(The fungus is food for the ants.)*

USE CONTEXT CLUES Remind students that they can use context clues to clarify information.

Ask What is the sex of all the ants shown in the diagram? *(female)* How do you know this? *(In the caption about the queen, the text states that workers are nonreproducing females. All the ants in the diagram, besides the queen, are identified as types of workers. Therefore, all are female.)*

USE VISUALIZATION Have students think about a small town or village where people live.

Ask What are some similarities between human societies and ant societies? *(Sample answers: live in colonies with many living spaces connected by roads; put trash in dumps; have specialized jobs that help the members of the society survive and reproduce)*

Speed Bump

Ask What are three examples of resources that animals protect when they defend their territories? *(Sample answers: food, water, nesting sites, shelter, mates)*

Ask What kinds of information do pheromones communicate? *(the boundaries of an animal's territory and an animal's readiness to mate)*

During Reading continued

Active Reading

MAKE A GRAPHIC ORGANIZER Pair students and have each pair make a graphic organizer to help them organize the information they learned about the different forms of animal communication. Students should reread the Communication section. For each blue head, they should list the main idea of the paragraph. Under each main idea, they should list important supporting details.

KEY QUESTION Read aloud the question and answer.

Ask What does *communicate* mean? *(to pass information from one individual to another)*

Speed Bump

Ask What types of signals do animals use to communicate? *(chemical, visual, and sound signals)*

Ask In what two ways is the use of language to communicate different from the use of signals? *(Language is a complex system, and it uses rules.)*

Post Reading

ASSESSMENT TIPS Review the highlighted words with students before they begin the assessment.

For Question 6, review the meanings of *cause* and *effect*. Point out that in the question, membership in a society is the cause, and evolutionary fitness is the effect.

MYSTERY CLUE After students read the clue, remind them that animals communicate in ways besides vocalization. Encourage students to think about what an elephant might sense by putting its trunk on the ground.

Wrap-Up Activity

STUDY ANIMAL BEHAVIOR Have students work individually or in pairs to observe the behaviors of a variety of animals, such as insects or fish in a terrarium or aquarium, or birds and mammals in a park, nature preserve, or zoo. **Caution students not to touch or get too close to wild animals.** Ask students to write detailed descriptions of the observed behaviors. Then, have them classify the behaviors as social interactions and/or behavioral cycles. As a class, discuss students' classifications.

BUILD Vocabulary

language a system of communication that combines sounds, symbols, and movements according to a set of rules about sequence and meaning, such as grammar and syntax

Visual Signals Many animals have well-developed eyes that can sense shapes and colors. These animals often use visual signals to communicate. In many species, males use colorful displays to signal when they are ready to mate. Squids can change their color to send a variety of signals. Some animals, such as fireflies, even send signals using light that is made within their bodies.

Sound Signals Most animals with good senses of hearing and the ability to make noises can communicate using sound. Dolphins use sound signals to communicate in the ocean. Elephants make sounds with their feet and vocal cords that can be used to identify individuals. Elephants, and some other animals, can send messages that can be felt rather than heard.

Language **Language** is a complex system of communication that uses rules to combine sounds, symbols, and movements. Many animals, including elephants, primates, and dolphins, seem to have "words"—calls with specific meanings, such as "lions on the prowl." However, untrained animals do not seem to use a set of rules for communication like those in human language.

Key Question How do animals communicate with others in their environments? **Animals may use a variety of signals to communicate with one another. Some can also use language.**

CHECK Understanding

Apply Vocabulary

Use the highlighted words from the lesson to complete each sentence correctly.

1. Behavioral cycles that happen daily are called ________.
2. ________ is seasonal movement from one environment to another.
3. Many animals defend a ________, which is an area that contains resources.

Critical Thinking

4. **Review** Name two ways in which animal behavior is related to environmental cycles.
5. **Review** List three types of social behavior.
6. **Relate Cause and Effect** How does membership in a society increase the evolutionary fitness of its members?
7. **Write to Learn** Answer the mystery clue.

MYSTERY CLUE

When threatened by predators, adult elephants form a circle around young elephants. What may have caused this behavior in the elephants at the watering hole? (**Hint:** See p. 703.)

BIOLOGY.com Search Lesson 29.2 GO • Lesson Assessment • Art Review

Check Understanding

1. circadian rhythms
2. Migration
3. territory
4. migration and circadian rhythms
5. Accept any three: courtship, territoriality, aggression, formation of societies
6. The evolutionary fitness of animals within a society is increased by the cooperative performance of tasks such as hunting, defense, and guarding young.
7. Elephants make sounds with their feet as well as their vocal cords. They can also feel messages that cannot be heard. Perhaps the elephant was able to feel a message when she touched her trunk to the ground.

Design Your Own Lab OPEN-ENDED INQUIRY

Pre-Lab: Termite Tracks

Problem How can you determine the type of stimulus that triggers a particular response?

Materials petri dishes, paper, scissors, termites, small paintbrushes, forceps, ballpoint pens, rollerball pens, felt-tip pens

Lab Manual Chapter 29 Lab

Skills Focus Form a Hypothesis, Design an Experiment, Draw Conclusions

Connect to the Big idea Animals react to their environments as they search for food, avoid predators, and look for a mate. The survival of a species can depend on these behaviors. Some behaviors are inherited and evolve over time due to natural selection. Over many generations, behaviors that help animals survive spread through a population. Behaviors that are not adaptive become less common. In this lab, you will observe an inherited behavior of termites. Then you will design an experiment to determine the type of stimulus that triggers the behavior.

Background Questions

a. **Review** How is a behavior usually defined?

b. **Explain** How do innate behaviors, or instincts, help animals survive?

c. **Infer** You see an ant walk across the ground. A minute later, you see another ant walk in the same exact line. What type of communication do you think is taking place?

Pre-Lab Questions

Preview the procedure in the lab manual.

1. **Control Variables** Why do you think the instructions ask you to draw a figure eight rather than a straight line?
2. **Draw Conclusions** How will you decide whether a termite has a positive reaction, a negative reaction, or no reaction to a stimulus?
3. **Predict** Read pages 703–704 of your textbook. Which of the signals described could be a stimulus for the termite in this lab? Explain your answer.

BIOLOGY.com Search Chapter 29 GO

Visit Chapter 29 online to test yourself on chapter content and to find activities to help you learn.

Untamed Science Video See how the latest technologies help the explorers from Untamed Science track the movements of different animals.

Tutor Tube Instinct? Sorting out everyday and scientific behavior terms.

Art Review Identify animal communication strategies.

Art in Motion Watch social behavior in a population.

CHAPTER LAB

Pre-Lab

Remind students that a hypothesis is a testable idea and that an experiment should test only one independent variable. To help students design their experiment, you might wish to have them record their hypothesis, independent variable, and dependent variable in a table like the one below.

Hypothesis (*question*)	Independent Variable (*what I change*)	Dependent Variable (*what I see happen*)

Review with students what a stimulus is.

Ask What stimulus wakes you up in the morning? *(Sample answers: alarm going off, sun shining in the room)*

Refer students back to the section on Communication. Have them skim the section and describe one example each of stimuli involving chemical signals, visual signals, and sound signals.

Review the Background Questions and Pre-Lab Questions as a class.

Lab

Tell students they will perform the chapter lab *Termite Tracks* described in **Lab Manual B**.

SAFETY Students should handle the termites and the sharp objects with care. Be sure students wash their hands after handling the termites.

Post-Lab

SCIENTIFIC MEETING Have students present their experiment to the class. Each presentation should include the hypothesis tested, the steps of the procedure, the data in an organized data table, and the conclusions drawn from the results.

Pre-Lab Answers

BACKGROUND QUESTIONS

a. A behavior is usually defined as the way an organism reacts to stimuli in its environment.

b. Sample answer: Innate behaviors allow animals to perform tasks that are essential for survival without having to learn the task.

c. Sample answer: The ants are using chemical signals, or pheromones, to mark the path.

PRE-LAB QUESTIONS

1. Sample answer: With the figure eight, it will be easier to tell whether the termite is responding to the drawn path.
2. Sample answer: I will classify a positive reaction as movement towards the stimulus, a negative reaction as movement away from the stimulus, and no reaction as no obvious change in movement.
3. Sample answer: The termite could respond to a visual signal or a chemical signal.

NATIONAL SCIENCE EDUCATION STANDARDS

UCP II

CONTENT C.6.b

INQUIRY A.1.a, A.1.b, A.1.d, A.1.f, A.2.a

CHAPTER SUMMARY

Chapter Summary

Think Visually

USE GRAPHICS Have students use information from this chapter to complete the following concept map.

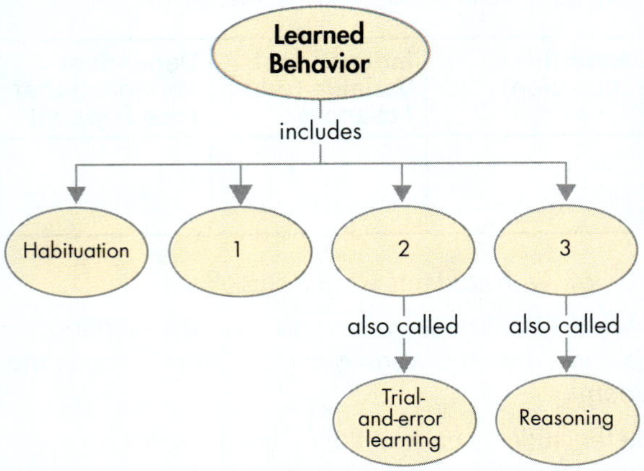

ANSWERS

1. Classical conditioning

2. Operant conditioning **3.** Insight learning

BIOLOGY.com Study Online

REVIEW RESOURCES

Tutor Tube Using actual bird songs as examples, this tutorial explains how both innate and learned behaviors are involved in complex behavior.

Data Analysis Students will analyze different kinds of animal migration data to derive migration routes and rates of travel.

Art in Motion Students explore various social behaviors, including courtship and aggression.

Art Review Students can drag and drop labels to identify different communication strategies.

Vocabulary Review The *Flash Cards* and *Crossword* are an interactive vocabulary review.

Editable Worksheets Editable pages of Study Workbook B, Lab Manual B, and the Assessment Resources Book are available online.

Lesson Overviews Lesson Overviews are based on Miller & Levine Biology. However, they are available on CD-ROM and can be edited for your class.

ASSESSMENT RESOURCES

Chapter Assessment Have students take an online version of the Chapter 29 Check Understanding.

Standardized Test Prep Students can take an online version of the Standardized Test Prep. You will receive their scores along with ideas for remediation.

Diagnostic and Benchmark Tests Use these tests to monitor your students' progress and supply remediation.

29 CHAPTER Summary

29.1 Elements of Behavior

- If an inherited behavior allows an animal to survive and reproduce, the behavior can spread through the population.
- An innate behavior is a behavior that an animal performs even though it has no previous experience with the stimulus to which it is responding. Innate behaviors are also called instincts.
- The four major types of learning are habituation, classical conditioning, operant conditioning, and insight learning.
- Many complex behaviors combine innate behavior with learning.

behavior (p. 696)
innate behavior (p. 697)
learning (p. 697)
habituation (p. 697)
classical conditioning (p. 698)
operant conditioning (p. 698)
insight learning (p. 698)
imprinting (p. 699)

29.2 Animals in Their Environments

- Many animals respond to the cycles in the environment with daily or seasonal cycles of behavior.
- Social behaviors can help animals to find and protect food, guard young, avoid predators, and find mates. Since these behaviors can increase an animal's survival and reproduction, they can improve its evolutionary fitness.
- Animals may use a variety of signals to communicate with one another. Some animals can use language.

circadian rhythm (p. 701)
migration (p. 701)
territory (p. 702)
aggression (p. 702)
courtship (p. 702)
society (p. 702)
communication (p. 703)
language (p. 704)

Assess the Big idea Evolution

Write an answer to the question below:

Q: How do animals interact with one another and their environments?

Constructed Response

Write an answer to each of the numbered questions below. The answer to each numbered question should be one or two paragraphs long. To help you begin, read the **Hints** below the questions.

1. **You are studying the behavior of a chimpanzee. You observe that the chimpanzee has learned how to use a key to open a box that contains food. How do you know whether the chimpanzee has learned this behavior using operant conditioning or insight learning?**

   Hint Look at the steps that show how a dog learns using operant conditioning on p. 698. Could a chimpanzee learn how to unlock a box using similar steps?

2. **Describe how the behaviors of animals in the northern United States help them survive when seasons change from fall to winter.**

   Hint What is the difference between the behavior of songbirds and the behavior of mammals when winter arrives?

3. **Some animals live in societies. What are the benefits of living in a group? What are the costs?**

   Hint Would resources be more available to an animal that lived in a group or less available?

UbD Transfer the Big Idea

TRANSFER Ask students to imagine that they are docents (volunteer educators) at a zoo or wildlife preserve. Tell them that their job is to talk to visitors about one of the animals in the zoo or preserve. Have each student choose one animal and prepare a talk about that animal's natural environment and behaviors. Direct students to research facts that would be of interest to a wide variety of people, including young children, teens, and adults. Encourage students to present their talks to the class.

29 CHECK Understanding

Foundations for Learning Wrap-Up

Use the table tents you made when you started the chapter to help you organize your thoughts about how animals interact with one another and with their environments.

Activity 1 Work together as a class. Write the seven main chapter topics on separate sheets of paper.

- Behavior and Evolution
- Innate Behavior
- Learned Behavior
- Complex Behaviors
- Behavioral Cycles
- Social Behavior
- Communication

Choose seven desks in the classroom. Put one of the sheets on each of the desks.

Have everyone place each of his or her table tents on the desk that matches its topic. Work together to make sure all the table tents are in the right place. Then, review each topic in small groups by looking at the table tents.

Activity 2 Divide up into pairs. Pick a topic from the chapter, but do not tell your partner what it is. Make a new table tent to illustrate that topic. Use mainly pictures on your table tent—just a few words. Be creative! Do not use the same pictures that are found in your textbook.

Tape the table tent together and give it to your partner. Have your partner guess what the table tent illustrates without looking at the textbook. Switch places and repeat the activity.

29.1 Elements of Behavior

Understand Key Concepts

1. The way an organism reacts to stimuli in its environment is called
   - **a.** behavior.
   - **b.** learning.
   - **c.** conditioning.
   - **d.** imprinting.
2. Animal behaviors can evolve through natural selection because
   - **a.** what an animal learns is added to its genes.
   - **b.** all behavior is completely the result of genes.
   - **c.** all behavior is completely the result of environmental influences.
   - **d.** genes that influence behaviors that increase fitness can increase in frequency from one generation to the next.
3. Which kind of behavior is shown below?

   - **a.** insight learning
   - **b.** imprinting
   - **c.** classical conditioning
   - **d.** operant conditioning
4. How can habituation contribute to an animal's survival?

Think Critically

5. **Infer** When a baby smiles, her mother picks her up and cuddles her. Over time, the baby smiles more and more often. Explain what type of learning the baby is showing.

Test-Taking Tip

Use Time Wisely When you are taking a test, answer all the questions you are sure about first. Put a mark next to the questions that you need to go back to. Skip any questions that you really cannot answer. Only go back to those questions if you have time.

Foundations for Learning Wrap-Up

Have students take out the table tents they prepared while reading the chapter. Provide additional sheets of paper and class time for the activities. Explain the activities, and have students complete them.

Assessment Tips

Pair students and have them work together to answer the questions on this page. Question 3 asks students to interpret a visual. Questions 5 and 12 require students to draw a conclusion based on given facts and their understanding of animal behavior. Question 6 requires students to synthesize information based on their knowledge. Questions 11 and 12 ask for interpretation of a graph.

Assess

Big idea Invite a volunteer to read the Big Idea and the question on the opposite page.

Ask How can behavior play a role in the evolution of an animal species? *(If a behavior helps an animal to survive and reproduce, the behavior can spread throughout the species.)*

Ask What is an example of animals interacting with other animals? *(Sample answers: living in societies, territoriality, aggression, courtship)*

Ask What is an example of animals interacting with their environment? *(Sample answers: building nests, hibernating, migrating, daily cycles)*

CONSTRUCTED RESPONSE Explain the activity and *Hints* carefully. Review how to make an outline to organize the topics to include in a paragraph.

Write Question 1 on the board and read it aloud. Underline *operant conditioning* and *insight learning*.

Read the hint.

Ask What does operant conditioning always involve? *(trial and error, or repetition)* What does insight learning involve? *(reasoning)*

Say The chimpanzee had to use a key to open the box.

Ask Is using a key a random action, like the dog brushing its tail against a bell? *(no)*

Model writing the first one or two sentences. *(Sample: The chimpanzee most likely used insight learning to open the box.)*

For Question 2, pair students and have them use the hints and your modeling to answer the question. Have students share their paragraphs with the class.

For Question 3, have students write the paragraphs independently for homework or as an anchor activity.

Answers

Lesson 29.1

UNDERSTAND KEY CONCEPTS

1. a
2. d
3. b
4. Habituation enables animals to ignore irrelevant stimuli, thus conserving energy for real threats.

THINK CRITICALLY

5. Operant conditioning; smiling is reinforced by the reward of cuddling.

ASSESSMENT

Lesson 29.2

UNDERSTAND KEY CONCEPTS

6. c

7. d

8. Sample answer: If a group searches for food, it is more likely to find it than if an individual searches; a large group is safer from predators than an individual is.

9. Animals exhibit aggression or threatening behavior to establish a territory and defend it when a rival tries to claim it.

THINK CRITICALLY

10. No; migration is seasonal movement from one environment to another. The movement of these animals is not related to changing seasons.

Connecting Concepts

11. The squirrel's body temperature fluctuates during winter months and is stable in warmer months.

12. In winter, it is in a state of dormancy.

CHAPTER MYSTERY After students have read the Chapter Mystery, discuss the elephants' use of vibrations to communicate.

Ask What kinds of information do elephants communicate with these vibrations? *(greetings, locations of food, possible danger)*

Ask How does living in a society benefit an elephant? *(Elephants help one another survive. They help one another defend against predators.)*

29 CHECK Understanding

29.2 Animals in Their Environments

6. Each year, a bird called the redstart travels from its winter home in South America to its nesting area in New York. This behavior is called
- **a.** imprinting.
- **b.** competition.
- **c.** migration.
- **d.** courtship.

7. Which of the following is NOT social behavior?
- **a.** courtship
- **b.** territoriality
- **c.** hunting in a pack
- **d.** operant conditioning

8. Describe two ways in which social behavior can benefit an animal.

9. Explain how aggression and territorial behavior are related.

Think Critically

10. Apply Concepts Because a highway has been constructed through a forest, many of the animals that once lived there have had to move to a different wooded area. Is this migration? Explain.

Connecting Concepts

In winter, some mammals enter into a state of dormancy. Dormancy is an energy-saving adaptation in which metabolism slows down, causing body temperature to become lower. The graph below tracks a ground squirrel's body temperature over the course of a year.

11. Explain Describe the pattern that you observe.

12. Infer What can you infer about the squirrel's behavior at different times of the year?

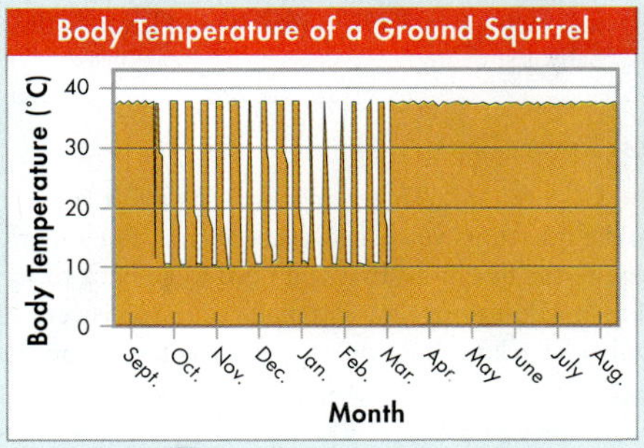

solve the CHAPTER MYSTERY

ELEPHANT CALLER ID?

Elephants can communicate with loud calls and vibrations that can travel long distances. Elephants send these vibrations by drumming on the ground with their feet. They also make low rumbling calls that contain very low-frequency vibrations. The air vibrations can travel up to 16 kilometers. The ground vibrations can travel up to 32 kilometers.

Elephants sense these vibrations with specialized pads of fat in their feet. They can also detect the vibrations with receptor cells in their feet and trunks. By standing still and pressing their feet and trunks to the ground, they can sense the vibrations better. Some patterns of vibrations can identify individuals—like caller ID for elephants!

The vibrations contain greetings, locations of food, and warnings of danger. The elephants at the watering hole had sensed a message from another herd of elephants: "Warning! Lions!"

1. Compare and Contrast How do vibrations in the ground compare with sounds in the air as a way to communicate over long distances?

2. Form a Hypothesis Sometimes researchers play sounds and vibrations from elephant groups that live far from Etosha. The elephants at Etosha do not always respond. Why might this be the case?

3. Connect to the Big Idea How do elephants' responses to low-frequency vibrations in their environment affect their survival?

UntamedScience **Never Stop Exploring Your World.** Finding the solution to the Elephant Caller ID? mystery is only the beginning. Take a video field trip with the ecogeeks of Untamed Science to see where the mystery leads.

CHAPTER MYSTERY ANSWERS

1. Ground vibrations are more effective than air vibrations as a means of long-distance communication because ground vibrations travel twice as far.

2. Sample answer: Since the groups are separated by a long distance, they are probably not related and have developed different signals to use in communication.

3. Elephants' responses to low-frequency vibrations help them survive possible attack by predators by allowing them to leave an area before the predators arrive.

Untamed Science To learn more about how scientists use remote sensing to study animal behavior, have students watch **When Animals Go Digital**.

Standardized Test Prep

Multiple Choice

1. A rat that learns to press a button to get food is exhibiting
   A insight learning. C classical conditioning.
   B operant conditioning. D habituation.

2. A dog that always salivates at the ringing of a bell is exhibiting
   A insight learning. C classical conditioning.
   B operant conditioning. D habituation.

3. A chimpanzee that stacks boxes in order to reach a banana hanging from the ceiling is showing
   A insight learning. C classical conditioning.
   B operant conditioning. D habituation.

4. A bird that stops responding to a repeated warning call when the call is not followed by an attack is showing
   A insight learning. C classical conditioning.
   B operant conditioning. D habituation.

5. Which kind of behavior does NOT involve learning?
   A habituation C imprinting
   B trial and error D instinct

6. A male three-spined stickleback fish will attack male red-bellied sticklebacks and models of fishes that have a red underside. It will not attack males or models lacking a red underside. What can you conclude from the three-spined stickleback's behavior?
   A The stimulus for an attack is a red underside.
   B The stimulus for an attack is aggression.
   C The stimulus for an attack is the presence of a fish with red fins.
   D The stimulus for an attack is the presence of a fish model.

7. Which of the following is NOT an innate behavior?
   A a dog looking for its food dish
   B a baby mammal sucking milk
   C a worm moving away from bright light
   D a spider spinning a web

Questions 8–9

A researcher observed sedge warblers during breeding season. She charted the number of different songs a male bird sang compared to the time it took him to pair with a mate. The graph shows her data.

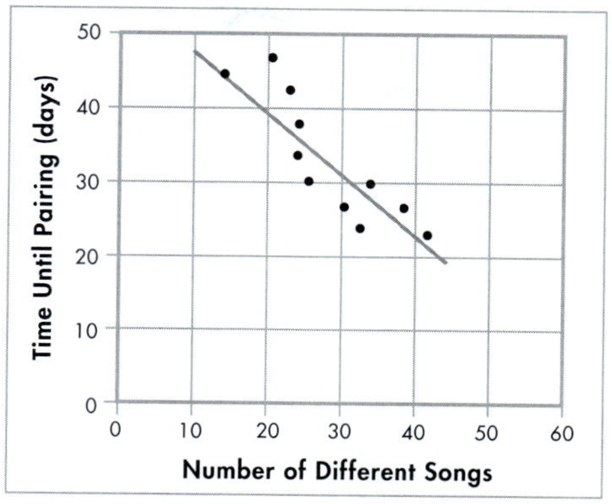

8. The researcher was trying to find out whether there is a correlation between
   A the number of a male bird's songs and the number of offspring.
   B the number of a male bird's songs and his attractiveness to females.
   C a male's age and when he mates.
   D a female's age and when she mates.

9. What can you conclude based on the graph?
   A Males prefer females that do not sing.
   B Females prefer males that do not sing.
   C Males prefer females with a larger number of songs.
   D Females prefer males with a larger number of songs.

Open-Ended Response

10. How does defending a specific territory benefit an animal?

If You Have Trouble With . . .

Question	1	2	3	4	5	6	7	8	9	10
See Lesson	29.1	29.1	29.1	29.1	29.1	29.1	29.1	29.2	29.2	29.2

Test-Taking Tip

READ THE TEST CAREFULLY

Remind students always to read the directions carefully while taking a test. Explain that certain words such as *except* and *not* can change the meaning of a sentence and that if such words are overlooked, students may select the wrong answer. Advise students to think about what the question is asking before reading each answer choice. Also advise eliminating from consideration answers students know are wrong before choosing the answer they think is correct. Recommend that students reread the question and the answer chosen as correct before moving on to the next question.

Before the Test

Review with students the Test-Taking Tip.

Point out the word *NOT* in Questions 5 and 7. Also point out that the answer to Question 9 can be found by studying the graph.

Remind students that an open-ended response means that they'll have to write a paragraph to answer the question. Have them paraphrase the question in their notebooks, and remind them about the steps in writing a paragraph.

Answers

1. B
2. C
3. A
4. D
5. D
6. A
7. A
8. B
9. D
10. Defending a territory reduces competition for resources within the territory, such as food, water, and shelter, and helps to ensure survival for the animal defending the area and its offspring.

Plan Ahead

Have students preview the Unit 7 Project a few days before they begin work on their exhibit. Suggest they "visit" zoos online, especially to see how exhibits have been designed to display a wide variety of animals. Consider having students work in small groups. Individual students can take responsibility for creating an exhibit of a particular group of animals, and then work with other group members to show similarities and differences among the animal groups.

Materials poster board, modeling clay, materials for papier-mâché, aluminum foil, cardboard, toothpicks, string, yarn, tape, glue, colored markers, additional art materials

Monitor the Project

Suggest students or groups begin their exhibit design by making lists of animal groups, specific animals, and life processes that should be included in the exhibit. As students work, ask groups or individuals questions that will help them organize and design their exhibits.

Ask What examples of animals are you including in your exhibit to represent different animal groups?

Ask How are you showing similarities and differences in the ways animals carry out basic life processes?

Project Assessment

Make sure students use the rubric and reflection questions to assess their work. Then, use the rubric to assign a final score. Note that it is important to value the creativity of students' work as well as the content when you score their projects. If desired, talk with students about any differences between their self-assessment scores and your assigned score.

UNIT 7 Animals

Unit Project

Zoo Exhibit

Have you ever been to the zoo? Going to the zoo is a great way to see and learn about living things from around the world. Imagine you work for a zoo in your local city and have been asked to set up a new exhibit entitled *The Diversity of Animals.*

Your Task Design an exhibit that highlights the different groups of animals in the animal kingdom and their characteristics.

Be sure to
- show the differences in shapes and forms of animals.
- show the similarities/differences in the ways in which groups of animals carry out basic life processes (circulation, respiration, excretion, movement, response, feeding).
- design the exhibit so that it is clear and engaging.

Reflection Questions

1. **Score your exhibit using the rubric below. What score did you give yourself?**
2. **What did you do well in this project?**
3. **What needs improvement?**
4. **What aspects of another group's exhibit did you like? Why?**

Assessment Rubric

Score	Scientific Content	Quality of Exhibit
4	Exhibit reveals a thorough understanding of animals and the ways they carry out life processes.	Exhibit is very organized, educational, and engaging to visitors.
3	Exhibit reveals an adequate understanding of animals and the ways they carry out life processes.	Exhibit is organized and clear, but it could be more engaging to visitors.
2	Exhibit reveals a limited understanding of animals and the ways they carry out life processes.	Exhibit needs improvement in clarity and organization. It is somewhat engaging.
1	Exhibit reveals significant misunderstandings about animals and the ways they carry out life processes.	Exhibit is very unclear and disorganized. It is not engaging.

21st Century Skills

To be successful in the 21st century, students need skills and learning experiences that extend beyond subject area mastery. The Unit 7 Project helps students build the following 21st Century Skills: *Communication Skills; Critical Thinking and Systems Thinking; Creativity and Intellectual Curiosity; Self-Direction;* and *Accountability and Adaptability.*

FOCUS ON INTERPERSONAL AND COLLABORATIVE SKILLS Extend this Unit Project by having small groups of students work together to design a contemporary zoo in which the animals are kept and exhibited responsibly. Ask students to discuss how representative animals from each of the major animal groups could be responsibly kept in captivity in ways that would allow them to thrive.

For more practice building 21st Century Skills, see The Chapter Mystery pages in **Study Workbook A.**

NATIONAL SCIENCE EDUCATION STANDARDS

UCP I, II, IV, V

CONTENT C.5.d, C.6.a, C.6.b

INQUIRY A.2.a

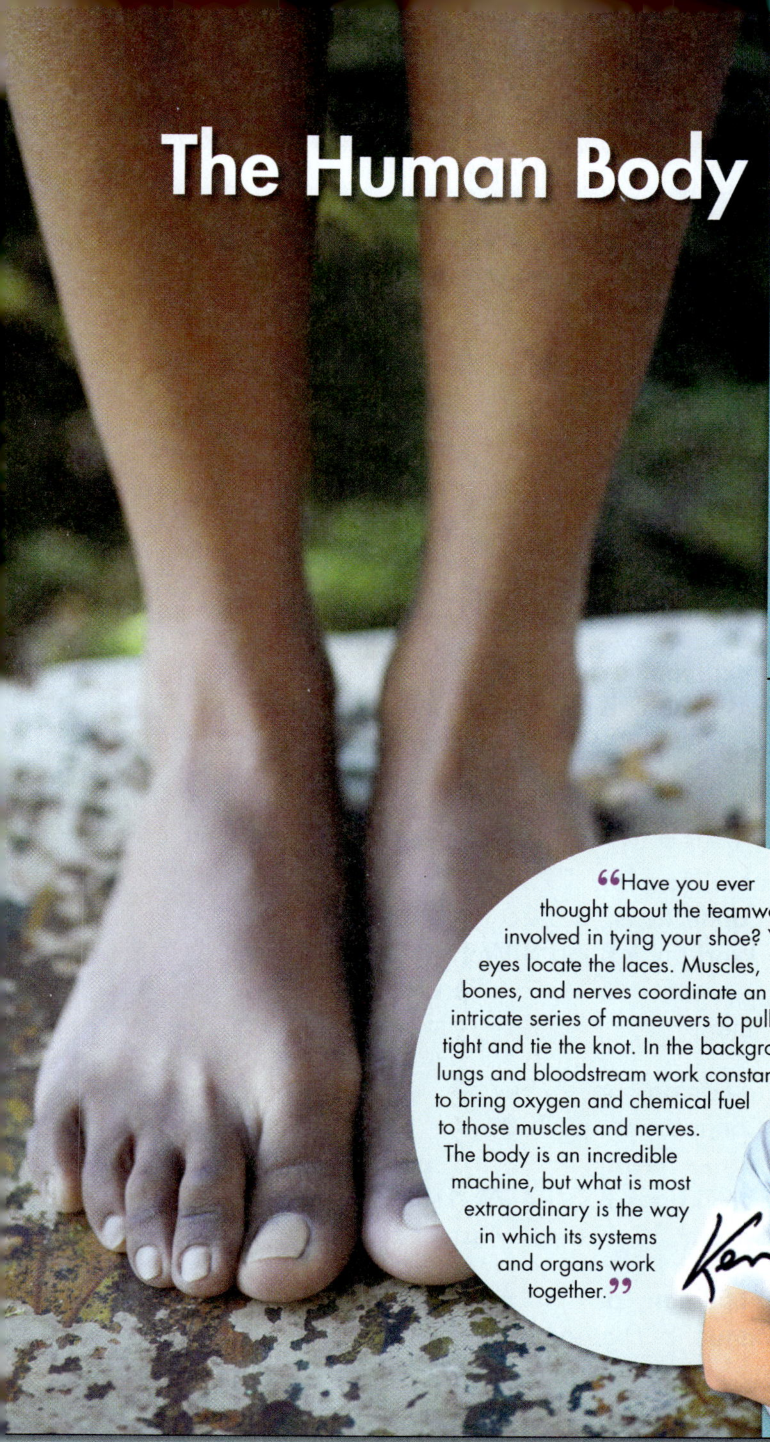

The Human Body

UNIT 8

Chapters

INTRODUCE the **Big ideas**

- **Homeostasis**
- **Structure and Function**

"Have you ever thought about the teamwork involved in tying your shoe? Your eyes locate the laces. Muscles, bones, and nerves coordinate an intricate series of maneuvers to pull them tight and tie the knot. In the background, lungs and bloodstream work constantly to bring oxygen and chemical fuel to those muscles and nerves. The body is an incredible machine, but what is most extraordinary is the way in which its systems and organs work together."

Ken Miller

Dear Colleague,

I still remember how it happened. I saw one of my teammates, took a step back, and looped a perfect pass over a defender right into his arms. It was only a touch football game, but it felt great to watch the ball spiral through the air—except for one little thing. One of guys on the other team tried to block the pass, and even though he failed, his arm collided with my wrist as I followed though.

It didn't hurt much at first. That night, I had to take a couple of aspirin tablets to dull the pain. But in the morning, I couldn't even move the hand. I had broken my navicular, a small bone in the palm just below the thumb. As he wrapped a cast around my hand, the doctor warned me that the navicular heals slowly. "Say goodbye to your hand," he said. "You won't see it for four months."

Unfortunately, he was right. But, during those four months, the most amazing things happened. The pain—my nervous system's warning that something was broken—subsided. Bone cells tore up the matrix on both sides of the break, and then began to lay down a new matrix joining both parts of the bone back together. The digestive system extracted calcium from my food, while the endocrine and excretory systems made sure that the right amount of that calcium was put into the circulatory system. My skeletal system used the calcium, together with phosphate, to build a new bone matrix. In short, the body healed itself. Even my nervous and muscular systems adjusted, getting used to the extra weight from the cast on my right hand.

Homeostasis is a complex process, but at its core is a simple idea. The body takes care of itself. It adjusts to changes in the environment, manages adversity, keeps things on an even keel, and finally, as I found out when the cast came off, it can even heal itself. If we can teach these things to our students, we can help them learn to appreciate one of the most remarkable things about their lives—the wonder of the human body and the role that homeostasis plays in keeping us alive.

Ken Miller

CHAPTER PLANNER 30 Digestive and Excretory Systems

Chapter Contents	NSES	Time	Core Resources
Chapter Preview			**Student Edition,** pp. 712–713 **Chapter Mystery,** p. 713 **Foundations for Learning,** p. 713
30.1 Organization of the Human Body Organization of the Body • Homeostasis	I, II, IV, V, C.1.a, C.1.f, C.5.d	1 period 1/2 block	**Student Edition,** pp. 714–718 **Build Connections,** p. 716 **Study Workbook B** 30.1 Worksheets **Biology.com** *Art Review:* Human Body Systems • *Tutor Tube:* Working Together to Stay Balanced • 30.1 Lesson Assessment
30.2 Food and Nutrition Food and Energy • Nutrients • Nutrition and a Balanced Diet	V, B.2, B.3, C.1.b, C.5.a, C.5.c, F.1	1/2 period 1/4 block	**Student Edition,** pp. 719–722 **Study Workbook B** 30.2 Worksheets **Biology.com** 30.2 Lesson Assessment
30.3 The Digestive System Functions of the Digestive System • The Process of Digestion • Absorption and Elimination	I, V, C.1.b, C.5.d	2 periods 1 block	**Student Edition,** pp. 723–728 **Build Connections,** p. 726 **Study Workbook B** 30.3 Worksheets **Biology.com** *Art in Motion:* Peristalsis • 30.3 Lesson Assessment
30.4 The Excretory System Structures of the Excretory System • Excretion and the Kidneys • The Kidneys and Homeostasis	I, III, IV, V, A.1.c, A.2.a, C.5.d, F.1	1 1/2 periods 3/4 block	**Student Edition,** pp. 729–733 **Build Connections,** p. 730 **Inquiry Into Scientific Thinking,** p. 733 **Study Workbook B** 30.4 Worksheets **Biology.com** 30.4 Lesson Assessment
Chapter Pre-Lab	A.1.b, A.2.a, C.4.c, C.5.d, F.1	1 period 1/2 block	**Student Edition,** p. 734 **Lab Manual B** *Digestion of Dairy Products*

Tools for ELL Instruction

ELL Handbook describes teaching strategies to make *Miller & Levine Biology: Foundation Edition* more accessible to ELL students.

Spanish Study Workbook is a Spanish translation of Study Workbook A.

Multilingual Glossary is the *Miller & Levine Biology* glossary translated into ten languages.

Tools to Challenge Students

Study Workbook A includes more challenging worksheets and activities such as Chapter Mystery activities that reinforce 21st Century Skills.

Lab Manual A labs are geared toward on-level students. The Lab Manual includes fifteen labs in addition to the Chapter Labs.

Biology.com Data Analysis activities offer students a way to practice data gathering and analyzing skills.

Visual Quizzes in the Assessment Resources Book are geared toward on-level students and use visuals from *Miller & Levine Biology.*

Additional Resources

Biology.com Untamed Science Video • Flash Cards

Study Workbook A 30.1 Worksheets
Lab Manual B *Hands-On Activity:* Maintaining Temperature
Spanish Study Workbook 30.1 Worksheets

Study Workbook A 30.2 Worksheets
Lab Manual B *Data Analysis:* The American Diet
Spanish Study Workbook 30.2 Worksheets
Biology.com *Data Analysis:* Balance Your Virtual Diet

Study Workbook A 30.3 Worksheets
Spanish Study Workbook 30.3 Worksheets
Assessment Resources Book Visual Quiz

Study Workbook A 30.4 Worksheets
Spanish Study Workbook 30.4 Worksheets
Assessment Resources Book Visual Quiz

Lab Manual A *Digestion of Dairy Products* • *Reducing Excess Gas*

Chapter Review

Student Edition Chapter Summary, p. 735
Study Workbook B Chapter 30 Review
Transparencies, pp. 332–341
Biology.com Untamed Science Video • Editable Worksheets of Study Workbooks A and B and Lab Manuals A and B • Chapter 30 Flash Cards and Match It

Untamed Science DVD • Classroom Resources CD (includes lesson presentations and editable worksheets)

Chapter Assessment

Student Edition Check Understanding, pp. 736–739
Study Workbook B Chapter 30 Taking a Standardized Test
Assessment Resources Book Chapter 30 Test B
Biology.com Chapter 30 Assessment • Editable Worksheets of Chapter 30 Visual Quizzes* and Chapter 30 Tests A and B

Exam*View* *Assessment Suite* • Classroom Resources CD (includes lesson presentations and editable worksheets)

Time: 1 period, 1/2 block

* Visual Quizzes are based on visuals from *Miller & Levine Biology.* Review carefully to make sure they are appropriate for your class.

Pressed for Time?

Preview the Chapter Preview the visuals Types of Tissue and The Digestive System and have students read the first Key Question for Lesson 30.4.

Cover the Chapter Quickly Have students read *Organization of the Body* in Lesson 30.1, all of Lesson 30.3, and *Structures of the Excretory System* and the first paragraph of *Excretion and the Kidneys* in Lesson 30.4.

Assess Assign Question 1 in the 30.1 Check Understanding, the 30.3 Check Understanding, and Question 1 in the 30.4 Check Understanding. In the Chapter 30 Check Understanding, assign Questions 1, 3, 4, 10–14, and 16–18.

CHAPTER 30

30 Digestive and Excretory Systems

Big idea **Homeostasis**

Q: How are the materials that enter and leave your body related to the processes that maintain homeostasis?

Food sellers display their goods at a floating food market on Dal Lake in northern India.

Get the Picture

Point out that some places in the world have large supermarkets or grocery stores where food is sold. However, in many places farmers may sell their goods directly to customers in markets, like this floating food market.

Ask What can you infer about the diet of people who buy the food in the photo? *(It is high in plant products, such as vegetables.)* How are the foods in the photo different from the foods in a supermarket in your neighborhood? *(Sample answers: There is less variety. There are no meat or dairy products.)* What do the foods in the photo have in common with foods in a supermarket? *(Sample answer: They contain many of the same nutrients, such as proteins and carbohydrates.)*

Activate Prior Knowledge

Ask What is needed in the habitat of any organism? *(food, water, shelter, and oxygen)*

Point out that food is essential to organisms. Have students recall that the chemical energy in food is changed to the energy needed for cellular processes during cellular respiration. Remind students that plants make their own food using sunlight.

UbD Connect to the Big Idea

Invite a volunteer to read aloud the chapter number and the chapter title. Read the Big Idea and the question. Emphasize the question, **How are the materials that enter and leave your body related to the processes that maintain homeostasis?**

Set the Purpose

PREVIEW AND PREDICT Draw the graphic organizer on the next page on the board. Guide students to the INSIDE feature on the next page, and invite a volunteer to read each lesson number and title.

Ask How do the chapter title and the Big Idea question relate to the lesson questions in the graphic organizer?

Remind students to search for the answers as they read and also to ask themselves what homeostasis is and why a healthy body depends on it.

NATIONAL SCIENCE EDUCATION STANDARDS

UNIFYING CONCEPTS AND PROCESSES

I, II, III, IV, V

CONTENT

B.2, B.3, C.1.a, C.1.b, C.1.f, C.4.c, C.5.a, C.5.c, C.5.d, F.1

INQUIRY

A.1.b, A.1.c, A.2.a

INSIDE:

CHAPTER MYSTERY

THE TELLTALE SAMPLE

All players had to get a physical on the first day of football practice. "Please provide me with a urine sample," the doctor said. Each student was handed a plastic cup and sent to the restroom. The players had no idea how much could be learned from their urine.

Philip and Seth handed in their samples. They were told right away to go home and drink plenty of water before practice the next day. The next day, Andrew was told he might have diabetes and should see his doctor. Several days later, another player was dropped from the team. He had violated the school's well-known antidrug policy. How was all this information learned from a urine sample?

Read for Mystery Clues As you read this chapter, look for clues to help you predict what can be learned by examining what leaves your body. Then, solve the chapter mystery.

FOUNDATIONS for Learning

Think of organs as building blocks for an organ system. Before you read the chapter, make two layered pinwheels, one for the digestive system and one for the excretory system. Use two sheets of paper. Write the name of the system at the top of the first sheet. To make the tabs of the pinwheel, fold the second sheet in half twice. Then cut the paper diagonally and along each fold, as shown below. Tape the pinwheel to the first sheet of paper. As you read the chapter, write the name of an organ from the system on each tab. Then write the functions of the organ under the tab. At the end of the chapter are two activities that use the pinwheels to help answer the question: How are the materials that enter and leave your body related to the processes that maintain homeostasis?

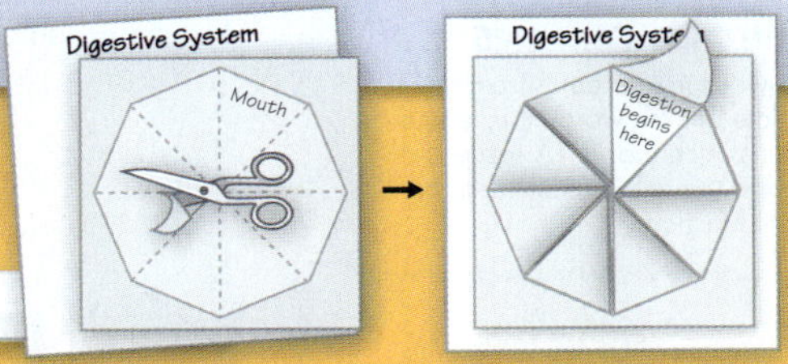

- Untamed Science Video
- Chapter Mystery

Chapter 30 Big Idea: Homeostasis

Chapter 30: How are the materials that enter and leave your body related to the processes that maintain homeostasis?

- **30.1:** How is the human body organized and regulated?
- **30.2:** Why are the essential nutrients important?
- **30.3:** How does the human body convert food into useful molecules?
- **30.4:** How does the human body get rid of wastes?

Foundations for Learning

You may want to do this activity as a class, or in small groups, to make sure students follow the directions correctly.

Ask What does each tab show you about the relationship between organs and organ systems? *(The pinwheels show how the organs in a system relate to each other.)*

As students read the chapter, have them think about how the information in the pinwheels can be arranged to also show the relationship between the digestive system and the excretory system.

In the video **Drat . . . I Smell Scat!** the Untamed Science crew gets to the bottom of a stinky mystery: What can we find out about animals from what they leave behind?

Have students read the Chapter Mystery. For each of the individual cases described in the Chapter Mystery, have students predict what might have been discovered in the urine sample. Then, have students predict in general how a urine sample could be used to learn about what has entered or has been formed by the body.

Digital Support

Extend your reach by using the digital assets offered at **Biology.com.**

LESSON 30.1

Pre-Reading

Objectives

30.1.1 Describe how the human body is organized.

30.1.2 Explain homeostasis.

Preview the Pages

1. Before reading about body organization, discuss other analogous situations where more specific items are organized into a whole. An example might be the produce section in a grocery store, where the section is divided into fruits and vegetables, and each of those categories is made up of specific types of fruits and vegetables.
2. Students might not know where the body parts mentioned in the text are located. Have a body chart or model available so that these parts can be identified. If one is not available, identify the parts using the Human Body Systems diagram.
3. Have students identify the vocabulary terms in the lesson and use context clues to develop operational definitions for the terms.

Activate Prior Knowledge

Elicit from students what they already know about the structures and functions of the circulatory, respiratory, muscular, and skeletal systems.

Build Understanding

PREVIEW VISUALS Tell students that the functions listed for each system in the Build Connections feature often provide clues to how systems interact with each other. Refer students to Workbook B to find out more about obtaining information from visuals.

NATIONAL SCIENCE EDUCATION STANDARDS

UNIFYING CONCEPTS AND PROCESSES

I, II, IV, V

CONTENT

C.1.a, C.1.f, C.5.d

For the corresponding lesson in ***Miller & Levine Biology,*** see pages 862–867.

30.1 Organization of the Human Body

Key Questions

How is the human body organized?

What is homeostasis?

BUILD Understanding

Previewing Visuals Study the Build Connections feature on p. 716. For each system, describe how it interacts with at least one other system.

In Your Workbook Go to your workbook for more about previewing visuals.

BUILD Vocabulary

epithelial tissue
a type of tissue that lines the interior and exterior body surfaces

connective tissue
a type of tissue that provides support for the body and connects its parts

nervous tissue
a type of tissue that transmits nerve impulses throughout the body

muscle tissue
a type of tissue that makes movements of the body possible

WORD ORIGINS

The word *tissue* comes from the Middle English word *tissu,* which means "rich fabric." A tissue is a "fabric" of cells that perform the same function.

Organization of the Body

How does a softball player field a ground ball? She uses her eyes and brain to figure out where to move. Her bones support her muscles as she moves. Her lungs absorb the oxygen her cells use during cellular respiration. Her brain monitors the ball and sends signals to guide her hand to catch it. An easy out, a routine play. How do all of these parts of the body work together to allow the softball player to field the ball?

Every cell in the human body works alone, yet also depends on the rest of the body. How can so many single cells and parts work together so efficiently? One way to answer this question is to study the organization of the human body.

Cells A cell is the basic unit of structure and function in living things. Remember that single cells in multicellular organisms are usually specialized. Bone cells, blood cells, muscle cells, and other types of cells each perform a specific function.

Tissues A group of cells that performs a single function is called a tissue. There are four basic types of tissue in the human body—epithelial, connective, nervous, and muscle.

▶ ***Epithelial Tissue*** The tissue that lines the interior and exterior surfaces of your body is **epithelial tissue.** Your skin and the lining of your stomach are examples of epithelial tissue.

▶ ***Connective Tissue*** The type of tissue that provides support for the body and connects its parts is called **connective tissue.** Fat cells, bone cells, and blood cells are types of connective tissue. Many connective tissues produce a protein called collagen. Collagen is a fiberlike protein that makes tissues strong and durable. Collagen helps the tissue keep its shape even when it is under pressure.

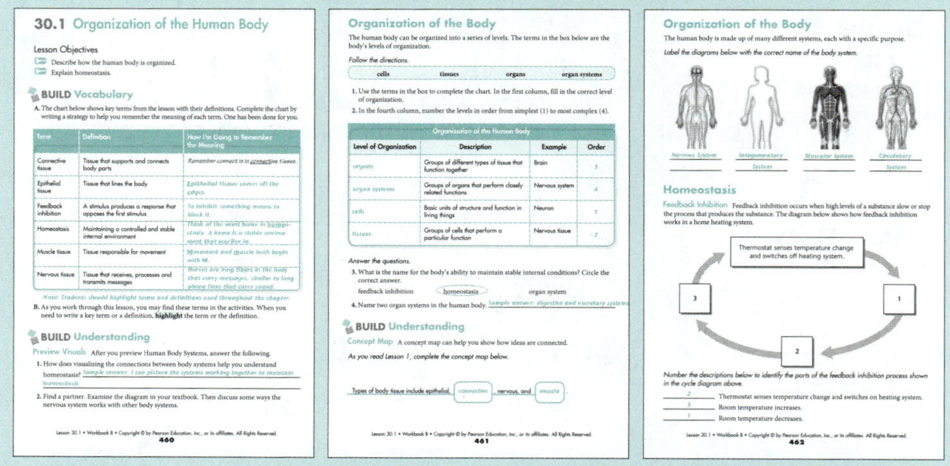

Types of Tissue The four major types of tissue are epithelial tissue, connective tissue, nervous tissue, and muscle tissue.

	Epithelial Tissue	Connective Tissue	Nervous Tissue	Muscle Tissue
FUNCTIONS	Protection, absorption, and excretion of materials	Binding of epithelial tissue to structures, support, and transport of substances	Receiving and transmitting nerve impulses	Voluntary and involuntary movements
LOCATIONS	Skin, lining of digestive system, certain glands	Under skin, surrounding organs, blood, bones	Brain, spinal cord, and nerves	Skeletal muscles, muscles surrounding digestive tract and blood vessels, the heart

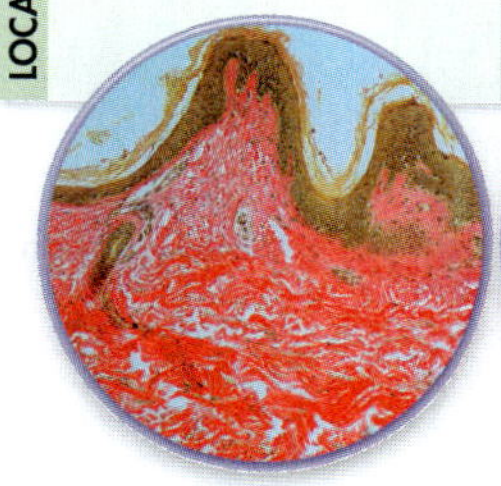
LM 65×

LM 280×

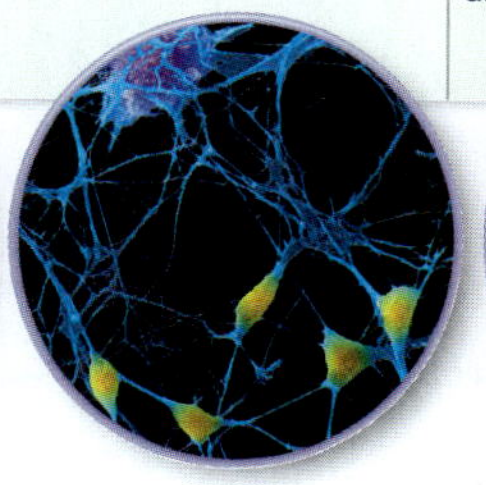
SEM 295×

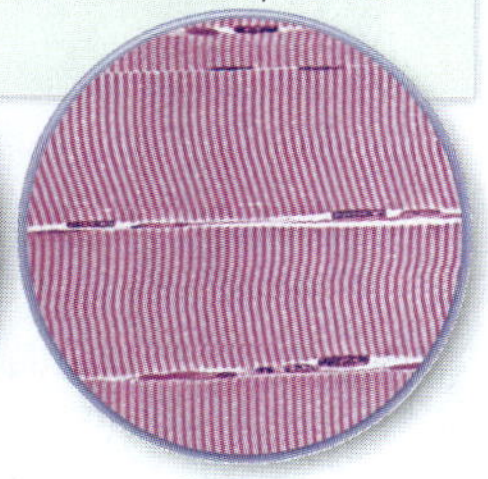
LM 275×

▶ ***Nervous Tissue*** Nerve impulses are transmitted all over the body by **nervous tissue.** Neurons and glial cells are examples of nervous tissue. Neurons carry the nerve impulses, while glial cells surround and protect the neurons.

▶ ***Muscle Tissue*** Movements of the body are possible because of **muscle tissue.** Some muscles you can control, some you cannot control. For example, you can control the muscles in your arms and legs. But you cannot control the muscles in the pupils of your eyes.

Organs An organ is a group of different types of tissue that work together. An organ may perform a single function or a group of related functions. Your eye is an organ that has all four types of tissues. All four tissues work together to enable us to see.

Organ System A group of different organs that perform closely related functions is called an organ system. The brain and spinal cord are organs of the nervous system. The organ systems work together to keep the body stable. The organ systems are shown on the next page.

Key Question How is the human body organized?
The human body is organized into four levels: cells, tissues, organs, and organ systems.

Hands-On Learning

OBSERVE CELLS Have small groups of students observe and compare cells.

1. Provide each group with a microscope and prepared microscope slides of tissues. Muscle and nerve cells are readily available and are good for comparison.
2. Review the safe, proper use of a microscope and slides with students.
3. Have students observe the slides, then develop group answers to the following questions.

   Ask What do you notice about all the cells in each type of tissue? *(The cells are alike or very similar.)*

   Ask How do the cells of one tissue compare to the cells of another tissue? *(The cells differ in shape and size.)*
4. Point out that cells differ in appearance from tissue to tissue because cells differ in function from tissue to tissue.

During Reading

Build Vocabulary

WORD ORIGINS Ask Why do you think the words *facial tissue* and *wrapping tissue* have *tissue* in their names? *(They are thin, fabric-like materials that perform a certain function.)*

Point out that even though the scientific use of the word *tissue* specifies a group of similar cells, both the scientific use and the common use of the word have the same word origin.

Active Reading

USE VISUALS Refer students to the visual Types of Tissues. Point out that the photos of tissue show examples of the type of tissue listed, but that the appearance of this tissue might vary, depending on its function and its location in the body.

Ask Do you think that muscle tissue in skeletal muscles and in the digestive system would look exactly the same? Why or why not? *(Sample answer: The tissues would probably look different because they perform different functions. Different kinds of muscles control voluntary and involuntary movement.)*

Ask What connective tissue listed shows that cells do not have to be connected to each other to form a tissue? *(blood)*

MULTIPLE MEANINGS Invite a volunteer to read the last sentence under the heading Muscle Tissue.

Say The word *pupil* can mean a student or the black center part of an eye.

Ask Which meaning of *pupil* is used in the text? *(the black center part of an eye)*

KEY QUESTION Have a student volunteer read the question and its answer.

Have students use a series of sentences to show how each level forms the following level. *(Sample answer: Cells that are alike form a tissue. Different tissues with related functions form organs. Different organs with related functions form organ systems.)*

Speed Bump

Ask What does it mean to say that cells are specialized? *(Each cell performs a specific function.)*

Digestive and Excretory Systems **715**

During Reading continued

Build Connections

VISUAL SUMMARY Even though the chapter deals primarily with the digestive and excretory systems, be sure students have some familiarity with all of the body systems. Use the following questions to review the information in the visuals on this page.

Ask What are the main structures of the circulatory system? *(heart, blood vessels, blood)*

Ask What is the function of the digestive system? *(It breaks down food, absorbs nutrients, and eliminates wastes.)*

Ask What function is shared by the digestive and the excretory systems? *(elimination of wastes)*

Ask What organs are shared by more than one system? *(Sample answer: lungs and skin)*

Ask For you to breathe, muscles related to the respiratory system must receive signals from what other system? *(the nervous system)*

Ask What function of the skin relates to the integumentary system and what function relates to the excretory system? *(In the integumentary system, the skin protects the rest of the body, forming a barrier to microorganisms. In the excretory system, the skin eliminates wastes through perspiration.)*

Ask How do you think each type of tissue mentioned in the visual on the previous page relates to the digestive system? *(Sample answer: Epithelial tissue lines the digestive system. Connective tissue surrounds the organs of the digestive system. Nervous tissue carries impulses from the brain to the digestive system. Muscle tissue makes up the muscles surrounding the digestive system.)*

Ask How do you think the reproductive system may interact with another system? *(Sample answer: The circulatory system carries nutrients to the cells of the reproductive system and carries wastes and hormones away from it.)*

Ask How might the respiratory system interact with the muscular system? *(Sample answer: The respiratory system provides oxygen for muscle cells and eliminates wastes produced by them. Muscles are necessary to inhale and exhale during breathing.)*

BUILD Connections

HUMAN BODY SYSTEMS

Each organ system shown here has a different set of functions. However, they all work together to maintain homeostasis in the body.

In Your Workbook Go to your workbook for an exercise about the functions of body systems.

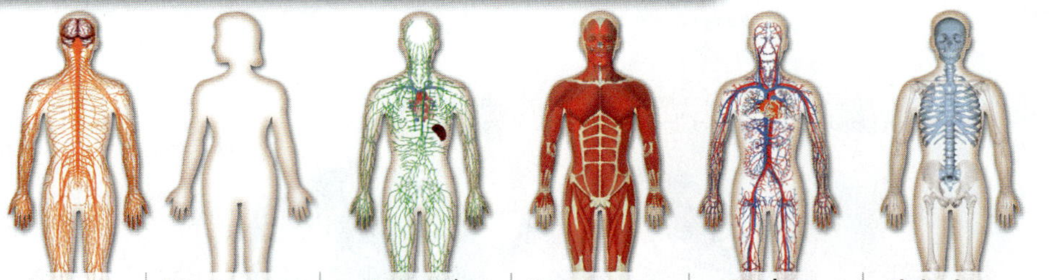

	Nervous System	Integumentary System	Immune/ Lymphatic Systems	Muscular System	Circulatory System	Skeletal System
STRUCTURES	Brain, spinal cord, nerves	Skin, hair, nails, sweat and oil glands	White blood cells, thymus, spleen, lymph nodes, lymph vessels	Skeletal muscle, smooth muscle, cardiac muscle	Heart, blood vessels, blood	Bones, cartilage, ligaments, tendons
FUNCTIONS	Recognizes and coordinates the body's response to changes in its internal and external environments	Guards against infection, injury, and ultraviolet radiation from the sun; helps to regulate body temperature	Helps protect the body from disease; collects fluid lost from blood vessels and returns it to the circulatory system	Works with skeletal system to produce voluntary movement; helps to circulate blood and move food through the digestive system	Transports oxygen, nutrients, and hormones to cells; fights infection; removes cell wastes; helps to regulate body temperature	Supports the body; protects internal organs; allows movement; stores mineral reserves; contains cells that produce blood cells

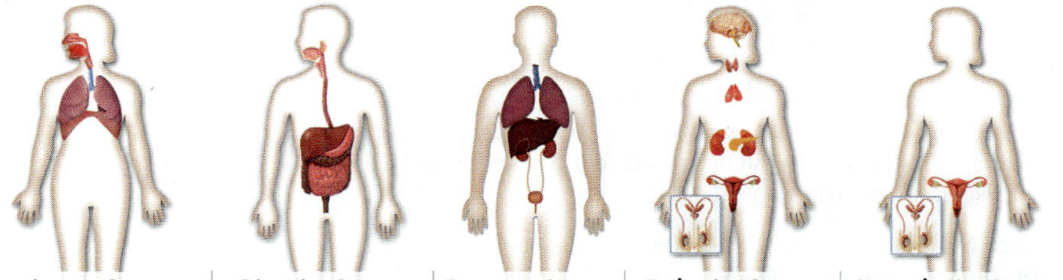

	Respiratory System	Digestive System	Excretory System	Endocrine System	Reproductive System
STRUCTURES	Nose, pharynx, larynx, trachea, bronchi, bronchioles, lungs	Mouth, pharynx, esophagus, stomach, small and large intestines, liver, pancreas, rectum	Skin, lungs, liver, kidneys, ureters, urinary bladder, urethra	Hypothalamus, pituitary, thyroid, parathyroids, adrenals, pancreas, ovaries (in females), testes (in males)	Testes, epididymis, vas deferens, urethra, and penis (in males); ovaries, Fallopian tubes, uterus, vagina (in females)
FUNCTIONS	Brings in oxygen needed for cellular respiration and removes excess carbon dioxide from the body	Breaks down food; absorbs nutrients; eliminates wastes	Eliminates waste products from the body	Controls growth, development, and metabolism; maintains homeostasis	Produces gametes; in females, nurtures and protects developing embryo

BIOLOGY.com Search Lesson 30.1 GO • Art Review

Focus on Academic Language

• system • structure • function

These words are used frequently throughout the chapter, but their meanings might not be familiar to students.

Ask Which word describes the purpose or use of something? *(function)*

Ask Which word describes a set of connected or related things that form a whole? *(system)*

Ask Which word describes how parts are arranged to make up a whole? *(structure)*

Point out that *structure* refers to how parts are arranged, but *structures* often refers to the parts themselves. Choose a system, and show students how the listed structures work together to form the structure of the system.

Homeostasis

Some things are easy to observe. When you run, swim, or even write, you can see your body at work. However, some things are not easy to see. Your body's systems work behind the scenes to keep your internal environment stable. This stable environment inside your body is called **homeostasis**. You cannot always see homeostasis, but to an organism it means life or death.

Feedback Inhibition Have you ever watched someone drive a car down a fairly straight road? Did you notice that the driver constantly moves the wheel left and right? Making small adjustments to the wheel helps keep the car in the middle of the road. Body systems also constantly make small adjustments. These adjustments keep your body's internal conditions within a certain range.

▶ ***A Nonliving Example*** One example of a system that automatically keeps conditions within a certain range is a home heating system. Many home heating systems use a furnace that burns oil or natural gas to provide heat. The system uses a device called a thermostat to monitor the home's air temperature. A sensor compares the air temperature to the one you set on the thermostat. When the air temperature is too cool, the thermostat turns the furnace on. Once the air temperature is warm enough, the thermostat turns the furnace off. The thermostat works to keep the air temperature within a narrow range.

The heating system is using a process called feedback inhibition to control the air temperature in the house. Your body uses **feedback inhibition** to stay in homeostasis. *Feedback* means "response." *Inhibit* means "to block." Feedback inhibition blocks the response to a stimulus. In the heating system, the warm air "feeds back" on the thermostat. The thermostat then "blocks" the release of more warm air by turning the furnace off.

▶ ***The Body's Thermostat—A Living Example*** One part of homeostasis is keeping your body temperature steady. Your body uses feedback inhibition to do this. The hypothalamus is the part of your brain that controls body temperature. The hypothalamus acts as your body's thermostat. Sensors in your body tell the hypothalamus when you are too cold or too hot.

BUILD Vocabulary

homeostasis
the maintenance of relatively constant internal physical and chemical conditions within an organism

feedback inhibition
a process that stops the response to a stimulus

ACADEMIC WORDS
The noun *inhibition* means "the act of blocking the action of." Therefore, feedback inhibition refers to a response that blocks further action of a stimulus.

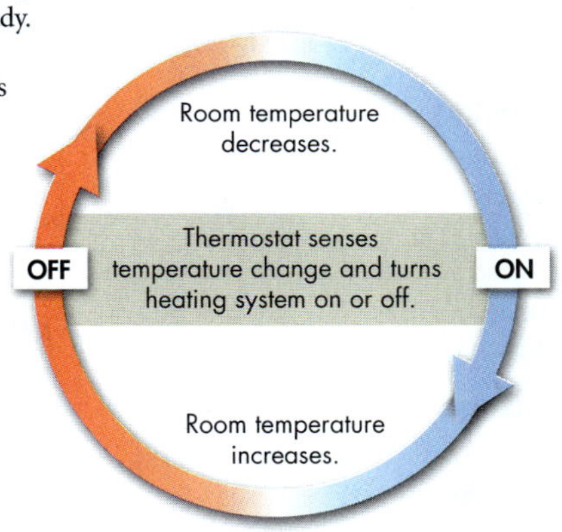

Feedback Inhibition A home heating system uses feedback inhibition. This keeps the environment in the house stable and comfortable.

ELL Focus on ELL: Build Background

INTERMEDIATE SPEAKERS Have students add water to a bowl until the water is near the top. Have them take turns carrying the bowl and observing how the water moves in the bowl. Have them use the model to demonstrate the meanings of *homeostasis, stable, adjustment,* and *feedback inhibition.*

CAUTION: *Have paper towels on hand and instruct students to clean up any spills immediately.*

Active Reading

USE ANALOGIES Refer students to the home heating system analogy in the text and in the diagram. Emphasize that the heating system is analogous to homeostasis in general and is not meant to be analogous to body temperature regulation only.

Ask What is a stable environment? *(Sample answer: an environment that stays about the same or changes only slightly)*

Ask Name a condition in the human body other than body temperature that needs to be stable. *(Sample answers: the amount of water in the body, blood glucose levels)*

Build Vocabulary

ACADEMIC WORDS Students might not fully understand the meanings of the words *maintain* and *stimulus* as they are used in the definitions of *homeostasis* and *feedback inhibition* respectively.

Ask What does it mean to maintain something? *(Sample answer: to keep something in the same shape)* Give an example of something that needs to be maintained in a home. *(the yard, trash taken out, clothes washed)*

Ask What is a stimulus? *(something that causes something else to happen)* What might be a stimulus for eating? *(Sample answers: hunger, ads for food, smelling food, seeing others eat)*

SCIENCE SUPPORT The heating system model describes feedback inhibition and its role in maintaining homeostasis. Explain that positive feedback also occurs in the human body.

Ask How do you think the human body responds to a positive feedback mechanism? *(The response does not resist the original stimulus; it reinforces it.)*

Tell students that two examples of positive feedback are the release of the pituitary hormone oxytocin during childbirth and chemicals released during blood clotting.

Speed Bump

Ask What is the relationship between feedback inhibition and a stimulus? *(Feedback inhibition blocks the response to a stimulus.)*

Ask If the body contains too much sodium, how does the body maintain homeostasis? *(The body eliminates sodium to keep a stable sodium level.)*

During Reading continued

Active Reading

COMPARE/CONTRAST Have each student choose an example or a model of homeostasis from the lesson. For their choice, have them compare and contrast the mechanisms that either raise or lower something to maintain homeostasis.

KEY QUESTION Have student pairs take turns reading the question and answer to each other.

Ask What is the difference between internal and external conditions? *(Internal conditions are inside the body. External conditions are outside the body.)*

Speed Bump

Ask What action does the hypothalamus stimulate in body cells when you are hot? *(to slow down activity)*

Post Reading

ASSESSMENT TIPS Have students review the vocabulary terms and their definitions before answering Questions 1–3.

Have students review the definition and examples of feedback inhibition before answering Question 5.

MYSTERY CLUE Students should infer that the substance found in Andrew's urine was some form of sugar (glucose) because sugar in the urine is an indication of diabetes.

Wrap-Up Activity

MODEL HOMEOSTASIS Divide the class into groups of three. Give one student warm water, one student crushed ice, and the third student, a container of either lukewarm or chilled water and a thermometer. Have the student place the thermometer in the water and read the temperature aloud immediately, then every two minutes. When the temperature is read, the other two students should use the warm water and/or the ice to maintain the original temperature of the water. After the activity, have students compare this experiment to what happens in the body during temperature regulation.

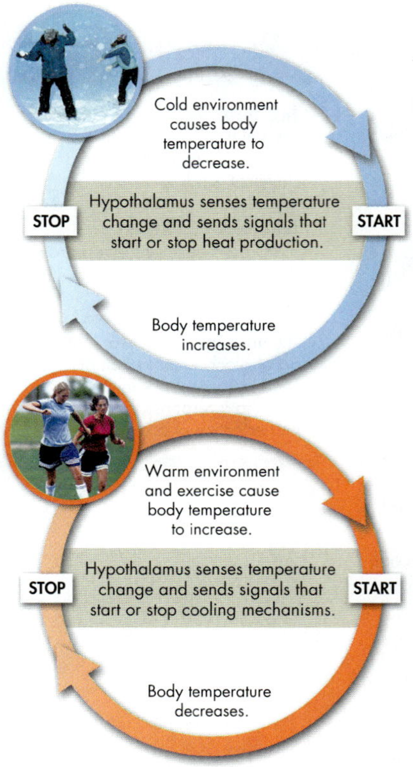

Getting Warm and Staying Cool Your hypothalamus uses feedback inhibition to control your body's temperature.

If you are too cold, the hypothalamus tells cells to speed up their activities. The extra activity produces heat to warm your body. When you are warm enough, the hypothalamus tells the cells they can slow down again. If you are too hot, your hypothalamus tells your sweat glands to produce more sweat. The sweat evaporates, cooling your body surface. The hypothalamus also tells your cells to slow down to produce less heat. This is one reason you feel tired and slow on a hot day.

The Liver and Homeostasis The liver, a part of the digestive system, is important in homeostasis. The liver breaks down toxic chemicals in the blood such as ammonia and drugs. The liver converts ammonia to urea, which is less toxic.

Another function of the liver is to control the amount of glucose in the blood. When blood sugar rises, as it does after a meal, the liver removes extra glucose and stores it. High blood glucose levels can damage the eyes, kidneys, heart, and immune system. If levels in the blood fall too far, the nervous system slows down, and you may even pass out. The liver solves this problem by releasing stored sugar back into the blood. In this way, the liver keeps blood sugar from rising or falling beyond normal levels.

Key Question What is homeostasis?
Homeostasis is a state of relative balance in an organism's physical and chemical conditions. Homeostasis continues even when internal and external environments change.

CHECK Understanding

Apply Vocabulary

1. The type of tissue that lines the inner and outer surfaces of the body is called ______ tissue.

2. When the body is in ______, there is a relatively constant level of glucose in the blood.

3. The hypothalamus uses ______ to keep body temperature constant.

Critical Thinking

4. **Explain** What are two roles of the liver in maintaining homeostasis?

5. **Apply Concepts** Do you think that feelings of hunger and fullness are an example of feedback inhibition? Explain.

6. **Write to Learn** Answer the question in the clue to the mystery below.

MYSTERY CLUE

The lab test of Andrew's urine sample showed an abnormal amount of a certain substance. What substance do you think it was? (**Hint:** Look ahead to p. 731.)

BIOLOGY.com Search Lesson 30.1 GO • Lesson Assessment

Check Understanding

1. epithelial
2. homeostasis
3. feedback inhibition
4. The liver breaks down toxic substances, such as ammonia and drugs, and it helps control blood sugar levels.
5. Sample answer: Yes, because the stimuli hunger and fullness are opposing stimuli. Hunger stimulates a person to eat. Fullness then stimulates a person to stop eating, which opposes the original action.
6. The substance is glucose. Abnormally high glucose levels can cause damage to cells in the eyes, kidneys, heart, and other parts of the body.

30.2 Food and Nutrition

Food and Energy

Have you ever wondered why you feel weak when you are hungry? Or, for that matter, why you need food in the first place? One answer is energy. You need energy to climb stairs, run, and even to think. Just as a car does, your body needs fuel, and it feels weak without it. Food is that fuel.

Energy The energy available in food can be measured in a laboratory. The way it is measured is surprisingly simple—the scientist burns it! When food is burned, most of the energy in the food is converted to heat. Heat is measured in calories. One calorie of heat raises the temperature of 1 gram of water by 1 degree Celsius. The "Calories" you have heard about in food are actually dietary Calories. Notice that a capital *C* is used. One dietary **Calorie** is equal to 1000 calories, or 1 kilocalorie (kcal). Remember that the energy stored in food molecules is released during cellular respiration. This energy is used to produce ATP molecules. ATP powers cellular activities.

Raw Materials Chemical pathways can get energy from almost any type of food. Why, then, does the type of food matter? Food supplies more than energy. It also supplies the raw materials used to build and repair body tissues. These raw materials are the building blocks needed to make molecules like enzymes and DNA. Food contains at least 45 substances that the body needs but cannot make. A healthy diet allows your body to get all of these required substances.

Key Question Why do we need to eat?
Molecules in food contain chemical energy. Cells use this energy to produce ATP. Food also supplies raw materials your body needs to build and repair tissues.

Food Supplies Raw Materials
Your body cannot make all of the raw materials it needs to build important molecules. Food contains at least 45 of the substances your body needs but cannot make.

Key Questions

Why do we need to eat?

What nutrients does your body need?

What is meant by the term "balanced diet"?

BUILD Understanding

Two-Column Chart Use a two-column chart to take notes about main ideas and details. List the main ideas in the left column. In the right column, write details that explain the main ideas.

In Your Workbook Go to your workbook for help with using a two-column chart.

BUILD Vocabulary

Calorie
the measure of energy in food; 1 Calorie = 1000 calories, or 1 kilocalorie (kcal)

carbohydrate
A class of nutrients that are the main source of energy for the human body

ACADEMIC WORDS
The word *carbohydrate* means "carbon with water." Substances that are "hydrated" contain water. For each carbon atom, carbohydrates contain roughly one oxygen and two hydrogens, the same proportions found in water (H_2O).

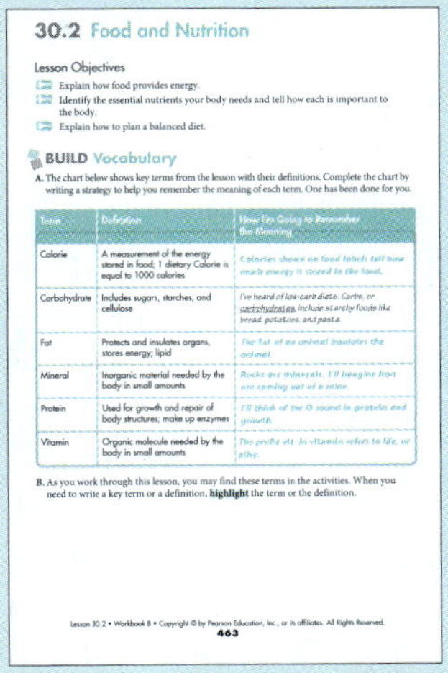

30.2 Food and Nutrition

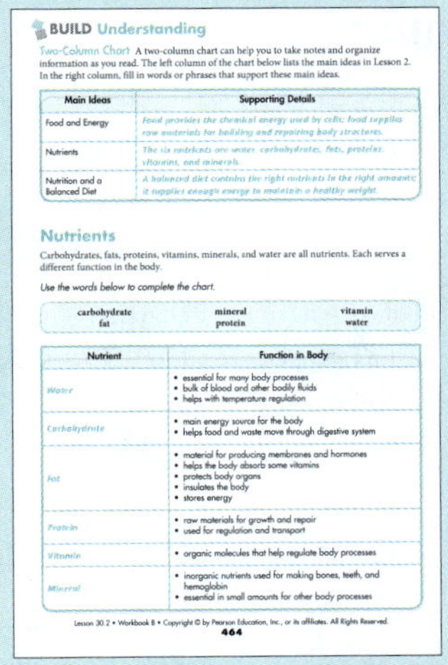

BUILD Understanding

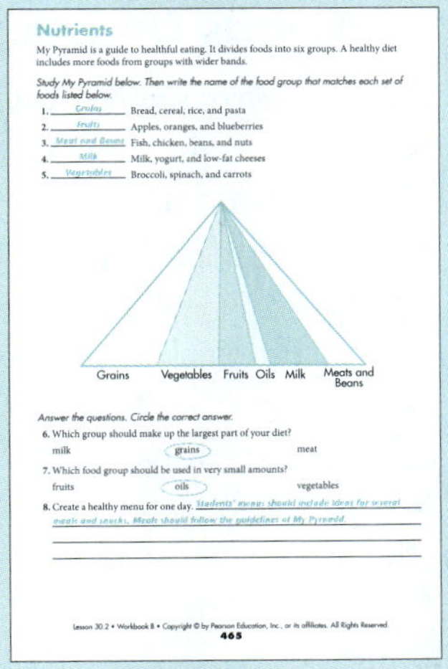

Nutrients

LESSON 30.2

Pre-Reading

Objectives

30.2.1 Explain how food provides energy.

30.2.2 Identify the essential nutrients your body needs and tell how each is important to the body.

30.2.3 Explain how to plan a balanced diet.

Preview the Pages

1. Point out the words *calorie* and *Calorie* in the Energy paragraph on this page. Explain that the capital *C* makes a difference in the word's meaning.
2. Have students examine the visuals that show foods containing carbohydrates, fats, and proteins. Have them compare the types of foods shown with what they ate yesterday.
3. Review from Chapter 9 how energy from food is released by cellular respiration.

Build Understanding

TWO-COLUMN CHART Direct students to Workbook B, where they will find a partially completed two-column chart. Have them fill it in. Remind them that they can use the completed chart as a study aid.

Speed Bump

Ask Why are Calories instead of calories used to refer to the energy in food? *(The number of Calories is easier to write and understand.)*

Ask What kind of energy is stored in food? *(chemical energy)*

NATIONAL SCIENCE EDUCATION STANDARDS

UNIFYING CONCEPTS AND PROCESSES

V

CONTENT

B.2, B.3, C.1.b, C.5.a, C.5.c, F.1

For the corresponding lesson in *Miller & Levine Biology*, see pages 868–873.

During Reading

Build Vocabulary

ACADEMIC WORDS Draw students' attention to the Build Vocabulary feature on the previous page. Emphasize that the carbon and the water form a compound—a carbohydrate—that has its own properties and not the properties of either carbon or water. Then, point out the *n* in the formula for a carbohydrate. Explain that *n* stands for a number and that this number is multiplied by the numbers for the C, H, and O. If a number is not visible it is understood to be 1.

Ask What is the chemical formula for glucose, a carbohydrate that contains six carbon atoms? *($C_6H_{12}O_6$)*

Active Reading

KEY QUESTION Have students cover the answer to the Key Question on the previous page. Then, have a volunteer read the question.

After listing students' answers, have students compare them to the answer given in the text.

USE CONTEXT CLUES Ask a volunteer to read the second paragraph under Water.

Ask Which word in this paragraph contains the word part *hydrate*? *(dehydration)* What does *hydrate* mean? *(water)* What does *dehydration* mean? *(having lost water)* Which clues in the paragraph helped you know the meaning of the word? *(Sample answers: the words* lost water *and* not replaced*)*

Say You could figure out the meaning of *dehydration* using your knowledge of word parts or using context clues.

CLARIFY WORD USAGE Invite a volunteer to read the first paragraph under Carbohydrates. Then, point out that the word *saccharide* means "sugar molecule." The prefix *mono-* means "one;" *di-* means "two;" *poly-* means "many." The prefixes refer to how many sugar molecules are in a carbohydrate.

Ask How many sugar molecules are in a disaccharide? *(two)* How many sugar molecules are in simple sugars? *(one or two)* How many sugar molecules are in starches? *(many)*

Speed Bump

Ask What happens to excess sugar in the blood? *(It is converted to glycogen or fat.)*

Ask What does the body do with fiber? *(moves it through the digestive system as bulk)*

BUILD Vocabulary

fat
nutrient needed by the body for cell membranes, vitamin absorption, regulation, insulation, and energy

protein
a macromolecule that contains carbon, hydrogen, oxygen, and nitrogen; needed by the body for growth and repair

vitamin
an organic nutrient that the body needs in very small amounts

mineral
an inorganic nutrient the body needs in very small amounts

MULTIPLE MEANINGS
In biology, the word *minerals* refers only to inorganic elements like sodium and magnesium. In geology, minerals are the basic units of rocks, and can be elements or compounds.

Nutrients

Nutrients are substances in food that supply energy and raw materials. Your body needs the energy and raw materials for growth, repair, and maintenance. There are six main types of nutrients—water, carbohydrates, fats, proteins, vitamins, and minerals.

Water Every cell in the human body needs water, which is the most important nutrient. Body processes such as chemical reactions are carried out in water. Large amounts of water are found in blood, extracellular fluid, and other bodily fluids. Water is lost from the body in urine and in sweat. You also lose water with every breath you exhale.

Humans need to drink at least 1 liter of fluid each day. Dehydration can occur if the lost water is not replaced. Dehydration leads to problems with many body systems. Under extreme conditions it can be fatal.

Carbohydrates **Carbohydrates** are major sources of food energy for the body. Simple carbohydrates are either monosaccharides, like glucose, or disaccharides, like sucrose, ordinary table sugar. Complex carbohydrates are polysaccharides, like starch and glycogen. Your digestive system breaks down starches into simple sugars. Sugars are absorbed into the blood and carried to the body's cells. Excess blood sugar is converted into glycogen. Glycogen is stored in the liver and in skeletal muscles. Excess sugar can also be converted to fat and stored in the body.

Another complex carbohydrate is cellulose. Cellulose is often called fiber. Your digestive system cannot break down fiber, but you need it in your diet for bulk. Bulk helps muscles move food and waste through your digestive system. Fiber may also reduce the risk of heart disease and Type II diabetes.

Milk is one of the few nonplant sources of carbohydrates. Milk is rich in a sugar called lactose.

Carbohydrates All of these foods are rich in carbohydrates. Pastas and cereals are also rich in carbohydrates. Simple carbohydrates do not have to be digested or broken down. Complex carbohydrates, like those found in whole grains, must be broken down before they can be used by the body.

Fruits, honey, and sugar cane contain sugars. Fruits are also rich in fiber.

Breads, grains, potatoes, and vegetables contain starch, a complex carbohydrate. Whole grains are also rich sources of fiber. Products made from whole grains have more fiber than products made from processed grains.

Hands-On Learning

DEMONSTRATE BREAKDOWN OF STARCHES Have students try this simple demonstration to observe the breakdown of starches into simple sugars.

1. Check with students about possible food allergies to wheat or gluten.
2. Give a saltine cracker to each non-allergic student.
3. Have students hold their cracker in their mouth without swallowing while observing the initial taste. *(The cracker should taste starchy, not sweet.)*
4. After students have sucked on their cracker for several seconds, ask them to describe any change in the taste of the cracker. *(The cracker should taste increasingly sweet.)*
5. Ask students to explain the results. *(The starch in the cracker breaks down into simple sugars, which taste sweet.)*

Fats Our bodies need fats, or lipids, as part of a healthy diet. **Fats** help the body absorb fat-soluble vitamins. Fats are found in cell membranes, nerve cells, and certain hormones. Body fat protects and insulates organs and stores energy.

Fats are made by combining glycerol molecules with fatty acids. Fats with only single bonds between carbons are called saturated. Unsaturated fats contain at least one double bond and fewer hydrogen atoms than saturated fats. Fats with more than one double bond are polyunsaturated. Trans fats are made by adding hydrogen atoms to unsaturated fats. Trans fats have been linked to health problems such as heart disease.

Proteins Proteins have many roles in the body. **Proteins** are used for growth and repair of tissues. Proteins called enzymes speed up chemical reactions in cells. Some proteins regulate different functions. Still others are transport molecules. Proteins can sometimes be used for energy.

Proteins are polymers of amino acids. The body is able to make 12 of the 20 amino acids used to make proteins. The other eight are essential amino acids. They must come from food. Animal sources usually provide all eight essential amino acids, but most plant foods do not.

Vitamins Organic molecules that the body needs in very small amounts are called **vitamins.** They help perform chemical reactions. Small amounts of the fat-soluble vitamins A, D, E, and K can be stored in body fat. The B vitamins and vitamin C are water-soluble. They dissolve in water and cannot be stored by the body. They must be included in the foods you eat each day.

A diet lacking in certain vitamins can cause serious health problems. Eating a variety of foods will meet the daily vitamin needs of most people.

Minerals **Minerals** are inorganic nutrients that the body needs, usually in small amounts. Calcium and iron are two examples of minerals needed by the body. Minerals are needed in the diet to replace those lost in sweat and bodily wastes.

Key Question What nutrients does your body need?
The body needs water, carbohydrates, fats, proteins, vitamins, and minerals.

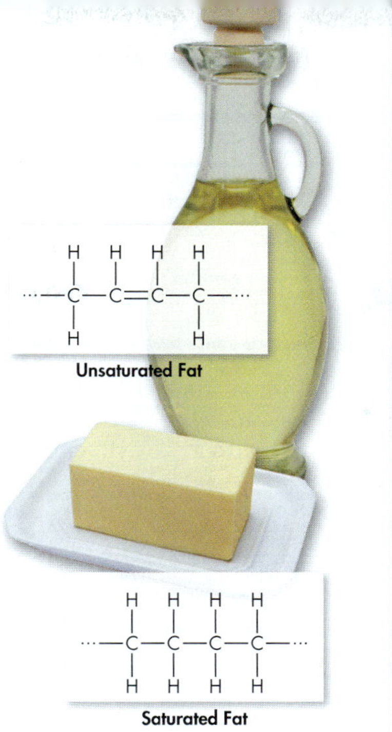

Fats Most saturated fats are solid at room temperature. Butter contains saturated fats. Most unsaturated fats are liquid at room temperature. Vegetable oils, like this olive oil, contain polyunsaturated fats.

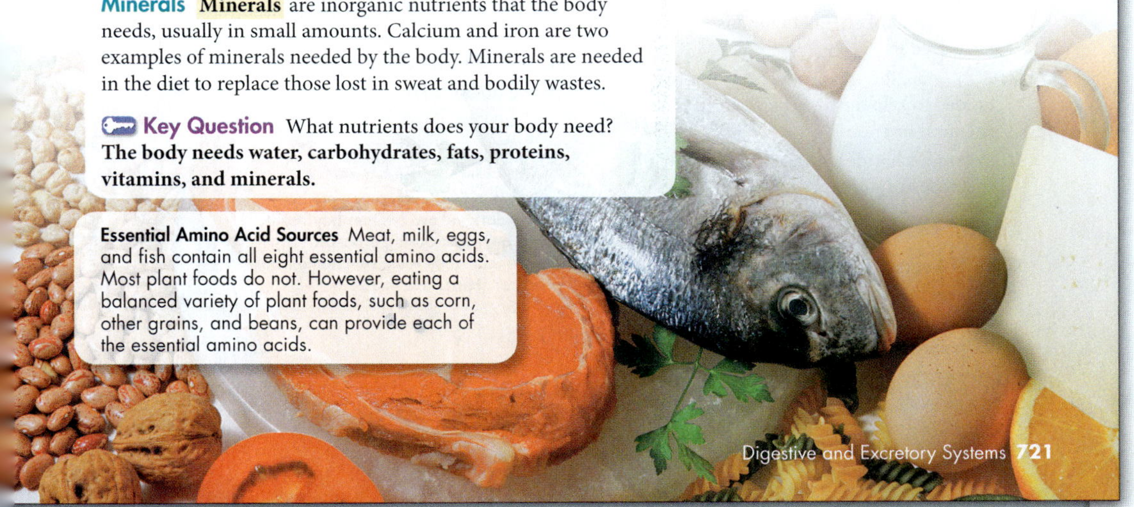

Essential Amino Acid Sources Meat, milk, eggs, and fish contain all eight essential amino acids. Most plant foods do not. However, eating a balanced variety of plant foods, such as corn, other grains, and beans, can provide each of the essential amino acids.

ELL Focus on ELL: Access Content

BEGINNING AND INTERMEDIATE SPEAKERS To help English learners organize lesson content, have students make a graphic organizer or other visual representation that shows the six types of nutrients. Visuals might include a cluster concept map or a circle sliced into six equal parts.

Beginning speakers can illustrate their graphic organizer by drawing sketches or pasting pictures of foods from a magazine.

For intermediate speakers, have students allow room in the visual to list how each nutrient is obtained and the function of the nutrient in the human body.

Active Reading

DRAW CONCLUSIONS Have students use the following questions to help them draw conclusions. Then, have students write additional questions they can use to draw conclusions.

Ask Why might people who eat no animal products have difficulty meeting their protein needs? *(Animal sources usually provide all essential proteins, but plant sources do not.)*

Ask What might be some consequences of not consuming enough fat? *(Sample answer: Not enough fat would be available for the body to produce cell membranes, nerve cells, and hormones.)*

USE GRAPHIC ORGANIZERS On the board, draw a Venn diagram with three overlapping circles. Label one Carbohydrates, the second Fats, and the third Proteins. Use the Venn diagram to show that some foods contain more than one of these nutrients. Add foods to the diagram in the appropriate circle(s), placing them in the overlapping areas as needed. Start by naming a food and having students predict where it should be placed. Then, have students name foods to be placed. Advise them to limit their suggestions to a single food, such as cheese or broccoli, rather than pizza, which has many ingredients. If possible, have a resource book available that shows how much of each nutrient is in each food.

Build Vocabulary

MULTIPLE MEANINGS Refer students to the Build Vocabulary feature on the opposite page. Pass around an empty container for a multivitamin that contains both vitamins and minerals. Make a list on the board of the different minerals contained in the multivitamin.

Ask What would be in the container if we were using the geological meaning of the word *mineral*? *(Sample answer: pieces of rock)*

KEY QUESTION Have a student volunteer read the question. Then, have volunteers each name a nutrient until all of them are named.

Speed Bump

Ask What is the difference between a saturated fat and an unsaturated fat? *(Unsaturated fats have at least one double bond. Saturated fats have only single bonds and are more harmful to the body.)*

Ask Why are some amino acids called essential? *(The eight amino acids not made by the body are called essential amino acids; the body must take in these amino acids.)*

LESSON 30.2

During Reading continued

Active Reading

KEY QUESTION Have student pairs read the question and the answer to each other. Ask the pairs to provide an example of one day's balanced diet.

Speed Bump

Ask What is the difference between a vitamin and a mineral? *(Sample answer: Vitamins are organic, and minerals are inorganic. Vitamins are frequently named by letters, such as A and B.)*

Post Reading

ASSESSMENT TIPS Before students answer Question 1, review the difference between a Calorie and a calorie.

Review the difference between vitamins and minerals before answering Question 2.

MYSTERY CLUE The urine of both boys was dark yellow because there was not enough water in the urine to dilute the wastes. The boys were dehydrated and needed to drink more water.

Wrap-Up Activity

NUTRIENTS IN FOOD Provide student groups with food labels from at least two different common cereals or snacks. Ask students to locate the amount of Calories, carbohydrates (or sugars), proteins, fats, vitamins, and minerals in one serving of the food and to compare the nutritional value of the foods. Then, ask students to justify which food is a better choice for a healthy, balanced diet.

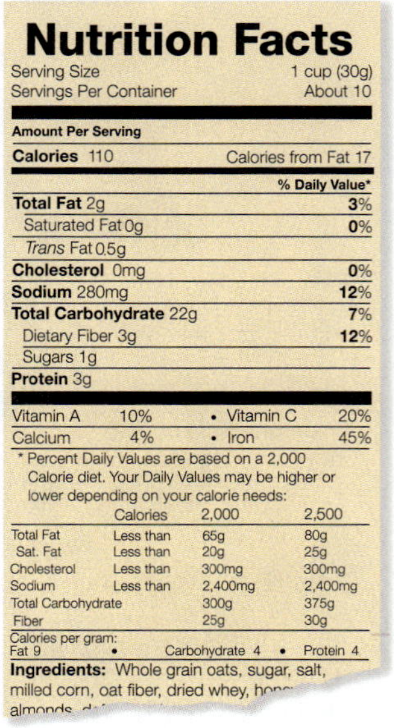

Food Label Reading food labels can help you track the amount of Calories and nutrients that you eat and drink.

Nutrition and a Balanced Diet

Nutrition is the study of food and its effects on the body. Nutrition tries to determine how food helps the body meet all its various needs. Nutritionists have developed many tools to help people plan healthful diets.

Balancing Your Diet How do you know which foods to eat for a balanced diet? Food labels can be used to choose healthful foods. You can use the labels to decide if you are consuming enough of some of the important vitamins and minerals. Nutrient needs are affected by age, gender, and lifestyle. People who are more active than average have greater energy needs. A person who stops growing or is less active has a lower need for energy.

Maintaining a Healthful Weight Exercising 30 minutes a day and eating a balanced diet helps maintain a healthful weight. Regular physical activity burns excess Calories and strengthens the heart, bones, and muscles.

Controlling fat intake is also important. Foods that contain high amounts of fat are high in Calories. Diets high in saturated and trans fats increase the risk of heart disease, Type II diabetes, or both.

Key Question What is meant by the term "balanced diet"?
A balanced diet provides the nutrients and energy needed for a person to maintain a healthful weight.

CHECK Understanding

Apply Vocabulary

Use the highlighted words from the lesson to complete each sentence correctly.

1. The amount of energy in food is measured in ______.
2. ______ are the organic nutrients that are needed to help the body perform chemical reactions.
3. Monosaccharides, disaccharides, and cellulose are all types of ______.

Critical Thinking

4. **Infer** Foods that contain many Calories but few raw materials are said to contain empty Calories. What do you think the term *empty Calories* means?
5. **Compare and Contrast** Saturated and unsaturated fats both are made of glycerol and fatty acids. How is a saturated fat different from unsaturated fat?
6. **Write to Learn** Answer the second clue of the mystery. Why do you think the color of the urine led the doctor to his diagnosis?

MYSTERY CLUE

Philip's and Seth's samples were both a very dark yellow. Neither boy drank water before or during practice. Why do you think they were sent home from practice? (**Hint:** See p. 720.)

BIOLOGY.com Search Lesson 30.2 GO • Lesson Assessment

Check Understanding

1. Calories
2. Vitamins
3. carbohydrates
4. Sample answer: The phrase *empty Calories* refers to foods that supply Calories without also supplying raw materials the body needs.
5. Unsaturated fats contain double bonds and fewer hydrogen atoms than saturated fats do.
6. The dark color of the urine showed that the urine was concentrated and that the boys were dehydrated. Urine becomes concentrated when the body conserves water, which it does when someone does not drink enough water.

30.3 The Digestive System

Functions of the Digestive System

When you are hungry, your whole body needs food. But the only system in the body that food actually enters is the digestive system. How does food get to the rest of the body after you digest it? The digestive system's job is to convert the food into useful molecules. Digestion happens in four phases—ingestion, digestion, absorption, and elimination.

Ingestion The first phase in digestion is getting food into the system. Your mouth is the opening of the digestive tract. The process of putting food into your mouth is called *ingestion.*

Digestion As food passes through the digestive tract, it is broken down in two ways—by mechanical digestion and chemical digestion. Mechanical digestion is the physical breakdown of large pieces of food into smaller pieces. These smaller pieces can be swallowed and accessed by digestive enzymes. Chemical digestion is the chemical breakdown of food into smaller pieces. During chemical digestion, enzymes break down food into small molecules that the body can use.

Absorption Once food has been broken into small molecules, it can be absorbed by the small intestine. During absorption, molecules enter the circulatory system from the digestive tract. The circulatory system carries food molecules throughout the body.

Elimination The digestive system cannot digest and absorb all the food that enters your body. Substances that cannot be absorbed travel through the large intestine and are eliminated from the body as feces. Cellulose is one example of a material your body cannot digest.

Key Question What are the functions of the digestive system?
The digestive system breaks down food into molecules the body can use in its cells. Food is processed in four phases—ingestion, digestion, absorption, and elimination.

Key Questions

What are the functions of the digestive system?

What occurs during digestion?

How are nutrients absorbed and wastes eliminated?

BUILD Understanding

Flowchart Make a flowchart that shows the steps in the process of digestion.

In Your Workbook Refer to your workbook for suggestions about how to use a flowchart to organize your notes.

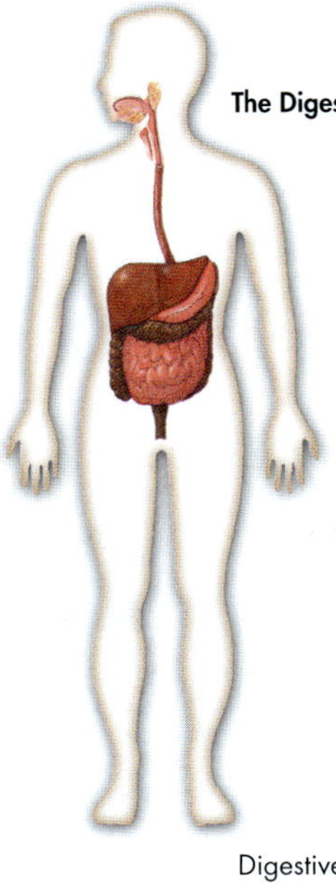

The Digestive System

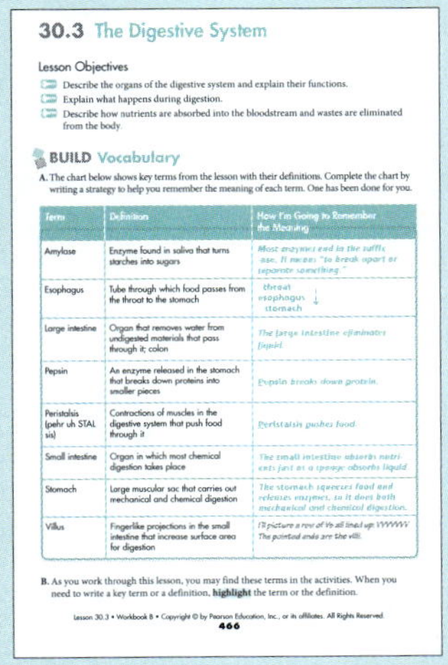

30.3 The Digestive System

Lesson Objectives
- Describe the organs of the digestive system and explain their functions.
- Explain what happens during digestion.
- Describe how nutrients are absorbed into the bloodstream and wastes are eliminated from the body.

BUILD Vocabulary

A. The chart below shows key terms from the lesson with their definitions. Complete the chart by writing a strategy to help you remember the meaning of each term. One has been done for you.

Term	Definition	How I'm Going to Remember the Meaning
Amylase	Enzyme found in saliva that turns starches into sugars	Most enzymes end in the suffix -ase. It means "to break apart or separate something."
Esophagus	Tube through which food passes from the throat to the stomach	throat esophagus stomach
Large intestine	Organ that removes water from undigested materials that pass through it; colon	The large intestine eliminates liquid.
Pepsin	An enzyme released in the stomach that breaks down proteins into smaller pieces	Pepsin breaks down protein.
Peristalsis (pehr uh STAL sis)	Contractions of muscles in the digestive system that push food through it	Peristalsis pushes food.
Small intestine	Organ in which most chemical digestion takes place	The small intestine absorbs nutrients just as a sponge absorbs liquid.
Stomach	Large muscular sac that carries out mechanical and chemical digestion	The stomach squeezes food and releases enzymes, so it does both mechanical and chemical digestion.
Villus	Fingerlike projections in the small intestine that increase surface area for digestion	I'll picture a row of Vs all lined up: VVVVVV The pointed ends are the villi.

B. As you work through this lesson, you may find these terms in the activities. When you need to write a key term or a definition, **highlight** the term or the definition.

Lesson 30.3 • Workbook B • Copyright © by Pearson Education, Inc., or its affiliates. All Rights Reserved.
466

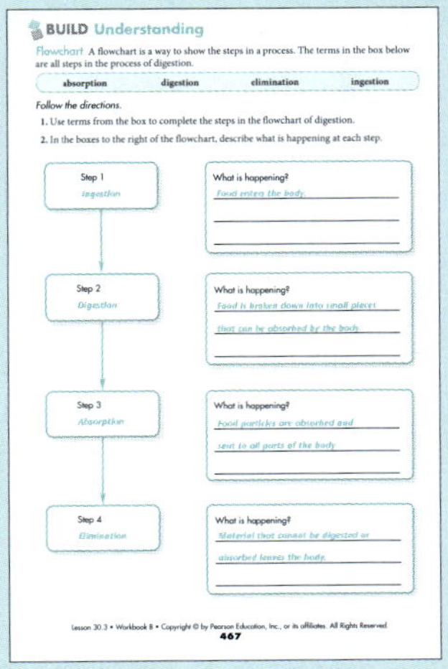

BUILD Understanding

Flowchart A flowchart is a way to show the steps in a process. The terms in the box below are all steps in the process of digestion.

absorption digestion elimination ingestion

Follow the directions.

1. Use terms from the box to complete the steps in the flowchart of digestion.
2. In the boxes to the right of the flowchart, describe what is happening at each step.

Step 1 Ingestion — What is happening? Food enters the body.

Step 2 Digestion — What is happening? Food is broken down into small pieces that can be absorbed by the body.

Step 3 Absorption — What is happening? Food particles are absorbed and sent to all parts of the body.

Step 4 Elimination — What is happening? Material that cannot be digested or absorbed leaves the body.

Lesson 30.3 • Workbook B • Copyright © by Pearson Education, Inc., or its affiliates. All Rights Reserved.
467

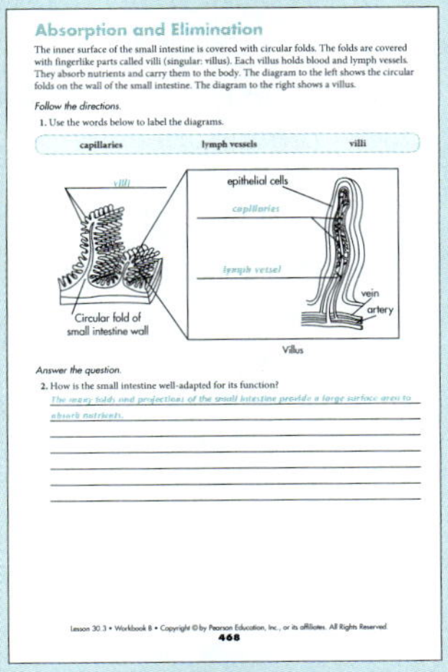

Absorption and Elimination

The inner surface of the small intestine is covered with circular folds. The folds are covered with fingerlike parts called villi (singular: villus). Each villus holds blood and lymph vessels. They absorb nutrients and carry them to the body. The diagram to the left shows the circular folds on the wall of the small intestine. The diagram to the right shows a villus.

Follow the directions.

1. Use the words below to label the diagrams.

capillaries lymph vessels villi

Answer the question.

2. How is the small intestine well-adapted for its function?

Lesson 30.3 • Workbook B • Copyright © by Pearson Education, Inc., or its affiliates. All Rights Reserved.
468

LESSON 30.3

Pre-Reading

Objectives

30.3.1 Describe the organs of the digestive system and explain their functions.

30.3.2 Explain what happens during digestion.

30.3.3 Describe how nutrients are absorbed into the bloodstream and wastes are eliminated from the body.

Preview the Pages

1. Have students use the flowchart from the Build Understanding activity to guide them through the steps in the digestive system.
2. Have students preview the visuals related to digestive processes without reading the captions and describe what they think is happening. Have them then compare what they think is happening to the information in the captions.

Activate Prior Knowledge

Students probably have some knowledge of the digestive system. Divide the class into an even number of groups. Have half the groups develop lists of what they already know about the digestive system. Then, have the other groups develop lists of questions they have about the system. Have the groups compare lists and determine if answers to them are known by another group.

Build Understanding

FLOWCHART Refer students to Workbook B for an example of a flowchart. Remind them that a flowchart can be used to show a sequence of events.

NATIONAL SCIENCE EDUCATION STANDARDS

UNIFYING CONCEPTS AND PROCESSES

I, V

CONTENT

C.1.b, C.5.d

For the corresponding lesson in *Miller & Levine Biology,* see pages 875–881.

LESSON 30.3

During Reading

Active Reading

KEY QUESTION Read the Key Question on the previous page to the class. Have students read the answer aloud as a group.

Ask The mechanical breaking down of food occurs in what two phases? *(ingestion and digestion)*

USE GRAPHIC ORGANIZERS Have students use the Build Understanding flowchart to make four cluster charts. Have them place each step in the center of one cluster. As students read through the lesson, have them add details to each cluster.

Ask Eating would be part of which cluster? *(ingestion)*

USE VISUALS Have students model peristalsis as shown in the visual. Mix cornstarch with water and make a small lump of mud-like consistency. Place the lump into the open end of an uninflated long balloon. Have students try different motions to move the lump a certain distance through the balloon.

Ask What motion was most effective? *(Answers may vary, but the motion mentioned should be one that first squeezes at the top of the lump, then moves the squeezing motion down.)*

Speed Bump

Ask Incisors, cuspids, bicuspids, and molars are all what kind of body part? *(teeth)*

Ask What might be more likely to happen in your body if saliva did not contain lysozyme? *(You might have more bacterial infections.)*

The Mouth Digestion begins in the mouth. The tongue, teeth, and saliva form food into a moist lump that can be swallowed.

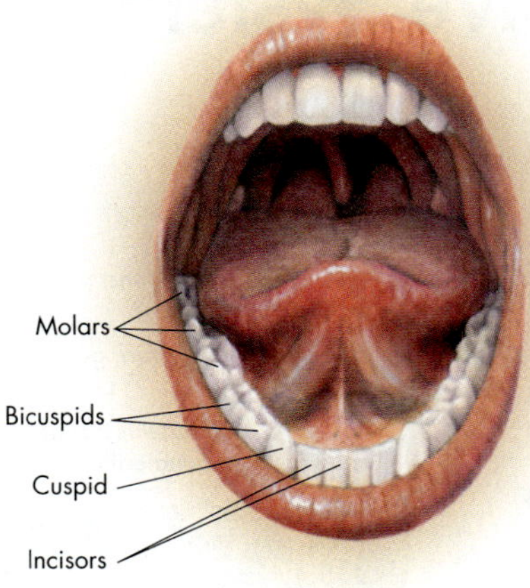

The Process of Digestion

The human digestive system is built around an alimentary canal. An alimentary canal is a one-way tube that passes all the way through the body. The entrance to this tube is the mouth.

Digestion Starts in the Mouth Once food enters your mouth, your teeth begin mechanical digestion. The incisors, cuspids, and bicuspids cut and tear the food. Then the molars grind the food into a paste that you can swallow. Your tongue moves the food around so that all of it can be chewed. While you are chewing, glands in your mouth release *saliva*. Saliva moistens the food and makes it easier to chew.

Saliva contains an enzyme called **amylase** that begins chemical digestion by breaking chemical bonds in starches to release sugars. Another enzyme, called lysozyme, fights infection. Lysozyme digests the cell walls of many bacteria that may enter the mouth with food.

Once food is chewed it moves into the pharynx, which is the back of the throat. As you swallow, a flap of tissue called the epiglottis closes over the opening to the trachea to keep food out. The trachea is the tube that leads from the pharynx to the lungs.

The Esophagus The food paste forms into a clump as it is swallowed. This clump, called a bolus, enters the **esophagus** from the pharynx. The esophagus is a muscular tube that connects the pharynx to the stomach. Muscle contractions push the bolus down the esophagus into the stomach. This process, called **peristalsis,** is shown below. A ring of muscle called the cardiac sphincter closes the esophagus after the bolus enters the stomach. This prevents food from moving back into the esophagus.

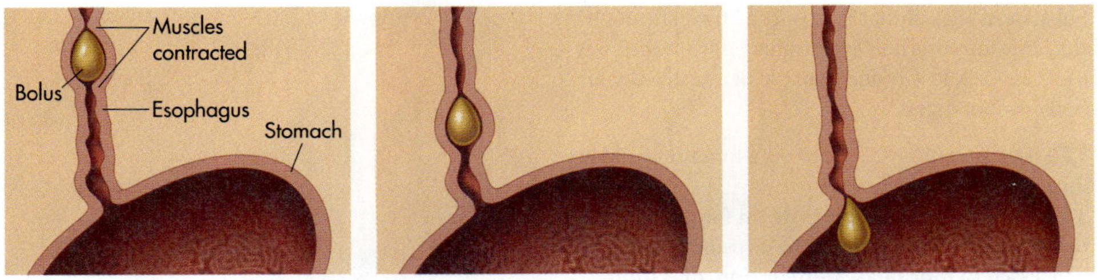

Peristalsis Muscles in the walls of the esophagus contract in waves. Each wave pushes the chewed clump of food, or bolus, in front of it. Eventually, the bolus is pushed into the stomach.

 BIOLOGY.com Search Lesson 30.3 GO • Art in Motion

Students Say

"At the beginning of a new chapter, our teacher had us read the title, all of the headings and subheadings, and only those. We couldn't read the chapter. Then, we had to present a commercial, live, in front of our classmates—10 seconds to sell them on what we thought we would learn in the chapter. Like, "Take a trip in the digestive system this summer. See exotic places like the esophagus and the stomach." Then we were allowed to read only the first paragraph under each heading, and then we'd discuss things, like whose ad made the most sense or was the most interesting."

Digestion in the Stomach Chemical and mechanical digestion continues in the **stomach.** Tiny glands in the stomach lining release hydrochloric acid, **pepsin,** and mucus. Pepsin is an enzyme that breaks down proteins into smaller pieces, and works best in acidic conditions. Mucus is a fluid that protects the stomach lining from the acid and lubricates the food. Smooth muscle layers in the stomach contract to mix the food with the hydrochloric acid and pepsin. The mixture that leaves the stomach is called chyme.

Digestion in the Small Intestine Chyme enters the upper part of the **small intestine** through the pyloric valve. This area of the small intestine is called the duodenum (doo oh DEE num), and is where most of the digestion and absorption of food occurs. The chyme mixes with enzymes and digestive fluids from the pancreas, liver, and the lining of the duodenum.

▶ ***Pancreas*** The pancreas is a small gland located behind the stomach. The pancreas produces enzymes, sodium bicarbonate, and hormones. The enzymes break down carbohydrates, proteins, lipids, and nucleic acids in the small intestine. Sodium bicarbonate neutralizes the acid in the chyme, protecting the small intestine's lining. It also gives the enzymes the right conditions to work. The hormones, as part of the endocrine system, regulate blood sugar levels.

▶ ***The Liver and Gallbladder*** The liver helps the pancreatic enzymes with fat digestion. The liver produces bile, a fluid loaded with lipids and salts. Bile is stored in the gallbladder until fat enters the duodenum. Fats glob together, making it hard for enzymes to digest them. Bile breaks the fat globs into smaller droplets. This makes it easier for enzymes to digest the fat.

Key Question What occurs during digestion? **During digestion, food travels through the mouth, esophagus, stomach, and small intestine. Mechanical digestion and chemical digestion break food down into molecules that can be absorbed.**

BUILD Vocabulary

amylase
the enzyme in saliva that begins breaking down starches to simple sugars

esophagus
the muscular tube that connects the throat to the stomach

peristalsis
the contractions of smooth muscle that move food through the digestive system

stomach
the organ where chemical and mechanical digestion continues after food leaves the mouth

pepsin
the stomach enzyme that begins protein digestion; it works best in acidic conditions

WORD ORIGINS
The prefix *amyl-* refers to starch. It has both Greek (*amylan*) and Latin (*amylum*) origins. The suffix *–ase* is commonly used to indicate that a substance is an enzyme. Amylase is an enzyme that breaks down starch.

Effects of Digestive Enzymes
Digestive enzymes speed up the breakdown of foods. This makes nutrients from the foods available to the body.

Effects of Digestive Enzymes

Active Site	Enzyme	Effect on Food
Mouth	Salivary amylase	Breaks down starches into disaccharides
Stomach	Pepsin	Breaks down proteins into large peptides
Small intestine (released from pancreas)	Pancreatic amylase	Continues the breakdown of starch
	Trypsin	Continues the breakdown of protein
	Lipase	Breaks down fat
Small intestine	Maltase, sucrase, lactase	Breaks down remaining disaccharides into monosaccharides
	Peptidase	Breaks down dipeptides into amino acids

LESSON 30.3

Active Reading

FIND THE MAIN IDEA As students read this page, have them find and write down the main idea of each paragraph. Under each main idea, have them list supporting details found in the text. To begin, ask the following questions about the first paragraph.

Ask What is the main idea in the paragraph on digestion in the stomach? *(Chemical and mechanical digestion starting in the mouth continues in the stomach.)* What details in the paragraph support this main idea? *(Student responses should include key ideas from each of the sentences following the main idea.)*

SCIENCE SUPPORT Because the cell is the basic unit of life, students may think that all substances in the human body form cells. Remind students that enzymes are proteins, not cells. Hormones, bile, and mucus are other examples of substances in the body that are not cells.

Build Vocabulary

WORD ORIGINS Read each vocabulary term aloud, then have students repeat it.

Ask Do the names of all enzymes end in *-ase*? How do you know? *(Sample answer: No, the name of the enzyme* pepsin *does not end in* -ase.*)*

Ask What do their names tell you about lipase and maltase? *(They are enzymes because they end in* -ase. *Some students may infer that lipase breaks down lipids.)*

KEY QUESTION Write the question on the board. Have students trace the path of digestion on a diagram of the human body or one of the digestive systems.

Ask Where on the diagram does mechanical digestion occur? *(Mechanical digestion occurs mostly in the mouth but also in the stomach.)* Where does chemical digestion occur? *(Chemical digestion occurs in the mouth, stomach, and small intestine.)*

Speed Bump

Ask Food travels through most parts of the digestive system. Which parts of the digestive system does food not travel through? *(the pancreas, the liver, and the gallbladder)*

ELL Focus on ELL: Extend Language

INTERMEDIATE AND ADVANCED SPEAKERS Prepare a set of flash cards that contains terms from this lesson that may be unfamiliar to English language learners. Write the term on one side of each card and its definition or description on the other side. Do not use vocabulary terms. Examples include: mucus, hydrochloric acid, valve, duodenum, and chyme.

As students read the lesson, have them use the cards to clarify meanings. Provide blank cards so that students can add additional terms.

Students can use the cards to clarify while reading the lesson or as a review at the end.

During Reading continued

Build Connections

VISUAL SUMMARY Have students use the visual to trace the path of food through the digestive system. Ask them to work in pairs to talk about what happens to the food as it goes through or by each digestive structure.

Ask What are the four functions of the digestive system? *(ingestion, digestion, absorption, elimination)*

Ask How is each labeled structure involved in one or more of these functions? *(The mouth is involved in ingestion. All labeled parts except the large intestine are involved in digestion. Absorption occurs in steps 4 and 5. Elimination occurs from the large intestine.)*

Ask Why do you think it takes 3–5 hours for food to pass through the small intestine? *(Sample answer: Food is released slowly into the small intestine. It takes a while for bile and enzymes to digest food there. If food passed through the small intestine too quickly, many nutrients would be eliminated instead of being absorbed, so that they can be used by the body.)*

Ask What is the function of the gallbladder in digestion? *(It stores bile from the liver until it is needed in the small intestine to aid in digestion of fats.)*

Explain to students that the duodenum takes up only about 25 cm of the small intestine. The part of the small intestine that absorbs nutrients is approximately 3 m long.

Ask Why is it important that most of the small intestine is used for absorption? *(The longer the area is available for absorption, the more nutrients are absorbed.)*

BUILD Connections

THE DIGESTIVE SYSTEM

Food travels through many organs as it is broken down into nutrients your body can use. The time needed for each organ to perform its role varies based on the type of food consumed.

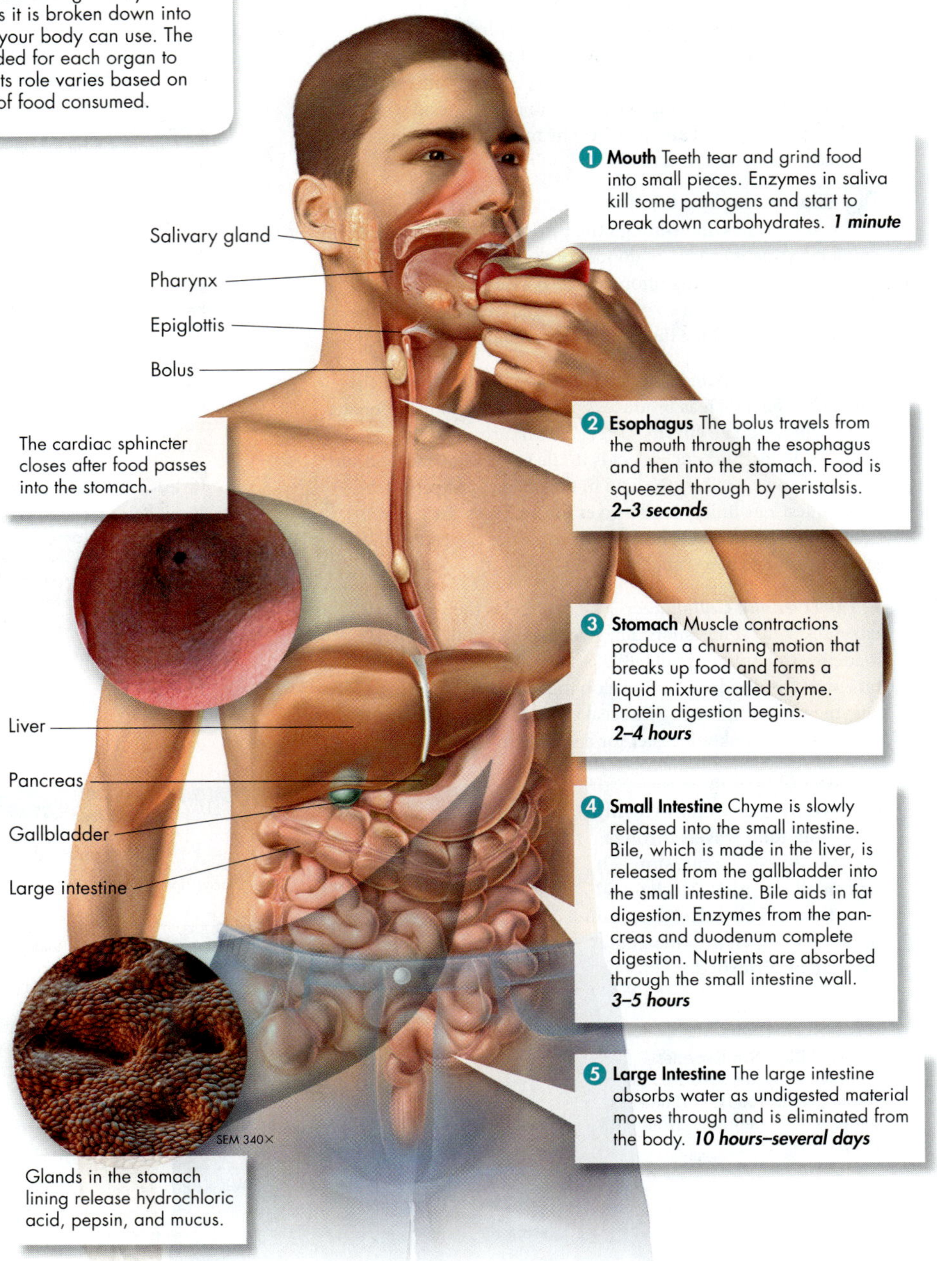

Hands-On Learning

DEMONSTRATE FUNCTION OF MUCUS Tell students that they will demonstrate one function of mucus production in the stomach. Divide the class into small groups. Have each group follow these steps:

1. Place a small piece of chalk in a transparent plastic cup.
2. Coat a second piece of chalk with petroleum jelly, and place it in a second cup.
3. Add vinegar or lemon juice to each cup and observe what happens.

Students should observe the acid react with the unprotected chalk but not with the chalk protected by the petroleum jelly. Point out that the liquid added is acidic, as is gastric juice. Tell students that gastric juice is much more acidic and would react even more strongly than the liquid they added.

The chalk represents the stomach lining. Like chalk, the stomach lining will react with acids it comes in contact with. The petroleum jelly serves the same purpose as mucus; it protects against exposure to acids.

Absorption and Elimination

Once digestion is finished in the small intestine, nutrients need to be absorbed from the alimentary canal. Absorption occurs in the intestines.

Absorption From the Small Intestine Muscles in the walls of the small intestine contract in waves. These contractions move the chyme down the intestine. By this time, most of the chemical digestion is complete. The chyme is now made of small and medium molecules that can be absorbed.

The structure of the small intestine is specialized for absorption of nutrients. The folded surface of the small intestine is covered in small structures called **villi** (singular *villus*). The villi are covered in tiny microvilli that absorb nutrients from the chyme. Most of the carbohydrates and proteins are absorbed into the capillaries in the villi. Most fats and fatty acids are absorbed by lymph vessels.

Chyme has almost no nutrients by the time it is ready to leave the small intestine. Complex food molecules have been digested and absorbed. The chyme now contains water, cellulose, and undigestable substances.

BUILD Vocabulary

small intestine
the digestive organ where most chemical digestion occurs

villus
a small, finger-shaped structure lining the small intestine that absorbs nutrients

large intestine
the organ that absorbs water and some vitamins from undigested material; also called the colon

WORD ORIGINS
In Latin, the word *villi* means "shaggy hair." Under a microscope, the villi in the intestine look like tiny hairs projecting from the intestine wall.

Absorption in the Small Intestine The lining of the small intestine consists of many folds. These folds are covered with tiny finger-like structures called villi. Each villus is covered in microvilli that absorb nutrients. Blood capillaries and lymph vessels in the villus carry the nutrients away from the villus.

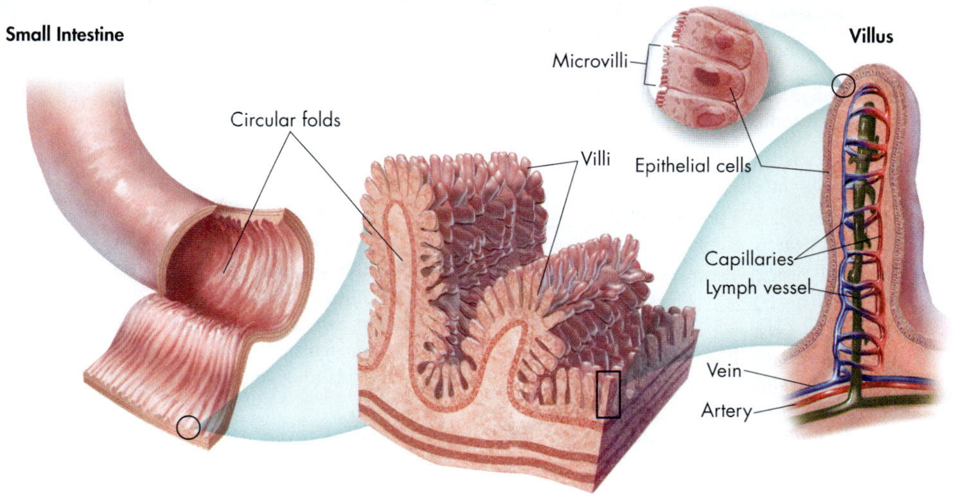

LESSON 30.3

Active Reading

COMPARE/CONTRAST Have students examine the loops on a terrycloth towel.

Ask How do the loops help the towel dry better? *(The loops increase the surface area of the towel, increasing the amount of liquid it can absorb.)*

Ask How do the loops on the towel compare to the villi in the small intestine? *(Both the loops and the villi increase the amount of surface area present, increasing the ability to absorb.)*

SELF-QUESTION Ask students to read about absorption in the small intestines. Then, have them reread the section, asking themselves questions about what they are reading. Their questions should include both those that can be answered from the text as well as any unanswered questions that arise. For example, when students reread the first paragraph, they might ask how the waves that transport chyme down the intestine move. Do they move through the entire intestine or only for a short length of the intestines? Do the same waves continue to move through the large intestine?

Build Vocabulary

WORD ORIGINS Explain that certain Latin terms ending in *-us* form plurals by changing the *-us* to *-i*. Villus follows this pattern. Another example of a word that follows this pattern is *alumnus (plural, alumni).*

Speed Bump

Ask What might happen if the cardiac sphincter does not close after food passes into the stomach? *(Sample answer: Some of the gastric acid might go into the esophagus, irritating it. Students might be familiar with gastric reflux and heartburn.)*

Ask What is absorbed in the small intestine? *(Most of the nutrients are absorbed in the small intestine.)*

Focus on Academic Language

• absorption • elimination • microvilli • undigestible

Write the words *absorption* and *elimination* on the board and link these nouns to the verbs *absorb* and *eliminate.* Underline *absorp* and *eliminat* in the nouns. Point out that *absorb* means "to take in" and *eliminate* means "to get rid of."

Also, point out that prefixes can make terms more descriptive. The prefix *micro-* means "very small." Thus, *microvilli* are very small villi. The prefix *un-* means "not." Thus, *undigestible* means "not able to be digested."

Ask students to point out uses of these words, including *absorb* and *eliminate,* in this section. As students read examples from the text aloud, rephrase each one by replacing the academic terms with their definitions.

LESSON 30.3

During Reading continued

Active Reading

COMPARE/CONTRAST Pronounce the words *diarrhea* and *constipation* for students. Then as a class compare and contrast the two conditions. Clarify by drawing a diagram on the board using arrows to show the movement of water into and out of the large intestine.

KEY QUESTION Read the question as two questions.

Ask How are nutrients absorbed? How are wastes eliminated?

For each question, have students read the part of the answer that applies to it.

Speed Bump

Ask How is the function of the large intestine related to the length of time it takes food to pass through it? *(Compared to the small intestine, the large intestine is not very long. If food did not pass slowly through it, enough water would not be absorbed and bacterial colonies would not produce vitamins.)*

Post Reading

ASSESSMENT TIPS Have students locate the vocabulary terms in the text and review their definitions before answering Questions 1–4.

Have students examine the visual and text about villi before answering Questions 4, 6, and 7.

Before answering Question 7, encourage students to look for more than one effect of celiac disease.

Wrap-Up Activity

THE ADVANTAGE OF VILLI Give groups of students a 10 cm × 10 cm square of a smooth cotton cloth and terrycloth toweling. Have students soak each piece of fabric in water, then remove it. Hold the fabric pieces by one corner for a minute until excess water has run out. Then, have them squeeze the cloths into two small, identical containers. Have them compare the amount of water each type of fabric absorbed. Finally, have students compare these results to having villi instead of smooth walls in the small intestines.

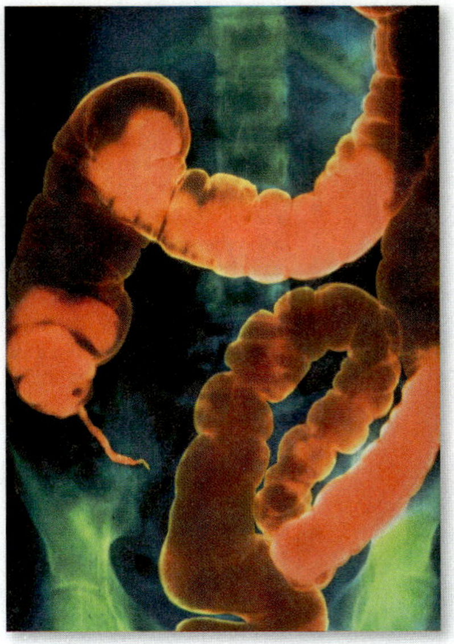

The Large Intestine This X-ray shows the large intestine and its contents.

Absorption from the Large Intestine Most nutrients have already been absorbed from the chyme before it enters the **large intestine.** The large intestine's job is to remove water from the chyme. The water is absorbed quickly, leaving the undigested material behind. Rich colonies of bacteria in the large intestine feed on this material. They produce compounds the body can use, such as vitamin K. Taking large doses of antibiotics for an infection can destroy these bacteria. When this happens, the body may not get enough vitamin K.

Elimination The waste material left after all of the water is absorbed is called feces. The feces move into the rectum and are eliminated from the body through the anus. If something happens to stop the large intestine from absorbing enough water, a condition called diarrhea occurs. If too much water is absorbed, a condition called constipation occurs.

Key Question How are nutrients absorbed and wastes eliminated?
Most nutrients are absorbed through the walls of the small intestine. The large intestine absorbs water and some vitamins, and prepares waste for elimination.

CHECK Understanding

Apply Vocabulary

Use the highlighted terms from the lesson to complete each sentence correctly.

1. ______ in your saliva begins breaking down starch while you chew.
2. When you swallow, a process called ______ moves the bolus down the esophagus to your stomach.
3. Hydrochloric acid is secreted into the stomach to help ______ digest proteins by lowering the pH.
4. The structures that line the folds in the small intestine are called ______. Blood and lymph vessels in these structures carry absorbed nutrients to the rest of the body.

Critical Thinking

5. **Compare and Contrast** What is the difference between mechanical digestion and chemical digestion?
6. **Explain** How are nutrients absorbed?
7. **Write to Learn** In people with celiac disease, many villi and microvilli are destroyed. How do you think this disease affects their ability to absorb nutrients? Write a short paragraph explaining your answer.

BIOLOGY.com Search Lesson 30.3 GO • Lesson Assessment

Check Understanding

1. Amylase
2. peristalsis
3. pepsin
4. villi
5. Mechanical digestion physically breaks down food into smaller particles. Chemical digestion uses enzymes to break these particles into molecules that can be absorbed.
6. Most nutrients are absorbed through the villi in the small intestines. Some water and some vitamins are absorbed in the large intestine.
7. The loss of villi and microvilli represent a loss not only of absorptive surfaces, but also of the blood and lymph vessels associated with the destroyed villi.

30.4 The Excretory System

Structures of the Excretory System

Every living thing produces chemical waste products. Some waste products are so toxic that they will cause death unless they are eliminated. The process of removing wastes to maintain homeostasis is called **excretion.**

The Skin The skin excretes excess water, salts, and a small amount of urea in sweat. Your skin is always releasing very small amounts of sweat.

The Lungs Carbon dioxide is a waste product of cellular respiration. Your blood transports carbon dioxide from your cells to your lungs. When you exhale, you excrete carbon dioxide and small amounts of water vapor.

The Liver The liver's job in the excretory system is to convert toxic ammonia into urea. The ammonia is produced when protein is broken down in the cells. Blood transports the urea to the kidneys where it can be eliminated from the body.

The Kidneys The major organs of excretion are the kidneys. The kidneys are a pair of fist-sized organs located on either side of the spinal column. The kidneys filter excess water, urea, and other wastes from the blood. The kidneys excrete these wastes as urine.

Key Question What is the main role of the structures of the excretory system?
The excretory system includes the skin, lungs, liver, and kidneys. The system excretes wastes from chemical processes in the body.

Excretion and the Kidneys

Nearly a million individual units process the blood in each kidney. These units are called **nephrons.** Each nephron filters out impurities and collects wastes, then sends the clean blood back into circulation. Blood is purified through two processes—filtration and reabsorption.

Key Questions

What is the main role of the structures of the excretory system?

How do the kidneys clean the blood?

How do the kidneys help maintain homeostasis?

BUILD Understanding

Preview Visuals Examine the visuals in this lesson. Think of questions to ask about the visuals.

In Your Workbook Refer to your workbook to help answer your questions.

BUILD Vocabulary

excretion
the process by which metabolic wastes are eliminated from the body

nephron
an individual unit in the kidney that processes wastes

WORD ORIGINS
The word *glomerulus* comes from the Latin word *glomus,* which means "ball of yarn." The twisted capillaries of a glomerulus resemble a ball of yarn.

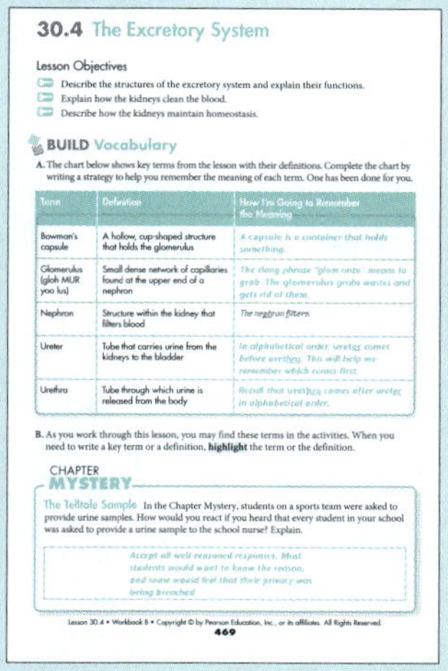
30.4 The Excretory System

Lesson Objectives
Describe the structures of the excretory system and explain their functions.
Explain how the kidneys clean the blood.
Describe how the kidneys maintain homeostasis.

BUILD Vocabulary

A. The chart below shows key terms from the lesson with their definitions. Complete the chart by writing a strategy to help you remember the meaning of each term. One has been done for you.

Term	Definition	How I'm Going to Remember the Meaning
Bowman's capsule	A hollow, cup-shaped structure that holds the glomerulus	A capsule is a container that holds something.
Glomerulus (gloh MUR yoo lus)	Small dense network of capillaries found at the upper end of a nephron	The slang phrase "glom onto" means to grab. The glomerulus grabs wastes and gets rid of them.
Nephron	Structure within the kidney that filters blood	The nephron filters.
Ureter	Tube that carries urine from the kidneys to the bladder	In alphabetical order, ureter comes before urethra. This will help me remember which comes first.
Urethra	Tube through which urine is released from the body	Recall that urethra comes after ureter in alphabetical order.

B. As you work through this lesson, you may find these terms in the activities. When you need to write a key term or a definition, **highlight** the term or the definition.

CHAPTER MYSTERY

The Telltale Sample In the Chapter Mystery, students on a sports team were asked to provide urine samples. How would you react if you heard that every student in your school was asked to provide a urine sample to the school nurse? Explain.

Accept all well-reasoned responses. Most students would want to know the reason, and some would feel that their privacy was being breached.

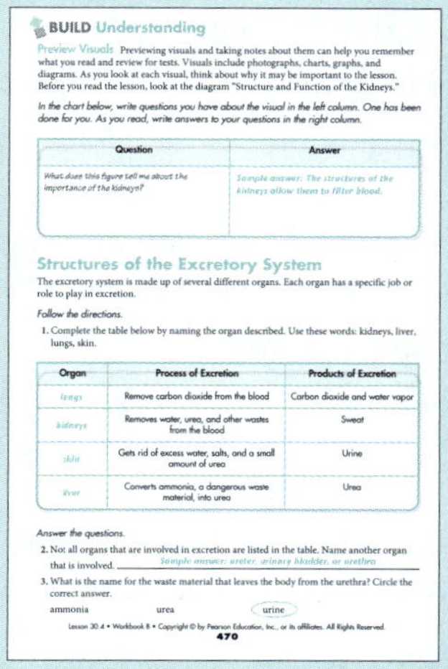
BUILD Understanding

Preview Visuals Previewing visuals and taking notes about them can help you remember what you read and review for tests. Visuals include photographs, charts, graphs, and diagrams. As you look at each visual, think about why it may be important to the lesson. Before you read the lesson, look at the diagram "Structure and Function of the Kidneys."

In the chart below, write questions you have about the visual in the left column. One has been done for you. As you read, write answers to your questions in the right column.

Question	Answer
What does this figure tell me about the importance of the kidneys?	Sample answer: The structures of the kidneys allow them to filter blood.

Structures of the Excretory System

The excretory system is made up of several different organs. Each organ has a specific job or role to play in excretion.

Follow the directions.

1. Complete the table below by naming the organ described. Use these words: kidneys, liver, lungs, skin.

Organ	Process of Excretion	Products of Excretion
lungs	Remove carbon dioxide from the blood	Carbon dioxide and water vapor
kidneys	Removes water, urea, and other wastes from the blood	Sweat
skin	Gets rid of excess water, salts, and a small amount of urea	Urine
liver	Converts ammonia, a dangerous waste material, into urea	Urea

Answer the questions.

2. Not all organs that are involved in excretion are listed in the table. Name another organ that is involved. Sample answer: ureter, urinary bladder, or urethra
3. What is the name for the waste material that leaves the body from the urethra? Circle the correct answer.
ammonia urea (urine)

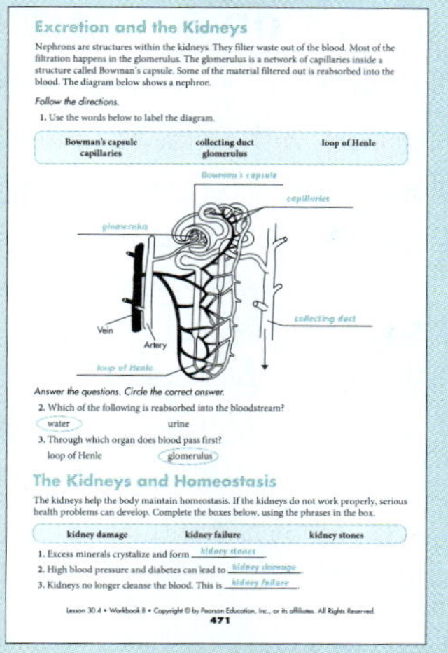
Excretion and the Kidneys

Nephrons are structures within the kidneys. They filter waste out of the blood. Most of the filtration happens in the glomerulus. The glomerulus is a network of capillaries inside a structure called Bowman's capsule. Some of the material filtered out is reabsorbed into the blood. The diagram below shows a nephron.

Follow the directions.

1. Use the words below to label the diagram.

Bowman's capsule, capillaries, collecting duct, glomerulus, loop of Henle

Answer the questions. Circle the correct answer.

2. Which of the following is reabsorbed into the bloodstream?
(water) urine
3. Through which organ does blood pass first?
loop of Henle (glomerulus)

The Kidneys and Homeostasis

The kidneys help the body maintain homeostasis. If the kidneys do not work properly, serious health problems can develop. Complete the boxes below, using the phrases in the box.

kidney damage, kidney failure, kidney stones

1. Excess minerals crystalize and form kidney stones.
2. High blood pressure and diabetes can lead to kidney damage.
3. Kidneys no longer cleanse the blood. This is kidney failure.

Pre-Reading

Objectives

30.4.1 Describe the structures of the excretory system and explain their functions.

30.4.2 Explain how the kidneys clean the blood.

30.4.3 Describe how the kidneys maintain homeostasis.

Preview the Pages

1. Locate each vocabulary term and read it and its definition aloud. Identify vocabulary terms that students have heard before.
2. Use section headings and the visuals to identify the main structures of the excretory system.
3. Ask students to predict how the excretory system affects homeostasis in the human body.

Build Understanding

PREVIEW VISUALS Ask a volunteer to read the caption for the visual Structure and Function of the Kidneys. Ask a second volunteer to read the labels on the image of the excretory system and a third to read aloud the numbered steps describing kidney function. Refer students to Workbook B, where they can record questions about the visual. They can fill in the answers as they read.

NATIONAL SCIENCE EDUCATION STANDARDS

UNIFYING CONCEPTS AND PROCESSES
I, III, IV, V

CONTENT
C.5.d, F.1

INQUIRY
A.1.c, A.2.a

For the corresponding lesson in *Miller & Levine Biology,* see pages 882–887.

During Reading

Active Reading

KEY QUESTION Invite a volunteer to read aloud the question and answer on the previous page. Review the labels on the visual of the excretory system.

Build Connections

VISUAL SUMMARY Be sure students understand that this visual is made up of three diagrams that show the structure and function of the kidneys in increasing detail.

Ask What does the top diagram show? *(the parts of the excretory system and how waste from the kidneys, collected in urine, is transported and then released from the body)*

Ask What does the middle diagram show? *(the structure of a kidney)*

Ask What does the bottom diagram show? *(how a nephron within the kidneys filters wastes from the blood and collects the wastes in urine)*

Ask If the diagram were to include one more level of increasing detail, what do you think it would show? *(Sample answer: The next level may focus on what happens in a glomerulus.)*

USE VISUALS Have students work in groups of three to review the bottom diagram, showing the process through which nephrons filter impurities from blood, collect wastes, and return purified blood to circulation. Have each student in a group choose a different step in the process. For the chosen step, have the student summarize what happens and relate this information to the rest of the group. Encourage groups to share their summaries with the rest of the class.

Ask How does the diagram help you explain how the kidneys clean blood? *(Each step can be traced in the diagram.)*

Speed Bump

Ask What are the major structures of the excretory system? *(skin, lungs, liver, kidneys)*

Ask Name two substances that the kidneys filter from the blood and then return to the blood. *(Possible answers are glucose, water, salt, vitamins, amino acids, and fats.)*

BUILD Connections

STRUCTURE AND FUNCTION OF THE KIDNEYS

Kidneys are made up of nephrons. Blood enters the nephrons, where impurities are filtered out and emptied into the collecting duct. Purified blood leaves nephrons through a vein.

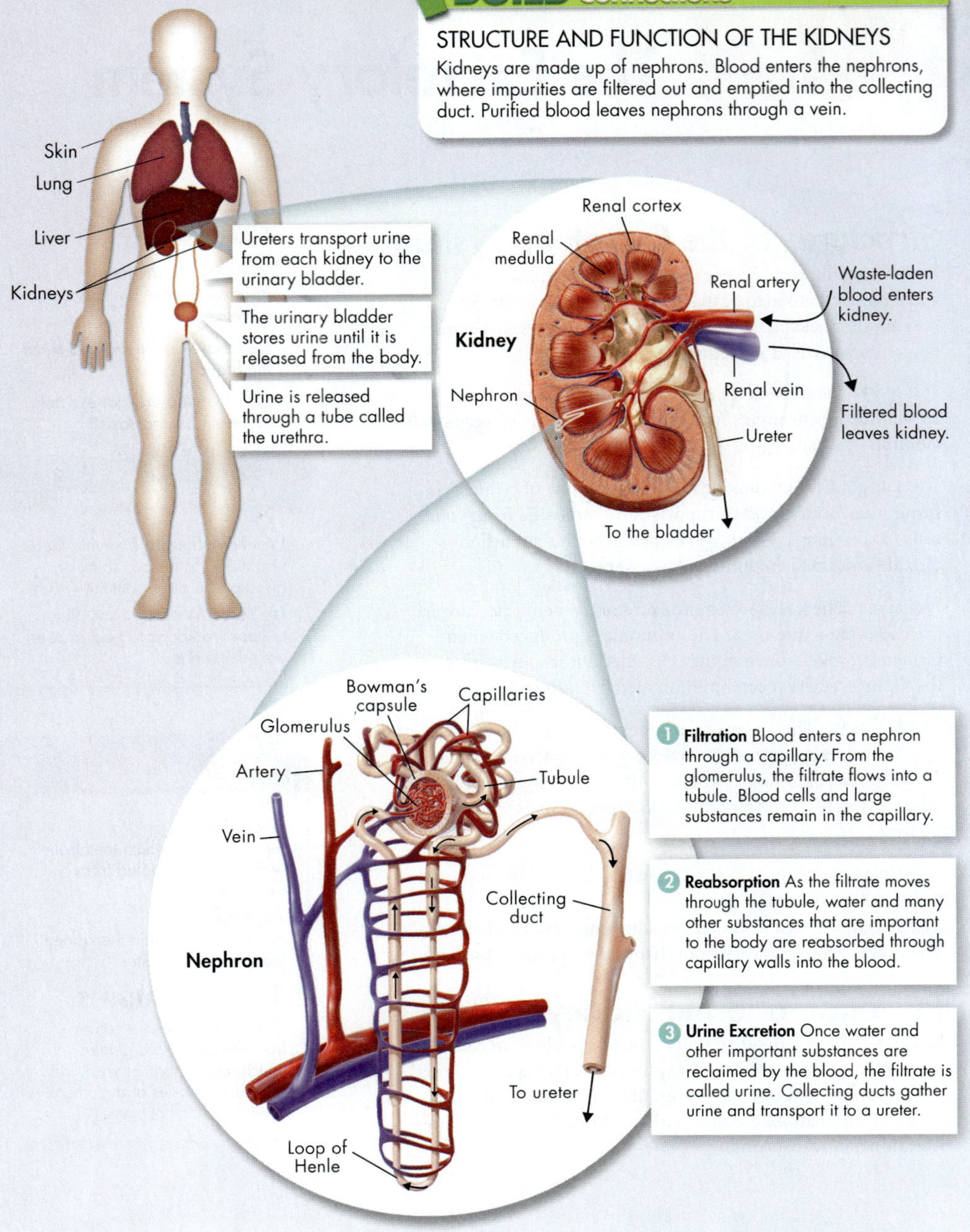

Focus on Academic Language

• filtration • bladder • reabsorption

Discuss with students these terms that relate to the excretory system.

Ask How do you use filtration in your daily life that is not related to body processes? *(Students are probably familiar with water filtration systems either in their homes or as large systems that help purify water in city water systems.)*

Say The gallbladder is part of the digestive system, and the urinary bladder is part of the excretory system.

Ask From the functions of these two body parts, what is a general definition for a bladder? *(a container that holds or stores materials)*

Ask What is the difference between reabsorption and absorption? *(Reabsorption involves absorption of something that had been previously absorbed.)*

Filtration The process of passing a gas or liquid through a filter is called **filtration.** Blood filtration takes place mainly in the **glomerulus** (gloh MUR yoo lus). A glomerulus is a very small, dense network of capillaries. The glomerulus is enclosed in a hollow, cup-shaped structure called Bowman's capsule in the nephron.

Blood is under pressure. This pressure helps fluid from the blood move easily through the walls of the capillaries and into Bowman's capsule. This fluid in the Bowman's capsule, called the filtrate, contains water, urea, glucose, salts, amino acids, and some vitamins. Large substances like proteins and blood cells are too large to move through the capillary walls.

Reabsorption The filtrate contains some waste materials, but most of the filtrate is not waste and must be returned to the blood. Most of the filtrate is returned to the blood through **reabsorption.** Salts, vitamins, amino acids, fats, and glucose are removed from the filtrate by active transport. Water follows these substances into the blood through osmosis.

Almost 99 percent of the water that was filtered out of the blood is reabsorbed in the **loop of Henle.** The loop of Henle is responsible for conserving water and minimizing the amount of filtrate. The filtrate remaining in the tubule is called urine and empties into a collecting duct.

Urine Excretion Urine flows from collecting ducts to the **ureter** of each kidney. The ureters carry urine to the **urinary bladder** for storage. Urine leaves the body through the **urethra.**

Key Question How do the kidneys clean the blood?
The kidneys filter the blood to remove urea, excess water and minerals, and other wastes.

The Kidneys and Homeostasis

The kidneys play an important role in homeostasis. Besides removing wastes, they maintain blood pH and regulate water content.

Control of Kidney Function Kidney function is controlled mainly by the blood's composition. If you have too much salt in your blood, your kidneys will excrete the excess. They will also excrete excess glucose. Glucose in the urine is a sign of diabetes.

Hormones also control kidney function. Your pituitary gland releases antidiuretic hormone (ADH) into your blood when you lose too much water. ADH tells the kidneys to reabsorb more water and excrete less in the urine. When blood has too much water, ADH secretion stops.

Urine Testing Urine tells a lot about your health. The color tells how hydrated you are. If urine is pale yellow, you are well hydrated. Darker urine means that the blood has too little water and your kidneys are conserving water. Protein or glucose in urine can be a sign of high blood pressure or diabetes. Since drugs, including steroids and illegal drugs, usually are not reabsorbed by the kidneys, they end up in the urine.

BUILD Vocabulary

filtration
the process of passing a liquid or gas through a filter to remove wastes

glomerulus
the network of capillaries inside the nephron

reabsorption
the process by which water and dissolved substances are taken back into the blood

loop of Henle
the section of nephron tubule that conserves water

ureter
a tube that carries urine from a kidney to the urinary bladder

urinary bladder
the muscular sac that stores urine until it is released

urethra
the tube that connects the urinary bladder to the outside of the body

WORD ORIGINS
The word *ureter* comes from the Greek word *ourētēr*. The word *urethra* comes from the Greek word *ourēthra*. *Ourētēr* and *ourēthra* share the same Greek root word, *ourein*. *Ourein* means "to urinate."

ELL Focus on ELL: Access Content

BEGINNING AND INTERMEDIATE SPEAKERS Have students draw a flowchart titled *"How do the kidneys clean blood?"* Have students label three squares in the flowchart *filtration, reabsorption,* and *urine excretion.* Ask students to write an explanation and add an illustration to each box. Allow for less-than-precise content words for beginning speakers.

LESSON 30.4

Active Reading

PREDICT Provide students with situations that can exist in the blood, and have them predict what the kidneys will do to maintain homeostasis.

Ask What happens when the blood contains excess water? *(The kidneys filter out some of the excess water.)*

Ask What do the kidneys do when the blood does not contain enough calcium? *(It does not filter out calcium.)*

SUMMARIZE Have students write three sentences that summarize how the kidneys maintain homeostasis. Have them discuss their summaries in small groups, evaluating why slightly different information might have been included in different summaries.

KEY QUESTION Have a student volunteer read the question and another volunteer read the answer.

Ask Why doesn't this filtration process remove blood cells? *(Blood cells are too large to pass through the walls of a glomerulus.)*

Build Vocabulary

WORD ORIGINS Point out the ureters and urethra on the visual of the Structure and Function of the Kidneys.

Ask Which other words in this section share the same root word? *(urine, urinary bladder)*

Tell students that another root word related to the kidneys is the Greek word *nephros,* which means "kidney."

Ask What vocabulary word from this lesson has *nephros* as a root word? *(nephron)*

Ask What kind of doctor is a nephrologist? *(a kidney specialist)*

Ask What kind of disease is nephritis? *(a disease of the kidneys)*

Speed Bump

Ask How does the color of urine relate to how much water a person's body has? *(Light colored urine indicates that the body has enough water. Dark colored urine indicates that the body is saving water.)*

During Reading continued

Active Reading

MAKE CONNECTIONS Ask students to share what they know about kidney disorders or transplants.

Ask Do you know anyone who has had a kidney transplant, kidney stones, or had to be on dialysis? *(Ask volunteers to share any experiences or reactions to the topics.)*

Ask How might high blood pressure cause kidney damage? *(The capillaries in the kidneys are quite fragile and can be permanently damaged by high blood pressure.)*

KEY QUESTION Ask students to cover the answer to the Key Question. Read the Key Question aloud. Ask volunteers to answer the question without looking at their books.

Review the function of the kidneys and the meaning of *homeostasis*.

Speed Bump

Ask What are three kidney disorders? *(kidney stones, kidney damage, and kidney failure)*

Ask How does dialysis help a person with kidney failure? *(The dialysis machine takes the place of the person's kidneys by filtering blood and returning it to the body.)*

Post Reading

ASSESSMENT TIPS Review the vocabulary terms and their definitions before answering Questions 1–3.

Allow students to work in pairs when answering Question 4. One student can trace the path of blood, and the other student can write the answer.

MYSTERY CLUE **Ask** How does the body respond when it needs to conserve water? *(It produces ADH.)*

Use students' answers to guide them to the conclusion that the boys' blood likely contains a high level of ADH.

Wrap-Up Activity

MODELING FILTRATION Add a liter of water to a small amount of a mixture of pebbles, sand, salt, and soil. Mix all the components together well. Pour the results through a strainer or a piece of window screen.

Ask How do the results compare to the filtration of blood by the kidneys? *(Water, the substances dissolved in the water, and very small particles pass through the strainer. Larger particles are too big to pass through.)*

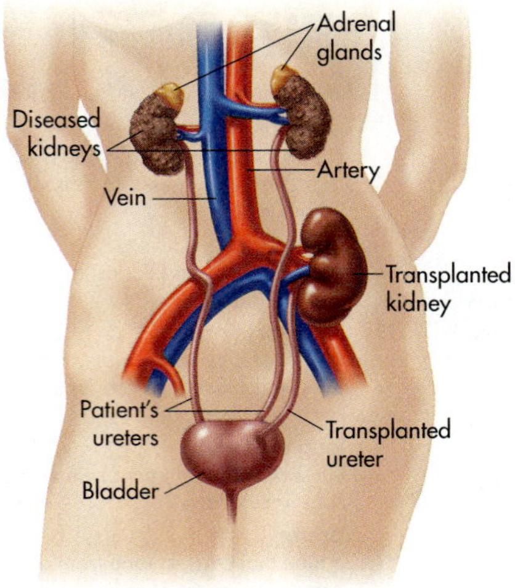

Kidney Transplantation A healthy kidney and ureter are transplanted from a donor. Diseased kidneys are left in place unless they are infected or causing high blood pressure.

Kidney Disorders If anything goes wrong with the kidneys, serious health problems will probably occur. Three of these problems are kidney stones, kidney damage, and kidney failure.

▶ ***Kidney Stones*** Sometimes minerals or uric acid salts in the urine form crystals. These crystals can grow into kidney stones. If the stones block a ureter, they cause great pain. Kidney stones can be broken into smaller pieces with sound waves. The pieces are then eliminated in the urine.

▶ ***Kidney Damage*** Many diseases, injuries, and dangerous substances can cause kidneys to function poorly. High blood pressure and diabetes are two of the leading causes of kidney damage in the United States.

▶ ***Kidney Failure*** Kidney failure occurs when the kidneys can no longer clean the blood and maintain homeostasis. Patients in kidney failure must have either dialysis or a kidney transplant. During dialysis, the blood is pumped through a machine that filters it. Dialysis is not painful. Dialysis patients must control their fluid intake and diet. In a kidney transplant, a person receives a kidney and ureter from a donor. The donor and the patient must have similar body chemistries. A person can survive with only one kidney.

Key Question How do the kidneys help maintain homeostasis? **The kidneys respond directly to blood composition and to hormones. Disruption of kidney function can lead to serious health problems.**

CHECK Understanding

Apply Vocabulary

Use the highlighted terms from the lesson to complete each sentence correctly.

1. The individual unit in the kidney where filtration and reabsorption takes place is called the _______.
2. The _______ is a tightly wound network of blood vessels that filters the blood.
3. Water reabsorption takes place in the _______.

Critical Thinking

4. **Interpret Visual** Refer back to the figure on page 730. List in order the structures that blood flows through in a kidney.
5. **Apply Concepts** Why do you think protein in the urine is a sign of kidney damage?
6. **Write to Learn** Answer the third clue of the mystery. What characteristics of the urine samples led you to your conclusion about ADH levels?

MYSTERY CLUE

Would Seth's and Philip's blood contain a high level or low level of ADH? (Hint: See p. 731.)

BIOLOGY.com Search Lesson 30.4 GO • Lesson Assessment

Check Understanding

1. nephron
2. glomerulus
3. loop of Henle
4. Waste-filled blood enters the kidney through the renal artery. Smaller arteries carry blood to the nephrons. Blood flows through the glomerulus. Blood flows out of the glomerulus through a small vein and re-enters the circulatory system through the renal vein.
5. Sample answer: In healthy kidneys proteins are too large to be filtered out of the blood. Protein in the urine means that the capillaries of the glomerulus are damaged and protein is entering the filtrate. Protein molecules are excreted because they are too large to be reabsorbed.
6. Their urine was a dark color, so the boys have high ADH levels. The kidneys are conserving water because blood water levels are low. ADH is secreted by the pituitary when blood does not contain enough water.

INQUIRY into Scientific Thinking

The Composition of Urine

The kidneys are selective filters. They remove urea, other impurities, and excess salts as blood passes through them. However, important substances such as water, protein and glucose stay in the blood. The collected waste products are excreted in urine. The table shows how the amounts of several important substances differ between the blood and the urine.

Study the data in the table. Notice that only glucose is lower in urine than in blood. All other substances have a higher concentration in the urine than in the blood.

How can you determine how much greater the concentration of a substance is in the urine than in the blood? You can compare the two amounts for a substance by using a ratio. Look at the calculation for potassium:

We want to know how much greater the amount of potassium is in the urine, so we will write our ratio as

$$\frac{\text{Concentration of potassium in urine}}{\text{Concentration of potassium in blood}} = \frac{0.20}{0.02}$$

To find out how many times more concentrated the potassium is in urine than in the blood, simplify the fraction:

$$\frac{0.20}{0.02} = \frac{10}{1}$$

For potassium, the fraction simplifies to 10/1. This means that potassium is ten times more concentrated in the urine than in the blood.

Concentrations of Selected Substances in Blood and Urine

Substance	Average Concentration in Blood (g/mL)	Average Concentration in Urine (g/mL)
Calcium	0.01	0.02
Glucose	0.10	0.00
Potassium	0.02	0.20
Sodium	0.32	0.60
Urea	0.03	2.00

Analyze and Conclude

1. Interpret Data

- Which substance has the highest concentration in the blood?
- Which substance has the lowest concentration in the blood?
- Which substance has the highest concentration in the urine?
- Which substance has the lowest concentration in the urine?

2. Calculate What is the ratio of urea in urine to urea in blood?

$$\frac{\text{Concentration of urea in urine}}{\text{Concentration of urea in blood}} = \frac{\quad}{\quad} = \frac{\quad}{1}$$

3. Predict Remember that urea is produced from ammonia, which is more toxic. Ammonia is produced when cells break down amino acids. Suppose someone eats a diet that is higher in protein than the person's usual diet. What do you think would happen to the urea concentrations in this person's blood and urine?

a. Urea increases in the blood and increases in the urine.

b. Urea increases in the blood and decreases in the urine.

c. Urea decreases in the blood and increases in the urine.

d. Urea decreases in the blood and decreases in the urine.

In Your Workbook Get help for this activity in your workbook.

Analyze and Conclude

1. sodium, calcium, urea, glucose

2. $\frac{\text{concentration of urea in urine}}{\text{concentration of urea in blood}} = \frac{2.00}{0.03} = 66.7$

The ratio is about 67 to 1.

3. a

LESSON 30.4

Inquiry Into Scientific Thinking

COMPARE/CONTRAST Remind students that a ratio is a comparison of two amounts and does not show a definite quantity. For example, a ratio of 8:3 for apples and oranges tells you that there are eight apples for every three oranges. It does not tell you how many apples or oranges are present.

Ask If you know a ratio and the number or amount of one item, can you find the number or amount of the other item? Give an example. *(Yes; If the ratio of sodium in urine to sodium in blood is 2:1, and you know there is 0.60 g of sodium in urine, you know there is about 0.30 g of sodium in blood.)*

MULTIPLE MEANINGS Be certain students understand the use of the term *concentration* in the table.

Ask In this table, what does the term *concentration* mean? (Concentration *relates to the amount of a substance dissolved in a certain volume of solvent. In this case, the solvent is water.)*

Ask What other meaning does *concentration* have? *(intense focusing on an object or a process)*

INTERPRET CHARTS AND TABLES Review the information in the table. Point out that the units used in this table (g/mL) are those commonly used for a solid dissolved in a liquid. Explain that all the substances listed are solids unless they are dissolved.

Ask If the blood and urine came from a group of people who all had diabetes, how would the concentration numbers change? *(The glucose levels for both blood and urine would be higher.)*

Ask Other than glucose, why are the concentrations of the substances higher in urine than in blood? *(The substances have been filtered out by the kidneys and the most of the water they were dissolved in is still in the blood.)*

Lead a Discussion

Discuss with students how the concentrations in the table relate to overall health.

Ask Could the concentration readings of a healthy person differ from the readings in the table? *(Yes; The concentrations listed are averages, which means that most readings are slightly higher or lower than the average. Desired concentrations are usually listed as a narrow range of readings.)*

Pre-Lab

Begin by having students identify the independent and dependent variables in this experiment. *(independent variable: whether or not the milk samples have milk-digestion aid added to them; dependent variable: the presence or absence of glucose in the milk samples)*

Ask If the purpose of the experiment is to evaluate the effect of milk-digestion aid on milk samples, why does one milk sample receive no milk-digestion aid? *(By testing a sample of pure milk, we can tell if a change has really taken place in the sample with milk-digestion aid.)*

Say A control is something used as a standard of comparison in evaluating experimental results.

Review the Background Questions and the Pre-Lab Questions as a class.

Lab

Tell students that they will perform the chapter lab *Digestion of Dairy Products* described in **Lab Manual B.**

SAFETY Students should wear goggles and aprons, and they should wash their hands thoroughly after completing the lab. Remind students to handle breakable items carefully and never to eat or drink any substance in the laboratory.

Post-Lab

DESIGN AN EXPERIMENT Challenge students to design an experiment that would determine the minimum dose of milk-digestion aid needed to break down 100 percent of the lactose in a glass of milk within 10 minutes.

NATIONAL SCIENCE EDUCATION STANDARDS

CONTENT C.4.c, C.5.d, F.1

INQUIRY A.1.b, A.2.a

Real-World Lab

GUIDED INQUIRY

Pre-Lab: Digestion of Dairy Products

Problem How can an enzyme deficiency affect digestion?

Materials spot plate, sheet of paper, glucose solution, milk, milk-digestion aid, toothpicks, glucose test strips, timer or clock

Lab Manual Chapter 30 Lab

Skills Focus Control Variables, Infer, Draw Conclusions

Connect to the Big idea Food is both a source of raw materials and a source of energy for your body. First the food must pass through your digestive system, where mechanical and chemical processes break the food down into smaller molecules. Enzymes play an essential role in chemical digestion. Different enzymes are needed to digest proteins, fats, and carbohydrates. In this lab, you will explore the role of enzymes in the digestion of milk and other dairy products.

Background Questions

a. **Review** What is an enzyme? Why are enzymes necessary for maintaining homeostasis?

b. **Review** Where do most digestive enzymes enter the digestive system?

c. **Compare and Contrast** Use glucose and sucrose to explain the difference between a monosaccharide and a disaccharide.

Pre-Lab Questions

Preview the procedure in the lab manual.

1. **Design an Experiment** What is the purpose of the glucose solution?
2. **Control Variables** What is the control in this lab?
3. **Communicate** Read the instructions on the package of glucose test strips. Then, briefly describe how you will test your samples for the presence of glucose.

BIOLOGY.com Search Chapter 30 GO

Visit Chapter 30 online to test yourself on chapter content and to find activities to help you learn.

Untamed Science Video Hold your noses as you join the Untamed Science crew to learn what scientists can discover about animals by investigating their scat.

Tutor Tube Tune in to Tutor Tube for another perspective on how homeostasis works in the human body.

Art Review Review your understanding of human body systems.

Art in Motion Watch peristalsis in action in the esophagus.

734 Chapter 30 • Pre-Lab

Pre-Lab Answers

BACKGROUND QUESTIONS

a. An enzyme is a protein that speeds up the rate of a biological reaction. Without enzymes, many biological reactions in cells could not take place (or at least not quickly enough to be useful).

b. Most digestive enzymes enter the duodenum, which is the upper part of the small intestine.

c. Monosaccharides, such as glucose, consist of single sugar molecules. Disaccharides contain two simple sugars joined together. The two molecules that join to form sucrose are glucose and fructose.

PRE-LAB QUESTIONS

1. The glucose solution will show what a positive test for glucose looks like.
2. The sample of milk without the milk-digestion aid is the control.
3. Depending on the type of test strips used, students will either dip a fresh strip into each solution or use toothpicks to place a bit of each solution on a single strip.

30 CHAPTER Summary

30.1 Organization of the Human Body

- The human body is organized into four levels: cells, tissues, organs, and organ systems.
- Homeostasis is a state of relative balance in an organism's physical and chemical conditions. Homeostasis continues even when internal and external environments change.

epithelial tissue (p. 714)
connective tissue (p. 714)
nervous tissue (p. 715)
muscle tissue (p. 715)
homeostasis (p. 717)
feedback inhibition (p. 717)

30.2 Food and Nutrition

- Molecules in food contain chemical energy. Your cells use this chemical energy to produce ATP. Food also supplies raw materials your body needs to build and repair tissues.
- The body needs water, carbohydrates, fats, proteins, vitamins, and minerals.
- A balanced diet provides the nutrients a person needs to maintain a healthful weight.

Calorie (p. 719)
carbohydrate (p. 720)
fat (p. 721)
protein (p. 721)
vitamin (p. 721)
mineral (p. 721)

30.3 The Digestive System

- The digestive system breaks down food into molecules the body can use in its cells. Food is processed in four phases—ingestion, digestion, absorption, and elimination.
- During digestion, food travels through the esophagus, stomach, and small intestine. Mechanical digestion and chemical digestion break food down into molecules that can be absorbed.
- The small intestine absorbs most of the nutrients released during chemical digestion. The large intestine absorbs water and some vitamins, and prepares wastes for eliminaion.

amylase (p. 724)
esophagus (p. 724)
peristalsis (p. 724)
stomach (p. 725)
pepsin (p. 725)
small intestine (p. 725)
villus (p. 727)
large intestine (p. 728)

30.4 The Excretory System

- The excretory system includes the skin, lungs, liver, and kidneys. This system excretes the wastes from chemical processes in the body.
- The kidneys filter the blood to remove urea, excess water and minerals, and other wastes.
- The composition of the blood helps determine what the kidneys remove during filtration. Hormones also tell the kidneys what substances to remove or keep. If the kidneys do not function properly, serious health problems can occur.

excretion (p. 729)
nephron (p. 729)
filtration (p. 731)
glomerulus (p. 731)
reabsorption (p. 731)
loop of Henle (p. 731)
ureter (p. 731)
urinary bladder (p. 731)
urethra (p. 731)

BIOLOGY.com Search Lesson 30 GO • Match It • Chapter Assessment

UbD Transfer the Big Idea

TRANSFER Explain to students that in the field of computer science, there is a common phrase, **garbage in, garbage out.** The phrase summarizes the idea that, no matter how good a computer or computer program is, it will not produce reliable results unless the data fed into the computer are accurate and reliable. Ask students to write a paragraph that explains how this phrase could be applied to the human body. Tell students to consider the role of materials entering the body in the maintenance of homeostasis, how the materials that enter the body are processed, and how the materials that leave the body reflect what enters the body and how well the body is functioning.

Chapter Summary

Think Visually

USE GRAPHICS Have students construct a flowchart that shows the path of a glucose molecule through a healthy nephron.

ANSWERS
Students' flowcharts should show a glucose molecule moving from the glomerulus into the Bowman's capsule as it is filtered out of the blood, then back into the capillary as it is reabsorbed.

BIOLOGY.com Study Online

REVIEW RESOURCES

Art Review Students can drag and drop labels to help them learn about the different structures and functions of human body systems.

Tutor Tube Online tutorials provide information about the nature of human body systems and homeostasis.

Data Analysis Students analyze their own Calorie and nutritional needs and plan a virtual meal that will help them meet those needs.

Art in Motion Students can watch an animation of peristalsis in the esophagus.

Vocabulary Review The *Flash Cards* and *Match It* provide an interactive way to review chapter vocabulary.

Editable Worksheets Pages of Study Workbook B, Lab Manual B, and the Assessment Resources Book are available online. These documents can be easily edited using a word-processing program.

Lesson Overviews Lesson Overviews are based on Miller & Levine Biology. However, they are available on CD-ROM and can be edited for your class.

ASSESSMENT RESOURCES

Chapter Assessment Have students take an online version of the Chapter 30 Check Understanding.

Standardized Test Prep Students can take an online version of the Standardized Test Prep. You will receive their scores along with ideas for remediation.

Diagnostic and Benchmark Tests Use these tests to monitor your students' progress and supply remediation.

ASSESSMENT

Assess

Big idea Ask a student volunteer to read the Big Idea question. On the board, write:

materials in —homeostasis→ materials out

Ask What types of materials are taken into the body? *(Answers include the six nutrients: water, carbohydrates, proteins, fats, vitamins, and minerals.)*

Ask What types of materials come out of the body? *(Answers include digestive wastes and the materials in urine, exhaled air, and perspiration.)*

Ask How is homeostasis involved in changing materials in to materials out? *(Accept any correct answer. Encourage students to focus on the involvement of homeostasis in the digestive and excretory systems.)*

CONSTRUCTED RESPONSE Paragraphs will vary. Explain the activity carefully. Be sure that students understand that the *Hints* should be used as a guide to help them answer the questions. Remind students that paragraphs have a main idea and details that support that main idea. When writing paragraphs, it is helpful to write the main idea and then fill in the details.

Say Question 1 refers to how systems work together to maintain balance in the human body. Although the question focuses on the digestive and excretory systems, the question assumes that other body systems still affect homeostasis.

Read the first hint.

Ask How do these nutrients affect the excretory system and homeostasis? *(Accept any answers that can be justified.)*

Read the second hint.

Ask Where do these wastes come from? *(the digestive system and chemical processes in the body)*

Model writing the first two or three sentences using the hints on the board and the text. *(Sample: The digestive and excretory systems work together to help maintain homeostasis. As food passes through the digestive system, nutrients that are needed by the rest of the body are absorbed into the bloodstream.)*

30 CHECK Understanding

Assess the Big idea Homeostasis

Write an answer to the question below.

Q: How are the materials that enter and leave your body related to the processes that maintain homeostasis?

Constructed Response

Write an answer to each of the numbered questions below. The answer to each question should be one or two paragraphs. To help you begin, read the **Hints** below the questions.

1. **How do the digestive and excretory systems work together to help the body maintain homeostasis?**

   Hint The digestive system absorbs nutrients needed to keep the body's cells healthy.

   Hint The excretory system removes wastes and regulates the body's water levels.

2. **Why is mechanical digestion important?**

   Hint Mechanical digestion breaks food into smaller particles.

   Hint Mechanical digestion mixes food with digestive juices.

3. **How could a failure to respond to ADH be dangerous to an athlete?**

   Hint The glomerulus filters water from the blood and the tubules reabsorb what is needed.

   Hint The pituitary releases ADH when water levels in the body are too low.

Foundations for Learning Wrap-Up

The pinwheels you prepared when you started the chapter are tools you can use to arrange your thoughts about the organization of the human body.

Activity 1 Working with a partner, go through the organs on each pinwheel. Discuss how the organ on each tab relates to the other organs and to the body system to which it belongs.

Activity 2 Working in small groups, create a flowchart that shows, in order, the organs through which food passes as it moves through the digestive system.

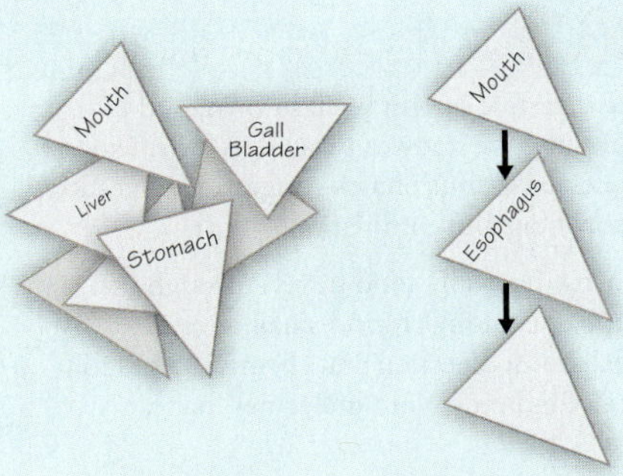

For Question 2, pair students and have them use the hints and your modeling to answer the question. Have students share their paragraphs with the class.

For Question 3, have students write the paragraphs independently for homework or as an anchor activity.

Foundations for Learning Wrap-Up

Have students use the pinwheel they completed as they progressed through this chapter. Remind students that the information written on the back of the pinwheel is available to help answer the questions.

Have student pairs complete the activities. Remind them that they also can refer to the Body System diagram in Lesson 30.1.

30.1 Organization of the Human Body

Understand Key Concepts

1. What do all types of tissue have in common?
   a. They are all made of connective tissue.
   b. They are all made of cells.
   c. They are all found in every organ.
   d. They are all made of organs.
2. Why is it important for an organism to maintain homeostasis?
3. Name the four types of tissues and describe one characteristic of each.

Think Critically

4. **Explain** Would you classify blood as a cell, tissue, or an organ? Explain.

30.2 Food and Nutrition

Understand Key Concepts

5. Inorganic nutrients that your body needs, usually in small amounts, are called
   a. vitamins.
   b. minerals.
   c. proteins.
   d. amino acids.

Test-Taking Tip

Watch for Qualifiers Sometimes questions have qualifiers that are clues to which answers are wrong. A qualifier gives you a limit, such as an amount. In question 5, you are told that the nutrients are usually needed in *small* amounts. Proteins and amino acids are needed in large amounts. So **c** and **d** are wrong answers. In this question you are also given the qualifier *inorganic*. So you can eliminate **a**, because vitamins are organic nutrients.

6. Energy in food is measured in
   a. ATP.
   b. fats.
   c. Calories.
   d. disaccharides.
7. Which nutrients provide the body with energy?
8. In what three ways are proteins important to the body?

Think Critically

9. **Infer** Many food manufacturers have replaced trans fats in their foods with other types of fats. The replacement fats have a lower level of heart disease risk. Some nutritionists fear that people will think foods such as French fries and doughnuts are healthful if they are not made with trans fats. Explain why these foods are still not healthful choices.

30.3 The Digestive System

Understand Key Concepts

10. Where does mechanical digestion begin?
    a. the esophagus
    b. the large intestine
    c. the mouth
    d. the small intestine
11. An enzyme in saliva that can break the chemical bonds in starch is
    a. pepsin.
    b. bile.
    c. amylase.
    d. chyme.
12. Explain why swallowed food does not normally enter the airway leading to the lungs.
13. Describe the functions of the pancreas.

Think Critically

14. **Relate Cause and Effect** Suppose that your doctor prescribed an antibiotic for an infection. The antibiotic killed all the bacteria in your body. What effect would this have on your digestive system?

30.4 The Excretory System

Understand Key Concepts

15. Urine is excreted from the body through the
    a. ureter.
    b. urinary bladder.
    c. urethra.
    d. renal vein.

Check Understanding 737

Assessment Tips

Have students review body organization, from cells to body systems, before answering Questions 3 and 4.

Before students answer Questions 5–9, have them list the six nutrients necessary for a healthy body.

Have students examine the full-page diagram of the digestive system before answering Questions 10 and 12.

Answers

Lesson 30.1

UNDERSTAND KEY CONCEPTS

1. b

2. Maintaining homeostasis provides a stable internal environment, which allows vital body processes, such as the functioning of enzymes, to occur. An organism that does not maintain homeostasis will die.

3. Sample answer: Epithelial tissue lines the inner and outer surfaces of the body. Connective tissue provides support for the body and connects its parts. Nervous tissue transmits nerve impulses. Muscle tissues are responsible for voluntary and involuntary body movements.

THINK CRITICALLY

4. Blood is a tissue because it is a group of cells that perform a function.

Lesson 30.2

UNDERSTAND KEY CONCEPTS

5. b

6. c

7. fats, carbohydrates, and proteins

8. Proteins supply raw materials for growth and repair of body structures; some proteins make up enzymes; and some proteins function in regulatory and transport functions.

THINK CRITICALLY

9. French fries and doughnuts supply a great number of Calories without providing many of the nutrients the body needs. These foods provide carbohydrates and fats, but very few raw materials for growth and repair of tissues. They also do not have a good balance of vitamins and minerals.

Lesson 30.3

UNDERSTAND KEY CONCEPTS

10. c

11. c

12. A flap of tissue called the epiglottis closes over the airway when a person swallows. This keeps food out of the airway.

13. The pancreas produces hormones that regulate blood sugar. It produces enzymes that break down carbohydrates, proteins, lipids, and nucleic acids. It also produces sodium bicarbonate, a base that neutralizes stomach acids.

THINK CRITICALLY

14. An antibiotic that killed all the bacteria in your body would destroy the beneficial bacteria in the large intestine. A vitamin K deficiency could result because these bacteria produce vitamin K.

Lesson 30.4

UNDERSTAND KEY CONCEPTS

15. c

16. a

17. The skin excretes water, salts, and small amounts of urea.

18. Water, urea, glucose, salts, amino acids, and some vitamins are filtered from the blood. Blood cells and large substances such as proteins remain in the blood.

THINK CRITICALLY

19. Without the kidneys to filter wastes and excess water from the blood, these substances can accumulate to toxic levels.

Connecting Concepts

20. hydrochloric acid, protein, fat

21. In the presence of hydrochloric acid, the pancreatic secretions are mostly bicarbonate, with a small amount of digestive enzymes. In the presence of fat, the reverse is true. The secretions are mostly digestive enzymes with a small amount of bicarbonate.

CHAPTER MYSTERY After students have read through the Chapter Mystery, discuss how the body's waste products reflect what is put into the body. Ask students to explain how urine can be used to identify diseases and disorders that disrupt homeostasis in the body.

Ask How could a physical examination of urine reveal that a person is dehydrated? *(If urine is dark yellow, it indicates that a person's kidneys are conserving water.)*

Ask What problems can be revealed by microscopic examination of urine? How do these problems relate to homeostasis? *(Kidney stones, metabolic problems, and infection can all be detected by microscopic examination of urine. Each of these problems disrupts homeostasis in the body by disrupting the normal functions of the body.)*

Ask How does chemical examination of urine reveal the relationship between the materials that enter the body and those that leave the body? *(Materials that enter the body are processed, and their byproducts are eliminated as wastes. Chemical examination of urine, a waste that leaves the body, can reveal what has been taken into the body.)*

30 CHECK Understanding

16. Which of the following is the basic functional unit of the kidney?

a. nephron
b. glomerulus
c. Bowman's capsule
d. loop of Henle

17. What is the role of the skin in excretion?

18. What materials are filtered from the blood in the kidney? What materials are not filtered from the blood?

Think Critically

19. **Apply Concepts** Explain why kidney failure can be a fatal condition.

Connecting Concepts

Use Science Graphics

Pancreatic secretions contain sodium bicarbonate and enzymes. The graph shows the secretions of the pancreas in response to three different substances in chyme. Use the graph to answer questions 20 and 21.

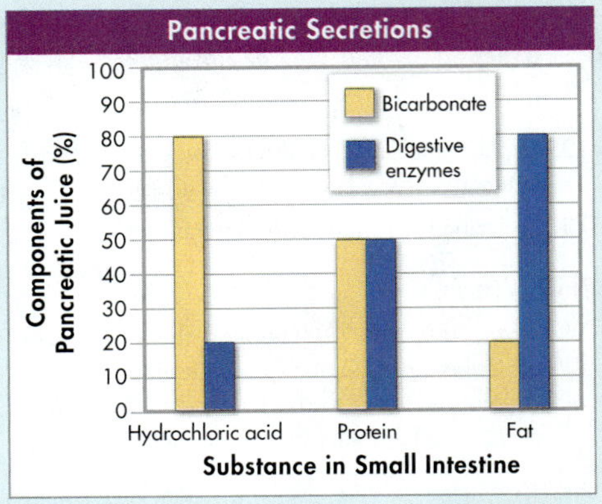

20. **Interpret Visuals** Each pair of bars represents the response of the pancreas to a different variable. What are the three variables?

21. **Compare and Contrast** Compare the composition of pancreatic secretions in the presence of hydrochloric acid and fat.

solve the CHAPTER MYSTERY

THE TELLTALE SAMPLE

For centuries, people have studied urine for clues to health and disease. For example, the Greeks knew that diabetics had too much sugar in their urine. They called this disease *diabetes mellitus. Mellitus* is the Greek word for "honey."

- **Physical Examination** First, doctors check the color of urine and look for cloudiness. The shade of yellow indicates the amount of water the kidneys are removing from the blood. Urine of a color other than yellow could indicate the presence of blood. Urine should be clear rather than cloudy.
- **Microscopic Examination** Mucus, white blood cells, or microorganisms in urine indicate a possible infection. Crystals may also cause cloudy urine. Crystals in the urine may indicate kidney stones or a problem with the chemical processes in the body.
- **Chemical Examination** Hundreds of chemical tests can be performed on urine. Chemical dipsticks that change color show that other chemicals are present. These tests can reveal a lot about kidney and liver function. They also reveal a lot about overall homeostasis in the body.

1. **Infer** How does urine reveal so much about the health of the human body?

2. **Form an Opinion** Most drug urine tests done for schools do not test for alcohol or tobacco. Why do you think this is the case? Do you agree or disagree? Explain.

Never Stop Exploring Your World. Finding the solution to the Telltale Sample mystery is only the beginning. Take a video field trip with the ecogeeks of Untamed Science to see where this mystery leads.

CHAPTER MYSTERY ANSWERS

1. Urine contains many waste products of normal reactions that occur in the body. It also contains excess materials that the body does not need. By examining its contents, a doctor can determine if many different body processes are occurring as they should.

2. Answers may vary. Accept responses supported by logical explanations.

Follow the Untamed Science crew in **Drat . . . I Smell Scat!** as they discover the treasure of information that can be unearthed from a most unlikely place: animal scat.

Standardized Test Prep

Multiple Choice

1. Which of the following is NOT a kind of tissue in the human body?
   A epithelial
   B connective
   C interstitial
   D nervous

2. Each of the following aids in the process of digestion EXCEPT
   A teeth
   B saliva
   C stomach
   D kidney

3. In the human body, hydrochloric acid is responsible for the low pH of the contents of the
   A kidney
   B gallbladder
   C stomach
   D liver

4. Which is NOT a function of the kidneys?
   A removal of waste products from the blood
   B maintenance of blood pH
   C regulation of water content of the blood
   D excretion of carbon dioxide

5. The main function of the digestive system is to
   A break down large molecules into smaller molecules.
   B excrete oxygen and carbon dioxide.
   C synthesize minerals and vitamins needed for a healthy body.
   D remove waste products from the blood.

6. In the kidneys, both useful substances and wastes are removed from the blood by
   A reabsorption.
   B excretion.
   C dialysis.
   D filtration.

7. Which of the following is NOT a role of fats in the body?
   A Deposits of fat act as insulation.
   B They are components of cell membranes.
   C They help with absorption of fat-soluble vitamins.
   D They are enzymes.

Questions 8–9

A student is studying the effect of temperature on the action of an enzyme in stomach fluid. The enzyme digests protein. An investigation was set up using five identical test tubes. Each tube contained 40 mL of stomach fluid and 20 mm of glass tubing filled with gelatin. Each tube was subjected to a different temperature. After 48 hours, the amount of gelatin digested in each tube was measured in millimeters. The results for the five test tubes are shown in the table.

Effect of Temperature on Enzyme Action

Test Tube	Temperature (°C)	Amount of Digestion After 48 Hours
1	2	0.0 mm
2	10	3.0 mm
3	22	4.5 mm
4	37	8.0 mm
5	100	0.0 mm

8. Which is the manipulated (independent) variable in this investigation?
   A gastric fluid
   B length of glass tubing
   C temperature
   D time

9. Another test tube was set up that was identical to the other test tubes and placed at a temperature of 15°C for 48 hours. What amount of digestion would you expect to occur in this test tube?
   A less than 3.0 mm
   B between 3.0 mm and 4.5 mm
   C between 4.5 mm and 8.0 mm
   D more than 8.0 mm

Open-Ended Response

10. Fad diets that boast of rapid weight loss often become popular. Many of these diets involve eating only a limited variety of foods. Explain why these diets are an unhealthful way to lose weight.

If You Have Trouble With . . .

Question	1	2	3	4	5	6	7	8	9	10
See Lesson	30.1	30.3	30.3	30.4	30.3	30.4	30.2	30.3	30.3	30.2

Test-Taking Tip

INTERPRET EXPERIMENTAL DATA

Tell students that when a question is based on experimental data, they should read the description of the experiment carefully to determine the steps that were followed. As students read the description, they should try to identify the independent and dependant variables, the hypothesis, and the controls that were used. Then, students should examine the data to identify trends that show relationships between the variables. These steps will help students correctly answer questions about the experimental procedure and results.

Before the Test

Review with students the Test-Taking Tip. Ask them to determine which questions on this page can be best solved by using the tip. *(8 and 9)*

Point out the word *NOT* in Questions 1, 4, and 7, and the word *EXCEPT* in Question 2.

Remind students that because open-ended questions do not have a specific answer, they can sometimes use facts based on personal experience for answers. Have any of them been on a rapid-loss diet? How did the diet affect their bodies?

Answers

1. C
2. D
3. C
4. D
5. A
6. D
7. D
8. C
9. B
10. Diets that include a limited variety of food may be unhealthy because they might not provide all of the nutrients needed for good health.

CHAPTER PLANNER 31 Nervous System

Chapter Contents	NSES	Time	Core Resources
Chapter Preview			**Student Edition,** pp. 740–741 **Chapter Mystery,** p. 741 **Foundations for Learning,** p. 741
31.1 The Neuron Functions of the Nervous System • Neurons • The Nerve Impulse	I, V, C.1.a, C.6.a, C.6.b	$1\frac{1}{2}$ periods $\frac{3}{4}$ block	**Student Edition,** pp. 742–746 **Build Connections,** p. 745 **Study Workbook B** 31.1 Worksheets **Biology.com** *InterActive Art:* The Nerve Impulse • *Visual Analogy:* The Moving Impulse • 31.1 Lesson Assessment
31.2 The Central Nervous System The Brain and the Spinal Cord • Addiction and the Brain	I, V, C.6.a, C.6.b, F.1	1 period $\frac{1}{2}$ block	**Student Edition,** pp. 747–750 **Build Connections,** p. 748 **Study Workbook B** 31.2 Worksheets **Biology.com** 31.2 Lesson Assessment
31.3 The Peripheral Nervous System The Sensory Division • The Motor Division	I, V, C.6.a, C.6.b	1 period $\frac{1}{2}$ block	**Student Edition,** pp. 751–753 **Study Workbook B** 31.3 Worksheets **Biology.com** *Art in Motion:* Reflex Arc • 31.3 Lesson Assessment
31.4 The Senses Touch and Related Senses • Smell and Taste • Hearing and Balance • Vision	I, V, A.1.c, A.2.a, C.6.a, C.6.b	1 period $\frac{1}{2}$ block	**Student Edition,** pp. 754–757 **Inquiry Into Scientific Thinking,** p. 755 **Study Workbook B** 31.4 Worksheets **Biology.com** *Tutor Tube:* The Physical Nature of Sound • *Art Review:* The Ear and the Eye • 31.4 Lesson Assessment
Chapter Pre-Lab	V, A.1.b, A.2.a, C.1.f, F.1	1 period $\frac{1}{2}$ block	**Student Edition,** p. 758 **Lab Manual B** *Testing Sensory Receptors for Touch*

Tools for ELL Instruction

ELL Handbook describes teaching strategies to make *Miller & Levine Biology: Foundation Edition* more accessible to ELL students.

Spanish Study Workbook is a Spanish translation of Study Workbook A.

Multilingual Glossary is the *Miller & Levine Biology* glossary translated into ten languages.

Tools to Challenge Students

Study Workbook A includes more challenging worksheets and activities such as Chapter Mystery activities that reinforce 21st Century Skills.

Lab Manual A labs are geared toward on-level students. The Lab Manual includes fifteen labs in addition to the Chapter Labs.

Biology.com Data Analysis activities offer students a way to practice data gathering and analyzing skills.

Visual Quizzes in the Assessment Resources Book are geared toward on-level students and use visuals from *Miller & Levine Biology.*

Additional Resources

Biology.com Untamed Science Video • Flash Cards

Study Workbook A 31.1 Worksheets
Spanish Study Workbook 31.1 Worksheets
Assessment Resources Book Visual Quiz

Study Workbook A 31.2 Worksheets
Spanish Study Workbook 31.2 Worksheets

Study Workbook A 31.3 Worksheets
Spanish Study Workbook 31.3 Worksheets
Assessment Resources Book Visual Quiz

Study Workbook A 31.4 Worksheets
Lab Manual B *Data Analysis:* Sound Intensity
Spanish Study Workbook 31.4 Worksheets
Biology.com *Data Analysis:* Looks Tasty: How Vision Affects Taste

Lab Manual A *Testing Sensory Receptors for Touch*

Chapter Review

Student Edition Chapter Summary, p. 759
Study Workbook B Chapter 31 Review
Transparencies, pp. 342–350
Biology.com Untamed Science Video • Editable Worksheets of Study Workbooks A and B and Lab Manuals A and B • Chapter 31 Flash Cards and Match It

Untamed Science DVD • Classroom Resources CD (includes lesson presentations and editable worksheets)

Chapter Assessment

Student Edition Check Understanding, pp. 760–763
Study Workbook B Chapter 31 Taking a Standardized Test
Assessment Resources Book Chapter 31 Test B
Biology.com Chapter 31 Assessment • Editable Worksheets of Chapter 31 Visual Quizzes* and Chapter 31 Tests A and B

Exam*View* *Assessment Suite* • Classroom Resources CD (includes lesson presentations and editable worksheets)

Time: 1 period, 1/2 block

* Visual Quizzes are based on visuals from *Miller & Levine Biology*. Review carefully to make sure they are appropriate for your class.

Pressed for Time?

Preview the Chapter Preview the visuals The Neuron and The Brain, and read the first Key Question in Lessons 31.1, 31.2, and 31.3.

Cover the Chapter Quickly Have students read *Functions of the Nervous System* and *Neurons* in Lesson 31.1. Go over the visuals The Moving Impulse and The Synapse. Assign *The Brain and the Spinal Cord* in Lesson 31.2, and *The Sensory Division* and the introduction to *The Motor Division* in Lesson 31.3. Discuss the visual Reflex Arc.

Assess Assign Questions 1 and 5 in the 31.1 Check Understanding, Question 4 in the 31.2 Check Understanding, and Questions 4 and 5 in the 31.3 Check Understanding. In the Chapter 31 Check Understanding, assign Questions 1, 2, 5, 6, 8, and 11–13.

Get the Picture

Use the photo of the baseball game to discuss how the human body gathers information about its surroundings and how it reacts to that information.

Ask What are the people looking at? *(Some are watching the game, others are looking at the ball.)*

Ask Why do the people in the center have their mitts in the air? *(They are going to try to catch the ball.)*

Ask What had to happen for these fans to respond in this way? *(Sample answer: They had to see the ball coming. They had to stand up and reach in the direction of the ball.)*

Say The subject of the chapter is the nervous system.

Ask How do you think the nervous system was involved in the fans' reaction to the ball? *(Answers may vary. Students should recognize that the senses, the brain, and directions to muscles probably played a role in the fans' actions.)*

Activate Prior Knowledge

Discuss some of the possible actions or reactions that might be involved in trying to catch the ball.

Ask Have you ever tried to catch something that was just out of your reach?

Point out that your hands and legs needed to coordinate so you would not get off balance. Your body needed to position your arms and hands near where you expected the object to fall. Maybe you needed to make adjustments for the object's motion and had to shift to the right or left.

Discuss how all of this seeing, thinking, and moving connects to the nervous system. Elicit words such as *nerves, senses, brain, signals,* and *muscles.*

NATIONAL SCIENCE EDUCATION STANDARDS

UNIFYING CONCEPTS AND PROCESSES

I, V

CONTENT

C.1.a, C.1.f, C.6.a, C.6.b, F.1

INQUIRY

A.1.b, A.1.c, A.2.a

31 The Nervous System

Big idea **Structure and Function**

Q: How does the structure of the nervous system allow it to regulate functions in every part of the body?

BIOLOGY.com Search Chapter 31 GO • Flash Cards

740

UbD Connect to the Big Idea

Invite a volunteer to read aloud the chapter number and title, the Big Idea, and the question. Emphasize the question, **How does the structure of the nervous system allow it to regulate functions in every part of the body?**

Set the Purpose

PREVIEW AND PREDICT Draw the graphic organizer on the board. Guide students to the INSIDE feature on the next page, and invite a volunteer to read each lesson number and title.

Ask How do you think this title relates to the questions in the graphic organizer? Remind students to search for the answers as they read the chapter. Remind them to look for how the structures and functions of the nervous system relate to one another.

INSIDE:

- 31.1 The Neuron
- 31.2 The Central Nervous System
- 31.3 The Peripheral Nervous System
- 31.4 The Senses

The sights, sounds, and smells at a ball game provide a fan's nervous system with a lot of stimulation.

- Untamed Science Video
- Chapter Mystery

CHAPTER MYSTERY

POISONING ON THE HIGH SEAS

From the middle to late 1700s, Captain James Cook commanded several voyages of discovery. During this time, he sailed to the South Pacific for Great Britain. His discovery of new lands brought him many riches. But the discoveries of new animals were not always pleasant.

September 7, 1774, was a remarkable day on the HMS *Resolution*. The ship's butcher died from a fall, and there was a solar eclipse. A clerk also traded some cloth for a freshly caught fish. Captain Cook ate only a few bites of the fish. But within a few hours, the captain felt an extraordinary weakness in his limbs. He also lost all sense of touch, and he could not sense the weight of objects. It took eleven days for the men who ate the fish to recover. A pig and dog that ate some of the fish's organs were dead by the next morning.

Read for Mystery Clues Look for clues as to how eating this fish could produce such deadly effects. Then, solve the mystery.

FOUNDATIONS for Learning

The nervous system is the control center for the body. Before you read the chapter, cut out about twenty 10 cm × 10 cm squares of paper. Punch small holes in each corner of the squares. As you read the chapter, write a structure or function of the nervous system on each square. Use the squares to keep track of important facts about each structure or function. At the end of the chapter are two activities that use these squares to help you answer the question: How does the structure of the nervous system allow it to regulate functions in every part of the body?

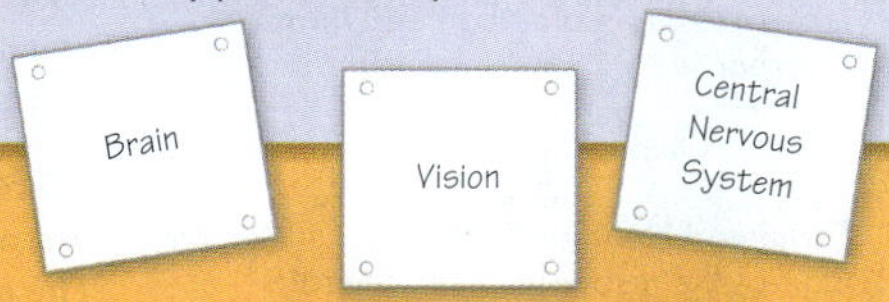

Chapter 31
Big Idea: Structure and Function

Chapter 31: How does the structure of the nervous system allow it to regulate functions in every part of the body?

- **31.1:** What is the basic function of the nervous system?
- **31.2:** What does the central nervous system do?
- **31.3:** How do messages travel between the central nervous system and the body?
- **31.4:** How do humans sense changes in their environment?

Foundations for Learning

Provide students with paper and hole punches and demonstrate how to make the paper squares. Students can also punch holes using a sharp pencil or a pen. Alternatively, pass out pre-cut, hole-punched paper squares.

Familiarize students with the activity by modeling one structure or function on one of the squares Make sure students understand the value of this activity. The cards give students an opportunity to write brief descriptions for future study aids. Students can sort the cards to look at how different parts of the nervous system are related.

Untamed Science

Students can take a video field trip to learn more about the effects of animal toxins in the video **Toxic Secretions**.

CHAPTER MYSTERY

Ask students if they know anyone who has gotten sick after eating a certain food. Invite a student to read aloud the Chapter Mystery. Have students predict how the mystery relates to the chapter title. Encourage them to predict what caused Captain Cook and the sailors to get sick and what killed the pig and the dog.

Tell students they will find clues in the chapter to answer the question, **How could eating fish have such deadly effects?**

Digital Support

BIOLOGY.com Extend your reach by using the digital assets offered at **Biology.com.**

LESSON 31.1

Pre-Reading

Objectives

31.1.1 Identify the functions of the nervous system.

31.1.2 Describe the function of neurons.

31.1.3 Describe how a nerve impulse is transmitted.

Preview the Pages

1. Preview the vocabulary words by having students find the words in the text, read the sentences in which the words appear, and define the words using context clues.
2. Preview the visuals with students. Point out that many of the visuals present a process or sequence of events. Stress to students that the order in which the events are shown is important. Use the sequence chart on this page as an example.
3. Review the following abbreviations with students: *V* for voltage, *mV* for millivolts, *Na* for sodium, *K* for potassium, and the + and – superscripts used to identify ions.

Build Understanding

FLOWCHART Direct students to Workbook B, where they will find a partially completed flowchart. Point out that each box in the chart describes an action, or step, in the process. Remind students that a flowchart will help them keep the steps in a process in the correct order.

NATIONAL SCIENCE EDUCATION STANDARDS

UNIFYING CONCEPTS AND PROCESSES

I, V

CONTENT

C.1.a, C.6.a, C.6.b

For the corresponding lesson in *Miller & Levine Biology,* see pages 896–900.

31.1 The Neuron

Key Questions

What are the functions of the nervous system?

What is the function of neurons?

How does a nerve impulse begin?

BUILD Understanding

Flowchart Create a flowchart to show information flow in the nervous system.

In Your Workbook Go to your workbook for help on creating the flowchart. Complete the flowchart in Lesson 31.1.

Functions of the Nervous System

We are all aware of the world outside our bodies. But how do you know about that world? When you opened the book to this page, how did you make this happen? How did the words that you are reading right now get into your mind? The answers to these questions are found in the nervous system.

The nervous system is our window on the world. The nervous system collects information about the body's internal and external environment. The system then processes that information and responds to it. These functions are accomplished by the peripheral nervous system and the central nervous system. The **peripheral nervous system** consists of nerves and supporting cells. This system collects information about the body's external and internal environment. The **central nervous system** consists of the brain and spinal cord. This system processes the information collected by the peripheral nervous system and creates a response. The response is delivered to the appropriate body part through the peripheral nervous system.

Think about what happens when you search for a pencil in your backpack. Your fingertips send information to your central nervous system about the objects you are touching. Your brain processes the information. It determines that the first object you touch is too square to be a pencil. Then your brain sends messages via your peripheral nervous system to the muscles in your hand. The messages command them to keep searching.

Information Flow in the Nervous System

Peripheral Nervous System
Gathers information and sends it to the central nervous system

↓ Input

Central Nervous System
Processes the information and forms a response

↓ Output

Peripheral Nervous System
Carries the response of the central nervous system to glands and muscles

Imagine the billions of messages that are sent throughout your body at any given moment. The messages may tell you to laugh at a funny joke. Or they may tell your brain that it's cold outside. These messages allow the different organs of the body to act together. They also allow the organs to react to conditions in the world around us. How does this communication occur?

Key Question What are the functions of the nervous system?

The nervous system collects information about the body's internal and external environment. The system then processes that information and responds to it.

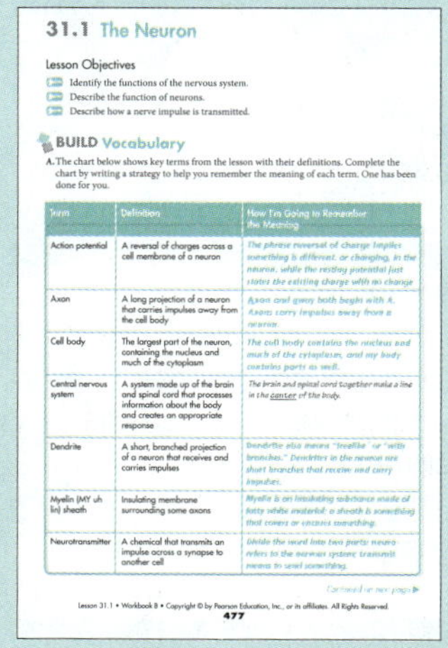

31.1 The Neuron

Lesson Objectives

Identify the functions of the nervous system.
Describe the function of neurons.
Describe how a nerve impulse is transmitted.

BUILD Vocabulary

A. The chart below shows key terms from the lesson with their definitions. Complete the chart by writing a strategy to help you remember the meaning of each term. One has been done for you.

Term	Definition	How I'm Going to Remember the Meaning
Action potential	A reversal of charges across a cell membrane of a neuron	The phrase reversal of charge implies something is different, or changing, in the neuron, while the resting potential just states the existing charge with no change
Axon	A long projection of a neuron that carries impulses away from the cell body	Axon and away both begin with A. Axons carry impulses away from a neuron.
Cell body	The largest part of the neuron, containing the nucleus and much of the cytoplasm	The cell body contains the nucleus and much of the cytoplasm, and my body contains parts as well.
Central nervous system	A system made up of the brain and spinal cord that processes information about the body and creates an appropriate response	The brain and spinal cord together make a line in the center of the body.
Dendrite	A short, branched projection of a neuron that receives and carries impulses	Dendrite also means "treelike" or "with branches." Dendrites in the neuron are short branches that receive and carry impulses.
Myelin (MY uh lin) sheath	Insulating membrane surrounding some axons	Myelin is an insulating substance made of fatty white material; a sheath is something that covers or encases something.
Neurotransmitter	A chemical that transmits an impulse across a synapse to another cell	Divide the word into two parts: neuro refers to the nervous system; transmit means to send something.

Continued on next page ▶

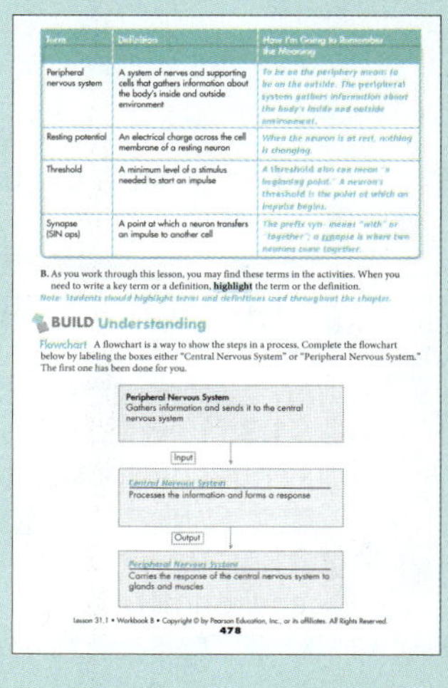

Term	Definition	How I'm Going to Remember the Meaning
Peripheral nervous system	A system of nerves and supporting cells that gathers information about the body's inside and outside environment	To be on the periphery means to be on the outside. The peripheral system gathers information about the body's inside and outside environment.
Resting potential	An electrical charge across the cell membrane of a resting neuron	When the neuron is at rest, nothing is changing.
Threshold	A minimum level of a stimulus needed to start an impulse	A threshold also can mean "a beginning point." A neuron's threshold is the point at which an impulse begins.
Synapse (SIN aps)	A point at which a neuron transfers an impulse to another cell	The prefix syn- means "with" or "together"; a synapse is where two neurons come together.

B. As you work through this lesson, you may find these terms in the activities. When you need to write a key term or a definition, **highlight** the term or the definition.
Note: Students should highlight terms and definitions used throughout the chapter.

BUILD Understanding

Flowchart A flowchart is a way to show the steps in a process. Complete the flowchart below by labeling the boxes either "Central Nervous System" or "Peripheral Nervous System." The first one has been done for you.

Peripheral Nervous System
Gathers information and sends it to the central nervous system

Input

Central Nervous System
Processes the information and forms a response

Output

Peripheral Nervous System
Carries the response of the central nervous system to glands and muscles

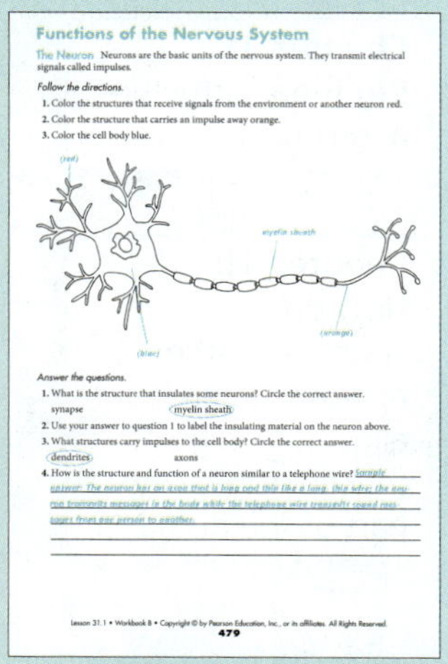

Functions of the Nervous System

The Neuron Neurons are the basic units of the nervous system. They transmit electrical signals called impulses.

Follow the directions.

1. Color the structures that receive signals from the environment or another neuron red.
2. Color the structure that carries an impulse away orange.
3. Color the cell body blue.

Answer the questions.

1. What is the structure that insulates some neurons? Circle the correct answer.
   synapse myelin sheath
2. Use your answer to question 1 to label the insulating material on the neuron above.
3. What structures carry impulses to the cell body? Circle the correct answer.
   dendrites axons
4. How is the structure and function of a neuron similar to a telephone wire? Sample answer: The neuron has an axon that is long and thin like a long, thin wire; the neuron transmits messages in the body while the telephone wire transmits sound messages from one person to another.

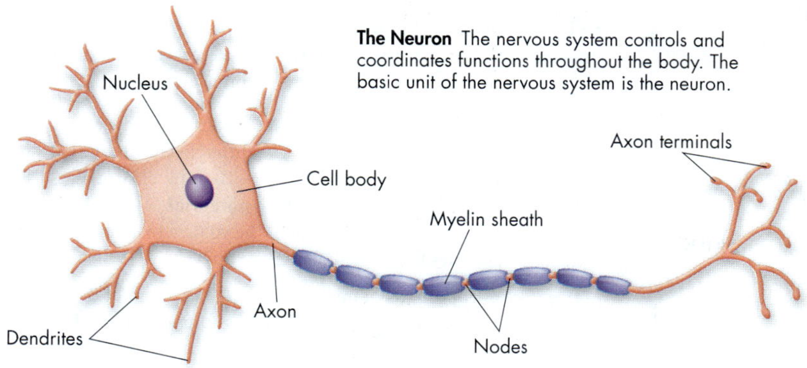

The Neuron The nervous system controls and coordinates functions throughout the body. The basic unit of the nervous system is the neuron.

Neurons

The messages carried by the nervous system are electrical signals called impulses. Impulses are carried by cells called neurons.

Types of Neurons Neurons can be classified into three types based on the direction in which an impulse travels. Sensory neurons carry impulses from the sense organs to the spinal cord and brain. Motor neurons carry impulses from the brain and the spinal cord to muscles and glands. Interneurons do the high-level work. They process information from sensory neurons. Then they send commands to other interneurons or to motor neurons.

Structure of Neurons Neurons come in many shapes and sizes, but they all share certain features. The largest part of a typical neuron is its **cell body.** The cell body contains the nucleus and much of the cytoplasm.

Short, branched extensions called **dendrites** spread out from the cell body. They receive impulses from other neurons and carry impulses to the cell body. The long fiber that carries impulses away from the cell body is the **axon.** The axon ends in a series of small swellings called axon terminals. A neuron may have dozens of dendrites, but usually only one axon. Axons and dendrites of different neurons form bundles of fibers called nerves.

The axons of some neurons are insulated by a membrane known as the myelin (MY uh lin) sheath. The myelin sheath around a long axon has many gaps, called nodes. Impulses move along the axon by jumping from one node to the next. Impulses travel faster along these axons than along axons without myelin.

Key Question What is the function of neurons?
Nervous system impulses are carried by cells called neurons.

BUILD Vocabulary

peripheral nervous system
a network of nerves and supporting cells that carries signals into and out of the central nervous system

central nervous system
a system that includes the brain and spinal cord; processes information and creates a response that it delivers to the body

cell body
the largest part of a typical neuron; contains the nucleus and much of the cytoplasm

dendrite
an extension of the cell body of a neuron that carries impulses from the environment or from other neurons toward the cell body

axon
a long fiber that carries impulses away from the cell body of a neuron

MULTIPLE MEANINGS
The dendrites on the cell body of a neuron give the neuron a treelike shape. In geology, a dendrite is a crystal with a treelike shape.

During Reading

Active Reading

MAKE CONNECTIONS Help students understand that sending and receiving messages to and from people around them is similar to how the body sends and receives messages to and from the brain.

Ask How do you send information to your friends? *(Sample answers: talking in person or on the phone, using e-mail, sending text messages)*

Ask How does the body transmit messages to and from the brain? *(through the nervous system)*

KEY QUESTION Ask a volunteer to read the Key Question and answer on the previous page aloud.

Ask What is the nervous system? *(the body system that collects information about the internal and external environment, processes that information, and responds to it)*

Ask Which part of the nervous system collects information? *(the peripheral nervous system)* Which part of the nervous system processes and responds to the information? *(the central nervous system)*

USE VISUALS Have students copy the neuron diagram onto a large sheet of paper. Ask them to label the neuron's parts as shown in the diagram and write a definition for each labeled part. Then, have each student exchange drawings with a partner and check the partner's work.

Build Vocabulary

MULTIPLE MEANINGS Ask students what the two meanings of the word *dendrite* have in common. *(they both refer to structures that are shaped like a tree)* Show students a map of a river and its tributaries. Explain that the tree-like pattern of a river system is called *dendritic.*

Have students find the word *impulse* on this page.

Ask What does *impulse* mean the way it's used on this page? *(an electrical signal)*

Ask What does *impulse* mean in this sentence? **He had to fight the impulse to run away.** *(Sample answer: a sudden urge to do something)*

KEY QUESTION Ask a volunteer to read the Key Question and answer on this page.

Speed Bump

Ask What are the three types of neurons? *(sensory neurons, motor neurons, and interneurons)*
Which neurons carry impulses from the sense organs to the brain? *(sensory neurons)*

ELL Focus on ELL: Extend Language

INTERMEDIATE AND ADVANCED SPEAKERS Prepare a list of words from the reading that can be used as different parts of speech. Examples include *process, commands,* and *signals.* Help students distinguish the different usages of these words by using each term in a sentence that reflects its meaning. *(Sample answers are shown below.)*

Interneurons *process* information from sensory neurons. *(verb)*

Steel is made by a special *process.* *(noun)*

Muscles receive *commands* from the central nervous system. *(noun)*

The president *commands* the entire military. *(verb)*

The spinal cord carries thousands of *signals* at one time. *(noun)*

A red light *signals* the need to stop. *(verb)*

During Reading continued

Active Reading

USE VISUALS Have students study the Resting Neuron diagram. Review the key with students.

Ask Which color dots represent Na^+ ions? *(gray)* Which color represents K^+ ions? *(purple)*

Have a volunteer read the first sentence in the second paragraph under the heading The Resting Neuron.

Ask Which part of the diagram relates to this sentence? *(the center section)*

Have another volunteer read the next sentence.

Ask Which part of the diagram relates to this sentence? *(the section at the far right that shows a K^+ ion moving through the cell membrane)*

Continue matching each sentence in the paragraph with the relevant part of the diagram.

MAKE CONNECTIONS Some students may have difficulty justifying the phrase *resting neuron* with the description of ion movement across the cell membrane.

Ask What do you do when you are at rest? *(Sample answer: breathe, sleep, listen, and so forth)*

Say When a neuron is not sending a signal, it is said to be at rest.

Ask What does a neuron do when it is at rest? *(It works to keep the inside of the cell negatively charged compared to the outside of the cell.)*

Ask What is the difference in the charge between the inside and outside of the neuron called? *(the resting potential)*

Build Vocabulary

ACADEMIC WORDS Show students a AA battery. Ask whether the battery or a neuron has more charge. *(the battery)* If students are not sure, explain that 1 V is equal to 1000 mV. Then, write the word *threshold* on the board.

Ask Where is there a threshold in this room? *(the doorway)* What is the definition of this kind of threshold? *(Sample answers: bottom part of a doorway; entrance to a room)* How do these meanings relate to the scientific meaning of *threshold*? *(They both refer to the start of something.)*

Speed Bump

Ask What is a nerve impulse? *(a sudden reversal of the resting potential)*

Ask How are a resting neuron and a moving impulse related? *(A moving impulse occurs when the neuron is no longer at rest.)*

BUILD Vocabulary

resting potential
an electrical charge across the membrane of a resting neuron

action potential
a reversal of charges across the cell membrane of a neuron; also called a nerve impulse

threshold
the minimum level of a stimulus that is required to cause an impulse

synapse
the point at which a neuron can transfer an impulse to another cell

ACADEMIC WORDS
In science, the word *potential* refers to a difference in electrical charge. Another word for *potential* is "voltage." In cells, *potential* is measured in millivolts (mV). In contrast, the *potential* of a typical AA battery is 1.5 volts.

The Nerve Impulse

Nerve impulses are a bit like the flow of an electric current through a wire. To see how this occurs, let's first examine a neuron at rest.

The Resting Neuron Like most cells, neurons have an electrical charge across their cell membranes. The difference in charge between the inside and the outside of a neuron is −70 millivolts (mV). This difference in charge is called the **resting potential.** The resting potential charge is about one twentieth the voltage in a flashlight battery. Where does this potential come from?

Active transport proteins pump sodium ions (Na^+) out of the cell and potassium ions (K^+) into the cell. Potassium ions also diffuse back through ungated potassium channel proteins in the neuron's membrane. These potassium channels allow K^+ ions to diffuse across the membrane more easily than Na^+ ions. The concentration of K^+ ions is greater inside the cell because of active transport. This greater concentration causes positively charged K^+ ions to move out of the cell. Then, the inside of the cell becomes negatively charged compared to the outside, producing the resting potential.

The Moving Impulse A neuron stays in its resting state until it receives an impulse because of a stimulus. An impulse begins when a neuron is stimulated by another neuron or by the environment. Once it begins, the impulse travels quickly down the axon toward the axon terminals. Impulses in myelinated axons move even more rapidly because they skip from one node to the next.

The impulse is a sudden reversal of the resting potential. At the leading edge of an impulse, "gated" sodium channels open. Positively charged Na^+ ions flow into the cell. The inside of the membrane temporarily becomes more positive than the outside. This charge reversal is the nerve impulse, or **action potential.**

The Resting Neuron The sodium-potassium pump uses ATP to pump Na^+ ions out of the cell. At the same time, it pumps K^+ ions in. A small amount of K^+ diffuses out of the cell through ungated channels. But gated channels block Na^+ from flowing into the resting neuron.

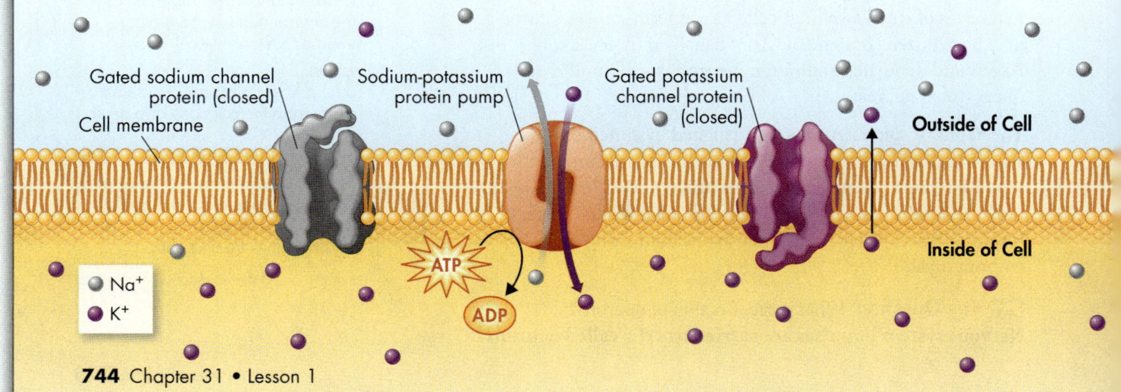

Focus on Academic Language

- transport
- flow

Point out that the word *transport* can be used as several parts of speech, including as a verb meaning "to move from one place to another" or as a noun meaning "movement from one place to another." It can also be used as an adjective.

Point out the sentence that begins, **The concentration of K^+ ions is greater . . .**

Ask Which part of speech is the word *transport* in this sentence? *(noun)* How can you tell? *(In this sentence, the word* transport *names a thing.)*

Point out the sentence that begins, **Active transport proteins pump sodium ions . . .**

Ask Which part of speech is *transport* in this sentence? *(adjective)* How can you tell? *(It's telling what kind of proteins pump sodium ions.)*

Have students find sentences on this page that include the word *flow.* Have students determine how *flow* is being used in each sentence.

BUILD Connections

A CHAIN REACTION

With a strong enough push, the fall of one domino leads to the fall of the next. An action potential moves along a neuron in a similar manner.

Once the impulse passes, sodium gates close and gated potassium channels open. The opening of these sodium channels allows K^+ ions to flow out, restoring the resting potential. The neuron is once again negatively charged on the inside. The sodium-potassium pump also keeps working, so the axon will be ready for more action potentials.

The flow of ions causes sodium channels just ahead of the point of impulse to open. The opening of these sodium channels allows the impulse to move rapidly along the axon. You could compare the flow of an impulse to the fall of a row of dominoes. As each domino falls, it causes the next domino to fall.

Threshold Not all stimuli can start an impulse in a neuron. The minimum level of a stimulus needed to cause an impulse is called the **threshold.** Any stimulus that is weaker than the threshold will not produce an impulse. A nerve impulse is an all-or-none response. Either the stimulus produces an impulse, or it does not produce an impulse.

The threshold principle can be illustrated by using a row of dominoes. If you were to gently press the first domino in a row, it might not move at all. A slightly harder push might make the domino wobble back and forth but not fall. A push may be strong enough to cause the first domino to fall into the second. If this push starts the whole row of dominoes falling, then the push is like a threshold stimulus.

All action potentials have the same strength. So how do we sense if a stimulus, like touch or pain, is strong or weak? The brain determines the strength of a stimulus from the frequency of action potentials. A weak stimulus might produce three or four action potentials per second, while a strong one might result in as many as 100 per second.

The Moving Impulse Once an impulse begins, it will continue down an axon until it reaches the end. In an axon with a myelin sheath, the impulse jumps from node to node.

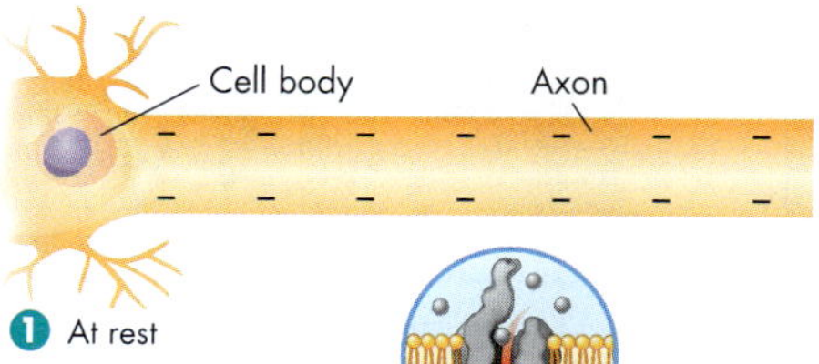

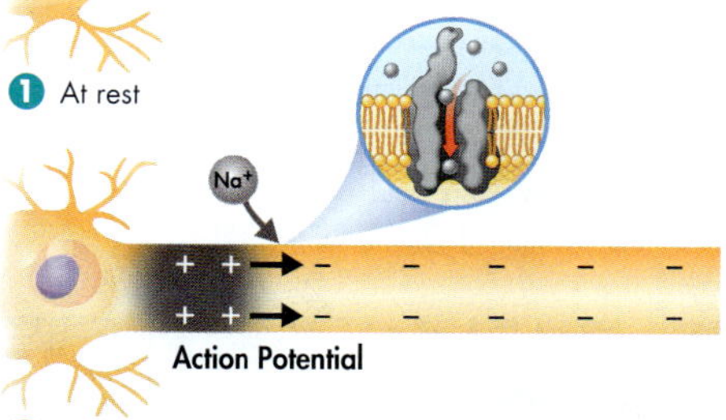

2 At the leading edge of the impulse, gated sodium channels open. Na^+ ions flow into the cell, reversing the potential between the cell membrane and its surroundings. This rapidly moving reversal of charge is called an action potential.

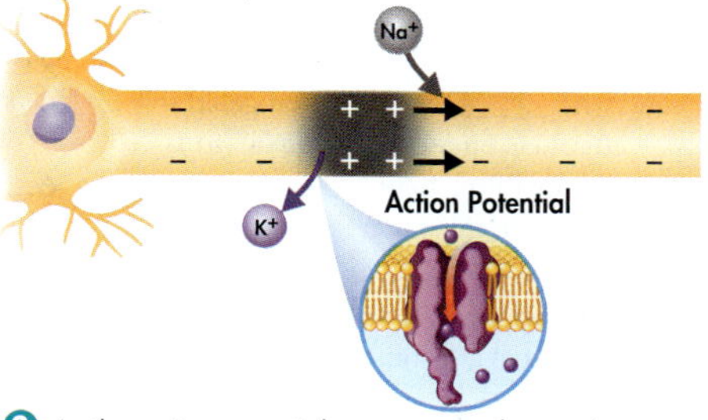

3 As the action potential passes, gated potassium channels open, allowing K^+ ions to flow out and restoring the resting potential inside the axon.

Build Connections

VISUAL ANALOGY Draw students' attention to the dominoes at the top of the page. Ask a volunteer to describe what is happening in the photo. *(Sample answer: Someone has tipped the first domino. As it falls, it hits the second domino. The second domino hits the next domino, and so on until all the dominos fall.)* Then, ask what the fall of a row of dominoes is being compared to. *(the movement of an action potential along a neuron)*

Active Reading

USE VISUALS Encourage students to study the diagram The Moving Impulse.

Ask What do the black arrows in parts 2 and 3 of the diagram represent? *(action potential)* What is another name for this? *(nerve impulse)*

Ask How does the gated sodium channel in this diagram compare to the gated sodium channel in a resting neuron? *(It's open during a moving impulse; it's closed in the resting neuron.)*

Ask What is a benefit for the resting potential to be restored after an impulse passes? *(This prepares the cell for the next impulse.)*

SEQUENCE EVENTS Write the following events in random order on the board:

(1) Impulse begins

(2) Impulse moves toward terminal

(3) Na^+ channels open

(4) Na^+ ions flow into the cell

(5) Inside membrane more positive than outside

(6) Impulse passes

(7) Na^+ gates close, K^+ channels open

(8) Resting potential is restored

(9) Inside membrane more negative than outside

Write the number *1* in front of *impulse begins.*

Ask What happens next? *(Impulse moves toward terminal)*

Continue until all the events are numbered. Suggest students write the events in order in their notebooks.

Speed Bump

Ask When do the gated sodium channels open? *(when the action potential reaches them)*

Ask When do the gated potassium channels open? *(right after the action potential passes them)*

Ask How is resting potential restored? *(The gated potassium channels open and K^+ ions flow out of the cell, making the inside of the cell more negative than the outside.)*

Hands-On Learning

MODEL THE FLOW OF AN IMPULSE Ask students to form a circle and touch hands. Tell them that each person will represent a neuron, with their left hand representing a dendrite and their right hand representing an axon.

Ask What will the rest of your body represent? *(the cell body)*

Choose a student in the circle. Have the student use his or her right hand to tap the left hand of the person on the right. Explain that this action models the beginning of a nerve impulse. Ask students to continue the impulse around the circle.

Ask How is the movement of the impulse in our circle similar to a row of dominoes falling? *(They are both chain reactions.)*

LESSON 31.1

During Reading continued

Active Reading

USE VISUALS Have students study the diagram of a synapse.

Ask What happens at a synapse? *(The neuron transfers an impulse to another cell.)*

Ask What are neurotransmitters? *(chemicals that carry an impulse across a synapse to another cell)*

KEY QUESTION Invite a volunteer to read both the Key Question and answer aloud.

Ask What happens to a nerve impulse once it begins? *(It moves along an axon until the end.)*

Ask What happens when an impulse reaches the end of the axon? *(Neurotransmitters carry it across the synaptic cleft.)*

Speed Bump

Ask If all action potentials, or nerve impulses, are the same strength, what makes the difference between slight pain and intense pain? *(Slight pain has a lower number of impulses per second, intense pain has a greater number of impulses per second.)*

Ask What happens to neurotransmitters once they have transmitted their message to another cell? *(Neurotransmitters are released from the receptors on the cell surface and then broken down by enzymes or taken up and recycled by the axon terminal.)*

Post Reading

ASSESSMENT TIPS Encourage students to review the diagrams and captions in this lesson before answering Question 4.

MYSTERY CLUE Discuss with students what they think might happen when Na⁺ ions cannot get into the cell and how that might affect muscle movement. Suggest students reread the section The Moving Impulse.

Wrap-Up Activity

ACTION NEWS Direct students to form groups of four to prepare a three-minute news segment on nerve impulses. Newsroom reporters can describe the resting neuron and the threshold, while on-the-scene reporters can describe the moving impulse and synapse. Encourage students to be imaginative in their reporting and remind them that visual aids would be helpful.

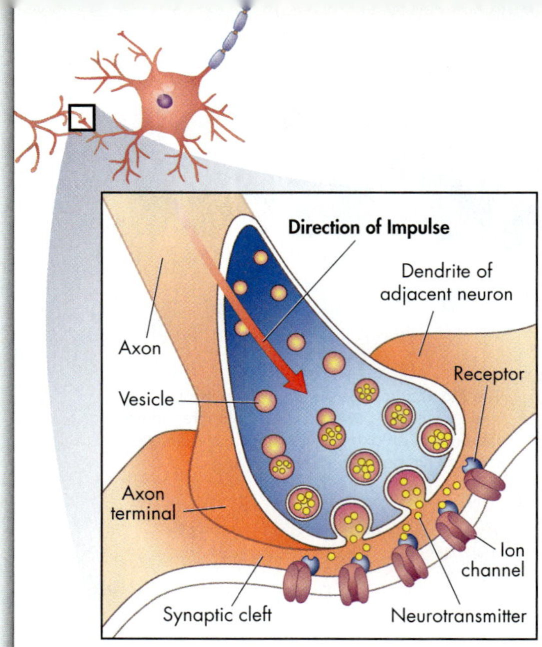

The Synapse When an impulse reaches the end of an axon, neurotransmitters are released into the synaptic cleft. The neurotransmitters bind to receptors on the membrane of an adjoining neuron.

The Synapse At the end of the neuron, the impulse reaches an axon terminal. The axon terminal may pass the impulse along to another cell. For example, a motor neuron may pass impulses to a muscle cell. These impulses cause the muscle cell to contract. The point where a neuron transfers an impulse to another cell is called a **synapse** (SIN aps). A space called the synaptic cleft separates the axon terminal from the cell.

The axon terminal at a synapse contains tiny sacs filled with neurotransmitters. Neurotransmitters are chemicals that transmit an impulse across a synapse to another cell. When an impulse arrives at the synapse, the axon releases neurotransmitters. The neurotransmitters diffuse across the synaptic cleft and bind to receptors on the receiving cell's membrane. These receptors then open ion channels in the membrane of the receiving cell. If the stimulation exceeds the cell's threshold, a new impulse begins.

Once their work is done, the neurotransmitters are released from the receptors on the cell surface. Then they are broken down by enzymes in the synaptic cleft. Or they may be taken up and recycled by the axon terminal.

Key Question How does a nerve impulse begin? **An impulse begins when a neuron is stimulated by another neuron or by the environment.**

CHECK Understanding

Apply Vocabulary

1. A(n) ______ is the part of a neuron that receives impulses from another neuron and carries it to the cell body.
2. A(n) ______ occurs when gated ion channels open, allowing Na^+ ions to flow into a neuron and reverse the electrical charge across its cell membrane.
3. The beginning of a nerve impulse is a sign that a stimulus has reached the neuron's ______ level.

Critical Thinking

4. **Explain** What happens when a neuron is stimulated by another neuron?
5. **Apply Concepts** Describe how your peripheral nervous system and your central nervous system were involved in a simple activity you performed today.
6. **Write to Learn** Answer the first clue of the mystery.

MYSTERY CLUE

The toxin found in this fish binds to gated sodium channels. This blocks the flow of Na^+ ions into a cell. How do you think this might affect muscle movement? (**Hint:** See p. 744.)

BIOLOGY.com Search Lesson 31.1 GO • Lesson Assessment

746 Chapter 31 • Lesson 1

Check Understanding

1. dendrite
2. action potential
3. threshold
4. When a neuron is stimulated by another neuron, an impulse travels down the axon toward the axon terminals.
5. Answers may vary. Students should indicate an understanding of how the peripheral and central nervous systems work and how they interact.
6. Answers may vary. If sodium ions cannot flow into nerve cells, a nerve impulse cannot be generated, and it cannot flow down the axon to stimulate a muscle. Without stimulation, muscles cannot move. Captain Cook had extraordinary weakness in his limbs. This suggests that the toxin was affecting his ability to move his muscles.

31.2 The Central Nervous System

The Brain and the Spinal Cord

The nervous system contains billions of neurons, each capable of carrying impulses and sending messages. What keeps them from sending impulses everywhere and acting like an unruly mob? Is there a central place where information is processed, decisions are made, and order is enforced?

Brain The control point of the central nervous system is the brain. The major areas of the brain are the cerebrum, cerebellum, and brain stem. Each of these areas is responsible for processing and relaying information. Like the central processing unit of a computer, information processing is the brain's primary task.

Most organs in the body function to maintain homeostasis. But the brain itself is constantly changed by its interactions with the environment. Sensory experience changes many of the patterns of neuron connections in the brain. Stem cells in the brain also produce new neurons throughout life. Many of these new cells form in regions connected with learning and memory. These changes make the brain highly flexible.

Spinal Cord Most of the neurons that enter and exit the brain do so in the spinal cord. The spinal cord is a large cluster of neurons and other cells. The spinal cord is the main communication link between the brain and the rest of the body. It is a bit like a major telephone line. It carries thousands of signals at once between the central and peripheral nervous systems. Thirty-one pairs of spinal nerves branch out from the spinal cord. These pairs connect the brain to different parts of the body. Certain kinds of information, including many reflexes, are processed directly in the spinal cord. A reflex is a quick, automatic response to a stimulus. Pulling your hand back quickly when pricked by a pin is an example of a reflex.

Key Question Where does processing of information occur in the nervous system?

The major areas of the brain are the cerebrum, cerebellum, and brain stem. Each of these areas is responsible for processing and relaying information. The spinal cord is the main communication link between the brain and the rest of the body.

Key Questions

Where does processing of information occur in the nervous system?

How do drugs change the brain and lead to addiction?

BUILD Understanding

Concept Map As you read, make a concept map that shows how the structures of the central nervous system are related to each other.

In Your Workbook Go to your workbook for help making a concept map. Complete the concept map in Lesson 31.2.

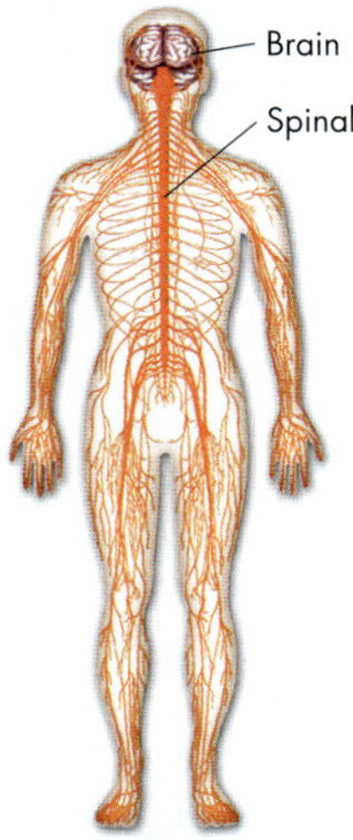

The Central Nervous System The central nervous system consists of the brain and spinal cord.

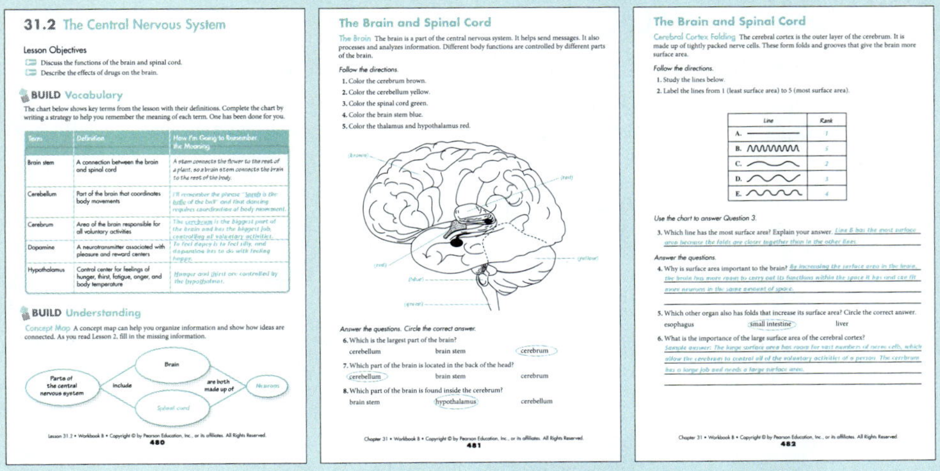

31.2 The Central Nervous System

Lesson Objectives
- Discuss the functions of the brain and spinal cord.
- Describe the effects of drugs on the brain.

BUILD Vocabulary

The chart below shows key terms from the lesson with their definitions. Complete the chart by writing a strategy to help you remember the meaning of each term. One has been done for you.

Term	Definition	How I'm Going to Remember the Meaning
Brain stem	A connection between the brain and spinal cord	A stem connects the flower to the rest of a plant, so a brain stem connects the brain to the rest of the body.
Cerebellum	Part of the brain that coordinates body movements	I'll remember the phrase "Sarah is the belle of the ball" and that dancing requires coordination of body movement.
Cerebrum	Area of the brain responsible for all voluntary activities	The cerebrum is the biggest part of the brain and has the biggest job, controlling all voluntary activities.
Dopamine	A neurotransmitter associated with pleasure and reward centers	To feel dopey is to feel silly, and dopamine has to do with feeling happy.
Hypothalamus	Control center for feelings of hunger, thirst, fatigue, anger, and body temperature	Hunger and thirst are controlled by the hypothalamus.

BUILD Understanding

Concept Map A concept map can help you organize information and show how ideas are connected. As you read Lesson 2, fill in the missing information.

Parts of the central nervous system — include — Brain, Spinal cord — are both made up of — Neurons

Lesson 31.2 • Workbook B • Copyright © by Pearson Education, Inc., or its affiliates. All Rights Reserved.
480

The Brain and Spinal Cord

The Brain The brain is a part of the central nervous system. It helps send messages. It also processes and analyzes information. Different body functions are controlled by different parts of the brain.

Follow the directions.
1. Color the cerebrum brown.
2. Color the cerebellum yellow.
3. Color the spinal cord green.
4. Color the brain stem blue.
5. Color the thalamus and hypothalamus red.

(brown) (red) (red) (yellow) (blue) (green)

Answer the questions. Circle the correct answer.
6. Which is the largest part of the brain?
cerebellum brain stem (cerebrum)
7. Which part of the brain is located in the back of the head?
(cerebellum) brain stem cerebrum
8. Which part of the brain is found inside the cerebrum?
brain stem (hypothalamus) cerebellum

Chapter 31 • Workbook B • Copyright © by Pearson Education, Inc., or its affiliates. All Rights Reserved.
481

The Brain and Spinal Cord

Cerebral Cortex Folding The cerebral cortex is the outer layer of the cerebrum. It is made up of tightly packed nerve cells. These form folds and grooves that give the brain more surface area.

Follow the directions.
1. Study the lines below.
2. Label the lines from 1 (least surface area) to 5 (most surface area).

Line	Rank
A.	1
B.	5
C.	2
D.	3
E.	4

Use the chart to answer Question 3.
3. Which line has the most surface area? Explain your answer. Line B has the most surface area because the folds are closer together than in the other lines.

Answer the questions.
4. Why is surface area important to the brain? By increasing the surface area in the brain, the brain has more room to carry out its functions within the space it has and can fit more neurons in the same amount of space.
5. Which other organ also has folds that increase its surface area? Circle the correct answer.
esophagus (small intestine) liver
6. What is the importance of the large surface area of the cerebral cortex?
Sample answer: The large surface area has room for vast numbers of nerve cells, which allow the cerebrum to control all of the voluntary activities of a person. The cerebrum has a large job and needs a large surface area.

Chapter 31 • Workbook B • Copyright © by Pearson Education, Inc., or its affiliates. All Rights Reserved.
482

LESSON 31.2

Pre-Reading

Objectives

31.2.1 Discuss the functions of the brain and spinal cord.

31.2.2 Describe the effects of drugs on the brain.

Preview the Pages

1. Make sure students understand that the visual The Brain shows a brain as if it were sliced into its hemispheres. Stress that living brain tissue is a pinkish color on the outside, mostly white on the inside. The colors of the image are added to help students discern the different areas of the brain.
2. Preview vocabulary words by having students find the words in the text. Have students read the text and state a definition in their own words to a partner. **Ask** Which of these words should you add to your word cards? *(cerebrum, hypothalamus, cerebellum, brainstem)*

Build Understanding

CONCEPT MAP Direct students to Workbook B, where they will find a partially completed concept map. Or, have students create their own concept maps showing how the structures of the central nervous system are related.

NATIONAL SCIENCE EDUCATION STANDARDS

UNIFYING CONCEPTS AND PROCESSES

I, V

CONTENT

C.6.a, C.6.b, F.1

For the corresponding lesson in *Miller & Levine Biology,* see pages 901–904.

LESSON 31.2

During Reading

Active Reading

KEY QUESTION Have a volunteer read the Key Question and its answer on the previous page.

Ask What is the brain's main task? *(to process information from the peripheral nervous system and produce responses)*

Build Connections

VISUAL SUMMARY Invite a pair of students to the front of the class. Cut out a 2 in. × 2 in. hole in the center of an index card. Have one student hold the card up to his or her eyes as if it were a camera.

Ask Where would you need to stand in relation to another person to take a picture that would match the drawing of the brain's hemispheres? *(behind a person)*

Ask Where would you need to stand to take a picture that would match the drawing of the brain's lobes? *(next to a person, on the person's left)*

FIND THE MAIN IDEA Invite volunteers to take turns reading each section about the cerebrum. After each section, stop the reading and stress the main idea of that section. Have students write the main ideas in their notebooks. Main ideas are given below.

Introductory Paragraph—The cerebrum is responsible for the voluntary, or conscious, activities of the body.

Hemispheres—Sensations from one side of the body go to the opposite hemisphere. Each hemisphere is divided into four lobes.

Cerebral Cortex—The outer layer of the cerebrum is made up of densely packed nerve cells that process information from sense organs and control body movements. This layer is called the cerebral cortex.

White Matter—White matter is made of bundles of axons that connect the cerebral cortex and the brain stem.

USE ANALOGIES Tell students that an analogy can help us learn from experience.

Say An analogy compares two items that seem to be different and shows how they can be similar.

Ask How is our brain like a computer? *(They both process and store information.)*

Speed Bump

Ask Where are long-term memories formed in the brain? *(in the limbic system)*

Ask What is the cerebral cortex? *(the outer layer of the cerebrum)*

BUILD Connections

THE BRAIN

The brain contains billions of neurons and other supporting tissue. The brain processes, relays, and forms responses to an unbelievable amount of information every moment.

In Your Workbook Explore the parts of the brain by completing the activity in Lesson 31.2.

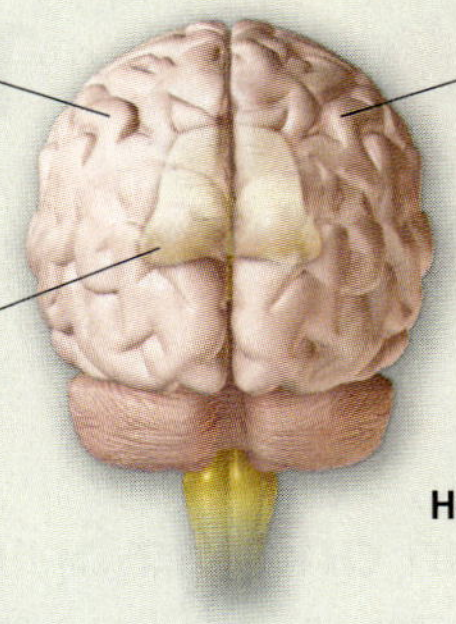

Cerebrum

The largest region of the human brain is the cerebrum. The **cerebrum** is responsible for the voluntary, or conscious, activities of the body. It is the site of intelligence, learning, and judgment.

Hemispheres A deep groove from front to back divides the cerebrum into right and left hemispheres. A band of tissue called the corpus callosum connects the hemispheres.

Remarkably, each hemisphere deals mainly with the opposite side of the body. Sensations from the left side of the body go to the right hemisphere. Sensations from the right side go to the left hemisphere. Commands to move muscles are generated in the same way.

Each hemisphere is divided into four regions called lobes. The lobes are named for the skull bones that cover them. Each lobe is associated with different functions.

Cerebral Cortex The cerebrum has two layers. The outer layer is called the cerebral cortex. It consists of densely packed nerve cell bodies known as gray matter. The cerebral cortex processes information from the sense organs and controls body movements. It is also where thoughts, plans, and learning are processed. Folds and grooves on the outer surface of the cerebral cortex greatly increase its surface area.

White Matter The inner layer of the cerebrum is known as white matter. Its whitish color comes from bundles of axons with myelin sheaths. These bundles may connect different areas of the cerebral cortex or they may connect the cerebrum to different areas of the brain.

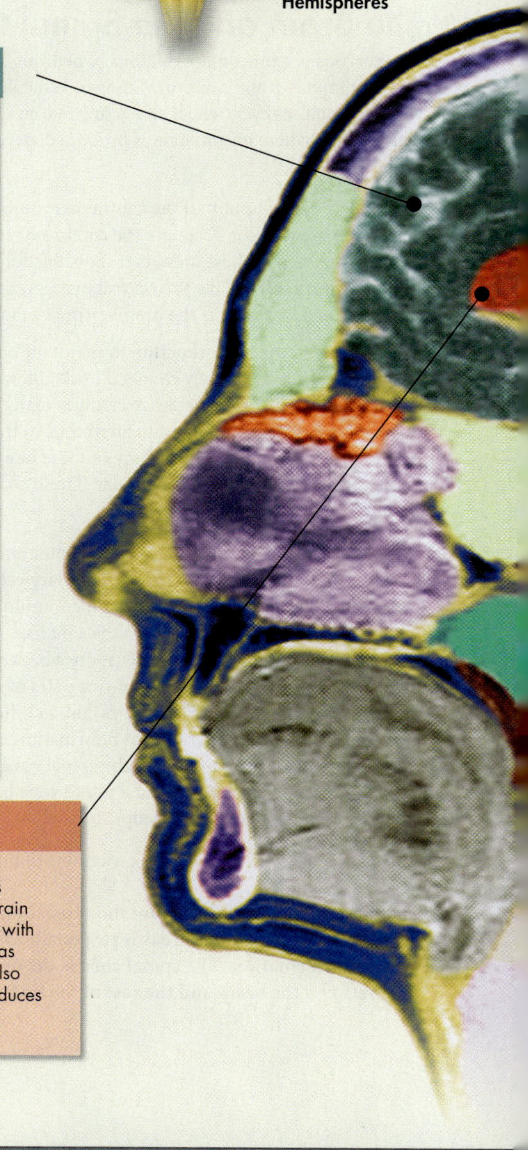

Limbic System

Many important functions have been linked to the structures that make up the limbic system. A region deep within the brain called the amygdala (uh MIG duh luh), has been associated with emotional learning. This includes fear and anxiety, as well as the formation of long-term memories. The limbic system is also associated with the brain's pleasure center. This region produces feelings of satisfaction and well-being.

ELL Focus on ELL: Access Content

BEGINNING SPEAKERS Information in the illustration is dense and may be confusing for less proficient readers. Use the following questions and sentence frames to help students focus on important information. Write the sentences on the board and read them aloud, having students supply the missing terms.

Where is the site of intelligence and learning in the brain?
The _____ *(cerebrum)* is the site of intelligence and learning.

Which part of the brain is responsible for coordinating muscle movements?
Coordination of muscles occurs in the _____ *(cerebellum)*.

What keeps the body functioning when you are asleep?
The _____ *(brain stem)* keeps the body's functions going when you are asleep.

Which system is associated with emotions, behavior, and memory?
The _____ *(limbic system)* is associated with emotions, behavior, and memory.

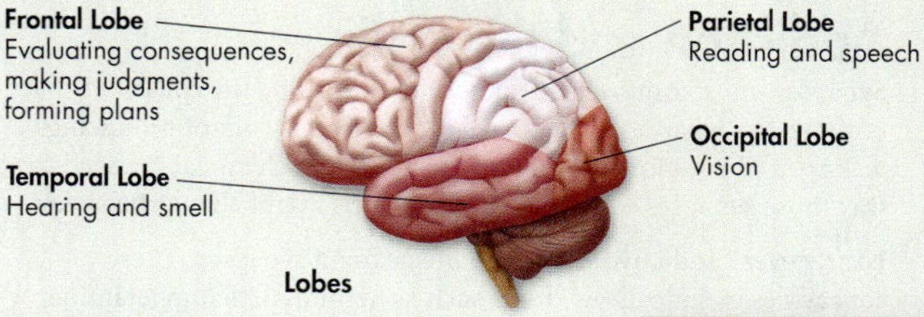

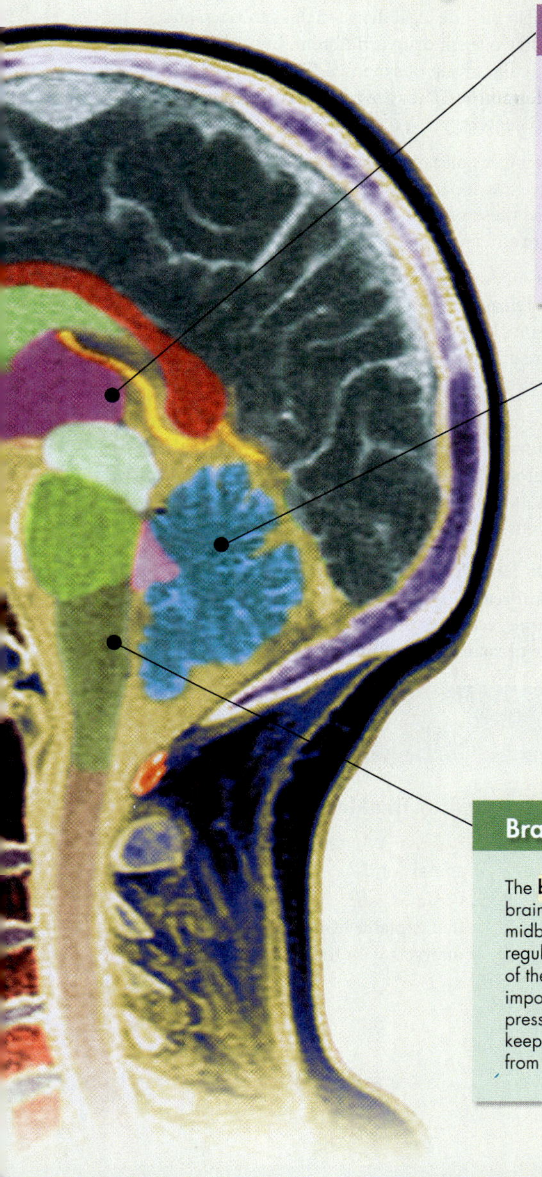

Thalamus and Hypothalamus

The thalamus and hypothalamus are found between the brain stem and the cerebrum. The thalamus receives messages from sensory receptors throughout the body. It then relays the information to the proper region of the cerebrum for further processing. Just below the thalamus is the hypothalamus. The **hypothalamus** recognizes and analyzes hunger, thirst, fatigue, anger, and body temperature. The hypothalamus also helps to coordinate the nervous and endocrine systems.

Cerebellum

The second largest region of the brain is the **cerebellum.** Information about muscle and joint position, and other sensory inputs, are sent to the cerebellum. The commands to move muscles come from the cerebral cortex. The sensory information allows the cerebellum to coordinate and balance the actions of these muscles. This ability to coordinate muscle actions allows the body to move gracefully and efficiently. When you begin any new activity involving muscle coordination, the cerebellum actually learns the movements. It then coordinates the actions of many individual muscles when the movement is repeated.

Brain Stem

The **brain stem,** located just below the cerebellum, connects the brain and spinal cord. The brain stem includes three regions—the midbrain, the pons, and the medulla oblongata. Each region regulates the flow of information between the brain and the rest of the body. The brain stem controls some of the body's most important functions. These functions include regulation of blood pressure, heart rate, breathing, and swallowing. The brain stem keeps the body functioning even when you lose consciousness from sleep or injury.

Focus on Academic Language

- coordinate

Students unfamiliar with English may have difficulty pronouncing *coordinate.* If they try to sound it out, they may pronounce the first two syllables as the single syllable, *coor* (koor). Have students look up *coordinate* in the dictionary.

Ask How is the word *coordinate* broken into syllables? *(co or di nate)*

Write the syllable breakdown on the board. Point out that students can use the syllables as a clue to the word's pronunciation (koh-OR-duhn-ayt).

Ask What is the syllable breakdown for *cooperation*? *(co op er a tion)*

Ask What does it tell you about the word's pronunciation? *(It's not KOOP-er-ay-shuhn but koh-ah-puh-RAY-shuhn.)*

Active Reading

READ, COVER, REMEMBER, RETELL Direct students to read the three paragraphs on this page. After reading, have them close their books and try to remember what they read.

Ask What area of your brain receives messages from your entire body? *(the thalamus, between the brain stem and the cerebrum)*

Ask What does the hypothalamus do? *(recognizes hunger, thirst, body temperature, anger, and fatigue; coordinates the nervous and endocrine systems)*

Ask What part of the brain receives information about muscle and joint position? *(cerebellum)*

Ask Which part of the brain is responsible for controlling blood pressure, heart rate, breathing, and swallowing? *(brain stem)*

MAKE CONNECTIONS Help students recognize which brain regions help them complete specific activities.

Ask How do the cerebral cortex and the cerebellum work together to help you learn a new dance step or a tennis serve? *(The cerebral cortex commands muscles to move while the cerebellum coordinates and balances the movements.)*

Ask What part of the brain works to keep you alive when you are unconscious? *(the brain stem)* What does it do to keep you alive? *(controls blood pressure, heart rate, and breathing)*

Ask Which part of your brain makes you want to reach out and pet a cute puppy? *(limbic system)*

Ask Which part of your brain is working hard to learn this information about the brain? *(cerebrum)*

CLARIFY WORD USAGE Students might confuse the terms *cerebrum* and *cerebellum.* Help students match the terms with their respective locations in the brain. Have students work individually or in pairs to create a rule or a mnemonic device to help them distinguish between the two terms. Encourage students to share these learning aids with the class for others to use.

Speed Bump

Ask Where does the thalamus send the messages it receives? *(the cerebrum)*

Ask What parts of the nervous system does the brain stem connect? *(the brain and the spinal cord)*

LESSON 31.2

During Reading continued

Active Reading

USE GRAPHIC ORGANIZERS Draw a cause-and-effect graphic organizer on the board. Write *addictive drugs* in the Cause square with three arrows pointing to three boxes. Ask students to fill in the Effect boxes. *(Sample answers: keep dopamine in synapse longer, release a flood of dopamine, stimulate other receptors to release dopamine)* Use this graphic organizer to review the content on this page.

KEY QUESTION Ask a volunteer to read both the Key Question and its answer.

Ask What is dopamine? *(a neurotransmitter)* What is a neurotransmitter? *(a chemical that transmits an impulse across a synapse to another cell)*

Build Vocabulary

PREFIXES Direct students to the visual of the brain.

Ask How can you use your knowledge of the prefix *hypo-* to find the hypothalamus? (Hypo- *means "under"' so the hypothalamus is likely under the thalamus.)*

Say The prefix *cerebr-* or *cerebro-* means "brain."

Ask What does *cerebral* mean in the term *cerebral cortex*? *(relating to the brain)*

Post Reading

ASSESSMENT TIP Refer students to the visual The Brain to help answer Questions 1–3. For Question 5, suggest that students re-read the paragraphs about the brain on the first page of this lesson.

Wrap-Up Activity

YOU AND YOUR BRAIN Have students form five groups. Explain that each group will prepare a booklet titled, *You and Your Brain.* Each group can be responsible for one of the five structures presented in the text. Information should include the structure's location and function. Suggest students include diagrams or graphic organizers where helpful. Have students then collate the reports into a single booklet.

BUILD Vocabulary

cerebrum
the part of the brain responsible for voluntary activities of the body; the "thinking" region of the brain

hypothalamus
the portion of the brain that acts as a control center for recognition and analysis of hunger, thirst, fatigue, anger, and body temperature

cerebellum
the part of the brain that coordinates movement and controls balance

brain stem
a structure that connects the brain and spinal cord; includes the medulla oblongata and the pons

dopamine
a neurotransmitter that is associated with the brain's pleasure and reward centers

PREFIXES
The prefix *hypo-* means "under or below." The hypothalamus is located below the thalamus in the brain.

Addiction and the Brain

Synapses transfer messages from cell to cell, doing the conscious work of thinking. Synapses also do the less-conscious work of producing feelings and emotions. What would happen if a chemical changed the way those synapses worked? Such chemicals might change behavior.

Nearly every addictive substance affects brain synapses. These substances include illegal drugs such as heroin, methamphetamine, and cocaine. They also include legal drugs such as tobacco and alcohol. The chemistry of each drug is different. But they all produce changes in one particular group of synapses. These synapses use the neurotransmitter **dopamine.** They are associated with the brain's pleasure and reward centers.

When an activity brings us pleasure, neurons in the hypothalamus and the limbic system release dopamine. Dopamine molecules stimulate other neurons across these synapses. This stimulation produces the sensation of pleasure and a feeling of well-being.

Addictive drugs act on dopamine synapses in several ways. Methamphetamine releases a flood of dopamine, producing an instant "high." Cocaine keeps dopamine in the synaptic region longer, intensifying pleasure and suppressing pain. Nicotine and alcohol cause an increase in the release of dopamine.

The brain reacts to excessive dopamine levels by reducing the number of dopamine receptors. Normal activities no longer produce the sensations of pleasure they once did. Addicts feel depressed and sick without their drugs. It takes greater amounts of addictive chemicals to get the same high with fewer receptors. This addiction is hard to break.

Key Question How do drugs change the brain and lead to addiction? **The brain reacts to too much dopamine by reducing the number of dopamine receptors. As a result, normal activities no longer produce the pleasurable feelings they once did.**

CHECK Understanding

Apply Vocabulary

1. The part of the brain that processes information from the eyes, ears, and skin is the ______.
2. The ______ is the part of the brain that recognizes when you are tired and hungry.
3. The part of the brain that learns muscle movements needed for playing a musical instrument or playing a sport is the ______.

Critical Thinking

4. **Explain** What is the role of the spinal cord?
5. **Write to Learn** Write a paragraph that explains how the brain interacts with the environment.

BIOLOGY.com Search Lesson 31.2 GO • Lesson Assessment

Check Understanding

1. cerebral cortex
2. hypothalamus
3. cerebellum
4. The role of the spinal cord is to be the main communication link between the brain and the rest of the body.
5. Sample answer: The brain is changed by interactions with the environment—neuron connection patterns change and stem cells produce new cells in areas associated with learning and memory. The changes allow the brain to be flexible.

31.3 The Peripheral Nervous System

The Sensory Division

It's all about input and output. A computer isn't useful unless it can accept input from the world around it. And the result isn't meaningful unless there's a way to output it. The central nervous system faces the same issues. Can you guess what it uses for input and output devices?

The peripheral nervous system consists of all the nerves and associated cells that are not part of the brain or spinal cord, including cranial nerves and spinal nerves. Cranial nerves go through openings in the skull. They stimulate regions of the head and neck. Spinal nerves stimulate the rest of the body. The cell bodies of cranial and spinal nerves are arranged in clusters called ganglia.

The peripheral nervous system is our link with the outside world. It has two major divisions—the sensory division and the motor division. The sensory division transmits impulses from sense organs to the central nervous system. The motor division transmits impulses from the central nervous system to the muscles and glands.

Sensory receptors are cells that transmit information about changes in the internal and external environments. These changes are called stimuli. Sensory receptors are grouped by the type of stimuli to which they respond. When they are stimulated, sensory receptors transmit impulses to sensory neurons. Sensory neurons then transmit impulses to the central nervous system.

Key Question How does the central nervous system receive sensory information?

The peripheral nervous system's sensory division transmits impulses from sense organs to the central nervous system.

Sensory Receptors
Sensory receptors react to a specific stimulus such as light or sound by sending impulses to sensory neurons.

Sensory Receptors

Type	Responds to	Some Locations
Chemoreceptor	Chemicals	Mouth, nose, blood vessels
Photoreceptor	Light	Eyes
Mechanoreceptor	Touch, pressure, vibrations, and stretch	Skin, hair follicles, ears, ligaments, tendons
Thermoreceptor	Temperature changes	Skin, hypothalamus
Pain receptor	Tissue injury	Throughout the body

Key Questions

How does the central nervous system receive sensory information?

How do muscles and glands receive commands from the central nervous system?

BUILD Understanding

Venn Diagram As you read, make a Venn diagram that compares the somatic nervous system and the autonomic nervous system.

In Your Workbook Go to your workbook to complete the Venn diagram in Lesson 31.3.

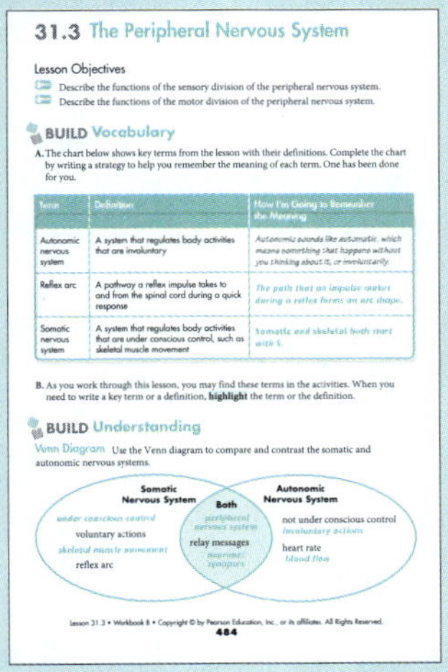

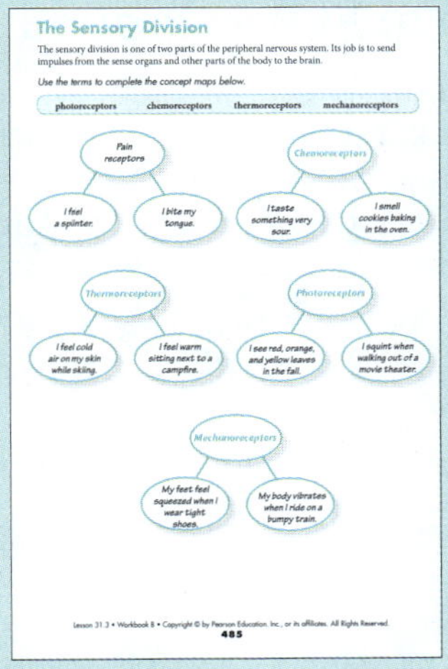

Pre-Reading

Objectives

31.3.1 Describe the functions of the sensory division of the peripheral nervous system.

31.3.2 Describe the functions of the motor division of the peripheral nervous system.

Preview the Pages

1. Preview the vocabulary words in the text. Focus on the pronunciations of *somatic* and *autonomic.* Ask students what *autonomic* sounds like *(probably* automatic*).* Invite students to conjecture how these words might be related.
2. Read each column heading in the Sensory Receptors table aloud and ask what each one means. *(*Type *refers to type of sensory receptor;* Responds to *refers to specific stimuli that activate the receptor, and* Some Locations *refers to where that kind of receptor can be found.)*
3. Review previous vocabulary words: *peripheral nervous system, stimulus, homeostasis,* and *neuron.*

Build Understanding

VENN DIAGRAM Direct students to Workbook B, where they will find a partially completed Venn diagram for the somatic and autonomic nervous systems. Remind students that a Venn diagram is a useful way to note the difference and similarities of two or more things.

NATIONAL SCIENCE EDUCATION STANDARDS

UNIFYING CONCEPTS AND PROCESSES

I, V

CONTENT

C.6.a, C.6.b

For the corresponding lesson in *Miller & Levine Biology,* see pages 906–908.

LESSON 31.3

During Reading

Active Reading

KEY QUESTION Invite a student to read the Key Question and its answer on the previous page.

Ask What is the central nervous system? *(the brain and spinal cord)* What is sensory information? *(information from the sensory receptors, or senses)*

INTERPRET DIAGRAMS Direct students' attention to the Reflex Arc diagram. Point out the connection between the numbered sentences in the text and the numbers on the diagram. Assist students in understanding the parts of the diagram.

Ask What does the red line represent? *(the nerve impulse path from the sensory receptors through the sensory neurons)* What about the blue line? *(the nerve impulse path from the motor neurons to a muscle)*

Ask What makes up the path of the nerve impulse along the red line? *(a sensory neuron)* along the blue line? *(a motor neuron)*

Ask What type of neuron is entirely housed in the spinal cord? *(interneuron)*

DRAW CONCLUSIONS Help students draw a conclusion about the benefits of reflex arcs.

Say When you draw a conclusion, you connect to types of knowledge—what you know from experience and what you have learned from the text.

Ask Why is it beneficial to the body to have reflex arcs? *(Sample answer: The reflex arc is a shorter route for impulses to travel. This results in a quicker response to possible danger, preventing some pain.)*

DRAW A DIAGRAM Suggest students draw a diagram showing the neuron path for a voluntary muscle movement. Check drawings for accuracy.

Ask How is the path for a voluntary movement different than the path for a reflex movement? *(A voluntary movement starts in the brain and travels to the muscle. The reflex movement starts with a sensory receptor, travels to the spinal cord, and then to the muscle.)*

Speed Bump

Ask The somatic nervous system and the autonomic nervous system belong to which division of the nervous system? *(the motor division)*

Ask The sensory division and the motor division belong to which part of the nervous system? *(the peripheral nervous system)*

The Motor Division

The nervous system maintains homeostasis by gathering and processing information from other systems and organs. Then it sends commands to muscles or glands through the peripheral nervous system's motor division. This division has two parts—the somatic nervous system and the autonomic nervous system.

Somatic Nervous System The **somatic nervous system** regulates body activities that are under conscious control. These activities include movement of skeletal muscles. But when your body is in danger, the central nervous system may take over.

▶ ***Voluntary Control*** When you wiggle your toes, you use motor neurons of the somatic nervous system. Impulses from the brain are carried through the spinal cord, synapsing with motor neuron dendrites. Axons from these motor neurons extend from the spinal cord, carrying the impulses directly to muscles. These impulses cause muscle contractions that produce voluntary movements.

Reflex Arc When you step on a tack, sensory receptors stimulate a sensory neuron. The neuron relays the signal to an interneuron within the spinal cord. The signal is then sent to a motor neuron. This neuron stimulates a muscle that lifts your leg.

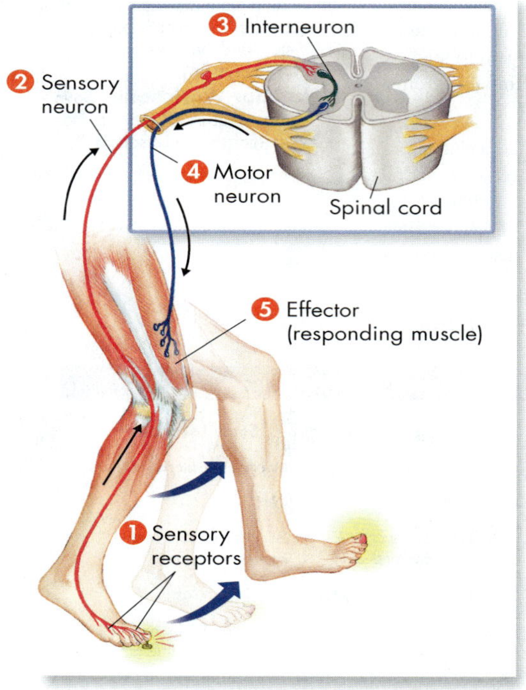

▶ ***Reflex Arcs*** The somatic nervous system is generally considered to be under conscious control. But some actions of the system occur automatically. Suppose you accidentally step on a tack with your bare foot. Your leg may pull back before you even notice the pain. This rapid response is a reflex. It is caused by impulses that travel a pathway called a **reflex arc.** 1 Sensory receptors react to the feeling of the tack, sending an impulse to sensory neurons. 2 Sensory neurons relay the information to the spinal cord. 3 An interneuron in the spinal cord processes the information and forms a response. 4 A motor neuron carries impulses that stimulate a muscle (its effector). 5 The muscle contracts, moving your leg. At the same time, impulses carry information about the injury to your brain. But by the time your brain interprets the pain, your leg and foot have already moved.

The spinal cord does not control all reflexes. Many reflexes that involve structures in your head are controlled by the brain. For example, blinking and sneezing are controlled by the brain.

ELL Focus on ELL: Build Background

ALL SPEAKERS English language learners are probably familiar with reflexes, although they may not know the steps outlined or vocabulary used to describe them. Have students work in groups to generate a list of reflexes they are familiar with. Examples include blinking, shivering, pulling your hand away from a hot light bulb, moving your leg when a doctor taps the knee, and so on.

Review the meaning of the following verbs as they apply to students' examples: *stimulate (a sensory neuron), relay (information), process (information), stimulate (a muscle), contract. (a muscle)*

Then, have students use these terms and the five steps outlined on this page to describe the reflexes they listed.

Autonomic Nervous System The **autonomic nervous system** regulates involuntary activities. These activities are not under conscious control. For instance, when you start to run, the autonomic nervous system speeds up your heart rate. It increases blood flow to the skeletal muscles and stimulates the sweat glands. It also slows down the contractions of smooth muscles in the digestive system. You may not be aware of some of these activities. But all of them help you run faster and farther.

The autonomic nervous system has two equally important parts, the sympathetic and parasympathetic nervous systems. In general, these systems have opposite effects on each organ they influence. The two systems produce a level of fine control that coordinates organs throughout the body.

For example, heart rate is increased by the sympathetic nervous system. But it is decreased by the parasympathetic nervous system. In general, the sympathetic system prepares the body for intense activity. When it is stimulated, blood pressure increases and energy-rich sugar is released into the blood. Activities not related to the body's "fight or flight" response to stress also shut down. But the parasympathetic system causes what might be called the "rest and digest" response. It lowers heart rate and blood pressure and activates digestion. It also activates pathways that store food molecules in the tissues of the body.

Key Question How do muscles and glands receive commands from the central nervous system? **The peripheral nervous system's motor division transmits impulses from the central nervous system to muscles or glands.**

BUILD Vocabulary

somatic nervous system
the part of the peripheral nervous system that carries signals to and from skeletal muscles

reflex arc
the sensory receptor, sensory neuron, motor neuron, and effector that are involved in a quick response to a stimulus

autonomic nervous system
the part of the peripheral nervous system that regulates activities that are involuntary, or not under conscious control; made up of the sympathetic and parasympathetic subdivisions

ROOT WORDS
The word *autonomic* comes from the word *autonomy*, which means "the state of being self-governing." The autonomic nervous system is the "self-governing" part of the peripheral nervous system.

CHECK Understanding

Apply Vocabulary

Use the highlighted words from the lesson to complete each sentence correctly.

1. The sensory receptors and effector are parts of a ________.
2. The heart, lungs, and muscles of the digestive tract are regulated by the ________.
3. When you reach for a glass of water, your hand and arm muscles are being controlled by the ________.

Critical Thinking

4. **Explain** Is a reflex a part of the central nervous system, the peripheral nervous system, or both?
5. **Apply Concepts** Give three examples of stimuli that your sensory division is responding to right now.
6. **Write to Learn** Answer the second clue to the mystery.

MYSTERY CLUE

Think about Captain Cook's symptoms of weakness. What part of his nervous system did consumption of the fish affect the most? (**Hint:** See p. 752.)

BIOLOGY.com Search Lesson 31.3 GO • Lesson Assessment • Art in Motion

Check Understanding

1. reflex arc
2. autonomic nervous system
3. somatic nervous system
4. A reflex is part of both the central and peripheral nervous systems because it involves sensory and motor neurons of the peripheral nervous system and is processed by either the brain or the spinal cord.
5. Sample answers: eyes responding to the light in the room, the feel of the book in the student's hands, ears picking up sounds or conversations in the classroom, the feel of the chair that the student sits on, and so on.
6. Captain Cook's weakness was in his limbs. This means his muscles were not responding to the somatic nervous system, a division of the motor division of the peripheral nervous system. The somatic nervous system relays impulses from the central nervous system to muscles and glands.

Active Reading

SUMMARIZE Remind students that summarizing can help them remember the important points about a topic.

Say When you summarize, include the main ideas of the passage. Summarize the information on the autonomic nervous system. *(Summaries should state that the autonomic system includes the sympathetic and parasympathetic nervous systems, which regulate involuntary activities.)*

KEY QUESTION Ask a volunteer to read the Key Question and its answer.

Ask What is the function of the peripheral nervous system? *(It transmits impulses from sense organs to the central nervous system.)*

Ask What kinds of neurons make up the peripheral nervous system? *(sensory neurons, motor neurons)*

Build Vocabulary

ROOT WORDS Ask students how the autonomic nervous system is autonomous. *(It directs activities that are not under conscious control.)* Then, tell students the root *neur* or *neuro* means "nerve."

Ask What words contain the root *neur*? *(Sample answers: neuron, interneuron, neurotransmitter)*

Speed Bump

Ask Which of the body's nervous systems regulates involuntary activities? *(autonomic nervous system)*

Ask Which part of the autonomic nervous system controls the "fight or flight" response? *(sympathetic nervous system)* Which part controls the "rest and digest" response? *(parasympathetic nervous system)*

Post Reading

ASSESSMENT TIPS For Question 4, review with students the Reflex Arc diagram.

MYSTERY CLUE Ask students to describe Captain Cook's symptoms. **Ask** What parts of his body were affected? What part of his nervous system controlled those parts?

Wrap-Up Activity

BUILD A CONCEPT MAP Have pairs write the following terms on seven index cards: *peripheral nervous system, sensory division, motor division, somatic nervous system, autonomic nervous system, sympathetic nervous system,* and *parasympathetic nervous system.* Students can use string to show the relationship among terms. Have pairs present their concept map and discuss it with the class.

Pre-Reading

Objectives

31.4.1 Discuss the sense of touch and identify the various types of sensory receptors in the skin.

31.4.2 Explain the relationship between smell and taste.

31.4.3 Identify the parts of the ears that make hearing and balance possible.

31.4.4 Describe the major parts of the eye and explain how the eye enables us to see.

Preview the Pages

1. Preview vocabulary words with students. Have them make a T-chart in their notebooks, listing familiar words on the left and unfamiliar on the right. They can add definitions as needed.
2. Use the Sound Levels graph as an opportunity to review the elements of a bar graph. Make certain students are able to read the sound level for each of the sounds the graph represents.

Activate Prior Knowledge

Ask volunteers to describe what happens to their senses of smell and taste when they have a cold.

Ask What happens to your sense of smell when you have a cold? *(Congestion makes it less sensitive.)*

Ask How does this affect your sense of taste? *(Food seems to have less taste when I have a cold.)*

Build Understanding

PREVIEW VISUALS Read aloud Preview Visuals. Ask if students have questions about the illustration. Have them record the questions in Workbook B.

NATIONAL SCIENCE EDUCATION STANDARDS

UNIFYING CONCEPTS AND PROCESSES

I, V

CONTENT

C.6.a, C.6.b

INQUIRY

A.1.c, A.2.a

For the corresponding lesson in *Miller & Levine Biology,* see pages 909–913.

31.4 The Senses

Key Questions

How does the body sense touch, temperature, and pain?

How are the senses of smell and taste similar?

How do the ears and brain process sounds and maintain balance?

How do the eyes and brain produce vision?

BUILD Understanding

Preview Visuals Before reading, preview the figure The Eye. Write down at least two questions you have about the information in the figure.

In Your Workbook Go to your workbook for a chart for previewing visuals.

Taste Buds The surface of the tongue contains many tiny projections. Taste buds line the tops of some of these and line the sides of other projections (LM 80×).

Touch and Related Senses

Different sensory receptors in the body respond to touch, temperature, and pain. Your skin has all three receptors, but some are also found in other areas. Nearly all regions of your skin are sensitive to touch.

Touch Human skin contains at least seven types of sensory receptors. Stimulation of these receptors creates the sensation of touch. But not all parts of the body are equally sensitive. For example, there is a higher density of touch receptors on your fingers than on your back.

Temperature Thermoreceptors are sensory cells that respond to heat and cold. They are found throughout the skin. They are also found in the hypothalamus, the part of the brain that senses blood temperature.

Pain Pain receptors are found throughout the body, but not in the brain. Many tissues also have pain receptors that respond to chemicals released during infection or inflammation.

Key Question How does the body sense touch, temperature, and pain? **Different sensory receptors in the body respond to touch, temperature, and pain.**

Smell and Taste

Your senses of smell and taste involve the ability of sensory cells that detect chemicals. These cells, called chemoreceptors, are located in the nose and mouth. They send impulses to the brain that are interpreted as sensations of smell and taste.

Smell Your sense of smell produces thousands of different sensations. Much of what we usually call the "taste" of food and drink is actually smell. Try eating a few bites of food while holding your nose. It seems to have little taste until you release your nose and breathe freely.

Taste The sense organs for taste are the taste buds. Most taste buds are on the tongue. Their sensory cells respond to salty, bitter, sweet, and sour foods. Scientists recently discovered that sensory cells also respond to the amino acid glutamate. This taste is called *umami* (Japanese for "savory").

Key Question How are the senses of smell and taste similar? **Sensations of smell and taste are both the result of impulses sent to the brain by chemoreceptors.**

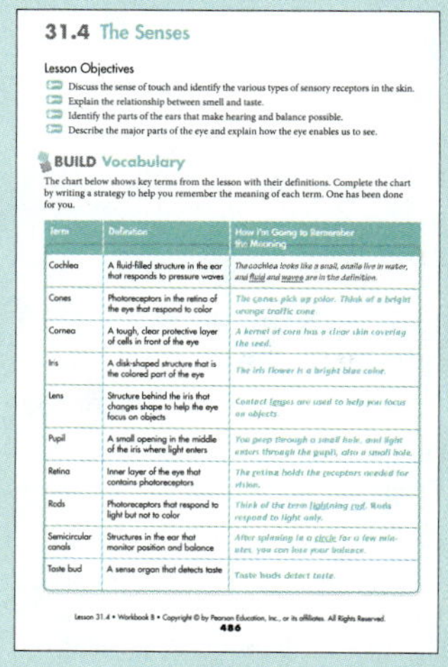

31.4 The Senses

Lesson Objectives

- Discuss the sense of touch and identify the various types of sensory receptors in the skin.
- Explain the relationship between smell and taste.
- Identify the parts of the ears that make hearing and balance possible.
- Describe the major parts of the eye and explain how the eye enables us to see.

BUILD Vocabulary

The chart below shows key terms from the lesson with their definitions. Complete the chart by writing a strategy to help you remember the meaning of each term. One has been done for you.

Term	Definition	How I'm Going to Remember the Meaning
Cochlea	A fluid-filled structure in the ear that responds to pressure waves	The cochlea looks like a snail, snails live in water, and fluid and waves are in the definition.
Cones	Photoreceptors in the retina of the eye that respond to color	The cones pick up color. Think of a bright orange traffic cone.
Cornea	A tough, clear protective layer of cells in front of the eye	A kernel of corn has a clear skin covering the seed.
Iris	A disk-shaped structure that is the colored part of the eye	The iris flower is a bright blue color.
Lens	Structure behind the iris that changes shape to help the eye focus on objects	Contact lenses are used to help you focus on objects.
Pupil	A small opening in the middle of the iris where light enters	You peep through a small hole, and light enters through the pupil, also a small hole.
Retina	Inner layer of the eye that contains photoreceptors	The retina holds the receptors needed for vision.
Rods	Photoreceptors that respond to light but not to color	Think of the term lightning rod. Rods respond to light only.
Semicircular canals	Structures in the ear that monitor position and balance	After spinning in a circle for a few minutes, you can lose your balance.
Taste bud	A sense organ that detects taste	Taste buds detect taste.

Lesson 31.4 • Workbook B • Copyright © by Pearson Education, Inc., or its affiliates. All Rights Reserved.
486

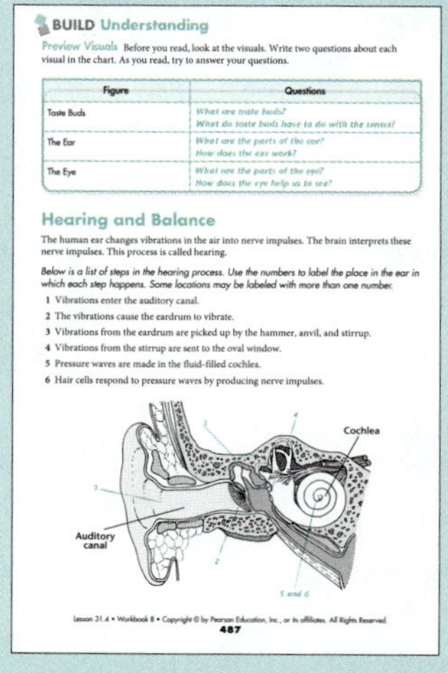

BUILD Understanding

Preview Visuals Before you read, look at the visuals. Write two questions about each visual in the chart. As you read, try to answer your questions.

Figure	Questions
Taste Buds	What are taste buds? What do taste buds have to do with the senses?
The Ear	What are the parts of the ear? How does the ear work?
The Eye	What are the parts of the eye? How does the eye help us to see?

Hearing and Balance

The human ear changes vibrations in the air into nerve impulses. The brain interprets these nerve impulses. This process is called hearing.

Below is a list of steps in the hearing process. Use the numbers to label the place in the ear in which each step happens. Some locations may be labeled with more than one number.

1 Vibrations enter the auditory canal.
2 The vibrations cause the eardrum to vibrate.
3 Vibrations from the eardrum are picked up by the hammer, anvil, and stirrup.
4 Vibrations from the stirrup are sent to the oval window.
5 Pressure waves are made in the fluid-filled cochlea.
6 Hair cells respond to pressure waves by producing nerve impulses.

Lesson 31.4 • Workbook B • Copyright © by Pearson Education, Inc., or its affiliates. All Rights Reserved.
487

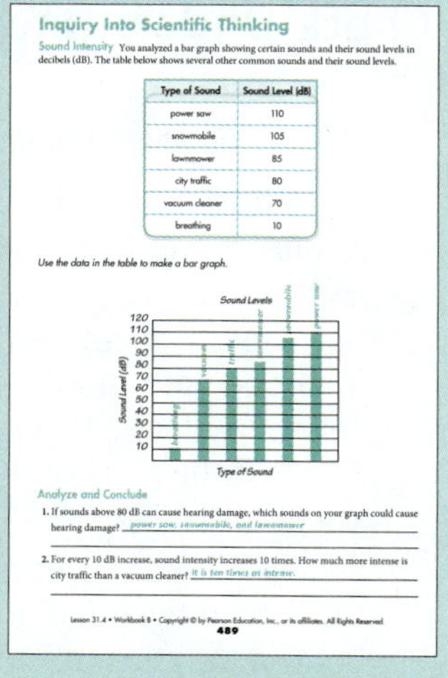

Inquiry Into Scientific Thinking

Sound Intensity You analyzed a bar graph showing certain sounds and their sound levels in decibels (dB). The table below shows several other common sounds and their sound levels.

Type of Sound	Sound Level (dB)
power saw	110
snowmobile	105
lawnmower	85
city traffic	80
vacuum cleaner	70
breathing	10

Use the data in the table to make a bar graph.

Analyze and Conclude

1. If sounds above 80 dB can cause hearing damage, which sounds on your graph could cause hearing damage? power saw, snowmobile, and lawnmower
2. For every 10 dB increase, sound intensity increases 10 times. How much more intense is city traffic than a vacuum cleaner? It is ten times as intense.

Lesson 31.4 • Workbook B • Copyright © by Pearson Education, Inc., or its affiliates. All Rights Reserved.
489

Hearing and Balance

The human ear has two sensory functions—hearing and balance. Mechanoreceptors found in parts of the ear transmit impulses to the brain. The brain translates the impulses into sound and information about changes in position.

Hearing Sound is vibrations moving through the air around us. The ears can distinguish both the pitch and loudness of those vibrations.

Vibrations enter the ear through the auditory canal and cause the tympanum (TIM puh num), or eardrum, to vibrate. Three tiny bones (the hammer, anvil, and stirrup) transmit these vibrations to a membrane called the oval window. Its vibrations create pressure waves in the fluid-filled **cochlea** (KAHK lee uh) of the inner ear. The pressure waves push the cochlea's tiny hair cells back and forth. In response, the hair cells send nerve impulses to the brain, which processes them as sounds.

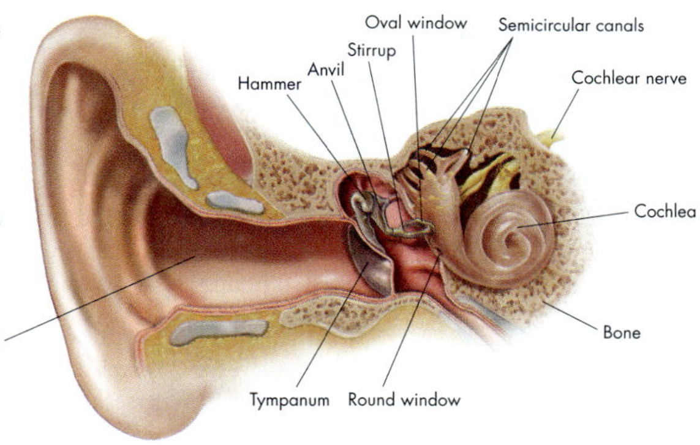

The Ear The structures in the ear transmit sound and allow a person to hear.

INQUIRY into Scientific Thinking

Sound Intensity

Sound intensity, or loudness, is measured in units called decibels (dB). The threshold of hearing for the human ear is 0 dB. For every 10 dB increase, the sound intensity increases ten times.

Based on the graph, how much more intense is a jet engine than a hair dryer?

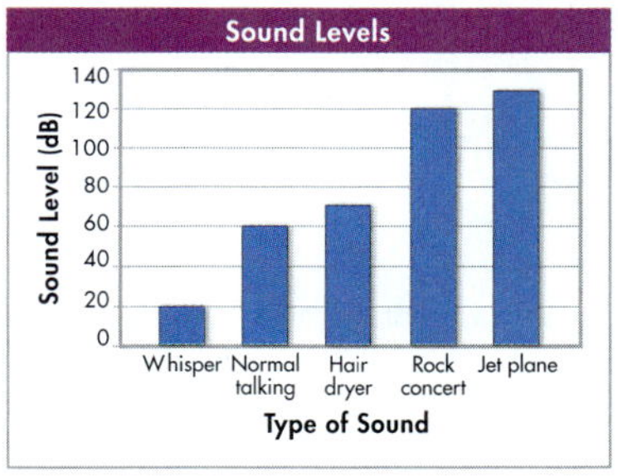

- The difference between the two sound intensities is

$$130 \text{ dB}_{jet} - 70 \text{ dB}_{hair\ dryer} = 60 \text{ dB}$$

- The jet engine is 60 dB louder than a hair dryer. This means it is 10^6 or 1,000,000 times louder than the hair dryer.

Loud noises can permanently damage the hair cells in the cochlea. These hair cells are the cells that sense vibrations. Exposure to sounds above 80 dB for several hours at a time can damage hearing. Exposure to sounds about 120 dB for even a few seconds can damage hearing.

Analyze and Conclude

1. Calculate How much more intense is normal talking than whispering?

2. Infer Repeated exposure to portable music devices set at high volume can cause hearing damage. Why do you think this damage might not reveal itself for many years?

In Your Workbook Get more help for this activity in your workbook.

The Nervous System **755**

Analyze and Conclude

1. Normal talking has an intensity that is 10,000 times greater than a whisper.
2. Sample answer: Hearing damage is cumulative. It would likely take a long time for a significant number of hair cells to become damaged. A person may not notice the damage until it accumulates to a certain level.

During Reading

Active Reading

ORGANIZING YOUR READING Have students use the blue headings on the previous page to organize information about touch and related senses.

Ask What three kinds of receptors does your skin have? *(touch, temperature, and pain)* How do the headings help you organize your reading and understand what you read? *(Sample answer: The headings help me see how the different topics are related and what bigger idea they are part of.)*

KEY QUESTION Have a volunteer read aloud the first Key Question and answer from the previous page.

Ask What does a sensory receptor do? *(It transmits information about changes in the internal and external environments.)*

Ask What are the names of the sensory receptors that sense touch, temperature, and pain? *(mechanoreceptor, thermoreceptor, pain receptor)*

KEY QUESTION Invite a different volunteer to read the second Key Question and its answer aloud.

Ask What do chemoreceptors sense? *(chemicals)*

Inquiry Into Scientific Thinking

MAKE CONNECTIONS Why do people working outside at airports wear special ear gear? *(Just a few seconds of sound above 120 dB can damage hearing. A jet makes a sound of about 130 dB.)*

Lead a Discussion

Explain that a bar graph compares data. In this case, each bar represents a different sound. Have a volunteer explain how to find the sound level, in decibels, of a jet. Guide students through the math necessary to determine that a jet is 10 times louder than a rock concert.

Ask Is a rock concert twice as loud as normal talking? Explain. *(No, for every 10 dB increase the sound intensity increases 10 times. The rock concert is 10^6, or 1,000,000 times louder than normal talking.)*

Ask How much louder than the rock concert is a jet engine? *(130 dB – 120 dB = 10 dB. The jet engine is 10 times louder than the rock concert.)*

Speed Bump

Ask How are the senses of taste and smell related? *(Both smell and taste are the result of information collected by chemoreceptors.)*

LESSON 31.4

During Reading continued

Active Reading

USE VISUALIZATION Ask students to imagine that they have about a half inch of water in a glass. Have them further imagine that they are rotating the glass so that the water is swirling around the bottom. Now have them visualize what will happen if they suddenly stop rotating the glass.

Ask What happens to the water in the glass? *(It will splash about before it stops swirling.)*

MAKE INFERENCES Have students use this visualization along with information from the text to make an inference about what causes the feeling of dizziness.

Ask Why do you get dizzy when you spin around and suddenly stop? *(Sample answer: When a person spins and then stops, the liquid in the semicircular canals continues to move and stimulate hair cells, sending confusing information to the brain.)*

KEY QUESTION Invite a volunteer to read the Key Question and its answer aloud. Ask students to look back at the sensory receptor chart in Lesson 31.3.

Ask What kinds of sensations do mechanoreceptors respond to? *(touch, pressure, vibration, stretch)*

Ask Which of these do mechanoreceptors in the ear sense? *(vibration)*

Build Vocabulary

WORD ORIGINS Reinforce the idea that many words in biology have Greek origins.

Say The word *tympanum* comes from the Greek *týmpanon*, which means "drum."

Ask What instrument do you think a timpanist in an orchestra plays? *(drum)*

Explain that many structures in the ear are named for how they look or behave. The tympanum is named for how it vibrates like a drum. The cochlea is named for its shape.

Ask Why do you think the bones called the hammer, anvil, and stirrup were so named? *(They look like these objects.)* Have pictures of an anvil and stirrup on hand for students to compare to the bones.

Speed Bump

Ask The cornea, iris, pupil, and lens work together. What do they do? *(They allow light into the eye and focus it on the retina.)*

Ask Which part opens and closes to control the amount of light that enters the eye? *(the pupil)*

Ask Which part opens and closes the pupil? *(the iris)*

BUILD Vocabulary

cochlea the fluid-filled part of the inner ear; contains nerve cells that detect sound

semicircular canal one of three structures in the inner ear that monitors the position of the body in relation to gravity

cornea a tough transparent layer of the eye through which light enters

iris the colored part of the eye

WORD ORIGINS

The word *cochlea* comes from the Greek word *kochlias*, which means "snail shell." The cochlea in the human ear is shaped like a snail shell.

Balance Your ears help your central nervous system maintain your balance. There are three tiny canals just above the cochlea called **semicircular canals.** These canals and two tiny sacs located behind them monitor your body's position.

The canals and sacs are filled with fluid that changes position as the head moves. These movements cause the hair on the hair cells lining the canals and sacs to bend. The brain uses the impulses created by this action to determine body motion and position.

Key Question How do the ears and brain process sounds and maintain balance? **Mechanoreceptors found in parts of the ear transmit impulses to the brain. The brain translates the impulses into sound and information about balance.**

Vision

The world around us is bathed in light that our eyes detect. Vision occurs when photoreceptors in the eyes transmit impulses to the brain. The brain translates these impulses into images.

Structures of the Eye Light enters the eye through a tough transparent layer of cells called the **cornea.** The cornea helps to focus the light. The light passes through a chamber filled with a fluid called aqueous (AY kwee uhs) humor. At the back of the chamber is a disk-shaped structure called the **iris.** The iris is the colored part of the eye.

The Eye The eye is a complicated sense organ. The sclera, choroid, and retina are three layers of tissues that form the inner wall of the eyeball.

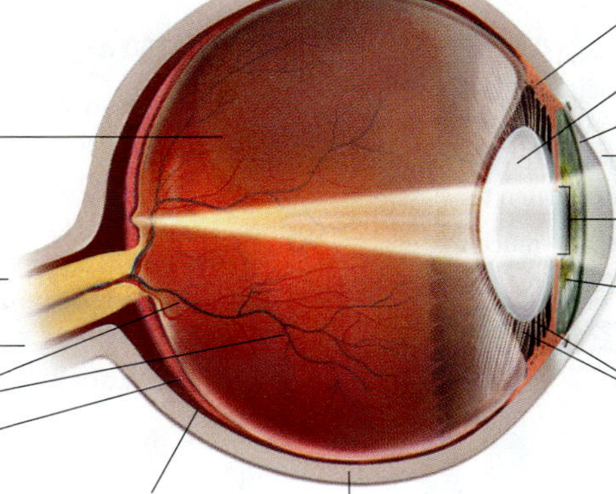

BIOLOGY.com Search Lesson 31.4 GO • Art Review

Hands-On Learning

TASTE AND SMELL Ask students to form groups of two or three and to wash their hands. Then, distribute a blindfold and small slices of onion, apple, and cooked potato to each group. **CAUTION:** *Be sure students are not allergic to any of these foods. Also be sure foods do not come in contact with lab equipment or surfaces.*

1. Have one student in each group wear the blindfold and hold his or her nose. A second student should then place a piece of potato on the partner's tongue.
2. The blindfolded student should describe the taste of the food. *(Apples, onions, and potatoes taste about the same absent their smell.)*
3. Repeat steps 2 and 3 for the onion and the apple.
4. Have students wash their hands after handling the food. Tell students to write a brief summary explaining what they learned about taste and smell.

The small opening in the middle of the iris is the **pupil.** Tiny muscles in the iris adjust the pupil's size. The pupil controls the amount of light that enters the eye. The pupil becomes smaller in bright light and larger in dim light.

Just behind the iris is the **lens.** Small muscles attached to the lens change its shape. This adjusts the eyes' focus to see near or distant objects clearly. The large chamber behind the lens is filled with a transparent, jellylike fluid called vitreous (VIH tree uhs) humor.

How You See The lens focuses light onto the **retina,** the inner layer of the eye. Photoreceptors in the retina convert light energy into nerve impulses. There are two types of photoreceptors: rods and cones. Rods are extremely sensitive to light, but they do not respond to colors. They only allow us to see black and white. Cones are less sensitive than rods. But they do respond to different colors, producing color vision. Cones are concentrated in the fovea, the site of sharpest vision.

The impulses gathered by this complicated layer of cells leave each eye through the optic nerve. This nerve carries the impulses to the appropriate regions of the brain. There are no photoreceptors where the optic nerve passes through the back of the eye. This lack of photoreceptors produces a blind spot in each image sent to the brain. The brain fills in the holes of the blind spot with information when it processes the impulses.

If the eye just took photographs, the images would be blurry and incomplete. The images we actually see are much more detailed. The reason is the complicated way in which the brain processes and interprets visual information.

Key Question How do the eyes and brain produce vision?
Vision occurs when photoreceptors in the eyes transmit impulses to the brain. The brain translates these impulses into images.

BUILD Vocabulary

pupil a small opening in the iris that admits light into the eye

lens a structure in the eye that focuses light rays on the retina

retina the innermost layer of the eye; contains photoreceptors

WORD ORIGINS
The word *retina* comes from the Middle English word *rethina*. *Rethina* is believed to come from the Latin word *rete*, which means "net."

CHECK Understanding

Apply Vocabulary

Use the highlighted words from the lesson to complete each sentence correctly.

1. Hair cells in the _______ send information about your body's position to the brain.
2. Muscles in the _______ control the amount of light that enters the eye.
3. Photoreceptors in the _______ convert light energy into nerve impulses.
4. The _______ is a fluid-filled structure with cells that detect sound.

Critical Thinking

5. **Infer** Some people suffer from night blindness. Which type of photoreceptor is likely not functioning correctly?
6. **Write to Learn** Answer the third clue of the mystery.

MYSTERY CLUE

Based on Cook's symptoms, which of his senses was greatly affected by the toxin? Explain. (**Hint:** See p. 754.)

BIOLOGY.com Search Lesson 31.4 GO • Lesson Assessment

Check Understanding

1. semicircular canals
2. iris
3. retina
4. cochlea
5. Sample answer: Rods are very sensitive to light, and are responsible for night vision. People who suffer from night blindness have rods that cannot detect the faint amount of light available at night.
6. Sample answer: Because Captain Cook's sense of touch was greatly affected by the toxin, his ability to feel temperature and pain might also have been affected.

LESSON 31.4

Active Reading

USE VISUALS Focus students' attention on the illustration of the eye on the opposite page. Have them trace the path of light as it moves through the eye. Write *Light enters the eye* on the board. Beneath it, list the following structures as shown and ask students to number them based on the order in which light passes through or strikes them.

- *(2)* aqueous humor
- *(6)* light shines on the retina
- *(1)* cornea
- *(5)* vitreous humor
- *(3)* pupil
- *(4)* lens

KEY QUESTION Invite a volunteer to read the Key Question and its answer.

Ask What are photoreceptors? *(sensory receptors that respond to light)* Where are they found? *(in the retina)*

Build Vocabulary

WORD ORIGINS Explain that the word *optic* comes from the Greek word *optikos*. *Optikos* is thought to come from the Greek word *optos* which means "seen."

Ask What do you think the words *optometrist* and *optometry* mean? *(Answers may vary but should refer to an eye specialist and the study of the eye.)*

Speed Bump

Ask What kind of photoreceptors respond to color? *(cones)*

Ask Why is there a blind spot at the back of the eye? *(There are no photoreceptors in this area where the optic nerve passes through the back of the eye.)*

Post Reading

ASSESSMENT TIPS Encourage students to review the highlighted vocabulary terms before they begin the assessment.

MYSTERY CLUE Discuss which kind of sensory receptors would be affected if someone lost his sense of touch. Help students determine if these receptors are part of the central or the peripheral nervous system.

Wrap-Up Activity

MEMORY AND MORE Distribute ten index cards. Have pairs of students write a fill-in-the-blank sentence about the senses on each. On a second set of cards have them write the term that correctly completes each sentence. Have pairs exchange cards, shuffle, arrange them face down, then take turns matching terms to the appropriate sentence.

CHAPTER LAB

Pre-Lab

Have students discuss what they know about touch receptors. Invite them to speculate whether touch receptors are evenly distributed throughout the body.

Ask What might be the advantage of having more touch receptors in some locations and fewer receptors in other locations? *(Sample answer: Some locations might be more prone to injury or damage. It might be helpful to have more receptors in these places.)*

Review the Background Questions and the Pre-Lab Questions as a class.

Lab

Tell students they will perform the chapter lab *Testing Sensory Receptors for Touch* described in **Lab Manual B.**

SAFETY Remind students to handle sharp objects carefully, keeping them away from the face and eyes. Request that students follow any of your specific safety precautions during the lab.

Post-Lab

DESIGN AN EXPERIMENT Ask students to design an experiment that would allow them to compare the density of touch receptors on the back of the hand and on the palm of the hand.

NATIONAL SCIENCE EDUCATION STANDARDS

UCP V

CONTENT C.1.f, F.1

INQUIRY A.1.b, A.2.a

Real-World Lab

GUIDED INQUIRY

Pre-Lab: Testing Sensory Receptors for Touch

Problem What factors affect a person's ability to sense gentle pressure on skin?

Materials bent paper clips, metric ruler

Lab Manual Chapter 31 Lab

Skills Focus Measure, Analyze Data, Draw Conclusions

Connect to the Big idea Your nervous system coordinates your response to stimuli from outside your body and inside your body. Sensory receptors react to stimuli by sending impulses to sensory neurons. Each receptor can detect only one type of stimulus. Receptors are classified by the type of stimuli to which they respond. Some respond to light, some to pain, some to chemicals, and so on. Mechanoreceptors are cells that respond to touch, pressure, vibrations, and stretch.

In this lab, you will investigate the mechanoreceptors in your skin that respond to gentle touch. You will compare the relative density of these receptors in three areas of your skin. You will also identify other factors that could affect a person's response to touch.

Background Questions

a. **Review** Which division of the peripheral nervous system transmits signals from receptors in your skin to your brain?

b. **Relate Cause and Effect** List two reasons why a touch might not produce a nerve impulse.

c. **Infer** People who are visually impaired use their fingertips to read books that are printed in Braille. In Braille, each letter of the alphabet is represented by a unique pattern of dots. What feature of the dots allows a reader to distinguish one set of dots from another?

Pre-Lab Questions

Preview the procedure in the lab manual.

1. **Predict** Which area will have the highest density of receptors for gentle pressure—your fingertips, the back of your hand, or your forearm?
2. **Control Variables** Why must you have your eyes closed while your partner touches your skin with the bent paper clip?
3. **Predict** Will you and your partner have the same density of touch receptors in a given area of skin? Give a reason for your prediction.

BIOLOGY.com Search Chapter 31 GO

Visit Chapter 31 online to test yourself on chapter content and to find activities to help you learn.

Untamed Science Video Hold on tightly as the Untamed Science crew takes you on a quick tour of how animal toxins affect the body.

Art Review Review your understanding of the structures in the eyes and ears with this drag-and-drop activity.

InterActive Art Watch a nerve impulse move down a neuron.

Visual Analogy Compare an action potential moving along a neuron to a row of falling dominoes.

Art in Motion View a short animation that shows how a reflex arc works.

Pre-Lab Answers

BACKGROUND QUESTIONS

a. the sensory division

b. Sample answer: There may not be a receptor for touch at the location being touched. The stimulus might not reach the threshold needed to produce a response.

c. The dots are raised so that their pattern will stimulate sensory receptors for touch in the reader's fingertips.

PRE-LAB QUESTIONS

1. Sample answer: Fingertips will have the highest density.
2. Sample answer: If I could see the paper clip touch my skin, I might assume that I felt two touches when I did not.
3. Because most physical traits vary between people, students may assume that the density of receptors can vary too.

758 Chapter 31 • Pre-Lab

31 CHAPTER Summary

31.1 The Neuron

- The nervous system collects information about the body's internal and external environment. The system then processes that information and responds to it.
- Nervous system impulses are transmitted by cells called neurons.
- An impulse begins when a neuron is stimulated by another neuron or by the environment.

peripheral nervous system (p. 742)
central nervous system (p. 742)
cell body (p. 743)
dendrite (p. 743)
axon (p. 743)
resting potential (p. 744)
action potential (p. 744)
threshold (p. 745)
synapse (p. 746)

31.2 The Central Nervous System

- The major areas of the brain are the cerebrum, cerebellum, and brain stem. Each of these areas is responsible for processing and relaying information. The spinal cord is the main communication link between the brain and the rest of the body.
- The brain reacts to too much dopamine by reducing the number of dopamine receptors. As a result, normal activities no longer produce the pleasurable feelings they once did.

cerebrum (p. 748)
hypothalamus (p. 749)
cerebellum (p. 749)
brain stem (p. 749)
dopamine (p. 750)

31.3 The Peripheral Nervous System

- The peripheral nervous system's sensory division transmits impulses from sense organs to the central nervous system.
- The peripheral nervous system's motor division transmits impulses from the central nervous system to muscles or glands.

somatic nervous system (p. 752)
reflex arc (p. 752)
autonomic nervous system (p. 753)

31.4 The Senses

- Different sensory receptors in the body respond to touch, temperature, and pain.
- Sensations of smell and taste are both the result of impulses sent to the brain by chemoreceptors.
- Mechanoreceptors found in parts of the ear transmit impulses to the brain. The brain translates the impulses into sound and information about balance.
- Vision occurs when photoreceptors in the eyes transmit impulses to the brain. The brain translates these impulses into images.

cochlea (p. 755)
semicircular canal (p. 756)
cornea (p. 756)
iris (p. 756)
pupil (p. 757)
lens (p. 757)
retina (p. 757)

UbD Transfer the Big Idea

TRANSFER Have students compare and contrast the structures and functions of a computer with the structures and functions of the human nervous system. Remind students that a graphic organizer can help them organize their thoughts. Encourage students to think about how the functions of a computer are similar to and different from the functions of the human nervous system.

Function	Computer	Human Nervous System
input	keyboard, mouse	sensory receptors
processor	hard drive	the brain
output	monitor	interneurons and motor neurons

Chapter Summary

Think Visually

USE GRAPHICS Ask students to complete the following graphic organizer to show the relationship between the divisions of the nervous system.

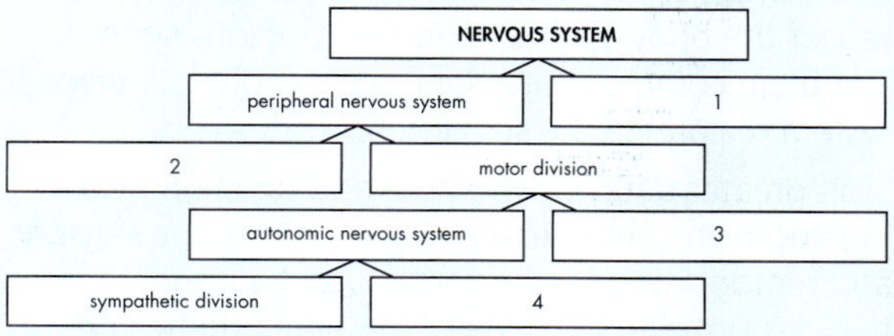

ANSWERS

1. central nervous system.
2. sensory division
3. somatic nervous system
4. parasympathetic division

BIOLOGY.com Study Online

REVIEW RESOURCES

InterActive Art Students view an animation of a nerve impulse.

Visual Analogy This visual analogy explores how the propagation of a nerve impulse is similar to the fall of a row of dominoes.

Art in Motion This short animation shows the steps in a reflex arc.

Art Review Students can drag and drop labels to help them learn the structures of the ears and eyes.

Data Analysis Students can analyze data to learn how the appearance of food can impact people's perception of flavor.

Vocabulary Review The *Flash Cards* and *Match It* are an interactive vocabulary review.

Editable Worksheets Editable pages of Study Workbook B, Lab Manual B, and the Assessment Resources Book are available online.

Lesson Overviews Lesson Overviews are based on Miller & Levine Biology. However, they are available on CD-ROM and can be edited for your class.

ASSESSMENT RESOURCES

Chapter Assessment Have students take an online version of the Chapter 31 Check Understanding.

Standardized Test Prep Students can take an online version of the Standardized Test Prep. You will receive their scores along with ideas for remediation.

Diagnostic and Benchmark Tests Use these tests to monitor your students' progress and supply remediation.

ASSESSMENT

Assess

Big idea Invite a volunteer to read the Big Idea and the question.

The minimum information students should provide in their answers is that the central nervous system acts as a control center, which communicates with the rest of the body via the peripheral nervous system. The brain communicates with the peripheral nervous system mainly via the spinal cord.

Elicit greater detail of structure and function by asking about the peripheral nervous system, the sensory and motor divisions, the somatic and autonomic nervous systems, and finally, the sympathetic and parasympathetic nervous systems.

CONSTRUCTED RESPONSE Paragraphs will vary. Explain the activity carefully. Be sure that students understand that the *Hints* should be used as a guide to help them answer the questions. Review how to write a paragraph.

Write Question 1 on the board, along with the two hints. Read the question aloud and underline *sodium-potassium pumps* and *neuron*.

Read the first hint.

Ask Do you remember in which direction the pump moves the sodium ions? *(Sodium ions leave the cell.)*

Ask What happens to the potassium ions? *(Potassium ions enter the cell.)*

Ask What effect does this have on the cell's potential? *(The inside of the cell is negatively charged compared to the outside of the cell.)*

Ask What is this called? *(resting potential)*

Read the next hint.

Ask *Reverse* means "opposite to a normal condition." What is the opposite of *negative charge*? *(positive charge)*

Ask If during resting potential the inside of a cell is negative compared to the outside, and the nerve impulse reverses resting potential, then what must be true about the potential during a nerve impulse? *(The inside of a cell is positive compared to the outside.)*

Model writing the first two or three sentences using the hints on the board and the text. *(Sample: The sodium-potassium pump moves sodium ions out of the neuron and potassium ions into the neuron. This maintains a resting potential for the neuron. It keeps the inside of the cell negatively charged compared to the outside of the cell.)*

31 CHECK Understanding

Assess the Big idea: Structure and Function

Write an answer to the question below.

Q: How does the structure of the nervous system allow it to regulate functions in every part of the body?

Constructed Response

Write an answer to each of the numbered questions below. The answer to each question should be one or two paragraphs. To help you begin, read the **Hints** below the questions.

1. What is the purpose of sodium-potassium pumps in a neuron?

   Hint The sodium-potassium pump moves sodium and potassium ions across the cell membrane.

   Hint Nerve impulses are reversals of resting potentials, which are electrical charges across the cell membrane.

2. Why must neurotransmitters be broken down or removed after a cell responds to a nerve impulse?

   Hint Axon terminals release neurotransmitters during a nerve impulse.

   Hint Neurotransmitters must bind to receptors on a cell to stimulate a response from the cell.

3. How do hair cells in the ear help transmit nerve impulses?

   Hint Hair cells are found in the cochlea, the semicircular canal, and the sacs associated with the semicircular canal.

   Hint Hair cells are mechanoreceptors.

Foundations for Learning Wrap-Up

Use the note squares you prepared when you started the chapter as tools to organize your thoughts about the nervous system.

Activity 1 Working with a partner, use your note squares to quiz each other about the structures and functions of the nervous system.

Activity 2 Working as a group, create idea "nervous systems" from your note squares by connecting the corners of the squares with brass fasteners. Use main ideas to create the "brain" and "spinal cord." Use details to create the "peripheral nerves." How do the "peripheral nerves" explain the main ideas in the "central nervous system"?

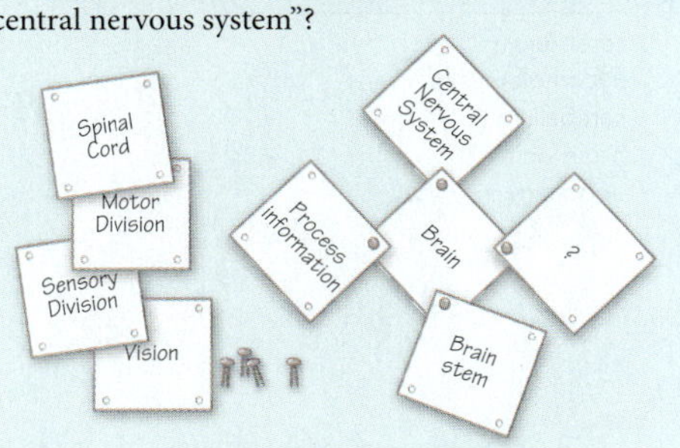

760 Check Understanding

For Question 2, pair students and have them use the hints and your modeling to answer the question. Have students share their paragraphs with the class.

For Question 3, have students write the paragraphs independently for homework or as an anchor activity.

Foundations for Learning Wrap-Up

Before starting the activities, students should make sure they have identified the structures and the functions for parts of the nervous system.

For Activity 2, suggest students follow the diagram at the beginning of the chapter to determine the placement of ten of the paper squares. Allow them to work in pairs to add to the construction so that there is a minimum of sixteen squares in the diagram.

31.1 The Neuron

Understand Key Concepts

1. The basic units of structure and function in the nervous system are
   a. neurons.
   b. axons.
   c. dendrites.
   d. neurotransmitters.
2. In the diagram below, the letter A is pointing to a(n)
   a. myelin sheath.
   b. axon.
   c. dendrite.
   d. cell body.

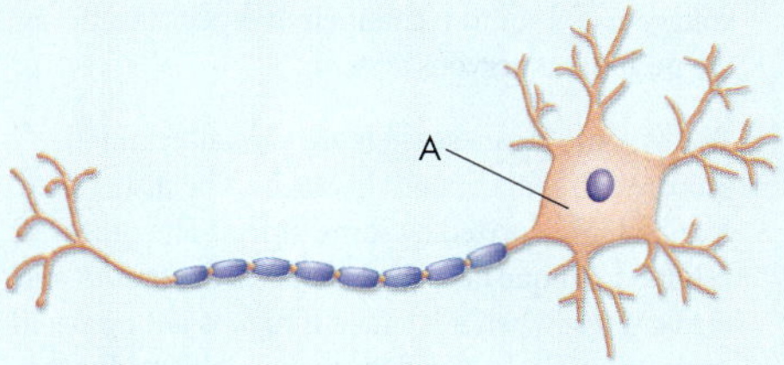

Test-Taking Tip

Interpret Graphics When you are asked to identify a structure in a figure, it is helpful to first review and label the figure. Question 2 includes the figure of a neuron. First, identify all the structures in the answer choices. The myelin sheath surrounds the axon. The axon is a long fiber that extends from the cell body. Dendrites are the short extensions of the cell body. The cell body contains the nucleus. So, A is pointing to the cell body.

3. The place where a neuron transfers an impulse to another cell is the
   a. synapse.
   b. dendrite.
   c. myelin sheath.
   d. receptor.
4. Why can an action potential be described as an all-or-none event?

Think Critically

5. **Infer** Suppose part of an axon is cut so that it does not connect to its cell body. How would that affect the transmission of impulses?

31.2 The Central Nervous System

Understand Key Concepts

6. The central nervous system consists of the
   a. sense organs.
   b. reflexes.
   c. brain and spinal cord.
   d. sensory and motor neurons.
7. Voluntary activities, or conscious activities of the body, are controlled primarily by the
   a. medulla oblongata.
   b. cerebrum.
   c. cerebellum.
   d. brain stem.
8. Describe the relationship between the brain stem and the spinal cord.

Think Critically

9. **Infer** Why might the folding of the cerebral cortex be important?

31.3 The Peripheral Nervous System

Understand Key Concepts

10. The sympathetic nervous system and parasympathetic nervous system are specific divisions of the
    a. peripheral nervous system.
    b. central nervous system.
    c. somatic nervous system.
    d. autonomic nervous system.
11. How do reflexes protect the body from injury?
12. What is the function of the sensory division of the peripheral nervous system?

Think Critically

13. **Apply Concepts** A routine examination by a doctor usually includes a knee-jerk reflex test. What is the purpose of this test?

Assessment Tips

Remind students that for multiple-choice questions, choosing the correct answer may be easier if students eliminate any answer(s) they know to be incorrect.

For Question 4, be sure students understand that an "all-or-none" event is one in which something either happens or it doesn't—not one in which something partly happens.

Tell students that Questions 5 and 9, which ask them to infer, require that they apply information they have learned to different situations.

Question 17 asks students to sequence the steps along a path. Point out that the steps must be in the correct order from first to last.

Answers

Lesson 31.1

UNDERSTAND KEY CONCEPTS

1. a
2. d
3. a
4. An action potential is an all-or-none event because it will not occur unless the stimulus reaches the threshold needed to cause the reversal of charges across the membrane.

THINK CRITICALLY

5. If the axon is cut, impulses would no longer be able to travel from the cell body to the axon terminals. The impulse would not be transmitted.

Lesson 31.2

UNDERSTAND KEY CONCEPTS

6. c
7. b
8. The brain stem connects the rest of the brain to the spinal cord. It regulates the flow of information between the brain and the rest of the body.

THINK CRITICALLY

9. Sample answer: The folding of the cerebral cortex allows the surface area to be much larger than a smooth surface. The greater the surface area, the greater the number of neurons and neuron networks that the cortex can hold. If the same amount of surface area occurred in a smooth surface, the overall size of the cerebrum, and thus the skull, would be much greater.

Lesson 31.3

UNDERSTAND KEY CONCEPTS

10. d
11. Sample answer: Reflexes protect the body from injury by allowing it to react more quickly to danger than a reaction that requires impulses to travel to the brain and be processed.
12. The sensory division transmits impulses from the sense organs to the central nervous system.

THINK CRITICALLY

13. The purpose of the knee-jerk test is to determine whether the person's reflexes are normal. The absence of a response could indicate a disorder of one of the components of the reflex arc.

ASSESSMENT

Lesson 31.4

UNDERSTAND KEY CONCEPTS

14. b

15. Rods and cones are photoreceptors, which send a nervous impulse when they are stimulated by light. Rods are very sensitive to light, and allow us to see black and white. Cones respond to different colors, producing color vision.

16. sweet, salty, sour, bitter, and umami

THINK CRITICALLY

17. Air vibrations make the tympanum vibrate. These vibrations are transmitted by the hammer, anvil, and stirrup bones to the oval window. Here, the vibrations create pressure waves in the fluid of the cochlea. The pressure waves move the hair cells in the cochlea. The hair cells change these movements into nerve impulses that are sent to the brain.

Connecting Concepts

18. b

19. Because alcohol depresses respiration and heart rate, it is most likely stimulating the parasympathetic nervous system.

CHAPTER MYSTERY After students have read through the Chapter Mystery, discuss the origin of the toxins and how they affect the central nervous system.

Ask What organism produces the toxins associated with the Silverstripe blaasop? *(bacteria that live in the fish's liver, gonads, intestines, and skin)*

Ask How does the toxin affect the nervous system? *(It binds to and blocks voltage-gated sodium channels. If sodium cannot enter the channels, nerve impulses cannot travel along the axons.)*

31 CHECK Understanding

31.4 The Senses

Understand Key Concepts

14. The senses of taste and smell involve sensory receptors called

a. photoreceptors.
b. chemoreceptors.
c. thermoreceptors.
d. mechanoreceptors.

15. What are the functions of the rods and cones?

16. What are the five basic tastes?

Think Critically

17. **Sequence** Trace the path of sound through the ear.

Connecting Concepts

Use Science Graphics

Use the illustration to answer questions 18 and 19.

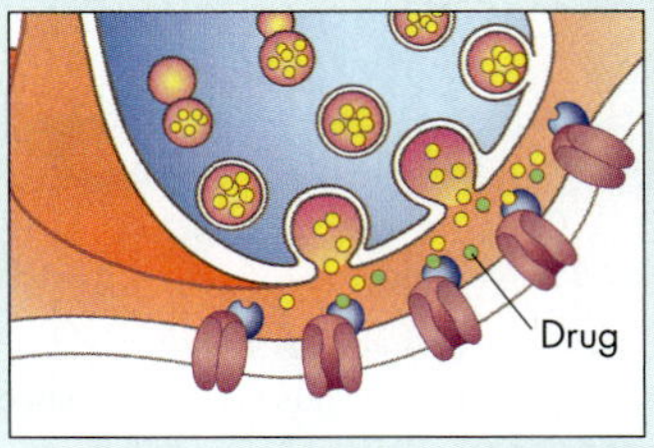

18. **Interpret Visuals** This illustration shows a drug

a. interfering with enzymes that break down a neurotransmitter at a synapse.
b. mimicking a neurotransmitter.
c. stimulating an axon terminal.
d. stimulating enzymes that break down a neurotransmitter at a synapse.

19. **Apply Concepts** Alcohol is called a depressant, because it depresses (slows) respiration and heart rate. Which part of the autonomic nervous system does alcohol most likely stimulate?

solve the CHAPTER MYSTERY

POISONING ON THE HIGH SEAS

A naturalist named Georg Forster traveled with Captain Cook. One of his sketches suggests that the crew ate a fish called *Tetraodon lagocephalus sceleratus*. This fish is also known as the Silverstripe blaasop. Bacteria that live in its liver, gonads, intestines, and skin produce a poison called tetrodotoxin. The poison can remain in the fish even after cooking the fish at high temperatures. Tetrodotoxin binds to and blocks voltage-gated sodium channels. It especially affects the peripheral nervous system.

Today, some Japanese chefs are specially trained to prepare fish that contain this toxin. The dish, called fugu, is highly prized by some diners. The prepared fish has a unique taste. It also makes the mouth and throat tingle when it is eaten. If fugu is not prepared correctly, it can lead to dangerous problems for the diner. Death is even possible.

Obviously, tetrodotoxin doesn't poison the fish that produce it. Studies of the fish's genome have revealed a mutation in an important gene. This gene codes for the structure of sodium channel proteins. The mutation changes the surface shape of the channel. This change prevents the toxin from binding to it.

1. **Infer** Some fish fillets are cooked and served with the skin left on. Would this be a safe way to prepare and serve fugu?

2. **Apply Concepts** Describe in your own words why the fish are not affected by their own toxin.

UntamedScience™ **Never Stop Exploring Your World.** Finding the solution to this mystery is just the beginning. Take a video field trip with the ecogeeks of Untamed Science to see where this mystery leads.

CHAPTER MYSTERY ANSWERS

1. Fugu would not be safe to prepare and serve with the skin left on because this is one of the places that the bacteria which produce the toxin are located.

2. Sample answer: The fish are not affected by their own toxin because a gene mutation has altered the shape of their gated sodium channels. This shape prevents the toxin from binding to the channel. Therefore, sodium can enter the cell when necessary, allowing nerve impulses to travel down the axons without any problem.

Untamed Science™ The Untamed Science crew explores the scary chemicals emitted by cone snails in **Toxic Secretions.**

Standardized Test Prep

Multiple Choice

1. The largest and most prominent part of the human brain is the
   A cerebrum. C thalamus.
   B cerebellum. D brain stem.

2. The point of connection between two neurons is called a
   A threshold. C neurotransmitter.
   B synapse. D dendrite.

3. The part of a neuron that carries impulses away from the cell body is called a(n)
   A axon. C vesicle.
   B dendrite. D synapse.

4. The minimum stimulus level that will cause a neuron to produce an action potential is called the
   A resting potential. C threshold.
   B impulse. D synapse.

5. The part of the brain responsible for collecting sensory input from the body and relaying it to appropriate brain centers is the
   A limbic system. C cerebellum.
   B thalamus. D cerebrum.

6. The major function of the spinal cord is
   A emotional learning and memory storage.
   B control of voluntary muscle movements.
   C fine control of detailed muscle movement.
   D a principal communication path between the brain and the rest of the body.

7. Involuntary activities carried out throughout the body are the primary responsibility of the
   A somatic nervous system.
   B autonomic nervous system.
   C spinal cord.
   D limbic system.

8. The part of the eye which contains photoreceptor cells is the
   A cornea. C retina.
   B iris. D optic nerve.

Questions 9–10

Blood alcohol concentration (BAC) is a measure of the amount of alcohol in the blood per 100 mL of blood. In some states, if a driver has a BAC of 0.08 percent, he or she is considered legally drunk. The table below lists an average BAC as alcohol consumption increases. Use the information in the table to answer the questions.

Blood Alcohol Concentration (Percent)

Drinks in One Hour	Body Mass					
	45 kg	54 kg	63 kg	72 kg	81 kg	90 kg
1	0.04	0.03	0.03	0.02	0.02	0.02
2	0.07	0.06	0.05	0.05	0.04	0.04
3	0.11	0.09	0.08	0.07	0.06	0.06
4	0.14	0.12	0.10	0.09	0.08	0.07
5	0.18	0.15	0.13	0.11	0.10	0.09
6	0.21	0.18	0.15	0.14	0.12	0.11
7	0.25	0.21	0.18	0.16	0.14	0.13
8	0.29	0.24	0.21	0.18	0.16	0.14

9. How many drinks in one hour would cause a 63 kg person to have a BAC of 0.08 percent?
   A 1 C 5
   B 3 D 7

10. If a 54 kg person had 3 drinks in one hour, what would his or her BAC percentage be?
    A 0.06 C 0.09
    B 0.08 D 0.11

Open-Ended Response

11. How do the parasympathetic and sympathetic nervous systems work together in the body?

If You Have Trouble With . . .

Question	1	2	3	4	5	6	7	8	9	10	11
See Lesson	31.2	31.1	31.1	31.1	31.2	31.2	31.3	31.4	31.2	31.2	31.3

Test-Taking Tip

REVIEW ANSWERS

Tell students to use any time remaining after they have answered all of the questions to review their answers. Point out that they should change only those responses they are sure are incorrect. If the answer booklet is separate from the test questions booklet, have students check to make certain that the question numbers correctly correspond with their responses in the answer booklet.

Before the Test

Review with students the Test-Taking Tip. Remind them that if they skip a question they are unsure of, they need to also skip the answer space.

Remind students that an open-ended response means that they'll need to write a paragraph to answer the question. Have them paraphrase the question in their notebooks, and remind them about modeling a paragraph, including topic sentences, main idea, and details. Caution them to leave enough time to read over their finished paragraph to make certain that it answers the question.

Answers

1. A
2. B
3. A
4. C
5. B
6. D
7. B
8. C
9. B
10. C
11. Sample answer: The sympathetic and parasympathetic nervous systems have opposite effects on the same organ systems, producing a level of fine control that coordinates organs throughout the body.

CHAPTER PLANNER 32 Skeletal, Muscular, and Integumentary Systems

Chapter Contents	NSES	Time	Core Resources
Chapter Preview			**Student Edition,** pp. 764–765 **Chapter Mystery,** p. 765 **Foundations for Learning,** p. 765
32.1 The Skeletal System The Skeleton • Bones • Joints	I, II, V, C.1.f, C.5.d, F.1	1 period 1/2 block	**Student Edition,** pp. 766–769 **Build Connections,** p. 767 **Study Workbook B** 32.1 Worksheets **Biology.com** *Visual Analogy:* The Skeleton • *InterActive Art:* Freely Movable Joints • 32.1 Lesson Assessment
32.2 The Muscular System Muscle Tissue • Muscle Contraction • Muscles and Movement	I, II, V, A.1.b, A.2.a, C.1.a, C.1.b, C.1.d, C.1.f, F.1	1 1/2 periods 3/4 block	**Student Edition,** pp. 770–774 **Build Connections,** p. 771 **Inquiry Into Scientific Thinking,** p. 774 **Study Workbook B** 32.2 Worksheets **Biology.com** *Art in Motion:* The Sliding Filament Model • *Tutor Tube:* Using Analogies to Understand Muscle Contraction • 32.2 Lesson Assessment
32.3 Skin—The Integumentary System Integumentary System Functions • Integumentary System Structures • Skin Problems	I, II, V, C.1.f, F.1	1 period 1/2 block	**Student Edition,** pp. 775–779 **Study Workbook B** 32.3 Worksheets **Biology.com** *Art Review:* The Structure of the Skin • 32.3 Lesson Assessment
Chapter Pre-Lab	V, A.1.b, A.2.a, C.1.f, C.3.c	1 period 1/2 block	**Student Edition,** p. 780 **Lab Manual B** *Comparing Limbs*

Tools for ELL Instruction

ELL Handbook describes teaching strategies to make *Miller & Levine Biology: Foundation Edition* more accessible to ELL students.

Spanish Study Workbook is a Spanish translation of Study Workbook A.

Multilingual Glossary is the *Miller & Levine Biology* glossary translated into ten languages.

Tools to Challenge Students

Study Workbook A includes more challenging worksheets and activities such as Chapter Mystery activities that reinforce 21st Century Skills.

Lab Manual A labs are geared toward on-level students. The Lab Manual includes fifteen labs in addition to the Chapter Labs.

Biology.com Data Analysis activities offer students a way to practice data gathering and analyzing skills.

Visual Quizzes in the Assessment Resources Book are geared toward on-level students and use visuals from *Miller & Levine Biology.*

Additional Resources

Biology.com Untamed Science Video • Flash Cards

Study Workbook A 32.1 Worksheets
Spanish Study Workbook 32.1 Worksheets

Study Workbook A 32.2 Worksheets
Spanish Study Workbook 32.2 Worksheets
Assessment Resources Book Visual Quiz

Study Workbook A 32.3 Worksheets
Lab Manual B *Data Analysis:* Rising Rate of Melanoma
Spanish Study Workbook 32.3 Worksheets
Biology.com *Data Analysis:* Sunlight and Skin Cancer
Assessment Resources Book Visual Quiz

Lab Manual A *Comparing Limbs* • *Modeling Breathing*

Chapter Review

Student Edition Chapter Summary, p. 779
Study Workbook B Chapter 32 Review
Transparencies, pp. 351–361
Biology.com Untamed Science Video • Editable Worksheets of Study Workbooks A and B and Lab Manuals A and B • Chapter 32 Flash Cards and Match It

Untamed Science DVD • Classroom Resources CD (includes lesson presentations and editable worksheets)

Chapter Assessment

Student Edition Check Understanding, pp. 780–783
Study Workbook B Chapter 32 Taking a Standardized Test
Assessment Resources Book Chapter 32 Test B
Biology.com Chapter 32 Assessment • Editable Worksheets of Chapter 32 Visual Quizzes* and Chapter 32 Tests A and B

Exam*View* *Assessment Suite* • Classroom Resources CD (includes lesson presentations and editable worksheets)

Time: 1 period, 1/2 block

* Visual Quizzes are based on visuals from *Miller & Levine Biology.* Review carefully to make sure they are appropriate for your class.

Pressed for Time?

Preview the Chapter Discuss the first Key Question listed for each lesson, and preview the visuals The Skeleton; Skeletal, Smooth, and Cardiac Muscle; and Structure of the Skin.

Cover the Chapter Quickly Have students read Lesson 32.1 and *Muscle Tissue* in Lesson 32.2. In Lesson 32.3, have students read *Integumentary System Functions* and *Integumentary System Structures.*

Assess Assign the 32.1 Check Understanding, Question 4 in the 32.2 Check Understanding, Questions 1–3 in the 32.3 Check Understanding, and the Chapter 32 Standardized Test Prep except Question 4.

CHAPTER 32

Get the Picture

Direct students to study the picture showing a gymnast performing a back flip. Inform students that their skeletal and muscular systems must work together to perform this and most other activities involving movement.

Ask How does the gymnast's skeletal system and muscular system help her perform the back flip correctly? *(The skeletal system provides support and structure. The muscular system allows the gymnast to move.)*

Ask What evidence do you see that the gymnast's body systems work to support her body during the activity? *(Students should note that the gymnast stands and moves as she performs. Students may observe that the gymnast's body is also performing other necessary activities, such as making her heart beat or breathing, during the activity.)*

Activate Prior Knowledge

Ask What are the parts of your skeletal system? *(Students are likely to respond that it includes their skeleton and/or all the bones in their body.)* Why do you think your skeletal system is important? *(Students may say that their skeleton gives them form or structure, or helps them stand up.)*

Ask What activities require you to use muscles? *(Sample answers: walking, running, riding a bicycle, any activity—even breathing—that involves the movement of parts of the body)*

Ask What are some parts that cover and protect your body? *(Sample answers: skin, hair, nails.)*

NATIONAL SCIENCE EDUCATION STANDARDS

UNIFYING CONCEPTS AND PROCESSES

I, II, V

CONTENT

C.1.a, C.1.b, C.1.d, C.1.f, C.3.c, C.5.d, F.1

INQUIRY

A.1.b, A.2.a

32 Skeletal, Muscular, and Integumentary Systems

Big idea Structure and Function

Q: What systems form the structure of the human body?

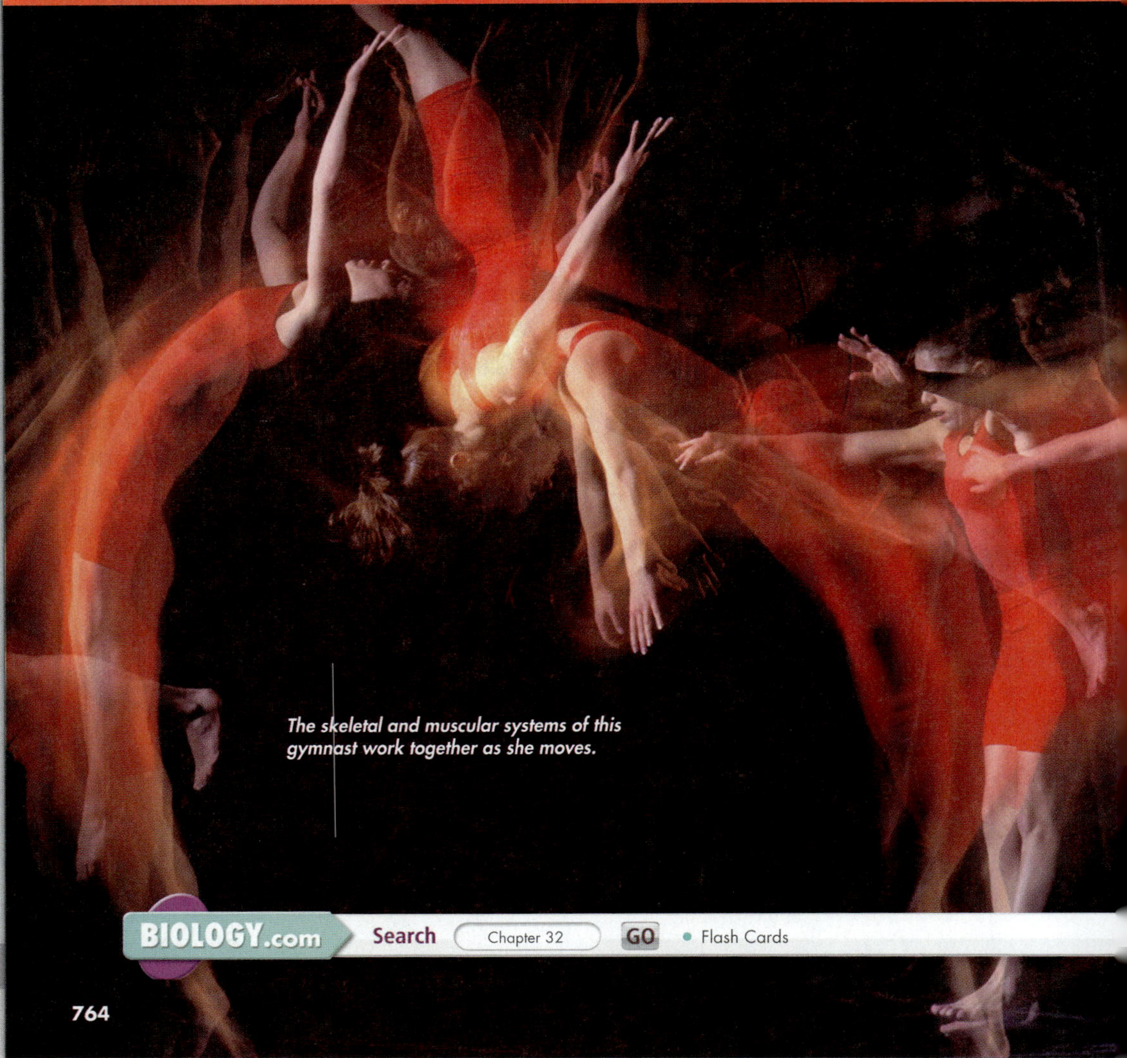

The skeletal and muscular systems of this gymnast work together as she moves.

BIOLOGY.com Search Chapter 32 GO • Flash Cards

UbD Connect to the Big Idea

Invite a volunteer to read aloud the chapter number and the chapter title. Read the Big Idea and question. Emphasize the question, **What systems form the structure of the human body?**

Set the Purpose

Draw the graphic organizer on the board. Guide students to the INSIDE feature on the next page, and invite a volunteer to read each lesson number and title.

Ask How do you think this title relates to the questions in the graphic organizer? Remind students to search for the answer as they read and to also ask themselves how their skeletons, muscles, and skin give structure to their bodies.

INSIDE:

- 32.1 The Skeletal System
- 32.2 The Muscular System
- 32.3 Skin—The Integumentary System

- Untamed Science Video
- Chapter Mystery

CHAPTER
MYSTERY

THE DEMISE OF A DISEASE

In the early 1900s, many children suffered from a disease called "rickets." Children with rickets had very soft, weak bones. These children often had bowed legs, twisted wrists, and many other problems with their bones. Many children with rickets were poorly nourished. Most of them lived in northern cities of the United States. Doctors were puzzled as to why few poorly nourished children living in the south had rickets.

No one knew the cause of rickets or how to cure the disease. Some people thought that cod liver oil cured rickets. Scientists wanted to find answers. What was the connection between rickets and northern cities? And could cod liver oil really be a cure?

Read for Mystery Clues As you read this chapter, look for clues that helped scientists form ideas about the cause of rickets. Then, solve the mystery.

FOUNDATIONS for Learning

Draw and label a three-column chart as shown below. As you read, list the functions in the second column. Cut the third column into flaps. Use the flaps to cover the second column as you study. At the end of the chapter are two activities that you can do with the chart to help answer the question: What systems form the structure of the human body?

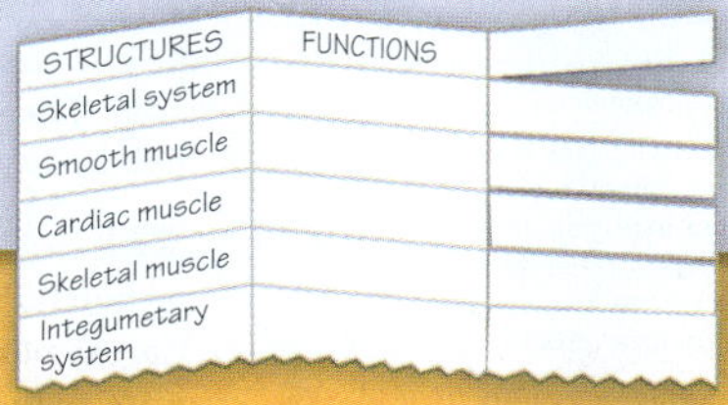

Chapter 32
Big Idea: Structure and Function

Chapter 32: What systems form the structure of the human body?

- **32.1:** How does the structure of the skeletal system allow it to function properly?
- **32.2:** How do muscles help you move?
- **32.3:** Why is the integumentary system a necessary organ system?

Foundations for Learning

Draw the table on the board and have students draw their own tables in their notebooks. Help students understand the importance of this activity.

Ask What does the word *structure* mean in the table? *(Sample answers: part or body part)* What does the term *function* mean? *(Sample answers: job, role, or purpose)*

Ask How is the table organized to reflect the Big Idea? *(The table organizes the three systems being studied into structures and functions.)*

Ask How will using this table to compare and contrast the three systems help you learn? *(Making notes side by side allows you to compare and contrast the systems and helps you see how they work together.)*

Remind students that they will use this table as a study aid as they read this chapter.

Untamed Science — Students can travel to NASA to learn how space travel affects bones in **Skeletons in Space**.

CHAPTER MYSTERY — Ask students if they have ever heard of the disease called rickets.

Ask What parts of the body does it affect? *(bones)*

Invite a volunteer to read aloud the Chapter Mystery. Have students predict why children living in northern cities were affected by rickets more than children living in southern cities. Tell students that they will find clues in the chapter to check their predictions. Tell students that they also will learn how one disease, such as rickets, can affect the functions of the three different body systems they are studying: the skeletal, muscular, and integumentary systems.

Digital Support

BIOLOGY.com — Extend your reach by using the digital assets offered at **Biology.com.**

LESSON 32.1

Pre-Reading

Objectives

32.1.1 List the structures and functions of the skeletal system.

32.1.2 Describe the structure of a typical bone.

32.1.3 List the different kinds of joints and describe the range of motion of each.

Preview the Pages

1. Review the vocabulary words with students before reading. Relate the term *axial skeleton* to the word *axis* and the term *appendicular skeleton* to the word *appendage.* Point out the body's axis and appendages.
2. Have students use the visuals to identify parts of their own skeletal system. For example, students can view the graphic on movable joints and demonstrate the movements of each type of joint on their own body. Review the difference between the meanings of the words *immovable* and *movable.*

Build Understanding

T-CHART Direct students to Workbook B, where they will find a partially filled-in T-chart. Remind students to use the T-chart to answer the questions they wrote from the headings in the lesson.

NATIONAL SCIENCE EDUCATION STANDARDS

UNIFYING CONCEPTS AND PROCESSES

I, II, V

CONTENT

C.1.f, C.5.d, F.1

For the corresponding lesson in *Miller & Levine Biology,* see pages 922–927.

32.1 The Skeletal System

Key Questions

What are the functions of the skeletal system?

What is the structure of a typical human bone?

What is the role of joints?

BUILD Understanding

T-Chart As you read the lesson, make a T-chart. Start with a two-column chart. In the left column, rewrite the headings in this lesson as questions. In the right column, write answers to the questions.

In Your Workbook Go to your workbook to learn more about making a T-chart. Complete the T-chart for Lesson 32.1.

FUNCTIONS OF THE SKELETON

Support The bones of the skeleton support and give shape to the human body.

Protection Bones protect the delicate internal organs of the body. For example, the ribs form a basketlike cage around the heart and lungs.

Movement Bones work with muscles to produce movement.

Mineral Storage Bones store minerals, including calcium. Minerals are important to many body processes. When blood calcium levels are low, some calcium is released from bones.

Blood Cell Formation Many kinds of blood cells are formed in soft tissue in the middle of some bones

The Skeleton

An animal's skeleton is so durable that it can often be recognized thousands of years after the animal's death. Because bones are so hard, it's easy to think of them as lifeless. If that were true, what would happen if a bone broke? As you know, broken bones can heal. What does that tell you about our skeletons?

Structure of the Skeleton To keep their shape, all organisms need structural support. Cells have cytoskeletons that support their shapes. The cytoskeleton also makes it possible for cells to move. Vertebrates have internal skeletons called endoskeletons. The vertebrate skeleton supports the body. It also makes it possible for the body to move.

An adult person has 206 bones. Some of these bones are in the axial skeleton. Others are in the appendicular skeleton. The skull, backbone, and ribs form the **axial skeleton.** The bones of the arms and legs, pelvis, and shoulder form the **appendicular skeleton.**

Functions of the Skeleton The skeleton holds the body up. It protects the internal organs. It helps the body move. The skeleton stores minerals. Blood cells form within the skeleton.

Key Question What are the functions of the skeletal system? **The skeleton holds the body up and protects the internal organs. The skeleton helps the body move. It stores minerals and is the site of blood cell formation.**

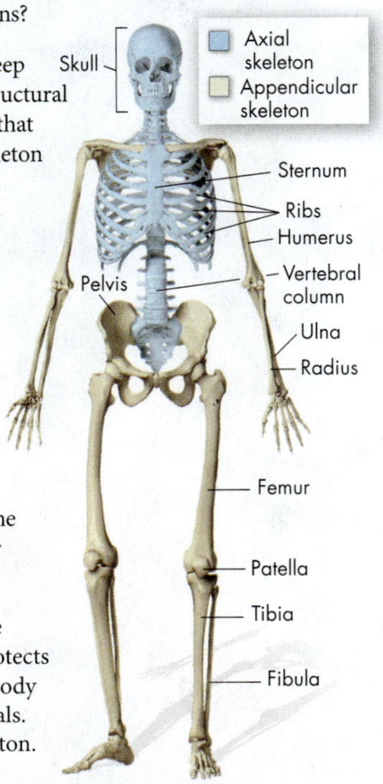

The Skeleton The human skeleton is made of living tissue that has many functions.

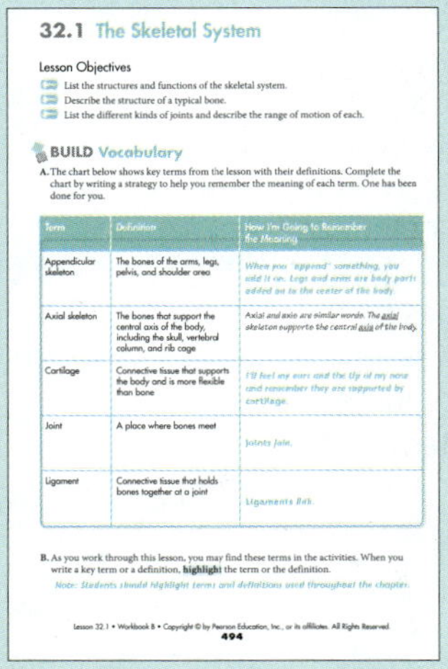

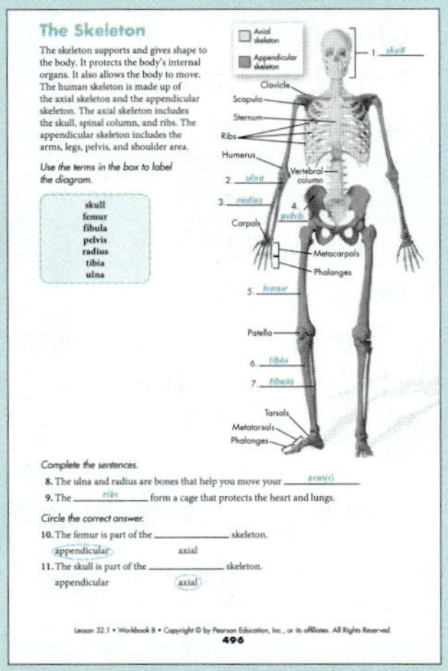

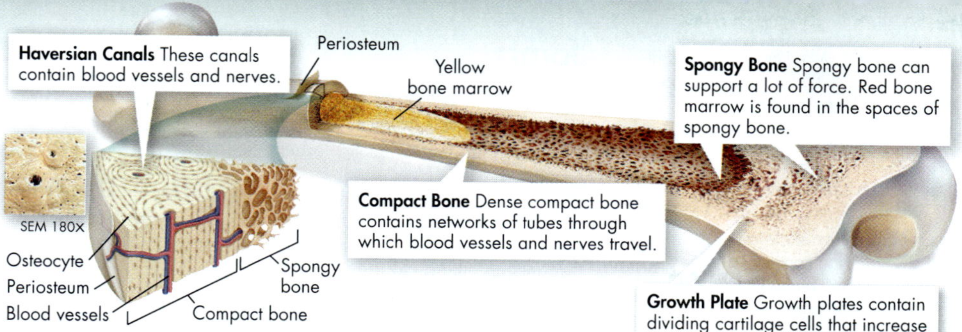

Structure of a Bone A long bone such as the femur is surrounded by the periosteum. Below the periosteum, Haversian canals run through the bone. Nerves and blood vessels run through the canals.

Bones

The mass of a bone is mostly mineral salts. Most of those salts are calcium and phosphorus. This composition makes it easy to think that bones are not really living tissue. However, bones are very much alive.

Structure of Bones Bones are made up of living cells and protein fibers. Calcium salts surround the fibers. The outer layer of bone is called periosteum (pehr ee AHS tee um). Nerves and blood vessels run through the bone in tubes. These tubes are called Haversian (huh VUR zhun) canals. Many bones have a tissue called bone marrow inside them. Yellow bone marrow stores fat. Red bone marrow contains stem cells that develop into blood cells.

Development of Bones Most of the skeleton of a human embryo is made of cartilage. **Cartilage** (KAHR tl ij) is softer and more flexible than bone. As an embryo grows, bone slowly takes the place of the cartilage. This process is called ossification (ahs uh fih KAY shun) and begins up to seven months before birth. It continues throughout life. Cells called osteoblasts (AHS tee oh blasts) add minerals that replace the cartilage. Most osteoblasts become osteocytes. Osteocytes help keep bones healthy.

Bone Remodeling and Repair Some osteoblasts continue to build bone tissue throughout a person's life. Osteoclasts are cells that break down bone tissue. Together, osteoblasts and osteoclasts allow bones to get stronger after exercise and injuries.

Osteoporosis Osteoporosis is a serious disorder in which bone breaks down more quickly than it is rebuilt. It weakens the bones of many older people to the point where the bones may break easily. Getting plenty of calcium in the diet can help prevent osteoporosis. Also, weight-bearing exercises such as walking help prevent osteoporosis.

Key Question What is the structure of a typical human bone? **Bones are made up of living cells and protein fibers. The fibers are surrounded by calcium salts.**

BUILD Connections

A person could not stand up without a bony skeleton, just as a house could not stand without its wooden frame.

In Your Workbook Go to your workbook to label the parts of the skeleton.

BIOLOGY.com Search Lesson 32.1 GO • Visual Analogy

ELL Focus on ELL: Extend Language

INTERMEDIATE AND ADVANCED SPEAKERS Ask students if they have heard the word *osteoporosis*, and invite them to share what they know. If students do not have experience with this term, show them ads for foods or dietary supplements that mention osteoporosis. Have students look up the word part *osteo-* in the dictionary. Help students understand that the word part *osteo-* refers to bones. Have students find other words with this word part on this page and discuss how their meanings are all related to bones.

During Reading

Active Reading

KEY QUESTION Read aloud the Key Question and answer on the previous page. Ask students what structure protects the brain from injury. *(the skull)*

INTERPRET DIAGRAMS Have students examine the diagram of bone structure. Point out the cut-away and enlarged views of compact bone. Read the captions in the four boxes aloud while students locate each part in the diagram.

Ask What are the two types of bone? *(compact and spongy)* Where are blood vessels and nerves found? *(Within canals in the compact bone.)* What is the outside covering of bone called? *(periosteum)*

KEY QUESTION Read aloud the Key Question and answer. Ask students why getting enough calcium in the diet is important. *(Bones contain calcium. Without it, bones can become weak and break.)*

Build Connections

VISUAL ANALOGY Remind students that an analogy is a kind of comparison: It takes two things that are different and shows how they can be similar.

Ask How does a house depend on its frame? *(The parts of the house are held up and supported by the frame.)* What might happen to a house if its frame is weak or incomplete? *(Sample answer: All or part of the house might fall down.)*

Have students make another analogy for the skeletal system using one of its other functions. Ask students to tell how their analogy works.

USE GRAPHIC ORGANIZERS Draw a flowchart that links the following terms from this page: *cartilage, ossification,* and *hard bone*.

Ask What are the bones of a baby like? *(They are softer and more flexible than the bones of an adult.)* Encourage students to feel the cartilage that makes up their nose and ears for comparison.

Ask What happens to bones as a child grows? *(Bones become harder and larger.)* What is this process called? *(ossification)*

During Reading continued

Active Reading

FIND THE MAIN IDEA Check students' T-charts that they began at the beginning of the lesson to make sure they are filling them in correctly. If students are having difficulty, demonstrate how to create main idea questions from the lesson headings and then search for the answers in order to fill in the T-chart. For example, students might write "What are joints?" Guide students to answer this question by looking for the yellow highlighted word *joint* in the first paragraph on this page.

Build Vocabulary

WORD ORIGINS Tell students that thinking about the origins of words can help them remember what the words mean and distinguish them from related words such as *ligament* and *tendon.*

Ask How can thinking about the origin of *ligament* help you tell it apart from *tendon*? *(The origin of* ligament *is a word that means "to tie," so ligaments are tissues that tie or hold bones together.)*

SELF-QUESTION Have students read the first paragraph on the structure of joints on the next page. Ask them to identify the areas in the paragraph that they have difficulty understanding. Have students reread these selections again slowly, using the graphic The Knee to help them understand. Finally, have students rewrite those selections using their own words.

Speed Bump

Show students a picture or model of the human skeleton. As you point to major joints, ask them to identify which type of joint it is and to explain their answer.

BUILD Vocabulary

axial skeleton
the skeleton that supports the central axis of the body; consists of the skull, vertebral column, and the rib cage

appendicular skeleton
the bones of the arms and legs along with the bones of the pelvis and shoulder area

cartilage
a type of connective tissue that is softer and more flexible than bone. In adults, cartilage supports both the nose and the ears.

joint
the place where one bone attaches to another bone

ligament
a tough connective tissue that holds bones together in a joint

WORD ORIGINS
The word *ligament* comes from the word *ligare*, a Latin word meaning "to tie." Ligaments "tie" two bones together.

Joints

A place where two or more bones meet is called a **joint.** Connective tissues hold the bones together. Joints let bones move without damaging each other.

Types of Joints Some joints, such as those of the shoulders, allow a lot of movement. Others, like the joints of the fully developed skull, allow no movement at all. There are three kinds of joints—immovable, slightly movable, and freely movable.

▶ ***Immovable Joints*** Immovable joints do not allow movement. The bones grow together until they are fused. The places where the bones in the skull meet are immovable joints.

▶ ***Slightly Movable Joints*** Slightly movable joints allow a small amount of movement. In immovable joints, the bones join and grow together. In slightly movable joints, the bones are separated from each other. The joints between the two bones of the lower leg are slightly movable joints. The joints between vertebrae are also slightly movable joints.

▶ ***Freely Movable Joints*** Freely movable joints allow movement in two or more directions. There are several kinds of freely movable joints. The types of joints are based on the amount of movement allowed by the joint.

Freely Movable Joints Freely movable joints make actions possible.

Hands-On Learning

MODEL JOINT MOVEMENTS Pass around objects or photos representing different types of joints for students to examine. Possibilities include a door hinge to model a hinge joint, a door knob to model a pivot joint, and a towing ball and hitch to model a ball-and-socket joint. Ask students to use the Freely Movable Joints visual to help them identify which joint each object or photograph represents. Then ask them to model the movements using their own joints.

Structure of Joints In freely movable joints, cartilage covers the surfaces where two bones come together. This protects the bones from damage as they move against each other. A joint capsule also surrounds the joint. The joint capsule helps hold the bones together while still allowing for movement.

The joint capsule has two layers. The outer layer forms strips of tough connective tissue called ligaments. The ligaments are connected to the membranes around the bones. **Ligaments** hold the bones in a joint together. The inner layer of the joint capsule is called the synovial (sih NOH vee uhl) cavity. This cavity holds cells that make synovial fluid. Small pockets of synovial fluid are called bursae (BUR see; singular: bursa). Synovial fluid allows the surfaces of the bones to slide over each other smoothly. Bursae also act as tiny shock absorbers.

Key Question What is the role of joints? **Connective tissue in joints holds bones together. Joints allow bones to move without damaging each other.**

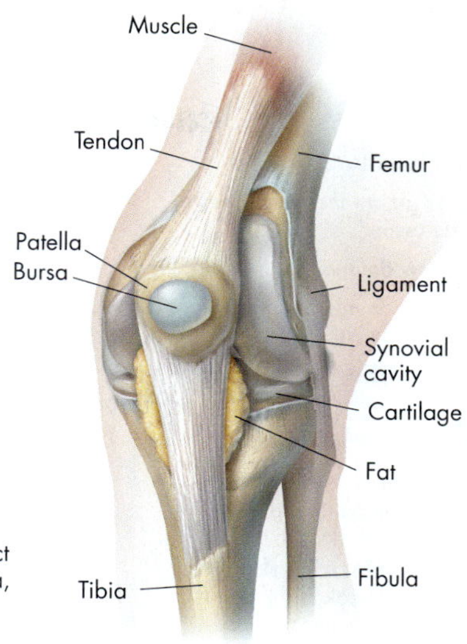

The Knee Cartilage and bursae protect the knee joint. The tibia, femur, patella, and fibula make up the knee joint. Ligaments hold these bones together.

CHECK Understanding

Apply Vocabulary

Use the highlighted words from the lesson to complete each sentence correctly.

1. Connective tissues within a _______ hold two or more bones together.
2. The ______ forms the central axis of the skeleton.
3. _____ supports the body and is more flexible than bone.

Critical Thinking

4. **Compare and Contrast** Explain how a human knee joint is similar to a door hinge. How is it different from a door hinge?
5. **Relate Cause and Effect** Wear and tear over the years often leads to a condition called osteoarthritis. The cartilage in the joints of fingers, knees, and hips begins to break down. What is the likely effect on the bones?
6. **Write to Learn** Answer the first clue of the mystery.

MYSTERY CLUE

Children with rickets have soft, cartilagelike bones that may bend under their own weight. Why might their bones be so soft?
(Hint: See p. 767.)

BIOLOGY.com Search Lesson 32.1 GO • Lesson Assessment • InterActive Art

Check Understanding

1. joint
2. axial skeleton
3. Cartilage
4. Most of the movement in a knee is in a back and forth direction, which is very similar to a door hinge. However, the ligaments in the knee also allow some side-to-side and pivoting motion.
5. Cartilage covers the surfaces of bones in a joint. The cartilage protects the bones as they rub together. Without the cartilage, the bones will become damaged.
6. Their bones may be soft because the osteoblasts are not depositing enough minerals.

Active Reading

INTERPRET DIAGRAMS Have students examine The Knee diagram. Tell them to list each part of the knee labeled in the diagram and write a brief description of the function in their own words.

KEY QUESTION Have students read and answer the Key Question.

Ask What would happen if a joint were damaged or did not function properly? *(The bones might not move easily or might damage each other as they move.)*

Speed Bump

Ask How are the tissues in cartilage and ligaments alike? *(They are both flexible tissues.)* How do their roles in a joint differ? *(Cartilage protects bones from damage as they move against each other in a joint. Ligaments hold bones together in a joint.)*

Post Reading

ASSESSMENT TIPS Review the components of the skeletal system with students before they begin the assessment.

For Question 4, refer students to the Freely Movable Joints picture.

For Question 5, explain that "wear and tear" is what happens as you use something a lot. For example, blue jeans may become lighter blue and thinner as a person wears them. Point out that Question 5 refers to what happens to a person's joints as a person uses them over many years.

MYSTERY CLUE Ask students to think about how cartilage turns to bone and what materials make bones hard. Students should realize that the bones of children with rickets do not undergo ossification as they would in healthy children.

Wrap-Up Activity

QUIZ A PARTNER Assign students to work in pairs. Have partners take turns asking each other a question about the lesson. Tell students that the students answering the question can use their text for help only if they get an answer wrong first.

LESSON 32.2

Pre-Reading

Objectives

32.2.1 Describe the structure and function of each of the three types of muscle tissue.

32.2.2 Describe the mechanism of muscle contraction.

32.2.3 Describe the interaction of muscles, bones, and tendons to produce movement.

Preview the Pages

1. Point out the visuals of the three muscle types covered in the lesson. Be sure students visually connect the enlarged images of the three muscle types with the organs colored in the human drawings.
2. Help students organize muscle structures from largest to smallest as they review skeletal muscle structure.
3. Preview the vocabulary with students. Remind them to look for these terms in the visuals.

Activate Prior Knowledge

Lead a discussion about the importance of the muscular system. Tell students to think about what they know about the muscular system. Encourage them to give specific examples of what muscles do.

Build Understanding

CONCEPT MAP Review with students the partially completed concept map in Workbook B. Explain how a concept map will help them visualize how the concepts they learn are connected to each other.

NATIONAL SCIENCE EDUCATION STANDARDS

UNIFYING CONCEPTS AND PROCESSES

I, II, V

CONTENT

C.1.a, C.1.b, C.1.d, C.1.f, F.1

INQUIRY

A.1.b, A.2.a

For the corresponding lesson in *Miller & Levine Biology,* see pages 928–933.

32.2 The Muscular System

Key Questions

What are the main kinds of muscle tissue?

How do muscles contract?

How do muscle contractions cause movement?

BUILD Understanding

Concept Map As you read, make a concept map that shows the relationship of ideas in this section.

In Your Workbook Go to your workbook for help in completing the concept map. Complete the concept map for Lesson 32.2.

Muscle Tissue

You know that bones provide support, but it's muscles attached to your bones that make movements possible. Did you know that about one third of the mass of a person's body is muscle? Large muscles in your arms and legs help your body move. These large muscles are part of the muscular system. However, this system is also made up of thousands of tiny muscles. These tiny muscles help move blood through the body and move food through the digestive system. Muscles power every movement of the body. There are three kinds of muscle: skeletal, smooth, and cardiac. Each kind of muscle performs a different function.

Skeletal Muscles Skeletal muscles are often connected to bones. They help the body make voluntary movements. Dancing and waving are voluntary. You decide to dance or wave. When viewed under a microscope, skeletal muscle looks striped. The stripes in skeletal muscle are called striations.

Skeletal muscle cells are large and have many nuclei. The cells are not always the same length. However, they are often long and thin. Skeletal muscle cells are often called **muscle fibers.**

Smooth Muscles Smooth muscle cells do not have striations. They look smooth under the microscope. These cells usually have only one nucleus. Smooth-muscle movements are usually involuntary. This means that you can't control these muscles. Smooth muscles form part of the walls of hollow parts of your body such as the stomach, blood vessels, and intestines. Smooth muscles move food through your digestive tract. Smooth muscles control the way blood flows through your circulatory system. Smooth muscles even change the size of the pupils of your eyes in bright light.

Skeletal, Smooth, and Cardiac Muscle Skeletal muscle cells appear striped under the microscope. Smooth muscle cells appear to be smooth. Cardiac muscle cells, like skeletal muscle cells, look striped.

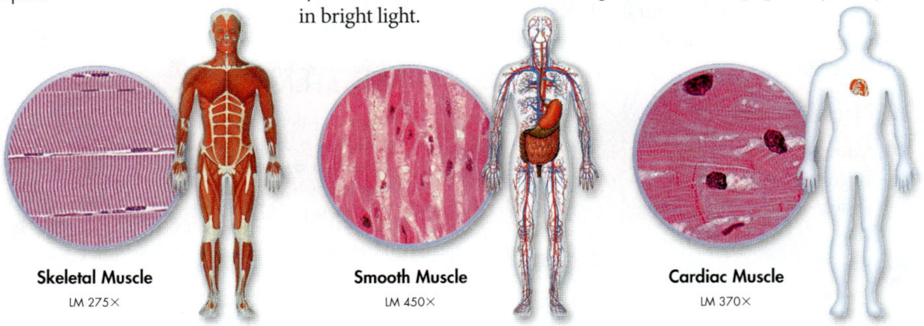

Skeletal Muscle LM 275×

Smooth Muscle LM 450×

Cardiac Muscle LM 370×

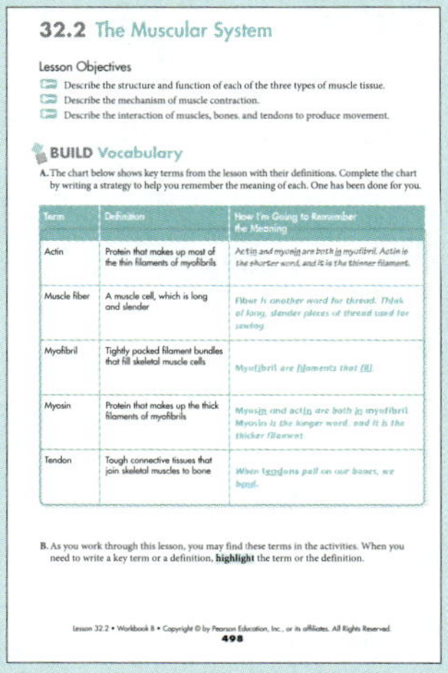

32.2 The Muscular System

Lesson Objectives

- Describe the structure and function of each of the three types of muscle tissue.
- Describe the mechanism of muscle contraction.
- Describe the interaction of muscles, bones, and tendons to produce movement.

BUILD Vocabulary

A. The chart below shows key terms from the lesson with their definitions. Complete the chart by writing a strategy to help you remember the meaning of each. One has been done for you.

Term	Definition	How I'm Going to Remember the Meaning
Actin	Protein that makes up most of the thin filaments of myofibrils	Actin and myosin are both in myofibril. Actin is the shorter word, and it is the thinner filament.
Muscle fiber	A muscle cell, which is long and slender	Fiber is another word for thread. Think of long, slender pieces of thread used for sewing.
Myofibril	Tightly packed filament bundles that fill skeletal muscle cells	Myofibril are filaments that fill.
Myosin	Protein that makes up the thick filaments of myofibrils	Myosin and actin are both in myofibril. Myosin is the longer word, and it is the thicker filament.
Tendon	Tough connective tissues that join skeletal muscles to bone	When tendons pull on our bones, we bend.

B. As you work through this lesson, you may find these terms in the activities. When you need to write a key term or a definition, **highlight** the term or the definition.

498

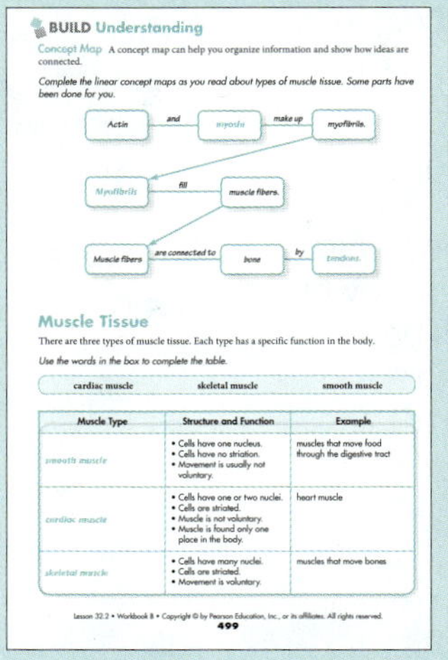

BUILD Understanding

Concept Map A concept map can help you organize information and show how ideas are connected.

Complete the linear concept maps as you read about types of muscle tissue. Some parts have been done for you.

Muscle Tissue

There are three types of muscle tissue. Each type has a specific function in the body.

Use the words in the box to complete the table.

cardiac muscle | skeletal muscle | smooth muscle

Muscle Type	Structure and Function	Example
smooth muscle	• Cells have one nucleus. • Cells have no striation. • Movement is usually not voluntary.	muscles that move food through the digestive tract
cardiac muscle	• Cells have one or two nuclei. • Cells are striated. • Muscle is not voluntary. • Muscle is found only one place in the body.	heart muscle
skeletal muscle	• Cells have many nuclei. • Cells are striated. • Movement is voluntary.	muscles that move bones

499

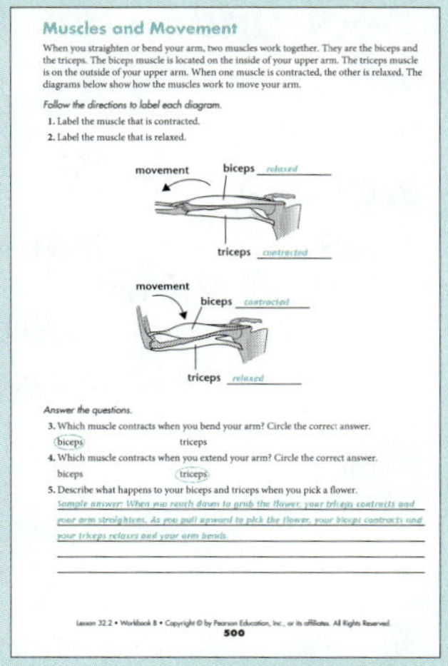

Muscles and Movement

When you straighten or bend your arm, two muscles work together. They are the biceps and the triceps. The biceps muscle is located on the inside of your upper arm. The triceps muscle is on the outside of your upper arm. When one muscle is contracted, the other is relaxed. The diagrams below show how the muscles work to move your arm.

Follow the directions to label each diagram.

1. Label the muscle that is contracted.
2. Label the muscle that is relaxed.

Answer the questions.

3. Which muscle contracts when you bend your arm? Circle the correct answer.
   biceps triceps
4. Which muscle contracts when you extend your arm? Circle the correct answer.
   biceps triceps
5. Describe what happens to your biceps and triceps when you pick a flower.
   Sample answer: When you reach down to grab the flower, your triceps contracts and your arm straightens. As you pull upward to pick the flower, your biceps contracts and your triceps relaxes and your arm bends.

500

The cells in smooth muscles are connected to one another by gap junctions. Gap junctions allow the electrical impulses that trigger contractions to move from one muscle cell to the next.

Cardiac Muscle Cardiac muscle is found only in the heart. Cardiac muscle is striated like skeletal muscle, but its cells are smaller. Cardiac muscle cells almost always have only one or two nuclei. Cardiac muscle is not under voluntary control. Like smooth muscle cells, cardiac muscle cells are also connected by gap junctions.

Key Question What are the main kinds of muscle tissue?
The three kinds of muscle tissue are skeletal muscle, smooth muscle, and cardiac muscle.

Muscle Contraction

Muscles cause parts of the body to move by getting shorter, or contracting, from end to end. Two kinds of filaments work together to cause contractions. These two filaments are called myosin and actin.

Muscle Fiber Structure Muscle fibers are filled with tightly packed filament bundles called **myofibrils** (MY uh FI bruhlz). Each myofibril is made up of two kinds of filaments. The thick protein filaments are called **myosin** (MY uh sin). The thin protein filaments are called **actin.** These filaments overlap within the myofibril. This pattern forms the striations that can be seen through a microscope. Actin filaments are bound together in areas called Z lines. Two Z lines and the filaments between them make up a unit called a sarcomere.

BUILD Vocabulary

muscle fiber
a long slender skeletal muscle cell

myofibril
a tightly packed filament bundle found within skeletal muscle fibers

myosin
a thick filament of protein found in skeletal muscle cells

actin
a thin filament of protein found in muscle cells

PREFIXES
The prefix *myo-* in *myofibril* and *myosin* means "muscle."

BUILD Connections

SKELETAL MUSCLE STRUCTURE

Skeletal muscles are made up of bundles of muscle fibers. The muscle fibers are made up of myofibrils. Each myofibril contains actin and myosin.

In Your Workbook Can you identify the characteristics of the different types of muscles? Go to your workbook to find out.

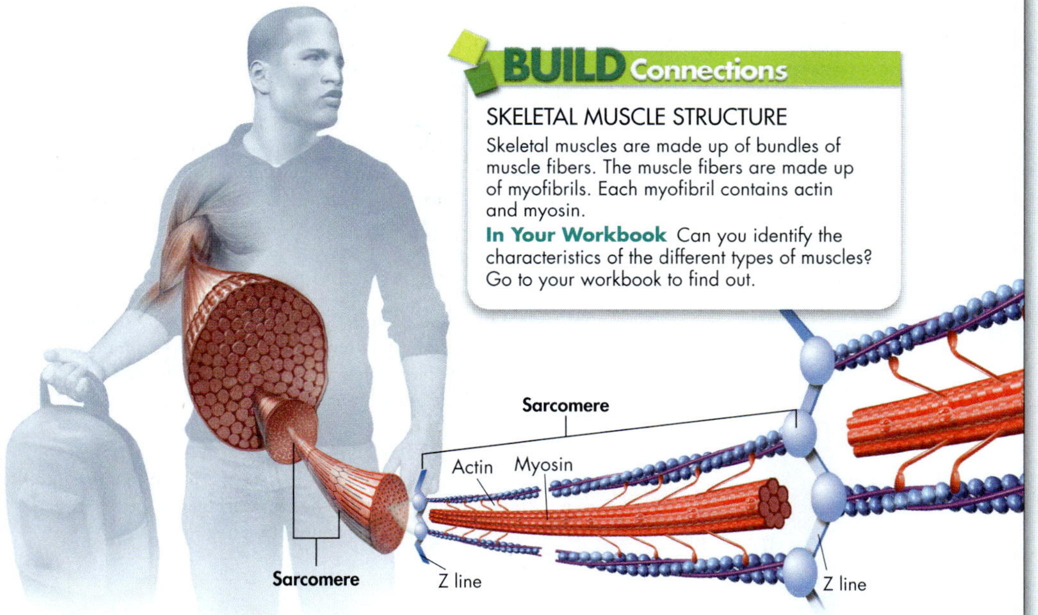

LESSON 32.2

During Reading

Active Reading

DRAW A DIAGRAM Show students how to make a Venn diagram to compare and contrast the three types of muscles.

Ask How is cardiac muscle similar to smooth muscle? *(Both types of muscle are involuntary.)* How are they different? *(Cardiac muscles are striated, while smooth muscle is not.)*

KEY QUESTION Read the Key Question aloud in class and list the answers on the board. Ask students to give examples of where each muscle is located as you list them.

Ask What kind of muscle is found in the heart? *(cardiac)* What kind of muscle is found in the stomach and blood vessels? *(smooth muscle)* What kind of muscle moves your arms and legs? *(skeletal muscle)*

Build Connections

VISUAL SUMMARY Encourage students to read the description of skeletal muscle and think of analogies to its structures. For example, students may be able to describe how a rope is made up of thin woven threads or other materials that are combined for strength.

Speed Bump

Ask In what ways are the three muscle types different from each other? *(Muscles differ in the size of the cells, the presence or absence of striation, the shape of the cells, and whether they are voluntary or involuntary.)*

ELL Focus on ELL: Access Content

ALL SPEAKERS Use a spring to link a muscle contraction with shortening. Press the ends of the spring together.

Ask How does the spring change when I press its ends together? *(It gets shorter.)*

Point out that when a muscle contracts, it also gets shorter.

Ask What will happen if I stop pressing on the spring? *(Students may say the spring will become longer again.)*

Slowly release the ends of the spring. Explain that when a muscle relaxes it returns to its original length just as the spring does. Use the spring to reinforce the vocabulary *contract, contracted muscle, relax, relaxed muscle.*

During Reading continued

Active Reading

INTERPRET DIAGRAMS Have students take turns explaining what is happening in each sequence of the Sliding-Filament Model diagram. Encourage them to make their own drawings or models to interpret the activity.

Ask Why are there two sets of pictures for each step in the diagram? *(The first picture shows the myofibril close up. The second picture shows several myofibrils together.)*

Point out that the red myosin cross-bridge binds to a different actin binding site in step 2 and step 3.

Ask What happens to the myofibril as the myosin moves down the actin? *(It pulls the actin closer to the center of the sarcomere, contracting the myofibril.)*

Speed Bump

Ask How does the binding of myosin to actin change the length of the muscle fibers as the muscle contracts? *(The muscle fibers shorten.)*

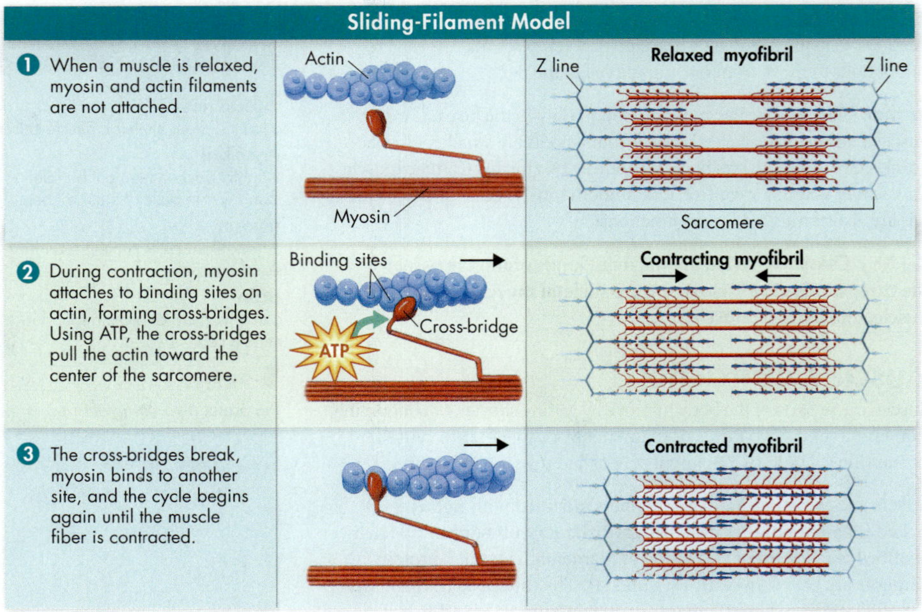

Sliding-Filament Model During muscle contraction, interaction between myosin and actin causes a muscle fiber to contract.

The Sliding-Filament Model Myosin and actin are like tiny engines that make your muscles move. When muscles are relaxed, the actin and myosin are not connected. During a muscle contraction, myosin connects to actin. The connections are called cross-bridges. The cross-bridges change shape and pull the actin filaments toward the center of the sarcomere. This action makes the fiber shorter. Then the cross-bridge breaks. The myosin connects to another place along the actin filament. As cross-bridges are made and the filaments slide past each other, the fiber gets shorter. The energy for this process comes from ATP. The sliding-filament model of muscle contraction shows how this process takes place. These contractions allow you to run, lift a heavy box, or even turn a page in a book.

Control of Muscle Contraction When you want to move your arm, the nervous system sends an impulse to certain muscles. Motor neurons carry the impulses directly to the muscle cells. The place where a motor neuron and muscle cell meet is called a neuromuscular (noo roh MUS kyoo lur) junction. Motor neurons release acetylcholine (as ih til KOH leen) to trigger the contraction. Acetylcholine (ACh) causes calcium to move into the muscle cell. Calcium causes the cross-bridges to form between actin and myosin, and the contraction begins.

Focus on Academic Language

- junction

Point out that the place where a motor neuron and a muscle cell meet is called a *neuromuscular junction.* Ask students to break this term down into the three parts they can identify. (Neuro- *refers to the nervous system;* muscular *refers to the muscular system; and a* junction *is a place where two things meet.)*

A muscle cell contracts until the motor neuron stops releasing ACh. Any remaining ACh is destroyed. The muscle cell pumps calcium back into storage. The cross-bridges stop forming, the contraction ends, and the muscle relaxes.

Key Question How do muscles contract?
During a muscle contraction, myosin filaments form cross-bridges with actin filaments. The cross-bridges change shape and pull the actin filaments toward the center of the sarcomere. This action shortens the fiber.

Muscles and Movement

Muscles can cause movement only by contracting in one direction. Yet, you know from experience that you can use your muscles in many directions. You can push and pull. How is this possible? Muscles work together to move your bones.

How Muscles and Bones Work Together Bones and joints act as levers. A lever is a machine that works around a fixed point. The fixed point is called the fulcrum. A seesaw is an example of one kind of lever. Skeletal muscles are connected to bones by tough tissue called **tendons.** By pulling on bones, tendons make bones work like levers.

The joint acts as the fixed point of the lever. Most of the time, several muscles pull in different directions around a joint to move the lever. When one muscle contracts, the other relaxes. In this way, muscles work in opposing pairs. Consider the muscles in your upper arm. When your biceps muscle contracts, it pulls the lower part of your arm. Your arm bends at the elbow joint. When the triceps muscle contracts, it opens the elbow joint. A controlled movement, such as holding a tennis racket, requires both muscles.

Exercise and Health Regular exercise is important for muscle strength and flexibility. Healthy muscles stay firm. Exercise increases the number of filaments in muscle cells. The large number of filaments increases the strength and size of the muscles.

Key Question How do muscle contractions cause movement? **Skeletal muscles cause movement by pulling on body parts as the muscles contract.**

BUILD Vocabulary

tendon
a tough connective tissue that connects skeletal muscles to bones

WORD ORIGINS
The word *tendon* comes from the Latin word meaning "to stretch."

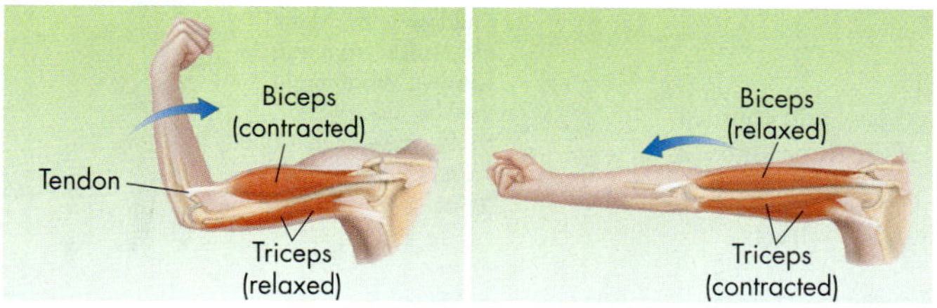

Opposing Muscle Groups By contracting and relaxing, the biceps and triceps in the upper arm enable you to bend or straighten your elbow.

Students Say

"We'd go on a 'Chapter Scavenger Hunt'. Pairs would write the page number down where we found things that were on the list such as, 'bold word that is a parasite' or 'graph that shows percent of water content.' After 10 minutes, a prize was awarded to the pair that found the most items. Then, we'd discuss the answers we found."

Build Vocabulary

WORD ORIGINS Remind students that the purpose of a tendon is to attach bones to muscles.

Ask How are tendons related to stretching? *(Tendons must be able to stretch in order to allow someone to stretch a part of the body.)*

Active Reading

MAKE INFERENCES Tell students that the muscular system helps other systems work.

Ask What causes food to move through your digestive system? *(Smooth muscle contracts.)* What causes blood to move through your body? *(Cardiac muscle contracts, causing your heart to pump blood through your circulatory system.)*

KEY QUESTION Ask a student volunteer to read the Key Question and answer aloud.

Ask What role do actin and myosin play in muscle contraction? *(They form cross-bridges, which shorten the muscle fiber.)* What is the muscle like before and after contraction? *(It is relaxed at full length with less tension.)*

MAKE CONNECTIONS Tell students that the visuals in this lesson show different details and levels of organization of the muscular system, but they all basically show the same thing. Have students practice comparing the visuals and describing how each one is related to the others. Encourage students to use their own words to describe these relationships.

KEY QUESTION Read aloud the Key Question and answer.

Ask Why do skeletal muscles exist in pairs? *(to allow movement in both directions)*

Speed Bump

Ask What kind of force moves muscles? *(a pulling force)* How are muscles able to work in both directions? *(Muscles work in pairs. One muscle pulls in one direction, while the other muscle pulls in the opposite direction.)*

Inquiry Into Scientific Thinking

COMPARE AND CONTRAST Ask students to observe the different structures that make up the chicken wing as they dissect it.

Ask How many muscles can you see in the chicken wing? *(Students should be able to see several different muscles.)* How can you tell muscle from bone? *(Muscle is soft and fleshy, while bone is hard and stiff.)* Describe the structure of the tendon. *(The tendon is strong, but flexible, and is stringy.)*

MAKE INFERENCES Have students pull on the tendon and observe what happens.

Ask What causes this movement? *(The student is pulling on the tendon.)* What causes this movement in a living chicken? *(The muscle in the chicken wing contracts and pulls on the tendon.)* Why is more than one muscle needed in the chicken wing? *(to move the wing in different directions)*

Lead a Discussion

Talk about how the skeletal and muscular systems work together to move the chicken wing.

Ask How does the bone structure affect the way the wing moves? *(The bones give the wing its shape. The type of joint enables the wing to move in specific ways.)* How does the muscle structure affect the way the wing moves? *(The size and placement of the muscles determines the force the muscles can exert and the direction in which movement can occur.)*

Post Reading

ASSESSMENT TIPS Students will need to understand the visuals of the muscular system and what they represent in order to answer the questions. Have students refer back to the figures as they answer the questions.

For Question 5, refer students to the Sliding-Filament Model diagram.

MYSTERY CLUE Have student pairs review how muscles contract normally. Then have them discuss what could go wrong with this process that might cause uncontrolled muscle movement.

Wrap-Up Activity

VOCABULARY TERMS Have student pairs create flash cards of the vocabulary terms. Tell students to take turns, with one student showing a card while the other student defines the term and explains its function.

INQUIRY into Scientific Thinking

What Do Tendons Do?

In Lesson 32.2, you learned how the biceps and triceps work together to make your arm move. In this activity, you will explore how the biceps muscle and one of its tendons make a chicken wing move.

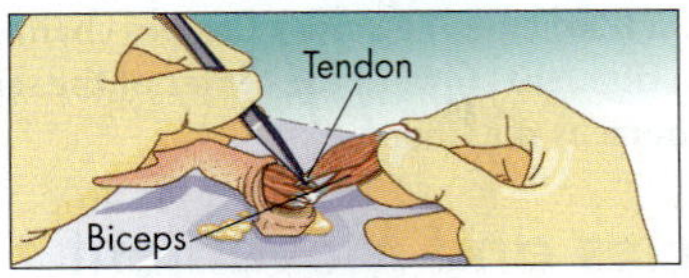

1. Put on gloves and an apron. Place a chicken wing on a paper towel. Peel back or cut away the skin and fat of the largest wing segment to expose the biceps. **CAUTION:** *Do not touch your face with your hands during the lab.*
2. Find the tendon that attaches the biceps to the bones of the middle segment of the wing.
3. Use forceps to pull on the tendon of the biceps and observe what happens to the chicken wing.
4. Your teacher will tell you how to clean your tools and dispose of the chicken wing and gloves. Wash your hands.

Analyze and Conclude

1. Observe What happened when you pulled on the tendon? In a live chicken, what structure would pull on the tendon to move the wing?

2. Compare and Contrast Observe the back of your hand as you move each of your fingers in turn. How is the way the wing moves similar to the way your fingers move?

In Your Workbook Go to your workbook for more help with this activity.

CHECK Understanding

Apply Vocabulary

Use the highlighted words from the lesson to complete each sentence correctly.

1. A muscle is made up of groups of ______.

2. Skeletal muscles are joined to bones by tough tissue called ______.

3. Tightly packed bundles of actin and myosin filaments are called ______.

Critical Thinking

4. Compare and Contrast Compare and contrast the structure and function of the three kinds of muscle tissue.

5. Explain Describe how a muscle contracts.

6. Write to Learn Answer the second clue of the mystery. Write a paragraph that includes the terms *actin*, *myosin*, and *muscle fiber*.

MYSTERY CLUE

Children with rickets may suffer from muscle spasms. What might they be lacking that could cause uncontrolled muscle movements? (Hint: See p. 772.)

BIOLOGY.com Search Lesson 32.2 GO • Lesson Assessment

Materials

raw chicken wing treated with bleach, paper towels, forceps, scissors or scalpel

Analyze and Conclude

1. I moved the wing by pulling the tendon. In a live chicken, the biceps muscle would pull the tendon.

2. A tendon moves the bones around a joint in my hand and in the wing.

Check Understanding

1. muscle fibers **2.** tendons

3. myofibrils

4. Skeletal: striated, large cells, many nuclei; voluntary. Cardiac: striated, small cells, one or two nuclei; involuntary. Smooth: not striated, one nucleus; involuntary

5. Myosin forms a cross-bridge with actin that pulls actin toward the center of the sarcomere.

6. The children may be lacking something needed to control muscle contractions such as ACh or calcium.

32.3 Skin—The Integumentary System

Integumentary System Functions

What's the largest organ in your body? No, it's not your ears or stomach, or even your lungs or heart. Would you be surprised to learn that the largest organ is the skin? You may think of the skin only as the outside of your body. However, the skin has many functions besides just covering your body. Skin is part of the integumentary system. The integumentary system is made up of the skin, hair, nails, and several glands.

Protection The most important function of the skin is to protect the inside of your body. The skin keeps germs and dirt from entering the body. Hairs keep dirt and germs from entering the nose, ears, and eyes. The skin keeps the body from drying out and protects against the sun's rays. Nails, which are produced by the skin, protect the tips of fingers and toes.

Body Temperature Regulation Body cells generate heat. The skin releases the extra heat, keeping the body temperature stable. Hair helps stop heat loss from the head.

Excretion Your sweat glands are always releasing a little sweat. Sweat takes waste, such as urea and salts, out of the body.

Information Gathering The skin helps you sense what is happening in the outside environment. The skin receives information about pressure, heat, cold, and pain and sends the information to the nervous system.

Vitamin D Production The skin makes vitamin D. This is one of the skin's most important functions. The body uses vitamin D to absorb calcium and phosphorus from the small intestine. The skin needs sunlight in order to make vitamin D.

Key Question What are the main functions of the integumentary system?
The integumentary system protects the body from dirt and germs. It also protects the body from the sun's rays and keeps the body from drying out. The integumentary system works to maintain body temperature. It also removes wastes, gathers information, and produces vitamin D.

Key Questions

What are the main functions of the integumentary system?

What are the structures of the integumentary system?

What are some problems that affect the skin?

BUILD Understanding

Preview Visuals Before you read, preview the figure Structure of the Skin on the next page. Make a two-column table. In the first column, list structures labeled in the figure. As you read, fill in the function of each structure in the second column.

In Your Workbook Go to your workbook for help in completing the table. Complete the table for Lesson 32.3.

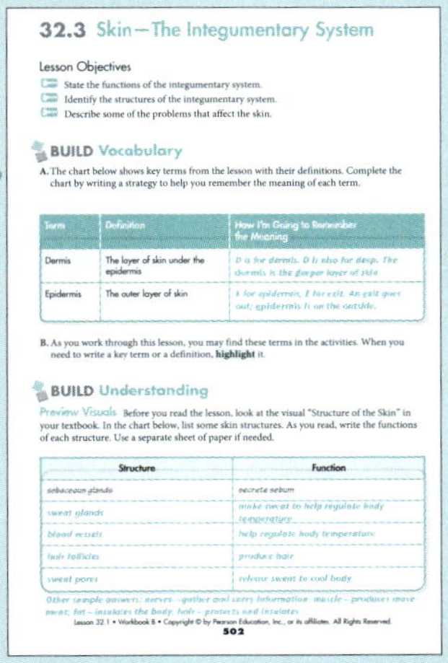

32.3 Skin—The Integumentary System

Lesson Objectives

- State the functions of the integumentary system.
- Identify the structures of the integumentary system.
- Describe some of the problems that affect the skin.

BUILD Vocabulary

A. The chart below shows key terms from the lesson with their definitions. Complete the chart by writing a strategy to help you remember the meaning of each term.

Term	Definition	How I'm Going to Remember the Meaning
Dermis	The layer of skin under the epidermis	D is for dermis. D is also for deep. The dermis is the deeper layer of skin
Epidermis	The outer layer of skin	E for epidermis, E for exit. An exit goes out; epidermis is on the outside.

B. As you work through this lesson, you may find these terms in the activities. When you need to write a key term or a definition, **highlight** it.

BUILD Understanding

Preview Visuals Before you read the lesson, look at the visual "Structure of the Skin" in your textbook. In the chart below, list some skin structures. As you read, write the functions of each structure. Use a separate sheet of paper if needed.

Structure	Function
sebaceous glands	secrete sebum
sweat glands	make sweat to help regulate body temperature
blood vessels	help regulate body temperature
hair follicles	produce hair
sweat pores	release sweat to cool body

Other sample answers: nerves – gather and carry information; muscle – produces movement; fat – insulates the body; hair – protects and insulates

Lesson 32.1 • Workbook B • Copyright © by Pearson Education, Inc., or its affiliates. All Rights Reserved.
502

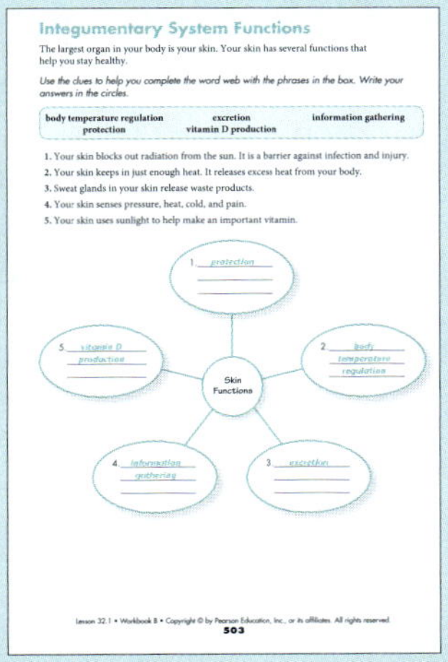

Integumentary System Functions

The largest organ in your body is your skin. Your skin has several functions that help you stay healthy.

Use the clues to help you complete the word web with the phrases in the box. Write your answers in the circles.

body temperature regulation **excretion** **information gathering** **protection** **vitamin D production**

1. Your skin blocks out radiation from the sun. It is a barrier against infection and injury.
2. Your skin keeps in just enough heat. It releases excess heat from your body.
3. Sweat glands in your skin release waste products.
4. Your skin senses pressure, heat, cold, and pain.
5. Your skin uses sunlight to help make an important vitamin.

Skin Functions
1. protection
2. body temperature regulation
3. excretion
4. information gathering
5. vitamin D production

Lesson 32.1 • Workbook B • Copyright © by Pearson Education, Inc., or its affiliates. All rights reserved.
503

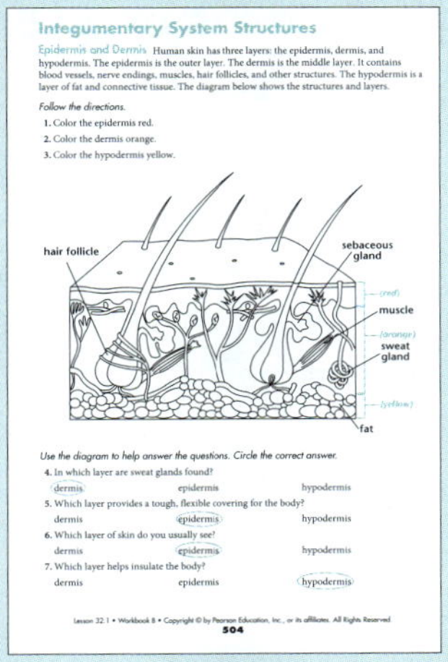

Integumentary System Structures

Epidermis and Dermis Human skin has three layers: the epidermis, dermis, and hypodermis. The epidermis is the outer layer. The dermis is the middle layer. It contains blood vessels, nerve endings, muscles, hair follicles, and other structures. The hypodermis is a layer of fat and connective tissue. The diagram below shows the structures and layers.

Follow the directions.

1. Color the epidermis red.
2. Color the dermis orange.
3. Color the hypodermis yellow.

Use the diagram to help answer the questions. Circle the correct answer.

4. In which layer are sweat glands found?
   (dermis) epidermis hypodermis
5. Which layer provides a tough, flexible covering for the body?
   dermis (epidermis) hypodermis
6. Which layer of skin do you usually see?
   dermis (epidermis) hypodermis
7. Which layer helps insulate the body?
   dermis epidermis (hypodermis)

Lesson 32.1 • Workbook B • Copyright © by Pearson Education, Inc., or its affiliates. All Rights Reserved.
504

LESSON 32.3

Pre-Reading

Objectives

32.3.1 State the functions of the integumentary system.

32.3.2 Identify the structures of the integumentary system.

32.3.3 Describe some of the problems that affect the skin.

Preview the Pages

1. Have students read the green heading at the top of each page of this section. Point out the relationship between the green headings and the blue headings underneath each one. For example, tell students that the blue headings on the first page form a list of the integumentary system functions.
2. Have students examine the Structure of the Skin diagram. Explain that this is a 3-D cross-section of the skin. Point out the three layers.

Build Understanding

PREVIEW VISUALS Direct students to the partially filled in two-column table in Workbook B. Have students look at the Structure of the Skin diagram and list the terms that go in column 1: sweat pores, hair follicles, blood vessels, sweat glands, sebaceous glands. Pronounce each term for the students.

NATIONAL SCIENCE EDUCATION STANDARDS

UNIFYING CONCEPTS AND PROCESSES

I, II, V

CONTENT

C.1.f, F.1

For the corresponding lesson in Miller & Levine Biology, see pages 935–939.

During Reading

Active Reading

KEY QUESTION Read aloud the Key Question and answer on the previous page.

Ask Which function of the integumentary system is the most important? *(protection)* Why? *(It keeps moisture in while keeping germs and the harmful effects of the sun away from the inside of the body.)*

MAKE CONNECTIONS Most students will not be familiar with the term *integumentary.* Point out that an integument is a natural outer covering.

Ask What covers your body? *(skin)* What covers your head? *(skin and hair)* What covers your hands and feet? *(skin and nails)*

Build Vocabulary

WORD ORIGINS Ask students to use the information about the root word *dermis* to guess the meaning of the word *hypodermis.*

Ask If *epi-* refers to the outside of skin in *epidermis,* what might *hypo-* refer to in *hypodermis*? (Hypo- *means "under" or "beneath," so* hypodermis *means "under the skin.")*

KEY QUESTION Have the class read the Key Question and answer to themselves.

Ask What are your body's outer coverings? *(skin, hair, and nails)* What glands are found in the skin? *(sebaceous and sweat glands)*

Speed Bump

Ask What are the three layers of skin in order from outer to inner? *(epidermis, dermis, and hypodermis)*

BUILD Vocabulary

epidermis
the outer layer of the skin

dermis
the layer of skin found beneath the epidermis

WORD ORIGINS
The prefix *epi-* in *epidermis* comes from the Greek word meaning "on" or "upon." *Dermis* comes from the Greek *derma,* meaning "skin."

Integumentary System Structures

The skin, hair, and nails work together to carry out the functions of the integumentary system. Skin has three layers—the epidermis, dermis, and hypodermis.

Epidermis The outer layer of the skin is the **epidermis.** The epidermis has two layers. Cells in the inner layer of epidermis divide quickly. The newer cells push older cells to the surface of the skin. These older cells produce a tough protein called keratin. As the cells fill up with keratin, they die, forming the outer layer of the skin.

The epidermis contains a dark brown pigment called melanin. Melanin protects the skin from the sun. Melanin is made by cells called melanocytes (MEL uh noh syts). These cells make more melanin in people with darker skin.

Dermis The **dermis** is the middle layer of skin, below the epidermis. The dermis helps keep the body temperature stable. On a cold day, the blood vessels in the dermis narrow. This keeps heat from escaping through the skin. On hot days, the blood vessels widen and sweat glands release sweat. Sweat takes heat away from your body as it evaporates.

Sebaceous (suh BAY shus) glands in the dermis make an oily substance called sebum. Sebum helps to keep the epidermis flexible and waterproof. It can also kill bacteria on the skin.

Hair and Nails Hair and nails are made mostly of keratin. Nails cover and protect the tips of the fingers and toes. Hair on the head protects the scalp from the sun's rays. It also keeps the head warm. Eyelashes and hairs in the nostrils and external ear canals keep dirt from entering the body. Hair is made by cells at the base of hair follicles. Hair follicles are pockets of epidermal cells that grow into the dermis.

Key Question What are the structures of the integumentary system? **The skin, hair, nails, and several glands make up the integumentary system.**

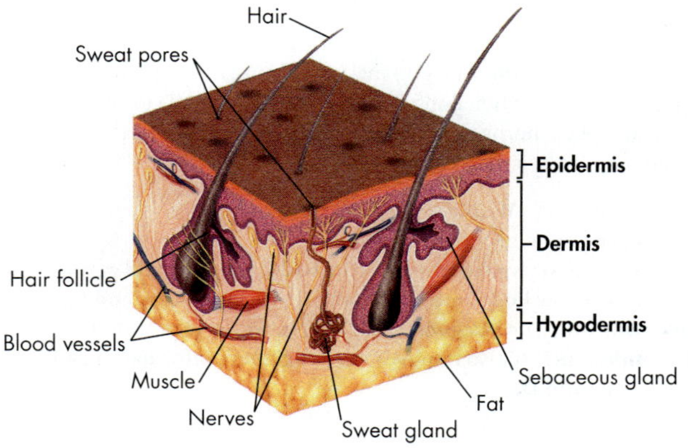

Structure of the Skin The skin has three layers. The outer layer is called the epidermis. The middle layer is called the dermis. The deepest layer is called the hypodermis.

 BIOLOGY.com Search Lesson 32.3 GO • Art Review

ELL Focus on ELL: Build Background

ALL SPEAKERS Have students form small groups, and assign each group one of the skin problems presented in the text on the next page. Have each group make a KWL chart in which they brainstorm what they already know about the skin problem. Then have students share their charts with the other groups.

Before students begin to read about skin problems in the text, have them discuss and record what they want to learn.

After reading, have students fill in the last column.

Skin Problems

The skin is always interacting with the environment. This interaction can cause many different problems. These problems can be as simple as a small scrape or as serious as skin cancer.

Acne Sometimes, sebum and dead skin cells form plugs in hair follicles. Trapped bacteria can then cause infection. Acne results from a bacterial infection of the sebaceous glands. Severe acne can leave scars on your skin for the rest of your life. A person with severe acne should see a dermatologist, a doctor who treats the skin.

Hives People with allergies sometimes get hives. During an allergic reaction, histamine causes small blood vessels to get bigger. Fluid can ooze from the vessels and move into the skin. This fluid causes swelling that leads to hives.

Skin Cancer Excessive tanning, outside or in a tanning bed, can cause skin cancer. Melanoma is the most dangerous form of skin cancer. Over 60,000 people are diagnosed with melanoma every year in the United States. As many as 8000 people die from it each year in the United States.

You can help protect yourself from skin cancer. Avoid tanning beds. Wear a hat and sunglasses whenever you plan to be outside for a long time. Always use a sunscreen that protects against UV-A and UV-B rays.

Key Question What are some problems that affect the skin?
Acne, hives, and skin cancer are some problems of the skin.

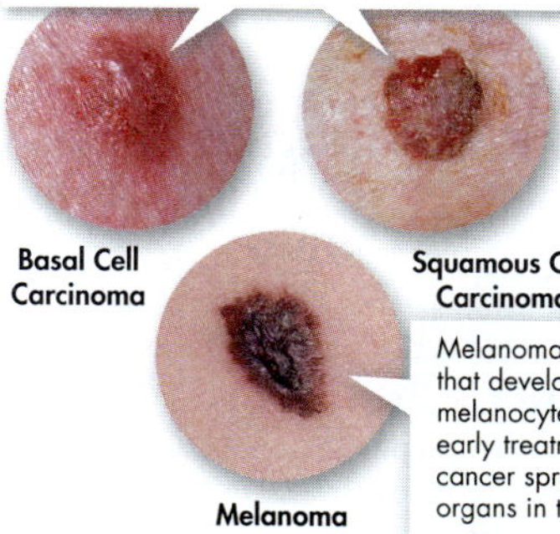

Skin Cancer Early detection is important in treating skin cancer. A sore that does not heal or a sudden change in a mole may be a sign of skin cancer. A mole that is larger than 6 mm or has irregular borders should be checked by a doctor. A mole that is an odd color should also be checked by a doctor.

CHECK Understanding

Apply Vocabulary

Use the highlighted words from the lesson to complete each sentence correctly.

1. New cells in the inner layer of the ______ push older cells out to the surface of the skin.
2. The middle layer of the skin is called the ______.

Critical Thinking

3. **Apply Concepts** Explain two ways that the skin can help remove excess heat from the body.
4. **Sequence** Explain the events that lead to acne.
5. **Write to Learn** Answer the third clue of the mystery. How is sunlight important to the skin's functions?

MYSTERY CLUE

How do you think the knowledge of the effect of sunlight on skin could have helped scientists unravel the rickets mystery? (**Hint:** See p. 775.)

BIOLOGY.com Search [Lesson 32.3] GO • Lesson Assessment

Check Understanding

1. epidermis
2. dermis
3. blood vessels widen and sweat glands produce sweat
4. Sometimes, sebum and dead skin cells block hair follicles. If bacteria are also trapped, infection occurs. This causes acne.
5. The body needs sunlight to produce vitamin D. Children in colder climates were probably outside less than children in warmer climates. Therefore, the children in the colder climates were not producing enough vitamin D.

Active Reading

USE VISUALS Point out the 6 mm reference in the caption. Ask volunteers to hold up their fingers to show how wide 6 mm is. Have students check their understandings by using a metric ruler. Have a volunteer show the class the actual width of 6 mm.

DRAW CONCLUSIONS Have students read the paragraphs on skin cancer.

Ask Should I draw the conclusion that I should avoid being outside on sunny days? *(no)* Then what is a conclusion I should draw? *(Sample answer: If I am going to be outside on a sunny day for long time, I should protect myself from skin cancer by wearing a hat, sunglasses, and sunscreen.)*

KEY QUESTION Read aloud the Key Question and answer.

Ask What causes acne? *(bacteria trapped in the skin around a hair follicle)* What are hives? *(red bumps on the skin caused by an allergic reaction)* How can you help prevent skin cancer? *(protect skin from the sun and avoid tanning beds)*

Post Reading

ASSESSMENT TIPS Tell students that in Question 3 the word *excess* means "extra" or "too much."

For Question 4, tell students that a sequence is a series of steps in the order in which they occur. Encourage students to begin to answer Question 4 by making notes in a numbered order and then using those notes to write their answer in paragraph form.

MYSTERY CLUE Encourage students to compare the levels of sunlight for healthy children and for children with rickets. Then have students review the effects of sunlight on the skin.

Wrap-Up Activity

APPLY CONCEPTS Have students form small groups to create a public service brochure or ad for healthy skin. Their material should describe the benefits of healthy skin as well as strategies for maintaining it.

CHAPTER LAB

Pre-Lab

Demonstrate how opposing muscles produce movement by stepping forward slowly and pointing out how the thigh muscles and hamstring muscles allow such movement.

Review the Background Questions and the Pre-Lab Questions as a class.

Lab

Tell students they will perform the chapter lab *Comparing Limbs* described in **Lab Manual B**.

SAFETY Have students wear proper protective equipment before beginning. Students should handle the dissection scissors with extra care. Make sure students dispose of the chicken wing and wash their hands and work area thoroughly after completing the lab. Remind students that raw chicken may contain bacteria that can cause food poisoning.

Post-Lab

Have students work in small groups and use sticks, tape, and string to construct a working model of a pair of muscles. Remind students that a muscle works by pulling and that a pair of muscles must pull a bone in different directions.

NATIONAL SCIENCE EDUCATION STANDARDS

UCP V

CONTENT C.1.f, C.3.c

INQUIRY A.1.b, A.2.a

Skills Lab

GUIDED INQUIRY

Pre-Lab: Comparing Limbs

Problem How is the structure of skeletal muscles and bones related to the functions of these body parts?

Materials disposable plastic gloves, chicken wing, disposable dissection tray, dissecting scissors, forceps, colored pencils or markers

Skills Focus Observe, Infer, Compare and Contrast

Connect to the Big idea The structure of your bones reflects the different functions of your skeleton. For example, your bones must be strong enough to support your body and protect internal organs. Your bones must also be rigid so that they provide a system of levers on which skeletal muscles can act.

Skeletal muscles have a structure that enables them to move bones around fixed points called joints. In skeletal muscles, the cells are long and narrow, which is why these cells are also called muscle fibers. When muscle fibers contract, they pull on the bone to which a muscle is attached. This force causes the bone to move in the direction of the contraction.

In this lab, you will observe and compare the structure and function of a human arm and leg. You will also compare the arm with a chicken wing.

Background Questions

a. Review What motion does a hinge joint allow? What motion does a pivot joint allow? Which of these joints are found in your elbows and knees?

b. Review What role does cartilage play in freely movable joints?

c. Explain How is it possible for bones to move in more than one direction around a joint?

d. Compare and Contrast How are a ligament and a tendon similar? How are they different?

Pre-Lab Questions

1. **Observe** How will you observe the structure and function of your elbow and knee joints?
2. **Relate Cause and Effect** Why is it important to wear goggles and disposable gloves while examining the chicken wing?
3. **Predict** Will the arrangement of bones and muscles in a chicken wing be similar to the arrangement in a human arm? Why or why not?

BIOLOGY.com Search Chapter 32 GO

Visit Chapter 32 online to test yourself on chapter content and to find activities to help you learn.

Untamed Science Video Hold onto your seats as the Untamed Science crew whisks you to NASA to learn about the effect space travel has on an astronaut's bones.

Tutor Tube Watch an analogy to help you learn the sliding-filament model of muscle contraction.

Art Review Review your understanding of the structures of the skin.

InterActive Art Watch how the various joints in the body move.

Art in Motion Watch the process of muscle contraction.

Visual Analogy How is the skeleton like the framework of a house?

Pre-Lab Answers

BACKGROUND QUESTIONS

a. A hinge joint allows back-and-forth motion. A pivot allows a bone to rotate, or turn, around another bone. Both the knee and elbow are hinge joints. The elbow is also a pivot joint.

b. Cartilage covers the surfaces where two bones meet and protects the bones from damage as they move against each other.

c. Most skeletal muscles work in opposing pairs. When one muscle in a pair contracts, the other muscle relaxes.

d. Both a ligament and a tendon are connective tissue. A ligament holds bones together in a joint. A tendon attaches a muscle to a bone.

PRE-LAB QUESTIONS

1. Sample answer: I will use diagrams to observe the structure of the elbow and knee joints. To observe function, I will feel each joint as I move my forearm and lower leg.
2. Raw chicken may contain salmonella bacteria, which can cause an infection.
3. Students may predict that the arrangement will be similar because humans and chickens share a common vertebrate ancestor. Students may predict that the arrangement will be different because wings have a different function than do arms.

32 CHAPTER Summary

32.1 The Skeletal System

- The skeleton holds the body up and protects the internal organs. The skeleton helps the body move. It stores minerals and is the site of blood cell formation.
- Bones are made up of living cells and protein fibers. The fibers are surrounded by calcium salts.
- Connective tissue in joints holds bones together. Joints allow bones to move without damaging each other.

axial skeleton (p. 766)
appendicular skeleton (p. 766)
cartilage (p. 767)
joint (p. 768)
ligament (p. 769)

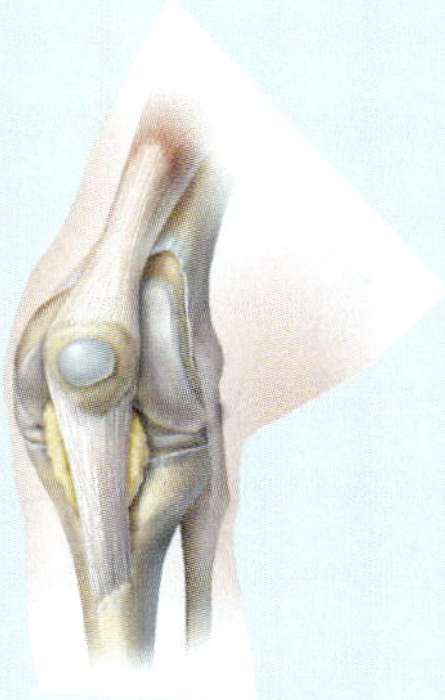

32.2 The Muscular System

- The three kinds of muscle tissue are skeletal muscle, smooth muscle, and cardiac muscle.
- During a muscle contraction, myosin filaments form cross-bridges with actin filaments. The cross-bridges change shape and pull the actin filaments toward the center of the sarcomere. This action shortens the fiber.
- Skeletal muscles cause movement by pulling on body parts as the muscles contract.

muscle fiber (p. 770)
myofibril (p. 771)
myosin (p. 771)
actin (p. 771)
tendon (p. 773)

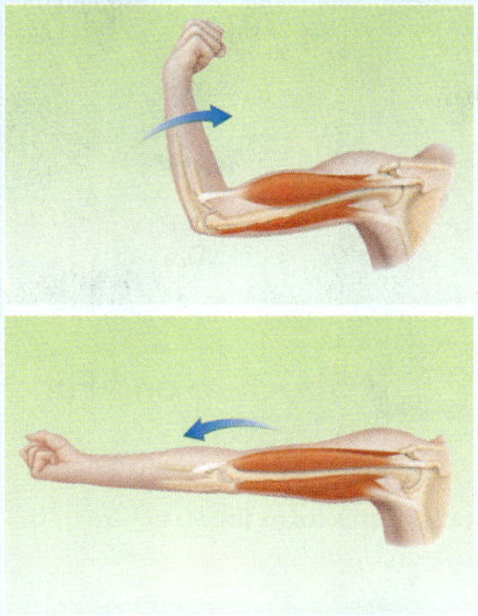

32.3 Skin—The Integumentary System

- The integumentary system protects the body from dirt and germs. It also protects the body from the sun's rays and keeps the body from drying out. The integumentary system works to maintain body temperature. It also removes wastes, gathers information, and produces vitamin D.
- The skin, hair, nails, and several glands make up the integumentary system.
- Acne, hives, and skin cancer are some problems of the skin.

epidermis (p. 776)
dermis (p. 776)

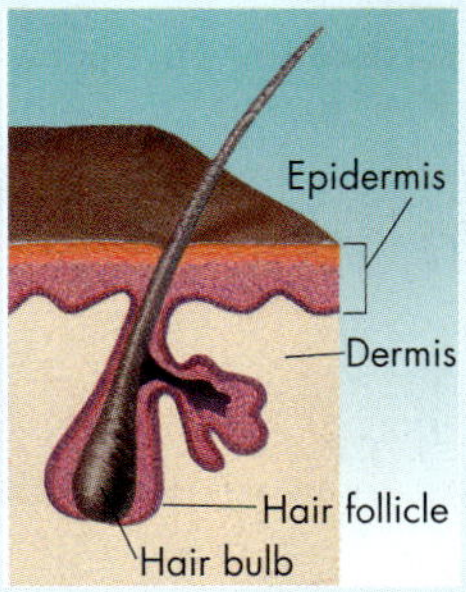

BIOLOGY.com Search Chapter 32 GO • Match It • Chapter Assessment

UbD Transfer the Big Idea

TRANSFER Have students make an illustrated study guide of their own for the chapter. Tell them to review the chapter and pick out the main ideas and the Big Idea. Have them illustrate these ideas along with notes that explain the concepts. If you wish, tell students they will be able to use their study guides while taking a chapter test.

Chapter Summary

Think Visually

USE GRAPHICS Have students develop a concept map to show the relationships between the different systems discussed in this chapter.

ANSWERS
Students' concept maps should show the relationships between the skeletal, muscular, and integumentary systems.

BIOLOGY.com Study Online

REVIEW RESOURCES

Visual Analogy Students compare a skeleton to the frame of a house.

InterActive Art Animated art shows how different joints move and how bones and muscles interact to provide this motion.

Art In Motion Students can further explore the interaction of actin and myosin as modeled by the sliding-filament model.

Tutor Tube Students will learn how to use analogies to help them understand difficult concepts like muscle contraction.

Art Review Drag-and-drop labels help students identify structures in a cross-sectional diagram of skin.

Data Analysis Students will analyze skin cancer epidemiology data to evaluate risk factors such as latitude, exposure to sunlight, and nutrition.

Editable Worksheets Pages of Study Workbook B, Lab Manual B, and the Assessment Resources Book are available online. These documents can be easily edited using a word-processing program.

Lesson Overviews Lesson Overviews are based on Miller & Levine Biology. However, they are available on CD-ROM and can be edited for your class.

Vocabulary Review The *Flash Cards* and *Match It* provide an interactive way to review chapter vocabulary.

ASSESSMENT RESOURCES

Chapter Assessment Have students take an online version of the Chapter 32 Check Understanding.

Standardized Test Prep Students can take an online version of the Standardized Test Prep. You will receive their scores along with ideas for remediation.

Diagnostic and Benchmark Tests Use these tests to monitor your students' progress and supply remediation.

ASSESSMENT

Check Understanding

Assess

Big idea Invite a volunteer to read the question aloud. Then draw a table on the board with two columns. Write *System* as the heading for the first column and *How it gives the body structure* as the heading for the second column. Lead the class in filling out the table as a way of answering the Big Idea question.

Ask What are the three systems you studied in this chapter?

Fill out the table as students respond.

Ask How does each system give the body structure? *(The skeletal system gives the body its frame. Muscles fill out the frame. Skin covers the bones and muscles.)*

Fill out the rest of the table as a class.

CONSTRUCTED RESPONSE Tell students that they will be using the *Hints* to write one or two paragraphs that answer each question completely. Students' paragraphs will vary, but should answer each question. Emphasize that each paragraph should include a main idea and supporting statements.

For Question 1, either display a copy of the Sliding-Filament Model diagram for the class, or have the class turn to that page in the textbook.

Say This question is about the sliding-filament model. Remind students of the diagram of the model in the chapter.

Read the first hint.

Ask How does the connection of the actin and myosin filaments affect the rest of the muscle? *(It shortens the muscle.)* When are actin and myosin connected in the model? *(They are connected when the muscle is stimulated to contract.)*

Check to make sure that students write their answer in paragraph form. Tell students to think about how well their paragraphs could be used to teach the concepts to someone who has not read this chapter. Explain that their goal is to write paragraphs that could be understood by someone who has not been taking this class.

32 CHECK Understanding

Assess the Big idea Structure and Function

Write an answer to the question below.

Q: What systems form the structure of the human body?

Constructed Response

Write an answer to each of the numbered questions below. The answer to each numbered question should be one or two paragraphs long. To help you begin, read the **Hints** below each of the questions.

1. **How does the sliding-filament model explain how muscles contract?**

   Hint When muscles are relaxed, the actin and myosin filaments are not connected.

2. **How do opposing muscle groups make a joint act as a lever?**

   Hint The part of the lever that does not move is called the fulcrum.

   Hint When one muscle in the group contracts, the other relaxes.

3. **Explain the statement, "In many ways, a bone is never finished growing."**

   Hint Osteoclasts break down bone minerals. Osteoblasts build bone tissue.

Foundations for Learning Wrap-Up

Use the table you prepared at the start of the chapter as a tool to help you organize your thoughts about the structures and functions of body systems.

Activity 1 Working with a partner, fold your paper so that only the structures are visible. Take turns quizzing each other about the structures and functions listed.

Activity 2 Your lists are separated into structures and functions. In a small group, discuss how the structure of each organ or system relates to its functions. For example, how do different structures of the skeleton help it perform its many functions? You may want to add rows for this information.

STRUCTURES	FUNCTIONS	
Skeletal system	• Support, • Protection • Movement • Mineral storage • Blood cell formation	
Smooth muscle	• Involuntary movement • Form part of the wall of blood vessels, intestines, stomach	
Cardiac muscle	• Involuntary movement • Most of the heart	
Skeletal muscle	• Voluntary movement • Move bones around joints	
Integumentary system	• Protects the body from dirt and germs • Protects from the sun • Keeps the body from drying out • Maintains body temperature • Removes wastes • Gathers information • Produces vitamin D	

For Question 2, pair students and have them use the hints and your modeling to answer the question. Have students share their paragraphs with the class.

For Question 3, have students write the paragraphs independently for homework or as an anchor activity.

Foundations for Learning Wrap-Up

Make sure that students have completed the table they started at the beginning of the chapter. Students should have been able to use headings in the chapter in their table. Pair students or assign them to small groups, and explain the activities.

32.1 The Skeletal System

Understanding Key Concepts

1. The network of tubes that runs through compact bone is called the
   a. periosteum.
   b. joint.
   c. Haversian canals.
   d. marrow.

> **Test-Taking Tip**
>
> **Eliminate Incorrect Answers** If you are having trouble answering a multiple-choice question, eliminate answers you know are not correct. This will help you find the correct answer. In question 1, answer **a** is the thin outer layer of the bone. Answer **b** is the place where two or more bones meet. Answer **d** is the soft tissue inside some bones where fat is stored and blood cells are formed. Once you have eliminated these answers, you see that answer **c** is the correct answer. Nerves and blood vessels run through the Haversian canals.

2. Ossification begins up to seven months before birth. What happens during ossification?
   a. Bones lose minerals and mass.
   b. Cartilage is replaced by bone.
   c. Vitamin D is made.
   d. Bones fracture more easily.
3. Which connective tissue joins two bones at a joint?
   a. bursae
   b. ligaments
   c. tendons
   d. cartilage
4. Draw a diagram of a long bone and label the structures.
5. Which type of freely movable joint allows for the greatest range of motion?
6. **Infer** Disks of rubbery cartilage are found between the individual bones in the backbone. What function do you think these disks serve?

Think Critically

7. **Interpret Visuals** Look at the images below. One shows a healthy vertebra. The other vertebra shows signs of osteoporosis. Which vertebra do you think shows signs of osteoporosis, choice *a* or choice *b*? Explain.

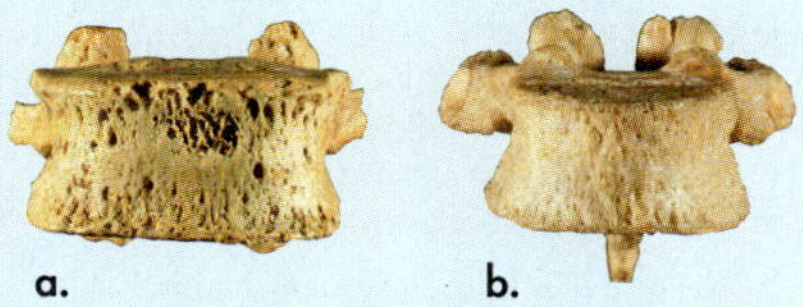

a. **b.**

8. **Use Models** Imagine that you want to build a robotic arm that works the way the human elbow works. Describe or sketch the elbow that you could use in your model.

32.2 The Muscular System

Understand Key Concepts

9. Which kind of muscle tissue is striated and has cells that have only one or two nuclei?
   a. cardiac
   b. skeletal
   c. smooth
   d. voluntary
10. Which two protein filaments are involved in muscle contraction?
    a. sarcomere and myofibril
    b. actin and myosin
    c. periosteum and cartilage
    d. ATP and acetylcholine
11. Describe the primary function of each of the three types of muscle tissue.
12. Describe how acetylcholine affects a muscle cell.
13. Explain this statement: "Most skeletal muscles work in opposing pairs."

Think Critically

14. **Relate Cause and Effect** Certain bacteria make a toxin that affects motor neurons. The toxin prevents motor neurons from releasing acetylcholine. Explain what effect this toxin is likely to have on the body.

Answers

Lesson 32.1

UNDERSTAND KEY CONCEPTS

1. c
2. b
3. b
4. Answers will vary. Correct answers should include the periosteum, Haversian canals, and blood vessels.
5. ball and socket
6. The cartilage between the vertebrae protects the bones as they move.

THINK CRITICALLY

7. *a* shows signs of osteoporosis. The holes in the bone make it less dense.
8. Models should include solid components that act as the bones above and below the joint. The robot will need pieces that act like tendons, pulling on each part of the arm when it is signaled to move.

Lesson 32.2

UNDERSTAND KEY CONCEPTS

9. a
10. b
11. cardiac: involuntary; powers the heart
    smooth: lines body cavities; involuntary
    skeletal: voluntary; moves bones around joints
12. Acetylcholine molecules cause the muscle cells to release calcium ions. These ions allow the cross-bridges to form. Muscle fibers contract until ACh is no longer released.
13. Several muscles surround a joint. When one muscle contracts, another muscle relaxes. One muscle causes the joint to open. Another muscle causes the joint to bend.

THINK CRITICALLY

14. Bacteria that prevent the release of acetylcholine will prevent the production of impulses in the cell membranes of muscle fibers. Without these impulses, calcium ions will not be released within the fibers and myosin will not be able to form cross-bridges with actin. Skeletal muscles, and many other muscles, will not be able to contract.

Assessment Tips

Have students work individually or in pairs to answer the questions. For Question 4, refer students to the Structure of a Bone diagram in Lesson 32.1.

For Question 5, refer students to the Freely Moveable Joints diagram and ask them to think about the objects in the diagram that represent each kind of joint.

For Question 14, suggest that students set up the question as a cause-and-effect graphic organizer. Help students get started by writing *toxin* in the Cause box. Then ask for suggestions for what to write in the first Effect box. *(Motor neurons can't release acetylcholine.)* Draw another Effect box beneath the first and explain that this is what students should fill in to answer the question.

ASSESSMENT

Answers

Lesson 32.3

UNDERSTAND KEY CONCEPTS

15. c

16. c

17. a

18. Any three of the following answers are acceptable: The skin stops germs and dirt from entering the body. The skin keeps the body from drying out. Nails protect the tips of fingers and toes. Hairs keep dirt and germs from entering the eyes, nose, and ears. The skin protects the body from the sun's rays.

19. Any two of the following answers are acceptable: Hair on the scalp keeps the head warm. When the body is cold, blood vessels narrow, preventing heat loss. When the body is warm, blood vessels widen. Sweat also releases heat as it evaporates.

Connecting Concepts

20. The hand on the left, which has the largest clear areas between the shaft and ends of the bones, belongs to the youngest person.

CHAPTER MYSTERY After students have read through the chapter mystery, discuss the impact of the skeletal, muscular, and integumentary systems on the health and function of the body as a whole.

Ask How did adding vitamin D to milk help lower the incidence of rickets? *(Children with rickets lacked vitamin D. Adding vitamin D to milk helped children avoid rickets.)*

Ask How did rickets affect the skeletal system's ability to protect the body's internal organs? *(Rickets weakens the structure of bones. Weaker bones are less effective in providing support and protection for internal organs.)*

Ask How does the example of rickets show how different human body systems are connected? *(Sample answer: In the example of rickets, the integumentary system is unable to produce vitamin D for the body. The skeletal system is unable to function properly without vitamin D. Muscles have spasms when vitamin D levels and, thus, calcium levels are low.)*

32 CHECK Understanding

32.3 Skin—The Integumentary System

15. What is the outer layer of skin called?
a. dermis
b. keratin
c. epidermis
d. melanin

16. Which of the following is a function of the eyelashes and hairs in the ears and nose?
a. They keep the body warm.
b. The help in the production of vitamin D.
c. They stop dirt and germs from entering the body.
d. They produce keratin.

17. In what layer of the skin do newer cells push older cells to the surface of the body?
a. the epidermis
b. the hypodermis
c. the dermis
d. the sebum

18. Describe three ways the integumentary system protects the body.

19. Describe two ways the integumentary system helps regulate body temperature.

Connecting Concepts

Use Science Graphics

20. **Infer** Cartilage does not appear on X-ray film. Instead, it is seen as a clear area between the shaft and the ends of bones. Examine the X-rays shown below. Which hand belongs to the youngest person? Explain.

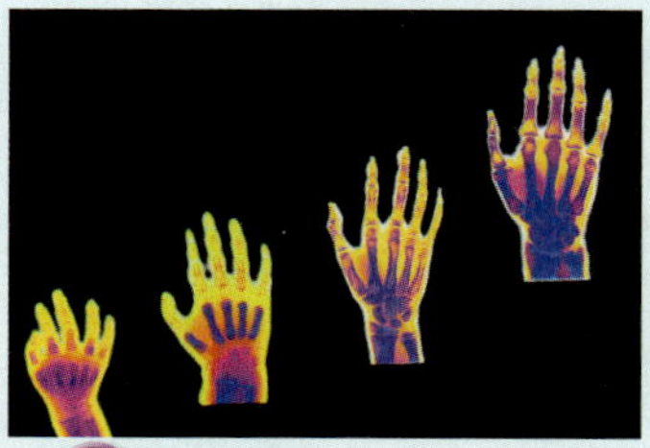

solve the CHAPTER MYSTERY

THE DEMISE OF A DISEASE

The research on the cause and a cure for rickets found two results. Both cod liver oil and exposure to ultraviolet light could prevent and cure rickets.

Cod liver oil has a nutrient involved in bone health. Starting in the 1930s, many parents in the United States gave their children a daily dose of cod liver oil.

The research also suggested that the sun influences bone health. This explained why children in colder climates were more likely to have rickets. They had little sun exposure during winter months.

But scientists still did not understand how such different treatments could have the same outcome.

Through the work of many scientists, we now know that both treatments provide vitamin D. Cod liver oil gives the body vitamin D. And, when exposed to the sun's rays, the skin helps the body make vitamin D. Vitamin D also helps the body absorb calcium and phosphorus from the digestive system.

Today, children are no longer given cod liver oil. Instead, vitamin D is added to milk. Rickets is now rare in the United States.

1. **Explain** Why were children in southern cities less likely to develop rickets?

2. **Compare and Contrast** Describe the structure of the bones of a healthy child in comparison to the bones of a child who developed rickets.

Untamed Science Solving the Demise of a Disease mystery is just the beginning. Take a video field trip with the ecogeeks of Untamed Science to see where the mystery leads.

CHAPTER MYSTERY ANSWERS

1. Children in southern cities were exposed to the sun's rays during more of the year than children in northern cities. The sun's rays helped the skin make vitamin D. Rickets is caused when children get too little vitamin D.

2. Vitamin D helps the body absorb calcium and phosphorus. Calcium salts make the bones hard and strong. Children with rickets have soft bones that are much like cartilage.

Untamed Science For more on what bones need to be healthy and strong suggest students watch **Skeletons in Space.**

Standardized Test Prep

Multiple Choice

1. What determines differences in skin color among individuals?
- **A** number of melanocytes
- **B** amount of melanin produced by each melanocyte
- **C** amount of keratin in the skin
- **D** amount of sebum produced

2. Smooth muscle is found in the
- **A** walls of blood vessels.
- **B** heart.
- **C** neuromuscular junctions.
- **D** joints.

3. All of the following are important roles of the skeletal system EXCEPT
- **A** protection of internal organs.
- **B** facilitation of movement.
- **C** storage of mineral reserves.
- **D** regulation of body temperature.

4. Which of the following supplies the energy required for muscle contractions?
- **A** myosin
- **B** ATP
- **C** acetylcholine
- **D** actin

5. The tough layer of connective tissue surrounding each bone is called
- **A** tendon.
- **B** ligament.
- **C** periosteum.
- **D** cartilage.

6. Joints that allow one bone to rotate around another are
- **A** gliding joints.
- **B** ball-and-socket joints.
- **C** hinge joints.
- **D** pivot joints.

7. Which of the following is NOT found in skin tissue?
- **A** keratin
- **B** collagen
- **C** marrow
- **D** melanin

8. What is a function of sebum?
- **A** It moistens the hair and skin.
- **B** It gives skin its color.
- **C** It insulates the body.
- **D** It makes nails and hair rigid.

Questions 9–10

As people age, the mineral content of their bone decreases. People who fail to build enough bone in adolescence and young adulthood or who lose bone at a faster than normal rate are at risk for developing osteoporosis. The graph below shows typical bone mass of men and women through most of the life span.

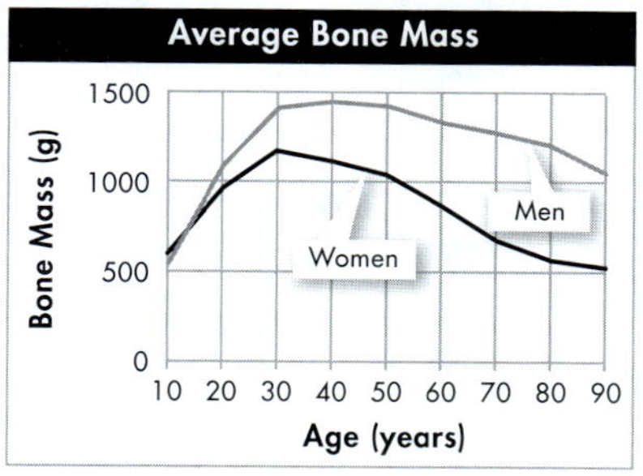

9. Between which ages do both men and women gain bone mass at the highest rate?
- **A** 10–20 years
- **B** 20–30 years
- **C** 30–40 years
- **D** 50–60 years

10. A valid conclusion that can be drawn from this graph is that, on average,
- **A** women lose more bone mass as they age than men do.
- **B** men lose more bone mass as they age than women do.
- **C** women and men lose the same bone mass as they age.
- **D** men continue to gain bone mass as they age.

Open-Ended Response

11. Doctors recommend that women eat calcium-rich foods and get plenty of exercise during adolescence and early adulthood. How could this help prevent osteoporosis later in life?

If You Have Trouble With . . .

Question	1	2	3	4	5	6	7	8	9	10	11
See Lesson	32.3	32.2	32.1	32.2	32.1	32.1	32.3	32.3	32.1	32.1	32.1

Test-Taking Tip

READ ALL THE ANSWER CHOICES

Remind students that when they are evaluating multiple choice answers, they should carefully read all of the choices, even when the first choice appears to be correct. Explain that evaluating and comparing all of the answer choices before selecting one can make it more likely that they will select the best choice.

Before the Test

Review the Test-Taking Tip with students.

Point out the word *EXCEPT* in Question 3.

Point out the word *NOT* in Question 7.

Explain that the answers to Questions 9 and 10 are found in the graph.

Remind students that an open-ended response means that they'll have to write a paragraph to answer the question. Have them paraphrase the question in their notebooks. Remind them to identify the meaning of key words before starting to write. Tell students to focus on the relationship between the recommendations and the osteoporosis.

Answers

1. B
2. A
3. D
4. B
5. C
6. D
7. C
8. A
9. A
10. A
11. Eating calcium-rich foods and exercising during adolescence and early adulthood maximizes bone mass. Having a high bone mass early in life can help prevent or minimize health problems related to loss of bone mass later in life.

33 Circulatory and Respiratory Systems

Chapter Contents	NSES	Time	Core Resources
Chapter Preview			**Student Edition,** pp. 784–785 **Chapter Mystery,** p. 785 **Foundations for Learning,** p. 785
33.1 The Circulatory System Functions of the Circulatory System • The Heart • Blood Vessels	I, V, C.5.d, F.1	$1\frac{1}{2}$ periods $\frac{3}{4}$ block	**Student Edition,** pp. 786–789 **Build Connections,** p. 786 **Study Workbook B** 33.1 Worksheets **Biology.com** *Visual Analogy:* A City's Transportation System • *Art in Motion:* Heartbeat • 33.1 Lesson Assessment
33.2 Blood and the Lymphatic System Blood • The Lymphatic System • Circulatory System Diseases • Understanding Circulatory Disease	I, V, A.2.a, C.1.a, C.5.d, F.1	1 period $\frac{1}{2}$ block	**Student Edition,** pp. 790–795 **Study Workbook B** 33.2 Worksheets **Biology.com** 33.2 Lesson Assessment
33.3 The Respiratory System Structures of the Respiratory System • Gas Exchange and Transport • Breathing • Smoking and the Respiratory System	I, V, A.1.b, A.1.c, A.2.a, C.5.d, F.1	$1\frac{1}{2}$ periods $\frac{3}{4}$ block	**Student Edition,** pp. 796–801 **Inquiry Into Scientific Thinking,** p. 797 **Build Connections,** p. 798 **Study Workbook B** 33.3 Worksheets **Biology.com** *Art Review:* The Respiratory System • *InterActive Art:* Breathing • *Tutor Tube:* The Role of the Diaphragm in Respiration • 33.3 Lesson Assessment
Chapter Pre-Lab	A.1.a, A.1.b, A.1.c, A.2.a, A.2.d, F.1	1 period $\frac{1}{2}$ block	**Student Edition,** p. 802 **Lab Manual B** *Tidal Volume and Vital Capacity*

Tools for ELL Instruction

ELL Handbook describes teaching strategies to make *Miller & Levine Biology: Foundation Edition* more accessible to ELL students.

Spanish Study Workbook is a Spanish translation of Study Workbook A.

Multilingual Glossary is the *Miller & Levine Biology* glossary translated into ten languages.

Tools to Challenge Students

Study Workbook A includes more challenging worksheets and activities such as Chapter Mystery activities that reinforce 21st Century Skills.

Lab Manual A labs are geared toward on-level students. The Lab Manual includes fifteen labs in addition to the Chapter Labs.

Biology.com Data Analysis activities offer students a way to practice data gathering and analyzing skills.

Visual Quizzes in the Assessment Resources Book are geared toward on-level students and use visuals from *Miller & Levine Biology.*

Additional Resources

Biology.com Untamed Science Video • Flash Cards

Study Workbook A 33.1 Worksheets
Spanish Study Workbook 33.1 Worksheets
Biology.com *Data Analysis:* Electrocardiography
Assessment Resources Book Visual Quizzes

Study Workbook A 33.2 Worksheets
Lab Manual B *Data Analysis:* Blood Transfusions
Spanish Study Workbook 33.2 Worksheets

Study Workbook A 33.3 Worksheets
Lab Manual B *Hands-On Activity:* Modeling Breathing
Spanish Study Workbook 33.3 Worksheets
Assessment Resources Book Visual Quiz

Lab Manual A *Tidal Volume and Lung Capacity*

Chapter Review

Student Edition Chapter Summary, p. 803
Study Workbook B Chapter 33 Review
Transparencies, pp. 362–372
Biology.com Untamed Science Video • Editable Worksheets of Study Workbooks A and B and Lab Manuals A and B • Chapter 33 Flash Cards and Match It

Untamed Science DVD • Classroom Resources CD (includes lesson presentations and editable worksheets)

Chapter Assessment

Student Edition Check Understanding, pp. 804–807
Study Workbook B Chapter 33 Taking a Standardized Test
Assessment Resources Book Chapter 33 Test B
Biology.com Chapter 33 Assessment • Editable Worksheets of Chapter 33 Visual Quizzes* and Chapter 33 Tests A and B

Exam*View Assessment Suite* • Classroom Resources CD (includes lesson presentations and editable worksheets)

Time: 1 period, 1/2 block

* Visual Quizzes are based on visuals from *Miller & Levine Biology.* Review carefully to make sure they are appropriate for your class.

Pressed for Time?

Preview the Chapter Read the Key Questions for Lesson 33.1 and preview the vocabulary for Lesson 33.3.

Cover the Chapter Quickly In Lesson 33.1, have students read *Functions of the Circulatory System,* the Heart Structure and Circulation sections of *The Heart,* and the Arteries, Capillaries, and Veins sections of *Blood Vessels.* Discuss the visuals The Heart and Structure of Blood Vessels. Assign *Structures of the Respiratory System* and *Breathing* in Lesson 33.3.

Assess Assign Questions 1–3 and 5 in the 33.1 Check Understanding, Questions 1 and 2 in the 33.3 Check Understanding, and Questions 1–6 and 13–16 in the Chapter 33 Check Understanding.

CHAPTER 33

Get the Picture

Ask students to look at the picture. Invite volunteers to demonstrate how they might breathe as they swim. Or, demonstrate several cycles of the crawl stroke, tilting your head to take in a breath of air.

Ask What happens to breathing when a swimmer goes under the water? *(Sample answer: The swimmer controls breathing by holding his or her breath.)* Contrast the controlled breathing during swimming with breathing in the classroom.

Ask Can you think of another time when a person might control their breathing? *(Sample answers: when a person is singing or when there is a bad smell)*

Activate Prior Knowledge

Ask Why is breathing important? *(Humans need the oxygen in air to live.)*

Ask Which parts of the body help us breathe? *(nose, mouth, lungs)* To prompt a response or if students are unsure, have them take a deep breath. Trace the path of air from the nose to the lungs.

Ask What happens to air after it enters the lungs? *(Sample answers: Oxygen is removed from the air, it is breathed out again.)*

Say All of the cells of the body need oxygen, which is in air. The respiratory system brings oxygen into the body. The circulatory system then delivers oxygen to the rest of the body.

Ask What are some parts of the circulatory system? *(Sample answers: heart, blood, blood vessels)* Explain that the circulatory system includes the blood, heart, and blood vessels, which transport materials throughout the body. Students will learn how the respiratory system and circulatory system work together.

NATIONAL SCIENCE EDUCATION STANDARDS

UNIFYING CONCEPTS AND PROCESSES

I, V

CONTENT

C.1.a, C.5.d, E.2, F.1

INQUIRY

A.1.a, A.1.b, A.1.c, A.2.a, A.2.d

33 Circulatory and Respiratory Systems

Big idea Structure and Function

Q: How do the structures of the circulatory and respiratory systems allow for their close functional relationship?

Usually, we are not conscious of breathing, but we can control it during activities, such as swimming.

BIOLOGY.com Search Chapter 33 GO • Chapter Mystery

784

UbD Connect to the Big Idea

Invite a volunteer to read aloud the chapter number and the chapter title. Read the Big Idea and question. Emphasize the question, **How do the structures of the circulatory and respiratory systems allow for their close functional relationship?**

Set the Purpose

PREVIEW AND PREDICT Draw the graphic organizer on the board. Guide students to the INSIDE feature on the next page, and invite a volunteer to read each lesson number and title.

Ask How does the graphic organizer show a connection between the circulatory and respiratory systems? Encourage students to describe how the systems interact. If necessary, explain that the graphic organizer describes the Big Idea in more detail and then branches out to explain different parts of the idea.

CHAPTER 33

INSIDE:
- 33.1 The Circulatory System
- 33.2 Blood and the Lymphatic System
- 33.3 The Respiratory System

CHAPTER **MYSTERY**

IN THE BLOOD

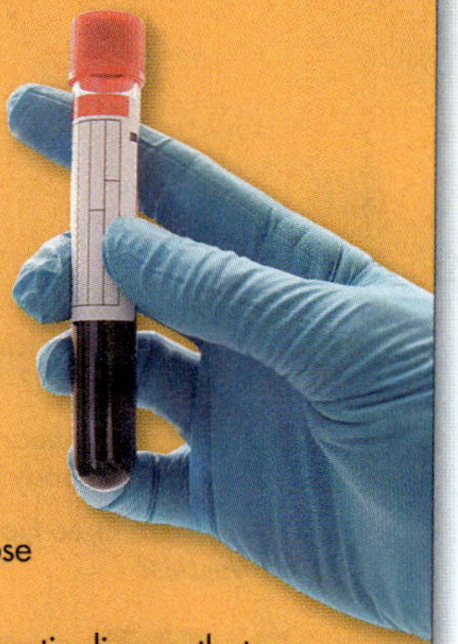

At the age of 60, John had surgery to send blood around blocked vessels in his heart. Since then, he has eaten less fat and stuck to an exercise program. Still, today he is meeting with his doctor. Fatty deposits are re-forming in his heart's vessels.

Down the hall, 6-year-old Lila is seeing her doctor today, too. Her vessels are also clogged. She is dangerously close to having a heart attack.

Both of these patients suffer from a genetic disease that affects a substance in their blood. What is the disease? Why did it affect them at such different ages?

Read for Mystery Clues As you read this chapter, look for clues about this disease. Then, solve the mystery at the end of the chapter.

FOUNDATIONS for Learning

Before you read this chapter, fold and cut a piece of paper as shown below to make a table tent. Then, label each "flap" with the words *circulatory*, *lymphatic*, and *respiratory*. As you read, take notes about the structures and functions of each body system on the inside of the flap. At the end of the chapter are two activities that use the table tent to help answer the question: How do the structures of the circulatory and respiratory systems allow for their close functional relationship?

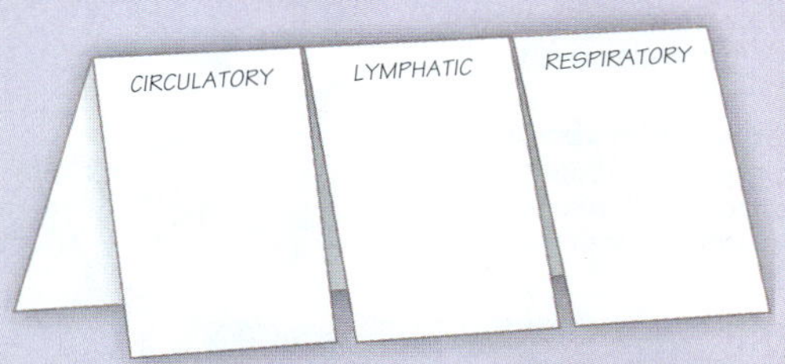

- Untamed Science Video
- Flash Cards

Chapter 33 Big Idea: Structure and Function

Chapter 33: How do the structures of the circulatory and respiratory systems allow for their close functional relationship?

- **33.1:** What structures transport substances throughout the human body?
- **33.2:** What are the roles of blood and the lymphatic system in the body?
- **33.3:** How are oxygen and carbon dioxide exchanged between humans and the environment?

Foundations for Learning

Show students how to construct the table tent. Tell them to use this tool to take notes on the structures they learn about as they read the chapter.

Ask How does organizing information this way help you learn? *(Organizing information as you learn it helps you think about what you are learning.)*

Ask How do structures in the body work together? *(Students should explain that structures in a system perform different functions. Each structure and function contributes to the overall function of the body system.)*

Tell students that thinking about the structures in the circulatory and respiratory systems will help them understand how the systems work together for the proper functioning of the entire body.

Untamed Science Some like it cold! By watching **Chillin' in the Cold,** students can take a video field trip that explores adaptations in animals that live in Earth's coldest environments.

CHAPTER MYSTERY Invite a volunteer to read aloud the Chapter Mystery. Have students brainstorm some reasons why John and Lila have the same symptoms even though they have different ages and medical histories. As a hint, suggest that students reread the Big Idea and the chapter question.

Have students predict which reason they think best explains John and Lila's condition. As they read the chapter, students should gather information that supports or disproves their predictions.

Digital Support

BIOLOGY.com Extend your reach by using the digital assets offered at **Biology.com.**

LESSON 33.1

Pre-Reading

Objectives

33.1.1 Identify the functions of the human circulatory system.

33.1.2 Describe the structure of the heart and explain how it pumps blood through the body.

33.1.3 Name three types of blood vessels in the circulatory system.

Preview the Pages

1. Have students preview the visuals for this lesson. Have students visually place The Heart diagram in the Circulatory Pathways diagram so that students can see where the arrows in The Heart diagram are coming from and going to.
2. Write the *oxygen-rich* and *oxygen-poor* on the board. Have students look for these terms in the text and infer their meanings. Then, link these terms and their meanings to the color-coding in the graphics: oxygen-rich blood is red and oxygen-poor blood is blue.
3. Point out that the word *atrium* has an unusual plural ending; the *-um* changes to *-a*. We say *one atrium,* but *two atria*.

Build Understanding

PREVIEW VISUALS Direct students to Workbook B where they will find directions for completing a two-column chart with information from The Heart visual. Encourage students to use the chart to study by covering the Function column and recalling as much information as they can.

NATIONAL SCIENCE EDUCATION STANDARDS

UNIFYING CONCEPTS AND PROCESSES

I, V

CONTENT

C.5.d, F.1

For the corresponding lesson in *Miller & Levine Biology,* see pages 948–953.

33.1 The Circulatory System

Key Questions

What are the functions of the circulatory system?

How does the heart pump blood through the body?

What are three types of blood vessels?

BUILD Understanding

Preview Visuals Before you read, look at the visual, The Heart. Draw a two-column chart. In the left column, write the structures listed in the image. In the right column, write down the functions of each structure.

In Your Workbook Go to your workbook for help completing the chart.

BUILD Connections

A CITY'S TRANSPORTATION SYSTEM

The human circulatory system is like the highways and streets of a large city.

In Your Workbook Go to your workbook to compare the needs of a person living in a large city with the needs of a cell in the body.

Functions of the Circulatory System

More than 1 million Americans suffer a heart attack each year. Of those, more than one-third die. This sad fact shows how important the heart and circulatory systems are to life.

Some animals have only a few cells. These cells are in direct contact with the outside of the animal's body. These animals are able to get all of the oxygen and nutrients they need through active transport and diffusion. They also remove wastes through active transport and diffusion. The human body, on the other hand, has millions of cells that are not on the outside of the body. Because it has so many cells, the human body needs a circulatory system. This system carries oxygen, nutrients, and other things the body needs throughout the body. It also removes wastes.

You can think of the human body as a large city. The cells are like people who live in the city. The people need things that are made elsewhere in the city. They need food. They need to get rid of garbage, and they need to move around the city. Your cells need similar things. How are these needs met in a city? The needs are met by the city's streets, highways, and subway or train lines. In the human body, the circulatory system meets these needs. This system is made up of the heart, blood vessels, and blood.

Key Question What are the functions of the circulatory system?
The circulatory system carries oxygen, nutrients, and other substances throughout the body. It also removes wastes.

786 BIOLOGY.com Search Lesson 33.1 GO • Visual Analogy

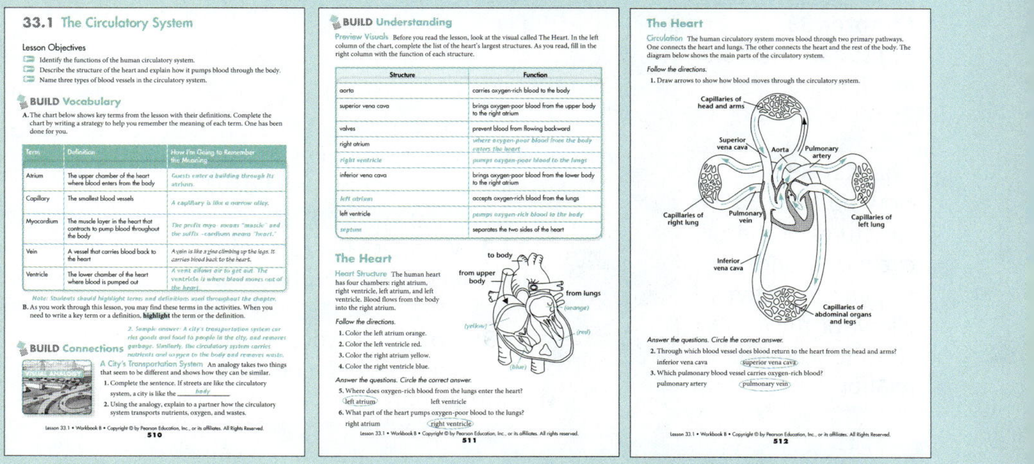

The Heart

Most of the time, you probably do not even notice your heart at work. But you notice it beating when you exercise.

Heart Structure An adult's heart beats about 72 times a minute. It beats more quickly during exercise. Your heart is about the size of your fist. It is almost all muscle. The muscles begin contracting before you are born. They stop only when you die. The layer of muscle in the walls of your heart is called the **myocardium.** Strong contractions of this muscle pump blood through the circulatory system.

The heart has four chambers. Each of the two upper chambers is called an **atrium** (plural: atria). Blood enters the heart through the atria. The lower chambers are called **ventricles.** They pump blood out of the heart.

Blood Flow Through the Heart Blood from the body enters the heart through the right atrium. Blood from the lungs enters the heart through the left atrium. Blood moves from the atria into the ventricles. Valves between the atria and ventricles keep the blood from flowing in the wrong direction. There are also valves at the exits of each ventricle. This system of valves keeps blood moving in one direction through the heart, similar to traffic on a one-way street.

BUILD Vocabulary

myocardium
the thick middle muscle layer of the heart

atrium
an upper chamber of the heart that receives blood from the rest of the body

ventricle
a lower chamber of the heart that pumps blood out of the heart to the rest of the body

WORD ORIGINS
The word *cardiac*, the prefix *cardio-*, and the suffix *-cardium* are all based on the Greek word *kardia*, which means "heart."

The Heart The human heart has four chambers: the right atrium, the right ventricle, the left atrium, and the left ventricle. Valves prevent blood from flowing backward through the heart.

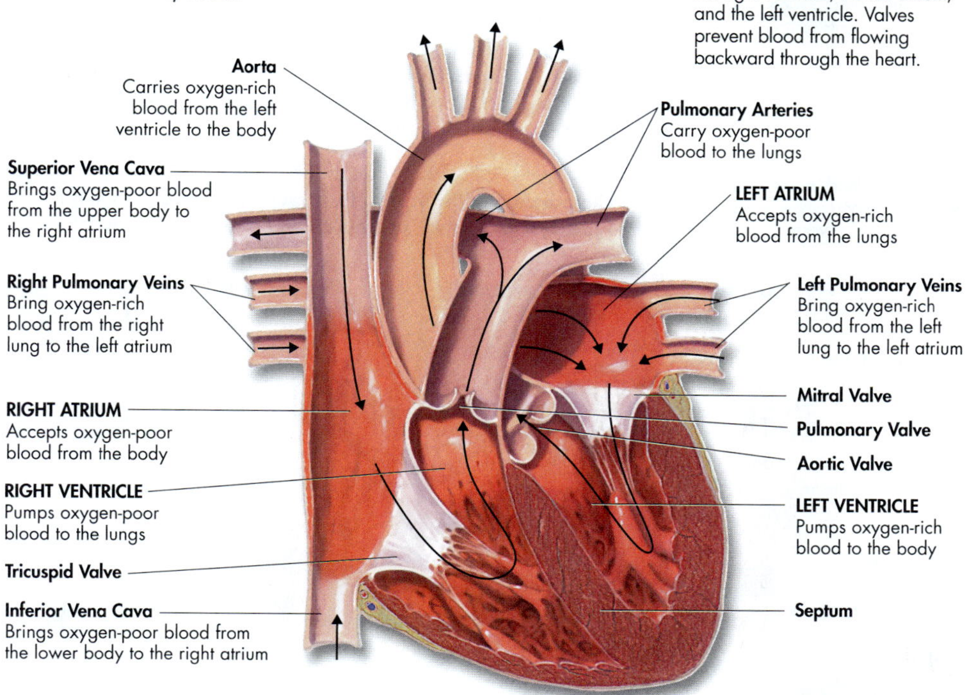

ELL Focus on ELL: Build Background

ALL SPEAKERS Ask students to define the word *chamber*. Students may be familiar with this word as meaning "a room." Explain that the chambers of the heart are similar to different rooms in a house.

Have students form small groups and make simple models of the heart using the form of a two-story house as a guide. Models can be drawn on paper or formed out of clay, cardboard boxes, or other material. Have each group divide their house into four main rooms and use the descriptions of the chambers of the heart to write similar descriptions on the corresponding rooms in their models. Tell students to use their own words. Encourage them to write descriptions that pertain to both the house and the heart.

During Reading

Build Connections

VISUAL ANALOGY Tell students to look at the Build Connections feature on the opposite page. Tell them that a visual analogy can help them visualize a difficult concept. By using an image that is comparable, but easy to think about, they can understand how something more complex works.

Ask If the circulatory system is like a city's transportation system, what do the roads represent? *(blood vessels)* What about the cars? *(blood cells)*

Active Reading

KEY QUESTION Have students read the Key Question and answer from the opposite page aloud.

Ask What does the circulatory system deliver to cells inside the body? *(oxygen and nutrients)*

USE DIAGRAMS Read the labels and caption of The Heart diagram aloud as students point to each structure.

Say Use your finger to trace the flow of blood through the heart as I read about it.

Read aloud the paragraph Blood Flow Through the Heart. Trace the pathway of blood as you read, and show students.

Ask Why do you think the muscle around the left ventricle is so much thicker than the other parts of the heart? *(Students may reason that this part of the heart must pump blood to the rest of the body. Therefore, it is thicker and stronger.)*

Point out that the names of the right and left chambers and veins seem reversed because they are named from the point of view of the heart's owner, not the view of the diagram.

Build Vocabulary

WORD ORIGINS Have students list and define words that contain the prefix -*cardio*. *(cardiac, cardiology, cardiovascular, cardiologist)*

Ask How do word origins help us understand new terms? *(Word origins can give us clues as to the meaning of the words, even if we are unfamiliar with the term. By using word origins, meanings can be built into the term itself.)*

Speed Bump

Ask What is the difference between the role of a ventricle and the role of an atrium? *(A ventricle pumps blood out of the heart. An atrium receives blood that enters the heart and pumps blood into the ventricles.)*

During Reading continued

Active Reading

USE DIAGRAMS Point out that the Circulation Pathways diagram is generalized and does not show the detail of the heart or blood vessels that is shown in other diagrams. For example, point out that the aorta does not show the three separate blood vessels that carry oxygen-rich blood to the head and arms.

Have students trace the pathway of blood in the diagram. Help students understand how blood moves through the body in two different loops.

Ask What is the difference between the movement of blood in the pulmonary and systemic circulations? *(Blood moves between only the heart and the lungs in pulmonary circulation. Blood moves from the heart to the rest of the body during systemic circulation.)*

Ask What happens to blood when it reaches the capillaries? *(Oxygen, nutrients, carbon dioxide, and wastes diffuse between the blood and the tissues.)*

KEY QUESTION Invite a student to read aloud the Key Question and answer.

Ask What does *contract* mean? *("squeeze or come closer together")* What is the myocardium? *(the thick middle muscle layer of the heart)*

Speed Bump

Ask What is the function of pulmonary circulation? *(to send blood to the lungs where it drops off carbon dioxide and picks up oxygen)*

Ask What is the function of systemic circulation? *(to carry oxygen to the rest of the body and pick up carbon dioxide from tissues)*

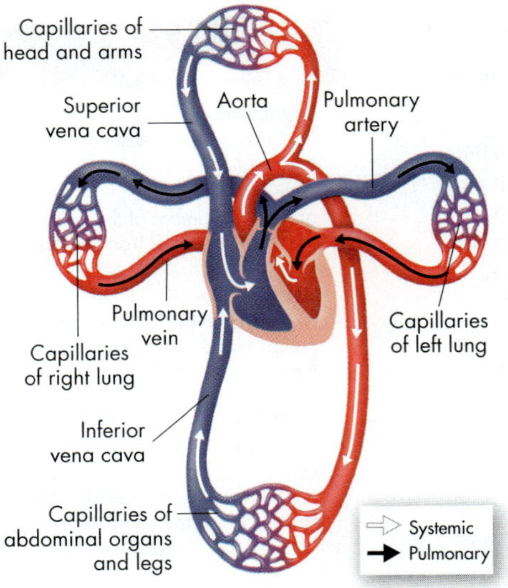

Circulation Pathways The circulatory system is divided into two pathways: pulmonary circulation and systemic circulation. Notice that the blue in this image represents oxygen-poor blood. The red represents oxygen-rich blood.

The Heart's Blood Supply The heart itself gets very little oxygen and nutrients from the blood it pumps through its chambers. Instead, a pair of vessels called *coronary arteries* run through the heart tissue. These provide blood to the heart muscle.

Circulation The heart acts as two pumps. One pump pushes blood to the lungs. The second pump pushes blood to the rest of the body.

▶ ***Pulmonary Circulation*** The first pump is the right side of the heart. It sends oxygen-poor blood from the heart to the lungs. This path is called pulmonary circulation. The blood picks up oxygen and gets rid of carbon dioxide in the lungs. Then the blood flows to the left side of the heart.

▶ ***Systemic Circulation*** The left side of the heart pumps oxygen-rich blood to the rest of the body. This path is called systemic circulation. The blood brings oxygen to the body's cells.

Heartbeat The heartbeat must be controlled. First the atria contract and then the ventricles contract. A small group of cardiac muscle fibers, called the SA node, is found in the right atrium. The SA node is sometimes called the pacemaker because it "sets the pace" of the heartbeats. The SA node sends an electrical impulse to the atria, and they contract. The impulse then goes to another group of muscle fibers called the AV node. The impulse is delayed for a moment, giving the atria time to contract. Blood flows into the ventricles, which contract. The ventricles pump blood out of the heart.

Key Question How does the heart pump blood through the body? **Contractions of the myocardium cause the atria and then the ventricles to contract, pumping blood through the body.**

Structure of Blood Vessels The structure of blood vessel walls contributes to the vessels' functions.

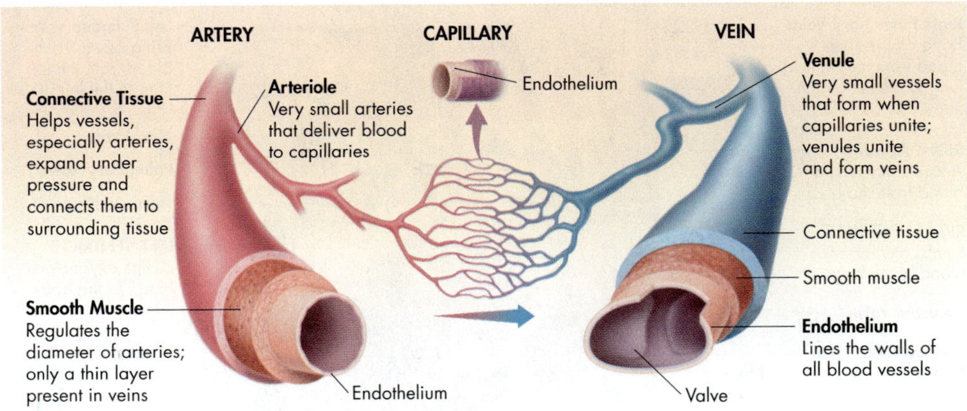

 BIOLOGY.com Search Lesson 33.1 GO • Art in Motion

Focus on Academic Language

• circulation • pulmonary • systemic

Point out that the form and meaning of the word *circulation* is similar to those of the words *circle* and *circulate*. Explain that *circulation* refers to moving in a loop or circle, as with traffic and blood.

Ask Where does the pulmonary circulation go? *(to the lungs and back to the heart)* Tell students that *pulmonary* refers to the lungs. Tell them that knowing this term can help them remember the difference between pulmonary and systemic circulation.

Ask What does the word *systemic* mean? *("pertaining to a system")* How might this be interpreted in the term *systemic circulation*? *(moving in a system, such as the body)*

Ask What do you think a *pulmonary disease* affects? *(the lungs)* What about a *systemic disease*? *(It affects the whole body.)*

Blood Vessels

Oxygen-rich blood leaving the left ventricle passes into the aorta. The aorta is the first of many vessels that carry blood through the body and back to the heart.

Arteries, Capillaries, and Veins As blood flows through the circulatory system, it moves through three kinds of blood vessels—arteries, capillaries, and veins. **Arteries** carry blood from the heart. Their thick walls can handle the pressure of the blood as it is pumped through them. **Capillaries** are tiny blood vessels. Many are so narrow that the blood cells pass through them in single file. Oxygen and nutrients diffuse through their thin walls into tissues. Carbon dioxide and other wastes move from tissues into capillaries. The blood returns to the heart through **veins.** When the skeletal muscles around many veins contract, they squeeze the veins. This squeezing action helps push the blood back toward the heart. Many veins also have valves to keep blood from flowing backward.

Blood Pressure Have you ever had your blood pressure taken at the doctor's office? Blood pressure that is too high or too low can cause big problems throughout the body. Skeletal muscles relax or contract if blood pressure is too high or too low. The kidneys help regulate blood pressure by controlling the amount of water in blood.

Key Question What are three types of blood vessels?
The three types of blood vessels are arteries, capillaries, and veins.

BUILD Vocabulary

artery
a large blood vessel that carries blood away from the heart to the tissues of the body

capillary
the smallest blood vessel; brings nutrients and oxygen to the tissues and absorbs carbon dioxide and waste products

vein
a blood vessel that carries blood from the body back to the heart

MULTIPLE MEANINGS

The word *artery* is often used to refer to major roads. This use of the term is making the same analogy between a city or town's transportation system and the body's circulatory system that was made at the beginning of this lesson.

CHECK Understanding

Apply Vocabulary

Use the highlighted words from the lesson to complete each sentence correctly.

1. Powerful contractions of the _______ pump blood through the circulatory system.
2. Oxygen-rich blood is pumped out of the heart to the body through the left _______.

Critical Thinking

3. **Apply Concepts** Why do humans need a circulatory system?
4. **Relate Cause and Effect** How would damage to the SA node affect the heart's function?
5. **Sequence** Through which blood vessels does blood flow on its path from the heart, to the rest of the body, and back to the heart?
6. **Write to Learn** Answer the first mystery clue. What is the function of the coronary arteries?

MYSTERY CLUE

The coronary arteries and vessels that branch from them are relatively narrow. Why is the heart especially susceptible to a disease that narrows blood vessels? (**Hint:** See p. 788.)

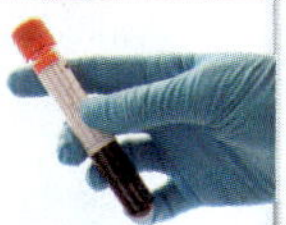

BIOLOGY.com Search Lesson 33.1 GO • Lesson Assessment

Check Understanding

1. myocardium
2. ventricle
3. The human body has millions of cells that are not in contact with the outside environment. A circulatory system is needed to transport oxygen and nutrients and remove wastes from these cells.
4. The SA node controls the contractions of the atria and ventricles. If the SA node were damaged, heartbeat would be irregular or it would stop.
5. Blood flows through the aorta, arteries, capillaries, and then veins.
6. Sample answer: The coronary arteries supply blood to the heart muscle. Because these vessels are already narrow, something that causes them to become even more narrow will quickly affect their ability to transport oxygen and nutrients to the heart muscle.

Active Reading

COMPARE AND CONTRAST Have students use the Structure of Blood Vessels diagram on the opposite page to make a chart that compares and contrasts different blood vessels.

Ask How are veins, arteries, and capillaries alike? *(They are all blood vessels.)*

Ask How are they different from one another? *(They are different sizes, move materials in different directions, and have different functions.)*

KEY QUESTION Invite a volunteer to read the question aloud. Then list the answers.

Ask How are veins and arteries connected to each other? *(through capillaries)*

Build Vocabulary

MULTIPLE MEANINGS Show students a map of your school.

Ask Which routes on the map might be considered arteries? *(Answers may vary. Help students identify hallways or sidewalks that connect to large rooms such as the gym or cafeteria.)*

Ask How are the arteries in a town similar to the arteries in your body? *(Both carry many things from one place to another.)*

Speed Bump

Ask Which blood vessels carry blood away from the heart? *(arteries)*

Ask What might be the effect of very low blood pressure? *(Cells may not receive oxygen and nutrients when they are needed.)*

Post Reading

ASSESSMENT TIPS Tell students to use the visuals as they answer the vocabulary questions.

For Question 4, tell students to recall the function of the SA node and then use it to figure out the answer.

MYSTERY CLUE Have students answer this question by first defining the role of coronary arteries. *(supply blood to heart muscle)*

Wrap-Up Activity

JEOPARDY Have students create a Jeopardy-type game. Tell them to think of five categories with five answers and point totals for each. Then, contestants have to choose the category and points to reveal the answer and state a correct question. Students can play in groups of four with three contestants and a host.

LESSON 33.2

Pre-Reading

Objectives

33.2.1 Explain the functions of blood plasma, red blood cells, white blood cells, and platelets.

33.2.2 Describe the role of the lymphatic system.

33.2.3 List three common circulatory diseases.

33.2.4 Describe the connection between cholesterol and circulatory disease.

Preview the Pages

1. Preview the headings in the lesson. Ask students to think about what they already know about topics covered under each heading.
2. Encourage students to scan the diagrams in the lesson. Tell students that diagrams with numbered steps indicate a sequence of events. These diagrams should be read in numerical order without skipping any of the illustrations.
3. Point out the diagram of the lymphatic system. Have students form some preliminary questions about the lymphatic system as they view the diagram. Ask them to think about the question "What does the lymphatic system do?" as they read that section.

Build Understanding

CONCEPT MAP Have the class read the Build Understanding box. Explain that a concept map helps organize information by linking ideas. You may wish to demonstrate how to make a simple concept map using non-science concepts so that students understand the process. Then, refer students to the concept map for Lesson 33.2 in Workbook B.

NATIONAL SCIENCE EDUCATION STANDARDS

UNIFYING CONCEPTS AND PROCESSES

I, V

CONTENT

C.1.a, C.5.d, F.1

INQUIRY

A.2.a

For the corresponding lesson in *Miller & Levine Biology,* see pages 954–961.

33.2 Blood and the Lymphatic System

Key Questions

What is the function of each component in blood?

What is the function of the lymphatic system?

What are three common circulatory diseases?

What is the connection between cholesterol and circulatory disease?

BUILD Understanding

Concept Map As you read, draw a concept map to organize the information in this lesson.

In Your Workbook Go to your workbook for help completing the concept map. Complete the concept map for Lesson 33.2.

Blood

Most likely, when you think about body tissues, you think of something with a definite shape, like muscle or skin. But blood is a tissue, too. It just happens to be in a liquid form. Blood has many important functions. It carries oxygen. It helps your body fight disease. It carries sources of energy such as sugars and fats. The different parts of blood also help keep body temperature stable and form clots to close wounds. How does this unusual tissue perform so many functions?

Plasma The human body contains 4 to 6 liters of blood. More than half of the body's blood is made up of a straw-colored fluid called **plasma.** Plasma is about 90 percent water. The other 10 percent is made up of dissolved gases, salts, nutrients, enzymes, hormones, waste products, plasma proteins, cholesterol, and other important compounds.

The water in plasma helps to control body temperature. Some proteins in plasma carry substances throughout the body. Others help the body fight infection or form blood clots.

Red Blood Cells The most numerous cells in blood are red blood cells. The main function of **red blood cells** is to carry oxygen through the body. They are red because of the iron in hemoglobin. Hemoglobin is a protein that binds to oxygen in the lungs. Hemoglobin releases the oxygen in capillary networks in other parts of the body. Red blood cells also carry carbon dioxide to the lungs.

Red blood cells are disks that are thinner in their center than along their edges. They are made in bone marrow. As they mature and fill with hemoglobin, they lose their nuclei and other organelles. Red blood cells stay in the blood for about 120 days. The cells are then destroyed in the liver and spleen.

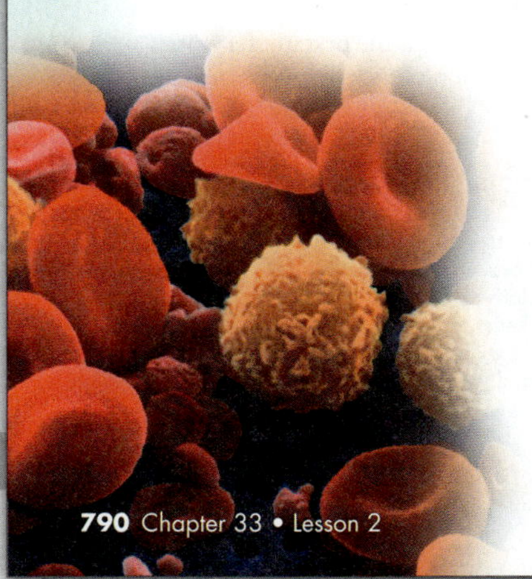

Blood Cells The red disks in this micrograph are red blood cells. The gold spheres are white blood cells. The pink fragments are platelets. (SEM 1866×)

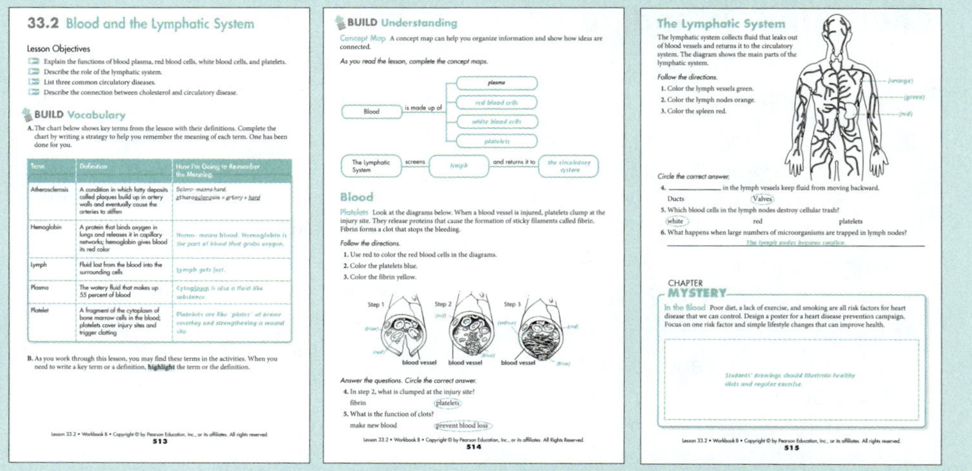
33.2 Blood and the Lymphatic System

BUILD Understanding

The Lymphatic System

White Blood Cells The main function of the **white blood cells** is to protect the body against infection, fight parasites, and attack bacteria. For example, B lymphocytes produce antibodies that provide immunity. When the body is fighting an infection, it makes more white blood cells. White blood cells do not always stay in blood vessels. Many white blood cells can move through capillary walls to fight pathogens. A pathogen is anything that can make you sick.

In a healthy person, there are many more red blood cells than white blood cells. Like red blood cells, white blood cells are made in bone marrow. Unlike red blood cells, white blood cells keep their nuclei and can live for years.

Platelets Losing too much blood can kill a person. Fortunately, a minor scrape or cut usually stops bleeding quickly. Why? Such minor wounds stop bleeding quickly because blood clots. The cytoplasm of some bone marrow cells divides into thousands of pieces. These pieces break away and enter the blood as **platelets.** Platelets are surrounded by cell membranes. Platelets and plasma proteins form blood clots. Platelets become sticky when they touch a broken blood vessel. They cluster around the wound, and the blood begins to clot.

BUILD Vocabulary

plasma
the straw-colored liquid portion of the blood

red blood cell
a blood cell containing hemoglobin that carries oxygen

white blood cell
a type of blood cell that guards against infection, fights parasites, and attacks bacteria

platelet
a cell fragment released by bone marrow that helps in blood clotting

SUFFIXES
The suffix *-let* in *platelet* means "little." Under the microscope, platelets look like small, broken plates.

Key Question What is the function of each component in blood?
Plasma controls body temperature, carries substances through the body, and fights infection. Red blood cells carry oxygen. White blood cells fight infection. Platelets form blood clots.

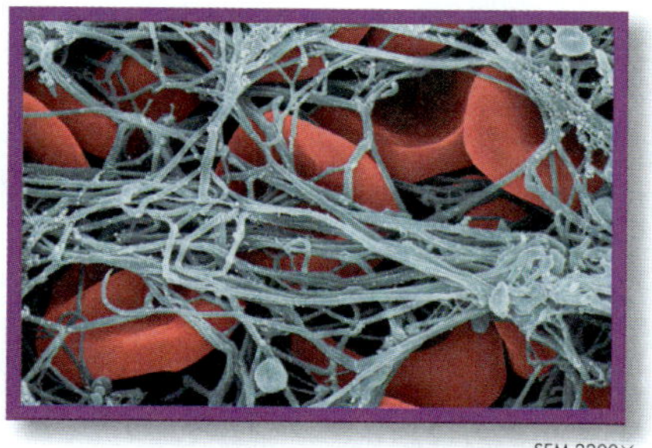

How Blood Clots Form When the clot is formed, filaments develop a net that prevents blood from leaving the damaged vessel.

SEM 2200×

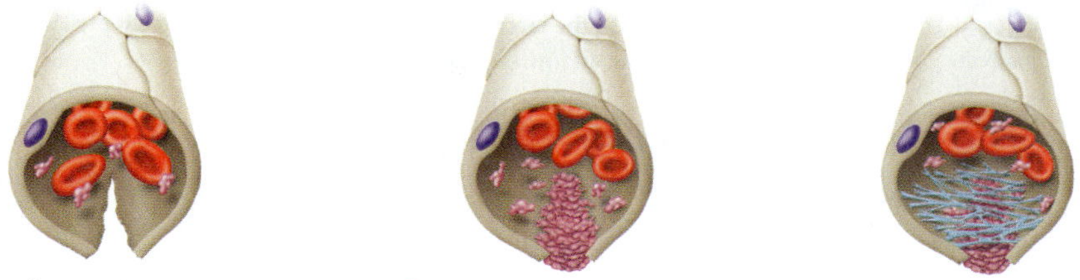

1 **Capillary Wall Breaks** A blood vessel is injured by a cut or scrape.

2 **Platelets Take Action** Platelets clump at the site, which triggers a series of reactions.

3 **Clot Forms** An enzyme called thrombin causes sticky fibrin filaments to form the clot. The clot seals the damaged area and prevents further loss of blood.

ELL Focus on ELL: Access Content

BEGINNING AND INTERMEDIATE SPEAKERS Have students form pairs and review the illustrated sequence on clotting. One person should read and describe a step in the clotting sequence. Then, the other person should rephrase the step using his or her own words and the graphic. Finally, have each pair create a simplified illustration for each step in the process. At the end of the exercise, have student pairs exchange their drawings with other pairs and review them.

During Reading

Build Vocabulary

SUFFIXES Tell students that suffixes, such as *-let*, are helpful because they allow us to change a word slightly instead of creating a whole new word.

Ask Using the definition of the suffix, what would you say is the definition of a *piglet*? *("a small, or young, pig")* Using the suffix, what word would describe a short book used to give some brief information? *(a booklet or pamphlet)*

Active Reading

CLARIFY WORD USAGE Invite a volunteer to read the Platelets paragraph. Tell the class to pay particular attention to the words *blood clot* and *blood clots*.

Ask What two parts of speech is the word *clot* or *clots* in this paragraph? *(noun and verb)* Compare the use of *clots* as a verb in *Because blood clots* with the use of *clot* or *clots* as a noun.

Ask In what other sentence is *clot* used as a verb? *(the last sentence of the paragraph)*

USE VISUALIZATION Invite students to think of other situations in which a net or similar structure prevents the passage or movement of something. *(Sample answers: straining something through a sieve while cooking, fish caught in a net, the goal net in soccer or hockey, tangle of grass roots holding soil in place)*

Ask How can thinking about these examples help you visualize how blood clots? *(Sample answer: They help me understand how a thing can get clogged up in a net even though the net has holes.)*

DRAW A DIAGRAM Have students construct a simple table that can be used to compare and contrast the different parts that make up blood.

Say Use the table to help you understand what the different parts of blood do. Write descriptions using your own words. Use simple drawings if you wish.

KEY QUESTION Read the Key Question aloud to the class. Then have a volunteer read the answer.

Ask Can you use the table you made to answer the Key Question? If not, go back and add information to your table so that it helps you answer the question.

Speed Bump

Ask What are the four main parts of blood? *(plasma, red blood cells, white blood cells, and platelets)*

During Reading continued

Active Reading

MAKE CONNECTIONS Tell students that the lymphatic system filters blood and removes germs.

Ask What other types of items use filters? *(Sample answers: drinking-water filter, coffee maker, air conditioner, swimming pool)* If possible, have some filters, or a sieve, on hand to show students.

Ask How do filters work? *(They hold back certain materials and let other materials flow through.)* Encourage students to compare these examples to the way the lymphatic system works.

KEY QUESTION Have a volunteer read the Key Question aloud without looking at the answer.

Ask What are the headings under Lymphatic System on this page? *(Role in Circulation, Role in Nutrition Absorption, and Role in Immunity)*

Then read the answer to the Key Question aloud and have volunteers find and state where the answer appears in the text.

Speed Bump

Ask What do lymph nodes do? *(Lymph nodes act as filters.)* Where are lymph nodes in the body? *(They are located throughout the body.)*

Ask What are swollen lymph nodes a sign of? *(This is a sign that the person is fighting an infection.)*

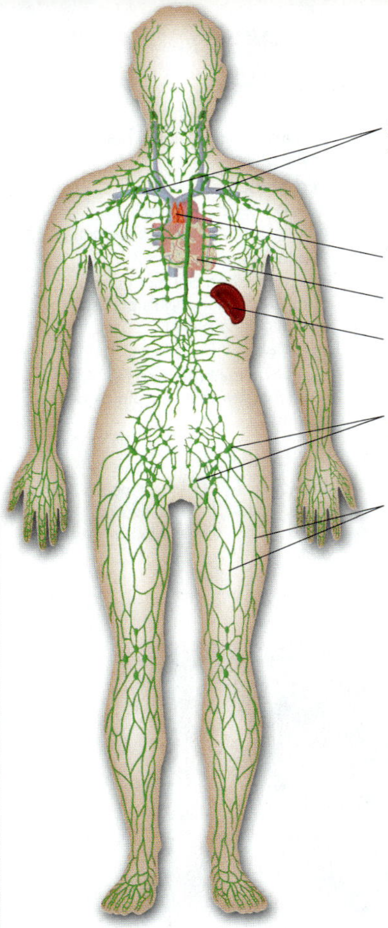

The Lymphatic System The lymphatic system is a network of vessels, nodes, and organs. This system recycles fluids from tissues and plays a role in nutrient absorption and immunity.

The Lymphatic System

As blood passes through the capillaries, some cells and parts of plasma leave the blood vessels. The cells and plasma form a fluid called **lymph.** Lymph carries nutrients, oxygen, and salts. Most of the lymph moves back into the capillaries. The rest becomes part of the lymphatic system. The lymphatic system collects the lymph, "screens" it for pathogens, and returns it to the circulatory system.

Role in Circulation Lymph collects in a system of lymphatic capillaries. It slowly moves into larger vessels. There is no pump in the lymphatic system. Instead, valves keep lymph from moving backwards. Skeletal muscles help move lymph through the system. Lymph is returned to the blood through openings in veins just below the shoulders.

Role in Nutrient Absorption The lymphatic system helps the body absorb nutrients. Lymph vessels running along the intestines pick up fats and fat-soluble vitamins. The vessels carry these nutrients into the blood.

Role in Immunity There are hundreds of lymph nodes throughout the body. These lymph nodes act as filters. They trap pathogens, damaged cells, and debris as lymph flows through them. White blood cells in the lymph nodes destroy the pathogens, damaged cells, and debris. When large numbers of pathogens are trapped in the lymph nodes, the lymph nodes may swell. When you are sick, you may notice "swollen glands" in your neck. You are really feeling swollen lymph nodes.

The thymus and spleen are organs of the lymphatic system. Certain white blood cells, called T lymphocytes, mature in the thymus. The spleen filters blood in much the same way that lymph nodes filter lymph. The spleen also removes old or damaged blood cells and stores platelets.

Key Question What is the function of the lymphatic system? **The lymphatic system collects lymph, filters it for pathogens, and returns it to the circulatory system. Lymph vessels also pick up fats and fat-soluble vitamins from the intestines.**

Circulatory System Diseases

Three common and serious diseases of the circulatory system are heart disease, stroke, and high blood pressure. Damage to the heart or brain can be fatal.

Heart Disease Heart disease is the leading cause of death in the United States. Heart muscle needs a constant supply of oxygen. Yet, the heart gets blood through only two coronary arteries and their smaller branches. Most forms of heart disease are caused when it becomes difficult for blood to flow through these vessels.

Hands-On Learning

LOOKING AT LYMPH ORGANS Write the word *tonsil* on the board. Invite volunteers to share their experience with tonsils, tonsillitis, and/or tonsillectomies with the class.

Explain that tonsils are lymph organs at the back of the throat. Point out that tonsils trap germs in the throat.

Tell students they can see these lymph organs. Provide volunteer pairs of students with a small hand mirror and a small flashlight. Ask students to look in the mirror, stick out their tongue, and look for two fleshy lumps on the sides of the back of the throat. Walk around the class helping students identify tonsils and linking them to the lymphatic system.

One type of heart disease is **atherosclerosis** (ath ur oh skluh ROH sis). Atherosclerosis is a condition in which fatty deposits called plaque build up in artery walls. The arteries become stiff. The plaque can make it difficult for blood to pass through the artery. The cap on a plaque can also rupture and cause a blood clot to form. The clot can block the artery. A heart attack happens as heart muscle cells are damaged and die from a lack of oxygen.

Symptoms of a heart attack include nausea, shortness of breath, and chest pain. People also often complain of pain in the neck, jaw, or left arm. These symptoms need *immediate* medical attention in order to save heart muscle.

Stroke When the blood supply to the brain is interrupted, some brain cells die. This event is called a stroke. Sometimes, a blood clot blocks a blood vessel in the brain and causes a stroke. A stroke may also be caused when a weak blood vessel breaks and causes bleeding in the brain. Symptoms include dizziness, severe headache, numbness, and trouble seeing or speaking. Some strokes cause death. Other strokes may cause paralysis or loss of speech.

High Blood Pressure A blood pressure reading above 140/90 is considered high. Doctors also call this condition hypertension. The heart is struggling to push blood through blood vessels. This pressure may damage the heart and the blood vessels. High blood pressure can lead to heart attack, stroke, and kidney damage.

Key Question What are three common circulatory diseases? **Three common circulatory diseases are heart disease, stroke, and high blood pressure.**

BUILD Vocabulary

lymph
a fluid that is filtered out of the blood

atherosclerosis
a condition in which fatty deposits called plaque build up on the inner walls of the arteries

WORD ORIGINS

Atherosclerosis comes from the Greek words *athero*, meaning "gruel" or "paste" and *sclerosis*, meaning "hardness." Atherosclerosis is the hardening of arteries that results from fatty deposits.

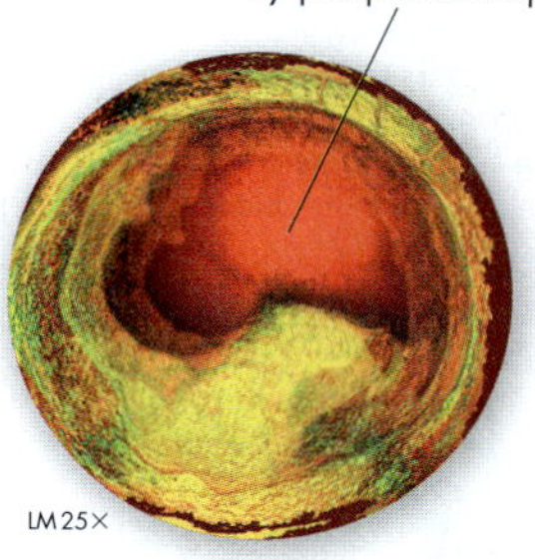

Atherosclerosis Most heart attacks happen when a plaque bursts in a coronary artery. A clot forms. The clot may block the artery. The clot could also break off and block a smaller artery.

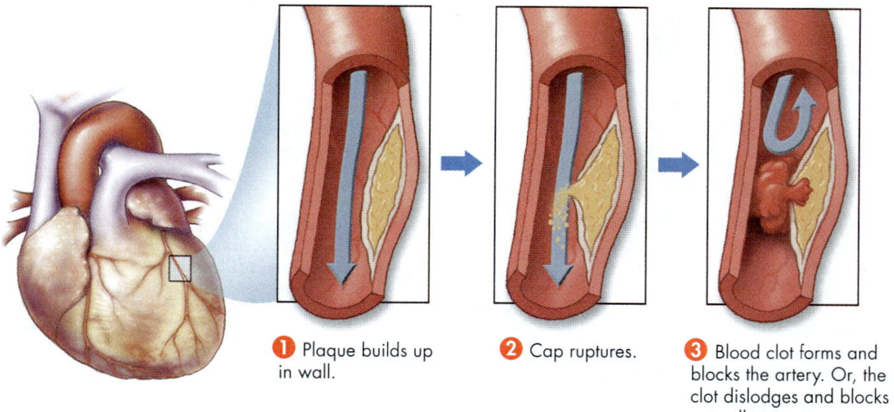

Active Learning

INFER Have students review the sections on heart disease and stroke.

Ask Heart attacks and strokes occur when the heart or brain do not receive enough of what? *(oxygen)*

INTEPRET DIAGRAMS Have students examine the diagram sequence of atherosclerosis.

Ask What is plaque? *(a fatty deposit that builds up in the walls of arteries)*

Ask How can the breaking of a plaque deposit lead to a heart attack or stroke? *(The plaque is in the wall of the artery. If the cap on the plaque breaks, a blood clot can form that blocks the flow of blood through the artery.)*

Build Vocabulary

WORD ORIGINS Ask students to read the origin of the word *artherosclerosis*. Show students a paste, such as school paste or toothpaste, and have them compare it to plaque. Place a sample of the wet paste on a paper towel and allow it to dry overnight. Allow students to touch the dried paste and compare it to the hardened fatty deposits associated with atherosclerosis.

KEY QUESTION Read the Key Question and answer aloud.

Ask What can plaque buildup lead to? *(blockages in blood flow, which can cause heart disease, stroke, and high blood pressure)*

Speed Bump

Ask What are three common serious diseases of the circulatory system? *(heart disease, stroke, and high blood pressure)*

Ask Imagine you are at a school event and you suspect that someone sitting next to you may be having a heart attack. What kind of symptoms would lead you to conclude this? *(nausea, shortness of breath, chest pain or pain in the neck, jaw, or left arm)*

Hands-On Learning

MODEL OF CLOGGED ARTERIES Give each student a plastic drinking straw. Tear up a small piece of paper into small pieces and place them on their desks. Have students blow through the straw and move the pieces of paper across their desk. Then use modeling clay to block approximately 50% of the opening of the straw. Have students blow through the straw to move the pieces of paper again. Ask them to describe the difference. Tell them that blowing through the partly blocked straw is similar to the heart trying to pump blood through clogged arteries.

During Reading continued

Active Reading

DRAW A DIAGRAM Have students reread the first paragraph on this page. Tell them to organize the risk factors into a two-column table. In the first column, students should list controllable risk factors. In the second, they should list risk factors that are not controllable.

Ask How can people reduce their risk for heart disease and stroke? *(by controlling the factors that are related to behavior, such as diet and exercise)*

Be sure students understand that they have the power, at least to a degree, to control behavior-related risk factors.

Ask Which risk factors cannot be controlled? *(age, family history, sex)* Distinguish these factors from those that can be controlled.

INTERPRET DIAGRAMS Have students study the graphic that demonstrates how LDL receptors work. Then, have volunteers read aloud each numbered step for cells with normal and then for cells with defective LDL receptors.

Ask How does the liver help remove LDL cholesterol from the blood? *(LDL cholesterol binds to receptor sites on liver cells and is taken in and destroyed.)* How is this different in liver cells with defective receptors? *(In liver cells with defective receptors, the LDL cholesterol binds to receptors but cannot be taken into the cell.)*

MAKE INFERENCES Which type of cholesterol do you think is sometimes called the "good cholesterol"? Why? *(HDL, because it generally helps the body remove more cholesterol.)*

Speed Bump

Ask Other than density, what is one difference between HDL and LDL cholesterol? *(LDL cholesterol can contribute to heart attack and stroke. HDL cholesterol helps remove harmful cholesterol from the blood.)*

Ask What are two ways that defective LDL receptors on liver cells cause LDL cholesterol levels to increase? *(1. LDL is not removed from the blood and instead builds up. 2. Because the liver cells are not taking in LDL, they continue to produce more LDL cholesterol and release it into the blood.)*

Understanding Circulatory Disease

Many factors affect a person's risk for heart disease and stroke. Some risk factors can be controlled and others cannot. Risk factors that cannot be controlled include a person's age and family history. Sex is another uncontrollable risk factor. Men have more heart attacks than women. Some controllable risk factors include choosing a healthy diet, exercising regularly, and not smoking. High blood pressure and high cholesterol are risk factors that can sometimes be controlled by medications. Researchers have learned a lot about cholesterol and how it is related to heart disease.

What Is Cholesterol? Cholesterol is part of animal cell membranes. The body uses it to make bile, vitamin D, and some hormones. Two substances carry cholesterol in the blood—low-density lipoprotein (LDL) and high-density lipoprotein (HDL). LDL becomes part of plaque. HDL generally helps the body remove cholesterol.

The liver makes cholesterol, which then moves through the blood to other body tissues. People also get cholesterol when they eat meat, eggs, dairy foods, and most fried foods.

Cholesterol and Atherosclerosis High cholesterol levels can lead to atherosclerosis and a higher chance of heart attack. Researchers Michael Brown and Joseph Goldstein received a Nobel Prize in 1985 for their study about cholesterol.

Cholesterol and the LDL Receptor Brown and Goldstein found LDL receptors on liver cells. When blood cholesterol levels are high, liver cells take the LDL from the blood and do not make more cholesterol. When cholesterol levels are low, liver cells make more cholesterol.

LDL Receptors Liver cells with normal LDL receptors take up LDL and use it or store it. But liver cells with defective LDL receptors cannot remove cholesterol from the blood.

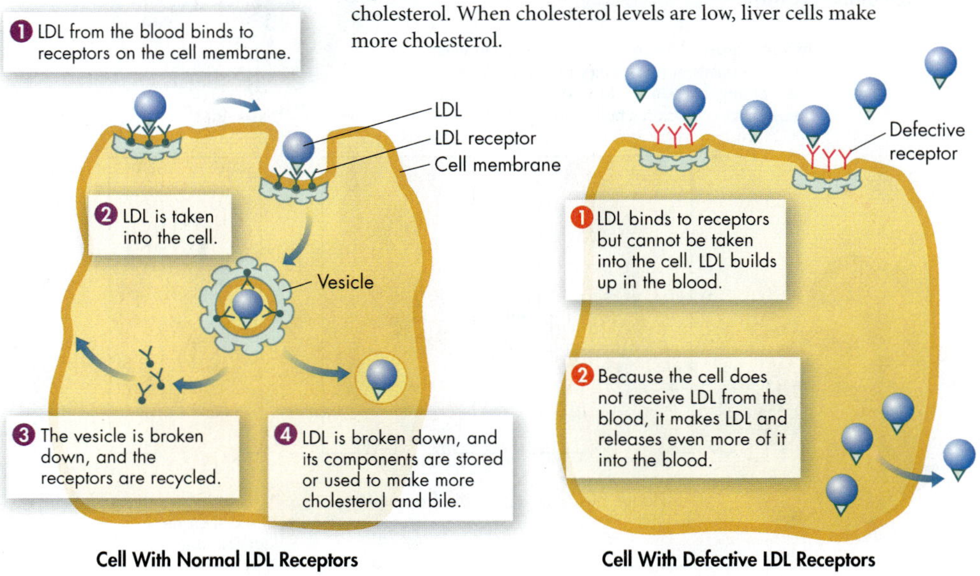

Hands-On Learning

NUTRITION LABELS Pass around several food packages or invite students to bring in empty packages of foods they eat, such as those for cereals, chips, granola/fruit/protein bars, or other snacks. Show the class how to find the amount of cholesterol in one serving. Encourage students to look for patterns in identifying high-cholesterol foods and low-cholesterol foods. Students may notice that high-cholesterol foods come from animal products.

Brown and Goldstein also found that some people carry genes that make defective LDL receptors. In these people, the liver cells cannot remove cholesterol from the blood. Also, the liver cells do not receive the signal to stop making cholesterol.

From Genetic Disease to the Public The genetic disorder helps us understand high cholesterol in people who do not have the disorder. People who eat high-fat diets store cholesterol in their liver cells. The liver cells then stop making LDL receptors. The liver cells stop removing cholesterol from blood. All of the excess cholesterol remains in the blood vessels. Therefore, a high-fat diet can cause symptoms that are similar to a genetic disorder.

Brown and Goldstein's work led to the development of drugs that can help people with high cholesterol. Some of these drugs stop the liver cells from making cholesterol. This causes the liver to make more LDL receptors. The LDL receptors remove extra cholesterol from the blood.

Keeping Your Circulatory System Healthy It is much easier to prevent heart disease than to cure it. Healthy habits when you are young help protect your heart. By eating a healthy diet, exercising regularly, and not smoking, you can help keep your circulatory system healthy.

Key Question What is the connection between cholesterol and circulatory disease? **High cholesterol levels can lead to atherosclerosis and a higher chance of heart attack and stroke.**

CHECK Understanding

Apply Vocabulary

Use the highlighted words from the lesson to complete each sentence correctly.

1. The main function of ______ is to carry oxygen throughout the body.
2. Blood clotting is made possible by plasma proteins and cell fragments called _______.
3. The lymphatic system collects ______, removes pathogens from it, and returns it to the circulatory system.
4. When plaque builds up in artery walls and causes them to stiffen, the condition is called ______.

Critical Thinking

5. **Infer** Hemophilia is a genetic disorder. One of the proteins in the clotting pathway does not work properly. What do you think happens to a person with hemophilia who has a minor cut?
6. **Relate Cause and Effect** Why do you think atherosclerosis may lead to high blood pressure?
7. **Write to Learn** Answer the second clue of the mystery. Remember that you inherit genes from both of your parents. Look back at Chapter 14 to review inheritance patterns.

MYSTERY CLUE

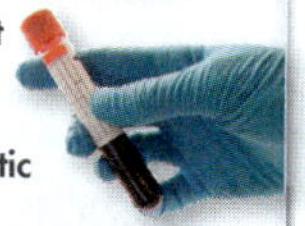

What genetic defect do both mystery patients carry? Can you think of a genetic reason why Lila's symptoms are so much worse than John's? (**Hint:** See p. 794.)

BIOLOGY.com Search Lesson 33.2 GO • Lesson Assessment

Active Reading

FIND THE MAIN IDEA Ask students to read and find the main idea of each paragraph under From Genetic Disease to the Public.

Ask What are the two main causes of defective LDL receptors? *(genetic reasons and high-fat diets)*

Ask What are the two main ways to treat high cholesterol? *(certain drugs and a change in lifestyle)*

KEY QUESTION Have a volunteer read the Key Question and answer aloud. Review the Atherosclerosis diagram and relate the buildup of plaque to heart attack and stroke.

Speed Bump

Tell students that some cholesterol drugs work by reducing LDL cholesterol, while other drugs work by increasing HDL cholesterol.

Ask Why do you think these different methods might both be effective? *(both lower LDL in the blood)*

Ask Why is it important to have your cholesterol levels checked regularly? *(to help prevent stroke and heart attack)*

Ask What are some examples of healthy habits that can keep your circulatory system healthy? *(choose a healthy diet, exercise regularly, avoid smoking)*

Post Reading

ASSESSMENT TIPS Point out to students that in order to relate cause and effect in Question 6, they must first define *atherosclerosis* and describe its effects. Then, they must connect one of the effects to high blood pressure.

MYSTERY CLUE Encourage students to review the genetic defects discussed in this lesson. Remind students that Lila is much younger than John although her symptoms are more severe. Help students realize that Lila may have inherited defective LDL receptor genes from *both* parents. Because John's symptoms are less severe, he may have inherited one defective gene and one normal gene. Review the patterns of inheritance.

Wrap-Up Activity

PRACTICE TEST Divide the class into groups. Have each group work together to make a test that covers this lesson's objectives. When they are finished, have them exchange tests with another group. Then, have students complete the tests, making note of the questions they miss to guide their studying.

Check Understanding

1. red blood cells
2. platelets
3. lymph
4. atherosclerosis
5. If a person's blood does not clot properly, a minor cut can become very serious. Bleeding will not stop as quickly or at all.
6. The heart has to work harder to force the blood through the narrow, stiff arteries.
7. John and Lila both have defective LDL receptors. Lila's condition would be worse if she inherited two defective copies of the gene and John inherited only one defective copy.

LESSON 33.2

LESSON 33.3

Pre-Reading

Objectives

33.3.1 Identify the structures of the respiratory system and describe their functions.

33.3.2 Describe gas exchange.

33.3.3 Describe how breathing is controlled.

33.3.4 Describe the effects of smoking on the respiratory system.

Preview the Pages

1. Show students the large graphic of the respiratory system. Ask them to look for parts of the circulatory system they may recognize, and remind them that connections between the two systems are part of the Big Idea.
2. Have students scan the headings in the lesson. Point out that the lesson is organized to first describe the parts of the respiratory system, and then to describe its functions.
3. Write the words *bronchus* and *bronchiole* on the board. Point out that the *ch* sounds like a /k/.

Activate Prior Knowledge

Ask a volunteer to tell why humans need oxygen. *(Our cells use oxygen for cellular respiration, the process that releases energy from food.)* Review Lesson 33.1 and how the circulatory system carries oxygen to cells in all parts of the body.

Build Understanding

FLOWCHART Tell students that a flowchart helps show the order of a series of events. Make sure that students understand how a flowchart works, using the flowchart in Workbook B before they begin.

NATIONAL SCIENCE EDUCATION STANDARDS

UNIFYING CONCEPTS AND PROCESSES

I, V

CONTENT

C.5.d, F.1

INQUIRY

A.1.b, A.1.c, A.2.a

For the corresponding lesson in *Miller & Levine Biology,* see pages 963–969.

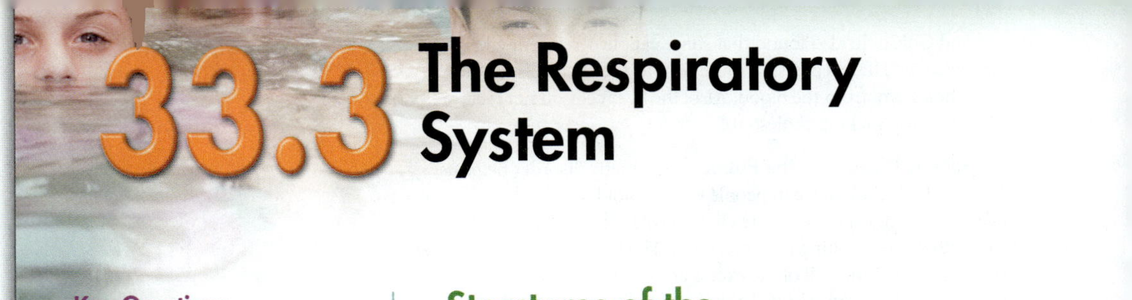

33.3 The Respiratory System

Key Questions

What is the function of the respiratory system?

How are oxygen and carbon dioxide exchanged and transported throughout the body?

What mechanisms are involved in breathing?

How does smoking affect the respiratory system?

BUILD Understanding

Flowchart As you read, make a flowchart that shows the path of air through the respiratory system.

In Your Workbook Go to your workbook to learn more about making a flowchart. Complete the flowchart for Lesson 33.3.

BUILD Vocabulary

pharynx
a tube at the back of the mouth that serves as a passageway for both air and food; also called the throat

larynx
a structure in the throat that contains the vocal cords

trachea
a tube that connects the larynx to the bronchi; also called the windpipe

WORD ORIGINS

The word *trachea* comes from the Greek phrase that means "rough artery." Rings of cartilage form the trachea and give it a rough surface.

Structures of the Respiratory System

Why do we need to breathe? All cells in our body, especially brain cells, need oxygen all of the time. Without oxygen, many cells die in minutes. The respiratory and circulatory systems work together to get oxygen to our cells.

For organisms, rather than single cells, *respiration* means the exchange of gases between a body and the environment. When you breathe in, your respiratory system picks up oxygen from the air. When you breathe out, your respiratory system lets carbon dioxide out into the air. The respiratory system is made up of the nose, pharynx, larynx, trachea, bronchi, and lungs.

Nose The lungs are made of some of the most delicate tissues in the body. Before entering the lungs, air must be filtered, moistened, and warmed. Hairs lining the inside of the nose trap large particles from the air. The air is warmed in the nasal cavity and sinuses. Mucus in these spaces moistens the air and catches even more dust particles.

Pharynx, Larynx, and Trachea Air moves from the nose to the throat, or **pharynx.** The pharynx is an area at the back of the mouth. Air and food move through the pharynx. When you swallow, a flap of tissue called the epiglottis at the back of the pharynx protects the airway. The epiglottis makes certain that the food moves into the esophagus.

Air moves from the pharynx to the larynx. The **larynx** contains the vocal cords. When muscles pull the vocal cords together, the air moving between them causes the cords to vibrate and produce sounds.

Air moves from the larynx to the **trachea,** or windpipe. Mucus in the trachea traps more particles from the air. Cilia push the particles and mucus away from the lungs toward the pharynx. From there, the mucus and particles can be swallowed or spit out.

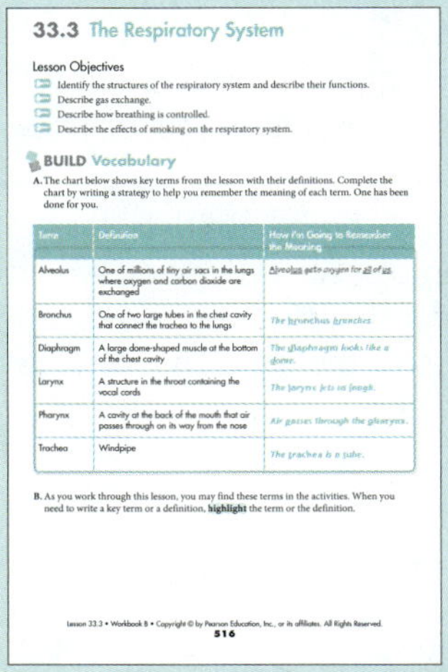

33.3 The Respiratory System

Lesson Objectives

- Identify the structures of the respiratory system and describe their functions.
- Describe gas exchange.
- Describe how breathing is controlled.
- Describe the effects of smoking on the respiratory system.

BUILD Vocabulary

A. The chart below shows key terms from the lesson with their definitions. Complete the chart by writing a strategy to help you remember the meaning of each term. One has been done for you.

Term	Definition	How I'm Going to Remember the Meaning
Alveolus	One of millions of tiny air sacs in the lungs where oxygen and carbon dioxide are exchanged	Alveolus gets oxygen for all of us.
Bronchus	One of two large tubes in the chest cavity that connect the trachea to the lungs	The bronchus branches.
Diaphragm	A large dome-shaped muscle at the bottom of the chest cavity	The diaphragm looks like a dome.
Larynx	A structure in the throat containing the vocal cords	[illegible]
Pharynx	A cavity at the back of the mouth that air passes through on its way from the nose	Air passes through the pharynx.
Trachea	Windpipe	The trachea is a tube.

B. As you work through this lesson, you may find these terms in the activities. When you need to write a key term or a definition, **highlight** the term or the definition.

516

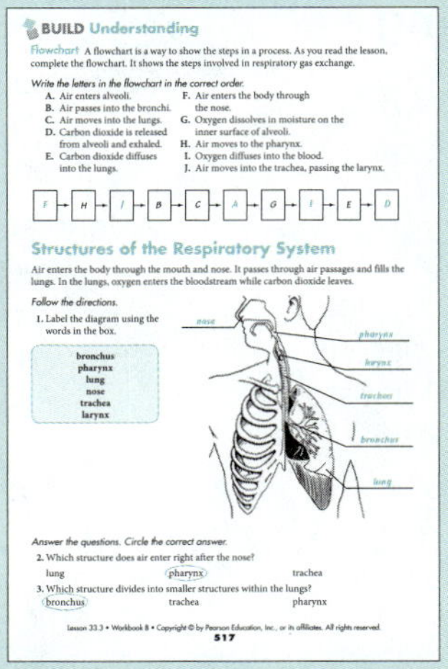

BUILD Understanding

Flowchart A flowchart is a way to show the steps in a process. As you read the lesson, complete the flowchart. It shows the steps involved in respiratory gas exchange.

Write the letters in the flowchart in the correct order.

A. Air enters alveoli.
B. Air passes into the bronchi.
C. Air moves into the lungs.
D. Carbon dioxide is released from alveoli and exhaled.
E. Carbon dioxide diffuses into the lungs.
F. Air enters the body through the nose.
G. Oxygen dissolves in moisture on the inner surface of alveoli.
H. Air moves to the pharynx.
I. Oxygen diffuses into the blood.
J. Air moves into the trachea, passing the larynx.

F → H → J → B → C → A → G → I → E → D

Structures of the Respiratory System

Air enters the body through the mouth and nose. It passes through air passages and fills the lungs. In the lungs, oxygen enters the bloodstream while carbon dioxide leaves.

Follow the directions.

1. Label the diagram using the words in the box.

bronchus
pharynx
lung
nose
trachea
larynx

Answer the questions. Circle the correct answer.

2. Which structure does air enter right after the nose?
lung pharynx trachea
3. Which structure divides into smaller structures within the lungs?
bronchus trachea pharynx

517

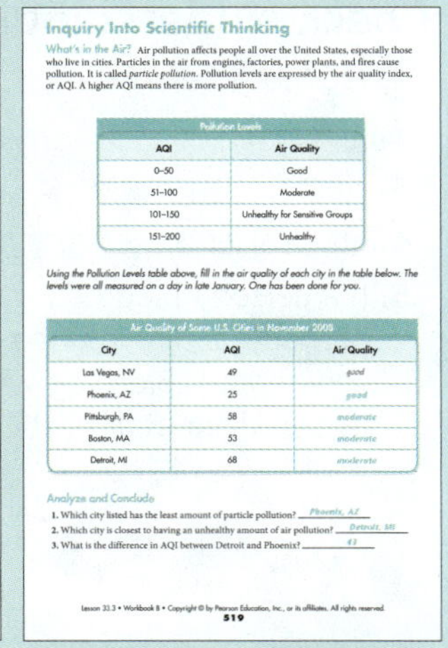

Inquiry Into Scientific Thinking

What's in the Air? Air pollution affects people all over the United States, especially those who live in cities. Particles in the air from engines, factories, power plants, and fires cause pollution. It is called *particle pollution.* Pollution levels are expressed by the air quality index, or AQI. A higher AQI means there is more pollution.

Pollution Levels

AQI	Air Quality
0–50	Good
51–100	Moderate
101–150	Unhealthy for Sensitive Groups
151–200	Unhealthy

Using the Pollution Levels table above, fill in the air quality of each city in the table below. The levels were all measured on a day in late January. One has been done for you.

Air Quality of Some U.S. Cities in November 2008

City	AQI	Air Quality
Las Vegas, NV	49	good
Phoenix, AZ	25	good
Pittsburgh, PA	58	moderate
Boston, MA	53	moderate
Detroit, MI	68	moderate

Analyze and Conclude

1. Which city listed has the least amount of particle pollution? Phoenix, AZ
2. Which city is closest to having an unhealthy amount of air pollution? Detroit, MI
3. What is the difference in AQI between Detroit and Phoenix? 43

519

Lungs Air moves from the trachea into two large tubes in the chest. These tubes are called **bronchi** (singular: bronchus). Each bronchus leads to a lung. Inside the lung, each bronchus divides into smaller and smaller passageways. These passageways are called bronchioles. These and the bronchi are surrounded by smooth muscles. These muscles are controlled by the autonomic nervous system. The contracting and relaxing of these muscles changes the size of the bronchi and bronchioles.

The bronchioles continue to divide until they reach a series of dead ends. The tiny air sacs at the end of the bronchioles are called **alveoli** (al VEE uh ly) (singular: alveolus). Alveoli are grouped in bunches, like grapes. Many capillaries surround each alveolus.

Key Question What is the function of the respiratory system?
The respiratory system picks up oxygen from the air we breathe in. It releases carbon dioxide back into the air when we breathe out.

BUILD Vocabulary

bronchus
one of two large tubes in the chest cavity that leads from the trachea to the lungs

alveolus
the tiny air sac at the end of a bronchiole in the lungs that provides surface area for gas exchange to occur

INQUIRY into Scientific Thinking

GUIDED INQUIRY

What's in the Air?

Your respiratory system has several structures that catch particles from the air before they can reach the lungs. For example, hairs in the nose trap large particles. In this activity, you will observe some of the particles in the air you breathe.

1. Trace the outline of a microscope slide onto a piece of graph paper. Cut out the outline and tape it to the slide.
2. Repeat step 1 to make four more slides.
3. Choose five places to put your slides. Some of your spots should be inside. Some should be outside.
4. On each piece of graph paper, write your initials, the date, and where you plan to place the slide.
5. Cover the front of each slide with a thin coat of petroleum jelly.
6. Leave your slides in your chosen spots for at least 24 hours.
7. Collect the slides and look at them under a microscope. Count the number of particles in ten of the squares on each slide. Record your results.

Analyze and Conclude

1. **Observe** Where had you placed the slide that had the most particles? Where had you placed the slide with the fewest particles?
2. **Draw Conclusions** Were you surprised by the results? Why or why not?
3. **Apply Concepts** List the structures in your body that prevent most of these particles from entering your lungs.

In Your Workbook Get more help for this activity in your workbook.

Materials

graph paper, microscope slides, petroleum jelly, microscope

Analyze and Conclude

1. Answers will vary depending on the locations of the slides.
2. Answers will vary. Accept all reasonable responses.
3. hairs lining the inside of the nose, and cilia in the trachea

During Reading

Build Vocabulary

WORD ORIGINS Tell students that scientists often name structures using Latin or Greek words that describe the structure's appearance. Compare the trachea to a vacuum cleaner hose in terms of the rough texture.

Active Reading

COMPARE AND CONTRAST Write these words on the board: *alveolus, bronchus, bronchiole*.

Say As air goes into the lungs, it passes through smaller and smaller structures.

Ask In which order does air move through these structures? *(bronchi, bronchioles, alveoli)* Which structure is smallest? *(alveoli)*

KEY QUESTION Read the question aloud and ask a volunteer to read the answer.

Ask What does the respiratory system do with the air it takes in? *(It takes oxygen from the air so that it can be carried to cells.)*

Speed Bump

Ask Why are alveoli referred to as dead ends? *(Once air reaches the alveoli, it does not go any farther in the respiratory system.)*

Inquiry Into Scientific Thinking

MAKE CONNECTIONS Ask students to think about how the location of the slide will affect what they find. Then ask them to think about how location makes a difference in the health of a person's respiratory system.

Lead a Discussion

Talk about the results as a class. Have students take turns showing and describing what they found.

Ask What is happening when you cough or sneeze? *(Your respiratory system is getting rid of particles that you have inhaled.)*

LESSON 33.3

During Reading continued

Build Connections

VISUAL SUMMARY Have students examine the figure of the respiratory system and trace the flow of air. Remind students to fill out their flowcharts and encourage them to use the figure for guidance.

Ask Where does air enter the respiratory system? *(through the nose or mouth)*

Ask How does this diagram show that it is possible to breathe in through your mouth? *(Both the mouth and nose are connected to the pharynx.)*

Ask How does the epiglottis protect a person from choking on food or liquid? *(When a person swallows food or liquid, the epiglottis closes off the trachea, so the substances will enter the esophagus.)*

Ask How many bronchi are in the respiratory system? *(two)*

Ask How does mucus in the trachea help with respiration? *(Mucus in the trachea traps dirt and particles and keeps them from entering the lungs.)* What happens to mucus and its trapped dirt particles? *(The mucus and particles are swept upward by cilia and either swallowed or spit out of the body.)*

Ask What happens to air when it enters the lungs? *(It moves into bronchioles and then into alveoli.)*

Ask Which structures of the circulatory system are in contact with alveoli? *(capillaries that lead to pulmonary veins)*

BUILD Connections

THE RESPIRATORY SYSTEM

Air moves through the nose, pharynx, larynx, trachea, and bronchi into the lungs.

In Your Workbook Go to your workbook for more practice with the respiratory system.

1 **Nose** Air enters the body through the nose, where it is filtered, moistened, and warmed.

2 **Pharynx, Larynx, and Trachea** From the nose, air moves into the pharynx. Then, it passes through the larynx, which contains the vocal cords, and through the trachea.

3 **Lungs** From the trachea, air moves into the bronchi. Each bronchus leads to one lung. The bronchi divide into bronchioles, which eventually end at alveoli.

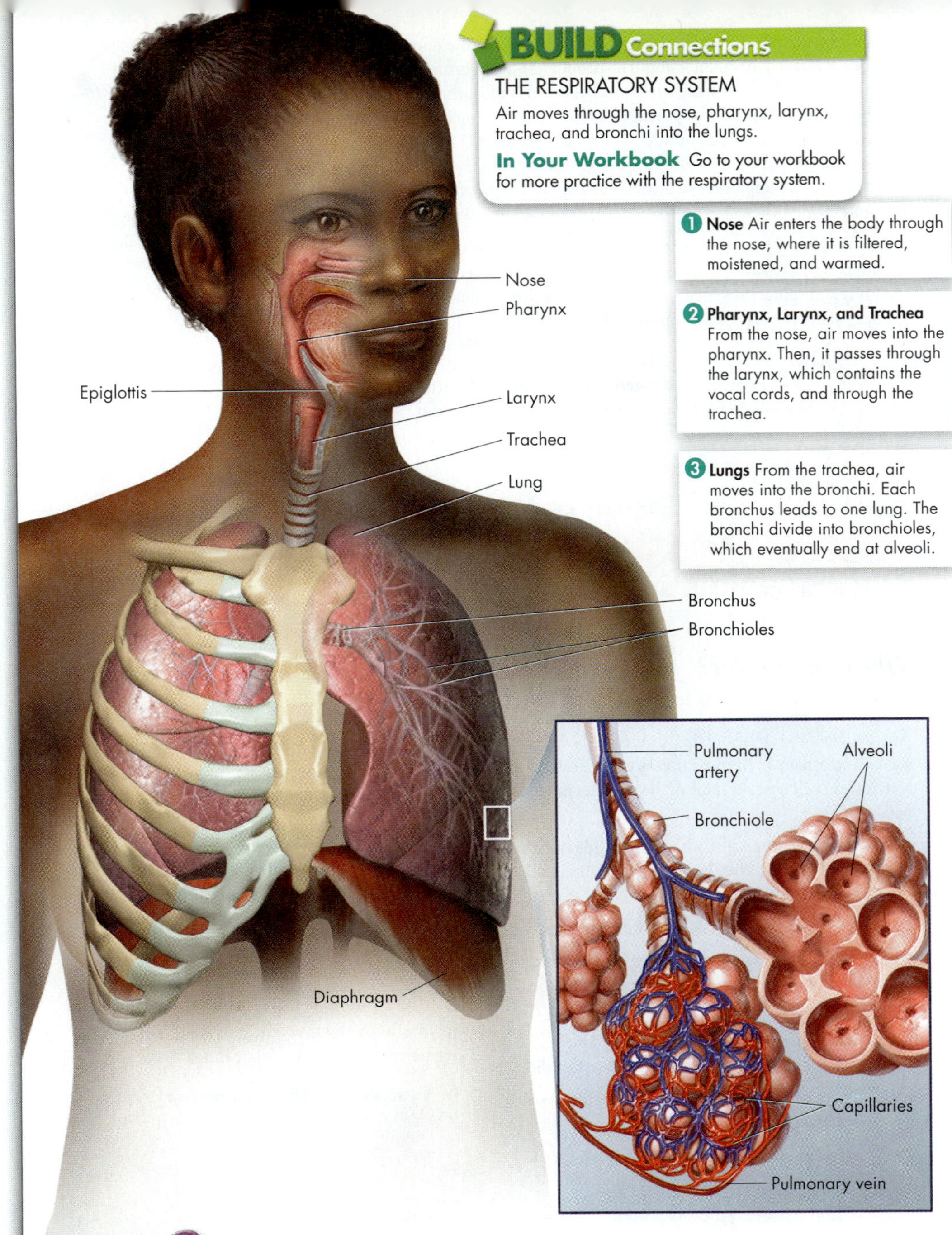

 BIOLOGY.com Search Lesson 33.3 GO • Art Review

ELL Focus on ELL: Extend Language

ALL SPEAKERS Because students have learned about cellular respiration, they may be confused when the term *respiration* is also applied to the respiratory system. Explain that cells need oxygen in order to carry out cellular respiration and the respiratory system delivers oxygen to cells throughout the body. Have students make two drawings: one showing cellular respiration at the cellular level and one showing respiration at the organism level. The illustrations do not need to be detailed, but they should include labels and indicate how the two processes are related.

Gas Exchange and Transport

Each healthy lung has about 150 million alveoli. The alveoli provide an enormous amount of surface area for gas exchange. Oxygen and carbon dioxide are exchanged across the alveoli and capillaries.

Gas Exchange The inner walls of alveoli are moist. When air enters the alveoli, oxygen dissolves in the moisture. There is more oxygen in the alveoli than in the blood inside the capillaries. So, the oxygen diffuses across thin capillary walls into the blood. At the same time, carbon dioxide diffuses into the alveoli.

Transport Oxygen diffuses passively from alveoli into capillaries. But diffusion stops if the oxygen concentration in the blood and the alveoli are the same. That's why hemoglobin in red blood cells is important. Hemoglobin binds to dissolved oxygen and removes it from the plasma. This helps to keep the concentration in the blood low.

Carbon dioxide diffuses from body tissues to capillaries in three ways. Most carbon dioxide combines with water in the blood. This combination forms carbonic acid. The rest dissolves in plasma or binds to proteins, including hemoglobin, in blood. When the blood reaches the lungs, the carbon dioxide is released into alveoli and exhaled.

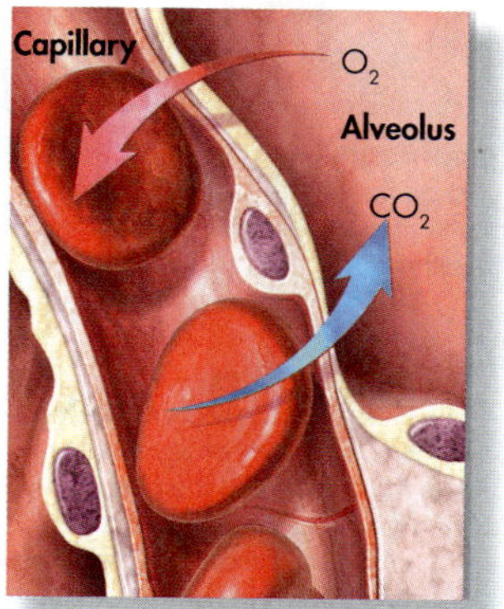

Gas Exchange Inside the lungs, the bronchi divide into bronchioles, which eventually end at alveoli. Carbon dioxide and oxygen diffuse across the walls of capillaries and alveoli.

Key Question How are oxygen and carbon dioxide exchanged and transported throughout the body?
Oxygen and carbon dioxide diffuse across the thin walls of alveoli and capillaries. Hemoglobin carries oxygen and some carbon dioxide. Carbon dioxide also dissolves in plasma or reacts with water to form carbonic acid.

Breathing

There are no muscles connected to our lungs. Breathing is driven by air pressure, the diaphragm, and muscles attached to the ribs.

Inhalation At the bottom of the chest cavity is a large muscle called the **diaphragm.** When you inhale, the diaphragm and muscles raise the rib cage. Because the lungs are sealed, this forms a small vacuum. Air rushes in to fill the space.

Exhalation Usually exhalation is passive. The diaphragm and rib cage relax. This relaxation causes air to be pushed out of the body. But to blow out a candle, sing, or yell, you need more force. The muscles between the ribs and the abdominal muscles can contract to push the air out.

This system works only because the chest cavity is sealed. If a wound punctures the chest, air may leak into the chest cavity. This leaking air makes breathing impossible.

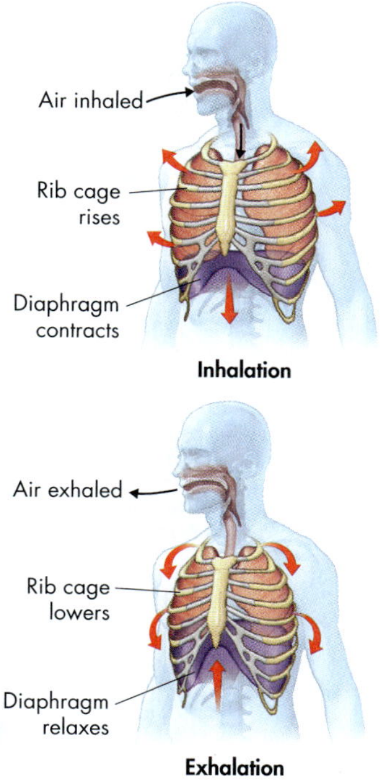

Breathing During inhalation, the rib cage rises and the diaphragm contracts. This increases the size of the chest cavity. During exhalation, the rib cage lowers and the diaphragm relaxes. This decreases the size of the chest cavity.

BIOLOGY.com Search Lesson 33.3 GO • InterActive Art

799

Active Reading

SUMMARIZE Have students read this page, and then write a summary in their own words for each heading. Make sure students focus on finding and expressing the main idea under each heading.

USE CONTEXT CLUES Tell students that they can gain clues to help their understanding and make connections as part of their reading. For example, have students read about oxygen transport.

Say Hemoglobin binds to oxygen, which keeps oxygen levels low in the blood.

Ask How does this process also help bring oxygen into the blood from the lungs? *(Lowering levels of oxygen in the blood causes oxygen to continue to diffuse across the capillary walls into the blood.)*

KEY QUESTION Read the Key Question and answer aloud to the class. Show students how to break the question into two parts.

Ask How are oxygen and carbon dioxide exchanged? *(Oxygen diffuses across the thin walls of the alveoli and capillaries into the bloodstream. Carbon dioxide diffuses from the bloodstream across the thin walls of the capillaries and alveoli to be exhaled.)*

Ask How are oxygen and carbon dioxide transported throughout the body? *(Hemoglobin carries oxygen to the body's tissues and carbon dioxide back to the lungs. Carbon dioxide also dissolves in plasma or reacts with water to form carbonic acid.)*

USE DIAGRAMS Refer students to the Breathing diagram. Use the labels and arrows to help explain the process of breathing. Tell students to trace the flow of air in and out of the body.

Ask What happens to the rib cage and diaphragm when you inhale? *(The rib cage rises and the diaphragm contracts.)* What does this do to the chest cavity? *(It increases the size of the chest cavity.)* Then what happens? *(Air rushes in to fill the larger space.)*

Ask What happens to the rib cage and diaphragm when you exhale? *(The rib cage lowers and the diaphragm relaxes.)* What does this do to the chest cavity? *(It decreases the size of the chest cavity.)* Then what happens? *(Air is pushed out of the body.)*

Speed Bump

Ask How does oxygen travel through the bloodstream? *(Oxygen binds to hemoglobin in the blood.)*

Ask Why might injury to the ribs cause pain when breathing? *(Muscles attached to the ribs help control breathing, and the ribs move during breathing.)*

Students Say

"When we read in groups of three or four, while one person reads out loud, the others read silently. At the end of each paragraph, the silent readers say 'CLINK' if they understood and 'CLUNK' if they didn't. The paragraph is explained by someone who said 'CLINK.' If no one said 'CLINK,' we try rereading. If it's still a 'CLUNK,' we raise our hands to get the teacher."

During Reading continued

Build Vocabulary

MULTIPLE MEANINGS Show students the position of the diaphragm under the chest.

Say The diaphragm controls the amount of air in the lungs by changing the size of the chest cavity.

Ask What happens when the diaphragm moves down? *(The lungs expand and air enters them.)* What happens when the diaphragm moves up? *(The lungs get smaller and air moves out of them.)*

Say A camera's diaphragm controls the amount of light by becoming smaller or larger, too.

Active Reading

KEY QUESTION Read the Key Question aloud, and have a volunteer read the answer. Then clarify the answer by listing the components and actions of breathing in a flowchart.

USE VISUALS Have students compare and contrast the photograph of the healthy lungs with the photograph of the smoker's lungs.

Ask What differences do you notice in the smoker's lungs? *(The smoker's lungs are black; the healthy lungs are pink.)* What caused the differences? *(Particles and chemicals from tobacco smoke damaged the smoker's lungs.)*

Speed Bump

Ask What triggers the breathing process? *(carbon dioxide levels in the blood)*

Ask Is it possible to suffocate by holding your breath? *(No, after a certain point, you cannot prevent your body from breathing.)*

Ask What three things in cigarettes are dangerous to your health? *(nicotine, carbon monoxide, and tar)* Which of these causes cancer? *(tar)*

BUILD Vocabulary

diaphragm
a large flat muscle at the bottom of the chest cavity that helps with breathing

MULTIPLE MEANINGS
In a camera, a diaphragm controls the amount of light that enters the camera. In the chest cavity, the diaphragm is a muscle that controls the amount of air that enters the body.

Breathing and Homeostasis You can control your breathing almost anytime you want. Yet, breathing is not purely voluntary, which is why people who drown have water in their lungs. When they lose consciousness, they "breathe" water into their lungs.

There is a breathing center in the brain that helps to control breathing and maintain homeostasis. The breathing center receives signals about the levels of carbon dioxide in the body. When carbon dioxide levels in the blood go up, the brain signals the diaphragm and chest muscles to contract. If carbon dioxide levels reach a certain point, you cannot keep yourself from breathing.

Key Question What mechanisms are involved in breathing? **Movements of the diaphragm and rib cage change the size of the chest cavity. Air rushes in during inhalation and is pushed out during exhalation.**

Smoking and the Respiratory System

The respiratory system filters out many particles that could damage the lungs. But some particles and chemicals can still reach the lungs. Chemicals in tobacco smoke damage structures throughout the respiratory system.

Effects on the Respiratory System Three of the most dangerous things in tobacco smoke are nicotine, carbon monoxide, and tar. Nicotine is addictive. It increases heart rate and blood pressure. Carbon monoxide blocks hemoglobin from binding with oxygen, which means that the blood cannot carry as much oxygen. Tar holds at least 60 compounds that cause cancer.

Tobacco smoke also damages cilia in the trachea. This damage allows more particles to enter the lungs. It also traps smoke-filled mucus along the passageways. Trapped particles and mucus lead to smoker's cough.

Effects of Smoking on the Lungs Chemicals in the smoke damage cilia in the lungs. The damage that smoking can cause is visible in the photograph on the right.

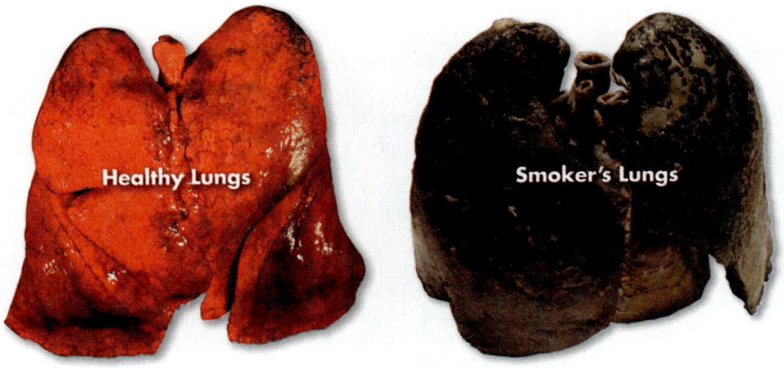

Hands-On Learning

WARNING LABELS Find examples of the warnings that are printed on the packages of cigarettes. Print the warnings and pass them around for students to read. Tell students that cigarette companies are required by law to print these warnings on their packages. If possible, have students research to find examples of warnings that other countries, such as Canada, put on their cigarettes. Have students share their findings with the class.

Diseases Caused by Smoking Smoking can cause permanent damage to the respiratory system. The damage from smoking can lead to chronic bronchitis, emphysema, and lung cancer. Smoking lowers life expectancy.

In chronic bronchitis, the bronchi become inflamed and clogged with mucus. Affected people often find simple activities, like climbing stairs, very hard. Emphysema (em fuh SEE muh) makes it hard to breathe because the lung tissue is damaged. People with this condition cannot get enough oxygen into the body. They cannot rid the body of extra carbon dioxide.

By the time lung cancer is found, it usually has spread to other parts of the body. Few people with lung cancer live more than five years after their diagnosis. About 87 percent of lung cancer deaths are due to smoking.

Other Effects of Smoking Smoking also hurts the circulatory system. It raises blood pressure, making the heart work harder to get oxygen to the body. Nonsmokers are also affected by secondhand smoke. Studies now show that children of smokers are twice as likely as children of nonsmokers to have asthma or other respiratory problems. Babies of women who smoked during pregnancy face many complications. Some of these can lead to lifelong problems.

Key Question How does smoking affect the respiratory system? **Chemicals in tobacco smoke damage structures throughout the respiratory system.**

CHECK Understanding

Apply Vocabulary

Use the highlighted words from the lesson to complete each sentence correctly.

1. The cavity at the back of the mouth called the ______ serves as a passageway for both air and food.
2. When you inhale, the ______ contracts and the rib cage rises.

Critical Thinking

3. **Explain** Describe the process of gas exchange in the lungs. What is hemoglobin's role in gas exchange?
4. **Relate Cause and Effect** Carbon monoxide gas binds to hemoglobin more easily than oxygen does. Based on this information, why do you think carbon monoxide alarms in homes have saved many lives?
5. **Apply Concepts** People with emphysema cannot exhale as much carbon dioxide as people with healthy lungs can. Why do you think this leaves them short of breath?
6. **Write to Learn** Answer the third clue of the mystery.

MYSTERY CLUE

John's doctor told him that if he hadn't stopped smoking, he probably would not have lived past the age of 50. Explain the doctor's reasoning. (**Hint:** See p. 801.)

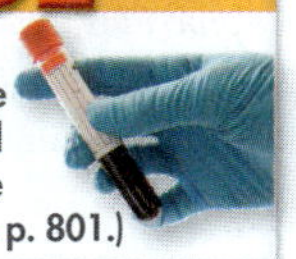

BIOLOGY.com Search Lesson 33.3 GO • Lesson Assessment

Check Understanding

1. pharynx
2. diaphragm
3. Oxygen diffuses from the alveoli to the blood, and carbon dioxide diffuses from the blood to the alveoli. Hemoglobin in red blood cells carries the oxygen. When hemoglobin binds with oxygen, the oxygen concentration in the blood drops. This allows diffusion to continue.
4. If carbon monoxide enters the lungs, it will bind to hemoglobin in place of oxygen. Carbon monoxide detectors alert people that there is a high level of the gas in their home.
5. Sample answer: If too much carbon dioxide remains in the alveoli, then the body cannot inhale enough oxygen to meet its needs.
6. Smoking raises blood pressure. John has high cholesterol, which also can lead to high blood pressure. These factors could lead to premature death.

Active Reading

MAKE INFERENCES Have students read the paragraphs under Diseases Caused by Smoking.

Say Life expectancy is the age to which a person is likely to live. Life expectancy varies based on when a person was born and where they live.

Ask How does smoking affect life expectancy? *(It lowers it.)*

Ask Do you have to smoke in order to experience the negative effects of smoking? Why or why not? *(No, people who breathe in smoke from smokers can also have respiratory problems. Children of smokers are at higher risk for asthma or respiratory problems. The babies of women who smoke during pregnancy often have health problems.)*

KEY QUESTION Read the Key Question and answer aloud. Then, have the class work together to produce a T-chart that lists the diseases caused by smoking and the respiratory structures they affect.

Speed Bump

Ask How can smoking affect the circulatory system? *(It can raise blood pressure.)*

Ask What is the impact of smoking on lung cancer cases? *(Approximately 87 percent of lung cancer deaths are caused by smoking.)*

Post Reading

ASSESSMENT TIPS For Question 4, point out to students that in order to answer the question, they need to explain what might happen to the body if a person were to breathe carbon monoxide.

MYSTERY CLUE Ask students to describe how smoking affects the circulatory system. *(It raises blood pressure.)* Remind students that John could already suffer from high blood pressure due to his condition.

Wrap-Up Activity

PUBLIC SERVICE ANNOUNCEMENT Have students work in pairs to write a public service announcement, or PSA, to encourage young people not to start smoking. Students can write their PSA for radio or TV broadcast. Encourage groups to consider a variety of formats, from simple statements and visuals to sound effects and live-action video. Have groups read, play, or show their PSAs to the class to evaluate which are the most effective.

CHAPTER LAB

Pre-Lab

Review the function of the diaphragm with the class before beginning the lab. Ask students to describe the connection between the muscles movements and changes in the lungs. Students should understand that the diaphragm changes lung volume by moving up and down.

Review the Background Questions and the Pre-Lab Questions as a class.

Lab

Tell students they will perform the chapter lab *Tidal Volume and Lung Capacity* described in **Lab Manual B**.

SAFETY Students should not start their experiments until their plans ave been approved. Necessary safety precautions will vary based on students' experimental design. Do not let students with latex allergies handle the balloons.

Post-Lab

COMIC STRIP Have students draw a comic strip that shows the sequence of events they tested. Encourage students to be creative but accurate.

NATIONAL SCIENCE EDUCATION STANDARDS

CONTENT F.1

INQUIRY A.1.a, A.1.b, A.1.c, A.1.d, A.2.a, A.2.d

Design Your Own Lab

GUIDED INQUIRY

Pre-Lab: Tidal Volume and Lung Capacity

Problem What factors can affect lung capacity?

Materials round balloons, metric ruler, meterstick

Lab Manual Chapter 33 Lab

Skills Focus Measure, Form a Hypothesis, Design an Experiment, Interpret Graphs

Connect to the Big idea Your lungs and circulatory system work together to provide the oxygen your cells need for cellular respiration. In your lungs, oxygen diffuses from the air you inhale into your blood. Carbon dioxide, a waste product of cellular respiration, diffuses from your blood into the inhaled air. Your lungs must have a large enough volume, or capacity, to supply all your cells with the oxygen they need.

Most of the time your lungs do not fill to capacity. But they can take in more air when you want to dive underwater or when you want to sing a long phrase without having to take another breath. In this lab, you will measure the volume of air you exhale when you are breathing normally and the volume of air you exhale after you take a deep breath.

Background Questions

a. **Sequence** List in order, from exterior to interior, the parts of the respiratory system that air passes through as you inhale.

b. **Review** Why does oxygen diffuse from inhaled air in the alveoli into the capillaries?

c. **Compare and Contrast** What is the difference between respiration and cellular respiration?

Pre-Lab Questions

Preview the procedure in the lab manual.

1. **Control Variables** What is the one difference between the procedures in Part A and Part B?

2. **Design an Experiment** Why must you use round balloons for this experiment?

3. **Predict** Which do you think will be greater—your estimated vital capacity or your measured vital capacity? Why?

BIOLOGY.com Search Chapter 33 GO

Visit Chapter 33 online to test yourself on chapter content and to find activities to help you learn.

Untamed Science Video Bundle up as the Untamed Science crew journeys to cold climates to show us how some animals handle extreme environments.

Art in Motion View a short animation that shows the beating of the heart as well as the transmission of impulses from the SA and AV nodes.

Art Review Review your understanding of the different parts of the respiratory system.

InterActive Art Watch an animation that shows the process of breathing and the production of sound.

Visual Analogy Compare the structure and function of the circulatory system to a system of highways and secondary roads.

Pre-Lab Answers

BACKGROUND QUESTIONS

a. nose, pharynx, larynx, trachea, bronchioles, alveoli

b. The concentration of oxygen is greater in the inhaled air in the alveoli than it is in the blood.

c. Respiration is the process of gas exchange in the lungs. Cellular respiration releases energy from food in the presence of oxygen.

PRE-LAB QUESTIONS

1. In Part A, the subject breathes normally. in Part B, the subject takes a deep breath before exhaling as much air as possible.

2. Sample answer: for accurate measurement

3. Students may say that the estimated vital capacity will be larger because they will not be able to capture all the exhaled air in the balloon. Some students may expect their measured capacity to be greater because they have done considerable aerobic training.

33 CHAPTER Summary

33.1 The Circulatory System

- The circulatory system carries oxygen, nutrients, and other substances throughout the body. It also removes wastes from body tissues.
- Heart action spreads through the myocardium, as first the atria and then the ventricles contract, pumping blood through the body.
- Arteries, capillaries, and veins are three kinds of blood vessels.

myocardium (p. 787)
atrium (p. 787)
ventricle (p. 787)
artery (p. 789)
capillary (p. 789)
vein (p. 789)

33.2 Blood and the Lymphatic System

- Plasma carries heat and a number of important substances through the body.
- Red blood cells carry oxygen. Hemoglobin in red blood cells binds to oxygen in the lungs and releases it in the capillaries.
- White blood cells fight infection.
- Platelets and plasma proteins form blood clots.
- The lymphatic system collects lymph, filters it for pathogens, and returns it to the circulatory system. It also picks up fat and fat-soluble nutrients from the intestines.
- Three common circulatory diseases are heart disease, stroke, and high blood pressure.
- High cholesterol levels can lead to atherosclerosis and a higher chance of heart attack and stroke.

plasma (p. 790)
red blood cell (p. 790)
white blood cell (p. 791)
platelet (p. 791)
lymph (p. 792)
atherosclerosis (p. 793)

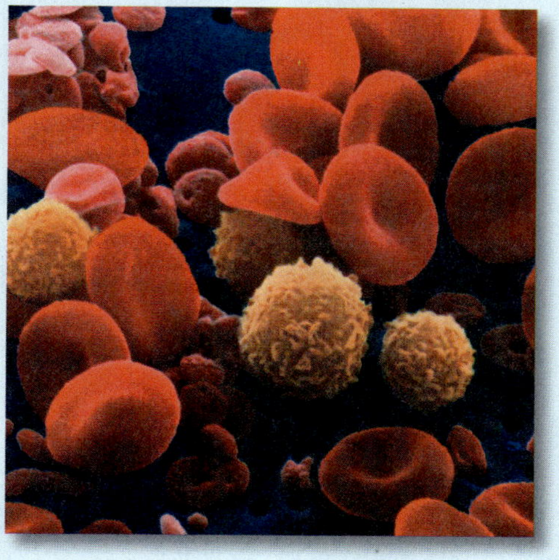

33.3 The Respiratory System

- The human respiratory system picks up oxygen from the air when you inhale. It releases carbon dioxide back into the air when you exhale.
- Oxygen and carbon dioxide diffuse across the thin walls of alveoli and capillaries. Hemoglobin carries oxygen and some carbon dioxide. Carbon dioxide also dissolves in plasma or reacts with water to form carbonic acid.
- Movements of the diaphragm and rib cage change the size of the chest cavity. Air rushes in during inhalation and is pushed out during exhalation.
- Chemicals in tobacco smoke damage structures throughout the respiratory and circulatory systems.

pharynx (p. 796)
larynx (p. 796)
trachea (p. 796)
bronchus (p. 797)
alveolus (p. 797)
diaphragm (p. 799)

BIOLOGY.com Search Chapter 33 GO • Match It • Chapter Assessment

803

UbD Transfer the Big Idea

TRANSFER Have students form small groups to create two full-page magazine advertisements titled *Take Care of Your Circulatory System* and *Take Care of Your Respiratory System*. Tell students that these ads are meant to be used in a nationwide advertising campaign to inform people about their respiratory and circulatory systems and how to keep them healthy. Have students work as a team to write material, design the pages, and choose photographs and illustrations. Tell students they should aim the ads at people who have very little science knowledge.

Chapter Summary

Think Visually

USE GRAPHICS Have students make a two-column table. They should title the first column *Structure* and the second column *Function*. Tell them to fill in the table with the structures described in this chapter—from both circulatory and respiratory systems—and their functions.

ANSWERS

Students' tables should include in the left column structures of the circulatory and respiratory systems, such as arteries, veins, aorta, heart, lungs, larynx, trachea, pharynx, and alveoli. In the right column, the functions of each structure should be described.

BIOLOGY.com Study Online

REVIEW RESOURCES

Visual Analogy Explore ways in which the human circulatory system is similar to an urban transportation network.

Art in Motion Have students use this animation to watch how blood flows through the heart.

Data Analysis This activity allows students to use electrocardiography to diagnose heart conditions.

Art Review Use this activity to help students understand the structures of the respiratory system.

InterActive Art Students can use this interactive activity to learn about breathing.

Vocabulary Review The *Flash Cards* and *Match It* provide an interactive way to review chapter vocabulary.

Editable Worksheets Pages of Study Workbook B, Lab Manual B, and the Assessment Resources Book are available online. These documents can be easily edited using a word-processing program.

Lesson Overviews Lesson Overviews are based on Miller & Levine Biology. However, they are available on CD-ROM and can be edited for your class.

ASSESSMENT RESOURCES

Chapter Assessment Have students take an online version of the Chapter 33 Check Understanding.

Standardized Test Prep Students can take an online version of the Standardized Test Prep. You will receive their scores along with ideas for remediation.

Diagnostic and Benchmark Tests Use these tests to monitor your students' progress and supply remediation.

ASSESSMENT

Assess

Big idea Read the Big Idea and question together as a class. Then ask students to take out their two-column tables listing the structures and functions of the circulatory and respiratory systems. Take a few moments to review the table as a class.

Ask Do these body systems share any functions? What are they? *(Yes, both systems are involved in the movement of oxygen and carbon dioxide in the body.)*

Ask How do capillaries and alveoli allow the two systems to function together? *(The thin walls of alveoli allow the respiratory system to take in oxygen from the environment into the body and take wastes from the body and remove it to the environment. The capillaries of the circulatory system allow the system to transport oxygen throughout the body and collect wastes from throughout the body. The thin walls of alveoli and capillaries allow them to transfer oxygen and wastes between them and bridge these connections between the cells and the environment.)*

CONSTRUCTED RESPONSE Paragraphs will vary. Explain the activity carefully. Be sure that students understand that the *Hints* should be used as a guide to help them answer the questions. Review how to make a main ideas and details chart. Point out that this chart can help them organize their paragraphs.

Write Question 1 on the board and read it aloud. Circle the words *structure* and *functions*.

Say This question asks you to make connections between structures and functions. Most of the time, a difference in structure also indicates a difference in function.

Read the first hint.

Ask What do you recall about the difference in functions and structures of arteries, capillaries, and veins? *(Answers may vary. Students may need to review the definitions or diagrams for each.)*

Read the next hint.

Ask What causes blood to move in each of these situations? *(Blood in arteries is pumped by the heart. Blood moving in veins is under less direct pressure and is often assisted by skeletal muscle movement.)*

Read the next hint.

Ask How are capillaries different from veins and arteries? *(Capillaries are much smaller and interact with individual cells.)*

33 CHECK Understanding

Assess the Big idea: Structure and Function

Write an answer to the question below.

Q: How do the structures of the circulatory and respiratory systems allow for their close functional relationship?

Constructed Response

Write an answer to each of the questions below. The answer to each question should be one or two paragraphs long. To help you begin, read the **Hints** below each of the questions.

1. **Explain how the structure of blood vessel walls contributes to the vessels' functions.**

   Hint As blood flows through the circulatory system, it moves through three kinds of blood vessels—arteries, capillaries, and veins.

   Hint Arteries carry blood away from the heart. Veins carry blood back to the heart.

   Hint Capillaries are the smallest blood vessels.

2. **Compare and contrast the functions of the lymphatic system and the rest of the circulatory system.**

   Hint The circulatory system includes the heart, blood vessels, and blood.

   Hint The lymphatic system is part of the circulatory system. The lymphatic system is a network of vessels, nodes, and organs including the spleen and thymus.

3. **Compare and contrast cellular respiration and respiration at the level of an organism.**

   Hint Cellular respiration is the process that releases energy from food, such as a sugar, when there is oxygen present. Review cellular respiration in Chapter 9.

   Hint For organisms, rather than single cells, *respiration* means gas exchange between a body and the environment.

Foundations for Learning Wrap-Up

Use the table tent you made while reading the chapter to help you organize your thoughts about the respiratory and circulatory systems.

Activity 1 Working with a partner, review the structures and functions of the respiratory, lymphatic, and circulatory systems. Add any structures and functions you may have missed. Then, quiz each other about the structures and functions of each system.

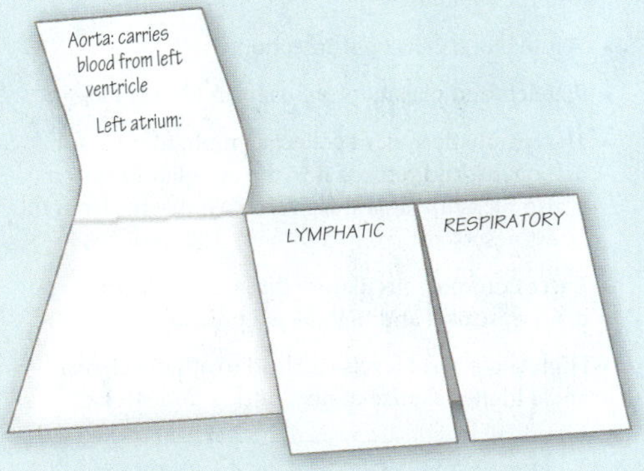

Activity 2 Work in a group. One person reads aloud any function in any system and anyone in the group can identify the structure. For every correct answer, add a tick mark to your table tent. Take turns. The person with the most tick marks at the end is the winner.

Extension: Working with a partner, read aloud the name of any structure in any system. Your partner has to spell the structure correctly. Take turns.

For Question 2, pair students and have them use the hints and your modeling to answer the question. Have students share their paragraphs with the class.

For Question 3, have students write the paragraphs independently for homework or as an anchor activity.

Foundations for Learning Wrap-Up

Have students take out the table tents they made. Be sure that students have included accurate descriptions of all structures. Either assign groups or pair students, and explain the activities.

33.1 The Circulatory System

Understand Key Concepts

1. Which of the following is NOT a type of blood vessel?
   a. bronchus
   b. artery
   c. capillary
   d. vein
2. The circulatory system includes the
   a. lungs, heart, and brain.
   b. lungs, blood vessels, and heart.
   c. heart, blood, and blood vessels.
   d. heart, trachea, and alveoli.

Test-Taking Tip

Choose Among Similar Answers As you read each answer choice, notice what they have in common. In question 2, every answer choice includes *heart*. So, the heart must be part of the circulatory system. The heart pumps blood through the body, so the circulatory system includes "blood." Only answer **c** includes blood.

3. Which of the following describes the function of the circulatory system?
   a. It digests food.
   b. It carries oxygen and nutrients to the body's tissues and removes wastes from the tissues.
   c. It lets a person breathe.
   d. It responds to light and sound.
4. Compare the size and structure of arteries, capillaries, and veins.
5. Compare pulmonary circulation and systemic circulation.

Think Critically

6. **Apply Concepts** Some large veins have one-way valves, which keep blood flowing in only one direction. Why don't arteries need similar valves?

33.2 Blood and the Lymphatic System

Understand Key Concepts

7. Cells that protect the body by engulfing foreign cells or by producing antibodies are
   a. red blood cells.
   b. plasma.
   c. platelets.
   d. white blood cells.
8. The process shown below is made possible by plasma proteins and cell fragments called
   a. hemoglobins.
   b. thrombins.
   c. platelets.
   d. lymphocytes.

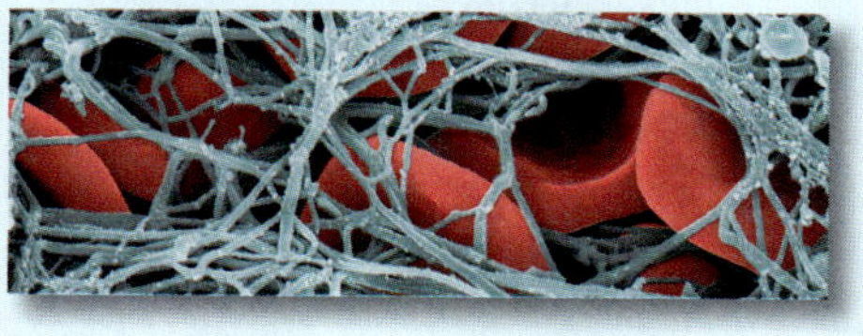

9. What are the primary functions of the lymphatic system?
10. List three common circulatory diseases.

Think Critically

11. **Apply Concepts** Why would a person with a low red blood cell count feel tired?
12. **Compare and Contrast** Explain how high blood cholesterol develops in a person with a genetic disorder versus someone who eats a high-fat diet.

33.3 The Respiratory System

Understand Key Concepts

13. The tiny hollow air sacs in the lungs where gas exchange takes place are called
    a. alveoli.
    b. lymph nodes.
    c. capillaries.
    d. bronchioles.
14. Which structure of the respiratory system contains the vocal cords?
    a. pharynx
    b. diaphragm
    c. trachea
    d. larynx

Assessment Tips

Answer questions to Lesson 33.1 as a class. After reading Question 2, read the Test-Taking Tip. Then, continue with the other questions. Ask students to keep an eye out for examples of similar answers.

When students reach Question 6, ask them to read the question and think about it. Tell them they are being asked to think of the differences between arteries and veins and then choose the difference that relates to valves.

Have students complete the questions to Lesson 33.2 and 33.3 on their own. If they have trouble answering the multiple choice questions, review the vocabulary and encourage students to use the visuals.

For Question 12, make sure students recall the different ways that cholesterol gets into the blood. Many students will think that cholesterol must come from the diet.

Tell students to read the questions in Lesson 33.3 carefully. For example, in Question 17, the question asks students to explain how the effects of tobacco contribute to smoker's cough. Caution students to find the main question and use the details to help answer it.

Answers

Lesson 33.1

UNDERSTAND KEY CONCEPTS

1. a
2. c
3. b
4. Arteries are wide vessels with thick walls. Capillaries are narrow vessels with walls just one cell thick. Veins are wide vessels with walls that are thicker than capillary walls but thinner than artery walls.
5. Pulmonary circulation carries blood between the heart and the lungs. Systemic circulation carries blood between the heart and the rest of the body.

THINK CRITICALLY

6. The powerful pressure produced when the heart contracts keeps blood moving in one direction in the arteries.

Lesson 33.2

UNDERSTAND KEY CONCEPTS

7. d
8. c
9. The primary functions of the lymphatic system are to collect fluid lost by the blood and return it to the circulatory system, to filter bacteria and other pathogens from the fluid, and to absorb fat and fat-soluble vitamins from the digestive tract and carry them to the circulatory system.
10. heart disease, stroke, and high blood pressure

THINK CRITICALLY

11. A person with a low red blood cell count has fewer red blood cells to transport oxygen to cells. Without enough oxygen, cellular respiration slows down. The cells do not get the energy they need.
12. In an individual with a genetic disorder, the liver cells cannot take up LDL from the blood. Therefore, they are not signaled to stop producing cholesterol. In an individual who eats a high-fat diet, excess cholesterol builds up in the liver cells. These cells stop making LDL receptors and no longer remove cholesterol from the blood.

Lesson 33.3

UNDERSTAND KEY CONCEPTS

13. a
14. d

ASSESSMENT

15. b

16. The respiratory system picks up oxygen from the air when we breathe in. It releases carbon dioxide into the air as we breathe out.

THINK CRITICALLY

17. Without the white blood cells, the respiratory system gets filled with debris. This causes the smoker to cough, as the body tries to get rid of debris.

Connecting Concepts

18. Student A: minute 6; Student B: minute 7

19. Student A is in better physical condition than Student B because his or her heart beats fewer times per minute to maintain the same level of effort as Student B. Student A's heart returns to its normal rate much more quickly than Student B's heart does. This also suggests a more efficient and healthier circulatory system.

20. During exercise, you would expect blood pressure to rise because blood is moving more quickly through the circulatory system. Also, breathing rate increases to keep pace with the increase in the amount of gas exchange taking place in the muscles of the body.

CHAPTER MYSTERY After students have read through the Chapter Mystery, discuss the role of cholesterol in the circulatory system.

Ask How does this gene defect on chromosome 19 disrupt the normal function of liver cells? *(It causes liver cells to produce defective LDL receptors. The cells cannot remove LDL from the blood.)*

Ask Why is this condition more severe in Lila than in John? *(John has one normal gene and one defective gene. Having one normal gene allows him to produce some normal LDL receptors in addition to the defective ones. This reduces his symptoms.)*

Ask How does the disease disrupt homeostasis? *(LDL receptors cannot take up LDL and so cholesterol builds up in the blood. The excess LDL is deposited in blood vessels and possibly other body sites.)*

33 CHECK Understanding

15. The large flat muscle that moves up and down and changes the volume of the chest cavity is the

a. pharynx. **c.** trachea.
b. diaphragm. **d.** larynx.

16. What is the function of the respiratory system?

Think Critically

17. **Infer** Tobacco smoke can kill white blood cells in the respiratory tract. The white blood cells help keep the respiratory system clean by consuming debris. How do you think this contributes to the development of smoker's cough?

Connecting Concepts

Use Science Graphics

The following graph is based on pulse rates taken each minute for two students doing the same exercises. The exercises begin at minute 1 and end at minute 8. Use the graph to answer questions 18–20.

Pulse Rate and Exercise

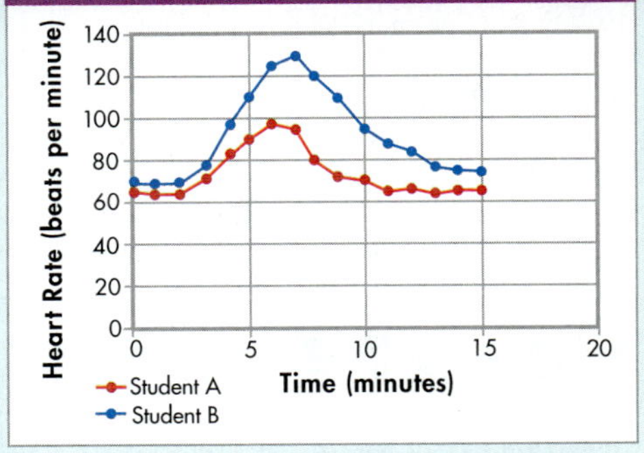

18. **Interpret Graphs** At about which minute did each student reach his or her highest heart rate?

19. **Draw Conclusions** Which of the two students is most likely in better physical condition? What evidence from the graph supports your answer?

20. **Predict** What other changes in the circulatory and respiratory systems would you expect to take place in the time interval shown?

solve the CHAPTER MYSTERY

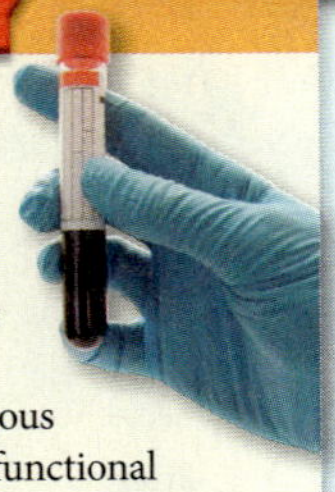

IN THE BLOOD

Both John and Lila have a genetic disease called familial hypercholesterolemia. This disease is caused by a gene defect on chromosome 19. John is heterozygous for the disorder. He inherited one functional copy of the gene and one defective copy. Although his liver cells make a mixture of normal and defective LDL receptors, his blood cholesterol levels were so high that he had serious atherosclerosis by age 35. Most people with this disease have had a heart attack by age 60.

Lila is homozygous for the defective allele—a very rare condition. Her liver cells do not produce any functional LDL receptors. Her atherosclerosis became apparent when she was only 4 years old. Fatty deposits can be seen in the corneas of her eyes and beneath the skin near her elbows and knees.

Research on this genetic defect helped uncover the role of liver cell LDL receptors in regulating blood cholesterol. Researchers then applied that information to cases of high cholesterol among the general public. The result was the development of several new classes of drugs that are helping many people live longer.

1. **Predict** If Lila were to have a child, what is the likelihood that she would pass on the allele for familial hypercholesterolemia? How likely is it that John passed on the allele to his son?

2. **Infer** Most patients with one defective allele can control their LDL levels with medication that keeps their liver from making cholesterol. But these medications generally do not lower the LDL levels of patients with two defective alleles. Why do you think that is so?

UntamedScience™ **Never Stop Exploring Your World.** Finding the solution to the In the Blood mystery is just the beginning. Take a video field trip with the ecogeeks of Untamed Science to see where the mysteries of the circulatory system lead.

CHAPTER MYSTERY ANSWERS

1. 100%; 50%

2. Because homozygous patients have no LDL receptors, medications to keep their liver cells from producing cholesterol do not help the body control the cholesterol that is consumed in food.

Untamed Science To learn how some animals survive and thrive in cold climates, suggest students watch **Chillin' in the Cold.**

Standardized Test Prep

Multiple Choice

1. In the human heart, oxygen-rich blood would be found in the
- **A** right atrium and the right ventricle.
- **B** right atrium and the left atrium.
- **C** left atrium and the left ventricle.
- **D** right ventricle and the left ventricle.

2. Which statement BEST describes an interaction between the circulatory system and the respiratory system that helps maintain homeostasis?
- **A** Blood plasma transports salts, nutrients, and proteins through the body to keep it healthy.
- **B** The diaphragm and rib cage work together to move air into and out of the lungs.
- **C** Lymph nodes filter out bacteria that could cause disease.
- **D** Blood cells pick up and carry oxygen from the lungs to the body's cells.

3. A heartbeat begins with an impulse from the
- **A** nervous system.
- **B** sinoatrial (SA) node.
- **C** atrioventricular node.
- **D** aorta.

4. All of the following are components of human blood EXCEPT
- **A** plasma.
- **B** mucus.
- **C** phagocytes.
- **D** platelets.

5. Nicotine in tobacco
- **A** is not addictive.
- **B** lowers blood pressure.
- **C** blocks the transport of oxygen.
- **D** increases heart rate.

6. Antibodies are produced by
- **A** red blood cells.
- **B** platelets.
- **C** B lymphocytes.
- **D** hormones.

Questions 7–10

Use the diagram below to answer the questions that follow.

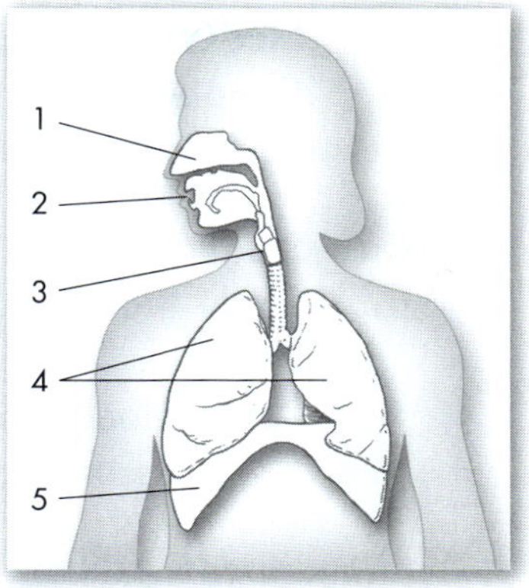

7. Which structure's primary function is to warm and moisten inhaled air?
- **A** 1
- **B** 3
- **C** 4
- **D** 5

8. Which structure contains the vocal cords?
- **A** 1
- **B** 2
- **C** 3
- **D** 4

9. Damage to which structure can lead to emphysema?
- **A** 2
- **B** 3
- **C** 4
- **D** 5

10. Which structure contains alveoli?
- **A** 2
- **B** 3
- **C** 4
- **D** 5

Open-Ended Response

11. Explain why the risk factors for heart disease and strokes are similar.

If You Have Trouble With . . .

Question	1	2	3	4	5	6	7	8	9	10	11
See Lesson	33.1	33.3	33.1	33.2	33.3	33.2	33.3	33.3	33.3	33.3	33.2

Before the Test

Review with students the Test-Taking Tip.

Point out the word *EXCEPT* in Question 4.

Remind students that an open-ended response means that they'll have to write a paragraph to answer the question. Have them paraphrase the question in their notebooks, and remind them about the steps in writing a paragraph.

Answers

1. C
2. D
3. B
4. B
5. D
6. C
7. A
8. C
9. C
10. C
11. Heart disease and strokes can both be caused by narrowing of the arteries—a result of the plaque buildup within the arteries. Therefore, both illnesses have similar risk factors.

Test-Taking Tip

INTERPRET VISUALS

Tell students to watch for questions, such as Questions 7–10 on this page, that require them to identify structures in a diagram. Explain that when they are asked about structures shown in a diagram, they should first identify each structure and think about its function. Then, they should answer the questions about the diagram.

CHAPTER PLANNER 34 Endocrine and Reproductive Systems

Chapter Contents	NSES	Time	Core Resources
Chapter Preview			**Student Edition,** pp. 808–809 **Chapter Mystery,** p. 809 **Foundations for Learning,** p. 809
34.1 The Endocrine System Hormones and Glands • Hormone Action	I, IV, V, C.1.a, C.1.c, C.1.d	1/2 period 1/4 block	**Student Edition,** pp.810–812 **Study Workbook B** 34.1 Worksheets **Biology.com** *Art Review:* Major Endocrine Glands • *Art in Motion:* Steroid and Nonsteroid Hormones • 34.1 Lesson Assessment
34.2 Glands of the Endocrine System The Human Endocrine Glands • Control of the Endocrine System	I, IV, V, C.1.d, C.5.d, F.1	1 period 1/2 block	**Student Edition,** pp. 813–816 **Study Workbook B** 34.2 Worksheets **Biology.com** 34.2 Lesson Assessment
34.3 The Reproductive System Sexual Development • The Male Reproductive System • The Female Reproductive System • Sexually Transmitted Diseases	I, V, A.1.c, A.2.a, C.2.b, F.1	1 1/2 periods 3/4 block	**Student Edition,** pp. 817–823 **Build Connections,** p. 821 **Inquiry Into Scientific Thinking,** p. 822 **Study Workbook B** 34.3 Worksheets **Biology.com** 34.3 Lesson Assessment
34.4 Fertilization and Development Fertilization and Early Development • Later Development	I, V, C.1.f, C.2.b, C.5.d, F.1	1 1/2 periods 3/4 block	**Student Edition,** pp. 824–829 **Study Workbook B** 34.4 Worksheets **Biology.com** 34.4 Lesson Assessment
Chapter Pre-Lab	I, A.2.b, F.1	1 period 1/2 block	**Student Edition,** p. 830 **Lab Manual B** *Diagnosing Endocrine Disorders*

Tools for ELL Instruction

ELL Handbook describes teaching strategies to make *Miller & Levine Biology: Foundation Edition* more accessible to ELL students.

Spanish Study Workbook is a Spanish translation of Study Workbook A.

Multilingual Glossary is the *Miller & Levine Biology* glossary translated into ten languages.

Tools to Challenge Students

Study Workbook A includes more challenging worksheets and activities such as Chapter Mystery activities that reinforce 21st Century Skills.

Lab Manual A labs are geared toward on-level students. The Lab Manual includes fifteen labs in addition to the Chapter Labs.

Biology.com Data Analysis activities offer students a way to practice data gathering and analyzing skills.

Visual Quizzes in the Assessment Resources Book are geared toward on-level students and use visuals from *Miller & Levine Biology.*

Additional Resources

Biology.com Untamed Science Video • Flash Cards

Study Workbook A 34.1 Worksheets
Spanish Study Workbook 34.1 Worksheets

Study Workbook A 34.2 Worksheets
Spanish Study Workbook 34.2 Worksheets
Biology.com *Data Analysis:* Diabetes
Assessment Resources Book Visual Quiz

Study Workbook A 34.3 Worksheets
Lab Manual B *Data Analysis:* Menstrual Cycle
Spanish Study Workbook 34.3 Worksheets
Assessment Resources Book Visual Quizzes

Study Workbook A 34.4 Worksheets
Lab Manual B *Hands-On Activity:* Growing Up
Spanish Study Workbook 34.4 Worksheets

Lab Manual A *Diagnosing Endocrine Disorders*

Chapter Review

Student Edition Chapter Summary, p. 831
Study Workbook B Chapter 34 Review
Transparencies, pp. 373–381
Biology.com Untamed Science Video • Editable Worksheets of Study Workbooks A and B and Lab Manuals A and B • Chapter 34 Flash Cards and Crossword

Untamed Science DVD • Classroom Resources CD (includes lesson presentations and editable worksheets)

Chapter Assessment

Student Edition Check Understanding, pp. 832–835
Study Workbook B Chapter 34 Taking a Standardized Test
Assessment Resources Book Chapter 34 Test B
Biology.com Chapter 34 Assessment • Editable Worksheets of Chapter 34 Visual Quizzes* and Chapter 34 Tests A and B

Exam*View Assessment Suite* • Classroom Resources CD (includes lesson presentations and editable worksheets)

Time: 1 period, 1/2 block

* Visual Quizzes are based on visuals from *Miller & Levine Biology.* Review carefully to make sure they are appropriate for your class.

Pressed for Time?

Preview the Chapter Read the Key Questions for all the lessons in this chapter.

Cover the Chapter Quickly Have students read *Hormones and Glands* in Lesson 34.1 and go over the Major Endocrine Glands visual. Assign *The Human Endocrine Glands* in Lesson 34.2. In Lesson 34.3, have students read *The Male Reproductive System* and *The Female Reproductive System* and go over the Male Reproductive System and Female Reproductive System visuals. Assign the Fertilization section of *Fertilization and Early Development,* and have students look over the visuals showing different stages of human embryonic development in Lesson 34.4.

Assess Assign Questions 1–4 in the 34.1 Check Understanding, Questions 1–4 in the 34.2 Check Understanding, Questions 1–2, and 5 in the 34.3 Check Understanding, and Question 2 in the 34.4 Check Understanding. Assign the "Think Visually" activity on p. 831, and Questions 2 and 4–15 in the Chapter 34 Check Understanding.

Get the Picture

Tell students to look at the photograph and imagine themselves as the painter walking on a cable of the Golden Gate Bridge.

Ask How would you feel if you were the painter in the picture? *(Sample answers: nervous, stressed, fearful, excited)* What physical reactions do you think your body would experience as you walked high on the bridge cable? *(Sample answers: sweating, increased heart rate, increased breathing rate)*

Say Such reactions are caused by signals inside your body that are sent by your endocrine system. That's one of the systems you will read about in this chapter.

Activate Prior Knowledge

Have students recall the functions of the nervous system.

Ask Can you give an example of a time when you sweated and your breathing sped up, but you were not doing anything that involved physical activity? *(Students will likely relay experiences that made them nervous, stressed, fearful, or excited.)*

Ask Why do you think you were sweating and/or breathing fast? *(My endocrine system sent messages telling my body how to respond to the situation.)*

Tell students while they are reading to look for the different types of messages the endocrine system sends.

34 Endocrine and Reproductive Systems

Big idea Homeostasis

Q: How does the body use chemical signals to maintain homeostasis?

BIOLOGY.com Search Chapter 34 GO • Chapter Mystery

808

NATIONAL SCIENCE EDUCATION STANDARDS

UNIFYING CONCEPTS AND PROCESSES

I, IV, V

CONTENT

C.1.a, C.1.c, C.1.d, C.1.f, C.2.b, C.5.d, F.1

INQUIRY

A.1.c, A.2.a, A.2.b

UbD Connect to the Big Idea

Invite a volunteer to read aloud the chapter number and the chapter title. Read the Big Idea and question. Emphasize the question, **How does the body use chemical signals to maintain homeostasis?**

Set the Purpose

PREVIEW AND PREDICT Draw the graphic organizer on the board. Guide students to the INSIDE feature on the next page. Invite a volunteer to read each lesson number and title.

Ask Why do you think homeostasis in the endocrine and reproductive systems is important? *(keeps the systems in balance)*

Tell students to search for answers to the questions in the graphic organizer as they read.

INSIDE:

- 34.1 The Endocrine System
- 34.2 Glands of the Endocrine System
- 34.3 The Reproductive System
- 34.4 Fertilization and Development

This painter may have had sweaty palms and a racing heart his first day on the job. His endocrine system is partly responsible for these reactions.

CHAPTER MYSTERY

OUT OF STRIDE

Lisa trained hard during spring track and over the summer. But as the new school year approached, she wasn't satisfied. She felt she needed to be faster for her team to win the state cross-country championship. A teammate suggested she lose a few pounds. Lisa had already lost weight over the summer, but she decided to lose some more.

Lisa continued her exhausting workouts. She also stopped snacking before practice and avoided high-calorie foods. She did lose weight. But she was always tired. She also noticed that she had not had a menstrual period in four months. The week before the championship meet, she collapsed in pain at practice. She had suffered a stress fracture to her lower leg. Her season was over.

Read for Mystery Clues Lisa's doctor told her that all of her symptoms were related to a single cause. As you read this chapter, look for clues to explain why too much exercise and dieting had these effects on Lisa. Then, solve the mystery.

FOUNDATIONS for Learning

Think of the Big Idea as a collection of facts that are connected to each other. Before you read the chapter, make an envelope from the pattern below. Fold the flaps in order along the lines. Glue or tape flap 3 to flaps 1 and 2. As you read the chapter, write important facts about each lesson or draw a picture to help you remember important facts on a slip of paper. Put your slips of paper in your envelope. At the end of the chapter are two activities that use the facts to help answer the question: How does the body use chemical signals to maintain homeostasis?

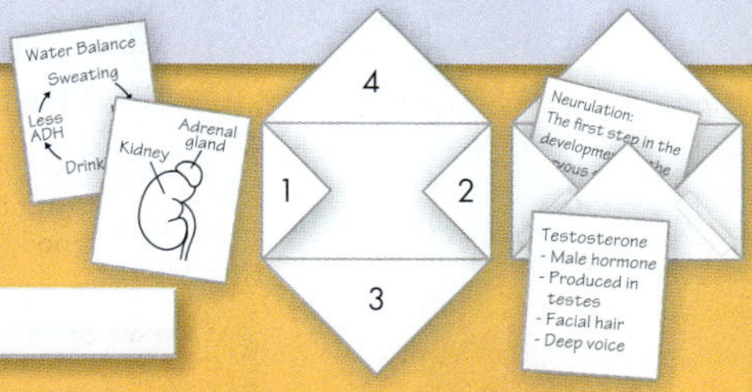

- Untamed Science Video
- Flash Cards

Chapter 34 Big Idea: Homeostasis

Chapter 34: How does the body use chemical signals to maintain homeostasis?

- **34.1:** How does the body send and receive chemical signals?
- **34.2:** What life processes are regulated by hormones?
- **34.3:** What body structures enable humans to produce offspring?
- **34.4:** How does a human develop from a single cell to a newborn baby?

CHAPTER 34

Foundations for Learning

If you have a source for inexpensive envelopes, you could use them to save time. Have students make a dozen or so blank slips and insert them in the envelope. The ready availability of slips will encourage students to write down the facts as they read. Demonstrate the use of the slips by developing the idea of a chemical signal.

Ask What is a chemical? *(Sample answer: A chemical is a certain substance, such as salt, water, or sugar.)* What is a signal? *(Sample answers: message, sign)*

Lead students to understand that a chemical signal in the body is a message carried by a specific substance from one place in the body to another. Have them write this description on one of the slips of paper and put it in the envelope. Remind students that they can use the slips as study aids throughout the chapter as well as at the chapter's end.

Untamed Science™ Why do you stand tall or run scared when presented with danger? Follow the Untamed Science crew as they explore the answer to this question in the short online video **Fight or Flight!**

CHAPTER MYSTERY Invite a volunteer to read aloud the Chapter Mystery. Have students review the meaning of *homeostasis*. Write the class definition on the board. Then, ask students to list the physical problems Lisa had experienced.

Ask How does the definition of *homeostasis* relate to Lisa's problems? *(Students may suggest that homeostasis was disrupted in Lisa's body in ways that caused her problems.)*

Digital Support

BIOLOGY.com Extend your reach by using the digital assets offered at **Biology.com.**

Pre-Reading

Objectives

34.1.1 Describe the components of the endocrine system.

34.1.2 Explain how hormones work.

Preview the Pages

1. Preview the vocabulary words by writing them on the board and pronouncing them. Then, have students find the words in the text, pronounce them, and read their definitions aloud.
2. Preview the visuals and read the captions and labels. Explain that the Major Endocrine Glands diagram shows both the female (ovaries) and male (testes) reproductive glands, with the male glands encircled. All the other glands shown are found in males and females.

Activate Prior Knowledge

Ask Can you recall how the prefixes change the meanings of the words *endoskeleton* and *exoskeleton*? *(endo-: inside; exo-: outside)*

Ask Based on their prefixes, how do you think that *endocrine* and *exocrine* differ in meaning? *(Guide students to conclude that the two words must relate to being inside and outside.)*

Build Understanding

COMPARE/CONTRAST TABLE Direct students to Workbook B, where they will find a compare/contrast table. Remind students that a compare/contrast table will help them quickly see the similarities and differences between two things.

NATIONAL SCIENCE EDUCATION STANDARDS

UNIFYING CONCEPTS AND PROCESSES

I, IV, V

CONTENT

C.1.a, C.1.c, C.1.d

For the corresponding lesson in *Miller & Levine Biology,* see pages 978–981.

34.1 The Endocrine System

Key Questions

What are the components of the endocrine system?

How do hormones affect cells?

BUILD Understanding

Compare/Contrast Table
As you read, make a table that compares and contrasts the two different types of hormones.

In Your Workbook
Refer to your workbook to learn how to make a compare/contrast table. Complete the compare/contrast table for Lesson 34.1.

Hormones and Glands

Your nervous system works like a telephone, with neurons acting as wires. Nerve impulses transmit messages directly from one cell to another cell. Your endocrine system works more like a radio station. It "broadcasts" chemical messages to all the cells in the body. The chemicals that carry these messages are called **hormones.** Hormones are released in one part of the body and travel through the blood to other parts of the body. Hormones can act on almost every cell in the body.

Hormones Hormones act by binding to specific chemical receptors on cell membranes or inside cells. Cells that have receptors for a particular hormone are called *target cells.* A cell must have receptors for a particular hormone for the hormone to affect it.

The body's responses to hormones are slower and last longer than its responses to nerve impulses. It may take several minutes, hours, or days for a hormone to have its full effect on its target cells. A nerve impulse may take only a fraction of a second to act on its target cells.

Glands A gland is an organ that produces and releases a secretion. **Exocrine glands** release their secretions through tube-like structures called ducts. Exocrine secretions are released either out of the body or directly into the digestive system. Exocrine glands include those that release sweat, tears, and digestive juices. **Endocrine glands** usually release their secretions (hormones) directly into the blood. Blood transports the secretions throughout the body. Bones, fat tissue, the heart, and the small intestine also produce and release hormones.

The Endocrine System
The endocrine system acts like a radio station. It broadcasts messages to the body's cells using hormones.

 BIOLOGY.com Search Lesson 34.1 GO • Art Review

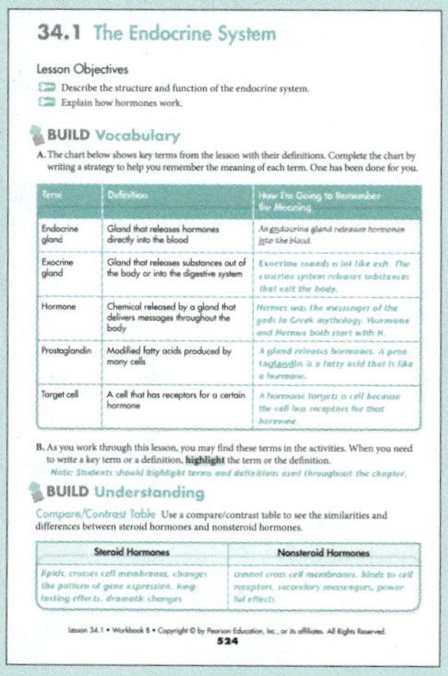

34.1 The Endocrine System

Lesson Objectives

Describe the structure and function of the endocrine system.

Explain how hormones work.

BUILD Vocabulary

A. The chart below shows key terms from the lesson with their definitions. Complete the chart by writing a strategy to help you remember the meaning of each term. One has been done for you.

Term	Definition	How I'm Going to Remember the Meaning
Endocrine gland	Gland that releases hormones directly into the blood	An endocrine gland releases hormones into the blood.
Exocrine gland	Gland that releases substances out of the body or into the digestive system	Exocrine sounds a lot like exit. The exocrine system releases substances that exit the body.
Hormone	Chemical released by a gland that delivers messages throughout the body	Hermes was the messenger of the gods in Greek mythology. Hormone and Hermes both start with H.
Prostaglandin	Modified fatty acids produced by many cells	A gland releases hormones. A prostaglandin is a fatty acid that is like a hormone.
Target cell	A cell that has receptors for a certain hormone	A hormone targets a cell because the cell has receptors for that hormone.

B. As you work through this lesson, you may find these terms in the activities. When you need to write a key term or a definition, **highlight** the term or the definition.

Note: Students should highlight terms and definitions used throughout the chapter.

BUILD Understanding

Compare/Contrast Table Use a compare/contrast table to see the similarities and differences between steroid hormones and nonsteroid hormones.

Steroid Hormones	Nonsteroid Hormones
lipids, crosses cell membranes, changes the pattern of gene expression, long-lasting effects, dramatic changes	cannot cross cell membranes, binds to cell receptors, secondary messengers, powerful effects

Lesson 34.1 • Workbook B • Copyright © by Pearson Education, Inc., or its affiliates. All Rights Reserved.
524

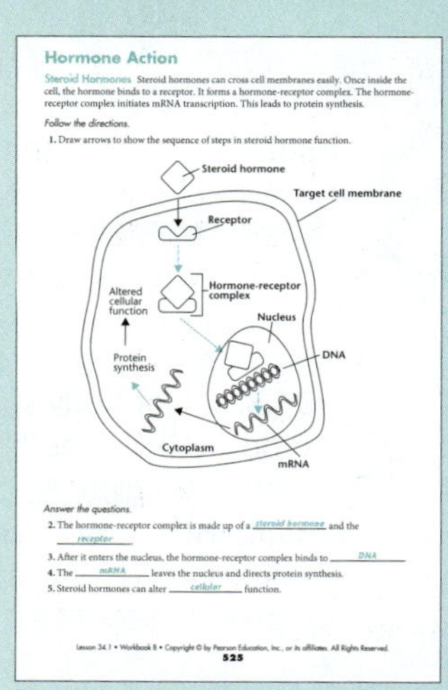

Hormone Action

Steroid Hormones Steroid hormones can cross cell membranes easily. Once inside the cell, the hormone binds to a receptor. It forms a hormone-receptor complex. The hormone-receptor complex initiates mRNA transcription. This leads to protein synthesis.

Follow the directions.

1. Draw arrows to show the sequence of steps in steroid hormone function.

Answer the questions.

2. The hormone-receptor complex is made up of a steroid hormone and the receptor
3. After it enters the nucleus, the hormone-receptor complex binds to DNA
4. The mRNA leaves the nucleus and directs protein synthesis.
5. Steroid hormones can alter cellular function.

Lesson 34.1 • Workbook B • Copyright © by Pearson Education, Inc., or its affiliates. All Rights Reserved.
525

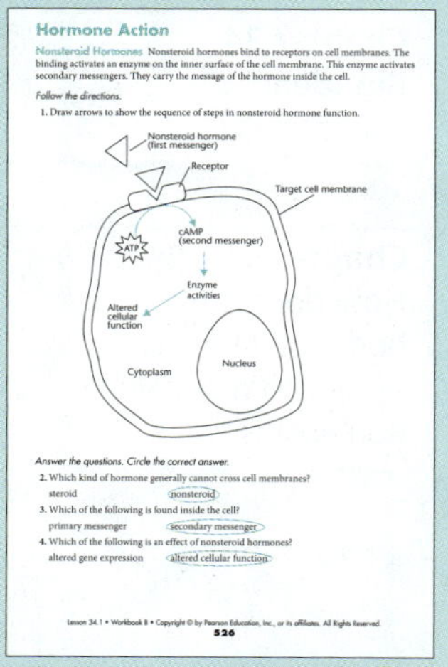

Hormone Action

Nonsteroid Hormones Nonsteroid hormones bind to receptors on cell membranes. The binding activates an enzyme on the inner surface of the cell membrane. This enzyme activates secondary messengers. They carry the message of the hormone inside the cell.

Follow the directions.

1. Draw arrows to show the sequence of steps in nonsteroid hormone function.

Answer the questions. Circle the correct answer.

2. Which kind of hormone generally cannot cross cell membranes?
   steroid nonsteroid
3. Which of the following is found inside the cell?
   primary messenger secondary messenger
4. Which of the following is an effect of nonsteroid hormones?
   altered gene expression altered cellular function

Lesson 34.1 • Workbook B • Copyright © by Pearson Education, Inc., or its affiliates. All Rights Reserved.
526

Prostaglandins Endocrine glands were once believed to be the only organs that produced hormones. However, nearly all cells produce small amounts of hormonelike substances called **prostaglandins** (prahs tuh GLAN dinz). Prostaglandins are modified fatty acids that are produced by a wide range of cells. They are sometimes called "local hormones" because they generally affect only nearby cells and tissues.

Key Question What are the components of the endocrine system?
The endocrine system is made up of glands that release hormones into the blood. Hormones deliver messages throughout the body.

Major Endocrine Glands
Endocrine glands produce hormones that affect many parts of the body.

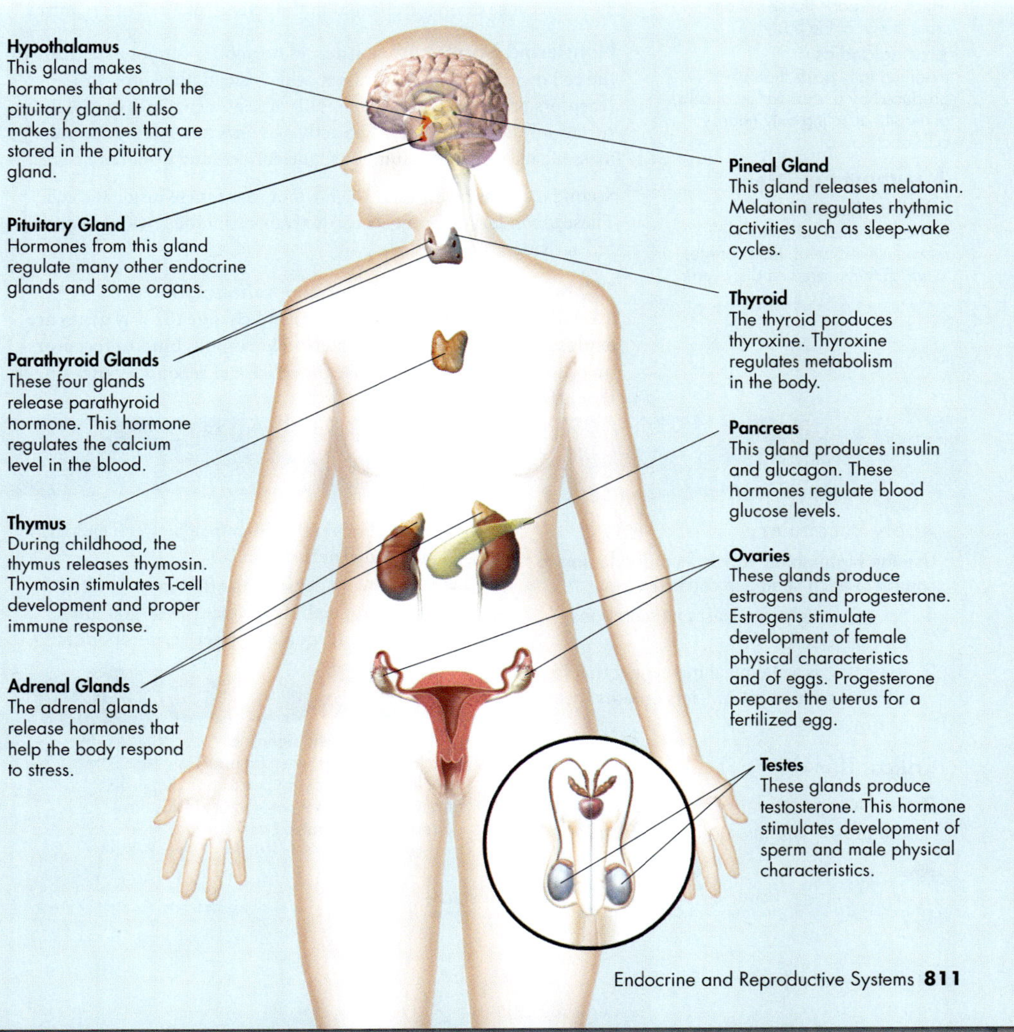

Focus on Academic Language

• regulate • control • stimulate

Write the words on the board. Ask students if they know how the words *regulate* and *control* are related. *(They are synonyms.)* Then ask volunteers to give examples that illustrate what it means to regulate or control something. *(Sample answer: A spigot controls, or regulates, the flow of water from a faucet.)*

Write the following sentences on the board:

The pancreas produces insulin and glucagon, which regulate the level of glucose in the blood.

The thymus releases thymosin, which stimulates T cell development.

Have students share ideas about how the meanings of *regulate* and *stimulate* differ. *(When you regulate something, you control it. When you stimulate something, you cause it to happen.)* If students are unsure, have them look up *stimulate* in a dictionary and use the word in a sentence.

During Reading

Active Reading

VISUAL ANALOGY Make sure students understand the analogy that compares the endocrine system to a radio station.

Ask Why does the image on the previous page show a radio station tower? *(The image is supporting an analogy between the endocrine system and a radio station.)*

Ask How are a radio station and the endocrine system alike? *(They both broadcast messages to a specific "audience.")*

USE VISUALS Write on the board a list of the glands shown in the Major Endocrine Glands diagram. Point to the name of each gland, pronounce it, and have students locate it in the diagram. Ask a volunteer to read the brief description of each gland. Help students with terms within the descriptions.

Ask Which endocrine gland helps the body respond to stress? *(adrenal gland)* Which endocrine gland produces hormones that regulate the amount of glucose in the blood? *(pancreas)*

Suggest students refer back to this diagram as they read about the glands in the following lessons.

KEY QUESTION Invite a volunteer to read aloud the question and its answer.

Ask What are glands? *(organs that produce and release secretions)* What are hormones? *(chemical messengers)*

Speed Bump

Ask What are the secretions released by endocrine glands called? *(hormones)*

Ask How do hormones travel to different parts of the body? *(through the blood)*

Ask If a cell has receptors for a particular hormone, what is the cell called? *(a target cell)*

During Reading continued

Active Reading

DRAW A DIAGRAM Be sure students complete their Workbook table that compares and contrasts steroid and nonsteroid hormones. Then, ask volunteers to use the information in their tables to draw two diagrams on the board that show the steps by which the two types of hormones affect their target cells.

Ask Why are steroid hormones able to cross cell membranes but nonsteroid hormones are not? *(Steroid hormones are lipids, but nonsteroid hormones are not. Lipids can pass through the cell membrane's lipid bilayer.)*

KEY QUESTION Invite a volunteer to read aloud the question and its answer.

Build Vocabulary

WORD ORIGINS Tell students that the word part *-crine* in the words *endocrine* and *exocrine* comes from a Greek word that means "to separate."

Ask By looking at the definitions for the two words, can you see what they have to do with the definition "to separate"? *(Sample answer: When the glands release secretions, they separate that matter from the rest of their contents.)*

Post Reading

ASSESSMENT TIPS For Question 3, have students make a compare/contrast table for hormones and prostaglandins.

For Question 4, have students review the meanings of the prefixes *endo- (inside)* and *exo- (outside)*.

MYSTERY CLUE **Ask** Do you think Lisa's fat tissue sent signals to her hypothalamus? *(Yes, because her body fat percentage decreased significantly.)* How might this be related to homeostasis? *(Sample answer: If fat reserves are low, the body may not be able to maintain homeostasis.)*

Wrap-Up Activity

MAKE A CHART Have student pairs make an endocrine system chart by dividing a blank sheet into three columns and labeling them Glands, Hormones, and Functions, respectively. Students can complete the chart with information from the Major Endocrine Glands diagram. Direct students to leave space between the rows so that they can add information as they continue with the chapter.

BUILD Vocabulary

hormone
a chemical produced in one part of an organism that affects another part of the same organism

exocrine gland
a gland that releases its secretions, through tube-like structures called ducts, directly into an organ or out of the body

endocrine gland
a gland that releases its secretions (hormones) directly into the blood, which transports the secretions to other areas of the body

prostaglandins
modified fatty acids that are produced by a wide range of cells; generally affecting only nearby cells and tissues

WORD ORIGINS
Prostaglandins get their name from a gland in the male reproductive system, the prostate, in which they were first discovered.

Hormone Action

Hormones fall into two general groups—steroid hormones and nonsteroid hormones. Steroid hormones are produced from a lipid called cholesterol. Nonsteroid hormones include proteins, small peptides, and modified amino acids. Each type of hormone acts on a target cell in a different way.

Steroid Hormones Because steroid hormones are lipids, they can easily cross cell membranes. Inside the cell, a steroid hormone binds to a receptor found only in the hormone's target cells. The hormone and receptor form a hormone-receptor complex that enters the cell nucleus.

In the nucleus, the hormone-receptor complex regulates gene expression. Steroid hormones can turn whole sets of genes on or off. This can cause dramatic changes in the activity of a cell or organism.

Nonsteroid Hormones Nonsteroid hormones cannot pass through the cell membranes of their target cells. Instead, they bind to receptors on the cell membrane. This activates enzymes on the inner surface of the membranes. The activated enzymes release secondary messengers such as calcium ions, nucleotides, and even fatty acids.

Secondary messengers relay the hormone's message inside the cell. These secondary messengers can activate or inhibit a wide range of cell activities.

Key Question How do hormones affect cells?
Steroid hormones enter the nucleus and change the way genes are expressed in a target cell. Nonsteroid hormones bind to receptors on cell membranes. They cause the release of secondary messengers that affect cell activities.

CHECK Understanding

Apply Vocabulary

Use the highlighted words from the lesson to complete each sentence correctly.

1. An ______ gland releases its secretions through ducts.
2. A ______ is a chemical messenger that is released from a cell and acts on other cells nearby.

Critical Thinking

3. **Compare and Contrast** How are hormones and prostaglandins similar? Different?
4. **Explain** Explain the difference between endocrine glands and exocrine glands.
5. **Write to Learn** Answer the first mystery clue. Write a paragraph that explains why you think maintaining adequate fat reserves is important.

MYSTERY CLUE

Fat tissue may send signals to the hypothalamus when fat reserves are low. Lisa's body fat percentage dropped from 17 percent to 9 percent. Could this have affected such signals?

BIOLOGY.com Search Lesson 34.1 GO • Lesson Assessment

Check Understanding

1. exocrine
2. prostaglandin
3. Hormones and prostaglandins are chemical messengers that change a target cell's activity. Hormones are released by glands and travel through the blood to affect target cells in other parts of the body. Prostaglandins are secreted by cells, but do not travel far—they act on cells surrounding the cells that release them.
4. Endocrine glands release their secretions directly into the blood. Exocrine glands release their secretions outside the body or into the digestive tract through ducts.
5. Answers should indicate that students understand that fat tissue produces and releases hormones, and therefore a drastic drop in fat reserves could have affected the signals sent from fat tissue to the brain.

34.2 Glands of the Endocrine System

The Human Endocrine Glands

The endocrine system is different from other organ systems. Most of the glands in the endocrine system are scattered throughout the body. However, they still work as a single system, regulating a wide variety of activities.

Pituitary Gland The pituitary gland is a bean-sized gland that hangs under the base of the brain. A slender stalk of tissue attaches it to the brain. It is divided into two parts: the anterior pituitary and the posterior pituitary. The pituitary gland secretes some hormones that directly regulate many body functions. It secretes other hormones that control the actions of other endocrine glands.

Hypothalamus The hypothalamus links the central nervous system to the endocrine system. It directly and indirectly controls the secretions of the pituitary gland.

Neurons connect the posterior pituitary to the hypothalamus. The cell bodies in the hypothalamus produce two hormones, antidiuretic hormone (ADH) and oxytocin. These hormones are stored in the axons, which extend into the pituitary. When the cell bodies are stimulated, the axons release the hormones into the blood.

The hypothalamus indirectly controls the anterior pituitary with **releasing hormones.** These hormones are secreted into blood vessels that lead to the pituitary. Each releasing hormone controls a different pituitary hormone.

Key Questions

What are the functions of the major endocrine glands?

How are endocrine glands controlled?

BUILD Understanding

Concept Map As you read, develop a concept map that shows the relationship between the human endocrine glands.

In Your Workbook Go to your workbook for help in completing your concept map for Lesson 34.2.

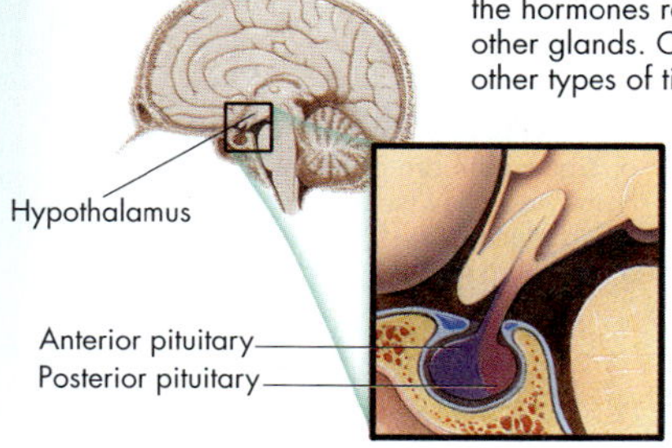

Pituitary Gland The pituitary gland is located below the hypothalamus in the brain. Some of the hormones released by the pituitary control other glands. Other hormones released affect other types of tissues.

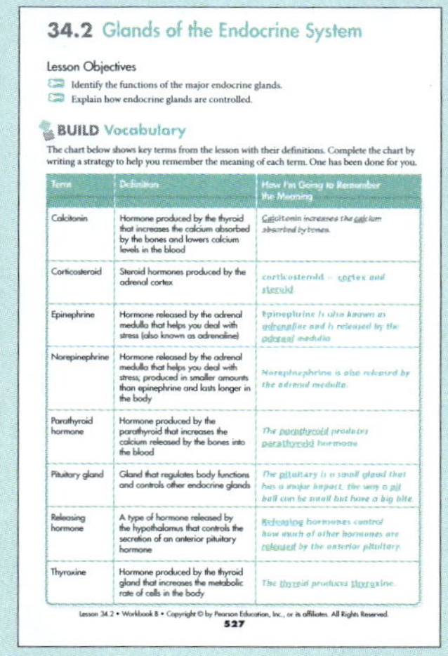

34.2 Glands of the Endocrine System

Lesson Objectives

Identify the functions of the major endocrine glands.

Explain how endocrine glands are controlled.

BUILD Vocabulary

The chart below shows key terms from the lesson with their definitions. Complete the chart by writing a strategy to help you remember the meaning of each term. One has been done for you.

Term	Definition	How I'm Going to Remember the Meaning
Calcitonin	Hormone produced by the thyroid that increases the calcium absorbed by the bones and lowers calcium levels in the blood	Calcitonin increases the calcium absorbed by bones.
Corticosteroid	Steroid hormones produced by the adrenal cortex	corticosteroid – cortex and steroid
Epinephrine	Hormone released by the adrenal medulla that helps you deal with stress (also known as adrenaline)	Epinephrine is also known as adrenaline and is released by the adrenal medulla.
Norepinephrine	Hormone released by the adrenal medulla that helps you deal with stress; produced in smaller amounts than epinephrine and lasts longer in the body	Norepinephrine is also released by the adrenal medulla.
Parathyroid hormone	Hormone produced by the parathyroid that increases the calcium released by the bones into the blood	The parathyroid produces parathyroid hormone.
Pituitary gland	Gland that regulates body functions and controls other endocrine glands	The pituitary is a small gland that has a major impact, the way a pit bull can be small but have a big bite.
Releasing hormone	A type of hormone released by the hypothalamus that controls the secretion of an anterior pituitary hormone	Releasing hormones control how much of other hormones are released by the anterior pituitary.
Thyroxine	Hormone produced by the thyroid gland that increases the metabolic rate of cells in the body	The thyroid produces thyroxine.

527

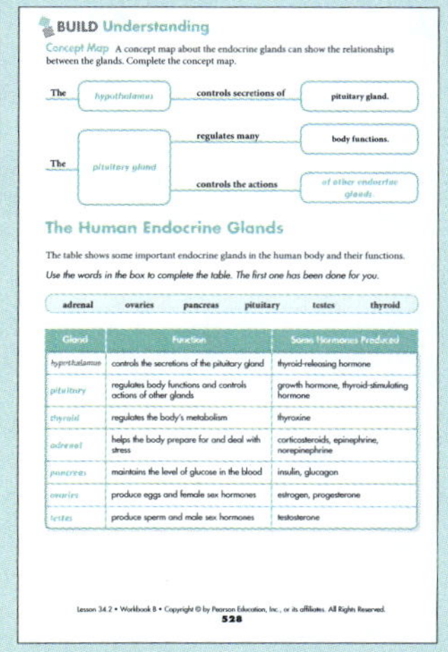

BUILD Understanding

Concept Map A concept map about the endocrine glands can show the relationships between the glands. Complete the concept map.

The [hypothalamus] controls secretions of [pituitary gland.]

The [pituitary gland] regulates many [body functions.]

The [pituitary gland] controls the actions [of other endocrine glands.]

The Human Endocrine Glands

The table shows some important endocrine glands in the human body and their functions.

Use the words in the box to complete the table. The first one has been done for you.

adrenal ovaries pancreas pituitary testes thyroid

Gland	Function	Some Hormones Produced
hypothalamus	controls the secretions of the pituitary gland	thyroid-releasing hormone
pituitary	regulates body functions and controls actions of other glands	growth hormone, thyroid-stimulating hormone
thyroid	regulates the body's metabolism	thyroxine
adrenal	helps the body prepare for and deal with stress	corticosteroids, epinephrine, norepinephrine
pancreas	maintains the level of glucose in the blood	insulin, glucagon
ovaries	produce eggs and female sex hormones	estrogen, progesterone
testes	produce sperm and male sex hormones	testosterone

528

Pre-Reading

Objectives

34.2.1 Identify the functions of the major endocrine glands.

34.2.2 Explain how endocrine glands are controlled.

Preview the Pages

1. Have students read aloud the blue headings under the first green heading. Guide them to note that they are the names of endocrine glands. Tell students that they will learn more about these glands and that they can add this information to the Glands, Hormones, and Functions chart they made for the Wrap-Up Activity in the previous lesson.
2. Have students preview the visuals. Relate the drawings of the pituitary, thyroid, and parathyroid glands to the Major Endocrine Glands diagram in the previous lesson.
3. This lesson includes several terms that students have studied in previous chapters, including *organ, system, sympathetic nervous system, glucose, glycogen, metabolism, gonads,* and *feedback mechanism.* List these terms on the board and have volunteers define or describe each, using the glossary as necessary.

Build Understanding

CONCEPT MAP Direct students to Workbook B, where they will find a concept map designed for this lesson. Students should fill in the concept map that shows the relationship between glands.

NATIONAL SCIENCE EDUCATION STANDARDS

UNIFYING CONCEPTS AND PROCESSES

I, IV, V

CONTENT

C.1.d, C.5.d, F.1

For the corresponding lesson in *Miller & Levine Biology,* see pages 982–987.

During Reading

Active Reading

SET PURPOSE FOR READING Students' understanding in this lesson depends strongly on their recall of terms and concepts studied earlier. Students may need to review carbohydrate metabolism and storage, digestion, and kidney function.

Ask What happens to glucose that is absorbed during digestion? *(It is broken down in cells, releasing energy.)* What can happen to glucose that is not needed immediately for energy? *(It is stored as glycogen or fat.)*

CAUSE AND EFFECT Make sure students understand the cause-and-effect relationships between food, insulin, glucagon, and glucose.

Say Look closely at the Blood Glucose Control diagram.

Ask What causes blood glucose levels to rise? *(eating food)* Why does the pancreas release glucagon between meals? *(the glucose levels have dropped)*

Ask How does this cause-and-effect situation relate to the question for this chapter? *(The cause and effect situation is in place to maintain homeostasis of blood glucose levels.)*

Build Vocabulary

PREFIXES Tell students that the prefix *epi-* means "above," and the root *-nephrine* refers to nephron, or kidney. So, the word *epinephrine* means "above the kidney."

Ask Which endocrine gland produces epinephrine? *(adrenal gland)* Where are the adrenal glands located? *(above the kidneys)*

Speed Bump

Ask What two hormones does the adrenal medulla release when the body is under stress? *(epinephrine and norepinephrine)* How do these hormones affect the body's heart rate and blood pressure? *(They cause both heart rate and blood pressure to increase.)*

Ask What two hormones does the pancreas release? *(insulin and glucagon)* What effect does each hormone have on blood glucose levels? *(Insulin causes blood glucose levels to decrease; glucagon causes them to increase.)*

BUILD Vocabulary

releasing hormone
a hormone produced by the hypothalamus that makes the anterior pituitary secrete hormones

corticosteroid
a steroid hormone produced by the adrenal cortex

epinephrine
a hormone released by the adrenal glands that increases heart rate and blood pressure and prepares the body for intense physical activity; also called adrenaline

norepinephrine
a hormone released by the adrenal glands that increases heart rate and blood pressure and prepares the body for intense physical activity

PREFIXES
The prefix *nor-* means "parent of." Scientists use this prefix to indicate when one compound is made from another compound. This prefix tells us that the body makes epinephrine from norepinephrine.

Adrenal Glands The adrenal glands, which sit on top of the kidneys, have two layers. The outer layer is the adrenal cortex. It produces a group of steroid hormones called **corticosteroids** (kawr tih koh STEER oydz). Aldosterone (al DAHS tuh rohn) regulates blood volume and pressure. The adrenal cortex releases aldosterone when blood volume decreases. Cortisol helps control the metabolism of carbohydrates, fats, and proteins. The adrenal gland releases cortisol during physical stress, such as intense exercise.

The inner part is the adrenal medulla. It produces **epinephrine** and **norepinephrine.** When you are under stress, the sympathetic nervous system sends impulses to the adrenal medulla. The adrenal medulla releases large amounts of these hormones, causing heart rate and blood pressure to rise. Epinephrine and norepinephrine also widen air passageways in the lungs and cause the release of extra glucose.

Pancreas The pancreas is an exocrine gland that releases digestive enzymes to help break down food. The pancreas is also an endocrine gland. It has clusters of cells that release hormones into the blood. These clusters, called the "islets of Langerhans," contain both beta cells and alpha cells. Beta cells secrete the hormone insulin. Alpha cells secrete the hormone glucagon.

▶ ***Blood Glucose Regulation*** Insulin and glucagon work together to keep blood glucose levels stable. Blood glucose levels rise after you eat, which causes the pancreas to release insulin. Insulin causes cells to take glucose out of the blood. Liver cells, skeletal muscle cells, and fat tissue store the glucose. Storing glucose in these cells keeps blood glucose from rising too rapidly. Liver and muscle cells store the glucose as glycogen, while the glucose in fat is converted to lipids.

Blood glucose levels drop within one or two hours after a person has eaten. As a result, the pancreas releases glucagon. Glucagon stimulates the liver and skeletal muscle cells to break down glycogen and release glucose into the blood. Glucagon also causes fat cells to break down fats and convert them to glucose. These actions help raise the blood glucose to normal levels.

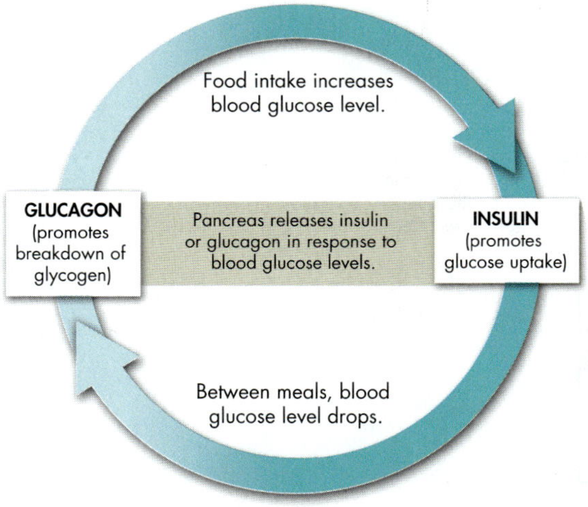

Blood Glucose Control Insulin and glucagon are opposing hormones. They make sure that blood glucose levels stay within a normal range.

ELL Focus on ELL: Access Content

ALL SPEAKERS On small sticky notes, have students write words on this page that they are having trouble reading. Have them place their sticky notes in the margin across from where each word appears. Then have students work in pairs to restate sentences in their own words, using the words they had trouble with.

▶ ***Diabetes Mellitus*** Diabetes mellitus occurs when the body fails to produce or properly respond to insulin. Diabetes causes very high blood glucose levels. Very high blood glucose levels can damage almost every system and cell in the body.

There are two types of diabetes mellitus. Type I diabetes is an autoimmune disorder that destroys beta cells. Type I diabetes usually develops in people before age 15. People with Type I secrete little to no insulin. Type I diabetics must follow a strict diet and get a daily dose of insulin.

People with Type II diabetes produce insulin, but it does not bind to insulin receptors properly. Type II usually starts in people after age 40. However, it is increasing rapidly in young people due to childhood obesity. In its early stages, Type II diabetes can often be controlled through diet and exercise.

Thyroid and Parathyroid Glands The thyroid gland is located in the neck and wraps around the upper part of the trachea. It plays a major role in regulating the body's metabolism and blood calcium levels. It produces two hormones: **thyroxine** and *calcitonin*. Thyroxine stimulates cells to become more active, use more energy, and produce more heat. Iodine is needed to make thyroxine. An iodine deficiency causes the thyroid to enlarge. This is called a goiter.

Calcitonin reduces blood calcium levels by signaling the kidneys to reabsorb less calcium from filtrate. It also inhibits calcium absorption from the small intestine and promotes absorption into bones. Its opposing hormone is parathyroid hormone (PTH), which is produced by the parathyroid glands. These glands are located on the back side of the thyroid. PTH increases the release of calcium from the bone. It also increases calcium reabsorption in the kidneys, and calcium uptake in the digestive system.

The Reproductive Glands The gonads—ovaries and testes—are the body's reproductive glands. The gonads serve two important functions: the production of gametes and the secretion of sex hormones. You will learn more about the gonads and their hormones in the next lesson.

Key Question What are the functions of the major endocrine glands?

The pituitary gland secretes hormones that directly regulate many body functions or control other endocrine glands. The hypothalamus controls the secretions of the pituitary gland. The thyroid gland helps regulate the body's metabolic rate. The gonads produce gametes and secrete sex hormones.

BUILD Vocabulary

thyroxine
a hormone produced by the thyroid gland, which increases the metabolic rate of cells throughout the body

ACADEMIC WORDS
Hormones such as thyroxine have very long official names based on their chemical structure. Scientists sometimes give these hormones shorter names for everyday use. The official chemical name for thyroxine is *O*-(4-Hydroxy-3, 5-diiodophenyl)-3,5-diiodo-L-tyrosine.

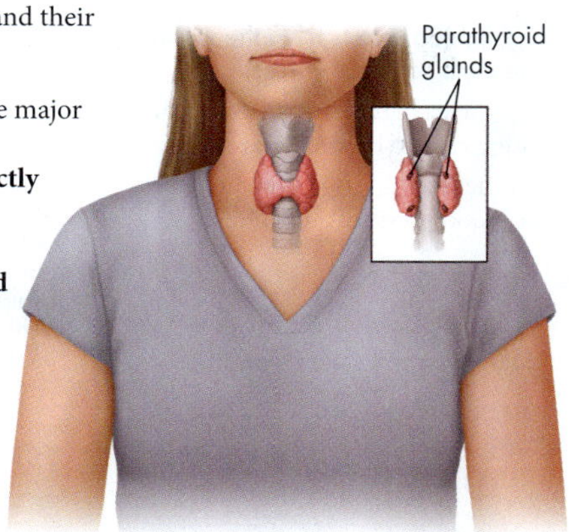

Thyroid Gland The thyroid makes thyroxine, which speeds up metabolism and helps control growth and development.

ELL Focus on ELL: Build Background

ALL SPEAKERS Some students may have a family member or friend who has Type II diabetes. Discuss with students the recommended diet for a person with diabetes. Point out that exercise and proper diet can help a person avoid diabetes.

Ask How does eating high-Calorie foods, such as chips, cookies, cakes, and ice cream, affect the blood glucose level of a person with Type II diabetes? *(Blood glucose increases.)*

Ask Why does the blood glucose level of a person with Type II diabetes remain high after eating rather than decreasing to normal levels? *(The insulin secreted by the pancreas does not properly bind to the insulin receptors.)*

Active Reading

COMPARE/CONTRAST As students read about diabetes, have them record the similarities and differences between Type I and Type II diabetes in a compare/contrast table.

MAKE INFERENCES After students read about goiter, point out that very few cases of goiter occur in the United States.

Ask Why do you think goiter is uncommon in the United States? *(People get enough iodine in their diets.)*

INTERPRET VISUALS In the drawing showing the thyroid and parathyroid glands, point out that the parathyroid glands lie on the back side of the thyroid pictured in the girl's throat.

KEY QUESTION Read the Key Question aloud. Have students use their Glands, Hormones, and Functions chart to follow along as a volunteer reads the answer.

Build Vocabulary

ACADEMIC WORDS Point out to students that the names of some hormones are shortened even further if the names consist of more than two words. Have students look for examples on this page and the next.

Ask What is the shortened name for parathyroid hormone? *(PTH)* For antidiuretic hormone? *(ADH)*

Speed Bump

Ask Which type of diabetes can be controlled with diet and exercise? *(Type II)*

Ask Which thyroid hormone stimulates the activity of cells? *(thyroxine)*

Ask How do calcitonin and parathyroid hormone work in opposing ways to control blood calcium levels? *(Calcitonin causes blood calcium levels to decrease. Parathyroid hormone causes blood calcium levels to increase.)*

LESSON 34.2

During Reading continued

Active Reading

MAKE AN ANALOGY Remind students of the analogy in Chapter 30 that compares feedback mechanisms in the body to a home-heating system.

Say A room thermostat is similar to the hypothalamus in that it senses changes and responds to them.

Ask How does a thermostat respond to changes? *(A thermostat senses a rise or fall in room temperature and sends a signal to the furnace to turn on or off.)*

Ask How is this similar to the action of the hypothalamus to maintain water balance? *(Sample answer: When the hypothalamus senses that water levels are low or high, it signals the kidneys, via the posterior pituitary, to regain water balance.)*

KEY QUESTION Invite a volunteer to read aloud the question and its answer. Have students study the Water Balance feedback loop for an example of how the hypothalamus and posterior pituitary regulate water balance.

Ask What are two ways that the body reacts to excessive water loss? *(Sample answer: thirst and decreased kidney action)*

Post Reading

ASSESSMENT TIPS For Questions 1 and 2, have students refer to their Glands, Hormones, and Functions chart.

For Question 3, remind students that *compare and contrast* means that they should tell how two things are similar as well as different.

MYSTERY CLUE Suggest that they reread the Adrenal Glands section to recall details relating to *cortisol*.

Ask When does the adrenal cortex release cortisol? *(during physical stress, such as intense exercise)* How is the release of cortisol helpful to a person who is exercising vigorously? *(More calcium is available for the person's skeletal muscles.)*

Wrap-Up Activity

COMPOSE A TEST Have students form groups of four. Direct groups to write a multiple-choice or fill-in-the blank question for each of the glands named in the blue headings. Have groups exchange and complete the tests. Tell them to return completed tests to check answers. Suggest that they discuss any questions or answers that are unclear.

Control of the Endocrine System

Like most other body systems, the endocrine system is regulated by feedback mechanisms, which help the body maintain homeostasis.

When you exercise heavily, you lose water as sweat. If water loss continued, you would become dehydrated. As you lose water, the blood becomes more concentrated. The hypothalamus responds by signaling the posterior pituitary to release antidiuretic hormone (ADH). ADH signals the kidneys to remove less water from the blood. Later, you experience a sensation of thirst. Thirst is a signal to replace lost water.

Water Balance
The hypothalamus and posterior pituitary use a feedback mechanism to regulate water balance.

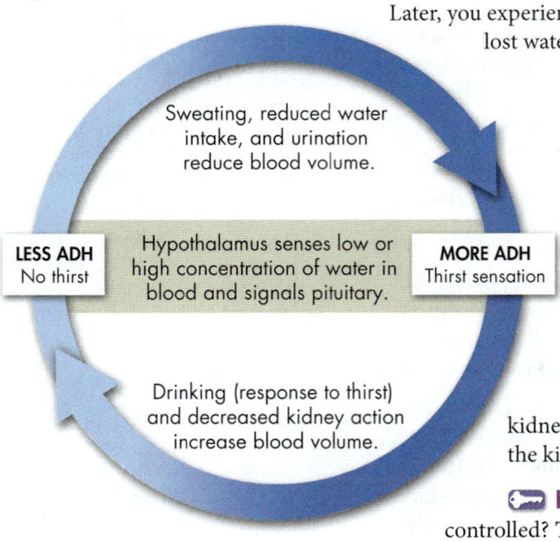

When you finally drink, you might take in a liter of water. Most of the water is absorbed by the blood. This rapid absorption could dilute the blood too quickly. Large amounts of water would diffuse out of the blood into body tissues. Body cells would swell with the excess water. The hypothalamus prevents this by signaling the pituitary to release less ADH when the blood becomes dilute. The kidneys respond by removing more water from the blood. So, the homeostatic system sets limits for water in the blood. When blood has too little water, ADH is released and the kidneys conserve water. An oversupply of water causes the kidneys to eliminate excess water in the urine.

Key Question How are endocrine glands controlled? **The endocrine system is regulated by feedback mechanisms that function to maintain homeostasis.**

CHECK Understanding

Apply Vocabulary

Use the highlighted words from the lesson to complete each sentence correctly.

1. ________ helps regulate metabolism by stimulating cells to become more active.
2. Aldosterone is a ________ that helps regulate blood pressure and volume.

Critical Thinking

3. **Compare and Contrast** Compare and contrast the two types of diabetes.
4. **Explain** Describe how the hypothalamus controls the secretions of the posterior pituitary gland.
5. **Write to Learn** Answer the second mystery clue.

MYSTERY CLUE

One effect of cortisol is the release of calcium from bones into the blood. This makes the calcium available for skeletal muscles. How could this effect of cortisol have contributed to Lisa's condition? (**Hint:** See p. 814.)

BIOLOGY.com Search Lesson 34.2 GO • Lesson Assessment

Check Understanding

1. Thyroxine
2. corticosteroid
3. Both result in difficulty regulating blood sugar levels. Type I diabetes occurs when the immune system destroys beta cells, preventing the pancreas from making insulin. The disease usually begins before age 15 and daily doses of insulin are required. In Type II diabetes, the insulin receptors do not respond properly to insulin. It usually begins after age 40 and can often be controlled by diet and exercise.
4. Axons of some neurons in the hypothalamus extend into the posterior pituitary. ADH and oxytocin are produced in the cell bodies and stored in the axons. When the cell bodies are stimulated, the hormones are released from the posterior pituitary.
5. Answers should indicate an understanding that cortisol would have made Lisa's bones weak and at risk for stress fractures.

34.3 The Reproductive System

Sexual Development

The reproductive system may be the only system in the body that isn't necessary for the survival of an individual. But without the reproductive system, the human species itself would not survive. In some ways, this makes the reproductive system the most important system in the body.

Male and female human embryos look almost identical until they are about seven weeks old. Then male gonads begin secreting testosterone, causing the male reproductive system to develop. In females, the gonads produce estrogens instead of testosterone. Estrogens cause the female reproductive system to develop.

Testosterone and estrogens are steroid hormones that are primarily produced in the gonads. They control sexual development of the human embryo. They also act on other cells and tissues. Testosterone causes facial hair to grow and increases muscular development in males. It also causes the male voice to deepen. Estrogens stimulate breast development and widening of the hips in females.

The testes and the ovaries cannot produce active reproductive cells until puberty. Puberty is a period of rapid growth and sexual maturation. The reproductive system becomes fully functional during puberty, which usually occurs between ages 9 and 15. Puberty begins in the brain, when the hypothalamus signals the release of two hormones from the pituitary. These hormones are follicle-stimulating hormone (FSH) and luteinizing hormone (LH).

Key Question What effects do estrogens and testosterone have on females and males?
Estrogens stimulate breast development and widening of the hips in females. Testosterone stimulates the growth of facial hair, muscular development, and deepening of the voice in males.

The Male Reproductive System

The release of LH stimulates cells in the testes to produce increased amounts of testosterone. Testosterone causes the male physical changes associated with puberty. Testosterone also works with FSH to stimulate sperm development. When puberty is finished, the reproductive system can produce and release active sperm.

Key Questions

What effects do estrogens and testosterone have on females and males?

What are the main functions of the male reproductive system?

What are the main functions of the female reproductive system?

What are some of the most commonly reported sexually transmitted diseases?

BUILD Understanding

Preview Visuals Before you read, preview the diagram called Sperm. Think about how you would describe this diagram.

In Your Workbook Go to your workbook for help in previewing this visual.

34.3 The Reproductive System

Lesson Objectives

- Describe the effects the sex hormones have on development.
- Name and discuss the structures of the male reproductive system.
- Name and discuss the structures of the female reproductive system.
- Describe some of the most common sexually transmitted diseases.

BUILD Vocabulary

A. The chart below and on the next page shows key terms from the lesson with their definitions. Complete the chart by writing a strategy to help you remember the meaning of each term. One has been done for you.

Term	Definition	How I'm Going to Remember the Meaning
Corpus luteum	The ruptured follicle, which turns yellow and releases progesterone after ovulation	*Corpus luteum means "yellow body" in Latin, and after cars go at a green traffic light, the light turns yellow.*
Epididymis	Place where sperm mature and are stored	*I'll picture a storage shed with the word epididymis on the door.*
Menstrual cycle	Regular sequence of events involving the ovaries, lining of the uterus, and the endocrine system	*The menstrual cycle is a series of events that includes menstruation.*
Menstruation	Phase of the menstrual cycle in which the lining of the uterus, the unfertilized egg, and blood are discharged through the vagina	*Menstruation is a part of the menstrual cycle.*
Ovary	One of the primary female reproductive organs	*An O is egg-shaped. I can remember the ovary is where the egg comes from.*
Ovulation	Release of an egg from the ovary	*Ovulation is when the egg is released from the ovary.*
Puberty	Time of rapid growth when the reproductive system becomes fully functional	*Puberty happens during the teen years. The teen years are when people begin to mature physically.*
Scrotum	External sac that contains the testes	*The first two letters of scrotum are S and C and it is a sac.*

Continued on next page ▶

Term	Definition	How I'm Going to Remember the Meaning
Semen	Combination of sperm and seminal fluid	*Semen includes seminal fluid.*
Seminiferous tubule	One of hundreds of tiny tubes in which sperm are produced	*Tubule sounds like a tiny tube.*
Sexually transmitted disease	Disease that spreads through sexual contact	*This is any disease spread by sex.*
Testis	One of the primary male reproductive organs (plural: testes)	*Males have two testes.*
Vas deferens	Tube that extends from the scrotum to the abdominal cavity	*Sometimes men have a vasectomy to avoid having children, because the vas deferens is what takes sperm out of the body.*

B. As you work through this lesson, you may find these terms in the activities. When you need to write a key term or a definition, **highlight** the term or the definition.

BUILD Understanding

Previewing Visuals The visuals in a lesson provide helpful information. Visuals include photographs, maps, charts, graphs, illustrations, and timelines. Before you begin reading Lesson 3, preview the diagram entitled sperm. Then fill in the chart below.

Visual	Description
Sperm	*A sperm has three major parts: a head, a midpiece, and a tail. The head contains the nucleus in which genetic material is found.*

The Male Reproductive System

Sperm are produced in the testes and mature in the epididymis. To leave the body, they travel through the vas deferens and the urethra. Glands, including the seminal vesicles, produce seminal fluid that nourishes and protects the sperm.

Follow the directions.

1. Use the words in the box to label the diagram.

epididymis	penis	seminal vesicle
testis	urethra	vas deferens

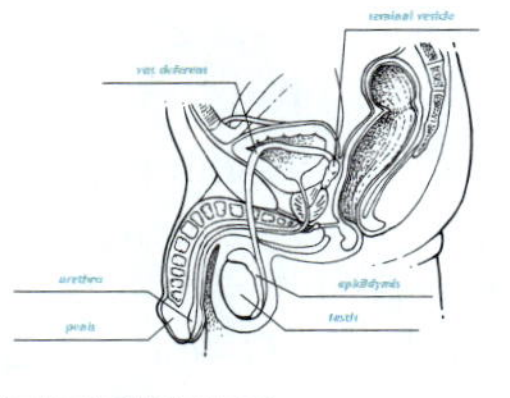

Answer the questions. Circle the correct answer.

2. Through what structure does the urethra pass?
testis — penis (circled)

3. Through which structure do sperm pass?
seminal vesicle — vas deferens (circled)

LESSON 34.3

Pre-Reading

Objectives

34.3.1 Describe the effects the sex hormones have on development.

34.3.2 Name and discuss the structures of the male reproductive system.

34.3.3 Name and discuss the structures of the female reproductive system.

34.3.4 Describe some of the most common sexually transmitted diseases.

Preview the Pages

1. Some students may find it difficult to maintain appropriate behavior during discussions of the reproductive system. Tell students you will use the phrase "think like a scientist" to remind them to use appropriate vocabulary and behavior during discussions of the reproductive system.
2. Preview the visuals and read the captions and labels. Be sure students recognize that the Male Reproductive System and the Female Reproductive System visuals show both front and side views and that the photographs showing ovulation and the bacteria that cause chlamydia are microscopic views.

Build Understanding

PREVIEW VISUALS Direct students to Workbook B, where they will find a preview visuals chart. Students are asked to preview the Sperm visual and fill in a description. You may want them to add to this chart with descriptions of other visuals in the lesson.

NATIONAL SCIENCE EDUCATION STANDARDS

UNIFYING CONCEPTS AND PROCESSES

I, V

CONTENT

C.2.b, F.1

INQUIRY

A.1.c, A.2.a

For the corresponding lesson in ***Miller & Levine Biology,*** **see pages 988–994.**

During Reading

Active Reading

KEY QUESTION Read aloud the Key Question and answer from the previous page.

Ask Where are estrogens and testosterone primarily produced? *(Estrogens are produced in the ovaries. Testosterone is produced in the testes.)*

INTERPRET DIAGRAMS Have students compare the front and side views of the male reproductive system.

Ask What does the side view show about the vas deferens and the urethra that is not visible in the front view? *(The vas deferens joins with the urethra.)*

MAKE INFERENCES Ask a volunteer to read aloud the first paragraph.

Say Some couples who want a baby have trouble getting pregnant. Sometimes doctors advise the man to wear loose clothing that is lightweight.

Ask Why might a doctor give this advice? *(so that the testes will not be as warm, thereby allowing more sperm to develop properly)*

Build Vocabulary

MULTIPLE MEANINGS Tell students that other words on this page with multiple meanings include *organ*, *cell*, and *cavity*. Have students find the words and look up their various meanings in a dictionary. Ask volunteers to make a sentence for each meaning.

Speed Bump

Ask What reproductive organs produce sperm? *(testes)* Where are the sperm stored while they mature? *(in the epididymis)*

Ask Through which tube does sperm travel from the epididymis to the urethra? *(the vas deferens)*

Ask What is semen? *(combination of sperm and seminal fluid)*

Ask Through which tube does semen travel through the penis? *(urethra)*

BUILD Vocabulary

testis
the primary male reproductive organ; produces sperm

ovary
in animals, the primary female reproductive organ; produces eggs

MULTIPLE MEANINGS
In plants, an ovary is a structure that surrounds and protects seeds.

Testes The **testes** (singular: testis) are the primary male reproductive organs. Just before or just after birth, they descend from the abdomen into an external sac called a *scrotum*. This external sac is necessary because sperm must be cooler than body temperature (37° Celsius) to develop properly.

Sperm Development Inside each testis are clusters of hundreds of tiny tubules called seminiferous (sem uh NIF ur us) tubules. Sperm develop inside these tubules from specialized diploid cells. The diploid cells divide by meiosis to form the haploid nuclei found in mature sperm. A haploid cell contains only a single set of chromosomes.

Sperm move from the seminiferous tubules into the epididymis (ep uh DID ih mis) to mature and be stored. Some of the sperm move from the epididymis into another tube called the vas deferens. The vas deferens extends upward from the scrotum into the abdominal cavity. Here it merges with the urethra, which leads to the outside of the body through the penis.

Glands lining the reproductive tract produce seminal fluid. This fluid provides nutrients for the sperm. It also protects them from the acidity of the female reproductive tract. The combination of seminal fluid and sperm is called *semen*.

Sperm Release When the male is sexually aroused, the penis becomes erect. Contractions of smooth muscles in the reproductive tract then eject semen from the penis. This process is called ejaculation. Ejaculation is regulated by the autonomic nervous system, so it is not completely voluntary. If ejaculation takes place in or near the female reproductive tract, sperm may enter and successfully fertilize an egg.

Male Reproductive System The main structures of the male reproductive system produce and deliver sperm.

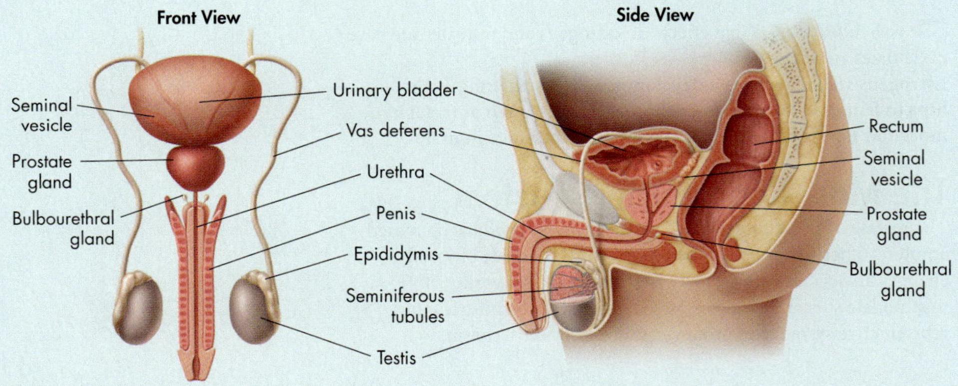

Focus on Academic Language

- testosterone • scrotum • meiosis • urethra • sperm • puberty
- diploid cell

Point out to students that one way to remember the meaning of a word is to classify it. For example, the *testis* is an organ, the *vas deferens* is a tube, and so on.

Write the following classifications on the board: Hormone, Process, and Structure. Discuss with students what each classification means. Then ask students to classify the following words: *testosterone (H), scrotum (S), meiosis (P), urethra (S), sperm (S), puberty (P),* and *diploid cell (S)*. Invite a volunteer to write each word under the appropriate heading.

Sperm Structure The head of a mature sperm cell contains a highly condensed nucleus. Its midpiece is packed with mitochondria that provide energy to the sperm. The tail (flagellum) moves the sperm forward. The tip of the head has a small cap that contains enzymes needed for fertilization.

Key Question What are the main functions of the male reproductive system?

When puberty is complete, the male reproductive system is fully functional, meaning that the male can produce and release active sperm.

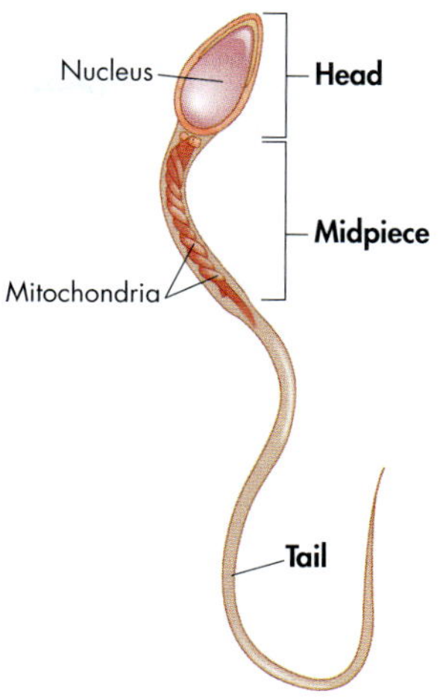

Sperm Sperm have large numbers of mitochondria. The mitochondria provide power for a sperm cell's trip through the female reproductive tract. If a sperm reaches an egg, enzymes in its cap can break down the egg's outer layer.

The Female Reproductive System

The **ovaries** are a female's primary reproductive organs. When puberty starts, FSH stimulates cells inside the ovaries to produce more estrogens. FSH also signals the ovaries to start producing egg cells.

Female Reproductive Structures Each ovary contains up to 400,000 primary follicles. A follicle is a cluster of cells surrounding a single egg, or ovum. The follicle helps the egg mature. Then the egg is released into the reproductive tract, where it can be fertilized by a sperm. Only about 400 mature eggs are released in a female's lifetime.

Other structures in the female reproductive system include the Fallopian tubes, uterus, cervix, and the vagina.

Female Reproductive System The main function of the female reproductive system is to produce eggs. The ovaries are the main organs of the female reproductive system.

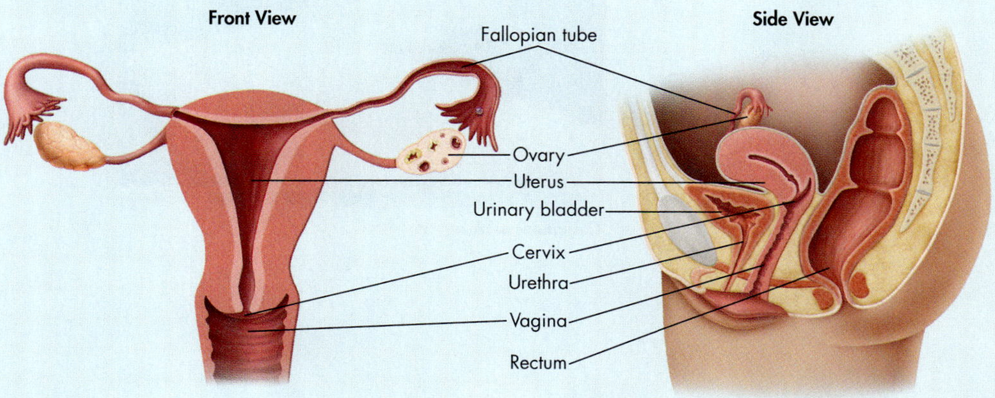

LESSON 34.3

Active Reading

DRAW CONCLUSIONS Ask a student volunteer to read aloud the caption describing a sperm cell. Direct students' attention to the shape of the cell.

Ask How is the shape helpful to the sperm cell? *(Its long, thin shape allows it to move easily through the female reproductive tract.)*

KEY QUESTION Invite a volunteer to read aloud the question and its answer.

Ask What do the testes produce once a male has reached puberty? *(sperm)*

Speed Bump

Ask What is the function of a sperm cell's tail? *(move the sperm forward)*

Ask What is the benefit of a sperm cell having a large number of mitochondria? *(to provide enough energy to move the sperm through the female reproductive tract)*

Active Reading

USE DIAGRAMS As students read the section Female Reproductive Structures, have them locate the structures in the diagram. Help students pronounce the names of the structures.

Speed Bump

Ask Where are eggs produced in a female? *(in the ovaries)*

Ask What is a follicle? *(cluster of cells surrounding an egg in an ovary)*

Focus on Academic Language

• follicle

Students may be familiar with the term *hair follicle* and may be confused by the use of the term *follicle* in connection with cells in the ovary. Tell students that the word *follicle* means "little bag."

Say A hair follicle is a small depression in the skin. The follicle acts like a little bag that surrounds, contains, and protects the hair's root.

Ask Why is the cluster of cells in the ovary called a follicle? *(because it surrounds, contains, and protects the egg)*

LESSON 34.3

During Reading continued

Active Reading

USE VISUALS The structures and processes described on this page relate to the diagram of the female reproductive system on the previous page and the diagram of the menstrual cycle on the next page.

As students read about ovulation, have them look at the Female Reproductive System diagram on the previous page.

Say Put your finger on an ovary. Trace your finger from the ovary to the end of the Fallopian tube.

Ask When an egg leaves the Fallopian tube, where does it go? *(into the uterus)*

Speed Bump

Ask What is the name of the cycle during which an egg is produced and released from an ovary? *(menstrual cycle)*

Ask What are the four phases of the menstrual cycle? *(follicular phase, ovulation, luteal phase, and menstruation)*

Ask What happens during ovulation? *(An egg is released from the ovary into the Fallopian tube.)*

BUILD Vocabulary

menstrual cycle
a regular sequence of events in which an egg develops and is released from the body

ovulation
the release of a mature egg from the ovary into one of the Fallopian tubes

The Menstrual Cycle One mature egg is usually produced and released from an ovary every 28 days or so. The process of egg production and release is part of the **menstrual cycle.** A menstrual cycle involves the ovaries, the lining of the uterus, and the endocrine system. Hormones from the hypothalamus, the pituitary, and the ovaries regulate this cycle.

During this cycle, an egg develops and is released from the ovary. If the egg is not fertilized, it leaves the body along with the lining of the uterus. If the egg is fertilized, then the menstrual cycle stops and embryonic development begins. The menstrual cycle includes the follicular phase, ovulation, the luteal phase, and menstruation.

▶ ***Follicular Phase*** Blood estrogen levels are low on day 1 of the menstrual cycle. The hypothalamus reacts to the low levels by producing a releasing hormone. The releasing hormone causes the anterior pituitary to release FSH and LH. These hormones stimulate the ovaries, causing a follicle to mature.

As the follicle develops, the cells surrounding the egg grow larger. The cells release more and more estrogens, causing FSH and LH levels to drop. The estrogens also cause the lining of the uterus to thicken so it can receive a fertilized egg. An egg matures in about 12 days.

▶ ***Ovulation*** The follicle releases more and more estrogens as it grows until the estrogens reach a certain level. This increased estrogen level triggers a burst of FSH and LH, causing the follicle to rupture. This rupture releases the egg into the Fallopian tube, a process called **ovulation.** Eggs are stalled in metaphase of meiosis II when they are released. Eggs remain in this phase of meiosis unless they are fertilized. Cilia push the egg down the Fallopian tube toward the uterus.

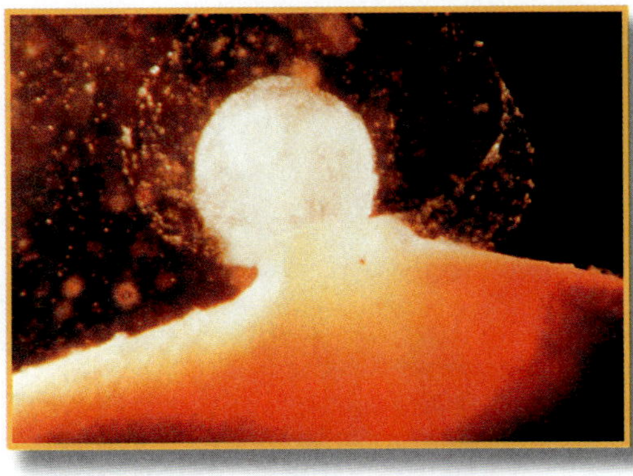

Ovulation (LM 160×)

Hands-On Learning

INTERPRET DIAGRAMS Give each pair of students a copy of a one-month calendar with space provided to record events for each day. Also, make sure each pair has three pencils or markers of different colors.

Tell students they will make a schedule using information from the Menstrual Cycle diagram. Model how to block out the first through the twelfth day with an arrow labeled *Follicular Phase.* Direct students to use a single color to record the follicular, ovulation, and luteal phases on their calendars. Tell students to use a second color to label days on which each hormone reaches its highest level. Finally, have students use a third color to indicate the days on which menstruation takes place. Suggest each student pair exchange calendars with another pair to check for clarity and accuracy.

▶ ***Luteal Phase*** Immediately after ovulation, the cells of the ruptured follicle change. The follicle turns yellow and is now called the *corpus luteum* (which means "yellow body"). The corpus luteum is still releasing estrogens, but it also begins releasing the steroid hormone, progesterone. Progesterone causes the uterine lining to develop. These hormones slow down the release of FSH and LH. This blocks the growth of more follicles during the cycle.

If the egg remains unfertilized, LH levels fall. This decrease in LH levels causes the corpus luteum to collapse. The collapse causes estrogen levels to decrease, and the follicular phase begins again.

▶ ***Menstruation*** When the new follicular phase starts, estrogen levels drop, causing the uterine lining to loosen. The tissue, blood, and the unfertilized egg are discharged through the vagina. This phase, called **menstruation,** lasts three to seven days. Estrogen production declines when a female is in her late forties to early fifties. This decrease in estrogen production stops ovulation and menstruation permanently. This phase is called menopause.

BUILD Connections

THE MENSTRUAL CYCLE

The menstrual cycle includes several phases. Notice the changes in hormone levels in the blood, follicle development, and the uterine lining.

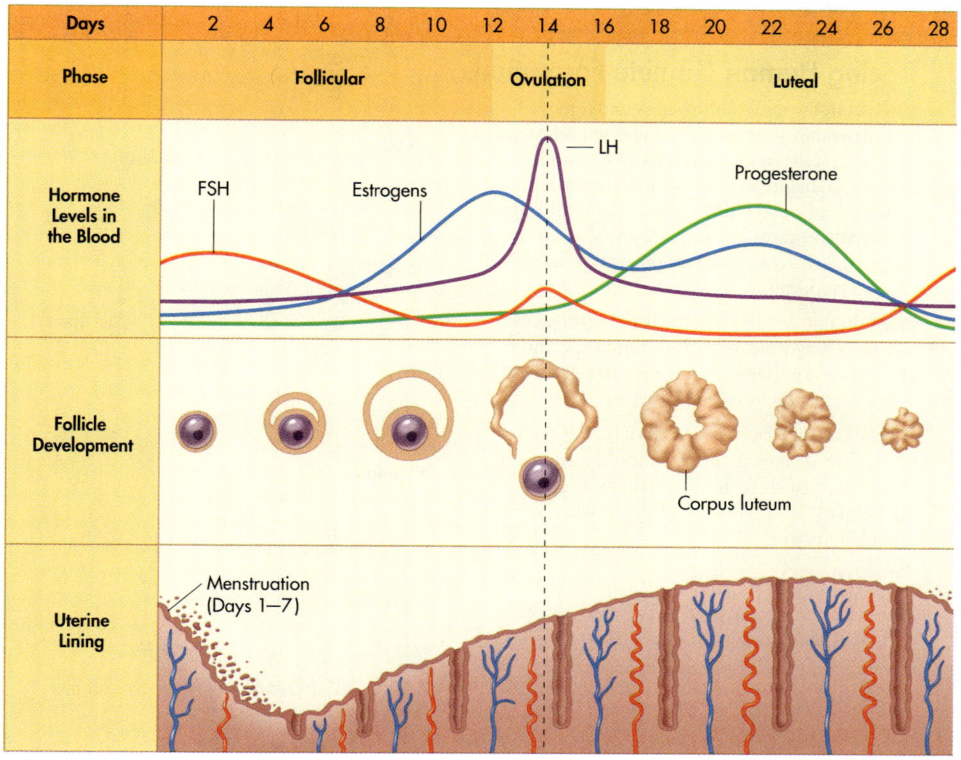

ELL Focus on ELL: Extend Language

INTERMEDIATE AND ADVANCED SPEAKERS Most students recognize *-ing* as a verb ending, as in walk*ing*, sleep*ing*, and so on. Tell students that there are times when this word form functions as if it were an adjective modifying a noun. Write *walking shoes*. Point out that, in this case, the *-ing* word limits the noun it modifies. Explain that *walking shoes* are not just any shoes; they are shoes used only for walking.

Have the class follow along as a volunteer reads the first paragraph under Follicular Phase on the previous page. Tell students to watch for the word *releasing*. Direct students to raise their right hands if they think *releasing* is acting as a verb and raise their left hands if they think it is acting as an adjective. *(*Releasing *acts as an adjective modifying the noun* hormone.*)* Repeat the activity with the first paragraph under Luteal Phase. *(*Releasing *appears twice as a verb.)*

Build Connections

VISUAL SUMMARY Some students may have difficulty interpreting the diagram showing the daily progress of the menstrual cycle, especially the curves showing hormone levels at the top of the diagram.

Say The hormone levels shown at the top of the diagram control the development of the follicle shown in the center and the lining of the uterus shown at the bottom.

Ask What does the diagram show about how the follicle develops? *(the days on which the follicle matures, the egg is released, and the corpus luteum forms and collapses)* Why does the top of the lining of the uterus have a curving shape in the diagram? *(The thickness of the lining changes on different days of the cycle.)*

Ask On what day of the cycle is the level of FSH the highest? *(day 2)* Why is the level of FSH low except at the beginning of the cycle? *(If FSH levels stayed high, another egg may start to mature.)* What happens to the lining of the uterus as the level of estrogen increases? *(The lining grows thicker.)* When does ovulation occur? *(at about day 14)*

Say Look at what is happening to the hormone levels from day 24 through day 28.

Ask What hormones are responsible for keeping the lining of the uterus in place? *(estrogens and progesterone)*

Ask What happens when the levels of these two hormones drop? *(The uterine lining loosens.)* Then what happens? *(Menstruation begins as the lining and the unfertilized egg are discharged through the vagina.)*

Active Reading

DRAW CONCLUSIONS Make sure students realize that the body reacts differently depending on whether or not the egg is fertilized.

Say Read the Luteal Phase section.

Ask What do you think would happen to the LH level and the corpus luteum if the egg were fertilized? *(The LH level would not fall and the corpus luteum would not collapse.)*

Speed Bump

Ask What pituitary hormones control the menstrual cycle? *(FSH and LH)* What hormones from the ovaries are part of the menstrual cycle? *(estrogens and progesterone)*

Ask What happens if the egg remains unfertilized during the luteal phase? *(A new follicular phase starts and menstruation begins.)*

During Reading continued

Active Reading

SCIENCE SUPPORT Some students may think fertilization occurs in the uterus. Refer them to the Female Reproductive System diagram. Point out that fertilization occurs in the Fallopian tube. By the time the egg has reached the uterus, it usually cannot be fertilized.

Ask Through which parts of the female reproductive system must sperm travel before fertilization can occur? *(vagina, cervix, uterus, and Fallopian tube)*

Say The fertilized egg then travels down the Fallopian tube into the uterus, where it attaches to the uterine lining and begins to develop into an embryo.

KEY QUESTION Invite a volunteer to read aloud the Key Question and its answer.

Ask What is the name of the cycle during which a female produces an egg and her uterus undergoes changes? *(menstrual cycle)*

Speed Bump

Ask What are two events that either end or disrupt the menstrual cycle? *(menopause and pregnancy)*

Inquiry Into Scientific Thinking

USE CONTEXT CLUES Read Step 1 to students.

Ask What is a replica? *(an exact copy of something)*

Say To replicate means to make a copy of something.

Ask What happens when a chromosome replicates? *(It makes an exact copy of itself.)*

INTERPRET DIAGRAMS Remind students that the chromosomes in the cell at the top of each diagram undergo replication before meiosis begins.

Ask How many chromosomes are in the cell just as meiosis begins? *(46 × 2 = 92)* How many chromosomes are in each daughter cell after meiosis I? *(46)* How many chromosomes are in each gamete? *(23)*

Lead a Discussion

Talk about how meiosis relates to human reproduction. Draw students' attention to the relative number of sperm cells and egg cells produced as a result of meiosis.

Ask What is the advantage of males producing large numbers of sperm? *(to increase the probability that a sperm cell reaches an egg cell and fertilizes it)*

Pregnancy The menstrual cycle stops if a woman becomes pregnant. An egg is most likely to be fertilized during the first two days after ovulation. A fertilized egg immediately completes meiosis and undergoes mitosis. It divides several times, forming a ball of cells that implants into the uterine lining. Soon, the uterus and embryo begin to release hormones that keep the corpus luteum functioning. This allows the uterine lining to nourish and protect the embryo for several weeks. It also blocks the menstrual cycle from starting again.

Key Question What are the main functions of the female reproductive system?
The main function of the female reproductive system is to produce eggs. It also prepares the female's body to nourish a developing embryo.

INQUIRY into Scientific Thinking

Tracing Human Gamete Formation

1. Cells in the testes and ovaries undergo meiosis as they form gametes—sperm and eggs. In meiosis, cells start out as diploid (2N) cells. Remember, before meiosis begins, chromosomes replicate, but the cell does not divide. If the cell starts with 46 chromosomes (23 pairs), how many total chromosomes are present in the cell after this replication?
2. At the end of meiosis I, the two daughter cells each have two copies of each chromosome. But the cells do not have homologous pairs. Are the daughter cells 2N or N?
3. Look at the diagrams to the right. For each letter, indicate how many chromosomes are in the cells at that stage. Is the cell diploid (2N) or haploid (N)? Answers *a.* and *e.* have been provided for you.

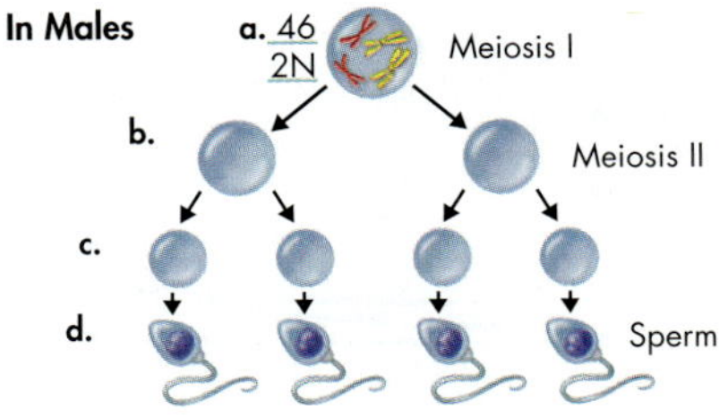

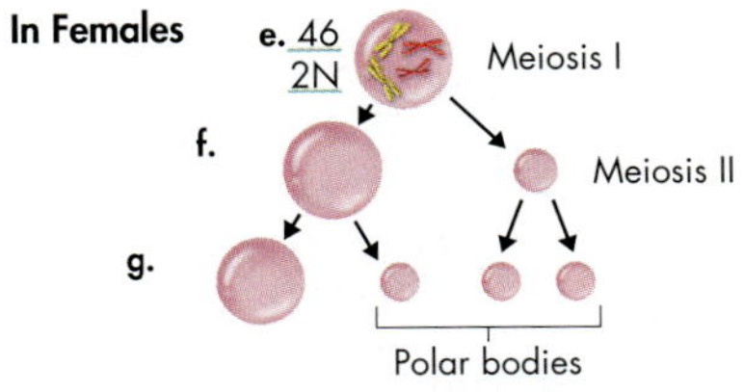

Analyze and Conclude

1. **Interpret Visuals** For every cell that undergoes meiosis in a male or female, how many more sperm than eggs are produced?
2. **Infer** What percentage of sperm cells will contain a Y chromosome?

In Your Workbook Get more help for this activity in your workbook.

Analyze and Conclude

1. 4 sperm to 1 egg, so 3 more sperm
2. 50%

Sexually Transmitted Diseases

Some diseases are spread by sexual contact. These diseases are called **sexually transmitted diseases (STDs).** STDs can be caused by bacteria or by viruses. Bacterial STDs include chlamydia, gonorrhea, and syphilis. Viral STDs include hepatitis B, genital herpes, and genital warts. Unlike bacterial STDs, viral STDs cannot be treated with antibiotics. Some viral STDs can be fatal. Tens of thousands of people die from AIDS each year. Human papillomavirus (HPV), which causes genital warts, is a major cause of cervical cancer.

STDs can be avoided. The safest way is to abstain from sexual contact before marriage. The next safest way is to use a condom, but even condoms do not provide 100 percent protection.

Key Question What are some of the most commonly reported sexually transmitted diseases?

Some of the most common bacterial STDs are chlamydia, gonorrhea, and syphilis. Some of the most common viral STDs are hepatitis B, genital herpes, genital warts, and AIDS.

BUILD Vocabulary

menstruation
the discharge of blood and the unfertilized egg from the body

sexually transmitted disease (STD)
a disease that is spread from person to person by sexual contact

WORD ORIGINS
The word *menstruation* comes from the Latin word *mensis*, meaning "month."

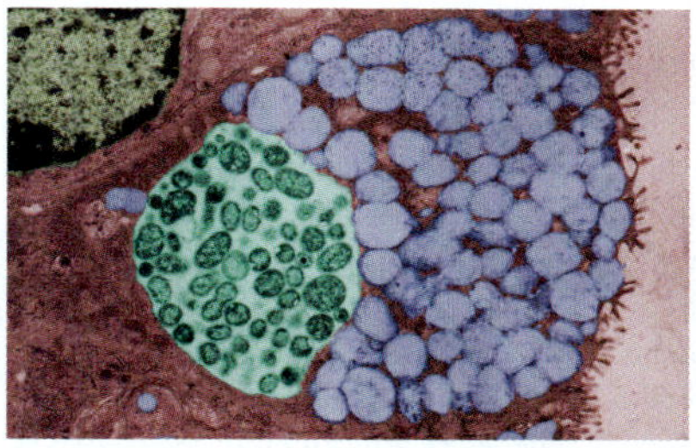

Infection This electron micrograph shows a group of *Chlamydia trachomatis* bacteria in green. The bacteria are growing inside a cell within the female reproductive tract. The bacteria will eventually cause the cell to burst, and the infection will spread.

CHECK Understanding

Apply Vocabulary

Use the highlighted words from the lesson to complete each sentence correctly.

1. ______ are the primary male reproductive organs.

2. The ______ is characterized by a follicular phase, ovulation, and a luteal phase.

Critical Thinking

3. Evaluate Why do you think young people are especially at risk for STDs?

4. Compare and Contrast Compare and contrast the sexual development of male and female embryos.

5. Sequence Explain how sperm develop.

6. Write to Learn Answer the third clue of the mystery.

MYSTERY CLUE

Low fat reserves in women are related to low FSH and LH levels. Tests showed that Lisa's blood had very low levels of these hormones. How might this have affected her menstrual cycle? (**Hint:** See p. 820.)

BIOLOGY.com Search Lesson 34.3 GO • Lesson Assessment

Check Understanding

1. Testes

2. menstrual cycle

3. Sample answer: Young people are at particular risk for STDs because they may be more prone to risky behaviors, such as unprotected sex. Young people may also be less likely to go to the doctor for treatment, and therefore, may be more likely to spread the disease.

4. In the male embryo, the gonads secrete testosterone. Testosterone causes the embryo to develop a male reproductive system. In the female embryo, the gonads secrete estrogens, causing the development of a female reproductive system.

5. Diploid (2N) stem cells in the lining of the seminiferous tubules undergo meiosis and yield the haploid (N) nuclei of sperm. Sperm move into the epididymis for further maturation.

6. FSH and LH are needed to stimulate follicular development, the secretion of estrogens, and ovulation, which are steps that are important in keeping the menstrual cycle going.

Active Reading

SUMMARIZE Have students make a T-chart that summarizes the information about STDs on this page. Tell them to list the names of bacterial STDs in one column and viral STDs in the other. Help students pronounce the names of the STDs aloud.

KEY QUESTION Read aloud the Key Question. Have students refer to their T-chart as you read the answer.

Speed Bump

Ask What is the safest way to avoid STDs? *(Abstain from sexual contact.)* What is the next safest? *(Use a condom.)*

Post Reading

ASSESSMENT TIPS For Question 1, have students refer to the Male Reproductive System diagram.

For Question 5, explain that sequencing means putting events in order from earliest to latest.

MYSTERY CLUE Have students reread the section of the lesson that describes the follicular phase of the menstrual cycle.

Ask What role do FSH and LH play in stimulating the menstrual cycle? *(They stimulate the ovaries, which causes a follicle to mature.)* What do you think would happen if levels of FSH and LH remained too low to stimulate the ovaries? *(The menstrual cycle might not happen.)*

Wrap-Up Activity

CARD SHUFFLE Have students work in groups of four. Provide ten index cards per group. Have two students in each group write on each of five cards a description or definition *(but not the name)* of one male structure, process, or hormone necessary for reproduction. The other two students can repeat the process for five female structures, processes, or hormones. Collect cards from all groups and shuffle. Distribute ten shuffled cards to each group. Have groups name the structure, process, or hormone described on each card and write it on the back of the card.

LESSON 34.4

Pre-Reading

Objectives

34.4.1 Describe fertilization and the early stages of development.

34.4.2 Identify the major events of later stages of development.

Preview the Pages

1. Preview the vocabulary words and their pronunciations with students. Have them find the words in the Fertilization and Implantation diagram and the Placenta diagram.
2. Have students look at the microscopic view of the sperm and egg on this page and compare their sizes.
3. Tell students that all the structures of the human body derive from three layers of cells. Have students follow the formation of these three layers in the Gastrulation diagram and their changes in the Neurulation diagram.

Activate Prior Knowledge

Initiate a discussion of human development.

Ask About how long is a normal human pregnancy? *(nine months)* What kinds of changes does the developing baby undergo? *(Sample answers: size, formation of body parts, increased mobility)*

Build Understanding

FLOWCHART Direct students to Workbook B, where they will find a flowchart designed for this lesson. Remind them that a flowchart will help them follow a sequence of events.

NATIONAL SCIENCE EDUCATION STANDARDS

UNIFYING CONCEPTS AND PROCESSES

I, V

CONTENT

C.1.f, C.2.b, C.5.d, F.1

For the corresponding lesson in *Miller & Levine Biology,* see pages 995–1001.

34.4 Fertilization and Development

Key Questions

What takes place during fertilization and the early stages of human development?

What important events occur during the later stages of human development?

BUILD Understanding

Flowchart As you read, draw a flowchart that shows the steps from fertilized egg to newborn baby.

In Your Workbook Go to your workbook to learn more about making a flowchart. Complete the flowchart for Lesson 34.4.

Fertilization and Early Development

Is anything more amazing than the formation of a new human from a single cell? The story of human development begins with the gametes. Gametes are sperm produced in the testes and egg cells produced in the ovaries. The process of development begins when a sperm and an egg cell fuse into a single cell.

Fertilization Hundreds of millions of sperm are released when semen is ejaculated into the vagina through the penis. Semen is generally released just below the cervix. The cervix is the opening that connects the vagina to the uterus. Sperm swim through the cervix and uterus into the Fallopian tubes. If an egg is in one of the Fallopian tubes, its chances of being fertilized are good.

The egg has a protective outer layer. A sperm can attach to binding sites on this layer. Its head then releases powerful enzymes that break down this protective layer. The sperm and egg are haploid (N). The sperm nucleus enters the egg, and their chromosomes are brought together. The two haploid nuclei fuse to form a single diploid (2N) nucleus. The nucleus contains a single set of chromosomes from each parent cell. The fertilized egg is called a **zygote,** or embryo.

The egg contains a layer of particles just inside its outer layer. It releases the contents of these particles when the sperm enters the egg. The material in the particles forms a barrier that prevents other sperm from attaching to, or entering, the egg.

Key Question What takes place during fertilization?
The fusion of a sperm and an egg is called fertilization.

Sperm Meet Egg Many sperm usually reach an egg. But only one sperm can successfully break through the egg's protective layer (SEM 650×).

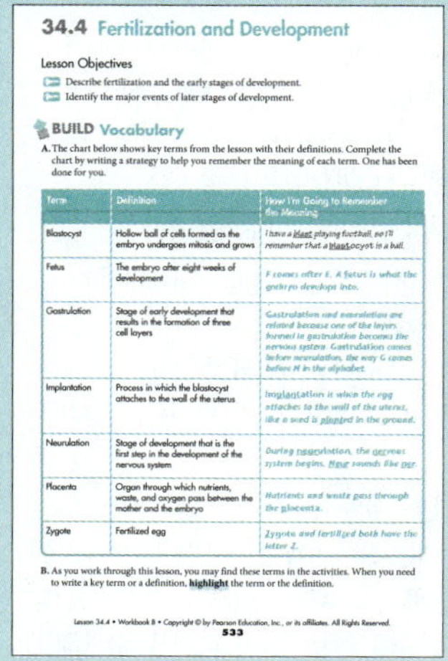

34.4 Fertilization and Development

Lesson Objectives

Describe fertilization and the early stages of development.

Identify the major events of later stages of development.

BUILD Vocabulary

A. The chart below shows key terms from the lesson with their definitions. Complete the chart by writing a strategy to help you remember the meaning of each term. One has been done for you.

Term	Definition	How I'm Going to Remember the Meaning
Blastocyst	Hollow ball of cells formed as the embryo undergoes mitosis and grows	I have a blast playing football, so I'll remember that a blastocyst is a ball.
Fetus	The embryo after eight weeks of development	F comes after E. A fetus is what the embryo develops into.
Gastrulation	Stage of early development that results in the formation of three cell layers	Gastrulation and neurulation are related because one of the layers formed in gastrulation becomes the nervous system. Gastrulation comes before neurulation, the way G comes before N in the alphabet.
Implantation	Process in which the blastocyst attaches to the wall of the uterus	Implantation is when the egg attaches to the wall of the uterus, like a seed is planted in the ground.
Neurulation	Stage of development that is the first step in the development of the nervous system	During neurulation, the nervous system begins. Neur sounds like ner.
Placenta	Organ through which nutrients, waste, and oxygen pass between the mother and the embryo	Nutrients and waste pass through the placenta.
Zygote	Fertilized egg	Zygote and fertilized both have the letter Z.

B. As you work through this lesson, you may find these terms in the activities. When you need to write a key term or a definition, **highlight** the term or the definition.

Lesson 34.4 • Workbook B • Copyright © by Pearson Education, Inc., or its affiliates. All Rights Reserved.
533

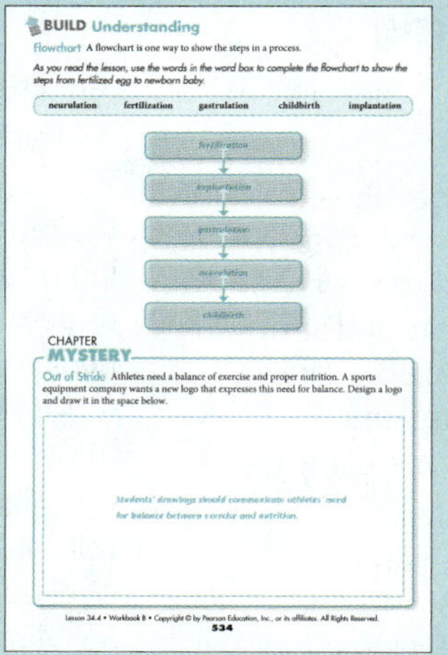

BUILD Understanding

Flowchart A flowchart is one way to show the steps in a process.

As you read the lesson, use the words in the word box to complete the flowchart to show the steps from fertilized egg to newborn baby.

neurulation fertilization gastrulation childbirth implantation

fertilization → implantation → gastrulation → neurulation → childbirth

CHAPTER MYSTERY

Out of Stride Athletes need a balance of exercise and proper nutrition. A sports equipment company wants a new logo that expresses this need for balance. Design a logo and draw it in the space below.

Students' drawings should communicate athletes' need for balance between exercise and nutrition.

Lesson 34.4 • Workbook B • Copyright © by Pearson Education, Inc., or its affiliates. All Rights Reserved.
534

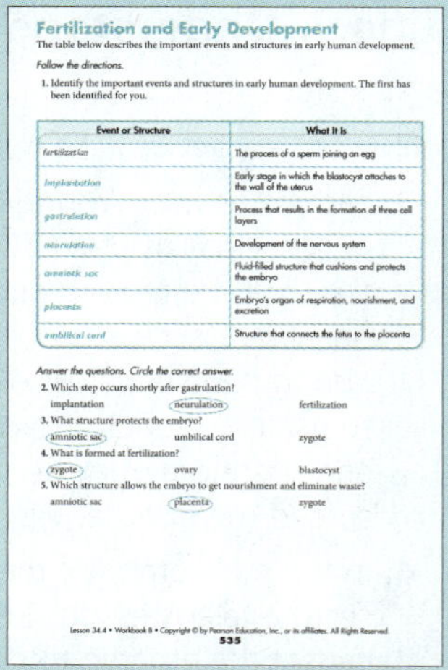

Fertilization and Early Development

The table below describes the important events and structures in early human development.

Follow the directions.

1. Identify the important events and structures in early human development. The first has been identified for you.

Event or Structure	What It Is
fertilization	The process of a sperm joining an egg
implantation	Early stage in which the blastocyst attaches to the wall of the uterus
gastrulation	Process that results in the formation of three cell layers
neurulation	Development of the nervous system
amniotic sac	Fluid-filled structure that cushions and protects the embryo
placenta	Embryo's organ of respiration, nourishment, and excretion
umbilical cord	Structure that connects the fetus to the placenta

Answer the questions. Circle the correct answer.

2. Which step occurs shortly after gastrulation?
implantation neurulation fertilization
3. What structure protects the embryo?
amniotic sac umbilical cord zygote
4. What is formed at fertilization?
zygote ovary blastocyst
5. Which structure allows the embryo to get nourishment and eliminate waste?
amniotic sac placenta zygote

Lesson 34.4 • Workbook B • Copyright © by Pearson Education, Inc., or its affiliates. All Rights Reserved.
535

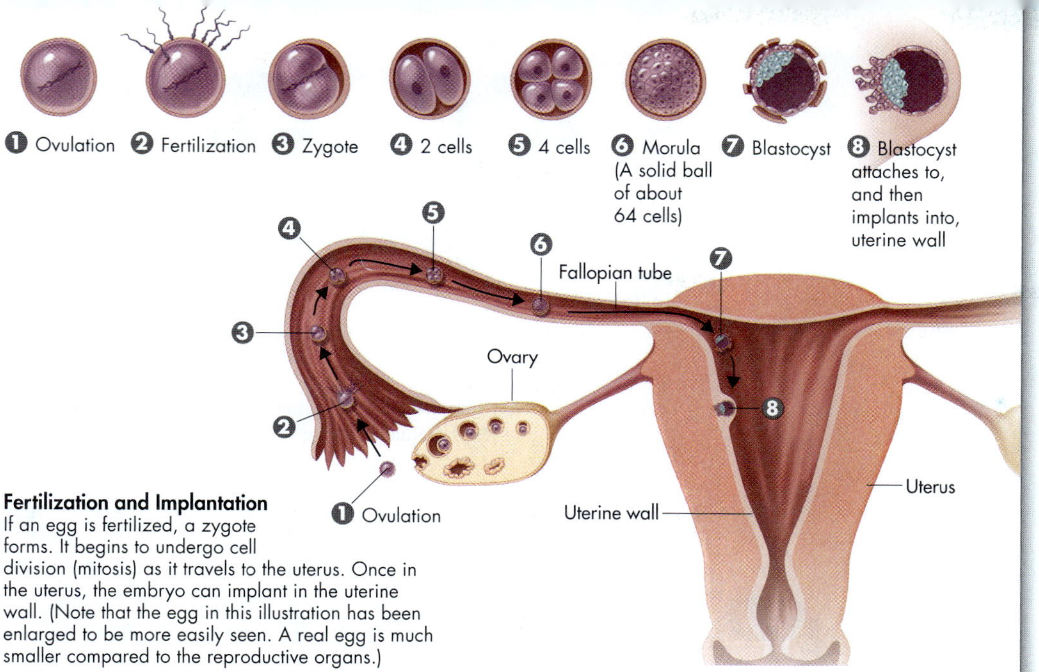

Fertilization and Implantation
If an egg is fertilized, a zygote forms. It begins to undergo cell division (mitosis) as it travels to the uterus. Once in the uterus, the embryo can implant in the uterine wall. (Note that the egg in this illustration has been enlarged to be more easily seen. A real egg is much smaller compared to the reproductive organs.)

Multiple Embryos Sometimes two eggs are released during the same menstrual cycle. If both are fertilized by different sperm, fraternal twins may result. Fraternal twins are not identical in appearance and may even be different sexes. If a single zygote splits apart, it produces two genetically identical embryos. These two identical embryos are called identical twins. Because they result from the same fertilized egg, identical twins are always the same sex.

Implantation The zygote begins to undergo mitosis inside the Fallopian tube. As the embryo grows, the ball of cells develops a cavity in the center. This hollow ball of cells is called the **blastocyst.** About six or seven days after fertilization, it attaches to the uterus wall. The blastocyst then grows into the tissues of the mother, a process called implantation.

At this point, cells in the blastocyst begin to specialize. This process is called differentiation. Differentiation causes the various types of body tissues to develop. A cluster of cells develops inside the inner cavity of the blastocyst. The body of the embryo will develop from this cluster. The remaining cells will differentiate into tissues that support and protect the embryo.

BUILD Vocabulary

zygote
a fertilized egg

blastocyst
a stage of early development in mammals that consists of a hollow ball of cells

placenta
a specialized organ in placental mammals through which respiratory gases, nutrients, and wastes are exchanged between the mother and her developing young

fetus
a human embryo after eight weeks of development

SUFFIXES
A *cyst* is a fluid-filled sac that forms in animal tissues. A structure in the body that has a fluid-filled cavity often has a name ending in *-cyst*. A blastocyst is a hollow ball of cells with a cavity in the center that contains fluid.

ELL Focus on ELL: Build Background

BEGINNING SPEAKERS English learners may have prior experience with the word *cavity.* Help them use that experience to better understand the structure of a blastocyst.

ASK Where have you heard the word *cavity* before? *(Answers may vary, but students probably associate the word* cavity *with the dentist's office.)* What is one meaning of *cavity*? *(Sample answer: a small hole, or area of decay, in a tooth)*

Tell students that *cavity* has a more general meaning that doesn't relate to teeth. Explain that the word *cavity* and the word *cave* both come from the same Latin root, *cavus,* meaning "hollow."

ASK What is a cave? *(a hollow space, or chamber, under the ground)*

SAY Like a cave, a cavity is a hollowed-out space in a mass. A blastocyst has a cavity, or hollowed-out space, in its center. In that cavity, cells grow that will eventually develop into the body of the embryo.

During Reading

Active Reading

FIND THE MAIN IDEA Read aloud the paragraph on Multiple Embryos.

Ask What is the main idea of this paragraph? *(Fraternal twins and identical twins form differently.)*

Ask What is the difference in the way the two types of twins form? *(Fraternal twins form from two different fertilized eggs. Identical twins form from one fertilized egg that splits apart.)*

INTERPRET DIAGRAMS Direct students' attention to the Fertilization and Implantation diagram.

Ask What is the relationship between the eight small pictures at the top of the diagram and the numbers in the Fallopian tube and uterus? *(The eight small pictures are blow-ups of stages of the fertilized egg on the way to the uterus and during implantation. The numbers show the location of each of the blow-ups.)*

MAKE CONNECTIONS Make sure students realize that during fertilization and implantation, the menstrual cycle continues.

Ask How can you tell that the wall of the uterus is preparing for implantation before the blastocyst arrives? *(We learned in the last lesson that the uterine lining begins developing right after ovulation. When the blastocyst arrives, the lining is ready to accept it.)*

Build Vocabulary

SUFFIXES Write the word *nematocyst* on the board and underline the word part *-cyst.*

Say A nematocyst is an organelle in jellyfish that contains a stinger.

Ask What do you think is the structure of a nematocyst? *(A nematocyst is a fluid-filled sac with a stinger inside.)*

Speed Bump

Ask How many sperm can fertilize a single egg? *(one)*

Ask Why are identical twins genetically identical? *(They result from the same fertilized egg.)*

Ask What does a blastocyst do during implantation? *(It attaches to the uterus and grows into the tissues of the mother.)*

LESSON 34.4

During Reading continued

Active Reading

INTERPRET DIAGRAMS Help students realize that the steps shown in the Neurulation diagram occur after those in the Gastrulation diagram at the top of the page. Students can follow the history of the three types of tissue—ectoderm, mesoderm, and endoderm—by referring to the different colors that represent the tissues in the diagrams.

Say In the diagrams, ectoderm is blue, endoderm is yellow, and mesoderm is red.

Ask Where does the mesoderm come from? *(cells of the ectoderm that migrate and form a middle layer)* From what layer of cells does the notochord form? *(mesoderm)* What about the neural tube? *(ectoderm)*

APPLY ACADEMIC VOCABULARY Remind students that *ecto-*, *meso-*, and *endo-* mean outside, middle, and inside, respectively.

Ask What layer of cells forms the skin? *(ectoderm)* What cell layer forms most of the body's structures? *(mesoderm)* What layer forms the linings of organs of the digestive, respiratory, and excretory systems? *(endoderm)*

MAKE CONNECTIONS Write *neur-*, *neurulation,* and *neural tube* on the board.

Ask What other words do you know that begin with *neur-*? *(Sample answers: neuron, neurotransmitter)* What does the word part *neur-* refer to? *(nerves or nervous system)* What happens in the embryo during neurulation? *(The nervous system begins to develop.)* What eventually happens to the neural tube formed during neurulation? *(It becomes the brain and spinal cord.)*

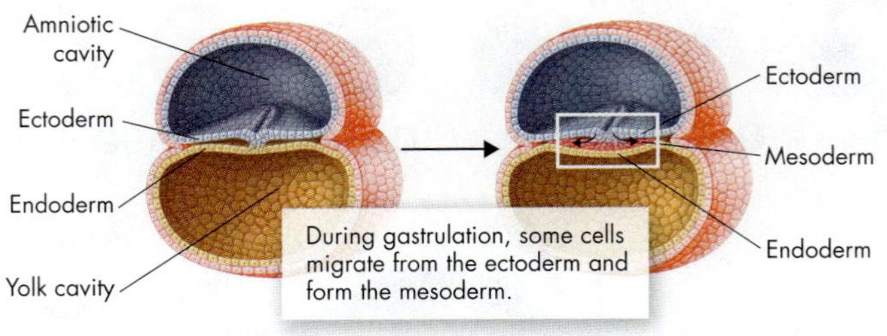

Gastrulation All of the organs and tissues of the embryo are formed from these three layers.

Gastrulation As development continues, the embryo begins a series of dramatic changes. These changes will produce the key structures and tissue layers of the body. One key event in early development is gastrulation (gas troo LAY shun). During gastrulation, three cell layers—the ectoderm, mesoderm, and endoderm—form. The ectoderm and endoderm form first. Then cells from the ectoderm migrate to form the mesoderm.

The ectoderm will develop into the skin and the nervous system. Mesoderm cells will differentiate into tissues such as bones, muscle, blood cells, and gonads. Endoderm forms the linings of organs in the digestive system, respiratory system, and excretory system.

Neurulation Gastrulation is followed by neurulation (NUR uh lay shun), an important step in the development of the nervous system. First, mesodermal tissue differentiates into the notochord. Ectoderm near the notochord also thickens, forming the neural plate. The raised edges of the neural plate form the neural crest and the neural folds. The folds gradually move together to form the neural tube. This will become the brain and spinal cord. Neural crest cells migrate to other areas, becoming nerve cells, skin pigment cells, and other structures.

If the neural tube does not close completely, a serious birth defect can result. Most cases of this defect, called spina bifida, can be prevented by the vitamin folic acid. Neurulation usually occurs before a woman knows she's pregnant. So, folic acid is an important nutrient in any woman's diet.

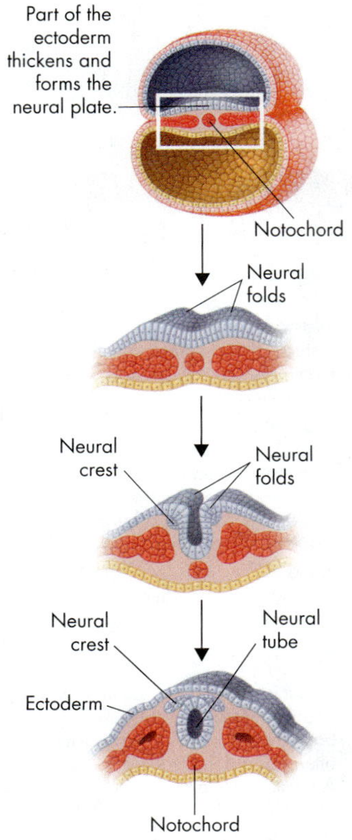

Neurulation During neurulation, changes in the ectoderm lead to formation of the neural tube. The neural tube will develop into the brain and spinal cord. Cells from the neural crest will develop into many types of nerves.

Focus on Academic Language

• dramatic • key • important • serious

Write these words on the board and have students find each one in the text.

Say Synonyms are different words that have similar meanings and can often be used to replace each other without changing what a sentence means. Many English words have several synonyms that are interchangeable. These four words on the board can be used as synonyms for each other.

Ask What do these words tell you about the nouns they modify? *(Sample answers: They are special; They are more than ordinary.)*

Have students rewrite each of the sentences in the text, substituting their own words for *dramatic, key, important,* and *serious.* Direct student pairs to exchange sentences. Allow time for discussion for any unclear or confusing substitutions. On the board, add to the list of possible synonyms.

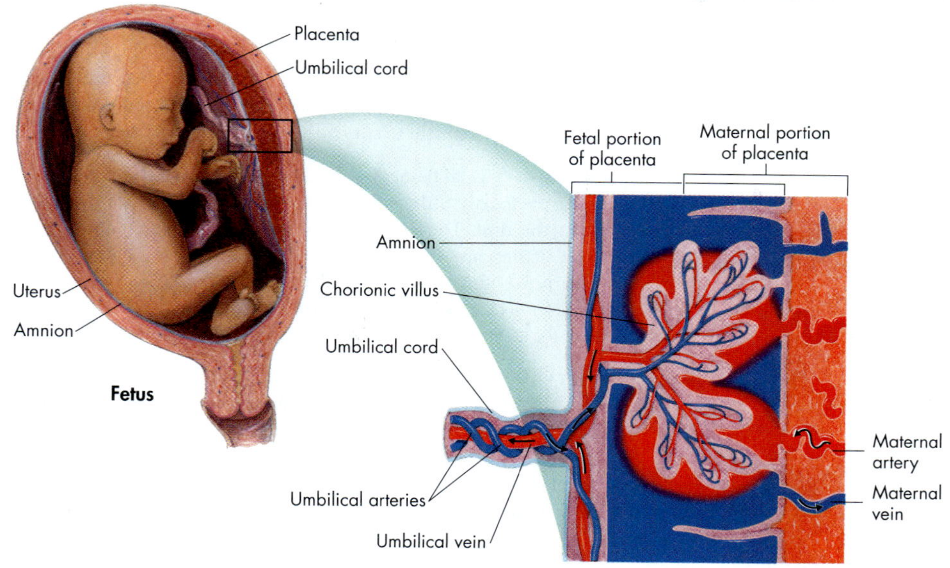

The Placenta The placenta is the connection between the mother and the embryo or fetus. The embryo gets oxygen and nutrients through the placenta. It sends wastes and carbon dioxide back to the mother through the placenta as well. The overlapping brackets show how the chorionic villi extend into the uterine lining.

The Placenta Specialized membranes form to protect and nourish the embryo as it develops. A sac called the amnion surrounds the embryo. It is filled with amniotic fluid, which cushions and protects the embryo. A sac called the chorion forms outside the amnion. Small, fingerlike projections called chorionic villi form on its outer surface.

The villi extend into the uterine lining, forming a vital organ called the **placenta.** The placenta connects the mother and embryo. The mother's blood and the embryo's blood flow past each other, separated by chorionic tissue. Gases, nutrients, and wastes diffuse across this tissue. Two arteries and a vein form the umbilical cord that connects the embryo to the placenta.

After eight weeks, the embryo is about 8 cm long and is called a **fetus.** Most major organs and tissues are fully formed by three months. The fetus may begin to move and show signs of reflexes.

Key Question What takes place during the early stages of human development?
Key events in early development are gastrulation and neurulation. Gastrulation produces the three cell layers of the embryo. Neurulation leads to the formation of the nervous system.

Active Reading

INTERPRET DIAGRAMS Direct students' attention to the fetal portion of the Placenta diagram. Point out that the box just above the fetus's hands is shown enlarged in the lower right portion of the diagram.

Ask What do the red and blue colors of the blood vessels represent? *(Red represents blood flowing from the mother to the fetus. Blue represents blood flowing from the fetus to the mother.)* Refer students to the arrows shown in the arteries and veins.

Ask In what direction do oxygen and nutrients move through the placenta? *(from the mother's arteries to the fetus's umbilical vein)* In what direction do carbon dioxide and wastes move through the placenta? *(from the fetus's umbilical arteries to the mother's veins)*

KEY QUESTION Read the Key Question aloud. Write, in random order, the words *fertilization, implantation, gastrulation* and *neurulation* on the board. Ask students to identify the correct order of these stages. Remind students that they can quickly confirm the order by reading the blue headings on the previous pages.

Speed Bump

Ask During what stage of development do the ectoderm, mesoderm, and endoderm form? *(gastrulation)*

Ask During what stage of development does the nervous system begin to form? *(neurulation)*

Ask Through what organ are gases, nutrients, and wastes exchanged between the mother and the fetus? *(placenta)*

Hands-On Learning

MAKE A MODEL Prepare a model of the amnion and amniotic fluid by placing an uncooked egg in a one-gallon plastic storage bag. Fill the bag with water and seal the bag. Invite pairs of students to try to break the egg without removing it or the water from the bag. Have students work over a pail or sink in case the bag breaks. Remind students that although the amnion and amniotic fluid do protect the developing fetus during the mother's daily life, they are not indestructible.

Ask What is protecting the egg—the bag or the water? *(Both the water and the bag protect the egg.)*

Ask In this model, what represents the fetus? *(the egg)* What does the plastic bag represent? *(the amnion)* What does the water represent? *(the amniotic fluid)*

Invite student to write a description of how the amnion and the amniotic fluid protect a developing fetus.

During Reading continued

Active Reading

USE GRAPHIC ORGANIZERS Draw on the board a Main Idea and Details chart. Have students draw it in their notebooks. Help them identify the main idea in the Months 4–6 paragraph. Have them write the main idea in the appropriate column. *(Sample answer: Fetal tissues become more complex and specialized.)* Ask students to identify one detail in their own words and record it in the Details column. Work with students to add three or more details to the organizer. *(Sample answers: Heart can be heard with stethoscope. Bone replaces cartilage. Skin is covered with hair.)*

Encourage pairs of students to add to the organizer, using the Months 7–9 paragraph. *(Sample main idea: Organ systems mature. Sample details: Organs change so they can work after birth. Fetus can control body temperature. Lungs and central nervous system are fully developed.)*

KEY QUESTION Read aloud the Key Question and answer. Tell students to check their graphic organizers to see if they have included all the details that are given in the answer. If not, have them add any they do not already have.

Speed Bump

Ask What is one change that occurs in a fetus during months 4 to 6 of its development? *(Sample answers: The heart increases in size. Bone replaces cartilage. Hair covers the skin.)*

Ask What is one change that occurs during months 7 to 9? *(Sample answers: The fetus doubles in mass. It can regulate its body temperature. The lungs and central nervous system finish developing.)*

Later Development

Most tissues and organs of the embryo have been formed after three months of development. But many systems will need about six more months of development to survive outside the uterus.

Months 4–6 Fetal tissues become more complex and specialized during the fourth, fifth, and sixth months. They also begin to function during this time. The fetal heart becomes large enough to be heard with a stethoscope. Bone continues to replace the cartilage that forms the early skeleton. A layer of soft hair grows over the skin of the fetus. As the fetus increases in size, the mother's abdomen swells to make room for it. The mother also begins to feel the fetus moving.

Months 7–9 The organ systems of the fetus mature during the last three months before birth. The fetus also doubles in mass. The lungs and other organs go through a series of changes needed for life outside the uterus. The fetus can now regulate its own body temperature. The central nervous system and lungs finish developing, too. The pictures below show an embryo and a fetus at different stages of development.

A fetus needs about nine months to develop fully. Babies born before eight months of development are called premature babies. They often have severe breathing problems because their lungs are not fully developed.

Key Question What important events occur during the later stages of human development?
During the fourth, fifth, and sixth months, tissues become more complex and specialized. They also begin to function. During the last three months before birth, organ systems mature. The fetus grows in size and mass.

Human Development At 7 weeks, most organs have begun to form and the heart (large, dark, rounded structure) is beating. By 14 weeks, the hands, feet, and legs have reached their birth proportions. The eyes, ears, and nose are well developed. At 20 weeks, muscles are more developed, and eyebrows and nails have grown in. At full term, a fetus can live on its own.

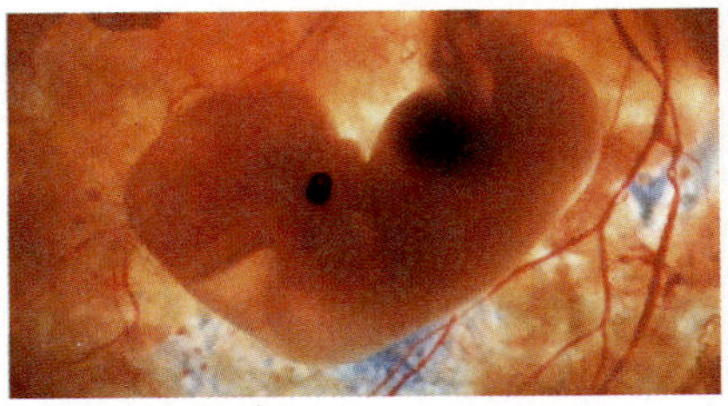
Embryo at 7 Weeks

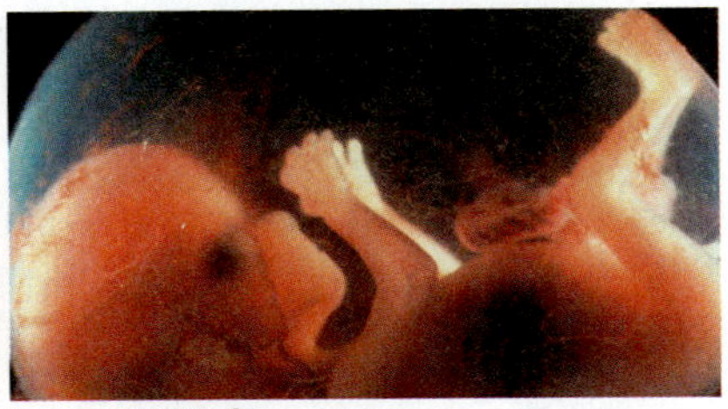
Fetus at 14 Weeks

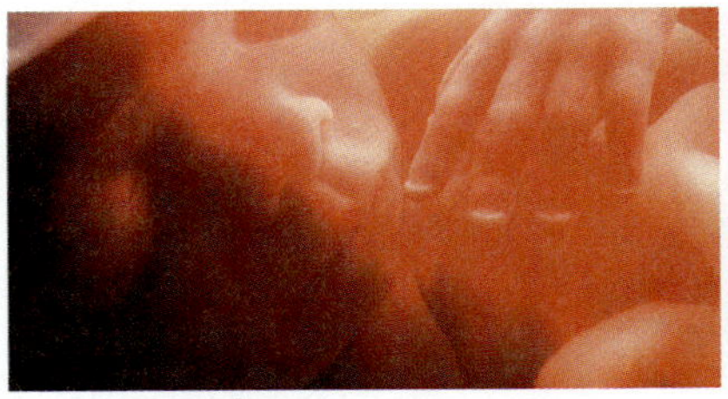
Fetus at 20 Weeks

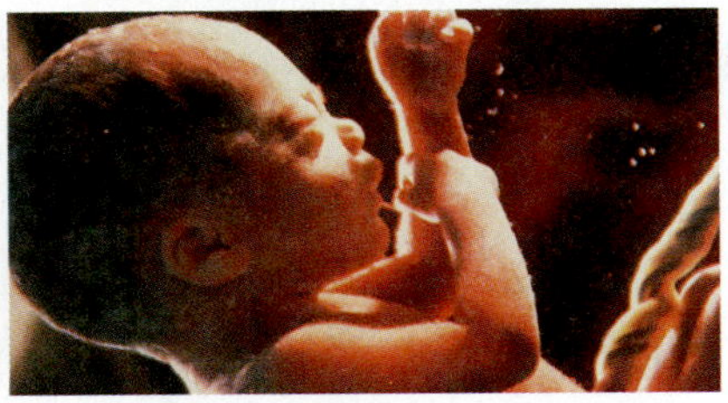
Fetus at Full Term

Students Say

"We'd 'Rate the Visual'—Was it big enough? Clear enough? What was it showing us? Did it add information? Go along with what was written on the page? Rating it made me look closer at the picture or chart, because sometimes I'd just kind of skip those things."

Childbirth Birth occurs about nine months after fertilization. Many factors trigger birth. One is the release of the hormone oxytocin from the mother's posterior pituitary gland. Oxytocin causes a series of rhythmic contractions, known as labor, to begin. During labor, the cervix expands until it is large enough for the head of the baby to pass through. The amniotic sac breaks, and the fluid it contains rushes out of the vagina. Soon, contractions force the baby out through the vagina. Most babies are delivered head first.

Babies often cough or cry when they emerge from the vagina. These actions help clear fluid from the baby's lungs, allowing the baby to start breathing almost immediately. The umbilical cord is clamped and cut. A small piece remains attached to the baby. This piece soon dries and falls off, leaving a scar called the navel. Now the newborn is independent. It can supply its own oxygen, excrete its own wastes, and maintain its body temperature.

A final series of uterine contractions expels the placenta and the empty amniotic sac. This is called the afterbirth. Within a few hours, the mother's pituitary releases the hormone prolactin. Prolactin signals the mother's breast tissues to produce milk. The mother's milk contains all the nutrients the baby needs for growth and development during the first few months of life.

Infant and Maternal Health The placenta acts as a barrier, but some diseases and chemicals can still cross it. Fetuses can be infected with AIDs. The virus that causes rubella (German measles) can cause birth defects. Alcohol, heroin, cocaine, and nicotine can also harm a fetus. A pregnant woman must behave responsibly and get proper medical care to protect the life inside her.

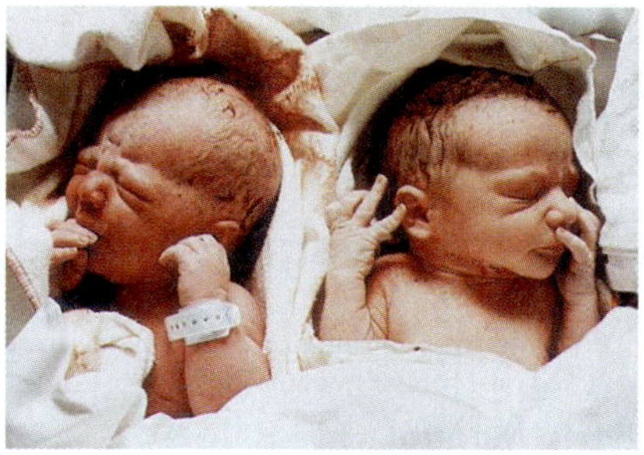

Newborns Twins, ten minutes after birth, adjusting to life outside the uterus

CHECK Understanding

Apply Vocabulary

Use the highlighted words from the lesson to complete each sentence correctly.

1. The hollow ball of cells that attaches to the uterine wall is the _______.
2. After eight weeks of development, an embryo is called a _______.

Critical Thinking

3. **Explain** What is the role of the placenta?
4. **Relate Cause and Effect** How do the results of gastrulation and neurulation contribute to human development?
5. **Quick Write** Doctors often recommend that pregnant women avoid many medications. Write one paragraph explaining why you think this is important.

BIOLOGY.com Search Lesson 34.4 GO • Lesson Assessment

Check Understanding

1. blastocyst
2. fetus
3. The placenta is the organ that connects the mother and the embryo. Nutrients, wastes, and gases are exchanged between the mother and embryo through the tissues of the placenta.
4. Gastrulation forms the ectoderm, endoderm, and mesoderm. Tissues and body systems form from these three layers. Neurulation forms the structures that will become the brain, spinal cord, nerve cells, skin pigment cells, and other structures in the body.
5. Answers will vary. Answers should indicate that the student understands that many chemicals can be transferred from the mother to the fetus across the placenta, and that these chemicals may cause birth defects or developmental defects in the nervous system. A fetus may also develop an addiction to prescription medication.

LESSON 34.4

Active Reading

USE GRAPHIC ORGANIZERS Draw a five-step flowchart graphic organizer on the board, and have students draw it in their notebooks. Title the organizer Steps that Occur During Childbirth. Help students fill in the chart by listing the first entry: *Oxytocin is released*. Encourage students to complete the chart by listing four more steps that occur just before childbirth. *(Oxytocin causes contractions. The cervix expands. The amniotic sac breaks. Contractions force the baby out.)*

Speed Bump

Ask During childbirth, what forces the fetus out through the mother's vagina? *(contractions)*

Ask Why is it important that a pregnant woman act responsibly? *(Diseases and harmful chemicals can pass from the mother to the fetus through the placenta and harm the fetus.)*

Post Reading

ASSESSMENT TIPS For Questions 1 and 2, have students refer to the Fertilization and Implantation diagram.

For Question 3, refer students to the caption for the Placenta diagram.

For Question 4, remind students that the effect of one event may be the cause of a later event.

For Question 5, suggest students write a rough draft of their paragraph before writing the final version.

Wrap-Up Activity

BUILD A TIME LINE Have students work in pairs to build a time line that includes the major steps of development from fertilization to birth. As a starting point, students can use the flowchart they developed in their workbooks, showing the steps from fertilized egg to newborn baby. Have students add additional steps and details as they develop their time lines.

CHAPTER LAB

Pre-Lab

Review with students the concept of homeostasis and the roles of the hypothalamus and pituitary glands in maintaining homeostasis. Use the Build Vocabulary table in the lab to make sure students understand what the words *normal, symptom, diagnose,* and *confirm* mean. Review the Background Questions and Pre-Lab Questions as a class.

Lab

Tell students they will perform the chapter lab *Diagnosing Endocrine Disorders* described in **Lab Manual B.**

Post-Lab

Have students add a fourth column to the Glands, Hormones, and Functions chart they made at the end of Lesson 34.1. If there is no room on the sheet to add a fourth column, students can tape a strip of paper to the right side of the sheet to extend the chart. Direct them to label the fourth column Possible Disorders.

Tell students to read over the information they have recorded in each row of the chart, paying particular attention to the functions they listed. At the end of each row, have students describe possible disorders a person might experience if a problem developed with the gland or hormone listed in that row.

Forensics Lab — GUIDED INQUIRY

Pre-Lab: Diagnosing Endocrine Disorders

Problem Can you diagnose an endocrine disorder based on a patient's symptoms?

Lab Manual Chapter 34 Lab

Skills Focus Analyze Data, Draw Conclusions, Relate Cause and Effect

Connect to the Big idea Organs of the endocrine system secrete hormones into the blood. Each hormone triggers a response in specific cells. Almost every cell in the body is affected by at least one hormone. The endocrine system regulates important processes such as growth, metabolism, and water balance. If one part of the endocrine system is not working properly, the body will be thrown off balance. If the imbalance is severe, it could threaten the health, or even the life, of a person.

Endocrinologists are medical doctors who diagnose and treat disorders of the endocrine system. The clues these doctors use to solve their mysteries are a patient's symptoms and the results of lab tests. In this lab, you will model the process of diagnosing endocrine disorders.

BIOLOGY.com Search Chapter 34 GO

Visit Chapter 34 online to test yourself on chapter content and to find activities to help you learn.

Untamed Science Video The Untamed Science crew helps us better understand the role epinephrine plays in regulating our response to fear and danger.

Art Review Review your understanding of the major endocrine glands in the body.

Art in Motion Watch how steroid and nonsteroid hormones act differently on cells.

Background Questions

a. Review Why doesn't every hormone affect every cell in the body?

b. Review Use a flowchart to describe the feedback loop for regulating the metabolic rate.

c. Use Analogies How are the hormones that regulate the level of glucose in the blood similar to the muscles that bend and straighten an arm?

Pre-Lab Questions

Preview the procedure in the lab manual.

1. Interpret Tables When patients complain of fatigue they are usually referring to a lack of energy or motivation. Which conditions listed in the data table have fatigue as a symptom?

2. Apply Concepts Why do doctors typically use blood tests to diagnose endocrine disorders?

Pre-Lab Answers

BACKGROUND QUESTIONS

a. A cell must have receptors for a given hormone for the hormone to have an effect on the cell's activities.

b. Low level of thyroxine in blood, hypothalamus secretes TRH; anterior pituitary gland secretes TSH, thyroid secretes thyroxine; increased level of thyroxine stops secretion of TRH.

c. Just as an opposing pair of muscles are needed to bend and straighten the arm, an opposing pair of hormones (insulin and glucagon) are needed to decrease and increase the level of glucose in blood.

PRE-LAB QUESTIONS

1. Type I diabetes, hypothyroidism, Addison's disease, hyperparathyroidism

2. They use blood tests because hormones and many of the substances that hormones regulate circulate in the blood.

NATIONAL SCIENCE EDUCATION STANDARDS

UCP I

CONTENT F.1

INQUIRY A.2.b

34 CHAPTER Summary

34.1 The Endocrine System

- The endocrine system is made up of glands that release hormones into the blood. Hormones deliver messages throughout the body.
- Steroid hormones can easily cross cell membranes. After they enter the nucleus, they change the way genes are expressed in target cells.
- Nonsteroid hormones bind to receptors on cell membranes. They cause the release of secondary messengers inside the cell. Secondary messengers then affect cell activities.

hormone (p. 810)
exocrine gland (p. 810)
endocrine gland (p. 810)
prostaglandin (p. 811)

34.2 Glands of the Endocrine System

- Hormones secreted by the pituitary gland directly regulate many body functions. Pituitary hormones also control the actions of other endocrine glands.
- The hypothalamus controls the secretions of the pituitary gland.
- The adrenal glands release hormones that help the body prepare for—and deal with—stress.
- Insulin and glucagon help to keep the blood glucose level stable.
- The thyroid gland has a major role in regulating the body's metabolism.
- The two functions of gonads are the production of gametes and the secretion of sex hormones.
- The endocrine system is regulated by feedback mechanisms that function to maintain homeostasis.

releasing hormone (p. 813)
corticosteroids (p. 814)
epinephrine (p. 814)
norepinephrine (p. 814)
thyroxine (p. 815)

34.3 The Reproductive System

- In females, the sex hormones cause breast development and a widening of the hips. In males, they cause the growth of facial hair and increased muscular development. They also cause deepening of the voice.
- The main functions of the male reproductive system are to produce and release active sperm.
- The main function of the female reproductive system is to produce egg cells. The system also prepares the female's body to nourish an embryo.
- Common bacterial STDs include chlamydia, gonorrhea, and syphilis. Common viral STDs include hepatitis B, genital herpes, genital warts, and AIDS.

testis (p. 818)
ovary (p. 819)
menstrual cycle (p. 820)
ovulation (p. 820)
menstruation (p. 821)
sexually transmitted disease (STD) (p. 823)

34.4 Fertilization and Development

- Fertilization occurs when a sperm and an egg cell fuse to form one cell.
- Gastrulation produces the three cell layers of the embryo. Neurulation leads to the formation of the nervous system.
- Fetal tissues become more complex and specialized during the fourth, fifth, and sixth months.
- The organ systems of the fetus mature during the last three months before birth. The fetus also doubles in mass.

zygote (p. 824)
blastocyst (p. 825)
placenta (p. 827)
fetus (p. 827)

BIOLOGY.com Search Chapter 34 GO • Crossword • Chapter Assessment

UbD Transfer the Big Idea

TRANSFER Suggest students role-play the following dialog:

Student 1: What's wrong?

Student 2: I feel awful today. It's hormones. They're wrecking my life.

Student 1: I agree. We'd all be better off without hormones.

Student 3: I disagree. Hormones are important.

Have pairs of students work together to list at least three examples that support Student 3's statement. Direct student pairs to exchange and evaluate each other's lists, adding appropriate examples if necessary. Finally, have students work in small groups to discuss their examples and evaluations.

Chapter Summary

Think Visually

USE GRAPHICS Have students trace the outline of the figure in Lesson 34.1 that shows the major endocrine glands.

Say Without referring to the figure in the lesson, label as many endocrine glands as you can.

ANSWERS

Students should label their outlines with the names of the endocrine glands shown in the figure. Suggest students refer to the figure to check and to revise their work. Encourage them to add to their outlines as needed.

BIOLOGY.com Study Online

REVIEW RESOURCES

Art Review Students study with a drag-and-drop version of the Major Endocrine Glands figure.

Art in Motion Using this feature, students can watch a short animation of steroid hormone and nonsteroid hormone pathways.

Data Analysis The real-world issues of blood glucose control, diabetes onset, and treatment of diabetes help students explore the important role of the endocrine system in maintaining homeostasis.

Vocabulary Review The *Flash Cards* and *Crossword* provide an interactive way to review chapter vocabulary.

Editable Worksheets Pages of Study Workbook B, Lab Manual B, and the Assessment Resources Book are available online. These documents can be easily edited using a word-processing program.

Lesson Overviews Lesson Overviews are based on Miller & Levine Biology. However, they are available on CD-ROM and can be edited for your class.

ASSESSMENT RESOURCES

Chapter Assessment Have students take an online version of the Chapter 34 Check Understanding.

Standardized Test Prep Students can take an online version of the Standardized Test Prep. You will receive their scores along with ideas for remediation.

Diagnostic and Benchmark Tests Use these tests to monitor your students' progress and supply remediation.

ASSESSMENT

Assess

Big idea Write the word *homeostasis* on the board.

Ask What is the meaning of this word? *(balance)*

Ask How is the meaning of *homeostasis* different from the meaning of *constant*?

(Homeostasis *means "balance," whereas* constant *means "never changing.")*

Ask What is the name of one of the body systems that is responsible for homeostasis? *(endocrine system)* What is the name of the chemical signals the body uses to maintain homeostasis? *(hormones)*

Ask How do hormones deliver messages throughout the body? *(They travel through the blood.)*

Ask What do hormones do once they reach their target cells? *(They bind to specific chemical receptors on cell membranes or inside cells.)*

CONSTRUCTED RESPONSE Paragraphs will vary. Explain the activity carefully. Be sure students understand that the *Hints* should be used as a guide to help them answer the questions. Review how to include details to support the main idea of a paragraph.

For Question 1, write the question on the board and read it aloud. Underline the words *hypothalamus* and *pituitary gland*.

Say This question asks how two different glands, the hypothalamus and the pituitary, work together. Remind students of the broadcast analogy in Lesson 34.1.

Read the first hint.

Ask How does the hypothalamus control the pituitary gland directly? *(Two hormones from the hypothalamus are stored in axons in the pituitary gland. These hormones are released directly into the blood.)*

How does the hypothalamus control the pituitary gland indirectly? *(Releasing hormones from the hypothalamus travel through blood vessels to the pituitary gland where they control pituitary hormones.)*

Read the next hint.

Say Here, the word *regulate* means the same as *control*. Do you remember what the hormones from the pituitary gland control?

Write *other endocrine glands* and *many body functions* on the board.

Model writing the first one or two sentences using the hints on the board and in the text. *(Sample sentences: The pituitary gland secretes some hormones that directly control many body functions. It secretes other hormones that control the actions of other endocrine glands.)*

34 CHECK Understanding

Assess the Big idea Homeostasis

Write an answer to the question below.

Q: How does the body use chemical signals to maintain homeostasis?

Constructed Response

Write an answer to each of the numbered questions below. The answer to each question should be one or two paragraphs. To help you begin, read the **Hints** below the questions.

1. **How do the hypothalamus and pituitary gland work together to help the body maintain homeostasis?**

   Hint The hypothalamus has direct and indirect control over the pituitary.

   Hint Pituitary hormones regulate other endocrine glands as well as other tissues and organs.

2. **Why is it important for the menstrual cycle to stop during pregnancy?**

   Hint The uterine lining becomes part of the placenta during pregnancy.

   Hint Estrogen levels fall at the end of the luteal phase.

3. **Why is it important that the egg cell prevents more than one sperm from fertilizing it?**

   Hint A sperm and egg cell each contain a haploid (N) nucleus.

Foundations for Learning Wrap-Up

The facts you collected as you read the chapter are tools you can use to arrange your thoughts about the endocrine and reproductive systems.

Activity 1 Work with a partner. On a separate sheet of paper, draw and label a chart with the following categories: Hormones, Endocrine glands, and Processes. Invite your partner to choose one card from your envelope and place it in the correct category on the chart. You then have to tell your partner everything you remember about the fact. Take turns. As you sort your facts, you may need to make up your own categories.

Activity 2 Working in small groups, create chains of related facts by pulling facts from each of the categories you created with your partners in Activity 1. Identify how the body uses chemical signals to maintain homeostasis in each chain of facts.

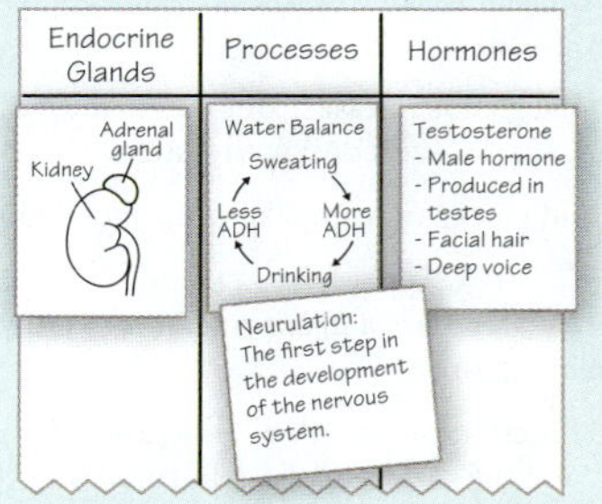

For Question 2, pair students and have them use the hints and your modeling to answer the question. Have students share their paragraphs with the class.

For Question 3, have students write the paragraphs independently for homework or as an anchor activity.

Foundations for Learning Wrap-Up

Have students take out the slips of paper they wrote on while reading the chapter. Be sure students have written information about how the hypothalamus, the pituitary gland, other endocrine glands, and hormones work together to maintain homeostasis. Slips should also contain information about specific hormones, such as insulin, epinephrine and calcitonin. Explain the activities, and assign either groups or student pairs to complete the activities.

34.1 The Endocrine System

Understand Key Concepts

1. Which is a chemical messenger that can directly influence gene expression?
a. nonsteroid hormone **c.** ATP
b. steroid hormone **d.** receptor

Test-Taking Tip

Find Key Words in the Question As you read the question, look for key words that can help you answer it. In question 1, the key words are *chemical messenger.* A chemical messenger is a hormone, so you know that answer **a** or **b** is correct. Read the question again. Another key word is *directly*. Only a steroid hormone can cross cell membranes and enter the nucleus. It affects the cell directly. The correct answer is **b**.

2. A modified fatty acid that is released by a cell and affects local cells and tissues is probably a(n)
a. nonsteroid hormone. **c.** prostaglandin.
b. steroid hormone. **d.** exocrine secretion.

3. Many body functions are influenced by the action of two hormones with opposing effects. Why are such pairs of hormones useful?

Think Critically

4. Infer After a hormone is secreted by a gland, the circulatory system transports it throughout the body. Why doesn't every cell respond to the hormone?

34.2 Glands of the Endocrine System

Understand Key Concepts

5. A hormone that helps regulate blood calcium levels is produced by the
a. posterior pituitary. **c.** thyroid.
b. thymus. **d.** pancreas.

6. Which hormone influences a person's rate of metabolism?
a. PTH **c.** thyroxine
b. aldosterone **d.** calcitonin

7. How does the secretion of epinephrine prepare the body to handle emergencies?

8. What happens if blood glucose levels are not properly regulated?

Think Critically

9. Apply Concepts The heartbeat of a swimmer was found to increase significantly both before and during a swim meet. Explain why this could happen.

34.3 The Reproductive System

Understand Key Concepts

10. The diagram shows the female reproductive system. Which structure is located at the X?
a. uterus
b. Fallopian tube
c. ovary
d. cervix

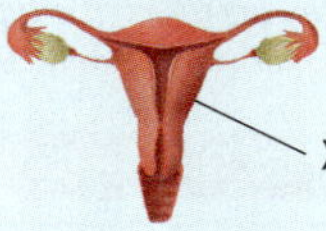

11. Which male reproductive structure releases sperm into the urethra?
a. epididymis **c.** prostate gland
b. vas deferens **d.** testis

12. Which two hormones stimulate the gonads to produce their hormones?

Think Critically

13. Apply Concepts Sperm cells contain many mitochondria. Explain how mitochondria might influence sperm activity.

34.4 Fertilization and Development

Understanding Key Concepts

14. Another name for a fertilized egg is a
a. gastrula. **c.** zygote.
b. placenta. **d.** blastocyst.

Answers

Lesson 34.1

UNDERSTAND KEY CONCEPTS

1. b

2. c

3. Pairs of hormones allow feedback mechanisms to maintain homeostasis.

THINK CRITICALLY

4. Not every cell responds because not every cell has receptors for every hormone.

Lesson 34.2

UNDERSTAND KEY CONCEPTS

5. c

6. c

7. Epinephrine increases heart rate and blood pressure, causes air passageways to widen, and stimulates the release of extra glucose into the blood. These actions prepare the body to respond quickly.

8. If blood glucose levels are not regulated, diabetes mellitus may occur. Very high blood glucose levels can cause serious complications or death.

THINK CRITICALLY

9. Before the swim meet, adrenaline is released as the swimmer anticipates the race, causing the heart rate to increase. During the race, the heart rate is increased due to oxygen demand.

Lesson 34.3

UNDERSTAND KEY CONCEPTS

10. a

11. b

12. FSH and LH

THINK CRITICALLY

13. Mitochondria provide energy for cells. Sperm need a lot of mitochondria so that they have enough energy to "swim" through the female reproductive system.

Lesson 34.4

UNDERSTAND KEY CONCEPTS

14. c

Assessment Tips

Pair students and have them work together to answer the questions.

For Questions 1–4, encourage students to review an overview of the endocrine system. Also, for Question 4 and for Question 18, remind students that *infer* means to use information in the text as a base for a new idea.

For Questions 5–9, tell students to use information about specific glands and the hormones they produce.

For Questions 19 and 20, tell students to think about the structures that form during pregnancy.

Tell students that after they answer a question requiring a written response, they should reread the question to make sure they have answered it completely.

ASSESSMENT

15. c

16. All of the tissues and organs of the embryo develop from the three layers that form during gastrulation. The mesoderm forms many internal structures. The endoderm forms the lining of the digestive, respiratory, and excretory systems. The ectoderm forms the skin and the nervous system.

17. The placenta, which contains maternal and fetal tissues, serves as the connection between the mother and fetus and is the fetus's organ of respiration, nutrition, and excretion.

THINK CRITICALLY

18. The Fallopian tube does not provide the fetus with enough room to grow; the tube will eventually rupture.

Connecting Concepts

19. 3 hours

20. The blue line represents someone without diabetes; the red line may represent someone with diabetes. The red line indicates high levels of blood glucose, an indicator of diabetes.

CHAPTER MYSTERY After students have read the Chapter Mystery, discuss female athlete triad disorder.

Ask How did Lisa have disordered eating, and what were its effects? *(Lisa continued exhausting workouts, but she stopped snacking before practice and avoided high-calorie foods. As a result, she lost weight and suffered from low energy.)*

Ask What is amenorrhea? *(three or more months without a menstrual cycle)* What caused Lisa's amenorrhea? *(Her low energy levels did not signal the release of FSH and LH. Without these hormones, her menstrual cycle stopped.)*

Ask What were the direct causes of Lisa's weakened bones? *(calcium loss due to high cortisol and low estrogen levels)*

34 CHECK Understanding

15. Fertilization usually occurs in the

a. uterus. **c.** Fallopian tube.
b. vagina. **d.** ovary.

16. Explain the importance of the three layers that form during gastrulation.

17. What is the function of the placenta?

Think Critically

18. Infer Occasionally, a zygote does not move into the uterus. Instead, it attaches to the wall of a Fallopian tube. Why might this be a very dangerous situation for the mother?

Connecting Concepts

Use Science Graphics

The graph shows the levels of glucose in the blood of two people during a five-hour period immediately following the ingestion of a typical meal. Use the graph to answer questions 19 and 20.

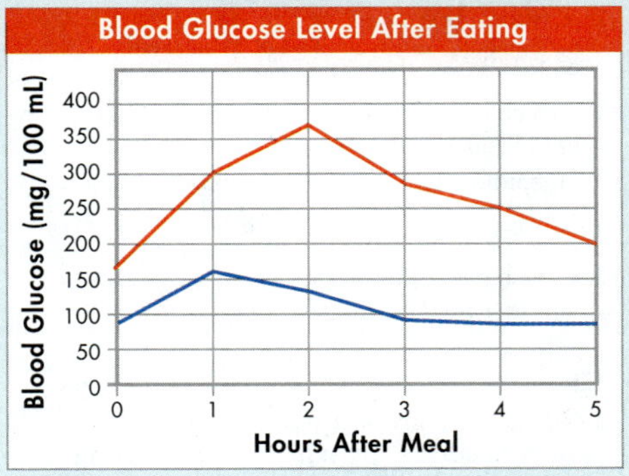

19. Interpret Graphs How long does it take the blood glucose level of the person represented by the blue line to return to a homeostatic value?

20. Draw Conclusions Which line represents a person who may have diabetes? Which line represents a person who does not have diabetes? Explain your answer.

solve the CHAPTER MYSTERY

OUT OF STRIDE

A healthful diet and exercise contribute to maintaining a healthy body. But a balance between the two is important. Lisa lost this balance. The reactions from her endocrine system led to a disorder known as the female athlete triad. The triad consists of three factors:

- **Disordered Eating** Lisa did not provide her body with enough nutrients and energy to support all of its functions.
- **Amenorrhea** Lack of menstrual cycles for three or more months is called amenorrhea. Lisa's hypothalamus detected her low energy levels. As a result, it did not signal the pituitary to release FSH and LH. This caused her menstrual cycle to stop and her estrogen levels to drop.
- **Weakened Bones** Lisa's bones lost more calcium than normal because of high cortisol and low estrogen levels. She also did not consume enough calcium because of her poor diet. This made her bones weak. Weak bones are at risk for stress fractures.

The problems associated with the female athlete triad are related to poor nutrition. Lisa used more energy and nutrients than she took in. The reaction of her endocrine system was normal, but it had negative effects on her health.

1. Relate Cause and Effect Explain why the menstrual cycle cannot continue without FSH and LH.

2. In your own words, explain what led to Lisa's stress fracture. What are some ways that Lisa can prevent this from happening again?

UntamedScience™ **Never Stop Exploring Your World.** Finding the solution to the Out of Stride mystery is only the beginning. Take a video field trip with the ecogeeks of Untamed Science to see where this mystery leads.

CHAPTER MYSTERY ANSWERS

1. The menstrual cycle cannot continue without FSH and LH because FSH and LH are necessary for follicles to mature. If the follicles do not reach maturity, the cells surrounding the eggs will not be triggered to release estrogens that cause the uterine lining to thicken.

2. Sample answer: Lisa had a stress fracture because she had a poor diet, a low estrogen level, and a high cortisol level. To prevent another stress fracture, Lisa should consume a balanced diet that supplies enough nutrients and energy for her body based on her activity level.

Untamed Science™ For more on the endocrine system, suggest students watch the Untamed Science video **Fight or Flight!**

Standardized Test Prep

Multiple Choice

1. Which sequence correctly describes the route sperm take through the human male reproductive system?
A vas deferens, urethra, epididymis
B epididymis, vas deferens, urethra
C vas deferens, epididymis, urethra
D urethra, epididymis, vas deferens

2. Each of these terms refers to a stage in the human menstrual cycle EXCEPT
A ovulation. C corpus phase.
B luteal phase. D follicular phase.

3. Which of the following is NOT an endocrine gland?
A pituitary gland C sweat gland
B parathyroid gland D adrenal gland

4. During which stage of embryonic development does the neural tube form?
A implantation C neurulation
B gastrulation D fertilization

5. Which of the following is a cluster of cells that surround a developing egg?
A follicle C ovary
B blastocyst D ovum

6. The structure(s) in the male reproductive system that stores mature sperm until they are released by the male reproductive system is (are) the
A vas deferens. C seminiferous tubules.
B penis. D epididymis.

7. Which statement best describes the relationship between the hypothalamus and the pituitary gland?
A The anterior pituitary gland makes hormones that are released by the hypothalamus.
B The hypothalamus produces releasing hormones that promote the release of particular hormones from the anterior pituitary.
C The hypothalamus produces releasing hormones that promote the release of particular hormones from the posterior pituitary.
D The posterior pituitary sends nervous signals to the hypothalamus to prompt the release of hormones.

Questions 8–11

The diagram below shows the female endocrine system. Use the diagram to answer the questions.

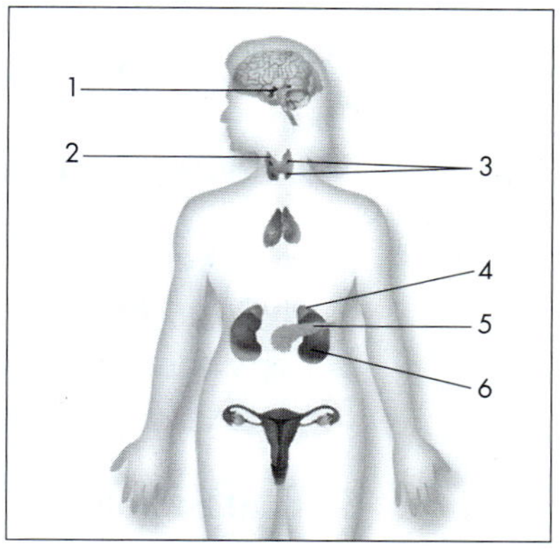

8. Which gland helps the body prepare for and deal with stress?
A 1 C 4
B 2 D 6

9. Which gland is both an endocrine and an exocrine gland?
A 2 C 4
B 3 D 5

10. Which gland secretes ADH?
A 1 C 3
B 2 D 5

11. Which gland secretes thyroxine?
A 1 C 3
B 2 D 4

Open-Ended Response

12. In a paragraph, describe the difference between the origins of fraternal and identical twins.

If You Have Trouble With . . .

Question	1	2	3	4	5	6	7	8	9	10	11	12
See Lesson	34.3	34.3	34.1	34.4	34.3	34.3	34.2	34.2	34.2	34.2	34.2	34.4

Test-Taking Tip

WATCH FOR QUALIFIERS

Explain to students that some questions contain qualifiers, such as *NOT* or *EXCEPT*. For these questions, students should first rule out any answer choices that fit the characteristic in question. Using this approach rules out three of the four choices. Then, to check the answer, a student should confirm that the choice does not fit the characteristic in question.

Before the Test

Review with students the Test-Taking Tip.

Point out the word *EXCEPT* in Question 2.

Point out the word *NOT* in Question 3.

Remind students that an open-ended response means that they will have to write a paragraph to answer the question. Have them paraphrase the question in their notebooks, and remind them about the steps you took when you modeled writing a paragraph.

Answers

1. B
2. C
3. C
4. C
5. A
6. D
7. B
8. C
9. D
10. A
11. B
12. Fraternal twins occur when two eggs are released and each is fertilized. Fraternal twins are not genetically identical. Identical twins occur when a single zygote splits and becomes two embryos. Identical twins are genetically identical.

Immune System and Disease System

Chapter Contents	NSES	Time	Core Resources
Chapter Preview			**Student Edition,** pp. 836–837 **Chapter Mystery,** p. 837 **Foundations for Learning,** p. 837
35.1 Infectious Disease Causes of Infectious Disease • How Diseases Spread	I, II, V, A.2.a, C.4.c, C.6.b, F.1	1/2 period 1/4 block	**Student Edition,** pp. 838–840 **Study Workbook B** 35.1 Worksheets **Biology.com** *Art Review:* Agents of Disease • 35.1 Lesson Assessment
35.2 Defenses Against Infection Nonspecific Defenses • Specific Defenses: The Immune System • The Immune System in Action	I, V, A.1.c, A.2.a, C.1.a, C.1.d, F.1	1 1/2 periods 3/4 block	**Student Edition,** pp. 841–845 **Build Connections,** p. 844 **Inquiry Into Scientific Thinking,** p. 845 **Study Workbook B** 35.2 Worksheets **Biology.com** 35.2 Lesson Assessment
35.3 Fighting Infectious Disease Acquired Immunity • Public Health and Medications • New and Re-Emerging Diseases	II, IV, A.2.a, A.2.b, C.1.a, C.4.e, F.1, F.5, F.6, G.3	1/2 period 1/4 block	**Student Edition,** pp. 846–848 **Study Workbook B** 35.3 Worksheets **Biology.com** 35.3 Lesson Assessment
35.4 Immune System Disorders When the Immune System Overreacts • HIV and AIDS	I, V, C.1.a, F.1	1 period 1/2 block	**Student Edition,** pp. 849–851 **Build Connections,** p. 850 **Study Workbook B** 35.4 Worksheets **Biology.com** *Art in Motion:* HIV Infection • 35.4 Lesson Assessment
Chapter Pre-Lab	II, A.1.b, A.2.a, A.2.b, A.2.c, F.1, F.5	1 period 1/2 block	**Student Edition,** p. 852 **Lab Manual B** *Detecting Lyme Disease*

Tools for ELL Instruction

ELL Handbook describes teaching strategies to make *Miller & Levine Biology: Foundation Edition* more accessible to ELL students.

Spanish Study Workbook is a Spanish translation of Study Workbook A.

Multilingual Glossary is the *Miller & Levine Biology* glossary translated into ten languages.

Tools to Challenge Students

Study Workbook A includes more challenging worksheets and activities such as Chapter Mystery activities that reinforce 21st Century Skills.

Lab Manual A labs are geared toward on-level students. The Lab Manual includes fifteen labs in addition to the Chapter Labs.

Biology.com Data Analysis activities offer students a way to practice data gathering and analyzing skills.

Visual Quizzes in the Assessment Resources Book are geared toward on-level students and use visuals from *Miller & Levine Biology.*

Additional Resources

Biology.com Untamed Science Video • Flash Cards

Study Workbook A 35.1 Worksheets
Spanish Study Workbook 35.1 Worksheets

Study Workbook A 35.2 Worksheets
Lab Manual B *Data Analysis:* Immune System "Memory"
Spanish Study Workbook 35.2 Worksheets
Assessment Resources Book Visual Quizzes

Study Workbook A 35.3 Worksheets
Lab Manual B *Hands-On Activity:* How Do Diseases Spread?
Spanish Study Workbook 35.3 Worksheets
Biology.com *Data Analysis:* Society and Immunity

Study Workbook A 35.4 Worksheets
Lab Manual B *Data Analysis:* Food Allergies
Spanish Study Workbook 35.4 Worksheets

Lab Manual A *Detecting Lyme Disease* • *Modeling Disease Transmission*

Chapter Review

Student Edition Chapter Summary, p. 853 • Unit Project, p. 858
Study Workbook B Chapter 35 Review
Transparencies, pp. 382–390
Biology.com Untamed Science Video • Editable Worksheets of Study Workbooks A and B and Lab Manuals A and B • Chapter 35 Flash Cards and Crossword

Untamed Science DVD • Classroom Resources CD (includes lesson presentations and editable worksheets)

Chapter Assessment

Student Edition Check Understanding, pp. 854–857
Study Workbook B Chapter 35 Taking a Standardized Test
Assessment Resources Book Chapter 35 Test B • Unit 8 Test B
Biology.com Chapter 35 Assessment • Editable Worksheets of Chapter 35 Visual Quizzes,* Chapter 35 Tests A and B, and Unit 8 Tests A and B

Exam*View* *Assessment Suite* • Classroom Resources CD (includes lesson presentations and editable worksheets)

Time: 1 period, 1/2 block

* Visual Quizzes are based on visuals from *Miller & Levine Biology*. Review carefully to make sure they are appropriate for your class.

Pressed for Time?

Preview the Chapter Preview the Key Questions and the vocabulary for Lesson 35.2.

Cover the Chapter Quickly Have students read Lesson 35.2 and go over the Inflammatory Response and Specific Immune Response visuals. Then, have students read *Acquired Immunity* in Lesson 35.3.

Assess Assign the 35.2 Check Understanding, Questions 1–3 in the 35.3 Check Understanding, and Questions 5–9 in the Chapter 35 Check Understanding.

CHAPTER 35

Get the Picture

Explain that the people wearing white suits in the picture are German soldiers who are going to investigate a possible outbreak of bird flu. Tell students that bird flu is a potentially deadly disease that can spread from birds to humans.

Say The white jumpsuits with hoods keep germs away from a person's skin, hair, and clothing.

Ask What other protective items are the soldiers wearing? *(goggles, masks, gloves, and boots)*

Ask What is a germ? *(something that causes a disease)* Tell students the germ that causes bird flu is a virus.

Activate Prior Knowledge

Help students recall that many common diseases pass from one person to others.

Ask What is a disease? *(A disease is an illness or sickness.)*

Say Colds and the flu are common human diseases. You may have heard others say "I caught a cold" or "I caught the flu."

Ask What does it mean for a disease to be "caught"? *(to get a disease because it was passed from another person)*

Introduce the concept that microorganisms are responsible for passing many diseases from person to person and also between animals and people.

NATIONAL SCIENCE EDUCATION STANDARDS

UNIFYING CONCEPTS AND PROCESSES

I, II, IV, V

CONTENT

C.1.a, C.1.d, C.4.c, C.4.e, C.6.b, F.1, F.5, F.6, G.3

INQUIRY

A.1.b, A.1.c, A.2.a, A.2.b, A.2.c

35 Immune System and Disease

Big idea **Homeostasis**

Q: How does the body fight against invading organisms that may disrupt homeostasis?

German soldiers on their way to a possible bird flu outbreak

BIOLOGY.com Search Chapter 35 GO • Flash Cards

836

UbD Connect to the Big Idea

Invite a volunteer to read aloud the chapter number and the chapter title. Read the Big Idea and question. Emphasize the question, **How does the body fight against invading organisms that may disrupt homeostasis?**

Set the Purpose

PREVIEW AND PREDICT Draw the graphic organizer on the board. Guide students to the INSIDE feature on the next page. Invite a volunteer to read each lesson number and title.

Ask What are two ways the body can protect itself from an agent that will be harmful if it gets inside the body? *(Sample answer: Keep the harmful agent from entering the body, and attack and kill the agent if it enters.)*

Remind students to search for specific answers to the question above as well as the questions in the graphic organizer as they read.

INSIDE:

- 35.1 Infectious Disease
- 35.2 Defenses Against Infection
- 35.3 Fighting Infectious Disease
- 35.4 Immune System Disorders

- Untamed Science Video
- Chapter Mystery

CHAPTER MYSTERY

THE SEARCH FOR A CAUSE

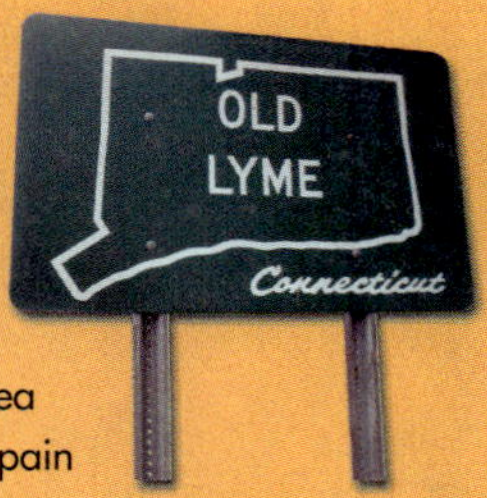

In 1975, researcher Allen Steere faced a medical mystery. Thirty-nine children and several adults living in one small area in Connecticut were suffering from joint pain and inflammation. The children's symptoms looked like a rare form of childhood arthritis. The adults' symptoms seemed like age-related arthritis. But why would so many cases of arthritis appear in such a small population in such a short time?

Steere thought that there must be some other reason that so many people were getting sick. The patients all lived in small towns and rural places. Their symptoms all started at more or less the same time of year. Could all of these people be suffering from a disease no one had discovered yet?

Read for Mystery Clues As you read this chapter, look for clues about this disease. Then, solve the mystery at the end of the chapter.

FOUNDATIONS for Learning

Before you read this chapter, make an undercover vocabulary tool as shown below. As you read, write each vocabulary word from the chapter in the first column. Write the definition in your own words in the second column. Then cut and fold the third column into tabs to cover the definitions. You may need to make more than one undercover vocabulary tool. At the end of the chapter there are two activities that use the undercover vocabulary tool to help you answer the question: How does the body fight against invading organisms that may disrupt homeostasis?

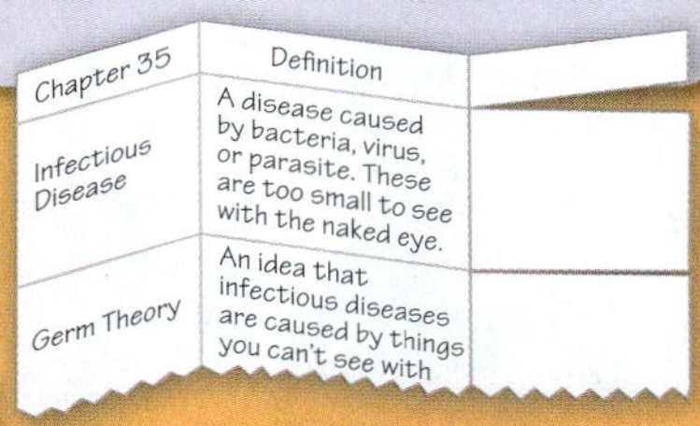

Chapter 35 Big Idea: Homeostasis

Chapter 35: How does the body fight against invading organisms that may disrupt homeostasis?

- **35.1:** How do people contract infectious diseases?
- **35.2:** How does the body defend against infection?
- **35.3:** How do humans prevent and fight the spread of disease?
- **35.4:** What happens when the immune system does not function properly?

CHAPTER 35

Foundations for Learning

Suggest that students use lined notebook paper. Emphasize that students are to record each vocabulary word and its definition as they read the chapter. Encourage them to write thorough definitions and to use additional sheets of paper, if needed. Explain that they will produce an undercover vocabulary tool.

Ask Why is it important for the body to maintain homeostasis? *(Sample answer: Life processes require certain conditions. The body cannot continue to live if vital processes do not function properly.)*

Encourage students to think about how each concept identified by a vocabulary term relates to homeostasis in the human body.

Ask How could you use the finished undercover vocabulary tools? *(Students may suggest working alone, in pairs, or in small groups to test their understanding of the vocabulary terms.)*

Untamed Science™ It's not just work that your students are allergic to. Follow the Untamed Science crew as they investigate what truly triggers allergic reactions in **A Reason for Sneezin'.**

CHAPTER MYSTERY Invite a volunteer to read aloud the Chapter Mystery.

Say Arthritis is usually caused by wear and tear in a person's joints over a long time.

Ask What could cause arthritis-like symptoms in both young and old people in a short period of time? *(Accept all reasonable explanations.)* Write students' responses on the board.

Have students predict which explanation is most likely to be true. As they read through the chapter, have students gather information that shows if their prediction was correct or incorrect.

Digital Support

BIOLOGY.com Extend your reach by using the digital assets offered at **Biology.com.**

LESSON 35.1

Pre-Reading

Objectives

35.1.1 Identify the causes of infectious disease.

35.1.2 Explain how infectious diseases are spread.

Preview the Pages

1. Preview the vocabulary terms by having students find them in the text and read each one aloud. Encourage students to discuss how knowing about each of these terms is important to their health.
2. Preview the visuals and read the captions. Remind students of previous chapters that contain information relating to each kind of disease-causing agent.
3. Preview Koch's postulates. Explain that these are steps in a scientific investigation that is conducted to determine the cause of an infectious disease.

Activate Prior Knowledge

Invite volunteers to share what they know about how diseases spread. Write their ideas in a list on the board and then return to the list after the lesson to correct any misconceptions.

Build Understanding

TWO-COLUMN CHART Direct students to Workbook B, where they will find a two-column chart. Encourage students to complete the chart as they read with descriptions of the ways that diseases are spread.

NATIONAL SCIENCE EDUCATION STANDARDS

UNIFYING CONCEPTS AND PROCESSES

I, II, V

CONTENT

C.4.c, C.6.b, F.1

INQUIRY

A.2.a

For the corresponding lesson in *Miller & Levine Biology,* see pages 1010–1013.

35.1 Infectious Disease

Key Question

What causes infectious disease?

How are infectious diseases spread?

BUILD Understanding

Two-Column Chart Use a two-column chart to list the ways diseases are spread and describe each way.

In Your Workbook Go to your workbook for a two-column chart to complete.

Causes of Infectious Disease

For thousands of years, people believed that sicknesses were caused by evil spirits, steam rising from dirty marshes, or dead plants and animals. This is not all that surprising. Until people had microscopes, they were not able to see most causes of disease!

About 150 years ago two scientists worked out a new explanation for how people get sick. Based on their observations and experiments, French chemist Louis Pasteur and German bacteriologist Robert Koch discovered that **infectious diseases** are caused by microorganisms. Infectious diseases cause changes to the body that disrupt normal body functions. The theory that microorganisms cause infectious diseases was called the **germ theory of disease.** Microorganisms used to be called germs. But today, the word *germ* has no scientific meaning.

Pathogens and Parasites If *germ* is not a scientific term, how should we describe the causes of infectious disease? Infectious diseases can be caused by viruses, bacteria, "protists," fungi, and parasites. Anything that can cause an infectious disease is called a pathogen.

Key Question What causes infectious disease?
Infectious diseases are caused by microorganisms. Diseases can be caused by viruses, bacteria, fungi, "protists," or parasites.

Examples of Agents of Disease Infectious diseases are caused by pathogens and parasites. These organisms invade a body and disrupt its normal functions.

Viruses
Characteristics: nonliving, replicate by inserting their genetic material into a host cell and taking over many of the host cell's functions
Diseases Caused: common cold, influenza, chicken pox, warts

▼ ***Influenza Virus,* Strain taken from a Beijing 1993 epidemic** (TEM 120,000×)

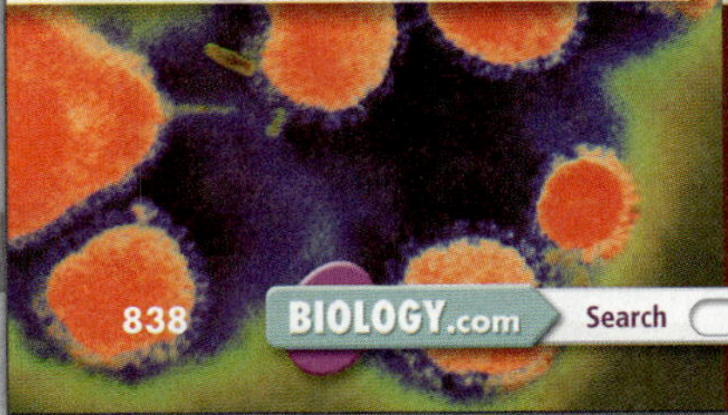

Bacteria
Characteristics: break down the tissues of an infected organism for food, or release toxins that interfere with normal activity in the host
Diseases Caused: streptococcus infections, diphtheria, botulism, anthrax

▼ ***Mycobacterium* causes tuberculosis** (SEM 10,600×)

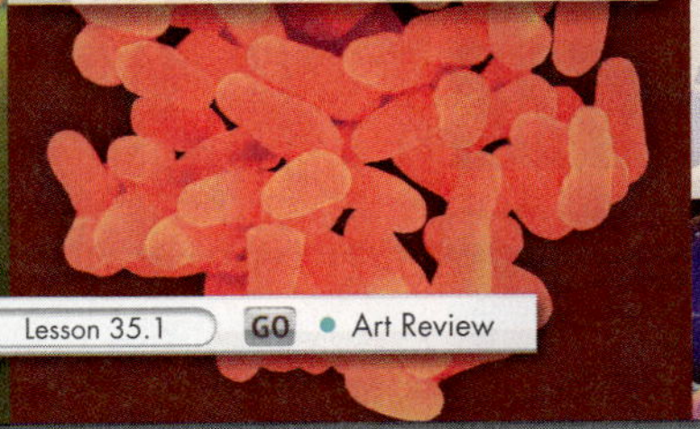

Fungi
Characteristics: cause infections on the surface of the skin, mouth, throat, fingernails, and toenails; dangerous infections may spread from lungs to other organs
Diseases Caused: ringworm, thrush

▼ ***Trichophyton interdigitale* causes athlete's foot** (SEM 2800×)

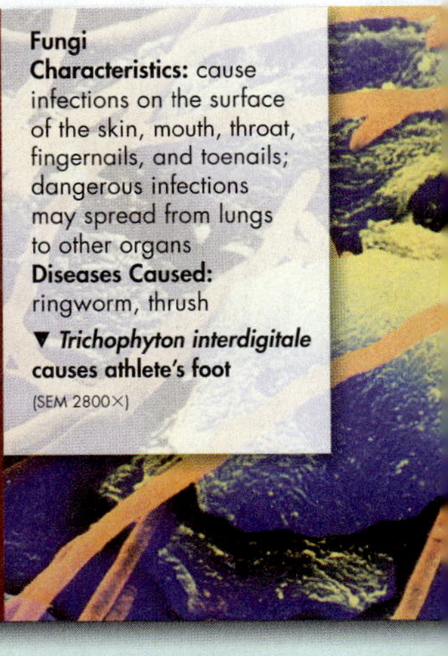

BIOLOGY.com Search Lesson 35.1 GO • Art Review

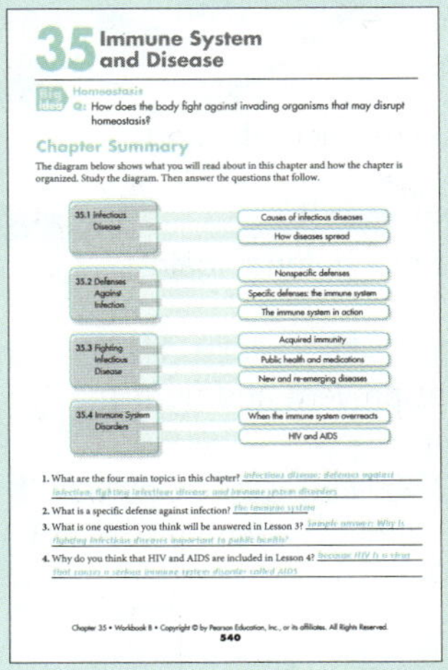

35 Immune System and Disease

Homeostasis
Q: How does the body fight against invading organisms that may disrupt homeostasis?

Chapter Summary

The diagram below shows what you will read about in this chapter and how the chapter is organized. Study the diagram. Then answer the questions that follow.

- 35.1 Infectious Disease: Causes of infectious diseases; How diseases spread
- 35.2 Defenses Against Infection: Nonspecific defenses; Specific defenses: the immune system; The immune system in action
- 35.3 Fighting Infectious Disease: Acquired immunity; Public health and medications; New and re-emerging diseases
- 35.4 Immune System Disorders: When the immune system overreacts; HIV and AIDS

1. What are the four main topics in this chapter? *infectious disease; defenses against infection; fighting infectious disease; and immune system disorders*
2. What is a specific defense against infection? *the immune system*
3. What is one question you think will be answered in Lesson 3? *Sample answer: Why is fighting infectious diseases important to public health?*
4. Why do you think that HIV and AIDS are included in Lesson 4? *Because HIV is a virus that causes a serious immune system disorder called AIDS*

Chapter 35 • Workbook B • Copyright © by Pearson Education, Inc., or its affiliates. All Rights Reserved.
540

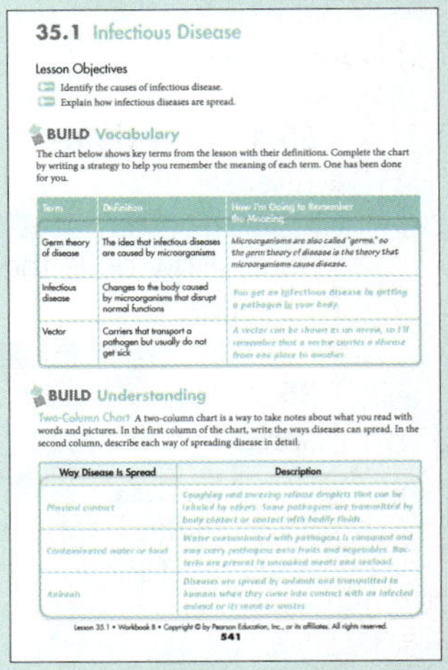

35.1 Infectious Disease

Lesson Objectives
Identify the causes of infectious disease.
Explain how infectious diseases are spread.

BUILD Vocabulary

The chart below shows key terms from the lesson with their definitions. Complete the chart by writing a strategy to help you remember the meaning of each term. One has been done for you.

Term	Definition	How I'm Going to Remember the Meaning
Germ theory of disease	The idea that infectious diseases are caused by microorganisms	*Microorganisms are also called "germs," so the germ theory of disease is the theory that microorganisms cause disease.*
Infectious disease	Changes to the body caused by microorganisms that disrupt normal functions	*You get an infectious disease by getting a pathogen in your body.*
Vector	Carriers that transport a pathogen but usually do not get sick	*A vector can be shown as an arrow, so I'll remember that a vector carries a disease from one place to another.*

BUILD Understanding

Two-Column Chart A two-column chart is a way to take notes about what you read with words and pictures. In the first column of the chart, write the ways diseases can spread. In the second column, describe each way of spreading disease in detail.

Way Disease Is Spread	Description
Physical contact	*Coughing and sneezing release droplets that can be inhaled by others. Some pathogens are transmitted by body contact or contact with bodily fluids.*
Contaminated water or food	*Water contaminated with pathogens is consumed and may carry pathogens onto fruits and vegetables. Bacteria are present in uncooked meats and seafood.*
Animals	*Diseases are spread by animals and transmitted to humans when they come into contact with an infected animal or its meat or wastes.*

Lesson 35.1 • Workbook B • Copyright © by Pearson Education, Inc., or its affiliates. All rights reserved.
541

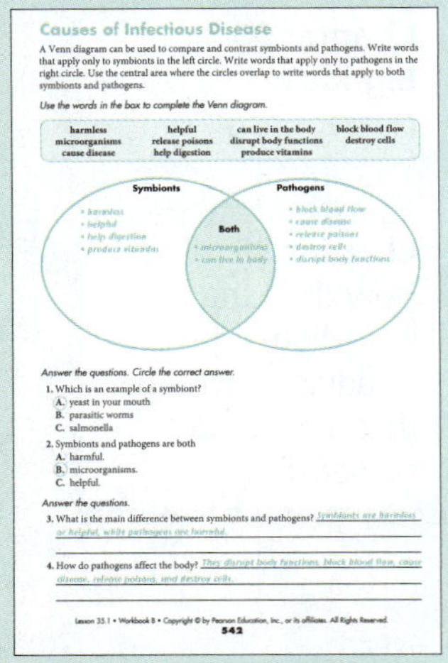

Causes of Infectious Disease

A Venn diagram can be used to compare and contrast symbionts and pathogens. Write words that apply only to symbionts in the left circle. Write words that apply only to pathogens in the right circle. Use the central area where the circles overlap to write words that apply to both symbionts and pathogens.

Use the words in the box to complete the Venn diagram.

harmless, microorganisms, cause disease, helpful, release poisons, help digestion, can live in the body, disrupt body functions, produce vitamins, block blood flow, destroy cells

Answer the questions. Circle the correct answer.

1. Which is an example of a symbiont?
   A. yeast in your mouth
   B. parasitic worms
   C. salmonella
2. Symbionts and pathogens are both
   A. harmful.
   B. microorganisms.
   C. helpful.

Answer the questions.

3. What is the main difference between symbionts and pathogens? *Symbionts are harmless or helpful, while pathogens are harmful.*
4. How do pathogens affect the body? *They disrupt body functions, block blood flow, cause disease, release poisons, and destroy cells.*

Lesson 35.1 • Workbook B • Copyright © by Pearson Education, Inc., or its affiliates. All Rights Reserved.
542

Koch's Postulates Based on his studies, Koch made a list of steps a scientist must follow to help decide which pathogen causes a specific disease. His steps are known as Koch's postulates.

1. The pathogen must always be found in the body of a sick organism and cannot be found in the body of a healthy organism.
2. The pathogen must be isolated from the sick organism and grown in the laboratory.
3. When the pathogen grown in the lab is given to a healthy organism, that organism must get the same illness.
4. The pathogen must be isolated from the second organism. The pathogen must be exactly the same as the original pathogen.

Koch was given a Nobel Prize in 1905. Today, we know that there are exceptions to his rules. Yet they are still very important steps to follow in order to learn the causes of new and emerging diseases.

Symbionts vs. Pathogens Many microorganisms live in our bodies. Most of them are symbionts. They are either harmless or they help us. Yeast and bacteria grow in the mouth and throat without causing trouble. Bacteria in the large intestine help with digestion and make vitamins.

What is the difference between harmless microorganisms and pathogens? The harmless microorganisms get nutrients, grow, and reproduce without harming normal body functions. Pathogens, on the other hand, cause problems in many ways. Some pathogens kill the cells of their host. Others may block blood flow or take up the host's nutrients.

How Diseases Spread

Pathogens cause disease. But how are pathogens spread from one person to the next? Some diseases spread through coughing, sneezing, or physical contact. Others spread through body fluids, or contaminated food or water. Some diseases can be spread to humans from other animals.

BUILD Vocabulary

infectious disease
a disease caused by a microorganism that disrupts normal body functions

germ theory of disease
the idea that infectious diseases are caused by microorganisms

MULTIPLE MEANINGS
The word *isolate* often means "to set apart from others." In biology and medicine, however, *isolate* means "to remove microorganisms from a person's body and then grow the microorganisms."

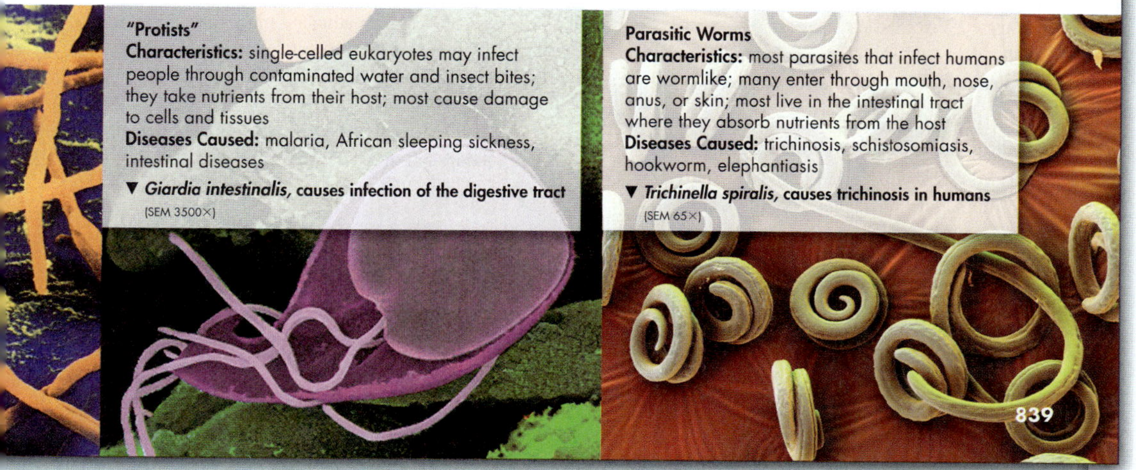

"Protists"
Characteristics: single-celled eukaryotes may infect people through contaminated water and insect bites; they take nutrients from their host; most cause damage to cells and tissues
Diseases Caused: malaria, African sleeping sickness, intestinal diseases
▼ ***Giardia intestinalis*, causes infection of the digestive tract**
(SEM 3500×)

Parasitic Worms
Characteristics: most parasites that infect humans are wormlike; many enter through mouth, nose, anus, or skin; most live in the intestinal tract where they absorb nutrients from the host
Diseases Caused: trichinosis, schistosomiasis, hookworm, elephantiasis
▼ ***Trichinella spiralis*, causes trichinosis in humans**
(SEM 65×)

839

Focus on Academic Language

- single-celled

Point out to students how hyphenated words can be used as modifiers.

Ask What does the hyphen in *single-celled* do? *(It links the two words in a compound adjective so that the reader knows the two words represent one adjective.)*

Ask Why is *single-celled* used in front of *eukaryotes* in the label for protists in the Examples of Agents of Disease image? *(Single-celled clarifies that these eukaryotes consist of one cell.)*

Ask What hyphenated word could you use to describe pathogens that cause diseases? *(disease-causing)*

Discuss with students other purposes for hyphens. *(Sample answers: to divide a word at the end of a line of text, to write out numbers, in expressions of time, to show syllabication)*

During Reading

Build Vocabulary

MULTIPLE MEANINGS Have students find sentences on this page that use the noun *host*. Ask volunteers to read the sentences aloud.

Ask What is a common meaning of the noun *host*? *(a person who invites guests to a social event; an emcee on a radio or television show)*

Ask What does *host* mean in biology? *(a living thing on which or in which another organism lives)*

Active Reading

FIND THE MAIN IDEA Invite a volunteer to read aloud the Pathogens and Parasites paragraph.

Ask What is the main idea of this paragraph? *(Answers may vary.)* Write some student responses on the board. Then, help students refine the responses into one main idea. *(The many different kinds of microorganisms and parasites that cause disease are called pathogens.)*

KEY QUESTION Read aloud the Key Question and answer from the previous page.

Students may not remember the characteristics of viruses, bacteria, fungi, "protists," and parasites. Encourage them to use the Examples of Agents of Disease image at the bottom of these two pages to make a comparison chart.

SUMMARIZE Have volunteers read Koch's postulates. After each postulate is read, help students compose a simplified version of the postulate and write it on the board. Encourage students to copy the list of simplified postulates in their notebooks. *(Sample answers: 1. Find pathogen in sick organism. 2. Grow pathogen in laboratory. 3. Put pathogen in healthy organism. 4. Find pathogen in now sick organism.)*

MAKE CONNECTIONS Have a volunteer read the first paragraph of Symbionts vs. Pathogens.

Say Yogurt contains bacteria that are symbionts.

Ask How do you think eating yogurt could help a person's digestive system? *(Eating yogurt may add bacteria to the digestive tract that can help with digestion and make vitamins.)*

Speed Bump

Ask How do viruses and bacteria differ in the way they cause diseases? *(Viruses take over cell functions. Bacteria break down tissues.)*

Ask What does a pathogen get from the organism it infects? *(food and a place to live and reproduce)*

LESSON 35.1

During Reading continued

Active Reading

USE VISUALS Draw students' attention to the photograph.

Ask Would you want to be near this person during the sneeze? Explain. *(No, because I might catch a disease from pathogens in the droplets.)*

MAKE INFERENCES Ask students to infer what would have happened if the sneezer had covered his mouth with a sleeve or tissue? *(Fewer droplets would have been released into the air.)*

KEY QUESTION Read aloud the Key Question and answer.

Ask If a disease is spread by coughing or sneezing, what must the droplets of fluid that come out of the body carry? *(the pathogen that causes the disease)*

Build Vocabulary

WORD ORIGINS Explain that *infectious* comes from a Latin word meaning "to dip into" or "to stain." Discuss how these meanings relate to *infectious.* *(Sample answer: They refer to unwelcome contact.)*

Speed Bump

Ask What are some ways that pathogens get inside the body? *(They are breathed in, swallowed with liquids or foods, injected under the skin by bites or needles, or transferred by touch or sexual activity.)*

Post Reading

ASSESSMENT TIPS For Question 3, encourage students to include at least two sentences in their answers.

For Question 4, point out the hint and be sure students understand how viruses replicate.

MYSTERY CLUE After students read the clue, suggest they locate the key term *vector* on this page and reread the information.

Wrap-Up Activity

MODEL SPREADING GERMS Give each group of four students a wooden block, white glue, and a different color of glitter. Tell students to smear a thin layer of glue on one side of the block, sprinkle glitter on it, pass it around their group, and then pass it to another group. Continue for five minutes. Have students note the colors of glitter on their hands. Discuss how this activity models the passing of pathogens. Upon completion, be sure students wash their hands and the blocks.

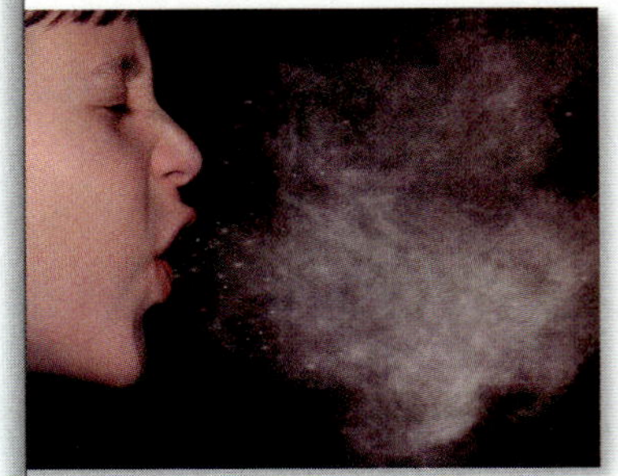

Sneezing Some infectious diseases are spread from person to person by sneezing. Thousands of pathogen particles can be released in a sneeze.

BUILD Vocabulary

vector
an animal that transports a pathogen to a human

WORD ORIGINS
Vector comes from the Latin word for "bearer" or "carrier." A vector carries the pathogen or parasite to the host.

Coughing, Sneezing, and Physical Contact Coughing and sneezing send thousands of tiny droplets flying into the air. Other people can breathe these into their bodies. Some droplets may land on things other people touch. Some pathogens can be spread through almost any body-to-body contact and through towels and some sports equipment. The best way to slow the spread of these pathogens is by washing your hands well and often. If you are ill, cover your mouth when you cough or sneeze.

Exchange of Body Fluids Some pathogens are spread only through specific kinds of direct contact. Sexually transmitted diseases, including chlamydia and AIDS, are spread through sexual contact. Some, including AIDS, may also be spread when drug users share needles. Diseases that are transmitted through blood and body fluids can be completely prevented only by avoiding sexual activity and drug use.

Contaminated Water or Food Many pathogens are spread through water contaminated with feces from infected people or other animals. People may drink the water. The water may also carry pathogens onto fruits or vegetables. Pathogens may also be present in seafood and uncooked meat. These foods must be stored and cooked properly.

Animals Diseases that can be spread from other animals to people are called zoonoses. Mad cow disease, Lyme disease, and Ebola are all zoonoses. An animal that carries one of these diseases from an animal host to a human host is called a **vector.** Vectors do not usually get sick themselves. A person can also pick up pathogens from an animal if the person is bitten or comes in close contact with a sick animal.

Key Question How are infectious diseases spread?
Infectious diseases can be spread through coughing, sneezing, or physical contact. Others spread through body fluids, contaminated food or water, or through bites or contact with other animals.

CHECK Understanding

Apply Vocabulary
Use the highlighted words from the lesson to complete each sentence correctly.

1. The _______ states that any infectious disease is caused by microorganisms.
2. West Nile virus is transferred from birds to humans by mosquitos. Mosquitos are _______.

Critical Thinking

3. **Explain** What are some ways that pathogens can cause disease in their hosts?
4. **Apply Concepts** Many pathogens make their hosts very sick without killing the host. Why do you think this is a beneficial adaptation? (**Hint:** Think about how viruses replicate.)
5. **Write to Learn** Answer the first clue of the mystery. Write a paragraph that uses the term *vector*.

MYSTERY CLUE

Many of the sick children remembered getting strange insect bites that summer. The bites developed into rashes. What clue did this give Steere? (Hint: See above.)

BIOLOGY.com Search Lesson 35.1 GO • Lesson Assessment

840

Check Understanding

1. germ theory of disease
2. vectors
3. Some pathogens kill the cells of their host. Others may block blood flow or take up the host's nutrients.
4. Sample answer: The pathogen often needs the host to stay alive in order for the pathogen to reproduce or replicate. For example, viruses replicate inside living host cells. Pathogens also depend on changes in host behavior, such as coughing and sneezing, to spread and infect new hosts.
5. This probably led Steere to think that an insect may have been involved in spreading this disease. Insects that can spread pathogens from animal hosts to humans are called vectors. Perhaps a mosquito or other insect acted as a vector in this case.

35.2 Defenses Against Infection

Nonspecific Defenses

There are pathogens all around us. Yet we are not usually sick. How do we stay healthy? How do we get better when we are sick? Our bodies have several amazing lines of defense.

First Line of Defense The skin blocks pathogens from entering the body. Cilia in the respiratory tract push pathogens away before they enter the body. Tears, saliva, and mucus have chemicals that break down bacterial cell walls.

Second Line of Defense If a pathogen gets past the skin, the body's second line of defense swings into action.

▶ ***The Inflammatory Response*** When a pathogen enters the body, the **inflammatory response** causes the infected area to become red and painful. Cells called mast cells release histamines. Histamines are chemicals that increase the flow of blood and other fluids to the area around the wound. This fluid can cause swelling. White blood cells, or phagocytes, move from the blood vessels into the tissue. Some of the white blood cells surround and destroy bacteria.

Key Questions

What are the body's nonspecific defenses against pathogens?

What are the functions of the immune system's specific defenses?

What are the body's specific defenses against pathogens?

BUILD Understanding

Concept Map Use the headings in this lesson to make a concept map. Add details to your map as you read.

In Your Workbook Go to your workbook for a partially completed concept map for Lesson 35.2.

Inflammatory Response The inflammatory response is a nonspecific defense against pathogens. When pathogens enter the body, white blood cells move into the area and destroy the pathogens.

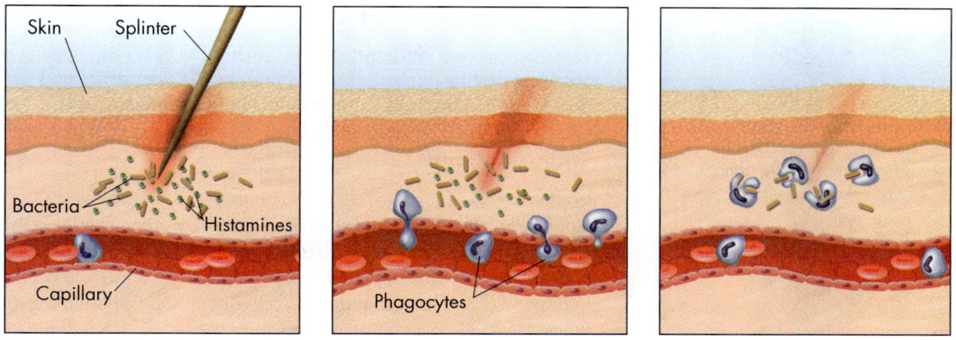

1 In response to the wound and invading pathogens, histamines increase blood flow to the area.

2 Local blood vessels open. Fluid leaves the capillaries and causes swelling. White blood cells move into the tissue.

3 White blood cells called phagocytes destroy the bacteria and damaged cells.

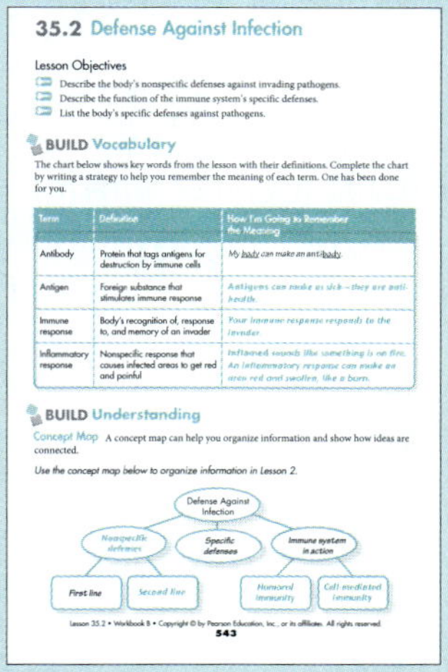

35.2 Defense Against Infection

Lesson Objectives

- Describe the body's nonspecific defenses against invading pathogens.
- Describe the function of the immune system's specific defenses.
- List the body's specific defenses against pathogens.

BUILD Vocabulary

The chart below shows key words from the lesson with their definitions. Complete the chart by writing a strategy to help you remember the meaning of each term. One has been done for you.

Term	Definition	How I'm Going to Remember the Meaning
Antibody	Protein that tags antigens for destruction by immune cells	*My body can make an anti-body.*
Antigen	Foreign substance that stimulates immune response	*Antigens can make us sick—they are anti-health.*
Immune response	Body's recognition of, response to, and memory of an invader	*Your immune response responds to the invader.*
Inflammatory response	Nonspecific response that causes infected areas to get red and painful	*Inflamed sounds like something is on fire. An inflammatory response can make an area red and swollen, like a burn.*

BUILD Understanding

Concept Map A concept map can help you organize information and show how ideas are connected.

Use the concept map below to organize information in Lesson 2.

Defense Against Infection → Nonspecific defenses (First line, Second line); Specific defenses; Immune system in action (Humoral immunity, Cell-mediated immunity)

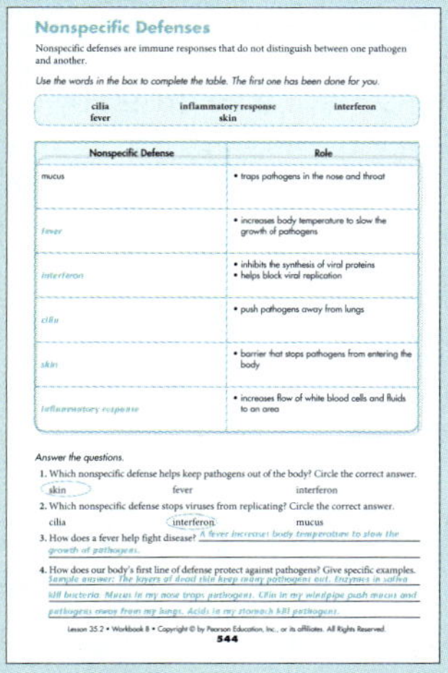

Nonspecific Defenses

Nonspecific defenses are immune responses that do not distinguish between one pathogen and another.

Use the words in the box to complete the table. The first one has been done for you.

cilia, inflammatory response, interferon, fever, skin

Nonspecific Defense	Role
mucus	• traps pathogens in the nose and throat
fever	• increases body temperature to slow the growth of pathogens
interferon	• inhibits the synthesis of viral proteins • helps block viral replication
cilia	• push pathogens away from lungs
skin	• barrier that stops pathogens from entering the body
inflammatory response	• increases flow of white blood cells and fluids to an area

Answer the questions.

1. Which nonspecific defense helps keep pathogens out of the body? Circle the correct answer.
   skin (circled) fever interferon
2. Which nonspecific defense stops viruses from replicating? Circle the correct answer.
   cilia interferon (circled) mucus
3. How does a fever help fight disease? *A fever increases body temperature to slow the growth of pathogens.*
4. How does our body's first line of defense protect against pathogens? Give specific examples. *Sample answer: The layers of dead skin keep many pathogens out. Enzymes in saliva kill bacteria. Mucus in my nose traps pathogens. Cilia in my windpipe push mucus and pathogens away from my lungs. Acids in my stomach kill pathogens.*

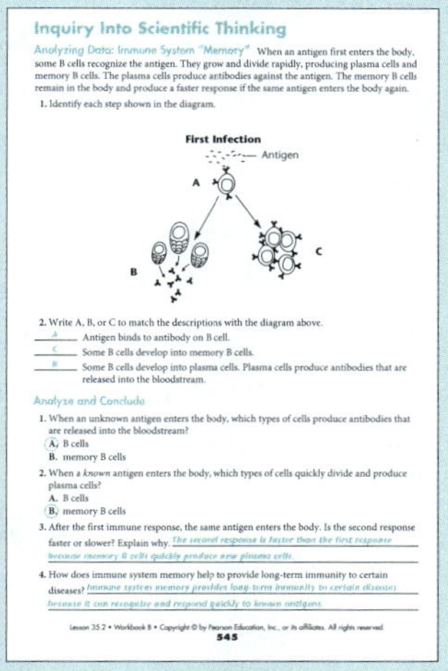

Inquiry Into Scientific Thinking

Analyzing Data: Immune System "Memory" When an antigen first enters the body, some B cells recognize the antigen. They grow and divide rapidly, producing plasma cells and memory B cells. The plasma cells produce antibodies against the antigen. The memory B cells remain in the body and produce a faster response if the same antigen enters the body again.

1. Identify each step shown in the diagram.

First Infection — Antigen — A — B — C

2. Write A, B, or C to match the descriptions with the diagram above.
   A Antigen binds to antibody on B cell.
   C Some B cells develop into memory B cells.
   B Some B cells develop into plasma cells. Plasma cells produce antibodies that are released into the bloodstream.

Analyze and Conclude

1. When an unknown antigen enters the body, which types of cells produce antibodies that are released into the bloodstream?
   A. B cells (circled)
   B. memory B cells
2. When a *known* antigen enters the body, which types of cells quickly divide and produce plasma cells?
   A. B cells
   B. memory B cells (circled)
3. After the first immune response, the same antigen enters the body. Is the second response faster or slower? Explain why. *The second response is faster than the first response because memory B cells quickly produce new plasma cells.*
4. How does immune system memory help to provide long-term immunity to certain diseases? *Immune system memory provides long-term immunity to certain diseases because it can recognize and respond quickly to known antigens.*

LESSON 35.2

Pre-Reading

Objectives

35.2.1 Describe the body's nonspecific defenses against invading pathogens.

35.2.2 Describe the function of the immune system's specific defenses.

35.2.3 List the body's specific defenses against pathogens.

Preview the Pages

1. Preview the vocabulary terms by having students find each term in the text and read it aloud. Encourage students to add the terms to the undercover vocabulary begun in Lesson 35.1.
2. Preview the visuals and read the captions and labels. Have students note the appearance of the different kinds of cells shown individually.

Activate Prior Knowledge

Ask What happens to the skin around a two-day-old cut that is really sore? *(Sample answer: The skin around the cut gets red, swollen, and warm.)*

Ask What do you think causes these symptoms to happen? *(Pathogens get into the person's body through the cut.)*

Build Understanding

CONCEPT MAP Direct students to Workbook B, where they will find a partially filled-in concept map that summarizes the defenses the body uses to protect itself from pathogens. Remind them to fill in the concept map as they read about each defense.

NATIONAL SCIENCE EDUCATION STANDARDS

UNIFYING CONCEPT AND PROCESSES

I, V

CONTENT

C.1.a, C.1.d, F.1

INQUIRY STANDARDS

A.1.c, A.2.a

For the corresponding lesson in *Miller & Levine Biology,* see pages 1014–1019.

During Reading

Build Vocabulary

WORD ORIGINS Ask students what the words *antibody* and *antigen* have in common. *(Both have the word part* anti-*.)* Explain that *anti* is Greek and means "against."

Ask What do antibodies work against? *(Antibodies work against invading pathogens.)*

Say In the word *antigen,* the word part *anti-* is a shortened form of *antibody.* The suffix *-gen* means "producing" or "generating." So, the combined meaning of the two word parts is "producing antibodies."

Ask Does that meaning make sense? Why? *(yes, because antigens are substances that trigger an immune response, including the production of antibodies)*

Active Reading

KEY QUESTION Invite a volunteer to read aloud the first Key Question and answer on this page.

Ask What do nonspecific defenses respond to? *(any pathogens that enter the body)*

DRAW CONCLUSIONS Have volunteers read aloud the paragraphs Recognizing "Self" and Recognizing "Nonself."

Ask What adjective is used to describe an immune system that recognizes self and nonself? *(healthy)* How, then, might you describe an immune system that does not always recognize self and nonself? *(Sample answers: unhealthy, flawed, deficient, defective)*

PREDICT Have students think about how people identify and recognize objects.

Ask What do the makers of CDs and brand-name clothes and foods use to help you recognize their products? *(Possible answers: labels, tags, markers)*

Say Your body also has ways of labeling parts that belong to it.

Have students read the Antigens paragraphs.

Ask What are the names of the labels or markers your body recognizes as nonself? *(antigens)*

KEY QUESTION Invite a volunteer to read aloud the second Key Question and answer on this page.

Ask How does the immune system recognize the cells and proteins that belong in the body as "self"? *(The body's cells and proteins carry chemical markers that tell the immune system not to attack.)*

BUILD Vocabulary

inflammatory response
the nonspecific defense reaction to tissue damage caused by injury or infection

immune response
the body's specific recognition of, response to, and memory of a pathogen attack

antigen
a substance that triggers an immune response

antibody
a protein that either attacks antigens directly or produces antigen-binding proteins

WORD ORIGINS
The word part *flamm-* comes from the Latin word that means "to burn." The inflammatory response often causes the area around the wound to become warm to the touch.

▶ ***Interferons*** When a virus enters the body, the body makes interferons. These proteins make it harder for viruses to make their own proteins. In this way, interferons slow down the infection.

▶ ***Fever*** Body temperature may rise during an infection. Fever may slow down or stop the growth of some pathogens. It also speeds up several parts of the immune response.

Key Question What are the body's nonspecific defenses against pathogens? **Nonspecific defenses include skin, hair, tears, the inflammatory response, interferons, and fever.**

Specific Defenses: The Immune System

The body's specific defenses attack particular pathogens with powerful weapons.

Recognizing "Self" A healthy immune system recognizes cells and proteins that belong in the body as "self." Your body's cells and proteins carry chemical markers that act like a secret password. That password tells the immune system "Don't attack me! I belong here!"

Recognizing "Nonself" The immune system's main job is to recognize anything that does not belong in the body. When the system notices something that is "nonself," it attacks. This reaction to specific pathogens is the **immune response.** If the same pathogen enters the body again, the body "recognizes" and attacks it even more quickly.

Antigens The immune response is triggered by certain molecules on the outer surfaces of bacteria, viruses, and parasites. These molecules are called **antigens.** The immune system responds to antigens in two ways. Certain body cells may attack the pathogen directly. Other cells make proteins called antibodies.

The shape of each **antibody** allows it to bind to only one antigen. Some antibodies are attached to immune cells. Other antibodies float around in plasma. Antibodies that bind to antigens act like signal flags. Those signal flags tell the immune system to destroy pathogens.

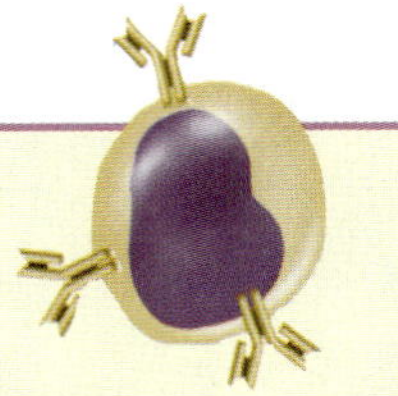

B Lymphocytes The antibodies on the surface of a B lymphocyte recognize one specific antigen.

T Lymphocytes T lymphocytes recognize infected body cells. Each T lymphocyte recognizes only one antigen.

Lymphocytes The main cells in the immune response are white blood cells called B lymphocytes (B cells) and T lymphocytes (T cells). B cells are made in red bone marrow. T cells are also made in red bone marrow, but mature in the thymus. Each B cell and T cell is able to respond to *one* specific antigen. B cells discover antigens in body fluids. T cells discover infected cells.

Key Question What are the functions of the immune system's specific defenses? **The specific defenses distinguish between "self" and "nonself" and respond to specific pathogens.**

ELL Focus on ELL: Extend Language

BEGINNING SPEAKERS Students may be confused by the usage of *marker* (on this page), which has multiple meanings, and *tag* (in the Build Connections caption), which can be used as either a noun or a verb. Explain that in this lesson both terms are used in explaining the functions of antigens and antibodies.

Ask What are some meanings of *marker*? *(Possible answers: a felt-tip pen, a sign that shows where or what something is)*

Ask What does *marker* mean on this page? *(a sign that shows where or what something is)* Help students understand that an antigen is like a sign that can be read by immune cells.

Ask What does *tag* mean when used as a noun? *(a small label, a game of chase)*

Ask What does *tag* mean when it's used as a verb? (fasten something onto) Point out that *tag* is used as a verb in this lesson that describes what antibodies do.

The Immune System in Action

There are two major parts of the specific immune response. They are humoral immunity and cell-mediated immunity.

Humoral Immunity Sometimes antibodies on B cells bind directly to the antigens on the surface of a pathogen. This part of the immune response is called humoral immunity. An antibody is shaped like the letter Y. It has two antigen-binding sites. Both sites recognize the same specific antigen. When the antigen and antibody on the B cell bind, T cells stimulate the B cell to grow and divide quickly. Two kinds of B cells are formed: plasma cells and memory B cells.

▶ ***Plasma Cells*** Plasma cells make antibodies and put them into the bloodstream. Antibodies bind to antigens that are free-floating or attached to pathogens. Then many immune system cells and proteins respond and attack the pathogen.

▶ ***Memory B Cells*** Once the pathogen has been "beaten," the plasma cells die. But memory B cells remain. If the same pathogen enters the body again, the memory B cells make plasma cells quickly. This secondary response happens very quickly and gives us long-term immunity to certain diseases.

Cell-Mediated Immunity Some pathogens infect the body's cells. This kind of infection is fought by T cells and macrophages. The response is called cell-mediated immunity. T cells also fight the body's own cells if they become cancerous.

Macrophages are B cells that engulf pathogens. When this happens or when a cell has been infected by a pathogen, the cell displays part of the antigen on its surface. This signals the helper T cells to divide and make more helper T cells. These activate B cells and cytotoxic T cells. Helper T cells also make memory T cells. Cytotoxic T cells kill infected body cells. Memory T cells help the immune system respond quickly if the same pathogen enters the body again.

Key Question What are the body's specific defenses against pathogens?
The body's specific defenses are humoral and cell-mediated immunity.

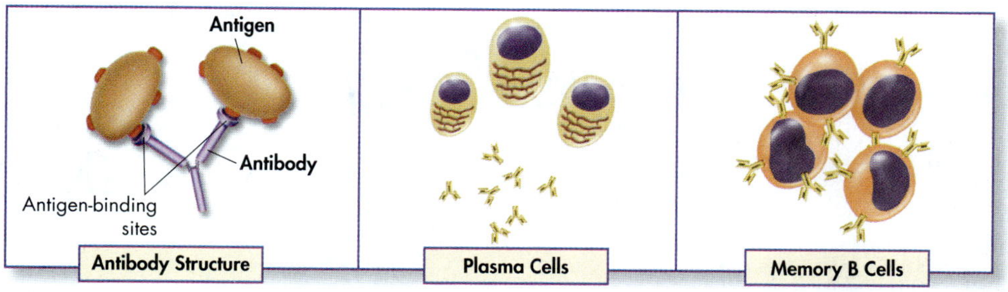

Focus on Academic Language

- cancerous
- destruction

Students might not recognize words they know after suffixes have been added.

Say If you look carefully at a word, you might see that you know a simpler form of the word and can use the simpler form to figure out the larger word.

Ask When you look at the word *cancerous,* do you see a simpler word you know? *(cancer)*

Say Some simpler words are not as easy to pick out because the whole word is not used in the bigger word.

Ask When you look at the word *destruction,* cover up the *–uction.* Do the remaining letters remind you of a word you know? *(destroy)*

Say Once you notice the simpler words within larger words, you can often infer the pronunciations and the meanings of the larger words.

Active Reading

READ, COVER, REMEMBER, RETELL Have students work in pairs to read and discuss the text on this page. Suggest that they take turns reading one sentence at a time. Explain that after a student reads a sentence, he or she should cover it and use his or her own words to tell the partner what the sentence meant.

VISUALIZE/DIAGRAM Divide the class into groups of three or four students. Instruct the groups to each draw a two-column chart with several empty rows. Tell groups to look for causes and their effects in each paragraph on this page and to write them in their charts. If necessary, provide questions such as the following to help groups identify causes and effects.

Ask What causes B cells to divide into plasma cells and memory B cells? *(the binding of antibodies to antigens)*

Ask What causes a body cell to display a pathogen's antigens? *(infection of a body cell by a pathogen or engulfing of a pathogen by a macrophage)*

Ask What causes the production of helper T cells and memory T cells? *(the display of an antigen by a macrophage or a body cell)*

KEY QUESTION Have a volunteer read aloud the Key Question and answer.

Ask Where, specifically, do the cells and antibodies of humoral immunity do their work? *(in the bloodstream and in other body fluids)*

Speed Bump

Ask What type of blood cells are B cells and T cells? *(lymphocytes)*

Ask How many kinds of antigens can each B cell or T cell respond to? *(only one)*

Ask Which types of lymphocytes gives you long-term immunity to certain diseases? *(memory B cells and memory T cells)*

Ask Which type of lymphocyte fights cancer cells? *(T cells)*

During Reading continued

Build Connections

VISUAL SUMMARY Have volunteers take turns locating the lesson's vocabulary terms in the Build Connections on this page. Instruct students to tell what the vocabulary term is, where it is located, and what it means. Be sure that students point out *immune response* (one time), *antibody* (three times), and *antigen* (four times).

Say The Build Connections can be broken into two halves in two ways. Notice the different background colors on the left and right sides.

Ask What type of immunity does the left side of the diagram describe? *(humoral immunity)* The right side? *(cell-mediated immunity)*

Say Now, notice the dashed line that separates the lower one-third of the diagram from the upper two-thirds.

Ask What does the top two-thirds of the diagram show? *(the primary immune response)* The lower one-third? *(the secondary immune response)*

Ask In the diagram, what symbols are used for antibodies? *(yellow Y-shaped structures)* What symbols are used for antigens? *(tiny purple triangles)*

Ask How does the diagram indicate the activation of T cells and B cells? *(with blue shapes that project from a T cell or B cell and that connect at and surround a purple triangle)* What do these connections show? *(that helper T cells recognize and bind to foreign antigens displayed by a pathogen or an infected cell)*

Speed Bump

Ask What type of immunity involves antibodies, antigens, and B cells? *(humoral immunity)*

Ask What type of immunity involves antigens and T cells? *(cell-mediated immunity)*

Ask What is the advantage of the secondary response? *(It responds more quickly to a pathogen than the primary response responded to the same pathogen.)*

BUILD Connections

SPECIFIC IMMUNE RESPONSE

In humoral immunity, antibodies bind to antigens in body fluids and tag them for destruction by other parts of the immune system. In cell-mediated immunity, body cells that contain antigens are destroyed.

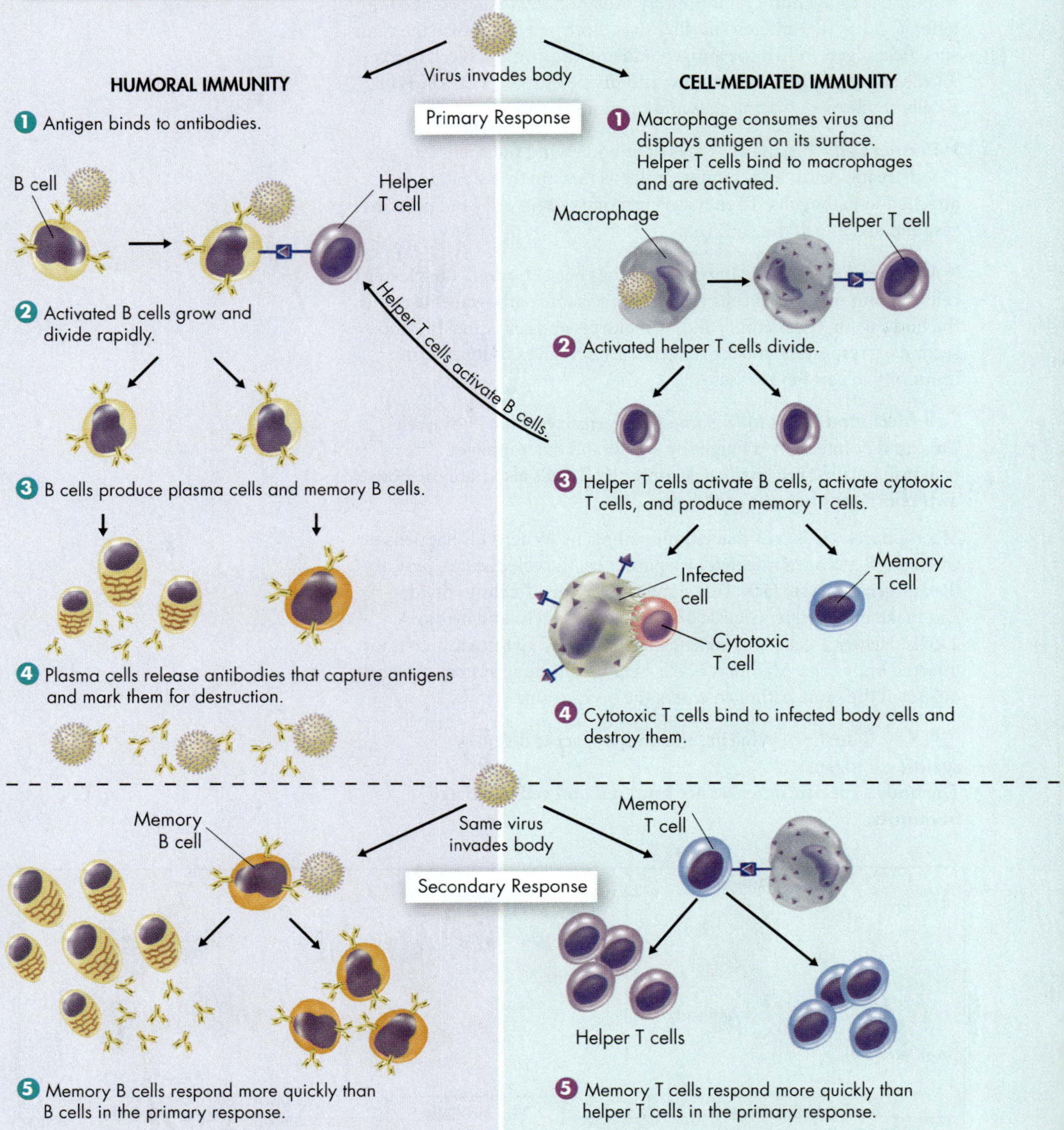

844 Chapter 35 • Lesson 2

Hands-On Learning

MODELING AN IMMUNE RESPONSE Pair students and hand out these materials: two small balloons in each of three colors—pink, blue, and yellow; pink, blue, and yellow adhesive notes

1. Tell students to partially inflate and seal the balloons. Explain that the inflated balloons represent pathogens and that each color represents the antigens on one type of pathogen.
2. Tell students that the adhesive notes represent antibodies and that each color corresponds to an antigen.
3. Have students use the balloons and adhesive notes to model the binding of antibodies to specific antigens. *(Students should attach adhesive notes to the balloons of the matching color.)*

Ask Which type of immunity does this activity model? *(humoral immunity)*

INQUIRY into Scientific Thinking

GUIDED INQUIRY

Immune System "Memory"

Your body is able to recognize pathogens and parasites as "nonself." Cells in your immune system recognize and try to destroy pathogens. During the initial infection, your body makes memory B cells. These cells allow your body to respond even more quickly if you are infected by the same pathogen again. Memory B cells make new plasma cells very quickly. Plasma cells then make and release antibodies to respond to the antigens on the surfaces of the pathogen.

The number of antibodies in a person's blood reveals the difference between the first and second immune response. Look at the graph. Assume the person was first exposed to Antigen A on Day 1. On Day 28, the person was exposed to Antigen A again and to Antigen B for the first time. The red line shows the levels of antibodies to Antigen A. The blue line shows the levels of antibodies to Antigen B. Use the graph to answer the questions that follow.

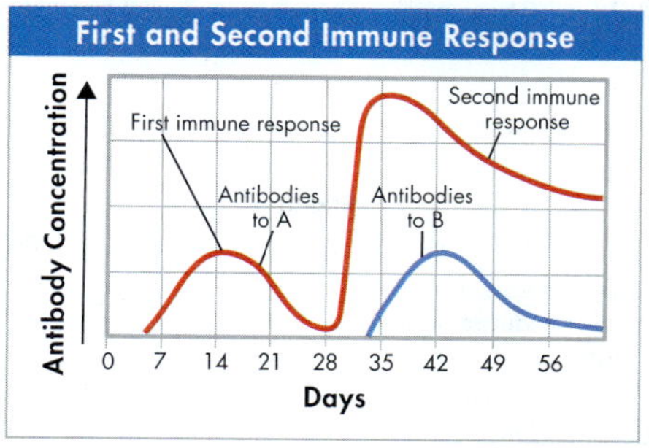

Analyze and Conclude

1. Interpret Graphs About how long does it take for your body to make antibodies after the first exposure to an antigen? About how long does it take for antibody levels to increase after the second exposure to an antigen?

2. Infer What could explain the significant increase in antibodies to Antigen A seen after Day 30?

In Your Workbook Get more help for this activity in your workbook.

CHECK Understanding

Apply Vocabulary

Use the highlighted words from the lesson to complete each sentence correctly.

1. Histamines are released during a(n) ________.

2. Pathogens have specific _______ that cause an immune response.

3. The body's specific recognition of, response to, and memory of a pathogen is called the ________.

Critical Thinking

4. Sequence Describe the steps of the inflammatory response.

5. Apply Concepts Although cytotoxic T cells are helpful in the immune system, they also make organ transplantation more difficult. How do you think cytotoxic T cells affect organ transplantation?

6. Write to Learn These two T cells are attached to a cancer cell. What kind of immune response are these cells a part of?

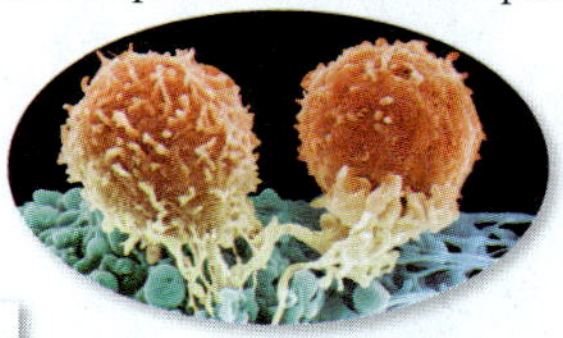

BIOLOGY.com Search Lesson 35.2 GO • Lesson Assessment

Inquiry Into Scientific Thinking

MAKE CONNECTIONS Explain that in this activity, students will compare numbers of a certain part of the immune response.

Ask What part of the immune response was measured for these data? *(antibodies)*

INTERPRET GRAPHICS Have students examine the graph and its labels.

Ask What does the red line in the graph represent? *(the number of antibodies that bind with antigen A)*

Ask How is the meaning of the blue line similar to the meaning of the first part of the red line? *(They both show what happens to antibodies in response to a specific antigen.)*

Lead a Discussion

Discuss the graph with the class to be sure they interpret it correctly.

Ask What do the two lines in the graph compare? *(number of antibodies produced in response to two different antigens over a period of time)*

Ask Based on the graph, is the response to antigen B a first immune response or a second immune response? *(first)*

Post Reading

ASSESSMENT TIPS For Question 4, be sure students understand that *sequence* means to arrange a number of steps in order from first to last. You might suggest they develop sequencing graphic organizers for their answers.

For Question 5, verify that students understand what an organ transplantation is.

As a hint for Question 6, you might ask students to think about whether the cells are attacking another cell or binding to antigens.

Wrap-Up Activity

MODELING Remind students that the immune system's specific defenses rely on the interaction of antigens and antibodies, which fit together like puzzle pieces. Provide modeling clay to small groups of students. Explain that the clay will be used to model antibodies. Have students use common small objects with different shapes, such as pencils, erasers, and keys, to model antigens. Tell students to mold clay around an object to make an antibody. Ask them how well their antibodies fit on their antigens and how this relates to the immune response. *(Students should recognize that the antibodies fit perfectly since they are made for a particular antigen.)*

Analyze and Conclude

1. about five days; about two days

2. This increase is the result of the secondary immune response. Memory B cells respond more quickly to the same pathogen invading the body than B cells in the primary response.

Check Understanding

1. inflammatory response

2. antigens

3. immune response

4. Mast cells release histamines, which increase the amount of fluid around the wound and cause swelling. White blood cells move into the tissue and some destroy bacteria. The area may become red and sore.

5. The immune system of the transplant patient is likely to recognize the donated organ as "nonself." Cytotoxic T cells would attack and kill the cells of the donated organ.

6. T cells are involved in cell-mediated immunity. These T cells are cytotoxic T cells, because they are bound to the cancer cell.

LESSON 35.3

Pre-Reading

Objectives

35.3.1 Distinguish between active immunity and passive immunity.

35.3.2 Describe how public health measures and medications fight disease.

35.3.3 Describe why patterns of infectious disease have changed.

Preview the Pages

1. Preview the vocabulary terms by having students find each term in the text and read it aloud. Point out that active immunity and passive immunity are related in some way because they are under the same main heading.
2. Preview the visuals and read the captions and labels. Point out to students that the first caption describes a way to prevent a disease, and the other captions describe sources of diseases.
3. Explain that the numbered red circles on this page represent steps in a simple process that may have prevented them from getting several life-threatening diseases.

Build Understanding

VENN DIAGRAM Direct students to Workbook B, where they will find a Venn diagram that they can use to compare active immunity and passive immunity. Remind them to fill in the Venn diagram as they read about each type of immunity.

NATIONAL SCIENCE EDUCATION STANDARDS

UNIFYING CONCEPTS AND PROCESSES

II, IV

CONTENT

C.1.a, C.4.e, F.1, F.5, F.6, G.3

INQUIRY

A.2.a, A.2.b

For the corresponding lesson in *Miller & Levine Biology,* see pages 1020–1022.

35.3 Fighting Infectious Disease

Key Questions

How do vaccines and externally produced antibodies fight disease?

How do public health measures and medications fight disease?

Why have patterns of infectious diseases changed?

BUILD Understanding

Venn Diagram Make a Venn diagram that compares and contrasts active and passive immunity.

In Your Workbook Go to your workbook to learn more about making a Venn diagram. Complete the Venn diagram for Lesson 35.3.

Acquired Immunity

More than 200 years ago, English physician Edward Jenner noticed that milkmaids who got a mild disease called cowpox did not get smallpox. At the time, smallpox killed many people.

Jenner wondered if people could be protected from smallpox by infecting them with cowpox. Jenner tested his hypothesis on a young boy named James Phipps. His experiment is outlined below:

1. Jenner put fluid from a cowpox sore into a small cut he made on James Phipps's arm.
2. Phipps got cowpox.
3. Two months later, Jenner injected Phipps with fluid from a smallpox infection.
4. Phipps did not get smallpox.

Ever since that time, doctors have used **vaccinations** to produce immunity. Doctors inject patients with a weakened form of a pathogen or a similar but less dangerous pathogen.

Active Immunity Today, we understand how vaccination works. The immune system makes memory B cells and memory T cells in response to antigens carried by the pathogen. This kind of immunity is called **active immunity.** People may develop active immunity after they are vaccinated. People may also develop active immunity after they fight an infection.

Passive Immunity Disease can be prevented in another way. Antibodies made against a pathogen by other individuals or animals can be used to give temporary immunity. If these antibodies are introduced into a person's blood, the result is **passive immunity.** Passive immunity lasts only a short time, because the immune system will destroy the foreign antibodies.

The First Vaccination Edward Jenner used the cowpox virus to vaccinate James Phipps against smallpox.

846

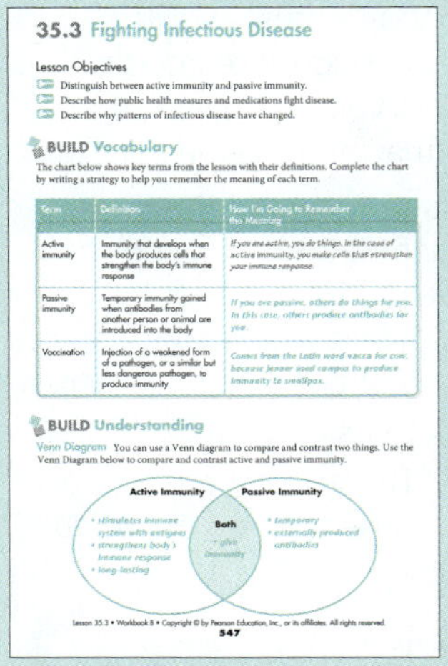

35.3 Fighting Infectious Disease

Lesson Objectives

- Distinguish between active immunity and passive immunity.
- Describe how public health measures and medications fight disease.
- Describe why patterns of infectious disease have changed.

BUILD Vocabulary

The chart below shows key terms from the lesson with their definitions. Complete the chart by writing a strategy to help you remember the meaning of each term.

Term	Definition	How I'm Going to Remember the Meaning
Active immunity	Immunity that develops when the body produces cells that strengthen the body's immune response	If you are active, you do things. In the case of active immunity, you make cells that strengthen your immune response.
Passive immunity	Temporary immunity gained when antibodies from another person or animal are introduced into the body	If you are passive, others do things for you. In this case, others produce antibodies for you.
Vaccination	Injection of a weakened form of a pathogen, or a similar but less dangerous pathogen, to produce immunity	Comes from the Latin word vacca for cow, because Jenner used cowpox to produce immunity to smallpox.

BUILD Understanding

Venn Diagram You can use a Venn diagram to compare and contrast two things. Use the Venn Diagram below to compare and contrast active and passive immunity.

Lesson 35.3 • Workbook B • Copyright © by Pearson Education, Inc., or its affiliates. All rights reserved.
547

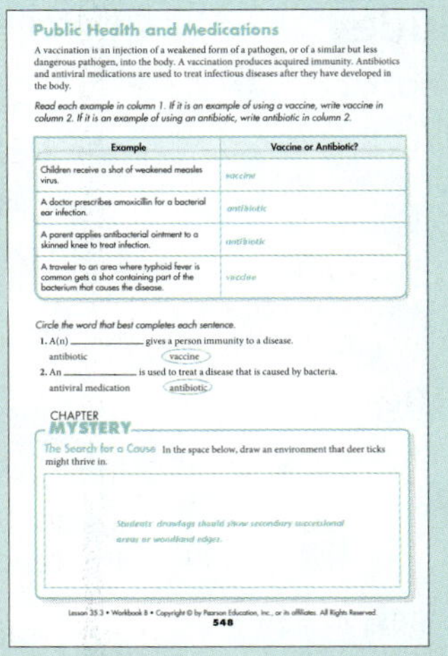

Public Health and Medications

A vaccination is an injection of a weakened form of a pathogen, or of a similar but less dangerous pathogen, into the body. A vaccination produces acquired immunity. Antibiotics and antiviral medications are used to treat infectious diseases after they have developed in the body.

Read each example in column 1. If it is an example of using a vaccine, write vaccine in column 2. If it is an example of using an antibiotic, write antibiotic in column 2.

Example	Vaccine or Antibiotic?
Children receive a shot of weakened measles virus.	vaccine
A doctor prescribes amoxicillin for a bacterial ear infection.	antibiotic
A parent applies antibacterial ointment to a skinned knee to treat infection.	antibiotic
A traveler to an area where typhoid fever is common gets a shot containing part of the bacterium that causes the disease.	vaccine

Circle the word that best completes each sentence.

1. A(n) ______________ gives a person immunity to a disease.
   antibiotic   vaccine
2. An ______________ is used to treat a disease that is caused by bacteria.
   antiviral medication   antibiotic

CHAPTER MYSTERY

The Search for a Cause In the space below, draw an environment that deer ticks might thrive in.

Students' drawings should show secondary successional areas or woodland edges.

Lesson 35.3 • Workbook B • Copyright © by Pearson Education, Inc., or its affiliates. All Rights Reserved.
548

Passive immunity can also happen naturally. A pregnant woman passes antibodies to her fetus. An infant gets antibodies through breast milk. In some cases, people can be given antibodies like a vaccination. For example, people who have been bitten by a rabid animal are injected with antibodies for the rabies virus.

Key Question How do vaccines and externally produced antibodies fight disease?
The immune system makes memory B cells and memory T cells in response to the antigen in the vaccination. Externally produced antibodies can be used to give temporary immunity.

Public Health and Medications

In 1900, more than 30 percent of deaths in the United States were caused by infectious disease. In 2005, less than 5 percent of deaths were caused by infectious disease. Public health measures and the development of medications helped to make this happen.

Public Health Measures When people live in large groups, behavior, cleanliness of food and water supplies, and sanitation all influence the spread of disease. The field of public health offers services and advice that help keep living conditions healthy. Public health measures help keep the food and water supplies clean. Encouraging vaccinations and recommending ways to avoid getting sick are also public health measures. Rules that make parents get vaccinations for their children, and education about hand washing, have greatly reduced the spread of many diseases.

Medications It is not possible to prevent every infectious disease. Medications are other weapons that can fight pathogens. Antibiotics can kill bacteria. Some antiviral medications can slow down viral activity.

The term *antibiotic* means something that kills bacteria without harming the host. In 1928, Alexander Fleming was the first scientist to discover an antibiotic. Fleming noticed that the mold *Penicillium notatum* seemed to make something that slowed bacterial growth. Research showed that this "something" was a compound Fleming named penicillin. Researchers learned to make large amounts of penicillin just in time to save the lives of thousands of World War II soldiers. Since then, dozens of antibiotics have saved millions of people.

Antibiotics have no effect on viruses. However, antiviral drugs have been made that fight certain viral infections. These drugs generally make it more difficult for viruses to get into the body's cells or to multiply once inside cells.

Key Question How do public health measures and medications fight disease?
Public health measures help keep living conditions healthy. Medications can be used to fight pathogens after someone is sick.

BUILD Vocabulary

vaccination
an injection of a weakened, or a similar but less dangerous, pathogen to produce immunity

active immunity
immunity that develops as a result of natural or deliberate exposure to an antigen

passive immunity
temporary immunity that develops as a result of natural or deliberate exposure to an antibody

WORD ORIGINS
The word *vaccination* comes from the Latin word *vacca,* meaning "cow," as a reminder of Jenner's work.

Broad Street Pump In 1854, Dr. John Snow interviewed residents and mapped cases to determine the source of a London cholera epidemic. He discovered that a water pump like this replica was the source. This is a major event in the history of public health.

ELL Focus on ELL: Access Content

ALL SPEAKERS Pair English learners of different ability levels. Have partners work together to read the second and third paragraphs under Medications and make cluster diagrams of each. Tell them to use one piece of paper for each paragraph. On each page, suggest that they begin with a center circle for the main idea and add radiating lines for the details. Point out that they can use as many or as few words as they would like to present each idea, as long as they are able to explain what they have written. Invite groups to share their cluster diagrams.

For a description of cluster diagrams, see Strategy 19 in the Workbook B Appendix.

During Reading

Build Vocabulary

WORD ORIGINS Explain that *immunity* comes from the Latin word *immunis,* meaning "free from something unpleasant."

Ask How does this meaning relate to the meaning of *immunity? (Someone with immunity is free from the disease or having to treat the disease.)*

Active Reading

KEY QUESTION Read aloud the first Key Question and answer on this page.

Ask What does *temporary* mean? *(lasting only a short time)*

SUMMARIZE Have a volunteer read the Public Health Measures section aloud. Then help students summarize the paragraph.

Ask According to the text, what two major activities are carried out by public health workers? *(They provide services and offer advice that help people keep living conditions healthy.)*

Say Now, read the caption above the photo on this page.

Ask What kind of activity does the caption describe? *(an investigation)*

Be sure students understand that public health agencies investigate the causes and sources of infectious disease as well as provide services and advice for keeping living conditions healthy.

KEY QUESTION Have a volunteer read the second Key Question and answer aloud.

Ask What are two examples of public health measures to prevent disease? *(Possible answers: purifying water, requiring vaccinations, educating the public, washing hands frequently, inspecting and cleaning food)* Clarify that public health measures should be practiced by everyone, not just by agencies.

Speed Bump

Ask What can be credited for the drop in percentage of deaths in the United States due to infectious disease? *(the development of public health measures and medications)*

Ask What type of organism did Alexander Fleming think had produced a substance that slowed the growth of bacteria? *(a mold)*

Ask How do antiviral drugs fight viral diseases? *(Possible answers: by making it harder for viruses to enter cells or to reproduce)*

During Reading continued

Active Reading

SELF-QUESTION Remind students that they can ask themselves questions to help them think through what they are reading and to develop a purpose for reading on. Demonstrate how they could ask questions about the text in the Changing Interactions With Animals paragraph. Read the first sentence.

Say When I read that sentence, the first thing I wonder is "What animals?" My self-questioning encourages me to read on to find an answer.

Read the next sentence.

Ask What is the answer to my self-question? *(Sample answer: Oh, I see. The animals are exotic animals that we do not normally come in contact with.)*

Say As you read, ask yourself questions and then look for the answers.

KEY QUESTION Read the Key Question and answer aloud.

Ask What are four examples of emerging diseases? *(Sample answers: AIDS, hantavirus, monkeypox, West Nile virus, Ebola virus, avian influenza)*

Speed Bump

Ask What has brought many emerging diseases to the United States? *(the exotic animal trade)*

Ask Why are diseases, such as measles, that were common in the past becoming more common again? *(People stopped getting vaccinations.)*

Post Reading

ASSESSMENT TIPS For Question 3, remind students that *contrast* means to list differences between things, and that the question does not ask for ways the two types of immunity are alike.

For Question 4, remind students that the answer to an *infer* question is not stated in the text.

MYSTERY CLUE After students read the clue, remind them that *anti-* means "against" and *biotic* refers to living things. Suggest they reread the information about antibiotics in the Medications section.

Wrap-Up Activity

GALLERY WALK Set up several microscopes with slides of different types of human white blood cells. Have small groups of students circulate from one microscope to another to view the cells. Have students draw what they see.

Causes of Emerging Disease
Illegally imported animals can lead to the spread of emerging disease. In 2003, dormice and other rodents from Africa spread monkeypox to prairie dogs in the United States. Humans were soon infected. The spread of SARS has been associated with the wild animal trade.

New and Re-Emerging Diseases

By 1980, many people thought that medicine had won the fight against infectious disease. Vaccination and other public health measures had wiped out polio in the United States. Smallpox had been wiped out around the world. Antibiotics seemed to have bacterial diseases under control. Researchers thought that epidemics would soon be history. They were wrong.

Recently, several new diseases have started making people sick. These include AIDS, SARS, hantavirus, monkeypox, West Nile virus, Ebola, and avian influenza ("bird flu"). Other diseases that people thought were gone are coming back, or re-emerging, and spreading to new places. What's going on?

Changing Interactions With Animals As people clear new land and as environments change, people come in contact with different animals and pathogens. Exotic animal trade, for pets and food, has also given pathogens new chances to jump from animals to humans. Both monkeypox and SARS are thought to have started this way.

Misuse of Medications Misuse of medications has led to the re-emergence of diseases that many people thought were under control. For example, many strains of the pathogens that cause tuberculosis and malaria are becoming resistant to many antibiotics and other medications. In addition, diseases such as measles are coming back because some people are not getting vaccinations.

Key Question Why have patterns of infectious diseases changed?
Through development and the exotic animal trade, people are coming into contact with new animals and pathogens. The misuse of medications has led to the re-emergence of certain diseases.

CHECK Understanding

Apply Vocabulary
Use the highlighted words from the lesson to complete each sentence correctly.

1. The injection of a weakened form, or of a similar but less dangerous form, of a pathogen to give the patient immunity is called a(n) ________.
2. Vaccination provides ________ by exposing a person to an antigen.

Critical Thinking

3. **Compare and Contrast** Describe the difference between active and passive immunity.
4. **Infer** In the past few decades, it has become much easier to travel around the world. How do you think this has affected the spread of emerging diseases? Explain.
5. **Write to Learn** Answer the second clue of the mystery. Think about what the term *antibiotic* means.

MYSTERY CLUE
Steere's patients were helped by antibiotics. What clue did this give him about the disease's pathogen? (Hint: See p. 847.)

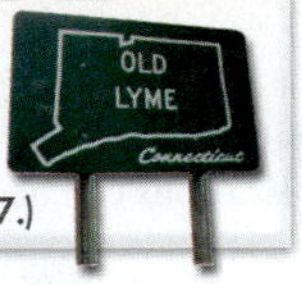

BIOLOGY.com Search Lesson 35.3 GO • Lesson Assessment

848

Check Understanding

1. vaccination
2. active immunity
3. Active immunity results from natural or deliberate exposure to an antigen. It is long-term immunity. Passive immunity results from natural or deliberate exposure to an antibody made by a different animal or person. This provides only temporary immunity.
4. Sample answer: As people travel, they come into contact with new pathogens. If they return home while they are infected, they may bring the pathogen with them and expose more people.
5. The term *antibiotic* means something that kills bacteria without harming the host. Antibiotics can kill bacteria, but they do not affect viruses. Therefore, the mystery disease's pathogen must have been a bacterium.

35.4 Immune System Disorders

When the Immune System Overreacts

Sometimes, the immune system fights the body's own cells. Other times, the immune system itself is weakened by sickness. Then what happens? A strong immune response to harmless antigens can cause allergies, asthma, and autoimmune disease.

Allergies Antigens on things like pollen, dust mites, and mold can cause allergic reactions. The overreaction of the immune system to these harmless antigens is called an **allergy.** Allergies cause mast cells to release histamines.

Asthma Allergic reactions in the respiratory system can cause asthma. **Asthma** is a disease that causes air passages to narrow. This causes wheezing and coughing. It also makes it difficult for the person to breathe. Asthma attacks can be triggered by infections, exercise, stress, and some medications. Cold or dry air, smoke, pollen, and dust can also cause an attack.

Asthma can be life-threatening. Patients need to take their medication properly. If they do not, an attack can cause long-term harm to the lungs. There is no cure, but some medications can relax the muscles around the airways and relieve symptoms.

Autoimmune Diseases Sometimes the immune system does not properly recognize "self." When this happens, it may attack cells or compounds in the body as if they were pathogens. This kind of illness is called an autoimmune disease. Type I diabetes and lupus are autoimmune diseases. Some autoimmune diseases can be treated with medications that lessen the symptoms. Others are treated by lowering the immune response. These also lower the normal immune response. The patient must be watched carefully.

Key Question How can misguided immune responses cause problems?
A strong immune response to harmless antigens can cause allergies, asthma, and autoimmune disease.

Pet Dander Pet dander is dead skin shed from cats and dogs. It often causes allergies.

Key Questions

How can misguided immune responses cause problems?

What causes AIDS and how is it spread?

BUILD Understanding

Anticipation/Reaction Guide
Before you read this lesson, try to answer the two key questions in your own words. As you read, modify your answers.

In Your Workbook Go to your workbook to complete the Anticipation/Reaction guide for Lesson 35.4.

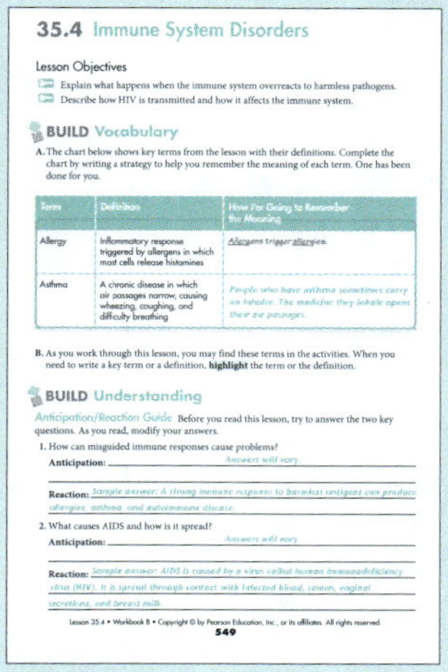

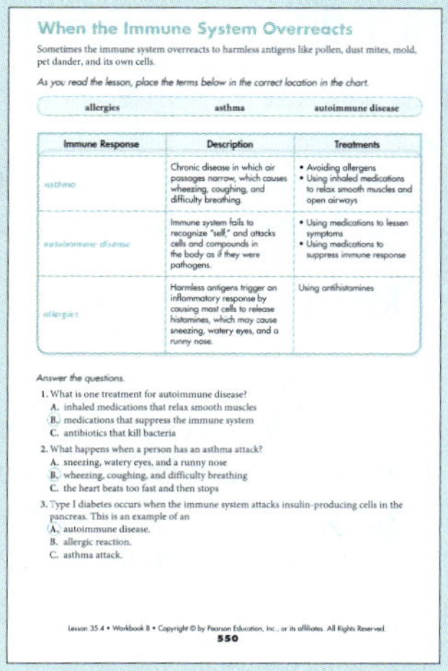

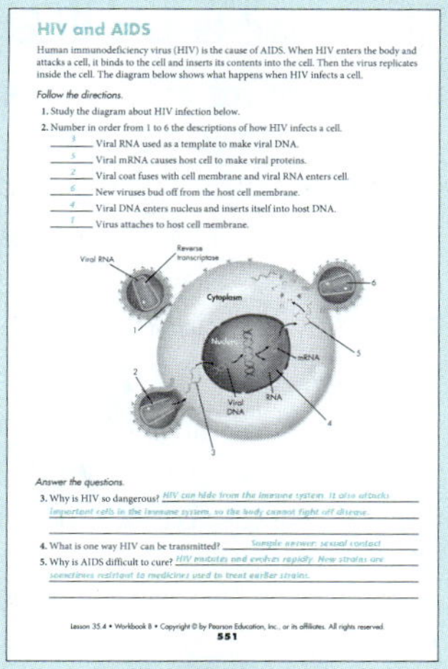

LESSON 35.4

Pre-Reading

Objectives

35.4.1 Explain what happens when the immune system overreacts to harmless pathogens.

35.4.2 Describe how HIV is transmitted and how it affects the immune system.

Preview the Pages

1. Preview the Pet Dander visual on this page. Be sure students realize what is shown in the microscopic view of the cat's fur. *(blue: tiny pieces of dead skin; yellow: cat's fur)*
2. Preview the HIV Infection diagram. Have volunteers read aloud the caption and labels. Identify the major parts of the diagram.
3. Preview the AIDS graph by reading the title, the axes labels, and the caption. Point out that the zigzag line means the numbers 1000 through 5000 are not included.
4. Have students preview the main ideas and details by reading the green headings and blue subheadings.

Build Understanding

ANTICIPATION/REACTION GUIDE Direct students to Workbook B, where they will find an anticipation/reaction guide for this lesson. Point out to students that this exercise will help them discover what they already know about immune system disorders. Encourage students to revise their initial answers as they read the lesson.

NATIONAL SCIENCE EDUCATION STANDARDS

UNIFYING CONCEPTS AND PROCESSES

I, V

CONTENT

C.1.a, F.1

For the corresponding lesson in *Miller & Levine Biology*, see pages 1024–1027.

LESSON 35.4

During Reading

Active Reading

USE CONTEXT CLUES Before students begin reading, have a volunteer read aloud the green heading on the previous page.

Ask What does the word *overreact* mean? *(to react with unnecessary or inappropriate force or emotion)*

Have volunteers give examples of overreactions from everyday life. *(Sample answer: a person may overreact when he finds his wallet is missing, only to find it later in his jacket pocket)*

Have students read the section When the Immune System Overreacts, looking for context clues that help answer the question, **What happens when the immune system overreacts?** Then have students work in pairs to write a brief response. *(Sample answer: When the immune system overreacts, it reacts with unnecessary force to things that are not pathogens. In some cases, the immune system attacks cells in the body as if they were pathogens.)*

KEY QUESTION Have a volunteer read aloud the Key Question and answer on the previous page.

Ask What is the connection between asthma and allergies? *(Asthma is a narrowing of air passages, sometimes caused by allergies.)*

Build Vocabulary

WORD ORIGINS Tell students that *asthma* is a Greek word that means "panting." Point out that this meaning relates to a common asthma symptom.

Build Connections

VISUAL SUMMARY Tell students that the diagram shows a cell that becomes infected by HIV and how the virus replicates by taking over the cell. Point out the cell's nucleus and DNA as well as the parts of the virus and its RNA. Have volunteers read the captions in order and review the meanings of *transcription* and *translation* they learned in Chapter 13.

Ask What part of HIV enters a cell? *(its RNA and reverse transcriptase)*

Ask What is the role of the reverse transcriptase? *(to make DNA using the viral RNA as a template)*

Make sure students go to Workbook B to find out if they understand all of the steps of HIV infection.

Speed Bump

Ask What change in blood cells signals the beginning of AIDS? *(a drop in the T cell count)*

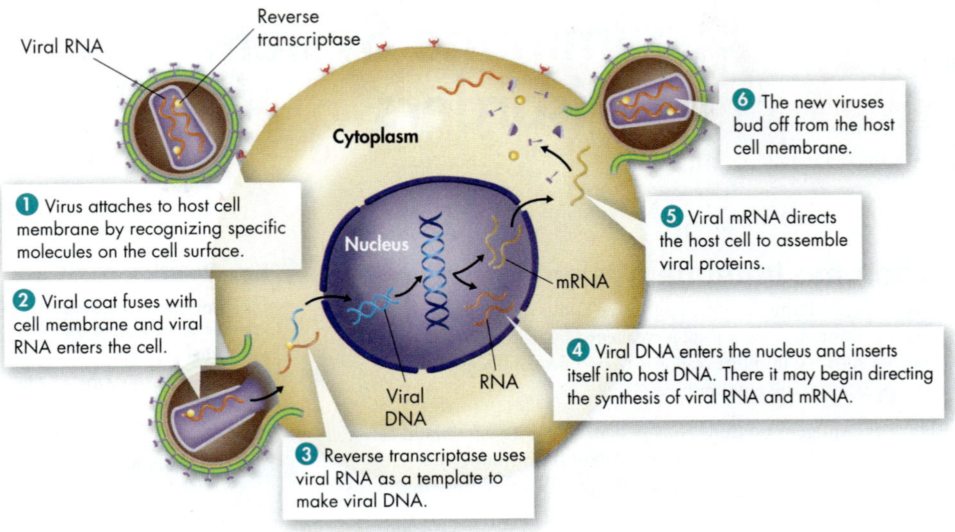

BUILD Connections

HIV INFECTION

HIV travels through the blood, where it binds to receptors on helper T cells. Inside the cell, the viral DNA directs the cell to make many new viruses. These new viruses are quickly released back into the blood, where they infect more cells.

In Your Workbook Do you know all of the steps of HIV infection? Go to your workbook to find out.

BUILD Vocabulary

allergy an overreaction of the immune system to an antigen

asthma a chronic respiratory disease in which air passages narrow, causing wheezing, coughing, and difficulty breathing

WORD ORIGINS

Allergy comes from the Greek words that mean "different or strange activity." An allergy is a strange or unusual reaction to something that should be harmless to the immune system.

HIV and AIDS

During the late 1970s, doctors began to report that some people were getting very sick from microorganisms that did not usually make people sick. Doctors concluded that these people must have weakened immune systems. Diseases that attack a person with a weakened immune system are called opportunistic diseases. Researchers concluded that these people all had the same disorder. They called it acquired immunodeficiency syndrome (AIDS). Eventually, it was discovered that this "syndrome" was caused by a pathogen new to science.

HIV Researchers discovered the pathogen that causes AIDS in 1983. They called it the human immunodeficiency virus (HIV). HIV is deadly for two reasons. It can hide from the immune system, and it can attack key cells in the immune system. This leaves the body with little protection from other pathogens.

Target: T Cells One of HIV's main targets is helper T cells. Helper T cells are important to the specific immune response. HIV destroys T cells. Eventually, the immune system is unable to fight HIV and other pathogens. When the patient's T cell count drops to about one sixth the normal level, he or she is said to have AIDS.

HIV Transmission HIV is deadly. But it does not pass from one person to the next easily. It cannot be passed through coughing, sharing clothes, or any casual contact. HIV can be spread only through contact with infected blood, semen, vaginal secretions, or breast milk.

ELL Focus on ELL: Build Background

INTERMEDIATE AND ADVANCED SPEAKERS Write the terms *acquired immunodeficiency syndrome* and *opportunistic diseases* on the board. Have students copy the terms on the board into their notebooks. Invite them to break the words down into understandable parts or to write words they know that are similar to the unfamiliar words.

Ask What two words can you see within the word *immunodeficiency*? *(immune and deficiency)*

Say Use a dictionary to find definitions for each unknown word in the two terms. Then, using your own words, write definitions for the two complete terms. *(Sample answers: acquired immunodeficiency syndrome—a health condition that a person is not born with and that is characterized by a non-working immune system; opportunistic diseases—diseases that take the opportunity of a weakened system to attack)*

Preventing HIV Infection You can choose behaviors that reduce your risk of becoming infected with HIV. The only way to be safe from HIV infection is to avoid sexual activity and intravenous drug use. Before 1985, HIV was transmitted to some patients through transfusions of infected blood or blood products. Now that doctors know the cause of HIV, such cases have been virtually eliminated. The blood supply is screened for HIV antibodies. Also, people who engage in certain activities are discouraged from donating blood.

Can AIDS Be Cured? Right now, there is no cure for AIDS. New drugs make it possible to survive HIV infection for years. However, HIV mutates and evolves rapidly. There are many strains that are resistant to most drugs.

The only way to control the virus is to use several expensive drugs. These drugs interfere with some of the virus' enzymes. These drugs are allowing more people in the United States to live with HIV rather than to die from it. However, these drugs are not available in many parts of Africa and Asia. Because HIV can now be treated (but not cured), some people have the dangerous misconception that HIV infection is not serious.

Key Question What causes AIDS and how is it spread? **AIDS is caused by HIV, a virus that attacks the immune system. It is spread though contact with infected blood, semen, vaginal secretions, or breast milk.**

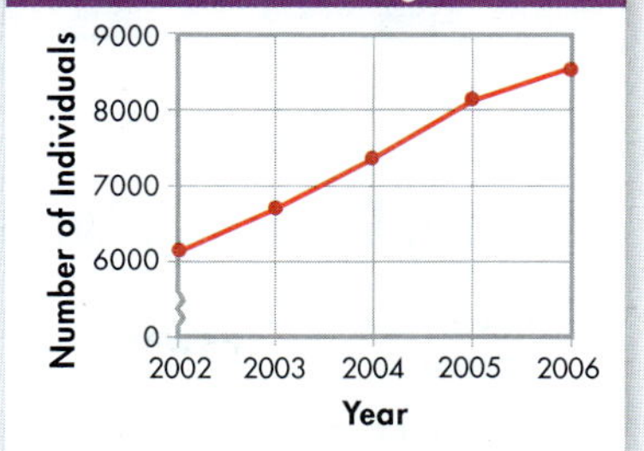

13–24-Year-Olds Living With AIDS The number of adolescents and young adults infected with HIV in the United States has been rising steadily since 2002.

CHECK Understanding

Apply Vocabulary

Use the highlighted words from the lesson to complete each sentence correctly.

1. An unusually strong immune response to an antigen is called a(n) ________.
2. A chronic respiratory illness that involves wheezing and narrowed air passages is called ________.

Critical Thinking

3. **Apply Concepts** When a person first visits a doctor about asthma symptoms, the doctor is likely to ask the patient to list times and places his or her asthma attacks tend to happen. Why do you think doctors do this?
4. **Infer** Why is it hard for a person with HIV to fight off other infections?
5. **Draw Conclusions** Look again at the graph at the top of this page. What are two conclusions that you can draw regarding the increasing number of adolescents living with AIDS?
6. **Write to Learn** Answer the third clue in the mystery. Consider what the pathogen might do to the immune system.

MYSTERY CLUE

As with HIV, blood tests of patients with the mystery disease showed that patients were making antibodies to a pathogen. What could be a reason the immune system cannot overcome the disease? (**Hint:** See p. 850.)

BIOLOGY.com Search Lesson 35.4 GO • Lesson Assessment

Check Understanding

1. allergy
2. asthma
3. Sample answer: The doctor is trying to learn which specific antigens trigger the asthma attacks. Once the antigens are identified, the doctor and patient can plan ways to prevent future attacks.
4. HIV attacks the person's helper T cells. The helper T cells are responsible for stimulating the immune response to foreign antigens. Without them, the immune system is not as able to respond to infection.
5. Sample answer: Either more adolescents are contracting HIV or adolescents are living longer with AIDS than in the past.
6. The mystery disease may attack key cells in the immune system, weakening the immune response to the disease.

Active Reading

USE GRAPHIC ORGANIZERS Have each student make a T-chart that lists the methods of HIV transmission in one column and ways to prevent transmission by each method in the other column.

After students have finished individual T-charts, conduct a class discussion and build a summary T-chart on the board. Encourage students to add as much information to their charts as possible.

KEY QUESTION Have a volunteer read the Key Question and answer aloud.

Ask When do you use the term *AIDS* and when do you use the term HIV? *(You use* AIDS *when referring to a disease of the immune system. You use* HIV *when referring to the virus that causes AIDS.)*

Speed Bump

Ask What are four body fluids through which HIV can be spread? *(blood, semen, vaginal secretions, and breast milk)*

Ask How are HIV infections treated? *(Several expensive drugs control the virus but cannot eliminate it from an infected person's body.)*

Post Reading

ASSESSMENT TIPS For Question 3, remind students that several different things can cause an asthma attack and that a doctor's questions can eliminate some possible causes.

For Question 4, remind students that answers to *infer* questions are not directly given in the text. To answer this question, suggest they think about what HIV infection does.

For Question 5, invite pairs of students to analyze the graph and discuss what it shows.

MYSTERY CLUE After students read the clue, remind them that antibodies themselves usually do not destroy pathogens but mark them for destruction by other cells. Suggest students reread the paragraph under the green heading When the Immune System Overreacts.

Wrap-Up Activity

JIGSAW REVIEW Divide the class into five learning circles. Assign each student in the learning circle a number. Then have all of the 1's, 2's, and so on, form their group. Each group should be assigned a different blue heading under HIV and AIDS and write a summary. Then, have students return to their original learning circle and take turns reading their summaries.

CHAPTER LAB

Pre-Lab

Review the importance of controlling variables.

Ask What is a variable? *(A variable is any factor that could affect the results of an experiment.)*

Say In a controlled experiment, you vary one factor, called the independent variable. And you measure changes in another factor, called the dependent variable, to see how it changes in response to variations in the independent variable. All other variables must be controlled, or kept the same.

Ask Why is it important to control all variables other than the independent variable? *(If more than one variable changes in an experiment, there is no way to know which variable explains the results.)*

Review the Background and Pre-Lab Questions.

Lab

Tell students they will perform the chapter lab *Detecting Lyme Disease* described in **Lab Manual B.**

SAFETY Students should wear goggles and disposable plastic gloves, take care with glassware, and wash their hands with warm, soapy water afterwards.

Post-Lab

ROUNDTABLE DEBATE Divide the class in half to debate Koch's postulates versus this procedure as methods for identifying infectious diseases.

Say You will debate the pros of the method you are defending and the cons of the other team's method.

Model some pros and cons.

Say For example, applying Koch's postulates is time consuming but very specific, while using ELISA is quick but may be less specific.

Forensics Lab

GUIDED INQUIRY

Pre-Lab: Detecting Lyme Disease

Problem How can a blood test be used to detect Lyme disease?

Materials well plate, permanent marker, white paper, 400-mL beaker, 100-mL beaker, distilled water, micropipettes, test solutions

Lab Manual Chapter 35 Lab

Skills Control Variables, Interpret Data, Draw Conclusions

Connect to the Big idea To maintain homeostasis, your immune system must defend against invasions by harmful pathogens. Some invaders enter the body through bites from insects. For example, a tiny deer tick can infect you with the bacterium that causes Lyme disease. As a precaution, you should avoid areas where deer ticks are active. If you visit a location where ticks are active, wear clothing that covers the skin and check for ticks.

Symptoms for Lyme disease can vary widely, but many people develop a bull's-eye rash at the location of the bite. People who suspect that they have been exposed to the bacteria that cause Lyme disease should consult a medical professional. Blood tests are used to diagnose Lyme disease. In this lab, you will model one of these tests.

Background Questions

a. **Review** What is an antigen?

b. **Review** How does the immune system respond to antigens?

c. **Explain** Why does the presence of antibodies in blood suggest that a person was exposed to an antigen?

Pre-Lab Questions

Preview the procedure in the lab manual.

1. **Sequence** Use a flowchart to show the order in which the solutions will be added to the well plate.
2. **Infer** What is the advantage of having a control for a positive test and a control for a negative test?
3. **Control Variables** Why must you rinse the micropipette with distilled water before adding a different solution to the well plate?

BIOLOGY.com Search Chapter 35 GO

Visit Chapter 35 online to test yourself on chapter content and to find activities to help you learn.

Untamed Science Video Be careful what you touch as you follow the Untamed Science crew on a journey through human allergies.

Art Review Review characteristics and examples of the different types of pathogens and parasites with this drag-and-drop activity.

Art in Motion View an animation of HIV infecting a cell.

Pre-Lab Answers

BACKGROUND QUESTIONS

a. any foreign substance that can stimulate an immune response

b. The immune system responds to antigens by increasing the production of cells that attack the antigen and cells that produce antibodies.

c. Antibodies are produced in response to specific antigens. The antibodies would not be present unless the person was exposed to a particular antigen.

PRE-LAB QUESTIONS

1. The flowchart should show the solutions being added in the following order: antigen, controls, donor samples, secondary antibody, and substrate.
2. Seeing the expected results for both a positive test and a negative test will make it easier to draw conclusions about the donor samples.
3. Rinsing keeps the contents of a well from being contaminated with the contents of a different well.

NATIONAL SCIENCE EDUCATION STANDARDS

UCP II

CONTENT F.1, F.5

INQUIRY A.1.b, A.2.a, A.2.b, A.2.c

35 CHAPTER Summary

35.1 Infectious Disease

- Infectious diseases are caused by microorganisms. Diseases can be caused by viruses, bacteria, fungi, "protists," and parasites.
- Infectious diseases can be spread through coughing, sneezing, or physical contact. Others spread through body fluids, contaminated food or water. Some diseases can be spread to humans from other animals.

infectious disease (p. 838)
germ theory of disease (p. 838)
vector (p. 840)

35.2 Defenses Against Infection

- Nonspecific defenses against pathogens include skin, hair, tears, and the inflammatory response. Interferons and fever play an important role in the inflammatory response.
- The immune system's specific defenses distinguish between "self" and "nonself" and respond to specific pathogens.
- The body's specific defenses against pathogens are humoral and cell-mediated immunity.

inflammatory response (p. 841)
immune response (p. 842)
antigen (p. 842)
antibody (p. 842)

35.3 Fighting Infectious Disease

- The immune system makes memory B cells and memory T cells in response to the antigen in the vaccination. Vaccination provides active immunity.
- Antibodies produced against a pathogen by other individuals or animals can be used to give temporary passive immunity.
- Public health measures help keep living conditions healthy. Medications can be used to fight pathogens after someone is sick. Antibiotics kill bacteria. Some antiviral medications can slow down viral activity.
- Through human development and the exotic animal trade, people are coming into contact with new animals and pathogens. The misuse of medications has led to the re-emergence of certain diseases.

vaccination (p. 846)
active immunity (p. 846)
passive immunity (p. 846)

35.4 Immune System Disorders

- A strong immune response to harmless antigens can cause allergies, asthma, and autoimmune disease.
- AIDS is caused by HIV, a virus that attacks the immune system. It is spread though contact with infected blood, semen, vaginal secretions, or breast milk.

allergy (p. 849)
asthma (p. 849)

BIOLOGY.com Search Chapter 35 GO • Crossword • Chapter Assessment

UbD Transfer the Big Idea

TRANSFER Have students work in small groups to write a breaking news report about the invasion of an unknown pathogen into a body and the immune system's response to that invasion. The report should be written from the point of view of a radio news reporter.

Say Your news report should accurately answer the following questions: Who? What? When? Where? How? Your report should also include the state of homeostasis within the body.

Ask What descriptive words might you use in your report? *(descriptive words, such as* invade, attack, latch onto, *and* go after, *that help listeners visualize the invading pathogens and the cells and molecules involved in the body's defense)*

Have groups deliver their news reports to the class.

Chapter Summary

Think Visually

USE GRAPHICS Have students make a flowchart that shows what occurs during the specific immune response. (Hint: Refer to the Specific Immune Response diagram.)

ANSWERS
Students' flowcharts should begin with the first step, Virus invades body. Flowcharts should then have separate tracks for humoral immunity and cell-mediated immunity. Each of the two tracks should include all of the steps shown in the Specific Immune Response diagram, with an additional step—Same virus invades body—before the final step in each track.

BIOLOGY.com Study Online

REVIEW RESOURCES

Art Review Students drag and drop labels to review different groups of pathogens.

InterActive Art A short animation shows students the steps in the primary and secondary immune response.

Art in Motion Students watch an animation of the process of HIV infection.

Vocabulary Review The *Flash Cards* and *Crossword* provide an interactive way to review chapter vocabulary.

Editable Worksheets Pages of Study Workbook B, Lab Manual B, and the Assessment Resources Book are available online. These documents can be easily edited using a word-processing program.

Lesson Overviews Lesson Overviews are based on Miller & Levine Biology. However, they are available on CD-ROM and can be edited for your class.

ASSESSMENT RESOURCES

Chapter Assessment Have students take an online version of the Chapter 35 Check Understanding.

Standardized Test Prep Students can take an online version of the Standardized Test Prep. You will receive their scores along with ideas for remediation.

Diagnostic and Benchmark Tests Use these tests to monitor your students' progress and supply remediation.

ASSESSMENT

Assess

Invite a volunteer to read the Big Idea and the question.

Ask What does the word *homeostasis* refer to in biology? *(It refers to the stable internal conditions of an organism that are maintained through constant change.)*

Ask What word means the opposite of health and homeostasis for an organism? *(disease)*

Ask What state do pathogens interrupt when they cause disease? *(homeostasis)*

Ask What body system guards the body's health by fighting pathogens? *(the immune system)*

CONSTRUCTED RESPONSE Paragraphs will vary. Explain the activity carefully. Be sure students understand that the *Hints* should be used as a guide to help them answer the questions. Review how to organize the topics to include in a paragraph by making an outline.

Write Question 1 on the board and read it aloud. Underline *virus, salivate,* and *beneficial.*

Say This question asks you to tell how symptoms caused by the rabies virus could help the virus.

Read the first hint.

Ask What happens when an animal with rabies bites another animal? *(Some of the rabid animal's saliva gets left in the bite wound.)*

Read the next hint.

Say Any viruses in the saliva can go directly into the bitten animal's bloodstream.

Ask Why does a virus need living cells? *(because viruses can only replicate when they are inside living cells)*

Model writing the first one or two sentences using the text and the hints. *(Sample: In order to produce more viruses, the virus must enter more hosts. The virus is probably in the saliva. The tendency of the infected animal to bite other animals is in the virus's favor.)*

For Question 2, pair students and have them use the hints and your modeling to answer the question. Have students share their paragraphs with the class.

For Question 3, have students write the paragraphs independently for homework or as an anchor activity.

35 CHECK Understanding

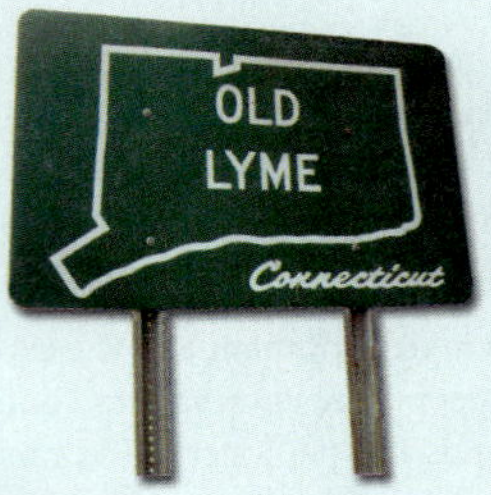

Assess the Big idea: Homeostasis

Write an answer to the question below.
Q: How does the body fight against invading organisms that may disrupt homeostasis?

Constructed Response

Write an answer to each of the questions below. The answer to each question should be one or two paragraphs long. To help you begin, read the **Hints** below each of the questions.

1. **Animals infected with the virus that causes rabies often salivate a lot and are more likely than healthy animals to bite other animals. Explain how these symptoms may be beneficial to the virus.**

   Hint Some pathogens are transferred by the exchange of bodily fluids.

   Hint Many viruses cannot live outside a host's body.

2. **Describe several ways that infectious diseases may be spread.**

   Hint Infectious diseases are caused by microorganisms, including bacteria, viruses, and fungi. These microorganisms are called pathogens.

   Hint Pathogens are often spread by symptoms of the disease they cause.

3. **Describe the steps in HIV infection.**

   Hint HIV targets helper T cells.

   Hint HIV uses an enzyme called reverse transcriptase to change the functioning of the cell.

Foundations for Learning Wrap-Up

Use the undercover vocabulary study guide you prepared before reading the chapter as a tool to help you organize your thoughts about homeostasis and the immune system.

Activity 1 Working with a partner, review the definitions you wrote on your undercover vocabulary study guide. If necessary, fill in any missing information and correct any errors. Then, quiz each other about each of the definitions.

Activity 2 Cut apart the vocabulary words and definitions to form two piles of "cards." Then shuffle the cards together. Working with a partner, try to match vocabulary words with their definitions.

Extension: Working with a partner, use the "cards" to make a flowchart describing how the body responds to infection. Challenge yourselves to use as many of the cards as possible.

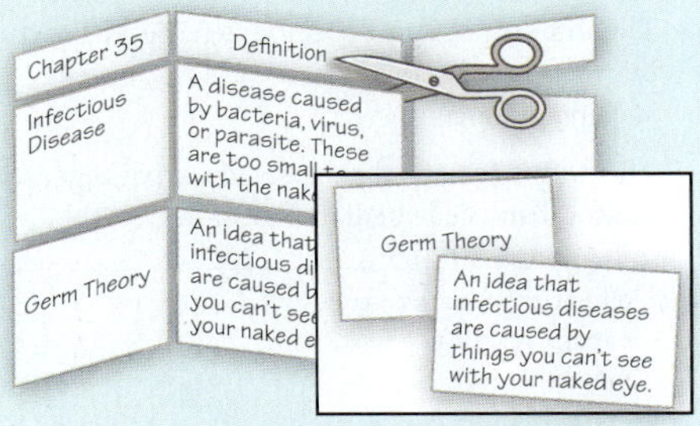

Foundations for Learning Wrap-Up

Have students take out their undercover vocabulary tools. Be sure students have recorded all of the vocabulary words from the chapter, along with definitions written in their own words. Pair students, and explain the activities.

35.1 Infectious Disease

Understand Key Concepts

1. Anything, including bacteria, viruses, and fungi, that can cause disease is called a(n)
   a. antibody. **c.** pathogen.
   b. antigen. **d.** vector.
2. Which of the following describes a zoonosis?
   a. any disease that can be transmitted through the exchange of bodily fluids
   b. any disease that is caused by microorganisms
   c. any disease that can be transmitted from animals to humans
   d. any disease that disrupts the body's normal functions

Test-Taking Tip

Rephrase the Question You might understand a question better if you rephrase it in your own words. Question 2 asks you to describe *zoonosis.* Rephrase the question to: "Define *zoonosis.*" Zoonosis is a disease that can be spread from animals to humans. Answer **c** is the correct choice.

3. What are some things that can prevent the spread of disease?

Think Critically

4. **Compare and Contrast** Compare and contrast harmless microorganisms that live in or on the human body to pathogens.

35.2 Defenses Against Infection

Understand Key Concepts

5. Which of the following structures is NOT involved in the body's first line of defense against infectious disease?
   a. skin **c.** T cells
   b. hair **d.** tears
6. A nonspecific defense reaction to tissue damage caused by injury or infection is known as
   a. the inflammatory response.
   b. the immune response.
   c. active immunity.
   d. passive immunity.
7. Distinguish between humoral immunity and cell-mediated immunity.

Think Critically

8. **Compare and Contrast** How does the secondary response to an antigen differ from the primary response to an antigen?

35.3 Fighting Infectious Disease

Understand Key Concepts

9. Antibodies made by animals are sometimes injected into a human to provide temporary immunity. This kind of immunity is called
   a. active immunity. **c.** vaccination.
   b. passive immunity. **d.** zoonosis.
10. Who discovered the first antibiotic and how did he discover it?
11. List two public health measures that have reduced the spread of infectious disease.

Think Critically

12. **Apply Concepts** How are human activities such as land-clearing related to the spread of new and re-emerging diseases?

35.4 Immune System Disorders

Understand Key Concepts

13. The main target cells of HIV are
   a. mast cells. **c.** B cells.
   b. T cells. **d.** red blood cells.

Assessment Tips

Pair students and have them work together to answer the questions.

Remind students to pay attention to the red subheadings. Point out that Questions 4, 8, 12, 16, and 17 require students to analyze and apply information.

Point out that when questions end with the word *Explain,* students need to both give an initial answer and then explain that answer.

Answers

Lesson 35.1

UNDERSTAND KEY CONCEPTS

1. c **2.** c

3. Sample answer: frequent thorough hand washing; avoiding contaminated water, improperly stored or cooked meats and seafood, close contact with animals' wastes or fluids, being bitten by infected animals, and eating infected meat

THINK CRITICALLY

4. Sample answer: They are similar in that both survive well within the human body. They are different in that harmless microorganisms may even be helpful to humans, while pathogens cause diseases.

Lesson 35.2

UNDERSTAND KEY CONCEPTS

5. c

6. a

7. Humoral immunity is the immune response in which antibodies defend against antigens in body fluids. Cell-mediated immunity involves T cells defending the body against pathogens inside body cells.

THINK CRITICALLY

8. The immune system responds more quickly during the secondary response as memory B cells make plasma cells more quickly than B cells in a primary response. This secondary response provides long-term immunity to certain pathogens.

Lesson 35.3

UNDERSTAND KEY CONCEPTS

9. b

10. Alexander Fleming was the first to discover an antibiotic. He noticed that the mold *Penicillium notatum* seemed to make something that slowed bacterial growth. Later, scientists found out the mold made a chemical that is now called penicillin.

11. Rules requiring parents to get their children vaccinated and education about the importance of hand washing have reduced the spread of infectious disease.

THINK CRITICALLY

12. Sample answer: Development in natural habitats brings humans and animals into closer contact. These animals may carry pathogens that cause previously unknown diseases.

Lesson 35.4

UNDERSTAND KEY CONCEPTS

13. b

14. b

15. HIV attacks helper T cells. Without the helper T cells, the body is unable to fight HIV or other pathogens.

THINK CRITICALLY

16. The body's response to a second infection by the same antigen is stronger than the first. The allergic response is a strong immune response to an otherwise harmless antigen. The second response to that antigen will be stronger and faster than the first.

Connecting Concepts

17. The pump on Broad Street was most likely the source. Most of the deaths caused by cholera happened very close to that pump.

18. Sample answer: Such a graph could be used if, for example, the source of the food poisoning were goods from a grocery store in a neighborhood where many people became sick with food poisoning.

CHAPTER **MYSTERY** After students have read the Chapter Mystery, discuss bacteria, vectors, and research on human diseases.

Ask Why did Steere think the bacterium he found caused Lyme disease? *(It could infect tissues related to arthritis.)*

Ask How can people avoid getting Lyme disease? *(They can stay away from areas where deer and deer ticks live or wear protective clothing if they do go near those areas.)*

35 CHECK Understanding

14. A strong response by a person's immune system to a harmless antigen is called

a. cell-mediated immunity.

b. an allergy.

c. an antibiotic.

d. an autoimmune disease.

15. Describe how HIV makes an infected person unable to fight off other infections.

Think Critically

16. **Apply Concepts** Why is a second bee sting more dangerous than the first for a person who is allergic to bee stings?

Connecting Concepts

Use Science Graphics

John Snow made a map similar to the one below to help him determine the source of the cholera outbreak in London in 1854. The dots represent the locations of people who died of cholera. The Xs represent pumps. Use the map to answer questions 17 and 18.

17. **Infer** Which pump do you think Snow determined was most likely the source of the cholera outbreak? Explain.

18. **Apply Concepts** Do you think a map such as this one could be used to discover the source of a food poisoning outbreak? Explain.

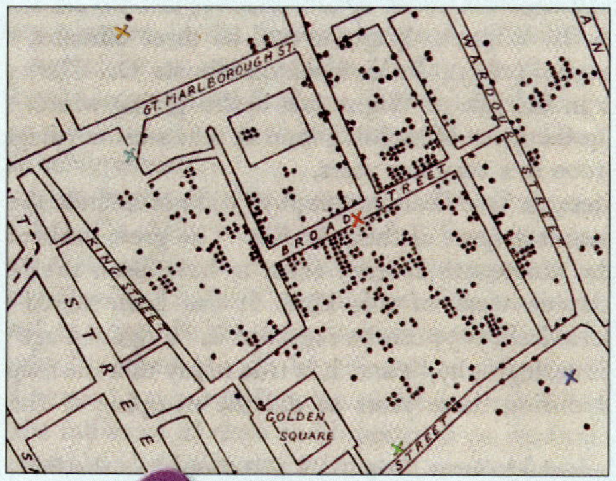

solve the CHAPTER MYSTERY

THE SEARCH FOR A CAUSE

Many of the patients suffering from this new disease lived in Lyme, Connecticut. The new disease was named Lyme disease, after the town. A researcher found a bacterium called *Borrelia burgdorferi* in deer ticks. These ticks had been captured in wooded areas near the patients' homes. Steere also found this bacterium in the patients. Could this bacterium be the cause of Lyme disease?

Steere infected healthy mice with the bacterium in his laboratory. The infected mice showed signs of the disease. Steere took bacteria from the sick mice and injected them into healthy mice. This second set of mice also became sick.

Now researchers know that a bite from a deer tick carrying *B. burgdorferi* can spread Lyme disease. The bacteria "swim" through tissues around the tick bites, causing the spreading rash. The bacteria can infect many types of cells. Some of the proteins made by the bacteria look like proteins around human nerve cells. This may cause an autoimmune response that leads to arthritis and other problems.

1. **Explain** What set of rules did Steere use to determine if *B. bergdorferi* was the pathogen responsible for Lyme disease?

2. **Infer** Deer and deer ticks thrive in wooded areas that grow back after the areas have been cleared and at the edges of woodlands. How might suburban development contribute to an increase in Lyme disease?

UntamedScience **Never Stop Exploring Your World.** Finding the solution to this medical mystery is only the beginning. Take a video field trip with the ecogeeks of Untamed Science to see where this mystery leads.

CHAPTER MYSTERY ANSWERS

1. Koch's postulates

2. Sample answer: Suburban development sometimes takes place at the edges of woodlands. Also, suburban development often involves clearing land for building houses. After the building is complete, small wooded areas re-emerge. Both deer and deer ticks thrive in environments found at the edges of woodlands and in woodland regrowth.

Untamed Science Join the crew of Untamed Science as they investigate human allergies in **A Reason for Sneezin'.**

Standardized Test Prep

Multiple Choice

1. All of the following prevent pathogens from entering the human body EXCEPT
   A red blood cells.
   B tears.
   C mucus.
   D skin.

2. Which of the following is NOT part of the inflammatory response?
   A White blood cells rush to infected tissues.
   B Blood vessels near the wound shrink.
   C Phagocytes engulf and destroy pathogens.
   D The wound becomes red.

3. What is the role of a vector in the spread of disease?
   A A vector is an inanimate object, such as a doorknob, where pathogens may collect.
   B A vector must infect a host for its life cycle to continue.
   C Vectors usually do not suffer from the infection, they just spread it from host to host.
   D A vector is a pathogen.

4. Which type of lymphocyte produces antibodies that are released into the bloodstream?
   A cytotoxic T cells
   B helper T cells
   C phagocytes
   D plasma cells

5. Which of the following is NOT a white blood cell?
   A interferon
   B macrophage
   C cytotoxic T cell
   D lymphocyte

6. Which is an example of naturally occurring passive immunity?
   A vaccination
   B exposure to a disease
   C an infant consuming antibodies in breast milk
   D antibodies injected from another person

7. How do medications help a person with asthma?
   A They counteract the effects of histamines.
   B They suppress the immune system.
   C They increase mucus production in the lungs.
   D They relax smooth muscles around airways.

Questions 8–9

A researcher measured the concentrations of HIV and T cells in 120 HIV-infected patients over a period of 10 years. Her data are summarized in the graph.

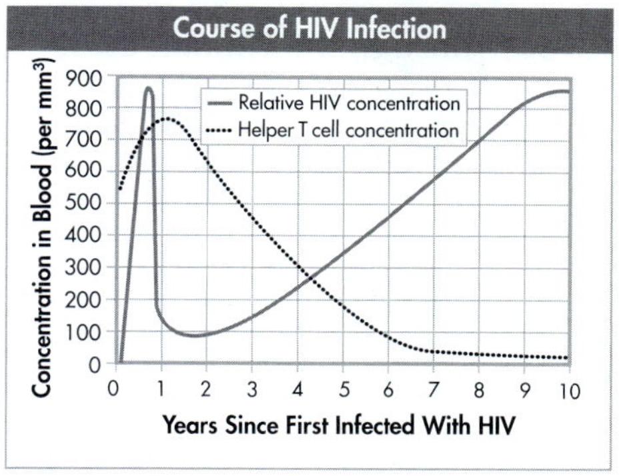

8. What happened to the HIV concentration over years 2 through 9?
   A It stayed about the same, then suddenly increased.
   B It stayed about the same, then suddenly decreased.
   C It steadily increased.
   D It steadily decreased.

9. What is probably responsible for the change in HIV concentration during the first year?
   A immune response
   B inflammatory response
   C passive immunity
   D HIV stopped replicating

Open-Ended Response

10. Explain why some symptoms of disease such as coughing and sneezing are advantageous to the pathogen that causes the disease.

If You Have Trouble With . . .

Question	1	2	3	4	5	6	7	8	9	10
See Lesson	35.2	35.2	35.1	35.2	35.2	35.3	35.4	35.4	35.4	35.1

Test-Taking Tip

READ ALL THE ANSWER CHOICES
Remind students that when evaluating multiple-choice answers, they should be sure to read all the answer choices, even if the first answer choice seems to be the correct one. By reading all the choices, students can make sure the answer chosen is the best answer to the question.

Before the Test

Review with students the Test-Taking Tip.

Point out the word *EXCEPT* in Question 1 and the word *NOT* in Questions 2 and 5.

Remind students that open-ended response means that they'll have to write a paragraph to answer the question in their notebooks. Have them paraphrase the question in their notebooks, and remind them about the steps in writing a paragraph.

Answers

1. A
2. B
3. C
4. D
5. A
6. C
7. D
8. C
9. A
10. Sample answer: Some symptoms of disease such as coughing and sneezing are advantageous to the pathogen that causes the disease because coughing and sneezing promote further spreading of the pathogen to new hosts. These symptoms release thousands of tiny pathogen-containing droplets that can be inhaled by other people or settle on objects.

UNIT 8

Plan Ahead

Have students read through the description of the Unit 8 Project a few days before they create their travel brochures to give them time to consider the task. Suggest they think about interesting ways they could describe and illustrate structures, organs, and organ systems inside the human body. Provide students with examples of travel brochures to stimulate their thinking and provide ideas.

Materials colored pens or pencils, paper

Monitor the Project

Suggest students start by making a basic outline of what they plan to include in their brochure. Tell them a good way to proceed is by making a page-by-page mockup that shows where headings, illustrations, and descriptions will be placed within the brochure. As students work, circulate through the classroom, asking questions that might help them in designing a good brochure.

Ask What attraction are you including for the muscular system? the excretory system?

Ask What dangers might tourists encounter as they travel through the body?

Project Assessment

Make sure students use the rubric and reflection questions to assess their work. Then, use the rubric to assign a final score. Note that it is important to value the creativity of students' work as well as the content when you score their projects. If desired, talk with students about any differences between their self-assessment scores and your assigned score.

Unit Project

A Tour Through the Human Body

Have you ever imagined what it would be like to shrink down to microscopic size and tour the inside of the human body? What interesting things would you see and hear? What "dangers" might you encounter?

Your Task Create a travel brochure in which you persuade a person to "visit" the human body. Discuss the various "attractions" inside the body to show how well you understand the systems.

Be sure to

- include at least one attraction for each organ system covered in this unit.
- design the brochure so that it is clear and easy to follow.
- be creative!

Reflection Questions

1. **Score your brochure using the rubric below. What score did you give yourself?**
2. **What did you do well in this project?**
3. **What needs improvement?**
4. **What "attractions" would you choose to visit? Why?**
5. **Exchange travel brochures with a partner. What attractions did he/she include?**

Assessment Rubric

Score	Scientific Content	Quality of Brochure
4	Brochure reveals an exceptionally thorough understanding of the human organ systems.	Brochure is clear, informative, and creative.
3	Brochure reveals a solid understanding of the human organ systems.	Brochure effectively conveys information about various attractions.
2	Brochure reveals a limited understanding of the human organ systems.	Brochure could be more clear and creative. It needs some editing.
1	Brochure reveals significant misunderstandings about the human organ systems.	Brochure is unclear and needs significant editing.

21st Century Skills

To be successful in the 21st century, students need skills and learning experiences that extend beyond subject area mastery. The Unit 8 Project helps students build the following 21st Century Skills: *Communication Skills; Creativity and Intellectual Curiosity; Interpersonal and Collaborative Skills; Self-Direction;* and *Accountability and Adaptability.*

FOCUS ON SOCIAL RESPONSIBILITY Extend this Unit Project by having small groups of students use their brochures to write a play about teens who take a tour through the human body. In the play, these young tourists should encounter thrills and attractions, but also problems in the various systems caused by smoking, drinking alcohol, using illegal drugs, and eating an unhealthful diet.

For more practice building 21st Century Skills, see The Chapter Mystery pages in **Study Workbook A.**

NATIONAL SCIENCE EDUCATION STANDARDS

UCP I, II, V

CONTENT C.1.f, C.5.d

INQUIRY A.2.a

858 Unit 8 Project

A Visual Guide to
The Diversity of Life

▲ *The Chambered Nautilus, found today in the Pacific Ocean, is one of the few living representatives of a group that once flourished in ancient seas 265 million years before the dinosaurs evolved. This Visual Guide will give you a glimpse of life's great variety and evolutionary history.*

A Visual Guide to
The Diversity of Life

CONTENTS

HOW TO USE THIS GUIDE

Use this visual reference tool to explore the classification and characteristics of organisms, including their habitats, ecology, behavior, and other important facts. This guide reflects the latest understandings about phylogenetic relationships within the three domains of life. Divided into six color-coded sections, the Visual Guide begins with a brief survey through the Bacteria and Archaea domains. It next discusses the major groups of protists, fungi, and plants. The final section provides information on nine animal phyla.

1. See how the group of organisms relates to others on the tree of life.

2. Learn about the general characteristics that all members of the group share.

3. Discover the members of the group and learn about their traits.

Animals

Cnidarians

KEY CHARACTERISTICS

Cnidarians are aquatic, mostly carnivorous, and the simplest animals to have specialized tissues (outer skin and lining of the gastrovascular cavity) and body symmetry (radial). Their tentacles have stinging cells called nematocysts used in feeding.

Feeding and Digestion Predatory, stinging prey with nematocysts; digestion begins extracellularly in gastrovascular cavity and is completed intracellularly; indigestible materials leave body through single opening; many, especially reef-building corals, also depend on symbiotic algae, or zooxanthellae.

Circulation No internal transport system; nutrients typically diffuse through body.

Respiration Diffusion through body walls

Excretion Cellular wastes diffuse through body walls.

Response Some specialized sensory cells: nerve cells in nerve net, statocysts that help determine up and down, eyespots (ocelli) made of light-detecting cells

Movement Polyps stationary, medusas free-swimming; some, such as sea anemones, can burrow and creep very slowly; others move using muscles that work with a hydrostatic skeleton and water in gastrovascular cavity; medusas such as jellyfish move by jet propulsion generated by muscle contractions.

Reproduction Most—alternate between sexual (most species by external fertilization) and asexual (polyps produce new polyps or medusae by budding)

Sea Nettle

Eco • Alert

Coral Symbionts

The color of this plate coral is caused by zooxanthellae algae living within it.

Reef-building coral animals depend on symbiotic algae called zooxanthellae for certain vital nutritional needs. In many places, reef-building corals live close to the upper end of their temperature tolerance zone. If water temperatures rise too high, the coral-zooxanthellae symbiosis breaks down, and corals turn white in what is called "coral bleaching." If corals don't recover their algae soon, they weaken and die. This is one reason why coral reefs are in grave danger from global warming.

1066

GROUPS OF CNIDARIANS

There are more than 9000 species of cnidarians.

HYDROZOA: Hydras and their relatives

Hydras and their relatives spend most of their time as polyps and are either colonial or solitary. They reproduce asexually (by budding), sexually, or they alternate between sexual and asexual reproduction. Examples: hydra, Portuguese Man-of-War

A Portuguese Man-of-War is actually a colony of polyps.

ANTHOZOA: Corals and sea anemones

Corals and sea anemones are colonial or solitary polyps with no medusa stage. The central body is surrounded by tentacles. They reproduce sexually or asexually. Examples: reef corals, sea anemones, sea pens, sea fans

Sea Anemone

SCYPHOZOA: Jellyfishes

Jellyfishes spend most of their time as medusas; some species bypass the polyp stage. They reproduce sexually and sometimes asexually by budding. Examples: lion's mane Jellyfish, Moon Jelly, Sea wasp

Black Sea Nettle

Jellyfishes such as this sea nettle are beautiful to us but deadly to their prey. The stinging cells on their tentacles can kill prey instantly and can ruin a human swimmer's day at the beach!

Diversity of Life 1067

4. Investigate current news and interesting facts about the group.

5. See photographs of representative animals within each group.

THE TREE OF LIFE

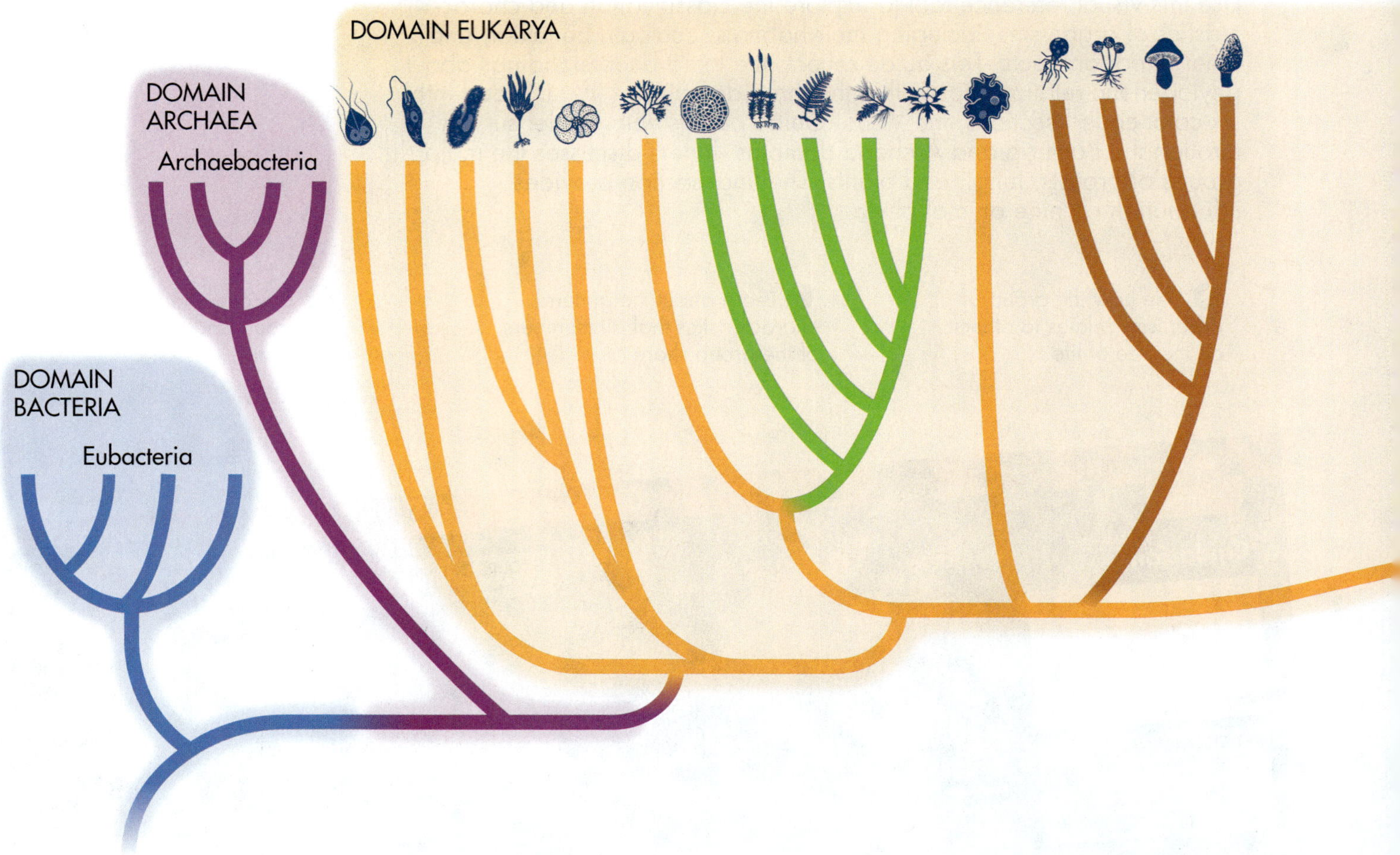

Before you begin your tour through the kingdoms of life, review this big picture from Chapter 18. The pages that follow will give you a glimpse of the incredible diversity found within each of the "branches" shown here.

DOMAIN BACTERIA

Members of the domain Bacteria are unicellular and prokaryotic. The bacteria are ecologically diverse, ranging from free-living soil organisms to deadly parasites. This domain corresponds to the kingdom Eubacteria.

DOMAIN ARCHAEA

Also unicellular and prokaryotic, members of the domain Archaea live in some of the most extreme environments you can imagine, including volcanic hot springs, brine pools, and black organic mud totally devoid of oxygen. The domain Archaea corresponds to the kingdom Archaebacteria.

DOMAIN EUKARYA

The domain Eukarya consists of all organisms that have cells with nuclei. It is organized into the four remaining kingdoms of the six-kingdom system: Protista, Fungi, Plantae, and Animalia.

THE "PROTISTS"

Notice that the branches for the kingdom Protista are not together in one area, as is the case with the other kingdoms. In fact, recent molecular studies and cladistic analyses have shown that "eukaryotes formerly known as Protista" do not form a single clade. Current cladistic analysis divides these organisms into at least six clades. They cannot, therefore, be properly placed into a single taxon.

FUNGI

Members of the kingdom Fungi are heterotrophs. Most feed on dead or decaying organic matter. The most recognizable fungi, including mushrooms, are multicellular. Some fungi, such as yeasts, are unicellular.

PLANTS

Members of the kingdom Plantae are autotrophs that carry out photosynthesis. Plants have cell walls that contain cellulose. Plants are nonmotile—they cannot move from place to place.

ANIMALS

Members of the kingdom Animalia are multicellular and heterotrophic. Animal cells do not have cell walls. Most animals can move about, at least for some part of their life cycle.

Bacteria

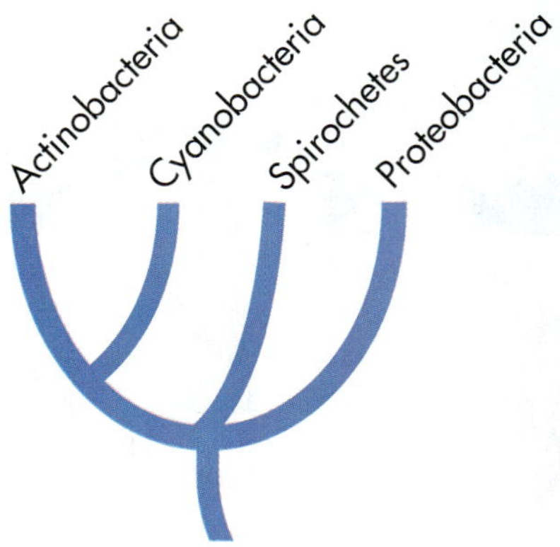

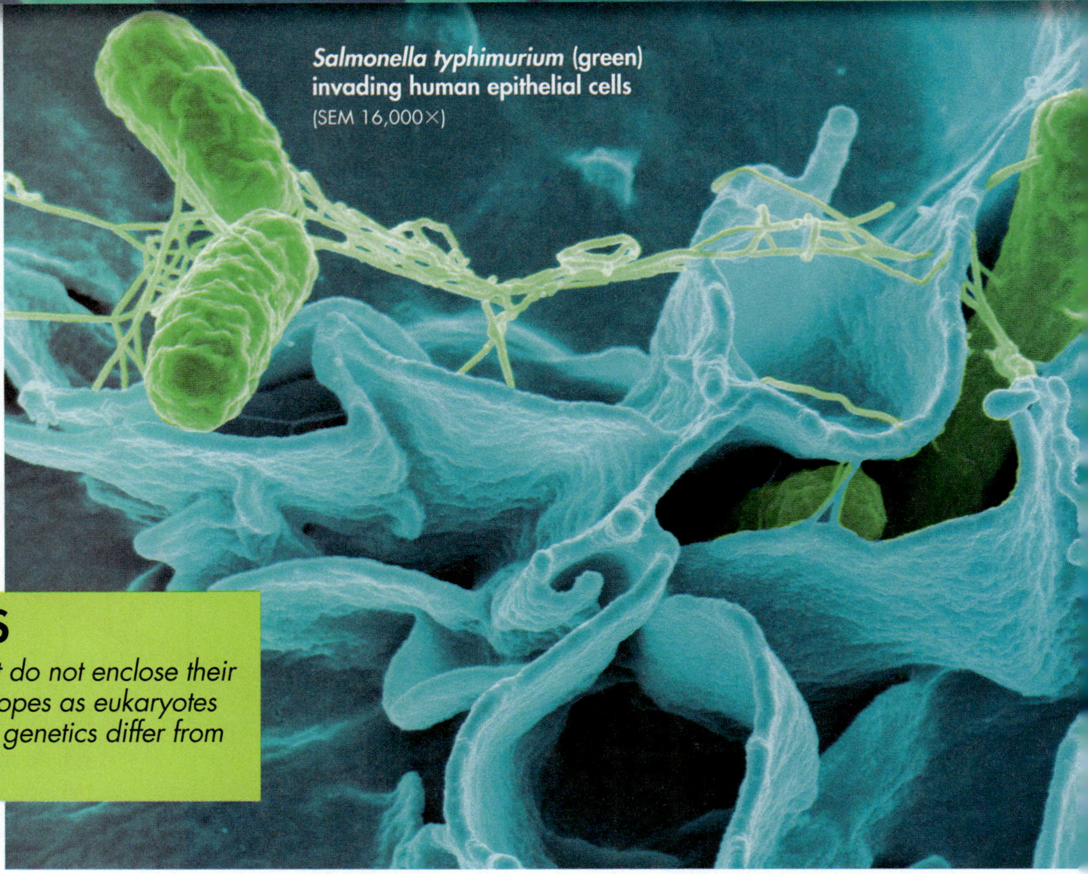

Salmonella typhimurium (green) invading human epithelial cells (SEM 16,000×)

KEY CHARACTERISTICS

Bacteria are prokaryotes—cells that do not enclose their DNA in membranous nuclear envelopes as eukaryotes do. Many details of their molecular genetics differ from those of Archaea and Eukarya.

Cell Structure Variety of cell shapes, including spherical, rodlike, and spiral; most have cell walls containing peptidoglycan. Few if any have internal organelles. Some have external flagella for cell movement.

Genetic Organization All essential genes are in one large DNA double helix that has its ends joined to form a closed loop. Smaller loops of DNA (plasmids) may carry nonessential genes. Simultaneous transcription and translation; introns generally not present; histone proteins absent

Reproduction By binary fission; no true sexual reproduction; some achieve recombination by conjugation.

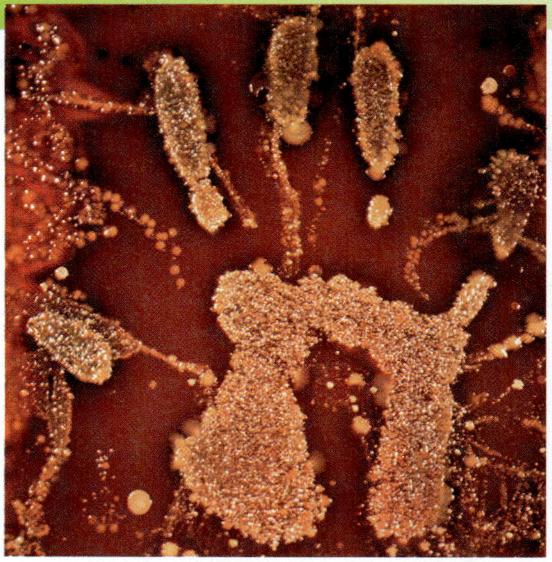

Did You Know?

A World of Bacteria

Putting Bacteria in Proper Perspective

"Planet of the Bacteria" was the title of an essay by the late Stephen Jay Gould. He pointed out that the dominant life forms on planet Earth aren't humans, or animals, or plants. They are bacteria. They were here first, and they inhabit more places on the planet than any other form of life. In fact, bacteria make up roughly 10 percent of our own dry body weight! In terms of biomass and importance to the planet, bacteria truly do rule this planet. They, not we, are number one.

◀ *The bacterial colonies shown here are growing in the print of a human hand on agar gel.*

GROUPS OF BACTERIA

There is no generally agreed phylogeny for the bacteria. Included here are some of the major groups within the domain.

PROTEOBACTERIA

This large and diverse clade of bacteria includes *Escherichia* (*E. coli*), *Salmonella*, *Helicobacter*, and the nitrogen-fixing soil bacterium *Rhizobium*.

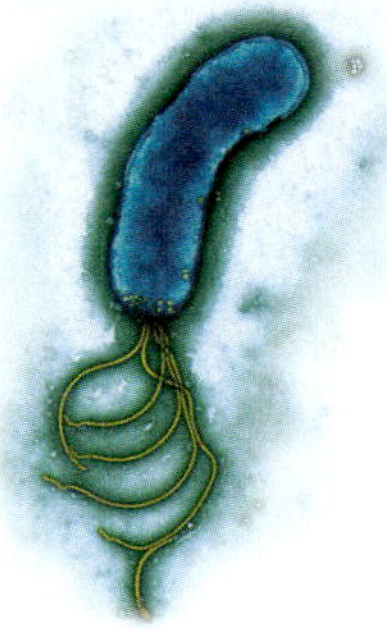

◀ *Helicobacter pylori is rod-shaped and has several flagella used for movement. This bacterium infects the stomach lining and causes ulcers in some people.* (TEM 7100×)

SPIROCHAETES

The spirochaetes (SPY roh keets) are named for their distinctive spiral shape. They move in a corkscrew-like fashion, twisting along as they are propelled by flagella on both ends of the cell. Most are free-living, but a few cause serious diseases, including syphilis, Lyme disease, and leptospirosis.

The spiral-shaped bacterium that causes syphilis is Treponema pallidum. (SEM 10,000×) ▼

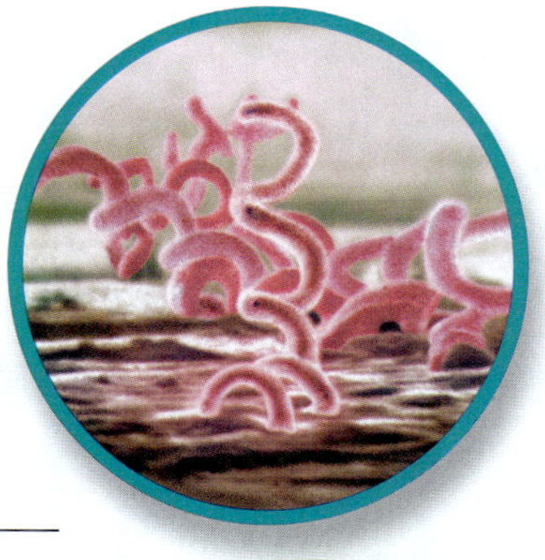

ACTINOBACTERIA

A large number of soil bacteria belong to this group. Some form long filaments. Members include the *Streptomyces* and *Actinomyces*, which are natural producers of many antibiotics, including streptomycin. A related group is the *Firmicutes*. The *Firmicutes* include *Bacillus anthracis* (anthrax), *Clostridia* (tetanus and botulism), and *Bacillus thuringensis*, which produces a powerful insecticide used for genetic engineering in plants.

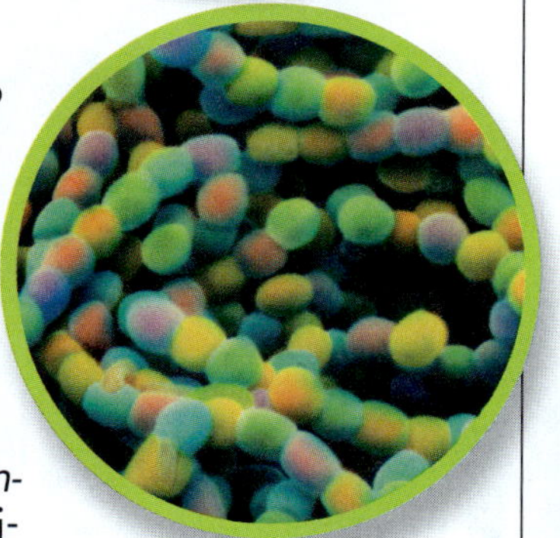

▲ *Chains of spores of soil bacteria, genus Streptomyces* (SEM 3400×)

CYANOBACTERIA

The cyanobacteria are photosynthetic prokaryotes that were once called "blue-green algae." They are among the oldest organisms on Earth, having been identified in rocks dating to more than 3 billion years ago. They are found in salt water and fresh water, in the soil, and even on the surfaces of damp rocks. They are the only organisms on Earth that are able to fix carbon and nitrogen under aerobic conditions, and this enables them to play critical roles in the global ecosystem, where they serve as key sources of carbon and nitrogen.

▼ *Many cyanobacteria form long filaments of attached cells, like those shown here (genus Lyngbya,* SEM 540×*).*

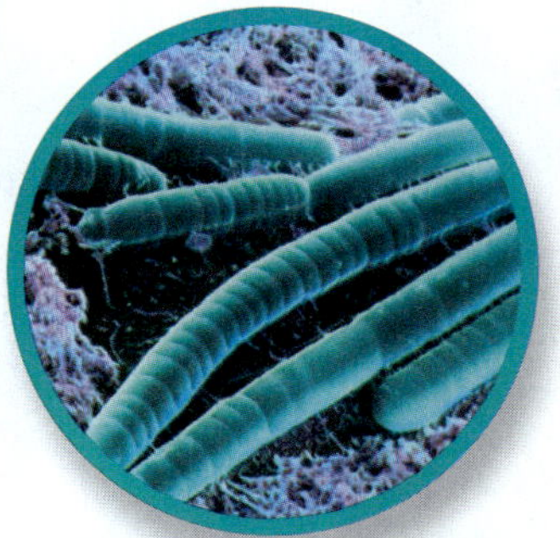

• A Closer Look

The Gram Stain

A Microbiologist's Quick Diagnostic

The Gram stain, developed by the nineteenth-century Danish physician Hans Christian Gram, allows microbiologists to categorize bacteria quickly into one of two groups based on their cell wall composition. Gram-positive bacteria lack a membrane outside the cell wall and take up the stain easily. Gram-negative bacteria, on the other hand, have an outer membrane of lipids and carbohydrates that prevents them from absorbing the gram stain. Many gram-negative bacteria are found among the proteobacteria. On the other hand, actinobacteria are mostly gram-positive.

Gram-positive bacteria appear purple after staining, while gram-negative bacteria appear pink. (LM 1000×) ▶

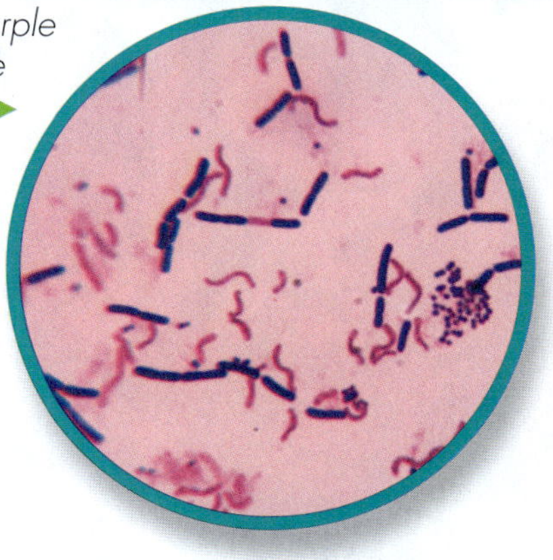

Archaea

KEY CHARACTERISTICS

Archaea are prokaryotes that differ from bacteria in so many details of structure and metabolism that they are viewed as a different domain than bacteria. Genetically, they have more in common with eukaryotes than with bacteria. Their cell walls do not contain peptidoglycan.

Cell Structure Cells similar to those of bacteria in appearance; many have flagella that are different in structure and biochemical composition from bacterial flagella. Cell membrane lipids also different from those of bacteria; few internal organelles

Genetic Organization As in bacteria, all essential genes are in one large DNA double helix that has its ends joined to form a closed loop. Proteins responsible for transcription and translation are similar to those of eukaryotes. Also like eukaryotes, most species contain introns, and all species contain DNA-binding histone proteins.

Reproduction By binary fission; no true sexual reproduction, but some achieve recombination by conjugation.

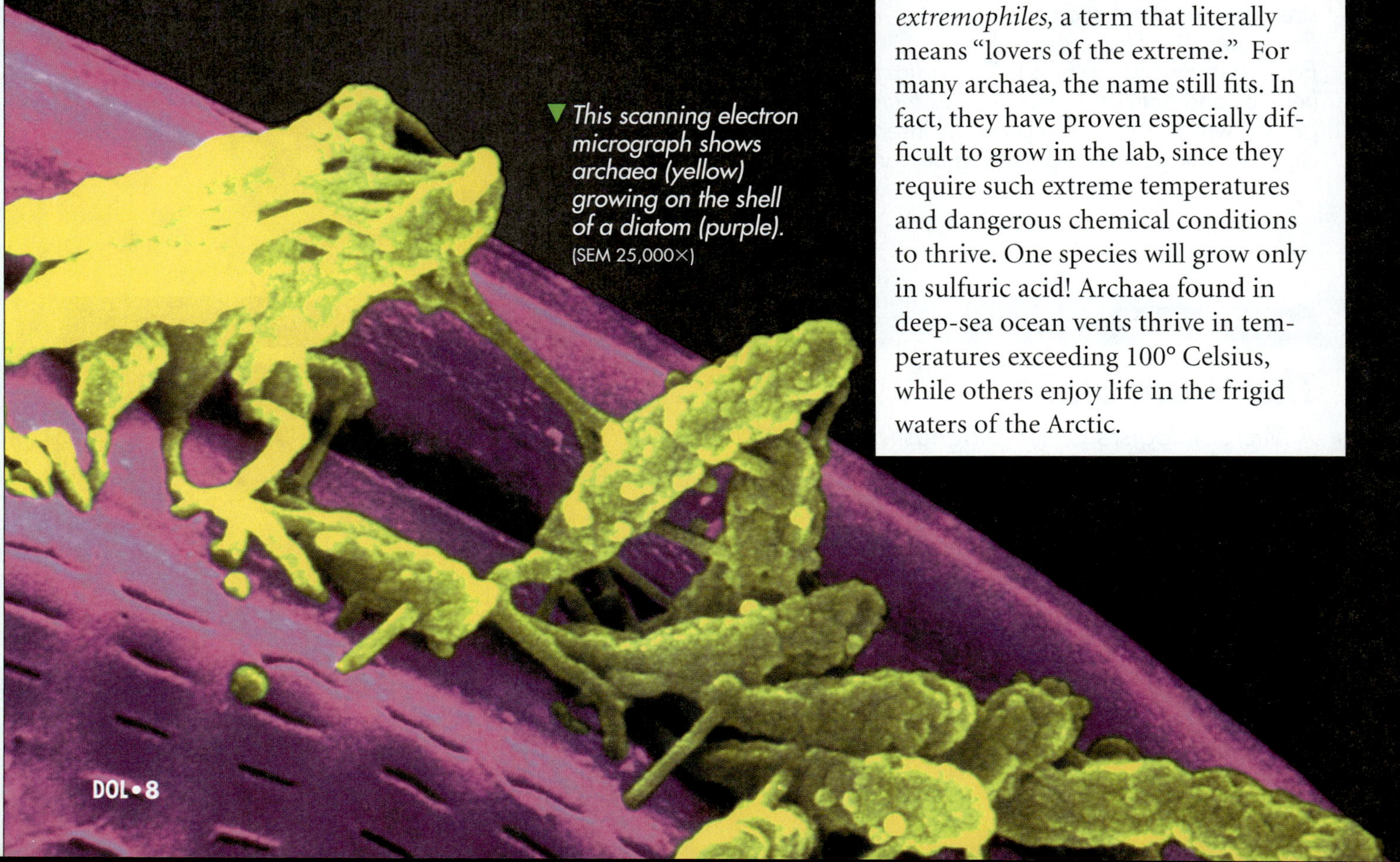

▼ *This scanning electron micrograph shows archaea (yellow) growing on the shell of a diatom (purple).* (SEM 25,000×)

▲ *The volcano Solfatara, near Naples, Italy, is home to many archaea in the genus* Sulfolobus.

•Did You Know?

Hot Enough for You?

The Original Extremists

Way before extreme sports and extreme reality TV shows came the archaea—the original and ultimate extremists. When archaea were first discovered, biologists called them *extremophiles,* a term that literally means "lovers of the extreme." For many archaea, the name still fits. In fact, they have proven especially difficult to grow in the lab, since they require such extreme temperatures and dangerous chemical conditions to thrive. One species will grow only in sulfuric acid! Archaea found in deep-sea ocean vents thrive in temperatures exceeding 100° Celsius, while others enjoy life in the frigid waters of the Arctic.

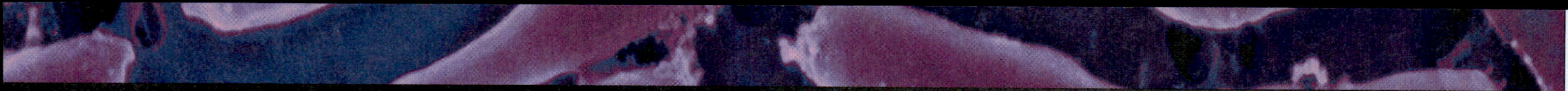

GROUPS OF ARCHAEA

To date, four major clades of archaea have been identified. Biologists continue to debate how these clades are related to one another.

CRENARCHAEOTES

The crenarchaeotes (kren AHR kee ohts) include organisms that live in the hottest and most acidic environments known. Most of the known species have been isolated from thermal vents and hot springs—the prefix *cren-* means "spring." Some species grow using organic compounds as energy sources, but others fix carbon from carbon dioxide, using hydrogen or sulfur to provide chemical energy.

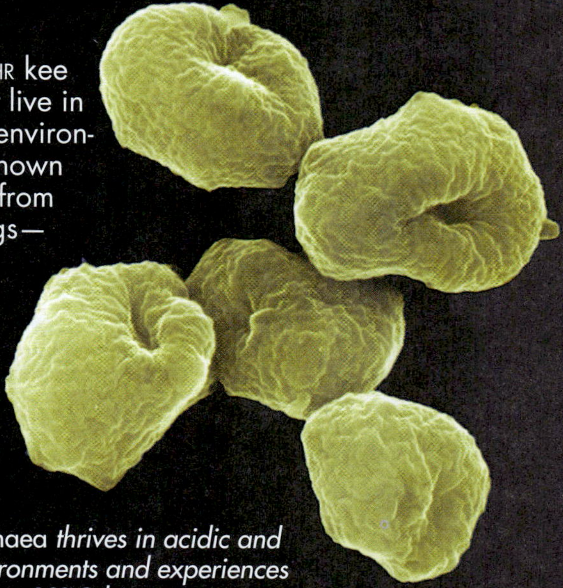

▶ Sulfolobus archaea *thrives in acidic and sulfur-rich environments and experiences optimal growth at 80° Celsius.* (SEM 33,200×)

KORARCHAEOTES

Scientists recently discovered the korarchaeote (kawr AHR kee oht) lineage in Obsidian Pool, Yellowstone National Park, and have since discovered more species in Iceland. Their DNA sequences place them apart from other archaea. The korarchaeotes may in fact be one of the least-evolved lineages of modern life that has been detected in nature so far.

▲ *Korarchaeotes from Obsidian Pool are shown in a lab culture with other microbes from their community.* (SEM 6000×)

NANOARCHAEOTES

Only a single species of this group has been discovered, in 2002, attached to a much larger crenarchaeote! Nanoarchaeotes (na noh AHR kee ohts) grow in hot vents near the coastal regions of the ocean and show definite molecular differences from other archaea. More research is needed to characterize this group, but what is known is that they have the smallest known genome of any organism.

▼ *The newly discovered* Nanoarchaeum equitans *(smaller cells) is shown attached to its host, genus* Ignicoccus *(larger cells).* (LM 2000×)

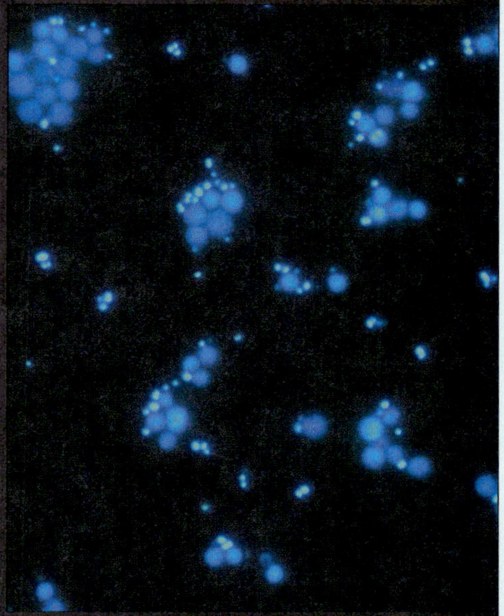

▼ Colony of *Methanosarcina mazei* (SEM 40,000×)

EURYARCHAEOTES

The euryarchaeotes (yoor ee AHR kee ohts) are a very diverse group of archaea, living in a broad range of habitats. The prefix *eury-* comes from a Greek word meaning "broad." The methanogens are a major group of euryarchaeotes that play essential roles in the environment. They help to break down organic compounds in oxygen-poor environments, releasing methane gas in the process. Another group, the *Halobacteria*, are found in salt ponds, where the concentration of sodium chloride approaches saturation.

Protists

KEY CHARACTERISTICS

A protist is a eukaryote, generally single-celled, that does not fit into any of the other major taxonomic groups. The protists do not make up a true kingdom.

Organization Great diversity of cell organelles and organization: some have cell walls, some have chloroplasts, most have mitochondria or organelles related to mitochondria; those that are multicellular have relatively little differentiation into tissues.

Movement Some move by cilia or flagella.

Reproduction Most reproduce by cell division; many have sexual phases to their life cycle; some exchange genetic material by conjugation.

▲ *This freshwater protist, a heliozoan in the clade Amoebozoa, has numerous thin pseudopods. It harbors symbiotic, photosynthetic algae called zoochlorellae.* (LM 730×)

•Did You Know?

The Kingdom That Isn't

The Challenges of Classifying Protists

Biologists traditionally classified protists by splitting them into funguslike, plantlike, and animal-like groups. This seemed to work for a while, but when they studied protists more carefully with new research tools, including genome-level molecular analysis, this traditional system simply fell apart.

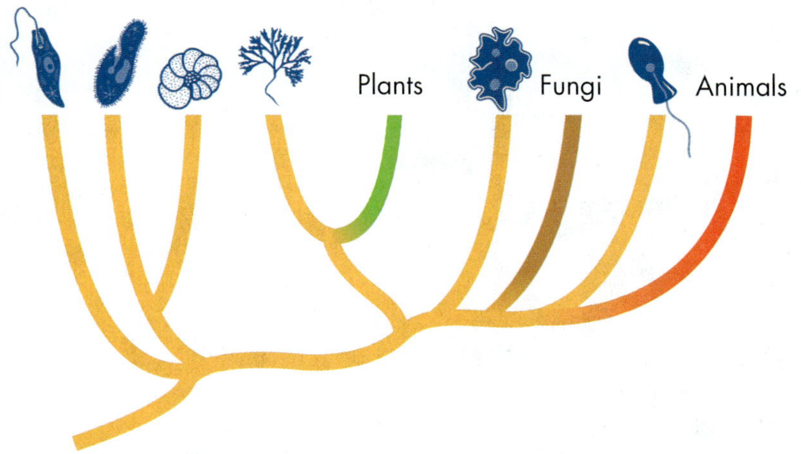

Biologists now think that protists shouldn't be classified as a kingdom at all. In fact, when scientists look for the deepest and most fundamental divisions among eukaryotes, they find that all of those divisions are within the protists themselves, not between protists and other eukaryotes. Starting over, biologists could simply use those divisions to define newer, more accurate "kingdoms," but that might cause new problems. For one thing, it would lump two of the traditional kingdoms (animals and fungi) together, and it would leave a handful of kingdoms that contain only unicellular organisms. There is no perfect solution to this problem. Here, "protists" are considered a kingdom for the sake of convenience, but keep in mind that their differences are really too great for any single kingdom to contain.

Excavates

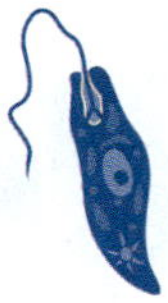

KEY CHARACTERISTICS

Excavates (EKS kuh vayts) have a characteristic feeding groove, usually supported by microtubules. Most have flagella. A few lack mitochondria and are unable to carry out oxidative phosphorylation, although they do possess remnants of the organelle.

GROUPS OF EXCAVATES

The excavates include a wide diversity of protists, from free-living photosynthesizers to some of humankind's most notorious pathogens.

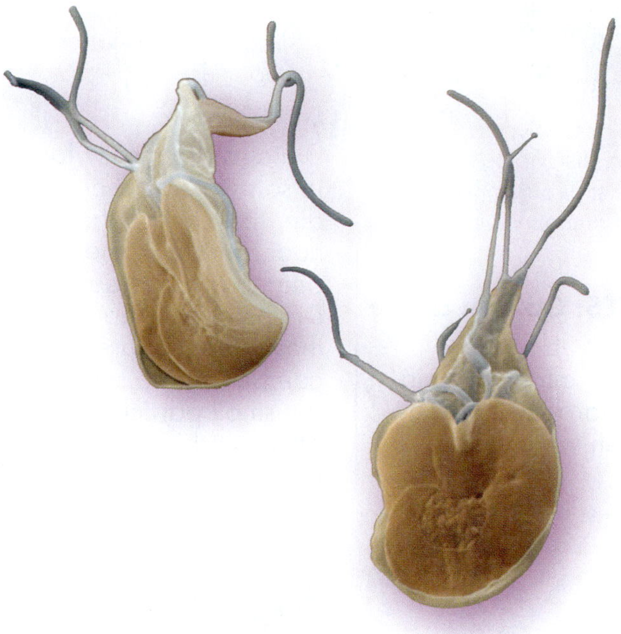

▲ *The diplomonad* Giardia *is a dangerous intestinal parasite that frequently contaminates freshwater streams.* Giardia *infections are common in wildlife and pet dogs and cats.* (SEM 1800×)

DIPLOMONADS

These organisms get their name from the fact that they possess two distinct and different nuclei (from Greek, diplo = double). The double nuclei probably derived from an ancient symbiotic event in which one species was engulfed by another. Cells contain multiple flagella, usually arranged around the body of the cell. Most species of diplomonads are parasitic.

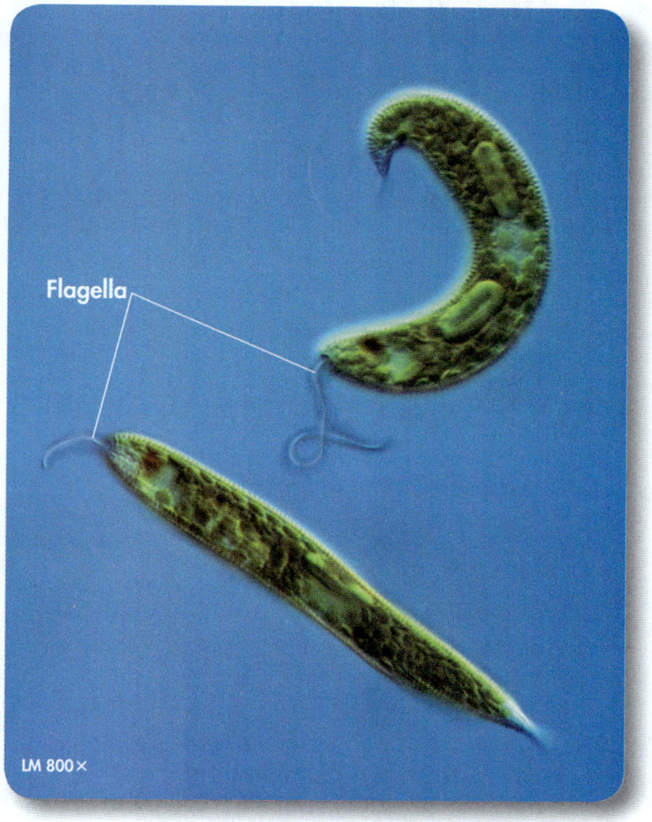

▲ *Photosynthetic* Euglena spirogyra *is commonly found in lakes and ponds.*

DISCICRISTATES

Discicristates (disk ee KRIS tayts) are named for the disc-shaped cristae present in their mitochondria. Some species are photosynthetic and free-living, such as *Euglena*, while others are dangerous parasites.

▼ *The ribbonlike cells of* Trypanosoma brucei *cause African sleeping sickness. The parasitic protist is transmitted by tsetse flies to humans, where it infects the blood, lymph, and spinal fluid. Severe nervous system damage and death are the usual result.* (SEM 6700×)

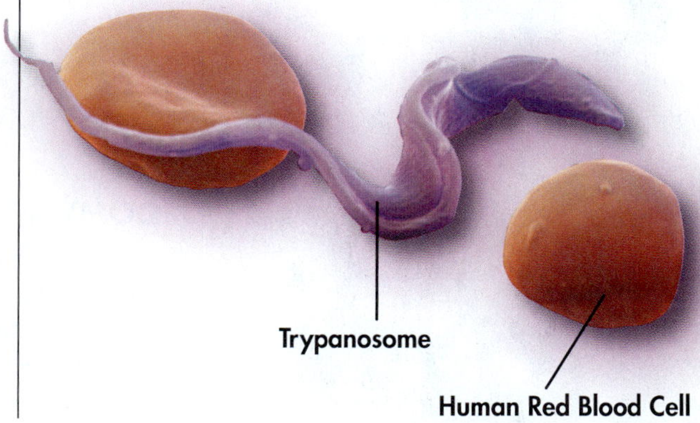

Chromalveolates

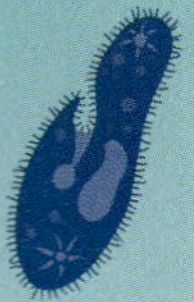

KEY CHARACTERISTICS

Chromalveolates (krohm AL vee uh layts) get their name from alveoli, flattened vesicles that line the cell membrane. The prefix chromo-, meaning "pigment," reflects evidence that members of this clade share a common ancestor that had accessory pigments used in photosynthesis.

SEM 280×

SEM 1000×

▲ *Diatoms often produce intricate shells made from silicon dioxide that persist long after they die.*

GROUPS OF CHROMALVEOLATES

The chromalveolates are one of the largest and most diverse groups of eukaryotes.

PHAEOPHYTES: Brown algae

Phaeophytes (FAY uh fyts) are mostly found in salt water. They are some of the most abundant and visible of the algae. Most species contain fucoxanthin, a greenish-brown pigment from which the group gets its common name. The multicellular brown alga known as giant kelp can grow as large as 60 meters in length.

▼ *Brown algae in genus* Fucus *are commonly found in tidepools and on rocky shorelines of the Pacific Coast of the United States.*

SEM 650×

▲ *This species, in genus* Synura, *is a colonial alga. In this photograph, you can see the scales that cover the cell surfaces of the individuals making up the colony.*

CHRYSOPHYTES: Golden algae

Chrysophytes (KRIS oh fyts) are known for colorful accessory pigments in their chloroplasts. Most are found in fresh water and are photosynthetic.

DIATOMS

Diatoms are mostly found in salt water. When they die, they sink to the ocean floor, and their shells pile up in large deposits. Diatomaceous earth, as these deposits are known, can be used to screen out small particles, and is often used in swimming pool filters.

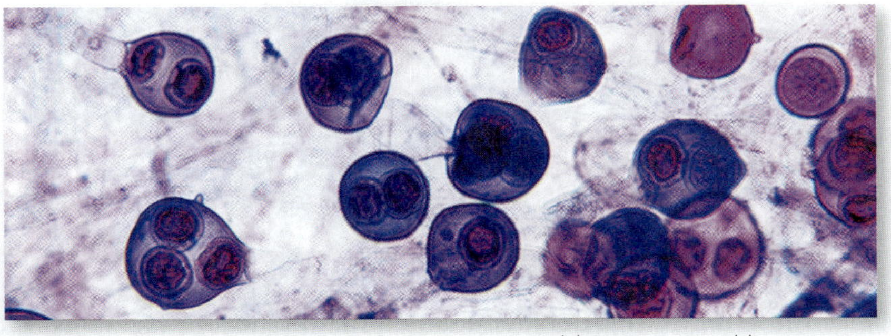

▲ *Water molds in genus* Achlya (LM 140×)

OOMYCETES: Water molds

These nonphotosynthetic organisms are often confused with fungi. Oomycetes (oh uh MY seed eez) typically produce fuzzy mats of material on dead or decaying animals and plants. Oomycetes are also responsible for a number of serious plant diseases, including potato blight, sudden oak death, and ink disease, which infects the American chestnut tree.

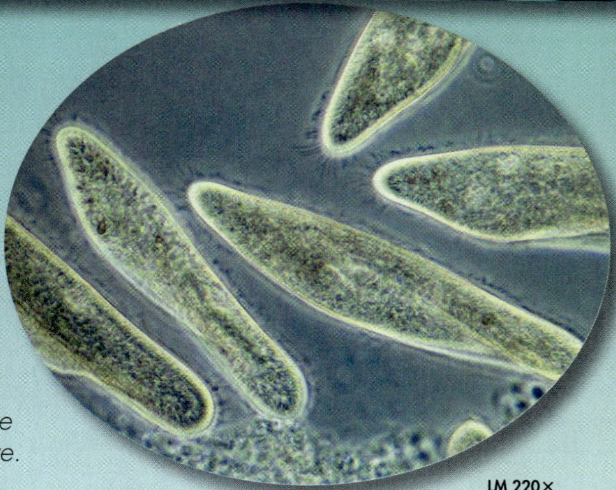

Paramecium multimicronucleatum *is the largest paramecium, with cells that are visible to the naked eye.*

CILIATES

These common organisms may contain hundreds or even thousands of short cilia extending from the surface of the cell. The cilia propel the ciliate through the water, and may sweep food particles into a gullet. Ciliates are large compared to other protists, with some cells exceeding 1 mm in length.

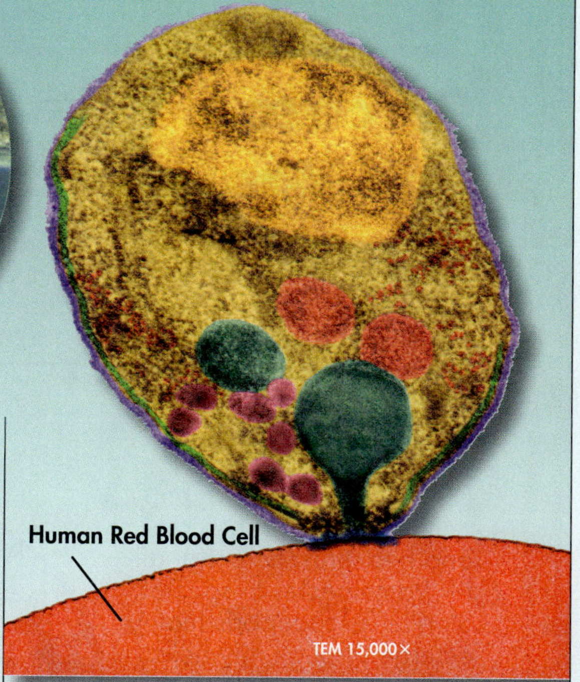

Apicomplexans in genus Plasmodium *are mosquito-borne parasites. Shown here is the sporozoite stage attached to a human red blood cell.*

DINOFLAGELLATES

Dinoflagellates are photosynthetic protists found in both fresh and salt water. Their name comes from their two distinct flagella, usually oriented at right angles to each other. Roughly half of dinoflagellate species are photosynthetic; the other half live as heterotrophs. Many dinoflagellate species are luminescent, and when agitated by sudden movement in the water, give off light.

The two flagella of dinoflagellates originate in grooves within thick plates of cellulose that resemble a cross shape, as shown here (genus Peridinium*).*

APICOMPLEXANS

The apicomplexans (AYP ih kum plek sunz) are named for a unique organelle near one end of the cell known as the apical complex. This structure contains vesicles with enzymes that allow apicomplexans to enter other cells and take up residence as parasites.

Eco • Alert

Toxic Blooms

Dangerous Dinoflagellates

Great blooms of the dinoflagellates *Gonyaulax* and *Karenia* have occurred in recent years on the East Coast of the United States, although scientists are not sure of the reason. These blooms are known as "red tides." *Gonyaulax* and *Karenia* produce a toxin that can become amplified in the food chain when filter-feeding shellfish such as oysters contentrate it in their tissues. Eating shellfish from water affected by red tide can cause serious illness, paralysis, and even death.

A red tide containing toxic dinoflagellates

Cercozoa, Foraminiferans, and Radiolarians

There is no single morphological characteristic that unites this diverse trio, but many have extensions of cytoplasm called pseudopods and many produce protective shells. The grouping together of Cercozoa, Foraminifera, and Radiolaria is based almost entirely on molecular analyses and not on morphology.

FORAMINIFERANS

Foraminifera (fawr uh min IF uh ra) produce intricate and beautiful shells that differ from species to species. Slender pseudopods that emerge through tiny holes in the shell enable them to capture food, including bacteria. As many as 4000 species exist.

Peneroplis plantus has a spiral-shaped shell.

Radiolarian shells are composed of silica or strontium sulfate.

RADIOLARIANS

These organisms have an intricate structure in which the nucleus is found in an inner region of the cell known as the endoplasm. The outer portion of the cell, known as the ectoplasm, contains lipid droplets and vacuoles. These organisms sometimes form symbiotic relationships with photosynthetic algae, from which they obtain food.

CERCOZOA

Members of this clade are common in soil, where they feed on bacteria as well as decaying organic matter. Many have flagella, and some produce scales made of silica that protect their surfaces.

• A Look Back in Time

Foraminiferan Fossils

Ancient Climates Revealed

Abundant fossils of foraminiferans have been found in sediments dating to the Cambrian period (560 million years ago). For decades, oil companies have taken advantage of these ancient fossils to locate the sediments most likely to contain oil, but now there is another use for them—measuring the sea temperature of ancient Earth. Foraminiferans take dissolved oxygen from seawater to make the calcium carbonate ($CaCO_3$) in their shells, and when they do so, they take up two isotopes of oxygen, ^{16}O and ^{18}O. Because water made from ^{16}O is less dense, more of it evaporates into the atmosphere when the seas are warm—increasing the amount of ^{18}O in the remaining seawater, and in the fossil shells. The ratio between ^{16}O and ^{18}O in these fossils allows scientists to study the history of seawater temperature, as shown in the graph above.

Foraminiferan Isotope Ratios and Climate Change

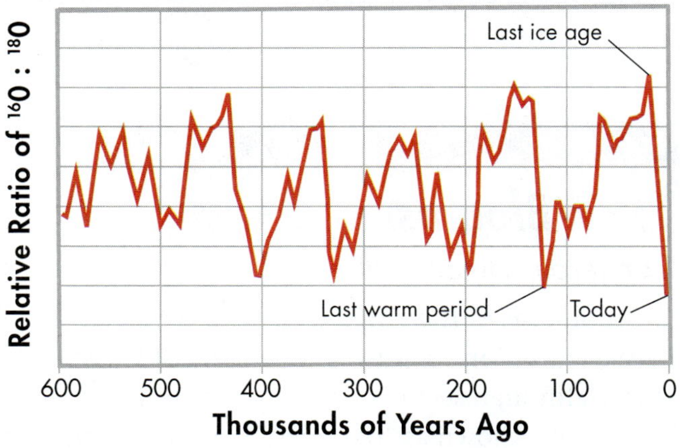

Rhodophytes

Also known as the red algae, these organisms get their name (from Greek, *rhodo* = red and *phyte* = plant) from reddish accessory pigments called phycobilins (fy koh BIL inz). These highly efficient pigments enable red algae to grow anywhere from the ocean's surface to depths as great as 268 meters. Most species are multicellular. Rhodophytes are the sister group to kingdom Plantae.

▼ Antithamnion plumula *is a marine rhodophyte that lives attached to coastal rocks.* (LM 35×)

Amoebozoa

Members of the Amoebozoa (uh MEE boh zoh ah) are amoebalike organisms that move by means of cytoplasmic streaming, also called amoeboid movement, using pseudopods.

◀ *Slime molds live as single microscopic amoebas in the soil, but aggregate into a colony when conditions are right, forming a multicellular fruiting body.*

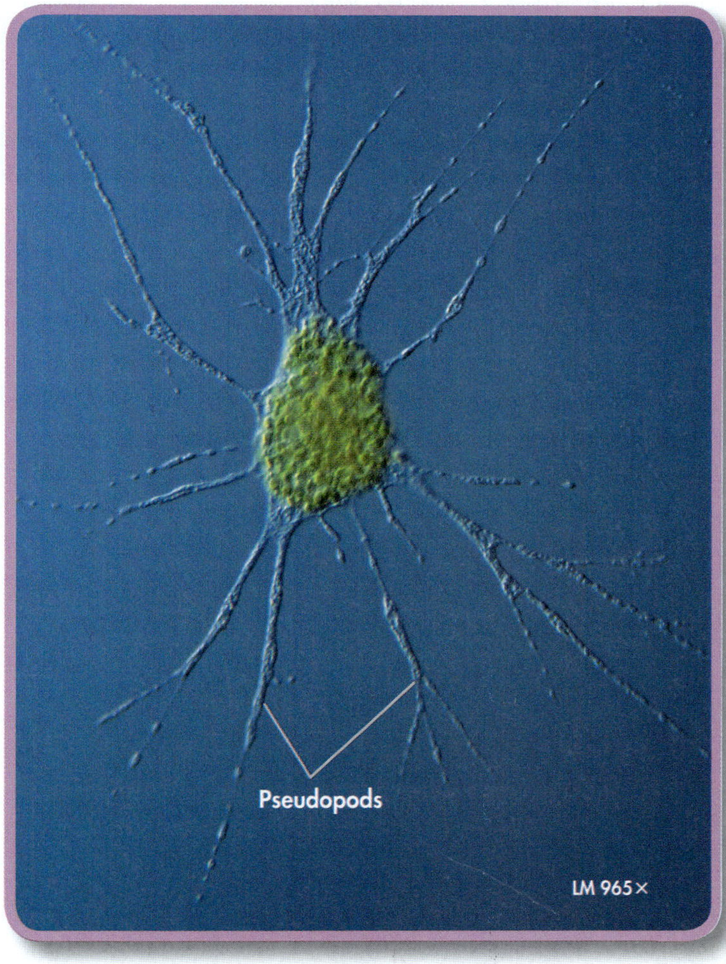

◀ *This solitary amoeba,* Penardia mutabilis, *has very slender pseudopods.*

Choanozoa

Members of the clade Choanozoa (koh AN uh zoh uh) can be solitary or colonial and are found in aquatic environments around the world. This clade is the sister group to kingdom Animalia.

Choanoflagellates are a major group in the clade Choanozoa. They get their name from a collar of cytoplasm that surrounds their single flagellum (form Greek, *choano* = collar.) Many species trap food within the collar and ingest it.

Fungi

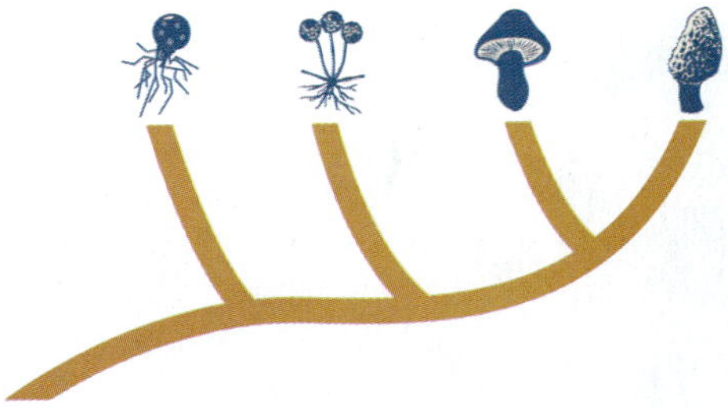

KEY CHARACTERISTICS

Fungi are heterotrophic eukaryotes with cell walls that contain chitin. Fungi were once thought to be plants that had lost their chloroplasts. It is now clear, however, that they are much more closely related to animals than to plants. More than 100,000 species of fungi are known. Distinctions among the phyla are made on the basis of DNA comparisons, cell structure, reproductive structures, and life cycles.

▲ *Stinkhorn fungus (genus* Dictyophora*)*

Organization Some are unicellular yeasts, but most have a multicellular body called a mycelium that consists of one or more slender, branching cells called hyphae.

Feeding and Digestion Obtain food by extracellular digestion and absorption

Reproduction Most have sexual phases to their life cycle and are haploid at most points during the cycle. Most produce tough, asexual spores, which are easily dispersed and able to endure harsh environmental conditions. Asexual reproduction by budding and splitting is also common.

• A Closer Look

▲ *Fly Agaric (*Amanita muscaria*) is poisonous to humans.*

Consumers Beware!

Edible and Inedible Mushrooms

Many types of fungi have long been considered delicacies, and several different species of mushrooms are cultivated for food. You may have already tasted sliced mushrooms on pizza, feasted on delicious sautéed portobello mushrooms, or eaten shiitake mushrooms. When properly cooked and prepared, domestic mushrooms are tasty and nutritious.

Wild mushrooms are a different story: Although some are edible, many are poisonous. Because many species of poisonous mushrooms look almost identical to edible mushrooms, you should never pick or eat any mushrooms found in the wild. Instead, mushroom gathering should be left to experts who can positively identify each mushroom they collect. The result of eating a poisonous mushroom can be severe illness, or even death.

Basidiomycetes

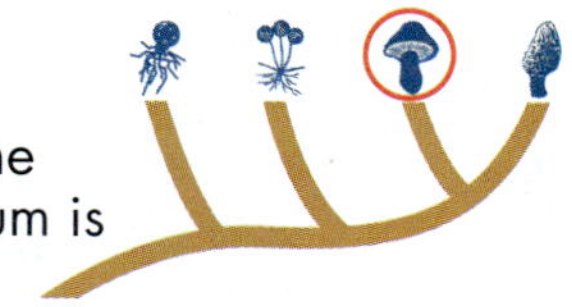

The basidiomycetes, or club fungi, are named for the basidium (buh SID ee um; plural: basidia). The basidium is a reproductive cell that resembles a club.

Life Cycle Basidiomycetes undergo what is probably the most elaborate life cycle of all the fungi, shown below.

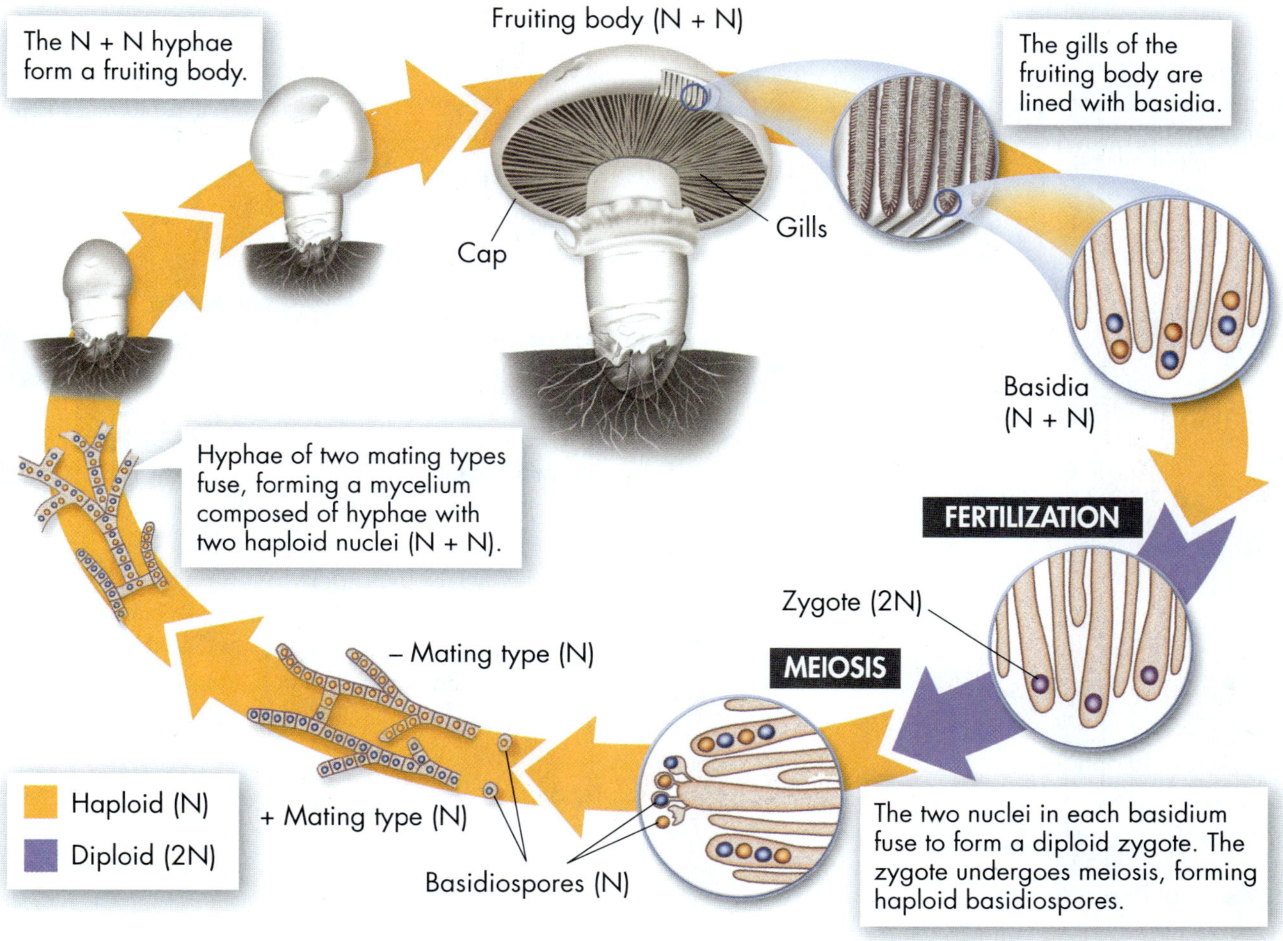

Diversity More than 26,000 species of basidiomycetes have been described, roughly a third of all known fungal species. Examples include the stinkhorn and fly agaric mushrooms shown on the previous page, and the shelf fungus and puffball at right.

► *Shelf fungi (Polypore family) often grow on the sides of dead or dying trees.*

▼ *A puffball releases its spores in an explosive cloud.*

Fungi

Ascomycetes

The ascomycetes, or sac fungi, are named for the ascus (AS kus), a saclike reproductive structure that contains spores.

Life Cycle The ascomycete life cycle includes an asexual phase, in which haploid spores are released from structures called conidiophores, and a sexual phase.

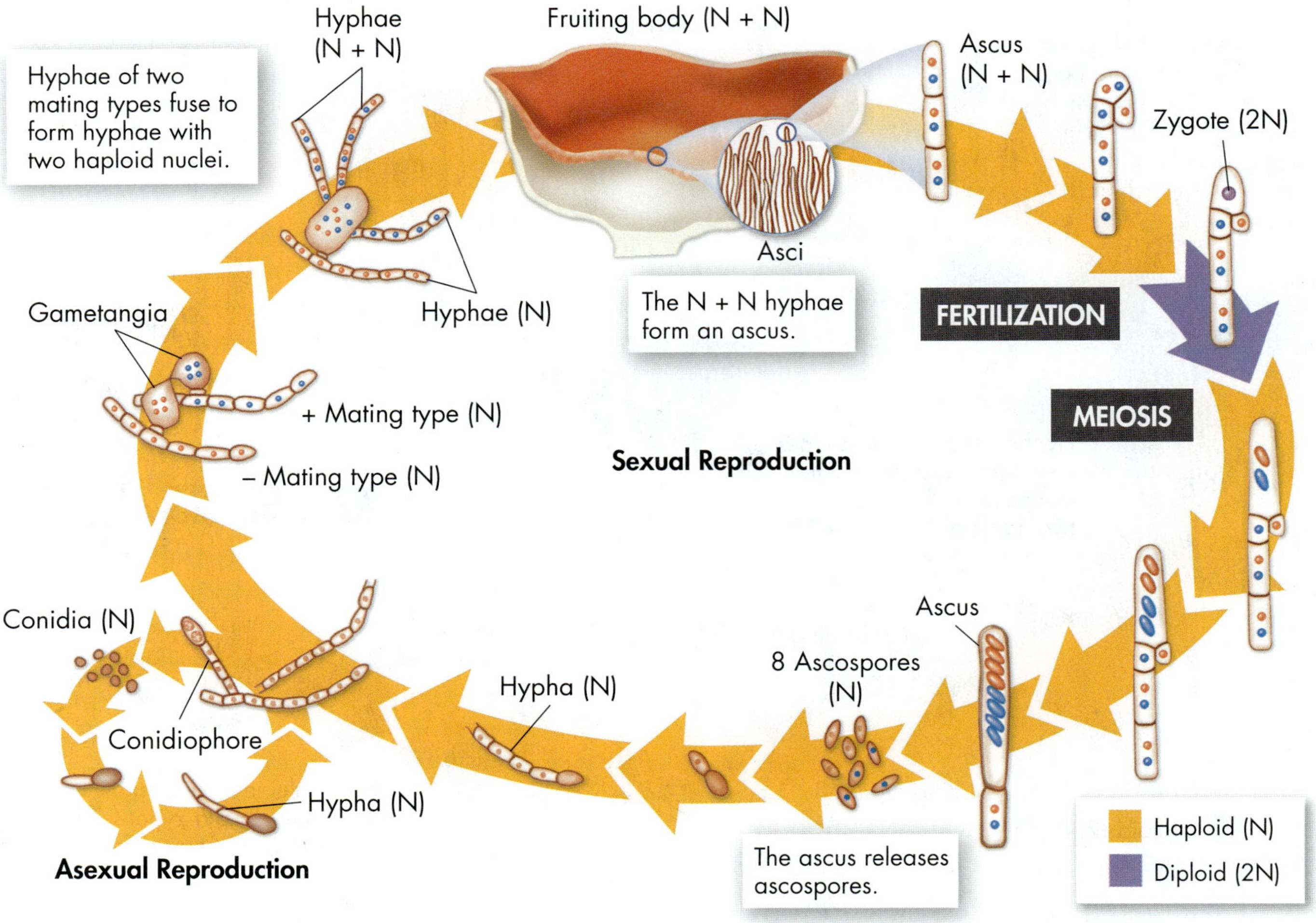

Diversity There are more than 50,000 species of ascomycetes, making it the largest phylum of the Fungi. Some ascomycetes, such as morels and cup fungi, are large enough to be visible when they grow above ground. Others, such as the common yeasts used for baking bread, are microscopic.

▼ *An edible morel in genus* Morchella

▼ *A cup fungus (genus* Cookenia*)*

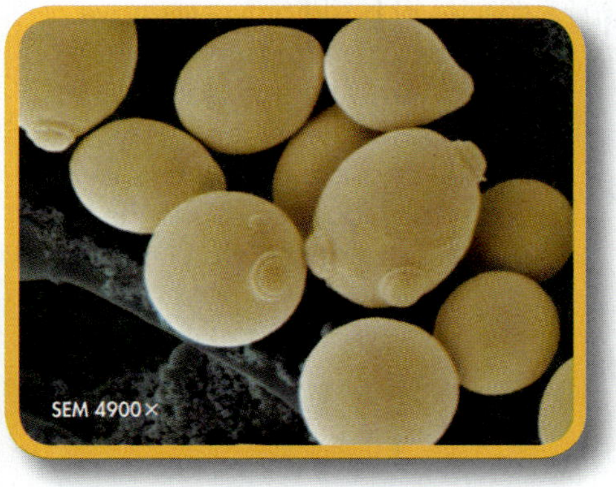

▲ Saccharomyces cerevisiae, *the yeast used to raise bread dough, is a unicellular ascomycete that reproduces asexually by budding.*

Zygomycetes

The hyphae of zygomycetes generally lack cross walls between cells. Zygomycetes get their name from the sexual phase of their reproductive cycle, which involves a structure called a zygosporangium that forms between the hyphae of two different mating types. One group within the zygomycetes, the Glomales, form symbiotic mycorrhizae (my koh RY zee) with plant roots.

◄ *The fruiting body of the common black bread mold,* Rhizopus stolonifer (SEM 450×)

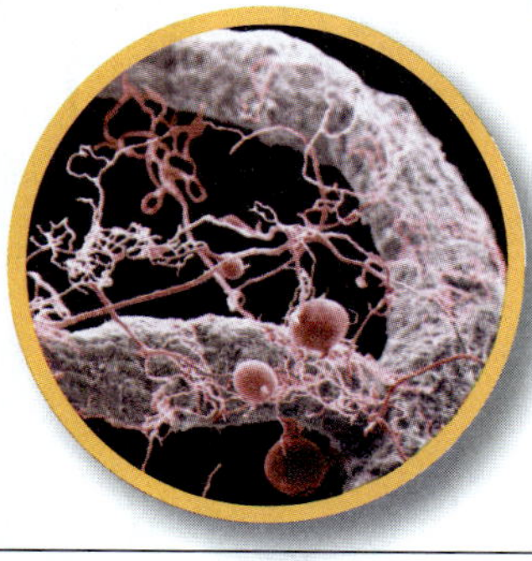

◄ *This micrograph shows mycorrhizal fungi in symbiosis with soybean roots. The soybean plant provides nutrient sugars to the fungus, while the fungus provides water and essential minerals to the plant.* (SEM 200×)

Chytrids

► *Chytriomyces hyalinus* (LM 500×)

Members of this phylum live in water or moist soil. Their reproductive cells have flagella, making them the only fungi known to have a motile stage to their life cycle. Chytrids are especially good at digesting cellulose, the material of plant cell walls—some live in the digestive systems of cows and deer, helping them to digest plant matter. Others are pathogens—certain chytrids have recently been associated with the decline of frog populations around the world. About 1000 species are known, many of them recently discovered.

Eco • Alert

▲ *Lichen-covered oak trees in Shenandoah National Park, Virginia*

Look to the Lichens

Lichens as Bio-Indicators

Lichens are mutualistic associations between a fungus, usually an ascomycete, and a photosynthetic organism, usually an alga. They are incredibly durable, and have even been reported to survive in the vacuum of space. However, they are also incredibly sensitive indicators of the state of the atmosphere. In particular, when sulfur dioxide is released into the atmosphere, it often reacts with water to form acids (including sulfuric acid) that pollute rainfall. Lichens can be severely damaged by acidic rainfall, although the degree of damage depends on the substrate upon which they grow. Lichens disappear first from the bark of pine and fir trees, which are themselves somewhat acidic. Lichens on elms, which have alkaline bark, are the last to go. By careful monitoring of the health of lichen populations of various trees, scientists can uses these remarkable organisms as low-tech monitors for the health of the environment.

Plants

KEY CHARACTERISTICS

Plants are eukaryotes with cell walls composed of cellulose. Plants carry out photosynthesis using the green pigments chlorophyll a and b, and they store the products of photosynthesis as starch.

▶ *A banana plant in bloom*

▼ *A typical plant life cycle*

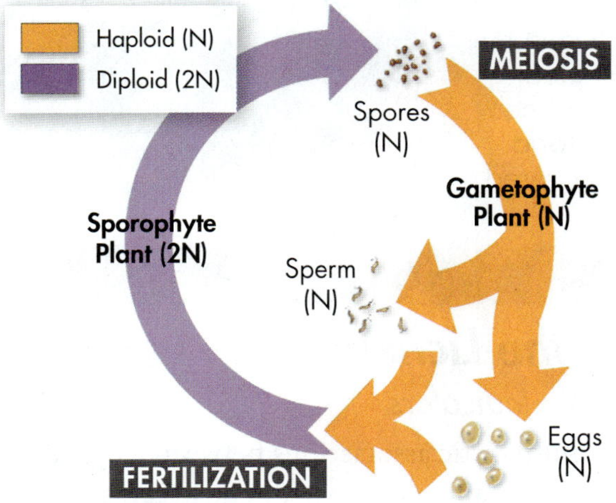

• A Closer Look

Prokaryotes Within

The Origins of Chloroplasts

Chloroplasts, which contain their own DNA, are found in all green plants, but where did they come from? In 1905, the Russian botanist Konstantin Mereschkowsky, noticing the similarities between chloroplasts and cyanobacteria, proposed that these organelles originated from a symbiotic relationship formed with the ancestors of today's plants.

This hypothesis still holds up very well today. New DNA studies suggest that all chloroplasts are descended from a single photosynthetic prokaryote, closely related to today's cyanobacteria.

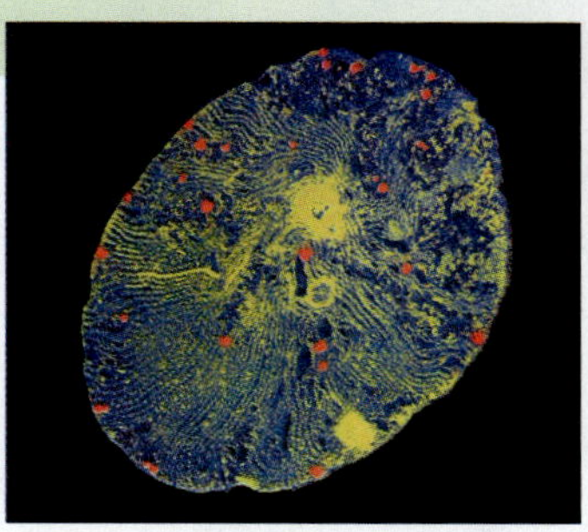

The photosynthetic membranes (shown in green) visible in this thin section of a cyanobacterium resemble the thylakoid membranes of plant cell chloroplasts. (TEM 14,000×)

Green Algae

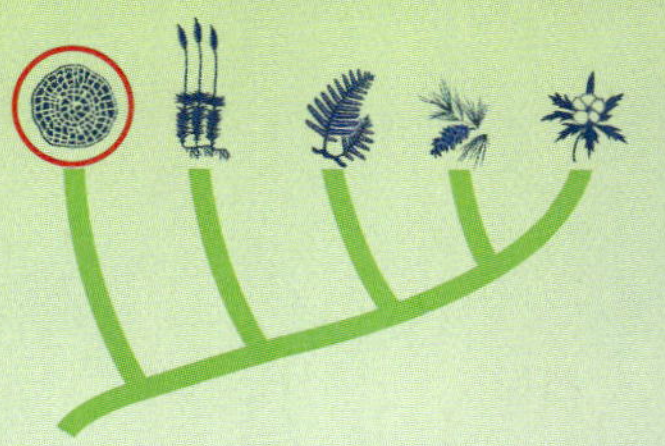

KEY CHARACTERISTICS

The green algae are plants that do not make embryos. All other plants form embryos as part of their life cycle. The green algae include both unicellular and multicellular species, and they are primarily aquatic.

Organization Single cells, colonies, and a few truly multicellular species

Movement Many swim using whiplike flagella.

Water Transport Water diffuses in from the environment.

Reproduction Asexual and sexual, with gametes and spores; some species show alternation of generations.

▲ *Clumps of* Spirogyra, *a filamentous green alga, are commonly called water silk or mermaid's tresses.*

GROUPS OF GREEN ALGAE

The three most diverse groups of green algae are profiled below.

CHLOROPHYTES: Classic green algae

These algae usually live as single cells, like *Chlamydomonas*, or in colonies, like *Volvox*. They are found in both fresh and salt water, and some species are even known to live in arctic snowbanks.

▶ Chlamydomonas *is a unicellular green alga. Each cell has two flagella, which are used in movement.* (SEM 3000×)

ULVOPHYTES: Sea lettuces

The ulvophytes are large organisms composed of hundreds or thousands of cells. Most form large, flattened green sheets and are often simply called seaweed. They show both haploid and diploid phases in their life cycle, but in many species, such as the common sea lettuce, *Ulva*, it is difficult to tell the two phases apart.

▼ *Ulva lactuca*

CHAROPHYTES: Stoneworts

Among the green algae, the charophytes (KAHR uh fyts) are the closest relatives of more complicated plants. They are mostly freshwater species. Their branching filaments may be anchored to the substrate by thin rhizoids.

◀ Chara *with antheridia (sperm-producing structures) visible*

Bryophytes

KEY CHARACTERISTICS

Bryophytes (BRY oh fyts), found mostly on land, are multicellular plants that lack true vascular tissue. This lack of vascular tissue limits their height to just a few centimeters and restricts them to moist soils.

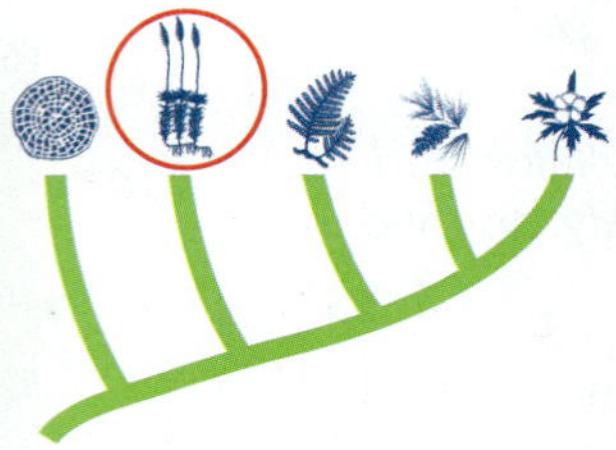

▲ *Mosses thrive in shady, damp locations, such as along the banks of this Oregon creek.*

Organization Complex and specialized tissues, including protective external layers and rhizoids

Movement Adults stationary; male gametes swim to egg cells using flagella.

Water Transport Diffusion from cell to cell; in some mosses, water flows through specialized tissue.

Reproduction All reproduce sexually with alternation of generations, producing gametes and spores. Most reproduce asexually, too. The gametophyte stage is dominant, with the sporophyte stage dependent on the gametophyte.

GROUPS OF BRYOPHYTES

Although they are listed together here, the three major groups of bryophytes are now considered to have evolved independently from each other.

MOSSES: Classic bryophytes

Mosses are found on damp, well-shaded soil, and occasionally along the sides of tree trunks.

LIVERWORTS

Liverworts are flat, almost leaflike plants that grow on the damp forest floor. Some species are shaped almost like the liver, from which they get their name.

HORNWORTS

Hornworts get their name from their sporophytes, tiny green structures resembling horns. Like other bryophytes, hornworts are found mostly in damp, well-shaded areas. Only about 100 species are known.

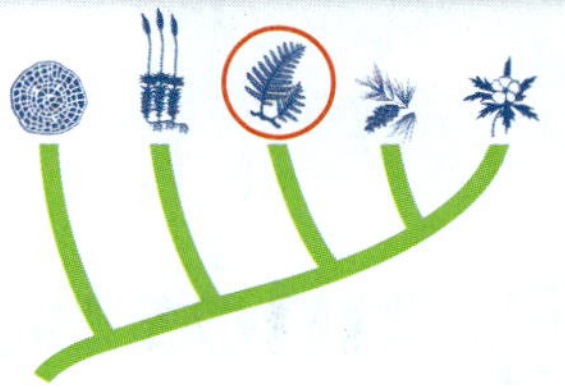

Seedless Vascular Plants

KEY CHARACTERISTICS

This informal grouping lumps together all the plants that have true vascular tissue but lack seeds. Vascular tissue is a key adaptation to life on land. By carrying water and food throughout plant structures, vascular tissue permitted the evolution of roots and tree-size plants, and it allowed plants to spread into dry areas of land.

Organization Complex and specialized tissues, including true roots, stems, and leaves

Movement Adults stationary; male gametes swim to egg cells using flagella.

Water Transport Through vascular tissue

Reproduction Alternation of generations, producing spores, eggs, and swimming sperm; the sporophyte stage is dominant, but the sporophyte is not dependent on the gametophyte as it is in bryophytes.

GROUPS OF SEEDLESS VASCULAR PLANTS

Besides the flowering plants, these organisms make up the most diverse collection of land plants, with more than 10,000 known species.

FERNS

Ferns are common and abundant. Because they need standing water to reproduce, ferns are generally found in areas that are damp at least part of the year. The sporophyte phase of the life cycle is dominant. Spores are produced in prominent clusters known as sori (SOH ry) on the undersides of leaves.

▼ *Polypodium vulgare*

CLUB MOSSES

Not really mosses, these vascular plants are also called lycopods (LY koh pahdz). These plants were especially abundant during the Carboniferous Period 360 to 290 million years ago, when they grew as large as trees. Today, their remains make up a large part of coal deposits mined for fuel.

▼ *The small club moss known as* Lycopodium *can be found growing on the forest floor throughout the temperate regions of North America. They look like tiny pine trees at first glance, but they are, in fact, small, seedless plants.*

HORSETAILS

Only a single living genus of horsetails is known, *Equisetum* (ek wi SEET um). These plants were thought to resemble horses' tails; their name is derived from this perception. Today, only 25 species are known, confined to wet areas of soil. But horsetails were once much more diverse, larger in size, and abundant. Abrasive silica, found in many horsetails, was used in colonial times as a scouring powder to help clean pots and pans.

▼ *Equisetum*

Gymnosperms

KEY CHARACTERISTICS

Gymnosperms are seed-bearing vascular plants whose seeds are exposed to the environment, rather than being enclosed in a fruit. The seeds are usually located on the scales of cones.

Organization True roots, stems, and leaves

Movement Adults stationary; within pollen grains, male gametophytes drift in air or are carried by animals to female structures, where they release sperm that move to eggs.

Water Transport Through vascular tissue

Reproduction Sexual; alternation of generations; the sporophyte stage is dominant. Female gametophytes live within the parent sporophyte. Pollen grains carry sperm to eggs, so open water is not needed for fertilization.

Some bristlecone pines are thousands of years old, like this one growing in Nevada.

Did You Know?

Rising From the Ashes

Fire's Role in Seed Germination

We generally think of forest fires as being natural disasters, and that's typically true. Some gymnosperm species, however, are so well adapted to the arid conditions of the American West that they actually depend upon such fires to spread their seeds.

The best-known example is the Jack Pine, *Pinus banksiana*. Its seed cones are thick and heat resistant. When engulfed in a fire, its seeds escape damage. The fire's high heat helps to open the outer coat of the cone, enabling the seeds to pop out afterward. As a result, Jack Pines are among the very first plants to repopulate a forest that has been damaged by fire.

The high heat of a forest fire opens the cones of the Jack Pines, releasing their seeds. The inset shows a Jack Pine seedling growing in the charred remains of the fire.

GROUPS OF GYMNOSPERMS

There are four groups of gymnosperms, representing about 800 species in total.

CONIFERS

Conifers are by far the most diverse group of living gymnosperms, represented by nearly 700 species worldwide. They include the common pine, spruce, fir, and redwood trees that make up a large share of the forests in the temperate regions of the world. Conifers have enormous economic importance. Their wood is used for residential building, to manufacture paper, and as a source of heat. Compounds from their resins are used for a variety of industrial purposes.

▲ *Most conifers retain their leaves year-round.*

CYCADS

Cycads (SY kads) are beautiful palmlike plants that have large cones. Cycads first appeared in the fossil record during the Triassic Period, 225 million years ago. Huge forests of cycads thrived when dinosaurs roamed Earth. Today, only nine genera of cycads exist. Cycads can be found growing naturally in tropical and subtropical places such as Mexico, the West Indies, Florida, and parts of Asia, Africa, and Australia.

▶ *A Sago Palm,* Cycas revoluta

▲ *Ginkgoes are often planted in urban settings, where their toughness and resistance to air pollution make them popular shade trees.*

GINKGOES

Ginkgoes (GING kohs) were common when dinosaurs were alive, but today the group contains only one species, *Ginkgo biloba*. The living *Ginkgo* species looks similar to its fossil ancestors—in fact, *G. biloba* may be one of the oldest seed plant species alive today.

GNETOPHYTES

About 70 present-day species of gnetophytes (NET oh fyts) are known, placed in just three genera. The reproductive scales of these plants are clustered in cones.

▶ Welwitschia mirabilis, *an inhabitant of the Namibian desert in southwestern Africa, is one of the most remarkable gnetophytes. Its huge leathery leaves grow continuously and spread across the ground.*

Angiosperms

KEY CHARACTERISTICS

Angiosperms are plants that bear seeds in a closed ovary. The ovary is part of a reproductive organ known as a flower. Seeds are formed in a double fertilization event, which forms a diploid embryo and a triploid endosperm tissue. As seeds mature, ovaries develop into fruits that help to disperse the seeds.

Organization True roots, stems, and leaves

Movement Adults stationary; within pollen grains, male gametophytes drift in air or are carried by animals to female structures, where they release sperm that move to eggs.

Water Transport Through vascular tissue

Reproduction Sexual, with alternation of generations; also asexual. The sporophyte stage is dominant. Female gametophytes live within the parent sporophyte. Pollen carries sperm to eggs, so open water is not needed for fertilization.

▶ *A Southern Long-Nosed Bat pollinates the Saguaro Cactus,* Carnegia gigantea, *while collecting nectar from its blossoms.*

A Closer Look

Whatever Happened to Monocots and Dicots?

Traditionally, flowering plants have been divided into just two groups, monocots and dicots, based on the number of seed leaves in their embryos. Today, however, molecular studies have shown that the dicots aren't really one group. Some of the most primitive flowering plants (like *Amborella*) are dicots, and so are some of the most advanced flowering plants, while the monocots fall right in between. So, while monocots are indeed a single group, the term *dicots* is now just an informal, though still useful, grouping.

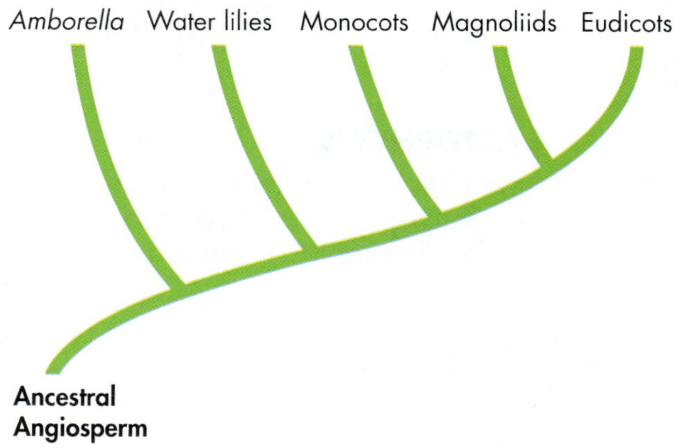

GROUPS OF ANGIOSPERMS

The great majority of plant species—over 260,000—are angiosperms.

▲ *Water lilies are aquatic plants that produce flowers and leaves, which float on the surface of the water.*

NYMPHAEACEAE: Water lilies

About 50 species of water lilies are known, and they are of special interest to plant taxonomists. Their DNA and flower structure suggest that they are, along with *Amborella*, one of the earliest groups to have split off from the main line of flowering plant evolution. Examples of water lilies are found throughout the world.

MAGNOLIIDS: Magnolia trees and others

The most famous genus of these plants is *Magnolia*, which includes nearly 200 species. Laurels and tulip poplars are also magnoliids (mag NOH lee ids). Because of their flower structure, magnoliids were once thought to be nearly as primitive as water lilies. Genetic studies now suggest that they split off from the rest of the angiosperm line after monocots and, therefore, do not represent the earliest flowering plants.

▼ *The Tulip Poplar is a long, straight tree often used as wood for telephone poles. Its flowers are greenish and shaped like tulips.*

► *Magnolia trees produce conspicuous flowers, which contain multiple stamens and multiple pistils.*

AMBORELLA

Amborella does not represent a group of plants but instead just a single species found only on the island of New Caledonia in the South Pacific Ocean. DNA studies show that *Amborella* is equally separated from all other flowering plants living today, suggesting that it is descended from plants that split off from the main line of flowering plant evolution as long ago as 100 million years.

▲ *The flowers of* Amborella trichopoda *are simpler than those of most other plants, and the species has a number of features that place it at the very base of flowering plant evolution.*

GROUPS OF ANGIOSPERMS CONTINUED...

MONOCOTS

The monocots include an estimated 65,000 species, roughly 20 percent of all flowering plants. They get their name from the single seed leaf found in monocot embryos, and they include some of the plants that are most important to human cultures. Monocots grown as crops account for a majority of the food produced by agriculture. These crops include wheat, rice, barley, corn, and sugar cane. Common grasses are monocots, as are onions, bananas, orchids, coconut palms, tulips, and irises.

▲ Many orchid species are grown by enthusiasts for their rare beauty. Notice the aerial roots on this specimen, which grows as an epiphyte in its natural environment.

▲ *Onions are just one of many examples of monocot crop species.*

▲ *This African hillside is dotted with clumps of Wild Pampas Grass.*

◀ *This sugar cane in Vietnam has been bundled for sale.*

Eco • Alert

Coevolution: Losing the Pollinators

The successes of flowering plants are clearly due to coevolution with their insect pollinators. Common honey bees are among the most important of these, gathering nectar from the flowers of hundreds of plant species and spreading pollen from plant to plant as they go.

Unfortunately, beekeepers around the world, including the United States, are facing a serious crisis. "Colony collapse disorder," as beekeepers describe it, causes bees to fly away from the hive and either never return, or return only to weaken and die. The disease threatens to affect scores of important crops, which depend upon bees to produce fruit and seeds. Suspicion has centered on a fungus or a virus that might spread from colony to colony, but at this point there is no definitive cause or cure.

EUDICOTS: "TRUE DICOTS"

Eudicots (YOO dy kahts) account for about 75 percent of all angiosperm species. The name means "true dicots," and these plants are the ones usually given as examples of dicot stem, leaf, and flower structure. Eudicots have distinctive pollen grains with three grooves on their surfaces, and DNA studies strongly support their classification in a single group. They include a number of important subgroups, five of which are described here.

▲ Clusterhead Pinks

Ranunculales

The ranunculales subgroup (ruh NUNH kyu lay les) includes, and is named after, buttercups (genus *Ranunculus*). Also included in this subgroup are a number of well-known flowers such as columbines, poppies, barberries, and moonseed.

▶ Rocky Mountain Columbine

Caryophyllales

Cacti are probably the most well-known plants in the caryophyllales subgroup (KAR ee oh fy lay les). Pinks and carnations, spinach, rhubarb, and insect-eating plants, such as sundews and pitcher plants, are also members.

▲ Orange

Rosids

The rosids include, as you might expect, the roses. However, this subgroup also includes many popular fruits, such as oranges, raspberries, strawberries, and apples. Some of the best-known trees, including poplars, willows, and maples, are also members.

Saxifragales

Plants in the saxifragales (SAK suh frij ay les) subgroup include peonies, witch hazel, gooseberries, and coral bells.

◀ Peony

Asterids

The nearly 80,000 asterid species include sunflowers, azaleas, snapdragons, blueberries, tomatoes, and potatoes.

▼ *The flower heads in a field of sunflowers all track the sun as it moves across the sky; thus, they all face the same direction.*

Animals

Snow Leopard

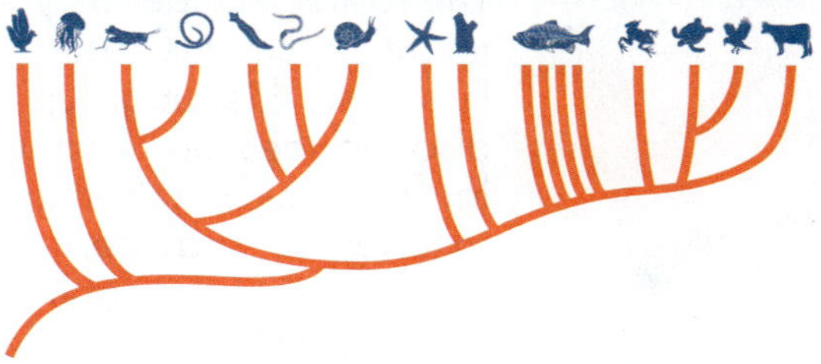

KEY CHARACTERISTICS

Animals are multicellular, heterotrophic, eukaryotic organisms whose cells lack cell walls.

A Closer Look

A Common Ancestor

Recent molecular studies and cladistic analyses recognize the clade Choanozoa to be the true sister group to all Metazoa—multicellular animals. Choanozoa is one group of organisms formerly called "protists" and is named for choanoflagellates (art and photo right), single-celled, colonial organisms that look like certain cells of sponges and flatworms. Current thinking suggests that the choanoflagellates alive today are the best living examples of what the last common ancestor of metazoans looked like.

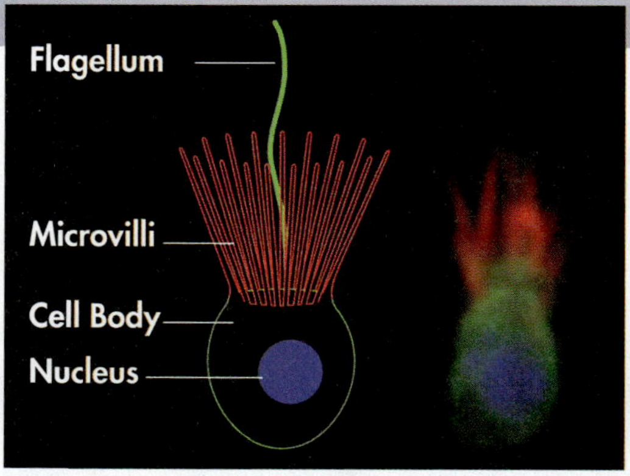

Porifera (Sponges)

KEY CHARACTERISTICS

Sponges are the simplest animals. They are classified as animals because they are multicellular, heterotrophic, lack cell walls, and have some specialized cells. They are aquatic, lack true tissues and organs, and have internal skeletons of spongin and/or spicules of calcium carbonate or silica. Sponges have no body symmetry.

Feeding and Digestion Filter feeders; intracellular digestion.

Circulation Via flow of water through body

Respiration Oxygen diffuses from water into cells as water flows through body.

Excretion Wastes diffuse from cells into water as water flows through body.

Response No nervous system; little capacity to respond to environmental changes.

Movement Juveniles drift or swim freely; adults are stationary.

Reproduction Most—sexual with internal fertilization; water flowing out of sponge disperses sperm, which fertilizes eggs inside sponge(s); may reproduce asexually by budding or producing gemmules.

GROUPS OF SPONGES

There are more than 5000 species of sponges; most are marine. Three major groups are described below.

DEMOSPONGIAE: Typical sponges

More than 90 percent of all living sponge species are in this group, including the few freshwater species. They have skeletons made of spongin, a flexible protein. Some species have silica spicules. Examples: Yellow Sponge, bath sponges, Carnivorous Mediterranean Sponge, tube sponges

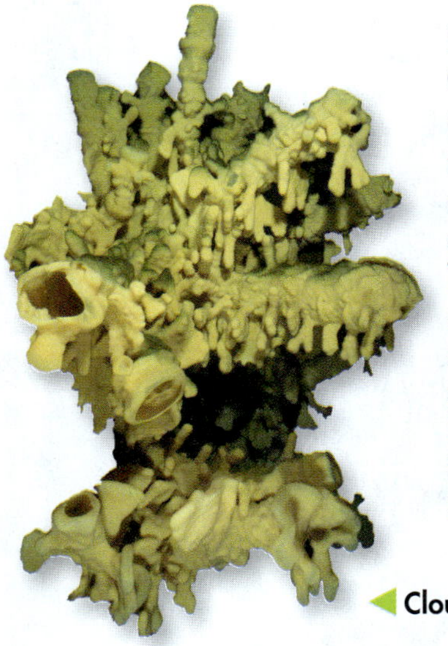

Cloud Sponge

HEXACTINELLIDA: Glass sponges

Glass sponges live in the deep ocean and are especially abundant in the Antarctic. They are called "glass" sponges because their skeletons are made of glasslike silica spicules. Examples: Venus's Flower Basket, Cloud Sponge

Elephant Ear Sponge

CALCAREA: Calcareous sponges

Calcareous sponges live in shallow, tropical marine waters and are the only sponges with calcium carbonate spicules. Example: *Clathrina*

Yellow Tubular Sponge

Cnidarians

▲ Sea Nettle

KEY CHARACTERISTICS

Cnidarians are aquatic, mostly carnivorous, and the simplest animals to have specialized tissues (outer skin and lining of the gastrovascular cavity) and body symmetry (radial). Their tentacles have stinging cells called nematocysts used in feeding.

Feeding and Digestion Predatory, stinging prey with nematocysts; digestion begins extracellularly in gastrovascular cavity and is completed intracellularly; indigestible materials leave body through single opening; many, especially reef-building corals, also depend on symbiotic algae, or zooxanthellae.

Circulation No internal transport system; nutrients typically diffuse through body.

Respiration Diffusion through body walls

Excretion Cellular wastes diffuse through body walls.

Response Some specialized sensory cells: nerve cells in nerve net, statocysts that help determine up and down, eyespots (ocelli) made of light-detecting cells

Movement Polyps stationary, medusas free-swimming; some, such as sea anemones, can burrow and creep very slowly; others move using muscles that work with a hydrostatic skeleton and water in gastrovascular cavity; medusas such as jellyfish move by jet propulsion generated by muscle contractions.

Reproduction Most—alternate between sexual (most species by external fertilization) and asexual (polyps produce new polyps or medusae by budding)

The color of this plate coral is caused by zooxanthellae algae living within it.

Eco • Alert

Coral Symbionts

Reef-building coral animals depend on symbiotic algae called zooxanthellae for certain vital nutritional needs. In many places, reef-building corals live close to the upper end of their temperature tolerance zone. If water temperatures rise too high, the coral-zooxanthellae symbiosis breaks down, and corals turn white in what is called "coral bleaching." If corals don't recover their algae soon, they weaken and die. This is one reason why coral reefs are in grave danger from global warming.

GROUPS OF CNIDARIANS

There are more than 9000 species of cnidarians.

A Portuguese Man-of-War is actually a colony of polyps.

HYDROZOA: Hydras and their relatives

Hydras and their relatives spend most of their time as polyps and are either colonial or solitary. They reproduce asexually (by budding), sexually, or they alternate between sexual and asexual reproduction. Examples: hydra, Portuguese Man-of-War

ANTHOZOA: Corals and sea anemones

Corals and sea anemones are colonial or solitary polyps with no medusa stage. The central body is surrounded by tentacles. They reproduce sexually or asexually. Examples: reef corals, sea anemones, sea pens, sea fans

Sea Anemone

Black Sea Nettle

SCYPHOZOA: Jellyfishes

Jellyfishes spend most of their time as medusas; some species bypass the polyp stage. They reproduce sexually and sometimes asexually by budding. Examples: Lion's Mane Jellyfish, Moon Jelly, Sea wasp

Jellyfishes such as this sea nettle are beautiful to us but deadly to their prey. The stinging cells on their tentacles can kill prey instantly and can ruin a human swimmer's day at the beach!

Arthropods

KEY CHARACTERISTICS

Arthropods are the most diverse of all multicellular organisms. They have segmented bodies and jointed appendages. They are supported by tough exoskeletons made of chitin, which they periodically shed as they grow. Arthropods are coelomate protostomes with bilateral symmetry.

Eco•Alert

Beetle Damage

You probably know that some insects can seriously damage crop plants. But insects affect plants in natural habitats, too. One example is the mountain pine beetle, which is dramatically extending its range. Global warming appears to be enabling the beetle to survive farther north, and at higher altitudes, than it used to. The new beetle infestation is causing extensive damage to northern and high-altitude forests in North America. The death of millions of acres of trees has resulted in the release of large amounts of carbon dioxide, a greenhouse gas, into the atmosphere. You can see the sort of damage the beetles cause in the photo at right.

▲ *Mountain pine beetle damage to pine trees in White River National Forest, Colorado*

Feeding and Digestion Extremely diverse: herbivores, carnivores, detritivores, parasites, bloodsuckers, scavengers, filter feeders; digestive system with two openings; many feeding specializations in different groups

Circulation Open circulatory system with heart and arteries

Respiration Terrestrial—tracheal tubes or book lungs; aquatic—gills or book gills (horseshoe crabs)

Excretion Terrestrial—Malpighian tubules; aquatic—diffusion into water

Response Well-developed nervous system with brain; sophisticated sense organs

Movement Muscles attached internally to jointed exoskeletons

Reproduction Usually sexual, although some species may reproduce asexually under certain circumstances; many undergo metamorphosis during development

Most animals, including this land crab, are arthropods.

GROUPS OF ARTHROPODS

Phylum Arthropoda contains more known species than any other phylum. Scientists have identified more than 1,000,000 arthropod species, and some scientists expect there are millions yet to be identified. Arthropods are classified based on the number and structure of body segments and appendages.

▲ Lobster

CRUSTACEA: Crustaceans

There are crustacean species in almost every habitat, but most are aquatic, and most of these are marine. They have two or three body sections, two pairs of antennae, and chewing mouthparts called mandibles. Many have a carapace, or "shell," that covers part or all of the body. Examples: crabs, lobsters, crayfish, pill bugs, water fleas, barnacles

---

CHELICERATA: Chelicerates

Living chelicerates include horseshoe crabs and arachnids. (Their extinct relatives include trilobites and giant "sea-scorpions.") Most living chelicerates are terrestrial. The body is composed of two parts—the cephalothorax and abdomen. The first pair of appendages are specialized feeding structures called chelicerae. Chelicerates have no antennae.

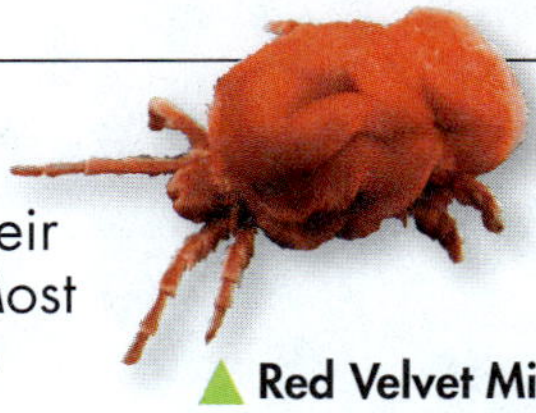

▲ Red Velvet Mite

Horseshoe crabs are actually more closely related to spiders than to crabs!

Merostomata: Horseshoe crabs

The class Merostomata once included many species, but only four species of horseshoe crab survive today. All are marine. They have five pairs of walking legs and a long, spinelike tail.

Arachnida: Arachnids

The vast majority of arachnids are terrestrial. They have four pairs of walking legs and no tail. Examples: spiders, ticks, mites, scorpions, daddy longlegs

▲ Mexican Beauty Tarantula

UNIRAMIA: Uniramians

Most uniramians are terrestrial, although some are aquatic for all or part of their lives. They have one pair of antennae, mandibles, and unbranched appendages. Uniramians include at least three fourths of all known animal species!

Uniramians include centipedes, millipedes, and insects—more than three fourths of all known animal species, including this Elephant Hawk Moth.

Praying Mantis

Insecta: Insects

There are more than 1,000,000 insect species in more than 25 orders. An insect body is divided into three parts—head, thorax, and abdomen. Insects have three pairs of legs and usually one or two pairs of wings attached to the thorax. Some insects undergo complete metamorphosis. Examples: termites, ants, beetles, dragonflies, flies, moths, grasshoppers

The Death's-Head Hawk Moth is named for the skull-like shape on the adult's head (above). Like many insects, this moth undergoes complete metamorphosis, during which the larva (below), or caterpillar, turns into a pupa, and, eventually, an adult.

Giant Millipede

Diplopoda: Millipedes

Millipedes have a long body composed of many segments. Each segment bears two pairs of legs. Most millipedes are herbivorous.

Chilopoda: Centipedes

Centipedes have a long body composed of many segments. Each segment bears one pair of legs. They are carnivorous and have poison claws for capturing prey.

Centipede

Insecta (continued)

Many "bugs" benefit humans. For example, ladybugs (which are not all "ladies") eat pests, and bees pollinate plants. Insects such as praying mantises, katydids, flies, moths, beetles, and ants also have important roles in ecosystems.

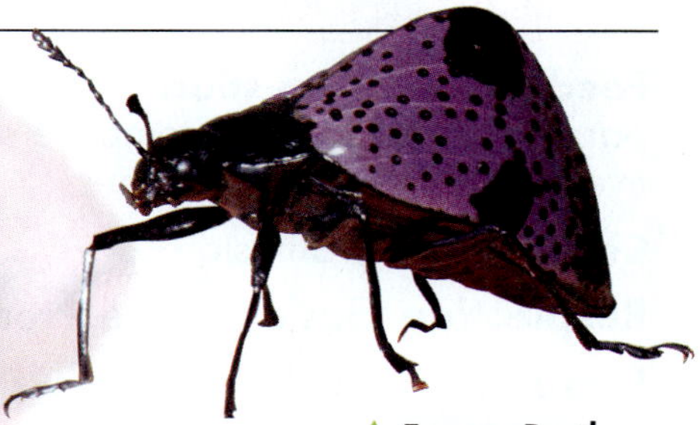

Fungus Beetle

Mint Beetle

Honeybee

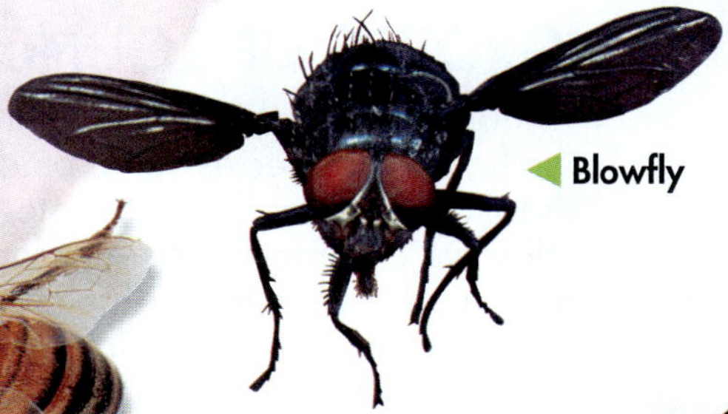

Blowfly

Ladybug

Desert Ant

Lobster Katydid

Nematodes (Roundworms)

"Hooks" in the mouth of a hookworm attach the worms to their hosts so that they can drink the host's blood or ingest their digested foods.

▲ *Foleyella*
(SEM 130×)

KEY CHARACTERISTICS

Nematodes, or roundworms, are unsegmented worms with a tough outer cuticle, which they shed as they grow. This "molting" is one reason that nematodes are now considered more closely related to arthropods than to other wormlike animals. Nematodes are the simplest animals to have a "one-way" digestive system through which food passes from mouth to anus. They are protostomes and have a pseudocoelom.

Feeding and Digestion Some predators, some parasites, and some decomposers; one-way digestive tract with mouth and anus

Circulation By diffusion

Respiration Gas exchange through body walls

Excretion Through body walls

Response Simple nervous system consisting of several ganglia, several nerves, and several types of sense organs

Movement Muscles work with hydrostatic skeleton, enabling aquatic species to move like water snakes and soil-dwelling species to move by thrashing around.

Reproduction Sexual with internal fertilization; separate sexes; parasitic species may lay eggs in several hosts or host organs.

GROUPS OF ROUNDWORMS

There are more than 15,000 known species of roundworms, and there may be half a million species yet to be described. Free-living species live in almost every habitat imaginable: fresh water, salt water, hot springs, ice, soil. Parasitic species live on or inside a wide range of organisms, including insects, humans, and many domesticated animals and plants. Examples: *Ascaris lumbricoides*, hookworms, pinworms, *Trichinella, C. elegans*

•A Closer Look

A Model Organism?

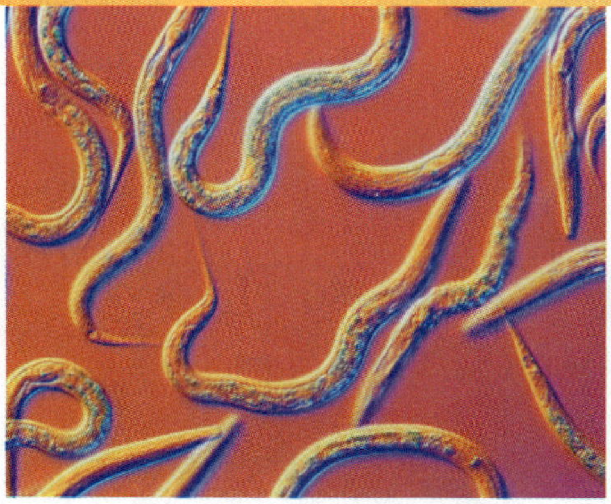

Caenorhabditis elegans is a small soil nematode. Fifty years ago, this species was selected as a "model organism" for the study of genetics and development. We can now chart the growth and development of *C. elegans*, cell by cell, from fertilization to adult. This information is invaluable in understanding the development in other species—including many other nematodes that cause serious disease.

◄ ***C. elegans*** (LM 64×)

Platyhelminthes (Flatworms)

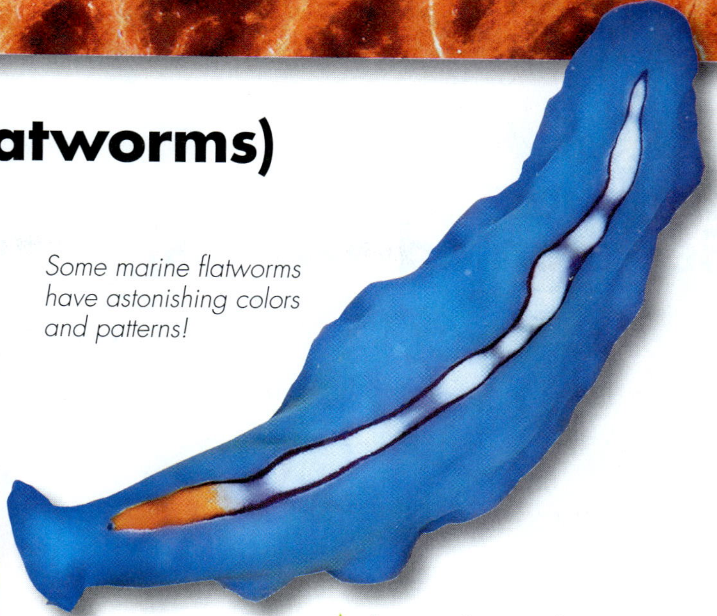

Some marine flatworms have astonishing colors and patterns!

▲ Blue Pseudoceros Flatworm

KEY CHARACTERISTICS

Flatworms are soft worms with tissues and internal organ systems. They are the simplest animals to have three embryonic germ layers, bilateral symmetry, and cephalization. They are acoelomates.

Feeding and Digestion Free-living—predators or scavengers that suck food in through a pharynx and digest it in a system that has one opening. Parasitic—feed on blood, tissue fluids, or cell pieces of the host, using simpler digestive systems than free-living species have. Tapeworms, which absorb nutrients from food that the host has already digested, have no digestive system.

Circulation By diffusion

Respiration Gas exchange by diffusion

Excretion Some—flame cells remove excess water and may remove metabolic wastes such as ammonia and urea. Many flame cells are connected to tubules that release substances through pores in the skin.

Response Free-living—several ganglia connected by nerve cords that run through the body, along with eyespots and other specialized sensory cells; parasitic—simpler nervous system than free-living forms have

Movement Free-living—using cilia and muscle cells.

Reproduction Free-living—most are hermaphrodites that reproduce sexually with internal fertilization; parasitic—commonly reproduce asexually by fission but also often reproduce sexually

GROUPS OF FLATWORMS

Flatworms are an amazingly diverse group of worms that include more than 20,000 species. They have historically been placed into three classes, but these taxa now appear not to be true clades, and will probably change.

TREMATODA: Flukes

Most flukes are parasites that infect internal organs of their hosts, but some infect external parts such as skin or gills. The life cycle typically involves more than one host or organ. Examples: *Schistosoma*, liver fluke

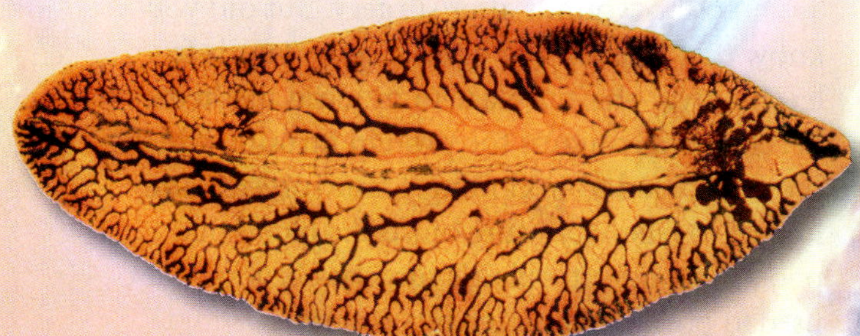

▲ Liver Fluke

TURBELLARIA: Turbellarians

Turbellarians are free-living aquatic and terrestrial predators and scavengers. Many are colorful marine species. Examples: planarians, polyclad flatworm

CESTODA: Tapeworms

Tapeworms are very long intestinal parasites that lack a digestive system and absorb nutrients directly through their body walls. The tapeworm body is composed of many repeated sections (proglottids) that contain both male and female reproductive organs.

Annelids (Segmented Worms)

Peacock worms, whose feather-shaped gills look somewhat like peacock feathers, are marine annelids, or polychaetes.

KEY CHARACTERISTICS

Annelids are coelomate protostome worms whose bodies are composed of segments separated by internal partitions. The annelid digestive system has two openings.

Feeding and Digestion Filter feeders, carnivores, or parasites; many obtain food using a muscular pharynx, often equipped with "teeth"; widely varied digestive systems—some, such as earthworms, have complex digestive tracts.

Circulation Closed circulatory system with dorsal and ventral blood vessels; dorsal vessel pumps blood like a heart.

Respiration Aquatic—gills; terrestrial—skin

Excretion Digestive waste exits through anus; nitrogenous wastes eliminated by nephridia

Response Nervous system includes a rudimentary brain and several nerve cords; sense organs best-developed in free-living saltwater species

Movement Hydrostatic skeleton based on sealed body segments surrounded by longitudinal and circular muscles; many annelids have appendages that enable movement.

Reproduction Most—sexual, some through external fertilization with separate sexes, but others are simultaneous hermaphrodites that exchange sperm; most have a trochophore larval stage

Did You Know?

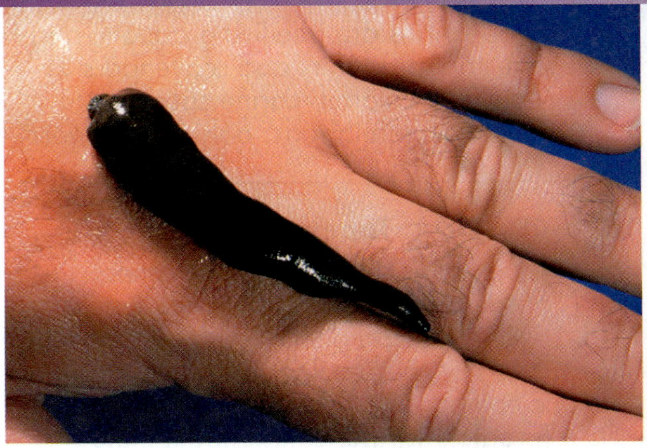

▲ *Leech* (Hirudo medicinalis) *drawing blood from a hand*

Not-So-Modern Medicine

You may have heard that medieval healers used leeches to remove "excess" blood from patients and to clean wounds after surgery. But did you know that leeches—or at least compounds from leech saliva—have a place in modern medicine? Leech saliva contains the protein hirudin, which prevents blood from clotting. Some surgeons use leeches to relieve pressure caused by blood that pools in tissues after plastic surgery. Hirudin is also used to prevent unwanted blood clots.

Feather-Duster Worms

Giant Earthworm

GROUPS OF ANNELIDS

There are more than 15,000 species of annelids.

HIRUDINEA: Leeches

Most leeches live in fresh water. They lack appendages. Leeches may be carnivores or blood-sucking external parasites. Example: medicinal leech *(Hirudo medicinalis)*

POLYCHAETA: Polychaetes

Polychaetes live in salt water; many move with paddle-like appendages called parapodia tipped with bristle-like setae. Examples: sandworms, bloodworms, fanworms, feather-duster worms

The white, bristle-like structures on the sides of this bearded fireworm are setae.

OLIGOCHAETA: Oligochaetes

Oligochaetes live in soil or fresh water. They lack appendages. Some use setae for movement but have fewer than polychaetes. Examples: *Tubifex*, earthworms

Mollusks

KEY CHARACTERISTICS

Mollusks have soft bodies that typically include a muscular foot. Body forms vary greatly. Many mollusks possess a hard shell secreted by the mantle, but in some, the only hard structure is internal. Mollusks are coelomate protostomes with bilateral symmetry.

Feeding and Digestion Digestive system with two openings; diverse feeding styles—mollusks can be herbivores, carnivores, filter feeders, detritivores, or parasites

Circulation Snails and clams—open circulatory system; octopi and squid—closed circulatory system

Respiration Aquatic mollusks—gills inside the mantle cavity; land mollusks—a saclike mantle cavity whose large, moist surface area is lined with blood vessels.

Excretion Body cells release ammonia into the blood, which nephridia remove and release outside the body.

Response Complexity of nervous system varies greatly; extremely simple in clams, but complex in some octopi.

Movement Varies greatly, by group. Some never move as adults, while others are very fast swimmers.

Reproduction Sexual; many aquatic species have free-swimming trochophore larval stage.

Colossal Squid

Did You Know?

The Colossal Squid

The Colossal Squid, the largest of all mollusks, has the largest eyes of any known animal. One 8-meter-long, 450-kilogram specimen of the species *Mesonychoteuthis hamiltoni* had eyes 28 centimeters across—larger than most dinner plates! The lens of this huge eye was the size of an orange.

GROUPS OF MOLLUSKS

Mollusks are traditionally divided into several classes based on characteristics of the foot and the shell; specialists estimate that there are somewhere between 50,000 and 200,000 species of mollusks alive today.

Giant Clam

BIVALVIA: Bivalves

Bivalves are aquatic. They have a two-part hinged shell and a wedge-shaped foot. They are mostly stationary as adults. Some burrow in mud or sand; others attach to rocks. Most are filter feeders that use gill siphons to take in water that carries food. Clams have open circulatory systems. Bivalves have the simplest nervous systems among mollusks. Examples: clams, oysters, scallops, mussels

Garden Snail

GASTROPODA: Gastropods

There are both terrestrial and aquatic gastropods. Most have a single spiral, chambered shell. Gastropods use a broad, muscular foot to move and have a distinct head region. Snails and slugs feed with a structure called a radula that usually works like sandpaper. Some species are predators whose harpoon-shaped radula carries deadly venom. They have open circulatory systems. Many gastropod species are cross-fertilizing hermaphrodites. Examples: snails, slugs, nudibranchs, sea hares

Chambered Nautilus

CEPHALOPODA: Cephalopods

Cephalopods live in salt water. The cephalopod has a highly developed brain and sense organs. The head is attached to a single foot, which is divided into tentacles. They have closed circulatory systems. Octopi use beaklike jaws for feeding; a few are venomous. Cephalopods have the most complex nervous systems among mollusks; octopi have complex behavior and have shown the ability to learn in laboratory settings. Examples: octopi, squids, nautilus, cuttlefish

Nudibranchs, such as this Blue Dorid, are marine gastropods without shells. They breathe through gills (the yellow structures) on their backs.

Animals

Echinoderms

KEY CHARACTERISTICS

Echinoderms are marine animals that have spiny skin surrounding an endoskeleton. Their unique water vascular system includes tube feet with suction-cuplike ends used in moving and feeding. The water vascular system also plays a role in respiration, circulation, and excretion. Echinoderms are coelomate deuterostomes. Adults exhibit 5-part radial symmetry.

Feeding and Digestion Method varies by group—echinoderms can be filter feeders, detritivores, herbivores, or carnivores.

Circulation Via fluid in the coelom, a rudimentary system of vessels, and the water vascular system

Respiration Gas exchange is carried out by surfaces of tube feet, and, in many species, by skin gills.

Crinoid fossil, about 400 million years old

Living modern crinoid (feather star)

• A Look Back in Time

Crinoids Then and Now

Echinoderms have a long fossil record that dates all the way back to the Cambrian Period. Although these animals have been evolving for millions of years, some fossil crinoids look a great deal like living crinoids.

Excretion Digestive wastes released through anus; nitrogenous cellular wastes excreted as ammonia through tube feet and skin gills.

Response Minimal nervous system; nerve ring is connected to body sections by radial nerves; most have scattered sensory cells that detect light, gravity, and chemicals secreted by prey.

Movement In most, tube feet work with endoskeleton to enable locomotion.

Reproduction Sexual, with external fertilization; larvae have bilateral symmetry, unlike adults.

You can't miss the 5-part radial symmetry of this candy-cane sea star moving across a sea anemone.

GROUPS OF ECHINODERMS

There are more than 7000 species of echinoderms.

Sea star

CRINOIDEA: Crinoids

Crinoids are filter feeders; some use tube feet along feathery arms to capture plankton. The mouth and anus are on the upper surface of the body disk. Some are stationary as adults while others can "walk" using short "arms" on the lower body surface. Examples: sea lily, feather star

Feeding crinoid

ASTEROIDEA: Sea stars

Sea stars are bottom dwellers whose star-shaped bodies have flexible joints. They are carnivorous—the stomach pushes through the mouth onto the body tissues of prey and pours out digestive enzymes. The stomach then retracts with the partially digested prey; digestion is completed inside the body. Examples: crown-of-thorns sea star, sunstar

Basket star feeding on Orange Finger Sponge

OPHIUROIDEA: Ophiuroids

Ophiuroids have small body disks, long, armored arms, and flexible joints. Most are filter feeders or detritivores. Examples: brittle star, basket star

ECHINOIDEA

Echinoids lack arms. Their endoskeleton is rigid and boxlike and covered with movable spines. Most echinoids are herbivores or detritivores that use five-part jawlike structures to scrape algae from rocks. Examples: sea urchin, sand dollar, sea biscuit

Red sea urchins and purple sea urchins feeding on kelp

Sea cucumber

HOLOTHUROIDEA: Sea cucumbers

Sea cucumbers have a cylindrical, rubbery body with a reduced endoskeleton and no arms. They typically lie on their side and move along the ocean floor by the combined action of tube feet and body-wall muscles. These filter feeders or detritivores use a set of retractable feeding tentacles on one end to take in sand and detritus, from which they glean food.

Nonvertebrate Chordates

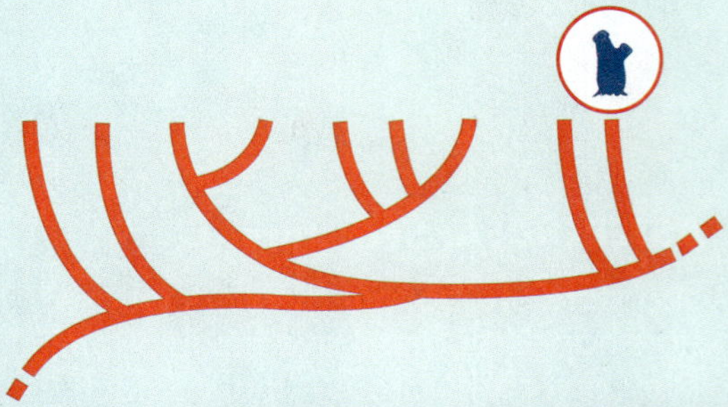

Tunicates are chordates named for the colorful tunic-like covering the adults have. As larvae, tunicates have all the characteristics of chordates, as well as bilateral symmetry, but as adults, they look very, very different.

KEY CHARACTERISTICS

The nonvertebrate chordates are the only chordates that lack a backbone. Like other chordates, they have a nerve cord, notochord, pharyngeal pouches, and a tail at some point during development. They are coelomate deuterostomes. The two subphyla, tunicates and lancelets, differ significantly.

Feeding and Digestion Filter feeders; tunicates—in most, water carrying food particles enters through an incurrent siphon; food is strained out in the pharynx and passed to the digestive system; lancelets—mucus in the pharynx catches food particles carried in by water, which are then carried into digestive tract

Circulation Closed; tunicates—heart pumps blood by "wringing out," and flow periodically reverses direction; lancelets—no heart, but blood vessels pump blood through body in one direction

Respiration Tunicates—gas exchange occurs in the gills and across other body surfaces; lancelets—through pharynx and body surfaces

Excretion Tunicates—most through excurrent siphon; lancelets—flame cells in nephridia release water and nitrogenous wastes into the atrium and out through an opening called an atriopore

Response Cerebral ganglion, few specialized sensory organs; tunicates—sensory cells in and on the siphons and other internal surfaces help control the amount of water passing through the pharynx; lancelets—a pair of eyespots detect light

Movement Tunicates—free-swimming larvae, but most are stationary as adults; lancelets—no appendages: they move by contracting muscles paired on either side of the body

Reproduction Tunicates—most sexual and hermaphroditic with external fertilization, but some reproduce by budding; most have free-swimming tadpole-like larvae that metamorphose into adults; lancelets—sexual with external fertilization

Eco • Alert

Out-of-Control Tunicates

You've never heard of them, but Asian stalked tunicates are disrupting marine ecosystems in Washington State; Prince Edward Island, Canada; and elsewhere. Tunicate larvae are carried in the ballast water of freight ships and discharged wherever the ships make port. There, away from their usual predators, the tunicates grow out of control, smothering shellfish beds and covering boats, docks, and underwater equipment. Researchers are still trying to figure out how to control them.

GROUPS OF NONVERTEBRATE CHORDATES

There are two major groups of nonvertebrate chordates: tunicates and lancelets (sometimes called amphioxus).

Two lancelets, Branchiostoma lanceolatum, poking out of sand.

CEPHALOCHORDATA: Lancelets

Lancelets are fishlike animals that have bilateral symmetry and live in salt water. They are filter feeders and have no internal skeleton. Example: *Branchiostoma*

▼ Pastel Sea Squirt

UROCHORDATA: Tunicates

Tunicates are filter feeders that live in salt water. Most adults have a tough outer covering ("tunic") and no body symmetry; most display chordate features and bilateral symmetry only during larval stages. Many adults are stationary; some are free-swimming. Examples: sea squirts, sea peaches, salps

▼ Sea Squirts

Animals

Fishes

KEY CHARACTERISTICS

The word fish is used informally to describe aquatic vertebrates that look similar even though they belong to several different clades, because all are adapted to life in water. Most vertebrates we call fishes have paired fins, scales, and gills.

Feeding and Digestion Varies widely, both within and between groups: herbivores, carnivores, parasites, filter feeders, detritivores; digestive organs often include specialized teeth and jaws, crop, esophagus, stomach, liver, pancreas

Circulation Closed, single-loop circulatory system; two-chambered heart

Respiration Gills; some have specialized lungs or other adaptations that enable them to obtain oxygen from air.

Excretion Diffusion across gill membranes; kidneys

Response Brain with many parts; highly developed sense organs, including lateral line system

Movement Paired muscles on either side of backbone; many have highly maneuverable fins; the largest groups have two sets of paired fins; some have a gas-filled swim bladder that regulates buoyancy.

Reproduction Methods vary within and between groups: external or internal fertilization; oviparous, ovoviviparous, or viviparous

• A Look Back in Time

Live Birth in Devonian Seas

▲ *Artist's conception of* Materpiscis *giving birth*

You might think that live birth is a recent addition to chordate diversity. Guess again. Recent fossil finds of fishes from the Devonian Period show that at least one group of fishes was already bearing live young 380 million years ago. Two incredibly well preserved fossils, including that of the fish *Materpiscis*, show the remains of young with umbilical cords still attached to their mother's bodies. This is the earliest fossil evidence of viviparity in vertebrates.

GROUPS OF FISHES

Fishes are the largest group of vertebrates, including more than 30,000 species. Evolutionary classification of these animals is still a work in progress; many traditional groups are now known not to be clades. "Fishes" actually represent several ancient clades, one of which includes tetrapods, or four-limbed vertebrates. Fishes, as we treat them here, include two groups of jawless fishes (hagfishes and lampreys), cartilaginous fishes, and bony fishes.

"Sweetlips" are, despite their funny faces, easily recognizable as fish.

"JAWLESS FISHES"

Hagfishes and lampreys make up separate clades, but their bodies share common features that distinguish them from other fishes. They have no jaws, lack vertebrae, and their skeletons are made of fiber and cartilage.

PETROMYZONTIDA: Lampreys

Lampreys are mostly filter feeders as larvae and parasites as adults. The head of an adult lamprey is taken up almost completely by a circular, tooth-bearing, sucking disk with a round mouth. Adult lampreys typically attach themselves to fishes. They hold on to their hosts using the teeth in their sucking disk and then scrape away at the skin with a rasping tongue. Lampreys then suck up their host's tissues and body fluids. Because lampreys feed mostly on blood, they are called "vampires of the sea."

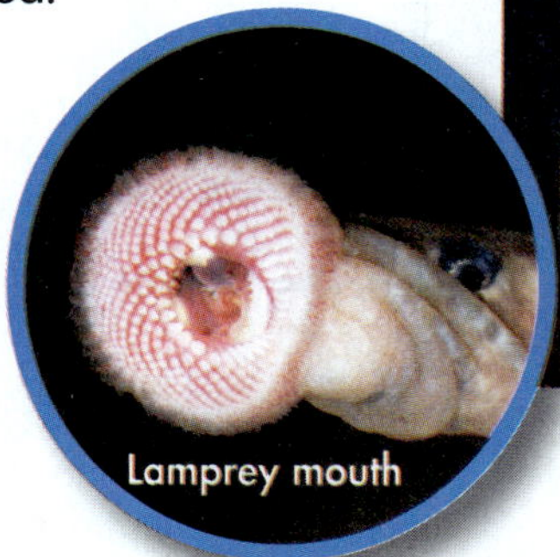

Lamprey mouth

▲ Lamprey

▲ Pacific Hagfish

MYXINI: Hagfishes

Hagfishes have pinkish gray wormlike bodies and four or six short tentacles around their mouths. They retain notochords as adults. Hagfishes lack image-forming eyes, but have light-detecting sensors scattered around their bodies. They feed on dead and dying animals using a rasping tongue that scrapes away layers of flesh.

Tiger Shark

CHONDRICHTHYES: Cartilaginous Fishes

Members of this clade are considered "cartilaginous" because they lack true bone; their skeletons are built entirely of cartilage. Most cartilaginous fishes also have tough, scales, which make their skin as rough as sandpaper.

Holocephalans: Chimaeras

Chimaeras have smooth skin that lacks scales. Most have just a few platelike, grinding teeth and a venomous spine located in front of the dorsal fin. Examples: ghostfish, ratfish, rabbitfish

Spotted Ratfish

Elasmobranchii: Sharks, skates, and rays

Sharks, skates, and rays are very diverse, but all have skin covered with toothlike scales known as dermal denticles. Elasmobranchii make up the vast majority of living cartilaginous fish species.

Dermal denticles on shark skin reduce drag, helping the shark to swim faster. (SEM 40×)

Galeomorphi: Sharks

Most of the 350 or so shark species have large, curved asymmetrical tails, torpedo-shaped bodies, and pointed snouts with a mouth underneath. Predatory sharks, such as the great white, have many teeth arranged in rows. As teeth in the front rows are worn out or lost, new teeth replace them. Some sharks go through 20,000 teeth in their lifetime! Other sharks are filter feeders, and some species have flat teeth for crushing mollusk and crustacean shells. Examples: Great White Shark, Whale Shark, Hammerhead Shark

Hammerhead Shark

Squalomorphi: Skates and rays

Skates and rays have diverse feeding habits. Some feed on bottom-dwelling invertebrates by using their mouths as powerful vacuums. Others filter-feed on plankton. When not feeding or swimming, many skates and rays cover themselves with a thin layer of sand and rest on the ocean floor. Example: stingray

Blue-Spotted Ribbon-Tailed Ray

OSTEICHTHYES: Bony fishes

The skeletons of these vertebrates are made of true bone. This clade includes the ancestors and living members of all "higher" vertebrate groups—including tetrapods.

Rainbow Trout

Actinopterygii: Ray-finned fishes

Almost all living bony fishes belong to a huge group called ray-finned fishes. The name *ray-finned* refers to the slender bony rays that are connected to one another by a layer of skin to form fins.

Coelacanth

Sarcopterygii: Lobe-finned fishes

Seven living species of bony fishes, including lungfishes and coelacanths, are classified as lobe-finned fishes. Lungfishes live in fresh water, but coelacanths are marine. The fleshy fins of lobe-finned fishes are supported by strong bones rather than rays. Some of these bones are homologous to the limb bones of land vertebrates. Examples: lungfish, coelacanths

This clade includes the ancestors of tetrapods, so, technically, all living tetrapods (including us!) are Sarcopterygians! As a result, the bony-fish clade includes almost half of all chordate species!

Amphibians

Marsupial Frog

KEY CHARACTERISTICS

The word amphibian *means "double life," an apt name for these vertebrates, most of which live in water as larvae and on land as adults. Most adult amphibians breathe with lungs, lack scales and claws, and have moist skin that contains mucous glands.*

Feeding and Digestion Tadpoles—usually filter feeders or herbivores with long, coiled intestines to digest plant material; adults—carnivores with shorter intestines for processing meat

Circulation Double-loop system with three-chambered heart

Respiration Larvae breathe through skin and gills; most adult species have lungs, though a few use gills; lungless salamanders breathe through their mouth-cavity lining and skin.

Excretion Kidneys produce urine.

Response Well-developed nervous and sensory systems; organs include protective nictitating membrane over moveable eyes, tympanic membranes, lateral line system

Movement Larvae have tails; adults have limbs (except caecilians); some have specialized toes for climbing.

Reproduction Most lay eggs without shells that are fertilized externally; most undergo metamorphosis from aquatic tadpole larvae that breathe with gills to land-dwelling adults, which usually have lungs and limbs.

Eco • Alert

Red-Eyed Treefrog

The Frogs Are Disappearing!

For several decades, scientists have noticed that amphibian populations worldwide have been decreasing, and a number of species have become extinct. Scientists have not yet pinpointed a single cause for this problem. It is, however, becoming clear that amphibians are susceptible to a variety of environmental threats, including habitat loss, ozone depletion, acid rain, water pollution, fungal infections, and introduced aquatic predators.

To better understand this decline, biologists worldwide have been focusing their efforts and sharing data about amphibian populations. One amphibian-monitoring program covers all of North America.

GROUPS OF AMPHIBIANS

The three orders of amphibians include more than 6000 species, roughly 5000 of which are frogs and toads.

Red Eft

URODELA: Salamanders and newts

Salamanders and newts have long bodies and tails. Most also have four legs. All are carnivores. Adults usually live in moist woods, where they tunnel under rocks and rotting logs. Some salamanders, such as the mud puppy, keep their gills as adults and live in water all their lives. Examples: Barred Tiger Salamander, Red Eft

American Toad

ANURA: Frogs and toads

Adult frogs and toads are amphibians without tails that can jump. Frogs tend to have long legs and make long jumps, whereas toads have shorter legs that limit them to shorter hops. Frogs are generally more dependent on bodies of fresh water than toads, which may live in moist woods or even deserts. Examples: treefrogs, Leopard Frog, American Toad, spadefoot toads

APODA: Caecilians

The least-known and most unusual amphibians are the legless caecilians. They have tentacles, and many have fishlike scales embedded in their skin—which shows that not all amphibians fit the general definition. Caecilians live in water or burrow in moist soil or sediment, feeding on small invertebrates such as termites. Examples: Ringed Caecilian, Yellow-Striped Caecilian

Ringed Caecilian

▶ *Because amphibian eggs must develop in water, most live in moist climates. Some, such as this Alpine Newt, live on cool, rainy mountain slopes.*

Reptiles

Saltwater crocodiles, such as this young one, are the largest living reptiles and sometime reach 6 meters long. But that's still only half as long as their famous dinosaur ancestor, T. rex!

KEY CHARACTERISTICS OF REPTILES

Living reptiles, traditionally classified in the class Reptilia, are ectothermic vertebrates with dry, scaly skin; lungs; and amniotic eggs. Modern evolutionary classification now recognizes a larger clade Reptilia that includes living reptiles, extinct dinosaurs, and birds—the living descendants of one dinosaur group.

Feeding and Digestion Feeding methods vary by group; digestive systems—herbivores have long digestive systems to break down plant materials; carnivores may swallow prey whole

Circulation Two loops; heart with two atria and one or two ventricles

Respiration Spongy lungs provide large surface area for gas exchange; lungs operated by muscles and moveable ribs

Excretion Kidneys; urine contains ammonia or uric acid

Response Brain; well-developed senses including, in some species, infrared detectors that can spot warm-bodied prey in the dark

Movement Strong limbs (except snakes)

Reproduction Internal fertilization via cloaca; amniotic egg with leathery shell

Eco • Alert

Calling Doctor 'Gator!

You might think of alligators mostly as killing machines, but their blood may soon provide medicines that can save lives. An alligator's immune system works quite differently from our own. Proteins in their white blood cells can kill multidrug resistant bacteria, disease-causing yeasts, and even HIV. Remarkably, these proteins work against pathogens to which the animals have never been exposed. Researchers are currently sequencing the genes for these proteins and hope to develop them into human medicines in the near future.

GROUPS OF REPTILES

There are nearly 9000 species of reptiles (not including birds).

SPHENODONTA: Tuataras

The tuatara, found only on a few small islands off the coast of New Zealand, is the only living member of this group. Tuataras resemble lizards in some ways, but they lack external ears and retain primitive scales.

Tuatara

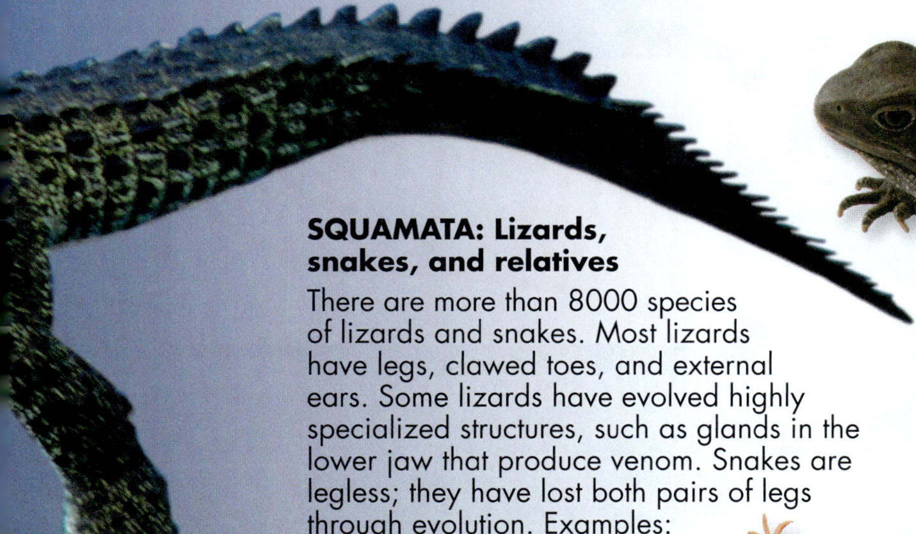

SQUAMATA: Lizards, snakes, and relatives

There are more than 8000 species of lizards and snakes. Most lizards have legs, clawed toes, and external ears. Some lizards have evolved highly specialized structures, such as glands in the lower jaw that produce venom. Snakes are legless; they have lost both pairs of legs through evolution. Examples: iguanas, Milk Snake, Coral Snake

Leopard Gecko

ARCHOSAURS: Crocodilians; pterosaurs and dinosaurs (extinct); and birds

This clade includes some of the most spectacular animals that have ever lived. The extinct dinosaurs and pterosaurs (flying reptiles), whose adaptive radiations produced some of the largest animals ever to walk Earth or fly above it, are the closest relatives of birds. Living crocodilians are short-legged and have long and typically broad snouts. They are fierce carnivorous predators, but the females are attentive mothers. Crocodilians live only in regions where the climate remains warm year-round. We discuss birds separately. Examples: extinct types: *Tyrannosaurus*, *Pteranodon*; living types: alligators, crocodiles, caimans, and birds (see following pages)

Paraguay Caiman

Leopard Tortoise

TESTUDINE: Turtles and tortoises

Turtles and tortoises have a shell built into their skeleton. Most can pull their heads and legs into the shell for protection. Instead of teeth, these reptiles have hornlike ridges covering their jaws equipped with sharp beaklike tips. Strong limbs can lift their body off the ground when walking or, in the case of sea turtles, can drag their body across a sandy shore to lay eggs. Examples: snapping turtles, green sea turtles, Galápagos tortoise

Animals

Birds

Today, only birds have feathers. These delicate, intricately interlocking and beautiful structures keep birds warm and cool and enable most to fly.

Common Kingfisher

KEY CHARACTERISTICS OF BIRDS

Birds, once placed in a class of their own, are now recognized as endothermic reptiles with feathers and hard-shelled, amniotic eggs that are descended from dinosaurs. Birds have two scaly legs and front limbs modified into wings, which enable most species to fly.

Feeding and Digestion No teeth; bills adapted to widely varied foods, including insects, seeds, fruits, nectar, fish, meat; organs of the digestive system include crop, gizzard, cloaca

Circulation Two loops with four-chambered heart; separation of oxygen-rich and oxygen-poor blood

Respiration Constant, one-way flow of air through lungs and air sacs increases the efficiency of gas exchange and supports high metabolic rate

Excretion Kidneys remove nitrogenous wastes from blood, converting them to uric acid, which is excreted through cloaca

Response Brain with large optic lobes and enlarged cerebellum; highly evolved sense organs including, in some species, eyes that can see ultraviolet light

Movement Skeleton made up of lightweight, hollow bones with internal struts for strength; powerful muscles; most fly

Reproduction Internal fertilization via cloaca; amniotic egg with hard, brittle shell; depending on species, newly hatched young may be precocial—downy-feathered chicks able to move around and feed themselves, or altricial—bare-skinned and totally dependent on their parents

• A Look Back in Time

Birds of a Feather

Fossils recently discovered in lake beds in China have greatly expanded our understanding of bird evolution. One exciting discovery was that of a four-winged dinosaur named *Microraptor gui* from about 125 million years ago. *Microraptor gui*, which was related to *Tyrannosaurus rex*, had feathers on both its wings *and* its legs, so some researchers hypothesize that it flew like a biplane! This and other fossils show that several lineages of dinosaurs and ancient birds evolved various kinds of feathers over millions of years.

Artist's conception of Microraptor gui

GROUPS OF BIRDS

Evolutionary classification of living birds is still a work in progress, as different techniques and analyses produce different results. There are about 10,000 species. The groups described below illustrate some of the diversity of birds.

PALEOGNATHAE: Ostriches, emus, kiwis, and relatives

This group represents an early branch of the bird family tree that is separate from all other living birds. This clade includes the largest birds alive today. Ostriches can be 2.5 meters tall and weigh 130 kilograms! Kiwis, however, are only about the size of chickens. Roughly a dozen living species are scattered throughout the Southern Hemisphere. All are flightless, but the larger species can run very fast. They generally eat a variety of plant material, insects, and other small invertebrates. Examples: Ostrich, emus, Brown Kiwi, Greater Rhea, Dwarf Cassowary

SPHENISCIDAE: Penguins

These flightless birds of the Southern Hemisphere are adapted to extreme cold and hunting in water. Though they cannot fly, they use their wings as flippers when they swim. Penguins have more feathers per square centimeter than any other bird; this density allows them to repel water and conserve heat effectively. Some species form large colonies. Examples: Emperor Penguin, Chinstrap Penguin, King Penguin

ANATIDAE: Ducks, geese, and swans

These birds spend much of their time feeding in bodies of water. Webbed feet enable them to paddle efficiently across the surface of the water. Most fly well, however, and many species migrate thousands of kilometers between breeding and resting locations. Examples: Redhead, Ross's goose, Trumpeter Swan

Galápagos Hawk

FALCONIDAE AND ACCIPITRIDAE: Falcons, eagles, and hawks

These fierce predators, often called raptors, typically have powerful hooked bills, large wingspans, and sharp talons. Raptors have powerful flight muscles and keen eyesight, enabling them to see prey at a distance. Examples: Eurasian Kestrel, Golden Eagle, Galápagos Hawk

---

Black Woodpecker with chicks

PICIDAE AND RAMPHASTIDAE: Woodpeckers and toucans

Woodpeckers are tree-dwelling birds with two toes in front and two in back. (Most birds have three in front and one in back; the two-and-two arrangement makes moving up and down tree trunks easier.) Woodpeckers are typically carnivores that eat insects and their larvae. Toucans usually use their huge, often colorful bills to eat fruit. Examples: Black Woodpecker, Keel-Billed Toucan

Keel-Billed Toucan

PASSERIFORMES: Passerines

Also called perching birds, this is by far the largest and most diverse group of birds, with about 5000 species. Most are songbirds. Examples: flycatchers, mockingbirds, cardinals, crows, chickadees, and finches.

Summer Tanager

Great Crested Flycatcher

Mountain Bluebird

Hooded Warbler

Lark Sparrow

Blue Grosbeak

Mammals

KEY CHARACTERISTICS

Mammals are endothermic vertebrates with hair and mammary glands that produce milk to nourish their young.

Feeding and Digestion Diet varies with group; foods range from seeds, fruits, and leaves to insects, fish, meat, and even blood; teeth, jaws, and digestive organs are adapted to diet

Circulation Two loops; four-chambered heart; separation of oxygen-rich and oxygen-poor blood

Respiration Lungs controlled by two sets of muscles.

Excretion Highly evolved kidneys filter urea from blood and produce urine.

Response Most highly evolved brain of all animals; keen senses

Movement Flexible backbone; variations in limb bones and muscles enable wide range of movement across groups: from burrowing and crawling to walking, running, hopping, and flying

Reproduction Internal fertilization; developmental process varies with group (monotreme, marsupial, placental)

Did You Know?

Platypus: Mix-and-Match Genome

The duckbill platypus has such an odd mix of reptile and mammal features that some scientists thought the first specimens were hoaxes produced by sticking parts of different animals together! Recent genome studies have revealed an equally odd mix of reptilian and mammalian genes. Genes for reptile-like vision, the production of egg yolk, and the production of venom link the platypus to reptiles. Genes for the production of milk link it to other mammals. The evidence provides confirmation that this monotreme represents a truly ancient lineage, one from the time close to that at which mammals branched off from reptiles.

GROUPS OF MAMMALS

The three living groups of mammals are the monotremes, the marsupials, and the placentals. There are about 5000 species of mammals, usually divided into about 26 orders, most of which are placentals. There is only one order of monotremes.

This black-backed jackal pup is enjoying a moment of independence from its family group. Mammals provide intensive parental care to their young.

Short-Beaked Echidna (Albino)

MONOTREMATA: Monotremes

Monotremes—egg-laying mammals—share two important characteristics with reptiles. First, the digestive, reproductive, and urinary systems of monotremes all open into a cloaca similar to that of reptiles. Second, monotreme development is similar to that of reptiles. Like a reptile, a female monotreme lays soft-shelled eggs incubated outside her body. The eggs hatch in about ten days. Unlike reptiles, however, young monotremes are nourished by mother's milk, which they lick from pores on the surface of her abdomen. Only five monotreme species exist today, all in Australia and New Guinea. Examples: Duckbill Platypus, echidnas

MARSUPIALIA: Marsupials

Marsupials bear live young at an extremely early stage of development. A fertilized egg develops into an embryo inside the mother's reproductive tract. The embryo is then "born" in what would be an embryonic stage for more familiar mammals. It crawls across its mother's fur and attaches to a nipple that, in most species, is located in a pouch called the marsupium. The embryo spends several months attached to the nipple. It continues to nurse until it can survive on its own. Examples: kangaroos, wallabies, wombats, opossums

Wombat

PLACENTALIA: Placental Mammals

Placental mammals are the mammals with which you are most familiar. This group gets its name from a structure called the placenta, which is formed when the embryo's tissues join with tissues within the mother's body. Nutrients, gases, and wastes are exchanged between embryo and mother through the placenta. The placenta allows the embryo to develop inside the mother longer so that placental young are born at a later stage of development than other mammals are. Development may take as little as a few weeks (mice), to as long as two years (elephants). After birth, most placental mammals care for their young and provide them with nourishment by nursing. Examples: Mice, cats, dogs, seals, whales, elephants, humans

Chiroptera: Bats

These are the only mammals capable of true flight. There are more than 900 species of bats! They eat mostly insects or fruit and nectar, although a few species feed on the blood of other vertebrates. Examples: fruit bats, Little Brown Myotis, Vampire Bat

Epauletted Bat, roosting

Lioness attacking Greater Kudu

Carnivora: Carnivores

Many members of this group, such as tigers and hyenas, chase or stalk prey by running or pouncing, then kill with sharp teeth and claws. Dogs, bears, and other members of this group may eat plants as well as meat. Examples: dogs, cats, skunks, seals, bears

Sirenia: Sirenians

Sirenians are herbivores that live in rivers, bays, and warm, coastal waters scattered throughout the world. These large, slow-moving mammals lead fully aquatic lives. Examples: manatees, dugongs

Manatee mother and nursing calf

African Hedgehog mother and baby

Insectivora: Insectivores

These insect eaters have long, narrow snouts and sharp claws that are well suited for digging. Examples: shrews, moles, hedgehogs

Perissodactyla: Hoofed, odd-toed mammals

This group is made up of hoofed animals with an odd number of toes on each foot. Like artiodactyls, this group contains mostly large, grazing animals. Examples: horses, zebras, rhinoceroses

Tapir hoof

Central American Tapir

Artiodactyla: Hoofed, even-toed mammals

These large, grazing, hoofed mammals have an even number of toes on each foot. Examples: cattle, sheep, pigs, hippopotami

Giraffe hooves

Maasai Giraffe

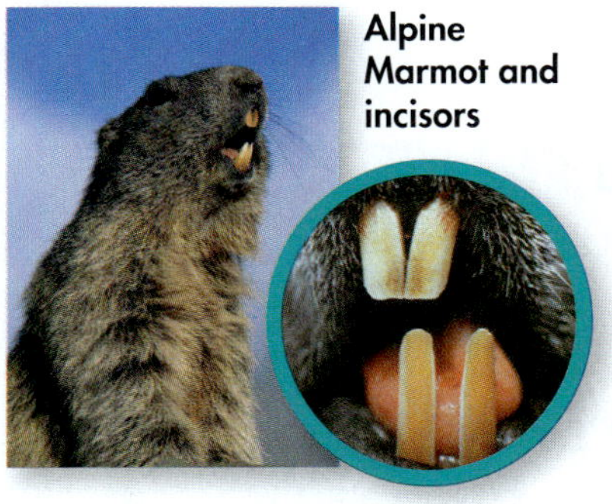

Alpine Marmot and incisors

Rodentia: Rodents

Rodents have a single pair of long, curved incisor teeth in both their upper and lower jaws, used for gnawing wood and other tough plant material. Examples: rats, squirrels, porcupines

Cetacea: Cetaceans

Like sirenians, cetaceans—the group that includes whales and dolphins—are adapted to underwater life, yet must come to the surface to breathe. Most cetaceans live and breed in the ocean. Examples: whales, dolphins

Atlantic Spotted Dolphin

Black-Tailed Jackrabbit

Lagomorpha: Rabbit, hares, and pikas

Lagomorphs are entirely herbivorous. They differ from rodents by having two pairs of incisors in the upper jaw. Most lagomorphs have hind legs that are adapted for leaping.

Northern Tamandua

Xenarthra: Edentates

The word *edentate* means "toothless," which refers to the fact that some members of this group (sloths and anteaters) have simple teeth without enamel or no teeth at all. Armadillos, however, have more teeth than most other mammals! Examples: sloths, anteaters, armadillos

Proboscidea: Elephants

These are the mammals with trunks. Some time ago, this group went through an extensive adaptive radiation that produced many species, including mastodons and mammoths, which are now extinct. Only two species, the Asian Elephant and the African elephant, survive today.

Asian Elephant and calf

Primates: Lemurs, monkeys, apes, humans, and relatives

Members of this group are closely related to ancient insectivores but have a highly developed cerebrum and complex behaviors.

Sifaka

Tarsier

Langur

Baboon and baby

Orangutan

Gorilla

Chimpanzee

Appendix A Science Skills

Data Tables and Graphs

How can you make sense of the data from a science experiment? The first step is to organize the data. You can organize data in data tables and graphs to help you interpret them.

Data Tables

You have gathered your materials and set up your experiment. But before you start, you need to plan a way to record what happens during the experiment. By creating a data table, you can record your observations and measurements in an orderly way.

Suppose, for example, that a scientist conducted an experiment to find out how many kilocalories people of different body masses burned while performing various activities for 30 minutes. The data table below shows the results.

Notice in this data table that the independent variable (body mass) is the heading of the first column. The dependent variable (for Experiment 1, the number of kilocalories burned while bicycling for 30 minutes) is the heading of the next column. Additional columns were added for related experiments.

Calories Burned in 30 Minutes

Body Mass	Experiment 1: Bicycling	Experiment 2: Playing Basketball	Experiment 3: Watching Television
30 kg	60 Calories	120 Calories	21 Calories
40 kg	77 Calories	164 Calories	27 Calories
50 kg	95 Calories	206 Calories	33 Calories
60 kg	114 Calories	248 Calories	38 Calories

Bar Graphs

A bar graph is useful for comparing data from two or more distinct categories. In this example, pancreatic secretions in the small intestine are shown. To create a bar graph, follow these steps.

1. On graph paper, draw a horizontal, or *x*-axis, and a vertical, or *y*-axis.
2. Write the names of the categories (the independent variable) along one axis, usually the horizontal axis. You may put the categories on the vertical axis if that graph shape better fits on your page. Label the axis.
3. Label the other axis with the name of the dependent variable and the unit of measurement. Then, create a scale along that axis by marking off equally spaced numbers that cover the range of the data values.
4. For each category, draw a solid bar at the appropriate value. Then, fill in the space from the bar to the axis representing the independent variable. Make all the bars the same width.
5. Add a title that describes the graph.

Pancreatic Secretions

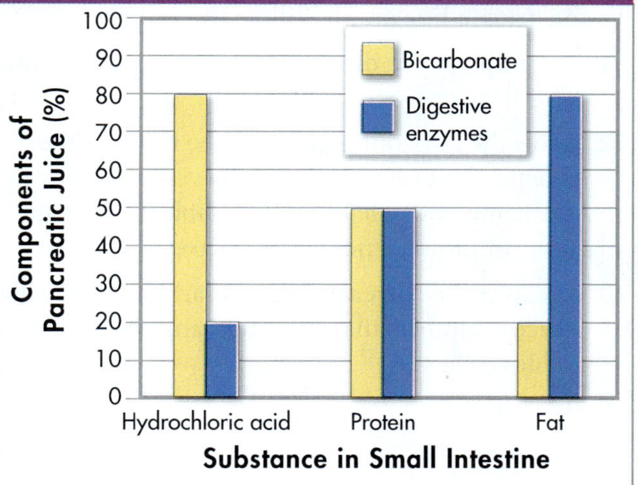

Line Graphs

A line graph is used to display data that show how the dependent variable changes in response to manipulations of the independent variable. You can use a line graph when your independent variable is continuous, that is, when there are other points between the ones that you tested. For example, the graph below shows how the growth of a bacterial population is related to time. The graph shows that the number of bacteria approximately doubles every 20 minutes. Line graphs are powerful tools because they also allow you to estimate values for conditions that you did not test in the experiment.

Number of Bacteria Cells

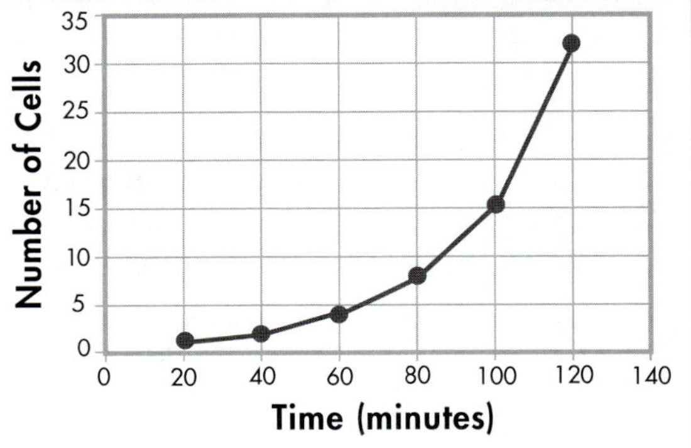

To construct a line graph, follow these steps.

1. On graph paper, draw a horizontal, or *x*-axis, and a vertical, or *y*-axis.
2. Label the horizontal axis with the name of the independent variable. Label the vertical axis with the name of the dependent variable. Include the units of measurement on both axes.
3. Create a scale on each axis by marking off equally spaced numbers that cover the range of the data values collected.
4. Plot a point on the graph for each data value. To do this, follow an imaginary vertical line extending up from the horizontal axis for an independent variable value. Then, follow an imaginary horizontal line extending across from the vertical axis at the value of the associated dependent variable. Plot a point where the two lines intersect. Repeat until all your data values are plotted.
5. Connect the plotted points with a solid line. Not all graphs are linear, so you may discover that it is more appropriate to draw a curve to connect the points.

The data in the graph at the left fit neatly on a smooth curve. But if you were to connect each data point on the graph below, you would have a mess that yielded little useful information. In some cases, it may be most useful to draw a line that shows the general trend of the plotted points. This type of line is often called a line of best fit. Such a line runs as closely as possible to all the points and allows you to make generalizations or predictions based on the data. Some points will fall above or below a line of best fit.

The Effect of Rainfall on Plant Productivity

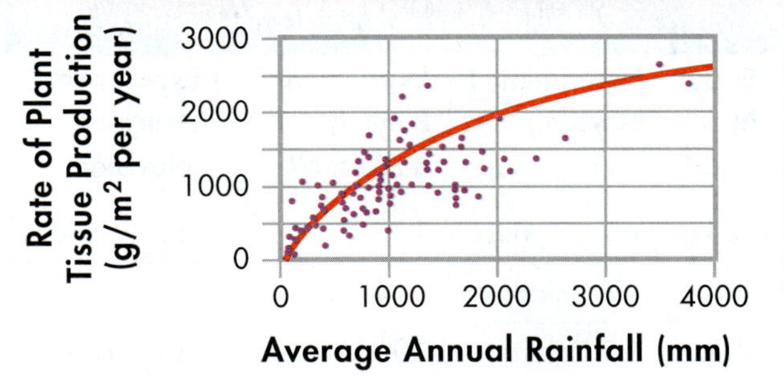

Circle Graphs

Circle graphs, or pie charts, display data as parts of a whole. Like bar graphs, circle graphs can be used to display data that fall into separate categories. Unlike bar graphs, however, circle graphs can only be used when you have data for all the categories that make up a given group. The circle, or "pie," represents 100 percent of a group, while the sectors, or slices, represent the percentages of each category that make up that group. The example below compares the different blood groups found in the U.S. population.

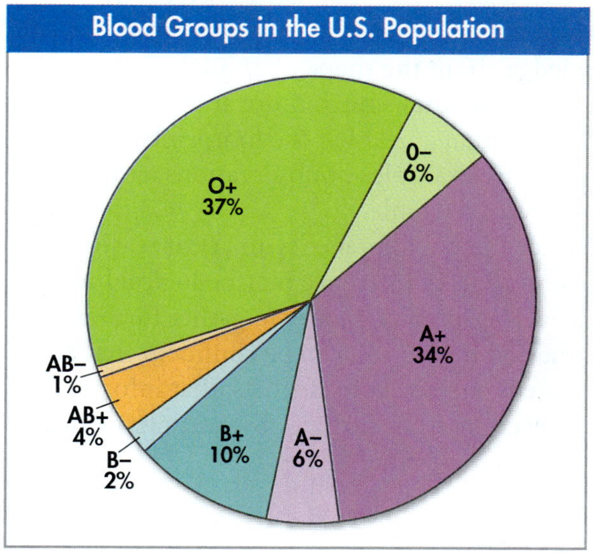

To construct a circle graph, follow these steps.

1. Draw a circle and mark the center. Then, draw a radius line from the center to the circle's edge.
2. Determine the size of a sector of the graph by calculating the number of degrees that correspond to a percentage you wish to represent. For example, in the graph shown, B^+ makes up 10 percent of all blood groups; 360 degrees $\times$ 0.10 = 36 degrees.
3. With a protractor fixed at the center of the circle, measure the angle—in this example, 36 degrees—from the existing radius, and draw a second radius at that point. Label the sector with its category and the percentage of the whole it represents. Repeat for each of the other categories, measuring each sector from the previous radius so the sectors don't overlap.
4. For easier reading, shade each sector differently.
5. Add a title that describes the graph.

Reading Diagrams

In scientific figures showing a cut-away of a structure, the diagram or photograph is showing the structure from a particular angle. Look for clues throughout this book that will help you interpret the view being shown.

Cross Sections

A cross section shows a horizontal cut through the middle of a structure. This icon will help you locate cross sections.

Longitudinal Sections

A longitudinal section shows a vertical cut through the middle of a structure. This icon will help you locate longitudinal sections.

Appendix A Science Skills

Basic Process Skills

During a biology course, you often carry out short lab activities as well as lengthier experiments. Here are some skills that you will use.

Observing

In every science activity, you make a variety of observations. Observing is using one or more of the five senses to gather information. Many observations involve the senses of sight, hearing, touch, and smell. On rare occasions in a lab—but only when explicitly directed by your teacher—you may use the sense of taste to make an observation.

Sometimes you will use tools that increase the power of your senses or make observations more precise. For example, hand lenses and microscopes enable you to see things in greater detail. Rulers, balances, and thermometers help you measure key variables. Besides expanding the senses or making observations more accurate, tools may help eliminate personal opinions or preferences.

In science, it is customary to record your observations at the time they are made, usually by writing or drawing in a notebook. You may also make records by using computers, cameras, videotapes, and other tools. As a rule, scientists keep complete accounts of their observations, often using tables to organize their observations.

Inferring

In science, as in daily life, observations are usually followed by inferences. Inferring is interpreting an observation or statement based on prior knowledge.

For example, suppose you're on a mountain hike and you see footprints like the ones illustrated below. Based on their size and shape, you might infer that a large mammal had passed by. In making that inference, you would use your knowledge about the shape of animals' feet. Someone who knew much more about mammals might infer that a bear left the footprints. You can compare examples of observations and inferences in the table.

Notice that an inference is an act of reasoning, not a fact. An inference may be logical but not true. It is often necessary to gather further information before you can be confident that an inference is correct. For scientists, that information may come from further observations or from research done by others.

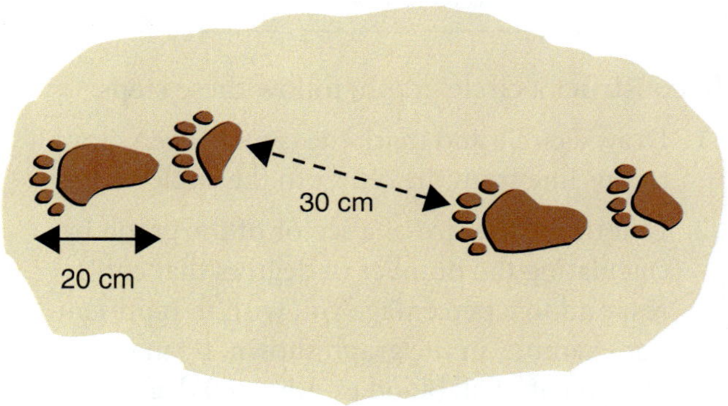

Comparing Observations and Inferences

Sample Observations	Sample Inferences
The footprints in the soil each have five toes.	An animal made the footprints.
The larger footprints are about 20 cm long.	A bear made the footprints.
The space between each pair of footprints is about 30 cm.	The animal was walking, not running.

As you study biology, you may make different types of inferences. For example, you may generalize about all cases based on information about some cases: *All the plant roots I've observed grow downward, so I infer that all roots grow downward.* You may determine that one factor or event was caused by another factor or event: *The bacteria died after I applied bleach, so I infer that bleach kills bacteria.* Predictions may be another type of inference.

Predicting

People often make predictions, but their statements about the future could be either guesses or inferences. In science, a prediction is an inference about a future event based on evidence, experience, or knowledge. For example, you can say, *On the first day of next month, it will be sunny.* If your statement is based on evidence of weather patterns in the area, then the prediction is scientific. If the statement was made without considering any evidence, it's just a guess.

Predictions play a major role in science because they provide a way to test ideas. If scientists understand an event or the properties of a particular object, they should be able to make accurate predictions about that event or object. Some predictions can be tested simply by making observations. At other times, carefully designed experiments are needed.

Classifying

If you have ever heard people debate whether a tomato is a fruit or a vegetable, you've heard an argument about classification. Classifying is the process of grouping items that are alike according to some organizing idea or system. Classifying occurs in every branch of science, but it is especially important in biology because living things are so numerous and diverse.

You may have the chance to practice classifying in different ways. Sometimes you will place objects into groups using an established system. At other times, you may create a system of your own by examining a variety of objects and identifying their properties.

Classification can have different purposes. Sometimes it's done just to keep things organized, to make lab supplies easy to find, for example. More often, though, classification helps scientists understand living things better and discover relationships among them. For example, one way biologists determine how groups of vertebrates are related is to compare their bones. Biologists classify certain animal parts as bone or muscle and then investigate how they work together.

Using Models

Some cities refuse to approve any new buildings that could cast shadows on a popular park. As architects plan buildings in such locations, they use models that can show where a proposed building's shadow will fall at any time of day in any season of the year. A model is a mental or physical representation of an object, process, or event. In science, models are usually made to help people understand natural objects and processes.

Models can be varied. Mental models, such as mathematical equations, can represent some kinds of ideas or processes. For example, the equation for the surface area of a sphere can model the surface of Earth, enabling scientists to determine its size. Physical models can be made of a huge variety of materials; they can be two dimensional (flat) or three dimensional (having depth). In biology, a drawing of a molecule or a cell is a typical two-dimensional model. Common three-dimensional models include a representation of a DNA molecule and a plastic skeleton of an animal.

Physical models can also be made "to scale," which means they are in proportion to the actual object. Something very large, such as an area of land being studied, can be shown at 1/100 of its actual size. A tiny organism can be shown at 100 times its size.

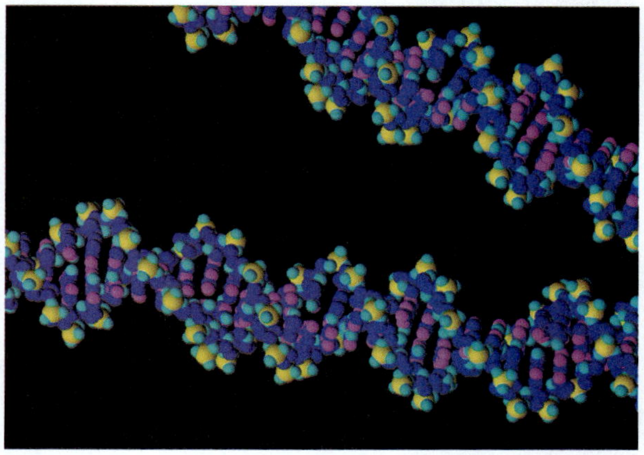

Organizing Information

When you study or want to communicate facts and ideas, you may find it helpful to organize information visually. Here are some common graphic organizers you can use. Notice that each type of organizer is useful for specific types of information.

Flowcharts

A flowchart can help you represent the order in which a set of events has occurred or should occur. Flowcharts are useful for outlining the steps in a procedure or stages in a process with a definite beginning and end.

To make a flowchart, list the steps in the process you want to represent and count the steps. Then, create the appropriate number of boxes, starting at the top of a page or on the left. Write a brief description of the first event in the first box, and then fill in the other steps, box by box. Link each box to the next event in the process with an arrow.

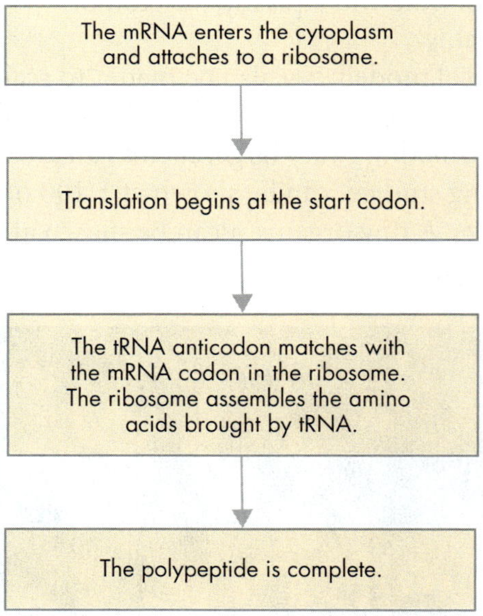

Concept Maps

Concept maps can help you organize a topic that has many subtopics. A concept map begins with a main idea and shows how it can be broken down into specific topics. It makes the ideas easier to understand by presenting their relationships visually.

You construct a concept map by placing the concept words (usually nouns) in ovals and connecting the ovals with linking words. The most general concept usually is placed at the top of the map or in the center. The content of the other ovals becomes more specific as you move away from the main concept. The linking words, which describe the relationship between the linked concepts, are written on a line between two ovals. If you follow any string of concepts and linking words down through a map, they should sound almost like a sentence.

Some concept maps may also include linking words that connect a concept in one branch to another branch. Such connections, called cross-linkages, show more complex interrelationships.

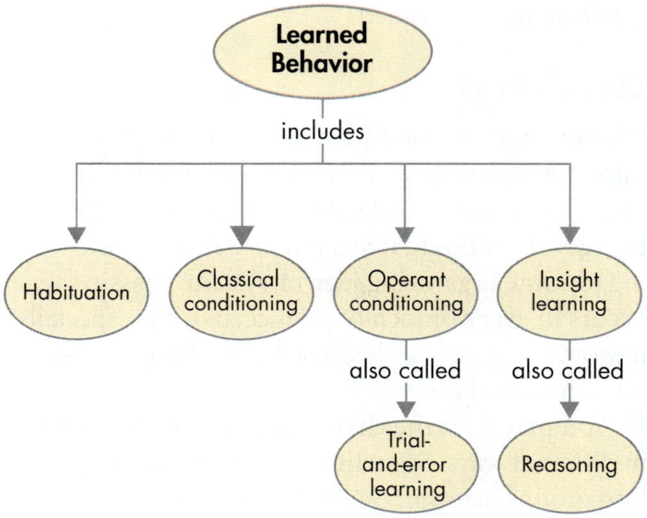

Compare/Contrast Tables

Compare/contrast tables are useful for showing the similarities and differences between two or more objects or processes. The table provides an organized framework for making comparisons based on specific characteristics.

To create a compare/contrast table, list the items to be compared across the top of the table. List the characteristics that will form the basis of your comparison in the column on the left. Complete the table by filling in information for each item.

Comparing Cellular Respiration and Fermentation

Characteristic	Cellular Respiration	Fermentation
Starting reactants	Glucose	Glucose, oxygen
Pathways involved	Glycolysis, several others	Glycolysis, Krebs cycle, electron transport
End products	CO_2 and alcohol *or* CO_2 and lactic acid	CO_2, H_2O
Number of ATP molecules produced	2	36

Venn Diagrams

Another way to show similarities and differences between items is with a Venn diagram. A Venn diagram consists of two or more ovals that partially overlap. Each oval represents a particular object or idea. Characteristics that the objects share are written in the area of overlap. Differences or unique characteristics are written in the areas that do not overlap.

To create a Venn diagram, draw two overlapping ovals. Label them with the names of the objects or the ideas they represent. Write the unique characteristics in the part of each oval that does not overlap. Write the shared characteristics within the area of overlap.

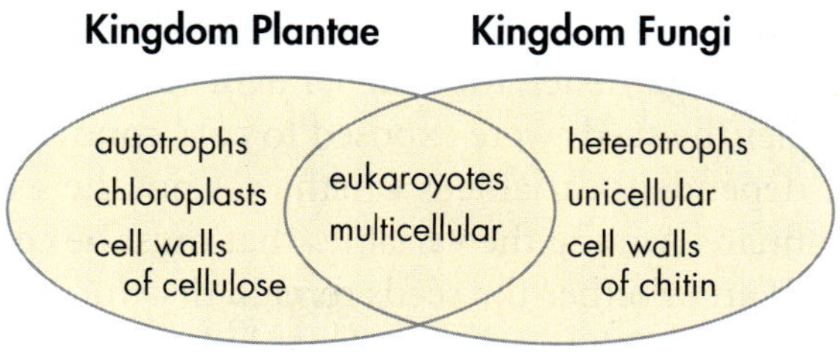

Cycle Diagrams

A cycle diagram shows a sequence of events that is continuous, or cyclical. A continuous sequence does not have a beginning or an end; instead, each event in the process leads to another event. The diagram shows the order of the events.

To create a cycle diagram, list the events in the process and count them. Draw one box for each event, placing the boxes around an imaginary circle. Write one of the events in an oval, and then draw an arrow to the next oval, moving clockwise. Continue to fill in the boxes and link them with arrows until the descriptions form a continuous circle.

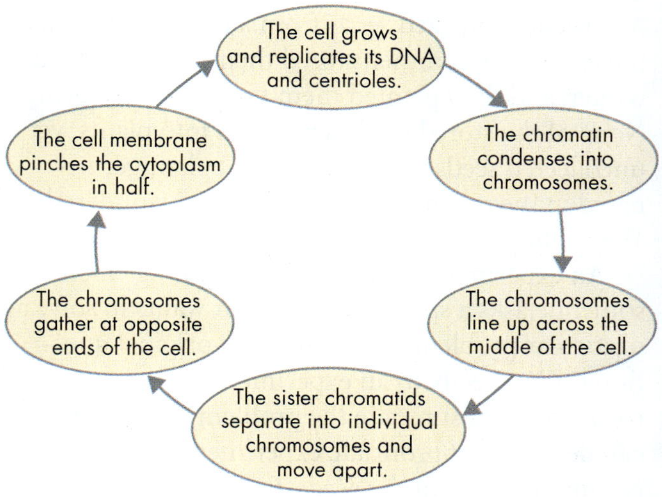

Appendix B Lab Skills

Conducting an Experiment

A science experiment is a procedure designed to test a prediction. Some types of experiments are fairly simple to design. Others may require ingenious problem solving.

Starting With Questions or Problems

A gardener collected seeds from a favorite plant at the end of the summer, stored them indoors for the winter, and then planted them the following spring. None of the stored seeds developed into plants, yet uncollected seeds from the original plant germinated in the normal way. The gardener wondered: *Why didn't the collected seeds germinate?*

An experiment may have its beginning when someone asks a specific question or wants to solve a particular problem. Sometimes the original question leads directly to an experiment, but often researchers must restate the problem before they can design an appropriate experiment. The gardener's question about the seeds, for example, is too broad to be tested by an experiment, because there are so many possible answers. To narrow the topic, the gardener might think about related questions: *Were the seeds I collected different from the uncollected seeds? Did I try to germinate them in poor soil or with insufficient light or water? Did storing the seeds indoors ruin them in some way?*

Developing a Hypothesis

In science, a question about an object or event is answered by developing a possible explanation called a hypothesis. The hypothesis may be developed after long thought and research, or it may come to a scientist "in a flash." How a hypothesis is formed doesn't matter; it can be useful as long as it leads to predictions that can be tested.

The gardener decided to focus on the fact that the nongerminating seeds were stored in the warm conditions of a heated house. That premise led the person to propose this hypothesis: *Seeds require a period of low temperatures in order to germinate.*

The next step is to make a prediction based on the hypothesis, for example: *If seeds are stored indoors in cold conditions, they will germinate in the same way as seeds left outdoors during the winter.* Notice that the prediction suggests the basic idea for an experiment.

Designing an Experiment

A carefully designed experiment can test a prediction in a reliable way, ruling out other possible explanations. As scientists plan their experimental procedures, they pay particular attention to the factors that must be controlled.

The gardener decided to study three groups of seeds: (1) some that would be left outdoors throughout the winter, (2) some that would be brought indoors and kept at room temperature, and (3) some that would be brought indoors and kept cold.

Controlling Variables

As researchers design an experiment, they identify the variables, factors that can change. Some common variables include mass, volume, time, temperature, light, and the presence or absence of specific materials. An experiment involves three categories of variables. The factor that scientists purposely change is called the independent variable. An independent variable is also known as a manipulated variable. The factor that may change because of the independent variable and that scientists want to observe is called the dependent variable. A dependent variable is also known as a responding variable. Factors that scientists purposely keep the same are called controlled variables. Controlling variables enables researchers to conclude that the changes in the independent variable are due exclusively to changes in the dependent variable.

For the gardener, the independent variable is whether the seeds were exposed to cold conditions. The dependent variable is whether or not the seeds germinate. Among the variables that must be controlled are whether the seeds remain dry during storage, when the seeds are planted, the amount of water the seeds receive, and the type of soil used.

Interpreting Data

The observations and measurements that are made in an experiment are called data. Scientists usually record data in an orderly way. When an experiment is finished, the researcher analyzes the data for trends or patterns, often by doing calculations or making graphs, to determine whether the results support the hypothesis.

For example, after planting the seeds in the spring, the gardener counted the seeds that germinated and found these results: None of the seeds kept at room temperature germinated, 80 percent of the seeds kept in the freezer germinated, and 85 percent of the seeds left outdoors during the winter germinated. The trend was clear: The gardener's prediction appeared to be correct.

To be sure that the results of an experiment are correct, scientists review their data critically, looking for possible sources of error. Here, *error* refers to differences between the observed results and the true values. Experimental error can result from human mistakes or problems with equipment. It can also occur when the small group of objects studied does not accurately represent the whole group. For example, if some of the gardener's seeds had been exposed to an herbicide, the data might not reflect the true seed germination pattern.

Drawing Conclusions

If researchers are confident that their data are reliable, they make a final statement summarizing their results. That statement, called the conclusion, indicates whether the data support or refute the hypothesis. The gardener's conclusion was this: *Some seeds must undergo a period of freezing in order to germinate.* A conclusion is considered valid if it is a logical interpretation of reliable data.

Following Up an Experiment

When an experiment has been completed, one or more events often follow. Researchers may repeat the experiment to verify the results. They may publish the experiment so that others can evaluate and replicate their procedures. They may compare their conclusion with the discoveries made by other scientists. And they may raise new questions that lead to new experiments. For example, *Are the spores of fungi affected by temperature as these seeds were?*

Researching other discoveries about seeds would show that some other types of plants in temperate zones require periods of freezing before they germinate. Biologists infer that this pattern makes it less likely the seeds will germinate before winter, thus increasing the chances that the young plants will survive.

Appendix B Lab Skills

The Metric System

The standard system of measurement used by scientists throughout the world is known as the International System of Units, abbreviated as SI (Système International d'Unités, in French). It is based on units of 10. Each unit is 10 times larger or 10 times smaller than the next unit. The table lists the prefixes used to name the most common SI units.

Common SI Prefixes

Prefix	Symbol	Meaning
kilo-	k	1000
hecto-	h	100
deka-	da	10
deci-	d	0.1 (one tenth)
centi-	c	0.01 (one hundredth)
milli-	m	0.001 (one thousandth)

Commonly Used Metric Units

Length To measure length, or distance from one point to another, the unit of measure is a meter (m). A meter is slightly longer than a yard.

Useful equivalents:

1 meter = 1000 millimeters (mm)
1 meter = 100 centimeters (cm)
1000 meters = 1 kilometer (km)

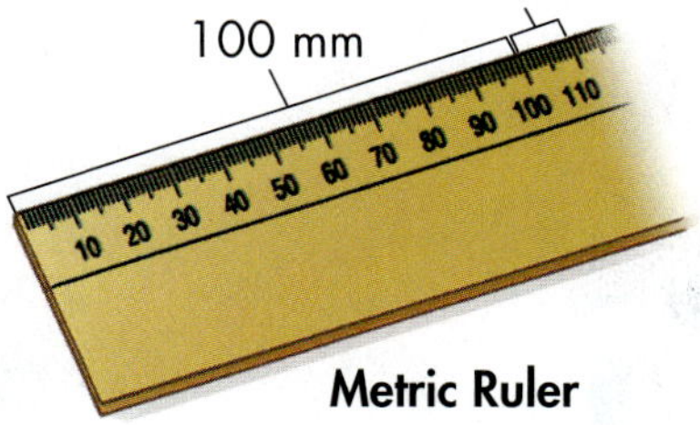

Metric Ruler

Volume To measure the volume of a liquid, or the amount of space an object takes up, the unit of measure is a liter (L). A liter is slightly more than a quart.

Useful equivalents:

1 liter = 1000 milliliters (mL)

Mass To measure the mass, or the amount of matter in an object, the unit of measure is the gram (g). A paper clip has a mass equal to about one gram.

Useful equivalents:

1000 grams = 1 kilogram (kg)

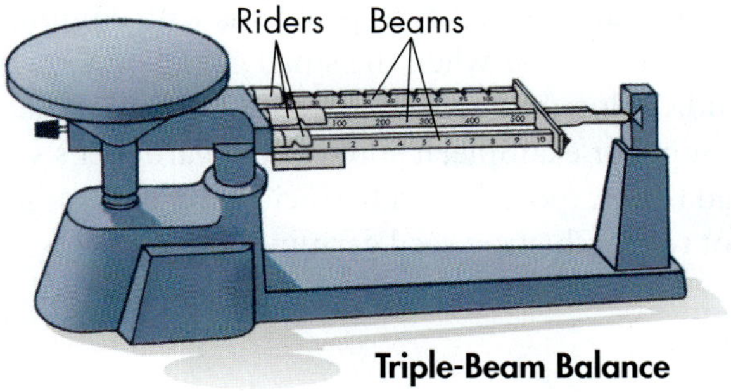

Triple-Beam Balance

Temperature To measure the hotness or coldness of an item, or its temperature, you use the unit degrees. The freezing point of water is 0°C (Celsius). The boiling point of water is 100°C.

Metric-English Equivalents

2.54 centimeters (cm) = 1 inch (in.)
1 meter (m) = 39.37 inches (in.)
1 kilometer (km) = 0.62 miles (mi)
1 liter (L) = 1.06 quarts (qt)
236 milliliters (mL) = 1 cup (c)
1 kilogram (kg) = 2.2 pounds (lb)
28.3 grams (g) = 1 ounce (oz)
°C = 5/9 × (°F – 32)

Safety Symbols

These symbols appear in laboratory activities to alert you to possible dangers and to remind you to work carefully.

Safety Goggles Always wear safety goggles to protect your eyes during any activity involving chemicals, flames or heating, or the possibility of flying objects, particles, or substances.

Lab Apron Wear a laboratory apron to protect your skin and clothing from injury.

Plastic Gloves Wear disposable plastic gloves to protect yourself from contact with chemicals or organisms that could be harmful. Keep your hands away from your face, and dispose of the gloves according to your teacher's instructions at the end of the activity.

Breakage Handle breakable materials such as thermometers and glassware with care. Do not touch broken glass.

Heat-Resistant Gloves Use an oven mitt or other hand protection when handling hot materials. Heating plates, hot water, and glassware can cause burns. Never touch hot objects with your bare hands.

Heating Use a clamp or tongs to hold hot objects. Do not touch hot objects with your bare hands.

Sharp Object Scissors, scalpels, pins, and knives are sharp. They can cut or puncture your skin. Always direct sharp edges and points away from yourself and others. Use sharp instruments only as directed.

Electric Shock Avoid the possibility of electric shock. Never use electrical equipment around water or when the equipment or your hands are wet. Be sure cords are untangled and cannot trip anyone. Disconnect equipment when it is not in use.

Corrosive Chemical This symbol indicates the presence of an acid or other corrosive chemical. Avoid getting the chemical on your skin or clothing, or in your eyes. Do not inhale the vapors. Wash your hands when you are finished with the activity.

Poison Do not let any poisonous chemical get on your skin, and do not inhale its vapor. Wash your hands when you are finished with the activity.

Flames Tie back loose hair and clothing, and put on safety goggles before working with fire. Follow instructions from your teacher about lighting and extinguishing flames.

No Flames Flammable materials may be present. Make sure there are no flames, sparks, or exposed sources of heat present.

Fumes Poisonous or unpleasant vapors may be produced. Work in a ventilated area or, if available, in a fume hood. Avoid inhaling a vapor directly. Test an odor only when directed to do so by your teacher, using a wafting motion to direct the vapor toward your nose.

Physical Safety This activity involves physical movement. Use caution to avoid injuring yourself or others. Follow instructions from your teacher. Alert your teacher if there is any reason that you should not participate in the activity.

Animal Safety Treat live animals with care to avoid injuring the animals or yourself. Working with animal parts or preserved animals may also require caution. Wash your hands when you are finished with the activity.

Plant Safety Handle plants only as your teacher directs. If you are allergic to any plants used in an activity, tell your teacher before the activity begins. Avoid touching poisonous plants and plants with thorns.

Disposal Chemicals and other materials used in the activity must be disposed of safely. Follow the instructions from your teacher.

Hand Washing Wash your hands thoroughly when finished with the activity. Use soap and warm water. Lather both sides of your hands and between your fingers. Rinse well.

General Safety Awareness You may see this symbol when none of the symbols described earlier applies. In this case, follow the specific instructions provided. You may also see this symbol when you are asked to design your own experiment. Do not start your experiment until your teacher has approved your plan.

Appendix B Lab Skills

Science Safety Rules

Working in the laboratory can be an exciting experience, but it can also be dangerous if proper safety rules are not followed at all times. To prepare yourself for a safe year in the laboratory, read the following safety rules. Make sure that you understand each rule. Ask your teacher to explain any rules you don't understand.

Dress Code

1. Many materials in the laboratory can cause eye injury. To protect yourself from possible injury, wear safety goggles whenever you are working with chemicals, burners, or any substance that might get into your eyes. Avoid wearing contact lenses in the laboratory. Tell your teacher if you need to wear contact lenses to see clearly, and ask if there are any safety precautions you should observe.
2. Wear a laboratory apron or coat whenever you are working with chemicals or heated substances.
3. Tie back long hair to keep it away from any chemicals, burners, candles, or other laboratory equipment.
4. Before working in the laboratory, remove or tie back any article of clothing or jewelry that can hang down and touch chemicals and flames.

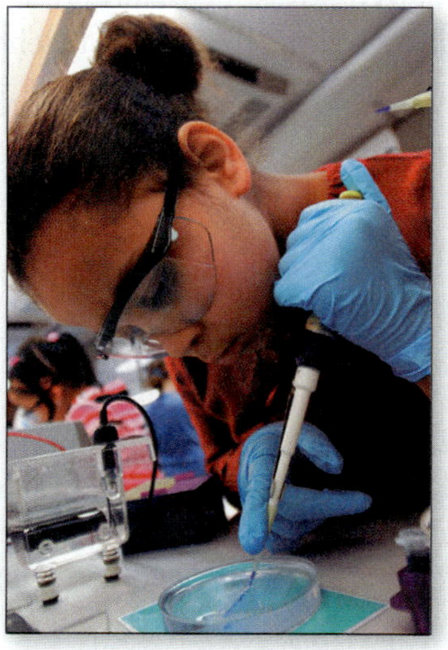

General Safety Rules and First Aid

5. Read all directions for an experiment several times. Follow the directions exactly as they are written. If you are in doubt about any part of the experiment, ask your teacher for assistance.
6. Never perform investigations your teacher has not authorized. Do not use any equipment unless your teacher is in the lab.
7. Never handle equipment unless you have specific permission.
8. Take care not to spill any material in the laboratory. If spills occur, ask your teacher immediately about the proper cleanup procedure. Never pour chemicals or other substances into the sink or trash container.
9. Never eat or drink in, or bring food into, the laboratory.
10. Immediately report all accidents, no matter how minor, to your teacher.
11. Learn what to do in case of specific accidents, such as getting acid in your eyes or on your skin. (Rinse acids off your body with lots of water.)
12. Be aware of the location of the first-aid kit. Your teacher should administer any required first aid due to injury. Your teacher may send you to the school nurse or call a physician.
13. Know where and how to report an accident or fire. Find out the location of the fire extinguisher, fire alarm, and phone. Report any fires to your teacher at once.

Heating and Fire Safety

14. Never use a heat source such as a candle or burner without wearing safety goggles.
15. Never heat a chemical you are not instructed to heat. A chemical that is harmless when cool can be dangerous when heated.
16. Maintain a clean work area and keep all materials away from flames. Be sure that there are no open containers of flammable liquids in the laboratory when flames are being used.
17. Never reach across a flame.

18. Make sure you know how to light a Bunsen burner. (Your teacher will demonstrate the proper procedure for lighting a burner.) If the flame leaps out of a burner toward you, turn the gas off immediately. Do not touch the burner. It may be hot. Never leave a lighted burner unattended!

19. When you are heating a test tube or bottle, point the opening away from yourself and others. Chemicals can splash or boil out of a heated test tube.

20. Never heat a closed container. The expanding hot air, vapors, or other gases inside may blow the container apart, causing it to injure you or others.

21. Never pick up a container that has been heated without first holding the back of your hand near it. If you can feel the heat on the back of your hand, the container may be too hot to handle. Use a clamp or tongs when handling hot containers or wear heat-resistant gloves if appropriate.

Using Chemicals Safely

22. Never mix chemicals for "the fun of it." You might produce a dangerous, possibly explosive substance.

23. Many chemicals are poisonous. Never touch, taste, or smell a chemical that you do not know for certain is harmless. If you are instructed to smell fumes in an experiment, gently wave your hand over the opening of the container and direct the fumes toward your nose. Do not inhale the fumes directly from the container.

24. Use only those chemicals needed in the investigation. Keep all container lids closed when a chemical is not being used. Notify your teacher whenever chemicals are spilled.

25. Dispose of all chemicals as instructed by your teacher. To avoid contamination, never return chemicals to their original containers.

26. Be extra careful when working with acids or bases. Pour such chemicals from one container to another over the sink, not over your work area.

27. When diluting an acid, pour the acid into water. Never pour water into the acid.

28. If any acids or bases get on your skin or clothing, rinse them with water. Immediately notify your teacher of any acid or base spill.

Using Glassware Safely

29. Never heat glassware that is not thoroughly dry. Use a wire screen to protect glassware from any flame.

30. Keep in mind that hot glassware will not appear hot. Never pick up glassware without first checking to see if it is hot.

31. Never use broken or chipped glassware. If glassware breaks, notify your teacher and dispose of the glassware in the proper trash container.

32. Never eat or drink from laboratory glassware. Thoroughly clean glassware before putting it away.

Using Sharp Instruments

33. Handle scalpels or razor blades with extreme care. Never cut material toward you; cut away from you.

34. Notify your teacher immediately if you cut yourself when in the laboratory.

Working With Live Organisms

35. No experiments that will cause pain, discomfort, or harm to animals should be done in the classroom or at home.

36. Your teacher will instruct you how to handle each species that is brought into the classroom. Animals should be handled only if necessary. Special handling is required if an animal is excited or frightened, pregnant, feeding, or with its young.

37. Clean your hands thoroughly after handling any organisms or materials, including animals or cages containing animals.

End-of-Experiment Rules

38. When an experiment is completed, clean up your work area and return all equipment to its proper place.

39. Wash your hands with soap and warm water before and after every experiment.

40. Turn off all burners before leaving the laboratory. Check that the gas line leading to the burner is off as well.

Appendix B Lab Skills

Use of the Microscope

The microscope used in most biology classes, the compound microscope, contains a combination of lenses. The eyepiece lens is located in the top portion of the microscope. This lens usually has a magnification of 10×. Other lenses, called objective lenses, are at the bottom of the body tube on the revolving nosepiece. By rotating the nosepiece, you can select the objective through which you will view your specimen.

The shortest objective is a low-power magnifier, usually 10×. The longer ones are of high power, usually up to 40× or 43×. The magnification is marked on the objective. To determine the total magnification, multiply the magnifying power of the eyepiece by the magnifying power of the objective. For example, with a 10× eyepiece and a 40× objective, the total magnification is $10 \times 40 = 400\times$.

Learning the name, function, and location of each of the microscope's parts is necessary for proper use. Use the following procedures when working with the microscope.

1. Carry the microscope by placing one hand beneath the base and grasping the arm of the microscope with the other hand.
2. Gently place the microscope on the lab table with the arm facing you. The microscope's base should be resting evenly on the table, approximately 10 cm from the table's edge.
3. Raise the body tube by turning the coarse adjustment knob until the objective lens is about 2 cm above the opening of the stage.
4. Rotate the nosepiece so that the low-power objective (10×) is directly in line with the body tube. A click indicates that the lens is in line with the opening of the stage.
5. Look through the eyepiece and switch on the lamp or adjust the mirror so that a circle of light can be seen. This is the field of view. Moving the lever of the diaphragm permits a greater or smaller amount of light to come through the opening of the stage.
6. Place a prepared slide on the stage so that the specimen is over the center of the opening. Use the stage clips to hold the slide in place.
7. Look at the microscope from the side. Carefully turn the coarse adjustment knob to lower the body tube until the low-power objective almost touches the slide or until the body tube can no longer be moved. Do not allow the objective to touch the slide.
8. Look through the eyepiece and observe the specimen. If the field of view is out of focus, use the coarse adjustment knob to raise the body tube while looking through the eyepiece. **CAUTION:** *To prevent damage to the slide and the objective, do not lower the body tube using the coarse adjustment while looking through the eyepiece.* Focus the image as best you can with the coarse adjustment knob. Then, use the fine adjustment knob to focus the image more sharply. Keep both eyes open when viewing a specimen. This helps prevent eyestrain.

1. **Eyepiece:** Contains a magnifying lens.
2. **Arm:** Supports the body tube.
3. **Low-power objective:** Provides a magnification of 10x.
4. **Stage:** Supports the slide being observed.
5. **Opening of the stage:** Permits light to pass up to the eyepiece.
6. **Fine adjustment knob:** Moves the body tube slightly to adjust the image.
7. **Coarse adjustment knob:** Moves the body tube to focus the image.
8. **Base:** Supports the microscope.
9. **Illuminator:** Produces light or reflects light up toward the eyepiece.
10. **Diaphragm:** Regulates the amount of light passing up toward the eyepiece.
11. **Stageclips:** Hold the slide in place.
12. **High-power objective:** Provides a magnification of 40x.
13. **Nosepiece:** Holds the objectives and can be rotated to change the magnification.
14. **Body tube:** Maintains the proper distance between the eyepiece and the objectives.

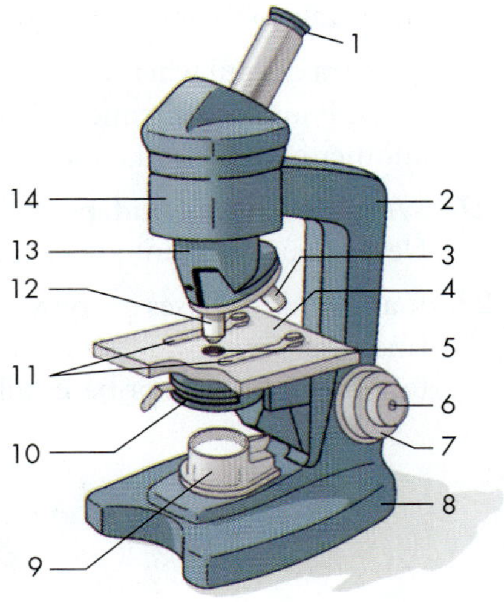

9. Adjust the lever of the diaphragm to allow the right amount of light to enter.
10. To change the magnification, rotate the nosepiece until the desired objective is in line with the body tube and clicks into place.
11. Look through the eyepiece and use the fine adjustment knob to bring the image into focus.
12. After every use, remove the slide. Return the low-power objective into place in line with the body tube. Clean the stage of the microscope and the lenses with lens paper. Do not use other types of paper to clean the lenses; they may scratch the lenses.

Preparing a Wet-Mount Slide

1. Obtain a clean microscope slide and a coverslip. A coverslip is very thin, permitting the objective lens to be lowered very close to the specimen.
2. Place the specimen in the middle of the microscope slide. The specimen must be thin enough for light to pass through it.
3. Using a dropper pipette, place a drop of water on the specimen.

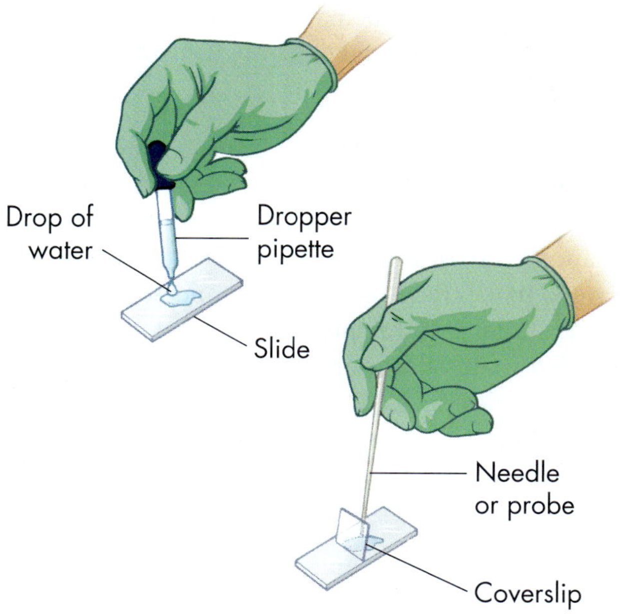

4. Lower one edge of the coverslip so that it touches the side of the drop of water at about a 45° angle. The water will spread evenly along the edge of the coverslip. Using a dissecting needle or probe, slowly lower the coverslip over the specimen and water as shown in the drawing. Try not to trap any air bubbles under the coverslip. If air bubbles are present, gently tap the surface of the coverslip over the air bubble with a pencil eraser.
5. Remove any excess water around the edge of the coverslip with a paper towel. If the specimen begins to dry out, add a drop of water at the edge of the coverslip.

Staining Techniques

1. Obtain a clean microscope slide and coverslip.
2. Place the specimen in the middle of the microscope slide.
3. Using a dropper pipette, place a drop of water on the specimen. Place the coverslip so that its edge touches the drop of water at a 45° angle. After the water spreads along the edge of the coverslip, use a dissecting needle or probe to lower the coverslip over the specimen.
4. Add a drop of stain at the edge of the coverslip. Using forceps, touch a small piece of lens paper or paper towel to the opposite edge of the coverslip, as shown in the drawing. The paper causes the stain to be drawn under the coverslip and to stain the cells in the specimen.

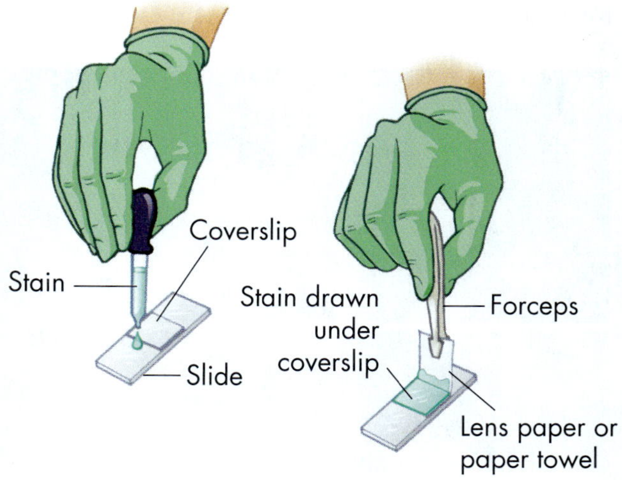

Appendix C Technology & Design

Engineers are people who use scientific and technological knowledge to solve practical problems. To design new products, engineers usually follow the process described here, even though they may not follow these steps in the exact order.

Identify a Need

Before engineers begin designing a new product, they must first identify the need they are trying to meet. For example, suppose you are a member of a design team in a company that makes toys. Your team has identified a need: a toy boat that is inexpensive and easy to assemble.

Research the Problem

Engineers often begin by gathering information that will help them with their new design. This research may include finding articles in books, in magazines, or on the Internet. It may also include talking to other engineers who have solved similar problems. Engineers also often perform experiments related to the product they want to design.

For your toy boat, you could look at toys that are similar to the one you want to design. You might do research on the Internet. You could also test some materials to see whether they would work well in a toy boat.

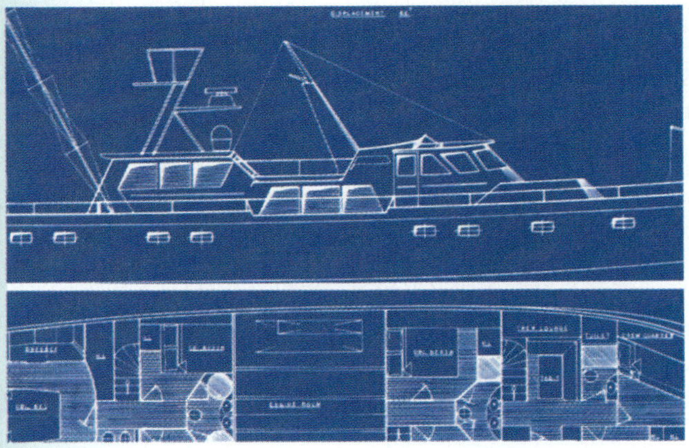

Design a Solution

Research gives engineers information that helps them design a product. When engineers design new products, they usually work in teams.

Generating Ideas Often, design teams hold brainstorming meetings in which any team member can contribute ideas. Brainstorming is a creative process in which one team member's suggestions can spark ideas in other group members. Brainstorming can lead to new approaches to solving a design problem.

Evaluating Constraints During brainstorming, a design team will often come up with several possible designs. The team must then evaluate each one.

As part of their evaluation, engineers consider constraints. Constraints are factors that limit or restrict a product design. Physical characteristics, such as the properties of materials used to make your toy boat, are constraints. Cost and time are also constraints. If the materials in a design cost a lot, or if the design takes a long time to make, it may be impractical.

Making Trade-offs Design teams usually need to make trade-offs. A trade-off is the acceptance of the benefits of one design aspect at the cost of another. In designing your toy boat, you will have to make trade-offs. For example, suppose one material is sturdy but not fully waterproof. Another material is more waterproof, but breakable. You may decide to give up the benefit of sturdiness in order to obtain the benefit of waterproofing.

Build and Evaluate a Prototype

Once the team has chosen a design plan, the engineers build a prototype of the product. A prototype is a working model used to test a design. Engineers evaluate the prototype to see whether it works well, is easy to operate, is safe to use, and holds up to repeated use.

Think of your toy boat. What would the prototype be like? Of what materials would it be made? How would you test it?

Troubleshoot and Redesign

Few prototypes work perfectly, which is why they need to be tested. Once a design team has tested a prototype, the members analyze the results and identify any problems. The team then tries to troubleshoot, or fix the weaknesses in the design. For example, if your toy boat leaks or wobbles, the boat should be redesigned to eliminate those problems.

Communicate the Solution

A team needs to communicate the final design to the people who will manufacture the product. To do this, teams may use sketches, detailed drawings, computer simulations, and written descriptions.

Activity

You can use the technology design process to design and build a toy boat.

Research and Investigate

1. Visit the library or go online to research toy boats.
2. Investigate how a toy boat can be powered, including wind, rubber bands, or baking soda and vinegar.
3. Brainstorm materials, shapes, and steering methods for your boat.

Design and Build

4. Based on your research, design a toy boat that
   - is made of readily available materials
   - is no larger than 15 cm long and 10 cm wide
   - includes a power system, a rudder, and a cargo area
   - travels 2 meters in a straight line while carrying a load of 20 pennies
5. Sketch your design and write a step-by-step plan for building your boat. After your teacher approves your plan, build your boat.

Evaluate and Redesign

6. Test your boat, evaluate the results, and identify any technological design problems in your boat.
7. Based on your evaluation, redesign your toy boat so it performs better.
8. As a class, compare the test results for each boat. Choose the model which best meets the needs of the toy company.

Appendix D Math Skills

Scientists use math to organize, analyze, and present data. This appendix will help you review some basic math skills.

Formulas and Equations

Formulas and equations are used in many areas of science. Both formulas and equations show the relationships between quantities. Any numerical sentence that contains at least one variable and at least one mathematical operator is called an equation. A formula is a type of equation that states the relationship between unknown quantities represented by variables.

For example, Speed = Distance ÷ Time is a formula, because no matter what values are inserted, speed is always equal to distance divided by time. The relationship between the variables does not change.

Example

Follow these steps to convert a temperature measurement of 50°F to Celsius.

1. Determine the formula that shows the relationship between these quantities.
   °F = (9/5 × °C) + 32°F
2. Insert values you know into the formula.
   50°F = (9/5 × °C) + 32°F
3. Solve the resulting equation.
   50°F – 32°F = (9/5 × °C)
   18°F = 9/5 × °C
   18°F × 5/9 = 10°C

Applying Formulas and Equations

There are many applications of formulas in science. The example described below uses a formula to calculate density.

Example

Follow these steps to calculate the density of an object that has a mass of 45 g and a volume of 30 cm^3.

1. Determine the formula that shows the relationship between these quantities.
   Density = Mass/Volume
2. Insert values you know into the formula.
   Density = 45 g/30 cm^3
3. Solve the resulting equation.
   Density = 1.5 g/cm^3

Mean, Median, and Mode

The mean is the average, or the sum of the data divided by the number of data items. The middle number in a set of ordered data is called the median. The mode is the number that appears most often in a set of data.

Example
A scientist counted the number of distinct songs sung by seven different male birds and collected the data shown below.

Male Bird Songs							
Bird	A	B	C	D	E	F	G
Number of Songs	36	29	40	35	28	36	27

To determine the mean number of songs, find the sum of the songs sung by all the male birds and divide by the number of male birds.

$$\text{Mean} = 231/7 = 33 \text{ songs}$$

To find the median number of songs, arrange the data items in numerical order and identify the number in the middle.

27 28 29 35 36 36 40

The number in the middle is 35, so the median number of songs is 35.

The mode is the value that appears most frequently. In the data, 36 appears twice, while every other item appears only once. Therefore, 36 is the mode.

Estimation

An estimate is a reasonable approximation of a numerical value. Estimates are made based on careful assumptions and known information.

Scientists use estimates in biology for two primary reasons: when an exact count or calculation cannot be made or is impractical to make, and to make reasonable approximations of answers that will be calculated or measured later.

One method for estimation used in biology is sampling. In sampling, the number of organisms in a small area (a sample) is multiplied to estimate the number of organisms in a larger area.

Example
Follow these steps to use sampling to estimate the total number of birds in the photo.

1. Count the birds in the highlighted area of the photo. In the highlighted area of the photo, there are 36 birds.
2. Determine the portion of the entire photo represented by the highlighted area. In this case, the highlighted area is 1/6 of the total area.
3. Calculate your estimate by multiplying the number of birds in the sample area by 6 (because the entire photo is 6 times as large as the sample area). A reasonable estimate of the total number of birds is 36×6, or 216 birds.

HINT: Estimates and calculated answers are rarely exactly the same. However, a large difference between an estimated answer and a calculated answer indicates there may be a problem with the estimate or calculation.

Using Measurements in Calculations

Density is an example of a value that is calculated using two measurements. Density represents the amount of mass in a particular volume of a substance. The units used for density are grams per milliliter (g/mL) or grams per cubic centimeter (g/cm^3). Density is calculated by dividing an object's mass by its volume.

Example
Follow these steps to calculate the density of an object.

1. Measure and record the mass of an object in grams.
2. Measure and record the volume of an object in mL or cm^3.
3. Use the following formula to calculate density:

$$\text{Density} = \text{Mass/Volume}$$

Effects of Measurement Errors

Density is calculated using two measured values. An error in the measurement of either mass or volume will result in the calculation of an incorrect density.

Example
A student measured the mass of an object as 2.5 g and its volume as 2.0 cm^3. The actual mass of the object is 3.5 g; the actual volume is 2.0 cm^3. What is the effect of the measurement error on the calculation of density?

Follow these steps to determine the effect of a measurement error on calculation.

1. Determine the density using the student's measurements.
   Density = Mass/Volume
   Density = 2.5 g/2.0 cm^3
   Density = 1.25 g/cm^3
2. Determine the density using the actual values.
   Density = Mass/Volume
   Density = 3.5 g/2.0 cm^3
   Density = 1.75 g/cm^3
3. Compare the calculated and the actual values.

In this case, a measurement of mass that was less than the actual value resulted in a calculated value for the density that was less than the actual density.

Accuracy

The accuracy of a measurement is its closeness to the actual value. Measurements that are accurate are close to the actual value.

Both clocks on this page show a time of 3:00. Suppose, though, that these clocks had not been changed to reflect daylight savings time. The time shown on the clocks would be inaccurate. On the other hand, if the actual time is 3:00, these clocks would be accurate.

Precision

Precision describes the exactness of a measurement. The clocks shown on this page differ in precision. The analog clock measures time to the nearest minute. The digital clock measures time to the nearest second. Time is measured more precisely by the digital clock than by the analog clock.

Comparing Accuracy and Precision

There is a difference between accuracy and precision. Measurements can be accurate (close to the actual value) but not precise. Measurements can also be precise but not accurate. When making scientific measurements, both accuracy and precision are important. Accurate and precise measurements result from the careful use of high-quality measuring tools.

Significant Figures

Significant figures are all of the digits that are known in a measurement, plus one additional digit, which is an estimate. In the figure below, the length of a turtle's shell is being measured using a centimeter ruler. The ruler has unnumbered divisions representing millimeters. In this case, two numbers can be determined exactly: the number of centimeters and the number of millimeters. One additional digit can be estimated. So, the measurement of this turtle's shell can be recorded with three significant figures as 8.80 centimeters.

Rules for Significant Digits
Follow these rules to determine the number of significant figures in a number.

All nonzero numbers are significant.

Example: 3217 has four significant digits.

Zeros are significant if

- They are between nonzero digits. Example: 509
- They follow a decimal point and a nonzero digit. Example: 7.00

Zeros are not significant if

- They follow nonzero digits in a number without a decimal. Example: 7000
- They precede nonzero digits in a number with a decimal. Example: 0.0098

Calculating With Significant Figures

When measurements are added or subtracted, the precision of the result is determined by the precision of the least-precise measurement. The result may need to be rounded so the number of digits after the decimal is the same as the least-precise measurement.

Example
Follow these steps to determine the correct number of significant figures when adding 4.51 g, 3.27 g, and 6.0 g.

1. Determine which measurement is reported with the least degree of precision. In this case, the least-precise measurement, 6.0 g, has one digit after the decimal point.
2. The result must be rounded so that it also has one digit after the decimal point. After rounding, the result of this calculation is 13.8 g.

When measurements are multiplied or divided, the answer must have the same number of significant figures as the measurement with the fewest number of significant figures.

Example
Follow these steps to determine the correct number of significant figures when multiplying 120 m by 6.32 m.

1. Determine the number of significant figures in each of the measurements. In this case, the measurement 120 m has two significant figures; the measurement 6.32 m has three significant figures.
2. The result must be rounded to have only two significant figures. After rounding, the result of this calculation is 760 m^2.

Scientific Notation

In science, measurements are often very large or very small. Using scientific notation makes these large and small numbers easier to work with.

Using scientific notation requires an understanding of exponents and bases. When a number is expressed as a base and an exponent, the base is the number that is used as a factor. The exponent tells how many times the base is multiplied by itself. For example, the number 25 can be expressed as a base and an exponent in the following way:

$$25 = 5 \times 5 = 5^2$$

In the example above, 5 is the base and 2 is the exponent. In scientific notation, the base is always the number 10. The exponent tells how many times the number 10 is multiplied by itself.

A number written in scientific notation is expressed as the product of two factors, a number between 1 and 10 and the number 10 with an exponent. For example, the number 51,000 can be expressed in scientific notation. To find the first factor, move the decimal to obtain a number between 1 and 10. In this case, the number is 5.1. The exponent can be determined by counting the number of places the decimal point was moved. The decimal point was moved four places to the left. So, 51,000 expressed in scientific notation is 5.1×10^4.

Numbers that are less than one can also be expressed in scientific notation. In the case of numbers less than one, the decimal point must be moved to the right to obtain a number between 1 and 10. For example, in the number 0.000098, the decimal point must move five places to the right to obtain the number 9.8. When the decimal point is moved to the right, the exponent is negative. So, 0.000098 expressed in scientific notation is 9.8×10^{-5}.

Calculating With Scientific Notation

Numbers expressed in scientific notation can be used in calculations. When adding or subtracting numbers expressed in scientific notation, the first factors must be rewritten so the exponents are the same.

Example

Follow these steps to add $(4.30 \times 10^4) + (2.1 \times 10^3)$.

1. Move the decimal point in one of the expressions so the exponents are the same.
   $(43.0 \times 10^3) + (2.1 \times 10^3)$
2. Add the first factors, keeping the value of the exponents the same.
   $(43.0 \times 10^3) + (2.1 \times 10^3) = 45.1 \times 10^3$
3. Move the decimal point so the first factor is expressed as the product of a number between and 1 and 10 and an exponent with base 10.
   $45.1 \times 10^3 = 4.51 \times 10^4$

When numbers expressed in scientific notation are multiplied, the exponents are added. When numbers expressed in scientific notation are divided, the exponents are subtracted.

Example

Use the following steps to determine the area of a rectangular field that has a length of 1.5×10^3 meters and a width of 3.2×10^2 meters.

1. Write down the expressions to be multiplied.
   $(1.5 \times 10^3 \text{ m})(3.2 \times 10^2 \text{ m})$
2. Multiply the first factors, add the exponents, and multiply any units.
   $= (1.5 \times 3.2)(10^{3+2}) \text{ m} \times \text{m}$
   $= 4.8 \times 10^5 \text{ m}^2$

Dimensional Analysis

Scientific problems and calculations often involve unit conversions, or changes from one unit to another. Dimensional analysis is a method of unit conversion.

Suppose you were counting a pile of pennies. If there were 197 pennies in the pile, how many dollars would the pennies be worth? To determine the answer, you need to know the conversion factor between pennies and dollars. A conversion factor simply shows how two units are related. In this case, the conversion factor is 100 pennies = 1 dollar. Determining that 197 pennies is equal to $1.97 is an example of a unit conversion.

In dimensional analysis, the conversion factor is usually expressed as a fraction. Remember that the two values in any conversion factor are equal to one another. So, the two values form a fraction with the value of 1. Look at the example below to see how dimensional analysis can be applied to an everyday problem.

Example
A student walked 1.5 kilometers as part of a school fitness program. How many meters did the student walk?

1. 1.5 km = _?_ m
2. 1 km = 1000 m
3. 1000 m/1 km
4. 1.5 km × 1000 m/1 km = 1500 m (cross out "km" in two places); 1.5 km = 1500 m

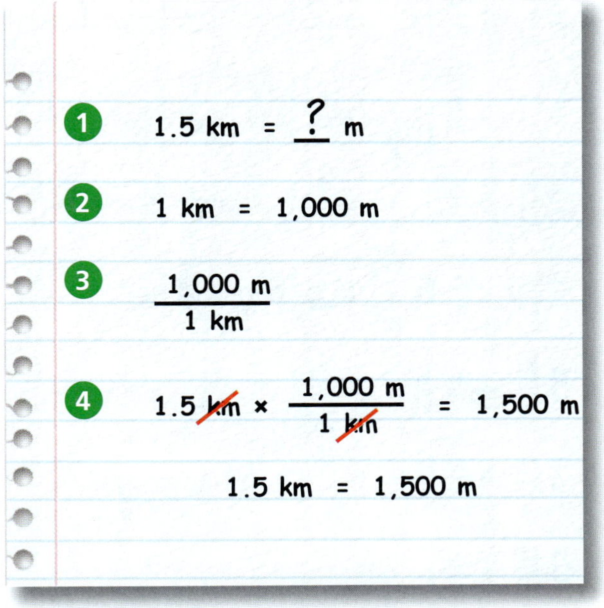

Applying Dimensional Analysis

There are many applications of dimensional analysis in science. The example below demonstrates the use of dimensional analysis to convert units.

Example
The average teenage girl needs about 2200 kilocalories of energy from food each day. How many calories is this equivalent to?

Use the following steps to convert kilocalories to calories.

1. Determine the conversion factor that relates the two units.
   1 kilocalorie = 1000 calories
2. Write the conversion factor in the form of a fraction.
   1000 calories/1 kilocalorie
3. Multiply the measurement by the conversion factor.
   2200 kilocalories × 1000 calories/1 kilocalorie = 2,200,000 calories

Appendix E Periodic Table

Periodic Table of the Elements

Representative Elements
- Alkali Metals
- Alkaline Earth Metals
- Other Metals
- Metalloids
- Nonmetals
- Noble Gases

Transition Elements
- Transition Metals
- Inner transition metals

C Solid
Br Liquid
He Gas
Tc Not found in nature

Key: 14 (Atomic number); 2 8 4 (Electrons in each energy level); Si (Element symbol); Silicon (Element name); * 28.086 (Average atomic mass)

*The atomic masses in parentheses are the mass numbers of the longest-lived isotope of elements for which a standard atomic mass cannot be defined.

Each cell: atomic number, symbol, name, average atomic mass, electrons in each energy level.

1 1A	2 2A	3 3B	4 4B	5 5B	6 6B	7 7B	8	9 8B	10	11 1B	12 2B	13 3A	14 4A	15 5A	16 6A	17 7A	18 8A
1 H Hydrogen 1.0079 (1)																	2 He Helium 4.0026 (2)
3 Li Lithium 6.941 (2 1)	4 Be Beryllium 9.0122 (2 2)											5 B Boron 10.81 (2 3)	6 C Carbon 12.011 (2 4)	7 N Nitrogen 14.007 (2 5)	8 O Oxygen 15.999 (2 6)	9 F Fluorine 18.998 (2 7)	10 Ne Neon 20.179 (2 8)
11 Na Sodium 22.990 (2 8 1)	12 Mg Magnesium 24.305 (2 8 2)											13 Al Aluminum 26.982 (2 8 3)	14 Si Silicon 28.086 (2 8 4)	15 P Phosphorus 30.974 (2 8 5)	16 S Sulfur 32.06 (2 8 6)	17 Cl Chlorine 35.453 (2 8 7)	18 Ar Argon 39.948 (2 8 8)
19 K Potassium 39.098 (2 8 8 1)	20 Ca Calcium 40.08 (2 8 8 2)	21 Sc Scandium 44.956 (2 8 9 2)	22 Ti Titanium 47.90 (2 8 10 2)	23 V Vanadium 50.941 (2 8 11 2)	24 Cr Chromium 51.996 (2 8 13 1)	25 Mn Manganese 54.938 (2 8 13 2)	26 Fe Iron 55.847 (2 8 14 2)	27 Co Cobalt 58.933 (2 8 15 2)	28 Ni Nickel 58.71 (2 8 16 2)	29 Cu Copper 63.546 (2 8 18 1)	30 Zn Zinc 65.38 (2 8 18 2)	31 Ga Gallium 69.72 (2 8 18 3)	32 Ge Germanium 72.59 (2 8 18 4)	33 As Arsenic 74.922 (2 8 18 5)	34 Se Selenium 78.96 (2 8 18 6)	35 Br Bromine 79.904 (2 8 18 7)	36 Kr Krypton 83.80 (2 8 18 8)
37 Rb Rubidium 85.468 (2 8 18 8 1)	38 Sr Strontium 87.62 (2 8 18 8 2)	39 Y Yttrium 88.906 (2 8 18 9 2)	40 Zr Zirconium 91.22 (2 8 18 10 2)	41 Nb Niobium 92.906 (2 8 18 12 1)	42 Mo Molybdenum 95.94 (2 8 18 13 1)	43 Tc Technetium (98) (2 8 18 14 1)	44 Ru Ruthenium 101.07 (2 8 18 15 1)	45 Rh Rhodium 102.91 (2 8 18 16 1)	46 Pd Palladium 106.4 (2 8 18 18)	47 Ag Silver 107.87 (2 8 18 18 1)	48 Cd Cadmium 112.41 (2 8 18 18 2)	49 In Indium 114.82 (2 8 18 18 3)	50 Sn Tin 118.69 (2 8 18 18 4)	51 Sb Antimony 121.75 (2 8 18 18 5)	52 Te Tellurium 127.60 (2 8 18 18 6)	53 I Iodine 126.90 (2 8 18 18 7)	54 Xe Xenon 131.30 (2 8 18 18 8)
55 Cs Cesium 132.91 (2 8 18 18 8 1)	56 Ba Barium 137.33 (2 8 18 18 8 2)	71 Lu Lutetium 174.97 (2 8 18 32 9 2)	72 Hf Hafnium 178.49 (2 8 18 32 10 2)	73 Ta Tantalum 180.95 (2 8 18 32 11 2)	74 W Tungsten 183.85 (2 8 18 32 12 2)	75 Re Rhenium 186.21 (2 8 18 32 13 2)	76 Os Osmium 190.2 (2 8 18 32 14 2)	77 Ir Iridium 192.22 (2 8 18 32 15 2)	78 Pt Platinum 195.09 (2 8 18 32 17 1)	79 Au Gold 196.97 (2 8 18 32 18 1)	80 Hg Mercury 200.59 (2 8 18 32 18 2)	81 Tl Thallium 204.37 (2 8 18 32 18 3)	82 Pb Lead 207.2 (2 8 18 32 18 4)	83 Bi Bismuth 208.98 (2 8 18 32 18 5)	84 Po Polonium (209) (2 8 18 32 18 6)	85 At Astatine (210) (2 8 18 32 18 7)	86 Rn Radon (222) (2 8 18 32 18 8)
87 Fr Francium (223) (2 8 18 32 18 8 1)	88 Ra Radium (226) (2 8 18 32 18 8 2)	103 Lr Lawrencium (262 (2 8 18 32 32 9 2)	104 Rf Rutherfordium (261) (2 8 18 32 32 10 2)	105 Db Dubnium (262) (2 8 18 32 32 11 2)	106 Sg Seaborgium (263) (2 8 18 32 32 12 2)	107 Bh Bohrium (264) (2 8 18 32 32 13 2)	108 Hs Hassium (265) (2 8 18 32 32 14 2)	109 Mt Meitnerium (268) (2 8 18 32 32 15 2)	110 Ds Darmstadtium (269) (2 8 18 32 32 17 1)	111 Rg Roentgenium (272) (2 8 18 32 32 18 1)	112 *Uub Ununbium (277) (2 8 18 32 32 18 2)		114 *Uuq Ununquadium				

Elements 104–114 are the transactinide elements.

*Name not officially assigned.

Lanthanide Series													
57 La Lanthanum 138.91 (2 8 18 18 9 2)	58 Ce Cerium 140.12 (2 8 18 20 8 2)	59 Pr Praseodymium 140.91 (2 8 18 21 8 2)	60 Nd Neodymium 144.24 (2 8 18 22 8 2)	61 Pm Promethium (145) (2 8 18 23 8 2)	62 Sm Samarium 150.4 (2 8 18 24 8 2)	63 Eu Europium 151.96 (2 8 18 25 8 2)	64 Gd Gadolinium 157.25 (2 8 18 25 9 2)	65 Tb Terbium 158.93 (2 8 18 27 8 2)	66 Dy Dysprosium 162.50 (2 8 18 28 8 2)	67 Ho Holmium 164.93 (2 8 18 29 8 2)	68 Er Erbium 167.26 (2 8 18 30 8 2)	69 Tm Thulium 168.93 (2 8 18 31 8 2)	70 Yb Ytterbium 173.04 (2 8 18 32 8 2)
Actinide Series													
89 Ac Actinium (227) (2 8 18 32 18 9 2)	90 Th Thorium 232.04 (2 8 18 32 18 10 2)	91 Pa Protactinium 231.04 (2 8 18 32 20 9 2)	92 U Uranium 238.03 (2 8 18 32 21 9 2)	93 Np Neptunium (237) (2 8 18 32 22 9 2)	94 Pu Plutonium (244) (2 8 18 32 24 8 2)	95 Am Americium (243) (2 8 18 32 25 8 2)	96 Cm Curium (247) (2 8 18 32 25 9 2)	97 Bk Berkelium (247) (2 8 18 32 27 8 2)	98 Cf Californium (251) (2 8 18 32 28 8 2)	99 Es Einsteinium (252) (2 8 18 32 29 8 2)	100 Fm Fermium (257) (2 8 18 32 30 8 2)	101 Md Mendelevium (258) (2 8 18 32 31 8 2)	102 No Nobelium (259) (2 8 18 32 32 8 2)

Glossary

A

abiotic factor: physical, or nonliving, factor that shapes an ecosystem (58)
factor abiótico: factor físico, o inanimado, que da forma a un ecosistema

abscisic acid: plant hormone that inhibits cell division and, therefore, growth (589)
ácido abscísico: hormona vegetal que inhibe la división celular y, por ende, el crecimiento

acetylcholine: neurotransmitter that produces an impulse in a muscle cell (772)
acetilcolina: neurotransmisor que produce un impulso en una célula muscular

acid: compound that forms hydrogen ions (H^+) in solution; a solution with a pH of less than 7 (36)
ácido: compuesto que en una solución produce iones hidrógeno (H^+); una solución con un pH inferior a 7

acid rain: rain containing nitric and sulfuric acids (136)
lluvia ácida: lluvia que contiene ácido nítrico y ácido sulfúrico

actin: thin filament of protein found in muscles (771)
actina: microfilamento de proteína que se halla en los músculos

action potential: reversal of charges across the cell membrane of a neuron; also called a nerve impulse (744)
potencial de acción: inversión de las cargas a través de la membrana de una neurona; también llamado impulso nervioso

activation energy: energy that is needed to get a reaction started (43)
energía de activación: energía necesaria para que comience una reacción

active immunity: immunity that develops as a result of natural or deliberate exposure to an antigen (846)
inmunidad activa: inmunidad que se desarrolla a consecuencia de la exposición natural o deliberada a un antígeno

adaptation: heritable characteristic that increases an organism's ability to survive and reproduce in an environment (388)
adaptación: característica heredable que aumenta la capacidad de un organismo de sobrevivir y reproducirse en un medio ambiente

adaptive radiation: process by which a single species or a small group of species evolves into several different forms that live in different ways (459)
radiación adaptativa: proceso mediante el cual una sola especie o un grupo pequeño de especies evoluciona y da lugar a diferentes seres que viven de diversas maneras

adenosine triphosphate (ATP): compound used by cells to store and release energy (192)
trifosfato de adenosina (ATP): compuesto utilizado por las células para almacenar y liberar energía

adhesion: force of attraction between different kinds of molecules (34, 568)
adhesión: fuerza de atracción entre diferentes tipos de moléculas

aerobic: process that requires oxygen (214)
aeróbico: proceso que requiere oxígeno

age structure: number of males and females of each age in a population (109)
estructura etaria: número de machos y de hembras de cada edad en una población

aggression: threatening behavior that one animal uses to exert dominance over another animal (702)
agresión: comportamiento amenazador que emplea un animal para ejercer control sobre otro animal

algal bloom: increase in the amount of algae and other producers that results from a large input of a limiting nutrient (510)
florecimiento de algas: aumento de la cantidad de algas y otros productores debido a una gran entrada de un nutriente limitante

allele: one of a number of different forms of a gene (263)
alelo: cada una de las diversas formas de un gen

allele frequency: number of times that an allele occurs in a gene pool compared with the number of alleles in that pool for the same gene (406)
frecuencia alélica: número de veces que aparece un alelo en un caudal genético, comparado con la cantidad de alelos en ese caudal para el mismo gen

allergy: overreaction of the immune system to an antigen (849)
alergia: reacción exagerada del sistema inmune ante un antígeno

alternation of generations: life cycle that has two alternating phases—a haploid (N) phase and diploid (2N) phase (507, 530)
alternancia de generaciones: ciclo vital con dos fases que se alternan, una fase haploide (N) y una fase diploide (2N)

alveolus (pl. alveoli): one of many tiny air sacs at the end of a bronchiole in the lungs that provides surface area for gas exchange to occur (652, 797)
alvéolos: pequeños sacos, ubicados en las terminaciones de los bronquiolos pulmonares, que proporcionan una superficie en la que tiene lugar el intercambio gaseoso

amino acid: compound with an amino group on one end and a carboxyl group on the other end (40)
aminoácido: compuesto que contiene un grupo amino en un extremo y un grupo carboxilo en el otro extremo

Glossary *(continued)*

amniotic egg: egg composed of shell and membranes that creates a protected environment in which the embryo can develop out of water (682)
huevo amniota: huevo formado por una cáscara y membranas que crea un ambiente protegido en el cual el embrión puede desarrollarse en un medio seco

amylase: enzyme in saliva that breaks the chemical bonds in starches (724)
amilasa: enzima de la saliva que fragmenta los enlaces químicos de los almidones

anaerobic: process that does not require oxygen (214)
anaeróbico: proceso que no requiere oxígeno

analogous structures: body parts that share a common function, but not structure (393)
estructuras análogas: partes del cuerpo que tienen la misma función, mas no la misma estructura

anaphase: phase of mitosis in which the chromosomes separate and move to opposite ends of the cell (242)
anafase: fase de la mitosis en la cual los cromosomas se separan y se desplazan hacia los extremos opuestos de la célula

angiosperm: group of seed plants that bear their seeds within a layer of tissue that protects the seed; also called flowering plant (536)
angiospermas: grupo de plantas con semillas, que están protegidas con una capa de tejido. Se conocen también como plantas que florecen.

anther: flower structure in which pollen grains are produced (579)
antera: estructura de la flor en la cual se generan los granos de polen

anthropoid: primate group made up of monkeys, apes, and humans (634)
antropoide: grupo de primates constituido por monos, simios y humanos

antibiotic: group of drugs used to block the growth and reproduction of bacterial pathogens (490)
antibiótico: grupo de drogas utilizadas para bloquear el desarrollo y la reproducción de organismos patógenos bacterianos

antibody: protein that either attacks antigens directly or produces antigen-binding proteins (842)
anticuerpo: proteína que ataca directamente a los antígenos o produce proteínas que se unen a los antígenos

anticodon: group of three bases on a tRNA molecule that are complementary to the three bases of a codon of mRNA (313)
anticodón: grupo de tres bases en una molécula de ARN de transferencia que son complementarias a las tres bases de un codón de ARN mensajero

antigen: any substance that triggers an immune response (842)
antígeno: cualquier sustancia que provoca una respuesta inmune

aphotic zone: dark layer of the oceans below the photic zone where sunlight does not penetrate (96)
zona afótica: sección oscura de los océanos donde no penetra la luz solar, situada debajo de la zona fótica

apical dominance: phenomenon in which the closer a bud is to the stem's tip, the more its growth is inhibited (588)
dominancia apical: fenómeno por el cual cuanto más cerca de la punta del tallo está un brote, más se inhibe su crecimiento

apical meristem: group of unspecialized cells that divide to produce increased length of stems and roots (555)
meristemo apical: grupo de células no especializadas que se dividen para producir un aumento en la longitud de tallos y raíces

apoptosis: process of programmed cell death (246)
apoptosis: proceso de muerte celular programada

appendage: structure, such as a leg or antenna, that extends from the body wall (625)
apéndice: estructura, como una pierna o una antena, que se proyecta desde la superficie corporal

appendicular skeleton: the bones of the arms and legs along with the bones of the pelvis and shoulder area (766)
esqueleto apendicular: los huesos de los brazos y de las piernas junto con los huesos de la pelvis y del área de los hombros

aquaculture: raising of aquatic organisms for human consumption (146)
acuicultura: cría de organismos acuáticos para el consumo humano

aquaporin: water channel protein in a cell (177)
acuaporina: proteína que canaliza el agua en una célula

Archaea: domain consisting of unicellular prokaryotes that have cell walls that do not contain peptidoglycan; corresponds to the kingdom Archaebacteria (439)
Arqueas: dominio formado por procariotas unicelulares cuyas paredes celulares no contienen peptidoglicano; corresponden al reino de las Arqueabacterias

artery: large blood vessel that carries blood away from the heart to the tissues of the body (789)
arteria: vaso sanguíneo grande que transporta la sangre desde el corazón a los tejidos del cuerpo

artificial selection: selective breeding of plants and animals to promote the occurrence of desirable traits in offspring (387)
selección artificial: cría selectiva de plantas y animales para fomentar la ocurrencia de rasgos deseados en la progenie

asexual reproduction: process of reproduction involving a single parent that results in offspring that are genetically identical to the parent (15, 236)
reproducción asexual: proceso de reproducción que involucra a un único progenitor y da por resultado descendencia genéticamente idéntica a ese progenitor

asthma: chronic respiratory disease in which air passages narrow, causing wheezing, coughing, and difficulty breathing (849)
asma: enfermedad respiratoria crónica en la cual las vías respiratorias se estrechan, provocando jadeos, tos y dificultad para respirar

atherosclerosis: condition in which fatty deposits called plaque build up inside artery walls and eventually cause the arteries to stiffen (793)
arteriosclerosis o ateroesclerosis: enfermedad en la cual se acumulan depósitos de grasa llamados placas en el interior de las paredes arteriales que, con el tiempo, causan un endurecimiento de las arterias

atom: the basic unit of matter (28)
átomo: unidad básica de la materia

ATP synthase: cluster of proteins that span the cell membrane and allow hydrogen ions (H^+) to pass through it (199)
ATP sintasa: complejo de proteínas unidas a la membrana celular que permiten el paso de los iones de hidrógeno (H^+) a través de ella

atrium (pl. atria): upper chamber of the heart that receives blood from the rest of the body (654, 787)
aurícula: cavidad superior del corazón que recibe sangre del resto del cuerpo

autonomic nervous system: part of the peripheral nervous system that regulates activities that are involuntary, or not under conscious control; made up of the sympathetic and parasympathetic subdivisions (753)
sistema nervioso autónomo: parte del sistema nervioso periférico que regula las actividades involuntarias, o que son independientes de la conciencia; está compuesto por las subdivisiones simpática y parasimpática

autosome: chromosome that is not a sex chromosome; also called autosomal chromosome (335)
autosoma: cromosoma que no es un cromosoma sexual; también llamado cromosoma autosómico

autotroph: organism that is able to capture energy from sunlight or chemicals and use it to produce its own food from inorganic compounds; also called a producer (60, 194)
autótrofo: organismo capaz de atrapar la energía de la luz solar o de las sustancias químicas y utilizarla para producir su propio alimento a partir de compuestos inorgánicos; también llamado productor

auxin: regulatory substance produced in the tip of a growing plant that stimulates cell elongation and the growth of new roots (588)
auxina: sustancia reguladora producida en la punta de una planta en crecimiento que estimula el alargamiento celular y el crecimiento de raíces nuevas

axial skeleton: skeleton that supports the central axis of the body; consists of the skull, vertebral column, and the rib cage (766)
esqueleto axial: esqueleto que sostiene al eje central del cuerpo; consiste en el cráneo, la columna vertebral y la caja torácica

axon: long fiber that carries impulses away from the cell body of a neuron (743)
axón: fibra larga que lleva los impulsos desde el cuerpo celular de una neurona

B

bacillus (pl. bacilli): rod-shaped prokaryote (486)
bacilo: procariota con forma de bastón

background extinction: extinction caused by slow and steady process of natural selection (457)
extinción de fondo: extinción causada por un proceso lento y continuo de selección natural

Bacteria: domain of unicellular prokaryotes that have cell walls containing peptidoglycan; corresponds to the kingdom eubacteria (439)
Bacteria: pertenece al dominio de los unicelulares procariota cuyas paredes celulares contienen peptidoglicano; corresponde al reino de las Eubacterias

bacteriophage: kind of virus that infects bacteria (289, 481)
bacteriófago: clase de virus que infecta a las bacterias

bark: tissues that are found outside the vascular cambium, including the phloem, cork cambium, and cork (563)
corteza: tejidos que se hallan fuera del cámbium vascular, incluidos el floema, el cámbium suberoso y el corcho

base: compound that produces hydroxide ions (OH^-) in solution; solution with a pH of more than 7 (36)
base: compuesto que en una solución produce iones hidróxido (OH^-); una solución con un pH superior a 7

Glossary *(continued)*

base pairing: principle that bonds in DNA can form only between adenine and thymine and between guanine and cytosine (295)
apareamiento de bases: principio que establece que los enlaces en el ADN sólo pueden formarse entre adenina y timina y entre guanina y citocina

behavior: manner in which an organism reacts to changes in its internal condition or external environment (696)
comportamiento: manera en que un organismo reacciona a los cambios que ocurren en su condición interna o en el medio ambiente externo

behavioral isolation: form of reproductive isolation in which two populations develop differences in courtship rituals or other behaviors that prevent them from breeding (415)
aislamiento conductual: forma de aislamiento reproductivo en la cual dos poblaciones desarrollan diferencias en sus rituales de cortejo o en otros comportamientos que evitan que se apareen

benthos: organisms that live attached to or near the bottom of lakes, streams, or oceans (96)
bentos: organismos que viven adheridos al fondo, o cerca del fondo, de lagos, arroyos u océanos

bias: particular preference or point of view that is personal, rather than scientific (12)
parcialidad: preferencia especial o punto de vista que es personal en lugar de ser científico

bilateral symmetry: body plan in which a single imaginary line can divide the body into left and right sides that are mirror images of each other (611)
simetría bilateral: diseño corporal en el cual una línea imaginaria divide al cuerpo en dos lados, izquierdo y derecho, que son imágenes reflejas una del otra

binary fission: type of asexual reproduction in which an organism replicates its DNA and divides in half, producing two identical daughter cells (487)
fisión binaria: tipo de reproducción asexual en la cual un organismo replica su ADN, se divide por la mitad y produce dos células hijas idénticas

binocular vision: ability to merge visual images from both eyes, providing depth perception and a three-dimensional view of the world (634)
visión binocular: capacidad de fusionar las imágenes visuales provenientes de ambos ojos, lo cual proporciona una percepción profunda y una visión tridimensional del mundo

binomial nomenclature: classification system in which each species is assigned a two-part scientific name (428)
nomenclatura binaria: sistema de clasificación en el cual a cada especie se le asigna un nombre científico que consta de dos partes

biodiversity: total of the variety of organisms in the biosphere; also called biological diversity (138)
biodiversidad: totalidad de los distintos organismos que se hallan en la biósfera; también denominada diversidad biológica

biogeochemical cycle: process in which elements, chemical compounds, and other forms of matter are passed from one organism to another and from one part of the biosphere to another (68)
ciclo biogeoquímico: proceso en el cual los elementos, los compuestos químicos y otras formas de materia pasan de un organismo a otro y de una parte de la biósfera a otra

biogeography: study of past and present distribution of organisms (392)
biogeografía: estudio de la distribución pasada y presente de los organismos

bioinformatics: application of mathematics and computer science to store, retrieve, and analyze biological data (344)
bioinformática: aplicación de las matemáticas y de la informática para almacenar, recuperar y analizar información biológica

biological magnification: increasing concentration of a harmful substance in organisms at higher trophic levels in a food chain or food web (134)
bioacumulación: concentración creciente de sustancias perjudiciales en los organismos de los niveles tróficos más elevados de una cadena o red alimentaria

biology: scientific study of life (13)
biología: estudio científico de la vida

biomass: total amount of living tissue within a given trophic level (67)
biomasa: cantidad total de tejido vivo dentro de un nivel trófico dado

biome: a group of ecosystems that share similar climates and typical organisms (57)
bioma: un grupo de ecosistemas que comparten climas similares y organismos típicos

biosphere: part of Earth in which life exists including land, water, and air or atmosphere (17, 56)
biósfera: parte de la Tierra en la cual existe vida, y que incluye el suelo, el agua y el aire o atmósfera

biotechnology: process of manipulating organisms, cells, or molecules, to produce specific products (355)
biotecnología: proceso de manipular organismos, células o moléculas con el fin de obtener productos específicos

biotic factor: any living part of the environment with which an organism might interact (57)
factor biótico: cualquier parte viva del medio ambiente con la cual un organismo podría interaccionar

bipedal: term used to refer to two-foot locomotion (635)
bípedo: término utilizado para referirse a la locomoción sobre dos pies

blastocyst: stage of early development in mammals that consists of a hollow ball of cells (249, 825)
blastocisto: etapa temprana del desarrollo de los mamíferos que consiste en una bola hueca formada por una capa de células

blastula: hollow ball of cells that develops when a zygote undergoes a series of cell divisions (612)
blástula: esfera hueca de células que se desarrolla cuando un cigoto atraviesa una serie de divisiones celulares

bone marrow: soft tissue found in bone cavities (767)
médula ósea: tejido blando que se halla en las cavidades de los huesos

bottleneck effect: a change in allele frequency following a dramatic reduction in the size of a population (411)
efecto cuello de botella: un cambio en la frecuencia alélica que resulta cuando el tamaño de una población reduce drásticamente

Bowman's capsule: cuplike structure that encases the glomerulus; collects filtrate from the blood (731)
cápsula de Bowman: estructura en forma de taza que encierra al glomérulo; recoge los filtrados provenientes de la sangre

brain stem: structure that connects the brain and spinal cord; includes the medulla oblongata and the pons (749)
tronco cerebral: estructura que conecta al cerebro con la médula espinal; incluye el bulbo raquídeo y el puente de Varolio

bronchus (pl. bronchi): one of two large tubes in the chest cavity that leads from the trachea to the lungs (797)
bronquio: cada uno de los dos conductos largos ubicados en la cavidad torácica que parten desde la tráquea y llegan a los pulmones

bryophyte: group of plants that have specialized reproductive organs but lack vascular tissue; includes mosses and their relatives (533)
briofitas: grupo de plantas que tienen órganos reproductores especializados pero carecen de tejido vascular; incluyen a los musgos y sus congéneres

bud: plant structure containing apical meristem tissue that can produce new stems and leaves (560)
yema o gema: estructura de las plantas que contiene tejido del meristemo apical y puede producir nuevos tallos y hojas

buffer: compound that prevents sharp, sudden changes in pH (36)
solución amortiguadora: compuesto que evita cambios bruscos y repentinos en el pH

C

calcitonin: hormone produced by the thyroid that reduces blood calcium levels (815)
calcitonina: hormona producida por la tiroides que reduce los niveles de calcio en la sangre

Calorie: measure of heat energy in food; equivalent to 1000 calories (719)
Caloría: medida de la energía térmica de los alimentos, equivalente a 1000 calorías

calorie: amount of energy needed to raise the temperature of 1 gram of water by 1 degree Celsius (212)
caloría: cantidad de energía necesaria para elevar la temperatura de 1 gramo de agua en 1 grado Celsius

Calvin cycle: light-independent reactions of photosynthesis in which energy from ATP and NADPH is used to build high-energy compounds such as sugar (201)
ciclo de Calvin: reacciones de la fotosíntesis independientes de la luz en las cuales se utiliza la energía del ATP y del NADPH para elaborar compuestos con alto contenido energético, como el azúcar

cancer: disorder in which some of the body's cells lose the ability to control growth (247)
cáncer: enfermedad en la cual algunas de las células del cuerpo pierden la capacidad de controlar su crecimiento

canopy: dense covering formed by the leafy tops of tall rain-forest trees (92)
dosel forestal: cubierta densa formada por las copas de los árboles altos del bosque tropical

capillary: smallest blood vessel; brings nutrients and oxygen to the tissues and absorbs carbon dioxide and waste products (789)
capilar: más pequeño de los vaso sanguíneo más pequeño; lleva nutrientes y oxígeno a los tejidos y absorbe dióxido de carbono y productos de desecho

capillary action: tendency of water to rise in a thin tube (568)
capilaridad: tendencia del agua a ascender en un tubo delgado

capsid: protein coat surrounding a virus (481)
cápsida: cobertura de proteínas que rodea a un virus

carbohydrate: compound made up of carbon, hydrogen, and oxygen atoms; type of nutrient that is the major source of energy for the body (38, 720)
hidrato de carbono: compuesto formado por átomos de carbono, hidrógeno y oxígeno; tipo de nutriente que es la fuente principal de energía para el cuerpo

carnivore: organism that obtains energy by eating animals (62)
carnívoro: organismo que obtiene energía al comer otros animales

Glossary *(continued)*

carpel: innermost part of a flower that produces and shelters the female gametophytes (579)
carpelo: parte interna de una flor que produce y alberga los gametofitos femeninos

carrying capacity: largest number of individuals of a particular species that a particular environment can support (111)
capacidad de carga: mayor cantidad de individuos de una especie en particular que un medio ambiente específico puede mantener

cartilage: type of connective tissue that supports the body and is softer and more flexible than bone (628, 767)
cartílago: tipo de tejido conectivo que sostiene al cuerpo y es más blando y flexible que el hueso

Casparian strip: waterproof strip that surrounds plant endodermal cells and is involved in the one-way passage of materials into the vascular cylinder in plant roots (558)
banda de Caspary: banda impermeable que rodea a las células endodérmicas de las plantas y participa en el transporte unidireccional de las sustancias hacia el interior del cilindro vascular de las raíces de las plantas

catalyst: substance that speeds up the rate of a chemical reaction (44)
catalizador: sustancia que acelera la velocidad de una reacción química

cell: basic unit of all forms of life (160)
célula: unidad básica de todas las formas de vida

cell body: largest part of a typical neuron; contains the nucleus and much of the cytoplasm (743)
cuerpo celular: parte más grande de una neurona típica; que contiene el núcleo y gran parte del citoplasma

cell cycle: series of events in which a cell grows, prepares for division, and divides to form two daughter cells (240)
ciclo celular: serie de sucesos en los cuales una célula crece, se prepara para dividirse y se divide para formar dos células hijas

cell division: process by which a cell divides into two new daughter cells (236)
división celular: proceso por el cual una célula se divide en dos células hijas nuevas

cell membrane: thin, flexible barrier that surrounds all cells; regulates what enters and leaves the cell (162)
membrana celular: barrera flexible y delgada que rodea a todas las células; regula lo que entra y sale de la célula

cell theory: fundamental concept of biology that states that all living things are composed of cells; that cells are the basic units of structure and function in living things; and that new cells are produced from existing cells (160)
teoría celular: concepto fundamental de la Biología que establece que todos los seres vivos están compuestos por células; que las células son las unidades básicas estructurales y funcionales de los seres vivos; y que las células nuevas se producen a partir de células existentes

cell wall: strong, supporting layer around the cell membrane in some cells (171)
pared celular: capa resistente que sirve de sostén y está situada alrededor de la membrana celular de algunas células

cell-mediated immunity: immune response that defends the body against viruses, fungi, and abnormal cancer cells inside living cells (843)
inmunidad celular: respuesta inmune que desde las células defiende al cuerpo contra virus, hongos y células anormales cancerígenas

cellular respiration: process that releases energy by breaking down glucose and other food molecules in the presence of oxygen (213)
respiración celular: proceso que libera energía al descomponer la glucosa y otras moléculas de los alimentos en presencia de oxígeno

central nervous system: includes the brain and spinal cord; processes information and creates a response that it delivers to the body (742)
sistema nervioso central: incluye el cerebro y la médula espinal; procesa información y genera una respuesta que es enviada al cuerpo

centriole: structure in an animal cell that helps to organize cell division (167, 242)
centríolo: estructura de una célula animal que contribuye a organizar la división celular

centromere: region of a chromosome where the two sister chromatids attach (242)
centrómero: región de un cromosoma donde se unen las dos cromátidas hermanas

cephalization: concentration of sense organs and nerve cells at the anterior end of an animal (613)
cefalización: concentración de órganos sensoriales y células nerviosas en el extremo anterior de un animal

cerebellum: part of the brain that coordinates movement and controls balance (671, 749)
cerebelo: parte del encéfalo que coordina el movimiento y controla el equilibrio

cerebral cortex: outer layer of the cerebrum of a mammal's brain; center of thinking and other complex behaviors (748)
corteza cerebral: capa externa del cerebro de un mamífero; centro del raciocinio y otros comportamientos complejos

cerebrum: part of the brain responsible for voluntary activities of the body; "thinking" region of the brain (671, 748)
cerebro: parte del encéfalo responsable de las actividades voluntarias del cuerpo; región "pensante" del encéfalo

chemical digestion: process by which enzymes break down food into small molecules that the body can use (723)
digestión química: proceso por el cual las enzimas descomponen los alimentos en moléculas pequeñas que el cuerpo puede utilizar

chemical reaction: process that changes, or transforms, one set of chemicals into another set of chemicals (42)
reacción química: proceso que cambia, o transforma, un grupo de sustancias químicas en otro grupo de sustancias químicas

chemosynthesis: process in which chemical energy is used to produce carbohydrates (61)
quimiosíntesis: proceso en el cual la energía química se utiliza para producir hidratos de carbono

chitin: complex carbohydrate that makes up the cell walls of fungi; also found in the external skeletons of arthropods (514)
quitina: hidrato de carbono complejo que forma las paredes celulares de los hongos; también se halla en los esqueletos externos de los artrópodos

chlorophyll: principal pigment of plants and other photosynthetic organisms (195)
clorofila: pigmento fundamental de las plantas y de otros organismos fotosintéticos

chloroplast: organelle found in cells of plants and some other organisms that captures the energy from sunlight and converts it into chemical energy (170)
cloroplasto: orgánulo de las células de las plantas y de otros organismos que captura la energía de la luz solar y la convierte en energía química

chordate: animal that has, for at least one stage of its life, a dorsal, hollow nerve cord, a notochord, a tail that extends beyond the anus, and pharyngeal pouches (606)
cordado: animal que, al menos durante una etapa de su vida, tiene un cordón nervioso hueco y dorsal, un notocordio, una cola que se prolonga más allá del ano y bolsas faríngeas

chromatid: one of two identical "sister" parts of a duplicated chromosome (242)
cromátida: una de las dos partes "hermanas" idénticas de un cromosoma duplicado

chromatin: substance found in eukaryotic chromosomes that consists of DNA tightly coiled around histones (239)
cromatina: sustancia que se halla en los cromosomas eucarióticos y que consiste en ADN enrollado apretadamente alrededor de las histonas

chromosome: threadlike structure of DNA and protein that contains genetic information; in eukaryotes, chromosomes are found in the nucleus; in prokaryotes, they are found in the cytoplasm (239)
chromosoma: estructura larga de ADN y proteina, con foma de hilo, que posee información genética; en los eucariotas, los chromosomas están dentro del núcleo; en los procariotas, los cromosomas están en el citoplasma

chyme: mixture of enzymes and partially digested food (725)
quimo: mezcla de enzimas y alimentos parcialmente digeridos

cilium (pl. cilia): short hairlike projection that produces movement (506)
cilio: pequeña prolongación parecida a un pelo que produce movimiento

circadian rhythm: behavioral cycles that occur daily (701)
ritmo circadiano: ciclos conductuales que ocurren diariamente

clade: evolutionary branch of a cladogram that includes a single ancestor and all its descendants (433)
clado: rama evolutiva de un cladograma que incluye a un único ancestro y a todos sus descendientes

cladogram: diagram depicting patterns of shared characteristics among species (433)
cladograma: diagrama que representa patrones de características compartidas entre especies

class: in classification, a group of closely related orders (430)
clase: en la clasificación, un grupo de varios órdenes relacionados estrechamente

classical conditioning: type of learning that occurs when an animal makes a mental connection between a stimulus and some kind of reward or punishment (698)
condicionamiento clásico: tipo de aprendizaje que ocurre cuando un animal realiza una conexión mental entre un estímulo y algún tipo de recompensa o castigo

climate: average year-to-year conditions of temperature and precipitation in an area over a long period of time (82)
clima: promedio anual de las condiciones de temperatura y precipitación en un área durante un largo período de tiempo

clone: member of a population of genetically identical cells produced from a single cell (361)
clon: miembro de una población de células genéticamente idénticas producidas a partir de una célula única

closed circulatory system: type of circulatory system in which blood circulates entirely within blood vessels that extend throughout the body (653)
sistema circulatorio cerrado: tipo de sistema circulatorio en el cual la sangre circula completamente dentro de los vasos sanguíneos que se extienden por todo el cuerpo

coccus (pl. cocci): spherical prokaryote (486)
coco: procariota de forma esférica

cochlea: fluid-filled part of inner ear; contains nerve cells that detect sound (755)
cóclea: parte del oído interno llena de fluidos; contiene las células nerviosas que detectan el sonido

codominance: situation in which the phenotypes produced by both alleles are completely expressed (271)

Glossary *(continued)*

codon: group of three nucleotide bases in mRNA that specify a particular amino acid to be incorporated into a protein (311)
codón: grupo de tres bases de nucleótidos en el RNA mensajero que especifican la incorporación de un aminoácido en particular en una proteína

coelom: body cavity lined with mesoderm (612)
celoma: cavidad corporal revestida de mesodermo

coevolution: process by which two species evolve in response to changes in each other over time (460)
coevolución: proceso por el cual dos especies evolucionan en respuesta a cambios mutuos en el transcurso del tiempo

cohesion: attraction between molecules of the same substance (34)
cohesión: atracción entre moléculas de la misma sustancia

collenchyma: in plants, type of ground tissue that has strong, flexible cell walls; helps support larger plants (553)
colénquima: en las plantas, tipo de tejido fundamental que tiene paredes celulares fuertes y flexibles; contribuye a sostener las plantas más grandes

commensalism: symbiotic relationship in which one organism benefits and the other is neither helped nor harmed (87)
comensalismo: relación simbiótica en la cual un organismo se beneficia y el otro ni se beneficia ni sufre daño

communication: passing of information from one organism to another (703)
comunicación: traspaso de información desde un organismo a otro

community: assemblage of different populations that live together in a defined area (56)
comunidad: conjunto de varias poblaciones que viven juntas en un área definida

companion cell: in plants, phloem cell that surrounds sieve tube elements (554)
célula anexa: en las plantas, célula del floema que rodea a los vasos cribosos

competitive exclusion principle: principle that states that no two species can occupy the same niche in the same habitat at the same time (86)
principio de exclusión competitiva: principio que afirma que dos especies no pueden ocupar el mismo nicho en el mismo hábitat al mismo tiempo

compound: substance formed by the chemical combination of two or more elements in definite proportions (30)
compuesto: sustancia formada por la combinación química de dos o más elementos en proporciones definidas

cone: in the eye, photoreceptor that responds to light of different colors, producing color vision (757)
cono: en el ojo, receptor de luz que responde a la luz de diferentes colores, produciendo la visión a color

coniferous: term used to refer to trees that produce seed-bearing cones and have thin leaves shaped like needles (93)
coníferas: término utilizado para referirse a los árboles que producen conos portadores de semillas y que tienen hojas delgadas con forma de aguja

conjugation: process in which paramecia and some prokaryotes exchange genetic information (487, 507)
conjugación: proceso mediante el cual los paramecios y algunos procariotas intercambian información genética

connective tissue: type of tissue that provides support for the body and connects its parts (714)
tejido conectivo: tipo de tejido que proporciona sostén al cuerpo y conecta sus partes

consumer: organism that relies on other organisms for its energy and food supply; also called a heterotroph (61)
consumidor: organismo que depende de otros organismos para obtener su energía y su provisión de alimentos; también llamado heterótrofo

control group: group in an experiment that is exposed to the same conditions as the experimental group except for one independent variable (6)
grupo de control: en un experimento, grupo que está expuesto a las mismas condiciones que el grupo experimental, excepto por una variable independiente

controlled experiment: experiment in which only one variable is changed (6)
experimento controlado: experimento en el cual sólo se cambia una variable

convergent evolution: process by which unrelated organisms independently evolve similarities when adapting to similar environments (460)
evolución convergente: proceso mediante el cual organismos no relacionados evolucionan independientemente hacia caracteres similares cuando se adaptan a ambientes parecidos

cork cambium: meristematic tissue that produces the outer covering of stems during secondary growth of a plant (562)
cámbium suberoso: tejido del meristemo que produce la cubierta exterior de los tallos durante el crecimiento secundario de una planta

cornea: tough transparent layer of the eye through which light enters (756)
córnea: membrana dura y transparente del ojo a través de la cual entra la luz

corpus luteum: name given to a follicle after ovulation because of its yellow color (821)
cuerpo lúteo: nombre dado a un folículo después de la ovulación debido a su color amarillo

cortex: in plants, region of ground tissue just inside the root through which water and minerals move (557)
corteza radicular: en las plantas, región de tejido fundamental situada en el interior de la raíz a través de la cual pasan el agua y los minerales

corticosteroid: steroid hormone produced by the adrenal cortex (814)
corticosteroide o corticoide: hormona esteroídica producida por la corteza de las glándulas adrenales

cotyledon: first leaf or first pair of leaves produced by the embryo of a seed plant (542)
cotiledón: primera hoja o primer par de hojas producidas por el embrión de una planta fanerógama

courtship: type of behavior in which an animal sends out stimuli in order to attract a member of the opposite sex (702)
cortejo: tipo de comportamiento en el cual un animal emite estímulos para atraer a un miembro del sexo opuesto

covalent bond: type of bond between atoms in which the electrons are shared (31)
enlace covalente: tipo de enlace entre átomos en el cual se comparten los electrones

crossing-over: process in which homologous chromosomes exchange portions of their chromatids during meiosis (276)
entrecruzamiento: proceso por el cual los cromosomas homólogos intercambian partes de sus cromátidas durante la meiosis

cyclin: one of a family of proteins that regulates the cell cycle in eukaryotic cells (246)
ciclina: un componente de la familia de proteínas que regulan el ciclo celular de las células eucariotas

cytokinesis: division of the cytoplasm to form two separate daughter cells (241)
citocinesis: división del citoplasma para formar dos células hijas separadas

cytokinin: plant hormone produced in growing roots and in developing fruits and seeds (589)
citoquinina: hormona vegetal que se genera en las raíces en crecimiento y en los frutos y semillas en desarrollo

cytoplasm: in eukaryotic cells, all cellular contents outside the nucleus, in prokaryotic cells, all of the cells' contents (164)
citoplasma: en una célula eucariota, todo el contenido cellular fuera del núcleo; en las células procariotas, todo el contenido de las células

cytoskeleton: network of protein filaments in a eukaryotic cell that gives the cell its shape and internal organization and is involved in movement (167)
citoesqueleto: en una célula eucariota, red de filamentos proteínicos que otorga a la célula su forma y su organización interna y participa en el movimiento

D

data: evidence; information gathered from observations (6)
datos: evidencia; información reunida a partir de observaciones

deciduous: term used to refer to a type of tree that sheds its leaves during a particular season each year (92)
caduco: término utilizado para referirse a un tipo de árbol que pierde sus hojas cada año durante una estación en particular

decomposer: organism that breaks down and obtains energy from dead organic matter (62)
descomponedor: organismo que descompone y obtiene energía de la materia orgánica muerta

deforestation: destruction of forests (133)
deforestación: destrucción de los bosques

demographic transition: change in a population from high birth and death rates to low birth and death rates (118)
transición demográfica: en una población, cambio de índices de nacimiento y mortalidad altos a índices de nacimiento y mortalidad bajos

demography: scientific study of human populations (118)
demografía: estudio científico de las poblaciones humanas

dendrite: extension of the cell body of a neuron that carries impulses from the environment or from other neurons toward the cell body (743)
dendrita: prolongación del cuerpo celular de una neurona que transporta impulsos desde el medio ambiente o desde otras neuronas hacia el cuerpo celular

denitrification: process by which bacteria convert nitrates into nitrogen gas (71)
desnitrificación: proceso por el cual las bacterias del suelo convierten los nitratos en gas nitrógeno

density-dependent limiting factor: limiting factor that depends on population density (112)
factor limitante dependiente de la densidad: factor limitante que depende de la densidad de la población

density-independent limiting factor: limiting factor that affects all populations in similar ways, regardless of the population density (114)
factor limitante independiente de la densidad: factor limitante que afecta a todas las poblaciones de manera similar, sin importar la densidad de la población

deoxyribonucleic acid (DNA): genetic material that organisms inherit from their parents (14)
ácido desoxirribonucleico (ADN): material genético que los organismos heredan de sus padres

dependent variable: variable that is observed and that changes in response to the independent variable; also called the responding variable (6)
variable dependiente: variable que está siendo observada y cambia en respuesta a la variable independiente; también llamada variable de respuesta

derived character: trait that appears in recent parts of a lineage, but not in its older members (434)
carácter derivado: rasgo que aparece en los descendientes recientes de un linaje, pero no en sus miembros más viejos

dermis: layer of skin found beneath the epidermis (778)
dermis: capa de la piel situada debajo de la epidermis

desertification: lower land productivity caused by overfarming, overgrazing, seasonal drought, and climate change (132)
desertificación: disminución de la productividad de la tierra debido al cultivo y al pastoreo excesivo, a la sequía estacional y al cambio climático

detritivore: organism that feeds on plant and animal remains and other dead matter (62)
detritívoro: organismo que se alimenta de restos animales y vegetales y demás materia orgánica muerta

deuterostome: group of animals in which the blastopore becomes an anus, and the mouth is formed from the second opening that develops (613)
deuteróstomos: grupo de animales en los cuales el blastoporo se convierte en ano y la boca se forma a partir del desarrollo de una segunda abertura

diaphragm: large flat muscle at the bottom of the chest cavity that helps with breathing (799)
diafragma: músculo plano y grande ubicado en la parte inferior de la cavidad torácica que participa en la respiración

dicot: angiosperm with two seed leaves in its ovary (542)
dicotiledónea: angiosperma con dos cotiledones (hojas embrionarias) en su ovario

differentiation: process in which cells become specialized in structure and function (248, 323)
diferenciación: proceso en el cual las células se especializan en estructura y función

diffusion: process by which particles tend to move from an area where they are more concentrated to an area where they are less concentrated (176)
difusión: proceso por el cual las partículas tienden a desplazarse desde un área donde están más concentradas hacia un área donde están menos concentradas

digestive tract: tube that begins at the mouth and ends at the anus (648)
tracto digestivo: tubo que comienza en la boca y termina en el ano

diploid: term used to refer to a cell that contains two sets of homologous chromosomes (275)
diploide: término utilizado para referirse a una célula que contiene dos series de cromosomas homólogos

directional selection: form of natural selection in which individuals at one end of a distribution curve have higher fitness than individuals in the middle or at the other end of the curve (410)
selección direccional: forma de selección natural en la cual los individuos que se hallan en un extremo de la curva de distribución poseen una mayor capacidad de adaptación que los individuos que se hallan en el centro o en el otro extremo de la curva

disruptive selection: natural selection in which individuals at the upper and lower ends of the curve have higher fitness than individuals near the middle of the curve (410)
selección disruptiva: forma de selección natural en la cual los individuos que se hallan en los extremos superior e inferior de la curva poseen una mayor capacidad de adaptación que los individuos que se hallan cerca del centro de la curva

DNA fingerprinting: tool used by biologists that analyzes an individual's unique collection of DNA restriction fragments; used to determine whether two samples of genetic material are from the same person (365)
prueba de ADN: herramienta utilizada por los biólogos mediante la cual se analiza el conjunto de los fragmentos de restricción de ADN exclusivo de cada individuo; utilizada para determinar si dos muestras de material genético pertenecen a la misma persona; también llamada huella genética o análisis de ADN

DNA microarray: glass slide or silicon chip that carries thousands of different kinds of single-stranded DNA fragments arranged in a grid. A DNA microarray is used to detect and measure the expression of thousands of genes at one time (364)
chip de ADN: superficie de vidrio o chip de silicona que contiene miles de diferentes tipos de fragmentos de ADN de una sola cadena dispuestos en una cuadrícula. Un chip de ADN se utiliza para detectar y medir la expresión de miles de genes a la vez

DNA polymerase: principal enzyme involved in DNA replication (297)
ADN polimerasa: enzima fundamental involucrada en la replicación del ADN

domain: larger, more inclusive taxonomic category than a kingdom (438)
dominio: categoría taxonómica más amplia e inclusiva que un reino

dopamine: neurotransmitter that is associated with the brain's pleasure and reward centers (750)
dopamina: neurotransmisor que está asociado con los centros de placer y de recompensa del cerebro

dormancy: period of time during which a plant embryo is alive but not growing (586)
latencia: período de tiempo durante el cual un embrión vegetal está vivo pero no crece

double fertilization: process of fertilization in angiosperms in which the first event produces the zygote, and the second, the endosperm within the seed (581)
doble fertilización: proceso de fecundación de las angiospermas en el cual se produce, en el primer suceso el cigoto y en el segundo, el endospermo dentro de la semilla

E

ecological footprint: total amount of functioning ecosystem needed both to provide the resources a human population uses and to absorb the wastes that population generates (143)
huella ecológica: cantidad total de ecosistema en funcionamiento necesaria para proporcionar los recursos que utiliza una población humana y para absorber los residuos que genera esa población

ecological hot spot: small geographic area where significant numbers of habitats and species are in immediate danger of extinction (142)
zona de conflicto ecológico: área geográfica pequeña donde cantidades importantes de hábitats y especies se hallan en peligro de extinción inmediato

ecological pyramid: illustration of the relative amounts of energy or matter contained within each trophic level in a given food chain or food web (66)
pirámide ecológica: ilustración de las cantidades relativas de energía o materia contenidas dentro de cada nivel trófico en una cadena o red alimenticia dada

ecological succession: series of gradual changes that occur in a community following a disturbance (88)
sucesión ecológica: serie de cambios graduales que ocurren en una comunidad después de una alteración

ecology: scientific study of interactions among organisms and between organisms and their environment (56)
ecología: estudio científico de las interacciones entre organismos y entre los organismos y su medio ambiente

ecosystem: all the organisms that live in a place, together with their nonliving environment (57)
ecosistema: todos los organismos que viven en un lugar, junto con su medio ambiente inanimado

ecosystem diversity: variety of habitats, communities, and ecological processes in the biosphere (138)
diversidad de ecosistemas: variedad de hábitats, comunidades y procesos ecológicos que existen en la biósfera

ectoderm: outermost germ layer; produces sense organs, nerves, and outer layer of skin (612)
ectodermo: capa embrionaria más externa; desarrolla órganos sensoriales, nervios y la capa exterior de la piel

ectotherm: animal whose body temperature is determined by the temperature of its environment (685)
animal de sangre fría: animal cuya temperatura corporal está determinada por la temperatura de su medio ambiente

electron: negatively charged particle; located in the space surrounding the nucleus (28)
electrón: partícula con carga negativa; ubicada en el espacio que rodea al núcleo

electron transport chain: series of electron carrier proteins that shuttle high-energy electrons during ATP-generating reactions (199)
cadena de transporte de electrones: serie de proteínas transportadoras que llevan electrones de alta energía, durante las reacciones generadoras de ATP

element: pure substance that consists entirely of one type of atom (29)
elemento: sustancia pura que consiste íntegramente en un tipo de átomo

embryo: developing stage of a multicellular organism (248)
embrión: una de las etapas de desarrollo de un organismo multicelular

embryo sac: female gametophyte within the ovule of a flowering plant (580)
saco embrionario: gametofito femenino dentro del óvulo de una planta que produce flores

emerging disease: disease that appears in the population for the first time, or an old disease that suddenly becomes harder to control (492)
enfermedad emergente: enfermedad que aparece en una población por primera vez o una enfermedad antigua que de pronto se vuelve más difícil de controlar

emigration: movement of individuals out of an area (109)
emigración: desplazamiento de individuos fuera de un área

endocrine gland: gland that releases its secretions (hormones) directly into the blood, which transports the secretions to other areas of the body (685, 810)
glándula endocrina: glándula que vierte sus secreciones (hormonas) directamente en la sangre, para ser transportadas a otras áreas del cuerpo

endoderm: innermost germ layer; develops into the linings of the digestive tract and much of the respiratory system (612)
endodermo: capa embrionaria más interna, a partir de la cual se desarrollan los revestimientos del tracto digestivo y gran parte del sistema respiratorio

endodermis: in plants, layer of ground tissue that completely encloses the vascular cylinder (557)
endodermis: en las plantas, un capa de tejido fundamental que envuelve completamente al cilindro vascular

endoplasmic reticulum: internal membrane system found in eukaryotic cells; place where lipid components of the cell membrane are assembled (168)
retículo endoplasmático: sistema de membranas internas de las células eucariotas; lugar donde se reúnen los componentes lipídicos de la membrana celular

endoskeleton: internal skeleton; structural support system within the body of an animal (675)
endoesqueleto: esqueleto interno; sistema estructural de sostén dentro del cuerpo de un animal

endosperm: food-rich tissue that nourishes a seedling as it grows (581)
endospermo: tejido nutritivo que alimenta a una plántula a medida que crece

endospore: structure produced by prokaryotes in unfavorable conditions; a thick internal wall that encloses the DNA and a portion of the cytoplasm (487)
endospora: estructura producida por los procariotas en condiciones desfavorables; una gruesa pared interna que encierra al ADN y a una parte del citoplasma

endosymbiotic theory: theory that proposes that eukaryotic cells formed from a symbiotic relationship among several different prokaryotic cells (464)
teoría endosimbiótica: teoría que propone que las células eucariotas se formaron a partir de una relación simbiótica entre varias células procariotas distintas

endotherm: animal whose body temperature is regulated, at least in part, using heat generated within its body (686)
endotermo: animal cuya temperatura corporal se regula, al menos en parte, utilizando el calor generado dentro de su cuerpo

enzyme: protein catalyst that speeds up the rate of specific biological reactions (44)
enzima: proteína catalizadora que acelera la velocidad de reacciones biológicas específicas

epidermis: in plants, single layer of cells that makes up dermal tissue (553); in humans, the outer layer of the skin (778)
epidermis: en las plantas, única capa de células que forma el tejido dérmico; en los seres humanos, la capa exterior de la piel

epididymis: organ in the male reproductive system in which sperm mature and are stored (818)
epidídimo: órgano del sistema reproductor masculino en el cual el esperma madura y se almacena

epinephrine: hormone released by the adrenal glands that increases heart rate and blood pressure and prepares the body for intense physical activity; also called adrenaline (814)
epinefrina: hormona liberada por las glándulas adrenales que aumenta la frecuencia cardíaca y la presión sanguínea y prepara al cuerpo para una actividad física intensa; también llamada adrenalina

epithelial tissue: type of tissue that lines the interior and exterior body surfaces (714)
tejido epitelial: tipo de tejido que reviste el interior y el exterior de las superficies del cuerpo

era: major division of geologic time; usually divided into two or more periods (454)
era: división principal del tiempo geológico; usualmente dividida en dos o más períodos

esophagus: tube connecting the mouth to the stomach (724)
esófago: tubo que conecta la boca con el estómago

estuary: kind of wetland formed where a river meets the ocean (98)
estuario: tipo de humedal que se forma donde un río se une al océano

ethylene: plant hormone that stimulates fruits to ripen (589)
etileno: hormona vegetal que estimula la maduración de los frutos

Eukarya: domain consisting of all organisms that have a nucleus; includes protists, plants, fungi, and animals (440)
***Eukarya* (eucariontes):** dominio compuesto por todos los organismos que tienen un núcleo; incluye a los protistas, las plantas, los hongos y los animales

eukaryote: organism whose cells contain a nucleus (162)
eucariota: organismo cuyas células contienen un núcleo

evolution: change over time; the process by which modern organisms have descended from ancient organisms (380)
evolución: cambio en el transcurso del tiempo; el proceso por el cual los organismos actuales se derivaron de los organismos antiguos

excretion: process by which metabolic wastes are eliminated from the body (656, 729)
excreción: proceso por el cual se eliminan del cuerpo los residuos metabólicos

exocrine gland: gland that releases its secretions, through tubelike structures called ducts, directly into an organ or out of the body (810)
glándula exocrina: glándula que vierte sus secreciones directamente a un órgano o al exterior del cuerpo a través de estructuras tubulares denominadas conductos

exon: expressed sequence of DNA; codes for a protein (310)
exón: secuencia expresada de ADN; codifica una porción específica de una proteína

exoskeleton: external skeleton; tough external covering that protects and supports the body of many invertebrates (674)
exoesqueleto: esqueleto externo; cubierta externa dura que protege y sostiene el cuerpo de muchos invertebrados

exponential growth: growth pattern in which the individuals in a population reproduce at a constant rate (109)
crecimiento exponencial: patrón de crecimiento en el cual los individuos de una población se reproducen a una tasa constante

extinct: term used to refer to a species that has died out and has no living members (450)
extinto: término utilizado para referirse a una especie que ha desaparecido y de la que ninguno de sus miembros está vivo

extracellular digestion: type of digestion in which food is broken down outside the cells in a digestive system and then absorbed (647)
digestión extracelular: tipo de digestión en la cual el alimento es degradado fuera de las células dentro de un sistema digestivo y luego se absorbe

F

facilitated diffusion: process of diffusion in which molecules pass across the membrane through cell membrane channels (177)
difusión facilitada: proceso de difusión en el cual las moléculas atraviesan la membrana a través de los canales de la membrana celular

family: in classification, group of similar genera (430)
familia: en la clasificación, grupo de géneros similares

fat: lipid; made up of fatty acids and glycerol; type of nutrient that protects body organs, insulates the body, and stores energy (721)
grasa: lípido; compuesto de ácidos grasos y glicerina; tipo de nutriente que protege a los órganos del cuerpo, actúa como aislante térmico y almacena energía

feedback inhibition: process in which a stimulus produces a response that opposes the original stimulus; also called negative feedback (608, 717)
inhibición de la retroalimentación: proceso en el cual un estímulo produce una respuesta que se opone al estímulo original; también llamada retroalimentación negativa

fermentation: process by which cells release energy in the absence of oxygen (223)
fermentación: proceso por el cual las células liberan energía en ausencia de oxígeno

fertilization: process in sexual reproduction in which male and female reproductive cells join to form a new cell (262)
fecundación: proceso de la reproducción sexual en el cual las células reproductoras masculinas y femeninas se unen para formar una célula nueva

fetus: a human embryo after eight weeks of development (827)
feto: un embrión humano después de ocho semanas de desarrollo

fever: increased body temperature that occurs in response to infection (842)
fiebre: temperatura corporal elevada que se produce como respuesta a una infección

filtration: process of passing a liquid or gas through a filter to remove wastes (731)
filtración: proceso de hacer pasar un líquido o un gas a través de un filtro para quitar los residuos

fitness: how well an organism can survive and reproduce in its environment (388)
aptitud: capacidad de un organismo para sobrevivir y reproducirse en su medio ambiente

flagellum (pl. flagella): structure used by protists for movement; produces movement in a wavelike motion (506)
flagelo: estructura utilizada por los protistas para desplazarse; produce un desplazamiento con un movimiento semejante al de una onda

food chain: series of steps in an ecosystem in which organisms transfer energy by eating and being eaten (63)
cadena alimenticia: serie de pasos en un ecosistema, en que los organismos transfieren energía al alimentarse y al servir de alimento

food vacuole: small cavity in the cytoplasm of a protist that temporarily stores food (510)
vacuola alimenticia: pequeña cavidad situada en el citoplasma de los protistas que almacena alimentos por algún tiempo

food web: network of complex interactions formed by the feeding relationships among the various organisms in an ecosystem (63)
red alimenticia: red de interacciones complejas constituida por las relaciones alimenticias entre los varios organismos de un ecosistema

forensics: scientific study of crime scene evidence (365)
ciencias forenses: estudio científico de las pruebas en la escena del crimen

fossil: preserved remains or traces of ancient organisms (382)
fósil: restos conservados o vestigios de organismos antiguos

founder effect: change in allele frequencies as a result of the migration of a small subgroup of a population (411)
efecto fundador: cambio en las frecuencias alélicas como consecuencia de la migración de un subgrupo pequeño de una población

frameshift mutation: mutation that shifts the "reading frame" of the genetic message by inserting or deleting a nucleotide (317)
mutación de corrimiento de estructura: mutación que cambia el "marco de lectura" del mensaje genético insertando o eliminando un nucleótido

fruit: structure in angiosperms that contains one or more matured ovaries (541)
fruto: estructura de las Angiospermas que contiene uno o más ovarios maduros

fruiting body: reproductive structure of a fungus that grows from the mycelium (515)
cuerpo fructífero: estructura reproductora de los hongos que se desarrolla a partir del micelio

Glossary *(continued)*

G

gamete: sex cell (264)
  gameto: célula sexual

gametophyte: gamete-producing plant; multicellular haploid phase of a plant life cycle (530)
  gametofito: planta que produce gametos; fase haploide multicelular del ciclo vital de una planta

ganglion (pl. ganglia): group of interneurons (670)
  ganglio nervioso: grupo de interneuronas

gastrovascular cavity: digestive chamber with a single opening (647)
  cavidad gastrovascular: cámara digestiva con una sola apertura

gastrulation: process of cell migration that results in the formation of the three cell layers—the ectoderm, the mesoderm, and the endoderm (826)
  gastrulación: proceso de migración celular que da por resultado la formación de las tres capas celulares—el ectodermo, el mesodermo y el endodermo

gel electrophoresis: procedure used to separate and analyze DNA fragments by placing a mixture of DNA fragments at one end of a porous gel and applying an electrical voltage to the gel (342)
  electroforesis en gel: procedimiento utilizado para separar y analizar fragmentos de ADN colocando una mezcla de fragmentos de ADN en un extremo de un gel poroso y aplicando al gel un voltaje eléctrico

gene: sequence of DNA that codes for a protein and thus determines a trait; factor that is passed from parent to offspring (263)
  gen: secuencia de ADN que contiene el código de una proteína y por lo tanto determina un rasgo; factor que se transmite de un progenitor a su descendencia

gene expression: process by which a gene produces its product and the product carries out its function (314)
  expresión génica: proceso por el cual un gen produce su producto y el producto lleva a cabo su función

gene pool: all the genes, including all the different alleles for each gene, that are present in a population at any one time (406)
  caudal de genes: todos los genes, incluidos todos los alelos diferentes para cada gen, que están presentes en una población en un momento dado

gene therapy: process of changing a gene to treat a medical disease or disorder. An absent or faulty gene is replaced by a normal working gene. (364)
  terapia genética o génica: proceso en el cual se cambia un gen para tratar una enfermedad o una afección médica. Se reemplaza un gen ausente o defectuoso con un gen de funcionamiento normal.

genetic code: collection of codons of mRNA, each of which directs the incorporation of a particular amino acid into a protein during protein synthesis (311)
  código genético: conjunto de codones del ARN mensajero, cada uno de los cuales dirige la incorporación de un aminoácido en particular a una proteína durante la síntesis proteica

genetic diversity: sum total of all the different forms of genetic information carried by a particular species, or by all organisms on Earth (138)
  diversidad genética: suma de todas las distintas formas de información genética portadas por una especie en particular, o por todos los organismos de la Tierra

genetic drift: random change in allele frequency caused by a series of chance occurrences that cause an allele to become more or less common in a population (411)
  tendencia genética: alteración al azar de la frecuencia alélica causada por una serie de acontecimientos aleatorios que hacen que un alelo se vuelva más o menos común en una población

genetic equilibrium: situation in which allele frequencies in a population remain the same (412)
  equilibrio genético: situación en la cual las frecuencias alélicas de una población se mantienen iguales

genetic marker: alleles that produce detectable phenotypic differences useful in genetic analysis (360)
  marcador genético: alelos que producen diferencias fenotípicas detectables, útiles en el análisis genético

genetics: scientific study of heredity (262)
  genética: estudio científico de la herencia

genome: entire set of genetic information that an organism carries in its DNA (334)
  genoma: todo el conjunto de información genética que un organismo transporta en su ADN

genomics: study of whole genomes, including genes and their functions (344)
  genómica: estudio integral de los genomas, incluyendo los genes y sus funciones

genotype: genetic makeup of an organism (267)
  genotipo: composición genética de un organismo

genus: group of closely related species; the first part of the scientific name in binomial nomenclature (428)
  género: grupo de especies relacionadas estrechamente; la primera parte del nombre científico en la nomenclatura binaria

geographic isolation: form of reproductive isolation in which two populations are separated by geographic barriers such as rivers, mountains, or bodies of water, leading to the formation of two separate subspecies (414)

aislamiento geográfico: forma de aislamiento reproductivo en el cual dos poblaciones están separadas por barreras geográficas como ríos, montañas o masas de agua, dando lugar a la formación de dos subespecies distintas

geologic time scale: time line used to represent Earth's history (453)

escala de tiempo geológico: línea cronológica utilizada para representar la historia de la Tierra

germ theory of disease: idea that infectious diseases are caused by microorganisms (838)

teoría microbiana de la enfermedad: idea de que las enfermedades infecciosas son causadas por microorganismos

germination: resumption of growth of the plant embryo following dormancy (586)

germinación: reanudación del crecimiento del embrión de la planta después de la latencia

giberellin: plant hormone that stimulates growth and may cause dramatic increases in size (589)

giberelina: hormona de las plantas que estimula el crecimiento y puede causar aumentos significativos de tamaño

gill: feathery structure specialized for the exchange of gases with water (651)

branquia: estructura tegumentaria especializada en el intercambio de los gases con el agua

global warming: increase in the average temperatures on Earth (147)

calentamiento global: aumento del promedio de temperatura en la tierra

glomerulus: small network of capillaries encased in the upper end of the nephron; where filtration of the blood takes place (731)

glomérulo: pequeña red de capilares encerrados en el extremo superior del nefrón; donde tiene lugar la filtración de la sangre

glycolysis: first set of reactions in cellular respiration in which a molecule of glucose is broken into two molecules of pyruvic acid (216)

glicólisis: primer conjunto de reacciones en la respiración celular, en las cuales una molécula de glucosa se descompone en dos moléculas de ácido pirúvico

Golgi apparatus: organelle in cells that modifies, sorts, and packages proteins and other materials from the endoplasmic reticulum for storage in the cell or release outside the cell (169)

aparato de Golgi: orgánulo de las células que modifica, clasifica y agrupa las proteínas y otras sustancias provenientes del retículo endoplasmático para almacenarlas en la célula o enviarlas fuera de la célula

gradualism: the evolution of a species by gradual accumulation of small genetic changes over long periods of time (458)

gradualismo: evolución de una especie por la acumulación gradual de pequeños cambios genéticos ocurridos en el transcurso de largos períodos de tiempo

grafting: method of propagation used to reproduce seedless plants and varieties of woody plants that cannot be propagated from cuttings (584)

injerto: método de propagación utilizado para reproducir plantas sin semillas y algunas variedades de plantas leñosas que no pueden propagarse a partir de esquejes

gravitropism: response of a plant to the force of gravity (590)

geotropismo: respuesta de una planta a la fuerza de la gravedad

green revolution: development of highly productive crop strains and use of modern agriculture techniques to increase yields of food crops (594)

revolución verde: el desarrollo de variedades de cultivos altamente productivos y el uso de técnicas agrícolas modernas para aumentar el rendimiento de los cultivos

greenhouse effect: process in which certain gases (carbon dioxide, methane, and water vapor) trap sunlight energy in Earth's atmosphere as heat (83)

efecto invernadero: proceso mediante el cual ciertos gases (dióxido de carbono, metano y vapor de agua) atrapan la energía de la luz solar en la atmósfera terrestre en forma de calor

growth factor: one of a group of external regulatory proteins that stimulate the growth and division of cells (246)

factor de crecimiento: una de las proteínas del grupo de proteínas reguladoras externas que estimulan el crecimiento y la división de las células

guard cell: specialized cell in the epidermis of plants that controls the opening and closing of stomata (565)

célula de guarda (o célula oclusiva): célula especializada de la epidermis vegetal que controla la apertura y el cierre de los estomas

gullet: indentation in one side of a ciliate that allows food to enter the cell (510)

citofaringe: hendidura a un costado de un ciliado que permite que los alimentos entren a la célula

gymnosperm: group of seed plants that bear their seeds directly on the scales of cones (536)

Gimnospermas: grupo de plantas fanerógamas que tienen sus semillas directamente sobre las escamas de los conos

H

habitat: area where an organism lives, including the biotic and abiotic factors that affect it (85)

hábitat: área donde vive un organismo, incluidos los factores bióticos y abióticos que lo afectan

Glossary *(continued)*

habitat fragmentation: splitting of ecosystems into pieces (140)
fragmentación del hábitat: la ruptura, o separación en partes, de los ecosistemas

habituation: type of learning in which an animal decreases or stops its response to a repetitive stimulus that neither rewards nor harms the animal (697)
habituación: tipo de aprendizaje en el cual un animal disminuye o cancela su respuesta ante un estímulo repetido que no recompensa ni castiga al animal

hair follicle: tubelike pockets of epidermal cells that extend into the dermis; cells at the base of hair follicles produce hair (776)
folículo piloso: sacos tubulares de las células epidérmicas que se prolongan hacia el interior de la dermis; las células situadas en la base de los folículos pilosos, producen pelo

half-life: length of time required for half of the radioactive atoms in a sample to decay (452)
vida media: período de tiempo requerido para que se desintegre la mitad de los átomos radiactivos de una muestra

haploid: term used to refer to a cell that contains only a single set of genes (275)
haploide: tipo de célula que posee un solo juego de cromosomas

Hardy-Weinberg principle: principle that states that allele frequencies in a population remain constant unless one or more factors cause those frequencies to change (412)
principio de Hardy-Weinberg: el principio que afirma que las frecuencias alélicas de una población permanecen constantes a menos que uno o más factores ocasionen que esas frecuencias cambien

Haversian canal: one of a network of tubes running through compact bone that contains blood vessels and nerves (767)
conducto de Havers: uno de los tubos de una red que recorre longitudinalmente el hueso compacto y contiene vasos sanguíneos y nervios

heart: hollow muscular organ that pumps blood throughout the body (653)
corazón: órgano muscular hueco que bombea la sangre a todo el cuerpo

heartwood: in a woody stem, the older xylem near the center of the stem that no longer conducts water (562)
duramen: en un tallo leñoso, el xilema más viejo situado cerca del centro del tallo que ya no conduce agua

hemoglobin: iron-containing protein in red blood cells that binds oxygen and transports it to the body (790)
hemoglobina: proteína de los glóbulos rojos que contiene hierro, fija el oxígeno y lo transporta al organismo

herbaceous plant: type of plant that has smooth and nonwoody stems; includes dandelions, zinnias, petunias, and sunflowers (542)
planta herbácea: tipo de planta que tiene tallos blandos y no leñosos; incluye dientes de león, cinias, petunias y girasoles

herbivore: organism that obtains energy by eating only plants (62)
herbívoro: organismo que obtiene energía alimentándose solo de plantas

herbivory: interaction in which one animal (the herbivore) feeds on producers (such as plants) (86)
herbivorismo: interacción en la cual un animal (el herbívoro) se alimenta de productores (como las plantas)

heterotroph: organism that obtains food by consuming other living things; also called a consumer (61, 194)
heterótrofo: organismo que obtiene su alimento consumiendo otros seres vivos; también llamado consumidor

heterozygous: having two different alleles for a particular gene (267)
heterocigota: que tiene dos alelos diferentes para un gen dado

histamine: chemical released by mast cells that increases the flow of blood and fluids to the infected area during an inflammatory response (841)
histamina: sustancia química liberada por los mastocitos que aumenta el flujo de la sangre y los fluidos hacia el área infectada durante una respuesta inflamatoria

homeobox gene: The homeobox is a DNA sequence of approximately 130 base pairs, found in many homeotic genes that regulate development. Genes containing this sequence are known as homeobox genes, and they code for transcription factors, proteins that bind to DNA, and they also regulate the expression of other genes. (324)
gen homeobox: el homeobox es una secuencia de ADN de aproximadamente 130 pares de bases, presente en muchos genes homeóticos que regulan el desarrollo. Los genes que contienen esta secuencia se denominan genes homeobox y codifican los factores de transcripción, las proteínas que se adhieren al ADN y regulan la expresión de otros genes

homeostasis: relatively constant internal physical and chemical conditions that organisms maintain (15, 181, 717)
homeostasis: las condiciones internas, químicas y físicas, que los organismos mantienen relativamente constantes

homeotic gene: a class of regulatory genes that determine the identity of body parts and regions in an animal embryo. Mutations in these genes can transform one body part into another (324)

gen homeótico: tipo de genes reguladores que determinan la identidad de las partes y regiones del cuerpo en un embrión animal. Las mutaciones de estos genes pueden transformar una parte del cuerpo en otra

hominine: hominoid lineage that led to humans (635)

homínino: linaje hominoide que dio lugar a los seres humanos

hominoid: group of anthropoids that includes gibbons, orangutans, gorillas, chimpanzees, and humans (635)

homínido: grupo de antropoides que incluye a los gibones, orangutanes, gorilas, chimpacés y seres humanos

homologous: term used to refer to chromosomes in which one set comes from the male parent and one set comes from the female parent (275)

homólogos: término utilizado para referirse a los cromosomas en los que un juego proviene del progenitor masculino y un juego proviene del progenitor femenino

homologous structures: structures that are similar in different species of common ancestry (393)

estructuras homólogas: estructuras que son similares en distintas especies que tienen un ancestro común

homozygous: having two identical alleles for a particular gene (267)

homocigota: que tiene dos alelos idénticos para un gen dado

hormone: chemical produced in one part of an organism that affects another part of the same organism (588, 810)

hormona: sustancia química producida en una parte de un organismo que afecta a otra parte del mismo organismo

Hox gene: a group of homeotic genes clustered together that determine the head to tail identity of body parts in animals. All hox genes contain the homeobox DNA sequence. (324)

gen Hox: grupo de genes homeóticos agrupados en un conjunto que determinan la identidad posicional de las partes del cuerpo de los animales. Todos los genes Hox contienen la secuencia de ADN homeobox

humoral immunity: immunity against antigens in body fluids, such as blood and lymph (843)

inmunidad humoral: inmunidad contra los antígenos presentes en los fluidos corporales, como la sangre y la linfa

humus: material formed from decaying leaves and other organic matter (93)

humus: material formado a partir de hojas en descomposición y otros materiales orgánicos

hybrid: offspring of crosses between parents with different traits (263)

híbrido: descendencia del cruce entre progenitores que tienen rasgos diferentes

hybridization: breeding technique that involves crossing dissimilar individuals to bring together the best traits of both organisms (355)

hibridación: técnica de cría que consiste en cruzar individuos diferentes para reunir los mejores rasgos de ambos organismos

hydrogen bond: weak attraction between a hydrogen atom and another atom (33)

enlace de hidrógeno: atracción débil entre un átomo de hidrógeno y otro átomo

hydrostatic skeleton: skeleton made of fluid-filled body segments that work with muscles to allow the animal to move (674)

esqueleto hidrostático: esqueleto constituido por segmentos corporales llenos de fluido que trabajan con los músculos para permitir el movimiento del animal

hypertonic: when comparing two solutions, the solution with the greater concentration of solutes (178)

hipertónica: al comparar dos soluciones, la solución que tiene la mayor concentración de solutos

hypha (pl. hyphae): one of many long, slender filaments that makes up the body of a fungus (515)

hifa: uno de muchos filamentos largos y delgados que componen el cuerpo de un hongo

hypothalamus: structure of the brain that acts as a control center for recognition and analysis of hunger, thirst, fatigue, anger, and body temperature (749)

hipotálamo: estructura del cerebro que funciona como un centro de control para el reconocimiento y el análisis del hambre, la sed, la fatiga, el enojo y la temperatura corporal

hypothesis: possible explanation for a set of observations or possible answer to a scientific question (5)

hipótesis: explicación posible para un conjunto de observaciones o respuesta posible a una pregunta científica

hypotonic: when comparing two solutions, the solution with the lesser concentration of solutes (178)

hipotónica: al comparar dos soluciones, la solución que tiene la menor concentración de solutos

I

immigration: movement of individuals into an area occupied by an existing population (109)

inmigración: desplazamiento de individuos a un área ocupada por una población ya existente

immune response: the body's specific recognition, response, and memory to a pathogen attack (842)

respuesta inmune: reconocimiento, respuesta y memoria específicos que tiene el cuerpo respecto al ataque de un organismo patógeno

implantation: process in which the blastocyst attaches to the wall of the uterus (825)

implantación: proceso en el cual la blástula se adhiere a la pared del útero

Glossary *(continued)*

imprinting: type of behavior based on early experience; once imprinting has occurred, the behavior cannot be changed (699)
impronta: tipo de comportamiento basado en las primeras experiencias; una vez que ocurre la impronta, el comportamiento no puede cambiarse

inbreeding: continued breeding of individuals with similar characteristics to maintain the derived characteristics of a kind of organism (355)
endogamia: la cría continua de individuos con características semejantes para mantener las características derivadas de un tipo de organismo

incomplete dominance: situation in which one allele is not completely dominant over another allele (271)
dominancia incompleta: situación en la cual un alelo no es completamente dominante sobre otro alelo

independent assortment: one of Mendel's principles that states that genes for different traits can segregate independently during the formation of gametes (269)
distribución independiente: uno de los principios de Mendel que establece que los genes para rasgos diferentes pueden segregarse independientemente durante la formación de los gametos

independent variable: factor in a controlled experiment that is deliberately changed; also called manipulated variable (6)
variable independiente: en un experimento controlado, el factor que se modifica a propósito; también llamada variable manipulada

index fossil: distinctive fossil that is used to compare the relative ages of fossils (451)
fósil guía: fósil distintivo usado para comparar las edades relativas de los fósiles

infectious disease: disease caused by microorganism that disrupts normal body functions (838)
enfermedad infecciosa: enfermedad causada por un microorganismo que altera las funciones normales del cuerpo

inference: a logical interpretation based on prior knowledge and experience (5)
inferencia: interpretación lógica basada en la experiencia y en conocimientos previos

inflammatory response: nonspecific defense reaction to tissue damage caused by injury or infection (841)
respuesta inflamatoria: reacción defensiva no específica al daño causado a los tejidos por una herida o una infección

innate behavior: type of behavior in which the behavior appears in fully functional form the first time it is performed even though the animal has had no previous experience with the stimuli to which it responds; also called instinct (697)
comportamiento innato: tipo de comportamiento en el cual la conducta aparece en forma completamente funcional la primera vez que se lleva a cabo, aunque el animal no tenga ninguna experiencia previa con los estímulos a los que responde; también llamado instinto

insight learning: type of behavior in which an animal applies something it has already learned to a new situation, without a period of trial and error; also called reasoning (698)
aprendizaje por discernimiento: tipo de comportamiento en el cual un animal aplica algo que ya ha aprendido a una situación nueva, sin un período de ensayo y error; también llamado razonamiento

interferon: one of a group of proteins that help cells resist viral infection (842)
interferón: un tipo de proteína que ayuda a las células a combatir las infecciones virales

interneuron: type of neuron that processes information and may relay information to motor neurons (669)
interneurona: tipo de neurona que procesa información y la puede transmitir para estimular las neuronas

interphase: period of the cell cycle between cell divisions (241)
interfase: período del ciclo celular entre las divisiones celulares

intracellular digestion: type of digestion in which food is digested inside specialized cells that pass nutrients to other cells by diffusion (647)
digestión intracelular: tipo de digestión en la cual los alimentos se digieren dentro de células especializadas que pasan los nutrientes a otras células mediante difusión

intron: sequence of DNA that is not involved in coding for a protein (310)
intrón: secuencia de ADN que no participa en la codificación de una proteína

invertebrate: animal that lacks a backbone, or vertebral column (606)
invertebrado: animal que carece de columna vertebral

ion: atom that has a positive or negative charge (30)
ion: átomo que tiene una carga positiva o negativa

ionic bond: chemical bond formed when one or more electrons are transferred from one atom to another (30)
enlace iónico: enlace químico que se forma cuando uno o más electrones se transfieren de un átomo a otro

iris: colored part of the eye (756)
iris: parte coloreada del ojo

isotonic: when the concentration of two solutions is the same (178)
isotónica: cuando la concentración de dos soluciones es la misma

isotope: one of several forms of a single element that contains the same number of protons but different numbers of neutrons (29)
isótopo: cada una de las diferentes formas de un único elemento, que contiene la misma cantidad de protones pero cantidades distintas de neutrones

J

joint: place where one bone attaches to another bone (675, 768)
articulación: sitio donde un hueso se une a otro hueso

K

karyotype: micrograph of the complete diploid set of chromosomes grouped together in pairs, arranged in order of decreasing size (334)
cariotipo: micrografía de la totalidad del conjunto diploide de cromosomas agrupados en pares, ordenados por tamaño decreciente

keratin: tough fibrous protein found in skin (776)
queratina: proteína fibrosa y resistente que se halla en la piel

keystone species: single species that is not usually abundant in a community yet exerts strong control on the structure of a community (86)
especie clave: especie que habitualmente no es abundante en una comunidad y sin embargo ejerce un fuerte control sobre la estructura de esa comunidad

kidney: an organ of excretion that separates wastes and excess water from the blood (656)
riñón: órgano excretor que separa los residuos y el exceso de agua de la sangre

kingdom: largest and most inclusive group in Linnaean classification (431)
reino: grupo más grande e inclusivo del sistema de clasificación inventado por Linneo

Koch's postulates: set of guidelines developed by Koch that help identify the microorganism that causes a specific disease (839)
postulados de Koch: conjunto de pautas desarrollado por Koch que ayuda a identificar al microorganismo que causa una enfermedad específica

Krebs cycle: second stage of cellular respiration in which pyruvic acid is broken down into carbon dioxide in a series of energy-extracting reactions (218)
ciclo de Krebs: segunda fase de la respiración celular en la cual el ácido pirúvico se descompone en dióxido de carbono en una serie de reacciones que liberan energía

L

language: system of communication that combines sounds, symbols, and gestures according to a set of rules about sequence and meaning, such as grammar and syntax (704)
lenguaje: sistema de comunicación que combina sonidos, símbolos y gestos según un conjunto de reglas sobre la secuencia y el significado, como la gramática y la sintaxis

large intestine: organ in the digestive system that removes water from the undigested material that passes through it; also called the colon (728)
intestino grueso: órgano del sistema digestivo que extrae el agua del material no digerido que pasa por él; también llamado colon

larva: (pl. larvae): immature stage of an organism (627)
larva: etapa inmadura de un organismo

larynx: structure in the throat that contains the vocal cords (796)
laringe: órgano situado en la garganta que contiene las cuerdas vocales

learning: changes in behavior as a result of experience (697)
aprendizaje: cambios en el comportamiento a consecuencia de la experiencia

lens: structure in the eye that focuses light rays on the retina (757)
cristalino: estructura del ojo que enfoca los rayos luminosos en la retina

lichen: symbiotic association between a fungus and a photosynthetic organism (518)
liquen: asociación simbiótica entre un hongo y un organismo fotosintético

ligament: tough connective tissue that holds bones together in a joint (675, 769)
ligamento: tejido conectivo resistente que mantiene unidos a los huesos en una articulación

light-dependent reactions: set of reactions in photosynthesis that use energy from light to produce ATP and NADPH (197)
reacciones dependientes de la luz: en la fotosíntesis, conjunto de reacciones que emplean la energía proveniente de la luz para producir ATP y NADPH

Glossary *(continued)*

light-independent reactions: set of reactions in photosynthesis that do not require light; energy from ATP and NADPH is used to build high-energy compounds such as sugar; also called the Calvin cycle (197)
reacciones independientes de la luz: en la fotosíntesis, conjunto de reacciones que no necesitan luz; la energía proveniente del ATP y del NADPH se emplea para construir compuestos con gran contenido energético, como el azúcar; también llamado ciclo de Calvin

lignin: substance in vascular plants that makes cell walls rigid (554)
lignina: sustancia de las plantas vasculares que hace rígidas a las paredes celulares

limiting factor: factor that causes population growth to decrease (112)
factor limitante: un factor que hace disminuir el crecimiento de la población

limiting nutrient: single essential nutrient that limits productivity in an ecosystem (72)
nutriente limitante: un solo nutriente esencial que limita la productividad de un ecosistema

lipid: macromolecule made mostly from carbon and hydrogen atoms; includes fats, oils, and waxes (39)
lípido: macromolécula compuesta principalmente por átomos de carbono e hidrógeno; incluye las grasas, los aceites y las ceras

lipid bilayer: flexible double-layered sheet that makes up the cell membrane and forms a barrier between the cell and its surroundings (171)
bicapa lipídica: lámina flexible de dos capas que constituye la membrana celular y forma una barrera entre la célula y su entorno

logistic growth: growth pattern in which a population's growth slows and then stops following a period of exponential growth (111)
crecimiento logístico: patrón de crecimiento en el cual el desarrollo de una población se reduce y luego se detiene después de un período de crecimiento exponencial

loop of Henle: section of the nephron tubule that is responsible for conserving water and minimizing the volume of the filtrate (731)
asa de Henle: una sección del túbulo de nefrón responsable de conservar el agua y minimizar el volumen del material filtrado

lung: respiratory organ; place where gases are exchanged between the blood and inhaled air (651)
pulmón: órgano respiratorio; lugar donde se intercambian los gases entre la sangre y el aire inhalado

lymph: fluid that is filtered out of the blood (792)
linfa: fluido procedente de la sangre

lysogenic infection: type of infection in which a virus embeds its DNA into the DNA of the host cell and is replicated along with the host cell's DNA (482)
infección lisogénica: tipo de infección en la cual un virus inserta su ADN en el ADN de la célula huésped y se replica junto con el ADN de dicha célula huésped

lysosome: cell organelle that breaks down lipids, carbohydrates, and proteins into small molecules that can be used by the rest of the cell (166)
lisosoma: orgánulo celular que descompone los lípidos, los hidratos de carbono y las proteínas en moléculas pequeñas que pueden ser utilizadas por el resto de la célula

lytic infection: type of infection in which a virus enters a cell, makes copies of itself, and causes the cell to burst (482)
infección lítica: tipo de infección en la cual un virus penetra una célula, hace copias de sí mismo y provoca la ruptura o muerte celular

M

macroevolutionary patterns: changes in anatomy, phylogeny, ecology, and behavior that take place in clades larger than a single species (456)
patrones de macroevolución: cambios que ocurren en la anatomía, filogenia, ecología y comportamiento de clados que abarcan a más de una especie

Malpighian tubule: structure in most terrestrial arthropods that concentrates the uric acid and adds it to digestive wastes (658)
túbulo de Malpighi: estructura de la mayoría de los artrópodos terrestres que concentra el ácido úrico y lo incorpora a los residuos digestivos

mammary gland: gland in female mammals that produces milk to nourish the young (682)
glándula mamaria: glándula de las hembras de los mamíferos que produce leche para alimentar a las crías

mass extinction: event during which many species become extinct during a relatively short period of time (457)
extinción masiva: suceso durante el cual se extinguen muchas especies durante un período de tiempo relativamente corto

matrix: innermost compartment of the mitochondrion (218)
matriz: compartimento más interno de la mitocondria

mechanical digestion: physical breakdown of large pieces of food into smaller pieces (723)
digestión mecánica: descomposición física de grandes pedazos de comida en pedazos más pequeños

meiosis: process in which the number of chromosomes per cell is cut in half through the separation of homologous chromosomes in a diploid cell (276)

meiosis: proceso por el cual el número de cromosomas por célula se reduce a la mitad mediante la separación de los cromosomas homólogos de una célula diploide

melanin: dark brown pigment in the skin that helps protect the skin by absorbing ultraviolet rays (776)

melanina: pigmento marrón oscuro de la piel que contribuye a protegerla al absorber los rayos ultravioletas

melanocyte: cell in the skin that produces a dark brown pigment called melanin (776)

melanocito: célula de la piel que produce un pigmento marrón oscuro llamado melanina

menstrual cycle: regular sequence of events in which an egg develops and is released from the body (820)

ciclo menstrual: secuencia regular de sucesos en la cual un huevo se desarrolla y se elimina del cuerpo

menstruation: discharge of blood and the unfertilized egg from the body (821)

menstruación: descarga de sangre y del huevo no fertilizado del cuerpo

meristem: regions of unspecialized cells responsible for continuing growth throughout a plant's lifetime (554)

meristemos: regiones de células no especializadas responsables del crecimiento continuo de una planta durante su vida

mesoderm: middle germ layer; develops into muscles, and much of the circulatory, reproductive, and excretory systems (612)

mesodermo: capa embrionaria media; se desarrolla para dar lugar a los músculos y gran parte de los sistemas circulatorio, reproductor y excretor

mesophyll: specialized ground tissue found in leaves; performs most of a plant's photosynthesis (564)

mesófilo: tejido fundamental especializado que se halla en las hojas; realiza la mayor parte de la fotosíntesis de una planta

messenger RNA (mRNA): type of RNA that carries copies of instructions for the assembly of amino acids into proteins from DNA to the rest of the cell (309)

ARN mensajero: tipo de ARN que transporta copias de las instrucciones para el ensamblaje de los aminoácidos en proteínas, desde el ADN al resto de la célula

metabolism: the combination of chemical reactions through which an organism builds up or breaks down materials (15)

metabolismo: la combinación de reacciones químicas a través de las cuales un organismo acumula o desintegra materiales

metamorphosis: process of changes in shape and form of a larva into an adult (681)

metamorfosis: proceso de cambios en la estructura y forma de una larva hasta que se convierte en adulto

metaphase: phase of mitosis in which the chromosomes line up across the center of the cell (242)

metafase: fase de la mitosis en la cual los cromosomas se alinean a través del centro de la célula

microclimate: environmental conditions within a small area that differs significantly from the climate of the surrounding area (82)

microclima: condiciones medioambientales de un área pequeña que difieren significativamente del clima del área circundante

migration: seasonal behavior resulting in the movement from one environment to another (701)

migración: comportamiento estacional que da por resultado el desplazamiento desde un medio ambiente a otro

mineral: inorganic nutrient the body needs, usually in small amounts (721)

mineral: nutriente inorgánico que el cuerpo necesita, usualmente en pequeñas cantidades

mitochondrion: cell organelle that converts the chemical energy stored in food into compounds that are more convenient for the cell to use (170)

mitocondria: orgánulo celular que convierte la energía química almacenada en los alimentos en compuestos más apropiados para que la célula los use

mitosis: part of eukaryotic cell division during which the cell nucleus divides (241)

mitosis: fase de la división de las células eucariotas durante la cual se divide el núcleo celular

mixture: material composed of two or more elements or compounds that are physically mixed together but not chemically combined (34)

mezcla: material compuesto por dos o más elementos o compuestos que están mezclados físicamente pero no están combinados químicamente

molecular clock: method used by researchers that uses mutation rates in DNA to estimate the length of time that two species have been evolving independently (417)

reloj molecular: método de investigación que emplea las tasas de mutación del ADN para estimar el lapso de tiempo en que dos especies han evolucionado independientemente

molecule: smallest unit of most compounds that displays all the properties of that compound (31)

molécula: la unidad más pequeña de la mayoría de los compuestos que exhibe todas las propiedades de ese compuesto

molting: process of shedding an exoskeleton and growing a new one (675)

muda: proceso de desprendimiento de un exoesqueleto y el crecimiento de uno nuevo

monocot: angiosperm with one seed leaf in its ovary (542)

monocotiledónea: angiosperma con un cotiledón (hoja embrionaria) en su ovario

monoculture: farming strategy of planting a single, highly productive crop year after year (129)
monocultivo: estrategia agrícola que consiste en plantar año tras año un único cultivo altamente productivo

monomer: small chemical unit that makes up a polymer (38)
monómero: pequeña unidad química que forma un polímero

monophyletic group: group that consists of a single ancestral species and all its descendants and excludes any organisms that are not descended from that common ancestor (433)
grupo monofilético: grupo que consiste en una especie con un único ancestro y todos sus descendientes y excluye a todos los organismos que no descienden de ese ancestro común

monosaccharide: simple sugar molecule (38)
monosacárido: molécula de azúcar simple

motor neuron: type of nerve cell that carries directions from interneurons to either muscle cells or glands (669)
neurona motora: tipo de célula nerviosa que lleva las instrucciones provenientes de las interneuronas a las células musculares o las glándulas

multiple alleles: a gene that has more than two alleles (272)
alelos múltiples: un gen que tiene más de dos alelos

multipotent: cell with limited potential to develop into many types of differentiated cells (250)
multipotentes: células con potencial limitado para generar muchos tipos de células diferenciadas

muscle fiber: long slender skeletal muscle cells (770)
fibra muscular: células largas y delgadas de los músculos esqueléticos

muscle tissue: type of tissue that makes movements of the body possible (715)
tejido muscular: tipo de tejido que hace posibles los movimientos del cuerpo

mutagen: chemical or physical agents in the environment that interact with DNA and may cause a mutation (318)
mutágeno: agentes físicos o químicos del medioambiente que interaccionan con el ADN y pueden causar una mutación

mutation: change in the genetic material of a cell (316)
mutación: cambio en el material genético de una célula

mutualism: symbiotic relationship in which both species benefit from the relationship (87)
mutualismo: relación simbiótica en la cual ambas especies se benefician

mycelium (pl. mycelia): densely branched network of the hyphae of a fungus (515)
micelio: la red de filamentos muy ramificados de las hifas de un hongo

mycorrhiza (pl. mycorrhizae): symbiotic association of plant roots and fungi (519)
micorriza: asociación simbiótica entre las raíces de las plantas y los hongos

myelin sheath: insulating membrane surrounding the axon in some neurons (743)
vaina de mielina: membrana aislante que rodea al axón de algunas neuronas

myocardium: thick middle muscle layer of the heart (787)
miocardio: capa media, gruesa y musculosa del corazón

myofibril: tightly packed filament bundles found within skeletal muscle fibers (771)
miofibrilla: manojos de filamentos muy apretados que se hallan dentro de las fibras de los músculos esqueléticos

myosin: thick filament of protein found in skeletal muscle cells (771)
miosina: filamento grueso de proteína que se halla en las células de los músculos esqueléticos

N

NAD^+ (nicotinamide adenine dinucleotide): electron carrier involved in glycolysis (216)
NAD^+ (dinucleótido de nicotinamida adenina): transportador de electrones que participa en la glucólisis

$NADP^+$ (nicotinamide adenine dinucleotide phosphate): carrier molecule that transfers high-energy electrons from chlorophyll to other molecules (196)
$NADP^+$ (fosfato de dinucleótido de nicotinamida adenina): molécula transportadora de electrones que transfiere electrones de alta energía desde la clorofila a otras moléculas

natural selection: process by which organisms that are most suited to their environment survive and reproduce most successfully; also called survival of the fittest (389)
selección natural: proceso por el cual los organismos más adaptados a su medioambiente sobreviven y se reproducen más exitosamente; también llamada supervivencia del más apto

nephridium (pl. nephridia): excretory structure of an annelid that filters body fluid (658)
nefridio: estructura excretora de los anélidos que filtra el fluido corporal

nephron: blood-filtering structure in the kidneys in which impurities are filtered out, wastes are collected, and purified blood is returned to the circulation (729)
nefrón: estructura filtradora de la sangre en los riñones, en la cual se filtran las impurezas, se recogen los desechos y la sangre purificada se devuelve a la circulación

nervous tissue: type of tissue that transmits nerve impulses throughout the body (715)
  tejido nervioso: tipo de tejido que transmite los impulsos nerviosos por el cuerpo

neuromuscular junction: the point of contact between a motor neuron and a skeletal muscle cell (772)
  unión neuromuscular: el punto de contacto entre una neurona motora y una célula de un músculo esquelético

neuron: nerve cell; specialized for carrying messages throughout the nervous system (668)
  neurona: célula nerviosa; especializada en conducir mensajes a través del sistema nervioso

neurotransmitter: chemical used by a neuron to transmit an impulse across a synapse to another cell (746)
  neurotransmisor: sustancia química utilizada por una neurona para transmitir un impulso a otra célula a través de una sinapsis

neurulation: the first step in the development of the nervous system (826)
  neurulación: primer paso en el desarrollo del sistema nervioso

niche: full range of physical and biological conditions in which an organism lives and the way in which the organism uses those conditions (85)
  nicho: toda la variedad de condiciones biológicas y físicas en las que vive un organismo y la manera en la que dicho organismo utiliza esas condiciones

nitrogen fixation: process of converting nitrogen gas into nitrogen compounds that plants can absorb and use (71)
  fijación de nitrógeno: el proceso por el cual el gas nitrógeno se convierte en los compuestos nitrogenados que las plantas pueden absorber y utilizar

node: part on a growing stem where a leaf is attached (560)
  nudo: parte de un tallo en crecimiento donde está adherida una hoja

nondisjunction: error in meiosis in which the homologous chromosomes fail to separate properly (341)
  no disyunción: error que ocurre durante la meiosis, en el que cromosomas homólogos no logran separarse adecuadamente

nonrenewable resource: resource that cannot be replenished by a natural process within a reasonable amount of time (131)
  recurso no renovable: recurso que no se puede reponer mediante un proceso natural dentro de un período de tiempo razonable

norepinephrine: hormone released by the adrenal glands that increases heart rate and blood pressure and prepares the body for intense physical activity (814)
  norepinefrina o noradrenalina: hormona liberada por las glándulas adrenales que aumenta la frecuencia cardíaca y la presión sanguínea y prepara al cuerpo para realizar actividad física intensa

notochord: long supporting rod that runs through a chordate's body just below the nerve cord (606)
  notocordio: extenso bastón de apoyo que se extiende a lo largo del cuerpo de los cordados, justo por debajo del cordón nervioso

nucleic acid: macromolecules containing hydrogen, oxygen, nitrogen, carbon, and phosphorus (40)
  ácido nucleico: macromoléculas que contienen hidrógeno, oxígeno, nitrógeno, carbono y fósforo

nucleotide: subunit of which nucleic acids are composed; made up of a 5-carbon sugar, a phosphate group, and a nitrogenous base (40)
  nucleótido: subunidad que constituye los ácidos nucleicos; compuesta de un azúcar de 5 carbonos, un grupo fosfato y una base nitrogenada

nucleus: the center of an atom, which contains the protons and neutrons (28); in cells, structure that contains the cell's genetic material in the form of DNA (162)
  núcleo: el centro de un átomo, contiene los protones y los neutrones; en las células, la estructura que contiene el material genético de la célula en forma de ADN

nutrient: chemical substance that an organism needs to sustain life (70)
  nutriente: sustancia química que un organismo necesita para continuar con vida

nymph: immature form of an animal that resembles the adult form but lacks functional sexual organs (681)
  ninfa: forma inmadura de un animal que se parece a la forma adulta, pero carece de órganos sexuales funcionales

O

observation: process of noticing and describing events or processes in a careful, orderly way (5)
  observación: el método de percibir y describir sucesos o procesos de manera atenta y ordenada

omnivore: organism that obtains energy by eating both plants and animals (62)
  omnívoro: organismo que obtiene energía alimentándose de plantas y animales

open circulatory system: type of circulatory system in which blood is only partially contained within a system of blood vessels as it travels through the body (653)
  sistema circulatorio abierto: tipo de sistema circulatorio en el cual la sangre, cuando fluye por el cuerpo, está solo parcialmente contenida dentro de un sistema de vasos sanguíneos

operant conditioning: type of learning in which an animal learns to behave in a certain way through repeated practice, to receive a reward or avoid punishment (698)
  acondicionamiento operante: tipo de aprendizaje en el cual un animal aprende a comportarse de cierta manera mediante una práctica repetida, para recibir una recompensa o evitar un castigo

operator: short DNA region, adjacent to the promoter of a prokaryotic operon, that binds repressor proteins responsible for controlling the rate of transcription of the operon (321)
operador: pequeña región de ADN, adyacente al promotor del operón de una procariota, que une las proteínas represoras responsables de controlar la tasa de transcripción del operón

operon: in prokaryotes, a group of adjacent genes that share a common operator and promoter and are transcribed into a single mRNA (320)
operón: en las procariotas, grupo de genes adyacentes que comparten un operador y un promotor en común y que son transcritas a un solo ARN mensajero

opposable thumb: thumb that enables grasping objects and using tools (635)
pulgar oponible o prensible: un pulgar que permite aferrar objetos y utilizar herramientas

order: in classification, a group of closely related families (430)
orden: en la clasificación, un grupo de familias relacionadas estrechamente

organ: group of tissues that work together to perform closely related functions (182)
órgano: grupo de tejidos que trabajan juntos para realizar funciones estrechamente relacionadas

organ system: group of organs that work together to perform a specific function (182)
sistema de órganos: grupo de órganos que trabajan juntos para realizar una función específica

organelle: specialized structure that performs important cellular functions within a eukaryotic cell (164)
orgánulo: estructura especializada que realiza funciones celulares importantes dentro de una célula eucariota

osmosis: diffusion of water through a selectively permeable membrane (177)
ósmosis: la difusión de agua a través de una membrana de permeabilidad selectiva

osmotic pressure: pressure that must be applied to prevent osmotic movement across a selectively permeable membrane (178)
presión osmótica: la presión que debe aplicarse para evitar el movimiento osmótico a través de una membrana de permeabilidad selectiva

ossification: process of bone formation during which cartilage is replaced by bone (767)
osificación: el proceso de formación de hueso durante el cual el cartílago es reemplazado por hueso

osteoblast: bone cell that secretes mineral deposits that replace the cartilage in developing bones (767)
osteoblasto: célula ósea que secreta depósitos minerales que reemplazan al cartílago de los huesos en desarrollo

osteoclast: bone cell that breaks down bone minerals (767)
osteoclasto: célula ósea que degrada los minerales óseos

osteocyte: bone cell that helps maintain the minerals in bone tissue and continue to strengthen the growing bone (767)
osteocito: célula ósea que ayuda a conservar los minerales en el tejido óseo y continúa fortaleciendo al hueso en crecimiento

ovary: in plants, the structure that surrounds and protects seeds (540); in animals, the primary female reproductive organ; produces eggs (819)
ovario: en las plantas, la estructura que rodea a las semillas y las protege; órgano reproductor femenino fundamental en los animales; produce huevos

oviparous: species in which embryos develop in eggs outside a parent's body (680)
ovíparo: especie animal en la cual los embriones se desarrollan en huevos fuera del cuerpo del progenitor

ovoviviparous: species in which the embryos develop within the mother's body but depend entirely on the yolk sac of their eggs (680)
ovovíparo: especie animal en la cual los embriones se desarrollan dentro del cuerpo de la madre, pero dependen completamente del saco vitelino de sus huevos

ovulation: the release of a mature egg from the ovary into one of the Fallopian tubes (820)
ovulación: liberación de un huevo maduro desde el ovario a una de las trompas de Falopio

ovule: structure in seed cones in which the female gametophytes develop (538)
óvulo: estructura de las semillas coníferas donde se desarrollan los gametos femeninos

ozone layer: atmospheric layer in which ozone gas is relatively concentrated; protects life on Earth from harmful ultraviolet rays in sunlight (145)
capa de ozono: capa atmosférica en la cual el gas ozono se encuentra relativamente concentrado; protege a los seres vivos de la Tierra de los perjudiciales rayos ultravioletas de la luz solar

P

pacemaker: small group of cardiac muscle fibers that maintains the heart's pumping rhythm by setting the rate at which the heart contracts; the sinoatrial (SA) node (788)
marcapasos: grupo pequeño de fibras musculares cardíacas que mantiene el ritmo de bombeo del corazón estableciendo la frecuencia a la que se contrae el corazón; el nodo sinusal

paleontologist: scientist who studies fossils (450)
paleontólogo: científico que estudia los fósiles

palisade mesophyll: layer of cells under the upper epidermis of a leaf (564)
mesófilo en empalizada: capa de células situada bajo la epidermis superior de una hoja

parasitism: symbiotic relationship in which one organism lives on or inside another organism and harms it (87)
parasitismo: relación simbiótica en la cual un organismo vive sobre otro organismo o en su interior y lo perjudica

parathyroid hormone (PTH): hormone produced by parathyroid gland that increases calcium levels in the blood (815)
hormona de la paratiroides: hormona producida por la glándula paratiroides que aumenta los niveles de calcio en la sangre

parenchyma: main type of ground tissue in plants that contains cells with thin cell walls and large central vacuoles (553)
parénquima: tipo principal de tejido fundamental de las plantas que contiene células con paredes celulares delgadas y vacuolas centrales grandes

passive immunity: temporary immunity that develops as a result of natural or deliberate exposure to an antibody (846)
inmunidad pasiva: inmunidad transitoria que se desarrolla a consecuencia de una exposición natural o deliberada a un anticuerpo

pathogen: disease-causing agent (489)
patógeno: agente que causa una enfermedad

pedigree: chart that shows the presence or absence of a trait according to the relationships within a family across several generations (336)
árbol genealógico: diagrama que muestra la presencia o ausencia de un rasgo de acuerdo con las relaciones intrafamiliares a través de varias generaciones

pepsin: enzyme that breaks down proteins into smaller polypeptide fragments (725)
pepsina: enzima que descompone las proteínas en fragmentos de polipéptidos más pequeños

period: division of geologic time into which eras are subdivided (454)
período: división del tiempo geológico en la que se subdividen las eras

peripheral nervous system: network of nerves and supporting cells that carries signals into and out of the central nervous system (742)
sistema nervioso periférico: red de nervios y células de apoyo que transporta señales hacia y desde el sistema nervioso central

peristalsis: contractions of smooth muscles that provide the force that moves food through the esophagus toward the stomach (724)
peristalsis: contracciones de los músculos lisos que proporcionan la fuerza que hace avanzar los alimentos a través del esófago hacia el estómago

permafrost: layer of permanently frozen subsoil found in the tundra (94)
permacongelamiento: capa de subsuelo congelado en forma permanente que se halla en la tundra

pH scale: scale with values from 0 to 14, used to measure the concentration of H^+ ions in a solution; a pH of 0 to 7 is acidic, a pH of 7 is neutral, and a pH of 7 to 14 is basic (35)
escala del pH: escala con valores de 0 a 14, utilizada para medir la concentración de iones H^+ en una solución; un pH de 0 a 7 es ácido, un pH de 7 es neutro y un pH de 7 a 14 es básico

pharyngeal pouch: one of a pair of structures in the throat region of a chordate (607)
bolsa faríngea: cada una de las dos estructuras situadas en la región de la garganta de los cordados

pharynx: tube at the back of the mouth that serves as a passageway for both air and food; also called the throat (796)
faringe: tubo situado a continuación de la boca que sirve de conducto para que pasen el aire y los alimentos; también llamada garganta

phenotype: physical characteristics of an organism (267)
fenotipo: características físicas de un organismo

phloem: vascular tissue that transports solutions of nutrients and carbohydrates produced by photosynthesis through the plant (534)
floema: tejido vascular que transporta por toda la planta las soluciones de nutrientes e hidratos de carbono producidos en la fotosíntesis

photic zone: sunlight region near the surface of water (96)
zona fótica: región cerca de la superficie del mar en la que penetra la luz solar

photoperiodism: plant response to the relative lengths of light and darkness (591)
fotoperíodismo: la respuesta de una planta a los tiempos relativos de luz y oscuridad

photosynthesis: process used by plants and other autotrophs to capture light energy and use it to power chemical reactions that convert carbon dioxide and water into oxygen and energy-rich carbohydrates such as sugars and starches (60, 194)
fotosíntesis: proceso empleado por las plantas y otros organismos autótrofos para atrapar la energía luminosa y utilizarla para impulsar reacciones químicas que convierten el dióxido de carbono y el agua en oxígeno e hidratos de carbono de gran contenido energético, como azúcares y almidones

photosystem: cluster of chlorophyll and proteins found in thylakoids (199)
fotosistema: conjunto de clorofila y proteínas que se hallan en los tilacoides

phototropism: tendency of a plant to grow toward a light source (590)
fototropismo: la tendencia de una planta a crecer hacia una fuente de luz

Glossary *(continued)*

phylogeny: the evolutionary history of a lineage (433)
filogenia: historia evolutiva del linaje

phylum (pl. phyla): in classification, a group of closely related classes (431)
filo: en la clasificación, un grupo de clases estrechamente relacionadas

phytoplankton: photosynthetic algae found near the surface of the ocean (63)
fitoplancton: algas fotosintéticas que se hallan cerca de la superficie del océano

pigment: light-absorbing molecule used by plants to gather the sun's energy (195)
pigmento: moléculas que absorben la luz, empleadas por las plantas para recolectar la energía solar

pioneer species: first species to populate an area during succession (88)
especies pioneras: las primeras especies en poblar un área durante la sucesión ecológica

pistil: single carpel or several fused carpels; contains the ovary, style, and stigma (579)
pistilo: un único carpelo o varios carpelos unidos; contiene el ovario, el estilo y el estigma

pituitary gland: small gland found near the base of the skull that secretes hormones that directly regulate many body functions and controls the actions of several other endocrine glands (813)
glándula pituitaria: pequeña glándula situada cerca de la base del cráneo que secreta hormonas que regulan directamente muchas funciones corporales y controla las acciones de varias otras glándulas endocrinas

placenta: specialized organ in placental mammals through which respiratory gases, nutrients, and wastes are exchanged between the mother and her developing young (683, 827)
placenta: órgano especializado de los mamíferos placentarios a través del cual se intercambian los gases respiratorios, los nutrientes y los residuos entre la madre y su cría en desarrollo

plankton: microscopic organisms that live in aquatic environments; includes both phytoplankton and zooplankton (97)
plancton: organismos microscópicos que viven en medios ambientes acuáticos; incluye el fitoplancton y el zooplancton

plasma: straw-colored liquid portion of the blood (790)
plasma: parte líquida de la sangre de color amarillento

plasmid: small, circular piece of DNA located in the cytoplasm of many bacteria (359)
plásmido: pequeña porción circular de ADN ubicada en el citoplasma de muchas bacterias

plasmodium: amoeboid feeding stage in the life cycle of a plasmodial slime mold (511)
plasmodio: etapa de alimentación ameboide del ciclo vital de los mohos mucilaginosos

plate tectonics: geologic processes, such as continental drift, volcanoes, and earthquakes, resulting from plate movement (454)
tectónica de placas: procesos geológicos, como la deriva continental, los volcanes y los terremotos, que son consecuencia de los movimientos de las placas

platelet: cell fragment released by bone marrow that helps in blood clotting (791)
plaqueta: fragmento celular liberado por la médula espinal que interviene en la coagulación de la sangre

pluripotent: cells that are capable of developing into most, but not all, of the body's cell types (249)
pluripotentes: células capaces de convertirse en la mayoría de células del cuerpo, pero no en todas

point mutation: gene mutation in which a single base pair in DNA has been changed (317)
mutación puntual: mutación genética en la cual se ha modificado un único par de bases en el ADN

pollen grain: structure that contains the entire male gametophyte in seed plants (536)
grano de polen: la estructura que contiene a todo el gametofito masculino en las plantas fanerógamas

pollen tube: structure in a plant that contains two haploid sperm nuclei (538)
tubo polínico: en una planta, estructura que contiene dos núcleos espermáticos haploides

pollination: transfer of pollen from the male reproductive structure to the female reproductive structure (536)
polinización: transferencia de polen desde la estructura reproductora masculina hacia la estructura reproductora femenina

pollutant: harmful material that can enter the biosphere through the land, air, or water (134)
contaminante: material nocivo que puede ingresar en la biósfera a través de la tierra, el aire o el agua

polygenic trait: trait controlled by two or more genes (272, 408)
rasgo poligénico: rasgo controlado por dos o más genes

polymer: molecules composed of many monomers; makes up macromolecules (38)
polímero: molécula compuesta por muchos monómeros; forma macromoléculas

polymerase chain reaction (PCR): the technique used by biologists to make many copies of a particular gene (358)
reacción en cadena de la polímerasa (PCR): técnica usada por los biólogos para hacer muchas copias de un gen específico

polypeptide: long chain of amino acids that makes proteins (311)
polipéptido: cadena larga de aminoácidos que constituye las proteínas

polyploidy: condition in which an organism has extra sets of chromosomes (319)
poliploidía: condición en la cual un organismo tiene grupos adicionales de cromosomas

population: group of individuals of the same species that live in the same area (56)
población: grupo de individuos de la misma especie que viven en la misma área

population density: number of individuals per unit area (108)
densidad de población: número de individuos que viven por unidad de superficie

predation: interaction in which one organism (the predator) captures and feeds on another organism (the prey) (86)
depredación: interacción en la cual un organismo (el predador) captura y come a otro organismo (la presa)

prehensile tail: long tail that can coil tightly enough around a branch (635)
cola prensil: cola larga que puede enrollarse apretadamente alrededor de una rama

pressure-flow hypothesis: hypothesis that explains the method by which phloem sap is transported through the plant from a sugar "source" to a sugar "sink" (568)
teoría de flujo por presión: teoría que explica el método por el cual la savia del floema recorre la planta desde una "fuente" de azúcar hacia un "vertedero" de azúcar

primary growth: pattern of growth that takes place at the tips and shoots of a plant (561)
crecimiento primario: patrón de crecimiento que tiene lugar en las puntas y en los brotes de una planta

primary producer: first producer of energy-rich compounds that are later used by other organisms (60)
productor primario: los primeros productores de compuestos ricos en energía que luego son utilizados por otros organismos

primary succession: succession that occurs in an area in which no trace of a previous community is present (88)
sucesión primaria: sucesión que ocurre en un área en la cual no hay rastros de la presencia de una comunidad anterior

principle of dominance: Mendel's second conclusion, which states that some alleles are dominant and others are recessive (263)
principio de dominancia: segunda conclusión de Mendel, que establece que algunos alelos son dominantes y otros son recesivos

prion: protein particles that cause disease (493)
prión: partículas de proteína que causan enfermedades

probability: likelihood that a particular event will occur (266)
probabilidad: la posibilidad de que ocurra un suceso dado

product: elements or compounds produced by a chemical reaction (42)
producto: elemento o compuesto producido por una reacción química

prokaryote: unicellular organism that lacks a nucleus (162, 485)
procariota: organismo unicelular que carece de núcleo

promoter: specific region of a gene where RNA polymerase can bind and begin transcription (310)
promotor: región específica de un gen en donde la ARN polimerasa puede unirse e iniciar la transcripción

prophage: bacteriophage DNA that is embedded in the bacterial host's DNA (482)
profago: ADN del bacteriófago que está alojado en el interior del ADN del huésped bacteriano

prophase: first and longest phase of mitosis in which the genetic material inside the nucleus condenses and the chromosomes become visible (242)
profase: primera y más prolongada fase de la mitosis, en la cual el material genético dentro del interior del núcleo se condensa y los cromosomas se hacen visibles

prostaglandin: modified fatty acids that are produced by a wide range of cells; generally affect only nearby cells and tissues (811)
prostaglandina: ácidos grasos modificados que son producidos por una amplia gama de células; generalmente afectan solo a las células y tejidos cercanos

protein: macromolecule that contains carbon, hydrogen, oxygen, and nitrogen; needed by the body for growth and repair (40, 721)
proteína: macromolécula que contiene carbono, hidrógeno, oxígeno y nitrógeno; necesaria para el crecimiento y reparación del cuerpo

protostome: animal whose mouth is formed from the blastopore (613)
protóstomo: animal cuya boca se desarrolla a partir del blastoporo

pseudocoelom: body cavity that is only partially lined with mesoderm (612)
pseudoceloma o falso celoma: cavidad corporal que está revestida sólo parcialmente con mesodermo

Glossary *(continued)*

pseudopod: temporary cytoplasmic projection used by some protists for movement (505)
seudópodo: prolongación citoplasmática transitoria utilizada por algunos protistas para moverse

puberty: period of rapid growth and sexual maturation during which the reproductive system becomes fully functional (817)
pubertad: período de crecimiento rápido y de maduración sexual durante el cual el sistema reproductor se vuelve completamente funcional

pulmonary circulation: path of circulation between the heart and lungs (788)
circulación pulmonar: recorrido de la circulación entre el corazón y los pulmones

punctuated equilibrium: pattern of evolution in which long stable periods are interrupted by brief periods of more rapid change (458)
equilibrio interrumpido: patrón de evolución en el cual los largos períodos de estabilidad se ven interrumpidos por breves períodos de cambio más rápido

Punnett square: diagram that can be used to predict the genotype and phenotype combinations of a genetic cross (267)
cuadro de Punnett: un diagrama que puede utilizarse para predecir las combinaciones de genotipos y fenotipos en un cruce genético

pupa: stage in complete metamorphosis in which the larva develops into an adult (681)
pupa: etapa de la metamorfosis completa en la cual la larva se convierte en un adulto

pupil: small opening in the iris that admits light into the eye (757)
pupila: pequeña abertura en el iris que deja pasar la luz al ojo

Q, R

radial symmetry: body plan in which any number of imaginary planes drawn through the center of the body could divide it into equal halves (611)
simetría radial: diseño corporal en el cual cualquier número de ejes imaginarios dibujados a través del centro del cuerpo lo dividirá en mitades iguales

radiometric dating: method for determining the age of a sample from the amount of a radioactive isotope to the nonradioactive isotope of the same element in a sample (452)
datación radiométrica: método para determinar la edad de una muestra a partir de la cantidad de isótopo radioactivo en relación a la de isótopo no radiactivo del mismo elemento en dicha muestra

reabsorption: process by which water and dissolved substances are taken back into the blood (731)
reabsorción: proceso por el cual el agua y las sustancias disueltas regresan a la sangre

reactant: elements or compounds that enter into a chemical reaction (42)
reactante: elemento o compuesto que participa en una reacción química

receptor: on or in a cell, a specific protein to whose shape fits that of a specific molecular messenger, such as a hormone (183, 588)
receptor: proteína específica que puede encontrarse en la membrana celular o dentro de la célula, cuya forma se corresponde con la de un mensajero molecular específico, por ejemplo una hormona

recombinant DNA: DNA produced by combining DNA from different sources (359)
ADN recombinante: ADN producido por la combinación de ADN de orígenes diferentes

red blood cell: blood cell containing hemoglobin that carries oxygen (790)
glóbulo rojo: célula sanguínea que contiene hemoglobina y transporta oxígeno

reflex: quick, automatic response to a stimulus (747)
reflejo: respuesta rápida y automática a un estímulo

reflex arc: the sensory receptor, sensory neuron, motor neuron, and effector that are involved in a quick response to a stimulus (752)
arco reflejo: el receptor sensorial, la neurona sensorial, la neurona motora y el efector que participan en una respuesta rápida a un estímulo

relative dating: method of determining the age of a fossil by comparing its placement with that of fossils in other rock layers (451)
datación relativa: método para determinar la edad de un fósil comparando su ubicación con la de los fósiles hallados en otras capas de roca

releasing hormone: hormone produced by the hypothalamus that makes the anterior pituitary secrete hormones (813)
hormona liberadora: hormona producida por el hipotálamo que hace que la glándula pituitaria anterior secrete hormonas

renewable resource: resource that can be produced or replaced by healthy ecosystem functions (131)
recurso renovable: recurso que se puede producir o reemplazar mediante el funcionamiento saludable del ecosistema

replication: process of copying DNA prior to cell division (296)
replicación: proceso de copia de ADN previo a la división celular

reproductive isolation: separation of a species or population so that they no longer interbreed and evolve into two separate species (414)

aislamiento reproductor: separación de una especie o de una población de tal manera que ya no pueden aparearse y evolucionan hasta formar dos especies separadas

resource: any necessity of life, such as water, nutrients, light, food, or space (85)

recurso: todo lo necesario para la vida, como agua, nutrientes, luz, alimento o espacio

response: specific reaction to a stimulus (669)

respuesta: reacción específica a un estímulo

resting potential: electrical charge across the cell membrane of a resting neuron (744)

potencial de reposo: carga eléctrica que pasa a través de la membrana celular de una neurona en reposo

restriction enzyme: enzyme that cuts DNA at a sequence of nucleotides (342)

enzima restrictiva: enzima que corta el ADN en una secuencia de nucleótidos

retina: innermost layer of the eye; contains photoreceptors (757)

retina: membrana más interna del ojo; contiene receptores susceptibles a la luz

retrovirus: RNA virus that contains RNA as its genetic information (483)

retrovirus: ARN viral cuya información genética está contenida en el ARN

ribonucleic acid (RNA): single-stranded nucleic acid that contains the sugar ribose (308)

ácido ribonucleico (ARN): hebra única de ácido nucleico que contiene el azúcar ribose

ribosomal RNA (rRNA): type of RNA that combines with proteins to form ribosomes (309)

ARN ribosomal: tipo de ARN que se combina con proteínas para formar los ribosomas

ribosome: cell organelle consisting of RNA and protein found throughout the cytoplasm in a cell; the site of protein synthesis (168)

ribosoma: orgánulo celular formado por ARN y proteína que se halla en el citoplasma de una célula; lugar donde se sintetizan las proteínas

RNA interference (RNAi): introduction of double-stranded RNA into a cell to inhibit gene expression (323)

ARN de interferencia: introducción de un ARN de doble hebra en una célula para inhibir la expresión de genes específicos

RNA polymerase: enzyme that links together the growing chain of RNA nucleotides during transcription using a DNA strand as a template (309)

ARN polimerasa: enzima que enlaza los nucleótidos de la cadena de ARN en crecimiento durante la transcripción, usando una secuencia de ADN como patrón o molde

rod: photoreceptor in the eyes that is sensitive to light but can't distinguish color (757)

bastoncillo: receptor ubicado en los ojos que es susceptible a la luz, pero que no puede distinguir el color

root cap: tough covering of the root tip that protects the meristem (557)

cofia: cubierta dura de la punta de las raíces que protege al meristemo

root hair: small hairs on a root that produce a large surface area through which water and minerals can enter (557)

pelo radicular: pelos pequeños sobre una raíz que producen una superficie extensa a través de la cual pueden penetrar el agua y los minerales

rumen: stomach chamber in cows and related animals in which symbiotic bacteria digest cellulose (648)

panza: cavidad del estómago de las vacas y otros rumiantes en la cual las bacterias simbióticas digieren la celulosa

S

sapwood: in a woody stem, the layer of secondary phloem that surrounds the heartwood; usually active in fluid transport (562)

albura: en un tallo leñoso, la capa de floema secundario que rodea al duramen; participa usualmente en el transporte de fluidos

sarcomere: unit of muscle contraction; composed of two z-lines and the filaments between them (771)

sarcómero: unidad de contracción muscular; compuesto por dos líneas "z" y los filamentos que hay entre ellas

scavenger: animal that consumes the carcasses of other animals (62)

carroñero: animal que consume los cadáveres de otros animales

science: organized way of gathering and analyzing evidence about the natural world (4)

ciencia: manera organizada de reunir y analizar la información sobre el mundo natural

sclerenchyma: type of ground tissue with extremely thick, rigid cell walls that make ground tissue tough and strong (553)

esclerénquima: tipo de tejido fundamental con células extremadamente rígidas y gruesas que lo hacen fuerte y resistente

scrotum: external sac that houses the testes (818)

escroto: bolsa externa que contiene a los testículos

sebaceous gland: gland in the skin that secretes sebum (oily secretion) (776)

glándula sebácea: glándula de la piel que secreta sebo (secreción oleosa)

Glossary *(continued)*

secondary growth: type of growth in dicots in which the stems increase in thickness (562)
crecimiento secundario: tipo de crecimiento de las dicotiledóneas en el cual los tallos aumentan su grosor

secondary succession: type of succession that occurs in an area that was only partially destroyed by disturbances (88)
sucesión secundaria: tipo de sucesión que ocurre en un área destruida sólo parcialmente por alteraciones

seed: plant embryo and a food supply encased in a protective covering (536)
semilla: embrión vegetal y fuente de alimento encerrada en una cubierta protectora

seed coat: tough covering that surrounds and protects the plant embryo and keeps the contents of the seed from drying out (537)
envoltura de la semilla: cubierta dura que rodea y protege al embrión de la planta y evita que el contenido de la semilla se seque

segregation: separation of alleles during gamete formation (264)
segregación: separación de los alelos durante la formación de gametos

selective breeding: method of breeding that allows only those organisms with desired characteristics to produce the next generation (354)
reproducción selectiva o selección artificial: método de reproducción que sólo permite la producción de una nueva generación a aquellos organismos con características deseadas

selectively permeable: property of biological membranes that allows some substances to pass across while others cannot; also called semipermeable membrane (173)
permeabilidad selectiva: propiedad de las membranas biológicas que permite que algunas sustancias pasen a través de ellas mientras que otras no pueden hacerlo; también llamada membrana semipermeable

semen: the combination of sperm and seminal fluid (818)
semen: combinación de esperma y de fluido seminal

semicircular canal: one of three structures in the inner ear that monitor the position of the body in relation to gravity (756)
canal semicircular: una de las tres estructuras ubicadas en el oído interno que controlan la posición del cuerpo en relación con la fuerza de la gravedad

seminiferous tubule: one of hundreds of tubules in each testis in which sperm develop (818)
túbulo seminífero: uno de los cientos de túbulos situados en cada testículo, en los cuales se produce el esperma

sensory neuron: type of nerve cell that receives information from sensory receptors and conveys signals to central nervous system (668)
neurona sensorial: tipo de célula nerviosa que recibe información de los receptores sensoriales y transmite señales al sistema nervioso central

sex chromosome: one of two chromosomes that determines an individual's sex (334)
cromosoma sexual: uno de los pares de cromosomas que determina el sexo de un individuo

sex-linked gene: gene located on a sex chromosome (336)
gen ligado al sexo: gen situado en un cromosoma sexual

sexual reproduction: type of reproduction in which cells from two parents unite to form the first cell of a new organism (15, 236)
reproducción sexual: tipo de reproducción en la cual las células de dos progenitores se unen para formar la primera célula de un nuevo organismo

sexual selection: when individuals select mates based on heritable traits (412)
selección sexual: cuando un individuo elige a su pareja sexual atraído por sus rasgos heredables

sexually transmitted disease (STD): disease that is spread from person to person by sexual contact (823)
enfermedad de transmisión sexual (ETS): enfermedad que se transmite de una persona a otra por contacto sexual

sieve tube element: continuous tube through the plant phloem cells, which are arranged end to end (554)
tubo crivoso: tubo continuo que atraviesa las células del floema vegetal, que están puestas una junto a otra

single-gene trait: trait controlled by one gene that has two alleles (408)
rasgo de un único gen (monogénico): rasgo controlado por un gen que tiene dos alelos

small intestine: digestive organ in which most chemical digestion and absorption of food takes place (725)
intestino delgado: órgano digestivo en el cual tiene lugar la mayor parte de la digestión química y la absorción de los alimentos

smog: gray-brown haze formed by a mixture of chemicals (136)
esmog: neblina marrón grisácea formada por una mezcla de compuestos químicos

society: group of closely related animals of the same species that work together for the benefit of the group (702)
sociedad: grupo de animales de la misma especie, estrechamente relacionados, que trabajan juntos para el beneficio del grupo

solute: substance that is dissolved in a solution (34)
soluto: sustancia que está disuelta en una solución

solution: type of mixture in which all the components are evenly distributed (34)
solución: tipo de mezcla en la cual todos los compuestos están distribuidos de forma homogénea

solvent: dissolving substance in a solution (34)
disolvente: sustancia que disuelve una solución

somatic nervous system: part of the peripheral nervous system that carries signals to and from skeletal muscles (752)
sistema nervioso somático: parte del sistema nervioso periférico que conduce señales hacia y desde los músculos esqueléticos

speciation: formation of a new species (414)
especiación: formación de una nueva especie

species: a group of similar organisms that can breed and produce fertile offspring (414)
especie: un grupo de organismos similares que pueden reproducirse y producir una descendencia fértil

species diversity: number of different species that make up a particular area (138)
diversidad de especies: número de especies diferentes que forman un área determinada

spirillum (pl. spirilla): spiral or corkscrew-shaped prokaryote (486)
espirilo: procariota con forma helicoidal o espiral

spongy mesophyll: layer of loose tissue found beneath the palisade mesophyll in a leaf (564)
mesófilo esponjoso: capa de tejido suelto situado debajo del mesófilo en empalizada de una hoja

sporangium (pl. sporangia): spore capsule in which haploid spores are produced by meiosis (507)
esporangio: cápsula en la cual se producen las esporas haploides mediante meiosis

spore: in prokaryotes, protists, and fungi, any of a variety of thick-walled life cycle stages capable of surviving unfavorable conditions (506)
espora: en los procariotas, los protistas y los hongos, cada una de las células que, en un momento de su ciclo de vida, produce una membrana gruesa y resistente capaz de sobrevivir en condiciones desfavorables

sporophyte: spore-producing plant; the multicellular diploid phase of a plant life cycle (530)
esporofito: planta productora de esporas; la fase diploide multicelular del ciclo vital de una planta

stabilizing selection: form of natural selection in which individuals near the center of a distribution curve have higher fitness than individuals at either end of the curve (410)
selección estabilizadora: forma de selección natural en la cual los individuos situados cerca del centro de una curva de distribución tienen mayor aptitud que los individuos que se hallan en cualquiera de los extremos de la curva

stamen: male part of a flower; contains the anther and filament (579)
estambre: parte masculina de una flor; contiene la antera y el filamento

stem cell: unspecialized cell that can give rise to one or more types of specialized cells (249)
célula troncal: célula no especializada que puede originar uno o más tipos de células especializadas

stigma: sticky part at the top of style; specialized to capture pollen (579)
estigma: parte pegajosa situada en la parte superior del estilo; especializado en atrapar el polen

stimulus (pl. stimuli): signal to which an organism responds (14, 668)
estímulo: señal a la cual responde un organismo

stoma (pl. stomata): small opening in the epidermis of a plant that allows carbon dioxide, water, and oxygen to diffuse into and out of the leaf (564)
estoma: pequeña abertura en la epidermis de una planta que permite que el dióxido de carbono, el agua y el oxígeno entren y salgan de la hoja

stomach: large muscular sac that continues the mechanical and chemical digestion of food (725)
estómago: gran bolsa muscular que continúa la digestión mecánica y química de los alimentos

stroma: fluid portion of the chloroplast; outside of the thylakoids (195)
estroma: parte fluida del cloroplasto; en el exterior de los tilacoides

substrate: reactant of an enzyme-catalyzed reaction (44)
sustrato: reactante de una reacción catalizada por enzimas

suspension: mixture of water and nondissolved material (35)
suspensión: mezcla de agua y material no disuelto

sustainable development: strategy for using natural resources without depleting them and for providing human needs without causing long-term environmental harm (131)
desarrollo sostenible: estrategia para utilizar los recursos naturales sin agotarlos y para satisfacer las necesidades humanas sin causar daños ambientales a largo plazo

symbiosis: relationship in which two species live close together (87)
simbiosis: relación en la cual dos especies viven en estrecha asociación

synapse: point at which a neuron can transfer an impulse to another cell (746)
sinapsis: punto en el cual una neurona puede transferir un impulso a otra célula

systematics: study of the diversity of life and the evolutionary relationships between organisms (430)
sistemática: estudio de la diversidad de la vida y de las relaciones evolutivas entre los organismos

systemic circulation: path of circulation between the heart and the rest of the body (788)
circulación sistémica: recorrido de la circulación entre el corazón y el resto del cuerpo

Glossary *(continued)*

T

taiga: biome with long cold winters and a few months of warm weather; dominated by coniferous evergreens; also called boreal forest (94)
taiga: bioma con inviernos largos y fríos y pocos meses de tiempo cálido; dominado por coníferas de hojas perennes; también llamada bosque boreal

target cell: cell that has a receptor for a particular hormone (588, 810)
célula diana o célula blanco: célula que posee un receptor para una hormona determinada

taste bud: sense organs that detect taste (754)
papila gustativa: órgano sensorial que percibe los sabores

taxon (pl. taxa): group or level of organization into which organisms are classified (430)
taxón: grupo o nivel de organización en que se clasifican los organismos

telomere: repetitive DNA at the end of a eukaryotic chromosome (297)
telómero: ADN repetitivo situado en el extremo de un cromosoma eucariota

telophase: phase of mitosis in which the distinct individual chromosomes begin to spread out into a tangle of chromatin (242)
telofase: fase de la mitosis en la cual los distintos cromosomas individuales comienzan a separarse y a formar hebras de cromatina

temporal isolation: form of reproductive isolation in which two or more species reproduces at different times (415)
aislamiento temporal: forma de aislamiento reproductivo en la cual dos o más especies se reproducen en épocas diferentes

tendon: tough connective tissue that connects skeletal muscles to bones (676, 773)
tendón: tejido conectivo resistente que une los músculos esqueléticos a los huesos

territory: a specific area occupied and protected by an animal or group of animals (702)
territorio: área específica ocupada y protegida por un animal o un grupo de animales

testis (pl. testes): primary male reproductive organ; produces sperm (818)
testículo: órgano reproductor masculino fundamental; produce esperma

tetrad: structure containing four chromatids that forms during meiosis (276)
tétrada: estructura con cuatro cromátidas que se forma durante la meiosis

tetrapod: vertebrate with four limbs (630)
tetrápode: vertebrado con quatro membros

thalamus: brain structure that receives messages from the sense organs and relays the information to the proper region of the cerebrum for further processing (749)
tálamo: estructura cerebral que recibe mensajes de los órganos sensoriales y transmite la información a la región adecuada del cerebro para su procesamiento ulterior

theory: well-tested explanation that unifies a broad range of observations and hypotheses, and enables scientists to make accurate predictions about new situations (11)
teoría: explicación basada en pruebas que unifica una amplia gama de observaciones e hipótesis; permite que los científicos hagan predicciones exactas ante situaciones nuevas

thigmotropism: response of a plant to touch (590)
tigmotropismo: respuesta de una planta al tacto

threshold: minimum level of a stimulus that is required to cause an impulse (745)
umbral: nivel mínimo que debe tener un estímulo para causar un impulso

thylakoid: saclike photosynthetic membranes found in chloroplasts (195)
tilacoide: membranas fotosintéticas con forma de bolsa situadas en los cloroplastos

thyroxine: hormone produced by the thyroid gland, which increases the metabolic rate of cells throughout the body (815)
tiroxina: hormona producida por la glándula tiroides que aumenta el metabolismo de las células de todo el cuerpo

tissue: group of similar cells that perform a particular function (182)
tejido: grupo de células similares que realizan una función en particular

tolerance: ability of an organism to survive and reproduce under circumstances that differ from their optimal conditions (85)
tolerancia: capacidad de un organismo de sobrevivir y reproducirse en circunstancias que difieren de sus condiciones óptimas

totipotent: cells that are able to develop into any type of cell found in the body (including the cells that make up the extraembryonic membranes and placenta) (249)
totipotentes: células capaces de convertirse en cualquier tipo de célula del cuerpo (incluidas las células que forman las membranas situadas fuera del embrión y la placenta)

trachea: tube that connects the larynx to the bronchi; also called the windpipe (796)
tráquea: tubo que conecta a la laringe con los bronquios

tracheid: hollow plant cell in xylem with thick cell walls strengthened by lignin (534)
traqueida: célula vegetal ahuecada del xilema con paredes celulares gruesas, fortalecida por la lignina

trait: specific characteristic of an individual (262)
rasgo: característica específica de un individuo

transcription: synthesis of an RNA molecule from a DNA template (309)
transcripción: síntesis de una molécula de ARN a partir de una secuencia de ADN

transfer RNA (tRNA): type of RNA that carries each amino acid to a ribosome during protein synthesis (309)
ARN de transferencia: tipo de ARN que transporta a cada aminoácido hasta un ribosoma durante la síntesis de proteínas

transformation: process in which one strain of bacteria is changed by a gene or genes from another strain of bacteria (288)
transformación: proceso en el cual una cepa de bacterias es transformada por uno o más genes provenientes de otra cepa de bacterias

transgenic: term used to refer to an organism that contains genes from other organisms (360)
transgénico: término utilizado para referirse a un organismo que contiene genes provenientes de otros organismos

translation: process by which the sequence of bases of an mRNA is converted into the sequence of amino acids of a protein (312)
traducción (genética): proceso por el cual la secuencia de bases de un ARN mensajero se convierte en la secuencia de aminoácidos de una proteína

transpiration: loss of water from a plant through its leaves (565)
transpiración: pérdida del agua de una planta a través de sus hojas

trochophore: free-swimming larval stage of an aquatic mollusk (627)
trocófora: estado larvario de un molusco acuático durante el cual puede nadar libremente

trophic level: each step in a food chain or food web (66)
nivel trófico: cada paso en una cadena o red alimenticia

tropism: movement of a plant toward or away from stimuli (590)
tropismo: movimiento de una planta hacia los estímulos o en dirección opuesta a ellos

tumor: mass of rapidly dividing cells that can damage surrounding tissue (247)
tumor: masa de células que se dividen rápidamente y pueden dañar al tejido circundante

U

understory: layer in a rain forest found underneath the canopy formed by shorter trees and vines (92)
sotobosque: en un bosque tropical, la capa de vegetación que se halla bajo el dosel forestal, formada por árboles más bajos y enredaderas

ureter: tube that carries urine from a kidney to the urinary bladder (731)
uréter: conducto que transporta la orina del riñón a la vejiga urinaria

urethra: tube through which urine leaves the body (731)
uretra: conducto por donde la orina sale del cuerpo

urinary bladder: saclike organ in which urine is stored before being excreted (731)
vejiga urinaria: órgano en forma de bolsa en el cual se almacena la orina antes de ser excretada

V

vaccination: injection of a weakened, or a similar but less dangerous, pathogen to produce immunity (846)
vacunación: inyección de un patógeno debilitado o similar al original, pero menos peligroso, para producir inmunidad

vaccine: preparation of weakened or killed pathogens used to produce immunity to a disease (490)
vacuna: preparación hecha con organismos patógenos debilitados o muertos que se utiliza para producir inmunidad a una enfermedad

vacuole: cell organelle that stores materials such as water, salts, proteins, and carbohydrates (166)
vacuola: orgánulo celular que almacena sustancias como agua, sales, proteínas e hidratos de carbono

valve: flap of connective tissue located between an atrium and a ventricle, or in a vein, that prevents backflow of blood (787)
válvula: pliegue de tejido conectivo ubicado entre una aurícula y un ventrículo, o en una vena, que impide el retroceso de la sangre

vas deferens: tube that carries sperm from the epididymis to the urethra (818)
conducto deferente: tubo que transporta el esperma desde el epidídimo a la uretra

vascular bundle: clusters of xylem and phloem tissue in stems (561)
hacecillo vascular: manojo de tejidos del xilema y del floema en los tallos

vascular cambium: meristem that produces vascular tissues and increases the thickness of stems (562)
cámbium vascular: meristemo que produce tejidos vasculares y aumenta el grosor de los tallos

Glossary *(continued)*

vascular cylinder: central region of a root that includes the vascular tissues—xylem and phloem (557)
cilindro vascular: región central de una raíz que incluye a los tejidos vasculares xilema y floema

vascular tissue: specialized tissue in plants that carries water and nutrients (533)
tejido vascular: tejido especializado de las plantas que transporta agua y nutrientes

vector: animal that transports a pathogen to a human (840)
vector: animal que transmite un patógeno a un ser humano

vegetative reproduction: method of asexual reproduction in plants, which enables a single plant to produce offspring that are genetically identical to itself (583)
reproducción vegetativa: método de reproducción asexual de las plantas que permite que una única planta produzca descendencia genéticamente idéntica a sí misma

vein: blood vessel that carries blood from the body back to the heart (789)
vena: vaso sanguíneo que transporta la sangre del cuerpo de regreso al corazón

ventricle: lower chamber of the heart that pumps blood out of heart to the rest of the body (654, 787)
ventrículo: cavidad inferior del corazón que bombea la sangre fuera del corazón hacia el resto del cuerpo

vertebrate: animal that has a backbone (607)
vertebrado: animal que posee una columna vertebral

vessel element: type of xylem cell that forms part of a continuous tube through which water can move (554)
elemento vascular (o vaso): tipo de célula del xilema que forma parte de un tubo continuo a través del cual el agua puede desplazarse

vestigial structure: structure that is inherited from ancestors but has lost much or all of its original function (393)
estructura vestigial: estructura heredada de los ancestros que ha perdido su función original en gran parte o por completo

villus (pl. villi): fingerlike projection in the small intestine that aids in the absorption of nutrient molecules (727)
vellosidad: proyección en forma de dedo en el intestino delgado que contribuye a la absorción de las moléculas nutrientes

virus: particle made of proteins, nucleic acids, and sometimes lipids that can replicate only by infecting living cells (480)
virus: partícula compuesta por proteínas, ácidos nucleicos y, a veces, lípidos, que puede replicarse sólo infectando células vivas

vitamin: organic molecule that helps regulate body processes (721)
vitamina: molécula orgánica que ayuda a regular los procesos corporales

viviparous: animals that bear live young that are nourished directly by the mother's body as they develop (680)
vivíparo: animal que da a luz crías vivas que se nutren directamente dentro del cuerpo de la madre mientras se desarrollan

W

weather: day-to-day conditions of the atmosphere, including temperature, precipitation, and other factors (82)
tiempo: condiciones diarias de la atmósfera, entre las que se incluyen la temperatura, la precipitación y otros factores

wetland: ecosystem in which water either covers the soil or is present at or near the surface for at least part of the year (97)
humedal: ecosistema en el cual el agua cubre el suelo o está presente en la superficie durante al menos una parte del año

white blood cell: type of blood cell that guards against infection, fights parasites, and attacks bacteria (791)
glóbulo blanco: tipo de célula sanguínea que protege de las infecciones, combate a los parásitos y ataca a las bacterias

woody plant: type of plant made primarily of cells with thick cell walls that support the plant body; includes trees, shrubs, and vines (542)
planta leñosa: tipo de planta constituida fundamentalmente por células con paredes celulares gruesas que sostienen el cuerpo de la planta; en este tipo se incluyen los árboles, arbustos y vides

X, Y, Z

xylem: vascular tissue that carries water upward from the roots to every part of a plant (534)
xilema: tejido vascular que transporta el agua hacia arriba, desde las raíces a cada parte de una planta

zoonosis (pl. zoonoses): disease transmitted from animal to human (840)
zoonosis: enfermedad transmitida por un animal a un ser humano

zygote: fertilized egg (277, 612, 824)
cigoto: huevo fertilizado

Index

Page numbers for vocabulary terms are printed in **boldface** type.

A

Index *cont'd*

B

C

Index *cont'd*

D

Index *cont'd*

E

Index *cont'd*

H

Index *cont'd*

I

J

K

L

M

Index *cont'd*

N

O

Q

R

S

Index *cont'd*

Credits

Staff Credits
Jennifer Angel, Amy C. Austin, Laura Baselice, Neil Benjamin, Peggy Bliss, Diane Braff, Daniel Clem, Glen Dixon, Alicia Franke, Julia Gecha, Ellen Granter, Katherine Immel, Anne Jones, Stephanie Keep, Beth Kun, Kimberly Fekany Lee, Ranida Touranont McKneally, Anne McLaughlin, Rich McMahon, Laura Morgenthau, Debbie Munson, Deborah Nicholls, Michelle Reyes, Rashid Ross, Laurel Smith, Lisa Smith-Ruvalcaba, Ted Smykal, Amanda M. Watters, Merce Wilczek, Berkley Wilson

Additional Credits
Bryan Cholfin, Lisa Furtado Clark, Sharon Donahue, Amy Hamel, Courtenay Kelley, Hilary L. Maybaum, Anakin S. Michele, Julia Osborne, Lisa Redmond, Kelly Rizk, Jan Van Aarsen, Rachel Youdelman

Front Cover, Spine, and Title Page: © Ralph A. Clevenger/Corbis.
Back Cover: © Pete Oxford/Minden Pictures.

iii: T and B, Stew Milne. **iii:** L, Courtesy of Grant Wiggins; R, Jen-Yi Wu. **ix:** TL, Pete Oxford/Minden Pictures; MR, Piotr Naskrecki/Minden Pictures. **x:** Ed Rescheke/Peter Arnold, Inc. **xi:** Andrew Syred/Photo Researchers, Inc. **xii:** T, Colin Keates/DK Images; B, Colin Keates/Dorling Kindersley, Courtesy of the Natural History Museum, London. **xiii:** TR, Southhampton General Hospital/Science Photo Library/Photo Researchers, Inc.; B, Bkgrnd, Neil Lucas/npl/Minden Pictures; BL, Georgette Douwma/ Getty Images; ML, Peter Chadwick/DK Images; **xiv:** T, Ben Twist/istockphoto.com; B, Nature Picture Library/Alamy. **xv:** L, Ingo Arndt , MINDEN PICTURES; R, Nigel J. Dennis; Gallo Images/CORBIS. **xix:** Courtesy of Beng Chiak Tang. **xx:** Stew Milne. **1:** ©Keren Su/CORBIS. **2–3:** ©Steve Wursta/Photopress Washington/Corbis. **3:** TR, Comstock/Jupiter Images. **4:** B, Flip Nicklin/Minden Pictures/Getty Images; T, ©Photopress Washington/Corbis Sygma. **7:** Michael Melford/National Geographic/Getty Images. **8:** Comstock/Jupiter Images. **9:** Photopress Washington/Corbis Sygma. **12:** Comstock/Jupiter Images. **13:** B, Masa Ushioda/Image Quest Marine; T, ©Photopress Washington/Corbis Sygma. **14–15:** Bkgrnd, Martin Rugner/Age Fotostock. **14:** BR, Visuals Unlimited/Corbis; BL, Dwight Kuhn. **15:** TR, John Marshall/Corbis; BL, Visuals Unlimited/Corbis; BR, ©G. Thomas Bishop/Custom Medical Stock ; Lower MR, Dwight Kuhn; Upper MR, Microscan/Phototake USA. **16:** Klein/Peter Arnold, Inc. **17:** Louie Psihoyos/CORBIS. **18:** T, Michael Nichols/ National Geographic/Getty Images; B, Tek Image/Photo Researchers, Inc. **19:** B, Comstock/Jupiter Images; T, Daniel J. Cox/ Corbis. **21:** TR, Brand X/Jupiter Images. **22:** Photopress Washington/ Corbis Sygma. **24:** Comstock/Jupiter Images. **26–27:** Hans Strand/ CORBIS. **27:** TR, Flip Nicklin/Minden Pictures. **28:** Hans Strand/ CORBIS. **31:** Flip Nicklin/Minden Pictures. **33:** Hans Strand/ CORBIS. **34:** BM, Richard Megna/Fundamental Photos. **35:** B, Noam Armonn/istockphoto.com; BM, fanelie rosier/istockphoto. com; TM, Milos Luzanin/istockphoto.com; T, Nancy Louie/ istockphoto.com. **36:** Flip Nicklin/Minden Pictures. **37:** Hans Strand/CORBIS. **42:** Hans Strand/CORBIS. **45:** M, Thomas A. Steitz, Howard Hughes Medial Institute, Yale University. **45:** Flip Nicklin/Minden Pictures. **48:** Hans Strand/CORBIS. **50:** Flip Nicklin/Minden Pictures. **52:** Keren Su/CORBIS. **53:** ML, ©Pete Oxford/Minden Pictures. **54–55:** Stan Osolinski/Oxford Scientific/ Jupiter Images. **55:** TR, Jeff Rotman/www.jeffrotman.com. **56:** BL, Tim Graham/Getty Images; BM, Arnold Streifengnu Shah/Peter Arnold Inc.; BR, Yva Momatiuk & John Eastcott/Minden Pictures; T, Stan Osolinski/Oxford Scientific/Jupiter Images. **57:** BR, NASA; BL, Dmitri Kessel/Time Life Pictures/Getty Images; BM, Planetary Visions Ltd./Photo Researchers, Inc. **59:** B, Jeff Rotman/www. jeffrotman.com; T, Frans Lanting/Minden Pictures. **60:** B, Piotr Naskrecki /Minden Pictures. T, Stan Osolinski/Oxford Scientific/ Jupiter Images. **61:** Vincenzo Lombardo/Photodisc/Getty Images. **62:** B, Jeff Rotman/www.jeffrotman.com; Inset,TR, Pete Oxford/ MINDEN PICTURES/Getty Images; Inset,TL, Kevin Schafer/Peter Arnold ; Inset, M, Roy Toft/National Geographic Creative/Getty Images; Inset, BR, Devon Graham/Project Amazonas, Inc.; Inset, BL, Wolfgang Kaehler/Alamy; Bkgrnd, Will & Deni McIntyre/ Stone/Getty Images; Inset, ML, Carol Farneti Foster/Oxford Scientific/PhotoLibrary. **63:** Stan Osolinski/Oxford Scientific/ Jupiter Images. **67:** Jeff Rotman/www.jeffrotman.com. **68:** Stan Osolinski/Oxford Scientific/Jupiter Images. **73:** Jeff Rotman/ Jeffrotman.com. **75:** Piotr Naskrecki/Minden Pictures. **76:** Stan Osolinski/Oxford Scientific/Jupiter Images. **78:** Jeff Rotman/www. jeffrotman.com. **81–82:** Anup Shah/ npl/Minden Pictures. **81:** George McCarthy/Nature Picture Library. **82:** Anup Shah/ npl/ Minden Pictures. **85:** Anup Shah/ npl/Minden Pictures. **87:** George McCarthy/Nature Picture Library. **88:** Anup Shah/ npl/Minden Pictures. **90:** MR, © Macduff Everton/CORBIS; B, ©Hallmark Institute /Photolibrary Group; ML, © Reuters/CORBIS; T, © Gary Braasch/CORBIS. **91:** TL, Anup Shah/ npl/Minden Pictures. **92:** T, © Galen Rowell/CORBIS.; M, ©Staffan Widstrand/Nature Picture Library; B, © Michele Burgess/SuperStock. **93:** T, © Darrell Gulin/ CORBIS; TM, ©Jim Brandenburg/Minden Pictures; M, © Debra Behr/Alamy; BM, © Natural Selection Jerry Whaley/Design Pics/ Corbis; B, © Stuart Westmorland/CORBIS. **94:** T, age fotostock/ SuperStock ; B, ©2008 Jim D. Barr/AlaskaStock.com. **95:** George McCarthy/Nature Picture Library. **96:** B, © Corbis/SuperStock ; T, Anup Shah/npl/Minden Pictures. **97:** MR, © Pacific Stock/ SuperStock; R, David Noton/Nature Picture Library; L, Philippe Clement/Nature Picture Library; ML, ©MICHAEL GADOMSKI/ Animals Animals - Earth Scenes; M, ©GORDON & CATHY ILLG/ Animals Animals - Earth Scenes. **98:** TL, ©Cindy Ruggieri; TR, © age fotostock/SuperStock; B, ©Norbert Wu/Minden Pictures. **99:** George McCarthy/Nature Picture Library. **102:** Anup Shah/npl/ Minden Pictures. **103:** Anthony Bannister; Gallo Images/CORBIS. **104:** George McCarthy/Nature Picture Library. **106–107:** Jean-Paul Ferrero/Auscape/Minden Pictures. **107:** TR, Eric Wanders/Foto Natura/Minden Pictures. **108:** T, Jean-Paul Ferrero/Auscape/ Minden Pictures; B, Agricultural Research Service, USDA/Stephen Ausmus. **109:** SciMAT/Photo Researchers, Inc. **110:** Ingo Arndt/ Minden Pictures. **111:** Eric Wanders/ Foto Natura/Minden Pictures. **112:** TR, Joke Stuurman-Huitema/ Foto Natura/Minden Pictures; TL, Jean-Paul Ferrero/ Auscape/Minden Pictures. **113:** D. Robert & Lorri Franz/CORBIS. **114:** Les Stocker/Photolibrary Group. **115:** T, AP Photo/Xinhua Photo, Jiang Yi; B, Eric Wanders/ Foto Natura/Minden Pictures. **117:** Jean-Paul Ferrero/ Auscape/ Minden Pictures. **119:** image100/Photolibrary. **121:** R, AP Photo/ Xinhua Photo, Jiang Yi; L, Ingo Arndt/Minden Pictures. **122:** Jean-Paul Ferrero/Auscape/Minden Pictures. **124:** Eric Wanders/ Foto Natura/Minden Pictures. **126–127:** NASA/NSSDC. **127:** SIME s.a.s/eStock Photo. **128:** BL, Douglas Peebles/eStockPhoto ; BR, Douglas Peebles/eStockPhoto ; T, NASA/NSSDC. **129:** Patrick Bennett/Superstock. **130:** Inset, The Wetlands Initiative archives; Bkgrnd, By Gary Sullivan, The Wetlands Initiative. **131:** SIME s.a.s/ eStock Photo. **132:** B, AP Photo; T, NASA/NSSDC. **135:** Maxphotos/Newscom. **136:** Jason Lee/Reuters Pictures. **137:** SIME s.a.s/eStock Photo. **138:** B, AP Photo/Penn State University, S. Blair Hedges); TL, NASA/NSSDC. **139:** M, Noel Hendrickson/GETTY IMAGES; R, Marcos G. Meider/Superstock; L, Paolo Aguilar/epa/ Corbis. **140:** B, Creatas/Photolibrary Group. **141:** Katherine Feng/ Globio/Minden Pictures. **142:** BR, SIME s.a.s/ eStock Photo; T, Barbara Walton/epa/Corbis. **143:** NASA/NSSDC. **145:** T, Images by Greg Shirah, NASA Goddard Space Flight Center Scientific Visualization Studio; B, Jerry Mason/Photo Researchers, Inc. **146:** T, Jeffrey L. Rotman/CORBIS; B, BRIAN J. SKERRY/National Geographic Image Collection. **147:** Armin Rose/Shutterstock. **148:** Sumio Harada/Minden Pictures. **149:** Inset, T, UPPA/Photoshot/ Newscom; Inset, B, Realimage/Alamy; Bkgrnd, Charmagne Leung/ California Academy of Sciences. **151:** Patrick Bennett/Superstock. **152:** NASA/NSSDC. **154:** SIME s.a.s/eStock Photo. **157:** Clouds Hill Imaging Ltd./CORBIS **158–159:** Dr. Robert Berdan. **159:** simple stock shots/Age Fotostock. **160:** B, ©Dwight Kuhn; T, Dr. Robert Berdan; M, ©Grafissimo /iStockphoto.com. **161:** BL, Michael Abbey/Photo Researchers, Inc.; BR, SciMAT/Photo Researchers, Inc.; BM, Dr. Gopal Murti/Visuals Unlimited. **163:** simple stock shots/Age Fotostock. **164:** Dr. Robert Berdan. **166:** ISM/Phototake; R, Eric Grave/Science Photo Library/Photo Researchers, Inc. **167:** Dr. Torsten Wittmann/Photo Researchers, Inc. **170:** Biophoto Associates/Photo Researchers, Inc. **171:** Ed Reschke/Peter Arnold, Inc. **172:** ISM/Phototake USA. **176:** Dr. Robert Berdan. **180:** B, © simple stock shots/Age Fotostock; T, NIBSC/Photo Researchers, Inc. **181:** T, Dr. Robert Berdan. B, Volker Steger/Christian Bardele/SPL/Photo Researchers, Inc. **183:** Dennis Kunkel/Phototake USA. **185:** B, Volker Steger/Christian Bardele/SPL/Photo Researchers, Inc. **186:** Dr. Robert Berdan. **188:** Alan Bailey/Rubberball/Getty Images. **190–191:** Perennou Nuridsany/Photo Researchers, Inc. **191:** TR, Watering can (earthenware), English School, (16th century)/Museum of London, UK,/The Bridgeman Art Library International. **192:** Perennou Nuridsany/Photo Researchers, Inc. **194:** T, Anup Shah/ Minden Pictures; B, Watering can (earthenware), English School, (16th century)/Museum of London, UK,/The Bridgeman Art Library International. **195:** B, Andy Small/GAP Photos/Getty Images; T, Perennou Nuridsany/Photo Researchers, Inc. **197:** Watering can (earthenware), English School (16th century)/ Museum of London, UK/The Bridgeman Art Library International. **199:** Perennou Nuridsany/Photo Researchers, Inc. **201:** Imagewerks/Getty Images. **203:** T, Richard Cummins/Corbis; B, Watering can (earthenware), English School, (16th century)/ Museum of London, UK/The Bridgeman Art Library International. **204:** Pearson Education. Photo by Russ Lappa. **205:** Anup Shah/Minden Pictures. **208:** Watering can (earthenware), English School, (16th century)/©Museum of London, UK,/ TheBridgeman Art Library International. **210–211:** Professor Pietro M. Motta/Photo Researchers, Inc. **211:** Flip Nicklin/Minden Pictures. **212:** T, Professor Pietro M. Motta/Photo Researchers, Inc., B, Stock Food Creative/Getty Images. **215:** Flip Nicklin/ Minden Pictures. **216:** T, Professor Pietro M. Motta/Photo Researchers, Inc.; B, Pete Oxford/MINDEN PICTURES/Getty Images **220:** Jim McIsaac/Getty Images. **222:** Flip Nicklin/Minden Pictures. **223:** Professor Pietro M. Motta/Photo Researchers, Inc. **225:** T, Brand X Pictures/Photo Library ; B, Flip Nicklin/Minden Pictures. **227:** Brand X Pictures/Photo Library. **228:** Professor Pietro M. Motta/Photo Researchers, Inc. **230:** Michael Nolan/Peter Arnold. **232–233:** Michael Abbey/Photo Researchers, Inc. **233:** Joe McDonald/CORBIS. **234:** Michael Abbey/Photo Researchers, Inc. **236:** R, ©Konrad Wothe/Science Faction; L, ©CNRI/Photo Researchers, Inc.; M, ©Oxford Scientific Films/Photolibrary. **237:** Joe McDonald/CORBIS. **239:** Michael Abbey/Photo Researchers, Inc. **243:** T, DR GOPAL MURTI/SCIENCE PHOTO LIBRARY/ Photo Researchers; M, Biodisc/Visuals Unlimited ; B, © Joe McDonald/CORBIS. **244:** All Photos, Ed Rescheke/Peter Arnold, Inc. **245:** BL, Paul Aresu/Getty Images; BR, Scott Camazine/Photo Researchers, Inc.; T, Michael Abbey/Photo Researchers, Inc. **246:** L, Inga Spence/Visuals Unlimited ; R, Nina Lampen/Phototake. **247:** Joe McDonald/CORBIS. **248:** BL, Scientifica/Visuals Unlimited ; B, ML, Jim Haseloff/The Wellcome Trust Medical Photographic Library; B, MR, Dr. John D. Cunningham/Visuals Unlimited ; BR, Neil Fletcher © Dorling Kindersley; TL, Michael Abbey/Photo Researchers, Inc. **251:** Joe McDonald/CORBIS. **253:** L, ©Konrad Wothe/Science Faction; R, ©DR GOPAL MURTI/SCIENCE PHOTO LIBRARY/Photo Researchers. **254:** ©Michael Abbey/ Photo Researchers, Inc. **256:** TR, © Joe McDonald/CORBIS. **259:** BSIP LAURENT/PhotoResearchers Inc. **260–261:** M, © Biosphoto/ Labat J.-M. & Rouquette F./Peter Arnold Inc. **261:** blickwinkel/ Alamy. **262:** ©Biosphoto/Labat J.-M. & Rouquette F./Peter Arnold Inc. **265:** blickwinkel/Alamy. **266:** B, Brand X/Jupiter Images; T, © Biosphoto/Labat J.-M. & Rouquette F./Peter Arnold Inc. **270:** Maximilian Weinzier/Alamy. **271:** BL, Christopher Burrows/ Alamy; BR, Christopher Burrows/Alamy; in grid, Cubo Images/ Alamy; T, © Biosphoto/Labat J.-M. & Rouquette F./Peter Arnold Inc. **273:** TL, Stepan Bartos/Photo Researchers, Inc.; TR, Stepan Bartos/Photo Researchers, Inc. B, blickwinkel/Alamy **275:** Biosphoto/Labat J.-M. & Rouquette F./Peter Arnold Inc. **279:** blickwinkel/Alamy. **282:** Biosphoto/Labat J.-M. & Rouquette F./ Peter Arnold Inc. **283:** Michael Quinton/Minden Pictures. **284:** blickwinkel/Alamy. **286–287:** Charles C. Benton/Kite Aerial Photography. **287:** T, Pearson Education. Photo by Heather Wright; B, ©aliciahh/istockphoto.com. **288:** Charles C. Benton/ Kite Aerial Photography. **292:** Charles C. Benton/Kite Aerial Photography. **293:** T, by Courtesy of the National Portrait Gallery, London; Inset, R, Courtesy Ava Helen & Linus Pauling Papers/ Special Collections/Oregon State University; Inset, L, With permission of the University Archives, Columbia University in the City of New York. **294:** T, A. Barrington Brown/Photo Researchers, Inc. ; B, Wellcome Library, London. **295:** T, Pearson Education. Photo by Heather Wright; B, aliciahh/istockphoto.com. **296:** B, Dr. Gopal Murti/Science Photo Library/Photo Researchers, Inc.; T, Charles C. Benton/Kite Aerial Photography . **299:** T, Pearson Education. Photo by Heather Wright; B, aliciahh/istockphoto.com. **301:** Courtesy Ava Helen & Linus Pauling Papers/Special Collections/Oregon State University. **302:** Charles C. Benton/Kite Aerial Photography. **304:** T, Pearson Education. Photo by Heather Wright; B, aliciahh/istockphoto.com. **306–307:** Konrad Wothe/ Minden Pictures. **307:** EYE OF SCIENCE/SPL/Photo Researchers, Inc. **308:** Konrad Wothe/Minden Pictures. **311:** Konrad Wothe/ Minden Pictures. **315:** ML, EYE OF SCIENCE/SPL/Photo Researchers, Inc. **316:** BL, Bob Gibbons/Photo Researchers, Inc.; T, Konrad Wothe/Minden Pictures; BR, David Scharf/Science Faction Images. **318:** Eye of Science/Photo Researchers, Inc. **319:** travismanley/istockphoto.com. **320:** B, G. Murti/Photo Researchers, Inc.; T, Konrad Wothe/Minden Pictures. **325:** EYE OF SCIENCE/SPL/Photo Researchers, Inc. **327:** Bob Gibbons/Photo Researchers, Inc. **328:** Konrad Wothe/Minden Pictures. **330:** EYE OF SCIENCE/SPL/Photo Researchers, Inc. **332–333:** Digital Vision. **333:** Dr. Stanley Flegler/Visuals Unlimited. **334:** T, Digital Vision; B, CNRI/Photo Researchers, Inc. **336:** Dave King,Dorling Kindersley. **337:** Dr. Stanley Flegler/Visuals Unlimited. **338:** Pearson. **339:** Digital Vision. **341:** Dr. Stanley Flegler/Visuals Unlimited. **342:** Digital Vision. **345:** B, Dr. Stanley Flegler/Visuals Unlimited; TR, Reprinted by permission from Macmillan Publishers Ltd: Nature, February 15, 2001, Volume 409, Number 6822. Copyright © 2001; TL, From *Science,* Volume 291, No. 5507, February 16 2001. Reprinted with permission from AAAS. Image © Ann Cutting. **346:** AP Images. **347:** Dave King,Dorling Kindersley. **348:** Digital Vision. **349:** Dennis Kunkel/Phototake. **350:** Dr. Stanley Flegler/Visuals Unlimited. **352–353:** Gyeongsang National University/HO/epa/Corbis. **353:** David Parker/Photo Researchers, Inc. **354:** TR, Thinkstock/Jupiter Images; B, Corbis Royalty Free/Jupiter Images; TL, ©Gyeongsang National University/HO/epa/Corbis. **356:** Jose F. Poblete/Corbis. **357:** B, Wernher Krutein/Photovault; T, ©Gyeongsang National University/ HO/epa/Corbis. **359:** Torunn Berge/Photo Researchers, Inc. **361:** David Parker/Photo Researchers, Inc. **362:** Goat, Jan L. Carson/ University of California at Davis; Girl, ©istockphoto.com/Katrina Brown; T, ©Gyeongsang National University/HO/epa/Corbis. **363:** UPI/Digital Railroad.net. **366:** T, Look at Science/Phototake USA; B, David Parker/Photo Researchers, Inc. **367:** B, © Charles Smith/ Corbis; T, ©Gyeongsang National University/HO/epa/Corbis. **368:** N. Cobbing./Still Pictures/Peter Arnold, Inc. **369:** T, www.glofish. com; B ©David Parker/Photo Researchers, Inc. **372:** Gyeongsang National University/HO/epa/Corbis. **374:** David Parker/Photo Researchers, Inc. **377:** Tui De Roy/ Minden Pictures. **378–379:** Chip Clark. **379:** Chris Johns/National Geographic Image Collection. **380:** R, AKG/Photo Researchers, Inc.; L, Chip Clark. **381:** DK IMAGES. **382:** T, Tui De Roy/ Minden Pictures; B, Tui De Roy/ Minden Pictures. **383:** B, Chris Johns/National Geographic Image Collection; T, Courtesy The Field Museum. **384:** B, Momatiuk - Eastcott/Corbis; T, © Chip Clark. **385:** Tom Vezo/Minden Pictures. **386:** Snark/Art Resource, NY. **387:** M, Portrait of Pouter Pigeon (coloured engraving), Wolsenholme, D. (fl.1862)/Down House, Kent, UK, /The Bridgeman Art Library International; R, Portrait of a Beard Pigeon (coloured engraving), Wolsenholme, D. (fl.1862)/Down House, Kent, UK, /The Bridgeman Art Library International; L, Portrait of a Carrier Pigeon (coloured engraving), Wolsenholme, D. (fl.1862)/Down House, Kent, UK, /The Bridgeman Art Library International. **388:** Chip Clark. **389:** TL, Hal Beral/V&W/The Image Works; BL, Steve Kaufman/Peter Arnold; BR, E.R. DEGGINGER/Animals Animals - Earth Scenes; TR, Markus Varesvuo/Nature Picture Library. **391:** B, CHRIS JOHNS/National Geographic Image Collection; T, Cambridge University Library. **392:** Chip Clark. **397:** CHRIS JOHNS/National Geographic Image Collection. **400:** Chip Clark. **402:** CHRIS JOHNS/National Geographic Image Collection. **404–405:** R. Hoelzl/Peter Arnold Inc. **405:** Pearson Education. Photo by Heather Wright. **406:** T, R.Hoelzl/Peter Arnold Inc. **407:** Brand X/ SuperStock. **408:** Pearson Education. Photo by Heather Wright. **409:** B, Charlie Ott/Photo Researchers, Inc.; T, R.Hoelzl/Peter Arnold Inc. **412:** Ashley Cooper/Corbis. **413:** Pearson Education. Photo by Heather Wright. **414:** B, Rick & Nora Bowers/Alamy; M, Tom & Pat Leeson/DRKphoto.com; T, R.Hoelzl/Peter Arnold Inc. **416:** Pearson Education. Photo by Heather Wright. **417:** R.Hoelzl/ Peter Arnold Inc. **422:** R.Hoelzl/Peter Arnold Inc. **424:** B, James Cavallini/Custom Medical Stock Photo 2002; T, Pearson Education. Photo by Heather Wright. **426–427:** Chip Clark. **427:** BR, Mark Hamblin/age fotostock; TR, Kevin Schafer/age fotostock. **428:** B, Hans Strand/Age Fotostock; T, Chip Clark. **429:** T, Matthew Ward /DK Images; Chart, T, Matthew Ward/DK Images; Chart, Upper ML, Matthew Ward/DK Images; BL, Matthew Ward/DK Images; Chart, Lower MR, Matthew Ward/DK Images; BR, Matthew Ward/DK Images; Chart, Upper MR, Peter Chadwick/ DK Images. **430:** Portrait of Carl von Linnaeus (1707–78) (oil on canvas), Roslin, Alexander (1718–93)/Chateau de Versailles, France, Lauros/Giraudon /The Bridgeman Art Library International. **431:** W. Perry Conway/CORBIS. **432:** TL, 2006 Kåre Telnes/Image Quest Marine; ML, age fotostock/SuperStock; MR, Piotr Naskrecki/ Minden Pictures; B, Mark Hamblin/age fotostock. **433:** Chip Clark. **436:** T, Joe McDonald/Animals Animals - Earth Scenes; M, B.K. Wheeler/VIREO; B, Chris Zwaenepoel/istockphoto.com. **437:** Upper MR, Gerry Ellis/Minden Pictures; Upper ML, Corbis/ SuperStock; Upper L, Raven Regan/robertstock.com; Upper R, Patricio Robles Gil/ Sierra Madre/Minden Pictures; B, Mark Hamblin/age fotostock. **438:** Chip Clark. **439:** Eye of Science/ Photo Researchers Inc. **443:** B, Science Source/Photo Researchers, Inc.; T, Portrait of Carl von Linnaeus (1707–78) (oil on canvas), Roslin, Alexander (1718–93)/Chateau de Versailles, France, Lauros/Giraudon /The Bridgeman Art Library International. **444:** Chip Clark. **446:** T, Kevin Schafer/age fotostock; B, Mark Hamblin/ age fotostock. **448–449:** Natural History Museum, London. **449:** Marvin Dembinsky Photo Associates/Alamy. **450:** B, Jason Edwards/National Geographic/Getty Images; T, Natural History Museum, London. **455:** Marvin Dembinsky Photo Associates/ Alamy. **456:** Natural History Museum, London. **461:** T, Frans Lanting/Corbis; B, Marvin Dembinsky Photo Associates/Alamy. **462:** T, Natural History Museum, London.; B, age fotostock/ SuperStock. **466:** ML, ©Chase Studio/Photo Researchers, Inc.; TL, Colin Keates/DK Images; BL, Colin Keates/DK Images; BR, Colin Keates/DK Images; MR, DK Images; TR, Colin Keates/DK Images.

467: ML, Publiphoto/Photo Researchers, Inc.; BR, Laurie O'Keefe/Photo Researchers, Inc. ; Upper MR, Interfoto Pressebildagentur/Alamy ; TL, Luis Rey/DK Images; BL, © Kevin Schafer/CORBIS; Lower MR, Colin Keates/DK Images; TR, Kathie Atkinson/Oxford Scientific Films/PhotoLibrary. **468:** MR, DeAgostini Picture Library/Getty Images; TM, John Cancalosi/Peter Arnold; BL, Oxford Scientific/PhotoLibrary ; BR, John Downs/DK Images; ML, Ibis Bildagentur/Peter Arnold; TL, Colin Keates/DK Images; MB, John Cancalosi/Peter Arnold; TR, Louie Psihoyos/Corbis. **469:** MR, Corbis/PhotoLibrary; TR, Colin Keates/DK Images; TL, Robin Carter ©Dorling Kindersley; BL, Colin Keates© Dorling Kindersley, Courtesy of the Natural History Museum, London; BR, ©2003 Brochard/AgenceImages. **472:** Natural History Museum, London. **474:** Marvin Dembinsky Photo Associates/Alamy. **477:** Studio Eye/Corbis. **478–479:** Ted Horowitz/CORBIS. **479:** Peter Cade/Getty Images. **480:** Ted Horowitz/CORBIS. **484:** Peter Cade/Getty Images. **485:** B, CNRI/Science Photo Library/Photo Researchers, Inc; T, Ted Horowitz/CORBIS. **486:** L, David Scharf/Peter Arnold; M, Stone/Getty Images; R, Scott Camazine/Photo Researchers, Inc. **487:** Tony Freeman/PhotoEdit, Inc. **488:** Wally Eberhart/Visuals Unlimited, Inc. **489:** Ted Horowitz/CORBIS. **490:** Scott Camazine/Alamy. **491:** B, Juergen Berger/Science Photo Library/Photo Researchers, Inc.; T, Ted Horowitz/CORBIS. **491:** Bkgrnd, Bettmann/Corbis; BL, Pearson Education Curriculum Group; BR, Saturn Stills/Science Photo Library/Photo Researchers, Inc.; Inset, L, Dr. Hans Gelderblom/Visuals Unlimited/Getty Images. **493:** Peter Cade/Getty Images. **495:** R, Pearson Education Curriculum Group; L, David Scharf/Science Faction/Getty Images. **496:** Ted Horowitz/CORBIS. **498:** B, Image:LambdaPlaques.jpg; T, ©Peter Cade/Getty Images. **500–501:** Patricio Robles Gil/Sierra Madre/Minden Pictures, Inc. **501:** Nigel Cattlin/Visuals Unlimited. **502:** BL, AlaskaStock/Photolibrary; BR, Steve Gschmeissner/Photo Researchers, Inc.; T, Patricio Robles Gil/Sierra Madre/Minden. **503:** T, M. I. Walker/Photo Researchers, Inc.; B, Eye of Science/Photo Researchers, Inc. **504:** ML, AlaskaStock/Photolibrary; MR, Steve Gschmeissner/Photo Researchers, Inc.; L, M. I. Walker/Photo Researchers, Inc.; R, Eye of Science/Photo Researchers, Inc. **505:** BL, Michael Abbey/Visuals Unlimited; BM, Michael Abbey/Visuals Unlimited ; BR, Michael Abbey/Visuals Unlimited ; T, Patricio Robles Gil/Sierra Madre/Minden. **508:** ©Nigel Cattlin/Visuals Unlimited. **509:** BL, Brandon D. Cole; BR, Birgitte Wilms/Minden Pictures, Inc.; T, Patricio Robles Gil/Sierra Madre/Minden Pictures, Inc. **510:** Eric V. Grave/Photo Researchers, Inc. **511:** Eric V. Grave/Photo Researchers, Inc. **512:** TR, Eye of Science/Photo Researchers, Inc.; BL, Eye of Science/Photo Researchers, Inc.; BR, Andrew Paul Leonard/Photo Researchers, Inc.; TL, John McConnico/AP Images. **513:** Nigel Cattlin/Visuals Unlimited. **514:** M, Alamy Images; B, Michael Hutchinson/Minden Pictures, Inc.; T, Patricio Robles Gil/Sierra Madre/Minden. **517:** T, Inga Spence/Photo Researchers, Inc.; B, Michael Fogden/DRK Photo. **518:** Alamy Images. **519:** T, Science VU/Visuals Unlimited ; B, Nigel Cattlin /Visuals Unlimited. **521:** Steve Gschmeissner/Photo Researchers, Inc. **522:** Patricio Robles Gil/Sierra Madre/Minden. **524:** M, Andrew Syred/SPL/Photo Researchers, Inc.; T, Nigel Cattlin/Visuals Unlimited; B, B. Runk/S. Schoenberger/Grant Heilman Photography. **526–527:** Craig Tuttle/Corbis. **527:** Vienna Report Agency/Sygma/Corbis. **528:** Craig Tuttle/Corbis. **531:** Vienna Report Agency/Sygma/Corbis. **532:** BL, Perennou Nuridsany/Photo Researchers, Inc.; BR, Roland Birke/Peter Arnold Inc.; T, Craig Tuttle/Corbis. **534:** Bkgrnd, Howard Rice/Photolibrary Group; Inset, David T. Webb, Botany, Univ. of Hawaii. **535:** Vienna Report Agency/Sygma/Corbis. **536:** Craig Tuttle/Corbis. **537:** TL, M.Gabriel/Peter Arnold Inc.; BL, BARRY RUNK/Grant Heilman Photography; TM, Walter H. Hodge/Peter Arnold Inc.; TR, Manfred Kage/Peter Arnold Inc.; Upper MR, Peter Chadwick/DK Images; BM, Runk/Schoenberger/Alamy; Lower MR, CMSP; BR, Runk/Schoenberger/Alamy. **538:** Perennou Nuridsany/Photo Researchers, Inc. **539:** Vienna Report Agency/Sygma/Corbis. **540:** BR, Herve Conge/ISM/Phototake; T, Craig Tuttle/Corbis. **541:** T, all, Peter Chadwick/DK Images; BR, ©Georgette Douwma/Getty Images; **543:** TL, Sangtae Kim, University of Florida. Photo courtesy the Botanical Society of America.: T, ML, Pavel Bortel/istockphoto.com; TM, The Garden Picture Library/Alamy; TR, Skip ODonnell/istockphoto.com; T, MR, Julien Grondin/istockphoto.com; B, Vienna Report Agency/Sygma/Corbis. **545:** L, Peter Chadwick/DK Images; R, ©Peter Chadwick/DK Images. **546:** Craig Tuttle/Corbis. **548:** T, Photo Archives of the South Tyrol Museum of Archaeology; B, WILLIAM MANNING/Grant Heilman Photography. **550–551:** Perennou Nuridsany/Science Photo Library/Photo Researchers, Inc. **551:** Ted Mead/PhotoLibrary **552:** Perennou Nuridsany/Science Photo Library/ Photo Researchers, Inc. **555:** T, Ray F. Evert, University of Wisconsin; M, Ray F. Evert, University of Wisconsin; B, Ted Mead/PhotoLibrary. **556:** BL, Runk Schoenberger/Grant Heilman Photography; BR, Derek Croucher/Alamy; T, Perennou Nuridsany/Science Photo Library **559:** Ted Mead/Photo Library. **560:** Perennou Nuridsany/Science Photo Library/ Photo Researchers, Inc. **563:** T, Manfred Kage/Peter Arnold; B, Ted Mead/PhotoLibrary. **564:** TL, Perennou Nuridsany/Science Photo Library/ Photo Researchers, Inc. **566:** ML, Kjell B. Sandved/Visuals Unlimited; TR, Peter Miller/Getty Images; MR, Muriel Hazan/BIOS/Peter Arnold; TL, David Kirkland/PhotoLibrary; B, Ted Mead/PhotoLibrary. **567:** Perennou Nuridsany/Science Photo Library/ Photo Researchers, Inc. **569:** Ted Mead/PhotoLibrary. **571:** L, Runk Schoenberger/Grant Heilman Photography; R, Derek Croucher/Alamy. **572:** Perennou Nuridsany/Science Photo Library/Photo Researchers, Inc. **574:** Ted Mead/PhotoLibrary. **576–577:** 2006 Bob Sacha and Jacob Louis Mey (AMNH). **577:** Martin Perez/istockphoto.com. **578:** 2006 Bob Sacha and Jacob Louis Mey (AMNH). **581:** R, ©Nature Production/Minden Pictures; L, Wermter, C./Peter Arnold Inc. **583:** TL, Ed Reschke/Peter Arnold Inc.; Bkgrnd, Richard Cummins/SuperStock; BL, Michael P. Gadomski/Photo Researchers, Inc. **584:** TL, National Trust Photographic Library/Stephen Robson/The Image Works; TR, BIOS Bios - Auteurs Lafranchis Tristan/Peter Arnold Inc.; B, Martin Perez/istockphoto.com. **585:** BL, Westend61/Alamy; BM, Eisenhut & Mayer/Stockfood/Getty Images; MR, Corbis/SuperStock ; TL, 2006 Bob Sacha and Jacob Louis Mey (AMNH). **586:** L, Thomas Mangelsen/Minden Pictures; R, Peter Lilja/age Fotostock; M, Werner H. Muller/Peter Arnold Inc. **587:** Martin Perez/istockphoto.com. **588:** 2006 Bob Sacha and Jacob Louis Mey (AMNH). **592:** Bkgrnd, T, Mark Bolton/Corbis; Inset, T, GRANT HEILMAN/Grant Heilman Photography; B, Martin Perez/istockphoto.com. **593:** B, Keren Su/CORBIS; T, 2006 Bob Sacha and Jacob Louis Mey (AMNH). **594:** Nicolle Rager Fuller, National Science Foundation. **595:** BR, © Brand X/SuperStock; BM, ©Mats Tooming/istockphoto.com; BL, © Christophe Boisvieux/Corbis; TL, © Ingram Publishing/SuperStock; TR, Photo by Heather Wright. ©Pearson Education. **598:** 2006 Bob Sacha and Jacob Louis Mey (AMNH). **600:** Martin Perez/istockphoto.com. **603:** Martin Harvey/Corbis **604–605:** Jeff Hunter/The Image Bank/Getty Images. **605:** Bkgrnd and Inset: Kelly McGann Hebert **606:** Jeff Hunter/The Image Bank/Getty Images. **608:** Tierbild Okapia/Photo Researchers, Inc. **610:** T, ©M. FOGDEN/OSF/Animals Animals - Earth Scenes; B, Bkgrnd and Inset: Kelly McGann Hebert **611:** Jeff Hunter/The Image Bank/Getty Images. **616:** M, BIO Klein & Hubert/Peter Arnold, Inc. ; T, Michio Hoshino/Minden Pictures; B, Bkgrnd and Inset: Kelly McGann Hebert **619:** BR, Peter Parks/Image Quests Marine; TR, Bkgrnd and Inset: Kelly McGann Hebert; TL, Jeff Hunter/The Image Bank/Getty Images. **622–623:** Editions Xavier Barral. Photo © Patrick Gries. **623:** O. LOUIS MAZZATENTA /National Geographic Society. **624:** B, Photo courtesy of Shuhai Xiao at Virginia Tech; T, Editions Xavier Barral. Photo © Patrick Gries. **625:** B, Pearson Education /Patrice Rossi Calkin. **626:** T, Waterframe/Alamy; Upper M, Fred Bavendam/Minden Pictures; B, Kim Taylor/Dorling Kinderesley; Lower M, Ben Twist/istockphoto.com. **627:** T, DAVID M. DENNIS/Animals Animals - Earth Scenes; Upper M, JOHN ANDERSON/Animals Animals - Earth Scenes; M, Dr. David Wachenfield/Minden Pictures; Lower M, Roger Steene/imagequestmarine.com; B, O. LOUIS MAZZATENTA /National Geographic Society. **628:** B, Simon Conway Morris, University of Cambridge; T, Editions Xavier Barral. Photo © Patrick Gries. **630:** Chip Clark. **631:** Anne Ryan/Polaris Images. **633:** M, DeAgostini/Getty Images; T, Carolina Biological Supply Company/Phototake. B, O. LOUIS MAZZATENTA/National Geographic Society. **634:** Editions Xavier Barral. Photo © Patrick Gries. **637:** SuperStock, Inc./SuperStock. **639:** B, Chip Clark; T, Profimedia/SuperStock. **640:** Editions Xavier Barral. Photo © Patrick Gries. **642:** O. LOUIS MAZZATENTA /National Geographic Society. **644–645:** Richard Du Toit/Minden Pictures. **645:** Anastasiya Maksymenko/istockphoto.com. **646:** L, Richard Du Toit/Minden Pictures; R, Radius Images/Photolibrary Group. **647:** BL, Norbert Wu/Minden Pictures; BR, Fred Bavendam/Minden Pictures; TR, PHONE PHONE - Auteurs Danna Patricia/Peter Arnold Inc.; TL, Hiroya Minakuchi/Minden Pictures. **650:** Richard Du Toit/Minden Pictures. **653:** Richard Du Toit/Minden Pictures. **655:** T, David Kjaer/Minden Pictures; B, Anastasiys Maksymenko/istockphoto.com. **656:** Richard Du Toit/Minden Pictures. **658:** Tom Vezo/Minden Pictures. **659:** Anastasiya Maksymenko/istockphoto.com. **660:** Fred Bavendam/Minden Pictures. **661:** B. RUNK/S. SCHOENBERGER/Grant Heilman Photography. **662:** Richard Du Toit/Minden Pictures. **664:** Anastasiya Maksymenko/istockphoto.com. **666–667:** David Tipling/Minden Pictures. **667:** Masa Ushioda/Image Quest 3-D. **668:** B, Peter Blackwell/ npl/Minden Pictures; T, ©David Tipling/Minden Pictures. **671:** Antti Leinonen /Jupiter Images. **672:** TR, Roger Steene/Image Quest Marine; BR, John F White/SuperStock ; BL, Larry West/Photo Researchers, Inc.; BM, David Scharf/Photo Researchers, Inc.; TL, Oxford Scientific / Photolibrary Group. **673:** Dale C. Spartas/CORBIS. **674:** BL, Dwight Kuhn; BR, Dwight Kuhn; T, David Tipling/Minden Pictures. **677:** TL, Stephen Dalton/Minden Pictures; TR, ©Greg Lasley Nature Photography; B, MC DONALD WILDLIFE PHOTOG./Animals Animals Enterprises. **678:** B, DAVID M. DENNIS/Animals Animals - Earth Scenes; T, David Tipling/Minden Pictures. **680:** Nature Picture Library/Alamy. **683:** B, 2002 Masa Ushioda/Image Quest 3-D; T, Mitsuaki Iwago/Minden Pictures; MR, Thomas Marent/Minden Pictures; ML, Tom Brakefield/CORBIS. **684:** B, Nigel J. Dennis; Gallo Images/CORBIS; T, David Tipling/Minden Pictures. **685:** Michael & Patricia Fogden/Minden Pictures. **686:** Simon King/npl/ Minden Pictures. **687:** Stephen J. Krasemann/Photo Researchers, Inc. **689:** Dale C. Spartas/CORBIS. **690:** David Tipling/Minden Pictures. **692:** 2002 Masa Ushioda/Image Quest 3-D. **694–695:** TIM FITZHARRIS /Minden Pictures. **695:** Ariadne Van Zandbergen/africanpictures.net. **696:** R, Polka Dot Images/SuperStock; L, TIM FITZHARRIS /Minden Pictures. **699:** M, LAZAR, TOM/Animals Animals Enterprises; L, Markus Botzek/zefa/Corbis; R, Konrad Wothe /Minden Pictures. **700:** TL, Thomas Wiewandt ; TM, www.operationmigration.org; TR, 2005 Jason Hahn Photography; B, Ariadne Van Zandbergen/africanpictures.net. **701:** TIM FITZHARRIS /Minden Pictures. **702:** Corbis/SuperStock. **704:** T, Flip Nicklin/Minden Pictures; B, Ariadne Van Zandbergen/africanpictures.net. **706:** ©TIM FITZHARRIS /Minden Pictures. **708:** TR, Ariadne Van Zandbergen/africanpictures.net. **711:** ML, © PNC/Blend Images/Corbis. **712–713:** Reuters/Corbis. **713:** Walter Smith/Workbook Stock/Jupiter Images. **714:** Reuters/Corbis. **715:** L, Image Source/Photolibrary; ML, Manfred Kage/Peter Arnold; MR, Visuals Unlimited; R, © ISM/Phototake. **718:** T, ©Greg Ceo/Taxi/Getty Images; M, Caroline Woodham/Stock Connection; B, Walter Smith/Workbook Stock/Jupiter Images. **719:** T, Reuters/Corbis; B, Jupiter Images/Pixland/Alamy Images. **720:** BL, Corel Professional Photos CD-ROM™; TL, Photodisc/Getty Images; R, ©Gord Horne/istockphoto.com. **721:** M, Image Source Pink/Jupiter Images; T, ©Elena Schweitzer/istockphoto.com; B, Emilio Ereza/Alamy. **722:** Walter Smith/Workbook Stock/Jupiter Images. **723:** Reuters/Corbis. **726:** B, Meckes/Otawa/SPL/Photo Researchers, Inc. ; T, Dr. David M. Martin/Photo Researchers, Inc. **728:** Photo Researchers, Inc. **729:** Reuters/Corbis. **732:** Walter Smith/Workbook Stock/Jupiter Images. **736:** Reuters/Corbis. **740–741:** Jonathan Daniel/Getty Images. **741:** Private Collection/Bridgeman Art Library. **742:** Jonathan Daniel/Getty Images. **746:** Private Collection/Bridgeman Art Library. **747:** Jonathan Daniel/Getty Images. **748–749:** BSIP/Photo Researchers, Inc. **751:** Jonathan Daniel/Getty Images. **753:** Private Collection/Bridgeman Art Library. **754:** B, Dr. Cecil H. Fox/Photo Researchers, Inc.; T, Jonathan Daniel/Getty Images. **757:** Private Collection/Bridgeman Art Library. **760:** Jonathan Daniel/Getty Images. **762:** Private Collection/Bridgeman Art Library. **764–765:** Joe McNally/Getty Images News. **765:** The Schlesinger Library/Radcliffe. **766:** Joe McNally/Getty Images News. **768:** Photo Alto/Tips Images. **769:** The Schlesinger Library/Radcliffe. **770:** BL, ISM/Phototake; BM, Oxford Scientific Films/PhotoLibrary ; BR, Ed Reschke/Peter Arnold, Inc. ; T, Joe McNally/Getty Images News. **774:** The Schlesinger Library/Radcliffe. **775:** Joe McNally/Getty Images News. **777:** TM, ISM/Phototake; TL, Dr. P. Marazzi/Science Photo Library/Photo Researchers, Inc.; TR, Dr. P. Marazzi/Science Photo Library/Photo Researchers, Inc.; B, The Schlesinger Library/Radcliffe. **780:** TL, Joe McNally/Getty Images News. **781:** R, ISM/Phototake; L, ISM/Phototake. **782:** T, The Schlesinger Library/Radcliffe.; B, © Salisbury District Hospital/Science Photo Library/Photo Researchers, Inc. **784–785:** Somos Images/Corbis. **785:** Libby Chapman/istockphoto.com. **786:** Somos Images/Corbis. **789:** Libby Chapman/istockphoto.com. **790:** B, National Cancer Institute/Science Photo Library/Photo Researchers, Inc.; T, Somos Images/Corbis. **791:** Visuals Unlimited/Corbis. **793:** ISM/Phototake USA. **795:** Libby Chapman/istockphoto.com. **796:** Somos Images/Corbis. **800:** L, Clark Overton/Phototake/Alamy; R, Nancy Kaszerman/ZUMA/Corbis. **801:** Libby Chapman/istockphoto.com. **803:** National Cancer Institute/Science Photo Library/Photo Researchers, Inc. **804:** Somos Images/Corbis. **805:** Visuals Unlimited/Corbis. **806:** Libby Chapman/istockphoto.com. **808–809:** Roger Ressmeyer/CORBIS. **809:** Nicholas Monu/istockphoto.com. **810:** B, Alamy Images; T, © Roger Ressmeyer/CORBIS. **812:** Nicholas Monu/istockphoto.com. **813:** Roger Ressmeyer/CORBIS. **816:** Nicholas Monu/istockphoto.com. **817:** Roger Ressmeyer/CORBIS. **820:** Claude Edelman/Photo Researchers, Inc. **823:** B, ©Nicholas Monu/istockphoto.com; T, David M. Phillips/Photo Researchers, Inc. **824:** B, Yorgos Nikas/Stone/Getty Images; T, Roger Ressmeyer/CORBIS. **828:** TL, Petit Format/Photo Researchers, Inc.; TR, Claude Edelmann/Photo Researches, Inc.; BL, Neil Bromhall/Photo Researchers, Inc.; BR, Petit Format/Photo Researchers, Inc. **829:** Simon Puschmann/Bilderberg/Aurora Photos. **832:** Roger Ressmeyer/CORBIS. **834:** Nicholas Monu/istockphoto.com. **836–837:** AP Photo/Thomas Haentzschel. **837:** TR, Christopher Wheeler. **838:** BL, NIBCS/SPL/Photo Researchers, Inc.; BM, Dennis Kunkel/Phototake, USA; BR, Biophoto Associates/Photo Researchers, Inc. **839:** L, Science Photo Library/Photo Researcher's Inc.; R, Eye of Science/Photo Researcher's Inc. 8**40:** T, William Radcliffe/Science Faction/Getty Images; B, Christopher Wheeler. **841:** AP Photo/Thomas Haentzschel. **845:** Steve Gschmeissner/Photo Researchers, Inc. **846:** B, Mary Evan Picture Library/Photo Researchers, Inc.; T, AP Photo/Thomas Haentzschel. **847:** Peter Vinton-Johansen, et al., Cholera, Chloroform, and the Science of Medicine: A Life of John Snow (2003), 289; and www.matrix.msu.edu/~johnsnow. **848:** T, Wang Haiyan/China Features/Corbis; M, Frank Greenaway/DK/Getty Images; B, Christopher Wheeler. **849:** BL, ©Eric Isselée/istockphoto.com; BR, Volker Steger/Photo Researchers, Inc.; T, AP Photo/Thomas Haentzschel. **851:** Christopher Wheeler. **853:** Wang Haiyan/China Features/Corbis. **854:** Christopher Wheeler. **856:** L, Royal Geographic Society Picture Library: TR; ©Carolina Biological Supply Company/Phototake. **DOL 1:** Header Bar, Harald Theissen/Photolibrary Group; M, 2003 James D. Watt/Image Quest 3-D. **DOL 2:** FRED BAVENDAM/Minden Pictures. **DOL 6:** T, Science Source/Photo Researchers, Inc.; B, Science Pictures Ltd./Photo Researchers, Inc. **DOL 6–7:** Header Bar, Dennis Kunkel/Phototake. **DOL 7:** TR, BIOMEDICAL IMAGING UNIT, SOUTHAMPTON /Photo Researchers, Inc. GENERAL HOSPITAL/SCIENCE PHOTO LIBRARY; TM, Science Source/Photo Researchers, Inc.; MR, Microfield Scientific Ltd/Photo Researchers, Inc.; M, Microfield Scientific Ltd/Photo Researchers, Inc.; BR, Jack Bostrack/Visuals Unlimited. **DOL 8:** TR, Stephen & Donna O'Meara/Photo Researchers, Inc.; B, LookatSciences/Phototake. **DOL 8–9:** Header Bar, Dennis Kunkel/Phototake. **DOL 9:** TM, EYE OF SCIENCE/SCIENCE PHOTO LIBRARY/Photo Researchers, Inc.; BL, Getty Images/Visuals Unlimited; M, Dr. Gerhard Wanner/Ludwig-Maximilans Universitat, Munich; MR, Karl Stetter, University of Regensberg. **DOL 10–11:** Header Bar, Dennis Kunkel/Phototake Inc./Oxford Scientific Images. **DOL 10:** TL, M. I. Walker/Photo Researchers, Inc. **DOL 11:** ML, Eye of Science/Photo Researchers, Inc.; BR, Eye of Science/Photo Researchers, Inc.; TR, Wim van Egmond/Visuals Unlimited. **DOL 12:** M, Dr. Dennis Kunkel/Visuals Unlimited; TR, Steve Gschmeissner/Photo Researchers, Inc.; BL, David Wrobel/Visuals Unlimited; BR, Biodisc /Visuals Unlimited; M, Dennis Kunkel/Phototake; MR, Susumu Nishinaga/Photo Researchers, Inc. **DOL 13:** BR, Karen Gowlett-Holmes/Photolibrary New York; TM, Michael Abbey/Photo Researchers, Inc.; TR, Dr. Dennis Kunkel/Visuals Unlimited. **DOL 14:** M, Wim van Egmond/Visuals Unlimited ; TR, Dr. Dennis Kunkel/Visuals Unlimited. **DOL 15:** MR, Dennis Kunkel/Phototake; BL, Wim van Egmond/Visuals Unlimited; TR, Wim van Egmond/Visuals Unlimited. **DOL 16:** TR, Glen Threlfo/Minden Pictures, Inc.; BL, Jaime Plaza Van Roon/Minden Pictures, Inc. **DOL 16–17:** Header Bar, Dennis Kunkel/Phototake Inc./Oxford Scientific Images. **DOL 17:** BM, Fletcher & Baylis/Photo Researchers, Inc.; BR, Bill Beatty/Visuals Unlimited. **DOL 18:** BL, Scott Camazine/Photo Researchers, Inc.; BM, Thomas Marent/Minden Pictures, Inc.; BR, Custom Medical Stock Photo, Inc. **DOL 19:** TR, Biophoto Associates/Photo Researchers, Inc.; TL, Science VU/M.F. Brown/Visuals Unlimited ; MR, John Taylor/Visuals Unlimited ; BL, Carr Clifton/Minden Pictures, Inc. **DOL 20:** MR, Naturfoto Honal/Corbis; BR, A. Barry Dowsett/Photo Researchers, Inc. **DOL 20–21:** Header bar, Creatas/Photolibrary Group. **DOL 21:** M, Dr. Richard Kessel/Visual Unlimited/Getty Images; BL, Lee W. Wilcox; BR, LAURIE CAMPBELL/NHPA/Photoshot; TR, Dr. Jeremy Burgess/Photo Researchers, Inc. **DOL 22:** BL, Peter Chadwick/DK Images; BM, B. RUNK/S. SCHOENBERGER/Grant Heilman Photography; BR, Lee W. Wilcox; TR, Craig Tuttle/CORBIS. **DOL 23:** BM, B. RUNK/S. SCHOENBERGER/Grant Heilman Photography; BL, GARDENWORLD IMAGES/L THOMAS; BR, Dwight Kuhn; Inset, GardenWorldImages/DR GARY BROWN. **DOL 24:** TR, Rob Blakers/Photolibrary; BR, Jim Brandenburg/Minden Pictures; BM, Bettmann/CORBIS. **DOL 25:** BR, Gerald Hoberman/Photolibrary Group; ML, Alan and Linda Detrick/Photo Researchers, Inc.; MR, G. R. Roberts/Omni-Photo Communications, Inc.; TR, Peter Chadwick/Dorling Kindersley. **DOL 26:** Dr. Merlin D. Tuttle/Photo Researchers, Inc. **DOL 27:** MR, Sangtae Kim. Sangtae Kim, Mi-Jeong Yoo, Victor A. Albert, James S. Farris, Pamela S. Soltis and Douglas E. Soltis. 2004. Phylogeny and diversification of B-function MADS-box genes in angiosperms; evolutionary and functional implications of a 260–million-year-old duplication. *American Journal of Botany* 91:2102-2118. Used by permission of the Botanical Society of America.; TL, xyno6/istockphoto.com; BR, Florida Images/Alamy; BL, HUBAUT Damien/Age Fotostock. **DOL 28:** M, Mafuta/istockphoto.com; TR, Trevor Sims/GardenWorld Images; ML, LARRY LEFEVER/Grant Heilman; MR, HOWARD BUFFETT/Grant Heilman Photography; BL, KirsanovV/istockphoto.com; BR, Kim Taylor/npl/Minden Pictures.

Credits *cont'd*

DOL 29: B, Creatas/SuperStock ; M, WILLIAM MANNING/Grant Heilman Photography; TR, Martin Siepmann/Age Fotostock; ML, Michael Valdez/istockphoto.com; MR, Stephen Hayward/DK IMAGES. **DOL 30:** T, ZSSD/ Minden Pictures; BR, Stephen Fairclough, King Lab. **DOL 30–31:** Header Bar, Corbis/SuperStock. **DOL 31:** BL, Norbert Wu/Minden Pictures; MR, Brandon Cole; BR, Daniel L. Geiger/SNAP/Alamy. **DOL 32:** TR, Michael Durham/ Minden Pictures; BL, Alex Kerstitch/Visuals Unlimited/ Getty Images. **DOL 33:** B, Michael Durham/Minden Pictures; M, Image by: Wernher Krutein/Photovault.com; TL, Norbert Wu/ Minden Pictures; TR, Hiroya Minakuchi/Minden Pictures. **DOL 34:** MR, AP Photo/Ed Andrieski; B, age fotostock/ SuperStock. **DOL 35:** T, Justine Gecewicz/istockphoto.com; MR, Mark Moffett/Minden Pictures; BR, Kim Taylor/Minden Pictures; BL, Colin Keates/DK Images. **DOL 36:** T, Stephen Dalton/ Minden Pictures; MR, Kim Taylor/npl/Minden Pictures; BR, Joke Stuurman-Huitema/ Foto Natura/Minden Pictures; BL, Cisca Castelijns/ Foto Natura/Minden Pictures. **DOL 36–37:** Bkgrnd, Stephen Dalton/Minden Pictures. **DOL 37:** Centipede, DAVID M. DENNIS/Animals Animals - Earth Scenes; Millipede, Kim Taylor/ npl/Minden Pictures; Desert Ant, Piotr Naskrecki/Minden Picture; Honeybee, Kim Taylor/npl/ Minden Pictures; Ladybug, Ingo Arndt/ Minden Pictures; Blowfly, Heidi & Hans-Jurgen Koch/ Minden Pictures; Katydid, Piotr Naskrecki/Minden Pictures; Mint Beetle, Kim Taylor/npl/Minden Pictures; Fungus Beetle, Michael and Patricia Fogden/ Minden Pictures. **DOL 38:** BL, Sinclair Stammers/Photo Researchers, Inc.; TR, Dr. Richard Kessel & Dr. Gene Shih/Getty Images. **DOL 38–39:** Bkgrnd, Sinclair Stammers/ Photo Researchers, Inc. **DOL 39:** TR, 2007 Jez Tryner/Image Quest Marine; BL, Volker Steger/Photo Researchers, Inc. **DOL 40:** B, Leslie Newman & Andrew Flowers/ Photo Researchers, Inc.; T, Kåre Telnes/Image Quest Marine. **DOL 41:** T, Carlos Villoch 2004/ Image Quest Marine; BL, Brandon Cole; BR, FABIO COLOMBINI MEDEIROS/Animals Animals - Earth Scenes. **DOL 42:** TR, Peter Herring/Image Quest 3-D; Bkgrnd, Flip Nicklin/Minden Pictures. **DOL 43:** B, Norbert Wu/Minden Pictures; TL, Ingrid Visser/Foto Natura/Minden Pictures; TR, Flip Nicklin/Minden Pictures; TM, superclic/Alamy. **DOL 44:** B, Norbert Wu/Minden Pictures; TL, Kaj R. Svensson/Photo Researchers, Inc.; TR, Norbert Wu/Minden Pictures. **DOL 45:** TR, FRANS LANTING/Minden Pictures; MR, Brandon Cole; ML, Fred Bavendam/ Minden Pictures; BL, Roger Steene/Image Quest Marine; TL, Norbert Wu/Minden Pictures. **DOL 46:** T, Jurgen Freund/npl/Minden Pictures; B, age fotostock/ SuperStock. **DOL 47:** B, T.Burnett-K.Palmer/V&W/Image Quest Marine; T, Heather Angel/Natural Visions; M, 2002 Roger Steene/ Image Quest 3-D. **DOL 48–49:** T, Norbert Wu/Minden Pictures. **DOL 48:** B, Nature Magazine. **DOL 49:** BM, B. RUNK/S. SCHOENBERGER/Grant Heilman Photography; BR, Gerald & Buff Corsi/Getty Images; BL, Peter Batson/imagequestmarine.com. **DOL 50:** TR, Norbert Wu/Minden Pictures; BL, Tui De Roy/ Minden Pictures; BR, Georgette Douwma/npl/Minden Pictures; ML, EYE OF SCIENCE/SCIENCE PHOTO LIBRARY/Photo Researchers, Inc.; TL, James D. Watt/ImageQuest 3D. **DOL 51:** B, Tom McHugh/Photo Researchers, Inc.; T, Tim Fitzharris/Minden Pictures. **DOL 52:** T, Pete Oxford/Minden Pictures; B, Michael Durham/Minden Pictures. **DOL 53:** TL, DAVID M. DENNIS/ Animals Animals/Earth Scenes; TR, Kevin Snair/SuperStock ; M, Pete Oxford/Minden Pictures; B, Rene Krekels/ Foto Natura/ Minden Pictures. **DOL 54 - 55:** M, Mike Parry. **DOL 55:** M, Michael Durham/Minden Pictures; BL, Winfried Wisniewski/Foto Natura/Minden Pictures; TR, ©Karl Shone/DK IMAGES; BR, Tui De Roy/Minden Pictures. **DOL 56:** T, BERNARD CASTELEIN/ Nature Picture Library; B, Journal Nature/Portia Sloan/Getty Images. **DOL 57:** T, age fotostock/SuperStock ; BL, Michael Durham/Minden Pictures; BR, Danny Ellinger/Foto Natura/ Minden Pictures. **DOL 58:** T, Tui De Roy/Minden Pictures; BR, Michael & Patricia Fogden/Minden Pictures; BL, Duncan Usher/ Foto Natura/Minden Pictures. **DOL 59:** ML, Tim Fitzharris/ Minden Pictures; TL, Tom Vezo/Minden Pictures; MR, TomVezo. com/MINDEN PICTURES; TR, Tom Vezo/Minden Pictures; M, Tom Vezo/Minden Pictures; B, Tom Vezo/Minden Pictures. **DOL 60–61:** TM, Suzi Eszterhas/Minden Pictures. **DOL 60:** B, Dave Watts/Nature Picture Library. **DOL 61:** BL, Mitsuaki Iwago/ Minden Pictures; BR, Pete Oxford/ Minden Pictures. **DOL 62:** TR, Gallo Images/CORBIS; BR, ZSSD/Minden Pictures; TL, Ingo Arndt/Nature Picture Library; BL, Todd Pusser/npl/Minden Pictures. **DOL 63:** ML, Ingo Arndt/Minden Pictures; TR, ZSSD/ Minden Pictures; Inset, TR, Richard Du Toit/Minden Pictures; B, Reinhard Dirscherl/age fotostock; TL, Creatas/PhotolibraryGroup; Inset, ML, PETER BAUMANN/Animals Animals/Earth Scenes; Inset, TL, Jim Clare/Nature Picture Library. **DOL 64:** TM, Michael and Patricia Fogden/Minden Pictures; TR, Lockwood & Dattatri/ Nature Picture Library; TL, Thomas Marent/Minden Pictures; BR, Creatas/Photolibrary; Center, ML, Thomas Marent/Minden Picutres; BL, Gerry Ellis/Minden Pictures; Center, R, Elio Della Ferrera/Nature Picture Library; BM, Prisma/SuperStock; Center, L, Pete Oxford/Minden Pictures; Center, MR, Anup Shah/npl/ Minden Pictures. **A-4:** Flip Nicklin/Minden Pictures/Getty Images. **A-5:** Kenneth Eward/Biografs/Photo Researchers, Inc. **A-9:** Pearson Education/PH School Division. **A-12:** Peter Casolino/Alamy. **A-16:** Taunton Yachts. **A-17:** Richard Haynes. **A-19** ©Kevin Fleming/ CORBIS. **A-20:** B, DK Image Library; T, Tom Pantages. **A-21:** Richard Haynes.